D1289106

AMERICAN HANDBOOK OF PSYCHIATRY

Volume One

AMERICAN HANDBOOK OF PSYCHIATRY

Silvano Arieti, EDITOR-IN-CHIEF

Volume One
The Foundations of Psychiatry
EDITED BY SILVANO ARIETI

Volume Two
Child and Adolescent Psychiatry, Sociocultural and Community Psychiatry
EDITED BY GERALD CAPLAN

Volume Three
Adult Clinical Psychiatry
EDITED BY SILVANO ARIETI AND EUGENE B. BRODY

Volume Four
Organic Disorders and Psychosomatic Medicine
EDITED BY MORTON F. REISER

Volume Five
Treatment
EDITED BY DANIEL X. FREEDMAN AND JARL E. DYRUD

Volume Six
New Psychiatric Frontiers
EDITED BY DAVID A. HAMBURG AND H. KEITH H. BRODIE

AMERICAN HANDBOOK OF PSYCHIATRY

SECOND EDITION

Silvano Arieti · Editor-in-Chief

VOLUME ONE

The Foundations of Psychiatry

SILVANO ARIETI · *Editor*

BASIC BOOKS, INC., PUBLISHERS · NEW YORK

Second Edition

Library of Congress Catalog Card Number: 72-89185
SBN: 465-00147-5
Manufactured in the United States of America
76 77 78 10 9 8 7 6 5 4 3 2

CONTRIBUTORS

Ian Alger, M.D.
Visiting Associate Clinical Professor of Psychiatry, Albert Einstein College of Medicine, New York; Clinical Assistant Professor of Psychiatry and Training Psychoanalyst, New York Medical College, New York.

Heinz L. Ansbacher, Ph.D.
Professor Emeritus of Psychology, University of Vermont; Editor, Journal of Individual Psychology.

E. James Anthony, M.D.
Blanche F. Ittleson Professor of Child Psychiatry, and Director, Eliot Division of Child Psychiatry, Washington University School of Medicine, St. Louis.

Kenneth Appel, M.D.
Professor Emeritus, Department of Psychiatry, University of Pennsylvania, Philadelphia, Pennsylvania.

Silvano Arieti, M.D.
Clinical Professor of Psychiatry, New York Medical College, New York; Training Analyst and Supervisor, William Alanson White Institute of Psychiatry, Psychoanalysis and Psychology, New York.

Jack Bemporad
Rabbi, Temple Emanu-El, Dallas, Texas; Past Director of the Commission on Worship of the Union of American Hebrew Congregations.

Jules R. Bemporad, M.D.
Director of Pediatric Psychiatry, Babies Hospital, Columbia-Presbyterian Medical Center, New York; Faculty Member, Psychoanalytic Division, New York Medical College, New York.

Therese Benedek, M.D.
Senior Staff Member, The Institute for Psychoanalysis, Chicago, Illinois.

Viola W. Bernard, M.D.
Former Clinical Professor of Psychiatry, Columbia University; Staff Psychoanalyst, Psychoanalytic Clinic for Training and Research, Columbia University, New York.

Donald A. Bloch, M.D.
Director, Nathan W. Ackerman Family Institute, New York; Editor, Family Process.

Henry Brill, M.D.
Director, Pilgrim State Hospital, West Brentwood, New York; Clinical Professor of Psychiatry, Down-State Medical Center, Stony Brook, New York.

Robert N. Butler, M.D.
Research Psychiatrist and Gerontologist, The Washington School of Psychiatry, Washington, D.C.; Associate Clinical Professor, Howard University School of Medicine, Washington, D.C.

Rachel Dunaway Cox, Ph.D.
Professor Emeritus of Psychology and Education, and Director, Longitudinal Study of Adult Personality, Bryn Mawr College, Bryn Mawr, Pennsylvania.

Nancy Datan, Ph.D.
Assistant Professor of Psychology, Department of Psychology, West Virginia University, Morgantown, West Virginia.

Katrina De Hirsch, F.C.S.T.
Former Lecturer in Pediatrics, College of Physicians and Surgeons, Columbia University; Director of Pediatric Language Disorder Clinic, Columbia Presbyterian Medical Center, New York.

Marvin G. Drellich, M.D.
Associate Clinical Professor of Psychiatry and Assistant Director of the Division of Psychoanalysis, New York Medical College, New York.

Abraham Edel, Ph.D.
Distinguished Professor of Philosophy, City University of New York Graduate School; Professor of Philosophy, The City College, New York.

Leon Edel, Ph.D.
Citizens' Professor of English, University of Hawaii; Member, Academy of Arts and Letters.

Henri F. Ellenberger, M.D.
Professor, Department of Criminology, University of Montreal; Psychiatrist, Hotel-Dieu Hospital, Montreal, Canada.

James L. Foy, M.D.
Associate Professor, Department of Psychiatry, Georgetown University School of Medicine, Washington, D.C.

Eugene A. Friedmann, Ph.D.
Professor of Sociology and Chairman, Department of Sociology and Anthropology, Kansas State University, Manhattan, Kansas.

Richard A. Gardner, M.D.
Assistant Clinical Professor of Child Psychiatry, Columbia University, College of Physicians and Surgeons, New York; Faculty Member, William Alanson White Institute of Psychiatry, Psychoanalysis and Psychology, New York.

Kurt Goldstein, M.D.
Deceased.

Ralph R. Greenson, M.D.
Clinical Professor of Psychiatry, University of California, Los Angeles, School of Medicine; Chairman, Scientific Advisory Board, Foundation for Research in Psychoanalysis, Beverly Hills, California.

Harry Guntrip, Ph.D.
Fellow of the British Psychological Society; Psychotherapist and Lecturer, Department of Psychiatry, Leeds University, England.

Joseph L. Henderson
Assistant Clinical Professor of Neuro-Psychiatry, Stanford University Medical School, California; President of the C. G. Jung Institute of San Francisco.

John W. Higgins, M.D.
Late Associate Professor Clinical Psychiatry, St. Louis University, St. Louis, Missouri.

Irene M. Josselyn, M.D.
Training Analyst, Southern California Psychoanalytic Institute; Member of Child Analysis Committee, American Psychoanalytic Society.

Bernard Kaplan, Ph.D.
University Professor and Professor of Psychology, Clark University, Worcester, Massachusetts.

Kenneth Keniston, Ph.D.
Professor of Psychology (Psychiatry), Yale University School of Medicine; Chairman and Executive Director, Carnegie Council on Children, New Haven, Connecticut.

Nolan D. C. Lewis, M.D.
Former Head, Department of Psychiatry, College of Physicians and Surgeons, Columbia University, New York.

Theodore Lidz, M.D.
Professor of Psychiatry, Yale University School of Medicine, New Haven, Connecticut; Career Investigator, National Institute of Mental Health.

Harold I. Lief, M.D.
Professor of Psychiatry and Director, Division of Family Study, University of Pennsylvania School of Medicine, Philadelphia, Pennsylvania.

Bruce Mazlish, Ph.D.
Professor of History, Massachusetts Institute of Technology, Cambridge, Massachusetts.

Rev. James R. MacColl, III, D.D.
Former President, Academy of Religion and Mental Health; Chairman of Executive Committee, Institutes of Religion and Health.

R. E. Money-Kyrle, Ph.D.
Full Member, British Psychoanalytical Society.

George Mora, M.D.
Assistant Clinical Professor of Psychiatry, New York Medical College, New York; Research Associate, Department of History of Science and Medicine, Yale University, New Haven, Connecticut.

Wendell S. Muncie, M.D.
Associate Professor of Psychiatry, Johns Hopkins University, Baltimore; Consultant, Springfield State Hospital, Sykesville, Maryland.

Simon H. Nagler, M.D.
Clinical Professor of Psychiatry, Director of Psychiatric Residency Training, and Director of Psychoanalytic Division, Department of Psychiatry, New York Medical College, New York.

Bernice L. Neugarten, Ph.D.
Professor, Human Development, and Chairman, Committee on Human Development, University of Chicago, Chicago, Illinois.

Daniel Offer, M.D.
Associate Director, Institute for Psychosomatic and

Psychiatric Research and Training, Michael Reese Hospital and Medical Center; Professor of Psychiatry, Pritzker School of Medicine, University of Chicago, Chicago, Illinois.

Harold L. Orbach, Ph.D.
Associate Professor of Sociology, Kansas State University, Manhattan, Kansas; President, Midwest Council for Social Research in Aging, Kansas City, Missouri.

E. Mansell Pattison, M.D.
Associate Professor and Vice-Chairman, Department of Psychiatry and Human Behavior, University of California at Irvine; Deputy Director, Training and Manpower Development Division, Orange County Department of Mental Health, Santa Ana, California.

Isidore Portnoy, M.D.
Associate Dean, American Institute for Psychoanalysis; Attending Psychoanalyst, Karen Horney Clinic, New York.

John D. Rainer, M.D.
Professor of Clinical Psychiatry, College of Physicians and Surgeons, Columbia University; Chief of Psychiatric Research (Medical Genetics) New York State Psychiatric Institute, New York.

Anatol Rapoport, Ph.D.
Professor of Psychology and Mathematics, University of Toronto, Canada.

Paul Ricœur
Professor of Philosophy, University of Chicago and University of Paris.

Alfred H. Rifkin, M.D.
Clinical Associate Professor of Psychiatry, New York University School of Medicine, New York; Associate Clinical Professor of Psychiatry, New York Medical College, New York.

Alan K. Rosenwald, Ph.D.
Chairman, Department of Psychology, Illinois State Psychiatric Institute, Chicago, Illinois.

Joseph F. Rychlak, Ph.D.
Professor of Psychology, Department of Psychological Sciences, Purdue University, West Lafayette, Indiana.

Melvin Sabshin, M.D.
Professor and Chairman, Department of Psychiatry, Abraham Lincoln School, University of Illinois, College of Medicine, Chicago, Illinois; Medical Director-designate, American Psychiatric Association.

David E. Schecter, M.D.
Director of Training, William Alanson White Institute of Psychiatry, Psychoanalysis and Psychology, New York.

Edward C. Senay, M.D.
Director, Illinois Drug Abuse Programs; Associate Professor, Department of Psychiatry, University of Chicago, Chicago, Illinois.

Natalie Shainess, M.D.
Faculty Member, William Alanson White Institute of Psychiatry, Psychoanalysis and Psychology, New York; Lecturer in Psychiatry, Columbia University, College of Physicians and Surgeons, New York.

William M. Sheppe, Jr., M.D.
Clinical Associate Professor, Department of Psychiatry, University of Virginia School of Medicine, Charlottesville, Virginia; Medical Director, Central Virginia Psychiatric Associates, Inc., Charlottesville, Virginia.

Cyril Sofer, Ph.D.
Fellow of Queens' College and University Reader in Industrial Management, Cambridge, England.

John A. Sours, M.D.
Private practice; Assistant Clinical Professor of Psychiatry, Columbia University, Psychoanalytic Clinic for Training and Research, New York.

Rose Spiegel, M.D.
Training Analyst and Supervisor, William Alanson White Institute of Psychiatry, Psychoanalysis and Psychology, New York; Associate Psychiatrist, New York State Psychiatric Institute, New York.

Ian Stevenson, M.D.
Carlson Professor of Psychiatry, Department of Psychiatry, University of Virginia School of Medicine, Charlottesville, Virginia.

Alberta Szalita, M.D.
Clinical Professor of Psychiatry and Training and Supervising Analyst, College of Physicians and Surgeons, Columbia University, New York; Visiting Professor, Tel-Aviv University, Tel-Aviv, Israel.

Ludwig von Bertalanffy, Ph.D.
Late Faculty Professor, Natural and Social Sciences, State University of New York at Buffalo.

Eilhard Von Domarus

Late Fellow and Member of the Board of Directors, Association for the Advancement of Psychotherapy; Late Supervisor, Postgraduate Center for Psychotherapy, New York.

Joseph B. Wheelwright, M.D.

Professor of Psychiatry, Langley Porter Clinic, San Francisco, California; Past President of the International Association of Analytical Psychology.

Earl Witenberg, M.D.

Director, William Alanson White Institute of Psychiatry, Psychoanalysis and Psychology, New York.

Milton Wittman, D.S.W.

Chief, Social Work Training Branch, Division of Manpower and Training Programs, National Institute of Mental Health, Alcohol, Drug Abuse, and Mental Health Administration, Department of Health, Education and Welfare, Rockville, Maryland.

Joseph Wolpe, M.D.

Professor of Psychiatry, and Director of the Behavior Therapy Unit, Temple University Medical Center and Eastern Pennsylvania Psychiatric Institute, Philadelphia, Pennsylvania.

PREFACE

Among the disciplines which are growing at the most accelerated pace is psychiatry, and understandably so. Intimately involved with general biology, neurology, psychology, sociology, cultural changes of any kind, it is constantly enriched by these fields and many others, which all converge toward a greater knowledge of man in health and mental illness.

When one looks at the first edition of this Handbook several thoughts emerge: how much of its content has preserved its validity; how little is no longer acceptable; but, most of all, how much is missing from it which we are now in a position to include. And yet the initial two volumes of the first edition appeared in 1959; the third in 1966. The time elapsed is short, but the work done in the field in these intervening years is enormous. A new and larger edition has thus become necessary. The success of the first edition, not only in the United States but throughout the world, gave the inspiration and provided the energy necessary for the preparation of this second edition.

When this new edition was planned it soon became evident that I could no longer be the only editor, as I had been for the 1959-1966 edition. It is increasingly difficult for one person to have the necessary competence in the many areas of rapidly expanding psychiatry. The need for section editors became a necessity. I am responsible for the planning, general organization, and division of the work; but competent colleagues have joined me in editing the various volumes. Only Volume One, "The Foundations of Psychiatry," is edited by me exclusively. Dr. Gerald Caplan is the editor of Volume Two, which deals with "Child and Adolescent Psychiatry, Sociocultural and Community Psychiatry." Volume Three, devoted to "Adult Clinical Psychiatry," is edited by Dr. Eugene B. Brody and myself. Volume Four, dealing with "Organic Disorders and Psychosomatic Medicine," is edited by Dr. Morton F. Reiser. Drs. Daniel X. Freedman and Jarl E. Dyrud have edited Volume Five, dealing with "Treatment," and Drs. David A. Hamburg and H. Keith Brodie have edited Volume Six, "New Psychiatric Frontiers."

The basic philosophy and guidelines of the first edition will be recognized in the second. All orientations, all schools, all respectable methods of treatment are represented in these volumes. Some of these approaches converge and integrate, others diverge. They actually represent in the field of psychiatry the spirit of innovation and the state of flux of modern man. At the present stage of our knowledge any attempt to integrate some approaches and to exclude others would not lead to higher syntheses or to a consistent view of psychiatry or of man, but to one or another brand of reductionism.

Completeness was preferred to simplification, representation of contrasting views and reliability were preferred to uniformity and attempts toward consistency. Each author was requested to cover his special field; he was free to express his personal point of view, but he was asked in almost all instances to present alternative conceptions and to reduce his private terminology to a minimum or to define it immediately. The chapters of these volumes differ from those of usual textbooks in being more complete, more analytical, and more authoritative without reaching monographic proportions. Except for Volume Six, no attempt was made to put emphasis on new or fashionable developments. Currently accepted or classical concepts are presented together with new findings and innovating ideas.

The authors who were invited to participate in this project are in most instances recognized authorities in their respective fields. In several instances, the task of selecting the author was difficult because many people had made important contributions in the same field. The search for recognized authorities did not lead us to discriminate against young age. Several chapters are written by promising young authors. Some colleagues whom we would have liked to have joined us, could not participate because of heavy involvement in other work.

Repeating some of the words of the preface to the first edition, I wish to say that this book is offered as representative of American psychiatry today. "American," used in this context, refers specifically to the receptivity to all possible approaches which is characteristic of psychiatry in the United States. Although most of the great psychiatric contributions originated in Europe, in no country other than the United States is there such willingness to listen to, try out, and evaluate all theories, methodologies, and techniques, and to absorb and find a place for all or many of them. We are also happy to have in our midst a representative number of authors from other countries.

When the first edition was published, it was the first time that a work of this kind appeared in the United States. It was also the first time that some of the topics and issues appeared in a textbook of psychiatry published anywhere. That this Handbook, within a short time, was imitated by other editors in the United States and abroad was proof of its value and a source of rejoicing for us. Perhaps this will happen again, as we hope that the usefulness of the innovations of the second edition will be evident.

It is apparent from the organization of the Handbook that the views expressed in the various chapters are the responsibility of the respective authors and do not necessarily represent those of the editors or the publisher. On the other hand, the single contributors are not responsible for the general editorial policies, with which, in a few instances, they may have disagreed.

As I have already mentioned, this first volume deals specifically with the foundations of psychiatry. Foundations are of various kinds, and they will all be represented in this book. They include the historical background, the very basic notions from which the discipline emerges, the methodology with which we examine and classify the patient, and the knowledge of the normal life cycle and its common vicissitudes. They also include the theoretical foundations on which the various schools of psychiatry have based their work.

It is not feasible to express gratitude individually to the hundreds of people who, in various ways, have helped prepare this project. Their work will unfold in these six volumes and will speak for itself. I shall mention only that for many years the staff of Basic Books has faithfully cooperated and made it possible to transform into reality what was once only a hopeful expectation.

SILVANO ARIETI

New York, June 1973

CONTENTS

Volume One

PART ONE: *History*

PART TWO: *Basic Notions*

PART THREE: *The Life Cycle and Its Common Vicissitudes*

PART FIVE: *Contributions to Psychiatry from Related Fields*

PART SIX: *Classification and Assessment of Psychiatric Conditions*

PART ONE

History

CHAPTER 1

PSYCHIATRY FROM ANCIENT TO MODERN TIMES

Henri F. Ellenberger

Introduction

MODERN PSYCHIATRY, like the other branches of science, is continually changing. Today's discovery will soon be made obsolete by tomorrow's discovery. At present certain medical papers mention only the literature of the past five years; the rest is almost as antiquated as Hippocrates. This being the case, one may wonder about the interest of the history of psychiatry and inquire into its meaning.

No science can progress if it lacks a solid theoretical fundament. There is no theory of science without a knowledge of the history of science, and no theory of psychiatry without a knowledge of the history of psychiatry. This becomes evident as soon as one ponders the basic principles of modern psychiatry. Why should we treat mental patients humanely? Is any kind of healing possible without a rational body of knowledge? Is there a difference between an empirical body of knowledge and a scientific, experimental one? Does psychiatry belong to medicine, or is it a science in its own right, or perhaps no science at all? In order to solve such problems philosophical cogitation is not enough: one needs a great deal of data, and these data can be secured only through historical inquiry. However, this implies in turn that historical inquiry be conducted not in an amateurish fashion, but by means of a scientific methodology.

Unfortunately medical history is a very young branch of science. It is not enough to say that there are wide gaps in our knowledge of Greco-Roman, Arabian, and medieval medicine (not to speak of Indian or Chinese medicine). The truth is that we possess only extremely fragmentary data, so that any reconstruction attempted on the basis of these data is doomed to be artificial. In regard to the last two or three centuries, the difficulty often stems from the immense accumulation of data, so that the trees hide the forest.

Today scientists are keenly aware that no progress can be called definitive, that any discovery can at any moment be displaced by a new one. But we should not overlook another, complementary viewpoint: the progress of today might also get lost through the regression of tomorrow. Science implies not only a striving toward progress but also a constant effort to maintain the permanent acquisitions of yesterday. These are two among the fruitful lessons that a psychiatrist may learn from the history of science.

The Roots of Psychiatry

Although scientific psychiatry is hardly more than one or two centuries old, it is in a large measure the outcome of notions and procedures that are perhaps as ancient as mankind itself. It would seem that from the beginning among all populations of the earth a special attention was bestowed upon certain conspicuous types of behavior or other abnormalities. Individuals whose behavior was deemed to be abnormal were dealt with in three possible ways. Many of them were treated in a downright inhuman, cruel way. Some others were treated in a human but nonmedical way (for instance, those psychotics whose utterances were taken for prophetic inspiration). A third group were submitted to a variety of healing procedures that we today understand to be anticipations of modern psychiatry.

The birth of scientific psychiatry was a long-range process. It started with the discarding of nonmedical ways of dealing with mental patients and the perfecting of primitive healing procedures. The next step was the constitution of systematic bodies of knowledge by medicine men (and later temple healers) and also by lay healers. The third step was the foundation of a rational medical art, severed from religion and superstition; this great revolution is symbolized by the legendary figure of Hippocrates. But whatever its merits, the rational medicine of the Greeks and Romans still remained on a prescientific level (to use Bachelard's terminology). The foundation of a properly scientific medicine did not occur before the seventeenth century, and that of psychiatry as a branch of medicine in its own right came still later.

In this chapter I shall attempt to retrace these steps as briefly as so complex a matter will allow. It has been, indeed, a long way from nonmedical attitudes to primitive healing, from primitive healing to rational prescientific medicine, and from there to scientific medicine and modern psychiatry.

Nonmedical Ways of Coping with Deviant Behavior and Abnormalities

For countless centuries a great number of those persons who would indisputably be recognized today as mental patients were the victims of social attitudes in direct opposition to the principles of a humane and scientific psychiatry. Feeble-minded infants, with or without bodily defects, were often killed shortly after their birth without further ado. Even among the culturally enlightened Greeks such children were exposed in the wilderness. Among certain Western and Central European rural populations there was a lasting belief in the existence of "changelings." Goblins or demons were supposed to steal newborn children and replace them with their own ugly progeny. Children believed to be changelings were often left without nourishment so that they would die of hunger; this was no crime since these beings allegedly did not belong to humankind. Those feeble-minded children who survived often were mercilessly exploited by their families, or led a miserable life, being the butt of their community's jokes.

Those senilely demented were often ruthlessly treated. Koty has shown that among the populations of the earth certain peoples are kind and considerate with old people and cripples; others harsh and cruel. But even in the best case these individuals were likely to be sacrificed in the eventuality of famine or other calamities.

Psychotics frequently fled from, or were rejected by, their communities. Some of them, the stronger, would escape to the woods or desert places, living as "wild men," feeding themselves on roots and berries. An example is the demoniac of Gadara described in the Gospel: he had been tied up several times but had managed to break his chains and escape; he lived in tomb vaults and frightened the surrounding population. When these psychotics attacked passers-by or isolated homesteads, they were outlawed so that they could be killed by anyone without compunction. Less severe psychotics would survive as vagrants and live off the charity of individuals and communities. Accounts from the Renaissance period tell of "fools" roving from place to place; in certain communities they were provided with some food or money and gently led to the border; in other places they were driven away with a whip; no doubt they paid a heavy toll in fatal illnesses or accidents. There are also stories of "fools" being set in a little boat on the river, so that the stream would drift them along to another place.

A frequent plight of chronic psychotics was a prolonged confinement in dungeons, cellars, or other dark places, either in a prison, a monastery, or a private home. Even after mental hospitals had been established, certain families preferred to have their patients secretly confined in their home, rather than institutionalized. A famous case was that of the *séquestrée de Poitiers* in 1901. Following an anonymous letter, the police discovered that a catatonic woman had been confined by her respected and well-to-do family in a dark room for 24 years in an indescribably filthy condition; neither neighbors nor friends knew of the existence of the woman.

Mild psychotics, especially those with delusions of grandeur, were often the plaything of children and adults, sometimes made the victims of practical jokes or utilized for nefarious purposes. Philo tells the story of a lunatic who lived in Alexandria in the first century A.D. and was the laughingstock of the children. It happened that Herod Agrippa, king of the Jews and protected by Emperor Caligula, visited Alexandria. A group of people of that city, out of hatred for the Jews, staged an insulting mockery by bringing this "fool" into the gymnasium, dressing him like a king, and greeting him with royal honors; this was the onset of an anti-Jewish riot and pogrom. It would be easy to collect numerous instances of disturbed individuals who were victimized in the worst fashions. Sir John Lauder, a young Scotsman who sojourned in France in 1665–1666, related that in the town of Montpellier a "fool" proclaimed that he had discovered a universal antidote that could nullify the effect of any poison. He proposed to try it on himself. Although it should have been clear that the man suffered from delusions, the pharmacists of the town prepared a poisonous beverage that the man drank together with his own draught, whereupon he died miserably within a few hours.

The worst fate was probably that incurred by the "furious," the agitated and aggressive psychotics. They usually were chained, sometimes with iron fetters, or mercilessly beaten until they quieted down—that is, fell into complete exhaustion. Sometimes they were sandwiched between two mattresses tightly bound with ropes, where they often died from choking. Herodotus relates that Cleomenes, King of Sparta, "was smitten with downright madness" and went striking every Spartan he met with his scepter. He was imprisoned by his kindred and his feet put in the stocks. He asked for a knife, and the servant who kept him did not dare to refuse his order. Cleomenes then cut gashes in his flesh along his legs, thighs, hips, loins, and his belly until he died. If this could happen to a king, one may imagine what would be the condition of a commoner befallen with a similar condition.

As to psychotic criminals, many of them were tried and sentenced as if they had been ordinary, nonpsychotic criminals. But the crime itself that brought the man to trial was all too often the product of the patient's delusions, or of mass suggestion exerted upon him, if not of false accusations and torture, as happened to thousands of unfortunate women during the witch psychosis of the sixteenth and seventeenth century.

On the other hand, it could happen that the social response was favorable to the patient. Various types of disturbed behavior were tolerated in many cultural settings. Certain mental patients were well cared for by the community. Such was the kind attitude of the poor mountaineers in the high valleys of the Alps toward the feebleminded afflicted with cretinism, a condition that was endemic on these mountains until the middle of the nineteenth century. In certain cultural settings psychotics with delusions of grandeur were honored as prophets; sometimes they gathered followers and launched a psychic epidemic.

One should not be too shocked by such "primitive" attitudes toward mental patients. In our own century did we not see the inmates of mental hospitals being systematically exterminated in certain civilized nations, and left to perish miserably of hunger in other countries? The humane and scientific care of mental patients has been a hard-won, delayed achievement in the history of mankind, a conquest that will always need to be defended against the powers of obscurantism and oppression.

Primitive Healing

As far as we can go back into the past we find evidence that a more privileged group of mental patients benefited from certain healing procedures. About healing in prehistoric times we know almost nothing. Numerous skulls from the neolithic era show the marks of trepanning that had been performed on living individuals and followed with cicatrization. Since even in recent times the same operation was performed by medicine men of various primitive populations as a cure for certain nervous conditions, it is likely that the rationale was the same among our prehistoric ancestors. A few pictures of prehistoric art point to the existence of magic and wizards. Such is the well-known picture of a sorcerer, his head adorned with deer's antlers, in the "Cave of the Three Brothers" in southern France, a picture believed to have been painted about 15,000 B.C. and to be the oldest known representation of a healer.

We assume that a continuity exists between prehistoric and primitive healing, such as it is known from ethnological inquiry. Unfortunately many primitive populations disappeared before any serious study could be conducted, and of those that survived many retained only distorted remnants of their former medical lore. However, the systematic study of primitive medicine, undertaken by Buschan, Bartels, and their followers, provides us with a fairly accurate knowledge of the main features of primitive medicine.

According to Sudhoff,[34] primitive medicine always and everywhere distinguished several kinds of disease, of treatment, and of healer. Certain conditions, obviously the effect of accidents, parasites, or poisons, suggested the use of rational, empirical treatments (this was *natural medicine*). Instances of death rapidly following hemorrhage or asphyxia led to the belief that the source of life was in the blood or the breath; this was the starting point of *speculative medicine*. Acute sickness or death occurring in an unexpected, inexplicable way was attributed to the action of evil spirits or wizards; this was the foundation of *magical medicine*. Hysterical or epidemic possession incited efforts to expel the mysterious intruder from the soul, and this was *demonological medicine*. Natural medicine was the realm of the "lay healer"; magical and demonological medicine the preserve of the medicine man, later of the priest; speculative medicine was to become the field of election of the rational prescientific physician.

The medicine man plays an essential role in his community. He is not only healer but often a dreaded wizard and one of the leaders of his tribe, along with the priest and the chief. He is a "man of high degree" (as Elkin termed the Australian medicine man), and he has undergone a long and difficult training that often includes the experience of an initiatory illness. His personality is the principal agent of the cure, provided that the patient, the healer himself, and the community are all convinced of his healing power. Thus, his healing methods are essentially psychological in nature. Primitive healing is almost always a public procedure, a ceremony conducted within a well-structured group.

Ethnologists distinguish several basic disease theories, each one linked to a specific healing procedure: these include the loss of the soul and its recovery by the shaman, the intrusion of a supposed disease object and its extraction, the intrusion of an evil spirit and its expulsion (mainly in the form of exorcism), the breach of a taboo and its propitiation, the pathogenic effect of magic and its cure through countermagic. These procedures are of great interest for transcultural psychiatry, comparative psychotherapy, and medical history. A direct continuity can be shown from exorcism to magnetism, magnetism to hypnotism, and from hypnotism to the newer dynamic therapies.

On the other hand, historians of medicine have convincingly shown that the true ancestor of the modern physician is the lay healer, whereas the medicine man is the ancestor of the priest, who was the physician's antagonist for centuries. Thus, the discontinuous line of evolution led from primitive healing to priestly healing, from there to rational prescientific therapy, and eventually to scientific psychiatry.

Priestly and Religious Healing

Around 4,000 B.C. the first kingdoms and empires were founded in Asia and in Egypt. This implied the advent of a new type of social organization with a large administrative system and of religions with colleges of priests and elaborate rituals. The medicine man gave way to the priest, whereas the lay healer became the physician, although the separation was not always very sharp. For many centuries the healing priest and the physician lived side by side, the physician more concerned with natural therapy (massage, dietetics, hot baths, nonmagical drugs), and the priest with psychological healing.

The healing powers of the priest were enhanced by the fact that, in addition to the attributes of the medicine man, he acquired the awesome prestige of being the representative of a healing god. A few healers came to be considered supernatural beings and gods. Such is the story of Imhotep in Egypt. He was born around 3,000 B.C. and became vizier of Pharaoh Zoser and chief of ritual. Special honors were conferred upon him after his death; later he was worshiped as a demigod; around 600 B.C. he had reached the status of a god of medicine. Numerous wonderful cures were reported to occur at his shrines, and it seems that a medical teaching was provided in his great temple at Memphis. Until about 500 A.D. he remained one of the most popular gods in Egypt.

It would seem that a similar evolution took place in Greece with Asclepius (Aesculapius) though we lack reliable information about his life. In the *Iliad* he is referred to as a physician and father of two physicians. A few centuries later he was a god and numerous patients flocked to his shrines, the *Asclepeia.* Most famous were his temples in Epidaurus, Pergamon, and Cos. The beautiful site, the prestige of the place, the stories of wonderful cures, the journey, the period of waiting, all affected the patient. An *Asclepeion* combined the holiness of a place of pilgrimage like Lourdes with the enjoyments of a fashionable health resort. After careful screening the patient had to undergo a period of purification and to perform preliminary rites. The highlight was the *incubation,* that is, the night of sleep in the sanctuary, either on the ground or on a couch called the *kline.* Then the patient might experience an *epiphania* (apparition of the god), or receive an oracle, or have a therapeutic dream, that is, a specific kind of dream that would in itself bring the cure. The cult of Aesculapius did not recede before the advent of rational medicine. As time went on, the number of his shrines increased throughout the Hellenistic and the Roman world, and for some time Aesculapius was a great rival of Christ. After the triumph of the Church, patients sought healing at the new Christian shrines. It is significant that in 1893 the skeptical Charcot wrote a paper *On Faith Healing,* declaring that he had seen patients cured at Lourdes after medical treatment had failed, so that the existence of powerful, unknown healing agents must be assumed.

The influence of religious healing extended over the patients' perception of their own illness. Depressive conditions became linked with the idea of sin; this was particularly marked in the Egyptian and Assyro-Babylonian worlds. Healing could be obtained through confession, propitiation, that is, reconciliation with the gods, and acceptance of the cosmic order. Concepts of the nature of man, as taught by the priests, also influenced the clinical pictures. One instance is the *Dialogue of a Life-Weary Man with His Soul*, an Egyptian writing of about 2,000 B.C. that might be considered the oldest known document on the psychology of suicide.

Rational Prescientific Medicine

A decisive step in the history of medicine was taken with the rise of autonomous schools of medical practice and teaching, dominated by astute clinicians and creative thinkers. In Egypt the Ebers Papyrus (of about 1,500 B.C.) is one of the oldest documents of this type of medicine. Similar developments took place in Assyro-Babylonia, Persia, India, China, Japan, and even in Mexico among the Aztecs; however, our knowledge of the history of medicine in these countries is extremely imperfect. Modern scientific medicine is predominantly the tributary of Greek prescientific medicine, even in its terminology.

Ancient Greek medicine was based on dogmatic concepts that survived for about 25 centuries. In the same way as Greek priestly medicine was dominated by the mythical figure of Aesculapius, Greek rational prescientific medicine is dominated by the hardly less mythical figure of Hippocrates.

All we know of Hippocrates (about 460 to 377 B.C.) is that he was one among several reputed physicians of his time; the rest is legend. The treatises collected under his name about three centuries later were divergent in style, dialect, and content. As Werner Jaeger[16] says, "The result of one century of research is that there is not one page in the Hippocratic collection which we can, with certitude, ascribe to Hippocrates himself." Nor do we have his portrait; but his name is traditionally attached to a noble bust that is, in Singer's[33] words, "an idealized representation of what the Greek would wish his physician to be." Whereas the god Aesculapius stood high above mankind, Hippocrates remained the more accessible figure of the "physician-philosopher." To quote Singer[33] again:

> His figure, gaining in dignity what it loses in clearness, stands for all time as the ideal physician. . . . Calm and effective, human and observant, prompt and cautious, at once learned and willing to learn, eager alike to get and give knowledge, unmoved save by the fear lest his knowledge may fail to benefit others . . . incorruptible and pure in mind and body. . . . In all ages he has been held by medical men in a reverence comparable only to that which has been felt towards the founders of the great religions by their followers.

Hippocrates is customarily called the Father of Medicine. Before him, however, Greek medicine had at least two centuries of intensive development in the form of autonomous schools of philosophical medicine, and many medical treatises, now lost, had been written. To quote Neuburger:[26] "We stand as in a devastated town, where only *one* building is extant, and we see only the rough outline of the streets." Among these schools the Cnidian strove to diagnose diseases and localize their seats, whereas the school of Cos was more concerned with the prognosis, with the organism as a whole within the environmental setting, and with the concept of the "healing power of Nature."

It would be irrelevant to speak of a "Hippocratic psychiatry." Throughout the whole Greek and Roman literature, mental disorders are described in one among several other chapters of medicine; nothing is known about that could resemble a textbook of psychiatry. The Hippocratic writings describe or mention three main mental conditions: phrenitis (acute mental disorders with fever), mania (mental disorders with agitation without fever), and melancholia (chronic mental disorders without fever and agitation). A variety of organic paroxysmal conditions in women are related to the uterus (in Greek, *hystera*, hence the adjective "hysterical" which meant "uterine" and did not have the connotation of a neurosis). A description of the "Scythian disease" refers to an instance of transsexuality in the setting of a particular culture.

The Hippocratic writings are often considered the beginning of scientific medicine. To be sure, they discard religion and superstition and strive to rely upon clinical observation, experience, and sound judgment. In this sense it is a rational medicine. However, we are still very far from the principles of scientific medicine. Hippocratic thinking belongs to the prescientific level of medicine. Robert Joly[17] has shown conclusively how even the best treatises of the Hippocratic collection are pervaded with "substantialism," "numerology," and other irrational elements that often interfere with and blur the otherwise acute clinical vision of the writer.

Greek medicine worked with the principles of physical qualities, the four humors, the pneuma, the faculties of the soul, and organic localization. These principles had to be correlated to each other and with the clinical pictures offered by medical practice.

Greek philosophers had widely discussed polarities of opposite qualities. The two main polarities finally chosen were those of dry-moist and warm-cold. Philosophers had also argued endlessly about the physical elements until Empedocles settled the question by compromise, proclaiming as the fundamental elements water, air, fire, and earth. Similarly physicians decided there were four humors of the human body: phlegm, blood, bile, and "black bile." Like the four elements, the four humors were correlated to the physical qualities (phlegm moist-cold, blood moist-warm, bile dry-warm, black bile dry-cold). Morbid conditions were explained by the resulting imbalance from the excess of one of the humors. Thus, *melancholy* was thought to result from an excess of black bile. However, it soon became necessary to involve also changes in the qualities of the humors, disturbances in their circulation, and to distinguish varieties among each one of the humors.

Meanwhile, primitive medicine's concept of breath (or pneuma) as a principle of life had found its way into philosophical medicine. Stoic philosophers and the Pneumaticist school of medicine distinguished three kinds of pneuma, physical, vital, and psychic, and interpreted certain diseases as resulting from a lack or alteration of the pneuma.

Greek concepts of the soul, originating in primitive and priestly medicine, were developed by the philosophers. Aristotle distinguished three main "souls" of the human psyche: vegetative soul (common to plants, animal, and man), animal soul (common to animals and man), and rational soul (property of man only). To each one he attributed several "powers" (or faculties). The use of this framework led to associating mental states with lack, excess, or disturbance of the various "souls" or "powers." Passions were considered excessive activities of the animal soul, or the absence of control over the animal soul by the rational soul. Certain mental conditions were supposed to result from a deficiency of the rational soul (anoia).

Greek philosophers and physicians argued about the seat of the soul and of its faculties. The ancients had thought that the diaphragm (*phrenes*) was the seat of the soul, hence the word "phrenitis" for acute fever delirium (this condition later was correlated to an inflammation of the brain or its coverings). As to the seat of the intellect, Empedocles, Aristotle, and Diocles said it was the heart, whereas Alcmaeon, certain Hippocratic writers, and the Alexandrian, Galen, held for the brain. The next step was the localization of the various "souls" and "powers."

We can see how heterogeneous were the theoretical assumptions underlying Greek psychopathology. The word "melancholia" was derived from humoral pathology, "anoia" from psychological concepts, "phrenitis" from supposed anatomic views, whereas the word "mania" (probably a colloquial word) simply meant "madness" or "fury." Epilepsy, a purely clinical designation, expressed the main symptom of the convulsive attack. The Hippocratic writers had inherited this terminology from their unknown forebears, and the whole development of Greek medicine was a ceaseless effort to integrate these various conflicting concepts and clinical pictures into a coherent system.

The legacy of ancient Greek rational medicine is thus a twofold one. Greek medicine gave to the world the ideal figure of the Father of Medicine, Hippocrates. On the other hand, it gave also an abstruse, increasingly artificial system of dogmatic medicine, which predominated throughout Hellenistic, Roman, Arabian, and Western medicine and eventually became an intolerable burden that impeded the progress of human thinking.

Psychiatry in the Greco-Roman World

In the Greek and Roman worlds there was no such thing as "psychiatry" as we know it today, but only a few scattered elements of that which was to become this science much later.

We have discussed the prescientific concepts utilized by Greek rational medicine. The Greek physician's way of thinking was not less different from ours. The modern scientist starts with observation and quantification, draws inductions and hypotheses that are tested by experimentation and according to the findings accepted or rejected. Not so with the Greeks, who used axiomatic thinking, started with statements considered self-evident, then drew deductions that they extended with the help of analogies. Conflicting statements were rarely submitted to experimental testing, but mostly settled by theoretical discussions and compromises without much regard to facts.

In Greek science there was no unified framework universally accepted and adopted by all scientists. Instead, there were a number of schools based on particular philosophical systems, each one professing its own dogmatic teaching incompatible with those of the other schools. Greek medicine remained inseparable from philosophical speculation. Each philosophical revolution had its counterpart in medicine: the Pneumatists based their system on Stoicism, the Methodists on Epicureanism, the Empiricists on Skepticism. The movement toward philosophical syncretism was paralleled by the medical eclecticism of Galen.

It would be fascinating if we could closely follow the evolution and vicissitudes of these medical schools. Unfortunately our knowledge is extremely fragmentary. Out of a thousand books written by the Greeks, hardly more than one or two have survived.

Among the oldest medical schools known to us were those of Sicily, of Cnidos, and of Cos, the latter two well-represented in the so-called Hippocratic writings. Then in the fifth century B.C. Athens became the center of Greek culture. Aristotle exerted a great influence on medicine. He edited an encyclopedic collection of monographs covering the whole field of contemporary learning; there was, for instance, a *History of Medicine* by his pupil Menon, of which fragments have survived. Diocles of Carystos, who was associated with that school, was considered one of the greatest Greek physicians; unfortunately his works are lost. However, his description of hypochondriac disturbances was to be taken over by Galen.

After the Macedonian conquest, the Eastern Mediterranean world was Hellenized and a new cultural center flourished in Alexandria, with its famous museum and library. Among the great Alexandrian physicians were Herophilus and Erasistratus. Both of them made discoveries in the field of anatomy, especially in regard to the nervous system. It is generally assumed that with these two physicians Greek medicine reached its highest scientific level, even though their theories were not free from speculation. Undoubtedly their works, had they survived, would hold great interest for the history of neurology and psychiatry.

The Empiricists school, whose philosophy was inspired by the Skeptics, was represented mostly by Heraclid of Tarent. They were good clinicians who strove to define diseases according to their seats and described them *a capite ad calcem*, that is, going from scalp and hair down to the heels and adding a chapter for general diseases. Mental diseases were classified among the diseases of the head, following those of the scalp and the skull. This type of classification was traditional up to the seventeenth century.

Inspirer of the Methodist school was Asclepiades (first century B.C.), a fashionable Greek physician and prolific writer who practiced in Rome. His works have been lost. He seems to have been much interested in mental diseases and to have applied a variety of treatments: hydrotherapy, gymnastics, massage, suspended bed, music therapy—obviously a therapy accessible only to a few wealthy patients. Another great Methodist physician, Soranus of Ephesus (first half of the second century A.D.), discussed the treatment of mental diseases along the same lines as Asclepiades. Most of his writings have perished, but something of his work is known thanks to a Latin adaptation by Caelius Aurelianus.

Another school, the Pneumaticists, basically followed the philosophy of Stoicism and emphasized the doctrine of the pneuma. Their great proponent was Archigenes; his works are lost but were utilized by Aretaeus who flourished around 100 A.D. Aretaeus gave clear descriptions of the traditional disease entities, phrenitis, mania, and melancholia, and mentioned that mania and melancholia could turn into each other. This simply meant that a chronic psychosis could begin or be interspersed with acute episodes.

The Romans did not found a school of medicine, but compiled encyclopedias embracing the whole field of knowledge. Those by Cato and Varro have perished, but we have the medical part of A. Cornelius Celsus' encyclopedia (first century A.D.) with a chapter on mental diseases. Among the treatments were fettering, flogging, starving, or terrifying the patient, and suddenly pouring cold water over his head; however, rocking the patient in a suspended bed was used in other cases.

The eclectic trend was personified by Claudius Galen (ca. 138–201 A.D.), an outstanding clinician, good investigator in the field of physiology, and passionate systematizer. He borrowed his philosophical principles from Aristotle, the Stoicists, and the Hippocratic writers. A prolific author, he is credited with writing about 400 treatises, of which about 80 are extant. Because the works of the most prominent physicians before and after him have been lost, Galen towers as a mighty isolated genius. He endeavored to reconcile his clinical and experimental findings with the traditional doctrines of the qualities, the humors, the pneuma, the "powers" of the soul, and the localizations. Galen laid great emphasis on the doctrine of the four basic humors, which had been much developed since the Hippocratic writers. Disease resulted from the excess of one humor and the resulting dyscrasia (imbalance), or from mixtures and alterations of the various humors, or from their accumulation in certain organs, or from the ascension of "vapors" from the stomach or other organs to the brain. Galen's humoral doctrine covers also the field of the innate constitutions or temperaments. Galen distinguished two kinds of black bile. The word "melancholia" could mean either a specific variety of innate character or one among a wide range of diseases caused by the action of the two kinds of black bile. The Galenic doctrine extended to the theory of drugs and their therapeutic indications.

On Melancholia, a treatise under Galen's name described three main forms of this condition: (1) general melancholia (from an excess of black bile in the whole body), (2) brain melancholia (from an excess of black bile in the brain), and (3) hypochondriac melancholia (from the ascension to the brain of vicious vapors from the stomach). The author refers also to constitutional melancholia, melancholia from a one-sided diet, from the "adustion" of yellow bile, from the suppression of hemorrhoidal or menstrual flux, from precipitating emotional factors. Finally he describes particular clinical types of melancholia, one of them being *lycanthropia*, that is, the delusion of being transformed into a wolf. This treatise standardized for the following fifteen or sixteen centuries the concept of melancholia. It was the model and prototype of a tradition that was to culminate with Robert Burton's *Anatomy of Melancholy* in 1621.

It is usually assumed that Greco-Roman medicine underwent a swift decline after Galen. This might partly be an illusion resulting from the fact that the works of the most original minds of the following few centuries have been lost: Posidonius (second half of the fourth century A.D.) seems to have been eminent in the fields of neurological and mental diseases. But the Roman Empire was crumbling under its inner weakness and the repeated assaults of the barbarians, and medical progress was hampered. Thus, Galen was resorted to as the infallible oracle who had said the final word about medicine.

One may wonder to what extent the patients actually benefited from the kind of medical care expounded by all these authors. Greco-Roman physicians were mostly interested in acute diseases; they gave up the care of chronic patients as soon as these appeared to be incurable. Medical care was mostly given to patients of the upper classes in the cities and to the military. According to George Rosen,[29] the family and friends of the Greek and Roman patients were expected to provide for them according to the accepted customs, and the condition of the insane poor was extremely miserable.

There were apparently a few privileged forms of mental disorders. Plato distinguished "divine madness" from natural madness. Divine madness included the four varieties of prophetic madness (given by Apollo), religious madness (given by Dionysus), poetic madness (inspired by the Muses), and erotic madness (inspired by Aphrodite and Eros). An unclear problem is the feigned madness of men such as Meton, Solon, and Brutus: it would seem that it was a kind of ceremonialized eccentric behavior conspicuously displayed in certain extraordinary circumstances in order to focus attention upon an impending public danger.

Throughout the ancient world there were a few practices that could deserve the name of psychotherapy. Specific techniques of mental training were associated with the philosophical schools. The Pythagorians practiced exercises in self-control, memory recall, and memorization for recitation. The Stoics learned the control of the emotions and practiced written and verbal exercises in meditation. The Epicureans resorted to an intensive memorization of a compendium of maxims that they recited ceaselessly, aloud or mentally. Psychological training could be individualized as evidenced by Galen's treatise *On the Passions of the Soul.* This method consisted of unceasing effort to control one's passions with the help of a wise mentor who would point out one's defects and dispense advice, then the gradual reduction of one's standard of living, attaining serenity and freedom from affects, until one was able and ready to help others in a similar way.

Lain Entralgo[21] has pointed out the importance of the "therapy by the word" among the ancient Greeks. There was a cathartic, a dialectic, and a rhetorical use of the spoken word. One particular practice was the consolation, a letter, sometimes a poem, written to a person suffering grief and intended to help him recover peace of mind. Several of these consolations have become literary classics, such as Plutarch's *Consolation* to his wife for the death of their child, or his *Consolation* to his friend Apollonius for the death of his son. This kind of supportive psychotherapeutic intervention was to revive in Western Europe in the sixteenth and seventeenth century.

The Romans are credited with founding public hygiene. They devised elaborate systems of water adduction and sewage. A slow development of public medicine took place: there were town physicians paid by city authorities and a rudiment of medical teaching. But their only hospitals were the *valetudinaria* for the recuperation of wounded soldiers, gladiators, or slaves. Their moral callousness, their predatory economy, the use of vast masses of miserable slaves, the cruel games of the circus, all these were the antipode of any kind of mental hygiene.

Christianity no doubt introduced a new spirit. Its founder imposed the care of the sick as a religious duty. Charity institutions and hospitals gradually developed, parallel to the flourishing of monasteries. Christianity influenced much of the philosophical and legal thinking, and the practice of religious confession stimulated the development of introspection: this is well illustrated by Saint Augustine's *Confession.*

Psychiatry in the Middle Ages

After the ruin of the ancient world, Greco-Roman culture split into three parts: the Byzantine Empire, the Arabia, and Western Europe, each with its vernacular language: Greek, Arabic, and Latin. Each developed its own civilization and medical tradition, but they were bound by certain mutual influences.

Although the barbarians had destroyed the Western Roman Empire, the Eastern Roman (or Byzantine) Empire survived for ten centuries, and its capital, Constantinople, became the cultural center of the world. In the sixth century A.D., under Emperor Justinian, many charitable institutions were founded, such as orphanages, homes for the aged, and hospitals, among which were a few *morotrophia* for the mentally sick. Unfortunately we do not know how these institutions functioned and what vicissitudes they underwent between the sixth and the fifteenth century. What could be the way of life of an insane person in Constantinople is shown by the descriptions we have of the *saloi* or "sacred fools": these men led an utterly marginal life, living off the charity of good people, being an object of ridicule, but enjoying the privilege of telling the truth to anybody. The Greek medical tradition was prolonged by men such as Aetius of Amida and Alexander of Tralles (sixth century), Paul of Egina (seventh century), and Johannes Actuarius (fourteenth century). Contributions of psychiatric interest could probably also be found among the writings of Byzantine theologians and writers: to give only one instance, Eustathius of Thessalonica (twelfth century) wrote a treatise, *On Simulation*, analyzing role playing in the theater and in life and giving a psychology of histrionism. But numerous precious works perished in the catastrophe that engulfed Byzantine civilization in 1453.

In the seventh century the Islamic invasion destroyed Hellenistic civilization in Asia Minor, Syria, and Egypt. Then, after the Arabs had conquered an empire stretching from Persia to Spain, they underwent the influence of the conquered countries, and by the late eighth century and during the ninth century a brilliant Islamic civilization flourished in Baghdad, Cairo, and Spain. It was an era of widespread development for a medicine based on the teachings of Hippocrates, Galen, Aretaeus, and the earlier Byzantines. But here, too, it is difficult to appraise the real originality of that period. Out of a thousand books written by Arabian physicians, very few have survived; among these few most either remained in manuscript or were not translated into modern Western languages.

Among the most outstanding names one should mention Rhazes (864–925), a Persian distinguished in all fields of medicine. Among his many treatises is one translated into English under the title, *The Spiritual Physick of Rhazes*: a classification of psychopathological disorders is given, according to the failure or excess in each one of the three "souls" (vegetative, animal, and rational). Rhazes also discusses drunkenness (a topic somewhat neglected in extant Greek medical books) and its motivations: to dispel anxiety, to meet situations requiring particular courage and cheerfulness. Arabian medicine culminated in the work of Avicenna (989–1037), a Persian poet, philosopher, and physician, whose *Canon of Medicine* was used as a medical textbook for centuries in the Islamic as well as in the Western European world. The *Canon* systematized the highly artificial concepts enounced by Galen. The diseases are described in the traditional order "from head to heels," mental diseases being included in the chapter on head diseases. Arabian literature, like the Byzantine, contains nonmedical works that are of interest for the psychiatrist, for instance, books by the mystics, the moralists, the philosophers, books on physiognomy, on the interpretation of dreams. But of all this wealth, the greatest part has perished, and very little of value is available from the rest.

Throughout the Islamic world, many mental hospitals were founded by pious benefactors, especially in the thirteenth century. Among these were *moristans*, most probably inspired by Byzantine *morotrophia*. We have wondrous descriptions of the great hospital in Cairo in the thirteenth century, and especially of the *moristan* of Adrianople (Edirne) founded in the fifteenth century. In the midst of paradise gardens was a luxurious marble building with fountains in the courts, summer and winter rooms; the patients lay on silk cushions, were nourished with the finest food, were treated with a combination of drug, music, and perfume therapy, received friendly visits from the beauties of the town; their chains were gilded with gold and silver. In sharp contrast with such a description, we read in a poem of Djelal-Eddin (thirteenth century): "In the corner of a dungeon sat a raving insane; his neck tied with a rope." The French traveler Chardin, who visited Persia in the seventeenth century, describes the *moristan* of Ispahan as a kind of cloister around a garden; there were about 80 cubicles, but only seven or eight lunatics, all of them in the most miserable condition, lying on straw, fettered by arms, body, and neck. The staff consisted of one doctor, one pharmacist, one *molla* (priest), one cook, one doorman, and one cleaner. Not much better is the account given by Edward Lane of the *moristan* in Cairo in the middle of the nineteenth century: there were two courts, one for male and one for female patients. Around the first court were 17 very small cells with grated windows. The patients were chained by the neck to a wall, had only straw to sleep upon, and wore scarcely any clothing. It was customary for each visitor to give them pieces of bread, so that as soon as they saw a stranger enter, they made a great clamor. Obviously the Cairo *moristan* had declined since the thirteenth century. Actually we know very little about the history and the functioning of the Islamic mental institutions.

After the ruin of the ancient world, medicine in Western Europe was in a precarious state. There was a long period of destruction and decline. The privileged class of city aristocracy for whom Greco-Roman medicine was meant had disappeared, the writings of ancient authors were lost, medical tradition was interrupted. However, as we have seen, Christianity forwarded the care of the sick. Saint Benedict's rule proclaims that "the care of the sick is to be placed above and before every other duty." There were infirmaries in the monasteries, where monks made use of medical compilations written by obscure authors.

Meanwhile, the Church slowly gathered a certain amount of psychiatric knowledge of its own. In monasteries and in the practical experience of the confessors, certain specific conditions had been observed: the *acedia* (boredom) of the monks, the *scrupulositas* (excessive scruples) of certain penitents. Later the distinction between genuine and false mysticism and other observations made by moral theologians were the objects of valuable inquiries. Not until our time, however, were they incorporated, as a section of "pastoral psychology," in the general body of psychiatry. On the other hand, demonological concepts were taken for granted by the Church and accepted by medicine. In view of the general belief in demonology, many cases of neurotic or psychotic disturbances took the form of devilish possession, and conversely many patients could be cured by exorcism. Suggestive healings also took place at the shrines of certain saints; patients could sometimes remain there for a prolonged time; this was the origin of the psychiatric colony in Gheel, Belgium, which still exists and can be traced back to the thirteenth century.

Gradually there was a revival of lay medicine and a rise of medical schools. The thread with ancient Greco-Roman medicine was found again. After the recovery of Galen's works, his system became in its whole complexity and artificiality the official doctrine of Western European medicine.

Another great event took place in the thirteenth century. As in the first centuries of Christianity, the cure of the sick was placed in the foreground of Christian duties, and many hospitals were created. There are reports on hospitals for the mentally sick scattered in several European countries; however, the main development took place in Spain. In Granada an asylum for the insane had been founded by the Moorish King Mohammed V in 1365–1367, and the plan of the older Spanish asylums copied the plan of the Arabian *moristans*. The first one was opened in Valencia in 1407 under the name of *los Desamparados* (the abandoned) by Father Jofré, and operated by an association of 100 clerks, 300 laymen, and 300 laywomen. Then came the foundation of the asylums of Saragossa in 1425, Seville in 1436, Barcelona in 1481, and Toledo in 1483. Spanish historians, notably Juan Delgado Roig,[8] have published the acts of foundation and official documents of these hospitals, but many details about their functioning are still unclear.

Psychiatry in the Renaissance Period

Appraisals of the Renaissance have varied. According to Jacob Burckhardt, it was the period of "coming into awareness of the human personality, of its nature and place in the universe." Man's image of the world was immeasurably widened following the geographic explorations and the rediscovery of ancient Greco-Roman culture. Man became able to perceive reality instead of what he had been taught to see. G. de Morsier,[9] a historian of brain anatomy, has shown that fourteenth- and fifteenth-century surgeons who dissected human bodies were unable to see anything except what Galen and the Arabs had taught; the ability to perceive the anatomy of the brain as it is, and not as one believes that it should be, appeared in the early 1500s almost simultaneously and independently from each other in men born in Italy, France, Belgium, and Holland. The death blow to Galenic anatomy was given by Vesalius in 1543. The same transformation occurred in botany, astronomy, physics, and other sciences. These innovations, however, met much resistance from the adherents of the traditional doctrines. In addition to these great advances, the Renaissance was also the period when slavery (which had been abolished since the fall of the Roman Empire) was reestablished, at least in the newly conquered colonies, when widespread genocide took place in the Caribbeans and Central America, when judicial torture was reinforced under the impact of resurgent Roman law, and when the witch psychosis underwent an unprecedented development. Another negative feature of the Renaissance was its contempt for the vulgar, the illiterate, and the "fool." But there was much interest in mental illness and in the multiform manifestations of *imaginatio*, a notion that included all that we call today suggestion and autosuggestion as well as many creations of the unconscious mind.

Medical authors of the Renaissance are much better known than those of previous periods because the invention of printing gave their works a better chance to survive. In Italy Arturo Castiglioni[7] praises the psychiatric writings of G. B. da Monte, Gerolamo Mercuriale, Prospero Alpino (all three mainly interested in melancholia), and Jerome Cardan (a precursor of the concept of moral insanity). One should also mention Jean Fernel in France, Johannes Schenck in Germany, and Timothy Bright in England. The latter's *Treatise of Melancholie* (1586) is considered the first psychiatric book written by an Englishman. His theory of melancholia was also highly successful in the literate world, and this may explain the great popularity of Robert Burton's *Anatomy of Melancholy* (1621). Although Burton admits that the distinction of the various subforms of melancholia is "a labyrinth of doubts and errors," he describes them in detail. Melancholy became the fashionable disease in England. The type of character that was moody, cold, bitterly ironic, eccentric, misanthropic, disgusted with life, and predisposed to suicide became identified with the term "melancholy," later called spleen or hypochondriasis, and on the continent it was sometimes considered to be the typical English character.

The break with the tyrannical domination of the Galenic system in medicine was effected mainly by two great pioneers, Paracelsus and Platter. Theophrastus Bombastus von Hohenheim, called Paracelsus (1493–1541) is perhaps the most problematic figure in the history of medicine. Many episodes of his wandering life are tinged with legend. He has been considered a quack, a paranoid schizophrenic, a genius, a mystic philosopher, an initiate to secret sciences, a precursor of the psychology of the unconscious, an eminent psychotherapist, a pioneer of modern medicine, or a mixture of all these. Few authors are more difficult to read and more untranslatable. Sigerist[32] says that "perhaps no one but a German can really understand him." His writings are an incoherent succession of abstruse philosophical concepts, abuse against his enemies, affirmations based on his belief in astrology, alchemy, witchcraft, and other superstitions, interspersed with good medical insights or striking aphorisms: "The highest foundation of medicine is love"; "What the Greeks have told is not true for us; every truth originates in its own country"; "Do not hold the effect of curses as nonsense, ye physicians, ye have no inkling of the power and might of the will"; "The child needs no star or planet: his mother is his planet and his star." On the positive side are Paracelsus' teaching of a new pharmacology based on the use of mineral substances (sulphur, antimony, mercury, and others), his emphasis on the effects of suggestion, his belief in the efficacy of healing springs, his interest in occupational diseases of the miners, his concern for cretins and other feeble-minded persons ("Fools are our brethren; like us they were saved by Christ"). Paracelsus died prematurely, leaving few adherents and no organized movement; a great part of his writings were lost or remained unpublished. He had the good luck to find almost four centuries after his death an admirer, Karl Sudhoff,[34] who published his extant works. But Paracelsus had already become a legendary figure and was made the hero of novels and theatricals. The (true or fictitious) story of his spectacular burning of Galen's works on the public place in Basel in 1527 acquired a symbolic value and was paralleled with Luther's burning of the papal bull on the public place in Wittenberg. It is ironic that Basel, where Paracelsus had failed, was the home of another medical reformer, Felix Platter (1536–1614), who carried out some of the reforms vainly attempted by his predecessor.

Platter's life is better known than most physicians' because he left his autobiography, his almost lifelong diary, his correspondence, the catalogue of his collections, and all of his writings. He studied medicine in Montpellier and lived the rest of his life in Basel. He became a brilliant practitioner who was called as a consultant to kings and princes; as a teacher he made the University of Basel famous and wrote standard textbooks on medicine; as a scientist he advanced Vesalian anatomy, clinical medicine, and psychiatry. In his medical works Platter described diseases exclusively as he had seen them in his practice, never quoting any author, never concealing his ignorance or his doubts, and giving case histories of his own patients as illustrations. This might seem a matter of course today, but at that time physicians used to follow the descriptions of the old masters, with case histories borrowed from the classics. As Paracelsus had done before him, Platter gave up the old classification of diseases "from head to heels" and introduced a new principle. Platter's descriptions are models of clarity. Each condition is subdivided under three headings: *genera, causae, curatio* (symptoms, causes, treatment). With Platter practical medicine acquires the precision of a scientific discipline; medicine is detached from philosophy and becomes a branch of natural science.

Certain historians of medicine have pointed out that the Renaissance was a period of great suffering for the mentally ill. According to Walsh[35] the hospitals were much better and the insane more humanely treated in the thirteenth century than later. Kirchhof[19] writes that in the Middle Ages great help was provided to the insance by monasteries, brotherhoods, and pilgrimages. Wherever such institutions were abolished by the Reformation, mental patients were abandoned to the harsh treatment of public civilian authorities.

The evil culminated with the witch psychosis of the sixteenth and seventeenth century. The belief spread that witches committed crimes in alliance with the devil and that they had organized a universal conspiracy to destroy mankind. In order to eradicate the alleged evil, extraordinary forms of prosecution and trial were used. The *Malleus Maleficarum*, a guide for the Inquisitors conducting witch trials, gives a sinister picture of the delusion; its third part amounts to a textbook on the technique of brainwashing. Thousands of women of all ages and walks of life confessed the most improbable crimes. Some of them undoubtedly were neurotic or psychotic, but a great many were normal individuals who accused themselves under the impact of mental and physical torture. Witch hunts, inaugurated by the Inquisition, were pursued actively by lay civilian authorities, Protestant and Catholic alike. Among the few who dared openly to defend the incriminated was Johann Weyer (1515–1588). Though sharing himself the contemporary belief in witchcraft, Weyer tried to demonstrate that the supposed witches were victims of the demon rather than his allies. His book contains many interesting clinical cases, but lacks psychiatric systematization. Later Weyer came to be considered a great pioneer of psychiatry and to be called, not without exaggeration, the promoter of the "First Psychiatric Revolution."

Those asylums for the humane treatment of the insane that had been founded in Spain in the fifteenth century were isolated institutions, supported by rich patrons or local authorities. A crucial development began with Juan Ciudad Duarte (1495–1550). A pious and eccentric merchant in Granada, he suffered an acute psychotic episode and was treated in the local hospital with merciless flogging. After his recovery he founded a hospital where patients were treated humanely; he operated it with the help of a few volunteers. His organization grew; after the founder's death it was declared the Order of the Hospitalers, and Juan Ciudad was canonized as Saint John of God. This Order created and operated general and mental hospitals. All institutions of the Order worked according to the same rules and were run by men trained at the same place, so that experience could be collected and transmitted. These institutions spread over Spain, Italy, France, and other countries. Later they served as models to Philippe Pinel and Jean-Etienne Dominique Esquirol when they devised modern mental hospitals. It is not surprising that the first mental hospital episodes to be found in European literature appear in Cervantes' *Don Quixote*. (Until then mental disease had often been described in poetry, novels, or on the stage, but not the mental hospital.) In one episode of the novel an inmate of the Seville asylum protests to the archbishop that he is unduly retained, whereupon the archbishop sends his chaplain to examine the patient. In Avellaneda's continuation of the first part, Don Quixote is the victim of numerous cruel practical jokes and is treacherously committed in the Toledo asylum; the reader is left to suppose that he will stay there for the rest of his life. This probably brought Cervantes to write the second part of the novel: the hero recovers his sound mind on his death bed; far from being just the victim of ridiculous delusions, Don Quixote now appears as a personification of the human condition.

Early Scientific Psychiatry

The advent of modern science was a slow and gradual process, and it took a long time until it extended to psychiatry. During the Renaissance man had learned to perceive reality as it is and not as it should be; now man attempted to fathom the depths of nature and elicit her laws. This implied the use of new methods, such as systematically conducted experiments and measurement techniques, and of new instruments, such as the microscope. In natural history the principle of specificity came to the foreground. This evolution started at the end of the sixteenth century and developed during the seventeenth and the eighteenth centuries. Among its theoreticians were Francis Bacon and Descartes; among its pioneers Copernicus, Kepler, Galileo, Newton, and Vesalius and Harvey in medicine. But new systems arose, founded partly on science and partly on speculation. Such were iatromechanicism (Borelli, Baglivi), iatrochemistry (Van Helmont, Sylvius), and the "animism" of Georg Ernst Stahl (1660–1734). The latter was to exert a great influence upon German Romantic psychiatrists. Actually seventeenth- and eighteenth-century medicine was a curious mixture of genuine science, loyalty to antiquated doctrines, and new irrational systems—all three often in the same man.

Psychiatry was not yet an independent discipline. However, it underwent a noteworthy evolution. To the extent that Galenism was overthrown, mental diseases ceased to belong to humoral pathology and were correlated with disturbances of the nervous system; what was formerly attributed to "black bile" began to be labeled as nervous ailments.

Several of the new pioneers of medicine remained staunch believers in Galenism. In Italy Paolo Zacchias (1584–1659), a man eminent both as a physician and a lawyer, gave in his *Quaestiones medico-legales* (1621–1635) an inexhaustible mine of documents and expounded a new system of psychiatric diagnosis. He is generally considered the founder of forensic psychiatry. In England Thomas Willis (1622–1675) was, in spite of his belief in Galenism, a keen clinician and indefatigable experimenter who made outstanding discoveries in the anatomy and physiology of the nervous system. He integrated neurology and the study of the neurovegetative system into psychiatry.

Among the new trends was that of the systematists. Sydenham had emphasized that infectious diseases are specific entities: they exist in their own right independently of the sick individuals they affect. Efforts were made to work out an allover classification of diseases in natural species, genera, and classes. The first attempt was probably that of François Boissier de Sauvages (1706–1767), who taught botany and medicine in Montpellier. An outline of his system appeared in 1732 and was considerably enlarged in his *Nosologia methodica* published in Latin in 1763 and in French in 1770. Diseases were divided into ten classes. The bulk of mental diseases belonged to Class VIII, divided into 4 orders and 23 genera. Among the latter, melancholia contained 14 species. Each species was illustrated by short clinical descriptions, either from Boissier's practice or from other authors. Clinical documents were gathered from every possible source, even from travelers, so that the Malayan "running amok" was incorporated into a European textbook. Boissier's work was far from being a tedious catalogue of diseases; there were lucid introductions to the whole book and to each part; it was written in a clear and flowing style and filled with interesting facts. No wonder it inspired a long series of successors, among them the Swedish naturalist Carl Linnaeus, the English physician William Cullen (who coined the term "neurosis"), and eventually Philippe Pinel. In spite of the artificial character of their classifications, the systematists exerted a useful influence on the development of medicine (including psychiatry). They advanced the practice of looking more attentively for new diseases and facts and organizing material in a more accurate and comprehensive way. It sufficed to replace the terms "classes," "genera," and "species" with "parts," "chapters," and "paragraphs" to obtain the pattern of the nineteenth-century textbooks.

A powerful trend toward rationalism developed under the impact of the cultural movement called the Enlightenment. It forwarded the cult of reason, the concept of society as created for man, and the belief in science. Typical is Voltaire's definition of the word "folie" (insanity) in his *Dictionnaire philosophique*: "A brain disease that keeps a man from thinking and acting as other men do. If he cannot take care of his property he is put under tutelage; if his behavior is unacceptable, he is isolated; if he is dangerous, he is confined; if he is furious, he is tied." We see here mental illness equated to brain disease, its clinical picture to intellectual disorders, its treatment to the protection of society. In view of the opposition of rationalism to any kind of superstition, the belief in demons and witches gradually receded; demonomania ceased to be considered a preternatural condition and was understood as a man's delusion of being possessed; the term "obsession" lost its meaning of being assaulted by evil spirits and took on its present meaning. La Mettrie, in his book, *L'Homme machine* (1748), contended that certain hard criminals (later called the "moral insane") should be considered sick individuals.

The concept of the unconscious (if not the term) appeared with Leibniz's discussion of the "small perceptions." However, there had already been many observations on the unconscious activities of the mind, even among rationalist philosophers. Descartes wrote in his treatise, *Les Passions de l'ame* (1649), that unexplained aversions may originate from occurrences in early childhood: the event is forgotten, the aversion remains. In another place Descartes told of his propensity for falling in love with cockeyed women and how, thinking about it, he remembered that as a child he had been in love with a young girl who had that defect; after he had understood the connection his predilection disappeared. However, it did not occur to him nor to anybody else that such observations might be utilized for mental healing. Some kind of rational psychotherapy existed in the guise of consolations, exhortations, and admonishments. There also were various other attempts. Diderot relates how the wife of one of his friends was afflicted with the vapors (the fashionable neurosis of eighteenth-century ladies). He advised the husband to simulate the symptoms of his wife's illness. The husband did so, and the wife, who was very much in love with him, took so much care of him that she forgot her own ailments and recovered.

Many psychological problems were known and debated by Catholic priests. For a long time the only objective classifications and descriptions of sexual deviations were to be found in the treatises of moral theology. In Germany certain Protestant ministers practiced with success "Cure of Souls" (*Seelsorge*); they effected what amounted to a kind of brief psychotherapy by relieving the person of the burden of a pathogenic secret and helping to overcome the situation. In England the famous preacher John Wesley practiced medicine, opened dispensaries, and wrote a book, *Primitive Physics*, in order to give people a basic knowledge of medicine and hygiene; he was one of the first to use electricity as a therapeutic agent and had a profound understanding of many emotional problems. The Quakers included the insane among the unfortunates for whom they were concerned.

During the seventeenth and the eighteenth centuries the number of mental institutions in Europe increased. They derived from two opposite models: the prison and the monastery. Institutions of the incarcerative type have been the subject of many descriptions. Who has not heard of the horrible conditions in Bicêtre before Pinel conducted his reform? Michel Foucault contends that the modern philosophy of the mental institution was a product of the Age of Reason, that is, a reaction against "unreason": since the sight of the "fools" had become intolerable, they were hidden from the rest of mankind and incarcerated in large institutions together with the criminals, the cripples, and the beggars.

The other type of mental institutions derived from the monastery. Saint Vincent de Paul (1576–1660) founded institutions for delinquents where they received humane treatment. A glimpse of the treatment methods may be found in one episode of Abbé Prévost's novel, *Histoire de Manon Lescaut et du Chevalier des Grieux* (1731). As a punishment for his misbehavior, the Chevalier was committed to the Maison de Saint-Lazare in Paris; he was treated with firm kindness, isolation, silence, reading of serious books, and frequent conversations with tactful religious men. Actually we know very little about the results of this kind of therapy. But we are better informed about one of the institutions of the Order of the Hospitalers of Saint John of God, the Charité of Senlis, not far from Paris. The archives of that institution were preserved and utilized as the basis for an excellent monograph by Dr. Hélène Bonnafous-Sérieux.[4] The inmates were classified into four groups: the *insensés* (mental defectives), the *fous* (psychotics), the *libertins* (amendable behavior disorders, and the *scélérats* (dangerous psychopaths). According to their behavior the inmates were placed in one of three sections: the *force* (maximum security), the *semi-liberté* (medium security), the *liberté* (inmates enjoying certain privileges. The treatment was individualized. Emphasis was laid on the religious character of the institution, and inmates were invited to join the morning prayers and Mass. Every new inmate was given a pseudonym under which he was henceforth known; he had to wear a uniform that was more religious than prisonlike in character. The "director of the inmates" had a personal interview with each one of the inmates every day. There was a library, but no work therapy. There was no question of chains, whips, or cages and no demonology. Patients benefited from the medications known at that time and from a kind of rational psychotherapy. The registers show that many patients recovered, and it was reported that certain *libertins* improved their behavior.

We thus see that by the end of the eighteenth century there existed a small number of monastic institutions for a minority of privileged patients and a large number of incarcerative institutions where the poor lived in dreadful conditions. Meanwhile, a trend toward "sensitiveness" had appeared and a concern for the disadvantaged and rejected members of the great human family. An Englishman, John Howard, made a systematic inquiry into the state of the prisons all over Europe. Institutions were opened and methods devised for the education of the blind, the deaf-mute, later of the idiots and feebleminded. Humane treatment was introduced in a few public mental hospitals, notably by Vincenzo Chiarugi in Florence, by Joseph Daquin in Savoie, and by Philippe Pinel (1745–1826) in Paris.

Few pioneers of psychiatry reached a fame comparable to Pinel's. However, recent historical research has shown that the current accounts of his life are covered with legend. For many years Pinel lived in Paris as a medical journalist and translator with a small medical practice, working upon a *Nosographie philosophique* in the style of Boissier de Sauvages. In 1793 he was appointed as physician to the medical wards of the hospital of Bicêtre. The administration was in the hands of the "Governor" Pussin, an able man who was much concerned with the welfare of the patients and had already achieved several improvements. Pinel benefited from Pussin's experience and collaboration and gave his medical authority to the reforms. In 1795 Pinel was appointed to the Salpêtrière hospital; he introduced the same reforms there, too. In 1801 he published his *Medico-Philosophical Treatise on Insanity*, expounding a clear and humane system for the care of mental patients. Contemporaries depict Pinel as a small, shy, unassuming, often absent-minded man with speech difficulties, but kind and good-humored. In later years a number of young physicians became his enthusiastic pupils. Pinel acquired the prestige of a hero: he had achieved his humane reforms in the notorious Bicêtre (also a prison and house of arrest for vagabonds) during the Terror, the worst period of the Revolution. Thus, he in his later years was a legendary figure, and after his death he became the patron saint of several generations of French alienists. Actually he had achieved a decisive step toward the foundation of a scientific and humane psychiatry.

Modern Scientific Psychiatry

With the nineteenth century began a new era of psychiatry. Within the framework of medicine—now a fully scientific discipline—psychiatry became an autonomous specialty (as were already surgery and ophthalmology). Attempts were made to give psychiatry a firm basis, either in brain anatomy or physiology, or even in a general "science of man" (one example was Cabanis's *Rapports du physique et du moral de l'homme* in 1802). Old Galenism and speculative systems were discarded, but new semiscientific systems based on a mixture of speculation and sound observation appeared on the periphery of psychiatry. Such were phrenology and mesmerism: their adherents elaborated fantastic speculations, but also made useful discoveries and contributed in their own way to the development of psychiatry.

The evolution of modern scientific psychiatry may be divided into three periods, which largely overlap each other:

1. From 1800 to 1860 the center of psychiatric activity was the mental hospital. Its main concern was the description and classification of mental diseases, and the devising of a "moral therapy," together with the study of brain anatomy.

2. From 1860 to 1920 the center of psychiatric activity was the university psychiatric clinic. The trend was toward the elaboration of great psychiatric systems. A striking development occurred

in the study of neuroses and culminated in the creation of schools of dynamic psychiatry.

3. After 1920 came a kind of psychiatric explosion, with an almost boundless widening of the field of psychiatry and its division into a multitude of subspecialties.

Asylum Psychiatry (1800–1860)

The main feature of the period of asylum psychiatry was the appearance of a new type of physician who devoted his whole time and activity to the mentally sick. These men sought to found, operate, and reform "asylums" satisfying the demands of science and humanity and open to all categories of mental patients. The old Charités institutions were sometimes taken as models, but the developments took different forms in various countries.

The main school of alienists flourished in France among a group of physicians who had been Pinel's pupils or proclaimed themselves his successors. Most prominent among them was Jean-Etienne Dominique Esquirol (1772–1840). "At the beginning of the continuous development of psychiatry stands the outstanding personality of Esquirol," wrote Karl Jaspers. Esquirol devised a system of clear psychopathological concepts and gave their present definition to terms like illusion and hallucination, remission and intermission, dementia and idiocy. He inaugurated the use of asylum statistics for the study of the causes, course, and prognosis of mental disease. He proclaimed that "an insane asylum is a therapeutic instrument in the hands of an able physician, and our most powerful weapon against mental illness." He laid the foundation of a collective therapy, analyzing the influence exerted by the patients upon each other and grouping them accordingly. He wrote the first really scientific textbook on mental diseases, which was also the first to have illustrations, a book remarkable for its style and clear clinical descriptions. He became the director of the National Asylum of Charenton, which had been rebuilt according to his own plans on the site of a former Charité of the St. John of God Hospitalers. He visited every place where mental patients were kept in France, "house by house, hospital by hospital, prison by prison." These inquiries resulted in his magnificent illustrated treatise on the construction of asylums. Esquirol was also foremost in forensic psychiatry and the author of the French law on the insane (1838).

A young French physician, Laurent-Jessé Bayle, discovered in 1822 that "general paresis" was not a complication—as believed until then—but a specific mental disease with constant cerebral lesions. This encouraged the alienists eagerly to study cerebral pathology with the hope of discovering new specific entities. Their efforts, however, brought fewer rewards than did the clinical observation of mental patients. Thus, Etienne Georget distinguished from idiocy and dementia a third condition, "stupidity" (transient, reversible impairment of intelligence). It was observed that mania and melancholia could succeed each other and alternate in such a way that they must constitute one identical condition. This discovery was published simultaneously by Jean-Pierre Falret and by Jules Baillarger in 1854. Ulysse Trélat gave fine descriptions of the *folie lucide* and Charles Lasègue of the delusions of persecution.

A new trend was open by Jacques-Joseph Moreau de Tours's book on hashish (1845), showing that this drug produced a state of delusions and hallucinations that—he believed—was for a few hours identical with the manifestations of insanity and afforded an experimental approach to its study. On this basis he came to consider the dream as the key to the knowledge of mental life, and the *désagrégation* (in today's language, "regression") as the basic psychopathological process.

Bénédict-Augustin Morel (1809–1873) published in 1852 a description of *démence précoce*, not as a disease entity, but as a peculiar form of rapid evolution of mental illness toward a state of severe emotional impairment (the term was later misunderstood to mean "dementia at an early age"). He also described phobias under the name of *délire émotif* (emotional delusions). Morel inaugurated the psychiatric application of physical anthropology and genealogy. He photographed patients, measured their skulls, took plaster casts of their heads, investigated their lives and the lives of their parents and ancestors. This was the basis for his theory of degeneracy. Degeneracy, Morel said, was a gradual process; its manifestations became worse from one generation to the next, until the descendants were sterile, which automatically ended the process. But Morel also discussed the ways of checking the progress of degeneracy and making it reversible; in these studies were the beginnings of social anthropology and mental prophylaxis. Morel also introduced the psychiatric use of etherization, that is, studying a patient's reactions under the administration of ether as a means of diagnosing between hysteria and organic disease (as today with narcoanalysis).

In the meantime German psychiatry, starting from a very different background, had undergone its own evolution. It seems that the German asylum tradition issued more from the prison than the monastery, and this is, according to Kirchhof,[19] the reason for their tremendous use of coercive measures. On the other hand, German psychiatry underwent the strong influence of Romanticism and of Schelling's philosophy of nature. Whereas French alienists based their nosology on the empirical study of clinical cases, their German colleagues often tried to isolate abstract types deduced from a psychological frame of reference. German psychiatry soon developed into two trends: the *Physiker* (organicists) and the *Psychiker* (who emphasized the psychological origin and treatment of mental illness).

Among the *Physiker* were Johannes Friedreich, also known as one of the first historians of psychiatry, and Maximilian Jacobi, author of a treatise on the architecture of mental hospitals.

The *Psychiker* took over from Stahl the notion of the role of emotions in the etiology of mental illness. Each one of them developed his own views about the human mind and the nature of mental illness with great originality and audacity. Johann Christian Reil (1759–1813), though also a brain anatomist, expounded in his *Rhapsodien* (1803) an extraordinary program of psychotherapy for mental illness, including the use of a "therapeutic theater." Johann Gottfried Langermann (1768–1832) is considered a pioneer of occupational therapy. Johann Christian Heinroth (1773–1843) pointed out the role of "sin" (in modern language, guilt feelings) in psychopathology; he, too, devised an extensive program of psychotherapy for mental illness. Karl Wilhelm Ideler (1795–1860) emphasized the effects of frustrated sexual drives in the psychogenesis of mental diseases. Heinrich Wilhelm Neumann (1814–1884) devised an original system of medical psychology. He discussed the relationship between anxiety and frustrated drives, and he described the masked manifestations of sexuality among psychotic patients. All these men were skeptical in regard to the current psychiatric nosologies; they shared the belief in the emotional causation of mental illness and the possibility of psychotherapy for even severe psychotics. They anticipated much of what Eugen Bleuler, C. G. Jung, and the psychoanalysts were to teach as great novelties several generations later.

One belated representative of German asylum psychiatry was Karl Ludwig Kahlbaum (1828–1899), who spent his career in remote asylums and whose publications were ignored for a long time. His main contribution to psychiatry was his descriptions of hebephrenia and catatonia (the former delegated to his collaborator, Ewald Hecker, in 1871; the latter in his own monograph, 1874). The validity of Kahlbaum's concepts was accepted later by Emil Kraepelin who made hebephrenia and catatonia two subforms of dementia praecox.

In Belgium Joseph Guislain taught that almost all mental diseases had emotional causes and were rooted in an all-pervading, underlying feeling of anxiety. In his institution in Ghent he organized an excellent system of work therapy. Wonderful accounts of moral therapy were also published by those who visited Pietro Pisani's institution in Palermo, Sicily.

In England William Tuke's pioneer work in the "Retreat" near York was pursued by descendants and their collaborators. One of the latter, M. Allen, founded his own institution at High Beach and wrote a highly instructive account of his therapeutic methods. To give only one instance: patients with violent agitation, accompanied by an attendant, were sometimes allowed to ramble about and scream in a forest for a whole day, with a subsequent lessening of their agitation. Henry Maudsley (1835–1918) began his *Physiology and Pathology of the Mind* (1867) with a chapter about the unconscious and the impossibility of exploring the human mind through introspection. Maudsley advocated the use of a "physiological method" that would take into account the "plan of development" of the mind (in the animal, the infant), the study of its "degeneration" (in the dream, delirium, insanity), and of its "progress and regress" throughout history.

In the United States Benjamin Rush (1745–1813) introduced "moral treatment" at the Pennsylvania Hospital in Philadelphia. Other mental institutions operated along similar lines. Isaac Ray (1807–1881) founded American forensic psychiatry, and Thomas Kirkbride (1809–1883) is famous for his book on the construction of mental hospitals.

Throughout the period of asylum psychiatry, a noteworthy discrepancy is often found between the theoretical views of the alienists and the real state of their mental hospitals, as pictured by visitors. The term "moral treatment" was given many conflicting meanings. In France François Leuret (1797–1851) used to treat delusional patients with icy cold showers until they came to recognize their "errors"; he called that method "moral treatment."

Outside asylum psychiatry ran a kind of underground current centered around the use of "magnetic sleep," later called hypnosis. Mesmer, Puységur, and a long succession of followers treated many neurotic and psychosomatic patients. They developed a full-fledged First Dynamic Psychiatry, which was later to give birth to the modern schools of dynamic psychiatry.

University Clinic Psychiatry (1860–1920)

The historical examples of Morel and Kahlbaum showed how difficult it was to achieve important creative work in a mental asylum. For that reason the center of scientific activity gradually moved to the psychiatric university clinic, a threefold institution for treatment, teaching, and research. For some reason German psychiatry outdistanced France around 1860 (with the exception of the field of neuroses), but as time went on, psychiatry became more international.

A new type of psychiatrist arose, the university psychiatrist who worked with a team of collaborators and pupils and was provided with well-equipped laboratories. Techniques of brain anatomy were improved, new biological methods and experimental psychology were resorted to. The field of psychiatry expanded swiftly. Among the new clinical pictures were alcoholism, morphinism, cocainism, neurasthenia, and traumatic neuroses.

Wilhelm Griesinger (1817–1868) is usually considered the founder of German university psychiatry. Possessing an eclectic mind, he introduced Herbart's psychological concepts into psychiatry, but his contention that "mental diseases are brain diseases" was taken as a slogan by the adherents of organicism. At that time Rokitansky and Virchow were laying the foundations of cellular anatomo-pathology, which appeared to be the one firm basis of all medicine. This encouraged psychiatrists to rebuild psychopathology on a similar basis. Theodor Meynert (1833–1892) supplemented his objective anatomical findings with hypotheses on the functional opposition between the brain cortex and brain stem and on the role of physiological brain disturbances. However, he was not so exclusively organicist as he is usually depicted; he pointed to the psychogenesis of homosexuality and other sexual deviations. Carl Wernicke (1848–1905) attempted a brilliant synthesis of brain anatomy, physiology of the nervous reflex bow, and associationist psychology. He considered the brain an associative organ and the soul as being "the sum of all possible associations." Only the most elementary mental functions are localized in the cortex, he said; the higher ones, including consciousness, are products of the associative activity. Unfortunately these men and many of their colleagues fell into "brain mythology," the tendency to explain all mental phenomena in terms of real or fictitious brain structures.

The credit for overcoming "brain mythology" as well as the current nosological confusion belongs to Emil Kraepelin (1856–1926). An eclectic researcher, Kraepelin combined neuroanatomical research with experimental psychology and a thorough investigation of the patients' life history. In the sixth edition of his textbook (1899) he defined the two great endogenous psychoses: manic-depressive psychosis (the old "circular illness") and dementia praecox. The latter comprised three subforms: hebephrenia, catatonia, and paranoia. Contemporaries felt that Kraepelin had shed definitive light on the chaos of psychiatric nosology. His classification was widely adopted in Europe and America.

In Zurich Auguste Forel (1848–1931) started with discoveries in brain anatomy but shifted toward a dynamic concept of the mind. Among his pupils were Eugen Bleuler and Adolf Meyer. Eugen Bleuler (1857–1930) sought a deeper psychological understanding of psychotic patients; he revolutionized the concept of dementia praecox, which he called schizophrenia. He introduced many new terms such as autism, ambivalence, and schizoidism.

In France Valentin Magnan (1835–1916) extended the theory of mental degeneracy so widely that this diagnosis covered almost the whole realm of psychiatry. His fame, however, was the result of his clinical studies on alcoholism. But the main French contribution to psychiatry was the work of a neurologist, Jean-Martin Charcot (1825–1893), and of an internist, Hippolyte Bernheim (1840–1919). These two men gave the official stamp of approval to the use of hypnosis, thus making it a respectable procedure, though their theoretical concepts and therapeutic use of the phenomenon differed widely. A large part of the teaching of the old magnetists and hypnotists was rediscovered. From about 1880 to 1900 hypnotism and suggestion were widely investigated and utilized.

In England probably the most original contribution to psychiatry was made by a neurologist, Hughlings Jackson (1834–1911). He conceived of man's neurological and mental structure as a hierarchy of functions, resulting from differentiation and perfection through evolution. Nervous and mental disorders were explained by "dissolution" (regression) to inferior, older levels of functions. In such disorders Jackson distinguished negative symptoms resulting from the loss of a higher function and positive ones resulting from the activation of functions of a lower level. Jackson's theory had a great impact on many neurologists and psychiatrists.

In the United States George Beard (1839–1882) described "neurasthenia," a state of physical and mental exhaustion, as a neurosis of modern, and especially American, life. Beard's ideas met with great success. S. Weir Mitchell (1829–1914) devised a standard method for the treatment of neurasthenia by means of rest, isolation, feeding, and massage. After the turn of the century the main American contribution to psychiatry came from Adolf Meyer (1866–1950). In his individualized, psychobiological approach, Meyer interpreted mental disorders as being faulty reactions to life situations, which he tried to understand with the help of a thorough study of the patient's life history and development.

Gradually several branches of psychiatry expanded and tended to become sciences in their own right. Thus, the field of sexual pathology had been explored by forensic psychiatrists. Now a brilliant systematization was effected by Richard von Krafft-Ebing (1840–1902). His *Psychopathia Sexualis* (1886) was the starting point of the flourishing new discipline of sexology. Another forensic psychiatrist, Cesare Lombroso (1836–1909) founded a school of criminal anthropology in Italy. He conceived of it as being a branch of biological medicine, but when sociologists and lawyers joined his group the newer criminology became an autonomous science.

We have seen that Charcot and Bernheim had incorporated into medicine a part of the teachings of the First Dynamic Psychiatry. Their attempts, however, were short-lived. The first man who undertook to build a new dynamic psychiatry was Pierre Janet (1859–

1947). Janet was able to correlate hysterical symptoms with "subconscious fixed ideas" and bring a cure by means of a "psychological analysis." He then elaborated a new theory of the neuroses, of mental energy, of the "hierarchy of tendencies," and eventually a vast conceptual model that illuminates in some way virtually every phenomenon of the mind. But whereas Janet had kept closely within the bounds of traditional psychiatry, Freud and the founders of the newer dynamic schools openly broke with official medicine.

The work of Sigmund Freud (1856–1939) is so widely known that it is hardly necessary to recall its main features. Among his theories are those of the dynamic unconscious and the deciphering of symptoms, the interpretation of dreams, repression and other defense mechanisms, the development, fixations, and regressions of libido, child sexuality, the superego, the Oedipus complex, the sexual origin of neuroses. Freud introduced a new approach to the unconscious by means of free associations, and analysis of resistance and transference: he developed a new approach to therapy by means of a development and resolution of a transference neurosis. From the beginning Freud made of psychoanalysis a movement, then an organization of a new type that had no parallel in modern times, that is, a close, tightly structured organization with its official doctrine and rules for membership, including a long initiation in the form of a didactic analysis. Gradually psychoanalysts extended their interpretations to sociology, anthropology, literature, art, religion, education, and all manifestations of cultural life. Some of its adherents had the feeling that the history of psychiatry was now divided in two periods: before and after Freud.

Psychoanalysis was the starting point of a vast development of dynamic psychiatry. Among Freud's disciples were orthodox and deviant groups. Both Alfred Adler (1870–1937) and Carl Gustav Jung (1875–1961) developed a system that was basically different from psychoanalysis. Adler's individual psychology is a pragmatic or concrete system centered around the dialectic of the individual's striving for superiority versus his community feeling. Jung's analytic psychology centers around the concepts of the collective unconscious, with the archetypes, and the process of individuation.

The new dynamic psychiatric systems grew outside the realm of "official" medicine. Their theories extended far beyond the traditional limits of psychiatry. This was a manifestation of a new trend that could be called the "psychiatric explosion."

The Psychiatric Explosion (1920–)

After World War I it became obvious that the psychiatric university clinic could no longer remain the only center of psychiatric progress. Kraepelin founded in Munich the first *Forschungsanstalt* (psychiatric research institute). It was discovered that considerable impetus could be given to scientific progress if specialists could be liberated from daily treatment and administrative work and devote their whole time to carefully planned research projects. Similar institutes were founded in Germany, Russia, America, and later in other countries.

From that time on, the field of psychiatry continually expanded. Former branches became autonomous disciplines (for instance, child psychiatry, sexual psychopathology, clinical criminology), and new branches were founded. Among the latter were genetics (Rüdin, Luxemburger), biotypology (Pende, Kretschmer, Sheldon), reflexology (Pavlov) electroencephalography (Berger), psychiatric endocrinology (Manfred Bleuler), intelligence testing (Binet and Simon, Terman), projective testing (Rorschach, Murray, Szondi), organo-dynamic psychiatry (Henri Ey), general psychopathology (Jaspers), phenomenology (Minkowski), existential analysis (Binswanger). There also was a great development in social psychology, transcultural psychiatry, family psychiatry, psychosomatic medicine, and a variety of approaches to community psychiatry.

Among the new therapeutic methods we may mention perfected forms of ergotherapy

(Herman Simon), malaria therapy (Wagner von Jauregg), narcotherapy (Klaesi), narcoanalysis (Horsley), cardiazol therapy (Meduna), insulin shock therapy (Sakel), electric shock therapy (Cerletti), psychosurgery (Egas Moniz), and last but not least, the pharmacotherapy of mental diseases (Laborit and many others). In psychotherapy a great variety of new methods were introduced, in addition to the adaptation of psychoanalysis to the treatment of children and psychotics. We may mention Rogers's nondirective therapy, Frankl's existence therapy, Schultz's autogenous training, and various methods of behavior therapy. Not less numerous are the varieties of group therapy, therapeutic communities, and psychodrama.

The description of all these clinical and therapeutic developments will be an absorbing and arduous task for psychiatric historians of the future.

Meantime, the traditional principles that had ruled psychiatry for a long time were shaken. Henri Ey had warned that the boundless expansion of psychiatry—which he called "panpsychiatry"—would inevitably lead to a reaction in the form of the dissolution of psychiatry. This is exactly what present-day psychiatry is witnessing. A number of "antipsychiatrists" now proclaim that mental illness does not exist; what is called by that name is nothing but the artificial product of social repression; psychiatric treatment is nothing but a disguised form of punishment, of social violence; a sojourn in a mental hospital can never cure a patient but only make his condition worse; mental hospitals should be closed and psychiatrists take up another profession. As Henri Ey had predicted, "panpsychiatry" has led to "antipsychiatry."

Thus, the outcome of 25 centuries of efforts is that the basic principles of psychiatry are in need of careful revision. This is not unique to psychiatry and reminds one of what Bachelard once wrote: "All scientific knowledge must be, at every moment, reconstructed. . . . In the work of science only, one can love that which one destroys; one can continue the past while denying it; one can honour one's master while contradicting him."

Bibliography

1. ACKERKNECHT, E., "Problems of Primitive Medicine," *Bull. Hist. Med.*, *11*:503–521, 1942.
2. ———, *A Short History of Psychiatry*, rev. ed., Hefner, New York, 1968.
3. ALEXANDER, F., and SELESNICK, S., *The History of Psychiatry*, Harper & Row, New York, 1966.
4. BONNAFOUS-SÉRIEUX, H., *La Charité de Senlis*, Presses Universitaires de France, Paris, 1936.
5. BROWNE, E., *Arabian Medicine*, Cambridge University Press, Cambridge, 1921.
6. CAMPBELL, D., *Arabian Medicine and Its Influence in the Middle Ages*, 2 vols., Kegan Paul, London, 1926.
7. CASTIGLIONI, A., *A History of Medicine*, Knopf, New York, 1941.
8. DELGADO ROIG, J., *Fundaciones psiquiatricas en Sevilla y Nuevo Mundo*, Editorial Paz Montalvo, Madrid, 1948.
9. DE MORSIER, G., *Essai sur la genèse de la civilisation scientifique actuelle*, Georg, Genève, 1965.
10. ELLENBERGER, H., *The Discovery of the Unconscious: The History and Evolution of Dynamic Psychiatry*, Basic Books, New York, 1970.
11. FISCHER-HOMBERGER, E., *Hypochondrie, Melancholie bis Neurose*. Huber, Bern, 1970.
12. HARMS, E., *Origins of Modern Psychiatry*, Charles C Thomas, Springfield, Ill., 1967.
13. HUNTER, R., and MACALPINE, I., *Three Hundred Years of Psychiatry*, Oxford University Press, New York, 1963.
14. HURRY, J. B., *Imhotep*, Oxford University Press, New York, 1926.
15. JACKSON, S. W., "Galen—On Mental Disorders," *J. Hist. Behav. Sci.*, *5*:365–384, 1969.
16. JAEGER, W., *Diokles von Karystos. Die griechische Medizin und die Schule des Aristoteles*, De Gruyter, Berlin, 1938.
17. JOLY, R., *Le niveau de la science hippocratique*, Les Belles Lettres, Paris, 1966.
18. KARCHER, J., *Felix Platter, Lebensbild des Basler Stadtarztes*, Helbling & Lichtenhahn, Basel, 1949.

19. KIRCHHOF, T., *Geschichte der Psychiatrie*, Franz Deuticke, Leipzig, 1912.
20. LAIGNEL-LAVASTINE, M., and VINCHON, J., *Les malades de l'esprit et leurs médecins du XVIe au XIXe siècle*, Maloine, Paris, 1930.
21. LAIN ENTRALGO, P., *Historia de la medicina*, Ed. Cientifico Medico, Barcelona, 1954.
22. LECHLER, W. H., *Neue Ergebnisse in der Forschung über Philippe Pinel*, Diss med., Munich, 1960.
23. LEIBBRAND, W., *Heilkunde. Eine Problemgeschichte der Medizin*, K. A. Freiburg, Munich, 1953.
24. ———, and WETTLEY, A., *Der Wahnsinn. Geschichte der abendländischen Psychopathologie*. Alber, Freiburg, 1961.
25. LESKY, E., *Die Wiener medizinische Schule im 19. Jahrhundert*, Graz, Böhlau, 1965.
26. NEUBURGER, M., *Handbuch der Geschichte der Medizin*, 3 vols., Fischer, Jena, 1902–1905.
27. PAGEL, W., *Paracelsus*, Karger, Basel and New York, 1958.
28. RIESE, W., *A History of Neurology*, M.D. Publications, New York, 1959.
29. ROSEN, G., *Madness in Society*, University of Chicago Press, Chicago, 1968.
30. SEMELAIGNE, R., *Les Pionniers de la psychiatrie française*, 2 vols., Baillière, Paris, 1930–1932.
31. SIGERIST, H. E., *The Great Doctors*, Norton, New York, 1933.
32. ———, *A History of Medicine*, 2 vols., Oxford University Press, New York, 1951, 1956.
33. SINGER, C., *A Short History of Medicine*, Clarendon Press, Oxford, 1928.
34. SUDHOFF, K., *Essays in the History of Medicine*, Medical Life Press, New York, 1926.
35. WALSH, J. J., *Medieval Medicine*, Black, London, 1920.
36. ZILBOORG, G., *A History of Medical Psychology*, Norton, New York, 1941.

CHAPTER 2

AMERICAN PSYCHIATRY FROM ITS BEGINNINGS TO WORLD WAR II

Nolan D. C. Lewis

PSYCHIATRY HAS DEVELOPED in combination and association with the other branches of medicine to the extent that its borders, as a specialty, in certain areas are not sharply outlined, and, moreover, most of its particular concepts and aims are exceedingly difficult to trace accurately in terms of their chronological appearance.

Medicine is seen within a matrix of economic, political, social, and cultural elements. It is but one aspect of the state of general civilization of a period, and its particular philosophy and practice are determined to a considerable extent by the current cultural conditions of its locality and time.

To attempt to trace the complex growth of psychiatric thought is an interesting experience. The most satisfactory way to understand a complicated phenomenon is to follow its pathway of genesis, but to cover the span of time indicated by the title of this chapter, one must resort to a condensed style of communication. The topic includes more than a hundred years, during the early parts of which are revealed only a few records of basic importance in our particular field; however, there are some outstanding persons and events that appear as bright promises of progress to come in later years.

Since the aim in this chapter is to indicate some of the most important trends of thought that constitute the background of present-day psychiatry in this country, no attempt has been made to present a complete history of psychiatry in America, and thus many workers and their accomplishments have been omitted. In psychiatry as well as in other branches of medicine, one can find pioneers, men whose energy and curiosity were so great as to lead them to devote their lives to the promotion and extension of knowledge beyond the existing outposts of what was accepted and practiced. Benjamin Rush (1745–1813), "America's first psychiatrist," was one of these men.

At the termination of the American Revolution in 1783, Dr. Benjamin Rush joined the staff of physicians at the Pennsylvania Hospital in Philadelphia, which event is often referred to as being the actual beginning of American psychiatry. Although at this time Rush was only 38 years old, he was already a very famous physician. He had been a precocious student, having graduated from the New Jersey College (now Princeton) at the age of 15. He received his doctor's degree from the University of Edinburgh at the age of 22. While at Edinburgh he studied under William Cullen (1710–1790), who was one of the most famous professors of medicine of his time, with an interest in the nervous system and in behavior phenomena. Cullen was a follower of Locke, but he originated a new doctrine of physiological significance. In his system irritability of the nervous system was of special importance, and the causes of mental disorders were within the individual, that is, endogenous. Insanity was a pathological condition of the mind and not some force entering from without, and, therefore, its expressions should be interpreted in the light of normal psychological functions. His *First Lines of the Practice of Physik* (London, 1777) described paranoid forms of mental disorder that he called "vesaniae." Rush was introduced to Cullen by a letter from Benjamin Franklin, a fortunate occurrence since Cullen exerted a strong influence on Rush's subsequent career, which was thus oriented in the theory and therapy then prevalent in Scotland and England.

Dr. Rush, equipped with a keen mind and the best training available at the time, attacked the problems of mental disorder, a subject that was infiltrated with superstition and ignorance as well as characterized by neglect and actual brutality. He was apparently the first American physician to approach the investigation and treatment of mental disorders from a scientific viewpoint, the first American teacher to propose an original systematization of psychiatry, and the first to write a general treatise on psychiatry in America. This book, published in 1812 under the title *Medical Inquiries and Observations upon Diseases of the Mind*, remained the only American book of its kind for 70 years.*

Rush's principal remedies were purgatives, emetics, and bloodletting, of which he was a particularly strong advocate. The condition of the blood circulation in the brain was a prominent focus for his theories and therapy. He devised two curious instruments, the gyrator, based on the principle of centrifugal action to increase cerebral circulation, and the tranquilizer (1810) "to obviate these evils of the 'strait waistcoat' and at the same time to obtain all the benefit of coercion." These instruments, which were certainly rough on the patient, were decidedly "shock" measures—a form of shock therapy, and, like all shock therapies including the modern ones, were interpreted as effecting cures.

Benjamin Rush was one of the earliest, if not the first, to recommend labor and prison sentences instead of capital punishment for criminal offenses, and he consistently advocated a modification of the harsh and cruel elements in the system of criminal jurisprudence. In the minds of many medical historians, he did for America what Pinel and Esquirol did for France. His biographers are in accord that he, like many originators and creators, was a strongly opinionated man, difficult to get along with, frankly intolerant of anyone who did not agree with his theories and practices, and at times actively quarrelsome. He was a very learned man who not only served the poor of his community but also was a leading patriot and a signer of the Declaration of Independence. His was an overwhelming personality, fully justifying the title by which he is known today, "The Father of American Psychiatry."

Dr. Adolf Meyer (1866–1950) in 1945 made the following pertinent comment:

> Had Benjamin Rush had a successor akin to his own spirit American psychiatry may have had the

* Benjamin Rush is also claimed by professional chemists as the "Father of American Chemistry" on the basis that he was the first professor of chemistry in America, his post being the College of Philadelphia (1769–1789).

lead over the European and its actually colonial development. It took a hundred years before the step towards a culture for responsible life began to assert itself. . . . I do not know what the names and teachings of the immediate successors were. . . . Psychiatry may see the influence of the 1840's in J. P. Gray.[28]

Psychiatry has always been more hospital-centered in England and other countries than it has here in America, but there were some important early, as well as later, developments of this nature that are of historical significance. For example, the old San Hipolito, established in Mexico in 1566 by the philanthropist Bernardo Alvares, was the first hospital in the Americas for the study and care of mental disorders and, of course, one of the earliest for this purpose in the whole world. At first this institution was not devoted exclusively to the treatment of mentally disordered patients, but before long the increase in the number of these cases made it necessary to restrict the admissions to nervous and mental patients.

It was nearly 200 years before the first institution to care for the mentally disordered in the American states was organized. In 1752 the Pennsylvania Hospital in Philadelphia received insane patients, and this was followed by a few other public and private institutions. In 1773 the first public hospital exclusively for mental patients was established at Williamsburg, Virginia, then the capital of the colony. For half a century mental patients were kept at the Pennsylvania Hospital in cells in different parts of the building, but in 1796 a new wing was provided for these patients, and at the end of another 50 years a separate building was constructed for them. The Bloomingdale Asylum was another of the early institutions for the mentally disordered. It was opened in the New York City area in 1821 as a separately managed hospital, although it had its beginnings in the earlier founding of the New York Hospital. It continues today in White Plains, New York, as the Westchester Division of the New York Hospital.

Out of a meeting held in 1844 by a group called the "original thirteen" came the organization known as the Association of Medical Superintendents of American Institutions, which was the first national society of physicians in the country. From 1893 to 1921 it was known as the American Medico-psychological Association, and is now the American Psychiatric Association, with a large and ever growing membership. In the year of its origin (1844) the *American Journal of Insanity*, later to become *The American Journal of Psychiatry*, the official organ of the Association, was established under the editorship of Amariah Brigham (1798–1849), one of the original thirteen superintendents. The *Journal* has continued to be a leader in the field by virtue of the management of a line of distinguished editors.

In their day the "original thirteen" made a lot of psychiatric history, both as a group and as individuals. It is well to note some of their particular contributions as pioneer workers.

Samuel B. Woodward (1787–1850) was the first president of the Association of Medical Superintendents of America and a strong force in its organization. He aided in establishing the Retreat for the Insane at Hartford, Connecticut. In 1832 Massachusetts was the first state in New England to build a state hospital. Woodward supervised the construction of this hospital and became its first superintendent. He believed in specialized medical care for alcoholics, and in 1838 his volume entitled *Essays on Asylums for Inebriates* was published. He was elected to the Connecticut state legislature in 1830.

Isaac Ray (1807–1881) was an avid student of scientific matters, both in America and in Europe, and a prolific writer. He was one of the most eminent and versatile of the group. In 1841 he became medical superintendent of the State Hospital for the Insane at Augusta, Maine, and later of the famous Butler Hospital in Providence, Rhode Island. He wrote a book on mental hygiene in 1863 and was interested particularly in the relationship between legislation and mental disease. As a matter of fact, the first book in the English language concerning forensic psychiatry was published by Ray in 1838; this book made a profound impression and established him as a pioneer and leader in medicolegal matters. The title of this book was *Treatise on the*

Medical Jurisprudence of Insanity. It was used as an authoritative text for court work for nearly 50 years.

Amariah Brigham (1798–1849), founder of the *American Journal of Insanity*, was a distinguished author, teacher, and hospital administrator whose influence for the better understanding and care of the mentally sick was widespread. He wrote lucidly on such diverse subjects as cholera, religion, clinical neurology, and clinical psychiatry. At one time he taught anatomy at the College of Physicians and Surgeons in New York City. In 1842 he was appointed superintendent of the State Lunatic Asylum at Utica, New York, the first New York State public institution for the mentally disordered. This hospital, under his inspiration and effort, became an outstanding training center for medical superintendents. His book entitled *Remarks on the Influence of Mental Cultivation upon Health* had a remarkably wide American and European distribution.

Pliny Earle (1809–1892), in 1844, was appointed superintendent of the Bloomingdale Asylum in New York. In 1863 he became professor of psychological medicine at the Berkshire Medical Institution in Pittsfield, Massachusetts. This may have been the first chair of psychiatry in a medical school in this country, and it was certainly one of the very early additions of psychiatry to a medical curriculum in the United States. From 1864 to 1885 he was superintendent of the State Lunatic Hospital at Northampton, Massachusetts.

Dr. Earle was an able hospital administrator and an outstanding contributor to the literature on mental disease. He was well versed in the European psychiatry of the day, introducing new ideas from Germany and France. He published many statistical papers, psychiatric reviews, and historical essays, and lectured widely. In 1877 he published a statistical treatise entitled *The Curability of Insanity.* He had a critical mind and a far-reaching vision, recommending small hospitals for intensive treatment and separate ones for the incurable patients. He was also an early advocate of family care and occupational therapy.

Thomas S. Kirkbride (1809–1883), after receiving his medical degree from the University of Pennsylvania in 1832, became resident physician at Friends Asylum, followed by a residency at the Pennsylvania Hospital, and later by the appointment as superintendent of the new Pennsylvania Hospital, a position that he held for 43 years during which time he was a progressive advocate of many reforms in the care and therapy of mental patients and served as an active participant in community affairs.

Dr. Kirkbride published many important books and articles. In 1847 he published an article on hospital construction that made such an impression that for at least 50 years the "Kirkbride type" of hospital building was adopted by many institutions. A notable number, if not most, of the "Kirkbride type" buildings are still in active use today in the older mental hospitals. Dr. Earl Bond[4] has written a most informative book, *Dr. Kirkbride and His Mental Hospital*, about this remarkable personality and his achievements.

William M. Awl (1799–1876), after some years in general practice, settled in Columbus, Ohio, in 1833, where he became active in the organization of a mental hospital, the Ohio State Asylum for the Insane. He became its first superintendent in 1838 and remained at this post for 12 years, when he was removed by political factions. He then returned to private practice and also became physician at the Ohio Institute for the Blind.

Although Dr. Awl possessed keen mental abilities, he wrote very little on mental disorders. He devoted much time and thought to the organization of the state asylum and the institute for the blind, and was concerned with the organization of the Ohio Medical Society.

Luther V. Bell (1806–1862) became superintendent of the McLean Asylum in Somerville, Massachusetts, when it had been in operation for 19 years. He received his medical degree from Dartmouth at the early age of 20, was in general practice for a time in New Hampshire, and also served in the state legislature, as well as participating in establishing the State Asylum at Concord before taking the post at McLean Asylum.

Dr. Bell wrote on a wide range of medical subjects, including an original description of a form of acute mania, which was named after him (Bell's disease). During his active professional lifetime he received many awards and honors and was a participant in a variety of local and national civic activities. He was the only one of the "original thirteen" to have served in the Civil War, during which he died, in 1862, in an army camp near Washington, D.C., where he was a medical inspector.

John S. Butler (1803–1890), after receiving his medical degree from the Jefferson Medical College in Philadelphia in 1828, entered private practice in Worcester, Massachusetts, until 1839, when he was appointed superintendent of the Boston Lunatic Hospital. Here he soon instituted various reforms and striking improvements in the hospital treatment of patients. In 1843 Dr. Butler was elected superintendent of the Retreat for the Insane at Hartford, Connecticut, to succeed Dr. Amariah Brigham, who had gone to Utica, New York. Dr. Butler remained at the Retreat for nearly 30 years, during which time he demonstrated outstanding leadership and a progressive spirit. He remained active in medical matters for another ten years after retirement from hospital service.

Nehemiah Cutter (1787–1859) received his medical degree from Yale University in 1817 and entered private practice in Pepperell, Massachusetts, where he gradually built up a specialty of treating the mentally ill to the point of organizing a private hospital known as the Pepperell Private Asylum. This hospital was successful despite the prejudice extant against private mental institutions. The hospital burned down in 1853 and was not rebuilt; however, Dr. Cutter continued his successful career as a practicing physician.

Charles H. Stedman (1805–1866) graduated from the Harvard Medical School in 1828, and in 1830 was appointed resident surgeon at the United States Marine Hospital at Chelsea, Massachusetts, where he worked for ten years. In 1834, being keenly interested in the theories of Gall and Spurzheim, he edited a translation of Spurzheim's book on brain anatomy. Dr. Stedman started a private practice in Boston in 1840, and two years later he became superintendent of the Boston Lunatic Hospital, following the resignation of Dr. Butler. Here he proved to be an exceptionally able administrator and student of mental disorders. He was also particularly active as a staff member and consultant in medicine and surgery in various other hospitals in the city.

Samuel White (1777–1845) opened an office for private practice at Hudson, New York, in 1797 at the age of 20. He apparently never attended a medical school but gained his professional education by means of the apprentice system, which was possible in those early times. In 1830 he opened a private hospital, called Hudson Lunatic Asylum, where he functioned successfully until his death. From the historical fragments that have been recorded about his life and ideas, one may assume that he was progressive in thought, was accepted as an authority by his contemporaries, and that he utilized the best methods known in practice.

Francis T. Stribling (1810–1874) obtained his medical degree from the University of Pennsylvania in 1830. He was in private practice for a time in Staunton, Virginia, his native town, and at the age of 26 was appointed physician to the Western Lunatic Asylum of Virginia. Dr. Stribling left very little published work, but he was an able clinician and was one of the early advocates of training for attendants, of prompt treatment for acute cases, and of occupational therapy.

John M. Galt (1819–1862) was a young man of 22 when he received his medical degree from the University of Pennsylvania in 1841, and in this year he was made the first superintendent of the Williamsburg Asylum, which had been established many years before (1773). He was an unusually erudite man who read many languages and was thus able to bring important psychiatric literature to the attention of American physicians. In 1846 he published a book entitled *The Treatment of Insanity*, which contained a summary of about all of what was known of mental disorders. Other books and many articles came from his pen, and, although his life span was comparatively short, his was a powerful influence for progress since he advocated the proper keeping of records, the performance of autopsies, the value of research, and the need for changes in the legal aspects of psychiatry. He utilized in his hospital several of the adjunct therapies that are in vogue today.

Of these founders of the Association of Medical Superintendents of American Institutions for the Insane, six were in charge of state institutions, five directed incorporated hospitals, and two were the proprietors of private hospitals. From the intelligence of this remarkable group of men and the examples they set in their respective areas have developed many of our modern attitudes and endeavors in the specialty. At the time they formed their national association, European psychiatry, although still relatively young as a medical discipline, had progressed to a point where the mentally sick were no longer treated as monsters or criminals but as patients suffering from very serious disorders.

Another important powerful force in the evolution of American psychiatry, for many years after 1841, was Dorothea Lynde Dix (1802–1887) of Boston. This remarkable woman, a retired schoolteacher, became "shocked" by how she saw mental patients cared for in certain places, where they were locked up in filthy cells and were not only neglected but often treated brutally as well. Through her widespread activities not only in the United States but also in Canada, the British Isles, and on the European continent, reforms were instituted successfully through various government agencies. She is credited with having been directly responsible for creating or extending the facilities of a total of 32 hospitals. In this country 20 states built or enlarged mental hospitals as a result of her personal efforts. The Government Hospital for the Insane (now St. Elizabeths Hospital) in Washington, D.C., is among those that she aided in establishing.

On the second day of the second annual meeting of the American Neurological Association held at the College of Physicians and Surgeons in New York City in 1876, with 16 members present, Dr. George M. Beard (1839–1883) presented a paper entitled "The Influence of Mind in the Causation and Cure of Disease and the Potency of Definite Expectation." This paper was remarkable in its originality as it dealt clearly with what is now known as psychosomatic medicine. Such a presentation was new and startling, in fact so startling that the famous neurologist, Dr. William A. Hammond, remarked during the discussion that if the doctrine advanced by Beard was to be accepted, he "should feel like throwing his diploma away and joining the theologians."

According to the original report in the *Transactions of the Association*, "Dr. Beard maintained that disease might appear and disappear without the influence of any other agency than some kind of emotion. Mental qualities, like drugs, could neutralize therapeutics and they could also increase the effect of drugs. Fear, terror, anxiety and care, grief, anger, wonder and expectation were regarded as the most likely to produce disease." To the criticisms referring to mental therapeutics as being nothing new, Beard replied that he had not implied that psychotherapy was new, but that it had not been carried out in a systematic manner by the profession. Although there was much vigorous critical discussion and skepticism, the brilliant Beard, pioneer in the field, defended his position with confidence and ability. When Beard's famous paper entitled "Neurasthenia (Nervous Exhaustion) and Morbid Fears as a Sympton of Nervous Disease" appeared (1869), his term "neurasthenia" was gradually adopted over the whole world as "Beard's disease," and he was duly recognized as a great pathfinder.

Among the prominent American alienists of the 1880s were Allan McClane Hamilton, Walter Channing, John Chapin, John P. Gray, W. A. Hammond, and E. C. Spitzka. Dr. Hammond, usually considered as a neurologist, also taught psychiatry and wrote a book, *Insanity in Its Medical Relations*, that came into active use as a text. Dr. Spitzka, in 1883, published the *Manual of Insanity*, which included about all that was known and taught of the subject. Although the term "paranoia" (which means "mental disorder" in the original Greek) was introduced into medicine by Vogel in 1764 after which it had rather widespread applications, it was Dr. Spitzka of New York who, out of a welter of conditions and names, differentiated it as a specific disorder. It is therefore an American-named psychosis, as we use and understand the term today.

One of the strongest influences to bear upon American psychiatry was the work of Emil Kraepelin (1856–1926). He was professor of psychiatry successively at Dorpat, Heidelberg, and Munich and is generally considered to have been the greatest of all descriptive psychiatrists whose concepts increased and enriched the whole field. He was a systematizer who, following periods of active analysis of data, attempted a synthesis of scattered facts and ideas. Kraepelin had a large amount of clinical and laboratory material with which to work, and he made his studies according to the natural history method, in that he followed the course of mental disorder in the individual cases from the earliest symptoms on through to the termination. By means of this method it became possible for him to make a new grouping of mental disorders on which some basis for prognosis, therapy, and prevention could be ascertained. He was after the causes, the course, and the outcome of mental disorders with reference to any pathological lesions and processes, and he was interested in any mental trends and clinical pictures that tended to terminate in various degrees of deterioration.

In 1883 Kraepelin published the first edition of his *Psychiatrie*, which subsequently passed through several revisions and translations into other languages. The English translation of the book changed the whole aspect of classification of mental disorders in America. His outstanding contributions to the description and delineations of dementia praecox states and manic-depressive conditions have held a central position in psychiatric thought and practice from the time they were first introduced to the present day. His clinical descriptions of the psychoses can still be read with a great deal of profit by the student of psychiatry.

The concept of dementia praecox was later elaborated and expanded into schizophrenia by Eugen Bleuler (1857–1939) in 1911. Among this eminent Burghölzli professor's numerous original contributions that have vitalized psychiatry in all civilized areas are the psychology of dementia praecox (schizophrenia), including the phenomena of ambivalence and of autistic thinking, and the delineation of the schizoid and syntonic personalities, and some later biophilosophical theories.

About one year after the birth of Kraepelin, a man was born who was not only to revolutionize psychiatry but to influence the philosophic thought of the whole world, namely Sigmund Freud (1856–1939). After a period of interest in neurology, Freud went to Paris to study under Charcot, who was the outstanding clinical neurologist of his day, and who had developed a system of hypnosis for the treatment of hysteria. After returning home to Vienna, Freud carried out some studies of his own, and in 1895 the *Studien ueber Hysterie* was published, followed by the *Traumdeutung* in 1900. These studies represented a new approach to the investigation of human motivation and to the phenomena of the unconscious. Freud had replaced hypnosis with his method of free thought association and dream interpretation.

Freud's technique made it possible, for the first time, for a physician to devote his time to attempting to understand the innermost working of the patient's mind rather than directing the patient's life in matters that were obscure to everyone involved. For a number of years Freud worked alone, but later, when he was well on his way to success despite the great amount of opposition from critical contemporaries, students of the mind began to take an interest in the concepts, and a number of now well-known names appear in the list of his special pupils, supporters, and co-workers. Among these are Karl Abraham, Sandor Ferenczi, Carl Jung, Alfred Adler, Wilhelm Stekel, Otto Rank, Hans Sachs, Paul Federn, and Ernest Jones.

Jung, Adler, Rank, and Stekel later developed some special modifications of psychoanalysis contrary to Freud's wishes. He insisted that these modifications were not really psychoanalytic and were of such a nature that they necessitated his breaking off professional relationship with these workers. However, these doctrines all found followers in America and elsewhere, as have the later modifications of some of Freud's theories and techniques such as those of Sandor Rado, Franz Alexander, Karen Horney, and Harry Stack Sullivan among others in this country.

In 1909 Freud made a personal appearance in America to lecture under the auspices of Clark University in Worcester, Massachusetts. It may be said that practically everything that Freud wrote has at one time or another been translated into the English language, but in those earlier years America became rather thoroughly acquainted with the Freudian literature through the efforts of Abraham A. Brill, who first translated Freud's works, and of Smith Ely Jelliffe and William A. White, both of whom were outstanding leaders in psychiatric thought and freely disseminated such knowledge through their prolific writings of books and articles. Psychoanalysis soon gained a wide acceptance and became gradually incorporated into psychiatric thought and procedure.

Psychoanalytic teaching and practice were organized and standardized by the establishment of psychoanalytic institutes and societies directed and supported by highly trained specialists in several large cities in America as well as abroad. The psychoanalytic literature that has accrued over the years reveals a considerable diversity of trends, but in the main they fall into two general categories—(1) that which includes investigations aimed at extending the field of metapsychology, where the unconscious structure of the superego, the material gained by hypnosis, and the pregenital infantile frustrations are the principal topics, and (2) that which attempts to adapt current theories to clinical confirmation. There are now an ever increasing number of able workers interested in testing the validity of the libido theory and other concepts pertaining to various psychiatric conditions, in understanding the physiological manifestations of mental attitudes by means of combined studies in cooperation with modern biological laboratory disciplines, and in establishing an interpretation of neurotic symptoms in terms of somatobiological activity.

A group of Fellows of the American Psychiatric Association formed a Section in Psychoanalysis within the Association in 1933, and a year later a symposium was given in this section on the relation of psychoanalysis to psychiatry, which marked the beginning of a new era in the history of psychoanalysis in America. The conclusions were to the effect that the psychoanalytic formulations of Freud were established as indispensable to the procedures of treatment and research, and that requirements of technique, the investigation of unclear or incomplete knowledge, and the application of therapeutic procedures to selected clinical groups were being continually studied and that it was to be expected that modifications of psychoanalytic conceptions would take place to a considerable extent.

Another prominent psychiatric development took place under the leadership of Adolf Meyer (1866–1950). He came to America from Switzerland as a young man, well trained in neuroanatomy, neuropathology, psychiatry, and philosophy. Following services as a pathologist in the state hospitals of Kankakee, Illinois, and Worcester, Massachusetts, he became director of the New York State Psychiatric Institute on Wards Island where, as early as 1906, he was working on his concept of integration with the total individual as a unit. From the beginning he emphasized the pathological evolution of the symptoms of psychiatric disorders in terms of all of the presenting facts. His was a dynamic system.

Later in 1912 Dr. Meyer became director of the Henry Phipps Psychiatric Clinic and Professor of Psychiatry at the Johns Hopkins Medical School. He formulated through the subsequent years the reaction type of behavior and his concepts of social adjustment. He created a fresh, unique attitude toward the individual as a totality and brought the term "psychobiology" into the foreground. By training many psychiatric teachers in this discipline, he created a school of thought—particularly the concept of psychobiology—that has had a notable influence on psychiatric developments, particularly in this country and Great Britain.

Early in the 1900s Elmer E. Southard (1876–1920), another great ferment in psychiatry, developed the psychopathic hospital idea and also brought into the foreground the training of social workers in psychiatry. Adolf Meyer had brought social work into psychiatry as early as 1904, but in 1912 it was Southard who expressed clearly the social aspects of psychiatry and extended its interests into the training of social workers and into industrial hygiene. His famous book *The Kingdom of Evils*, with Mary C. Jarrett as co-author, was published in 1922 after Southard's death. This book dealt with social case work, presenting 100 case histories together with a classification of social divisions of evil. In 1919 he said, "Social psychiatry is far from the whole of mental hygiene, for mental hygiene includes also the far more difficult and intriguing topic of the individual as related to himself and his organ processes." Here is reflected the keen interest he always had in the organic reactions of the body and brain, which were the principal foci of his original researches.

It seems that the term "mental hygiene," in its generally accepted meaning, first appeared in the literature in 1843 in a book entitled *Mental Hygiene or an Examination of the Intellect and Passions Designed to Illustrate Their Influence on Health and Duration of Life* by Dr. W. C. Sweetser, Professor of Therapy and Practice of Physic at the University of Vermont. Dr. Isaac Ray also wrote a book ten years later entitled *Mental Hygiene*. He defined it as the art of preserving the health of the mind, and he emphasized that all methods of preserving mental health came under it. However, many years were to pass before the mental hygiene movement became a force for psychiatric progress. In 1908 Clifford Beers published his book, *The Mind That Found Itself*, which created a tremendous interest. By his efforts, together with those of Adolf Meyer and others, the National Committee for Mental Hygiene was organized in 1908. Dr. Meyer suggested the name "mental hygiene" for this committee, which has now been renamed the National Committee for Mental Health. It began with twelve charter members, among whom were three prominent psychiatrists, August Hoch, Adolf Meyer, and Frederick Peterson; the famous psychologist, William James; and the Johns Hopkins internist, Lewellys Barker. In 1912 Dr. Thomas W. Salmon (1876–1927) was appointed director.

The chief aims of the Committee were clearly stated: (1) to work for the protection of the mental health of the public; (2) to strive to raise the standard of care for the mentally ill and for those in danger of becoming mentally disordered; (3) to promote the study of all types of mental illness and to disseminate information concerning their causes, therapy, and prevention; (4) to obtain from every available source reliable data regarding methods of dealing with mental disorders; (5) to enlist the help of the federal government to the extent desirable; and (6) to coordinate the existing agencies and to aid in the organization, in each state, of an allied but independent society for mental hygiene.

The dynamic concepts of Freud, Pierre Janet (1859–1947), and Adolf Meyer strongly emphasized psychotherapy. The early workers in America as well as elsewhere had considered psychotherapy in terms of moral treatment, occupational therapy, recreational activities, and similar procedures, but the followers of the exponents of dynamics, such as A. A. Brill, S. E. Jelliffe, and W. A. White, among others, basing their teachings on the concept that mental symptoms are the expressions or symbols of meanings that are obscure and hidden in the unconscious, made a widespread impact on psychotherapeutic practice, which proceeded along the lines of what could be learned from a study of what was going on in the deeper recesses of the patient's mind.

Under the influence of some of Janet's ideas and formulations, but contributing a number of concepts of his own, was the Boston psychiatrist Dr. Morton Prince (1854–1929) of Tufts Medical School. He worked with what he termed "co-conscious phenomena," adding much to the knowledge of hysteria, multiple personalities, and hallucinations. He never fully accepted the significance of the unconscious in the Freudian sense of its meaning, and he disagreed with Freud's theory of the libido. He founded the *Journal of Abnormal Psychology*. His famous case of a girl with multiple personality was studied for over seven years and published in 1908 under the title *Dissociation of a Personality*; it is probably the most complete report ever made of this phenomenon. Another interesting contribution to the literature was his *Experimental Study of Visions*. He applied psychotherapy within the framework of his personal concepts.

Trigant Burrow (1875–1951), originally a Freudian, organized his own special group of students in what he termed "phyloanalytic" investigations of human behavior and social adjustment. He stressed the defensive distortions of man's natural reactions through the dominance of visual-verbal part functions resulting in internal tensional disorders called neuroses. He developed a system of psychotherapy aimed at the reduction of these tensions.

Paul Schilder (1886–1940), coming to America after having had a distinguished career in Vienna, contributed to the knowledge of several areas of the psychiatric and neurologic fields. His concepts of integration and his development of the body-image significance in various mental states are among the subjects with which he enriched the literature. The work of Harry Stack Sullivan (1892–1949) and his associates systematically developed the theme of interpersonal relationships, which Freud had taken into consideration but did not develop or emphasize in the same way. The basis of some of Sullivan's conclusions stemmed from the assumption that parataxic illusions are derived from earlier anxiety experiences. To Karen Horney and Sullivan neuroses were products of cultural environment and distorted interpersonal relationships rather than based on the preformed instincts, as Freud thought. Sullivan organized and published the journal called *Psychiatry*, which still serves as a useful medium for psychiatric thought.

In the field of therapy as well as those of pathology and diagnosis, a significant event was the obtaining in pure culture of the *Treponema pallidum* in 1911 by Hideyo Noguchi (1876–1928), and in 1913 Joseph W. Moore (1879–1957), of New York. Noguchi demonstrated the spirochete in the brain tissue of patients dying of general paresis, thus proving that this mental disorder was precipitated by a form of syphilis. Following this laboratory demonstration, general paretics were treated by the arsenicals and other drugs used for syphilitic disorders, but without marked success. In 1918 Julius Wagner von Jauregg (1857–1940) of Vienna announced his now well-known malarial therapy for general paresis. This treatment, strikingly successful as compared to other known procedures, promptly came into general use, and the first to apply it in America was the staff of the St. Elizabeths Hospital in Washington, D.C., from which the first paper in English on the results of this treatment was published, as far as I am informed.[25]

During World War I psychiatry was brought more prominently into the foreground and received a long-neglected consideration among the other medical specialties. At the beginning of the war the government called upon the National Committee for Mental Hygiene for aid, and this organization was exceedingly fortunate to obtain and recommend the services of Dr. Thomas W. Salmon (1876–1927), who had a remarkable organizing ability. Through his efforts it was demonstrated that psychiatry could contribute a great deal to the efficiency of the armed services by recognizing and treating the war neuroses. Salmon and his associates in the Army, with its numerous civilian as well as military personnel problems, were instrumental in pointing out the relationship of these problems to medical and social disciplines. By extraordinary patience and hard work, Salmon overcame seemingly insurmountable obstacles and problems. He had confidence in the future of psychiatry and was always alert to defend it against any unfair criticism. He profoundly influenced the growth and course of psychiatry in America, and it may be pointed out that what he did for the Army has stood the test of time, as his methods are still more or less in force and, in the minds of some, are still the best available, having been so proved in World War II and in the Korean campaign.

One of the most significant and practically important phases of the mental hygiene movement to the credit of American psychiatry was the creation of child guidance clinics, which started with William Healy (1869–) in 1909 in the institute devoted to the study of juvenile problems in Chicago. This was the pioneer child guidance clinic, although the term was not coined until 1922. The Boston Psychopathic Hospital, opening in 1912, accepted children from the first, as did the Phipps Psychiatric Clinic in Baltimore where a special outpatient children's clinic was active as early as 1913. The Allentown State Hospital in Pennsylvania was also a pioneer in this field, with a clinic under way in 1915. Child guidance clinics developed rapidly in other cities and areas under such well-known specialists as Lawson Lowrey (1890–1957), David Levy (1892–), Frederick Allen (1890–1969), J. S. Plant (1890–1946), and Douglas Thom (1887–1954) and are continuing to grow in size and distribution, as the needs for the study and correction of the behavior disorders of childhood are still greater than the available facilities.

Researches of many types have been carried on from early times, particularly on nervous system anatomy and functions, but intensive organized research, as such, on mental disorders is of comparatively recent development. For example, in 1884 Dr. I. W. Blackburn of Philadelphia, an excellently trained pathologist with a particular interest in the relationship between the brain and mental diseases, was appointed special pathologist at the Government Hospital for the Insane (now St. Elizabeths Hospital). He was the first full-time pathologist in a mental institution in America (at least I can find no record of an earlier full-time appointment of this kind). He published 25 papers and books during a period of 25 years, many of them illustrated by his own hand, since he was a skilled artist as well as a scientist.

Probably the first large-scale stride toward organized systematic psychiatric research in this country was made in 1895 by the establishment of the Pathological Institute of the New York State Hospitals as an integral part of the state hospital system. Located at first in New York City with Dr. Ira van Gieson (1865–1913), a neuropathologist, as director, this institute was later constructed on the grounds of the Manhattan State Hospital on Ward's Island where it remained for many years under the successive directorships of Adolf Meyer, August Hoch (1868–1919), and George Kirby (1875–1935). In 1929 the old institute was replaced by a modern building constructed as a unit in the Columbia Presbyterian Medical Center and named the New York State Psychiatric Institute and Hospital. During the past 60 years this institute has been unique since it has been supported for the sole purpose of research.

The Boston Psychopathic Hospital built in 1912 under the direction of Elmer Southard (1876–1920), the State Psychopathic Hospital under Dr. Albert M. Barrett (1871–1936) constructed in 1906, and later the Colorado Psychopathic Hospital under Franklin Ebaugh (1895–1972) are among the earliest centers with active research units employing full-time research workers. With these beginnings acting as stimuli, there have been many similar developments in university and hospital centers. Moreover, the philanthropic foundations have devoted large sums of money to train young research workers in the psychiatric field, to support investigators already in key positions, and to finance laboratories where original studies could be carried out. In 1934 the Scottish Rite Masons of the Northern Masonic Jurisdiction began their support of research projects in schizophrenia, which has continued to the present day with accomplishments that have served as an example of the progress that may be expected with the growth of research interests and financial aid.

The work of Sir Charles Sherrington (1857–1952) of England on the physiology of the nervous system and of Walter Cannon (1871–1945) of America on the relationship between the emotions and the vegetative nervous system, as well as the investigations of the followers of Pavlov's (1849–1936) concepts and to a lesser extent those of Kretschmer (1888–1971) on bodily conformation, have served as a background upon which the structure of "psychosomatic" or "comprehensive" or "total" medicine has been erected and expanded in recent years. The term "psychosomatic medicine" has been known for at least 100 years, but in the sense of the modern clinical application and theories in psychopathology and medical problems it is rather distinctly an American development to which Adolf Meyer and Smith Ely Jelliffe contributed a great deal.

Here it is well to remember that George Beard of New York described a syndrome in 1868 that he called "neurasthenia" and under which he included pathological fatigues and so-called nervous exhaustion. He said, "While modern nervousness is not peculiar to America, yet there are special expressions of nervousness that are found here only: and the relative quantity of nervousness and of nervous diseases that spring out of nervousness are far greater here than in any other nation in history and it has a special quality." These disorders, regardless of whether there were more of them in America than elsewhere, were the disorders to which Silas Weir Mitchell (1829–1914) of Philadelphia devoted much thought and therapy. He ascribed them to deficiencies in fat, blood, and other elements and recommended complete rest as the cure. From the general interest in so-called functional disorders came one of the special roots now known as psychosomatic medicine.

Forensic psychiatry, as such, has developed slowly in America through the years despite the fact that there have been outstanding pioneers with constructive concepts and writings. Beginning with Isaac Ray's book in 1838, one finds interesting and pertinent references to the relationship between psychiatry and the law and recommendations for dealing with these complicated problems. In more recent years the publications and personal influence of such authorities as L. Vernon Briggs (1863–1942), William A. White (1870–1937), Bernard Glueck (1884–1973), Winfred Overholser (1892–1964), and Gregory Zilboorg (1890–1958) have made excellent contributions to the subject in ways that have favorably modified some of the practical legal procedures.

Since World War I industrial psychiatry, school and college psychiatry, and other penetrations of psychiatry into public affairs have made notable strides that have led to a deeper understanding of emotional phenomena. Clinical psychology, occupational therapy, and social service activities, all of which are valuable psychiatric auxiliaries, have shown remarkable progress. In the field of clinical psychology the use of the Rorschach and other projective techniques became established and were freely applied to psychiatric problems. The social worker, from the first, proved to be practically indispensable in investigating the home situation, in the problems of hospital patients and of the child guidance field, and in interpreting the nature of the problem to the family as well as aiding in the readjustment of the improved patient in the community.

Occupational therapy, in the broad sense of its original definition ("any activity, mental or physical, definitely prescribed and guided to hasten recovery from disease or injury"), has always accompanied enlightened, humanitarian treatment, especially in the field of mental illness. It may have begun in the United States with the opening of the Pennsylvania Hospital in 1752, when the board of managers provided wool-carding and spinning equipment "to employ such Persons as may be capable of using the same." Activities ranging from music, games, and arts and crafts to industrial or housekeeping tasks and education were utilized as an essential feature of the nineteenth-century "moral treatment" as mental hospitals opened in Philadelphia, New York, and Boston. Nursing personnel frequently supervised these activities, though no formal preparation for such leadership was available until Susan E. Tracy established a course for student nurses at Adams Nervine Hospital. By 1917, when the Society for the Promotion of Occupational Therapy (later the American Occupational Therapy Association) was formed, two more training centers were functioning. Three other schools, originated during World War I to train "reconstruction aides" for Army hospitals, continued to grow and were among those accredited in 1938 by the American Medical Association. Graduate therapists, like their predecessors, worked in various hospitals and rehabilitation centers, using a wide range of activities to promote patients' recovery from physical as well as mental illness.

As late as the 1930s it was advocated by many that psychiatry was not a science but an art, since the personality of the psychiatrist bulks so large in the practice of mental medicine, and psychiatry could never free itself from its philosophic elements. Regardless of what was thought and said, an "American school" was developing, and by 1937 it had the following characteristics, most of which have persisted.

The American school had started to use the new shock therapies that later became so widespread in application, and it had also recognized the great possibilities of sociology and cultural anthropology as contributors to psychiatric knowledge. However, this American movement was not merely a composite eclecticism, but it was beginning a positive frontal attack upon the problems of mental medicine that was new, frank, and critical. It advocated treating all behavior, thought, and feeling of an individual as real or actual performance or, in other words, as a personality experience. The differentiation of human performances and capacities is just as much a real dynamic phenomenon as is any feature of somatobiological processes or of social culture. It also emphasized proceeding along the following general lines in dealing with a particular psychiatric problem: (1) the correction of all recognized defects in the soma of the patient; (2) the establishment of adequate rapport between physician and patient; (3) the careful study and evaluation of familial, economic, and social situations; (4) the detailed investigation of the personality problem; (5) the attitude toward and adjustment to reality; (6) the ventilation of conflicts and desensitization of the patient; (7) the institution of re-educational training; (8) the formation of an adequate philosophy of life; and (9) the desirability of follow-up studies.

The American school recognized, and attempted with considerable success to correct, the lack of adequate instruction and training in psychiatry in many of the medical schools; in fact, in most of the medical centers psychiatry was inadequately represented. However, despite these educational handicaps, a greater number of medical students sought postgraduate training, and the number of women physicians entering psychiatry became notably larger since earlier strong prejudices against them had subsided considerably. It is a matter of interest to note that in 1890 a law was passed by the New York State legislature authorizing the appointment of at least one woman physician in each state hospital. Obviously we have come a long way during the last half century. The lack of sufficient knowledge of mental hygiene and the psychic components of disease on the part of the general practitioners and the lack of proper psychiatric information among the clergy and the population in general were also well recognized, and movements were started to try to correct these deficiencies.

In recent years, and particularly since World War II, more adequate methods and techniques suitable for attacking vital problems have appeared and are still in various stages of evolution. In scrutinizing the past in an attempt to perceive the simple beginnings and origins of the present-day science of man, one should remember that each worker, regardless of his individuality, is in some degree dependent on the thought and practice of his time, uses to some extent formulations suggested by others, and is stimulated and influenced by the particular interests of his friends and colleagues.

Bibliography

1. ACKERLY, S., "Thirty Years of Child Psychiatry," *Am. J. Psychiat., 110:*567, 1953.
2. AMDUR, M. K., "Dawn of Psychiatric Journalism," *Am. J. Psychiat., 100:*205, 1943.
3. ———, "Psychiatry a Century Ago," *Am. J. Psychiat., 101:*18–28, 1945.
4. BOND, E. D., *Dr. Kirkbride and His Mental Hospital,* Lippincott, Philadelphia, 1947.
5. ———, "Psychiatry in Philadelphia in 1844," *Am. J. Psychiat., 101:*16–17, 1945.
6. ———, "Therapeutic Forces in Early American Hospitals," *Am. J. Psychiat., 113:*407, 1956.
7. BRACELAND, F. J., "Kraepelin: His System and His Influences," *Am. J. Psychiat., 113:*871, 1957.
8. CARLSEN, E. T., "Amariah Brigham: I. Life and Work; II. Psychiatric Thought and Practice," *Am. J. Psychiat., 112:*831; *113:* 911, 1956.
9. COLLIER, G. K., "History of the Section on Convulsive Disorders and Related Efforts," *Am. J. Psychiat., 101:*468–471, 1945.
10. DEUTSCH, A., *The Mentally Ill in America: A History of Their Care and Treatment from Colonial Times,* Doubleday, New York, 1937.
11. DIAMOND, B. L., "Isaac Ray and Trial of Daniel McNaugton," *Am. J. Psychiat., 112:*651, 1956.
12. DRUKURS, R., and CORSINI, R., "Twenty Years of Group Therapy," *Am. J. Psychiat., 110:*567, 1954.
13. EBAUGH, F. G., "The History of Psychiatric Education in the United States from 1844 to 1944," *Am. J. Psychiat., 101:*151–160, 1945.
14. FARR, C. B., "Benjamin Rush and American Psychiatry," *Am. J. Psychiat., 101:*1–15, 1945.
15. HADDEN, S. B., "Historic Background of Group Psychotherapy," *Int. J. Psychother., 5:*162–168, 1955.
16. HALL, J. K., *et al.* (Eds.), *One Hundred Years of American Psychiatry,* Columbia University Press, New York, 1944.
17. HASKELL, R. H., "Mental Deficiency over a Hundred Years," *Am. J. Psychiat., 101:* 107–118, 1945.
18. "Historical Note: C. K. Clarke (1857–1957)," *Am. J. Psychiat., 114:*368, 1957.
19. KANNER, L., "The Origins and Growth of Child Psychiatry," *Am. J. Psychiat., 101:* 139–143, 1945.
20. KARWIN, E., "Contribution of Adolf Meyer and Psychobiology to Child Guidance," *Ment. Hyg., 29:*575, 1945.
21. LAZELL, E. W., "The Group Treatment of Dementia Praecox," *Psychoanal. Rev., 8:* 168–179, 1921.
22. LEBENSOHN, Z. M., "Contributions of St. Elizabeths Hospital to a Century of Medico-Legal Progress," *M. Ann. District of Columbia, 24:*469–477, 542–550, 1955.
23. LEWIS, N. D. C., "Review of the Scientific Publications from St. Elizabeths Hospital During the Past 100 Years," *M. Ann. District of Columbia, 25:*143–147, 1956.
24. ———, *A Short History of Psychiatric Achievement,* Norton, New York, 1941.
25. ———, HUBBARD, L. D., and DYAR, E. G., "Malarial Treatment of Paretic Neurosyphilis," *Am. J. Psychiat., 4:*175, 1924.
26. MAPOTHER, W., "Impressions of Psychiatry in America," *Lancet, 1:*848, 1930.
27. MEYER, A., "Objective Psychology or Psychobiology with Subordination of the Medically Useless Contrast of Mental and Physical," *J.A.M.A., 65:*860–862, 1915.
28. ———, "Revaluation of Benjamin Rush," *Am. J. Psychiat., 101:*433–442, 1945.
29. ———, "Thirty-Five Years of Psychiatry in the United States and Our Present Outlook" (Presidential Address, American Psychiatric Association), *Am. J. Psychiat., 8:*1, 1928.
30. MORENO, S. R., "History of First Psychopathic Institute of the Great American Continent," *Am. J. Psychiat., 99:*194, 1942.
31. OBERNDORF, C. P., "Psychiatry at Ward's Island 40 Years Ago," *Psychiat. Quart.,* suppl., *24:*35, 1950.
32. RIESE, W., "An Outline of History of Ideas in Psychotherapy," *Bull. Hist. Med., 25:*5, 1951.
33. RUGGLES, A. H., "Clifford Beers and American Psychiatry," *Am. J. Psychiat., 101:*98–99, 1945.
34. RUSSELL, W. R., "From Asylum to Hospital —A Transition Period," *Am. J. Psychiat., 101:*87–97, 1945.
35. SHRYOCK, R. H., "The Psychiatry of Benja-

min Rush," *Am. J. Psychiat.*, *101*:429–432, 1945.

36. SILVERMAN, M., "Julius Wagner von Jauregg (1857–1957)," *Am. J. Psychiat.*, *113*:1057, 1957.
37. STERNS, A. W., "Isaac Ray: Psychiatrist and Pioneer in Forensic Medicine," *Am. J. Psychiat.*, *101*:573, 1945.
38. STEVENSON, G. S., "The Development of Extramural Psychiatry in the United States," *Am. J. Psychiat.*, *101*:147–150, 1945.
39. STONE, S., "Psychiatry in New Hampshire, First 100 Years," *New Eng. J. Med.*, *228*: 595, 1943.
40. THOMPSON, G., "The Society of Biological Psychiatry," *Am. J. Psychiat.*, *111*:389, 1954.
41. TUCKER, B. R., "Development of Psychiatry and Neurology in Virginia," *Virginia Med. Month.*, *69*:480, 1942.
42. ———, "Silas Weir Mitchell," *Am. J. Psychiat.*, *101*:80–86, 1945.
43. WALKER, W. B., "Medical Education in the Nineteenth Century," *J. Med. Educ.*, *31*: 765, 1956.
44. WITTELS, F., "The Contribution of Benjamin Rush to Psychiatry," *Bull. Hist. Med.*, *20*: 157–166, 1946.
45. ZILBOORG, G., "Eugen Bleuler and Present-Day Psychiatry," *Am. J. Psychiat.*, *119*: 289–303, 1957.

CHAPTER 3

RECENT PSYCHIATRIC DEVELOPMENTS (SINCE 1939)

George Mora

Introduction: General and Methodological Issues

A SUCCINCT PRESENTATION of the development of psychiatry in the last three decades is not an easy task, especially in view of the great deal of progress made in this field since the end of World War II, of the difficulties involved in carrying on meaningful research in this area, and of the lack of adequate historical perspective to evaluate properly the events related to this progress.

The fact remains, however, that in the period under consideration psychiatry has gained acceptance in the overall realm of medicine[370] and, even more, in the American culture by and large. Psychiatry has now reached the point of being able to look comfortably at its present situation and to draw from the past inspiration for the future.

In this country, following Albert Deutsch's *The Mentally Ill in America*[149] and Gregory Zilboorg's *A History of Medical Psychology*,[618] a number of general histories of psychiatry,[4,8,12,96,336,522,581] biographical studies[7,63,79,603] (mainly the thorough, yet biased, study on Freud by E. Jones),[306] histories of diseases,[296,572] of institutional care, [137,262,506,510] and of basic concepts and trends related to psychiatry[74,136,142,165,227,274,430,439,502] (such as the important volume on the development of the unconscious by H. Ellenberger)[175] have appeared.

Moreover, the emphasis on newly published primary sources—for example, Freud's correspondence with pupils and admirers—and on the historical dimension of sociological attitudes toward the mentally ill has become significant. Even the American Psychiatric Association, which has taken many initiatives through its Committee on History,[13,275] recognized the importance of its own development by republishing the presidential addresses of the last quarter-century on the occasion of the 125th anniversary of its foundation in 1969. In the introduction to this publication,[426] as well as elsewhere,[423,424,425,427] I myself have discussed many important points related to the history of psychiatry, to which the interested reader is referred.

The historical dimension, when presented from a broad cultural perspective, can help to predict future developments and to introduce optimism into the study of certain phenomena —such as apparent manifestations of collective psychopathology and widespread use of "drugs of the mind"—that find antecedents in similar episodes in the past.

Because of space limitations only topics relevant to the main areas of psychiatry have been considered here. Events related to special and collateral fields of psychiatry, included in the first edition of this *Handbook*, have been omitted. Also, in discussing specific points, full source information has not been provided in the text or in the bibliography, as this will be done in the following chapters of this work.

National Recognition of Psychiatry

Shortly after the first edition of this *Handbook* was published in 1959, *Action for Mental Health*, a milestone in American psychiatry, appeared. The main psychiatric events that took place in the 25 years from the end of World War II to then can be seen as progressively leading to the realization of such an important document.

American psychiatry had originated from the British and continental impetus toward moral treatment in the first part of the nineteenth century, followed by the emphasis on institutionalization of a large number of mentally ill in isolated settings in the second part of the century. The emphasis on the organic etiology of mental disorders has been slowly superseded by Meyer's psychobiology, a convergence of the new European psychodynamic theories and the optimistic American view of environmental forces.

In the late thirties early attempts were made to improve the treatment of hospitalized patients through the development of aftercare programs in the community, better administrative policies in state institutions, and better training of personnel at every level. The introduction of shock therapies around that time helped to focus attention once again on psychotic patients and on organic psychopathology and research in general; in the two previous decades, at the beginning of the child guidance movement, attention had shifted to neurotic and antisocial patients.

The child guidance movement, originating from the desire to prevent juvenile delinquency, had eventually based its platform on a combination of individual psychodynamics and environmental behaviorism. By the third decade of this century research on new ideas —child psychiatry, criminology, alcoholism— advanced and new clinical techniques, such as projective tests, were introduced. Under the influence of the many psychoanalysts who emigrated from Europe to this country as at the beginning of the forties to escape the Nazi persecution of the Jews, new impetus was given to individual psychotherapy in its various modalities and, to a lesser extent, to research in the new areas of psychosomatic medicine and experimental neuroses.

By that time World War II had revealed the magnitude of psychiatric problems: syndromes of acute breakdown in relation to combat, a large number of inductees rejected for psychiatric reasons, and rehabilitation of many veterans suffering from psychiatric disorders. In a short time the problem of mental illness came to be recognized at a national level. Especially urgent was the need to train a great number of professionals and nonprofessionals, far above the few psychiatrists hitherto trained with the assistance of some private organizations, such as the Rockefeller Foundation and the Commonwealth Fund.

The Vocational Rehabilitation Act (1942), the National Mental Health Act (1946)[89] leading to the creation of the National Institute of Mental Health (1949)—for research, training, and assistance in developing mental health programs—the National Governors' Conference on Mental Health, the establishment of separate departments of mental hygiene on mental health in many states, these and many other developments can be seen as steps along a continuum of increasing national concern about mental illness.

In 1955 the Mental Health Study Act was passed, providing for the creation of the Joint Commission on Mental Illness and Health. This commission, composed of outstanding leaders from many organizations, in five years of intensive work produced the above-mentioned *Action for Mental Health*, essentially geared to shift the emphasis from institutional to community care of the mentally ill. Eventually in 1963 under President Kennedy the Community Mental Health Act was passed, providing for the creation of a network of community mental health centers able to offer a comprehensive program of prevention, treatment, and rehabilitation throughout the entire nation. Finally in 1965 funds for staffing these centers were allocated through an amendment to the act.

Regardless of the effectiveness of these legislative actions, they point unequivocally to the recognition of the problem of mental illness at a national level and to a concerted effort to deal with it.

Research and Methodology

In this country research was first carried on systematically at the Pathological Institute (now New York Psychiatric Institute) founded in 1896,[330,525] and at the Henry Phipps Psychiatric Clinic of Johns Hopkins University inaugurated in 1919 by Adolf Meyer. Early research projects tended to focus on histopathology, genetics, endocrinology, and neurophysiology; later on the themes of juvenile delinquency, psychosomatic disorders, and emotional deprivations became prominent.

Research in psychiatry is vitiated by certain methodological drawbacks related to the difficulties in measuring psychological phenomena, in reaching agreement on symptoms and diagnoses, in producing animal experimentation meaningful for human beings, especially by interdisciplinary teams composed of scientists having different ideas and biases.[193,263,300,474,475,509,545] Moreover, most of the research tends to be supported by the federal government, which relies on a relatively small number of experts, who are inclined to favor certain projects and themes. Efforts by the government to encourage young psychiatrists to carry on research through mental health careers, investigation careers, and career development awards have met with limited success.[266]

In addition, psychiatrists arrive at the end of their training in their thirties, when the process of creativity is already in decline; in the course of their training they do not receive adequate preparation in research methodology. Consequently research designs tend to be carried on by psychologists, more interested in proper methodology than in creativity, and often in themes peripheral to the mainstream of psychiatry. It is a fact that most of the "discoveries" in psychiatry—from psychoanalysis to shock treatment, psychopharmacology, and community innovations—have been made mostly in Europe by individual psychiatrists often working in poorly equipped settings. The typical pattern has been for American research teams to thoroughly investigate and critically assess discoveries achieved somewhere else through a variety of methodology including double-blind designs, use of placebo, selection of cohorts of patients for comparative purposes, follow-up of patients, "research alliance" between researcher and patient, and so forth.

Twenty years ago Lawrence Kubie, an outstanding American psychiatrist, wrote that "research in psychiatry is starving to death."[343] Since then a considerable amount of research has been carried on. Recently, however, the issue of research has become more complex than ever as the result of the upsurge of community psychiatry, which requires a difficult type of multidisciplinary research, involving clinical, statistical, and sociological dimensions and not immune from ethical and political pressures.[484] The fear of losing the uniqueness of the doctor-patient relationship, so close to the core of psychiatry, has been voiced by some.

Classification: Normality and Mental Disorders, Epidemiology Statistics

Pathology is meaningful only vis-à-vis normality.[127,594] Yet normality has not been the focus of psychiatric research up until the last decade or so;[480,530] possibly the new attention on normality is a response to the need of assessing large groups of people in the context of the new community mental health movement. Among the recent pertinent publications mention should be made of *Current Concepts of Positive Mental Health* (edited by M. Jahova), which emphasizes individual self-actualization; *Normality and Pathology in Childhood* (by A. Freud), which is based on the psychoanalytic developmental perspective; and *Normality: Theoretical Concepts of Mental Health* (by D. Offer and M. Sabshin), which is based on the four dimensions of health, utopia, average, and process. In very recent years the pseudoissue of a "supernormality," that is, "expansion of consciousness" achieved with the help of certain drugs, has been brought forward. Although most psychiatrists reject this notion as absurd, it has a relationship with centuries-old techniques of mastering the body through the mind—from Yoga to Zen—used in the Far Eastern cultures.

Regardless of all this, in psychiatry, like medicine, the urge to classify reflects the fundamental antithesis of looking for what is different while, at the same time, trying to find what is common.[544,562] Some years ago H. Ellenberger[177] pointed to the biases of psychiatric nosology in terms of the nature and kind of classifications, the concept of nature, the projection of intellectual schemata, and the unconscious position of the researcher. Yet the history of psychiatry coincides with the history of psychiatric classifications.[408,573] Since the beginning of our century Kraepelin's notion of the rigid pattern of mental diseases has been superseded by Freud's developmental views and, in this country, by A. Meyer's emphasis on mental diseases as "reaction types"—a position that was accepted in the 1952 official classification of the American Psychiatric Association.[493,546]

Since then many psychiatrists have stressed the increasingly "dull" aspects of psychiatric symptoms, up to the point of simple boredom or "alienation"; others have emphasized the trend from a cross-sectional to a longitudinal dimension and from "outer" behavior to "inner" feeling;[282] the poorly differentiated "borderline syndrome" (thoroughly described by R. Grinker)[261] and "social breakdown syndrome" (E. Gruenberg)[268] have been described; and finally mental illness has been considered as a "myth" supported by the psychiatric establishment (T. Szasz).[558]

In the light of all this the *Diagnostic and Statistical Manual of Mental Disorders* (DSM-II) published by the American Psychiatric Association in 1968[297] seems a rather conservative document, attempting to fit into the general scheme of the eighth edition of the International Classification of Diseases (ICD-8), published by the World Health Organization in 1966. At this point it is uncertain what future awaits bold new attempts[310] (such as *New Approaches to Personality Classification*, edited by A. Mahrer) to base classification systems on new parameters (for example, having the patient himself participate in his own evaluation).

Rather, in the light of the community mental health movement, the importance of carrying on research on epidemiology has become evident.[313] Following the methodological clarification of the notions of "incidence rate," "prevalence rate," and "system analysis," epidemiological studies have focused on the incidence of mental disorders in a certain area in toto (Hollingshead and Redlich in New Haven, Rennie, Srole, and collaborators in

Manhattan, the Leighton's group in Stirling County, Dohrenwend, in the Washington Heights district of New York City), or in regard to hospitalized mental patients (Malzberg in New York, Dayton in Massachusetts), to particular ethnic groups (Eaton and Weil on the Hutterite), to samples of brain damaged and retarded children (Pasamanick, *et al.*, in Baltimore).[280]

The impetus toward epidemiological research[298]—in this country by E. Gruenberg and others under the sponsorship of the Milbank Foundation[118]—has resulted in the establishment of psychiatric registers (for example, in Rochester, N.Y., by the Tri-County in Raleigh, N.C., and by the National Clearing House of the NIMH in Bethesda, Md.). Computers, first introduced at the Institute of Living in Hartford, Conn., in 1962, have later been used in other places (New York State Department of Mental Hygiene, New York State Psychiatric Institute, Department of Computer Science of the Stanford University Medical Center in Palo Alto, Calif., Missouri Institute of Psychiatry in St. Louis, the Reiss-Davis Child Study Center in Los Angeles, and a few others).[241] The statistical refinement brought about by the use of computers is outweighed in the minds of some by their dehumanizing aspect, which runs counter to the very essence of psychiatry.[55,504] As a senior psychiatrist, Carl Binger, has put it, if the computer takes over, the psychiatric role may be "reduced to machine-processed data of being pushed around like pawns on a chessboard of science."[64]

Psychopathology

Biological Research

America's original contribution to psychiatry can be traced back to Cannon's concepts of homeostasis and of the autonomous reactions of the organism under stress in the mid-twenties, influenced by the dynamic theory of personality and by behaviorism. Such a typical expression of American melioristic philosophy of life was challenged by the great depression in the thirties and by the spread of psychoanalytic ideas in the forties.

The biological trend in psychiatry, which had been prominent in the late nineteenth century, gained momentum again at the end of World War II, for example, the founding of the Society of Biological Psychiatry.[563] Among the main research themes were: stress reactions in Air Force servicemen (R. Grinker and J. Spiegel), theory of emotions (J. Papez), visceral brain (P. MacLean), reticular system (G. Moruzzi and H. Magoun), theory of cell assembly (D. Hebb), stimulation of cerebral cortex (W. Penfield), functions of the frontal lobes (J. Fulton), "stress syndrome" and "general adaptation syndrome" (H. Selye), theory of bodily defensive reaction (H. G. Wolff), experimental neuroses (J. Masserman), and psychosomatic conditions (F. Alexander and collaborators at the Chicago Psychoanalytic Institute).

The tendency of psychiatric research in the forties, under the influence of the psychoanalytic movement, to rely more on personal intuition than on scientific methodology, was superseded by the discoveries in psychopharmacology in the fifties, which represented a return to the philosophy of biological psychiatry.[87,566] Of the many topics mention should be made here at least of: the genetic role of the DNA molecule and the transmission of coded messages through the RNA; conditioning responses leading to behavior forms of therapy; clinical recognition of positive spike phenomena in EEG and of REM signs in sleep and dreaming; clinical aspects of sensory deprivation of psychominetic agents, of drug tolerance and abuse; functions of lymbic and reticular systems, of the temporal lobe, and of the neurotransmitters in psychiatric syndromes and in relation to chemotherapy; finally inborn errors of metabolism, enzymatic defects, and chromosexual abnormalities (Down's, Turner's, and Klinefelter's syndromes), leading to new investigating techniques (sex chromation determination, amniocentesis, cytogenetic study of criminal individuals) and to concern for family planning based on genetic counseling.

Such research, mainly carried on by teams of experts from the fields of neuroanatomy, neurochemistry, electrophysiology, and heredity, is largely supported by the National Institute of Mental Health, the National Institute of Child and Human Development, and the National Science Foundation. Even if one disregards Freud's prediction that the ultimate cause of mental disorders will be found one day in biological processes, the fact remains that nowadays a comprehensive view of the personality has to take into consideration the importance of biological research. The collaborative volume *Psychiatry as a Behavioral Science* (published under the auspices of the Committee on Science and Public Policy of the National Academy of Science and the Problems and Policy Committee of the Social Science Research Council) represents an excellent survey of this field.[277]

Anxiety and Related States

Anxiety is intrinsically related to the human condition of longing for eternity but having to accept death: a theme that can be traced from St. Augustine and Pascal to Kierkegaard and the contemporary existentialists.[403]

In psychiatry the role of anxiety became paramount in Freud's various formulations; his final conceptualization of anxiety as a signal of danger from within (1926) led to the notions of defense mechanisms and of ego psychology. In addition to the semantic difficulty of differentiating between the philosophical and the psychopathological meaning of the term,[372] anxiety has been seen as both "positive" (that is, facilitating purposeful behavior) and "negative" (that is, interfering with such behavior) from time to time.

In this country from the forties on, many studies have been devoted to "separation anxiety"; Hans Selye has described the abovementioned "stress syndrome" and "general adaptation syndrome" as a global reaction of the organism resulting from the interplay of two opposite endocrinological constellations; S. Wolf and H. Wolff have differentiated fear from anxiety on the basis of experimental studies on gastric secretion; others (D. Funkenstein) have differentiated unexpressed anger (anger-in) from expressed anger (anger-out) on the basis of the mechanism of action of epinephrine and norepinephrine; still others have studied anxiety in experimentally induced neuroses.

These studies have been criticized on various grounds, such as the difficulty of translating animal into human behavior. From a more clinical perspective Sandor Rado has attempted to explain the notion of emergency behavior as related to the level of organization of emotional and unemotional thought; and cultural anthropologists, followed by the neo-Freudians, have insisted on the importance of societal factors in the causation of anxiety, which from a defense may turn into a symptom (K. Horney). Finally the advent of psychopharmacology in the mid-fifties has brought the focus back on the symptomatic factors of anxiety regardless of the total personality. No matter what perspective one adheres to, anxiety remains a highly complex subject, as shown in the two monographs recently published by S. Lesse and by W. Fischer.

Unconscious, Dreams, Sexuality

The existence and main characteristics of the unconscious appeared to be established beyond doubt when the first edition of this *Handbook* was published a decade ago. Freud's overwhelming emphasis on the unconscious, based on solid data on hypnosis, dreams, and countertransference, and its acceptance in psychiatric circles signified the final outcome of a long tradition beginning with the Greeks, running through the Romantics in the early nineteenth century, and leading to study of hypnosis by the French schools in the latter part of that century.[601]

Although from the forties on the importance of the unconscious came to be reduced, with the emphasis on defense mechanisms and ego psychology, no one questioned its existence. As a matter of fact, fresh research on dreams, sensory deprivation, and posthypnotic phenomena seemed to add new evidence to the classical notion of the unconscious.

Yet in the last decade or so the advocates of the unconscious have found themselves on the defensive.[144,468] Supporters of the new "behavior therapy" and of other related approaches question, not the validity of the notion of the unconscious, but its relevance for psychiatric treatment, which they view as solely based on learning theory and on conditioning. Whether such a threat to the notion of the unconscious represents a temporary fad or will signify a persistent trend remains to be seen.

It is well known that dreams, like the unconscious, have a centuries-old tradition that can be traced back to ancient Middle East cultures, the Greeks, and the Middle Ages.[46,271] Such a tradition was given scientific form for the first time in Freud's *Interpretation of Dreams* (1899), which, by introducing a new methodology in psychiatry, initiated the "royal road" to psychoanalysis.

For several decades the importance of dreams was not questioned, although different emphases were placed on their interpretation at variance with Freud's insistence on their sexual and aggressive aspects: compensation for social inferiority feelings (Adler); manifestation of collective unconscious and archetypal images (Jung); expression of ego's thrust for synthesis (Hartman and Kris); attempt to unify past and future in the light of ego identity and of a life plan (Erikson); struggle to achieve personal self-awareness and responsibility (neo-Freudians); finally a mode of personal existence (existentialists).

In recent years a revolutionary event has taken place: a new methodology based on the discovery of regular periods of eye motility during sleep (E. Aserinsky and N. Kleitman)[27] and their relation to EEG patterns and content of dreams (C. Fisher, W. Dement, M. Ullman, F. Snyder, and others). An imposing amount of data has been gathered on the relation between rapid eye movements (REMPs) and nonrapid eye movements (N-REM), instinctual versus teleological role of sleep, clinical importance of "dream deprivation," interplay of neocortex and basal centers, new meaning of enuresis, somnambulism, nightmares, and so forth.

Interestingly enough this experimental research proves the erroneousness of some of Freud's basic tenets (such as dreams as protecting sleep and the connection between dreaming and psychotic states), while at the same time it points to the important role of interpersonal relationships, cultural aspects, and historically bound factors in the understanding of dreams. As a result, dreams are seen today from the threefold perspective of their neurobiological substratum, of their psychotherapeutic value, and of their relation to the preconscious level of artistic creativity. The recently established Association for the Psychophysiological Study of Sleep (APSS) is especially concerned with the first of these three aspects. Many interesting points are discussed in the literature of the last two decades, such as Fromm's *Forgotten Language*, Boss's *The Analysis of Dreams*, Tauber and Green's *Pre-logical Experience*, Bonime's *The Clinical Use of Dreams*, French and Fromm's *Dream Interpretation*, and Hall and Van de Castle's *The Content Analysis of Dreams*.

Expressions of sexuality easily can be traced back to every culture, from the early times on, in the literary as well as in the figurative fields. This is true even during periods of severe sexual repression, such as in the Middle Ages, when sex aberrations were expressed in the context of the witchcraft mania.

Early in our century Freud and his disciples faced the manyfold psychological aspects of sexuality in a candid way through the use of a new verbal technique. Thus for the first time the centuries-old intuitions concerning the relationship of sexuality to psychopathology were given systematic form in terms of individual development. Freud's rather rigid model of the progression from the oral to the anal to the genital stage has remained valid to our day, though modified by some: for example, Alexander has attributed the sexual urge of adults to the surplus of energy after growth is completed at adolescence, while the neo-Freudians have stressed the cultural components of the sexual instinct.

Under the influence of the Freudian school

the development of sexuality came to be studied directly in children, rather than in retrospect in adults, from the comprehensive perspective of anatomy, heredity, endocrinology, ethology, and sociology, and research was conducted also in non-Western cultures to test the universality of the psychoanalytic postulates. Eventually even the validity of a notion as basic as that of the oedipal complex came to be questioned.[123]

Undoubtedly a more liberal view of sexual expressions has become noticeable concomitantly to the increasing acceptance of psychiatry.[91,315,453] Without dealing with the issue of cause-and-effect relationship between these two phenomena, mention should be made here of: the two Kinsey reports on the sexual behavior of the male (1949) and of the female (1953), indicating a wide range of sexual activity in the American culture; the bold methodological approach introduced by R. Masters and V. Johnson in the psychiatric treatment of sexual disorders; the slowly gaining view of considering homosexuality from the psychological rather than from the moral perspective; finally the rapid reassessment of the role of the female vis-à-vis that of the male in this country as well as in other Western nations.

Formal or Structural Mechanisms, Cognitive Functions, the Intrapsychic Self

The various concepts listed in the heading of this section have come to acquire significance as the result of revisions of traditional Freudian notions and of the new fruitful integration of data acquired from psychoanalysis, developmental psychology, and other fields. It is the special merit of Silvano Arieti to have focused on the neglected area of the cognitive aspects of the personality in a series of publications that span more than two decades. According to Arieti, this new orientation has the following background: the pioneering eighteenth-century writings on prelogical thinking by the Italian philosopher Giambattista Vico; the differentiation between "abstract attitude" and "concrete attitude" in brain-damaged and schizophrenic patients (K. Goldstein); the essence of identity in paleological thinking being based on identical predicates rather than on identical subjects as in mature reasoning (Von Domarus); and finally the various models of the genetic development of the mind presented by H. Werner in *Comparative Psychology of Mental Development* and by J. Piaget in his many monographs.

In the past some attempts had been made to establish a relationship between psychotic symptoms and formal mechanisms of dreams, languages, and other human expressions (for example, by the Swiss A. Storch).[551] However, traditionally the emphasis has been on the study of the content rather than the form of psychopathology. As Arieti put it: "The study of formal mechanisms reveals *how* we think and feel. The study of dynamics of psychoanalytic mechanisms reveals *what* we think and feel and the motivation of our thinking and feeling. Both the formal and the dynamic approaches are necessary if we want to understand psychological phenomena fully."[18]

In his early writings, mainly in his *Interpretation of Schizophrenia*,[18] Arieti has described in detail the mechanisms of dreams, verbal associations, and infantile and paleological thinking.[17,23] Later on he has tried to overcome the shortcomings of Freud's positivistic model of the mind and the culturalistic model of the neo-Freudians by asserting the role of the intellect: a position that, historically, represents the grafting of contemporary concepts on the Western intellectual tradition. In his *The Intrapsychic Self: Feeling, Cognition and Creativity in Health and Mental Illness*,[19] he has described the fundamental stages of human development as a succession of the three categories of primary symbolic cognition (phantasmic stage of inner reality followed by the endocept of the preverbal level and the preconceptual level of thinking), secondary conceptual thinking, and tertiary thinking or creativity. Slowly Arieti has arrived at a phenomenological view of the human personality that has considerable relevance for future developments.[24]

Schizophrenia

Schizophrenia has remained to the present the most studied, yet the most baffling, of the various psychiatric syndromes, whether it is considered as a "disease" according to the European tradition (Kraepelin, 1896, Bleuler, 1911) or as a "reaction" according to Adolf Meyer's philosophy; that is, whether the emphasis is put on its difference from neurosis or, conversely, on a continuum of the neurotic process.[70,72,388,550,559] Some genetic studies, especially those on twins carried on at the New York Psychiatric Institute by F. Kallman and then by J. Rainer, have thrown some light on this issue, at least in terms of a premorbid personality due to genic factors developing into a schizophrenic process under the influence of environmental factors.

In the late thirties the empirical approach of shock therapies was emphasized, followed shortly thereafter by the psychosurgical procedures (in this country especially by W. Freeman and J. Watts). This overshadowed the therapeutic method of "total push," geared at a massive utilization of all the patient's resources in the context of the hospital setting, as well as the biological research carried on in a few places, notably at the Worcester State Hospital by R. Hoskins and associates.[293] Such a trend was resumed in the mid-fifties, following the introduction of chemotherapy, resulting in a variety of studies and hypotheses (disturbance of the catecholamines, faulty epinephrine metabolism, serotonine blockade, pathological transmethylation, taraxein or blood protein factor, presence of the plasma protein alpha-2 globulin, finally urinary discharge of dimethoxyphenylethamine [DMPEA]).

On the psychological side the American contribution has focused on the notions of ego integration, "pseudoneurotic schizophrenia" (A. Hoch and P. Polatin)[287] or "ambulatory schizophrenia" (G. Zilboorg),[617] "early infantile autism" (L. Kanner),[312] "symbiotic psychosis" (M. Mahler),[391] other child schizophrenic syndromes (L. Bender, L. Despert, B. Rank, W. Goldfarb, L. Eisenberg, and others). In the above-mentioned *Interpretation of Schizophrenia*, S. Arieti—on the basis of some fundamental notions on paleological thinking enunciated by E. Von Domarus—has indicated that the schizophrenic way of thinking is based on "primary classes" (that is, Freud's primary process) instead of "secondary classes" (that is, the secondary process of the Aristotelian logic).[20,21] Examples of such primary thinking are the spontaneous productions of patients (as in the famous Schreber's "Memoirs"),[45] which are increasingly assessed today from the intrinsic perspective of the psychopathological process rather than from their difference from the normal process of thinking.

In recent years many studies in this country have focused on the issue of the faulty ego development of the patient in relation to his family,[419] already anticipated years ago by the notions of "pseudocommunity" (N. Cameron) and loss of "consensual validation" (H. Sullivan) in paranoid patients. From 1956 on the so-called Palo Alto group (G. Bateson, D. Jackson, J. Haley, and J. Weaklan) has concentrated on the "double-bind" theory of schizophrenia, based on the ambiguous message that the schizophrenic receives from his family members, and on the concept of "pseudomentality," that is, the similarities between the disturbed logic of the schizophrenic and the disturbed interpersonal patterns of his family. Along similar lines T. Lidz and associates at Yale have found evidence of deficiencies of ego nurturing in schizophrenic patients.

The literature of the last decade—for example, *The Origins of Schizophrenia* by J. Romano, *Family Process and Schizophrenia* by Mischler Waxler, *The Meaning of Madness* by C. Rosenbaum and co-workers, and *The Schizophrenic Reactions* by R. Cancro,—has been influenced by the above-mentioned concepts. The various books of the British psychiatrist R. D. Laing have had considerable resonance in this country (so to justify their mention here). Laing's interests have shifted from the phenomenological discussion of the inner process of schizophrenia (*The Divided Self,*

1960), to the dynamics of the communication patterns (*The Self and Others*, 1961; *Sanity, Madness and the Family*, 1964), and recently to a metapsychological position calling for major social and political reforms to make possible the reinsertion of the schizophrenic into society (*The Politics of Experience*, 1967; *The Politics of the Family*, 1969).[86]

Depression

Like schizophrenia, depression has been considered either as a disease (Kraepelin) or as a progressive worsening of a neurotic condition up to a "reaction type" of personality (Meyer's school). No matter what concept psychiatrists adhered to, they came to be increasingly influenced by Freud's famous paper on "Mourning and Melancholia" (1917), until in the forties the attention shifted to the treatment of depression by means of shock therapies.

The emphasis of the British psychoanalytic school (M. Klein) on a normal depressive position in the earlier stages of life never had too much following in this country. American contributions, instead, centered on "anaclitic depression" (R. Spitz)[543] resulting from severe emotional deprivation in infancy and on "bereavement" (E. Lindemann)[378] as a critical condition conducive to maladjustment. Since the introduction of chemotherapy in the mid-fifties, a great deal of research has focused on the biochemical aspect of depression, such as the antidepressant activity of the inhibitors of monamine oxidase (IMAE), the "catecholamine hypothesis" related to deficiency of noradrenaline, and the effectiveness of lithium carbonate in the treatment of the manic phase of the depressive condition.

All this should not overshadow the advances made in the psychodynamic understanding of depression.[51] In particular, Arieti, aside from the common form of self-blaming depression, has described the "claiming type" of depression, in which there is a loss of the "dominant other," that is, of the idealized parental figure toward whom the patient is dependent. Moreover, these types of patients present a "fear of autonomous gratification"—that is, independent of external approval—and their severe ego defect neurosis makes them unable to transform the interpersonal into the intrapsychic; thus they remain quite vulnerable to the loss of sources of self-esteem.[22]

Worth mentioning are the monographs published on manic-depressive psychosis by L. Bellak, on depression by R. Grinker and associates, and on pharmacotherapy of depression by A. Hordern, J. Cole, and J. Wittenborn. From the perspective of epidemiology and public health—which has recently received a great deal of attention along with the biochemical and the dynamic orientations—measures to deal with acute depressions (0.5 to 2 percent of the general population) include 24-hour emergency assistance in many American cities and a variety of recommendations by the Center for the Study of Suicidology, established at the National Institute of Mental Health in Bethesda, Md.

Psychosomatic Medicine

In the thirties a number of psychoanalysts "rediscovered" the centuries-old belief in the influence of the mind upon the body in the wake of Freud's original concept of somatic compliance in the mechanism of hysteria.[395] Certain diseases—peptic ulcer, asthma, rheumatoid arthritis, colitis, dermatitis, hypertension, and hyperthyroidism—were considered as mainly psychosomatic (the ample monograph by F. Dunbar in 1943 is typical).

Under the overall influence of Freud's theory of anxiety (1926), F. Alexander and pupils from 1932 carried on a great deal of research on psychosomatic conditions, which led to the concept of "specificity," that is, of a definite correlation between each one of these conditions and a particular emotional conflict (for example, repressed hostility in hypertension). No matter how meaningful the study of psychological factors has been (such as the correlation between dreams and the biological phases of the menstrual cycle by T. Benedek), the presence of a particular preexisting organ vulnerability (constitutional factor "X") under conditions of stress has been assumed by practically everyone (for example, by J. Mirsky and associates in cases of duodenal ulcer by measuring the secretion of serum pepsinogen).

Other researchers in the field of psychosomatic medicine have made use of Cannon's emergency theory, of Selye's stress theory, of Schur's resomatization concept, of the metapsychological postulates of Hartmann and of Rapaport, of hypnosis, of projective techniques, of verbal behavior in particularly structured interviews (Deutsch's "associative anamnesis," open-end medical interview of the Rochester group), or simply of casual happenings (such as the famous case of a gastric fistula illustrated by G. Engel and associates). All this research has brought up a number of issues: mechanism of expression of the "body language," interplay of voluntary and involuntary innervated systems, alternation of psychosomatic and psychotic conditions, and so forth.

Among the important publications in this field are F. Alexander's *Psychosomatic Medicine*, T. Benedek's *Psychosexual Functions in Women*, F. Deutsch's *On the Mysterious Leap from the Mind to the Body*, A. Garma's *Peptic Ulcer and Psychoanalysis*, and G. Engel's *Comprehensive Psychological Development in Health and Disease*. From the historical perspective two factors stand out: (1) the methodological approach has shifted from the exclusive psychoanalytic to an interdisciplinary one, inclusive of biochemists, internists, and behavioral scientists; and (2) the various theoretical models of psychosomatic disorders based on a closed system appeared to be superseded by models based on an open system, as that presented by the general system theory.

The Psychoanalytic School

The study of the life and work of Freud has continued to be the subject of a number of studies in the last two decades,[61,67,132,387,392,434,485,486,494,523,533,548,596] typically the three-volume monograph by E. Jones[306] completed in 1957. Such a monograph typified the mythical representation of Freud's message to which he himself unconsciously gave a prophetic character. Since then, with the help of newly published material—such as Freud's correspondence with some pupils and friends[214–219] and the minutes of the early Vienna Psychoanalytic Society[437]—important historical studies have appeared by various authors.[104,490] Controversies have arisen concerning Freud's academic career[171,239] and his involvement in the suicide of V. Tausk, author of the classical paper on the "influencing machine" in schizophrenia (1919).[172,492]

All these studies have been overshadowed by the monographs by E. Erikson on Luther (1958) and on Gandhi (1970), which presented an entire historical period from the perspective of the development of one person, thus opening the new field of psychohistory.[101,107,393,394,404,413] This has signified a new advance in the application of psychoanalytic insight to literature and the figurative arts, which has had a long tradition in Europe as well as in this country.[78,169,170,189,320,321,340,578]

Clinical Developments

The most important event in the history of psychoanalysis has been the shift of emphasis from the unconscious to the ego (mainly A. Freud's *The Ego and the Mechanisms of Defense* and H. Hartmann's *Ego Psychology and the Problem of Adaptation*), which occurred in the late thirties, shortly before the exodus of a large number of psychoanalysts from central Europe to this country.[196,301] Freud's anticipation at the occasion of his lectures at Clark University in 1909 that psychoanalysis would receive such a great acceptance in the United States to the point of losing its identity appeared to be confirmed.

Actually the bulk of the psychoanalytic movement remained faithful to Freud's traditional teaching based on the integration of empirical therapeutic procedures and theoretical notions of the structural, economic, genetic, and topographical aspects of the mind, as in Fenichel's classical *Psychoanalytic Theory of Neurosis* (1945). The above-mentioned studies on ego psychology, as well as the research on psychosomatic medicine carried on at the Chicago Psychoanalytic Institute under F. Alexander and the new perspectives presented by the neo-Freudians (C. Thompson, E. Fromm, H. Sullivan), did not have a significant impact until the fifties.

Here mention at least should be made of the outstanding American representatives of the psychoanalytic movement and their particular area of interest: Helene Deutsch for psychology of women; Theresa Benedek for psychosexual disorders of women; Franz Alexander, Carl Binger, Flanders Dumbar, Thomas French, and Roy Grinker for psychosomatic medicine; Felix Deutsch and Maurice Levine for integration of medicine and psychoanalysis; Jules Masserman for experimental neuroses; Spurgeon English, David Levy, Gerald Pearson, and Emmy Sylvester for emotional disturbances in childhood; Erich Lindemann and Nathan Ackerman for family dynamics; Frieda Fromm Reichmann and Gustav Bychowski for psychotherapy of psychosis; Kurt Eissler for psychotherapy of delinquents; Abram Kardiner, Kenneth Appel, Greta Bibring, Phyllis Greenacre, Ives Hendrick, Robert Knight, Lawrence Kubie, Bertrand Lewin, Sandor Lorand, Karl and William Menninger, Herman Nunberg, Clara Thompson, Gregory Zilboorg, and many others for various clinical matters.

In particular, a few words should be said regarding the work of Franz Alexander,[6] which spanned three decades of uninterrupted creativity. Particularly important are *Psychoanalytic Therapy* (1946), *Studies in Psychosomatic Medicine* (1948) for its many innovating techniques, and *Our Age of Unreason* and *Western Mind in Transition*, dealing with broad cultural issues; his various papers collected under *The Scope of Psychoanalysis* (1961) focus especially on the three dynamic principles of homeostasis, economy, and surplus energy. Among the other important American contributors, Thomas French has attempted to represent psychoanalysis as a process of progressive adaptation to achieve integration.

Theoretical Developments: Ego Psychology, Life Cycle

The emphasis on the ego that became prominent in the forties[471,472] left unsolved the issue of its genesis. In the fifties H. Hartmann, E. Kris, and R. Loewenstein tried to explain the development of the ego from an undifferentiated state of id-ego under the influence of (1) congenital ego characteristics, (2) primary instinctual drive, and (3) external realities conducive to ego development.

The main advance related to the concept of adaptation, which was anticipated by H. Nunberg and thoroughly investigated by D. Rapaport, who defined it as the balance of ego autonomy from the id and ego autonomy from the environment. The adaptive point of view was integrated with the genetic one in Hartmann's definition of the ego as the matrix of the personality, mastering the apparatus of internal and external motility and the perception, contact with reality, and inhibition of primary instinctual drives.

Other theoretical developments deal with the role of the somatic ego related to the body image, the concepts of "ego strength," of "area of the ego free of conflicts," and of "neutralized energy" (that is, desexualized and aggression-free energy), and the role of introjection and identification in the formation of the ego.[236] All this has come to signify historically a rapprochement between psychoanalysis and genetic psychology, this latter represented mainly by H. Werner and J. Piaget. In fact, Piaget's books became significant for American psychiatry in the last two decades coincidentally with the advent of ego psychology, leading to attempts to compare psychoanalysis with the school of Geneva (P. Wolff, J. Anthony, and others).

Aside from this the main innovation consists of the concepts developed in this country by the Danish-born and Viennese-trained E. Erikson, a highly creative personality imbued with literary gifts. Erikson views the personality from the perspective of a comprehensive life cycle, in which the "normal" or "normative" rather than the pathological acquires preeminence. Erikson's two most celebrated books, *Young Man Luther* (1956)[181] and *Gandhi's Truth* (1969), represent the best example of an entire historical period viewed in the light of the individual dynamics of an important figure.[178,179,180] In other publications (*Identity and Life Cycle*, 1959; *Insight and Responsibility*, 1964; *Identity: Youth and Crisis*, 1969), all stemming from his basic *Childhood and Society* (1950), he has brought to the fore the identity crisis of adolescence in American society and the basic stages of the life cycle. These latter (in succession "hope," "will," "purpose," "skill," "fidelity," "love," "care," "wisdom"), which he has called "basic virtues," can be easily connected with the fundamental "virtues" of the Judeo-Christian tradition. His views on adolescence have come to be very relevant in light of the growing impact of youngsters in the American cultural scene. Erikson's work has influenced many others (as typically represented by the important volume *The Person* by T. Lidz) and transcends the field of psychoanalysis proper, so one may justifiably question whether Erikson belongs to this school.

New Psychoanalytic Trends

The difficulty of separating theoretical from practical issues has been a constant one in the psychoanalytic school. By and large, however, most American psychoanalysts have remained faithful to the basic principles established by Freud.

As the early generation of European-born psychoanalysts is slowly disappearing, the relevance for psychoanalysis of the biological sciences, on the one hand, and of the social sciences, on the other hand, is being recognized.[290,317,318,396,469,470] The American Academy of Psychoanalysis, established about 15 years ago, has represented this new trend, as evidenced by the proceedings of the meetings edited by J. Masserman under the title *Science and Psychoanalysis*. In the collaborative volume *Modern Psychoanalysis: New Directions and Perspectives* edited by J. Marmor, the views of the main exponents (R. Grinker,[254] J. Ruesch, etc.) of the integration of psychoanalysis—conceived of as an open system—with biological and social sciences are clearly stated: use of findings from the fields of communication theory, electrical engineering, cybernetics, information theory, automation, and computing; consideration of adaptional aspects derived from information, self-regulatory, and transactional systems and consideration of the fields of forces in which the therapeutic relationship takes place; in general, replacement of a "closed system" based on the death instinct, the narcissistic drives of the ego, and the isolation of the psychotherapeutic relationship with an "open system" (von Bertalanffy, 1962) as a "reciprocal and reverberating process," a "transaction, rather than self-action or interaction, which is the effect of one system on another, is the relationship of two or more systems within a specific environment which includes both, not as specific entities, but only as they are in relation to each other within a specific space-time field" (Grinker, 1968).

It is to be hoped that the introduction of biological and social dimensions in the reformulation of psychoanalytic principles will result in a much needed clarification of concepts, elimination of tautologies, and improvement of the communication between psychoanalysts and scientists from other fields.[249] As an example of this, the analysis of the symbolic-linguistic system, which is basic in the psychotherapeutic relationship, is being investigated with the help of a new methodology.

Research

As mentioned above, the American contribution to psychoanalysis has mainly consisted in the clarification and, if possible, the measurement of some of the classical findings by the early psychoanalysts. Among the first examples of this trend, quite often represented by psychologists, is the survey of psychoanalytic data published by R. Sears in 1945.[527] The difficulties inherent to research in psychoanalysis—especially impressionistic bias by the observers, the problem of experimenting with human subjects, and the question of confidentiality—have not been overcome even in the last two decades, when advances were made in the methodological approach to psychiatric research.[128,464,582]

From the developmental perspective, in the last 25 years three main areas have become prominent in psychoanalytic research: assessment of psychoanalytic tenets in various experimental situations, observation of child development in terms of psychoanalytic theory, and measurement of the results of psychoanalytic therapy. In the first area, the research on experimental neurosis—originally introduced by J. Masserman and then by J. Dollard and N. Miller in their classical *Frustration and Aggression*—has been extended more recently to the areas of sleep, hypnosis, sensory deprivation, and mother-child relationship by a number of scientists (H. Middell, H. Harlow, and others), resulting in a rapprochement between psychoanalysis and conditioning. In the second area, a number of centers on child development and treatment have undertaken research projects; perhaps the most important one (following the early studies in the forties by D. Levy and by R. Spitz) is that carried on by A. Freud and coworkers at the Hampstead Child Therapy Clinic in London.[213,516] In the third area falls the Menninger Psychotherapy Research Project, initiated in 1954 and still in progress; it is hoped that this will result in findings more meaningful than those presented in the Report of the Ad Hoc Committee on Central Fact Gathering of the American Psychoanalytic Association.[278]

Present State of the Psychoanalytic School

When the first edition of this *Handbook* was published in 1959, this historical chapter was presented from a perspective in which psychoanalysis occupied a prominent position.[258] Today this is no longer the case, as it has become clear that psychoanalysis is going through a progressive decline following the high peak it reached in the mid-fifties.[36,48,250,260] This is reflected in the expectation of some segments of the population, in the acceptance of psychoanalytic modes of treatment, and ultimately in the change of self-image of the young psychiatrist, who does not identify any longer with the typical sophisticated, reserved, and well-to-do psychoanalyst.

This is not to say that the psychoanalytic tenets based on the development of the personality from the unconscious matrix in the context of family relationships have been replaced by other more relevant systems. Rather, the attitude of many toward these tenets, especially when not sufficiently proven, has become increasingly critical, and attempts are made to view them from a broader interdisciplinary perspective.[259,260] Even before the advent of community psychiatry, criticism of the official position of the psychoanalytic association controlled by Freud and his disciples was very vehement, resulting in a number of secessions: in this country, for instance, the founding of K. Horney's American Institute for Psychoanalysis, of W. A. White's Institute, and of T. Reik's Society for Psychoanalytic Psychology. The other two major issues of the integration of psychoanalysis into medical schools[397] (for example, at Columbia University,[244] the New York Medical College, and the Downstate Medical Center in Brooklyn) and of the exclusion of lay analysts from official recognition (with the exception of very few, such as E. Kris, B. Bornstein, B. Rank, and especially E. Erikson) were particularly debated by the American Psychoanalytic Association, which, founded in 1911, acquired autonomy from the International Psychoanalytic Association in 1938, at the time of the immigration of a large number of analysts from Europe.

While for many years, as it will be shown in the next section of this chapter, the question centered around the acceptance, or rejection, of the official position of the psychoanalytic group, today the main issue concerns the very relevance of psychoanalysis in view of the spread of the community psychiatry movement, which has received massive political and financial support. The reaction of psychoanalysis to this movement has been far from consistent and uniform; it has been punctuated by criticism of community psychiatry for disregarding the basic dyad patient-doctor relationship and training in long-term psychotherapy in order to follow ill-defined methods of treatment and community approaches.[27,37] All this does not mean that psychoanalysis is dying, as some sensational journalistic reports seem to indicate,[369] but rather that it is increasingly seen as a specific technique instead of a general philosophy of treatment.[237,319]

Other Schools and Trends of Psychoanalytic Derivation: Jung, Rank, Adler, Reik, Reich, Klein

Aside from some basic notions presented early in his career and absorbed into the mainstream of the psychoanalytic school (mainly introversion and extroversion, complex and collective unconscious), the work of the Swiss Carl Gustav Jung (1875–1961)[307,592] has received very little notice here. One reason is that his pupils, being of non-Jewish extraction, did not have to emigrate to this country. Particularly ignored is his late production, which has been exceedingly influenced by mystical, esoteric, and religious concepts not very palatable to the pragmatic American mind. Despite the availability of his main writings in this country for many years (and the current publication of his complete works by the Bollingen Foundation), Jung's ideas have found followers mainly in Switzerland and in England (especially M. Fordham, F. Fordham, Y. Jacobi, and A. Jaffe) and in artistic rather than in psychiatric circles.[124,202]

In contrast to Jung, Otto Rank (1884–1939),[559] a brilliant and favored pupil of Freud, has had a considerable following in this country. His influence, however, results not so much from his original concepts—birth trauma, birth of the hero, Doppelgang, and other literary and artistic themes (presented in the journal *Imago*, which he directed)—but from the so-called functional school of social work that he established at the University of Pennsylvania and that was continued by his pupils V. Robinson and J. Taft.

In regard to Alfred Adler (1870–1937),[83,440 445,603] his basic notions of inferiority feelings and of organ inferiority have become universally accepted, in spite of his bitter separation from Freud's school in 1911, followed by the founding of the Society for Individual Psychology. In this country, aside from a few pupils (A. Ansbacher,[14,15] R. Dreikurs, and others), important aspects of his work—notably his application of psychoanalysis to education, resulting in the child guidance movement—have been almost entirely forgotten, probably because of his rather unassuming personality, unconcerned with academic recognition, and because of the poorly organized style of his writings, directed to the general public rather than to professionals.

Other well-known analysts who worked in this country for many years following their arrival from Europe include Theodore Reik (1888–1969), whose books on various clinical aspects of psychoanalysis have met with success, and Wilhelm Reich (1897–1957),[478] who early in his career introduced the innovating notion of "character analysis" (which anticipated ego psychology), then attempted an integration of psychoanalysis and Marxism[442] (very recently brought to the fore again),[479] and eventually became involved in the controversial "discovery" of "orgone" energy as the basis of life.

Finally the "English" school of the German-born Melanie Klein (1882–1960) has had very little impact in this country. Her main views on the development of the superego in infancy (anticipated by her teacher, K. Abraham), on the crucial role of the introjection of "good" and "bad" objects, and on a normal "depressive position" early in life have been considered too overdetermined by the American mentality concerned with environmental influences. M. Klein's main contribution lies in her pioneering use of play therapy in the twenties and in some anticipations of the psychoanalytic therapy of psychotic children.

Original American Contributions. Neo-Freudians, Cultural, and Interpersonal Schools: Rado, Horney, Sullivan, Fromm

Rado, Horney, Sullivan, and Fromm typify the original American contribution to the psychoanalytic movement. Although all of them, with the exception of Sullivan, were European-born, their work has taken place almost exclusively in this country. They have all been influenced by the Swiss-born Adolf Meyer (1866–1950),[373] the founder of the school of psychobiology, which had considerable impact on American psychiatry—probably because of its optimistic view of human nature in contrast to Freud's pessimism—and which was represented by a large number of pupils who eventually acquired leading academic positions in this country as well as abroad. Aside from its eclectic orientation and broad acceptance, psychobiology is generally considered to have facilitated the introduction of the psychoanalytic movement in this country.

The "adaptational psychodynamics" of Sandor Rado (1890–1972), based, like Meyer's philosophy, on an eclectic methodology and on an evolutionary biological orientation, aims at describing the hierarchical levels of central integration and control of the organism's motivation and behavior.

The German-born Karen Horney (1885–1952), an early pupil of Franz Alexander in Chicago, became widely known for her many books directed to the general public, in which she stressed the importance of environmental influences at variance with the rigid aspect of Freud's doctrine of the instincts and of family dynamics (for example, the dominant male role). Her explanation of neurosis as "moving toward," "moving away," and "moving against" and of defenses as "self-effacement," "expansiveness," and "resignation" can be viewed as an anticipation of today's clinical pictures of alienation and lack of emotional involvement.

Harry Stack Sullivan (1892–1949) is unquestionably the most original and significant representative of neo-Freudianism. Under the influence of social scientists (R. Benedict, M. Mead, E. Sapir, H. Lasswell, and others) he departed from Freud's rigid notions of the individual development of the personality (that is, stages of libido, oedipus complex) and elaborated notions based on various modalities of experiencing interpersonal relationship ("prototaxic," "parataxic," and "syntaxic") and on the central role of anxiety as experienced disapproval from others, leading to the appearance of substitutive neurotic and disintegrative psychotic symptoms. Today Sullivan is especially remembered in American psychiatry for his pioneering attempt to view psychotherapy as a mutual learning experience between patient and doctor and to consider even psychoses as treatable through a correction of distorted processes of communication.[560]

Also geared to the general public are the many volumes of the German-born Erich Fromm (b. 1900), in the past associated with the William Alanson White Institute in New York City. His presentation of personality types ("receptive," "exploitative," "hoarding," "marketing," and "productive") reflects his dramatic view of man in conflict between individual aspirations and dehumanizing collective forms of life. Fromm's writings, in which he has paid tribute to both Freud and Marx, have an appealing and engaging style, but are rather peripheral to the central theme of psychiatry proper. The same can be said of the philosopher Herbert Marcuse (b. 1898), whose humanistic defenses of man from the Marxist perspective have been taken as a symbol by the New Left.

It is too early to pass judgment on the historical significance of Fromm, Marcuse, and others. For the neo-Freudians, instead, the comprehensive works by R. Munroe, C. Thompson, and others[100] are available.

Existentialist Schools

Since the existentialist movement pertains essentially to philosophy and developed mainly in Europe, it is important to state that, like the rest of this chapter, this presentation deals exclusively with the *American* developments of existentialism in relation to psychiatry and presupposes a basic knowledge of its main tenets. The matter is complicated by the vagueness of the core and boundaries of existential psychiatry, which inherently defies any attempt at categorization into a definite school with clearly established teaching.

Rather, it is generally accepted that existentialism, a fundamental theme of human existence from the Greeks on, tends to become significant at times of general insecurity and weakening of social institutions, leading to a defense of the uniqueness of the individual person.[489] The sources of the existential movement (mainly Kierkegaard, Dilthey, Husserl, Buber, and Heidegger in Europe and W. James in this country)[192,602] have been well established and presented in comprehensive form, especially in the monograph by H. Spiegelberg.[542] Also it has been said that the essence of the psychotherapeutic relationship includes an existential motive.

The fact remains that in this country, probably in relation to the awareness of new social dimensions (poverty, alienation, racial conflicts, and so forth), existential psychiatry came to the fore in the mid-fifties. Various works by European exponents of existential psychiatry (K. Jaspers,[305,362,380] L. Binswanger,[65] M. Boss,[81] and others)[252] became available in translation; a comprehensive volume on this field (*Existence: A New Dimension in Psychiatry and Psychology*, edited by R. May, E. Angel, and H. Ellenberger, 1958) was published; three journals were founded with the support of G. Allport, C. Rogers, E. Weigert, C. Bühler, H. Murray, and others. The original American contributions worth mentioning are R. May's *The Meaning of Anxiety* (1950), P. Tillich's *The Courage to Be* (1953), and A. Maslow's *Toward a Psychology of Being* (1962); Maslow is a representative of a "third force" in psychology,[242] between behaviorism, on the one hand, and psychoanalysis, on the other hand.

Today, a decade later, it safely can be said that the original impetus of the existential movement has subsided.[415] Even the publications of Erwin Strauss, a distinguished European-born existentialist who has been active in Lexington, Ky., for more than two decades, have received very little notice.[552,553,554] It appears that the reaction of many disenchanted with the traditional American style of life has taken the form of "irrational" group expressions (such as the acceptance of Marxism, the spread of collective movements from the hippies to encounter sessions, the refuge into all kinds of beliefs from occultism to Far Eastern practices), rather than of an individual response like in Europe.

New Trends: Ethology, General System Theory, Ecology, Structuralism

These various trends, though apparently heterogeneous, have in common two main aspects: (1) their appearance in the last two decades or so, in response to dissatisfaction with current concepts of human behavior as not relevant to the new needs of man's changing role under the pressure of collective systems; (2) their interdisciplinary approach, from comparative neuroanatomy to anthropology, sociology, electrical engineering, and environmental planning, aimed at facing today's overwhelming problems of population explosion, environmental contamination, and rise of underdeveloped nations by defending the humanistic core of the individual without escaping from the world, to the point of constituting a sort of "new utopia" (W. Boguslaw).[74] Also two of them, ethology and structuralism, are of European origin and based on innate and congenital postulates; the other two, general system theory and ecology, are of American origin and based on environmental and behavioristic postulates.

Their relevance to psychiatry can be summarized in a few points. Ethology,[443,614] initiated by Lorenz, Tinbergen, and others, is mainly related to the findings of developmental psychology (R. Spitz, P. Wolff) and comparative development (H. Harlow). General systems theory (L. Bertalanffy) based on the notions of homeostasis, transactional relationship, and communication and information processes,[62] has resulted in works by R. Grinker (*Toward a Unified Theory of Human Behavior*), by K. Menninger (*The Vital Balance*, a new classification of mental disorders based on this theory), by S. Arieti (who has stressed that mental dysfunction is a system disturbance rather than a loss of single functions, especially in schizophrenia), and by others (J. Ruesch's concern with the human aspects of systems, J. Spiegel's notion of foci in a transactional field, L. Frank's views on organized complexity, J. Miller's behavioral theory as having relevance for community mental health), all presented in the recent volume *General Systems Theory and Psychiatry*, edited by W. Gray, F. Duhl, and N. Rizzo. Ecology, rapidly seen as important for psychology (for example, in *Environmental Psychology: Man and His Physical Setting*, edited by H. Proshansky, W. Ittelson, and L. Rivlin) has resulted in the new ecological model of mental illness and treatment based on the interplay between the individual and his environment (E. Auerswald).[30] Finally structuralism (mainly founded on the writings of the French anthropologist C. Lévi-Strauss[283]) is still too new for its relevance for psychiatry to be seen,[168,354,359,433,576] but it has definite connections with psycholinguistics, communication theory, and transcultural psychiatry, as well as psychology (for this latter, mainly in the book recently published on structuralism by J. Piaget).[449]

Overall Development of Psychiatric Treatment: From Hospital to Community

The expression "community psychiatry" has become increasingly popular in the last ten years. From the historical perspective of this chapter, two points are important in relation to this issue: the developments that led to the preeminence of community psychiatry and the definition of its core and boundaries vis-à-vis other collateral fields.

In regard to the first point, throughout history the mentally ill have been seen in different ways, from being possessed by devils to being emissaries of gods, and consequently worshiped, tortured, or simply neglected. Recent historical studies (mainly by E. Ackerknecht,[2] G. Rosen, I. Galdston, M. Foucault, and others)[42] have attempted to investigate the social and cultural dimensions underlying these various attitudes. In this country a progression can be followed from the emphasis on institutionalization during the late nineteenth century to the recognition of the value of treating the patient in his own environment, to the awareness of prevention of mental disorders, and finally to the ambitious plan of making psychiatry available to everyone at the community level.[74] Some historical presentations of community psychiatry (by J. Ewalt and P. Ewalt,[184] J. Brand,[90] W. Barton,[41] A. Freedman,[211] W. Ryan,[511] I. Galdston[226] and A. Rossi[505]) are available.

In regard to the second point, community psychiatry has to be differentiated from social psychiatry. This latter, first defined in this country by T. Rennie[482] in 1956 as concerned with individual and collective forces in relation to adaptation and psychopathology, has been from time to time especially interested in environmental (F. Redlich and M. Pepper),[476] sociocultural (A. Leighton), transcultural (E. Wittkower),[607] ecological (J. Ruesch)[508] and interdisciplinary (N. Bell and J. Spiegel)[47] aspects. In general, the focus of social psychiatry is on theory and research in relation to sociological theories and ecological models.

Instead, the focus of community psychiatry —concretely represented through the concept of "catchment area" as an area of 50,000–75,000 people identifiable for common ethnic, social, and cultural dimensions—is on treatment and on prevention. Treatment is carried on in a variety of ways—also due to the different ethnic, cultural, and religious backgrounds of various groups in this country—justifying the criticism of being "a movement without a philosophy." Prevention relies heavily on the fields of epidemiology and public health and has achieved recognition especially through the many studies published by G. Caplan, who is responsible for the subdivisions of primary, secondary, and tertiary prevention. It is likely that the various schools of community psychiatry now operating in this country (especially important are those at Harvard, Columbia, and the University of California) will contribute in time to the clarification of this new field.

The Movement toward Community Psychiatry: "Action for Mental Health"

The movement of community psychiatry, which was officially initiated in 1961 with the publication of *Action for Mental Health*, represented the culmination of a long period of incubation and the convergence of various trends that can be followed for a considerable period.

Prior to World War II some developments anticipated themes central to community psychiatry: opening of outpatient clinics for adults and then for children, organization of psychiatric social work, A. Meyer's pioneering views of today's concept of "catchment area,"[412] interdisciplinary input by sociology[498] and anthropology, lay involvement in the mental hygiene movement, and new therapeutic optimism derived from collective forms of treatment.

It is well known that World War II emphasized the magnitude of psychiatric disorders and the need for a national program to adequately face this issue. In succession a series of steps were taken: establishment of vast facilities for treatment and training by the Veterans' Administration; passing of the Hill-Burton Act for federal assistance to allocate psychiatric beds in general hospitals (which now number more than 30,000); foundation of the National Institute of Mental Health in 1949, which, under the long leadership of R. Felix,[195] developed a Community Services Branch; participation in research on community aspects of mental health by some foundations, notably the Milbank Memorial Fund.

Meanwhile, in the psychiatric field the social structure of the mental hospital was first described by A. Stanton and M. Schwartz in their pioneering study *The Mental Hospital* (1954), carried out at Chesnut Lodge in Rockville, Md.; this received ample recognition and was followed by others (for example, *From Custodial to Therapeutic* and *The Patient and the Mental Hospital*, both edited by M. Greenblatt, *et al.*). Also the important research on the sociological aspects of psychiatric treatment by A. Hollingshead and F. Redlich and the "Mid-Town Study" on psychiatric epidemiology in Manhattan by L. Srole, *et al.*, were published.

All this, as well as other developments, eventually had an impact on the political scene. Community Mental Health Acts to assist local community programs were approved (first in New York State in 1954), and the annual Governors' Conference on Mental Health offered the impetus for passing adequate legislation. As a result of the Mental Health Study Act of 1955, the Joint Commission on Mental Illness and Health was established under the leadership of K. Appel and L. Bartemeier to make an assessment of the system of treatment and care of the mentally ill, identify needs, and propose recommendations.

At the end of five years of work[40,185] the Commission's Chairman J. Ewalt and collaborators found that the 13,000 psychiatrists then available were largely insufficient to take care of the large number of people in need of assistance; the 1,250 state institutions, where the great majority of the 700,000 mental patients were, tended to be overcrowded and understaffed; moreover, less than one million people were treated as outpatients, although statistics showed that 10 per cent of the general population were affected by nervous and mental illness.

The Commission's recommendations centered around three points: (1) improvement in the utilization of manpower by gearing psychiatrists toward community mental health and relying on the help of other professionals and nonprofessionals; (2) opening of many new facilities, such as clinics, psychiatric wards in general hospitals, and centers for rehabilitation; (3) provision of adequate funds at local, state, and federal levels.

Aside from the widely distributed and comprehensive volume, *Action for Mental Health*, nine other books were published on the following topics: concepts and public images of mental health, economics, manpower, community resources, epidemiology, role of schools and churches in mental health, new perspectives on mental patient care, and research resources in mental health. A basic suggestion for the implementation of this new approach was to convert large state hospitals into units of no more than 1,000 patients and to provide a mental health clinic for each 50,000 people.

With the support of many professionals and laymen (such as Mary Lasker and Mike Gorman, executive director of the National Committee against Mental Illness), as well as legislators (mainly Senator L. Hill, Congressmen J. Priest, O. Harris, and J. Fogarty, and A. Ribicoff, then Secretary of Health, Education, and Welfare), proposals to implement the recommendations of the Commission were introduced in Congress. Significant impetus toward the success of this endeavor was provided by the late President Kennedy; on February 5, 1963, in his memorable message to the 88th Congress on mental illness and retardation, he indicated that what was needed was "a national mental health program to assist in the inauguration of a wholly new emphasis and approach to care for the mentally ill—which will return mental health to the mainstream of American medicine, and at the same time upgrade mental health services."[458]

Eventually this political action (described in detail in *Politics of Mental Health*, edited by R. Connery)[130] resulted in the passing of the Community Mental Health Act in 1963, which provided for the establishment of a community mental health center for each "catchment area" of about 75,000 people. Any center had to offer five types of services: inpatient, outpatient, partial hospitalization, emergency and consultation, and education. Other services, such as diagnostic, vocational, training, and research were also recommended, but not mandated. In addition to the $150 million over a three-year period to finance these centers, an amendment to the Act was signed by President Johnson in 1965 to provide federal funds also for staffing. By 1970 more than 400 centers were in operation: some received construction grants, others staffing grants or both. The NIMH budget for that year was $348 million, while moneys allocated by the states reached $2½ billion.

Developments in the Pattern of Delivery of Services: Mental Hospitals, Outpatient Clinics, Community Mental Health Centers

Historically services for the mentally ill have developed according to the order followed in the above heading. Mental hospitals first appeared at the end of the eighteenth century in the process of differentiating the mentally ill from all other outcasts of society; the philosophy of "moral treatment" based on the paternalistic approach of the superintendent was carried on successfully in the small, homogeneous private mental hospitals in the early nineteenth century; later on, with the arrival of many immigrants and the expansion of the frontier, the mentally ill were increasingly institutionalized in large state institutions (many built under the impetus of D. Dix's crusade)[398] where treatment became more impersonal and custodial.[110]

Around the second decade of our century, under the influence of several currents (progressivism, psychoanalysis, behaviorism, and others) outpatient treatment for many people affected with emotional disturbances gained momentum. However, the practice persisted of keeping the mentally ill anonymously in large institutions away from the community. At times it reached the point of neglect and despair, as portrayed in *The Shame of the States* by A. Deutsch and in *The Snake Pit* by M. Ward.

Only in the mid-fifties two concomitant developments, the introduction of chemotherapy on a large scale and a more accepting attitude toward mental illness on the part of many, resulted in a substantial improvement in the delivery of services to mental patients. The first important step in this direction was the "therapeutic community" described by M. Jones in England in 1953. As a result of clarification of structures, roles, and role relationship reached in mental hospitals through T groups, sensitivity training, crisis situations, and face-to-face confrontations, there was an improvement in staff-patient interaction and increased participation by patients in the therapeutic program (on such issues as confidentiality, authority, decision making, and limit setting.)[37]

In time other modalities of treatment were introduced, first in Europe and then in this country: "day hospitals" for patients not needing full hospitalization; family care[452] and aftercare services; ex-patient clubs; use of volunteers; assignment of patients from the same geographical area to a "unit" in the state hospital to facilitate contacts with their community. All these pioneering endeavors came to be named the "open-door policy" (M. Jones), that is, a shift from a custodial to a therapeutic setting and from a closed to an open system. "Therapeutic community" has also become a very commonly used expression, not only in terms of the mental hospitals, but also in terms of the community at large. As one would expect, these developments have resulted in a progressive decrease—for the first time in the last 150 years—in the number of hospitalized patients from about 559,000 in 1953 to less than 425,000 fifteen years later.

Recently the notion of "revolving door" has been introduced to signify the flexible approach both of the hospital and of the community, as quite often the problems of the mentally ill cannot be properly met simply by transferring the responsibility for the patient from the institution to his family. Worth mentioning also is the significant role that private mental hospitals (about 170 caring for almost 17,000 patients and involved in the National Association of Private Psychiatric Hospitals) have played in the above-mentioned developments.

The literature on all these events is quite extensive. Among the most valuable works are *The Therapeutic Community* by M. Jones, *The Psychiatric Hospital as a Small Society* by W. Caudill, *Day Hospital* by B. Kramer, *The Prevention of Hospitalization* by M. Greenblatt, *Partial Hospitalization for the Mentally Ill* by Glasscote, *et al.*, *The Day Treatment Center* by Meltzoff and Blumenthal, *The Treatment of Family in Crisis* by Langsley and Kaplan, *Community as Doctor* by R. Rapaport, *Social Psychiatry in Action: A Therapeutic Community* by H. Wiener, and *The Psychiatric Hospital as a Therapeutic Community* by A. Gralnick.

Outpatient clinics have an important tradition in this country, which can be traced back to the convergence of various movements of social work, voluntary agency, welfare programs, settlement houses, and others early in this century. The original philosophy of these clinics was eclectic and depended largely on community resources. A number of clinics (281 out of a total of 373 in 1935) served patients discharged from mental hospitals, and most of them were located in the five states of New York, Massachusetts, Pennsylvania, New Jersey, and Michigan.

In the forties and fifties, under the influence of the psychoanalytic school, clinics came to be geared toward long-term treatment of intrapsychic problems by members of various disciplines (mainly psychology and social work) that identified with the psychotherapeutic role of the physicians.[139] Increasingly the philosophy of treatment tended to favor young, intelligent, and sophisticated patients whose values were similar to that of the staff, while the contact with community agencies and schools became negligible.

Around the mid-fifties the country suddenly became aware of the conditions of poverty, neglect, and rejection of a considerable segment of the population (M. Harrington, F. Riessman, M. Deutsch, and others). The fact that middle-class people tended to be treated in clinics while low-class people ended in mental hospitals was well documented by A. Hollingshead and F. Redlich.[220] Slowly many sicker patients, no longer in need of institutional care because of the success of psychopharmacological treatment and the above-mentioned open-door policy of mental hospitals, came to be treated by outpatient clinics with the help of new techniques, such as family and group therapy.

However, the controversies concerning the role of the approximately 2,000 clinics now existing (organized under POCA, Psychiatric Outpatient Centers of America) are far from over. One of the sharpest critics, G. Albee, has written that "the psychiatric clinics in the United States are treating the wrong people; they are using the wrong methods; they are located in the wrong places; they are improperly staffed and administered; and they require vast and widespread overhaul if they are to continue to exist as a viable institution."[5]

In the context of the community mental health movement, the philosophy of the clinics tends to be influenced by social factors: new therapeutic modalities aimed at treating low-income and culturally disadvantaged groups, as well as patients in critical need of treatment (adolescents, alcoholics, drug addicts, etc.), are developed; efforts are made to open clinics in rural areas and in Midwestern and Southern states with the substantial help of public moneys and, to a less extent, of insurance coverage.

The main issue remains the identity of the outpatient clinic vis-à-vis the community mental health movement. This is colored by considerable ambivalence:[57] on the one hand, the nostalgic feeling toward the traditional small clinic whose staff was quite involved with the patients; on the other hand, the commitment to serve as many people from all backgrounds as possible, in the context of the network of medical, social, educational, and rehabilitation services in the community.

Community mental health centers, being only less than a decade old, are difficult to assess from the historical perspective.[238,267,507,613] The complexity of any one of such centers, composed of various agencies staffed by an heterogeneous group and located in areas culturally different, contributes greatly to such difficulty.

Yet ten years from *Action for Mental Health*, some trends concerning the development of community mental health have emerged. It is unquestionable that the great expectations raised initially that this movement would constitute a "third" (N. Hobbs)[286,600] or a "fourth" (L. Linn)[379] psychiatric revolution are not accepted by many. At best it is accepted that this movement helped considerably to create a climate of more favorable acceptance toward emotional disorders and a more optimistic outlook toward their treatment.

However, it is increasingly recognized that the mental health movement is not a panacea for gigantic social problems, from the Vietnam War to drug addiction to changes in the traditional values of this country (J. Seeley). Too often existing services, no matter how labeled, have remained unchanged and therapeutic modalities have remained unaffected by this movement; the role of the so-called paraprofessionals or "activators," torn between their commitment to treating the underprivileged and their identification with the professionals, has become controversial.

From an overall perspective the dilemma of the psychiatrist toward the patient, or toward the community forces tending to control the patient, has been brought forward, notably in a dramatic form by T. Szasz. Moreover, criticism has been expressed in leftist quarters, mainly in nonpsychiatric literature, toward the "psychiatrization" of social conflicts[9] (for instance, turning delinquency and youth unrest into an illness); on the opposite side conservative groups have seen the mental health movement as a plot against patriotism by government infringement on the mental health of the citizens.[29,39,225,512]

Among the professionals it is commonly accepted that this movement, while making good use of principles of epidemiology and of public health, lacks a conceptual foundation, to the point of being called "a movement in search of a theory" (J. Newbrough)[436] or "the newest therapeutic bandwagon" (H. Dunham).[160] From the psychoanalytic viewpoint the approach stressed by this movement has been characterized as "a retreat from the patient" (L. Kubie),[341,344] and a strong defense of the "medical" (that is, "dyadic") model of the doctor-patient relationship[256] over the "social" model has been voiced by some (R. Kaufman,[316] L. Kolb).[331]

All this should not deter anyone from recognizing the moral implications of a movement that, in line with the American democratic tradition, attempts to bring help to the largest possible number of people in need. Unquestionably some positive results have been achieved: a more flexible use of professionals reached through a slow reorientation of goals and functions; the integration of many nonprofessionals in the work of each community mental health center; the new pattern of cooperation with social agencies, schools, institutions, and other facilities in the community; the increasing integration of health and mental health services; the progressive acceptance of responsibility toward the emotionally disturbed on the part of local, state, and federal agencies; and last but not least, the more accepting attitude toward mental illness by many segments of the population.

These points, and others, have been brought forward in a number of publications, such as the *Handbook of Community Psychiatry* (edited by L. Bellak), *Perspectives in Community Mental Health* (edited by A. Bindman, R. Williams, and L. Ozarin), *Progress in Community Mental Health* (edited by L. Bellak and H. Barten), as well as in special journals (mainly the *Community Mental Health Journal* and *Hospital and Community Psychiatry*).

Thus far most of the impetus toward community psychiatry has taken place in the states of the East Coast, in California, and in some Midwestern states. The realignment of national priorities related to the slight economic recession and other social problems of this country indicates that local communities will have to assume most of the responsibilities for the community mental health movement. How this will affect the success of this movement in the long run remains to be seen.

¶ Therapy

Any attempt at modifying the mental functioning of a person has to be viewed from the perspective of the theory of mind and body prevailing in each culture at a particular period. In the Western tradition the centuries-old Aristotelian notion of a body-mind unity was replaced by Descartes' splitting of body and mind in the seventeenth century. Consequently mental disorders, which were traditionally considered in the light of that unity, came to be "discovered" from that time on.

In regard to therapy, the decades between the end of the eighteenth century and the beginning of the nineteenth century saw the rise of mesmerism for neurotic patients and of moral treatment for psychotic patients; later on for several decades therapy was influenced by the organogenic notion of the ascendancy of the body over the mind; psychoanalysis reversed this situation by emphasizing the characteristics of mental functions and the treatment of neurotic disorders. The psychodynamic trend has persisted to our days, although organic theories of the mind became prominent again in the thirties in connection with the introduction of shock therapies and in the fifties with the discovery of chemotherapy.[567]

While all this justifiably has aroused in many the urge to reach a new unitary concept of body and mind, the orientation of most psychiatrists with the psychogenic or organogenic tradition makes this goal unattainable at present. The need to replace today's hybrid eclectism with a comprehensive formulation of body and mind—perhaps based on the new ecological framework of the general systems theory—may very well constitute the challenge of the seventies.

Organic Therapies

SHOCK THERAPIES

The notion that sudden and unexpected events (such as a loud noise, an unpredictable shower, or a fall into the water) may alter the mental status of a person is very old; it was used empirically by some German psychiatrists in the early nineteenth century.

In the late thirties shock therapies were introduced in a matter of a few years, first in Europe and then in the United States.[514] M. Sakel (1900–1957) initiated insulin coma in Berlin and Vienna; J. Meduna (1896–1964) cardiazol shock in Budapest; V. Cerletti (1877–1963) and L. Bini (1908–1964) electric convulsions in Rome. The first two moved to this country shortly thereafter, so that their therapies became an intrinsic part of American psychiatry, while electric shock was imported here by a few European psychiatrists (mainly L. Kalinowski, R. Almansi, and D. Impastato).

From the historical perspective the significance of shock therapies has been to bring new optimism to the treatment of psychiatric conditions, which had been largely missing in the psychodynamic schools, both in the patients and in the professionals. After the wave of enthusiasm shock therapies came to be limited mainly to electric shock, because of its easy use and safety, and to the treatment of forms of depression. The efforts of many to find the explanation of the intrinsic mechanism of action of shock therapies (based on biochemical abnormalities or on other notions) have been unsatisfactory. Also limited have been the technical improvements, such as use of anesthesia and various substances to achieve relaxation. Worth mentioning also are the psychological implications of shock therapies, that is, the patient's regression and dependence on the staff coupled with symbolic death and rebirth. In the last two decades, in connection with the rise of psychopharmacology, the literature on shock treatment has decreased considerably. The classic book on the subject is *Shock Treatment, Psychosurgery and Other Somatic Procedures in Psychiatry* by L. Kalinowski and P. Hoch, continually brought up-to-date.

PSYCHOSURGERY

Archaeological remnants of past civilizations bring evidence that skull trepanation for the purpose of liberating epileptic as well as mental patients from the alleged possession by evil spirits was extensively practiced. Surgical interventions on the brain are recorded in Roman, Byzantine, and Arabian medicine, while, later on, caution prevailed in connection with the discoveries of the delicate functions of the central nervous system.

By the thirties some knowledge had been gathered on the relationship between cortical and subcortical functions on the basis of data obtained from ablation of frontal lobes in monkeys (J. Fulton, C. Jacobsen), from electroencephalography, and from stimulation and inhibition of cortical areas. The Portuguese Nobel Prize winner Egas Moniz (1874–1955) was the first to perform a successful lobotomy in 1936. His technique was imported to this country and widely used for a number of psychiatric conditions in the forties, mainly by W. Freemann and Y. Watts in Washington, D.C., and elsewhere.

Regardless of the new surgical techniques—topectomy, thalathomy, cingulectomy, and others—the opposition to psychosurgery has mounted in professional quarters in regard to the indications for selection of patients and the postoperative impairment of intellectual functioning and will. Lay and religious groups, from the Catholic Church to Soviet Russia, have condemned psychosurgery on moral grounds. The controversies about psychosurgery have decreased considerably because of its decline and the corresponding rise of psychopharmacology, which can result in a sort of "functional" lobotomy without producing personality changes and moral conflicts. Very recently some of these issues have been raised again in connection with the research by J. Delgado at Yale University on modification of psychotic behavior through electrodes implanted in various areas of the brain.[146]

PSYCHOPHARMACOLOGY

The field of psychopharmacology, less than two decades old, has become one of the most important in psychiatry.[108] Today's extensive use of "drugs of the mind" has brought forward many similarities between them and a variety of drugs employed in magic-religious ceremonies of healing in preliterate cultures.[190,272,291,346,587,588] Comparative research has been carried on by some, often working in interdisciplinary teams of pharmacologists, psychiatrists, anthropologists, and others.

In the history of Western medicine, aside from hellebore in Greek times, the list of the "drugs of the mind" includes antimony, belladonna, hyoscyamus, cannabis indica, quinina, followed in the nineteenth century and later on by opium, bromides, chloral hydrate, paraldehyde, and finally barbiturates.[50,163] Mescaline was isolated in 1896 by the German L. Lewin (who published a famous book on the drugs of the mind)[369] and later synthetized, and its hallucinogenic effects were described in the German literature in this century.[56]

Later on the therapeutic importance of diphenylhydantoin for some forms of epilepsy was proved by the neurologist T. Putnam. In 1938 lysergic acid diethylamide (LSD 25) was discovered by the Swiss chemist A. Hoffmann, and in the forties amphetamines (benzedrine and dexedrine) were studied by G. Alles, while serotonin was isolated by I. H. Page in 1945. By that time extensive use had been made during World War II of sodium amytal and similar compounds in narcocatharsis, narcoanalysis, and narcosynthesis for the intensive treatment of acute breakdowns. On a more theoretical basis research on hormonal substances was carried on in some centers, notably at the Worcester Foundation by H. Hoagland and associates.

The credit for first having used chlorpromazine in psychotic and agitated patients is attributed to the French P. Deniker, H. Leborit, and J. Delay early in the fifties. This opened the way to a great step forward in psychiatry, that is, to a more optimistic view of mental illness on the part of professionals and patients and ultimately of the community. In rapid succession a number of other important drugs were discovered and used: meprobamates by F. Berger and B. Ludwig (1950); LSD 25 for clinical purposes by J. Elkes; reserpine by the Swiss H. Bein (1956); the antidepressing imipramine by the Swiss R. Kuhn (1957); butyrophenone by the Belgian P. Janssen (1958); chlorprothixene (Taractan) by the Danish R. Ravn (1959); benzodiazepines (Librium) by I. Cohen at the University of Texas (1960); finally the antimanic effects of lithium by the New Zealander J. Cade in the sixties.

Regardless of national boundaries, great rapidity has characterized the use of these new drugs, be these "tranquilizers" (a term first used by F. Yonkman) or "psychic energizers" (a term coined by N. Kline) or "neuroleptic drugs," which is the term commonly used in Europe (originally introduced by Delay and Deniker). Impetus toward research and practical application of psychopharmacology has resulted from the establishment of research centers (mainly the Psychopharmacological Service of the NIMH and the one at St. Elizabeths Hospital in Washington), from the sponsoring of a number of international symposia with the help of private organizations (such as the Macy Foundation), from the foundation of the Collegium Internationale Neuropsychopharmacologicum in 1957, and from the publications of important serial volumes (such as *Recent Advances in Biological Psychiatry*, edited by J. Wortis). Also worth mentioning from the historical perspective is the monograph containing the proceedings held at Taylor Manor Hospital in Baltimore in 1970 by the discoverers of psychopharmacology.[321] From the lively account of the participants one learns about the creative process of discovery, the interplay of pure research and the interests of supporting drug companies, legal aspects in various nations, and the continuity of the tradition of "drugs of the mind" from preliterate cultures to our civilization.

No matter to what psychiatric school one adheres, he cannot dispute the value of psychopharmacology in alleviating many emotional conditions—especially some that previously required hospitalization. On the other hand, exaggerated expectations about the uncritical use of "psychomimetics" (mainly LSD 25) in the treatment of mental disorders and especially about the power of some drugs to enlarge the field of consciousness and provide new philosophical and religious insights are unrealistic. Concretely many use chemotherapy in conjunction with psychotherapy, regardless of the psychodynamic aspects of the administration of drugs, mainly the orally dependent and the suggestive effects (as proved by some research on placebo). From a broader perspective psychopharmacology has resulted in a much more enlightened attitude toward mental illness on the part of many general practitioners and other physicians and especially the general population at large.

Psychological Therapies

Since the first edition of this book was published, psychological therapies have also undergone a considerable process of reassessment from the historical perspective. Their importance was unquestionably brought forward by Freud's basic concept of the one-to-one relationship. Following the widespread acceptance of Freud's ideas, for a number of years such psychoanalysts as G. Roheim investigated healing practices for mental disorders carried on in past or present preliterate cultures.

The innovation that has taken place recently consists in the new methodological approach toward such healing practices by researchers well versed in psychiatry and anthropology (for example, G. Devereux, C. Kluckhohn, A. Leighton,[206,364] J. Frank, M. Opler). Considerable light has been thrown on the causes of mental disorders in preliterate cultures, be these nonphysical events (that is, power of devils or ancestors and action of words and deeds) or events attributed to the person himself (that is,[77,361] disregard for certain taboos). Also in the last decade or so, methods of psychological healing stemming from the Greek tradition and their relation to present methods have been made the subject of thorough studies by W. Riese, P. Lain Entralgo, H. Ellenberger, and others.[222,435]

Psychotherapy has unquestionably become more accepted in this country in recent years,[273,568] as evidenced by the foundation of the American Academy of Psychotherapy in 1959 and by the annual publication *Progress in Psychotherapy*, edited by J. Masserman since 1956. However, a critical view of the psychotherapeutic process has been advanced by some: for example, J. Ehrenwald's[166] notion of "doctrinal compliance" to explain the tendency of the therapist to fit everything into his own system. Moreover, the boundaries of psychological therapies have been loosened considerably, not only by the success of nondyadic modalities, such as family and group therapy, but by the spread of new approaches, from brief therapy and crisis intervention to encounter groups, and by the inclusion of nonprofessionals among the therapists.

Underlying many of these developments appears to be a basic conflict between the traditional Freudian approach based on the doctor-patient relationship in a stable cultural context and the new collective approaches resulting from the urge toward action brought forward by the pressing social problems of this country.

Individual Psychotherapy: Psychoanalytic Psychotherapy and Psychoanalytically Oriented Psychotherapy

For the historian it is intriguing to investigate the causes of a major shift in regard to psychoanalytic therapy that has taken place in the last dozen years. In 1959, when this book first appeared, this chapter was written from the perspective of the preeminence of the psychoanalytic doctrine in the overall field of psychiatry, in terms of expectations from the sophisticated self-image of the psychiatrists, methods of psychiatric training, and doctrinal adherence of most of the psychiatrists in teaching positions.

As a result of many currents, such as the well-documented study by A. Hollingshead and F. Redlich showing that psychotherapy was available only to middle- and upper-class patients while lower-class patients tended to receive somatic therapies in institutions, *Action for Mental Health* (1961) offered a nationwide guideline for delivery of psychiatric services to all the citizens of the nation. Increasingly the traditional psychoanalytic methods have become diluted by emphasizing symptoms at the expense of the basic personality and healthy potential rather than pathology, to the point of fully justifying Freud's prediction that psychoanalysis would become so accepted in this country that it would lose its identity.

This is paralleled by the trend among professionals, either well-trained psychoanalysts or psychiatrists, to make wide use of psychoanalytically oriented psychotherapy and to restrict classical psychoanalysis to few patients in need of such procedure, which, in addition, is costly and available only in some urban areas. Moreover, the tendency toward integration of psychoanalysis in the training curriculum of some medical centers has been the subject of bitter controversies between those inclined to a dogmatic defense of Freud's message and those open to a dialogue with adherents of other schools.

In the attempt to introduce clarity and specificity in psychoanalytic therapy, many new nomenclatures have been presented in the literature. Aside from the loose distinction of deep versus superficial, insight versus supportive, verbal versus active therapy, descriptions have been offered of listening, clarification, confrontation, interpretation, suggestion, prohibition, and manipulation, and therapeutic techniques have been listed as suggestive, abreactive, clarifying, interpretative, suppressive, expressive, supportive, exploratory, educational, up to paternal and maternal. To the classic "correctional emotional experience" of Alexander and French have been added the intensive psychotherapy of psychosis (Fromm-Reichmann), the "sector therapy" (F. Deutsch), the "anaclitic therapy" (Margolin and Lindemann), the "diatrophic relationship" (Gitelson), the "working alliance" (Greenson), and the "therapeutic alliance" (Zetzel).

Essentially what these esoteric denominations have in common are: the emphasis on ego psychology and analysis of defenses (A. Freud, Hartmann, Rapaport, Erikson, Lowenstein, Kris) and on current developmental crises rather than exploration of the unconscious; a more modest view of the healing role of the psychoanalyst; and, historically, a return to some themes brought forward in the early psychoanalytic literature and then forgotten.

Even so, dissatisfaction toward the psychoanalytic movement in toto is mounting, and a "generation gap" is emerging between classical psychoanalysts and young therapists open to eclectic and unorthodox approaches and more attuned to present social realities. While the basic clinical postulates of psychoanalysis will remain valid in the future, it is difficult to predict the outcome of the theoretical foundations of this movement. At the moment the attempts to graft them on a broader and more relevant context, such as Grinker's "transactional" views, seem the most promising and fruitful.

Hypnosis

Hypnosis, scientifically practiced in the late nineteenth century by the Salpêtrière School (Charcot) and by the Nancy School (Liébeault and Bernheim), is at the root of the early psychotherapeutic treatment of neuroses practiced by Freud and Breuer in cases of hysteria in 1893 and 1895. As the psychodynamic school gained momentum, the historical roots of hypnosis, traceable to Mesmer and, further back, to the Greeks and preliterate cultures, were illustrated by some.[3,564]

It is well known that Freud rejected hypnosis after a few years and that its use for anesthetic and surgical purposes, introduced in the mid-nineteenth century, was soon forgotten, also as a result of the theatrical use of hypnosis. Only in the forties, following some pioneer work by C. Hull (*Hypnosis and Suggestibility,* 1933), was hypnosis scientifically investigated by M. Erickson, L. Mecron, J. Schneck, L. Wolberg, and others. By that time considerable experience had been gathered in hypnotherapy and narcoanalysis during World War II. Eventually the Society for Clinical and Experimental Hypnosis (1949) and later the American Society of Clinical Hypnosis, which publishes *The American Journal of Clinical Hypnosis,* were founded.

In recent years the theories of play acting to please the hypnotist (M. Orne) and of archaic oral-dependent relationship between subject and therapist (M. Gill and M. Brennan) have been postulated. Also the connection between hypnosis and depth and extension of the field of consciousness achieved through the use of particular drugs (LSD 25, mescaline, and others) and the relationship of hypnosis to behavior modifications have been investigated. Despite these developments the future of hypnosis remains vague at this point, although its value in conjunction with psychotherapy is well established.

CLIENT-CENTERED THERAPY

Among the various schools of psychotherapy the only one with a fully American origin was developed by the psychologist Carl Rogers (b. 1902) first in Ohio (1940–1945) and then in Illinois (1945–1950). This school focuses on the genuine, understanding, involved, yet supposedly neutral attitude of the therapist, who continually reflects his feelings toward his client. Historically the lay analyst Otto Rank early in the century advocated the analysis of the therapist's feelings and respect for the patient.

Client-centered therapy appears to have elicited an ambivalent reaction from American psychiatry: on the one hand, psychologists, counselors, and other members of nonmedical groups have mainly used this form of treatment for young and sophisticated clients suffering from problems rather than definite clinical entities (thus nonpatients in the medical sense); on the other hand, Rogers and his pupils have done important research on psychotherapy with the help of purposely designed inventories and tests. In essence this school represents a combination of the humanistic defense of the person and the scientific approach to psychotherapy.

GROUP PSYCHOTHERAPY

Group psychotherapy results from the confluence of many trends originally independent from psychiatry. The American inclination toward collective gatherings of different types (intellectual, political, religious) historically appears to be an attempt to overcome the feelings of isolation resulting from the loss of the support provided by each society from which the immigrant groups came. Moreover, in the European societies centuries-old cultural and religious ceremonies contributed to the release of emotions.

In this country great release of feelings was achieved by some religious sects—notably the Christian Science movement in the nineteenth century. With the progress of civilization and concomitant urbanization, as well as decline of the sources of support provided by traditional values, today's man tends to feel alienated, or "other-directed" in the sense of Riesman (*The Lonely Crowd*, 1950), and influenced by "groupism" in the sense of Whyte (*The Organization Man*, 1956).

As a result of this situation groups of patients of different kinds were formed by some: "classes" of tuberculous patients by J. Pratt in Boston (1905), lecture classes for mental patients by E. Lazell at St. Elizabeths Hospital in Washington, D.C. (1919) and by T. Burrows and by L. Marsh in the New York City area, the "impromptu theater" by J. Moreno also in New York (already practiced in Vienna early in our century). While these various groups were composed for different reasons and the emotional outlet was only coincidental, later on psychoanalytic concepts tended to prevail in groups formed for psychiatric purposes: mainly, analytic group therapy introduced by L. Wender and P. Schilder, activity groups for disturbed children practiced by S. Slavson at the Jewish Board of Guardians, and, to a less extent, "psychodrama," which makes use of particular techniques (auxiliary ego, mirror, double, and role reversal), practiced by J. Moreno, all in the New York City area.

Moreno was the first one to attempt to conceptualize his methods by founding the journal *Sociometry: A Journal of Interpersonal Relations* in 1937. However, aside from psychoanalytic notions, only with the advent of K. Lewin's "field theory" in the thirties (first at Harvard University and then at the University of Michigan in Ann Arbor) were the theoretical foundations of group dynamics able to offer a much needed scientific basis to the field of group psychotherapy. Since then this field has acquired a more stable image with the help of professional meetings and journals, especially *Group Psychotherapy* (1947) and *International Journal of Group Psychotherapy* (1951), founded by the homonymous associations.

As a result of all this, progress has been made in identifying goals, in selecting proper patients, and in structuring the role of the leader in the formation of groups.[326] However, some basic issues, such as the conceptual definition of group psychotherapy and the modalities of training for leaders, are still clouded by uncertainty. In recent years the rise of all kinds of new groups (from Alcoholic Anonymous to encounter groups, sensitivity training, and others) has put the professional movement of group psychotherapy on the defensive. Proper historical perspective may be found in the surveys by J. Klapman, R. Dreikurs and R. Corsini,[159] G. Bach and J. Illing,[35] and others.[519]

Family Therapy

Family therapy, scarcely mentioned in the first edition of this work, has become prominent in the last dozen years, to the point of being considered as the treatment of choice by some.[264,515] For the historian the reason for this rapid success, over and above the field of psychiatry, has to be found in family therapy's attempt to strengthen an institution that traditionally has contributed a great deal to the prevention and treatment of mental disorders and that is now affected by some basic problems: decrease of family ties, virtual loss of the extended family, and rise of the divorce rate.

As in group psychotherapy, various trends stemming from psychoanalysis and from group dynamics have contributed to the development of family therapy. No one has been more instrumental in fostering the field of family therapy than N. Ackerman (1908–1971), through his many writings (especially *The Psychodynamics of Family Life*, 1958), lectures, courses, and, eventually, formal training at the Family Institute in New York City founded by him in 1960.

In time the early themes of family therapy based on psychoanalytic notions (family secrets, emergence of a scapegoat, and so forth) have been replaced by notions acquired through study of the group dynamics of the family process either at the research level (J. Spiegel, J. Bell, and others) or from the practical perspective (G. Bateson, D. Jackson, and V. Satir of the so-called Palo Alto Group, which has published *Family Process* since 1962; T. Lidz and A. Cornelison of the Yale Group; Boszormenyi-Nagy, J. Minuchin, C. Sager, as well as R. D. Laing in England, and others).

Essentially the emphasis is on the threefold perspective of the patient's dynamics and role, the sociological approach, and the cultural dimension; this is clinically manifested through identification of areas of health and pathology, focus on communicating and sharing, and movement toward the ecological model of community psychiatry (E. Auerswald). All this constitutes a considerable departure from the classic dyadic relationship of orthodox psychoanalysis, in which the family was taken for granted as a solid institution. This image of the family is so altered today in American society to be a cause of concern for many. The historical consideration that the family has remained a landmark in all cultures at all times may help overcome this pessimistic view. Within the limits of psychiatry proper, there is no question that the field of family psychiatry will continue to develop, at the expense not only of individual psychotherapy but also of child and adolescent psychotherapy.

Behavior Therapy

Behavior therapy is so recent that it was not even mentioned in the first edition of this work. In a matter of a few years this school has gained a great deal of interest, if not of acceptance, partially as a result of the caustic criticism of psychoanalytic therapy by some, notably H. Eysenck at the Maudsley Hospital in London.[188]

The foundations of behavior therapy have to be related to the two schools of conditioning (or conditional) learning theory and of reinforcement theory, respectively.[210,612] The first school is identified with the Russian I. Pavlov (1849–1936),[34,598] who derived some of his concepts from his fellow countryman I. Sechenov (1829–1905). To a less extent, even the research on "reflexology" by V. Bekhterev (1857–1927) in Leningrad, resulting in the so-called rational therapy—a mixture of medical and environmental regime—is pertinent. Pavlov's notions of conditioning were used in this country first by behaviorists in the twenties, then by animal researchers, and finally by clinicians, notably W. Gantt (b. 1893), who studied under him in Russia and who founded the Pavlovian Laboratory at Johns Hopkins University in 1930 and, more recently, the Pavlovian Society and the *Conditional Reflex and Soviet Psychiatry Journal*.[230] A basic criticism of this movement is the difficulty of translating animal into human behavior, that is, to make the higher nervous activity relevant to clinical issues. The other school of reinforcement theory based on learning, anticipated by E. Thorndike's instrumentalism and by Hull's stimulus-response theory, led to some clinical studies (notably on frustration and aggression by J. Dollard and N. Miller) and to the controversial operant conditioning by the Harvard psychologist B. Skinner.

Behavior therapy, so named by R. Lazarus and H. Eysenck in the late fifties, is especially identified with the work of the psychiatrist J. Wolpe (b. 1915), first at the University of Witwatersrand in South Africa and then at Temple University in this country. His main tenets, presented in detail in Chapter 43 of this volume, are based on desensitization and reciprocal inhibition, positive and negative reinforcement, aversive conditioning, extinction, and other techniques.

Some reasons for the success of behavior therapy are the dissatisfaction with psychodynamic therapies, its apparent measurability consonant with the English empirical tradition, and, perhaps, the increased status of the psychologist functioning as a therapist. Within the movement of behavior therapy there are considerable internal contrasts between those inclined to emphasize the theoretical assumptions (J. Wolpe) and those inclined to emphasize the empirical applications (R. Lazarus), as well as between the researchers and the practitioners in the psychological profession.

The attitude of most psychiatrists toward this movement is ambiguous:[311] the therapeutic successes are often considered only symptomatic, superficial, and transient; nevertheless, this approach may be useful in treating large numbers of unsophisticated patients, especially in the context of the community mental health movement. In the light of all this, it is difficult to pass judgment on the long-range importance of this school. It is likely, however, that common points—mainly relevance of symptoms, role of the therapist, and the doctor-patient relationship—between the behavioral and psychodynamic schools, rather than areas of disagreement, will be stressed in the future.[1,535]

NEW AREAS OF PSYCHOTHERAPY

Psychotherapy of Schizophrenia. Even from the scanty reports of treatment of very disturbed patients carried on during the period of "moral treatment" in the nineteenth century, as well as by other systems at other times, it is clear that psychotherapy of psychoses was at times successful.[481] For a number of years after the advent of psychoanalysis, psychotherapy of psychoses was rejected almost universally (K. Abraham was perhaps the only exception) on the basis that the patient was unable to develop a transference neurosis.

In the forties H. Sullivan illustrated concrete cases of psychotherapy of schizophrenia carried out by him at the Sheppard Pratt Hospital in Towson, Md., on the basis of his interpersonal theory of behavior. While his empirical efforts influenced many in this country and in Europe, his theoretical formulations never gained popularity. Much more accepted was F. Fromm-Reichmann's *Principles of Intensive Psychotherapy* (1950), in which the concept of the "schizophrenogenic mother" was stressed. By that time considerable controversy had been elicited by the technique of direct analysis introduced by J. Rosen, first in New York City[503] and then at the Institute for Direct Analysis (founded in 1956 at Temple University in Philadelphia);[292] direct analysis is similar to the so-called symbolic realization described by M. Sechehaye in Switzerland.

In recent years sound attempts have been made by some—notably S. Arieti—to correlate each technique of treatment of schizophrenia with a particular stage of individual development: for example, lack of maturational development in early childhood (H. Sullivan, J. Rosen, L. Hill) or, conversely, compensatory defenses of the second stage in terms of reestablishing disturbed communications between the patient and his family (G. Bateson, D. Jackson, T. Lidz, L. Wynne).

Historically, similar to the neuroses, the emphasis appears to have shifted from individual treatment to the treatment of the patient in the context of his family and community. Thus far, psychotherapy of psychoses has tended to be carried on by a few therapists endowed with strong personality and keen intuition. The need in the future is for methodologically sound research.

Brief Psychotherapy. The expression "brief psychotherapy" has appeared in the literature in recent years, probably as a result of the need to offer treatment to the large number of patients brought forward by the community mental health movement (P. Castelnuovo-Tedesco, 1962).[115] While the focus on this approach is new, its use in some cases goes back to Freud himself (particularly in the famous treatment of the conductor Bruno Walter)[547] and to some of the early psychoanalysts. Alfred Adler practiced short-term therapy for many low-income patients in whom he had considerable interest; S. Ferenczi and O. Rank advocated a limited number of sessions to avoid unnecessary dependency and regression in the patient in their volume *The Development of Psychoanalysis* (1923); during World War II brief psychotherapy for the treatment of combat neuroses was widely used, as reported by R. Grinker[253] and A. Kardiner; finally shortening of psychotherapy was advocated by F. Alexander and T. French in *Psychoanalytic Therapy* (1946) on the basis of their research at the Chicago Institute for Psychoanalysis.[460]

In spite of all this, psychoanalytic therapy has tended to be taught and practiced from the perspective of long-term treatment. What is new is the emphasis by many today on considering brief psychotherapy as the treatment of choice on the basis of diagnostic considerations and practical aspects (cost, waiting list, availability of staff, community attitudes, and so forth). Clearly, brief psychotherapy focuses on crises, traumata, emergencies, and, in general, acute decompensations, rather than on personality disturbances, and its success is partly related to the patient's and the therapist's expectations.

The main criticism of brief psychotherapy centers on its lack of a theoretical dimension, as it is the result of the amalgamation of all kinds of practices derived from systems as diverse as psychoanalysis and behavior therapy (as illustrated in the recent book by L. Small, *The Briefer Psychotherapies*). This is certainly a considerable weakness, though certainly not unique in the field of psychotherapy, in the light of the unproven assumptions of long-term psychoanalytic therapy. Nevertheless, in the context of the community mental health movement, brief psychotherapy will foreseeably acquire importance, necessitating proper theoretical formulations and methods of teaching.

Psychotherapy according to Interpersonal Theory, Cognitive and Volitional School, Communication Theory, and Transactional Analysis. Common to all these various trends —most of them already mentioned somewhere in this chapter—are their American origin (though not devoid of indirect European influences) under the impact of the melioristic approach of sociology in this country, their defense of the individual against dehumanizing forces, and their modest view of the role of the therapist as a mediator or interpreter in contrast to the omnipotent image of the psychoanalytic tradition. Moreover, their tenets have been illustrated by their various originators in a rather unsystematic way without the support of rigid organizations, justifying calling them trends rather than schools.

The *interpersonal theory* of behavior is mainly represented by K. Horney, H. Sullivan, and E. Fromm, Horney, influenced by character analysis (H. Schulz-Henke and W. Reich) and the American concept of the "self" (G. Mead), in a number of popular books illustrated the clinical implications of self, self-image and relationship of the self to others on the basis of the experience gathered at the Psychoanalytic Institute in Chicago and then at her own American Institute of Psychoanalysis in New York City. While these notions have been relevant to American psychiatry, especially in the treatment of asymptomatic characterological patterns, her attempt to replace didactic analysis with self-analysis has been rejected.

Sullivan's lasting contribution to therapy, originated with the above-mentioned treatment of schizophrenic patients, lays in his emphasis on the role of the patient as "participant-observer" and the patient-staff interaction in the mental hospital, while his nomenclature of psychiatric conditions is now almost completely forgotten.

Fromm's numerous volumes, primarily addressed to the intellectual and progressive elite, stem from an attempt to combine Marx's notions (mainly the pathos of the "alienated" man in today's collective society) with psychoanalytic insight. Though very appealing, their relevance for psychiatry is quite limited.

The main thrust of the *cognitive and volitional school*, a new trend represented by psychiatrists and psychologists (A. Beck, J. Barnett, J. Bemporad, and others), has been clearly stated by S. Arieti, its originator, in Chapter 40C of this volume: "A stress on cognition and volition does not imply that affects are not major agents in human conflicts and in conscious and unconscious motivation. It implies, however, that at a human level all feelings except the most primitive are consequent to meaning or choice. In their turn they generate new meanings and choice."[25] In essence primary consideration is given here to the cognitive dimension in its relation both to conflict and creativity in the light of its unique importance for the human condition.

The *communication theory* of behavior, influenced by philosophical analysis of language, linguistics, and notions of physical field and psychological field theory, has focused attention on the process of verbal communication through which psychotherapy occurs. General systems theory, mainly through the efforts of R. Grinker, has been particularly interested in the transactional perspective.

Finally Eric Berne (1910–1970) has described, in his popular and successful book *Games People Play*, his system of *transactional analysis* in which exteropsychic, neopsychic, and archaeopsychic ego states (colloquially called parent, adult, and child) are combined in the different forms of basic human interactions.

Psychotherapeutic Borderlines: Sensitivity Training, Encounter, and Marathon Groups

The various movements listed in this section are quite recent and all of American origin. Their rapid success in some quarters, mainly outside the realm of psychiatry proper, appears to derive from the increasing isolation and alienation felt by many; paradoxically these feelings are reinforced by the close and impervious atmosphere of the individual psychotherapeutic relationship.

Historically the psychological aspects of the behavior of crowds were studied by French sociologists at the beginning of our century, followed by the development of social psychology and, in the forties, by forms of collective therapy. The founding of sensitivity training is attributed to the importance given to group self-evaluation during sessions for training of community leaders dealing with racial problems by three educational and social psychologists, L. Bradford, R. Lippitt, and K. Berne in 1946.[54] Out of this initial effort, supported by the Gould Academy in Bethel, Maine, resulted the work of the National Training Laboratories.[143,247] Eventually in the fifties the emphasis shifted toward individual self-actualization and in the sixties toward anti-intellectual and nonconforming techniques.

A variety of issues have risen in connection with this movement: the transitional nature of their methods based on the "here and now"; their appeal mainly to sophisticated people (not "patients") of the East and West Coast; the call for "honesty" on the part of every participant and leader (and the consequent ambiguous role of the leader, perceived as a peer but different); the possibility of bringing underlying psychopathology to the surface through these sessions; and consequently the ethical aspect of the leader's responsibility.

Similar issues have been raised in regard to nonverbal techniques emphasizing the unity of body and mind and grouped under the term of Gestalt therapy. Common to all of them (initiated by the neo-Reichian A. Lowen with his bioenergetic group therapy) is the postulate that release of tension, which expresses itself through emotional disturbances or peculiarities of muscular posture, can be achieved just as successfully through physical activities as through verbal psychotherapy. The techniques developed at the Esalen Institute in California since 1964 by F. Perls, B. Schutz, and B. Gunther center around the manifestations of body language in a variety of ways.

For the historian it is too early to pass judgment on the significance of this movement, since it is not clear at this point whether it constitutes a momentary fad or the beginning of a new orientation in psychotherapy. It is a fact that systems of healing of classical Greece and of Eastern civilizations took into consideration the body as well as the mind, as pointed out in some publications (such as *Asian Psychology*, edited by G. Murphy and L. Murphy;[429] *Psychotherapy East and West*, by A. Watts;[589] and a number of monographs on Zen Buddhism).[221,590,621] Those therapeutic methods, however, represented an expression of their own culture, while the methods described in this section appear to be isolated attempts to counteract the dehumanizing trends of our civilization.

Research in Psychotherapy

Reference to research in regard to therapy, including psychotherapy, has already been made in some scattered passages in this chapter. However, the matter deserves more consideration in view of the importance of psychotherapy.

Two points stand out, in some way related to each other. In the first place psychiatrists appear to have been unconcerned traditionally with the careful assessment of the results and follow-up of psychotherapy. This attitude may have been due to a number of reasons:[585] the discouraging number of variables intervening in the psychotherapeutic process; the lack of adequate training in research methodology for medical students; the exclusion of psychologists and other research-oriented nonmedical professionals from psychoanalytic associations; the adherence of each therapist to a particular school, which interferes with the objectivity necessary for research; the empirical attitude of the American mentality, coupled with a humanistic defense of the individual obviously opposed to the quantifying orientation of research; and finally an exaggerated concern about the confidentiality of psychotherapeutic scenes.[459]

In the second place, partly as a result of this situation, partly as an attempt to show their vital role in the field of psychotherapy, psychologists have taken most of the initiatives in regard to research on psychotherapy. This is substantiated by a perusal of the most important publications in this field such as *Methods and Research in Psychotherapy* edited by L. Gottschalk and H. Auerbach, *Research in Psychotherapy* by J. Meltzoff and M. Kornreich, preceded by the monographs *Research in Psychotherapy* (1959, 1962) published by the American Psychological Association.

Since most psychiatrists appear to disregard psychological literature, the introductory statement to a thorough review of the entire field of research in psychotherapy by H. Strupp and A. Bergin seems justified: "Thus far, research in psychotherapy has failed to make a deep impact on practice and technique."[556] The few issues dedicated to this topic by the *American Journal of Psychotherapy*,[182,561] the small monograph *Psychotherapy and the Dual Research Tradition* (1969) by the Group for the Advancement of Psychiatry, and the above-mentioned *Psychiatry as a Behavioral Science* edited by D. Hamburg do not essentially alter this statement.

This state of things, which accurately portrays the present situation, may, however, change in the future in connection with some developments: the decrease of the omnipotent image of the therapist and of the charismatic role of therapy; the influence of behavior models of psychotherapy conducive to measurement; and the overall political and social situation, which calls for a more thorough justification of the use of public funds in the field of mental health, even in psychotherapy.

Milieu Therapy

The influence of the environment in the care and treatment of the mentally ill in any institutional setting may have been overlooked, but it has certainly been present from early times on. Evidence of the importance of environmental factors can be easily found in the early psychiatric literature at the beginning of the nineteenth century and then in the period of moral treatment. Eventually, with the prevalence of large mental hospitals in the latter part of the century, the environment became more custodial and impersonal.

The advent of the psychodynamic approach did not alter this situation as the focus came to be on neurotic, nonhospitalized patients. The same situation persisted with the introduction of shock therapies in the late thirties. Even the interpersonal theories of behavior that developed in this country, while reducing the omnipotent role of the therapist to more modest proportions, did not modify the essentially obscure role of the environment.

A substantial change, in the sense of the environment itself becoming the focus of attention, took place only in the mid-fifties in connection with some important studies: notably, *The Therapeutic Community* (1953) by M. Jones in England and *The Mental Hospital* (1954) by Stanton and Schwartz, based on their experience at Chesnut Lodge, under the influence of F. Fromm-Reichmann, H. Sullivan, W. Menninger, and others.

Since then many studies carried on in mental hospitals have attempted to define the characteristics of the physical setting, roles and role relationships, authority and control, communication and culture in general. Much effort has been made to prove the underlying assumption that a stable environment, in which the operating forces are known, contributes to the reinforcement of the ego of the patient.

This has resulted in attempts at defining such forces in terms of conflict-free areas of the ego and adaptation (H. Hartmann), in terms of clarification and learning of roles (T. Parsons, G. Mead), on the basis of the notion of the field of forces (K. Lewin's "lifespace"), and from the overall perspective of Western democratic society. Up until now the best attempt to conceptualize the therapeutic areas of the environment remains *Ego and Milieu Therapy* by J. Cumming and E. Cumming. It is likely that in the future the therapeutic milieu will increasingly be considered as an open system from the viewpoint of the general systems theory and will be concretely affected by the movement of community psychiatry.

¶ Psychiatric Education

From the broad perspective of systems of healing and attitudes toward the mentally ill in preliterate societies, it is clear that methods of apprenticeship for medicine men have been in use from earliest times (for instance, in the training of shamans in many cultures). Among their common characteristics were the "call" of the candidate through dreams, a period of isolation under the close guidance of an experienced healer, and finally the return to the community, which had definite expectations of supernatural powers in the new medicine man, to be manifested through rigidly established rituals.[164] All this, of course, bears resemblance to the training of today's psychotherapists.

However, the awareness of these similarities is only recent.[487] Traditionally psychiatric education has been traced back to the early nineteenth century, when young physicians underwent a highly personalized, yet unstructured, system of apprenticeship around moral treatment in the early mental hospitals in the Western countries. The decline of such a philosophy of treatment late in the century coincided with the rise of "scientific" psychiatry, mainly in German universities, based on the belief in the ultimate neuropathological etiology of mental disorders.

In this country, instead, psychiatric training (as well as research) continued to take place very empirically in mental hospitals, thus justifying the famous critical address given by the neurologist W. S. Mitchell in 1894 at the occasion of the fiftieth anniversary of the American Psychiatric Association.[420] Eventually a new spirit conducive to better training was introduced at the beginning of this century at the Worcester State Hospital by A. Meyer (who first developed there his "life chart" of the individual development of the patient). Later on the Henry Phipps Clinic at Johns Hopkins University also opened under Meyer's leadership, following the model of the Kaiser Wilhelm Institute for Psychiatry in Munich, the so-called Kraepelin Institute.

Since then much has happened in the field of psychiatric training in this country. The American Board of Psychiatry was established in 1934 to set up standards of training and certify physicians in the specialty of psychiatry.[111] Around the same period teaching in psychiatry came to be organized, both at the undergraduate and at the postgraduate levels, in a number of hospitals and outpatient facilities with the help of private foundations (especially the Commonwealth Fund and the Rockefeller Foundation) and a few interested professionals (such as A. Gregg, F. Ebaugh, and C. Rymer, authors of *Psychiatry in Medical Education*, and H. Witmer, who published *Teaching Psychotherapeutic Medicine*).

World War II, by emphasizing the great need for psychiatry, led to a vast program of psychiatric training by the Veterans' Administration. From the early fifties on, such a program has been undertaken almost exclusively by the National Institute of Mental Health through its division of training and manpower. By the time the First Conference on Psychiatric Education was held at Cornell University in 1951, the report on medical education sponsored by the Group for the Advancement of Psychiatry (1948) had identified the main areas of training as personality development, unconscious motivations, and dynamic comprehension of the individual case; a program of "comprehensive medicine" inclusive of psychological development and case work principles was developed in some medical schools (Western Reserve, Colorado, Pennsylvania, Tennessee, and Harvard); and departments of "behavioral sciences" and courses in "human ecology" were offered by others (Syracuse and North Carolina, respectively).

In the fifties, with the rise of psychoanalysis, the issue of psychoanalytic training as an essential aspect of the training of the psychiatrist, to be carried on independently in psychoanalytic institutes or in the framework of medical schools, became outstanding (for example, in the 1954 symposium in the *International Journal of Psychoanalysis* and in F. Alexander's *Psychoanalysis and Psychotherapy*). Despite the many controversies, this issue is still unsettled today, but it has been essentially bypassed by the events related to the decline of the psychoanalytic movement and the parallel rise of the community mental health movement. Instead, what has remained of the psychoanalytic influence has been the system of supervision,[521] defined in its threefold aspect of patient-centered, process-centered, and trainee-centered, especially in the well-known study by R. Ekstein and R. Wallerstein.[173]

In the last two decades most of psychiatric training has taken place with the support of the federal government: the original program of NIMH began with a little more than 200 grants annually and now is in the realm of 10,000. This has had, on the one hand, the advantage of fostering a certain degree of homogeneity and of maintaining high standards; on the other hand, it has tended to favor large and well-known centers located in urban areas, thus interfering with the aim of making psychiatry available even in less populated areas of the country.

Serious efforts have been made toward offering a comprehensive type of training, including experience in state institutions, outpatient clinics, special facilities (for example, for children or for delinquents), and presenting a manifold philosophical orientation (genetic, organicistic, psychodynamic, and epidemiological) and various therapeutic approaches (chemotherapy, individual as well as group and family therapy, and others). In view of the vastness of each new field of psychiatry, such an ambitious program can be realistically carried on in very few places.

Likewise, limited success has characterized the various efforts to influence practicing physicians to take a more progressive attitude toward psychiatry and emotional disturbances in general, either at the undergraduate level or at the postgraduate level, with the help of federally supported seminars. In spite of the widespread belief of the pervasive influence of mental on physical pathology in many patients (as represented in the appealing volume *Man, Mind and Medicine*, edited by the distinguished surgeon, Oliver Cope)[131], only slow gains toward the acceptance of psychiatry have been made thus far among the already established professionals. The young physicians recently graduated from medical schools seem to be more open-minded toward psychiatry, even in regard to some new controversial therapeutic group approaches.[325]

Recent developments in the field of psychiatric education include the tendency toward specialization in psychiatry during training in medical school and toward training in a subspecialty of psychiatry during the residency period (initiated at Yale);[374] the concern for providing adequate training in community psychiatry, especially in regard to urban problems; the controversies related to discontinuing the internship before the psychiatric residency (as a result of the Mill's Citizen's Commission on Graduate Medical Education in 1966); finally the programs for continuing education for psychiatrists (mainly under the leadership of the late W. Earley) and the self-assessment project sponsored by the American Psychiatric Association.[38,97,161,162,375,499,586]

The future of psychiatric manpower remains far from bright[68] although between 10 to 20 percent of medical students decide now to embrace psychiatry, which has become the third largest specialty. Also a cause of concern is the shortage of psychiatrists actively involved in research, in spite of the efforts made by the National Institute of Mental Health.

In recent years it has become apparent that many young psychiatrists are interested in humanistic medicine, that is, in the comprehensive approach to patients of any social class and their families and communities.[223] Unfortunately the training of the physician and, subsequently, of the psychiatrist is so long as to discourage a number of young men from entering this field. This has given further impetus to the proposal—advanced by L. Kubie[342] years ago—to establish a doctorate in medical psychology. In view of the traditional determination of professional psychiatric and psychoanalytic organizations to limit training in psychotherapy to physicians, this has remained a very debatable issue up until now. But efforts in this direction continue to be made, as evidenced by the recently published book *New Horizons for Psychotherapy, Autonomy as a Profession*, edited by R. Holt.

The future of psychiatric education, though promising in terms of innovations and increased flexibility, has recently been clouded by the financial restrictions applied to the budget of the National Institute of Mental Health. This may result in a broader support of training at the state and local levels, whose impact in psychiatry is difficult to assess at the present moment.

American Psychiatry in the Context of Psychiatry in Other Countries

Within the limits of this chapter it is possible only to present some general trends concerning the interplay of psychiatry in the United States and in other countries, the reciprocal influence of models of the mind and therapeutic methods, and the foreseeable developments in the future. Since no comprehensive publication on psychiatry throughout the world exists at the moment, the following remarks are the result of a broad perusal of pertinent literature scattered in many publications. The lack of a comprehensive view on this subject here is partially compensated for by the various references to theoretical and practical aspects of psychiatry in other countries made in other sections of this chapter.

From the overall historical perspective the relevance of discussing psychiatry abroad for a better comprehension of American psychiatry is unquestionable, at least on three grounds: (1) the manyfold cultural traditions from Europe, Africa, and, to a less extent, other continents that are the basis of American society; (2) the persistent interest that this country has had in supporting scientific and humanitarian projects abroad, especially in Western countries damaged by World War II, some countries of Latin America, and newly developing Afro-Asian areas; (3) the convergence of practical methods of healing carried on in underdeveloped countries and approaches to community mental health recently introduced in this country.

Even a succinct discussion of the above points calls, however, for some preliminary considerations. The methodological perspective in the comprehension of other countries has undergone a tremendous change in the course of the century and a half of development of modern psychiatry. During a good part of the nineteenth century American psychiatry tended to follow European psychiatry, first in regard to the practical aspects of moral treatment, then in regard to the clinical orientation of the French school and the theoretical research on neuropathology of German universities. The acceptance of the Freudian message, prepared for by the emphasis on environmental factors in the etiology of emotional disturbances brought forward by various trends (progressivism, behaviorism, Meyer's psychobiology), led to the tendency to assess mental pathology and attitudes toward the mentally ill in our country from the almost exclusive perspective of psychoanalysis up until recently. Finally, parallel to the spread of the community mental health movement and to the progress in communication systems (easy transportation, international meetings, speedy translations), approaches to mental illness and its treatment carried on in other countries are becoming more known even in the United States.

Concomitant with this has been a better definition of the three growing fields of: (1) *trans*cultural psychiatry, in which scientific observation is extended to non-Western practices; (2) *cross*-cultural psychiatry, which focuses on comparative and contrasting dimensions of psychiatry in various countries; (3) *international* psychiatry, which emphasizes teaching, training, and, in general, organizational aspects in the different political entities of the world. These three fields overlap to a certain extent, and, furthermore, a thorough discussion of the first two will be offered in other sections of this handbook. Consequently here the presentation will be limited mainly to the developmental aspects of international psychiatry, which is not covered elsewhere.

In spite of the tendency toward the spread of Western practices all over the world, resulting in the transformation of "pure" cultures into cultures at different levels of acculturation, for didactic purposes the following discussion is divided into: (1) countries of the Western tradition, that is, mainly Europe (including Russia), Canada, Australia, South Africa, Israel, Latin America; (2) countries of the Far Eastern tradition, that is, India, Japan, China, and some others; and (3) international psychiatry. The developments of psychiatry in these various countries will be presented essentially in their relevance to American psychiatry.

Psychiatry in Countries of the Western Tradition

On the one hand, in view of the different ethnic, cultural, economic, and political aspects of the many countries of the Western tradition, it is impossible to present the developments of psychiatry in a compact and unitary form. On the other hand, the interchange of ideas and people among these countries has been so great that it is impossible to conceive of the developments of psychiatry in each one of them independently from the others. Yet some general characteristics have emerged in most of these countries, and the discussion will center around them.

To begin with, a few main aspects stand out clearly on the basis of some important publications by G. Allport,[10] H. Ellenberger,[174] L. Bellak,[49] and a few others. There are many similarities between British and American trends, at variance with continental trends: the empirical tradition (derived from

Locke), which emphasizes environmental forces, social interaction, and optimistic expectations; a positivistic approach to the formulation of "brain models"; the study of traits, attitudes, and motivation rather than the total personality; the tendency to consider mental disturbances as "reactions" rather than symptoms; the inclination toward pragmatic use of therapeutic methods developed somewhere else regardless of theoretical speculations; finally the interest in experimentally testing the above methods by interdisciplinary teams of scientists. The main difference between Great Britain and the United States consists, of course, in the uniformity of the ethnic and cultural scene of the former vis-à-vis the racial and social pluralism of the latter, with definite repercussions on psychiatry. This is reflected in the mobility of professionals in psychiatry and collateral fields, leading to rapid spread of ideas, homogeneity in clinical practices, and eclectic approaches.

In contrast, Continental Europe, with the notable exception of Russia and, to a less extent, of other communist countries, has been characterized by the rational tradition (derived from Leibnitz and Kant), which has resulted in a more philosophical view of mental life; the persistent concern with the "whole man," which stems from the Greek tradition and which is somehow responsible for a humanistic (and humanitarian) orientation; a pessimistic orientation toward life in general, probably influenced by the long-term experience of wars, migrations, famines, exterminations, and other horrible events, as evidenced by Freud's preoccupation with death and by the dramatic aspects of the existentialist movement; a concern with the individual rather than with the social dimensions of the personality; a preoccupation with symptomatology and diagnostic categories at the expense of therapeutic efforts; a succession of brilliant "discoveries," from psychoanalysis to shock treatment, psychopharmacology, and community psychiatry, by isolated scientists; finally a tendency toward fragmentation of psychiatric theory and practices into definite "schools," each one led by an academician and followed by his pupils (and represented by special journals and publications), quite often bitterly opposed to each other.

In general, it is very difficult to follow the developments connected with the interrelationship of American and European psychiatry. Regardless of differences due to language barriers, personality patterns, and social customs, the traditional influence of European on American psychiatry underwent a significant change after World War II; such an influence persisted in regard to discoveries and introduction of new therapeutic systems (for example, in community psychiatry), but was paralleled by an opposite influence of American on European psychiatry, especially in the fields of psychoanalysis, child psychiatry, research, and training.

Historically it would appear that certain periods have been characterized by a common psychiatric approach in Europe and the United States: (1) moral treatment in the early nineteenth century; (2) organicistic philosophy in the late nineteenth century; (3) psychoanalytic influence in the early twentieth century, followed by an interest in shock treatment later on; (4) psychopharmacological approach and methods of community psychiatry in the last two decades. For the purpose of this chapter the discussion will be necessarily limited to this last point, as advances in chemotherapy have already been discussed above, also from the international perspective.

Interest in community psychiatry began to be noticeable in this country in the mid-fifties, following the important publications in England on the "open-door policy" in mental hospitals, on day hospitals (J. Bierer), and on the therapeutic community (M. Jones), as well as the successful program of public mental health introduced in Amsterdam by A. Querido.

Shortly thereafter, the volume *Impressions of European Psychiatry* (edited by W. Barton, *et al.*)[43] appeared in 1959 under the sponsorship of the American Psychiatric Association. It then became evident what the main positive aspects of European psychiatry consisted of: a better reciprocal respect between patients and staff in institutions, probably because of the humanitarian tradition, systems of education, and division into rigid class systems; a more supportive role by physicians (paternal), by nurses, at times religious (maternal), and by community resources of all types (this latter related to the social stability of the population), resulting in better systems of communication, involvement with families, partial hospitalization, and follow-up; a flexible therapeutic approach, based on chemotherapy and short-term psychotherapy, and carried on by a dedicated staff more interested in their professional vocation than in financial rewards. On the negative side were the uneven and loose systems of training, the lack of systematic support for research even to dedicated scientists, the resistance of professionals to work in interdisciplinary teams (especially in areas such as child psychiatry where this approach is most valid), the dependence of psychiatry on neurology in some Latin countries, the tendency to centralize the decision-making process in psychiatry (be this service, training, or research) on a few people.

Interesting enough from the historical perspective is the fact that European psychiatry, which appeared to be backward when seen from the American psychoanalytic viewpoint of the fifties, was seen as advanced from the viewpoint of community psychiatry of the sixties.[224] Somewhat related to a more tolerant attitude toward the mentally ill on the part of society, which is basic to community psychiatry, is the pervasive European belief that mental illness is *occurring to* an individual rather than being *synonymous with* him (that is, a patient "has" a schizophrenic illness, but he is not "a schizophrenic").

In *Great Britain*[285,357,363,371,537] the brilliant psychiatric tradition, initiated with the "moral treatment" and the movement of "no restraint" in the early nineteenth century, was later influenced by the evolutionary concepts of C. Darwin, T. Huxley, and H. Jackson. At the beginning of this century, under the impact of Meyer's psychobiology and Freud's notions, psychoanalysis received great impetus there, mainly through the work of E. Jones. Eventually a good number of psychoanalysts preferred to accept the concepts advanced by M. Klein and even to follow ideas put forward by C. Jung. The trend of organicistic psychiatry has remained very strong, however, in line with the excellent neurological tradition.[518] Training and research are mainly carried on at the Tavistock Institute of Human Relations[154] and at the University of London Institute of Psychiatry, linked to Bethlem Royal Hospital and Maudsley Hospital, while the most important periodical is the *British Journal of Psychiatry* (which superseded the *Journal of Mental Science*). Many of the 3,350 practicing psychiatrists hold the Diploma in Psychological Medicine and gather around the Royal Medico-Psychological Association, founded in 1841. Among the most prominent psychiatrists are (or have been) D. Henderson, D. Hill, J. Bowlby, E. Miller, A. Lewis, D. Winnicott, W. Fairbain, W. Sargant,[517] D. Leigh, J. Howells, and G. Carstairs. Since 1930, when the Mental Health Act was passed, the social dimension has continually gained importance in psychiatry:[584] in 1948 the National Mental Health Service Act placed the care and treatment of the mentally ill under the Ministry of Health on a regional basis; the Mental Health Act of 1959 removed any legal distinction between patients in psychiatric and in general hospitals.[124] The great majority of psychiatrists work in the context of social medicine and private practice is very limited.

In *France*[44,187,450] the pioneering work of Pinel[488] was followed by his favored pupil Esquirol and by many others mainly interested in the clarification of clinical symptoms, in the connection with neurology and in legal-psychiatric matters in the mid-nineteenth century. Later on the advances made in neurophysiology (C. Bernard), psychopathology (T. Ribot), experimental psychology (A. Binet), and social psychology (E. Durkheim and others) were overshadowed by the clinical application of hypnosis to neurotic disorders by the school of Salpêtrière in Paris (J. M. Charcot) and by the rival school of Nancy (H. Bernheim and A. Liébeault), both of which influenced Freud directly as well as P. Janet (1859–1947).[303,304] Through Janet and some of his early disciples the psychoanalytic movement developed on a limited scale,[428] also with the support (in contrast to the belief of many) of some progressive Catholic quarters.[451] Aside from some psychotherapeutic innovations[84] (such as the technique of "directed daydream" by R. Désoille) the most important event has been the secession from the Psychoanalytic Society of Paris by the French Psychoanalytic Society, led by D. Lagache and J. Lacan; the latter is the author of a famous paper (1953) in which he identified the structure of the unconscious with the structure of the language.[491] Another important group of psychiatrists has centered around the journal *Evolution Psychiatrique*, mainly dominated by F. Minkowsky,[416,417,418] a pioneer in the field of phenomenology (represented by M. Merleau-Ponty)[411] and not immune from existentialist influences (J.-P. Sartre). In recent years the two psychiatrists J. Delay[145] and H. Ey,[186] both authors of many publications aiming at the integration of organic and dynamic concepts, have acquired prominence. Recognition has also been given to projects in community psychiatry (especially in the thirteenth arrondissement in Paris).[446,447] The new trend of structuralism—mainly represented by the anthropologist C. Lévi-Strauss and the philosopher M. Foucault[205] (the latter is the author of an important historical study of psychiatry in the age of the Enlightenment)[204]—may also influence psychiatry considerably. Finally mention should be made of the traditional international role played by France, which has contributed to the communication of ideas; in Paris were held both the First International Congress of Child Psychiatry (1937) and the First World Congress of Psychiatry (1950).[457]

Belgium, the country where family care of the mentally ill was continued uninterruptedly at Gheel from the thirteenth century to the present, has essentially followed the French tradition. At the Catholic University of Louvain, where psychology was initiated by Cardinal Mercier, efforts to combine psychoanalysis with a spiritualistic conception of man were made by A. Michotte and J. Nuttin.[438] The *Netherlands*, where J. Wier, the sixteenth-century pioneer of modern psychiatry was born, has lately become known for the important school of phenomenology (J. Buytendijk,[105,106] H. Van den Berg, H. Rümke, and others)[113,534] and for advancements in community psychiatry (A. Querido).[465,466,467]

In *Germany*[289] psychiatry appears divided into many schools, historically explainable on the basis of the political and academic independence of each region. The controversy between the "mentalists" and the "somatists" in the early nineteenth century resulted in the predominance of the latter, represented by W. Griesinger and later by the school of anatomopathology and histopathology (C. Wernicke, T. Meynert, O. Vogt, and others). Early in this century, preceded by the clinical contributions of E. Hecker and K. Kahlbaum, E. Kraepelin established the fundamental dicotomy of manic-depressive psychosis versus dementia praecox, which influenced nosology everywhere. It was followed by the constitutional school of personality of E. Kretschmer and by the important trend of genetic psychiatry, mainly represented by E. Rudin and his pupil, F. Kallmann; the latter was active for many years at the New York Psychiatric Institute. It is regrettable that in Germany some adherents to genetic psychiatry attempted to offer scientific justification for the Nazi persecution of the Jews and the alleged superiority of the Aryan race.[597] In the thirties for a few years the Berlin Psychoanalytic Institute, where didactic analysis was first introduced, acquired prestige. After World War II the various psychiatric trends have all been influenced by the existentialist movement, anticipated by the philosophers Husserl and Heidegger: the clinical contributions by H. Grühle,[269] V. Gebsattel, V. Weizsäcker, K. Kolle,[334] as well as the so-called neopsychoanalytic movement of H. Schultz-Henke, the current centered around the journal *Psyche* (edited by A. Mitscherlich),[421] the yearly "Lindauer Psychotherapeutic Week" organized by Speer and others.[270,337,366] Mention should also be made of J. Schultz-Henke's "autogenic training," which combined Western and Eastern healing practices,[385] the contributions to ethology by K. Lorenz and others (N. Tinbergen, C. Schiller,[520] H. Hass[281]), and the recent studies dealing with sociological (including Marxist[233]) aspects of psychology,[167] preceded by the pioneering work by T. Adorno in this country and then in Germany (especially his study on the authoritarian personality). Valuable historical studies on psychiatry have also appeared there.[156,333,444]

In *Austria*[289] basic concepts of mental hygiene anticipating psychodynamics advanced by E. von Feuchtersleben (*Textbook of Medical Psychology*, 1845) and, later on, studies by the organicistic school (T. Meynert, R. Krafft-Ebing) received impetus through the renown of the medical school of Vienna and the extension of the Austrian empire. For the past two decades or so the developments of Freud's ideas and of the psychoanalytic movement have been the subject of many studies in the United States. In Austria, however, Freud's role remained quite limited among his contemporaries, when compared with the success achieved by the Nobel Prize winner Wagner-Jauregg, discoverer of malaria therapy (1917) and, later on, by M. Sakel who introduced insulin therapy in the thirties. Almost all of Freud's pupils from Austria (as well as the sympathizer P. Schilder, whose studies on the body image and on clinical applications of psychoanalysis have become largely known in this country) emigrated to the United States, with the exception of A. Aichhorn, who did pioneering work on the psychoanalytic treatment of juvenile delinquents.[191] In recent years, aside from academic psychiatry (mainly represented by A. Stransky and H. Hoff),[288] recognition has been accorded to logotherapy (V. Frankl)[208,209] and so-called personalistic psychoanalysis (I. Caruso, W. Dain), which emphasizes values and the purpose of life rather than gratification of instincts.[155]

Switzerland[53,176,483] has gained an important place in the development of modern psychiatry, probably owing to its geographical location, traditional neutrality, and multilingual background. In Geneva psychology was cultivated at the Institut Rousseau in succession by T. Flournay, E. Claparède, and J. Piaget (whose studies on the development of the child have become universally known and have increasingly been compared with psychoanalysis). Except for the method of "sleep therapy" introduced by J. Klaesi in Bern in 1922, psychiatry, instead, has traditionally flourished in Zurich, mainly at the mental hospital of Burghhölzli, directed in succession by A. Forel, E. Bleuler (who coined the term "schizophrenia"),[69] and then M. Bleuler.[71] There A. Meyer received his first psychiatric training before moving to this country; the clinical application of psychoanalysis was first introduced by C. Jung (who wrote his famous works on dementia praecox and word associations early in this century);[308] and H. Rorschach developed his test that soon became known the world over. Aside from Jung's followers such as C. Meier,[417] C. Kereny, and others, gathered around the Jung Institute in Zurich and the so-called Eranos meetings in Ascona, worth remembering are the names of the psychoanalysts R. De Saussure, C. Baudouin, O. Pfister, A. Maeder, E. Oberholzer (who moved to New York City, where he introduced the Rorschach test in the twenties), and H. Zulliger. In recent years existential analysis—mainly represented by L. Binswanger,[66] M. Boss, G. Bally and a few others, all originally influenced by psychoanalysis—has acquired momentum. Aside from biological studies relevant to psychosomatic medicine and psychodynamics (R. Brun,[103] A. Portmann[454,455]) and characterological studies (L. Klages, L. Szondi), two main trends have emerged in Switzerland: child psychiatry, mainly represented by A. Répond, M. Tramer, L. Bovet, J. Lutz, and H. Hanselmann (who initiated the movement of "Heilpädogogik" on therapeutic education), and known through the international journal *Acta Paedopsychiatrica*; and the psychotherapeutic treatment of schizophrenia, which, related to the pioneering contributions by C. Jung, E. Bleuler, and A. Storch, has been applied first by M. Sechehaye with her technique of "symbolic realization" and then by C. Müller in Bern and by the Italian-born G. Benedetti[52] in Basel. Finally mention should be made of the outstanding contributions made by some chemical companies of Basel to the development of psychopharmacology and of the role played by Switzerland in sponsoring international meetings (notably, the First International

Congress on Therapeutic Education in 1939[279] and the Second World Congress of Psychiatry in 1957, both held in Zurich[129]).

In *Scandinavian countries*[347,355] psychiatry has been dependent on the German tradition up to World War II and on the English tradition thereafter. The psychodynamic influence has been limited; it is important mainly at the Erica Foundation for Child Guidance in Stockholm and at the Therapeia Foundation recently established in Helsinki by the Swiss-trained M. Siirala. In view of the stability of the social situation and the long-term absence of great military conflicts, research on hereditary factors in mental illness has received a great deal of attention (T. Sjogran, G. Langfeldt) in the context of a strongly supported organicistic framework in psychiatry, as evidenced by the studies published in the important journal *Acta Psychiatrica et Neurologica*. The broad-minded social legislation enacted in Scandinavian countries for some time has brought about successful developments in community mental health.[571]

Italy, Spain, and Portugal are presented together here because of close ethnic, religious, and cultural affinities that justifiably can be extended to psychiatry. In these countries care of the mentally ill has had an illustrious, centuries-old tradition, psychiatry has been predominantly organicistic and dependent on neurology, and psychodynamics has been opposed by the Fascist and Franchist political regimes on various grounds. In *Italy*[119] the pioneering reform in the treatment of mental patients introduced by V. Chiarugi in Florence at the end of the eighteenth century was soon forgotten. Instead more recognition was accorded to the histoneurological studies by the Nobel Prize winner C. Golgi and the pathographic studies by C. Lombroso—the founder of criminal anthropology—early in this century. At the University of Rome S. De Sanctis's notions on child psychosis (*Dementia praecocissima*, 1906) were soon overlooked, while electric shock introduced by U. Cerletti and L. Bini in the late thirties was applied everywhere.[120] Recently interest in psychoanalysis[140] and existentialism has increased.

Spain, considered "the cradle of psychiatry" because of the foundation of some pioneering mental hospitals in the fifteenth century, later on went through a long period of decline. Only at the beginning of this century ample recognition was granted to the studies on histopathology of the nervous system by Ramon y Cajal and his school. In recent years psychodynamic and existentialist concepts have gained momentum there (J. Lopez-Ibor, R. Sarro), and psychiatry has been given more recognition, as evidenced[348,349,382,383] by the Fourth World Congress of Psychiatry held in Madrid in 1966.[381] Important studies on the history of psychiatry have been published there. *Portugal* has become known in psychiatry because of the introduction of frontal lobotomy there in 1936 by E. Moniz, who eventually received the Nobel Prize.

Among the various countries of the British Commonwealth, *Canada* has traditionally been in the unique position of economic and cultural dependence on the United States. In fact, developments in Canadian psychiatry have been often considered in conjunction with American psychiatry, and the interchange of professionals has always been very high (for example, C. Farrar from Toronto was for many years editor of the *American Journal of Psychiatry*, and E. Cameron was elected president of the American Psychiatric Association). The psychoanalytic movement, first introduced by E. Jones in Toronto early in the century and then dependent on the British association, gained autonomous status in the fifties. It has been heavily influenced by Catholic philosophy in the French province of Quebec[248] (for example, by K. Stern in his *The Third Revolution* and by N. Mailloux at his Centre d'Orientation for delinquent adolescents in Montreal), where religious orders have taken care of institutionalized mental patients for the past two centuries. Academic psychiatry has been particularly cultivated in Toronto (mainly at the Toronto General Hospital and at Clarke Institute founded in 1966) and in Montreal, where the Allen Memorial Institute at McGill University (opened by E. Cameron in 1944) has acquired prominence through the work by E. Wittkower on psychosomatic medicine and transcultural psychiatry,

R. Cleghorn on neuroendocrinology, H. Lehmann on psychopharmacology, as well as through the participation of fine neurologists (W. Penfield, H. Jasper, F. Gibbs, E. Gibbs, and H. Selye, who described the general adaptation syndrome). Since the 1,600 practicing psychiatrists are mainly located on the east and west coasts, efforts have been made (for example, through the report *More for the Mind* published in 1963 by the Canadian Mental Health Association under the leadership of J. Griffin) to establish a network of services of community psychiatry throughout the whole nation. Such an endeavor, also supported by the Canadian Psychiatric Association (founded in 1951), has already led to considerable progress in some provinces, notably in Saskatchewan, as reported in the *Canadian Psychiatric Association Journal* published since 1956. The Third World Congress of Psychiatry was held in Montreal in 1961.[462]

In *Australia*[82,98,135,550] the 500 practicing psychiatrists are increasingly involved in a vast program of mental health services (following the project initially developed by the State of Victoria, where a Mental Health Research Institute has been operating under A. Stoller since 1955), while relatively few take part in the psychoanalytic movement (at the Melbourne Institute of Psychotherapy).

In *South Africa, Hong Kong,* and other countries of the Commonwealth, aside from development of the Western practice of psychiatry, important research on transcultural psychiatry has been carried on.[76,356]

Israel, a small and young nation, offers a great deal of interest for psychiatry. Even before that country officially became independent in 1948, some psychoanalysts practiced there: M. Eitington from Berlin, P. Wolff from Russia, and E. Neumann of the Jungian school. Since its founding Israel has become a fertile ground for psychiatry in three main areas: (1) coping with psychiatric problems of migration and acculturation, related to the rapid conglomeration of people from many parts of the world (for example, *Migration and Belonging* by A. Weinberg);[593] (2) establishment of a decentralized program of community psychiatry (in general hospitals, special centers, "therapeutic communities," and "work villages") facilitated by the lack of a tradition of institutional psychiatry; (3) relevance for personality development and psychopathology of raising children in the collective form of the kibbutz, made the subject of many studies even by American authors. In the academic field psychiatry is cultivated at Tel Aviv University and Hadassah University in Jerusalem, and research is published in the *Israel Annals of Psychiatry and Related Disciplines.*

Psychiatry in *Latin America* offers a very complex and varied impression owing to marked differences in the ethnic composition (especially Indians), geographical and cultural situation, economic and social level in each country, in addition to the unstable political scene. Yet certain common trends can definitely be seen everywhere: Western psychiatry has tended to be practiced mainly in urban areas under the strict leadership of a university professor; the prevailing psychiatric orientation was dependent on the French tradition in the mid-nineteenth century, later on the German tradition, and after World War II on the American tradition; in many countries methods of indigenous psychological healing continued to be practiced in preliterate cultures, though increasingly influenced by Western civilization.[58,83,473,575]

The efforts made toward improvement in the field of psychiatry by the Inter-American Council of Psychiatric Associations—including the American, the Canadian, and the Latin American Associations—have led to rather meager results thus far.[94] In general, most of the broad long-range projects initiated in Latin America tend to achieve limited results because of social and political difficulties.[294]

Puerto Rico is in a particular situation, since a considerable percentage of its population has relocated itself in the United States since the end of World War II. From the psychiatric perspective two main areas are thus identifiable: (1) the psychiatric problems presented by those who decided to resettle in American communities, related to the dynamics of separation from their background and difficulties in acculturation (language barriers, loss of the extended family, stress of urbanization and industrialization, and so forth) and resulting in a colorful psychopathology characterized by the so-called Puerto Rican syndrome[197] (a sort of hysteric attack highlighted by sudden loss of control, falling to the floor, and hyperkinetic movements of various kinds), frequent suicide attempts, and a high percentage of psychosomatic (especially asthmatic) disorders, for which short-term psychotherapy and chemotherapy carried on in emergency services and storefront facilities appear most successful; (2) the psychiatric problems of the population remaining in Puerto Rico, taken care of by about 100 psychiatrists. Most work in institutional settings, and others are connected with the Puerto Rican Institute of Psychiatry, reorganized in 1958 with American help but functioning under local leadership (E. Maldonado, R. Fernandez, and others).

In *Mexico* the American influence has been particularly strong, both in the overall field of psychiatry (mainly represented by A. Millán, R. de LaFuente, J. Velasco Alzaga, and others) and in psychoanalysis (E. Fromm has been active since 1951), as evidenced by the Fifth World Congress of Psychiatry held in Mexico City at the end of 1971. Knowledge of the family and cultural background underlying individual psychopathology[114] (extended family, masculinity of the man and dependence of the woman, formalistic expressions of social behavior, etc.) has been increased through the anthropological books of O. Lewis (*Life in a Mexican Village*, 1951).

In *Cuba* the traditional American influence on psychiatry has been drastically reduced by the advent of the Marxist regime. A magazine on transcultural psychiatry *Revista de Psiquiatría Transcultural* is edited there by the leading psychiatrist, J. Bustamante. Likewise, it is difficult to assess the position of psychiatry in *Chile*, where psychodynamic concepts were introduced in the past (mainly by J. Matte-Blanco and C. Nassar), after the rise to power of the leftist government of the physician S. Allende, who was himself interested in mental health in the thirties.

In contrast, the situation has remained more conservative in *Colombia*, where C. León is the most active psychiatrist, and in *Peru*, where psychodynamic notions were introduced first by H. Delgado and later by C. Seguin and where an Institute of Social Psychiatry was founded in 1967 at the University of San Marcos in Lima, the oldest medical school in the American continent.[528,529]

In *Brazil* psychiatry has developed mainly in the coastal cities, while in the interior—as in other Latin American countries—systems of healing are still carried on in the traditional framework of preliterate societies. In Rio de Janeiro after World War II the German-born W. Kemper introduced psychoanalysis and the Spanish-born E. Mira was active in various areas of psychiatry. More recently impetus for psychiatry has been provided in Sao Paulo by A. Pacheco e Silva.

Finally in *Argentina*[59] psychiatry and, in particular, psychoanalysis have acquired a great deal of acceptance, probably as a result of the immigration there of some Jews from Central Europe in the thirties, followed by many other immigrants in the late forties. All this overshadowed the fact that important pioneering work in psychiatry by some pupils of the Italian Lombroso had taken place there early in the century. The Argentine Psychoanalytic Association, founded in 1942, publishes the important *Revista de Psicoanálisis* of eclectic orientation (that is, influenced by Kleinian, Adlerian, and Jungian trends). Recently some have followed Pavlovian concepts. Among the professionals recognition has been won by A. Garma for his studies on peptic ulcer and by E. Krapf for his influence in the academic field. A National Institute of Mental Health, aimed at providing better facilities for the mentally ill throughout the whole country, was founded in 1957.

Russia and the communist countries offer a rather complex and varied picture from the historical perspective. Russia and Yugoslavia present a manifold background in terms of ethnic, linguistic, and religious dimensions; in most communist countries, with the exception of Czechoslovakia and, to a less extent, of Russia, an agricultural economy prevails. Also their allegiance to the Marxist doctrine is not homogeneous. In terms of psychiatry proper,

Western influences have always been prominent, first the German organicistic school in the mid-nineteenth century, then the French clinical school of hypnosis at the end of the century, followed by a short-term Freudian orientation in the twenties, and, from then on, by the Pavlovian doctrine.[619]

Russian psychiatry remained virtually unknown to the United States because of the reciprocal attitude of diffidence and lack of contact. Since foreign travelers have been admitted to communist countries in the last decade, a number of reports (including an official one by a special mission of the American Psychiatric Association in 1967) have become available.[33,198,199,245,246,327,360] It has been found that Russia has developed a network of services for mental patients, first at the level of the polyclinic (one for every 5,000 people), then the neuropsychiatric dispensary (one for every 500,000 people), and finally the district mental hospital, in all of which extensive work on prophylaxis, diagnosis, and rehabilitation takes place. Such work is possible because of the large number of physicians (more than 600,000, mostly women), nurses, and medical technicians (that is, paraprofessionals, called "feldshers") available in community facilities as well as in mental hospitals (where the ratio of physicians to patients is 1:16). Therapy centers around social readaptation, by keeping patients as "vertical" as possible and by the eclectic use of short-term supportive relationship,[605] chemotherapy, suggestion (up to hypnosis), and an extensive program of day care (initiated in 1930) and sheltered workshops, often in conjunction with industries.[284,384] In contrast to the United States, the conception of psychiatry from the narrow "medical" perspective (reinforced by the view of psychodynamics as linked to the bourgeois system),[200] with the consequent complete absence of the collateral fields of clinical psychology and social work, the limited research facilities[386] (mainly at two centers in Moscow and Leningrad), and the recently established complacency of some psychiatric hospitals in certifying as mentally ill enemies of the regime[405] have naturally been seen in a very negative light.

Despite this, Soviet psychiatry, in the past considered backward from the psychoanalytic viewpoint, has recently elicited a favorable or at least an interested attitude in this country for two reasons: from the perspective of therapy, because of the spread of behavior therapy, whose philosophy has been heavily influenced by Pavlovian notions;[448,579] from the perspective of community psychiatry, because of the apparent success achieved in providing services for the mentally disturbed at the community and district level.[26,31,332,497,565,615,616] Aside from the 20-year-old monograph on Soviet psychiatry by J. Wortis,[610] the recent historical study by J. Brozek and D. Slobin,[102] and some reports by Russian authors translated into English,[236,456] much more direct knowledge of psychiatry, especially in relation both to the scientific attitude and the political dimension, is needed.[11,95,390] It looks as if both the United States and Russia may gain from the reciprocal observation of their psychiatric systems.

The other *communist countries* present considerable differences from Russia because of the reasons mentioned above, in spite of the efforts made by them to introduce a national system of psychiatric services early in the fifties. *East Germany, Hungary, Czechoslovakia,* and the *Croatian part of Yugoslavia* tend to be heavily influenced by Western psychiatric concepts, which had a long tradition in each one of them. In *Poland* much of the care for the institutionalized mentally ill is still in the hands of Catholic orders, while in *Bulgaria* and *Rumania,* as in the rest of the communist countries, systems based on Pavlovian principles are reinforced under the pressure of political forces. Even more than for Russia, firsthand reports[309] (with the exception of the books edited by A. Kiev[323] and by J. Masserman,[400] respectively) are lacking.

Psychiatry in the Countries of the Far Eastern Tradition

From an overall historical perspective—and taking into consideration the need to simplify matters in a short historical presentation—the countries of the Far Eastern tradition can roughly be divided into three types: (1) those that developed as part of the British Commonwealth (India, Ceylon, etc); (2) those in which psychiatry consisted of the amalgamation of autochthonous practices and Western concepts (Japan, Philippines, Taiwan); (3) and those in which indigenous practices were only very limitedly influenced by foreign notions.

In *India* official psychiatry was once represented by the Indian Division of Royal Medico-Psychological Association, which in 1947 became the Indian Psychiatric Society. Today there are about 200 psychiatrists for a population of more than 500 million. Psychiatric facilities tend to be undeveloped, and many mentally ill receive minimal care in their communities.[531,532] Worth mentioning is the rapprochement between traditional Hindu concepts of the mind (yoga and others) and some contemporary Western systems.[299]

In *Japan* academic psychiatry was in the past heavily influenced by the German school, while, in general, the attitude toward the mentally ill was rather punitive or neglectful (with some exceptions, such as the system of community care practiced at Iwakura, near Kyoto).[345] In the late twenties the psychoanalytic movement had a number of followers there, especially in the large cities, although in retrospect it appears that Freudian notions were basically modified by local customs—childrearing, role of the woman, servile attitude toward the elders and, in particular, the Emperor, and, in general, ambivalence toward Western progress, admired but also hated (J. Moloney's *Understanding the Japanese Mind*, 1954). Around the same time the so-called Morita therapy (named after the Tokyo psychiatrist S. Morita) was introduced, with a moderate degree of success that has persisted to the present. Essentially intended for patients suffering from neuroses and compulsions (common in Japan because of the tendency of people to internalize conflicts), this method consists of a period of complete isolation in bed, followed by progressive activity up to reinsertion in the community, carried on in a rigidly established situation of complete dependency by the patient on the doctor and the nurse (which has prompted the psychoanalytic view of Morita therapy as regression followed by "corrective ego experience"). As a result of the American involvement with Japan during World War II, firsthand anthropological studies were made on the whole Japanese culture by R. Benedict (*The Chrysanthemum and the Sword*, 1946), followed more recently by social psychological studies on attitudes and practices toward the mentally ill (mainly by W. Caudill,[116,117] T. Doi, C. Schooler, and others). Since then Japanese psychiatry has been greatly influenced by American trends, also through joint meetings (such as the one between the American and Japanese Psychiatric Associations in 1963). Today many of the 4,000 psychiatrists tend to follow eclectically the organicistic and psychodynamic schools. Regardless of differences in orientation, the importance of cultural factors stands out: close, almost symbiotic, relationship between mother and child; intense repression of feelings; conflicts between individuality and collaterality of family members; and especially conflicts between allegiance to traditional values and identification with Western mores.[557]

In *China*,[121,335] a gigantic country relatively little influenced by Western civilization, attitudes toward the mentally ill depend heavily on autochthonous cultural beliefs. Proper behavior is related to Tao, mainly based on reverence to the ancestors and on the public image of each individual; disregard for Tao may cause an imbalance between the two basic forces Yin and Yang, to which are subjected the various organs and channels connecting the inside to the periphery and the five constituents of the body, that is, earth, fire, water, wood, and metal. Since early times (such as in *The Yellow Emperor's Classic of Internal Medicine*, about 1,000 B.C.) the treatment of mental illness consisted of acupuncture and moxibustion (application of needles and of ignited substances at the surface of the body) to facilitate the flow of Yin and Yang along the proper channels.[574] In the coastal cities of Canton, Shanghai, and a few others, Western practices were introduced a century ago mainly by American missionaries, so that the few psychiatrists came to be influenced by Meyer's psychobiology. Dr. K. Bowman, sent there in 1947 by the United Nations to help organize the National Neuropsychiatric Institute, reported that there were about 50 psychiatrists and 6,000 psychiatric beds.[85] Since the communists took over in 1949, psychiatry has come to be seen from the Pavlovian perspective, although Mao Tse-Tung's writings are important in terms of prevention, as they stress priority of services to the masses, combination of mental hygiene and public health, and amalgamation of local and Western systems. Eventually the Chinese Society of Neurology and Psychiatry and the *Chinese Neurological and Psychiatric Journal* came into existence, psychiatric training was introduced in all 50 medical schools, and to the regularly trained physicians were added many paraprofessionals ("barefoot doctors" or "peasant-scientists"), especially in the rural areas. Regardless of the theoretical emphasis[122] (strictly Pavlovian in the fifties when politically China was very close to Russia, more a combination of indigenous and Western practices from the mid-sixties on), milieu therapy focused on group sessions directed toward ideological discussions has received the primary emphasis, in addition to physical treatment and chemotherapy. This may be justified from a perspective that considers mental disorders essentially as social problems, but it leaves unanswered some basic questions related to the individual's inability to express feelings openly, even in his own family, and to the emotional transferal of areas of personal life to a society that is based on austerity and purposefulness of common ideals.[577,611] These various points have many implications for mental health from the Western psychodynamic perspective and need further assessment.[358]

Psychiatry at the International Level

As mentioned above, cross-cultural and transcultural psychiatry will be presented in detail in other parts of this work. Here it is enough to mention the change in the methodological approach to the study of non-Western culture that has taken place during the period considered in this chapter, from the traditional psychoanalytic model of "culture and personality" (for example, E. Sapir, R. Benedict, M. Mead, C. DuBois, M. Opler, A. Kardiner, A. Kluckhohn and, among the opponents of psychoanalysis, B. Malinowski and A. Krober [314,328]) to a broader multidimensional perspective by interdisciplinary teams.[147,148,157,201,203,228,302,322,329,399,410,414,608,620] As an example of this switch, the volume by Carothers on the African mind (1953), which he considered incapable of reaching the level of Western sophistication, is today already obsolete.[112] Impetus toward research on mental disorders in other countries has come from the rapidly increasing contacts among people of different nations since the end of World War II and, even more, from the rise of many independent nations in the Afro-Asian areas.[376]

In connection with this latter event governments all over the world are involved in making plans for prevention, treatment, and rehabilitation of the mentally ill, often relying on a combination of Western and local practices.[570] The role of the United States has been so prominent in helping these various governments in this endeavor as to justify the separate presentation of this matter in the present section.

However, some essential preliminary points have to be mentioned on the basis of recent studies[92] (such as the comprehensive survey by A. Kiev):[324] the scientific approach attempts to explain *how*, folk attitudes *why* (e.g., influence of evil spirits) mental disorders occur; the incidence of severe psychopathology is constant everywhere, but the content (e.g., hallucinations or delusions) and the form (e.g., unusual syndromes due to altered states of consciousness) are different in each culture, according to its meaning (compensatory or pathoplastic) and methods of healing (e.g., cathartic); depending on the social expectation of the role played by the mentally ill (G. Devereux),[150,151,152] in each culture certain diagnoses are emphasized or not (e.g., alcoholism, homosexuality) and occur more often than others (e.g., frequent acute schizophrenic breakdown versus rare depressive conditions in Africa); in each culture are to be found healers of mental disorders, whose methods are based on a mixture of exorcism, drugs, and particular rituals, such as the interpretation of dreams; rapid social changes occurring mainly in Afro-Asian countries tend to result in stress that facilitates the rise of messianic and superstitious cults.

Taking these various points into consideration, it becomes understandable why it is difficult to establish common criteria to obtain epidemiological data on mental disorders in underdeveloped countries, where hospital facilities are rare, life expectancy is shorter, and migrations as well as political and social conflicts (urbanization, industrialization, etc.) are frequent.[536] It also becomes understandable why in many countries the tendency has prevailed to combine local and Western practices in the handling of psychiatric disorders with the help of nurses and paraprofessionals and the support of indigenous leaders and groups, resulting in efficient types of care, for instance, at the Aro Mental Hospital in Abeokuta, Nigeria (opened in 1954 and described by T. Lambo and others),[28,351,352,353] where a high percentage of patients live in villages under the supervision of trained personnel.

Three institutions have been particularly active in the organizational field of mental health: the World Federation for Mental Health, the World Health Organization, and the World Psychiatric Association.

Founded in 1948 in London (in connection with the Third International Congress in Mental Health),[461] the World Federation for Mental Health, under the dedicated leadership of few professionals (first the British J. Rees, then the Swiss A. Répond, the Americans F. Freemont-Smith, O. Klineberg, G. Stevenson, J. Millet, and others), has been instrumental in providing publications, seminars, workshops, lectures, and research on various aspects of mental hygiene.[538,539,540]

Likewise, meetings on many topics related to psychiatry have been sponsored by the World Health Organization (established as an agency of the United Nations in Geneva in 1949 and composed of five regional offices, one for each continent) with the support of an expert advisory panel and various study groups. In addition, the WHO has published important monographs on juvenile delinquency (by L. Bovet, 1951) on maternal care and mental health (by J. Bowlby, 1952), and on other subjects.[338,339,402]

The World Psychiatric Association was founded in 1961, at the time of the Third World Congress of Psychiatry held in Montreal; the late E. Cameron was elected the first president.[109] Through the dedication of many psychiatrists (mainly the French J. Delay, H. Ey, and P. Sivadon, the Spanish J. Lopez-Ibor, the Swiss M. Bleuler, the British D. Leigh and J. Wing, and the American H. Tompkins, D. Blain, and F. Braceland), it has fostered the dissemination of professional information among 72 national psychiatric organizations, representing more than 63,000 practitioners throughout the world (as evidenced by the new *Directory of World Psychiatry*, edited by J. Gunn), and has organized technical sections, symposia, and regional meetings on various subjects.

Among the other worldwide organizations are the European Association of Child Psychiatrists, the International Society for Social Psychiatry, the International Association of Child Psychiatry, and the recently established Association of Psychiatrists in Africa (actually Pan-African Psychiatric Conferences have been held there since 1961). The *International Journal of Psychiatry* has been edited by J. Aronson in New York City since 1963, while, among the developing nations, the periodical *Psychopathologie Africaine* has appeared in Dakar since 1965.

Epilogue

In 1959, at the conclusion of the first edition of this chapter, the question of the responsibility of American psychiatry toward the mental health goals of the future was raised. That same year Karl Menninger, a senior psychiatrist, in his Academic Lecture entitled "Hope,"[407] reminded his audience of the words pronounced by Ernest Southard[232], a pioneer in broadening the scope of psychiatry, in 1919:

"May we not rejoice that we [psychiatrists] . . . are to be equipped by training and experience better, perhaps, than any other men to see through the apparent terrors of anarchism, of violence, of destructiveness, or paranoia—whether these tendencies are showing in capitalists or in labor leaders, in universities or in tenements, in Congress or under deserted culverts. . . . Psychiatrists must carry their analytic powers, their ingrained optimism and their strength of purpose not merely into the narrow circle of frank disease, but, like Seguin of old, into education; like William James, into the sphere of morals; like Isaac Ray, into jurisprudence; and, above all, into economics and industry. I salute the coming years as high years for psychiatrists!"

It took half a century before American psychiatry woke up to its responsibility toward all the citizens of the nations.[463] For the historian it is important to determine the main currents that have created this situation, beginning with the discussion on transcultural, cross-cultural, and international psychiatry in the previous section.

There it was established that American psychiatry had played an increasingly important role at the international level, first through research on transcultural and cross-cultural dimensions, followed by forms of assistance to European countries affected by World War II and, later on, to newly emerging Afro-Asian countries. Actually interest in international matters on the part of this country can be traced back to the establishment of the International Committee on Mental Hygiene in 1918 (preceded by the National Association for Mental Health, 1909), which organized the First International Congress for Mental Health in Washington, D.C., in 1930, and to involvement on the part of the American Psychiatric Association (through joint meetings with other national associations and the work of various committees), aside from the substantial support given to the United Nations.[229]

Direct psychiatric influence at the international level can also be related to the pre-World War II practice of obtaining training in European facilities (mainly England) by a few American psychiatrists, and vice versa. Following the end of the war, with the help of various organizations, a good number of European physicians received training in this country in the fifties; in fact, some of them eventually settled permanently in this country. In the sixties the majority of the foreign physicians in training came from developing Far Eastern nations (India, Korea, Philippines); in fact, during the academic year 1967–1968 more than 30 percent of the psychiatric trainees were graduates of foreign medical schools. This puts this country in a position of ambiguous responsibility: on the one hand, it allows foreign graduates to provide necessary manpower to poorly staffed public facilities; on the other hand, America has a commitment to developing countries to help train staff for their own programs.[377,422]

The issue of the responsibility of American psychiatry at the international level increasingly has become related to the developments that have taken place in this country.[500,580,591] Through a convergence of important studies (mainly J. Galbraith's *The Affluent Society*, 1958, and M. Harrington's *The Other America*, 1963), widely publicized campaigns (bus boycotts, ghetto riots, and the Poor People's March), and some important judicial and political actions (the 1954 Supreme Court decisions to outlaw segregation in public education, President Johnson's War on Poverty, etc.), this country has become aware, as never before, of large areas of poverty and deep racial and social conflicts. As a result of this new awareness, the community mental health movement, discussed in detail in another section of this chapter, has been taking place.

Such a movement, however, cannot be properly implemented only through legislation or allocation of funds, but requires the active participation of many people at various levels eager to modify the traditionally middle-class oriented psychiatric philosophy to be of more relevance to large segments of low-income population. Some modest attempts have already been made, such as introducing proper therapeutic modalities (E. Auerswald, S. Minuchin, F. Riessman, and others), assessing community-based facilities in some areas (for example, in a section of New York City by L.

Kolb and associates in *Urban Challenge to Psychiatry*, 1969), and encouraging well-motivated people to become paraprofessionals (or "indigenous workers or mental health expediters").

In view of all this, the situation of psychiatry in many other parts of the world, where poverty and social conflicts are endemic, has become quite relevant to this country. While, from the psychoanalytic perspective of the fifties, psychiatry in practically all the rest of the world was seen as inferior to American standards, from the perspective of the community mental health movement of the sixties, systems of care and treatment of the mentally ill used in underdeveloped countries as well as countries of the communist bloc may very well acquire a great deal of importance for American psychiatry. This ranges from a more tolerant attitude toward the mentally disturbed by the general population, to the development of community-based facilities, to the extensive use of paraprofessionals of all types.[133,569] Therefore, the mission grandiosely held by American psychiatry on the wave of the military victory of World War II, of disseminating everywhere psychodynamic principles leading to intensive therapeutic relationships, has been replaced by a more modest philosophy of individual treatment and, by a thorough commitment toward prevention and help to many more people throughout the country. Moreover, Western psychodynamic principles, seen from the broad cross-cultural perspective, have been found to share many similarities with indigenous forms of treatment based on public expression of feelings practiced in many countries by recognized healers; and, likewise, research in the context of other cultures (for example, on the phenomenology of the person in the Yoruba society by I. Laleye[350] and on the African Oedipus by M. Ortigues and E. Ortigues)[441] has pointed to the relativity of some dynamic notions traditionally held to be universal. Finally from the historical perspective the tendency toward internationalization of many issues (from youth revolt[401] to drug abuse[73]) and toward world-wide cooperation (e.g., in space exploration and in ecological projects) cannot but affect also the field of psychiatry.

In reference to history, the thesis recently presented in the volume *A Social History of Helping Services* (1970) by M. Levine and A. Levine[368] that periods of prevailing "intrapsychic models of help" (that is, psychodynamic) in eras of political conservatism alternate to periods of prevailing "situational modes of help" (that is, community mental health) in eras of social reforms, may be debatable. The fact remains, however, that historically a definite change is taking place in the role of psychiatry from the viewpoint of the psychiatric profession itself, of the patient, and of the public at large.[501]

In regard to the psychiatric profession,[141,194,431,595] the aristocratic image presented by psychiatrists in the past (R. Holt's *Personality Patterns of Psychiatrists*, 1958) is being challenged by the notion that there are few differences between psychoanalysts, psychiatrists, psychologists, and social workers (W. Henry, *et al.*, *The Fifth Profession*, 1971), especially if psychotherapy is seen as "the purchase of a friendship" (W. Schofield).[524] Most psychiatrists tend to be influenced by the organicistic, the individual, or the community model (E. Strauss, 1964),[555] quite often in a rather narrow way (W. Freeman, 1968).[212] In reality the first two models are the most pervasive, while thus far community action has not attracted many, as shown in the thorough study by A. Rogow.[496] Claims of psychoanalytic contributions to community action (in D. Milman and G. Goldman's *The Psychoanalytic Contributions to Community Psychiatry*, 1971) or of substantial progress achieved through preventive programs (in *Crisis in Child Mental Health Challenge for the 1970's*, 1970, which is the report of the Joint Commission of Mental Health of Children)[134] are far from being substantiated.

Very little has been said by patients in regard to their treatment. F. Redlich has been among the few who has attempted to answer the question of how a person finds a psychiatrist (*Harper's*, 1960).[477] The results of his own research (with Hollingshead, 1959), that low-income people tend to receive organic treatment in institutions and middle-class people psychotherapy in outpatient clinics, have not been substantially modified by the follow-up study by J. Myers and L. Beam (1968). There is a need for studies along the lines of H. Strupp's *Patients View Their Psychotherapy* (1969).

From the broad perspective of the public several studies are available,[409] all pointing to the fact that the majority of people tend to turn for help to other kinds of healers (mainly clergymen) before going to psychiatrists (M. Krout's *Psychology, Psychiatry and the Public Interest*, 1956; J. Nunnally's *Popular Conceptions of Mental Health*, 1961; Elison's *Public Image of Mental Health Services*, 1967; C. Kadushin's *Why People Go to Psychiatrists*, 1969).

It is questionable whether the above trend depends on the shortage of psychiatrists or rather on the widespread ambiguity toward psychiatry. The shortage of psychiatric manpower is still grave,[432] although psychiatry has become the third most frequent choice of specialty by young physicians after medicine and surgery, and there has been massive federal support of training programs and ingenious attempts to encourage students to enter this field (for example, with the booklet *Careers in Psychiatry*, published by the National Commission on Mental Health Manpower, 1968). Efforts to train paraprofessionals are certainly very laudable, although in reality their effectiveness is handicapped by the conflict between their identification with the values of the professionals and their allegiance to their own values.

The issue of values needs to be mentioned at this point. In the past psychiatry, as a field of medicine, has been seen from the traditional perspective of medical ethics.[88] The early psychoanalytic movement, stemming from a highly homogeneous patriarchal society, beginning with Freud assumed that values were not relevant to psychiatry. This notion was later challenged by many who became aware of the unconscious identification of the patient with the therapist's own values. In the fifties, during the short period of the rise of existentialism, values in psychiatry constituted the subject of considerable discussion[80] (for example, C. Bühler's *Values in Psychotherapy*, 1962).

In the last decade the main issue in regard to values has shifted from medical ethics to a much broader perspective involving the responsibility of the psychiatrist as a professional and as a citizen.[255,389,604] This shift has been influenced by several events: in the mental health field, the controversies generated by the many publications of T. Szasz indicating that psychiatrists perpetuate the "myth of mental illness" by supporting attitudes that place certain individuals in the role of the mentally ill; in the academic field, the bipolarity between Skinner's behavioristic model of personality[183] and Allport's[234] and Maslow's humanistic psychology; on a larger scale, the ambiguous image offered by psychiatry in relation to matters such as professional assessment of political figures (especially at the time of the 1964 Presidential election), the Vietnam War, the youth unrest, the spread of drug addiction, the fight against poverty, the ethnic conflicts, and the epidemics of violence.

The fact that many of these latter problems transcend the boundaries of our nation should not deter American psychiatrists from meaningful involvement, taking into consideration, of course, the possibility of conflict between confidentiality to the patient and service to the community.[99] This is the position officially taken by the Group for the Advancement of Psychiatry in *Psychiatry and Public Affairs*, 1966.[265]

Deep-seated attitudes are difficult to modify, even in psychiatrists,[60,513] as recently shown, for instance, in J. Kovel's *White Racism: A Psychohistory* (1970) and in T. Thomas and J. Sillen's *Racism and Psychiatry* (1972). There is evidence, however, that a new breed of young psychiatrists is emerging, committed to alleviating people's miseries at every level quite at variance with the traditional cliché of the psychoanalyst exclusively involved with a sophisticated clientele. The publications by R. Coles on underprivileged children, by M. Dumont (*The Absurd Healer*, 1969), by R. Leifer (*In the Name of Mental Health*, 1969), and by others on the uncertain role of the psychiatrist at present are expressions of this trend. Also notable are the liberal attitude of some psychiatrists toward the use of drugs and toward sexual behavior (especially homosexuality), as evidenced by the research of R. Masters and V. Johnson at the Reproductive Biology Research Foundation in St. Louis. The American Psychiatric Association has sponsored studies on violence[126,138,207,240,445,541,609] and its current president, A. Freedman, has been elected with the support of the Committee for Concerned Psychiatrists, a newly formed group directed toward social action.

All this points to the fact that in a matter of a few years a generation gap appears to have developed between the traditionally oriented psychiatrists, who still control many academic positions, and many socially committed psychiatrists. Certainly, on the one hand, the latter should realize that psychiatry (called "the uncertain science" in a thorough survey in a popular magazine, 1968)[365] cannot be the answer to all problems,[153,231] especially after J. Seeley has convincingly showed that increasingly psychiatry is expected to take over the roles left by the decline of traditional social and religious institutions.[527] On the other hand, inactivity vis-à-vis urgent issues is in itself a decision, as cogently pointed out by S. Halleck in *The Politics of Therapy* (1971).[276]

As stressed in an exceedingly stimulating paper by the distinguished historian, Stuart Hugues, at the 1969 convention of the American Psychiatric Association, society, rather than the patient, appears to be sick today. Yet the irrational expressions of many, up to despair, should not deter us from reason: "Sooner or later," he concluded, "the soft voice of reason will be heard once again."[295]

These words appear to echo the prophetic statement made in 1944 by Alan Gregg, a great mentor of our profession, on the occasion of the centenary of the American Psychiatric Association: "Psychiatry, along with the other natural sciences, leads to a life of reason. . . . Psychiatry gives us a sort of oneness-with-others, a kind of exquisite communion with all humanity, past, present and future. . . . Psychiatry makes possible a kind of sincere humanity and naturalness. . . . Psychiatry makes it possible *to bring to others* these things I have mentioned. . . . Also it makes one able to receive these same gifts."[251]

Perhaps looking back at history may represent for American psychiatry a source of confidence in meeting the great challenge of the future. This interest in history may be essential for the rise of a new humanism to which psychiatry, imbued with science and humanity, can validly contribute.

¶ Bibliography

In view of the historical and general character of this chapter, the following principles have been followed in the preparation of the bibliography:

1. Well-known books are cited in the text only, together with the name of their authors. They are not listed in the bibliography because the complete reference is given in other chapters dealing specifically with the topic to which the book refers.
2. Full reference is given in the bibliography to the books and articles that belong to the following categories:
    a. Historical material, especially if concerned with broad issues.
    b. Surveys of a particular topic that offer a view of its development.
    c. Specific articles that historically represent the beginning of a new field.
    d. Contributions by American authors.

As bibliographical guides to the psychiatric literature, the two following comprehensive books are available:

ENNIS, B., *Guide to the Literature in Psychiatry*, Partridge, Los Angeles, 1971.

MENNINGER, K., *A Guide to Psychiatric Books in English*, 3rd ed., Grune & Stratton, New York, 1972.

1. ABROMS, G. M., "A New Eclecticism," *Arch. Gen. Psychiat.*, 20:514–523, 1969.
2. ACKERKNECHT, E., *Medicine and Ethnol-*

ogy: *Selected Essays*, Johns Hopkins Press, Baltimore, 1971.

3. ———, "Mesmerism in Primitive Societies," *Ciba Symposia*, 9:826–831, 1948.
4. ———, *A Short History of Psychiatry*, 2nd ed., Hafner, New York, 1968.
5. ALBEE, G. W., "A Specter Is Haunting the Out-Patient Clinic," in Tulipan, A. B., and Feldman, S. (Eds.), *Psychiatric Clinics in Transition*, Brunner-Mazel, New York, 1969.
6. ALEXANDER, F., and ROSS, H. (Eds.), *Twenty Years of Psychoanalysis*, Norton, New York, 1953.
7. ———, EISENSTEIN, S., and GROTJAHN, M. (Eds.), *Psychoanalytic Pioneers*, Basic Books, New York, 1966.
8. ———, and SELESNICK, S., *The History of Psychiatry*, Harper & Row, New York, 1966.
9. ALINSKY, S., "The War on Poverty-Political Pornography," *J. Soc. Issues*, *21*:41–47, 1965.
10. ALLPORT, G. W., "European and American Theories of Personality," in David, H. P., and Von Bracken, H. (Eds.), *Perspectives in Personality Theory*, pp. 3–26, Basic Books, New York, 1957.
11. ALT, H., and ALT, E., *Russia's Children*, Bookman Associates, New York, 1969.
12. ALTSCHULE, M. D., *Roots of Modern Psychiatry: Essays in the History of Psychiatry*, Grune & Stratton, New York, 1957.
13. *American Journal of Psychiatry, Centennial Anniversary Issue, 1844–1944*, Washington, D.C., 1944.
14. ANSBACHER, H. L., "Alfred Adler: A Historical Perspective," *Am. J. Psychiat.*, *127*:777–782, 1970.
15. ———, and ANSBACHER, R. R. (Eds.), *The Individual Psychology of Alfred Adler*, Basic Books, New York, 1956.
16. ARIES, P., *Centuries of Childhood*, Alfred A. Knopf, New York, 1963.
17. ARIETI, S., "Conceptual and Cognitive Psychiatry," *Am. J. Psychiat.*, *122*:361–366, 1965.
18. ———, *Interpretation of Schizophrenia*, Brunner, New York, 1955.
19. ———, *The Intrapsychic Self: Feeling, Cognition, and Creativity in Health and Mental Illness*, Basic Books, New York, 1967.
20. ———, "New Views on the Psychodynamics of Schizophrenia," *Am. J. Psychiat.*, *124*:453–458, 1967.
21. ———, "The Psychodynamics of Schizophrenia: A Reconsideration," *Am. J. Psychother.*, *22*:366–381, 1968.
22. ———, "The Psychotherapeutic Approach to Depression," *Am. J. Psychother.*, *16*:361–366, 1962.
23. ———, "Some Elements of Cognitive Psychiatry," *Am. J. Psychother.*, *21*:723–736, 1967.
24. ———, "The Structural and Psychodynamic Role of Cognition in the Human Psyche," in Arieti, S. (Ed.), *The World Biennial of Psychiatry and Psychotherapy*, Vol. I, pp. 3–33, Basic Books, New York, 1970.
25. ———, *The Will to Be Human*, Quadrangle, New York, 1972.
26. ARONSON, J., and FIELD, M. G., "Mental Health Programming in the Soviet Union," *Am. J. Orthopsychiat.*, *34*:913–924, 1964.
27. ASERINSKY, E., and KLEITMAN, N., "Regularly Occurring Periods of Eye Motility and Concomitant Phenomena during Sleep," *Science*, *118*:273–274, 1953.
28. ASUNI, T., "Aro Hospital in Perspective," *Am. J. Psychiat.*, *124*:763–770, 1967.
29. AUERBACH, H., "The Anti-Mental Health Movement," *Am. J. Psychiat.*, *120*:105–112, 1963.
30. AUERSWALD, E., "Interdisciplinary versus Ecological Approach," *Fam. Process*, *7*:202–215, 1968.
31. AUSTER, S. L., "Impressions of Psychiatry in One Russian City," *Am. J. Psychiat.*, *124*:538–542, 1967.
32. AYD, F. J., and BLACKWELL, B. (Eds.), *Discoveries in Biological Psychiatry*, Lippincott, Philadelphia, 1970.
33. BABAYAN, E. A., "The Organization of Psychiatric Services in the USSR," *Int. J. Psychiat.*, *1*:31–35, 1965.
34. BABKIN, B. P., *Pavlov: A Biography*, University of Chicago Press, Chicago, 1949.
35. BACH, G. E., and ILLING, H., "Historische Perspektive zur Gruppenpsychotherapie," *Zeitschrift für Psycho-somatische Medizin*, *2*:132–147, 1956.
36. BAK, R. C., "Psychoanalysis Today," *J. Amer. Psychoanal. A.*, *18*:3–23, 1970.
37. BANDLER, B., "The American Psychoanalytic Association and Community Psy-

chiatry," *Am. J. Psychiat.*, *124*:1037–1042, 1968.

38. BANDLER, S., "Current Trends in Psychiatric Education," *Am. J. Psychiat.*, *127*: 585–590, 1970.
39. BARAHAL, H. S., "Resistances to Community Psychiatry," *Psychiat. Quart.*, *45*: 333–347, 1971.
40. BARTEMEIER, L. H., "The Future of Psychiatry: The Report of the Joint Commission on Mental Illness and Health," *Am. J. Psychiat.*, *118*:973–981, 1962.
41. BARTON, W., "Historical Perspective in the Delivery of Psychiatric Services," in Stokes, A. B. (Ed.), *Psychiatry in Transition, 1966–1967*, pp. 3–18, University of Toronto Press, Toronto, 1967.
42. ———, "Prospects and Perspectives: Implications of Social Change for Psychiatry," *Am. J. Psychiat.*, *125*:147–150, 1968.
43. ———, FARRELL, M. J., LENEHAN, F. T., and MCLAUGHLIN, W. F., *Impressions of European Psychiatry*, American Psychiatric Association, Washington, D.C., 1961.
44. BARUK, H., *La psychiatrie française de Pinel à nos jours*, Presses Universitaires de France, Paris, 1967.
45. BATESON, G. (Ed.), *Perceval's Narrative: A Patient's Account of His Own Psychosis, 1830–1832*, Stanford University Press, Stanford, 1961.
46. BECKER, R. DE, *The Understanding of Dreams: Their Influence on the History of Man*, Hawthorn, New York, 1968.
47. BELL, N. W., and SPIEGEL, J. P., "Social Psychiatry: Vagary of a Term," *Arch. Gen. Psychiat.*, *14*:337–345, 1966.
48. BELLAK, L., "The Role of Psychoanalysis in Contemporary Psychiatry," *Am. J. Psychother.*, *24*:470–476, 1970.
49. ———, "Some Personal Reflections on European and American Psychiatry," in Bellak, L. (Ed.), *Contemporary European Psychiatry*, pp. vii–xxvi, Grove Press, New York, 1961.
50. BELLONI, L., "Dall'elleboro alla reserpina," *Archivio di psicologia, neurologia e psichiatria*, *17*:115–148, 1956.
51. BEMPORAD, J. R., "New Views on the Psychodynamics of the Depressive Character," in Arieti, S. (Ed.), *The World Biennial of Psychiatry and Psychotherapy*, Vol. I, pp. 219–243, Basic Books, New York, 1971.
52. BENEDETTI, G., *Neuropsicologia*, Feltrinelli, Milan, 1969.
53. ———, and MÜLLER, C., "Switzerland," in Bellak, L. (Ed.), *Contemporary European Psychiatry*, pp. 325–359, Grove Press, New York, 1961.
54. BENNE, K. D., "History of the T-Group in the Laboratory Setting," in Bradford, L. P., Gibb, J. R., and Benne, K. D. (Eds.), *T-Group Theory and Laboratory Method: Innovations and Re-Education*, Chap. 4, John Wiley, New York, 1964.
55. BENNETT, C. L., and GRUENBERG, E., "A Substitute for Clinical Thinking," *Int. J. Psychiat.*, *9*:621–627, 1970.
56. BERINGER, K., *Der Meskalinrausch*, Springer, Berlin, 1927.
57. BERLIN, I. N., "Transference and Countertransference in Community Psychiatry," *Arch. Gen. Psychiat.*, *15*:165–172, 1966.
58. BERMAN, G., "Mental Hygiene in Latin America," in Shore, M. J. (Ed.), *Twentieth Century Mental Hygiene*, pp. 403–428, Social Sciences Publishers, New York, 1950.
59. ———, *La salud mental y la asistencia psiquiatrica en la Argentina*, Paidos, Buenos Aires, 1965.
60. BERNARD, V. W., "Interracial Practice in the Midst of Change," *Am. J. Psychiat.*, *128*: 978–984, 1972.
61. BERNFELD, S., "Freud's Scientific Beginnings," in Lorand, S. (Ed.), *The Yearbook of Psychoanalysis*, Vol. 6, pp. 24–50, International Universities Press, New York, 1950.
62. BERTALANFFY, L. VON, *General System Theory*, Braziller, New York, 1968.
63. BINGER, C., *Revolutionary Doctor: Benjamin Rush, 1746–1813*, Norton, New York, 1966.
64. ———, *Two Faces of Medicine*, Norton, New York, 1967.
65. BINSWANGER, L., *Being-in-the-World: Selected Papers*, Basic Books, New York, 1963.
66. ———, *Schizophrenie*, Neske, Tübingen, 1957.
67. ———, *Sigmund Freud: Reminiscences of a Friendship*, Grune, New York, 1957.
68. BLAIN, D., POTTER, H., and SALOMON, H., "Manpower Studies with Special Refer-

ence to Psychiatrists," *Am. J. Psychiat.*, *116*:791–797, 1960.

69. BLEULER, E., *Dementia Praecox or the Group of Schizophrenias*, International Universities Press, New York, 1950.
70. BLEULER, M., "Conception of Schizophrenia within the Last Fifty Years and Today," *Proc. Roy. Soc. Med.*, *66*:945–952, 1963; repr. in *Int. J. Psychiat.*, *1*:501–523, 1965.
71. ———, *Endokrinologische Psychiatrie*, Thieme, Stuttgart, 1954.
72. ———, "Research and Changes in Concepts in the Study of Schizophrenia, 1941–1950," *Bull. Isaac Ray Med. Lib.* (Providence, R. I.), *3*:1–132, 1955.
73. BLUM, R. H., *et al.*, *Society and Drugs*, 2 vols., Jossey-Bass, San Francisco, 1970.
74. BOCKOVEN, S., *Moral Treatment in American Psychiatry*, Springer, New York, 1963, 2nd ed., *Moral Treatment in Community Mental Health*, Springer, New York, 1972.
75. BOGUSLAW, W., *The New Utopians*, Prentice-Hall, Englewood Cliffs, N.J., 1965.
76. BOHANNAN, P. (Ed.), *African Homicide and Suicide*, Princeton University Press, Princeton, 1960.
77. BOLMAN, W. M., "Cross-Cultural Psychotherapy," *Am. J. Psychiat.*, *124*:1237–1244, 1968.
78. BONAPARTE, M., *The Life and Works of Edgar Allan Poe*, Imago, London, 1949.
79. BOND, E., *Dr. Kirkbride and His Mental Hospital*, Lippincott, Philadelphia, 1947.
80. BOOTH, G., "Values in Nature and Psychotherapy," *Arch. Gen. Psychiat.*, *8*:22–32, 1963.
81. BOSS, M., *Psychoanalysis and Daseinsanalysis*, Basic Books, New York, 1963.
82. BOSTOCK, J., *The Dawn of Australian Psychiatry*, Australian Medical Association, Mervyn Archdall Medical Monograph Number 4, Sidney, 1968.
83. BOTTOME, P., *Alfred Adler*, Vanguard, New York, 1957.
84. BOUTONNIER, J., *L'angoisse*, Presses Universitaires de France, Paris, 1949.
85. BOWMAN, K. M., "Psychiatry in China," *Am. J. Psychiat.*, *105*:70–71, 1948.
86. BOYERAS, R. (Ed.), *R. D. Laing and Anti-Psychiatry*, Harper & Row, New York, 1971.
87. BRACELAND, F., "Biological Research Developments in Psychiatry," in Talkington, P. C., and Bloss, C. L. (Eds.), *Evolving Concepts in Psychiatry*, pp. 52–64, Grune & Stratton, New York, 1969.
88. ———, "Historical Perspectives of the Ethical Practice of Psychiatry," *Am. J. Psychiat.*, *126*:230–237, 1969.
89. BRAND, J. L., "The National Mental Health Act of 1946: A Retrospect," *Bull. Hist. Med.*, *39*:231–245, 1965.
90. ———, "The United States: A Historical Perspective," in William, R. H., and Ozarin, L. D. (Eds.), *Community Mental Health. An International Perspective*, pp. 18–43, Jossey-Bass, San Francisco, 1968.
91. BRECHER, E. M., *The Sex Researchers*, Little, Brown, Boston, 1969.
92. BRODY, E. B., "Cultural Exclusion, Character and Illness," *Am. J. Psychiat.*, *122*:852–857, 1966.
93. ———, (Ed.), *Mental Health in the Americas*, Report of the First Hemisphere Conference, Washington, D.C., American Psychiatric Association, 1969.
94. ———, "Psychiatric Education in the Americas," *Am. J. Psychiat.*, *127*:1479–1484, 1971.
95. BROFENBRENNER, U., *Two Worlds of Childhood: U.S. and U.S.S.R.*, Russell Sage Foundation, New York, 1970.
96. BROMBERG, W., *Man above Humanity: A History of Psychotherapy*, Lippincott, Philadelphia, 1954.
97. BROSIN, H. W., "The Changing Curriculum," in Stokes, A. B. (Ed.), *Psychiatry in Transition, 1966–1967*, pp. 39–55, University of Toronto Press, Toronto, 1967.
98. BROTHERS, C. R. D., *Early Victorian Psychiatry, 1835–1905*, Mental Health Authority, Melbourne, 1962.
99. BROWN, B. S., "Psychiatric Practice and Public Policy," *Am. J. Psychiat.*, *125*:141–146, 1968.
100. BROWN, J. A. C., *Freud and the Post-Freudians*, Penguin Books, Baltimore, 1961.
101. BROWN, N., *Life against Death: The Psychoanalytic Meaning of History*, Wesleyan University Press, Middletown, 1959.
102. BROZEK, J., and SLOBIN, D. I., *Psychology in the USSR: An Historical Perspective*,

International Arts and Science Press, White Plains, N.Y., 1972.

103. BRUN, R., *General Theory of Neurosis.*, International Universities Press, New York, 1951.

104. BURNHAM, J. C., *Psychoanalysis and American Medicine, 1894–1918*, International Universities Press, New York, 1967.

105. BUYTENDIJK, F. J. J., *Attitudes et mouvements*, Desclée de Brouwer, Paris, 1957.

106. ———, *Pain*, Hutchinson, London, 1961.

107. CAHNMAN, W. J., and BOSKOFF, A. (Eds.), *Sociology and History: Theory and Research*, Free Press, New York, 1964.

108. CALDWELL, A. E., *Origins of Psychopharmacology: From CPZ to LSD*, Charles C Thomas, Springfield, Ill., 1970.

109. CAMERON, D. E., "The World Psychiatric Association," *Am. J. Psychiat.*, *123*:1439–1441, 1967.

110. CAPLAN, R. (in collaboration with CAPLAN, G.), *Psychiatry and the Community in Nineteenth-Century America*, Basic Books, New York, 1969.

111. CARMICHAEL, H. T., "Perspectives on the American Board of Psychiatry and Neurology," *Arch. Gen. Psychiat.*, *8*:405–417, 1963.

112. CAROTHERS, J. C., *The African Mind in Health and Disease: A Study in Ethnopsychiatry*, World Health Organization, Geneva, 1953.

113. CARP, E. A. D. E., *The Affective Contact*, Proceedings of the International Congress for Psychotherapeutics, Leiden, Sept. 5–10, 1951, Strengholt, Amsterdam, 1952.

114. CARVALHO-NETO, P., *Folklore y Psicoanalisis*, Mortiz, Mexico City, 1956.

115. CASTELNUOVO-TEDESCO, P., *The Twenty-Minute Hour: A Guide to Brief Psychotherapy for the Physician*, Little Brown, Boston, 1965.

116. CAUDILL, W., and DOI, L. T., "Psychiatry and Culture in Japan," in David, H. P. (Ed.), *International Trends in Mental Health*, pp. 97–108, McGraw-Hill, New York, 1966.

117. ———, and SCHOOLER, C., "Symptom Patterns and Background Characteristics of Japanese Psychiatric Patients," in Caudill, W., and Lin, T. (Eds.), *Mental Health Research in Asia and the Pacific*, pp. 114–147, East-West Center Press, Honolulu, 1969.

118. *Causes of Mental Disorders: A Review of Epidemiological Knowledge, 1959*, Proceedings of a Round Table, Arden House, Harriman, N.Y., October 27–28, 1959, Milbank Memorial Fund, New York, 1961.

119. CERLETTI, U., "Italy," in Bellak, L. (Ed.), *Contemporary European Psychiatry*, pp. 185–216, Grove Press, New York, 1961.

120. ———, "Old and New Information about Electroshock," *Am. J. Psychiat.*, *107*:87–94, 1950.

121. CERNY, J., "Chinese Psychiatry," *Int. J. Psychiat.*, 229–247, 1965.

122. CHIN, R., and CHIN, A. S., *Psychological Research in Communist China: 1949–1966*, MIT Press, Cambridge, 1969.

123. CHODOFF, P. "A Critique of Freud's Theory of Infantile Sexuality," *Am. J. Psychother.*, *20*:507–517, 1966.

124. CLARK, D. H., MONRO, A. B., SAINSBURY, P., and DICKS, H. V., "The British National Health Service and Psychiatry," *Am. J. Psychiat.*, *125*:1218–1238, 1969.

125. CLARK, R. A., "Analytic Psychology Today," *Am. J. Psychother.*, *15*:193–204, 1961.

126. ———, "Psychiatrists and Psychoanalysts on War," *Am. J. Psychother.*, *19*:540–557, 1965.

127. CLINARD, M. B., *Sociology of Deviant Behavior*, Rinehart & Co., New York, 1957.

128. COLBY, K. M., *An Introduction to Psychoanalytic Research*, Basic Books, New York, 1960.

129. *Congress Report of the 2nd International Congress for Psychiatry*, Zürich, 1–7 September 1957, 4 vols., Füssli, Zürich, 1959.

130. CONNERY, R. H. (Ed.), *The Politics of Mental Health*, Columbia University Press, New York, 1968.

131. COPE, O., *Man, Mind and Medicine: The Doctor's Education*, A Chairman's View of the Swampscott Study on Behavioral Science in Medicine, 23 Oct.–4 Nov. 1966, Lippincott, Philadelphia, 1968.

132. COSTIGAN, G., *Sigmund Freud: A Short Biography*, Macmillan, New York, 1965.

133. COWEN, E. L., GARDNER, E. A., and ZAX, M., *Emergent Approaches to Mental*

Health Problems, Appleton-Century-Crofts, New York, 1967.

134. *Crisis in Child Mental Health: Challenge for the 1970's*, Report of the Joint Commission on Mental Health of Children, Harper & Row, New York, 1970.
135. CUNNINGHAM, D. E., "Psychiatry in Australia," *Am. J. Psychiat.*, *124*:180–186, 1967.
136. DAIN, N., *Concepts of Insanity in the United States, 1789–1865*, Rutgers University Press, New Brunswick, 1964.
137. ———, *Disordered Minds: The First Century of Eastern State Hospital in Williamsburg, Va., 1766–1866*, University of Virginia Press, Charlottesville, 1971.
138. DANIELS, D. N., GIGULA, M. F., and OCHBERG, F. M., *Violence and the Struggle for Existence*, Little, Brown, Boston, 1970.
139. DAVID, K., "Mental Hygiene and the Class Structure," *Psychiatry*, *1*:55–65, 1938.
140. DAVID, M., *La psicoanalisi nella cultura italiana*, Boringhieri, Turin, 1966.
141. DAVIDSON, H. A., "The Image of the Psychiatrist," *Am. J. Psychiat.*, *121*:329–334, 1964.
142. DAVIES, J., *Phrenology, Fad and Science*, Yale University Press, New Haven, 1955.
143. DAY, M., "The Natural History of Training Groups," *Int. J. Group Psychother.*, *17*:436–446, 1967.
144. DEAN, S. R., "Is There an Ultraconscious beyond the Unconscious?" *J. Canad. Psychiat. A.*, *15*:57–62, 1970.
145. DELAY, J., *The Youth of André Gide*, University of Chicago Press, Chicago, 1963.
146. DELGADO, J. M. R., *Physical Control of the Mind*, Harper & Row, New York, 1969.
147. DENKO, J. D., "How Preliterate Peoples Explain Disturbed Behavior," *Arch. Gen. Psychiat.*, *15*:398–409, 1966.
148. ———, "The Role of Culture in Mental Illness in Non-Western Peoples," *J. Am. Med. Women's A.*, *19*:1029–1044, 1964.
149. DEUTSCH, A., *The Mentally Ill in America*, 2nd ed., Columbia Univerity Press, New York, 1959.
150. DEVEREUX, G., "Cultural Factors in Psychoanalytic Therapy," *J. Am. Psychoanal. A.*, *1*:629–655, 1953.
151. ———, "Cultural Thought Models in Primitive and Modern Psychiatric Theories," *Psychiatry*, *21*:359–374, 1958.
152. ———, *Mohave Ethnopsychiatry and Suicide: The Psychiatric Knowledge and the Psychic Disturbances of an Indian Tribe*, Smithsonian Institution, Bureau of American Ethnology, Bulletin 175, Government Printing Office, Washington, D.C., 1961.
153. DEVOTO, A., *La tirannia psicologica. Studi di psicologia politica*, Sansoni, Florence, 1960.
154. DICKS, H. V., *Fifty Years of the Tavistock Clinic*, Routledge & Kegan Paul, London, 1970.
155. DIENELT, K., *From Freud zu Frankl*, Osterreichischer Bundesverlag für Unrerricht, Wissenschaft und Kunst., Vienna, 1967.
156. DORNER, K., *Bürger und Irre. Zur Sozialgeschichte und Wissenschaftsoziologie der Psychiatrie*, Europäische Verlaganstalt, Frankfurt am Main, 1969.
157. DOUGLAS, M. (Ed.), *Witchcraft, Confessions and Accusations*, Tavistock Publications, London, 1970.
158. DREIKURS, R., "Early Experiments with Group Psychotherapy," *Am. J. Psychother.*, *13*:882–891, 1959.
159. ———, and CORSINI, R., "Twenty Years of Group Psychotherapy," *Am. J. Psychiat.*, *110*:567–575, 1954.
160. DUNHAM, W., "Community Psychiatry: The Newest Therapeutic Bandwagon," *Arch. Gen. Psychiat.*, *12*:303–313, 1965.
161. EARLEY, L. W. (Ed.), *Teaching Psychiatry in Medical School*, Proceedings of the Conference on Psychiatry and Medical Education, Atlanta, March 6–10, 1967, American Psychiatric Association, Washington, D.C., 1969.
162. EBAUGH, F. G., "The Evolution of Psychiatric Education," *Am. J. Psychiat.*, *126*:97–100, 1969.
163. EBIN, D. (Ed.), *The Drug Experience*, Grove Press, New York, 1961.
164. EDSMAN, C. M. (Ed.), *Studies in Shamanism*, Proceedings of the Symposium on Shamanism, Abo, Sept. 6–8, 1962, Almquist & Wiksell, Stockholm, 1967.
165. EHRENWALD, J., *From Medicine Man to Freud: An Anthology*, Dell, New York, 1956.
166. ———, *Psychotherapy: Myth and Method. An Integrative Approach*, Grune & Stratton, New York, 1966.
167. EHRHARDT, H., PLOOG, D., and STUTTE, H.

(Eds.), *Psychiatrie und Gesellschaft*, Huber, Bern, 1958.

168. EHRMANN, J. (Ed.), *Structuralism*, Doubleday, New York, 1970.

169. EISSLER, K. R., *Goethe: A Psychoanalytic Study*, 2 vols. Wayne State University Press, Detroit, 1963.

170. ———, *Leonardo da Vinci: Psychoanalytic Notes on the Enigma*, International Universities Press, New York, 1962.

171. ———, *Sigmund Freud und die Wiener Universität*, Bern and Stuttgart, 1966.

172. ———, *Talent and Genius: The Fictitious Case of Tausk contra Freud*, Quadrangle, New York, 1971.

173. EKSTEIN, R., and WALLERSTEIN, R. S., *The Teaching and Learning of Psychotherapists*, 2nd ed., Basic Books, New York, 1972.

174. ELLENBERGER, H., "A Comparison of European and American Psychiatry," *Bull. Menninger Clin.*, *19*:43–52, 1955.

175. ———, *The Discovery of the Unconscious: The History and Evolution of Dynamic Psychiatry*, Basic Books, New York, 1969.

176. ———, "La psychiatrie Suisse," *Evol. Psychiat.*, *16*:321, 619, 1951; *17*:139, 369, 593, 1952; *18*:298, 719, 1953.

177. ———, "Les illusions de la classification psychiatrique," *Evol. Psychiat.*, pp. 221–242, 1963.

178. ERIKSON, E. H., "Autobiographic Notes on the Identity Crisis," *Daedalus*, *99*:730–759, 1970.

179. ———, "On the Nature of Psycho-Historical Evidence: In Search of Gandhi," *Daedalus*, *97*:695–730, 1968.

180. ———, "Psychoanalysis and Ongoing History: Problems of Identity, Hatred and Non-Violence," *Am. J. Psychiat.*, *122*:241–253, 1965.

181. ———, *Young Man Luther*, Norton, New York, 1958.

182. "Evaluation of the Results of the Psychotherapies," Second and Third National Scientific Meetings of the Association for the Advancement of Psychotherapy, *Am. J. Psychother.*, *20*:1–202, 1966.

183. EVANS, R. I., *R. F. Skinner: The Man and His Ideas*, Dutton, New York, 1968.

184. EWALT, J. R., and EWALT, P. L., "History of the Community Psychiatry Movement," *Am. J. Psychiat.*, *126*:43–52, 1969.

185. ———, SCHWARTZ, M. S., APPEL, K. E., BARTEMEIER, L. H., and SCHLAIFER, C., "Joint Commission on Mental Illness and Health," *Am. J. Psychiat.*, *116*:782–790, 1960.

186. EY, H., *La conscience*, Presses Universitaires de France, Paris, 1968.

187. ———, "Trends and Progress in French Psychiatry," *J. Clin. & Exper. Psychopath.*, *15*:1–8, 1954.

188. EYSENCK, H. J. (Ed.), *Behavior Therapy and the Neuroses*, Pergamon Press, New York, 1960.

189. FABER, M. D., *The Design Within: Psychoanalytic Approaches to Shakespeare*, Science House, New York, 1970.

190. FANCHAMPS, A., "Des drogues magiques des Aztèques à la thérapie psycholytique," *Acta Psychotherapeut.*, *10*:372–384, 1962.

191. FARAU, A., *Der Einfluss der österreischen Tiefenpsychologie auf die amerikanische Psychotherapie der Gegewart*, Sexl, Vienna, 1953.

192. FARBER, M., *The Foundation of Phenomenology*, Harvard University Press, Cambridge, 1943.

193. FELDMAN, P. E., "The Personal Element in Psychiatric Research," *Am. J. Psychiat.*, *113*:52–54, 1956.

194. FELIX, R. H., "The Image of the Psychiatrist: Past, Present and Future," *Am. J. Psychiat.*, *121*:318–322, 1964.

195. ———, *Mental Illness: Progress and Prospects*, Columbia University Press, New York, 1967.

196. FERMI, L., *Illustrious Immigrants: The Intellectual Migration from Europe, 1930–1941*, University of Chicago Press, Chicago, 1968.

197. FERNANDEZ-MARINA, R., "The Puerto Rican Syndrome: Its Dynamics and Cultural Determinants, *Psychiatry*, *24*:47–82, 1961.

198. FIELD, M. G., "Approaches to Mental Illness in Soviet Society: Some Comparisons and Conjectures," *Soc. Problems*, *7*:277–297, 1960.

199. ———, "The Institutional Framework of Soviet Psychiatry," *J. Nerv. & Ment. Dis.*, *138*:306–322, 1964.

200. ———, "Psychiatry and Ideology: The Official Soviet View of Western Theories

and Practices," *Am. J. Psychother.*, *22:* 602–615, 1968.

201. FIELD, M. J., *An Ethno-Psychiatric Study of Rural Ghana*, Northwestern University Press, Evanston, 1960.
202. FORDHAM, M. (Ed.), *Contact with Jung*, Lippincott, Philadelphia, 1963.
203. FOSTER, E. B., "The Theory and Practice of Psychiatry in Ghana," *Am. J. Psychother.*, *16:*7–51, 1962.
204. FOUCAULT, M., *Madness and Civilization: A History of Insanity in the Age of Reason*, Pantheon Books, New York, 1965.
205. ———, *The Order of Things: An Archaeology of the Human Sciences*, Pantheon Books, New York, 1971.
206. FRANK, J. D., *Persuasion and Healing: A Comparative Study of Psychotherapy*, Johns Hopkins Press, Baltimore, 1961.
207. ———, *Sanity and Survival: Psychological Aspects of War and Peace*, Random House, New York, 1967.
208. FRANKL, V. E., *The Doctor and the Soul*, Alfred A. Knopf, New York, 1955.
209. ———, *Psychotherapy and Existentialism: Selected Papers on Logotherapy*, Washington Square Press, New York, 1967.
210. FRANKS, C. M., *Behavior Therapy: Appraisal and Status*, Chap. 1, McGraw-Hill, New York, 1970.
211. FREEDMAN, A. M., "Historical and Political Roots of the Community Mental Health Centers Act," *Am. J. Orthopsychiat.*, *37:* 487–494, 1967.
212. FREEDMAN, W., *The Psychiatrists, Personalities and Patterns*, Grune & Stratton, New York, 1968.
213. FREUD, A., "Clinical Studies in Psychoanalysis (Research Project at the Hampstead Child Therapy Clinic)," *Proc. Roy. Soc. Med.*, *51:*937–974, 1958.
214. FREUD, E. L. (Ed.), *The Letters of Sigmund Freud and Arnold Zweig*, Harcourt, Brace & World, New York, 1970.
215. FREUD, S., *The Freud Journal of Lou Andreas-Salome*, ed. by Leavy, S. A., Basic Books, New York, 1964.
216. ———, *Letters of Sigmund Freud*, Basic Books, New York, 1960.
217. ———, *The Origins of Psychoanalysis. Letters to Wilhelm Fliess, Drafts and Notes: 1887–1902*, Basic Books, New York, 1954.
218. ———, *Psychoanalysis and Faith: The Letters of Sigmund Freud and Oskar Pfister*, ed. by Meng, H., and Freud, E. L., Basic Books, New York, 1963.
219. ———, *A Psycho-Analytic Dialogue: The Letters of Sigmund Freud and Karl Abraham, 1907–1926*, ed. by Abraham, H. C., and Freud, E. L., Basic Books, New York, 1965.
220. FRIED, M., "Effects of Social Change on Mental Health," *Am. J. Orthopsychiat.*, *34:*3–28, 1964.
221. FROMM, E., SUZUKI, D. T., and DE MARTINO, R., *Zen Buddhism and Psychoanalysis*, The Zen Studies Society, New York, 1960.
222. FROMM-REICHMANN, F., "Notes on the History and Philosophy of Psychotherapy," in Fromm-Reichmann, F., and Moreno, J. L. (Eds.), *Progress in Psychotherapy*, pp. 1–23, Grune & Stratton, New York, 1956.
223. FUNKENSTEIN, D. H., "A New Breed of Psychiatrist?" *Am. J. Psychiat.*, *124:*226–228, 1967.
224. FURMAN, S. S., *Community Mental Health Services in Northern Europe, Great Britain, Netherlands, Denmark and Sweden*, PHS Pub. No. 1407, U.S. Government Printing Office, Washington, D.C., 1965.
225. ———, "Obstacles to the Development of Community Mental Health Centers," *Am. J. Orthopsychiat.*, *37:*758–765, 1967.
226. GALDSTON, I., "Community Psychiatry, Its Social and Historic Derivations," *J. Canad. Psychiat. A.*, *10:*461–473, 1965.
227. ——— (Ed.), *Historic Derivations of Modern Psychiatry*, McGraw-Hill, New York, 1967.
228. ——— (Ed.), *The Interface between Psychiatry and Anthropology*, Brunner Mazel, New York, 1971.
229. ———, "International Psychiatry," *Am. J. Psychiat.*, *114:*103–108, 1957.
230. GANTT, W. H., PICKENHAIN, L., and ZWINGMANN, C. (Eds.), *Pavlovian Approach to Psychopathology: History and Perspectives*, Pergamon, New York, 1970.
231. GARBER, R. S., "The Presidential Address: The Proper Business of Psychiatry," *Am. J. Psychiat.*, *128:*1–11, 1971.
232. GAY, F. P., *The Open Mind: Ernest Southard, 1876–1920*, Normandie House, Chicago, 1938.

233. GENTE, H. P., *Marxismus. Psychoanalyse. Sexpol*, Fischer, Frankfurt am Main, 1970.
234. GHOÛGASSIAN, J. P., *Gordon W. Allport's Ontopsychology of the Person*, Philosophical Library, New York, 1972.
235. GILL, M., "The Present State of Psychoanalytic Theory," *J. Abnorm. & Soc. Psychol.*, *58*:1–8, 1959.
236. GILYAROVSKI, V. A., "The Soviet Union," in Bellak, L. (Ed.), *Contemporary European Psychiatry*, pp. 281–323, Grove Press, New York, 1961.
237. GITELSON, M., "On the Identity Crisis in American Psychoanalysis," *J. Am. Psychoanal. A.*, *12*:451–476, 1964.
238. GLASSCOTE, R. M., *et al.*, *The Community Mental Health Center: An Analysis of Existing Models*, American Psychiatric Association, Washington, D.C., 1964.
239. GLICKHORN, J., and GLICKHORN, R., *Sigmund Freuds Akademische Laufbahn im Lichte der Dokumente*, Urban & Schwarzenburg, Vienna, 1960.
240. GLIDDEN, H. W., "The Arab World," *Am. J. Psychiat.*, *128*:984–988, 1972.
241. GLUECK, B. C., JR., and STROEBEL, C. F., "The Computer and the Clinical Decision Process. II," *Am. J. Psychiat.*, *125*, Suppl. 1–7, 1969.
242. GOBLE, F. G., *The Third Force: The Psychology of Abraham Maslow*, Grossman, New York, 1970.
243. GOFFMAN, E., *Asylums: Essays on the Social Situation of Mental Patients and Other Inmates*, Doubleday, New York, 1961.
244. GOLDMAN, G., "Retrospect and Prospect: Review and Evaluation of Developments at Columbia Psychoanalytic Clinic," in Goldman, G. S., and Shapiro, D. (Eds.), *Developments and Psychoanalysis at Columbia University*, pp. 335–347, Proceedings of the 20th Anniversary Conference, Psychoanalytic Clinic for Training and Research, Columbia University, October 30, 1965, Hafner, New York, 1966.
245. GORMAN, H., "Soviet Psychiatry and the Russian Citizen," in Freeman, H. (Ed.), *Progress in Mental Health*, pp. 85–98, Proceedings of the Seventh International Congress on Mental Health, Grune & Stratton, New York, 1969.
246. GORMAN, M., BARTON, W. E., VISOTSKY, H. M., SIROTKIN, P., MILLER, A. D., BAZELON, D. L., and YOLLES, S. R., "Impressions of Soviet Psychiatry," *Am. J. Psychiat.*, *125*:638–674, 1968.
247. GOTTSCHALK, L. A., and PATTISON, E. M., "Psychiatric Perspectives on T-Groups and the Laboratory Movement: An Overview," *Am. J. Psychiat.*, *126*:823–839, 1969.
248. GRATTON, H., *Psychoanalyses d'hier et d'aujourd'hui*, Cerf, Paris, 1957.
249. GREENFIELD, N. S., and LEWIS, W. C. (Eds.), *Psychoanalysis and Current Biological Thought*, University Wisconsin Press, Madison, 1965.
250. GREENSON, R. R., "The Origin and Fate of New Ideas in Psychoanalysis," *Int. J. Psychoanal.*, *50*:503–516, 1969.
251. GREGG, A., "Critique of Psychiatry," *Am. J. Psychiat.*, *101*:285–291, 1944.
252. GRENE, M., *Approaches to a Philosophical Biology*, Basic Books, New York, 1965.
253. GRINKER, R. R., "Brief Psychotherapy in Psychosomatic Medicine," *Psychosom. Med.*, *9*:78–103, 1947.
254. ———, "Conceptual Progress in Psychoanalysis," in Marmor, J. (Ed.), *Modern Psychoanalysis: New Directions and Perspectives*, pp. 19–43, Basic Books, New York, 1968.
255. ———, "The Continuing Search for Meaning," *Am. J. Psychiat.*, *127*:725–731, 1970.
256. ———, "Emerging Concepts of Mental Illness and Models of Treatment: The Medical Point of View," *Am. J. Psychiat.*, *125*:865–876, 1969.
257. ———, "Identity or Regression in American Psychoanalysis?" *Arch. Gen. Psychiat.*, *12*:113–125, 1965.
258. ——— (Ed.), *Mid-Century Psychiatry: An Overview*, Charles C Thomas, Springfield, Ill., 1953.
259. ———, "Psychiatry Rides Madly in All Directions," *Arch. Gen. Psychiat.*, *10*:228–237, 1964.
260. ———, "The Sciences of Psychiatry: Fields, Fences and Riders," *Am. J. Psychiat.*, *122*:367–376, 1965.
261. ———, WERBLE, B., and DRYE, R. C., *The Borderline Syndrome: A Behavioral Study of Ego-Functions*, Basic Books, New York, 1968.

262. GROB, G., *The State and the Mentally Ill: A History of the Worcester State Hospital in Massachusetts, 1830–1920*, University of North Carolina Press, Chapel Hill, 1966.
263. Group for the Advancement of Psychiatry, *Clinical Psychiatry: Problems of Treatment, Research and Prevention*, Science House, New York, 1967.
264. ———, *The Field of Family Therapy*, Report #78, New York, 1970.
265. ———, *Psychiatry and Public Affairs*, Aldine, Chicago, 1966.
266. ———, *The Recruitment and Training of the Research Psychiatrist*, Report #65, April 1967.
267. GRUENBERG, E. M. (Ed.), *Evaluating the Effectiveness of Community Mental Health Services*, Proceedings of a Round Table at the Sixth Anniversary Conference of the Milbank Memorial Fund, April 5–7, 1965, Mental Health Materials Center, New York, 1966.
268. ———, "The Social Breakdown Syndrome: Some Origins," *Am. J. Psychiat.*, *123:* 1481–1489, 1967.
269. GRÜHLE, H. W., *Verstehen und Einfühlen, Gesammlete Schriften*, Springer, Berlin, 1953.
270. ———, JUNG, R., MAYER-GROSS, W., and MÜLLER, M., *Psychiatrie der Gegenwart*, 3 vols., Springer, Berlin, 1961–1967.
271. GRUNEBAUM, G. E. VON, and CAILLOIS, R. (Eds.), *The Dream and Human Societies*, University of California Press, Berkeley, 1966.
272. GUERRA, F., *The Pre-Columbian Mind*, Seminar Press, New York, 1971.
273. HADFIELD, J. A., *Introduction to Psychotherapy: Its History and Modern Schools*, Allen & Unwin, London, 1967.
274. HALE, N. G. JR., *Freud and the Americans: The Beginnings of Psychoanalysis in the United States, 1876–1917*, Oxford University Press, New York, 1971.
275. HALL, J. (Ed.), *One Hundred Years of American Psychiatry*, Columbia University Press, New York, 1944.
276. HALLECK, S. L., *The Politics of Therapy*, Science House, New York, 1971.
277. HAMBURG, D. A. (Ed.), *Psychiatry as a Behavioral Science*, Prentice-Hall, Englewood Cliffs, N.J., 1970.
278. HAMBURG, D. A., *et al.*, "Report of the Ad Hoc Committee on Central Fact-Gathering Data of the American Psychoanalytic Association," *J. Am. Psychoanal. A.*, *15:* 841–861, 1967.
279. HANSELMANN, H., and SIMON, T. (Eds.), *Bericht über den 1. Internationalen Kongress für Heilpädagogik*, Leemann, Zürich, 1939.
280. HARE, E. H., and WING, J. K. (Eds.), *Psychiatric Epidemiology*, Proceedings of the International Symposium Held at Aberdeen University, July 22–25, 1969, Oxford University Press, New York, 1970.
281. HASS, H., *The Human Animal: The Mystery of Man's Behavior*, Putnam's, New York, 1970.
282. HAVENS, L., "Main Currents of Psychiatric Developments," *Int. J. Psychiat.*, *5:*288–345, 1968.
283. HAYES, E. N., and HAYES, T. (Eds.), *Claude Levi-Strauss: The Anthropologist as Hero*, MIT Press, Cambridge, 1970.
284. HEIN, G., "Social Psychiatric Treatment of Schizophrenia in the Soviet Union," *Int. J. Psychiat.*, *6:*346–362, 1968.
285. HENDERSON, D. K., *The Evolution of Psychiatry in Scotland*, Levingstone, Edinburg, 1964.
286. HOBBS, N., "Mental Health's Third Revolution," *Am. J. Orthopsychiat.*, *34:*822–833, 1964.
287. HOCH, P. H., and POLATIN, P., "Pseudoneurotic Forms of Schizophrenia," *Psychiat. Quart.*, *23:*248–255, 1949.
288. HOFF, H., *Lehrbuch der Psychiatrie*, 2 vols., Schwabe, Basel, 1965.
289. ———, and ARNOLD, O. F., "Germany and Austria," in Bellak, L. (Ed.), *Contemporary European Psychiatry*, pp. 43–142, Grove Press, New York, 1961.
290. HOFLING, C. K., and MEYERS, R. W., "Recent Discoveries in Psychoanalysis: A Study of Opinion," *Arch. Gen. Psychiat.*, *26:*518–523, 1972.
291. HOLMSTEDT, B., "Historical Survey," in Efron, D. (Ed.), *Ethnopharmacological Search for Psychoactive Drugs*, Public Health Service Publication No. 1645, U.S. Government Printing Office, Washington, D.C., 1967.
292. HORWITZ, W. A., POLATIN, P., KOLB, L. C., and HOCH, P., "A Study of Cases of Schizophrenia Treated by "Direct Analy-

sis," *Am. J. Psychiat.*, *114*:780–783, 1958.

293. HOSKINS, R. G., *The Biology of Schizophrenia*, Norton, New York, 1946.

294. HUDGES, R. W., *et al.*, "Psychiatric Illness in a Developing Country: A Clinical Study," *Am. J. Pub. Health, 60*:1788–1805, 1970.

295. HUGHES, H. S., "Emotional Disturbance and American Social Change, 1944–1969," *Am. J. Psychiat.*, *126*:21–28, 1969.

296. HUNTER, R., and MACALPINE, I., *Three Hundred Years of Psychiatry, 1535–1960*, Oxford University Press, London, 1963.

297. JACKSON, B., "The Revised Diagnostic and Statistical Manual of the American Psychiatric Association," *Am. J. Psychiat.*, *127*:65–73, 1970.

298. JACO, E. G., *The Social Epidemiology of Mental Disorders*, Russell Sage Foundation, New York, 1960.

299. JACOBS, H., *Western Psychotherapy and Hindu-Sadhana*, International University Press, New York, 1961.

300. JACOBS, L., and KOTIN, J., "Fantasies of Psychiatric Research," *Am. J. Psychiat.*, *128*: 1074–1080, 1972.

301. JAHODA, M., "The Migration of Psychoanalysis: Its Impact on American Psychology," in Flemming, D., and Bailyn, B. (Eds.), *The Intellectual Migration: Europe and America, 1930–1960*, pp. 420–445, Harvard University Press, Cambridge, 1969.

302. ———, "Traditional Healers and Other Institutions Concerned with Mental Illness in Ghana," *Int. J. Soc. Psychiat.*, *7*: 245–268, 1961.

303. JANET, P., *The Major Symptoms of Hysteria*, Fifteen Lectures Given in the Medical School of Harvard University, Macmillan, New York, 1907.

304. ———, *Psychological Healing*, 2 vols., Macmillan, New York, 1925.

305. JASPERS, K., *General Psychopathology*, University of Chicago Press, Chicago, 1968.

306. JONES, E., *The Life and Work of Sigmund Freud*, 3 vols., Basic Books, New York, 1953–1957.

307. JUNG, C. G., *Memories, Dreams, Reflections*, Basic Books, New York, 1963.

308. ———, *The Psychology of Dementia Praecox*, Journal of Nervous Disease Pub. Co., New York, 1909.

309. KALINOWSKY, L. B., "Practices of Psychiatry in the Communist Countries of Eastern Europe," *Am. J. Psychother.*, *16*:301–307, 1962.

310. KANFER, F. H., "Behavioral Analysis: An Alternative to Diagnostic Classifications," *Arch. Gen. Psychiat.*, *14*:529–539, 1965.

311. ———, and PHILLIPS, J. S., "Behavior Therapy: A Panacea for All Ills or a Passing Fancy?" *Arch. Gen. Psychiat.*, *15*:114–128, 1966.

312. KANNER, L., "Autistic Disturbances of Affective Contact," *Nerv. Child*, *2*:217–230, 1943.

313. KAPLAN, B. H. (Ed.), *Psychiatric Disorder and the Urban Environment: Report of the Cornell Social Science Center*, Behavioral Publications, New York, 1971.

314. KARDINER, A., and PREBLE, E., *They Studied Man*, Secker & Warburg, London, 1962.

315. KARLEN, A., *Sexuality and Homosexuality: A New View*, Norton, New York, 1971.

316. KAUFMAN, M. R., "Psychiatry: Why 'Medical' or 'Social' Model?" *Arch. Gen. Psychiat.*, *17*:347–360, 1967.

317. ———, "Psychoanalysis and American Psychiatry," *Psychiat. Quart.*, *43*:301–318, 1969.

318. KELMAN, H., "The Changing Image of Psychoanalysis," *Am. J. Psychoanal.*, *26*: 169–176, 1966.

319. KEPECS, J. G., "Psychoanalysis Today: A Rather Lonely Island," *Arch. Gen. Psychiat.*, *18*:161–167, 1968.

320. KIELL, N., *Psychiatry and Psychology in the Visual Arts and Aesthetics*, University of Wisconsin Press, Madison, 1965.

321. ———, *Psychoanalysis, Psychology and Literature: A Bibliography*, University of Wisconsin Press, Madison, 1963.

322. KIEV, A. (Ed.), *Magic, Faith and Healing: Studies in Primitive Psychiatry Today*, Free Press, New York, 1964.

323. ———, *Psychiatry in the Communist World*, Science House, New York, 1968.

324. ———, *Transcultural Psychiatry*, Free Press, New York, 1972.

325. KLAGSBRUN, S. C., "In Search of an Identity," *Arch. Gen. Psychiat.*, *16*:286–289, 1967.

326. KLAPMAN, J. W., *Group Psychotherapy: Theory and Practice*, Ch. 1, Grune & Stratton, New York, 1947.

327. KLINE, N. S., "The Organization of Psy-

chiatric Care and Psychiatric Research in the Union of Soviet Socialist Republics," *Ann. N.Y. Acad. Sci.*, *84*:147–224, 1960.

328. KLUCKHOHN, C., "The Influence of Psychiatry on Anthropology in America during the Past One Hundred Years," in Hall, J. K. (Ed.), *One Hundred Years of American Psychiatry*, pp. 589–617, Columbia University Press, New York, 1944.
329. ———, *Navaho Witchcraft*, Beacon Press, Boston, 1944.
330. KOLB, L., "The Institutes of Psychiatry: Growth, Development and Future," *Psychol. Med.*, *1*:86–95, 1970.
331. ———, "The Presidential Address: American Psychiatry, 1944–1969 and Beyond," *Am. J. Psychiat.*, *126*:1–10, 1969.
332. ———, "Soviet Psychiatric Organization and the Community Mental Health Center Concept," *Am. J. Psychiat.*, *123*:433–440, 1966.
333. KOLLE, K. (Ed.), *Grosse Nervenaerzte*, 3 vols., Thieme, Stuttgart, 1956–1963.
334. ———, *An Introduction to Psychiatry*, Philosophical Library, New York, 1963.
335. KORAN, L. M., "Psychiatry in Mainland China: History and Recent Status," *Am. J. Psychiat.*, *128*:970–978, 1972.
336. KRAEPELIN, E., *One Hundred Years of Psychiatry*, Citadel, New York, 1962.
337. KRANZ, H., and EINRICH, K., *Psychiatrie im Uebergang*, Thieme, Stuttgart, 1969.
338. KRAPF, E. E., "The Work of the World Health Organization in the Field of Mental Health," *Ment. Hyg.*, *44*:315–338, 1960.
339. ———, and MOSER, J., "Changes of Emphasis and Accomplishments in Mental Health Work, 1948–1960," *Ment. Hyg.*, *46*:163–191, 1962.
340. KRIS, E., *Psychoanalytic Explorations in Art*, International Universities Press, New York, 1952.
341. KUBIE, L. S., "Pitfalls of Community Psychiatry," *Arch. Gen. Psychiat.*, *18*:257–266, 1968.
342. ———, "The Pros and Cons of a New Profession: A Doctorate in Medical Psychology," *Tex. Rep. Biol. & Med.*, *12*:125–170, 1954.
343. ———, "Research in Psychiatry Is Starving to Death," *Science*, *116*:239–243, 1952.
344. ———, "The Retreat from Patients," *Arch. Gen. Psychiat.*, *24*:98–106, 1971.
345. KUMASAKA, Y., "Iwakura: Early Community Care of the Mentally Ill in Japan," *Am. J. Psychother.*, *21*:666–675, 1967.
346. LA BARRE, W., "The Peyote Cult," *Yale University Publications Anthropology*, No. 19, 1938.
347. LACHMAN, J. H., "Swedish and American Psychiatry: A Comparative View," *Am. J. Psychiat.*, *126*:1777–1781, 1970.
348. LAIN ENTRALGO, P., *Doctor and Patient*, McGraw-Hill, New York, 1969.
349. ———, *The Therapy of the Word in Classical Antiquity*, Yale University Press, New Haven, 1971.
350. LALEYE, I. P., *La conception de la personne dans la pensée traditionelle Yoruba, "approche phénomenologique,"* Lang, Berne, 1970.
351. LAMBO, T. A., "Experience with a Program in Nigeria," in William, R. H., and Ozarin, L. O. (Ed.), *Community Mental Health. An International Perspective*, pp. 97–110, Jossey-Bass, San Francisco, 1968.
352. ———, "Patterns of Psychiatric Care in Developing African Countries: The Nigerian Village Program," in David, H. P. (Ed.), *International Trends in Mental Health*, pp. 147–153, McGraw-Hill, New York, 1966.
353. ———, "Socioeconomic Change, Population Explosion and the Changing Phases of Mental Health Programs in Developing Countries," *Am. J. Orthopsychiat.*, *36*:77–83, 1966.
354. LANE, M. (Ed.), *Introduction to Structuralism*, Basic Books, New York, 1970.
355. LANGFELDT, G., "Scandinavia," in Bellak, L. (Ed.), *Contemporary European Psychiatry*, pp. 217–279, Grove Press, New York, 1961.
356. LAUBSCHER, B. J. F., *Sex, Custom and Psychopathology*, Routledge & Sons, London, 1937.
357. LAUGHLIN, H. P., "Psychiatry in the United Kingdom," *Am. J. Psychiat.*, *126*:1790–1794, 1970.
358. LAZURE, D., "Politics and Mental Health in New China," *Am. J. Orthopsychiat.*, *34*:925–933, 1964.
359. LEACH, E., *Claude Levi-Strauss*, Viking, New York, 1970.
360. LEBENSOHN, Z. M., "Impressions of Soviet

Psychiatry," *Arch. Neur. Psychiat.*, *80*: 735–751, 1958.

361. LEDERER, W., "Primitive Psychotherapy," *Psychiatry*, *22*:255–265, 1959.
362. LEFEBRE, L. B., "The Psychology of Karl Jaspers," in *The Philosophy of Karl Jaspers*, pp. 467–497, Tudor, New York, 1959.
363. LEIGH, D., *The Historical Development of British Psychiatry*, Pergamon, London, 1961.
364. LEIGHTON, A. H., PRINCE, T., and MAY, R., "The Therapeutic Process in Cross-Cultural Perspective: A Symposium," *Am. J. Psychiat.*, *124*:1171–1183, 1968.
365. LEMON, R., "Psychiatry: The Uncertain Science," *Saturday Evening Post*, *241*: 37–54, 1968.
366. LENZ, H., *Vergleichende Psychiatrie. Eine Studie über die Beziehung von Kultur, Soziologie und Psychopathologie*, Maudrich, Vienna, 1964.
367. LEO, J., "Psychoanalysis Reaches a Crossroad," *New York Times*, August 4, 1968.
368. LEVINE, M., and LEVINE, A., *A Social History of Helping Services: Clinic, Court, School and Community*, Appleton-Century-Crofts, New York, 1970.
369. LEWIN, L., *Phantastica, Narcotic and Stimulant Drugs*, Kegan Paul, London, 1931.
370. LEWIS, A., "Empirical or Rational? The Nature and Basis of Psychiatry," *Lancet*, *2*:1–9, 1967.
371. ———, "Great Britain," in Bellak, L. (Ed.), *Contemporary European Psychiatry*, pp. 145–183, Grove Press, New York, 1961.
372. ———, "Problems Presented by the Ambiguous Word 'Anxiety' as Used in Psychopathology," *The Israel Annals of Psychiatry and Related Disciplines*, *5*: 105–121, 1967. Reprinted in *Int. J. Psychiat.*, *9*:62–79, 1970.
373. LIDZ, T., "Adolf Meyer and the Development of American Psychiatry," *Am. J. Psychiat.*, *123*:320–332, 1966.
374. ———, "Toward Schools of Psychiatry," in Lidz, T., and Edelson, M. (Eds.), *Training Tomorrow's Psychiatrist: The Crisis in Curriculum*, Yale University Press, New Haven, 1970.
375. ———, and EDELSON, M. (Eds.), *Training Tomorrow's Psychiatrist*, Yale University Press, New Haven, 1970.
376. LIN, T., "Community Mental Health Services: A World View," in William, R. H., and Ozarin, L. O. (Eds.), *Community Mental Health: An International Perspective*, pp. 3–17, Jossey-Bass, San Francisco, 1968.
377. ———, *et al.*, "Psychiatric Training for Foreign Medical Graduates: A Symposium," *Psychiatry*, *34*:233–257, 1971.
378. LINDEMAN, E., "Psychosexual Factors as Stress Agents," in Tanner, J. (Eds.), *Stress and Psychiatric Disorders*, Blackwell, Oxford, 1960.
379. LINN, L., "The Fourth Psychiatric Revolution," *Am. J. Psychiat.*, *124*:1043–1048, 1968.
380. LOEWENBERG, R. D., "Karl Jaspers on Psychotherapy," *Am. J. Psychother.*, *9*:502–513, 1951.
381. LOPEZ IBOR, J. J. (Ed.), *Proceedings of the Fourth World Congress of Psychiatry*, 4 vols., Excerpta Medica, Amsterdam, 1968.
382. LOPEZ PINERO, J. M., *Origines historicas del concepto de neurosis*, Industrias Graficas ECIR, Valencia, 1963.
383. ———, and MORALES MESEGUER, J. M., *Neurosis y psicoterapia: Un estudio historico*, Espasa-Calpe, Madrid, 1970.
384. LUSTIG, B., *Therapeutic Methods in Soviet Psychiatry*, Fordham University (Monographs in Soviet Medical Sciences, No. 3), New York, 1963.
385. LUTHE, W., "Autogenic Training: Method, Research and Application in Medicine," *Am. J. Psychother.*, *17*:174–195, 1963.
386. LYNN, R., "Russian Theory and Research on Schizophrenia," *Psychol. Bull.*, *60*:486–498, 1963.
387. MCGLASHAN, A. M., and REEVE, C. J., *Sigmund Freud: Founder of Psychoanalysis*, Praeger, New York, 1970.
388. MCKNIGHT, W. K., "Historical Landmarks in Research on Schizophrenia in the United States," *Am. J. Psychiat.*, *114*: 873–881, 1958.
389. MCNEIL, J. N. M., LIEWELLYN, C. E., and MCCOLLOUGH, T. E., "Community Psychiatry and Ethics," *Am. J. Orthopsychiat.*, *40*:22–29, 1970.
390. MADISON, B. Q., *Social Welfare in the Soviet Union*, Stanford University Press, Stanford, 1968.
391. MAHLER, M. S., "On Child Psychosis and

Schizophrenia: Autistic and Symbiotic Infantile Psychoses," in *The Psychoanalytic Study of the Child*, Vol. 7, pp. 286–305, International University Press, New York, 1952.

392. MANNONI, O., *Freud*, Pantheon, New York, 1971.
393. MANUEL, F. E., "The Use and Abuse of Psychology in History," *Daedalus, 100:* 187–213, 1971.
394. MARCUSE, H., *Eros and Civilization: A Philosophical Inquiry into Freud*, Beacon Press, Boston, 1955.
395. MARGETTS, E. L., "Historical Notes on Psychosomatic Medicine," in Wittkower, E., and Cleghorn, J. (Eds.), *Recent Developments in Psychosomatic Medicine*, pp. 41–69, Lippincott, Philadelphia, 1954.
396. MARMOR, J., "Current Status of Psychoanalysis in American Psychiatry," in Galdston, I. (Ed.), *Psychoanalysis in Present-Day Psychiatry*, pp. 1–16, Brunner/Mazel, New York, 1969.
397. ———, "The Reintegration of Psychoanalysis into Psychiatric Practice," *Arch. Gen. Psychiat., 3:*569–574, 1960.
398. MARSHALL, H., *Dorothea Dix: Forgotten Samaritan*, University of North Carolina Press, Chapel Hill, 1937.
399. MARWICK, M. (Ed.), *Witchcraft and Sorcery: Selected Readings*, Penguin, Baltimore, 1970.
400. MASSERMAN, J. H., *Psychiatry: East and West. An Account of Four International Conferences*, Grune & Stratton, New York, 1968.
401. ——— (Ed.), *Youth: A Transcultural Psychiatric Approach*, Grune & Stratton, New York, 1969.
402. MAY, A. R., "Development of Mental Health Services in Europe," in Freeman, H. (Ed.), *Progress in Mental Health*, Proceedings of the Seventh International Congress on Mental Health, pp. 77–84, Grune & Stratton, New York, 1969.
403. MAY, R., "Historical Roots of Modern Anxiety Theories," in Hoch, P. H., and Zubin, J. (Eds.), *Anxiety*, Proceedings of the 39th Annual Meeting of the American Psychopathological Association, pp. 3–16, Grune & Stratton, New York, 1950.
404. MAZLISH, B. (Ed.), *Psychoanalysis and History*, Prentice-Hall, Englewood Cliffs, N.J., 1963.
405. MEDVEDEV, Z. A., and MEDVEDEV, R. A., *A Question of Madness*, Alfred A. Knopf, New York, 1971.
406. MEIER, C. A., *Ancient Incubation and Modern Psychotherapy*, Northwestern University Press, Evanston, 1967.
407. MENNINGER, K., "Hope" (The Academic Lecture), *Am. J. Psychiat., 116:*481–491, 1959.
408. ———, (with Mayman, M., and Pruyser, P.), *The Vital Balance*, Viking, New York, 1963.
409. *Mental Health Education: A Critique*, Pennsylvania Association for Mental Health, Philadelphia, 1960.
410. MERING, O. VON, and KASDAL, L. (Eds.), *Anthropology and Behavioral and Health Sciences*, University of Pittsburgh Press, Pittsburgh, 1970.
411. MERLEAU-PONTY, M., *Phenomenology of Perception*, Humanities Press, New York, 1962.
412. MEYER, A., "Organizing the Community for the Protection of Its Mental Life," *Survey, 34:*557–560, 1915.
413. MEYERHOFF, H., "On Psychoanalysis and History," *Psychoanal. & Psychoanalyt. Rev., 49:*2–20, 1962.
414. MIDDLETON, J. (Ed.), *Magic, Witchcraft and Curing*, The Natural History Press, Garden City, N.Y., 1967.
415. MILLER, M. H., WHITAKER, C. A., and FELLNER, C .H., "Existentialism in American Psychiatr. Ten Years Later," *Amer. J. Psychiat., 125:*1112–1115, 1969.
416. MINKOWSKI, E., *Lived Times*, Northwestern University Press, Evanston, 1970.
417. ———, *La schizophrénie*, 2nd ed., *Desclée de Brouwer*, Paris, 1953.
418. ———, *Traité de psychopathologie*, Presses Universitaires de France, Paris, 1966.
419. MISCHLER, E. G., and WAXLER, N. E., "Family Interaction Processes and Schizophrenia: A Review of Current Theories," *Int. J. Psychiat., 2:*375–413, 1966.
420. MITCHELL, W. S., "Address before the 50th Annual Meeting," *J. Nerv. & Ment. Dis., 21:*413–437, 1894.
421. MITSCHERLICH, A., *Society without the Father: A Contribution to Social Psychology*, Basic Books, New York, 1969.
422. MITTEL, N. S., "Training Psychiatrists from

Developing Nations," *Am. J. Psychiat.*, *126*:1143–1149, 1970.

423. MORA, G., "The Historiography of Psychiatry and Its Development: A Re-Evaluation," *J. Hist. Behav. Sci.*, *1*:43–52, 1965.

424. ———, "The History of Psychiatry: A Cultural and Bibliographical Survey," *Psychoanal. Rev.*, 52:299–328, 1965.

425. ———, "The History of Psychiatry: The Relevance for the Psychiatrist," *Am. J.* France, Paris, 1961.

426. ———, "Introduction," in *New Directions in American Psychiatry, 1944–1968*, Presidential Addresses of the American Psychiatric Association over the past Twenty-Five Years, American Psychiatric Association, Washington, D.C., 1969.

427. ———, and BRAND, J., *Psychiatry and Its History: Methodological Problems in Research*, Charles C Thomas, Springfield, Ill., 1971.

428. MOSCOVICI, S., *La psychanalyse. Son image* 1964.

429. MURPHY, G., and MURPHY, L. (Eds.), *Asian Psychology*, Basic Books, New York, 1968.

430. ———, and ——— (Eds.), *Western Psychology*, Basic Books, New York, 1969.

431. MYERS, J. M., "The Image of the Psychiatrist," *Am. J. Psychiat.*, *121*:323–328, *et son public*, Presses Universitaire de *Psychiat.*, *126*:957–967, 1970.

432. National Institute of Mental Health, *The Nation's Psychiatrists*, Public Health Service Pub. #1885, Chevy Chase, Md., 1969.

433. NAYES, E. J., and NAYES, T. (Eds.), *Claude Levi-Strauss: The Anthropologist as Hero*, MIT Press, Cambridge, 1970.

434. NELSON, B. (Eds.), *Freud and the 20th Century*, Meridian Books, New York, 1957.

435. ———, "Self-Images and Systems of Spiritual Directions in the History of European Civilization," in Klausner, S. Z. (Ed.), *The Quest for Self-Control: Classical Philosophies and Scientific Research*, pp. 49–103, Free Press, New York, 1965.

436. NEWBROUGH, J. R., "Community Mental Health: A Movement in Search of a Theory," in *Symposium on Community Mental Health*, Ninth Inter-American Congress of Psychology, PHS Publications No. 1504, U.S. Government Printing Office, Washington, D.C., 1966.

437. NUNBERG, H., and FEDERN, E., *Minutes of the Vienna Psychoanalytic Society*, vol. 1, *1906–1908*, vol. 2, *1909–1911*, International University Press, New York, 1962, 1967.

438. NUTTIN, J., *Psychoanalysis and Personality*, Sheed & Ward, New York, 1962.

439. OBERNDORF, C., *A History of Psychoanalysis in America*, Grune, New York, 1953.

440. ORGLER, H., *Alfred Adler: The Man and His Work*, Liveright, New York, 1963.

441. ORTIGUES, M. C., and ORTIGUES, E., *Oedipe Africain*, Plon, Paris, 1966.

442. OSBORN, R., *Freud and Marx: A Dialectic Study*, Gollancz, London, 1937.

443. OSTOW, M., "Psychoanalysis and Ethology," *J. Am. Psychoanal. A.*, *8*:526–534, 1960.

444. PANSE, F., *Das psychiatrische Krankenhauswesen*, Thieme, Stuttgart, 1964.

445. PAPANEK, H., and PAPANEK, E., "Individual Psychology Today," *Am. J. Psychother.*, *15*:4–26, 1961.

446. ———, PAUMELLE, P., and LEBOVICI, S., "An Experience with Sectorization in Paris," in David, H. P. (Ed.), *International Trends in Mental Health*, pp. 97–108, McGraw-Hill, New York, 1966.

447. ———, and WILLIAM, R. H., "Issues on a Pilot Program" (Paris) in William, R. H., and Oxarin, L. O. (Eds.), *Community Mental Health: An International Perspective*, pp. 157–166, Jossey-Bass, San Francisco, 1968.

448. PAYNE, T. R., *S. L. Rubinstein and the Philosophical Foundations of Soviet Psychology*, Humanities Press, New York, 1968.

449. PIAGET, J., *Structuralism*, Basic Books, New York, 1970.

450. PICHOT, P., "France," in Bellak, L. (Ed.), *Contemporary European Psychiatry*, pp. 3–39, Grove Press, New York, 1961.

451. PLE, A., *Freud et la réligion*, Editions du Cerf, Paris, 1968.

452. POLLOCK, H. M. (Ed.), *Family Care of Mental Patients: A Review of Systems of Family Care in America and in Europe*, State Hospital Press, Utica, N.Y., 1936.

453. POMEROY, W. B., *Dr. Kinsey and the Institute of Sex Research*, Harper & Row, New York, 1972.

454. PORTMANN, A., *Animals as Social Beings*, Viking, New York, 1961.
455. ———, *New Paths in Biology*, Harper & Row, New York, 1964.
456. PORTNOV, A., and FEDOTOV, D., *Psychiatry*, Mir Publications, Moscow, 1969.
457. *Premier Congrès Mondial de Psychiatrie. Compte Rendus des Séances (18–27 Septembre, Paris)*, 7 vols., Hermann, Paris, 1952.
458. "President Kennedy's Message," *Am. J. Psychiat.*, *120*:729–737, 1964.
459. "Privacy and Behavioral Research," *Science*, *155*:535–538, 1967.
460. *Proceedings of the Brief Psychotherapy Council*, Institute for Psychoanalysis, Chicago, 1942, 1944, 1946.
461. *Proceedings of the International Conference on Mental Health, London, 1948*, Lewis, London, 1949.
462. *Proceedings of the Third World Congress of Psychiatry (Montreal, 4–10 June 1961)*, 3 vols., University of Toronto Press, Toronto, 1963.
463. "The Psychiatrist, the APA and Social Issues: A Symposium," *Am. J. Psychiat.*, *128*:677–687, 1971.
464. PUMPIAN-MINDLIN, E. (Ed.), *Psychoanalysis as Science*, Basic Books, New York, 1952.
465. QUERIDO, A., "Early Diagnosis and Treatment Services: Elements of a Community Mental Health Program," in Proceedings of a Round Table at 1955 Annual Conference, Milbank Memorial Fund.
466. ———, *The Efficiency of Medical Care*, Kroeser, Leiden, 1963.
467. ———, "Mental Health Programs in Public Health Planning," in David, H. P. (Ed.), *International Trends in Mental Health*, pp. 32–54, McGraw-Hill, New York, 1966.
468. RABKIN, R., "Is the Unconscious Necessary?" *Am. J. Psychiat.*, *125*:313–319, 1968.
469. RANGELL, L., "Psychoanalysis: A Current Lock," *Bull. Am. Psychoanal. A.*, *23*: 423–431, 1967.
470. ———, "Psychoanalysis and Neuropsychiatry: A Look at Their Interface," *Am. J. Psychiat.*, *127*:125–131, 1970.
471. RAPAPORT, D., "A Historical Survey of Psychoanalytic Ego Psychology," *Bull. Philadelphia A. Psychoanal.*, *8*:105–120, 1958.
472. ———, *Organization and Pathology of Thought*, Columbia University Press, New York, 1951.
473. RASCOVKY, A., "Notes on the History of the Psychoanalytic Movement in Latin America," in Litman, R. E. (Ed.), *Psychoanalysis in the Americas: Original Contributions from the First Pan-American Congress for Psychoanalysis*, pp. 289–299, International Universities Press, New York, 1966.
474. REDLICH, F. C., "Research Atmosphere in Departments of Psychiatry," *Arch. Gen. Psychiat.*, *4*:225–236, 1961.
475. ———, and BRODY, E. B., "Emotional Problems of Interdisciplinary Research in Psychiatry," *Psychiatry*, *18*:233–239, 1955.
476. ———, and PEPPER, M., "Are Social Psychiatry and Community Psychiatry Subspeciality of Psychiatry?" *Am. J. Psychiat.*, *124*:1343–1349, 1968.
477. ———, and PINES, M., "How to Choose a Psychiatrist," *Harper Magazine*, *220*:33–40, 1960.
478. REICH, I. O., *Wilhelm Reich: A Personal Biography*, Saint Martin's Press, New York, 1970.
479. REICH, W., *The Mass Psychology of Fascism*, Farrar, Straus & Giroux; New York, 1970.
480. REIDER, N., "The Concept of Normality," *Psychoanal. Quart.*, *19*:43–51, 1950.
481. REIK, L. E., "The Historical Foundations of Psychotherapy of Schizophrenia," *Am. J. Psychother.*, *10*:241–249, 1956.
482. RENNIE, T. A. C., "Social Psychiatry: A Definition," *Int. J. Soc. Psychiat.*, *1*:5–13, 1955.
483. REPOND, A., "Mental Hygiene in Switzerland," in Shore, M. J. (Ed.), *Twentieth Century Mental Hygiene*, pp. 383–402, Social Sciences Publishers, New York, 1950.
484. "Research Aspects of Community Mental Health Centers: Report of the APA Task Force," *Am. J. Psychiat.*, *127*:993–998, 1971.
485. RICOUER, P., *Freud and Philosophy: An Essay on Interpretation*, Yale University Press, New Haven, 1970.
486. RIEFF, P., *Freud: The Mind of a Moralist*, Viking, New York, 1959.
487. ———, *The Triumph of the Therapeutic:*

Uses of Faith after Freud, Harper & Row, New York, 1966.

488. RIESE, W., *The Legacy of Philippe Pinel*, Springer, New York, 1969.
489. ———, "Phenomenology and Existentialism in Psychiatry. An Historical Analysis," *J. Nerv. & Ment. Dis.*, *132*:469–484, 1961.
490. ———, "The Pre-Freudian Origins of Psychoanalysis," in Masserman, J. (Ed.), *Science and Psychoanalysis*, vol. 1, pp. 29–72, Grune & Stratton, New York, 1958.
491. RIFFLET-LEMAIRE, A., *Jacques Lacan*, Dessart, Brussels, 1970.
492. ROAZEN, P., *Brother Animal: The Story of Freud and Tausk*, Alfred A. Knopf, New York, 1969.
493. ROBBINS, L. L., "A Historical Review of Classification of Behavior Disorders and One Current Perspective," in Eron, L. D. (Ed.), *The Classification of Behavior Disorders*, pp. 1–37, Aldine, Chicago, 1966.
494. ROBERT, M., *The Psychoanalytic Revolution: Sigmund Freud's Life and Achievement*, Harcourt, Brace & World, New York, 1966.
495. ROGERS, R. R., "The Emotional Climate in Israel Society," *Am. J. Psychiat.*, *128*: 987–992, 1972.
496. ROGOW, A. A., *The Psychiatrists*, Putnam's, New York, 1970.
497. ROLLINS, N., "Is Soviet Psychiatric Training Relevant in America?" *Am. J. Psychiat.*, *128*:622–627, 1971.
498. ROMAN, P. M., "Labeling Theory and Community Psychiatry (The Impact of Psychiatric Sociology on Ideology and Practice in American Psychiatry)," *Psychiatry*, *34*:378–390, 1971.
499. ROMANO, J., "The Teaching of Psychiatry to Medical Students: Past, Present and Future," *Am. J. Psychiat.*, *126*:1115–1126, 1970.
500. ROME, H. P., "Psychiatry and Foreign Affairs: The Expanding Competence of Psychiatry," *Am. J. Psychiat.*, *125*:725–730, 1968.
501. ———, *et al.*, "Psychiatry Viewed From the Outside: The Challenge of the Next Ten Years," *Am. J. Psychiat.*, *123*:519–530, 1966.
502. ROSEN, G., *Madness in Society: Chapters in the Historical Sociology of Mental Illness*, University of Chicago Press, Chicago, 1968.
503. ROSEN, J. N., "The Treatment of Schizophrenic Psychosis by Direct Analysis," *Psychiat. Quart.*, *21*:117–131, 1947.
504. ROSENBERG, M., and ERICSON, R. P., "The Clinician and the Computer: Affair, Marriage or Divorce?" *Am. J. Psychiat.*, *125*: Suppl. 28–32, 1969.
505. ROSSI, A. M., "Some Pre-World War II Antecedents of Community Mental Health Theory and Practice," *Ment. Hyg.*, *46*:78–98, 1962. Reprinted in Bindman, A. J., and Spiegal, A. D. (Eds.), *Perspectives in Community Mental Health*, Aldine, Chicago, 1969.
506. ROTHMAN, D. J., *The Discovery of the Asylum: Social Order and Disorder in the New Republic*, Little, Brown, Boston, 1971.
507. RUBIN, B., "Community Psychiatry: An Evolutionary Change in Medical Psychology in the United States," *Arch. Gen. Psychiat.*, *20*:497–507, 1969.
508. RUESCH, J., "Social Process," *Arch. Gen. Psychiat.*, *15*:577–589, 1966.
509. ———, "The Trouble with Psychiatric Research," *Arch. Neurol. & Psychiat.*, *77*: 93–107, 1957.
510. RUSSELL, W., *The New York Hospital: A History of the Psychiatric Service, 1771–1936*, Columbia University Press, New York, 1945.
511. RYAN, W., "Community Care in Historical Perspective," *Canada's Mental Health*, *17*: Suppl. 60, 1969.
512. SABSHIN, M., "The Anti-Community Mental Health Movement," *Am. J. Psychiat.*, *125*:1005–1012, 1969.
513. ———, DIESENHAUS, H., and WILKERSON, R., "Dimensions of Institutional Racism in Psychiatry," *Am. J. Psychiat.*, *127*: 787–793, 1970.
514. SACKLER, A. M., *et al.* (Eds.), *The Great Physiodynamic Therapies in Psychiatry*, Harper, New York, 1956.
515. SAGER, C. J., "The Development of Marriage Therapy: An Historical Review," *Am. J. Orthopsychiat.*, *36*:458–467, 1966.
516. SANDLER, J., "Research in Psychoanalysis: The Hampstead Index as an Instrument of Psychoanalytic Research," *Int. J. Psychoanal.*, *43*:287–291, 1962.

517. SARGANT, W., *Battle for the Mind*, Doubleday, New York, 1957.
518. ———, "Psychiatric Treatment Here and in England," *Atlantic Monthly*, *214*:88–95, 1964.
519. SCHEIDLINGER, S., "Group Psychotherapy in the Sixties," *Am. J. Psychother.*, *22*:170–184, 1968.
520. SCHILLER, C. H. (Ed.), *Instinctive Behavior: The Development of a Modern Concept*, International Universities Press, New York, 1957.
521. SCHLESSINGER, N., "Supervision of Psychotherapy: A Critical Review of the Literature," *Arch. Gen. Psychiat.*, *15*:129–139, 1966.
522. SCHNECK, J., *A History of Psychiatry*, Charles C Thomas, Springfield, Ill., 1960.
523. SCHOENWALD, R., *Freud, the Man and His Mind, 1856–1956*, Alfred A. Knopf, New York, 1956.
524. SCHOFIELD, W., *Psychotherapy: The Purchase of a Friendship*, Prentice-Hall, Englewood Cliffs, N.J., 1964.
525. SCHULMAN, J., *Remaking an Organization: Innovation in a Specialized Psychiatric Hospital*, State University of New York Press, Albany, 1969.
526. SEARS, R. R., "Survey of Objective Studies of Psychoanalytic Concepts," *Soc. Sci. Res. Comm. Bull.*, No. 51, 1945.
527. SEELEY, J. R., *The Americanization of the Unconscious*, Lippincott, Philadelphia, 1967.
528. SEGUIN, C. A., "The Theory and Practice of Psychiatry in Peru," *Am. J. Psychother.*, *18*:188–211, 1964.
529. ———, "Twenty-Five Years of History of Psychiatry," in Heseltine, G. F. D. (Ed.), *Psychiatric Research in Our Changing World*, pp. 75–81, Excerpta Medica, Amsterdam, 1969.
530. SELLS, S. B. (Ed.), *The Definition and Measurement of Mental Health*, U.S. Department of Health, Education and Welfare, Washington, D.C., 1968.
531. SETHI, B. B., SACHDEV, S., and NAG, D., "Sociocultural Factors in the Practice of Psychiatry in India," *Am. J. Psychother.*, *19*:445–454, 1965.
532. ———, THACORE, V. R., and GUPTA, S. C., "Changing Patterns of Culture and Psychiatry in India," *Am. J. Psychother.*, *22*:46–54, 1968.
533. SHAKOW, D., and RAPAPORT, D., *The Influence of Freud on American Psychology*, International Universities Press, New York, 1964.
534. SITUATION, *Contributions to Phenomenological Psychology and Psychopathology*, Vol. 1, Spectrum, Utrecht, 1954.
535. SLOANE, R. B., "The Converging Paths of Behavior Therapy and Psychotherapy," *Am. J. Psychiat.*, *125*:877–885, 1969.
536. SMARTT, C. G. F., "Mental Maladjustment in the East African," *J. Ment. Sci.*, *102*:441–466, 1956.
537. SODDY, K., "Mental Hygiene in Great Britain," in Shore, M. J. (Ed.), *Twentieth Century Mental Hygiene*, pp. 361–382, Social Sciences Publishers, New York, 1950.
538. ———, and AHRENFELDT, R. H. (Eds.), *Mental Health in a Changing World*, Vol. 1, Lippincott, Philadelphia, 1965.
539. ———, and ——— (Eds.), *Mental Health in Contemporary Thought*, Vol. 2, Lippincott, Philadelphia, 1967.
540. ———, and ——— (Eds.), *Mental Health in the Service of the Community*, Vol. 3, Lippincott, Philadelphia, 1967.
541. SPIEGEL, J. P., "Psychosocial Factors in Riots: Old and New," *Am. J. Psychiat.*, *125*:281–285, 1968.
542. SPIEGELBERG, H., *The Phenomenological Movement*, 2 vols, 2nd ed., Nijhoff, The Hague, 1969.
543. SPITZ, R. A., and WOLF, K. M., "Anaclitic Depression: An Inquiry into the Genesis of Psychiatric Conditions in Early Childhood," in *The Psychoanalytic Study of the Child*, Vol. 2, pp. 313–342, International Universities Press, New York, 1946.
544. STAINBROOK, E., "Some Historical Determinants of Contemporary Diagnostic and Etiological Thinking in Psychiatry," in Hoch, P., and Zubin, J. (Eds.), *Current Problems in Psychiatric Diagnosis*, pp. 3–18, Grune & Stratton, New York, 1953.
545. STEIN, M., "Psychiatrist's Role in Psychiatric Research," *Arch. Gen. Psychiat.*, *22*:481–489, 1970.
546. STENGEL, E., "Classification of Mental Disorders," *Bull. WHO*, *21*:601–663, 1960.
547. STERBA, R., "A Case of Brief Psychotherapy

by Sigmund Freud," *Psychoanal. Rev.*, *35*:75–80, 1951.

548. STEWART, W. A., *Psychoanalysis: The First Ten Years, 1888–1898*, Macmillan, New York, 1967.
549. STIERLING, H., "Bleuler's Concept of Schizophrenia: A Confusing Heritage," *Am. J. Psychiat.*, *123*:996–1001, 1967.
550. STOLLER, A., "Public Health Aspects of Community Mental Health" (Australia), in William, R. H., and Ozarin, L. O. (Eds.), *Community Mental Health, An International Perspective*, pp. 123–139, Jossey-Bass, San Francisco, 1968.
551. STORCH, A., *The Primitive Archaic Forms of Inner Experiences and Thought in Schizophrenia*, Nervous and Mental Diseases Publishing Co., New York, 1924.
552. STRAUS, E. W., *Phenomenological Psychology*, Basic Books, New York, 1966.
553. ———, *The Primary World of Senses*, Free Press, New York, 1963.
554. ———, and GRIFFITH, R. M. (Eds.), *Phenomenology of Memory*, Duquesne University Press, Pittsburgh, 1970.
555. STRAUSS, E., *et al.* (Eds.), *Psychiatric Ideologies and Institutions*, Free Press, New York, 1964.
556. STRUPP, H. H., and BERGIN, A. E., "Some Empirical and Conceptual Bases for Coordinated Research in Psychotherapy: A Critical Review of Issues, Trends and Evidence," *Int. J. Psychiat.*, *7*:18–90, 118–168, 1969.
557. SUK CHOO CHANG, "The Cultural Context of Japanese Psychiatry and Psychotherapy," *Am. J. Psychother.*, *19*:593–606, 1965.
558. SZASZ, T. S., *The Myth of Mental Illness*, Harper, New York, 1961.
559. TAFT, J., *Otto Rank*, Julian Press, New York, 1958.
560. TAUBER, E. S., "Sullivan's Conception of Cure," *Am. J. Psychother.*, *14*:666–676, 1960.
561. *Techniques and Statistics in the Evaluation of the Results of the Psychotherapies*, Proceedings of the First National Scientific Meeting of the Association for the Advancement of Psychotherapy, *Am. J. Psychother.*, *18:* Suppl., 1964.
562. TEMKIN, O., "The History of Classification in the Medical Sciences," in Katz, M. J., Cole, J. O., and Barton, W. E. (Eds.), *The Role and Methodology of Classification in Psychiatry and Psychopathology*, Proceedings of a Conference in Washington, D.C., November 1965, Department of Health, Education and Welfare, Washington, D.C., 1968.
563. THOMPSON, G. W., "The Society of Biological Psychiatry," *Am. J. Psychiat.*, *111:* 389–391, 1954.
564. TINTEROW, N. M., *Foundations of Hypnosis: From Mesmer to Freud*, Charles C Thomas, Springfield, Ill., 1970.
565. TORREY, E. F., "Emergency Psychiatric Ambulance Services in the USSR," *Am. J. Psychiat.*, *128*:153–157, 1971.
566. TOURNEY, G., "History of Biological Psychiatry in America," *Am. J. Psychiat.*, *126*:29–42, 1969.
567. ———, "A History of Therapeutic Fashions in Psychiatry, 1800–1966," *Am. J. Psychiat.*, *124*:784–796, 1967.
568. ———, "Psychiatric Therapies: 1880–1968," in Rothman, T. (Ed.), *Changing Patterns in Psychiatric Care*, pp. 3–44, Crown, New York, 1970.
569. TREFFERT, D. A., "Psychiatry Revolves as It Evolves," *Arch. Gen. Psychiat.*, *17:* 72–94, 1967.
570. TSUNG-YI, L., "Community Mental Health Services: World View," in William, R. H., and Ozarin, L. D. (Eds.), *Community Mental Health: An International Perspective*, pp. 3–17, Jossey-Bass, San Francisco, 1968.
571. VAIL, D. J., *Mental Health Systems in Scandinavia*, Charles C Thomas, Springfield, Ill., 1965.
572. VEITH, I., *Hysteria: The History of a Disease*, University of Chicago Press, Chicago, 1965.
573. ———, "Psychiatric Nosology: From Hippocrates to Kraepelin," *Am. J. Psychiat.*, *114*:385–391, 1957.
574. ———, "Psychiatric Thought in Chinese Medicine," *J. Hist. Med.*, *10*:261–268, 1955.
575. VELASCO-ALZAGA, J. M., *La salud mental en las Americas*, Scientific Publications No. 81, pp. 8–28, Pan American Health Organization, Washington, D.C., 1963.
576. VIET, J., *Les méthodes structuralistes dans les sciences sociales*, 2nd ed., Mounton, The Hague, 1969.
577. VOGEL, E., "A Preliminary View of Family and Mental Health in Urban Communist

China," in Caudill, W., and Lin, T. (Eds.), *Mental Health Research in Asia and the Pacific*, pp. 393–404, East-West Center Press, Honolulu, 1969.

578. VOLMAT, R., and WIART, C. (Eds.), *Art and Psychopathology*, Proceedings of the Fifth Congress of the International Society of Art and Psychopathology, Paris, June 7–10, 1967, Excerpta Medica Foundation, Amsterdam, 1969.
579. VYGOTSKY, L. S., *The Thought and Language*, MIT Press, Cambridge, 1962.
580. WAGGONER, R. W., "The Future of International Psychiatry," *Am. J. Psychiat., 126:*1705–1710, 1970.
581. WALKER, N., *A Short History of Psychotherapy*, Routledge, London, 1957.
582. WALLERSTEIN, R. S., "The Challenge of the Community Mental Health Movement to Psychoanalysis," *Am. J. Psychiat., 124:*1049–1056, 1968.
583. ———, "The Current State of Psychotherapy: Theory, Practice, Research," *J. Am. Psychoanal. A., 14:*183–225, 1966.
584. WALTON, H. J., "Psychiatry in Britain," *Am. J. Psychiat., 123:*426–433, 1966.
585. WARD, C. H., "Psychotherapy Research: Dilemmas and Directions," *Arch. Gen. Psychiat., 10:*596–622, 1964.
586. ———, and RICKELS, K., "Psychiatric Residency Training: Changes over a Decade," *Am. J. Psychiat., 123:*45–54, 1966.
587. WASSON, R. G., *Soma: Divine Mushrooms of Immortality*, Harcourt, Brace & World, New York, 1969.
588. WASSON, V. P., and WASSON, R. G., *Mushrooms, Russia and History*, 2 vols., Pantheon, New York, 1957.
589. WATTS, A. W., *Psychotherapy East and West*, Pantheon, New York, 1961.
590. ———, *The Way of Zen*, Pantheon, New York, 1957.
591. WEDGE, B., "Training for a Psychiatry of International Relations," *Am. J. Psychiat., 125:*731–736, 1968.
592. WEHR, G., *Portrait of Jung*, Herder, New York, 1971.
593. WEINBERG, A. A., *Migration and Belonging: A Study of Mental Health in Israel*, Nijhoff, The Hague, 1961.
594. WEINBERG, S. K. (Ed.), *The Sociology of Mental Illness: Analyses and Readings in Psychiatric Sociology*, Aldine, Chicago, 1967.
595. WEINSHEL, E. M., *et al.*, "The Changing Identity of the Psychiatrist in Private Practice: A Symposium," *Am. J. Psychiat., 126:*1577–1587, 1970.
596. WEISS, E., *Sigmund Freud as a Consultant*, Intercontinental Medical Book Corp., New York, 1970.
597. WEITBRECHT, H. J., *Psychiatrie in der Zeit des National-sozialismus*, Hanstein, Bonn, 1968.
598. WELLS, H. K., *Pavlov and Freud*, 2 vols., International Publishers Co., New York, 1956.
599. WENDER, P. H., "Dementia Praecox: The Development of the Concept," *Am. J. Psychiat., 119:*1143–1151, 1963.
600. WHITTINGTON, H., "The Third Psychiatric Revolution, Really?" *Comm. Ment. Health J., 1:*73–80, 1963.
601. WHYTE, L., *The Unconscious before Freud*, Basic Books, New York, 1960.
602. WILD, J., "William Janes and Existential Authenticity," *J. Existentialism, 5:*244–256, 1965.
603. WILDER, J., "Alfred Adler in Historical Perspective," *Am. J. Psychother., 24:*450–460, 1970.
604. ———, "Values and Psychotherapy," *Am. J. Psychother., 23:*405–415, 1969.
605. WINN, R. B. (Ed.), *Psychotherapy in the Soviet Union*, Philosophical Library, New York, 1961.
606. WITTELS, F., *Freud and His Time*, Liveright, New York, 1931.
607. WITTKOWER, E. D., "Transcultural Psychiatry," *Arch. Gen. Psychiat., 13:*387–394, 1965.
608. ———, "Transcultural Psychiatry in the Caribbean: Past, Present and Future," *Am. J. Psychiat., 127:*162–166, 1970.
609. WOLIN, S. S., "Violence in History," *Am. J. Orthopsychiat., 33:*15–28, 1963.
610. WORTIS, J., *Soviet Psychiatry*, Williams & Wilkins, Baltimore, 1950.
611. YANG, C. K., *The Chinese Family in the Communist Revolution*, Harvard University Press, Cambridge, 1959.
612. YATES, A. J., *Behavior Therapy*, Ch. 1, John Wiley, New York, 1970.
613. YOLLES, S. F., "Past, Present and 1980: Trend Projections," in Bellak, L., and Barten, H. H. (Eds.), *Progress in Com-*

munity Psychiatry, vol. 1, pp. 3–23, Grune & Stratton, New York, 1969.

614. ZEGANS, L. S., "An Appraisal of Ethological Contributions to Psychiatric Theory and Research," *Am. J. Psychiat.*, *124*:729–739, 1967.

615. ZIFERSTEIN, I., "The Soviet Psychiatrist's Concept of Society as a Therapeutic Community," in Masserman, J. H. (Ed.), *Current Psychiatric Therapies*, vol. 6, pp. 367–373, Grune & Stratton, New York, 1966.

616. ———, "The Soviet Psychiatrist: His Relationship to His Patients and to His Society," *Am. J. Psychiat.*, *123*:440–446, 1966.

617. ZILBOORG, G., "Ambulatory Schizophrenia," *Psychiatry*, *4*:149–155, 1941.

618. ———, *A History of Medical Psychology*, Norton, New York, 1941.

619. ———, "Russian Psychiatry: Its Historical and Ideological Background," *Bull. N.Y. Acad. Med.*, *19*:713–728, 1943.

620. ZUBIN, J. (Ed.), *Field Studies in the Mental Disorders*, Proceedings of the Conference sponsored by the American Psychopathological Association, February 15–19, 1959, Grune & Stratton, New York, 1961.

621. ZUZUKI, D., *Zen Buddhism*, Doubleday, New York, 1956.

PART TWO

Basic Notions

CHAPTER 4

A GENERAL ASSESSMENT OF PSYCHIATRY

Alfred H. Rifkin

TO ASSESS is to determine the amount or worth of, to appraise, to evaluate, to take the measure of. We do not ordinarily consider assessing fields like cardiology, ophthalmology, or pediatrics in the sense of determining their worth, although we may inquire into the efficacy of a particular procedure or a medication. Yet it seems appropriate to seek a general assessment of psychiatry. Why? The question reflects a certain uneasiness, a need to clarify the scope of psychiatry, to fix the proper limits of its concern, to determine the nature of the problems to which it should address itself, and to study the conceptual and technical tools fashioned for the solution of problems. What is to be measured and evaluated? Shall it be the diagnostic scheme and the criteria for health or illness, the "cure" rate achieved by different treatments, the incidence and prevalence of specified disorders, or the various theories and schools of thought? These questions do not have merely academic or theoretical interest; they are of direct and immediate concern in determining many matters of public policy; in assigning investigative priorities; in allocating resources, funds, and personnel; and in establishing programs of education and training for the mental health professions.

A proper assessment of psychiatry requires that the subject be viewed in historical perspective. The physical sciences, dealing with relatively discrete phenomena amenable to direct observation or experimentation, and aided by the powerful tools of mathematics, were the first to emerge from speculative philosophy. Medicine as a whole was slower to emerge, but its scientific foundations were laid in the sixteenth and seventeenth centuries by anatomical investigations, the discovery of basic physiological functions, and the gradual delineation of clinical pictures. Psychiatry lagged behind as the understanding of human behavior long remained the province of metaphysics and theology. In general, "schools" of medicine faded as cellular pathology and bacteriology established a firm basis for understanding disease processes and their treatment. Today there are hardly any remnants of the great controversies that raged around bloodletting and purgation, or homeopathy versus allopathy. But it was not until 1824–1825 that J. F. Herbart, a philosopher and educator, published his *Psychology as Science, Newly Based on Experience, Metaphysics, and Mathematics*.[25] The steps forward as well as the burdens of the past are evident in this title. It was several decades later before Gustav Fechner demonstrated that psychological functions are amenable to experimental measurement, and only in 1879 did Wilhelm Wundt establish the first psychological laboratory. Mechanism and vitalism are no longer urgent issues, but psychogenesis, the modern descendant of the ancient body-mind dilemma, remains a subject of lively contention in psychiatry. Deeply rooted schisms abound that are not differences in emphasis or variations in technique but fundamental divergences in conceptualization.

Practical limitations require that an assessment of psychiatry must be selective rather than exhaustive. Ideally it should deal with trends and lines of development insofar as these can be discerned, rather than with specific syndromes or procedures. The most significant and most difficult task is to identify central themes and problems, and the strategies devised to explore them.

Schools of Thought

The first matter that calls for attention and appraisal is the prominence of schools of thought. There are many lines of cleavage that take on the qualities of slogans: organic versus psychogenic, heredity and constitution versus life experiences, instinct versus culture, psychotherapy versus chemotherapy or psychosurgery, depth analysis versus "behavioristic" symptom elimination, the medical model versus game theory and social learning. Such profound differences in approach have few, if any, parallels in other branches of medicine. It is often said that theories flourish where ascertainable facts are few. In principle a theory is not provable in any absolute sense; it merely becomes more plausible and perhaps more useful as observations consistent with it accumulate. Furthermore, a theory must be of such form and content that it may be refuted by some crucial observation. The problem for psychiatry is that these conditions are difficult to fulfill. On the contrary, observations may often be marshaled in support of several competing theories, while none is conclusively refuted. These strictures apply to most psychiatric theorizing, but are most often raised against psychodynamic, especially psychoanalytic, theories. Shall we then accept Jaspers'[28] view that ". . . in psychopathology there are no proper theories as in the natural sciences"?

Concerned as it is with disorders of human behavior, psychiatry must take into account a truly overwhelming diversity of factors and variables. To deal with the vast array, to apply the available modes of reasoning and investigation, this complexity must be simplified and ordered. This is the basic function of schools of thought. Each school of thought offers a body of theory that provides a framework upon which to arrange the data of experience, observation, and experiment. Theory directs attention to certain phenomena in preference to others, determines what will be considered important or relevant and what will be disregarded or dismissed. Theory is also limiting. Applying Sapir's viewpoint we may posit that each school of thought establishes its own language, which classifies, organizes, and to a significant degree predetermines experience for its users. Theory projects meaning into experience and imposes certain modes of observation and interpretation.

It is hardly possible to make a full inventory of the various theories of human behavior, but some useful categories may be delineated. Every individual is part of many interlocking and interacting systems or levels of integration. Within himself he is a delicately balanced set of chemical, physiological, and psychological subsystems. Each individual is also part of several systems of social interaction, extending from those with whom he is in closest contact to the widening circles comprised of groups of differing size, composition, and complexity to which he has some relationship. Psychiatry encompasses at one and the same time the scientific study of man on the biological level, the psychological study of the individual, and the sociological and anthropological study of human groups on a small and large scale. Not only is psychiatry concerned with these various levels of integration, but also it is a composite of all the levels. Collectively designated the behavioral sciences, these traditional disciplinary subdivisions reflect the course of historical development in the study of human behavior. They bring portions of the subject matter within manageable bounds, but at the same time fragment our understanding.

Recognition of this fragmentation has had several significant consequences in psychiatry. In recent years there have been increasing efforts to make critical comparisons among different viewpoints as well as to develop "unified" theories. The rapid accumulation of information that characterizes the current era makes it less and less possible for any one individual to attain competence in several diverse fields, or even the subdivisions of any one discipline. As a result research efforts, journal articles, and books are more often group undertakings. On the other hand, professional training, institutional affiliation, and the scientific literature are still largely disciplinary.

The educational and organizational issues posed by this state of affairs will be discussed below in connection with some speculations about the future of psychiatry. The more immediate need is to find ways to amalgamate information from the parochial fields of knowledge in usable fashion, given the present structure of the behavioral sciences. Several steps may be envisaged. The first and most obvious is exemplified by the multidisciplinary conferences and projects already mentioned. What can be expected of these? At the very least the results should include mutual enrichment of information, clarification of definitions and problems, and a deeper understanding of other viewpoints. The disciplines may be using different terminology for the same phenomena, but they may also be applying the same terminology to fundamentally different phenomena.*

Beyond this it may be possible to identify related events at several levels of integration. Simple examples would take the form of parallelism or concomitance of chemical events and psychological states, or correlation between social class status and symptom patterns. Such studies usually fall within the newly delineated fields of psychosomatic medicine and social psychiatry.

Psychosomatic medicine is confronted with "the mysterious leap from mind to body"[19] and the reverse. Two major theoretical formulations have emerged: (1) certain bodily disorders are the consequence of physiological concomitants of psychological (mainly emotional) states; (2) certain bodily disorders are the symbolic expression of psychological states (mainly conflict or the need to communicate). Placing the "psycho" first seems to have meant for most users of the term that the psychological state is primary or causal. Recent findings and more detailed analysis indicate that the interrelationship is more complex. Thus Knapp[31] formulates the issue as a series of feedback interactions that include social as well as psychological and physiological factors. Weiner,[59] relying on other experimental data, sees concomitance that does not necessarily imply causality in either direction.

For social psychiatry the problem of interrelationships is likewise complex. Human survival requires that societal forms be consistent with basic biological needs. The well-documented variation encountered in different societies makes it clear that there is no simple translation of biological needs into cultural institutions. There are evidently many ways in which these needs can be met. The most direct formulations focus on ways in which society facilitates or hinders the expression or satisfaction of needs, with emphasis usually on the repressive aspects of the social system. These formulations are oversimplifications. Societies also create needs apparently as im-

* Published proceedings of conferences and compilations of papers on various subjects are too numerous to warrant singling out particular ones for specific mention.

perative as those ordinarily considered basic and biological. Furthermore, the impact of society on individual needs is not merely to hinder or facilitate. Bell,[10] surveying the literature, lists some of the interrelationships postulated between individual and social variables: one step direct, one step indirect, two step direct, and two step indirect. Even this description is too schematic. Examination of various studies suggests there are many interweaving processes and variables that operate to a greater or lesser degree simultaneously, such as frustration, disorganization, social change versus stability, definition versus ambiguity of role, and others. The problem in each instance is to specify the variables and the mode of interactions.

Many comprehensive formulations have been made. Knapp[31] presents evidence for a "transactional model" in the etiology of bronchial asthma. Another sophisticated example is the investigation of genetic factors in schizophrenia originating in the National Institute of Mental Health. From a very carefully designed and executed series of studies summarized by Pollin,[42] it appears that genetic factors are significant in this disorder and may consist of predispositions to abnormal metabolism of catecholamines and indolamines due to peculiarities in the inducibility of certain enzymes. Affected individuals therefore suffer a hypothesized hyperarousal state of the nervous system that renders them specially vulnerable to stress. What is stressful, however, depends largely on the life history and experiences of the individual within his familial and sociocultural milieu.

Thus, we have pictured a sequence which involves an external event, the potential stressor—its resultant psychological meaning and signal, which is based on previous experience, role within the family and within the social structure—the resultant biochemical and physiologic response on the organ and cellular level, its extent determined in part by genetically controlled enzyme activity, and in part by changes in enzyme levels induced by previous experience.

In summary, experiential factors are seen as operating in four distinct ways. (1) They form the dictionary with which an individual translates the meaning and significance of any given current experience; and the yardstick by which he automatically measures the amount of threat, ie, stress, it constitutes for him. (2) Their residue determines the quantity and style of defenses and coping abilities with which the individual will attempt to deal with a stressful current life-situation. The determinants include such varied processes as the clarity of intrafamilial communication, the nature and extent of intrafamilial alliances and identifications, and the pattern of perceptions and coping mechanisms associated with one's location in the social class matrix. (3) Prior experience determines, in part, the extent of biochemical response to stress by influencing the level of enzymes available, through enzyme induction. (4) It seems likely that certain events, possibly concentrated in the intrauterine period, are relevant because of a direct slight effect on CNS structure, ie, neonatal anoxia causing minimal brain changes that later appear as subtle decrease in the capacity to maintain focal attention. (Pollin, pp. 35–36.)

This formulation is comprehensive and flexible enough to accommodate new findings at any level, including, for example, detailed investigations of formal thought disturbances, family constellations, or social class position. For the purposes of this general assessment it is not necessary to review the evidence for or against the various aspects of Pollin's thesis. Admittedly some portions are speculative. The significant advance is that investigations at different levels are not posed against one another as competing etiologies but are woven into a coherent whole. The concept of stress provides the framework for relating the diverse disciplinary approaches. It also permits relatively simple organization and statistical analysis of data, mainly by treating stress as an intervening variable and relying on twins to equate genetic factors. Such simplifications are essential.

The underlying logic of most investigations depends on one or more of J. S. Mill's classical canons for discovery of causal relationships in which possibly relevant factors are made to vary one at a time. Twin studies are an elegant application of Mill's canons; the genetic factor is taken as constant so that the effects of other variables can be studied.

Even in the case of identical twins it is difficult fully to satisfy Mill's canons and keep factors truly constant. Differences of personality in identical twin pairs have been noted repeatedly along with many similarities. Pollin cites a twin pair only one of whom was schizophrenic. From an early age there were marked differences in personality. In a further investigation of differences between identical twins, Pollin[43] and his co-workers point out that intrauterine and birth experiences may differ, leading to constitutional but not necessarily genetic differences. Such differences may be reflected in marked differences in personality and behavior at all ages, which, in turn, may evoke different treatment from the parents. Throughout life there is a complex interplay among heredity, constitution, the interpersonal twin relationship, subsequent life events, and extrafamilial relationships.

In general, human behavior must be taken as the resultant of complex systems of factors that vary and interact in many different ways. Any assertion of cause and effect depends on a condition that is usually unstated: "other things being equal." Since other things cannot ordinarily be taken as equal, demonstration of cause and effect requires procedures like matching and randomization, supplemented by statistical tests to cancel out factors other than those under investigation within some limit of allowable error. In real life situations there are rarely single causes, but rather a multiplicity of factors better conceived of as causal chains or networks. Changes in any portion of the network are associated with changes in other portions, and these, in turn, may affect other variables by feedback. Under such circumstances the classical definitions and demonstrations of cause and effect no longer suffice. Mathematical techniques for dealing with chains and networks are less advanced and more time-consuming. Mathematics aside, it is difficult to think in terms of networks without using simplifications that possibly vitiate the line of reasoning. An interesting example of ways in which such problems may be approached comes from the field of economics.[4] In this area, because numerical measures are available for many variables, complex interactions can be represented by systems of simultaneous equations. It then is possible to identify a hierarchical ordering of variables, which is the basis for defining cause and effect. The procedures are formidable, but in principle it is possible by these methods to derive important conclusions about the behavior of the real economic system even though the available information may be incomplete or approximate. Furthermore, subsystems of variables exhibit predictable degrees of stability and change within stated limits of probability.

These theoretical considerations have a bearing on the matter of schools of thought. In a multifactorial process, modification of any subsystem will produce measurable change in the system. In practice this will mean that a variety of approaches will find some support in terms of demonstrable effects. Competing theories may each find an acceptable degree of confirmation simultaneously because each affects some portion of the causal network. By directing attention to different subsystems, several investigations may be able to demonstrate what appear to be different valid cause-and-effect relationships. In the schizophrenia studies cited above the conceptual framework makes clear that there is no contradiction between biochemical or genetic etiologies and the competing family structure and social class etiologies. Knapp[31] proposes a similar framework for the etiology of bronchial asthma. Leighton[34] developed a comprehensive "Outline for a Frame of Reference" of similar scope to serve as the basis for testable hypotheses in the well-known Stirling County Study. Competing schools of thought will no doubt persist until sufficient investigation establishes unified conceptual frameworks. As matters now stand, multidisciplinary studies appear to offer the best prospect toward this goal.

Models—Medical and Otherwise

Psychiatry is said by some to be in the throes of an identity crisis epitomized as a challenge to "the medical model." The challenge comes from many quarters and ranges from mild criticism to outright rejection. The sharpest attack comes from Szasz, who denies there is any mental illness, suggesting, instead, that there are only "problems of living," which are moral and ethical, not medical.[56] * Similar criticisms of the medical model are voiced by Adams,[1] Albee,[2,3] Becker,[8] Laing,[32] Leifer,[33] Sarbin,[48] Scheflen.[51] Somewhat less stringent criticism is offered by Cowen,[15] Ellis,[20] Mowrer,[36] Reiff.[44] Virtues and drawbacks are attributed to the medical model by Ausubel,[5] Cohen,[14] Crowley,[16] Halleck,[24] Sarason and Ganzer.[47] Related criticisms from a sociological viewpoint are made by Goffman,[22] Scheff,[50] Spitzer and Denzin.[52] Rejoinders to Szasz and the criticisms of the medical model are offered in a sequence of papers and discussions by Davidson, Slovenko, Rome, Donnelly, Weihofen,[17] and by Begelman,[9] Brown and Long,[12] Brown and Ochberg,[13] Glaser,[21] Grinker,[23] Kaufman,[30] Reiss,[45] Thorne.[57]

Models are devices to make thinking about complex subjects easier. In a formal sense a model is an abstract logical system whose elements correspond to a set of events or things in the real world.[11,37] If this model is well chosen it may be possible to perform "thought experiments" with the model and draw conclusions that might not be feasible with the real system in the external world. There are also dangers. The usefulness of a model depends on the degree of correspondence between its elements and the "real" system in the external world. Valid conclusions can be drawn only to the extent of such correspondence. In other respects the model may not behave like the "real" system at all. In the discussion of schools of thought in the preceding section of this chapter the economic system was "modeled" by systems of simultaneous equations. It must be determined empirically how well the model corresponds to the real economic system, and transpositions from model to reality must remain within these limits. Several steps may be involved in using models. Light may first be compared with (modeled by) waves in a real medium (water or air), or the atom is modeled by electrons revolving around a nucleus after the fashion of the solar system. Then a mathematical model is devised to represent the properties of waves or rotating bodies.

It is clear that the medical model being criticized is not of this formal variety. The term is used to describe the explanation, mechanism, or process involved. The challenge is to the medical model. Is there such a model, one model that can properly be labeled the medical model?

Concern with illness is evidently as old as mankind. Explanations and theories about causes and what the process consists of have reflected the state of knowledge in each historical period. Primitive peoples could conceive only of processes that they themselves experienced directly. A man could inflict pain, so pain of unknown origin was attributed to spirits or demons. The medical model consisted of anthropomorphized creatures of the primitive imagination. The Greeks attributed illness to excesses or deficiencies of the four humors, or to excess blood. Galen left a theory of natural spirits, vital spirits, and animal spirits that survived through the Renaissance, gradually giving way before the advancing knowledge of anatomy, physiology, and chemistry following the Renaissance. During the seventeenth and eighteenth centuries there were several medical models: (1) iatrophysics—the body conceived as a machine operating in accordance with the principles of physics and mechanics then being elucidated; (2) iatrochemistry—based on investigations of acids, alkalis, and fermentation; (3) vitalism—the living body is governed by special laws of its own, not those of the chemistry and physics of inanimate objects. Bloodletting was widely employed by all, along with numerous herbal folk remedies. Cellular pathology and the role of microorganisms became the medical model of the latter half of the nineteenth century. Subsequently other facets were added: toxins and antitoxins, immune and hypersensitivity reactions, allergies, autoimmune reactions, nutritional deficiencies, and more recently, enzyme defects. Many general principles have

* Citations in this section are to recent publications, which give references to each author's earlier writings as well as bibliographies.

been formulated: patterned reflex reaction, host resistance and defense, homeostasis, generalized stressors (Selye) to name a few. As infectious diseases have receded in frequency, the emphasis has been shifting to neoplasia, to "wear and tear"—the processes of aging and degeneration, and to the external circumstances of life and psychological stress. If from this overview a medical model can be discerned, it would seem to be merely that the processes and mechanisms of illness are to be investigated, using previous information as fully as possible but advancing to whatever new views are justified by the facts discovered.

The major criticisms of the medical model are directed against specific features attributed to that model and may be summarized as follows:

1. The medical model assumes the existence of a disorder of the mind that is like disorders of the body. Brain pathology produces specific neurological disorders, but psychiatry deals with functional disorders in which no structural or chemical alteration has been demonstrated. It is usually asserted that mind is not an organ and in principle cannot be reduced to chemical, electrical, or other physiological processes.

2. The locus of the disorder is within the affected person, and the disorder is to be corrected or removed by the physician as is the case with known diseases of the body. This is misleading because the real locus of the disorder may be, and usually is, outside of the individual, in the social system or his interaction with the social system.

3. The medical model fosters a superior, authoritarian attitude in the physician and a dependent, subservient attitude in the patient. Both are antithetical to successful treatment, which requires that the patient achieve a greater degree of independence or autonomy.

4. Not only are there no demonstrable structural or physiological changes, but also there are no objective criteria for disturbed behavior. Mental illness can only be inferred from behavior that the illness is then supposed to explain, an obvious circularity in reasoning. Furthermore, judgment of behavior is subjective, tied to the value systems of the culture and open to various kinds of bias.

5. Because it depends on deviation from some norm, the medical model fosters conformity and stifles originality and creativity.

Some corollary criticisms may be added. The medical model focuses on pathology that is to be removed rather than on growth, development, and maturation. Many aspects of disordered behavior are not deviations from a norm, but rather normal responses to external conditions. Pathology should be ascribed to these external conditions, not to the individual reaction. The medical model requires vast numbers of highly trained personnel to deal with existing problems. Adequate care for all who need it cannot be attained within any foreseeable time; the model must therefore be replaced.

Leifer and Szasz carry the argument further. In their view a psychiatric diagnosis is intended to derogate and destroy anyone who deviates from accepted societal norms. Once labeled, the patient is stigmatized and victimized while the power and prestige of the psychiatrist are enhanced.

Szasz's objections to the medical model begin with the definition of mental illness. His experience is that at best mental illness is a metaphor that likens personal unhappiness and socially deviant behavior to symptoms and signs of bodily ailments. Signs and symptoms are caused by specifiable disorders of bodily organs, which can be stated with some precision in anatomical or physiological terms. Disorders of the brain cause neurological defects. But the term "disorders of the mind" refers to a false substantive, mind, which cannot be a cause. Personal unhappiness is subjective and cannot be stated in precise terms. These definitional and philosophical issues are significant but not crucial. Psychiatry may be quite adequately defined without reference to mind, and many textbooks do not use the term, or if they do, specify that it is a collective designation for certain functional activities of the organism rather than a metaphysical entity. Although Szasz states that he rejects the ancient body-mind dualism in favor of a hierarchical scheme of levels of integration, his argument treats mental illness as a metaphysical entity and thus appears to resurrect the philosophical dilemma.

A related aspect of the metaphysical problem is the contention that one cannot "have" a mental disease in the sense that one can "have" diabetes. (The quotation marks indicating special meaning are terms of the critics.) The argument is that diabetes refers to an entity that is "real," while mental disease does not. What is at issue is the meaning of terms like "have, entity," and "real." "Entity" is a metaphysical abstraction usually used to designate a hypothetical property separate and apart from the tangible and experiential aspects of an object. It is an abstraction that has meaning only as a part of a philosophical system that postulates the existence of such properties. Diabetes is not an entity in this sense. Diabetes designates a class defined by certain characteristics. A class name is neither real nor unreal; it may be more or less useful. Whether a class corresponds to anything in the "real" external world is an epistemological problem. A person does not "have" diabetes in the sense of possessing a material object. Saying a person has diabetes merely places him in the class of individuals who show certain defining biochemical characteristics. The class falls within the larger class of diseases, which are defined as states of discomfort and/or impairment of function. There is no logical necessity to restrict the range of discomforts or impairments to be included, although the nature and basis of the discomfort or impairment are legitimate subjects for investigation. In common usage the class of discomfort or impairment is extended beyond the individual. Thus the dictionary definition of disease includes "a derangement or disorder of the mind, moral character and habits, *institutions*, the *state*, etc." Physiological alterations or the mechanisms of the signs and symptoms are characteristics added to the definition of specific diseases as knowledge about them expands. Diabetes was recognized as a disease and named on the basis of one of its conspicuous and easily observable features long before there was any information about its pathological physiology. In other instances older descriptions had to be revised and new classifications added to accommodate newly acquired knowledge. Some behavioral aberrations that are today called mental diseases were recognized and described as far back as the period of classical Greek antiquity. Szasz proposes that these aberrations be renamed "problems in living" because of restrictions that he believes should be imposed on the category "disease."

The crucial criticism of the medical model focuses on deviancy. The outline of the argument is deceptively simple. Deviancy is a departure from a norm. Norms are either evaluative or statistical. If evaluative, they are arbitrary, culture-bound, and probably biased to reflect a predominant ideology. If statistical, then the distribution of any trait will have extremes, and it is illogical to label the extremes as abnormal. Careful review of the extensive writings of Szasz and the other critics of the medical model reveals several additional elements implicit in the argument. The most important one is that nothing other than a departure from a norm is involved. Mere deviancy, without further qualification or limitation as to its nature, is said to call forth a social reaction of rejection that is institutionalized under the pseudoscientific label of mental illness. It is difficult to accept this very general thesis. Only certain aberrations are labeled in any particular culture. In Szasz's view the problem of mental illness is the right to be different in the face of societal demands for conformity. This view involves at least two further assumptions: (1) social roles and behavior are imposed against an inherent resistance; (2) conformity or nonconformity is the only relevant dimension and nonconformity is more desirable. Similar views are expressed by Leifer,[33] Parsons,[39] and Scheff.[50] Laing[32] regards psychoses as a superior and creative nonconformity in response to pressures of living in a world he considers irrational.

This line of argument has great appeal. It is addressed to the established tradition of individuality and freedom of expression. It takes advantage of the difficulty of arriving at a satisfactory definition of mental illness, which has vexed even those who accept the concept. It invokes images of arbitrary and capricious restrictions by malevolent control agents, the institutional psychiatrists.

This line of argument is open to question on many grounds. It represents a very incomplete statement of the problem of normality and abnormality in the medical model. Mere deviation from norms does not constitute disease; rather it calls for attention and further investigation into the significance of the deviation. The precision associated with bodily disease has to do only with measurements of certain indicators of physical processes. Norms for judgment are not different in principle from those applied in many psychological processes. To take the example of diabetes again, the measurement of blood sugar is precise and objective. The dividing line between normal and diabetic is arbitrary. It is based on the relative probabilities that particular levels of blood sugar will be associated with the other alterations characteristic of diabetes. If the blood sugar is near the arbitrary borderline, judgment may be difficult and additional information may be required. In behavioral deviation the case is not different. Certain gross aberrations have been regarded as abnormal, while lesser aberrations may fall into the area of uncertainty. The final judgment is made on the basis of additional information—for example, whether behavioral deviation is associated with changes in mood or thinking processes or whether it is functionally disabling. Deviations are judged in the context of such other indicators and with due regard for background factors such as culture.

The medical model does not take all hallucinations as indicators of schizophrenia any more than it would take all elevations of blood sugar as indicators of diabetes. In a culture where hallucinations are accepted and even approved, the occurrence of hallucinations will not necessarily indicate psychosis. Yet it is possible in such a culture to make a judgment about pathology even though in specific instances the case may be borderline. The most stringent critics of the medical model seem on the whole to be describing individuals who exhibit minor deviations and give little evidence of subjective distress. An adequate model, however, must include those who are severely depressed, markedly agitated, or paralyzed by irrational fears.

Another major objection to the medical model is that it is based on conditions in which there are demonstrable anatomical or physiological alterations. On the other hand, psychiatry is for the most part concerned with the so-called functional disorders in which no consistent physiological changes have as yet been found despite many investigations. Invoking the concept of levels of integration, the critics contend that psychological processes by their very nature can never be "reduced" to a physiological level. Yet, unless one accepts the possibility of a nonmaterial spirit, psyche, or mind, psychological processes must in some way be connected with, and not separate from, the physiological level. The nonreductionist position can only be accepted as an assertion of belief subject to revision in the light of further investigation. Consider an individual who arrives at the incorrect conclusion that two plus two equals five. The error may be due to ignorance or mental defect. Suppose these causes are excluded and the error persists for reasons that are not known but are presumed to be "psychological." The nonreductionist position would hold that these psychological reasons comprise the full description of the error within the appropriate level of integration. Recent investigations by John[29] indicate that there are demonstrable differences in the electrical activity of the brain when correct and incorrect choices are made. It is possible also to demonstrate electrical patterns associated with "psychological" processes like stimulus generalization and abstraction. If such studies can be extended, it may be possible to resolve the reductionist objection and arrive at physiological measures for normality and abnormality in psychological functions.

Many critics of the medical model assert that it fosters superior, authoritarian attitudes in physicians and dependent, subservient attitudes in patients. No doubt there are physicians and patients about whom these assertions are correct. As a generalization, however, more tangible evidence would be required than is now offered by the critics. As the matter stands, the argument is an appeal to prejudice. Patients are given "orders" by their physicians, and they "depend" on his technical knowledge and skills. But "orders" and "depend" in this context do not have the pejorative meaning ascribed by the critics. Authoritarianism or dependency are individual qualities in no way inherent in the medical model and not necessarily characteristic of patients or physicians. Szasz holds that the medical model obscures moral and ethical problems that must be confronted. There is no reason to believe that his position is any less likely than the medical model to foster attitudes of superiority.

A major objection to the medical model is that it requires facilities and personnel far beyond what could conceivably be made available in the foreseeable future. A new model must therefore be developed. What alternatives are offered?

Mowrer suggests that the concept of sin has been too hastily excluded. Better sin than sickness in his view. If the responsibility for sin is acknowledged there is at least the prospect of redemption. Szasz does not propose any specific alternatives. He insists that all involuntary treatment and hospitalization be abolished, but is silent about what is to be done with those now under care. Presumably he anticipates that they can all be returned to the community; therefore, he does not attempt to devise any other program. For a select few he offers an austere and forbidding version of psychoanalysis he names autonomous psychotherapy. Most of the other critics refer in general terms to nonmedical psychotherapy, group therapy, day and night hospitals, behavioral and conditioning therapies, and a variety of environmental services. Albee, among the staunchest critics of the medical model, acknowledges the magnitude of the problem of providing such personnel in sufficient numbers. His proposal is a social learning theory of mental disorder.

> Once it is finally recognized and accepted that most functional disorders are learned patterns of deviant behavior, then the institutional arrangement which society evolves to deal with these problems probably will be *educational* in nature. . . . It is quite possible that they will be combinations of present day-care centers recast as small tax-supported state schools with a heavy emphasis on occupational therapy, reeducation and rehabilitation. . . . It will take several generations, perhaps a century, to replace the illness model . . .[3] (pp. 71–72.)

Albee's new institutions are strongly reminiscent of ideal hospitals envisaged by psychiatrists save only that he omits all mention of medication or other treatments. Elsewhere in his writings he has already dismissed all biochemical, neurophysiological, and genetic factors as irrelevant.[2] Surely this is too one-sided and premature a view to serve as the basis for professional or public policy in the mental health field. In any event the personnel and financing required would hardly be less than under the medical model.

The most general alternative to the medical model is the social model, which attributes the major portion of mental illness to the impact of social and economic factors. This viewpoint is more concerned with prevention than with caring for those now afflicted, although there is reasonable ground to expect that improvements in social and economic circumstances might assist current patients by reducing the rate of recurrences and rehospitalizations.

What conclusions can be drawn from this overview of the challenge to *the* medical model? What is *the* medical model? On the whole it appears that the medical model is whatever critics attribute to it. The model used by most psychiatrists accepts the notion that the disability of the mental patient is real and represents a dysfunction, but is not specific about the nature of the dysfunction. The concept of cause in the mental realm requires redefinition and will probably turn out to be a complex network of factors. The medical model does not require that treatment be directed internally even if the dysfunction is within the individual. Treatment can be directed at alteration of external circumstances or can consist of combinations of modalities.

These are matters for investigation. Given the present state of knowledge, most of the alternative models offered represent hypotheses subject to investigation in experimental or pilot programs. The most urgent need is for a truly comprehensive model. The closest approach at this time is the kind of formulation for schizophrenia offered by the researchers at the National Institute of Mental Health.

The Future of Psychiatry

If assessment of the present status of psychiatry is subjective, time-bound, and difficult, what can be said of the future? Probably the surest prophecy is that today's speculations will soon prove to be an acute embarrassment, for history seldom moves as predicted. On the other hand, plans for the future are necessary, and some risk must therefore be undertaken. Prediction is so much a matter of personal evaluation that I will in this section depart from the customary style of handbooks and use the pronoun "I." I will attempt to anticipate the future in three time spans—near, intermediate, and long range—and three aspects—theory, education, and delivery.

For the immediate future I believe psychiatry will have to be primarily oriented to the sudden expansion of demand engendered by the extension of health care to all segments of the population under some form of insurance. At this time there is some official hesitation about providing psychiatric services because of the anticipated high costs. Ways will have to be devised to supply the needs, and individual psychiatrists and professional organizations will have to take on roles of active advocacy. It has already been demonstrated that short-term (not brief) ambulatory treatment is insurable.[7] Further explorations are under way. An essential feature of any system of care will be provision of a full range of services and employment of all forms and modalities of treatment. Psychiatrists must resist efforts to limit services to the least expensive treatments. The aim must be to provide a basis for evaluating the relative efficacy as well as the range of applicability and limitations of different treatments. Heretofore the distribution of treatment has followed economic lines. Only when the various modalities are available to all regardless of economic status will it be possible to determine the indications for each modality. Evaluation of formal treatments will also have to be correlated with the effects of environmental measures, such as improved housing, education, recreation, and social services generally, to help determine their relative influence.

Theory must play a dual role. On the one hand, theory must supply clues for new approaches. These, in turn, will modify and enrich theory. The outcome should be in the form of "unified" theories, drawing upon all levels of integration, as illustrated by the work of the NIMH group previously cited. The most active development should be on the biochemical-neurophysiological level and on the social psychiatry level.

Psychiatric education will have to undergo the most drastic change. I foresee several interrelated pressures. The first is the already discussed challenge to the medical model. The overall impact of this trend has been to move psychiatry away from the rest of medicine. The outcome will depend on the competition between psychotherapies, behavioral therapies, and environmental modification, on the one hand, and chemotherapies, on the other. The results of this competition will not emerge early, carrying the issue into the intermediate time span. In the meantime there will be mounting pressure from the various subprofessions to achieve independent status. An example of this trend is the move toward autonomous schools of psychotherapy,[26] which will no doubt be accompanied by corresponding professional organizations. The new categories of mental health workers, the indigenous paraprofessionals, will undergo increasing professionalization with gradually increasing educational requirements. At the same time some psychiatrists and associated workers in the basic sciences will be pursuing their investigations.

I see in these various anticipated lines of development a trend to fragment rather than to unify the mental health field. Specialization, I believe, is inevitable, yet the need is for integration. A possible solution is to move in the direction of institutes of behavioral science rather than toward autonomous schools of psychotherapy, social work, or psychiatry. These institutes would be the base for both clinical and research activities and should be located within a university that has a medical school. The core of such institutes would be departments of psychiatry, psychology, and social work, that is, the disciplines most closely involved in treatment. Each institute should also be a major center for research and should include departments of biochemistry, neurophysiology, and experimental psychology, as well as psychoanalysis and the behavioral and conditioning therapies. In addition to the foregoing, the behavioral science institute should have strong representation from anthropology, sociology, political science, economics, and social psychology.

In an institute organized along these lines the major subdivisions would probably be around treatment and research, but the unified framework would help minimize the current coolness between clinicians and researchers so often encountered. The department of psychiatry would still be concerned with the overall integration of activity and information aimed at treatment, and would centralize and coordinate the delivery of services and evaluation of efficacy.

The relationship of the department of psychiatry to the medical school requires separate attention. If both the medical school and the behavioral sciences institute are based in a university, the pressure to separate psychiatry from the rest of medicine will be minimized. But I also foresee changes in medical education, which is also beset by the trend to specialization and the need for integration. Several medical schools have already introduced a system of "tracks." This arrangement groups the preclinical sciences and basic general clinical training in the first two to two- and one-half years. The student can then choose a track leading toward specialization or continue his general medical training. Psychiatry is one such track.[35] If this track could be located in the institute of the behavioral sciences, interested medical students would have early access to all the disciplines relevant to psychiatry, such as psychology, anthropology, and sociology, and could at the same time maintain their connection with the rest of medicine. Other students in the behavioral sciences institute might find their primary base in one of the related disciplines and take clinical work in the psychiatry track. The content of the M.D. degree may also change. Ph.D.'s are now granted in a specific discipline. The trend toward earlier and more intensive specialization may lead to an M.D. in a specialty—M.D. in surgery or gynecology and obstetrics—or perhaps the M.D. will be the basic medical degree, to be followed by a Ph.D. or Doctor of Medical Science in a specialty if desired. A major advantage of a behavioral institute is that it would promote adequate comparative studies and facilitate mutual enrichment of information.

These changes in medical education will certainly not be completed in the immediate future, but will extend into the intermediate time span.

The intermediate term should see the resolution of many current uncertainties. The delivery systems should be functioning with relative efficiency and the roles of the various professionals should have been clarified. Indications for different treatments should be well established and etiological factors should be more accurately defined.

Predictions for the long term are, I am afraid, mainly affirmations of faith. I therefore affirm my belief in the ultimate perfectibility of mankind, at least to the point that men will be able to live harmoniously with other men. I see no basic antithesis between instinct and civilization, so I do not believe that either analyses or psychiatry are interminable. We do not yet know whether some individuals may be so predisposed, say to schizophrenia or manic-depressive disease, that no amelioration of societal pressures will spare them from their disease. If in the intermediate term this should prove to be the case, I do not doubt that ways will be found to improve the genetic stock of the race. When all this has been done, then perhaps psychiatrists will no longer be concerned with mental diseases but only with problems of living.

Bibliography

1. ADAMS, H., "Mental Illness or Interpersonal Behavior," *Am. Psychol.*, *19*:191–197, 1964.
2. ALBEE, G., "Emerging Concepts of Mental Illness and Models of Treatment: The Psychological Point of View," *Am. J. Psychiat.*, *125*:870–876, 1969.
3. ———, "The Relation of Conceptual Models to Manpower Needs," in Cowen, E., Gardner, E., and Zax, M. (Eds.), *Emergent Approaches to Mental Health Problems*, Appleton-Century-Crofts, New York, 1967.
4. ANDO, A., FISHER, F., and SIMON, H., *Essays on the Structure of Social Science Models*, MIT Press, Cambridge, 1963.
5. AUSUBEL, D., "Personality Disorder *Is* Disease," *Am. Psychol.*, *16*:69–74, 1961.
6. ———, "Relationships between Psychology and Psychiatry: The Hidden Issues," *Am. Psychol.*, *11*:99–105, 1956.
7. AVNET, H., *Psychiatric Insurance*, Group Health Insurance, Inc., New York, 1962.
8. BECKER, E., *The Revolution in Psychiatry*, Free Press, Glencoe, 1964.
9. BEGELMAN, D., "Misnaming, Metaphors, the Medical Model and Some Muddles," *Psychiatry*, *34*:38–58, 1971.
10. BELL, N., "Models and Methods in Social Psychiatry," in Zubin, J., and Freyhan, F. (Eds.), *Social Psychiatry*, Grune & Stratton, New York, 1968.
11. BRAITHWAITE, R., *Scientific Explanation*, Cambridge University Press, Cambridge, 1955.
12. BROWN, B., and LONG, E., "Psychology and Community Mental Health: The Medical Muddle," *Am. Psychol.*, *23*:335–341, 1968.
13. ———, and OCHBERG, F., "The Medical Muddle," *Int. J. Psychiat.*, *9*:22–25, 1970–1971.
14. COHEN, L., "Health and Disease: Observations on Strategies for Community Psychology," in Rosenblum, G. (Ed.), *Issues in Community Psychology and Preventive Mental Health*, Behavioral Publications, New York, 1971.
15. COWEN, E., "Emergent Approaches to Mental Health Problems," in Cowen, E., Gardner, E., and Zax, M. (Eds.), *Emergent Approaches to Mental Health Problems*, pp. 389–455, Appleton-Century-Crofts, New York, 1967.
16. CROWLEY, R., "The Medical Model in Psychoanalysis," in Masserman, J. (Ed.), *Science and Psychoanalysis*, Vol. 12, pp. 1–11, Grune & Stratton, New York, 1968.
17. DAVIDSON, H., "The New War on Psychiatry," *Am. J. Psychiat.*, *121*:528–548, 1964.
18. DAVIS, K., "The Application of Science to Personal Situations: A Critique of the Family Clinic Idea," *Am. Sociol. Rev.*, *1*:238–247, 1936.
19. DEUTSCH, F., *On the Mysterious Leap from the Mind to the Body*, International Universities Press, New York, 1959.
20. ELLIS, A., "Should Some People Be Labeled Mentally Ill?" *J. Consult. Psychol.*, *31*: 435–446, 1967.
21. GLASER, F., "The Dichotomy Game: A Further Consideration of the Writings of Dr. Thomas Szasz," *Am. J. Psychiat.*, *121*: 1069–1074, 1965.
22. GOFFMAN, E., *Asylums: Essays on the Social Situation of Mental Patients and Other Inmates*, Doubleday, New York, 1961.
23. GRINKER, R. R., SR., "Emerging Concepts of Mental Illness and Models of Rx: The Medical Point of View," *Am. J. Psychiat.*, *125*:865–869, 1969.
24. HALLECK, S. L., *The Politics of Therapy*, Science House, New York, 1971.
25. HERBART, J. F., Cited in Boring, E., *A History of Experimental Psychology*, Appleton-Century-Crofts, New York, 1950.
26. HOLT, R. R. (Ed.), *New Horizon for Psychotherapy: Autonomy as a Profession*, International Universities Press, New York, 1971.
27. JAHODA, M., *Current Concepts of Positive Mental Health*, Basic Books, New York, 1959.
28. JASPERS, K., *General Psychopathology*, University of Chicago Press, Chicago, 1963.
29. JOHN, E. R., "Where is Fancy Bred?" in

M. Hammer, *et al.* (Eds.), *Psychopathology*, John Wiley, New York, 1972.

30. KAUFMAN, M. R., "Psychiatry: Why 'Medical' or 'Social' Model," *Arch. Gen. Psychiat.*, *17*:347–361, 1967.
31. KNAPP, P., "Revolution, Relevance and Psychosomatic Medicine: Where the Light Is Not," *Psychosom. Med.*, *33*:363–374, 1971.
32. LAING, R. D., *The Politics of Experience*, Penguin, Harmondsworth, 1967.
33. LEIFER, R., "The Medical Model as Ideology," *Int. J. Psychiat.*, *9*:13–21, 31–35, 1970–1971.
34. LEIGHTON, A., "Psychiatric Disorder and Social Environment," in Bergen, B. J., and Thomas, C. (Eds.), *Problems in Social Psychiatry*, Charles C Thomas, Springfield, Ill., 1966.
35. LIDZ, T., and EDELSON, M., *Training Tomorrow's Psychiatrist*, Yale University Press, New Haven, 1970.
36. MOWRER, O. H., " 'Sin' the Lesser of Two Evils," *Am. Psychol.*, *15*:301–304, 1960.
37. NAGEL, E., *The Structure of Science*, Harcourt, Brace & World, New York, 1961.
38. OFFER, D., and SABSHIN, M., *Normality*, Basic Books, New York, 1966.
39. PARSONS, T., "Definitions of Health and Illness in the Light of American Values and Social Structure," in Jaco, E. G. (Ed.), *Patients, Physicians and Illness*, Free Press, Glencoe, 1958.
40. PHILIPS, D. L., "Identification of Mental Illness: Its Consequences for Rejection," *Com. Ment. Health J.*, *3*:262–266, 1967.
41. ———, "Rejection: A Possible Consequence of Seeking Help for Mental Disorders," *Am. Sociol. Rev.*, *28*:963–972, 1963.
42. POLLIN, W., "The Pathogenesis of Schizophrenia," *Arch. Gen. Psychiat.*, *27*:29–37, 1972.
43. ———, and STABENAU, J. R., "Biological, Psychological and Historical Differences in a Series of Monozygotic Twins Discordant for Schizophrenia," *J. Psychiat. Res.*, 6 (Suppl. 1):317–332, 1968.
44. REIFF, R., "Mental Health Manpower & Institutional Change," in Cowen, E., Gardner, E., and Zax, M. (Eds.), *Emergent Approaches to Mental Health Problems*, pp. 74–78, Appleton-Century-Crofts, New York, 1967.
45. REISS, S., "A Critique of Szasz's 'Myth of Mental Illness,' " *Am. J. Psychiat.*, *128*: 1081–1086, 1972.
46. SANDER, F., "Some Thoughts on Thomas Szasz," *Am. J. Psychiat.*, *125*:1429–1431, 1969.
47. SARASON, I., and GANZER, V. J., "Concerning the Medical Model," *Am. Psychol.*, *23*:507–510, 1968.
48. SARBIN, T. R., "On the Futility of the Proposition That Some People Be Labeled 'Mentally Ill,' " *J. Consult. Psychol.*, *5*:447–453, 1967.
49. ———, and JUHASZ, J., "The Historical Background of the Concept of Hallucination," *J. Hist. Behav. Sci.*, *3*:339–357, 1965.
50. SCHEFF, T. J., "Role of the Mentally Ill and the Dynamics of Mental Disorder: A Research Framework," *Sociometry*, *26*:436–453, 1963.
51. SCHEFLEN, A. E., "An Analysis of a Thought Model Which Persists in Psychiatry," in Bergen, B. J., and Thomas, C. (Eds.), *Issues and Problems in Social Psychiatry*, Charles C Thomas, Springfield, Ill., 1966.
52. SPITZER, S., and DENZIN, N., *The Mental Patient—Studies in Social Deviance*, McGraw-Hill, New York, 1968.
53. SZASZ, T. S., *The Ethics of Psychoanalysis*, Basic Books, New York, 1965.
54. ———, *Ideology and Insanity*, Doubleday-Anchor, New York, 1970.
55. ———, *The Manufacture of Madness*, Dell, New York, 1970.
56. ———, *The Myth of Mental Illness: Foundations of a Theory of Personal Conduct*, Hoeber-Harper, New York, 1961.
57. THORNE, F. L., "An Analysis of Szasz's 'Myth of Mental Illness,' " *Am. J. Psychiat.*, *123*: 652–656, 1966.
58. TURNER, R. J., and CUMMING, J., "Theoretical Malaise and Community Mental Health," in Cowen, E., Gardner, E., and Zax, M. (Eds.), *Emergent Approaches to Mental Health Problems*, Appleton-Century-Crofts, New York, 1967.
59. WEINER, H., "Some Comments on the Transduction of Experience by the Brain: Implications for Our Understanding of the Relationship of Mind to Body," *Psychosom. Med.*, *34*:355–380, 1972.

CHAPTER 5

THE GENETICS OF MAN IN HEALTH AND MENTAL ILLNESS

John D. Rainer

In no branch of medical science have there been more rapid, more significant, and more fascinating advances in recent years than in the understanding and application of genetics. Once the research tools and conceptual frameworks became available, psychiatry, too, finally accepted genetics as one of its foundation stones. Gene-borne and gene-external influences can now be more readily conceived as continuously interacting in human development rather than as mutually exclusive, and the study of the machinery, the processes, and the consequences of this interaction can now be seen as essential to the understanding, the control, and even the treatment of psychiatric disorder.

In the interdisciplinary setting of modern psychiatry, genetics may be thought of as an area on a map surrounded by the research fields of biochemistry, pharmacology, physiology, experimental psychology, and ethology, and the clinical pursuits of diagnosis, treatment, community and social psychiatry, and developmental and psychodynamic understanding. In the past decade and a half, major advances in cytological and molecular genetics have captured the imagination of psychiatrists and suggested a basis for the possible mechanisms of genetic influence. New approaches to the study of the genetics of behavior have provided data on strain differences in species from drosophila to dogs, and problems of population growth and ecology have drawn attention to crucial aspects of population genetics, evolution, and the human gene pool. Meanwhile, clinicians have become more aware of individual differences in patients of all ages, in the genetic aspects of family patterns, and in hereditary differences in metabolism and response to drugs. All of these developments have opened the borders on the map described above and have made for productive communication and cooperation between genetics and the other basic and clinical sciences in psychiatry.

This, of course, was not always the case; the nature-nurture controversy, which still smoulders and breaks out occasionally, divided behavioral scientists into biologically and psychologically minded groups, and heredity and environment were thought of as separate and distinct forces. In the United States oversimplified views of one side or the other prevailed, with psychogenic or environmental forces usually given the dominant role. The behaviorism of J. B. Watson and the psychobiology of Adolf Meyer emphasized the role of external forces in molding behavior and life style; and in the transfer of psychoanalysis from Europe before and after World War II, the attention of Freud[24] and Ernest Jones[41] to inborn differences was largely ignored. A few psychoanalysts of that generation preserved the unitary approach; Hartmann[27] studied inborn characteristics of the ego, and Rado[80] considered psychodynamics to be established on the bases of genetics and physiology. In clinical psychiatry a few American family and twin studies of the major psychoses were reported by Rosanoff,[91] and Pollock and Malzberg[78] in the attempt to assess the genetic contribution, but it was the publication in 1938 of Franz Kallmann's *The Genetics of Schizophrenia*,[45] based on his Berlin family study, and in 1946 of his paper "The Genetic Theory of Schizophrenia,"[46] based on his New York State twin investigation, that for a while divided and then aroused American psychiatry to the importance of genetic factors. During his influential career Kallmann devoted his attention to many areas,[47] including schizophrenia, manic-depressive psychosis, homosexuality, mental deficiency, aging and longevity, tuberculosis, early total deafness, cytogenetics, and genetic counseling.

Throughout the description of the various applications of genetics to psychiatry, it is well to think of the influences and interactions as taking place developmentally in time at various levels of biological organization; the atomic and molecular, the cellular and systemic, the metabolic and neurophysiological, the psychodynamic, the familial, the demographic, and the social. The psychiatric status of the organism at any given time is defined by its interaction with its surroundings, with an interplay of effective forces whose various loci are genes, chromosomes, cytoplasm; enzymes, hormones, metabolites; brain and nervous system; infectious and toxic agents and diet; parents, families, groups, and society. In the largest sense the aim of psychiatric genetics as a scientific approach is to understand and control the etiological mechanisms both in psychiatric illness and in normal, adaptive behavior.

Interaction of Genetic and Experiential Determinants

Before any details of basic genetics or clinical application are presented, it is of prime importance to consider the problem of illustrating, conceptualizing, and describing the interaction of all types of molding forces, and in particular that between the information carried by genes and chromosomes and that provided by the natural and man-created environments. The activation or repression of genes or entire chromosomes, so obviously important in providing for cell differentiation and smooth physiological function in response to the needs of tissue and organism, is being seen more and more in terms of feedback-interaction; one model is provided by the mechanism of regulator genes outlined for bacteria by Jacob and Monod,[34] another by hormonal control.[106] Another example of orderly interaction at the molecular level is the production by the cells of a specific antibody in response to antigenic stimulation; less adaptive, at least to the individual, are environmentally induced gene mutations or chromosome breaks, or disruption of genetic function by viral intrusion.

At the other extreme on the scale of time and number, the mechanisms of population change and evolution are being carefully re-evaluated; positive and negative selective forces in the environment, the workings of chance, and the complexity of the genetic structures of the individual and the population, all contribute to the dynamic patterns of adaptation and change.

These various basic mechanisms, by no means completely understood, but essential to the very substance of modern genetics, can make it easier to conceive of the way genetics may function in individual development and psychopathology by serving as prototypes in some kind of general systems approach. If interaction is more than co-action, the need for two forces to combine, but rather a unique process with mutual feedback, spiral development, and critical stages, proper language must be found to describe the interworking of genetic and nongenetic factors as they come to light. Meanwhile, a few examples of clinical efforts in this direction may be presented.

Psychodynamics and Genetics

The science of psychodynamics has been considered by many psychoanalysts to be part of a total biological conception of man. In Rado's scheme, for example, by the psychodynamic approach to an understanding of motivation and emotional control, it was considered possible to describe behavior problems and character disorders that would require for their complete explanation investigating the role of genetic transmission.[80] Such an approach was certainly implicitly and explicitly formulated in the writings of Freud,[24] who considered both the constitutional and accidental causes of neurotic disease, and who spoke of primary congenital variations in the ego and in the defense mechanisms an individual selects. Hartmann[28] also, as part of his interest in twins, wrote of personality structure as the result of interaction between heredity and environment. He considered character anlagen that differentiate into character traits in the course of development, and he felt that twin studies might throw light on the possible substitution of one trait for another. Genetics may thus act as a unifying principle in future psychodynamic work, providing an opportunity to recognize fundamental genetic differences among individuals and to correlate these differences with various forms of developmental interaction.

Studies of Infants and Children: Reaction Patterns

As science progresses by the interplay of theory and observation, new formulations in ego psychology go hand in hand with fresh approaches to investigation. Verifying the observations of almost any mother, child psychiatrists have turned their attention to individual differences among children, found them to be as important as maternal attitudes in shaping subsequent behavior patterns, and begun to study the details of the interaction in a longitudinal framework. As Anna Freud[23] wrote, "inherent potentialities of the infant are accelerated in development, or slowed up, according to the mother's involvement with them, or the absence of it." The individual differences have been measured in various areas, among them sleep, feeding, and sensory responses, activity and passivity, motor behavior, and specific reaction patterns. Thomas, Chess, and Birch[104] have identified nine temperamental qualities in early infancy that tend to persist at least during the first two years. These are activity level, rhythmicity, approach-withdrawal, adaptability, intensity of reaction, threshold of response to stimulation, quality of mood, distractibility, and attention span and persistence. Various psychophysiological studies of neonates have also provided pertinent data.

The role of heredity in determining these constitutional variations requires further clarification, since the interaction process begins from the moment of conception and prenatal and perinatal influences come into play as well. Refined observational techniques coupled with methods of genetic analysis discussed below can be expected to clarify this intricate matter.

Interaction Models in Psychosomatic Disorders

The principle of the interlocking effects of genetic, nurtural, and social contributions in psychopathology is well illustrated in Mirsky's[63] studies in the etiology of peptic ulcer and Spitz's[100] formulations regarding infantile eczema. In the ulcer studies both neonates and healthy older persons were noted to vary in the degree of secretion of pepsinogen, as measured in urine and blood; this variation appears to be genetically determined. Under conditions of environmental stress, army training, for example, a large number of those who were rated as hypersecretors developed signs and symptoms of a duodenal ulcer, but none of those in the hyposecretor group did. In Mirsky's formulation infants who are genetically hypersecretors have intense oral needs that are insatiable by the average good mother; the mother is therefore perceived as rejecting, the dependent wishes persist, and they can be revived in later life when fears of the loss of security are mobilized by environmental stress. The resultant anxiety, by pathways as yet not certain (hypothalamic, autonomic; hyperexcretion of pepsinogen, hyperchlorhydria, hypermotility, hyperadrenalism), may precipitate the ulcer in the predisposed individual.

Similar observations were made by Spitz[100] on a group of infants institutionalized along with their unwed mothers. The infants who developed eczema differed from those who did not in a congenital predisposition and in a psychological factor originating in the mother-child relation. In particular, the infants who would develop eczema in the second six months of life showed a heightened set of responses at birth in the area of cutaneous reflexes. Psychologically their absence of eight-month anxiety meant to Spitz a retardation in affective development and led to the discovery of manifest anxiety and repressed hostility on the part of their mothers. The mothers did not like to touch their child or care for them, deprived them of cutaneous contact, and therefore refused to gratify the very need that already at birth had been shown to be increased in this group of infants. The pathways to the actual somatic lesion are again not clear, but the model for interaction provides the new kind of dynamic role for genetics in psychiatric thought.

Molecular Chemistry of Genetics

Classical Transmission Genetics

The rediscovery of Gregor Mendel's garden pea experiments at the turn of the century established the fact that single hereditary traits are determined by paired particles (later called genes) that are unchanged throughout life, independently separate during gamete formation, and are transmitted via the ovum and sperm respectively to combine again in the zygote.[61,103] The sum total of the genetic constitution of an individual, as thus established in the zygote and duplicated in every somatic cell, is referred to as the genotype, and his appearance and physiological state at any given time as his phenotype. It is possible for an individual to receive the same (or an essentially indistinguishable) gene from each parent at a given locus, and under proper environmental conditions such a homozygote will exhibit the characteristics associated with the action of that gene. On the other hand, each parent may contribute a different gene (allele) at the given locus; such a heterozygote may exhibit traits intermediate to those associated with the homozygote for either gene, or he may display the same trait as a homozygote, in which case the trait is known as dominant. A trait that is not expressed in an easily discernible fashion in a heterozygote is known as recessive. A mutated gene arising with predictable frequency in the population, if not immediately lethal, will usually have a pathological effect expressed against the genetic background of many other factors. If dominant it will be transmitted in pedigrees in the direct line of descent through the affected parent in each generation, appearing in approximately 50 per cent of that parent's offspring. Recessive traits require inheritance from both parents, who, since they are usually heterozygotes, are rarely affected themselves. A child of two heterozygotes has a 25 per cent risk of being a homozygote and of being affected under usual environmental conditions. While heterozygotes, or "carriers," for a given gene are relatively frequent in the population, the chance of two such individuals mating under conditions of random choice is much less common; when the gene is rare, many of such matings will be represented by consanguineous marriages. Transmission of the gene is chiefly along collateral lines, from heterozygote to heterozygote.

Since a gene produces a given trait only via a long series of steps that may be modified at every level by both prenatal and postnatal environmental forces and requirements, as well as by the action of other genes, the effects described above may vary from complete expression to total lack of penetrance. In many traits, particularly in variations within the range of normalcy in such characteristics as height or intelligence but also in graded pathological syndromes, the genetic contribution is made by the resultant effect of many genes, either all having minor intermediate effect (multifactor inheritance), or modifying the effect of single major genes.

Replication, Transcription, and Translation

For a half century this empirical body of knowledge was related to what was known of chromosome structure and processes of cell division, to inborn errors of metabolism, to the study of mutations, usually radiation-produced, and to the experimental, mathematical, and statistical investigation of populations and evolutionary processes. In the 1950s a new level of understanding was achieved through major advances in molecular genetics and cytology. For some time it had been known that the genetic material transmitting information from generation to generation was deoxyribonucleic acid (DNA). In 1953 the structure of this molecule was determined by Watson and Crick.[109,110] Two sugar (deoxyribose)-phosphate-sugar chains are twisted about each other, forming a double helix. To each sugar is attached a nucleotide base—a purine (adenine or guanine) or pyrimidine (cytosine or thymine). From a sugar on one chain to the corresponding sugar on the other, there is a hydrogen bond linkage that may only take place between either adenine on one chain and thymine on the other, or guanine on one and cytosine on the other. A DNA molecule may consist of thousands of such nucleotide linkages. Replication of the DNA molecule during cell division preserves the sequence of nucleotide bases, since separation of the strands, breaking the linkages between the nucleotides, is followed by the construction by each half of a new sister chain, with the sequence complementary to the original half chain. There result two double helical chains, each identical to the original. This model made possible an unparalleled explosion of investigation and knowledge.

The importance of the base sequence in the DNA molecule was soon realized, as the process of protein manufacture by the cell was elucidated. In very schematic form the code represented by the sequence in the double helix is transcribed to a single stranded molecule, messenger ribonucleic acid (mRNA), which is then found in the cytoplasm attached to structures known as ribosomes that move along its length. Here the code is translated, directing the sequence of amino acid assembly into a polypeptide chain. In this process specialized molecules of soluble or transfer RNA (sRNA) carry one by one in turn specific amino acids to specific codons, sequences of three nucleotide bases that code for one amino acid alone. Once formed, the chain assumes the spatial configuration of a specific protein molecule.

Mutation

Gene mutations represent a change in the genetic instruction. In the light of the above scheme, they arise from a change in the nucleotide sequence in the DNA, an error during replication possibly resulting from the action of certain chemicals or from radiation. This change will affect the transcription-translation process so that a modified polypeptide or protein will be produced. Since this product is either an enzyme or a structural protein such as hemoglobin, the result may be disordered function or structure—for example, the loss of enzyme activity responsible for phenylketonuria, or the altered hemoglobin structure found in sickle-cell anemia.

Control of Gene Action

In the normal functioning of the above process, it is obvious that there must exist some method of control or regulation. With every cell containing the same genes, some of these must be inactive, their message never transcribed, to account for the differentiation of cell functions. In all cells there must be some feedback mechanism whereby the production of enzymes and other products is regulated at any given time according to the needs of the entire organism. There are various theories to account for such control. Jacob and Monod[34] have studied the action in bacteria of regulator genes whose products may switch on or off the action of structural (enzyme-producing) genes depending on the presence or quantity of the substrate on which the given enzyme acts, or the end product of such action. In this operon theory genes are responsive to their surroundings; this is the most basic example of the process of interaction so central to the modern conception of genetics. In higher organisms more complex mechanisms, including the action of hormones, may play a role in the regulatory process.

Human Cytogenetics

While these advances in the molecular chemistry of genetics were being made, techniques were being developed that led to a corresponding explosion in cytogenetics, the visualization and study of human chromosomes in health and disease. As in many fields of knowledge, dramatic discoveries concerning human chromosomes followed quickly upon the perfection of new techniques. Some long-held hypotheses had to be discarded while a few careful observations came to be explained in a definitive manner.

For many decades the study of human chromosomes had been neglected. Even the number of chromosomes and the mechanism of sex determination had been matters of controversy. For a long time most textbooks perpetuated the erroneous statement that the diploid number of chromosomes in man was 48. It was known that females had two X chromosomes and males one X and one Y, but by analogy with drosophila maleness was thought to be due to the presence of only one X chromosome, insufficient to balance masculinizing genes on the autosomes. There were no satisfactory methods of correlating chromosomal or nuclear patterns with clinical abnormalities.

Sex Chromatin and "Nuclear Sex" Determination

In retrospect the first breakthrough in modern human cytogenetics occurred in 1949, when Barr and Bertram[4] accidentally discovered an important morphological difference between female and male cells. Examining neurons of the cat, they found, in some animals only, a densely staining chromatin mass applied to the inner surface of the nuclear membrane. It was soon realized that it was female cats that demonstrated this nuclear chromatin. This phenomenon was then demonstrated in other mammals and in man, where from 30 to 60 per cent of the cells of a normal female show the dark-staining mass, referred to as sex chromatin or a Barr body. In 1954 a corresponding sex difference was discovered in polymorphonuclear leucocytes, where an additional small lobe is found in about one in 40 cells in females and in less than one in 500 cells in males.[17] Because of its shape, this small lobe was called a "drumstick."

Since the sex chromatin body was found in

females, who have two X chromosomes, and not in males, who have one, it was originally believed to be made up of the dark-staining regions of two such chromosomes. Later it was established that a Barr body consists of all or part of one X chromosome, one that was modified, suppressed, or partially or totally "inactivated." According to Lyon's[60] hypothesis, this "inactivation" of one X chromosome in each cell takes place early in the development of the female embryo.

Sex chromatin determination in humans may be accomplished by examining any cellular tissue; in practice it is most convenient to study smears made of oral mucosal cells. After swabbing the inside of the cheek gently with a gauze pad, the roughened surface is scraped with the edge of a narrow metal spatula. The buccal material is spread onto a glass slide, which is fixed in 95 per cent ethyl alcohol for 30 minutes. It is then washed and stained by a nuclear stain such as thionin or cresyl violet, cleared, and mounted; between 100 and 200 cells are then examined.

It was not long before the process of "nuclear sexing," as it was then called, was applied to certain cases of human infertility and abnormal sexual development. In 1954 female patients with gonadal aplasia (Turner's syndrome) were reported to be chromatin-negative, that is, to lack the nuclear chromatin body.[76,111] Such females tend to be of short stature and sexually infantile, and their ovaries are reduced to streaks of connective tissue. Associated defects such as webbed neck and increased carrying angle of the arms are often present. Occurring with an estimated frequency of 0.2 to 0.3 per 1,000 females, Turner's syndrome may include slight intellectual impairment but not severe mental deficiency. Because of the absence of nuclear chromatin, early investigators considered such patients as "chromosomal males" with some form of sex reversal. Although this interpretation was not generally accepted, it was finally corrected only by later findings, which are described below.

A second classical example of nuclear sex anomaly was Klinefelter's syndrome, characterizing males with small atrophic testes, sterility, gynecomastia, and increased excretion of gonadotropin. Occurring in about one in 500 male births, it accounts for about 1 per cent of male mentally retarded patients in institutions. In 1956 a high proportion of patients with this clinical syndrome were found to be chromatin-positive;[74] they were for a time incorrectly termed "chromosomal females," presumably with two X chromosomes. Further elucidation of the etiology of this disorder also awaited study of the chromosomes themselves and direct observation of the patients' sex chromosome constitution.

Technical Advances In Human Cytogenetics

The next set of studies in human cytogenetics followed the perfection of methods for observing human chromosomes microscopically. Earlier investigations were done mostly on testicular biopsies, but degenerative changes as well as clumping and overlapping of chromosomes led to inaccurate counts. The introduction of tissue culture methods made available a better source of cells in the process of mitotic division, the stage in which the chromosomes are distinguishable. Treatment with colchicine arrested dividing cells in this stage, and the use of hypotonic solutions swelled the nuclei, spreading the chromosomes and separating them for examination before final staining of the cells in slide preparations. Photography under the oil immersion lens revealed the chromosomes as X-shaped bodies, each representing a single chromosome in the process of dividing. First applied to skin biopsy material and bone marrow, these techniques have been variously modified and at present are applied most practically to lymphocytes of peripheral blood.[6]

Suspended in their plasma and stimulated to divide by the addition of phytohemagglutinin, such cells are protected against bacterial contamination by antibiotics. After three days of incubation at 37°C, they are ready for harvesting after being arrested at the cell division stage by adding colchicine or a derivative of it. The addition of hypotonic sodium citrate solution is followed by fixation, slide preparation, staining, clearing, and mounting.

Normal Human Karyotype

In 1956 these new techniques resulted in the demonstration of 46 chromosomes as the correct number in man.[105] It became possible to study clear and easily examined preparations of human chromosomes, to photograph them, and to enlarge, cut out, pair, and mount them in order of decreasing length. Such an arrangement is called a karyotype; a diagrammatic representation thereof is called an idiogram.

Chromosomes may be distinguished from one another by their length and by the position of the constriction called the centromere, the point at which the two parts of the dividing chromosome still remain joined. The centromere may be median (metacentric), submedian (submetacentric), or subterminal (acrocentric) in location; in many chromosomes of the latter type satellite bodies may be seen projecting from their short arm.

When arranged in order of decreasing size, the human chromosomes fall into seven groups, which were standardized at a historic meeting in Denver in 1960.[18] The groups may be clearly distinguished, and chromosomes within a group can often be recognized with a high degree of probability. In line with the system proposed at the Denver meeting, the description of the chromosomes is as follows:

Group 1–3. (A) Large chromosomes with approximately median centromeres. The three chromosomes are readily distinguished from each other by size and centromere position.

Group 4–5. (B) Large chromosomes with submedian centromeres. The two chromosomes are difficult to distinguish, but chromosome 4 is slightly longer.

Group 6–12. (C) Medium-sized chromosomes with submedian centromeres. The X chromosome resembles the longer chromosomes in this group, especially chromosome 6, from which it is difficult to distinguish. This large group presents the major difficulty in identification of individual chromosomes.

Group 13–15. (D) Medium-sized chromosomes with nearly terminal centromeres (acrocentric chromosomes). Chromosome 13 has a prominent satellite on the short arm. Chromosome 14 has a small satellite on the short arm. No satellite has been detected on chromosome 15.*

Group 16–18. (E) Rather short chromosomes with approximately median (in chromosome 16) or submedian centromeres.

Group 19–20. (F) Short chromosomes with approximately median centromeres.

Group 21–22. (G) Very short, acrocentric chromosomes. Chromosome 21 has a satellite on its short arm. The Y-chromosome is similar to these chromosomes.

Further conferences were held in London[58] in 1963, in Chicago[14] in 1966, and in Paris in 1971, with the aim of increasing the accuracy of chromosome identification. The presence of secondary constrictions and the determination by radiographic means (incorporation of radioactive thymidine into replicating chromosomes) of the time sequence of chromosome replication were among the methods suggested at the London conference; in Chicago a standard nomenclature was adopted for describing the human chromosome complement. More recently new staining techniques were reported that promise to make each chromosome individually recognizable by disclosing banded regions;[29] among the distinguishing characteristics are fluorescent patterns seen after exposure to quinacrine derivatives and deeply colored bands using various modifications of Giemsa staining methods.

Chromosomal Abnormalities

Within three years after the correct determination of the human karyotype, and even before the numbering system was standardized, important discoveries were made correlating abnormalities in the karyotype and clinical syndromes. The first group of these abnormalities arose from the presence of an abnormal number of chromosomes and resulted from a process known as nondisjunction

* Subsequent investigations revealed that satellites are carried by all the acrocentric chromosomes, except the Y, although they cannot always be seen; therefore, the use of satellites to distinguish between acrocentric chromosomes is probably not reliable.

in the formation of the gamete (sperm or ovum). Instead of the two members of a given pair of chromosomes separating, each one going into a separate sperm cell (in the male), or into the ovum and polar body, respectively (in the female), they both go into the same gamete, producing two kinds of abnormal gametes, those with two chromosomes of a given pair and hence with one chromosome too many, or with none of that pair and hence one too few. When united with a normal sperm or ovum, such a gamete yields a fertilized ovum (zygote) with either 47 or 45 chromosomes in all. Similar processes of nondisjunction at an early cleavage stage of the zygote, resulting in cells with even more than 47 chromosomes, have been found in conjunction with the X and Y chromosomes. It is also possible in such cases that cells of more than one kind are formed, each one of which perpetuates its line, giving rise to individuals with cells of two or more types. This phenomenon is known as mosaicism.

The first condition found to be associated with a chromosomal anomaly was mongolism, preferentially referred to as Down's syndrome. The association of this syndrome in the preponderance of cases with children born to older mothers, and the one-egg twin concordance rate of close to 100 per cent, foreshadowed an early germinal, possibly a chromosomal, defect as the responsible agent. Early in 1959 tissue culture methods showed that patients with Down's syndrome had 47 instead of 46 chromosomes and that the additional chromosome was one of the smallest chromosomes, no. 21 in the conventional numbering system.[11,36,56] Arising by nondisjunction (especially in the female, as is suggested by the association of the condition with advancing maternal age), the extra chromosome, small as it is, causes widespread morphological and biochemical abnormalities.

Some cases of Down's syndrome, especially those born to younger mothers or those with a familial concentration, were assumed for a while to have a normal chromosome complement. However, upon closer examination of the karyotype, one of the chromosomes of the 13 to 15 group was found to have attached to it the long arm of an extra chromosome 21.[75] Thus there was a translocation of chromosomal material, a situation that may be transmitted from generation to generation. If two normal chromosomes 21 are present, in addition to the 15/21 translocation chromosome, the individual is affected; if only one is present, the individual is normal. In the latter case, however, he or she may be a "carrier," that is, may bear affected children, by producing a gamete with both the translocation chromosome and a normal chromosome 21. Translocations have been found also between chromosomes 21 and 22. From a practical point of view the translocation mechanism for producing this syndrome is very important.[12] For most mothers of a child with Down's syndrome, the chance of having a second such child is very small, hardly greater than for any other woman of the same age (1 in 600 altogether, up to 1 in 50 in mothers over 45). A mother with a 15/21 translocation has a much higher risk (about 10 per cent) of having a second affected child, apart from that of having children who are carriers. If the father has the translocation, the empirical risk of having an affected child is lower (2 to 3 per cent).[12] Microscopic chromosome examination is therefore strongly indicated in counseling such families.

Other conditions with trisomies of autosomes have been described that are rare in live-born infants and usually result in early death. One of the conditions, involving a chromosome of the D group, is characterized by symptoms that include microphthalmus, harelip, cleft palate, and polydactyly, frequently accompanied by congenital heart defects,[69] another, in which an E group chromosome is involved, is characterized by micrognathia, low-set ears, overlapping of fingers, and other skeletal and cardiac defects.[19]

Abnormalities Involving Sex Chromosomes

After interest had been aroused in sexual abnormalities with anomalous chromatin patterns, it was not long before the techniques of studying human karyotypes were applied to these conditions. Chromatin-negative Turner's syndrome was found to be associated with only 45 chromosomes, one of the sex chromosomes being missing.[22] The resulting sex chromosome constitution was first called the XO condition, "O" indicating simply the absence of a chromosome normally present; in the Chicago nomenclature it is called 45, X. Apparently this condition may originate by nondisjunction in the formation of either the sperm or the egg. Thus, while normal sperm cells have either an X or Y chromosome (in addition to 22 autosomes), the nondisjunction phenomenon may yield sperm cells having both X and Y (XY) with 24 chromosomes in all, or neither (O) with 22 chromosomes in all. Similarly, while normal ova all have one X chromosome, the abnormal ones may have two (XX) or none (O). Fertilization of an O gamete by a normal X gamete results in the karyotype of Turner's syndrome.

In similar fashion patients with Klinefelter's syndrome are known to have 47 chromosomes with an extra X chromosome (XXY), brought about by the union of an abnormal XY sperm with a normal ovum, or an abnormal X ovum with a normal Y-bearing sperm.[37] A third classic example of chromosomal abnormality in the human is represented by females, sexually infantile and amenorrheic, with an XXX chromosome pattern.[35]

The sex chromatin pattern in these anomalies can now be shown to be related in a simple way to the chromosome structure. The number of nuclear chromatin patches is always one less than the number of X chromosomes found—hence Turner's syndrome has none, Klinefelter's has one, and the triple X female has two. In terms of the Lyon hypothesis, this finding can be explained by the inactivation of all X chromosomes in excess of one. At the same time it has been established that the presence of a Y chromosome is necessary and sufficient for maleness, since the XO individual is female with only one X chromosome but no Y, while the XXY, possessing two X chromosomes but also possessing a Y, is male.

These sex chromosomal anomalies are the prototypes of more complex pictures that have been reported. Nondisjunction during later cell divisions after the zygote is formed may produce such abnormalities as XXXY or XXXXY individuals (males similar to Klinefelter's syndrome but with severe mental deficiency) and many varieties of mosaicism. It has been estimated that one in every 150 newborn infants has a chromosome aberration identifiable by present techniques.[87]

Chromosomal Abnormalities and Psychiatric Syndromes

Aside from the importance of human chromosome analysis for diagnosis and counseling, its value in unraveling the processes of gene action is only beginning to be exploited. For one example, the correlation of specific disorders with syndromes associated with chromosome abnormalities may serve to localize the genes involved on particular chromosomes or the metabolic (enzymatic) pathways of gene action and may also help to distinguish nosological subgroups in major diagnostic categories.

In the major syndromes of psychiatric importance chromosomal findings have been equivocal.[48,86] Only a small proportion of schizophrenic patients seem to display a sex chromosome abnormality; conversely a few of the mentally defective patients with abnormal karyotype show psychotic or schizophreniclike behavior.

Klinefelter's syndrome, with 47,XXY haryotype and its variants, leads to degrees of eunuchoid habitus and weak libido, low marriage rate, often mental deficiency, and personality disorders ranging from inadequate personality and delinquency to symptoms of schizophrenia. This trisomic condition has been found three times as frequently in mental hospitals as among newborns. If this figure is upheld it will still have to be clarified whether the patients have psychotic-like syndromes resulting from the chromosome imbalance or whether the chromosomal syndrome precipitates an otherwise determined psychosis by adding an additional biological or social burden. Similar considerations apply to the extra X chromosome in the female, the 47,XXX syndrome. Once misnamed "superfemales," those who have been found in clinics and hospitals are often mildly retarded and socially withdrawn; they seem to be found more often than expected in mental hospitals, but their incidence at birth (1 in 1,000) is only a rough estimate. Schizophrenia in women with the XXX karyotype has recently been studied by Vartanyan and Gindilis.[107]

Turner's syndrome, typically a short and sexually undeveloped female, is associated with the absence of the second sex chromosome, leaving 45 chromosomes with a single X. If emotional reaction, sexual immaturity, and body defects play any role in the extra X syndromes described above, they do not seem to result in any disturbance here, for these girls and women have been described as resilient to adversity, stable in personality, and maternal in temperament. They are not mentally retarded except in some nonverbal skills centering about space-form appreciation and constructional skills.[64] This finding is significant as one of the few examples of a specific intellectual correlate of a karyotypic defect.

A great deal of interest has been generated in the 47,XYY anomaly. Although noted in the general population, males with this chromosome constitution were found in seeming excess in correctional institutions where they were noted to be tall, mentally deficient, and prone to aggressive behavior patterns.[16] The specificity of this syndrome has been questioned, especially since the frequency of XYY males in the total population is not yet accurately known.[65] Fluorescent staining techniques in buccal epithelial cells[71] should be able to provide better data both on the incidence of the karyotypic abnormality and its possible consequences in the course of development. Cases studied so far, including a pair of adolescent twins,[84] suggest that at least some of these men are usually passive, withdrawn, inadequate, and docile, but under stress they may exhibit impulsive behavior. There is also suggestive evidence that these periodic episodes have a seizurelike course and that the electroencephalogram may be abnormal.

Further information on the relationship between chromosome imbalance and mental symptoms may one day elucidate some of the genetic mechanisms in behavior disorders. An analogy has been provided in studies of the fruit fly by the correlation of behavioral response with contributions from specific chromosomes.[20]

Methods of Investigation in Psychiatric Genetics

Pedigree Method

The study of individual families or pedigrees has often been used to suggest forms of genetic transmission for larger-scale investigation or to provide tentative material on rare, well-defined pathological conditions. They have to be supplemented by the study of more representative samples with proper attention to statistical methods of correcting for ascertainment.

Census Method

In geographically confined populations, small enough to be visited individually, and with good medical and demographic records, it has been possible to study the population and family distribution of diagnosable psychiatric disorders; this method has provided some excellent reports in certain Scandinavian areas.[55]

Family Risk Studies

An extension of the pedigree method, studies of family risk (sometimes referred to as contingency methods) have as their goal comparing the expectancy risk for a given condition in the relatives of affected individuals with that in the general population. A number of pitfalls must be avoided. The affected individuals must be diagnosed according to uniform criteria, and they must represent a consecutive series of patients (or a random sample thereof). They are designated then as index cases, or probands. All relatives in the desired categories must then be located and diagnosis made. Since one is interested not in the incidence (new cases) or even the prevalence (total cases) but rather in the expectancy rate (cases that may arise during the lifetime of the relatives), it is necessary either to wait for many years or more practically to apply a correction method. A simplified method often used is the Weinberg abridged method. A manifestation age interval is derived clinically; in schizophrenia, for example, it is usually taken at 15 to 45 years. The number of cases found among a given category of relatives (the numerator) is related to a denominator that is not the total number of relatives in this category, but rather that number diminished by all of those who have not reached the earliest manifestation age and half of those still within the manifestation period. This method yields a risk figure, representing the expectancy of developing the given condition for those who will live through the manifestation period. This figure can then be compared with risk figures for other groups of relatives and for the general population. The latter have been determined in many cases by total or sample population studies, or by studies of the relatives of control patients.

Twin Family Method

An extension of the family risk method to families containing twins makes possible a series of further approaches to understanding the complementary roles of nature and nurture, heredity and environment.[97,98] First used by Galton, the method depends upon the occurrence of two types of twins, those derived from a single fertilized ovum that has split and developed as two separate individuals ("identical," monozygotic, or one-egg twins) and those derived from two ova, fertilized by two different spermatozoa during the same pregnancy ("fraternal," dizygotic, or two-egg twins). In the first case the twins are always of the same sex (barring a rare sex chromosome loss early in embryonic division, which may result in a pair of one-egg twins made up of a normal male and a female with Turner's syndrome). On the average dizygotic twins are one-quarter of the time both male, one quarter of the time both female, and half the time of opposite sex. Zygosity determination can be made with a high degree of accuracy by similarity of somatic characteristics. The accuracy is further increased by comparison of fingerprints and blood groups, while immunological methods such as reciprocal skin grafting may provide certainty in individual pairs where intensive study requires and warrants such procedures. Monozygotic twins are born with the same genotype; dizygotic twins are as similar genetically as any pair of sibs.

In the twin family method the index cases are twins. Expectancy rates can be compared between monozygotic twin partners, dizygotic twin partners of the same sex, dizygotic twin partners of opposite sexes, full sibs, half sibs, and step sibs. This method provides a graded series of genetic relationships and a differently graded series of environmental ones; the extent to which the risk in the various categories of relatives corresponds to the genetic similarity bears a relation to, or is a measure of, the genetic contribution to the etiology of the condition. For example, if the expectancy in monozygotic twins is higher than that in dizygotic twins of the same sex (despite the environmental similarities) and that in dizygotic twins is close to that in full sibs (to which they are genetically equivalent), there is a measurable genetic contribution. Another way of separating genetic and environmental influences would be to study identical twins reared apart from birth, preferably in homes with disparate social and psychological settings, but such pairs who also present a particular syndrome are rare.

There are a number of cautions and misunderstandings about twin studies to be noted. Before using the twin concordance method, it is necessary, for example, to determine that a condition to be investigated is no more prevalent in twins than in single-born individuals. In doing a study the sample of twins should be complete, with the expected proportion of monozygotic and dizygotic pairs, and the diagnosis of zygosity and that of illness should be made by separate persons without knowledge of the other's findings. In interpreting the results of twin studies, the assumption of comparable environmental patterns for monozygotic twins and dizygotic twins, at least those of the same sex, has been questioned; yet observation of twins and their families has often shown similar family dynamics and role assignment—twin dependency, parental need to distinguish, and so forth—with both types of twins. Indeed, monozygotic twins are often more disparate in size and vigor at birth than dizygotic ones. Finally it should be realized that it is not necessary to have 100 per cent concordance in monozygotic twins to postulate a genetic contribution; provided they are sufficiently higher than the dizygotic rates, monozygotic concordance rates of under 50 per cent, for example, can be shown by analysis of variance techniques to indicate a high heritability value.[26]

Co-Twin Control Studies

The use of longitudinal data from selected pairs of monozygotic twins, particularly those discordant or dissimilar with respect to the overt expression of a behavioral trait or syndrome, may serve to point out important developmental factors. Among the reported case histories of twins and other sets of persons of multiple birth, the study of Rosenthal's[92] group on the aptly named Genain quadruplets is particularly notable because of its thoroughness and its adherence to the concept of heredity-environment interaction. Although all members of the set were found to be schizophrenic, their psychoses varied in symptomatology as well as in intensity.

Other investigations of this type include the series of homosexually dissimilar twins studied by Rainer, Mesnikoff, Kolb, and Carr.[51,85] Obtaining data through psychiatric anamnesis, family interviews, and free association, these studies paid particular attention to: (1) the prenatal fantasies and wishes of each parent about the sex of the expected child; (2) the attitude of the parents toward the twins' birth; (3) the family significance of the naming, both as established before birth and as modified by the birth of the twins; (4) the parental efforts at differentiating the twins; (5) the occurrence of physically distinguishing features in the twins; (6) the emotional connotations of such features to the parents and the extended family; (7) the effect of such bodily distinctions on the differential mode of mothering for each twin; (8) the differing object relations of the twins from birth onward; (9) the attitude of each twin to his body and to his self, as perceived and as seen ideally; (10) the fantasy life of the twins, particularly in the sexual area; and (11) the superego growth of the twins, particularly in relation to sexually allowed and prohibited forms of activity. Benjamin[7] studied a pair of male twins concordant for asthma but dissimilar with respect to a number of important psychological features, in the context of seeking to "conceptualize most complex behaviors in terms of different sorts of interactions between innate and experiential constants and variables."

In a pair of twins concordant for the XYY chromosome anomaly, but dissimilar in the degree of control over episodic impulsive behavior, Rainer, Abdullah, and Jarvik[84] noted the mother's search for personality differences at birth, and the occurrence of petit mal seizures in one twin at the age of three. It was suggested that these early events may have set a double pattern—focusing of the mother's concern on one twin as being psychiatrically and neurologically abnormal, and a protective and caring response on the other child's part, which helped him to strengthen his control and responsibility over himself. As in all of these studies, similarities in underlying personalities were noted, particularly in projective testing, with differences in overt behavior then attributable suggestively to patterns of family dynamics as well as to physiological differences.

In the study of identical twin pairs discordant for schizophrenia by Pollin[77] and his co-workers, again there were physiological differences, namely, lower birth weight for the subsequently schizophrenic twin, that directly or in interaction with family dynamic patterns may have influenced their development.

These studies help one to understand that genes actually determine a norm of reaction, the exact expression of which depends on many prenatal, perinatal, and postnatal interactions. The pathways are labyrinthine that lead in twins from the identical molecular structures that constitute the genic patterns to the manifestations in later life of behavior traits; minor shifts in the dynamic process at nodal points may lead to wide phenotypic divergence. Preconceived ideas on the locus of such nodal points should be regarded with caution. A spiral-like development toward marked behavioral dissimilarity may arise not only from postnatal influences on the twins but also in the prenatal stages of development. Infant, childhood, adolescent, and adult experiences may thereafter precipitate a single gene effect or shape the expression of a polygenic trait in a unique way.

Adoption Studies

The physical separation from their biological parents of children reared in adoptive homes offers a valuable method of weighting genetic and environmental influences. Some studies involving adopted children have compared those with affected biological parents to those with unaffected biological parents, while others have compared biological and adoptive parents and other relatives of affected adopted children. Some of the findings in these kinds of studies are discussed in Volume Three, Chapter 25, of this handbook in connection with genetic studies in schizophrenia (Rosenthal).

Longitudinal Studies

Longitudinal studies of high-risk infants and children are of the greatest potential value in learning about the role and vicissitudes of genetic predisposition in mental illness. Since there is evidence for increased risk with greater genetic loading in a given syndrome, children of one or of two affected parents may be followed, comparing them with their sibs and with control children, searching for early behavioral, biochemical, neurological, psychophysiological, and clinical signs, for patterns of development, and for protective factors.

Other Methods of Genetic Investigation

Of course, the most direct way of establishing genetic etiology would be to identify errors in the genetic code itself; the closest available approach is through identification of a specific enzyme deficiency or modification, particularly in the brain or nervous system. This may be done by electrophoretic methods or by studying products of intermediate metabolism. Loading techniques—tolerance tests—may make such deficiencies more obvious in the case of heterozygotes. Correlation of psychiatric symptoms with chromosome anomalies has been discussed; more subtle chromosomal changes will be detected with the perfection of new staining techniques, as well as with techniques of cell hybridization.

A method for establishing homogeneous genetic categories is the study of genetic linkage; a syndrome may be associated with the effect of a single gene if it appears to stem from a locus on the same chromosome with, and at a given distance from, the locus of another known gene. In principle linkage is suggested by the detection of pedigrees in which there is a reduced frequency of recombination between genes at two particular loci; for example, depression and color blindness, occurring in the same family, may be found usually together or usually separately.

The correlation of psychiatric syndromes with certain physical characteristics—blood types, dermatoglyphics, constitutional habitus—may be a lead to the presence of a genetic basis. Finally, in cross-cultural studies, if variations in the frequency of a syndrome across different populations are determined only by population-genetic variables, such as consanguinity, genetic drift, or selection factors, the syndrome is more likely to have a genetic basis; if, however, there are strong differences in prevalence that are determined by cultural and environmental factors and that are not due to differences in diagnostic procedures, the genetic contribution, although it may still be present, will be more difficult to detect.

Genetics and Psychopathology

Schizophrenia

For over 40 years family and twin studies, studies of adopted children, and longitudinal studies, both retrospective and prospective, have yielded data consistent with the presumption of a necessary, although not sufficient, hereditary basis for schizophrenia.[93] Chapter 25, Volume Three, of this handbook is devoted to the genetics of schizophrenia. From the point of view of the present discussion, four themes may be noted in appraising this area of research today.[83] The first is the need to define more precisely the significant genetic and environmental influences; the second is to find better ways to describe their interaction; the third is the increasing evidence that genetic determination seems to vary with the severity of the illness and to bear a relation to newly delineated clinical types; and the fourth is the emergence of longitudinal studies that promise to throw light on developmental mechanisms and the natural history of the condition.

Because of modern treatment and community care methods, the marital rates of schizophrenic patients have increased, resulting in a higher number of fertile marriages and higher overall reproductive rates;[21] therefore, the need for better understanding of genetic data for use in counseling and child-care programs is evident.

Affective Disorder

Studies of families and twins have played an important role in both the nosology and etiology of affective disorder. Earlier reports have focused on manic-depressive illness. In his classic investigation of this syndrome, Kraepelin[53] observed large numbers of relatives who had the same illness, but he did not find an increase in dementia praecox among these relatives. In most European and American populations the general rate for manic-depressive psychosis varies between 0.4 per cent and 1.6 per cent,[115] though in a few special situations, such as isolated populations, rates of as low as 0.07 per cent[10] and as high as 5 or 7 per cent[30,108] have been found. An excess of females has been consistently observed. In the case of parents, sibs, and children of manic-depressive index cases, the expectancy rates are much higher, with a morbidity risk in first-degree relatives of about 20 per cent reported in earlier studies.[2,73,96,102] With both parents affected, the morbidity risk in children has been found to be as high as 40 per cent.[102,113]

In the largest twin family study that has been reported, Kallmann[44] located 27 one-egg and 58 two-egg pairs. In this series the expectancy of manic-depressive psychosis varied from 16.7 per cent for half sibs to 22.7 and 25.5 per cent for sibs and two egg co-twins, respectively, and 100 per cent for one egg co-twins. Parents of index cases showed a rate of 23.4 per cent. Since only patients admitted to a mental hospital, and hence the most severe cases, were included as index cases, the apparently perfect concordance rate of 100 per cent for one-egg twins was considered an artificial maximum value. Other twin studies corroborated these data.

Recent investigations have tended to subdivide affective disorders into two subgroups, bipolar illness with periods of mania, and unipolar illness with symptoms limited to recurrent depression.[2,57,73,112] Family studies conducted in these two groups have shown different morbidity risk in relatives, with the bipolar cases showing more genetic loading than the unipolar cases or than the earlier combined groups. A 34 per cent risk was found, for example, in the first-degree relatives of bipolar patients studied by Winokur.[113] Moreover, both bipolar and unipolar illness have been found in the first-degree relatives of bipolar cases, while families of unipolar probands have shown only unipolar illness.[2]

These data seem to indicate that bipolar and unipolar illness are different with regard to their genetic component, although the nature of the genotype of the unipolar relatives found in bipolar families remains in question. Much current research is directed toward exploring further genetic heterogeneity in the bipolar group of affective disorders. Asano[3] attempted to delineate subgroups of manic-depressive illness on the basis of clinical symptoms, with a "typical" group showing manic-depressive symptoms in a relatively pure form, and an "atypical" group showing clouding of consciousness, or schizophrenic or severe autonomic symptoms. The typical forms were found to have a greater and more specific genetic loading than the atypical forms. Some studies have divided the group according to time of onset, with earlier onset associated with more severe genetic loading, later onset with a negative family history;[32,49,62] while others have classified patients on the basis of their response to pharmacological treatment. In the latter category is the proposal by Pare and Mack[68] to differentiate genetically between patients and affected family members who respond only to MAO inhibitors and those who respond only to tricyclic antidepressants. Similar studies are in progress with relation to lithium response.

Kallmann[44] and Stenstedt[102] favored an autosomal dominant form of inheritance in manic-depressive illness, with incomplete penetrance and variable expressivity. Rosanoff[90] in 1934 and currently Winokur and Tanna[114] support the presence of at least two dominant factors in the transmission of bipolar affective disorder, one X-linked, the other autosomal. Perris[73] has supported X-linked transmission only in unipolar cases. Other investigators have reported data on ancestral secondary cases that are consistent with a polygenic hypothesis.[72,99] None of these results are really contradictory, as one of the genes involved in polygenic inheritance may well be X-linked.

The X-linked hypothesis is based on the observation of an excess of females in the sex ratio of manic-depressive illness and the rarity of father-to-son transmission. Recently Reich, *et al.*,[89] reported on linkage studies between manic-depressive illness and two X-linked genetic markers—Xga blood group and color blindness.

From a psychodynamic point of view Rado[82] described the psychological characteristics of the individual predisposed to manic-depressive psychosis as a tendency to emotional overreaction from infancy, a persistent alimentary dependent state, a strong craving for gratification from without, and an intolerance to pain. He considered the depressive spell to be the final stage of an etiological progression beginning with the genotype. Further data regarding modification of symptoms by developmental factors were provided by the New York study of persons totally deaf since birth or early childhood.[1] Although the prevalence of affective illness did not differ from that among the normal hearing population, there was a decrease in symptoms of guilt and retarded depression, with paranoid symptoms or agitation taking their place. It was hypothesized that the tenuous nature of early object relationships played a role in this modification.

Involutional depression was considered by Kraepelin to belong to the large group of manic-depressive psychosis. Kallmann[47] found an elevated risk of schizophrenia in the relatives of probands with involutional melancholia, with no increased risk for manic-depresssive psychosis. He considered the syndrome to be pathogenetically complex and genetically associated more closely with the group of schizoid personality traits than with manic-depressive psychosis. Other investigators found an increased morbidity risk for affective illness in the relatives of involutional probands.[2,49,101] These discrepancies are probably due to differences in diagnostic criteria.

Criminality

Early studies of twins involving criminal behavior suffered from sample selectivity and heterogeneity; they tended to show high concordance rates for both monozygotic and same-sex dizygotic twins. This distribution of concordance rates led genetic investigators to suspect a large environmental role in the pathogenesis of criminal behavior. Individual pairs of twins have been described with divergent overt histories, and it would seem to be more appropriate to study specific personality traits leading to a life of crime rather than criminal behavior itself. The current status of the role of chromosomal abnormalities in criminal behavior has been considered earlier in this chapter.

Intelligence and Mental Defect

There is perhaps more controversy today about the role of heredity in intelligence than any other aspect of behavior genetics; the value of early childhood education, the nature of learning, the problem of underprivileged groups, and innumerable school policies have become involved with this issue. The question often narrows down to one of priorities in making very practical decisions. Underlying much of the divergence of conclusion is the basic difficulty of measuring intelligence apart from environmental influence. Intelligence scores, largely consisting of IQs, show remarkable correlation with genetic closeness; these similarities persist in longitudinal studies of twins, for example, and in twins reared apart since early life. It is not so clear yet what factors are measured by the various tests, or how their results are affected by such environmental factors as maternal health and nutrition, or communication and educational patterns at home and in school.

A valuable contribution to the role of communication in intelligence was made by the study of a group of 33 pairs of twins discordant for early total deafness.[94] In these pairs the verbal IQs showed significant difference that disappeared in the performance scales. The role of emotional factors in learning, both those associated with interpersonal channels of acceptance and identification and those related to security and motivation, preclude simple formulations.

In the field of mental defect some of the data are more clear-cut. Mental deficiency syndromes based on specific gene mutations and chromosomal aberrations are well known, although there are intelligence differences among persons with such syndromes that must be explained on other genetic or environmental grounds. According to Reed and Reed,[88] however, these specific genetic factors, together with obvious infectious or birth-traumatic incidents, account for barely half of persons with IQs below 70. These authors consider that polygenic inheritance is responsible for most of the rest, with social deprivation usually playing a secondary or modifying role. In any event they conclude that five-sixths of persons in the given lower IQ range have had at least one parent or an aunt or an uncle similarly retarded.

Aging and Life Span

In human populations there have been a number of studies relating the life span of offspring to that of their parents. In Pearl's work[70] longevity ("regarded as a single numerical expression of the graded effects of all the forces that operate upon the individual, innate and environmental") was investigated by comparing the ancestors of two groups of persons, one a group of persons still living at 90 years and above and the other a random group of individuals. The sum of the ages at death of the six immediate ancestors of the index cases was significantly greater in the longevous group than in the comparison group.

The most extensive and carefully followed investigation of aging has been that conducted in the Department of Medical Genetics of the New York State Psychiatric Institute since 1945 by Kallmann and various colleagues[43] and continued at the present time by Jarvik and associates.[38,39,40] These investigations started with 1,603 twin index cases over the age of 60 and followed the mortality patterns of 584 pairs where both twins qualified as index cases. The length of time between the death of the first twin and the death of the second twin has been consistently greater in the dizygotic pairs than in the monozygotic pairs. These studies also confirmed the relationship between the mean life span of the twins and their sibs and the age of death of their parents. The investigators felt that life span potential was demonstrated to have a genetic basis that could be assumed to follow the multifactor type of inheritance. Also studied were intrapair differences in psychometric test scores, which were larger for dizygotic than for monozygotic twins. There was some indication that stability of intellectual performance was associated with survival.[9,39]

In searching for possible biological concomitants of aging, Jarvik and her colleagues[40] undertook chromosome examinations of peripheral blood lymphocytes to determine whether there were increasing deviations from the normal chromosome number with increasing life span. Such losses in peripheral blood lymphocytes might reflect similar aneuploidy in glial cells. As compared with a group of young individuals, an excess of chromosome loss was found in the aged women but not in the aged men. The chromosome loss among aged women was random and not limited specifically to C group chromosomes; in the aged males, although they showed no greater frequency of overall chromosome loss than young males, the chromosomes that were lost were largely in the G group.

Finally a higher frequency of chromosome loss was found by Nielsen, *et al.*,[66] and by Jarvik and her associates[38] in the peripheral leukocytes of women with organic brain syndrome compared with other women of comparable age. In the New York study a significant increase in the frequency of chromosome loss was found in the women who had organic brain syndrome without evidence of cerebral arteriosclerosis, compared with women without organic brain syndrome. The relationship between chromosome loss and organic brain syndrome in males, however, appeared to be purely random. In women a positive association between chromosome loss and memory loss was also demonstrated by psychological testing.

Genetics of Psychosexual Development

Intersexual Conditions

Concerning the genetic aspects of intersexual conditions, recent findings related to chromosomal abnormalities have already been discussed. Not all sexual anomalies, however, are by any means associated with karyotypic changes. Most hermaphrodites with both testicular and ovarian tissues are found to have normal karyotypes, although some are mosaics. A few pseudohermaphroditic states, although distinguished by normal karyotypes, are apparently caused by gene-borne influences. Among them are the testicular feminization syndrome (male karyotype with testicular tissue, but female habitus and genitalia) and some instances of the adrenogenital syndrome (virilization in the female).

What one can say is that the chromosomal pattern determines the differentiation of the gonad, while the subsequent course of development—duct systems and genitalia—is hormonally controlled. More precisely the phenotypic sex (type of gonad) is the result of sex-determining genes on all the chromosomes. Male-determining genes are present strongly on the Y chromosome and probably on other chromosomes as well. Female-determining genes are probably on the X chromosome and may also exist on the autosomes. Normally the strong male determiners on the Y chromosome shift the balance in the direction of maleness. Without this chromosome female determiners tend to exert such a strong influence that the individual develops as a female.

Regarding the management of patients with doubtful sex, it is generally thought to be desirable to determine the most suitable sex as early as possible in life, considering the chromosome pattern as well as the anatomy, and, after surgical and hormonal treatment, if necessary, to rear the child in the chosen sex.

Male Homosexuality

In the case of homosexuality earlier hypotheses implied that some male homosexuals might have a female chromosome structure. They were based on the finding of a greater proportion of males among their sibs than would be expected normally.[54] These inferences were criticized on statistical grounds and could not be confirmed in chromatin studies on various series of male homosexuals, which showed negative sex chromatin patterns.[5,8,67,79] Nevertheless, in a group of 401 consecutive admissions of males diagnosed as homosexual at the Bethlem Royal and Maudsley hospitals, these patients showed a later birth order and a high maternal age, with a variance in the latter as great as that in Down's syndrome. These data were interpreted by Slater[95] as supporting a hypothesis of heterogeneity in the etiology of male homosexuality, with some of them possibly connected with a chromosomal anomaly, others associated with social and psychological causes. Moreover, a number of case reports linked homosexuality, transvestitism, and pedophilia in patients with Klinefelter's syndrome.[31] Recent reports of abnormal patterns of hormone excretion in homosexuals, if confirmed, raise questions regarding the role of endocrine imbalance as cause or effect.[52,59]

Kallmann's[42] investigation of a series of male twins with homosexuality was interpreted as suggesting a gene-controlled disarrangement between male and female psychosexual maturation patterns. Almost perfect concordance was found in a series of 40 monozygotic twin pairs, whereas in 45 dizygotic pairs the degree of concordance was no higher than that expected on the basis of Kinsey's statistics for the general population. In the explanation of these concordance data, sexual behavior was considered to be part of the personality structure rather than the gonadal or hormonal apparatus. Not ruling out the possibility that some male homosexuals, particularly the infertile ones, might have been intersexes, Kallmann generally put aside this special explanation, along with other theories of single-factor causation, in favor of a range of genetic mechanisms capable of disturbing the adaptational equilibrium between organic sex potentialities and psychosexual behavior.

It is of historical interest to quote in this connection the opinion expressed in 1931 by Goldschmidt,[25] a pioneer in the study of intersexuality in insects:

> As far as human homosexuality is concerned, the biologist must be extremely cautious in commenting on this much disputed field. I concede that during an earlier period (1916) I was less cautious and believed, on the basis of extensive studies of the literature, that it was justifiable to classify the clearly congenital form of homosexuality as an incipient form of intersexuality. At present I can no longer hold to this theory. What was discussed in the previous sections makes it very difficult to assign homosexuals of either sex a place in the series of intersexuals. Without pretending to be an expert in this field I should like to point out that what has been previously said about gynecomastia (an inherited change in the reactivity of the breast tissue to hormones) would also seem to be the most likely explanation for homosexuality, except that instead of the mammary gland *the brain* would be the end organ (italics added).

If the site of the divergent development in homosexuality is in the personality and ultimately, therefore, in the "mind" or in the brain or the nervous system, it becomes possible to study this deviation within the framework of the function of genetics in psychiatry, tracing the processes of interaction in the organism from its earliest origin and with its genetic potentialities.

Some observations made by Kinsey and his associates[50] regarding the biological basis for the usual heterosexual choice, for example, may provoke speculation regarding the areas of defect in homosexuality. The factors facilitating heterosexual choice were listed as follows: (1) greater aggressiveness in the male with a tendency to avoid sexual behavior with another individual of the same level of aggressiveness; (2) greater ease of intromission in heterosexual contact, with more of the "satisfactions which intromission may bring"; (3) olfactory and other anatomical and physiological characteristics differentiating the sexes; and (4) the conditioning effect of the "more frequently successful" heterosexual contacts.

In the heterosexual individual these factors would seem to require a normal rate of maturation of personality development, marked by the ability (1) to perceive and respond to biological sexual stimuli of a pleasurable nature; (2) to feel and recognize satisfaction and success; and (3) to utilize these experiences as integrating forces and guides to future action. A primary defect in perception or self-perception or response to pleasure may render an individual vulnerable to accidental or to family- or society-encouraged deviant behavior. Any vulnerability factors in homosexuality may well be of this nature, rather than what Rado[81] calls "counter-fragments" of the opposite sex.

Other theories based on evolutionary considerations have been advanced by Hutchison[33] and Comfort.[15] The former suggested that the genotype responsible for homosexuality may operate on the rates and extent of the development of the neurophysiological mechanisms underlying the identification processes and other aspects of object relationship in infancy. Comfort considered the evolutionary significance of the time in which castration anxiety begins, a phenomenon that may protect immature male animals from competitive harm between the onset of sexual maturity and the attainment of adequate size and strength. It was assumed that if this anxiety develops too strongly or too early, one reaction may be an avoidance of heterosexual behavior and a turning to homosexual behavior.

The plan used in observing those few one-egg twin pairs found with one clearly homosexual and one predominantly heterosexual partner has been referred to above. Psychoanalytic and other data on a pair of 30-year-old male twins classified as clinically discordant revealed important similarities, principally in psychological test findings.[13,85] Both twins were reported as having shown marked sexual confusion and body-image distortion. Taking these results as a first approximation to the similar underlying personality characteristics expected in identical twins, the interpretation of the developmental material was focused on divergent patterns of experience. In this case, as well as in some others studied subsequently, an important factor seemed to be a difference in the twins' relationship with their mother. The preferred and overprotected twin may have become frustrated in his heterosexual contacts and formed a poor masculine identification. It was assumed that the gene-influenced personality potentials were relatively vulnerable in both twins but led to homosexual symptom formation in only one. With this type of investigation aiming to pinpoint the crucial components of the developing character structure and the most sensitive periods at which they can be affected, it is obvious that many more cases must be studied before any detailed interactional synthesis can be achieved.

Genetic Counseling

From the clinical point of view the final common pathway for the application of genetic knowledge to patients and their families is the responsible practice of genetic counseling—the provision of scientifically accurate guidance in connection with problems of marriage and parenthood. Changing attitudes and policies regarding fertility regulation, parenthood planning and population growth, and concerns with birth control, voluntary sterilization, genetic diagnosis by amniocentesis, abortion, and adoption procedures, have created an interest in and a demand for genetic information and for help in understanding that information.

The psychiatrist is involved in genetic counseling in two ways. First, he is best qualified to diagnose behavioral syndromes, as well as to interview family members and interpret records and laboratory tests. Second, he will have special skill in assessing the emotional aspects of the search for genetic information, the motivations, rationalizations, and defenses of the persons coming for help, and the impact of the information provided. Many persons accept misinformation and superstition if it is in accordance with their wishes or fears. Some couples look for reasons to avoid marriage and parenthood, while others who have had an affected child seek to alleviate their shame or guilt. Marital problems may arise, such as hostility and sexual inhibition. Considered as a psychotherapeutic procedure, genetic counseling presents, therefore, a growing need for community mental health resources staffed by professional persons trained in modern genetic research methods, able to empathize with persons in need of guidance, and equipped with a sense of public responsibility. With proper concern the growing body of scientific and conceptual advances in genetics may be turned to human benefit, and psychiatrists may participate in the social planning that lies ahead as well as discharge their duty to foster responsible parenthood and genuinely healthy development in the families under their care.

Bibliography

1. ALTSHULER, K. Z., "Personality Traits and Depressive Symptoms of the Deaf," in Wortis, J. (Ed.), *Recent Advances in Biological Psychiatry*, Plenum Press, New York, 1964.
2. ANGST, J., and PERRIS, C., "Nosology of Endogenous Depression, A Comparison of the Findings of Two Studies," *Archiv für Psychiatrie und Zeitschrift für die Gesamte Neurologie, 210:*373–386, 1968.
3. ASANO, N., "Clinico-Genetic Study of Manic-Depressive Psychoses," in Mitsuda, H. (Ed.), *Clinical Genetics in Psychiatry*, Shoin, Tokyo, 1967.
4. BARR, M. L., and BERTRAM, E. G., "A Morphological Distinction between Neurons of the Male and Female, and the Behavior of the Nucleolar Satellite during Accelerated Nucleoprotein Synthesis," *Nature, 163:*676–677, 1949.
5. ———, and HOBBS, G. E., "Chromosomal Sex in Transvestites," *Lancet, 1:*1109–1110, 1954.
6. BARTALOS, M., and BARAMKI, T. A., *Medical Cytogenetics*, Williams & Wilkins, Baltimore, 1967.
7. BENJAMIN, J. D., "Some Comments on Twin Research in Psychiatry," in Tourlentes, T. T., Pollack, S., and Himwich, H. E. (Eds.), *Research Approaches to Psychiatric Problems*, Grune & Stratton, New York, 1962.
8. BLEULER, M., and WIEDEMANN, H. R., "Chromosomengeschlecht und Psychosexualität," *Archiv für Psychiatrie und Z. Neurologie, 195:*14–19, 1956.
9. BLUM, J. E., JARVIK, L. F., and CLARK, E. T., "Rate of Change on Selective Tests of Intelligence: A Twenty-Year Longitudinal Study of Aging," *J. Gerontol., 25:*171–176, 1970.
10. BÖÖK, J. A., "A Genetic and Neuropsychiatric Investigation of a North-Swedish Population with Special Regard to Schizophrenia and Mental Deficiency," *Acta Genet., 4:*1–139, 1953.
11. ———, FRACCARO, M., and LINDSTEN, J., "Cytogenetical Observations in Mongolism," *Acta Paediatrica, 48:*453–468, 1959.
12. BREG, W. R., "Family Counseling in Down's Syndrome," *Ann. N.Y. Acad. Sc., 171:*645–654, 1970.
13. CARR, A. C. (Ed.), *The Prediction of Overt Behavior through the Use of Projective Techniques*, Charles C Thomas, Springfield, Ill., 1960.
14. Chicago Conference, "Standardization in Human Cytogenetics," *Birth Defects Original Article Series*, 2, 1966.
15. COMFORT, A., "Sexual Selection in Man—A Comment," *Am. Naturalist, 93:*389–391, 1959.
16. COURT BROWN, W. M., Males with an XYY Sex Chromosome Complement," *J. Med. Genet., 5:*341–359, 1968.
17. DAVIDSON, W. M., and SMITH, D. R., "A Morphological Sex Difference in Polymorphonuclear Neutrophil Leucocytes," *Brit. M. J., 2:*6–7, 1954.
18. DENVER REPORT, "A Proposed Standard System of Nomenclature of Human Mitotic Chromosomes," *Am. J. Human Genet., 12:*384–388, 1960.
19. EDWARDS, J. H., HARNDEN, D. G., CAMERON, A. H., CROSSE, V. M., and WOLFF, O. H., "A New Trisomic Syndrome," *Lancet, 1:*787–790, 1960.
20. ERLENMEYER-KIMLING, L., and HIRSCH, J., "Measurement of the Relations between Chromosomes and Behavior," *Science, 134:*1068–1069, 1961.
21. ———, NICOL, S., RAINER, J. D., and DEMING, W. E., "Changes in Fertility Rates of Schizophrenic Patients in New York State," *Am. J. Psychiat., 125:*916–927, 1969.
22. FORD, C. E., JONES, K. W., POLANI, P. E., *et al.*, "A Sex-Chromosome Anomaly in a Case of Gonadal Dygenesis (Turner's Syndrome)," *Lancet, 1:*711–713, 1959.
23. FREUD, A., *Normality and Pathology in Childhood*, Hogarth, London, 1966.
24. FREUD, S. (1913), "The Predisposition to Obsessional Neurosis, a Contribution to

the Problem of the Option of Neurosis," in *Collected Papers*, Vol. 2, Hogarth, London, 1924.

25. GOLDSCHMIDT, R., *Die Sexuellen Zwischenstufen*, Springer, Berlin, 1931.
26. GOTTESMAN, I. I., and SHIELDS, J., "A Polygenic Theory of Schizophrenia," *Int. J. Ment. Health, 1*:107–115, 1972.
27. HARTMANN, H., "Comments on the Psychoanalytic Theory of the Ego," in *Essays on Ego Psychology*, International Universities Press, New York, 1964.
28. ———, "Psychiatric Studies of Twins," *Jahrbücher für Psychiatrie und Neurologie*, 50, 1934, and 51, 1935.
29. HECHT, F., WYANDT, H. E., and ERBE, R. W., "Revolutionary Cytogenetics," *New Eng. J. Med., 285*:1482–1484, 1971.
30. HELGASON, T., "Epidemiology of Mental Disorders in Iceland," *Acta Psychiatrica Scandinavica*, Suppl. *173*, 1964.
31. HOAKEN, P. C. S., CLARKE, M., and BRESLIN, M., "Psychopathology in Klinefelter's Syndrome," *Psychosom. Med., 26*:207–223, 1964.
32. HOPKINSON, G., "A Genetic Study of Affective Illness in Patients over 50," *Brit. J. Psychiat., 110*:244–254, 1964.
33. HUTCHINSON, G. E., "A Speculative Consideration of Certain Possible Forms of Sexual Selection in Man," *Am. Naturalist, 93*:81–91, 1959.
34. JACOB, F., and MONOD, J., "Genetic Regulatory Mechanisms in the Synthesis of Proteins," *J. Molec. Biol., 3*:318–356, 1961.
35. JACOBS, P. A., BAIKIE, A. G., COURT BROWN, W. M., MACGREGOR, T. N., MACLEAN, N., and HARNDEN, D. G., "Evidence for the Existence of the Human 'Super-Female,'" *Lancet, 2*:423–425, 1959.
36. ———, ———, ———, and STRONG, J. A., "The Somatic Chromosomes in Mongolism," *Lancet, 1*:710, 1959.
37. ———, and STRONG, J. A., "A Case of Human Intersexuality Having a Possible XXY Sex-determining Mechanism," *Nature, 183*:302–303, 1959.
38. JARVIK, L. F., ALTSHULER, K. Z., KATO, T., and BLUMNER, B., "Organic Brain Syndrome and Chromosome Loss in Aged Twins," *Dis. Nerv. System, 32*:159–170, 1971.
39. ———, KALLMANN, F. J., FALEK, A., and KLABER, M. M., "Changing Intellectual Functions in Senescent Twins," *Acta Genetica et Statistica Medica, 7*:421–430, 1957.
40. ———, and KATO, T., "Chromosome Examinations in Aged Twins," *Am. J. Human Genet., 22*:562–572, 1970.
41. JONES, E. "Mental Heredity," in *Essays in Applied Psychoanalysis*, Vol. 1, Hogarth, London, 1951.
42. KALLMANN, F. J., "Comparative Twin Study on the Genetic Aspects of Male Homosexuality," *J. Nerv. & Ment. Dis., 115*:283–298, 1952.
43. ———, "Genetic Factors in Aging: Comparative and Longitudinal Observations in a Senescent Twin Population," in Hoch, P., and Zubin, J. (Eds.), *Psychopathology of Aging*, Grune & Stratton, New York, 1953.
44. ———, "Genetic Principles in Manic-Depressive Psychosis," in Hoch, P., and Zubin, J. (Eds.), *Depression*, Grune & Stratton, New York, 1953.
45. ———, *The Genetics of Schizophrenia*, Augustin, New York, 1938.
46. ———, "The Genetic Theory of Schizophrenia," *Am. J. Psychiat., 103*:309–322, 1946.
47. ———, *Heredity in Health and Mental Disorder*, Norton, New York, 1953.
48. KAPLAN, A. R., "Association of Schizophrenia with Non-Mendelian Genetic Anomalies," in Kaplan, A. R. (Ed.), *Genetic Factors in Schizophrenia*, Charles C Thomas, Springfield, Ill., 1971.
49. KAY, B., "Observations on the Natural History and Genetics of Old Age Psychoses, A Stockholm Material, 1936–1937," *Proc. Roy. Soc. Med., 52*:29–32, 1959.
50. KINSEY, A. C., POMEROY, W. B., MARTIN, C. E., and GEBHARD, P. H., *Sexual Behavior in the Human Female*, Saunders, Philadelphia, 1953.
51. KOLB, L. C., RAINER, J. D., MESNIKOFF, A., and CARR, A., "Divergent Sexual Development in Identical Twins," in *Third World Congress of Psychiatry, Proceedings*, Vol. 1, University of Toronto Press, Montreal, 1961.
52. KOLODNY, R. C., MASTERS, W. H., HENDRYX, J., and RORO, G., "Plasma Testosterone and Semen Analysis in Male Homosexuals," *New Eng. J. Med., 285*:1170–1174, 1971.

53. KRAEPELIN, E., *Manic-Depressive Insanity and Paranoia*, E & S Livingstone, Ltd., Edinburgh, 1921.

54. LANG., T., "Die Homosexualität als Genetisches Problem," *Acta Geneticae Medicae et Gemellogiae*, 9:370–381, 1960.

55. LARSSON, T., and SJÖGREN, T., "A Methodological, Psychiatric, and Statistical Study of a Large Swedish Rural Population," *Acta Psychiatrica et Neurologica Scandinavica*, Suppl. 89, 1954.

56. LEJEUNE, J., GAUTIER, M., and TURPIN, R., "Etude des chromosomes somatiques de neuf enfants Mongoliens," *C. R. Academie Science Paris*, *248*:1721–1722, 1959.

57. LEONHARD, K., *Aufteilung der Endogenen Psychosen*, Akademie Verlag, Berlin, 1959.

58. London Conference on the Normal Human Karyotype, *Cytogenetics*, *2*:264–268, 1963.

59. LORAINE, J. A., ADAMAPOULOS, D. A., KIRKHAM, K. E., ISMAIL, A. A. A., and DOVE, G. A., "Patterns of Hormone Excretion in Male and Female Homosexuals," *Nature*, *234*:552–554, 1971.

60. LYON, M., "Sex Chromatin and Gene Action in the Mammalian X-Chromosome," *Am. J. Human Genet.*, *14*:135–148, 1962.

61. MCKUSICK, V. A., *Human Genetics*, Prentice-Hall, Englewood Cliffs, N.J., 1969.

62. MENDLEWICZ, J., FIEVE, R. R., RAINER, J. D., and FLEISS, J. L., "Manic Depressive Illness: A Comparative Study of Patients With and Without a Family History," *Brit. J. Psychiat.*, *120*:523–530, 1972.

63. MIRSKY, I. A., "Physiologic, Psychologic and Social Determinants of Psychosomatic Disorders," *Dis. Nerv. System, Monogr. Suppl.*, *21*:50–56, 1960.

64. MONEY, J., "Hormonal and Genetic Extremes at Puberty," in Zubin, J., and Freedman, A. M. (Eds.), *The Psychopathology of Adolescence*, Grune & Stratton, New York, 1970.

65. National Institute of Mental Health, *Report on the XYY Chromosomal Abnormality*, U.S. Government Printing Office, Washington, D.C., 1970.

66. NIELSEN, J., JENSEN, L., LINDHARDT, H., STOTTRUP, L., and SONDERGAARD, A., "Chromosomes in Senile Dementia," *Brit. J. Psychiat.*, *114*:303–309, 1968.

67. PARE, C. M. B., "Homosexuality and Chromosomal Sex," *J. Psychosom. Res.*, *1*: 247–251, 1956.

68. ———, and MACK, J. W., "Differentiation of Two Genetically Specific Types of Depression by Response to Antidepressent Drugs," *J. Med. Genet.*, *8*:306–309, 1971.

69. PATAU, K., SMITH, D. W., THERMAN, E., *et al.*, "Multiple Congenital Anomaly Caused by an Extra Autosome," *Lancet*, *1*:790–793, 1960.

70. PEARL, R., and PEARL, R. DEW., *The Ancestry of the Long-Lived*, Johns Hopkins Press, Baltimore, 1934.

71. PEARSON, P. L., BORROW, M., and VOSA, C. G., "Technique for Identifying Y Chromosomes in Human Interphase Nuclei," *Nature*, *226*:78–80, 1970.

72. PERRIS, C., "Abnormality on Paternal and Maternal Sides: Observations in Bipolar (Manic-Depressive) and Unipolar Depressive Psychoses," *Brit. J. Psychiat.*, *118*:207–210, 1971.

73. ———, "A Study of Bipolar (Manic-Depressive) and Unipolar Recurrent Depressive Psychoses," *Acta Psychiatrica Scandinavica*, Suppl. *194*, 1966.

74. PLUNKETT, E. R., and BARR, M. L., "Testicular Dysgenesis Affecting the Seminifuous Tubules, Principally with Chromatin Positive Nuclei," *Lancet*, *2*: 853–856, 1956.

75. POLANI, P. E., BRIGGS, J. H., FORD, C. E., *et al.*, "A Mongol Girl with 46 Chromosomes," *Lancet*, *1*:721–724, 1960.

76. ———, HUNTER, W. E., and LENNOX, B., "Chromosomal Sex in Turner's Syndrome with Coarctation of the Aorta," *Lancet*, *2*:120–121, 1954.

77. POLLIN, W., and STABENAU, J. R., "Biological, Psychological, and Historical Differences in a Series of Monozygotic Twins Discordant for Schizophrenia," in Rosenthal, D., and Kety, S. S. (Eds.), *The Transmission of Schizophrenia*, Pergamon Press, London, 1968.

78. POLLOCK, H. M., MALZBERG, B., and FULLER, R. G., *Heredity and Environmental Factors in the Causation of Manic-Depressive Psychoses and Dementia Praecox*, State Hospital Press, Utica, N.Y., 1939.

79. RABOCH, J., and NEDOMA, K., "Sex Chro-

matin and Sexual Behavior," *Psychosom. Med.*, *20*:55–59, 1958.

80. RADO, S., "Adaptational Psychodynamics: A Basic Science," in *Psychoanalysis of Behavior*, Grune & Stratton, New York, 1956.
81. ———, "An Adaptational View of Sexual Behavior," in Hoch, P. H., and Zubin, J. (Eds.), *Psychosexual Development*, Grune & Stratton, New York, 1949.
82. ———, "Hedonic Control, Action-Self, and the Depressive Spell," in *Depression, Proceedings of the Forty-Second Annual Meeting of the American Psychopathological Association*, Grune & Stratton, New York, 1954.
83. RAINER, J. D., "A Reappraisal of Genetic Studies in Schizophrenia," in Sankar, S. (Ed.), *Schizophrenia: Current Concepts and Research*, PJD Publications, Hicksville, N.Y., 1969.
84. ———, ABDULLAH, S., and JARVIK, L. F., "XYY Karyotype in a Pair of Monozygotic Twins: A 17-Year Life-History Study," *Brit. J. Psychiat.*, *120*:543–548, 1972.
85. ———, MESNIKOFF, A., KOLB, L. C., and CARR, A., "Homosexuality and Heterosexuality in Identical Twins," *Psychosom. Med.*, *22*:251–259, 1960.
86. RAPHAEL, T., and SHAW, M. W., "Chromosome Studies in Schizophrenia," *J.A.M.A.*, *183*:1022–1028, 1963.
87. REDDING, A., and HIRSCHHORN, K., "Guide to Human Chromosome Defects," *Birth Defects Original Article Series*, *4*, 1968.
88. REED, E. W., and REED, S. C., *Mental Retardation: A Family Study*, W. B. Saunders, Philadelphia, 1965.
89. REICH, T., CLAYTON, P., and WINOKUR, G., "Family History Studies: V. The Genetics of Mania," *Am. J. Psychiat.*, *125*:1358–1369, 1969.
90. ROSANOFF, A. J., HANDY, L. M., and PLESSET, I. R., "The Etiology of Manic-Depressive Syndromes with Special Reference to their Occurrence in Twins," *Am. J. Psychiat.*, *91*:726–762, 1935.
91. ———, HANDY, L. M., PLESSET, I. R., and BRUSH, S., "The Etiology of So-Called Schizophrenic Psychoses with Special Reference to Their Occurrence in Twins," *Am. J. Psychiat.*, *90*:247–286, 1934.
92. ROSENTHAL, D. (Ed.), *The Genain Quadruplets*, Basic Books, New York, 1963.
93. ———, *Genetic Theory and Abnormal Behavior*, McGraw-Hill, New York, 1970.
94. SALZBERGER, R. M., and JARVIK, L. F., "Intelligence Tests in Deaf Twins," in Rainer, J. D., Altshuler, K. Z., and Kallmann, F. J. (Eds.), *Family and Mental Health Problems in a Deaf Population*, New York State Psychiatric Institute, New York, 1963.
95. SLATER, E., "Birth Order and Maternal Age of Homosexuals," *Lancet*, *1*:69–71, 1962.
96. ———, "The Inheritance of Manic-Depressive Insanity," *Proc. Roy. Soc. Med.*, *29*:981–990, 1936.
97. ———, "Psychiatric and Neurotic Illnesses in Twins," *Medical Research Council, Special Report Series No. 278* Her Majesty's Stationery Office, London, 1953.
98. ———, and COWIE, V., *The Genetics of Mental Disorders*, Oxford University Press, London, 1971.
99. ———, MAXWELL, J., and PRICE, J. S., "Distribution of Ancestral Secondary Cases in Bipolar Affective Disorders," *Brit. J. Psychiat.*, *118*:215–218, 1971.
100. SPITZ, R. A., *The First Year of Life*, pp. 224–242, International Universities Press, New York, 1965.
101. STENSTEDT, A., "Involutional Melancholia," *Acta Psychiatrica Scandinavica*, Suppl. *127*, 1959.
102. ———, "A Study in Manic-Depressive Psychosis: Clinical, Social and Genetic Investigations," *Acta Psychiatrica et Neurologica Scandinavica*, Suppl. 79, 1952.
103. STERN, C., *Principles of Human Genetics*, W. H. Freeman, San Francisco, 1960.
104. THOMAS, A., CHESS, S., BIRCH, H. G., HERTZIG, M., and KORN, S., *Behavioral Individuality in Early Childhood*, New York University Press, New York, 1963.
105. TJIO, J. H., and LEVAN, A., "The Chromosome Number of Man," *Hereditas*, *42*:1–6, 1956.
106. TOMPKINS, G. M., and MARTIN, D. W., "Hormones and Gene Expression," *Annual Rev. Genet.*, *4*:91–106, 1970.
107. VARTANYAN, M. E., and GINDILIS, V. M., "The Role of Chromosomal Aberrations in the Clinical Polymorphism of Schizophrenia," *Int. J. Ment. Health*, *1*:93–106, 1972.

108. VON TOMASSON, H. "Further Investigations on Manic-Depressive Psychosis," *Acta Psychiatrica et Neurologica, Scandinavica*, *13*:517–526, 1938.

109. WATSON, J. D., *Molecular Biology of the Gene*, W. A. Benjamin, New York, 1970.

110. ———, and CRICK, F. H. C., "The Structure of DNA," *Cold Spring Harbor Symposia on Quantitative Biology*, *18*:123–131, 1953.

111. WILKINS, L., GRUMBACH, M. M., and VAN WYK, J., "Chromosomal Sex in Ovarian Agenesis," *J. Clin. Endocrin.*, *14*:1270–1271, 1954.

112. WINOKUR, G., and CLAYTON, P., "Family History Studies: 1. Two Types of Affective Disorders Separated According to Genetic and Clinical Factors," in Wortis, J. (Ed.), *Recent Adv. Biolog. Psychiat.*, *9*:35–50, 1967.

113. ———, and ———, *Manic-Depressive Illness*, C. V. Mosby, St. Louis, 1969.

114. ———, and TANNA, V. L., "Possible Role of X-Linked Dominant Factor in Manic-Depressive Disease," *Dis. Nerv. System*, *30*:89, 1969.

115. ZERBIN-RÜDIN, E., "Endogene Psychosen," in Becker, P. E. (Ed.), *Humangenetik*, Georg Thieme Verlag, Stuttgart, *2*:446–577, 1967.

CHAPTER 6

THE PERSONALITY

Joseph F. Rychlak

THIS CHAPTER will be limited to the strictly psychological views of personality, as opposed to the more properly psychiatric theories. All theories must deal with certain basic philosophical questions, however, and we would like to review five such points of relevance before moving on to see how they have been answered by theorists interested in the question, "What is man?"

The Classical Philosophical Questions

What Is a Cause?

It is frequently overlooked that today's conception of a cause is only one of several used earlier in history. In particular, based upon Aristotle's theory of knowledge, one could once speak of at least four causes. The first he called the material cause. In describing a chair we can say that we know it is a chair because like most chairs it is made of wood, or metal, or the like. Another cause of the chair is the fact that it was assembled by someone or something (a machine). This Aristotle termed the efficient cause. Chairs also meet our blueprint conceptions of what chairs "look like." This usage Aristotle termed the formal cause.

Finally Aristotle noted that there is often a purpose in events, a "that for the sake of which" something like a chair is made to come about. The "sake" for which a chair is constructed might be termed "utility" in eating, writing, and so forth. Of course, the chair does not itself decide to "come about." It is the human being who obtained the wood (material cause) and made it (efficient cause) into a chair matching his physical requirements (formal cause) so that he might live more comfortably (final cause) who may be said to have a purpose or an intention.

Are Theoretical Meanings Bipolar or Unipolar?

When we theorize we essentially deal in meanings. The early Greeks viewed the world as consisting of "many" meanings, tying into one another by way of opposition. Just as to know "left" is to know "right," so, too, did men like Socrates and Plato assume that all meanings were at some point united through bipolar opposites. That which is "error" is tied oppositionally to that which is "truth." To split up this totality of knowledge a method termed the dialectic was employed (see Rychlak,[40] p. 256). Aristotle eventually countered this reliance on oppositional discourse and dialectical reasoning as organon by arguing that when one begins in error he ends in error. One cannot extract truthful conclusions from premises that are false to begin with, dialectically or otherwise. Only through premises of a "primary and true" or factual nature could science advance. This demonstrative strategy in reasoning laid emphasis on the unipolarity of meaning, in which the "law of contradiction" (A is not not-A) separated sense from nonsense.

What Is a Scientific Explanation?

Although Aristotle wanted to move the scientist out of his armchair into the world of facts, he was not above employing final causes in his description of nature. For example, in his *Physics* Aristotle[6] theorized that leaves exist for the "purpose" of providing shade for the fruit on trees (pp. 276–277). In helping to forge modern scientific methods, Bacon[7] later waged a spirited attack on this Aristotelian use of teleological explanation in nature. Since his time the natural scientist has made conscious effort to explain events in only material and/or efficient cause terms, with modest use of formal causality, but *no* use of the final cause.

John Locke then followed in this British empiricist tradition to say that man's basic reasoning capacities are entirely demonstrative as well, consisting of small units of unipolar meanings (simple ideas), which added up to more involved combinations of meaning (complex ideas) in quasi-mathematical fashion. Meanings thus "issued from below," and man's "tabula rasa" intellect was passively molded via input influences from the external environment.

In Continental philosophy, on the other hand, we have the more Kantian model of intellect taking root. Kant stressed man's "categories of the understanding," which were like intellectual spectacles (formal causes) framing in meaning "from above." Whereas Locke felt that we could not—as human beings—subdivide, frame, or invent one "new" simple idea in mind, Kant recognized that through exercise of a "transcendental dialectic" in free thought man could and often did see the *opposite* implication of these Lockean inputs. This led to alternative implications for meanings were again taken as bipolar, and hence man *could* be said to influence his relationship to "reality" in a way impossible to conceive of on the Lockean model.

Are All Theories Written from the Same Meaningful Perspective?

Natural science explanation was thus to be written in material and especially efficient cause terms, utilizing a Lockean model of summative structures or, in the case of mentality, "inputs." This placed the theoretical account at a "third person" or extraspective perspective. The extraspectionist writes his theory about "that, over there," the object or organism under empirical observation. On the other hand, with the rise of psychiatry as a science of man, we see a more introspective or "first person" theory being written. The introspectionist writes about "this, over here," the individual or subject under study in a "personal" way. In this case a more Kantian formulation is possible as we consider the intellectual spectacles as a "point of view" for the sake of which (final cause) an organism may be said to behave. This shift in theoretical perspective is probably the main alteration in scientific procedure brought about by the rise of psychiatry.

What Is Proper Evidence for Belief in a Theory?

The final aspect of knowledge that science was to affect has to do with the nature of proof. What should we require as evidence before we believe the truth value of a proposition? If one believes a theoretical account because of its intelligibility, consistency with common-sense knowledge, or its implicit self-evidence, he uses as grounds for his conviction procedural evidence. On the other hand, science was to raise the status of validating evidence. In the latter case we believe a theoretical proposition only after having submitted it to "control and prediction," which involves an observable succession of events that have been designed to test a prediction about the effect of one (predictor) variable on another (criterion) variable. Here again psychiatry has been the focal point for these interplaying vehicles for the exercise of evidence.

The Major Schools of Personality Theory in Historical Overview

One could trace personality study back to early philosophy, but "modern" personality theory is usually dated from Freud's brilliant work beginning in the closing decades of the nineteenth century. We shall formulate the major intellectual traditions in terms of Lockean versus Kantian models as reflected in man's image.

Mixed Lockean-Kantian Models and the Psychoanalytic Tradition

It is not difficult to show that Freud is a twentieth-century dialectician, who tried to find his way within the strictures of a demonstrative science that did not quite meet his theoretical needs. Freud is the father of modern personality theory because he did—seemingly unknowingly—depart from the material and efficient causes of the "medical model" to assign formal and especially final causes to man's description. His concept of a rational unconscious, directing man from out of a region of wishes and desires, was entirely Kantian in formulation. Thanks to the use of dialectic the psyche is divided into subidentities, each with its own "that for the sake of which" it operated. This is what makes Freud's account so true to life, so *human* and familiar to us who are enacting a series of daily events that we know too well are crazy quilt patterns of contradiction and inconsistency. Freud made man intelligent and introspectively directing by seeing that his physical (hysterical) symptoms were—like our chair, above—*themselves* in existence "for the sake of" causes that lay behind them (intentionality). A symptom carried meaning, and even more complexly, such meanings were *always* compromises between two (bipolar) wishes: the repressing and the repressed!

Freud was instructed in science by Brücke and encouraged to write more scientifically by Fliess. Both of these men were uncompromising Lockeans in scientific commitment. It was under pressure from Fliess that Freud[15] began writing his ill-fated *Project for a Scientific Psychology*. This is the clearest Lockean formulation in all of Freud, but one that he could not complete and in later years tried to have destroyed unpublished. Whereas in his very first theoretical account Freud[16] had actually referred to antithetic ideas and counter wills (pp. 117–128), in the ill-fated *Project* he was to speak of "quantitatively determinate states of specifiable material particles" (p. 295).[15] The former constructs are clearly on the side of formal and final causes, whereas the latter are on the side of efficient and material causes. Fliess had pushed Freud over the line to scientific respectability, but within three months' time Freud[15] could honestly say to his friend: "I no longer understand the state of mind in which I concocted the psychology [*the Project*]" (p. 134).

What Freud *did* do in time was to introduce his libido theory, as a kind of efficient cause (thrust) and possibly material cause (does libido "exist?") translation of introspective mental mechanisms into a pseudo-Lockean frame of reference. It is for this reason that we classify the psychoanalytical tradition as a mixed Lockean-Kantian model. Freud most surely wanted to be "scientific" in his approach, and he did *not* wish to be called a dialectician, an appellation he identified with "sophist." But anyone with the proper grasp of history can see that he made his energies behave dialectically. Drive power always issues from a dialectical ploy of some sort, oriented teleologically for goals in conflict (lust versus propriety, and so on), and then rephrased in energy terms (efficient causes) after the implications are clear.

Although he retained the essentially "reductive" tactic of Freud's Lockean substrate energies, Jung[20] moved his concept of libido even more teleologically over to a direct parallel with Bergson's *elan vital* and Schopenhauer's concept of *Will*, both of which are teleological constructs (p. 147). Jung[21] also clearly recognized that his therapeutic approach was dialectical in nature (p. 554). Adler[2] was to prove the most teleological of the original founders of analytical thought, rejecting quasi-physical energies altogether in favor of an emphasis on the "natural" tendency for movement to occur in human behavior without having to be propelled (p. 41). Moreover, this movement was always fixed by some goal (telos) as embodied in a life plan, prototype, or life style (formal causes that, when exercised, permitted a "that for the sake of which" purposiveness in Adlerian thought). Let Adler retained a healthy respect for the tough-minded approach to theoretical description. His Lockeanism is reflected in a basic distrust of the idealism that our Kantian "spectacles" suggest, and Adler[1] was adamant in his rejection of the dialectical ploy (p. 145).

It was Sullivan[43] more than any other person who shifted the locus of psychoanalysis to an interpersonal rather than an intrapersonal frame of reference. This was a decidedly Lockean shift, bringing about a more extraspective formulation in behavioral description. His concept of energy was far less teleological in connotation, and, indeed, Sullivanian conceptions of behavior are the most compatible of all analytical formulations with the more "mechanistic" theories of what are called in academic circles the "behavioral sciences." Sullivan relied heavily on the formal cause in his concept of the *dynamism*, or patterned energic distribution. Dynamisms were viewed as akin to the Lockean building blocks, as unipolar identities that might then combine into more complex assemblages constituting a person or a society (p. 103).

In more recent years Arieti has attempted to offset the heavy reliance of the Sullivanians on the interpsychic as opposed to the intrapsychic aspects of behavior. Whereas Sullivan had adapted Cooley and Mead's "looking-glass self" conceptions to say that man is what his environment makes him, Arieti[5] stated flatly that "the self is not merely a passive reflection" (p. 370). Psychic identities are not *tabula rasa*, but rather the individual has a certain cognitive contribution to make to his ultimate personality. Thus we find in Arieti[4] a return to the more balanced, mixed models of the main psychoanalytical stream. The Kantian spectacles are now "levels" of cognitive development, moving across the age span to bring the individual to higher and higher stages of symbolic organization.

There are many other theorists who might be cited in this tradition, such as Horney, Fromm, Rank, and so forth, but we must refer the reader to the other chapters of this volume for a more thorough coverage of such views. We wish now to move into another major tradition having implications for theories of personality.

Lockean Models: The Behavioristic Tradition

Important as psychoanalysis was in the framing of man's image in the twentieth century, this theory was never popular as a formal position in the psychology departments of American academia. The difficulty lay in Freud's having to rely upon the procedural evidence of his patients. If he made an interpretation that "struck home" and aided his client's subsequent psychological adjustment, Freud naturally assumed that his theory had validity—for this client, and quite likely for all humans as well. Unfortunately, as events were to demonstrate both within and without the analytical camp, there are many such therapeutic insights to proffer, therapists to proffer them, and clients to be healed by the knowledge so garnered. If they all work equally well to heal, which "insight" is the "true" one, reflecting what is actually taking place in the personality—a Freudian's, Jungian's, Adlerian's, or Sullivanian's?

From its very beginnings in the academic centers of Germany, psychology has sought to be a scientific discipline. Two of its important founding fathers, Helmholtz and Wundt, were dedicated Lockeans who argued that not until a behavioral phenomena had been traced back to "simple forces" (Helmholtz) and "motion" (Wundt) could it be said that a complete account of its nature was rendered (Cassirer, pp. 86, 88).[10] We can see here the substrate efficient cause "reduction" that had been the hallmark of sound science since the days of Bacon. It remained for John B. Watson to pull together the demonstrative, extraspective, Lockean tenets of this style of "natural science" description and press them upon the study of man in his school of behaviorism. Here was a "truly scientific" rendering of human activity, one that academic psychology could embrace and further through experimentation.

As Watson[45] said of the behaviorist: "The rule, or measuring rod, which the behaviorist puts in front of him always is: Can I describe this bit of behavior I see in terms of 'stimulus and response'?" (p. 6). We recognize in this stimulus-response conception the *sine qua non* of efficient causality. As behaviorists we stand "over here" and describe the behavior of others "over there" in efficient cause terms. There is no attempt to speculate about the "inside" of "that" person "over there." Such introspective efforts were effectively discredited by Watson. Hence teleological positions resting upon final causation are literally impossible in the behaviorist's world view. All that can transpire is an "input" (stimulus) and/or an "output" (response), each of which is extraspectively observable, hence subject to scientific manipulation. Indeed, the behaviorist typically equates his research terms (independent variable and dependent variable) with his theoretical terms (stimulus and response.) The world of reality is "out there," and the behaviorist makes no bones about his job being that of mapping it "as discovered."

What then leads to regularities in behavior? Here the behaviorists have differed over the years. Watson relied upon Pavlov's conditioned reflex and the Pavlovian-Thorndikian conception of a reinforcement that supposedly cemented this connection of stimulus and response by way of some kind of physical process. This theory has been termed a "drive reduction" view, based essentially on a belief in the efficacy of inborn needs such as hunger, thirst, sex, and so forth to establish regularities in behavior entirely outside of awareness, much less intentionality. Indeed, when he voiced his initial call for behaviorism in 1913, Watson[46] made it clear that this approach "recognizes no dividing line between man and brute" (p. 158). This resulted in an entirely mechanical conception of human behavior, with reflexes being combined à la Lockean building blocks into habits, and congeries of habits leading to higher level behaviors at the social level (not unlike the Sullivanian conception referred to above). But nowhere was there the dialectical clash, the internal jockeying, or the self-deceiving aspects of behavior so characteristic of psychoanalysis.

Not all behaviorists were to foster such an exclusive reliance upon extraspective theory. Tolman,[44] for example, drew inspiration from the Gestalt theorists and proposed that animals (including man) approached life in terms of a Kantianlike *sign-gestalt* (p. 135), which acted as a sort of road may along which behavior could be directed by the individual—even in terms of his expectancies. We see here more of a mixed model and the hint of final causality, although Tolman formally rejected teleology in the best tough-minded tradition. Hull[18] doubtless raised behaviorism and so-called learning theory to its highest level of expression, continuing in the Watsonian style of a drive reduction to account for habitual behavior. Mowrer[30] added the significant idea that a "reduction in anxiety" could act as a potent reinforcement, leading to the stamping in of abnormal responses. Although the response is self-destructive in the long run, leading to neurotic symptoms, the fact that it reduces anxiety over the short run tends to maintain it (see Dollard and Miller section, below).

Skinner was to alter drive reduction thinking by proposing what has often been termed an "empirical law of effect" position. In his concept of operant conditioning Skinner argued that "whatever" leads to a recurrence of the response following that response's "operation" on the environment may be termed a reinforcement. Thus, if a bear lumbering through the forest turns over a log (operant response) and is rewarded thereby with a rich lode of insects, it follows that he will be seen turning over logs on subsequent occasions. But it is not the "hunger" drive which is being reduced that leads to the later logrolling. At least, argues Skinner, it adds nothing to speculate on such "unobservable" obscurities as needs or drives "within the organism." As he once said to Evans:[13] "I don't see any reason to postulate a need anywhere along the line. . . . As far as I'm concerned, if a baby is reinforced by the sound made by a rattle, the sound is just as useful as a reinforcer in accounting for behavior as food in the baby's mouth" (p.10). As operant conditioning behaviorists, we look extraspectively outward and keep our theories empirically pure.

Skinner does retain the language of stimuli and responses, of course. And it is this fundamental attempt to account for all of behavior in efficient cause terms that stamps a man as a behaviorist, neobehaviorist, or whatever. In fact, the cybernetic account of behavior is comparably built on an efficient cause conceptualization.[48] Whether we call them stimuli and responses, or inputs and outputs, whether mediations between stimuli and responses, or feedback circuits, the tie binding all such "mechanistic" accounts of man is their fundamental Lockeanism. Although Wiener[48] has drawn parallels between man and machine, it has never properly occurred to the cyberneticist that man has any other than a demonstrative power of reason. Yet, as Freud properly grasped, whereas the Ten Commandments fed into a machine would "teach it" or "communicate information" to it in a unidirectional (unipolar) sense, so that these proscriptions would never be violated (their premises would never be challenged), the same commands fed into a man would *of necessity* teach him "ten possible sins!" But in Skinner's[42] world not only are all such one-sided "controls" possible, but also they are desirable and of the essence of existence. For that is the nature of behavior; it is seen as obeying determinate laws, functioning with complete efficient-cause predictability. The trick is to find how best to direct this flow of impetus factors across time.

Kantian Models: Phenomenology, Gestalt Psychology, and Existential Psychology

It was precisely the "wooden" conception of man the Helmholtzian-Wundtian and the behavioristic formulations led to that moved men like Köhler,[24] Koffka,[23] and Wertheimer[47] to found a reactionary form of scientific approach they termed Gestalt psychology. This school took root about the time Watsonian behaviorism was emerging in the second decade of this century. But the Gestaltists were not the only voices rising in opposition to natural science descriptions of man. Throughout the 1920's, 1930's, and 1940's, and particularly following World War II, a rising tide of criticism was voiced by the existentialists, men whose philosophical antecedents went back to Kierkegaard and Nietzsche. What both Gestalt psychology and existentialist psychology have in common is their conviction that traditional natural science theory applied to man somehow robs him of that spontaneous sense of subjective experience that he knows as reality. This is termed the "phenomenal" realm of experience.

It was Kant who carefully showed how reality was constituted of *phenomena* (sensory knowledge, as via seeing, hearing, and so on) and *noumena* (the presumed underlying "stuff" of "things in themselves"). Put in terms of our causes, Kant was saying that material causation was always dependent upon an assumption that "things are really there, even though all I can know about palpable events is what my senses tell me." And when we now consider the objectivities of natural science, it is obvious that they all rely on a kind of "intersubjective agreement" between individuals who are themselves functioning within their own, private, subjective phenomenal field of awareness. Both Gestaltists and existentialists seek to say something about this phenomenal field, which, in turn, pitches their theories to the introspective perspective—making it difficult for a sensitive communication to take place with the exclusively extraspective theories of behaviorism.

There is also a kind of "hope" in the line of theoretical descent now under consideration that is yet to be realized. It concerns a new method of arriving at scientific proofs to rival the "control and prediction" tactic of extraspective validation. Building on the theme of alienation first introduced by Hegel, and then popularized in the writings of Kierkegaard, the existentialists argue that man has been alienated from his true (phenomenal) nature by science's penchant for objective measurement, control, and stilted, nonteleological description. It was Husserl[19] who first pointed to the need for such a variant form of scientific method. Although he worked to lay down the principle of just how this method might be conceived, the actual process was never crystallized.

Binswanger's[8] existential analysis, or *daseinsanalyse*, is conceived in terms roughly equivalent to the phenomenological method of Husserl—as the full description of an individual's experience without "scientific" prejudice or bias, even in the sense of presuming that certain experience is normal, other abnormal, and so forth (p. 110). Hallucinations are thus phenomenally as true as are perceptions of a more "objective" cast. Through *daseinsanalyse* Binswanger essentially hopes to trace back the individual's present conceptual schemes (attitudes, beliefs, personality predilections) to what might be termed the "existential *a priori*" that conditions them, or determines their nature, much as a major premise directs the ultimate conclusions drawn in the syllogism. The Kantian emphasis here is obvious. We find the spectacles, or the "world designs" to use Binswanger's language, that frame in experience for the individual and in this way come to see *his* reality (dasein) from his subjective perspective.

What this phenomenological method has come down to again and again is 100 percent reliance upon procedural evidence. Although it may indeed be based upon intersubjectivity, and doubtless the resultant account leaves out much that is rich in a subjective sense, validating evidence does at least point to objective generalizations. With nothing else to go on, the scientist *can* state this objective "probability" as a body of knowledge without having to haggle over the details of "what do we know about phenomenon X?" And this has been the great indictment of phenomenological efforts. Not that they are wrong as to theoretical statement, but that they have lacked the methodological or evidential support to be taken as authoritative rather than simply as literary accounts. It must not be overlooked that *all* so-called clinical accounts suffer from the same problem. Freud, as we know, took psychoanalysis to be a valid scientific method, as do many analysts today. Yet not the least of the reasons that he found it so difficult settling disputes with students and colleagues can be traced to the exclusive reliance that clinical analyses must make upon procedural evidence.

Even though Gestalt psychology was to meet the strictures of validating evidence by proposing a series of remarkably creative experiments, the lock that behaviorism has on academia never really permitted this more Kantian approach to flourish. One annually hears of the complete demise of Gestalt psychology, but the truth is that it makes rather frequent rebounds under the guise of so-called cognitive psychology. The Gestaltists, too, were advocates of phenomenology, which Koffka[23] once essentially defined as the attempt "to look naïvely, without bias, at the facts of direct experience" (p. 73). Gestalt psychology is best known for its supposed attempt to prove that "the whole is greater than the sum of its parts." What has not often been made clear is that this conception of the relations between the "one" and the "many" has philosophical precedents dating back to the pre-Grecian philosophers (see Rychlak,[40] pp. 257–267). And invariably over the course of the centuries it was the more idealistic, dialectically oriented philosopher who argued this point. Plato, for example, viewed knowledge as "one," as having "many" facets, but each of these latter aspects were configurated into a single, overriding totality. By beginning at any point with a given (thesis), reasoning dialectically to its opposite (antithesis), and resolving the inner contradictions thus implied, the student (and teacher) could arrive at a higher state of totality (synopsis, later "synthesis" à la Hegel).

Although they did not press a dialectical formulation, basing their studies on sensory mechanisms such as the eye, ear, and so forth, the Gestaltists did show again and again that man contributes something to reality by way of his innate equipment (analogical to Kantian "spectacles"). Knowledge is not simply a question of information input. There are certain "laws of organization" that result in perceptual constancy, rules of memory and thought that make certain "total organizations" likely to result in one phenomenal experience while the same factors slightly reorganized result in another. The Gestaltists were theorizing introspectively, describing what was taking place "over here," as an observer looking out onto the world studies his own processes. Their behaviorist counterparts found this sort of talk almost spiritual, harking back to the Middle Ages. There was a certain truth in this, of course, because the major factor of a "totality" is that it has organization; that is, it is a formal cause. And, assuming now that the individual may be said to behave "for the sake of" this total experience rather than simply responding to inputs, it follows that we begin taking on the meaning of a final cause in our theoretical accounts. Little wonder, therefore, that tensions were to arise. What is probably most regrettable in this academic confrontation is that the issues separating behaviorists from Gestaltists have never been made clear. Often *ad hominems* have been substituted for rational discourse, and nowhere does one find analyses in terms of causation, theoretical perspective, or philosophical presumptions. Actually the Gestaltists would welcome such discourse, but the behaviorists find it just another example of the tenderminded theorist's mania for obfuscation and cheap verbal triumphs.

Some Examples of Classical Answers to the Classical Questions

Having now reviewed the three major intellectual traditions that have made an impact on psychological theory, we might review a number of theoretical constructs that have been proffered within these lines of descent. Since, as we noted above, the psychoanalytical tradition has been slighted in the formal outlooks of academicians, it will be clear to the reader that psychologists have placed greatest emphasis on the behavioral and the phenomenological aspects of human behavior. It is often possible to show that these constructs were directed at some clearly philosophical issue, such as determinism versus teleology. A theorist's name and his major construct will be given in the title to each of the subsections.

Sheldon's "Morphogenotype"

William H. Sheldon[41] has continued and furthered the style of theorizing about man that dates back through Kretschmer, Lomboroso, and others to the very founder of such speculations, the father of medicine: Hippocrates. The emphasis here is on material and efficient causation, which—along with formal causality in the syndrome picture—has been the mainstay of all medical models of illness, including the psychiatric. Hereditary concepts are of this nature, and the transmission of various characteristics that might be seen in overt behavior is along a "chance" line, sketched entirely in terms of a Lockean model. The point here is that "genes" are "primary and true" items of physical structure that combine in various ways entirely due to physicochemical laws (efficient causes). Natural selection (Darwin) has determined the final result, for there has been no teleological advance in this descent of man. At least there has been no deity teleology or natural teleology at work. Whether there has been a human teleology—and how this is to be conceived—is a question that has not yet been settled.

Sheldon's theory rests on the assumption that an hereditary factor termed the morphogenotype (gene-induced bodily form) is in operation. Analogizing to the in utero development of the human embryo, Sheldon argued that the morphogenotype selectively works in physicochemical terms to emphasize the development of the ectodermal (nervous system, sense organs, etc.), endodermal (visceral and digestive organs, etc.), or mesodermal (muscles, bones, blood vessels, etc.) layers of the developing organism. The resulting bodily structure at birth is predominantly ectomorphic (thin, linear, delicate), endomorphic (rotund, often corpulent), or mesomorphic (muscular, large-boned, strong). Components of each of these dimensions are to be seen in every human form, and Sheldon has devised a series of ratings to score individuals along these "primary components of physique."

Sheldon reviewed the literature on personality and devised what he termed were the "primary components of temperament," as follows: viscerotonia (people who love physical comfort, are socially outgoing, complacent, amiable, and love to eat and drink in the company of others); somatotonia (assertive, physically active people, who love risk-taking, competition, and the leader role); and cerebrotonia (people with restraint in action and emotion, a love of privacy, and a hypersensitivity to pain). Empirical study of the relationship between the physique and temperament established that endomorphy was related to viscerotonic personality tendencies, ectomorphy to cerebrotonic traits, and mesomorphy to somatotonic behaviors. Although there are obvious problems here of the "chicken-egg" variety, not to mention the effects of diet on bodily developments, we see here one fine example of a theory of personality relying upon purely "natural science" explanation. It is not a very thorough explication of the human pattern, but neither does it presume to explain all things. Sheldon has extended his study to include individual differences among delinquents, the sexes, and so forth.

Allport's "Functional Autonomy"

One of the more challenging issues put to men who considered themselves students of personality was the question of just how behavior in the present was sustained. Addressing himself directly to the behaviorist's conception of a stimulus-response sequence (efficient cause) that had been stamped into habit by a reinforcement (material cause in physical satiation), Gordon Allport[3] proposed that some behaviors become functionally autonomous of such "drive reductions." Although a child might have initially studied his school books because his parents showed him love and gave him financial rewards for good grades, the mature adult can in fact acquire along the way a love of knowledge per se. The "conditioning" process of earlier years is not irrelevant to the now functionally autonomous motivation, that is, free from the tie to specific reinforcements—that the activity itself has taken on. Behaviorists would term all such "function pleasures" to be extensions of the basic drive reductions (the physical caressing from parents) as so-called secondary reinforcements (pleasure in reading, acquiring knowledge, and so forth).

But Allport was trying in his own way to break personality description free of "yesterday's" blind directedness. He rejected not only behaviorism in this regard but also psychoanalysis—where he felt that man's higher behaviors were invariably reduced to yesterday's "fixations" having no real value for the behavior as witnessed. Allport was a transitional figure in psychology, accepting the merits of stimulus-response psychology even as he tried to conceptualize human behavior in less mechanistic and hence more teleological ways. Rather than speak of behavioral habits, Allport[3] took as his basic unit of study the construct of a trait, which he defined as: ". . . a generalized and focalized neuropsychic system (peculiar to the individual), with the capacity to render many stimuli functionally equivalent, and to initiate and guide consistent (equivalent) forms of adaptive and expressive behavior" (p. 295). Note the tie given here to the physical structure of the central nervous system (material cause). A trait is a formal cause notion, since it implies a self-bearing "style" of behavior.

Allport was thus hoping to point out that man's behavior is not blindly habitual, responsive to stimulus input *only*, but also to some degree self-directing and stylized. Man could behave "for the sake of" an interest, a fascination, a freely operating desire that he had learned over time was worthy of perpetuating in its own right. Allport popularized Windleband's distinction between the nomothetic and idiographic sciences in psychology. Psychology, he contended, must be like history—a study of the trend line of development over the course of life. One can learn a good deal about the nature of digestion by studying the alimentary canals of thousands of animals, from lower to higher, within the confines of an isolated laboratory. But one cannot learn why the French value one style of living while the Italians another without considering the respective histories of these nations. In the same way personality as a compendium of traits must be seen uniquely evolving across time. And to immerse oneself in the first five years, as Freud had done, or to believe that only through base reinforcement does man find the motivation to approach life, as Watson had done, was for Allport a common error in theoretical formulation. Man can be functionally autonomous from such base reinforcements, as he can be functionally autonomous from the fixations of toilet training.

Murphy's "Canalization"

Another historically important attempt to counter the more "mechanical" formulations of stimulus-response psychology was made by Gardner Murphy,[31] when he distinguished between conditioning and canalization—a term used earlier by Pierre Janet in a different sense. A conditioned response, said Murphy, was indeed a mechanical sequence of events (efficient causes). But these movements were simply preparatory; they oriented the animal for eventual gratification that was itself more in the nature of an anticipated achievement leading to goal realization (p. 193). As the restaurant waitress approaches our table to take our order, there are various mechanical, unthinking conditioned responses that we make in ordering our meal; for example, we arrange our plate and table utensils, tap nervously on the menu, and so forth. But the consummatory act, the *goal* of eating what we choose to eat, is not itself a conditioned response or a class of such responses. What a person comes to prefer, comes to work for and select in life is not simply conditioned—it is canalized. Behavioral patterns thus become fixed through active, purposive, self-directed attempts on the part of the individual to channel (canalize) his behavior in terms of something he has personally discovered to be satisfying. Conditioned responses are passive and routine. Canalizations are active and selective.

Thus man is passively shaped in the Watsonian sense, but he is also actively self-created through a process of anticipation and achievement (p. 170). One can see here an effort to bring some modicum of "that for the sake of which" into personality theory. As Allport would have it, man passes through life with an intellect open to the possibility in things. He is being demonstratively "input" with experience, but he has this capacity to evaluate such input, to take delight as well as to respond with simple animal satisfaction. At some point the taking delight begins selectively to direct what is satisfying. This directing or canalizing is entirely on the side of a final cause, possibly also including the formal cause as a kind of plan, wishful design, and so forth. Murphy thus did not deny the behavioral findings of the rat laboratory. He simply wished to point out that there was more to human behavior than this routine, blindly repetitive series of responses. Choice and purpose were "in" the picture.

Murray's "Regnancy of a Need"

H. A. Murray was to solve the problem of directed behavior in somewhat more physical terms than either Allport or Murphy. The behaviorists had referred to needs as specific tissue deprivations of some sort, as in the case of hunger or thirst. Murray[32] was to broaden the scope of this term, using it much as Allport had used traits to include the purely psychological aspects of behavior. He viewed the need as a hypothetical property of force, presumably a force in the brain region that organizes behavior in a directional sense (p. 123). Needs generate action, which, in turn, eventuates in some counteracting environmental force termed a *press*. The individual with a great need for achievement might well find that there are forces in his experience that counter an easy access to wealth, the attainment of accolades for athletic prowess, or the professional recognition from colleagues. There are all kinds of environmental pressures against which the individual must struggle in order to gain the satisfaction of his needs.

The typical fashion in which people go about meeting their needs in the face of counteracting pressures from their life's milieu Murray called the *thema*. This entire theoretical account is essentially an analogy from Murray's[29] projective test, the Thematic Apperception Test (TAT). In the same way that the clinician analyzes a "hero" figure in a TAT story, so, too, does the personality theorist assess the individual in his life circumstance.

Murray proposed a series of need terms to be used in the description of behavior, including aggression, achievement, affiliation, exhibition, order, and so forth. Precisely what an individual is like in personality could now be assessed in terms of the particular combinations of his unique needs, their level of satiation or deprivation, and the life circumstances (press) that faced him. The highly affiliative person, for example, after a period of time in which he might be forced to be alone, could well begin appearing highly frustrated and even abnormal simply because of his rising need state and the circumstances of his life milieu. Murray retained a physical tie-in to the functioning of the body, making an effort to resolve the dualisms of Freudian or Jungian formulations. Dualisms have never fared well in psychological academic circles. The behaviorists had argued that habits took on a hierarchical arrangement—from lower-level, simple stimulus-response connections, to higher-level complexities of a more global nature (we see here a Lockean model being pressed). Murray[32] now builds on this conception by arguing that certain needs were prepotent (overriding force) to other needs. Such needs demand answering in the "now"; they dominate our brain process as *regnancies* (predominant physiological reactions), assuring that some form of behavior will be undertaken to seek the goal that can satisfy the condition of a rising motivation (p. 45).

Since the need concept is now a psychological one, we can say that Murray has effectively resolved the dualism of classical analysis by claiming "wishes" or "cathexes" are regnant brain processes. The man under regnant brain processes heralding a prepotent need for achievement is to be seen driving himself forward across life's way to success at all costs. This theoretical usage puts a kind of directionality (final cause) in the account without actually making it seem that way, because all needs are either given at birth in physical constitution or they are "learned" (input influences à la Lockean model) over the years of development. Hence the image of man here is more introspective than classical behaviorism would have it, but we still do not find that heavy aura of the "internal world" that Freud and Jung provide us with. Murray has milked the dialectical side of man out of his theory. Man no longer takes in an input meaning, reasons to its opposite meaning-implication, and thence directs himself via a true choice to some alternative or compromise creation all his own. Man is directed by regnancies emanating from physicochemical forces in the brain, forces that have been planted there by nature or by the social milieu. What quasi-dialectical clash there is takes place between these internal forces (needs) and the external counterforces in the milieu (press). But man *qua* man is not "in the middle" as Freud believed him to be, with an ego identity struggling internally to compromise the wishes of the id and superego identities.

Dollard and Miller's "Anxiety"

An even more thorough job of taking the dialectical capacities of man out of his theoretical conceptualization was accomplished by Dollard and Miller.[12] Their motives were laudable, in that they hoped to cement laboratory theory of a Hullian cast with the insights of the consulting room and thereby unite psychology in a way it had never been united previously. Since stimulus-response theory is more abstract than clinical formulations, this translation could have been performed on Adler, Jung, Sullivan, and so forth. But Freud was selected, and the procedure adopted was to rewrite Freudian terminology into the more abstract Hullian terminology of cue, drive, response, reinforcement, habit, and so forth. This is a drive reduction theory. A "drive" is a strong stimulus impelling action, which, when reduced (material cause), acts as a reinforcement of the stimulus-response regularity (efficient cause) that preceded it. Freud's sexual concept is interpreted as such a drive. There are basic drives (such as pain) and drives of a secondary cast, which can be easily attached (learned) to stimuli that do not ordinarily bring about a basic drive arousal.

One such secondary drive is anxiety, which is interpreted as a learned drive having the properties of fear, except that in the case of anxiety the source of the threat is vague (p. 63). We fear a train bearing down on us, but we are anxious knowing why. The point of importance for learning theory, however, is that when such vaguely stimulated anxieties are reduced—when we flee the elevator situation (claustrophobia)—this return to a normal emotional level acts as a reinforcement. It samps in the flight response, in relation to the elevator stimulus. Hence the next time we face an elevator situation we will be sure to flee in order to reduce the anxiety that has once again mounted due to inexplicable reasons.

Now this theoretical treatment of anxiety has become extremely important in psychology. For example, it lies at the heart of the so-called behavioral approaches to psychotherapy (see, for example, Wolpe).[49] It has proved very popular because, just like Murray's regnancies, we have reduced the Freudian dualisms of mental events versus bodily drives to a single, hence monistic, formulation. This is all very much in the traditions of natural science, and the essence of this tactic is to say that a physically based drive (material cause) brings about and sustains behavior (efficient cause) entirely "on its own." The neurotic's symptom is sustained because he reduces anxiety by performing it. His grasp of why he does this is vague, thanks in part to the fact that he has not paid sufficient attention to his life circumstances in the past. For example, when he was frightened as a child in an elevator, he "stopped thinking" as a response ("repression") and therefore never really knew what it was about the elevator that actually set off his fear (a loud noise, a frightening passenger, the fact that he was being taken to the dentist, and so forth). So far as learning theory is concerned, the actual reason for the symptom is unimportant, or at least quite secondary to the fact that a symptom now is coming about and must be removed.

Although many psychoanalysts today would agree with the statement "neurotics behave as they do to avoid anxiety," the actual translation of an introspective, Kantian-Lockean (Freudian) model into an extraspective, entirely Lockean (behaviorism) model *has* lost something in the translation. In the first place Freud's concept of repression was not one of a passive "stopping thought," or overlooking possible cues in the environment. The unconscious mind, according to Freud, knew only too well what it feared or lusted or hated. Second, Freud definitely did not want anxiety to take over the motivation properties of the personality structure. That was the job of libido—an entirely mental construction of force or drive. This is why he stressed that only the ego could experience anxiety! As is well known, Freud[14] relegated anxiety to an instrumental role as a warning sign (pp. 57–59). The warning to consciousness in neurotic anxiety was something to the effect: "watch out, your lustful desire for mother and your wish to kill father is going to pop up here in a moment and then you will have consciously to admit that you are a rapacious, incestuous pig and murderer." The conscious aspect of the ego is "inoculated" with a modicum of anxiety so that, rather than consciously dealing with such incestuous and murderous intentions or "wishes" (final causes), it deals with an unpleasant physical sensation (material and efficient causes). But to say that symptoms are aimed at avoiding anxiety is completely to misconstrue the meanings of Freudian theory. Neurotics behave as they do not to avoid anxiety, but to avoid the awareness of their completely psychic, unacceptable intentions!

Hence the wedding of Freudian and behavioral theory must and has altered the image of man being described. Theorists are dualists (mind-body) for reasons, and when one alters the necessary teleological implications of this dualism to meet the strictures of monistic scientific thought, he violates the reason impelling dualistic formulations from the outset. Freud had his Fliess, his "natural science conscience," and he gave the Lockean model a sincere effort in the *Project*. But he could not forego the meanings he was trying to convey by conforming exclusively to the style of description that Dollard and Miller were later to employ. Hence we must count the latter's laudable efforts as only partially successful, through no real fault of their own. Some meanings simply *must* stand as framed.

Cattell's "Source Traits"

Raymond B. Cattell must surely be the foremost theorist to take a measurement approach to the assessment of personality. There is much of the Allport and Murray tactic in Cattell, for he begins with the assumption that behavior is constituted of traits, and that these Lockean building blocks combine to form the personality superstructure. Rather than a hierarchy Cattell speaks of a "dynamic lattice" in which some traits "subsidiate" (take precedence over and hence enter into) others. The unique twist that Cattell[11] gives to the Lockean model is that he sees both surface and source traits in behavioral operation. Surface personality traits are the apparent manifestation of individual differences, superficial assessments that we make as observers because we have no way of directly viewing the commonalities lying beneath. Source traits, on the other hand, "promise to be the real structural influences underlying personality" (p. 27). Thus by using the trait designation Cattell captures a formal cause meaning, but his "source" adaptation gives us a kind of analogy to the reductive explanations of natural science. We get a quasi-material and quasi-efficient cause meaning here through rough analogy, if nothing else.

What one must do to jump the gap between a surface and a source trait manifestation is carefully to measure overt behavior and then submit these crude empirical measures to the statistical refinement of factor analysis. One finds in this way a common "factor" accounting for various overt manifestations. In time a series of reference factors having universal relevance to behaviors will be empirically identified and carefully validated. This collection of source traits (universal index) can then be applied by the psychologist much as the chemist makes use of his periodic table of elements. So much of source factor A, combined with so much of source factors B and C, results in what we call superficially the (surface) trait of "leadership," and so on. Although an oversimplification this portrayal of Cattell's approach is basically accurate, and we can see in it the substrate notions of Lockean "simple" structures combining to form the higher order, "complex" structures. The specifics of just how personalities got to be the way they are "now" constituted would depend upon the typical "input" notion (efficient cause) of environmental influence, including conditioning and hereditary explanations.

Skinner's "Contingency"

We have already noted above that B. F. Skinner was an important figure in the "empirical law of effect" interpretation of reinforcement. Rather than attributing behavioral regularities to such "inside the organism" concepts as needs, wishes, aspirations, intentions, and so on, Skinner argued that only those responses that "operate" on the environment to bring about rewarding events are retained by an organism. At least, most of an animal's response and virtually all of human responding is of this nature. Skinner termed this an operant response, and that "something" in the environment that serves as a reinforcement of such operants he termed a contingency. What are the contingencies of reinforcement available to an organism in the environment? If we know this, then we can easily predict what behavior it will *emit*, for behavior is always under the control of some class of empirically demonstrable reinforcing contingencies.

Although Skinner[42] is not precisely a "personality theorist," surely his image of man has been given enough serious consideration by specialists in this area to rank him at the very top of contributors to the study of personality. And what he constantly emphasizes in all of his characterizations of man's behavior is that the environment and not man is the selective agent in behavioral control. He specifically rejects the concept of "autonomous man" (p. 67). Skinner, more than any other modern psychologist, is an uncompromising classicist in his image of man as exclusively an efficiently caused succession of events. He has acknowledged a debt to British empiricism (Lockean model) by noting to Evans[13] that he "short-circuited" Kantian formulations (p. 15).

An even more remarkable observation on his theoretical stance is reflected in the following: "Operant behavior, as I see it, is simply a study of what used to be dealt with by the concept of purpose. The purpose of an act is the consequences it is going to have" (Evans,[13] p. 19). We find here an almost startling preempting of the final by the efficient cause meaning. Through viewing man exclusively on extraspective terms, and fixing on the consequences of "that" behavior "over there," Skinner can assess the consequences (contingent reinforcements) of "that" behavior to see which consequences perpetuate it and which do not. Once he determines empirically what such contingencies entail, he can exert what he takes to be a form of efficient cause control over it. But what if the organism "over there" is considered introspectively and judged to behave "for the sake of" personally held intentions, after all? What if, in the case of man, rather than his being under the control of the extraspective manipulator, he is actually simply conforming or cooperating with what he takes to be the manipulator's purposes?

Well this would make no essential difference to the Skinnerian formulation. For example, psychologists have shown to general satisfaction that "being aware" of the response-reinforcement contingency, or, in other terms, knowing that verbal behavior X will lead to reinforcement Y, decidedly facilitates the efficiency of learning verbal behavior X. Some psychologists claim that there is very little, if any, verbal learning without such an awareness on the part of the subject. This could easily be taken as evidence in support of a final cause view of behavior, with the response-reinforcement contingency interpreted as "that advantage, clue, goal, or plan for the sake of which" behavior is acquired and perpetuated. Yet such questions are not thought worthy of serious theoretical consideration in the Skinnerian world view because what is fixed upon is *only* the flow (impetus, efficient cause) of events across time. This is taken as "behavior." Just so long as it can be shown that certain contingencies lead to behavioral pattern A and other contingencies lead to behavior pattern B, this is all the Skinnerian feels obliged to deal with as he perfects his ability to "control" such patterns—from A to B and back again.

Rotter's "Expectancy"

When Tolman was working out his variant brand of "purposive behaviorism," he emphasized that input stimuli are rarely translated directly into output responses, because as Woodworth had observed, there is a certain "mediation" of the organism in between. This Kantian notion of a "cognitive map" was central to the Gestaltist theorists who inspired Tolman, but in accounting for the continuing influence that an organism (rat, man) has upon his experience, Tolman was to speak of what have ever since been called "mediators" in learning theories. Dollard and Miller have made considerable use of this mediation construct in their translations of Freud. Higher mental processes (thoughts, words, language) all come down to an operation of some such mediating "cue stimulus" or "anticipatory goal response." Animals learn to begin responding even before they see their reward (reinforcement), said Hull. Anthropomorphizing, we might say that the dog "knows" his dinner is waiting ahead, for he salivates noticeably and breaks into a more rapid run as he sees his master's house ahead. But actually the dog anticipates nothing at all. He has simply been trained to respond to certain "antedating" cues, so that over time his salivation response began moving ahead in time, from his food dish, to the door leading from outdoors, to the silhouette of the entire house ahead, and so forth.

Hull was the behaviorist to develop this line of theory most creatively, and he was obviously trying to account for what in other contexts might be termed intentional or purposive behavior without resorting to a teleology. In personality study Adler and Lewin had been developing conceptions of human behavior based on what has since been called the level of aspiration. Adler made no bones about this being a teleological conception, claiming that people laid down a definite plan (prototype, life plan) "for the sake of which" they then aspired to further their advantages in living. Lewin's conception was also teleological, but he was not quite so outspoken because he was trying to meet some of the natural science objections to final cause description.

It remained for Julian B. Rotter,[39] a theorist who was influenced by Hull, Adler, and (to a lesser extent) Lewin, to raise this conception of aspiration level to what is probably its most thorough and well-rounded expression. Rotter changed the descriptive label to expectancy, but it has the same meaning of "that for the sake of which" an individual may be influencing his behavior. The child who expects to earn (aspires to) school grades at the A level is crushed with a grade of B, whereas the child expecting C's is elated with the same achievement. One's life circumstance cannot be entirely circumscribed by the "simple facts" of reality. Rotter also added the concept of reinforcement value to say, for example, that even some things that are easy to attain in life are not valued. If we wish to predict behavior we must know not only what the person is expecting, but how much he values that which is upcoming. The child who does not value education will give little effort to it even if achieving good grades is relatively easy for him.

Although he has moved his descriptions over to the introspective perspective, and his account is far less mechanistic than the classical behaviorist's, Rotter's psychology remains heavily Lockean in tone. He is clearly in the line of descent we have been reviewing to this point. Expectancies amount to past "inputs," learned through conditioning in experience on the basis of an empirical law of effect and functioning in the present as special kinds of mediators. The value of reinforcement is also a function of past reinforcement. Man is not viewed as capable of reasoning to the opposite of what is given, drawing out an expectancy-aspiration of some other possibility, and then aspiring to what was *never* known, much less reinforced, in the past. Yet Rotter's theory is probably best classified as a mixed Kantian-Lockean model, and he forms a nice bridge to the more clearly Kantian approaches we now turn to.

Lewin's "Life Space"

Although he was not an orthodox Gestalt theorist, there can be little doubt that Kurt Lewin received considerable stimulation from the work of Wertheimer, Koffka, and particularly Köhler—all of whom were his colleagues for a time at Berlin University before he came to America.[26] Lewin took the Gestalt concept of a perceptual phenomenal field and drew it out into a view of the life space, or the total psychological environment that each of us experiences subjectively. This (formal cause) construct embraced needs, goals, unconscious influences, memories, and literally anything else that might have an influence on one's behavior.[25] Rather than seeing behavior as an incoming process of stimulus-to-response, Lewin constantly stressed that behavior takes on field properties as an ongoing process of organization and interpretation following Gestaltlike principles.

The course of behavior follows paths or pathways between one's present location in his life space and the goal or goal region (level of aspiration) that attracted him. Other goal regions might repel the individual (negative valence), and it is the sum total of all the field forces (efficient causes) entering into an overall pattern (Gestalt, formal cause) that led to behavior (locomotion) within the life space. Hence behavior was directed, and although Lewin might be said to have introduced a modicum of teleology in this more introspective account, the directedness of this behavior was put extraindividually in the sense that the "person" or "personality" is merely one organized subportion of the entire life space. Motions within the field (efficient causes) can be induced by any portion of this life space.

Lewin's handling of teleology is therefore quite unique and rather moderate in relation to the more extreme final cause formulations that can be tied to man's image. The life space is, of course, a derivative concept of the Kantian categories or predicating "spectacles" that are the major contributor to behavior in this theoretical style. But Lewinian psychology is not a complete idealism. Lewin accepted what might be termed a noumenal world on the "other side" of the phenomenal life space. Such influences on behavior as the fact that a path under our foot is slippery, or a roof over our head leaks water, were influences emanating from the foreign hull. As a permeable membrane the life space (formal cause) interacted with such foreign hull influences (material causes), making such alterations in its organization as were called for and mutually altering the status of the foreign hull by reciprocal influences.

Kelly's "Personal Constructs"

One of the most clearly Kantian formulations in the literature is to be seen in George A. Kelly's[22] "Psychology of Personal Constructs." Unlike Rotter, who viewed expectancies as past input influences based upon a reinforcement principle, Kelly ascribed an active intellect to man, one that construed experience rather than passively took it in. For Kelly an expectancy is to be thought of in terms of both a formal and a final cause meaning. It is a stylized meaning through which or "for the sake of which" the individual advances on life daily. Of course, Kelly's actual term is that of the personal construct rather than the expectancy. Just as Kant had argued that freely created thought is dialectical in its essence, so, too, did Kelly view the process of construing as bipolar in nature (p. 304). When one affirms the commonality of events that he has observed recurring over time, he must also negate some other aspect of that recurring experience. To say "Redheads tend to be hotheads" is also to say "Nonredheads tend to be level-headed."

Thought is only possible, said Kelly, because man can dichotomize elements of experience into similarities and contrasts (p. 62). The products of thought, or constructs, state in either clear or highly nebulous terms how "two elements are similar and contrast with a third" (p. 61). Constructs are working hypotheses, predictions, appraisals, and even pathways of movement, for they frame in our meaningful experience like transparent templets (Kantian spectacles), and hence predetermine just what is possible for us to do. Man is determined mechanically only when he construes himself in this fashion. Constructs are either permeable and capable of change, or they are impermeable, rigid, and frozen into a form of thought Kelly termed pre-emptive. To change behavior we must change the constructs determining that behavior. The philosophy that expresses a strong faith in man's capacity to do precisely this Kelly termed "constructive alternativism." Although constructs are ultimately highly subjective or "personal" in nature, they can be understood introspectively if we make serious efforts to see things from the slant of the personality under study.

Constructs can also be objective in that many men can understand the meanings implied in the same set of constructs. To further a clinician's ability to identify the constructs of his clients Kelly formulated the "Role Construct Repertory Test," or, more simply, the "Rep Test."[22] A role construct is one that defines the individual's more important interpersonal behavior; for example, when he perceives another individual as also a construer, and to that extent enters into an interpersonal relationship with him. By contrasting and comparing how various figures in his life (mother, father, best friend, admired teacher, disliked associate, and so forth) were like and yet different from a third figure, Kelly was able to fashion a list of core personal constructs that he then used to see the world from his client's eyes. "How are your mother and ex-girl friend alike, and yet different from your wife?" This would be a typical example of the way in which role constructs are evoked. The individual is free to select his own terms. Kelly devised a nonparametric procedure for factor analyzing these many different constructs to find the very heart of the individual's construct system. This idiographic manner of factor analyzing data is quite different from that of Cattell's, and the image of man that results is, of course, diametrically opposed to the Lockean formulations of the more nomothetic approach.

Maslow's "Third Force Psychology"

Abraham Maslow coined the phrase "The Third Force" in psychological theory by which he meant an approach in the traditions of people like Allport and Rogers.[17] He seems to have identified this approach as "humanistic psychology," a phrase that has achieved considerable prominence in the post-World War II era. After passing through a period of fascination with Watsonian behaviorism, Maslow[27,28] moved on to emphasize such concepts as self-actualization, human potential, and peak experience. These terms attest to man's capacity for teleological advance, based upon a hierarchy of lower-to-higher needs that rest upon one another, yet are fundamentally independent of each other. Maslow thus picks up the conception of need developed by Murray, as dealing with both physical and psychological necessities. Physiological needs lie at the base of the hierarchy of needs, with higher-level needs such as love, esteem, the need to grow and self-actualize coming into the organization as kind of emergents. The important point is that one cannot find the meaning of higher-level needs by reducing them to the lower-level needs. Further, it is inevitable that as the lower-level, physical needs are being met, the more humanistic needs will begin to seek expression and gain satisfaction.

Hence, just as neo-Darwinian theorists speak of emergents in the evolutionary processes of nature, so, too, does Maslow rely upon this tactic to modify the Lockean hierarchy that held that the lower levels constitute higher levels and to know the latter we must deal with the formed. This is a clear Gestalt or holistic infusion, a tempering of the more mechanistic features of Lockean thought while striving to retain continuity with the physical aspects of nature. As a theoretical device it is comparable to Freud's uniting of body (physical) and mind (psychological) through use of the instinct concept. Maslow actually based much of his thought on biological conceptions, feeling that there was a "growing tip" to the advance of organismic life (natural teleology). If we want to get a sense of the higher life that evolution is making possible, we should investigate our more self-actualized human life histories. Maslow did just that, isolating the factors of important historical figures like Lincoln and Einstein, whom he judged to be self-actualized individuals.

Maslow claimed that self-actualized individuals see life more clearly than other people. They are more decisive and can take a stand with greater confidence, for they are prepared to name what is right and what is wrong about life. They have a childlike simplicity and usually admit their lack of knowledge in an area of what is clearly their expertise. Though very confident they are humble and more open in their general approach to others. Without exception they have some worthy task to which they commit themselves completely—a career, duty, or vocation that presses on them, fascinates them, and gives them a sense of fulfillment even though it is not always easy or pleasurable to accomplish. They are, above all, spontaneous and creative in their behavior, willing to "be themselves" for they lack pretense and defensiveness. Maslow coined the term *Eupsychian* to describe the society that a group of such self-actualized individuals would form if left to their own devices—say, on a secluded island. Presumably the society would reflect their common tendencies: a biological utopia of our very best, the "growing tip" clipped off and transplanted to flower as all utopias do—apart and unmolested by the common foliage.

Piaget's "Schemata"

Although he worked for years in relative obscurity, Jean Piaget has assumed major importance in the outlook of many psychologists during the post-World War II years. Piaget[34] has a concept of the schema that is reminiscent of Kellyian constructs, but it takes on a developmental frame of reference in that presumably we are dealing here with innately prompted constructions. Thus Piaget argues that schemata are first brought into play on the basis of reflexive activity, as when the infant first employs his sucking reflex, bringing it to bear on the mother's nipple. This process of aligning a patterned behavior (schema) to a proper stimulus Piaget termed accommodation. Once fixed in this fashion the experience of sucking takes on meaning to the child, although of course, the extent of meaningful grasp is limited due to the lack of language. One sees here a decidedly introspective formulation of what is a formal cause term (schema, pattern). The nature of human maturation is now a question of extending and in time patterning various schemata into more and more meaning. The child begins to notice and suck other objects—his fingers, a blanket, and so forth—coming to enlarge this already accommodated schema. This process of enlarging and enriching schemata Piaget termed assimilation.

The essence of Piagetian motivation theory is that the child and then the adult tries to keep his schema relevant and applicable to experience. Schemata that are not assimilable are meaningless by definition, so it is essential to human intelligence that a continuing growth takes place. Much of Piaget's empirical work has involved the study of maturing children, tracing how this process of continuing, expanding, and changing schemata takes place. For example, he early found that the natural experience of reality for the child is anthropomorphic.[33] The child perceives natural events of all sorts, including rain, wind, and so on in terms of intentions and willful acts. The five year old says that the sun's rays push the wind into activity or organize the clouds to look pretty. Only by about age eight or ten does the child completely divest the physical world of human qualities and view it in purely physical, mechanical terms. Piaget called this early phase "precausal" thinking, but we can see here the time-honored issue of final-formal versus material-efficient causes manifesting itself. The basic question is: do children think "primitively" or do they think entirely "naturally," so that their teleological formulations are actually the phenomenal truth? Piaget leans in the former direction, and thus he departs from existentialism.

Rogers' "Wisdom of Organic Evidence"

It is well and good for science to lay down its principles of explanation, viewing animistic explanations such as children proffer to be "precausal" or "primitive," but does this change what is taking place? The child does, after all, experience intentionality phenomenally. Who is to say that this experience is not therefore just as vital to the meaning of existence as the so-called scientific laws that presumably are the "real cause" of such experience? Although anthropomorphic experience may be recast in the efficient cause substrate of stimulus-response psychology, this does not mean that the causes that propel the individual through his phenomenal field are being identified. What if teleological considerations are at play? This line of argument takes us deeply into the phenomenological-existentialistic or "cognitive" theoretical sphere, and a foremost spokesman here has been Carl R. Rogers.

Rogers[37] is widely known for his expounding of a phenomenal field construct, which is similar to, although more subjective than, Lewin's life space concept (p. 97). Man's physiological-biological and psychological experience combines to funnel into his organismic experience by way of the phenomenal world. Distinctions of body-mind are thus dropped for all practical purposes since, in essence, man comes to know as much by way of his sensory feelings as he does by way of his conscious symbols. Indeed, says Rogers,[35] there is "a discriminating evaluative physiological organismic response to experience, which may precede the conscious perception of such experience" (p. 507). This organic valuing process is important to the individual. It has a sense of what is worthy and true phenomenally even before rational justification might be given symbolically in words. Hence there is in a sense wisdom within the organic evidence of feeling tones. Extending this, Rogers[38] literally comes to a "naturalistic ethic," for he claims that people all over the world have a common base of organismic valuing, stemming essentially from man's common base in organic evolution. As he summarizes it: "The suggestion is that though modern man no longer trusts religion or science or philosophy nor any system of beliefs to *give* him values, he may find an organismic valuing base within himself which, if he can learn again to be in touch with it, will prove to be an organized, adaptive, and social approach to the perplexing value issues which face all of us" (p. 441).[38]

Hence modern man must be unafraid to be "what he is." He cannot allow science to define him or to control him. To be what Rogers calls the fully functioning person (Maslow's self-actualized individual), the human being must trust to his feelings and to the feelings of others. Science may tell the person that he is under the control of outwardly determined natural laws, but what does he subjectively perceive (phenomenally) if not a sense of personal decision and self-direction? Rogers[36] moved from individual therapy to a concern with group encounter based on this naturalistic ethic. The point of sensitivity training is to make one aware of his personal contributions to the phenomenal reality of others. By turning in on himself and discovering a pattern of natural feelings very similar to others, the individual can drop the façades of social niceties and the masks of social defenses. The person as enacted in overt behavior can become "one" or congruent with the feeling tones he has been ignoring or denying in the past. With everyone in the group 100 per cent in tune with their sincere feelings, a higher level of phenomenal living is achieved.

In one sense Rogers has avoided the dualism of mind-body in his uniting phenomenal field construct; but in another he has brought on a second dualism of the "feeling versus symbolizing" variety. Although clearly a Kantian and having the typical existentialistic-phenomenologistic approach to man, Rogers comes back to a firm basis for ethics in the physical reactions of the body. Material causes (feelings) somehow clue us to what is "best" (final cause) through a patterning of sensations (formal cause). They tell us when we feel this way or that, and no further "reason" —reduction to substrate stimulus-responses, or Freudian fixations—is needed for a more healthy pattern to emerge. If everyone listened to their feelings and behaved genuinely, in time behaviors would seek their level. The bully would acknowledge his hostility and the coward would express his fears in a way never before possible. The result would be a more genuine, sincere, fully functioning life for all.

Boss's "Meaning Disclosing" Dasein

In developing our phenomenological tradition we have made it appear that all theorists in this line are clearly Kantian, that they take on some such "spectacles" notions as Binswanger's world designs, Kelly's constructs, or Piaget's schemata. Actually there are positions that are not this easy to classify within the phenomenological camp. Medard Boss[9] is an excellent case in point. Although both he and Binswanger were stimulated by the philosophy of Heidegger, Boss's interpretation of the latter's philosophy seems closer to accuracy. Heidegger—at least in his later writings—was trying to avoid the separation of Dasein (existence, experience) into the *a priori* (Kantian spectacles) and the *a posteriori* (the resultant existence as phenomenally gleaned). The meanings of Dasein for Binswanger and other classical Kantian views are "endowed" by the world designs that frame in experience. For Boss, on the other hand, Dasein is always "disclosed." Boss[9] liked to speak of it as *luminating* or shining forth, disclosing itself to man's awareness rather than vice versa (p. 39).

This has the practical effect of making Boss's existentialism appear more Lockean in the sense that Dasein is issuing toward awareness in a quasi-input sense. Actually, of course, this is not the intent of Boss's construction. What he is emphasizing here is the completely free and unbiased nature of phenomenal experience. Even a Kantian "category of understanding" is to that extent forming experience. It is pressing on experience something that is not part and parcel of that experience, much in the way that science presses its arbitrary efficient causes onto teleological behavior or Freud presses his infant analogues onto mature behavior. We must not reduce one level to another in a truly phenomenological-existential approach.

This "purity criticism" can be taken back in the history of modern existentialism to Kierkegaard's ridicule of Hegelian logic (see Rychvlak, p. 390).[40] Hegel had concocted a logic that was a brilliant example of how the mind can create "a position." But when Hegel now took this to be "the" position he made himself ridiculous on the fact of things, said Kierkegaard. In like fashion the very heart of existentialistic positions has always been to undermine the pat, the set, the rigid certainties of contrived experiences. What it seeks is "pure" experience, as immediately luminated to the mind's eye. The anthropomorphizing child is therefore not "primitive"; he is entirely "human" (see Piaget, above). Man is teleological in his essence, and he must therefore take responsibility for what he does. Existentialistic catchwords like commitment, engagement, and confrontation flow from this philosophical premise.

The Final Question: What Is Personality?

It should be clear now that the reason it was impossible to write this chapter around "the" personality is because of the interlacing classical issues framed by our opening questions. It does seem that a personality term borrows greatest meaning from the pattern or style of behavior witnessed among people. This would make it predominantly a "formal cause" term. Such styles are usually first identified in terms of the "total" person, so that the personality scheme is likely to begin as a typology. We study individuals and construe an oral or anal personality picture. Then, as surely as anything, the type gradually develops into a trait theory as we begin seeing signs of orality and anality in other and then all people "more or less." This is not to say that trait theories cannot begin as such; it is to suggest that the usual historical pattern has been to move from typologies to trait theories.

It is when we begin to explicate why there are such "individual differences" among people that our other causes come into play. A classical solution here is to fall back on genetic-hereditary explanations, as in the morphogenotype of Sheldon. There is almost no rebuttal to this proclamation that people differ because they are born that way. Enlarging upon this we can say that people differ because they have different needs, or that their needs (instincts, drives, etc.) have been differentially met. We can now extend this to include learned needs, and even Rotterian expectancies that serve to influence behavior in one direction rather than another. The behavioral approach of stimulus-response psychology has been fundamentally opposed to the study of individual differences in behavior. Individual differences are merely differential controls being exerted on the basic organism that itself follows common (basic) laws. Needs are reinforcers when met, and insofar as the term "personality" has any meaning at all, it refers to the habit hierarchy that results when these reinforced behaviors have been fashioned or "shaped" (formal cause) by experience.

The social milieu is extremely important to the behaviorist's formulations, since it must be assumed that the inputs that fashion behavior are prompted by one's culture, social group, and so forth. In the final analysis every behaviorist psychologist is a social psychologist. Cattell's test-based conceptions of personality are no different, for he would think of personality as a set of source characteristics employed by a psychologist to predict how person X behaves in situation Y. This is a fairly general attitude among laboratory psychologists, who have moved away from the more classical personality grand theory formulations to rely increasingly on measurement and methodological test to substantiate a more restricted area of study, which is sometimes called a miniature theory. But the upshot is that individual differences and uniqueness in the study of personality have given ground to the more common formulations of behavior. This "common" behavior is well-captured in efficient cause terms, such as the S-R construct. But what of the style of such behaviors? And how much influence does the individual himself have on this styling of his personal behavior?

Those theorists who reject the more passive, Lockean conceptions of the human condition argue that behavior is also shaped by the individual, who formulates personal constructs (Kelly) or furthers schemata (Piaget) over his lifetime. Man is a potentially higher animal for some (Maslow), and hence we can not find his unique humanity in the reductive common substrate of a lower form of behavior. The phenomenologists and existentialists take this a good deal further, and staunchly defend the thesis that only man can be the measure for man.

When we now focus on that most human of all animals we must of necessity ponder the corollary to our present question: "What is human?" And here it would seem that, by science's own standards, if the world of natural events is not to be anthropomorphized, then the anthrop- is not to be naturalized! Existentialism is most eloquent in this argument, but we must surely see in every attempt to account for human behavior a kind of teleology being espoused; or at least a good deal of theoretical effort put into a substitute for this kind of description. Continuing in this vein, it would seem to us that to be human is to be responsive to final causation.

But how is this to be conceptualized? Is it not unquestionably certain that man responds to his past antecedents? Does not learning fashion his present behavioral predilections? Without denying the fact of antecedent events we must point out that antecedents can be such things as the Adlerian "life plans," which are, when put into effect, done so "for the sake of" their intended goals. But are these not simply mediators, plans themselves put into the so-called human being just as mediating stimuli are programmed into a rat? The final rejoinder here is: "No, not if meanings are bipolar and man as a human animal can reason dialectically. If this is true then that state of 100 per cent control from outside the organism that theorists like Skinner speak about is flatly impossible." Hence the heuristic device that a theorist *must* fall back on at some point if he is to ascribe humanity to man is the dialectic. One cannot program an animal that reasons by opposites!

So, in closing this chapter, it would be our argument that the term "personality" is superfluous without the predicate assumptions of a humanly dialectical intelligence, which can take in meaningful experience, consider alternatives by way of opposite implication in this experience, and then project a plan, hypothesis, goal, aim, intention, purpose—call it what you will!—"for the sake of which" it behaves. This behavioral pattern is not simply the result of control, nor is it even mediated behavior. Rather it is created, conformed to, and premised upon. Here again the reader is under the persuasion—*not* the control—of a neo-Kantian theorist. He can reason to the opposite of the argument now being summarized, and—we firmly believe—come up with an alternative that will best suit him. But if he can do that, he can also "be" a distinct, unique person. Terms used to describe him in these creative efforts are properly thought of as personality concepts.

Bibliography

1. Adler, A., *The Education of Children*, Allen and Unwin, London, 1930.
2. ———, *The Practice and Theory of Individual Psychology*, Littlefield, Adams, & Co., Totowa, N.J., 1968.
3. Allport, G. W., *Personality: A Psychological Interpretation*, Holt, New York, 1937.
4. Arieti, S., *The Intrapsychic Self: Feeling, Cognition and Creativity in Health and Mental Illness*, Basic Books, New York, 1967.
5. ———, "The Psychodynamics of Schizophrenia: A Reconsideration," *Am. J. Psychother.*, *22*:366–381, 1968.
6. Aristotle, "Topics," and "Physics," in Hutchins, R. M. (Ed.), *Great Books of the Western World*, pp. 143–223, pp. 257–355, Encyclopedia Britannica, Chicago, 1952.
7. Bacon, F., "Advancement of Learning," in Robert M. Hutchins (Ed.), *Great Books of the Western World*, pp. 1–101, Encyclopedia Britannica, Chicago, 1952.
8. Binswanger, L., *Being-In-The-World*, Basic Books, New York, 1963.
9. Boss, M., *Psychoanalysis and Daseinsanalysis*, Basic Books, Inc., New York, 1963.
10. Cassirer, E., *The Problem of Knowledge*, Yale University Press, New Haven, 1950.
11. Cattell, R. B., *Personality: A Systematic, Theoretical, and Factual Study*, McGraw-Hill, New York, 1950.
12. Dollard, J., and Miller, N. E., *Personality and Psychotherapy: An Analysis in Terms of Learning, Thinking and Culture*, McGraw-Hill, New York, 1950.
13. Evans, R. I., *B. F. Skinner: The Man and His Ideas*, Dutton, New York, 1968.
14. Freud, S., *The Ego and the Id, and Other Works* in Strachey, J. (Ed.), *Standard Edition*, Vol. 19, Hogarth, London, 1961.
15. ———, *The Origins of Psycho-Analysis, Letters to Wilhelm Fliess, Drafts and Notes: 1887–1902*, Basic Books, New York, 1954.
16. ———, "Pre-Psycho-Analytic Publications and Unpublished Drafts," in Strachey, J. (Ed.), *Standard Edition*, Vol. 1, Hogarth, London, 1966.
17. Goble, F., *The Third Force: The Psychology of Abraham Maslow*, Grossman Publishers, New York, 1970.
18. Hull, C. L., *Principles of Behavior*, Appleton-Century-Crofts, New York, 1943.
19. Husserl, E., *Phenomenology and the Crisis of Philosophy*, Harper & Row, New York, 1965.
20. Jung, C. G., *Civilization in Transition*, Pantheon Books, New York, 1964.
21. ———, *Psychology and Religion: West and East*, Pantheon Books, New York, 1958.
22. Kelly, G. A., *The Psychology of Personal Constructs*, Vols. 1 & 2, Norton, New York, 1955.
23. Koffka, K., *Principles of Gestalt Psychology*, Harcourt, Brace & Co., New York, 1935.
24. Köhler, W., *The Task of Gestalt Psychology*, Princeton University Press, Princeton, 1969.
25. Lewin, K., *A Dynamic Theory of Personality*, McGraw-Hill, New York, 1935.
26. Marrow, A. J., *The Practical Theorist: The Life and Work of Kurt Lewin*, Basic Books, New York, 1969.
27. Maslow, A. H., *Motivation and Personality*, Harper & Row, New York, 1954.
28. ———, *Toward a Psychology of Being*, Van Nostrand, New York, 1962.
29. Morgan, C. D., and Murray, H. A., "A Method for Investigating Phantasies," *Arch. Neurol. & Psychiat.*, *34*:289–306, 1935.
30. Mowrer, O. H., "A Stimulus-Response Analysis of Anxiety and Its Role As a Reinforcing Agent," *Psychol. Rev.*, *46*:553–566, 1939.

31. MURPHY, G., *Personality: A Biosocial Approach to Origins and Structure*, Harper, New York, 1947.
32. MURRAY, H. A., *et al.*, *Explorations in Personality*, Oxford University Press, New York, 1938.
33. PIAGET, J., *The Child's Conception of Physical Causality*, Kegan Paul, London, 1930.
34. ———, *The Origins of Intelligence in Children*, International Universities Press, New York, 1952.
35. ROGERS, C. R., *Client-Centered Therapy*. Houghton Mifflin, Boston, 1951.
36. ———, "The Process of the Basic Encounter Group," in Hart, J. T., and Tomlinson, T. M. (Eds.), *New Directions in Client-Centered Therapy*, pp. 292–313, Houghton Mifflin, Boston, 1970.
37. ———, "A Theory of Therapy, Personality, and Inter-Personal Relationships, as Developed in the Client-Centered Framework," in Koch, S. (Ed.), *Psychology: A Study of a Science*, pp. 184–256, McGraw-Hill, New York, 1959.
38. ———, "Toward a Modern Approach to Values: The Valuing Process in the Mature Person," in Hart, J. T., and Tomlinson, T. M. (Eds.), *New Directions in Client-Centered Therapy*, pp. 430–441, Houghton Mifflin, Boston, 1970.
39. ROTTER, J. B., *Social Learning and Clinical Psychology*, Prentice-Hall, Englewood Cliffs, N.J., 1954.
40. RYCHLAK, J. F., *A Philosophy of Science for Personality Theory*, Houghton Mifflin, Boston, 1968.
41. SHELDON, W. H., with STEVENS, S. S., and TUCKER, W. B., *The Varieties of Human Physique: An Introduction to Constitutional Psychology*, Harper, New York, 1940.
42. SKINNER, B. F., *Beyond Freedom and Dignity*, Alfred A. Knopf, New York, 1971.
43. SULLIVAN, H. S., *The Interpersonal Theory of Psychiatry*, Norton, New York, 1953.
44. TOLMAN, E. C., *Purposive Behavior in Animals and Men*, Appleton-Century-Crofts, New York, 1960.
45. WATSON, J. B., *Behaviorism*, Norton, New York, 1924.
46. ———, "Psychology as the Behaviorist Views It," *Psychol. Rev.*, *20*:158–177, 1913.
47. WERTHEIMER, M., *Productive Thinking*, Harper, New York, 1945.
48. WIENER, N., *The Human Use of Human Beings*, Houghton Mifflin, Boston, 1954.
49. WOLPE, J., *Psychotherapy by Reciprocal Inhibition*, Stanford University Press, Stanford, 1958.

CHAPTER 7

THE FAMILY OF THE PSYCHIATRIC PATIENT

Donald A. Bloch

Introduction

THIS CHAPTER will consider the relationship between a set of events usually thought of as being characteristic of a single individual, his psychiatric status, and events taking place in a natural social group, his family.

Interest in the family *of* the psychiatric patient has blended in recent years with an interest in the family *as* the psychiatric patient. As a therapeutic modality, family therapy originated in child psychiatry, drew important technical skills from play therapy, group therapy, and psychodrama, with more recent additions from the encounter and training group fields. On the theoretical side it has drawn much from psychoanalytic theory, from anthropological and sociological studies of the family and of small groups, and, more recently, from cybernetics, general systems theory, linguistics, and kinesics. While the family field has burgeoned in the last two decades as a therapeutic and social movement, it is conceptually diverse and varied. As to practice, there are no generally accepted standards of certification or accreditation, and there is no national organization of family therapists; teaching in the field goes on in many different settings: medical schools, family institutes, freestanding workshops, in-service training programs of social work and mental health facilities, and graduate programs of departments of psychology.

This diversity can be found at all levels of theorizing, research, clinical conceptions, and therapeutic management. In this survey we will attempt to present the main elements of this array. In doing so we will focus on the following issues:

1. The Concept of the Psychiatrically Relevant Family
2. Categories of Definition of the Relation of the Family to Psychiatric Disorder
    a. Psychiatric disorder as an exogenous stress on the family
    b. Psychiatric disorder as caused by pathogenic relationships in the family

c. The family as a pathogenic culture carrier
d. Psychiatric disorder as an expression of the systems properties of the family
3. Family Characteristics Related to Specific Psychiatric syndromes
4. The Psychiatric Future of the Family

The Psychiatrically Relevant Family

An anchoring concept used in this essay is that of the *psychiatrically relevant family*, a term used to delimit a natural social system, based upon, but not limited to, the nuclear family. The psychiatrically relevant family may be defined along two principal dimensions, structure and function. By the first term we refer to the *structural units of the family* held in any given instance to be of psychiatric interest, for example, the mother-child relationship, the oedipal triangle, the nuclear family, the household, the three-generational family, the kin network. This is essentially a temporo-spatial definition. By function we refer to those aspects of *family interaction or process* considered of psychiatric interest, for example, role definition, power distribution, communication patterning, identity formation, intergenerational relationships, anxiety sources, defensive strategies. Naturally structure and function here, as elsewhere, are interrelated aspects of the same phenomenon.

The concept is necessary because of the wide variation in definitions of the family from culture to culture, from science to science, within any single science, and from one to another psychiatric problem. Thus relevance necessarily expresses the *ad hoc* nature of present psychiatric studies of the family. Target symptoms, syndromes, attitudes, or traits are used as the defining elements in regard to the disorder being considered, as well as to define related characteristics of the family. As Leichter[105] says, ". . . the family unit may shift according to the purpose of analysis." It follows that the psychiatrically relevant family should be thought of as a distinctive social unit that may or may not overlap other defined families. Legal, sociological, and anthropological definitions of the family, among others, are useful as definitional starting points in any culture, but such "families" rarely are congruent with the family of interest to the psychiatric theoretician or practitioner.

The starting place for most definitions is the nuclear family* (often referred to as the isolated nuclear family); that is, biological parents and children, generally considered in Western society to occupy the same household until the departure of the children in young adulthood to establish new families of their own. Although this is what is most often referred to as the family, it is unsuitable as a definition of the psychiatrically relevant family in many instances, since it may not conform to the group actually dealt with by clinicians; it does not take into account the wide variation in family and household composition, even in Western industrial societies; nor does it indicate the extent of that family that is psychologically meaningful to the individual.

A structural delimitation of the boundaries of the family may begin by noting the *temporal* aspects of such a definition. The outer limit of personal psychosocial time probably extends backward for four generations and into the future for three generations. Even for the neonate the family extends backward in time; there is a specific *familial* inheritance that is part of the psychosocial matrix into which the new human is born. Naturally the more remote in time, the less influential persons and events are; death and other dislocations alter or obliterate their idiosyncratic influence, and they are finally perceived as part of the general cultural inheritance. With the development in the individual of a sense of a

* In general terms family is a biopsychosocial system found in all cultures that acts: (1) to reproduce other families (Families reproduce reproductive units, namely families. It is the human biopsychosocial system that uniquely recruits new components by biological means, i.e., sexual reproduction); (2) to nurture and acculturate young humans; and (3) to care for *some* of the biopsychosocial needs of its members.

personal future, the family comes to have a meaning as an extension forward of these unfolding processes. Although the extensive branching network of progenitors and descendants can hardly ever be assessed in detail, nevertheless, it is in this matrix that family themes, myths, identity positions, conflicts, and defensive positions are elaborated and synthesized. In connection with the temporal aspect of the family, its constantly changing character over time must be mentioned. Not only does the family have a past and a future, but also it is eternally flowing into new forms, monitoring and processing the developmental changes of its members and of the systems exterior to it.

Thus the psychiatrically relevant family is historical in nature, developmental and changing in character, multigenerational in structure. In addition, it has varying structural boundaries at any single point in time.

The most common subunit of the family of clinical interest is the marital pair. Without regard to the nature of the presenting problem, the husband and wife are most often apt to be considered as the primary unit of interest. If the group is enlarged, one child may be added, usually the identified patient, or the entire nuclear family may be included. Astute clinicians[161] are sensitized to the defensive exclusion of a particular family member, often one child, from the therapeutic enterprise on the grounds that this person does not need to be involved, or will be damaged by exposure to secret or painful material, and are apt to insist that all members be included.

The issue may be conceptualized differently so as to include as "the family" all members of the household; practically this often indicates who is actually available for therapeutic work. This definition is particularly useful where cultures do not adhere to the nuclear model, and in situations where other persons have been regular and meaningful members of the household (a housekeeper or maiden aunt, for example).

Kin other than the nuclear family are often considered as important and within the bounds of relevance. Most commonly the parents and the siblings of the parents of the nuclear family are included. Many theories of transmission of pathology within families depend on a three-generational model, regarding as critical the necessities imposed on any one generation by the need to mediate and deal with issues originating in previous generations. Ackerman,[1,3] for example, has emphasized the issue.

Other components of the extended kin network may be included as part of the psychiatrically relevant family as well. Bell[17] discusses four aspects of extended kin articulation with nuclear families. In the first two, extended families serve as countervailing forces, or as continuing stimulators of conflict; here the dynamics of intergroup relations are dealt with. In the second dimension the extended families act as screens for the projections of conflicts, or as competing objects of support and indulgence; here the social psychological qualities of the relationships are being dealt with.

Finally there is a kind of conceptual enlargement of the family system that extends the concept of relevance to other systems with which the family interacts and of which it is a part. The important instances of this are network therapy and ecological therapy. In the first instance relevance is extended to include members of the social networks of which the family is a part; friends, extended family, neighbors, and involved professionals may meet as a group with the therapist to consider the psychiatric problems of one member of that network (Speck[162,163] and Attneave).[7]

Other workers are more systems-oriented in their conception of the boundaries of relevance. Auerswald, in this connection, speaks of the ecological approach. Typically such therapists include representatives of all social institutions that are believed to have power or to provide an important maintenance function for the family, such as the school, hospital, court, and welfare department. Thus in these approaches the nuclear family is considered not only as a system in its own right, or as part of an extended family system, but also in relation to its external boundary with other natural social institutions.

Categories of Definition of the Relations of the Family to Psychiatric Disorder

It is possible to classify views of the relations between psychiatric disorder and the family into several distinctively different categories. Each of these categories suggests a differing causal or etiological connection between the two and implies quite different management or therapeutic strategies. The variation between these categories is so great as to suggest profoundly different conceptions of the nature of the phenomenon of psychiatric disorder. Fully coherent and integrated statements of these positions do not exist, nor will exponents of one view hold to it in all circumstances. Depending on the purposes of the moment and on the material at hand, there may be some shifting back and forth between orientations. Nevertheless, the positions are distinctively different; those preferring one view are unlikely to vacate it easily.

Psychiatric Disorder as an Exogenous Stress on the Family

This view of the relation between the family and psychiatric disorder holds that in a particular case there is no substantial connection between the disorder and the family except insofar as it is a misfortune accidentally impinging on that particular context. An analogous situation would be a physical illness that in no way is related to the character structure of the family members or its health practices—a brain tumor, for example. There is taken to be no etiological connection, and the family issues to be dealt with are thought of entirely as reactive to the event.

The ability of the family to cushion the blows of adversity upon its members is an essential capacity. Stresses originating at other systems levels may impinge upon a family as crises that are dealt with successfully. A death adequately mourned, a disaster coped with or endured, a developmental crisis integrated through personal and group growth—these are all instances of the family healing its own members and protecting its own structure in the face of external pressures.

As a man in his seventies lay dying of cancer, he requested a physician nephew to visit him. Barely able to speak, the dying man asked the physician in his presence to inform the wife and only daughter that his condition was hopeless. For the first time they were able to openly weep together and begin the process of leave-taking and mourning. In this instance, the extended family could be used to help heal the nuclear family by facilitating available restitutive processes.

Some of the complexities of this model can be seen when one considers how psychiatric disorder itself is to be understood. Is it a stress upon the family system that requires adapting to; is it to be understood as an adaptation in and of itself to strains originating elsewhere; or is a blend of both views perhaps most appropriate.

Early psychiatric interest in the family assumed the disorder to be accidental, to arise endogenously in the patient, and, in its own turn, to pose problems for the family growing out of the need to manage the ensuing difficulties. The era of modern clinical study of the family began when this view was challenged as being universally applicable.

Psychiatric Disorder as Caused by Pathogenic Relationships in the Family

The bulk of research aimed at discovering the causal relationship between family issues and psychiatric matters has been carried out in terms of a model of linear causality. The underlying assumption is that a sufficiently homogeneous psychiatric entity can be delimited and that specific parental attitudes, traits, or practices can be located, most often in the mother, that cause this. The earliest versions of such studies indicted a blend of physical and social heredity as being pathogenic; for example, Dugdale's study of the Jukes[43] and Goddard's study of the Kallikak Family.[69]

Psychoanalytic theory, at about the same time —that is, at around the turn of the century— added the important dimensions of relationship and developmental vulnerability as being pathogenically critical. Freud's cases of Dora[63] and Little Hans[62] are discussed below. Subsequent psychoanalytic literature has richly developed these themes.

In the 60 years from 1910 to 1970 an extensive literature has developed in this field. Excellent reviews are to be found in Spiegel and Bell[165a] and Aldous and Hill.[6] More recently Walter and Stinnett[181] reviewed the literature of the 1960's on parent-child relationships. They found "a distrust of simplistic explanations concerning the direction of causality in explaining the nature of parent-child relationships." This includes a greater interest in contingent explanations and in qualified generalizations; even the direction of assumed causality is questioned. They quote Kysar[100] as to the possibility that disturbed children may produce impaired functioning in their parents, rather than the other way around. The authors note a shift in the studies reviewed from an earlier almost exclusive concern with mother-child relationships to greater interest in fathers, although they note that the father-son relationship is explored much more than that of fathers and daughters.

They find that boys seem to be more susceptible than girls to parental influence, that "parental warmth is a factor which influences occupational choices among children as well as their academic achievement, leadership, and creative thinking. Poor parent-child relations on the other hand, are related to aggressive, antisocial behavior, and a tendency for children to be involved in disciplinary action. Extreme parental restrictiveness, authoritarianism, and punitiveness, without acceptance, warmth and love, tend to be negatively related to a child's positive self-concept, emotional and social development."

The principal feature of the model being used is that a parental trait or attitude (Levy's[108] maternal overprotection or Bowlby's maternal deprivation are good examples) is damaging because of some direct linear effect it produces on the child. The model is that of medical pathogenicity. The maternal overprotectiveness, for instance, is abstracted from the systems of which it is a part, such as a particular cultural attitude toward women and motherhood, a mode of mate selection, a multigenerational context, a pattern of parental relatedness, the mutual circular reinforcing programming of mother and child, and dealt with as if it existed in isolation and *caused* the pathological response to appear in the child.

If the wrong attitudes, traits, and practices can cause psychological illness in the child, it is reasonable to hope that one can discover these wrong attitudes and, having discovered them, teach parents how to do better. Out of such hopes has grown the broad field of parental attitude and child development research, as well as the influential child study movement in the United States, which, in its political impact and in molding social attitudes, has done much to create the current climate of childrearing practices. While still an active social influence, this movement seems to be reducing its impact. The research yield has decreased considerably as well although the research model is attractive to students, because it lends itself to simple designs and straightforward quantification.

The Family as a Pathogenic Culture Carrier

As might be expected there have been heavy sociological inputs to this field, and investigators and theorists alike have attempted to use the language and research methods of sociology to elucidate the origins of psychiatric disorder in the context of the family. Bales,[9] Bell,[17] Bell and Vogel,[18] Cottrell,[40] Foote and Cottrell,[60] Handel,[82,83] Handel and Hess,[84] Hare, *et al.*,[85] Parsons and Bales,[136] Spiegel,[165] and Spiegel and Klyckhohn[166] may be mentioned as seminal. The field is vast, and it is only possible here to indicate the general direction research and theorizing takes.

As part of the larger Stirling County Study, Cleveland and Longaker,[38] using the research setting of a rural clinic, investigated neurotic patterns of interaction in a single family. Essentially the approach is sociological and consists of an attempt to relate the problems of the individual, "to certain important features of his personality (needs), to behavior expected of him by the society (role) and/or to difficult and conflicting patterns of cultural prescription (value)." At the same time the cultural patterns of childrearing and modes of socialization are considered, with particular emphasis on interpersonal devaluation as a preferred socialization mode incompatible with the needs, role, and value system requirements of the culture. The authors' view is that the family mediates these constraints and that individuals are "in some sense trapped by the conflicting tenets of these two broad paths of life" (p. 171).[38]

The predominant analytic model here is sociological; the family for the most part is seen as a straightforward social transmission conduit, with emphasis on role and value conflict and the use of a learned socialization technique—disparagement in this case—as a way of coping with these conflicts. In this view neurosis in an individual is precipitated as an immediate response to excess stress, such as organic illness, that makes untenable a previous integration of needs, role, and value. In analysis of this sort the family is seen primarily as an acculturator and socializer. The larger society provides a range of options as to how these functions can be accomplished. Presumably a set of constraints, deriving from the nature of man in general and his procreative, nurturant, developmental, characteristics in particular, operate to determine the final course of these events.

It is necessary, of course, to take into account the particular features of each homogeneous subculture. Vassiliou[175] has attempted to deal with the cross-cultural issue by the use of the concept of milieu specificity. In an article on this subject Greek culture and the historical development of the modern Greek family is reviewed, leading to a picture of the contemporary Greek family. "Man-woman relations are characterized by superordination of the man, mutual ambivalence, and ingratiating and contemptuous attitudes of women toward men." There is a great emphasis on "proper" marriage, with clear role ascription. Milieu specificity refers to the particular way in which needs or wishes, presumably panhuman, tend to play themselves out in the specific cultural situation. Attempting to combine the psychoanalytic and sociological points of view, the author uses such terms as penis envy and castration anxiety to indicate these needs and points out how the specific interaction within Greek family life modifies their unfolding.

Miller and Westman[124] describe a mode of analysis of marriage and family that is based, in the beginning, on the partner's identities, subidentities, roles, and social positions. Identity is defined in terms of both self-identity, that is, how the person views himself, and the objective public identity, that is, how he is viewed by others. The sociological term "altercasting" is used to describe the pressures exerted on others to assume a particular subidentity, presumably complementary in nature. A general theory of pathology is developed based on the degree to which there is compatibility between experience and subidentities. Thus the effeminate husband may be expected to show masculine initiative on his job and be subservient at home. A newborn child may act as a stabilizer, or may tip the balance toward instability, because his presence "requires the development of new sub-identities in parents and siblings."

The family as acculturator may induct its members into roles, value systems, and perceptual-cognitive modes that deviate from those of the larger culture. The dominant culture may define these deviant ways as "sick," "criminal," or "defective"; this labeling, in turn, may be regarded by the subordinate culture as political scapegoating. A recent instance may be found in the controversy over the Moynihan Report. This report contended that an important proportion of black families were enmeshed in a particular version of the culture of poverty, growing out of the heritage of slavery days. Moynihan believed that these black families were caught up in a self-perpetuating pattern of female dominance, female economic power, together with male impotence and extrusion from the family. This was related to the systematic attack on the masculine role in the black family in slavery, with the development of a protective matriarchy. Discriminatory social, educational, and employment practices in industrial society were held to undermine further the position of the black male, with the consequence that a vicious generational cycle was established and maintained. Critical to this conception is the notion that the self-image for both sexes is distorted and that family relationships and child-rearing practices are so structured in a hostile white world as to maintain the pattern.

Critics[171] of the report note that there is no such thing as a "homogeneous and perdurable" white family unrelated to class and culture; that even if there were it is questionable if it should serve as a paradigm, or if it "actually possesses the virtues attributed to it. . ." Blackness is minimized as etiological in these critiques and poverty emphasized.

Aside from the substantive merits of these arguments one must note the political importance attached to allegedly scientific assessments of family functioning.

Psychiatric Disorder as an Expression of the Systems Properties of the Family

A promising recent theoretical approach to understanding the family comes from general systems theory. Originally this approach was developed by von Bertalanffy, a biologist, who pioneered the field in the 1940's. Systems approaches consider phenomena as part of networks of circular causality rather than in terms of linear causality. It is an organismic approach where self-regulating mechanisms are the essential building blocks. Thus it is uniquely suited to dealing with living systems. To think of the family in this way is to ask how it can be conceptualized as a set of structures organized so as to maintain patterned integrated functioning of the entire system.

Roy R. Grinker, Sr.,[72] notes that "general systems theory includes concepts of integration and process by which integration is maintained in all open living systems inescapable from their environment. Each system is composed . . . of sub-systems under control and regulation within specified gradients. Information exchange occurs among sub-systems, and between systems, at their interfaces by means of reverberating circular transactions" (p. 135). These words felicitously introduce the notion of information exchange.

All energy transformations can be conceptualized as information exchanges as well and understood in terms of the thermodynamic laws. The family, in this sense, is an open system in that there are transactions across its boundaries of information, or energy. The family as a system makes up a set of interfaces with other social institutions and can be characterized then as to the nature of these interfaces and the transactions taking place across them. Among these social institutions would be other parts of the larger kin network, other families, educational systems, medical care systems, and mental health intervention systems.

Internal subsystems of the family include generations, the parental pair, and sex splits, among others. In all of these instances the same principles of interface and information exchange across it apply.

Perhaps the keystone systems concept used by family therapists is that of homeostasis, the tendency of the organism or system to dynamically and actively re-equilibrate itself. Homeostasis, or homeodynamics as Ackerman calls it, has been a conceptual tool of considerable importance to the clinician in the field of family studies. It serves as a reliable guide to understanding mechanisms resisting therapeutic change. Simple examples include roles or identity positions that must be filled in the family. Speck[162] notes that when a suicidal patient gives up his self-destructive ideation, his role in the family will be replaced by the appearance of a depression in another family member. It is as if the equilibrium in the family can only be maintained by the presence of some expression of this affect. Perry, *et al.*,[139] observed that families in a disaster acted as a unit to maintain one parent in the strong protective role and suffered severe anxiety when this no longer could be accomplished. Rashkis[145] noted the importance of depression in maintaining family homeostasis, and numerous other observers, Jackson[92] most importantly, have spoken on this issue.

It is possible to conceptualize new family formation in systems terms in part as an effort by each partner to reproduce in dynamic terms the configurations and interchanges experienced in the family of origin. Tomas[173] has called attention to the importance of sibling position in dynamic terms as a factor in mate selection. Napier[133] speaks of this process as cross-generational complementarity. He describes the matrix of complementary expectations to be filled by the mate, matrices that have developed in the context of the homeostatic requirements of the family of origin. Finally Erlich and Bloch[49] describe therapy in terms of the transaction of two family systems, that of the patient and therapist, as each struggles to integrate old themes (parameters) in the new therapeutic relationship. They point out the similarity, in systems terms, between the new pair formation of therapist and patient and mate selection.

Family Characteristics Related to Specific Psychiatric Syndromes

Investigators attempting to relate dimensions of family life to pathology have approached the issue in ways that reflect their personal biases, professional technologies, as well as the special nature of the problem being studied. Clinicians in recent years have considered almost all of the well-defined psychiatric entities from the family point of view. In a number of instances rigorous, quantified research has also been conducted.

Schizophrenia

Understandably families with schizophrenic members have attracted great interest on the part of both clinicians and researchers. In the same spirit that has led other investigators to search for a unitary cause for this "disease," family researchers have tried to define the uniquely pathogenic qualities of the family with a schizophrenic member. While the hope of finding the specific family etiology of the disorder seems clearly illusory, there have been a number of significant contributions to a better understanding of the syndrome.

Overall these studies have become increasingly sophisticated, developing more meaningful typologies of families with schizophrenic members based on the sex, premorbid history, quality of thought disorder, and social class of the index patient.

One group of studies may be broadly characterized as having a psychoanalytic orientation. Alanen,[6] for example, offers a developmental conception beginning with "the defective initiation into important object relations, above all that with the mother." Davis[42] attempts to get at the oedipal contribution to the etiology of schizophrenia in males. In acute schizophrenic breakdown he claims that anxiety arises out of the relationship with the mother and "the illness begins when a change in this relationship increases the anxiety. The frustration of incestuous wishes contributes to this anxiety." This is associated with "a failure of identification with the father, a poor relationship between mother and father. . . ."

The most important studies of this genre are those of the Lidz group at Yale,[57,58,111,112,114] which reports on investigations of middle-class, structurally intact families with schizophrenic young adult members. Middle-class families were chosen for study so as to avoid the effects of social disorganization under economic stress. There were many clinical contacts with the families and with kin, friends, servants, and so on. In addition to clinical interviews at home and in the hospital, a complete battery of psychological tests was administered. To summarize their findings, all families "were severely disturbed, distorted by conflict, and beset by role uncertainties by family members other than the patient."

Patients' mothers appear severely disturbed, often bordering on the psychotic, but without a single predominant personality type, rather showing a wide range of disturbance. The authors underscore the importance of the fact that these women were "paired with husbands who would either acquiesce to any irrational and bizarre notions . . . or who would constantly battle with and undermine an already anxious and insecure mother."

The fathers in half the families were described as paranoid, "paired with a submissive acquiescing spouse. . . ." Again there is no "characteristic type of disturbed father," but a general inability to play the paternal role successfully for a variety of reasons. The suggestion is that if they had married more supportive wives, they might have done better as fathers and husbands. The Lidz group's findings seem comparable with those of Cheek,[35,36] who says, "The profile of the father of the schizophrenic differed less from that of the father of the normal than the profile of the mother of the schizophrenic differed from the profile of the mother of the normal." Cheek agrees with Parsons[136] to the effect that "schizophrenia may be related to a lesser differentiation of parental sex roles in the family." However, he notes, "Our study does not tell us whether or not this distortion of parental roles has been a cause of, or a reaction to, the schizophrenia."

In the Lidz studies two types of families are described. Schismatic families are "beset by chronic strife and controversy, primarily between the parents. . . . The parents undermine each other's worth, despising each other as man or woman, depriving each other of much needed support. . . ." This is particularly important in that it raises severe identity problems for the children.

The second type of family they describe is the skewed family, which may be peaceful on the outside, but where the peace is maintained because the parents have "overtly or covertly reached a compromise concerning a serious personality defect in one or the other." (Wynne's[192] pseudomutuality and Laing's[102] mystification are related concepts.)

Considering the difference between families in which a male rather than a female becomes schizophrenic, they note, among other things, that same sex siblings seem to be more disturbed than opposite sex siblings. Three issues appear to be critical: "(1) the faulty model for identification provided by the parent of the same sex as the patient; (2) the impediments to proper resolution of the patient's oedipal attachments, created by the disturbed parental interaction; and (3) the failure of the parents to maintain proper generation boundaries between themselves and the patient, either by being seductive with a child or by being more like a rivalrous sibling than a parent, with both patterns sometimes occurring in the same family."

Considering the influence of parental homosexual tendencies, Lidz and Lidz[111] points out, "The de-erotization of the child-parent relationship is one of the cardinal functions of the family," and in a series of cases they illustrate the failure to accomplish this in the mother-daughter relationship. In these families this failure increased the insecurities of the child, who focused on a complex and perplexing tie to the mother as a mode of maintaining security.

Turning to the central problem in schizophrenia, the nature of the thought disorder, they say, "We are following the hypothesis that the schizophrenic patient escapes from an untenable world in which he is powerless to cope with insoluble conflicts by the device of imaginatively distorting his symbolization of reality."[112]

The mother of a schizophrenic son is seen as needing the son to complete her frustrated life. She demands the impossible from him, feels only he understands her, and excuses all difficulties as being the fault of some other person or institution. A hopeless enmeshment is created for the child, nor is there any exit from this because of the qualities of the father.

The Lidz group speaks of folie à famille where the life of the family centers around the distorted beliefs of one parent, usually the father, and lives according to a social pattern quite different from the larger social context. The parents hold a rigid and fantastic conception of the environment and the family into which the children must fit without regard to reality or their own needs. These parents are described as being "impervious," unable to hear the child's emotional needs, and, in general, much given to projective identification, the attribution to the child of the parental needs, principally in the service of maintaining the parent's picture of himself and of simplifying a complex environment.

Lyman Wynne and his co-workers[116,155,192,193] set themselves the formidable task of trying "to develop a psychodynamic interpretation of schizophrenia, that takes into conceptual account the social organization of the family as a whole." They build their theory on the need of all humans to be related and on the striving to develop a sense of personal identity.[48,60] The attempt to deal with these isues may lead in three directions: mutuality, nonmutuality, and pseudomutuality. The latter they regard as occurring widely and consisting of a "predominant absorption in fitting together, at the expense of the differentiation of the identities of the persons in the relation." The essential difficulty in differentiation of identity characterizing the schizophrenic is put this way: "There is a characteristic dilemma: divergences are perceived as leading to disruption of the relationship . . . but if divergence is avoided growth of the relation is impossible." Under these circumstances noncomplementarity, that is, individuation and separation, must be avoided at all costs since it is the most severe threat in these families. Thus, in terms of the development of the schizophrenic modes of life, important motivational systems must be excluded in the spirit of maintaining the appearance of mutuality.

For a variety of reasons, including the painful experience of separation, there is held to be a strong drive toward maintaining a *sense* of relation. The inevitable discontinuity of relatedness in situations of even minor conflict or disagreement cannot be tolerated. Genuine mutuality leading to the growth of the relationship through the exploration and enlargement of alternative modes of interacting cannot be achieved because of the underlying rigidity and fragility. The authors emphasize the role rigidities of these families as well as the absence of imaginative play or affective release in the interactions of family members. Noncomplementarity is seen as being the most serious threat.

Shared mechanisms operate within the family to maintain this pseudomutuality, including the nonrecognition or delusional reinterpretation of any deviations from the rigid family role structure. A consequence is that differentiation and individuation are severely impaired and that it is extremely difficult for the members of these families to distinguish the family boundary. The term used for this by Wynne is "the rubber fence," an "unstable but continuous boundary, with no recognizable openings, surrounding the schizophrenic family system, (which) stretches to include that which can be interpreted as complementary and contracts to extrude that which is interpreted as noncomplementary."

According to Wynne and his co-workers, "The fragmentation of experience, the identity diffusion, the disturbed modes of perception and communication . . . are to a significant extent derived, by processes of internalization, from characteristics of the family's social organization." In this regard it is possible to see the similarities between the Lidz and Wynne approaches to these issues. Assisting in the formulation of these theories is the subjective experience of the therapist in direct work with these families (Schaffer, *et al.*),[155] an experience that "tended informally to be reflected in the use of words such as maddening, enraging, bewildering, and exhausting. . . ." The sense is that a pattern is established of idiosyncratic meanings within the family that are destructive of meaning to those unfamiliar with the premises and logical interrelationships.

Wynne and Singer[193] attempt to relate schizophrenic thinking disorders to family experience in more detail by a more refined assessment of the nature of the thinking disorder. Clinical observations were "concerned with the styles of focusing attention and communicating in these families, which may have a disorganizing, complementary effect, or may provide models which are internalized in the same form into the ego structure of the growing offspring."

Wynne and Singer's study used parents of childhood schizophrenics, adult schizophrenics, acting-out children, and withdrawn, neurotic children. When the Rorschach and TAT tests of the parental pair were compared, blind ranking of these tests showed high correlation with the diagnosis of the child. Moreover, it was possible to relate the distinctions between the groups to a cognitive style and experiential mode that might be understandably related to the child's pathology. Thus on the TAT parents of autistic children "have clear percepts; depict people, events, feelings, consequences clearly, compared to parents of adult schizophrenics," who "have people, events, feelings, consequences remaining global, abstract, overly general, attention appears fragmented or amorphous." This would support other findings (Goldfarb[71] and Meyers and Goldfarb,[123] as well as those of Reiss on cognition),[147,148,149,150] distinguishing families of autistic children as being comparatively less pathogenic.

Laing[104] has advanced the view that pathology lies, as it were, in the eye of the beholder. As he sees it, this is true for individual psychopathology, and even more so for the concept of family pathology: "It extends the unintelligibility of individual behavior to the unintelligibility of the group."[101,102,103,104] With various co-workers Laing has constructed an experiential approach to individual and family psychodynamics. Dyads, triads, and more complex networks, of which each individual is a part, are related by mutual percepts and in layers of further complexity, the perception of the perception of the perception. A unique reality is constructed thereby, which is true for the particular individual and family; this reality is to be discovered in the therapeutic work, especially the processes whereby contradictions and illogicalities are reconciled.

One such process of key importance in Laing's thinking is *mystification*. Laing[103] emphasizes that mystification is both an *act* and a *state* and points out that the person being mystified may not know it or feel it. It is a process operating within the family (and elsewhere for that matter) in which the person is systematically forced to deny data from one realm in the interest of maintaining a percept or relationship. Relating mystification to the double-bind concept Laing[103] notes, "The double-bind would appear necessarily to be mystifying, but mystification need not be a complete double-bind" (p. 353).

An extensive case report embodying this point of view can be found in Schatzman.[156] In this paper the Schreber case, which played a key part in the development of Freudian psychoanalytic theories of paranoia, is reviewed, with emphasis on the use of mystification to cover over deeply conflicted issues in the relationship between Schreber and his father. Schatzman concludes that Schreber was indeed persecuted by his father and mystified in addition, by being told the persecution was loving. His delusions then represented an effort to make sense out of this total configuration.

Bateson, Haley, Jackson, and Weakland[10,11,12,77,78,79,91,186] aim to develop a communicational theory of schizophrenia. The cornerstone propositions are that learning occurs in a context that has formal characteristics and that this context always occurs within a wider context called a metacontext, indeed, within a series of these.

Bateson[11] says, "Even more shocking is the fact that there may be infinite regress of such relevant contexts." There is a relationship between context and metacontext that may be congruent or incongruent. Metamessages tell the receiver what the relationship is between messages of different levels, with the possibility that "in human relations another sort of complexity may be generated; e.g., messages may be emitted forbidding the subject to make the meta-connection."

Bateson, Jackson, *et al.*,[12] discuss the relationship between family homeostasis and schizophrenia in communicational terms. The essential feature of this relationship is the *double-bind*, described as a repeated experience in which a person (called a victim) is subjected to messages containing a primary negative injunction, such as "do not do so and so, or I will punish you," together with a secondary injunction at a more abstract level, which conflicts with the first. It is essential that the "victim" also be prevented from escaping from the field. The authors suggest that an individual repeatedly exposed to this mode of communication will suffer a "breakdown in his . . . ability to discriminate between logical types, . . . in a double-bind situation, a shift to metaphorical statement brings safety." This use of metaphor often corresponds to what is called psychotic behavior.

Specifically, in regard to the family situation in which these events occur, the authors suggest: "1. The child's very existence has a special meaning to the mother, which arouses her anxiety . . . when she is in danger of intimate contact with the child. 2. The mother's way of denying this anxiety and hostility is to express overt loving behavior. 3. There is no strong insightful person available to the child to support it in the face of the contradictions." The emphasis throughout is on the child's being forbidden to discriminate the metacommunicative messages, which are inherently contradictory. As the authors note, the child "must deceive himself about his own internal state in order to support mother in her deception."

Analyzing the communicative pattern of schizophrenics in this context, Haley[78] emphasizes that various kinds of *disqualification* occur in messages; for example, that the person is not really the source of the message, that the words are action and not really the message, that the receiver is not really being talked to (this emphasizes the self-negating quality of paranoid statements). The emphasis here is on the schizophrenic's effort to extricate himself from the family communicational pattern and on the conceptualization of the schizophrenic symptoms in those terms. "It can be argued that psychotic behavior is a sequence of messages which infringe a set of prohibitions but which are qualified as not infringing them." Haley emphasizes the degree to which the disqualification process avoids the risk of accepting being governed by other people and that this, in turn, destroys relationships.

Reiss[147,148,149,150] performed a series of experiments aimed at distinguishing the cognitive styles of families with normals, character disorders, and schizophrenics. Five families in each group were tested by a series of experiments; a chief feature of which was to distinguish the degree to which intrafamily cues and ideas were attended to rather than those coming from outside the family. Schizophrenics were distinguished for their attention to intrafamilial cues. This finding is commensurate with the unique and idiosyncratic information environment within the family containing a schizophrenic person.

In an effort to distinguish efforts due to socioeconomic status, the relation Rorschach,[116] in which several family members are asked to compose a response to the Rorschach card, was used to study 17 black schizophrenics, 11 white schizophrenics, and 11 lower-class black control families.[15] The transcripts of the tests done in the homes were studied by raters who did not know the families. The results "indicate the communication and interaction patterns of lower-class families with a schizophrenic differ in a significant way from these patterns in families whose class background is similar, but which do not contain a schizophrenic." Interestingly a distinction could *not* be made in regard to the white and black schizophrenic groups.

Many studies[13,32,44,110,157,174] have investigated the relation between the spouses of schizophrenic persons. Becker,[14] for example, found that the husbands of schizophrenic wives generally came from materially and emotionally deprived backgrounds, with duty-bound, close ties to harsh, demanding fathers and affectionate mothers. It often seemed as if these husbands did not find it possible to separate from their families of origin in order to make strong ties with their wives. This led to frequent neglect, increasing with the birth of new children, which often, in turn, led to schizophrenic breakdown in the wives.

DuPont and Grunebaum[44] point out that paranoid women tend to marry men who are walled off, passive, and unable to express hostile or erotic feelings. In some instances the wife's psychosis is precipitated by the husband's reduction in sexual activity. They point out that the husband's "eagerness to reunite with his wife and to exclude others was reflected in his failure to support her therapeutic alliance with her psychiatrist and his failure to consider divorce."

Phobias

The case of "little Hans,"[62] first published by Freud in 1909, presaged later psychoanalytic theories of pathology in the family as well as techniques of family therapy. The identified patient, a five-year-old boy, is treated through the agency of his father, of whom Freud says, "the special knowledge, by means of which he was able to interpret the remarks made by his five-year-old son, was indispensable . . . and (treatment was possible) . . . because the authority of a father and a hand of a physician were united in a single person. . . ."

Strean[168] discusses this case as an early example of family diagnosis and treatment. In the same spirit the celebrated case of Dora[63] can be read as an instance of symptoms appearing in one member of the family as a result of events occurring elsewhere in the family system. Here the identified patient is an 18-year-old girl, diagnosed as hysterical, with a nervous cough, aphonia, easy fatigability, and suicide threats. The precipitating event of her illness was her father's involvement with the wife of a business friend; the friend had on one occasion made a sexual approach to Dora. Freud uses both of these cases to demonstrate the unfolding of psychosexual development; they are equally useful as early statements about the relationship between family interaction and psychopathology.

Repeatedly anxious, phobic patients are viewed as expressing affect present in other members of the family, often a husband or wife. As Fry[65] has observed, "The spouses of the patients in this group . . . are typically negativistic, anxious, compulsive, and show strong withdrawal tendencies." Upon careful study the spouses reveal "a history of symptoms closely resembling, if not identical to, the symptoms of the patient. Usually they are reluctant to reveal this history." It is as if the symptoms of one (more severe) make it unnecessary to face the symptoms of the other. Frequent collusive involvement is seen. For example, the spouse will volunteer to stay with the anxious patient, even though the patient does not need or wish it.

Fry also notes, as does Carek[33] and Miller,[125] that symptoms seem to break out at a point when the spouse has some change—usually for the better—in his life; there is a constant dual supervision and control, and the couple are frequently held together by the symptoms.

Johnson, *et al.*,[96] writing in 1941, described a pattern of generational transmission of school phobias, in which the mothers of children express "an inadequately resolved dependency relationship to their mothers, with intense repressed resentments." In turn, their own children are pulled into a pattern of dependency, which is then shifted to the school, because of the differing patterns of authority there. "When the teacher, as a more consistent disciplinarian, frustrates the child, she arouses his rage." The child's rage toward the mother is inhibited and the teacher becomes the phobic object. In explanation a three-generational structure is invoked as well as comments about the specificity of interaction in the child's involvement with the school. The minimum systems model seems to include the child at an interface between his generational family and the school.

Ackerman[3] reports a case of phobia in a four-and-one-half-year-old boy. He relates this to the mother's inability to establish close contact with the boy, her guilt, domination, inconsistency, overindulgence, and eventually his "frightened retreat from her oral, devouring attitude" (p. 119). The concept of homeostasis, as used by Ackerman, really concerns much more the homeostatic equilibrium of the self, rather than of the family.

Psychosomatic Disorders

Some psychosomatic illness[3,27,32,126] seem clearly connected to aspects of family life. The connection has been made, for example, in certain cases of severe childhood asthma. Purcel, *et al.*,[141] note, "It is clear that substantial numbers of asthmatic children admitted to hospitals lose their symptoms rather rapidly, while others show relatively little symptom change in response to institutionalization and separation from home." An effort was made to distinguish the two groups. An imaginative procedure was used, whereby the child stayed physically in his own home although all parental and sibling figures were removed, and a substitute parent was provided. The carefully controlled study showed that children with a history of emotionally precipitated asthmatic attacks were improved by removing the people from the house—even though the allergenic situation was unchanged.

Meissner,[122] reviewing the problem of the connection between family processes and psychosomatic illnesses, notes that it is necessary to suggest "a link between emotional dysfunction related to psychosomatic illness and discernible patterns of family dynamics." The issue is not whether there is a connection between emotional stress and somatic disorder, rather whether there is some unique connection between family patterns and such disorder. Although research in this area has been relatively unproductive thus far, it is our conviction that several modes of investigation will fuse in the near future to improve our understanding of these relationships. Specifically it seems likely that kinesthesiologic studies of family interaction from the viewpoint of operant conditioning, using highly responsive physiological measures, will reveal the subtle modes whereby families teach somatic responses to their members.

Titchener,[172] in a case study of a family in which one son developed ulcerative colitis, suggests, "The rigid and confining patterns of object relations were not only formed in the binary mother-child symbiosis, but were conditioned by the multi-dimensional matrix of object relations constituting the field in which his personality developed." Jackson and Yalom[94] describe such families as being very restrictive with limited interaction, affectivity, and contact outside the family, and with a communicative restrictiveness through the generations.

Learning Difficulties

Grunebaum, *et al.*,[76] studied the families of boys with neurotic learning inhibitions. They identified patterns of interaction between the parents that prevented the boys from adequately identifying with their fathers or from perceiving their fathers to be competent or worthy.

Similarly Miller and Westman[125] noting that the mothers in these families exercise the real power, "postulate that parents and children resist change in the reading disability, because it contributes to the family's survival." The main line of analysis concerns the support for an identity based on the parent's projections. In addition, "The symptom and subidentity act as governors on aggression between the parents." When the symptom changes, there is increasing disturbance elsewhere in the family—evidence of the homeostatic function of the behavior.

Delinquency

Ferreira[52] explores the concept of the double-bind as it relates to delinquency and speaks specifically of "the split double-bind," where "the victim is caught in a sort of *bipolar message* in which A emanates from father, for instance, and B (a message about Message A) from mother." He notes the special importance for delinquency of *punishment* as a consequence of the efforts of the child victim to avoid the double-bind; each half of the parental pair threatens a series of punishments as the consequence for not obeying his half of the self-canceling message.

Addictive States

Ewing,[50] using concurrent group therapy for wives and husbands, notes that "marriage to an alcoholic is no accident. . . ." The wife seems to need a weak, dependent husband and sabotages efforts to improve his strength or accept her own dependent needs. One case is described in which there is a constant interplay between the husband's drinking bouts and the wife's depression, which requires hospitalization.

Homosexuality

Bieber, *et al.*,[19] conducted a study of 206 male psychoanalytic patients, of whom 106 were homosexuals. Brown's[26] findings in his study of Air Force personnel, in regard to mother-son and father-son relationships, are confirmed here as well. The authors consider, too, the "triangular system," involving both parents and the son. The basic mode of analysis consists of "constructing classifications of mother-son, father-son, and interparental power-affect parameters." They find, "The classical homosexual triangular pattern is one where the mother is CBI (Close Binding Intimate) and is dominant and minimizing towards a husband who is a detached father, particularly a hostile-detached one. . . . Chances appear to be high that any son exposed to this parental combination will become homosexual, or develop severe homosexual problems" (p.172).[19]

Depression and Suicide

Various writers have related suicide and depression to characteristics of family systems, noting their transactional nature. Thus Goldberg and Mudd[70] develop a categorization based on the notion that "in married persons suicidal behavior may be regarded as a transaction between the suicidal individual and his spouse." Essentially the threats and plans are seen as a move to maintain domination and control of the partner.

Speck[162] is concerned with the homeostatic equilibrium of the family that is maintained by the suicidal behavior and speaks of families who seem to hold to the dictum, "better mad than bad," citing case material where improvement in one member, the suicidal patient in this instance, produces an increase in seclusiveness and withdrawal in other family members.

Whitis,[191] speaking of a child's suicide, observed, "The suicide served to spotlight the apartness each member of this family felt in this inability to communicate with one another in a helpful or constructive way."

([The Psychiatric Future of the Family

Any assessment of the relation between psychiatric status and the family must recognize the influence on this issue of the rapid process of social change presently characteristic of the family in the Western world. As standards of normative behavior change, diversify, and broaden, the reference points used for judgments about deviation change. The social vantage point, i.e., bias, from which a psychiatric judge views phenomena determines the meaning of the behaviors being assessed. Also the very processes of rapid social change themselves, operating on families over time, have an effect equivalent to exposure to markedly different cultural sets and expectations. There is little reason to expect these factors to abate in the future; more likely their effects will intensify.

As a consequence it becomes even more difficult for the psychiatrist confidently to assume that he knows the directions family life should take; yet it is not possible for him to abandon or deny his own value systems since they are the points of reference he must use in order to define behavior as psychiatrically relevant. This is especially so when he is dealing with cultures other than his own. The white view of the black family is notably deficient in this regard, as is that of the male psychiatrist judging the "proper" role for women.

The psychiatric future of the family cannot therefore be separated in any way from the broad processes at work to create change in society at large. A few of the more important of these processes may be noted. First, there is the rapid change in the status of women vis-à-vis men. Associated with this are profound revisions of the ideal and normative expectations of both sexes with regard to such family functions as childrearing, domestic work, and sexual intimacy. Patterns of pair bonding are changing rapidly; premarital sexual liaisons are openly accepted in many quarters, and the age of permissible sexuality is lowered on one hand and raised on the other. Adolescents and youth are expected to have active sexual lives prior to marriage, while the sexual needs and capabilities of the aged are better understood and accepted.

The contractual patterns of family life are changing, again in the direction of greater permissible variation. Illegitimacy as a status is becoming increasingly obsolete, the single-parent family more common. Not only is divorce ubiquitous, less opprobrious, and easier to secure, but also unconventional household and childrearing arrangements such as group or homosexual marriages are becoming steadily more common and more frequently legitimized. Thus participation in a nonstandard family arrangement is no longer prima facie evidence of psychiatric disorder.

Demographic patterns are changing rapidly, again with profound impact on the normative expectations for the family. Deliberately childless couples are more common, and the birth rate has recently dropped (1972) to the lowest point in U.S. history, a rate approximating that needed to achieve the goal of zero population growth. Abortion laws are increasingly liberalized in the direction of abortion on demand; these, together with changed contraceptive technology and social attitudes, have markedly reduced the number of children available for adoption, increasing the movement into various forms of single-parent adoptions, cross-race and cross-religion adoptions, and the adoptions of children previously unplaceable by virtue of age or physical handicap.

Another demographic change of consequence for the family is the increased number of aged persons in the population. The generational splitting, age-grading, and emphasis on youth that has characterized much of recent U.S. social history has, when combined with patterns of increasing longevity, produced many serious problems of social isolation in the aged. The primary breakdown in ties to kin networks results in an epidemic psychiatric public health problem. Differential mortality rates between men and women accentuate this; a pattern of marriage of women to men two or more years senior, when coupled with this differential longevity, statistically guarantees an increasing pool of older, isolated, sexually active single women.

There are no base lines established for rates of social experimentation in new modes of intimate relatedness. Thus it is only an impression that these are proliferating at a great rate; the impression may in part be due to greater acceptance of public discussion of issues formerly barred. Constantine and Constantine,[39] studying group and multilateral marriages, have been impressed with the large number of persons considering establishing such relationships in comparison with the small number actually doing so. Despite this it seems certain that the family of the future, and thus the psychiatrically relevant family of the future, will refer less often to the modal nuclear family of the past and with increasing frequency to a nonstandard pattern of relatedness. Perhaps most important in terms of the generational model that family psychiatry has found to be so helpful, the natural social system into which a person is born, the natural social system in which he lives his adult life and procreates, and the natural social system in which his children (if any) live their lives, all will be different.

Bibliography

1. ACKERMAN, N. W., "Adolescent Problems: A Symptom of Family Disorder," *Fam. Process, 1*:202–213, 1962.
2. ——— (Ed.), *Family Therapy in Transition*, Little, Brown, Boston, 1970.
3. ———, *Psychodynamics of Family Life, Diagnosis and Treatment in Family Relationships*, Basic Books, New York, 1958.
4. ———, BEATMAN, F. L., and SHERMAN, S. (Eds.), *Exploring the Base for Family Therapy*, Family Service Association of America, New York, 1961.
5. ALANEN, Y., "Some Thoughts on Schizophrenia and Ego Development in the Light of Family Investigations," *Arch. Gen. Psychiat., 3*:650–656, 1960.
6. ALDOUS, J., and HILL, R., *International Bibliography of Research in Marriage and the Family 1900–1964*, University of Minneapolis Press, Minneapolis, 1967.
7. ATTNEAVE, C. L., "Therapy in Tribal Settings and Urban Network Intervention," *Fam. Process, 8*:192–210, 1969.
8. BAILEY, M. B., "Alcoholism and Marriage," *Quart. J. Stud. Alcohol., 22*:81–97, 1961.

9. BALES, R. F., *Interaction Process Analysis*, Addison-Wesley, Cambridge, 1950.
10. BATESON, G., "The Biosocial Integration of Behavior in the Schizophrenic Family," in Ackerman, N. W., Beatman, F. L., and Sherman, S. (Eds.), *Exploring the Base for Family Therapy*, Family Service Association of America, New York, 1961.
11. ———, "Minimal Requirements for a Theory of Schizophrenia," *Arch. Gen. Psychiat.*, *2*:477–491, 1960.
12. BATESON, G., JACKSON, D. D., HALEY, J., and WEAKLAND, J., "Toward a Theory of Schizophrenia," *Behav. Sc.*, *1*:251–264, 1956.
13. BAXTER, J. C., "Family Relations and Variables in Schizophrenia," in Cohen, I. M. (Ed.), *Psychiatric Research Report No. 20*, American Psychiatric Association, Washington, D.C., 1966.
14. BECKER, J., "'Good Premorbid' Schizophrenic Wives and Their Husbands," *Fam. Process*, *2*:34–51, 1963.
15. BEHRENS, M., ROSENTHAL, A. J., and CHODOFF, P., "Communication in Lower-Class Families of Schizophrenics," *Arch. Gen. Psychiat.*, *18*:689–696, 1968.
16. BELL, J., *The Family in the Hospital*, National Institute of Mental Health, Chevy Chase, Md., April 1969.
17. BELL, N., "Extended Family Relations of Disturbed and Well Families," *Fam. Process*, *1*:175–193, 1962.
18. ———, and VOGEL, E. S. (Eds.), *A Modern Introduction to the Family*, Free Press, Glencoe, 1961.
19. BIEBER, I., *et al.* (Eds.), *Homosexuality: A Psychoanalytic Study*, Basic Books, New York, 1962.
20. BILLINGSLEY, A., *Black Families in White America*, Prentice-Hall, Englewood Cliffs, N.J., 1968.
21. BIRDWHISTELL, R. L., "An Approach to Communication," *Fam. Process*, *1*:194–201, 1962.
22. BOSZORMENYI-NAGY, I., and FRAMO, J. L. (Eds.), *Intensive Family Therapy: Theoretical and Practical Aspects*, Harper & Row, New York, 1965.
23. BOTT, E., *Family and Social Network*, Tavistock, London, 1957.
24. BOWEN, M., "A Family Concept of Schizophrenia," in Jackson, D. D. (Ed.), *The Etiology of Schizophrenia*, Basic Books, New York, 1960.
25. BRODEY, W., "Image, Object and Narcissistic Relationships," *Am. J. Orthopsychiat.*, *31*:69–73, 1961.
26. BROWN, D. G., "Homosexuality and Family Dynamics," *Bull. Menninger Clin.*, *27*:227–232, 1963.
27. BRUCH, H., "Psychological Aspects of Obesity," *Psychiatry*, *10*:373, 1947.
28. BUCK, C. W., and LADD, K. L., "Psychoneurosis in Marital Partners," *Br. J. Psychiat.*, *3*:587–590, 1965.
29. BUELL, B., *et al.*, *Community Planning for Human Services*, Columbia University Press, New York, 1952.
30. BURGESS, E. W., "The Family and Sociological Research," *Soc. Forces*, *26*:1, 1947.
31. ———, "The Family as a Unity of Interacting Personalities," *The Family*, *7*:3, 1926.
32. BURSTEN, B., "Family Dynamics, The Sick Role, and Medical Hospital Admissions," *Fam. Process*, *4*:206–216, 1965.
33. CAREK, D. J., HENDRICKSON, W. J., and HOLMES, D. J., "Delinquency Addiction in Parents," *Arch. Gen. Psychiat.*, *4*:357–362, 1961.
34. CHANCE, E., "Measuring the Potential Interplay of Forces within the Family during Treatment," *Child Development*, *26*:241, 1955.
35. CHEEK, F., "Family Interaction Patterns and Convalescent Adjustment of the Schizophrenic," *Arch. Gen. Psychiat.*, *13*:138–147, 1965.
36. ———, "The Father of the Schizophrenic: The Function of a Peripheral Role," *Arch. Gen. Psychiat.*, *13*:336–345, 1965.
37. ———, PERATZ, D., and ZUBIN, J., *Social Factors in Prognosis of Mental Disorders*, Project No. 387, Columbia University Council for Research in the Social Sciences, New York.
38. CLEVELAND, E. J., and LONGAKER, W. D., "Neurotic Patterns in the Family," in Leighton, A. H., Clausen, J. A., and Wilson, R. N. (Eds.), *Explorations in Social Psychiatry*, Basic Books, New York, 1957.
39. CONSTANTINE, L. L., and CONSTANTINE, J. M., "Group and Multilateral Marriage: Definitional Notes, Glossary, and Anno-

tated Bibliography," *Fam. Process, 10:* 157–176, 1971.

40. COTTRELL, L. S., "New Directions for Research on the American Family," *Soc. Casework, 34:*54, 1953.
41. COUNTS, R., "Family Crisis and the Impulsive Adolescent," *Arch. Gen. Psychiat., 17:*64–74, 1967.
42. DAVIS, D. R., "The Family Triangle in Schizophrenia," *Br. J. Med. Psychol., 34:* 53–63, 1961.
43. DUGDALE, R. L., *The Jukes: A Study in Crime, Pauperism, Disease and Heredity,* Putnam, New York, 1910.
44. DUPONT, R., and GRUNEBAUM, H., "Willing Victims: The Husbands of Paranoid Women," *Am. J. Psychiat., 125:*151–159, 1968.
45. EASSON, W. M., and STEINHILBER, R. M., "Murderous Aggression by Children and Adolescents," *Arch. Gen. Psychiat., 4:*27–35, 1961.
46. EHRENWALD, J., "Neurosis in the Family: A Study of Psychiatric Epidemiology," *Arch. Gen. Psychiat., 3:*232–242, 1960.
47. EISENSTEIN, V. (Ed.), *Neurotic Interaction in Marriage,* Basic Books, New York, 1956.
48. ERIKSON, E. H., *Childhood and Society,* Norton, New York, 1950.
49. ERLICH-AMITAI, H. S., and BLOCH, D. A., "Two Families: The Origins of a Therapeutic Crisis," *Fam. Process, 10:*37–52, 1971.
50. EWING, J. A., LONG, V., and WENZEL, G. G., "Concurrent Group Psychotherapy of Alcoholic Patients and Their Wives," *Int. J. Group Psychother., 11:*329–338, 1961.
51. FERBER, A., KLIGLER, D., ZWERLING, I., and MENDELSOHN, M., "Current Family Structure," *Arch. Gen. Psychiat., 16:*659–667, 1967.
52. FERREIRA, A. J., "The 'Double-Bind' and Delinquent Behavior," *Arch. Gen. Psychiat., 3:*359–367, 1960.
53. ———, "Psychosis and Family Myth," *Am. J. Psychother., 21:*186–197, 1967.
54. ———, and WINTER, W. D., "Family Interaction and Decision-Making," *Arch. Gen. Psychiat., 13:*214–223, 1965.
55. FISHER, S., and MENDELL, D., "The Communication of Neurotic Patterns over Two and Three Generations," *Psychiatry, 19:*41–46, 1956.
56. ———, and ———, "The Spread of Psychotherapeutic Effects from the Patient to His Family," *Psychiatry, 21:*133–140, 1958.
57. FLECK, S., "Family Dynamics and Origin of Schizophrenia," *Psychosom. Med., 22:* 333–344, 1960.
58. ———, LIDZ, T., and CORNELISON, A., "Comparison of Parent-Child Relationships of Male and Female Schizophrenic Patients," *Arch. Gen. Psychiat., 8:*1–7, 1963.
59. FONTANA, A. F., "Familial Etiology of Schizophrenia: Is a Scientific Methodology Possible?" *Psychol. Bull., 66:*214–227, 1966.
60. FOOTE, N., and COTTRELL, L. S., *Identity and Interpersonal Competence: New Directions in Family Research,* University of Chicago Press, Chicago, 1955.
61. FRAMO, J. L., "The Theory of the Technique of Family Treatment of Schizophrenia," *Fam. Process, 1:*119–131, 1962.
62. FREUD, S., "Analysis of a Phobia in a Five-Year-Old Boy," in *Collected Papers,* Vol. 10, pp. 5–149, Hogarth, London, 1955.
63. ———, "Fragment of an Analysis of a Case of Hysteria," in *Collected Papers,* Vol. 7, pp. 7–124, Hogarth, London, 1955.
64. FRIEDMAN, A. S., "Delinquency and the Family System," in Pollak, O., and Friedman, A. S. (Eds.), *Family Dynamics and Female Sexual Delinquency,* Science & Behavior Books, Palo Alto, Calif., 1969.
65. FRY, W. F., JR., "The Marital Context of an Anxiety Syndrome," *Fam. Process, 1:* 245–252, 1962.
66. GEHRKE, S., and KIRSCHENBAUM, M., "Survival Patterns in Family Conjoint Therapy," *Fam. Process, 6:*67–80, 1967.
67. GETZELS, J. W., and JACKSON, P. W., "Family Environment and Cognitive Style: A Study of the Sources of Highly Intelligent and of Highly Creative Adolescents," *Am. Sociol. Rev., 26:*351–359, 1961.
68. GLICK, I., and HALEY, J., *Family Therapy and Research—An Annotated Bibliography,* Grune & Stratton, New York, 1971.
69. GODDARD, H. H., *The Kallikak Family: A Study in the Heredity of Feeble-Mindedness,* Macmillan, New York, 1912.
70. GOLDBERG, M., and MUDD, E., "The Effects

of Suicidal Behavior upon Marriage and the Family," in Resnick, H. L. P. (Ed.), *Suicidal Behaviors*, Little, Brown, Boston, 1968.

71. GOLDFARB, W., "The Mutual Impact of Mother and Child in Childhood Schizophrenia," *Am. J. Orthopsychiat.*, *31*:738–747, 1961.

72. GRAY, W., DUHL, F. J., and RIZZO, N. D. (Eds.), *General Systems Theory and Psychiatry*, Little, Brown, Boston, 1969.

73. GREIG, A. B., "The Problem of the Parent in Child Analysis," *Psychiatry*, *3*:539, 1940.

74. GROTJAHN, M., *Psychoanalysis and the Family Neurosis*, Norton, New York, 1960.

75. GROVES, E. R., "Freud and Sociology," *Psychoanal. Rev.*, *3*:241, 1916.

76. GRUNEBAUM, M. G., HURWITZ, I., PRENTICE, N. M., and SPERRY, B. M., "Fathers of Sons with Primary Neurotic Learning Inhibitions," *Am. J. Orthopsychiat.*, *32*:462–472, 1962.

77. HALEY, J., "Experiment with Abnormal Families," *Arch. Gen. Psychiat.*, *17*:53–63, 1967.

78. ———, "The Family of the Schizophrenic: A Model System," *J. Nerv. & Ment. Dis.*, *129*:357–374, 1959.

79. ———, "Observation of the Family of the Schizophrenic," *Am. J. Orthopsychiat.*, *30*:460–467, 1960.

80. ———, "Speech Sequences of Normal and Abnormal Families with Two Children Present," *Fam. Process*, *6*:81–97, 1967.

81. ——, and HOFFMAN, L., *Techniques of Family Therapy*, Basic Books, New York, 1967.

82. HANDEL, G., "Psychological Studies of Whole Families," in Handel, G. (Ed.), *The Psychosocial Interior of the Family*, Aldine, Chicago, 1967.

83. ——— (Ed.), *The Psychosocial Interior of the Family*, Aldine, Chicago, 1967.

84. ———, and HESS, R. D., "The Family as an Emotional Organization," *Marriage & Fam. Living*, *18*:99–101, 1956.

85. HARE, A. P., BORGATTA, E. F., and BALES, R. F. (Eds.), *Small Groups—Studies in Social Interaction*, Alfred A. Knopf, New York, 1965.

86. HENRY, J. "Family Structure and the Transmission of Neurotic Behavior," *Am. J. Orthopsychiat.*, *21*:800, 1951.

87. HILGARD, J., and NEWMAN, M., "Parental Loss by Death in Childhood as an Etiological Factor among Schizophrenic and Alcoholic Patients Compared with a Non-Patient Community Sample," *J. Nerv. & Ment. Dis.*, *137*:14–28, 1963.

88. HILL, R., "A Critique of Contemporary Marriage and Family Research," *Soc. Forces*, *33*:268, 1955.

89. ———, "Methodological Issues in Family Development Research," *Fam. Process*, *3*:186–206, 1964.

90. HOOVER, C., "The Embroiled Family: A Blueprint for Schizophrenia," *Fam. Process*, *4*:291–310, 1965.

91. JACKSON, D. D. (Ed.), *The Etiology of Schizophrenia*, Basic Books, New York, 1960.

92. ———, "The Question of Family Homeostasis," *Psychiat. Quart. Suppl.*, Part I, *31*:79, 1957.

93. ———, "The Study of the Family," *Fam. Process*, *4*:1–20, 1965.

94. ———, and YALOM, I., "Family Research on the Problem of Ulcerative Colitis," *Arch. Gen. Psychiat.*, *15*:410–418, 1966.

95. JENSEN, G. D., and WALLACE, J. G., "Family Mourning Process," *Fam. Process*, *6*:56–66, 1967.

96. JOHNSON, A. M., FALSTEIN, E. I., SZUREK, S. A., and SVENDSEN, M., "School Phobia," *Am. J. Orthopsychiat.*, *11*:702, 1941.

97. KAFKA, J. S., "Critique of Double Bind Theory and Its Logical Foundation (an Alternative Formulation)," paper read to the Double Bind Symposium at Annual Meeting of American Psychological Association, Washington, D.C., September 1969.

98. KATZ, M., "Agreement on Connotative Meaning in Marriage," *Fam. Process*, *4*:64–74, 1965.

99. KEMPLER, W., IVERSON, R., and BEISSER, A., "The Adult Schizophrenic Patient and His Siblings," *Fam. Process*, *1*:224–235, 1962.

100. KYSAR, J. E., "Reactions of a Professional to Disturbed Children and Their Parents," *Arch. Professional Psychiat.*, *19*:562–570, 1968.

101. LAING, R. D., *The Divided Self*, Tavistock, London, 1960.
102. ———, "Mystification, Confusion and Conflict," in Boszormenyi-Nagy, I., and Framo, J. L. (Eds.), *Intensive Family Therapy*, Harper & Row, New York, 1965.
103. ———, and ESTERSON, A., *Sanity, Madness and the Family*, Vol. 1, *Families of Schizophrenics*, Tavistock, London, 1964.
104. ———, PHILLIPSON, H., and LEE, A., *Interpersonal Perception—A Theory and a Method of Research*, Tavistock, London, 1966.
105. LEICHTER, H. J., "Boundaries of the Family as an Empirical and Theoretical Unit," in Ackerman, N. W., Beatman, F. L., and Sherman, S. (Eds.), *Exploring the Base for Family Therapy*, Family Service Association, New York, 1961.
106. LERNER, P., "Resolution of Intrafamilial Role Conflict in Families of Schizophrenic Patients. I. Thought Disturbance," *J. Nerv. & Ment. Dis.*, *141*:342–351, 1965.
107. ———, "Resolution of Intrafamilial Role Conflict in Families of Schizophrenic Patients. II. Social Maturity," *J. Nerv. & Ment. Dis.*, *145*:336–341, 1967.
108. LEVY, D. M., *Maternal Overprotection*, Columbia University Press, New York, 1943.
109. LEVY, J., and EPSTEIN, N. B., "An Application of the Rorschach Test in Family Investigation," *Fam. Process*, *3*:344–376, 1964.
110. LEWIS, V. S., and ZEICHNER, A. N., "Impact of Admission to a Mental Hospital on the Patient's Family," *Ment. Hyg.*, *44*:503–509, 1960.
111. LIDZ, R., and LIDZ, T., "Homosexual Tendencies in Mothers of Schizophrenic Women," *J. Nerv. & Ment. Dis.*, *149*:229–235, 1969.
112. LIDZ, T., CORNELISON, A., FLECK, S., and TERRY, D., "Intrafamilial Environment of Schizophrenic Patients. II. Marital Schism and Marital Skew," *Am. J. Psychiat.*, *114*:241–248, 1957.
113. ———, "Schism and Skew in the Families of Schizophrenics," in Bell, N. W., and Vogel, E. F. (Eds.), *A Modern Introduction to the Family*, Free Press, Glencoe, 1960.
114. ———, "Intrafamilial Environment of the Schizophrenic Patient: VI. The Transmission of Irrationality," *Arch. Neurol. & Psychiat.*, *79*:305–316, 1958.
115. LOMAS, P. (Ed.), *The Predicament of the Family—A Psychoanalytical Symposium*, International Psychoanalytical Library 71, Hogarth, London, 1967.
116. LOVELAND, N. T., WYNNE, L. C., and SINGER, M. T., "The Family Rorschach: A New Method for Studying Family Interaction," *Fam. Process*, *2*:187–215, 1963.
117. LU, Y.-C., "Contradictory Parental Expectations in Schizophrenia," *Arch. Gen. Psychiat.*, *6*:219–234, 1962.
118. ———, "Mother-Child Role Relations in Schizophrenia: A Comparison of Schizophrenic Patients with Non-schizophrenic Siblings," *Psychiatry*, *24*:133–142, 1961.
119. MACHOTKA, P., PITTMAN, F. S., and FLOMENHAFT, K., "Incest as a Family Affair," *Fam. Process*, *6*:98–116, 1967.
120. MCCORD, W., PORTA, J., and MCCORD, J., "The Familial Genesis of Psychoses," *Psychiatry*, *25*:60–71, 1962.
121. MALMQUIST, C., "School Phobia: A Problem in Family Neurosis," *J. Am. Acad. Child Psychiat.*, *4*:293–319, 1965.
122. MEISSNER, W. W., "Family Dynamics and Psychosomatic Processes," *Fam. Process*, *5*:142–161, 1966.
123. MEYERS, D., and GOLDFARB, W., "Studies of Perplexity in Mothers of Schizophrenic Children," *Am. J. Orthopsychiat.*, *31*:551–564, 1961.
124. MILLER, D. R., and WESTMAN, J. C., "Family Teamwork and Psychotherapy," *Family Process*, *5*:49–59, 1966.
125. ———, "Reading Disability as a Condition of Family Stability," *Fam. Process*, *3*:66–76, 1964.
126. MILLER, H., and BARUCH, D. W., "Psychosomatic Studies of Children with Allergic Manifestations. I. Maternal Rejection: A Study of 63 Cases," *Psychosom. Med.*, *10*:275, 1948.
127. MINUCHIN, S., MONTALVO, B. GUERNEY, B. G., ROSMAN, B. L., and SHUMER, F., *Families of the Slums: An Exploration of Their Structure and Treatment*, Basic Books, New York, 1967.
128. MISHLER, E. G., and WAXLER, N. E., "Family Interaction Processes and Schizo-

phrenia," *Int. J. Psychiat.*, *2*:375–430, 1966.

129. ———, *Interaction in Families*, John Wiley, New York, 1968.

130. MORRIS, G., and WYNNE, L., "Schizophrenic Offspring, Parental Styles of Communication," *Psychiatry*, *28*:19–44, 1965.

131. MOWRER, H. R., "Alcoholism and the Family," *J. Crim. Psychopath.*, *3*:90, 1941.

132. MURRELL, S., and STACHOWIAK, J., "Consistency, Rigidity, and Power in the Interaction Patterns of Clinic and Non-Clinic Families," *J. Abnorm. Psychol.*, *72*:265–272, 1967.

133. NAPIER, A. Y., "The Marriage of Families: Cross-Generational Complementarity," *Fam. Process*, *10*:373–395, 1971.

134. NOVAK, A. L., and VAN DER VEEN, F., "Family Concepts and Emotional Disturbance in the Families of Disturbed Adolescents with Normal Siblings," *Fam. Process*, *9*:157–171, 1970.

135. OTTO, H. A., "Criteria for Assessing Family Strength," *Fam. Process*, *2*:329–338, 1963.

136. PARSONS, T., and BALES, R. F., *Family Socialization and Interaction Process*, Free Press, Glencoe, 1955.

137. PAUL, N. L., and GROSSER, G. H., "Family Resistance to Change in Schizophrenic Patients," *Fam. Process*, *3*:377–401, 1964.

138. ———, "Operational Mourning and Its Role in Conjoint Family Therapy," *Com. Ment. Health*, *1*:339–345, 1965.

139. PERRY, S. E., SILBER, E., and BLOCH, D. A., *Children in a Disaster: A Study of the 1953 Vicksburg Tornado*, National Academy of Science—National Research Council, Committee on Disaster Research, Washington, D.C., 1955.

140. PITTMAN, F., LANGSLEY, D., and DEYOUNG, C., "Work and School Phobias: A Family Approach to Treatment," *Am. J. Psychiat.*, *124*:1535–1541, 1968.

141. PURCEL, K., *et al.*, "The Effect on Asthma in Children of Experimental Separations from the Family," *Psychosom. Med.*, *31*:144–164, 1969.

142. RABKIN, L. Y., "The Patient's Family: Research Methods," *Fam. Process*, *4*:105–132, 1965.

143. RABKIN, R., *Inner and Outer Space*, Norton, New York, 1970.

144. RAPOPORT, R., "Normal Crises, Family Structure, and Mental Health," *Fam. Process*, *2*:68–80, 1963.

145. RASHKIS, H., "Depression as a Manifestation of the Family as an Open System," *Arch. Gen. Psychiat.*, *19*:57–63, 1968.

146. REISS, D., "Individual Thinking and Family Interaction. I. Introduction to an Experimental Study of Problem Solving in Families of Normals, Character Disorders and Schizophrenia," *Arch. Gen. Psychiat.*, *16*:80–93, 1967.

147. ———, "Individual Thinking and Family Interaction. II. A Study of Pattern Recognition and Hypothesis Testing Families of Normals, Character Disorders and Schizophrenics," *J. Psychiat. Res.*, *5*:193–211, 1967.

148. ———, "Individual Thinking and Family Interaction. III. An Experimental Study of Categorization Performance in Families of Normals, Those with Character Disorders and Schizophrenics," *J. Nerv. & Ment. Dis.*, *146*:384–403, 1968.

149. ———, "Individual Thinking and Family Interaction. IV. A Study of Information Exchange in Families of Normals, Those with Character Disorders, and Schizophrenics," *J. Nerv. & Ment. Dis.*, *149*:473–490, 1969.

150. ———, "Varieties of Consensual Experience. I. A Theory for Relating Family Interaction to Individual Thinking. II. Dimensions of a Family's Experience of Its Environment," *Fam. Process*, *10*:1–36, 1971.

151. RICHARDSON, H. B., *Patients Have Families*, Commonwealth Fund, Harvard, Cambridge, 1945.

152. ROSEN, V. H., "Changes in Family Equilibrium Through Psychoanalytic Treatment," in Eisenstein, V. W. (Eds.), *Neurotic Interaction in Marriage*, Basic Books, New York, 1956.

153. ROSENTHAL, A. J., BEHRENS, M. I., and CHODOFF, P., "Communication in Lower Class Families of Schizophrenics," *Arch. Gen. Psychiat.*, *18*:464–470, 1968.

154. RUESCH, J., and BATESON, G., *Communication, the Social Matrix of Psychiatry*, Norton, New York, 1951.

155. SCHAFFER, L., WYNNE, L. C., DAY, J., RYCKOFF, I. M., and HALPERIN, A., "On the Nature and Sources of the Psychia-

trist's Experience with the Family of the Schizophrenic," *Psychiatry*, *25*:32–45, 1962.

156. SCHATZMAN, M., "Paranoia or Persecution: The Case of Schreber," *Fam. Process*, *10*:177–212, 1971.

157. SCHWARTZ, C. G., "Perspectives on Deviance—Wives' Definitions of Their Husbands' Mental Illness," *Psychiatry*, *20*: 275, 1957.

158. SINGER, M. T., and WYNNE, L. C., "Differentiating Characteristics of Parents of Childhood Schizophrenics, Childhood Neurotics, and Young Adult Schizophrenics," *Am. J. Psychiat.*, *120*:234–243, 1963.

159. SLUZKI, C., "Transactional Disqualification: Research on the Double Bind," *Arch. Gen. Psychiat.*, *16*:494–504, 1967.

160. ———, and VERÓN, E., "The Double-Bind as a Universal Pathogenic Situation," *Fam. Process*, *10*:397–411, 1971.

161. SONNE, J. C., SPECK, R. V., and JUNGREIS, J. C., "The Absent-Member Maneuver as a Resistance in Family Therapy of Schizophrenia," *Fam. Process*, *1*:44–62, 1962.

162. SPECK, R. V., "Family Therapy of the Suicidal Patient," in Resnick, H. L. P. (Ed.), *Suicidal Behaviors*, Little, Brown, Boston, 1968.

163. ———, and ATTNEAVE, C. L., "Network Therapy," in Haley, J. (Ed.), *Changing Families*, Grune & Stratton, New York, 1971.

164. SPEER, D. C., "Family Systems: Morphostasis and Morphogenesis, or 'Is Homeostasis Enough?'" *Fam. Process*, *9*:259–278, 1970.

165. SPIEGEL, J. P., "A Model for Relationships among Systems," in Grinker, R. R. (Ed.), *Toward a Unified Theory of Human Behavior*, Basic Books, New York, 1956.

165a. ———, and BELL, N. W., "The Family of the Psychiatric Patient," *Am. Hbk. Psychiat.*, *1*:114–149, 1959.

166. ———, and KLYCKHOHN, F. R., *Integration and Conflict in Family Behavior*, Report No. 27, Group for the Advancement of Psychiatry, Topeka, Kansas, 1954.

167. STACHOWIAK, J., "Decision-Making and Conflict Resolution in the Family Group," in Larson, C., and Dance, F. (Eds.), *Perspectives on Communication*, Speech Communication Center, University of Wisconsin, Milwaukee, 1968.

168. STREAN, H. S., "A Family Therapist Looks at 'Little Hans,'" *Fam. Process*, *6*:227–234, 1967.

169. TERRILL, J. M., and TERRILL, R. E., "A Method for Studying Family Communication," *Fam. Process*, *4*:259–290, 1965.

170. THOMAS, A., "Simultaneous Psychotherapy with Marital Partners," *Am. J. Psychother.*, *10*:716, 1956.

171. ———, and SILLEN, S., *Racism and Psychiatry*, Brunner/Mazel, New York, 1972.

172. TITCHENER, J. L., RISKIN, J., and EMERSON, R., "The Family in Psychosomatic Process: A Case Report Illustrating a Method of Psychosomatic Research," *Psychosom. Med.*, *22*:127–142, 1960.

173. TOMAS, W., *The Family Constellation: Its Effects on Personality and Social Behavior*, Springer, New York, 1969.

174. TOWNE, R., MESSINGER, S. L., and SAMPSON, H., "Schizophrenia and the Marital Family: Accommodations to Symbiosis," *Fam. Process*, *1*:304–318, 1962.

175. VASSILIOU, G., "Milieu Specificity in Family Therapy," in Ackerman, N. W. (Ed.), *Family Therapy in Transition*, Little, Brown, Boston, 1970.

176. VOGEL, E. F., "The Marital Relationship of Parents of Emotionally Disturbed Children: Polarization and Isolation," *Psychiatry*, *23*:1–12, 1960.

177. VOGEL, E. F., and BELL, N. W., "The Emotionally Disturbed Child as the Family Scapegoat," in Bell, N. W., and Vogel, E. F. (Eds.), *A Modern Introduction to the Family*, Free Press, Glencoe, 1960.

178. VON BERTALANFFY, L., *General System Theory*, George Braziller, New York, 1968.

179. ———, and RAPOPORT, A. (Eds.), *General Systems Yearbook of the Society for General Systems Research*, Vols. 1–16, Society for General Systems Research, Washington, D. C., 1955–1971.

180. WAHL, C. W., "Antecedent Factors in Family Histories of 392 Schizophrenics," *Am. J. Psychiat.*, *110*:668, 1954.

181. WALTER, J., and STINNETT, N., "Parent-Child Relationships: A Decade Review of Research," *J. Marriage & Fam.*, *33*: 70–111, 1971.

182. WATZLAWICK, P., "A Review of the Double

Bind Theory," *Fam. Process*, 2:132–153, 1963.

183. ———, BEAVIN, J. H., and JACKSON, D. D., *Pragmatics of Human Communication*, Norton, New York, 1967.

184. WEAKLAND, J. H., "The Double-Bind Hypothesis of Schizophrenia and Three-Party Interaction," in Jackson, D. D. (Ed.), *The Etiology of Schizophrenia*, Basic Books, New York, 1960.

185. ———, "Family Therapy as a Research Arena," *Fam. Process*, 1:63–68, 1962.

186. ———, and FRY, W. F., "Letters of Mothers of Schizophrenics," *Am. J. Orthopsychiat.*, 32:604–623, 1962.

187. WEBLIN, J., "Communication and Schizophrenic Behavior," *Fam. Process*, 1:5–14, 1962.

188. WELLDON, R. M. C., "The 'Shadow of Death' and Its Implications in Four Families, Each with a Hospitalized Schizophrenic Member," *Fam. Process*, 10: 281–302, 1971.

189. WELLISCH, D. K., GAY, G. R., and MCENTEE, R., "The Easy Rider Syndrome: A Pattern of Hetero- and Homosexual Relations in a Heroin Addict Population," *Fam. Process*, 9:425–430, 1970.

190. WESTLUND, N., and PALUMBO, A. Z., "Parental Rejection of Crippled Children," *Am. J. Orthopsychiat.*, 16:271, 1946.

191. WHITIS, P. R., "The Legacy of a Child's Suicide," *Fam. Process*, 7:159–169, 1968.

192. WYNNE, L. C., RYCKOFF, I. M., DAY, J., and HIRSCH, S. I., "Pseudo-Mutuality in the Family Relations of Schizophrenics," *Psychiatry*, 21:205–220, 1958.

193. ———, and SINGER, M. T., "Thought Disorder and Family Relations of Schizophrenics. I. A. Research Strategy. II. A Classification of Forms of Thinking," *Arch. Gen. Psychiat.*, 9:191–206, 1963.

194. ZILBACH, J. J., "Crisis in Chronic Problem Families," in Belsasso, G. (Ed.), *Psychiatric Care of the Underprivileged*, Little, Brown, Boston, 1971.

195. ZUK, G. H., "On the Pathology of Silencing Strategies," *Fam. Process*, 4:32–49, 1965.

196. ———, "On the Theory and Pathology of Laughter in Psychotherapy," *Psychotherapy*, 3:97–101, 1966.

CHAPTER 8

THE CONCEPT OF NORMALITY

Daniel Offer and Melvin Sabshin

Introduction

WHY HAS THE STUDY of normality and health been a relative newcomer to psychiatry? One obvious reason is that it is a result of training, since training produces not only a set of skills but also what Kaplan[30] calls a trained incapacity. Trained to recognize the abnormal, the psychiatrist and his teacher have had difficulty in recognizing the normal. Recently there has been a surge of interest among mental health professionals in studying normality. In part it is important to study normality because it serves as a base line for all behavior whether pathological or not. Reference to the normal also gives empirical validation to theoretical constructions concerning any kind of behavior. That is not to say, however, that the concept of normality is a clear and concise one. On the contrary, the concept is ambiguous, has a multiplicity of meanings and usages, and is burdened by being value-laden.

In the past, within the behavioral sciences in general and specifically within psychiatry, there has been an implicit understanding that mental health could be defined as the antonym of mental illness. Given such an assumption, the absence of gross psychopathology was often equated with normal behavior. A number of recent trends have cast doubt on the usefulness of this assumption and have made it increasingly important for psychiatrists to become concerned with providing more precise concepts and definitions of mental health and normality.

As psychiatrists began to move out of their consulting rooms and hospital wards into the community, they began to come into contact with segments of the population not previously seen in their more traditional role functions. The broader acceptance of preventive, social, and community psychiatry has necessitated a re-examination of preventing *what* in *whom*. Psychiatrists have also become increasingly involved in agency consultation, where they are called upon to make decisions about who is healthier rather than about who is too sick for various positions. Interest in evaluating the outcome of psychiatric therapeutic endeavors has also brought the issue of what is mental health or normality into focus. Indeed, assessing therapeutic outcome has been handicapped by lack of clarity regarding the concepts of normality and mental health.

This chapter is written with the intent of clarifying some of the conceptual issues related to normality and mental health. We cannot provide a definitive answer to the question, "What is mental health or normality?" since such an answer must eventually evolve out of new research and new experiences. Because cultural and personal values and biases are so intimately tied to one's conception of normality, it is doubtful whether, even in the long run, we will have *one* definition of normality. We shall attempt, however, to delineate the perspectives of normality, evaluate some of the current research on normal populations, and point to newer directions that promise to elucidate the issues still further.

The Four Perspectives of Normality

In our review of the literature on the theoretical and clinical concepts of normality, we were able to define what we called the four "functional perspectives" of normality.[40] Although each perspective is unique and has its own definition and description, the perspectives do complement each other, so that together they represent the total behavioral and social science approach to normality. The four perspectives are (1) normality as health, (2) normality as utopia, (3) normality as average, and (4) normality as process. Let us now define these four perspectives in more detail.

The first perspective, *normality as health*, is basically the traditional medical psychiatric approach to health and illness. Most physicians equate normality with health and view health as an almost universal phenomenon. As a result behavior is assumed to be within normal limits when no manifest psychopathology is present. If we were to transpose all behavior upon a scale, normality would encompass the major portion of the continuum and abnormality would be the small remainder. This definition of normality correlates with the traditional model of the doctor who attempts to free his patient from grossly observable signs and symptoms. To this physician the lack of signs or symptoms indicates health. In other words, health in this context refers to a *reasonable* rather than an *optimal* state of functioning. In its simplest form this perspective is illustrated by Romano,[45] who states that a healthy person is one who is reasonably free of undue pain, discomfort, and disability.

The second perspective, *normality as utopia*, conceives of normality as that harmonious and optimal blending of the diverse elements of the mental apparatus that culminates in optimal functioning. Such definition emerges clearly when psychiatrists or psychoanalysts talk about the ideal person or when they grapple with a complex problem such as their criteria of successful treatment. This approach can be dated directly back to Freud[10] who, when discussing normality, stated, "A normal ego is like normality in general, an ideal fiction." While this approach is characteristic of a significant segment of psychoanalysts, it is by no means unique to them. It can also be found among psychotherapists of quite different persuasion in the field of psychiatry and psychology (for example, Rogers).[43]

The third perspective, *normality as average*, is commonly employed in normative studies of behavior and is based on the mathematical principle of the bell-shaped curve. This approach conceives of the middle range as normal and *both* extremes as deviant. The normative approach that is based on the statistical principle describes each individual in terms of general assessment and total score. Variability is described only within the context of groups and not within the context of one individual. Although this approach is more commonly used in psychology and biology than in psychiatry, recently psychiatrists have been using pencil and paper tests to a much larger extent than in the past. Not only do psychiatrists utilize results of IQ tests, Rorschach tests, or T.A.T. but they also construct their own tests and questionnaires. Conceptually the normality as average perspective is similar to Kardiner's[31] "basic personality structure" for various cultures and subcultures. In developing model personalities for different societies, one assumes that the typologies of character can be statistically measured.

The fourth perspective, normality as process, stresses that normal behavior is the end result of interacting systems. According to this perspective, temporal changes are essential to a complete definition of normality. In other words, the normality as process perspective stresses changes or processes rather than cross-sectional definitions of normality. Investigators who subscribe to this approach can be found in all the behavioral and social sciences. Most typical are Grinker's[18] thesis of a unified theory of behavior encompassing polarities within a wide range of integration and Erikson's[8] conceptionalization of the seven developmental stages that are essential for mature adult functioning. The recent interest in general system theory (Von Bertalanffy,[51] Gray, Duhl, and Rizzo)[16] further stressed the general applicability of general system research for psychiatry. Normality as a system has been recently outlined by Grinker;[18] variables from the biological, psychological, and social fields all contribute to the functioning of a viable system over time. The integration of the variables into the system and the loading (or significance) assigned to each variable will have to be more thoroughly explored in the future.

Illustrative Research Strategies

Studies on normal population have to a large extent concentrated on normality as health and normality as utopia perspectives. The epidemiological studies of Srole[49] focused on disturbed signs and symptoms that interfered with the person's functioning and behavior. Leighton[35] and his colleagues concentrated on gross psychopathology. On the other hand, studies of superior or very competent individuals have been undertaken by Silber, *et al.*,[48] White,[54] Cox,[6] Heath,[24] Westley and Epstein.[53] There have been very few clinical or longitudinal studies illustrating the normality as average or normality as process perspectives. In the section below we have chosen two research strategies that illustrate the usefulness of the latter two perspectives in empirical studies of normality and health.

Normality as Average—The Modal Adolescent Project

The overall aim of the Modal Adolescent Project has been to study intensively and over a long period of time one representative group of "typical," "average," or "normal" adolescents. A modal student was defined as one whose performance on a specially constructed Self-Image Questionnaire (Offer and Diesenhaus)[39] was within one standard deviation from the mean in nine out of ten scales. As such it was our aim to study one population of average students by eliminating both extremes; the disturbed adolescent as well as the teen-ager who functions on a superior level. Since the above study was undertaken in two suburbs of Chicago, the adolescents represent average middle-class suburban teen-agers. Data were collected in order to provide a base line for comparative data collected on other populations both in our culture and cross-culturally as well as in future generations. It will become part of a data pool that will give scientists a better understanding concerning the totality of the phenomena of adolescence.

At the onset of our project we wanted to select our modal group from the widest possible spectrum of teen-agers living in a particular community, so a natural choice was a high school attended by all teen-agers in the community. We chose two different high schools that were limited to middle-class populations but, nevertheless, represented the full range of the middle class. For our selection procedure we devised a Self-Image Questionnaire that we administered to the two total entering freshmen classes.[39] "Normal" or modal subjects were chosen on the basis of a statistical approach, normality as average. This method was selected in order to attempt to eliminate the extremes of psychopathology and of superior adjustment.

Our research strategy helped us understand the kind of problems typical teen-agers have, the way they cope with them, and the reasons behind their successes and failures in the coping process. Our findings are summarized below:

1. There was a history of almost complete absence of gross psychopathology or severe physical illness.
2. There was a history of mastery of previous developmental tasks without serious setbacks.
3. Our subjects demonstrated ability to experience affects flexibly and actively bring their conflicts to a reasonable resolution. In general, our subjects coped well with anxiety, depression, shame, guilt, and anger.
4. Our subjects had relatively good peer relationships as well as a capacity to relate to and identify with adults. During high school the father is seen as reliable and the mother as understanding. The adolescents feel closer emotionally to the mother. After high school the subject gets closer emotionally to the father.
5. Twenty-two per cent of our subjects experimented with delinquency between the ages of 12 and 14. They had a capacity to learn from their experiences and did not become chronic delinquents. During high school they handled their aggressive impulses mainly by channeling them into sports. After high school and particularly during the first year away from home, we noted some mild depression among our subjects.
6. There was evidence of discomfort but no marked conflict with the increasing strength of the sexual impulses. Our subject moves slowly in the direction of heterosexuality. There is no evidence of a "sexual revolution" in behavior among our subjects. Nevertheless, in their attitudes they seem open and willing to discuss sex.
7. There was no evidence for the existence of serious "adolescent turmoil." The lack of serious "adolescent turmoil" was not a reflection of denial or closed off communication with the self. Indeed, introspection was prominent and communication with the self is present. The adolescents show some awareness of conflicts with parents and authority figures, but the conflicts do not spin out of control. They have surprisingly realistic self-images and are actively bringing their conflicts to a successful resolution.
8. There was a great deal of continuity between their values, those of their parents, and those of the social milieu of their community.

We discussed our findings concerning one group of normal or typical teen-agers in detail to illustrate how such a research strategy can be useful. It tells us how a specific group of adolescents feels and behaves. It serves to establish the beginning of behavioral norms for one group of adolescents; thus when future groups of teen-agers are studied, especially from the same sociocultural backgrounds, they can be meaningfully compared with the group described above.

Similar studies on modal populations have to be performed on a variety of populations throughout the life cycle in order for us to have a better understanding of the factors that contribute to adaptation and health as well as those that contribute to the development of psychopathology and maladaptation.

Normality as Process— Parental Anticipatory Mourning

In the previous discussion we have indicated that a process conception of normality involves an investigation of multiple systems of adaptation transacting over a period of time. In contrast to cross-sectional perspectives, a process analysis emphasizes temporal progression of coping techniques and assumes the continuity of change in behavioral patterns. While this particular perspective has been espoused by many mental health professionals on a verbal level, the empirical evidence supporting this understanding has been less convincing in the behavioral sciences than in several of the biological sciences (for example, embryology and evolutionary theory). In part the understanding has been hampered by a lack of empirical data so that the discussions of the process perspective have often appeared to be abstruse and divorced from concrete behavior. Consequently we have selected a particular research program to illustrate a more comprehensible picture of normality as process. The illustration is intended to summarize the formulation of a process analysis derived from empirical data in a longitudinal study.

The particular work that we have selected as an illustrative paradigm for normality as process is the work during the past five years by Futterman, Hoffman, and their colleagues[12,13,14] in the area of family adaptation to the fatal illness of a child. In this study each of the four perspectives is utilized to a certain degree. However, the overriding interest of the investigators is in studying coping behavior over time. The specific interactions of the various defensive structures (or subsystems) and their contributions to cluster behavior are being investigated. Their method is similar to the one utilized by Vaillant[50] in his 30-year follow-up study of healthy adults. The central question is what were the contributing factors that helped bring about successful adaptation.

These investigators have studied the coping mechanisms of parents with leukemic children. They have interviewed and observed a normative sample of parents and their children from the initial communication about the fatal illness up to, and in some cases following, the death of the child. To carry on such studies is in itself a most formidable task considering its extreme poignancy. Indeed, recent work by Paykel[41] has indicated that the death of a child is considered to be the most upsetting event conceivable to an adult population of psychiatric patients and their families. Nevertheless, several research teams (Friedman, *et al.*,[11] Richmond and Waisman,[42] Bozeman, *et al.*,[4] Natterson and Knudson,[37] Hamovitch,[23] Binger, *et al.*,[2]) have carried on investigations in this area with the realization that much can be learned by studies of a normative population undergoing stress of high intensity over a substantial period of time.

Futterman and Hoffman[12,13,14] conducted formal interviews with 24 sets of parents supplemented by informal contacts with many other families, in collaboration with the Department of Pediatrics at the University of Illinois. From the very beginning the observers were impressed with the wide repertoire of coping strategies utilized by the involved families. While the exact criteria for successful coping with this dreadful stress remain to be worked out, the research team has focused upon the modal processes of adaptation in their sample. Examination of the data after two and a half years of observations indicated that each set of parents had to cope with the following adaptational dilemmas:

1. To trust the caregivers (physicians, *et al.*) and yet to accept their limitations.
2. To be active in mastery of the chronic stress and yet to delegate appropriately to the caretakers.
3. To acknowledge helplessness periodically and yet to maintain their own sense of worth.
4. To cherish and cling to the leukemic child and yet to allow the child to separate from them and continue to grow.
5. To invest in the leukemic child and yet to invest also in the child's siblings.
6. To accept medical reality and yet to maintain a certain amount of hope.
7. To develop acceptable outlets for anger and yet to avoid excessive bitterness at others or persistent self-blame.
8. To mourn the impending loss and yet to maintain investment in the child's further development until death becomes imminent.
9. To focus on the immediate crises and yet to plan for the future of the family.
10. To protect significant others (relatives, friends, teachers) from too great a stress and yet to prepare them also for the child's death.
11. To maintain strategies of continued existence and yet face feelings of hopelessness, helplessness, and grief.

After analyzing these dilemmas the investigators further postulated the following interdependent five areas as major adaptational tasks of parents with leukemic children:

1. Maintaining family integrity.
2. Maintaining confidence (worth, mastery,

trust) in themselves and the child's physicians.
3. Maintaining meaningful value orientations (for example, causality, meaning of life and death, hope, and so forth).
4. Managing their awareness process and feeling states.
5. Anticipatory mourning.

Each of the above adaptational tasks was further analyzed into its various components, and efforts were made to follow these adaptational subtasks throughout the course of the child's illness and even beyond the death of the child.

The concept of parental anticipatory mourning was developed empirically from the data as a series of adaptive part-processes interwoven throughout the trajectory of the illness (Futterman, Hoffman, and Sabshin).[14] Preliminary inspection of the data showed that certain components emerged and reached prominence earlier in the child's illness while others peaked in the latter stages of the disease. The part-processes emerging from qualitative analysis of the data included the following:

1. Acknowledgment: development of explicit awareness (Glaser and Strauss)[15] of the fatal prognosis; becoming convinced that the child's death is inevitable.
2. Grief and grieving: experiencing and expressing the emotional impact of the anticipated loss and the physical, psychological, and interpersonal turmoil associated with it.
3. Reconciliation: coming to terms with the child's expected death in a manner that preserves a sense of appreciation of the value of the child's life and of life generally.
4. Detachment: withdrawal of emotional investment from the child as a growing being with a real future.
5. Memorialization: development of a relatively fixed internal image of the child with investment in his memory replacing investment in the real child, occurring before the death of the child.

These phases were postulated after the data were examined qualitatively; they are being checked currently by quantitative analysis to assess the frequency of their occurrence, their precise phase-specific relationships, and the various deviations from modal patterns. Several of these variables have been discussed by other investigators of parental adaptation to fatal childhood illness. However, other authors have not treated the course of parental anticipatory mourning as a total process, nor have they described the progression of each of its part-processes and their dynamic interactions over time. In addition, a number of specific constructs, among them "reconciliation" and "memorialization," are formally introduced by Futterman, *et al.*,[12,13,14] for the first time as integral aspects of anticipatory mourning.

Parental anticipatory mourning is still a relatively unexplored field. In this chapter we do not wish to focus on it as a clearly defined entity; rather we wish to illustrate empirical research in which the processes of coping over time are studied in a sample of individuals who do not specifically seek psychiatric help. We envisage a variety of such longitudinal studies as helpful in elucidating processes of adaptation so that normality as process becomes much more understandable as a perspective complementing the others discussed in this chapter.

Coping and Adaptation

Normality and health have often been associated with terms such as adaptation, competence, and most recently coping. We have to keep in mind that these concepts, like normality and health, are complex theoretical constructs that have not been adequately explained on the basis of any past or present single theory or hypothesis. One of the major problems that has plagued investigators in the behavioral sciences has been the great difficulty in making successful predictions about the long-term future behavior of an individual, or even of a majority within a group. Current research endeavors in the behavioral sciences have, therefore, tended to identify operationally clusters of traits and behavior that describe the variety of healthy or normal populations. (See, for example, Silber, *et al.*,[48] Grinker, *et al.*,[19] White,[54] Heath,[24] Westley and Epstein,[53] Cox,[6] Offer,[38] Beiser,[1] and Vaillant).[50]

The interest of behavioral scientists in studies of coping and adaptation has increased rapidly in the past two decades.[5,7] A variety of definitions of coping has been offered in the literature. Hamburg and Adams[22] define coping as "seeking and utilizing of information under stressful conditions." Heath[24] defines adaptation as "to so regulate behavior as to optimize simultaneously both the stability of the self structures and their accommodation to environmental requirements." Coping is defined by Lazarus, *et al.*,[34] as consisting of problem-solving efforts made by an individual when the demands he faces have a potential outcome of a high degree of relevance for his welfare (that is, a situation of great jeopardy or promise), and particularly when these demands tax heavily his adaptive resources. Competence, according to White,[54] means "fitness or ability. The competence of an organism means its fitness or ability to carry on those transactions with the environment which result in its maintaining itself, growing or flourishing."

Ability to cope is another way of assessing the type of character of the defenses utilized by individuals; the "healthier" the defenses the better the coping abilities. (See, for example, Grinker[19] and Vaillant.)[50] A recent review by Weinshel[52] includes a discussion of the direct correlation between ego strength and health and normality. Here again the stronger (or healthier) the ego the easier it is for the individual to adapt to his internal as well as to his external environments.

From a physiological point of view studies of coping mechanisms tended to agree that: "It is not the stimulus that is specific but the response. Response specificity clearly is based on phenotypic patterns based on combinations of genic and experiential factors."[17] Recognizing the importance of response specificity, the most significant investigations of coping behavior have involved analysis of variations in responses to stressful situations. The fact that a large variety of stimuli from either the external environment or the internal environment (for example, affects such as anxiety, anger, fear, or depression) all have a common physiological pathway has made it easier to study the somatic responses to stress. It has not, however, made it any easier to study the coping behavior of people from a psychological point of view. The individualized response, which is based on constitutional, experiential, familial, social, and cultural factors, is not uniform. Hence the 20 universal stressful situations* outlined by Hamburg[21] are coped with in a variety of ways rather than having a uniform pattern of "response specificity." Thus what causes one person to be unable to cope (for example, death of a close relative or friend) *at a particular time* will not necessarily cause the same response at a different developmental state. A knowledge of the individual's background can tell us what defense mechanism (or coping behavior) an individual will choose to combat a particular stress. However, psychiatrists have not been too successful in the past in predicting who will cope successfully under stressful conditions.

In the section on parental anticipatory mourning above, we have described a specific study that deals with one of the stressful situations outlined by Hamburg. The physiological system that has been activated by the parents in acute stages of grief is not intrinsically different from the system that the soldier activates in battle. The psychosocial response, however, is less specific. Hamburg and Adams[22] state that there are basically four stages in relieving stress of major proportions: "1. Personality attributes that tend to facilitate involvement in and mastery of the new situa-

* (1) Separation from parents in childhood; (2) displacement by siblings; (3) childhood experiences of rejection; (4) illness and injuries in childhood; (5) illness and death of parents; (6) severe illnesses and injuries of the adult years; (7) the initial transition from home to school; (8) puberty; (9) later school transitions, for example, from grade school to junior high school to college; (10) competitive graduate education; (11) marriage; (12) pregnancy; (13) menopause; (14) necessity for periodic moves to a new environment; (15) retirement; (16) rapid technological and social change; (17) wars and threats of wars; (18) migration; (19) acculturation; (20) social mobility.

tion or task; 2. Ego processes that serve to develop an adequate self-image in regards to the new task or situation; 3. Ego processes that help to maintain otherwise distressful affects within manageable limits; and 4. Personality processes that tend to maintain and/or restore significant inter-personal relationships."

The basic assumption still is that the royal route to studying psychosocial coping is through deviancy, be it psychopathology (Jones)[27] or unusual (that is, stressful) situations or conditions (Hamburg and Adams,[22] Grinker and Spiegel,[20] and Lazarus).[34]

To many investigators the term "coping" implies that there is something negative to which one has to attend. This view of coping is conceptually similar to the psychoanalytic theory of defense mechanisms—the defenses are erected to protect the person against the upsurge of powerful aggressive and sexual instinctual demands from within and to protect him from undue pressure from without. From this perspective the best way to study coping is by observing populations that are experiencing stressful reactions in highly conflictual situations.

Erikson[8] presents human growth from the point of view of conflict, crisis, and stress: "For man, in order to remain psychologically alive, constantly resolves these conflicts just as his body increasingly combats the encroachment of physical deterioration" (p. 91). He postulates that during the eight specific crisis periods one can study the human condition better because both the potential for growth and the emergence of pathological defense mechanisms are more evident. It remains unclear whether Erikson extends the crisis periods to include all substages or phases of the eight crisis periods. The concepts that delineate the beginning and end of each crisis period have not as yet been operationally defined. Erikson assumes that crisis, like stress, elicits emergency defenses; hence the study of crisis serves as a way to decipher how well a person is able to cope.

An alternative perspective to sequential life crises is one that assumes periods of relative equilibrium and disequilibrium. The modal adolescent described above goes through adolescence without reacting to it *in toto* as a stressful situation or even as a series of *major* crises. Coping can, therefore, be studied in this context by observing quantitatively different phenomena, for example, the most ordinary life situations of one group of teen-agers. Adolescence is certainly not conflict-free, and the adolescent must disengage himself from parental domination. He can do this without total renunciation of parental values, but rather through conflicts or minor issues that have been endowed with major importance for the adolescent's own growth and development. At least in the modal adolescents the reactions do not give rise to the storm and stress described by investigators studying intensive experiences of adolescence and experiences such as combat conditions (for example, *Men under Stress*).[20] To describe the adolescent period as a period of great inner turmoil, rebellion, storm and stress is true for a segment of the adolescent population, but is not universally observed.[38] Like any period in life it is highly stressful for some individuals. For the latter group their coping abilities are often inadequate, and hence they commonly seek help in the psychiatrists' offices. The adolescents who cope successfully with the period are an enigma to many clinicians. We need new terminology that will describe the nuances in coping behavior and that will take into account the tremendous variability in reactions of individuals to stress and crisis. It will allow us to view a continuum of stressful situations from most stressful to the least stressful, with crisis being at one end of the continuum and successful coping with everyday tasks at the other end.

Studies of adaptation and coping would enable us to formulate what Jahoda[26] called "positive aspects of mental health." While psychiatrists have always been quick to extrapolate from psychopathology to normative theory, it has not been simple for psychiatrists to focus upon adaptive aspects of human behavior. Until very recently our theoretical concepts have facilitated the examination of nuances and details in studying psychopathology. In addition, psychiatrists are adept in perceiving coping behavior correctly, but still clothing it in the garments of psychopathology.* With the developing sophistication of ego psychology, it has become increasingly obvious that there is a need to investigate the correlation between the behavior and specific coping ability of people and their overall psychological health and normality.

Returning full circle now to our original question: What is normality and health? We believe that there are a number of adaptational routes open to individuals. Based on a complex interaction of biopsychosocial variables in a system that we call an individual, a person will have developed a coping style unique to him early in life. It is our opinion that in a majority of individuals the psychological system that an individual erects in order to cope with crisis, stress, and everyday life will by and large remain relatively constant through life, since the repertoire of adaptive mechanisms, including defenses, is relatively fixed in early childhood. Some individuals will undergo *Sturm und Drang* at every major crossroad through the life cycle. Others will go through life with little turmoil. We postulate that the healthier individuals will have a variety of coping techniques available to them, will have a larger array of defenses at their disposal, will utilize the more adaptive and object-seeking defenses, and will show more flexibility in dealing with internal as well as external events. Whether the ones who show less overt (behavioral) turmoil are the ones who are optimally healthier cannot be determined without intimate knowledge of the biopsychosocial variables that contribute to the behavior. In other words, behavior has to be studied in the context of the individual, the family, the group, and the society. Only then can it be fully understood.

It is imperative that we study the behavior, psychodynamics, and coping styles of a large segment of the population in order to understand the contributing factors to adaptation or maladaptation. Recent trends in social and community psychiatry afford us some new opportunities to achieve part of this goal.

* For example, in diagnosing or labeling precision work as "compulsive."

¶ Social and Community Psychiatry

The issue of normality has already become a visible question in the development of social and community psychiatry. Although some investigators have clearly recognized that conceptual dilemmas regarding the definition of normality exist when one conducts epidemiological studies or attempts primary preventive programs (Mechanic[36] and Bolman),[3] the problem is rarely acknowledged. There is reason to expect, however, that a series of events will force even greater consideration by community mental workers of perspectives of normality.[46,47] Discrepancies in prevalence and incidence rates of psychiatric illness across subcultural and geographic boundaries are determined, at least in part, by variations in the operational definitions of normality or health. Furthermore, efforts at evaluation of preventive programs (primary as well as secondary prevention) must include criteria for successful as well as negative outcomes. Comparative analysis of such results in community mental health programs across the nation will necessitate consideration of the operational definitions of illness and health, and there will be increasing pressures to improve upon our gross definitions of psychopathology and healthy adaptation.

There is another reason why normality will emerge as a central research and pragmatic issue in social and community psychiatry. The development of community mental health programs affords an excellent opportunity to begin a series of investigations of normative behavior not heretofore possible. Although the image of communities as laboratories for research has brought about resistance as well as consternation in some areas, there is reason to believe that the efforts to conduct investigations *with* communities will, after some hesitation, supplant our tendency to do research *on* communities. Evidence already exists indicating the advantage to communities of increased understanding of its population's behavior as compared to stereotyping and its attendant distortions. Furthermore, communities will be interested in evaluation of intervention programs within their borders, and they will cooperate with investigations that facilitate such evaluation.

One of the best illustrations of *access* to a normative sample and community *sanction* for research in the area of psychological adaptation and maladaptation has been the pioneering investigations by Kellam and Schiff.[32,33] They have demonstrated the feasibility of such studies and have also contributed significantly to our understanding of behavior in a large sample of schoolchildren. It is to be hoped that over time many more populations will be investigated so that our empirical evidence will be considerably augmented.

Summary

Concepts of health and normality are not universal or all-encompassing. They tend to reflect the particular orientation or values of the investigator and are often tied to the social context in which the research is undertaken. We have described four major perspectives on normality: normality as health, normality as utopia, normality as average, and normality as process. We further described two research strategies that can be seen as illustrating two of the four perspectives on normality.

The trend toward increased study of normative population by mental health professionals can be well documented in a variety of other mental health areas. It is important to note that convergences have already begun to take place, as, for example, in the convergence of community psychiatric studies and human development within an epidemiological framework. We have seen in the past decade a dramatic increase in investigators studying the factors contributing to coping under stress, as well as the psychological processes relative to successful adaptation in everyday life. Many more studies will be necessary before a useful integration of the concepts of healthy adaptation, coping, and normality will be possible.

Bibliography

1. BEISER, M., "A Psychiatric Follow-Up Study of 'Normal' Adults," *Am. J. Psychiat.*, *127:* 1464–1473, 1971.
2. BINGER, C. M., ABLIN, A. R., FEUERSTEIN, R. C. KUSHNER, J. H., ZOGER, S., and MIKKELSON, C., "Childhood Leukemia: Emotional Impact on Patient and Family," *New Eng. J. Med.*, *280:*414–418, 1969.
3. BOLMAN, W. M., "Theoretical and Empirical Bases of Community Mental Health," *Am. J. Psychiat.*, *129:*4, 1967, Suppl.
4. BOZEMAN, M. F., ORBACH, C. E., and SUTHERLAND, A. M., "Psychological Impact of Cancer and Its Treatment, III. The Adaptation of Mothers to the Threatened Loss of Their Children through Leukemia: Part I," *Cancer*, *8:*1–19, 1955.
5. Conference on Normal Behavior, *A.M.A. Arch. Gen. Psychiat.*, *17:*258–330, 1967.
6. COX, F. D., *Youth into Maturity*, Mental Health Materials Center, New York, 1970.
7. "The Definition and Measurement of Mental Health: A Symposium," National Center for Health Statistics, U.S. Public Health Service, September 1967.
8. ERIKSON, E. H., *Identity: Youth and Crisis*, Norton, New York, 1968.
9. FREUD, A., *Normality and Pathology in Childhood*, International Universities Press, New York, 1965.
10. FREUD, S. (1937), "Analysis, Terminable and Interminable," in *Collected Papers*, Vol. 5, Basic Books, New York, 1959.
11. FRIEDMAN, S. B., CHODOFF, P., MASON, J. W., and HAMBURG, D. A., "Behavioral Observations on Parents Anticipating the Death of a Child," *Pediatrics*, *32:*610–625, 1963.
12. FUTTERMAN, E. H., and HOFFMAN, I., "Shielding from Awareness: An Aspect of Family Adaptation to Fatal Illness in Children," *Arch. Thanatol.*, *2:*23–24, 1970.
13. ———, and ———, "Crisis, Adaptation and Anticipatory Mourning in Families with a Leukemic Child, in Anthony, E. J. and Koupernik, C. (Eds.), *The Child in His Family*, Vol. 2, pp. 127–144, Wiley, New York, 1973.

14. ———, ———, and SABSHIN, M., "Parental Anticipatory Mourning," in Schoenberg, B., *et al.* (Eds.), *Psychosocial Aspects of Terminal Care*, Columbia University Press, New York, 1972.
15. GLASER, B. G., and STRAUSS, A. L., *Time for Dying*, Aldine, Chicago, 1968.
16. GRAY, W., DUHL, F. J., and RIZZO, N. D., *General Systems Theory and Psychiatry*, Little, Brown, Boston, 1969.
17. GRINKER, R. R., SR., "Brief Impressions of the Conference on Coping," paper presented at the Conference on Coping and Adaptation, Stanford University, March 1969.
18. ———, *Towards a Unified Theory of Human Behavior*, Basic Books, New York, 1956.
19. ———, with GRINKER, R. R., JR., and TIMBERLAKE, J., "A Study of 'Mentally Healthy' Young Males (Homoclites)," *A.M.A. Arch. Gen. Psychiat.*, *6*:405–453, 1962.
20. GRINKER, R. R., and SPIEGEL, J., *Men under Stress*, Blakiston, Philadelphia, 1948.
21. HAMBURG, D. A., "A Perspective on Coping," paper presented at the Conference on Coping and Adaptation, Stanford University, March 1969.
22. ———, and ADAMS, J. E., "A Perspective on Coping Behavior," *Arch. Gen. Psychiat.*, *18*:277–284, 1967.
23. HAMOVITCH, M. B., *The Parent and the Fatally Ill Child*, Delmar Publishing Co., Los Angeles, 1964.
24. HEATH, D. H., *Exploration of Maturity*, Appleton-Century-Crofts, New York, 1965.
25. HOFFMAN, I., and FUTTERMAN, E. H., "Coping with Waiting," *Comprehensive Psychiat.*, *12*:67–81, 1971.
26. JAHODA, M., *Current Concepts of Positive Mental Health*, Basic Books, New York, 1958.
27. JONES, E., Cited in Birnbach, M., *Neo-Freudian Social Philosophy*, Stanford University Press, Stanford, 1961.
28. ———, "The Concept of a Normal Mind," in *Papers in Psychoanalysis*, Williams & Wilkins, Baltimore, 1948.
29. KAGEN, J., and MOSS, M. A., *Birth to Maturity: A Study in Psychological Development*, John Wiley, New York, 1962.
30. KAPLAN, A., "A Philosophical Discussion of Normality," *A.M.A. Arch. Gen. Psychiat.*, *17*:325–330, 1967.
31. KARDINER, A., *The Individual and His Society*, Columbia University Press, New York, 1939.
32. KELLAM, S. G., and SCHIFF, S. K., "The Origins and Early Evolution of an Urban Community Mental Health Center in Woodlawn," in Duhl, L., and Leopold, R. (Eds.), *Casebook on Community Psychiatry*, Basic Books, New York, 1967.
33. ———, and ———, "The Woodlawn Mental Health Center: A Community Mental Health Center Model," *Soc. Service Rev.*, *40*:255–263, 1966.
34. LAZARUS, R. S., AVERILL, J. R., and OPTION, E. M., JR., "Assessment of Coping," paper presented at the Conference on Coping and Adaptation at Stanford University, March 1969.
35. LEIGHTON, A. H., *My Name Is Legion*, Basic Books, New York, 1959.
36. MECHANIC, D., *Mental Health and Social Policy*, Prentice-Hall, Englewood Cliffs, N.J., 1969.
37. NATTERSON, J. M., and KNUDSON, A. G., "Observations Concerning Fear of Death in Fatally Ill Children and Their Mothers," *Psychosom. Med.*, *22*:456–465, 1960.
38. OFFER, D., *The Psychological World of the Teen-ager: A Study of Normal Adolescent Boys*, Basic Books, New York, 1969.
39. ———, and DIESENHAUS, H., "The Self-Image Questionnaire for Adolescents," Special Publication, Chicago, 1969.
40. ———, and SABSHIN, M., *Normality: Theoretical and Clinical Concepts of Mental Health*, Basic Books, New York, 1966.
41. PAYKEL, E. S., "Life Events and Acute Depression," paper presented as part of a symposium on "Separation and Depression: Clinical and Research Aspects" at the Annual Meeting of the American Association for the Advancement of Science, Chicago, Illinois, December 26–30, 1970.
42. RICHMOND, J. B., and WAISMAN, H. A., "Psychological Aspects of Management of Children with Malignant Diseases," *A.M.A. Am. J. Dis. Child.*, *89*:42–47, 1955.
43. ROGERS, C. R., "A Theory of Therapy, Personality and Interpersonal Relationships as Developed in Client-Centered Framework," in Koch, S. (Ed.), *Psychology: A*

Study of a Science, Vol. 3, McGraw-Hill, New York, 1959.

44. ROGLER, L. H., and HOLLINGSHEAD, A. B., *Trapped: Families and Schizophrenia*, John Wiley, New York, 1965.
45. ROMANO, J., "Basic Orientation and Education of the Medical Student," *J.A.M.A.*, *143*:409, 1950.
46. SABSHIN, M., "The Boundaries of Community Psychiatry," *Soc. Service Rev.*, *40*:3, 246–254, 1966.
47. ———, "Toward More Rigorous Definitions of Mental Health," in Roberts, L. M., Greenfield, N. S., and Miller, M. H. (Eds.), *Comprehensive Mental Health: The Challenge of Evaluation*, pp. 15–27, University of Wisconsin Press, Madison, 1968.
48. SILBER, *et al.*, "Adaptive Behavior in Competent Adolescents," *A.M.A. Arch. Gen. Psychiat.*, *5*:359–365, 1961.
49. SROLE, L., *et al.*, *Mental Health in the Metropolis*, McGraw-Hill, New York, 1962.
50. VAILLANT, G. E., "Theoretical Hierarchy of Adaptive Ego Mechanisms," *A.M.A. Arch. Gen. Psychiat.*, *24*:107–118, 1971.
51. VON BERTALANFFY, L., *General Systems Theory*, George Braziller, New York, 1968.
52. WEINSHEL, E. M., "The Ego in Health and Normality," *J. Am. Psychoanal. A.*, *18*:682–735, 1970.
53. WESTLEY, W. A., and EPSTEIN, N. B., *Silent Majority*, Jossey-Bass, San Francisco, 1970.
54. WHITE, R. W., "Competence and the Psycho-Sexual Stages of Development," in *Nebraska Symposium on Motivation*, pp. 97–138, University of Nebraska Press, Lincoln, 1960.
55. ZINBERG, N. W., and KAUFMAN, I., *Normal Psychology of the Aging Process*, International Universities Press, New York, 1963.

CHAPTER 9

THE CONCEPT OF PSYCHOLOGICAL MATURITY

Rachel Dunaway Cox

The Meaning of Maturity

Since the early 1950's maturity as a term and a concept has come into widespread use in psychiatry, psychology, social work, and education. Its facile employment in both formal presentations and informal discussions might suggest that the parameters of maturity have been systematically, unambiguously established. Actually this is by no means the case; although the state of maturity, often assumed to be synonymous with psychological health, is described in a growing body of literature, the number of empirical studies is still relatively small.

Psychological maturity is, in fact, a difficult subject for scientific inquiry and scholarly discourse. The hazard in treating it stems in no small measure from the tacit assumption that everybody—professional and lay people alike—knows what is meant by the term. The seeming general agreement about meaning may well arise out of the vagueness of the concept itself and the looseness of the way it is discussed. The inexactness and hence the slipperiness of the ideas clustering around psychological maturity is no doubt characteristic of any emerging concept and its lexicon, which will in time give way to greater precision. Meanwhile, the usefulness of the notion of psychological maturity is matched, perhaps exceeded, by its opposite—immaturity. Both concepts have come into rapid currency because they seem to describe aptly an important unitary something about persons. The great serviceability of the idea and the terminology has thus led to an accretion of meaning that makes for a certain level of communication. Indeed, maturity has become an all-purpose word, vague enough to fit into any context where a desirable, healthy, balanced commendable state is implied. Helen Lynd[76] perceptively points out an unacknowledged but implicit meaning that the term has come to carry—a meaning that nonjudgmental professionals would probably disclaim in referring to attitudes and behavior. The words "good" and "bad," she says, "have been replaced by mature and immature, productive and unproductive, socially adjusted and maladjusted. And when these words are used by the teacher, counselor or therapist they carry the same weight of approbation and disapprobation as the earlier good and bad," and constitute a code as specific as that they have replaced. If she is right, certain moral overtones have crept unbidden into estimates of maturity and immaturity.

In the field of education the concept is also found serviceable. Unhappily, teachers sometimes resort to the term "immature" when they either feel it is inadvisable to tell parents the hard truth or are themselves very unclear about what ails a failing or near failing pupil. In this situation confusion rather than understanding is communicated. Only one thing is certain when a teacher tells a parent her child is immature. This is that the teacher regards the child as unsatisfactory. In what ways and to what degree and with what hope of change for the better are often obscure to teacher and parent alike.

A frontal approach by research on psychological maturity or, to use the interchangeable name, psychological health has been long delayed. Preoccupied with pathology, psychiatry has until quite recently directed its attention away from health. Committed to precision of methodology and narrowly defined variables, academic psychology has explored isolated bands of behavior in experimental settings or limited itself to infinitely detailed and thorough descriptions of child growth. The Early Childhood Education movement, borrowing freely from both psychoanalysis and academic child psychology, has historically reached a more profound understanding than any other branch of psychology of how body, intellect, emotion, and social interaction meet and fuse with environmental opportunity to produce successive levels of maturity. In their close association with young children developing at a very rapid pace, nursery school teachers long ago saw that growth must be understood as a holistic phenomenon, indivisible into its separate components. Important and perceptive as its work has been, Early Childhood Education has remained for the most part a self-contained applied field. So-called basic research and theory building has been on the whole left to other branches of the discipline.

Thus for a variety of reasons the systematic exploration of the concept and status now known as psychological maturity has only fairly recently become a topic to be researched. Heath[63] who has produced one of the more thorough studies, acknowledges that the construct "maturity" has been criticized by psychologists impatient with inexact and subjective evaluations. He insists, however, that the results of his research on college men show *how* persons subjectively judged mature differ from those judged immature. In fact, the findings, not only of Heath but also of the body of empirical research on psychological maturity —scant though that research is—are in substantial agreement about the manifest traits and behavior of the so-called psychologically healthy person. This general unanimity points toward what Heath calls an underlying psychological reality.

Even while the loose and sometimes confused use of the terminology of maturity has gained momentum, research, theory, and expository discussion have produced a stimulating and significant, if numerically limited, literature. Many streams of thought mingle in present-day conceptualizations of psychological maturity. Contributions from the mental measurement movement, from the psychology of cognition, from sociology, anthropology, and social psychology, from both psychoanalytic and academic personality theory can be discerned in most scholarly and workaday references to the idea and the terminology. The significant point is that in its present use psychological maturity is a concept relatively new to science. While it unquestionably has a legacy from moral philosophy and religion, that rigor is softened by known facts about how development is rooted in genetic potential interacting with environmental offerings sometimes generous, sometimes niggardly. The concept includes the idea that physical soundness supports the totality of growth, but does not accept the notion that poor health forecloses the possibility of mature personality. The quality has been studied almost exclusively by using subjects who are intellectually well endowed, highly educated, and well above the poverty line economically; but the certainty remains in the minds of expert and laity alike that many individuals only moderately bright, poorly educated, and in the lower ranges of income actually do acquit themselves exceedingly well.

What, then, is psychological maturity? A consensus of experts on human behavior coincides reasonably well with the popular use of the term. This holds that mature persons are clearly aware of reality and that they do not run head on to violate it, though they do what they can to shape it to positive ends. Reaching out in trust and warmth to other persons, they are able to give and accept affiliation. Being essentially at peace with themselves, they are attentive to the needs, joys, and griefs of others. Enjoying productivity, they work in accordance with their gifts and tend to grow steadily toward higher levels of competence. These characteristics are attainable at any time of life by individuals of any status. They constitute what might be called the invariants of maturity.

It may be that the aptitude to achieve the somatic, psychic, social unity required has a genetic core. This is a question as yet scarcely formulated. It cannot be denied that good intellect, a basic minimum of economic well-being, social and vocational opportunity that protects against intolerable frustration, and, most critically, good physical health are circumstances most often associated with psychological maturity in researched groups. Yet the possibility of an overall effective level of being that satisfies the criteria of the reality-oriented, productive, and loving way of life is achieved by persons who have relatively little to work with. How such individuals cope with their situations, preserve their intactness as persons, and keep to a steady course is a research area as yet wholly unexplored. Locating and getting the cooperation of subjects and the choice of feasible methodology present problems so formidable that the hurdles in work with the highly endowed are child's play by comparison. Nevertheless, it is to be hoped that qualified investigators will make the attempt. Meanwhile, the overview of the field undertaken in this chapter must perforce deal with the research and discussion now available.

Work dealing specifically with psychological maturity has concentrated almost exclusively upon adult subjects, most of them in their twenties and thirties when full biological and social maturation has presumably been attained. Important early work was carried on by Murray[90] and by a group under medical auspices at Harvard (Hooton,[65] Heath[42]) before World War II. The years following the war brought a succession of studies (Barron,[13] Bond,[18,19] Brown[22] Cox[31,33] Sanford[115] Terman),[132,133] all of them beamed upon the adult years. On exploring what was essentially uncharted territory, these investigations had to adapt old or invent new methodologies that the individual researcher believed would produce relevant data. As publication dates show, much of this work was carried on simultaneously, each project having slightly different if similar aims and its own theoretical basis. As a result most of these investigations seem to begin *de novo*. It is captious to complain that studies did not at this early stage build one upon another. Timing and differences in research situations and resources as well as various aims made replication either impossible or inadvisable. The more significant fact is that work independently pursued and from different starting points came out with pictures of the psychologically healthy young adult that are in essential agreement from study to study. Furthermore, though the findings of the empirical work have given added dimension to the description of personality at its healthiest proposed by theorists and clinicians, they rarely contradicted it. Longitudinal studies did reveal some surprises. This had to do with the severity of the stresses encountered by presumably mature young subjects in their later experience. Some of the stresses were environmental and could be said to be of the "act of God" variety. Others were psychological and developed out of conflicts and pain unique to the individual's psychic organization. Nevertheless, one of the most consistently appearing trends in the psychologically mature subject was the ability to recover from stress. Faced with disappointment or loss or disillusion that shakes him profoundly, he regains his equilibrium as if by some inner "righting mechanism" and resumes his course. (Note especially Barron,[12] Cox,[33] and Heath).[63]

The steps by which the "righting" comes about are of particular interest. Cox, for example, found that continuing in an accepted role was supportive as was the less tangible framework implicit in a self-ideal. Ruff's[113] work with the astronauts revealed that their command of specific skills that gave them a range of appropriate ideas of what steps might correct a system failure enabled them to control anxiety in space flights.

From all the in-depth research on criterion groups comes evidence that mature people are sensitively aware of the persons and events around them and recognize and accept the claim of others upon them. The mature individual is usually in harmony with people and with the basic institutions of his society; but if his own value system and his own clear sense of what he is and what he wants to become run counter to the ideas and events he confronts in persons or in society, this tranquillity may be disturbed but not fatally disrupted. His work, as expected, is well suited to his capabilities and is pursued with satisfaction to himself and for the enrichment of his society.

The strong and resilient people who have been subjects are given various descriptive names or descriptive phrases by the investigators who studied them. They appear in the literature as self-actualizing, sound, psychologically healthy, or normal more often than as specifically mature. The different titles or names appear to carry essentially the same content and will be treated here as bearing on the same psychological phenomenon. In actual fact the subjects of this particular part of the literature have been biologically and socially adult.

Although the major research efforts have not denied the possibility that there is such a thing as psychological maturity before and after the blossoming of the early adult years, they have either not entertained the idea or have given it only passing mention. However, the present-day widespread use of the term and the idea of mature and immature levels of development in the growing years make imperative the recognition that there is not one but many levels of maturity—each appropriate to particular points in the life cycle.

Long before specific studies of psychological maturity under various labels began to appear, research in developmental psychology had produced a vast body of information about infants and young and school age children who were functioning in accord with their good potential. Terman[132] and Hollingworth,[64] Witty[141] and others studied gifted youngsters and revealed a complex of desirable personality traits as well as excellent physical endowment accompanying remarkable intellectual gifts. These researches did not set out to delineate psychological maturity as an entity in the latency years, but one of the significant by-products was just that. Contrary to belief current at the time that highly intelligent individuals were physically puny, socially withdrawn, and emotionally unstable, the gifted children were found as a group to be superior in every way, enjoying robust health, well-coordinated physically, socially well-adjusted and acceptable to their peers. Their behavior and personal qualities would seem to be precursors of the modes that much more recent research has found in adults who meet criteria of being notably well-functioning persons.

In numerous intensive cross-sectional and longitudinal studies of children appear evidences of individuals whose level of behavior reflects unusual capability in dealing with the experiences of their time of life. Retrospectively it seems clear that these children were working and loving in a way appropriately sensitive, complex, and effective as well as intellectually appropriate for their chronological age. Follow-up studies found them after infancy and earliest childhood maintaining not only their high I.Q.'s but, more significantly, their competent way of life. These findings suggest that psychic maturity can be validly viewed as a process, a continuously evolving state that can be attained early in life but that moves unceasingly toward later revisions as changing tasks and circumstances require. I shall have more to say presently of the concept of the norm from which this idea of psychological maturity derives.

Over the last four or five decades two threads of psychological investigation, following different theoretical paths and employing entirely different methodology, have simultaneously made important contributions to the concept of maturity. On the one hand, psychoanalysis, and the elaboration of its theory in ego psychology, has traced affective and intellectual growth toward full stature. On the other hand, more recently the psychology of cognition in children has become a powerful influence on the way maturity is conceptualized. Its single most significant thrust has been its explication of the orderly progress of the genetically rich but cognitively naïve infant toward the extraordinary complexity of abstract thought in the adolescent. These two approaches to the maturation process have been, and in large measure remain, separate despite the urgent need on the part of psychiatry, psychology, and education for a synthesis. Such a synthesis is by no means in the immediate offing, but the theories, empirical investigation, and expository statements of the analytic and the cognitive approaches seem to be moving together into much greater harmony.

Leading the way for psychoanalysis, Hartman,[57] as early as 1939, declared for an autonomous ego, active in "conflict-free" spheres of memory, imagination, and maturity. "I think," he wrote again in 1964, "we have the right to assume that there are, in man, inborn apparatuses which I have called primary autonomy, and that these primary autonomous apparatuses of the ego and their maturation constitute one foundation of the relation to reality. Among these factors originating in the hereditary core of the ego, there are those which serve postponement of discharge, that is, which are of an inhibitory nature."[58] He goes on to say that although the inhibitory functions derive from the conflicts between the ego and id, not all ego activities are devoted to such struggles. The ego is able to achieve on its own strength and to secure those achievements against reversibility.

The significant difference made in analytic theory by the postulation of an autonomous ego has only gradually been realized in the precincts of educational and descriptive child psychology where for so long psychoanalysis has been either rejected or disregarded. The revision, nevertheless, makes an incalculable difference in the acceptability of analytic theory in those circles and may eventually make synthesis possible. It is, however, Erik Erikson's[39,40,41] particular interpretation of psychoanalysis that has created a climate in which nonanalytic physicians, teachers, and academic child psychologists can approach the theory with lessened wariness. Erikson accepts the psychosexual stages of historical psychoanalysis and traces progress from the orality of infancy to full genitality at the dawn of adult life. When the individual falters on one step in the process, all that follows is in some degree compromised. The developmental sequence by which the ego eventually is able to deal effectively with the id and superego, and at the same time attain mastery of relevant actualities of existence, constitutes, he says, the history of maturation. Implicit in Erikson's view is the possibility of personal growth that can match the exigencies of unfolding experience, moving ahead unceasingly through early, middle, and declining years. He sees polarities of achievement and failure in succeeding epochs of the individual life cycle. These polarities state the central problem whose solution at each particular stage is necessary for the continuing development and health.

The polarity of basic trust versus basic mistrust in the early months of life is followed by autonomy versus shame and doubt in early childhood. In turn, the polarities of industry versus inferiority, identity versus diffusion, intimacy versus isolation, generativity versus stagnation lead on to the final stage in advanced age of integrity versus despair. Recognizing learning as related to these steps in growth, and weaving into his theoretical fabric a consideration of social identity, Erikson speaks with a persuasive voice to many schools of thought and has done much to lower the barriers that have so long compartmentalized descriptive, experimental, and learning psychology from the psychoanalytic and clinical. The massive resurgence in the last decade of the experimental impulse in child psychology and the rise of the conditioning therapies take the science in a direction that seems for the moment to negate the rapprochement of recent years. Yet neither experimental nor neo-Pavlovian work has remained immune to what Erikson and like-minded psychologists have been doing. The new research differs markedly from the laboratory work with infants and children before the publications of Ribble,[108] Goldfarb,[51] and Provence,[103] among many others.

Maturity of Cognitive Processes

The contributions of the psychology of cognition to the concept of psychological maturity are of a different order. Beginning in the early 1950's research and theory building directed toward the discovery of ways of knowing in infancy, childhood, and adolescence took a vigorous leap forward in America. This was due in large part to the impact of the work of Jean Piaget.* His prolific publications have continued to appear in a steady stream since the late 1920's, when his interest was first aroused by the ways in which children's comprehension of the world differs from the adult mode. He pointed out—and his observations and researches over half a century support his conclusion—that many fundamental ways of thinking usually assumed to be part of inborn human equipment are actually the upshot of maturation, of discovery and learning on the part of the infant and young child.† Equally important for both theory and practical affairs is his demonstration that cognitive mastery awaits both the biological maturation coming with the passage of time and the assimilation of experiences made accessible to the developing organism by its own maturing structures and the offerings from the environment.

The somewhat cool reception accorded Piaget's early work in the United States seems to have been due to his overreliance upon verbal responses from child subjects whose ideation, American psychologists believed, was far more sophisticated than their language could express. Piaget himself now concedes the point. The further fact that most of his work was available only in French limited his hearing in America. His later studies, now translated, on his own three young children and his ingenious experiments with larger groups of different ages have aroused enormous enthusiasm here, as elsewhere, and stirred a growing company of psychologists to replicate and extend his approach and methods. In general, his ideas about the stages of cognitive development have been supported by investigations in Europe and America (Elkind,[36] Smedslund,[122] Wohlwill).[143] These stages will be described in more detail later in this chapter.

* The work of Heinz Werner,[138,139] though in many ways paralleling that of Piaget, has until quite recently been less well known. His theory of mental development evolved over 40 years ago into an abstract formulation in which he confronts the problem of stability in the presence of ongoing change. His teaching at Clark University profoundly influenced his students. Piaget's enormous impact on American psychology has in a sense made developmental psychologists more aware of Werner.

† Stimulus-response learning theory and methodology have produced a vast literature that bears a tangential relationship to cognitive psychology and through cognition to the holistic interpretation taken in our approach to psychological maturity. S-R learning work dealing with small bands of behavior emitted by subjects responding to strictly controlled stimuli stands, however, in sharp contrast to the genetic approach, which has proved more useful in the understanding of persons as functioning wholes. White's[140] excellent review of the S-R learning theory tradition in relation to developmental psychology is illuminating.

Operant conditioning as a therapeutic method is, of course, an application of S-R learning theory, as is Skinner's contribution to education in machine teaching. These topics are dealt with more appropriately in another chapter of the *handbook*.

He describes the development of intellect as taking place through the processes of assimilating experience and accommodating the existing structures or schemata to the unfamiliar information. This, in turn, creates increased readiness for assimilating new incoming experience and the elaboration of schemata to the new state of affairs.

This formulation actually constitutes a fresh statement of the vexing nature-nurture problem that reached stalemate some 30 years ago. As such it speaks with particular relevance to whole societies as well as to professionals concerned with maximizing human potential. In this context maturation is seen as an open-ended proposition dependent neither wholly upon genetic constitution nor upon socially proffered nurture, but upon some as yet loosely defined, dynamic interweaving of their respective contributions. Concurrent evidence from animal laboratories showing how early deprivation stunts and distorts a wide variety of growth patterns has provided an extension of data from the experimental sector.

The importance of Piaget's work for therapy, teaching, and social policy can scarcely be overstated. There are, of course, levels of deficit that no amount of skill in the helping process or enrichment of nurturance can overcome, but the necessity of continual provision of stimulation, highlighted by Piaget's work, as a requisite to further growth provides solid theoretical foundation for invigorated social and individual efforts on behalf of the socially deprived as well as the meagerly endowed. His work has also given additional urgency to social planning for early childhood since he stresses that later abstract levels of thinking rest upon foundations of schemata developed in early successive stages of growth. The critical importance of the capacity for abstract thought in developing a citizenry capable of democratic self-government is not actually stressed by Piaget, but it is of vital concern to American educators and is likely to be influential in shaping teacher education in the last quarter of the century.

Although firm in his view that the sequence of steps in intellectual maturation is invariant, Piaget is less positive about what happens in the long run when there is a time lag in attaining any given level of growth. He surmises, however, that there is an optimal rate of progress from stage to stage. If the organism is delayed in a new phase, the ability to achieve the next level organization of experience may be unfavorably affected since the "new organizations cannot be indefinitely postponed . . . since they would lose their power of internal combination."[97]

Piaget is, of course, concerned with the theory of cognition, but the principles and the stages he delineates also have applicability to treatment situations. The capacity to think, transpose, understand, and generalize can be, and indeed are, fruitfully adapted to the affective sphere. Thinking abstractly is supportive of the reorganization not only of cognition but also of behavior and emotion. The almost universal reluctance to make psychotherapy available to mentally retarded individuals has some justification in the fact that such patients, being handicapped in abstract thought, make little progress in the usual type of psychotherapy.*

A concern, which Piaget sees as typically American, with whether intellectual maturation might be accelerated in extremely bright children by "forced feeding" appears to him ill-considered, and he sounds a warning on what the upper limit of expectation had better be.

A review of the range of work on psychological maturity reveals that the cognitive factor is a critical issue. As Heath[63] points out, "A theme that weaves through most developmental theories is the centrality of man's maturing cognitive processes in assisting his adaptation." Both ego and cognitive psychology recognize this. The development by the two separate theoretical systems of the concept of stages—each resting upon that which goes before and

* The neglect of the retarded by psychiatry is currently being remedied to some degree by the implementation of comprehensive mental health programs. Public and private education and the multifaceted program of the National Association for Retarded Children have undertaken the habilitation—as distinct from rehabilitation—of this group and will probably continue to do so under comprehensive mental health care with more adequate support and overall coordination.

laying the foundation of what is to follow—suggests that both schools of thought discern currents of growth that proceed along similar developmental paths. The fact that one system has addressed itself primarily to the affective sphere and the other to the cognitive makes the parallelism the more interesting.

As the foregoing discussion indicates, psychoanalytic, developmental, and clinical psychologies espouse quite similar assumptions about an interacting somatic, psychic, social indivisibility with predictable and stable sequences related in a general way to the sequences of life experience. Two other lines of investigation are of special importance to an understanding of the nature of maturity. The first of these is the concept of age-appropriate development. This idea has grown out of the mental measurement movement. Originating early in the twentieth century, mental measurement now finds many of its assumptions taken for granted in lay as well as in professional thinking. Perhaps the most important of its working postulates is the use of the norm or central tendency as the reference point in evaluating performance and attitudes of many kinds. The second line of investigation derives from the self-psychology of academic personality theory.

The ideas, the methodology, and the extensive testing programs associated with mental measurement and normative research have always been strongly practical, even though important theories about the nature of ability based upon statistical manipulation of test scores have been a notable spin-off. Self-psychology, on the other hand, has been very largely theoretical and speculative, buttressed by both clinical experience and insight. The literatures of the two fields overlap chiefly where empirical studies of personality employ psychological tests. It is true also that the notion of the norm is implicit, even when unexpressed, in most evaluative approaches to living organisms. In the field of measurement the existence of an underlying integrating unity in persons, a major concern for self-psychology, is taken for granted. Ways to measure ego strength, which in many discussions seems to be used interchangeably with personal integration or self-stability, is of interest to theorists and clinicians alike. Personality inventories such as the MMPI and several of the projective techniques have been used by self, or ego, researchers sufficiently sophisticated in these methods to use them effectively. (See especially Barron,[13] Heath).[63] It is probably not unduly pessimistic to say, however, that the clear delineation of norms for overall psychological maturity, and the measurement of self or ego strength, are still in the pioneering stage.

Age-Appropriate Maturity

Early in this century Alfred Binet developed the seemingly simple, but extraordinarily seminal, idea of age-appropriate mental development. Binet's very practical application to educational purposes of Cattel's earlier work on individual differences is based upon the recognition that a child's relative brightness must be judged by relating what he is able to do to his own chronological age, and that within broad limits his performance at any given age must be evaluated by comparing it with that of other children of the same life age. This device, later extended, with certain technical modifications by other workers, to adult subjects (Wechsler,[137] and the testing programs of the armed services in World War I and World War II), has led to the elaboration of a vast and varied structure of norms in the form of quantitative statements of expectable, average performance on tests of scholastic aptitude and achievement, manual skills, attitudes and states of emotional health.

The age-appropriate idea leads at once to the proposition that the status called psychological maturity—that is, the ability to deal effectively and resiliently with experience—cannot be conceptualized as a state achieved only in biological and sociological maturity, coextensive with the early and middle years of adult life. It is true, of course, that almost without exception the specific research on so-called psychological maturity has been done on adult subjects. This is probably so because intensive study of normal and supernormal persons has only fairly recently attracted research talent. It can be argued, nevertheless, that there are as many levels of psychological maturity as there are steps in the life cycle.

The point of view advanced in this chapter is that for each age span certain developmental tasks are implicit, and that psychological maturity is attained when these undertakings are accomplished in phase with chronological age and, in some degree of synchrony, with the expectations of the society of which the individual is a member. In infancy and childhood the rapid growth of the biological basis of the person increases and elaborates the requirements of age-appropriate maturity at dizzy speed. With every widening of the visual field, every extension of physical mobility, every proliferation of human contacts, every possibility for responding to social incitement, the complexity of required attainment increases. Thus, to maintain a once achieved relative level of psychological maturity, the individual must undertake continued revision.

At no time of life is static maturity possible since new situations demand new adaptation. Furthermore, at certain critical junctures accelerated alterations of overt behavior and inner growth are required. Separation experiences occurring in new contexts at different points in the life cycle, the taking on of new roles and heavier responsibilities, the elaboration of human relationships as extension of the social network calls for increased subtlety and sensitive responses to hierarchies of loyalty—these and countless other passages change the constellation comprising mature feeling, thought, and action. Similarly in the decades following the middle years, receding physical resilience, the withdrawal in retirement of the social consensus concerning the individual's meaning and worth, and the loss of significant others make unprecedented demands.

As Scott[119] has aptly said, "Maturation is more than ripening or growing older; it involves organization and differentiation as well." Of psychological maturity this is especially true. Equally true is the fact that only if reorganization proceeds unceasingly can equilibrium be maintained.

Psychological Maturity in Infancy and Childhood

To be in phase with appropriate overall maturation in the earliest days of life depends in major proportion upon sound biological equipment capable of advancing through normative processes of organic growth, myelination, and differentiation. This is not to say that individuals born with crippled bodies or damaged nervous systems cannot possibly achieve integrated and effective lives. Evidence provided by occasional remarkable personalities shows that fully flowering personalities can develop in tragically damaged bodies. Nevertheless, very early somatic deviations, especially those involving the central nervous system, will slow and run great danger of permanently compromising healthy personality. Since manifestations of psychological adequacy in infancy are both experienced and expressed in somatic terms, being in phase with chronological age at that time of life depends upon soundness of body. This is especially true in the early stages of extrauterine life when the infant's grasp upon his world can be secured only by his contact with the space, sounds, touch, smells, and temperature of that world.

The interaction of the growing infant with his milieu in reaching toward comprehension of his world has been richly documented over the last hundred years by students of childhood as diverse as Darwin,[34] Preyer,[101,102] Shinn,[121] Pratt,[100] Carmichael,[27] McGraw,[78] Gesell,[49] Bayley,[15] Bühler,[24] Escalona,[42] Kessen,[69] to name but a few from an innumerable list. The most favored approach to the study of maturation in the growing years has been the naturalistic or genetic one. Using longitudinal methods with fairly small groups, and cross-sectional approaches with much larger numbers, developmental psychology has minutely catalogued somatic structure, speech, ideation, social behavior, and the like from birth through adolescence. Norms of growth have been identified at closely spaced points along the continuum of chronological age with a competence that will probably never be excelled. From the normative picture of different age levels, the characteristics and features of outstandingly effective functioning in the growing years can be spelled out.

As noted above, Jean Piaget has infused the genetic approach with new vitality. He cannot really be classified as concerned with norms in the usual sense, because his commitment to exploring the epistemology of cognition gives his work a theoretical and philosophic depth and breadth scarcely found in descriptive accounts of growth. Nevertheless, in setting out a sequence and in giving the steps approximate time boundaries, he does contribute to the notion of age-appropriate cognitive development.*

The importance of the infancy and early childhood work for our understanding of the maturation of the total person can scarcely be overstated. From the earliest baby biographies down to the present, these researches throw into bold relief the basic interrelatedness of physical, emotional, and intellectual factors. Since World War II a series of studies have highlighted the blighting effect upon the growing organism where resources for wholesome growth are not available. The benchmark studies of Ribble[108,109] and Spitz[124] were followed by numerous investigations that sustained the early finding that infants deprived of ongoing contact with supportive emotional and social nurturance fall behind on all dimensions of development—including alimentation, neuromuscular coordination, cognitive and social learning. The vehement critics of the Ribble and Spitz work have tended to fall silent as evidence has accumulated sustaining the central theses of their reports. The meticulously documented studies of Provence and Lipton[103] demonstrate how lack of development merges into pathology—the marked opposite of a robustly normal state. More recently Rheingold[106] has focused attention on the springs of language acquisition in the preverbal interchanges between the infant and his caretaker. Acting as a mother surrogate with a small group of institutionalized infants for a period of weeks, she offered them opportunities for vocal exchange such as mothers carry on with their own babies. Later evaluators, uninfluenced by knowledge of which infants had received social-vocal stimulation and which belonged to a control group, found the experimental infants notably more socially reactive and, indeed, somewhat more competent on baby tests than the controls. Rheingold's finding is in line with that of others who report that language acquisition is especially retarded by the absence of stimulation. Her experiment gives, however, the positive side of the coin in demonstrating that the introduction of preverbal stimulation makes a significant difference in the infant's own preverbal outreach.

Tracing the progress of age-appropriate affect has been the special province of psychoanalysis. According to that theory, the neonate is—and must be—an oral being, since to fail in the task of sucking, as some newborns with seriously impaired nervous systems do, is to incur grave hazard to life itself. But as Erikson has pointed out, the infant takes in much more than physical nurture in this oral stage. He knows the world and interprets it in accordance with his oral experience. According to Erikson, the basic emotional achievement at this stage is the development of a belief in the readiness of his world to care about and for him. At three, or thereabouts, he will have come to terms with society's initial interference with his reflexive and instinctual life and begun to channel and control his intense loves and hatreds.

The ego of the psychologically mature child has gathered considerable strength by the time he is of school age. The conquest of his oedipal strivings has laid the foundations for identification and social feeling. His control of his own body, his command of information and skills, his growing awareness of the "rules" and expectations of society, as well as his capacity for satisfaction in knowing those rules and rising to those expectations, are impressive. At this point his faith in the world and his confidence in his own powers ready him for the further complications of life beyond the reach of his mother's arm.

* Editor's Note: For a description of Piaget's approach to cognition, see Chapter 15 by Jules Bemporad.

Latency, as postulated by traditional psychoanalysis, is relatively free of the somatic and emotional struggle around sexuality, although Kinsey's retrospective reports on childhood showed that the years from six to twelve are by no means asexual and that interest in sex and sporadic sexual activity are neither perverse nor abnormally precocious. However, the most telling indicator of psychological maturity at this stage is the style and energy with which the child gets on with the expansion of scope and skill in the ego apparatuses. Erikson's depiction of the latency period as characterized, in emotional health, by industry speaks most congenially to developmental, social, and educational psychology. The industry that he proposes as the central enterprise of normal latency is expected to undertake school learning, to confirm a biologically suitable sexual identity, and to come to identify the self as a member of small and larger social groups. The opposed polarity, inferiority, is expressed in failure to achieve these goals. As teachers know, lag in the mastery of the various forms of symbolic learning does not merely constitute inconvenient deficits in the tools of school work. Such lacks are felt by the child and the society as deficits in the individual himself. The shock waves of these failures, reverberating into every aspect of the child's life, damage even the most tender social relationships, undermine self-esteem, and deteriorate self-image. This being so, it is inevitable that concern be keen that the young person "keep up" and remain close to suitably advanced school achievement.

Criteria of psychological maturity in middle childhood thus must include adequate progress in classroom learning. In deciding what is adequately "mature" in such matters for any child at a particular point in time, educators lean upon age and grade norms where these are available, as they are for most school subjects. The dangers inherent in the rigid use of norm tables in dealing with individuals who differ markedly from the majority of schoolchildren in cultural nurture or personal idiosyncrasy is fortunately well known today. At the same time humane use of norms as guidelines for teaching, and occasional and judicious resort to them as a motivating force, can be helpful to both the child and his mentors. They serve to keep the realities of the child's school situation in view. School learning is, after all, a pressing reality for the child.

In the more subtle matter of becoming the sort of person the family, the school, and the society demand, landmarks are far more vague. There are no quantified norms for being an acceptable boy or girl, a normative son or daughter, a good team member. Although rarely spelled out, standards are nonetheless communicated constantly and are correctly understood by the latency child to be fairly unbending. He learns by apprenticeship, as it were, how to act and what roles to carry out. The histories of adults studied as notably mature reveal that these men and women were capable schoolchildren, not only successful in school learning but also in the varied and more inclusive roles and functions of their personal lives. There is little solid evidence on whether children who fail badly over an extended period in the tasks of latency can later achieve a high level of personal integration. We do know that adults who have been studied in criterion groups have not been failures in the past. Research on the psychological status of grown-ups who were earlier diagnosed as dyslexic or otherwise severely out of phase in school learning might throw much needed light on the question.

Psychological Maturity at Adolescence

As childhood draws to a close, changes associated with puberty accelerate changes in psychological status. Intensifying demands of school curricula enforce decisions entailing long-range consequences. Educational choices made at this point are, happily, more reversible in the United States than in most societies, but even here the child must begin to think with a time perspective he has hardly known before. What he does—or more stringently, what he does not do—in his school tasks can be recouped only by unusual exertion. In his acceptance of these hard facts he expresses his developmental maturity. As to sex role, the individual has, of course, never had a choice, and awareness of his correct role has been emphasized since early childhood. In Western cultures, however, equalizing of opportunities and diversification in sociological roles has in the last half century tended to focus attention upon the samenesses in being human at the expense of attention to sex differences. With puberty, however, the female, especially, is confronted inescapably by the physical aspects of her femininity along with all its implications for emotional and sociological specialization. While physical maturation is less likely to be viewed as a mixed blessing by the young male, the sociological challenge to his male aggressiveness, toughness, and responsibility is formidable.

Both male and female must now relinquish old certainties and accustomed protections. Erikson[41] describes adolescence as a period of natural uprootedness. Part of the maturation process is the questioning of values until now taken for granted. As this evaluation goes on the youth retires into his inmost self, questioning what he is really like and what he may become. This state of affairs is reflected in his Rorschach protocol, which now shows a decided swing toward introversion. This is surprising in view of the strong group orientation of teen-agers. But as Cox has pointed out, the teen-age clique serves as a kind of public relations agent for the adolescent while he, at his deepest levels, goes into executive session with himself. Meanwhile, observing the social scene, seeking clues about himself in the ways his fellows relate to him, trying out attitudes and roles in which he only half believes, he moves toward a revised individuality. So much is opening up for the adolescent that anything may seem possible to him. Fantasy and enthusiasm together with his newfound powers to entertain the hypothetical entice him at times beyond himself. The line between adolescent dreams and unreality may, even in the healthiest, sometimes become shadowy.

Erikson[39] sees as the critical task the securing of the emerging identity against diffusion. In historical periods such as ours, rapid cultural change intensifies for all age groups beyond earliest childhood the problems of who one is and where he shall stand. For the adolescent whose self-revision is already creating so many inner uncertainties, the chaotic social scene presents almost intolerable confusion about values, means, and ends. Emotional balance at this difficult time is best maintained, so clinical work indicates, by individuals who are able to preserve continuity with the past self while reorganizing and revising standards and goals in terms compatible with the realities of their present situation and with what the society beyond the family will expect of them and permit them to do and be.

With the upsurge of genital capacity young persons who have come through earlier phases of emotional growth and sex identification successfully now turn toward members of the opposite sex for recognition and confirmation of acceptability. The flowering of the capacity to love and to be loved by a sexually different being, and more especially to commit oneself unreservedly in trust to another, is the critical next step for the genitally mature. By the end of the adolescent years the emotionally mature man and woman are able to enter into faithfulness in love and the competence in work that can undergird the long-term relationship and the responsibility of nurturing a new generation.

Psychological Maturity in the Adult Years

Age-appropriate psychological maturity in the adult years has been more explicitly researched than that in any other period. The qualities of personality found in actual persons comprising criterion groups have been set forth in psychological test terms and in verbal descriptions by a number of investigators (Barron,[13] Bond,[18,19] Brown[23] Cox,[33] Golden,[50] Heath,[62,63] Hooten,[65] Sanford,[114] Maslow).[86,87] Informed reflections, growing out of clinical work, teaching and other human encounters, about the personality of the mature adult are found scattered through the literature of personality theory. (Note particularly Allport,[6,7] Fromm,[48] Lynd,[76] Rogers,[110,111] Schactel[22]). The mature adult's qualities, as noted in the first section of this chapter, fully support Freud's surmise that in psychological health the individual would be able to love and to work.

Cox,[33] taking functioning as an indicator of maturity, inquired into the adequacy of her subjects in the five spheres she deemed most important: preparation for vocational future, effectiveness in work, relationship to spouse and children, relationship to aging parents, and management of the practical realities of the home and money. Despite the generally effective functioning that this study revealed, her findings agree with Heath's[63] that psychological maturity does not bring, even in these years of greatest emotional fulfillment, exemption from conflict and pain. The heightened awareness that their well-developed intellect and emotional comprehension bring, their openness to sensory impressions and intuitions, their essentially object-oriented approach in human relationships, and their sensitive involvement in the social issues of their time and place make them vulnerable. Their emotional health lies in their assimilation of these enriching experiences and their use of them to reach still higher levels of personality organization.

The capacity to love deeply is not, as we have seen, reserved to the span beyond adolescence. But during the adult years it takes on greater depth and scope. The mature love experienced in lasting heterosexual commitments opens realms of feeling and action new even to those who have been able to love sincerely and securely in childhood and youth. Maslow found such relationships at their best to be singularly spontaneous, free of pretense and defensiveness, generously giving, seeking the comfort and enhancement of the beloved. Such a characterization is essentially that reported by the fairly limited number of other investigations that have inquired into the nature of love among adults in sound emotional health.*

Erikson, placing emphasis upon the production and care of a rising generation, gives to generativity the central role in emotional maturity of the adult years; and Saul[117] sees the balance between receiving and giving to be more heavily weighted on the giving side.

* Psychology as a science has shown marked reluctance to study tender, affiliative relationships. Allport[7] notes that Ian Suttle sees this as a "flight from tenderness" spurred by the fear of seeming sentimentality. Despite the general taboo, investigators of psychological health have found love relationships to be a central theme in the lives of their subjects.

Later Maturity and Old Age

The dimensions of psychological maturity in the last three decades of life have been explored very little by formal research. There is, of course, abundant evidence on the intellectual status of late adulthood and old age. In a long series of studies, beginning with the cross-sectional research of Jones and Conrad and culminating in the standardization of the Wechsler-Bellevue Test of Adult Intelligence, the curve of growth from childhood to late adult life has been exhaustively documented. When the summed scores on standardized psychological tests are used to represent intellectual growth, the curve reaches its highest point in the early twenties, then enters upon a plateau that it maintains with relatively minor loss for two or three decades. Thereafter, the curve declines, reaching at length in the late sixties and early seventies the level attained by the average 14 year old. When the total score was analyzed and subtest scores scrutinized, it was found that the decline differs from one kind of ability to another. Verbal facility is maintained or increases right up to the onset of senility; tasks requiring completely new associations and those calling for rapid eye-hand coordination lose ground early and at a relatively rapid rate; abstract thinking, particularly as measured by dealings with spatial relations, occupies a middle range.

Later studies using longitudinal methods have thrown further, more optimistic light on the problem of aging. Intelligence tests repeated after 20 years on the same subjects did not yield falling scores. The earlier results seem to have been produced in part at least by an artifact of poorer schooling in older subjects. Yet when younger subjects today whose schooling is more recent and presumably more adequate are compared with older subjects, the young still outscore the older.

On the broader, more important, issue of emotional maturity unique to the closing years of life, there is very little solid evidence. Much is written and said about the unhappy plight of the very old—their loneliness, their sense of being forgotten, their sense of disintegration as persons because there are so few messages of reassurance from the non-self world. It is true that individuals selected as outstandingly mature in their earlier lives enjoyed then remarkably good physical health, generous recognition in their work, and satisfaction in their close love relationships. Does it follow that if and when these supports are withdrawn, as they inevitably are for most of the aging, there is no possibility for further growth or even for maintenance of already achieved levels of psychological maturity? Or is it reasonable to hope that in the presence of an intact central nervous system, the values, the manner of relationship, and the certainty of what matters, developed over a lifetime, can sustain and carry the individual onward to some hitherto unattained upland? Few psychological and social questions press so urgently for an answer. In this cause it is important that longitudinal studies already in progress be extended into the later lives of the criterion subjects.

Meanwhile, gerontology is discovering that not only in late maturity but also in advanced age, men and women can be restored and maintained in health by a vigor of physical activity once thought to be surely lethal.* The productivity of unusually endowed persons into their eighties suggests that continued work spurred by expectations on the part of the community supports emotional health and, indeed, the totality of the person.†

Erikson believes that the later years offer the opportunity for the fully developed person to affirm the worthwhileness of the human enterprise and his own particular part in it. In rounding out this final phase of his growth, the individual may arrive at wisdom that maintains and conveys the integrity of experience despite the decline of bodily and mental functions. He concludes: "Potency, performance and adaptability decline, but if vigor of mind combines with the gift of responsible renunciation, some old people can envisage human problems in their entirety." Such old people can, he says, "represent to the coming generation a living example of the 'closure' of a style of life."[41] Such acceptance and responsibility might well challenge human beings to further levels of maturity until the very end of life.

In sum then, the presence or absence in any individual of the quality of maturity is expressed by the way he bears himself in the tasks and roles of his particular time of life.

The Mature Self

The question of how an organism develops a style peculiarly its own, while keeping within the biological template of its species, continues to challenge science. Genetic endowment and the stage of physical and cognitive growth, together with the input from the environment, set limits. The level of actual achieved maturity is, however, better conceived as the outcome of the ways in which these contributing components are marshaled in the service of the needs and purposes of the total organism. The synthesizing of resources, the selective responding, the directing of energy, and the harmonizing of many levels of thought and action in their totality are known as integration. The tendency or entity that makes integration possible appears to have its origins early in the life process. Since it unquestionably makes its appearance soon after birth and since its mode of relating to experience shows a certain consistency even then, it is surmised that this embryonic tendency toward a unique style may have an hereditary core. This entity, which becomes clearer cut and more individualized as experiences and the memories of them accumulate, is referred to by some theorists as the self. Others call it the ego. Although the terms are not synonymous, both desig-

* Research on very vigorous physical exercise by subjects in their eighties has produced very promising results at the Lankenan Hospital, Philadelphia.

† Note particularly such people as Pablo Casals and Picasso in the arts, Hilda Smith in public service, and Lily Taylor in classical scholarship.

nate a central, powerful aspect of human personality—its capacity to behave as a coherent unity over time, in a wide range of situations. A substantial portion of the theoretical and empirical literature on psychological maturity is devoted directly or indirectly to the nature of this unifier and undertakes to spell out its dimensions and the conditions of its development.

The present-day interest of academic psychology in the self represents a significant return to a topic that the science largely abandoned after 1890 or thereabouts. Some psychologists believe that the revitalized interest in the self is evidence of a maturing process in the science itself since the trend means that there is in some quarters less disposition than formerly to treat as nonexistent an issue for which current conceptualizations and methods were inappropriate or inadequate. Historically, when psychology broke away from philosophy late in the nineteenth century, the notion of a unifier became increasingly uncongenial. William James suggested that there was no need for what might be called a supervising entity. For him the self was only the sum total of experience. "Each moment of consciousness," he said, "appropriates each previous moment and the knower is thus embedded in what is known. The thoughts themselves are the thinker" (Allport).[3] A few psychologists resisted this view, insisting that a unifying principle held the parts together and was the overriding phenomenon (Allport),[3] but they fought a losing battle. Uneasiness among the proponents of the new theory and methodology may have arisen, Allport[3] surmises, out of their fear that psychology would excuse itself from its chosen tasks by assigning what was not readily understood in psychic behavior to "a mysterious central agency."

Psychoanalysis, however, going its separate way, never relinquished its interest in a unifier. Indeed, as Allport[3] points out, "Freud played a leading, if unintentional, role in preserving the concept of ego from total obliteration throughout two generations of strenuous positivism" in psychological science. Freud's own ideas about the ego shifted over time and the concept has been further elaborated by Anna Freud.[45] The ego psychologists, as indicated above, introduced into psychoanalytic theory important modifications, giving the ego its own autonomy from the instinctual drive sources postulated by analytic theory.

Meanwhile, the turn of the wheel in academic psychology brought into being a point of view and a body of research that bears directly upon the idea and meaning of psychological maturity. This relatively recent development makes it possible to entertain the notion that there is such a psychic phenomenon as the self and to ask what a mature self is like. In his 1955 account of the renewed interest in the self by academic psychology, Allport[3] said that many psychologists "had commenced to embrace what two decades ago would have been considered a heresy. They have reintroduced self and ego unashamedly and, as if to make up for lost time, have employed ancillary concepts, such as self image, self actualization, self affirmation, phenomenal ego, ego involvement, ego striving and many other hyphenated elaborations which to experimental positivists still have a slight flavor of scientific obscenity." A rising generation of psychologists has been less bound by the self-imposed restrictions of their predecessors and have been able to assert that an indispensable something was missing from the positivists' account of the human psyche. Whether the new work on the self takes theory beyond the point where the road forked in the nineteenth century remains to be seen. What is crucial, however, is that when some members of the psychological community turned away from the molecular approach, they returned to the field, as it were, and thus may be in a position to make progress in the understanding of the self.

Relevant to our discussion of psychological maturity is the fact that in most treatments of maturity the idea of the self is recurrent. The notion that the psychologically healthy person has a solid sense of himself that he effortlessly affirms most of the time is substantiated by research and informed discussion. Like so many aspects of health, the robustness of the self is taken for granted unless it is belatedly or imperfectly established, or, having been partially developed, becomes unsteady. Then its keystone position in personality structure is unmistakable.

Of course, a unifying biological principle is operating in every living organism; but at the human level it masters extraordinary complexity, with the psyche not only maintaining its own consistency, capable of planning activities and developing values, but also standing apart from and regarding itself. It is symptomatic of the profound disturbance of the present era that uncertainty about the self is so widespread. It may well be, as Erikson has suggested, that uncertainty about identity is endemic in our time. On every hand are heard reports of young people who are vague and bewildered. They "do not know who they are" or what they are all about. The drug culture appears to be one of the malignant symptoms of the confusion. In the idiom of academic personality theory the unhappy state results from a poorly developed self. In psychoanalytic terms the egos of these individuals are weak, brittle, shaky.

Psychological maturity, on the other hand, is characterized by a sense of inner cohesion that makes it unnecessary for the individual to question who he is. Except at natural choice points—as in career and mate selection—he is pretty clear about where he wants to go and about how his gifts are matched to the journey. Very early he develops a life style that he maintains over time, undertaking those inner revisions and those manipulations of milieu that enable him to realize his own potential and to use environmental givens fruitfully. In this vein Maslow[87] describes self-actualizing persons. Fromm[48] says that "the mature and productive individual derives his feeling of identity from the experience of himself as the agent who is one with his powers." Lecky insists that self-consistency is a necessary component of health. Heath,[63] writing of the mature person whom he uses as coordinate with the term "mature self," finds him to be stably organized, able to resist disturbing information, well-integrated, allocentric in his thought and feeling, and notably accurate in his self-image. Barron[12] describes ego strength, a significant characteristic of the mature self, as a function of intelligence since, he says, the ability to comprehend experience, to store it in memory, and use it in positive ways depends upon the quality of the brain. He goes on to say, however, that intelligence alone does not account for ego strength. No less important are flexible controls that use repression selectively but at the same time keep the person "optimally open to experience." On the basis of his data on personal soundness in a criterion group of graduate students, he concludes that within that population of subjects of ordinary physical and psychological integrity, soundness is by no means exclusively determined by circumstances but may be considered in the nature of an unintended—perhaps largely unconscious—personal achievement. Of his subjects, he said, "They are sound largely because they bear with their anxieties, hew to a stable course and maintain some sense of the ultimate worthwhileness of their lives."[12] Through these descriptions of psychological maturity runs the thread of a clearly defined, well-maintained, purposive self moving toward judgmentally selected goals.

To recognize and even to describe the mature self is not the same as understanding it or divining its secret. Research on groups of mature persons gives an account of the nature of maturity and its antecedent circumstances: adequate intellect, stable childhood home, at least minimal economic provision, parents devoted to the child's welfare, educational and vocational opportunity. These favorable factors show high positive correlation with judgments of maturity in selfhood. Yet most of the studies also report that exceptional individuals become psychologically mature in the absence of one or several of these supports (Barron,[13] Bond,[18] Cox,[33] Heath).[63] Cox found that, in general, the more numerous the misfortunes and deprivations the greater the likelihood of loss of the personal equilibrium established at an earlier time. This agrees with the findings of the large sample study of Langner and Michael,[73] but Cox insists, as does Barron, that this trend is contravened in some subjects who grow to full stature despite difficulty.

Thus, the possibility that maturation may actually be accelerated and confirmed by felt difficulty is indicated by empirical research. Psychoanalysis long ago proposed that the ego grows through frustration and denial. Nagera,[92] following the path laid out by Sigmund and Anna Freud, asserts that "some aspects of human development are due to frustration of primary needs of the newborn infant. Such frustrations," he adds, "are inevitable even in the presence of ideal conditions and ideal mothering." He adds that disturbance of the basic homeostatic equilibrium acts as a spur toward recovery of the equilibrium and concludes that "complete satisfaction would have made development from the original state unnecessary." This statement of the dynamics of ego growth recalls the growth of cognitive processes as Piaget outlines it. The development of cognition comes about when assimilation of new information demands accommodation of existing structures. This, in turn, facilitates the development of more complex schemata. The beneficent effects of stress are, of course, circumscribed by the capacity of the individual to absorb it, and if, as in the case of institutionalized infants, the deprivation is extreme, development is blocked.

Some years ago Claparède[29] inquired where subjects felt the self to reside: in the right hand, in the viscera, in the center of the forehead, behind the eyes, and so forth. This somewhat whimsical approach no longer engages psychology. The notion of body image, however, is influential in clinical practice and has inspired substantial research (Albee and Hamlin,[2] Berman,[17] Elkisch,[37] Machover,[79,80] Modell).[89] Drawing of a person, whether designated as a self-portrait or simply as a man (or as a woman for female subjects), is believed to be an expression of the total self, with the inner sense of one's special strengths, weaknesses, and one's unique style expressed in line quality, relationship of parts to whole, size of drawing, and placement on the page. The self can, the clinician believes, be read from the body image expressed in the drawing. Whether this is true remains controversial, but there is enough supporting evidence to lead clinicians to use the drawings as one of several personality instruments.

Allport proposes what he calls the "proprium" to include all the regions of our life that we regard as peculiarly ours and reviews those aspects of personality that make for inward unity. In his eightfold schema of the proprium* he gives special priority to the bodily sense. The centrality of the body in psychological health is affirmed by all investigators. If psychological maturity is conceived—as we do here—to be expressed by the relationship between chronological age and psychic development, the bodily basis for the self is especially important. During the growing years rapid and continuous revisions of the body might be expected to challenge integration. Rarely, however, does healthy physical change disturb the self. On the contrary, increasing strength and stature, improved neuromuscular control, the development of bodily skills, quite as much as the growth of cognition, extend and elaborate the infant self. During the latency years the self is supported and spelled out significantly in bodily terms.

In the beginning, as work with infants shows, the bodily self is learned. The way in which the young child discovers that the boundaries of his body have special relevance to all the rest of him and establishes his physical dimensions was vividly chronicled by Wilhelm Preyer in 1882.[101,102] He recounted how his infant son learned by biting his fingers painfully that they belonged to the nearer parts of him. Later when the baby was standing in his crib, Preyer asked him to "give the shoe"—which lay on the crib beside the child. The little son complied. A moment later, Preyer, preparing to put the shoe on the child, requested, "Give me the foot." The child's response was to stoop and tug at his own foot in the same way he had reached for and lifted the shoe. At this point the knowledge that the

* Allport's[3] eightfold schema for the proprium—or the self—includes the following: bodily sense, self-identity, ego enhancement, ego extension, rational agent, self-image, propriate striving, and the proprium as knower (pp. 41–62).

foot belonged to the self in a way quite different from the shoe had not been mastered.* Preyer reported a series of explorations the little boy made discovering the outlines of his own head, the difference between his hand and the tray of his highchair. Fifty-five years later, Piaget's published studies of his three young children gave an account of the emergence of the idea of the object as a continuing entity apart from the perceiver. He concludes that the notion of the self as an object among objects was one of the achievements of the sensorimotor period.

The awareness that oneself is not merely one other object but has a special status is also attained in the early months of life. The toddler recognizes his own reflection in a mirror as a baby to whom he responds with interest as if to another child. By the time he reaches 18 months, however, he gives clear indication that the baby in the mirror is himself. This may come before he has enough language to express his understanding verbally, but his self-conscious little smile and his gestures make his meaning unmistakable. Thus, for a child even as young as this, one of the most complicated phenomena of psychic life has come into being (Stutzman).[128] For a child at this time of life this demonstrates the maturation of self.

Recent research has inquired into the very earliest appearance of the style that will become the self. Mahler[84] traces the gradual emergence of the individual from the autistic shell by which the newborn has been sheltered from excessive stimulation into "a protective but also receptive and selective positively cathected stimulus shield" that envelops a child-mother unity in the early normal symbiosis. After the healthful and necessary symbiotic stage is past, the infant gradually becomes aware, from two to nine months old, of the world outside the protecting shell and can attend increasingly to the world about him. The "holding behavior" of the mother, so indispensable for the protection of the neonate and young infant, can now loosen as the infant becomes ready to "hatch" from the symbiotic orbit smoothly and gradually without undue strain on his own resources. During the early infancy period a circular interaction of cues goes on between the infant and mother in which the child gradually alters his behavior in a characteristic way. His emergent mode, Mahler writes,[84] grows out of his own endowment and the mother-child relationship. Already "certain overall qualities of the child's personality" are discernible. What we seem to see here is the birth of the child as an individual. In the mutual cuing of mother and child the groundwork is laid for the "infant's becoming the child of his particular mother."

Mahler recounts the processes of separation and individuation, noting periods of heightened vulnerability. Optimal maturation will bring the young child to readiness for both physical and psychic distance from the mother. This stage is reached when ego functions have reached a fairly advanced stage and, equally important, when he has attained the level of object constancy in which the object is intrapsychically available. His memory of the mother, his cognitively held assurance of her existence and availability even when she is physically invisible, sustains him in his separateness. Clearly the timing of this maturation cannot be dated precisely, for the intricate network of competencies that enable the child, or older person, to exist without panic is made up of many strands and can be weakened by the failure of any one of them. Yet some children exhibit very early in life—even in infancy—an unusually strong thrust toward mastery. Mahler's reference to innate endowment is well-taken, for a characteristic level of directed energy is immediately interactive with environmental circumstances. This is apparently an original given. A strong core of self thus seems present almost at birth.

Before nursery school age individual differences in the strength and clarity of self are apparent. The three year old with a firmly developed self can tolerate brief separation from his mother without great anxiety and, in

* A series of baby biographies and other less extensive reports touch upon these same phenomena, although few prior to Piaget have done so with details as telling as Preyer's.

achieved inner cohesiveness, is able to grasp the opportunities offered by the nursery school. As he advances into the school years, his life style is elaborated and his mold as a person is more firmly established. His achievements and failures, his acceptances and rejections are assimilated. The psychologically mature self during latency, building upon an already developed identity, is modified to include awareness that the self has at its command both a range of capabilities and the power to master increasing difficulty and scope as occasions require.

With each epoch in the growth cycle, changing biological structures, functions, and social roles call for reorganization of the existing self. Evidence on adult life for periods as long as 15 to 20 years indicates that the level of relative maturity and the trajectory of purpose is maintained. The extent to which life style remains essentially unchanged over the entire life span is still to be determined, but the evidence from longitudinal work encourages the hypothesis that those who have over the years been mature personalities will be so to the end (Cox).[31]

The concepts and terms clustering around the self have generated research and stirred extensive discussion. Particularly germane to our purposes here are the ideas expressed by self- or ego identity, self-esteem, and self- or ego ideal.

The Self Known to Itself

The establishment of an identity recognizable to and accepted by the self as its own is a facet of the development of all persons and is not a property of only the psychologically mature. As in so many other facets of his makeup, the notably healthy individual is clearer about his identity and maintains it more steadfastly than the immature one does. Heath[63] experimentally induced mild uncertainty about self-image in his immature and mature subjects and found the mature better able to withstand disturbing evidence.

Identity, nonexistent at birth, evolves as the infant defines himself in terms of his genetic idiom and also in terms of the responses he gets from others. Soon after extrauterine life begins the child identifies the self as male or female, and later still as a good or poor game participant, as a quick or slow learner. Similarly he understands his meaning in belonging to a particular family, race, social and national group. Only much later does he grow into his basic identity as a member of the family of man.

To be sure, there are certain normal and inevitable crises in the life cycle that shake identity. These are known to be associated with puberty, childbirth, and the climacteric as well as the socially produced trauma of vocational retirement and the dislocations following revolutions and wars. In such experiences complex supports of identity change drastically or are lost. One of the tests of maturity is the way the assurance of what one really is holds up.

A distinction is drawn in some of the research between the private and the social self.* In health the self is not wholly dependent upon the way one is conceptualized by others, but there is evidence that in the mature person there is congruence between what one feels himself to be and how he believes others view him. Furthermore, he is pretty accurate in his estimate of the way he appears to others (Heath).[63] Here both integration and realism operate.

Self-Esteem

Capacity to love unreservedly in the most intimate relationships and readiness to turn outward in disinterested generosity to the whole range of human relationships is said to have its roots in love of self. This sentiment is given various names by personality theorists: self-esteem, self-regard, self-love. It is necessary, of course, to be clear that self-love is not the same as selfishness or self-centeredness. Fromm has been at some pains to make this distinction. Maslow found self-love to be "sy-

* Identity is regarded by some investigators to be so deeply related to social factors that it has no existence apart from them. They see it as a personal-social phenomenon. This would lead to the conclusion that there is no such thing as a private identity.

nergic with rather than antagonistic to love of others." Beneath the outreach to others that psychologically mature persons without exception show, we find a basic peace with the self (Cox,[33] Fromm,[48] Heath,[63] Maslow).[87] Pride in achievement, conviction of one's worth, and confidence that one is worthy of love enable the individual to extend the largesse of his disinterested care beyond the bounds of his own particular concerns. To entertain these feelings about oneself does not go hand-in-hand with smugness. Rather this attitude might be likened to that of the good strict parent described by Susan Isaacs—accepting and warm but not uncritical nor negligent about responsibility and future growth. In sum, the psychologically mature are found to be favorably disposed and cherishing toward the totality of what the self has been able to make of its quantum of life.

Self-Ideal

The significance of the self-ideal as a magnetizing force in moving the person toward as yet unachieved goals and levels of being is emphasized by various theorists. Lynd[76] is especially clear about the place of this dynamic in personality: "It is readily apparent that the ideal of who one might or desires to be or become has important bearing on one's feeling about who one is." The self-ideal by definition reaches outside of and beyond the achieved state or self-image but is included in the sense of who one is. To aspire to excellence, to strength, to saintliness is to claim kinship with these qualities and to borrow some of their effulgence. At the same time, unless there is intention and effort toward bringing the ideal self to reality, the ideal can become a substitute for striving. We see this in poorly integrated children who, when asked to set a goal, place it so far beyond anything they could possibly attain that they neutralize the pull of the goal (Cox).[32] However, in realistic and responsible persons, from childhood onward the idea of what one wishes and tries to become exercises significant influence and is, indeed, an inalienable part of self.

The Mature Person

In the course of change and in the maintenance of stability, certain characteristics mark the psychologically mature individual, distinguishing him from the immature. Although the list of characteristics could be indefinitely extended, I would like to suggest the following trends or themes as most nearly universal and timeless: firm anchorage in reality, warmth and caring for other persons from an increasingly giving posture, productivity in work suited to ability, responsibility toward the small and the large social group, secure sense of self, development of a value system, and resilience under stress.

Bibliography

1. ACKERMAN, N. W., *The Psychodynamics of Family Life*, Basic Books, New York, 1958.
2. ALBEE, G. W., and HAMLIN, R. M., "Judgement of Adjustment from Drawings: the Application of Rating Scale Methods," *J. Clin. Psychol.*, *6*:363–365, 1950.
3. ALLPORT, G. W., *Becoming: Basic Considerations for a Psychology of Personality*, Yale University Press, New Haven, 1955.
4. ———, "The Ego in Contemporary Psychology," *Psychol. Rev.*, *50*:451–478, 1943.
5. ———, *The Individual and His Religion*, Macmillan, New York, 1950.
6. ———, *Pattern and Growth in Personality*, Holt, Rinehart and Winston, New York, 1961.
7. ———, *Personality and Social Encounter*, Beacon, Boston, 1960.
8. ANASTASI, A., *Psychological Testing*, 3rd ed., Macmillan, New York, 1963.
9. ANDERSON, H. H. (Ed.), *Creativity and Its Cultivation*, Harper & Row, New York, 1959.
10. ANDERSON, T. B., and OLSEN, L., "Ideal Self and Occupational Choices," *Personnel & Guidance J.*, *44*:171–176, 1965.

11. ARIETI, S. (Ed.), *American Handbook of Psychiatry*, 3 vols., Basic Books, New York, 1959, 1966.
12. BARRON, F., *Creativity and Psychological Health*, Van Nostrand, New York, 1963.
13. ———, *Personal Soundness in University Graduate Students: An Experimental Study of Young Men in the Sciences and Professions*, University of California Publications Personality Assessment and Research, No. 1, 1954.
14. ———, "What Is Psychological Health?" *California Monog.*, *68*:22–55, 1957.
15. BAYLEY, N., "Development of Mental Abilities," in Mussen, P. H. (Ed.), *Carmichael's Manual of Child Psychology*, Vol. 1, John Wiley, New York, 1970.
16. BENNETT, V. D., "Combinations of Figure Drawing Characteristics Related to the Drawer's Self Concept," *J. Projective Techniques & Personality Assessment*, *30*:192–196, 1966.
17. BERMAN, S., and LAFFAL, J., "Body Type and Figure Drawing," *J. Clin. Psychol.*, *9*:368–370, 1953.
18. BOND, E. D., "The Student Council Study: A Preliminary Report," *Am. J. Psychiat.*, *107*:271–273, 1950.
19. ———, "The Student Council Study: An Approach to the Normal," *Am. J. Psychiat.*, *109*:11–16, 1952.
20. BOWLBY, J., *Attachment and Loss*, Vol. 1, Basic Books, New York, 1969.
21. ———, *Child Care and the Growth of Love*, 2nd ed., Penguin Books, Harmondsworth, England, 1968.
22. BROWN, D. R., "Non-Intellecting Qualities and the Perception of the Ideal Student by College Faculty," *J. Educ. Sociol.*, *33*:269–278, 1960.
23. ———, "Personality, College Environment, and Academic Productivity," in Sanford, N. (Ed.), *The American College*, John Wiley, New York, 1961.
24. BÜHLER, C., *The First Year of Life*, John Day, New York, 1930.
25. BUTLER, R. N., "Aspects of Survival and Adaptation in Human Ageing," *Am. J. Psychiat.*, *123*:1233–1243, 1967.
26. CABAK, V., and NAJDANVIC, R., "Effects of Undernourishment in Early Life on Physical and Mental Development," *Arch. Dis. in Childhood*, *40*:532–534, 1965.
27. CARMICHAEL, L., "The Onset of Early Development of Behaviour," in Mussen, P. H. (Ed.), *Carmichael's Manual of Child Psychology*, Vol. 1, John Wiley, New York, 1970.
28. CASLER, L., "The Effects of Extra Tactile Stimulation on a Group of Institutionalized Infants," *Genet. Psychol. Monog.*, *71*:137–175, 1965.
29. CLAPAREDE, E., "*Note sur la localisation du Moi*," *Arch. of Psychol.*, *19*:172–182, 1924.
30. COOPERSMITH, S., *Antecedents of Self Esteem*, W. H. Freeman, San Francisco, 1967.
31. COX, R. D., "The Normal Personality: An Analysis of Rorschach and Thematic Apperception Test Responses of a Group of College Students," *J. Projective Techniques*, *20*:70–77, 1956.
32. ———, "Verbal Expression of Level of Aspiration as a Defense Against Anxiety," Unpublished case study of Robert.
33. ———, *Youth into Maturity*, Mental Health Materials Center, New York, 1970.
34. DARWIN, C., "Biographical Sketch of an Infant," *Mind*, *2*:285–294, 1877.
35. DUNN, H. L., "Dynamic Maturity for Purposeful Living," *Geriatrics*, *21*:205–208, 1966.
36. ELKIND, D., *Children and Adolescents: Interpretative Essays on Jean Piaget*, Oxford University Press, New York, 1970.
37. ELKISCH, P., "Children's Drawings in a Projective Technique," *Psychol. Monog.*, *58*, 1945.
38. ERB, E. D., and HOOKER, D., *The Psychology of the Emerging Self: An Integrated Interpretation of Goal-Directed Behaviour*, F. A. Davis, Philadelphia, 1967.
39. ERIKSON, E., *Childhood and Society*, 2nd ed., Norton, New York, 1963.
40. ———, *Growth and Crisis in Healthy Personality*, Monograph 1, Vol. 1, Psychological Issues, International Universities Press, New York, 1959.
41. ———, *Insight and Responsibility*, Norton, New York, 1964.
42. ESCALONA, S. K., "Emotional Development in the First Year of Life," in Senn, M. J. E. (Ed.), *Problems of Infancy and Childhood*, Josiah Macey, Jr., Foundation, New York, 1953.
43. FLAVELL, J. H., *The Developmental Psy-*

chology of Jean Piaget, Van Nostrand, Princeton, 1963.

44. FREUD, A., "The Concept of Developmental Lines," in *The Psychoanalytic Study of the Child*, Vol. 18, pp. 245–265, International Universities Press, New York, 1946.
45. ———, *The Ego and the Mechanisms of Defense*, International Universities Press, New York, 1946.
46. FREUD, S. (1923), *The Ego and the Id*, Norton, New York, 1960.
47. ———, *New Introductory Lectures on Psychoanalysis*, Norton, New York, 1933.
48. FROMM, E., *Man for Himself*, Rinehart & Co., New York, 1947.
49. GESELL, A. L., and AMATRUDA, C. S., *Developmental Diagnosis*, 2nd ed., Holber, New York, 1960.
50. GOLDEN, J., MANDEL, N., GLUECK, B. C., JR., and FEDER, Z., "Summary Description of Fifty Normal White Males," in Haimowitz, M. L., and Haimowitz, N. (Eds.), *Human Development*, Thomas G. Crowell, New York, 1966.
51. GOLDFARB, W., "Variations in Adolescent Adjustment of Institutionally Reared Children," *Am. J. Orthopsychiat.*, *17:* 449–457, 1947.
52. GORNEY, J. E., and TOBIN, S. S., "Experiencing among the Aged," *Proceedings of the 20th Annual Meeting of the Gerontological Society*, *34*, 1967.
53. GREENACRE, P., "Early Physical Determinants in the Development of the Sense of Identity," *J. Am. Phychoanal. A.*, *6:* 612–627, 1958.
54. GURIN, G., VEROFF, J., and FELD, S., *Americans View Their Mental Health*, Basic Books, New York, 1960.
55. HARRIS, D. B., *Children's Drawings as Measures of Intellectual Maturity*, Harcourt Brace World, New York, 1963.
56. HARTMAN, H., "Comments on the Psychoanalytic Theory of the Ego," in *The Psychoanalytic Study of the Child*, Vol. 5, pp. 74–96, International Universities Press, New York, 1950.
57. ——— (1939), *Ego Psychology and the Problem of Adaptation*, International Universities Press, New York, 1958.
58. ———, *Essays on Ego Psychology: Selected Problems in Psychoanalytic Theory*, International Universities Press, New York, 1964.
59. ———, "Psychoanalysis and Developmental Psychology," in *The Psychoanalytic Study of the Child*, Vol. 5, pp. 7–17, International Universities Press, New York, 1950.
60. HAVIGHURST, R. J., *Developmental Tasks and Education*, 2nd ed., David McKay, New York, 1952.
61. ———, *Human Development in Education*, Longman Green, London, 1953.
62. HEATH, C. W., *What People Are: A Study of Normal Young Men*, Harvard University Press, Cambridge, 1946.
63. HEATH, D., *Explorations of Maturity*, Appleton-Century-Crofts, New York, 1965.
64. HOLLINGWORTH, L., *Children above 180 I.Q. Stanford Binet*, World, New York, 1942.
65. HOOTEN, E., *Young Man, You Are Normal: Findings from a Study of Students*, G. P. Putnam's Sons, New York, 1945.
66. INHELDER, B., and PIAGET, J., *The Growth of Logical Thinking from Childhood to Adolescence*, Basic Books, New York, 1958.
67. JAHODA, M., *Current Concepts of Positive Mental Health*, Basic Books, New York, 1958.
68. JONES, M. C., "Psychological Correlates of Somatic Development," *Child Dev.*, *36:* 899–911, 1965.
69. KESSEN, W., *The Child*, John Wiley, New York, 1965.
70. KOHLBERG, L., and KRAMER, R., "Continuities and Discontinuities in Childhood and Adult Moral Development," *Human Dev.*, *12:*93–120, 1969.
71. KOTKOV, B., and GOODMAN, M., "The Draw-a-Person Tests of Obese Women," *J. Clin. Psychol.*, *9:*362–364, 1953.
72. LANGER, J., "Werner's Theory of Development," in Mussen, P. H. (Ed.), *Carmichael's Manual of Child Psychology*, John Wiley, New York, 1970.
73. LANGNER, T. S., and MICHAEL, S. T., *Life Stress and Mental Health: The Midtown Manhattan Study*, Vol. 2, Free Press, Glencoe, 1963.
74. LIDZ, T., *The Person: His Development throughout the Life Cycle*, Basic Books, New York, 1968.
75. LORBER, N. M., "Concomitants of Social

Acceptance: A Review of Research," *Psychology, 6*:53–59, 1969.

76. LYND, H. M., *On Shame and the Search for Identity*, Science Editions, New York, 1961.
77. MACCOBY, E. E. (Ed.), *The Development of Sex Differences*, Stanford Studies in Psychology, Tavistock, London, 1967.
78. MCGRAW, M. B., *Growth: A Study of Johnny and Jummy*, Appleton Century, New York, 1935.
79. MACHOVER, K., "Drawing of the Human Figure: A Method of Personality Investigation," in Anderson, H. H., and Anderson, G. L. (Eds.), *An Introduction to Projective Techniques*, Pp. 341–369, Prentice-Hall, New York, 1951.
80. ———, *Personality Projection in the Drawing of the Human Figure*, Charles C Thomas, Springfield, Ill., 1949.
81. MACKINNON, D. W., "The Highly Effective Individual," *Teachers College Record, 61*:367–378, 1960.
82. MADDOX, G. L., "Disengagement Theory: A Critical Evaluation," *The Gerontologist, 4*:80–82, 103, 1964.
83. ———, "Fact and Artifact: Evidence Bearing on Disengagement Theory from Duke Geriatrics Project," *Human Dev., 8*:117–130, 1965.
84. MAHLER, M. S., *On Human Symbiosis and the Vicissitudes of Individuation*, Vol. 1, International Universities Press, New York, 1968.
85. MARMOR, J., "The Crisis of Middle-age," *Am. J. Orthopsychiat., 37*:336–337, 1967.
86. MASLOW, A. H., *Motivation and Personality*, Harper & Row, New York, 1954.
87. ———, "Self-actualizing People: A Study of Psychological Health," *Personality*, Symposium No. 1, 11–34, 1950.
88. ———, *Toward a Psychology of Being*, 2nd ed., Van Nostrand Reinhold, New York, 1968.
89. MODELL, A. H., "Changes in Human Figure Drawings by Patients Who Recover from Regressed States," *Am. J. Orthopsychiat., 21*:584–596, 1951.
90. MURRAY, H. A., *Explorations in Personality: A Clinical and Experimental Study of Fifty Men of College Age*, Oxford University Press, New York, 1938.
91. MUSSEN, P. H., CONGER, J. J., and KAGAN, J., *Child Development and Personality*, 3rd ed., Harper & Row, New York, 1969.
92. NAGERA, H., *Early Childhood Disturbances, The Infantile Neuroses and Adulthood Disturbances*, Monograph Series of the Psychoanalytic Study of the Child, International Universities Press, New York, 1966.
93. NASS, M. L., "The Superego and Moral Development in the Theories of Freud and Piaget," in *The Psychoanalytic Study of the Child*, Vol. 21, pp. 51–68, International Universities Press, New York, 1966.
94. NEUGARTEN, B. (Ed.), *Personality in Middle and Late Life*, University of Chicago Press, Chicago, 1968.
95. OLIM, E., "The Self-Actualizing Person in the Fully Functioning Family: A Humanistic Viewpoint," *Family Co-ordinator, 17*:141–148, 1968.
96. PIAGET, J. (1937), *The Construction of Reality in the Child*, Basic Books, New York, 1954.
97. ——— (1936), *The Origins of Intelligence in Children*, International Universities Press, New York, 1952.
98. ———, "Piaget's Theory," in Mussen, P. H. (Ed.), *Carmichael's Manual of Child Psychology*, Vol. 1, John Wiley, New York, 1970.
99. ——— (1947), *The Psychology of Intelligence*, Harcourt Brace World, New York, 1950.
100. PRATT, K. C., "The Organization of Behaviour in the New Born Infant," *Psychol. Rev., 44*:470–490, 1937.
101. PREYER, W., *The Mind of the Child*, Pt. I, *The Senses and the Will*, Appleton, New York, 1888.
102. ———, *The Mind of the Child*, Pt. II., *The Development of the Intellect*, Appleton, New York, 1893.
103. PROVENCE, S., and LIPTON, R. C., *Infants in Institutions*, International Universities Press, New York, 1962.
104. RABINOWITZ, M., "The Relationship of Self Regard to the Effectiveness of Life Experiences," *J. Counsel. Psychol., 13*:139–143, 1966.
105. RHEINGOLD, H. L., "The Effects of Environmental Stimulation upon Social and Exploratory Behaviour in the Human Infant," in Foss, B. M. (Ed.), *Determi-*

nants of Infant Behaviour, Pp. 143–171, John Wiley, New York, 1961.

106. RHEINGOLD, H. L., GEWIRTZ, J. L., and ROSS, H. W., "Social Conditioning of Vocalization in the Infant," *J. Comp. Physiol. Psychol.*, 52:68–73, 1959.

107. RHUDICK, P. J., and GORDON, C., "Test-Retest I.Q. Changes in Bright Ageing Individuals," *Proceedings of the 20th Annual Meeting of the Gerontological Society*, 34, 1967.

108. RIBBLE, M. A., "Infantile Experience in Relation to Personality Development," in Hunt, J. McV. (Ed.), *Personality and the Behaviour Disorders*, Ronald, New York, 1944.

109. ———, *The Personality of the Young Child*, Columbia University Press, New York, 1955.

110. ROGERS, C. R., *On Becoming a Person*, Houghton Mifflin, Boston, 1961.

111. ———, "Toward a Theory of Creativity," in Anderson, H. H. (Ed.), *Creativity and Its Cultivation*, Harper & Row, New York, 1959.

112. RUBIN, I. M., "Increased Self-Acceptance: A Means of Reducing Prejudice," *J. Pers. & Soc. Psychol.*, 5:233–238, 1967.

113. RUFF, G., and SHELDON, J. K., "Adaptive Stress Behaviour," in Appleby M. H., and Trumbull, R. (Eds.), *Psychological Stress*, Appleton-Century-Crofts, New York, 1967.

114. SANFORD, N. (Ed.), *College and Character*, John Wiley, New York, 1964.

115. ———, "Personality Development during the College Years," *J. Soc. Issues*, *12*:1–70, 1956.

116. SANTOSTEFANO, S., "Relating Self-Report and Overt Behaviour: The Concept of Levels of Modes for Expressing Motives," *Perceptual and Motor Skills*, *21*: 940, 1965.

117. SAUL, L., *Emotional Maturity*, 2nd ed., J. B. Lippincott, Philadelphia, 1960.

118. SCHILDER, P., *Image and Appearance of the Human Body*, Kegan Paul, London, 1935.

119. SCOTT, J. P., "The Genetic and Environmental Differentiation of Behaviour," in Harris, D. B. (Ed.), *The Concept of Development*, University of Minnesota Press, Minneapolis, 1957.

120. SHACHTEL, E., *Metamorphosis*, Basic Books, New York, 1959.

121. SHINN, W. M., *The Biography of a Baby*, Houghton Mifflin, New York, 1900.

122. SMEDSLUND, J., "The Effect of Observation on Children's Representation of the Spatial Orientation of a Water Surface," *J. Genet. Psychol.*, *102*:195–201, 1963.

123. SPENCE, D. L., "The Role of Futurity in Ageing," *Gerontologist*, *8*:180–183, 1968.

124. SPITZ, R. A., *The First Year of Life: A Psychoanalytic Study of Normal and Deviant Development of Object Relations*, International Universities Press, New York, 1965.

125. SMITH, M. B., "Mental Health Reconsidered: A Special Case of the Problem of Values in Psychology," *Am. Psychol.*, *16*:299–306, 1961.

126. SROLE, L., LANGNER, T., MICHAEL, S. T., OPLEY, M. K., MARVIN, R., and THOMAS, A. C., *Mental Health in the Metropolis: The Midtown Manhattan Study*, Vol. 1, McGraw-Hill, New York, 1962.

127. STEWART, R. A., "Transcendence of Opposites," *Psychology*, *6*:62–64, 1969.

128. STUTZMAN, R., *Mental Measurement of Preschool Children with a Guide for the Administration of the Merriel-Palmer Scale of Mental Tests*, World Book, Yonkerson-Hudson, N.Y., 1931.

129. TAFT, J., *The Dynamics of Therapy in a Controlled Relationship*, Dover, New York, 1962.

130. TANNER, J. M., *Growth at Adolescence*, 2nd ed., Blackwell, Oxford, 1962.

131. ———, "Physical Growth," in Mussen, P. H. (Ed.), *Carmichael's Manual of Child Psychology*, Vol. 1, John Wiley, New York, 1970.

132. TERMAN, L. M., and ODEN, M. H., *The Gifted Child Grows Up: 25-Year Follow-up of a Superior Group*, Stanford University Press, Stanford, 1948.

133. ———, *The Gifted Group in Mid-life: 35-Year Follow-up of the Gifted Child*, Stanford University Press, Stanford, 1960.

134. TIPPET, J. S., and SILFER, E., "Autonomy of Self Esteem: An Experimental Approach," *Arch. Gen. Psychiat.*, *14*:372–385, 1966.

135. VAN DEN DEALE, L., "A Developmental

Study of the Ego Ideal," *Genet. Psychol. Monog.*, *18*:191–256, 1968.

136. VISPO, R. H., "On Human Maturity," *Perspectives Biol. & Med.*, 9:586–602, 1966.

137. WECHSLER, D., *The Measurement and Appraisal of Adult Intelligence*, 4th ed., Williams & Wilkins, Baltimore, 1958.

138. WERNER, H., *Comparative Psychology of Mental Development*, International Universities Press, New York, 1948.

139. ———, "The Concept of Development from a Comparative and Organismic Point of View," in Harris, D. B. (Ed.), *The Concept of Development*, University of Minnesota Press, Minneapolis, 1957.

140. WHITE, S., "The Learning Theory Approach," in Mussen, P. H. (Ed.), *Carmichael's Manual of Child Psychology*, John Wiley, New York, 1970.

141. WITTY, P., (Ed.), *The Gifted Child* (American Association for Gifted Children) Heath, Boston, 1951.

142. WOLFF, K., "Personality Type and Reaction toward Ageing and Death: A Clinical Study," *Geriatrics*, *21*:189–192, 1966.

143. WOHLWILL, J. F., and LOWE, R. C., "An Experiment Analysis of the Development of the Conservation of Number," *J. Child Dev.*, 33:153–167, 1962.

144. ZILLER, R. C., and GROSSMAN, S. A., "A Developmental Study of the Self-Social Constructs of Normals and the Neurotic Personality," *J. Clin. Psychol.*, 23:15–21, 1967.

145. ———, ———, HAGEY, J., SMITH, M., and LONG, B. H., "Self Esteem: A Self-Social Construct," *J. Consult. & Clin. Psychol.*, *33*:84–95, 1967.

146. BENTZ, V. J., "A Test-Rerest Experiment on the Relationship between Age and Mental Ability," *Am. Psychol.* *8*:319–320, 1953.

147. OWENS, W. A., "Age and Mental Abilities: A Longitudinal Study," *Genet. Psychol. Monog.*, *48*:3–54, 1953.

PART THREE

The Life Cycle and Its Common Vicissitudes

CHAPTER 10

THE LIFE CYCLE: INTRODUCTION

Theodore Lidz

The psychodynamic understanding of the personality and its disorders rests heavily upon the study of the life cycle. The commonalities and similarities in the course of all lives make possible the generalizations and abstractions necessary for the scientific study of the personality. Although no two persons are identical and no life stories are the same, the basic themes are limited and it is the variations upon them that are infinite and inexhaustible. Behind their individual uniqueness all persons are born with physical endowments that are essentially alike and with similar biological needs that must be met. Like all living things, they go through a cycle of gestation, maturation, maturity, decline, and death. In common with all humans, each individual starts life totally dependent on others and remains immature and dependent for many years, during which he forms intense emotional ties to those who nurture him—bonds that must be loosened sufficiently to enable him to live without them and to form new meaningful relationships. He requires many years not only to mature sexually and physically but also to learn the adaptive techniques he needs to survive and guide his life; and he depends upon a society to provide his essential environment. He possesses the unique human capacities for language and tool bearing, and he depends upon verbal communication for collaboration with others, upon thought and foresight, and upon his ability to change his environment to meet his needs. Alone among animals he is aware of death and his position in his course from birth to death. Man's life is never static, for the passage of time of itself changes his functioning. Each phase of the life cycle brings new potentialities and closes off old ones; the opportunity and challenge must be met and surmounted for the individual to be prepared for the next phase.

Phasic Nature of the Life Cycle

The development of the personality and the course of the life cycle proceed phasically, not at a steady pace. The child goes through periods of relative quiescence during which his progress seems slight and then undergoes marked changes as he enters a new phase that opens new potentialities, provides new areas to explore, and sets new tasks that require the acquisition of new skills and abilities to master. Thus, the ability to walk, which must await maturation of the infant's nervous system as well as sufficient practice, changes the limits of the child's world and his perspective of his surroundings, as well as his relationships with his parents. He quickly gains new opportunities to explore and learn, but he also requires more delimitation from those who take care of him. Similarly the hormonal shifts that precede and accompany the advent of puberty will move the child into a new phase of his life by rapidly altering his size and contours and by initiating unfamiliar sexual impulsions.

The phasic nature of the life cycle derives from several interrelated factors.

Physical Maturation

The acquisition of certain capacities must await the maturation of the organism. The infant cannot become a toddler until the pyramidal tracts in the spinal cord that permit voluntary discrete movements of the lower limbs become functional around the tenth month. Even after maturation allows the acquisition of a new attribute, gaining the skill and knowledge to develop it can require considerable time and practice. Simple skills must be mastered adequately before they can be combined with others and incorporated into more complex activities.

In a related manner phasic shifts in the physiological balance of the organism initiate new phases in the life cycle. The metamorphosis of puberty that ushers in adolescence provokes changes without regard to prior developmental progress, and the menopause is likely to produce a basic reorientation in a woman's life.

Changes in the Individual's Cognitive Development and Decline

The child's cognitive development does not progress at an even pace. Qualitatively different capacities emerge in rather discrete stages and influence the child's capacity to assume responsibilities and direct his own life.

The Society

The roles the society establishes for persons of various ages set expectations that promote shifts in ways of living. The time of weaning and bowel training, for example, are markedly influenced by societal norms. Becoming a married person or a parent involves socially set expectations such as rescinding areas of independence to care for and consider the needs of a spouse or a child. Yet for a society to remain viable, the roles and expectations it establishes must be compatible with a person's abilities and needs at each phase of life. A child is moved into the role of schoolchild, with its many demands and privileges, at the age of five or six partly because primary socialization is usually completed and partly because his cognitive capacities have reached the stage of "concrete operations."[10]

The Passage of Time

Not only does the individual move into age-appropriate roles, but time brings changes in physical makeup that require changes in attitudes, as when a person reaches middle life and realizes his life is approaching a climax, or when diminished abilities lead to retirement.

The Epigenetic Principle

The epigenetic principle maintains that the critical tasks of each developmental phase must be met and surmounted at the proper time and in the proper sequence to assure healthy personality development. The principle was adopted by psychoanalysis from embryology. The birth of a normal infant depends upon each organ's arising out of its fetal anlage in the proper sequence, with each development depending upon the proper unfolding of the preceding phase. Personality development is, however, not as rigidly set as embryonic maturation, and even though development is impeded when a developmental phase is not properly mastered, compensations are possible and deficiencies can sometimes be turned into strengths. It is clear that a child who does not gain adequate autonomy from his mother prior to going to school will have difficulties in attending school, learning, and relating to peers, but failures are usually partial and not productive of an irremediable anomaly as in the embryo.

Progression, Fixation, Regression

Every life contains a series of developmental crises that arise from the need to meet the new challenges that are inherent in the life cycle. The individual gains new strength and self-sufficiency through surmounting these crises. Similarities exist in the ways in which different people meet similar developmental problems, and there is likely to be something repetitive about the ways in which the same person surmounts certain crises in his life.

Often there is a pause or delay before the child finds the confidence to attempt to face the strange needs of a new phase in his life. The need for emotional security sets limits on the pace of development. He is often prey to opposing motivations. There is an impetus toward expansion and mastery of new skills and situations and the child wishes to emulate parental figures, become more grown up, and achieve greater independence; but the new situations and demands bring insecurity, failures create frustrations, and greater independence requires renunciation of the comforts of dependency. The anxieties that are aroused can lead the child to seek the security of known ways and to renounce temporarily further forward movement, or even to fall back upon increased dependency.

A child requires guidance and support to progress properly. At times he may need to be checked from the unrestrained use of new capacities as when he begins to walk or when he first matures sexually, but at other times he may need support or even some prodding to move forward as when he is reluctant to leave the familiar and protected home to attend school. Developmental hazards exist on both sides; too much support may leave the child overly dependent and fearful of venturing forward; too little can leave the child struggling to keep afloat.

The failure to cope with and master the essential tasks of a developmental phase leaves the child unprepared to move forward into the next phase. The child gives up, or more commonly he moves ahead in some areas and remains stuck in others. He squanders energy in coping with old problems instead of moving on. A child who has never felt adequately secure at home continues to seek maternal protection and affection when his peers are secure with one another. Such arrests are termed *fixations*. Movement backward to an earlier developmental phase in which the child felt secure is termed *regression*. Paradoxically regression is a normal aspect of the developmental process, for every child will regress at times in order to regain security or to reestablish his equilibrium after feeling defeated. The small child tends to progress securely when he feels that he can find parental protection at the center of his expanding world should he need it.

Even though fixations and regressions are important means of maintaining or regaining emotional security, they create insecurities in turn if they are not simply temporary expediencies. The child remains improperly prepared to meet the developmental challenges of the next phase of the life cycle and is unable to accept the opportunities afforded him. Even though the child is pulled in two directions, the motivations to move forward are normally more powerful. He is carried onward by his growth, by impulsions for the stimulation of new experiences, by his drives, by needs for the affection and approval of his parents, by desire to go along with his peers, by the yearning of his body for another, by the needs of survival, by the roles provided by society, by the desire for progeny, and by other such influences.

The Divisions of the Life Cycle

The life cycle has been divided into stages or phases somewhat differently by various students of the developmental process; some of the differences reflect differing purposes in making these divisions, and others reflect differing theoretical orientations and conceptualizations of human development. However, despite differences in theory and terminology, there is considerable overlap in where the dividing lines are placed. The description and study of the salient features of these stages have evolved largely from four rather different approaches to understanding the phasic emergence of essential attributes of the personality: those of Freud,[5] Sullivan,[12] Erikson, and Piaget.[10,11] As the contributions of Freud and Sullivan are described in detail in other chapters, only Erikson's and Piaget's approaches will be discussed here.

Erikson's Phases of Psychosocial Development

Adhering to a more classical psychoanalytic framework than Sullivan, Erikson superimposed an epigenesis of psychosocial development upon the psychosexual phases.[3] He designated what he considered to be the critical psychosocial task of each phase that the individual must surmount in order to be properly prepared to meet the opportunities and tasks of the next stage. He also went beyond the traditional psychosexual phases that end with the "genital phase." In particular, he emphasized the critical moment of late adolescence, when the personality must gel and a person achieve an ego identity. He then continued to consider the developmental problems of adult life. He has formulated eight stages of psychosocial development, focusing upon the specific developmental tasks of each stage and examining how differing societies help the individual to cope by providing essential care, promoting independence, offering roles, and by having institutionalized ways of assuring survival, proper socialization, and emotional stability. Meeting and surmounting the developmental task of a period leads to the acquisition of a fundamental trait essential for further stable development, whereas failure leads to an enduring deficiency.

As the infant is almost completely dependent upon others to satisfy his vital needs and to keep him comfortable, the basic psychosocial task of the oral phase concerns the achievement of a basic trust in others and also in the self, with failures leading to varying degrees of basic mistrust; the "basic" conveys that the trust is not conscious but blends into the total personality and forms an inherent component of it. The emphasis of the anal phase or the second year of life is upon the attainment of muscular control in general rather than upon bowel control in particular. In learning self-control during this phase, the child properly gains a lasting sense of autonomy; on the other hand, failure to achieve self-control often leads to a pervasive sense of doubt and shame. The resolution of the oedipal conflict during the phallic phase leads to a heightening of conscience; it is the time, too, when the child needs to develop the prerequisites for either masculine or feminine initiative or become prey to a deep and lasting sense of guilt. In the latency period the child starts school where he finds that gaining admiration, approval, and affection depends upon achievement, and now he must acquire a capacity for industry or become subject to a pervasive sense of inferiority. Then, instead of emphasizing the relationship between genital sexuality and emotional maturity, Erikson focuses upon the need to attain an ego synthesis by the end of adolescence that affords a sense of ego identity, "the accrued confidence that one's ability to maintain an inner sameness and continuity is matched by the sameness and continuity of one's meaning for others."[3] If this cannot be attained the person is subject to identity diffusion. Then, after the young adult has achieved a sense of identity, he can move to relate to another with true intimacy and have the concomitant capacity to distantiate the self from persons or forces whose essence is dangerous to his own; failure to gain the capacity for intimacy almost inevitably leads to self-absorption. The next phase of adult life has as its critical issue the interest in producing members of the next generation and guiding and laying foundations for it; that is, a capacity for generativity, with stagnation as the negative outcome of the phase. The final phase of the life cycle concerns the achievement of mature dignity and integrity through maturely accepting "one's own and only life cycle" and taking responsibility for how it has turned out, whereas despair involves the feeling that this one chance has been wasted and, in essence, has been worthless.

The specific dichotomies that Erikson has utilized to characterize the critical issues of each developmental phase sharpen his emphasis upon the need to cope with tasks rather than simply to pass through a phase without suffering traumata that cause fixations. Still, the critical issues selected neglect other developmental tasks and personality attributes that would seem just as significant.

Piaget and the Epigenesis of Cognitive Development

The theory that we have been considering is not explicitly concerned with the person's linguistic and cognitive development, even though these capacities are essential to what is uniquely human in personality development. Fortunately Piaget, because of his interest in epistemology, sought to uncover the psychological foundations of knowing. The development of intellectual functioning, including language, reasoning, conceptualizing, and categorizing, have been examined, and concomitant studies of the child's moral development have also provided new insights. Cognitive development is conceptualized in strictly epigenetic terms. Piaget and his co-workers have traced the ever increasing scope of the child's abilities through the constant process of adaptation of the existing state of the organism to new experiences. The child cannot utilize experiences that his cognitive schema are not yet ready to assimilate. The foundations for experiencing emerge step by step through the expansion and reorganization of prior capacities as they take in new experiences, thus becoming prepared to react to and utilize more complex experiences. The process is a very active one in which the organism is, so to speak, ever reaching out to incorporate new experiences within the limits permitted by its organization at that moment in its development. Piaget's observations are of great importance to any conceptualization of the life cycle because they provide guides to how the child and adolescent regard the world and can think about it at each stage of development.

Piaget has divided cognitive development into four major periods, each of which is subdivided into various stages and substages.[10] The *sensorimotor* period lasts from birth through the first eighteen to twenty-four months and is essentially concerned with the child's preverbal development. The child's ways of interacting with the world are traced step by step from the neonate's reflex movements through the time when the child uses internalized visual and motoric symbols to invent very simple means of solving problems. In the *preoperational* period, which lasts until about the time the child enters school, the child becomes capable of using symbols and language. He will not yet have the ability to adapt what he says to the needs of the listener, to note contradictions, or to construct a chain of reasoning. During this period he moves beyond his static ways of thinking as he gains experience and as words begin to symbolize categories. The period of *concrete operations* approximates the "latency" period, lasting from about the start of schooling to the onset of puberty. The child now possesses a coherent cognitive system with which he can understand his world and into which he can fit new experiences in a rather ordered fashion. He can carry out simple classifications, but he is not yet capable of abstract conceptualizations. The period of *formal operations* starts in early adolescence when the youth becomes capable of thinking in propositions, using hypotheses, and carrying out operations that are abstracted from concrete examples. Formal operations may require considerable formal education, for only exceptional persons appear to reach this stage on their own.

A Comprehensive Orientation to the Phases of the Life Cycle*

It is not always feasible to sum up the various tasks of a developmental phase under a common rubric, nor is it always wise, for it can convey an oversimplification of a very complex process. Various aspects of the personality develop at differing tempos, and it is essential to study each developmental line separately. The interrelationships between the development of such essentials as cognitive abilities, ethical concepts, object relationships, self-concepts, and gender identity require continuing study. Still, dividing the life cycle into a series of rather natural phases permits the comparison of the various developmental lines, and how an individual's development may be globally or partially impeded, fixated, or regressed at a given phase in development.

Infancy

The period of infancy approximates the first fifteen months of life when the baby is almost completely dependent upon others to care for his essential needs, to provide a sense of security and the stimulation required for his proper emotional and cognitive development. The psychoanalytic term "oral phase" emphasizes that the infant's life centers upon taking in nutriment through sucking but also that his emotional well-being rests upon the assimilation of feelings of security from those who nurture him. If his essential needs are filled and untoward tensions do not repeatedly arise within him, he will have established at the core of his being a basic trust in the world and those who inhabit it. It is upon this trust in others that confidence in himself and in his capacities to care for himself will later develop. Still, a great deal more than gaining a sense of basic trust and a satisfaction of oral needs must take place if the infant is to develop properly. These are a very long fifteen months during which the child undergoes a greater physical and developmental transformation than at any other time in life. So much occurs that we may doubt the wisdom of considering it as a single developmental period. Indeed, there are profound differences between the first and second halves of infancy. During the first half, particularly during the first four months, the baby's physical maturation is of dominant importance, and the care of the child's physical needs takes precedence. Relative neglect of socializing and affectional care can probably be neutralized by later efforts, but such needs cannot be neglected during the second half of infancy without producing permanent effects. With adequate security and proper stimulation from those who raise him, the child will complete most of his sensorimotor development to gain increasing control over his movements, to differentiate himself from his environment, and to experiment with ways of manipulating it. He learns to crawl, and as he begins to toddle and as his babbled jargon turns into words, he moves into a new phase of development.

* See T. Lidz, *The Person*.[8]

The Toddler

As the baby begins to walk and talk, he enters the new phase in which the crucial problems involve the imbalance between his newfound motor skills and his meager mental capacities. He is impelled to use his new abilities to explore his surroundings, but limits must be set upon his activities for his own safety. Because of his limited verbal and intellectual capacities reasons cannot be explained to him, and the baby is only beginning to tolerate delays and frustrations. The delimitations and the increasing expectations for self-control almost inevitably lead to some conflict with the nurturing persons and can provoke stubbornness, negativism, withholding, and ambivalent feelings toward parental figures. The phase is critical to establishing a basic trust in the self and a sense of initiative. Too much delimitation and fostering a fear of exploration can stifle initiative and the development of self-confidence. Mothers who overestimate the toddler's capacities to conform and who cannot tolerate disorder can easily instill a pervasive sense of guilt or shame. The struggle over conformity often focuses upon bowel training and thereby can involve anal erotic fixations. During this period the child requires subtle guidance in the complex process of acquiring language and a trust in the utility of language. Now the child is also gaining a firm sense of being a boy or a girl, and after the age of three the child's gender identity cannot be reversed without leaving considerable sense of confusion. During this phase the child's life interdigitates so closely with his mother's that prolonged separation from her affects him profoundly, causing physiological and depressive disturbances. When the child is about three years old, the period comes to a close; a reasonably good equilibrium has been established between his motor skills and his ability to comprehend and communicate verbally. If progress through the period has gone well, he is ready to gain a more definite autonomy from his mother and find his or her place as a boy or girl member of the family.

The Oedipal or Preschool Child

Some time around the age of two and a half or three, the child ceases to be a baby and becomes a preschool child and begins to go through a period of decisive transition. The infant and toddler had required a close and erotized relationship with his mother, but now the erotic aspects of the relationship must be frustrated to enable the child to gain a more realistic and less egocentric grasp of his place in his mother's life. The transition depends partly upon his mother's ability gradually to frustrate her child's attachment to her, but also upon other changes that are taking place. This phase is critical to the eventual achievement of autonomy. At the end of it he will have completed the tasks of primary socialization, accepted his place as a child member of the family, and be prepared to invest his energies and attention in schooling and in activities with peer groups. Early in this developmental phase the boy experiences his father as an intruder into his relationship with his mother, develops hostile feelings toward his father, and fears reciprocal hostility from him. He properly regains an emotional equilibrium by rescinding his primacy with his mother and seeking instead to become a man like his father, a person capable of gaining the love of his mother. The girl, in giving up her basic attachment to her mother, usually finds a new primary love object within the family in her father and forms a close attachment that will be rescinded about the time of puberty. The child is now gaining the experience as well as the cognitive tools he needs for improved reality testing, and an important aspect of the period is the differentiation of fantasy from reality. He also properly gains a trust in the utility and validity of verbal communication as a help in solving problems and in collaborating with others. His ego function, his ability to guide himself, thus progresses considerably and is further strengthened as he internalizes parental directives. By the age of five or six the child will have gained considerable organization of his personality, and the major lines of his future development will have been laid down—his ways of relating to others and reacting emotionally, having been patterned within his family setting. The resolution of the oedipal situation terminates early childhood, and the precise manner in which the oedipal situation or "family romance" is worked through sets a pattern that will be relived in later interpersonal relationships. During these few years the child has taken a giant step toward becoming an independent and self-sufficient person even though he has done so through appreciating the long road ahead before he can attain adult prerogatives.

The Juvenile

The equilibrium the child gained within his family is disrupted as he goes off to school, where he will be judged by his achievements rather than by ascription,[9] and as he spends increasing amounts of his time with his neighborhood peer group, where he must also find his place on his own. As Erikson has noted, unless the child now gains a sense of industry, he can develop pervasive feelings of inferiority and inadequacy. The trait of industry is encompassed, on the one side, by the danger of habitual compulsive striving to excel competitively and, on the other, by defeatist trends with unwillingness to face meaningful challenges. Other traits also have their roots in this time of life. A sense of belonging develops, the feeling that one is an accepted integral part of the group as contrasted with the feeling of being an outsider; it is a sense that not only concerns social ease as contrasted with social anxiety, but ultimately affects the person's identification with the society in which he lives and his commitment to its values and ethics, as contrasted with a deep-seated feeling of alienation. It is now that a sense of responsibility begins to develop, which involves the willingness to live up to the expectations that one arouses in others. The child's self-concept begins to crystallize in reaction to the way his teachers and peers relate to him. At about the time the child enters school, there is a notable shift in his cognitive processes as he moves into Piaget's stage of concrete operations; and his moral judgment matures from a morality of constraint that is based upon adult edicts to a morality of cooperation.[11] Although psychoanalytic theory considers this time of life as the latency period, it is apparent that these early school years are critical to the child's socialization and the development of many personality characteristics as he begins to move beyond the shelter of the home into the broader world of the schoolroom and his peer groups.

Adolescence

Adolescence may be properly defined as the period between pubescence and physical maturity, but in considering personality development we are interested primarily in the transition from childhood to the attainment of adult prerogatives, responsibilities, and integration. Adolescence involves the discrepancy between sexual maturation and the physical, emotional, and social unpreparedness for commitment to intimacy and the responsibilities of parenthood. Usually it involves an inner struggle to overcome dependency upon parents, which is reflected in an outer struggle to assert independence from parental authority. The adolescent goes through the difficult process of gaining freedom from the earlier repressive ban upon sexual expression while redirecting his choice of love objects outside of the family circle. During these years his cognitive capacities change, and he enters Piaget's stage of formal operations; he becomes capable of using propositional logic, can think about concepts abstracted from reality, and is able to utilize hypothetical-deductive thinking. Ethical or moral capacities also develop as the adolescent becomes interested in ideas, ideals, and ideologies, and his behavior is apt to be directed by the values of the social system rather than simply by interpersonal relationships and the values stemming from them.[6] At the start of adolescence the individual is still a child at play, dependent on his parents and with an amorphous future; at the end he becomes responsible for himself, the direction his life will take is fairly well settled, and his personality has gelled into a workable integrate. As Erikson has properly emphasized, the primary task of adolescence concerns the attainment of an ego identity, but usually, particularly for a girl, the attainment of an ego identity also requires the gaining of a capacity for intimacy.

The lengthy period of adolescence can usefully be divided into three subphases. Early adolescence includes the reorientation required by the prepubertal spurt of growth as well as the physical changes and altered impulsions and feelings that come with puberty. Midadolescence, which is likely to begin twelve to eighteen months after pubescence, is an expansive period. It includes movement toward the opposite sex, which leads to a realignment of intimate friendships. Characteristically it is a time of revolt and conformity, revolt from parental standards and authority and conformity to peer group standards and loyalties. Late adolescence is usually a time when delimitation is required as the young person seeks to find and define his own identity and when cravings for intimacy with a specific individual become important.

Youth

Keniston has suggested that we are currently experiencing the emergence of a new developmental phase among some young people, which he suggests we designate as "youth."[7] It may seem strange to insert a new phase in the life cycle, but the course of the life cycle varies with the social and cultural setting. Aries[1] has pointed out that during the Middle Ages childhood stopped at the age of six or seven. Adolescence may not have been a major developmental stage until secondary school education became common early in this century. Among many preliterate peoples the child becomes an adult shortly after puberty by means of initiation ceremonies. Currently the inability to depend upon traditional ways of adapting in a rapidly changing world has created new developmental problems for some highly educated youths.

Youth includes the period that usually has been included as part of late adolescence and early adult life. It concerns the tensions between the self and society, with a reluctance, or a refusal, to accept either conventional societal standards or a conventional role in society—the "social contract" cannot be accepted if it conflicts with a personal morality based upon higher principles. Critical to this stage of life may be the new awareness of how greatly a person's traits and potentialities depend upon the social environment in which he has been raised as well as upon the ethical standards he embraces. The period is marked by a struggle against acceptance of a conventional way of life and sometimes by a reluctance to accept commitment to a single way of life that will make one an adult. The dangers quite clearly lie in the extremes of alienation—alienation from society, instead of reasonable acceptance of it, or alienation from the self with a pathological self-absorption and withdrawal from considerations of reality.

The Young Adult

The young adult is at the height of his physical and mental vigor, and his energies and interests can now be directed beyond his own growth and development. He has become reasonably independent of his natal family, and having overcome blocks to intimacy with a member of the opposite sex, he is ready to establish a new interdependence with another person and to find his place in the social system. Now commitments are made, and if a person cannot make them he will lack the opportunity of being meaningful to others and having others become meaningful to him. Occupational choice and marital choice, if not made previously, are critical issues, and both will greatly influence the further course of his life cycle. Choice of an occupation determines much of the physical and social environment in which a person will live; it selects out traits that are utilized and emphasized; it provides social roles and patterns for living; and by determining the sort of persons with whom one spends much of his life, it markedly influences values and ethical standards. The choice of a spouse constitutes the major decision that can complement or alter the personality makeup before the production of a new generation. In marrying, one gains a partner who shares, supplements, and supports and upon whom one can rely because the well-being of each partner is bound up with the fate of the other. Marriage also helps individual integration by providing a way of life for which there are traditional directives and a place in the social system. Marital adjustment provides a major test of personality integration and organization, for it requires the malleability to interdigitate one's drive satisfaction, way of life, and ethical standards with a person raised in another family. Marriage not only tests the success of the earlier oedipal transition but also many other facets of the developmental process.

Parenthood

The position of parenthood in the life cycle is not fixed, but for most it comes after commitment to marriage and when the process of self-creation is more or less completed. Spouses are transformed into parents and their marriage into a family. The birth of the first child often alters their life patterns and perspectives more than had their marriage. The person who becomes a parent usually gains a new sense of achievement and completion by the act of creativity. Women, in particular, now feel fulfilled, for to many their biological purpose seems to require completion through conceiving, bearing, and nurturing children and strong cultural directives had added moment to such desires. The woman's creativity as a mother becomes a central matter that provides meaning and balance to her life; and for many years her child or children will be a prime focus in her life.[2] The birth of a child also brings increased self-esteem and sense of responsibility to the father. Parenthood often provides a severe test of the integration of a parent's personality and capacities to accept responsibility for dependents. The child forms and will long continue to form a bond between the parents, a source of shared interest and identification, but he can also be a divisive influence, and in each marriage may in different proportions and ways be a unifying and separating force. For those who have achieved the necessary capacities, the new responsibilities are offset by the fulfillment that comes with experiencing the other side of the parent-child relationship; the self-realization that derives from being needed and loved by a child to whom the parent is so very important; and the constant renewal that comes with the changes in the offspring.

The Middle Years

The passage over the crest of life is a particularly critical period. It is a time of stock-taking concerning the manner in which the person's one and only life is passing—the gratifications attained from marriage and career, and an assessment of prospects for the future. The turn into middle life involves a state of mind rather than some specific bodily change or some clearly demarcated shift in life roles, although awareness of wear and tear on the body and increasing incidence of chronic illness, incapacitation, and death among peers influences the state of mind. The new phase starts with a persistent awareness that it will soon be too late to attempt to change one's course through life. For women in particular middle age may start when the children leave home, or at least no longer require a major share of her attention. Middle age for the woman contains the menopausal "change of life," but she is usually well into middle age when it occurs. A sense of integrity comes with the feeling that a life has been meaningful, that relationships with those who are significant are happy. A negative balance sheet can lead to changes in career or marital partners and frantic efforts for sexual gratification. Dignity, perhaps shaded by resignation, can ward off feelings of despair when life has not produced meaningful relationships or the prospect of achieving goals. As the person passes over the peak, he finds himself one of the older, responsible generation. Accepting responsibilities that one has achieved or has thrust upon one leads to further growth in contrast to stagnation. It is a time of fruition for some, but for others it is a time of regret and disillusion that often leads to resentments toward those who seem to have frustrated.

Old Age

Currently old age is somewhat arbitrarily considered to start at about the age of sixty-five, the time when most men retire from an active occupation. Many of the contemporary problems of the aged involve the difference between a desirable, unharried, and dignified closure of life and a hollow survival in which the person feels useless, unneeded, and burdensome. It is, or should be, a period of surcease from striving and a need to prove one's worth. A sense of equanimity requires the acceptance of failures as well as accomplishments. Sooner or later the aged person becomes increasingly dependent, and his contentment rests upon the reliability of those upon whom he must depend. In some respects there is a reversal of the attainments of adolescence: bodily strength and sexual desire decrease, secondary sexual characteristics and cognitive capacities wane, and concomitantly the pride and confidence derived from such attributes diminish.

It is useful to consider three phases of old

age even though not all persons go through them. The elderly person is not greatly changed from middle life and considers himself capable of taking care of himself and his affairs. Eventually physical changes or alterations in life circumstances make him dependent upon others and he is considered *senescent*. The final phase, which many are spared, is *senility*, when the brain is no longer sufficiently intact to serve its essential function as an organ of adaptation.

Death

Death is the end of the life cycle and an inevitable outcome that brings closure to every life story. Because man is aware of this eventuality from an early age, it profoundly influences how he lives his life. The desire for some type of continuity into the future is pervasive, but the ways of seeking some semblance of immortality are diverse. The finite character of life provides delimitation by directing the individual to specific objectives, by countering diffuse and unbridled strivings, and by bringing a desire to provide a proper ending to a life. Death lends incisiveness to the meaning of experiences, sharpening appreciation of the transitory and the beautiful. Above all, it heightens the preciousness of those one loves because of their mortality. To those who have obtained some wisdom in the process of reaching old age, death often assumes meaning as the proper outcome of life. It is nature's way of assuring much life and constant renewal. Times and customs change but the elderly tire of changing; it is time for others to take over, and the elderly person is willing to pass quietly from the scene.

Panphasic Influences

The developmental process is also profoundly affected by influences that are not phasic. Children usually have the same parents throughout the first decades of their lives, and the parents' personalities and their ways of relating to each other as well as to the child pervade all phases of the child's early development. While a parent may relate more salubriously or deleteriously at a given phase of a child's development, many such influences are panphasic. The parental models for identification, as well as the intrafamiliar milieu and the socioeconomic setting in which the family exists, can affect the individual throughout all developmental stages. Some of these panphasic influences will be presented in Chapter 11.

Bibliography

1. ARIES, P., *Centuries of Childhood*, Knopf, New York, 1962.
2. BENEDEK, T., "Parenthood as a Developmental Phase: A Contribution to the Libido Theory," *J. Am. Psychoanal. A.*, 7:389–417, 1959.
3. ERIKSON, E., "Growth and Crises of the 'Healthy Personality,'" *Psychol. Issues*, Vol. 1, No. 1, Monograph No. 1, International Universities Press, New York, 1959.
4. FLAVELL, J., *The Developmental Psychology of Jean Piaget*, Van Nostrand, Princeton, N.J., 1963.
5. FREUD, S. (1905), "Three Essays on the Theory of Sexuality," in Strachey, J. (Ed.), *Standard Edition, Complete Psychological Works*, Vol. 7, Hogarth, London, 1953.
6. INHELDER, B., and PIAGET, J., *The Growth of Logical Thinking from Childhood to Adolescence* (Tr. by Parsons, A., and Milgram, S.), Basic Books, New York, 1958.
7. KENISTON, K., "Youth: A 'New' Stage of Life," *Am. Scholar*, 39:631–654, 1970.
8. LIDZ, T., *The Person*, Basic Books, New York, 1968.
9. PARSONS, T., "The School Class as a Social System: Some of Its Functions in American Society," in *Social Structure and Personality*, Free Press, New York, 1964.
10. PIAGET, J., *Judgment and Reasoning in the Child* (Tr. by Warden, M.), Harcourt Brace, New York, 1928.
11. ———, *The Moral Judgment of the Child* (Tr. by Gabain, M.), Free Press, Glencoe, 1948.
12. SULLIVAN, H. S. (1946–1947), *The Interpersonal Theory of Psychiatry* (Ed. Perry, H., and Gawel, M.), Norton, New York, 1953.

CHAPTER 11

THE FAMILY: THE DEVELOPMENTAL SETTING

Theodore Lidz

Studies of personality development and maldevelopment have been seriously impeded by a dearth of understanding of the central role of the family in directing the developmental process. The critical early stages of the life cycle, upon which all later development and the stability of the personality rest, take place in the nidus of the family. The stability and satisfaction of the lives of most adults depend greatly upon their marital and parental transactions within their families of procreation. A very large proportion of the work in any dynamic psychotherapy is concerned with the reevaluation of parental influences and the reorganization of patient's reactions to them, so he can be freed from deleterious internalizations and transferences that interfere with his interpersonal relationships and his own evaluation of himself. The various childhood phases of the life cycle unroll favorably or unfavorably not so much because of innate characteristics as because of the manner in which the parental figures and the intrafamilial transactions guide the child through the phase. Attempts to study the young child's development independently of the family setting distort even more than they simplify, for they leave out essential factors of the process.

Importance of the Family

Because the family is ubiquitous it has, like the air that we breathe, been very much taken for granted and many of the vital functions it subserves have been overlooked. Indeed the human being is so constructed that the family is an essential correlate of his biological makeup. It is the basic institution that permits his survival and his development into an integrated person by augmenting his inborn adaptive capacities. Man, after all, is virtually unique among animals in depending upon two endowments: he has both a genetic inheritance and a cultural heritage. His genetic endowment transmits his physical structure and his physiological makeup, which, as in all other animals, permits survival within a relatively narrow range of environmental conditions. Many of his critical adaptive techniques are not inborn; he is born with a unique brain that permits him to acquire language and thereby to acquire from those who raise him the instrumentalities that his society has developed for coping with the environment and for living with one another. This permits him to develop a personality suited to that specific society in which he grows up.[5] The human mechanisms for survival and adaptation are vastly different from those possessed by any other organism, and we can never understand human development and functioning properly unless we take full cognizance of man's dual heritage.

Everywhere the family must meet two requisites: the biological nature and needs of man, and the requirements of the particular society in which it exists and which it subserves by preparing children to live in it. Thus, wherever families exist they will have certain essential features in common even while handling similar problems in differing ways in accordance with the needs of a specific society. In this chapter, I shall seek to designate the essential functions of the family, particularly for childrearing, and the requisites for carrying them out.

The family is an essential correlate of man's biological endowment, for it is the basic social system that mediates between the child's genetic and cultural endowments, provides for his physical needs while instilling societal techniques, and stands between the individual and society, offering a shelter against the remainder of society. Because the child must remain dependent on others for many years, it is important that he be raised by persons to whom his well-being is as important as their own. His dependency upon them and his prolonged attachment to them provide major motivations and directives for his development. As the family forms the earliest and most pervasive influence that encompasses the still unformed infant and small child, the family's ways are *the* ways of life for the child, the only ones he knows. All subsequent interpersonal experiences are perceived, consciously and unconsciously understood, and reacted to according to patterns laid down within the family. These family patterns and the child's reactions to them become so thoroughly incorporated in the child that they are difficult to differentiate from genetically determined factors with which they interrelate. This difficulty greatly complicates the study of the child's physical and personality development. Later influences will modify those of the family, but they can never undo nor fully reshape these early core experiences.

The Family's Primary Functions

The family is usually considered essential because of its childrearing functions, but we cannot properly understand either why the family is omnipresent or how it rears children unless we appreciate that it also subserves essential needs of the spouses and of the society. It not only fills a vital societal need by carrying out the basic enculturation of its children, but the family also constitutes the fundamental social unit of virtually every society: it forms a grouping of individuals that the society treats as an entity; it helps stabilize a society by creating a network of kinship systems; it constitutes an economic unit in all societies and a major economic unit in some; and it provides roles, status, motivation, and incentives that affect the relationships between individuals and the society. In addition, the nuclear family completes and stabilizes the lives of the spouses who formed it. These three sets of functions of the family—for the society, for the parents, and for the children—are interrelated, and it is likely that no other institution could simultaneously fill these three functions without radical change in our social structure and probably not without grave consequences. It is even highly probable that these functions essential to human adaptation cannot be met separately at all except under very special circumstances, but must be fused in the family. Nevertheless, these functions can also conflict, and some conflict between them seems virtually inevitable. Fulfilling parental roles obviously often conflicts with a person's functions as a spouse, and society's demands can obviously conflict with the needs of both the spouses and children, as when the husband is taken from the home into military service.

The Purposes of Marriage

In order properly to grasp the nature of the family setting, it seems essential to examine briefly why people marry and form new families. Such considerations have particular pertinence at the present juncture in history when the value of the family is being challenged as part of the broader questioning of existing institutions and mores. Although people marry for many reasons—love, passion, security, status, to escape from the parental home, to have children, to legitimize a child—marriage is a basic institution in virtually all societies primarily because of man's biological makeup and the manner in which he is brought up to reach maturity.

In growing up in a family, a person forms an essential bond to those who raise him, and he assimilates and internalizes their ways and their attributes. However, an individual cannot achieve completion as an adult within his family of origin. Minimally some degree of frustration must occur because he cannot become a parent with the prerogatives of parenthood and because sexual gratification cannot be united with his affectional relationships. Within his natal family, however, he has enjoyed the security of being a member of a mutually protective unit in which his welfare, at least theoretically, has been of paramount importance to his parents. When he leaves his family of origin, his emotional attachments to it remain unresolved, and he has strong conscious and unconscious motivations to bring closure to these emotional imbalances that move him toward a new union with a person who seems to fill the image of the desired complementary figure sufficiently to be transformed into it. He hopes through marriage to regain the security afforded by a union in which his well-being and needs are again of paramount importance to another—and in marriage the spouses' well-being and security are intimately if not irrevocably interconnected.

The division of the human species into two sexes has created another major impetus for marriage. Men and women are drawn together not only by sexual impulsions but also because the two sexes complement one another in many different ways. Males and females are subjected to gender-linked role training from earliest childhood, which gives them differing skills and ways of relating to people and regarding the world even if such differences are instigated by anatomical, genetic, and hormonal factors. Speaking broadly, neither a man nor a woman can be complete alone. The two sexes are raised to divide the tasks of living and to complement and complete one another as well as to find common purposes sexually and in raising children.

In a marriage the husband and wife can assume very differing types of role relationships and find very diverse ways of achieving reciprocity with one another provided they are satisfactory to both, or simply more satisfactory than separating. The variant ways in which marital couples live together are countless. However, when the birth of a child turns a marriage into a nuclear family, the spouses' ways of relating to one another must not only shift to make room for the children, but limits are also set upon how they can relate to one another if they are also to provide a proper developmental setting for their children.

The Family as a Small Group

Even though a marriage relationship is a very complicated matter, it can be studied and understood in terms of a dyadic interaction, including the influence of other persons and other situations upon the two marital partners. A family, in contrast, cannot be grasped simply in interactional terms, for it forms a true small group with a unity of its own. The family has the characteristics of all true small groups, of which it is the epitome: the action of any member affects all; unless members find reciprocally interrelating roles, conflict or the repression of one or more members follows; to function properly the group requires unity of objectives and leadership toward these objectives; the maintenance of group morale requires each member to give some precedence to the needs of the group over his own desires; it has a tendency to divide up into dyads that exclude others from significant relationships and transactions. These and still other characteristics of small groups are heightened in the family because of the intense and prolonged interdependency of its members, which requires the family, in particular, to have structure, clarity of roles, and leadership to promote the essential unity and to minimize divisive tendencies. The family, moreover, is a very special type of group with characteristics that are determined both by the biological differences of its members and also by the very special purposes it serves. A designation of these characteristics can lead to an appreciation of why the structure of the nuclear family must meet certain requirements.

Generational Differences

The nuclear family is composed of persons of two generations, and the members of each have different needs, prerogatives, obligations, and functions in the family. The parents who have grown up in two different families seek to merge themselves and their backgrounds into a new unit that satisfies the sexual and emotional needs of both and helps bring completion to their personalities in a relationship that seeks to be permanent for them. The new relationship requires the intrapsychic reorganization of each spouse to take cognizance of an alter ego. Wishes and desires of a spouse that can be set aside must be differentiated from needs that cannot be neglected. Although individuals, as parents they function as a coalition, dividing the tasks of living and childrearing. They are properly dependent on one another, but parents cannot be dependent on immature children without distorting the children's development. They provide nurturance and give of themselves so that the children can develop, serving as guides, educators, and models for their offspring even when they are unaware of it. As objects of identification and as basic love objects for their children, how the parents behave and how they interrelate with one another, and not simply what they do to and for their child, are of utmost importance to the child's personality development.

Children, in contrast to parents, receive their primary training in group living and in socialization within the family and are properly dependent upon their parents for many years, forming intense bonds to them while developing through assimilation from the parents and the introjection of their characteristics. Yet the children must so learn to live within the family that they can eventually emerge from it into the broader society, or at least be capable of starting families of their own as members of the parental generation.

Gender Differences

The nuclear family is also composed of persons of two genders with complementary functions and role allocations as well as anatomical differences. The primary female role derives from woman's biological makeup and is related to the nurturing of children and the maintenance of the home needed for that purpose, which has led women to have a particular interest in interpersonal relationships and the emotional harmony of family members—an expressive-affectional role. The male role, also originally related to physique, traditionally is concerned with the support and protection of the family and with establishing its position within the larger society—an instrumental-adaptive role.[7]

Intrafamiliar Bonds

The family relationships are held firm by erotic and affectional bonds. As the marriage is expected to be permanent, the parents are not only permitted but expected to have sexual relationships. Conversely all direct sexual relationships within the family are prohibited to the children; and even the erogenous gratification from parental figures that properly accompanies nurturant care must be progressively frustrated lest the bonds to the family become too firm and prevent the child's investments of interests, energy, and affection beyond the family. The de-erotization of the child's relationships to other family members is a primary task of the family.

The Family as a Shelter

The family forms a physical and emotional shelter for its members within the larger society. However, the family must reflect and transmit the society's techniques of adaptation to its children including the culture's systems of meanings and logic, its ethos and ethics, to assure that the children will be able to function when they emerge from the family into the broader society.

These characteristics of the nuclear family, and corollaries derived from them, set requisites for the parents and for their marital relationship if the family they form is to provide a suitable setting for the harmonious development of their offspring and for directing their development into reasonably integrated adults.

¶ The Parental Requisites

In considering the family's essential functions in regard to the development of its children, it is of critical importance to recognize that the child does not grow up to attain a mature, workable personality simply through the nurturance of inborn directives and potentialities; he does not simply develop into an integrated and adaptable person unless fixations occur because of some innate tendency, some emotional trauma, or some flaw in maternal nurturance during the early years of his development. The child requires positive direction and guidance in a suitable interpersonal environment and social system. The positive molding forces have largely been overlooked because they have been built into the institutions and customs of all workable societies and particularly into the family, which has everywhere knowingly or unknowingly carried out the task of shaping the child's development. The family fosters and organizes the child's development by carrying out a number of interrelated functions, which I shall consider under the headings of *nurture, structure,* and *enculture*. We must examine the nature of these essential functions to understand human personality development and its aberrations properly.

Nurture

The parental nurturant function must meet the child's needs and supplement his immature capacities in a different manner at each phase of his development. This is the one function of the family that has been specifically recognized by most developmental theories. As it has been the focus of intensive study, it does not require elaboration here. It concerns the nature of the nurturance provided from the total care given to the newborn to how the parents foster adolescent movement toward independence from them. It involves filling not only the child's physical needs but also his emotional needs for security, consistency, and affection; and it includes furnishing opportunities for the child to utilize new capacities as they unfold. Proper nurturance requires the parents to have the capacities, knowledge, and empathy to alter their ways of relating to the child in accord with his changing needs. The capacity to nurture or to be maternal is not an entity. Some mothers can nurture a child properly so long as he is almost completely dependent but become apprehensive and have difficulty as soon as he becomes a toddler and cannot be fully guarded from the dangers in his surroundings; some have difficulties in permitting the child to form the erotized libidinal bonds essential for the proper development of the preoedipal child, whereas others have difficulty in frustrating the child's erotized attachment during the oedipal phase. However, unstable parents and grossly incompatible parents are often disturbing influences throughout all of the child's developmental years, and such panphasic influences are often more significant in establishing personality traits or disturbances in children than are fixations at a specific developmental phase. While the mother is usually the primary nurturant figure to the child, particularly when the child is small, and though she is usually the family expert in childrearing and child care, her relationship with the child does not transpire in isolation but is influenced by the total family setting. The mother's capacity to care for her child properly is influenced greatly by her marital interaction with her husband, by the demands of other children, and by the relationships between her children as well as by her husband's relationship with each child. The quality and nature of the nurture that a child receives profoundly influences his emotional development. It affects his capacities to differentiate from the mother and the emotional context of his relationships to others; it affects his vulnerability to frustration and the anger, aggressiveness, anxiety, hopelessness, and helplessness he experiences under various conditions. As Erikson has emphasized, it influences the quality of the basic trust a child develops—the trust he has in others, in himself, and in the world in which he lives.[3] It contributes to the child's self-esteem as a member of the male or female sex. It lays the foundations for trust in the reliability of collaboration with others and in the utility of communicating verbally as a means of solving problems. A person's physiological functioning can be and perhaps always is permanently influenced by the way in which the parental nurturing figures respond to his physiological needs.[2]

From these brief comments it is apparent why so much attention has properly been paid to the parental nurturant activities and how profoundly they influence the development of a person; but they are but one aspect of what a child requires from his parents and his family.

Structure

Let us now turn to consider the relationship between the dynamic organization of the family and the personality integration of its offspring. Although the family organization differs from one society to another and with social class and ethnic group within a society, it seems likely that the family everywhere follows certain organizational principles both because of its biological makeup and because of the specific functions it subserves. As noted above, the family members must find reciprocally interrelating roles or distortions in the personalities of one or more members will occur. The division of a family into two generations and two sexes lessens role conflicts and tends to provide an area free from conflict into which the immature child can develop, and directs him or her to grow into the proper gender identity, which forms the cornerstone of a stable ego identity. While all groups require unity of leadership, the family contains two leaders—the father and the mother—with different but interrelated functions that permit them to form the required coalition that permits unity of leadership. We may hazard that in order for the family to develop a structure that can properly direct the integration of its offspring, *the spouses must form a coalition as parents, maintain the boundaries between the generations, and adhere to their respective gender-linked roles.* These requirements may sound rather simple until we explore their ramifications.

The Parental Coalition

As has been noted, all small groups require unity of leadership, but the family has a dual leadership. The mother, no matter how subjugated, is the expressive-affectional leader; the out his or her cardinal functions. The mother, father, the instrumental leader. A coalition between these leaders is necessary not only to provide unity of direction but also to afford each parent the support essential for carrying for example, can better delimit her erotic investment in the small child to maternal feelings when her sexual needs are being satisfied by her husband. The family is less likely to break up into dyads that create rivalries and jealousies if the parents form a unity in relating to their children; and, particularly, a child's tendency to possess one or the other parent for himself alone—the essence of the oedipal situation—is more readily overcome if the parental coalition is firm and the child's egocentric fantasies are frustrated and redirected to the reality that requires repression of such wishes. If the parents form a coalition both as parents and as a married couple, then the child, who is provided with adult models that treat one another as alter egos, each striving for the partner's satisfaction as well as for his own, is very likely to grow up valuing marriage as an institution that provides emotional satisfaction and security, thus gaining a long-range goal.

The child properly requires two parents: a parent of the same sex, with whom he can identify and who forms an object of identification to follow into adulthood, and a parent of the opposite sex, who serves as a basic love object and whose love and approval is sought by identifying with the parent of the same sex. However, a parent fills neither role effectively for a child if denigrated, despised, or treated as a nonentity, or even as an enemy, by the spouse. Parents who are irreconcilable in reality are likely to become irreconcilable introjects in the child, causing confused and contradictory internal directives. It is possible for parents to form a reasonable coalition for their children despite marital discord and to some extent even despite separation; they can agree about how children should be raised and support their spouses to the children as worthwhile persons and as good parents even if their ways and ideas differ.

A variety of difficulties can follow upon failures of the parental coalition. The growing child may invest his energy and attention in supporting one or the other parent or in seeking to bridge the gap between them, rather than utilizing his energies for his own development. Sometimes the child becomes a scapegoat with his problems magnified into the major source of dissent between the parents, and he comes to feel responsible for their difficulties. A child may willingly oblige and assume the role of villain in order to mask the parental discord, thereby retaining the two parents he needs. The child may also be caught in an impossible situation in which any attempt to please one parent elicits rebuff or rejection from the other. When the parents fail to achieve a coalition, there are many ways in which the child becomes subject to conflicting motivations, directives, and standards that interfere with the development of a well-integrated personality.

The Generation Boundaries

The division of the nuclear family into two generations lessens the danger of role conflict and furnishes space free from competition with a parent into which the child can develop. The generational division is a major factor in providing structure to the family. The parents are the nurturing and educating generation and provide adult models and objects of identification for the child to emulate and internalize. The child requires the security of dependency to be able to utilize his energies in his own development, and his personality becomes stunted if he must emotionally support the parents he needs for security. A different type of affectional relationship exists between parents than between parent and child. However, the situation is complicated because of the intense relationship heightened by erogenous feelings that properly exists between the mother and each preoedipal child and by the slow differentiation of the child from his original symbiotic union with his mother. The generational division aids both mother and child to overcome the bond, as is essential to enable the child to find a proper place as a boy or girl member of the family and then to invest his energies in peer groups and schooling, as well as in gaining his own identity. The generation boundaries can be breached by the parents in various ways, such as by the mother failing to establish boundaries between herself and a son; by the parent using a child to fill needs unsatisfied by a spouse; by the parent acting as a rival to a child; by the father seeking to be more of a child than a spouse. Incestuous and near incestuous relationships in which a parent overtly or covertly gains erotic gratification from a child form the most obvious disruptions of generation lines. When the child is used by one parent to fill needs unsatisfied by the other, the child can seek to widen the gap between his parents and insert himself into it; and by finding an essential place in completing a parent's life he need not—and perhaps cannot—turn to the extrafamilial world for self-completion. The resolution of the oedipal situation thus depends for its proper completion upon having a family in which the parents are primarily reliant upon one another or at least upon other adults. Further, if a parent feels excluded by a child, the child's fears of retribution and retaliation may not be simply projections of his own wishes to be rid of a parent, but may derive from the reality of having a jealous and hostile parent.

Confusions of the generation boundaries within the nuclear family together with the ensuing role conflicts can distort the child's development in a variety of ways, some of which have already been indicated. The child's proper place within the family is invaded; rivalry with parents absorbs energies and fosters internalized conflicts; a parent's dependency upon a child occupies and preoccupies a child prematurely with problems of completing the life of another rather than with his own development. Aggressive and libidinal impulses directed toward the parents become heightened, rather than undergoing repression and gradual resolution, and are controlled only through strongly invested defensive mechanisms.

The Sex-Linked Roles

Security of gender identity is a cardinal factor in the achievement of a stable ego identity; and of all factors entering into the formation of personality characteristics, the child's sex is probably the most decisive. Confusions concerning sexual identity and dissatisfactions with one's sex can contribute to the etiology of many neuroses and character disorders as well as to perversions; and probably all schizophrenic patients are seriously confused concerning their sexual identity. A child does not attain proper sex-linked attributes simply by being born a boy or girl, but through gender role allocation that starts at birth and then develops through role assumptions and identifications as he grows older. The maintenance of appropriate gender-linked roles by the parents is one of the most significant factors in guiding the child's proper development as a boy or girl. Clear-cut reversals in the parents' sex-linked roles obviously distort the child's development, either when they are in the sexual sphere, as when a parent is overtly homosexual, or when they concern the divisions of tasks in maintaining the family. While it is clear that a child whose father performs the mothering functions, both tangibly and emotionally, while the mother supports the family will usually gain a distorted image of masculinity and femininity, the common problem is usually more subtle: the inability of the mother to fill an expressive-affectional role or of the father to provide instrumental leadership for the family. As Parsons and Bales have pointed out, a cold and unyielding mother is more deleterious than a cold and unyielding father, whereas a weak and ineffectual father is more damaging than a weak and ineffectual mother.[7] Still more explicitly, a cold and aloof mother may be more detrimental to a daughter who requires experience in childhood with a nurturant mother in order to attain feminine and maternal characteristics, whereas an ineffectual father may be more deleterious to a son who must overcome his initial identification with his mother, as well as his early dependency upon her, to gain security in his ability to provide for a wife and family. Although the sharing of role tasks has become a necessity in most contemporary families, which leads to some blurring of the parental roles, there is still a need for the parents to maintain their primary gender-linked roles and to support one another in them.

The child's identification with the parent of the same sex is likely to be seriously impeded when this parent is unacceptable to the other whose love the child seeks, a difficulty that can be heightened by the homosexual tendencies of a parent. The mother may be basically unacceptable to a father with homosexual proclivities simply because she is a woman; and the daughter may respond by seeking to be boyish, by gaining the father's affection by being intellectual, or through some other means that do not threaten him by feminine appeal. If a mother, on the other hand, is consciously or unconsciously rivalrous with all men, her son may readily learn that masculinity will evoke rebuff from her, and fears of engulfment or castration by the mother become more realistic sources of anxiety than fears of retaliatory rejection or castration by the father.

Of course, other problems can create difficulties for a child in gaining a secure gender identity, such as when parents convey the wish that the boy had been born a girl or vice versa, or the need to avoid incestuous involvements; still, when parents adequately fill their own gender-linked roles and accept and support the spouse in his or her role, a general assurance of a proper outcome is provided.

The relationship between the family structure and the integration of the offspring's ego development is a topic that is only beginning to be studied. Still, a little consideration leads us to realize that the provision of proper models for identification, motivation toward the proper identification, security of sexual identity, the transition through the oedipal phase, the repression of incestuous tendencies before adolescence, and many other such matters are profoundly affected by the family's organization. Unless the parents can form an adequate coalition, maintain proper boundaries between the generations, and provide appropriate gender role models by their behavior, conflicts and role distortions will interfere with the proper channeling of the child's drives, energies, and role learning.

The Family and Enculturation

The proper enculturation of the child within the family may be more properly divided into the process of socialization and the process of enculturation. Under socialization we may somewhat arbitrarily subsume how the child learns basic roles and institutions through interactions between family members; and enculturation concerns that which is transmitted symbolically from generation to generation rather than through societal transactions. However, there is considerable overlap, and the two functions cannot always be clearly distinguished.

The form and function of the family evolves with the culture and subserves the needs of the society of which it is a subsystem. The family is the first social system that the child knows, and simply by living in it he properly gains familiarity with the basic roles as they are carried out in the society in which he happens to live: the roles of parents and child, of boy and girl, of man and woman, of husband and wife. He also learns how these roles of the family members interact with the broader society. Whereas roles are properly considered units of the social system rather than of the personality, they also are important in personality development through directing behavior to fit into roles and by giving cohesion to the personality functioning. Individuals generally do not learn patterns of living entirely on their own, but in many situations learn roles and then modify them to their specific individual needs.

The child also learns from his intrafamilial experiences about a variety of basic institutions and their values, such as the institutions of family, marriage, economic exchange, and so forth; and societal values are inculcated by identification with parents, superego formation, teaching, example, and interaction. The wish to participate in or avoid participating in an institution—such as marriage—can be a major motivating force in personality development. It is the function of the family to transmit to the offspring the prescribed, permitted, and proscribed values of the society and the acceptable and unacceptable means of achieving such goals. Within the family the child is involved in a multiplicity of social phenomena that permanently influence his development, such as the value of belonging to a mutually protective unit; the rewards of renouncing one's own wishes for the welfare of a collectivity; the acceptance of hierarchies of authority and the relationship between authority and responsibility. The family value systems, role definitions, and patterns of interrelationship enter into the child through the family behavior far more than through what he is taught or even what the parents consciously appreciate.

The process of enculturation concerns the acquisition of the major techniques of adaptation that are not inherited genetically but that are assimilated as part of the cultural heritage that is a filtrate of the collective experiences of man's forebears. In a complex industrial and scientific society such as ours, the family obviously can transmit only the basic adaptive techniques to its offspring and must rely upon schools and other specialized institutions to teach many of the other instrumentalities of the culture.

The cultural heritage includes such tangible matters as agricultural techniques and food preferences, modes of housing and transportation, arts and games, as well as many less tangible matters such as status hierarchies, religious beliefs, ethical values and behavior that are accepted as divine commands or axiomatically as the only proper way of doing things and are defended by various taboos. The transmission of language is a primary factor in the inculcation of both techniques and values because the totality of the enculturation process depends so greatly upon it. After the first year of life the acquisition of almost all other instrumental techniques depends to a greater or lesser extent upon language; and the collaborative interaction with others, which is so critical to human adaptation, depends upon the use of a shared system of meanings. Indeed, the capacity to direct the self, to have any ego functioning at all, depends upon having verbal symbols with which one constructs and internalizes a symbolic version of the world that one can manipulate in imaginative trial and error before committing oneself to irrevocable action. As virtually all intact children learn to speak, we are apt to overlook the complexities of the process of learning language as well as its central importance to ego functioning. It required the linguistic anthropological studies of Sapir[8] and Whorf[11] to illuminate how profoundly the specific language that a person utilizes influences how he perceives, thinks, and experiences. The studies of schizophrenic patients and their parents illustrated how greatly faulty and distorted language usage can affect personality development and functioning.[6] The inculcation of a solid foundation in the language of the culture is among the most crucial tasks carried out by the family.

Speaking very broadly, the child learns language through attempting to solve problems. Meanings are established rapidly or slowly, with clarity or vagueness, in accord with how effectively and consistently the proper usage of words gains objectives for the child. The process depends upon reciprocal interaction between the child and his tutors, the consistency among his teachers, the cues they provide, the words to which they respond or remain oblivious, the meanings that they reward consistently or sporadically, or that they indicate are useless, ineffectual, undesirable, repugnant, or punishable. Many other factors are also involved in the child's attainment of language, but it is clear that the family plays a very important part in the process. The categorizing of experiences through the abstraction of common attributes, the labeling of categories by words, and the sharpening of the meaning of words by grasping the critical attributes designated by the word are essential for both ego development and proper ego functioning.

In contrast to the commonly accepted dictum expressed by Hartmann,[4] the infant is not born adapted to survive in an average, expectable environment. The range of environments in which he is physiologically capable of surviving are relatively limited, but every viable society develops a set of instrumental techniques and institutions that take the infant's essential needs into account and modify the environment to the child's capacities. Then, very largely through the use of language, the child learns the culture's techniques of adaptation more or less adequately, and gains an ability to delay the gratification of basic drives, to internalize parental attributes, directives, and teachings, and to be motivated by future security as well as by drive impulsions. Further, the world in which he lives, the behavior of others, and his own needs gain some degree of order and predictability through the categories provided by the language.

Upon consideration it seems apparent that the parental styles of behaving, thinking, and communicating, as well as their specific patterns of defenses, are critical factors in the development of various personality traits in their children, both through direct example and the internalization of such traits by the children as well as through the reactions that such styles produce in the child. When Bateson and Jackson[1] formulated their double-bind hypothesis of schizophrenia; when Lidz, Fleck, et al.[6] noted how schizophrenic patients had been taught to misperceive, to deny the obvious, to be suspicious of outsiders; and when Wynne and Singer[9,10,12,13] documented that virtually all schizophrenic patients have parents who have either amorphous or fragmented styles of communicating, a new dimension was added to the study of personality development as well as to the study of psychopathology. In the study of the obsessional character, for example, instead of focusing primarily upon what occurred in the patient's "anal" phase of development, we begin to note that the parents of such persons are usually obsessional themselves, unable to tolerate direct expressions of hostility in themselves or in their children. They are very likely to teach the use of isolation, undoing, and reaction formation as a defense against the expression of hostility both through their own example and through what they approve and disapprove in their children. Such obsessional parents are likely to use rigid bowel training and to limit the young child's autonomy and thus foster ambivalence, stubbornness, shame, and undoing defenses in many ways other than simply through the way they direct the child's bowel training.

Conclusions

Personality development cannot be properly studied or understood abstracted from the family matrix in which it takes place. The major foci of attention—the childrearing techniques and the emotional quality of the nurturant care provided the child—while clearly very important do not encompass the topic. The child's development into an integrated individual is guided by the dynamic organization of his family, which channels his drives and directs him into proper gender and generation roles. The child must grow into and internalize the institutions and roles of the society as well as identify with persons who themselves have assimilated the culture. The child acquires characteristics through identification but also by reactions to parental figures and through finding reciprocal roles with them. His appreciation of the worth and meaning of both social roles and institutions is markedly affected by the manner in which his parents fill their roles, relate to one another, and behave in other contexts. The perceived reliability of the verbal tools that are necessary for collaboration with others and for thinking and self-direction depend greatly upon the tutelage within the family and on the parents' styles of communicating. In the study of personality development and in our search for proper guidelines to help provide stable emotional development, the emphasis upon what parents should or should not do to the child, for the child, and with the child in each phase of his development has often led to neglect of other more significant familial influences. Who the parents are; how they behave and communicate; how they relate to one another as well as to the child; and what sort of family they create including that intangible factor, the atmosphere of the home, are of paramount importance. Numerous sources of deviant personality development open before us when we consider the implications of the approach to personality development presented in this chapter.

Bibliography

1. BATESON, G., JACKSON, D., HALEY, J., and WEAKLAND, J., "Toward a Theory of Schizophrenia," *Behav. Sci.*, *1*:251–264, 1956.
2. BRUCH, H., "Transformation of Oral Impulses in Eating Disorders: A Conceptual Approach," *Psychiat. Quart.*, 35:458–481, 1961.
3. ERIKSON, E., *Childhood and Society*, Norton, New York, 1950.
4. HARTMANN, H. (1939), *Ego Psychology and the Problem of Adaptation*, International Universities Press, New York, 1958.
5. LIDZ, T., *The Family and Human Adaptation*, International Universities Press, New York, 1963.
6. LIDZ, T., CORNELISON, A., TERRY, D., and FLECK, S., "The Transmission of Irrationality," in Lidz, T., Fleck, S., and Cornelison, A., *Schizophrenia and the Family*, International Universities Press, New York, 1965.
7. PARSONS, T., and BALES, R., *Family, Socialization and Interaction Process*, Free Press, Glencoe, 1955.
8. SAPIR, E., *Selected Writings of Edward Sapir in Language, Culture and Personality*, University of California Press, Berkeley, 1949.
9. SINGER, M., and WYNNE, L., "Thought Disorder and Family Relations of Schizophrenics: III. Methodology Using Projective Techniques," *Arch. Gen. Psychiat.*, *12*:187–200, 1965.
10. ———, and ———, "Thought Disorder and Family Relations of Schizophrenics: IV. Results and Implications," *Arch. Gen. Psychiat.*, *12*:201–212, 1965.
11. WHORF, B., *Language, Thought, and Reality: Selected Writings of Benjamin Lee Whorf* (Ed. by Carroll, J.), M.I.T. and J. Wiley & Sons, New York, 1956.
12. WYNNE, L., and SINGER, M., "Thought Disorder and Family Relations of Schizophrenics: I. A Research Strategy," *Arch. Gen. Psychiat.*, 9:191–198, 1963.
13. ———, and ———, "Thought Disorder and Family Relations of Schizophrenics: II. A Classification of Forms of Thinking," *Arch. Gen. Psychiat.*, 9:199–206, 1963.

CHAPTER 12

INFANT DEVELOPMENT

David E. Schecter

THE FRAME OF REFERENCE of the behavioral sciences, including psychiatry, is increasingly expanding to include psychohistorical and intergenerational considerations, especially with the realization that the needs of child, youth, and adult are, to a large degree, mutually interdependent—each group having needs to confirm and be confirmed by the other. We mean to emphasize in our discussion of infancy—approximately the first fifteen months of life—the reciprocal relatedness between child and caretaker, whether the caretaker is parent, professional, or extended family. With new possibilities in societal and childrearing structures (the daycare center, the kibbutz, the commune), it is more important than ever to understand the nature of the infant and his dependency on his caretakers, even if the child were to be conceived in a test tube. For practical purposes we will assume that the family—or a variant thereof—is still viable and still a rather universal matrix in which children are reared, notwithstanding the influences of other institutions.* Traditional parental roles—and the security that comes with these roles—have already broken down to varying degree in the Western nuclear family, with an ensuing search for new forms of childrearing that may be adaptive to a relatively unknown and unpredictable future world. Hence the widespread phenomenon of acute and chronic parental doubting concerning a range of problems connected with child and adolescent rearing, with life style, and with basic value orientation.

Although traditionally concerned with alleviating symptoms and altering deviant behavior, psychiatry has moved into community concerns, recognizing its potential contribution to the fostering of mental and emotional well-being as well as to the prevention and treatment of suffering and destructive behavior among the people of the world. If we can understand the nature of human development in its various sociocultural forms, we increase the possibilities of knowing the conditions under which "healthy" development can be facilitated. Since the human organism has a wide range of adaptability, the issues of "health" and adaptation are closely related. "Adaptation to what" involves matters of

* See Murdock[53] on the issue of the universality of the family.

human values and goes beyond the usual boundaries defined by a narrow scientific approach that attempts to remain value-free or—more accurately—value-blind. For example, if a family or a society's goal is to encourage self-control in a child, there are many routes toward this behavioral end. One end of the spectrum would rely on providing a milieu that encourages self-control by example and encouragement as well as by a clear setting of limits; the other direction would make use of techniques weighing most heavily on fear, coercion, or shaming. Although the surface behaviors in each instance may have a similar appearance, the different psychological structures involved in these examples would have completely different implications for the child's total development and for the social relations of which he becomes a part.

The Birth of a Family

The biological helplessness of the infant demands nurturance from caretakers who have a high stake in the infant's growth and development. Traditionally the family has been entrusted with the functions of fostering the child's biological, social, and cultural development. This implies that from the time of conception parents will try to influence the new organism in directions largely dictated by their personal and socially shared values. However, the freedom to nurture optimally a completely dependent fetus and infant depends a good deal on the level of psychological maturity attained by the parents.

Parenthood can be regarded as a developmental phase, incorporating several substages, with stage-specific tasks, stresses, and opportunities for growth. The young adult usually comes to parenthood while he is still undergoing a series of individuation experiences in which he has attempted to liberate himself psychologically from his family of origin through an intense inner struggle to establish his—or her—own sense of identity. Simultaneously a need develops for a relatively enduring, intimate relationship, which often involves the formation of a new social unit, a "couple."* In such a setting we can observe fluctuations from states of relative psychological separateness to states of greater fusion or loosening of ego boundaries. The latter is experienced not only during sexual excitement and orgasm but in the not infrequent expectation that the one partner have *identical* wishes, tastes, and values as the other. With the relative sense of exclusive possessiveness seen in some new couples, there are temptations toward regressions, including increased dependency and fusion that evoke affects associated with earlier parent-child experiences. In part this accounts for the cyclical moving toward and away from one another in any intimate relationship.

Against this background of relatively exclusive couplehood, a pregnancy can potentially come to signify an intrusion into the new unit. At the very least, even when the pregnancy and newborn are consciously and unconsciously welcomed, there will be a marked disequilibrium and need for new homeostasis in the new family unit. The capacity for the dyad to grow into a relatively harmonious triad is one of the essential developmental tasks of parenthood, the outcome of which will have enormous impact on the child's and the family's development.

We know from clinical as well as from direct family observations that pregnancy and infancy set up new strains in both parents. Aside from the demandingness of the new infant, there is a shift of emotional investment (cathexis) in the mother—especially after quickening—when she becomes aware of a new being that is inside and part of her. If the husband needs his wife at the same level of intensity as in prepregnancy, he may experience some deprivation, which can be overcome—if he is mature enough—especially with the feeling of a new pride in his role as father. In families we have studied in weekly observation for the first two years of life, we have noted in some fathers clear signs of dep-

* For discussions of parenthood that include developmental and intergenerational points of view, see Erikson's[27] eight stages of man, Lidz,[47] and R. Blanck and G. Blanck.[13]

rivation, jealousy, and feelings of being excluded.*

Since there tends to be a social idealization of infancy, demanding an unambivalently blissful family atmosphere into which the infant is supposed to be received, ambivalent affects are generally suppressed or repressed, and investigation of these feelings has been hampered by a sense of taboo surrounding this issue. Also, from clinical experience with parents in various settings, one would suspect that there may be a widespread incidence of new families suffering in a quiet spirit of desperation.†

Although discussion of father's role is often absent in consideration of early family development, such considerations are manifest in mythology—probably indicating the rather universal nature of these problems. There is a whole other side to the Oedipus-type myths that Freud did not emphasize. This perspective involves predictions that the newborn (son) would preempt and perhaps kill the father of the new family. This prophecy of the Delphic oracle provoked Laius and Jacosta to attempt infanticide by tying Oedipus's feet and leaving him to die in the fields in order to save their marriage and kingdom.§

The observations that a father may feel that his wife has "the inside track" with their son and that the mother feels despairing about her lack of abundance to provide for both child and husband should not obscure the fact that these conflicts are often resolved in a positive direction; the couple transcends its former structure to make room for—and, indeed, feel enriched by—the common pleasures and tasks involved in the child's growth. With this capacity of the parent to move from exclusive couplehood to a communal triad, there develops a sense of growing into generative adulthood (Erikson)[26] in which one's potency and surplus vitality is expressed in the sharing of the care and concern for another whose importance is experienced as at least equal to one's own.

Parent-Child Developmental Fit: Individual Differences and Sociocultural Factors

The many variables that are involved in the adequate growth of a family include the character and maturity of the parents, the constitutional difficulties and demands posed by a newborn, and his "match" with what his parents—and by implication their sociocultural group—can tolerate and respond to. The work of Bridger,[18] Birns,[12] Chess, Thomas, *et al.*,[23] and Escalona[31] reveals the early appearance of individual differences in infantile reactivity and temperament and the complex relation of these factors to parental expectations and responsiveness. The matter of parental expectations and hopes for their infant is of paramount importance in determining how the

* The observations referred to above have been written up in an unpublished manuscript, "Some Early Developments in Parent-Child Interaction."[70] The undertaking was part of a project "Studies in Ego Development" at the Albert Einstein College of Medicine (New York) and was made possible by Grant #HD 01155–01 provided by the National Institute of Child Health and Human Development.

† The public health possibilities in helping newly developing families are enormous, although complex. One can imagine parent and child caretaker groups being formed under skilled leadership on a widespread basis with the function of helping parents with problems that have been considered unique, unshareable, and taboo. If we can further develop the professional knowledge and skills in this area of family development, such feelings can emerge step by step from considerations of the practical matters of infant rearing to the more personal issues that beset all parents.

§ We presume that common myths are part dramatization and part "answer" to certain universal human concerns. Although Laius's action can be linked to his jealous, "immature," authoritarian character, there is yet another side to his mythic action that functions as a cultural expression of the incest taboo. In ontogenetic terms this taboo expresses the superego imperative that derives, in part, from Laius's own boyhood oedipal strivings.

This type of analysis of the Oedipus myth reveals another motive for Oedipus's death wish, namely, one of *revenge* against the father authority whose son's growth was viewed as an unpardonable threat to the father's exclusive power and possessions. For a further discussion of the "triangular" parental affects in the preoedipal phase, see Fromm's[38] presentation of the issue of authority in the oedipal complex and my own discussion of these matters.[68]

parents will evaluate the infant's progress and their own worth and "goodness" as parents.*

Our own observations of infant-parent interaction impressed us with the intensity of parental expectation and hope for their infant's achievements in accordance with a kind of idealized developmental schedule. If one of these goals—for example, to be independent, to play by himself, to reach and grasp—was not achieved by a certain age, there could ensue a sense of disappointment and failure leading to blame of self, child or spouse, and to mounting family tensions, whose origins the parents would soon become unaware of in the complex mesh of secondary interpersonal stress. We observed a two-month-old infant with colic shrieking for relief, but when his mother carried him "too long" in order to soothe him she was accused by her husband—and eventually accused herself—of "spoiling" the child. The fear of spoiling the infant proved quite prominent in a number of parents and was one of the sources of doubt and dulling of spontaneity in the parents' attitudes toward their child. Behind the issue of spoiling is a fear of inducing overdependence and an omnipotence of will in the infant. At times these matters could be amusing, but they also signified a damaging form of patterning when many spontaneous behaviors in a very young infant were assigned the significance of willfulness and defiance.† Thus, a three-month-old infant was spanked by her father because she "refused" to go to sleep; on another occasion, using unfamiliar strident vocal tones and threatening facial gestures, her mother severely reprimanded her when the infant's protruding of her tongue was interpreted as a sign of disrespect. A further example of the fear of giving "too much attention" was seen in the mother's decision to withhold bodily contact with the baby by propping the bottle during most feedings. We must emphasize that although these were not the most sensitive of parents, they were also not too unrepresentative in many respects of what we saw in lower-class and lower-middle-class homes.§

Messiness and *dirtiness* in connection with feeding proved to be another source of maternal anxiety and consequent scolding or punishment in the early months of life. In this way important early autonomy striving such as self-feeding by finger, spoon, or cup can bc discouraged by a mother whose sensitivity to messing or whose need to control overrides the child's readiness for certain masteries. Spock[78] has written about how critical the period of five to six months is for the development of increasing autonomy and self-reliance through the potential mastery of cup feeding at this time.

We saw the beginnings of the "power struggle" from the parents' side much earlier than the classic autonomy phase usually ascribed to the infant in his second and third year of life. Even though we were impressed with the observations of how maternal attitudes and behaviors were influenced by the infant's behavior, we were even more impressed with the limitations of this proposition.‡ Parental char-

* There is more than analogy to the above model in the major transference elements that are brought into the psychotherapeutic situation. The therapist is frequently cast in the role of magic helper by the patient, but he has also been cast into this role—in varying degree—by his own motivations in becoming a therapist. Understanding and working through these mutual needs between therapist and patient constitute a major part of the work of intensive psychoanalytic psychotherapy. If we continue to follow the ensuing issues of family development in infancy, we will see that there is probably not a single dimension —whether it be attachment behavior, separation or stranger anxiety—that fails to be represented in the psychotherapeutic situation, especially if this situation is analyzed in some depth. Hence an understanding of infancy and its salient developmental issues enriches the psychotherapist's work with any age group.

† The whole problem of child abuse (including the "battered child syndrome") can only be mentioned here. The rage leading to violence has been connected by the attacking parents to the inability—as they perceive it—to bend the will of the child to obedience. It is not only the child's *actual* autonomous, defiant, or negativistic behavior that provokes attack but spontaneous behaviors—including crying—that are experienced by the parent as willful and defiant.

§ We had less opportunity to observe upper-middle-class and upper-class families.

‡ Coleman, Kris, and Provence[24] have described in some detail how parental attitudes and unconscious fantasies are continuously influenced by the child's growth and development. More recent research, for example, Moss,[52] has been even more specific about the fact that different variables such as sex, age, and "state" contribute to the shaping of maternal response.

acter—in some aspects individually unique, but in its basic dimensions determined by sociocultural patterning—was seen to exert a powerful influence on the threshold, intensity, quality, and flexibility of parental response in relation to the infant's behavior, including the developmental changes in his behavior organization.

Despite the warps of development that can follow from inappropriate parental standards and expectations, it is clear that standards are—in varying degree of flexibility—common to all societies. Parental hopes for their offspring are universal and have been expressed in the form of the messianic ideal; this theme runs through various religious, mythological, and artistic motifs throughout history. The intense affective investment in the child increases the chances of species and individual survival as well as cultural continuity, even though it lays the groundwork for potential disappointment, disillusion, blame, and resentment in family relationships. Even so, despite the inevitability of parental inner conflict and guilt in Western culture,* one can observe parents within this framework who show remarkable sensitivity and skill in facilitating and enabling opportunities for the infant's movement toward an optimal balance of interdependence and autonomy.

Development of the Human Bond

Of all the developments in the first year of life, that of the human bond between the infant and his caretaker—or caretakers—is probably the most fateful for his future life. We have presumed that the unfolding of the child's subsequent interpersonal relationships derives heavily from the patterning of the first social relationships. I have avoided the use of the word "mother" here, lest we assume that it is only with one's actual mother that the primary social bonds can be formed. Originally psychological, including psychoanalytic, formulations concerning the mother-child relationship were largely reconstructive or theoretical. Only in recent years have there been direct and specific kinds of researches to help elucidate the exact nature of the unfolding of the first human relationship.

There are several types of theories concerning the process by which the child becomes socially related. One type is concerned with the formation of the social bond, largely through secondary psychological dependency deriving from repeated cycles of gratification through the reduction of need-tension, primarily oral. This type of theory—largely involving the precepts of social learning—is in essence the one formulated by Freud, who properly emphasized the helpless nature of the newborn infant whose survival depends on the ministrations and need gratifications by his caretaker. However, Freud[37] also postulated a primary instinctual sucking drive that was anaclitic in its nature since it "leaned on" what he then referred to as the self-preservative ego instincts. The object of the sucking drive was seen as the breast—the social bond to mother being developed largely through the secondary psychological dependency described above.

Bowlby[17] and others (Balint,[4] Fairbairn[32]) have emphasized a primary object seeking tendency in the infant from the time of birth. Bowlby postulates mechanisms—derived from ethological models—by which the primary attachment to the mother is mediated. These mechanisms, referred to as "component instinctual responses," are made up of species-specific behavior patterns, determined by heredity and emerging within specific developmental periods during the first years of life. The five "instinctual responses" suggested consist of sucking, clinging, following behavior (both visual and locomotor), crying, and smiling. Bowlby himself stresses that his theoretical model was intended to retain but update Freud's original schema of component instincts. From an evolutionary point of view, Bowlby considered the instinctual responses as having evolved with the adaptive function of

* See R. Levy's[45] exposition of the proposition of the inevitability of guilt provocation in Western cultures, which are constantly aspiring to new standards, thus making it very unclear what—at any given moment—may be "good" or "bad."

eliciting nurturing behavior on the part of the mother. As Yarrow[87] points out, a third theoretical framework, central to psychoanalytic theory—that of "object relationships"—is probably the broadest one, with a clearly developmental orientation that takes into account different kinds of social responses at different developmental stages. The controversy that has ensued between primary and secondary attachments (Bowlby)[15] is largely spurious because in human development it is, at least at the present time, practically impossible to separate out the primarily innate from the experiential since early infantile experience tends to become organized and patterned, and presumably immediately begins to have its effects on later development. Nevertheless, this does not rule out the fact that certain maturations must occur before certain kinds of experience can be undergone and organized by the infant. Many ethologists have largely abandoned the idea of an entirely innate origin of instinct-based behavior; Schneirla,[71] for example, in his studies of cats, has shown the influence of early learning in complex mother-kitten interaction on the eventual "bio-psychological" mother-child relationship.

Oral and Feeding Behavior

That the oral and feeding experience of the infant constitutes one of the important basic roots of social attachment is not in question. However, historically, because of the obvious power of the sucking drive and its crucial connection with satisfaction of hunger and with survival itself, it was seen as the dominating experience that mediated the attachment to the mother. Indeed, the whole period of infancy was conceptualized as "the oral phase" of libidinal development. In nature, however, the oral experience involves other sensory modalities such as tactile, auditory, visual, and olfactory stimulation. Nevertheless, even with the exciting new discoveries of the importance of the visual modality—described elsewhere in this chapter—we should not underestimate the critical quality of the feeding experience. We know, for example, that mothers who respond to their child's cry and discomfort almost exclusively by offering the breast or the bottle condition their infants in such a way that oral craving is experienced and oral satisfaction may be more usually sought out when distress is felt.*

A number of authors, including Erikson,[26] Sullivan,[80] and Brody,[20] have emphasized the building up of a quality of interpersonal mutuality through the feeding experience; a number of fine manipulations and adaptations must be made by both partners in order to achieve a reciprocally gratifying feeding experience.

The significance of *contact comfort* and *tactile gratification* early in the life of the infant is dramatically demonstrated when the crying infant quiets upon being picked up and held, at first by any caretaker, but after a few months usually by the preferred mother. We assume that tactile, pressure, thermal, olfactory, and kinesthetic stimulation (Mason)[50] have an ongoing impact on infantile experience. Harlow's[43] work dramatized the importance of tactile experience in infant macaques, who apparently preferred artificial terry cloth mother surrogates to wire mesh lactating surrogates. Although Harlow had reason to conclude that contact comfort was more important than feeding as an antecedent to social

* Psychoanalysis has systematized the positive and negative oral character traits that presumably derive from the period of infancy. Oral optimism is seen as a consequence of having been adequately gratified in this area. On the other hand, such traits as excessive longing and a compulsive need for acquisition and intake of various kinds have been seen as consequences of either overly or underly gratified oral experience. There are as yet no definitive studies (which would have to be of a direct observational longitudinal nature) to indicate whether there is a definite relationship between oral patterning and such problems as obesity (Bruch),[21] vulnerability to drug addiction, alcoholism, and cigarette smoking.

J. Bruner's[58] work with infants has taught us that four to six week olds can learn to alter their rate of sucking (for example, to suck in longer bursts "to produce" a clearer focus in a projected picture). Moreover, by reversing the conditions the infant can even learn to desist from sucking on his nipple in order to obtain a consequently clearer picture. This kind of work is indicative of the tremendous range of learning that begins to take place in the early weeks of life—even in an area as "drive-oriented" as the oral zone.

attachment, I believe this is a false kind of distinction because this type of competitive choice between these two particular modalities does not occur in nature as it does in Harlow's experiments. Also, in general, any inferences to humans from infrahuman species carry a risk, although the relevance of such inferences for early infantile development may be of a somewhat higher order. There is, however, initial evidence that experimental stimulation of institutional infants—exclusively in the tactile modality—contributes to significant developmental gains.[5]

The evidence for the existence of important individual differences in the intensity of oral drive and the pleasure experienced in close physical contact is rather impressive. For example, Schaffer and Emerson[66] studied a group of infants who could be differentiated into "cuddlers" and "noncuddlers." The authors concluded that the noncuddlers were not suffering primarily from maternal contact deprivation; rather, as a group, they were more advanced and more active motorically, and at the same time they tended to resist restraint of movement, including the restraint consequent to close physical contact.*

Visual and Auditory Modalities

In recent years research has revealed that the visual apparatus is relatively ready to function soon after birth. Tauber[81] demonstrated the optokinetic nystagmus reaction movement in newborns; Wolff and White[86] observed visual pursuit of objects with conjugate eye movements in three to four day olds; Fantz[34] described more prolonged visual fixation upon more complex visual patterns as against simpler ones in the early days and weeks of infancy.

The normal face is similarly "preferred" to comparable head shapes with scrambled features in infants from one to six months.[33] Spitz[77] and Ahrens[1] in separate, very detailed analyses were able to demonstrate that the infant appeared to smile in response to a "sign gestalt," at first centering around the two eyes and later becoming more differentiated to include the mouth. A number of observers, including Wolff[85] and Robson,[63] have noted the development of preferential visual fixation upon, and following of, the human face from the early weeks of life.†

Wolff,[85] through careful observation and experimentation, discovered that as early as the third week in the infant's life the specifically human stimulus of a high-pitched voice elicits a smile more consistently than any other stimulus at that time. The voice also served to reduce the infant's fussiness as well as evoke a smile. Our own experience, as well as that of others, indicates that a most effective way of evoking a smile in an infant from the second month onward is by a "social approach," consisting of a smiling, nodding face, with accompanying musical vocalizations—that is, with the cumulative potency of various modalities.

Research observations would point to the probability that the infant is equipped innately with the capacity for a smiling response, a capacity that is evoked by a set of key "releaser" stimulus configurations—such as

* This is the kind of individual difference that is most important to psychiatry and psychotherapy because we tend to assume that people have more or less the same degree and quality of need in various modalities, be they oral or contact stimulation. The fact of individual difference by no means diminishes the fundamental importance of the tactile modality to the formation and maintenance of the social bond throughout the life cycle. Witness the emphasis on the use of touch and kinesthetic experience—as attempts to overcome individual alienation—in the encounter group phenomena and the "human potential movement."

† D. Stern[79] discovered through a film microanalysis that by the age of three months stable patterns of eye-to-eye contact and eye aversion between mother and infant have already been developed and tend to remain stable for a number of months thereafter. If this work is validated, the implications seem far-reaching for the understanding of the patterning of interpersonal relationships. Such poorly understood, but crucial, phenomena as "empathic" communication (Sullivan)[80] and "contagion" of affect (Escalona)[28] may be better understood through the microsignaling visual "ballet" that Stern describes as occurring between mother and child. More speculatively, patterns of visual aversion may also constitute one of the first anlagen of later classical ego defenses, including denial and possibly repression. Stern's later work appeared after this chapter was written.

the human face gestalt—which become effective at certain phases of development. This point of view clearly does not exclude a complementary one that regards the smiling response, once elicited, as being immediately open to various influences of learning, including conditioning and the increasing emotional investment in the recognition of familiar persons.*

Social and Playful Interaction

There is by now a mounting volume of evidence† that the crucial variables in determining the outcome of social responsiveness in the potentially healthy infant are the patterned social stimulations and responsiveness of the significant persons in the environment. Without adequate social (including perceptual) stimulation—as, for example, in blind and institutionalized infants—deficits develop in emotional and social relationships, in language, in abstract thinking, and in inner controls. Barring social traumata and deprivation of varying degree, nature and culture seem to guarantee reciprocal responsiveness by the fact that healthy adults, especially those who are intensely invested in their infants, find the infant's smiles and vocalizations irresistible; they apparently must respond unless the caretakers are particularly depressed or disturbed.

By five to seven months the infant who is being enjoyed by his parents spends a good part of his day in social interactions involving mutual regard, sometimes with intense eye-to-eye contact, and mutual smiling and vocalizations; these may include tactile and kinesthetic stimulations, modalities that are all combined in various ongoing patterns of interaction. Many of these patterned exchanges become idiosyncratically personal to the mother-infant couple, while others represent traditional social play, such as presemantic vocal "conversations" and repetitive social approaches and responses, which may involve nuzzling, jiggling or jouncing, postural games of lifting and lowering, "upside down" and "airplane" and—most dramatically—the game of peekaboo.§

By the end of the first half year of life, mother and infant have developed important patterns of social interaction. In some couples the baby is given maximal opportunity to actively respond and initiate; in others he is coerced into the position of a relatively passive recipient of stimulation that may excite him to the point of painful stress. Some of the variables involved in a systematic study of the patterns of reciprocity include: the infant's and mother's sensitivity and activity levels, their initiatory tendencies, mother's need to dominate rather than facilitate her infant, the nature of her personification of her infant (is he to be docile or actively initiatory), mother's anxiety level, her fear of "spoiling" the child through play, and so on. A prominent feature of maternal style includes the mother's *capacity to enjoy and respond* to her infant's activity, including his developmental progressions. The maternal variables are stressed here for the moment, since social reciprocity in the infant is given largely as a potentiality and, to a great degree, must be induced and sustained by the significant adults.

In the earliest months of life, the mother responds to physiologically based needs (hunger, cold, sleep). She functions as a protector from excessive inner and outer stimulation as well as a provider of perceptual stimulation. As the infant develops a repertoire of recipro-

* Several workers have experimentally demonstrated that one can reinforce the infant's smiling response by responding to his smile with a smile, or tend to extinguish the smile by failing to respond to it. Rheingold[62] and Weisberg[83] similarly demonstrated that an infant's vocalizations can be markedly increased by the adult's social responsiveness, in contrast to conditioning by contingent nonsocial responses such as a door chime.

† See the classic studies of Spitz,[74] Bowlby,[16] Goldfarb,[40] and the more recent observations of Provence and Lipton,[59] Schaffer and Emerson,[65] and Rheingold.[61]

§ See Kleeman's excellent description and analysis of this particular form of play. Kleeman[44] does not reduce peekaboo to the mastery of separation anxiety or to tension reduction, but sees it also in its own right as a "form of interaction, play, a social game pleasurable to infant and adult." This kind of playful interpersonal exchange can often take precedence over the activity of nursing.

cal playful experiences, he comes to anticipate and learn that he can evoke a social response even when he is not hungry, cold, wet, or in pain. *With this realization develops a new sense of "social potency" and trust that is qualtatively different from urgent need tension relief.* The child can now obtain not only reduction of tension but also positively stimulating and playful patterns of response in relation to a human partner, as well as with objects.*

Social playfulness, perhaps more than any other modality, constitutes a remarkably easy vehicle for the mutual exchange of affectionate and exuberant affects. Since play is characterized by the quality of continuing improvisation and hence by the availability of novel elements of experience, it operates as a powerful motivator of learning. As Piaget[56,57] has demonstrated, the development of learning structures proceeds by the assimilation of novel inputs in the infant's experience, followed by appropriate accommodations. *Reciprocal play* appears to involve the utmost of focal attention and absorption of the two partners—a kind of sacred ritual that one dare not intrude upon. When the caretaker—adult or adolescent—experiences this with the infant, the latter gains a new degree of human status, now being perceived as a psychological and social as well as physiological being. At the same time a new kind of parental pride appears; the mother's self-esteem is validated by her feeling that she has succeeded in helping her baby become socially human. It is at this point in development that fathers—often for the first time—experience themselves as a meaningful part of the infant-parent relationship. Playfulness requires special conditions, for instance, an appropriate level of stimulation and an absence of coercion and domination. In this sense we can see that mutual playfulness is a model of freedom and spontaneity in human relatedness. It helps prepare the individual and group for communication, language, and collaboration and provides a means for overcoming destructiveness through playful aggression.†

The internalization of "good" reciprocal relationships comes to be organized as part of "good me" and "good mother" and contributes eventually to the sense of one's self-esteem. With the confidence that he can evoke a response, the child is freed to "be alone in the presence of mother," as Winnicott[84] states it, a phase that prepares him for separations from mother's physical presence for longer periods without undue anxiety. This is a crucial step in the development of the infant's autonomy. If parents are depleted emotionally for any reason (depression, social deprivation, and hardship), one of the first qualities of a relationship to fall away is playfulness since this depends on a surplus of emotional well-being. When we speak of emotional deprivation in infancy, this refers not only to the gross kinds of deprivation seen in such situations as institutionalization and obvious parental psychopathology but also to the more subtle quality of the interaction between parent and child.§ Moreover, we must keep in mind not only the quality of parenting but also the individual characteristics of the infant that determine the nature of the stimulation he requires;[30] for example, a passive, low-energy infant who cannot actively send out signals that will bring response is more in need of stimulation that is initiated by the parents than a more active, self-initiatory infant.

* We suggest the hypothesis that with deprivation of relatively enduring reciprocal social relationships, including playfulness, children and adults will appeal for response by re-creating and communicating the urgent need tensions that had been successful in bringing about a response. Hence, hunger, pain, and later in life various expressions of anxiety, hypochondriac fears, psychosomatic conditions, acting out, and compulsive activity can be understood, at least in part, as *an appeal for responsiveness that has had no alternatively stable and successful interpersonal pathway.*

† Genuine play, whether with words, metaphors, ideas, sounds, or design, is an important basis for the formation of culture. I have elaborated on the theme of social playfulness elsewhere,[69] stressing the quality of lack of immediate purposiveness, which frees the partners to improvise and explore new forms of action, symbolism, and relatedness. The structural development of play in infancy has been studied by Piaget,[57] and more recently a stimulating review of the subject has been offered by Galenson.[39]

§ A fuller discussion of "masked deprivation"[60] is included in the section of this chapter titled "Separation and Individuation."

Developmental Stages of Social Attachment

When looked at closely the development of social attachment can be described as a complex series of steps in achieving a meaningful special relationship to other persons. Because of different rates of development and the diversity of research definitions and criteria in studying social attachment, we cannot expect to find a fixed age at which a given level of attachment is achieved. Furthermore, the observer can only use behavioral reactions and from them draw inferences about qualities and levels of social attachment. The infant is in no position to verify or contradict these inferences.

What is the significance to psychiatry of detailed developmental studies of social attachment? It is nothing less than the foundation of all human relatedness and of personality development. To study carefully the different levels of social relationship allows the student of human behavior to recognize, for example, at what level of relatedness a particular person —or group—may be operating at a given time and also at what level a person may have been arrested in his development. Such knowledge will allow for a reconstructive viewpoint in attempts at individual and social change. If we know there tends to be a relatively invariant sequence of stages A, B, C, D in the formation of a human attachment, we will not expect or demand D level behavior if step C has never been achieved. This developmental viewpoint alters the conception of individual and social therapies that have been modeled largely on the issue of conflict, without too much regard for the structural elements of personality that are needed for a certain level of interpersonal behavior and conflict. For a person who has not achieved the capacity for close specific social attachment in infancy, one would not expect to find the higher level oedipal-type conflicts that already presume a capacity for specific social attachment.

From the work of a number of investigators, including Spitz,[73] Benjamin,[6] Schaffer and Emerson,[65] and Yarrow[87] we can summarize the various stages by which social attachment between infant and mother is achieved during infancy.

1. *Undifferentiated presocial phase.* The infant in the early days or weeks of his life may fail to discriminate social and nonsocial objects.

2. *Indiscriminate social responsiveness.* The infant now discriminates social and nonsocial objects but responds without apparent discrimination among various social objects. It is at this stage that Spitz's[73] comment applies: the mother is a function and not yet a face.

3. *Selective responsiveness* to familiar versus unfamiliar people. One type of selective responsiveness involves the recognition of the mother as revealed by a series of behavioral signs, including selective concentration on the mother, excitement and approach movements at the sight of her, as well as differential crying, smiling, and vocalization. Mother is no longer merely a function; she has a face.

There seems to be agreement that perceptual discrimination precedes the possibility of recognition of the mother as a specific person, and that recognition, in turn, is a prerequisite for stage 4.

4. *Specific social attachment.* According to Schaffer and Emerson,[65]* there was a peak of specific social attachment at 10 to 11 months, followed by a slow decline. At 18 months there seemed to be an increase in the attachment curve, reaching its previous peak that had been found at 10 months. It is likely that the development of specific attachments and expectations toward the mother or other significant persons depends on certain perceptual and cognitive developments, including the beginning concept of "object permanence" (Pi-

* The criterion used by these authors to assess the level of achievement of specific social attachment to the mother was that of protest upon separation from her. The assumption here was that the infant had a need for proximity, at least on a visual level; when he suffered a cutoff of such visual contact, he expressed his protest in affecto-motor sounds and movements, including crying.

aget)—the mental representation of objects when they are outside the infant's immediate perceptual field. It is important to note that in Piaget's experiments on object permanence, the child *begins* to retain an image of the disappeared object and seeks it out under a napkin beginning around nine months of age, completing the mastery of the complexities of object permanence around 18 months. In nature, of course, one cannot separate the cognitive from the affective-social bonds; these dimensions are abstracted from a unified gestalt experience in the infant's life.

5. *The Confidence relationship.* This higher level of interpersonal relationship—derived by Yarrow from Benedek—involves the development of specific expectations toward the mother and overlaps significantly with Erikson's concept of basic trust. However, since complex inferences about the meaning of behavior are necessary, the development of behavioral criteria for "confidence" or "trust" is extremely difficult. Yarrow chose, as one criterion, "the expectation of soothing when in distress," and he found that about half the infants had developed this relationship of confidence to the mother by age three months and 56 per cent at age six months. (His study did not go beyond eight months.) Yarrow concluded that the development of confidence in the mother is not simply a maturationally determined development, but undoubtedly influenced by many environmental as well as idiosyncratic factors, including the patterns of maternal care, "such as the depth of the relationship, the consistency with which mother responds to the child as well as the general level of predictability of the environment based on recurring and predictable sequences of gratification."[87]

From our own and others' observations of infants, we know that the increasing capacity to anticipate and wait for specific responses in the mother increases with age in a good relationship and, indeed, is an early sign of what is meant psychoanalytically by the concept of object constancy.

6. *Object constancy.* As Fraiberg[35] has indicated, the criteria for achievement of object constancy vary a great deal with different authors, so that its achievement is placed at ages ranging from eight months until after the second year.[49] In any case what is meant by this concept is that not only can the child discriminate and selectively value his mother but also he has begun to represent her mentally with qualities of increasing permanence and objectivity. Even in the face of frustration or cruelty or during a limited absence, the mother usually continues to be preferred and central to the child's life.

The evidence for achievement of object constancy in the psychoanalytic sense—in contrast to Piaget's purely cognitive concept of "object permanence"—is not on certain grounds empirically, but there are behaviors that would indicate the mother is represented mentally and invested with intense affect.*

The child, for example, will call for his mother by whatever "call sounds" he has developed to summon her to himself; when she is absent he will verbally refer to her or to her possessions and he will miss her grievingly. Even nonverbally, his beginning dramatic play indicates that he is developing the capacity to mentally represent himself and mother in a mobile symbolic act—for instance, when he places a doll to sleep during a play sequence. His mental operations have progressed beyond immediate imitation to what Piaget calls deferred imitation, and then to the formation of more lasting identifications with the mother. Evi-

* Reconstructive data derived from psychiatric and psychoanalytic histories indicate that the achievement of object constancy—in the meanings described above—is essential to later mental health. Whether there is a "critical period" for its achievement and whether there is the possibility for compensation are discussed under the topic "Separation and Individuation." Many psychiatric disturbances are associated with the failure to develop stable interpersonal relationships; schizophrenic, schizoid, and sociopathic persons particularly suffer such incapacity. These warps of interpersonal development—as Sullivan refers to them—derive in large part from a failure to achieve a level of object constancy with one or more significant persons early in life, for a complex variety of reasons. The achievement of object constancy can be unstable and—as we know from work with all age groups—subject to breakdown under stress.

dence for these identifications are revealed in the toddler's play, whose content is in part concerned with parental attitudes and the roles of provider, helper, protector, and comforter (Schecter).[67]

During the second year of life, the child's very special relationship to the mother, in a nuclear family, becomes increasingly complex and elaborated. The child shares his inner and outer world with his mother, verbalizing fantasies and fears, bringing her objects, naming them, and expecting an affirming response from her. Even as the child increasingly individuates, he becomes capable of sharing a rather private world—a shared "mythology"—in the sense that there are idiosyncratic words for special objects; there are frequent recapitulations of memory experiences that both have shared together; and there's repetitive playing of games that are bodily, kinesthetic, verbal, musical, that is, play involving almost every modality. This is probably the period of bliss frequently represented in Renaissance art in the idealized version of mother and cherub—for many, a period to remain imbued with paradisic feeling.

Intensity and Breadth of Social Attachments

Schaffer and Emerson[65] found that factors increasing the seeking of proximity to the mother include:

1. Pain, teething, illness, fatigue, and fear
2. A period of the mother's absence (which corroborates Bowlby's finding of greater clinging and demandingness after separation)
3. The habituation effect of a period of great stimulation such as occurs during the visit of a doting relative
4. When the infant enters a strange environment

All these factors are relevant to later development, including adulthood. We have already indicated the importance of understanding the factors that contribute to a predisposition to habituation and addictions of various kinds. More intensive longitudinal study of habituation levels in infancy and childhood could test their correlation with personality outcomes characterized by needs for strong stimulation and input of various kinds.

An important topic under discussion currently is that of the "breadth of attachments" of infants to significant persons. We can infer from the work of Schaffer and Emerson—as well as from observations of societies with multiple child caretakers—that a single person is not necessarily the first step in forming a specific social attachment. In Schaffer and Emerson's[65] research almost one-third of the infants showed attachments to multiple persons in the phase of specific social attachments. However, their work does not disprove the possibility that intensive early mothering may have been a prerequisite to the broadening of specific attachment. In most cases (62 per cent) fathers were found to be specific objects of social attachment after the onset of the phase of specific attachment. In fact, for 4 per cent of the infants in their sample, the father was the only specific object of attachment at seven months. Hence we find the possibility of a "hierarchy of object persons," the most intense attachment being shown to the principal object person, who is not necessarily the mother.*

* Some other important findings that are relevant to contemporary issues involving new forms of childrearing include the following: (1). High intensity of attachment correlated with the degree of stimulation by the mother. However, such a conclusion poses a problem of what is cause and what is effect, since certain individual differences in babies may demand higher levels of stimulation. (2). High intensity of attachment also correlated with mothers who respond quickly to demands. Again, the nature and intensity of infant demandingness—probably innate in part—may influence the mother's behavior as well as vice versa. (3). High intensity of attachment was found mostly in families with fewer caretakers.

The "selection" of the infant's principal object of attachment correlated closely with the particular adult's responsiveness to the infant's crying and with the amount of interaction between the significant adult and the child. The authors concluded that the breadth of social attachments is related to the opportunity of interacting with people who will offer relevant stimuli, especially socializing and caretaking functions in the widest sense.

For a discussion of these issues from a cultural anthropological point of view, see Mead,[51] Spiro,[72] and Bettelheim,[9] who have studied new forms of multiple caretaking; the latter two authors have examined the kibbutzim in Israel.

Origins of Infantile Anxiety

As Benjamin[7,8] points out, a satisfactory design to tease out the relative contributions of hereditary, intrauterine, birth, and early postnatal factors has been impossible to actualize until now. Even though Greenacre[41,42] re-presented Rank's idea of the birth trauma in a far more sophisticated form, it is still useful mainly as a concept of a single variable in a predisposition to anxiety. More recently the possibility of learning from experience in the early weeks of life adds a new variable both to the predisposition and to the idea of actual anlagen to anxiety experience. So much depends on how anxiety is defined in infancy that it may be more fruitful to describe the various critical periods of its development—leaving open the theoretical question of what constitutes anxiety,* in contrast to infantile fear or undifferentiated negative affect, which we see, for example, in reaction to overstimulation of various kinds.

Benjamin's observations led him to postulate a "critical period" at age three to four weeks when a rapid rate of neurophysiological maturation accounts for an increased capacity to register internal and external stimulation. Benjamin hypothesized that in the ensuing weeks the quality of mothering in protecting the infant from this new source of stimulation might contribute to the subsequent predisposition to anxiety. The relatively undifferentiated negative affect expressed upon being left alone can be seen as a consequence of a form of habituation to a certain level and quality of stimulation rather than to the experience of loss of a truly discriminated mother.†

Before the appearance of infantile stranger anxiety proper, Benjamin postulates an innate fear of the strange as is seen in the two- to four-month-old infant's negative reaction to strange objects or sounds—or to being handled in an unaccustomed manner. An apprehensive response may also be aroused in the young infant by altering an anticipated gestalt pattern through the addition of unfamiliar elements or by the omission of some apparently crucial familiar element. We have noted, for example, that some infants respond with a look of apprehension at the appearance of a smiling, nodding adult face that is presented without the accustomed vocal accompaniment. When vocalizations are added the infant relaxes and smiles, giving the impression of closure of the anticipated familiar gestalt.§ In a similar vein a humming or falsetto voice lacking visual presentation of the human face could produce fearful reactions that disappeared once the face was brought into view.[6] The infant may respond with apprehension to a variety of alterations of the facial gestalt, such as the placing of pads over the eyes, the forbidding expression with vertical—in contrast to horizontal—forehead creasing,[73] or changes in the mother's appearance when she wears a new hat, glasses, or hair curlers. However, it is the expression around the adult's eyes that seems to have particular significance for the infant; this fits with Wolff's[85] observation that the infant tends to search out the eye area and make eye-to-eye contact before smiling at the presentation of a face.‡ These signs of increasing perceptual discrimination predate and constitute a necessary precondition

* S. Brody and S. Axelrad[20] have attempted to describe the development of infantile anxiety in relation to ego formation from a psychoanalytic frame of reference, using direct observations of infants.

† More recent work is revealing that perceptual discrimination of the mother—if not the specific social attachment to her—is developed in the early months of life. The fact that caretakers would appear to be interchangeable does not contradict the observations that certain signs of discrimination are appearing concurrently very early in life.

§ The experimental observations noted here have not been carried out systematically on a sufficiently significant number of infants to allow any solid conclusion about how characteristic these reactions may be.

‡ H. S. Sullivan[80] took pains to point out the understandably magical significance that many people—most particularly schizophrenics—attribute to the power of the eyes, to being looked at, and, we would add, to intense eye-to-eye contact. One of the most comprehensive reviews of the significance of early eye-to-eye contact has been written by Robson.[63] The most thorough exploration of communication by facial signs will probably be achieved through intensive film studies of interpersonal behavior.

for the development of infantile stranger anxiety.

The infantile stranger reaction has been properly distinguished from separation anxiety by Benjamin.[6] Freud[36] and Spitz[76] considered the "eight-month anxiety" in reaction to the stranger to be based on the same dynamic found in separation anxiety, namely, the fear of object loss. Benjamin found that although stranger and separation anxieties are related dynamically and even positively correlate statistically, nevertheless, there are babies showing high levels of separation anxiety but low stranger anxiety and vice versa. Stranger anxiety can occur whether or not the mother is present, whereas separation reactions occur in the absence of the mother whether or not anyone else is present. Moreover, the average and peak time of onset are different for each type of anxiety, occurring somewhat earlier for stranger anxiety.* Phenomenologically the reaction to strangers runs a gamut from no apparent reaction, visual concentration without apparent affect, reserved friendliness after initial wariness to sobering with mild apprehension, inhibition of motor behavior ("freezing"), aversion of visual gaze, clinging, withdrawal, active protest, screaming, and panic-like behavior.

Aside from its own intrinsic significance, this description of the reaction to strangers is also offered to help elucidate another poorly understood form of anxiety central to H. S. Sullivan's[80] theory of interpersonal relations—that is, the anxiety induced in the infant by the anxiety of the mothering one through as yet unknown mechanisms that Sullivan referred to as "empathic linkage." Various observers have noted, for example, that a baby would suffer feeding disturbances when fed by mothers who were high-strung and excitable while accepting the same formula from another feeder.[29] It is our hypothesis that when the mother is anxious or distressed, she can be experienced as both familiar and strange by her infant. We assume that from his very early discriminations of familiar persons as well as from familiar ways of being handled the baby comes to learn and anticipate behavioral signs connected with "good mothering." These signs probably include cues from all the various sense modalities. When the mother is anxious her facial configuration is altered by a frown or tight lips, her vocalizations become tense and strident, her handling becomes less graceful and smooth, and it is conceivable that there may be olfactory-sensitive changes in her odors as well. We suggest that there is a shock of "strangeness" in such a situation after the infant has learned to anticipate a pleasurably "good" gestalt of experience. We would also speculate that the *anxious smile*[6]—which we have observed in seven- and eight-month-old infants—may represent in effect a smile of recognition contaminated with the expression of the tension of anxiety. Once infants become mobile—by crawling or walking, or by the early use of mechanical walkers—they are commonly subjected to a multitude of anxiety-ridden no's. At such times parents become aware of a momentous change in the previously "innocent" relationship, once the socially disapproving modalities come into operation—especially when there are conflict and anxiety in clashing with the child's real or imagined "will." Sullivan and other analysts suggest that "good" and "bad" feeling experiences become organized and grouped in a polarized way, leading to the symbol formations associated with "good-me" and "bad-me" and with "good-mother" and "bad-mother." *Learning through the experiencing and avoidance of anxiety becomes one of the most powerful means through which socialization may then take place.*

One of the mother's—and the father's—major functions in facilitating separation and individuation is to help render the outside unfamiliar world available for exploration. Aside from the practical aspects involved in this function, the parent mediates for her child the new and strange objects, sounds, and people in the environment. We see the origins of

* Compare Schaffer and Emerson's apparently opposite but differentiating results, presumably due to the use of slightly different criteria for each form of anxiety. Yarrow[87] reserves the term "stranger anxiety" for those infants who manifest active protest or withdrawal.

what might be called the "magical blessing" when mother, for example, allays the child's fear of receiving and exploring a new toy from a stranger by simply handling the toy and offering it to the child herself. One has the impression that by such mediations the mothering one can detoxify strange, anxiety-laden elements of the environment. She does this, in part, by helping the child cope with the frighteningly strange aspects of his world in ways that allow them to be experienced as engagingly novel or even as partly familiar.*

Separation and Individuation†

Separation anxiety would seem to be a ubiquitous phenomenon in infancy and remains a lifelong vulnerability at any stage of the life cycle. Yarrow[87] found that by the age of eight months 100 per cent of his sample of infants suffered both mild and severe signs of separation anxiety. He also found that the greater the discrepancy in patterns of maternal care between the first and second caretaker the greater was the postseparation disturbance. Spitz's and Bowlby's pioneering work in this area has already greatly changed our sensitivity to the problems of separation in infancy and childhood to the point of emphasizing the importance of parental presence, if feasible, during a child's hospitalization.

In his studies of attachment, loss, and grief in childhood, Bowlby[15,16,17] has revealed what appears to be a frequent sequential pattern of reaction to separation, especially in children over six months and under three years of age. At first the child—especially if he has had a close relation to the mother—reacts with *protest*, crying and searching for the missing mother as if he expects her return. This reaction is followed by a phase of *despair* characterized by intermittent crying, inactivity, and withdrawal, indicating increasing hopelessness and what Bowlby believes to be equivalent to a state of mourning. A third stage of *detachment* follows which is often welcomed as a sign of recovery, although when the mother visits there is a striking lack of attachment behavior toward her—as if the child had selectively lost interest in the mother. If there is a series of losses of mothering figures, the child will commit himself less and less to each succeeding figure and will develop rather superficial relationships in which people come to be experienced as sources of supplies rather than as special people in their own right. In the extreme of the neglected institutionalized child, one can observe the deterioration of almost every area of functioning, including the development of language, cognition, motoric control, autoerotic activity as well as adequate affective interpersonal relationships.§ Hence the enormously important public health issue of providing an adequate stable nurturing environment for a child who does not have this environment available in the traditional family setting. *The propagation of defects of early development constitutes one of the widespread human crises of our time* since the relation between such defects and subsequent personal and social pathology is more than merely speculative at this stage in our knowledge.

From the work of Spitz, Bowlby, and more recently Tennes and Lampl,[82] one can postulate that a number of reactions to separation represent prototypical precursors of major human defensive and coping systems throughout life. The infant's reactions to separation include visual avoidance (of the strange substitute caretaker), inhibition of activity, and withdrawal as well as active attempts at mastery, for example, attempting to follow a

* The psychotherapist has a similar task in fostering the patient's movement from the familiar and "embedded" (Schachtel)[64] into new areas of experience that had been avoided because of their association with anxious affect. The therapist in this sense is also a mediator who looks with the patient into dark, unknown, dissociated areas and through the therapeutic alliance gradually helps to detoxify both the traumatic and untried areas of living.

† M. S. Mahler[48] has made a significant contribution to the process of separation-individuation.

§ See the classic work of Spitz[74,75] and Provence and Lipton.[59] The latter authors found that some of the institutionalized children subsequently placed in families, despite some improvement, still revealed serious ego deficiencies such as an incapacity for delay, failure in generalization from learning, overly concrete thinking, and a continuing superficiality in social relationships.

mother who is disappearing through the door. When active attempts at mastery are thwarted, the experience of futility and affects of hopelessness and helplessness ensue. Although we are only referring to research involving overt separation, one can postulate the importance of these reaction patterns to character development in less obvious experiential patterns, involving what might be called "affective separation" or "masked deprivation"[60]—which can occur when a mothering figure is physically present but not adequately stimulating or responsive.*

In so-called disadvantaged children, defensive character detachment and precocious pseudoindividuation, with an implicit loss of hope for good relationships, are frequently observable; they partly account for the lack of richness of experience and the failure of adequate cognitive-affective development when "enriched" environments are subsequently made available. Even though human development does not seem to proceed in the rather rigidly defined "critical periods" seen in various other animals, nevertheless, there would seem to be optimal periods during which certain experiences are most productive to the cognitive and social-affective development of the growing child. Deficits in the various stages of infancy described here can be "made up" or compensated for only to a limited degree. This statement is certainly no argument for not attempting later developmental compensations; quite the reverse. Nevertheless, it is clear that our social focus should be on prevention of deficit as well as on attempts to find adequate methods of compensation.†

The direction of psychic development in infancy is from symbiotic fusion to individuation with increasing differentiation and structuralization of the ego;[48] this, in turn, permits interpersonal relationships on an increasingly higher level of reciprocity. There are signs of a dawning sense of self in the first year of life as the infant remembers and anticipates experience and comes to discriminate his self, his mother, and others. As maturation and experience—including environmental facilitation—make this possible, the child begins to do for himself and for others what had been done for him: he feeds himself, he manipulates objects and toys, he transports himself, finally in the upright posture. He decides on a course of action even if this means opposing or negating those closest to him. He learns, largely by identification, a gesture and word—"no"—to express semantically his autonomous strivings.[73,76] It is through decision making, goal setting, and goal mastery that the sense of self is experienced in its most heightened intensity.

With the development of language and locomotion in the second year, we begin to consider the infant as entering a new phase of development, the "toddler" or "autonomous" stage, which is ushered in around 15 months of age. By this time the infant has begun to share his experience with his parents who are—if all is going well—delighted with his humanoid capacities to walk, talk, and begin to communicate his needs and experience.

The very achievement of a sense of self exposes the young child to an awareness of being observed and evaluated, giving rise to a

* We note in our discussion that hypotheses concerning character development have proceeded largely from research connected with trauma and psychopathology. This is due, in part, to the fact that as yet there are few direct observational longitudinal studies reported in depth to connect patterns of experience with character formation. Moreover, we wish to emphasize that with deprivational and traumatic experience, defensive patterns are evoked in an unbalanced or extreme form, whereas the relatively "normal" range of experience of frustration or periodic separation is—when in proper dosage and at the appropriate stage of development—assumed to contribute to the formation of "healthy" coping capacities and "ego strength." There are, of course, many other sources of character development considered in detail in the psychoanalytic literature, including identification and a whole range of coping mechanisms and ego defenses. See Nagera,[55] Murphy,[54] and Schecter.[67] Learning—in all its forms—and cognitive styles contribute heavily to ego development. The whole topic of learning and cognition will be considered formally in Chapter 14 of this volume.

† For a discussion of attempts at compensatory work with deprived children, see Lichtenberg and Norton's[46] review of the research literature and Deutsch's *The Disadvantaged Child*.[25] For a more extensive review of the effects of maternal deprivation, see Ainsworth's work.[2] Birch[10,11] has carefully studied the devastating consequences on development of early malnutrition.

Caldwell[22] offers an excellent broadly ranging review of the entire subject of infant care.

self-consciousness and a fateful subject-object split in the self that lays the groundwork for shame and doubt.[27] The infant and toddler becomes all too aware that a socially disapproved act will bring a disapproving signal or more subtly, but not less potently, a withdrawal of parental behaviors that have the power to reduce anxiety and induce security. In the child's new stage of awareness of his separateness and vulnerability to loss of self-esteem, it becomes crucial that his induction into the social world proceed with a net balance allowing for zestful enjoyment of activity, mastery, autonomy, and initiative since this is a time when there are increasingly necessary limitations—physical and social—on the child's spontaneous activities.

Bibliography

1. AHRENS, R., "A Contribution to the Development of the Recognition of Physiognomy and Mimicry," *Zeitschrift fur Experimentelle und Angewandte Psychologie, 11:* 412–494, 599–633, 1954.
2. AINSWORTH, M., *et al.*, "Deprivation of Maternal Care: A Reassessment of Its Effects," Public Health Papers No. 14, World Health Organization, Geneva, 1962.
3. ALDRICH, R., "Preface," in Hellmuth, J. (Ed.), *The Exceptional Infant*, Vol. 1, Brunner/Mazel, New York, 1967.
4. BALINT, M., *Primary Love and Psychoanalytic Technique*, Liveright, New York, 1953.
5. BARSCH, R., "The Infant Curriculum—A Concept for Tomorrow," cites Casler, L., "The Study of the Effects of Extra Tactile Stimulation on the Development of Institutionalized Infants," in Hellmuth, J. (Ed.), *The Exceptional Infant*, Vol. 1, Brunner/Mazel, New York, 1967.
6. BENJAMIN, J., "Further Comments on Some Developmental Aspects of Anxiety," in Gaskill, H. (Ed.), *Counterpoint: Libidinal Object and Subject*, International Universities Press, New York, 1963.
7. ———, "The Innate and the Experiential in Development," in Brosin, H. (Ed.), *Lectures in Experimental Psychiatry*, pp. 19–42, Pittsburgh Press, Pittsburgh, 1961.
8. ———, "Some Developmental Observations Relating to the Theory of Anxiety," *J. Am. Psychoanal. A.*, 9:652–668, 1961.
9. BETTELHEIM, B., *The Children of the Dream*, Macmillan, New York, 1969.
10. BIRCH, H., "Malnutrition and Early Development," in Grotberg, E. (Ed.), *Day Care: Resources for Decisions*, pp. 340–372, Office of Economic Opportunity, Office of Planning, Research, and Evaluation, Experimental Research Division, 1971.
11. ———, and GUSSOW, J., *Disadvantaged Children: Health, Nutrition and School Failure*, Harcourt Brace/Grune & Stratton, New York, 1970.
12. BIRNS, B., "Individual Differences in Human Neonates' Responses to Stimulation," *Child Dev.*, *36:*249–256, 1965.
13. BLANCK, R., and BLANCK, G., *Marriage and Personal Development*, Columbia University Press, New York, 1968.
14. BOWLBY, J., *Attachment and Loss*, Vol. 1, Basic Books, New York, 1969.
15. ———, "Grief and Mourning in Infancy and Early Childhood," in *The Psychoanalytic Study of the Child*, Vol. 15, International Universities Press, New York, 1960.
16. ———, *Maternal Care and Mental Health*, 2nd ed., Monograph Series No. 2, World Health Organization, Geneva, 1952.
17. ———, "The Nature of the Child's Tie to His Mother," *Internat. J. Psychoanal.*, *39:* 350–373, 1958.
18. BRIDGER, W., BIRNS, B., and BLANK, M., "A Comparison of Behavior Ratings and Heart Rate Measurements in Human Neonates," *Psychosom. Med.*, *27:*123, 1965.
19. BRODY, S., *Patterns of Mothering*, International Universities Press, New York, 1956.
20. ———, and AXELRAD, S., *Anxiety and Ego Formation in Infancy*, International Universities Press, New York, 1970.
21. BRUCH, H., and TOURAINE, G., "Obesity in Childhood: The Family Frame of Obese Children," *Psychosom. Med.*, *2:*141–206, 1940.
22. CALDWELL, B., "The Effects of Infant Care," in Hoffman, M., and Hoffman, L. (Eds.), *Child Development Research*, pp. 9–89, Russell Sage Foundation, New York, 1964.
23. CHESS, S., THOMAS, A., BIRCH, H., *et al.*, *Behavioral Individuality in Early Child-*

hood, New York University Press, New York, 1963.

24. COLEMAN, R., KRIS, E., and PROVENCE, S., "The Study of Variations of Early Parental Attitudes," in *The Psychoanalytic Study of the Child*, Vol. 8, International Universities Press, New York, 1953.
25. DEUTSCH, M., *et al.*, *The Disadvantaged Child*, Basic Books, New York, 1967.
26. ERIKSON, E., *Childhood and Society*, Norton, New York, 1950.
27. ———, "Identity and the Life Cycle," *Psychol. Issues*, *1*:1, 1959.
28. ESCALONA, S., "Emotional Development in the First Year of Life," in Senn, M. (Ed.), *Problems of Infancy and Childhood*, Josiah Macy, Jr., Foundation, New York, 1953.
29. ———, "Feeding Disturbances in Very Young Children," *Am. J. Orthopsychiat.*, *15*:76, 1945.
30. ———, "Patterns of Infantile Experience and the Developmental Process," in *The Psychoanalytic Study of the Child*, Vol. 18, p. 241, International Universities Press, New York, 1963.
31. ———, *The Roots of Individuality*, Aldine Chicago, 1968.
32. FAIRBAIRN, R., *Psychoanalytic Studies of the Personality*, Tavistock, London, 1952.
33. FANTZ, R., "The Origin of Form Perception," *Scientific American*, *204*:66–72, 1961.
34. ———, "Pattern Vision in Young Infants, *Psychol. Rec.*, *8*:43, 1958.
35. FRAIBERG, S., "Libidinal Object Constancy and Mental Representation," in *The Psychoanalytic Study of the Child*, Vol. 24, International University Press, New York, 1969.
36. FREUD, S. (1925–1926), "Inhibitions, Symptoms and Anxiety," in Strachey, J. (Ed.), *Standard Edition, Complete Psychological Works*, Vol. 29, p. 169, Hogarth, London, 1964.
37. ——— (1905), "Three Essays on the Theory of Sexuality," in Strachey, J. (Ed.), *Standard Edition*, Vol. 7, Hogarth, London, 1953.
38. FROMM, E., *The Forgotten Language*, pp. 196–231, Rinehart, New York, 1951.
39. GALENSON, E., "A Consideration of the Nature of Thought in Childhood Play," in McDevitt, J., and Settlage, C. (Eds.), *Separation-Individuation: Essays in Honor of Margaret Mahler*, International Universities Press, New York, 1971.
40. GOLDFARB, W., "Variations in Adolescent Adjustment of Institutionally-Reared Children," *Am. J. Orthopsychiat.*, *17*:449, 1947.
41. GREENACRE, P., "The Biological Economy of Birth," in *The Psychoanalytic Study of the Child*, Vol. 1, pp. 31–51, International Universities Press, New York, 1945.
42. ———, "The Predisposition to Anxiety," *Psychoanal. Quart.*, *10*:66, 1941.
43. HARLOW, H., "The Nature of Love," *Am. Psychol.*, *13*:675, 1958.
44. KLEEMAN, J., "The Peek-a-Boo Game," in *The Psychoanalytic Study of the Child*, Vol. 22, International Universities Press, New York, 1967.
45. LEVY, R., "Tahiti, Sin and the Question of the Integration between Personality and Sociocultural Systems," in Muensterberger, W., and Esman, A. (Eds.), *The Psychoanalytic Study of Society*, International Universities Press, New York, 1971.
46. LICHTENBERG, P., and NORTON, D., "Cognitive and Mental Development in the First Five Years of Life," pp. 83–98, National Clearinghouse for Mental Health Information, Maryland, 1970.
47. LIDZ, T., *The Person*, Basic Books, New York, 1968.
48. MAHLER, M., *On Human Symbiosis and the Vicissitudes of Individuation*, International Universities Press, New York, 1968.
49. ———, "On the Significance of the Normal Separation Individuation Phase," in Schur, M. (Ed.), *Drives, Affects, Behavior*, Vol. 2, pp. 161–169, International Universities Press, New York, 1965.
50. MASON, W., "Early Social Deprivation in the Non-Human Primates: Implications for Human Behavior," in *Proceedings of Conference on Biology and Behavior: Environmental Influences*, Rockefeller University, New York, 1967.
51. MEAD, M., "A Cultural Anthropologist's Approach to Maternal Deprivation," in Ainsworth, M., *et al.* (Eds.), *Deprivation of Maternal Care: A Reassessment of Its Effects*, World Health Organization, Geneva, 1962.
52. MOSS, H., "Sex, Age and State as Determinants of Mother-Infant Interaction," in Chess, S., and Thomas, A. (Eds.), *Annual*

Progress in Child Psychiatry and Child Development, pp. 73–91, Brunner/Mazel, New York, 1968.

53. MURDOCK, G., "The Universality of the Nuclear Family," in Bell, N., and Vogel, E. (Eds.), *A Modern Introduction to the Family*, Free Press, Glencoe, 1962.
54. MURPHY, L., *The Widening World of Childhood*, Basic Books, New York, 1962.
55. NAGERA, H., *Early Childhood Disturbances, The Infantile Neurosis and the Adult Disturbances*, International Universities Press, New York, 1966.
56. PIAGET, J., *The Construction of Reality in the Child*, Basic Books, New York, 1959.
57. ———, *Play, Dreams and Imitation in Childhood*, Routledge Kegan Paul, London, 1962.
58. PINES, M., "Infants Are Smarter than Anybody Thinks (A Report on Some of the Work of Jerome Bruner, Harvard University Center for Cognitive Studies)," *New York Times Magazine*, November 29, 1970.
59. PROVENCE, S., and LIPTON, R., *Infants in Institutions: A Comparison of Their Development with Family Reared Infants during the First Year of Life*, International Universities Press, New York, 1952.
60. PRUGH, D., and HARLOW, R., "Masked Deprivation," in Ainsworth, M., *et al.* (Eds.), *Deprivation of Maternal Care: A Reassessment of Its Effects*, World Health Organization, Geneva, 1962.
61. RHEINGOLD, H., "The Modification of Social Responsiveness in Institutional Babies," *Monog. Soc. Res. Child Dev.*, *21*:63, 1956.
62. ———, GEWIRTZ, J., and ROSS, W., "Social Conditioning of Vocalizations in the Infant," *J. Comp. Physiol. Psychol.*, 52:68, 1959.
63. ROBSON, K., "The Role of Eye-to-Eye Contact in Maternal-Infant Attachment," in Chess, S., and Thomas, A. (Eds.), *Annual Progress in Child Psychiatry and Child Development*, pp. 92–108, Brunner, New York, 1968.
64. SCHACHTEL, E., *Metamorphosis*, pp. 48–49, Basic Books, New York, 1959.
65. SCHAFFER, H., and EMERSON, P., "The Development of Social Attachments in Infancy," *Monog. Soc. Res. Child Dev.*, *29*: 3, 1964.
66. ———, and ———, "Patterns of Response to Physical Contact in Early Human Development," *J. Child Psychol. & Psychiat.*, *5*:1–13, 1964.
67. SCHECTER, D., "Identification and Individuation," *J. Am. Psychoanal. A.*, *16*:48–80, 1968.
68. ———, "The Oedipus Complex: Considerations of Ego Development and Parental Interaction," *Contemp. Psychoanal.*, *4*: 111–137, 1968.
69. ———, "On the Emergence of Human Relatedness," in Witenberg, E. (Ed.), *Interpersonal Explorations in Psychoanalysis: New Directions in Theory and Practice*, Basic Books, New York, 1971.
70. ———, and CORMAN, H., "Some Early Developments in Parent-Child Interaction," 1971 (unpublished mss.).
71. SCHNEIRLA, T., and ROSENBLATT, J., "Behavioral Organization and Genesis of the Social Bond in Insects and Mammals," *Am. J. Orthopsychiat.*, *31*, 1961.
72. SPIRO, M., *Children of the Kibbutz*, Schocken, New York, 1966.
73. SPITZ, R., *The First Year of Life*, International Universities Press, New York, 1965.
74. ———, "Hospitalism: A Follow-up Report," in *The Psychoanalytic Study of the Child*, Vol. 2, p. 113, International Universities Press, New York, 1946.
75. ———, "An Inquiry into the Genesis of Psychiatric Conditions in Early Childhood," in *The Psychoanalytic Study of the Child*, Vol. 1, International Universities Press, New York, 1945.
76. ———, *No and Yes*, p. 54, International Universities Press, New York, 1957.
77. ———, "The Smiling Response: A Contribution to the Ontogenesis of Social Relations," *Genet. Psychol.*, *34*:57–125, 1946.
78. SPOCK, B., "The Striving for Autonomy and Regressive Object Relationships," in *The Psychoanalytic Study of the Child*, Vol. 18, International Universities Press, New York, 1963.
79. STERN, D., "A Micro-Analysis of Mother-Infant Interaction: Behavior Regulating Social Contact between a Mother and Her 3½ Month-Old Twins," *J. Am. Acad. Child Psychiat.*, *10*:501–518, 1971.
80. SULLIVAN, H., *The Interpersonal Theory of Psychiatry*, Norton, New York, 1953.
81. TAUBER, E., and KOFFLER, P., "Optomotor

Response in Human Infants to Apparent Motion: Evidence of Innateness," *Science*, *152*:383, 1966.

82. TENNES, K., and LAMPL, E., "Defensive Reactions to Infantile Separation Anxiety," *J. Am. Psychoanal. A.*, *17*:1142–1162, 1969.
83. WEISBERG, P., "Social and Non-Social Conditioning of Infant Vocalizations," *J. Child Dev.*, *34*:377–388, 1963.
84. WINNICOTT, D. (1958), "The Capacity to Be Alone," in Winnicott, D., *The Maturational Processes and the Facilitating Environment*, International Universities Press, New York, 1965.
85. WOLFF, P., "Observations on the Early Development of Smiling," in Foss, B. (Ed.), *Determinants of Infant Behavior*, Vol. 2, Methuen, London, 1963.
86. ———, and WHITE, B., "Visual Pursuit and Attention in Young Infants," *J. Child Psychiat.*, *4*, 1965.
87. YARROW, L., "The Development of Focused Relationships during Infancy," in Hellmuth, J. (Ed.), *The Exceptional Infant*, Vol. 1, Brunner/Mazel, New York, 1967.

CHAPTER 13

GROWTH AND DEVELOPMENT DURING THE TODDLER YEARS

John A. Sours

THE TODDLER YEARS commence at about 10 to 14 months, with the beginning of creeping and crawling movements, and extend to the third year, when the child has developed the motor skills, language, cognitive activities, and defenses sufficient for separation-individuation and the autonomy needed for the nursery school experience.* This developmental interval is also referred to as the anal-muscular phase; developmentally it lies between the oral phase and the phallic-oedipal phase.

During the anal-muscular phase the child grows from lap babyhood to toddlerhood. Erikson refers to this phase as one of autonomy versus shame and guilt.[10] The erogenous zone has shifted from the mouth to the anus-rectum with the development of anal erotism[29] and a conflicting biological mode—namely, fecal retention and elimination.[10] The phase is characterized by vigorous self-assertion in the service of separation-individuation and autonomy.[42] The toddler strives for independence as a separate individual with his own identity in the family group.[5,28] The child acquires a sense of autonomy to combat his sense of doubt and shame. His physical, psychological, and social dependency, however, fosters doubts about his capacity and freedom to assert himself. His urge to prove muscular strength and mobility is ever present. It is hard for him to stay in one activity or space. He wants to explore and accomplish new feats and skills. He is now increasingly capable of controlling his anal and urinal sphincters. His diet has been changed so that his stools are harder, and they are easier to control.

* See references 3, 7, 8, 9, 10, 38, 42, 46, 53, 54, and 55 in the Bibliography at the end of this chapter.

The child in this stage of development experiences intense self and object ambiguity and ambivalence. He wishes to hold on and to let go, not only with his hands, but also with his mouth, eyes, and sphincters. His muscular control is more refined.* Movement has developed to the point where he is able to creep, crawl, stand, and walk and also to hold on, clutch, and release. He now has a strong desire to manipulate objects. As the child's locomotion increases, the mother begins to say "no" to him. The "no" is a prohibiting gesture. The mother is afraid that he will hurt himself. The child understands the mother's prohibition through identification and imitates the negative headshaking movements; this is the child's first abstract concept.[3,9,10,38,42] How is this concept acquired? Spitz has suggested that the sign of negation evolves from the neurophysiological rooting reflex. At six months visual and muscular coordination makes rooting unnecessary for sucking. Sucking is then changed to withdrawal and refusal when the child is sated. At 15 months the motor behavior has taken on an ideational content. Then the head movement means a refusal. On the other hand, the prototype of the affirmative gesture is different. Head nodding is not derived from rooting. The young infant's neck muscles are not strong enough for extension. From three to six months of age the infant is able to support his head. Visual orientation is then possible. The affirmative head nodding is based on the vertical approach movement of the head to the nipple. The affirmative nodding retains its affirmative function. At two years it takes on an expressive meaning after the negative is acquired.

Early oppositional behavior can be noted in infants. A mother will refer to her baby who is slow to wake up for feeding as stubborn. She sees her child as resistive to being fed. Stranger anxiety at eight months may be confused with shyness and stubbornness.[39,41] At ten to eleven months displays of independence through rejection of the offered spoon are seen. The child can also spit out food he does not like, an early manifestation of the development of independence, which increases in children up to two years of age. Negativism appears when the child feels a need for protecting this developmental process. Oppositional behavior includes mutism, food refusal, bowel and bladder incontinence, and willfulness. Oppositionalism is not necessarily a displacement from anal erotism; it is intrinsic to the developmental process. Resistance to external influences is the most characteristic aspect of this behavior, but individual differences are noteworthy. The capacity to resist external influence is the toddler's first flowering of self-determination and autonomy. Oppositional behavior fosters separation, protects against compliance, energizes the child to overcome ego immaturities, and helps him to develop and use his own controls.

The toddler faces separation-individuation with a rate determined in part by neuromuscular and language development.[3,6,7,9] Central nervous system development leads to increased myelination and functioning of the stretch reflex in the gamma afferent loop systems. At 12 months the child is able to pull himself to the standing position, at 14 months to stand alone, and at 15 months to walk awkwardly. His new locomotion gives him a great deal of pleasure but also causes conflicts and struggles with his parents. The maturation of motor function seems independent of learning and experience.

This is not true, however, of language, where maturation and learning work hand in hand. Language hastens the child's integration into the family and allows him to start using secondary process thinking and expression. Under one year he has used principally vowels. From one to two years he switches more to consonants and begins using single words and some grammatical structure.[54] The vowel-consonant reversal after age one occurs because of his introduction to solid foods, the appearance of dentition, the decrease in buccal fat pads, the increase in size of the mouth, and the attainment of erect posture. Language

* Mussen, Conger, and Kagan have described in detail the toddler's progressive development in movement, language, perception, and cognition. P. H. Mussen, J. J. Conger, and J. Kagan, *Child Development and Personality* (New York: Harper & Row, 1963).

development is fostered by secondary reward and generalization learning. Identificatory learning is also important in this aspect of development. Girls surpass boys consistently in language until age three, when boys' articulation improves.

Toilet training becomes part of the overall communication between the child and the mother.[53] Reward and punishment are critically involved in this behavioral acquisition. Many children, however, are not trainable until one and a half to two years. And frequently children are not dry at night until well after two years.

The child begins to play at age two, but shows little interest in reciprocal play and affiliation with his peers. Later he is able to play cooperatively in the preschool period from three to five years. In his play he engages in exploratory behavior and learns a safe range of autonomy. He manipulates objects and is able to affect changes in his world. From walking, exploring, and manipulating come some feelings of confidence. His play activity satisfies his intrinsic need to deal with the environment. If a self-initiated response is rewarded, then it is permissible to be curious.

By six months the child tries playing with many playthings in his microsphere or thing-world.[10] In early infancy playthings are soft and cuddly, reproducing the mother's body and providing a transitional object for the child's security at bedtime. By nine months the child is able to play pat-a-cake and peekaboo with the mother, turning passivity into activity in order to assimilate the anxiety of object loss. Such symbolic play expresses separation from and retrieval of the mother.[56] In the toddler stage movable toys with strings extend this separation play. In addition, the toddler uses socially acceptable substances for symbolic excrement play involving filling, emptying, and messing.

Galenson[30] has extensively reviewed cognitive aspects of nonverbal play. The use of symbols with no fixed meaning and the lack of negation are aspects of primary process thinking as well as presentational symbolism, both apparent in play behavior and its manifestations as nonverbal thinking.

Bowel-Bladder Pleasure and Control and Transformations in Personality Development

With myelination of the neural pathways for micturition and defecation, the toddler is ready to assume control of his sphincters.[7,9,11,42] At the same time the bowels and bladder become associated with physiological tension and pleasure. The mother imposes her adult standards on the child to force him to give up pleasurable retention and elimination, as well as play with the feces. She cannot control his sphincters as she controlled his rhythmic nutritive and non-nutritive needs. She now has to negotiate with the child by pleading, begging, and cajoling. This interaction, once thought to be the most important aspect of toddlerhood before the separation-individuation subphases were recognized, is still central to this phase of development.

The child may retain feces to prolong the pleasurable sensation of evacuation. He may retain to defy the mother, an act that may mark the onset of stubbornness or be a part of his need for mastery. He may retain feces to incur the mother's pleasure. He may let go to defy the mother. Soiling and smearing are hostile acts and are often precursors to anal, urethral, and phallic profanity. The child may urinate and move his bowels to please his parents. The feces may be regarded by the child as valuable possessions, his own creation. The mother may regard the feces as a laudatory product or, on the other hand, as dirt and filthiness. Frequently both attitudes are expressed at one time or another. The child may view defecation as getting rid of something bad or dangerous.

Urination for boys provides an aggressive and libidinal pleasure owing to the powerful stream. This play may evoke envy in little girls. If a child is frustrated too much by toilet training, his frustration may result in chronic anger or in timidity and submission. For boys stimulation of the genitals during defecation often leads to erection, retraction of the testicles, and simultaneous contraction of the sphincter anuses. The stimulation may lead to confusion about whether the fecal mass or the genitals are dropping off during defecation.[4]

In order to understand bowel-bladder pleasure and control and their transformations into personality development, it is necessary to trace the development of anal erotism and body awareness. Anal erotism refers to the drive qualities of the anal phase of development when the erogenous zone has shifted to the anus and rectum. The erotism is experienced during defecation and in connection with anal masturbation and anal retention.[2] Stimulation of the mucus membranes lining the anus and rectal canal leads to erotism.[25] The pleasure is autoerotic. In this way the child negotiates with the mother over sphincter control so that the erotism becomes object related. With the appearance of teeth, the increase in musculature and strength of sphincteric functions, anal and sadistic impulses emerge. As in oral erotism, there are two distinguishable stages to anal erotism. The first is characterized by destructive impulses. The second is reflected in impulses to keep and possess.

Freud first referred to the anal phase in a letter to Fliess in 1897.[8,11,56] Freud commented on how money and excrement have long been associated in history. In the 1905 edition of the "Three Essays"[29] Freud described excitation of the anal zone. He indicated that children enjoy anal erotogenic stimulation in holding back stools to the point where the accumulation results in a violent contraction. In addition, he indicated that anal masturbatory pleasures are derived from stimulation of the anal zone. In his paper on "Little Hans"[18] he referred to Hans's fantasy of taking big children to the toilet to make them "widdle" and to "wipe their behinds." Freud felt that the fantasy indicated that Little Hans had been looked after as an infant in much the same way and this had been a source of pleasurable sensations to him. Freud suspected that Little Hans had been a child who enjoyed retaining his feces so that he could enjoy "a voluptuous sensation from their evacuation." In his paper on "Character and Anal Erotism"[19] Freud indicated a connection between anal erotism and character traits of orderliness, parsimony (connected with an interest in the product of defecation), and obstinacy (which is associated with the act of defecation), as well as the child's unwillingness to comply with the mother's wishes. Later he indicated that anal erotic aims are unserviceable culturally and must undergo sublimation. Later in the analysis of the "Rat Man"[27] Freud revealed ways in which anal erotism appeared in the man's illness. For the "Rat Man" the rat had many meanings such as money, worms that bury in the anus, and dirty animals feeding upon excrement. In "Disposition to Obsessional Neurosis"[20] Freud indicated that there is a pregenital sexual organization, thus using for the first time the term "pregenital." At this point of development he thought that the component instincts have already come together for the choice of an object and that the primacy of the genital zones has not yet been established. He believed that the component instincts that dominated the pregenital organization of sexual life are the anal erotic and sadistic ones.[20] He also indicated the mechanism of reaction formations against anal erotic and sadistic impulses and the possibility of regression in instinctual life to the pregenital sadistic and anal erotic stages.

Freud characterized the sexual aim of anal erotism as two in number, active and passive. Activities are supplied by instinctual mastery. The passive trend "is fed by anal erotism whose erotogenic zone corresponds to the old undifferentiated cloaca."[20] In "Three Essays"[29] Freud stated that "the activity is put into operation by the instinct for mastery through the agency of the somatic musculature; the organ which, more than any other, represents the passive sexual aim is the erotogenic mucus membrane of the anus. Both of these currents have objects which however are not identical."

In 1917 in "The Transformation of Instincts with Special Reference to Anal Erotism"[28] Freud remarked on the complex unconscious relationship between the ideas of feces, child, and penis. He later elaborated on this in his study of the "Wolf Man"[21] who preferred copulation from behind and was attracted to the female buttocks, all part of his anal erotic disposition. And in "Three Essays"[29] Freud further elaborated on the unconscious meaning of feces, suggesting that they are at first for a child a "gift" and later acquire the meaning of "baby," one of the sexual theories of children.[24,26] In his study of the "Wolf Man" Freud also concluded that the fecal mass stimulates the erotogenic mucus membranes of the male and is the active organ just as the penis stimulates the vaginal mucus membrane. In this sense the fecal mass acts as though it were a forerunner during the cloacal period. Freud felt that giving up feces was a prototype of castration, the first time a child is made to part with a piece of his body in order to gain the love of an object. Feces, baby, penis thus all form a unity, "the little ones" that can become separate from one's body.

In another paper Freud remarked that the child must decide between a narcissistic and an object-loving attitude. As he makes the choice of autoerotic satisfaction and becomes defiant, his attitude comes from his narcissistic clinging to anal erotism.[29] He remarked that a large part of anal erotism is carried over to the cathexis of the penis. He further remarked that fantasies of patients were filled with the equation of feces-money-gift-baby-penis. Interest in the vagina occurs later and springs from anal erotic sources. He quoted Andreas-Salome, who referred to the vagina as "taken on lease" from the rectum; this is most commonly seen in the fantasies of homosexuals, where the vagina is represented by the rectal canal.[26]

Thus Freud viewed the anal phase as the second phase of libidinal development in which the discharge of dammed up instinctual energy is connected with the act of defecation and the accompanying pleasurable sensations. The child's interest at first lies in retention and expulsion of feces, which he regards as a valuable creation, a prized personal possession, that unfortunately he must eventually relinquish. Finally the child repudiates his anal wishes and sublimates the energy into activities such as playing in the sand, staying clean, and painting. If the child is not permitted to give up his anal investments slowly and is forced into a traumatic situation, he is apt to use the mechanisms of repression and reaction formation in an attempt to rid himself immediately of such interests. Repressed unconscious anal wishes may continue to play a large part in his life and produce various warped defense mechanisms.

The Creation of the Representational World

The child creates his representational world out of images of the external world that he comes to experience as the "external" world.[52] The representational world also contains the child's own body sensations as part of his body schema, along with representations of instinctual drives as need and affect representations.[49,52] Piaget has studied the maturational process by which images are formed into lasting representations during toddlerhood (about 16 months).[13,43] At this stage object constancy is perceptual, followed at about 36 months by libidinal object constancy.[14] From body representations arise self-representations, which are enriched by activities, experiences, identifications, and, with the resolution of the Oedipus complex, introjection of parental authority.[49,50]

From the development of sensorimotor and representational intelligence the toddler's representational world arises. Piaget has described this process over a period of some years.[13,43]

The neonatal cycles of tension—quiescence, sleep, wakefulness—continue through the first year. A fatigued and satisfied infant goes to sleep until hungry and restless. If not immediately fed, he then begins sucking his finger, which gives him experience in eye and mouth activity, the basis of primary circular reactions. This activity leads to tension reduction by mental images that are satisfying for a while until hunger increases, and the infant cries and becomes more restless. The mother then responds with milk. The infant then lies in an alert, active state, playing and exploring to the point that he again becomes fatigued and falls asleep. Thus the sleep-waking cycle is completed. Sensorimotor intelligence is dependent for its development on gratification of instinctual needs.

Whenever the infant's oral needs are satisfied, the baby uses sensorimotor schemata for explorations and short excursions by which he acquires increasing knowledge of his environment. He explores his world until fatigue sets in and he is forced to go to sleep.

During the first two years of life the child goes through six stages of cognitive development. The basic biological processes whereby he acquires knowledge are assimilation and accommodation. By assimilation Piaget means that the organism takes in aliments from the world about him. He makes constant perceptual and motor adjustments to the world on several levels. At the biological level there are physiochemical incorporations into the organic structure of the body. At a primitive psychological level the organism incorporates sensory and motor components of behavior into endogenous reflex schemata that are then activated in later psychological development. The mind incorporates ideas about the external world and products of mental activity into already existing schemata of sensorimotor activity. By accommodation Piaget means a process whereby what is assimilated is so changed as to bring about an adjustment within the environment. Accommodation modifies the experience of new stimuli or conditions. Intelligence is a capacity to distinguish objects from the self in respect to space, time, and form. It is a capacity to manipulate intrapsychically symbols and schemata to solve problems of the environment. The child's cognitive development is facilitated when his instinctual needs are satisfied and he is provided with a sense of safety by the mother.

The first stage in the development of intelligence (birth to four weeks) is made up of reflexes needed for survival, such as sucking, rooting, and grasping. These reflexes are the building blocks to intelligence. At this stage of development the baby is able to differentiate the nipple from the blanket. He can recognize satisfying or unsatisfying objects. Successful repetition of the reflexes results in an assimilation in the neural apparatus. The assimilated pattern of behavior is called a schema, which may be a mental experience and at a later stage an image. It is an adjustment to the environment.

At the second stage the primary circular stage (one to four months) the infant practices rhythmic mouthing and hand movements. He watches his mother and visually pursues her. The sensory stimuli in his environment provide the aliments for activity. The human face is a visual schema. The human voice is an auditory schema. At four months sucking becomes adjusted to a new type of schemata. The assimilation of grasping and visual pursuit leads to new accommodations. The child is now able to grasp what he sees. He now is capable of intersensory coordination, but he is still not able to identify things in space as having identity. For instance, if a toy is hidden under a cover, the toy is totally lost to him. The image is no longer a stimulus when the object is gone.

The third stage is the secondary circular reaction (four to eight months). It is the first time the infant performs an intentional act. The baby acts upon things and uses relationships as a means toward an end. For instance, he will grasp a suspended rattle, strike it, or swing it. If he is supine he will swing at the rattle rather than grasp it. He will practice with pleasure. He may make the rattle swing by shaking the crib. Thus the schemata are reciprocally assimilated. In following the mother, the baby will use these gestures to perpetuate a variety of spectacles. He may induce another person to make the rattle move. Still objects have no independence outside of the movement of the baby in respect to them. For example, if an adult hides a ball under a cover, the baby will look for it as an extension of his own visual environment, or he will look into the adult's hand. There is now some kind of permanence of images for him at this stage. Space has become a function of the images of prehension and motor activity; which exist in a trajectory of visual and motor activity. This marks the beginning of interest in causal relations. Imitation begins as a means of inducing a person in the environment to repeat interesting acts.

The fourth stage, coordination of sensory schemata and their application to new situations (eight to twelve months), is a stage in which the infant assigns to external objects a consistency independent of himself, but without permanence. He pursues ends not immediately attainable. For example, if an object is hidden, he will remove the object with one motion and then take hold of it with another. He has the capacity to keep in mind his goal. He uses instruments as intermediary means to an end. For instance, he will push the adult's hand away. Signs and symbols indicate specific events. At this stage the infant will turn toward the sound of footsteps. He will explore objects as though trying to understand them. He will grasp relations between size and distance. He is now well aware of three dimensionality. In this stage eighth-month anxiety appears with the recognition of the mother's face. He can now only tolerate short separations.

At the next stage, tertiary circular reactions (12 to 18 months), the baby discovers new means to active experimentation by adapting himself to unfamiliar situations not only by using schemata acquired earlier but also by seeking and utilizing new means to do it. He no longer is satisfied with producing familiar results. If he has a new toy, he will feel it, pick it up, drop it. His interest is taken up with what is new and different about the toy. He may also, for instance, take a ball and roll it and delight with the new activity he has given the object. He will search for novelty and be overjoyed if he finds an entity that can be made to do different things. He may use a stick or string to bring things nearer. Spatial and causal relations are extended. For instance, if a ball is hidden at point A, he will find it. If the ball is then hidden successively under B, C, and D, he will go immediately to D. He is able to take into account visual displacement. Objects have an identity and a relationship to continued perceptual cues that reach the baby's mind.

The child is capable of discovering and utilizing complex relations among objects. This is possible through extensive experimental study of distance as well as space. He studies movements of objects from place to place. The exploration of equilibrium, rotation, reversals, positions, and content is possible, yet knowledge of the relationships does not transcend perception.

In the sixth stage of development (beginning at 18 months) the invention of new means through mental combinations marks the start of true intelligent behavior. The child is able to conceive of objects as entities with dimension, qualities, spatial, temporal, and causal relations, even when the objects are no longer perceived. Thus the child has attained conceptual object constancy. He forms mental representations that include the object not only as it is actually perceived but also as it might appear under other circumstances. External objects are identical to the self, separate from the self and his actions, regardless of how displaced or hidden. The child will search for hidden objects or displacements not visible to him. He is able to imagine displacement in space and to internalize cause-and-effect relationships. The child's body can be subjectively experienced as an independent object, differentiated from inanimate objects and from persons. He now has the capacity for mental representations of the self.

At the end of the 24 months the child enters a new phase of intelligence. He has left the sensorimotor stage of intelligence and is now in the period of preparation for the organization of concrete operations. Now the child is capable of crude symbolizations. Piaget has divided this phase of development into two parts. In the preoperational representation phase, from two to four years, there is the beginning of representational thought. Then from four to five and a half simple reproductions or intuitions occur, followed from five and a half to seven by articulated representations or intuitions. At age seven the child enters the stage of concrete operations. He is now capable of conceptual organization that takes on stability and coherence by series of groupings. After age 11 to about 15 the child is in the period of formal operations. New structures, visuomorphic to groups and lattices of logical algebra are now possible.

Separation-Individuation during the Infant-Toddler Years

Through her early studies of young psychotic children, Mahler[39] has identified and described the process of separation-individuation during the first three years of life. Many of her concepts relate to or interdigitate with those of Piaget and Sandler. She has helped develop the view of the life cycle as involving progressive stages in distancing from and internalizing the symbiotic mother and negotiating the eternal longing for the actual or fantasized "ideal state of self"—a symbiotic fusion with the "all-good" symbiotic mother, who at one point in infancy is part of the self, as the core to the blissful state of well-being.[39,48,49]

During the neonatal period the infant attempts to achieve homeostasis, at first through the help of the mother and later through the development of increasingly protective neurophysiological mechanisms. The infant begins to differentiate between pleasure and pain through expulsive activity and maternal gratification.

By the second month of infancy he begins to have an awareness of need-satisfying objects. Mother and infant are a dual unity with a common boundary. At this point there is no difference between the "I" and "not-I." The inside and outside are experienced together. At this point the infant has formed a maternal symbiosis that strengthens the inborn instinctual stimulus barrier and protects the child from premature phase-specific strain.[41] The symbiosis is an hallucinatory somatopsychic omnipotent fusion with the representation of mother and the sense of common boundaries. It is necessary that displacement of libido take place from the inside to the outside of the body.

During the normal autistic phase there is an absolute primary narcissism. The infant is unable to differentiate between "self" and "not-self." By the third month, however, there is increasing awareness of need satisfaction coming from part objects. Still the infant views the outside as continuous with an omnipotent, symbiotic dual unity. Gradually there is demarcation of the representation of the body ego from the symbiotic matrix.[34] The body ego becomes increasingly differentiated. There is a shift of proprioceptive and enteroceptive cathexis to sensoriperceptive cathexes of the outside world, which is essential to body ego formation. In addition, there is rejection by projection of destructive aggressive energy beyond the body's self boundaries. The infant begins to develop an inner sensation for the core of the self. Feelings of the self form the beginnings of a sense of identity. The sensoriperceptive organ further demarcates the self. Object relations start to develop as the infant differentiates from the mother-infant dual unity and enters a normal phase of human symbiosis. There then is a transition from lap babyhood to toddlerhood; this progresses through steps of separation-individuation only made possible, on the one hand, by autonomous ego development and, on the other hand, by a host of identificatory mechanisms. The process of growing away from the baby-mother symbiosis involves a mourning process, which, as part of every step toward independent functioning, is associated with the threat of object loss. This is especially prominent in the separation-individuation process from the fourth to fifth month to the thirtieth to thirty-sixth month. The separation-individuation phase has been divided by Mahler into four subphases: differentiation, practicing, rapprochement, and "on the way to libidinal object constancy."

The first phase, differentiation, takes place from four to five months of age at the peak of symbiosis. In the symbiotic months the infant comes to identify the mother as the specific object toward whom he forms a specific bond. The symbiotic unit is created through entero-proprioceptive and contact-perceptual information. Sensoriperception attention-cathexis expands his perception beyond the symbiotic orbit and fosters a perceptual conscious system and a permanently alert sensorium during periods of wakefulness. From six to seven months the child is fascinated with the mother's face and body, which he explores with his available sensory modalities. In addition, he plays peekaboo games. He is now able to recognize his mother and check out unfamiliar objects against the already familiar; Brody calls this activity "customs inspection." He attempts for the first time to break away from his passive lap babyhood. He does this in miniscule ways by pushing back from his mother's hold, sliding down her lap, and sitting at her feet, all providing him with increasing distance from the mother and new views of the maternal object. As he sees new objects distinctly different from his mother, he experiences "stranger reactions." He is curious about and full of awe at every new strange thing he sees. If he is trustful of his mother, anxiety is manageable.

The second phase, practicing, overlaps the phase of differentiation to some degree. The second phase occurs from seven to ten months and lasts up to fifteen to sixteen months of age. Activity in this period justifies dividing this phase into two parts: (1) the early phase, which overlaps differentiation and is marked by the infant's early ability to push away physically from the mother through crawling, righting himself, and climbing, yet at the same time holding on; and (2) the practicing phase proper, which is characterized by free, upright, bipedal locomotion. The child's increasing awareness of separateness is involved in rapid body differentiation from the mother, the establishment of a specific bond with her, and the growth and functioning of the autonomous ego apparatus in close proximity to the mother.

Because of his new relationship with the mother he can now look beyond her at other human and inanimate objects. He can play with toys of all sorts and explore them with his eyes, nose, hands, and mouth. One or more of these objects are apt to become transitional objects. Part-objects of the mother are used to assuage anxiety during her absence at night. During this phase of development the infant must move and explore. As he does so he makes brief sorties away from the mother, returning periodically to "home base" in order to refuel through physical contact. In this early practicing subphase there are brief periods of separation anxiety. The infant tries not to lose sight of his mother. Much depends on the mother's comfort in seeing the child moving away from her. Walking has for the mother, as for the child, an enormous symbolic meaning. If she is able to believe that the child is capable of moving progressively away from her, she contributes to his feeling of safety, and perhaps as Sandler has indicated, he gives up some of his magical omnipotence for autonomy and developing self-esteem.

In the practicing subphase proper, once the child attains upright locomotion, he has, as Greenacre has said, a "love affair with the world." Libidinal cathexis goes into the service of the growing autonomous ego and its functions. The child appears to be ecstatic with his new faculties and the wonders of his new world. He seems to regard himself as indestructible by injury. He acts as if he could not care less about his mother's presence. His relationship with his new faculties and world is also a defensive maneuver to avoid absorption into the mother's protective sphere. The toddler has now gone from a passive peekaboo game to an active game that aims at avoiding reengulfment by the mother. By fleeing from her he is able to retain her actively, as she eventually catches him and reminds him of her potential comfort. He plays these games and activities in various moods. Although he gives a semblance of indifference to the mother's presence, he nevertheless returns periodically, seeking a sense of physical proximity. The toddler's moods become, as Mahler has phrased it, "low-key" only when he is aware that the mother is absent. His motility diminishes as does his interest in his surroundings, and he appears to be preoccupied with inwardly concentrated attention, which has been called "imagining." The low-keyness has been compared to a miniature anaclitic depression. If the child is separated from the mother and a different person returns to comfort him, he is apt to burst into tears. When he is reunited with the mother, his low-keyness diminishes. Mahler believes that such behavior results from the child's effort to hold onto a state of mind that Joffe and Sandler have termed the "ideal state of self."

The third subphase of separation-individuation takes place from 16 to 25 months and commences with the mastery of upright locomotion and a diminishing excitement in that ego function. The toddler is now increasingly aware of physical separation. He is now not quite so resistive to frustration; neither is he as seemingly indifferent to the mother's presence. Increasing separation anxiety and awareness of object loss is apparent. His behavior becomes much more active with the mother. He now wants his mother to share with him all his new skills and experiences. Mahler has referred to this subphase of separation-individuation as the period of rapprochement.[40] The earlier refueling type of contact is now replaced by a quest for constant interaction of the toddler with the mother, most markedly evident in both vocal and nonvocal communication, as well as symbolic play. The toddler now begins to experience the impediments to his anticipated conquest of the world. With the marked increase in autonomous ego functions, there is an increasingly clear differentiation between intrapsychic representation of the object and self-representation. At this point the ego has created a representational world. As he reaches toward a greater height of mastery, it becomes apparent to the toddler that the world is not quite his oyster; he realizes that he must learn to cope with increasing independence even though he is still a relatively small, helpless individual unable to command the world by voice, gesture, and feeling since he has relinquished much of his omnipotence. One finds the child now actively wooing his mother; this is often perplexing to her because he should now be more independent and self-assertive. Frequently mothers during this phase find the child's demandingness unbearable and inexplicable. Other mothers find it difficult, even if the child does return to woo her, to view the child as increasingly independent and separate. In the third subphase individuation proceeds very rapidly just as the child is becoming increasingly aware of his separateness and uses a number of ploys to avoid separation from the mother. He finds that he must use verbal communication more and more with his mother. Gestural communication and preverbal empathy between mother and child no longer provide the same sense of satisfaction and well-being that they used to.[36] The junior toddler realizes and must accept that his parents are separate individuals with their own individualistic tendencies and interests. He is forced to give up the delusion of his own omnipotence and grandeur. This Mahler has termed the rapprochement crisis, the crossroad where three basic anxieties of early childhood come together: namely, the fear of object loss, the fear of loss of love, and signs of castration anxiety. It is the crossroad between fusion and isolation, part of the lifelong struggle with separation-individuation.

The fourth subphase of separation-individuation is that of increasing attainment of object constancy in the Hartmann sense. The child is "on the way to object constancy." This subphase occurs from 25 to 36 months. At the beginning the child is gradually able to accept once again separation from the mother as he did during the practicing subphase. The verbal communication that had begun during the third subphase develops very rapidly during the fourth phase and replaces to some degree gestural language, affective motility and primary process language. In this regard the mother must often play along with the toddler, providing him with secondary process language and ways of viewing the world. In addition, the toddler in this subphase shows a more constructive and purposeful type of play, often associated with fantasy, role playing, and make-believe. The child is increasingly interested in adults other than the mother. His sense of time develops along with a capacity to tolerate delay of gratification. He is better able now to endure separation at a time when individuation is also progressing. Oppositional behavior is more prominent than during the rapprochement phase. Now he actively resists the demands of adults and at times is quite negativistic in his behavior.

As the child is "on his way to object constancy," he obtains a unified representation of the parental object, which then becomes intrapsychically available to him as a love object. Now separation and individuation are more closely connected to developmental processes. Yet if development is disturbed, they can proceed divergently as a consequence of the lag in one or the other. If a rapprochement crisis occurs, it may result in unresolved interpsychic conflict, which is then established as a fixation point interfering with later oedipal development.

Oral, anal, and early genital forces and conflicts are found at the time of the rapprochement crisis. Symbiotic omnipotence must be renounced in the face of increasing awareness of body and its sensations and pressures at zonal points. Awareness of bowel and urinary sensations during the toilet training process can heighten the child's reaction to the discovery of anatomical sex differences, causing him prematurely to experience castration anxiety.

The father's role in separation-individuation is not as clearly defined, partly because of his lesser importance during the early infancy months.[11] The baby's relationship with the father begins during the symbiotic phase, as manifested in the infant's smiling response to the father. At this time the father becomes somewhat familiar to the infant and is no longer just an object toward whom he may experience stranger anxiety. From the very beginning of infancy the infant is more reactive to the mother and even to siblings than to the father. In the differentiation subphase attachment to the father occurs, but it becomes pronounced only at the beginning of the practicing subphase, at which time the father is truly the second parent.[1] The father is now the more interesting object, in space away from the mother and readily available for exploration. The infant's elevated mood seen during this period is most conspicuous with the father, with whom exuberant play takes place.

It has been observed that girls attach themselves to the father earlier than boys and are more cautious with strange men and with all unfamiliar people. Boys are able to approach male objects earlier, a fact probably related to their greater interest in distance, space, and inanimate objects. Even at the age of two years girls tend to be passive in seeking affection from the father and his substitutes. Boys, on the other hand, are more distant and show a dichotomy between the symbiotic and the nonmaternal world.

Toward the end of the practicing subphase toddlers seek out the attention of more than one adult. Children of this age are quite competitive, manifesting considerable envy and aggression toward one another. Competition toward the father is not evidenced at this point, but the child's relationship to the mother shows increasing ambivalence during the rapproachement crisis. After the beginning of the rapprochement crisis the paternal object appears in the fantasy world of the toddler as a more powerful parent. This image of the father may be essential for the attenuation of the child's ambivalence to the mother during the rapprochement position. The child at this stage has three mental representations, those of the self and parental objects, which provide the necessary triadic elements for the Oedipus complex.

Roiphe and Galenson[44,45] have found that in the separation-individuation phase, particularly between 15 and 24 months, there is increasing genital manipulation, masturbatory activity, and curiosity about and reactions to anatomical sex differences. They suspect that there is at this time an increase in endogenous genital sensations, associated with bowel and bladder activity, that is probably independent of toilet training. They suggest that this part of separation-individuation be considered an early genital phase, necessary for consolidation of self and object representations and the establishment of a primary schematization of the body and genital outline. They have found that castration anxiety is apparent after awareness of anatomical differences for some children who have not formed stable body schema. Castration anxiety is most apparent in those children who have suffered physical illnesses, surgery, or birth defects or have lost a parent. Their observations come from nursery school experiences, but analytic child and adult data also indicate much the same thing.[10,31,32,47] During the rapprochement phase the vulnerability of this subphase is heightened if the child experiences excessive castration anxiety. The anxiety leads to a regressive breakdown of toilet training, with massive enuresis and negativism. The child's increasing ambivalence results in splitting of the good and bad mother images and turning of aggression against the self, with depressive mood and collapse of self-esteem. Castration anxiety weakens ego functions and brings about inhibition of curiosity and play.

Studies of the early genital phase indicate that children are well aware of genital parts and are readily upset by becoming aware of genital differences. It is not unusual for toddlers, if given the opportunity, to examine each other's genitals. A little girl may reach out to touch a boy's penis in order to see whether the penis is really there or to give vent to an incompletely inhibited aggressive impulse to take the penis.

Toilet training has traditionally been viewed as an activity imposed upon the child by the parents, with the expressed aim of giving the child autonomy, but at the same time making the child struggle for independence. The toddler's sexual curiosity, interest, and activities are also typical of this period and have strong influence on the entire toilet training process. Castration reactions at this phase can disrupt toilet training and, if part of the rapprochement crisis, deflate the child's self-esteem at a time when he is already bereft with the loss of his omnipotence and the uncertainty about the presence and availability of his mother. With dissolution of the toilet training routine, the child is put into a further crisis with the mother and is more apt to develop a hostile, dependent, paralyzing relationship with the mother.

With the attainment of bowel-bladder control and libidinal object constancy, the child enters firmly into the triadic world of his family. Growth and development during the phallic-oedipal phase is discussed in Chapter 14.

Bibliography

1. ABELIN, E. L., "The Role of the Father in the Separation-Individuation Process," in McDevitt, J. B., and Settlage, C. F. (Eds.), *Separation-Individuation*, pp. 229–252, International Universities Press, New York, 1971.
2. ABRAHAM, K. (1921), "Contribution to the Theory of Anal Character," in Abraham, K., *Selected Papers on Psychoanalysis*, pp. 370–393, Hogarth, London, 1948.
3. BALDWIN, A. L., *Theories of Child Development*, John Wiley, New York, 1967.
4. BELL, A., "Some Observations on the Role of the Scrotal Sac and Testicles," *J. Am. Psychoanal. A.*, 9:261–286, 1961.
5. BLOS, P., *On Adolescence: A Psychoanalytic Interpretation*, Free Press, New York, 1962.
6. BRACKBILL, Y. (Ed.), *Infancy and Early Childhood*, Free Press, New York, 1967.
7. BRECHENRIDGE, M. E., and VINCENT, E. L. *Child Development*, W. B. Saunders, Philadelphia, 1965.
8. BRENNER, C., *An Elementary Textbook of Psychoanalysis*, Doubleday, New York, 1957.
9. CAMERON, N., *Personality Development and Psychopathology*, Houghton Mifflin, Boston, 1963.
10. ERIKSON, E., *Childhood and Society*, pp. 182–218, Norton, New York, 1950.
11. FENICHEL, O., *The Psychoanalytic Theory of Neurosis*, Norton, New York, 1943.
12. FERENCZI, S., "The Ontogenesis of the Interest in Money," *Selected Papers*, Vol. 1, pp. 319–331.
13. FLAVELL, J. H., *The Developmental Psychology of Jean Piaget*, pp. 83–142, Van Nostrand, New York, 1963.
14. FRAIBERG, S., "Libidinal Object Constancy and Mental Representation," in *The Psychoanalytic Study of the Child*, Vol. 24, pp. 9–47, International Universities Press, New York, 1969.
15. FREUD, A., *The Ego and the Mechanism of Defense*, Hogarth, London, 1936.
16. ———, *Normality and Pathology in Childhood*, International Universities Press, New York, 1963.
17. ———, "Observations on Child Development," in *The Psychoanalytic Study of the Child*, Vol. 6, pp. 18–30, International Universities Press, New York, 1951.
18. FREUD, S. (1909), "Analysis of a Phobia in a Five-Year-Old Boy," in Strachey, J. (Ed.), *Standard Edition*, Vol. 10, pp. 3–152, Hogarth, London, 1961.

19. ——— (1908), "Character and Anal Erotism," in Strachey, J. (Ed.), *Standard Edition*, Vol. 9, pp. 167–176, Hogarth, London, 1961.
20. ——— (1913), "The Disposition to Obsessional Neurosis," in Strachey, J. (Ed.), *Standard Edition*, Vol. 12, pp. 322–326, Hogarth, London, 1961.
21. ——— (1918), "From the History of an Infantile Neurosis," in Strachey, J. (Ed.), *Standard Edition*, Vol. 17, pp. 3–124, Hogarth, London, 1961.
22. ——— (1923), "The Infantile Genital Organization of the Libido," in Strachey, J. (Ed.), *Standard Edition*, Vol. 19, pp. 141–148, Hogarth, London, 1961.
23. ——— (1926), "Inhibitions, Symptoms and Anxiety," in Strachey, J. (Ed.), *Standard Edition*, Vol. 20, pp. 77–178, Hogarth, London, 1961.
24. ———, "The Interpretation of Dreams," in Strachey, J. (Ed.), *Standard Edition*, Vol. 5, pp. 354–356, Hogarth, London, 1961.
25. ——— (1905), "Jokes and Their Relation to the Unconscious," in Strachey, J. (Ed.), *Standard Edition*, Vol. 8, pp. 3–249, Hogarth, London, 1961.
26. ———, "New Introductory Lectures on Psychoanalysis," in Strachey, J. (Ed.), *Standard Edition*, Vol. 22, pp. 3–250, Hogarth, London, 1961.
27. ——— (1909), "Notes upon a Case of Obsessional Neurosis," in Strachey, J. (Ed.), *Standard Edition*, Vol. 10, pp. 133–330, Hogarth, London, 1961.
28. ——— (1917), "On Transformations of Instinct as Exemplified in Anal Erotism," in Strachey, J. (Ed.), *Standard Edition*, Vol. 17, pp. 125–134, Hogarth, London, 1961.
29. ——— (1903), "Three Essays on the Theory of Sexuality," in Strachey, J. (Ed.), *Standard Edition*, Vol. 7, pp. 123–248, Hogarth, London, 1961.
30. GALENSON, E., "A Consideration of the Nature of Thought in Childhood Play," in McDevitt, J. B., and Settlage, C. F. (Eds.), *Separation-Individuation*, pp. 41–59, International Universities Press, New York, 1971.
31. GREENACRE, P. "The Childhood of the Artist: Libidinal Phase Development and Giftedness," in *The Psychoanalytic Study of the Child*, Vol. 12, pp. 47–72, International Universities Press, New York, 1957.
32. ———, "Perversions: General Considerations Regarding Their Genetic and Dynamic Background," in *The Psychoanalytic Study of the Child*, Vol. 23, pp. 47–62, International Universities Press, New York, 1968.
33. HENSCHER, J. E., *A Psychiatric Study of Fairy Tales*, Charles C Thomas, Springfield, Ill., 1963.
34. JACOBSON, E., *The Self and the Object World*, International Universities Press, New York, 1964.
35. JOFFE, W. G., and SANDLER, J., "Notes on Pain Depression and Individuation," in *The Psychoanalytic Study of the Child*, Vol. 20, pp. 394–424, International Universities Press, New York, 1965.
36. ———, and ———, "Some Conceptual Problems Involved in the Consideration of Disorders of Narcissism," *J. Child Psychother.*, *2*:52–71, 1967.
37. LENNEBERG, E., *Biological Foundation of Language*, pp. 127–142, John Wiley, New York, 1967.
38. LIDZ, T., *The Person*, Basic Books, New York, 1968.
39. MAHLER, M. S., "On the First Three Subphases of the Separation-Individuation Process," *Int. J. Psychoanal.*, *53*:333–338, 1972.
40. ———, "Rapprochement Subphase of the Separation-Individuation Process," *Psychoanal. Quart.*, *41*:487–506, 1972.
41. ———, and FURER, M. "Observations on Research Regarding the Symbiotic Syndrome of Infantile Psychosis," *Psychoanal. Quart.*, *29*:317–327, 1960.
42. MAIER, H. W., *Three Theories of Child Development*, Harper, New York, 1965.
43. PIAGET, J., and INHELDER, B., *The Psychology of the Child*, Basic Books, New York, 1969.
44. ROIPHE, H., "On the Early Genital Phase with an Addendum on Genesis," in *The Psychoanalytic Study of the Child*, Vol. 23, pp. 348–365, International Universities Press, New York, 1968.
45. ———, and GALENSON, E., "Early Genital Activity and the Castration Complex," *Psychoanal. Quart.*, *41*:334–347, 1972.
46. RUTTER, M., "Normal Psychosexual Development," *J. Child Psychol. & Psychiat.*, *11*:259–283, 1971.

47. SACKS, L. J., "A Case of Castration Anxiety Beginning at Eighteen Months," *J. Amer. Psychoanal. A.*, *10*:329–337, 1962.
48. SANDLER, J., "The Background of Safety," *Int. J. Psychoanal.*, *41*:352–356, 1960.
49. ———, "On the Concept of the Superego," in *The Psychoanalytic Study of the Child*, Vol. 13, pp. 128–162, International Universities Press, New York, 1960.
50. ———, *et al.*, "The Ego-Ideal and the Ideal Self," in *The Psychoanalytic Study of the Child*, Vol. 18, pp. 139–158, International Universities Press, New York, 1963.
51. ———, and Joffe, W. G., "Notes on Obsessional Manifestations in Children," in *The Psychoanalytic Study of the Child*, Vol. 20, pp. 425–438, International Universities Press, New York, 1965.
52. ———, and ROSENBLATT, B., "The Concept of the Representational World," in *The Psychoanalytic Study of the Child*, Vol. 17, pp. 128–148, International Universities Press, New York, 1962.
53. SEARS, R. R., MACCOBY, E., and LEVIN, H., *Patterns of Child Rearing*, Row, Peterson & Co., Evanston, Ill., 1957.
54. STONE, L. J., and CHURCH, J., *Childhood and Adolescence*, Random House, New York, 1957.
55. WAELDER, R., *Basic Theory of Psychoanalysis*, International Universities Press, New York, 1960.
56. WATSON, R. I., *Psychology of the Child*, John Wiley, New York, 1963.
57. WOODWARD, M., "Piaget's Theory," in Howell, J. G. (Ed.), *Modern Perspectives in Child Psychiatry*, Charles C Thomas, Springfield, Ill., 1965.

CHAPTER 14

GROWTH AND DEVELOPMENT IN CHILDHOOD

John A. Sours

PSYCHOSEXUAL DEVELOPMENT has a central place in all phases of child development and personality formation. Its influences and transformations affect all aspects of the developmental process. From the standpoint of physical sexuality, however, there are no appreciable changes before puberty between the sexes.[95] Sex hormone production for both boys and girls involves small amounts of estrogens and androgens. Androgen production is increased in both sexes, more marked in boys,[95,115] at about age nine to ten, with subsequent sharp rises in adolescence. The excretion of estrogens gradually rises in both sexes from about the age of seven.

Psychosexual development, on the other hand, entails many more preadolescent changes than those of physical development. These developments are apparent in a child's increased sexual masturbatory activity and explorative interests, his awareness of genital differences, his preoccupation with theories about the creation of babies, his attraction to the contrasexual parent, and his castration anxiety. In the phallic-oedipal phase of psychosexual development numerous changes occur independently of hormonal change.[76,82,95]

The phallic-oedipal phase of development has also been called the nursery school years, the preschool years, and the stage of initiative versus guilt.[76,82] Although this phase of development occurs between the ages of three and six, there are variations in its chronology from child to child. In some respects many developmental issues during this phase overlap the age span of three to six years. This fluidity between psychosexual stages is clearly seen at both ends of the phallic-oedipal period. Mahler, Roiphe, and Galenson suggest that sexual arousal can occur before the second birthday; for some children, who observe differences between the sexes, distinct castration reaction can also occur this early.[73,74] Although many character traits are discernible by the age of five years and have some permanence, character structure is subject to future modification and emendation during latency and adolescent years before consolidation at the end of late adolescence occurs.

The phallic-oedipal phase has a number of component parts.[26,27,28,29,30,33,35,37,40] The phenomena of infantile sexuality include discovery of sex differences, erotic genital exploration, sexual curiosity, sexual play, attachment to the contrasexual parent with expectation of loss of love, injury, and retaliation from the isosexual parent, and castration anxiety.[2,6,24,47,51,108,118] Castration anxiety, however, is most nuclear of the developmental conglomeration and is a pivotal force in the creation and dissolution of the oedipal situation. During the anal-muscular phase of development, especially the subphases of differentiation and practicing, the toddler attempts to increase his experience and capacity and reaches libidinal object constancy at about 36 months.[73] With his entry into the phallic years he experiences another burst of energy, curiosity, and initiative. Erikson[19] has referred to the phallic-intrusive mode as the descriptive and dynamic hallmark of this phase of development. "Being on the make," "making," and pleasure in the "conquest" are colloquialisms for this phase of development.[2,77,82] This intrusive quality is more prominent in boys, but, nevertheless, it has its parallel in girls as far as "catching," reaching out, "being on the make," and being attractive and endearing.

At this point of development the phallic child is capable of conceptualizing his place in his family and his relationship to people in his immediate community. According to Mahler, at the end of the fourth subphase of separation-individuation (36 months) the child has attained object constancy in the Hartmann sense and has established mental representations of parents and self.[73] He now has a social role in his group that allows him, among other things, to delineate more clearly his self and gender representation. After the age of three or four years he is repeatedly exposed to environmental reminders that he is no longer a small child. On the other hand, his intrusiveness quickly brings him by way of frustrations to the realization that he is hardly an adult. Consequently he repetitively experiences defeat and humiliation in his competitions and in his everyday interactions with his family, nursery school teachers, and his peer group. To some extent his aspirations inevitably must be dampened. As he goes through the phallic-oedipal years, he fumbles his way through his frustrations and failures and tries to remain hopeful.

As his awareness of himself within the matrix of his family increases, he queries more and more his exact relationship to his parents. He starts wondering about pregnancy and birth and how he came to be born. Even if he has been instructed in sexual facts by his parents or has witnessed sexual intercourse, he still must, by virtue of his level of capacity and emotional life, strive to create his own theories of intercourse, pregnancy, and delivery. He wonders in many ways about pregnancy, curious about the possibility of his mother having more children; he raises questions about his own ability to have a baby, even with his father. He is now more inclined to ask his mother, not too sheepishly, whether he can marry her. Often he sees no incongruity in his intent, but whatever he fantasies and initiates in reality only reinforces his sense of inferiority, anxiety, and guilt. His wishes pale before reality. He is smaller and less strong than adults. He cannot run as fast and, in fact, is quite apt to trip and fall. His ability to throw a ball is most likely better than his sister's, but he cannot throw as well as his father. Now he tries to drive his father's car, but the best he can do is to sit on his father's lap, only to have the wheel taken from him as soon as the father notices an oncoming automobile. He wants to help his father cut the grass, shovel snow, and rake leaves. These tasks challenge his strength, and he becomes fatigued more quickly than his father. If his father has mechanized yard equipment, he is told to stay away lest his foot or hand be cut off. His father takes him out to show him how to play tennis or golf. It is bad enough that he cannot play these sports as well as his father. He is further humiliated in the locker room when he notices the smallness of his genitals. He wonders how he will ever be able to fill his father's role. All his experiences as a failure reinforce his feelings of smallness and his awareness of genital smallness. The phallic-oedipal struggle meets with defeat—much like his experience in toilet training—when he attempts to assert his power and autonomy. His is a hopeless fantasized situation that leads to guilt and resignation. The joys of his power fantasies turn to the horrors of his monster dreams.

The girl during phallic-oedipal years experiences no happier an existence. She, too, is frustrated in many ways and fails every day. She tries to help her mother with household chores, but she is always told that she cannot do sufficiently well. Often her job is said to be taking care of her little brother or sister, whom she deeply resents and is tempted to hurt surreptitiously.When she looks at her younger sibling, she cannot help but feel that she would be a better mother than her own mother. Often she makes invidious comparisons with her mother, noticing how inept the mother is in doing household chores and how lacking the mother is in her tastes and manners. The mother's cooking is hardly gourmet enough for the girl's father. The mother is getting older and older, so much so that no longer will she even reveal her age to her daughter. The girl at this time can hardly imagine why her father continues to show the mother so much attention and, even at times, love. Before the father comes home the girl quickly dresses up for him. She rushes to her closet to pull out a new dress that only that day she was able to buy over the objections of her mother. As the father comes home, she rushes to show him her new dress. His response is enthusiastic. He may even be adoring of her. Too quickly does the girl see that the father's attention is directed, instead, to his wife. He tells his wife about what happened during the day, his various business or professional successes and disappointments. The mother then regales the father with the events of her own day, with the result that the girl is ignored, left only with the prospect of having to help the mother set the table or feed her little brother. Thus the girl at this time feels increasingly annoyed with the mother. She views the mother as inadequate and lacking, wonders whether she too may be deficient in the same manner, and strives all the harder to make an impact upon her father. Nevertheless, regardless of her efforts, her charms, and her beauty, the father's attention to her is interrupted by his involvement with his wife and his preoccupation with his work and with social interests, which take both him and the girl's mother out of the house.

It is readily apparent, both in clinical work and direct observation of children, that in order to reach the phallic phase of psychosexual development and to proceed on to resolution of the oedipal complex, a remarkable degree of maturation[3] and learning[17,22,68] is necessary. In addition, the child must go through many socializing experiences within both the family and the peer group.[82]

In the age span from three to five years physical growth accelerates. The average boy grows about five inches and gains almost ten pounds. The average girl is slightly shorter and lighter. Physical growth is very rapid during the preschool years; in fact, by the fifth year 75 per cent of the weight gain is due to muscular development. However, the central nervous system develops most rapidly during the phallic period. When the child reaches age five, 90 per cent of the nervous system has attained an adult level of maturation.[82] Myelination is now almost completed in the higher brain centers, particularly the cortical and subcortical areas. Even though the genital organs are physically immature at this time, the neural organization involved in erotic excitation and orgasm is developed. Its threshold for stimulation, however, is comparatively high and remains so until androgens are secreted in puberty. Heightened androgen production lowers the threshold for erotic stimuli.[4,72]

Improvement in psychomotor coordination and physical dexterity are encouraging to the phallic child. By the age of four years the child is capable of smoother movements, faster running, and stronger broad jumps. His movements are finer, more synchronized, and less total. Nevertheless, the child still shows awkward movement in various activities. Overhand throwing is still rough and unpredictable in its accuracy. By the time the child is five his sense of balance is markedly increased, but, nevertheless, he is still not capable of hopping, skipping, and jumping smoothly. Fine digital movements, however, are quite good and are appreciated by the child as soon as he enters kindergarten and first grade. Now he is able to pick up pellets quite easily, draw straight lines, copy squares and triangles. As a result, he has many adult motor patterns that allow him to use tools and more complicated toys with some dexterity. He is a better competitor in sports, at least able now to stand at bat and make a valorous try at hitting a ball. With all these new motor skills he tries to play out various roles. He may be the engineer or the truckdriver or the athlete or some other model figure. He seems to enjoy the status of power and control in the child's world.

Language development has gone beyond basic phonemes to simple sounds or morphemes, which develop by the time the child is three. Morphological rules, necessary for the construction of words, and syntactical rules are increasingly grasped by the child when he reaches the age of five. Consequently, the five-year-old child can organize sentences with a good choice of vocabulary. Through the use of words the child is now not as dependent upon fantasy and ludic play. He is now better able to establish connections between thoughts and words. And with his improved communicative capacity he is able to make greater strides toward awareness of meaning in object relations. He can use language to describe, classify, and comprehend better the phenomena that he encounters every day in his world.

By the time he is three he uses words in an undifferentiated and syncretic way. Words stand for objects and events as well as actions and fantasies. For a three-year-old a dog is a class of all animals with four feet. A child syncretically thinks of words such as "eat" as meaning food and being fed, as well as the process of eating. By the age of five, however, words are differentiated into meanings to apply to specific objects and events. Classification of objects and things in common is now possible.

In his object relations the child remains egocentric up to the age of four to five. He cannot put himself empathicly in the place of others. As he passes through the phallic-oedipal period, his object relations become more sociocentric. His speech is more socially oriented. Fantastic and symbolic play is less needed as a child moves during the latency years into sociocentric communications and relationships with his peer group. The child now has less need for imaginary companions. There is less need for dramatization of different roles.

Perceptual capacities increase as the child enters the phallic-oedipal years.[3,20,48,59] A preschool child can better differentiate stimuli in his environment. Stimuli become much more distinct when specific language labels can be applied to them. By the time the child is age five, he can label the component parts of stimuli. He is now able to attend to both the whole and the parts. A five-year-old, however, has difficulty in detecting differences between shapes and mirror images. Only with increasing age can the child regard spatial organization of objects as a relevant dimension. This is due in large part to his learning the labels "right" and "left" as well as "up" and "down."

It is very difficult to separate perception and language from cognition in any study of intellectual development. Intellectual ability is dependent upon the acquisition of language, increasing memory capacity, the differentiation of perceptual experience, and the learning of rules of mathematics and logic and their application.[17,59,68] Between two and four years, according to Piaget, a child can form a representational world.[22] During the preconceptual phase of intelligence the child develops a symbolic representation of the world. He requires labels for the things that he now perceives. From the age of four to seven years he is better able to articulate simple representations. He now can construct more complicated images and more elaborate concepts.

Nevertheless, his understanding of a concept is still based on the perceptual aspect of the stimulus. As an example, at the age of four years the child thinks that beads in a tall cylinder jar are greater in quantity than the same number held in a short squat jar. If a round piece of clay is pressed out onto an elongated cylinder, a child of this age is apt to think that there is more clay in the latter form. From the age of five to seven, however, the child begins to understand that the amount of beads or clay remains constant regardless of changes in the shape of the containers or the clay. Thus the phallic-oedipal child adds thought to his perception and is better able to comprehend his world. This is what Piaget means when he says that the late phallic child develops the capacity of reversibility and conservation.[121]

The manifest content of children's dreams during the early phallic period shows an emphasis on narrations of events with global impressions using concrete simple attributes. Frequently phallic children have difficulty in differentiating between the dream as a private experience and the dream as a shared experience. The phallic-oedipal child is preoccupied with iconic images of monsters and dangerous animals. Death and physical damage are very much present in the dreams. Simple wish fulfillment dreams do occur, but they are far outnumbered by dreams of traumatic experiences.

Play during the phallic-oedipal years shows a great variety of activity. A child tends to play house, puts on adult clothing, and plays the roles of cowboy, doctor, ship captain, jet pilot, and nurse. His play reflects his anxiety about his smallness, with a compensatory fantasy aimed at the denial of the anxiety. The wish is to be big and to do the things that big people do. Dramatic play occurs during this time in many exfoliations. The child's energy, vitality, and inventiveness are most impressive. Oedipal play is replete with high feelings of triumph, invulnerability, invincibility, and happiness. Their open expression in play is one of the delightful aspects of oedipal play. Fantasies tend to be very rich and dramatic, centering primarily on the theme of "twins" and "family romance." Fairy tales embody many aspects of oedipal development, emphasizing the dynamic themes of this phase of development. *Jack and the Beanstalk* and *Cinderella* are two outstanding examples found in all developed cultures throughout the world.

Relationships with peers change as the child enters the phallic-oedipal years. Until age three his peers were relatively unimportant to him. Reciprocal play was minimal. His contact with children stylistically reflected his learning experiences at home. In the nursery school situation, however, the child discovers that many of the responses that his parents accepted and rewarded prove unacceptable elsewhere and, in fact, may even incur for him shame, punishment, and rejection. This awareness becomes increasingly acute and can bring about changes in his social behavior as he moves into the preschool years. The child goes from solitary to parallel play and then to cooperative play and finally to reciprocal play during the phallic-oedipal years. His play activity is vitally important, not just for the discharge of instinctual drive and motor energy, but also for the practice of new skills and the opportunity to try new roles and modes of behavior.

The phallic-oedipal phase of development is not just a psychosexual period but also an important psychosocial developmental epoch with enormous developmental overlay. For a child to end this phase of development and to resolve the Oedipus complex, he must attain a higher level of maturation both in the cognitive and perceptual spheres and in living and socializing within the family. Psychosocially the main emphasis is on awareness of goals, many of which can be gratified only in play and fantasy. Genital arousal and pleasure are obtainable only through fantasy, masturbation, and sexual exploration.

The boy wants to take over his mother. The girl wants to be like the mother, even though she may resent and denigrate her, in order to take over the father. The boy tries to do in his fantasies what his father can do in everyday life. And this provides a fertile soil for guilty feelings that can undercut the boy's sense of initiative. His guilt is much greater than the shame and doubt he experiences at the anal-muscular stage of development. The girl experiences guilt in her desire to usurp the mother's role, but her principal emotional burden is one of self-disappointment, frustration, shame, and inadequacy. She feels more and more that she is defective, and it is only on her mother that she can cast the blame. As she looks about from her position of injury and inferiority, she sees her father and brothers seemingly in a far superior position. They seem to have more fun, to be capable of more activity, and to be the recipients of more rewards. Before a boy and girl enter first grade they must have enough initiative left—after their multiple disappointments, shame, and guilt—so they can optimistically start school and commence the stage of industry in the latency years.

For each child the oedipal experience is unique, determined by earlier ego growth and experience, the instinctual and ego development of the phallic period, and interaction with his parents and family; death and illness, the birth of new siblings, and separation experiences are further influences on phallic-oedipal development.

Phallic Erotism

During the phallic-oedipal phase of development[28,29,30,31,33,34,35,37,38,40] libidinal impulses originate in the zone of the penis and clitoris. The vagina at this phase of development has relatively little psychic significance for the little girl. The phallic child now enters a second phase of infantile masturbation. In early infancy there is evidence of phallic excitation and masturbation. In the second phase of infantile masturbation the excitation is to the objects of the Oedipus complex, with the specific aim of penetration and procreation. The erotism of the phallic-oedipal period is clearly different from genital erotism, which appears developmentally only with puberty and its hormonal changes.[30,33] Thus phallic impulses during the phallic-oedipal years are impulses to penetrate but not to discharge semen.

Freud viewed the phallic-oedipal phase of development as one stage in the sequence of infantile libidinal development. He viewed the sexuality of the girl as primarily masculine in character, pointing out that the erotogenic zone of the female at this time is located in the clitoris, which is homologous to the masculine genital zone of the glans penis. For both sexes at this developmental phase only the male genital has prominence and primacy.

Associated with phallic primacy is the impulse to knock to pieces, to hit, to press in, to tear open, and to make a bull's-eye. If the child is made sexually knowledgeable by his parents, he is still puzzled in his attempt to understand phallic penetration in intercourse. As a consequence, he falls back on other ideas of sexual contact and pregnancy, ideas that come from his own experience and feelings. His wishes lead him to want to penetrate through the mouth, anus, navel and in some indefinable way to create a baby. The aim of phallic erotism is thus to penetrate and to procreate. The passive aim is to be penetrated and to bear a baby. Thus the passive wishes of the boy make up the negative Oedipus complex, in which the wish to be castrated becomes a necessary condition of being penetrated and is usually fantasied as being anally penetrated. The child's wish to have the phallic zone stimulated by another person is also regarded as a passive phallic aim.

The child attains sexual knowledge through his perceptions and fantasies.[30,33,34,38,39] At first the child believes there is only one sexual organ—the penis. But later he must face the perception that the female has no penis. This perception is not acceptable so he disavows it. Later, because of his increasing reality sense, he cannot maintain the belief that there is no perceptual difference between the sexes. Consequently he must elaborate a series of fantasies to account for it. Now the disavowal becomes more sophisticated.[25] ("She doesn't have a penis now, but it will grow back.") The child may go on to develop neurotic avoidance of the female genital or femininity in general. Normally these upsetting fantasies are repressed at this stage and the child resolves his Oedipus complex. The pervert, however, has no solution to the Oedipus complex. He is stopped between the disavowal of perception and denial through fantasy. Freud states in his paper on splitting of the ego[39] that the pervert is faced with the fact that his mother has no penis. To fill the gap he creates a fetish or a phobia.

Male Phallic Psychosexual Development

Early in the phallic stage the boy becomes aware of anatomical gender differences,[15] and his awareness of the male genitalia furthers his identification with his father, brothers, and male peers. Now he thinks that he can do what his father does. In addition, his view of the father's relationship with the mother is heightened, particularly in his fantasies of the father's lovemaking with the mother. Now the boy wants to exhibit his phallus to his mother and at the same time look at her breasts and genitals. He may request that she go to the bathroom with him, or he may blatantly exhibit himself at bedtime to his mother. He views his penis as a penetrating instrument and as a source of great pleasure—earlier, as one of several pleasurable body areas and, later, as a special pleasure structure of the body that he has become aware of through masturbation, exhibition, and sex play.

The boy views himself as a little man; his father is his adult rival. The mother becomes the object of all his pleasures, around whom he weaves a not too innocent "romance." The boy at this time tends to be very protective of his mother, imitating his father in many respects. On the other hand, he may downgrade his father or even at times openly attack him verbally. He may try to play games involving the father's role. He is apt to make up stories and fantasies about larger rivals whom he can easily overcome, like *Jack and the Beanstalk.* The boy realizes, however, that the father's size, dominance, and strength make the competition futile. So threatening is the father that he assumes for the little boy an aura of dread and expectant injury. This threat assumes the form of castration anxiety, which ushers in a new developmental theme. This threat can exist only in the child's fantasies and has no real roots in reality. Often enough, however, the boy has sensed the father's annoyance because of his persistent intrusions into the parents' relationship. Then, too, any punishments the father dispenses to his son during this period, for omissions and commissions irrelevant to the boy's phallic wishes, are seen as direct punishments for his phallic impulses. The conflicts centered around castration eventually lead to the repudiation of the boy's oedipal wishes. The suppression and repression shift the ego's defense organization. Evidence for this developmental pattern is aptly found in the everyday play of children, and as well as in the fantasies, fairy tales, and dreams of this period of development. Furthermore, anthropological studies of totemic animals, children's graphic productions, data from child analysis, and psychoanalytic reconstructive data are also supportive of these developmental facts.

When the mother indicates to the boy her unwillingness to respond to his wishes, he can react to her in a jealous rage, which either gives rise to or reinforces a wish to kill her and to be loved by the father in her place. The negative Oedipus complex also leads to fear of injury and castration because of passive aims. Both negative and positive wishes in the oedipal period, therefore, arouse castration anxiety.

The boy is quite prone to respond to the oedipal situation with fear of castration. There are many determinants to this fear. The erogenous phallus is an object of attention, pleasure, and fantasy. The boy's concern with uncontrollable phallic tumescence gives rise to his fears that his fantasies will be exposed to his parents. When he discovers the female genitalia, he concludes that the girl has lost her penis. Direct and indirect threats of punishment from either the father or the mother reinforce the fantasy that he, too, may lose his penis. In addition, his earlier experience of losing the breast in weaning and in losing feces in defecation are preoedipal antecedents to body loss and castration anxiety. The fantasy of castration involves loss of the penis and not necessarily the testicles. The fantasy is based on his view of loss of body part going back to his previous stage of development, where loss of the fecal "stick" suggested a body loss. The boy's castration anxiety is further manifested by overt concern with the body, excessive fear of injury, phobias, and nightmares. The boy must resolve his oedipal conflict by renouncing his wishes and by holding them in check through superego controls, available defensive mechanisms, and ego techniques for resolution of conflict. The boy relinquishes the libidinal object and replaces it by identification with the mother. There is also an intensification of the identification with the father. Identification with the father strengthens the masculinity of the boy's character. The boy's libidinal wishes are partly desexualized and sublimated and partly exhibited and transformed into impulses of affection.

Female Phallic Psychosexual Development

The girl's psychosexual development is in several respects more complicated and less well understood than the boy's.[5,58] Freud initially assumed that sexual development in boys and girls was similar, a view that he changed in his paper, "A Child Is Being Beaten."[29] He later realized that the girl has to change not only her sexual object from the mother to the father but also her erotogenic zone from the clitoris to the vagina. Psychosexually the girl is much like the boy, starting out with the mother as the main libidinal object. Her earliest erotic fantasies, like the boy's, involve the mother as object. Between the ages of two and three years the girl begins showing a preference for the father. Much like the boy, the girl increasingly masturbates, using the clitoris instead of the penis. It is here that a gender difference in psychosexual development occurs. Her desire to play the man does not flounder on castration fear. It is castration fear that propels the girl toward the Oedipus complex. She blames her mother for the genital difference and seeks out her father in the hope of repairing her body damage and becoming a man. Even though she may become aware in the phallic-oedipal period of her vagina, the clitoris remains the erogenous zone. Later in development, especially during adolescent years, she starts to switch from the clitoris to the vagina as a primary erogenous zone.

Decisive awareness of gender difference occurs when the girl shifts her libidinous strivings from her first object, the mother, to the father in the hope of achieving an adult heterosexual status. Turning from the mother to the father can be abrupt or quite gradual and can involve antagonism and hostility toward the mother. Often the girl catalogues her many complaints against her mother. The shift from the mother to the father occurs at the time when the girl has become aware of anatomical differences between herself and the male. Her first impulse is to repudiate the sexual difference. The fantasy of having been deprived of a penis by the mother furthers her antagonism toward the mother and increases her fear of more retaliation from the mother. The wish for the penis heightens her interest in her father and brothers. Penis envy expressed as shame, inferiority, jealousy, and rage is most evident at this time. The girl has a passive wish for the father's penis. She then turns to the father as her principal love object. She is soon rebuffed by the father and forced to renounce her oedipal wishes. The girl's feelings of castration—the female castration analogues—are jealousy and a sense of mortification as well as fear of genital injury, partly a consequence of the wish to be penetrated by the father.

The boy discovers his penis and the girl is quickly made aware of her deficiency. Both boys and girls during the phallic-oedipal years evidence marked narcissistic investment in their bodies. The phallic child displays body curiosity. He explores his body, checking every orifice. The body narcissism is manifested not only by preoccupation with body function but also by dread of injury. The slightest injury is regarded by the child as worthy of a bandaid.

The discovery of sex differences means to the boy that the girl has lost her penis.[15,51] He is frightened that he, too, may suffer the same fate. He resorts to ego-defensive maneuvers. He may believe that everybody is built like himself. He can try to convince himself that the girl has a penis that is hidden inside her vulva. Or he can believe that the girl's penis will grow back some day. He can openly exhibit himself as a way of proving that his penis is still there. He can play games in which he hides his penis between his legs and then allows it to pop forth. He can displace to other parts of his body his concern over castration. He can deprecate girls, even trying to hurt them, to further his image of them as injured. His anxiety can lead to increased masturbation aimed at reassuring himself that his penis is still intact. He can threaten other children with castration. He may do this by intimidating children or by actually trying to hurt them.

The girl, on the other hand, can deny the fact that she has no penis. She may feel that the mother took it away from her. She can repress her awareness of anatomical differences until she reaches adolescence. Often phallic girls, even from sophisticated backgrounds, can deny the vagina well into latency years. The phallic child's view of the parents' external genitalia can further intensify castration fears. The boy, in seeing his father's penis, is apt to feel even more inferior. The girl is frightened in seeing her father's penis. She can fear that her wish to touch his penis may come true. Or she can be alarmed by oral fantasies that aim to incorporate the father's penis. The phallic boy, in seeing his mother nude, may fantasize that her penis is inside her pubic hair.

Primal scene experiences are exciting, stimulating, and often very frightening to the phallic child. All children experience at least auditory primal scenes. Hearing the parents moving about in bed, awareness of the mother's giggles, the father's heavy breathing, and various noises at the time of orgasm are most stimulating to the child's fantasies. The child does not know who is doing what to whom. A number of children also see their parents having intercourse. Visual primal scenes during the toddler years make the child even more vulnerable to primal scenes during the phallic-oedipal period. Whether the child's primal scene is visual or auditory, the child tends to view parental intercourse as an act of aggression or a fact of exclusion.

The phallic child's awareness of the mother's pregnancy is another source of fantasies during this stage of development. The pregnancy stirs up the child's sibling rivalry feelings and forces the child to elaborate his sexual theories of intercourse, pregnancy, and birth.

Preoedipal Sexual Identification with Love Objects and the Negative Oedipus Complex

Both the male and female child have initially an undifferentiated primary identification with the mother.[7,11,12,53,57,62,67,102,107,108,122] Later the child internalizes parts of both parents.[49,98,99] At around two to three years the girl imitates the mother and carries out a flirtation with the father. At this time most of the little girl's investment is with the mother. And in this sense the little girl is experiencing her negative oedipal complex. With the first separation and individuation the little girl then proceeds to the positive oedipal complex in which she reaches out more affirmatively toward the father. She aggressively devalues the mother. For a girl the negative oedipal complex involves transfer of her aggression to the father. She attaches her libidinal drives to the oedipal mother. For the male at age two and three and a half, the identification shifts from the mother to the father. The boy in his negative oedipal feelings remains aggressively attached to the mother and regards her as engulfing, overprotective, and frequently nongiving. If he is able to resolve these feelings, he then makes a switch in his allegiances to the mother and transfers his aggression to the father. The mother is now valued and the father is seen as a competitor.

If the boy's development is to proceed in a phase-specific manner, his affectionate attachment to his father must intensify. He remains aggressively attached to the mother, with whom he identifies. The boy is then in a passive stage in which he subordinates his active wishes for his mother. During the time of the negative oedipal complex there are two possibilities. The boy can have sexual wishes for the father, wanting to put himself in the father's place and to play a passive role. Or the boy is apt to want to supplant the mother and to be loved by the father.

The Oedipus Complex

The Oedipus complex is regarded by Freud as the "cornerstone" of psychoanalysis. It is the central part of his psychosexual theory of development and is in many respects the phenomenological core of his theories of infantile sexuality. Rutter's[95] critique of infantile sexuality, for instance, reviews the empirical data that support many of Freud's phallic-oedipal formulations.

The developmental trends that lead to the Oedipus complex involve the child's obligatory dependency on his parents[21] and the child's need to be loved by and to love his parents—the first love objects. The Oedipus complex is a triadic developmental family phenomenon, involving the child's sexual strivings that bring him face to face with his erotic feelings and phallic wishes toward the parent of the opposite sex. The desire for both affection and stimulation from the contrasexual parent pushes the child into a competitive relationship with the isosexual parent. For the boy this results in fear of castration. For the girl this intensifies her antagonism to the mother, her wish to have a romance with her father, be like her father, and be part of the male world. The Oedipus complex is an apprenticeship for heterosexuality and a necessary developmental stage for male and female psychosexual identity. The boy's Oedipus complex develops out of the phase of phallic sexuality. Under the influence of castration anxiety he is compelled to abandon the Oedipal feelings at about the age of five or six, thereby repressing and sublimating his incestuous wishes and further identifying with his father. Castration anxiety, therefore, terminates the oedipal complex in boys. The little girl, however, must shift from the first object, the mother, to the father. This development occurs when she notices genital anatomical differences. She feels antagonistic and hostile toward her mother. Her first impulse is to repudiate the genital difference by attempting a masculine identification with the father. She masturbates more, with fantasies of having a penis. Because of guilt, however, she may give up masturbation for a while; or, on the other hand, her wish for possession of the father's penis may be strengthened. Rage and despair intensify the little girl's rivalry with the mother. She further turns to the father as her principal love object. She is then rebuffed in one way or another by her father, so that she must now renunciate and repress these oedipal wishes the best she can. She may attempt to remain a "pal" of the father, wanting to do things with him, particularly if she is competitive with her brother. In passing through the "tomboy" stage, she furthers her masculine identification with the father. The shift back to the mother only occurs during adolescence. Frequently, however, the shift is not completely made until the girl has married and has had a child of her own. Since no castration threat can make the girl relinquish her father, the girl's oedipal conflict is not resolved until adolescence or adulthood.

Freud viewed the infantile neurosis as a universal phenomenon in man's progression through the phallic-oedipal phase—apparent for a short time and manifest as a childhood neurotic disturbance, or not apparent at all, sometimes later appearing as neurotic symptoms in adulthood. Freud related the infantile neurosis to the Oedipus complex. Nagera[83] has reviewed the concept of infantile neurosis from the developmental standpoint. His view is that "the infantile neurosis is . . . an attempt to organize all the previous and perhaps manifold neurotic conflicts and developmental shortcomings with all the conflicts typical of the Phallic-Oedipal phase, into a single organization . . . into a single unit of the highest economic significance. . . . For these reasons—the 'Phallic-Oedipal' phase is in fact an essential turning point in human development."[83] The infantile neurosis is the first form of neurosis and can lead, with accretions from latency and adolescent years, to adult neurosis. Anna Freud[26] also emphasizes the necessity for certain developmental steps to have been completed before the infantile neurosis occurs. Early deprivations, absent objects, improper environmental handling, various forms of deprivations, and constitutional deficiencies do not allow for adequate identifications, complete internalizations, and normal structuralization of the personality. Thus the child remains in a preoedipal level of personality organization.

Neo-Freudian Formulations of the Oedipus Complex

Since Freud's formulation of the Oedipus complex, there have been a number of other postulations that take issue with Freud's view, particularly in regard to infantile sexuality.

Adler regarded the oedipal child as a pampered child.[81] He saw the normal attitude of the phallic child as that of an equal interest in the father and mother. While he agreed to the possibility of a boy's overstimulation by the mother, he regarded sexual pleasure as incidental to the quest for power. He saw sexual pleasure as incidental in the boy's quest for power and domination over the mother. For Adler the Oedipus complex is simply one of the many forms of child pampering, and in many respects the child eventually becomes the victim of his own fantasies toward the mother.

Jung saw the Oedipus complex as mainly involving issues of independence and autonomy.[81] He felt that if there is no freedom for the child, the Oedipus complex must inevitably lead to conflict. He introduced the term "Electra complex" to denote the girl's conflict between the fantasized infantile-erotic relationship with the father and her will to power. Failure to achieve more autonomy and the need to return to the relative security of the father are key factors in Jung's view of the Electra complex. Freud rejected Jung's Electra complex because it was meant to "emphasize the analogy between the attitudes of the two sexes." Furthermore, Freud felt that Jung's view added nothing.

Another neo-Freudian formulation of the Oedipus complex was made by Rank, who viewed the complex as the origin and destiny of man.[81] The average person, Rank believed, never really overcame his birth trauma. From the primal horde came the primal family, the group family, and eventually the matriarchal society. The Oedipus complex, for Rank, became a saga, a sociological phenomenon, a compromise between the wish to have no children and the necessity to renounce one's own immortality in favor of children. The Oedipus complex is a reaction to coercion by the human species, which requires marriage and fatherhood against the individual's will.

Horney attacked Freud's biological view of the Oedipus complex.[81] For her the family relationship is a force in the molding of the matrix of character formation. Anxiety comes from conflict between dependence on the parents and hostile impulses toward them, as well as the need to cling to parents for love and security. She doubted whether sexual wishes toward the parents had any relationship to the child's development.

The rebellion of the son against patriarchal authority is central to Fromm's concept of the Oedipus complex.[81] He does agree with Freud that there are sexual strivings in children, that the tie to the contrasexual parent is frequently not severed, and that the father-son conflict is characteristic of patriarchal society. But he seriously questioned whether the Oedipus complex is a result of sexual rivalry. Furthermore, Fromm questioned the universality of the Oedipus complex. He pointed to cultures in which there is no rivalry and no patriarchal authority. He, too, doubted whether the tie to the mother is primarily sexual; instead, the fixation to the mother is caused by maternal dominance.

Harry Stack Sullivan also took issue with Freud on the Oedipus complex.[81] He suggested that feelings of familiarity toward the child on the part of the isosexual parent foster an authoritarian attitude that produces resentment and hostility in the child. A parent treats the child of the opposite sex with more consideration because of a sense of the child's strangeness. A parent feels justified in dictating to a child who seems to be like himself. The feelings of strangeness deprive the parent of motives for control of the child's life. Consequently the child is treated more carefully. The freedom from pressure results in the child's feeling greater affection for the contrasexual parent and being more attracted to this parent.

Another formulation of the Oedipus complex comes from the adaptational view, which rejects infantile sexuality and the libido theory. This view was most strongly espoused by Rado.[81] He suggested an ontogenesis of family relations. In his view parents provide the child with conditions that foster the development of omnipotence. A child delegates these feelings to the parents during the anal-muscular phase of development. A rewarding relationship between the child and the parent generates tender affects and parental idealization. Thus the child's object choice is "learned"; it is not a reflection of inborn instinctual drives and their vicissitudes. The child's first sexual pleasure is from autoerotic manipulations. A subsequent shift to another object comes only after the child has established an affectionate tie to the mother. Now he has learned that he must share the mother with the father and siblings. This initially intensifies his need for exclusive possession of the mother as a dependency object. Consequently both boys and girls in the family are interested in the mother primarily as the affectionate object. In this sense there is no essential difference in sexual development between the sexes. The striking gender difference is in the attitude of the little girl toward her genital equipment. She attributes cultural privileges of greater masculine freedom in play and assertiveness to the fact that the boy possesses a penis. This is the root of the girl's penis envy, a repressed wish to castrate the boy as a means of resolving her dependency needs.

According to the adaptational view,[55] the boy's clinging to his mother results in overt disapproval from both parents, which then leads to an awareness of the incest taboo.[85] The child's response to parental intimidation is hostility to the father, which must be repressed because of fear of castration. The child must further protect himself by repressing libidinal desires for the mother. As the mother becomes more inaccessible he retreats autoerotically to his own genitals or indulges in sex play with other children. If these activities are not interfered with, the Oedipus complex is resolved initially by substitution of himself through masturbation and later by substitution of nonincestuous objects for his mother. In a society that is sexually inhibited the likelihood of parental intimidation exists. As a consequence all sexual gratification for the child must be given up. Thus the child falls back on the mother again as a sexual object since the inhibition of his independent executive action forces him to return to earlier dependency attachments. The Oedipus complex is thus perpetuated from one generation to another. Heterosexual objects are identified by the boy with the forbidden mother, which heightens his castration anxiety.

In the adaptational view—and this is true also in the classical view of the Oedipus complex—the girl's sexual development is much more complicated and in many respects not as well understood. For the girl the first sexual object is her mother. Later she turns to her father, not, according to the adaptational view, because of sexual differences, but because of subtle persuasion by both the mother and the father. The mother dampens the daughter's sexual interest in herself. She promotes the shift in the little girl's attention to the father through the many examples of her own behavior to the father. As a consequence she helps the girl identify with her, especially if the father sets up no objections. Through his own playful, affectionate attitude toward his daughter he facilitates her identification with the mother. The shift of the daughter to the father is reinforced by the social institutions of our culture, which emphasize attachment to a man as the highest goal for a woman. A contradiction soon arises for the girl. The parents become alarmed at the dramatic shift they have encouraged. Alarm occurs because of sexual overtones in the heightened relationship between the father and daughter, which then leads to sexual intimidation of the girl in much the same manner that this occurs for the boy. The little girl is now aware of genital differences. She unconsciously attributes her lack of a penis to castration by the mother because of her oedipal strivings toward the father.

In the adaptational formulation castration anxiety for the boy and the girl occurs after parental intimidation. It is not in any respect related to discovery of genital differences. The sexual development of the girl then proceeds in the same manner as that of the boy. This view challenges the view that castration anxiety ends the Oedipus complex in boys but initiates it in girls. In the adaptational view the relationship of castration anxiety is the same for both sexes. It sets in motion repression of the Oedipus complex. In addition, sexual inhibition for the woman is further enhanced by social institutions that limit the woman's freedom in her female role. The woman must present herself as virtuous, chaste, and dimly interested in intercourse. This view facilitates her repression of her sexuality. Her penis envy is exaggerated by the various contrapuntal cultural attitudes between men and women. Intercourse is viewed as an instrument in the battle for dominance-submission. Because of her position vis-à-vis man and society, she is vulnerable to injury; she can only assuage injury fantasies by reparative fantasies aimed at taking over the man's penis. Her sense of male hostility and retaliation is augmented by these reparative fantasies. In addition, penetration by the penis arouses repressed oedipal strivings and connected feelings of guilt and punishment by the mother, thereby adding another increment of retaliatory fear for the woman. Consequently the woman withdraws from men and is forced to deny her sexual yearnings for pleasure and comfort.

The Resolution of the Oedipus Complex

In the dissolution of the Oedipus complex there occur changes in the aims and objects of the instinctual drives. Progressive modification of the relationship with past objects (parents and siblings) is seen. Anaclitic and erotic aspects of objects are dissociated. Development of new object relations, free of the incest taboo, takes place.[78] Further sexual differentiation occurs. Object cathexis is given up and is replaced by identification with the authority of the father introjected into the ego as the authority of the superego, perpetuating the incest taboo and preventing the ego from returning to libidinal object cathexis. The libidinal components of the Oedipus complex are desexualized and sublimated; in part they are inhibited in their aim and transmuted into impulses of affection. Freud saw the process as more than repression—"destruction and an abolition of the complex."[30] "In boys . . . the complex is not simply repressed, it is literally smashed to pieces by the shock of the threatened castration. Its libidinal cathexis are abundant, desexualized and in part sublimated; its objects are incorporated into the ego, where they form the nucleus of the superego. . . . In normal, or, it is better to say, in ideal cases, the Oedipus complex exists no longer, even in the unconscious; the superego has become its heir."[30] In the *Ego and the Id* Freud, in discussing the loss of the love object resulting in a "setting up of the object in the ego," further comments about the dissolution of the Oedipus complex: "The broad general outcome of the sexual phase dominated by the Oedipus complex may, therefore, be taken to be the forming of a precipitate in the ego, consisting of these two identifications in some way united with each other. This modification . . . confronts the other contents of the ego as the ego-ideal or superego."[36]

The destruction of the child's phallic-genital organization results from the threat of castration and the experience of painful disappointments. The castration threat is in large part the result of the child's use of the mechanism of projection, but it is further strengthened by overt or covert punishment for masturbation and sex play and the boy's observation of sexual differences. When the child abandons the Oedipus complex, his identification with the father maintains the object relationship to the mother that has been linked to the positive Oedipus complex and at the same time replaces the object relationship to the father that has been part of the negative Oedipus complex. The same intrapsychic events take place in connection with the mother identification. Freud feels that the relative intensity of the two identifications is related to the preponderance of one or other of the two sexual dispositions. With the abandonment of sexual aims and the change of object cathexis into identifications, there results a sublimation that permits diffusion of the libidinal and aggressive components of the object cathexis. This freed aggression helps in the formation of the superego.

Because of the gender differences in the resolution of the Oedipus complex, Freud thought that the superego formation was quite different for the girl. Fear of castration plays no part in the breakup of the infantile genital organization for the girl. For the girl threats of loss of love threaten the Oedipus complex, but what eventually abolishes the Oedipus complex is her increasing awareness through later development that her wish to have a child from the father can never be fulfilled. As a consequence the woman's "superego is never so inexorable, so impersonal, so independent of its emotional origins as we require it to be in men."[30]

Jacobson[49] has analyzed the development of the superego from its earliest precursors to the superego consolidation at the end of adolescence. She has delineated the various types of self and object representations as they take part in the formation of the superego, as well as the role of superego introjection and ego identifications and their relationship to development and experience. Sandler[99] has developed these concepts further in his concept of the "representational world." The ego establishes the representational world, the center of which is the self-representation, an amalgam of integrated self-images, surrounded by object representations formed out of a synthesis of object images. Before the phallic period the superego schema reflects the idealized and desirable qualities of the parents and encourages the child to behave, which gains for the child a feeling of being loved. At this point the superego is a preautonomous schema.[98] It is only with the resolution of the Oedipus complex that introjection of the parents occurs and an autonomous structure is formed. Now the introject can substitute for the real object as a source of narcissistic gratification.

With the resolution of the Oedipus complex the child enters the phase of latency, which phenomenologically commences with the shedding of the first deciduous teeth and is dynamically associated with the resolution of the "family romance" and the diminution in the child's narcissistic preoccupation with his body, its functions and its orifices. Latency extends to the prepubertal growth spate of adolescence. During this time further sexual identification and ego differentiation are perceived. Peer interaction is much more common as the child spends less time with his family. In his relationship with other authoritarian figures, as well as in his relationship with peers, there is a strengthening of superego and ego ideals.

The triadic transaction in the Oedipus complex is not limited to the phallic-oedipal period of development. In adolescence there is a resurgence of many aspects of the Oedipus complex as the adolescent attempts to rework previous developmental issues. And in adulthood oftentimes prior to a son's marriage there is a resurgence of oedipal feelings on the part of the father vis-à-vis his future daughter-in-law. The same, of course, can occur for the mother in her feelings toward her new son-in-law. Later in life grandparents can experience a return of their rivalrous feelings toward a son-in-law or daughter-in-law after the birth of a grandchild; and toward the end of life, if aging has taken its toll on cognitive control and capacity for instinctual delay, a man may give vent to long repressed oedipal urges through molestation of young girls.

There are many reasons for unsuccessful resolution of the Oedipus complex.[18,21,70,71,76,86,93,103,106,107] If a child has had an unsustaining relationship with two parents or has had multiple deprivations during childhood, these experiences will result in a relative inability to relate to parental objects, with impairment in psychosexual development and sexual identity. As a result fragile control of sexual and aggressive impulses, along with little capacity for pleasure, can ensue.[56] Harlow's primate studies have demonstrated that female infants who are deprived of their mothers grow up to be inadequate mothers.[82] Males deprived of their mothers as infants have difficulty in expressing aggression and sexuality. If rather extreme frustration or extreme overindulgence at the preschool level has occurred, oedipal involvement may be impeded. If the contrasexual parent is removed at the beginning of the child's phallic development and remains absent, the child's contrasexual object relation will be strained. He can become overly attached to the parent of the same sex, making it more difficult for him to work out a heterosexual object choice. Many times, however, relatives, friends, and older siblings offer other identificatory models. The oedipal outcome also depends on whether the remaining parent takes the child as a symbolic substitute for the lost spouse.

The contrasexual parent can behave in an unduly seductive way toward the child, leading to overstimulation of the child, with intensification of any unresolved Oedipal feelings. Death, illness, injury, or desertion of the parent of the same sex during the oedipal period can foster conflicts since these events coincide with unconscious or thinly disguised aggressive wishes of the child toward the isosexual parent. Because of magical thinking, whereby a wish or fantasy is viewed as tantamount to an act, death or illness of a parent can be viewed by the child as a result of his aggression.

Ideally for the resolution of the Oedipus complex the isosexual parent should be nonpunitive, attractive, and strong enough to serve as a model for identification. Second, the contrasexual parent should not be unduly punishing, emotionally unpredictable, or seductive so that the child can place full confidence in objects of the opposite sex. Third, parents should show no indication or rejection of the child's genetic sex. Gender-role behavior should not be obscured by the parents in terms of teaching cross-sex-role responses. Fourth, the child should identify with a satisfactory view of marriage, seeing it as providing pleasure, comfort, and security so that he may subsequently view marriage as desirable and likely for himself.

Fathers who are harsh and punitive make oedipal resolution more difficult. Struggling with such a father, a boy may feel it necessary to give up all women and to assume a passive, compliant attitude toward his father. He may aspire to be the main object in his father's life and see his mother as a rival. Exaggerated character traits like passivity, compliance, ingratiation, and timidity can result. On the other hand, the boy may overtly identify with the father and develop very aggressive pseudomasculine traits, by way of identification with the aggressor. With such a father the girl can assume a masochistic attitude toward men and regard the penis as a sadistic weapon to be passively accepted. In this situation frigidity is common. If the mother is aggressive and a relatively masculine person and the father is passive and feminine in character, the boy must deal with his aggressive, punitive mother.

His inadequately masculine father cannot help him and cannot provide him with a satisfactory identificatory model. Thus the boy must follow in the father's footsteps in the passive, feminine, masochistic way.

A girl may feel that she was deprived by the mother during her preoedipal development. She may think that her feminine attitude has been depreciated by the mother. Thus she can adopt a rather passive attitude toward the mother, maintaining the mother as an object, depreciating the father as weak and withholding, and entering into a homosexual relationship later in adolescence. On the other hand, she can assume a maternally protective attitude toward the father and establish strong bonds with him that she is never able fully to relinquish. Sometimes older siblings of the opposite sex provide substitutes for the parents, and the oedipal problems may therefore be displaced onto these siblings. Sibling competition with a much older brother can be intensified by oedipal displacements.

Infantile Sexuality and Behavioral Science Research

The existence of infantile sexuality is thought to be substantiated on the basis of direct observation of children, sexual foreplay, sexual perversions, psychosis, the phenomena of regression, data of child analysis, and psychoanalytic reconstruction of infantile sexual life.[5,14] Freud has been challenged by many neo-Freudians who view infantile sexuality as of no importance compared to the effects of environmental factors or regard infantile sexuality as a result of pathological family interaction. Regardless of theoretical orientation there is unanimous agreement that further study of infantile sexuality, particularly phallic erotism and the Oedipus complex, must be done. With the marked emphasis, however, on methodology and electronic instrumentation for the study of infancy, the phallic-oedipal period and infantile sexuality in general have been relatively neglected. Children's fantasies, dreams, wishes, and other mental events have been comparatively ignored by research workers in child development. Castration fears, primal scene experiences, penis envy, birth fantasies, and oedipal and masturbatory fantasies are phallic phenomena that warrant research. In addition, fairy tales and myths provide rich sources of research material. The nature of infantile amnesia is poorly understood, partly because of our ignorance of memory mechanisms in general. Further study of the developmental nuances of the female Oedipus complex and its resolution is needed.[3,5,58] Many of the postulates of infantile sexuality, however, are not easily tested by empirical methods. Controlled research in this area is extremely difficult. Preschool children cannot easily report experiences. The use of play and direct observation can produce very misleading and at times naïve results. Play techniques are useful as modes of expression. Translations from play to language are necessary, but frequently distortion results.[69]

Child analytic work offers the most convincing evidence for infantile sexuality. Since the child analyst has the opportunity to observe the child over a period of time, he is able to see the multiple variations and transformations in instinctual drive and ego development that occur throughout development. The evidence from child analysis, however, is not always acceptable if statistical requirements are made.

Neurophysiological speculation attempting to relate orality and sexuality by way of contiguous neural limbic pathways does not at this time supply a physiological support to psychoanalytic concepts.[4,72] The connection of thumb-sucking with masturbation has been questioned by Wolfe, who views rhythmic infant activities as motor discharges rather than as autoerotic movements.[16] And the relevance of penile erections during sleep to infantile sexuality is no longer assumed by sleep researchers. It has been shown that penile erections occur principally during a period of rapid eye movement (REM) and may be merely manifestations of altered metabolic states.

The most useful behavioral science research has been applied to questions of gender role and gender identity. Hampson and Money have identified the variables of sex differentiation.[79,80] Genetic sex, hormonal sex, gonadal sex, and the morphology of both internal and external reproductive organs are the important variables in sex determination. Studies of pseudohermaphrodites have indicated that psychosexual differentiation takes place as an active process of editing and assimilating experiences that are gender-specific and is reinforced by the individual's genital appearance. The gender-role assignment and early gender learning are the most critical factors in psychosexual differentiation.[6,13] The sex of assignment, according to Money's research, must be clearly made by the eighteenth month of life in order to avoid psychosexual deviation in later years. The preoedipal process of sex-role determination is thought to be related to some type of early learning. Imprinting, a model taken from animal studies of learning, has been suggested as the possible cognitive activity that occurs at the time of the toddler stage.[80] It appears that the acquisition of gender role and sexual identity is quite similar to language in that for both there is a critical period of growth. Stoller's[111,112,113] research has corroborated much of Money's early work. He points out that gender identity depends in the beginning upon sex assignment and that it is established by the age of two and a half to three years. Gender identity comes from experiential rather than constitutional factors and occurs well before the appearance and resolution of the Oedipus complex. Stoller views this process as leading to the development of core gender identity and feels that biological forces silently augment the formation of identity. Therefore, gender identity is not entirely the result of fixation by conflict, but rather the effect of imprinting upon a passively developing identity at a critical period. Thus castration anxiety, penis envy, and the like change only the quality of gender identity.

Cognitive development plays an important role in sex-role behavior. The child cognitively organizes his social world along role dimensions. This patterning is based on the child's concept of physical things. The social world concept is cognitively organized in terms of universal physical dimensions. The organization of role perception and role learning depends on the child's concepts of his body and his role. The schemata that cement events together include his concepts of the body, of the physical and social world, and of general categories of relationships, including substantiality, causality, space, time, quantity, logical identity, and inclusion. The child develops an unchangeable sexual identity that has its analogues in his concepts of the invariable identity of physical objects. From the patterning arises a self-categorization as boy or girl that is a basic organizer of sex-role attitude.[63]

There is a paucity of developmental studies on memory in children.[101,120] These are needed to further our understanding of infantile amnesia. Repression at the phallic-oedipal stage is a countercathexis that impedes release of memory traces into consciousness.[28,30,33,37] But, in addition, memory in childhood is an immature and developing ego function. Memory is not only a function of previous experience and storage of information but also the capacity to organize information and reason. A child classifies experiential data differently from an adult. In all likelihood changes in cognitive ego style in the oedipal phase facilitate repression and allow for amnesia.

Central to the core of the classical formulation of the Oedipus complex is the fear of incest.[81,85,104] There are recent family studies of incest but only a meager number of developmental studies of children involved in incest.[91,119] From both a statistical and clinical standpoint the frequency of incest between mother and son, father and daughter, and between siblings is not known. Incest studies of father-daughter dyads, for instance, are not illuminating.

Freud's concept of the clitoral-vaginal transfer theory has been challenged by Masters' and Johnson's research on the female orgasm.[16,96] They suggest that the female sexual drive is based on clitoral eroticism and that it is impossible to distinguish the vaginal from the clitoral orgasm. No such thing as a vaginal orgasm exists; it is an orgasm of the circumvaginal venous chambers. In the latent phase there is increasing pelvic congestion with transudation and an increase in sexual desire. Psychosexual excitation increases the vasocongestive reaction during any phase of the menstrual cycle.

In support of their concept of the female orgasm, Masters and Johnson point out that genetic sex at fertilization is potent during the first six weeks through a number of forces. Initially the embryo is female with sexual bipotentiality, which changes to a unipotentiality as the testicular inductance substance forms and leads to the production of androgen between the seventh and twelfth weeks. In the male the "clitoris" becomes the penis. From the crura and bulb emerge the male external genitalia. Thus there is in the very beginning of life a constitutional bisexuality and by the phallic-oedipal phase a bisexuality in terms of identifications and introjects.

Freud's theories of infantile sexuality involve many levels of conceptualization. Fear of bodily injury, preoccupation with bodily functions and integrity, masturbation, behavioral responses to primal scene experience, and transactions of the child vis-à-vis isosexual and contrasexual parents are phenomenological. In his review of normal psychosexual development Rutter[95] points to a corpus of empirical data that child analysts encounter daily in their work. At another level of observation are dreams, fantasies, and memories, reportable by children through verbal and nonverbal means. At another level of conceptualization are postulates of the libido theory[21] and the theory of object relations.[49,57,99,109] It is in this area that child analysts[73,74,94] can provide data and evidence for the whole field of child development.

Latency: The Middle Childhood Years

When the child enters school he is no longer young. He has entered middle childhood, the latency years, or, in Erikson's terminology, the stage of industry versus inferiority. When the child begins shedding his first deciduous teeth, he enters a period of repressive "calm."[105] Between the dramatic changes of the preschool years, with the resolution of the oedipal situation and the diminution of his narcissistic preoccupation with his body and its various orifices and functions, and the prepubertal growth spurt of adolescence, with the increase in instinctual drives and the environmental confusion of that period, the latency child goes through a great deal of psychosexual and psychocultural development. Between the ages of three and five he is struggling with his oedipal relationships and his competitive dependency position vis-à-vis his siblings and his parents. During this time he lives out a happy illusion of being a miniature adult. From age five to seven, however, he is forced to come to terms with his illusion through the solution of the family "romance" and through fantasies that substitute for primary objects, as well as through reinforcement of his peer group behavior and through a new vision created by experience outside the boundaries of the home. He becomes aware of his relative position as a child in a world of adults. Perhaps this is part of the force behind the typical jokes and puzzles of latency children, remarks that mock adults' intelligence and for a brief period cast the latency child intellectually alongside the grown-up.

In latency the child is no longer satisfied with just play and make-believe. He identifies with people who know things and suggest to him a sense of competence. The danger that his success is illusive is always present. During the stage of industry, when workmanship is so important, the child forgets most of his earlier experiences. He tries to attenuate his family relations and deny what he wants from them. His past hopes and wishes are now less pressing on him. He turns at this time to tools, objects, and work from which he can win recognition by producing things. He learns with others to do competitive tasks with the idea of finishing things. Unlike the child during the stage of initiative, when the emphasis is on the goal but actually very little is done, the latency child aims at completion. During this time he learns to read and write. In primitive cultures he turns instead to gender-appropriate activities that are essential to the survival of the tribe. In most societies selected adults are designated as the teachers, and systematic instruction is provided.

In middle childhood the differentiation of male and female becomes more complete. Boys are taken away from their homes, given tools, instructed in the techniques of the culture, and encouraged to develop skills and workmanship. But some children in every culture want to retreat to earlier years. They are unable to enjoy the tool situation and often feel inadequacy and despair in regard to tools and skills. The period of inferiority can become overpowering for the child, particularly if his mastery of the preceding stage was not accomplished. Work contributes another step in identity development in that the child learns to do what counts. He has more than glimmerings of the fact that part of what he will be is determined by what he does. If mistrust, shame, and guilt are experienced during growing up, he is more likely to develop a negative image of himself.

Harry Stack Sullivan referred to latency as the juvenile era where the chief problems are competition and compromise; the period emphasizes social confidence. School and playground provide the backdrop for the competitive display. New roles are promoted and the child learns new modes of participating with his peers and adults.[105]

Sex-role identification and differentiation are advanced during latency because of peer group and school activities. Peer interaction enables the child to learn his role in society, obtain some familiarity with associative-cooperative play, develop new patterns of aggression including competition and mastery, develop the ability to subordinate his individual needs and wishes to the group and to reexamine his values in terms of his peer group. During latency he develops his academic skills, largely those involving numbers and words. There is also a strengthening of superego and ego ideals. The superego authority is introjected as part of the outcome of the oedipal conflict. Now for the first time it is available to assist the ego in controlling id impulses. These developmental issues are precursors to adolescent and adult development. Structures are modified and emended during middle childhood and the adolescent years.

In 1896 Freud referred to latency as the time of transition during which repression, a forerunner of defense, usually takes place. He thought that the period occurs from age eight to ten and is marked by sexual quiescence. The resolution of the Oedipus complex, Freud believed, leads to repression as the typical latency mechanism. Reaction formations, sublimations, and superego development are pivotal to this period. Sexual energy is displaced in its gratification since aims are inhibited. In 1905, however, Freud indicated that latency is from six to ten.[1] He underscored his belief that a sexual quiescence continues during these years until puberty. Sarnoff[100] points out in his review of latency theories that Freud continued to believe that phylogenesis produces the ego functions that permit what we know as latency.

Subsequent theories have characterized latency as a period in which sexuality is less observable, usually repressed, and most commonly suppressed. Few have continued to believe that latency is a biologically determined period in which drive activity is lessened. Regression to pregenital impulses is commonly seen, more so in boys. The child has to work out new defenses against the pregenital drives. Ego development and the establishment of defenses occur. There is an increase in the prominence of secondary process thinking. The reality principle becomes more firmly established, and positive object relations and identifications are consolidated.

Freud was less interested in latency partly because he felt that most of the personality development had occurred by the time of resolution of the Oedipus complex. He emphasized biological factors as most important in the resolution of the Oedipus complex and considered cultural and educational forces as a secondary importance. Freud's concept of latency led to a falsified understanding of the problems of control, stultified research in this area of development, and cast a pessimistic note on treatment of children during this developmental phase.

Latency is not found in any other mammal; it is unique to man. Chimpanzees do not demonstrate this development. It is thought more and more that latency is a cultural phenomenon. Sexual quiescence is hardly seen in middle childhood. Drive strengths are maintained.[60,61,89] Kinsey's data, as well as those of clinical experience, suggest that sexual experimentation and curiosity persist. Masturbatory activities and fantasies and sexually tinged play are quite common during latency. The distinguishing feature of sexuality during latency is that children are much more circumspect and socially aware about sexuality.

Structural development, particularly superego functions, increases at this time so that the child becomes more aware of what is socially acceptable. Those children who are cognitively impaired, such as mentally retarded children, demonstrate that without cognitive control sexuality during latency is very manifest. Furthermore, transcultural studies abundantly indicate that phenomenological aspects of sexuality are determined in large part by the degree of cultural freedom afforded the child.

The Phases of Latency

Since latency covers a period of years, it can only be understood if it is divided into two phases.[9,10] The first phase is from age six to eight. During this phase repression, suppression, and regression are more apparent. Masturbation is inhibited. The child is greatly involved with himself. The ego can temporarily regress to pregenitality with reaction formations of shame, disgust, and guilt. The superego is extremely strict and feels like a foreign body. In interpersonal relationships the superego is crude and ambivalent, especially toward the isosexual parent. Impulses that must be defended by the superego intensify the ambivalence. Parents cannot be viewed as sexual objects. Drive discharge is through fantasy. Masturbatory fantasies vary according to the child's fixation points. Masturbatory equivalents, such as nail-biting, scratching, and head banging, sadomasochistic to varying proportions, appear in latency. Compulsive talking with excitement is often part of an active fantasy life, which is the primary means of coping with stress. Castration anxiety can persist for the girl. Defenses against penis envy are usually quite apparent. Secondary process thinking waxes and wanes. Increased contact with reality furthers differentiation of secondary autonomous ego functions.

The later phase, age eight to ten, shows better cognitive functions and reality testing. Sublimation is now more successful. There is less suffering over masturbation. There is more gratification in the external world. Fantasy is less needed for drive discharge and by age eleven is used more for reality planning. The superego is less strict. The child in the second phase of latency shows decreasing omnipotence. Defense and affect are now closer to the ego than impulse. Children begin moving more in gangs without adult supervision. From eight to eleven years children begin to show homosexual and heterosexual curiosity and interest. Girls demonstrate an increasing ability to dramatize their feelings.

Between the ages of six to nine the child has to find out how to get along with others and how to compete successfully. Initially rules for the child are seen as absolute and unchangeable. The crudeness of the superego stems partly from cognitive immaturity. At the age of nine the child, however, has a concept of reciprocity. He now feels that it is fair to change the rules provided everybody agrees and new rules apply to all.[63] From age eleven to twelve he develops a sense of equity. He gives special concessions to handicapped players in games with his peer group. He is able to realize that another child's needs can be different and has thus transcended his egocentrism. He is able to shift his perspective and to be empathic.

The Ego of Late Latency

At the age of eight the child has a well-developed time perspective.[121] His concept of death as something permanent is attained. He has a sense of trust in the future and a sense of objectivity. He now is capable of some detachment from the narcissistic equation of his self and other objects. He is better able to delay gratification and has some tolerance for frustration. He has realistic, nonmagical perceptions of cause and effect. He now tries to master previous disappointments. He shows initiative in the capacity for taking responsibility. His quest for mastery is apparent in many things that he does. Self-esteem increases and is contingent on mastery of competition and competence. There is an expanding cognitive awareness of the world based on increasing trust, autonomy, initiative, and industry. With this cognitive expansion and shift toward the peer group, the latency child becomes more involved with his peer relations. He depends on them in part to point out external reality, limit inappropriate impulses, share skills and modes of coping, repair damage to self-esteem, offer social means of lessening tension, and reduce the danger to the balance between excitement and control. Peers remind one another of social goals and gratifications. The group helps to maintain its own controls and to resist seduction or provocation. It shows curiosity and praise for other's accomplishments. Thus the peer group provides an auxiliary ego to the latency child.[92]

Physical Growth

From age six onward there occurs a deceleration of growth during the latency interval. At age six the average child is 46 inches tall and weighs 48 pounds. By the time he is age 12, he is 60 inches tall and weighs from 95 to 100 pounds. Physically boys surpass girls in that they are slightly taller and heavier than girls, stronger, better integrated and coordinated neurologically, and have a faster speed of reaction. The gender differential in growth, however, generally ceases at the time of the pubescent growth spurt when girls become slightly taller than boys and remain so until about the age of fifteen.

School-Peer Group Socialization

The child becomes just another child in the school and neighborhood. He learns new roles and modes of behavior. In play he shares things in associative-cooperative play. There are new patterns of aggressive behavior that lead toward mastery and self-control. In peer groups the child must feel that to subordinate his individual needs and goals to the total group, he must strike a balance between the activities of his parents and those of his peer group. The peer group forces the child to reevaluate his internalized values. His means for coping with aggressive and sexual needs are constantly evaluated by his peer group and his teachers as well as by his parents. Performance is a standard by which the child is now evaluated. Internal shifts and accretions in superego and ego ideals occur, with some reduction in the strictness of the conscience between eight and ten years.

The partial transfer of parental roles to teachers and organized peer groups affects the school-age child in many ways. He becomes more critical of his parents and invests them with less delegated omnipotence. At times he is disillusioned with them. Generally he perceives them in more realistic terms. On entering school he is confronted with an organized peer group, and at the same time he experiences varying degrees of separation anxiety from his parents. In the absence of parents the child's ambivalence shows its hostile side in the form of aggressive feelings and fantasies toward his parents. His modes of defense against anxiety are now challenged by his ongoing experience with teachers and peers.

The school-age child must be more self-reliant in handling fears. His fears change from concrete symbolic representations such as ghosts, monsters, and animals to more generalized and less symbolic substitutions of his fears of parental rejection and disapproval. Concomitantly he develops more reality-based anxieties, particularly ones around performance. He remains in a vicarious balance between his old magical world and the new world of reality, often shifting back and forth in his cognitive appraisal of himself and the world. The child may defensively regress temporarily to an earlier stage where he can utilize old behavior in the service of emotional conflict. He then further develops reaction formations such as guilt, disgust, and shame against infantile impulses that during this growth period become part of the child's character development.

In his sex-role development the child seeks new models for identification. His parents are not enough since he can do so few of the things adults do. Thus girls tend to idealize prestigious and romantic figures. Yet in their play activities they will often act out the roles of mother and daughter. Boys, on the other hand, are prone to identify with people who have status in society—leaders who are strong and assertive. For both sexes peer group formation occurs on the basis of sex differences. In this respect there is a cleavage between the sexes. Popularity within the peer group depends in large part on the adequacy of the child's sex-typed activities and attitudes. Girls are more obsessive, boys, more compulsive, in keeping with the assigned, socially acceptable mode of the sex role.

Girls can more easily disguise neurotic traits because of their compliance to school situations and their overall need to display only proper social behavior. As in school, behavioral problems of girls are not as apparent as those of boys in the physician's office.

Models are selected for identification on the basis of the child's own skills and personality traits and the familial and peer group pressure toward the adoption of sex-appropriate roles. For example, a boy who is well coordinated and strong, or assertive and outgoing, will pick an athletic star for his model, more so if the parents indicate their preference and encourage him to develop masculine interests and activities. Pressure from the family varies inversely with social class. In other words, lower-class parents are concrete and more specific about what constitutes culturally sex-appropriate role behavior. Children perceive the masculine role as more aggressive than the feminine role. Both boys and girls view the male as both more competent in problem solving and more fear-inspiring. This dichotomous distinction persists through adulthood, still subject to the cultural variants of social class. One study indicated that by the age of 14 girls esteem aggressive-competitive personality characteristics in males as much as boys do.[82] This fact, among others, has been related to the more negative self-concept that the average girl possesses in our culture.

Sex-appropriate behavior is important in the development of children because it is one of the factors that shapes the attitude of peers toward the child. In addition, many personality traits seem to cluster around the adequacy of the gender role. Male-female dichotomy holds for assertive-aggressive and avoidance-withdrawal behavior. Although the child's fantasies can be at variance with his behavior, career choices are in consonance with gender role. Career choice, however, is a phenomenon of adolescence, but expressed ambitions and aspirations are phenomena of middle childhood. Kagan[52,53] has suggested that the most stable aspect of personality development is the sex-role identification, which gives both longitudinal direction and organization to the formation of the self-concept, self-esteem, and body image.

The ease in socialization is determined by the child's acceptance or rejection by the parents. If he is rejected by the parents, he is apt to be fearful on entering school and in seeking out relationships. If rejected by parents, he is apt to be aggressive. A second parental behavioral polarity is restrictiveness versus permissiveness. Restrictive parents frustrate the child's need for autonomy. Patterns of control can deviate in either direction. Domineering parents are apt to foster shyness, inhibition, and apathy in the child. Overprotective parents encourage children to be stubborn and irresponsible. Parental permissiveness tends to produce a more assertive, spontaneous child, often to aggressive proportions. Excessive instinctual gratification can lead to faulty integration of superego and impulse control. Limits for behavior must be set by the parent so that the child can make his choices and decisions. Complete freedom for the child is now an old-fashioned childrearing concept. Latency children still require consistent guidelines and disciplinary measures as signs of love and interest on the part of parents, particularly from the isosexual parent.

As the child moves from the home-centered world of preschool years, he begins to concentrate on objective reality. The child becomes more interested in knowledge, skills, and activities. He searches for general principles and is fascinated with the new, strange, and distant places and people of the world. New experiences with language enable him to communicate with symbols so that he can generalize and abstract. School is the playground in which new activities and skills are discovered and developed. Play is transformed into work. To a lesser extent, scout activities and organized recreational activities serve the same purposes. Parental teaching is supplemented and new experiences in coping with problems are provided. These socializing activities enable a child to initiate the formation of long-range goals.

The child's behavior in school is in large part a continuum of his behavior in the parent-child relationship. The child's adaptation to school depends on factors other than his emotional organization. Separation anxiety is a crucial determinant. Specific capabilities and maturational readiness to undertake the essentials of education are important. His position and social class also bear on his school behavior. Lower-class children adjust less readily to school because their parents are less encouraging and the peer group less reinforcing of the positive merits of the school experience. Furthermore, lower-class children are accustomed to immediate gratification. The academic curriculum for the middle-class child does not always suit the child in the lower social class.

The sex of the child is another important factor. Boys are apt to adjust less readily to the school experience because of feminization in teaching. Boys tend to react to women teachers with anxiety and tension. They are more hostile to school. Kagan indicates that if more men were to teach children, particularly in the lower primary grades, boys would be more apt to associate the act of acquiring knowledge with masculinity. For example, boys have far more reading and spelling difficulties than girls.[52] This has been explained in the past totally on the basis of a psychoneurological differential in brain damage at birth. Psychocultural factors, however, play an important part in this gender differential. Boys look upon the customary activities of kindergarten and first grade as feminine since those classroom activities tend to involve cutting out, pasting, and doing "pretty things." Furthermore, boys are much more comfortable in handling aggressive feelings if they are supervised by male teachers.

The IQ of the latency child can change during the course of his education.[82] Several studies have demonstrated that the correlation between IQ of children obtained from age six to ten is .70. At the Fels Institute 50 per cent of the children studied showed changes over the years, most markedly the boys. The drive to learn increased as the children became more assertive and independent at school. Those girls whose IQ changed over the years were closely attached to their mother with whom they identified and whom they perceived as valuing intellectual achievement. Latency children do not have fully developed cognitive functions until adolescence. When they first enter school, latency children tend to be concrete but less animistic and magical in thinking. Later the latency child begins to perform logical manipulation. At first the manipulation involves concrete objects, but toward the end of late latency deductive processes are available to the child. Piaget has serially documented the various stages in intellectual development, from the concrete to the abstract, during this growth period.[121]

Latency Peer Group Selection, Interaction, and Culture

There is a wide variation in peer group ego ideals and values existing within social classes.[14] A lower-class child may be poorly accepted by his peers if he excels in school. The heroes of the lower-class children are not upper-class authority figures. These children are apt to be more receptive to the "heroes" of the television world. The violence of television has more impact upon children from the lower social classes, in which there is less anxiety about aggressive behavior and a greater tendency on their part to identify with aggressive models provided in television programs.

Friendship patterns in middle childhood years are principally determined by the personality characteristics of the child. Nonintellectual traits are usually selected. Mutual friends will resemble each other in socioeconomic background and general intellectual level, but not in academic achievement. The accomplishment and status of the parents have no direct bearing on acceptance by the child's peer group. Aggressive boys will seek out aggressive friends. Friendships are made on the basis of gender grouping. The pattern of the friendships is unstable because mutual interests frequently shift. With increasing age general interests become crystallized and the friendships are more apt to be enduring. The peer group interaction has a very decided socialization function. It has a molding effect on the child's self-image and goals. If the values of the peer group are discordant with those of the child's parents, the child may be much more susceptible to peer influence, especially if the child has not made a strong identification with the parent of the same sex, or if he comes from a home where nurturance has been inadequate. This discordant value selection may also be made more likely if the child is not accepted by his peer group and is forced to seek out the group through compliant behavior.

Playing the games of the culture is most important for the latency child. Games with rules, Piaget's third category of play, come into prominence. Hobbies, halfway between play and work, are selected.[27] Some can lead to ego mastery and a sense of job competence. In general, the latency child plays down his deficiencies, plays up his achievements, plays up to his identificatory models, and tries to play fair with his peers.

Latency has its own special culture with its own rules, games, rhymes, and riddles, from which all grown-ups are scrupulously excluded. Children in middle childhood tell repetitiously the same old jokes. Their rites include obsessive counting, peculiar little superstitions, tongue twisters, and odd collections of objects that serve as talismen. Hobbies commence quite early and frequently are precursors to later work. Coded communications are used in the subsocieties of clubs, packs, and gangs, all of which pride themselves on the group's solidarity and individuality. Play is more realistic and imaginative. Peer groups are meticulously organized and founded on rules, oaths, and clandestine passwords. The rules, however, are made to be broken. The activities of latency groups are devoted more to planning than to actual accomplishment. The goals are primarily the exclusion of those children who do not fit into the special identity of the group. Girls organize their clubs for the express purpose of keeping other girls out and gaining strength to reject masculinity. On the other hand, boys are much more involved with the concrete, such as trips, games, and innumerable projects. Their trading of collections and prized objects is quite common. Barter is a boy's way of comparing assets and strength.

Pubescence

Pubescence or prepuberty is generally regarded as the two-year interval that precedes the onset of full puberty.[82] In many respects it is psychologically closer to the middle school years. Its onset is marked by a spurt in physical growth. Physical changes take place including both primary and secondary sexual changes. Pubescence can be short-lived, hardly noticed by either the child or the parents. The potentialities for physical growth are suddenly realized, with rapid spurts in height and weight and rapid growth of the arms, legs, and neck; changes in body structure, increased sexual impulses, and growth of the genitals occur. The triggers for pubescence are neuroendocrinological. Both physical and emotional changes are not fully understood. At puberty the biological changes reach a peak marked in the female by menarche and in the male by spermatogenesis. This period of development is often dubbed preadolescence, which is confusing because the term is often used to designate the middle school years as an epoch.

Pubescence should be regarded as part of early adolescence. It is a distinct interval in child development during which instinctual drives increase; sexuality blossoms, and physical growth pushes forward with breakneck speed. With the intensification of both aggressive and sexual drives, the onset of rapid growth, changing body concepts, and narcissistic body concern, the pubescent child is completing his shift in cognitive function from the concrete to the abstract.[82] Thus, at a time when he is developmentally capable of greater rationality, he is beset with more physical discomfort and emotional uneasiness. Temporary losses in the personality gains made during the middle childhood years often produce a picture of the dissocial or early psychotic adolescent. Regressive trends to both phallic and preoedipal phases are present and frequently ephemerally apparent. In early adolescence attitudes of cleanliness become converted to dirtiness, neatness to disorder, sociability to boorishness; moods swing from euphoria to depression; and there are frequent rumblings and alienation from parents. There occurs a resurgence of dependency and ambivalence, especially in boys. Girls seem to experience less regressive pull in their development at this time.

In general boys are more immature and more apt to regress than their more mature and sophisticated female counterparts. Boys often feel inferior. They fear any trace of femininity in themselves. They are concerned about growth changes, particularly ones involving the genitals. Both boys and girls are very much aware of body hair as an early indication of sexual maturation. Modesty, which was of little importance during the early middle childhood years, now becomes maximally important to shroud the child until he has "arrived."

Physical Changes in Adolescence

Physiological and morphological changes in early adolescence are striking from ages 11 to 15.[82] In the male from age 13 to 15 there occurs a growth spurt. Fourteen inches is maximal growth for boys, the average being about eight. The average boy gains roughly 40 pounds. Between 15 and 18 growth slows down and then ceases. The girl experiences her growth between 11 and 13. Her average gain is five to six inches. Her growth is thus two years earlier than that of the boy. From age 13 to 16 her growth slows down and then ceases. The popular notion that girls mature earlier than boys stems primarily from the fact that girls attain their adult height and weight about two years earlier than boys.

During adolescence body proportions change; this occurs several years later for boys. The "baby phase" of childhood begins to disappear. The forehead becomes higher and wider. The mouth widens and flat lips become fuller. The slightly receding chin of childhood begins to jut out. Relatively large head characteristics of childhood become smaller in proportion to the total body height, due to the fact that during this period the extremities are growing at a faster rate than the head. The reproductive system particularly increases in size, but the brain does not change noticeably during this period. There is a decrease in subcutaneous fat so that the proportion of body weight attributable to fat decreases, more so in boys than in girls. Basic changes in size, proportion, and shape are most pronounced in the bones. Ossification speeds up and is completed in the female by age 17 and in the male by age 19. There is a marked increase in muscle tissues in relation to total body weight. For the female between age 12 and 15 there is pronounced growth; for the male, between age 15 and 16. Motor performance, coordination, strength, speed, and accuracy in the female improve to age 14, in the male to age 17. Boys in many respects show better motor performance than girls.

Changes in the Reproductive System and Sex Characteristics

In the female the primary sex characteristics involve the genital urinary system. The secondary characteristics appear at age 11, with the development of breast buds usually appearing two years before menarche. This is followed by pubic and axillary hair and increased widening and deepening of the pelvis. When ovulation appears two to four years after menstruation, conception becomes possible. For the male primary sex changes occur from 12 to 14. Secondary sex characteristics are initially pubic with later axillary, facial and bodily hair and changes in voice.

The sequence for the male is increasing size of the testicles and scrotum, the appearance of pubic hair from 12 to 14, enlargement of the prostate gland, penile growth, and the occurrence of ejaculation one year after pubic hair. Spermatogenesis takes place between the age 15 and 16 and marks the beginning of nubility.

Tertiary sex characteristics for the female include a specific infrafemoral angle, arm swing, pelvic tip, and lateral ball throwing.[6] Menarche for the American girl appears on the average at the age of 13. In 1900 the average age was 14. This change is attributable to better health and nutrition. Ninety-seven per cent of girls have first menarche between 11 and 15, but the age range is from 9 to 20. The year following the year during which there has been maximal growth in height is frequently the year of onset of menarche.

Menstruation is a maturational crisis for the girl since it is a symbol of sexual maturity. In one study only 6 per cent of teen-age girls indicated any sense of pride in menarche. This predominantly negative attitude is influenced by the attitudes of parents and friends, who foster a negative feminine identification. This is found in other cultures such as Bali where menstruation is responded to with shame. The Manus and Arapesh view menstruation as a result of injury or an indication of dangerous powers.[82] Young girls are apt to view menstruation as equivalent to excretion, loss of control, fear of injury, evidence of genital damage, punishment for masturbation and incestuous desire, an indication of need for surgery, and a precipitant for depression.

Nocturnal emission is the male psychological counterpart to menstruation. It is equivalent in the sense that it frequently comes as a surprise and is a source of worry to the pubescent boy. Frequently there is minimal information given by parents to teen-age boys. Nocturnal emissions usually occur with erotic dreams. Fischer and Dement have demonstrated that during stage two REM sleep, penile erection occurs. A boy's sense of shame about nocturnal emissions may stem from his fear of instinctual vulnerability and his concern that oedipal wishes will be detected by his parents.

Central Nervous System Development

In puberty neurohumoral stimuli from the hypothalamus result in increased secretion of gonadotropin hormone from the anterior pituitary gland.[105] This leads to gonadal development with increased secretion of sex hormones, both gonadal and adrenal. In females the pituitary produces follicle-stimulating hormones (FSH). In nubility interstitial-cell-stimulating hormones (ICSH) become operative. The remaining follicular cells form corpora lutea. The luteotrophin from the anterior pituitary produces progesterone from the corpus luteum, which prepares the endometrium for implantation. Biological changes have profound influence on psychic development. Now there is a genital drive quality to psychic activity.

The psychic changes that take place in adolescence are part of the second separation-individuation phase,[8] which involves intrapsychic infantile objects rather than the "raw" parental objects of the toddler years. These aspects of adolescence are reviewed in Chapter 18 of this volume.

Bibliography

1. ALPERT, A., "The Latency Period," *Am. J. Orthopsychiat.*, *11*:126–132, 1941.
2. BALDWIN, A. L., *Theories of Child Development*, John Wiley, New York, 1967.
3. BARTON, D., and WARE, T. D., "Incongruities in the Development of the Sexual System," *Arch. Gen. Psychiat.*, *14*:614–623, 1966.
4. BEACH, F. A. (Ed.), *Sex and Behavior*, John Wiley, New York, 1965.
5. BENEDEK, T., "An Investigation of the Sexual Cycle in Women," *Arch. Gen. Psychiat.*, *8*:311–322, 1963.
6. BIRDWHISTELL, R. L., *The Tertiary Sexual Characteristics of Man*, AAAS, Montreal, 1964.
7. BLAUVELT, H., and MCKENNA, J., "Mother-Neonate Interaction: Capacity of the Human Newborn for Orientation," in Foss, B. M. (Ed.), *Determinants of Infant Behavior*, pp. 38–49, John Wiley, New York, 1961.
8. BLOS, P., "The Second Individuation Process of Adolescence," in *The Psychoanalytic Study of the Child*, Vol. 22, pp. 162–186, International Universities Press, New York, 1967.

9. BORNSTEIN, B., "Masturbation in the Latency Period," in *The Psychoanalytic Study of the Child*, Vol. 8, pp. 279–285, International Universities Press, New York, 1951.
10. ———, "On Latency," in *The Psychoanalytic Study of the Child*, Vol. 8, pp. 279–285, International Universities Press, New York, 1951.
11. BOWLBY, J., *et al.*, "The Effects of Mother-Child Separation: A Follow-up Study," *Brit. J. Med. Psychol.*, *29*:211–247, 1956.
12. BRENNER, C., *An Elementary Textbook of Psychoanalysis*, Doubleday, New York, 1957.
13. BROWN, D. G., "Sex-Role Development in a Changing Culture," *Psychol. Bull.*, *55*:232–242, 1958.
14. CAMPBELL, J. D., "Peer Relations in Childhood," in Hoffman, M. L., and Hoffman, L. W. (Eds.), *Review of Child Development Research*, Vol. 1, pp. 289–322, Russell Sage Foundation, New York, 1964.
15. CASUSO, G., "Anxiety Related to the Discovery of the Penis," in *The Psychoanalytic Study of the Child*, Vol. 12, pp. 169–174, International Universities Press, New York, 1957.
16. CHODOFF, P., "A Critique of Freud's Theory of Infantile Sexuality," *Int. J. Psychiat.*, *4*:35–48, 1967.
17. DECARIE, T. G., *Intelligence and Affectivity in Early Childhood*, International Universities Press, New York, 1965.
18. DICKES, R., "Fetishistic Behavior: A Contribution to Its Complex Development and Significance," *J. Am. Psychoanal. A.*, *11*:203–330, 1963.
19. ERIKSON, E. H., "Identity and the Life Cycle," *Psychol. Issues*, *1*:50–100, 1959.
20. FALKNER, F. (Ed.), *Human Development*, Saunders, Philadelphia, 1966.
21. FENICHEL, O., *The Psychoanalytic Theory of Neurosis*, Norton, New York, 1945.
22. FLAVELL, J. H., *The Developmental Psychology of Jean Piaget*, Van Nostrand, Princeton, 1963.
23. FORD, C. S., and BEACH, F. W., *Patterns of Sexual Behavior*, Ace Books, New York, 1951.
24. FRANK, R. L., "Childhood Sexuality," in Hoch, P., and Zubin, J. (Eds.), *Psychosexual Development*, pp. 143–158, Grune & Stratton, New York, 1949.
25. FREUD, A., *The Ego and the Mechanisms of Defense*, Hogarth, London, 1936.
26. ———, "The Infantile Neurosis: Genetic and Dynamic Considerations," in *The Psychoanalytic Study of the Child*, Vol. 26, pp. 79–90, Quadrangle Books, New York, 1972.
27. ———, *Normality and Pathology in Childhood*, International Universities Press, New York, 1965.
28. FREUD, S. (1909), "Analysis of a Phobia in a Five-Year-Old Boy," in Strachey, J. (Ed.), *Standard Edition*, Vol. 10, pp. 3–143, Hogarth, London, 1955.
29. ——— (1919), "A Child Is Being Beaten," in Strachey, J. (Ed.), *Standard Edition*, Vol. 17, pp. 175–204, Hogarth, London, 1961.
30. ——— (1924), "Dissolution of the Oedipus Complex," in Strachey, J. (Ed.), *Standard Edition*, Vol. 19, pp. 173–179, Hogarth, London, 1961.
31. ——— (1931), "Female Sexuality," in Strachey, J. (Ed.), *Standard Edition*, Vol. 21, pp. 223–246, Hogarth, London, 1961.
32. ——— (1927), "Fetishism," in Strachey, J. (Ed.), *Standard Edition*, Vol. 21, pp. 149–158, Hogarth, London, 1961.
33. ——— (1923), "The Infantile Genital Organization of the Libido," in Strachey, J. (Ed.), *Standard Edition*, Vol. 19, pp. 141–148, Hogarth, London, 1961.
34. ——— (1924), "The Loss of Reality in Neurosis and Psychosis," in Strachey, J. (Ed.), *Standard Edition*, Vol. 19, pp. 183–190, Hogarth, London, 1961.
35. ——— (1908), "On the Sexual Theories of Children," in Strachey, J. (Ed.), *Standard Edition*, Vol. 9, pp. 205–226, Hogarth, London, 1961.
36. ——— (1938), "An Outline of Psychoanalysis," in Strachey, J. (Ed.), *Standard Edition*, Vol. 23, pp. 141–208, Hogarth, London, 1961.
37. ——— (1917), The Sexual Life of Human Beings," in Strachey, J. (Ed.), *Standard Edition*, Vol. 10, pp. 3–148, Hogarth, London, 1955.
38. ——— (1925), "Some Psychological Consequences of the Anatomical Distinction between the Sexes," in Strachey, J. (Ed.),

Standard Edition, Vol. 19, pp. 243–260, Hogarth, London, 1961.

39. ——— (1940), "Splitting of Ego and Process of Defense," in Strachey, J. (Ed.), *Standard Edition,* Vol. 23, pp. 271–278, Hogarth, London, 1961.

40. ——— (1905), "Three Essays on the Theory of Sexuality," in Strachey, J. (Ed.), *Standard Edition,* Vol. 7, Hogarth, London, 1961.

41. FRIES, M. E., "Review of the Literature on Latency Period," *J. Hillside Hosp., 7:*3–10, 1958.

42. GELENSON, E., "Prepuberty & Child Analysis (Panel Report)," *J. Am. Psychoanal. A., 12:*600–609, 1964.

43. GIFFORD, S., "Repetition Compulsion," *J. Am. Psychoanal. A., 12:*632–649, 1964.

44. GOODMAN, S., "Current Status of the Theory of the Superego Report," *J. Am. Psychoanal. A., 13:*172–180, 1965.

45. HARTLEY, R. E., and HARDESTY, F. P., "Children's Perception of Sex Role in Childhood," *J. Genet. Psychol., 105:*43–51, 1964.

46. HARTMANN, H., "The Genetic Approach to Psychoanalysis," in *The Psychoanalytic Study of the Child,* Vol. 1, pp. 11–30, International Universities Press, New York, 1945.

47. HENDRICK, I., *Facts and Theories of Psychoanalysis,* Alfred A. Knopf, New York, 1958.

48. HOFFMAN, M. L., and HOFFMAN, L. N. W. (Eds.), *Review of Child Development Research,* Vol. 1, Russell Sage Foundation, New York, 1964.

49. JACOBSON, E., *The Self and the Object World,* International Universities Press, New York, 1964.

50. JESSNER, L., and PAVENSTEDT, E. (Eds.), "Some Observations on Children Hospitalized during Latency," in *Dynamic Psychopathology in Childhood,* pp. 257–266, Grune & Stratton, New York, 1959.

51. JONES, E., "The Phallic Phase," *Int. J. Psychoanal., 14:*1–33, 1933.

52. KAGAN, J., "Acquisition and Significance of Sex Typing and Sex Role Identity," in Hoffman, L. W., and Hoffman, M. L. (Eds.), *Review of Child Development Research,* Vol. 1, pp. 137–168, Russell Sage Foundation, New York, 1964.

53. ———, "The Concept of Identification," in Mussen, P. H., Conger, J. J., and Kagan, J. (Eds.), *Readings in Child Development and Personality,* pp. 212–224, Harper, New York, 1965.

54. KAPLAN, B., and WAPNER, S. (Eds.), *Perspectives in Psychological Theory,* International Universities Press, New York, 1960.

55. KARDINER, A., KARUSH, A., and OVESEY, L., "A Methodological Study of Freudian Theory," *J. Nerv. Ment. Dis., 129:*11–19, 133–143, 207–221, 341–356, 1959.

56. KATIN, A., "Distortions of Phallic Phase," in *The Psychoanalytic Study of the Child,* Vol. 15, pp. 208–214, International Universities Press, New York, 1960.

57. KERNBERG, O., "A Contribution to the Ego-Psychological Critique of the Kleinian School," *Int. J. Psychoanal., 50:*317–333, 1969.

58. KESTENBERG, J. S., "Vicissitudes of Female Sexuality," *J. Am. Psychoanal. A., 4:*453–477, 1956.

59. KIDD, A. H., and RIVOIRE, J. L. (Eds.), *Perceptual Development in Children,* International Universities Press, New York, 1966.

60. KINSEY, A. C., *et al., Sexual Behavior in the Human Female,* Saunders, Philadelphia, 1953.

61. ———, *Sexual Behavior in the Human Male,* Saunders, Philadelphia, 1948.

62. KOFF, R. H., "A Definition of Identification: A Review of the Literature," *Int. J. Psychoanal., 42:*362, 1961.

63. KOHLBERG, L., "A Cognitive-Developmental Analysis of Children's Sex-Role Conceptions and Attitudes," in Maccoby, E., (Ed.), *Development of Sex Differences,* pp. 82–172, Stanford University Press, Stanford, 1966.

64. KURTH, F., and PATTERSON, A., "Structuring Aspects of the Penis," *Int. J. Psychoanal., 49:*620–628, 1968.

65. LAMPL-DEGROOT, J., "Ego Ideal and Superego," in *The Psychoanalytic Study of the Child,* Vol. 17, pp. 94–106, International Universities Press, New York, 1962.

66. ———, "On Masturbation and Its Influence on General Development," in *The Psychoanalytic Study of the Child,* Vol. 5, pp. 153–174, International Universities Press, New York, 1950.

67. ———, "The Pre-Oedipal Phase in the De-

velopment in the Male Child," in *The Psychoanalytic Study of the Child*, Vol. 2, pp. 75–83, International Universities Press, New York, 1946.

68. LAURENDEAU, M., and PINARD, A., *Causal Thinking of the Child*, International Universities Press, New York, 1962.
69. LEVY, D., "Studies of Reaction to Genital Differences," *Am. J. Orthopsychiat.*, *10*: 755–762, 1940.
70. LICHTENSTEIN, J., "Identity and Sexuality," *J. Am. Psychoanal. A.*, *9*:179–260, 1961.
71. MCDOUGALL, J., "Primal Scene and Sexual Perversion," *Int. J. Psychoanal.*, *53*:371–384, 1972.
72. MACLEAN, T. D., "New Findings Relevant to the Evolution of Psychosexual Functions of the Brain," *J. Nerv. & Ment. Dis.*, *135*:289–301, 1962.
73. MAHLER, M. S., "On the First Three Subphases of the Separation-Individuation Process," *Int. J. Psychoanal.*, *53*:333–338, 1972.
74. ———, "Rapprochement Subphase of the Separation-Individuation Process," *Psychoanal. Quart.*, *41*:487–506, 1972.
75. MAIER, H. W., *Three Theories of Child Development*, Harper, New York, 1965.
76. MARMOR, J. (Ed.), *Sexual Inversion*, New York, Basic Books, 1965.
77. MEAD, M., "The Childhood Genesis of Sex Differences in Behavior," in Tanner, J. M., and Inhelder, B. (Eds.), *Discussions on Child Development*, Vol. 3, pp. 13–90, International Universities Press, New York, 1958.
78. MINUCHIN, D., "Sex-Role Concept and Sex Typing in Children as a Function of School and Home Environments," *Child Dev.*, *36*:1033–1048, 1965.
79. MONEY, J., "Psychosexual Differentiation," in Money, J. (Ed.), *Sex Research*, pp. 3–23, Holt, New York, 1965.
80. ———, HAMPSON, J. G., and HAMPSON, J. L., "Imprinting and the Establishment of Gender Role," *Arch. Neurol. & Psychiat.*,*77*:333–336, 1957.
81. MULLAHY, P., *Oedipus: Myth and Complex*, Hermitage Press, New York, 1948.
82. MUSSEN, P. H., CONGER, J. J., and KAGAN, J., *Child Development and Personality*, Harper, New York, 1965.
83. NAGERA, H., "Early Childhood Disturbances, the Infantile Neurosis, and the Adulthood Disturbances," in *The Psychoanalytic Study of Child*, International Universities Press, New York, 1966.
84. NOVICK, J., and NOVICK, K. K., "Beating Fantasies in Children," *Int. J. Psychoanal.*, *53*:237–242, 1972.
85. PARSONS, T., "The Incest Taboo in Relation to Social Structure and the Socialization of the Child," *Brit. J. Sociol.*, *5*:101–117, 1954.
86. PAULY, I. B., "Male Psychosexual Inversion: Trans-sexualism," *Arch. Gen. Psychiat.*, *13*:172–181, 1965.
87. PELLER, L., "Reading and Daydreams in Latency: Boy-Girl Differences," *J. Am. Psychoanal. A.*, *6*:57–70, 1958.
88. RADO, S., "An Adaptational View of Sexual Behavior," in *Psychoanalysis of Behavior*, Vol. 1, pp. 201–203, Grune & Stratton, New York, 1956.
89. RAMSEY, G. V., "The Sexual Development of Boys," *Am. J. Psychol.*, *56*:217–233, 1943.
90. RANGELL, L., "The Role of the Parent in the Oedipus Complex," *Bull. Menninger Clin.*, *19*:9–16, 1965.
91. RAPHING, D. L., CARPENTER, B. L., and DAVIS, A., "Incest: A Genealogical Study," *Arch. Gen. Psychiat.*, *16*:505–512, 1967.
92. RAUSH, H. L., and SWEET, B., "The Preadolescent Ego," *Psychiatry*, *24*:122–132, 1961.
93. REEVY, W. R., "Child Sexuality," in Ellis, A., and Abarbanel, A. (Eds.), *The Encyclopedia of Sexual Behavior*, Vol. 1, pp. 258–267, Hawthorn Books, New York, 1961.
94. ROIPHE, H., and GALENSON, E., "Early Genital Activity and the Castration Complex," *Psychoanal. Quart.*, *41*:334–347, 1972.
95. RUTTER, M., "Normal Psychosexual Development," *J. Child Psychol. & Psychiat.*, *11*:259–283, 1971.
96. SALZMAN, L., "Psychology of the Female," *Arch. Gen. Psychiat.*, *17*:195–204, 1967.
97. SANDLER, J., "On the Concept of the Superego," in *The Psychoanalytic Study of the Child*, Vol. 15, pp. 128–162, International Universities Press, New York, 1960.
98. ———, *et al.*, "The Ego Ideal and The Ideal Self," in *The Psychoanalytic Study of the Child*, Vol. 18, pp. 139–158, Inter-

national Universities Press, New York, 1963.
99. ———, and ROSENBLATT, B., "The Concept of the Representational World," in *The Psychoanalytic Study of the Child*, Vol. 17, pp. 128–148, International Universities Press, New York, 1962.
100. SARNOFF, C., "Ego Structure in Latency," *Psychoanal. Quart.*, *40*:387–414, 1971.
101. SCHACHTEL, E. G., *Metamorphosis*, Basic Books, New York, 1959.
102. SEGAL, H., *Introduction to the Work of Melanie Klein*, Basic Books, New York, 1964.
103. SEGAL, M. M., "Transvestitism as an Impulse and as a Defense," *Int. J. Psychoanal.*, *46*:209–217, 1965.
104. SLATER, M. K., "Etiological Factors in the Origin of Incest," *Am. Anthropol.*, *61*: 1042–1047, 1959.
105. SOURS, J. A., "Growth and Emotional Development of the Child," *Postgrad. Med.*, *40*:515–522, 1966.
106. ———, "Phallic-Oedipal Development: Deviations and Psychopathology of Sexual Behavior in Children," in Wolman, B. B. (Ed.), *Manual of Child Psychopathology*, McGraw-Hill, New York, 1972.
107. SPERLING, M., "Fetishism in Children," *The Psychoanalytic Quarterly*, *32*:374–392, 1963.
108. SPITZ, R. A., "The Derailment of Dialogue," *J. Am. Psychoanal. A.*, *12*:752–775, 1964.
109. ———, *The First Year of Life*, International Universities Press, New York, 1965.
110. ———, "A Note on the Extrapolation of Ethological Findings," *Int. J. Psychoanal.*, *36*:162–165, 1955.
111. STOLLER, R. J., "A Contribution to the Study of Gender Identity," *Int. J. Psychoanal.*, *45*:220–226, 1964.
112. ———, "The Hermaphroditic Identity of Hermaphrodites," *J. Nerv. & Ment. Dis.*, *139*:453–457, 1964.
113. ———, "Passing and the Continuum of Gender Identity," in Marmor, J. (Ed.), *Sexual Inversion*, pp. 190–210, Basic Books, New York, 1965.
114. ———, "The Sense of Maleness," *Psychoanal. Quart.*, *34*:207–218, 1965.
115. TANNER, J. M., *Growth at Adolescence*, 2nd ed., Blackwell, Oxford, 1962.
116. TENNES, R. H., and LAMPL, E. E., "Stranger and Separation Anxiety in Infancy," *J. Nerv. & Ment. Dis.*, *139*:247–254, 1965.
117. VELIKOVSKY, I., *Oedipus and Acklination*, Sidgwick & Jackson, London, 1960.
118. WAELDER, R., *Basic Theory of Psychoanalysis*, International Universities Press, New York, 1960.
119. WEINBERG, S., *Invest Behavior*, Citation Press, New York, 1965.
120. WHITTY, C. W. M., and ZANGWILL, O. L., *Amnesia*, Butterworth, London, 1966.
121. WOODWARD, M., "Piaget's Theory," in Howell, J. G. (Ed.), *Modern Perspectives in Child Psychiatry*, Charles C Thomas, Springfield, Ill., 1965.
122. YARROW, L. J., "Separation from Parents during Early Childhood," in Hoffman, M. L., and Hoffman, L. N. W. (Eds.), *Review of Child Development Research*, Vol. 1, Russell Sage Foundation, New York, 1964.

CHAPTER 15

COGNITIVE DEVELOPMENT

Jules R. Bemporad

Introduction

FOR DECADES the area of cognition was uniformly ignored by psychiatry, which concentrated on the energetic-dynamic model of psychoanalysis or the medically oriented "organic" approach to mental illness. This neglect, in fact, has caused one writer to call cognition the Cinderella of psychiatry.[4] As in the fairy tale, in recent years this neglected Cinderella has indeed been discovered by a number of psychiatric princes and has become a topic of great popularity.

The reasons for this sudden surge of interest in cognition, which for many years had been the domain of somewhat isolated academicians, are many and complex. On a theoretical level the existence of mind has again become fashionable in psychology, and the strict behaviorism that attempted to deal solely with stimulus and response has lost its former prominence. It can no longer be held as valid that an organism's response is simply determined by an internal copy of a physical stimulus. Any stimulus is cognitively modified, interpreted, and perhaps partially created by the organism, which then selects a response pattern. The ever increasing number of postulated "mediational processes" confirms the need to consider active, selective mental processes between the presented stimulus and the observed response. Similarly the stress on the greater abilities of the ego in psychoanalytic theory has shown that behavior is more than the simple transformation of unconscious drives. The id psychology of early psychoanalysis has been replaced by a theory that makes the ego and its cognitive structures central to the causes of behavior. While some "ego psychologists" such as Hartmann and Rapaport have tried to fit their innovations within the older framework of the energetic "economic" point of view, other psychoanalysts have elaborated core concepts such as identity, the self, and object-relations that are difficult to integrate with classical metapsychology and seem to imply cognitive rather than energetic structural models. Therefore, the importance of the cognitive functions of the mind has reasserted itself in terms of dealing with both external and internal stimulation.

On a more pragmatic level the coming of age of child psychiatry as an independent discipline has directed attention to the developmental process of cognitive growth. The actual observation of children rather than the former reliance on retrospective accounts of adults has underscored the need for an appreciation of cognitive development to understand the psychic world of childhood as well as the eventual psychic world of the adult. The more children were studied, the more it became apparent that their behavior could not be explained solely on the basis of accumulated S-R habits or on the basis of libidinal stages, but that the development of intellectual faculties, independent of other factors, held an important key to the understanding of childhood.* Furthermore, the recent movement of psychiatry into preventive and community mental health brought to light a great number of psychiatric casualties who were lacking in cognitive abilities. The effects of early failure of appropriate cognitive development in the poor and in minority groups mobilized a massive effort at remediation and prevention of these early deprivations.

It is apparent that from the theoretical, academic, and social points of view cognition and its ontogenetic development have become an important area of study in psychiatry. This has been a salutary change since, after all, in studying human behavior we are really studying cognitive processes. The human psyche, whether viewed in terms of repressed memories, current attitudes, or expectations of the future, is made up of cognitive constructs, and, therefore, the alterations and creation of these structures are quite justifiably the province of psychiatry and should be included in a psychiatric textbook. In the following pages some of the more pertinent issues, as well as the work of major theorists, in the area of cognitive development will be summarized.

Current Status of Cognitive Developmental Theory

The development of cognitive functions is predominantly viewed as a fluid process, with the emergence of novel capabilities at various steps of ontogenesis. Although the causes for the course of this development are unknown, current theory stresses a phase-specific interplay between innate and environmental factors, with appropriate stimulation needed to elicit or reinforce inborn patterns. There is also a great emphasis on the role of mental activity, of action, rather than simply passive reception of stimulation, as necessary for the growth of cognitive skills. Underlying this view of cognitive development as an evolving progression toward more increasing complexity is the concept of epigenesis, which states that mature abilities at each stage grow out of simpler forms of cognition and that these complex abilities are reliant on the mastery of more primitive tasks. The hallmark of current developmental theory is thus a conception of ontogenesis as a creative process, with the emergence of qualitatively new forms of behavior that require proper environmental stimulation as well as successful completion of earlier adaptations for the process to continue.

As such, most theorists see cognitive development as beginning with simple innate reaction patterns and passing through various phases to culminate in adult capabilities such as abstract ideas, relative independence from environmental events, and the establishment of self-regulatory systems. At each stage the mind is seen as *actively* structuring a world view, beginning with a primitive logic of action and building on each successive stage to transform its relationship to the environment toward greater autonomy and complexity.

* An actual incident may demonstrate the role of cognition in the behavior of children. A three-year-old girl whose father often traveled by airplane developed an enthusiasm for flying. When the family planned to take a vacation and to travel by air the girl was ecstatic. However, when they were ready to board the plane the young girl panicked and refused to go on the plane. After calming her down, her parents inquired about the reasons for her "plane phobia"; the girl quite simply and honestly replied that she didn't want to get on the plane because she didn't want to shrink. The girl had been watching planes taking off and diminish in size as they became airborne, and her immature concept of size constancy did not allow for such rapid and extreme changes. The point is that her fear was a result of her immature cognitive ability and not of underlying dynamic events or prior learned habits.

Historical Orientation: Idealism, Empiricism, and Interactionism

In the past theories of cognition usually stressed a specific philosophic position. Plato's doctrine of *Anemnesis* proposed that prior to earthly life the soul had knowledge of ideal forms whose recollection allowed man to structure his phenomenological world. Through development myriad concrete experiences re-evoked the dormant memory of the ideal forms. In modern dress Plato's myth stresses the dependence on predetermined, nonlearned cognitive modes for intellectual development. This idealist tradition thus holds that the sources of knowledge derive from innate structures that are independent of experience. Perhaps the most extreme adherent of this view was the philosopher Leibnitz who conceived of the mind as a "windowless monad" that had no contact with its environment and whose experience was the unfolding of a predetermined program.

At the other extreme empiricists have attempted to demonstrate that knowledge is totally reliant on experience and that cognitive development is the mind's gradual accumulation of sense data. As John Locke proposed, the mind at birth is a *tabula rasa*, a blank tablet, which has no inborn structure or organization, and upon which are imprinted the perceptions of sense data. The empiricist tradition had its greatest popularity in England and France in the eighteenth and nineteenth centuries, perhaps as a result of the impact of scientific methodology and the fascination with technology. Itard, a French educator and psychologist, tried to put empiricist theory to practical use in his attempts to educate the "wild boy" of Aveyron. Itard's[28] two monographs describing his procedures are classics, and his ingenious methods of pedagogy have enormously shaped the current educational practices with defective children. However, Itard's intent was to prove the correctness of the empiricist position, and he extensively quotes Condillac, one of its most extreme adherents. He believed that through proper stimulation he could develop the mind of a boy who had apparently lived alone in the forests of France until puberty. Ultimately Itard's experiment ended in failure: the child never attained the use of language, and Itard concluded that the wild boy had been born defective. It may well be, however, that too many years had passed and that it was too late for linguistic abilities to be formed.

Although the empiricist point of view experienced a revival of interest in this country with the work of Watson and the behaviorists, today few, if any, theorists would subscribe to either a strict idealism or empiricism in describing cognitive development. It is granted that both innate factors and specific experiences are necessary for the development of intellectual skills. It is less a question of how much is due to innate or learned abilities than how do these two sources of development interact to reinforce each other and optimally combine to exert their maximal effect. Associated problems that arise concern the specificity of eliciting certain capabilities at set stages of development as well as the possibility of reversing the effects of prior deprivation or overstimulation.

D. O. Hebb[23] has approached the problems of cognition from a neurophysiological standpoint and has demonstrated that much of the former controversies between psychologists regarding innate versus learned behavior may ultimately depend on the species of organism that had been studied. For example, insects appear able to perform complex functions immediately after birth, while higher organisms require long periods of nurturance and stimulation before showing even rudimentary forms of adult behavior. In primates and humans Hebb speculates that there occur two forms of learning. The first form utilizes a haphazard trial-and-error strategy and can be observed in very young children or in more mature organisms that were deprived of specific stimulation (such as children born with congenital cataracts or monkeys raised in total darkness). The second type of learning is characterized by insight, rapidity, and flexibility. Hebb comments that higher organisms that show this second form of learning also demonstrate a prolonged and less effective form of the primary type. It seems that as we go up the phylogenetic scale, there is an increase of type two learning but also that type one learning takes longer and longer. For example, a rat reared in darkness may require 15 minutes to an hour to learn a visual task after removal from deprivation; for a chimpanzee the time required to learn an analogous task may be weeks or months. Hebb attempts to explain this difference in terms of neuroanatomy, dividing the cortex into primary sensorimotor areas, which record external stimulation, and association areas, which combine and integrate these sense data. Through phylogenesis there is a gradual increase in the amount of association areas in comparison to sensorimotor areas, a relationship that Hebb calls the A/S ratio. Species with low A/S ratios show rapid primary learning since sensorimotor sequences are easily established because there is a paucity of association fibers to be organized. However, this same lack of association areas limits the amount of flexibility of behavior after the connections have been established. In contrast, organisms with high A/S ratios require a great deal of time for stimuli to organize the massive association areas. However, once this is accomplished the complexity of the connections allows for versatility of response as well as relative independence from the immediate situational stimulus.* The early stage of primary learning is still of paramount importance for higher organisms since it is during this period that the association areas are structured and thus cognitive relationships may be formed that persist throughout life. This aspect of Hebb's work strongly emphasizes the importance of early experience in all later behavior. In later writings Hebb seems to view the plasticity of the very young nervous system as so great that it is difficult to differentiate innate from learned behavior. Some support for Hebb's theories has come from Sapir's and Whorf's studies[53] in comparative linguistics showing that the structure of the language learned in childhood influences adult cognitive patterns.

Hebb's work is closely related to what has been termed "critical period" theory, which also attempts to deal with the manner of organism-environment interaction, stressing that there are "critical times" for certain behaviors to be learned or for specific stimulation to have its maximal effect. Critical period theory evolved from the field of ethology with the early experiments of Lorenz with newly hatched geese and partridges.[45] He found that although these chicks would normally follow their mother in normal development, they would learn to follow him if he presented himself first. This strong bond continued through maturation so that Lorenz speculated that as a result of this one-time presentation, his image had been "imprinted" in the chicks. Later experiments by Hess[26] showed that there were peak times for imprinting to occur and that after a certain amount of maturation imprinting could no longer be elicited. Scott's[46] work with puppies showed that even in higher species there were optimal times for the formation of an emotional bond to a trainer. These studies again point to the hypothesis that early experiences or the deprivation of such experiences may have a profound effect on future development.

Harlow's[21] work with rhesus monkeys showed that subjects raised without adequate mothering became sexually inadequate as adults and that the females were incapable of maternal behavior. Harlow found that monkeys raised in isolation for two years were grossly abnormal in play, defensive, and sexual behavior; those raised in isolation for six months were able to develop some play behavior; while those who returned to a natural setting after 80 days of age showed an almost total reversal of the isolation effects. There appeared to be a "critical period" time after which compensatory stimulation was ineffective.

* Hebb developed neurophysiological correlates of these types of learning. Very briefly, early simple percepts create "cell assemblies" or self-stimulating reverberating neuronal circuits in the brain. More complex impressions are stored in the form of "phase sequences" made up of a series of cell assemblies in a specific series. Although this aspect of this theory is important in neurophysiology, its detailed explanation in terms of anatomy is beyond the scope of this chapter.

Similar conclusions have been reached in work with deprived infants. Rene Spitz[47] found that infants who were separated from their mothers at six months of age and were not provided with an adequate substitute became listless and withdrawn, were retarded in their motor and intellectual development, and were prone to infectious disease. Spitz found that the effects of maternal deprivation were reversible only up to a period of three months, after which reunion with the mother did not completely nullify the effects of separation. Dennis and Najarian[10] came to similar conclusions following their studies of institutionalized children, although they interpreted the eventual developmental defects as a result of stimulus deprivation rather than the absence of mothering. Similarly Province and Lipton[40] found that institutionalized children showed significant delay in smiling responses, handling of toys, speech development as well as interpersonal relations. In follow-up studies Goldfarb[20] found that children who were placed in foster homes had higher IQs than children who remained in institutions.

The pressing questions raised by critical period theory are whether the effects of early stimulus deprivation are irreversible or if later remediation can be compensatory. Scott[45] believes that "organization inhibits reorganization," indicating that once patterns of cognitive and affective experience have been formed, it is difficult, if not impossible, to impose new patterns. Counterarguments are that most of the evidence for critical periods has been in lower species, which, as described by Hebb, have little capacity for complex neuronal interconnections in the association areas and possibly are thus less able to modify past learned behavior. Also most critical periods studied involve formation of affective rather than cognitive functions. The lack of hard evidence for the existence of critical periods in human cognitive development has caused Wolff[55] to criticize strongly the application of this formulation to child development, especially in the form of educational gadgets or "enrichment" toys that are promoted as necessary for proper intellectual growth. Wolff believes that the greater complexity of the human brain makes feasible much more flexibility than critical period theory allows. He believes there are optimal periods for the attainment of certain abilities such as the learning of a second language, but that on the whole the human mind can make compensations that are beyond lower species. In support of his argument he quotes recent studies[9,56] that seem to indicate that the effects of early deprivation are not as irreversible as previously thought. Another important point raised by Wolff is that there is little agreement about what critical periods exist in human cognitive development. Retardation may result from lack of stimulation or novelty, but it may also be a product of overstimulation and unpredictability, which prevents the child from assimilating his experience into meaningful categories. Finally the social milieu may actively discourage exploration or fantasy, negatively rewarding attempts to integrate and master experiences. Many of Wolff's arguments reflect the current controversy surrounding Head Start programs. Culturally disadvantaged children repeatedly have been shown to be deficient in language skills and abstract ability; however, the cause as well as the best methods for remediation are still hypothetical.

Cynthia Deutsch[11] has stressed difficulties in learning auditory discrimination as basic to later language disorders. The child is surrounded by meaningless "noise" with little meaningful auditory stimulation, so he does not develop the ability to identify nuances of sound.* Fineman[12] stresses the importance of mother-infant interplay in the development of an active inner fantasy life and ultimate cognitive ability. Pavenstedt[31,32] emphasizes the need for emotional tranquillity and regularity for any learning to take place.

Most remedial programs have found that intervention before age three is crucial,[8,29]

* In neurophysiological studies Hernandez-Peon, *et al.*,[25] have found that distraction tends to shut out information. Similarly Galambos, *et al.*,[19] have shown that stimuli that are not reinforced in terms of reward or punishment are extinguished and not recorded by the brain. The analogy is that unless language or sounds have meaning for the child they will not be integrated.

some suggesting that compensatory tutoring should begin prior to 14 months of age.[43] These findings reinforce the concept of epigenesis that unless early cognitive tasks are properly mastered, later development will be deficient. However, positive care during the first three years does not insure continued development; proper support and stimulation must continue throughout childhood. In fact, most of the gains made by Head Start programs have been rapidly lost once these programs were discontinued.[29] The difficulties that beset attempts to specify the causes of deficient cognitive development in disadvantaged children may serve as illustrative to the field of cognitive development in general.

The development of cognitive functions cannot be divorced from emotional factors such as a healthy mother-infant relationship in the early years, respect for autonomy and curiosity during early childhood, parental motivation and models during middle childhood, and the effect of peer pressure in the later stages. As systems of cognitive growth will be discussed below, it is important to keep in mind that such descriptions are really artificial since the development of intellectual functions cannot be separated from the total growth of the child. Recent studies have shown that children learn predominantly through action, that sensory stimulation by itself is insufficient unless it elicits a motor response that alters the environmental situation. There is a unity between stimulation, response, and alteration of the stimuli in the growth of behavior. Numerous experiments by Held[24] and others have shown that proper "stimulation" involves a process of motor response and resultant sensory feedback—a sequence that Held calls "reafference." A typical study involved a series of two kittens raised in darkness, but regularly placed in a circular drum whose interior was painted with vertical lines. One kitten was strapped to a harness connected to a central axis in the middle of the drum, but it was free to exercise. The other kitten was placed in a gondola directly opposite the "active" kitten, but it was unable to exercise. Each time the active kitten moved, he caused the "passive" kitten to be passively moved by a bar connecting the kittens. In this manner both kittens were exposed to the same stimulation. After an average of 30 hours in the apparatus the active member of each pair showed normal visual behavior in terms of averting collision with objects, not going off a "visual cliff," and so forth. After the same period the passive kitten failed to show these behaviors.

Toward a General Concept of Development

The preceding section has considered the sources of development as well as the question of interaction between maturation and environmental effects for optimal cognitive development. However, what is actually meant by "development?" It is well known that a 15 year old can perform mental tasks that are beyond the capacity of an 8 year old, who, in turn, can utilize concepts unavailable to a 2 year old. However, when a definition of what lies behind these obvious changes is sought, great difficulties are encountered. It appears easier to document static levels of achievement at different ages than to grapple with what is being altered in development. For example, most "intelligence tests" are based on purely empirical standardizations of certain given problems at various ages. These tests may give us evidence of the deviation of cognitive development from a standardized norm, but they offer little in terms of theory beyond the known fact that certain abilities increase with age.

Attempts to grasp *the essence of development*, its formal structure, rather than to document it, have come not from the fields of education or psychological measurement but rather have been spurred by the application of Darwinian evolutionary concepts to ontogenesis. One of the earliest and best known of these attempts is G. Stanley Hall's recapitulation theory. Hall speculated that human development presents in a condensed form all of the stages through which the species had traversed in evolution. The behavior of children could then be seen as atavistic remnants of former evolutionary periods. Hall's theory found its greatest influence in describing the play activities of children: the child's insistence on swinging from branches showed the influence of our simian ancestors just as the pleasure in outdoor group activities was seen as a vestige of primitive man's tribal existence. A very similar theory was espoused by Thorndike, who saw the behavior of the child as expressing those actions that had allowed the species to survive. Although avoiding Hall's Lamarckian interpretation of evolution, this theory as well neglects the effects of acculturation and seems to equate ontogeny with phylogeny. Today these two processes are seen as similar but far from identical so that the one cannot be used to explain the other, although both may follow analogous laws.

It was toward a discovery and elaboration of these laws of general development that the comparative psychologist Heinz Werner[49,50] applied himself in a series of articles and his major book, *The Comparative Psychology of Mental Development.* Werner believed that developmental psychology has two basic aims: (1) to grasp the characteristic pattern of each genetic level and the structure particular to it, and (2) to establish the genetic relationship between these levels in respect to the direction of development in order to discover any general tendency in development. Through his work Werner eventually proposed an "orthogenetic principle"[52] by which he hoped to be able to explain any open system in the process of evolutionary change. Toward the end of his career Werner had even discarded the dimension of time in his consideration of development, basing his studies on increasing complexity of certain factors in discovering the direction of movement for any open system. He thus antedated "general systems theory" and was a pioneer in the field of theoretical model-building for biological systems. Through this type of analysis Werner tried to encompass the evolution of cultures, the growth of language, the changes observed in psychopathology, as well as the normal development of the individual. In his book, in fact, Werner discusses many important aspects of ethology, anthropology, and psychopathology; however, for our purposes the discussion will exclude these other contributions and concentrate on his theory of human cognitive development.*

Werner criticized most psychological theories of development, with the exception of Gestalt psychology, as too mechanistic, considering development as the mere addition of new abilities. For Werner development is creative and organismic, bringing about totally new abilities and radically changing the developing entity. Each new stage represents a totally new organizational synthesis that must be understood through different criteria than previous or later stages. "Any level, however primitive it may be, represents a relatively closed, self-subsisting totality. Conversely, each higher level is fundamentally an innovation, and cannot be gained by merely adding certain characteristics to those determining the preceding level" (p. 22).[49] Werner attempted to find standards by which to compare and measure these differing stages of development. The two major changes that occur in development and that thus could be used to assess it were: (1) that development proceeds from a state of relative globality to a differentiation of parts, and (2) that there is an increasing integration of these parts into a hierarchical arrangement. Here again it must be stressed that Werner was interested in formulating the formal overall plan of development rather than a detailed descriptive analysis.†

In order to measure these two major dimensions, Werner formulated five sets of parameters that he applied to cognitive development. The first, that of *syncretic* versus *discrete* functioning, refers to the interpenetration of

* See Chapter 49 for an exposition of these aspects of Werner's theories.

† These concepts clearly have their roots in biological theory. For example, primitive embryonic cells are undifferentiated and similar. Through maturation these cells differentiate and become capable of performing specific functions of the various organ systems. Furthermore, the cellular structure ultimately forms a hierarchy so that certain cells control the activities of others and direct the organisms' behavior. These two principles, differentiation and hierarchization of parts, form the essence of Werner's definition of development.

functions in the child that become increasingly discrete in the adult. For example, the child often fuses together sensory, motor, and emotional components of experience. Young children exhibit synesthesia or the fusing of two sensory modalities, such as a three-year-old boy, reported by Werner, who stated that "a leaf smells green." This syncretic tendency of a child's thinking was experimentally demonstrated by Wapner[48] at Clark University. Children of increasing age were asked to adjust the speed of a series of pictures moving past an aperture in a board so that these pictures moved at the same speed as another series of pictures moving past another aperture in the same board. In comparing the adjustments made for a static picture (grazing horse) and a dynamic picture (running horse), it was found that younger children increasingly adjusted the dynamic picture series to a slower speed than the static picture series. The dynamic aspects of the presented form were syncretically fused with its actual speed for the younger children so that the running horse series seemed to be moving faster than the grazing horse series.

A further example of syncretic functioning is given by Werner and Kaplan[51] from their studies of linguistics, although their conclusions are equally applicable to child development. In studying the language of the Trobriand Islanders, they found that, in contrast to our own language, a specific word contained many exact yet unconnected meanings. The word "yam," for example, in addition to being an edible type of potato, denotes a specific degree of ripeness, bigness, and roundness. If one of these characteristics is missing, the object is no longer a yam—it is something else. Similarly the child infuses an object with a myriad of unessential qualities that to the adult seem separate from the object itself.

An important instance of the child's syncretic fusing of cognitive and emotional modes is expressed in Werner's concept of "physiognomic perception." Werner means that the child ascribes human qualities to all sorts of objects so that they appear animate and express some sort of inner life. A rock may be happy or a hat may be sad. Even adults retain this form of description in such phrases as an angry sea or melancholy sunset. However, while for the adult the imagery is metaphorical, the child actually ascribes feelings to inanimate objects, again fusing qualities that to the adult seem quite discrete. Werner also noted that in pathological states such as brain damage and schizophrenic psychosis there is a regression of formal thought operations to the syncretic mode.

A second parameter is that of *diffusion* versus *articulation*, indicating the transition from the global perception of wholes to the appreciation of parts in relation to the whole. Werner showed this tendency of young children to omit details and grasp only vague global characteristics by asking them to reproduce geometric figures. A square, as well as the letter "C," was reproduced as a circle, demonstrating a trend toward symmetry and an avoidance of nonglobal characteristics such as corners or breaks in contour. More complex figures were similarly simplified. A more crucial aspect of the diffusion-articulation continuum is the young child's tendency to equate a detail of the whole with the whole itself. Werner reports a boy who was afraid of spiders becoming upset when a hair stuck to his fingers. When the hair was removed the child asked, "Didn't the hair bite you?" showing that he equated qualities (biting) of the whole (spider) with the part (hair). In an analogous manner two situations may be equated if they share an element, however trivial, in common. Another child reported by Werner, who was picked up by an uncle wearing a rose in his lapel, expected to be nursed by the uncle because his wet nurse often wore violets. Here the sight or smell of flowers made the situation of being picked up by uncle identical to being fed by nurse. Again the use of part-symbols to represent wholes is often used by adults in allegorical works, but to the child the identity of the associated situations or objects is real rather than poetic or metaphorical. Werner termed this type of cognition *pars pro toto* (or part for whole) reasoning. This type of logical association has been subsequently found to occur in adult psychopathological states; in addition, it is similar to processes described by Freud in his interpretation of normal dreams. As will be described below, Arieti has discussed the same process as fundamental to "paleologic thought" and Piaget has coined the term "transduction" to describe association of wholes by their similar parts.

The remaining three parameters can be dealt with more briefly. The *indefinite-definite* continuum refers to the organization of goal-directed behavior and the ability to withstand desires to give way to momentary gratifications. The *rigid-flexible* continuum describes the transition from fixed, stereotyped responses to the ability to vary or modify behavior to meet the needs of a situation. Finally the *labile-stable* continuum refers to the distractibility of the child by external stimuli in contrast to adult behavior, which is characterized by persistence to a task despite environmental interruptions.

It is apparent from this very brief résumé of Werner's system that he was more concerned with formulating laws of development than in describing various developmental stages and in stating broad principles than in noting the abilities of children at various ages. His experiments and illustrations serve primarily to confirm his general hypothesis: that development is a fluid, evolving process that can be assessed by specific parameters. Wapner, who worked with Werner on numerous studies, has recently tried to postulate a series of stages of development as derived from the "organismic" point of view. Wapner describes four major stages in the cognitive development of the child. The first is the *biological-organismic* level, in which there is minimal separation of the self from the environment, physiological intraorganism mechanisms predominate, and behavior consists largely of innate reflexes. The next state is the *sensorimotor* level, characterized by a separation of the self from the environment and the willful manipulation of objects. The third stage is the *perceptual operations* level, in which thought and behavior are dominated by the most striking perceptual elements in the environment. The last stage is the *conceptual operations* level, in which behavior is a result of symbolic representations of objects and events rather than pure mirroring of reality. Wapner stresses, as did Werner, that as newer levels are reached, the older cognitive styles are not eliminated but relegated to a lower hierarchical status and that primitive forms of behavior may still emerge in the presence of stress or disease.

Despite these attempts at delineating steps in cognitive development, it must be reemphasized that Werner's major contribution seems to have been in trying to grasp the essence of the developmental process rather than in defining discrete stages. The more painstaking and perhaps thorough task of studying the child's cognitive behavior as it relates to developmental stages as well as the total process has been more closely actualized by Piaget and his collaborators, whose work will be considered next.

The Genetic Epistemology of Jean Piaget

Theoretical Introduction

Piaget has carefully termed his work "studies in genetic epistemology," emphasizing his concern with the way the individual constructs and organizes his knowledge of the world during his historical development.[18,37] At each stage of development problems in world construction are handled according to age-appropriate conceptions, with the earlier attempts forming the foundations for later capabilities, culminating in the logical thought of the adult.

Piaget[39] states, "psychologists eventually began to wonder whether the logic was innate or resulted from a gradual development. To solve problems of this kind they turned to the study of the child and in so doing promoted child psychology to the rank of genetic psychology. Genetic psychology becomes an essential tool of explicative analysis to solve the problems of general psychology."

In attempting to derive the adult modes of thought from childhood experience, Piaget shows his basic roots in biology and his great debt to Darwinian thought. In his later works[33] he acknowledges that our knowledge of reality is not a copy but an organization of the real world. But at the same time this knowledge cannot go beyond our actual experience as human beings enmeshed in nature. He discards any transcendental aspect of truth that goes beyond our biological makeup. Knowledge is seen as a biologically useful ability that regulates the individual's exchanges with the environment, and cognitive functions are conceived of as specialized organ systems that help higher organisms relate to their world.

Piaget specifies three types of intelligence: (1) innate abilities such as instincts; (2) learning that relates to the concrete physical world; and (3) logical mathematical structures that are utilized in nonrepresentational abstractions. Whereas most species are suitably adapted through instinct, in the higher animals there is a breakdown of instinct or hereditary programming in favor of new cognitive self-regulations. These take the forms of "reflective interiorization," meaning the ability to re-create internally segments of experience, and "experimental exteriorization," meaning the ability to vary the course of a behavioral response for maximum benefit. For Piaget, cultural or social interaction takes the place of previously fixed instinctual responses in protecting an organism in its exchange with nature. This Darwinian view of the origins of cognition bears a strong resemblance to the equally evolutionary theory of thought in recent psychoanalytic literature. Hartmann,[22] for example, states that as an organism ascends the phylogenetic scale, the id becomes increasingly estranged from its reality, giving rise to the need for greater ego functions. For psychoanalysis this estrangement of instinct from nature leads to inescapable problems in biological expression and fulfillment; for Piaget, however, this development of mind does not bring about any psychological disharmony.

The other major difference between Piaget and the classical epistemologists is his emphasis on the role of action in cognition. For Piaget an object is known to the extent that it is acted upon or forms part of an action process. Cognitive functions not only subserve action but also are forms of action themselves. Thinking is not only for action but also is action. What Piaget calls operational thinking is in essence internalized action sequences; thus the early motor action sequences constitute the source of all later cognitive operations. This stress on mind as action and as specifically concerned with solving problems that will insure adequate functioning helps explain Piaget's notion of the nature of intelligence and of some of the driving forces in cognitive growth. This view of thought as internalized action shows Piaget's roots in pragmatism and ultimately his seeming disinterest of the richness of inner life.

Growth as Sequential Equilibration

Piaget broadly equates intelligence with biological adaptation, which he sees as more complex than simply survival or the establishment equilibrium between organism and environment.[36,38] Adaptation is a process involving the reciprocal sequences of assimilation and accommodation. By assimilation is meant the process whereby "reality data are treated or modified in such a way as to become incorporated into the structure of the subject."[39] The environment is grasped or reacted to only if it can be fitted or transformed into the existing mental capabilities of the subject. For Piaget "intelligence is assimilation to the extent that it incorporates all the given data of experience within its framework."[39] The corollary process, accommodation, occurs when alterations in the environment cause the organism to broaden its mental framework so as to integrate the new data. Rather than altering or misperceiving the stimulus to fit pre-existing cognitive modes, the organism must modify itself to deal with this new experience. Each process is never pure since each assimilation introduces some new element that causes accommodation; and each accommodation, by creating new mental schemes, sets the stage for more sophisticated assimilation. It is this interplay between assimilation and accommodation that describes the progressive enlargement and development of mental structures or schemes. At each stage of mental development there is a constant sequential balance between the two processes. It is important to note Piaget's stress on the reciprocity of stimulus and response. The cyclic process can be found in all stages of development, from the neonate's spontaneous movements and reflexes to the adult's abstract formulations.

Although the equilibration of assimilation and accommodation are used to describe development, these are not the forces that cause cognitive growth; rather Piaget delineates four general factors that are important causes for mental development. The first factor is organic growth or maturation of the nervous and endocrine systems. A second factor is the role of exercise and acquired experience from action performed upon physical objects. This area of development relates directly to first-hand experience with the routine physical world. The third factor is social interaction and transmission. The fourth factor is an inherent striving toward equilibrium or self-regulation. This last factor is as far as Piaget seems willing to go in favor of any pre-established or teleological view of development. He quite specifically states, "in the development of the child there is no pre-established plan but a gradual evolution in which each innovation is dependent on the previous one.[39]

The Stages of Cognitive Development

Through the interlocking processes of assimilation and accommodation there is a progressive evolution in the mental structures that are utilized to adapt successfully to the environment. Piaget's experimental effort has been to investigate and describe the schemes or modes of mental organization at various stages of development. This goal helps explain his most frequently used experimental method called the "clinical method," in which children of various ages are asked to perform the same or similar tasks. Following the presentation of a task the child is asked about his understanding of his performance. Piaget is not so interested in the result of the child's behavior as in the types of mental procedure that underlie the child's performance at various stages of ontogenesis. Through these investigations Piaget has been able to document a series of stages that describe the major cognitive modes of childhood. What changes during development are the mental schemes, the mode of organizing and constructing experience.[34]

The earlier or more primitive schemes are concerned with action per se without ideational representation; they are characteristic of the sensorimotor period that extends from birth to roughly 24 months of age. Although behavior during this period is essentially devoid of any conceptual representation of the external environment, the behavior is generally adaptive and therefore "intelligent." At first the sensorimotor level is characterized by reflexes, which, however, should not be conceived of as pure automatism, but as requiring exercise for their continued usage and development. Thus after a few days the newborn "nurses with more and more assurance and finds the nipple more easily when it has slipped out of his mouth than at the time of his first attempts."[39] This stage of reflexes gives rise to the stage of acquired habits. The infant begins coordinating vision and prehension; that is, the child begins to manipulate freely objects in his visual field. This ability to act upon environmental objects initiates what J. M. Baldwin had called a "circular reaction," which implies the eliciting of a stimulus from the environment as a result of the infant's behavior, which, in turn, tends to affect further behavior. In this manner the response pattern tends to prolong itself and to stabilize its own existence. Through a positive feedback mechanism a primary circular reaction is the exercise and practice of innate reflexes such as thumb-sucking. The secondary circular reaction, however, is more akin to what has been called operant conditioning. The infant's behavior accidentally produces an environmental change that causes a repetition of the act that had fortuitously produced the change. This type of behavior appears around four months of age and ushers in the beginning of intentional adaptations, that is, the beginning of an act in the service of a desire, and with it a gradual differentiation of means and ends. Piaget describes how the child learns to pull a cord in order to cause the shaking of a rattle at this stage. At the same time the child begins to separate internal actions from external results. He begins to form the concept of an external world of permanent objects existing in continuous space. After the age of four months the child begins to react toward a partially hidden object, such as the nipple or a bottle, as he formerly did toward the whole bottle.[34] In the next stage of the sensorimotor level extending from 8 to 12 months, the child is able to utilize his action schemes in new situations to produce novel results and achieve new goals. The schemes have become an instrumental act that can be applied to varied situations.

There are correspondent advances in the child's concept of an object. At about one year of age the child can search for an object if it is placed behind a barrier. Thus the child has the concept of the permanence of that object that continues to exist even when it is not directly observed. However, if the object is removed from behind one barrier and in full view of the child placed behind a second barrier, the child will continue to search for it behind the first. The child has the idea of object permanence but not of the multiple displacements of the object. It has to exist where it belongs. It is not until a later stage extending from 12 to 18 months of age that the child appears to understand the multiple displacement of objects and to search for the object wherever it was last put. Yet at this stage the child still cannot cope with "invisible" movements. Thus if an object such as a ball is held in a clenched fist, the fist put behind a pillow and out of view of the child, and the ball released behind the pillow, the child will persistently search the hand for the ball, never considering that the ball may have been left behind the pillow. It is only with the last stage of the sensorimotor level of intelligence, extending from 18 to 24 months of age, that the child can conceive of actions that he does not witness directly and thus searches for the ball behind the pillow even though he did not see it put there.

The later stages of sensorimotor development are also characterized by tertiary circular reactions whose central attribute is the invention of new behaviors to achieve desired ends. In the secondary circular reactions the child was able to use established patterns in novel situations. Here the response pattern itself is original. This behavior is similar to trial-and-error learning in which alternate solutions are attempted until the correct one is accidentally stumbled upon. It is important to note that these schemes are not created *de novo* but evolve from acquired habits that, in turn, grew out of innate reactions.

The final stage of the sensorimotor level is marked by the beginning of internal representation, which is the transition to later forms of intelligence. The child is able to find new means, not only by "external or physical groping but also by internalized combinations that culminate in sudden comprehension or insight."[39] There is no longer a need to see objects in order to realize their permanence. The ability to picture objects mentally allows the child to search for them after multiple displacements. This understanding underlies the child's concept of spatial relations. The idea of an object being able to cross a path from multiple fixed positions is basic to seeing space as a continuous medium. At the same time the internal representation of space is necessary for the child to see himself in a space and to be able to follow a continuous map from one position to another. A possible limitation of Piaget's pragmatic view of thought as internalized action sequences is his relative neglect of the importance of internalized images during early childhood. The works of Melanie Klein and later Arieti have shown the crucial role that internalization of emotionally important personages and events during this stage plays in later life. Similarly, although Piaget has devoted an entire book to the subject of symbolism, he seems to view the symbolic functions of the mind in terms of their use for strategies and not important in their own right, as emphasized by Cassirer and Langer.

In summary, the sensorimotor level traces the development from birth to the beginnings of internalized thought. It shows the evolution of habits from reflexes and trial-and-error behavior, culminating in internalized action schemes that make possible the mental manipulations of objects that are not immediately present. With the advent of true representative cognitive structures, the child is no longer at the stage of sensorimotor behavior, yet, on the other hand, neither is he able to form "operative" schemes, which imply a separation of thought from the immediate aspects of reality. The stage of operational thinking consists of representational actions that are reversible in thought but not necessarily so in reality. Prior to this stage the schemes are "preoperational." They are composed of internalized actions but are still tied to perceptual rather than cognitive criteria.

Another characteristic of the preoperational level is "egocentric" thinking, which should be taken only in its epistemological sense: the child cannot put himself in the place of another person. For example, when asked to describe the view of a model of mountains from positions other than his own, the child at the preoperational level is unable to describe or to select out a proper perspective of the model other than what he sees from his own position. There is also a similar ascendancy of assimilation over accommodation. The child tends to fit his experience into his own categories rather than to expand these categories to give a more realistic grasp of his environment. This is evidenced by the preoperational child's great use of fantasy in play. The play is unhampered by external rules, which are altered as the situation demands, or by a strict adherence to the realistic nature of objects. These characteristics help explain another aspect of the preoperational level, that of ascribing a willful causality to most physical events. For example, children feel that night comes so that we can go to sleep or that if you put on a raincoat it will rain.

This level of intelligence can further be subdivided into two stages, that of preconceptual thought, which lasts from roughly age two to age four, and that of intuitive thought, which lasts from roughly age four to age seven. The former stage begins with the internalization of action, that is, true conceptual intelligence. The child shows deferred imitation, which gives evidence of the beginnings of this internalized representation. The transition from representation in action to representation in thought is reinforced through symbolic play and through the use of drawing and painting materials. Finally the development of language gives the child the usage of symbols by which to handle internal images and memories. Through language the child can represent a long chain of action sequences very rapidly (while at the sensorimotor level the child had to follow events without altering their speed). The reasoning of the young child, however, is still concrete and at the same time distorts reality. Piaget describes how the young child forms primitive classes based on particulars of objects that have no essential relevance. As stated above, this process, which Piaget calls transduction,[35] is similar Werner's *pars pro toto* reasoning. The particular of an object is equated with the whole in relating objects. The child has not as yet derived a system of concepts that separates details of objects from their essential nature. This centering on an unessential, although perhaps the most conspicuous, aspect of an object continues into the intuitive phase.

The second phase of preoperational intelligence is characterized by a growing capability of separating thought from action. The child is better able to group objects into classes, but only on a perceptual rather than a cognitive basis. The child has not achieved the primacy of thought over appearance. For example, Piaget presented children with a box containing 20 wooden beads, 18 of which were brown and 2 of which were white. The children were then asked whether a necklace made from the brown beads would be longer, shorter, or equal in length to a necklace made from wooden beads. Children at the intuitive level mostly replied the the brown beads would make the longer necklace. From this and similar experiments Piaget concluded that the child at this stage is incapable of thinking of two subordinate classes, that is, white and brown beads, at the same time as he is thinking of the whole class, that is, the total number of wooden beads. The child's attention is centered on the preponderance of brown beads in relation to white beads, the most conspicuous property, and he cannot simultaneously switch from the relation of "brown beads to white beads" to the relation of "brown beads to the total number of beads." The child's thought is guided by the perceptual aspect of reality. Children continue to conclude that a brown bead necklace would be longer even after they acknowledge that there are more wooden beads than brown beads. The point is that these two aspects of the same situation cannot be integrated at this level. At the level of concrete operations, stretching from roughly age seven to age eleven, however, the child immediately answers that the wooden bead necklace would be longer because "there are more wooden beads than brown beads." Here the child can simultaneously take into consideration the relation of brown beads to white beads (part to part) and brown beads to wooden beads (part to whole). Piaget asserts that at this later stage the child's thought is "decentered," that is, no longer exclusively focused on the perceptual, and is "reversible," that is, can move back and forth through a logical relational thought sequence.

These principles are also exemplified by Piaget's experiments on conservation. One of these experiments involved an equal amount of liquid in two identical beakers, A–1 and A–2, which the child acknowledged as the same. The liquid from beaker A–2 was then poured into two smaller beakers, B–1 and B–2, directly in front of the child. The child is then asked if the liquid in the beakers B–1 and B–2 was equal to the amount in the original beaker A–1. Children at the intuitive level of intelligence felt that the quantity of liquid had been altered when poured into the two smaller beakers. Similar alterations in quantity were ascribed to when the liquid was poured into different shape beakers, although here again the liquid was poured directly in view of the child. Piaget concludes that these interpretations are due to the child's lack of the schemes of reversibility and conservation; the child centers on what he sees and cannot disengage his thought from his perceptions so that he can mentally reverse the process and conclude that the amounts of water were originally the same. Because of this perception bound set the child cannot consider two aspects of one situation simultaneously, but can only examine one aspect at the expense of all the others. For example, when the child at this stage is shown a ball of clay that is rolled into a sausage shape, he will say that there is either more clay because the sausage is longer than the ball or less clay because the sausage is thinner than the ball. The child cannot conceive of the ball simultaneously becoming both longer, thus having more clay, and thinner, thus having less clay. It is this ability to attend to two aspects of the environment and to relate them in a coherent fashion that leads to the scheme of conservation and marks the beginning of the operational stage. Similarly with the bead experiment described above, the child in the preoperational stage cannot conceive that white beads and brown beads equal wooden beads. He is centered on the color and cannot conceptualize the two aspects of color and total number simultaneously.

It is with the onset of the operational period that perceptual appearances no longer predominate over thought processes. In this period thought becomes truly "operational," that is, independent of what is phenomenologically possible. The child can go through action steps in his mind that would be impossible in reality. With this enlargement of perspective or "decentering," the child at seven or eight becomes less egocentric. He can shift rapidly between his own views and the views of others, facilitating communication because there is also less idiosyncratic meaning given to words. There is an overall autonomy of the central cognitive processes from the immediacy of the situation. Thought begins to follow a logical, rather than a haphazard, perceptual structure. The child forms the schemes of reversibility and conservation despite perceptual evidence to the contrary. The schemes are seen to go from a quasi-logical order of groupings to a higher hierarchical level of classes.

Yet during the first part of the operational period children can only reason in this manner when they are actually manipulating concrete objects. Their abilities are not revealed on a verbal or nonrepresentational task. A child who can easily arrange a series of dolls according to height and match the dolls to different size sticks either in a progressive or reverse order cannot as yet answer questions such as "Jane is lighter than Nancy. Jane is darker than Lois. Who is the darkest of the three?" The child can reason independently of perceptual influences, but only when the elements of his cognitive behavior are concretely present. Thus the first part of the operational period is called that of concrete operations. To quote Piaget,[39] "The concrete operations relate directly to objects and to groups of objects (classes), to the relations between objects and the counting of objects. Thus the logical organization of judgments and arguments is inseparable from their content." The child can only reason with what is palpably before him. He cannot utilize nonrepresentational abstractions in his cognitive operations. There can be no hypotheses unless the elements of the propositions are directly present in reality. The eventual freeing of thought from content makes the advance to the stage of formal operations possible. In this final stage there is a disconnection of thought from concrete objects so that the mind is capable of dealing with relationships between things rather than with only the things themselves. The preadolescent becomes capable of forming classes according to abstract principles and similarities that defy concrete representation. This "combinatorial system" allows the mind to consider all possible alternatives of any given situation. Some ramifications of this advance are that the adolescent can conceive of the future and plan for it, can think not only about concrete things but also about his own thought.

Furthermore, this ability to consider all alternatives leads to an "experimental spirit." The adolescent, when presented with a problem, takes a preliminary inventory of all the factors and then varies each factor alone, keeping the others constant. At the level of concrete operations the child proceeds directly to action with little attempt at systematization. One of Piaget's and Inhelder's[27] experiments demonstrates the difference between concrete and formal operations quite clearly. They arranged a series of jars of colorless liquids and then showed the child that by adding a few drops from the last jar to an unknown mixture of the liquids a yellow color could be produced. The significant aspect of this problem was that there was no way for the child to figure out ahead of time which mixture of liquids would produce the yellow color once the indicator was added. Inhelder and Piaget then observed the manner by which children tried to combine liquids to arrive at the yellow color. The younger child attacked the problem by adding a few drops from the last jar to each of the others and then felt essentially defeated. From here on he proceeded in no particular order and usually did not think of mixing various liquids and then adding the drop from the last jar. The child at the stage of formal operations, however, solved the problem by systematically going through all combinations of liquids, often keeping notes to be sure that he could keep track of his experimentations. Eventually he not only solved the problem but also identified the different liquids as to their relationships. For example, one jar contained a substance that prevented the color from appearing. Therefore, in the stage of formal operations the person can generate theories about relationships and derive laws that will explain occurrences in the environment. These laws are not limited to their immediate content and can be applied to analogous events. To quote Piaget,[39] "indeed the essential difference between formal thought and concrete operations is that the latter are centered on reality whereas the former grasps transformations and assimilates reality only in terms of imagined or deduced events."*

* In a rigorous manner Piaget has been able to characterize formal operations as a system of four cognitive processes: i, identical transformations; n, inverse transformations; r, reciprocal transformations; and c, correlative transformations—each describing a process. Significant as this reduction is, a thorough exposition of it is beyond the scope of this summary.

Summary and Overview of Piaget's System

It will be obvious at this point that Piaget and his associates have derived an enormously rich and complex concept of development, any small segment of which could generate an enormous amount of research and study. However, it is important in reading Piaget to constantly keep in mind his overall system and his epistemological interest. Briefly Piaget believes that cognitive growth is best described as the progressive evolution of the schemes that underlie outward intelligent behavior. At the sensorimotor level schemes are pure action, but gradually this action is internalized into mental representation so that the schemes become truly cognitive. These cognitive schemes progress from representations of the environment to more sophisticated attempts to construct an understanding of the world. During the preoperational phase this construction of reality is haphazard and follows a loose, disconnected pattern. Objects and events form groupings rather than logical classes and are associated by idiosyncratic or egocentric linkages. The child centers on the most obvious perceptual elements of his environment. This dependence on what can be observed is gradually subsumed under what can be reasoned, exhibiting the supremacy of thought over perception. This decentering allows the mind to follow reason rather than appearance. Piaget thus differs from the Gestalt psychologist's attempt to describe perception and cognition by similar laws. For Piaget these functions are entirely different, with perception being an inferior and more rigid form of behavior. The mind conceives of schemes, such as reversibility, that are impossible in the real world. At first these operations are possible only when directly involved with concrete objects, but eventually operations involving abstract, nonrepresentational forms are evolved. Piaget thus views cognitive development as progressively freeing thought from its concrete surroundings, culminating in abstract schemes that are totally independent of concrete experience. At the same time, however, Piaget remains firmly rooted in biology, stressing that each advance in schemes serves the more pertinent purpose of organismic adaptation. Piaget seems most interested in the mind as a creator of strategies and as the solver of problems. In later years he has devised mathematical-logical models to describe the mind's processes at various stages. As such, he has been less interested in the irrational, poetic, and creative aspects of mind, and his effort has been one of a rigorous logician.

In recent years there has been a renewed interest in Piaget that might be best described as an explosive revival. Piaget has been discovered, not just by developmental psychologists, but by educators, philosophers, social theorists, and psychoanalysts. Some applications of Piaget's work have been extremely important and interesting. Inhelder has applied some of Piaget's concepts to the diagnosis of mental retardation, which might ultimately culminate in a new system of intelligence testing. Others have used Piaget's experiments to assess the effects of cultural differences. Golden and Sims, for example, found that children of varying social classes did equally well on experiments like Piaget's until two years of age, when lower socioeconomic groups of children began performing much more poorly than higher socioeconomic groups. Wolff[54] has attempted to integrate psychoanalytic findings of object relations with Piaget's work on object constancy in infancy. Odier[30] has utilized much of Piaget's work to show the childhood origins of adult symptoms when extreme anxiety causes a reemergence of prelogical forms of cognition. For example, Odier sees a resemblance between the child's reification of his own feelings and perceptions in believing in their objective reality and the adult's use of projection in pathological states. Similarly Freeman and McGhie[13] have attempted a description of schizophrenic thought according to Piaget's levels of cognitive development, and Anthony[1] has tried to apply some of Piaget's concepts to child psychiatry. Although these applications have been salutary, Piaget has warned against the overly pragmatic use of theory and what he calls "the American question," meaning the ways by which to speed up intellectual development. Piaget makes a strong point that development has to proceed at its own pace. Too much stimulation or instruction is simply not assimilated and eventually might be harmful.

Finally Piaget seems more interested in theory than in these pragmatic applications. In his summary works he quotes experiments that test other theoretical hypotheses rather than practical aspects of his own work. While much applied, and misapplied, to all sorts of educational and psychological endeavors, Piaget seems to wish to remain in the realm of philosophy and epistemology rather than applied science.

The Psychoanalytic Approach to Cognitive Development

Until recent years the field of cognitive development played a very limited role in psychoanalytic thought. Although certain cognitive concepts have always been a part of psychoanalysis, these have been used mainly in reference to symbolization in symptoms and distortions in dreams.[16] The study of cognitive life as recognized by academic psychologists was considered to be too superficial or too removed from basic motivating forces to exert any real effect on development. In addition, the early years of psychoanalysis were primarily devoted to the demonstration of the existence of unconscious life, infantile sexuality, and psychic determinism—against massive opposition—so that such topics as cognitive development had to await easier times.

Nevertheless, any system of psychology as far-reaching and comprehensive as psychoanalysis could not entirely ignore cognitive factors, and Freud did attempt to formulate a theory of thinking. According to this model, mental excitation flowed from sensory structures to the motor process that discharged the excitation. Along this basically S-R path memory traces were activated by the excitation, aiding in the choice of discharging action. When, however, the ability to discharge the energy through action was blocked, it reversed its flow back toward the sensory end, resulting in an internal psychic experience rather than action. This model sought to account for dreams, hallucinations, and psychiatric symptoms. The point of this model is that thought or conscious experience occurs when motor discharge of energy is blocked. In a later paper Freud[15] used the same model to account for the mental life of infants and the development of thought. The beginnings of thought or images appear when instinctual satisfaction, and thus the discharge of energy, is not possible because of the lack of a satisfying object. The instinctual energy then activates a memory trace of previous satisfaction as an attempt to partially discharge the energy, and this previous state of satisfaction is hallucinated by the child. Therefore, thought begins as an attempt at wish fulfillment and follows what Freud called the "pleasure principle." Ultimately, however, the hallucinated satisfaction does not really gratify the instinctual need, and the infant must come to grips with external reality as well as his internal wishes. There is, therefore, a transformation to the "reality principle." However, the mind is always ready, as in dreams or in psychosis, to revert to an hallucinatory world of pleasurable wish fulfillments. Freud's view of mind, his metapsychology, has been repeatedly criticized as being limited by the prevalent scientific modes of thought of his time. This early discription of the "psychic apparatus" leans heavily on energetics and resembles the hydrodynamic models of physics popular at the turn of the century.

Eventually, through the evolution of psychoanalytic theory, thought became more than simply an internal agency of wish fulfillment. As a result of revisions brought about by the "structural theory" and ego psychology, the ability to internalize action sequences into thought took on protective aspects and evolutionary significance. The ability to think was seen not just as a way of postponing actual satisfaction and temporarily discharging energy, but as a means of trial action that helped the individual adapt to his environment. Hartmann[22] especially stressed this pragmatic survival function of thought in psychoanalytic theory emphasizing its adaptive and "conflict-free" functions. Nevertheless, Hartmann attempted to reconcile his "new" view of the ego with the original economic theory of the "psychic apparatus" as an instrument to reduce tension or energy of the instincts.

Later psychoanalytic thought stated that the ego does not arise out of the id through conflict resolution, but that the ego and its functions develop independently if the environment is adequate. This significant innovation stresses that the ego (and thus cognitive functions) has an autonomous development that unfolds throughout ontogenesis and that development is not contingent on resolution of conflict. Anna Freud[14] has extensively documented various "developmental lines" that trace the path of ego functions through childhood, stressing mastery of such areas as eating, eliminatory functions, body management, as well as relationships to others.

Concomitant with these views on the purpose of thought in ontogeny and phylogeny, Freud postulated two great classes or modes of thought that he called the primary and secondary processes. The primary process was solely concerned with the discharge of energy through any means. Its goal was release of tension or pleasure, and it was not characterized by any strict logical structure. This primacy of instinctual discharge was described in terms of "loosely bound cathexes," which meant mobility and instability. Cognitively this meant that instinctual energy could be displaced onto neutral figures, or that an image could become a composite of many objects (condensation), if this facilitated tension reduction. The primary process was essentially the mode of the unconscious; it manifested itself in dreams, psychosis, parapraxis, and other symptoms that revealed the true instinctual desires of the individual. On the other hand, the secondary process is characterized by Aristotelian logic and typifies most of our everyday conscious behavior. This secondary process mode of thought is concerned with the individual's relationship to his environment rather than with the release of tension. In terms of development the infant's cognition follows the primary process and his main concern is the attainment of pleasure. Through repeated frustrations, however, the infant learns to cope with the environment, and through the attainment of the reality principle he develops secondary process modes of cognition.[15,17] This gradual transition from primary to secondary process and the development of ego controls over the biological drives that cannot be expressed because of environmental restrictions have formed the bulk of psychoanalytic studies on the development of cognitive structures in the child. David Rapaport,[41] who was perhaps the psychoanalytic theorist most interested in cognitive theory, stated quite clearly that the major concern of psychoanalytic formulations in the development of cognition was to trace the evolution of ego mastery in the perpetual conflict between satisfaction of internal drives and societal demands. Numerous psychoanalytic studies of child development repeat the theme of ego mastery and view cognitive growth only in the service of this primary aim.

Only in recent years have psychoanalysts turned their attention to cognitive processes as significant in their own right and not simply as better methods of defense against unconscious drives. One problem that is currently attracting a great deal of attention is the process of internalization of environmental figures and events into intrapsychic structures. Sandler,[42] in a number of articles, has stressed the importance of the "representational world" inside the mind as directing behavior. Schafer,[44] in an important monograph, has tried to refine the vagueness ascribed to terms such as introjection, identification, and incorporation that are commonly used in psychoanalytic parlance. Although these attempts are crucial in directing study to the child's cognitive experience and lessening the emphasis on the pure unfolding of instinctual processes, this area of research is still in its beginning stages and has not as yet arrived at definitive formulations. Other authors who have dealt with topics such as identity formation or the relation of the self to environmental figures without trying to integrate their theories into an energetic model have, on the other hand, refused to classify their concepts as basically cognitive as if this meant a betrayal of psychoanalytic principles.

Another avenue of psychoanalytic study has been the work of Arieti, who has dealt with the development of inner reality and motivational forces from a truly cognitive frame of reference. While praising the psychoanalytic emphasis on inner life, Arieti[2,6,7] disagrees with its stress on instinctual and biological forces. He believes that the individual is ultimately motivated by higher order concepts such as expectancies of others, demands from the self, and internalizations of significant others rather than discharge of energy or primitive biological needs. In a recent book Arieti[4] has attempted to describe the development of these inner concepts through ontogenesis. His concern has not been with the child's ability to manipulate the physical environment as much as with the transformation that occurs in intrapsychic life.

He describes the first months of life as dominated by "simple feelings," or "protoemotions," such as tension, appetite, fear, rage, and satisfaction. In general, these protoemotions are global, nonlocalized bodily experiences that are elicited by a stimulus directly present in the environment. These sensations require a minimum of cognitive ability and appear to be the felt components of autonomic reaction patterns.

Toward the end of the first year of life, according to Arieti, the infant begins to retain enduring mental representations of external objects, events, and relationships. The child can create an image—a memory trace that has psychical representation. The image becomes a substitute for the external object and can from then on direct behavior. With the formation of the image inner life truly begins. Cognition can no longer be considered a hierarchy of mechanisms but a psychological content that retains the power to affect its possessor, now and in the future. In contrast to Piaget and some of the stricter learning theorists, Arieti is concerned with the content as well as the processes of the child's psyche. This ability to create lasting internal images initiates the child into what Arieti[4] calls the *phantasmic stage of inner reality*. This stage is characterized by a higher level of emotional life since the child responds to internal mental constructs as well as to the stimuli of the environment. Arieti calls these feelings "second order emotions" because they are not elicited by immediate, external threats to homeostasis but by the anticipation of such a change; that is, by a mental event that predicts an external event. In this manner purely inner cognitive events begin to alter the life of the child. As early as 1947 Arieti[5] attempted to separate fear and anxiety on a cognitive basis: fear is a primitive reaction to an environmental object, while anxiety results from an anticipated image, that is, from a cognitive construct.

At this stage of development the child also constructs what Arieti has termed the "paleosymbol." By this he means a specific mental concept that represents something that truly exists in reality but whose meaning and value are highly private and personal; the meaning of the paleosymbol is not commonly shared by others. These highly idiosyncratic values given to paleosymbols may account for the young child's seemingly irrational likes or fears of common objects. According to parental handling, neutral objects such as a feeding table or baby stroller can become internally associated with various emotions so that as rudimentary symbols for these objects are internalized, specific meanings are given to their internal representations. For Arieti not only is the creation of symbols in the service of external adaptation, but also it is essential for the construction of inner life; they become the building blocks for psychic reality. Again there is an important emphasis on inner content and its emotional impact as well as on adaptation to external demands. In addition, however, at this phantasmic stage of development the child has difficulty in clearly distinguishing between internal and external reality. Much of what has only internal meaning is projected or acted out on the environment. The child readily mixes fantasy and fact. This flexibility may account for the richness of clinical material that may be obtained through symbolic play techniques with children. At this stage they literally act out their inner fantasies through the use of dolls and the like, losing themselves in the context of play so that it is no longer "play" in the adult sense of the word. It is only gradually that the child separates the two domains of internal and external reality. This final separation begins to be accomplished when the child is able to form true concepts and has completely mastered verbal means of expression.

Between the phantasmic level of inner experience and the eventual mature form of psychic reality, Arieti postulated the emergence of a mental construct that he calls the "endocept." By this concept Arieti means a primitive organization of memories and images that, however, are nonrepresentational. The endocept is a feeling state that cannot be expressed with the precision of language: it is "at times experienced as an 'atmosphere,' an intention, a holistic experience which cannot be divided into parts or words . . ." (p. 97),[4] comparable to what Freud has described as the "oceanic feeling." This experiential level can be relieved in adult life in aesthetic experiences or in certain empathic feelings toward others.

However, in regard to purely cognitive growth the phantasmic world of paleosymbols is replaced by a primitive attempt at logical thinking. Arieti calls this stage of experience "the paleological world," stressing both its primitive and its logical characteristics. During the first stage there is an attempt to structure experiences and to associate events into broader categories, but the associative linkages are highly arbitrary and idiosyncratic. Arieti describes this clustering of mental events as a "primary aggregate," appearing like a strange agglomeration of disparate things put together, as they are in a collage. At other times it may be recognized as an embryonic structure from which conceptual structures eventually emerge. Or in some cases it may even embrace a field that at higher levels of development corresponds to a highly abstract concept. The primary aggregate form of cognition is replaced by "paleologic thinking." While not as yet conforming to the adult standard of Aristotelian logic, this form of mentation is not haphazard; it does follow specific laws or organization in which elements of thought are arranged into classes. However, the characteristics by which this organization of elements takes place appear insignificant or unessential to the adult. Objects or events are seen as identical if they have one quality in common. This stage is comparable to Piaget's preoperational level, where objects were associated by their most prominent perceptual quality, as well as to Werner's *pars pro toto* functioning. It is also similar to what Freud had described as primary process thinking as universally manifested in dreams. For example, an individual may dream that he is in the presence of a king, and on analysis of the dream he will reveal that the king was representative of the dreamer's father. Here a similar element (exalted authority) had been used to create a paleologic identification. In view of the similarity of paleologic thinking to the primary process, Arieti terms its form of cognitive organization the "primary class." In contrast, what Arieti calls the "secondary class" (similar to Freud's secondary process) is a collection of objects or events that have elements in common, but these elements are seen as similarly modifying the event or object. Each of these similarities is seen as separate from the object that may be abstracted or conceived of in pure form. This type of organization that underlies adult thought does not deduce identity from similarity. However, in pathological states such as schizophrenia, there may occur a reversal to primary class formation, explaining the peculiarities of thought found in this disorder. This "teleological regression" to primitive forms of cognition is utilized to satisfy inner needs that cannot be gratified in reality.

While the phantasmic world was predominantly visual and static, the paleologic world is mainly auditory and utilizes language. Together with linguistic forms there is also an appreciation of sequence and causality. The paleologic world is not inhabited by copied images of reality, but by rudimentary abstractions. The child can begin to reason and to build on his experience: he can separate similar data from the manifold of objects and begin to organize these objects into classes. However, because this process of abstraction is far from complete, the part is often confused with the whole, or two dissimilar events can be conceived of as identical. In other words, reality can be severely misunderstood and as a result internally misrepresented. As a result children are prone to make generalizations that follow a primary class organization and are unfortunately retained into adult life. These early misrepresentations are all too often seen in the irrational behavior of adults.

Concomitant with this increased ability to organize life events, the child begins to consider problems of causality. Prior to this stage, according to Arieti, the child's world is acausal, events simply occur. As the child begins to give causes to events, however, he utilizes extremely teleological explanations in which most things occur because someone has so willed them. Arieti speculates that this explanatory model is chosen by the child because so much of his life is actually determined by the will of powerful adults. The pertinent aspect of this new dimension of thought for the emotional and cognitive development of the child is that events are caused by people and thus one is responsible for these events. This new insight is necessary for what Arieti calls "third order" emotions. These emotional states presuppose a knowledge that a person can have an emotional affect on another, that one can cause a feeling state in someone else. Typical of this highest class of emotions are depression, hate, love, and joy. Here again Arieti is stressing that affective states grow in correspondence to cognitive development.

When the child reaches the level of conceptual thought—that is, his internal world consists predominantly of consensually validated concepts, rather purely personal symbols or images—secondary process mechanisms truly prevail and concepts are organized into secondary classes. This type of cognition becomes perfected through childhood and reaches true prominence in adolescence. Arieti's use of the term "concept," however, encompasses more than is generally acknowledged by developmental psychologists. He is more interested in internal concepts such as one's self-image or one's view of the mothering figure. These concepts are in a state of change throughout development just as concepts of geometry or logic differ at various stages of childhood. Furthermore, according to Arieti, these concepts should be viewed as cognitive structures in their own right. Arieti believes that ultimately we are motivated and gratified through mental concepts rather than through instinctual energies. As the child develops, higher order concepts form his aspirations and fears. These mental structures should be seen as cognitive and not reduced to physiological drives. Here there is a blending and reinforcing of the affective and cognitive aspects of development. Even neurotic defenses should be viewed as cognitive configurations rather than solely emphasizing their affective or energetic basis. To quote one of Arieti's[7] recent articles:

> In a large part of psychiatric, psychoanalytic and psychological literature, concepts are considered static, purely intellectual entities separate from human emotions and unimportant in psychodynamic studies. I cannot adhere to this point of view. Concepts and organized clusters of concepts become depositories of emotions and originators of new emotions. . . . Not only does every concept have an emotional counterpart, but concepts are necessary for high emotions. In the course of reaching adulthood, emotional and conceptual processes become more and more intimately interconnected. It is impossible to separate the two. They form a circular process. The emotional accompaniment of a cognitive process becomes the propelling drive not only toward action but also toward further cognitive processes. [p. 23]

At each stage of development the evolution of cognitive structures that allow for a greater behavioral repertoire with the environment also creates a more sophisticated internal world with the emergence of higher level emotions and more abstract methods of mentation. Thus the motivation of the individual varies according to levels of development. Greater cognitive abilities give rise to higher order emotions, which, in turn, push toward growth of awareness, interpersonal relationships, and abstract processes.

In summary, Arieti has attempted to explore the growth of "inner reality" and the interlocking relationship of affective and cognitive structures in development. Furthermore, he has suggested that some psychic structures that are pertinent in everyday behavior, such as the self-concept, be viewed as cognitive entities that grow through ontogenesis. Arieti has stressed that the emotional aspect of development should not be neglected, and the reconstruction and understanding of the inner world of childhood has an important place in the study of cognitive development.

Conclusion

From the foregoing presentation of current theorists, it becomes apparent that the developmental process is directed at self-regulatory behavior and the relative independence of behavior from immediate stimulation. Following a period of primarily reflexive, innate behavior, most theorists describe a stage of internalization of the environment so that events in the world can be mentally represented. Once this internal world is created, the child begins to structure stimuli and select adaptive responses rather than reacting automatically. However, as has been shown, this early representation of the environment is impressionistic and prone to error. The child is still bound to an egocentric yet concrete view of his environment, in which events and objects are associated by primitive logical linkages, into "nestings" or "primary aggregates" rather than hierarchical and logical classes. When the child begins rationally to order experience, it appears to be on a perceptual rather than logical basis. He responds to what is most striking in his environment rather than through understanding.

The last step in cognitive development, that of reason over appearance, is what Piaget has termed operational thinking. This divorce of thought from appearance is perhaps the true hallmark of the human psyche and the very foundation of human functioning. The reliance on logic and the attempt to formulate cognitive laws to order the chaotic world of the senses are the ultimate goals of cognitive development. It is equally important, however, to realize that cognitive development does not occur in a vacuum, but is interwoven with the development of internal emotional life as well as increased sensory and motor capabilities.

For many years the study of the ontogenesis of thought was predominantly an academic, ivory tower preoccupation. Today, however, the complexity of our modern industrial society, greater than at any time in the history of civilization, requires of each individual cognitive abilities that in the past were the province of a very small privileged minority. Despite this requirement for more sophisticated levels of mentation for both economic and social survival, an inordinately large percentage of individuals do not achieve this ultimate aim and cannot share in the rewards or responsibilities of modern society. The study of how the mind develops, the forces that affect this development, and the ways to correct deficiencies due to both internal and external forces has thus become a desperately important scientific discipline, not simply for the satisfaction of intellectual curiosity, but for the greater benefit of all.

Bibliography

1. ANTHONY, E. J., "The Significance of Jean Piaget for Child Psychiatry," *Brit. J. M. Psychol.*, 29:20–34, 1956.
2. ARIETI, S., "Cognition and Feeling," in Arnold, M. B. (Ed.), *Feelings and Emotions*, Academic Press, New York, 1970.
3. ———, "Contributions to Cognition from Psychoanalytic Theory," *Science and Psychoanalysis*, 8:16–37, 1965.
4. ———, *The Intrapsychic Self*, Basic Books, New York, 1967.
5. ———, "The Process of Expectation and Anticipation," *J. Nerv. & Ment. Dis.*, *106*: 471–481, 1947.
6. ———, "The Role of Cognition in the Development of Inner Reality," in Hellmuth, J. (Ed.), *Cognitive Studies*, Brunner/Mazel, New York, 1970.

7. ———, "The Structural and Psychodynamic Role of Cognition in the Human Psyche," in Arieti, S. (Ed.), *The World Biennial of Psychiatry and Psychotherapy*, Vol. 1, Basic Books, New York, 1971.
8. BOBATH, B., "Very Early Treatment of Cerebral Palsy," *Develop. Med. Child Neurol.*, *9*:372–390, 1967.
9. CASSLER, L., "Maternal Deprivation: A Critical Review of the Literature," *Monogr. Soc. Res. Child Devel.*, *26*:11, 1961.
10. DENNIS, W., and NAJARIAN, P., "Infant Development under Environmental Handicap," *Psychol. Monogr.*, *71*, 1957.
11. DEUTSCH, C., "Auditory Discrimination and Learning Social Factors," *Merrill-Palmer Quart.*, *10*:277–296, 1964.
12. FINEMAN, J. A., "Observations on the Development of Imaginative Play in Early Childhood," *J. Am. Acad. Child Psychiat.*, *1*:167–181, 1962.
13. FREEMAN, T., and MCGHIE, A., "The Relevance of Genetic Psychology for the Psychopathology of Schizophrenia," *Brit. J. M. Psychol.*, *31*:176–187, 1958.
14. FREUD, A., *Normality and Pathology in Childhood*, International Universities Press, New York, 1965.
15. FREUD, S. (1911), *Formulations on the Two Principles of Mental Functioning*, in Strachey, J. (Ed.), *Standard Edition*, Vol. 12, Hogarth, London, 1958.
16. ——— (1900), *The Interpretation of Dreams*, in Strachey, J. (Ed.), *Standard Edition*, Vols. 4 & 5, Hogarth, London, 1953.
17. ——— (1905), *Three Essays on Sexuality*, in Strachey, J. (Ed.), *Standard Edition*, Vol. 7, Hogarth, London, 1953.
18. FURTH, H. G., *Piaget and Knowledge*, Prentice-Hall, Englewood Cliffs, N.J., 1969.
19. GALAMBOS, R., *et al.*, "Electrophysiological Correlates of a Conditioned Response in Cats," *Science*, *123*:376–377, 1955.
20. GOLDFARB, W., "Emotional and Intellectual Consequences of Psychological Deprivation in Infancy: A Re-evaluation," in Hoch, P., and Zubin, J. (Eds.), *Psychopathology of Childhood*, Grune & Stratton, New York, 1955.
21. HARLOW, H. F., and HARLOW, M. K., "Social Deprivation in Monkeys," *Scientific American*, *207*:136–146, 1962.
22. HARTMANN, H., *Ego Psychology and the Problem of Adaptation*, International Universities Press, New York, 1958.
23. HEBB, D. O., *The Organization of Behavior*, Science Ed. Inc., New York, 1961.
24. HELD, R., "Plasticity in Sensory-Motor Systems," *Psychobiology: Readings from Scientific American*, W. H. Freeman, San Francisco, 1967.
25. HERNANDEZ-PEON, R., *et al.*, "Modification of Electric Activity in Cochlear Nucleus during 'Attention' in Unanesthetized Cats," *Science*, *123*:331–332, 1956.
26. HESS, E., "Imprinting in Birds," *Science*, *146*, 1964.
27. INHELDER, B., and PIAGET, J., *The Growth of Logical Thinking From Childhood to Adolescence*, Basic Books, New York, 1958.
28. ITARD, J. M. G., *The Wild Boy of Aveyron* (Tr. Humphrey, G., and Humphrey, M.), Appleton-Century-Crofts, New York, 1962.
29. LICHTENBERG, P., and NORTON, D. G., *Cognitive and Mental Development in the First Five Years of Life*, Public Health Service Publications, No. 2057, 1971.
30. ODIER, C., *Anxiety and Magic Thinking*, International Universities Press, New York, 1956.
31. PAVENSTEDT, E., "A Comparison of Child-rearing Environment of Upper-Lower and Very Low-Lower Class Families," *Am. J. Orthopsychiat.*, *35*:89–98, 1965.
32. ——— (Ed.), *The Drifters*, Little Brown, Boston, 1967.
33. PIAGET, J., "Biologie et Connaissance," in Furth, H. G., *Piaget and Knowledge*, Prentice-Hall, Englewood Cliffs, N.J., 1969.
34. ———, *The Construction of Reality in the Child*, Basic Books, New York, 1954.
35. ———, *Play, Dreams and Imitation in Childhood*, Norton, New York, 1962.
36. ———, *The Origins of Intelligence in Children*, International Universities Press, New York, 1952.
37. ———, "Psychology and Philosophy," in Wolman, B. B., and Nagel, E., *Scientific Psychology*, Basic Books, New York, 1965.
38. ———, *The Psychology of Intelligence*, Routledge & Kegan Paul, London, 1947.
39. ———, and INHELDER, B., *The Psychology of the Child*, Basic Books, New York, 1969.
40. PROVINCE, S., and LIPTON, R., *Infants in Institutions*, International Universities Press, New York, 1962.

41. RAPAPORT, D., "Psychoanalysis as a Developmental Psychology," in *Collected Papers*, Basic Books, New York, 1967.
42. SANDLER, J., and ROSENBLATT, B., "The Concept of the Representational World," in *Psychoanalytic Study of the Child*, Vol. 17, pp. 128–145, International Universities Press, New York, 1962.
43. SCHAEFER, E. S., and AARONSON, M., "Infant Education Research Project: Implementation and Implication of a Home Tutoring Program," Mimeographed manuscript.
44. SCHAFER, R., *Aspects of Internalization*, International Universities Press, New York, 1968.
45. SCOTT, J. P., "Critical Periods in Behavior Development," *Science*, *138*:949–958, 1962.
46. ———, "Critical Periods in the Development of Social Behavior in Puppies," *Psychosom. Med.*, *20*:42–54, 1958.
47. SPITZ, R., *The First Year of Life*, International Universities Press, New York, 1966.
48. WAPNER, S., "Some Aspects of a Research Program Based on an Organismic Developmental Approach to Cognition: Experiments and Theory," *J. Am. Acad. Child Psychiat.*, *3*:193–230, 1964.
49. WERNER, H., *The Comparative Psychology of Mental Development*, International Universities Press, New York, 1948.
50. ———, "The Concept of Development from a Comparative and Organismic Point of View," in Harris, D. B. (Ed.), *The Concept of Development*, University of Minnesota Press, Minneapolis, 1957.
51. ———, and KAPLAN, B., "The Developmental Approach to Cognition. Its Relevance to the Psychological Interpretation of Anthropological and Ethnological Data," *Am. Anthropol.*, *58*:866–880, 1956.
52. ———, and ———, *Symbol Formation*, John Wiley, New York, 1964.
53. WHORF, B. L., *Language Thought and Reality*, John Wiley, New York, 1956.
54. WOLFF, P. H., "The Developmental Psychologies of Jean Piaget and Psychoanalysis," *Psychol. Issues*, *11*, 1960.
55. ———, and FEINBLOOM, R. I., "Critical Periods and Cognitive Development," *Pediatrics*, *44*:999–1007, 1969.
56. YARROW, L. J., "Separation From Parents during Early Childhood," in Hoffman, L. W., and Hoffman, M. L. (Eds.), *Review of Child Development Research*, Vol. 1, pp. 89–136, Russell Sage Foundation, New York, 1964.

CHAPTER 16

EARLY LANGUAGE DEVELOPMENT

Katrina de Hirsch

Introduction

FREUD[29] SAID THAT LANGUAGE brings into being a higher psychical organization. This chapter discusses some aspects of language acquisition, a process that is part and parcel of the young child's organismic growth.

The linguistic development of the child, according to Lewis,[50] is determined by the interaction of many factors: those that spring from within himself, those that derive from the interplay between him and his mother, and those that impinge upon him from the particular culture to which he belongs.

Consequently there are several ways of looking at language acquisition, each concerned with different aspects of verbal development.

The potential for language is laid down in the central nervous system. Thus one focus of interest is upon the biological and neurophysiological factors that make possible the development of language and its specific human characteristics: the ability to symbolize.

Another focus of interest is concerned with the affective aspect of communication, which develops in the matrix of the mother-child relationship. The role of language in the formation of the ego is related to this inquiry. It is by no means clear, according to Edelheit,[22] to what extent ego organization is reflected in the language system and to what extent the internalized language system constitutes the regulatory function of the ego.

An equally legitimate concern is the sociocultural determinants of language. As a result of the growing awareness of deprived children's linguistic and learning difficulties, which are closely related, this aspect is, at present, very much in the foreground.

A relatively new type of investigation is being undertaken by those linguists who investigate the developmental facets of the phonemic, semantic, and syntactical elements of the child's language.

It is far beyond the scope of this presentation to sketch more than the outlines of the different stages in language learning.

Stages of Language Development

Earliest Responses to Auditory Stimuli and Vocalizations

There is some evidence of fetal response to sounds.[2,21,40] According to Eisenberg,[23] the cochlea seems to be functional at about the twentieth week of intrauterine life, and the basic mechanism for coding intensity and frequency is probably operant by the thirtieth week of gestation. The neuronal mechanisms for processing sounds are fully mature at birth. Toriyama, *et al.*,[74] reported EEG responses to 60 db tones bursts three hours after delivery. Recent data sharply contradict previous views of the newborn as presenting an auditory *tabula rasa*. Eisenberg's[23] laboratory studies involving several hundred infants demonstrate that neonates not only respond to auditory stimuli but also respond selectively. Pure tones are far less effective in eliciting responses than are complex ones.[75]

The function of the birth cry is physiological, serving to establish normal respiration and oxygenation of the blood. Palmer[61] maintains that the cries of infants damaged prenatally or at birth differ in pitch, rhythm, and volume from those of normal babies. Karelitz and Fischelli[43] have found that certain crying contours may be predictive of mental retardation.

From his earliest days the baby vocalizes, utilizing the expiratory phase of the respiratory cycle. The air flow is modified by the vocal cords and the various parts of the peripheral speech mechanism, setting up sound waves that may be detected auditorily. The identical process holds for all vocal productions: the baby's cry, his cooing, and at a later time his speech. Lieberman[51] has shown that the wave forms of the baby's cry have the same characteristics as adult sentences. Long before the child uses words, alterations in breathing produce variations in intonation that are, in turn, instrumental in carrying messages.

Hunger, pain, and discomfort will make the neonate cry. These discomfort cries approximate low front vowels and are usually nasalized. Shortly afterward the active and alert baby will produce different kinds of vocalizations after feeding and during states of well-being. Such sounds consist of low front vowels combined with high back ones and a few consonants such as guttural /r/ and palatal /g/. Very early, then, the child uses the physiological mechanisms of respiration and phonation for expressive purposes, and for the mother each characteristic vocal pattern has a distinct meaning.

Quite early the baby distinguishes between male and female voices and between friendly and angry voices. By measuring changes in heart rate McCaffrey[53] and Moffit[58] demonstrated in two different laboratory studies that 11 to 20 eight-week-old infants distinguish between phonemic contrasts of both synthetic speech and the human voice. On the basis of their experiments Elmas, *et al.*,[24] concluded that perception in the linguistic mode may well be part of the biological makeup of the organism and must be operative at a surprisingly early age. That babies do a far greater amount of discriminative listening than has hitherto been shown is demonstrated by Friedlander's[30,31] work. The infant's experience with hearing spoken language, he believes, is an indispensable requisite for his eventual ability to speak it.

Long before children respond to the phonetic patterns of words, they respond to the prosodic features of language, to the variation in intonation and pitch that carry the emotional load of communication. The mother's vocalizations are at first experienced as part of the total constellation that includes her smiling face, the presence of caresses, and the satisfaction of bodily needs. Turnure[76] has shown that a three-month-old baby can be "tuned in" to a tape recording of his mother's voice, even in her absence.

Cooing

Before they are three months old babies engage in playful cooing interchanges with the mother when she imitates the baby's own sounds. This is probably a universal phenomenon. Piaget[65] observed such early forms of interaction from the first month on and called it "verbal contagion." These vocalizations show a preponderance of back rounded vowels and—at a somewhat later stage—glottal stops, in addition to clicks of several varieties. In the fourth month, with increasing chewing activities as the result of the introduction of solid foods and the baby's tendency to put all sorts of objects into his mouth, lip and tongue muscles become more active and the cooing interchanges become more varied and livelier.

Mother-Child Dialogue

The vocalizations that mothers use with their babies are tailored to the child's affective needs. They differ from those she uses in other situations in a number of features: in dynamic accent, in melody and rate, and in a marked tendency to rhythmic iterations. Snow[69] says that children who are learning language have available a sample of speech that is simpler, more redundant, and less confusing than adult speech.

Di Carlo[20] has made the pertinent observation that the mother caresses the child with her voice. Wyatt[80] calls the dialogue between the mother and her baby a process of mutual feedback. The learning of the "mother language," she says, is achieved through unconscious identification. Infants are continually talked to in situations that are essential to their well-being. The vocal and affect communication between mother and child during bathing, feeding, and dressing must of necessity play an important part in the child's communicative attitudes, of which language is but one aspect. The mother's voice is undoubtedly heavily cathected for the baby. The extent to which this phenomenon bears on subsequent listening attitudes has not been explored, but it is quite possible, as maintained by Eveloff,[26] that the first 18 months are crucial for symbolic language development and that serious flaws in the interaction between the baby and the mothering adult would have repercussions in terms of the child's linguistic and cognitive development.

We do know that normal maturational processes are delayed or disturbed in children brought up in institutions. This is convincingly shown in Provence and Lipton's study[66] of such infants, whose vocal and verbal development is often severely delayed. We have no systematic investigations comparing the early vocalizations of institutionalized infants with those of well-mothered babies, who have a repertoire of most of the vowels and half of the consonants by the end of the first year. Not all institutions are alike, of course. Tizard, *et al.*, found that the language comprehension scores of young children correlated significantly with both the quality of the speech directed to them and the way the residential nurseries were organized. A careful comparison of the vocalizations of infants even in "good" institutions with those of babies who have received ample sensory and affect stimulation might, nevertheless, reveal much earlier deficits in institutionalized babies than those shown by Brodbeck and Irwin,[10] who found that differences in verbal development between deprived youngsters and middle-class ones begin to show at 18 months. It is of interest that in a later study Irwin[37] proved that with very systematic stimulation beginning at 14 months, the phoneme frequency of stimulated babies was significantly higher than that of their control group.

Babbling

Between the fifth and the tenth month of life, usually around the time he attempts to turn over, to sit up, to drink from a bottle, the baby begins to babble.* There is no clear demarcation from expressive sounds to babbling. The various stages of language development overlap. From short cooing phrases the baby proceeds to the use of long strings of sounds that are no longer tied to specific situations. Now the baby lies in his crib, amuses himself with his fingers and toes, and plays with sounds that he produces spontaneously, obviously for the sheer pleasure of making them. These vocalizations are presymbolic. They are playful and probably autoerotic. There is no mistaking the intense gratification the baby derives from the mouthing of sounds and the enjoyment he gets from this repetitive playful experimentation and stimulation. It is thus not surprising that analysts are impressed by the role this particular oral activity plays in early psychosexual development.

During babbling the baby "discovers" how to mold the outgoing breath stream so as to produce a whole repertoire of sounds. At the same time an auditory feedback loop is being established. As sound-producing movements are being repeated over and over again, a strong link is being forged between tactual and kinesthetic impressions, on the one hand, and auditory sensations, on the other. A pattern of alternate hearing and uttering is set up. Sounds—and as time goes on they take on a more and more repetitive character—are no longer random but tend to resemble the phonetic and intonational configurations of the original utterance. One of the main functions of babbling, therefore, is an intensive kinesthetic-proprioceptive auditory learning experience, resulting in primitive sound-movement schemata. The baby not only produces a whole range of different sounds but also derives pleasure from those he himself produces. The self-stimulation the baby engages in during this activity may play an important role in his increasing awareness of linguistic organization.[78]

Every mother knows that she can stimulate long babbling conversations between herself and her child. It has been shown that nonsocial stimuli such as the sound of chimes do not increase vocalizations. Nor does the mere presence of the human being result in more copious babbling.[79] It is the human voice[52] that stimulates babbling, and babbling is reinforced by the presence of a significant adult.[34] Mother and child engage in what Piaget[65] calls "mutual imitation." The baby, stimulated by his mother's vocalizations, produces more of his own. Thus, even at this early stage, a corrective feedback mechanism is established that, according to Wyatt,[80] is a *sine qua non* for smooth language development. Of course this is possible only if the child has learned to store—if only for a limited time—and to retrieve the auditory-motor configurations the mother models for him. It is she who monitors the baby's vocalizations by providing the phonemes used in her own linguistic culture.

Normal children vary in the amount of babbling they do; some babble early, some much later. However, intactness of the central nervous system is a prerequisite for the auditory-motor feedback loop that stimulates babbling behavior. This loop does not become functional in babies who are deaf or hard of hearing. Deaf babies babble, but they discontinue doing so after a while because they lack the stimulation derived from hearing their own sounds as well as those of the environment. New investigations have demonstrated that with amplification at the babbling stage, that is to say, with the very early establishment of the auditory feedback loop, the language development of severely hard of hearing children approximates that of normal ones.

It has been assumed that babies from widely differing language backgrounds have an extensive repertoire of babbled sounds that

* Cruttenden[18] objects to the definition of babbling as the stage in which children utter sounds just for the pleasure of making them. He prefers a more objective definition and suggests that babbling be defined as the stage when the baby first produces "pulmonic-lingual" consonants, for example, dental and velar plosives, alveolar and palatal nasals, in addition to a growing variety of vowel-like sounds. The production of these sounds would imply "a new awareness" on the part of the child of how to combine tongue movements with breathing. This stage, he says, more or less ends with the emergence of the first words.

include not only those of their mother language but also many others that drop out later on.[5,13,33,59] Cruttenden,[18] in an interesting analysis of the babbling of his twin girls—admittedly an atypical sample—maintains that this claim is not borne out by the facts. He did not, for instance, find fricatives or affricatives* /th/, /s/, /z/, /sh/, or /ch/ among his children's babbled sounds. He concluded that the range of babbled consonants is not nearly as extended as is often implied. Nevertheless, he believes that the babbling stage is important not only as a period for experimentation but as one in which the child's repertoire shifts in the direction of the language to be learned. Weir[78] found that the sounds of five- to seven-month-old Chinese babies differed from those of Russian and American ones. Early and selective imitation of input, therefore, must play a significant role in babbling. Quantity and quality of babbling varies from one child to the next, but as babies advance in age, their babbled sounds become more differentiated and more clearly patterned.

Echolalia

Around the tenth month, as control over volume and pitch develops, duplications become more frequent, intonational contours more distinct, and the baby begins to imitate the more patterned utterances he hears in his environment. He starts repeating words heard, but not necessarily understood. It is the main accented word in a sentence, the word that is most heavily loaded with affect, that is selected for imitation, and it is the initial syllable of such a word that is usually chosen for duplication.

Echolalia is a normal phase in language development, and echolalic responses comprise a significant portion of children's speech up to 27 months.[27] Toddlers often repeat words or combinations of words for purposes of clarification, to try them on for size as it were. Echolalia is pathognomonic for mental retardation, autism, or aphasia only if it is the only response available to the child, if it has an intensely compulsive quality, or if it occurs in older children.

Listening and Language Comprehension

That effective listening is indispensable for the development of speech and language has been known for some time. Infants' systems for processing acoustic and linguistic input mature much faster than those for generating language. How do babies make some order out of the "buzzing confusion" that surrounds them? Friedlander[30] says they face a monumental task learning to order auditory and linguistic signals into more or less stable categories. In a series of highly ingenious experiments, he has shown that, among other achievements, eight- to fifteen-month-old infants can discriminate between backward and forward speech and between flat and bright intonations.†

Long before they decode linguistic patterns, children learn to detect certain regularities from the intonational contours and phoneme boundaries of the speech that surrounds them. An enormous amount of learning, therefore, must be going on during the seemingly passive listening done by very small children. This fact has important implications for early intervention and enrichment programs.

Crude processing of input precedes language comprehension. We assume that by a process of selecting, filtering, and transforming acoustic input, small children arrive at the decoding of linguistic information. What are the requisites for comprehension? A fun-

* In contrast, Leopold[48] asserts that in the case he studied, affricatives occurred very early during cooing and babbling but quite late—in the last third of the second year—in imitation of real words.

† Friedlander's[31] tape recordings of the children's "natural home language environment" and specifically of the language interaction between parents and children could provide important clues in terms of verbal competence at later ages.

damental requirement is that the child be able to make a modicum of sense out of the world that surrounds him, that he use some objects appropriately, and that he has a rudimentary sense of self, of being separate. Lewis[49,50] says that the child responds first to the intonational configurations that form part of specific situations. Slowly the phonemic pattern becomes intertwined with situational and intonational features, until very gradually, as further differentiation takes place, the phonemic pattern becomes dominant, irrespective of the situation. Around the age of 12 months the child shows signs of understanding simple phrases and commands, particularly those that are heavily loaded affectively.

Expressive Language

Although new research has thrown light on some of the factors that determine the ability to process verbal information, we do not really know what happens when the child *produces* his first meaningful utterances. Indeed, there is little agreement on what constitutes the first word. Is it the word the child uses in contexts other than imitation, the utterance that stands for an object, a person, an event? McCarthy[54] argues that there is a tremendous gap between the mere production of a word and the symbolic representational use of that word in an appropriate situation.

In recent years a lively controversy has arisen between the proponents of learning theory and nativistically oriented researchers regarding the emergence of the first meaningful words. Learning theorists argue that language is slowly shaped out of the multitude of babbling sounds by means of selective reinforcement on the part of the parents, who show their delight when the child produces sound combinations that resemble actual words. As a result of such reinforcement, the child gradually begins to associate such sound combinations with specific situations, people, or objects. Linguists such as Jakobson and Halle,[39] on the other hand, draw a sharp dividing line between babbling and representational speech. Brown,[11] in an intermediate position, suggests a "babbling drift" in describing the transition from babbling to language.

We know that at the age of 12 months the child is able to imitate the intonational contours of his mother's speech. Many preverbal babies utter long strings of sounds that resemble those of adults in melody and pitch. Mothers respond to these vocalizations as if they were definite messages, statements, commands, or questions. A message is clearly associated with the modulated strings of sounds the baby produces even before he can utter a single word. Advanced babbling *is* a sort of language, and we cannot, therefore, talk of the emergence of symbolic speech as a sudden event, but must look upon this development as an ongoing process that goes through a series of interrelated and overlapping phases, each new phase representing not merely an addition but a new and more highly integrated configuration.

Linguistic Dimensions

Several dimensions of language are described by linguists: (1) the phonological refers to elementary verbal forms and their combinations into words; (2) the semantic deals with word meanings; (3) the syntactical concerns the formal relations and underlying rules for processing and generating sentences. Developmental linguists attempt to construct a model to explain the child's unfolding ability to understand and produce an infinite number of sentences.*

* *Readings in Language Development*, edited by Lois Bloom,[7] is an excellent introduction to developmental linguistics.

The author is aware of the limitations of the section that follows; it only attempts to summarize the groundbreaking work done in this area. Not only does the scope of the chapter exclude a more penetrating summary, but also the original manuscript was submitted nearly two years ago and thus does not refer to much of the recent work. (The new book by Cazden, *Child Language and Education*,[14] is an example.) The author realizes that her presentation is not only incomplete but also probably biased in favor of the transformational school of linguistics.

Phonology describes the matrix of features that differentiates one speech sound from another by virtue of certain characteristics. It deals with contrasts between certain features, such as the one between voiced and voiceless sounds, for example, /p/ and /b/.

For the infant sounds are probably at first indistinguishable from one another. As children's perceptive faculties develop, they gradually learn to differentiate between coarser and finer shades of sounds. Research on the perceptual distinctions children are able to make between birth and the age of three years is relatively recent. Jakobson[38] postulated that successive perceptual contrasts are acquired in a more or less consistent sequence. One of the first is probably the distinction between vowels and consonants.*

There are considerable variations between children when it comes to production of sounds. Some, for example, acquire fricatives far earlier than others. Not only the timing but also the quality of sound reproduction varies. Many children have trouble not only with a single phonetic feature but with several of them. A child who substitutes "dar" for "car" misses not only the place of articulation—alveolar rather than velar—but also the voiced-nonvoiced contrast. Menyuk[57] found that Japanese children acquire mastery of the phonological system roughly at the same time and in the same order as do American youngsters. This suggests that the hierarchy of sound acquisition depends on the developing productive capacity of the child.

Not all of the 40-odd sounds of English are mastered simultaneously. The child's earliest utterances contain only two or three phonemes. At 18 months he may cope with from eight to twelve of them, including diphthongs. From then on his production of the sounds of his language increases by leaps and bounds as a result of further differentiation. Sudden spurts are followed by phases of consolidation. American children manage 80 per cent of all sounds by age five.[73]

As a rule children use only those sound sequences that are part of their native language. An English child will not, for instance, produce the cluster "pf," which is a common one in German (as in the word "pferd"), but he will use the cluster "pl," which is acceptable in English. Children do not usually need formal training to learn these rules. While it is obvious that the mastery of the phonological system of the language requires a degree of central nervous system maturation, the child primarily needs a model. It is the mother who usually serves as the primary language model and who literally feeds back to the child those sound sequences that are acceptable in his language environment. During certain critical stages of language learning, the model is of decisive importance in the mastery of the phonological system of the language.

The affective and communicative exchange between mother and child also plays an important part in the acquisition of vocabulary and in the growth of meaning, that is, in the *semantic* dimension of language.

Most children begin to use symbolic utterances between 12 and 20 months. They usually select those words that provide essential information or those that are strongly affectively colored. Words at this age reflect the magical universe of the small child. There is a close link between the name and the qualities of an object. Single words at this age may also stand for groups of experiences and configurations that are lumped together (holophrases). Many children cling to one-word sentences for a long period, using intonation to express a variety of meanings. The upward shift in pitch might indicate a question, for example, car?[12]

Brown[11] says that parents intuitively provide the names that reflect the structure of the child's world. The parents may say, "spoon" and not "silverware," using the more concrete referent in preference to the more abstract one. But they will say "fish" rather than "trout," using the general rather than the more specific referent. Children enlarge their vocabulary through a process of extension and

* Not discussed in this chapter is the "motor theory" of perception proposed by the Haskins group.[25]

restriction. Its growth is characterized by continuing modification of meaning, involving both the acquisition of new words and the expansion and refinement of word meanings previously learned. Increasing differentiation is reflected in more precise usage. "Doggie" is at first only the child's small, fluffy toy dog; a little later the word describes the real dog the child sees at a neighbor's, and finally it stands for all members of the species "dog." The meaning of the word has been expanded. The opposite process takes place when the child says, for instance, "He write a picture." When he learns that the word "write" refers only to those graphic activities that involve letters and numbers, he has restricted and refined the meaning of the word.

Some concepts and the words that represent them are learned earlier than others. Only 10 per cent of 240 two year olds in Palmer's group[61] understood the word "around." Concepts such as "backward" are difficult for three year olds."

The development of meaning, which belongs to the semantic aspect of language, cannot be explained by associative bonds. Meanings are not constant. They change throughout the individual's entire life. At age three the child uses 700 to 800 words, and between three and five years he adds at least 50 new words to his vocabulary each month. The growth of meaning depends on the verbal richness of the child's cultural background and probably on his linguistic endowment (although linguists do not agree, clinicians who work with language-delayed children know that this endowment may differ within social milieus and even within given families). This growth depends primarily on the mother's interest in and knowledge of words, on her ability to clarify shades of meaning, on the easy give and take between her and her child, and on the constant feedback provided by the significant people in his environment. Stodolsky's[71] study indicated that the best single predictor of Negro children's recognition vocabulary was the mother's vocabulary score on the Wechsler Intelligence Scale for Adults. The maternal teaching variable was the extent to which the mother defined tasks and named specific qualities of the environment. Thus the child's cultural and affective milieu play a vital part in fostering delight in the use of words and interest in shades of meanings. Children do need models, above all, for the acquisition of the lexical aspects of communication.

On the other hand, modern linguistic theory postulates that in the development of *syntax*, the set of rules that governs the organization of the more or less limitless number of possible utterances, imitation is not of primary significance. Children do need exposure in order to be able to deduce the underlying rules of language, and linguists admit that the very young child begins with repetition of bits of speech heard. Because of limitations in their auditory memory and because they cannot as yet program longer units, they repeat only two to four items. These repetitions are usually systematic reductions of the input perceived and are apparently processed as units rather than as lists of single words. (Children repeat those parts of speech that carry the significant information.)

Learning to generate complete sentences (as opposed to repeating) is a slow process and proceeds from primitive to more and more differentiated syntactical and morphological structures. Before or around their second birthday, children produce their first spontaneous, nonimitative two-word combinations. Brown[11] points out that the two single utterances "push" and "car" are not simply joined. The unit is programmed as a whole, and "push" is subordinated to "car" by means of lesser stress and lower pitch. The transition from the generating of single words to two-word units that are hierarchically organized is a tremendous step forward. Words can no longer be changed around like beads. The position of words in the string becomes relevant; the principle of word order, a powerful linguistic signal in English, is established.

Children's spontaneous two-word combinations, like their repetitions, consist of the most highly stressed and informative words—nouns, verbs, adjectives—content or "open class" words. This telegraphic speech is characterized by the omission of functor words, prepositions, auxiliaries, inflections, articles.

These constructions are slowly expanded by the use of modifiers such as "push pretty car" and are the building blocks of subsequent sentences.

Brown and Bellugi[12] describe the next level, the noun phrase, as an essential milestone in the acquisition of syntax. In standard English a noun phrase consists of a noun and assorted modifiers: "my girl," "more coffee," and the like. Two or more such modifiers can be combined with nouns, and since children construct interim grammars on their way to linguistic mastery, they slowly learn the privileges and constraints belonging to each modifier. The noun phrase can be moved to different positions in the string and can even be replaced by a pronoun. According to Chomsky's theory,[17] phrases are cohesive grammatical units. They reflect the child's discovery of basic phrase structure and lead to the formation of "kernel sentences," which are composed of at least two units: one functioning as subject, the other as predicate.

From such kernel sentences new forms evolve by additions, deletions, and rearrangements. Serialization is effected through the use of the word "and"; questions are formulated by prefacing sentences with "wh" words. Thus, beginning with the simplest two-word combinations, followed by noun phrases, one can trace the gradual emergence of elaborated structures. Complexity and length of units increase simultaneously. By 36 months most children's linguistic development is so advanced that they are able to produce all of the varieties of simple English sentences consisting of up to ten or eleven words.* It is important to realize, however, that what children learn are not specific combinations of words but the rules that govern these combinations. Chomsky[17] says that children internalize underlying grammatical rules on the basis of their exposure to limited numbers of correctly formed sentences. That language learning involves much more than the storing of longer and longer bits of utterances heard has been demonstrated by Berko.[3] In a series of ingenious experiments she used nonsense words to test children's ability to apply morphological rules: that is to say, tenses, plurals, possessives, and so forth. The examiner points to a doll, for instance, and says: "this is a child who knows how to 'wug,' today he is . . ." If the child supplies "wugging" although he has never heard the present participle of "wug," he has spontaneously applied the underlying grammatical rule that asks for the ending "ing" in this particular construction. A child may have memorized the plural form "witches," Lenneberg[47] says. But if he produces the plural "gutches," he shows that he has incorporated the rule that the plural of words ending with /ch/ is formed by adding /ez/. The very existence of "overgeneralizations" such as "mouses" (as in "houses") proves that the child has derived a set of abstract rules from the linguistic data to which he was exposed. These rules are basic to the decoding and encoding of language.

Mastery of syntax is a gradual process. A three year old understands the contrast between subject and object in the active but not in the passive voice; this distinction belongs to a chronologically older age.

* Bloom[6] feels that what is called "transformational grammar" describes only the most formal aspects of children's language development. Her own sophisticated formulations show that a child's identification and reproduction of a particular syntactical structure is intimately related to his interaction with the world of objects, events, and relationships.

Controversies

McNeill[55] and other nativists claim that children are born with a set of "linguistic universals" as part of their innate endowment and that they are neurologically so preprogrammed with a language acquisition device that they need only a minimum of stimulation for its realization. Language learning, they say, is reminiscent of embryonic development in the sense that it constitutes the unfolding of an inherent potential. A less radical position would be to postulate that children are born with certain propensities, "anlagen," that enable them to organize linguistic inputs into cohesive categories and structures. Goldstein[32] referred to the organism's drive to "actualize" inborn capacities. As far back as 1907 the Sterns[8] stressed that the child who learns to speak is neither a phonograph reproducing sounds nor a sovereign creator of concepts. His speech is the result of "convergence" of the continuous interaction of external impressions and internal systems.

Finally there is the behaviorist position, asserting that language learning depends on differential reinforcement on the part of adults who care for the child.*

The universality and the consistency of normal language development would seem to speak for the nativistic position. Penfield and Roberts[64] speak of a "biological timetable" for language development. However, while this holds for the majority of youngsters, there are, nevertheless, large variations in timing. It is the *sequencing* that is relatively stable. Most children pass through babbling, echolalic speech, two-word sentences, etc., on the road to fully developed verbal communication.

Stimulation

Lenneberg's[46] position that normal language learning proceeds without actual training is correct only to a point. An enormous amount of informal teaching, or corrective feedback, goes on in most families. Mothers respond to their children's telegraphic speech by "expansion"; that is, they modify their children's utterances by adding functor words, auxiliaries, prepositions, etc., and in so doing, present them with a model of the nearest correctly formed sentence.†

This is not a form of correction. Cazden[14,15] rightly points out that correction extinguishes communication. Expansion, according to Cazden, is most effective at "critical" stages of linguistic development when a growth in structure is taking place—an important fact to remember in terms of early intervention. What expansion does is clarify and enrich verbalization at the very moment the child is engrossed in a situation that holds his attention. Young children focus attention best when they are affectively engaged. If a youngster cries out, "Teddy table hurt," and the mother takes him on her lap and says, "Poor Teddy, yes, you bumped into the corner of the table, it hurts," she is involved with him in a highly meaningful experience. Mothers filter their own output: they seem to know which stage in language development the child is passing through at a given moment. With young children they use the kind of constructions the children themselves will use a number of months later. In terms of content mothers deal very much with the "here" and "now." The timing and the tuning in to the child's concerns are all-important. There are mothers who fail to cue in. They do not expand their children's telegraphic utterances, and they thus miss an opportunity to enrich communication. Or they swamp the child with long and complex verbalizations that, in his terms, are totally irrelevant. Wyatt[80] presents some excellent examples of the kind of affective and linguistic interaction that fosters comprehension and use of language and in the last instance presents the child with a view of the world.

* That the Skinnerian model of language acquisition is inappropriate has been shown in a brilliant review by Chomsky,[17] who more or less demolished the theories expounded in *Verbal Behavior*.[68] Nor can Braine's[9] contingency and transfer of training theory account for even fairly elementary phrase structure. Although it is true that certain aspects of language learning in children are more dependent on contact with the adult than others (parent-child interaction has more impact on semantic and phonological than on syntactical development), it is far too simplistic to think of such learning in terms of even highly sophisticated stimulus-response theory.[60] Children might learn words and even phrases by way of differential reinforcement on the part of the adult who "shapes" desirable verbal responses; they will not, however, acquire language—that is to say, the attitudes underlying communication, the investment in verbal exchange, and the joy of expressing subtle meaning by way of words.

† Brown[11] makes the interesting differentiation between the adult's imitation of the child's speech by means of expansion and the child's imitation of adult speech by means of reduction.

Sociocultural Factors

Hess and Shipman[36] look at the mother as a teacher, a programmer of experience in the preschool years. They discuss different styles in information processing, which is mediated through language. Nobody who is familiar with language stimulation programs involving young children and their mothers will deny that there are large differences within the same social group in the way mothers foster linguistic and conceptual development. Thus it is not social class alone that determines the child's exposure to adequate language models. Nevertheless, the experiments carried out by Hess and his group convinced them that the teaching style of many lower-class mothers is less explicit and therefore less effective than that of middle-class ones. Tulkin, *et al.*, found that working-class mothers often did not believe that their children were able to communicate and hence felt it was futile to attempt to interact with them verbally. The fact that these mothers' handling of their children, in general, tends to be more authoritarian and less based on verbal exchange than that of middle-class mothers would account for some of the intergroup differences.

It was Bernstein[4] who originally postulated that the very speech of lower-class individuals is characterized by rigidity of syntax and limited options for sentence organization. Lower-class speech, Bernstein said, is less elaborated and more restricted than middle-class speech, and it offers less opportunity for categorizing environmental stimuli. As a result it is relatively ineffective for conceptualization.

Bernstein's work, which is both interesting and controversial, has been under heavy attack by sociolinguists. They maintain that black dialect is by no means a more primitive language than standard English, but is rather a *different* language. Baratz,[1] Stewart,[70] Taylor,[72] and others have demonstrated that lower-class black children show the same proportion of grammatical transformations as do middle-class children of comparable age. Black youngsters incorporate the linguistic rules of *their* language environment. The use of the double negative, for instance, is an accepted feature in adult black speech and cannot, therefore, be considered to be defective in the black child. A recent paper by Quay[67] shows that much of the research discussing language deficits in black children is based on language expression rather than on comprehension, which clearly puts youngsters who speak a dialect at a disadvantage. Comprehension is a much more sensitive indicator of a child's linguistic maturity than is production.

Moreover, it is not legitimate on methodological grounds to evaluate the language of one subculture by means of norms derived from a different subculture. And it is not justifiable to use developmental language scales standardized on middle-class children as criteria for youngsters who do not speak standard English.

Cazden,[15] who has dealt extensively with the deficit-difference issue, feels that both concepts are inappropriate. For one thing they are probably too global. Middle-class children's vocabularies are richer and more differentiated than those of lower-class black youngsters, but the grammatical competence of lower-class black children is probably comparable to that of their middle-class peers.

What counts in the last analysis is children's communicative effectiveness. Thus Cazden[15] stresses the speaking situation and the children's ability to perceive and categorize a social situation and to adjust their speech to its requirements.

The literature is replete with statements referring to the superiority of girls' early language development. McCarthy[54] says: "Whenever groups of boys and girls are well matched in intelligence and socio-economic background . . . there appear slight differences in favor of girls." Not only length of sentences, but almost all other measures that show developmental trends with age—articulation, word usage, complexity of output, and grammatical competence—demonstrate slightly more rapid maturation in girls. Kagan's report[41] that, in contrast to boys, vocalizations of girls between four and fourteen months showed greater stability, independent of social class, is of interest in this context. Scores on an early vocalization index proved to be a far better predictor of subsequent cognitive development in girls than was the case with boys. One might speculate that the earlier neurophysiological maturation of girls as compared to boys is an important factor in the linguistic superiority of girls, at least during the earlier years.

Birth Order

Another striking fact is the marked linguistic superiority of only children. McCarthy[54] points out that this superiority is out of proportion to that which could be expected from their sex, mentality, and socioeconomic background, and she relates it to only children's far greater opportunity for linguistic and affect interaction with adults. Firstborns do better than younger children probably for the same reason. Twins and triplets, who have to share their mother's lap, are worse off then single babies. The study of the language development of identical twins is complicated by the fact that it involves not only neurophysiological factors (many twins are prematurely born) but also problems of identity. The tendency of twins to create a universe separate from the world that surrounds them may be another variable.

Language and the Organization of the Child's World

It is impossible to overrate the importance of language in children's development. In the young child speech is a means of expressing feelings and demands, a thing to play with, a device that substitutes for action, or an attempt to control the behavior of others. Kaplan[42] describes young children's speech as governed by affective needs, including play, and as such infused with highly personal meanings.*

The world in which we live is conceptually organized,[63] and language makes this organization possible. Language is our main tool in the construction of a comprehensible universe. The preverbal child probably has some schemata, some ways of ordering the world, but these schemata can be assumed to be amorphous. The ability to use verbal symbols permits the child to impose order on his experience and to arrive at some clarification of his outer and inner cosmos.

Language, Edelheit[22] states, is an obligatory component of human biological organization and plays a crucial role in the formation of the ego, which is the special human organ for adaptation. Ego development depends largely on an optimal level of verbal expression, and conversely ego defects of whatever origin are reflected in disturbances of language. As far back as 1936, Anna Freud[28] wrote: "The association of affects and drive processes with verbal signs is the first and most important step in the direction of mastery of instinctual drives. . . . The attempt to take hold of the drive processes by linking them with verbal signs which can be dealt with in consciousness is one of the most general, earliest and most necessary accomplishments of the ego, not as one of its activities."

The secondary process requires word representation. Ability to verbalize helps to delay action; it constitutes the "trial action" Freud[29] speaks of. Words are needed for purposes of impulse control. Postponing gratification is infinitely more difficult without the use of verbalization.

With the acquisition of language the distinction between self and nonself becomes clearer. The child probably has an emerging sense of self even before he begins to use words. But the distinction between self and nonself becomes much more differentiated

* No attempt was made to discuss the development of *inner* speech, which appears more or less around the time when "egocentric" speech disappears. For a classic discussion of both, see Vygotsky, *Thought and Language*.[77]

when categories such as "I" and "mine" emerge.[63] The two later stages of the separation-individuation phase[56] run parallel with the most rapid spurt in language development. One would assume that there is a reciprocal relationship between these events, one fostering the other. According to Despert,[19] "Since the appearance of the first person pronoun in language development follows that stage of individuation which corresponds to the child's consciousness as one, whole and apart from others, the importance of this sign cannot be overestimated."

Things acquire stability and permanence by being named. Names help to anchor the significant people in the child's environment; they reassure him that the mother will return even if she is gone for a while; and they thus contribute to what is called object constancy.

Language permits the child to externalize magical and omnipotent fantasies and thus renders them less dangerous. It makes possible the distinction between fantasies, on the one hand, and reality, on the other. The shift from a magical, prelogical world to a reality-oriented world is mainly accomplished by language. The labeling of inner states helps the distinction between the outer and inner cosmos, thereby assisting reality testing, one of the most essential functions of the ego.[35]

The formation of conscience is immensely facilitated when the child learns to internalize rules and laws. Such internalization remains precarious without the use of verbal concepts.

Learning to express verbally feelings of fear, pain, and sadness protects the child from being overwhelmed.

It is well known and fully accepted that language is an indispensable tool for cognition. There is thus no need for extensive elaboration of this point in the present chapter although it might be helpful to stress a few points. Luria[52] showed that language stabilizes perception and mediates also during nonverbal tasks. Kimbal and Dale[45] have shown that forming categories of color names facilitates recognition. By freeing the child from the here and now, by making possible operations such as classifying, serializing, and formulating, words allow the child to discover the common properties of perceptions and events and permit him to group single entities into larger wholes. Language helps in the shift from associative to cognitive levels of learning and paves the way for high order operations.

To quote a poem by O. Mandelstam cited by Vygotsky:[77] "I have forgotten the word I intended to say and my thought unembodied returns to the realm of shadows."

Bibliography

1. BARATZ, J. C., "Language and Cognitive Assessment of Negro Children: Assumptions and Research Needs," *ASHA*, *11*:87, 1969.
2. BENCH, J., "Sound Transmission to the Human Foetus through the Maternal Abdominal Wall," *J. Gen. Psychol.*, *113*:85, 1968.
3. BERKO, J., "The Child's Learning of English Morphology," *Word*, *14*:150, 1958.
4. BERNSTEIN, B., "Social Class and Linguistic Development: A Theory of Social Learning," in Halsey, A. J., Floud, H., and Anderson, C. A. (Eds.), *Education, Economy and Society*, Free Press, New York, 1961.
5. BERRY, M., *Language Disorders of Children*, Appleton-Century-Crofts, New York, 1969.
6. BLOOM, L., *Form and Function in Emerging Grammars*, M.I.T. Press, Cambridge, 1970.
7. ———, *Readings in Language Development*, Simon & Schuster, New York, 1970.
8. BLUMENTHAL, A. L., *Language and Psychology*, John Wiley, New York. Reference to Carla and William Stern, 1907.
9. BRAINE, M. D. S., "The Ontogeny of English Phrase Structure: The First Phase," *Language*, *39*:1, 1963.
10. BRODBECK, A. J., and IRWIN, O. C., "The Speech Behavior of Infants without Families," *Child Dev.*, *17*:145, 1946.
11. BROWN, R., "The Acquisition of Language," in Riach, D., and Weinstein, E. A. (Eds.), *Disorders of Communication, Res. Publ. Ass. Nerv. Ment. Dis.*, *42*:56, 1964.
12. ———, and BELLUGI, U., "Three Processes in the Child's Acquisition of Syntax," *Harvard Educ. Rev.*, *34*:133, 1964.
13. CARROLL, J. B., "Language Development in Children," in Saporta, S. (Ed.), *Psycho-*

linguistics: A Book of Readings, Holt, New York, 1961.

14. CAZDEN, C., *Child Language and Education*, Holt, Rinehart & Winston, New York, 1972.
15. CAZDEN, C. B., "Three Sociolinguistic Views of the Language and Speech of Lower-Class Children—With Special Attention to the Work of Basil Bernstein," *Dev. Med. Child Neurol.*, *10*:600, 1968.
16. CHOMSKY, N., *Aspects of the Theory of Syntax*, M.I.T. Press, Cambridge, 1965.
17. ———, "Review of 'Verbal Behavior' by B. F. Skinner," *Language*, *35*:26, 1959.
18. CRUTTENDEN, A., "A Phonetic Study of Babbling," *Brit. J. Disord. Comm.*, *5*:110, 1970.
19. DESPERT, L., Discussion of "Irrelevant and Metaphorical Language in Early Infantile Autism," *Am. J. Psychiat.*, *103*:242, 1946.
20. DI CARLO, S., Personal Communication.
21. DWORNICKA, B., JASIENSKA, J., SMOLARY, W., and WAWRYK, R., "Attempt of Determining the Foetal Reaction to Acoustic Stimulation," *Acta Otolaryng.*, *57*:571, 1964.
22. EDELHEIT, H., "Language and Ego Development," *J. Am. Psychoanal. A.*, *16*:13, 1968.
23. EISENBERG, R. B., "The Development of Hearing in Man," *ASHA*, *12*:119, 1970.
24. ELMAS, P. D., SIQUELAND, E. R., JUSCZYK, P., and VIGORITO, J., "Speech Perception in Infants," *Science*, 303–306.
25. ERVIN-TRIPP, S., and SLOBIN, D. I., "Psycholinguistics," *Ann. Rev. Psychol.*, *17*:435, 1966.
26. EVELOFF, H., "Some Cognitive and Affective Aspects of Early Language Development," *Child Dev.*, *43*:1895–1905, 1971.
27. FAY, W. H., and BUTLER, B. V., "Echolalia, IQ, and the Developmental Dichotomy of Speech and Language Systems," *J. Speech & Hearing Res.*, *11*:365, 1968.
28. FREUD, A., *The Ego and the Mechanisms of Defense*, International Universities Press, New York, 1936.
29. FREUD, S. (1911), "Formulations on the Two Principles of Mental Functioning," in Strachey, J. (Ed.), *Standard Edition*, Vol. 12, Hogarth, London, 1958.
30. FRIEDLANDER, B. Z., "Receptive Language Development in Infancy: Issues and Problems," *Merrill-Palmer Quart.*, *16*:7, 1970.
31. ———, "Time Sampling Analysis of Infants' Natural Learning Environment in the Home," *Child Dev.*, *43*:730–740, 1972.
32. FRIEDMAN, D. G., *Personality Development in Infancy*, Holt, Rinehart & Winston, New York, 1968, pp. 258–287 reference to Kurt Goldstein.
33. FRY, D. B., "The Development of the Phonological System in the Normal and Deaf Child," in Smith, F., and Miller, G. A. (Eds.), *The Genesis of Language*, M.I.T. Press, Cambridge, 1966.
34. GIBSON, E. J., *Principles of Perceptual Learning and Development*, Appleton-Century-Crofts, New York, 1969.
35. GROSSMAN, W., and BENNET, S., "Anthropomorphism," in *The Psychoanalytic Study of the Child*, Vol. 24, p. 78, International Universities Press, New York, 1969.
36. HESS, R., and SHIPMAN, V. C., "Influences upon Early Learning," in Hess, R., and Bear, R. H. (Eds.), *Early Education: Current Theory, Research and Action*, Aldine, Chicago, 1968.
37. IRWIN, O. C., "Development of Speech during Infancy," *J. Exp. Psychol.*, *36*:43, 1946.
38. JAKOBSON, R., *Selected Writings I*, Mouton and Co., The Netherlands, 1962.
39. ———, and HALLE, M., *Fundamentals of Language*, Mouton and Co., The Netherlands, 1956.
40. JOHANSSON, B., WEDENBERG, E., and WESTIN, B., "Measurement of Tone Response by the Human Foetus," *Acta Otolaryng.*, *57*:188, 1964.
41. KAGAN, J., "Continuity in Cognitive Development during the First Year," *Merrill-Palmer Quart.*, *15*:101, 1969.
42. KAPLAN, B., "The Study of Language in Psychiatry," in Arieti, S. (Ed.), *American Handbook of Psychiatry*, Vol. 3, Basic Books, New York, 1959.
43. KARELITZ, S., and FISCHELLI, V. R., "Infants' Vocalizations and Their Significance," *Clin. Proc. Children's Hospital, Washington, D.C.*, *25*:11, 1969.
44. KATAN, A., "Some Thoughts about the Role of Vocalization in Early Childhood," in *The Psychoanalytic Study of the Child*, Vol. 16, p. 184, International Universities Press, New York, 1961.
45. KIMBAL, M., and DALE, P., "The Relationship between Color Naming and Color Recognition Ability of Preschoolers," *Child Dev.*, *43*:1895–1905, 1971.

46. LENNEBERG, E., *The Biological Foundations of Language*, John Wiley, New York, 1967.

47. ———, "Speech as a Motor Skill with Special Reference to Nonaphasic Disorders," in Bellugi, U., and Brown, R. (Eds.), "The Acquisition of Language," *SRCD Monographs*, *29*:115, 1964.

48. LEOPOLD, W. F., "Patterning in Children's Language Learning," *Language Learning*, *54*:1, 1953.

49. LEWIS, M. M., *Infant Speech*, Humanities Press, New York, 1951.

50. ———, *Language, Thought and Personality in Infancy and Childhood*, Harrap, London, 1963.

51. LIEBERMAN, P., *Intonation, Perception and Language*, M.I.T. Press, Cambridge, 1967.

52. LURIA, A. R., *The Role of Speech in the Regulation of Normal and Abnormal Behavior*, Pergamon Press, New York, 1961.

53. MCCAFFREY, A., "Speech Perception in Infancy," Personal Communication.

54. MCCARTHY, D., "Language Development in Children," in Carmichael, L. (Ed.), *Manual of Child Psychology*, 2nd ed., John Wiley, New York, 1954.

55. MCNEILL, D., "Developmental Psycholinguistics," in Smith, F., and Miller, G. A. (Eds.), *The Genesis of Language*, M.I.T. Press, Cambridge, 1966.

56. MAHLER, M. S., "Thoughts about Development and Individuation," in *The Psychoanalytic Study of the Child*, Vol. 18, International Universities Press, New York, 1963.

57. MENYUK, P., "The Role of Distinctive Features in Children's Acquisition of Phonology," *J. Speech & Hearing Res.*, *11*:138, 1968.

58. MOFFITT, A. R., "Consonant Cue Perception by Twenty- to Twenty-Four-Week-Old Infants," *Child Dev.*, *42*:717, 1971.

59. NAKZIMA, S., OKAMOTO, N., MURAI, J., TANAKA, M., OKUNO, S., MAEDA, T., and SHIMIZU, M., "The Phoneme Systematization and the Verbalization Process of Voices in Childhood," *Shinrigan-Hyoron*, *6*:1, 1962.

60. OSGOOD, C. E., "Motivational Dynamics of Language Behavior," in Jones, M. R. (Ed.), *Nebraska Symposium on Motivation: 1957*, University of Nebraska Press, Lincoln, 1957.

61. PALMER, F. H., "Socioeconomic Status and Intellective Performance among Negro Preschool Boys," *Develop. Psychol.*, *3*:1, 1970.

62. PALMER, M. F., "The Speech Development of Normal Children," *J. Speech Disord.*, *5*:185, 1940.

63. PELLER, L. E., "Freud's Contribution to Language Development," in *The Psychoanalytic Study of the Child*, Vol. 21, p. 448, International Universities Press, New York, 1966.

64. PENFIELD, W., and ROBERTS, L., *Speech and Brain Mechanisms*, Princeton University Press, Princeton, 1959.

65. PIAGET, J., *The Origins of Intelligence in Children*, International Universities Press, New York, 1952.

66. PROVENCE, S., and LIPTON, R., *Infants in Institutions*, International Universities Press, New York, 1962.

67. QUAY, L. C., "Language Dialect, Reinforcement and the Intelligence-Test Performance of Negro Children," *Child Dev.*, *42*:5, 1971.

68. SKINNER, B. F., *Verbal Behavior*, Appleton-Century-Crofts, New York, 1957.

69. SNOW, C., "Mothers' Speech to Children Learning Language," *Child Dev.*, *43*:549–565, 1972.

70. STEWART, W., "Continuity and Change in American Negro Dialects," *Florida FL Reporter*, *7*:1, 1968.

71. STODOLSKY, S. S., "Maternal Behavior and Language and Concept Formation in Negro Pre-school Children: An Inquiry into Process," Ph.D. diss., University of Chicago, 1965.

72. TAYLOR, O., "Recent Developments in Sociolinguistics: Some Implications for ASHA," *ASHA*, *13*:341, 1971.

73. TEMPLIN, M., "The Study of Articulation and Language Development during the Early School Years," in Smith, F., and Miller, G. (Eds.), *The Genesis of Language*, M.I.T. Press, Cambridge, 1966.

74. TORIYAMA, M., MATSUZAKI, T., and HAYASHI, H., "Some Observations on Auditory Average Response in Man," *Int. Audiol.*, *5*:234, 1966.

75. TORKEWITZ, G., BIRCH, H., and COOPER, K., "Responsiveness to Simple and Complex Auditory Stimuli in the Human Newborn,"

Developmental Psychobiology, Vol. 5, pp. 7–19, John Wiley, New York, 1972.

76. TURNURE, O., "Response to Voice of Mother and Stranger by Babies in the First Year," paper presented at Society for Research in Child Development Meeting, Santa Monica, Calif., 1969.
77. VYGOTSKY, L. S., *Thought and Language*, M.I.T. Press, Cambridge, 1962.
78. WEIR, R. H., "Some Questions on the Child's Learning of Phonology," in Smith, F., and Miller, G. A. (Eds.), *The Genesis of Language*, M.I.T. Press, Cambridge, 1966.
79. WEISBERG, P., "Social and Nonsocial Conditioning of Infant Vocalization," *Child Dev.*, 34:377, 1963.
80. WYATT, G. L., *Language Learning and Communication Disorders in Children*, Free Press, New York, 1969.

CHAPTER 17

THE JUVENILE AND PREADOLESCENT PERIODS OF THE HUMAN LIFE CYCLE

E. James Anthony

The Juvenile Period

Definition

The juvenile period, outside its legal sense, was a term incorporated by Sullivan[24] into his developmental system, but others have classified the same chronological span of time in ways more reflective of their particular professional viewpoints. To the educator, for example, these are the years passed in grade school and bifurcated for good educational reasons into a primary triennium from grades one to three and an elementary triennium from grades four to six. For the developmental psychologist this is the era of middle childhood, a time of relative quiescence interposed between the turbulences of preschool and adolescence. This is the case both physically and psychologically. In the context of physical development there is a plateau extending from the preschool to the pubertal growth spurts; Stone and Church[22] have called this the "growth latency," suggesting that it may be a physiological counterpart of the Freudian sexual latency between the passing of the Oedipus complex and its reactivation in early adolescence. For the dentist the juvenile stage begins with the last of the baby teeth and ends with the complete eruption of the permanent set, minus the "wisdoms." In different theoretical systems of development, other prime characteristics have been accentuated. Piaget has pointed to the concrete operational mode of thinking, Sullivan to cooperation within interpersonal relationships, and Erikson to the opposing tendencies of industry and inferiority that can make or mar the school child. There would appear to be, therefore, some grounds for believing that the middle years of childhood represent a psychobiological moratorium in certain crucial and perhaps distracting areas that allows the child to restrict himself to the conflict-free spheres of ego "business" in order to learn as much as he can about the environment in which he lives and his own particular place in it. There is a growing flight from family life and especially from close parental contact, an increasing immersion in peer society, and a restless pursuit and exploration of a universe expanding in space and time with every year of development. The middle years form a self-contained world, irresponsible, irreflective, and halcyon, to which adults harken back with pleasure across the disturbing memories of adolescence. More recent investigations of latency have indicated that these golden recollections are to a large extent a sustaining, utopian illusion and that the miseries of the middle years are as real and as scarifying as any experienced in other parts of the life cycle.

Descriptive Development

Physical growth during the middle years is surprisingly regular and free from breaks or spurts as long as the environment is optimal. The evidence for the existence of a midgrowth spurt between five and eight years is dubious; until more exact longitudinal data are available, there is every indication of a growth latency apart from minor seasonal variations that may be superimposed on the rate-of-growth curve. Growth continues at a slower pace: the average child entering latency at six years measures 3½ feet and weighs 40 pounds and the average child leaving latency at 12 years measures 5 feet and weighs 80 pounds. His muscular strength more or less doubles during the same period. Latency girls, in general, keep ahead of the boys and enter puberty earlier. Very little is known about hormone excretion during the child's development, but there would seem to be no striking changes from infancy to puberty. Some parts of the pituitary, the adrenal cortex, and the gonads do not function until puberty, but others are secreted in proportion to the size of the growing child. The 11-oxysteroids of the adrenal cortex, believed to be concerned in the response to stress, are produced from birth onward at the same intensity as in the adult, per surface area. The 17-ketosteroids, mostly end products of androgenic hormones from the adrenal cortex and gonads, make a surprising appearance around about nine or ten years in both male and female children and some have termed this the "adrenarche." The nervous pattern of copulation and orgasm is complete by early childhood but not normally stimulated into action until the sex hormones are secreted and lower the threshold to stimuli. More is known about the electrophysiological developments in the brain during latency. The slow or delta activity declines sharply during the preschool years and is not a prominent feature of the juvenile record as far as amplitude and abundance of activity is concerned. According to Grey Walter,[14] it persists (statistically) in docile, manageable, and easily led children whom he terms "ductile." These children tend to drift inevitably into delinquency. The theta rhythms are characteristically prominent during early latency and are easily evoked with frustration. The decline in theta parallels the decline in temper outbursts in preschool children, and its association with the frequent annoyance and deprivation of the preschool period suggests the reason for its relative inactivity during the latency phase. Its presence has been linked to the level of emotionality in the individual. The alpha rhythms attain their adult distribution sometime between the ages of nine to eleven, but the variations may be considerable. There are juveniles who show no alpha rhythm at all and others in whom it persists throughout the waking day, whether the eyes are closed or open, and whether the child is concentrating or relaxed. The rise of alpha may have something to do with the child's new orientation to his environment during latency and his attempts to reduce its diversity to regular, recognizable, and familiar patterns.

Emotional development does not settle until later in latency. Gesell[13] depicts the emotionality of the younger juvenile and the marked disequilibrium between him and others in his environment. He is given to extremes of expression, to love and hate, and to extreme sensitivity to praise and blame. He is fearful during these early years, especially of the supernatural, of animals, of self-injury, and of something dreadful happening to his mother that may take her away from him. His dreams reflect these same concerns, and girls especially may fear and dream of men getting into their rooms and attacking them. In the later part of latency the emotions are brought more under control, and the juvenile is in much better equilibrium with others. Whereas in the early years he was somewhat humilious and unamenable to management with humor, later he is able to develop a good sense of humor and can even laugh at himself. The later fears tend to be realistic and concerned with school problems, competitive situations, and worries about failure, and although far fewer dreams are reported in the later part of the period, they mostly have to do with experiences regarding possessions, playmates, and personal difficulties or worries connected with these.

These are, of course, group profiles drawn by Gesell[13] of white middle-class children, and they say very little about children in general, children in different cultures, children under conditions of disadvantage, children with different constitutional propensities, and individual children. The self of this white middle-class juvenile undergoes changes highly reflective of the child-centered world to which it belongs. In early latency Gesell describes the self as the center of its universe, expansive, undiscriminating, self-willed, and demanding to be first, to be loved the best, to be praised and to win. At times the self may even seem to inhabit "another world" and to withdraw into this whenever circumstances are not propitious. There is also a self-consciousness about the body and the beginnings of modesty, especially in the girl. Shame is a powerful affect stemming from lapses of competence. Later, from nine onward, the child seems gradually to change for the better, to be more outgoing, more social, more interested in relationships especially with other children; more independent, self-sufficient, dependable, and trustworthy; more responsible and much busier and more actively interested in school work, hobbies, and a variety of tasks. At this point he becomes a "worker" and may even prefer work to play. He is increasingly self-aware about what he does, what he is, and what he can and cannot do. He is likely to complain a lot about life in general, his life in particular, and the way in which his body is malfunctioning. At the end of the period he may become markedly hypochondriacal and fearful of death and may even experience an existential crisis.

He now belongs almost completely to the society of children, and although living within adult society, he picks his way through it preoccupied with the concerns of childhood. He is part of a special, separate subculture with its own traditions, rules, regulations, values, and loyalties, and like primitive cultures it is handed down largely by word of mouth from child to child. In this proving ground the child learns to function apart from adults and practices a variety of roles both in play and in social interaction. For the younger child play especially represents a time out in which he allows himself to escape from the limitations of reality and to undergo, if only momentarily, in Sully's[23] term, a "transmutation of the self." He often plays as industriously and as seriously as he works, and since he is still reverberating from the emotions of preschool, he may use play further for catharsis, wish fulfillment, fostering anxiety, problem solving, circumventing the conscious, symbolizing conflict, stimulating new identifications, softening the impact of trauma by repetition, and elaborating new systems of defense. In the latter part of childhood participation in games teaches him to compete, to win and lose gracefully, to eschew cheating, and to cooperate with his peers. As a member of his group or gang he may take his place in the pecking order and construct a fairly valid concept of himself in terms of his usefulness to the group, capacity for leadership, and organizational skill. The membership carries with it the joys of belonging and the pains of exclusion or ostracism. Neighborhood groups are subdivided by age and sex, and in stable residential areas there may be direct continuity between early childhood groups and the street-corner society of preadolescence and adolescence. Some of the groups maintain their organizational identities while successive generations of children pass through them.

At this stage it is usual for the two sexes to play separately, although they may still share in the same interests and even play the same games. Few latency girls are likely to participate in predominantly male games, but practically no boys ever openly engage in what are thought of as female ones. However, the interest of the two sexes shows convergence, and the sex cleavage is no longer as absolute. In the gangs a good deal of sexual experimentation may take place, and heterosexual exploration at this age is also becoming commoner as parents countenance or promote mixed parties. The continuity of sex role, to use Benedict's[3] expression, is more manifest for girls who may spend a large part of latency rehearsing for motherhood, homemaking, and child care. Only a few adult occupations are continuous with boyhood culture.

There is no doubt that siblings provide an intense and crucial "corrective emotional experience" for the child and although the contacts are likely to be, according to Stone and Church,[22] marked by "baiting, bantering, bickering, battling, belittling, and bedlam" within the family, an extraordinary solidarity may prevail outside it. The children hand down clothes, toys, traditions, customs, and play patterns from one developmental stage to the next, and they model for one another in terms of appropriate sex and age behavior. The world of parents is gradually left further and further behind although they are still a central pivot in the child's life. He may boast of their achievements, demand their backing and enthusiasm in his pursuits, and still appeal to them as the final arbiters where sensitive matters of justice and fairness are concerned. Nevertheless, deidealization is gradually progressing, and the child is becoming increasingly aware of his parents as fallible and sometimes frail human beings.

Psychosexual Development (Freud)

The primary psychoanalytic concept relevant to the juvenile period is that of sexual latency. Latency begins with the resolution and repression of the Oedipus complex and ends with its reactivation at puberty; it is a kind of psychosexual moratorium in human development that allows the future adult to learn the technical skills necessary for a future work situation. This is different from the psychosocial moratorium given by some cultures to the adolescent so that he can find a place for himself in society. The first moratorium is brought about by mechanisms of repression and sublimation, and the second by suppression and prohibition. In our Western culture latency children manifest a great show of indifference and even antagonism toward members of the opposite sex, and such reaction formations are further rewarded by the attitude of society in general. The upsurge of intellectual curiosity observed in schoolchildren has been ascribed to sublimation, to transformations of repressed sexual energies, and also to disguised sexual curiosity. The rules and rituals of this stage are also seen as magic devices for controlling impulses and feelings, including the anxieties and hostilities derived from the oedipal conflict. Latency children, with their superstitions, their collecting and hoarding habits, their mysterious chants and formulas, show a slight resemblance to patients suffering from an obsessive-compulsive neurosis, and latency children will often complain of compulsions, thoughts, and snatches of music that they cannot be rid of, magical counting devices before test situations, and sometimes quite elaborate systems of checking and undoing. All these are still within the range of normative development and the symptoms disappear with the stage. The compulsions may be a means of giving the child some sense of security, certainty, order, and meaning following his escape from the chaotic and confusing experiences of the oedipal period. Since no conflicts can be perfectly repressed, some of the child's rituals represent a disguised, symbolic enactment of an impulse escaping the repression barrier. Therefore, the characteristic latency child is a mildly obsessional child whose compulsions are very likely to pass with the onset of puberty.

Questions have been raised about how latent the situation really is, and investigators such as Broderick[5] have pointed to the amount of heterosexual as well as homosexual interests and play in the preteen years. He feels that sexuality is still active but simply hidden from adult attention. Forbidden information, "smutty" magazines, crude drawings, dirty jokes, bad words, medical diagrams are passed around the underground with a great deal of nervous excitement and pleasure. Curiosity is probably stronger than craving at this age and the toilet function inseparable from the sexual one. In children who have been overstimulated or seduced, a greater degree of precocity may develop and at the same time provoke much anxiety. The main repression, according to psychoanalytic theory, is directed toward sexual feelings for the parent of the opposite sex, and it is this element that is universally latent. The Henrys[16] have furnished us with protocols from the doll play of children in an Indian tribe, and the record is replete with gross references to copulation. The latency children in many primitive societies spend a lot of time in sexual acting out under the permissive aegis of the elders, and Harlow[15] has pointed to the fact that sexual play among immature rhesus monkeys is essential to sexual competence in adult life. The rehearsal of sexual elements becomes increasingly proficient with practice.

The drives that made the preschool child dream and play undergo sublimation into a host of latency activities, the most important being the sharpening of the epistemological interest. The child becomes a glutton for knowledge as his sexual curiosity is transformed into scholarly inquisitiveness. Leftover anxieties are dealt with by a wide range of defenses characteristic of the latency child and manifest themselves in minor phobias, obsessions, compulsions, conversions, and depressions, all within the normal range. Shame in reference to real or imagined inferiorities and guilt in relation to masturbation and masturbatory fantasies may push the latency child ever closer to a manifest neurosis. Infantile sexual theories are further embroidered and changed following research in latency, but it is surprising, even with systematic sexual education, how fantastically the process of reproduction can be perceived. But as Freud pointed out, many erroneous ideas are invented by children in order to contradict older, better, but now unconscious and repressed knowledge. Furthermore, the latency child may admit that all parents, except his own, may copulate in the manner described by his instructor: "It is possible that your father and other people do such things, but I know for certain that my father would never do it."

Two fantasied offshoots of the Oedipus complex in latency are, first, the beating fantasy, which has its origin in the girl's incestuous attachment to the father or the boy's feminine attitude toward his father, and second, the family romance, which represents a stage in the estrangement of the child from his parents. In early childhood he may feel that he is not getting his full share of parental love and attention, and he may develop the idea that he is a stepchild or an adopted child. Somewhat later he may invent parents of higher social status. When he has become aware of the sexual determinants of procreation, the family romance undergoes a "peculiar curtailment"; the father becomes a stranger and exalted but the mother remains the same, although in dreams they remain king and queen. In one of his papers Freud[11] reproduces a letter from a motherless girl of 11 years who urgently asks her aunt the "truth." "We simply can't imagine how the stork brings babies. Trudel thought the stork brings them in a shirt. Then we wanted to know, too, how the stork gets them out of the pond, and why one never sees babies in ponds. And please will you tell me, too, how you know beforehand when you are going to have one." (The writer of this touching request has developed a compulsive neurosis full of obsessive questioning.)

Psychosocial Development (Erikson)

The central concept in Eriksonian developmental theory is the epigenesis of identity, the slow unfolding of the human personality through a succession of psychosocial crises or turning points. In this system the child enters latency, if all has gone well, with a requisite measure of basic trust, of confidence (Benedek),[2] of mutuality, of autonomy, and of initiative, all contributing to his overall rudimentary sense of identity. According to Erikson,[6] identity is one of the "indispensable coordinates" of the life cycle and emerges more firmly, more completely, and with an increasing sense of inner unity with each crisis that is encountered and mastered. Each turning point in the epigenetic ground plan represents a period of specific vulnerability, a concomitant shift in instinctual urges, and a new interpersonal perspective. The child who is a developmental success actively masters his environment and begins to perceive himself and others more correctly. When development is failing, the child comes into latency overwhelmed by mistrust, shame, doubt, and guilt. Trust is the cornerstone of the healthy personality and stems from transactions with a trustworthy mother, imbued with a sense of fairness and justice. She has also been able to inculcate a deep and almost "somatic" conviction that there is meaning to her management of the child. The psychosocial issue of trust to mistrust is a crucial quotient for the rest of the development. In latency the child is confronted with a widening social radius, expanding libidinal needs, a large number of developing capacities, and a specific psychosocial strength deriving from a favorable trust-mistrust ratio.

Erikson believes in the essential "wisdom of the ground plan" so that when the child reaches school age following the expansive period of the preschool years, he is more than ready to get down to work and discipline himself to perform well and to share in whatever is being constructed and planned. He becomes a manufacturer, a producer, a tool user and occupies himself industriously with making things as perfectly as he can and persevering with the job until it is fully done. When things go wrong, on the other hand, feelings of inferiority can develop, especially when he persists in comparing himself unfavorably with his father. A great deal depends on the parental surrogates who step into the picture to continue the work of the parents. The good enough teacher knows how to alternate work and play, recognize special efforts, and encourage special gifts. The good enough parents encourage their children to trust their teachers. The good enough school is not simply an extension of grim adulthood and the puritan ethic into the classroom or an extension of the expansive imagination of childhood without restriction, but something of both. The contribution, therefore, of the school age to the sense of identity can be expressed in the words: "I am what I can learn to make work." To many children this is also the end of identity and its consolidation around an occupational or technical capacity. This is a foreclosure of identity and abrogates the "higher" elements of identity.

Interpersonal Development (Sullivan)

Sullivan[24] constructed a heuristic classification of personality development ranging from infancy to adulthood and governed by the central principle of interpersonal cooperation. In this scheme the juvenile era begins with the need for playmates, defined as cooperative individuals of comparable status, and ends with the need for an intimate relationship with a single person of comparable status. Sullivan reiterates many times that this particular era is of special importance since it constitutes the first occasion when all the drawbacks of living in the family can be remedied in the world outside the home. It almost seems as if Sullivan is implying that one stage of development can be therapeutic for another, and if this is at all true it emphasizes the need to study the developmental stages in parallel in order to be able to exploit this ameliorating capacity to the full. Sullivan suggests that unless the juvenile era can alter the spoilt or self-effacing child for the better, he is going to be increasingly impossible to live with and work with in later life. (It is well to remember that Sullivan was particularly inclined to express himself in somewhat exaggerated terms and that most of his assertions need to be toned down to some extent.) At school the child is required to subordinate himself to new authority figures with whom he has often had minimal experience. In order to get along in the new milieu, the juvenile must learn some degree of social accommodation, which is mainly a principle of expediency. However, social accommodation has the effect of increasing social competence as well as providing the individual with a basic education in getting along, getting by, and getting away with things. With increasing social sophistication the juvenile is able to detoxify the frightening authority figures and reduce them to the proportion of people. Without such necessary de-idealization the parents and their homologues remain sacrosanct and the juvenile continues in servitude.

Therefore, the juvenile era is the time, as Sullivan[24] puts it, "when the world begins to be really complicated by the presence of other people" in relation to whom there must be cooperation, criticism, competition, compromise, concentration of attention, and real interpersonal communication. As a result of all these, the juvenile learns a good deal about how to create a secure environment for himself. The self-system at the core of Sullivanian personality development is a dynamism that organizes experience, controls the content of consciousness, and deals with the intrusion of anxiety. It is not synonymous with the psychoanalytic superego, although it sounds like it at times. Sullivan does little to clarify these overlapping terms and tends to avoid the issue. "It has been so many years since I found anything but headaches in trying to discover parallels between various theoretical systems that I have left that for the diligent and scholarly, neither of which includes me."

Another piece of learning that goes on during the juvenile era is "sublimatory reformulation," which consists in substituting acceptable behavior for unacceptable behavior for the sake of appeasing authority figures. Since this diminishes gratification, the unsatisfied need must be worked off in a dream or daydream.

Since life at this time is largely lived in groups, the question of belonging to outgroups or ingroups becomes a crucial factor for healthy development. It may be hard for the juvenile to understand the mysterious workings of segregation and ostracism, and no amount of wishful thinking can transform an outgroup into an ingroup. Self-esteem is one of the major casualties of the group organization of the juvenile world. In addition to grouping, the individuals at this stage also resort to stereotyping, and if this becomes fixed, it can give rise to real trouble later on. A great many juveniles arrive at preadolescence with quite rigid stereotypes about the classes and conditions of mankind. These troublesome stereotypes also invade the self-system and set up "supervisory patterns" in the processes and personifications that constitute the self-system. These imaginary people are always with one and can be hearers (judging the relevancy of what one is saying) or spectators (of what you showed to others). Since these act in an inhibitory way, they can be compared to the psychoanalytic superego and ego ideal system. The supervisory patterns come into being in the juvenile era and form a part of the elaborate organization for maintaining self-esteem and self-respect.

As the individual reaches the end of the juvenile era, he gains, whether he likes it or not, some sort of "reputation" in terms of popularity, sportsmanship, brightness, and other desirable or undesirable attributes. If one leaves the juvenile era with a bad reputation, it is likely to stick to one and prove a handicap during further development.

The individual is not only subjected to influences from the peer group but also from the parent group, and one of the less fortunate things that parents can do at this stage is to teach disparagement. As Sullivan[24] puts it in his usual vivid style, "the disparaging business is like the dust of the streets—it settles everywhere." It can be very disastrous for the individual if the only way he can maintain his self-esteem is by pulling down the standing of others. Security achieved in this way is not real security and can easily crumble.

By the end of the juvenile era, all being well, one becomes "oriented in living" or, put in another way, "well-integrated." The integration includes one's interpersonal relationships, their management with relatively little anxiety, and the capacity for postponing immediate gratification in favor of some eventual gain. The juvenile who is well oriented to living has a more or less assured future; the juvenile not so oriented is destined to be, to use some of Sullivan's favorite negative epithets, "unimportant," "troublesome," and a "lamentable nuisance."

Psychocognitive Development (Piaget)

The juvenile period, on Piaget's[19,20,21] scale, approximates the end of the representational period and the stage of concrete operational thinking. Egocentrism is on the wane throughout the stage. Symbolic play gradually disappears, to be replaced by games. The child's conception of the world is gradually approaching the adult one.

At first glance there would appear to be some similarities between Piaget's and Gesell's[13] notion of development. As Hunt[17] has pointed out, both employed the cross-sectional method of confronting children of various ages with situations and materials, at the same time observing how the children reacted and responded; both described behaviors typical of children at successive ages; both recognized an epigenetic system of change in the structure of behavior as essentially predetermined, although Piaget is more of an environmentalist and his conception of the organism includes a theory of adaptation governed by the input and adjustment mechanisms of assimilation and accommodation. As a "stage" psychologist, Piaget has postulated a definite order in development, a succession of more or less self-contained stages, with an intrastage hierarchical organization of attributes and transitions from one stage to the next. Anthony[1] has suggested that a system such as this with well-defined stages, transitional periods, and a dynamic adaptation theory could well provide a basis for a psychopathology.

Egocentrism is narcissism without Narcissus or self-awareness. The process of differentiation between the ego and the world is still incomplete. The child is at the center of the universe, and his perspective is subjective, absolute, and incapable of allowing for another viewpoint. Early in the juvenile period a revolution in perspective takes place, and the child sees himself as a thing among many other things in the universe and by no means the most important of them. This is something in the nature of a Copernican shift. From being highly subjective and absolute, the viewpoint becomes objective and relative, and for the first time the child becomes capable of getting himself in someone else's shoes and perceiving the world through the other's eyes. However, by six or seven years he is not wholly free from egocentrism, and his world is still alive in every part with animistic projections and magical-phenomenistic forces. The sun and moon follow him around; his dreams come in through a window at night; a stone will feel a prick if it is rolling; twisted string unwinds itself because it feels all twisted up; shadows are emanations from objects; and so forth.

As logical, operational thinking takes over, even if tied exclusively to the manipulation of concrete objects, this dynamic, primitively conceived world sobers down and becomes much more realistic and down-to-earth. The child gives up thinking by transduction and juxtaposition of ideas and is surprisingly capable of conceiving reciprocal relationships and reversing his thought processes. Reversability is the crucial cognitive achievement of the juvenile era. Not only does the child think more logically, but his previously egocentric speech, seldom more than a "collective monologue," becomes socialized and intended for communication. In fact, his whole approach to his environment becomes less amorphous and syncretistic and more precise and differentiated. His cognitions can now encompass a variety of mental transformations referred to as groupings (a logico-mathematical structure that Piaget finds useful for describing cognitive operations). As a result the juvenile can solve many problems that lie beyond the capacity of the preschool child. A child from seven to eleven years behaves as though his primary task consists of organizing and ordering what is immediately and concretely present before him. He is essentially a here-and-now child, and what is absent or what is potential is not his concern. He is bound to the phenomenal and engages the elements of his universe one by one in his efforts to master them. He learns about classification, serialization, conservation, inclusion, the dimensions of time and space, and basic notions of movement and velocity, all of which help to make him familiar with the workings of his environment.

What drives the child to all this accomplishment? According to Piaget, it is because he has to "nourish" his internal cognitive schemata by repeatedly incorporating nutriments from the environment. There is a need to function, which is the only motivation that Piaget recognizes. Furthermore, the child's feelings are not tied to his motivations but to his cognitions and undergo a parallel development that is simply the reverse side of the coin. The intellect furnishes the structure and the affect provides the energy. During the juvenile era, the emotional life is isomorphic to the logico-arithmetic organizations and is characterized by highly structured systems of values, concepts of justice and obligation, fairness in transactions, and interpersonal relations founded on reciprocity and individual autonomy. The "two moralities of childhood" stretch across the whole juvenile period. The first morality, during the earlier years, is authoritarian and based on constraint and unilateral respect. Games are governed by coercive rules that are absolute and inflexible, and moral judgment derives from external and heteronomous sources. There is a moral realism about crime and punishment and the need to fit them together literally, so that a child who breaks six glasses is twice as culpable as the child who breaks three and must be punished twice as harshly. Punishment is dictated by the *lex talionis*. No one can escape justice; it is immanent in the entire universe, so that a child who has been stealing apples may very well break a leg as he attempts to escape over a wall. In the second part of the juvenile period, morality is based on norms of reciprocity involving cooperation and mutual respect. Games are governed by rules agreed upon by the participants, and the sense of fairness may extend even to the point of making allowances for natural inequalities. For example, the child who has an injured knee will be given a handicap in the race. The moral sense is now internal and autonomous and functions even in the absence of authority figures. The justice meted out is "distributive," which means fairly apportioned in terms of responsibility.

The Possibilities and Problems of Synthesis

Taken in totality, the picture of juvenile development by Gesell,[13] Freud[9,10,11,12] Erikson,[6,7] Sullivan,[24] and Piaget,[19,20,21] combined with the empirical findings of academic developmental psychology, is rich, complex, and varied. The different systems do not articulate very closely, nor do they agree very often on the interpretation of similar behavior, but they do frequently seem to be saying similar things in different ways. The language of the systems is idiosyncratic, and this in itself creates a climate of difference. The juvenile child turns away from his family and from his egocentric self to a society of peers with whom he engages in various cooperative transactions that are governed by principles of reciprocity and equality. Because of the physical and psychosexual latency, he can use a lot of his energies for epistemological ends and gradually constructs an internal representation of the world that is fairly authentic and based on reality. He talks now in order to communicate, and his concrete thinking allows him to categorize his world, though to a mildly obsessional extent. He lives and works in groups and attempts to conform to the group's code of behavior; he is miserable when excluded from the group. He is on the way to developing a full and consistent identity and a sense of self that gives him a feeling of continuity throughout all the vicissitudes of developmental change. The self, which has private and public components and both true and false aspects, may emerge into self-consciousness suddenly and dramatically at the end of the juvenile era.

The Preadolescent Period

Definition

The end of childhood is indefinite since puberty may intervene almost any time after the tenth year. Educationally the largest number of preadolescents are concentrated in the junior high schools between the seventh and the ninth grades. Preadolescence can therefore be defined as an indeterminate period of variable duration, but mostly brief, between the eleventh and the fourteenth year. It is the transitional period between childhood proper and maturity. Many developmental psychologists ignore its existence; some give it a passing reference, while others, like Sullivan,[24] speak impressively of its significance. He refers to it as "an exceedingly important but chronologically rather brief period." For Piaget[20] formal or abstract operational thinking makes its first appearance during this era, and with it a new conception of the world unfolds.

The term "pubescence" refers to a period of about two years immediately before puberty and to the preparatory physical changes that take place in this time. Its onset is marked by a prepubertal growth spurt signaling the end of growth latency, and along with this comes changes in the facial and body proportions, the appearance of primary and secondary sexual characteristics, and a variety of other physical changes that may sometimes persuade the child that he is turning into a completely different person. In both sexes the extremities and neck grow at a faster rate than the head and trunk, giving the self-conscious youngsters the long-legged, gawky, and coltish look characteristic of this age. The shoulders broaden in the boy and the pelvis widens in the girl, who, to a greater extent than the boy, develops a layer of subcutaneous fat that rounds and softens the contours of the body. In general, the boys remain leaner and more angular. The growth of pubic, axillary, facial, and chest hair adds to the transformation. The skin becomes coarser, the pores larger, and the sebaceous glands more active, and as a result the young person becomes subject to the miseries of acne. The composition of the sweat also alters and the odor becomes stronger. This, together with the odor of menstruation, intensifies self-consciousness. In girls the areolas grow larger and become elevated and pigmented, and the breasts may attain almost full size prior to the menarche (according to Aristotle, menstruation tended to begin when the breast was two fingers abreast in size). A typical feature of pubescent physical development is asynchrony in the development of bodily parts. (Stone and Church[22] referred to this as "split growth.") The resulting asymmetry may further intensify the child's feeling of being "off kilter" and this, together with early and late maturing, may promote transient maladjustments in the preadolescent. (These effects may be more permanent than was previously supposed. Jones[18] has shown that personality differences between early and late maturing individuals may persist well into adult life. Early maturers as a group were poised, responsible, successful, and conventional, and late maturers were active, insightful, independent, and impulsive.)

The functioning of the endocrine glands during this period of growth and development provides an example of "feedback control," in which pituitary hormones stimulate adrenal hormones, which, in turn, stimulate gonadal hormones, which then inhibit the pituitary hormones and cause secretions to slow down. As a result of this, the pituitary-inhibiting secretions of the gonads also slow down and the pituitary once again becomes active. Why the pituitary becomes active at this particular time is not known, but it is thought to be part of a genetically programmed pattern of normal maturation.

Psychologically the preadolescent individual is more akin to the latency child than to the adolescent.

Psychosexual Development (Freud)

The new developments in the physical system create new feelings and new urges in the psychosexual sphere, with some increase in sexual tension. However, the mechanisms operating in the latter part of latency undergo a complementary increase in strength, so that the child's ego may become more heavily defended than at any other time during his development. Some analytic therapists insist that the child is more difficult to treat at this period than at any other and that transference work is particularly hard to carry out.

If one looks at the middle years in terms of defense, there are marked changes in the picture presented. During the first phase of latency the new superego is harsh, rigid, and, according to Bornstein,[4] "still a foreign body." Two different sets of defenses begin to operate. The first is directed against the genital impulses and leads to a temporary regression to pregenitality, which seems at first to be less dangerous, but later offers problems of its own and provokes a second defense against pregenitality that is largely reaction formation. This brings about the first character change in latency. As a result of increasing genital pressure and constant defense against it, the child becomes increasingly ambivalent and oscillates between obedience, rebellion, guilt, and self-reproach. There is little he can do during this "intermediate stage of superego development" (Anna Freud)[8] except to identify with the aggressor or project the guilt. The whole equilibrium is extremely precarious, and every now and then a crisis situation occurs that may manifest itself in acute symptoms such as animal phobias, fears of being hurt or dying, insomnia and nightmares, and sometimes a new wave of separation anxiety. In the second phase of latency things are not as smooth as often described. Adults especially, when recalling latency, seem to be conjuring up the "ideal of latency" during which the instinctual demands have been successfully warded off. There is some truth, however, in the assertion that infantile neurosis decreases during the second part of latency. Certainly the sexual demands seem less pressing, the superego less rigid, the ego more preoccupied with reality, and masturbation less frequent. The gradual de-idealization of the parents helps to attenuate conflict between ego and superego. Because of the resurgence of instinctual forces accompanying pubescent development, the defensiveness of the preadolescent has become a therapeutic byword. The child is caught between latency and pubertal struggles, between old and new sexual aims. The latency peer relationship is now reduced to a group of two with homosexual undertones. As a result mutual masturbation and other sexual practices may erupt from time to time, associated with shame and guilt. A recapitulation of the psychosexual development of the first five years may come into evidence toward the end of the preadolescent period.

Psychosocial Development (Erikson)

Erikson[6,7] has very little to say about preadolescence, and it does not figure as a separate "box" in his epigenetic diagram. It is difficult to explain this oversight on his part, since the child obviously does not jump from industry into the confusions and turbulences of adolescence without some transitional or prodromal experiences. He clearly believes that at each stage the child becomes a very different person with greater capacities for action, reaction, and interaction and that the psychosexual moratorium is there for the specific purpose of learning the basic grammar of the culture during the sexual lull when the polarities of industry and inferiority are operating and competence begins to emerge.

Interpersonal Development (Sullivan)

Sullivan[24] has much to say about preadolescence and what is for him the primary characteristic of the period, which is the preadolescent collaboration, different in every way from the cooperation, competition, and compromise of the juvenile era. The developmental force that brings this about is the integrating tendency, otherwise known as love or the need for interpersonal intimacy. This is a specific new type of interest in a *particular* member of the *same* sex that eventually develops into what Sullivan calls a "chumship." What he is saying is that real love begins in preadolescence and that all previous relationships during development with peers and parents are essentially exploitative rather than reciprocal. In the loving chumship of preadolescence each partner moves toward the other for the purpose of mutual satisfaction. This special sensitivity with regard to the other person and his feelings is the hallmark of this stage. As Sullivan[24] puts it: "Collaboration is a great step forward from cooperation." A change takes place from predominantly me-feelings to we-feelings. Because of the intimacy and the new capacity for seeing oneself through the other's eyes, the preadolescent phase of personality development is especially significant in correcting autistic and fantastic notions about oneself or others. Here again Sullivan[24] puts forward his fascinating idea of a developmental stage being therapeutic for disturbances inherited from other stages. "Development of this phase of personality is of incredible importance in saving a good many rather seriously handicapped people from otherwise inevitable serious mental disorder." Here he isn't allowing for his usual liberal use of superlatives. This idea would seem to be something that every therapist should keep in mind, involving, as it seems to do, a synergism between natural and artificial therapy. The essential therapeutic ingredient in the preadolescent phase is the "chumship."

The gang at this time is basically structured on the two-group interlocking with other two-groups. Another basic unit is the three-group, made up of a two-group and a preadolescent providing model qualities to the other two. Thus, there is a lot of mutual influencing in the gang. When the members are confronted with serious problems that overwhelm them, one of the "models" may assume the role of leadership for the purpose of solving the pressing issues. The preadolescent personality is therefore molded by this developing leadership-led system of relationships. Another important aspect of the gang is that it provides, for the first time during development, what Sullivan refers to as a "consensual validation of personal worth." The improved communication between chums helps to rectify the self-deceptions and mythologies. The intimacies of preadolescent socialization may also help to correct the leftover egocentrisms of the juvenile era or the marginal case, constantly on the fringe of ostracism, who may have been in the outgroup of juvenile society. Sullivan pooh-poohs the alleged dangers of homosexual development and feels that the chumship is a marvelously corrective emotional experience for the bad sports and spoiled children emerging out of the juvenile era. He is shocked by the idea that preadolescence could be considered as a training for an antisocial career because of the gang influence. He agrees that in bad circumstances and bad environments there are likely to be bad gangs, but even in these extreme cases the gang experience can be of value for socialization. So therapeutic is this phase in Sullivan's eyes that even juveniles entering preadolescence with what he calls a "malevolent transformation of personality" may have this ameliorated, reversed, or even cured. The juvenile who has a vicious disparaging tendency that constantly pulls people down may also have this somewhat mitigated by the preadolescent experience. Schizoid individuals may also benefit, although there is a chance that their self-isolating behavior may make it difficult to establish the type of intimacy required for a chumship, and puberty may be established before they can reap the benefits of the period. The persistent juvenile who tries to avoid growing up may eventually wend his way into an irresponsible gang unless he can be salvaged by a chumship. The handicapped child is particularly likely to suffer from the competitions and rivalries of the juvenile era, but he may come into his own with the development of preadolescence and the helping hand of a chum.

The two-group has its built-in disadvantages as well as advantages. If the relationship is too intense, fixation in preadolescence may occur or homosexual developments might take place. Once again Sullivan dismisses this risk and insists that his own hope would be that the preadolescent relationship was intense enough "for each of the two chums literally to get to know practically everything about the other one that could possibly be exposed in an intimate relationship." It is difficult to know exactly what he means by this, but he clearly is proclaiming his faith in the "great remedial effect of preadolescence." At times like this he does seem perilously close to pushing a pathological development, and it is possible that wide experience of the adult outcome has provided him with the confidence for such assertions. What is sometimes disastrous for the preadolescent chumship or gang is unevenness of development at this time, which puts the participants out of phase and sometimes out of relationship with each other. Bad couplings may result from preadolescents' linking with those still juvenile or those already pubertal, and in these situations homosexuality or schizophrenia are possible risks. The different time schedules of development bring about differences in attitudes, values, and behavior that carry with them the seeds of disruption.

In preadolescence, where real intimacy first begins, there can also be bitter experiences of ostracism and consequent loneliness. Without the protection of intimacy, loneliness can be terrible, and it is in this era that loneliness reaches its full significance and goes on relatively unchanged for the rest of life. The defense against loneliness is what shapes existence.

Psychocognitive Development (Piaget)

In Piaget's system a cognitive revolution takes place in the preadolescent. His reasoning frees itself from the concrete and begins to undertake such formal operations as deduction from hypotheses. A final state of equilibrium is reached at about 14 to 15 years. The preadolescent also starts to construct theories and make use of the ideologies that surround him. Piaget links these new capacities to the development of neurological structures and suggests an isomorphism with them, as do the Gestaltists. Nevertheless, he also is aware of social and cultural influences and feels that education may accelerate individual development. As he says, the maturation of the nervous system can only determine the totality of possibilities and impossibilities at a given stage, but the social environment remains indispensable for the realization of these possibilities. This constant circular process characterizes all exchanges between the nervous system and society, and the individual has to learn to adapt to both physical and social worlds.

The preadolescent differs from the juvenile in that he thinks beyond the present and commits himself to possibilities. Unlike the juvenile, he begins to build "systems" as well as to think systematically. The juvenile does not try to systematize his ideas and is not reflective or self-critical. The preadolescent can also make theories, although they tend to be oversimplified and awkward at first. He may also develop theories as part of his intellectual life with the gang. He is not only capable of thinking about the past, the present, and the future, the present and the absent, but he is also able to think about thinking, which is another leap forward in self-consciousness.

What Piaget calls the third form of egocentrism emerges at this stage and continues through adolescence as one of its most characteristic features. It is only when this is "decentered" that a true adult work orientation becomes possible. The egocentrism of the preadolescent still makes him try to adjust the environment to his ego rather than the other way about. As a result there is a relative failure to distinguish between his own point of view as an individual and the point of view of the group. The preadolescent society helps to mitigate and correct this perspective. This egocentrism allows the preadolescent to attribute an unlimited power to his own thoughts, so that he dreams of a glorious future of transforming the world through ideas or reforming the present sorry state of affairs. All this is generally fuzzy and contains elements of play rather than serious work.

One should point out, if it is not already sufficiently apparent, that Piaget does not have an adequate personality theory. In a somewhat oversimplified way he sees the ego as naturally egocentric and personality as a decentered ego. This is not enough to construct a personality theory and certainly not enough to make a psychopathology.

Conclusion

Preadolescence is, therefore, a gateway to adolescence; it is the crucial transitional era between the juvenile and adolescent stages; it is a time for high defensiveness, for intimate chumships, for therapeutic relationships, for the flowering of a new and important kind of human intelligence whose eventual purpose is to shape the future of the world. It is there to repair all the disastrous damages of development and prepare the person for the new and different world in front of him.

Bibliography

1. ANTHONY, E. J., "The Significance of Jean Piaget for Child Psychiatry," *Brit. J. Med. Psychol.*, 29:20–34, 1956.
2. BENEDEK, T., "Adaptation to Reality in Early Infancy," *Psychoanal. Quart.*, 7:200–214, 1938.
3. BENEDICT, R., "Continuities and Discontinuities in Cultural Conditioning," *Psychiatry*, 1:161–167, 1938.
4. BORNSTEIN, B., "On Latency," in *The Psychoanalytic Study of the Child*, Vol. 6, pp. 279–285, International Universities Press, New York, 1951.
5. BRODERICK, C. B., and FOWLER, S. E., "New Patterns of Relationship between the Sexes among Preadolescents," *Marriage & Fam. Living*, 23:27–30, 1961.
6. ERIKSON, E. H., *Childhood and Society*, Norton, New York, 1963.
7. ———, *Identity, Youth and Crisis*, Norton, New York, 1968.
8. FREUD, A., *The Ego and the Mechanisms of Defence*, Hogarth, London, 1942.
9. FREUD, S., "Family Romances," in *Collected Papers*, Vol. 5, pp. 74–78, Basic Books, New York, 1959.
10. ———, "On the Sexual Theories of Children," in *Collected Papers*, Vol. 2, pp. 59–75, Basic Books, New York, 1959.
11. ———, "The Sexual Enlightenment of Children," in *Collected Papers*, Vol. 2, pp. 36–44, Basic Books, New York, 1959.
12. ———, "Three Essays on Sexuality," in Strachey, J. (Ed.), *Standard Edition, Complete Psychological Works*, Vol. 7, pp. 135–243, Hogarth, London, 1953.
13. GESELL, A., and ILG, F., *Child from Five to Ten*, Harper & Row, New York, 1943.
14. GREY, WALTER W., "Electroencephalographic Development of Children," in Tanner, J. M. and Inhelder, B. (Eds.), *Discussions on Child Development*, International Universities Press, New York, 1971.
15. HARLOW, H. F., and HARLOW, M. K., "The Affectional Systems," in Schier, A. M., Harlow, H. F., and Stollnitz, F. (Eds.), *Behavior of Nonhuman Primates*, Academic Press, New York, 1968.
16. HENRY, J., and HENRY, L., "The Doll Play of Pilega Indian Children," *Am. J. Orthopsychiat. Res. Mono.*, 4, 1944.
17. HUNT, McV. J., "The Impact and Limitations of the Giant of Developmental Psychology," in Elkind, D., and Flavell, J. H. (Eds.), *Studies in Cognitive Development*, Oxford University Press, New York, 1969.
18. JONES, M. C., "Psychological Correlates of Somatic Development," *Child Dev.*, 36: 899–911, 1965.
19. PIAGET, J., *The Child's Conception of the World*, Harcourt Brace, New York, 1929.
20. ———, *The Growth of Logical Thinking from Childhood to Adolescence*, Basic Books, New York, 1958.
21. ———, *The Moral Development of the Child*, Free Press, Glencoe, 1932.
22. STONE, L. J., and CHURCH, J., *Childhood and Adolescence*, Random House, New York, 1968.
23. SULLY, J., *Studies of Childhood*, Longmans Green, London, 1895.
24. SULLIVAN, H. S., *The Collected Works of Harry Stack Sullivan*, Vol. 1, Norton, New York, 1953.
25. TANNER, J. M., and INHELDER, B., *Discussions on Child Development*, International Universities Press, New York, 1971.

CHAPTER 18

ADOLESCENCE

Irene M. Josselyn

Adolescence Defined

The term "adolescence" is ubiquitous; as a result its use creates confusion for the reader unless it is specifically defined in the context of the discussion offered. In this chapter the term "adolescence" is used to describe the psychological stage in personality and character development that follows the latency phase of childhood.*

Prior to latency the child has, through adaptive and defense mechanisms, achieved some integration of his internal needs, urges, and wishes within the relatively circumscribed reality provided by his primary love objects. During latency this integration pattern widens, with progressive adaptation to an expanded reality and a meaningful social milieu. The typical latency child is occupied primarily with external realities to which he responds with relative predictability and minimal disturbance over crises. It is not, as implied by the term "latency," a dormant period, but rather a span of time during which, optimally, the child's psychological gestalt is becoming enlarged and basically melded to provide a firm foundation for the next span of growth, adolescence.

During adolescence there is structured upon this foundation of latency what will become the architecture of the adult personality and character, as well as the crystallized potentiality for neuroses or psychoses. The analogy of the architectural structure is not chosen lightly. No total foundation is sound unless the bricks and mortar that are the fundamental components of that foundation are sound and discretely interrelated. Thus, latency is not optimally effective as a step toward maturation unless prelatency components have reached a constructive solution. Any seriously faulty solution of prelatency conflicts will result in the latency phase being poorly balanced; in addition, that lack of balance will distort what occurs as the superstructure develops and will effect the completion of the architectural design, adulthood. While adolescence is the preadult phase from which adulthood will evolve, adolescence is affected by the foundation upon which it is built. It is important to em-

* Editor's Note: Some authors distinguish also a juvenile and preadolescent period. See Chapter 17.

phasize that as the significance of the adolescent phase becomes more and more understood, it is not to be evaluated as independent of the past but rather, for good or bad, a superstructure upon that past.

As many have pointed out, adolescence had not been studied with any intensity until approximately the last 30 years. Even with the evolution of thinking regarding psychological development from birth on, for a long time the adolescent phase was ignored by most students of human psychological development. Outstanding exceptions to this were Aichhorn,[1] Bernfeld,[4] Healy,[15] Hall,[13] and in her early work, Anna Freud.[9] Prior to their writings there were protests by some against youth and wonderment or despair about what the youth of that day would do to the future. For the most part it was presumed that if adolescence was ignored the idiosyncrasies would end. There was little attempt to understand why the phase existed and what it achieved.

Adolescence as a Phase of Development

In recent years there has been increasing interest in exploring the manifestations and significance of adolescence. From this study has evolved two diametrically different points of view. There are those who describe typical adolescence as a period of turmoil and confusion.[9] This has implied to others a degree of pathology that they do not believe exists. The latter group believe that normal adolescence is not full of turmoil; they argue that those who indicate otherwise base their conclusions upon their study of manifestly disturbed individuals and therefore attribute to all the troubles of relatively few.*

The disagreement concerning the dynamics of adolescence may be related to terminology and emphasis rather than a basic contradiction. Every phase of psychological maturation has inherent in it new conflicting desires to be harmonized. During the process of resolving the conflict, there are transient episodes of clinging to the old; at the same time the strongest force in the normal individual is to thrust forward. There are also brief false steps toward the next phase.

Adolescence is a phase in development. No one would assume that the ten-year-old boy of today will be the same fifteen years from now; he will have "grown up." He will have grown from childhood to adulthood, having reevaluated many facets of himself and the outer world during adolescence. As he strives during adolescence to attain a satisfactory adult role, he does not just seek it in playacting as he did earlier; rather he seeks for an internal realignment that will be effectively automatic, self-expressive, and predictable, a new self gestalt. It is this striving to abandon childhood and attain adulthood that is adolescence, an inherent phase in maturation.

The Cultural Impact

These changes, while primarily self-initiated and the adolescent's own psychological task, are affected by the particular culture's definition of the adult's role. Many cultures define sharply the child's role and the adult's role; the path from childhood to adulthood is well trodden and clearly marked. In contrast, in Western culture the child is only familiar with vague directions. His position is comparable to that of the pioneers who explored the West; they knew the way led west, but the miles between their familiar homes and the anticipated riches at the end of their trek were uncharted except for the vague guidelines of the sun and the stars. Children of Western culture know that adulthood implies certain privileges, and certain obligations, among them the freedom to choose on their own, within broad limits. Before the individual makes his ultimate choices, he reevaluates many concepts that he previously had uncritically accepted.

* These apparently opposing points of view are summarized in Wiener's *Psychological Disturbance in Adolescence*.[20]

Problems in Evaluating the Adolescent

This reevaluation and ultimate mastery of an adaptation to a new internal and external reality may be manifested in a variety of ways. Many adolescents show no external evidence of psychological change. The process is gradual, and only when it is relatively completed do those who know him well realize that he is different from what he was as a child. Frequently the change is only apparent after a period of absence. A boy of this type goes away to college, and when vacations reunite the family he seems strange to his parents. He is, as the family vaguely describes it, more grown up. He also is aware of, and frequently uncomfortable about, the change, but he tolerates the stress until vacation ends.

Another group of adolescents cause concern for their parents and possibly their teachers. They are moody, at times manifesting a gratifying pleasure in living and at other times acting depressed, bored, or irritable. Any attempt to explore the latter difficulty frequently leads to the response, "I don't know, I just feel this way," or a curt, "Can't you let me alone?" School achievement may be less than in the past, or it may manifest the same sporadic shifts as the mood swings, with successes and failures occurring unpredictably. This group in particular are frequently still diagnosed as "just adolescent"; family and teachers are assured that "it will pass." This group is perhaps most secretive about its thoughts. For a variety of reasons the individual prefers to find his own solutions to the internal problems with which he struggles. Only when he feels overwhelmed by some aspect of his confusion does he suddenly confide in his parents or in an adult friend. When he does confide he may appear to be dramatizing and exaggerating his difficulties. This evaluation by another *may* be correct, but usually the normal adolescent is not dramatizing. He is actively aware of his discomfort and experiences transiently a sense of having entered a blind alley that suddenly appears to be sealed at both ends.

A third group of adolescents show the degree of their turmoil through their overt actions. It is this group that makes the news and arouses the apprehension of adult society. Its members seek to handle their own confusion by changing yesterday so that today will be the beginning of a future in utopia. They are inexperienced, naïve, and have too many contradictory goals to outline a program that adults can grasp or apply. At times they do find pseudoadults who, as confused as they are, may become their mentors. On the other hand, most of them are soon disillusioned by these pseudoadults, will talk with them but will not, except for a shorter or longer period of time, follow them. In the latter instance the adolescent has tried on for size the ideas the adult has expressed and finally, often after long consideration, decides those ideas do or do not fit.

There are adolescents that may or may not be a part of the third group mentioned above. Their behavior is grossly destructive. Whether an individual manifesting such behavior has a deep-seated core of faulty adaptation, and therefore a character defect or a serious psychological disturbance, only careful study over time will tell. They at least have a frustration tolerance that is too low for the frustrations they are experiencing. A certain number of them, either with help or by their own mastery of their inner turmoil, will leave this type of behavior behind them as they begin to resolve their phase-typical conflicts, and they will seek wiser ways to bring about constructive social, political, or personal change as adults. Others will undoubtedly become psychologically pathological adults.

The last paragraph illustrates the difficulty in evaluating "normal" adolescence. The behavior of adolescents, their verbalization, and often their developmental history does not give a means of quick diagnosis of the syndrome they present.[9] The first group described above may have among its members those who are preschizophrenic or actually schizophrenic, the symptoms masked by their slow, plodding attempts to deal with phase-characteristic adolescence. The second group may have among its members hysterical characters, incipient hysterics, or a manic-depressive constellation that will become manifest in adulthood. The third group may include those who in adulthood will still be psychologically young adolescents, remaining impulsive, intolerant of frustration, and unpredictable, or who will be sociopaths, schizophrenics, or delinquents. Any individual, irrespective of which group he appears to represent now, may develop later a mental illness that would appear more characteristic of a different adolescent type. Probably the soundest tentative diagnosis of normal adolescence is when, after a period of real communication with the individual, the adult becomes aware that the young person he is studying does not consistently fall into any of the above groups; rather he is shifting from one to the other and is at times immature, at other times strikingly mature, and, by adult standards, rational. This would suggest that the diagnosis "normal" is most valid when a diagnostician who knows adolescents, and who knows the prodromal symptoms of mental illness or of character defects, is confused about the proper diagnosis, the diagnosis being difficult because of the shifts that occur so readily in the constellation of behavior patterns and verbalizations—signs that the adolescent is seeking to find himself.

Psychosomatic Nature of the Adolescent Emotional State

From one viewpoint the psychosomatic implications of adolescence are very clear. The onset of psychological adolescence is closely related to the increased function of the glands of internal secretion, particularly the reproductive glands. (The significance of those glands in relation to the spurt of growth prior to puberty will be discussed later.) The impact of the reproductive glands upon the emotional state of the individual is well illustrated in adults by various studies, such as those of Benedek[3] concerning the menstrual cycle in women. The young adolescent of either sex, about the time that he attains physical puberty, becomes sensitive to things he either previously ignored, took for granted, or let pass. He is more easily hurt, more easily elated, more readily depressed, and more quickly angered over minor frustrations. His sensitivity also expresses itself in less egocentric ways. He reacts intensely to what is beautiful and to what is ugly, the former typically gratifying, the latter typically very painful.

He loves intensely and hates intensely. It is difficult for him to tolerate his ambivalent feelings. As a result he may love something or someone intensely until a minor defect becomes apparent; then, particularly in interpersonal relationships, he may hate as intensely as he loved. Similarly, if he finds a positive element in the person or object that he has hated, he may reverse his attitude and abandon his hostility and temporarily only feel love. He is not completely unaware of this shift from positive to negative feelings. He often, particularly with people he cares for a great deal, finds his inability to handle his love and his anger, or his inability to accept imperfections, very confusing. This is especially so when he is aware that those toward whom he is reacting in this way do not understand and respond to each of his opposing feelings as if each erased the other. If others cannot harmonize his feelings, how can he? This is readily observed in an adolescent who has a basically sound relationship with his parents. At times he feels hostility toward them and can't understand why they are not the people he would love. Then later he loves them and can't understand why they don't believe that love.

It is important to bear in mind that the adolescent will not always manifest this confusion in feelings or indicate the intensity of either positive or negative responses to the beautiful or the ugly. Some adolescents deny positive feelings for people or for experiences that one would anticipate they would enjoy. Such denial, if indicative of an established attitude, would suggest pathology. While an apparently insensitive adolescent may be a disturbed adolescent, this evaluation is not necessarily true, particularly among boys. In our present culture the all-American boy, by adult standards, is supposed to be above expression of tender feelings, intense investment in beauty, and other emotional reactions that are experienced by young adolescents. Boys therefore put up a veneer of indifference that is not necessarily how they really feel within themselves. One boy—who seemed in every aspect of his life to present a picture of a psychologically healthy adolescent—broke down crying after many hours of discussing his remoteness from his parents, "Gee, don't they realize I love them. I can't show it." If one accepted only his overt behavior of indifference to his parents, it was easy to see why they felt that he didn't love them.

This sensitivity in the adolescent is often the secret that they carefully shelter from the invasion of others. As disturbing as today's drug culture is to all adults, one thing can be learned from certain participants. Their underlying sensitivity to the meaning of interpersonal relationships and beauty, which is usually a well-kept secret, is revealed under the influence of drugs.

The psychological sensitivity that the adolescent is experiencing is new and therefore unfamiliar; he does not know how to channel it adequately into his future life. This sensitivity to his internal feelings explains in part why he works so hard to find a way to bring conflicting sensations into some sort of working order. Since such a process cannot be mastered by intellectualization alone, he is confused. As he recapitulates aspects of his early childhood, attempting to find new solutions, this sensitivity to his own confusion is always intermingled with his techniques for trying to attain a new self-image.

In his confusion society does not offer him a great deal of help; modern civilization offers very little sharply defined structure as a framework for the adolescent. Most of the advanced cultures of today are oriented to the importance of progress, not to a static society. This is particularly true of those societies based upon a democratic philosophy. Today a typical adolescent is confused by the teachings of democracy that stress the rights of the individual. While the adult recognizes there is a difference between freedom and license, many adolescents consider the differentiation quite hazy. It is difficult for them to understand why a statesman can make a statement that, if used in a school theme, would result in a threat of suspension from school! They cannot see the interrelationship of the total forest because they are able to grasp only one tree at a time.

The Ego Ideal and Superego during Adolescence

If adolescence is a normal step in maturation, what are the goals and conflicts characteristic of that developmental stage? Many explorers of this question have stressed the identity crisis as the chief characteristic.[7] Prior to adolescence the child had found his identity, but it was his identity as a child. During normal latency he lived and enriched that identity. With the advent of psychological adolescence that childhood identity begins to lose its effectiveness as a source of security that it once provided. Because of the pressure for maturation, a pressure that originates from an internal source as well as from the external world, a new identity must be sought. Self-identity is an internal experience, the result of the interlocking of multiple aspects of living with external and internal reality.[18] Important components of self-identity are the ego ideal and the superego. Together they assure a certain degree of self-predictability based upon the individual's own values. While the ego ideal and superego are typically so interrelated it is often difficult to tell them apart; during adolescence the demarcation between the two is frequently apparent and the source of confusion for the adolescent (as well as others).

The theoretical differentiation used in this discussion is based on Piers's[17] delineation of the difference between shame and guilt. Shame is related to a failure to live up to the individual's self-concept of what he wishes to be; guilt is the response to failure to live up to what is expected by those meaningful in interpersonal relationships, primarily the parents and the established social mores. The ego ideal defines the healthily narcissistic value system that assures self-appreciation. It is the formulation of a gestalt that indicates, "This is the kind of me *I* can love." The superego, when artificially separated from this, defines, "This is the kind of person who will not be punished, but will be loved by others." The reason for the typical interrelatedness of these two in a psychologically healthy individual is apparent. There is not much love for the self if punishment, including loss of love, by others is not avoided.

During adolescence the childhood ego ideal, which was based upon being a self-loving and lovable child, is partly abandoned. The lovable self must attain an identity as an adult, not only in a social, economic, and parental role, but as an integrated internal self-concept. The adolescent strives to find an ego ideal that will, as it becomes crystallized, result in respect for the self as an adult, not as a child.

During his latency the adolescent achieved a meaningful relationship with his peer group. He learned the rules of the game for living in the world of his peers; his earlier ego ideal had become enriched by including the requirement of being a good player in that game. His self-image thus had incorporated social standards as well as the standards of his family. Discarding this social ego ideal and alienating himself from his peer group threatens a basic component of what he seeks, a new ego ideal. As a result he is ashamed if he does not protest a parental demand as his peer group does, or claims they do. He is ashamed if he dresses differently from his companions and if he does not join them in violating a school dress code. He probably would not mind not wearing ragged blue jeans, and even would not object to wearing a necktie to school, except that to do so would be to abandon the self-confidence that has resulted from social acceptance by that world of which he had become a part during latency.

On the other hand, often the code of his peer group violates an aspect of a family-rooted superego. The superego that developed during childhood was primarily the result of an internalization of parentally imposed standards. As the superego takes charge, the responses become autonomous, providing guidelines not only to assure avoidance of punishment but, more importantly, to assure a continuity of the needed primary love of parental figures. If these internalized guidelines are not followed, guilt results.

Some of the values of the childhood superego relate to wise prohibitions relevant only to childhood; others relate to values applicable at all ages. It is not always an easy task for the adolescent to differentiate them. A simple example of a significant component of the childhood superego that must be modified if adulthood is to be attained is the child's internalization of the attitude that parents must be obeyed. There are cultures in which that aspect of the superego remains intact in adult life. On the whole, however, in our current culture the normal adult's attitude is based upon the concept of self-determination.

To attain mature status the individual must learn to differentiate between those superego standards that are universally applicable to society and those that are a remnant of a childhood role. While parents are often aware of the necessity for this shift and give verbal sanction to the change, the childhood superego has become so much a part of the individual that its origin in parental standards has become vague. While the parents' assurance that modifications are correct, because what applied in childhood no longer applies, does aid the adolescent in his attempt to attain a more mature superego, it does not automatically bring about the shift.

Helen, a college student, who was reared in an academically oriented family, had (except for one episode of serious illness), an unbroken record of class attendance through high school. This record was highly praised by the parents; any deflection from it because of a cold or a headache (she had migraines) would have been seriously criticized. She became a student at a college dedicated to independent study, where class attendance was determined by the student's own desire to learn what the professor was discussing at the time. In spite of her parents' enthusiasm for this educational approach, she could not fail to attend daily every class in which she was enrolled; she indicated she would feel too guilty if she didn't. As a result she could not find the time to pursue her independent study. Finally she solved the problem by transferring to a college that required class attendance and repressed any student initiative!

The typical adolescent, sometimes overtly, sometimes covertly, handles the conflict between his childhood superego and the pressure to modify it to comply with his image of an adult value system by rebelling, sometimes seriously, more typically in a token fashion, against his own superego. In doing so he frequently first projects that superego, using its original source as the object against which to rebel—namely, his parents or those he sees as rigid and authoritative people. (This may at times be a realistic technique!) He may reveal some behavior that he has tried to keep hidden from his parents, insisting they would not let him do it if they knew. A discussion with the parents may indicate that they would not condemn the act; rather they had wondered why he was not doing it. Pointing this out to the young person may relieve him; in some instances, however, it makes him uncomfortable. He has been deprived of a token enemy, the projection onto another of an internally established but inappropriate value system that he wishes but fears to abandon.

To the adolescent, rebellion against an internalized value system is at times frightening. The superego provides security, assuring him that his behavior will not be unpredictable. As a consequence the rebellion against the superego often may be abandoned temporarily and the early standards held more rigidly than previously. At such times there is a possibility of the evolvement of a strict, crippling superego in adulthood. In the typical individual this possibility results in a reactivation of the rebellion against the earlier value system and, usually through verbalization rather than action, a flaunting of imposed standards.

Currently there may be an interesting manifestation of this. Young people, and many adults, talk about the change in sexual mores among the young. Yet statistical studies of college students during the 1960's did not confirm this modification.[2,6,8,12,14] Why does this contradiction exist? It is possible that one aspect of the contradiction represents a wish and the other represents an actuality. Maybe the mores are not changing as much as the young people indicate they are. On the other hand, it is equally possible that the mores have changed. If so, to what extent does the acceptance of the change provide an opportunity for a wish to be expressed unconsciously through interest in the behavior of others rather than to be acted upon secretively and guiltily? Perhaps when the present generation are in their fifties we will know whether statistics or impressions were correct during this time span.

As alarming as the rebellion against the superego may appear both to adults and to the adolescent himself at times, it usually does not take a serious form. A well-structured superego developed during childhood results in the major aspects of the value system remaining intact. Those aspects that succumb to the adolescent rebellion are more typically relatively minor changes as far as social living or individual maturity is concerned. He may rebel against wearing socks to keep his feet clean, but he doesn't steal or murder!

As indicated previously, differentiation between the ego ideal and the superego is usually difficult because the two, which during childhood and hopefully when adulthood is attained are melded and interrelated, are frequently in conflict during adolescence. A patient described his conflict succinctly in one situation. He and a group of friends planned a trip to Europe that would involve a minimal expenditure for living expenses. Each individual of the group had earned the money for the limited budget they projected. His parents, not wishing him to be stranded in a foreign country, gave him traveler's checks to be used in an emergency. At first he was happy to have the extra security, but then he experienced an overwhelming sense of shame. His friends were acting as independent, mature individuals; he was secretly more secure because, under a veneer of independence, he was a child being taken care of by his parents. He became so uncomfortable under this burden that he returned the traveler's checks to his parents. When the parents' attempt to persuade and cajole him to accept the checks failed, they became angry, pointing out they would worry about him and their summer would be ruined. This created intolerable guilt in him; he almost abandoned the trip because of the discomfort resulting from the conflict between his ego ideal and his superego. Fortunately he solved the conflict by telling his friends about his "neurotic parents," and by plotting with his friends for a splurge with the traveler's checks during their last days in Europe.

Adults find it difficult to understand certain aspects of the adolescent's struggle to define an adult ego ideal and superego for himself; therefore, they are often unable to help him wisely in his psychological clumsiness. Part of what the adolescent seeks are an ego ideal and superego that will incorporate an occupational role he respects for himself and the training for and achievement of a work identity that will, through acceptance of obligations, achieve approval, not punishment, and a place in society for him. High school and college educators commonly indicate to parents that if their son or daughter are planning to go to college, the parents should not be concerned if the young person frequently shifts his or her vocational goals, or appears to have none; he or she will find that in college. Regardless of the practical aspects of ability, a beginning college student who has not mapped out his occupational future is still confused about one aspect of his adult ego ideal and superego; he has not as yet determined who, vocationally and as a part of a social structure, he wants to become.

Regardless of the healthy effectiveness of his childhood ego ideal and superego, the adolescent, in order to reach relatively healthy adulthood, must abandon those values that were appropriate only during childhood, or that were necessary in his situation as a child, and selectively retain those of value for all ages. While his selection is facilitated through the wise, not rigid, counseling of others, the selection must be *his* if he is to experience a true self-identity, not just one that is an image in a mirror, the image of someone else.

Adolescence as Rebirth

The ego ideal and superego do not develop independent of other steps in psychological growth. They are both the result of, and the stimulus for, that growth. This suggests another facet to explore in regard to the adolescent period. Blos[5] has referred to adolescence as a time of the second individuation process, which for brevity could be termed "rebirth." This conceptualization would seem the most broadly descriptive term of adolescent psychology. In seeking a new identity, the individual recapitulates the process of development that resulted in the integration achieved in his childhood. Although it would be simple if the repetition compulsion described by Freud[11] were not a part of human psychology, clinical evidence suggests that at least the human species does not abandon the past to start anew, but rather repeats the old in order, if possible, to find a new answer. During adolescence the individual seeking a new identity recapitulates the phases of development that resulted in his childhood identity. In contrast, however, to the clinically typical repetition compulsion, he does so with new tools and with new goals. To the extent that he mastered any earlier phase, the recapitulation during adolescence will be more easily handled, other things being equal, than if his mastery was incomplete or dealt with by the utilization of unfortunate defenses. It is because of this tendency to relive the past, and reevaluate it in terms of the future, that adolescence cannot be understood as either isolated from the past or as the beginning of attainment of the goals for the future. It is a rebirth in which neither the first birth can be ignored nor the future of the reborn person be assumed to be alien to the initial growth process.

Body Image in Adolescence

According to the concepts of this article, psychological adolescence typically begins at the time of the spurt in growth and the body changes associated with prepuberty and puberty.[19] This physical change results in the young adolescent experiencing an unfamiliarity with himself. We know a piece of glass is a mirror because when we look into it we see a body that is ours. Imagine what the experience of seeing his reflection would be for a thirteen-year-old boy, who had not seen himself in a mirror for a year, during which time he had grown several inches and his body contour had changed. Actually, in a less dramatic way, the adolescent's body, because of the typical body changes, is less familiar to him than in the past. As one boy who had grown quite rapidly half-humorously commented, "I wonder if I am farsighted; my feet are so far away when I stand up, but I can still see them."

Because children and adolescents have been encouraged to engage in physical activities that require good body coordination, the body awkwardness of 50 years ago is not as universally observed among the adolescents of today; during the period of rapid growth the young person is sufficiently involved in physical activity to coordinate more effectively on the whole. If an adolescent does show the physical awkwardness that used to be considered so typical of adolescence, and if his history indicates that prior to his spurt of growth he appeared well coordinated, it is possible to observe how an adolescent actually does, as a result of physical changes, feel unfamiliar with his body. His self-identity through body familiarity is undermined.

If adolescence is conceptualized as a rebirth, this manifestation resembles what occurred in the past as the infant began to become aware of his body. The infant lacked the development that enabled him to handle his body as he would later. In contrast to the adolescent, the infant did not know from where he came; he could not recall his fetal state before he had hands, arms, and legs. The adolescent, discovering a new encasement of himself, does know his physical past; he is often proud of the changes he sees in himself. But he does not necessarily feel at home in this new encasement.

As a consequence of the changes that have occurred, an adolescent is very conscious of his body, both its beauty and its malfunctioning. He is aware of body sensations that are often alarming to him because he doesn't understand them; he is fearful that abnormalities are developing that will make him an inadequate adult in the framework of his own sex. An example of this is the temporary development of a deposit of adipose tissue in the breasts of a boy, particularly if he is obese; he fears his body will be effeminate. The girl may be concerned because her breasts are not developing rapidly enough, and she fears she will have a masculine body while she is striving to see her body as indicative of her femininity. These responses to confusing patterns of physical development are significant in evaluating the "normal" adolescent. Thus, the boy who has a feminine identification may value his breast development; the girl may value her "boyish" figure because it denies her biological femininity. But concern about body development does not always indicate pathology.

Awareness of the possibility that his body encasement might not be perfect should make the adolescent very careful of his body. Although he will be careful at times, at other times he will act alarmingly to the contrary. He then may manifest many of the characteristics of the oral phase of development. When feeling lost in his interpersonal relationships and in his world, he may revert to the same comfort source that he found gratifying during infancy. He may overeat in spite of gaining too much weight, or he may eat foods that he knows the doctors believe cause acne. Food represents to him comforting that in infancy was provided by others but that he can now provide for himself.

The problem of differentiating between a normal conflict, for which a temporary solution is sought, and a serious conflict, which the adolescent cannot handle, becomes apparent in the case of adolescent eating patterns. As indicated above, if transiently insecure he may handle his longing for the support found during infancy by eating. He may also be obese or deny the doctor's recommendations about acne because being unattractive is protection against the gratification, particularly sexual, that he fears he would be tempted to achieve were he not physically unattractive.

Kate illustrated rather dramatically the interlocking between a transient and a deeper, more sustained conflict. With a lifelong history of a tendency toward overweight, she became a compulsive eater and rapidly gained weight during early adolescence. She was quite insecure during this period, an insecurity that appeared related to her confusion about her goals and to parental criticism of the lack of future plans. A rather sudden spurt in academic achievement resulted in increased self-confidence and adherence to a reducing diet. The latter was very effective, and in about two months she became an attractive, feminine adolescent. Then her compulsive eating patterns returned and with them her obesity. Her self-confidence appeared to remain intact. It became clear, however, that her compulsive eating returned at the time boys in her class changed their response to her. Previously she had been an asexual friend to them; after she lost weight they sought her as a "girl friend." She finally recognized that she had been frightened by this change in the boys' attitude. Her sexual conflict, which had aspects of a conflict deeper than her original insecurity, only then became clear.

Dependency during Adolescence

This reliving of early infancy becomes manifest in other broader psychological responses. Although it is often stressed that adolescence is a time of striving for independence—which it certainly is—the great dependency needs of the adolescent may be overlooked. As he attempts to become more independent and strives to find his own self in a new form, he may overextend himself, or fear that he has. This creates anxiety; he then appears to regress to a dependent state characteristic of a period when he really was helpless. He loses all confidence in his own ability to make decisions and seeks advice from parents or parent surrogates. He usually does not seek advice on deep philosophical questions, which adults would probably willingly give at great length. The advice frequently sought is whether to ask a certain girl for a date, what courses to take the next semester or, by the girl, what dress to wear, or whether her lipstick is really a good color for her. The normal adolescent, when he seeks such advice, may become angry with the advice that is given. From the standpoint of adolescent psychology this is easy to understand. If the advice is asked because of anxiety, the fact that advice is given relieves some of the anxiety; then a new discontent arises. The adolescent does not want to be dependent upon parents or even parent surrogates; he wants to have respect for himself as an individual. The only way he can achieve that, in his mind, is to refuse the advice given. He therefore says the advice is stupid and does the opposite. Fortunately, if the earlier child-parent relationship was positively meaningful, the adolescent is more apt to confuse his parents by seeking advice as if he wanted it primarily to do the opposite only in regard to relatively minor, token aspects of his living; the color of lipstick is really not that important.

There is another factor in his seeking advice on minor events in his life. If the advice is given it assures him that in times of real stress he can still depend on his parents for guidance, even if he is attempting to be independent. The mature adult does not wish to be independent to the extent that he will not utilize the expertise of others. It is not surprising then that the adolescent, while striving for his naïve concept of independence is frightened at times and seeks in a real or token way people to whom he can turn who are more experienced and who will offer a helping hand if it is needed.

In areas related to long-time, meaningful goals, the adolescent is more willing to listen to the adult because he does want external help in sorting out his confusion; the long-time goals are aspects of his life that he is mulling over but about which he does not have to make an immediate decision as he would if he had to decide whether to take a girl on a certain date. On important issues the normal adolescent listens and often stores away what he hears to think it over in his own secret way, even though on the surface he may seem to resent the advice and imply that he will do the opposite. Adults who say to the adolescent asking for advice, "Don't ask me because you will do as you please anyway," may lock a door that should be left open, irrespective of whether the adolescent walks through it at that point.

Redefining of Self-Object during Adolescence

Another manifestation of adolescence as a rebirth is the negativism observable in many psychologically healthy adolescents. Just as it does with a two year old, negativism plays a significant role in the maturation of the adolescent. The two year old, as he becomes aware of the strength inherent in being an individual instead of a compliant extension of another, gains increasing confidence in his separateness by saying "no." "Yes" is as if he were putty; "no" indicates a strength of a solid form. The infant, as he becomes aware of his separateness from others, assures himself of a new-found identity by being negativistic. Childrearing practices of any culture lead to a gradual remission of negativism paralleling a growing assurance of self-identity within socially imposed and reality limitations.

The young adolescent, since he senses that the identity he established as a child is not the adult identity that he now is seeking, utilizes again the mechanisms that first established his own self-concept. He also, in a different form, is negativistic. If society decreed that all men should have long hair, it could be anticipated that many long-haired adolescent boys of today would shear off their locks. When miniskirts became acceptable high school attire, the maxi skirt became the vogue. It is the new style accepted by the extremists but questioned by the greater majority of adults that often, if the protest of the latter is strong enough, provides an immovable object against which to assert the valued, irresistible force of negativism, becoming the adolescent *cause célèbre*. It is the way to be "different," not different from the peer group, but different in the eyes of those who established the childhood code the adolescent had accepted in the past.

The psychologically healthy adolescent does not intentionally carry his negativism beyond a point of no return unless he is driven to do so by the counternegativism of adults. Unfortunately the point of no return cannot always be delineated and can at times be evaluated only by a longitudinal study. Will an intelligent boy, who is negativistic in his rejection of education beyond high school and apparently deaf to all advice concerning the advantages of further education, ultimately invest himself in further education after he has effectively established his own self-realized identity, or will he cling to the identity his negativism provided him with? The answer may become apparent retrospectively after his adult configuration has become solidified. Fortunately most psychologically healthy adolescents do not go beyond the point of no return.

The Sexual Drive in Adolescence

With the advent of adolescence the sexual drive becomes intensified, partly because of the increased secretions of the reproductive glands. Less sexual confusion exists in simpler societies in which the increased drive can, within the framework of the mores of the social structure, find more readily a permitted pathway of expression with the opposite sex. For many reasons Western culture does not sanction such a pathway; instead, it complicates the early discharge of sexual tensions by fostering the ideal that mature sexuality represents a fusion of a biological urge and the emotional desire to love and be loved.

In Western civilization being loved and loving, at least consciously asexually, are freely expressed, optimal emotional experiences during childhood. Many theories of psychological growth patterns consider the preadolescent experience with love as the roots from which the ultimate heterosexual fusion of sexuality and love grows. The masturbatory fantasies of the child prior to adolescence indicate that the fusion, in an immature form, begins prior to the onset of adolescence. The intensified sexual feelings of the young adolescent are not yet directed solely toward an heterosexual goal. They readily become a part of any emotionally invested experience. They contribute to the intensity with which the young person responds to both pleasure and disappointment. These modes of discharge are often experienced without conflict.

In a desirable family milieu the child's primary love objects are his parents; the clearest pathway for expression of fused sexual response and love is thus the familiar one of love for the parents. Because of the increased intensity of sexual feelings, this leads to a reactivation of the oedipal conflict. The incest taboo became well imbedded in the psychological format of the individual as he struggled for a solution to the oedipal conflict in childhood. The prohibition is reinforced during adolescence because in addition to the incestuous implication, love for the parent represents remaining immature and a child in his own eyes and those of his parents. He thus must deny and repress the reactivated oedipal wish. He frequently denies it in his emphatic refusal to tolerate physical contact or any suggestion of seductive behavior by the parent of the opposite sex. Repression of the wish is not as easily achieved. Dreams of the adolescent often have a poorly disguised or undisguised incestuous nature. To add to the confusion, the most apparently rejecting son or daughter will have moments of being pleasantly and harmlessly seductive with the parent of the opposite sex.

Frequently the adolescent attempts to solve the oedipal conflict by a denial of affection for the parent of the opposite sex, and he experiences an intensification of positive feelings for the parent of the same sex. This offers a dual conflict again. To relate in this fashion to a parent of the same sex again represents a threat of returning to childhood. It also frightens the adolescent because he recognizes the possible implications of a homosexual orientation. Consequently he denies, or attempts to repress, his affection for the parent of the same sex and resents any implication of a strong emotional tie. The adolescent vacillates between this and a "man-to-man" or a "woman-to-woman" relationship with the parent of the same sex.

Parents contribute to the intensity of the reactivated oedipal conflict. The mother finds her son in his good moments a delightful male, and she utilizes her flirtatious manners that may have become ineffectual with her long-time mate. She avoids the conflict inherent in such a response to her son by being "motherly." When, for example, she kisses him and he protests, her answer is, "Why won't you let *your mother* kiss you?" He probably doesn't know why, but he does know that he is uncomfortable. A father likewise, when the son is in a good mood in which he feels free to express his affection in a manner that he cannot express it to his male friends, finds with him a gratifying relationship. The reverse situation exists with a girl; her father finds her budding femininity stimulating, and he, under the guise of being "after all her father," responds warmly to her. She, if needing to deny her affection for her father, is often quite resistant to his overtures. When she is not fighting any expression of her affection for her mother, she again provides the mother with a closeness that the mother does not know with friends.

These responses by parents, while increasing the young person's conflicts, are not, unless excessive because of the neurotic need of the parents or of the young person, an unfortunate response. A boy gains confidence in his own masculine attractiveness as a result of his mother's response. He also gains strength from his closeness with his father if the struggle for self-identity becomes too overpowering and external support is needed. The girl likewise gains confidence in her femininity from her father's response. At the same time her closeness to her mother, when it is needed, provides her with an opportunity to discuss her own femininity and to model herself after a loved woman. Just as the oedipal conflict is an essential part of maturation in childhood, it is an essential part of the maturational phase of adolescence. If parents are not available to play this role, appear too threatening, or are unwilling, other surrogates are frequently sought, as indicated in the ambivalent relationship with a basically loved teacher, or other adult, expressed by the "crush."

During the adolescent phase many solutions may be tried to resolve the oedipal tie, solutions that if they were to become a permanent part of the adult character structure would have unfortunate repercussions. For example, a boy, instead of turning toward his mother, may transiently seek a girl who will mother him; he may seek a girl who represents only a sexual object; he may deny any interest in any girl; or he may choose a girl who does not have a psychologically feminine orientation. He may turn to an older man or to some male peer whom he can love as he fears to love his father. He may renounce his masculinity, rejecting any identification with his father or any masculine figure. These intense, frequently ambivalent relationships are often brief. While this brevity may be evaluated by the adult world as evidence of fickleness and a lack of investment in any relationship, it may be indicative of a wise rejection of possible solutions to the oedipal conflict.

One of the disquieting manifestations of this attempt to solve the oedipal conflict—disquieting to both the young person and to adults—is the implication that adolescent sexual attachment toward a person of the same sex is the harbinger of an ultimate homosexual adult life pattern. In some this type of behavior may represent a fixation at a homosexual level. In a psychologically normal adolescent this is not true; it may, however, become a fixation at an adolescent homosexual stage if unwisely handled. Many adolescents who at times appear to be homosexual are in reality trying to find a substitute relationship with someone who will serve in a surrogate parental role as the struggle to be free of the child-parent relationship is being waged in spite of the anxiety such a separation arouses. As mentioned above, because the goal of sexual feelings is not yet crystallized, such feelings may become a direct, unsublimated component of any internally meaningful experience.

While the reintensification of the oedipal conflict is an important part of the confusion of the adolescent as he deals with the increased intensity of his sexual aims, it is by no means the sole source of sexual conflict. In most cultures he faces many other major problems, as he tries to attain mature heterosexuality, which is both biological and social. Particularly if his relationship with a parent figure (during childhood) has led to an identification with a mature person of his own sex, he sees adulthood as attained sexually by emulating the person of the same sex. In the past this adult had appeared to be a giant, he himself a pygmy. As a consequence his identification with the older person has not assured him that he is the same, particularly in sexual potential. He wishes to attain sameness; he is not confident he can.

The adolescent feels unsure of his sexual competence. Because he has doubts about his own sexual adequacy, he fears it will be seen by others; because of his tie to a person of the same sex he worries about homosexuality. To deny these concerns, he may talk or act out a genital sexual pattern compatible with heterosexuality. If he only talks it, he often is masking a deep fear that he can't act it out. If he acts out a pseudoheterosexual role, he either finds it unsatisfactory or is vaguely aware that the sexual act itself is a mutual masturbatory experience, far short of what he believed a true, mature heterosexual experience would be. The latter is particularly disturbing to the young adolescent girl, for she is often frigid. She interprets this as evidence of her failure in her feminine sexual identity. The fact that the reproductive glands do not reach full maturity with the increased functioning at puberty probably contributes to some adolescent confusion. There has not been sufficient study of this to state it as more than an hypothesis, but it may be that part of the difficulty in fulfilling the sexual goals of the young adolescent is related to this immaturity.

In today's culture sexual fulfillment is not evaluated only by the capacity for genital discharge of sexual tension with anyone of the opposite sex.[16] It is considered to be attained when love for the sexual partner and sexual discharge are one. Often the love an adolescent experiences for another person of the opposite sex is an expression of a need of the young girl or boy that is not readily melded with sexuality. Possibly such melding cannot occur until the sexual glands have attained a stable maturity, and emotional needs and responses have also attained relative maturity.

Social Meanings during Adolescence

Fortunately for parents, as well as for the adolescent, the normal adolescent has days or months in which he is reliving the latency period. He is interested in his school work, but now he orients it to the reality of his future, not just to the present. He works out in fantasy many possible roles that he can play in the social and economic structure when he becomes an adult. He is interested in sports, either as a participant or as an observer, and formalizes into a social philosophy for the future the standards of fair play and sportsmanship that he learned during his latency. Temporarily the mountain toward maturity that he has been struggling to climb has become a restful plateau, both for himself and for his parents.

There is one exception, however, to this plateau concept. During this period he often translates his philosophy into thinking about social issues, the status of minority groups or of others less fortunate than himself, political formulations or international relations. These ideas have intense affective meaning, but inevitably reflect a limited grasp of all the reality implications. His protest becomes verbal, and he may join with those who with similar idealistic goals verbalize their feelings, often expressing hostility toward those whom they consider barriers to the fulfillment of their goals, for example, the Establishment. The reliving and reevaluation of latency temporarily becomes a source of concern to others and may lead the young adolescent into activities that are basically incompatible with the philosophy he is expressing.

This behavior has many components. The social formulations of latency provide a way to escape complete self-absorption; the social structure, not the self, must find new rules by which to live. Lack of life experience leads often to naïve, but sincere concepts. At the same time, however, it is not only inexperience and the resultant naïveté that energizes the drive toward these social goals. The drive may be intensified by: the wish to find an area of adequacy; the need to find a securely dependent relationship in a group that provides a familylike unity that is obviously not the primary family; the need to rebel against the symbol of a childhood relationship with his family (at present the Establishment) in order to establish a confident respect for the self; or a wish to discharge sexual tension through love for a cause and for individuals who are abstractions and thus do not threaten intimacy.

"Regression" during Adolescence

The recapitulation during adolescence of childhood patterns of adaption and defense, and the firm establishment of defenses and adaption against ego-alien impulses, is described by many as regression in response to the intensification of inherent needs and drives. This concept would imply a temporary ego failure. In some cases even in relatively normal adolescents this is probably a valid assumption. An example of such regression would be when overeating to the extent of causing obesity occurs because food represents love (as experienced in infancy). In the normal adolescent, however, the way in which childhood is relived is not characteristic of the past, but rather is an attempt to correct the child-oriented solutions of psychological conflicts, thereby attaining, hopefully, an adult identity.

In many ways it parallels the latter part of a successful psychoanalysis; the patient, after regressing and coming to recognize the infantile manner in which problems have been handled, reconstitutes himself to attain in an adult fashion those goals that were sought in childhood in a childlike way. The normal adolescent does not have to uncover the unconscious use of childhood tools; he is experimenting with new tools that may more adequately serve his needs. Many of the former tools prove of value with or without some modification; even though at times he casts them aside, he ultimately uses them again. Other and new tools appear at times to be the only answer to reconstituting himself as an adult. Gradually he recognizes the importance of multiple and functionally interrelated approaches. As a consequence he interlaces his multiple needs, wishes, and drives, binding them together with adaptive and defense mechanisms that finally represent an identity that is himself as an adult.

Resolution of Adolescence

This process occurs throughout adolescence, but during the early phase there is a chaotic pattern that gradually becomes a discernible design in later adolescence. He achieves that ultimate design by finding, not always optimally, a solution to each recapitulated phase of early development.

He becomes at home in his new body and with his new physiology. This is facilitated by the completion of the physical and biological growth that has been initiated by prepuberty and puberty. He has a body image of himself that represents his external self-concept.

He learns by experimentation, evaluation, and from the behavior of others the wise limits of independence. He also, as he feels safer in his achievement of wise independence, finds that to be dependent on others is not necessarily to be a child. Dependence is childish only when it is sought as a child would. He becomes aware or unconsciously accepts that society is based upon the interdependence of its members—an adult form of dependency in the human species.

As he becomes increasingly familiar with, and secure in, his nonchild self-identity, he does not have to protest so emphatically that he is a person in his own right. What was negativism earlier becomes a capacity to think rationally about those issues that previously aroused his negativism. Hopefully he does not completely abandon his earlier naïve attitudes, but rather exposes them to a rational evaluation that enables him to see their weaknesses and potentials, and to outline a course toward fulfillment of their positive potentials.

He resolves the reactivated oedipal conflict by turning to a member of the opposite sex for sexual gratification, a nonincestuous object acceptable to himself. His sexual feelings, lacking goal direction in early adolescence, has become goal oriented; he has attained a mature heterosexual level of sexual maturation. With this forward step he can experience his nonsexual love for his parents without anxiety, accepting them and hopefully being accepted by them as an adult with virtues and shortcomings. A tie to his parents persists, well-established through the years of childhood. The tie does not bind, it only enriches. It contributes to the capacity of the young adult to be a parent.

Experience and biological maturation brings confidence in his (or her) sexual identity as a male (or female). A sense of sexual adequacy relieves the anxiety experienced earlier and frees the individual to enjoy that adequacy. At the same time there is a relief from the strain of multiple sexual goals, and gratification in meaningful relationships such as friendships that have now become asexual is found.

Whereas previously he had vacillated, he incorporates defined role patterns for himself with the development of familiarity and the acceptance of his own identity. He clarifies his role in his family of the past and, if married, his new family; he finds secure employment; and he accepts responsibilities in accordance with the society of which, as he now accepts, he is an integral part. He has become an adult. Later he may modify or change the roles, but he will not do so impulsively or confusedly, but only after careful exploration of the multiple facets of his familiar role and the new one by which he is tempted.

Obviously the preceding recapitulation of the resolution of adolescence formulates a theoretical termination that in reality is possibly never completely achieved by any individual. One of the frequent unfortunate consequences of failure to achieve an adequate resolution of adolescence is the loss of the emotional sensitivity and responsiveness of adolescence. As their naïveté is confronted with the negating aspects of reality, too many resolve the conflict by repressing the sensitivity and responsiveness rather than incorporating it into an approach to reality. Unfortunately this is not usually considered as pathology, but rather as evidence of growing up. To the extent that repression instead of maturation of this aspect of adolescence occurs, to that extent any cultural growth is retarded. That sensitivity and responsiveness, if it matures, results in the individual's capacity to care for, to be involved in, and to offer wise participation and leadership in areas beyond his immediate interpersonal relationships. That widened capacity protects society from his use of any unusual creativeness he may have solely egocentrically; instead, the creativeness will also be beneficial to those who are represented to him only through the abstractions of "society."

To summarize, during infancy and early childhood the individual struggles with the basic maturational conflicts that appear to be characteristic of the psychological development of at least the human species. With the attainment of adolescence these conflicts are reactivated, and the individual is "reborn" to face the psychological task of formulating an adult identity. What he ultimately formulates, or fails to formulate, will be significantly affected by his infancy and early childhood experiences, but not necessarily irrevocably determined by them. To the extent childhood conflicts are resolved constructively, to that extent adolescent conflicts will have a greater possibility of a constructive resolution. Irrespective of the childhood past, however, adolescence is a phase of rebirth that, if relatively achieved, brings forth an adult.

The preceding article is a description of the manifest behavior of the normal adolescent. To probe deeper into the psychodynamics of the adolescent phase, the following publications are suggested:

1. BLOS, P., *On Adolescence,* Free Press, New York, 1962.
2. ———, *The Young Adolescent,* Free Press, New York, 1970.
3. DEUTSCH, H., "Selected Problems of Adolescence," *The Psychoanalytic Study of the Child,* Monograph No. 3, International Universities Press, New York, 1967.
4. ERIKSON, E. H., *Identity and Youth Crisis,* Norton, New York, 1968.
5. FREUD, A., "Adolescence," in *The Psychoanalytic Study of the Child,* Vol. 13, pp. 255–278, International Universities Press, New York, 1958.

Bibliography

1. AICHHORN, A., *Wayward Youth,* Viking, New York, 1925.
2. BELL, R., and BUERKLE, J., "Mother and Daughter Attitudes to Pre-marital Sexual Behavior," *Marriage & Fam. Living, 23:* 390–392, 1961.
3. BENEDEK, T., *Studies in Psychosomatic Medicine: Psychosexual Functions in Women,* Ronald Press, New York, 1952.
4. BERNFELD, S., "Types of Adolescence," *Psychoanal. Quart., 7:*243–253, 1938.
5. BLOS, P., "The Second Individuation Process of Adolescence," in *The Psychoanalytic Study of the Child,* Vol. 22, pp. 162–186, International Universities Press, New York, 1967.
6. EHRMANN, W., *Premarital Dating Behavior,* Holt, New York, 1959.
7. ERIKSON, E. H., *Identity and Youth Crisis,* Norton, New York, 1968.
8. FREEDMAN, M. B., "The Sexual Behavior of American College Women," *Merrill Palmer Quart., 11:*33–48, 1965.
9. FREUD, A., "Adolescence," in *The Psycho-*

analytic Study of the Child, Vol. 13, pp. 255–278, International Universities Press, New York, 1958.

10. ———, *The Ego and the Mechanisms of Defense*, International Universities Press, New York, 1936.
11. FREUD, S. (1933), "New Introductory Lectures on Psychoanalysis," in Strachey, J. (Ed.), *Standard Edition, Complete Psychological Works*, Vol. 22, pp. 3–182, Hogarth, London, 1964.
12. GRINDER, R., and SCHMITT, S., "Coeds and Contraceptive Information," *J. Marriage & Fam.*, 28:471–479, 1966.
13. HALL, G. S., *Youth, Its Education, Regimen, and Hygiene*, D. Appleton & Co., New York, 1904.
14. HALLECK, S., "Sex and Mental Health on the Campus," *J.A.M.A.*, 200:684–690, 1967.
15. HEALY, W. H., *The Individual Delinquent: A Textbook of Diagnosis and Prognosis for All Concerned in Understanding Offenders*, Little, Boston, 1915.
16. LICHTENSTEIN, H., "Identity and Sexuality: A Study of Their Interrelationship," *J. Am. Psychoanal. A.*, 9:179–260, 1961.
17. PIERS, G., and SINGER, M., *Shame and Guilt*, Charles C Thomas, Springfield, Ill., 1953.
18. SCHAFER, R., *Aspects of Internalization*, International Universities Press, New York, 1968.
19. STUART, H. C., "Normal Growth and Development during Adolescence," *New Eng. J. Med.*, 234:666–672, 693–700, 732–738, 1946.
20. WEINER, I. B., *Psychological Disturbance in Adolescence*, Wiley-Interscience, New York, 1970.

CHAPTER 19

YOUTH AND ITS IDEOLOGY

Kenneth Keniston

MOST DISCUSSIONS of the life cycle assume a certain fixity of life stages.* Psychiatrists, social scientists, and laymen in the Western world take for granted that life can be divided into a definite series of "stages," beginning with early infancy, continuing with two or three further stages during the preschool years, followed by a stage of "childhood" that extends until puberty, then by adolescence, early adulthood, and so on until old age and death. These are, after all, the stages of life that are acknowledged, sanctioned, and institutionalized in modern industrial societies.

The Psychohistorical Contingency of Life Stages

It has gone relatively unnoticed until recently that other societies and other historical eras segment the life cycle in different ways. As Erikson[22] notes in his discussion of Indian concepts of the life cycle, there are indeed parallels between his own theory of human development and the traditional wisdom of Indian culture. But equally impressive are the differences—the "failures" of Indian tradition to recognize developmental milestones that are considered critical in Western society, the Indian emphasis on developmental issues that pass virtually without notice in the West. Similarly Eisenstadt,[20] in his discussion of age-grading in the life cycle, underlines the enormous variability between societies in the extent to which they place children in narrowly defined age-grades.

Historians, too, have recently pointed out that even in medieval European societies such apparently universal stages of life as childhood and adolescence were not recognized. For example, Ariès,[3] in his study of medieval concepts of the life cycle, shows that nothing like the modern notion of childhood existed. Instead, the medieval child at birth entered a stage of life called infancy, which lasted six or more years. After infancy he was simply considered a small adult, expected to dress and

* Throughout this essay I assume some familiarity with the works of Erik Erikson,[21–27] virtually all of which clarify the stage I term "youth." Portions of this essay were previously published in *The American Scholar*.[52] The research on which it is based is supported by a grant from the Ford Foundation.

act like what we would today consider a grown-up. Obviously, medieval Western society must have recognized that those whom we call children were in some ways different from those whom we call adolescents and young adults.[40] But this recognition was nowhere institutionalized or sanctioned. For example, insofar as schools existed at all, they were graded by ability, not age: the same medieval classroom might contain within it students from eight to 45, grouped on the basis of their mastery of the classical curriculum. Ariès argues that a "sentimental" view of childhood only began to emerge in the seventeenth and eighteenth centuries. In this new view children were thought to possess an "innocence" that required protection from the adult world; age-grading increasingly dominated the widening network of schools; and a stage of life corresponding to what we call "childhood" was gradually recognized and sanctioned.

Ariès's intent is largely to describe rather than to explain. But it is noteworthy that the recognition and sanctioning of childhood as a stage of life overlaps with social, technological, and economic changes that drastically altered medieval life. Specifically, childhood emerged and flourished alongside mercantile capitalism, which produced an increasingly large and prosperous bourgeoisie that could exempt its children from the work previously necessary for family economic survival. A mercantile society also required of an ever growing minority of the young a new fluency with written language and numbers that rarely could be taught at home. For the new clerks and merchants of the seventeenth century, education was no longer simply a kind of training for virtue through knowledge of the classical curriculum and the Scriptures. Increasingly it became a prerequisite for social mobility, wealth, and power. Furthermore, the development of an entrepreneurial society in Europe seems to have been impelled by the psychological qualities summarized in the concept of the Protestant ethic. And it may be that the development of these qualities requires a period of "protection" from adulthood that extends well beyond the medieval period of infancy. Clearly, then, the emergence of a concept of childhood is correlated with changes in the social and economic structure of Western Europe, which, in turn, transformed the matrix of opportunities and expectations within which children grew up.

If the widespread acceptance of childhood as a separate stage of life is no more than three centuries old in Western societies, the concept of adolescence is even more recent. Freud, writing at the turn of the century, spoke of the "transformations of puberty,"[32] but did not discuss adolescence as a separate stage. Indeed, before the twentieth century it is hard to find references to a stage of life like our contemporary adolescence. John and Virginia Demos,[17] reviewing nineteenth-century writings on childrearing and the family, note the virtual absence of mention of anything like modern adolescence until the latter part of the century. And even then those whom we would call adolescents were often discussed as a new problem, as what one writer called the "dangerous classes" who made up the street gangs that terrorized many nineteenth-century American cities. It was not until the twentieth century, and in large part because of the work of G. Stanley Hall,[37] that adolescence was finally recognized, sanctioned, and culturally approved as a separate stage of life.

Once again the discovery of a "new" stage of life corresponded with major social and economic changes. Industrialization in a half a dozen nations began to transform the social, economic, and familial structures of those societies. The old working family, where parents and children labored side by side in fields and factories, gradually began to disappear. Urbanism increasingly supplanted an agrarian way of life. The growing productivity of industrialized societies made it possible to "excuse" ever larger numbers of postpubescent men and women from the requirements of work. In addition, an industrial society required far more than rudimentary education from any young man or woman who aspired to status, respect, or wealth. Increasingly society demanded higher level skills that could rarely be taught at home, but that became the special province of the secondary schools that opened by the tens of thousands throughout

America in the nineteenth and early twentieth centuries. In all these ways the "discovery" of adolescence corresponded with major social and economic changes that drastically affected the lives of children as they grew up.[17]

The "discovery" of childhood and adolescence in Western societies during the last three or four centuries suggests that the segmentation of the life cycle is not a psychological or cultural constant, and that the way life stages are distinguished and defined is intimately related both to the social, economic, and technological conditions of a society and to the actual environment in which children grow up. To be sure, this correspondence between definitions of the life cycle, socioeconomic context, and the actual experiences of growing individuals is far from perfect. But the "discovery" of childhood and adolescence at times of major socioeconomic change suggests that concepts of life stages tend to correspond with, reflect, and sanction actual changes in the modal experience of young people as they grow up in any given society and historical era.

If these surmises be correct, they have far-reaching implications for the understanding of human development. We have tended to assume that the development of all men and women is in some fundamental respect similar to that of those middle-class, relatively prosperous, and advantaged young men and women who constitute the source material for most contemporary theories of development. We have seen the life cycle as a kind of escalator, onto which the child steps at birth and up which he continues willy-nilly until he has passed through all the stages until death. This escalator model is a tempting one, for it corresponds with what we know of physiological maturation, which, barring the grossest of insults to the organism, tends to be relatively invariant in sequence regardless of social and historical conditions. The escalator model also corresponds with the fact that every society, however it defines the life cycle, does segment life in some way: age-grading, accompanied by different expectations concerning those in each age-grade, is culturally universal.

But human development, though related to both maturation and age-grading, is identical to neither.[38,41] It entails qualitative changes in psychological functioning, presumably based upon what Anna Freud[31] calls the increasing "structuralization" of the personality or what Piaget[42] calls the development of ever more complex and inclusive schemata. The escalator model does not correspond with our growing knowledge concerning the contexts, environments, or matrices that promote human development, when development is distinguished from both physiological maturation and social age-grading.

Recent research has made clear that true psychological development is anything but automatic. On the contrary, development in every area, sector, or "developmental line" that has been examined turns out to be contingent upon what have been variously called environmental nutriments, supports, confirmations, or challenges. The most vivid example of developmental arrest in the absence of environmental facilitation comes from studies of institutionalized children who, if deprived of the necessary environmental supports and challenges during the first few years, suffer severe and probably irreversible retardations.[12,74,83] Furthermore, the utility of developmental concepts like fixation, foreclosure, arrest, retardation, and lag indicates the frequency of failures of development that occur either because of the absence of a facilitating environment or because of the presence of obstructions to psychological growth. At every stage of life and in every area of development, therefore, psychological development must be considered problematic rather than inevitable.

The contingency of human development may help us understand the large differences in the way different societies segment and organize the life cycle. For example, in many nonliterate and subsistence societies children are routinely expected to join their parents in adult work, at around the age of seven or eight. What little we know about the conditions that stimulate development in childhood suggests that such societies should have drastically different effects on the final cognitive and affective development of the typical adult, when compared with societies that insist on prolonged schooling during the same years.

Literacy alone appears to have massive effects on development. For example, Bruner and his colleagues,[14,35] working with nonliterate groups in Senegal, have shown that nonliterate children and adults cannot understand the question, "Why did you say that?" Bruner argues that only with literacy does there emerge a capacity to reflect upon language itself, to discuss propositions as objects of discourse.

If development is contingent, there may well be important developmental potentials that remain largely unactualized in some or most societies because these societies do not provide the needed developmental facilitations or because they place active obstructions in the way of development. Furthermore, we may discover that many definable adult "achievements" are not attained by many or most people in any given society. Such, indeed, seems to be the case. For example, the psychoanalytic concept of "genitality" refers to a psychosexual orientation that is often not realized. Similarly Kohlberg's[53-60] work on moral development (see below) has demonstrated empirically the existence of high level structures of moral reasoning that are not attained by most Americans. Hauser's[39] studies of non-college-bound, working-class black and white adolescents suggest that identity foreclosure rather than development may be the rule, especially among blacks. And other recent research suggests that a near majority of adult Americans lack certain aspects of the cognitive capacity for formal operations[63] that in Piaget's work is said to emerge around the age of puberty.

Other examples abound, but the general point is surely clear. If we consider development contingent, and view the life cycle in a developmental sense not as an escalator but as a pitted and problematic path, then it becomes more comprehensible why some societies should recognize stages that others ignore. It is, of course, possible that some societies are simply "blind" to real developmental changes that actually occur in their young. But an alternative hypothesis is also opened: developmental stages may commonly occur in some societies, but these same stages (or the developmental changes and issues that define them) may be infrequent or virtually absent in other societies. Thus, major changes in the social, economic, technological, and familial matrix within which children grow up may either encourage the foreclosure of development or stimulate development to higher levels in any of a number of sectors or developmental lines.

The existing evidence—anthropological, historical, and scientific—therefore supports the hypothesis that shifting historical definitions of the life cycle reflect not merely shifting cultural fashions but actual changes in the developmental matrix within which the typical child grows up, and, thus, real changes in the type and level of development actually experienced by typical children. It is unlikely that any stage will be recognized, approved, and sanctioned until a significant portion of the young have begun to pass through that stage. Furthermore, if we assume that developmental potentials are universal, while their actual unfolding is contingent, the variability of environments in all societies makes it likely that some rare individuals will pass through any given stage of life, even in a social context that makes entry into that stage very unlikely. Thus, once a stage is culturally recognized, we can look back in history and discover men and women who passed through it long before it was named and sanctioned. But the cultural "discovery" of a stage of life tends to coincide with a time when ever more young men and women are, in fact, entering this stage, exhibiting the visible hallmarks that accompany it, and developing—at least in certain areas—in different ways and perhaps to a greater extent than had previous generations.

From Adolescence to Youth

Today the concept of adolescence, first defined and disseminated by G. Stanley Hall, is unshakably enshrined in our view of human life. To be sure, the precise nature of adolescence still remains controversial. Some observers believe that Hall, like most psychoanalytic observers, vastly overestimated the inevitability of turbulence, rebellion, and upheaval in this stage of life.[1,19,34,69] But whatever the exact definition of adolescence, no one today doubts its existence. A stage of life that barely existed a century ago is now universally accepted as an inherent part of the human condition.

In the seven decades since Hall made adolescence a household word, American society has once again transformed itself. From the industrial era of the turn of the century, we have moved into a new era without an agreed upon name—it has been called postindustrial, technological, postmodern, the age of mass consumption, the technetronic age. And a new generation, the first born in this new era of postwar affluence, television, and the Bomb, raised in the cities and suburbs of America, socially and economically secure, is now coming to maturity. Since 1900 the average amount of education received by children has increased by more than six years. In 1900 only 6.4 per cent of young Americans completed high school, while today almost 80 per cent do, and more than half of them begin college. In 1900 there were only 238,000 college students: in 1970 there were more than seven million, with ten million projected for 1980.

These massive social transformations are today reflected in new public anxieties. The "problem of youth," "the now generation," "troubled youth," "student dissent" and "the youth revolt" are topics of extraordinary concern to most Americans. No longer is our anxiety focused primarily upon the teen-ager, upon the adolescent of Hall's day. Today we are nervous about new "dangerous classes"—those young men and women of college and graduate school age who can't seem to "settle down" the way their parents did, who refuse to consider themselves adult, and who often vehemently challenge the existing social order. In mid-1970 "campus unrest" was considered America's foremost problem.

The factors that have brought this new group into existence parallel in many ways the factors that produced adolescence: rising prosperity, the further prolongation of education, the enormously high educational demands of a postindustrial society. Behind these measurable changes lie other trends less quantitative but even more important: a rate of social change so rapid that it threatens to make obsolete all institutions, values, methodologies, and technologies within the lifetime of each generation; a technology that has created not only prosperity and longevity but power to destroy the planet through warfare or violation of nature's balance; a world of extraordinarily complex social organization, instantaneous communication, and constant revolution. The "new" young men and young women emerging today both reflect and react against these trends.

But if we search among the concepts of psychology for a word to describe these young men and women, we find none that is adequate. Characteristically they are referred to as "late adolescents and young adults"—a phrase whose very mouth-filling awkwardness attests to its inadequacy. Those who see in youthful behavior the remnants of childhood immaturity naturally incline toward the concept of "adolescence" in describing the unsettled twenty-four-year-old, for this word makes it easier to interpret his objections to war, racism, pollution, or imperialism as "nothing but" delayed adolescent rebellion. To those who are more hopeful about today's youth, "young adulthood" seems a more flattering phrase, for it suggests that maturity, responsibility, and rationality lie behind the unease and unrest of many contemporary youths.

But in the end neither label seems fully adequate. The twenty-four-year-old seeker, political activist, or graduate student often turns out to have been through a period of adolescent rebellion ten years before, to be all too formed in his views, to have a stable sense of himself, and to be much further along in his psychological development than his fourteen-year-old high school brother. Yet he differs just as sharply from "young adults" of twenty-four whose place in society is settled, who are married and perhaps parents, and who are fully committed to an occupation. What characterizes a growing minority of postadolescents today is that they have not settled the questions whose answers once defined adulthood: questions of relationship to the existing society, questions of vocation, questions of social role and life style.

Faced with this dilemma, some writers have fallen back on the concept of "prolonged" or "stretched" adolescence[5,6,8]—a concept that suggests that those who find it hard to settle down have failed the adolescent developmental task of abandoning narcissistic fantasies and juvenile dreams of glory. Thus, one remedy for "protracted adolescence" might be some form of therapy that would enable the young to reconcile themselves to abilities and a world that are rather less than they had hoped. Another interpretation of youthful unease blames society, not the individual, for the "prolongation of adolescence."[4] It argues that youthful unrest springs from the unwillingness of contemporary society to allow young men and women, especially students, to exercise the adult powers of which they are biologically and intellectually capable. According to this view, the solution would be to allow young people to "enter adulthood" and do "real work in the real world" at an earlier age.

Yet neither of these interpretations seems quite to the point. For while some young men and women are indeed victims of the psychological malady of "stretched adolescence," many others are less impelled by juvenile grandiosity than by a rather accurate analysis of the perils and injustices of the world in which they live. And plunging youth into the "adult world" at an earlier age would run directly counter to the wishes of many youths, who view adulthood with all of the enthusiasm of a condemned man for the guillotine. Far from seeking the adult prerogatives of their parents, they vehemently demand a virtually indefinite prolongation of their nonadult state.

If neither "adolescence" nor "early adulthood" quite describes the young men and women who so disturb most highly industrialized societies today, what can we call them? My answer is to propose that we are witnessing today the emergence on a mass scale of a *previously unrecognized stage of life*, a stage that intervenes between adolescence proper and adulthood. I propose to call this stage of life the stage of *youth*, assigning to this venerable but vague term a new and more specific meaning. Like Hall's adolescence, "youth" is in no absolute sense new: indeed, once having defined this stage of life, we can study its historical emergence, locating individuals and groups who have had a "youth" in the past.[16,33] But what is "new" is that this stage of life is today being entered, not by tiny minorities of unusually creative or unusually disturbed young men and women, but by millions of young people in all the nations of the world.

Like all stages youth is a stage of transition rather than of completion. In the remarks that follow I will attempt to define this emergent stage in several ways. First, I will comment on some of the major themes that dominate consciousness, development, and behavior during this stage. But human development rarely if ever proceeds on all fronts simultaneously: instead, we must think of development as consisting of a series of sectors or "developmental lines," each of which may be in or out of phase with the others. Thus, an account of youth must include an account of the more specific transformations in feeling, behavior, and personal relationships that occur during this stage. Third, youth is of all stages of life that in which issues of ideology, built on cognitive, intellectual, and ethical development, play the most visible and central role. Fourth, youth, like all stages of life, has its own stage-specific psychopathology, just as disturbances common to other stages of life may take a specific youthful form. And finally I will try to make clear what youth is not, in order to underline what it is. What follows, then, is a preliminary sketch of the issues that seem important to understanding youth as an emergent stage of life.

Major Themes in Youth

Perhaps the central conscious issue during youth is the *tension between self and society*. In adolescence young men and women tend to accept their society's definitions of them as rebels, truants, conformists, athletes, or achievers. But in youth the relationship between socially assigned labels and the "real self" becomes more problematic and constitutes a focus of central concern. The awareness of actual or potential conflict, disparity, lack of congruence between what one is (one's identity, values, integrity) and the resources and demands of the existing society increases. The adolescent is struggling to define who he is; the youth begins to sense who he is and thus to recognize the possibility of conflict and disparity between his emerging selfhood and his social order.

In youth *pervasive ambivalence* toward both self and society is the rule: the question of how the two can be made more congruent is often experienced as a central problem of youth. This ambivalence is not the same as definitive rejection of society, nor does it necessarily lead to political activism. For ambivalence may also entail intense self-rejection, including major efforts at self-transformation employing the methodologies of personal transformation that are culturally available in any historical era: monasticism, meditation, psychoanalysis, prayer, hallucinogenic drugs, hard work, religious conversion, introspection, and so forth. In youth, therefore, the potential and ambivalent conflicts between autonomous selfhood and social involvement, between the maintenance of personal integrity and the achievement of effectiveness in society, are fully experienced for the first time.

The effort to reconcile and accommodate these two poles involves a characteristic stance vis-à-vis both self and world, perhaps best described by the concept of the *wary probe*. For the youthful relationship to the social order consists not merely in the experimentation more characteristic of adolescence, but with more serious forays into the adult world, through which its vulnerability, strength, integrity, and possibilities are assayed. Adolescent experimentation is more concerned with self-definition than are the probes of youth, which may lead to more lasting commitments. This testing, exacting, challenging attitude may be applied to all representatives and aspects of the existing social order, sometimes in anger and expectation of disappointment, sometimes in the urgent hope of finding honor, fidelity, and decency in society, and often in both anger and hope. With regard to the self, too, there is constant self-probing in search of strength, weakness, vulnerability, and resiliency; constant self-scrutiny designed to test the individual's capacity to withstand or use what his society would make of him, ask of him, and allow him.

Phenomenologically youth is a time of alternating *estrangement and omnipotentiality*. The estrangement of youth entails feelings of unreality, absurdity, isolation, and disconnectedness from the interpersonal, social, and phenomenological world. Such feelings are probably more intense during youth than during any other period of life. In part they spring from the actual disengagement of youth from society; in part they grow out of the psychological sense of incongruence between self and world. Much of the psychopathology of youth involves such feelings, experienced as the depersonalization of the self or the derealization of the world.

Omnipotentiality[75] is the opposite but secretly related pole of estrangement. It is the feeling of absolute freedom, of living in a world of pure possibilities, of being able to change or achieve anything. There may be times when complete self-transformation seems possible, when the self is experienced as putty in one's own hands. At other times, or for other youths, it is the nonself that becomes totally malleable; then one feels capable of totally transforming another's life, or of creating a new society with no roots whatsoever in the mire of the past. Omnipotentiality and estrangement are obviously related: the same sense of freedom and possibility that may come from casting off old inhibitions, values, and constraints may also lead directly to a feeling of absurdity, disconnectedness, and estrangement.

Another characteristic of youth is the *refusal of socialization* and acculturation. In keeping with the intense and wary probing of youth, the individual characteristically begins to become aware of the deep effects upon his personality of his society and his culture. At times he may attempt to break out of his prescribed roles, out of his culture, out of history, and even out of his own skin. Youth is a time when earlier socialization and acculturation are self-critically analyzed, and when massive efforts may be made to uproot the now alien traces of historicity, social membership, and culture. Needless to say, these efforts are invariably accomplished within a social, cultural, and historical context, using historically available methods. Youth's relationship to history is therefore paradoxical. Although it may try to reject history altogether, youth does so in a way defined by its historical era, and these rejections may even come to define that era.

In youth we also observe the emergence of *youth-specific identities* and roles. These contrast both with the more ephemeral enthusiasms of the adolescent and with the more established commitments of the adult. They may last for months, years, or a decade, and they inspire deep commitment in those who adopt them. Yet they are inherently temporary and specific to youth: today's youthful hippies, radicals, communards, and seekers often recognize full well that, however reluctantly, they will eventually become older and that aging itself will change their status. Some such youth-specific identities may provide the foundation for later commitments; but others must be viewed in retrospect as experiments that failed or as probes of the existing society that achieved their purpose, which was to permit the individual to move on in other directions.

Another special issue during youth is the enormous value placed upon change, transformation, and *movement*, and the consequent abhorrence of *stasis*. To change, to stay on the road, to retain a sense of inner development or outer momentum is essential to many youths' sense of active vitality. The psychological problems of youth are experienced as most overwhelming when they seem to block change: thus, youth grows panicky when confronted with the feeling of "getting nowhere," of "being stuck in a rut," of "not moving."

At times the focus of change may be upon the self, and the goal is then to be moved. Thus, during youth we see the most strenuous, self-conscious, and even frenzied efforts at self-transformation, using whatever religious, cultural, therapeutic, or chemical means are available. At other times the goal may be to create movement in the outer world, to move others: then we may see efforts at social and political change that in other stages of life rarely possess the same single-minded determination. On other occasions the goal is to move through the world, and we witness a frantic geographic restlessness, wild swings of upward or downward social mobility, or a compelling psychological need to identify with the highest and the lowest, the most distant and apparently alien.

The need for movement and terror of stasis often are a part of a heightened *valuation of development* itself, however development may be defined by the individual and his culture. In all stages of life, of course, all individuals often wish to change in specific ways: to become wittier, more attractive, more sociable, or wealthier. But in youth specific changes are often subsumed in the devotion to change itself—to "keep putting myself through the changes," "not to bail out," "to keep moving." This valuation of change need not be fully conscious. Indeed, it often surfaces only in its inverse form, as the panic or depression that accompanies a sense of "being caught in a rut," "not being able to change." But for other youths change becomes a conscious goal in itself, and elaborate ideologies of the techniques of transformation and the *telos* of human life may be developed.

In youth, as in all other stages of life, *the fear of death* takes a special form. For the infant to be deprived of maternal support, responsiveness, and care is not to exist; for the four-year-old nonbeing means loss of body intactness (dismemberment, mutilation, castration); for the adolescent to cease to be is to fall apart, to fragment, splinter, or diffuse into nothingness. For the youth, however, to lose one's essential vitality is merely to stop. For some even self-inflicted death or psychosis may seem preferable to loss of movement; and suicidal attempts in youth often spring from the failure of efforts to change and the resulting sense of being forever trapped in an unmoving present.

The youthful *view of adulthood* is strongly affected by these feelings. Compared to youth, adulthood has traditionally been a stage of slower transformation when, as Erickson[23] has noted, the relative developmental stability of parents enables them to nurture the rapid change of their children. This adult deceleration of personal change is often seen from a youthful vantage point as concretely embodied in apparently unchanging parents. It leads frequently to the conscious identification of adulthood with stasis, and to its unconscious equation with death or nonbeing. Although greatly magnified today by the specific political disillusionments of many youths with the "older generation," the adulthood=stasis (=death) equation is inherent in the youthful situation itself. The desire to prolong youth indefinitely springs not only from an accurate perception of the real disadvantages of adult status in any historical era but from the less conscious and less accurate assumption that to "grow up" is in some ultimate sense to cease to be really alive.

Finally, youths tend to band together with other youths in *youthful countercultures*,[79] characterized by their deliberate cultural distance from the existing social order, but not always by active political or other opposition to it. It is a mistake to identify youth as a developmental stage with any one social group, role, or organization. But youth is a time when solidarity with other youths is especially important, whether the solidarity be achieved in pairs, small groups, or formal organizations. And the groups dominated by youth reflect not only the special configurations of each historical era but also the shared developmental positions and problems of youth. Much of what has traditionally been referred to as "youth culture" is, in the terms used here, adolescent culture; but there are also groups, societies, and associations that are truly youthful. In our own time, with the enormous increase in the number of those who are entering youth as a stage of life, the variety and importance of these youthful countercultures is steadily growing.

This compressed summary of themes in youth is schematic and interpretative. It omits many of the qualifications necessary to a fuller discussion, and it neglects the enormous complexity of development in any one person in favor of a highly schematic account. Specifically, for example, I do not discuss the ways the infantile, the childish, the adolescent and the truly youthful interact in all real lives. Perhaps most importantly, my account is highly interpretative, in that it points to themes that underlie diverse acts and feelings, to issues and tensions that unite the often scattered experiences of real individuals. The themes, issues, and conflicts discussed here are rarely conscious as such; indeed, if they all were fully conscious, there would probably be something seriously awry. Different youths experience each of the issues considered here with different intensity. What is a central conflict for one may be peripheral or unimportant for another. These remarks, then, should be taken as a first effort to summarize some of the underlying configurations that characterize youth as an ideal type.

Affective and Interpersonal Transformations in Youth

A second way of describing youth is by attempting to trace out the various psychological and interpersonal transformations that may occur during this stage. Once again only the most preliminary sketch of youthful development can be attempted here. Somewhat arbitrarily I will distinguish between development in several sectors or areas of life, noting only that, in fact, changes in one sector invariably interact with those in other sectors.

In pointing to the *self-society relationship* as a central issue in youth, I also mean to suggest its importance as an area of potential change. The late adolescent is only beginning to challenge his society's definition of him, only starting to compare his emerging sense of himself with his culture's possibilities and with the temptations and opportunities offered by his environment. Adolescent struggles for emancipation from external family control and internal dependency on the family take a variety of forms, including displacement of the conflict onto other "authority figures." But in adolescence itself, the "real" focus of conflict is on the family and all of its internal psychic residues. In youth, however, the "real" focus begins to shift: increasingly the family becomes more paradigmatic of society rather than vice versa. As relatively greater emancipation from the family is achieved, the tension between self and society, with ambivalent probing of both, comes to constitute a major area of developmental "work" and change. Through this work young people can sometimes arrive at a synthesis whereby both self and society are affirmed, in the sense that the autonomous reality-relatedness yet separateness of both is firmly established.

There is no adequate term to describe this "resolution" of the tension between self and society, but C. G. Jung's[44] concept of *individuation* comes close. For Jung the individuated man is a man who acknowledges and can cope with social reality, either by accepting it or by opposing it with revolutionary fervor. But he can do this without feeling his central selfhood overwhelmed. Even when most fully engaged in social role and societal action, he can preserve a sense of himself as intact, whole, and distinct from society. Thus, the "resolution" of the self-society tension in no way necessarily entails "adjusting" to the society, much less "selling out"—although many youths see it this way. On the contrary, individuation refers partly to a psychological process whereby self and society are differentiated internally. But the actual conflicts between men and women and their societies remain and, indeed, may become even more intense.

The meaning of individuation may be clarified by considering the special dangers of youth, which can be defined as extremes of *alienation*, whether from self or from society.[50] At one extreme is the total alienation from self that involves abject submission to society, "joining the rat race," "selling out." Here society is affirmed but selfhood denied. The other extreme is a total alienation from society that leads not so much to the rejection of society as to its existence being ignored, denied, and blocked out. The result is a kind of self-absorption, an enforced interiority and subjectivity, in which only the self and its extensions are granted reality, while all the rest is relegated to a limbo of insignificance. Here the integrity of the self is purchased at the price of a determined denial of social reality and the loss of social effectiveness. In youth both forms of alienation are often assayed, sometimes for lengthy periods. And for some whose further development is blocked, they become the basis for lifelong adaptations—the self-alienation of the marketing personality, the social alienation of the perpetual drop-out. In terms of the developmental polarities suggested by Erikson, we can define the central developmental tension of youth as individuation versus alienation.

Sexual development continues in important ways during youth. In modern Western societies, as in many others, the commencement of actual sexual relationships is generally deferred by middle-class adolescents until their late teens or early twenties: the modal age of first intercourse for American college males today is around twenty, for females about twenty-one. Thus, despite the enormous importance of adolescent sexuality and sexual development, actual sexual intercourse often awaits youth. In youth there may occur a major shift from masturbation and sexual fantasy to interpersonal sexual behavior, including the gradual integration of sexual feelings with intimacy with a real person. And as sexual behavior with real people commences, one sees a further working through, now in behavior, of vestigial fears and prohibitions whose origins lie in earlier childhood—specifically, of oedipal feelings of sexual inferiority and of oedipal prohibitions against sex with one's closest intimates. During youth, when these fears and prohibitions can be gradually worked through, they yield a capacity for genitality, that is, for mutually satisfying sexual relationships with another whom one loves.

The transition to genitality is closely related to a more general pattern of *interpersonal development*. I will term this the shift from "identicality" to mutuality. This development begins with adolescence* and continues through youth. It involves a progressive expansion of the early adolescent assumption that the interpersonal world is divided into only two categories: first, me and those who are identical to me (potential soul mates, doubles, and hypothetical people who "automatically understand everything"), and, second, all others. This conceptualization gradually yields to a capacity for close relationships with those on an approximate level of "parity" or similarity with the individual.

The phase of parity in turn gives way to a phase of "complementarity," in which the individual can relate warmly to others who are different from him, valuing them for their dissimilarities from himself. Finally the phase of complementarity may yield in youth to a phase of "mutuality," in which issues of identicality, parity, and complementarity are all subsumed in an overriding concern with the other as other. Mutuality entails a simultaneous awareness of the ways in which others are identical to oneself, the ways in which they are similar and dissimilar, and the ways in which they are absolutely unique. Only in the stage of mutuality can the individual begin to conceive of others as separate and unique selves and relate to them as such. And only with this stage can the concept of mankind assume a concrete significance as pointing to a human universe of unique and irreplaceable selves.

Relationships with elders may also undergo characteristic youthful changes. By the end of adolescence the hero worship or demonology of the middle adolescent has generally given way to an attitude of more selective emulation and rejection of admired or disliked older persons. In youth new kinds of relationships with elders become possible: psychological apprenticeships, then a more complex relationship of mentorship, then sponsorship, and eventually peership. Without attempting to describe each of these substages in detail, the overall transition can be described as one in which the older person becomes progressively more real and three-dimensional to the younger one, whose individuality is appreciated, validated, and confirmed by the elder. The sponsor, for example, is one who supports and confirms in the youth that which is best in the youth, without exacting an excessive price in terms of submission, imitation, emulation, or even gratitude.

Comparable changes continue to occur during youth with regard to *parents*. Adolescents commonly discover that their parents have feet of clay and recognize their flaws with great acuity. Childish hero worship of parents gives way to a more complex and often negative view of them. But it is generally not until youth that the individual discovers his parents as complex, three-dimensional, historical personages whose destinies are formed partly by their own wishes, conscious and unconscious, and partly by their historical situations. Similarly it is only during youth that the questions of family tradition, family destiny, family fate, family culture, and family curse arise with full force. In youth the question of whether to live one's parents' life, or to what extent to do so, becomes a real and active one. In youth one often sees what Ernst Prelinger[73] has called a "telescoped reenactment" of the life of a parent—a compulsive need to live out for oneself the destiny of a parent, as if to test its possibilities and limits, experience it from the inside, and (perhaps) free oneself of it. In the end the youth may learn to see himself and his parents as multidimensional persons, to view them with compassion and understanding, to feel less threatened by their fate and failings, and to be able, if he chooses, to move beyond them.

Every developmental stage tends to re-evoke, often in a highly selective fashion, themes and conflicts that date from earlier

* Obviously interpersonal development, and specifically the development of relationships with peers, begins long before adolescence, starting with the "parallel play" observed at ages two to four and continuing through many stages to the preadolescent same-sex "chumship" described by Sullivan.[84] But puberty in middle-class Western societies is accompanied by major cognitive changes that permit the early adolescent for the first time to develop hypothetical ideals of the possibilities of friendship and intimacy. The "search for a soul mate" of early adolescence is the first interpersonal stage built upon these new cognitive abilities.

stages. In each individual the weight and origin of the reevoked past will vary, depending on the idiosyncrasies of his experience. But beyond idiosyncratic variations common links seem to join postpubescent development with prepubescent development. For example, the stage of early adulthood involves entry for the first time into the world of work; it thus reevokes isomorphic issues of competence, industry, and inferiority that were experienced in childhood upon entry for the first time into the world of school.

If we differentiate adolescence from youth, we must reconsider those formulations that see revivified oedipal themes as central to adolescence. More common in adolescence, I believe, is the reawakening of preoedipal concerns focused on the "anal" stage, with its conflicts over autonomy, will, order, control and impulsivity, messiness and neatness. If we define adolescence, in oversimplified fashion, as involving, above all, emancipation from dependency on and control by the family, then the isomorphism between the anal phase and adolescence becomes clearer. Each stage entails, at a different level, an effort to move away from controlling and dependency-inducing others toward a more self-controlling and independent position. Thus, during adolescence, behavior and fantasy are commonly dominated by struggles for independence and thinly veiled wishes for dependency, by provocativeness and negativism, by sloppiness and compulsive neatness, by willfulness coupled with irresponsibility—all of which most resemble psychologically the toddler's early efforts to establish self-control and to avoid total control by his parents. Even the proverbial embarrassment of the adolescent, now focused upon his or her incipient adult sexuality, resembles the "shame" of the child during the anal stage.

In youth these anal issues recur, but they are increasingly overshadowed by oedipal themes. In Erickson's terms, the toddler establishes that he has a will of his own; during the oedipal period he learns to take initiative, using that will. In a parallel way the adolescent hopefully achieves emancipation and independence from his family; during youth he confronts the issue of how, where, and when his independent and autonomous selfhood is to be exercised. In the normal course of youthful development, a whole series of conflicts, complexes, and inhibitions whose origins lie in the oedipal phase are therefore reevoked. When youthful development falters, it is often because the inhibitions of the oedipal stage remain so powerful that the individual is compelled either to reenact them repetitively or to regress to modes of functioning (including cognitive functioning) more characteristic of the three- to five-year-old than of the stage of youth itself.

The tendency among youth to defer actual sexual relationships until the twenties also helps explain the centrality of oedipal themes in this stage of life. Although oedipal issues obviously recur in different forms throughout life, they tend to be reexperienced and reenacted with particular intensity at the point in life when the individual attempts to unite intimacy and sexuality in a real, instead of a fantasied, relationship. At this juncture the powerful, if usually unstated, taboos laid down within the child's family, prohibiting the joining of overtly erotic and affectionate ties, are powerfully reawakened. For this reason the first real sexual relationships are more often than not psychologically triadic rather than simply dyadic. Same-sex friends and rivals are often as much in the forefront of consciousness during the beginning of sexual relationships as is the sexual partner himself or herself. Indeed, the sexual partner may remain shadowy or vague in comparison to the dreaded and feared rivals with whom one competes for his or her affection; just as often, the initiation of sexual relationships appears to require a permissive contemporary, perhaps more sexually experienced or perceived as more "mature," to whom one reports in more or less graphic detail the sexual and interpersonal complexities of the new sexual relationship.

These triadic relationships often appear flagrantly neurotic, and can be, if they do not eventually yield to a more truly dyadic sexual relationship in which the real existence of the partner is central rather than peripheral. But the longitudinal clinical study of young men and women in the process of beginning sexual relationships suggest that such triadic "oedipal" relationships are so frequent as to be virtually normal in educated youth. Indeed it may be that this reenactment of triangular love relationships permits some youths to work through in actual behavior childhood feelings of inferiority in love relationships or childhood fears that heterosexual relationships will result in devastating reprisals from same sex rivals.

Having emphasized that these analytically separated lines of development are, in fact, linked in the individual's experience, it is equally important to add that they are never linked in perfect synchronicity. If we could precisely label one specific level within each developmental line as distinctively youthful, we would find that few people were "youthful" in all lines at the same time. In general, human development proceeds unevenly, with lags in some areas and precocities in others. One young woman may be at a truly adolescent level in her relationship with her parents, but at a much later level in moral development; a young man may be capable of extraordinary mutuality with his peers, but still be struggling intellectually with the dim awareness of relativism. Analysis of any one person in terms of specific sectors of development will generally show a simultaneous mixture of adolescent, youthful, and adult features. Once again the point is that the concept of youth here proposed is an ideal type, a model that may help understand real experience but can never fully describe or capture it.

Ideology in Youth

Of all stages of life, youth is the most ideological. The potential for the development of ideology rests upon affective and cognitive changes that may begin in adolescence but that extend far beyond it. In the ideological preoccupations of youth are focused, condensed, and expressed concerns of diverse origins: psychosexual and historical, interpersonal and societal, defensive and generational, affective and ethical. It is therefore arbitrary and misleading to separate ideology formation from those affective, psychosexual, and interpersonal transformations discussed in the previous section. All are intertwined in the effort, often vehement and agonized, to develop what Erikson, quoting George Bernard Shaw, termed "a clear view of the world in the light of an intelligible theory."[23]

As commonly used, "ideology" has two distinct meanings. Sociologists[2,67] and intellectual historians use it to refer to a public body of shared doctrine, belief, evaluation, and exhortation, transmitted from one individual or one generation to another and embodied in collective values, symbols, theories, and norms. For psychologists and psychiatrists, however, the term has a primarily individual referent: it points to the individual's developing world view, conscious or unconscious, shared or idiosyncratic, especially seen as a defensive and expressive system. Thus, on the one hand, the intellectual historian may interpret ideology only as a cultural cement that binds together a society, while the individually-oriented psychiatrist may consider ideology only a "projective system"[45] that expresses the complexes and conflicts of early childhood.

An adequate understanding of ideology during youth must include both perspectives. The sociologist is right in underlining that most of the ingredients of any individual's ideology are borrowed from others, from his historical tradition, from his culture. But the psychologist is right in emphasizing that the synthesis of these "borrowed" elements is built around enduring and often idiosyncratic themes, conflicts, defenses, and talents within the individual himself. The development of a world view is thus a battleground on which individual and sociohistorical issues come together and are sometimes resolved. Ideology lies at the juncture between the individual and his society and history; ideology formation entails at best a synthesis of elements that are uniquely personal with ideas, symbols, and values that connect the individual to his community and to his historical tradition.

The realm of abstract ideas, interpretative

principles, theories, and symbols, being farthest from observable reality, is most intimately related to the deepest themes of the psyche. Those who have studied the psychodynamics of individual ideologies[15,50,51,64] have shown that they integrate, synthesize, and to a degree "rationalize" intrapsychic issues ranging from deeply rooted conflicts to preferred adaptive styles. The assumptions and perspectives from which any individual perceives and interprets the world indeed perform important defensive functions of intellectualization and rationalization.[29,30] But to stress only the expressive or defensive element in ideology development neglects the roots of this process in cognitive, intellectual, and ethical developments that are themselves powerfully adaptive; it only ignores the functional role of ideology in orienting the individual in a world that might otherwise be overwhelming in its chaotic complexity. Ideology thus provides not only a defense against unacceptable impulses or a way of explaining away unattractive impersonal characteristics, but a framework within which the individual can relate himself simultaneously to his own personal history and to the history of his society and generation.

The development of an ideology that transcends the internalized prescriptions, prohibitions, and perspectives of childhood is made possible and necessary by cognitive, intellectual, and ethical restructurings of the personality that can occur after adolescence. The most crucial of these is what Inhelder and Piaget[42] term the capacity for formal operations—namely, the ability to perform hypothetico-deductive operations whereby the observable and concrete world of the real comes to be seen as a subset of the world of the possible. In Piaget's middle-class Swiss subjects this capacity is generally attained in rudimentary form around the time of puberty. But as Piaget notes, the full integration of the capacity for formal operations is often long deferred. At issue in this integration is a gradual "decentering," a transition from an exclusively egocentric mode of formal operations to an exercise of this capacity in a context that includes some acceptance of the gap between the real and the possible, the actual and the ideal.

Many of the ideological struggles of youth are built around the painful process of attaining a more integrated use of formal operations. It is as if the emergence of this capacity overwhelmed many adolescents with the terrifying awareness of desirable personal and societal possibilities that are not, and may never be, actualized. During youth this conflict between the ideal and the actual is worked on, and at best worked through, with regard to both self and society. Vis-à-vis the self, the conflict entails an awareness of all that one might conceivably be (and become) but is not (and will never be). It therefore involves a growing awareness and acceptance of one's human limitations. At the level of society the struggle involves confrontation with the gulf between societal creed and deed, a gulf probably widest in societies like America whose ideals are so frequently and nobly stated but so rarely achieved.[68] In most societies "mature adults" are ordinarily expected to have accepted the gap between the real and the ideal with regard to both self and society.

In youth such a "mature acceptance" of the nonideal nature of psychological and social reality has not been achieved. For this reason youth is the stage of life where the demand for both psychological and social change is most compelling. Nor should this demand be automatically dismissed as a sign of immaturity. On the contrary, one may ask whether the acceptance of the imperfection of self and society expected of adults does not constitute a real retrogression from the idealism of youth. In many adults this "acceptance" has been so excessive as to leave them with a lifelong fear that they have betrayed the best in themselves or "compromised" their own youthful vision of a better society. Paradoxically, how one later reacts to "youthful idealism" and its common subsequent waning depends largely on the ideological position attained in youth, for one of the central functions of ideology formation is to come to terms with the gap between the real and ideal. And how one does this determines where on the spectrum between revolutionary and organization man one is likely to spend the remainder of one's life.

In some youths there also occurs a further stage of cognitive development that goes beyond the integration and consolidation of formal operations. Jerome Bruner has suggested that beyond the stage of formal operations there may lie a further stage of "thinking about thinking."[13] In practice just such a stage can be observed in some intellectual and introspective youths. It involves a breaking away of thought from specific mental processes, a consciousness of consciousness, a separation of the phenomenological "I" from the contents of thought.

This breakaway of the phenomenological ego permits youth to engage in phenomenological games, intellectual tricks, and kinds of creativity that are rarely possible during adolescence itself. And as consciousness becomes an object of consciousness, it also becomes a potential target of manipulation, a state to be changed through transformations of self and environment. Throughout history there have always been some rare youths who deliberately sought to alter consciousness, using the instruments provided by their culture and their technology. Today this focus on alternation of consciousness involves not only agents like hallucinogenic drugs but a return to ancient forms of consciousness-changing like meditation, yoga, and Zen. Finally the consciousness of consciousness makes possible one of the most frightening states experienced in youth, the disappearance of the phenomenological ego in an endless regress of awareness of awareness of awareness.

Given a cognitive capacity for formal operations, regular stages of intellectual and moral development can be observed. William Perry[71,72] has recently provided a description of the stages of potential intellectual development that may occur during late adolescence and youth. Perry's work emphasizes the complex tradition from epistemological dualism to an awareness of multiplicity and to the realization of relativism. Relativism, in turn, may give way to a more "existential" sense of truth, culminating in what Perry terms "commitment within relativism." Thus, during youth we expect to see a passage beyond simple views of right and wrong, truth and falsehood, good and evil to a more perspectival view. And as youth proceeds, we may look for the development of commitment within a universe that remains epistemologically relativistic. The stages defined by Perry—dualism, multiplicity, relativism, and commitment within relativism—are a description of a developmental line that most individuals clearly do not complete to its end. In particular Perry underlines that the transition to the stage of relativism is fraught with difficulty, and that many young men and women, vaguely sensing the uncertainties of a relativistic universe, may temporize or return to earlier certainties.

Lawrence Kohlberg's[53-60] work on moral development provides another paradigm of changes that occur only in youth if they occur at all. Summarized oversimply, Kohlberg's theory of the development of moral reasoning distinguishes three broad stages. The earliest or *preconventional* stage involves relatively egocentric concepts of right and wrong as that which one can do without getting caught or as entailing primitive notions of reciprocity ("I'll scratch your back if you'll scratch mine"). This stage is followed, usually during later childhood, by a stage of *conventional* morality, during which good and evil are identified with the concept of a good boy or a good girl (that is, with conformity to role expectations), or, at a somewhat higher level, with standards of the community and concepts of law and order. In the conventional moral stage, which parallels Perry's dualistic stage, morality is perceived as objective, as existing "out there." Kohlberg's third and highest stage of moral reasoning is termed autonomous or *postconventional*; if development into this stage occurs at all, it is most likely to occur during youth. This stage involves more abstract moral reasoning that may lead the individual into conflict with conventional morality. The lower level of the postconventional stage involves the definition of right and wrong in terms of the long-term welfare of the community. This type of reasoning can be called the social contract level, in that it construes moral rules and moral behavior as ultimately determined by considerations of the well-being of all members of a society. Unlike conventional moral reasoning, this level views moral codes, rules, and laws not as objective and eternal, but as manmade, changeable, and amendable.

In the highest (and final) postconventional level, the individual affirms general moral principles that may transcend not only conventional morality but even the social contract: this level confronts the possibility that even rules made in the public interest may be unjust. At this level general precepts, stated at a high level of generality, are seen as personally binding, though not necessarily "objectively" true. However these precepts are stated—as the Golden Rule, the sanctity of life, the Kantian categorical imperative, the promotion of human development, or the concept of justice itself—they share their universalism and their search for moral structures that are applicable to all situations and persons regardless of time, place, society, station, and situation. At the level of personal principles the individual not only may find himself in conflict with existing concepts of law and order but also may at times consider democratically arrived at laws unjust if they violate his universalistic principles.

The work of Piaget, Perry, and Kohlberg is not intended as a comprehensive account of development, or even as an inclusive analysis of the development of ideology. Each sets itself a more limited task: the delineation of the sequences of development in a particular sector. For this reason the parallels between the three accounts are the more striking. All three accounts emphasize the centrality of the transition from the empirically, culturally, or morally "given" to a new mode of functioning that transcends the given. Stated more psychodynamically, all three suggest the limitations of "internalization" as a mechanism of learning beyond adolescence and point to the possibility of "synthetic" developments that go beyond and may be in conflict with inductive generalizations, received truths, or community moral standards.

Equally important, all three accounts emphasize the hazards that accompany the transition from the given to the synthetic, from the traditional to the formulative. Perry notes that the movement from dualism to relativism is fraught with special dangers that result from renouncing the certainty inherent in a dualistic perspective for the uncertainty of a universe where truth is perspectival, and where a commitment to a perspective requires an "existential" affirmation for which the individual must accept personal responsibility. Confronted with an imminent awareness of relativism, Perry argues, many young men and women temporize, stagnate, or retreat to earlier forms of dualism, now the more dogmatically held because dimly apprehended relativism threatens to undercut certainty. It follows that no dogmatism is more fierce than that of the man who has paused on the threshold of relativism to reassert the verity of the world view from which he was moving or the transcendental truth of some new ideology.

Kohlberg describes a similar process in moral development. In longitudinal studies of young American men, Kohlberg and his associates have observed a phenomenon they term the "Raskolnikov Syndrome"—a moral regression that may occur as the individual begins to question the adequacy of conventional moral structures. To be sure, many young men and women seem to make the transition to postconventional moral reasoning with little apparent difficulty. But among those who begin to challenge conventional moral codes, a number "regress" to preconventional moral structures rather than (or before) moving on to the affirmation of autonomous postconventional moral principles. In Perry's terms it is as if confrontation with the relativism, and to that degree the arbitrariness, of the moral structures accepted by the socializing community (and until then internalized) pushes some people back to a kind of "anything goes," preconventional moral framework. Both Perry and Kohlberg are thus describing, in different terms, a process whereby the transition from the "conformist" or "conscientious" position to a more "autonomous"[66] or "individuated" perspective is aborted, resulting not in further development but in a return to more familiar, primitive structures that also negate the validity of conventional moral norms and dualistically conceived truths.

Although the language in which this "cognitive regression" is described may be unfamiliar to the clinician, the process itself is observed in clinical practice and constitutes the cognitive core for much of the ideological and psychological turmoil seen in youth. Especially in times of cultural crisis and widespread exposure of injustice, abrupt confrontation with the arbitrariness and hypocrisy that exist in the social world (or in the individual) may "force" a youth back to the only nondualistic, nonconventional position he knows, namely, an earlier view of truth as that which is convenient and of morality as that which avoids punishment and increases pleasure. When this regression occurs in large numbers of young men and women, it tends to become ideologized at a collective level and institutionalized in youthful countercultures. The consequence is a scathing collective attack on conventional pieties, and an explicit sanctioning of behavior that scandalizes the most sacred norms and taboos of those who adhere to dualistic truths and conventional moralities. In a society that traditionally emphasizes the sanctity of property, for example, property is likely to be singled out for special attack by the destruction of property as an act of ideological affirmation, by the deliberate avoidance of possessions, or by a view of theft (the "rip-off") as a morally justified or noble act.

Accounts of the cognitive, intellectual, and moral aspects of ideology formation necessarily exclude all those other noncognitive conflicts and confrontations that serve as catalysts for and are intertwined with cognitive, intellectual, and ethical development in youth. But if we study real individuals in depth over an extended period, they rarely speak in exclusively cognitive, intellectual, or moral terms; and if they do, we rightly suspect that we are hearing only a part of the story, which serves as a "cover" for that which is left unmentioned. In practice, then, the cognitive, intellectual, ethical changes that underlie ideology formation are intertwined with the psychosexual, interpersonal, and affective developments discussed earlier. What psychologists conveniently separate into "affective" and "cognitive" are interdependent at all stages of life. For example, to imagine and strive toward a truly mutual relationship with another person presupposes the cognitive capacity of formal operations; but to achieve some degree of mutuality with a loved one, in turn modifies, "decenters," extends and consolidates that purely cognitive capacity.

As a rule the catalysts for intellectual and moral development in youth are found in realms other than the purely intellectual and moral: they may include an ambivalent realization that one's parents are limited, unhappy people; a prolonged immersion in an alien culture; falling in love with someone whose world view is very different from one's own; membership in a countercultural group devoted to an attack on conventional truths, policies, or moral codes; realizing (from friends, enemies, drugs, or psychotherapy) that one's pretensions to virtue are fraudulent; growing awareness of one's interpersonal, sexual, moral, or physical inhibitions; a concrete confrontation with the gap between public practice and public preaching; and so on. In the end human development is neither affective nor cognitive, but both; only as a matter of analytic convenience or research strategy can we separate its aspects—and then, hopefully, as a preface for showing their relationship.

To describe further the vicissitudes of the development of ideology during youth is simply to describe youth itself. Erikson's[23,24,25,27] insightful accounts of identity, ideology, and ideological innovators illustrate and expand the points made here: the interweaving of the uniquely personal with the generational, historical, and universal; the way in which the emergence of new cognitive, intellectual, and ethical capacities spurs new confrontations, advances, and regressions; the agonized and often prolonged shock that follows the "discovery" of the relativism of truth and the arbitrariness of conventional moral structures; the possibility of a return to old dogmatisms (or the creation of new ones) in the search for a world view to replace the dualistic security of childhood; the crucial role of ideology in uniting the members of youthful groups who are undergoing parallel developmental struggles.

The foregoing discussion of ideology development should be read in the light of my earlier comments on the contingency of psychological development, for there is clear evidence that the majority of young men and women between the ages of 18 and 30 do *not* pass through most of the stages I have outlined. For most adult men and women it is most precise to say that ideology is borrowed rather than developed. These are those whom William James[43] long ago termed the "once-born," who accept with little anguish or re-examination the cultural traditions learned in their early lives.

The evidence for the contingency of development is especially conclusive in the realm of ideology. Kohlberg and his associates[63] have found that only a minority of teen-agers possess the cognitive capacity for formal operations; and if higher levels of this capacity are measured, then a majority of all contemporary middle-class American adults lack it. Studies in a rural Turkish village indicate that virtually no adults posses this capacity. It follows, as Kohlberg notes, that the majority of chronological teen-agers in America are not cognitive adolescents—if, that is, we accept the prevailing view that cognitive development is both a companion and a catalyst for psychological adolescence. Since all the ideological developments of youth discussed here presuppose a highly developed capacity for formal operations, we must conclude that real ideology development occurs only in a minority of young Americans.

This conclusion is empirically substantiated in the work of both Perry and Kohlberg. Perry[71] found that only a few of the college students he studied had moved to his highest level of intellectual functioning, commitment within relativism. And Kohlberg,[53] whose research methods enable him to measure the individual's level of moral reasoning, finds that the majority of middle-class college students have not moved beyond conventional structures at the age of 24. Among working-class subjects (and their fathers) the proportion using postconventional moral reasoning is even lower. Other studies indicate that for both male and female college students, the modal moral position is the conventional position.[28,36]

Studies like these require us to define both adolescence and youth as "contingent" from a cognitive, intellectual, moral, and ideological point of view—and probably in other respects as well. They remind us again that "youth" as here defined is in no sense a universal stage of development, even among the advantaged, educated young men and women in economically advanced societies who occupy so much of psychiatrists' time. These studies further impel us to remember that it is possible and normal for men and women to live—and to live well, healthily, and happily—at the stage of concrete operations, with the dualistic view of truth, and with a conventional morality. It seems that most of mankind has always lived this way and that, as a result, most men and women have not experienced the full force of the ideological tensions that constitute a central part of youth. Youth involves the actualization of the usually latent human potential for departure from societal certainty, for questioning of conventional moral codes, for imagining a world different from that which exists, and for thinking about thought.

Psychopathology in Youth

The developmental component in psychological disorders is often neglected or discussed solely by noting that the origin of pathological symptoms often lies in crises, conflicts, or feelings that date from earlier developmental stages. But psychological disturbances are also related to, organized around, and precipitated by the psychological issues and conflicts of the developmental stage *during which they occur.* Thus, the relationship between development and psychopathology is at least fourfold. First, developmental "work" left undone or incompletely consolidated at one stage of life compromises later development and may produce significant psychopathology during later stages. Second, psychological disturbances necessarily occur during a stage of life, and their content is therefore shaped by stage-specific themes, changes, and conflicts; third, psychopathology is often triggered by blockings, foreclosures, or impasses in the developmental work faced by the individual at a given stage of his life; and finally, normal development entails what Anna Freud[31] and Peter Blos[8] have termed "regression in the service of development" and therefore involves a recapitulation, reenactment, and reexperiencing of the past[26] that, if it proves too deep and enduring, may be termed pathological. A comprehensive discussion of psychopathology during any stage of life should ideally consider all of these relationships.

During youth such protean human disturbances as *depression* are both precipitated by and organized around the themes that assume special prominence in this stage of life. I earlier noted the centrality to youth of continuing movement, the consequent abhorrence of stasis, and the unconscious identification of immobilization with death. Depressions during youth are strongly influenced by these themes. The immediate catalyst of many youthful depressions is a sense of development impasse, stasis, or stagnation, a feeling of being "stuck," "trapped," or "blocked"—whether spacially, temporally, or developmentally. Especially dysphoric is the experienced failure of efforts to move, however movement may be defined by the individual. For example, in those imbued with an achievement ethic, achievement itself may be equated with motion, and failures to achieve are therefore seen as portents of a life without further mobility. Similarly, when efforts at self-reform and self-improvement fail, they may produce depressions associated with a sense of being "locked into" an abhorred personality style or character structure. Such depressions far exceed "realistic" reactions to failure: they evoke deeper fears of immobilization and reawaken primitive anxieties associated with deprivation. The individual feels himself unresponsive, frozen, immobilized, effectively "dead."

Whatever the underlying etiology of youthful depressions, the most common response is to attempt to regain momentum. All types of movement are unconsciously equated: geographic movement and social mobility can be a substitute for psychological change; the effort to move others may be a response to the fear of being trapped oneself; the pressure for social reform and revolution may be animated by a sense of personal impasse and stagnation. Because movement is the quintessential youthful antidepressive maneuver, youthful depressions tend to be agitated, restless, jittery, and frantic, often accompanied by hypomanic ideation or activity. Today, with the ease of transportation, geographic motion—across the continent or around the world—is a favored and sometimes successful self-therapy. So, too, is rapid motion—the fast car, the racing motorcycle, the fleeting jet. And finally, the contemporary availability of a variety of drugs, from marijuana to amphetamines, which speed up experience or relieve the depressive flattening of affect and perception, provides yet another way of overcoming, at least temporarily, the loss of a sense of motion that is central to a feeling of vitality in youth.

Serious thoughts of *suicide*, suicidal gestures, and suicide attempts in youth are commonly related to these issues. Few young men and women pass through this stage of life without considering, whether distantly or concretely, the possibility of suicide. Serious suicidal thoughts or attempts are often animated by the fantasy that suicide will produce change, relieve an unbearable sense of being immobilized, and thus generate new movement. In working with severely depressed youths, it is, of course, necessary to explore the varied idiosyncratic meanings of suicide. But it is equally important to acknowledge that suicidal thoughts may be an (irrational) expression of the feeling that to remain "stuck" in one's present situation is untenable, that it is essential somehow to move. Indeed, the two most common causes of death among youth—accidents caused by speeding and suicide—must be understood not only as efforts to end life but as misguided attempts to regain a sense of inner momentum, and with it a feeling of "really being alive." If so, then it is the inner sense of already being psychologically dead that paradoxically impels many youths into activities that lead to physical death.

The special youthful potential for phenomenological estrangement also influences psychological disturbances during this stage. Especially for interpersonally isolated youths, estrangement may reach pathological proportions of acute *depersonalization and derealization*. The individual searches desperately for efforts to connect up with others, with himself, with meaningful groups, admired persons, or valued historical traditions. Even to the casual observer these efforts may seem frantic and driven not by affirmative developmental goals but by a growing sense of unreality and disconnection. The outcome of periods of acute estrangement depends mostly on the actual availability to the individual of others who are capable of understanding him, for the only lasting remedy for feelings of estrangement is the experience of real relationships. Those who are severely inhibited in their dealings with others therefore run a special danger of an acute sense of estrangement, sometimes culminating in periods of truly psychotic depersonalization.

The normal omnipotentiality of youth also may be incorporated into psychological disturbances. Driven by a fear of estrangement, a young man or woman may be impelled to *omnipotent fantasies of total transformation* of himself, others, or the world. Youth is capable of a special grandiosity that rests on the need to possess total control over one's destiny or over the fates of others. Adult recognition of limitations on personal ability and power tend to be absent from much youthful thought and planning. In working with youths who assume a kind of infantile omnipotence in dealings with the world, it is important not to undercut these fantasies too harshly, for these may be desperately needed to escape far more frightening feelings of estrangement, helplessness, and alienation.

In discussing ideology in youth, I emphasized the possibility of regression in ideological development. Clinicians are often too quick to apply pathological labels to youthful ideologies that seem strange, unfamiliar, or provocative. The search for a comprehensive world view may lead to ideas that seem bizarre, schizoid, or even psychotic to an outside observer. Such ideologies often bear little relationship to the "consensual reality" of most adults or to the "mature perspective" of most psychotherapists. Especially today, any attempt to define pathology in the realm of ideology must rest on a thorough knowledge of the prevailing variants of the youth culture. For example, an intelligent youth who in 1950 took astrology as a serious guide to life would have been considered psychotic; in the 1970's he is more likely to be simply a faithful member of the counterculture.

None of this should blind us to the existence of truly *disturbed ideology formation*, which may be part and parcel of bizarre, destructive, or self-destructive behavior. Just as the formation of ideology cannot be separated from other aspects of development during youth, so disorders in development, thought, and functioning are almost invariably reflected in ideology. Today the proliferation of exotic countercultural groups, cults, and communes allows many disturbed youths a more or less sanctioned ideological expression of their own conflicts, and at times a protected context where they can navigate an unusually troubled stage of development.

The greatest danger for clinicians who work with ideological youth, however, is the temptation to view ideology solely in its "psychological," expressive, and defensive role. It is indeed accurate to say that each youth "uses" the ideological alternatives of his world as an arena on which he can "act out" his developmental conflicts. But if we say no more than this, we neglect all the positive functions of ideology in individual as well as collective life. At an individual level the development of an ideology beyond dualism and conventional moralism makes possible, indeed demands, personal commitment to a world view that transcends the circumstances of childhood and of the previous generation. And as Erikson[27] has pointed out, ideology at a collective level can express and give meaning to widespread conflicts, common character traits, and a shared awareness of the disparity between the real and the ideal. Thus, the therapist who works with youth does well to recognize that, whatever the pathologies of ideology in a patient, the goal of therapy must not be the eradication of ideology, but support for those processes that will enable the individual to develop an ideology that both expresses his unique selfhood and unites him to the rest of mankind.

Perhaps the most common disturbance in youth lacks any diagnostic label: I will call it an *interpersonal imbroglio*. Characteristically involving at least three people, it fuses feelings of rivalry, identification, competition, symbiosis, sexuality, ambivalence, splitting, projection, and obsessionality in one embroiled relationship—all strongly influenced by oedipal fantasies and reenactments.

These interpersonal situations have a number of common features. They generally involve a core dyad locked in ambivalent intimacy and hostility, together with a third person who may be friend, rival, and competitor (or all of these things) of one or both of the core partners. All parties involved are likely to acknowledge that the situation is a "mess": the partners generally believe that they are temperamentally incompatible; the third party may be explicitly being "used" by both partners; or the couple itself may be organized primarily around shared fantasies about the "outsider," who is psychologically more important to them than their own relationship. A sense of doom hangs over such relationships, with everyone sensing that a rupture must eventually come. But even though the impossibility of the relationship is openly acknowledged, it tends to drag on interminably, with no one able to make (or sustain) a real break. Guilt, neediness, identification, genuine affection, fears of isolation, the hope of change, the fantasy of omnipotentiality, and a host of other factors are the glue that makes these triads so sticky.

When it is possible to explore in depth the fantasies that animate and are evoked by such imbroglios, oedipal themes predominate. A young woman who alternates between two lovers—one assertive and aggressive, the other quiet and sensitive—may be struggling to unify a split in her image of her father, all the while avoiding the guilt that would be aroused by a relationship with a "whole man." The young man who sustains an unhappy affair with a married woman often turns out to be far more preoccupied with his fantasies about the husband than with his lover. Finally these situations tend to be obsessionally preoccupying to everyone concerned, evoking and focusing conflicts at every psychological level from every stage of life. Sleepless nights, tears of pain, rage, and frustration, anxiety-laden reunions, and joyous if brief reconciliations are the stuff of which they are made.

Viewed developmentally, the interpersonal imbroglio can be a way station on the road to greater intimacy and mutuality. But so happy an outcome is never assured; and even at best, the participants tend to experience enormous anguish in the process. Because oedipal themes play so central a role in these relationships, those involved often persist in impossible relationships almost indefinitely, and if the relationship finally ends, they may compulsively seek out another equally entangled imbroglio. At times, however, working oneself out of such a relationship may also permit a working through of the inner conflicts enacted in it, enabling the individual to enter into less conflict-filled love relationships. Viewed several months or years later, then, such imbroglios (retrospectively recalled as "unhappy affairs") may be seen to have performed a positive role in freeing a youth from oedipal fears, inhibitions, and fantasies that stood in the way of any prolonged intimacy.

The most dramatic psychological disturbance in youth is the *acute psychotic break*. Psychotic episodes in youth often have a sudden onset in young men and women with no previous history of overt psychological disorder. A careful reconstruction of the events that led up to the psychotic break, however, often reveals a common pattern.[10,11] The beginning of disturbance dates from an experience of acute upheaval following a loss or separation. The individual reacts to the loss by isolating himself from anyone who might enable him to express or test his fantasies, ideas, and feelings. Upset and isolated, he becomes even more ideational; reads, writes, or introspects with extraordinary intensity; and is at times manically productive. As time passes, he experiences flights of ideas accompanied by a sense of "breakthrough"—of achieving extraordinary and often genuine insights into himself, others, or society. The sense of breakthrough, however, may lead to a loss of inner control: ideas come too fast, sleep is impossible, the mind races, the insights achieved cannot be verbalized. Often within a period of hours or days, there follows a more drastic sense of breakdown of previously established boundaries between self and world, inner and outer, conscious and unconscious, fantasy and reality. As these boundaries crumble, confusion mounts and the earlier sense of omnipotentiality, movement, breakthrough, and insight gives way to a feeling of confusion, disorientation, and even suicidal panic. The process may culminate in florid delusions, convincing hallucinations, and organized paranoid ideation.

An account of the prepsychotic experience of those who have acute schizophrenic episodes in youth is not an adequate account of the etiology of these disturbances.* But an effort to understand the immediate precursors of a psychotic episode may be vitally important in therapy, for the insights achieved in the prepsychotic period may contain important truths about the individual's problems and conflicts, however indirectly, symbolically, or incommunicably they were stated. The sense of "too much, too fast" in the prepsychotic phase may be a precise description of the experience of being overwhelmed by recollections and reexperiencings of episodes and fantasies that are crucial in his development. It is therefore important not simply to write off psychotic thoughts and behaviors as "crazy." If we study those many individuals for whom a "prepsychotic" phase does not culminate in a psychotic episode, one factor that differentiates them is their capacity to formulate, organize, and communicate the excitement and breakthroughs of the "prepsychotic" period. Even as the therapist explores the more distant antecedents of the psychotic episode, he should try to help the patient to unravel, decipher, and communicate whatever was valid in the psychotic experience itself.

Finally, no discussion of disorders in youth today can be complete without mentioning the interweaving of individual disturbances and youthful countercultures. As more and more of those between the ages of 18 and 30 enter this stage of life, we have witnessed the flowering of an enormous variety of youthful countercultures, often transient in ideology, membership, and existence. These countercultures, whose existence is rapidly disseminated by the mass media, provide a haven and at times a kind of therapy for many youths, including some whose development is highly disturbed. What Erikson[23,24] aptly termed a "psycho-social moratorium" has always been a characteristic of youths whose growth was uneven and troubled. But in the past such moratoriums tended to be individually negotiated and culturally unsanctioned (unless we consider higher education a form of moratorium, which it clearly still is for some). But today such moratoriums are increasingly collective rather than private. Countercultural groups and movements act as special magnets for two kinds of youths: those whose development is more "advanced" than that of their contemporaries, and those whose development requires a shorter or longer period of "regression," recapitulation, and reexperiencing, preferably in the company of others navigating a comparable developmental course.[25,36,49]

Contemporary countercultural groups tend to define themselves by opposition to or isolation from the surrounding culture. As a result they share many of the commonly observed tendencies of all small groups: encapsulation, lack of reality testing, rejection of outgroup members, the development of special characteristics of their youthful members, often producing at a group level behavior and ideology that would be considered pathological if it occurred at an individual level. To some outsiders many such groups appear to demand acceptance of a delusional group ideology, to encourage provocative, destructive, or danger-

* Research work with nonpatients suggests that the experiences I have termed "prepsychotic" are common among intellectual youth; that they are often precipitated by hallucinogenic drugs; and that they only occasionally lead to a truly psychotic episode.

ous behavior, and to promote pathological interpersonal relationships. Highly disturbed individuals are indeed drawn to such groups; but so, too, are youths who are in other contexts quite sane. The result may be a kind of group-induced "pathology," even in the absence of individual psychopathology in group members.

Immersion in a countercultural group obviously may be anything but helpful for some disturbed (or not so disturbed) youths. But for others these groups may in the long run serve an adaptive and developmental role. One study of young men and women involved in the East Village "hippie" community in the late 1960's[81] found that the average member stayed less than two years, that the majority eventually returned to their homes and reentered the educational process from which they had dropped out, and that only those who were highly disturbed or highly creative (or both) remained for longer periods. For the remainder the "hippie" world seems to have served as a collective psychosocial moratorium, a shared "developmental regression" that permitted them to work out and work through some of their inner conflicts. The rapid turnover in youthful communes, political groups, and the like suggests that these may perform a comparable developmental role for certain of their members.

I hope that these comments on the psychopathology of youth have made clear the fuzziness of the boundary between significant disorder and normal developmental turmoil during this stage of life. Much that would be considered ominous at other stages is routine and without prognostic implications during youth. At all stages of life, development proceeds from conflict, entails turmoil, and often requires the capacity to regress before moving forward. But during youth conflict, turmoil, and regression may last not for weeks or months but for years. Schizoid ideation, hypomanic locomotor restlessness, acute depression, an exaggerated sense of illumination, grandiose but unrealistic plans to change the world, serious suicidal thoughts, tormented interpersonal relationships—all are common. They themselves, their peers, and their elders do much harm to those in this phase of life by prematurely labeling as "pathological" routine developmental turmoil and retrogression. During youth, even more than during other stages, a step forward may have to be preceded by a step backward, and the absence of turmoil may be a symptom not of serenity but of foreclosure.

As a rule most disturbances during youth have a good prognosis; many and perhaps most work themselves out without special therapy. The study of youths who are not patients provides examples of young men and women whose self-prescribed "therapies," however remote from the traditional modalities of formal psychotherapy, prove highly successful. People were helping themselves and each other long before "psychotherapy" was invented; and they continue to do so, even today, in ways that often defy the explanatory power of our psychological theories.

Those who do work as therapists with troubled youths may therefore need certain special qualities: humility with regard to formal psychotherapy as the preferred modality of treatment; respect for the ultimate integrity of the patient, however disturbed his behavior may be; a capacity to focus on the stage-specific ingredients in the patient's disturbance along with its more distant etiology; a willingness to be fallible, direct, and honest with the patient; and, above all, an absence of any illusions of omniscience or omnipotence. Whatever the therapist's own ideological persuasion and techniques, it may be that he helps his youthful patients most by providing them with a real and honest human relationship with another person who has more or less successfully weathered the turmoil of youth.

What Youth Is Not

A final way to clarify the meaning of youth as a stage of life is to make clear what it is not. For one thing youth is not the end of development. I have described the belief that it is—the conviction that beyond youth lie only stasis, decline, foreclosure, and death—as a characteristically youthful way of viewing development, consistent with the observation that it is impossible truly to understand stages of development beyond one's own. On the contrary, youth is but a preface for further transformations that may (or may not) occur in later life. Many of these center around such issues as the relationship to work and to the next generation. In youth the question of vocation is crucial, but the issue of work—of productivity, creativity, and the more general sense of fruitfulness that Erikson calls generativity—awaits adulthood. The youthful attainment of mutuality with peers and of peerhood with elders can lead to further adult interpersonal developments by which one comes to be able to accept the dependency of others, as in parenthood. In later life, too, the relations between the generations are reversed, with the younger now assuming responsibility for the elder. Like all stages of life, youth is transitional. And although some lines of development, such as moral development, may be "completed" during youth, many others continue throughout adulthood.

It is also a mistake to identify youth with any one social group, role, class, organization, or position in society. Youth is a *psychological* stage; and those who are in this stage do not necessarily join together in identifiable groups, nor do they share a common social position. Not all college students, for example, are in this stage of life:[70] some students are psychological adolescents, while others are young adults—essentially apprentices to the existing society. Nor can the experience of youth as a stage of life be identified with any one class, nation, or other social grouping. Affluence and education can provide a freedom from economic need and an intellectual stimulation that may underlie and promote the transformations of youth. But there are poor and uneducated young men and women, from Abraham Lincoln to Malcolm X, who have had a youth, and rich, educated ones who have moved straightaway from adolescence to adulthood. And although the experience of youth is probably more likely to occur in the economically advanced nations, some of the factors that facilitate youth also exist in the less advanced nations, where comparable youthful issues and transformations are expressed in different cultural idioms.

Nor should youth be identified with the rejection of the status quo, or specifically with student radicalism. Indeed, anyone who has more or less definitively defined himself as a misanthrope or a revolutionary has moved beyond youthful probing into an "adult" commitment to a position vis-à-vis society. To repeat: what characterizes youth is not a definitive rejection of the existing "system," but an ambivalent tension over the relationship between self and society. This tension may take the form of avid efforts at self-reform that spring from acceptance of the status quo, coupled with a sense of one's own inadequacy vis-à-vis it. In youth the relationship between self and society is indeed problematical, but rejection of the existing society is not a necessary characteristic of youth.

Youth obviously cannot be equated with any particular age range. In practice most young Americans who enter this stage of life tend to be between the ages of 18 and 30. But they constitute a minority of the whole age-grade. Youth as a developmental stage is emergent; it is an "optional" stage, not a universal one. If we take Kohlberg's studies of the development of postconventional moral reasoning as a rough index of the "incidence" of youth, less than 40 per cent of middle-class (college-educated) men, and a smaller proportion of working-class men, have developed beyond the conventional level by the age of 24. Thus, "youths" constitute but a minority of their age group. But today those who are in this stage of life largely determine the public image of their generation.

Admirers and romanticizers of youth tend to identify youth with virtue, morality, and mental health.[9,76,79] But to do so is to overlook the special youthful possibilities for viciousness, immorality, and psychopathology. Every time of human life, each level of development, has its characteristic vices and weaknesses, and youth is no exception. Youth is a stage, for example, when the potentials for zealotry and fanaticism, for reckless action in the name of the highest principles, for self-absorption, and for special arrogance are all at a peak. Furthermore, the fact that youth is a time of psychological change also means inevitably that it is a stage of constant recapitulation, reenactment, and reworking of the past. This reworking can rarely occur without real regression, whereby the buried past is reexperienced as present and, one hopes, incorporated into it. Most youthful transformation occurs *through* brief or prolonged regression, which, however benignly it may eventually be resolved, constitutes part of the psychopathology of youth. And the special compulsions and inner states of youth—the euphoria of omnipotentiality and the dysphoria of estrangement, the hyperconsciousness of consciousness, the need for constant motion, and the terror of stasis—may generate youthful pathologies with a special virulence and obstinacy. In one sense those who have the luxury of a youth may be said to be "more developed" than those who do not have (or do not take) this opportunity. But no level of development and no stage of life should be identified either with virtue or with health.

Finally youth is not the same as the adoption of youthful causes, fashions, rhetoric, or postures. Especially in a time like our own, when youthful behavior is watched with ambivalent fascination by adults, the positions of youth become part of the cultural stock-in-trade. There thus develops the phenomenon of *pseudoyouth*: preadolescents, adolescents, and frustrated adults masquerade as youths, adopt youthful manners, and disguise (even to themselves) their real concerns by the use of youthful rhetoric. Many a contemporary adolescent, whether of college or high school age, finds it convenient to displace and express his battles with his parents in a pseudoyouthful railing at the injustices, oppression, and hypocrisy of the Establishment. And many an adult, unable to accept his years, may adopt pseudoyouthful postures to express the despairs of his adulthood.

To differentiate between "real" and pseudoyouth is a tricky, subtle, and unrewarding enterprise. For, as I have earlier emphasized, the concept of youth as defined here is an ideal type, an abstraction from the concrete experience of many different individuals. Furthermore, given the unevenness of human development and the persistence throughout life of active remnants of earlier developmental levels, conflicts, and stages, no one can ever be said to be completely "in" one stage of life in all areas of behavior and at all times. No issue can ever be said to be finally "resolved"; no earlier conflict is completely "overcome." Any real person, even though on balance we may consider him a "youth," will also contain some persistent childishness, some not outgrown adolescence, and some precocious adulthood in his makeup. All we can say is that, for some, adolescent themes and levels of development are *relatively* outgrown, while adult concerns have not yet assumed full prominence. It is such people whom one might term "youths."

¶ Social and Historical Implications

If the hypotheses stated in this essay are correct, we have experienced and will continue to witness a growth in the proportion of young men and women in their late teens and twenties who pass through that postadolescent stage I have called youth. If this be true, it will become the more important to recognize the differences between adolescence, youth, and early adulthood, viewing youth as neither prolonged adolescents nor young adults who have not quite been able to navigate the passage to adulthood. Clearly these comments on youth will require further qualification and elaboration. We must examine in greater detail the way sociohistorical change enables more young people to enter a period of youth, yet at the same time defines for them the alternatives available during that stage of life. We must learn to identify the factors that make each of the characteristic transformations of youth more or less probable. And we must learn to understand and cope better with the hazards, disturbances, and pathologies that inhere in the possibility of youth.

The emergence of an ever larger proportion of young people who exhibit the characteristics of youth will also have important social, cultural, and political implications. I earlier suggested that throughout human history the great majority of men and women have been concrete in their cognitive functioning, conventional in their moral judgments, and dualistic in their views of truth. It can be argued that cultural, social, and political stability, along with the secure transmission of culture from one generation to the next, has required these psychological characteristics. If so, then the emergence of an ever larger group of nondualistic, postrelativistic, postconventional youths will have major effects upon all of those institutions, from the family to the churches, from the political system to the schools, that have traditionally been charged with social stability and the transmission of culture.

Similarly that "cake of custom" that has, at a psychological level, bound societies together and insured their continuity may today be crumbling, with implications we begin to see in the sense of social and cultural malaise in all the industrially advanced nations. Psychological research suggests that, however widespread the ideal of the autonomous man and woman, this ideal has rarely if ever been historically actualized on a mass scale. It may be that modern conditions—the increase in cross-cultural contact, the age of affluence, the impact of the new media, the prolongation of education, the deferral of adult responsibilities, the rapidity of social change, and so on—are creating an ever larger minority (perhaps eventually a majority) of autonomous men and women. A society of such men and women would be vastly different from any we have known. It may not be too early to begin imagining how it could be a better society than ours.

Finally one established principle of cognitive psychology is that all individuals find it impossible to comprehend the mental processes of those whose cognitive development is much advanced beyond their own. For example, the conventional, law-and-order moralist described by Kohlberg finds it impossible really to understand autonomous, principled moral reasoning. Instead, he has no choice but to identify principled reasoning with the preconventional, egocentric, "anything goes" morality that he outgrew as a child. A similar principle appears to hold in other sectors of development. For example, the concept of interpersonal mutuality, defined as care for another person because of his or her irreplaceable individuality, tends to be incomprehensible to people who can care only for those whom they consider like themselves. Those who are related to their parents through bonds of authoritarian submission will tend to confuse a three-dimensional relationship to parents with childish rebellion and disrespect.

The general principle that higher levels of development are incomprehensible to those who have not experienced them has important implications for therapists, as for others who deal with youth. It is, of course, possible to understand empathically many aspects of another's experience that one has not experienced oneself: a therapist need not have been psychotic himself in order to understand a patient who has been or is. Furthermore, the common complaint of youths that their therapists do not understand them is often a true transference reaction that must be traced back to earlier feelings about their parents.

But the possibility remains that, just as able and complex high school students are often misunderstood by teachers less able and less complex than themselves, so some youthful patients may encounter therapists who misunderstand their behavior and thought processes, confusing higher levels of functioning with regressions. Psychotherapeutic training armors the therapist with powerful and often accurate ways of explaining away the patient's objections to his own interventions. In the end, however, the possibility remains that the patient may be right—that the therapist may not only have failed to understand but be incapable of understanding. There is no sure-fire safeguard against this possibility. But we can at least hope that therapists themselves will be relatively "developed" human beings, that they will not automatically equate their inability to understand with pathology in their patients, and that they will view the therapeutic process not only as a way of helping someone else to grow but as a human experience in which they, too, may develop.

Bibliography

1. ADELSON, J., "The Mystique of Adolescence," *Psychiatry, 1*:1, 1964.
2. APTER, D. E. (Ed.), *Ideology and Discontent*, Free Press, New York, 1964.
3. ARIÈS, P., *Centuries of Childhood: A Social History of Family Life*, Random House, New York, 1962.
4. BERGER, B. M., "The New Stage of American Man—Almost Endless Adolescence," *New York Times Magazine*, November 2, 1969.
5. BERNFELD, S., "Types of Adolescence," *Psychoanal. Quart.*, 7:243, 1938.
6. ———, "Über eine typische Form der männlichen Pubertät," *Imago*, 9, 1923.
7. BION, W. R., *Experiences in Groups*, Basic Books, New York, 1961.
8. BLOS, P., *On Adolescence*, Free Press, New York, 1962.
9. BOURNE, R.
10. BOWERS, M. B., "The Onset of Psychosis: A Diary Account," *Psychiatry*, 28:346, 1965.
11. ———, and FREEDMAN, D. X., "Psychedelic Experiences in Acute Psychoses," *Arch. Gen. Psychiat.*, 15:240–248, 1960.
12. BOWLBY, J., *Maternal Care and Mental Health*, World Health Organization, Geneva, 1951.
13. BRUNER, J. S., "A Psychologist's Viewpoint: Review of Inhelder and Piaget, 'The Growth of Logical Thinking,'" *Brit. J. Psychol.*, 50:363–370, 1959.
14. ———, OLVER, R. S., and GREENFIELD, P. M., *Studies in Cognitive Growth*, Chapters 11–14, John Wiley, New York, 1967.
15. DAVIES, A. F., *Private Politics: A Study of Five Political Outlooks*, Melbourne University Press, Melbourne, 1966.
16. DAVIS, N. Z., "The Reasons of Misrule: Youth Groups and Charivaris in Sixteenth-Century France," paper presented at Sir George William University, March 5, 1970.
17. DEMOS, J., and DEMOS, V., "Adolescence in Historical Perspective," *J. Marriage & Fam.*, 31:632–638, 1969.
18. DEUTSCH, H., *The Psychology of Women*, Vol. 1, Grune & Stratton, New York, 1944.
19. DOUVAN, E., and ADELSON, J., *The Adolescent Experience*, John Wiley, New York, 1966.
20. EISENSTADT, S. N., *From Generation to Generation*, Free Press, New York, 1956.
21. ERIKSON, E., *Childhood and Society*, 2nd ed., Norton, New York, 1963.
22. ———, *Gandhi's Truth: On the Origins of Militant Non-Violence*, Norton, New York, 1969.
23. ———, "Identity and the Life Cycle," *Psychol. Issues*, 1, 1959.
24. ———, *Identity: Youth and Crisis*, Norton, New York, 1968.
25. ———, *Insight and Responsibility*, Norton, New York, 1964.
26. ———, "Reflections on the Dissent of Contemporary Youth," *Daedalus*, 99:154–176, 1970.
27. ———, *Young Man Luther: A Study in Psychoanalysis and History*, Norton, New York, 1958.
28. FISHKIN, J., KENISTON, K., and MACKINNON, C., "Moral Reasoning and Political Ideology," *J. Pers. & Soc. Psychol.*, 27:109–119, 1973.
29. FREUD, A., "Adolescence," in *The Psychoanalytic Study of the Child*, Vol. 13, International Universities Press, New York, 1958.
30. ———, *The Ego and the Mechanisms of Defense*, International Universities Press, New York, 1946.
31. ———, *Normality and Pathology in Childhood: Assessments of Development*, International Universities Press, New York, 1965.
32. FREUD, S., "Three Essays on the Theory of Sexuality," in Strachey, J. (Ed.), *Standard Edition, Complete Psychological Works*, Vol. 7, Hogarth, London, 1953.
33. FEUER, L. S., *The Conflict of Generations*, Basic Books, New York, 1969.
34. FRIEDENBERG, E. Z., *The Vanishing Adolescent*, Dell, New York, 1962.
35. GREENFIELD, P. M., and BRUNER, J. S., "Culture and Cognitive Growth," in Goslin, D. A. (Ed.), *Handbook of Socialization Theory and Research*, Rand McNally, Chicago, 1969.

36. HAAN, N. BREWSTER SMITH, M., and BLOCK, J. H., "Moral Reasoning of Young Adults: Political-Social Behavior, Family Backgrounds, and Personality Correlates," *J. Pers. & Soc. Psychol., 10*:183–201, 1968.
37. HALL, G. S., *Adolescence: Its Psychology, and Its Relations to Psysiology, Anthropology, Sociology, Sex, Crime, Religion, and Education*, D. Appleton & Co., New York, 1904.
38. HARRIS, D. B. (Ed.), *The Concept of Development: An Issue in the Study of Human Behavior*, University of Minnesota Press, Minneapolis, 1957.
39. HAUSER, S. T., *Black and White Identity Formation*, Wiley-Interscience, New York, 1971.
40. HUNT, D., *Parents and Children in History: The Psychology of Family Life in Early Modern France*, Basic Books, New York, 1970.
41. INHELDER, B., "Some Aspects of Piaget's Approach to Cognition," *Monog. Soc. Res. Child. Dev., 27*:19–33, 1962.
42. ———, and PIAGET, J., *The Growth of Logical Thinking from Childhood to Adolescence*, (Tr. by Parsons, A., and Milgram, S.), Basic Books, New York, 1958.
43. JAMES, W., *The Varieties of Religious Experience*, Longmans, Green, New York, 1902.
44. JUNG, C. G., *Psychological Types: The Psychology of Individuation*, Random House, New York, 1962.
45. KARDINER, A., *The Psychological Frontiers of Society*, Columbia University Press, New York, 1945.
46. KATZ, J., *et al., No Time for Youth: Growth and Constraint in College Students*, Jossey-Bass, San Francisco, 1968.
47. KENISTON, K., "Postadolescence (Youth) and Historical Change," in Zubin, J., and Freeman, A. (Eds.), *The Psychopathology of Adolescence*, Grune & Stratton, New York, 1970.
48. ———, "Psychological Development and Historical Change," *J. Interdisciplinary Hist., 2*:329–345, 1971.
49. ———, "Student Activism, Moral Development and Morality," *Am. J. Orthopsychiat., 40*:577–592, 1970.
50. ———, *The Uncommitted: Alienated Youth in American Society*, Harcourt, Brace & World, New York, 1965.
51. ———, *Young Radicals: Notes on Committed Youth*, Harcourt, Brace & World, New York, 1968.
52. ———, "Youth as a Stage of Life," *Am. Scholar, 39*:631–654, 1970.
53. KOHLBERG, L., "The Child as a Moral Philosopher," *Psychology Today*, 25–30, 1968.
54. ———, "The Concept of Moral Maturity," paper read at NICHD Conference on the Development of Values, Washington, D.C., May 15, 1968.
55. ———, "The Development of Children's Orientations towards a Moral Order I: Sequence in the Development of Moral Thought," *Vita Humana, 6*:11–33, 1963.
56. ———, "Education for Justice: A Modern Statement of the Platonic View," Burton Lecture on Moral Education, Harvard University, Cambridge, 1968.
57. ———, "The Relations between Moral Judgment and Moral Action: A Developmental View," paper read at Institute of Human Development, Berkeley, 1969.
58. ———, "Stage and Sequence: The Cognitive-Developmental Approach to Socialization," in Goslin, D. A. (Ed.), *Handbook of Socialization Theory and Research*, Rand McNally, Chicago, 1969.
59. ———, and GILLIGAN, C., "The Adolescent as a Philosopher: The Discovery of the Self in a Postconventional World," *Daedalus, 100*:1051–1086, 1971.
60. ———, and KRAMER, R., "Continuities and Discontinuities in Childhood and Adult Moral Development," *Human Dev., 12*: 93–120, 1969.
61. KRAMER, R., "Changes in Moral Judgment Response Pattern during Late Adolescence and Young Adulthood," Ph.D. diss., University of Chicago, 1968.
62. KREBS, R. L., "Some Relationships between Moral Judgment, Attention and Resistance to Temptation," Ph.D. diss., University of Chicago, 1967.
63. KUHN, D., LANGER, J., and KOHLBERG, L., "The Development of Formal-Operational Thought: Its Relation to Moral Development," unpublished paper, Harvard University, 1971.
64. LERNER, M., "Personal Politics," Ph.D. diss., Yale University, 1970.
65. LIDZ, T., "Egocentric Cognitive Regression and a Theory of Schizophrenia," paper

presented at the Fifth World Congress of Psychiatry, Mexico City, December 1971.

66. LOEVINGER, J., "The Meaning and Measurement of Ego Development," *Am. Psychol.*, *21*:195–206, 1966.
67. MANNHEIM, K., *Ideology and Utopia*, Harcourt Brace, New York, 1936.
68. MYRDAL, G., *An American Dilemma*, Harper & Row, New York, 1944.
69. OFFER, D., *Psychological World of the Teenager: A Study of Normal Adolescent Boys*, Basic Books, New York, 1969.
70. PARSONS, T., and PLATT, G. M., "Higher Education, Changing Socialization," in Riley, M. W., Johnson, M. E., and Foner, A. (Eds.), *A Sociology of Age Stratification*, Volume III, *Of Aging and Society*, pp. 236–291, Russell Sage Foundation, New York, 1972.
71. PERRY, W. G., JR., *Forms of Intellectual and Ethical Development in the College Years*, Holt, Rinehart and Winston, New York, 1970.
72. ———, "Patterns of Development in Thought and Values of Students in a Liberal Arts College," Harvard University Bureau of Study Counsel, Cambridge, 1968.
73. PRELINGER, E., Personal Communication, 1969.
74. PROVENCE, S., and LIPTON, R. C., *Infants in Institutions*, International Universities Press, New York, 1962.
75. PUMPIAN-MINDLIN, E., "Omnipotentiality, Youth and Commitment," *J. Am. Acad. Child Psychiat.*, *4*:1–18, 1965.
76. REICH, C. A., *The Greening of America*, Random House, New York, 1970.
77. REST, J., "Developmental Hierarchy in Preference and Comprehension of Moral Judgment," Ph.D. diss., University of Chicago, 1968.
78. ———, TURIEL, E., and KOHLBERG, L., "Level of Moral Development as a Determinant of Preference and Comprehension of Moral Judgments Made by Others," *J. Pers. & Soc. Psychol.*, *37*:225–252, 1969.
79. ROSZAK, T., *The Making of a Counterculture*, Doubleday, New York, 1969.
80. SHAPIRO, R. L., "Adolescence and the Psychology of the Ego," *Psychiatry*, *26*:77–87, 1963.
81. SILVERSTEIN, H., "Study of East Village Youth," unpublished ms., prepared for Joint Commission of Mental Health of Children, 1968.
82. SLATER, P., *Microcosm: Structural, Psychological and Religious Evolution in Groups*, John Wiley, New York, 1966.
83. SPITZ, R., "Hospitalism," in *The Psychoanalytic Study of the Child*, Vol. 1, pp. 53–75, International Universities Press, New York, 1945.
84. SULLIVAN, H. S., *The Interpersonal Theory of Psychiatry*, Norton, New York, 1953.
85. WHITE, R .W., *Lives in Progress*, Dryden Press, New York, 1952.

CHAPTER 20

OCCUPATIONAL DIFFICULTIES IN ADULTHOOD

Cyril Sofer

Characteristics of Work in Advanced Industrial Societies

THE occupational system of advanced industrial societies has certain structural features that appreciably distinguish the work experiences of most persons from those of persons in other types of society. These structural features present the worker with particular types of opportunities. At the same time they confront him with particular types of problems or constraints that affect the quality of his everyday work experiences and relations and the way in which his work fits in with the rest of his personal life and social experience. This chapter is concerned with the social structural sources of strain deriving from occupations.

Work is physically segregated from the home or local residential area. It takes place in a designated workplace to which the worker has to travel because of the concentration there of machines and materials juxtaposed with each other in a planned set of arrangements; an appreciable part of his lifetime (for which no remuneration is received) may be consumed in the journey to work. There is typically little privacy in the work situation; the worker performs before a small "public" and is usually under continuous or intermittent surveillance designed to insure that he meets standards set by his superiors in the quantity and quality of his output. The work is subject to constraints of place, time, and standards set by others. The expression of strong emotion is discouraged.[53] Work is a social activity in which the worker typically interacts with peers, superiors, and subordinates. These exert different types of pressure on him, sometimes conflicting with each other, and he has to manage these in a way that makes his behavior acceptable to all parties. He will be subject to particular pressures to subscribe to the norms of his immediate colleagues in regard to output and to an ideology that attaches special value to their contribution to their employing organization and to society. He must cope with the stresses of dependence and interpersonal conflict, while maintaining enough of his idiosyncratic approach to life to retain his sense of personal identity.

Most work takes place in large bureaucratic organizations. This means that it is done in the name of objectives not necessarily identical with the personal aims of employees; rationalistic means are used to achieve goals; persons are treated as resources available for pursuit of those goals; and subgoals are differentiated into specialized tasks to be performed by specified departments. Tasks are assigned from the point of view of overall efficiency, not from that of the preferences of employees; creativity and innovation are the prerogatives of seniors; rewards tend to be of an extrinsic type (money, space, comfort), rather than being intrinsic to the performance of the task itself, and are related to observed contribution to the overall "organizational" task. Bureaucratic systems of organization facilitate the performance of large-scale tasks in mass societies, yield economic and social benefits from specialization and coordination, allocate work to employees systematically on the basis of qualifications, experience, and competence, and help to insure that favoritism and discrimination are minimized. They provide opportunities to many for self-fulfillment in an orderly and progressive career: for colleagueship, training, advice, and mutual aid. But the ethic of rationalism places the individual at the disposal of his employers, whose requirements may well shift during his work life at points not coincidental with his own needs as these develop.

It is widely assumed or suggested that bureaucracy suppresses individuality. This view has been challenged by recent studies of 3,100 men representative of employees throughout the United States working in similar jobs.[37,38] Taking organizational size and the number of formal levels of supervision as the main indices of bureaucracy, the results show that those in the more bureaucratic organizations were more likely to be open-minded and receptive to change and tended to value self-direction rather than conformity. They spent their leisure in more intellectually demanding activities. These findings applied to both the public and private spheres of employment and to blue-collar as well as white-collar workers. The researcher attributes the results to three of the occupational effects of bureaucracy: the greater job security of persons working in such organizations, their higher incomes compared to men of similar educational levels, and the more complex jobs they do. At the same time he notes that bureaucracies provide especially close supervision of staff and that, having hired educated men and given them complex jobs to perform, they may fail to give them as much self-direction as their educational attainments and the needs of the work allow.

When blue-collar workers are considered on their own, smaller subunits are characterized by higher job satisfaction, lower absence rates, lower turnover rates, and fewer labor disputes. The explanation for this is probably that as organizations increase in size and tasks become more specialized, it is more difficult to maintain high cohesiveness or a level of communication felt as satisfactory.[58] These findings do not necessarily imply that largeness of the overall organization in itself determines level of satisfaction, since one can assume, at least in some technologies, that the size of work groups can vary independently of this.

Workers do not choose randomly with respect to the type of firm they join. Those who have a very strong economic orientation have been shown to work for very large companies, and those who emphasize noneconomic factors select the smaller ones. This type of differential selection leads to similar turnover rates in the two types of firms.[32] A separate study of auto workers reports that the men deliberately gave priority to high economic returns at the understood cost of otherwise unrewarding and undemanding work. The majority said they would not be bothered by job changes that would move them away from their current colleagues. Nor did they look to their supervisors for social and psychological support in their work roles.[24]

Large, complex organizations appear to have certain typical forms of psychological stress associated with their functioning. Although most persons, asked whether they would continue to work if this were not financially necessary for them, say they would continue to do so, those in large organizations are especially prone to say they would prefer to do so in a different job or to express significant reservations about what they are doing. Two major stress-producing conditions appear to be the role conflict and role ambiguity, and these appear largely as an outcome of employees' being required to cross organizational boundaries, to produce innovative solutions to nonroutine problems, and to take responsibility for the work of others. No doubt processes of social selection take place that increase the likelihood that persons who can handle or who enjoy conflict and ambiguity fill roles where these conditions are especially likely to occur. But however they are handled, the endemic nature of these stressful conditions is incontrovertible. In a U.S. national survey of 725 persons representing that part of the labor force employed during the spring of 1961, about half the respondents reported that they were caught in the middle between two conflicting persons or factions: in 88 per cent of these cases at least one of the persons involved was an organizational superior of the respondent. Almost half of all respondents reported work overload, that is, conflict among tasks or problems in setting priorities. Four forms of ambiguity were reported as particularly troublesome: uncertainty about the way one's superior evaluates one's work, about opportunities for advancement, about scope for responsibility, and about the expectations of others regarding one's performance. The difficulties people had with their organizational roles increased as conflict and ambiguity increased.[35] On the other hand, ambiguities in such areas can be worked into defensive attitudes and ideologies that function to protect self-esteem.[50,64] In the study just described, personality dimensions mediated significantly the degree to which a given intensity of objective (situational) conflict was experienced as a strain by the respondent. The effects are also mediated by the helpfulness or otherwise of colleague relations. The authors distinguished between core problems located principally in the objective environment (on which they concentrated): mismatches between the requirements of a role and the capacities of the occupant; and difficulties that are primarily intrapsychic but are acted out in the work environment. But all these intermesh, as core problems are succeeded or elaborated by derivative problems, that is, those created by the individual's attempts to cope with the core problem. Tensions arising from difficulties in work roles are not necessarily expressed wholly in those roles but may manifest themselves elsewhere in the total array of roles we all fill—as husband, father; sibling, friend, or citizen.

Intergroup conflict is a standing feature of work situations in contemporary society. Organizations are arenas of constant struggle between individuals and groups for status, power, resources, and higher shares of the collective output, in the course of which the parties seek to overcome, injure, neutralize, or eliminate their rivals. In relations between managements and workers, strikes are the most conspicuous examples of such conflicts, but they appear also in such less organized forms as work limitation, slowdowns, waste, labor turnover, absenteeism or, on the management side, overstrict discipline and discriminatory dismissals. While vertical conflicts between managements and workers arouse most public attention, this type of opposition is only one of many. What we conventionally call management is divided into a variety of subcategories and subgroups whose interests, outlook, and aspirations differ from and compete with each other. Nor are employees a homogeneous general class. They are divided into many varieties of occupation that struggle against each other for rewards, recognition, and precedence, just as many of them struggle against management. In the large, modern, complex organizations a particularly prominent form of intergroup conflict is connected with division into groups of colleagues each responsible for a share of the work: the development of occupational identities based on life investment plays a key part in the acuteness of such conflicts. Conflict is typically contained in the work organizations of advanced industrial countries, rather than bursting out in its more destructive forms, because the parties usually have other overriding interests in common and because lines of cleavance are multiple, so that it is not in the interests of the parties to press any one division to its logical end.

Conflict can confer benefits on members of organizations and on those whom they serve by facilitating change, enabling alliances to be formed, providing groups with a sense of identity, bringing problems to the attention of higher authorities, clarifying expectation, correcting imbalances in reciprocity and power, mobilizing energy, clearing the air of simmering trouble, and allowing hostile impulses to be diverted outward. From a wide, societal point of view, conflict is not a pathological phenomenon to be understood as the desperate efforts of a deprived and discontented people, but a normal aspect of "antagonistic cooperation"* in a competitive society in which groups cooperate but at the same time use their power to influence the outcomes of that cooperation to their greater benefit.

At the same time conflict confronts groups and individuals with breakdowns in their capacity to cooperate and to control their impulses, substitutes warfare as a motive or basis for association, blunts judgment about the behavior of oneself and others, encourages unrealistic beliefs and ideologies, reinforces self-doubt, provides a rationale for dishonesty or opportunism, and results for some in loss of power, esteem, income, or even employment. Even though structural forces and social processes provide the foundations or causes of conflicts, these tend to become personalized in the form of criticisms of the integrity, competence, and motivations of the individuals actively concerned. Since conflict is endemic in social life and inescapable in the organizational and occupational affairs of complex economic systems, it is a source of constant or intermittent stress in the work experience of most adults.

The individual's relationship with his organization appears partly to be replacing or displacing personal contacts.[41] For many people that relationship comes to constitute one of the main sources of continuity in their life experience: while this can be argued to confer meaning and identity, there is clearly also a risk that people will become unduly dependent on their employing organizations and will play out on the organizational stage the strong emotions otherwise reserved for more private social relations.

The nature of work has changed appreciably during the past few decades. Physical tasks are increasingly being relegated to machines and physical labor is being displaced by mental effort. The occupational skills of a person may have to change often during his lifetime or may become altogether outdated.[3,63] Education and training, either formal or informal, are becoming continuing processes or matters of repeated injection/application as work processes and systems change. Employers have become active in the educational and training process. Large employers are involved in a process of secondary socialization of their members: personality and identity are not formed once and for all in childhood, and at least their external manifestations continue to be actively reshaped in adulthood, largely through the deliberate efforts of employers.

With the rapid introduction of new technologies, certain older occupations disappear and new ones emerge, for example, those of systems analyst and computer programmer. This process reduces certainty in occupational planning, introducing new occupational routes less standardized or predictable than the old and increasing the diversity of alternatives. New dangers and risks in occupational investments accompany the creation of new opportunities.

The bureaucratization of work and the rapid introduction of new occupations and techniques have contributed to changes (and possibly tensions) in the relations of younger persons to older. The career arrangements of bureaucracies place the young (even those with the latest technical skills) at the mercy of the older,[64,69] putting the power to set norms and make decisions into the hands of the middle-aged.

Occupational roles have attached differ-

* This phrase derives from W. G. Sumner.

ences in social status, with professional workers and senior businessmen ranking at the top and unskilled manual occupations at the bottom. The higher the level of the occupation, the higher the morale of the person.[58]

It is widely thought desirable to try to move up the occupational ladder, either during one's own lifetime (which is highly unusual) or from one's father's occupational status, and to help one's children move up from one's own occupation. Persons in a wide variety of occupational roles would prefer higher-status roles or may have aspired to higher-status roles that they have failed to achieve. Adaptation to such a situation, in the form of scaling down of ambition, can be expected in a large proportion of cases, partly because of the prevalence of this condition and the inevitable development of defensive ideologies that help to rationalize lack of achievement. But equally one can expect some less socially and personally "adjustive" reactions, such as deviant ways of gaining goals, retreat, or rebellion.[48]

When comparisons were made of the mental health of men at different occupational levels, from white-collar workers through skilled workers, semiskilled workers, and low-level production workers, mental health was found to vary consistently with level, with more problems reported at the lower levels.[39] It is not altogether clear, however, how far this is a direct result of factory work (and therefore possibly transient) or how far a consequence of prejob characteristics, such as low education and downward mobility, or of such factors as low income and poor housing.[44]

Differences in Occupational Cultures

Probably the most important dividing line between social categories in Western society today is that between blue-collar workers and white-collar workers.[42] These labels connote profound differences not only in occupational experience but also more generally in the quality of one's personal and social life, the opportunities one is likely to be able to make available to one's children, the values to which one adheres (including those involved in bringing up children), and the model of society that one holds.

The major factors that determine which occupational category a person will enter are the social class (largely occupational class) of his parents, the number of years of formal education he has had, the size of his family, and the attitudes of the family toward education. Of these factors the strongest is one's own education, although the effects on one's progress of both education and social origin decrease in importance over time relatively to what has happened in one's career. Important handicaps to career advance are having a broken family (which affects both the husband and male offspring), having many siblings, or coming from a family not favorably disposed toward education.[8]

Where mobility (either upward or downward) is appreciable, the mobile person experiences problems in his interpersonal relations and social identifications.[7,75] Mobile persons are not liable to be closely integrated with persons either in their former or present social class in regard to life style and reciprocal influence. Mobile persons are more apt to be prejudiced against minorities and more apt to be preoccupied with their health; these attitudes may reasonably be interpreted as evidence of felt insecurity.

Intergenerational mobility is apt to create sharper discontinuities, particularly between the type of culture experienced in childhood and adulthood. Because of the increased importance of professionalism, educational selection divides children into separate educational streams with different occupational potentialities and consigns them to different social fates and life styles. When the children of professionals are not successful enough in their education to achieve the occupational level of their parents, participation in a different culture, possibly disvalued and probably more localized, is their likely social destiny.[73]

Blue-Collar Workers

Operatives

Persons who enter blue-collar occupations have not normally had a great deal of systematic vocational guidance and are not well equipped with information to help in their occupational choices, for instance, in estimating the probable fate of various industries. They tend to enter occupations yielding relatively high short-term earnings but having relatively little opportunity for further learning or scope for advancement. This is particularly important in view of the binding character of most early occupational commitments. Many of the requisite skills are acquired in the course of the work itself and are derived from immediate work colleagues: allegiance to the work group and adherence to group norms and standards of output become a necessary condition for learning and for pleasant or tolerable social relations while at work.

A good deal of the work of persons in this sector is simple, monotonous, or boring and may consist of repetitive operations on physical objects. It may offer little scope for affirming personal competence (in a way that gives the worker feedback on his effectiveness); tasks may not be closely related to self-identity; and there may be little scope for achievement in the sense of recognition by others of the worth of one's contribution.

The context of the work may involve isolation from what is felt to be the heart of the organizational operation, a struggle to maintain or increase autonomy or control over working conditions, and effort to maintain and gain power vis-à-vis other occupational groups and particularly vis-à-vis managerial groups.

In an early industrial research classic, 180 representative workers were interviewed in their homes about their employment in one of the most modern automobile assembly lines in the world.[71] Ninety percent of them expressed a dislike of the many jobs in the plant characterized by mechanical pacing and repetitiveness. A large number wanted jobs that lacked these characteristics. The mass production characteristic most disliked was mechanical pacing. Workers in unskilled jobs tended to devaluate themselves. The work often demanded "surface mental attention"; that is, it required a high degree of attention but was not intellectually demanding, thus being conducive to boredom. The research workers found that work arrangements denied to all but a few opportunities for a team relationship with other workers. Ability to interact was also limited by noise, speed of the line, and the amount of physical energy demanded.

The combination of the content and context of manual work is considered often to be alienating in that working conditions may foster powerlessness (the feeling that one is an object controlled and manipulated by others or by an impersonal system); self-estrangement (occupation is not experienced as constituting personal identity in an affirmative way); isolation (the feeling of being in, but not of, society, a sense of remoteness from the large social order, an absence of loyalty to intermediate collectivities); and meaninglessness (individual roles are not seen as fitting into a total system of goals of a group, organization, community, or society).[9]

This is not a complete picture. There are many kinds of occupation and skill and varieties of working conditions in the blue-collar band. There is a large variety of technologies, each offering different opportunities for learning, effort, social relations, and satisfaction.[9]

Workers of all grades have a great capacity for adapting themselves to the levels of satisfaction actually available in their jobs[9] and for adapting even to the fact that few satisfactions (apart from income) may be available on the job and must be found elsewhere. In surveys relatively few describe themselves as acutely dissatisfied. Persons at the same occupational level vary widely in the extent to which their self-esteem is affected by the work they do; this depends on their relative investments in a wider range of human satisfactions.

Meaning can be introduced into jobs and

job contexts found boring or unsatisfactory by those performing them. This is done, for example, by elaborating the movements required, by unofficial exchanges of duties between workers, by working to targets defined by the worker himself or by groups of workers, by introducing social rituals or games, or by thwarting supervisors and managerial groups.[60,61,71] Colleagues often unwittingly collaborate with each other to use their relationships to express feelings and attitudes (both positive and negative) not strictly related to their work situation. Institutions can develop that act as defense mechanisms against tasks in a work system that arouse anxiety; an example is the depersonalization of patients, overelaborate division of labor, and ritualization of duties that occur in hospitals.[47]

Persons in manual occupations grow up in subcultures in which expectations of satisfaction from work may not be high and the instrumental aspects of work are emphasized. Satisfaction with a particular job does not depend purely on the quality of the work experience but on the relation between that experience, the expectations of the worker, and his assessment of the alternatives realistically open to him.

Once they have become established in their occupations the job aspirations of blue-collar workers are typically not high, and their definitions of what constitutes success have to be judged by job needs as they perceive them, that is, largely in terms of security of employment and reasonableness of hours, colleagueship, and supervision. On these criteria blue-collar jobs have their own typical sources of anxiety and stress. Such jobs are less secure than others; the posts held by an individual typically are not related to each other in an orderly sequence; and earnings do not rise concurrently with increasing family needs, being related more to strength, skill, and speed than to the improvements in judgment, experience, or to the capacity for looking after others that is generally accepted to come with age. The industrial worker is more subject to occupational vicissitudes than persons in more skilled or senior posts; is likely to experience or perceive randomness in occupational success; is likely to fear machines and technological improvements for their possible effects on him; is not likely to see a close connection between ability, effort, and success; and is prone to develop a philosophy of life that attributes such success to luck, pull, discrimination, or the power of others in a better position to manipulate society. A distinct ideology of the underdog or underprivileged worker expresses hopelessness, resignation, and unwillingness to believe that one matters as much as other people and that one's needs and wishes will be taken into account.[36,66,67]

In studies of the mental health of industrial workers it is reported that a large proportion suffer from "poor life adjustment." The mental health of these workers is reported to vary significantly with situational factors, chief among which is the opportunity that work offers for the use of their abilities and for associated feelings of interest, sense of accomplishment, personal growth, and self-respect. Mental health is apparently also affected, though less appreciably, by financial stress; pace, intensity, and repetitiveness of work; the quality of supervision and personal relations on the job; opportunities for advancement and improved social status. Mental health is apt to be poorer where plants are large, a high proportion of employees are at a low level of education and skill, and personnel policies and services are below average. Mental health tends to be poorer as job level falls. Men performing routine semiskilled work experience general dissatisfaction, have a narrow range of spare-time activities, and have relatively little devotion to larger social purposes. Self-esteem is especially low in the less skilled.[39]

In most advanced industrial countries a majority or substantial proportion of workers are affiliated to trade unions that seek to protect and advance their economic and related interests. The relationship of the individual with the union generally bulks larger as he comes to scale down his aspirations, settles for a steady job, and relies on the union to protect and advance his interests in the employment relationship and in competition with other occupational groups. In such cases the worker adjusts realistically to the fact that his mobility aspirations are more likely to be realized through his children. Ambivalence toward union membership and participation continues among those who remain ambitious to rise into the ranks of management.

The sheer size of modern industrial organizations combined with the existence of inevitably competing interests creates the occasion for the intervention of unions to deal with problems of disputes, fairness, and grievances; only in small firms it is possible to deal effectively with such matters through personal relationships. In their relations with unions, American workers appear to value most highly the job security and the protection from the dangers of wage reductions, arbitrary treatment, and deterioration of working conditions that they derive from membership.

American workers tend to have different expectations of the employer and the union. Acute conflicts in loyalty are most likely to occur in the organizing stage of unionism where ties of loyalty to the employer have to be broken for a union to be formed. In some cases the worker has little use for either employer or union, viewing both as evils to be tolerated and the whole work situation as unrewarding. But these workers appear to be greatly outnumbered in the United States by those who have no basic quarrel with the organization of industry and society and look to the union for a more equitable distribution of business income and for insurance against improper treatment.

Supervisors and Foremen

Supervisory positions have advantages over lower level jobs. Basic wage rates are higher; earnings are more regular since they do not depend on hours worked per week; status is higher in both factory and community; and promotion is known to be generally on the shop floor from among equals on the basis of demonstrated merit.

At its highest this role has been described as mediation between the rigidities and impersonalities of command and the infinite varieties of human nature, and as safeguarding of individual human dignity.[72] It is around the supervisor that the formal instrumental pressures of managerial systems and the informal "communal" aspects of shop floor work life have to mesh.[59]

A review of empirical studies has shown that greater job satisfaction exists among supervisory personnel than among workers.[58] In a research project all 55 production foremen in the same plant were interviewed. On the whole and in spite of rough times, difficulties, and frustrations, the foremen liked their jobs, their employing organization, their economic rewards, and the challenges of a dynamic work environment. But real dissatisfaction existed and real problems stemmed from the technical environment and the organization of work. The conveyor belt did not substitute mechanical compulsion for human supervision. It did not solve the foreman's major problems of maintaining quality and keeping the line manned. Combined with repetitiveness it created its own problems of adjustment and morale among subordinates. To be successful with his men the foreman had to absorb pressure from many sources and still not become a symbol of pressure to his men. The foremen felt they needed help from their managements, and this was the main criterion for evaluating managerial colleagues. The foreman's main way of keeping up productivity was by keeping the line manned, but absenteeism (partly prompted by the assembly line) often frustrated this aim. The foreman was exposed to the fact that those who experience compulsions associated with "nonhuman" or "mechanical" methods are apt to accuse the agents of those compulsions of being inhuman themselves. The essence of the foreman's job was doing something different every minute.[72]

Intermediate leaders are subject to different demands from above and below—typically demands for productivity (or its equivalent) from above and for human consideration from below—and hence are liable to actual or potential conflict in their behavior. In the U.S. Army during World War II, the noncommissioned officer was valued from above for "being a strong military leader" and from below for "being a good fellow."[66,67] Ratings of foremen by superiors and subordinates have been found to be inversely correlated, with a tendency for those who are well thought of by one group to be badly thought of by the other.[19] The supervisor must uphold the organizational standards while at the same time retaining the cooperative attitudes of subordinates. He is dependent on his subordinates both for his own reputation with his superiors and for the daily social interaction that makes work situations pleasurable or tolerable. If he does not collude with subordinates in circumventing unpopular rules, his personal position may become untenable. The supervisor may have to choose between "riding" his men or being "ridden" by his own superiors; he has to carry heavy responsibility; he has to make decisions.

In the vertical information flow of the organization, he has to keep his superiors informed about what is going on without reporting in a way that brings criticisms upon either himself or his subordinates. He is likely to be by-passed by subordinates talking to other supervisors or to union representatives about what is going on in his section. The foreman acts as the work group's intermediary with parallel work groups as well as with intermediaries. Given the demand of subordinates for consideration, one can assume ambivalence in the attention of the foreman to external tasks. He operates in a context in which he is expected to get results, but the results of his section depend on the cooperativeness of his subordinates and the largely independent activities of senior colleagues whom he does not control.

As a consequence of increased specialization, the foreman is surrounded today by a dozen (other) bosses, each a technical or staff man who has taken away a different segment of what used to be the supervisory job.

The contemporary supervisor is expected to be knowledgeable about a wide range of matters (including labor contracts and laws, company policy and regulations, technological processes, training methods) although he may well have come up the hard way and not have had much formal education. Nor may he be well prepared for the managerial side of his role: workers at the bottom of the corporate hierarchy learn to combat authority or to accept it, not how to exercise it or how to mediate between factions.[14]

For the majority of supervisors little lies ahead in the way of promotion, since they do not possess the educational requisites for white-collar or managerial rank. Desk work may not in any case be attractive, or the entry into "polite society" may be embarrassing.

White-Collar Workers

Within the white-collar band new entrants experience some similar difficulties to blue-collar workers in occupational search and job establishment, although their vocational guidance facilities at school are almost certainly better. They have insufficient knowledge of what is involved in different occupations. They experience difficulty in estimating the prospects of different occupations, organizations, and industries. On the other hand, their preoccupations lie more with deciding what to do than with getting an acceptable job.

They may have to build a view of their capacities and occupational life chances through a series of rejections at selection procedures. They spend an amount of time choosing between options that is disproportionate to the relatively small investment of time made in generating options.[17] Anxiety to get settled, to be an accepted member of the world of work leads to premature job crystallization, often with subsequent regret as the binding character of early commitments is realized. But compared with blue-collar workers, white-collar and executive workers in bureaucracies enter well-defined channels, and their steps thereafter are prescribed, visible, well-regulated, and supervised.[22] Executives are helped to develop in ways not normally available to other grades of workers.

Like blue-collar workers, the white-collar and executive occupational classes value security, continuity, interaction with others, autonomy, and duties that enable them to maintain and build self-respect, test and affirm their competence, and provide scope for achievement. Like these other categories, they are resources available to their senior management and experience the pressures of working to standards imposed by others. But they are usually able to take for granted the security, continuity, and adequacy of income that are problematic for the blue-collar man. Their problems lie rather in whether they can exercise a relatively high measure of autonomy, in whether their personal resources are being well used, or whether they can identify as their own and render visible to others the contributions they make to collective efforts in which the impact of any individual is diffuse and hard to specify. Work tasks are more central to personal identity in this group, so that expectations from the work situation are high and dissatisfaction or frustration especially painful. The incorporation of work tasks into the self and a high measure of identification with the employing organization makes life satisfaction highly dependent on the work situation and career progress.

While the blue-collar worker has to tolerate the risk that he may be regarded as expendable in times of economic stringency, the white-collar worker has to tolerate a different form of insecurity and potential strain, namely, that flowing from high dependence on one employer. His skills are only partly technical but are also largely his expertise in the administrative and political processes of his own organization. A common situation is that the executive becomes heavily invested in, dependent on, and committed to one employer and is in a poor position to seek alternative employment should he experience strain or come to dislike his work situation.

Persons at higher occupational status levels report greater satisfaction with their work than those at lower levels; satisfaction seems to arise from the very fact of being in an occupation that ranks higher than others, and this has a halo effect on other aspects of the occupation and the work it involves. White-collar workers have higher expectations of their work and, in general, more sources of satisfaction. The executive and the professional have more control than junior grades over their time, the pace at which they work, and their technological and social environment, and relative freedom from hierarchical authority, at least in the way of direct surveillance. They have the opportunities to extend themselves, to define or influence their roles in accordance with their capacities and inclinations, to exercise judgment in regard to priorities among tasks. Their work contributes to their self-esteem and can be integrated into their nonwork activities. This contrasts with the operative who may see no connection between what he does as a way of earning a living and what sort of person he is.[10,64]

Executives and technical specialists at a wide range of levels have their needs for security and for social interaction met, but satisfactions in esteem, autonomy, and self-actualization increase as rank rises. Among equals in managerial rank the content of duties apparently matters: line managers report more satisfaction in their jobs than staff managers, especially in regard to esteem and self-actualization.[57]

Substantial differences exist between blue-collar and white-collar definitions of success. Emphasis on achievement in one's job and career appears particularly in nonmanual occupations, and emphasis on economic security appears particularly in manual occupations.[55] Movement up the administrative hierarchy becomes in itself a criterion of success for the executive,[34] a criterion whose significance is apparently maintained for a high proportion of the work life even though objective opportunities for promotion actually decline.

Clerical Workers and Lower-Level Managers

The modern office and its staff are an outgrowth of large-scale industry. When industry was conducted on a small scale, offices were correspondingly small, there were relatively few office employees, they were appreciably better educated than blue-collar workers and in close contact with management. With the subsequent growth in size of organizations, in finance invested, in government controls and taxation of industry, in procedures for collecting personal taxes through employers, and in unionism, there has been a great growth of accountability, recording, and paper work. The numbers of office workers has increased greatly.

U.S. studies have emphasized that increased costs have led to the rationalization of office work and changed roles for the office worker.[62] The roles of bookkeeper, clerk, and secretary have been fragmented, become more routinized and to some extent more mechanized—in this way following somewhat the same course as shop floor roles. Nevertheless, important differences exist in the connection of the duties of the office worker with abstract symbols rather than physical materials, in the clerk's physical proximity and access to members of the managerial hierarchy, in cleaner and quieter physical working conditions, and in the opportunity to wear clothing also appropriate outside the workplace.

In regard to work satisfaction, one of the few exceptions to the rule that reported satisfaction decreases as level in the occupational pyramid lowers is that skilled manual workers are slightly more satisfied than rank-and-file white-collar workers. This is found in the United States, Germany, and the U.S.S.R.[33]

Over the past few decades the salaries of U.S. office workers and blue-collar workers have converged, partly through rationalization of work, partly through wider educational opportunities, partly through the greater strength of blue-collar unions. Office workers seem to have been put in a position of status ambiguity, a difference arising between their status claims and aspirations, on the one hand, and their objective position and the reluctance of blue-collar workers to concede their superiority, on the other.

Strains among office workers and the lowest levels of management are thought to arise from salaries that are too low to finance the middle-class aspirations characteristic of these employees and from relative status decline. To the extent that office work becomes mechanized one can expect increasing strains similar to that experienced by the operative confronted with technological advances.

In Britain the basic skill of literacy used to set the clerk apart from the working man and give him a foothold on the lowest rung of the middle-class ladder.[43] Although he lacked the income and security of the middle-class person, this is where his identification lay. Status considerations were powerful. But the connection with middle-class status has been altered by social and economic changes. Economic power has shifted toward organized labor and educated people have become less scarce. There has been a fall in the desirability and prestige of clerical work in the eyes of other occupational groups.

The clerk belongs neither to the middle class nor the working class. In social background, education, working conditions, proximity to authority, and opportunity for upward mobility, clerks in Britain can still perhaps claim a higher status than most manual workers. But using other criteria such as income and skill, they may be rated lower, or no different, by manual workers. Clerical workers are also divided among themselves or uncertain on the question of where they stand in the social hierarchy.

The position of clerical workers and lower-level managers has recently been particularly affected by the introduction of electronic data processing. Persons in these grades have tended to be especially security-minded, and many have worked under conditions more stable than those of either blue-collar workers or executives. The disorganization and reorganization associated with change to EDP may have found them less ready in personality and experience to meet the demands for adjustment. The nature of clerical jobs is changing with the automation of office processes. The most routine jobs are being eliminated and work pace is becoming more closely tied to machines, so that absence and lateness matter more. While the office worker still has more freedom than the blue-collar worker to leave his workplace or to vary his production level, more specific work quotas have been imposed. Opportunities for promotion may have been somewhat reduced, and some shift work has been introduced. There is lower tolerance for errors, which become transmitted far through the system and are more visible to superiors and customers. With the rationalization of the work system, each job gains significance in the continuity of the overall process and the individual becomes more accountable for his actions. With automation a proportion of jobs may be eliminated or changed radically. While new jobs are created, the persons displaced may not be able to fill them or may not be easily amenable to retraining.[3,45]

In a British study drawing on data from nine firms that had introduced computers, it was found that this and concomitant reorganization were experienced as a threat to staff security. There was much initial anxiety, even where little redundancy ensued. Staff often had to be redeployed. It proved difficult to avoid reducing the status and salary of staff in the 45 to 55 age group, particularly if they had held supervisory positions. In the clerical area there was a greater concentration of responsibility and a need for more staff of high intelligence: those associated with running computers had to be able to master complicated new skills. In some firms junior managerial roles also changed, sometimes only eliminating routine, but in other cases involving transfers from reduced departments and loss of status. Not all the results of the new computer installations were negative. Many of the younger clerical workers were attracted toward machine work, which seemed to promise greater interest and more promotion opportunities than the "elementary" clerical duties they were performing.[52]

Middle-Level Managers

A review of empirical studies has shown that median satisfaction with jobs is higher for managerial than nonmanagerial personnel and that satisfaction increases with echelon level within management.[58]

The middle manager's work is carried out in a highly "political" context; he is part of a constant struggle for relative power, autonomy, and influence over the way the overall enterprise is run, not only between individuals, but also between departments and occupational groups. The acuteness of these struggles derives largely from the investments made by the persons involved in their occupational training and experience and from the difficulty of radical occupational shifts in adulthood—particularly among those wishing to protect their status as higher-level managers or technical specialists.

The middle-level manager's situation is deeply affected by expectations (both his own and others) that he should climb up the administrative hierarchy. He has two tasks to perform simultaneously: that of carrying out organizational tasks in cooperation with colleagues and that of facilitating his upward rise. Most organizational tasks are carried out by groups that develop solidarity, shared definitions of their mission and status, and shared conceptions of how they should relate to other task groups, subordinates, and superiors. Therefore, the individual must be an acceptable colleague on easy terms with equals, prepared to give credit to their contributions and to refrain from pressing his own claims to special consideration and merit. But if he is to advance it is also necessary that he differentiate himself from his peers and convey his potential ability to distance himself enough to view their organizational contribution objectively or, if required, to control it from a position of greater authority. Both task performance and career building make high demands on interpersonal competence. Discussion of this topic usually emphasizes the need for the executive to conform, to represent himself as similar to others. But the task is more correctly viewed as a wider form of self-management, involving both a measure of conformity and a capacity for discreet and acceptable differentiation.

The character of a great deal of executive work is diffuse and is not subject to clear-cut criteria of adequacy. This makes assessment of the executive's competence dependent on evaluation by colleagues, of which evaluation by seniors is the most crucial. Because of the absence of objective evidence of performance and the difficulty, where there is an ascertainable output, of establishing that it is a direct consequence of a particular executive's input, evaluation of the executive tends to acquire a personal quality, so that he gets evaluated on such grounds as enthusiasm, confidence, assertiveness, patience, and cooperativeness alongside technical competence, knowledge, and capacity for problem solving. The personal component of this evaluation is accentuated by the fact that appraisal by superiors covers not only his performance but also his potential: an aspect that necessarily brings into issue the amount of flexibility and capacity for growth and change in his personality relative to the more senior roles in the organization. It is common practice for large organizations to inform their executives in appraisal interviews of their personal strengths and weaknesses in relation to their careers; the rationale is that this will increase the likelihood that the person will alter in a direction likely to bring greater organizational rewards, with resultant benefits to both the employer and himself. The extent of defensiveness in appraisees and the protective distortions that occur imply that the interviews are stressful events at least for the appraisee and possibly for the appraiser also. The executive faces a unique life situation in that he cannot escape or by-pass an organizational mirror that confronts him with the image held of him by his most important seniors.

He is commonly subject to personal dependence on one or more sponsors, persons with whom he has an interdependent relationship of reciprocal support. The executive's career becomes linked with that of his sponsor or sponsors, increasing his chances of rising as his sponsor gains status and influence, but simultaneously increasing his vulnerability as his sponsor's career goes through its own vicissitudes.

A systematic connection between rank, conflict, and tension has been found in a national survey, with the maximum of conflict occurring at the upper-middle levels of management. The research workers interpret this as partly a consequence of the still unfulfilled mobility aspirations of middle management, in contrast with the better actualized aspirations of top management. The greatest pressure is evidently directed to a person from others who are in the same department as himself, who are his superiors in a hierarchy and who are sufficiently dependent on his performance to care about his adequacy without being so completely dependent as to be inhibited in making their demands known.[35]

The tapering of the administrative hierarchy and the definition of success as movement combine to produce career disappointment in many executives. Some adapt fairly comfortably to their work situation or are helped to do so by the personnel institutions of their organizations. The disappointed person may be able to call on his individual defense mechanisms in the sense that he can tell himself that he has not been given a fair chance or that he would not stoop to the methods used by others to secure advancement. He is commonly helped by the existence of shared group ideologies, developed with colleagues in the same position as himself, whose central theme is that personal merit is not the main determinant of success.[64] Some are able to find compensations in other aspects of their lives. External compensatory roles may include devotion (or more devotion) to familial, recreational, political, or social service pursuits. Role segmentation may usefully insulate compartments of one's life against failure and disappointment in others.[23] One form of (negative) response is to turn "sour" in relations with the employing organization by withdrawing enthusiasm, good will, and vitality from one's work role, complying with the formal requirements of the role but withdrawing identification from it.[23]

A key factor in career disappointment is that persons come to base their self-identity on the assumption that they will in due course come to occupy certain organizational roles more senior than their own. A career disappointment can then come to constitute a disturbance in self-conception, involving the necessity to reassess or reclassify oneself, and an embarrassing withdrawal of the image of the prospective self that one may have held out to intimates. Seniors may be felt to have betrayed one. Personal and familial sacrifices may have been made in the vain pursuit of illusory occupational goals. Internal readjustment may be painful, and in seeking to justify to those closest to him his revised expectations and evaluations, the person may have to perform a stressful about-face and express himself in transparent rationalizations.

Technical Specialists

Line managers report themselves to be more satisfied in their jobs than staff managers, mainly because the latter feel they have to be directed by others.[58]

Technical specialists complain that they are subjected to administrative routines and controls that are alien to their notions of reciprocal control by equals; that they are overconstrained by considerations of practicality and profitability and have too little opportunity for work of long-term importance; that undue secrecy is placed on their work, on relations with colleagues in other organizations, and on publication; that they are appraised and managed by nonspecialists who do not understand their work; and that their rewards and promotion chances are inferior to those of line administrators. A distinction has been drawn between cosmopolitans and locals: this differentiates persons whose major commitment is toward their discipline and to colleagues wherever they are located from those whose more abiding interests and personal loyalty reside with their employing organization.[25] While important conflicts no doubt exist, most people are unlikely to lie at one extreme or another.

Accommodative mechanisms arise or are developed to reduce strain, to forestall and to settle actual or incipient conflict. In general, bureaucratic organizations are adapting themselves to the high proportion of technical specialists they now employ. Special roles are created for specialists in segregated departments, for example, research and development. Special administrative arrangements can be made to allow for colleague control in project teams whose leadership shifts in accordance with the dominant specialty involved. Specialist leaders are appointed who have qualifications and interests in both the disciplines involved and in "organizational" needs, and who can mediate between the technical specialists and superiors. Leave can be arranged to provide sufficient contact with outside colleagues. Dual career ladders have sometimes been created so that specialists can improve their prestige, income, and facilities without giving up their specialties. Those people who are most insistent on pure science and academic freedom are apt to avoid careers in the applied research departments of large organizations. Even in the most enthusiastic and tenacious research, worker compromise and adaptation will occur.

Industrial research provides appropriate positions for those technical specialists (who may be a substantial proportion) who are distinctly interested in their subject and feel it is an important part of their lives, but who are unwilling to have it as the dominating factor in their lives.

Some studies have concentrated on the organizational scientists who remain anxious to retain their full status as members of the scientific profession. They have to span two areas, since consolidation of their organizational positions is a prerequisite for gaining and maintaining recognition by external colleagues. Only a positive evaluation of their own work by their organizational superiors will give them the facilities and resources they need for successful research. At the same time they become dependent on having assistants and subordinates who become counters in their own career success.[21]

Senior Administrators

Taking various dimensions of job satisfaction separately, a review of U.S. research shows that within management security and social affiliation needs are equally met, but that the needs for esteem, autonomy, and self-actualization are better met at each higher level.[58] The phenomenon of greater job satisfaction at higher levels of management is not confined to the United States; in a cross-cultural investigation of managerial attitudes in 14 countries, higher levels of management on the whole reported that their needs were better satisfied in their jobs.[58]

Seniors are more exposed and vulnerable to criticism than middle-level executives or blue-collar workers. The less senior members of an organization are better protected by the narrower specification of their responsibilities and are in a better position to discharge their obligations by adhering to the instructions issued to them. Their errors are usually less visible and their rights more likely to be defended by formal collective associations or their informal equivalent.

Senior administrators may experience frustrations from the limits to their behavior set by the crystallization of organizational practices over time by laws, regulations, and agreements covering the roles involved in the organization, the means by which these must be filled or vacated, and the conditions under which personnel may be employed or dismissed.[16]

While it is one of the tasks of the senior administrative group to initiate changes, such changes are liable to alter the existing distribution of power and prestige. A tension can be set up between following out the logic of an organizational change and accepting the implications that the change will have for one's own position and behavior.[65]

In the earlier phases of his career the senior administrator will probably first have developed and practiced some form of specialist expertise, then have combined this with responsibility for a group of subordinates. At the most senior levels opportunities for direct development and practice of such special knowledge decrease, and usually the transition has to be made to being a generalist. The security of acknowledged specialist knowledge and techniques is given up for the exercise of judgment and wisdom on issues on which there is no clear-cut right and wrong and where one is more exposed to challenge.

The senior administrator has to steer a path between the advice of a variety of rival specialist colleagues. He has to show that he respects the views of each but may, nevertheless, have to act in ways that advance the interests of some and obstruct those of others. In order to maintain objectivity in his judgments he will find it necessary to maintain distance from previously close colleagues, with probable feelings of isolation or loneliness. His wife may also find herself in a similar position in relation to the families of her husband's colleagues. The administrator carries responsibility for large sums of money, large numbers of people, and far-reaching decisions but does so in a chronic state of uncertainty, since knowledge of the relevant facts, including future events, is always imperfect.

The roles that senior administrators play at the top of an organization have more scope for variation and for idiosyncratic interpretation than those lower down. This is largely because it is the function of the senior executive person or executive group to define organizational priorities. If, as is usual in large organizations, a number of people are involved, either formally or informally, some accommodation must take place between them in regard to relative power and duties. How things turn out will clearly depend on a combination of formal position, personality, traditions of colleagueship and popular support among subordinates and, perhaps, sponsors and shareholders. Interpersonal relations with close colleagues will bulk very large in the psychological life space. There will be a further division of labor in the psychological sense that each senior administrator will tend to perform particular psychological functions or have these projected on to him by subordinates. One might, for instance, symbolize instrumental purposes, another the supportive "maintenance" activities that also appear necessary in organizational functioning. A study of a hospital has identified the allocation of these roles among a colleague group of three and implied that, apart from objective task definition, problems may arise if administrators fit poorly into the psychological functions allocated to them by group processes, if they are too similar in personality, or if their own prior socialization has prepared them inadequately for these highly demanding social psychological roles.[31]

Bibliography

1. AVERY, R. W., "Enculturation in Industrial Research" excerpted in Glaser, B. G. (Ed.), *Organizational Careers*, Aldine, Chicago, 1968.
2. BARBER, B., "The Sociology of the Professions," *Daedalus*, Fall 1963.
3. BEDROSIAN, H., "Managerial Obsolescence in Banking," in Lazarus, H., and Warren, E. K. (Eds.), *The Progress of Management*, Prentice-Hall, Englewood Cliffs, N.J., 1968.
4. BELBIN, E., and BELBIN, R. M., "New Careers in Middle Age," in Neugarten, B. L., *Middle Age and Aging*, University of Chicago Press, Chicago, 1968.
5. BELL, D., *Work and Its Discontents*, Beacon, Boston, 1956.
6. BLACKBURN, R. M., *Union Character and Social Class*, Batsford, London, 1967.
7. BLAU, P. M., "Social Mobility and Interpersonal Relations," *Am. Sociol. Rev.*, *21*, 1956.
8. ———, and DUNCAN, O. D., *The American Occupational Structure*, John Wiley, New York, 1967.
9. BLAUNER, R., *Alienation and Freedom: The Factory Worker and His Industry*, University of Chicago Press, Chicago, 1964.
10. ———, "Work Satisfaction and Industrial Trends in Modern Society," in Galenson, W., and Lipset, S. M. (Eds.), *Labor and Trade Unionism: An Inter-disciplinary Reader*, John Wiley, New York, 1960.
11. BOROW, H. (Ed.), *Man in a World of Work*, Houghton Mifflin, Boston, 1964.
12. CAIN, L. D., "Life Course and Social Structure," in Faris, R. E. L. (Ed.), *Handbook of Modern Sociology*, Rand McNally, Chicago, 1964.
13. CENTERS, R., "Attitude and Belief in Relation to Occupational Stratification," *J. Soc. Psychol.*, *27*:159–185, 1948.
14. CHINOY, E., *The Automobile Worker and the American Dream*, Doubleday, New York, 1955.
15. COSER, L., *The Functions of Social Conflict*, Free Press, Glencoe, 1956.
16. CROZIER, M., *The Bureaucratic Phenomenon: An Examination of Bureaucracy in Modern Organizations and Its Cultural Setting in France*, University of Chicago Press, Chicago, 1964.
17. DILL, W. R., HILTON, T. L., and REITMAN, W. R., *The New Managers: Patterns of Behaviour and Development*, Prentice-Hall, Englewood Cliffs, N.J., 1962.
18. DOWN, E., and ADELSON, J., "Social Mobility in Adolescent Boys," *J. Abnorm. & Soc. Psychol.*, *56*, 1958.
19. FLEISHMAN, E. A., HARRIS, E. F., and BURT, H. E., *Leadership and Supervision in Industry: An Evaluation of a Supervisory Training Programme*, Ohio State University Press, Columbus, 1955.
20. GLASER, B. G. (Ed.), *Organizational Careers: A Source Book for Theory*, Aldine, Chicago, 1968.
21. ———, *Organizational Scientists: Their Professional Careers*, Bobbs-Merrill, Indianapolis, 1964.
22. ———, and STRAUSS, A., *Status Passage: A Formal Theory*, Aldine-Atherton, Chicago, 1971.
23. GOFFMAN, E., "On Cooling the Mark Out: Some Aspects of Adaptation to Failure," *Psychiatry*, *15*, 1952.
24. GOLDTHORPE, J. H., LOCKWOOD, D., BECHHOFER, F., and PLATT, J., *The Affluent Worker: Industrial Attitudes and Behaviour*, Cambridge University Press, Cambridge, 1968.
25. GOULDNER, A. W., "Cosmopolitans and Locals: Toward an Analysis of Latent Social Roles," *Admin. Sci. Quart.*, *2*, 1957.
26. ———, *Patterns of Industrial Bureaucracy*, Free Press, Glencoe, 1954.
27. ———, "The Unemployed Self," in Fraser, R. (Ed.), *Work*, Vol. 2, Penguin Books, London, 1969.
28. GROSS, E., *Industry and Social Life*, Brown, Dubuque, Iowa, 1956.
29. ———, *Work and Society*, Thomas Y. Crowell, New York, 1958.
30. HERZBERG, F., MAUSNER, B., and SNYDERMAN, B. B., *The Motivation to Work*, John Wiley, New York, 1959.
31. HODGSON, R. C., LEVINSON, D. J., and ZALEZNIK, A., *The Executive Role Constellation*, Harvard Graduate School of Business Administration, Cambridge, 1965.
32. INGHAM, G. J., *Size of Industrial Organisation and Worker Behaviour*, Cambridge University Press, Cambridge, 1970.

33. INKELES, A., "Industrial Man: The Relation of Status to Experience, Perception and Values," *Am. J. Sociol.*, *66*:1–31, 1960.
34. JENNINGS, E. E., *Routes to the Executive Suite*, McGraw-Hill, New York, 1971.
35. KAHN, R. L., WOLFE, D. M., QUINN, R. P., and SNOEK, J. D., *Organizational Stress: Studies in Role Conflict and Ambiguity*, John Wiley, New York, 1964.
36. KNUPPER, G., "Portrait of the Underdog," *Public Opinion Quart.*, *11*, 1947.
37. KOHN, M. L., "Bureaucratic Man," *New Society*, October 28, 1971.
38. ———, "Bureaucratic Man: A Portrait and an Interpretation," *Am. Sociol. Rev.*, *36*, 1971.
39. KORNHAUSER, A., *Mental Health of the Industrial Worker, A Detroit Study*, John Wiley, New York, 1965.
40. KORNHAUSER, W., *Scientists in Industry*, University of California Press, Berkeley, 1962.
41. LEVINSON, H., "Reciprocation: The Relationship between Man and Organisation," *Admin. Sci. Quart.*, *9*:370–390, 1964.
42. LIPSET, S. M., and BENDIX, R., *Social Mobility in Industrial Society*, University of California Press, Berkeley, 1959.
43. LOCKWOOD, D., *The Blackcoated Worker*, Unwin, London, 1958.
44. MACWHINNEY, W. R. H., and ADELMAN, S. R., "Mental Health of the Industrial Worker: An Analysis and Review," *Human Organization*, *25*, 1966.
45. MANN, F. C., and WILLIAMS, L. K., "Organizational Impact of White Collar Automation," *Proceedings of Eleventh Annual Meeting, Industrial Relations Research Association*, 1958.
46. MARCSON, S., *The Scientist in American Industry*, Harper, New York, 1960.
47. MENZIES, I. E. P., "A Case Study in the Functioning of Social Systems as a Defence against Anxiety: A Report of a Study of the Nursing Services of a General Hospital," *Human Rel.*, *13*, 1960.
48. MERTON, R. K., *Social Theory and Social Structure*, Free Press, Glencoe, 1957.
49. MILLS, C. W., *White Collar*, Oxford University Press, London, 1951.
50. MOORE, W. E., and TUMIN, M. M., "Some Social Functions of Ignorance," *Am. Sociol. Rev.*, *14*, 1949.
51. MORSE, N. C., and WEISS, R. S., "The Function and Meaning of Work and the Job," *Am. Sociol. Rev.*, *20*, 1955.
52. MUMFORD, E., *Living with a Computer*, Institute of Personnel Management, London, 1964.
53. NEFF, W. S., *Work and Human Behaviour*, Atherton Press, New York, 1968.
54. NEUGARTEN, B. L. (Ed.), *Middle Age and Ageing*, University of Chicago Press, Chicago, 1968.
55. PALMER, G. L., "Attitudes toward Work in an Industrial Community," *Am. J. Sociol.*, *63*, 1957.
56. PELLEGRIN, R. J., and COATES, C. H., "Executives and Supervisors. Contrasting Definitions of Career Success," *Admin. Sci. Quart.*, *1*, 1956–1957.
57. PORTER, L. W., "Job Attitudes in Management," *J. Appl. Psychol.*, *46*, 1962; *47*, 1963; *47*, 1963.
58. ———, and LAWLER, E. E., "Properties of Organization Structure in Relation to Job Attitudes and Job Behaviour," *Psychol. Bull.*, *64*, 1965.
59. ROETHLISBERGER, F. J., *Man in Organization*, Harvard University Press, Cambridge, 1968.
60. ROY, D. F., "'Banana Time': Job Satisfaction and Informal Interaction," *Human Organization*, *18*, 1959–1960.
61. ———, "Work Satisfaction and Social Rewards in Quota Achievement: An Analysis of Piecework," *Am. Sociol. Rev.*, *18*, 1953.
62. SCHNEIDER, E. V., *Industrial Sociology*, McGraw-Hill, New York, 1957.
63. SOFER, C., "Buying and Selling: A Study in the Sociology of Distribution," *Sociol. Rev.*, *13*, 1965.
64. ———, *Men in Mid-Career*, Cambridge University Press, Cambridge, 1970.
65. ———, *The Organisation from Within*, Tavistock Publications, London, 1961.
66. STOUFFER, S. A., LUMSDAINE, A. A., LUMSDAINE, M. H., WILLIAMS, R. M., BREWSTER SMITH, M., JANIS, I. L., STAR, S. A., and COTTRELL, L. S., *The American Soldier*, Vol. 2, *Combat and Its Aftermath*, Princeton University Press, Princeton, 1949.
67. ———, SUCHMAN, E. A., DE VINNEY, L. C., STAR, S. A., and WILLIAMS, R. M., *The American Soldier*, Vol. 1, *Adjustment during Army Life*, Princeton University Press, Princeton, 1949.

68. TURNER, A. N., and LAWRENCE, P. R., *Industrial Jobs and the Worker*, Harvard Graduate School of Business Administration, Cambridge, 1965.
69. VON MISES, L., *Bureaucracy*, Yale University Press, New Haven, 1946.
70. VROOM, V. H., *Work and Motivation*, John Wiley, New York, 1964.
71. WALKER, C. R., and GUEST, R. H., *The Man on the Assembly Line*, Harvard University Press, Cambridge, 1952.
72. ———, ———, and TURNER, A. N., *The Foreman on the Assembly Line*, Harvard University Press, Cambridge, 1956.
73. WATSON, W., "Social Mobility and Social Class in Industrial Communities," in Gluckman, M. (Ed.), *Closed Systems and Open Minds*, Oliver & Boyd, London, 1964.
74. WILENSKY, H. L., "Work, Careers and Social Integration," *Internat. Soc. Sci. J.*, 12, 1960.
75. ———, and EDWARDS, H., "The Skidder: Ideological Adjustments of Downward Mobile Workers," *Am. Sociol. Rev.*, 24, 1959.
76. WEISS, R. S., and RIESMAN, D., "Social Problems and Disorganization in the World of Work," in Merton, R. K., and Nisbet, R. A. (Eds.), *Contemporary Social Problems*, Harcourt Brace, New York, 1961.
77. ZYTOWSKI, D. G., *Vocational Behaviour: Readings in Theory and Research*, Holt, Rinehart, New York, 1968.

CHAPTER 21

MARRIAGE AND MARITAL PROBLEMS

Ian Alger

Introduction

WHATEVER its origins in various civilizations throughout recorded history, and whatever the joys and anguishes it may bring,[8] the fact that in our society marriage is the goal, both striven for and achieved, of the overwhelming majority of people insures it a high place in the hierarchy of "normal" maturational goals. At the beginning it should be made clear that this chapter intends to deal with marriage at this current time in history, and more specifically with middle-class American marriages.[17,67] This narrowing of focus is being undertaken because this type of marriage is the one about which most clinical psychiatric experience has been accumulated and, indeed, is the one about which over 90 per cent of the psychiatric literature is written. This is not to say that marriages in other classes of our society, among minority groups of various racial and ethnic backgrounds are not relevant and important. It is to say, however, that knowledge about them comes from a sociological and anthropological base, that even this knowledge is only gradually being developed with any completeness as survey and statistical studies improve, and that these marriages are by and large not the ones that have found referrals to psychiatrists.

Just as types of marriage other than the middle-class variety exist in our society, different types of marriage exist in other cultures and societies, not only now, but also back through history.[31,76,82,89] Again no major focus will be placed on these other varieties, except to underline the point that the institution of marriage is by no means universal in its characteristics, but differs from class to class, in different societies, and at different periods of history in the same societies.

The organization of institutions in a society can be explained on the basis of providing for the needs of that society and its people, and these needs can be described as biological and social. For example, needs for food, shelter, and sexual experience and needs for security, companionship, and child nurture may be seen as expressions of pure biological necessities, or pure social necessities, or in some instances a combination of both. The institution of marriage, depending on the details of its organization in any particular instance, can be seen as meeting necessities such as child care, companionship and security, and sexual fulfillment.[78] One theory explains the differentiation of the primal horde on the basis of sexual jealousy, which resulted in the males claiming the right to their own females. Will Durant[25] suggests another theory when he writes, "Some powerful economic motives must have favored the evolution of marriage. In all probability (for again we must remind ourselves how little we really know of origins) these motives were connected with the rising institutions of property. Individual marriage came through the desire of the male to have cheap slaves, and to avoid bequeathing his property to other men's children."

This theme finds an echo in the demands of today's women liberationists who call for equality and freedom from the enslavement of marriages in which women are treated as property.[21] Leaving aside for the time a discussion of the merits of this argument, it would seem true that marriage, especially as it becomes a legal contract, has great influence in the way in which power and wealth can be accumulated and then passed in a very discriminatory fashion from one generation to the next. Powerful families arrange marriages with other powerful families, and a king in one realm marries a princess from another. Monogamous marriage, with its nuclear organization, preserves and stabilizes the social organization and also provides for a continuity through one generation and on into the next. Thus, not only is power and property preserved, but also religious beliefs and social position, which, in turn, tend to preserve the economic and political structure of the society.[69,80]

With this introduction let us turn to an examination of marriage as it exists in middle-class America today.

The Marriage System

Marriage—The Majority Choice

Nearly all of us grew up in a family, in a nuclear family if you will. And although our experiences may have varied greatly, those many of us who experienced pain in that growing up, and who saw the frequent despair and anguish of our parents, nevertheless join with the large majority of our fellows and, despite any earlier protestations to the contrary, get married. But the tendency to marry is even more impressive than that. Over the past 40 years the divorce rate in this country, and not alone in this country, has risen markedly.* In the past decade the rate has shown another increase. Yet most of those who have divorced remarry, and, indeed, contrary to earlier hypotheses, the majority of those in second marriages claim that they have found greater satisfaction and happiness than in their first.† Thus, since 1940 for every unit of increase in the proportion of the population divorced, there have been five units of increase in the proportion married (p. v).[16]

Therefore, the almost inevitable route for an individual in our society is to become married, or to put it another way, to join a marriage system. Becoming a member of a marriage system is no small undertaking, for while it can be entered into fairly readily, the implications are enormous "from that moment on."

* Rates of final divorce decrees granted under civil law per 1,000 population: United States, 1932, 1.3; 1945, 3.5; 1960, 2.2; 1965, 2.5; France, 1932, 0.5; 1945, 0.6; 1950, 0.9; 1960, 0.7; Sweden, 1932, 0.4; 1945, 1.0; 1955, 1.2; 1960, 0.9; 1965, 1.2. Divorce rate per 1,000 married couples: United States, 1935, 7.8; 1964, 10.7; France, 1935, 2.3; 1963, 2.9; Sweden, 1934, 2.3; 1962, 5.0 (pp. 29, 31).[16]

† A very large proportion of the marriages that are dissolved within less than 20 years are followed by remarriages. For example, in 1960 over four-fifths of the white men who had first married some 18 years previously and had been divorced had remarried. The remarriage rate of nonwhite men was nearly as high; the corresponding rate for white women was seven-tenths, and that for nonwhite women was about five-eighths (p. 400).[16]

These implications involve not only the relationship between the two partners but also possibly myriad responsibilities to the larger society and to associate members of the marriage system, such as children, the couples' own parents, other relatives and assorted in-laws, and occasional friends. Furthermore, these responsibilities, at times cause for joy, at other times merely liabilities, may begin in random and erratic ways from the moment of the wedding vow and continue until the system closes with death or makes a transition into the system of divorce.

Society has always made entering the marriage system easier than leaving it. This commitment by the two partners holds both the promise of satisfaction in a close relationship with another human being through a continuity of mutual experience and the possibility of entrapment and disillusioned agony through a process of mutual erosion. We want to examine and understand this institution of marriage as we also look at the problems that arise in relationship to it, and while we then consider what therapies and remedies are, or may become, available to ease the suffering.

Why the Choice Is Made

Among other things marriage has been labeled an "institution," but I have chosen to talk of it as a "system" because I want to imply the features of change and dynamism, of interrelationships within the marriage and between the marriage system and other systems in the society. In addition, I wish to include the concept of the system existing and changing in time as other systems in the world also alter in time and relationship. When marriage is considered under this concept of flux, the impossibility of defining "marriage" becomes instantly apparent. Let us pursue the idea of the "marriage system" a little further, then. In this system two main partners voluntarily enter a relationship that actually is licensed, or certified, by the authority of society and that thereby becomes, whatever else it may be, a legal contractual agreement. The contract makes a demand, rare even in ordinary promises, much less legal vows, that each partner shall promise to love the other as long as both shall live. The impossibility of anyone's having sufficient control over feelings to make this promise realistically has, of course, been commonly recognized; and although for centuries hapless couples remained miserably together, the easing brought about by recent divorce legislation now removes the eternity from the promise and opens the option of certain release from such impossibly burdensome emotional obligation. However, rarely at the moment of making the vows do the parties to the contract consider the future shadow it may cast. Rather, they are more often eager and desirous of publicly declaring their affection and love for one another, and they look forward to the happiness that can come from the growing together of two people who care for each other, who share much with each other, and who long to share more. They anticipate the unique human fulfillment that can come from the intimacy of a physical and spiritual continuity of closeness between a man and a woman. In such hopes are found one of the basic forces that make marriage so appealing. Loneliness is an unhappy feeling, and although in one sense we must all exist alone, in another sense we all seek companionship. We developed our very humanness in this sharing of experience with others, with our own parents and those who reared and cared for us as infants and children. As we live we seek the constant renewal of that humanity in this sharing. For those who grew up deprived of this kind of human contact and love, relationships with others may be anticipated with anxiety and defensiveness. Still others may have grown up with such self-mistrust and low self-esteem that they continue to seek a chronically infantile dependency on others. More will be said later of the problems these and other developmental deprivations and distorting experiences can create in the marriage system. For the moment, however, let us emphasize again the compelling attractiveness of this anticipated future of love growing in a dyadic closeness.

In addition to this appeal, the marriage system also offers the couple the sanction of having their own children and moving from the role of children themselves to that of mother and father. Indeed, in these days when young people so often easily live together with no formal marriage vows, it is the wish to have children legitimated in society's eyes that leads many of them, if not to the altar, then to the desk of the county clerk. From another perspective the desire to legitimate the child premaritally conceived is the most telling motive in many teen-age marriages that are statistically most likely to end in early divorce.*

If the desire for children motivates the origin of many marriage systems, then the presence of the children binds the marriage more tenaciously than any other factor. Children conceived in love so often become the main reason for continuing the marriage when the love is gone. Although love for the children may often be a telling factor in one parent's decision not to leave, it is also true that the children can serve as a convenient rationalization for those who fear to move on in their lives. Some partners have become so dependent on the marriage for realistic or emotional reasons that they dare not attempt to make a life apart from their current marriage, feeling that such an effort would be doomed to failure.

A third reason for marriage has a less lofty tone, but nevertheless resounds with practicality. If suburbia has little to offer for single people, the rural climes have less, and urbanity notwithstanding, the big city can be a very lonely place as well. In other words, the larger society is compatible with the marriage system, and one might even surmise that the two were made for each other. It seems evident that coupled living is favored by all kinds of other systems in the society, from the income tax to the double occupancy. Two may not be able to live as cheaply as one, but in our society the establishment of a marriage system with two major partners is encouraged on every hand.†

A fourth reason for marriage has been deliberately left to last, not because it plays no part, but because it plays a diminishing role in the deliberation preceding marriage. It is the factor of sexual fulfillment. With the widespread availability of reliable and easy methods of oral contraception, and with all the other changed conditions that permit young people an easy and early intimate association, sexual experience before marriage has become more and more the usual course, and marriage is more rarely hurried along on the wings of intense but frustrated lust.[49,50] One might think that with increasingly liberal sexual attitudes and expanding sexual freedom the importance of romantic love as a central force in mate choice and marriage would diminish. But romantic love is actually not that closely associated with sex itself; rather, it grew from the eleventh- and twelfth-century courtly games of idealizing the "loved" one beyond all reality, and in the haze of daydreaming ecstasy imagining a life of superb joy lived in idyllic wonderment and delight in a never ending transcendence of rapture.[51,77] The myth of this impossibility spilled from the chambers of the court, beyond the aristocracy, and today the romanticism floods our lives, given special propulsion by the power of multimedia advertising. The promise of perpetual youthful glow, of poised and promising breast, of never fading allure sweeps through the airwaves and drowns us in a commercial gush. But still the seductive promise of a life of everlasting love, of the one and only, falsely leads the multitude to anticipate the continual glow of warmth and closeness. Alas, the romanticism has a short half-life, and the hopes of one who has depended on the idealized romanti-

* The divorce rates per 1,000 married persons, from four selected states having rates available, 1960–61: the total divorce rate for men under twenty was 24.8, and for women 29.0 (p. 57).[16]

† A great deal more is known about the divorced population than the separated . . . at the time of the 1960 census, over 2 million were separated and over 3 million were divorced. (p. 222).[16]

About 2.7 million men in 1967 were bachelors (statistically 35 years old or over who have never married), and 2.8 million women were spinsters (statistically 30 years old and over who have never married). (p. 298).[16]

cized image of his beloved are soon shattered by the day-to-day realities with which the marriage system has to engage.

Who Is Chosen

From considering why people choose to enter a marriage system, let us continue our exploration by looking at the partners people actually choose. In cultures where families arrange marriages the considerations are clear. Economic and social advantages play a large part, and if love grows between the marriage partners, all well and good. Of course, in our own society many marriages are still arranged, although the details of the arranging are often kept obscure. Nevertheless, marriage within the same class not only is common but is given much support, especially in the upper class, and when this tradition is broken, wills are frequently changed and the details of trust funds altered so that the passage of wealth and power can be controlled and channeled. It is true that in this country considerable marriage across vertical class lines take place, and certainly the trend to upward mobility favors this kind of intermarriage. In spite of this, however, and in spite of the apparently democratic idea of the freedom to marry the person one loves, the great majority of marriages occur between partners who share common cultural, ethnic, racial, and educational backgrounds.* For example 99.8 per cent of white spouses are married to white spouses, and 99 per cent of black spouses are married to black spouses. The least crossing of class occurs in the upper and politically powerful classes where the largest stakes in terms of money and power are in jeopardy. Royalty and the upper classes have always been keenly aware of the importance of marriage in maneuvers of political power, and although they have always had more sexual liberty and freedom of relationships than the rest of the population, they have been scrupulously careful in the way marriage contracts have stipulated the ownership and inheritance of property. Undoubtedly much of the age-old concern for women's chaste behavior has grown from men's fears that their wordly gains would be inherited by another man's children. This consideration, of course, has relevance to the status of women, to the esteem in which women are or are not held as persons; it will be explored further in the section dealing with the nature of the relationship between the partners in the marriage system and with the burgeoning women's liberation movement.

Reference was made to marriage across vertical class lines, but horizontal mobility is also an important factor in our current culture that affects not only the mate choice but also the quality of married and family life that develops. Large proportions of our population move every year, often to new areas thousands of miles away from the old home. With this constant shifting, kinship systems have greater difficulty maintaining any meaningful relationships, and the new marriage system, when it is established, must begin to function often with very little outside support from the families of either partner. Although many attributes of such newly formed couples may lie within narrow and similar boundaries, such as educational level, race, and religion, many other features may be quite unfamiliar, having to do, for example, with local customs, verbal and nonverbal communication patterns, and even general life outlook.[18,86]

Mate choice may also be influenced by personal manipulative factors, such as marrying the boss's daughter or marrying a doctor or

* Among couples who married during the 1950's and were still in their first marriages in 1960, the median educational attainment was 12.3 years for both husbands and wives.

About three-fourths of the married couples in the United States in 1960 comprised a husband and wife of the same national origin. English-speaking and German persons of foreign stock had high rates of outmarriage to persons of other origins including native Americans, whereas those of Polish, Russian, and Italian foreign stock—with a higher proportion of first-generation Americans—had lower rates of outmarriage. Persons of Russian foreign stock very rarely married persons of Irish or Italian foreign stock.

A cross section of married Protestants and Roman Catholic adults in the United States have a similar moderate amount of intermarriage; married Jewish adults have relatively little religious intermarriage by comparison with the levels for the other two groups.

The 1960 census showed that only 0.4 per cent of United States married couples had a different race reported for husband or wife (pp. 391–394).[16]

lawyer. And only the diehard romanticists would say that such marriages could not succeed. Although that sought after quality of close intimacy may be lacking, unions based on these self-serving yet solid motivations may endure and fulfill the lives of those so united in many ways. Other reasons for mate choice may fall into an increasingly neurotic category and be related more to the personal distortions and inadequacies of one or both of the partners than to cultural or other forces in the general social system. An immature person, for example, may seek a new family haven in a marriage at the point where other demands that may be experienced as terrifying are presented. Rather than go to work or continue in college, a young person may see marriage as a way to avoid the anxiety these other routes may evoke. If a boy or girl is insecure and frightened of social venturing, the offer of a sinecure may be overwhelmingly attractive. In a society where a woman's value as a marriage partner is distortedly based on her physical appearance and her sexual coefficient, many young women still caught in the trap of such a dehumanizing attitude may actually decide to accept a marriage proposal "now when there is a chance" rather than wait, because to most waiting may mean waiting forever, and to many in our culture this seems like a fate worse than married death.

The basis of neurotic attraction may be even more closely tied to the intimate interactions between the two people involved.[65] Many very comfortable arrangements work out between people who have vividly disparate characteristics. By no means is the attraction of opposites to be construed as neurotic in itself. Rather, the delight and stimulation, the excitement and adventure, the peace and comfort that can come from such intermingling of differences make for more richness in life's experiences. However, cases where the differences fulfill neurotic patterns—although they may create temporarily stable arrangements that allow greater freedom for both partners in the marriage system—may also result in vulnerable linkages that will be least able to withstand stresses of living. Then the individuals involved may not be able to adapt by themselves to a new life in which more independence is required.

Functioning in the Marriage System

All factors notwithstanding, the marriage system is entered into. Those who have the temerity to give advice may say that it should not be undertaken too young, and yet not too late, that marriage is best undertaken between those of different natures, and yet between those of the same background. Statistics show some convincing data on the marriages that have some better chance of surviving,* but other critics might question whether survival of a marriage is any valid criterion of its worthwhileness.[58] It would seem that when it comes to a specific case the task of advising a couple whether to marry is most difficult. What makes the prognosis even more uncertain is the very fact that marriage is a system and will evolve in relationship to all the other world and life forces; thus, the mass of factors

* In 1960 rates of dissolution were highest for those who married in their teens and lowest for those who married in their twenties. Persons with marriages "not intact" included the separated as well as the divorced and the widowed; among these persons the proportion of nonwhites was about twice as high as that for whites. The percentage of adults who were divorced was smaller among the foreign born than among the second generation.

The Northeast had by far the lowest percentage divorced. The West had uniformly the highest percentage divorced. The rural areas generally had the lowest percentage divorced and separated, and the large cities had the highest.

Well-educated white men had a much lower percentage of divorces than the less educated. The relationship was more complex for nonwhite men; the percentage divorced rose irregularly as education increased to the level of entrance into college and then fell as education increased further. Likewise among women, rising education up through the early college years was associated with a rising divorce rate; although the percentage of divorces was lower for college graduates, it reached its highest level for women with graduate school training.

Occupations in which especially small proportions of divorced men were found in 1960 included accountants, college professors, draftsmen, physicians, and high school teachers. Corresponding occupations for women included librarians, music teachers, and elementary and high school teachers (pp. 400–403).[16]

affecting the prediction lie mostly in the future and beyond the grasp of the most skilled analysts or readers of fortunes.[72]

Once the system is established it has a most complex evolution, indeed.[29] The course will be different depending on how old the partners are; what is their economic underpinning; whether it is a first, second, or third marriage; whether there are auxiliary members of the system at the start, such as children from earlier marriages or relatives who will be living in the same household. Obviously I cannot undertake consideration here of every possible combination of circumstances. To begin with, then, let us consider some of the challenges and situations that must be met by a newly formed marriage system in a somewhat uncomplicated setting with no auxiliary members of the system as yet.

The Newly Formed System

Earlier mention was made of the existential truth that in one sense each of us is alone in this life. And yet in another sense we are never alone, and our way of life can never be understood except in a social context, in relationship to others. In each system of which we are a member, we have a usual role or series of roles, and our social beings are defined by this multiplicity of roles. Our individuality, therefore, expresses itself in the number and complexity of roles we have in different contexts or systems. A sense of self, or person, runs through all the roles, but the role self is no less real because the behavior changes from role to role. The idea that in marriage "two become one" has a romantic beauty, but a practical pitfall. The individual who grew from a child in a family was part of that family system. But as he grew he joined other systems at school, in social groups, and at work. As he developed he usually organized less and less of his life in conjunction with the family system, and, indeed, he often had formed a separate existence in several ways by the time he was ready to enter marriage. The marriage system as a rule requires a more stringent "togetherness" than any other system, even the primary family. From being an individual involved in many systems, one suddenly merges in the identity of a "couple," and it is in the constricting implications of this coupling that the narrowness of many marriages begins to pinch. Of course, the renowned honeymoon period may so counter the disadvantages in the coupling system that weeks or months (or occasionally years) may pass with the spouses so involved in the pleasure of their relationship together, and with the possibilities of growth contained within it, that the limitations are not experienced as such. The newly formed system has much pleasure and challenge to offer. Since close living will be one of the hallmarks, there is a continuing communication on a verbal and nonverbal level, both in awareness and beyond awareness, during which basic rules of living are established.[6,10,45] The intricacy is such that rules must evolve about who makes rules, and in this way hierarchies are established in different areas. Such achievements are necessary for an orderly existence that is not always brewing anxiety in its confusion. Perhaps no area is more important for the establishment of a tolerable and workable marriage system than this one of communication. Greene[35] has noted in his categorization of problems most often brought to the counselor that problems in communication now rank first.

In the early period of the marriage, in addition to establishing modes and rules of communication, the couple also may be enjoying an opportunity for experience together, with just the two of them, that may be very sweet and quite rare in their previous experience. It is true that the trend now, especially in cities, is for joint living before marriage; as a result this kind of close personal living experience, together with working out some of the practical problems of living, occurs before the formality of marriage vows. An advantage of these informal arrangements is that one person may live with several different partners; thereby he learns that the experience can be very different with different people, and yet that some of the living situations that develop have to do specifically with the necessities deriving from the situation itself, and not from the particular people in it.

In the early stages of the marriage there are also many new and exciting tasks that can best be accomplished by the two working together, and from this kind of joint effort a good deal of satisfaction is inevitable. Usually a new living quarter is needed, which must be furnished.* Beyond this all the chores and everyday jobs must be accomplished, and with the development of the communication patterns, agreements and contracts are developed for handling these things. In the economic area agreements must also be made, either openly or covertly, about the patterns of earning, buying, and saving that will develop in the particular marriage system. Two people from similar backgrounds and having similar outlook will have less difficulty reaching agreement on such practical issues. Agreement, of course, also depends on the creation of adequate communication, and factors other than background determine this, varying from personal communication style, to personal prejudice, to neurotic or double-binding communication experience in the past.

The "Couple" Identity

Although the honeymoon period can be happy, productive, and very important in laying the basis for important patterns to be followed as the marriage continues, there still is the potential for much trouble deriving from the consideration of the individuals in the system as a "couple." The veneer of enchantment in being known as "that nice couple" can wear very thin. It may be true that entering a marriage system does not necessarily mean closing down other possibilities for personal growth, and, indeed, the thesis that marriage *need* not mean this will be pursued later in the chapter; but the experience of a multitude of married persons has been that entering the marriage system has meant embarking on a voyage down an increasingly constricted tunnel, chained by love and obligation that eventually erode to hate and resentful duty. The reasons for the erosion may actually relate less to the closeness than to the isolation that is so often associated with it. And the reasons for the isolation vary in each marriage system, yet may stem from personal characteristics, from the nature of the marriage system itself, or from the context of the larger society.

Reference has already been made to our highly mobile population; this mobility not only breaks up kinship systems but also disrupts friendship systems, so that no longer do groups of people easily maintain friendships over many years. When a family is going to live for a few years at most in one community, and then be relocated on the other side of the country for the next few years, the tendency to develop a self-sufficiency as a closed system is encouraged. In high-rise apartments families are often engulfed and overwhelmed by sheer closeness and numbers, and in the midst of the crowd become more isolated rather than less. Paradoxically as space is squeezed, and community areas for recreation and even local shopping where friends can meet are eliminated, families are increasingly isolated in their living arrangements. This is true as well in the suburbs where neighbors may live next to one another for years and yet never meet or even speak for more than a superficial exchange. Again the isolation is the key, with the resulting turning in. A further factor is work for members of the family. Places of employment may be far from the place where one lives. If both husband and wife work they may travel in different directions. This could lead to an independence that one would think might counter the effects of the too great closeness. But if both must return to be together after work, the advantages of the separateness cannot be exploited, and the strings of the marriage "bag" may be drawn more closely and tightly.

* During the first year of marriage one in every ten couples lives with relatives, according to data for 1960. For nonwhite couples the rate was more than twice that high.

Homeowners were twice as numerous as renters in 1960. Even among young couples married less than five years, every third couple had already started to pay for a home of their own (pp. 394–395).[16]

Each couple's marriage system is only one in a virtual sea of such relationships. The pressure, therefore, is to stay within each narrow system, not only because of factors in that particular marriage, but also because if one partner would like to move out more freely, he or she may find no welcoming for this, since the multitude of others are sequestered in their own togetherness. Social events tend to be organized for singles or for couples. Both partners in a marriage system are usually invited out for dinner, to a party, or to the theater. If independent activities are included, they most often are organized along sexually segregated lines, such as women's card groups or men's bowling clubs. If sexually mixed activities are sanctioned they most often are related to some specific task, such as educational classes, parent-teachers associations, political clubs, and the like. From these examples one can see that opportunities for one individual to develop himself or herself in a separate direction are to some degree available in spite of social pressures favoring couple togetherness.

The greater pressure, however, may come from within the structure of the marriage system itself, for in such a closed system expectations are set up that work against individual spontaneity and that tend to regulate the lives of the people in the system. Since so much of life is approached from the need to work in the system mutually, compromises are made and each partner tends to gear his life more and more in relationship to the other. Thus, if the wife is working during the day and taking classes for a degree some evenings, expectations will be set for the times she should be through and ready to spend time with her husband. Since he may have arranged his time to meet her after class, her option of making arrangements with someone later is limited. To expand this example, the pressure of expectation and accountability that can develop is enormous. In addition to the social and system aspects of the closeness, personal characteristics must also be considered. One of the most insidiously destructive forces in a marriage system is the neurotic dependency that develops and, indeed, is fostered by the very way in which the system now operates.[59] To be cherished, loved, and cared for by another person is truly a wonderful experience. To be able to turn to another person for help and to find them dependable in their concern for you is also a most gratifying human experience. But to define, especially unconsciously, your own value in terms of the way another human being behaves in relation to you is to be not only personally underdeveloped in self-esteem and maturity but also tremendously vulnerable to hurt. The corollary is that this neurotic dependency on another for one's own sense of self leads to excessive demands on the other person that disregard the other's own individuality and personal integrity and attempt to control the other person. The aftermath of such relating is a deluge of resentment, frustration, constriction, and possibly depression or rage.

In brief, then, a marriage in which the partners are free to grow as individuals and to relate themselves to other systems in a full and human way is not much encouraged by the nature of our society, is handicapped by the ubiquitous marriage system as it now exists, and is virtually impossible in any marriage system where either partner is personally so insecure that he operates in a neurotically dependent way, opposing the growth and expansion of his spouse as well as the development of himself.[20,53]

Women in the System

The particular place of women in our society must be considered here, because it plays such a crucial part in the marriage systems that are developed, and also because it is being challenged anew with a vigor that already has shown results.[30,64] That women have been discriminated against as a group seems beyond dispute. Documentation of the extent of the discrimination is easily found, and only relatively recently have women even begun to have equal rights under the law in civic matters as basic as the right to vote. But the effects of the discrimination are so widespread and so universal that the lives of women in our society are adversely affected either directly or indirectly every day in every place where women are to be found—the home, the school, work, the arts. In spite of revolutionary efforts on the part of women's rights advocates, there is a remarkable persistence of the conviction that a woman's place is in the home, that a girl should be raised primarily to be a good wife and mother, and that, indeed, if this does not continue the very institution of the home and family will be undermined and the future of the country itself placed in jeopardy. Raising a girl to believe that her basic goal is to become capable of attracting a man who will then marry her may make it almost impossible for her to achieve her full potential as an individual person in her own right. When such tremendous emphasis is placed on physical attributes, as is done in our culture, effectively forcing the woman into an image of a plasticized sex object, the damage to dignity and personality is likely to be extensive. A further crushing bind is placed on the girl when the idea is insinuated that to the extent she develops herself as a person and shows herself to be intelligent and capable she will be running the danger of engendering competitive and hostile feelings from threatened men, who will then relegate her to the role of a "castrating woman" and thereby diminish her chances of finding relationships with interesting men and eventually forming a marriage relationship. The poignancy of this problem can be no better illustrated than to note the incredible fact that in a test involving the completion of a story about a student who placed first in a medical school graduating class, the women being tested, more often than not, saw this honor as a disadvantage when the first place student was a woman; in contrast, when the top student was a man, the distinction only heralded a future of continuing benefits.[44]

Not only does the denigration of women as people have stifling effects on the growth of a maturing girl, but also the deleterious effects continue throughout the marriage and eventually lead to intense resentments and possibly tragedies.[15] The superior attitude of men fosters an atmosphere of competition and lack of equality that breeds chronic domestic warfare and alienation. In the later stages of marriage the woman is often more hopelessly trapped in an intolerable life because she lacks the grounding that could enable her to develop her own individuality and find a new and independent, if not separate, existence for herself. If the attitudes discussed encourage a battle of the sexes, small wonder that it is over sex that many of the battles are fought. Again it must be noted that many in the younger generation today have moved a long distance from the attitudes described. With the acceptance of equality between men and women has come the recognition of women's right to enjoy their own sexuality with the same freedom traditionally accorded to men, both married and single.[11] Modern contraception has provided much more freedom for women, and liberalized abortion laws, such as the one passed in 1971 in New York, has added further social sanction to the new sexual and personal independence of women. As women truly are given more equality, their degraded position as sexual objects loses relevance, and sexuality between men and women will also express their mutual humanity, and not be "the" prize either taken, or withheld, or proffered.

Early Phases of the System

In our review of the marriage system we have considered the newly married and some aspects of the way in which the partners develop rules of cooperative behavior and communication systems, as well as some of the tasks and goals related to this phase of their lives. Often both may work while one or both continue in school, and a period of adaptation ensues. Divorce is frequent at this early stage, especially in those who marry young. Probably the young have not each developed sufficiently as individual persons to allow a contract that has much likelihood of endurance. Before long, the growth in both partners may be such that disparities develop that can never be bridged. This, together with the adventuresomeness and energy of youth and the more liberal attitudes to divorce already present in our society, makes it more understandable why the rate of divorce among the very young is so high.[79]

For those who enter the marriage system and do not soon leave it for divorce, the next important phase may be that of childbearing and childrearing.* Although more mothers continue to work after the birth of a child, and although day-care centers are being established at an increasing rate, a large majority of mothers stay home during the years when the children are young. The advent of children makes a drastic change in the system, for while a marriage system remains, a family system is now established; and even though the marriage system can end by death or divorce, the family system does not, and the members of that system (mother, father, and siblings, plus all the extended family members) continue in their family relationships as long as they live. Children, let it be said again, profoundly affect the marriage system. If the system was locked in before, the ties that bind are now vastly multiplied. Responsibility for the children tie the parents to each other as well as to the children, and the love and affection for the children create some of the strongest bonds and devotion known. Not only do children raise the cohesion coefficient, but also they alter the intimacy factor in the marriage system. Earlier we mentioned that the close intimacy in the marriage often resulted in an isolation and alienation of the partners from other relationships. With children added to the system not only is the former isolation from others still present, but now the closeness between the husband and wife is encroached upon. Intimate times are difficult logistically, and when they are finally arranged the husband and wife (now also father and mother) may have little energy and incentive left. Such are the joys of parenthood.

* More women now than a generation ago are sharing in the process of bearing and rearing children. Now only about a tenth bear no children, as compared with a fourth of the older generation. A part of this change resulted from a drop from about one-twelfth to one-twenty-fifth in the proportion who remained single through the childbearing period. At the same time the average number of children has declined from about four to two and one-half (p. 394).[16]

The mother more often than not bears the brunt of the added work and strain at home, while the father increasingly becomes aware of his new financial responsibilities. The importance of maintaining and even increasing his income binds him further not only into his family situation but also into his work situation. Such phases in development are often referred to euphemistically as "learning the realities of life." The fact that marriage systems are organized as separate units contributes to the strain because sharing the tasks is difficult, and the continual yet often trivial chores are not relieved, even by the relatives, the aunts, grandmothers, and sisters who used to be available when the family was a larger conglomeration of generations. Wives often grow depressed, and withdraw from relationships with their husbands. The sexual relationship may be adversely affected, especially if it is experienced as a duty and an exploitation. During this period men are more likely to seek relationships with other women as relief from the tedium at home with its incessant demands and frequent sexual rebuff.

Stresses on the System

As we continue to examine the flow of the dynamics in the marriage system, we can see how the relationship between the man and woman fluctuates in response to a multiplicity of factors both in their own personal growth and in the marital situation, which, in turn, is affected by forces in the larger society as well as by the addition of children. When stress threatens to disrupt the system, adaptational changes are necessary, or the dysfunction engendered will be manifested in unusual behavior or symptoms in one or more members of the system, or in some unusual function of the entire system. Gross disruption of the system would include one member leaving the system entirely, or the collapse of one member, such as hospitalization, or violence among the members. Stress may be applied by forces outside the marriage system itself. For example, the husband may lose his job, not as a result of personal incompetence, but rather as a result of overall economic recession in the society.

Or severe illness may strike a child or the mother and require a mobilization of all other family members, with a reallocation of priorities and a major shifting of emphasis in everyone's lives. The children's development brings a series of crisis points, and these can be moments when the adaptational capacity of the family may be sorely tested. On the other hand, at such moments of stress the family has the greatest challenge and opportunity for growth, and each member of the family system may have a new occasion to redefine goals and relationships, so that not only may the family continue as a more integrated and successfully functioning system, but also each individual might have an enlarged personal capacity.

The problems that children bring in infancy and early childhood are often the source of much tension because the parents are aware of the literal helplessness of the youngsters, and therefore feel especially keenly the responsibility that falls on them as adults. In later childhood school and peer problems are often the focus of family stress. Then as adolescence approaches and the children increasingly move in their own directions and challenge the authorities at home and at school, a whole new series of tensions must be met that take on a more social aspect and that break into the insularity of the family. Here again a paradox is seen, because in our society family life has been constricted as many of the functions formerly undertaken in the family, such as education and recreation, are more and more being performed by other systems in the society. Education in schools now includes sex and family life, while the education brought into the homes through television completely overwhelms the attempts in many families to create some special and unique direction for that particular family. The tensions that began as children neared adolescence burst into clear-cut warfare among teen-agers and parents in many families, and the alliances and blame thrown back and forth may stir a turbulence that rocks the foundations of the family and the marriage. At those critical points—infancy and early childhood, preadolescence, and the teen-age years—the stresses in the family may become so severe that counseling or psychiatric help is sought. When family therapy is sought the issue most often resolves back to the marriage system, and the family adaptation is strengthened when the marital partners can redefine their goals and relationship and then cooperate in managing ways to meet the demands and responsibilities of the growing family, while at the same time not being engulfed to the extent that no life as individuals, or as man and woman, is available.

As a family grows the economic burden usually increases as well, and if college education is included in the children's future the economic strain and the sacrifice demanded may be quite severe. But assuming that a family manages to hold together through these stages, a new and most critical point in the marriage system is in the offing. At the time when the children have reached the point where they are more and more responsible for themselves, and certainly by the time some or all of the children are either in college or are working for themselves, the focus inevitably begins to turn again to the marriage relationship and to the fate of each of the two individuals in that relationship.

Middle Phases of the System

For the woman in many families, moments of personal decision may have been experienced when it was planned to begin a family, when the youngest child entered school, leaving more of the day free for other activities, when children reached their teens and were able to provide more care for themselves, and then when most of the children were old enough to leave home. At such moments of decision, many women resume earlier employment; some mothers may go back to teaching when their own children are in school, and others may reenter jobs in business. Other women, who had always wanted to further their education, begin to take courses, perhaps on a part-time basis, and a large number of women in their late twenties and early thirties become very aware that unless they begin to make progress in their own development they may not have the opportunity later, for it will be too late.

For these reasons, among others, many women reach a point of crucial decision in their thirties, when they review the lives they are living, and consider whether any changes may be possible. At this point women may feel that if they are ever going to make a change in their marital situation they would best do it now, for their youthfulness is still reflected in their appearance and their energy, and the possibility of establishing a new marriage is still a realistic one. In addition, many women now feel that an unmarried life provides greater possibilities of satisfaction than a married one, and so the fear of this different life style is not nearly as intimidating as it once was, nor is the social stigma of divorce and a single life nearly so powerful, or likely to be incurred.

At this time the husband may feel more pressure to increase his income, and many men may take on an extra job, or work overtime, or undertake further education in the evening. In addition, owing to the competitive business atmosphere, many men until the age of forty are engaged in an intensive struggle to reach a position of some strength and stability, because after forty further significant advancement will be unlikely, and the pattern for the future of the employment career will be fairly well settled. On this somewhat different timetable, men may experience their time of crisis between forty and forty-five, when they realize that this is their last chance for a major change in life style. At this point many men form new relationships with younger women and experience a resurgence of youthful feeling, with the desire to live through many of the fantasies that they may have suppressed in their adolescence.[27] In a sense a second adolescence may emerge, with the feeling that if the home, in this case the home of the secondary family, is not left at this time, the future will be closed, and the old life will continue until true death brings an end. At this time many men also make rather radical changes in career, or at the least make a major shift within the same field, possibly involving a geographical move as well.

In some instances a third possibility is chosen, when both husband and wife are able to change together and evolve a very new kind of life for themselves. When this does not occur, and when the resolution does not result in a continuation of the old life pattern, divorce ensues, and friends may wonder at the dissolution of an apparently successful marriage of over twenty years' duration.[91] However, the real surprise may be that the divorced individuals often are infused with a new excitement, and that both may go on to experience happier and more fulfilled lives. The course of the marriage, it would seem apparent, has to do not only with the external realities of children and finances but also with the intrinsic strengths of the marriage partners as independent individuals. This theme, then, keeps repeating itself; namely, that a crucial factor in determining the course of a marriage, as well as determining what happens after divorce, is the degree of personal individuation and personal security and development of each partner.

Later Phases in the System

When the marriage system continues after the critical period of middle age, new goals emerge. The possibility of more leisure time together may bring new enjoyment into the lives of the couple. The loss of friends becomes more common, and the facts of aging and death are increasingly evident. Illness may have been experienced before, but now the certainty is that illness and decline will be more and more an inevitable part of the agenda. Often families established by children may be a large source of interest and delight, but again the tendency for the breakup of the closer kinship systems in modern America affects the intensity of such relationships, so many older people may have to adjust themselves to distance and infrequent visiting. Although more affluent older people may be able to establish retirement ways of life, large numbers of the population are left increasingly to their own resources, which usually are meager; with their failing health and the inadequacy of publicly supported programs, many are doomed to an increasingly isolated and frighteningly lonely existence. The dependence of the married couple upon each other becomes more crucial and necessary than ever. It can be the source of tremendous resentment, as well as of severe anxiety, for roles of who will be dependent on whom may suddenly, after a lifetime, forcibly be reversed. Goldfarb[34] has noted that the crux of most marital problems in older people is their handling of dependency relationships.

Changing the System

Intrinsic Factors

In the discussion so far, we have explored some of the features of the marriage system. We have considered reasons why people may join the system, factors influential in the choice of a partner, problems in establishing a workable system, and an overall view of the goals and difficulties present at different stages in the evolution of life as the marriage system attempts to provide a workable arrangement for its members within the intertwining matrix of all the other systems in the society. At this point let us examine more closely certain intrinsic factors in the traditional system that reflect both its strengths and its shortcomings, and particularly let us examine the relationship of the individual in the marriage to the system itself, to himself, and to the larger world.

Monogamous marriage contains within it a dilemma, because the flourishing of love, which is one of its top priority goals, is discouraged by all the features that coerce and demand allegiance and affection. Compulsion breeds rebellion and resentment; duty and obligation discourage spontaneity and delight. Love cannot be forced by legal decree and formal contract, any more than it is engendered by blame or guilt. Thus, all the arrangements, formal and informal, that insure the binding of the marriage and that are justified by the necessity of protecting property rights, and the social good of preserving stable family situations for the nurturing and rearing of children, serve at the same time to entrap the partners in the marriage and provide a soil hostile to the development of mutual affection and love. The commitment sanctified in the promise to forsake all others becomes instead a captivity, and the added irony is that the spouse who promised to love now hosts a growing resentment, and at the same time through his own vow has cut himself off from the love of those he has forsaken. Nor in this system of values is much thought given to the pain of those who are forsaken.

The problem facing the individual partner in the marriage system is to establish a modus vivendi permitting his own and his partner's individual growth without jeopardizing the possibilities for an intimate relationship together; at the same time he must assure a cooperative and mutually respectful endeavor to meet all the housekeeping and other maintenance necessities, and fulfill, in addition, the requirements and obligations to children and members of the community at large. This task is almost too much to expect from any individual, particularly in a system that puts so many barriers in the way.

Cuber and Harroff[22] undertook an in-depth interview survey of 211 middle-class married people, and from the detailed reports concluded that there were five main marriage arrangements. The classification was based on the interview material of people whose marriages had lasted ten years or more and who said that they had never seriously considered divorce or separation. The first type is the "conflict-habituated." In this relationship there is considerable tension, although it is usually controlled. "There is private acknowledgment by both husband and wife as a rule that incompatibility is pervasive, that conflict is ever potential, and that an atmosphere of tension permeates the togetherness."[22] The authors note that some psychiatrists have speculated that the deep need to do psychological battle with one another constitutes the cohesive factor insuring the continuity of the marriage.

The second type of relationship is described as "devitalized." The authors write, "These people usually characterize themselves as having been 'deeply in love' during the early years, as having spent a great deal of time together, having enjoyed sex, and most importantly of all, having had a close identification with one another. The present picture . . . is in clear contrast—little time is spent together, sexual relationships are far less satisfying . . . and interests and activities are not shared."[22] Further comment is made that this type of "duty" relationship in marriage is very common, and that those in it often judge that "marriage is like this—except for a few oddballs or pretenders who claim otherwise."

The third described type is the "passive-congenial," which is much like the devitalized, except that the passivity pervading the association has been there from the start, giving the devitalized at least a more exciting set of memories. The authors speculate, "The passive-congenial life style fits societal needs quite well also, and this is an important consideration. The man of practical affairs, in business, government service, or the professions, quite obviously needs 'to have things peaceful at home' and to have a minimum of distractions as he pursues his important work. He may feel both love and gratitude toward the wife who fits this mode."[22]

The fourth category is in extreme contrast to the first three and is called the "vital" relationship. The essence of it is that the mates are intensely bound together psychologically in important life matters. As one of the interviewed subjects said, "The things we do together aren't fun intrinsically—the ecstasy comes from being *together in the doing*." The authors elaborate, "They find their central satisfaction in the life they live with and through each other—all else is subordinate and secondary."[22]

The final type outlined is the "total" relationship, which is described as like the vital with "the important addition that it is multifaceted . . . in some cases all the important life foci are vitally shared." Both these last two types are described as rare, both in marriage and out.

From the description of these five main types, it can be seen that factors other than the individual personality characteristics of the partners played an important part in the style of marriage arrangement. Relationships outside the marriage, whether these included sexual involvement or not, play a significant part in the lives of many people. The restriction of freedom of relationship can be the single most eroding factor in the marriage system. By having the freedom to develop other relationships, and to engage in activities with other people, a marriage partner can pursue his or her own development in a way that is not possible if most activities are restricted to those engaged in with one's marriage partner. Possibly one of the most destructive demands in the covert marriage contract is that which encourages the belief that one person can supply another person's needs in every area. A man or woman in this framework is expected to be a companion, a friend, an intellectual equal, a lover, a helper, a parent, a sharer in interests in sports, and on and on and on. No one person can meet the needs of another in all these ways, and, indeed, no one may be able to satisfy another fully in any one of these areas.

If the marriage system were not threatened by the inclusion of other people, it seems possible that much enrichment would then be made available for each of the partners. In turn, as each flourished with the addition of this new stimulation, he or she would return to the marriage relationship renewed and with more to offer the partner. In this light the idea grows that loving one person increases ones' capacity to love others, and that love is not diminished by the inclusion of others. In practice, however, the inclusion of other relationships frequently does threaten the marital relationship, and probably for this reason so many of these outside relationships are carried on clandestinely. This, in turn, poses a dilemma for many, who feel uncomfortable with this double role and who therefore suffer guilt and anxiety. As a result they either have periodic unsatisfying "flings," or live with a burdened feeling of uneasiness, or eventually break the doubleness by breaking the marriage, or by giving up the double life and settling for a more drab and unsatisfying continuity in their marriage.

Therapeutic Approaches

In a discussion of marital therapy, the first focus should be on the problems. In the chapter so far, problems in living of almost every sort have been discussed. Within the definition of the marriage system, it is evident that any problem that affects any individual in the system, or in the expanded family system, or any situation that influences the effective functioning of the marriage system can be considered a problem related to the marriage. In a sense, though, this means that almost every human problem can be seen as a problem related to marriage and the family. Indeed, there is much to commend this point of view. With the development of general systems theory and theories of communication, the idea that any human behavior can be understood outside the context of its occurrence has lost ground.[40] The increasing expansion of family theory and family therapy is a clear-cut application of these newer theories to actual clinical problems. More and more frequently, when an individual, be he a child, or the adult in a marriage or in a larger family system, is found exhibiting behavior that could be labeled "symptomatic," or in behaving in a disruptive manner, that particular "identified patient" is understood by clinicians as a member of a system, and his behavior is understood in the light of the behavior of those around him, and in the particular context of his living. So it is, more specifically, that when one member of a marital partnership has behavioral or symptomatic disturbances, the context of the marriage itself is now frequently included in the therapeutic effort. Some clinicians have defined family therapy, of which marriage therapy is a variety, as a method for treating the family, as opposed to treating the individual. However, Haley,[39,41] a pioneer in the field of family therapy, has noted that the concept of family therapy is inclusive and operates even when a single member of a family system is treated by a therapist. He makes the point that no one participant in a system can be influenced in his behavior without that influence having an effect on all other parts of the system. Briefly, then, he holds that family therapy is all that *has* ever been practiced, although often the practitioners were themselves unaware that this was the case.

Before continuing with any further discussion of therapy, let us return for the moment to a further appraisal of the problems. Leaving aside those behaviors that appear in individuals, such as depressions, acute psychotic episodes, anxiety states, alcoholism, and various behavior disorders, let us consider the complaints that overtly can be related to the marriage system and in fact are presented as such. Greene[35] has provided a recent listing of the major reasons couples give when they appeal for professional help with their marriages. Based on the responses of 750 couples, the survey shows that the most frequent complaint now is lack of communication. The next eleven specific marital complaints, in order of frequency, are constant arguments, unfulfilled emotional needs, sexual dissatisfaction, financial disagreements, in-law trouble, infidelity, conflicts about children, domineering spouse, suspicious spouse, alcoholism, and, finally, physical attack.

These, then, are the leading reasons why partners in marriage seek counsel and therapy. Since we have tried to look at marriage as a system, it would be consistent to consider the marriage therapist as a consultant to that system. This is possibly a different way to consider the role of the psychiatrist, but it has some advantages. For one, the therapist is not placed in the position of having to be a saving influence. Rather, the initiative is clearly left with the partners, who come to an expert for appraisal and consultation, who may be able to gain some further understanding of the reasons for the dysfunction of their system, and who may also take suggestions and advice from the consultant and in an experimental mode try out various new possibilities. In light of the previous discussion on the factors affecting the system at various moments in its evolution, one can see that the consultant would attempt to evaluate the strengths and weaknesses of the particular system, in terms of its members and of the supporting systems interacting with it, and he would further attempt to understand the nature of the stresses being applied in both general and specific ways. With this kind of background information he might then suggest various ways in which new operations in the system could be implemented, or in which certain tasks formerly attempted by the system might be given less important priority or abandoned altogether. For example, perhaps grandmother need only be visited once a year, or possibly one partner might cut down on committee activities in a local political club. Again, the consultant might suggest that problems in the way the two partners were communicating might be contributing to massive misunderstandings; by helping with the analysis and correction of such problems, the consultant may contribute to a new level of cooperation and satisfaction. In another problem area the consultant may provide direct help with sexual difficulties, utilizing approaches formulated by Masters and Johnson[61,62]—if not directly himself, then through referral to therapists trained in such methods.

Without further specific examples, it can be seen how the marriage therapist can act as consultant to the marriage system; and by considering the multitude of factors present in the system, and by utilizing his own special training and skills, he can engage the partners in their own therapeutic endeavor.

There is still a difference in approach among those who will see only the couple, those who will see only one of the partners, those who will see both partners, but only in separate sessions, and those who will vary their approach and see at times one, at others times both, and at other times each in separate sessions.[1,2,24,33,48,54,56,60,66,81,83,90] Perhaps some of the confusion can be resolved by realizing that different systems are involved. Each individual is in essence a system in himself, but never *only* a system in himself. The individual always relates to other people, and he relates to people in many different systems, such as his marriage system, his friendship systems, his work system, his community system, and so on. The dilemma for the therapist develops if he must choose between the individuality of the single patient or the continuity of functioning of one of the systems of which the patient is a participant member. This pivotal choice becomes the therapist's moment of truth, for it is at this juncture that he reveals himself as one who believes in the freedom of the individual to choose his own destiny, or as one who believes that the needs and merits of the particular system override those of any particular member, and thereby allies himself as therapist-facilitator to the continuation of the existing system.

Other Directions

In their struggle for growth people move in many different directions, make many false starts and return to start over again, and often never are able to achieve the growth they would like. A couple may seek marriage therapy, but find as they explore the possibilities that one or both do not want to continue either the therapy or the marriage. The variations are many. One or both may then enter therapy by themselves, to continue in the marriage or to break it. They may begin individual therapy and then combine again for marriage therapy. It is also important to realize that people in a marriage system have many other possibilities than therapy. Some, of course, stay stifled in a continuing unhappy and unproductive system. Some seek divorce. Only some seek therapy, which, as has been noted, may eventuate in several possible outcomes. More recently others have been seeking new styles of life that attempt to make over the old marriage system entirely.[14,27] Some have moved toward communal living. Others have experimented with group marriage.[19,42,74] Others have tried to break some of the constrictions in regular marriage by becoming "swingers," combining affairs and other relationships in an open contract with mutual participation.[68]

The divorce rate, and the amount of unfulfillment and unhappiness in marriages that continue, together with the increasing attempts by so many of our younger generation to find new possibilities for relationship, leads to the conclusion that at this time in history the traditional marriage system is having great problems in effectively fulfilling many of its stated functions.[46,70,85,88] Nevertheless, the system does not seem about to disappear; it is so intimately related to the basic structure of other systems in the total society that nothing short of a complete social revolution would lead to its early disappearance and replacement. Therefore, the continuing need for marriage therapists, or consultants, is apparent. Possibly the traditional system can continue to evolve. Perhaps the most promising hope lies in the direction of more openness in the system, of greater freedom for the individuals in the system to develop independently, while yet retaining the core cooperative relationship of intimacy and continuity. In the fostering of this evolution, it is hoped that the consultant-therapist can make a significant and helpful contribution.

Bibliography

1. ACKERMAN, N. W., "The Family Approach to Marital Disorders," in Greene, B. L. (Ed.), *The Psychotherapies of Marital Disharmony*, Free Press, New York, 1965.
2. ALGER, I., "Joint Psychotherapy of Marital Problems," *Curr. Psychiat. Ther.*, *7*:112–117, 1967.
3. ———, "Joint Sessions: Psychoanalytic Variations, Applications, and Indications," in Rosenbaum, S., and Alger, I. (Eds.), *The Marriage Relationship*, Basic Books, New York, 1968.
4. ———, "Therapeutic Use of Videotape Playback," *J. Nerv. & Ment. Dis.*, *148*:430–436, 1969.
5. ———, and HOGAN, P., "The Use of Videotape Recordings in Conjoint Marital Therapy," *Am. J. Psychiat.*, *123*:11, 1967.
6. ARD, B. N., JR., "Communication Theory in Marriage Counseling: A Critique," in Ard, B. N., Jr., and Ard, C. C. (Eds.), *Handbook of Marriage Counseling*, Science and Behavior Books, Palo Alto, Calif., 1969.
7. ASHLEY, M., "Love and Marriage in 17th Century England," *Hist. Today*, 8:667, 1958.
8. BALZAC, H., *Epigrams on Men, Women, and Love*, Peter Pauper Press, New York, 1959.
9. BANDURA, A., *Principles of Behavior Modification*, Holt, Rinehart and Winston, New York, 1969.
10. BATESON, G., JACKSON, D. D., HALEY, J., and WEAKLAND, J., "A Note on the Double Bind-1962," *Fam. Proc.*, *2*:154–162, 1963.
11. BERNE, E., *Sex in Human Loving*, Simon & Schuster, New York, 1970.
12. BIRD, H. W., and MARTIN, P. A., "Countertransference in the Psychotherapy of Marriage," *Psychiatry*, *19*:353–360, 1956.
13. BLINDER, M. G., and KIRSCHENBAUM, M., "The Technique of Married Couple Group Therapy," in Ard, B. N., Jr., and Ard, C. C. (Eds.), *Handbook of Marriage Counseling*, Science and Behavior Books, Palo Alto, Calif., 1969.
14. BRIEN, A., "Anyone for Polygamy?" *New Statesman*, *76*:787, 1968.
15. CADWALLADER, M., "Marriage as a Wretched Institution," *Atlantic*, *218*:62, 1966.
16. CARTER, H., and GLICK, P. C., *Marriage and Divorce: A Social and Economic Study*, Harvard University Press, Cambridge, 1970.
17. CAVAN SHONLE, R., *American Marriage*, Thomas Y. Crowell, New York, 1960.
18. CONE, J. D., JR., "Social Desirability and Marital Happiness," *Psychol. Rep.*, *21*:770–772, 1967.
19. CONSTANTINE, L. L., and CONSTANTINE, J. M., "Group and Multilateral Marriage: Definitional Notes, Glossary, and Annotated Bibliography," *Fam. Proc.*, *10*:157–176, 1971.
20. COOPER, D., *The Death of the Family*, Pantheon, New York, 1970.
21. CRONAN, S., "Marriage," The Feminists, New York, Unpublished ms.
22. CUBER, J. F., and HARROFF, P. B., *Sex and the Significant Americans*, Penguin Books, Baltimore, 1966.
23. DREIKURS, R., *The Challenge of Marriage*, Duell, Sloan and Pearce, New York, 1946.
24. DRELLICH, M. G., "Psychoanalysis of Marital Partners by Separate Analysts," in Rosenbaum, S., and Alger, I. (Eds.), *The Marriage Relationship*, Basic Books, New York, 1968.
25. DURANT, W., *Our Oriental Heritage*, Simon & Schuster, New York, 1954.

26. ———, and DURANT, A., *The Lessons of History*, Simon & Schuster, New York, 1968.

27. ELLIS, A., "Healthy and Disturbed Reasons for Having Extramarital Relations," *J. Human Rel.*, *16*:490–501, 1968.

28. ———, "A Plea for Polygamy," *Eros*, *1*:22–23, 1962.

29. FITZPATRICK, J. P. (S.J.), "The Structure of the Family in American Society," in BIER, W. C. (S.J.) (Ed.), *Marriage—A Psychological and Moral Approach*, Fordham University Press, New York, 1965.

30. FRIEDAN, B., *The Feminine Mystique*, Dell, New York, 1963.

31. FRUMKIN, R. M., and FRUMKIN, M. Z., "Sex, Marriage and the Family in the U.S.S.R.," in *J. Hum. Rel.*, *9*:254–264, 1961.

32. GEHRKE, S., and MAXON, J., "Diagnostic Classification and Treatment Techniques in Marital Counseling," *Fam. Proc.*, *1*:253–264, 1962.

33. GIOVACCHINI, P. L., "Treatment of Marital Disharmonies: The Classical Approach," in Greene, B. L. (Ed.), *The Psychotherapies of Marital Disharmony*, Free Press, New York, 1965.

34. GOLDFARB, A. I., "Marital Problems of Older Persons," in Rosenbaum, S., and Alger, I. (Eds.), *The Marriage Relationship*, Basic Books, New York, 1968.

35. GREENE, B. L., *A Clinical Approach to Marital Problems: Evaluation and Management*, Charles C Thomas, Springfield, Ill., 1970.

36. ———, "Introduction: A Multi-Operational Approach to Marital Problems," in Greene, B. L. (Ed.), *The Psychotherapies of Marital Disharmony*, Free Press, New York, 1965.

37. ———, BROADHURST, B. P., and LUSTIG, N., "Treatment of Marital Disharmony: The Use of Individual, Concurrent, and Conjoint Sessions as a 'Combined Approach,'" in Greene, B. L. (Ed.), *The Psychotherapies of Marital Disharmony*, Free Press, New York, 1965.

38. ———, and SOLOMON, A. P., "Marital Disharmony: Concurrent Psychoanalytic Therapy of Husband and Wife by the Same Psychiatrist, The Triangular Transference Transactions," *Am. J. Psychother.*, *17*:443–456, 1963.

39. HALEY, J., "Family Therapy: A Radical Change," in Haley, J. (Ed.), *Changing Families—A Family Therapy Reader*, Grune & Stratton, New York, 1971.

40. ———, "Marriage Therapy," *Arch. Gen. Psychiat.*, *8*:213–234, 1963.

41. ———, *Strategies of Psychotherapy*, Grune & Stratton, New York, 1963.

42. HALL, E., and POTEETE, R. A., "Do You Mary, and Anne, and Beverly, and Ruth, Take These Men . . . A Conversation with Robert H. Rimmer," *Psychology Today*, *5*:57–59, 1972.

43. HEFNER, H. M., "Excerpts from the Playboy Philosophy," in Ard, B. N., Jr., and Ard, C. C. (Eds.), *Handbook of Marriage Counseling*, Science and Behavior Books, Palo Alto, Calif., 1969.

44. HORNER, M., "A Bright Woman is Caught in a Double Bind. In Achievement-Oriented Situations She Worries Not Only About Failure but Also about Success," *Psychology Today*, *3*:37–40, 1969.

45. JACKSON, D. D., and BODIN, A. M., "Paradoxical Communication and the Marital Paradox," in Rosenbaum, S., and Alger, I. (Eds.), *The Marriage Relationship*, Basic Books, New York, 1968.

46. JEGER, L. M., "Why Not Marriage Reform?" *New Statesman*, *72*:219, 1966.

47. JOHNSON, N., "The Captivity of Marriage," *Atlantic*, *207*:38, 1961.

48. KEMPLER, W., "Experiential Psychotherapy with Families," in Fagan, J., and Shepherd, I. L. (Eds.), *Gestalt Therapy Now—Theory, Techniques, Applications*, Science and Behavior Books, Palo Alto, Calif., 1970.

49. KINSEY, A. C., POMEROY, W. B., and MARTIN, C. E., *Sexual Behavior in the Human Male*, W. B. Saunders, Philadelphia, 1948.

50. ———, ———, ———, and GEBHARD, P. H., *Sexual Behavior in the Human Female*, W. B. Saunders, Philadelphia, 1953.

51. KNIGHT, T. S., "In Defense of Romance," *Marriage & Fam. Living*, *21*:107–110, 1959.

52. KOHL, R. N., "Pathologic Reactions of Marital Partners to Improvement of Patients," *Am. J. Psychiat.*, *118*:1036–1041, 1962.

53. LAING, R. D., and ESTERSON, A., *Sanity, Madness and the Family*, Vol. 1, Basic Books, New York, 1964.

54. LAQUEUR, H. P., LABURT, H. A., and

MORONG, E., "Multiple Family Therapy: Further Developments," in Haley, J. (Ed.), *Changing Families—A Family Therapy Reader*, Grune & Stratton, New York, 1971.

55. LEDERER, W. J., and JACKSON, D. D., *Mirages of Marriage*, Norton, New York, 1968.
56. LEICHTER, E., "Group Psychotherapy of Married Couples Groups—Some Characteristic Treatment Dynamics," *Internat. J. Group Psychother.*, *12*:154–163, 1962.
57. LEWIS, C. S., "We Have No Right to Happiness," *Saturday Evening Post*, *10*:10–12, 1963.
58. LUCKEY, E. B., "Number of Years Married as Related to Personality Perception and Marital Satisfaction," *J. Marriage & Fam.*, *28*:44, 1966.
59. MARTIN, M. J., "Hostile Dependency in Marriage," Editorial, *Minnesota Med.*, *51*:511, 1968.
60. MARTIN, P. A., and BIRD, H. W., "An Approach to the Psychotherapy of Marriage Partners: The Stereoscopic Technique," *Psychiatry*, *16*:123–127, 1963.
61. MASTERS, W. H., and JOHNSON, V. E., *Human Sexual Inadequacy*, Little, Brown, Boston, 1970.
62. ———, and ———, *Human Sexual Response*, Little, Brown, Boston, 1966.
63. MEARES, A., *Marriage and Personality*, Charles C Thomas, Springfield, Ill., 1958.
64. MILLETT, K., *Sexual Politics*, Doubleday, New York, 1970.
65. MITTELMANN, B., "Complementary Neurotic Reactions in Intimate Relationships," *Psychoanal. Quart.*, *13*:479–491, 1944.
66. ———, "The Concurrent Analysis of Marital Couples," *Psychoanal. Quart.*, *17*:182–197, 1948.
67. MONTAGU, M. F. A., "Marriage—A Cultural Perspective," in Eisenstein, V. W. (Ed.), *Neurotic Interaction in Marriage*, Basic Books, New York, 1956.
68. O'NEILL, N., and O'NEILL, G., *Open Marriage*, M. Evans and Co., New York, 1972.
69. OSMOND, M. W., "Toward Monogamy: A Cross-Cultural Study of Correlates of Types of Marriages," *Soc. Forces*, *44*:8–16, 1965.
70. PARKE, R. JR., and GLICK, P. C., "Prospective Changes in Marriage and the Family," *J. Marriage & Fam.*, *29*:249, 1967.
71. RAVICH, R. A., "A System of Notation of Dyadic Interaction," *Fam. Proc.*, *9*:297–300, 1970.
72. REIDER, N., "Problems in the Prediction of Marital Adjustment," in Eisenstein, V. W. (Ed.), *Neurotic Interaction in Marriage*, Basic Books, New York, 1956.
73. RIMMER, R. H., *Harrad Experiment*, Bantam, New York, 1971.
74. ———, *Proposition 31*, New American Library, New York, 1968.
75. ROBBIN, B. S., "Psychological Implications of the Male Homosexual 'Marriage,'" *Psa. Rev.*, *30*:428–437, 1943.
76. RODMAN, H., "Marital Power in France, Greece, Yugoslavia, and the U.S.: A Cross-National Discussion," *J. Marriage & Fam.*, *29*:32, 1967.
77. ROSENBLATT, P. C., "Marital Residence and the Functions of Romantic Love," *Ethnology*, *6*:471, 1967.
78. RUSSELL, B., *Marriage and Morals*, Liveright, New York, 1929.
79. RYDER, R. G., "Dimensions of Early Marriage," *Fam. Proc.*, *9*:51–68, 1970.
80. SARWER-FONER, G. J., "Patterns of Marital Relationships," *Am. J. Psychother.*, *17*:31–44, 1963.
81. SATIR, V. M., "Conjoint Marital Therapy," in Greene, B. L. (Ed.), *The Psychotherapies of Marital Disharmony*, Free Press, New York, 1965.
82. SEIDENBERG, R., *Marriage in Life and Literature*, Philosophical Library, New York, 1970.
83. SOLOMON, A. P., and GREENE, B. L., "Marital Disharmony: Concurrent Therapy of Husband and Wife by the Same Psychiatrist," *Dis. Nerv. Sys.*, *24*:1–8, 1963.
84. TERESHKOVA, V., Chairman, Soviet Women's Committee, "The New Soviet Marriage Law," *New Times* (Moscow), July 17, 1968.
85. TOFFLER, A., *Future Shock*, Random House, New York, 1970.
86. TURNER, C., "Conjugal Roles and Social Networks: A Re-examination of an Hypothesis," *Human Rel.*, *20*:121–129, 1967.
87. VAN DE VELDE, T. H., *Ideal Marriage: Its Physiology and Technique*, Random House, New York, 1930.
88. VAN HORNE, H., "Are We the Last Married Generation?," *McCalls*, *96*:69, 1969.

89. WESTERMARCK, E. A., *The History of Human Marriage*, 5th ed., Allerton Book Co., 1922.
90. WHITAKER, C. A., "Psychotherapy with Couples," *Am. J. Psychother.*, *12*:18–23, 1958.
91. ———, and MILLER, M. H., "A Re-evaluation of 'Psychiatric Help' When Divorce Impends," in Haley, J. (Ed.), *Changing Families—A Family Therapy Reader*, Grune & Stratton, New York, 1971.
92. ZEHV, W., "Letter to the Editor. The American Sexual Tragedy: Critique of a Critique," *J. Marriage & Fam.*, *27*:417, 1965.

CHAPTER 22

THE EFFECT OF CHANGING CULTURAL PATTERNS UPON WOMEN

Natalie Shainess

IT IS GENERALLY AGREED that the changes of the last decade have been so great that their extent is hardly recognized. The scientific advances alone are said to equal or exceed those of the last century, which in itself was remarkable. All human beings have experienced massive changes, and any specific changes affecting women have also affected men. Man and woman are interdependent, and each affects the balance and movement of the other.[39]

Before considering the changes that appear to be specific to women, it is necessary to consider the major changes of the decade—the effect of technology and the mass media. In *The New People* Winick[61] has observed changes in personality and social life that reflect a "massive shift of human consciousness." He feels the "new people" have taken over the world, participating in the destruction of old concepts of identity, sexuality, and ways of living. He suggests that the most radical changes of all are in the areas of sexual identity and sex roles, and that the new tone of life—"a bitter, metallic existence"—may simply not be worth the price of living it. He comments that all extremes are becoming blurred into neuter—a tendency earlier sensed by both Freud and Erich Fromm in their consideration of Ibsen's *Peer Gynt* and the image of the button-molder. Winick also noted the significance of Playboy-fostered voyeurism, nongenital sex, and the general lack of sexual passion.

In Slater's *Pursuit of Loneliness*[53] Americans are described as hard, surly, bitter people, fanatically pursuing objects although already surfeited by them, and thus postponing (avoiding?) living. Slater describes the emotional and intellectual poverty of the housewife's role—one that often delivers the *coup de grâce* to her early promise.

Arieti has noted a new type of alienation,[2] which he distinguishes from the first type characteristic of the withdrawn schizophrenic individual, or the second, described by Horney, in which the person is cut off from his own feelings, although there is no other gross abnormality. In Arieti's "new alienation" the individual maintains effective contact with the outer environment and seeks ever more stimulation, but has undergone mutilation in his inner life: "Inner conflicts are denied, introspection is frowned upon . . . the person *bathes* in his environment . . . there are no long-range ideals or goals . . . and at times a new kind of depression occurs characterized by constant demands and claims on others." These claims utilize a projective style, blaming others, and are used by the young in particular to blame "society." Man, Arieti suggests, has become a robot—functioning, one might add, in an overstimulated and undercommitted society.

This conception of current personality and intrapsychic change can be understood, from a genetic vantage point, as relating to another concept of Arieti's, that of volition.[1] He views the inception of volition in the infant toward the end of the first year of life as the capacity for choice in pleasing self or mother—the inception of the capacity to defer instant gratification, leading to that maturational ability to choose the reality principle over the pleasure principle. Thus this viewpoint suggests that permissive childrearing,* a change of the last two decades or so, in which mothers—and to some extent, fathers—did not "care enough" to engage in the process of resolving infantile grandiosity, has left this generation grandiose, disconnected, undisciplined, doubtful of self-worth, and empty.

So dehumanization and impulsivity are the all-pervasive undercurrents to any changes being considered. Turning to more specific change affecting women, the contrast with the picture of even a decade ago is very great; of course, this refers to Western woman, and particularly the American woman. Traditional stereotypes of the passive, in some ways delicate, woman, preoccupied with her double role as wife and mother, taking on all of the gender tasks assigned her—cooking, cleaning, caring for her children, serving her husband in a multitude of ways, and working outside the home, in the majority of cases, only out of economic necessity—all these are passing.

Fostered by social and technological changes, including the perfecting of contraceptive techniques, the general rise in sexual preoccupation, and culminating in the women's liberation movement, a new woman is emerging. She rejects the traditional roles described, demanding individual satisfaction and success. Large families are essentially a thing of the past, and the one-child family, with or without a husband and father, is growing in numbers. The high divorce rate has resulted in women spending longer periods of life alone: most men divorce to remarry; women to get free of an unpleasant situation.

The sexual revolution—meaning, for the most part, free and casual sex—is a very real change, although many still tend to deny this. It is rather unique in terms of its genesis. Fostered in good measure very deliberately by SIECUS,* and by the mechanistic and (in my view) methodologically unsound and unproven[41,43,45,48] sex-in-the-laboratory studies and alienated therapy of Masters and Johnson,[19,20] varied changes have taken place, from lessened guilt and greater freedom for some, to a general tendency to a greater voyeurism and exhibitionism, loss of authenticity, and sexual alienation, on to the setting up of sexual norms that have created new and more damaging demands upon sexual participation, with "forced change."

While the old double standards have seemingly been abandoned, they are still present in new form, favoring the attractive young woman in her immediate sexual opportunities, but leaving long-range effects still to be appraised, and making marriage a rapidly dis-

* It confirms the unfortunate reality that *reaction* is not *improvement*, and that valuable change must come from a new parameter.

* Sex Information and Education Council of the United States.

appearing option. The distress for some has been great, noticeable at the university as well as beyond, and reported by patients in treatment.

The emphasis on youth and sexual desirability without regard to the *person* has created problems for the older woman. Whether changing values simply mean that women have to pay a price until new patterns are established, or whether it bodes ill for them ultimately, remains to be seen. Lederer,[16] in *The Fear of Women*, makes a significant and disturbing statement that suggests the latter: "Nuclear Age Woman has reason to be concerned about the threats to these (the family, and her value).... Men, glorifying in the new technology, visualizing an eventual mastery over nature, accomplished in their own exuberant feats of strength . . . are no longer part of the mother-goddess scheme of things."

These changes have been part of the ongoing blurring of gender distinctions leading to an androgynous society (no value valence is intended here, as there are many facets to change of this kind). The fading of gender boundaries has been accompanied by fashion change, in the phenomenon of "unisex" clothing and fetishistic styles (the exposed thigh between boot and "hot pants" is a current example) as well as the alienated "hard" and hardware-trimmed space-age type of clothing.[51] Greenson has also observed these changes.[11]

Perhaps toughened by some aspects of the new alienation and student activism, the women's liberation movement has been the greatest single force for change affecting women. It has also been greatly concerned about gender distinctions because these have interfered with woman's status as equals, and studies of gender-role stereotypes have examined the validity of rigid concepts of masculinity and femininity, biologically, psychologically, and socially.

Cultural change has thus affected every aspect of women's lives: the acceptance of gender-role assignments, living style and work, mode of sexual expression, heterosexual relationships including dating style, all aspects of the reproductive function and role including concepts of marriage, family, motherhood and child care, abortion and contraception, and finally gender identity and psychology.

Changing Gender Concepts and Identifications

With regard to changing concepts of gender and gender role today, including sexual role, Erikson[6] has described the "identity diffusion" of youth today—a lack of firm boundaries, or what might be called a failure of complete severance of the original symbiotic mother-child tie. Erikson notes that the process involved in the formation of identity lies at the core of the individual and also at the core of his communal culture. It involves what Sullivan termed "reflected appraisals"[57] and Erikson calls "simultaneous reflection and observation," the individual judging himself in the perceptions of others and molding himself along those lines. He observed that the media may take a major part in the identity-forming process, usurping the place of parents, and that traditional sources of identity-strength are now only fragments.

Erikson suggests that a new balance of male and female is presaged[6] in the contemporary changes in the relations of the sexes and in the wider awareness spread by the media and technology. He notes the great change between the youth of today and those of 20 years ago. It appears that for many the identity conflict, more hidden than apparent, no longer exists in the same way, but instead it is dramatically displayed in dress. It includes bisexual confusions in sexual identity or a diffusion in which gender distinctions of dress no longer exist: the "hairy, unisex style" is a blatant challenge to parents and authority. Yet there is also a very positive trend among the young to be more collaborative in household tasks, ignoring previously practiced false gender-role distinctions.

With what will the young identify today? This question is most disturbing. Sent early to day-care centers, perhaps never, even in their earliest moments, savoring the sweet experience of an exclusive loving relationship to their mother, they will start their identification process with the dispassionate day-care worker, the mechanical teacher—be it television or machine. Will they identify with the characters created to titillate or to sell learning, like the squawky Big Bird of "Sesame Street," or the colored numbers jumping up and down? Perhaps they will identify with the characteristics of power as revealed in the violence on TV, or the visions of protesters and drug addicts paraded on TV and watched repeatedly as a *pleasurable* immersion in our ills, or the violent TV detective shows in which the female human being is being replaced with an image of the "sex-parts female." Baby dolls have now been replaced with sex dolls, wearing bikinis and sexy clothes. Will their sexual identification be related to all of the pseudo-sex-education films—in or out of school or on TV—in which there is a programmed sexual style and response? Further, these exposures and images appear with ever increasing speed. All suggest the development of a hard, alienated woman who is primarily a sex object, unyielding in interpersonal transactions, and hardly likely to be a nurturant mother. All women will want to work, even as that work loses meaning in itself and is only a means to an end: passing time and obtaining more *things*.

Neutering Process

The neutering process, one manifestation of which has been considered in relation to the new alienation and identity, contains other elements. As a result of technological growth, overpopulation, the increased size of business organizations, the messages spread by the mass media, the concept of team play, the rebellion against authority, and many other processes, a hatred (not too strong a word) has developed for elitism. A competitive resentment at anyone who rises, in nonconforming ways, above the masses, the denial of expertise and its misinterpretation as authority—all of these represent on a more general level what can also be observed on a sexual level: the denial of difference, the comfort and protection in being identical members of society, like sheep in a flock. In some ways it seems a return to primitive tribalism; for example, rock music, with its insistent beat provoking the loosely paired dance of masturbatory exhibitionism, is a reflection of the neutering.

Fostered by Playboy voyeurism (and its hidden premise: "You can be a man without being potent"), the voyeurism of television-filled lives, and by the exhibitionism of the tidal wave of pornography, young men and women seem arrested in the infantile sexuality of the Freudian pregenital phase; that is, sexual expression has increasingly become what was formerly considered foreplay, including acts also considered perversions. These are fast becoming the total aim in sex; they offer the instant satisfaction of partial sex and masturbation to those who want to push a button and obtain a response, like buying something from a vending machine. The lack of anticipation, yearning, and passion, the too easy availability, the ocean of "sex-object sex" into which the individual is submerged and seems unwilling or unable to escape, especially the use of sexual innuendo in advertising every product from air travel to chutney, the alienation from the person's own sentience—all result in this neutering, eliminating gender difference, disguising anatomical difference, and denying qualitative difference in sexual action and experience.

Social Perceptual Change: The Demand for Freedom and Equality

Turning to the adult, the rising expectations of women have accelerated in recent years and are both cause and effect of the women's liberation movement. These expectations apply to the social, sexual, and marital spheres, as well as to work and economic considerations. Sociologist Komarovsky[15] observes that women have been hindered by "the old chestnuts" men have offered, and adds that men's "hostility and arrogance have been disguised as benevolent paternalism." She cites Marynia Farnham[8] as an example of a psychiatric "neo-antifeminist" (or Aunt Jane) who views feminine assertion as violating woman's deepest needs—the Freudian glorification of passivity and the child-penis equation supposedly characterizing feminine psychology. Komarovsky states that the housewife is discontented today because "*satisfaction* depends upon *aspiration*," and the aspiration level has shot up like the mercury in a hot thermometer. "A dozen times a day, events belie the sermons directed to women."

Rossi suggests that equality between the sexes is regarded as an "immodest proposal."[30] She feels that American society has been so inundated by psychoanalytic thinking that dissatisfaction of women with their role is invariably viewed as a personal problem. Ignored by society is the fact that girls often undergo a kind of cultural hazing process, keeping them locked in their assigned gender role at every level of development. Consider the reinforcing effect of a statement by Albert Einstein: "The center of gravity for creative activity is located in different parts of the body in men and women."*

Although woman's struggle for greater freedom and equality has continued from earliest recorded history, and was even expressed in Chaucer's *Canterbury Tales* (Tale of the Wife of Bath),[39] Betty Friedan's[9] outcry against the Freudian psychology of women and the bonds of the stereotyped feminine gender role gave new impetus and created an almost tangible new draught in the air. But how have the majority of women responded?

Erikson[7] has observed that the relative emancipation ceded to women by middle-class "self-made men" has been utilized in "gaining access to *limited* career competition, standardized consumership, and strenuous one-family homemaking . . . it has not led to equivalent or actual role in the game of power." He feels that the fashionable discussion about how women might become "fully human" is really a "cosmic parody," and that it is still amazingly hard for the vast majority of women to say clearly what they feel most deeply. In considering woman's psychic structure Erikson[7] rejects Freud's "wound concept" of female sexual psychology, substituting "inner bodily space"; this is a valid observation, yet insufficient in that it does not include the cultural.

Why the status quo is accepted by many women was strikingly revealed in a statement by a wealthy "professional wife" to a gathering that included a number of professional women. She said: "Taking care of my husband is *my* profession, and it gives me all the satisfaction I want." Its pointedness was made even clearer by the elaborate jewel she was wearing. But why men actively maintain the status quo is even clearer: few people voluntarily surrender power.

As a reflection of women's growing discontent with their position, they are expressing themselves more confidently, and researchers in the social and psychological fields have undertaken numerous studies of gender concepts. Notable among these is one by Rosenkrantz,[28] which concludes that different standards of emotional health are utilized for men and women, paralleling sex-role stereotypes; the very features that are considered healthy in men, such as aggressiveness, are judged unhealthy in women. The reverse also held: passivity, the standard for women, was considered sick for men. Other studies, such as one by Steinmann,[55] confirmed this.

This observation was also noted by Jean Baker Miller,[23] who studied man-woman relationships in terms of the processes between dominant and subordinate groups. She noted the self-criticism of the subordinate group ("identification with the aggressor") and suggested new political directions for women,[22] stating unequivocally that "anger is necessary." Indeed, one aspect of cultural change is the growth of overt anger and the demand for change by women as a group—a group widespread and seeking identity in a determined way, for perhaps the first time in history. The formation of the National Organization for Women was an indication of this change.

* *The Born-Einstein Letters*, reviewed by D. M. Locke, *Saturday Review*, September 11, 1971.

A curious side issue is the growing envy by men of certain aspects of the feminine gender-role stereotype, just as women themselves repudiate it. Not the breast-envy or womb-envy of an earlier time, it is now envy of the passive-dependent role and of the presumed idle time spent as the individual wishes, while being supported. Indications of this seem increasingly numerous of late. Women have observed that it is virtually impossible to take a taxi in a large city without listening to a monologue on this theme delivered by the driver, and generally including "how lucky you dolls are."

Sexual Expression and Dating Patterns

Studies have revealed that there is strong agreement in the young of both sexes on the inherent difference between the sexes in sexual self-concepts and in sexual stereotypes.[29] Yet there is no doubt about the vast change in concepts of sexuality and sexual behavior within the last five years. Effective and easy (though not danger-free) contraceptives, especially the birth-control pill, have contributed, as well as value change. Virginity, perhaps always of ambivalent value, has now come to be considered "something to be gotten rid of"—an alienated, impersonal term in general usage. A female college student echoed the statements of friends, who discussed, not whether they might be overcome (with passion) at the moment and yield to their boyfriends, but when and how they could "lose their virginity." Veryl Rosenbaum,[27] a lay psychoanalyst, has written a sensitive poem describing the current view of the first sexual encounter and the loss of virginity as a "button to be ripped off, to roll in the dust under a shabby bed."

Women seem to feel that frigidity, a symptom of a generation back, has in reality been an accusation against them, and they seem almost frantic in their efforts to overcome this and to live up to the norms set by sex researchers. In the women's liberation movement the concept of anything beyond clitoral orgasm is denied, with Anne Koedt's "The Myth of Vaginal Orgasm"[13] as a banner. Many psychiatrists have adopted this position as well.[32] But while basically rejecting Freud's feminine psychology, I still see Freud as correct in postulating a vaginal response.[47,52] I view the clitoral as a partial response, lending itself to "instant sex-on-demand"[50] from manual stimulation. Concern with orgasm has become enormous and omnipresent, leading to what has been described as "orgasm worship."[38] Noted along with this is the erotization of language and the use of sexual imagery to describe other problems.

Frequently dating patterns among young college students seem to be exercises in instant sex and nonrelationship, leading nowhere. A college senior had a blind date arranged with the brother of a girl she knew. He arrived at her apartment and within ten minutes announced that he was "horny." She thereupon felt it was incumbent upon her to relieve that state. She appeared as devoid of feeling as Camus' *Stranger*, who seemed to move through life without feeling, even without motivation, going like a vehicle along some predestined track, at a preordained pace, taking turns that he had no part in choosing. The Stranger attends his mother's funeral, even though he has no feeling for her, not even hate, meets a girl, goes to the beach, commits a senseless murder—and the reader is left wondering why, why? He is a new human among us, Camus' prophecy of the new alienation—a man of the Horney type of alienation on the verge of Arieti's third type: beginning to seek sensory experiences to submerge in. It is not by accident that the beach, a place for fun, is the setting for murder. But returning to the girl described, she *did* note the young man's failure to call her again—the failure of her perhaps manipulative use of "instant sex" to buy a relationship.*

As another example, a young professional

* Case illustrations will be anecdotal, since they are intended only to illustrate cultural trends rather than specific pathology.

woman, despairing of marriage prospects, decided to give up all restraints on sex. A friend arranged a holiday weekend date for her with a highly eligible young man. There was noticeable evidence of her self-esteem plummeting as she followed blindly her predetermined course. He departed the city without saying good-by, but had time to comment to the mutual friend: "Say, that was a superchick you fixed me up with!" The young woman's subsequent depression almost resulted in a literal enactment of her symbolically suicidal behavior.

The problem generally seems to be that rejection of the "good" or "bad" value in relation to sexual behavior has left young people bereft of any other standards. Most lack the ability to evaluate sensibly the so-called situational ethics. Here is need for sex education, in terms of value in the individual's development and in interpersonal relations.

Halleck[12] has noted that in spite of the sexual revolution, matters have not changed considerably for young women. The double standard, perhaps in a new form, persists because there cannot be equality in sex without total equality, especially in the distribution of power. "Women lose power and status when they become too indiscriminate in granting their sexual charms, or when their sexual attractiveness begins to wane."

Gloria Steinem[55] has commented on the fact that "men wise in the ways of power understand its sexual uses well." Perhaps the ugliness of sex in return for status and power on both sides of the equation is more clearly revealed in the involvement of glamorous young women of the entertainment world with top political figures—a long-enduring but recently accentuated pattern. Sex, of course, has always been a woman's ticket of admission to the social scene, but it has now become virtually the *only* means, as little else enters into relationships. It incurs dangerous new risks for women.

The "singles bar" is a relatively new development in the United States, appearing especially in large cities. A manifestation of the growing hold of drinking alcoholic beverages on social style, these bars are establishments where men and women go with the acknowledged purpose of meeting for quick sexual liaisons. A very personable young male patient reported on his escapades via this medium. Although he appeared the soul of innocence and safety, he was something less than that, and it became unavoidable to reflect upon the risks young women were taking, in leaving a bar and going off to the apartment of a total stranger after a couple of drinks, and without telling anyone of their destination, as is common practice. Although no murders, to my knowledge, have as yet been reported, it seems inevitable that they will appear. This approach to sex exemplifies the new alienation, in terms of instant precipitation into sensual experience devoid of inner meaning—the "drug kick" of the deadening of cognition through alcohol, and the instant immersion into sex with anyone. The new language of sexual request or invitation reflects the new alienated style. As reported in *The New York Times*, a man asked a woman he had just met: "Would you like a sensual visitor tonight?"*

Sexual liaisons of young women, whether with young or older men, seem to conform to a style of abandoning their own pursuits in good measure and going to live with the man; however, there seems to be a local style involved, and in some cities the reverse is true in location, but not in who makes the major life change. This self-abandonment, in the face of the current struggle for equality, seems another aspect of the new alienation. Seidenberg[33] has pointed out the problems of sexual inequality, especially for the gifted young woman of today who wants to share intellect as well as sex.

A new sexual aggressiveness in women has also manifested itself, visible in virtually assaultive passes at boyfriends in the street, who seem to endure rather than enjoy it. These have been in line with images fostered by the mass media as well as by sex researchers. In contrast—and perhaps astonishing to the older generation—is the practice of "bedding

* *New York Times*, January 23, 1972.

down," of college-age boys and girls sleeping together without anything sexual transpiring. It seems a mode of obtaining comfort, to alleviate what could be called a "babes-in-the-woods syndrome," and also a manifestation of the neutering process going on.

With greater experience women are sexually wiser, have expectations of men as lovers, and experience a growing discontent. Many men are finding these expectations hard to live up to, especially since they cannot as successfully disguise or deny their inadequacies. Perhaps the new insistence on the normality and desirability of sex play that was formerly considered perverse is an effort to compensate; again it suggests a sexual neutering process. To the author it appears that *passion* favors genital sex.

Along different lines an old, and yet new, sexual style has developed—new in the sense that the age disparity of incest-model dating has grown in sexual relationships of very young women with considerably older men. From reports—and to be observed on the streets—young women in their late teens are to be seen with men easily in their fifties or even sixties. The "anything goes" concept of sex, together with the new alienation, seems responsible for this, and young women do not seem aware that they may be "selling themselves short." Judith Viorst wrote[60] a delightful poem called "A Lot To Give Each Other," capturing the lack of commonality in anything but sex: "He worries about his prostate, and She worries about her acne, and He was born before television, and She was born after running boards—but they feel they've got a lot to give each other . . ." The reverse young-old pattern with women as the older partners, of which there are always a few instances among the wealthy, seems not to have caught hold. In the light of the aforementioned change, many men are living in a "captain's paradise," that is, they have a girlfriend (young) and a wife over a period of many years. But, of course, it is not always pure fun for such men, some of whom are compulsively trapped in a dual relationship, which serves deep neurotic needs and is at times quite punishing.

The Older Woman

For the older woman new problems seem to exist, and the new sexual freedom is proving to be a bad check. If married, and unless there has been an extraordinarily satisfying marriage, there is a drifting apart of all interests including the sexual, especially as the large corporation has made increasing demands on men and offered all kinds of "rewards." Infidelity has become almost a norm. Jessie Bernard has indicated that sex researchers have *promoted* the acceptance of infidelity.[3] An example of sexual image-making is to be seen in Neubeck's book, *Extramarital Relations*,[24] in which virtually the first words to meet the eye, in big bold type, are: "WE ARE HUMAN." It is an instance of incitement related to the indulge-yourself philosophy of the times, also expressed in a common phrase of late, especially by men: "After all, we're not going to pass this way again." Unverbalized is the concluding phrase: "so grab everything you can."

Although Bernard questions the hurt involved if infidelity is not known to the partner (and the impression is that it is *always* known at some level), she does note that it is much more commonly acted upon by men. Wolf[62] notes that in a patriarchal society there is less infidelity by women. He also observes the stress added to the wife's role by the image she must fill as an organization wife. Yet he feels there is less hypocrisy in marriage today. Quite aside from the fact that marriages that do not have common goals are in the process of destruction, Bernard fails to note the social and economic base to infidelity—in short, the power distribution makes infidelity by men more possible, while women have a history of accepting hurt, or settling for the most likely means of survival. My impression is that infidelity, whatever its basis, always causes a serious wound in either sex, however it is glossed over. When the interests of two partners diverge, marriage becomes a precarious living arrangement.

The problems of aging for men, aside from

the inherent psychological components of the experience, relate primarily to retirement at an age when still valuable powers and capacities exist, so that Goodman Ace has humorously suggested a "Used Man Lot" to deal with the problem of male obsolescence. However, the older woman's problems are even more serious. She has to contend with the menopause, which seems to punctuate the approaching end of her sexuality and her life, at the same time that she must cope with the "empty nest syndrome"—the end of her major life role as a mother. Where her stereotyped gender role, into which she has been molded, has not been tempered by any other interest or any activity bolstering self-esteem, her emptiness is experienced by her as a defect, and she is likely to "crack." She views with horror the long period of life—empty life—still ahead of her in the face of her tenuous relationship with her husband, whose interest is virtually nonexistent or turned elsewhere. One divorced woman in her forties reported a dream with a single visual image: an empty suitcase. In associating to the dream, she recalled the suitcase as an expensive one belonging to her former husband, usually stored in the closet, and now scratched and worn. She said: "I am alone and empty. I am unused as a woman. I am no longer valued, even by myself."

Fortunately women have begun to solve the problem belatedly by returning to study and work—although often at a level far below their neglected abilities. Yet this situation, compounded of "social maltraining" as well as personal difficulties, often results in serious depressions, as Rose Spiegel[54] and others have stated.

For the older woman alone, social isolation is a singularly punishing reality in a paired society, especially for the woman who sees herself as only capable of the passive feminine role for which society ordains her. The combined despair and anger this elicits was revealed in a Christmas card sent me by an attractive but faded widow in her early sixties, who had arranged a consultation because of depression. When it seemed little could be done for her, the unreality of her approach to life in "waiting for my Prince Charming to come" was pointed out, and it was suggested (without complete conviction, it must be acknowledged, since nothing compensates for the lack of a close relationship) that feelings of pleasure and reward would grow as she undertook some useful activity, the various possibilities of which were explored, although rejected by her. The card contained the following note: "Although I feel so disheartened because my contacting you was such a dismal failure, I am truly appreciative of your having seen me."

This predicament of the older woman alone has been well documented by Isabella Taves,[59] who observed her sexual as well as social dilemma: she must either engage in transient or (rarely) enduring sexual encounters with married men or live a celibate life. The divorced older men who remarry almost invariably seem to choose a considerably younger woman. As a professional man put it to a female colleague when discussing his own remarriage: "Why should I have undertaken the problems of marrying an older woman, when I can have an attractive, sexually appealing young woman who will wait on me hand and foot, be happy about what I can give her, and make few demands upon me? And it makes me feel young to start a family again." As logic it seems unassailable, but it spells a serious existential dilemma for the older woman, especially as her value as the *person she has become* seems nonexistent, and her sexual life must either become degrading or disappear. One 39-year-old attractive divorcée, who had several affairs with inadequate men out of desperation, asked: "Why aren't there men to appreciate the sterling qualities my friends insist I have?"

Marriage

In general, brittleness and unyielding self-centeredness of an almost militant nature seem to characterize marital attitudes of both sexes today. Perhaps the fact that for women, marriage with conventional gender-role stereotypes has often been a trap, a situation symbolized in Sartre's play *No Exit*, has contributed to this. Marmor[18] points out how psychoanalytic interpretations of a woman's rebellion in a marriage depends upon the analyst's point of view, and whether he espouses the view that passivity is normal for women. Seidenberg[34] also calls attention to the unfair expectations of woman's role in marriage, both in life and as portrayed in literature. And philosopher John Stuart Mill[21] has given us one of the most moving and clear statements about what is wrong with unequal marriage, and what marriage can achieve with true equality and communion between partners. His view, of course, particularly relates to intellectuals.

Symonds[58] has characterized the developmental restrictions of marriage for many women, and the symptomatic expression of the problem, as "Phobias after marriage: woman's declaration of dependence." She points out that phobias are the women's way of handling repressed anger within the marriage, and that with growth the marriage is often disrupted, as the fundamental nonverbalized premises are challenged. Yet up to now working and professional women have not greatly challenged these premises and still undertake the Herculean effort of doing what amounts to three or four jobs at once. Caroline Bird,[4] whose social study of the inequitable position of women broke ground for the women's liberation movement, recently pointed out the added advantage that men, and professional men in particular, have in wives who serve them personally and help them professionally. As professional women ruefully remark: what every successful woman needs is a *wife*![55]

Although the majority of women still seem satisfied with, or accept, the stereotyped view of marriage, as Komarovsky has indicated,[14,15] or even have a false view of marriage as freedom in the sense of escape from the childhood home, there is a gap between expectations and reality. Among blue-collar workers Komarovsky found that the male was dominant in the marriage and that the threat of violence was an important basis for this power. She also pointed out a very important consideration in marriage: the individual partner's bargaining position depends on the degree of emotional involvement, which is one factor explaining feminine subservience. Yet caught in the cross-currents of social change, women feel growing disappointment as some of the compensations of an earlier time have disappeared, and they find marriage less rewarding. As one woman, a professor of psychology, expressed it: "Marriage has become a living arrangement." Certainly less is heard of intimate relationship or of love, and more is heard of life style and sex.

Marriage as an institution is seriously embattled, and all sorts of alternatives are proposed, as Otto suggested,[25] including serial monogamy, marriage networks, and open marriage, among others. Greene[10] believes that the individual may be carrying out a neurotic repetition in sequential marriage. Yet the new choice may also be a good one, and the result of growth. Unfortunately women seem to have less opportunity for positive change by the time growth has occurred. Extreme woman liberationists support these alternative concepts, or deny that marriage can be viable in any form. From a psychoanalytic viewpoint, although undoubtedly some early choices of marital partner are predominantly neurotic mistakes, transference distortions and narcissism (self-centeredness) are still the major factors in marital failure, but the distribution of power in marriage is increasingly recognized as a determinant of workability.

Many express a belief that marriage and the nuclear family are failures, and some of the young have entered communal living arrangements and communal marriages. A famous illustration was that reported as a "course in communal living"—an experimental college set up at Columbia University.* Yet no change is ever monolithic, and many young college students are marrying legally after living together in experimental marriages. These are healthy marriages based on a satisfying relationship, not entered into out of fantasied

* *New York Post*, December 5, 1970.

expectations and sexual guilt. Many young couples write their own marriage ceremony, stressing equality and mutuality. Along with these changes sometimes goes the perhaps syn thetic type of back-to-nature ceremony, complete with barefoot bride in see-through dress without underclothes. Of course, some of these experimental marriages are legalized because of the young woman's pressure: the nest-building tendency is still a part of feminine nature—and a valid one. On a positive note Rostow[31] suggests that the newer concepts of marriage hold it to be a process in which the husband and wife cooperate on many levels to permit each separately and together to satisfy needs and achieve goals.

As part of the rejection of marriage, and while undoubtedly determined by other and unconscious forces, there has been a rise in lesbianism. Radical women seem to make a conscious effort in this direction, as part of an attempt to eliminate men from their lives. As one college girl, who was "trying out" lesbianism, put it, "If only my friend Vera had the head of Jane Fonda, and the body of a man!" Retreats and encounter weekends fostering lesbianism have become almost commonplace. This undoubtedly relates also to the growing reluctance to undertake motherhood.

Abortion, Mothering, and Child Care

In this aspect of women's lives some strange paradoxes are to be found. On the one hand, in spite of the range and availability of contraceptives, the one-parent family, consisting of unwed mother and child, is notably on the increase. There seems to be a kind of defiant pleasure in the pregnancy[46]—in the street many girls seem to take pains to be noticed as pregnant. One gets the impression that this is their weapon against mother and society. But there is a large faction who totally reject the idea of motherhood, which Betty Rollin described in a *Look Magazine* article as "as unnecessary as spaghetti." Yet many, married or otherwise, insist upon having children despite their reluctance to care for them, and they call for creches for infants and child-care centers as a regular aspect of child life from birth on. That is, they seem to choose biological, but not nurturing, motherhood.

On the other hand, women are demanding the right to total control over their reproductive function, including abortion. Their efforts, coupled with those of physicians, psychiatrists, and attorneys, have brought about great change—a subject too wide to document here. However, psychiatrists have been slow to realize the unfairness and unfortunate consequences of unwilling motherhood,[35,36,37,46] or to come to a conclusion that it has taken me a long time to clarify and state simply: there is absolutely no relationship between theological dogma and mythology and intrapsychic experience for women, who *must* resent an unwanted pregnancy, and who carry their resentment over *unalterably*, whether consciously or not, into their affective relationship and dynamisms with the child. In the past women have been expected to be Christlike, carrying their crucifix within; they have often been sanctimoniously counseled by some who have had rather casual attitudes toward the act that placed women in this position and little empathy for their circumstances. The plethora of research now going on to ascertain the effects of abortion or its refusal is belated and almost ludicrous, in the light of the years of cruel entrapment of women by refusal to recognize their needs and best interests—without benefit of research!

Women are beginning to see the relationship between social mythology and reality; and those freed from the "psychological set" of having to defer to men are very clear about it and are taking effective action. They have begun to realize that the refusal of abortion can also be seen as a power mode to restrict sexual expression, but only by *one* sex, or, politically, to enlarge a specific group. In any event the tide is turning.

How will women ever solve the career versus motherhood problem? This is a difficult dilemma. Women have an added burden if they wish to function as both human beings and females, and it is hard to avoid the feeling that they must give something extra to achieve their goals.[40] It does not seem in the best interests of children and society for women to attempt to live in exactly the same style as men.[44] Women's lives are perhaps more phase-oriented. It seems as if women should receive support from society to spend a few years with their young children, with the assurance that a place will be made for them when they return to their outside endeavors. Perhaps they need to be subsidized during the early mothering period, until child-care centers can take over. They also need more help from men, who have largely been absentee fathers, and who will gain from their wives' efforts. Already there is evidence among young couples of a change in the direction of a more participant father role.

My own solution, which I was fortunate in being able to carry out, was to stop my professional involvements—in all but a few limited instances—until my children were attending school full-time. The opportunities for any informal kind of study under such circumstances are great. Inevitably, of course, it does entail sacrifices and losses, certainly compensated in large measure by the evidence of happiness and growth in the child, as well as by the close relationship itself. Undue attention to the ease with which men direct their lives can be embittering. But as a corollary, men need to be more generous and more helpful in aiding women to do their best. For women whose work is an economic necessity when the children are small, or who flatly reject child care, obviously child-care centers are the answer.

Conclusion

Whether compounded of pessimism or a realistic appraisal, the impression persists that the direction in which the culture is moving is toward the creation of male and female "humanoids," becoming still more alienated, mechanical, unfeeling, compulsively cruel, asexual but sexually preoccupied, frantically trying successive modes of sensual experience including drugs, devoid of the capacity for deep relationships, and with an increasing will to power by any means. The society of the onrushing future, which is already today, has not only veered from the repressive to the expressive—it has lost its humanity along with its superego.

More specifically, the cultural factors affecting women are multidimensional, and in some areas they are in direct conflict. Women are more direct—and "harder." They demand social equality in every form, yet a majority are slow to seek or accept change, which has always been promoted by the young. They are sexually freer, yet also sexually compulsive and alienated. They have not gained too much through the "sexual revolution." They, as well as men, are less totally involved in the marital relationship. As a group they have considerably less interest in motherhood. And many are tending to doubt psychiatry's ability to help them.

Implications for Treatment

How will psychiatry function as the changes relating to alienation grow? The zenith of human understanding, peaking with the appearance of Freud, Adler, Sullivan—to select a limited few as representatives of aspects of thought about inner experience and relationships—seems already to have begun to decline. The behavioral therapists, the symptom modifiers, and the "containers" will deal with the *symptoms* of an expanding human illness: alienation and the loss of humanity, in some ways an even greater problem for women than for men. Perhaps social psychiatrists and psychologists will join them to deal with broader aspects of this illness, but there is little ground for optimism about constructive change.

On the other hand, the degree of change and its effect upon women within the last decade suggests something not sufficiently recognized: human beings are remarkably plastic, and great care should be taken before considering psychic and behavioral expressions innate. Perhaps the most significant fact of our time in relation to treatment generally, and with regard to women specifically, is the necessity to understand the social setting of problems as well as their interpersonal determinants, rather than regarding them as exclusively intrapsychic in genesis.

Social psychiatry applies to the individual as well as the group, and a new concept of preventive psychiatry and treatment has been described by Robert J. Lifton[17] as "Advocacy Psychiatry." Here he extends the view of Sullivan[57] that the therapist must be clear about his own values yet not impose them on the patient, to examining the integrity of values and attempting social change where indicated.

In this sense Freudian feminine psychology, which has been nibbled at by dissenters, but not challenged as a totality, *must* be supplanted. Ideas of feminine passivity as health and the necessity of deference to masculine interests must be jettisoned if women are really to be helped to mature and attain reasonably satisfactory lives. Mastery over their own bodies and reproductive lines is essential for women's emotional health and emancipation. The recognition that *all* humans are aggressive at times, that it is not a particularly desirable quality, but no worse in women than men, although more in conflict with male societal expectations, is another essential basis to good therapy for women. And again, as with all human beings, women must be helped to become as free of anxiety as possible and to learn collaborative modes of relationship as equals.

It has been suggested that there is psychologically *no* difference between men and women,[42,49] except perhaps as a remote component of their biological sexual representation, in terms of receptiveness in a woman, the quality of being penetrating in a man. But even these are distorted by other processes. For women any specific feminine psychology relates only to the significant nodal points of their developing reproductive function. These have yet to be properly organized into a body of psychological concepts, although the direction is suggested.[49]

May Romm[26] has asserted that "the thinking woman can no longer accept the demeaned and submissive role implied in the statement 'Anatomy is Destiny.'" She feels women must be brought into the category of homo sapiens. So what emerges with great clarity is that male therapists, and female, too, who have previously been either unquestioning of or insistent upon old theories, must scrupulously examine their own gender concepts and prejudices, *listen* carefully to women patients, and search for the *setting* of their discontent, depression, or sexual unresponsiveness—and, indeed, other kinds of symptomatology that have been used perhaps more accusatively than therapeutically. Here the choice of language is an important modality to note.

Lastly awareness of the three modes of alienation suggests a way of leading women back into contact with themselves—an important issue for men as well. Certainly it has been my experience that many a rebellious, angry, or destructive young woman has been led along the path of introspection to greater rationality and more rewarding or vital living, after she has had a chance to perceive that stereotyped, and sometimes demeaning, labels have not been pressed upon her.

Bibliography

1. ARIETI, S., "Volition and Value: A Study Based on Catatonic Schizophrenia," *Comprehensive Psychiat.*, 2:74–82, 1962.
2. ———, *The Will to Be Human*, Chap. 6, Quadrangle Books, New York, 1972.
3. BERNARD, J., "Infidelity: Some Moral and Social Issues," in Masserman, J. (Ed.), *Science and Psychoanalysis*, Vol. 16, Grune & Stratton, New York, 1970.
4. BIRD, C., *Born Female*, David MacKay, New York, 1968.
5. ———, "Woman Who Make Great Success . . . Have Husband Who Make Great Coffee," *New Woman*, November 1971.
6. ERIKSON, E., *Identity: Youth and Crisis*, Norton, New York, 1968.
7. ———, "Inner and Outer Space: Reflections on Womanhood," in Lifton, R. J. (Ed.), *The Woman in America*, Beacon Press, Boston, 1965.

8. FARNHAM, M. F., *Modern Woman: The Lost Sex*, Harper, New York, 1947.
9. FRIEDAN, B., *The Feminine Mystique*, Norton, New York, 1963.
10. GREENE, B. L., "Sequential Marriage: Repetition or Change? in Rosenbaum, S., and Alger, I. (Eds.), *The Marriage Relationship*, Basic Books, New York, 1968.
11. GREENSON, R., "Masculinity and Femininity in Our Time," in Wahl, C. W. (Ed.), *Special Problems of Diagnosis and Treatment in Medical Practice*, Free Press, New York, 1967.
12. HALLECK, S. L., "Sex and Power," *Med. Aspects Human Sexuality*, 3:8–24, 1969.
13. KOEDT, A., "Myth of the Vaginal Orgasm," *Notes from the First Year*, N. Y. Radical Women, June 1968.
14. KOMAROVSKY, M., *Blue Collar Marriage*, Random House, New York, 1962.
15. ———, *Woman in the Modern World*, Little, Brown, Boston, 1953.
16. LEDERER, W., *The Fear of Women*, Grune & Stratton, New York, 1968.
17. LIFTON, R. J., "Experiments in Advocacy Research," William V. Silverberg Memorial Award Lecture, Midwinter Meetings of the American Academy of Psychoanalysis, December 4, 1971.
18. MARMOR, J., "Changing Patterns of Femininity: Psychoanalytic Implications," in Rosenbaum, S., and Alger, I. (Eds.), *The Marriage Relationship*, Basic Books, New York, 1968.
19. MASTERS, W. H., and JOHNSON, V. E., *Human Sexual Inadequacy*, Little, Brown, Boston, 1970.
20. ———, *Human Sexual Response*, Little, Brown, Boston, 1968.
21. MILL, J. S., *On the Subjection of Women*, Fawcett, New York, 1971.
22. MILLER, J. B., *Social Policy: New Political Directions for Women*, International Arts & Science Press, White Plains, N.Y., 1971.
23. ——, and MOTHNER, I., "Psychological Consequences of Sexual Inequality," *Am. J. Orthopsychiat.*, 41:767–775, 1971.
24. NEUBECK, G., *Extramarital Relations*, Prentice-Hall, Englewood Cliffs, N.J., 1969.
25. OTTO, H. A., "Has Monogamy Failed?" *Saturday Review*, April 25, 1970.
26. ROMM, M., "Women and Psychiatry," *J. Am. Med. Women's A.*, 24:1–8, 1969.
27. ROSENBAUM, V., *Long Way from Home*, American Poet Press, Santa Fe, 1966.
28. ROSENKRANTZ, P., BROVERMAN, I., BROVERMAN, D. M., CLARKSON, F., and VOGEL, S. R., "Sex-Role Stereotypes and Clinical Judgments of Mental Health," *J. Consult. & Clin. Psychol.*, 34:1–7, 1970.
29. ———, VOGEL, S. R., BEE, H., BROVERMAN, I., and BROVERMAN, D. M., "Sexual Stereotypes and Self-Concepts in College Students," *J. Consult. & Clin. Psychol.*, 32: 287–295, 1968.
30. ROSSI, A., "Equality between the Sexes," in Lifton, R. J. (Ed.), *The Woman in America*, Beacon Press, Boston, 1965.
31. ROSTOW, E. G., "Conflict and Accommodation," in Lifton, R. J. (Ed.), *The Woman in America*, Beacon Press, Boston, 1965.
32. SALZMAN, L., "Sexuality in Psychoanalytic Theory," in Marmor, J. (Ed.), *Modern Psychoanalysis*, Basic Books, New York, 1968.
33. SEIDENBERG, R. R., "Is Sex without Sexism Possible?" *Sexual Behavior*, pp. 47, 48, 57–62, 1972.
34. ———, *Marriage in Life and Literature*, Philosophical Library, New York, 1970.
35. SHAINESS, N., "Abortion Is No Man's Business," *Psychology Today*, May 1970.
36. ———, in Schulder, D., and Kennedy, F. (Eds.), *Abortion Rap*, pp. 121–137, McGraw-Hill, New York, 1971.
37. ———, "Abortion: Social, Psychiatric and Psychoanalytic Perspectives," *N.Y. State J. Med.*, 68:3070–3074, 1968.
38. ———, "The Danger of Orgasm Worship," *Med. Aspects Human Sexuality*, 4:73–80, 1970.
39. ———, "Images of Woman: Past and Present, Overt and Obscured," *Am. J. Psychother.*, 23:77–97, 1969.
40. ———, "Is Motherhood Unnecessary?" *Marriage Magazine*, May 1972.
41. ———, "Is There a Normal Sexual Response?" *Psychiatric Opinion*, 5:27–30, 1968.
42. ———, "Is there a Separate Feminine Psychology?" *N.Y. State J. Med.*, 70:3007–3009, 1970.
43. ———, "Masters and Johnson Reconsidered," lecture, University of Syracuse Medical School Department of Psychiatry, Feb. 10, 1972, (published as: "How 'Sex

Experts' Debase Sex," *World*, Jan. 2, 1973).

44. ———, "Mother-Child Relationships: an Overview," in Masserman, J. (Ed.), *Science and Psychoanalysis*, Vol. 14, Grune & Stratton, New York, 1969.

45. ———, "The Problem of Sex Today," *Am. J. Psychiat.*, *124*:1076–1081, 1968.

46. ———, "Psychological Problems Associated with Motherhood," in Arieti, S. (Ed.), *American Handbook of Psychiatry*, Vol. 3, Basic Books, New York, 1966.

47. ———, "A Re-Assessment of Feminine Sexuality and Erotic Experience," in Masserman, J. (Ed.), *Science and Psychoanalysis*, Vol. 10, Grune & Stratton, New York, 1966.

48. ———, "Review: Human Sexual Inadequacy," *J.A.M.A.*, *231*:2084, 1970.

49. ———, "Toward a New Feminine Psychology," *Current Med. Dialog*, April 1972.

50. ———, "Women's Liberation and Liberated Woman," in Arieti, S. (Ed.), *World Biennial of Psychiatry and Psychotherapy*, Vol. 2, Basic Books, New York, 1973.

51. ———, "Viewpoints: Do Women Dress to Please Men or Other Women," *Med. Aspects Human Sexuality*, *4*:10, 1970.

52. ———, "Authentic Feminine Orgastic Response," in Kaplan, H. and Adelson, E. (Eds.), *Sexuality and Psychoanalysis Revisited*, Brunner/Mazel, New York, in press.

53. SLATER, P. E., *The Pursuit of Loneliness*, Beacon Press, Boston, 1970.

54. SPIEGEL, R., "Depressions and the Feminine Situation," in Goldman, G. D., and Milman, D. S. (Eds.), *Modern Woman*, Charles C Thomas, Springfield, Ill., 1969.

55. STEINEM, G., "Woman and Power," *Reflections*, *4*:15–26, 1969.

56. STEINMANN, A., "Male-Female Perceptions of the Female Role in the United States," *J. Psychol.*, *64*:265–276, 1966.

57. SULLIVAN, H. S., *Conceptions of Modern Psychiatry*, William A. White Psychiatric Foundation, Washington, D.C., 1946.

58. SYMONDS, A., "Phobias after Marriage: Woman's Declaration of Dependence," *Am. J. Psychoanal.*, *31*:144–152, 1971.

59. TAVES, I., *Women Alone*, Funk and Wagnalls, New York, 1968.

60. VIORST, J., "A Lot to Give Each Other," *New York Magazine*, June 22, 1970.

61. WINICK, C., *The New People*, Western Publishing Co., New York, 1969.

62. WOLF, A., "The Problem of Infidelity," in Rosenbaum, S., and Alger, I. (Eds.), *The Marriage Relationship*, Basic Books, New York, 1968.

CHAPTER 23

THE PSYCHOBIOLOGY OF PARENTHOOD

Therese Benedek

Introduction

THE PLEASURES AND PAINS, the gratifications and frustrations of parenthood are existential components in the adult life of humans. In spite of the ubiquity of its problems, the psychology of parenthood has not been studied systematically. Science progresses slowly. Generations of scientists labor arduously to build the foundation for an insight that a genius formulated long years before. I refer here to a statement of Darwin:[9] "The feeling of pleasure from society is probably an extension of the parental or filial affections, since the social instinct seems to be developed by the young remaining for a long time with their parents" (p. 6). It is obvious that Darwin, the naturalist, arrived at this insight from innumerable, seemingly unrelated observations. Today the verity of this generalization appears evident to students of behavior, whether the objects of observation are human, subhuman mammalians, or the lower levels of the evolutionary scale.

Psychoanalytic theory is (primarily) a biological approach to psychology. Psychoanalytic investigations of various aspects of behavior afforded the framework within which biology, psychology, and sociology as continuum can be explained.[12] This brief essay written in such a broad setting can serve only as an outline. At the same time it intends to show that parenthood is the focus in which biological, psychological, and cultural factors converge.

More than ever before, parenthood in our age, as it evolves in individuals reared in our culture under the pressure of a rapidly changing civilization, appears removed from its biological sources. Parental behavior, as a culturally molded pattern, and its individual variations, are focal points of psychological and psychiatric studies of children and adults, but the psychodynamic processes of normal parenthood have not been conceptualized, as if taken for granted. Since it was assumed that personality integration is achieved during adolescence, the genetic theory of psychoanalysis does not include the psychodynamic processes of reproduction and parenthood as drive motivations for further development. Yet investigation of the psychosexual functions of women has demonstrated that personality development continues beyond adolescence under the influence of reproductive physiology, and that "parenthood utilizes the same primary processes which operate from infancy in mental growth and development" (p. 389).[5]

Theoretical Considerations

The instinct of survival in the offspring assures the survival of the species. This instinct is considered the organizer of those complex species-specific behavioral patterns through which survival of the species is maintained. The drive organization has three consecutive phases: (1) the sexual drive, which motivates courtship and mating behavior; (2) reproductive physiology, which accounts for the maturation of the germ cells (gametes), sets in motion the processes that maintain the fertilized ovum, supports its maturation, and guides parturition; (3) the care of the offspring, which, although it takes place outside of the mother's body, is a part of the reproductive physiology. Strictly under hormonal control in all species, the care of the offspring in some nonmammalian species is the function of the mate. Mothering behavior of human parents can be modified and divided between the sexes, either by choice or by necessity.

The phasic evolution of female sexuality exposes to investigation the drive organization of the propagative function. In Chapter 28 each phase of the female propagative function is discussed. Here will be pointed out only those aspects of the female drive organization that elucidate the difference between the sexes.

Investigation of the woman's sexual cycle revealed the development of the sexual drive. The psychodynamic tendencies that characterize the phases of pregenital development are repeated in correlation with the evolution of the gonadal cycle. The pregenital tendencies (oral, anal, and pregenital) are integrated in the mature sexual drive, which reaches its peak at the time of ovulation and regresses again during the premenstrual-menstrual phase. The pregenital tendencies are manifestations of the *primary instincts* that maintain the homeostasis of the organism and secure its growth and maturation so that the *secondary, sexual instinct* can come to the fore at puberty.[15] The slow, phasic evolution of ovogenesis in woman exposes the integration of the sexual drive through the psychological manifestations that accompany the hormonal cycle. Although such investigation probably cannot be performed in men, one may assume that man's sexual drive is derived from the pregenital tendencies during development from infancy to puberty.

Characteristic of man's sexual drive is its plasticity. "Sexual energy," its "appetitive strength," and the "intensity of its consummatory behavior" (for the sake of brevity, ethologists' terms are useful) are constitutionally "given" individual characteristics molded by ontogenic development. The regulation of sexual need and activity is central; hormones induce changes in the nervous system by affecting those systems that coordinate arousal and mating behavior. While courtship and mating represent the most accurate coordination of hormones and behavior for all the vertebrate species, this is not wholly true of infrahuman primates. In regard to man, history, as well as current anthropological and cultural changes, masks the hormonal effects from the physiology of the procreative function.

Women became independent from the limitations of estrus. The factors that promoted this evolutionary fact continued to interact with intraorganismic and environmental conditions and increased the gap between sexuality and procreation. The characteristics of the sexual drive hold true for both sexes. They motivate and integrate the sexual act, which may or may not be in the service of procreation. In women the integration of the psychodynamic tendencies that accompany ovulation indicate that motherhood has an instinctual origin; thus one may speak of "mother instinct" in scientific terms.[7]

But what about man? Man's role is discharged in one act that does not involve tissue changes beyond the production and deposition of semen. This process is under hormonal control. The innate specificity of the procreative function expresses the fundamental difference between the sexes. The psychophysiological organization of the male serves one act, that of insemination; the psychophysiological organization of the female serves the function of pregnancy and motherhood beyond the mating behavior.

This raises the question, what about fatherhood? Are men trapped into the social (sociological) role of fatherhood just by the compelling desire for orgasmic discharge, or does there exist in man a primary instinctual tendency toward being a father, a provider? The biological root of fatherhood is the instinctual drive for survival. The drive organization of species survival has three phases differently employed in the sexes. In phase one both are equal; in phase two the function of the male lasts a short time; in phase three the male of all species is involved for the time necessary for the maturation of the offspring. In Homo sapiens man's biological function as provider reaches beyond the maturation of the children; it reaches even beyond the family; it is a source of socioeconomic organizations. Since the biological role of fatherhood is to protect and thereby to provide the territory that secures the survival of the pair bond and their offspring, is there a psychobiological source of the quality that we term fatherliness?

Fatherhood, fatherliness, and providing are parallel to motherhood, motherliness, and nurturing. Fatherhood and motherhood are complementary processes that evolve within the culturally established family structure to safeguard the physical and emotional development of the child. The role of the father and his relationship to his children are further removed from the instinctual roots that make his relationship with his children a mutual developmental experience.[3] Fatherliness, like motherliness, has two sources: one is the biological anlage; the other is rooted in developmental experience. Yet there are differences between the sexes regarding the evolution of these primary attributes of parental behavior. In the development of fatherliness, the biological bisexuality and the male infant's biological dependence on the mother are primary factors.[4]

Bisexuality is a biological attribute of both sexes. The propagative functions of nonmammalian vertebrates offer striking examples of the different distribution of courtship, preparatory activities, and especially care of the young. In many instances the male takes over the care of the deposited ova or the feeding of the young as the instinctual organization of the species requires. Even in mammals there are examples of the male's participation in the care of the offspring. Nature seems to be able to reach deep into the bisexual propensities to meet the need of adaptive processes in a species. Our knowledge of man's bisexuality is still very limited. Investigation has been impeded by cultural denial. Hormone chemistry has helped but little since androgenic and estrogenic hormones, even progesterone, are closely related compounds; they occur in both sexes; their function in relation to symptoms has not been clarified. In the last decade intensive research has been conducted in relevant fields in an effort to clarify the role of "normal bisexuality" in man. Some level of predisposition may be seen in the varying degress of aptitude men show in the performance of mothering functions with their own babies. One also observes a great variety in women's skill and aptitude for genuine motherliness. The inhibition of these primarily biological functions may be attributed to the bisexual anlage of woman. The behavioral manifestations of the biological anlage, however, are strongly influenced in both sexes by the developmental process, especially that of the oral-dependent phase.

Every man's earliest security, as well as his orientation to his world, has been learned through identifications with his mother during infancy. In the normal course of male development this early identification with and dependence upon the mother are surpassed by the developmental identification with the father directed by the innate maleness of the boy. This results not only in the sexual, oedipal competition with the father but also in multiple identifications with the various roles of the father as protector and provider. In the development of the girl the infantile identifications with the mother reinforce the gender anlage and facilitate the normal evolution of female sexuality.*

The primary drive organization of the oral phase, the prerequisite and consequence of the metabolic needs that sustain growth and maturation and lead to differentiation of the procreative function, is the origin of parental tendencies, of motherliness and fatherliness.

Parenthood

The term "parenthood" refers to a psychobiological status of great significance for the individual and for the society in which he lives. Becoming a parent means being a link in the chain of generations. "It is only Homo sapiens who has the distinction and the responsibility for raising children beyond that procreative cycle which produced the particular child to full maturity and adulthood" (p. 119),[2] and so to convey to their children not only what the parents inherited (with the genic code) but also the complex culture with its ethical restrictions and potential gratifications.

In this chapter parenthood is presented from two viewpoints. First, the parents' interactions will be characterized in relation to the phasic development of the child; second, parenthood will be viewed as a crucial experience during the life cycle.

Marriage and family structures evolved as the consequence of the lengthy dependence of the human child. The family is the psychological field in which the transactional processes between parents and between parents and their children take place. The core of this field is the husband and wife, who bring to their marriage particular personalities as they have developed from infancy in transaction with their own parents, siblings, and other significant persons and events in their environment. Manifold and often tenuous reciprocal adaptations occur in every marriage until the couple is welded biologically through parenthood. Heterosexual love alone is not a guarantee of a lasting relationship. The ability to maintain a lasting relationship, which secures the permanence of a marriage, depends on the total personality of each partner. It requires of each a self-organization that does not become discouraged by the changing aspects of love as an experience, since it invests the marriage as an institution with (narcissistic) libido. If such self-investment exists, the feedback of being married and being a parent supports the interpersonal relationship between the marital partners through the vicissitudes of marriage. Speaking not of happy but of enduring marriages, it should be emphasized that, stimulated by the ongoing psychodynamic interaction, the personality of each partner achieves another level of integration. The process of mutual maturation gains another dimension through parenthood, through relationship with the child, through communication with each other via the child.[2]

The carrier of nonverbal communication is empathy. Empathy can be defined in psychoanalytic terms as a psychic energy charge that directs attention, facilitates perception, and furthers integration within the psychic apparatus. In general, empathy enlarges the psychic field of any individual and enables him to encompass in his responsiveness everyone and everything to which he may relate. While empathy itself is unconscious, the empathic response usually appears as an intuitive, spontaneous reaction that often mobilizes affects and motivates responses. In our culture many individuals are so guarded against their intuitive feelings that they suppress their primary empathic responses even in the most intimate situations, such as sexual interaction, and even in their transactions with their children.

Closest to its biological source is the mother's empathy for her infant. This determines the quality of her motherliness and leads to competent, successful mothering. The adjective "competent" calls attention to another

* First emphasized by psychoanalysts, this fact has been confirmed by investigations of ethologists. Recently the studies by Harlow[11] are the most widely known.

level in the use of empathy. As the child develops and becomes more and more a person in his own right, parental empathy has to undergo intrapsychic elaboration. Empathic response is a direct instinctual or intuitive reaction to the child's need. Empathic understanding is arrived at by a preconscious process of self-reflection that leads the parent to an understanding of the motivations of the child's behavior and at the same time to an understanding of the motivations of his own reaction.

What is said about the empathy of the mother also holds true for the father. Although the father's empathic response to his infant cannot be related directly to his function in procreation, most men do exhibit genuine fatherliness. Fatherliness, like motherliness, is an instinctually rooted character trend that enables the father to act toward his child or toward all children with immediate empathic responsiveness. Fatherliness has early and differing manifestations. It seems to appear in the father's first smile greeting the newborn; it is expressed in his ability to cradle the infant securely; or later it is displayed in his participation in the care of the infant, in his patience and tolerance of the disturbance and difficulties that naturally arise in rearing a child.

The psychobiology of fatherhood seems to have evaded investigation as if it were hidden by the physiology of male sexuality and by the socioeconomic function of fathers as providers. While biology makes invariable the role played by the mother in the propagation of the species, the role of the father changes with cultural and socioeconomic conditions. Surprising as it may appear, the socioeconomic function of providing as well as the characterological quality of fatherliness are derivatives of the instinct for survival.

Only human parents have two sources of parental behavior. One is rooted in physiology as in any other creature; the other evolves as an expression of the personality that has developed under environmental influences that can modify motherliness and fatherliness. After the child outgrows his infancy, the mother becomes more independent of procreative physiology. Thus the motivational system of parental behavior becomes the same in both parents. Parental behavior is motivated by the response to the actual need of the child, by the situation in which the need arises. The unconscious motivations of parental behavior are rooted in the personality. This colors the meaning of the parental experience and stimulates anticipations that parents project onto their children often before and more concretely after they are born. Indeed, Freud[9,10] was right when he stated that the tender love of fond parents for their children originates in the narcissism of the parents, in the libido reservoir that maintains motherliness as well as fatherliness.

It is not possible to outline the normal range of motherliness and fatherliness in action. The limits change from culture to culture, from individual to individual. When we consider motherliness and fatherliness as developmental attainments in close contact with instinctual sources, we become aware of their oscillations. For motherliness and fatherliness appear to fluctuate under affect; they seem to regress and reintegrate in interaction with the child and his total environment.

Infants learn to anticipate the parents' responses faster than adults imagine. The significance of the child's anticipation of the parent's reaction to his behavior has been studied in detail from birth through adolescence. The balance between the child's confident expectation of gratification and his fear of frustration modifies his sense of security with his parents. The reciprocal process in the parent rarely has been studied. It is, or used to be, generally assumed that the adult parent's ego organization is not subject to change under the influence of his object relationship with his infant, with his growing child, and even with his grown-up child. Probably such self-secure, mostly authoritarian parents still exist in other civilizations. In our culture modern parents cannot even envy the security of the Victorian parent. Soon that generation will be parents whose grandparents were raised by Victorian parents.

The parent's emotional security toward the child, even when expressed as authority, has a double function. It protects the child and insures the parent against being unduly affected by the child's behavior. His authority helps him to repress or deny his fears and negative anticipations about the child and about his own ability to cope and love at the same time. The anticipation of negativistic attitudes in their children makes parents insecure, afraid, and often angry even before the child gives them cause. This mobilizes primitive behavior that, even though it may be appropriate, is followed by a sense of guilt in modern parents. The guilty feelings may increase the insecurity, and so a negative spiral evolves between parents and child.

Fearful insecurity is characteristic of young mothers, especially with their first child or with a child who is not healthy or normal. In patriarchal families fathers usually felt uninvolved with and not responsible for the care of the infant. In the young families of our age fathers feel involved and consider it a duty to help their wives. Not infrequently they prove themselves more secure in handling the newborn than the young mother. This, however, may have a negative effect on the wife, who, feeling inferior in performing this innate duty, may become depressed and alienated from the child. Such an incidence illustrates that the emotional balance of the family triangle depends on each of the participants, that to provide a satisfactory environment for the growth of the child, the parents' empathic understanding for each other is a prerequisite. Conflicts arising in the primary triangle—father-mother-child—that originate within the parents certainly influence, at least transiently, their behavior toward the child, but this seems secondary in regard to parenthood as a developmental process. This concept refers to those transactional processes between parent and child that, motivated by the phasic libidinal development of the child, reactivate in the parent old conflicts of the same period.

The transactional processes of early infancy can be easily conceptualized since in the mother and child they originate in primary biological needs; in the father they probably originate in the formidable adaptational task of becoming and being a father.

Based upon the model of reciprocal interaction between parent and child during the oral phase of development, we may generalize that the spiral of transactions in each phase of development can be interpreted on two levels of motivation in terms of each participant. The parent's behavior is determined unconsciously by his developmental past and consciously by his immediate reaction to the needs and behavior of the child. By incorporating the many traces of the parent's behavior, the infant learns and so acquires a past that enables him to anticipate the parent's response to his behavior. This introduces a third aspect into the motivational pattern, namely, anticipation of the emotional course of future experiences. This motivational pattern is not yet existent in the young infant, but it becomes noticeable early in the second quarter of the first year; from this time on it evolves to facilitate the child's orientation to and interaction with his environment. It is rarely observed how much irritability of the parents toward each other, stimulated by inefficiency of the mother or by unavailability of the father's help, influences the development of the infant; even less investigated is how much the infant's thriving compensates for the emotional stress between the parents.

Is there any psychoanalytic evidence that would support the thesis that the child, being the object of the parent's drive, has, psychologically speaking, a similar function in the psychic structure of the parent? Does the child, evoking and maintaining reciprocal intrapsychic processes in the parent, become instrumental in the further developmental integration of the parent? Observations and psychoanalytic investigations yield positive and negative examples of the intrapsychic processes of the parent in reaction to the child.

Imitation is a well-studied aspect of the parent's interaction with the child. The imitating child holds up a mirror image to the parent. Thus the parent may recognize and even say to himself or to the child, "This is your father;

this is me in you." If the child's imitative behavior shows the positive aspect of their relationship, the parent will like what he sees and consequently will feel that both child and parent are lovable. Imitation then reinforces the positive balance of identifications. It can also happen that the child shocks the parent by exposing the representation of hostile experience in the past or in the present. In this event the parent feels the child's rejection and withdraws from him, even if just for a moment, since the unloved self equals the unloved, unlovable child. Imitation externalizes what has been internalized from infancy. It exposes not only the child's identification with the "omnipotent" parent but also his anger because of frustrations imposed upon him by the parent. The parent's responses to the hostile imitation of the child is a record of his acceptance of the growing independence of the child with whom he identifies in the process. It should not be forgotten that any manifestation of the child's positive identification with a parent reassures that parent: "I am a good father"; "I am a good mother."

Normally the child's idealization of the parent gives the parent gratification. There is no need to describe how fathers respond to the admiration of their sons or to the flirtation of their three- to four-year-old daughters. Just as obvious is the mother's pleasure in her daughter's wish to become like her or in her son's promise to marry her because she is the best, the most beautiful mother.

Much has been written about the oedipal child, but, except for the actually seductive, pathogenic behavior of parents, very little has been written about the parents' participation in the development of the normal oedipal phase. This may be explained by many parents' restrained physical contact with the child of that age, by the tendency to hide, to forget, actually to repress libidinal impulses that were more freely expressed with the younger child. On the other side are those parents who, under cultural influences, assume that any sexual control is inhibiting to the child's psychosexual development. This mistaken rationalization allows them to expose their children to undue sexual stimulation, yet such parents often have to struggle with their own conflicts and with the psychological consequences of their laxity during later phases of their parenthood.

Psychoanalytic investigations have revealed that parents anticipate the child's failure in the area of their own developmental conflicts. Unconscious as the motivating conflict remains, the symptom is remembered. Well known to all of us is the parent who, because he or she was enuretic as a child, concentrates anxiously on the toilet training of the child. Even if this has been successfully achieved at an early age, such parents anticipate a relapse, especially when the child approaches the age at which their own relapse occurred. One may generalize that, unaware as parents usually are of the repressed conflicts of their childhood, the transactional processes evolve relatively smoothly until the child reaches the developmental level in which the parent becomes insecure in his response to the child's behavior. The anxious behavior of the parent is instrumental in conveying to the child the parent's own fixation.* The fear that a childhood symptom may be repeated by the child does not necessarily lead to anticipation that this will occur. Looking back at the childhood symptom from the security of his adulthood, the parent relives with the child his own conflict, now without fear. In the successful interactions with the child, the parent resolves his own developmental conflict, with an addition to his self-esteem: "I am a good parent."

Each parent has to deal in his own way with the positive as well as the negative revelations of himself in the child. "It is the individually varying degree of confidence in oneself and in the child which enables the parent not to overemphasize the positive and not be overwhelmed by the negative aspects of the self as it is exposed through the child" (p. 131).[2] With the help of the positive manifestations of the child's development, the parents' confidence in their child grows and with it grows the conviction that they are achieving

* Adelaide Johnson[13] described this process as the "etiology of fixation."

the goal of their existence. In terms of dynamic psychology this means that while the parents consciously try to help the child achieve his developmental goal, they cannot help dealing unconsciously with their own conflicts, and thus they achieve a new level of maturation themselves.

Parenthood during the Life Cycle*

Parenthood implies continual adaptation to physiological and psychological changes within the self, parallel to and in transaction with changes in the child and his expanding world. In discussing the limitless variations of conflicts recurring during the life cycle of the parents, it is helpful to conceptualize parenthood as a process that has an early, a middle, and a late phase. With each child all parents live through these three phases, which necessarily overlap. Parents can be in the late phase of parenthood with their oldest child and at the same time be young parents with their youngest child.

Parenthood as an experience is more in focus during the early phase, which Kestenberg[14] refers to as "total parenthood." The parent's involvement with and responsibility for the child is almost exclusive during infancy. The reciprocal psychodynamic interactions are most significant during infancy and the separation-individuation phase. But even in these early years there are exceptions to total parenthood. In many cultures mothers have helpers within the kinship; in our society the upper classes may have maids, nurses, or governesses who take over the duties of the mother; in other situations mothers go to work and therefore need helpers. Besides these, nursery school and kindergarten shorten the period of "total parenthood." But before school age the child's developmental needs for expansion are basically under parental surveillance.

The beginning of school in Western civilization coincides with the age and maturational level to which Freud attributed the end of the oedipal phase. With this the mental development achieves the ability to incorporate the expanding environment of classroom, teacher, and classmates. School, a socially regulated partial separation of parents from their children, facilitates the repression of the Oedipus complex; this induces the latency period. Kestenberg[14] states, "*Latency* stands out as a time of *part-time parenthood*!" (p. 305), meaning by this only the diminishing activities involved in childrearing. Yet it is worthwhile to mention that fortunately this separation evolves slowly, since otherwise it would activate fear or negativistic reactions in the parents against the growing independence of the child and against those who promote it. Usually the second and third child replenishes the libidinal supplies of the parents (more that of the mother than the father) when the first child reaches school age.

One could discuss the reciprocal developmental processes from the viewpoint of the parents' psychological separation from the child. Such conceptualization, however, does not cover parenthood during the life cycle. Conceptualizing parenthood according to its early, middle, and late phases affords the opportunity to organize the most frequent problems of parenthood as they change in time, keeping in focus the transactional processes between the generations.

Parents are total parents with each of their children and live through the early phase of parenthood with each of them until and through various stretches of their adolescence. The overlapping phases of parenthood, however, may cross the boundaries between generations. It is not infrequent that a young grandmother is at the same time a young mother. Paradoxes of family lineage may thus occur. The baby of the grandmother is the aunt or uncle of the child born to a son or daughter.

In the early phase of parenthood the mutu-

* Material in this section has been drawn extensively from my Chapter 8 with the same title in *Parenthood: Its Psychology and Psychopathology*, edited by E. J. Anthony and myself (Boston: Little, Brown, 1970), pp. 167–183.

ality of the ongoing processes of identification-separation dominates the psychic economy of the parent-child relationship. The shift in these processes pushes the child in the direction of separation, the parents toward holding on.

During the preoedipal and oedipal phase the evolution of the dominant libidinal conflict and the corresponding ego growth activate unconsciously motivated, characteristic responses in the parents. Normally these responses quickly disappear under the pressure of the newly arising developmental trends in the child. Secure in their love for the child, parents rarely feel responsible for his passing problems. All that happens seems to be open to the empathic understanding of the parents; therefore, they respond with the feeling that it is natural, that the child will outgrow it.

School age often disturbs the security of the parents. School represents authority for the parents as it did when they were children. School means to parents that their child's behavior, his performance at work and at play will be exposed to scrutiny, and thus the parents themselves feel exposed. In order to diminish their disconcerting feeling of responsibility and also their (probably) hurt narcissism, parents eagerly supervise the various sources of extrafamilial influences that their children experience. Their vigilance is often biased by prejudices and preconceived ideas. Playmates and neighbors are judged. Television programs are considered welcome entertainment; their influence upon the child—good or bad—usually cannot be assessed by the parents.

Yet parents observe with concern that their children are growing up faster than they did themselves. Very often they seem to want to slow down the tempo of externalization that characterizes the growth of the latency child. On the one hand, they would like to hang on to the past when they felt that they knew everything about the child; on the other hand, they have to weigh the child's competitive achievement with their ambition that he should perform on every level with adequate competence. But modern parents are wrought with apprehensions regarding educational aims. While they conscientiously strive to bring up independent, secure, and efficient individuals, they refrain from applying controls lest the child become inhibited through punishment and grow up to hate them.

These conflicting problems of modern parents are pointed out here to illustrate the difference in the educational tasks of parents of preschool children and parents of latency children and adolescents. The preschool child evokes in the parents empathic, affective, goal-directed responses to behavior that is a manifestation of a maturing individual. Whereas with young children the parents' developmental past refers prevalently to unconscious processes, in response to latency children and adolescents, parents remember their own behavior and its consequences. Conscientious parents' emotional responses to the problems set by their children of that age are motivated by reaction formation to the actual or psychological consequences of their own experience.

This model seems to apply mainly to those parents whose developmental past justifies their wish to provide their children with better conditions than they themselves had. However, there are many parents who, raised by permissive parents, grew up with the advantages of an affluent society. They, too, want to bring up their children to become productive, capable, contented adults. Can one formulate the educational task of such parents? It is beyond the scope of this chapter to discuss the transactional phenomena that set normative goals for the children of such parents by holding parental ambition within limits realistically measured by the capacity of the growing individual.

An essay on the psychobiology of parenthood should concentrate on parental reactions to the sexuality of their children. A half century ago one could have responded to such a request with a simple statement. The puritanical sexual mores invested in the Judeo-Christian tradition of Western civilization deny the existence of sexuality in human beings until marriage. Now parents observe the sexuality of their children and usually deal with it according to the state of their own conflicts.

They are not too disturbed to observe a young child playing with his genitals. The father who threatens a three-year-old child that he will "Cut it off" is becoming rare. During the boy's latency fathers usually are more concerned with the son's athletic ability and with his general manliness than with his sexual behavior. With the increasing sexual freedom among adolescents, fathers become concerned when their 16- to 18-year-old son is not sexually active. On the same basis they feel differently toward their daughter. As they suppress their own libidinal interest toward their daughters, they assume that girls are blissfully innocent. Mothers are different. Being more aware of their defensiveness against sexual impulses, they often become suspicious when children, even of the same sex, play together behind closed doors and are too quiet. Mothers are intent on protecting their children, the daughters more than the sons. Their worries become intensified when the period of dating begins and rarely cease until the daughter lands safely in marriage. Regarding their sons, their worries are different in degree, but not in kind.

In general, one may say that women who feel positively about their femininity and enjoy sexuality are usually less envious and less suspicious of their daughters' sexual lives. They trust their children since they trust themselves and their own experiences. In their intuitive confidence in the power of their own personalities, they feel they have conveyed to the children their own value system and what is in accordance with it is anticipated without anxiety. However, such mothers are also shocked sometimes by disappointment. Self-confident parents have such an intrinsic need to trust their children, to assume that "what should not happen cannot happen," that they are often blind to the obvious. There are parents who want to be even more modern; they convince themselves that this is "her life," and that they want her to "live it fully." They are shocked by the realization that they have deceived themselves; the daughter's pregnancy becomes their personal shame; the disappointment in the daughter is their own failure. The middle-aged mother might respond with a serious depression to an event whose emotional significance she has denied. Even parents who trust their children and enjoy their confidence are left in the dark about the most important experiences of their children. This probably must be so. If it is sincere, sexuality is a private experience between two individuals; it is least communicable to parents.

Mothers whose behavior toward their children is characterized by intrusive vigilance have usually repressed their own sexuality and had neurotic conflicts. Women having little or no confidence in their femininity usually convey their insecurity to their children. When the period of dating comes, they realize that their daughters are not popular, their sons are not going out with girls. The insecure mother relives her own adolescence with pangs of waiting, of being left out, of being alone. Such mothers suffer more than fathers from the inferiority feelings of their children. They begin to push their adolescent children; they advise and scheme in order to help. Painfully aware of the well-meant but unbearable concern, the adolescent tries to escape. For daughters as well as for sons, college or work away from home appears to be the best way to find relief. This often leads to emancipation of the daughter and the son; it permits new experiences away from the watchful eye at home.

Fathers are helpful by being more tolerant of the daughter's lack of popularity. The more usual complaint of daughters about their father is that he scrutinizes the boys too closely, criticizes them frankly, and often tries to scare them away. One can say that fathers have a double standard regarding their adolescent children's sexuality. They watch with Argus eyes over the virginity of their daughters, but unconsciously identify with their son's experience; they smile at the young girl with whom the son is in love as long as there are no consequences. For father, traditionally, propriety ends when the son comes home downhearted to announce that the girl is pregnant.

The variations of individual experiences of the *middle phase of parenthood* are many, but the main characteristic is the parent's involvement in and preoccupation with the children's sexual life. Whether this be traditional courtship under the watchful eyes of parents, or the now frequent series of love affairs leading to consecutive promiscuity before marriage, it is all to culminate in marriage. Whether the parents' marriage is happy, just tolerable, or a cauldron of explosive emotions, no matter how deeply the marital struggles of the parents influence their children, all of the past is forgotten and the future appears rosy in the light of a new marriage. Except when difference in race, religion, and social status seems irreconcilable, both sets of parents unite in their hope that the children will live happily ever after.

With a child's marriage the immediate responsibility of parenthood for the child discontinues. The parents cease to be closest of kin by law, since the new husband and wife, even if they have known each other for only a short time, become next of kin. Parents feel this first probably when the wedding is over and they come home to rest. Then they begin to feel and rationalize about their sudden sadness. These parents may still be young people, living in the unity of early parenthood with their younger children, yet they have entered the late phase of parenthood with the newly married child.

In cultures in which the young wife customarily has to leave her parental home to live in the parental home of her husband, the marriage of the daughter represents an almost complete separation. The mother has neither the right nor the opportunity to remain involved with her daughter. In our civilization it is still not uncommon that both sets of parents of the young couple live close to each other, are neighbors or friends, that the young people had known each other from childhood. Marriage in such a situation does not involve such sharp separation, does not require such a difficult adjustment. In our present culture, when neither social nor geographic boundaries restrict the choice of a mate, marriage often implies separation of parents from their children, which may activate a critical phase of parenthood.

In any case the marriage of a child represents a new adaptational task for the parents; they have to encompass the husband of the daughter or the wife of the son, not only in their own family, but also in their own psychic system as an object of their love. Psychologically this occurs through the identifications with their own child. The object relationship to the in-laws remains shaky for a time. The ambivalence easily flares up, rationalized by the parental concern for the happiness of their child. Just as when the child first went to school, parental narcissism makes them see the fault in the other rather than in their own child. Yet the young couple can "fight it out" and settle the differences more easily when the conflict remains their own problem and does not spread in circles like pebbles thrown in a pond. In this respect mothers are more often at fault than fathers.

Mothers often cannot relax their influence on their daughters; they identify with the married daughter or son and want to be involved in, or at least informed about, every detail of their life. Whether we see the problem from the point of view of the young husband or the wife, the mother is almost always the "in-law," the often feared, critical investigator of one partner of the marriage. Yet today one sees very definite changes in this respect. As long as daughters grew up in families in which they owed devoted dependence to their mothers, even when married, they accepted the mother's opinion with unquestioning deference. Such "good" mothers were the "feared mothers-in-law," the butt of jokes, ridicule, and hidden or open hatred. Now the more self-reliant woman's husband does not need to fear that his wife has to side with her mother. Mother-in-law jokes have almost disappeared from magazines, indicating a significant change in the structure of the family and in the relationship between the generations.

The example of mothers who cannot psychically separate from their married children shows the more universal psychological problem of the late phase of parenthood. The slogan "generation gap" is not affixed to the adaptational problems of the late phase of parenthood, for obvious reasons. But there are some factors that seem to justify a comparison. One of the characteristics of the late phase of parenthood is the emotional consequence of the married children's alienation from their parents. Whether the parents are middle-aged or older, the child who becomes a parent does not have the same psychological relationship with the parent; his psychic structure has changed. More than the deepening relationship with the spouse, parenthood does change the psychic structure of the young parent. The parents of married children have to adapt to their not being needed as they were before; this reduces the parents' self-esteem. This generation gap does not cause vehement upheaval, since it is not stirred by the maturation of adolescence, but by slow evolution of the late phase of parenthood. The gap between generations, which began with marriage and parenthood of the young generation, now deepens because of the physiological factors of aging. The psychophysiological reactions to "change of life" in women intensify the mother's emotional reactions and make her aware of all that which "hurts" in aging, even without severe, clinical depression. Yet her sensitivity increases the rift between the generations.

Fathers usually do not get into similar troubles of alienation at such a relatively early age as mothers. One reason for this is that fathers maintain more distance from the interpersonal problems of the family. As long as the father's ability to work is not diminished, aging has a mellowing effect on his attitude toward his children. In the disquieting experience of alienation from their married children, normal, healthy fathers frequently function as negotiators, trying to make peace and avoid a rift. The late phase of parenthood arrives later for fathers than for mothers, or it seems so because at that age level fathers become more interested and therefore more involved with their families than they were previously.

Before the last phase of actual "childless parenthood" arrives, late parenthood brings about the gratification of the life cycle, *grandparenthood.*

Psychoanalysis of both men and women whose married children are childless, whether voluntarily or because of infertility, reveals disappointment and frustration, and also the source of the anxiety caused by this condition. Sometimes these individuals have guilt feelings and blame themselves for wishing for something beyond their ken. The somatic correlations of such depressive states originate in the wish to survive in the grandchildren.

There is a noticeable difference in the attitude of the prospective grandparents toward the pregnancy of the daughter. Fathers do not identify with the experience of pregnancy as do mothers. Prospective grandmothers remember what their mothers told them about the pleasurable or frightening experience of delivery and lactation. The prospective grandmother, reliving her own pregnancies in identification with her daughter, in her wish to protect her daughter, may convey her anxiety to the pregnant woman. Such anxious overidentification of the prospective grandmother, however, often interferes with the actual bliss of grandparenthood.

Grandparenthood is parenthood one step removed. It is a new lease on life since grandmothers as well as grandfathers relive the memories of the early phases of their own parenthood in observing the growth and development of the grandchildren. Relieved from the immediate stress of motherhood and the responsibilities of fatherhood, grandparents appear to enjoy their grandchildren more than they enjoyed their own children. Since they do not have the responsibility for rearing the child toward an unknown goal, their love is not burdened by doubts and anxieties; they project the hope of the fulfillment of their narcissistic self-image to their grandchildren.

The indulgence of grandparents toward grandchildren has its psychodynamic (instinctual) motivation. If the relationship between the grandparents and the child's parents is not burdened by jealousy and hostility, open or suppressed, the grandparents can feel free to love their grandchildren. This does not mean just giving candy and toys or playing with them. The love of grandparents gives the child a sense of security, in being loved without always deserving it. What does the grandparent receive in return? A loving glance from a happy child, a trusting hand, an actual appeal for help; whatever it is, it is a message to the grandparent that he or she is needed, wanted, loved. Grandparents accept gratefully the reassurance from the child that they were, and still are, good parents.

Grandchildren, however, grow up and grow away from grandparents. As they reach adolescence, their attitude appears to reach that postambivalent phase of object relationship that Karl Abraham[1] described as characteristic of maturity. The ambivalence of adolescence, the rebellion of youth are directed toward the parents, who are the objects of their conflicting instinctual drives. The relationship with the grandparents is never so highly charged; therefore, the grandparents become the recipients of considerate and indulgent behavior by the maturing individuals, who, in the awareness of their strength, see the weakness of the doting grandparents even earlier than might be justified. The grandparents respond to the manifestations of the protective, somehow even condescending love of their grandchildren as balm for whatever wounds old age inflicts upon them.

Grandparenthood is, however, not the same for everyone. There are differences depending on the personalities of the interacting individuals belonging to three generations. The emotional content of grandparenthood and the expectations of the young parents in regard to them depend upon the cultural and socioeconomic changes in the family structure.

Of the many factors that influence the emotional meaning of grandparenthood, the chronological age of the grandparents is probably the most significant. Experience of grandparenthood has different emotional colorings if the grandparents are young, still in possession of their procreative capacities. It is obvious that the involvement of grandparents in such families overshadows their emotional need for grandchildren. This need seems urgent when the grandparents are well over their procreative period and they have had to wait a long time for grandchildren.

Old age, if not hastened by illness, arrives slowly, bringing with it the adaptive tasks of aging itself. From the multitude of these tasks, only those will be mentioned that influence intrafamilial relationships and consequently the status and function of the elderly parent in the family. In order to put this in a psychodynamic frame of reference, the overall psychodynamic character of each major phase of the life cycle is pointed out: (1) from infancy through adolescence the vector of metabolic and psychological processes is self-directed, i.e., receptive; (2) during the reproductive period the vector is expressed in the object-directed, expansive, giving attitude of parenthood; (3) as the supply of vital energies declines with aging, the positive, extraverted tendencies slowly become outweighed by the energy-conserving, restricting, self-directed tendencies of old age.[6] These unconscious factors bring about the psychological (often psychiatric) manifestations (symptoms) of old age; they motivate also the psychological processes that bring about the age-determined changes between parents and their children.

The style of aging depends more on the personality pattern than on chronological age. Since aging reduces the libidinal expansiveness of the individual, the hostile components of the character become more pronounced. This explains the domineering, know-it-all behavior of many aging mothers and grandmothers, who become embittered if the younger generation does not follow suit as they did earlier. When the pattern and course of the psychodynamic processes of the parent are known, it is not difficult to establish the distortions caused by the involutional processes of the parent and understand with sympathy the influence that old age exerts within the family.

There is no doubt about the specific blend of narcissism in the aged. Since it cannot draw upon fresh resources of libido, it enlarges the remaining resource by identification with the young and by rekindling the memories of past gratifications. Current frustrations increase preoccupation with memories of youthful experiences. Being engrossed in what one was often becomes irritating, even to the grandchildren, let alone to their parents. But this irritation means increased frustration and makes increasing demands in the senescent. The senescent person's ability for empathy with the younger generation diminishes. The defenses of the self-centered personality become more tenacious so that the younger generation's complaints about the egotism of the old are justified. The solace offered by the younger generation usually does not satisfy the senescent since he unconsciously wishes and, in some ways, consciously demands that his children and grandchildren remove the burdens of his age and make him unaware of his weakness. Many manifestations of "nonparenthood" with "nonchildren" can be described and explained; they all illustrate the complete turn of the cycle. As one time the parent was the need-fulfilling object of the child; now the "adult child" or the middle-aged child is the need-fulfilling object of the aged parent. The old parent, however, clings to the status of being a parent. Originating in the instinct for survival in the offspring, parenthood establishes a sense of identity that integrates the biological and social functions of the personality.

Being a parent is at the center of a normal parent's self-concept. In old age, removed from his procreative period by two generations, he clings to his adult children and seeks in them the psychic images they once had been and therefore will always remain, his children. Supplied by memories of past experiences, parenthood is timeless. In the sense of intrapsychic processes, parenthood ends when memory is lost and psychic images fade out.

Bibliography

1. ABRAHAM, K. (1924), "A Short Study of the Development of the Libido, Viewed in the Light of Mental Disorders," in *Selected Papers*, pp. 418–501, Basic Books, New York, 1955.
2. BENEDEK, T., "The Family as a Psychological Field," in Anthony, E. J., and Benedek, T. (Eds.), *Parenthood: Its Psychology and Psychopathology*, Chap. 4, Little, Brown, Boston, 1970.
3. ———, "Fatherhood and Providing," in Anthony, E. J., and Benedek, T. (Eds.), *Parenthood: Its Psychology and Psychopathology*, Chap. 7, Little, Brown, Boston, 1970.
4. ———, "On the Organization of the Reproductive Drive," *Int. J. Psychoanal.*, *41*:1–15, 1960.
5. ———, "Parenthood as a Developmental Phase," *J. Am. Psychoanal. A.*, *7*:389–417, 1959.
6. ———, "Personality Development," in Alexander, F., and Ross, H. (Eds.), *Dynamic Psychiatry*, pp. 63–113, University of Chicago Press, Chicago, 1952.
7. ———, "The Psychosomatic Implications of the Primary Unit: Mother-Child," *Am. J. Orthopsychiat.*, *19*:642–654, 1949.
8. DARWIN, C., *The Descent of Man*, Murray, London, 1871. Quoted from Kaufman, C., "Biologic Consideration of Parenthood," in Anthony, E. J., and Benedek, T. (Eds.), *Parenthood: Its Psychology and Psychopathology*, p. 6, Little, Brown, Boston, 1970.
9. FREUD, S. (1914), "On Narcissism: An Introduction," in Strachey, J. (Ed.), *Standard Edition*, Vol. 14, pp. 67–102, Hogarth, London, 1957.
10. ——— (1940), "An Outline of Psychoanalysis," in Strachey, J. (Ed.), *Standard Edition*, Vol. 23, pp. 141–207, Hogarth, London, 1964.
11. HARLOW, H. F., "Primary Affectional Patterns in Primates," *Am. J. Orthopsychiat.*, *30*:676–684, 1960.
12. HARTMANN, H., KRIS, E., and LOEWENSTEIN, R. M., in Wilbur, G. B., and Muensterberger, W. (Eds.), *Psychoanalysis and Culture* "Some Psychoanalytic Comments on 'Culture and Personality,'" pp. 3–31, International Universities Press, New York, 1951.
13. JOHNSON, A. M., "Factors in the Etiology of Fixations and Symptom Choice, *Psychoanal. Quart.*, *22*:475–496, 1953.
14. KESTENBERG, J., "The Effect on Parents of the Child's Transition into and out of Latency," in Anthony, E. J., and Benedek, T. (Eds.), *Parenthood: Its Psychology and Psychopathology*, Chap. 13, pp. 289–306, Little, Brown, Boston, 1970.
15. KUBIE, L., "Instincts and Homeostasis," *Psychosom. Med.*, *10*:15–20, 1948.

CHAPTER 24

PSYCHOLOGICAL ASPECTS OF DIVORCE

Richard A. Gardner

Of the 2,146,000 marriages reported in the United States in 1969,[47] one in three may reasonably be expected to end in divorce. This rather high proportion of failed marriages by no means represents a sudden collapse of the institution. According to Census Bureau figures,[39,47] the divorce rate in the United States has been rising since the turn of the century. In 1890 the divorce rate was 0.5 per 1,000 total population. In 1946 the rate was 4.6—a peak that is explained by the wholesale dissolution of ill-considered "war marriages." After 1946 the divorce rate gradually decreased to a low of 2.1 in 1958. Since then the rate has been climbing steadily; figures for recent years are: 1966, 2.5; 1967, 2.6; and 1968, 2.9.

In 1969 there were 660,000 divorces—a figure that represents a rate of 3.3 per 1,000 total population. Since 2,146,000 marriages were reported during this year, the ratio of divorces to marriages is 660,000/2,146,000 or 1/3.25. This ratio is the basis for the prediction that one in three marriages will end in divorce. This figure cannot, however, be considered an accurate projection. Since the number of reported marriages is increasing every year, there is no direct correspondence between 1969 marriages and 1969 divorces. The latter represent a chronological accumulation of failed marriages that began over a wide range of previous years, any one of which had fewer total marriages than 1969. Therefore, the possibility of a 1969 marriage ending in divorce is probably greater than one in three.

The latest Census Bureau figures on the duration of marriages that ended in divorce are for the year 1967.[40] The median duration was 7.1 years. The modal duration was between one and two years. Other generalizations regarding divorce may be drawn from the Census Bureau data. There is, for example, a greater risk of divorce for those who marry in their teens. The percentage of divorces involving children also appears to be increasing. In 1953, 45.5 per cent of the divorces in the United States involved children; in 1958 the figure was 55.1 per cent; and in

1963 the percentage was 61.1. Apparently the belief that marriage should be maintained "for the sake of the children" is losing its force.[39]

Within the United States there are significant regional variations in the divorce rate. In 1963 the rates were 0.9 per 1,000 total population in the Northeast, 2.2 in the North-Central states, 2.8 in the South, and 3.6 in the West.[39] These regional differences are in part related to varying degrees of permissiveness in state laws regarding divorce that encourage what the Census Bureau calls "migratory divorces." In 1967, for example, the divorce rate in New York was 0.4 per 1,000 total population while Nevada had a rate of 22.3.[40] The divorce rate in Nevada was thus 56 times the rate in New York. Far from being a statement about the relative stability of marriages in the two states, the figure primarily expresses the fact that a great many New Yorkers were going to Nevada to obtain their divorces. In spite of migratory divorces, however, the West has had a higher rate of divorce than the East since the beginning of the century.

How do divorce rates in the United States compare with those of other countries? Although almost all countries report their annual divorce totals to the Statistical Office of the United Nations, which publishes them in the *Demographic Yearbook*,[9] meaningful comparisons are difficult. A number of countries, including Argentina, Brazil, Chile, Colombia, Ireland, Malta, Paraguay, Peru, Philippines, and until 1970, Italy, do not provide legal means for the dissolution of marriage.

TABLE 24–1. **Number of Divorce Decrees per 1,000 Population Granted under Civil Law in 1968**

United States	2.91
U.S.S.R.	2.73
Southern Rhodesia	2.14
Hungary	2.07
Egypt	1.92
South Africa	1.72
East Germany	1.68
Czechoslovakia	1.49
Libya	1.41
Sweden	1.39
Austria	1.32
Bulgaria	1.16
Finland	1.15
Yugoslavia	1.02

SOURCE: *Demographic Yearbook*, United Nations, New York, 1969, pp. 671–674.

Although Table 24–1 lists the countries with the highest divorce rates for 1968, the order is somewhat selective. For example, the Falkland Islands reported a divorce rate of 2.50, which would make it third highest in the world. The population of the islands is so small, however, that only five divorces were necessary to achieve this rate. Factors other than population must be considered in comparing divorce rates between nations. Figures reported for Southern Rhodesia and South Africa, for example, include only the white population.

Psychosocial Factors in Marriage That Contribute to Divorce

If one marriage in three is doomed to failure, it is only to be expected that some would advance the opinion that monogamy is not congenial to man's basic personality structure. The ever rising divorce rate has led others to consider monogamy outmoded or dysfunctional in our highly mobile and specialized technological society.[42]

In opposition to the preceding theorists, Kardiner[25] has argued persuasively that man has tried practically every conceivable arrangement for marriage and childrearing and monogamy has proved to be the most effective, its deficiencies notwithstanding. It allows for the closest continuous contact between parent and child that all agree is essential if the child is to become a self-sufficient and contributing member of society.

In discussing alternative systems, such as the polyandrous Marquesan Islanders whom he studied intensively, Kardiner convinces the reader that the mother's incredibly complex interactions with her three to six husbands leave her but scant time for her children. Polygamous cultures, on the other hand, foster intense rivalries between the males that distract them from domestic involvements. A multiplicity of maternal or paternal figures confuses the child and lessens the likelihood of strong attachments.[25]

Bettelheim confirms Kardiner's point in his recent study of children raised in Israeli kibbutzim.[6] Although describing an encouragingly low incidence of juvenile delinquency, drug addiction, and severe emotional disturbance, he nevertheless concluded that the child raised in a setting with diffused parental figures tends to be more detached in his interpersonal relationships than children raised in more conventional settings.

Given the reasonably durable nature of the monogamous relationship, what is there in the Western variety, and that found in the United States in particular, that causes disequilibrium and divorce?

When two people marry today, "love" is taken to be the primary and only acceptable reason. While secondary considerations such as physical attractiveness, similar interests, status, and money may be admitted, anyone who states that he is marrying without "being in love" is branded with pejorative labels. He is "materialistic," "opportunistic," "sick," or "foolish." As with other psychological phenomena in the human repertoire, cultural influences have played an important role in the formation, manifestation, importance, and meaning of love.

The romantic love that Westerners deem so necessary to marriage is a legacy of the early French Renaissance and the chivalric tradition. The ancients had sung the joys and struggles of love, but it was not until the thirteenth century that the concept of courtly love gained acceptance and began to dictate such requirements and proofs of passion as the abdication of all selfish motives, complete fealty, and Platonic idealization of the beloved.[35] Without entirely excluding sensuality, the new love placed great emphasis on purity and virtue. As Huizinga[23] says, "Love now became the field where all moral and cultural perfection flowered." Such a marital relationship was considered to be far superior to those arranged for mundane considerations by parents and overlords.

People marrying today are undoubtedly freer to explore the pagan possibilities of their bodies than their forefathers. Nevertheless, they still feel strongly that to get married there must be a spiritual bond, euphoric feelings, and at least a measurable degree of fidelity. Some part of these feelings must be attributed to novelty. Whatever the composite origin, they do not seem to endure in marriage with its inevitable restrictions, frustrations, and inescapable confrontations not only with the partner's all too human defects but also with the simple realities of mundane cohabitation. In discussing these romantic expectations, Kubie[28] condemns them for exacerbating the major neurotic elements in marriage that are "inflated and reinforced by the romantic Western tradition which rationalizes and beatifies a neurotic state of obsessional infatuation. . . . It is an obsessional state which, like all obsessions, is in part driven by unconscious anger." The disenchantment that accompanies the waning of romantic euphoria is frequently associated with divorce. "We are no longer in love" is probably the most common reason given for divorce.

An adjunct of the chivalric inheritance is the notion propagated in our culture that marriage will increase one's personal happiness.[35] In the extreme, as Ackerman[1] describes it, "marriage is approached as a potential cure for whatever psychic ails a man may suffer." The failure of marriage to supply this elusive happiness plays an important role in divorce. This quest is a factor driving some from one marriage to another. Hunt[24] believes that "the wide use of divorce today is not a sign of a diminished desire to be married, but of an increased desire to be happily married."

Our Western society places a premium on youth and beauty. Many men display their wives in accordance with Veblen's principle of conspicuous consumption. A marriage based *primarily* on such attraction cannot but falter with the inevitable changes brought about by the years.

In American society an ideal, "happy" marriage is considered to be one in which there is an interlocking of needs and mutual gratifications in the higher areas of functioning. "Togetherness" is extolled to such a degree that those who wish to look elsewhere for some of their important satisfactions (not necessarily sexual) may consider their marriage defective. In no other relationship are such demands made. The "togetherness" a couple feels in college, for example, where interests are not only shared but also similar, may begin to evaporate when the husband begins to acquire the highly specialized skills necessary for success in our technological society. He can no longer communicate his major interests to his wife, and this breakdown in "togetherness" may contribute to the decision to divorce.

There can be little doubt that increased social mobility has contributed to a greater incidence of marriages between persons of different class and value systems. Montagu[35] considers such marriages to be intrinsically unstable because they lack the stabilizing influence of a shared kinship group.

Perhaps reflecting the social relaxation of restrictions and prohibitions, religious strictures against divorce have been eased, and religious obligations and commitments no longer impede the dissolution of an unsatisfactory marriage.[35]

While increased social equality for women has given them more power to extract themselves from a painful marriage, it has at the same time engendered professional interests that may conflict with their childbearing and homemaking desires and obligations. Some women dissolve their marriages in order to freely and unequivocally pursue their professional interests. Others remain married, but they are so guilt-ridden over role conflicts that their gratifications are markedly reduced.[28,35]

Kubie[27,28] suggests that one explanation for the increasing divorce rate may be the increasing life span. Marriages have always been fraught with difficulty, but the participants died before their years of agony could culminate in divorce. With an increase in divorce in the middle and late years of life, younger people have now had more exposure to the divorce experience. They are, Kubie feels, more likely to emulate their predecessors and feel more free to divorce.

Last, but certainly not least important, are the individual neurotic factors that contribute to marital discord and disillusionment.[33] Kubie[28] considers the neurotic difficulties to arise primarily from the discrepancy between the partners' conscious and unconscious desires in the relationship. Examples are legion. One woman unconsciously wants a father in her husband, and her spouse unconsciously wants a mother, although each vociferously professes the desire for an egalitarian relationship. Frustration mounts as their underlying demands are not met. Another woman may basically relate best to a man who is dependent on her. All may go well in the early years of marriage when he relies on her efforts while he builds his career. Once he has established himself, there is no longer any realistic need for dependency, and the woman's neurotic necessity to perpetuate his anaclisis may cause divorce.

Ackerman[1] emphasizes the factor of anxiety assuagement as a reason for getting married. The impotent man may marry a frigid woman to hide his deficiency. If either partner becomes more desirous of sexual activity, the neurotic equilibrium is disrupted, and marital discord becomes manifest.

Unfortunately divorce is rarely a solution to the damage and frustration caused by neurotic interaction in marriage. As Bergler[3] says, "Since the neurotic is unconsciously always on the lookout for his complementary type, the chances of finding happiness in the next marriage are exactly zero. . . . The second, third, and nth marriages are but repetitions of the previous experience." Monahan[34] concluded that second marriages are twice as likely to break up as first marriages, and that those who marry a third time are accepting an even greater risk. Specifically, if both spouses have been divorced two times or more, the probability of another failure is nearly five times greater than that for a first marriage.

The dire statistical projections of Bergler and Monahan are not open to dispute. Bernard,[4] however, presents a more optimistic outlook by suggesting that most first marriages lead to divorce not so much from neurotic factors but because of normal maladjustments and inexperience. "The experience of an unhappy first marriage, although it may constitute a high tuition fee, may nevertheless serve as a valuable educational prerequisite to a successful second marriage." Bernard does concur, however, with the finding that second marriages are 50 per cent more likely to fail than first marriages. In Goode's[21] study of remarried mothers 87 per cent described their second marriages as much better than their first. Such statements cannot, however, be taken as satisfactory evidence of better second marriages. Having failed in one marriage, these mothers are less likely to admit failure in a second—even to themselves.

Despite encouraging signs, it appears that the same neurotic needs that drive a person into his first unfortunate marital relationship remain to influence his future attempts. As a psychiatrist I would like to believe that treatment can lessen the likelihood of divorce in subsequent marriages, but I have not been able to find any studies that satisfactorily confirm or deny the efficacy of therapy in this regard.

Consulting with Patients Contemplating Divorce

Since omniscience is a prerequisite for predicting whether a particular patient will be better off married or divorced, it behooves the therapist to maintain a strictly neutral position regarding the question.

Generally, when the question of divorce arises, the therapist's efforts should be directed toward clarifying the issues and alleviating pathological behavior so that the patient may make healthier and more prudent choices. In all events the decision to divorce must be the patient's. He must feel that he took the risk on his own, that no one but himself is to blame if his decision turns out to be an unfortunate one, and that there is no one else to thank if his choice proves to be a judicious one.

The experienced therapist who speaks proudly of never having had a divorce in his practice is probably pressuring some of his patients into remaining married when both partners would be better off divorced. The therapist with a high frequency of divorces should consider the possibility that he may be inappropriately encouraging divorce when a more conscientious effort at working through the difficulties might have been preferable. When a therapist applies such inadvisable pressures, his own marital history often plays a role. If his own divorce resulted in a significant improvement in his life, he may tend to overstress the values of separation. If his own marriage is gratifying, he may strongly encourage working through when separation might be the more therapeutic course. And the therapist who has never been married, whatever his assets, is compromised in his ability to appreciate fully the problems and conflicts of marriage.

Therapeutic consultations regarding divorce fall into three major categories: (1) those in which a couple enters therapy with the express purpose of averting an impending divorce; (2) those in which a patient already in therapy finds himself facing a decision regarding divorce; and (3) those in which one partner only presents himself in order to forestall or work through the problems of an impending divorce.

I agree with Whitaker and Miller[46] that the ideal counseling situation is one in which the marital partners are seen conjointly by a therapist who has had no previous therapeutic experience with either partner. These circumstances facilitate the impartiality that is vital to such counseling. Not only does conjoint therapy allow the therapist to hear both sides of the story; it also permits him to observe the interaction between the couple. Further, if the outcome of the consultations is maintenance of the marriage, then the partners have had an experience in mutual inquiry that should serve them in good stead in their future relationship. If they decide to divorce, their sessions should leave them clearer about their reasons for separation. What they have learned may even help each avoid another unsatisfactory marriage.

In working with such couples the therapist must side only with health, supporting healthy and appropriate positions and discouraging pathological and inappropriate behavior, regardless of who professes it. If he acts as a benevolent participant observer, there is less likelihood that either spouse will accuse him of favoring the other even though in any given session one may get more criticism than the other.

The therapist must resist either partner's attempts to use him as a tool in neurotic manipulations. The husband, for example, may try to enlist the therapist's support in influencing his wife to stay with the marriage when she is strongly inclined to terminate it. The wife may attempt to get the therapist to pressure her husband into drinking less or spending more time at home when he has little real motivation to do so.

Even if the therapist maintains the most careful neutrality, motives will be imputed to the therapist that are really the projections of his patients. The wife, for example, may believe that the decision to divorce was encouraged by the therapist because she needs support and agreement for such an independent step. The husband may consider the therapist's failure to condemn his infidelity as sanction.

When a couple seeks consultation to avert divorce, they often claim that they want to stay together "for the sake of the children," an attitude that has both realistic and pathological elements. Studies suggest that *on the whole* there is less psychiatric disturbance in children from broken homes than in those from intact but unhappy homes.[29,37] Nevertheless, one still cannot predict which will be the better situation for any given child.[10] The realistic argument, therefore, that the spouses should remain together for the children is suspect; and I generally make it clear that one cannot know in advance whether or not the children will be better off.

Professions of concern for the children are often only rationalizations to buttress neurotic interactions. Sadomasochistic, overprotective, overdependent, symbiotic, or other pathological relationships may be serving as the basis of the marriage. The therapist should, insofar as it is possible, clarify these underlying issues for the couple while playing down the falsely benevolent considerations regarding the children's welfare.

Whitaker and Miller[46] further recommend including the couple's parents and children in the consultations. I have done this occasionally and found it helpful. Information is often obtained that would not otherwise become available. Certain issues, however, are more justifiably discussed in the more intimate interviews with the couple alone.

Individual therapy with married patients presents special problems when the possibility of divorce arises. If therapeutic work with a married patient is successful, his healthier adaptations may be most anxiety-provoking to his spouse. The latter may become frustrated attempting to maintain the pathological patterns of interaction. Sometimes the partner can form healthier patterns of relating, and the marriage may be continued along new lines. Often he cannot make these adjustments, and his only alternative is then to seek others who can provide him with the pathological gratifications that he craves. Therapy in such cases is, of course, instrumental in bringing about the divorce.

Even without healthier adaptations on the part of the patient in therapy, the intimacy that the therapist shares with the patient cannot help causing some feelings of alienation in the partner, and a marital schism may be widened as a result. In extreme cases the therapeutic relationship may precipitate divorce in a marriage that might otherwise have been realistically reconcilable.

The individual who presents himself for treatment because his spouse threatens divorce is a poor candidate for therapy. His motivation does not generally stem from an inner desire to change things within himself because of the personal pain his problems cause him. He comes, rather, with the hope of altering those aspects of his behavior that are alienating his spouse or with the intention of learning ways to manipulate his partner into staying.

Sometimes both partners are in therapy and one or both may refuse conjoint sessions because they would be in "foreign territory" with their spouse's therapist. In such situations the couple is deprived of the benefits to be derived from the adjunctive joint sessions.

On occasion a patient may need the therapist's meaningful involvement to make the divorce process more bearable. This need may, in fact, be a primary reason for entering therapy when separation impends, and it is particularly applicable to women who, in my experience, are more likely than men to institute divorce proceedings without being significantly involved with a third person. Sometimes a spouse who feels guilty about instituting the divorce may encourage treatment for the partner who is left behind in order to assuage his guilt over the "abandonment." The same guilt-alleviating mechanism may be operative in the departing partner's encouraging (either consciously or unconsciously) the remaining spouse to take a lover.

Therapeutic Implications of Divorce Litigation

In the United States most states adhere to the adversary system in divorce litigation. Divorce laws, therefore, are predicated upon concepts of guilt and innocence, punishment and restitution. The divorce is granted only when the complainant or petitioner has proven that he has been wronged or injured by the defendant or respondent. Acceptable grounds for divorce are narrowly defined and, depending on the state, include mental cruelty, adultery, abandonment, habitual drunkenness, and nonsupport. The law punishes the offending party by granting the divorce to the successful complainant. If the court finds both husband and wife guilty of marital wrongs, a divorce is usually denied. In actual practice, however, the attorneys negotiate a settlement that includes alimony, child support obligations, custody, and visitation privileges. Only a small percentage of divorce proceedings culminate in a contested trial.[12,22,26,42,44]

Since the adversary system is antithetical in spirit to the mutually cooperative inquiry vital to successful joint therapy, such consultations are rarely successful once litigation has been instigated. The lawyer advises his client to withhold information that might endanger his legal position and to gather whatever data he can that might strengthen his case. Patients who are naïve enough to think that anything therapeutically meaningful can be accomplished in such an atmosphere should not, in my opinion, be supported in this delusion by the therapist. I only accept couples for conjoint therapy on the condition that therapy will continue only as long as neither partner instigates legal action.

When a spouse instigates legal proceedings, the intent may not always be clear and the motivation is often fragile. For these reasons the lawyer's usual practice of recommending reconciliation at his initial meeting with his prospective client may have untoward results. It may have taken the prospective client many years to take a healthy step toward divorce, and the lawyer's implied moral condemnation may serve to perpetuate a pathological situation. Others want to be told by an authoritative figure to work out their marital problems, and the lawyer's advice may help them proceed. Quite frequently the initial legal consultation is used by the spouse as a warning that the marriage has seriously deteriorated, and more constructive efforts on the part of both partners may result.

Once divorce litigation has begun, the woman, more usually, may try to involve the lawyer for other than legal reasons. This involvement, which doesn't necessarily include sex, may help the woman compensate for the loneliness and loss of self-esteem caused by the divorce. More pathologically she may be seeking a substitute neurotic tie to replace the one that is being severed. While every divorce lawyer has experience with these involvements, not all are aware of their psychological implications.[43]

The therapist who must assume the role of King Solomon and offer recommendations about the custody of children is fortunate indeed when he meets parents who are genuinely seeking a solution that is best for the children. In practice, however, each parent usually pleads only his own cause, backed up by a lawyer who wants the psychiatrist's testimony only if it supports his client's case.

All too often custody discussions are distorted by exaggerated emotional claims by each parent—the child and his welfare are subordinated to the desire to wreak vengeance on the spouse by depriving him of a prized possession. Under such circumstances psychiatric evaluation is most difficult.

It is preferable, therefore, that custody consultations be conducted by one who has not been, and will not be, the therapist for any of the children involved. His decision cannot fail to alienate the parent who has "lost" the child in litigation, which, in turn, jeopardizes future therapy with the child. Any child's chances of being helped in therapy are markedly reduced if either of the parents is significantly hostile toward the therapist. The child senses his parent's antagonism toward his therapist (even when not overtly expressed) and is torn between the two—hardly a situation conducive to a good therapist-patient relationship. Divorced parents can have a good relationship with their child's therapist, but it is hardly possible if he has participated in the custody decision.

The Divorce Decision

Generally the decision to divorce takes months and years to mature, no matter how explosive the announcement itself may appear. The prospect of what divorce entails may be quite frightening, and it may safely be said that the divorce decision, tainted as it is by so many negative aspects, is much harder to arrive at than the original decision to marry. Inertia and the specters of loneliness and hardship plague both partners. Some may need a lover to help them bridge the gap. Ambivalent separations and reconciliations may be necessary, and the therapist must respect his patient's reactions, his need for desensitization and accommodation. Time itself can be very therapeutic, and it behooves the therapist not to pressure his patient into proceeding rapidly—even after the decision has been made and the divorce promises to be salutary.

Primarily because guilt is so pervasive at this point, the therapist is often asked "how to tell the children." I generally suggest that both parents together tell the children, describing the main issues in terms that are comprehensible to the child: "Mommy and Daddy don't love one another any more." "Daddy has had trouble drinking too much whiskey, and now Mommy is tired of his drinking and doesn't want to live with him any more." Withholding information is deleterious because it promotes an atmosphere of secrecy and dishonesty that the child senses and reacts to at a time when he most needs a trusting relationship with his parents.[2,16]

Specific details such as impotency, frigidity, and more extreme sexual problems need not be disclosed. The child may request information on such matters, and he may be told that just as he has certain matters that are private, so do his parents. The important thing is to encourage the parents to establish an atmosphere of open inquiry in which the child has the feeling that most of his questions will be answered. The child should, in fact, be encouraged to repeat his questions, for they are part of a process that is vital to his working through of the divorce.

Some parents hesitate to engage in open confrontations because they may get upset—a prospect they see as damaging to their children. On the contrary, such displays of emotion may be most salutary for they show the child acceptable ways to handle his own reactions.[7] If parents aren't free enough to honestly express their emotional reactions, they can hardly expect their children to do so.

The Child's Psychological Reactions to Divorce

Since the divorce rate is highest in the first few years of marriage,[39] the affected children tend to be young and, therefore, more vulnerable than older children to its deleterious effects. Bowlby's[8] extensive review of the literature demonstrates that parental deprivation (especially maternal) is particularly conducive to the development of psychopathology and that the younger the child at the time of the abandonment the more severe the psychiatric disorder. Symptoms indicative of deprivation in the infant include loss of appetite, depression, lack of responsivity, and in the extreme, marasmus and death. In the older child, the reactions run the gamut of psychiatric disorders.[5,30,31,32] It may be difficult to separate the effects of the divorce from those of the traumas and prolonged strains that have preceded it. The divorce can even be salutary for the child because it ends the years of bickering and misery that have contributed to his psychological disturbance.

When the child first learns about the divorce, he may respond with denial. Some children will react so calmly to the announcement of their parent's forthcoming departure that the parent may seriously question his child's affection and involvement. Even after the parent has left, and the child has been repeatedly and painstakingly told about the separation, he may quietly ask when the parent will return or why he hasn't yet come back. Or the child may intellectually accept the fact of the divorce but go through his daily routines as if there were absolutely no change in his household. Such a child is repressing the inevitable emotional reactions that are evoked by the divorce, and his repression may be consciously or unconsciously sanctioned by the parents. Stoic advice such as "Be brave" and "Big boys and girls don't cry" may bolster the child's repression of his emotional reactions.[22] The parents' decision not to express their own feelings in front of the children is another way in which the denial and repression reactions can be fostered.

Encouraging the child to express his grief is a far healthier reaction. It allows a piecemeal desensitization to the trauma. Play is an excellent medium through which some children can work through their grief reactions—in or out of the therapeutic situation.[32]

Children may also react to the divorce with symptoms of depression, withdrawal, apathy, insomnia, and anorexia. About one-third of the children of divorce studied by McDermott[31] were depressed and many exhibited accident prone behavior, unconcern with their safety, and suicidal fantasies. The depression is not simply reactive in many cases. Hostility redirected from the parents to the child himself and the feelings of self-loathing that the child feels because of what he considers to be an abandonment may contribute to the depression. The depression generally lasts about six to eight weeks.

Some children regress in an attempt to get more attention in compensation for that which has been lost. Overprotective parents and those who attempt to assuage the guilt they feel over the divorce by indulging the child may foster this adaptation.

Occasionally a child will run away from home—usually in an attempt to rejoin the departed parent. The enhanced attention that the act generates may also be a motivating factor. In addition, the hostile impulses that the child harbors toward the parents for the divorce can be gratified by his awareness of the worry and frustration that his absence causes them.

In working with such children it is important to help them accept their angry feelings so as to lessen the likelihood that it will be discharged through neurotic channels. It is also necessary to help them express their anger in constructive ways—ways that will help rectify the situations that are generating it.

Acting out the anger in an antisocial fashion is common. The divorce may leave the child weak and vulnerable, and he may gain compensatory power through his violent actions.[31] Observing his parents to be so flagrantly hostile to one another and so insensitive to each other's feelings may contribute to the superego deficiency that permits guiltless acting out.[31] It is not surprising then that a number of studies reveal a relationship between juvenile delinquency and divorce.[20,31,46]

The child may feel guilty and consider the divorce to have been his fault. He may believe that the departing parent can no longer stand his "badness" or that he has been too much of a financial burden. Often comments that the parents may have made during their altercations may be taken by the child as verifications of these ideas. Such guilt is complex. It may be related to oedipal problems. The boy may feel that his father's departure was caused by his own conscious or unconscious wishes. Girls may experience similar guilt when it is the mother who leaves. The guilt reaction is often related to the child's desire to gain control over this chaotic event in his life. Control is implied in the notion, "It's my fault."[17,19] The child may reason: "If they got divorced because I was bad, maybe they'll get married again if I'm good." The hostility the child feels toward the parents for having divorced may also contribute to his guilt feelings. Whatever the psychodynamics, to reassure the child that the divorce was not his fault is misguided; the fundamental issues that have brought about the guilt must be dealt with if it is to be alleviated.

On rare occasion the child may have contributed to the divorce. He may have been unwanted or he may suffer from a severe illness and the departing parent is unwilling to assume the burdens of his upbringing. Such children must be helped to appreciate that the real defect lies less within them and more with the parent who has left.

Anger is an inevitable reaction and it may be handled by a variety of mechanisms. Denial of it is common. The child has already lost one parent; he fears doing anything that might alienate the other. Some direct their anger toward the parent with whom they live, since the absent parent is not so readily available. Often the person who first instigated the divorce becomes the primary target—no matter how justified the initiation of divorce proceedings may have been.[15] Other children have temper tantrums. Some may utilize compulsive rituals for the symbolic discharge of hostility; others project their anger and then see themselves as innocently suffering at the hands of malevolent figures. Nightmares are a common manifestation of the repressed hostility. Some handle their anger through reaction formation: they become excessively concerned for the welfare of one or both parents and fear that they will be sick, injured, or killed. Some harbor the notion that their angry thoughts may harm the parent, and this produces guilt and fear.

Parental duplicity, often well-intentioned, may complicate the child's life. He may become confused over contradictions between what his parents say about their affection for him and how they act toward him. Father, he is told, still loves him although he never visits or sends support money. Mother is said to love him, yet she spends many nights and weekends away with strange men. His parents may adhere to the dictum that they should not speak unfavorably of one another to the child —lest his respect for the criticized parent be compromised. Here again confusion is engendered: the child can only ask, "If he was so perfect, why did you divorce him?" Such parental dishonesty (no matter how well-intentioned) can only create in the child distrust of his parents and confusion about what love is.

The child must be helped to perceive his parents as accurately as possible—as people with both assets and liabilities. This will lessen the likelihood that he will have unrealistic goals in his own marriage, and it will increase his chances of more realistic expectations from all people whom he encounters. In addition, he must be helped to appreciate that if there are deficiencies in the affection of one or both parents for him, this does not mean that he is unlovable. It means only that there is something wrong with a parent who cannot love his own child; and there is no reason why the child cannot obtain the love of others both in the present and future.

Many factors may contribute to the feelings of inadequacy that children of divorce almost invariably suffer. The child may consider the "abandonment" as proof that he is unworthy. He may feel that the parent with whom he has been left is as equally worthy of rejection as himself, further deepening his insecurity.[15] The divorce produces a basic feeling of the instability of human relationships. If one parent can leave him, what is to stop the second from doing so as well? If his mother and father (whom he once considered to be omnipotent) cannot solve their problems, the world must be a shaky place indeed. If his mother, for example, can get rid of his father so easily, what is to stop her from getting rid of him with equal impunity?[15] The parent with whom the child lives may attempt to make the child a confidant or force him to participate in decisions that he is ill-equipped to make. Although some children rise to the occasion and assume a maturity beyond their years, most become even more insecure as they observe themselves incapable of meeting the demands made upon them—demands that they may feel they should be capable of meeting. The parent with whom the child lives may become increasingly resentful of the child because of the greater responsibilities and restrictions placed upon him. This parent, in addition, may displace hostility toward the absent ex-spouse onto the child. Being the object of such hostility increases the child's feelings of worthlessness.

Some children develop severe separation anxieties. As mentioned above, having lost one parent they fear the other will leave as well. When the separation anxiety becomes severe (such as when a school phobia develops), other factors such as unconscious hostility and death wishes are usually operating. The child needs to be constantly by the side of his parent to be reassured that his death wishes have not been realized. Such anxieties over hostility may contribute to the formation of other phobic symptoms such as exaggerated fear of dogs, injections, or heights.

After death the absent parent is usually idealized; after divorce there may be an opposite tendency to devalue the parent who is no longer in the household. In either case exaggerations distort the child's view of his parent, and healthy identification is thereby impaired. The child of divorced parents has fewer opportunities to use reality testing to correct the misconceptions he may have about his absent parent. He may devalue the parent who has gone in order to protect himself from the painful feelings of having been abandoned. It is as if he were saying to himself: "No great loss. He wasn't such a good person anyway." Although he may derive some specious solace from this defense mechanism, its utilization deprives him of a model for emulation, identification, and superego formation. Excessive idealization is also common. Often it serves as reaction formation to the feelings of hate and detestation the child has toward the parent who has left. To admit these feelings might expose the child to guilt and self-loathing. Such idealization also hinders the formation of valid identifications.

Children whose parents are divorced are quite prone to the development of oedipal difficulties. The child may try to take over the role of the absent parent—especially when the child is the same sex as the departed parent. Such an adaptation may be encouraged by a seductive parent. The child, however, is seldom mature enough to assume the awesome responsibilities inherent in this attempt, and it may therefore entail significant anxiety. Such parental seduction need not serve sexual purposes. A mother, by getting her son to act like his departed father, may more readily justify the use of him as a scapegoat upon whom she can vent the rage she feels toward her former husband.[31] A female child, sensing her mother's continuing attachment to the absent father, may assume a male identity in order to insure her mother's affection. A male child, without a father to identify with, may also develop homosexual tendencies. These and other pathological oedipal resolutions are discussed by Mahler[30] and Neubauer.[36]

In some communities the child of divorced parents may be stigmatized. But even when this does not occur, he invariably feels different. Others live with two parents while he lives with only one. He may become ashamed to bring other children home and may even try to conceal the divorce from his friends. Hiding the fact of the divorce produces a continual fear of disclosure that only increases his difficulties. This duplicity also adds to the child's feelings of low self-regard. Children in nonbroken homes may feel threatened by the divorce of their neighbor and may reject the child of divorce. Or they may obsessively question the child about the details of his parents' divorce because the acquisition of such facts can be anxiety-alleviating to them.[15]

If the divorced parents are still fighting, the child may take advantage of their discord and try to play one against the other for his own gain. He may recognize that by fomenting their conflict, he is sustaining their relationship. Although the interaction is malevolent, it is better, as he sees it, than no relationship between them at all.

Some children become obsessed with effecting a reconciliation, and they may persist in this futile endeavor for years. When this occurs it usually reflects a failure by the child to obtain substitute relationships to compensate for the loss of the parent. This capacity, which is vital to the child's healthy adjustment to the divorce, reaches its extreme form when the child uses peers as parental surrogates. Freud and Burlingham[13,14] described this phenomena with English war orphans in World War II. It is an adaptation that is a true testament to the adaptability of the human psyche.

Rarely the child's reaction to the divorce is so severe that he exhibits psychotic decompensation—manifested by vague wandering, severe regression, detachment, and soiling.[32] The child who reacts in this manner, however, has probably suffered from significant psychopathology prior to the separation, and the divorce was probably the precipitating trauma.

With divorce practical problems arise that may not necessarily be related to any pathological processes in the child. He may come to view his visiting father as the "good guy" whose main purpose is to provide entertainment and his mother as "mean" because she always seems to be the one imposing restrictions on him. Because the child has already been traumatized, the parent may be hesitant to apply reasonable restrictions. The visitations may become a chore for both the father and child as each feels compelled to live up to the full allowance of time together as stipulated in the divorce contract. Actually both would be far better off if all would agree to a more flexible schedule in which the child could choose to skip an occasional visit, or to shorten the visit, or to bring a friend. Also it is not necessary that all siblings visit simultaneously. These arrangements can diminish the pressures on both the child and the visiting parent and insure more gratifying experiences on visitation days.

If the mother works, the child may exhibit angry and depressive reactions—especially if she has never worked before. Such absences can impose upon the child new jobs and responsibilities that he may resent, particularly if he must forego recreational activities that his peers have time to enjoy. These responsibilities can be maturing and ego-enhancing. Some children rise to the occasion, and the sense of mastery and accomplishment that they may enjoy from their new-found obligations can be salutary. Others, however, regress in the face of these new duties.

The parents' dating may arouse in the child reactions such as confusion, jealousy, anger, or denial. He may, on the one hand, try to get rid of each new date for fear that his privileged position with the parent with whom he lives will be jeopardized. On the other hand, he may approach each new date with a question about his marital intentions—much to the embarrassment of all adults concerned. A common reaction involves displacing the hostility that the child feels toward the absent parent onto the date. On the positive side a new date, or friend of the parent, is a potential stepparent and—proverbial stepparent hostility notwithstanding—can provide the child with a meaningful substitute relationship for the lost parent. The child who is jealous of his parent's new relationship should be reminded that the stepparent may once again make his home complete and provide him with vital gratifications. In addition, he should be told that if his divorced parent is happier through the new marital tie, he, too, will benefit through the happier state his parent is in. I have discussed these and some of the other more practical problems that children of divorce must deal with in a book written specifically to be read by children—either alone or along with a parent.[16]

Most children whose parents divorce are not in need of therapy. Although it is a traumatic experience, judicious and humane handling should enable most children to adjust adequately enough to avoid therapeutic intervention.[38,41] Those who do require therapy, in my opinion, generally have had problems before and the divorce has served merely as a precipitating event. Westman's[45] study reveals that those in treatment came primarily from homes in which there was a pathological postdivorce interaction or where there was total abandonment. On occasion a parent may bring the child to the therapist—not so much because he believes the child to be in need of treatment (although he may rationalize the necessity), but because he feels very guilty over his having left the child. By placing him in the hands of a therapist he hopes to lessen his guilt and insure that no further damage will be done. Putting the child in "good hands" also serves the parent's purpose of getting someone else to undo the psychological damage that he has done—either in reality or in fantasy.

Although many of the specific recommendations made in this chapter may be helpful, time itself is a potent healer—the child does not seem to be able to dwell long on calamities. One of the dangers, however, that the divorce holds for the child is that he will generalize from his experiences with his parents and when older will eschew marriage entirely because his view of the marital state is one of unpredictability, unreliability, and intense psychic trauma. Another danger is that he will reproduce in his own marriage the same pathological interactions that his parents have exhibited.

Postdivorce Pathological Interaction between Parents

Many couples, following the issuance of the divorce decree, remain bound together in neurotic ties that may persist for a lifetime. Even the remarriage of one or both may not break this pathological tie. The continuance of the malevolent relationship may become the primary obsession of each parent and have its toll on the children as well as subsequent spouses. About half of the 425 divorcées studied by Goode either wished to punish or remarry their former husbands.[21] Most therapists agree that the ideal to be attained is that the divorcé be able to relate to the former spouse without significant neurotic involvement in those areas that still require mutual cooperation. Generally this involves the children, but on occasion professional and social contacts may also be necessary.[18]

The one who has been left often considers the rejection a severe insult to his self-esteem. He or she may press for reunion, not so much out of love but in the misguided attempt to repair the ego defect that the abandonment has caused.

Each may become excessively involved with the child of the opposite sex, who may come to symbolize the absent spouse. Oversolicitous attitudes, indulgence, seductivity, and overprotection are manifestations of this adaptation, which may be a feeble attempt to gain love in compensation for that which has been lost. The parents may vie with one another to gain the preference of the child. (This is not only an attempt to make up for the feelings of being unloved that the divorce may have engendered but has hostile implications as well.)

The children may be used in many other ways as pawns in the parental conflict. The mother may express her hostility toward the father by refusing to let him see the children despite his legal right to do so. (He may then have to resort to litigation in order to see them —at no small expenditure of time, money, and energy.) Or, more passive-aggressively, she may structure the children's preparations for his visit in a way that frequently results in their being late. She may "forget" what day it is and not have them home when the father arrives to pick them up. The father may express his hostility by withholding funds, not showing up for the children after the mother has planned her day around their absence, or returning them at other than the arranged for times. The father may withhold support or alimony payments—often forcing his ex-wife into expensive and time-consuming litigation. In such cases the courts may be used as the weapons with which the parents continue their battles.

The child may be used as an informer to acquire information for parental neurotic gratification, or for the purpose of litigation. Such a child is placed in a terrible bind. Cooperating with his parents in these maneuvers produces guilt over his disloyalty. By refusing to "spy" and be a "tattletale," the child risks rejection at a time when he is extremely vulnerable. Worst of all, even the parent who encourages him to provide information cannot but distrust him—because he has already proven himself an informer.

The child may have to endure for years the parents' derogation of each other. Valid criticisms of one parent by the other can help the child gain a more accurate picture of his parents that can serve him well in the formation of future relationships. More often, however, he is exposed to diatribes, seething rage, and criticisms of such distortion that he becomes confused and his relationship with the vilified parent is undermined.

Vituperation and vengeance may become the way of life. There are women who claim they will not remarry because to do so would result in their having to give up the gratification they derive from knowing how much of a burden the alimony payments are to their former husbands. This is really another way of saying that the gratifications of the malevolent involvement with the former husband are more meaningful than a possibly more benevolent relationship with another man.

The therapist who treats a divorced person involved in such a tragic and wasteful struggle does his patient a great service indeed if he can help bring about its cessation. He must be aware that his patient's perception of the former spouse may be distorted; he should not take at face value all the criticisms that are presented to him; and he must try to help his patient look into his own contributions to the maintenance of the malevolent relationship. Some divorced patients try to elicit the pity of others as they bemoan their fates. The recitation of woes may, in fact, become their primary mode of relating. The therapist does his patient a disservice if he gets caught up in pitying his patient's plight rather than encouraging more constructive adaptations. The patient may have to be encouraged to consciously restrain himself from vengeful acts not only for the sake of the children but as a step toward extracting himself from the conflict. He must be helped to see that the mature and ultimately the most beneficial response to provocation is not necessarily retaliation. When father doesn't send support and alimony, mother need not retaliate by withholding the children from visitation. She may do her children and herself a greater service by earning her own money and resigning herself to the fact the funds will not be forthcoming. She may then be poorer, but she will not be expending her energies in futile endeavors or allowing herself to be tantalized.

The parent must be helped to appreciate that the most effective defense against the ex-spouse's vilification of him to the child is not to react in kind, or to point out the absurdities in each of the criticisms, but rather to exhibit behavior that appropriately engenders genuine respect and admiration in the child. Trying to present a perfect image to the child is also an ineffective way of countering the former spouse's slanderous remarks. Admitting one's weaknesses, when appropriate, is the more courageous course and is more likely to enhance the child's respect. Treatment of such patients also involves helping them come to terms with their new way of life, and the new kinds of relationships they will have to form—both with their former friends as well as their new-found ones.

Psychological Problems of the Divorced Mother

The divorced woman, feeling that she has been a failure as a wife, may try to compensate by proving herself a supercompetent mother. She may become overprotective of her children, and they may become the main focus of her life. Such involvement may also provide her with a rationalization for removing herself from adults. Hostility toward her ex-husband may be displaced onto her children, especially the male child whose rejection may even result in his being sent to a foster home or boarding school. Or the children may become more nonspecific scapegoats for the resentments and frustrations her new situation engenders in her.

For the first time in her life she may have to take a job. Guilt over exposing the children to further parental deprivation is common, and the frustrations of this added burden produce even further resentment and unhappiness.

The divorcée is usually faced with many sexual difficulties. Many men see her as easy prey, and wives may be threatened that she will be a lure to their husbands. Dating may present her with many difficulties. What will be the effects on her children of their seeing each new date? Will it raise up and then dash their hopes for her remarriage? If a man friend sleeps over, how will this affect her children and her reputation in the neighborhood? In some cases having a man sleep over may provide a litigious and vengeful ex-husband with grounds for having her declared unfit as a mother and thereby deprived of the custody of her children. Some may hide their dating from their children in order to protect themselves from the hostility that dating causes in the youngsters. Others may use the children's hostility as a rationalization for not dating at all.

Her whole way of life and her concept of herself must be altered. With her married friends she may feel out of place—like a "fifth wheel." She may now find herself more comfortable with divorced men and women. Forming new relationships and altering her whole *modus vivendi* in the middle of her life is a difficult task indeed.

Psychological Problems of the Divorced Father

The father, too, usually has to adjust to a whole new way of life. His may be the lonelier existence. His separation from his children may be particularly painful and guilt-evoking. The divorce may bring home to him for the first time just how important his children were to him. He may get feelings that he is superfluous to them. He is now deprived of involving himself in many important decisions regarding his own children: schooling, medical care, and others that are vital to the child's welfare.[11] He can only see his children by appointment under strict regulations defined in the divorce decree.

The visitations often present problems. Some fathers indulge the children in order to assuage the guilt they feel over having left them. Others do this to compete with their ex-spouses for the children's affection. The father may be hesitant to discipline appropriately lest the child become even more resentful. The days spent together are often contrived—fun and entertainment become forced and are considered to be the only acceptable activities on the agenda. Such fathers would do far better for themselves and their children if they would try to spend the day more naturally, combining both the usual day-to-day activities and the recreational ones. Relating to the child in activities that are *mutually* meaningful can be salutary. Many fathers primarily take the child's wishes into consideration when planning the visitation. The resentment they thereby feel when engaging in an activity that is boring or only tolerated cannot but be felt by the child, and so he is robbed of the enjoyment. Some fathers concentrate on spectator entertainment as a way of avoiding more directly relating to the child. Some will bring the child along on business while deluding himself into thinking that he is involving himself in a meaningful way. Most often the child is bored and resentful, but may fear expressing his feelings. Some are ashamed to bring their children to their new dwelling because it may compare so unfavorably with the old. When remarried the visitation with the children often is resented by the new wife, and the father may be placed in a difficult bind.

The father with custody may feel quite resentful of his extra responsibilities and vent his hostility on the child. The father who uses the child as a weapon against his former wife, or who withholds the child's support payments in an ongoing postdivorce battle, may compromise his feelings of self-worth—vengeful gratifications notwithstanding.

Concluding Comments

The question of prevention of divorce cannot be discussed without prior consideration of the whole issue of marriage: how satisfactory an arrangement it really is and how suited it is to men's and women's personality structures. Now, as in the past, we are experimenting with new arrangements. Perhaps the whole concept will be discarded, and then, of course, this discussion of divorce will be of only historical interest to future readers. If Kardiner[25] is correct in believing that humankind has already experimented many times over with all possible arrangements and monogamy still proves itself to be most consistent with his needs, then changes will certainly have to be made if the institution is to be improved or, as some might say, salvaged.

The trend among young people today to live with one another prior to marriage may ultimately play a role in lessening the divorce rate. The element of unfamiliarity that contributes to many divorces is thereby obviated. Greater sexual freedom, increasing availability of abortion, and lessened stigma over unwed motherhood may also lessen the number of poor marital relationships that will end in divorce. Young people today profess more vociferously than their predecessors concern with basic reality elements in society. "Tell it like it is" has become their byword. If, indeed, this trend proves to bring people into closer contact with reality, it may play a role in lessening man's predilection to utilize the kinds of denial and excessive euphoric fantasy formation seen in romantic love and to hold unrealistic expectations about others. Such developments, if they come to pass, may also increase the likelihood of more satisfactory marriages. It is to be hoped that education, and the beneficial effects that psychiatric understanding will ultimately provide society, may also be conducive to happiness in human interaction—be it in the married or nonmarried state.

Bibliography

1. ACKERMAN, N. W., *The Psychodynamics of Family Life*, Basic Books, New York, 1958.
2. ARNSTEIN, H. S., *What to Tell Your Child*, Pocket Books, New York, 1962.
3. BERGLER, E., *Divorce Won't Help*, Harper & Brothers, New York, 1948.
4. BERNARD, J., *Remarriage: A Study of Marriage*, Dryden Press, New York, 1956.
5. BERNSTEIN, NORMAN R., and ROBEY, J. S., "The Detection and Management of Pediatric Difficulties Created by Divorce," *Pediat.*, *30*:950–956, 1962.
6. BETTELHEIM, B., *The Children of the Dream*, Macmillan, New York, 1969.
7. BLAINE, G. B., JR., "The Effect of Divorce upon the Personality Development of Children and Youth," in Grollman, E. A. (Ed.), *Explaining Divorce to Children*, Beacon Press, Boston, 1969.
8. BOWLBY, J., *Maternal Care and Mental Health*, World Health Organization, Geneva, 1951.
9. *Demographic Yearbook*, United Nations, New York, 1969.
10. DESPERT, J. L., *Children of Divorce*, Doubleday, New York, 1953.
11. EGLESON, J., and EGLESON, J. F., *Parents without Partners*, Ace Books, New York, 1961.
12. FREEDMAN, H. C., "The Child and Legal Procedures of Divorce," in Grollman, E. A. (Ed.), *Explaining Divorce to Children*, Beacon Press, Boston, 1969.
13. FREUD, A., and BURLINGHAM, D. T., *Infants without Families*, International Universities Press, New York, 1944.
14. ———, and ———, *War and Children*, International Universities Press, New York, 1943.
15. GARDNER, G. E., "Separation of the Parents and the Emotional Life of the Child," *Ment. Hyg.*, *40*:53–64, 1956.
16. GARDNER, R. A., *The Boys and Girls Book about Divorce*, Science House, New York, 1970.
17. ———, "The Guilt Reaction of Parents of Children with Severe Physical Disease," *Am. J. Psychiat.*, *126*:636–644, 1969.
18. ———, "More Advice for Divorced Parents," *Psychiat. & Soc. Sco. Rev.*, *3(10)*:6–10, 1969.
19. ———, "The Use of Guilt as a Defense against Anxiety," *Psychoanal. Rev.*, *57*:124–136, 1970.
20. GLUECK, S., and GLUECK, E., *Unraveling Juvenile Delinquency*, Harvard University Press, Cambridge, 1950.
21. GOODE, W. J., *After Divorce*, Free Press, Glencoe, Illinois, 1956.
22. GROLLMAN, E. A., "Prologue," in Grollman, E. A. (Ed.), *Explaining Divorce to Children*, Beacon Press, Boston, 1968.

23. HUIZINGA, J., *The Waning of the Middle Ages*, Doubleday, New York, 1954.
24. HUNT, M. M., *The World of the Formerly Married*, McGraw-Hill, New York, 1966.
25. KARDINER, A., "A Psychological Understanding of Monogamy," in Rosenbaum, S., and Alger, I. (Eds.), *The Marriage Relationship*, Basic Books, New York, 1968.
26. KLING, S. G., *The Complete Guide to Divorce*, Parallax Publishing Co., New York, 1967.
27. KUBIE, L. S., "The Challenge of Divorce," *J. Nerv. & Ment. Dis.*, *138*:511–512, 1964.
28. ———, "Psychoanalysis and Marriage: Practical and Theoretical Issues," in Eisenstein, V. W. (Ed.), *Neurotic Interaction in Marriage*, Basic Books, New York, 1956.
29. LANDIS, J. T., "The Trauma of Children When Parents Divorce," *Marriage & Fam. Living*, *22*:7–13, 1960.
30. MAHLER, M. S., and RABINOVITCH, R., "The Effects of Marital Conflict on Child Development," in Eisenstein, V. W. (Ed.), *Neurotic Interaction in Marriage*, Basic Books, New York, 1956.
31. MCDERMOTT, J. F., "Divorce and Its Psychiatric Sequelae in Children," *Arch. Gen. Psychiat.*, *23*:421–427, 1970.
32. ———, "Parental Divorce in Early Childhood," *Am. J. Psychiat.*, *124*:1424–1432, 1968.
33. MITTLEMANN, B., "Analysis of Reciprocal Neurotic Patterns in Family Relationships," in Eisenstein, V. W. (Ed.), *Neurotic Interaction in Marriage*, Basic Books, New York, 1956.
34. MONAHAN, T. P., "The Changing Nature and Instability of Remarriages," *Eugenics Quart.*, *5*:73–85, 1958.
35. MONTAGU, M. F. A., "Marriage—A Cultural Perspective," in Eisenstein, V. W. (Ed.), *Neurotic Interaction in Marriage*, Basic Books, New York, 1956.
36. NEUBAUER, P. B., "The One-Parent Child and His Oedipal Development," in *The Psychoanalytic Study of the Child*, Vol. 15, pp. 286–309, International Universities Press, New York, 1960.
37. NYE, I. F., "Child Adjustment in Broken and and in Unhappy Unbroken Homes," *Marriage & Fam. Living*, *19*:356–361, 1957.
38. OBER, R., "Parents without Partners—with Children of Divorce," in Grollman, E. A. (Ed.), *Explaining Divorce to Children*, Beacon Press, Boston, 1969.
39. PLATERIS, A. A., *Divorce Statistics Analysis: United States—1963*, Public Health Service Publication No. 1000, Series 21, No. 13, U.S. Government Printing Office, Washington, D.C., 1967.
40. ———, *Increases in Divorces: United States—1967*, Public Health Service Publication No. 1000, Series 21, No. 20, U.S. Government Printing Office, Washington, D.C., 1970.
41. POLATIN, P., and PHILTINE, E. C., *Marriage in the Modern World*, J. B. Lippincott, Philadelphia, 1964.
42. SCHWARTZ, A. C., "Reflections on Divorce and Remarriage," *Soc. Casework*, *49*:213–217, 1968.
43. WATSON, A. S., "Psychoanalysis and Divorce," in Rosenbaum, S., and Alger, I. (Eds.), *The Marriage Relationship*, Basic Books, New York, 1968.
44. WELS, R. H., "Psychiatry and the Law in Separation and Divorce," in Eisenstein, V. W. (Ed.), *Neurotic Interaction in Marriage*, Basic Books, New York, 1956.
45. WESTMAN, J. C., CLINE, D. W., *et al.*, "Role of Child Psychiatry in Divorce," *Arch. Gen. Psychiat.*, *23*:416–420, 1970.
46. WHITAKER, C. A., and MILLER, M. H., "A Reevaluation of 'Psychiatric Help' When Divorce Impends," *Am. J. Psychiat.*, *126*:611–618, 1969.
47. *The World Almanac*, "Marriages, Divorces and Rates in the U.S.," Newspaper Enterprise Assoc., New York, 1971.

CHAPTER 25

ADOPTION

Viola W. Bernard

THROUGH ADOPTION society has established a type of family in which parenthood is based, not on having given birth, but rather on parental functioning and the ties of the child-parent relationship.* Adoptive family *formation* takes place for the most part during a circumscribed period of time: the children are still quite young and the parents are likely to be in their thirties or so. The adoptive family *process*, however, extends throughout successive stages of the life cycle for each of the participants, during which time they influence each other's existence fundamentally. It is from this dynamic of adoption that its powerful potential stems for preventive psychiatry—particularly for the children, whose still unfolding development is so critically at stake.

Specialists in the fields of child health, child development, and child welfare are in general agreement that, for dependent young children who are without parents, a permanent family of their own that has been socially created can offer optimal protection against the damaging effects of parental deprivation. To be sure, children born into the families that raise them are not thereby automatically insured against developmental hazard; but children who lack parents and family are clearly at great maturational disadvantage. They constitute the primary population at risk with which this chapter is concerned.

The number of such children in any one year far exceeds the number of those adopted. Thus in 1970, the last year for which national child welfare statistics were reported, about 89,200 children were adopted by nonrelatives, and another 85,800 by stepparents or other relatives.[45] Many thousands more, however, were at that time living out their childhood in long-term foster care or in child-care institutions;[39,40] still others, their numbers unknown, were being repeatedly shifted about, through informal arrangements, from one to another of a succession of temporary caretakers.

It is central to my orientation vis-à-vis adoption that any valid assessment of its psychosocial value must be measured against the yardstick of this massive unmet need; that

* This chapter deals only with the adoption of children by nonrelatives; almost half of the adoptions in this country each year are by stepparents or other relatives. The former are sometimes referred to in the literature as extrafamilial adoptions (EFA).

adoption is, on the evidence, the plan of choice to prevent or mitigate the destructive effects of such need; and that adoption is essentially an *affirmative* experience.

In this chapter I shall focus on a particular population—that of adoptive families—from the standpoint of primary, secondary, and tertiary prevention. That is to say, I shall explore how certain concepts, knowledge, and techniques of psychiatry may be applied, in concert with elements from other fields, at successive stages of adoptive family formation and process, so as to help strengthen the psychological health of this population group.

From the perspective, then, of mental health prevention at its three levels, a reciprocal relationship exists between adoption* and psychiatry. Adoption can contribute to the psychological health of its key participants, and psychiatry can contribute to the psychological success of adoption. Or, to put it another way, each of these can help prevent or reduce obstacles to the emotional well-being of a sizable and specific population of children and adults.

Although, as stressed above, adoption can have significant influence on mental health, many different sets of factors determine whether in any particular instance that influence is positive or negative. It has high potential for a positive impact in that adoption can be a means for: supporting sound infant-child development by preventing parentlessness; making the fulfillment of family life possible for couples who are unable to bear children, and whose longing and potential for parenthood would otherwise be thwarted; enhancing the growth-promoting potential of the family as a whole when parents adopt who already have children; finally, as a conflict solution, relieving pathogenic stress for many of those who, having given birth, are unable, for one or another reason, to function as parents.

Adoption services are mainly the responsibility of social workers in both voluntary and public welfare agencies, but they involve other professionals as well in various capacities: psychologists, pediatricians, psychiatrists, nurses, geneticists, lawyers, and others. (There is marked variation among agencies in this regard.) The rationale underlying adoptive practices is synthesized by drawing upon the theories, assumptions, and data of other fields, such as child development, psychoanalysis, and social science research. Prevailing sociocultural conditions at any given period, as well as factors of expediency, also influence adoptive service patterns and adoption policy.

In 1955 the Child Welfare League of America† convened a National Conference on Adoption, attended by members of allied professions, including this author, to examine and exchange ideas and information about the complexities of adoption.[54] Since then major changes have evolved, the more so in the past few years, so that adoption today is very different in many fundamental respects.

These changes have critical relevance to the mental health implications of adoption and to the ways in which the content of dynamic psychiatry and the work of psychiatrists can help maximize the psychic well-being of adoptive families.

* It should be noted that for convenience the single term "adoption" is used throughout this chapter to refer to any or all of its several meanings: a form of personal family experience; a social institution based on law; a specialized area of professional service and of investigation.

† Because of its recognized leadership role in the effort to improve adoptive service, the Child Welfare League of America should be known to anyone interested in the field of adoption. This footnote is by way of introducing it to those non-social worker readers who may not already be acquainted with it.

The League is a privately supported national organization with an accredited membership of over 300 member agencies that provide care and services for deprived, neglected, and dependent children. Its activities cover the entire child welfare field, of which adoption service is a part. As one of its range of services, the League develops and publishes standards, with continual updating, that are generally regarded as authoritative, for example, *Standards for Adoption Service* (revised in 1968).[13] In the form of monographs and books, as well as articles in its monthly journal, *Child Welfare*, the League's publications on adoption form an important part of the literature on that subject (several of its titles are included in this chapter's bibliography). It also sponsors research and conferences on adoption, and initiates special projects such as ARENA—the Adoption Resource Exchange of North America. As a totality the League has been a force, in the public as well as the voluntary agency sector, for positive change on behalf of children.

Steps and Stages in Adoption

For readers who are unfamiliar with the sequence of steps and stages in adoption, it may be useful to review these very briefly, since it is according to that sequence that considerations of preventive psychiatry will be explored here.

But first it should be explained that in this overview, and, indeed, throughout the chapter, only those placements that are arranged by authorized agencies, public and voluntary, are under discussion. It is true that many independent adoptions turn out well and that many agency adoptions turn out badly. On balance, however, agency placement offers by far the greater advantage and protection to the greatest number of those concerned, both children and adults. My stand on this question, discussed more fully elsewhere,[3] is in keeping with position statements by several medical and social welfare organizations.[1,21] Efforts to educate the public, plus stronger laws against the black market in babies, seem to have had some effect. In 1970, slightly more than three-quarters of the reported placements in the United States (for nonrelative adoption) were through agencies. This figure represents a steady rise since 1957, when the numbers of independent and agency placements were almost equal.[45]

Preplacement Phase

Couples who want to adopt and *agency workers in search of adoptive homes* for children in their care undertake a series of meetings and interviews. Through these the couples learn more about the realities of adoption; they explore and confront their own motives, conflicts, fears, and preferences; and they reassess their decision to adopt. If they do go on they have thereby become better prepared for the actuality of accepting a child as their own. The agencies utilize these interviews for the dual functions of assessment and of helping with problems about adoption through anticipatory guidance. The sessions are a way of screening out and screening in couples, on the basis of their potential as adoptive parents in general, for certain kinds of needy children in particular, and finally, if agencies and couple decide to go ahead, for a specific child.

Biological parents, of whom a majority are unwed mothers (88 per cent of the children adopted by nonrelatives in 1970 were born out of wedlock),[45] can be helped through counseling with their conflictual decision making about adoption and with the process of surrendering their baby to the agency, which then becomes the guardian pro tem—that is, until legal adoption has been consummated.

Society's responsibility for children whose parents have been found to neglect them is carried out by authorized child welfare agencies, through providing foster homes or institutional care for these children. It is important that such agencies, which are charged with strengthening family relationships for the children in their custody, reach out actively to help these parents—often themselves the victims of parental deprivation. In combination, psychiatric consultation, case work, and practical aid with social, economic, and medical problems may mobilize latent parental strengths and thereby prevent family breakup; they may, on the other hand, lead to the diagnostic conclusion, with or without parental concurrence, that family reunion is not feasible and that a permanent adoptive home for the child is the plan of choice.

Parents who have come to agree with this conclusion, through such a process of help and clarification, free their child for adoption through voluntary surrender of their parental rights. According to present laws, however, most of the children whose parents neither free them for adoption nor undertake their care are consigned year after year to the limbo of indefinite temporary care.

Yet there are situations in which courts may intervene despite the failure or refusal of parents to surrender a child. Thus, where parents are found to have abandoned the children or to have left them for long periods in foster care without any meaningful contacts or efforts to plan for the future, the courts are empowered to terminate parental rights and to give guardianship to the agency that has custody of a child, with authorization to place for adoption. Before such a determination is made, it is important that the aforementioned therapeutic effort and differential diagnosis on the part of the agency should have distinguished a case in which legal retention of family ties is in the best interests of the child from one in which it is not. The agency will thereby have become equipped to present evidence on which the court can base a considered and sound decision.[58]

For the children the agency's preplacement services differ in many respects, of course, as between infants and older children. In general, children are cared for in agency nurseries or temporary foster homes, where their special attributes are studied and their special needs met, and where their development is assessed. Infants are placed in adoptive homes as soon as one is available, depending on factors referable to the baby's condition and history, as well as the legal status of his surrender. In some instances it is feasible to transfer newborn infants directly from the hospital; older children, however, need various forms of specialized help in overcoming the effects of prior traumata, in psychological preparation for adoption, and in understanding and participating in the placement process as much as possible. Helping them to comprehend and to have a say about what is happening in their lives prevents or reduces painful and detrimental confusion and the sense of being a helpless pawn.

Between Placement and Legal Adoption

For the new parents and the children, whatever their age, the child's actual entry into the home as a family member is a critical, emotionally laden event in the adoptive sequence. Sensitivity in the way it is brought about, together with understanding support in the initial postplacement period, is especially important for all concerned. During the months before legal adoption can be consummated—state laws differ on the time requirement—agencies must maintain supervisory contact, based on their continuing legal responsibility for the child's welfare. The agency workers try to implement this contact by offering practical and psychological help. The parents' reactions to the evaluative function, however, tends to limit how fully they can, at that same time, accept and make use of the help. Group meetings of couples who are all going through this experience, for the discussion of their common problems, seem particularly useful, perhaps because they are less threatening than individual parent-worker contacts. This psychodynamically active period of family formation involves many different and complex ways in which the adoptive child and his new parents—and new siblings, too, in some cases—start to work through and establish their relationships.

For the biological parents, usually the mother, there is a period of time—it varies in duration, depending on local laws—after signing the surrender during which she can change her mind before the adoption is made legally final. This is a means of protecting her from hurried decisions made under the pressure of strain and anxiety. Appropriate practical and counseling help before and during her decision to surrender, and the availability of such help in the immediate weeks thereafter, can be of the utmost value in the protection of and planning for her own and her child's best interests.

Poignant court custody battles have dramatized the pressing need for adoption law reforms. (Anna Freud's seminars at Yale Law School have exemplified some of the potentialities for primary prevention between the broad fields of law and child development.)

Postadoptive Phase

Since the postadoptive period extends for an indefinite time after legal adoption, one may think of it in both short- and long-range terms.[8,10,50] A number of agencies offer adoptive parents individual counseling and group meetings at such key times as just before their children are ready for school. Agency provision of individual consultation and group counseling for adoptees at the time of adolescence is another instance of timely intervention. Aside from adoptive agency service, adopted children and parents utilize the full range of public and private social, health, and mental health services that are available generally.

Trends in Adoption Practice

Outstanding among the changes occurring in adoption has been the widened range of children who are considered adoptable. For a long time such children were almost exclusively white infants, with no detectable defects or deviations, physical or psychological. Increasingly in recent years placements are being made, and homes recruited, not only for these infants but also for older children, children from minority groups and of mixed racial and religious background, and children with various physical, emotional, and intellectual handicaps. These changes in adoptability have been paralleled by changes in eligibility criteria for adoptive parents, who are no longer drawn predominantly from among infertile couples. The psychic correlates of both these sets of changes profoundly affect every phase of adoptive family dynamics, and hence the role that preventive psychiatry can play in relation to them.

It has been customary to refer collectively to these more recent child entrants to adoption as "the hard to place" (agencies now refer to them as "children with special problems"). While descriptively this is true, such lumping together of children who differ so basically from one another tends to obscure their far more salient particularities. It is these that dictate the specifics for applying mental health concepts and methods on their behalf to adoptive service policies and practices.

Closely linked to the foregoing has come the recognition that it is the *well-being of the children* that is the primary purpose of adoption services. As an adaptive solution adoption has been a creative way of balancing the needs of adoptable children, of adoptive couples, and of parents who could not function as such. That balance has now shifted decisively in favor of the child. Every child who is capable of family living is seen as entitled to a permanent home of his own.[13]

Commitment to this philosophy has led to an enormous increase in the numbers of children for whom adoptive homes are now being sought. Since this has been coupled with the hugely escalating number of children born out of wedlock each year—about a third of them find their way into adoption—the need for homes has greatly outdistanced their availability, even though the number of applicants has also been increasing greatly throughout the past decade.

The relative decline in the applicant-child ratio does not hold true, however, for healthy white infants (nor recently for black infants either); on the contrary, the number of these that are available for adoption has been markedly decreasing. (Unfortunately this seems to be reviving a black market in such babies.) Apparently this is due to at least two factors: (1) liberalized abortion laws and wider use of contraceptives are reducing the number of unwanted babies who would otherwise be given up for adoption; and (2) many more white unwed mothers are now deciding (as have their nonwhite counterparts right along, largely as the result of there being fewer adoptive opportunities available to them) to keep and rear their infants.[15] One can surmise that more tolerant community attitudes toward illegitimacy, with lessening of stigma, is at least one determinant of this.

It does not seem appropriate to the purposes of this chapter, nor is it feasible, to try to describe the current picture of adoption with any degree of completeness.* In many respects the situation is in flux at the time of

* For some of the more up-to-date reviews in the social service literature of these changing trends, the reader is referred to the Child Welfare League's 1968 revision of *Standards for Adoption Service*,[13] as well as to overview articles by Chevlin,[12] Kadushin,[28] and Mech.[43]

this writing. The available statistics are not up-to-date enough to reflect adequately the rapidly moving situation. Indeed, it should be pointed out that much of the literature on which this chapter must of necessity draw is already, in a number of respects, out of keeping with many current actualities. Two trends, however, in addition to those that have already been referred to, do seem to warrant special mention from the standpoint of primary prevention.

For one thing there has been considerable gain in the social acceptance accorded adoption; this, of course, has been of value in terms of heightening the self-esteem of adoptive family members and lessening their need for concealment, with all its attendant emotional problems. In fact, many adoptive parents have formed highly vocal and visible organizations. The sharing of interests at both the personal and societal levels as well as the concerted approach to common problems, helps these parents to dispel feelings of isolation. Chapters of the National Council of Adoptive Parent Organizations are working actively for legislative reforms and for the improvement of public attitudes toward adoption. Also the Open Door Society, whose members are adoptive parents of minority group children, seeks to encourage that particular type of adoption.

Placements in Infancy

As of 1970, more children were being placed at earlier ages (two-thirds of all the children adopted in 1970 by unrelated persons were less than three months old when placed in an adoptive home).[45] This reflects the influence on adoptive practice of recommendations made by child development and mental health specialists that, in the adoption of infants, the most favorable time for placement, from the standpoint of primary prevention, is during the first few months of life. It is generally thought that such early placement can prevent or mitigate certain adverse effects on the child's development of prior maternal deprivation and of maternal separation. Moreover, such placement permits the early establishment of health-conducive patterns of mothering and parent-child relationships.

Yarrow's investigations are of particular value in this regard. Within the overall concept of maternal deprivation, he has distinguished four major types of deviation in early maternal care. Maternal separation is one of these;[61] and, with regard to it, six different kinds are identified, with correspondingly different effects on the child's subsequent development. The most lasting and damaging effects on mental health were found to occur when permanent separation from the parents was followed by repeated separations from subsequent foster home placements.[62]

Except for newborns who go directly into their adoptive homes, children are undergoing, at the time when they are being placed in their permanent adoptive homes, at least one separation—from whoever it was who had been taking care of them. But, as Yarrow's distinctions emphasize, separation experiences are not identical by any means. He and Goodwin[63] studied the effects on children who have been adopted as infants of the single separation from the agency's foster home, following which the new permanent mother figure had been immediately available. They reported finding few long-term personality disturbances, when the children were five years old, that could be attributed primarily to the early separation experience.

They did note, however, that there might be a critical period of immediate reactions to the separation. As early as three months infants were responding with disturbances; by six months there were fairly severe reactions. All those children placed after six months of age, however, showed some disturbance—for the most part quite marked. The infant's vulnerability to discontinuity of the mother figure may be related to his capacity at six months for focused attachment to her as a particular individual. Yarrow suggests that if the separation takes place when the "stranger reaction" is at its height, a change of the mother may be extremely disruptive to the infant who, at that time, is developmentally in an active phase of establishing an object relationship with the mother figure. Yarrow and his co-workers,[64] however, found no significant differences, when the children became ten years old, on a rating of overall adjustments between those who had left their foster mothers before and after six months of age.

According to their research, infants who are moved into adoptive homes from a prior temporary foster home tend to experience separation without the complications of maternal deprivation. For them the long-term effects of separation stress reactions, at placement, on cognitive and personality development are much less important than the quality of mother-infant interactions during the first year and the subsequent range of relationships and life experiences that serve to mitigate or reinforce the impact of the original separation stresses.

Because of the transactional nature of parent-infant relationships, early placement also favors arousal of emotional involvement in the baby on the part of the new mothers—and fathers—and thus stimulates parental capacities. This is all the more true of couples who are adopting their first child, particularly if they are infertile and have not gone through the parental role preparation of pregnancy.

As with all parenthood there is great variation, along multiple dimensions, among "good" patterns and styles of adoptive parenting. Aside from obvious external differences among infants who have been placed for adoption, they also differ from one another, of course, with regard to such individual attributes as temperament, innate constitution, predispositions, and latent vulnerabilities and talents. As yet we know far too little about which personality characteristics of the parents can be matched with which infant characteristics in order to enhance the chances for optimal adoptive family psychological outcomes. Agencies have by now discarded their former practices of attempting to "match" adoptive parents and children in terms of physical appearance. It was found to be needless for mutual identification; indeed, it can be detrimental, by playing into a need to deny the fact of adoption, based on unresolved feelings of discomfort with it. But research that can deal with the complexity of relevant variables in such a way as to arrive at a greater understanding of subtle types of parent-infant matching could have great importance for primary prevention.*

Patterns and Styles of Adoptive Parenting

The contrasting effects of two good but different kinds of mothering on the same adoptive child is well illustrated in Krugman's "A New Home for Liz: Behavioral Changes in the Deviant Child."[34] The facts of this case may be paraphrased and condensed as follows:

This child, who had been under the agency's care since birth, was placed at three months in an adoptive home where the same agency had placed another baby some years before, who had adjusted well, thus attesting to the couple's adoptive parental competence. Liz, however, showed markedly deviant development in her new home. Her mounting behavioral symptoms, panics, and distress became so intense that when she was 34 months of age, the agency's psychiatrist, psychologist, and social work staff decided on the drastic therapeutic step of a total change in home milieu. The child was therefore removed from the adoptive home and transferred to one of the agency's supervised boarding homes. There her behavior and adjustment improved so drastically that at five years of age she was replaced for adoption. In this second adoptive home she adjusted well, without any unusual efforts or special plans being made on her behalf. Her maturational progress continued till age six, when the case report stops.

Krugman attributes the child's remarkable

* I am participating in a prospective longitudinal study of adoption in infants that may shed some light on this issue. It is in process at the Child Development Center, New York City, under the direction of Dr. Peter B. Neubauer.

improvement in behavior and symptomatology in the main to the very different styles of response to Liz and her problems manifested by the adoptive and the special foster home mothers.

This case illustrates the fact that under certain circumstances it can be therapeutic to uproot a three year old from her familiar and apparently benign social and physical environment by moving her out of it into a totally strange new life situation. But how can we reconcile this with the findings, cited above, about the damaging deprivation and discontinuity of multiple placements? Would not such separations be all the more traumatic for an unstable child like Liz? As a pivotal criterion for distinguishing between traumatic and therapeutic separations, one needs to ascertain *from* what and *from* whom the child is separating, and *to* what and *to* whom he is going. Permanence is the unique attribute of adoption as a form of substitute family care; it is what gives it such power to prevent the ravaging effects of parental deprivation. Yet if the wrong child happens to be with the wrong parents, this very permanence can render such an adoption pathogenic by locking parents and child into lifelong destructive relationships.

In this instance, therefore, it was the agency's responsibility to evaluate and act before the finality of legal adoption took place. The appropriate emphasis for describing the move Liz made from her first adoptive home would seem to be that of *gaining* an environment in which her development could go forward, rather than that of *losing* a mother and a home that were actually imperiling her future sanity.

A more clear-cut case of secondary prevention, achieved by removing a nondeviant child from an unsuitable adoptive home prior to legal adoption, may be illustrated by the following case vignette from my own experience.

Sarah was already almost 11 years old when her pressing need for adoption became evident. At that time she was indeed "hard to place," even more so then—it was in the mid-1950's—than nowadays. When she was nine her mother, who was unmarried, had to enter the hospital, where she died four months later. Because of a series of deaths and rejections by relatives with whom she might have lived, Sarah spent one and a half years in a small children's institution. This deeply hurt child, who had unusual strengths of personality and intellect, was placed for adoption, following some case work help in the institution and psychiatric consultation at the adoption agency to evaluate her adoptability. The preadoptive study of the couple with whom she was placed when 11, as well as the placement process, were speeded up because even the institution was about to be closed down.

After the newly formed adoptive family had spent one summer together, a basic incompatibility became evident. The parents, especially the mother, sought my assurance as the psychiatrist for the agency that, in effect, everything about the adopted child could be changed to their specifications as the condition for keeping her (these specifications struck me as befitting a trained seal more than a daughter). Sarah, who had already been acutely sensitized to separation and bereavement, was terrified of losing this last semblance of a home. Consciously she desperately tried to please and to mold herself into what was wanted; unconsciously she reacted against such basic rejection with symptomatically disguised forms of protest, which the parents and the guilt-ridden child herself interpreted as further proof of her unacceptability.

My initial clinical objective was to help preserve this adoption by working with both the parents and the child. Protection of Sarah against still another abandonment seemed of overriding importance. Nevertheless, I was forced to the reluctant conclusion that for Sarah to remain in such an intractably pathogenic situation would be untenable. It was a momentous decision! This youngster was already showing psychic ill effects from all the discontinuities and misfortunes she had previously experienced; to remove her from this adoptive home was bound to entail trauma in the here and now, which could harm her further—perhaps seriously. Not to remove her, however, would destroy her chances, I felt sure, ever to reverse her emotional difficulties or to remotely fulfill her substantial potentialities. Upon concluding that these long-range dangers outweighed the short-term risks, I recommended, and the agency agreed to, her therapeutic removal. The adoptive parents not only agreed but were distinctly relieved.

To inform Sarah and explain this to her was one of the most painful tasks I have encountered. It was a psychiatric equivalent of radical surgery. Every ounce of her energy for survival was mobilized in fighting to stay in the placement. It took all the confidence that I could muster in my professional judgment to almost literally tear this child away from the loveless home to which she was clinging with all her might; it felt to her like the only alternative to an engulfing world of nothingness.

This is a vignette, not a full case report; its intent is to illustrate a limited aspect of secondary prevention in adoption. Suffice it to say, therefore, that four months later, after a period of special foster care, Sarah was once again placed for adoption. Legal adoption followed at age 13. During the interval between adoptive placements the case worker and I both worked intensively with the child, while the agency made a strenuous and at that time innovative home-finding effort on her behalf. By enlisting the cooperation of all the agencies in this and neighboring states, the number of adoptive applicants who could learn about and become interested in Sarah was greatly multiplied.* This made it possible to select from among quite a few families—a process in which Sarah took an active part.

Did that decision in 1955—to separate Sarah from her first adoptive home—turn out to have been so vital to her subsequent mental health? Because of the unusual opportunity I have had to follow the course of her development up to the present time, I can answer this with an unequivocal "Yes!"

Sarah and the members of her new permanent family had a turbulent period of initial adjustment; but they weathered it, thanks to the positive emotional qualities and effort each brought to their interrelationships. In her late teens, while at college, Sarah resumed contact with me (directly this time, not through the agency). She felt she needed psychotherapy, as, indeed, she did. She had matured and grown secure enough to want to work on her unfinished emotional business. In debating whether I should become her therapist, as she requested, I weighed the possible drawbacks of having intervened so actively in the realistic circumstances of her childhood against the advantages of my continuity between epochs in her life. I decided in favor of the latter, and it did prove to be a major therapeutic asset. Sarah achieved an excellent treatment result. I still see her from time to time, as when she came by to show me photographs of her husband and their little girl.

When the adoption agency first intervened —Sarah was just under 11—her prognosis for healthy development was poor, in the absence of special remedial measures. Removal from the first ill-chosen adoptive home was a crucial crossroads experience, prerequisite to the effectiveness of everything else that helped to undo the psychic injuries she was already then showing. It made her accessible to the emotional nourishment of the adoptive family process, which in synergistic combination with psychotherapy seemed to have made a superior adaptation possible for this particular young woman. I have learned a great deal from her, over a span of 18 years, about how to help others in comparable predicaments.

The cases of Liz and Sarah were cited to illustrate several points. Although in both these instances the first placement came to grief and was terminated, it would be incorrect for the reader to conclude that agencies frequently remove children once they have placed them for adoption. On the contrary, this seldom occurs.

* It is now generally recommended that statewide adoption resource exchanges be set up in order to increase the opportunities for both children and adoptive applicants. These exchanges, in turn, should work through ARENA, the Adoption Resources Exchange of North America, which has already been established by the Child Welfare League of America. Also at the international level recommendations for improving intercountry laws and practices were made by the 1971 First World Conference on Adoption and Foster Placements.

Criteria and Procedures for Parent Selection

Agencies have been very reluctant to disrupt placements already made. Instead, their efforts have been focused on preadoptive procedures of selecting and preparing particular parents and children for making a good life together. In recent decades as the weighting of factors has shifted with regard to what is considered desirable in adoptive parents—from affluence and religious and civic standing, for example, to psychological capabilities for parental competence and relationship—the concepts and methodologies of psychiatry and psychoanalysis have come to play a greater role in determining appropriate criteria and ways of evaluating adoptive applicants.

As psychiatric consultants to agencies, several of us have sought to contribute to this aspect of adoptive practice because of its obvious strategic importance to primary prevention.[3,4,11,26] It would certainly do violence to the true complexity of variables at issue to develop a check list, as it were, of qualifying and disqualifying psychological items for adoptive parenthood, or to seek a single personality stereotype or hypothetical paragon as a model among adoptive applicants. The basic personal qualities that have been sought as mental health assets for all adoptions, whatever the age and characteristics of the child—qualities that may be expressed through a diversity of life styles and personality patterns—concern the capacity for warm, mature love for a child as an individual in his own right, by each parent and by both as a unit* along with a stable compatible marriage and the flexibility to cope with life's unpredictable vicissitudes.

Criteria that are more specific to adoptive parenthood also enter into preadoptive assessments. Which infertile couples, for example, can or cannot feel comfortable enough about adoption as a substitute form of family formation really to accept such a child as their own? Have various emotional problems around their infertility been sufficiently resolved so as not to impair how they perceive and respond to the adoptive child? Or in regard to transracial adoptions, for instance, does a particular white couple want to adopt a nonwhite youngster as a way of making a sociopolitical statement, or is it primarily for love of the child? Can another couple, who would like to adopt an older child, allow him his memories and the sense of his past without feeling it as a threat to their sense of parenthood? And are still another couple, eager to adopt a handicapped child, too caught up in their own rescue fantasies to be able to know the child as the person he really is?

Evaluation of applicants' motivations to adopt—conscious and unconscious—requires a high degree of skill, sensitivity, and objectivity. One must know how to listen for meanings behind what is said, and to understand how these are likely to affect the adoptive family process. The screening out of those with manifestly neurotic motivations is a valuable means of preventing childhood maladjustment. Certain motivations are also generally deemed to be contraindications for adoption, at least at the time when the application is being made. The desire for an immediate "replacement child," for example, to relieve the bereavement of a couple whose own child recently died is almost bound to lead to misery: disappointment for the parents and rejection for the child. Instead, such couples should be invited to reapply, if they still want to adopt, after having gone through a mourning process for the child they lost. Nor should adoption ever be considered, of course, as a means of trying to hold together a faltering marriage.

However, unless relatively clear-cut reasons are uncovered why adoption would turn out to be unsound, psychodynamic insight in current adoptive practice is being utilized more for assisting applicant couples to become parents than for ruling them either in or out, by way of some diagnostic prediction, in advance of their actual experience with an adoptive child.

The rationale for this shift in emphasis be-

* Under certain conditions some single individuals are also now being approved as adoptive parents.

tween the intertwined processes of preadoptive appraisal and enablement in favor of the latter rests on certain factors. One of these has to do with greater awareness of the limits to accurate prediction of emotional capabilities for adoptive parenthood, in both the near and the far future, before the couple has even begun the process of becoming parents.

We know that feelings of parenthood develop as a process over time, and that this holds true for both biological and adoptive parenthood. The experience of pregnancy serves as a psychobiological preparation for motherhood that adopting mothers—at least those who are childless—lack. The interval between agency approval of a couple's application and placement of the child provides a compensatory opportunity in such instances to prepare emotionally for the mother role—a process that a psychologically sensitive case worker can facilitate. But for most couples the period between their application to adopt and its approval by the agency is a time of stress and uncertainty. Far from using this interval as an occasion to face and work out conflicts and anxieties that the much desired adoption may unconsciously evoke, they more often tend to defend themselves against the risk of agency turndown by not letting themselves believe that they will really become parents. Therefore, protracted agency preadoptive studies, intended to make sure of selecting "good couples," have in many instances defeated their purpose.

On a continuum of suitability for adoptive parenthood, most applicants are in the middle range. Selection among these entails such subtle differentiations that, in their case, agency overreliance on predictions about adoptive outcome, prior to placement, is not warranted. (A more feasible aim at that stage, in terms of reliability, is to identify and screen out, from the psychological standpoint, the more definitely unsuitable applicants, such as Sarah's first adoptive parents.) The conditions of agency practice add to the reasons for this. Such a diagnostic attempt demands unrealistic levels of skill on the part of most social workers; moreover, it has to be carried out within the relatively few interviews that comprise an adoptive home study. Furthermore, applicant couples understandably try to put their best foot forward and tend to conceal or misrepresent what they think might disqualify them. Another frequent source of error is the worker's own unconscious attitudes and prejudices; despite her best efforts these may influence her diagnostic judgments.

Lest this signify to the reader that preadoptive parental appraisal is useless, it should be noted that according to one study,[49] even those families with intervening "unpredictable" disruptions that seemed to be associated with the children's maladjustment have shown in retrospect certain signals of disturbance at the time of adoptive study. With keener diagnostic alertness to these clues in the first place, the problematic nature of these homes might have been predicted after all. If so, adoptive maladjustment might have been prevented either by the agency's not placing the child in the home or by appropriate interventions as indicated—not only before legal adoption, but also after it.

The possibility for such interventions goes to the heart of profound changes in emphasis with regard to adoptive services today: greater emphasis on facilitating the parental potential of applicants and on helping them with problems than on hyperselectivity; greater emphasis on services to parents and children throughout the adoptive experience, not only in the stages before legal adoption, but extending on after it. These changes follow from agency efforts to cope with new situations: in sharp contrast with the past, there is now a shortage of applicants relative to the increased numbers and kinds of children for whom adoption has become a recognized right. In response to the pressing need for homes, especially for nonwhite and older children, and in the light of improved research, many earlier requirements for adoptive parent eligibility have been revised, relaxed, or dropped. An intensified search for more homes, in relation to need, is also being carried on by a diversity of other changes in practice. Among these are the use of TV and public education programs for recruiting homes for children with special needs.

The magnitude of the problem with regard to nonwhite children may be conveyed by 1970 figures from 240 agencies, which show that 116 white homes were approved for every 100 white children available for adoption, while only 39 nonwhite homes were approved for every 100 nonwhite children reported as needing adoptive placement.[20] According to the latest published estimate (1970), the percentage of all children adopted who are black or who belong to other minority groups has not risen significantly, despite special efforts by social agencies to find adoptive homes for them. In 1969 children from these groups constituted an estimated 10 per cent of nonrelative adoptions. Published figures for 1970 reveal a 2 per cent increase. In that year 12 per cent of all nonrelative adoptions were of nonwhite children.

One of the most promising among the new ways of reducing this shortage of homes is a subsidy, through legislative provision, to parents who could not otherwise afford to adopt.[18] Not only does this enable a larger number of less affluent black couples to qualify for adoption, but also it makes it possible for black foster parents to convert their status to that of adoptive parents when their relationship with the child and his legal situation warrant this.

The Children's Aid Society of Pennsylvania has reported on a study of black adoption families who participated in a so-called quasi-adoption program.[36] Although these couples were interested in undertaking long-term care of foster children, they were ambivalent about assuming the financial and psychological responsibilities of adoption. They were subsidized as foster parents and also given active psychological and social help by the agency, with the understanding that this might lead to their deciding to adopt. Concurrently the agency worked with the children as well. The study compared 50 black families who adopted through the traditional methods of that agency and 50 who adopted through the subsidy program—a program that had doubled the number of homes available for the permanent placement of black children.

Transracial Adoption

Transracial adoptions, which are increasing, constitute another alternative for minority group children who would otherwise grow up without any family. American Indian, black, and Oriental children, as well as those of mixed racial backgrounds, including part Mexican-American children, have been placed in Anglo-American Caucasian homes at an accelerating rate in recent years. In discussing the psychodynamics of transracial adoptions, Marmor[42] has suggested that the psychological qualifications of such parents should include, in addition to the basic criteria, their relative freedom from ethnocentricity and their minimal dependence on family and community approval.

Between 1958 and 1967, 355 homeless American Indian children were placed for adoption with white families by agencies affiliated with the Child Welfare League of America, in a demonstration project that the League undertook in cooperation with the Bureau of Indian Affairs. Fanshel[14] has just completed a study of 96 such families, living in 15 states, to develop knowledge about the characteristics of those who adopted these children. He also sought to learn about the phenomenon of adoption across ethnic and racial lines, as well as to develop a five-year follow-up picture of the experiences encountered by these families and children. The children came from tribes in 11 states. He found that the adoptive parents, rather than being a homogeneous group, represented a cross-section of attitudes among Americans. Although their political and philosophical views do not appear to have been an important stimulus for the adoption, the parents of these Indian adoptees did evidence more independence of mind and a stronger civil libertarian view than seems to have been true for adoptive parents of white children—a finding that supports Marmor's suggested qualifications for transracial adopters.

On the whole the children seemed to be doing remarkably well from the standpoints of physical health, cognitive competence, personality development, behavior patterns, and "imbeddedness" in their adoptive families. With regard to the parents most of them appeared to be very positive about the adoption. Many of them recognized that there might be rough periods ahead for their youngsters around problems of dating and of their sense of personal identity and worth, especially when they reached their teens. They were planning ways of protecting the children against the ill effects of racial difference, in part by trying to foster in them a strong sense of the value of their own backgrounds.

As Fanshel points out, there have been important changes in the American Indian struggle against social injustice during the decade or so since this project began—changes that have made for conflicting attitudes among Indian groups about transracial adoption.

It might be added here that comparable conflicts are also manifest within black communities about the transracial adoption of black children by white parents. On the one hand, given the still current social circumstances that have deprived so many nonwhite children of families of their own race—biological or adoptive—the data show that transracial adoption does provide these children with a growing-up experience of far greater emotional security than the damaging life situations they would otherwise face. On the other hand, a segment of the black movement has renounced the goal of integration, at least for now, as a threat to their people's effort to achieve a positive sense of individual ethnic identity and group unity, which they regard as essential to their fight for equality. On this basis they sharply oppose the adoption of black children by white parents.

Close family attachments between parents who are white and their black or partly black children are epitomes of racial integration. It is ironic, therefore, that these transracial adoptions have come under fire from opposite directions—from white racism and from black militancy. Agencies, however, favor this kind of family formation as a here-and-now solution, at least so long as the relative scarcity of black adoptive homes persists. Most recently, thus not yet in published reports, it has become generally possible for agencies to find black homes for adoptable black infants; it is the older nonwhite for whom the shortage of homes is most acute.

Some guidelines for recruiting and selecting families that can enjoy transracial adoption and experience it constructively have been emerging from the experience to date. Agencies, for instance, recognize that some couples are able to accept American Indian or other nonwhite children, including at times children of mixed black and white backgrounds, but are unable to accept fully black children. Such applicants are encouraged to acknowledge and explore these attitudes, which are taken into account in determining the kind of child to be placed with them. Agencies are also tending to place black or interracial children with fertile couples who have already borne children, but who are interested in what is now referred to as "room for one more." Many of the case reports of these placements reveal how greatly those children who are already part of the family can help the newly arrived child—often in ways in which the parents cannot—with his initial adjustment and with the process of becoming a full-fledged family member. (This is generally true in the adoptive placement of older children, transracial or not.)

In addition to the professional literature on this subject—Fanshel's[14] book, incidentally, contains a very useful list of selected references on transracial adoptions—David C. Anderson[2] has written a sensitive and insightful book, *Children of Special Value*, from the viewpoint and experiences of white adopting parents. The journalist author and his wife have themselves adopted three children across racial lines. He tells the story of four families who between them have adopted ten such children. Racially these were black, American Indian, part black and part Indian, part black and part white, and Korean.

These highly personal stories provide a vivid, real-life picture, in which parents and children emerge as the differentiated individuals they actually are. Anderson writes about some of the influences leading up to such couples' decisions to adopt across racial lines, the reactions of their families and friends, their difficulties and their successes. He also reports on the different ways in which a strong sense of parenthood came into being for these mothers and fathers—sometimes quickly, sometimes gradually, yet influenced by how they felt about their particular child's characteristics and his ways of responding to them.

Anderson recognizes that interracial adoption is not for everyone, and that adopting parents will vary as to how well they raise their children for reasons that may or may not have to do with race. He is convinced that "adoptive parents develop a sense of parenthood for their children every bit as strong as that of biological parents, perhaps because the growth of any parent-child relationship depends far more on the behavior of the human beings involved than on biology."

As the title of his book conveys, Anderson sees the blackness and brownness of these children not as liabilities, but as assets for the white families who adopt them. As "children of special value" they make possible for their parents a human insight only rarely acquired by "conventional" parents. Parental perception of such children's special value, he believes, makes all the problems to be expected in raising them more manageable.

Other Children with Special Problems

Of course, the same child may be in several hard-to-place categories at once: he may be nonwhite, no longer an infant, and handicapped—physically, emotionally, or intellectually. In order to evaluate the adoption of older children per se, Kadushin[27] followed up 91 families who had adopted white healthy children between five and eleven years old. When adoptive outcomes were assessed on the basis of overall parental satisfaction, it was found that between 82 and 87 per cent were successful.

Clinical rather than statistical approaches, however, that have focused directly upon the older adoptive child have delineated certain psychological factors that agencies need to take into account in order to prevent and reduce problems for both the child and parents. This requires that agencies apply psychodynamic insight to their procedures and services before, during, and after placement, as well as following legal adoption.

Sometimes, for instance, between placement and legal adoption these children may suffer an unconscious conflict of which their new parents, and the agency workers, too, are unaware; in recognizing it case workers could obviate parent-child misunderstandings and the problems to which these may lead. One example of such conflict is between the child's overt fear that he *will not* be legally adopted and his simultaneous fear that he *will*. Already hypersensitive to rejection as a result of the experiences that made him adoptable, he is in terror lest his new family decide against keeping him "forever." This makes it all the more vital, he feels, to conceal his opposite fear both from them and from himself: that the finality of being legally merged into this new family unit—emotionally and often socioculturally strange to him—will dissolve his sense of continuity with himself and his past, and thus destroy his internalized attachments to previous parent figures. As a general principle, instead of older children *being adopted*, they and their parents really need to *adopt each other*.

Increasingly, as agency policies have changed, and more adoptive applicants have been willing to assume the risks of parenting *children with various medical impairments*, such children, once thought to be unadoptable, have been placed.[16] Study data have provided evidence that permanent placement of these youngsters should not be delayed until medical prognosis has been established or corrective treatment begun, since this "can inflict irreparable damage to mental health." The same study stresses the importance of subsidies, when required, so that the prospect of costly medical payments need not deter the permanent placement of some of these children.

Adoptive parents are also now being found for *mentally retarded children* when the degree of retardation is mild, and when the child is thought to be able to fit into family life and to become self-supporting. In discussing factors to be considered in effecting such adoptions, Gallagher[17] states: "For many children, mental retardation is a dynamic rather than a static condition, subject to change as the environment changes." That adoptive experience may, indeed, have a secondary as well as a tertiary preventive effect on some retarded children is borne out by a long-term follow-up study by Skeels.[56]

Single-Parent Adoptions

Under certain circumstances some agencies have been departing from traditional practice in recent years by arranging single-parent adoptions. This represents one further way of increasing the number of permanent homes for children who would otherwise face long-term foster care "careers." So far such placements are made infrequently and only when no suitable two-parent family can be found, because of a child's special needs. In practice nowadays these "special needs" usually mean that the child is black and of school age, or close to it. Despite its relative infrequency, however, single-parent adoption has aroused a great deal of interest among professionals because of its many implications for theories of child development and for principles of practice, as well as among the general public, in response to the attention given to this innovation by the news media.

Psychoanalytic case studies are the most intensive and detailed mode of psychological investigation on this question. Neubauer[46] reviewed the ten case reports in the literature—and presented one of his own—in which the absence of one parent had taken place before or during the oedipal phase. This was found to cause "oedipal deficiency," which, in turn, was related to the child's pathological and social development. Unlike children who had never known a father relationship and were then adopted by a single woman, the small sample of patients reported by Neubauer *had* experienced some relationship with the parent with whom they had lost partial or total contact in their early years.

As against such microscopic perspective is the awareness that from six to nine million children are currently growing up in one-parent homes in this country. A disproportionate number of these families, for the most part headed by women, are poor; of these a disproportionate number are black. Herzog and Sudia[22] have noted a widespread tendency to regard the one-parent family as a sick family, a nonfamily, or an unfamily—a tendency that affects the children adversely. The fact that one-parent families include millions of children and have produced many effective and apparently happy adults warrants, they believe, "recognition of the one-parent family as a family form in its own right—not a preferred form, but one which exists and functions. . . ." For single women who want to adopt, agency criteria are being formulated; especially careful exploration of such applicants' motives to adopt is indicated.[30]

It is still too early to evaluate the outcomes of such adoptions in any systematic way. Nevertheless, it is now being pondered whether it may not be preferable, from the mental health standpoint, to place older black children, for whom two-parent black homes are still too scarce, with single black women rather than with white couples. The outlook for transracial adoption, in terms of its emotional assets and drawbacks, would seem to hinge on the future course of the black movement. If the present vehement opposition to such adoption by some black groups and professionals becomes ascendant, adoption policy may favor intraracial, single-parent adoptions over transracial ones as less stressful alternatives for hard-to-place black children.

Adoptive Outcomes

In the foregoing pages adoption has been viewed as a means of preventing maladjustment and of strengthening the mental health potential of children at special risk. Although evidence in support of these basic hypotheses has been interspersed throughout the discussion, it seems worthwhile to consider further, however briefly, some of the research findings on adoptive outcomes and adoptee adjustment. We ask of this research two kinds of questions about adoption in relation to prevention. First, given the fact that adoption does prevent physical and legal homelessness for children, to what extent does it succeed or fail psychologically as a form of positive family experience? Second, can adoption provide a context for reinforcing a child's potential for healthy maturation, and if so, how can adoptive services increase such a prospect? Or is adoptive experience per se, because of certain built-in factors, conducive to psychopathology for adoptees (as suggested by some psychiatrists and refuted by others, including this author)? If so, how can adoptive service reduce such risk?

Because follow-up studies of adoptive outcome vary so greatly as to methodology, sampling, degree of complexity, outcome criteria, and general level of research sophistication, findings from particular studies are far from comparable. In overall terms, however, the results of all these studies, taken together, do confirm the preventive value of adoption for children who lack homes and who have often experienced antecedent deprivation.[23,24,25,27,41,49] Thus Kadushin[27] tabulated 14 adoptive outcome studies that had been reported between 1924 and 1968 under the following headings: size of study group and lapse of time between placement and study; number and percentages of subjects in each outcome category; the outcome criteria used for categorization (good, questionable, poor, and so forth); data used for follow-up assessment (interviews with adoptive children, adoptive parents, case records, and the like); and nature of the adoption being studied, whether agency or independent. In totaling these reported results, he found that 2,236 adoptees had been followed up; the adjustment of 74 per cent of these had been rated as unequivocally successful, 11 per cent as fairly successful or intermediate, and 15 per cent as unsatisfactory. Obviously many significant variables are washed out in this form of tabulation; nor can it do justice to the richness and complexity of the data or to the interpretations of their pertinence for practice.*

The research just referred to has been conducted by psychologists, social workers, and sociologists, singly or in teams, but without any direct participation by psychiatrists or psychoanalysts. Indeed, our contributions to the literature on adoption have mainly reflected our specialized concern and expertise with psychopathology.

Questions of Adoption and Emotional Disturbance

One major form of psychiatric and psychoanalytic contribution has been through reports of intensive case studies of emotionally disturbed adopted children and their parents. Such reports have been valuable for deepening understanding about psychodynamics and adoption-connected problems, such as have been found among these troubled individuals with at least some degree of regularity. Psychological data from disturbed members of adoptive families, however, cannot validly be assumed to apply *in toto* to that great number of well-adjusted adoptive families to whom clinicians generally lack access. The tendency to commit this error, some of us think, accounts in part for the conclusions reached by some clinicians—and challenged by others—that adoptive status per se has primary causal significance for child maladjustment, and that such risk is heightened if children who were placed as infants are told about their adoption before they have reached the latency phase of development.

In addition to case reports, the psychiatric literature on adoption of the past dozen years or so has consisted mainly of studies comparing the rates of emotional disturbance, and its symptomatology, of adopted and nonadopted children in the case loads of clinical settings. Much of the impetus for this series of investigations came from a 1960 paper by Schech-

* Several of the more outstanding investigations of adoptive outcome, covering a range of methodologies that help to supplement each other, are listed in the Bibliography at the end of this chapter.

ter,[51] in which he reported that about 13 per cent of the 120 children he had seen in private practice over a six-year span had been adoptees (EFA); he concluded that "this indicates a hundredfold increase of patients in this category . . . compared with what could be expected in the general (child) population."

Schechter attributed much of this adoptee pathology to the fact that these children were told about their adoption when they were between the ages of three and six. Such telling, he believed, leads to various problems of superego and ego ideal formation. Thus, to learn that they really do have dual parentage—by birth and by adoption—at an age when children still tend to fantasize two sets of parents, one "good" and one "bad," prevents adoptees from the subsequent normal fusing (so Schechter thinks) of the split between parental images. Information about adoption should therefore be postponed until after the conflictual oedipal phase of development. Postponement would also spare the child, thought Schechter, from the narcissistic injury and anxiety that would come from his learning, before his ego was mature enough to cope with such knowledge, that he had been "rejected" by his original parents. Similar warnings about early "telling" were advanced by Peller[48] soon after the Schechter article appeared. Agencies, by contrast, in consultation with psychiatrists,[3] had been advising parents to begin explaining the child's adoption to him very early—almost as soon as he understood language. This advice has been predicated on the awareness that such communication entails a gradual process over time, not a one-time event, and must therefore be sensitively attuned, in terms of content, to the growing child's emotional level and stage of comprehension.

Schechter's article was given wide publicity by mass media versions of its content, under headlines such as "The Truth Hurt Our Adopted Daughter" and "Why So Many Adoptions Fail." Understandably such "revelations" were extremely anxiety-arousing for adoptive family members, for child welfare workers, for natural parents considering surrender, and for prospective adoptive parents. There is no way of knowing how many children in need of homes were denied them because of the deterring effect on potential adopters of these articles. If adoption has, indeed, been discovered to be pathogenic and the early revealing of it to children a cause of maladjustment, then it would, of course, be necessary to effect appropriate changes in the light of this knowledge. If it has not, however, then the undermining of professional and public confidence in the primarily preventive value of adoption may be viewed as a calamity from the standpoint of public health and social policy.

Controversies among professionals that were set off by these initial articles have continued. In trying to resolve them, a succession of studies have been carried out.[9,19,33,55,57,60] Although none of these—including a second by Schechter, in collaboration with Carlson, *et al.*[52]—showed as high a percentage of adoptees among children referred for treatment as Schechter had first reported, they have confirmed the fact that EFA children are overrepresented to some extent in clinical settings.

Granted the agreement now that a disproportionate number of adopted children have appeared at clinical facilities, there is still wide disagreement about why this occurs and what its significance is. A number of investigators have shared the view of Schechter and his co-workers that it is the factors inherent in adoption that render adoptees more liable than other children to emotional disorder; another array of researchers, whose interpretations of the evidence seem far more convincing to me, do not regard the overrepresentation at clinics as a reliable indicator of the true incidence of adoptee disturbance.[29,33,38] We also challenge the appropriateness of the comparison groups and the accuracy of the base rates that were used in reaching such conclusions.

A number of psychiatrists who have stressed that the rate of disturbance is greater for adopted than for nonadopted children have also described the adoptee patients as showing more aggressive and sexual symptomatology than nonadopted clinic children. Again the tendency among psychiatrists to apply findings from patients to nonpatients has helped to create the impression that adopted children characteristically manifest aggressive and sexual behavior problems. Evidence to the contrary has been provided by studies that compare the adjustment of adopted and nonadopted children from nonclinic populations.

Mikawa and Boston,[44] for instance, studied two groups of "normal" children—20 of them had been adopted by nonrelatives and 20 were living with their parents; they found no significant differences between the adopted and nonadopted groups. From among other studies with similar results, that by Witmer, Herzog, *et al.*,[59] is unusually rigorous and comprehensive. A sample of 500 adoptions studied ten years after placement were matched with a control group of nonadoptive children. Very little difference was found between the adopted children and their schoolmate controls, in terms of social and emotional adjustment.

In taking issue with the type of incidence and prevalence studies discussed earlier, I do not mean to deny that adoptive children and parents are subject to special psychological strains, which we as mental health professionals need to recognize and help to mitigate. It is just that it seems regrettable that so much of the available psychiatric time and effort has had to be deflected from needed study of urgent questions about adoption by repeated testing of the issues that Schechter first raised —especially since I, at least, do not regard the general population of children who live with their own families as an appropriate control group for assessing the psychological well-being of adopted children.

To illustrate how studies of certain kinds of adoptee problems, however, can point to ways of preventing them, several investigations have found that a disquieting number of adopted children show some degree of cerebral damage.[31,47] This seems, in large part, referable to the prenatal period, and to the infrequency with which the natural mothers of these children obtained prenatal care before the final months of pregnancy, if at all. In order, therefore, to help prevent reproductive casualty among adoptees, provision of prenatal services attuned to the needs of unwed pregnant girls and women, though not directly a function of adoptive services, is relevant to preventing some of the neuropsychiatric problems that afflict adoptees.[7] A comparable preventive effort is being made by informing adoption workers more fully about medical genetics, insofar as this knowledge can be applied to the field of adoption.*[53]

Now that agencies have corrected their earlier policy of severing contact with adoptive couples and children after legal adoption, they can far better help to prevent and reduce the psychological problems that do, of course, arise. One area, for example, of helping adoptive parents stems from our recognition nowadays that it is the acknowledgment and acceptance of the differences between adoptive and other family forms, rather than their denial, that is the more conducive to successful adoptive experience. Kirk[32] has elaborated on this issue in his book *Shared Fate*.

Earlier mention was made about Schechter's and Peller's advice, on theoretical grounds, to postpone the "telling" of adoption until children had passed through the oedipal phase. Many of us have found that parental anxieties in facing the facts of adoptive differences are especially likely to be mobilized around the questions of whether, what, and when to tell their children about their adopted status. In postadoptive counseling of parents, for instance, this has proven to be one of the most regularly recurring sources of concern and uncertainty; it is also one on which the interactions of group process can exercise a very beneficial effect.[50]

Side by side with their genuine wish to protect the child from psychological harm and insecurity, parents often reveal an unconscious displacement of their own conflicts about

* Such application of knowledge from other fields to our topic exemplifies the reciprocal relationship that exists between adoptive concepts and practices, on the one hand, and the research uses of adoption for investigating broader basic issues, on the other. Psychiatric consultants to adoptive agencies can perform a useful function by extracting and integrating pertinent findings from each of these different but mutually supplementary research approaches, for application to agency policies and procedures.

adoption onto the more ego-acceptable fear of hurting the child emotionally by such disclosure. For parents with unresolved problems about their infertility, for instance, telling may mean exposing themselves as inadequate. For others an unrealistic sense of guilt—as if they had stolen the baby from its mother, or somehow did not feel entitled to be its parents—may underlie their discomfort in revealing his adoption to the child. Many parents, too, seem to fear that once their youngster learns he was adopted, he will stop loving them as much and wish for his original parents instead.

Most parents agree, intellectually at least, with the consensus of adoption experts—namely, that children do need to know that they were adopted and to learn that fact from their adoptive parents. To find it out from others or by accident—as is almost certain to occur at some point—can be extremely traumatic for the child and can also seriously impair parent-child relationships. With regard to the controversy among professionals stimulated by the arguments advanced by Schechter and Peller, a number of us feel that the disadvantages of postponement outweigh the advantages, at least for the majority of children who are not emotionally disturbed.

Of course, just how one defines the essential nature of parenthood has much to do with the kind of meaning that one gives to what the adoptive parents are really "telling." From an experiential point of view, they can justifiably feel and convey that they are their child's true parents, not only by law, but also by virtue of their parental role behavior and relationship ever since his adoption. According to a widely held biological orientation, however, they are informing him that the unknown beings who gave him birth are by virtue of that fact alone "more really" his parents than they themselves are—a troubling situation, indeed, for adoptive parents and children to have to live with.

Krugman,[35] in commenting on the biological orientation of both Schechter and Peller with regard to this question of where real parenthood lies, has noted the frequency with which, in his initial article, Schechter used the words "real" and "own" in connection with the biological parents, while neither adjective was used at all with the adoptive parents. She challenges these authors' assumptions that children who are adopted as infants do, in actual fact, have two sets of parents—the first, and more "real," set having rejected them—and that for children to learn about this, especially in early childhood, is a critical determinant of psychopathology.

In contrast to the case material presented by these and other therapists from adoptees as patients, Krugman reports on her experience, as consulting psychologist to an agency, in appraising the development of over 50 children who had *not* been referred because of problems in themselves or their families. They had been placed for adoption as infants, were between three and seven when studied, and had all been told about their adoption before she saw them. She could find no evidence of parental image diffusion or of any other distinguishing "signs" of adoptive living among them.

As we seek to improve the psychological understanding of adoption, it would, indeed, seem that we, as psychiatrists, cannot mainly rely on our customary problem-focused approaches—at least with regard to the large majority of adoptive family members who are not emotionally disturbed. A range of services, based on such understanding, does have potential for preventing and mitigating the effects, for adoptive parents and children, of the special stresses and strains entailed in adoptive family process. But the central reality of adoption is its power to prevent the misery and maldevelopment of children who lack homes of their own. From that perspective adoption is a repair of trauma, not a trauma in itself (except for unsuitable placements); it is not a losing or a taking away of what never was, but a mutual giving and gaining of affirmative family relationship.

Bibliography

1. American Academy of Pediatrics, *Adoption of Children*, rev. ed., American Academy of Pediatrics, Evanston, Ill., 1967.

2. ANDERSON, D. C., *Children of Special Value: Interracial Adoption in America*, St. Martin's Press, New York, 1971.
3. BERNARD, V. W., "Adoption," in *Encyclopedia of Mental Health*, Vol. 1, pp. 70–108, Watts, New York, 1963.
4. ———, "Application of Psychoanalytic Concepts to Adoption Agency Practice," in Heiman, M. (Ed.), *Psychoanalysis and Social Work*, International Universities Press, New York, 1953.
5. ———, "Community Mental Health Programming," *J. Am. Acad. Child Psychiat.*, *4*:226–242, 1965.
6. ———, and CRANDELL, D., "Evidence for Various Hypotheses of Social Psychiatry," in Zubin, J., and Freyhan, F. A. (Eds.), *Social Psychiatry*, pp. 183–196, Grune & Stratton, New York, 1968.
7. BERNSTEIN, B., "Deterrents to Early Prenatal Care and Social Services among Unwed Mothers," *Child Welfare*, *40*:6–12, 1961.
8. BISKIND, S. E., "Helping Adoptive Families Meet the Issues in Adoption," *Child Welfare*, *45*:145–151, 1966.
9. BORGATTA, E. F., and FANSHEL, D., *Behavioral Characteristics of Children Known to Psychiatric Outpatient Clinics*, Child Welfare League of America, New York, 1965.
10. BROWN, F. G., "Services to Adoptive Parents after Legal Adoption," *Child Welfare*, *38*:16–22, 1959.
11. CHESS, S., "The Adopted Child Grows Up," *Child Study*, *30*:6, 36, 1953.
12. CHEVLIN, M. R., "Adoption Outlook," *Child Welfare*, *46*:75, 1967.
13. Child Welfare League, *Standards for Adoption Service*, rev. ed., Child Welfare League of America, New York, 1968.
14. FANSHEL, D., *Far from the Reservation: The Transracial Adoption of American Indian Children*, Scarecrow Press, Metuchen, N.J., 1972.
15. FESTINGER, T. B., "Unwed Mothers and Their Decisions to Keep or Surrender Children," *Child Welfare*, *50*:254, 1971.
16. FRANKLIN, D. S., and MASSARIK, F., "The Adoption of Children with Medical Conditions," *Child Welfare*, *48*, Nos. 8, 9, 10, 1969.
17. GALLAGHER, U. M., "The Adoption of Mentally Retarded Children," *Children*, *15*, Jan., Feb., 1968.
18. GOLDBERG, H. L., and LINDE, L. H., "The Case for Subsidized Adoptions," *Child Welfare*, *48*, Feb., 1969.
19. GOODMAN, J. D., and MANDELL, W., "Adopted Children Brought to Child Psychiatric Clinic," *Arch. Gen. Psychiat.*, *9*: 451, 1963.
20. GROW, L. J., *A New Look at Supply and Demand in Adoption*, Child Welfare League of America, New York, 1970.
21. "A Guide for Collaboration of Physician, Social Worker, and Lawyer in Helping the Unmarried Mother and Her Child," *Child Welfare*, *46*:218–219, 1967.
22. HERZOG, E., and SUDIA, C. E., "Family Structure and Composition: Research Considerations," in *Race, Research and Reason: Social Work Perspectives*, National Association of Social Workers, New York, 1969.
23. HOOPES, J. L., SHERMAN, E. A., LAWDER, B. A., ANDREWS, R. G., and LOWER, K. D., "Post-Placement Functioning of Adopted Children," in Shapiro, M. (Ed.), *A Follow-up Study of Adoptions*, Vol. 2, Child Welfare League of America, New York, 1969.
24. JAFFE, B., "The Outcome of Adoption: Perceptions of Adult Adoptees and Their Adoptive Parents," unpublished ms., 1972.
25. ———, and FANSHEL, D., *How They Fared in Adoption: A Follow-up Study*, Columbia University Press, New York, 1970.
26. JOSSELYN, I. M., "A Psychiatrist Looks at Adoption," in Schapiro, M. (Ed.), *A Study of Adoption Practice*, Vol. 2, Child Welfare League of America, New York, 1956.
27. KADUSHIN, A., *Adopting Older Children*, Columbia University Press, New York, 1970.
28. ———, "Adoption," in Maas, H. S. (Ed.), *Research in the Social Sciences: A Five-Year Review*, National Association of Social Workers, New York, 1971.
29. ———, "Adoptive Parenthood: A Hazardous Adventure?" *Social Work*, July 1966.
30. ———, "Single-Parent Adoptions: An Overview and Some Relevant Research," *Soc. Service Rev.*, *44*:263, 1970.
31. KENNY, T., BALDWIN, R., and MACKIE, J. B., "Incidence of Minimal Brain Injury in Adopted Children," *Child Welfare*, *46*:24, 1967.
32. KIRK, H. D., *Shared Fate: A Theory of Adoption and Mental Health*, Free Press, New York, 1964.
33. ———, JONASSOHN, K., and FISH, A. D.,

"Are Adopted Children Especially Vulnerable to Stress?" *Arch. Gen. Psychiat.*, *14*:291, 1966.

34. KRUGMAN, D. C., "A New Home for Liz: Behavioral Changes in a Deviant Child," *J. Am. Acad. Child Psychiat.*, *7*:398, 1968.

35. ———, "Reality in Adoption," *Child Welfare*, *43*:349, 1964.

36. LAWDER, E. A., HOOPES, J. L., ANDREWS, R. G., LOWER, K. D., and PERRY, S. Y., *A Study of Black Adoption Families: A Comparison of a Traditional and a Quasi-Adoption Program*, Child Welfare League of America, New York, 1971.

37. ———, LOWER, K. D., ANDREWS, R. G., SHERMAN, E. A., and HILL, J. G., "Post-Placement Functioning of Adoption Families," in Schapiro, M. (Ed.), *A Follow-up Study of Adoptions*, Vol. 1, Child Welfare League of America, New York, 1969.

38. LAWTON, J. J., JR., and GROSS, S. Z., "Review of Psychiatric Literature on Adopted Children," *Arch. Gen. Psychiat.*, *11*:635, 1964.

39. MAAS, H. S., "Children in Long-Term Foster Care," *Child Welfare*, *48*:321, 1969.

40. ———, and ENGLER, R. E., JR., *Children in Need of Parents*, Columbia University Press, New York, 1959.

41. MCWHINNIE, A. M., *Adopted Children: How They Grow Up*, Routledge, London, 1967.

42. MARMOR, J., "Psychodynamic Aspects of Trans-Racial Adoptions," in *Social Work Practice*, Selected Papers, National Conference on Social Welfare, Columbia University Press, New York, 1964.

43. MECH, E., "Adoption: A Policy Perspective," in Caldwell, B. M., and Ricciuti, H. N. (Eds.), *Review of Child Development Research*, Vol. 3, University of Chicago Press, Chicago, 1972.

44. MIKAWA, J. K., and BOSTON, J. A., JR., "Psychological Characteristics of Adopted Children," *Psychiat. Quart.*, *42*:274, 1968.

45. National Center for Social Statistics, *Adoptions in 1970*, U.S. Government Printing Office, Washington, D.C., 1972.

46. NEUBAUER, P. B., "The One-Parent Child and His Oedipal Development," in *The Psychoanalytic Study of the Child*, Vol. 15, p. 286–309, International Universities Press, New York, 1960.

47. PASAMANICK, B., and KNOBLOCH, H., "Epidemiologic Studies on the Complications of Pregnancy and the Birth Process," in Caplan, G. (Ed.), *Prevention of Mental Disorders in Children*, Basic Books, New York, 1961.

48. PELLER, L., "Comments on Adoption and Child Development," *Bull. Philad. A. Psychoanal.*, *11*, No. 4, 1961; *13*, No. 1, 1963.

49. RIPPLE, L., "A Follow-up Study of Adopted Children," *Soc. Service Rev.*, *42*:479, 1968.

50. SANDGRUND, G., "Group Counseling with Adoptive Families after Legal Adoption," *Child Welfare*, *41*:248, 1962.

51. SCHECHTER, M. D., "Observations on Adopted Children," *Arch. Gen. Psychiat.*, *3*:21, 1960.

52. ———, CARLSON, P. V., SIMMONS, J. Q., III, and WORK, H. H., "Emotional Problems in the Adoptee," *Arch. Gen. Psychiat.*, *10*:100, 1964.

53. SCHULTZ, A. L., and MOTULSKY, A. G., "Medical Genetics and Adoption," *Child Welfare*, *50*:4, 1971.

54. SCHAPIRO, M. (Ed.), *A Study of Adoption Practice*, Vols. 1, 2, 3, Child Welfare League of America, New York, 1956.

55. SIMON, N. M., and SENTURIA, A. G., "Adoption and Psychiatric Illness," *Am. J. Psychiat.*, *122*:858, 1966.

56. SKEELS, H. M., *Adult Status of Children with Contrasting Early Life Experiences: A Follow-up Study*, Monographs of the Society for Research in Child Development, Serial No. 105, Society for Research in Child Development, Chicago, 1966.

57. SWEENY, D. M., *et al.*, "A Descriptive Study of Adopted Children Seen in a Child Guidance Center," *Child Welfare*, *43*:345, 1964.

58. U.S. Department of Health, Education, and Welfare, Children's Bureau, "Legislative Guides for the Termination of Parental Rights and Responsibilities and the Adoption of Children," Publication No. 394, Children's Bureau, Washington, D.C., 1961 (reprinted 1966, 1968).

59. WITMER, H. L., HERZOG, E., WEINSTEIN, E. A., and SULLIVAN, M. E., *Independent Adoptions: A Follow-up Study*, Russell Sage Foundation, New York, 1963.

60. WORK, H. H., and ANDERSON, H., "Studies in Adoption: Requests for Psychiatric Treatment," *Am. J. Psychiat.*, *127*:948, 1971.

61. YARROW, L., "Maternal Deprivation: Toward

an Empirical and Conceptual Re-evaluation," *Psychol. Bull.*, 58:459–490, 1961.

62. ———, "Separation from Parents during Early Childhood," in Hoffman, M., and Hoffman, L. (Eds.), *Review of Child Development Research*, Russell Sage Foundation, New York, 1964.

63. ———, and GOODWIN, M., "Some Conceptual Issues in the Study of Mother-Infant Interaction," *Am. J. Orthopsychiat.*, 35: 473, 1965.

64. ———, GOODWIN, M. S., MANHEIMER, H., and MILOWE, I. S., "Infancy Experiences and Cognitive and Personality Development at Ten Years," in Stone, L. J., Smith, H. T., and Murphy, L. B. (Eds.), *The Competent Infant: A Handbook of Readings*, Basic Books, New York, 1972.

CHAPTER 26

THE PSYCHOLOGY OF ABORTION

Edward C. Senay

Introduction

A WOMAN WHO SEEKS to abort a pregnancy she does not want presents us with a complex problem. The basic thesis of this chapter is that she is a human being in an acute psychological and social crisis. The definition of her situation is crucial. Some see her in moral terms; others see her in operant terms—for example, they view her as a "manipulator." Few appear to appreciate the psychological isolation inherent in her crisis because most tend to diminish their perceptions of her in order to deal with the abstract ethical dilemma she represents. Despite the recent partial liberalization of abortion laws, medical, legal, and political recognition of the true nature of her situation still leaves much to be desired. In some measure this is so because the psychology of her problem has not been understood and articulated clearly.

At this writing the woman in question remains an ethical dilemma for psychiatry. Professionally, with few exceptions, we have maintained a distance from the woman with an unwanted pregnancy. The lack of social sanction for our engagement with the problem, coupled with the possibility of legal reprisal, has served to create our collective attitude. Our response has also been hindered by a lack of knowledge, for it is only in the past two decades that we have started to study significant numbers of women with unwanted pregnancies. The work of Taussig,[30] Rosen,[24] Calderone,[2] and Tietze[31] has been highly important in the United States in this regard.

Despite our professional distance it is a fact that each year millions of women elect to solve the crisis of unwanted pregnancy by obtaining an abortion; they do so in every known religious group and social class in the Western world, and they do so and have done so in every known cultural and historical circumstance.[4] Despite the emergence of so-called liberal laws, many, if not most, women still abort in social and psychological circumstances that demean them.

Proponents of further liberalization of abortion laws argue that something precious in genuine human pride is destroyed by requesting women to register, or to subject themselves to psychiatric examination, or to undergo the dehumanizing experience frequently resulting from illegal abortions. Opponents of liberal abortion laws argue that something more precious in our general moral climate is destroyed by lowering the barrier protecting the potential life of the fetus.

As noted above, the basic hypothesis of this chapter is that unwanted pregnancy should be defined fundamentally in psychological and social terms. A second major hypothesis is that regular contact with women caught up in this intensely human crisis leads one inexorably to a position supporting abortion on demand. Most people, whatever their moral persuasion, cannot, on a face-to-face basis, regularly consign women to being demeaned, or to self-instrumentation with its attendant dangers of septicemia and death, or to referral for psychiatric resolution of their crisis when psychiatric facilities are either unacceptable, unavailable, ineffective, or nonexistent; nor can they deny a given woman access to all modern medical means to express her vision of what is moral. A corollary of this second hypothesis is that one can have an infinite set of positions with respect to the philosophical question of abortion. But the set will narrow rapidly if one will consider the real individuals involved.

The most complete expression of the abortion dilemma lies in the minor pregnant by paternal rape. Such cases, while not frequent, do occur and individually and professionally we are responsible for the alternatives in such a situation. Any serious discussion of the problem of abortion must deal with such a case; each of the options for the girl, for her family, for her potential child, and for society must be made explicit. These alternatives must be weighed in the light of modern findings that, for the large majority of women, abortion does not have negative long-term psychological effects. It is significant here to note that what is generally the best and most scholarly modern review of the question of abortion[3] nowhere considers the question of pregnancy by rape nor does it confront the issue of pregnancy in minors. Such omissions are inevitable when one moves from abstractions to people; therefore, the bias in this chapter is that one should create a guiding abstraction by starting from the experiences of people, not vice versa.

Psychological Findings in Unwanted Pregnancies

Clinical experience supported by recent research strongly suggests that almost every woman with an unwanted pregnancy has a variable subset of the following symptoms: high and persisting anxiety, somatic complaints, insomnia, depressive feelings, guilt, withdrawal from her usual relationships and pursuits, decrease in self-esteem, anorexia or bulimia, and, in probably three of four such women, at least transient thoughts of suicide. These symptoms represent the mildest form of crisis seen in association with unwanted pregnancy; the greater the precrisis level of personality problems or the more disturbed the family situation, the more severe the crisis picture. While its form may vary, women with unwanted pregnancies have significant and severe psychological pain without exception. Further, this pain can have destructive effects if the woman cannot surmount it.

Ordinary defensive styles usually are employed to cope with the crisis, but, as is true in so many areas of life, those with fewest supports and poorest adaptive styles have far the worst time of it. For the Aid-to-Dependent-Children mother with six or more children, perhaps living on the ninth floor of a public housing project where she fears to use the elevator because of robbery or gang harassment, the prospect of another child usually precipitates a more severe crisis than is seen in the married suburban mother who has a stable home, fewer children, and less social and economic pressure.

The girl prone to psychotic breakdown whose pregnancy represents an unsuccessful and desperate attempt to bind an object may present a florid psychosis; hypomanic episodes also occur in relation to unwanted pregnancy. Sociopathic women may, of course, present the same kinds of cynical manipulative defensive operations that they present in any other crisis, but to define the general problem of abortion by making these few women its symbol suggests naïveté or untenable prejudice.

The most constant psychological feature of unwanted pregnancy is depression, and this is so regardless of personality style or variation in social setting. The psychology of suicide then is a matter of clinical relevance. Suicide in relation to the abortion question is sufficiently controversial to be the focus of a section below.

The presence of a psychological crisis as represented by a complex of the symptoms described above involves some response from significant others in the patient's life as well as a response from the broad society of which the woman is a part. Clinical observation will confirm that in the majority of instances the response of significant others has negative effects on the woman involved, and in almost every instance—even in states with liberal laws—the social response as evidenced in the medical and legal communities is frankly destructive. Women with unwanted pregnancies are led to feel that they are bad, inferior, criminal, inhuman, and undeserving of the same consideration as those "genuinely" in need. It is human—particularly in time of crisis—to internalize such attitudes with the result that the crisis is usually compounded.

We have no systematic studies of the dreams of women with either wanted or unwanted pregnancies. Clinically, women seeking abortion on psychiatric grounds frequently report disturbing dreams with death and mutilation themes. This is so despite the fact that these same women deny being conscious of fantasies about the fetus within them.

Women who have been raped and have become pregnant as a consequence of the rape appear to have a uniquely difficult experience to overcome. Some such women have reported the fantasy that a residue of the brutality and viciousness of the rapist had been deposited within them, and they clearly identified the fetus with these frightening qualities. They fantasized that continuation of the pregnancy would result in the growth of these ugly potentials within them. Such fantasies, of course, form a powerful stimulus for anxiety. Many months were needed by one woman to work through such fears following an otherwise uneventful abortion.

Solutions to the Crisis of Unwanted Pregnancy

We are particularly concerned in this chapter with the option of abortion for the woman with an unwanted pregnancy. Space does not permit an exploration of other options, such as carrying the pregnancy to term and then attempting to become the social and psychological mother, or giving the child up for adoption, or giving the child to its grandparents for raising. All these options are apparently not grossly destructive from a psychological point of view for those who can choose them. There is no question that for certain women they serve well to resolve the crisis, but we have no systematic data from which we might make fully scientific judgments in this regard.

Clinical experience would suggest that more than one out of four pregnancies brought to term are unwanted for one reason or another. Thuwe and Forssman's[8] study suggests that the unwanted pregnancy may represent an even more consequential crisis for the fetus than for the mother, for they found that social and psychological outcomes at age 21 were poorer on all measures for unwanted children in comparison to wanted children.

The salient fact is that abortion is the option selected regularly by substantial numbers of women of every social class, intelligence level, religious, ethical, or cultural background; and, as far as can be determined, this choice represents an absolute historical continuity, for no human group of any time or place studied to date appears to have been abortion-free. Options other than abortion are considered and rejected by perhaps a plurality of women with unwanted pregnancies; to some the thought of separation from a child to whom they have given birth is almost unthinkable, and their moral revulsion is equivalent in all respects to that of persons who equate abortion with murder.

In Scandinavian countries when abortion was first legalized, utilization of the option of legal abortion, somewhat surprisingly, was greatest by married women. The commonest motivation of such a married woman was the entirely healthy desire to raise the level of care for her living children and to improve her chances of fulfilling her potential as a person. It is a grievous error to associate such strivings with criminality; it is a grievous error to do anything other than to nurture such aspirations; collective survival may depend upon it. It appears now in the United States that single women are beginning to use legal abortion as much or more than their married cohorts.

A major effect of the legalization of abortion anywhere short of abortion on demand is the creation of a new option that large numbers of women will elect; apparently they are women who will not use illegal avenues, for the rate of illegal abortions under such conditions does not appear to decrease.[12] Single Swedish girls, for example, go to Eastern Europe rather than face registration in their own country.

These facts suggest that there is great pressure in the female population for abortion services, and that many women, particularly the single and more youthful girl, will not use medical services that involve registration or the possibility of scrutiny by any specialists or committees not elected by or known to them. They will literally risk life—and the psychology of late adolescence lends itself particularly to such behavior—to avoid certain kinds of relationships with established medical facilities; but the evidence is incontrovertible that they will use medical services where their only relationship is with a physician of their choice and where no other parties are involved.

On clinical grounds it would appear that most women refused legal abortion in the United States use illegal channels. Scandinavian workers, however, have found that only 12 percent of women refused abortion then actually obtained illegal ones. The fact that married women predominate in Scandinavian samples needs to be borne in mind in this regard. Hook[11] concluded from a study of 249 women refused abortion that the mental status of many suffered from the fact of refusal. Hook also noted that nearly one-third of those who went ahead and gave birth were not functioning well in their maternal role. Clinical experience suggests that the burden of unwanted pregnancies and the resulting children contribute significantly to the chronic anxiety and depression suffered by many women. This observation in no sense contradicts the finding that there are significant numbers of women refused abortion who manage satisfactorily.[11,19]

¶ The Role of the Psychiatrist

Laws in various states create different problems for psychiatrists and for the women who seek them out on this question. Where rigid antiabortion laws prevail, one of the few paths open to women is through psychiatry. Usually suicide, psychosis, or impending severe behavior disorders furnish grounds upon which women can obtain complete modern medical care.

Under these conditions a dilemma is created for the psychiatrist who will consent to examine such patients. He is faced with the unacceptable task of deflecting women who do not meet stringent psychiatric criteria away from legitimate medical services to the dangers of the illegal market. The situation he faces is complex and deserves careful scrutiny.

The basic fact is that the intent of abortion laws cannot be determined; for example, "to preserve the life of the mother" can be interpreted variously. I do not equate life with regular cardiovascular or respiratory function. It is also relevant to note that no psychiatrist has ever been prosecuted successfully in connection with the abortion problem.

Given the fact of legal uncertainty, it is entirely legitimate and proper for the psychiatrist to be involved on the same grounds upon which he engages with any other patient in crisis and to interpret the laws according to his best determination of their meaning.[26] Indeed, in confused areas like this he has an obligation to stimulate society to consider the possibility of institutionalizing his special insights. In the event of a conflict between state and patient his reference values always should be medical and psychological.

Suicide and the Abortion Question

The most common sanction for abortion on psychiatric grounds is the judgment that a patient has some degree of risk of suicide if pregnancy is not terminated. Since such judgments are highly subjective, there is ample room for the psychiatrist who identifies with the predicament of the woman with an unwanted pregnancy to be liberal in his assessment of such risks; conversely the critics of such a psychiatrist, particularly those who equate abortion with murder, are inclined to read dishonesty into such judgments, and there is, of course, no scientific way of resolving the issue.

Sim,[27] Sloane,[29] and others have written that suicide is less frequent in pregnant than nonpregnant females and that suicide rates are low in the entire female population during the childbearing years, but clinical experience, supported by the recent research of Whitlock and Edwards,[33] suggests that suicide is equally frequent in pregnant and nonpregnant females and is by no means so rare as to be irrelevant.

The concerned but inexperienced psychiatrist may need an orienting position. This need may be satisfied by considering that the prevention or reduction of relatively unlikely risks is a perfectly proper medical-psychological undertaking; general anesthesia carries a slight risk, and this risk is considered frequently in surgical decision making without inciting the least comment. A similar degree of risk in the highly emotional question of abortion, however, generates much controversy.

As noted above, suicidal thoughts are common in women with unwanted pregnancies. Most of them do not have crystallized plans, but the psychology of suicide is prominent and must be a constant concern for the psychiatrist dealing with such patients. The acceptance of the uncertainties and risks of illegal abortion may have roots in depression and self-destructive tendencies. It is rare to encounter a woman with suicidal thoughts who needs immediate hospitalization, but such cases do occur just as successful suicides are known to every psychiatrist experienced in this area. The question of reporting false suicidal thoughts is frequently raised, usually by those without much exposure to the realities of the problem. Patt, *et al.*,[21] found that a few patients appeared to have misrepresented their symptoms. Similarly most experienced clinicians do not feel that falsification is frequent. When it occurs it needs to be confronted. There is usually no difficulty in identifying the fact that even these women are genuinely in crisis.

Outcome of Therapeutic Abortion

Myths to the effect that therapeutic abortion leads to sterility or to severe mental disturbance have been entirely disproved by the research carried out in the past 20 years. Ekblad's[5] work, the largest of the early Scandinavian efforts, demonstrated that three of four women legally aborted were content with their choice of abortion and had no negative sequelae of any kind. The balance felt, for limited periods of time, some self-reproach and guilt or they suffered from mild depression, but only 1 per cent was in real difficulty and these were women who would have had major problems with or without abortion.

Subsequent Scandinavian studies[16] and a recently growing number of American[32,34] and English[14] studies appear to confirm these early findings. As Fleck[6] has detailed, abortion in licensed hospitals, with full medical and psychological services available, is for most women a matter of no great psychological moment; indeed, Simon, *et al.*,[28] Levene and Rigney,[18] Peck and Marcus,[22] Meyerowitz, *et al.*,[20] and Ford, *et al.*,[7] have found in their studies that many patients were improved following abortion, and this was so despite the fact that disturbed women were overrepresented in their study samples.

The findings of improvement provides a new element in the basic philosophical debate about abortion, for now, in addition to the growing body of evidence suggesting that unwanted pregnancies are socially and psychologically detrimental to both mother and fetus, we have the criticism that antiabortion laws prevent some women from obtaining the psychological benefit that could accrue to them if they were free to follow their moral codes.

Jansson[13] compared rates of "psychic insufficiencies" after legal abortion, spontaneous abortions, and normal delivery and found that the respective rates were 1.92, 0.27, and 0.68. The greater rate of disorder following legal abortion no doubt reflects the fact that women predisposed to psychiatric problems were much more heavily represented in those having legal abortion than was the case in those having spontaneous abortions or normal deliveries. Psychological aspects of spontaneous abortions have not received the attention they deserve.

All workers find that single women are more prone to develop postabortion problems than married women,[5,13,26] and there is practically universal agreement that the more psychopathology seen before abortion, the more probable the observation of psychopathology following abortion. The real question here lies in before and after comparison in the single case, not in comparison of populations. Interestingly age does not appear to contribute much to variance with respect to postabortion psychiatric symptoms.

The crisis syndrome described in the early section of this chapter usually resolves rapidly once the patient learns that her pregnancy can be terminated legally. Anxiety decreases, depression lifts, sleep is restored, and for the married woman there may be an increase in family solidarity as the crisis is overcome.

Following the abortion itself, most women appear to suppress the entire episode and to accomplish an attenuated form of mourning without much difficulty. The fetus appears to be easier to mourn in these circumstances because these patients have inhibited the normal process of fantasy in pregnancy. Women seeking abortion regularly report that they have had no or few fantasies about the potential life within them.

In a study of the fantasies of 46 women—28 with wanted and 18 with unwanted pregnancies—I found that 16 of the 18 women with unwanted pregnancies reported no fantasies about the fetus, while none of the 28 women with wanted pregnancy reported such a complete inhibition.[25] Confirmation of the theory of mourning advanced here is apparent in the work of Kennell, *et al.*,[15] who observed that mourning of dead newborn infants was most difficult for mothers who were most pleased with their pregnancies.

If in the postabortion period one asks women specifically to focus on their feelings, some will experience transient anxiety or guilt, but such feelings will disappear rapidly.

As is true in other areas of psychiatry, active ego work by patients has desirable consequences. Women with ambivalence about their pregnancies who seek psychiatric consultation during which they examine alternatives open to them and who then elect abortion will have an increase in self-esteem and no symptoms whatsoever in the postabortion period. Such an outcome is even more likely to occur if the abortion is carried out in circumstances where psychological support is readily available and where there is staff acceptance of the idea that serving such patients is fully legitimate. On the other hand, passivity and dependence carry heavy costs. The patient who will experience persisting postabortion symptoms scores high on both these traits. Usually she also has a high degree of psychopathology combined with a low degree of social support. Her pregnancy represents a new stress that overloads her defenses to the point of new symptom formation or frank breakdown, but her request to be aborted is, in parallel with that of the normal housewife, a healthy striving and it is imperative that we recognize this and support it.

Occasionally, severe psychopathology is seen in a postabortion patient, but most observers feel that one never sees such pictures in women who were not severely disturbed before the abortion.[10] Examination of the histories of these women reveals that their pregnancies resulted from psychotic motives and once established were used, at least in part, for manipulative ends. Many psychotic women having therapeutic abortions appear to recognize with great clarity that they could not perform as mothers.

The Question of Indications

The grounds usually cited by psychiatrists—nowhere codified in law incidentally—for therapeutic abortion are threat of suicide, psychosis, or severe personality disorder. Previous postpartum psychoses are also sometimes mentioned. The GAP report on abortion[9] reviews such so-called indications before concluding that abortion should be removed completely from the domain of criminal law and that it should become a matter between the patient and physician with no third parties involved. Pfieffer,[23] who discriminates between psychiatric indication, for example, suicide, and psychiatric justification, for example, avoidable stress due to economic hardship, also feels that psychiatrists should be removed from their positions as "social decision makers" and should not have to be involved unless "hard" psychiatric indications are in question.

If one defines the unwanted pregnancy as a psychological and social crisis, the role for the psychiatrist is clear. He or she can be involved meaningfully because human crisis is a proper area for psychiatric intervention. If abortion is the option selected by the patient, we can play a proper consultative role, and we can contribute significantly to the welfare of our patients, as recent research has demonstrated. Our role vis-à-vis legal aspects of this kind of crisis should be one of advocacy for our patients.

A new problem has been created by the advances in amniocentesis. Early diagnosis of fetal abnormality is now possible, but in states where fetal indications are not accepted, women are, of course, probing the psychiatric route to abortion. The prospect of seven to nine months of living with the knowledge of carrying a deformed fetus certainly appears to be a prepotent stress, so we as psychiatrists are being sought out and we confront another dilemma. Again it should be stressed that one should make a distinction between acting as a professional psychiatrist and acting as a concerned human being. One's bias should be made explicit. It appears to me that public morality cannot be built upon the compulsory pregnancies[19] of psychotic minors who have been raped, or upon the suffering involved in carrying a fetus known to be deformed, or upon systematic frustration of the healthy psychological strivings of women with unwanted pregnancies.

The question of medical indications raises similar issues. Many internists take the position that any disease occurring in association with pregnancy can be managed without loss of the mother's life. Such a judgment equates life with respiration, and other definitions of life can be held that appear to be more humane. One can wonder at the morality of bringing a barely respiring mother through a complicated delivery to present her and her newborn with an extremely uncertain future. In reality women spare us such scenes, for the overwhelming majority of women with major medical problems turn to the illegal market when refused by licensed physicians and they risk their life by so doing.

Management

The management of women seeking abortion on psychiatric grounds is in most cases an elementary exercise in crisis intervention. Significant others always should be involved. Husbands usually passively or actively agree with their wife's decisions in these matters, but in the event that they do not, resolution of the difference is imperative. One can advance no rule of thumb for decision making in this infrequent event. My practice is to attribute primacy of decision to the woman, but sustained efforts should be made to obtain consensus either for or against abortion.

Women in the crisis of unwanted pregnancy profit from learning the results of the research of the past two decades. Commonly women—even those who are highly educated—appear to feel that abortion causes grave psychological damage. Simply learning that other women appear to improve following abortion can contribute significantly to successful coping.

Not surprisingly, many women develop negative feelings about the psychiatrist involved in this question, even though they may also have a strong positive transference. Women who feel that we should have abortion on demand are more likely to have such negative feelings, and women who feel high degrees of guilt about their situation also tend to develop negative transference. Obviously ventilation and working through such feelings can be significant if it is possible. Many such patients will not permit examination of these feelings, and they are best offered the continuous option of discussing them without pressure to do so.

Some women may use abortion to punish significant males in their lives. Again, if possible, such motives should be faced and worked through. Bernstein and Tinkham[1] report that older married women made good use of a group exploration centered on the abortion experience, while younger and single women tended to have more difficulty in using such a therapy. If preabortion preparation has been adequate there is usually no need for postabortion psychotherapy. As noted above, most women want to suppress the experience and, as one put it, "to get back to normal as soon as I can."

Contraindications

Abortion is contraindicated in the woman who is carrying out the wishes of someone else without engaging her ego with the problem.[26] Commonly such women are passive and dependent and will disregard their own wishes and feelings in order to maintain masochistic ties to a husband, mother, or adviser. The psychiatrist's role here calls for gentle insistence on separating out the patient's feelings and helping her to identify them. Minors of 12 years of age are no exception here; their feelings should be identified and they should be encouraged to make an active decision; the resolution of the crisis syndrome of unwanted pregnancy may be quite dramatic when such patients become active with regard to their problems.

Some of the few severe postabortion syndromes one will encounter are clearly related to failure to resolve the intense ambivalence associated with unwanted pregnancy. Proper preparation for therapeutic abortion can prevent such complications.

The Law and Abortion

Limitations of space preclude detailed review of the bewildering varieties of laws in various states and nations. The present differences in laws between Illinois and New York, for example, has the effect of creating discrimination against the poor and disadvantaged in Illinois. A woman with sufficient resources from this state can afford the trip to New York, but her poorer sister is restricted to the same set of options she inherited from the past.

Some socialist countries in Eastern Europe appear to have changed abortion laws according to economic needs; when low birth rates resulting from psychologically rational laws began to have economic consequences, the leaders of some of these nations reinstituted repressive laws. Such experiences suggest that unless there is concerted action by all nations on the population problem, it may be that there can be no sustained advance anywhere.

In closing we should observe that legalization of abortion on demand everywhere will not make psychiatry superfluous in the abortion question. A role is assured if only to identify those patients for whom the procedure is contraindicated. If the definition of this problem as a human crisis is correct, women will continue to seek us out, regardless of legal circumstance.

We are just beginning to explore the psychology of wanted and unwanted pregnancies, and we have just started to consider the social and political implications of the little we have learned. Psychiatry, therefore, has a role in both wanted and unwanted pregnancies into the foreseeable future.

Bibliography

1. BERNSTEIN, R., and TINKHAM, B., "Group Therapy Following Abortion," *J. Nerv. & Ment. Dis.*, *152*:303–313, 1971.
2. CALDERONE, M. S. (Ed.), *Abortion in the United States*, Hoeber-Harper, New York, 1958.
3. CALLAHAN, D., *Abortion: Law, Choice and Morality*, Macmillan, New York, 1970.
4. DEVEREUX, G., "A Typological Study of Abortion in 350 Primitive, Ancient, and Pre-industrial Societies," in Rosen, H. (Ed.), *Therapeutic Abortion*, Julian Press, New York, 1954.
5. EKBLAD, M., "Induced Abortion on Psychiatric Grounds: A Follow-up Study of 479 Women," *Acta Psychiatrica et Neurologica Scandinavia*, Suppl. 99–*102*:3–238, 1955.
6. FLECK, S., "Some Psychiatric Aspects of Abortion," *J. Nerv. & Ment. Dis.*, *151*:42–50, 1970.
7. FORD, C. V., TEDESCO, P. C., and LONG, K. D., "Abortion—Is It a Therapeutic Procedure in Psychiatry?" *J.A.M.A.*, *218*: 1173–1178, 1971.
8. FORSSMAN, H., and THUWE, I., "One-Hundred and Twenty Children Born after Application for Therapeutic Abortion Refused," *Acta Psychiatrica Scandinavia*, *42*:71–88, 1966.
9. Group for the Advancement of Psychiatry, "The Right to Abortion: A Psychiatric View," *7*:75, 1969.
10. HARDIN, G., "Abortion—or Compulsory Pregnancy," *J. Marriage & Fam.*, *30*:249, 1968.
11. HOOK, K., "Refused Abortion: A Follow-up Study of 249 Women," *Acta Psychiatrica Scandinavia*, 39 Suppl. *168*:1–156, 1963.
12. HULDT, L., "Outcome of Pregnancy When Legal Abortion Is Readily Available," *Lancet*, *1*: 467–468, 2 March, 1968.
13. JANSSON, B., "Mental Disorders after Abortion," *Acta Psychiatrica Scandinavia*, *41*: 108–110, 1965.
14. KAY, D. W., and SCHAPIRA, K., "Psychiatric Sequelae of Termination of Pregnancy," *Br. M. J.*, *1*:299–302, 1967.
15. KENNELL, J. H., SLYTER, H., and KLAUS, M. H., "The Mourning Response of Parents to the Death of a Newborn Infant," *New Eng. J. Med.*, *283*:344–349, 1970.
16. KOLSTAD, P., "Therapeutic Abortion: A Clinical Study Based upon 968 Cases from a Norwegian Hospital, 1940–1953," *Acta Obstetrica et Gynecologica Scandinavia*, Suppl. 6, *36*:72 pp., 1957.
17. KUMMER, J. M., "Post-Abortion Psychiatric Illness—A Myth?" *Am. J. Psychiat.*, *119*: 9982–9986, 1963.
18. LEVENE, H. I., and RIGNEY, F. J., "Law, Preventive Psychiatry, and Therapeutic Abortion," *J. Nerv. & Ment. Dis.*, *151*:51–59, 1970.
19. LINDBERG, B. F., "What Does the Applicant for Abortion Do When the Psychiatrist Has Said 'No'?" *Svenska Lak-Tidn*, *45*:1381–1391, 1948.
20. MEYEROWITZ, S., SATLOFF, A., and ROMANO, J., "Induced Abortion for Psychiatric Indication," *Am. J. Psychiat.*, *127*:73–80, 1971.
21. PATT, S. L., RAPPAPORT, R. G., and BARGLOW, P., "Follow-up of Therapeutic Abortion," *Arch. Gen. Psychiat.*, *20*:408–414, 1969.
22. PECK, A., and MARCUS, H., "Psychiatric Sequelae of Therapeutic Interruption of Pregnancy," *J. Nerv. & Ment. Dis.*, *143*: 425, 1966.
23. PFEIFFER, E., "Psychiatric Indications or Psychiatric Justification of Therapeutic Abortion?" *Arch. Gen. Psychiat.*, *23*:402–407, 1970.
24. ROSEN, H. (Ed.), *Therapeutic Abortion*, Julian Press, New York, 1954.
25. SENAY, E. C., and WEXLER, S., "Fantasies About The Fetus in Wanted and Unwanted Pregnancies," *J. of Youth and Adolescence*, *1*:333–337, 1972.

26. ———, "Therapeutic Abortion: Clinical Aspects," *Arch. Gen. Psychiat.*, 23:408–415, 1970.

27. SIM, M., "Psychiatric Sequelae of Termination of Pregnancy," *Brit. Med. J.*, 1:563–564, 1967.

28. SIMON, N. M., SENTURIA, A. G., and ROTHMAN, D., "Psychiatric Illness Following Therapeutic Abortion," *Am. J. Psychiat.*, 124:64, 1967.

29. SLOANE, R. B., "The Unwanted Pregnancy," *New Eng. J. Med.*, 280:1206–1212, 1969.

30. TAUSSIG, F., *Abortion—Spontaneous and Induced: Medical and Social Aspects*, C. V. Mosby, St. Louis, 1936.

31. TIETZE, C., "Therapeutic Abortion in the United States," *Am. J. Obstetrics & Gynecol.*, 101:784–787, 1968.

32. WHITE, R. B., "Induced Abortions: A Survey of Their Psychiatric Implications, Complications, and Indications," *Texas Reports on Biology and Medicine*, 24:531–538, 1966.

33. WHITLOCK, F. A., and EDWARDS, J. E., "Pregnancy and Attempted Suicide," *Comprehensive Psychiat.*, 9:11, 1968.

34. WHITTINGTON, H. G., "Evaluation of Therapeutic Abortion as an Element of Preventive Psychology," *Am. J. Psychiat.*, 126:1224–1229, 1970.

CHAPTER 27

SEXUAL FUNCTIONS IN MEN AND THEIR DISTURBANCES

Harold I. Lief

SEPARATING MALE FROM FEMALE SEXUALITY is an exercise in abstraction, perhaps even in frustration. Ordinarily one's sexuality is so much fashioned by the relationship between the sexes and its understanding so dependent on our appreciation of *the process of relating* that to study the sexual function of one's sex apart from the other is bound to be misleading. My assignment to deal with male sexuality requires me to concentrate my remarks on that sex, but of necessity constant reference will be made to the transactions between the sexes, especially since most sexual disturbances arise in the context of marriage or some other heterosexual relationship.

A frame of reference is necessary. Some general remarks about human sexuality will be followed by a discussion of the influence of culture on sexual behavior, after which sexuality as a system will be conceptualized and its components further defined in order to set the stage for a discussion of male disturbances in sexual functioning.

Introduction

Human sexuality, including sexual feelings and behavior, is so diverse and complex, its manifestations so protean, that its understanding requires the insights of artists as well as scientists and clinicians. Feelings include multiple bodily sensations, fantasies, emotions, and attitudes. Attitudes, in turn, are composed of belief systems and their related values. Important facets of sexual attitudes are linked with the image of oneself as male or female, masculine or feminine, even combinations of these identifications and identities. Highly significant as well are other internal states such as sexual drive or interest and the capacity for erotic arousal and responsivity. Sexual behavior goes far beyond "physical" sex, for it includes as well sex-typed behavior that not only varies enormously among individuals but also varies in different cultures, or in the same society in different historical periods.

Cultural Variations

Variations in sexual values and behavior in different historical periods in different cultures have been strikingly described by Marmor.[29] Attitudes toward masturbation, premarital sex, infidelity, abortion, homosexuality, and the relations between men and women are so different that they may change the sexual "climate of opinion" drastically.

> Even a cursory look at the recorded history of human sexuality makes it abundantly clear that patterns of sexual behavior and morality have taken many diverse forms over the centuries. Far from being "natural" and inevitable, our contemporary sexual codes and mores, seen in historical perspective, would appear no less grotesque to people of other eras than theirs appear to us. Our attitudes concerning nudity, virginity, fidelity, love, marriage, and "proper" sexual behavior are meaningful only within the context of our own cultural and religious mores. Thus, in the first millennium of the Christian era, in many parts of what is now Europe, public nudity was no cause for shame (as is still true in some aboriginal settings), virginity was not prized, marriage was usually a temporary arrangement, and extramarital relations were taken for granted. Frank and open sexuality was the rule, and incest was frequent. Women were open aggressors in inviting sexual intercourse. Bastardy was a mark of distinction because it often implied that some important person had slept with one's mother. In early feudal times new brides were usually deflowered by the feudal lord (jus primae noctis). In other early societies all the wedding guests would copulate with the bride. Far from being considered a source of concern to the husband, these practices were considered a way of strengthening the marriage in that the pain of the initial coitus would not be associated with the husband. [p. 165][29]

Malinowski,[28] Devereux,[7] Mead,[32] Kluckhohn,[24] Linton,[27] DuBois,[8] Kardiner,[19] Opler,[36] Ford and Beach[9] were among the behavioral scientists who demonstrated that sex roles were dependent on childrearing practices, which, in turn, were based on cultural institutions and their accompanying values.

An extraordinary variety in human sexual behavior exists. To quote Karlen,[20] "there are societies where modesty calls for hiding body and face, others where it insists that the male hide only his glans penis; where widows commit suicide, and where women have several husbands; where girls commence coitus at 11, and where they and their lovers are put to death for premarital intercourse; where relatively few women masturbate, and where they do so with a reindeer's leg tendon or with a live mink whose jaws are tied shut; where every male has experienced sodomy at some time in his life, and where one homosexual act may cause ostracism or even execution." (p. 475).

Despite this, several generalizations are possible: (1) no society exists in which there is unlimited sexual access to most potential partners; (2) some form of incest barrier is always found, involving at least some members of the nuclear family; (3) heterosexual coitus is the standard pattern for adults everywhere; no society makes homosexuality, masturbation, bestiality, or any other noncoital form of sexual activity the dominant form for adults.

Sexuality as a System

If sexuality refers to the totality of one's sexual being, its essence is even more what one "is" than what one "does." One's sense of being male and masculine, female and feminine and the various roles these self-perceptions engender or influence are important ingredients of sexuality. Cognitive, emotional, and physical sex each contribute to the totality.

Sexuality may be described in terms of a system analogous to the circulatory or respiratory system. The components of the sexual system include:

1. Biological sex—chromosomes, hormones, primary and secondary sex characteristics, and so forth
2. Sexual identity (sometimes called core gender identity)—sense of maleness and femaleness
3. Gender identity—sense of masculinity and femininity
4. Sexual role behavior—(a) gender behavior, behavior with masculine and feminine connotations; (b) sex behavior, behavior motivated by desire for orgasm (physical sex)

Biological Aspects

Chromosomal Abnormality

In the vast majority of children gender based on biological sex is unambiguous. However, abnormalities of the sex chromosomal patterns (or of hormonal secretions) result in mixed internal reproductive systems and ambiguous genitalia. These problems of intersexuality, if not corrected early in life, may lead to conflicts in sexual or gender identity. The sex assigned is, with rare exceptions, more important in determining gender identity than is biological sex, and change in sex is usually unwise after eighteen months.

Some of these people and some without evident biological defect have grave difficulties in developing a core gender identity of the same biological sex. These are transsexuals, mostly biological males who think of themselves as "a female trapped in a male body."[42]

Effect of Fetal Hormones

In the last decade particularly, evidence has been accumulating to indicate that fetal androgens, mainly testosterone, determine the organization of neural tissues that mediate sexual behavior.[38] If fetal androgen is absent during a critical period of fetal development, the fetus will develop as a female. However, the presence of androgen in the proper amounts during the critical period masculinizes the individual, whether genetic male or female. The "organizing action of testosterone" establishes a neural system so that differential sensitivity and responsiveness are built into the controlling mechanisms. The evidence seems to suggest that the hypothalamus plays a leading role in the control mechanisms. In normal development the genetic factors structure the endocrine environment, which, in turn, affects the psychosexual bias of the nervous system. Instead of having a bisexual constitution as theorized by Freud, the developing fetus, if there are no genetic or hormonal abnormalities, is organized in a male or female direction at very early stages of development. "The genetic code of XX or XY apparently determines only the differentiation of ovary or testes, after which fetal hormones from either gonad take over further differentiation."[11]

As Stoller[43] puts it, "The genital anatomic fact is that, embryologically speaking, the penis is a masculinized clitoris; the neurophysiological fact is that the male brain is an androgenized female brain."

Sexual Identity— Core Gender Identity

With relatively rare exceptions development of sexuality along usual developmental lines leads to a secure sense of maleness or femaleness that is generally complete by the age of three. If there are difficulties created by intersexuality and an ambiguous sex assignment, difficulties in sexual identity may be a consequence. Transsexualism, not always associated with known biological defect, is a special case of conflict over sexual identity. Even in ambiguous sexual development the sex of rearing can be established independently of biological sex. In the vast majority of cases the assignment of sex during the first years of life will prevail, even overriding an opposed biological sexual identity.[34] (In some rarer instances the opposite seems true; a disordered neural organization will lead to cross-sex attitudes and behavior despite unambiguous sex rearing.)

Of special interest is the implication from animal studies that "maleness and masculinity are more difficult to achieve and more vulnerable to disruption"[11] than are femaleness and femininity. Among primates male copulatory behavior is inhibited more readily by extraneous stimuli and is more subject to the effect of previous experience. Males, more than females, have to learn to copulate, and this capacity among males is more readily destroyed by cerebral cortical ablation experiments. In the experiments conducted by the Harlows,[16] the lack of peer group social and sexual play differentially affected the sexes, so that males, even more than females, failed to develop the capacity for adult sex behavior and successful copulation.

Gender Identity

Money and the Hampsons[34] defined gender identity as "all those things that a person says or does to disclose himself or herself as having the status of boy or man, girl or woman respectively." The unfolding of gender identity is affected mostly by the cultural and familial factors provided by all the transactional experiences, explicit and implicit, planned and unplanned, encountered in childhood and adolescence. Indeed, the process of gender identity continues into adult life, creating modifications of self-image through a variety of nonsexual as well as explicitly sexual encounters.

The inevitable stages of psychosexual development (oral, anal, phallic, and genital) postulated by Freud[10] are only gross approximations of the usual maturational processes, the nuances of which are very different in different individuals even in the same culture. For example, personality development during the anal stage is not related to an alleged anal eroticism, but, instead, is related to the nature of the interaction of the child with his parents and the conflict between submission to and defiance of parental demands for sphincter control. Similarly the latency period disappears in segments of society or in cultures where there is little sexual repression and perhaps even active encouragement of sexual, even coital, activity.

For the creation of the standard sexual identity, the child must have a parent or parent substitute of the same sex who is neither so punishing or weak as to make it impossible for the child to identify with him; a parent or parent substitute of the opposite sex who is neither so seductive, punishing, emotionally erratic, or withholding as to make it impossible for the child to trust members of the opposite sex; and parents who do not systematically reject the child's biological sex and attempt to force him into behavior more in keeping with the opposite sex.

Sexual Role Behavior

GENDER IDENTITY

The stages of psychosexual development postulated in classical psychoanalytic theory have been criticized on these grounds:

1. The notion that the stages are an inevitable unfolding of an instinctual energy within the child, suggesting that the child has the initiatory capacities within himself, is erroneous. It seems more likely that the initiation of sexual transactions comes from his parent.
2. This formulation fails to take into account the enormous influence of culture and its institutions in modifying early sexual learning.
3. The assumption that all contacts with or stimulation of the child's end organs have either a protosexual or completely sexual meaning right from the start ignores the quite different motivations surrounding oral or anal behavior. These criticisms should not obscure the importance of Freud's discovery that the child's sexual activity and interests were "an essential precursor and component of the development of the character structure of the adult."[12]

During the period three to seven years of age, the child gradually realizes that people are placed in two related categories, boys or girls, men or women, fathers or mothers. Certain cultural values and attitudes modifying masculine and feminine behavior are transmitted through clearly distinguishable cues including dress, bodily form and proportion, strength, distribution of hair, depth of voice, posture at the toilet, differential parental behavior with boys and girls, and characteristic sex-linked behavior in the kitchen, the garage, or the backyard. Physical differences between men and women, boys and girls, play a large part in the concept of masculinity and femininity learned at an early age; for example, a girl should be pretty and small, a boy large and strong. By the time the child is eight or ten, the primary sex-typed attribute for girls is having an attractive face, while for boys it is having a tall, muscular physique.

Although it is possible, as Margaret Mead[32] has demonstrated, for a culture to instill aggression in females and passive dependency in males, most cultures, including our own, promote aggressivity in the male and a passive-receptive stance in females. Many studies have demonstrated the connection between maleness and aggression, and the evidence is almost as strong for a greater dependency, conformity, and social passivity for females than for males at all ages.

There are class variations. Sex-role differentiation is sharper in lower-class families.[18] Lower-class mothers encourage sex-typing more consistently than do middle-class mothers, and the difference between middle class and lower class is especially sharp for girls. More flexibility is afforded the middle-class girl than the lower-class girl in choosing toys and in undertaking activities of the opposite sex.

Therefore, despite the rapid shift in societal values with regard to masculine and feminine behavior, children still continue to act as if aggression, dominance, and independence are more appropriate for males and passivity, nurturance, and feelings more appropriate for females.

In the development of sex-linked behavior, it is of interest that the girl must acquire reactions from other people since she cannot know whether she is attractive or poised or passive without frequent feedback from her environment. This promotes her dependency upon other people. The boy, on the other hand, develops many important sex-typed behaviors while alone. He can perfect his motor or mechanical skills by himself (shooting baskets, fixing his bicycle), and these will strengthen his convictions that he is acquiring masculine attributes.

The pressures from society and from families for stereotyping sex-linked behavior are enormous. The studious or artistic boy and the mathematically inclined girl may suffer humiliation at the hands of his or her family or peers. These stereotypic attitudes and values learned early in life generally persist into adult life and often come into sharp conflict with changing societal values, especially with regard to the role of women. Traditional masculine-feminine role behaviors, creating a sharp differentiation between men and women, are changing so rapidly that there is hardly a job or task that is completely absorbed by one sex alone. Sexual roles are now no longer assigned by tradition but are negotiated, creating a potent source of conflict between husband and wife. It is clear that efforts at reducing the perhaps excessive polarization of males and females have to start in the early stages of personality development.

Despite the changes, given our still current cultural emphases, by the time a boy is an adolescent a satisfactory sexual identity for him involves sexual experiences with girls. Thus, much sexual activity among males is as much motivated by the desire to strengthen a masculine identification as by strong erotic urges. For both boys and girls some concerns about their sexual identity and consequent inhibitions in sex-linked behavior seem inevitable. Whether this carries over to the area of physical sex itself depends on the intensity of anxiety, guilt, or hostility toward the opposite sex and on the types of reinforcing or extinguishing (corrective) stimuli the child receives during adolescence.

¶ Sexual Behavior

Sexuality in Infancy and Childhood

Infants of both sexes seem to experience pleasure from the stimulation of the genitals and other areas of the body that are commonly recognized as erogenous zones. In the early weeks of life male infants respond to internal or external stimuli with erections, and even at the time of delivery erections in the newborn are seen. This indicates the presence of a built-in sexual reflex. In adult life the basic biological nature of the reflex sexual response is seen in penile erections (and vaginal lubrication) during REM sleep. Orgasmic experiences without ejaculation occur throughout the preadolescent period. A study[23] of 700 four-year-olds in an English urban community found that 17 per cent of the children engaged in genital play at this age. Curiosity leads most children to engage in exploratory activities involving themselves and other male or female children. When children are scolded or obviously distracted from such activities, children often experience their first deep-seated feelings toward sex, namely, that sex is threatening and that sexual expressions are to be regarded with guilt and shame.

Despite Freud's notion that there is a period of latency—a phase of the child's psychosexual life and development when sexual interest and behavior come to a halt—evidence clearly shows that there is a steady rise in sexual activity involving nearly 80 per cent of preadolescent boys between 10 and 13.[3] At most, between the ages of 5 and 11, there is a relative decline in protosexual or sexual interest. During this same age period there is an enormous broadening of the child's range of interests. As the child learns many things about the world in which he lives and as he increases his contacts with his fellows, curiosity about his family, including intense sexual curiosity, abates. He leaves the protective canopy of his family for the wider world, but his sexual interests are still there, somewhat masked by his involvement and commitment to the exploration of his milieu. If that environment is stagnant or punitive or otherwise inhibits his move outward to enlarge his horizons, the child's sexual curiosity and behavior may become active, even florid. (In a St. Louis housing project prepubertal children actively engaged in coitus or in attempts at coitus. All around them older children and adults were engaged openly in sexual encounters of all types, in the hallways, elevators, lobbies, laundry rooms. Mothers even encouraged children to engage in adultlike sex play, for it kept them off the streets and out of trouble with the police.)[15] Broderick[3] and others have demonstrated that there is no period in which the majority of boys and the great majority of girls are not interested in the opposite sex. Four out of five pubertal boys and even a larger number of girls have fantasies of getting married someday.

During the prepubertal period, approximately from the ages of 9 to 12, boys are commonly interested in their own sex. Many of them go through a period of "hating girls." This is probably a time in which they are consolidating their gender identity, and it is probably a mistake to call this a "homosexual stage." There is a danger that parents will regard this same-sexed sexual activity as homosexual, creating a good deal of anxiety on the part of the boy who is participating in mutual sex.

Puberty

At puberty the increase in hormonal secretions brings with it the capacity for erotic arousal and responsivity, eventually leading to ejaculation in males. Preoccupation with sex causes many concerns at this time. Boys are concerned with masturbation or with what they assume to be a small penis.

Commonly puberty arrives for boys about two years later than for girls (11 to 18 for boys and 9 to 16 for girls). Whether the age of puberty has actually been getting lower for both sexes over the past half-century is still a controversial matter. There is a definite physical maturational sequence. In boys the order of pubescent phenomena is: (1) beginning growth of the testes and penis, (2) appearance of straight pigmented pubic hair, (3) early voice changes, (4) first ejaculation, (5) kinky pubic hair, (6) period of maximum growth, (7) auxillary hair, (8) marked voice changes, and (9) development of the beard.

In early adolescence masturbation becomes a central concern for most youth, especially for the white middle-class male. Masturbation is part of the standard sexual pattern and becomes a problem only if there is guilt or anxiety associated with it. While masturbation is now accepted as a perfectly natural part of sexual development, for the first time its positive features are being stressed. Aside from the pleasurable release of sexual tensions, masturbation leads to increased information about the adolescent's bodily responses and to a developing sense of mastery over newfound sexual capacities, and it helps in preparing for heterosexual relationships.

The cumulative incidence of masturbation is more than 90 per cent for males and above 60 per cent for females.[21,22] Most people who masturbate as adolescents continue to do so in adult life, at least occasionally. For many individuals masturbation constitutes virtually the only overt outlet.

In about three-fourths of males and about half of females, masturbation is accompanied by fantasies or daydreams, but with a difference between the sexes in the type of fantasy. Females usually fantasize about doing nonsexual things with someone who is admired, whereas males are more likely to have direct coital fantasies with someone of whom they are fond. Persons with more education fantasize more frequently; yet they are absent in about 12 per cent of college men and about 20 per cent of college women. Women are much less likely than men to fantasize having sex with a stranger or in a group. On the other hand, fantasies of being forced into sex are more common in women (rape fantasies are a way in which the woman decreases her sense of responsibility and of guilt).

Although about one out of three adolescent boys has a homosexual experience leading to orgasm (and about half that percentage of girls), in the majority this is a transient stage in developing their own gender identity. About 10 per cent of males and about 4 per cent of females become preferentially homosexual. A great majority of adolescent boys and girls find their dominant push toward heterosexual activity. Working out their gender identity, gaining control over their impulses, and learning how their bodies function are more important in the early phases than release of sexual tension. In this culture girls are more interested in interpersonal relations than in sexual gratification. Embarking on sexual activity, boys and girls commonly go through a series of stages involving necking, light petting, heavy petting (petting below the waist), petting to orgasm, and finally coitus. The standard pattern for high school seniors of petting to orgasm gradually shifts to coitus sometime during their college years. A minority of high school boys and girls have coital experience (the differences among social classes are highly significant), but the majority have engaged in coitus by the time they finish college. Studying a representative sample of the population (not restricted to college youth), researchers at Johns Hopkins[6] found that 44 per cent of 19-year-old unmarried girls reported having had sexual intercourse. The major shift over the past 30 years is in the sexual behavior of girls. A generation ago about 30 per cent of senior college girls had premarital intercourse. Now over 50 per cent report "going all the way." The predominant sexual norm both for high school and college students is "permissiveness with affection."[40] If affection is present there is little stigma attached to coital experience and behavior compared to the adolescents of a generation ago.

The complicated process of developing the capacity for intimacy, one of the major tasks of adolescence, involves the capacity for intimate sharing of pain as well as of pleasure with a loved person, in addition to the capacity to trust, to value fidelity, and to permit the loved person to develop his own unique style of living. With all the pressures impinging on the adolescent that involve his developing sense of self-reliance and sense of masculinity, it is often difficult for him to put aside his own satisfactions sufficiently long to develop genuine intimacy. This is a process that has its beginnings in early life, becomes a significant factor in adolescence, and continues with many vicissitudes into adult life. Although sex can be pleasurable in the absence of love or of intimacy, for most people sex is greatly enhanced in the presence of an intimate concern with another person who is returning affection and tenderness.

An intimate relationship usually leads to marriage. However, more and more young people are living together outside a formal marriage arrangement. The number of people below the age of 35 remaining single has gone up significantly between 1960 and 1970. There is greater concern with fidelity than with permanence. Nonetheless, over 90 per cent of people eventually marry.

Masters and Johnson[30] estimate that over half the married couples in the United States have a significant sexual problem, and they are referring only to sexual effectiveness and competence, not to conflicts over sex-linked behavior, conflicts over differing perceptions about what is appropriate for a husband and for a wife. The battle of the sexes more often involves battles over sexual roles than conflicts over foreplay and coitus. Nevertheless, as indicated above, difficulties in sexual performance are very frequent; indeed, they probably affect every couple at least at some point in their marriage.

Human Sexual Response

The four phases described by Masters and Johnson,[31] excitement, plateau, orgasm, and resolution, appear to be similar in males and females. One important difference is that men have a refractory period in which they are unable to repeat the ejaculation without an interval of some minutes at least, whereas females can have orgasms following each other at intervals of a few seconds.

Prior to the initial stage of excitement is a preparatory stage of sexual psychic arousal, sometimes called "the sexual motive state," a psychic readiness for sexual activity. It is affected by many factors such as the partner, the setting, the mood, and the positive or negative effect of emotion (love, anxiety). The details of the four stages of human sexual response have been described thoroughly by Masters and Johnson.

Sexual relationships not only can promote intimacy but also can serve as a means of satisfying nonsexual needs such as the need for reassurance, playfulness, companionship, and so forth. Unfortunately sex is often the setting for aversive feelings such as anger, revenge, dominance, and self-aggrandizement. Angry feelings may make sexual activities a veritable battleground for the expression and reception of enraged attacks. If these negative emotions are coupled with some degree of sexual incompetence, marital disharmony is assured. Thus, marital sexuality can be said to be bipotential, since it can serve as the expression of the most significant, tender, affectionate, and intimate feelings between two people or it can serve to release feelings of hate, revenge, and punitiveness.

Generally, a decline in sexual interest and in frequency of sexual intercourse takes place through the life cycle. In the forties the average frequency is closer to two times a week, whereas early in marriage it is about four times a week. Nonetheless, as people get older they can continue to have an active and satisfying sexual relationship. Over 50 per cent of men in their sixties and at least 25 per cent of men over 70[37] continue to have satisfactory sexual experiences, even though the male has to depend much more on direct stimulation than on psychic stimulation as the threshold for his "sexual motive state" increases with age. Since his ejaculatory needs have diminished, he is better able to control his "staying power," slowing down his ejaculation far more effectively than at an earlier age. Sexual capacity in older people depends on continued sexual expression and interest through middle life and the availability of a willing and interested partner.

In middle age many couples report waning sexual interest. This is either a reflection of a lack of variety and the routinization of sexual behavior or a reflection of other aspects of their relationship. Sexual gratification is more often a barometer of the total relationship than a result of an insufficient variety of partners or sexual behavior. If open communication and mutual participation in a range of interests have gone on throughout marriage, sex is likely to be fulfilling and enhancing. Clinics dealing in marital therapy report that about 75 per cent of couples coming for counseling have a significant sexual problem, but only 15 per cent have a sexual problem that is a primary cause of marital disharmony. In the other 60 per cent sexual problems are a consequence of disharmony in other areas of their relationship.

Areas of Sexual Dysfunction

Major disturbances of sexual functioning such as homosexuality, transvestitism, voyeurism, exhibitionism, fetishism, and sadomasochism are covered in other volumes of the *American Handbook of Psychiatry*. This chapter is meant to deal with more common vicissitudes of sexual functioning, most of which occur within the context of the marital or other sex-pair relationship. This chapter will contain discussions of impotence, premature ejaculation, ejaculatory incompetence, paradoxical orgasm, and, in addition, certain disturbances in relationships (created by ignorance or the influence of anxiety, guilt, or rage), conflicts over sex-linked behavior, difficulties in communication and perception, conflicts over frequency of sexual relations or choice of methods of sexual stimulation (such as oral-genital sex and coital positions), conflicts over the emotional components of sexuality, and infidelity. Finally this chapter will deal with special problems of the single male such as the fear of intimacy, the fear of marriage, and performance anxiety.

Comments on History Taking and Interviewing

Skills in history taking and interviewing are, of course, related to a person's general skills in interviewing. One must learn to be a good interviewer before one is competent to talk to a patient with a sexual problem. The usual dimensions of interviewing apply here as well as elsewhere. The interviewer should follow the patient's leads and not unduly structure the interview so as to cut off information and the flow of affect. Generally he should try to use nonstructured questions and "bridges" when changing topics. Obviously, here as elsewhere, the greater his mastery of the field the greater will be his competence in taking a history and in interviewing during following sessions. In general, the interviewer must develop a style that fits his own personality.

In taking a sexual history or in interviewing a patient with a sexual problem, special skills are required. Since sex evokes so many highly charged feelings, and anxiety and embarrassment are such frequent occurrences, the interviewer's attitude and manner are all-important. If he is uncomfortable he will communicate his discomfort to the patient whose embarrassment will increase. Conversely, if the interviewer is comfortable and relaxed, the patient will soon overcome his own anxiety. If the patient is clearly inhibited about discussing sexual matters, the therapist must take the initiative. Sometimes the decision whether to probe or wait until the sexual material comes up in a more appropriate context after the patient has worked through some of his resistances is a difficult one, and the interviewer will have to depend on his experience to make such a judgment. More often than not, gentle and tactful questioning turns out to be the most effective means of eliciting information and associated feelings.

In the opinion of many psychiatric educators, residents in psychiatry are not exempt from the almost universal anxiety and embarrassment found in other house officers in dealing with sexual problems of patients. Most of them have never learned this kind of interviewing when they were medical students, and little is done during residency training to teach the management of patients with sexual problems. As a consequence anxiety about one's lack of competence ("competence anxiety") occurs frequently. The only cure for this anxiety is to carry out interviewing under close supervision until the psychiatrist's growing skills make him more confident of his ability.

Sexuality is an area in which the physician's own unresolved problems can create particular countertransference difficulties. For example, he may be inhibited about questioning because of his own unconscious voyeurism, or his guilt about certain facets of sexuality may create much discomfort. Anxieties about potential homosexual behavior may create a blind spot in dealing with related aspects of sexual attitudes and behavior. Anxieties about his own sexual performance may lead either to inhibitions in interviewing or to a counterphobic brusqueness and aggressivity. His own sexual standards and life style may be in contrast to those of his patient, yet he has to learn not to be unduly influenced by his own sexual morality and preferences. One should not impose one's values on the patient. Condemnatory feelings about premarital or extramarital relations will certainly damage the relationship and interfere with communication between doctor and patient. Attempting to "liberalize" a patient before he is ready for the suggested behavior may be just as injudicious. As in any psychotherapeutic encounter, timing is all-important—timing of questions as well as of suggestions.

In general, there are certain technical maneuvers or "gambits" the interviewer can follow. He can move from less to more highly charged areas—for example, from a discussion of marital relations to sex within marriages; from a discussion of wet dreams to masturbation; from a discussion of dating to petting to intercourse. He can discuss the ubiquity or normalcy of behavior as a way of broaching the subject—for example, "The majority of married people engage in oral-genital sex, so tell me your feelings about this." One can assume normalcy when asking about masturbation or other facets of sexual behavior. Instead of asking "Did you ever masturbate?" it is more effective to ask "How young were you when you first masturbated?" Another help in sexual interviewing is to ask about attitudes before behavior. The physician may say, for example, "Research tells us that the majority of married men have one or more extramarital relations. What are your feelings about that?" It is usually easier for the patient to talk about his attitudes than about his personal experiences. Another general technique is to learn what expectations the patient may have entertained before actually undergoing certain experiences. "What were your expectations on your honeymoon?" followed by "How did your actual experiences match your expectations?"

The content of a sexual history can be subdivided into the following categories: identifying data, childhood sexuality, which should include family attitudes about sex, learning about sex, childhood sex activity, primal scene and childhood sexual theories or myths, the onset of adolescence, orgastic experiences before and after marriage, feelings about oneself as masculine (or feminine), sexual fantasies and dreams, dating, engagement, marriage, including premarital sex with partner, the wedding trip, sex in marriage, extramarital sex, sex when divorced or widowed, sexual deviations, certain effects of sex activities such as venereal disease or illegitimate pregnancy, and the use of erotic stimuli. It may be helpful to use a structured questionnaire such as the Sexual Performance Evaluation, a form developed at the Marriage Council of Philadelphia. This questionnaire permits the patient to answer questions about his own and his spouse's perception of a variety of sexual behaviors.

There are special advantages to interviewing husband and wife together, a technique called "conjoint interviewing." The physician has the opportunity of observing their interaction firsthand by the way they look at and talk with each other as well as by their bodily movements. In these cases he gets important information about their feelings and attitudes toward each other. Facial expressions and movements of the body may indicate concern, protectiveness, disdain, anger, and so forth. The physician is able to watch how a husband and wife attempt to control their relationship. If a wife complains that her husband gives her little affection, one can observe very directly whether the husband has the capacity to express affectionate feelings or even angry feelings.

In every marital situation there is a perceptual system with eight dimensions. The husband has a perception (1) of himself, (2) of his wife, (3) of his wife's perception of him, (4) of the marital relationship. This is likewise true for the wife, who has the complementary perceptual system. The interviewer is able to check this perceptual system and get immediate feedback. To take one example, the therapist may ask the wife, "How do you think your husband feels about the way you respond to him in bed?" (Her perception of his perception of her.) She will respond and the interviewer can then turn to the husband and say, "How does this fit in with the way you feel about her sexual responses?" (His perception of her.) In this fashion one can get immediate feedback about any of these perceptual areas or about communication in general. It is sometimes helpful to have a spouse repeat back what the other one has just said in order to make certain that one of them is not being "tuned out" by the other.

When one therapist is working with both the husband and wife, the usual format is to take a short sexual history from the husband and wife together to get some notion of their interaction and then to interview each of them separately, either on the same occasion or at the next interview. If a dual sex therapy team is doing the interviewing, as recommended by Masters and Johnson,[30] it is customary to have the husband interviewed by the male therapist and the wife by the female therapist separately and then switch at the next session so that each marital partner is being interviewed by the therapist of the opposite sex. Conjoint interviewing then takes place at the third session.

The use of a dual sex therapy team has many advantages. Better understanding is assured by the identification between the therapist and the patient of the same sex. Often the therapist of the same sex can articulate the feelings of that patient far more effectively than the two members of the opposite sex in the room. As in any group therapy situation, the co-therapists can aid each other in multiple ways such as clarification of points, making observations that the other therapist has missed, checking misdirected interventions by one therapist, and so forth. In addition, both transference and countertransference responses are diminished by dual sex conjoint interviewing.

Diagnosis or Appraisal

The therapist needs to define the sexual problem as precisely as possible. More than that, he must be able to appraise how the sexual problem fits into the fabric of the patient's life. The man who comes in complaining of impotence and wants treatment for this dysfunction presents a different problem to the therapist than does the patient who comes in for treatment of his depression, but who discloses occasional episodes of impotence when he suffers a blow to his self-esteem.

Because one's sexuality is so interwoven with one's interaction with others, it is rather difficult to categorize sexual problems without diminishing the unique and subtle aspects of an individual's life style. Yet some gross categories are aids to conceptualization. Sexual problems are either overt and out in the open in the initial interview or soon after, or they may remain masked by a variety of other symptoms such as depression, phobia, anxiety attacks, obsessive ruminations, and bodily complaints. Sexual problems may be a consequence of difficulties in the relationship, or they may have arisen almost entirely from difficulties prior to the relationship, especially influenced by early life experiences. Of course, the relationship may augment or exaggerate potentialities for dysfunction that the patient brought into the relationship. With some patients the sexual problem is critical in the sense that not only does it serve as an important source of unhappiness and lowered self-esteem, but also it is a significant threat to an important relationship. With other patients the sexual problem seems to be less significant and takes second place to other difficulties in adaptation and relationships. Further complications are created by different perceptions of the problem by the partners. A 37-year-old schoolteacher came for help because of impotence. Indeed, there had been no satisfactory coital experience for seven years. When his wife was interviewed, it turned out that she was less concerned about the sexual dysfunction than about her husband's lack of interest in her needs, his "selfishness" and passivity. It is usually impossible to treat the sexual problem without dealing with the total marital or sex-pair relationship.

In making the appraisal, the history can be organized around the following considerations: (1) the circumstances under which the symptom first occurred, (2) the patient's reaction to the symptom, (3) the wife or partner's reaction to the symptom, (4) the nature of the marital interaction, (5) situational components (the symptom may appear only at times or may occur only with the wife or with a mistress).

Based on the appraisal, the therapist must decide whether the patient requires conjoint marital therapy or individual treatment. These considerations will be taken up more thoroughly in the section on Therapeutic Considerations.

Impotence

The impotent man cannot achieve and/or maintain an erection sufficient for him to penetrate the vagina and maintain successful coitus. In *primary impotence* the man has never been able to complete coitus satisfactorily because of his failure to achieve an erection. In *secondary impotence* the male's rate of failure to complete successful sexual intercourse approaches 25 per cent of his opportunities, according to an arbitrary definition of Masters and Johnson.[30]

Primary Impotence

This condition is relatively rare. Masters and Johnson treated only 32 males with primary impotence over an 11-year period. The dominant feature in these men is the association of sex with sinfulness and dirtiness. Guilty fear is the "emergency" emotion found most frequently. There may have been frequent sexual, although not necessarily coital, experiences with the mother during adolescence or intense restrictions may have been placed on the adolescent boy during his dating and courting period, resulting in an almost complete absence of sexual encounters with girls. When some encounters take place, there is such a feeling of awkwardness resulting from his fumbling efforts at sexual contact that anxiety about his competence is added to the underlying feeling of wrongdoing. For their mates these men tend to select virginal wives whose own inexperience augments the male's awkward efforts. In some men a homosexual predilection, either overt or covert, leads to a failure in erection with a heterosexual partner. Humiliating and degrading experiences with prostitutes during the initial attempts at sexual intercourse are other contributing causes. Occasionally initial coital efforts under the influence of alcohol or drugs leaves a highly vulnerable young man with the impression that he is totally incompetent sexually. Initial failures, whatever their underlying cause, are then intensified by the strong anxiety about adequate performance.

Secondary Impotence

Over 90 per cent of cases of secondary impotence are psychogenic, yet the therapist must be aware of the possibility of some defect in the machinery of the body. The differentiation can be made readily. If a patient is able to have an erection from any source, whether from foreplay or during masturbation, or if he wakes up with morning erections or is aware of erections during nighttime dreaming, it is clear that there is no neuromuscular or circulatory disorder affecting his capacity to have an erection.

SYSTEMIC DISEASE

Of all the systemic diseases that are associated with secondary impotence, diabetes mellitus is the most frequent. Occasionally secondary impotence is one of the earliest symptoms of diabetes. The reasons for the relationship between impotence and diabetes are not clear, since one does not always find neuropathy in such cases. Other systemic illnesses in which impotence occasionally occurs are syphilis, multiple sclerosis and other degenerative diseases of the spinal cord, and even more infrequently endocrine dysfunction such as hypopituitarism and hypothyroidism. Infections and intoxications may produce temporary secondary impotence. Weakness and fatigue seem to be the significant factors.

LOCAL DISEASE

Local disease such as phimosis may, by its associated pain on erection, cause secondary impotence. The effects of castration are variable, although most adult males will not be troubled by impotence following castration. The effects of prostatic disease and prostatectomy are also variable, although again the vast majority of patients following prostatectomy do not demonstrate impotence. The specific method of prostatectomy does play some role in that there are more cases of impotence following the perineal method. However, impotence is not an inevitable consequence of even radical surgery.

Drugs

Many drugs taken in excessive quantities may cause impotence. This is particularly true of alcohol and the opiates. Even sedatives and tranquilizers may be implicated. While small doses of amphetamines may delay ejaculation in some men, larger doses may cause either impotence or ejaculatory abnormalities. The same is true of the psychoactive tranquilizers such as Mellaril.

Aging

As has been indicated earlier, aging is not an inevitable cause of impotence. If an active sexual life has been maintained through middle age and there is a willing and cooperative partner, men are often capable of having adequate erections in their seventies or even eighties. Because of the decrease in ejaculatory need, premature ejaculators in early years find that this disturbance disappears in midlife. The decreased need to ejaculate also affects the arousal state of the male. He is more dependent on local stimulation than he was in his early years, and psychic factors have less capacity to bring on quick arousal. A man may have satisfactory sexual intercourse without ejaculation, and if his partner understands this and does not demand ejaculation as part of a "normal" coital experience, the man's increased control may result in highly satisfactory coital experiences.

History Taking and Interviewing

The precise details of secondary impotence must be elicited. The difficulty may be situational, occurring at certain times in certain situations or with certain partners. It may follow a pattern in that the occurrences may follow some competitive defeat in work or in athletic competition. They may occur following angry encounters with the partner, and the anger itself may be a consequence of a specific pattern of interaction between the man and his wife. If it occurs, for example, almost every time the wife has been flirtatious at a cocktail party, this factor and its psychodynamic significance needs to be clarified. Sometimes impotence occurs only with ingestion of too much alcohol or drugs.

After establishing the pattern the history taking should be organized around the topic suggested earlier.

Circumstances under which Secondary Impotence First Occurred. The first episode of impotence often occurs after excessive ingestion of alcohol or attempted coitus when the patient is excessively fatigued, preoccupied, or distracted by some interruption of sexual activity such as childrens' voices, unexpected telephone calls, and the like. Other important causes in the first episode are anger at his wife, the failure of his wife to respond, or guilt toward his wife, perhaps for some extramarital relationship, real or fantasied. Occasionally the first episode occurs because of something that has nothing directly to do with the patient's sexual life. It may have followed a failure in work or in social life. Some humiliating experience outside the home may be responsible for the first episode. The interview should be directed toward determining as precisely as possible the circumstances surrounding the first episode.

The Patient's Reaction to the Symptom. The degree of embarrassment this causes varies from man to man. Sometimes it is so intense that the man's entire life is affected. He may become seriously depressed and unable to function. At any rate the most common response is a redoubled effort to perform properly, which in itself augments the problem since one cannot will an erection any more than one can control his breathing for any length of time. The more the man concentrates on his performance, the less able he is to have an erection or to maintain it. The other frequent response is an increasing avoidance of sexual encounters in order to avoid embarrassment.

The Wife's Reactions. Sometimes the wife is understanding, sympathetic, and supportive. This makes treatment much easier. However, more frequently the wife is frustrated and angry. She may increase her demands on the husband perhaps thinking that she is losing her attractiveness and wishing reassurance, or she may withdraw from sexual activity in order to avoid the frustration or to protect her husband from his own embarrassment and humiliation. At any rate an expression of disappointment or frustration on her part adds to the man's sense of humiliation. He may redouble his efforts to please his wife; the increased demand for performance can only have an adverse effect on his sexual competency.

Nature of the Marital Interaction Including the Sexual Relationship. In the majority of cases secondary impotence is a response to other difficulties in the marriage. Interviewing should be aimed at elucidating the nature of conflict areas in the marriage as well as those affiliative forces that may improve the prognosis. A detailed history of the marital couple's sexual interaction will be helpful to put the present difficulty in its historical context. The degree of responsivity of the wife is an important factor. Whether premature ejaculation preceded the development of impotence is another factor.

Situational Impotence. As has been indicated, a man may be impotent with his wife but fully potent with a mistress or the reverse may be true. Sometimes a man is potent with only one type of woman, such as one from a different ethnic background or social class. Misidentification of wife and mother is the most frequent instance of this. Because of repressed incestuous impulses toward his mother, he separates passion from love. Passion is evoked by a woman who reminds him least of his mother. Since sex is also associated with sinfulness and wrongdoing, he is most often passionate with a female whom he views as degraded, such as a prostitute. Women are put into two polarized categories, Madonnas or prostitutes, angels or whores, or as someone has said, Marys or Eves.

It is probably true that all men have some anxiety about maintaining potency especially with the advance of age. Most highly vulnerable are those men with a deficient sense of masculinity (gender identity) in whom castration anxiety is particularly strong. In these men the anxiety that follows the initial failure to achieve or maintain an erection is so intense that the fear of failure (performance anxiety) plunges the man into repeated episodes of impotence. In this connection I am indebted to Jules Masserman for a differentiation between "anxiety" and "panic." Anxiety occurs when for the first time a man is unable to achieve an erection twice, and it is panic when for the second time a man is unable to achieve an erection once.

CASE ILLUSTRATIONS OF SECONDARY IMPOTENCE

Case 1: Guilt over an Extramarital Relationship. A man in his mid-forties had been impotent for two years. Although occasionally troubled by premature ejaculation, he had had little trouble achieving or maintaining an erection in more than 20 years of marriage. Two years prior to his seeking help, influenced by stories of his business partner's extramarital adventures, on two occasions he had attempted intercourse with a young woman. He was impotent on both occasions. Feeling great remorse over his ineffectual attempts to be "one of the boys," he became impotent with his wife as well. Treatment ultimately involved both marital therapy and individual treatment for him.

Case 2: Impotence as a Result of a Wife's Bitter Reproaches. A man in his mid-thirties had been impotent for seven years. Raised in a family in which the males were given complete care and attention by the females, very little was asked in return except to be breadwinners. When he married he didn't have the faintest notion that he would have to give his wife any emotional support. He married a woman who had been previously married and divorced. She had had two difficult pregnancies and deliveries, fraught with considerable danger because of eclampsia. When she became pregnant for the third time (the first in his marriage), the husband became even more indifferent and virtually left her alone all during the labor. Her feelings of isolation were heightened by some careless and inconsiderate nursing care. She never really forgave her husband for his neglect. Consequently she began to withdraw from him emotionally, while she reproached him for his passivity and lack of concern. From premature ejaculation he began to ejaculate with semierections, and then finally he developed a complete case of secondary impotence. Conjoint marital therapy was the treatment of choice.

Case 3: The Man Who Liked Nymphets. A retired businessman in his late thirties had frequent episodes of impotence. Raised by a domineering mother, he fantasized that the vagina was a huge hole that would engulf him. As a consequence he preferred nymphets whom he could fondle and caress but with whom he avoided coitus. He enjoyed foreplay with his wife but would try as much as possible to avoid sexual intercourse. On the other hand, his wife's idea was that any kind of sex that did not quickly move on to coitus was perverse. Both of them were highly frustrated by this interaction. Again marital therapy was employed to resolve this dilemma.

Case 4: Impotence Resulting from a Fear of Marital Entrapment. A man in his late forties had been married for less than a year almost 15 years earlier. He had had no difficulty with potency prior to his marriage and during the first three months following the wedding. When his wife became pregnant she announced that this was all she wanted out of the marriage and refused him any further sex. Within the year they were divorced. Following this trauma, the patient had a phobia that he would impregnate a woman and then be trapped into another unfortunate marriage. He did not trust any contraceptive method. He had been impotent with scores of women during the ensuing years. On only one occasion was he potent. This followed his partner's discovery that his condom, into which he had ejaculated with a semierect penis, was still in her vagina. The patient began to tremble, to sweat profusely, and to feel faint, but within 20 minutes he had the best erection he had had in 17 years and was able to have successful coitus for the first and only time. Discussion of his feelings during this episode revealed that his fear of pregnancy had been completely relieved by the fact if he were going to impregnate the woman, he had already done so and no further harm was possible. By the time he came for treatment his phobia of impregnating a woman was so intense and so impervious to reason that he would not even consider the possibility of vasectomy.

Case 5: A Bisexual with Fear of Women. A man who had had extensive homosexual encounters in adolescence and early adulthood was married to a nonorgasmic female. Coitus had never been consummated. The patient had previously been married to a woman with whom he had had no erectile difficulties until he discovered that she was having an extramarital relationship. His mistrust of women, which had been submerged during his first marriage, came to the surface. He selected a woman who had a violent misconception of sex and who unconsciously sabotaged any possibility of overcoming the husband's fear and mistrust of women. This vicious cycle necessitated marital therapy.

Case 6: Revenge through Impregnation. Years ago, long before the pill, a man of 40 came for consultation because of impotence. By his count he had impregnated 14 women by putting holes in his condoms. His story was that at age 13 he had discovered his mother, to whom he had been closely attached, in bed with his father's employee. He chose a unique method of gaining revenge on all women by "knocking them up." The only way he could have intense sexual excitement was when there was the danger of impregnating his female partner. It was with great glee that he would receive the news that the girl was pregnant. As a crude jest he called himself "Jack the Dripper." Over time his pleasure had begun to wane, and he no longer got an intense thrill from his vindictive triumphs. Ultimately, in a somewhat unusual example of the law of the talion, "an eye for an eye and a tooth for a tooth," he became impotent. His depression was so intense that he required psychiatric hospitalization.

Premature Ejaculation

Premature ejaculation is difficult to define. If a man characteristically ejaculates prior to vaginal penetration, almost everyone would agree that this behavior is abnormal. When he ejaculates within one to two minutes after entrance, the pathological nature of his behavior is much less certain. Kinsey[22] reported that this is the average time; hence he found nothing abnormal about it. Since rapid ejaculation is found among animals of diverse species including the primates, Kinsey held that in the human rapid ejaculation was a natural expression of a biological phenomenon. Assuming this to be true, if the male is interested in bringing his partner to a climax, he must learn how to control ejaculation, for only a small minority of women seem to be able to respond within one or two minutes following intromission. (Highly responsive women can masturbate to orgasm within two to three minutes, but even these women are not usually that quickly responsive to coitus.) Whether a man is a premature ejaculator therefore depends on two factors: (1) the nature of the interaction between himself and his sexual partner, and (2) the norms established by society. It is clear that female satisfaction is a cultural demand in our society. Presumably far fewer men were labeled as premature ejaculators in Victorian England.

Given these two factors, an arbitrary and historically time-bound definition is inevitable. Presumably the one set forth by Masters and Johnson[30] is as workable as any. They consider "a man a premature ejaculator if he cannot control his ejaculatory process during intravaginal containment for time sufficient to satisfy his partner during at least 50 per cent of their coital exposures." However, they immediately point out that this definition is absurd "if the female partner is persistently nonorgasmic for reasons other than rapidity of the male's ejaculatory process."

It is impossible to ascertain the prevalence of premature ejaculation. For every male who presents himself to a therapist for treatment, presumably there are thousands who try to deal with their failure to control the speed of ejaculation as best they can without professional help. Since cultural expectations play a role, it is probable that there are important social class differences. Among many people in the lower social class, there may be no recognition that such a malady exists. If the husband is intent only on his pleasure and sheds responsibility for his wife's pleasure and if the female takes little pleasure in coitus and would just as soon get it over with, then neither will complain about premature ejaculation. Furthermore, it is probable that premature ejaculation is an almost ubiquitous phenomenon during adolescence. The cultural demand for speed, especially when sexual activity is being carried out in situations where the couple may be "caught" as in the back seat of a car or in the living room of the girl's parental home, predisposes the adolescent to the development of premature ejaculation. Anxiety about being caught is superimposed on anxiety about performance, which in itself is well-nigh universal. If, on top of this, there is also fear of impregnating the girl, along with the feeling that one is doing something wrong anyway, the combined anxiety will almost certainly produce premature ejaculation.

It is difficult to separate psychological from physiological factors; the intense erotic arousal in young males, especially when there has been infrequent contact with the opposite sex, and the heightened excitement of a new or relatively new experience, coexist with the anxieties cited previously. It is little wonder that premature ejaculation occurs so frequently. What is more remarkable is that the majority of men seem able to learn to control ejaculation either before or during marriage.

Essential to this control is a desire to please one's partner and the ability to recognize the subjective sensation just prior to ejaculatory inevitability. If this ability to recognize the penultimate subjective sensation is absent, there is nothing the man can do to prevent ejaculation. But if he learns to recognize this internal signal, he can stop thrusting or otherwise signal his partner to modify their pelvic movements in order to decrease excitement.

This is the rationale for the "squeeze technique" developed by James Semons and elaborated by Masters and Johnson. This will be discussed under Specific Forms of Treatment in Sex-Oriented Therapy.

In marriage or other close sex-pair relationships, feelings of resentment toward the partner may in themselves cause premature ejaculation. However, probably many thousands of men enter marriage with a marked fear of performance and fear of their partner's scorn and ridicule and continue the pattern of premature ejaculation for many years. Some of these men learn that frequent coital experiences will help them gain control since their ejaculatory demand is decreased. Similarly the decrease in ejaculatory demand with aging produces greater control.

The typical case coming to the therapist is that of a man in his late twenties or early thirties who, after five to ten years of marriage, has had a fairly constant pattern of premature ejaculation. His wife is also nonorgasmic at least through coitus, although she may be fully responsive with other methods or partners. The pattern of sexual disappointment and angry reproaches has heightened the demands for performance that only increase the man's anxiety and failure. This vicious cycle requires marital therapy for its solution.

Ejaculatory Incompetence

Ejaculatory incompetence is defined as the inability to ejaculate in the vagina during coitus, despite a good erection. Occasionally one finds a male who is even unable to masturbate to ejaculation, but most men who have ejaculatory incompetence have no such difficulty. The condition is rather rare. Masters and Johnson report having seen only 17 males with ejaculatory incompetence in an 11-year period. It probably is not as rare as the literature would lead one to believe. I have seen five cases in the past two years.

Psychodynamically, ejaculatory incompetence is related to premature ejaculation, even though physiologically they are the converse of each other. In many cases marked guilty fear seems to be the most prominent emotion. Severe reprisal for masturbation and even for nocturnal emissions may result in a marked fear of ejaculation. Ejaculation may mean a successful and pleasurable completion of masturbation that increases the guilt, or it may simply mean that detection is that much more possible since the soiled clothes or bed sheets may be a cue to repressive parents. On occasion a fear of impregnating a woman is also implicated in the etiology. Many cases turn up striking examples of psychic trauma. Masters and Johnson report the case of a young man who was caught at the point of ejaculation by the police while parked in a lover's lane. In another case they cite, children burst into the room just as the husband was ejaculating. Since these experiences must happen to thousands of people without the development of ejaculatory incompetence, it has to mean that the groundwork for ultimate dysfunction has been laid prior to the traumatic event.

Of special interest in some cases is the fantasy that the male will either soil or be soiled in the process of ejaculation; either the vagina is seen as a dirty hole, contaminating the penis, or the ejaculate itself is seen as a contaminant of the female partner. In a number of men the symptom merely indicates a "holding back," an inability to give of oneself to a partner who is either despised or hated. Either out of hostility to the woman or the fear of impregnating her, some men deliberately stop thrusting when they are about to ejaculate. If this is done over a period of time, unconscious mechanisms take over and the man becomes unable to ejaculate even if he consciously wants to.

An unusual case of ejaculatory incompetence was discovered in an 18 year old who could not masturbate to ejaculation. He had started to masturbate at about 11 but never could ejaculate. His first seminal emission occurred when he was 14. He masturbated frequently during the three-year period before the first seminal emission and established a pattern of masturbation without ejaculation, probably on a physiological basis to begin with. He also was intensely frightened by the first wet dream and thought that he had developed a serious illness. In three sessions a member of my staff, with reassurance and education, was able to cure a symptom that had been present for seven years.

Paradoxical Orgasm

This is a very rare condition in which male orgasm, generally with a semierect penis, occurs in inappropriate places and circumstances. Unlike the midstage between premature ejaculation and impotency in which ejaculation occurs with a semierect or flaccid penis, but in which the circumstances are highly appropriate to erotic arousal, paradoxical orgasm occurs in situations in which sexual arousal is entirely out of place. In most cases studied, orgasm takes place in a competitive situation, such as the conference room of a corporation during a business meeting or an argument over the telephone. On occasion paradoxical orgasm will take place in symbolic enclosures like a phone booth. But even here the conversation during the phone call is usually of a competitive, aggressive nature. Psychodynamically these men have markedly aggressive impulses, unconsciously murderous in nature, while at the same time they have a strong fear of retaliation. Symbolically they want to "put it out" but are afraid it will be chopped off. These men usually have pseudohomosexual anxieties, if not latent homosexual impulses. They are usually incompetent with their wives or sex partners, having either severe premature ejaculation or impotence.

Relationship Problems

Since the majority of sexual problems within marriage or other close sex-pair relationships arise out of conflict in nonsexual areas, with the sexual problem being a manifestation of these other areas of dysfunction, it is important to at least take passing notice of the kinds of problems that frequently create sexual difficulty. Inner conflicts in the male between passivity and self-assertion or between dependency and self-reliance lead to difficulties in assuming the masculine role. As compensation for passivity and dependency, the male may be overassertive, making his wife feel she is an object for use and manipulation rather than an equal co-partner. However the man may directly act out his passivity and dependency by being markedly underassertive, giving his wife the feeling that she has a baby to care for instead of a man to lean on when she needs him. All of these problems can lead not only to the difficulties cited above, but also to conflicts over frequency of sexual intercourse and arguments over types of foreplay or coital positions. Inner and interpersonal conflicts over the expression and control of emotions or words—in general, over communication—are almost invariably found in evaluating a marital unit with sexual difficulties. Feelings of resentment about the failure to communicate feelings or to express sufficient sexual interest are other frequent complaints. All of these can lead not only to dissatisfaction with the sexual aspect of the marriage, but also to infidelity.

Infidelity

The incidence of extramarital sex is increasing. Recent figures from the Institute of Sex Research at Indiana indicate that about 60 per cent of husbands (35 per cent of wives) have had at least one episode of extramarital sex by midlife. The number of people involved alone would tell us that not all of them are reacting to neurotic conflicts. Extramarital sex may result from a search for variety, sometimes even sanctioned by the spouse, who may be engaging in parallel affairs. In many cases no lasting harm is done to the marriage and on occasion it may be beneficial. If the wife is not available sexually because of physical or emotional handicaps, or is simply not interested in sex, extramarital sex may satisfy the man's erotic interest without endangering the marital relationship. This may be true as well if the wife is interested in sex but not responsive. The lack of response may diminish the husband's interest in marital sex while it awakens his interest in extramarital sex. Without outside interests the marriage would be threatened. Infidelity may be an alerting signal that something is wrong with the relationship, and it is possible for the partners to react constructively when both are aware that the marital difficulties have reached a point of crisis.

Since most people are acculturated to believe not only in permanency but in fidelity, extramarital sex is often destructive of the relationship. Diminished self-esteem, a consequence of feeling less preferred than the mistress or lover, may lead to depression or rage, with frequent marital quarrels, withdrawal from the relationship, or retaliative extramarital sex. Even when one partner seemingly accepts the situation with resignation, passive-aggressive techniques may eventually sabotage the marital relationship.

If the therapist is consulted because of infidelity, marital therapy is almost always the treatment of choice. If, however, it is a repetitive pattern, apparently indicative of neurotic conflict, individual psychotherapy may be preferable. Some of the frequent inner anxieties or conflicts leading to infidelity are an intense search to prove one's gender identity, the Madonna-prostitute complex, and a fear of intimacy.

Nontraditional Relationships

Cluster marriages in which two or three couples live together, a ménage à trois, and group sex are the most frequent types of nontraditional sexual styles found among married people. However, few people come for help because of living in a nontraditional relationship. Of these newer forms group sex has been the only one studied extensively, and that has been studied by participant observation[1] rather than by clinical experience with the participants. In group sex the emphasis is on sexual pleasure, not on the relationship; indeed, intimate, affectionate relationships are a threat to the lustful enjoyment of sex for its own sake. One of the striking findings in group sex is the marked increase in homosexuality, especially among females, that seems to take place in this setting. Group sex (swinging) either attracts people with the potential for homosexual or bisexual behavior, or it releases impulses that are widespread in the population but are held under check within traditional relationships. The former explanation is more likely than the latter.

Special Problems of the Young Man

The adolescent and young adult male frequently is torn by internal sexual conflicts. On the one hand, he has to manage intense feelings of sexual arousal, demanding release and gratification, augmented by the need to prove his masculinity. On the other hand, even if he does not have to deal with moral anxiety based on some association of sexuality with sinfulness, almost always he has to cope with the fear that he may not be a competent sexual performer. As has been indicated previously, these anxieties provide the basis for premature ejaculation and, occasionally, impotence.

The capacity to develop intimate relations develops slowly over a number of years, and the fear of intimacy is omnipresent. This may take the specific form of fearing to make a girl pregnant and the consequent premature commitment to a permanent relationship. Young men's frequent fear of marriage slows down the development of a capacity for intimacy and occasionally interferes with sexual competence itself. Another problem seen fairly often is the misconception of sex as an act of violence. This leads to a fear either of hurting the female or of being hurt in return. The fear of genital damage or diminution commonly referred to as castration anxiety may be highly significant in interfering with sexual performance. Any of these anxieties inhibiting appropriate dating and courtship behavior, compounded by a perceived personal rejection by the opposite sex, may create the fear that one is homosexual; this, in turn, augments the underlying fears about one's heterosexual competence and sets up a vicious cycle increasing the loss of pleasure and decreasing effective heterosexual adjustment.

Therapeutic Considerations

The selection of appropriate therapy is of prime importance. A patient should not be placed in a given modality of therapy merely because the therapist is skillful in that form of treatment. This means that the therapist has to become competent in a variety of forms of therapy or, failing that, he has to be able to refer patients to another therapist who possesses the requisite skills in the appropriate method of treatment. The modalities in the treatment of sexual problems fall into three main groups: (1) individual therapy, (2) marital therapy, or (3) a combination of these. Individual therapy may consist of reconstructive psychoanalysis, individual psychotherapy (psychotherapy with more limited goals than the reconstruction of personality), or behavior therapy. Elsewhere in the *American Handbook of Psychiatry* various forms of therapy are discussed in detail. Essentially behavior therapy has three basic techniques: (1) gradual desensitization, (2) flooding, (3) the assignment of tasks outside the treatment to overcome inhibitions. Another form of behavior therapy less frequently used is aversive therapy. Marital therapy may consist of conjoint therapy, the modality in most frequent use at this time, seeing the husband and wife separately, or a combination of the two. Conjoint marital therapy requires specialized training since the skills required are quite different from those involved in individual therapy. If marital therapy is generally the treatment of choice in problems of sexual dysfunction, as I believe, it means that relatively few psychiatrists have been sufficiently well trained to be competent sex counselors. Although 50 per cent of departments of psychiatry now claim that they teach some form of marital therapy, less than a handful have organized formal courses or programs in marital therapy apart from family therapy. A combination of individual psychotherapy or behavioral therapy and marital therapy is sometimes indicated.

What are the factors that determine the selection of treatment? The expectations of the patient are important. If he comes in for individual therapy and sees himself as the primary recipient of psychiatric care, this may either preclude marital therapy or the patient may require persuasion that marital therapy is the better choice. Put briefly and concisely, the issue is whether the relationship between the husband and wife or the intrapsychic conflicts of the patient will be the focus of therapy. If the relationship is to be the focus, the spouse's cooperation is mandatory. If the husband and wife come in together to deal with the problem, there is an explicit expectation that the relationship will be dealt with. Even in this situation one spouse may consciously or unconsciously sabotage the efforts of the therapist. If the patient has come in alone and the spouse's cooperation is required, efforts will have to be made to obtain her commitment to the therapeutic process. Most spouses can be brought in willingly if they are asked to cooperate in the treatment process by aiding the therapist to overcome the patient's difficulties. If the wife has any commitment to the marriage at all, this appeal generally is effective.

When marital therapy is the treatment of choice, the therapist has to decide whether his primary approach will be a "sex-oriented" therapy or a "marriage-oriented" therapy. In sex-oriented therapy the emphasis is on the sexual adjustment, and the various forms of marital disharmony are worked through as they form resistances to the treatment of the sexual dysfunction. In marriage-oriented therapy the emphasis is on the total relationship, and the sexual adjustment is a secondary consideration until enough of the problems in the marriage have been worked through so that the therapist can tackle the sexual problems directly. In the former approach the hostility of one spouse to the other might manifest itself as a refusal to follow the regimen set forth by the therapist. If coitus is interdicted in the first phase of treatment, he or she might attempt intercourse out of an unconscious wish to prevent the treatment from becoming successful. One partner may go through the motions of acting in accordance with the suggestions of the therapist and still fail to respond emotionally to the change in performance. When the marital relationship is the focus rather than the sexual problem, problems of communication, mutual hostility, and so forth are dealt with first in order to increase the motivation of the couple to try to work more effectively on the sexual problem.

Specific Forms of Treatment in Sex-Oriented Therapy

Impotence

For a full description of the technique developed by Masters and Johnson, the reader is referred to *Human Sexual Inadequacy*.[30] Basically their program involves three stages: (1) nongenital pleasuring, (2) genital pleasuring, and (3) nondemand coitus, in which the less involved spouse acts as "co-therapist." The essential ingredient of the treatment of all sexual dysfunctions is avoidance of the demand for performance. The impotent patient must be protected from the demand for penetration both from his spouse and himself. This decreases the fear of being incompetent. As a method of treating impotence, the need to avoid sexual intercourse and to decrease the patient's own demand for successful penetration has been known for a long time, at least since the time of John Hunter over 200 years ago. He claimed that he was able to treat impotent men successfully by insisting that they "refrain from coitus during six consecutive amatory experiences." This principle has been thoroughly developed by Masters and Johnson.

After the couple has learned the pattern of mutual pleasuring that Masters and Johnson called the "sensate focus," erotic arousal has been increased, and the male has had a series of successful erections, the wife (or partner surrogate), in a superior (on top) coital position in a place and at a time when she can control the decision for penetration, places her mate's penis in her vagina until the man can appreciate the sensation of vaginal containment without ejaculation. In this fashion his confidence is built up until he can generally last for 15 to 20 minutes before ejaculation. If his penis becomes soft, the woman withdraws, repeats the stimulation until erection is secured, and once again mounts him. This process is repeated until impotence is overcome altogether.

If the male is not married or has no partner whom he can bring in for sex-oriented marital therapy, individual therapy is the only form of treatment that can be applied. The therapist then has to choose between psychoanalysis, individual psychotherapy, or behavior therapy. Behavior therapists have concentrated their efforts on frigidity rather than on either impotence or premature ejaculation, yet there are isolated reports of the successful behavioral modification of both conditions. Desensitization, with or without the assignment of tasks outside of therapy to overcome inhibitions, is the prevalent mode of behavior therapy of male sexual dysfunction. If, for example, the hierarchy of feared situations involves anxiety about telephoning or making contact with a female, the patient may be assigned the task of carrying this out in addition to desensitization in the therapist's office. In successive stages the patient is brought through finding a partner, hand holding, petting with clothes on, undressing and light petting, heavy petting, and finally, coitus. In psychotherapy the decision to deal with the total personality or to deal primarily with symptom removal is critical. The approach may be primarily historical, dealing with psychic traumas and their effect on present functioning, or therapy may emphasize the anxieties in the patient's current life, hoping that a cognitive technique may decrease inhibitions sufficiently so that the patient begins to carry out in real life what he has thus far been unable to do.

Premature Ejaculation

A special sex-oriented technique to treat premature ejaculation has been devised by Masters and Johnson based on a technique for the individual male originally formulated by Semons. The "squeeze technique" involves the female placing her thumb and two fingers at the coronal ridge of the penis when she has been signaled by her mate or realizes through other cues that he is about to ejaculate. Firmly squeezing the penis in this fashion will prevent ejaculation and cause partial flaccidity. This is repeated many times until the male senses the sensation just prior to ejaculatory inevitability. Since he can stop stimulation at this point, this gives him greater control. Eventually he learns to signal his mate when he senses that he is about to ejaculate, and both of them can stop or decrease their efforts at mutual stimulation. In this way the patient acquires the necessary control to continue intromission for 15 to 20 minutes. The individual male who does not have a partner can learn to apply the squeeze technique himself (the original recommendation of Semons).

Ejaculatory Incompetence

Masters and Johnson suggest that the wife masturbate her partner to ejaculation. When this is successful it is rare that the patient fails to be able to ejaculate in the vagina. Ejaculating in the presence of the female seems to be the first necessary step in the process of overcoming this difficulty. In my experience inhibiting attitudes toward the mate have to be worked out. These almost invariably are feelings of hostility, sometimes deeply buried beneath a façade of agreeableness and a desire to please the female.

Special Forms of Marital Therapy

If marriage-oriented therapy is the treatment of choice, the relationship is dealt with until sufficient motivation is developed to begin the treatment of the sexual problem. When this motivation is high and the interfering emotional factors have been reduced, then the sex-oriented therapy just described can be used effectively. The dual sex therapy team, recommended by Masters and Johnson, has many advantages, some of which have been described earlier in this chapter in the section on Interviewing. Since most therapists do not have access to another therapist of the opposite sex, they should be reassured that they can handle these problems effectively although perhaps over a somewhat longer time than when a dual sex therapy team carries out the treatment.

A special form of treatment of sexual problems is the treatment of groups of couples with sexual problems. When this is done the use of audiovisual aids is very helpful. Showing groups of couples explicit pictures of sexual encounters developed by the National Sex Forum interwoven with small-group discussions, in which partners are separated, has been very useful in overcoming inhibitions and increasing the ease in communication between the husband and wife. In a similar fashion audiovisuals may be used with individual couples.

The psychotherapist who wishes to treat sexual dysfunction competently should learn a variety of techniques so that he can select the appropriate one for a particular patient. Since many of these techniques are not taught in residency or graduate training, some postgraduate training in the more specific treatment modalities described in this chapter seems necessary. Since sexual problems arise most often in the context of a marital relationship, and since, furthermore, in over half the couples the partner of the "presenting" patient also has a form of sexual inadequacy, marital therapy is the most important of these modalities.

Bibliography

1. BARTELL, G. D., *Group Sex*, Peter H. Wyden, New York, 1971.

2. BEACH, F. A. (Ed.), *Sex and Behavior*, Wiley, New York, 1965.
3. BRODERICK, C. B., "Heterosexual Interests of Suburban Youth," *Med. Aspects Human Sexuality*, *4*:83–103, 1971.
4. BROWN, D. G., and LYNN, D. B., "Human Sexual Development: An Outline of Components and Concepts," *J. Marriage & Fam.*, *28*:155–162, 1966.
5. CHRISTENSEN, H. T., and GREGG, C. F., "Changing Sex Norms in America and Scandinavia," *J. Marriage & Fam.*, *32*: 616–627, 1970.
6. Commission on Population Growth and the American Future, *Population and the American Future*, New American Library, New York, 1972.
7. DEVEREUX, G., "Institutionalized Homosexuality of the Mohave Indians," in Ritenbeck, Hendrick (Ed.), *The Problem of Homosexuality*, E. P. Dutton, New York, 1963.
8. DUBOIS, C., *The People of Alor*, Harvard University Press, Cambridge, 1960.
9. FORD, C., and BEACH, F., *Patterns of Sexual Behavior*, Ace Books, New York, 1951.
10. FREUD, S., *Three Contributions to the Theory of Sex*, 4th ed., Nervous and Mental Disease Publishing Co., New York, 1930.
11. GADPAILLE, W. J., "Research into the Physiology of Maleness and Femaleness," *Arch. Gen. Psychiat.*, *26*:193–206, 1972.
12. GAGNON, J. H., "Sexuality and Sexual Learning in the Child," *Psychiatry*, *28*:212–228, 1965.
13. GREEN, R., and MONEY, J. (Eds.), *Transsexualism and Sex Reassignment*, Johns Hopkins Press, Baltimore, 1969.
14. Group for the Advancement of Psychiatry, Committee on the College Student, *Sex and the College Student*, GAP Report No. 60, 1965.
15. HAMMOND, E. B., and LADNER, A. J., "Socialization into Sexual Behavior in a Negro Slum Ghetto," in Broderick, C. B., and Bernard, J. (Eds.), *The Individual, Sex and Society*, Johns Hopkins Press, Baltimore, 1969.
16. HARLOW, H., and HARLOW, M., "The Effect of Rearing Conditions on Behavior," in Money, J. (Ed.), *Sex Research: New Developments*, pp. 161–175, Holt, Rinehart and Winston, New York, 1965.
17. JOHNSON, W., *Masturbation*, SIECUS Study Guide No. 3, 1966.
18. KAGAN, J., "Acquisition and Significance of Sex Typing and Sex Role Identity," in Hoffman, M. L., and Hoffman, L. W. (Eds.), *Review of Child Development and Research*," Vol. 1, pp. 137–167, Russell Sage Foundation, New York, 1964.
19. KARDINER, A., *The Individual and His Society*, Ace Books, New York, 1951.
20. KARLEN, A., *Sexuality and Homosexuality: A New View*, p. 475, Norton, New York, 1971.
21. KINSEY, A. C., POMEROY, W. B., and MARTIN, C. E., *Sexual Behavior in the Human Female*, W. B. Saunders, Philadelphia, 1953.
22. ———, ———, and ———, *Sexual Behavior in the Human Male*, W. B. Saunders, Philadelphia, 1948.
23. KIRKENDALL, L. A., and RUBIN, I., *Sexuality and the Life Cycle*, SIECUS Study Guide No. 8, 1969.
24. KLUCKHOHN, C., "As an Anthropologist Views It," in Deutsch, A. (Ed.), *Sex Habits of American Men: A Symposium of the Kinsey Report*, Prentice-Hall, New York, 1948.
25. LIDZ, T., *The Person*, Basic Books, New York, 1968.
26. LIEBERMAN, B. (Ed.), *Human Sexual Behavior: A Book of Readings*, John Wiley, New York, 1971.
27. LINTON, R., *Culture and Mental Disorders*, Charles C Thomas, Springfield, Ill., 1956.
28. MALINOWSKI, B., *Sex and Repression in Savage Society*, Meridian Books, New York, 1955.
29. MARMOR, J., "'Normal' and 'Deviant' Sexual Behavior," *J.A.M.A.*, *212*:165–170, 1971.
30. MASTERS, W. H., and JOHNSON, V. E., *Human Sexual Inadequacy*, Little, Brown, Boston, 1970.
31. ———, and ———, *Human Sexual Response*, Little, Brown, Boston, 1966.
32. MEAD, M., *Male and Female*, New American Library, New York, 1955.
33. MONEY, J., *Sex Errors of the Body*, Johns Hopkins Press, Baltimore, 1968.
34. ———, HAMPSON, J. G., and HAMPSON, J. L., "Hermaphroditism: Recommendations Concerning Assignment of Sex, Change of Sex and Psychological Manage-

ment," *Bull. Johns Hopkins Hospital*, 97: 284–300, 1955.

35. OFFER, D., and SABSHIN, M., *Normality: Theoretical and Clinical Concepts of Mental Health*, Basic Books, New York, 1966.
36. OPLER, M., "Anthropological and Cross-Cultural Aspects of Homosexuality," in Marmor, J. (Ed.), *Sexual Inversion*, Basic Books, New York, 1965.
37. PFEIFFER, E., VERWOERDT, A., and DAVIS, G. C., "Sexual Behavior in Middle Life," *Am. J. Psychiat.*, 128:82–87, 1972.
38. PHOENIX, C. H., GOY, R. W., and RESKO, J. A., "Psychosexual Differentiation as a Reaction of Androgenic Stimulation," in Diamond, M. (Ed.), *Reproduction and Sexual Behavior*, Pp. 33–49, Indiana University Press, Bloomington, 1968.
39. POMEROY, W. B., and CHRISTENSON, C. V., *Characteristics of Male and Female Sexual Responses*, SIECUS Study Guide No. 4, 1967.
40. REISS, I. L., *The Social Context of Premarital Sexual Permissiveness*, Holt, Rinehart and Winston, New York, 1967.
41. SHILOH, A. (Ed.), *Studies in Human Sexual Behavior: The American Scene*, Charles C Thomas, Springfield, Ill., 1970.
42. STOLLER, R. J., *Sex and Gender*, Science House, New York, 1968.
43. ———, "The 'Bedrock' of Masculinity and Femininity: Bisexuality," *Arch. Gen. Psychiat.*, 26:207–212, 1972.
44. TAYLOR, D. L. (Ed.), *Human Sexual Development*, F. A. Davis Co., Philadelphia, 1970.

CHAPTER 28

SEXUAL FUNCTIONS IN WOMEN AND THEIR DISTURBANCES

Therese Benedek

THE AIM OF THIS CHAPTER is to present the interaction of physiological and psychological factors in the sexual function of women. One cannot discuss the sexual function of women without considering that "the mind-body problem looks different in the light of this century's thought and knowledge—of developmental processes, of physiology and psychology" (p. viii).[27] Although the following presentation is based on the empirical theory of psychosexual development of the personality, one cannot give a comprehensive overview of the interacting physiological and psychological factors and the modifying cultural influences without beginning with the basic theory of psychoanalysis, the instinct theory.

Theory of the Sexual Instinct

In Freud's time the concept of instinct did not need explanation, it was self-explanatory. It was based on the thesis that what is universal is self-evident truth for those who perceive the lawfulness of nature. Such self-evident, universal experience is conceptualized in "the instinct of self-preservation" and in "the instinct of preservation of the species." Freud[24] equated the former with "ego instincts," the latter with "sexual instincts." In his search for universal explanations of psychic conflicts, he assumed that the psychic representations of these biological tendencies are energized by forces that are in conflict with each other.

This hypothesis was useful in explaining behavior at a time when vigorous control and repression of sexuality was the cornerstone of childrearing and morals. Soon, however, Freud discovered that psychic development, the maturation of the ego, occurs in interaction with libido processes and that sexuality itself is dominated by tendencies belonging to ego instincts. Psychoanalysis revealed the intricate interplay of these forces, not as opponents, but as partners in living.

Instinct, as Freud[24] conceived of it, is "a concept on the frontier between the mental and the somatic" (p. 132). In the same paper he defines instinct according to its source, aim, object, and impetus, thus formulating the concept of drive. Instincts are not accessible to experience; only their effect on behavior can be observed. Drives, in contrast, are energy structures that, by responding to need, activate physiological processes to eliminate the need, thus securing survival. Freud conceived of the mental representations of the processes induced by drives as derivatives of instincts. Instincts, drives, and their mental representations are intrinsically interwoven processes. The instinct theory, based on the axiom that the mind originates in and is always dependent on the body, is the foothold of psychoanalysis among the natural sciences. Psychoanalysis as an empirical psychology arrived at its conclusions from observations of adults and children; this led, almost paradoxically, to the concept of pregenital sexuality.

Viewed from the point of view of biology, the term "pregenital sexuality" refers to manifestations of those innate, autonomous, self-differentiating patterns that evolve in any organism in the process of maturation. It was Freud's genius that he deduced from behavioral manifestations, from dreams, fantasies, and symptoms of adults that which could be observed in children. In these he and Abraham[1] recognized the sequence of the evolutionary processes that are repeated in the maturation of the human embryo. He also envisioned that the mind receives its energy, at least partially, from the source that in time would supply the drive for procreation.

The "libido theory," the genetic theory of personality development, although it originated in the instinct theory, did not deal with sexuality and its function in procreation. However, the feedback cycle of psychoanalytic discoveries is demonstrated in the method used in the investigation of the sexual cycle of woman.[12] Without the theory and technique of dream interpretation, without the aim-directed analyses of the multiple motivations of psychic events, it would not have been possible to establish the correlations between phases of the gonadal hormone cycle and psychodynamic tendencies.

From the point of view of general biology, instincts are regulatory principles that, functioning automatically, secure the survival of the organism.[34] Cannon[14] defined instincts as "coordinators of internal regulatory systems which maintain adaptive stabilization." The instinctual processes that serve the homeostatic regulatory functions of the organism are the vital, primary instincts. They serve the regulation of breathing, water balance, food intake, elimination, and maintenance of tissue substance. This sequence is indicatory of the difference in the urgency of the need that the instinct represents; it shows the physiological time interval between the need and its gratification.

The sexual instincts are secondary; they are in the service of procreation. Intake from the outer world seems to play no role in this instinct; sexual deprivation seems to be significant primarily as a psychological rather than a physiological deprivation. Since sexual instincts do not serve metabolic needs, the lengthening intervals between need and gratification allow for complex psychic elaborations and distortions.

Kubie[33] called attention to a seemingly neglected fact concerning primary instincts. "When instincts are viewed as serving the body by effecting the necessary interchange of substances with the outer world, the organ and the function around which all psychological representations focus must be the appropriate aperture" (p. 3). This would have been a good conceptualization of the insight that Freud and Abraham must have had when they considered the organ of intake, the mouth, and the organs of elimination, the anus and the urethra, the origins of the instinctual development during infancy.

The symbolic representations of primary instinctual functions center around the apertures of organs that serve primary instincts. The significance of this hypothesis cannot be overestimated regarding the genetic theory of personality development. It is even more important as a link in the unity of psyche and soma, since it indicates the way by which tensions originating on the level of psychological experience (whether conscious or unconscious) are converted into somatic processes. In the context of this study, however, the meaning is that it supports the method and the result of the investigation of the sexual cycle. It is gratifying to realize that the "apertural hypothesis" puts into the frame and language of biology the same idea that Franz Alexander[2] expressed in his vector concept and used for 30 years as the basis for his psychosomatic research. Alexander's hypothesis means that emotions deprived of their ideational content are expressions of one or more fundamental directions that characterize biological processes. The vector qualities expressed in the psychodynamic tendencies of intaking, retaining, and eliminating are psychic representations of instinctual processes and their related apertures. This expresses in a nutshell the application of the psychoanalytic method of interpretation in the investigation of the sexual cycle.[12]

In the investigation of the sexual cycle the recorded psychoanalytic material was analyzed; the motivations of dreams, fantasies, and behavior were reduced to the dominant psychodynamic tendency; from this the phase of the hormonal cycle was predicted. This prediction was compared with the independently established diagnosis of the state of the gonadal cycle based on the basal body temperature and the Papinacolou smear of the vaginal mucosa. The frequency of the correlations showed that the psychodynamic tendencies in correlation with low hormonal phases of the cycle are expressed by "apertural," pregenital motivations; at the peak of the hormonal cycle they are integrated on the genital level as actively receptive, heterosexual tendency. The biochemical component and the related psychodynamic tendency represent a unit that might be taken as the core of the sexual instinct, far distant, of course, from the complexity of those processes that account for sexual behavior.

What is sexuality? Sexuality is the complex manifestation of the sum total of the species-specific processes necessary for the survival of the species. These can be divided into two groups: in one belong those behavior patterns that bring about mating and conception, in the other belong the physiological processes of gestation, parturition, and the care of the offspring. To which group does the sexual cycle belong? In women the sexual cycle repeats monthly the developmental pattern of the preprocreative personality, but ovulation is a *sine qua non* of procreative sexuality. The "preparatory act"—coitus—is, however, relatively independent of the procreative goal of sexuality in the human female. The species is "free from estrus."

Beach[22] defines estrus as "the physiological condition of a female at the time when she can become pregnant as a result of copulation," and he adds, "Some authors differentiate behavioral from physiological estrus, using the former term to cover all periods in which the female will copulate, even though she may not be susceptible to impregnation" (p. 273). The manifestation of such "behavioral estrus" indicates that the freedom from estrus as an exact differentiating quality of human from infrahuman mammals is not borne out by observations.*

Investigation of the sexual cycle showed that the traces of estrus can be discovered by psychoanalysis; there is a psychodynamic peak of the hormonal cycle that occurs at ovulation. The investigation revealed also that the sexual instinct is rooted in the primary instincts that maintain the homeostasis of the organism and

* There are signs indicating the ways by which the estrus disappeared through the ages. Apes and monkeys are known to indulge in sexual intercourse in which they cannot be fertilized. Menstruation is not a distinction of the human female only, since menstruation occurs in all species of apes and in some kinds of monkeys, governed by the regularly alternating secretion of hormones.

secure its growth and maturation so that the secondary sexual instinct can come to the fore and preserve the species.

Beyond the universal characteristics of the sexual instinct, most significant for man is its plasticity. Having become independent of the limitations of estrus, from the obligation to serve the survival of the species, the factors that promote that basic evolutionary fact continue to interact with intraorganismic and environmental conditions and increase the gap between sexual instinct and behavior. Molded by psychological processes during the long time lag between infancy and maturity, by the network of instinctual derivatives, the secondary and tertiary elaboration of the instinct broke the boundaries of sexuality. No instinctual process is governed by the body's biochemistry alone; even respiration may become the nucleus of a complex psychological superstructure. In regard to the sexual instinct the process appears reversed. The body's physiological need is secondary; the activities of sexual behavior seem to be stimulated by psychological processes. Whether they are perceptual, as in visual impression, or intrapsychic motivations, it is generally known that the whole gamut of emotions, from love to hate, from tenderness to brutality, from happiness to sorrow, may find release, comfort, and consolation through sexual activity.

What is that which we call sexual drive in men and women? The answer is not simple since the drive is not energized by the sexual instinct alone. Sexual drive can be experienced; as a phenomenon it can be described in men more easily than in women. Freud[24] defined the difference by the basic direction of the drive: active in the male and passive in the female. It is assumed that sexual energy, its appetitive strength and the intensity of its "consummatory behavior," to use the ethologist's term, are constitutionally given, innate characteristics molded by ontogenic development.

What role, if any, do hormones play? According to Beach,[4] "Hormones induce changes in the nervous system which affect those systems that coordinate arousal and mating behavior" (p. 270). Although in many vertebrate species, courtship and mating activities are under hormonal control, this does not hold true with the same stringency for infrahuman primates. In regard to man, history as well as current anthropological and cultural changes screen the hormonal effects from the physiology of the procreative function. But arousal and mating behavior is preparatory to achieving the aim of the sexual instinct. Seeking the mate seems superficial, quasi the end of the chain of hormonal feedback circuits. The master gland, the hypophysis, controls the function of the pituitary gland. In women the pituitary produces the follicle-stimulating hormone (FSH) and the luteinizing hormone (LH) that maintain the cyclical changes of the ovarian hormones, estrogen and progesterone. Although the human species is free from the limitations of estrus regarding sexual behavior, the chain of hormonal reactions maintain the survival of the species, the aim of the sexual instinct.

The Organization of the Propagative Drive

There is one biological motivational system for the survival of the species divided between the sexes. The sexual drive is organized differently in the male and female in order to serve a specific function in procreation. The propagative function of the male under the control of one gonadal hormone, androgen, is discharged in one act, motivated by a compelling urge for ejaculation that may or may not have the consequence of inseminating (impregnating) the heterosexual mate. Female physiology is under the control of two alternating gonadal hormones: the follicular ripening hormone (FSH), estrogen, which brings about the maturation of the ovum, and the hormone of the corpus luteum (LH), lutein or progesterone, which prepares the uterus for nidation and maintains the pregnancy if impregnation occurs.

The psychoanalytic concept of sexual maturity is based on the organization of male sexuality. The term "genital primacy" refers to a drive organization that is consummated in heterosexual coitus, climaxing in orgasm. Female sexuality does not fit into this model since heterosexual intercourse, from the point of view of reproduction, is only a preparatory act. Pregnancy and lactation constitute the completion of the propagative cycle. Yet the drive organization that motivates pregnancy and lactation is not genital in the same sense as mating behavior.

The phasic nature of the female procreative physiology exposes the drive organization to investigation. The sexual cycle in monthly repetition of the correlated hormonal and psychological processes reveals the instinctual links between pregenital development and procreative physiology. It demonstrates that the primary homeostatic instincts that motivate growth and development prepare the evolution of mature sexuality. Ovulation and its psychodynamic superstructure disclose that motherhood originates in instincts; thus we may speak of the "mother instinct" in scientific terms.

Developmental Factors and Processes

The sex of the individual is determined at the time fertilization of the egg takes place, through the chromosomal makeup of the gametes (the ovum and the sperm nucleus). The embryo is thus endowed with the potentiality to develop toward one sex.[35] It was generally assumed that the structural differentiation stemmed from an undifferentiated condition. New observations contradict this assumption and demonstrate an initial stage of femaleness in every embryo, the duration of which differs according to the species. In the human embryo it lasts five to six weeks. Then the influence of the sex genes begins to be exerted. The primordial testes begin to produce androgens that induce the transformation of the female genital anlage into the male growth pattern. This inductor theory of primary sexual differentiation has been developed during the last two decades.*

It is beyond the scope of this chapter to give a summary of observations that demonstrate that sexual dimorphism is complete only in the area of procreative physiology. Experiments show that an abundance of opposite sex hormones in the fetal stage affect each sex, influencing development toward the potential for behavioral manifestations characteristic of the opposite sex. In the neuromuscular and psychic structures of both sexes, there remain rudimentary patterns of response and activity characteristic of the other sex. The biological anchorage of bisexuality, however, has not yet been safely established.

In the language of psychoanalysis the term "bisexuality" refers to a specific disposition to certain psychological reactions to environmental influences that exert control over the development of the personality. Our concept of bisexuality adds dimensions of psychology and social influence that broaden the understanding of human sexuality.

Infants of both sexes introject memory traces of responses to both parents. The complex "learning" during the oral phase occurs primarily through identification with the mother in boys and girls alike. This implies the basic assumption that children of both sexes have a biological predisposition to develop empathic responsiveness to individuals not only of their own but also of the opposite sex; without this, communication between the sexes would not be possible. Such communication is a prerequisite for establishing a relationship between the sexes, be it "love" or just "sex."

As far as our present knowledge goes, hormones play no discernible role in the psychic manifestations of pregenital development. Children of both sexes excrete in their urine both androgenic and estrogenic hormones. These remain on a low oscillating level until the age of six to eleven years when there is a fairly sharp increase in the hormone produc-

* For more detailed literature on the inductor theory of primary sexual differentiation, see Sherfey.[45]

tion in both sexes.[18] There are reports on the excretion of estrogenic and androgenic hormones at the age of two to three years, but no increasing hormone production was found in connection with the oedipal phase. Jones[30] found as early as 1927 that in the preoedipal "phallic" phase girls as well as boys might have the tendency to identify with the opposite sex as a defense against heterosexual impulses. In consequence the psychosexual development remains fixated in the phallic pregenital level of organization, motivating penis envy in girls and castration anxiety in boys. This accounts for the manifestations of bisexuality, which complicate the motivational system of the oedipal phase, with its ambivalent object relationship toward both parents. During the latency period the personality of the child enlarges its field of action externally and stabilizes its organization internally. One may speculate whether the extragonadal steroid hormones, estrogens and androgens (in synergy with corticosteroids), have some influence on the self-activation that characterizes the normal latency period.

Puberty is induced by the function of the anterior lobe of the pituitary gland, which secretes a gonadotrophic hormone. This stimulates the ovaries to produce estrogenic hormones and to promote the growth of the ovarian follicles. At about the age of 10 to 11 years girls show a marked rise in estrogen production but not in androgenic substances. The estrogenic hormone is the "hormone of preparation."[16] Produced from childhood on, at the time of early puberty its function is to stimulate the growth of the secondary sex characteristics and the genital organs. Estrogenic hormone levels continue to increase and later assume a cyclic variation as a consequence of the rhythmic ovarian activity that is brought about by ovulation. Estrogens thus prepare the uterus and maintain its readiness for the changes that will be imposed upon it by the corpus luteum, which produces progesterone, the "hormone of maturation."[16]

Menstruation is a central event in the girl's puberty, but not an unmistakable sign of her sexual maturity. Menstruation in the human female and also in subhuman primates may occur before ovulation. This brings about a period of adolescent sterility.[41,47]

Psychoanalytic theory equated menarche with a physiological sign of the castration of the female sex. As if menarche were a puberty rite cast upon women by nature itself, it is expected to be accompanied by intense emotional reactions and is surrounded with superstition.* The anticipation of menstruation as well as the conscious and unconscious responses to it are motivated not only by biological but also by cultural factors. In our present civilization education and hygiene prepare the girl for menarche in a sympathetic way. This seems not only to diminish the fear of menstruation but also to lessen its often painful accompaniments. Not educated in the fear of becoming women, girls expect menstruation with eagerness. This generalization, however, does not mean that girls are never disappointed by the experience. There are anatomical variations, not pathological per se, and hormonal conditions that cause menstrual cramps, lower backaches, and other discomforts.

Not only adolescent girls but also mature women often have anxiety dreams before the menstrual flow sets in that may be interpreted as fear of mutilation. Whether the anxiety dream originates in castration fear or in penis envy has to be investigated in each case. There are other causes of the woman's fear of being a woman. Menstruation forecasts the pain of defloration and, even more significant, injuries connected with childbearing. Menstruation enforces the awareness of the organ that until menarche is not part of the body image. It also activates the process of integration of the latent, partial identification with the mother toward the future task of childbearing. Fliess[21] formulated this process by calling attention to the womb (Gebär-mutter) as a symbol of mother. When menstruation stimulates the awareness of the uterus as an integral part of the body, "mother enters the body" (p. 216).

* Customs and rites surrounding the menstruating woman in different civilizations may indicate, however, man's fear of her. Blood coming from the female body may activate the dread of castration in men.[15]

The next step in the girl's development is to accept "mother-womb" as part of the self and with it to accept herself as a future mother and motherhood as a goal of her femininity. The maturation toward the female procreative goal depends upon the previous developmental identifications with the mother. If these identifications are not charged with reciprocal hostility, the girl is able to accept her heterosexual desire without anxiety and to regard motherhood as a desired goal. This, in turn, determines the girl's reaction to menstruation. If the feminine sexual function is desired, the girl accepts menstruation without undue protest. In contrast, if the bisexual (phallic) fixation dominates her psychosexual development, hostility toward the mother continues as a protest against menstruation and against other aspects of femininity.

In the study of the adolescent struggle to achieve psychosexual maturity, the question arises whether the culturally imposed restrictions on sexuality—the demands of the superego—have any influence on the maturation of the procreative physiology. The correlation is not a simple one. Although physiological maturation may be arrested because of inhibiting emotional factors, one cannot generalize and assume that the stronger the prohibitions and the related anxiety and guilt, the longer procreative maturation will be delayed. In this regard, constitutional, genetic, and intrauterine influences may determine the effect of interpersonal (experiential) factors. If all variations were to be arranged in a continuous series, at one end would be those cases in which the stability of the physiological processes is such that normal procreative maturation is not influenced by emotions, and at the other end would be those in which emotional factors can prematurely arrest maturation. The greatest number will lie between the extremes and present a variety of disturbances in coordination with the various phases of procreative physiology. Correct psychosomatic diagnosis tends to evaluate the proportionate relation and interaction of organic and psychological factors in every symptom. This general principle of psychosomatic pathology is pointed out here in connection with menstruation because menstruation is a cornerstone of female development, a focal point where physiological and psychological factors meet and forecast whether adaptation to sexuality and the tasks of motherhood will succeed or fail.

The Sexual Cycle

In woman the ebb and flow of the gonadal hormone production renders the interaction between endocrine functions and psychodynamic processes accessible to study. The first such investigation was published by the author in collaboration with Boris B. Rubenstein.[12] In spite of the complex and variable superstructure of human personality, it was demonstrated that the emotional manifestations of the sexual drive evolve in correspondence with the ovarian hormone production; the gonadal cycle and the psychodynamic response pattern represent a psychosomatic unit —*the sexual cycle.*

The cycle begins with the gradual production of estrogen; parallel with this the emotions are motivated by an active, object-directed, heterosexual tendency. The estrogenic phase is usually accompanied by a sense of well-being. The direct manifestations of the heterosexual tendency are recognized in dreams and fantasies as well as in behavior. If the desire is not satisfied, the emotional tension may increase parallel with the hormone; restlessness and irritability, even anger and anxiety, may be signs of thwarted sexual urge. The heterosexual tendency can be recognized also in the defenses against it; hostility toward men can express heterosexual tendency by covering it up. Defenses are simple or complex ego operations that, although disguised, reveal the psychodynamic tendency. They occur in any phase of the cycle, allowing for variations in the emotional manifestations of instinctual tendencies.

Shortly before ovulation a minimal progesterone production occurs. The heterosexual tendency, fused with the passive-receptive tendency directed toward the self, intensifies the need for sexual gratification. If frustration is anticipated, the increased physiological tension stirs up suppressed conflicts that may bring about the exacerbation of symptoms. The emotional manifestations of this hormonal state vary depending on the individual's age, emotional maturity, and also on the availability of gratification; a tense emotional state characterizes the preovulative state. After ovulation relief from preovulative tension follows. Then the increased receptive tendency fused with heterosexual tendency creates the highest level of psychosexual integration, that is, the biological and emotional readiness for conception.

Ovulation is a unique physiological event, accompanied by systemic reactions. The best known is the heightened basal body temperature. From the point of view of psychodynamics, the somatic preparation for pregnancy is accompanied by the introversion of libido; the woman's emotional interest shifts to her own body. What one may describe as "narcissistic" characterizes that phase of the cycle in which the woman is most desirous for intercourse. The heightened libidinal state is felt as a satisfaction with her body, an assurance that she is lovable and therefore she can let herself be loved. On the basis of psychoanalytic observations only, Helene Deutsch[17] made the generalization that "a deep-rooted passivity and a specific tendency toward introversion" are characteristic qualities of the female psyche. These propensities are intensified in correlation with the specifically female gonadal hormone—lutein-progesterone—in correspondence with ovulation. This observation justifies the assumption that the specific retentive and receptive tendencies are the psychodynamic correlates of the "mother instinct."

After ovulation the progesterone phase evolves. This is a high hormonal phase since both hormones, estrogen and progesterone, are produced. But the psychological manifestations of the heterosexual tendency appear masked, overshadowed by the emotional preparation for motherhood. The wish for pregnancy, or the fear of it, or hostile aggressive defenses against it may be on the surface of the psychoanalytic material. On a deeper level the biological learning process can be recognized. Introduced by the introversion of psychic energies, the increased receptive and retentive tendencies mobilize the memory traces of infancy, the oral dependent phase. While these appear "regressive," compared with the genital level of the psychological material during the estrogen phase of the cycle, they repeat and represent the vicissitudes of the developmental identifications of the girl with her mother through which "motherliness" evolves. In the psychophysiologically well-defined progesterone phase of the cycle are telescoped the oscillations of those processes by which the dominant tendencies of infancy—the need to be fed and taken care of—are replaced by an adult woman's ability to give, to succor, to be a mother. In adolescence the psychological material of the progesterone phase is often characterized by the repetition of conflicts related to the mother in various phases of development. Mood changes, overeating, or anorexia indicate the exacerbation of developmental conflicts during this phase of the cycle. As the physiological and emotional maturation progresses, the progesterone phase reflects the "reconciliation" with the mother, the reconciliation of the conflicting tendencies toward motherhood.[8]

Upon the decline of progesterone production, the *premenstrual* phase follows. With the diminishing progesterone, which in the later phase of the cycle inhibits estrogen production, the gonadal hormone level declines. Parallel with this process a regression of the psychosexual integration seems to take place; pregenital manifestations of urethral and genital eliminative tendencies motivate the psychodynamic trends. This regression and the increased general irritability of the sympathetic nervous system are manifestations of the moderate degree of ovarian deficiency of the premenstrual phase,[29] which is a significant factor in what Freud[15] termed the recurrent "neurosis of women."

Few women are completely free from mood changes and from discomfort during this part of the cycle. The symptom manifestations show great variation. Apprehension of what might happen to one's body, fear of pain and mutilation, recurrence of other infantile fantasies may lead to anxiety, excitability, and anger. In other cases fatigue, crankiness, sensitiveness to being hurt, and weeping spells characterize the "premenstrual depression." The low hormone phase is probably just one of the factors responsible for the fact that at this phase of the cycle all needs and desires appear imperative, all frustrations seem unbearable; the emotions are less controlled than at any other time during the cycle. This is characteristic of sexual behavior also. The heterosexual desire that occurs during the late premenstrual phase (stimulated probably by the beginning of the next cycle) often appears more intense and demanding than the sexual desire at the peak of the cycle.

The end of the sexual cycle is marked by menstrual flow that, ushered in by a sudden decrease of hormone production, continues for several days. Soon after the flow is established, the tense and fearful mood relaxes, the excitability decreases, and the depression lifts.

The course of the sexual cycle shows that the gonadal cycle forces the emotional processes of the adult woman into regulated channels. This was confirmed soon after the preliminary publication of the sexual cycle research[11] and by more recent investigators. Gottschalk, *et al.*,[26] observed the ovulative change, that is, the progesterone effect. Their investigation measured the variations in basic affects—anxiety and aggression—and found that the intensity of both diminished transiently about the time of ovulation; they found that data from multiple cycles show the effects of the menstrual cycle on the emotional responses measured. The large-scale investigation of Moos, *et al.*,[39] is based on the measurement of the levels in mood changes throughout several complete cycles of the same individual. Their findings are generally consistent with the findings of the first investigation (Benedek and Rubenstein)[11] of the sexual cycle. The variables measured in their investigation changed in cyclical fashion related to the menstrual cycle. These investigations show that neuroendocrinological influence is recognizable in such general manifestations as moods that transiently pervade the ego state.[39]

These investigations show the effect of hormones on the emotional processes. But affects and emotions exert influence upon the hormonal cycle also. The comparative study of a series of cycles in the same woman reveals the effects that stimulating and inhibiting emotional factors have upon the course of the gonadal cycle. It is well known that emotions may precipitate or delay the menstrual flow; less known is the fact that ovulation may vary under emotional influence. Although ovulation occurs, on the average, in the period from the eleventh to the seventeenth day of the cycle, the variation in the time of ovulation is such that probably no immutable period of infertility exists in the human species. Such condition occurs approximately the last week preceding menstruation. Besides the direct and immediate effects of emotions upon the hormonal cycle, psychic factors exert long-term influence on the gonads, as the symptom of psychogenic infertility indicates.

The comparative study of the sexual cycle of several individuals who were psychoanalyzed reveals that the pattern of the sexual cycle unfolds in interaction with the constitutional and environmental factors that determine the developmental organization of the personality. In the evolution of the gonadal cycle, estrogen is necessary for the completion of ovulation and for the ensuing progesterone phase. Similarly, in the integration of psychosexual maturity, the capacity for heterosexual love prepares the woman for acceptance of the feminine sexual role, not only the physiological processes of childbearing, but also the emotional manifestations in motherliness.

Just as psychosexual maturation is an interrelated sequence of physiological processes, so the phases of the sexual cycle represent a transactional sequence, one phase determining the course of the other. Although estrogen is produced in varying degrees from infancy on, progesterone appears only after puberty as a function of the ovum. If the individual reaches psychosexual maturity without fixating traumata in the pregenital phases, the relationship between the estrogen and progesterone phases of the cycle will be normal; this implies practically normal ovulation and average length of cycles. If, because of organic disposition or crippling traumata or both, fixation occurs on a pregenital level, the disturbance of the psychosexual maturation is reflected in the cycle. For example, in puerile, bisexual individuals the progesterone phase is short. Such a woman has short cycles. Women whose infantile fixation causes a prevalence of receptive tendencies (in cases of bulemia, obesity) usually have long, low-level progesterone phases. If the psychosexual development is even more defensive against the feminine reproductive goal, as it is in anorexia nervosa cases, the psychosomatic symptoms may lead to the complete suppression of the gonadal cycle, thus to sterility.

Orgasm and Frigidity

The investigation of Masters and Johnson[38] clarified the physiological process of "the human sexual response." They pointed out the interrelatedness and the differences between the sexes in this area of physiology. The four distinct phases through which coitus, or any sexual manifestation that moves toward orgasm, passes is terminated by ejaculation in men. When, through the excitement and plateau phases the process reaches the point of full testicular elevation, the sensation of impending ejaculation is uncontrollable, an autonomous, reflexive process. In women the clitoris and outer third of the vagina constitute a functional unit. Since the clitoris consists of erectile tissue, it builds up the orgasmic platform at the outer third of the vagina. Because of this, clitoral and vaginal orgasms are not separable in the sense of physiology. Experientially, however, women differentiate "clitoral" orgasm from "vaginal" orgasm; the former can be attained by any adequate stimulation; the latter often remains under the threshold of sensual awareness. The fact that there is a built-in physiology of orgasmic apparatus in women does not necessarily indicate that the process is identical in both sexes.

The female orgasm is attained through the tumescent reaction of the clitoris during the excitement phase with its consequent retraction into the clitoral hood (prepuce), which functions like a "miniature vagina" during the plateau phase (p. 78).[45] If stimulation is reduced the retracted clitoris descends, that is, regains its normal size. The retraction and descent of the clitoris can be repeated several times during the orgasmic process; the stimulation, however, has to be maintained until the orgasmic process is completed, since the woman's orgasmic process is interrupted at any point of the response pattern if stimulation ceases.[45] This is evidence that the woman's orgasm is not an autonomous process; the evolution of the response pattern depends on the male's ability to maintain effective stimulation.

The human sexual response is (not independent of but related to its physiology) an experience of the total personality. It is even more complex: human sexual experience is the outcome of a transactional process between two individuals, normally of different sex. Beginning with a surge of physiological and emotional "chemistry" that is experienced as sexual attraction, a flood of libido quickens the sexual response in man and woman. The heightened libidinal state is felt as a satisfying state of one's own body and gives rise to a libidinal sensation in the genitals—erection in man, and sensation of moisture in the vagina in woman. This sets in motion "modes of behavior" that maintain reciprocal stimulation until the initial reaction becomes a satiating orgasmic experience. Whatever sexual tactics and techniques human couples may employ to develop, enhance, prolong, and vary their experience, it begins with the man's tendency to take over and the woman's wish to submit. This quality of the sexual drive enhances the empathy of the lovers for the need or desire for the other. The "modes of behavior" express the emotions that make the technique of lovemaking effective. While the physiology of the sexual response is universal, the modes of behavior that bring about orgasm have innumerable variations. From these each man and woman adopt a few—some more, others less. Motivated by the unconscious representations of their developmental past, these behavior patterns are charged with the current flow of libido and help to arrive at the completeness of the mutual experience.

Is the experience of the male and the female identical or comparable? Much has been written about this, especially since orgasm as an equal birthright of woman has focused attention on the problematic nature of women's capacity for orgasm. Kinsey[31] assumed that frigidity is a biological characteristic of the female sex since the vagina has no sensory nerves; embryologically the vagina is homologous to the shaft of the penis; the sensory cells known as "genital corpuscles" are confined in the glans clitoridis in woman, as they are in the glans penis of man.[37]

The woman, however, not only perceives the orgasm that she excited in her sexual partner, but she also continues actively to add to the stimulation that she receives. The muscle contractions involved in preorgasmic tension and orgasmic release, which extend beyond the musculature of the pelvic organs, the buttocks and involve her whole body, add actively to her receptive gratification. Yet experience shows that women are so active only if their lovers desire it. Otherwise, women can relax in the flow of their receptive feelings, not worried whether the orgasm is complete or not. Observations indicate that women, especially young and unmarried women, consider their sexual experience satisfactory if their performance gratifies the partner and is praised by him.

The difference in the meaning of sexual experience is rooted in the biological difference in the aim of the sexual instinct, and its consequence in the personality organization of the woman. It is a biological characteristic of man that he be "dominant" and by achieving orgasm impregnate his mate. It is a cultural achievement that also requires men to be concerned with the sexual gratification of their women. The empathic, intuitive responsiveness makes the sexual response truly human. By achieving sexual gratification for himself, man also aspires to give gratification to the other, the object of his love. Probably this is the biological "reason" that such orgasmic experience elevates man's ego beyond what is the result of sexual gratification. It consolidates his sexual identity as it also forges the link of identification with the woman who enables him to have such an experience of his virility.

Women also experience a gratifying expansion of their feminine self as a result of orgasm. Yet the woman's sexual identity does not depend on her orgasm to the same degree that the man's does. Her sexual experience, stirred by the feeling of being loved, desired, and cherished, overflows to her sexual partner in sharing his experience. Women, however, generally do not consider the male orgasm as their own achievement. This kind of elated reaction to their own achievement that men experience after satisfactory, especially after mutual orgasm, women frequently attribute to parturition. Then the feeling of a job well done, having produced a miracle, flows over the woman's ego state and encompasses her infant in her psychic system. Of course, this sensation is not "orgasm," however strong the narcissistic gratification may be, but it indicates that woman's sexual identity is invested more in the aspiration to bear children than in orgasm.

The orgasm of women is "hidden," introverted, and diffuse in comparison with the orgasm of men. Our knowledge of the phasic course of the physiological orgasmic process does not clarify the variations in the sexual response of the same woman with the same mate (and even with different sexual partners). Our knowledge of the sexual cycle, the (gonadal) hormonal stimulation of psychodynamic (instinctual) tendencies, may help to account for variations in the desire for sexual intercourse; however, this in itself does not influence the intensity or the quality of the orgasm. The more we study the sexual experience of women, the further away we get from its physiological substratum, described by Masters and Johnson. The developmental integration of the sexual drive with the personality organization has been pointed out. Here it should be added that the ego in interaction with the drive experience may build up defenses in each phase of development that make free experience of orgasm impossible.[7] This is true for both sexes, but, in general, man's potency is less frequently affected than woman's capacity for orgasm. Biological and psychological factors interact in causing the vulnerability of woman's capacity for orgasm.

Since the sexual response of women is variable and diffuse, when is frigidity as a diagnostic label justified? Psychoanalytic theory considers frigidity a derivative of personality development, a neurotic symptom. Bisexuality and its consequent "penis envy" is the rock bottom[23] of the developmental conflicts that may (or may not) lead to frigidity. The term should be used with caution. Every woman experiences at some time the shrinking feeling of withdrawal that does not seem to tolerate the most tender touch. Indeed, passionate women may often experience such transient frigidity. Those women who are unaware of the vaginal participation in their clitoral orgasm are not "frigid" either. Really, there are many fewer frigid women than one used to assume.

Frigidity can be related to hormonal dysfunction only in rare cases of severe hypogonadism. Women may have all forms and degrees of frigidity and still have normal gonadal function.* Fear of being injured by the penis, fear of being impregnated, and hostility toward men often bring about inhibitions in the woman's capacity for orgasm. Since the social and emotional significance of frigidity is different from that of impotence, women often use the suppression of their own sexual needs as a weapon against their husbands, especially in times past when orgasm, or the desire for it, was considered unwomanly. The effect of the sexual mores that prescribe the "complete orgasm" as a necessary requisite of woman's fulfillment does not appear much healthier. This may increase in many women the anticipation of frustration. Because of this such women watch the course of the act with anxious impatience; thus they interfere with what they are seemingly eager to achieve.

Dyspareunia, painful intercourse in women, may be caused by local injury, by pathological and inflammatory lesions due to anatomical anomalies of the urogenital structure. But the Greek word dispareunia, means "badly mated."[3] It was known in Hippocratic times that pain or fear of pain can be utilized to render intercourse unbearable. *Vaginismus* is a spasm of the vaginal muscles. The spasm may be located in different areas of the vagina; it indicates extreme aversion to coitus against which the spasm is a somatic symptom of defense. This symptom, as well as the milder form of dyspareunia, may be cured by another mate. Above and beyond the fantasies that these symptoms may express, vaginismus achieves its unconscious goal by excluding the penis, by expelling it, or by painfully enclosing it. Since the vagina is a receptive organ, vaginismus can be considered as an expression of powerful incorporative tendencies; it seems to be the realization of men's fearful fantasy of the "vagina dentata." Vaginismus occurs usually in young women whose psychosexual makeup reveals, besides pregenital fixations, a general sexual infantilism.

Separating the interacting factors responsible for the somatic consequences of the serious forms of frigidity, one may assume that besides the constitutional predisposition, the primary factors are psychological, originating in the pregenital development. The chronic stress that evolves through each gonadal cycle may in time suppress the cyclical evolution of gonads, creating anovulatory cycles and infertility. The stagnation in the pelvic organs may involve the autonomic nervous system, causing vasodilatation, and further stasis in the pelvic and sexual tissues. Although frigidity may be a transient manifestation of moods and cannot be considered a symptom, it might have somatic consequences if caused by unresolved developmental conflicts that have interfered with the normal course of the sexual physiology of the woman.

* The opposite seems to be true in many instances. Women, even young girls, who have not had sexual relations, may have normal sexual appetite and enjoy heterosexual coitus even after total extirpation of the ovaries and uterus. This is evidence of the central organization of sexuality, independent of the gonads.

The Procreative Function and Its Decline

Woman's reproductive physiology is phasic. One phase, the monthly cycle that prepares for conception, is repeated from menstruation to menstruation; the other evolves after conception, regulated by the physiology of pregnancy, parturition, and lactation. It is a physiological characteristic of woman that her reproductive function—the progesterone phase, pregnancy, and lactation—goes hand in hand with an increase of receptive and retentive tendencies; they are the psychic representations of the need for fuel to supply energy for growth.

Pregnancy

When conception occurs the cyclical function of the ovaries is interrupted and is not reestablished with regularity until lactation is terminated. Because of the enhanced function of the corpus luteum, the psychodynamic processes of pregnancy can be understood as an intensification of the progesterone phase of the menstrual cycle. The metabolic processes, necessary to maintain the pregnancy, augment the vital energies of the mother. The pregnant woman's body abounds in libidinous feelings. This is the manifestation of the reciprocal processes between the mother and the fetus, of the biological symbiosis. Some women enjoy this narcissistic state of "vegetative" calmness; others, especially in the second trimester of pregnancy, find outlet for their physical well-being by enlarging their activities. As the metabolic and emotional processes of pregnancy replenish the primary narcissism of the woman, this becomes the wellspring of her motherliness. It increases her pleasure in bearing her child, stimulates her hopeful fantasies, and diminishes her anxieties. While the mother feels her growing capacity to love and to care for the child, she actually experiences a general improvement in her emotional state. Many women who suffer from neurotic anxiety are free from it during pregnancy; others, despite the discomforts of nausea and morning sickness, feel emotionally stable and have the "best time" during pregnancy. Thus healthy women demonstrate during pregnancy, as during the high hormone phase of the cycle, an increased integrative capacity of the ego.

Although pregnancy is biologically normal, still it is an exceptional condition that tests the physiological and psychological reserves of women. There are many attitudes, realistic fears, and neurotic anxieties that disturb the woman's desire for pregnancy. Some of these may be motivated by realistic situations. Bad marriage, economic worry, or conception out of wedlock may cause the pregnancy to be unwanted and the infant rejected before it is born. Even such pregnancies usually have a more or less normal course. Only if the psychosexual organization is laden with conflicts about motherhood do actual conditions stir up deeper conflict and disturb the psychophysiological balance of pregnancy.

Psychoanalytic observations trace the "existential anxiety" of a woman to the beginning of her life. The complex steps toward motherhood and motherliness begin in the mother's own oral phase of development. It seems that the memory traces of those experiences that were never conscious are repeated in the psychodynamic process of normal and pathological pregnancy. Passive dependent needs having been revived, the pregnant woman thrives on the solicitude of her environment. If her needs are unfulfilled, the frustration may activate the well-known symptoms of nausea, morning sickness, and perverse appetites. These symptoms seem to diminish in frequency and intensity under cultural influences. Such subjective symptoms of pregnancy are minimal in women who, becoming pregnant by mutual consent, can count on their husbands' care and loving participation in their great experience. Mood swings, however, occur in every pregnancy, especially during the first trimester, when the endocrinologic adjustment may cause fatigue, sleepiness, headaches, vertigo, and even vomiting. These symptoms are usually alleviated as the pregnancy progresses. During the last month the increasing bodily discomfort, disturbance of sleep, and fears connected with parturition may again interfere with the contented mood of pregnancy.

Self-centered and dominated by receptive and retentive tendencies, the pregnant woman is in a regressed, vulnerable ego state. The integrative task that the woman has to face physiologically, psychologically, and realistically is much greater than she has had to meet previously. In some cases the adaptive task appears greater with the first child. The physiological and emotional maturation of the first pregnancy makes motherhood generally easier with the second and third child. Yet it may happen that fatigued by the never ceasing labor of motherhood, women experience pathogenic regressions during later pregnancies.

Normally the psychodynamic regression of pregnancy and lactation serves as a source of growth and development; however, it harbors dangers for those whose ego cannot withstand the psychodynamic processes inherent in the procreative function. External circumstances, such as lack of love from the husband or other disappointments, may stir up the deep-seated conflicts of motherhood. Then the frustrated pregnant woman, like a hungry infant, cannot find gratification within herself; consequently the symbiosis evolves as a vicious circle. The frustrated, hostile pregnant woman becomes anxious since she senses her inability to love her child. The rejection of the pregnancy goes hand in hand with the hostility toward the self, the rejection of the self. Anorexia, toxic vomiting, and consequent severe metabolic disturbances such as eclampsia are "equivalents of depression," the psychosomatic symptoms of pathological pregnancy.

It is a well-established concept of psychoanalysis that regression to the oral phase is the psychodynamic condition of depression. Since such regression is inherent in pregnancy and lactation—even in the progesterone phase of the cycle—depression of varying severity and psychosomatic conditions of oral structure are the basic manifestations of the psychopathology of the female propagative function. Further elaboration of the primary pathology depends on the total personality of the woman. A detailed discussion of the psychopathology of pregnancy cannot be included here. It will suffice to say that the individually significant fantasy may be recathected during pregnancy, thus influencing its emotional course. For example, if the hostility is concentrated on the embryo, women may have fantasies of harboring a monster, a gnawing animal, a cancerous growth. This gives rise to panic or psychotic hypochondriasis. The aggressive impulses toward the embryo may be experienced as inadequacy in taking care of the child. Phobic reactions, even suicidal impulses, may represent defenses against hostility toward the content of the womb. There are many common infantile fantasies. In some instances the "missing penis," in others the envied beauty of the mother or envy of her pregnancy, is the nucleus of the fantasies projected on the embryo.

The fetus is a part of the mother's body. Normally the fetus is cathected with narcissistic libido.[25] This does not always mean pleasurable emotional sensations. There are many ambivalent tendencies in the mother's self-concept that may be projected to the unborn child. Thus the fetus can represent the "bad, aggressive, devouring self," engendering the fear of having a "monster." Many women identify the fetus with feces and relive during the pregnancy the ambivalent feelings and the mysteries of the "anal child." There are many instances in which the regressive fantasies do not interfere with the pleasurable expectation of pregnancy. The mother's object relationship to the unborn child becomes strongly hostile, however, when the fantasy that is projected to the child is charged with ambivalence. The anxiety and depression so caused may remain on the level of neurosis. It depends on the predisposition of the personality whether the anxiety activates a true psychosis with hallucinations, paranoid defenses, and schizophrenic reactions, endangering the woman and her fetus.[19]

Parturition

Parturition is not a growth process as pregnancy and lactation are. It is the physiological interruption of pregnancy that mobilizes, even if transiently, the "basic anxiety" of women. The fear of labor is universal. Even those women who do not anticipate labor with apprehension may suddenly experience cataclysmic anxiety during critical phases of parturition. Modern obstetricians are aware of the significance that sudden and violent emotions have on the course of labor, just as psychiatrists are aware of the influence of obstetrical traumata on the mother and her relationship to the child. A long period of severe labor pains may activate severe postpartum reactions. Pain is a subjective experience. Even if it is caused by organic processes, the experience of pain depends in a particular way on the experiencing ego. "Natural childbirth" is based on this fact. This form of mental hygiene of parturition consists in preparing the woman's ego for experiencing childbirth by reducing the fear-tension-pain circuit and thus reducing the intensity of pain.[43]

The exaggerated expectations of the effect of mental hygiene on parturition harbor psychiatric hazards. There may be unexpected obstetrical complications that interfere with the course of delivery. There are also psychological complications of labor. One of them is the woman's disappointment in herself if she cannot "behave" during labor as her self-respect requires. The sense of failure may then activate inferiority feelings, hidden under the cover of a perfectionistic ego ideal. Since "natural childbirth" permits the woman to feel responsible for what she has to endure, the obstetrical trauma may become complicated by a narcissistic blow. The psychopathological effect of this varies according to the personality of the mother and the maturity of her motherliness.

The Postpartum Period

The trauma of birth interrupts the biological symbiosis. The hormonal changes that induce and control parturition, the labor pains, and the excitement of delivery interrupt the emotional continuity of the mother-child unity. It is, indeed, a biological model of "separation trauma." During the last phase of labor the mother usually concentrates upon her own survival. After delivery the love for the newborn wells up in her as she first hears the cry of the baby. This sensation reassures the mother of the continuity of her oneness with the child. But it may not avoid spells of tearful sadness that often occur, with a pervading sensation of vulnerability, probably a result of hormonal imbalance. In a recent investigation Hamburg[28] compared the mood of postpartum women to premenstrual depression since both conditions are referable to a sharp drop in progesterone production. Incipient postpartum depressions are frequent, especially in primiparas who experience an emotional lag after delivery. Disappointed by a lack of feeling for the infant, they feel "empty" and cry frequently. The free flow tears is similar to the symptom of premenstrual depression. It depends on the personality of the woman whether her lack of immediate love for her child will activate a sense of guilt in her, make her insecure about her ability to be a good mother, and therefore cause her to withdraw from the infant. The incipient depression often goes unnoticed and is cured by the love and affection that the woman receives from her husband. This helps to overcome the fear of being alienated from her child. More serious postpartum reactions usually evolve during the first or second trimester; these conditions are beyond the scope of this chapter.

Lactation

The organism of the mother is not ready to give up the symbiosis after parturition; biologically and psychologically she prepares to continue the symbiosis through lactation.[9] Lactation is stimulated and maintained by prolactin, a hormone of the anterior lobe of the pituitary. This usually suppresses the gonadal function and induces an emotional attitude that is similar to that of the progesterone phase of the cycle.

During lactation, just as during the progesterone phase and pregnancy, receptive tendencies gain in intensity; they become the axis around which the mothering activities center. The woman's desire to nurse the baby, to be close to it bodily, represents the continuation of the biologic symbiosis, not only for the infant, but for the mother as well. While the infant incorporates the breast, the mother feels united with the baby. The identification with the baby permits her to "regress," to repeat (on the unconscious level) her own receptive incorporating wishes. Lactation permits a slow, step by step integration of motherliness. [P. 647][9]

Motherliness in women is not a simple response to hormonal stimulation brought about by pregnancy and the ensuing necessity to care for the young. Instead, motherliness in humans develops through the cyclical repetition of hormonal stimulation, which, interacting with other aspirations of the personality, reaches functional maturity through complex processes of personality development.

In each act of mothering we might differentiate two levels of motivation. One is dominated by the emotional manifestations of the passive, receptive-retentive tendencies; the other, more accessible to consciousness, mobilizes the loving, succoring activities of the mother. In simple societies the biological and ego aspirations of the woman are easily integrated in her ego ideal. Our culture, however, conveys to the woman an active, extraverted, "masculine" ego idea, which may conflict with the passive tendencies inherent in the procreative function. Consequently women may respond with defenses to the biological regression in motherhood. If they cannot permit themselves to be passive, to enjoy lactation, they deplete their source of motherliness. Such women respond with guilt and anxiety and often with a sense of frustration to their inability to live up to the biological functions of mothering. It may be that the mother-child symbiosis can best be maintained by maximum physiological participation such as in breast-feeding; in many instances, however, encouraging the mother to bottle-feed the infant is the better counsel since it may lessen the potentially disturbing ambivalence of the mother and improve the mother-child interaction.

In spite of hormonal control, lactation is influenced by emotional attitudes and may be suppressed by volition. This leads to such statements as "The ability to nurse is equivalent to the will to nurse" (p. 182).[32] This is, of course, an oversimplification of those processes through which emotions affect physiological processes. Galactorrhea, a free flow of the breast during pregnancy or after weaning, is caused by the structure of the gland, yet it seems to be under emotional influence also since it may occur if the woman sees another lactating.

Not only the attitude of masculine protest can motivate the psychopathology of mothering. There are women for whom the instinctual gratifications of mothering remain an everlasting temptation, who enjoy pregnancy, lactation, and caring for the infant as long as he is completely dependent. As if the procreative function of these women were completely under physiological controls, they are unable to develop an object relationship with the child beyond infancy. Psychologically such women do not mature through motherhood. They neglect each child as they become pregnant with the next, often abandoning their child because they can not cope with the responsibilities of motherhood for growing individuals.

Less known manifestations of postpartum pathology are those that, originating in the mother's fear of separation from the infant, intensify her receptive, incorporating need and consequently that of her child. Beata Rank[42] observed the "transmission" of conflicts by which infants become feeding problems. Escalona[20] deals with a broader concept of contagion. The present author interprets these processes in terms of the reciprocal processes of the postpartum symbiosis.[10]

The mother's receptive needs toward the child are not easily recognized. From the time of conception the offspring stimulates her receptive tendencies; after parturition the newborn remains the object of these instinctual tendencies. With her baby the mother feels whole, complete, but not without him. Many young mothers feel "emptiness" when they leave the baby, are compelled to eat, or become anxious and worried. This indicates an increased tension in their receptive needs. The affect hunger of the mother, her need for love and affection, her wish to reunite with her baby, to overprotect and overpossess him are all exaggerated manifestations of the very emotions that originate in the psychobiological processes of motherhood. As if the separation of parturition could be undone, mothering and each manifestation of love for the child may be fused with an urgent need to hold on to him. Characterizing such mothers, we speak of "overprotective, possessive" mothers. As the mother clings to her child, he becomes for her psychic economy (at least partially) the needed, loved, and hated object of her infancy. Reciprocally the child, affected by the incorporative need of the mother that does not allow self-differentiation,* responds to any attempt toward independence as if it meant separation, deprivation of love and protection. Thus the child remains dependent; the emotional symbiosis continues for an undue and often interminable period.

Psychosomatic Infertility

Infertility, if not caused by organic pathology, can be considered as a defense against the dangers inherent in the procreative function. There are few known reasons for absolute infertility, but reasons for relative infertility are manifold. It is generally assumed that besides delayed physiological maturation that may change with time and sexual experience, the psychosexual development is responsible for psychosomatic infertility. It is assumed that emotional factors may play a role in functional infertility, even when some organic pathology is present.[40]

Investigating the factors that bring about functional infertility, one has to take into account the constitutionally "given" differences in the fertility rates in women. It would be almost impossible, however, to establish a complementary series of fertility rates from one extreme to the other. The variables are too numerous. In some individuals fertility varies under the influence of interpersonal factors such as reliance on the mate's support in the case of pregnancy, his sexual potency, and, of course, his reproductive potential. It is necessary to emphasize that fertility (although in much less degree than the woman's capacity for orgasm) is a conjugal phenomenon.† The next category of causative factors lies in the developmental history of the woman. In general, investigations confirm that the psychogenetic course of the mother-daughter relationship is an essential factor in the development toward motherhood. But the concepts of developmental processes do not clarify the pathogenesis of this unique symptom. In the third category belong the cultural processes that, in interaction with the former factors, play a role in the organization of the personality and may modify the woman's attitude toward motherhood.[6]

The woman responds to the tasks of motherhood during the progesterone phase of the cycle from puberty on. Normally, through the monthly repetition, she overcomes her anxieties and absorbs the conflicts of childbearing. It is difficult for those women who as adolescents perceive the stimulation of the progesterone phase with signal anxiety. When developmental conflicts reinforce the signal anxiety, the primary anxiety that originates in the fear of pregnancy may be repressed, but the related conflicts exert a stress, a continuous and monthly intensified charge of defensive psychic energy that may lead to suppression of ovulation. Such a process may interfere with full maturation of the gonadal cycle, but it occurs often after full maturation, even after one or two full-term pregnancies.

The unconscious conflicts that may lead to infertility operate through interaction of neurohormonal and psychological processes; sometimes it appears as if they acted through reflexes activated by the primitive affects of fear and rage. Reflex action may play a role in infertility by a spasm of the fallopian tubes, which may be transitory or persistent. This form of infertility occurs in anxious, hysterical women. It seems justifiable to differentiate be-

* These are the processes that Mahler[36] termed "individuation."

† Westoff and Kiser[46] probably expressed the same idea: "The woman's ability to conceive would depend on the psychosomatic balances both within and between the would-be parents (p. 421).

tween the psychodynamics of infertility in these women and in those active, capable, and adaptable women who, except for functional infertility, are well-integrated, healthy individuals. In these instances infertility develops as a somatic counterpart of the characterological defenses.[44] Among infertile women there are many whose ego aspirations are in conflict with the propagative function.

"Relative infertility" may be the result of long-practiced birth control, not only with oral contraceptives, but with other methods also. Birth control may be achieved, of course, by inhibition of the heterosexual drive. Women who are afraid of men and marriage are not considered infertile from a diagnostic point of view, yet the psychoanalysis of unmarried women often reveals that fear of childbearing is the primary motivation for their sexual inhibition. Women may appear sterile if the desire for intercourse is suppressed during the fertile period and coitus takes place only during the infertile phases. This method of birth control is practiced voluntarily, but it is a "masked practice" of "rhythm." Another "masked practice" occurs in those cases of relative infertility in which ovulation occurs during menstruation when coitus usually does not take place.

The investigations of Seward, *et al.*,[44] confirm psychoanalytic observations that women, frustrated in their instinctual wish for motherhood, often clamor for pregnancy. Yet this does not exclude the possibility that the same women may avoid conception or would suffer severe anxiety if they conceived.

This condensation of the multiple factors that may lead to infertility does not clarify the psychopathology of the symptom. What is primary, the somatic or the psychological process? Its somatic core is the "inhibition" or suppression of gonadal hormone production that evolves in constitutionally predisposed individuals under intraorganismic "stress." In this context stress means the defensive cathexis against the signal anxiety that prevents the basic anxiety from breaking through. Infertility, a psychosomatic defense against the dangers of pregnancy, parturition, and motherhood, is a condition that fits into the "flight pattern" of reactions to anxiety.

Under the influence of a variety of experiences, functional sterility can end with a spontaneous cure. It is assumed that adoption often "works miracles" and facilitates successful childbearing. It would be easy to hypothesize that direct contact with an infant relieves the woman's anxiety about childbearing. Psychoanalytic explorations of such couples reveal that the conditions that facilitate conception, the sexual and emotional relationship between husband and wife, had improved before the adoption occurred. The therapeutic effect of psychoanalysis in cases of infertility can often be explained on the same basis; in other instances one may have evidence of the resolution of the conflict that impeded the woman's potential to become pregnant. The effects of hormone therapy with gonadotrophic hormones have been demonstrated, but not with gonadal hormones.

Pseudocyesis

This puzzling symptom simulates the signs of pregnancy such as cessation of menstruation, enlargement of the abdomen, breast changes consisting of swelling, tenderness, secretion of colostrum, even milk. It is not a "conversion hysterical" symptom that expresses the frustrated instinctual wish for pregnancy, since many organic factors must cooperate in simulating the course of normal pregnancy, sometimes throughout all nine months. No doubt the personality organization of women who undergo and exaggerate the unpleasant accompaniments of pregnancy without being pregnant may be "infantile" or "hysterical," may even have the qualities of an imposter, but the most characteristic quality of these women is a faulty reality testing, certainly in regard to pregnancy. In spite of the physiological processes that simulate pregnancy, there are many signs lacking and some are produced voluntarily; thus the woman could have realized that she is not pregnant. The symptom might occur in young women, in adolescents who have never had sexual intercourse. Indeed, the condition is well characterized by Brown and Barglow[13] in the subtitle of their paper, "A Paradigm for Psychophysiological Interactions."

The psychodynamic factors instrumental in inducing and maintaining pseudocyesis seem to be opposite to those that account for psychosomatic infertility. The latter is characterized by the lack of ovulation, thus by the lack of lutein (progesterone) production; in the sense of its psychodynamics, the symptom represents defense against the anxiety rooted in the danger of childbearing. Pseudocyesis, in contrast, is induced by the physiological and psychological signs of ovulation. Barglow[13] suggests that the "affect of depression has a crucial significance in the etiology of pseudocyesis" (p. 224). This is consistent with the observation that each phase of the procreative function of woman goes hand in hand with the increase of receptive-retentive tendencies. The psychodynamic correlates of the metabolic need that prepares for and maintains pregnancy may bring about depression in those women whose personality organization regresses under the regressive pull of the physiology of pregnancy. Briefly stated, the persistence of the corpus luteum accounts for the depression that seems crucial for pseudocyesis.

What accounts for the maintenance of the corpus luteum? The wish for pregnancy in itself is not enough. The hormonal complement is afforded by the reciprocal interaction of the pituitary gland and the corpus luteum. The few known steps in this interaction are: (1) the persistence of the corpus luteum suppresses follicular stimulation, thus interrupting the gonadal cycle, and menstruation ceases; (2) the pituitary gland, which produces lactogenic hormones, brings about breast changes. The cessation of menstruation and the breast changes suffice as evidence for pregnancy, especially during the first trimester. With this evidence the fantasy life of the woman may complete the picture, even if the attitude toward pregnancy is motivated by ambivalent emotions. Various "secondary" psychosomatic symptoms are nausea and vomiting to create another cycle of what sometimes are serious symptoms of pregnancy.

These alterations of the psychobiology of woman's procreative function have been included here to demonstrate the interaction between hormonal and psychological factors; that is, the manifestations of instinctual processes on a level of organization higher than that on which they were discussed in connection with the sexual cycle.

Climacterium

The procreative period in women averages 35 years. Its decline approaches gradually, marked by the cessation of the menstrual flow. Analogous with the terms "puberty and adolescence," the term "menopause" refers to the physiological process, the term "climacterium" to the period of adaptation to the definite "change of life." There are few women who pass through the transition without some psychosomatic symptoms. The increased irritability of the vegetative nervous system, especially the vascular apparatus, causes hot flashes, chills, palpitation, cardiac arrhythmia, and so forth. Mood swings, irritability, paresthesia, vertigo, and other signs of "nervousness" indicate a lowered tolerance for emotional tension. Since the symptoms usually respond to replacement therapy, it was assumed that the failure of the ovaries brings about the symptoms. The symptoms, however, are only temporary; they, therefore, are due rather to an imbalance of hormones than to simple deprivation. Hoskins[29] assumes that the overproduction of pituitary gonadotrophin—a characteristic feature of the climacteric—may be one factor in prolonging, if not initiating, the disturbance (p. 310).

There is evidence that menopause sets in earlier and often with more intense reactions in women who had never had children than in those who had several normal pregnancies. This is in harmony with the assumption that with complete sexual maturation and function, the regressive emotional manifestations (which characterize the low ebb of the premenstrual hormone state) become absorbed by the adaptional processes of development. On the basis of the woman's characteristic reactions to this phase of her sexual cycle, one might wager a guarded prediction on the course of her climacterium.

The course of climacterium, just as the girl's reaction to menarche, changes under cultural influence. The anticipation of the climacterium intensifies the reciprocal interaction between the neurohormonal symptoms and the emotional reactions to them. As long as menopause meant not only cessation of the reproductive capacity but also the end of sexual attractiveness, the end of the capacity to enjoy sexuality, menopause was a narcissistic blow. This led to the "mortification"[17] that menopause used to mean to many women. It implied that women had to repress sexuality or be ashamed of it. This was probably one of the motivations of the increased sexual excitation that often threatened to break down the previous sexual morals of women.

It is different for most women today. Following the menopause, healthy women respond to the gradual desexualization of their emotional household with an influx of extraverted energy. Their personalities, still flexible and relieved from the responsibilities of child-rearing, seek and find new aims for their psychic energy. As in childhood the repression of sexual impulses led to superego formation and socialization, so in climacterium the cessation of the gonadal function releases new impetus for socialization and learning. Thus when the gonadal stimulation subsides permanently, the healthy woman is not severely threatened by the loss. The accomplishments of the reproductive period, its capacity to love and give, to do for others, will sustain her personality when the woman faces the change of life. With the passing of physiological symptoms, the improvement in her physical and emotional health and her growing interest and activity indicate that the adaptational processes of climacterium may represent actual development in the life of many women.[6]

With aging many of the interpersonal attitudes of the woman change. She does not love with youthful ardor, but much of her ambivalence, jealousy, and insecurity have been overcome. Her love becomes more tolerant and shows more "postambivalent" qualities.[1] These evolve effortlessly toward her grandchildren. Identification with her pregnant daughter or daughter-in-law permits the aging woman to be a mother again. One step removed, she is now less involved; her love for her grandchildren is freer from the conflicts that as a mother she had toward her own children. Through the manifold gratifications within and outside her family, healthy women approach unobtrusively—and many years after climacterium—the physical and mental involution of senescence.

In spite of the positive aspects of climacterium, many women suffer neurotic, psychotic, and psychosomatic manifestations of varied severity. Because these symptoms occur during climacterium, they are usually attributed to the stresses of menopause. Yet the psychoanalytic study of such cases reveals that the symptoms that are aggravated during this period have existed before, or were preformed in the precarious organization of the personality. The trigger of the premenstrual symptoms is the regression that accompanies the low hormone level. As a result the integrative capacity of the ego declines, permitting repressed conflicts and affective reactions to emerge, requiring renewed efforts of repression. Similarly, but even more, the climacteric woman may suffer the permanent cessation of gonadal activity and the consequent decline of libido production. The sensation that indicates the deprivation of libido I designate as *internal frustration.** It refers to the inability of the woman to feel love, to feel satisfied with herself or with anything or anybody in her object world. Such a sensation of aridity expresses the lack of libido. The rising intrapsychic tension brings about defusion of the psychic energy. The freed aggression, manifest in the flow of hostility toward the self

* In psychoanalysis the term "frustration" implies that the drive is thwarted in attaining its goal. Thus frustration may be the result of external or of internal prohibitions of the superego that exclude the instinctual need from gratification.

and others, is the origin of the "agitated depression," the characteristic symptom of the aftermath of menopause.

The symptom that develops as a result of internal frustration depends on the organization of the personality and on the life experiences motivated by it. The interacting instinctual factors, motivating the pathology, are: (1) the bisexual component played a more than usual significance in the personality development; (2) the narcissistic ego strivings dominate the psychic economy throughout life; (3) these factors interfere with the primary gratifications of female sexuality. The narcissistic gratifications may protect the woman from awareness of her basic lack of fulfillment as long as everything goes well, as long as she is able to do well. Such personalities, however, are often unable to tolerate the disappointment that comes when the ego loses the power to perform according to their own standards. Whether it is a physical symptom that is unacceptable, or fantasies, emotions, or hostile impulses that are ego alien, the woman tries to defend herself with all her ego strength. "This can't happen to me" is a protest often heard. As the patient perceives her failure in maintaining the usual level of ego mastery, inferiority feelings and self-accusations mount, causing anxiety and depression. The anxiety brings to the fore the previously compensated pathological potentials of the personality organization. Oversensitivity, phobic reactions, or paranoid projections develop in the struggle to maintain the self. The process may lead to full-blown psychosis.

Summarizing the factors that cause serious psychopathology, one may generalize that severe psychiatric conditions occur in those individuals whose adaptation to internal and external stress occurred throughout their life at a cost of increased narcissistic defenses. The endocrine imbalance and the accompanying autonomic symptoms are precipitating agents; the causes of psychiatric symptoms are deeply embedded in the psychic organization. This is indicated by the fact that endocrine therapy affects the autonomic nervous system reactions and not the psychiatric symptoms.

Many women go through climacterium without any psychiatric disturbance, and then later develop more or less severe depression when physical or psychological manifestations of aging, especially in the area most significant to the individual, bring about a fear of helplessness and intensify defensive processes, as described above, even without endocrine imbalance.

Every single phase of life involves a shifting and reorganization of psychic structures in accordance with the supply of vital energies. As they decline the vector of psychic and somatic processes changes. The giving and expansive attitudes needed for the procreative period become outweighed, step by step, by the restricting, self-centered tendencies characteristic of old age. Yet this, too, is the result of the adaptive function of the ego, which tends to maintain the individual as she knew herself, as she used to be—to the last.

Bibliography

1. ABRAHAM, K., "A Short Study of the Development of the Libido, Viewed in the Light of Mental Disorders," in *Selected Papers*, pp. 418–501, Basic Books, New York, 1955.
2. ALEXANDER, F., "The Logic of Emotions and Its Dynamic Background," *Int. J. Psychoanal.*, *16*:399–413, 1935.
3. *American Medical Dictionary*, 13th ed., Saunders, Philadelphia, 1926.
4. BEACH, F. A., *Hormones and Behavior*, Hoeber, New York, 1948.
5. BENEDEK, T., "Climacterium: A Developmental Phase," *Psychoanal. Quart.*, *19*:1–27, 1950.
6. ———, "Infertility as a Psychosomatic Defense," *Fertility & Sterility*, *3*:527–537, 1952.
7. ———, "On Orgasm and Frigidity," Panel on Frigidity in Women, reported by Moore, B. in *J. Am. Psychoanal. A.*, *9*:571–584, 1961.
8. ———, "The Psychobiological Aspects of Mothering," *Am. J. Orthopsychiat.*, *21*:272–278, 1956.

9. ———, "The Psychosomatic Implications of the Primary Unit: Mother-Child," *Am. J. Orthopsychiat.*, *19*:642–654, 1949.
10. ———, "Toward the Biology of the Depressive Constellation," *J. Am. Psychoanal. A.*, *4*:389–427, 1956.
11. ———, and RUBENSTEIN, B. B., "Correlations between Ovarian Activity and Psychodynamic Processes, I. The Ovulative Phase; II. The Menstrual Phase," *Psychosom. Med.*, *1*:245–270; 461–485, 1939.
12. ———, and ———, *The Sexual Cycle in Women*, National Research Council, Washington, D.C., 1942.
13. BROWN, E., and BARGLOW, P., "Pseudocyesis, A Paradigm for Psychophysiological Interactions," *Arch. Gen. Psychiat.*, *24*:221–229, 1971.
14. CANNON, W. B. (1932), *The Wisdom of the Body*, Norton, New York, 1963.
15. CHADWICK, M., *The Psychological Effects of Menstruation*, Nervous and Mental Disease Publ. Co., New York, 1932.
16. CORNER, G. W., *The Hormones in Human Reproduction*, 2nd ed., Princeton University Press, Princeton, N.J., 1947.
17. DEUTSCH, H., *The Psychology of Women: A Psychoanalytic Interpretation*, 2 vols., Grune & Stratton, New York, 1944–1945.
18. DORFMAN, R. I., and SHIPLEY, R. A., *Androgens: Biochemistry, Physiology and Clinical Significance*, John Wiley, New York, 1956.
19. DUNBAR, H. F., *Emotions and Bodily Changes: A Survey of Literature on Psychosomatic Relationships, 1910–1953*, 4th ed., Columbia University Press, New York, 1954.
20. ESCALONA, S. K., "Emotional Development in the First Year of Life," in Senn, M. J. E. (Ed.), *Problems of Infancy and Childhood*, pp. 11–92, Transcript 6th Conference, Macy Foundation, New York, 1953.
21. FLIESS, R., *Erogeneity and Libido: Addenda to the Theory of the Psychosexual Development of the Human*, International Universities Press, New York, 1957.
22. FORD, C. S., and BEACH, F. A., *Patterns of Sexual Behavior*, Harper, New York, 1951.
23. FREUD, S., "Femininity," in Strachey, J. (Ed.), *Standard Edition*, Vol. 22, pp. 112–135, Hogarth, London, 1964.
24. ——— (1915), "Instincts and Their Vicissitudes," in Strachey, J. (Ed.), *Standard Edition*, Vol. 15, pp. 117–140, Hogarth, London, 1963.
25. ——— (1914), "On Narcissism: An Introduction," in Strachey, J. (Ed.), *Standard Edition*, Vol. 14, pp. 67–102, Hogarth, London, 1957.
26. GOTTSCHALK, L., KAPLAN, S., GLESER, G., and WINGET, C., "Variations in Magnitude of Emotion: A Method Applied to Anxiety and Hostility during Phases of the Menstrual Cycle," *Psychosom. Med.*, *24*:300–311, 1962.
27. GREENFIELD, N. S., and LEWIS, W. C., "Preface of the Editors," in *Psychoanalysis and Current Biological Thought,*" University of Wisconsin Press, Madison, Wisc., 1965.
28. HAMBURG, D. A., "Effects of Progesterone on Behavior," *Res. Publ. Ass. Nerv. & Ment. Dis.*, Vol. 43, *Endocrines and the Central Nervous System*, Williams & Wilkins, Baltimore, 1967.
29. HOSKINS, R. G., *Endocrinology: The Glands and Their Functions*, Norton, New York, 1950.
30. JONES, E. (1927), "The Phallic Phase," in *Papers on Psychoanalysis*, pp. 452–484, Williams & Wilkins, Baltimore, 1948.
31. KINSEY, A., *et al.*, *Sexual Behavior in the Human Female*, Saunders, Philadelphia, 1953.
32. KROGER, W. S., and FREED, S. C., *Psychosomatic Gynecology*, Saunders, Philadelphia, 1951.
33. KUBIE, L. W., "The Central Representation of the Symbolic Process in Psychosomatic Disorders," *Psychosom. Med.*, *15*, 1953.
34. ———, "Instinct and Homeostasis," *Psychosom. Med.*, *10*:15–29, 1948.
35. LOEB, J. (1912), *The Mechanistic Concept of Life*, Harvard University Press, Cambridge, Mass., 1964.
36. MAHLER, M. S., "Thoughts about Development and Individuation," in *The Psychoanalytic Study of the Child*, Vol. 18, pp. 307–324, International Universities Press, New York, 1963.
37. MARMOR, J., "Some Considerations Concerning Orgasm in the Female," *Psychosom. Med.*, *16*:240–245, 1954.
38. MASTERS, W. H., and JOHNSON, V. E., *Human Sexual Response*, Little, Brown, Boston, 1966.
39. MOOS, R. H., KOPPELL, B. S., MELGES, F. T.,

YALOM, I. D., LUNDE, D. T., CLAYTON, R. D., and HAMBURG, D. A., "Fluctuations in Symptoms and Moods during the Menstrual Cycle," *J. Psychosom. Res.*, *13*:37–44, 1969.

40. MORRIS, T. A., and STURGIS, S. A., "Practical Aspects of Psychosomatic Sterility," *Clin. Obstetrics & Gynecol.*, *2*:890–899, 1959.

41. PRATT, J. P., "Sex Functions in Man," in Allen, Edgar, *et al.*, *Sex and Internal Secretions*, pp. 1263–1334, Williams & Wilkins, Baltimore, 1959.

42. RANK, B., PUTNAM, M. C., and ROCHLIN, G., "The Significance of the 'Emotional Climate' in Early Feeding Difficulties," *Psychosom. Med.*, *10*:279–283, 1948.

43. READ, G. D., "Correlations of Physical and Emotional Phenomena of Natural Childbirth," *J. Obstetrics & Gynecol., Brit. Emp.*, *53*:55–61, 1946.

44. SEWARD, G. H., WAGERN, P. S., HENRICH, J. F., BLOCH, S. K., and MEYERHOFF, H. L., "The Question of Psychophysiologic Infertility: Some Negative Answers," *Psychosom. Med.*, *27*:533–545, 1965.

45. SHERFEY, M. J., "The Evolution and Nature of Female Sexuality in Relation to Psychoanalytic Theory," *J. Am. Psychoanal. A.*, *14*:28–128, 1966.

46. WESTOFF, C. F., and KISER, C. W., "Social and Psychological Factors Affecting Fertility; An Empirical Re-Examination and Intercorrelation of Selected Hypothesis Factors," *Milbank Mem. Fund Quart.*, *31*:421–438, 1953.

47. YOUNG, W. C., and YERKES, R. M., "Factors Influencing the Reproductive Cycle in the Chimpanzee; the Period of Adolescent Sterility and Related Problems," *Endocrinology*, *33*:121–154, 1943.

CHAPTER 29

THE MIDDLE YEARS

Bernice L. Neugarten and Nancy Datan

MIDDLE LIFE has received relatively little attention from students of the life cycle. The early part of life, paced by a biologically timed sequence of events, continues to attract major attention from behavioral scientists; and in the past few decades a large body of research has grown up about old age. The years of maturity, by contrast, are less understood. Only a few attempts have been made to develop a psychology of the life cycle, and the search for regular developmental sequences of personality change in adulthood is relatively new.[28,32,55,56,69] This neglect of middle life has led by default to the uncritical view on the part of many psychologists that personality is stabilized and the major life commitments completed in the period of youth and that nothing of great significance occurs for the long time until senescence appears. Clinicians, on the other hand, have shown considerable awareness of the potential hazards to mental health that arise in midlife.[9] For women the climacterium is thought to constitute one such crisis;[6,8,23,25] for men the perception of decline in sexual prowess is considered a serious crisis;[73,76] and the phenomenon of depression in middle age has received much attention.[68] All this has led to somewhat unbalanced views of middle age as either plateau or crisis.

The focus of this chapter is upon social psychological perspectives, rather than upon the psychodynamics of personality change. In taking note of salient biological, psychological, and social changes of middle age, we shall draw upon such empirical studies as exist and focus primarily upon investigations of so-called normal or nonclinical populations.*

The Changing Boundaries of Middle Age

The boundaries of middle age have been defined by various indices. Chronological age definitions are perhaps the most arbitrary. Typically the period being described is 40 to 60 or 65, but it is sometimes as broad as a 30 year span, 30 to 60, or sometimes only a single decade, the forties. There is no consensus that any single biological or social event constitutes

* Portions of the present chapter have appeared in earlier papers by Bernice Neugarten.[53,54]

the lower boundary of middle age. While it is often said that retirement constitutes the upper boundary for men, there is no agreed upon boundary for women. The major life events that characterize the middle part of the life span—reaching the peak of one's occupational career, the launching of children from the home, the death of parents, climacterium, grandparenthood, illness, retirement, widowhood—while they tend to proceed in a roughly predictable sequence, occur at varying chronological ages and are separated by varying intervals of time. Most behavioral scientists therefore concur that chronological age is not a meaningful index by which to order the social and psychological data of adulthood; and the individual's own awareness of entry and exit from middle age, as will be described in more detail below, seems to emerge from a combination of biological and social cues rather than from a fixed number of birthdays.[53,62,78] Since there are no clear boundaries, it is sometimes said that middle age should be described as a state of mind rather than as a given period of years. If so, it is a state of mind that has an important influence upon the individual's perceptions of himself and his strategies for managing his world.

For both observer and observed, middle age cannot meaningfully be separated from that which precedes and follows it in the life cycle, for the individual always assesses his present in terms of both his past and his future. Accordingly, perceptions of the life cycle are meaningful data in understanding the middle-aged.

Perceptions of the Life Cycle

Perceptions vary from one person to the next regarding the timing and rhythm of major life events and the quality of life at successive periods of adulthood, but important bases of consensus are to be found as well as consistent group differences. In one study of the views of men and women aged 40 to 70 (a community sample drawn from a metropolitan area), adulthood was generally seen as divided into four periods: young adulthood, maturity, middle age, and old age, each period having its unique characteristics. These major periods of life were recognized by all respondents, but there were differences in the views of men and women and differences according to socioeconomic level. Men saw a succession of minor dividing points and a relatively gradual progression from one period of life to the next. Women saw one major dividing point that outweighed the others in significance, and they often described adult life in terms of two somewhat disconnected lives, one before and one after 40. Among business executives and professionals, a man did not reach middle age until 50, nor old age until 70. For the blue-collar worker, on the other hand, life was paced more rapidly, and a man was described as middle-aged by 40 and old by 60.

The themes of life associated with middle age varied also with the social status of the respondent. For the upper-middle class, young adulthood (20 to 30) was described as a time of exploration and groping, of "feeling one's way" in job, marriage, and other adult roles, and as a period of experimentation. Maturity (30 to 40) was the time of progressive achievement and increasing autonomy. Middle age was described as the period of greatest productivity and of major rewards, the "prime of life." Women, while mentioning the adjustments required by the departure of children from the home, also described middle age as a period of mellowness and serenity. Old age was viewed as a period of relaxation, leisure, security, partial withdrawal, and resting on one's laurels.

The blue-collar worker had a different view. Both men and women described young adulthood as the period, not when issues are explored, but when they become settled, and when life's responsibilities loom up as inescapable. One becomes increasingly sensible, older, wiser, and quieter. Not only does middle age come early, but it is described in terms of decline—slowing down, physical weakening, becoming a has-been. Old age is the period of withdrawal and progressive physical decline and is described in pessimistic terms, the "old age, it's a pity" theme.[62]

In a related study of over 600 middle-aged men and women, there was a striking consistency in attitudes about growing older. While the largest proportion of responses were neutral, in those persons who expressed negative or contingent attitudes the fear of dependency was paramount. (Contingent attitudes were those in which the respondent said, "Growing old will be fine if my health stays good," or "I don't mind old age as long as I don't become a burden to anyone.") Dependency was always seen as having two sources, loss of income and loss of health. Fear of death was never expressed, nor fear of social isolation. Fear of dependency was the only theme to occur with any frequency, and it occurred approximately as often for men as for women and for people at all social class levels.[57]

The Changing Rhythms of the Life Cycle

Perceptions of the life cycle are influenced, of course, by social change. From a historical perspective there is documentation to show that not until the seventeenth and eighteenth centuries, with the growth of industrialization, a middle class, and formal educational institutions, did the view appear that childhood was a discernible period of life with special needs and characteristics.[2] The concept of adolescence can be viewed as essentially a twentieth-century invention.[24] Most recently a case has been made for a new stage called youth,[41] a stage that has appeared only in the last few decades when social change has become so rapid that it threatens to make obsolete all institutions and values within the lifetime of each generation and when a growing minority of young persons face the task of reconciling the self with the social order. It has been suggested that, in terms of the developmental polarities suggested by Erikson,[28] in which the central psychological issue of adolescence is identity versus diffusion, the central issue for youth is individuation versus alienation.

A parallel case can be made for the fact that the period of middle age is a recently delineated stage in the life cycle. In this instance the significant dimensions of social and technological change are the enormous increase in longevity that has occurred since the beginning of this century and the growth of leisure in the affluent postindustrial society. At the risk of overdoing these parallelisms, it might be ventured that the central psychological task for middle age relates to the use of time, and the essential polarity is between time mastery and capitulation.

From a somewhat narrower historical perspective, social change creates alterations in the rhythm and timing of life events, alterations that have their inevitable effects upon perceptions of what it means to be middle-aged. This can be illustrated from changes in the family cycle and the work cycle. Dramatic changes have occurred over the past several decades as age at marriage has dropped, as children are born earlier in the marriage and are spaced closer together, as longevity and consequently the duration of marriage has increased.[30]

Census data show, for instance, that in 1890, only 80 years ago, the average American woman left school at about age 14, married at 22, had her first-born child within two to three years and her last-born child when she was 32. Her husband died when she was only 53, her last child married when she was 55, and she herself lived to about 68. The average woman was widowed, then, before her last child left home.

In 1966 this picture was very different. Then the average woman left school at age 18; she married at 20; her first child was born within one year; her last child, by the time she was 26; and all her children were in school full-time when she was only 32. The projections for this group of women were that the last child would marry when she was 48, her husband would die when she was 64, and she herself could expect to live to almost 80. Thus, our average woman can now look forward to some 45 years of life after her last-born child is in school.

These trends are equally reflected, of

course, in the lives of men. Historically the family cycle has quickened for both sexes as marriage, parenthood, empty nest, and grandparenthood all occur earlier, and as the interval of time becomes extended (now some 16 years) when husband and wife are the remaining members of the household. (While the family cycle runs its course a few years later for men and women at higher social class levels, the general pattern of historical change is the same for both higher and lower levels.)

Parent-child relationships in middle age have been affected by the fact that parenthood has been coming earlier in life. Changes in parental behavior, with, for example, fathers becoming less authoritarian, may in part reflect this increasing youthfulness. It is the relative youth of both parents and grandparents, furthermore, that may be contributing to the complex patterns of help between generations that are now becoming evident, including the widespread financial help that flows from parents downward to their adult children.[75] In a study of three-generation families in which styles of behavior by grandparents were delineated, it was found that younger grandparents (those under 65 as compared with those over 65) more often followed the fun-seeking pattern.[63] The fun-seeker is the grandparent whose relation to the child is informal and playful and who joins the child for the purpose of having fun, somewhat as if he were the child's playmate. Grandchildren are viewed by these grandparents as a source of leisure activity, as a source of self-indulgence. Authority lines become irrelevant, and the emphasis is upon mutual gratification. Similarly, with grandparenthood coming earlier in life, there is the emergence of an extended family system that encompasses several generations. (In 1962, 40 per cent of all persons in the United States who were 65 or over had *great*-grandchildren.)

Changes in the work cycle are also changing rapidly, affecting the perceptions of middle age. With longer education and later entry into the labor force, and with earlier retirement, the proportion of the life span spent at work is diminishing for men. For women the trend is the opposite. While fewer women work, and fewer work full-time, the trend is to extend the proportion of the life span spent in the labor force. With more than half of all women aged 50 to 55 now in the labor force, middle-aged women have gained in status relative to men, and relationships are changing not only between the sexes but between the generations within the family. Not only the mother goes to work now, but also the middle-aged grandmother.[52,55]

While direct data are lacking, it is probable that these changes in the family and in the economy are contributing to major differences between the sexes in adaptations to middle age and in patterns of aging. For women, although there is an increased burden of caring for aged parents, lightened family responsibilities, the marriage of the last-born child, and the taking on of new economic and civic roles now tend to coincide with the biological changes of the climacterium, probably producing an increasingly accentuated new period of life and contributing to the new sense of freedom expressed by many middle-aged women.

The rhythms of the work career show great variability at different occupational levels, and there are large differences from one occupation to another in the timing of career stages, in the rewards that come at each stage, and in the relationships between younger and older participants in the work setting. Increased complexity of knowledge is required, not only of the practicing professional, but for a wide variety of occupations. Even more significant is the accompanying problem of obsolescence of skills and technical knowledge, a problem that characterizes most occupational groups. Age lines tend to become blurred and age-deference systems weakened in instances where a younger man's up-to-date knowledge has the advantage over the older man's experience. To take but one example, studies of business leaders show that the average age of the business elite rose steadily from 1870 to 1950.[33,79] The increase in size, complexity, and bureaucratization that characterizes big business today has been accompanied by a lengthening of the early phases of the career line. It has been taking longer than in earlier generations for a man to rise to the top of the administrative ladder, a fact that perhaps underlies some of the efforts now being made by business firms to push young men into leadership roles.

The changing technology, the changing job patterns, as well as the changing family cycle, are all influencing the experience of middle age.

The Subjective Experience of Middle Age

Middle-aged men and women, while they recognize the rapidity of social change and while they by no means regard themselves as being in command of all they survey, nevertheless recognize that they constitute the powerful age group vis-à-vis other age groups; that they are the norm bearers and the decision makers; and that they live in a society that, while it may be oriented toward youth, is controlled by the middle-aged. There is space here to describe only a few of the psychological issues of middle age as they have been described in one of our studies in which 100 highly placed men and women were interviewed at length concerning the salient characteristics of middle adulthood.[53] These people were selected randomly from various directories of business leaders, professionals, and scientists.

The enthusiasm manifested by these informants as the interviews progressed was only one of many confirmations that middle age is a period of heightened sensitivity to one's position within a complex social environment, and that reassessment of the self is a prevailing theme. As anticipated most of this group were highly introspective and verbal persons who evidenced considerable insight into the changes that had taken place in their careers, their families, their status, and in the ways in which they dealt with both their inner and outer worlds. Generally the higher the individual's career position, the greater was his willingness to explore the various issues and themes of middle age.

The Delineation of Middle Age

There is ample evidence in these reports, as in the studies mentioned earlier, that middle age is perceived as a distinctive period, one that is qualitatively different from other age periods. Middle-aged people look to their positions within different life contexts—the body, the career, the family—rather than to chronological age for their primary cues in clocking themselves. Often there is a differential rhythm in the timing of events within these various contexts, so that the cues utilized for placing oneself in this period of the life cycle are not always synchronous. For example, one business executive regards himself as being on top in his occupation and assumes all the prerogatives that go with seniority in that context, yet, because his children are still young, he feels he has a long way to go before completing his major goals within the family.

Distance from the Young

Generally the middle-aged person sees himself as the bridge between the generations, both within the family and within the wider contexts of work and community. At the same time he has a clear sense of differentiation from both the younger and older generations. In his view young people cannot understand nor relate to the middle-aged because they have not accumulated the prerequisite life experiences. Both the particular historical events and the general accumulation of experience create generational identification and mark the boundaries between generations. One 48 year old says,

> I graduated from college in the middle of the Great Depression. A degree in Sociology didn't prepare you for jobs that didn't exist, so I became a social worker because there were openings in that field. . . . Everybody was having trouble eking out an existence, and it took all your time and energy. . . . Today's young people are different. They've grown up in an age of affluence. When I was my son's age I was supporting my father's family. But my son can never understand all this . . . he's of a different generation altogether.

The middle-ager becomes increasingly aware of the distance—emotionally, socially, and culturally—between himself and the young. Sometimes the awareness comes as a sudden revelation:

I used to think that all of us in the office were contemporaries, for we all had similar career interests. But one day we were talking about old movies and the younger ones had never seen a Shirley Temple film. . . . Then it struck me with a blow that I was older than they. I had never been so conscious of it before.

Similarly, another man remarked:

When I see a pretty girl on the stage or in the movies, and when I realize she's about the age of my son, it's a real shock. It makes me realize that I'm middle-aged.

An often expressed preoccupation is how one should relate to both younger and older persons and how to act one's age. Most of our respondents are acutely aware of their responsibility to the younger generation and of what we called "the creation of social heirs." One corporation executive says,

I worry lest I no longer have the rapport with young people that I had some years back. I think I'm becoming uncomfortable with them because they're so uncomfortable with me. They treat me like I treated my own employer when I was 25. I was frightened of him. . . . But one of my main problems now is to encourage young people to develop so that they'll be able to carry on after us.

And a 50-year-old-woman says,

You always have younger people looking to you and asking questions. . . . You don't want them to think you're a fool. . . . You try to be an adequate model.

The awareness that one's parents' generation is now quite old does not lead to the same feeling of distance from the parental generation as from the younger generation.

I sympathize with old people, now, in a way that is new. I watch my parents, for instance, and I wonder if I will age in the same way.

The sense of proximity and identification with the old is enhanced by the feeling that those who are older are in a position to understand and appreciate the responsibilities and commitments to which the middle-aged have fallen heir.

My parents, even though they are much older, can understand what we are going through; just as I now understand what they went through.

Although the idiosyncrasies of the aged may be annoying to the middle-aged, an effort is usually made to keep such feelings under control. There is greater projection of the self in one's behavior with older people, sometimes to the extent of blurring the differences between the two generations. One woman recounted an incident that betrayed her apparent lack of awareness (or her denial) of her mother's aging:

I was shopping with mother. She had left something behind on the counter and the clerk called out to tell me that the "old lady" had forgotten her package. I was amazed. Of course the clerk was a young man and she must have seemed old to him. But I myself don't think of her as old.

Marriage and Family

Women, but not men, tend to define their age status in terms of the timing of events within the family cycle. Even unmarried career women often discuss middle age in terms of the family they might have had.

Before I was 35, the future just stretched forth, far away. . . . I think I'm doing now what I want. The things that troubled me in my thirties about marriage and children don't bother me now because I'm at the age where many women have already lost their husbands.

Both men and women, however, recognized a difference in the marriage relationship that follows upon the departure of children, some describing it in positive, others in negative terms, but all recognizing a new marital adjustment to be made.

It's a totally new thing. Now there isn't the responsibility for the children. There's more privacy and freedom to be yourself. All of a sudden there are times when we can just sit down and have a conversation. And it was a treat to go on a vacation alone!

It's the boredom that has grown up between us but which we didn't face before. With the kids at home, we found something to talk about, but now the buffer between us is gone. There are just the two of us, face to face.

A recent review of the literature on the family of later life points to the theme of progressive disenchantment with marriage from the peak of the honeymoon to the nadir of middle life, and to the fact that a number of studies indicate that middle-aged marriages are more likely than not to be unsatisfactory. There are other studies, however, that indicate the opposite.[77] In either case, and whatever the multiplicity of possible interpretations, middle age is not the point at which marriages characteristically break up. On the contrary, nationwide statistics show that divorce rates are highest for teen-age marriages, then drop steadily with age and duration of marriage.

One difference between husbands and wives is marked. Most of the women interviewed feel that the most conspicuous characteristic of middle age is the sense of increased freedom. Not only is there increased time and energy available for the self, but also a satisfying change in self-concept takes place. The typical theme is that middle age marks the beginning of a period in which latent talents and capacities can be put to use in new directions.

Some of these women describe this sense of freedom coming at the same time that their husbands are reporting increased job pressures or—something equally troublesome—job boredom. Contrast this typical statement of a woman,

I discovered these last few years that I was old enough to admit to myself the things I could do well and to start doing them. I didn't think like this before. . . . It's a great new feeling.

with the statement of one man,

You're thankful your health is such that you can continue working up to this point. It's a matter of concern to me right now, to hang on. I'm forty-seven, and I have two children in college to support.

or with the statement of another man, this one a history professor,

I'm afraid I'm a bit envious of my wife. She went to work a few years ago, when our children no longer needed her attention, and a whole new world has opened to her. But myself? I just look forward to writing another volume, and then another volume.

The Work Career

Men, unlike women, perceive the onset of middle age by cues presented outside the family context, often from the deferential behavior accorded them in the work setting. One man described the first time a younger associate held open a door for him; another, being called by his official title by a newcomer in the company; another, the first time he was ceremoniously asked for advice by a younger man.

Men perceive a close relationship between life line and career line. Middle age is the time to take stock. Any disparity noted between career expectations and career achievements—that is, whether one is "on time" or "late" in reaching career goals—adds to the heightened awareness of age. One lawyer said,

I moved at age forty-five from a large corporation to a law firm. I got out at the last possible moment, because if you haven't made it by then, you had better make it fast, or you are stuck.

There is good evidence that among men most of the upward occupational mobility that occurs is largely completed by the beginning of the middle years, or by age 35.[18,39] Some of the more highly educated continue to move up the ladder in their forties and occasionally in their fifties. On the other hand, some men, generally the less schooled, start slipping sometime in the years from 35 to 55. The majority tend to hold on, throughout this period, to whatever rung they managed to reach. Family income does not always reflect a man's job status, for by the time of the mid-forties such large numbers of wives have taken jobs that family income often continues to rise. For this and other reasons there is considerable variation in family income in middle age as compared to earlier periods in the family cycle.

The Changing Body

The most dramatic cues for the male are often biological. The increased attention centered upon his health, the decrease in the efficiency of the body, the death of friends of the same age—these are the signs that prompt many men to describe bodily changes as the most salient characteristic of middle age.

> Mentally I still feel young, but suddenly one day my son beat me at tennis.

Or,

> It was the sudden heart attack in a friend that made the difference. I realized that I could no longer count on my body as I used to do . . . the body is now unpredictable.

One 44 year old added,

> Of course I'm not as young as I used to be, and it's true, the refrain you hear so often in the provocative jokes about the decrease in sexual power in men. But it isn't so much the loss of sexual interest, I think it's more the energy factor.

A decrease in sexual vigor is frequently commented on as a normal slowing down: "my needs have grown less frequent as I've gotten older," or "sex isn't as important as it once was." The effect is often described as having little effect on the quality of the marriage.

> I think as the years go by you have less sexual desire. In fact, when you're younger there's a *need*, in addition to the desire, and that need diminishes without the personal relationship becoming strained or less close and warm. . . . I still enjoy sex, but not with the fervor of youth. . . . Not because I've lost my feelings for my wife, but because it happens to you physically.

Although a number of small-scale studies have appeared recently, the data are poor regarding the sexual behavior of middle-aged and older people. It would appear that sexual activity remains higher than the earlier stereotypes would indicate and that sexual activity in middle age tends to be consistent with the individuals' earlier behavior; but at the same time there is gradual decrease with age in most persons, and the incidence of sexual inadequacy takes a sharp upturn in men after age 50. Masters and Johnson, making many of the same points made earlier by Kinsey and others, point to the manifold physiological and psychological factors involved in sexual behavior in both sexes.[19,43,50] Most clinicians seem to agree with their view that in a high percentage of cases men can be successfully treated for secondarily acquired impotence, and that the diminution of both the male's sexual prowess and the female's responsiveness is the reflection of psychological rather than biological factors, mainly the boredom and monotony of repetitious sexual relationship, preoccupations with career or family pressures, mental or physical fatigue, health concerns, and, particularly for the male, fear of performance associated with any of these factors.

Changes in health and in sexual performance are more of an age marker for men than for women. Despite the menopause and other manifestations of the climacterium, women refer much less frequently to biological changes or to concern over health. Body monitoring is the term we used to describe the large variety of protective strategies for maintaining the middle-aged body at given levels of performance and appearance; but while these issues take the form of a new sense of physical vulnerability in men, they take the form of a rehearsal for widowhood in women. Women are more concerned over the body monitoring of their husbands than of themselves.

That widowhood is a critical concern for middle-aged women is borne out by other studies. Not only is there the grief and reorganization of established life patterns to face, but widows experience a drop in status with the death of their husbands.[46] The situation of many resembles that of a minority group who are singled out for unequal treatment and who regard themselves as objects of social discrimination. Many widows feel demeaned by limiting their socializing to other widows and feel that friends avoid them in the attempt to ignore the whole subject of death and grief. The rehearsal for widowhood is obviously reality-based. With 20 to 25 per cent of all women aged 55 to 65 living as widows, most married women number among their friends or relatives other women who have been recently widowed.

The Changing Time Perspective

Both sexes, although men more than women, talked of the new difference in the way time is perceived. Life is restructured in terms of time left to live rather than time since birth. Not only the reversal in directionality, but the awareness that time is finite, is a particularly conspicuous feature of middle age. Thus,

> You hear so much about deaths that seem premature. That's one of the changes that comes over you over the years. Young fellows never give it a thought.

Another said,

> Time is now a two-edged sword. To some of my friends, it acts as a prod; to others, a brake. It adds a certain anxiety, but I must also say it adds a certain zest in seeing how much pleasure can still be obtained, how many good years one can still arrange, how many new activities can be undertaken.

The recognition that there is "only so much time left" was a frequent theme in the interviews. In referring to the death of a contemporary, one man said,

> There is now the realization that death is very real. Those things don't quite penetrate when you're in your twenties and you think that life is all ahead of you. Now you know that death will come to you, too.

This last-named phenomenon we called the personalization of death: the awareness that one's own death is inevitable and that one must begin to come to terms with that actuality. Death rates over the life span show a sudden and dramatic rise at middle age. The rate for men aged 45 to 64 is six times as high as it is in the preceding 20-year period, and for women it is three times as high. A second factor that may be equally significant is that from childhood through early adulthood the leading cause of death is accidents, but for the age range 45 to 64, for both men and women, malignant neoplasms and heart disease account for nearly two-thirds of all deaths. To put this another way, in early life death is exceptional and accidental; but in middle age not only does death strike frequently, but it strikes from within.

The Prime of Life

Despite the new realization of the finiteness of time, one of the most prevailing themes expressed by middle-aged respondents is that middle adulthood is the period of maximum capacity and ability to handle a highly complex environment and a highly differentiated self. Very few express a wish to be young again. As one of them said,

> There is a difference between wanting to *feel* young and wanting to *be* young. Of course it would be pleasant to maintain the vigor and appearance of youth; but I would not trade those things for the authority or the autonomy I feel—no, nor the ease of interpersonal relationships nor the self-confidence that comes from experience.

The middle-aged individual, having learned to cope with the many contingencies of childhood, adolescence, and young adulthood, now has available to him a substantial repertoire of strategies for dealing with life. One woman put it,

> I know what will work in most situations, and what will not. I am well beyond the trial and error stage of youth. I now have a set of guidelines. . . . And I am practiced.

Whether or not they are correct in their assessments, most of our respondents perceive striking improvement in their exercise of judgment. For both men and women the perception of greater maturity and a better grasp of realities is one of the most reassuring aspects of being middle-aged.

> You feel you have lived long enough to have learned a few things that nobody can learn earlier. That's the reward . . . and also the excitement. I now see things in books, in people, in music that I couldn't see when I was younger. . . . It's a form of ripening that I attribute largely to my present age.

There are a number of manifestations of this sense of competence. There is, for instance, the 45 year old's sensitivity to the self as the instrument by which to reach his goals; what we have called a preoccupation with self-utilization (as contrasted to the self-consciousness of the adolescent):

> I know now exactly what I can do best, and how to make the best use of my time. . . . I know how to delegate authority, but also what decisions to make myself. . . . I know how to protect myself from troublesome people . . . one well-placed telephone call will get me what I need. It takes time to learn how to cut through the red tape and how to get the organization to work for me. . . . All this is what makes the difference between me and a young man, and it's all this that gives me the advantage.

Other studies have shown that the perception of middle age as the peak period of life is shared by young, middle-aged, and older respondents.[3] In one such study there was consensus that the middle-aged are not only the wealthiest, but the most powerful; not only the most knowledgeable, but the most skillful.[16]

There is also the heightened self-understanding that provides gratification. One perceptive woman described it in these terms:

> It is as if there are two mirrors before me, each held at a partial angle. I see part of myself in my mother who is growing old, and part of her in me. In the other mirror, I see part of myself in my daughter. I have had some dramatic insights, just from looking in those mirrors. . . . It is a set of revelations that I suppose can only come when you are in the middle of three generations.

In pondering the data on these men and women, we have been impressed with the central importance of what might be called the executive processes of personality in middle age: self-awareness, selectivity, manipulation and control of the environment, mastery, competence, the wide array of cognitive strategies. We are impressed, too, with reflection as a striking characteristic of the mental life of middle-aged persons: the stocktaking, the heightened introspection, and, above all, the structuring and restructuring of experience—that is, the conscious processing of new information in the light of what one has already learned and the turning of one's proficiency to the achievement of desired ends. These people feel that they effectively manipulate their social environments on the basis of prestige and expertise; and that they create many of their own rules and norms. There is a sense of increased control over impulse life. The middle-aged person often describes himself as no longer "driven," but as the "driver"—in short, "in command."

Although the self-reports quoted here were given by highly educated and successful persons, they express many of the same attitudes expressed less fluently by persons of less education and less achievement, persons who also feel in middle age the same increasing distance from the young, the stocktaking, the changing time perspective, and the higher degrees of expertise and self-understanding.

Furthermore, while these self-reports are taken from a single study, they demonstrate the salient issues identified by researchers who have studied middle age in various biological, social, and psychological contexts, and they are remarkably consistent with theoretical formulations of development in adulthood as set forth by psychologists and psychiatrists. They seem to us, for instance, to be congruent with the views of students of the life cycle such as Erikson,[28] Jung,[40] Buhler,[13,14] and others who conceive of midlife as a developmental period in its own right, a period in which the personality has new dilemmas and new possibilities for change; and when the personality is not to be understood as fixated upon the conflicts of childhood.[7,10,12,13,14,15,29] These self-reports are also congruent with the views of Havighurst,[36] Peck,[67] and others who have described the developmental tasks of middle-age, and with the insights of clinicians like Butler,[15] Levinson,[44] Gould,[31] and Soddy,[73] who have turned their attention to gathering data on nonclinical populations.

Is Middle Age a Crisis Period?

As mentioned earlier, middle age is often described in the psychiatric as well as in the popular literature as a period of crisis. The climacterium and the departure of children from the home are usually mentioned as trauma-producing events for women; and for men, health problems, sexual impotence, and career decline. The implication is often that early middle age is second only to adolescence as a period of stress and distress.

The Climacterium

Like puberty and pregnancy the climacterium is generally regarded as a significant turning point in a woman's psychosexual development; one that frequently reflects profound psychological as well as endocrine and somatic changes. Because it signifies that reproductive life has come to an end, it has often been described as a potential threat to a woman's feminine identity, and adaptation to the climacterium as one of the major tasks of a woman's life.[25] Benedek[7] is one of the few psychoanalysts who has taken a more optimistic view; she regards it as a developmental phase in which psychic energy previously used to cope with the fluctuations of the menstrual cycle and reproduction is now released for new forms of psychological and social expansion.

Although there is a large medical and popular literature on the climacterium, there is a conspicuous lack of psychological research with nonclinical samples. While an estimated 75 per cent of women experience some discomfort during the climacterium, only a small proportion receive medical treatment, suggesting that conclusions drawn from clinical observations cannot be generalized to the larger population.

For example, in one study of 100 working-class and middle-class women aged 43 to 53, all of them in good health, all with children of high school age, data were obtained on a large number of psychological and social variables, including both overt and covert measures of anxiety, life satisfaction, and self-concept.[61] It was found that these women minimized the significance of the menopause, regarding it as unlikely to produce much anxiety or stress. Among the aspects disliked most about middle age, only one woman of the 100 mentioned menopause, and even after considerable time given to the topic on two different interview occasions, only one-third could think of any way that a woman's physical or emotional health was likely to be adversely affected. Many welcomed menopause as relief from menstruation and fear of unwanted pregnancies. A majority maintained that any changes in health, sexuality, or emotional status during the climacteric period were caused by idiosyncratic factors or individual differences in general capacity to tolerate stress. "It depends on the individual. Personally I think if women look for trouble, they find it."

On a specially devised checklist of attitudes toward menopause, the large majority attributed no discontinuity in a woman's life to the climacterium. Three-fourths took the view that, except for the underlying biological changes, women even have a relative degree of control over their symptoms. Using a checklist of those menopausal symptoms most frequently reported in the medical literature (hot flushes, paresthesia, irritability, and so on), it was found that even those women with high symptom scores discounted the importance of the climacterium as a factor in their current morale. And using several different measures of psychological well-being, some based on standardized tests, others on projective tests, some on direct self-report, very little correlation was found between psychological well-being and measures of climacteric status, symptoms, or attitudes toward menopause. The study produced no evidence, in short, to support a crisis view of the climacterium.

In two related studies the attitudes-toward-menopause and the symptom checklists were administered to several hundred women who ranged in age from 13 to 65.[60,64] It was found that young women (under 40) had more negative and more undifferentiated views of menopause, but the middle-aged and older women saw it as only a temporary inconvenience. Highest frequency of symptoms occurred in the adolescents and in the menopausal women; but at adolescence the symptoms were primarily emotional, at menopause, primarily somatic. In only a few scattered instances were psychological symptoms reported more frequently by menopausal women than by any of the other age groups.

A similar picture among European-born women emerged in a cross-cultural study that included women of three Israeli subcultures: European immigrants, Near Eastern Jewish immigrants, and Israeli-born Muslim Arabs.[22,49] When European-born women were interviewed about their concerns at middle age, they seldom related them to climacteric change. This was not true, however, of the other two cultural groups, both of whom came from traditional settings in which the role of women had altered relatively little since Biblical times. For women in the latter groups climacteric changes were more salient and more closely related to their perceptions of major changes at middle age.

Nor did our data support the common view that the cessation of fertility is perceived as a major loss. Among the 100 American women just described, no regret over lost fertility was expressed: on the contrary, many women stated that they were happy to be done with childrearing.[61] A survey of nearly 1,200 middle-aged women of five Israeli subcultures that varied from traditionalism to modernity produced a parallel finding. Despite great differences in family size (an average of more than eight children among the most traditional women, and an average of two children among the most modern women), women in every cultural group emphatically welcomed the cessation of fertility.[20]

All this is not to gainsay the fact that while perhaps most middle-aged women attach only secondary significance to the climacterium, some experience considerable disturbance and should and do seek out treatment. Some clinicians have proposed that any physiological or psychological distress at climacterium is primarily due to temporary endocrine imbalance and should be treated by correcting the estrogen deficiency;[42] others have come to view the menopause as itself a hormone-deficiency disease, advocating estrogen maintenance from puberty to death;[22,80,81] but still others take a different view and see the distress as primarily due to psychodynamic factors or failure in psychosocial adaptation.[27] Therapeutic approaches vary accordingly, although it is usual for the therapeutic approach to be geared to both physiological and psychological dimensions.[17,74] In restricting the present chapter to the discussion of social psychological factors, we wish merely to point to the importance of biological and psychodynamic factors as well, but to leave this last topic to be treated at greater length elsewhere in this *Handbook*.

Climacterium in men is discussed by many writers; it is generally concluded that although there are exceptional occurrences of abrupt involution comparable to ovarian involution, and occasional reports of symptoms such as headaches, dizziness, and hot flushes, if there is any change at all for the majority of men other than the gradual involution of senescence, the change is neither abrupt nor universal.[68,73,76,78] It has been pointed out also that even if there is a decline in spermatogenesis the psychological consequences of such a change are unknown. Redlich and Freedman[68] are among those who suggest that the mild depressive symptoms seen among men in their fifties are perhaps related to a career decline rather than to intrinsic organic change. Soddy,[73] however, compares the climacteric period to adolescence, noting that although there is no definitive physiological marker among adolescent boys comparable to menarche in girls, the psychological changes are probably similar. He suggests that the logical inference is that there are very few differences between the sexes in attitude changes that are due directly to hormonal influences; sex differences are more likely due to cultural and social factors related to the more manifest climacteric changes in women.

The Empty Nest

With regard to another presumed crisis, the empty nest, available studies indicate that it, too, has been given exaggerated importance in terms of its consequences for mental health. One set of evidence comes from the women whose reactions to the menopause have just been described.[61] These women were divided according to family stage: the intact stage, in which none of the children had yet left home; the transitional stage, with one or more gone but one or more remaining; and the empty nest stage. Those women who had children under age 14 at home were also compared with those whose youngest child was 15 or older. Life styles or role orientations were identified, and the sample was separated into those who were primarily home-oriented, community-oriented, work-oriented, or mixed home-community-oriented. These women were studied also for *change* in the pattern of role activities, assessing the extent to which each woman had expanded or constricted her activities in family and in nonfamily roles over the past five to ten years. The women were grouped into "expanders," "shifters," "statics," and "constrictors." The relationships between these social role patterns and measures of psychological well-being were low. Rather than being a stressful period for women, the empty nest or postparental stage was associated with a somewhat *higher* level of life satisfaction than were the the other family stages. For at least the women in this sample, coping with children at home was presumably more stressful than seeing their children launched into adult society.

There is other evidence.[26] For example, a study of 54 middle-aged men and women whose youngest child was about to leave home showed that while some persons in the sample had serious problems, the problems were not related to the departure of the children.[47] The authors concluded that the confrontation of the empty nest, when compared with retrospections of the low points in the past and expectations of the future, is not of a nature to justify the use of the term "crisis."

Indices of Mental Health

Somewhat the same case can be made for the effects of other life events. While important alterations in the life space of the middle-aged person take place and may necessitate a certain degree of personal reorganization, it can be argued that the normal life events of middle age do not in themselves constitute emergencies for most people. This argument is supported by national statistics for age of first admission to mental hospitals (1967), which rise from childhood to a peak in the period 25 to 34—that is, during young adulthood—and then drop gradually through middle age, rising sharply after age 65. Similarly, in a study of 2,500 adults 21 and over, reports of past emotional crises were given by about a fifth in all age groups, but by a slightly higher proportion in the group aged 35 to 44, with the proportion then dropping steadily with age.[34] The same study also showed that self-reported symptoms of psychological distress rise somewhat with age, but show no tendency to peak in middle age. Finally U.S. suicide rates climb steadily with age for males but do not peak in middle age, while for females the rate remains low with a slight drop after age 60 to 69. Thus, neither self-reports nor gross external criteria such as these support the crisis view of the middle years.

We interpret all this as support for our view of the "normal, expectable life cycle." Adults develop a set of anticipations of the normal life events, not only what those events should be, but also when they should occur.[54] They make plans, set goals, and reassess those goals along a time line shaped by those expectations. Adults have a sense of the life cycle: that is, an anticipation and acceptance of the inevitable sequence of events that occur as men grow up, grow old, and die; and they understand that their own lives will be similar to the lives of others and that the turning points are inescapable. This ability to interpret the past and foresee the future, and to create for oneself a sense of the predictable life cycle, presumably differentiates the healthy adult personality from the unhealthy one. From this point of view the normal, expectable life events are seldom trauma-producing. Women in their forties and fifties regard the climacterium as inevitable; they know that all women survive it; and most women therefore take it in stride. Similarly men and women expect their children to grow up and leave home, just as they themselves did in their own youth, and their feelings of relief and pride are important parts of their mixed emotions.

The normal, expectable life event too superficially viewed as a crisis event can be illustrated for events at the upper boundary of middle age as well. To an increasingly large proportion of men, retirement is a normal, expectable event. Yet in much of the literature on the topic, the investigator conceptualizes it as a crisis, with the result that the findings from different studies are at variance, with some investigators unprepared for their discovery of no significant losses in life satisfaction or no increased rates of depression following retirement. The fact is that retirement is becoming a middle-aged phenomenon, with many workers now being offered and accepting the opportunity to withdraw from work at age 55. The latest national survey indicates that a surprisingly large proportion of workers in all industries are choosing to retire earlier and earlier, with the main, if not the single, determining factor being level of income—as soon as a man establishes enough retirement income, he chooses to stop working. A more recent study shows that nearly 70 per cent of persons who retired as *planned* were content in their retirement, compared with less than 20 per cent of the unexpected retirees, those who retired unexpectedly because of poor health or loss of job.[5] Even death becomes a normal, expectable event to the old, and there are various studies that describe the relative equanimity with which it is anticipated.[51] Judging from our many interviews with old people gathered in the course of large-scale studies of aging, the crisis is not death itself as much as the manner in which one will die. It appears that it is the prospect of dying in a nonnormal, unexpected circumstance (in an institutional setting, say, rather than in one's accustomed family setting) that creates the crisis.[45]

The situation with regard to widowhood is more equivocal; yet even here one study has shown that while somatic symptoms increased somewhat for both younger and older widows in the months following bereavement, consultation rates for psychiatric symptoms were very high for women under 65 but not for women over 65.[66] In a large-scale study of American widows, it has been pointed out that the hardest hit are those who experience widowhood earlier than their associates.[46] On the same point a review of epidemiological data regarding the widowed as a high-risk group in terms of mental illness, physical illness, and mortality indicate that there is excess risk at younger rather than at older ages for both sexes.[4] Is this because by the time a woman reaches old age the death of her husband, a husband usually several years her senior, moves into the category of the expected?

All these findings are not to deny that the expectable life event precipitates crisis reactions in some persons, especially in those who come to the attention of the mental health professional. But even for the minority it may more often be the *timing* of the life event, not its occurrence, that constitutes the problematic issue. This observation is not a denial of the fact that many of the major life events in middle age (and in old age) are losses to the individuals concerned and that grief is their accompaniment. It is to say, rather, that the events are anticipated and rehearsed, the grief work completed, the reconciliation accomplished without shattering the sense of continuity of the life cycle.

In drawing a distinction between illness and health, between the clinical and the normal, the psychology of grief is not synonymous with the psychology of mental illness. The relationships between loss, grief, physical illness, and mental illness are complex, and "loss" is itself a multidimensional factor. A recent study supporting this point revealed that persons who had experienced retirement or widowhood in the three years prior to being interviewed were not more frequently diagnosed by psychiatrists as mentally ill, nor had they found their way into mental hospitals with any greater frequency than others.[48] Mental illness, on the other hand, was associated with self-blame, with reports of having missed one's opportunities or having failed to live up to one's potentials; in short, with intrapunitiveness.

The distinction is worth making then, that it is the unanticipated, not the anticipated, that is likely to represent the traumatic event. Major stresses are caused by events that upset the sequence and rhythm of the expected life cycle, as when the death of a parent comes in adolescence rather than in middle age; when marriage does not come at its desired or appropriate time; when the birth of a child is too early or too late; when occupational achievement is delayed; when the empty nest, grandparenthood, retirement, major illness, or widowhood occur off time.

From this perspective it is an inaccurate view that middle age constitutes a crisis period in the life cycle any more than any other period of life. For most persons middle age brings with it the anticipated changes in family and work and health. Some of these changes are not necessarily interpreted as losses by the people who experience them. Whether perceived as losses or gains, the life events of middle age may produce new stresses for the individual, but they also bring occasions to demonstrate an enriched sense of self and new capacities for coping with complexity.

Bibliography

1. ANDERSON, J. E. (Ed.), *Psychological Aspects of Aging*, American Psychological Association, Washington, D.C., 1956.
2. ARIES, P., *Centuries of Childhood*, Random House, New York, 1962.
3. BACK, K. W., and BOURQUE, L. B., "Life Graphs: Aging and Cohort Effects," *J. Gerontol.*, 25:249–255, 1970.
4. BALER, L. A., and GOLDE, P., "Conjugal Bereavement: Strategic Area of Research in Preventive Psychiatry," in *Working Papers in Community Mental Health*, Vol. 2, No. 1, Harvard School of Public Health, Department of Public Health Practice, Cambridge, 1964.
5. BARFIELD, R., *The Automobile Worker and Retirement: A Second Look*, Institute for Social Research, University of Michigan, Ann Arbor, 1970.
6. BART, P., "Depression in Middle-Aged Women: Some Socio-Cultural Factors," Ph.D. diss., University of California, Los Angeles, 1967.
7. BENEDEK, T., "Climacterium: A Developmental Phase," *Psychoanal. Quart.*, *19*:1–27, 1950.
8. ———, "On the Psychic Economy of Developmental Processes," *Arch. Gen. Psychiat.*, *17*:271–276, 1967.
9. BIGELOW, N., "The Involutional Psychoses," in Arieti, S. (Ed.), *American Handbook of Psychiatry*, Vol. 1, pp. 540–545, Basic Books, New York, 1959.
10. BIRREN, J. E. (Ed.), *Relations of Development and Aging*, Charles C Thomas, Springfield, Ill., 1964.
11. BOTWINICK, J., *Cognitive Processes in Maturity and Old Age*, Springer, New York, 1967.
12. BRIM, O. G., and WHEELER, S., *Socialization after Childhood*, John Wiley, New York, 1966.
13. BUHLER, C., "The Course of Human Life as a Psychological Problem," *Human Dev.*, *11*:184–200, 1968.
14. ———, and MASSARIK, F. (Eds.), *The Course of Human Life*, Springer, New York, 1968.
15. BUTLER, R. N., "Toward a Psychiatry of the Life Cycle," *Psychiat. Res. Rep.*, *23*:233–248, 1968.
16. CAMERON, P., "The Generation Gap: Which Generation Is Believed Powerful versus Generational Members' Self-Appraisals of Power," *Develop. Psychol.*, *3*:403–404, 1970.
17. COHEN, S., DITMAN, K. S., and GUSTAFSON, S. R., *Psychochemotherapy: The Physician's Manual*, rev. ed., Western Medical Publications, Los Angeles, 1967.
18. COLEMAN, R., and NEUGARTEN, B. L., *Social Status in the City*, Jossey-Bass, San Francisco, 1971.
19. CRISTENSON, C. V., and GAGNON, J. H., "Sexual Behavior in a Group of Older Women," *J. Gerontol.*, *20*:351–357, 1965.
20. DATAN, N., "Women's Attitudes towards the Climacterium in Five Israeli Subcultures," Ph.D. diss., University of Chicago, 1971.
21. ———, MAOZ, B., ANTONOVSKY, A., and WIJSENBEEK, H., "Climacterium in Three Cultural Contexts," *Tropical & Geographical Med.*, *22*:77–86, 1970.
22. DAVIS, M. E., LANZL, L. H., and COX, A. B., "Detection, Prevention and Retardation of Menopausal Osteoporosis," *Obstetrics & Gynecol.*, *36*:187–198, 1970.

23. DE BEAUVOIR, S., *The Second Sex*, Knopf, New York, 1953.

24. DEMOS, J., and DEMOS, V., "Adolescence in Historical Perspective," *J. Marriage & Fam.*, *31*:632–638, 1969.

25. DEUTSCH, H., "The Climacterium," in Deutsch, H., *The Psychology of Women*, Vol. 2, *Motherhood*, Grune & Stratton, New York, 1945.

26. DEUTSCHER, I., "Socialization for Postparental Life," in Cavan, R. S. (Ed.), *Marriage and Family in the Modern World*, 2nd ed., Thomas Y. Crowell, New York, 1965.

27. DUNLOP, E., "Emotional Imbalance in the Premenopausal Woman," *Psychosom.*, *9*: 44–47, 1968.

28. ERIKSON, E. H., *Childhood and Society*, Norton, New York, 1950.

29. FRENKEL-BRUNSWICK, E., "Adjustments and Reorientations in the Course of the Life Span," in Neugarten, B. L. (Ed.), *Middle Age and Aging*, University of Chicago Press, Chicago, 1968.

30. GLICK, P. C., HEER, D. M., and BERESFORD, J., "Family Formation and Family Composition: Trends and Prospects," in Sussman, M. B. (Ed.), *Sourcebook in Marriage and the Family*, 2nd ed., Houghton Mifflin, Boston, 1963.

31. GOULD, R. L., "The Phases of Adult Life: A Study in Developmental Psychology," *Am. J. Psychiat.*, *129*:521–531, 1972.

32. GOULET, L. R., and BALTES, P. B., *Life Span Developmental Psychology*, Academic Press, New York, 1970.

33. GREGORY, F. W., and NEU, I. D., "The American Business Elite in the 1870's," in Miller, W. (Ed.), *Men in Business*, Harvard University Press, Cambridge, 1952.

34. GURIN, G., VEROFF, J., and FELD, S., *Americans View Their Mental Health: A Nationwide Interview Study*, Basic Books, New York, 1960.

35. GUTMANN, D. L., "An Exploration of Ego Configurations," in Neugarten, B. L., *et al.*, *Personality in Middle and Late Life*, Atherton Press, New York, 1964.

36. HAVIGHURST, R. J., "Changing Roles of Women in the Middle Years," in Gross, I. (Ed.), *Potentialities of Women in the Middle Years*, Michigan State University Press, Lansing, 1956.

37. ———, "The Social Competence of Middle-aged People," *Genet. Psychol. Monog.*, *56*:297–375, 1957.

38. JACQUES, E., "Death and the Mid-life Crises," *Internat. J. Psychoanal.*, *46*:502–514, 1965.

39. JAFFE, A. J., "The Middle Years," *Industrial Gerontology*, September 1971, special issue.

40. JUNG, C. G., "The Stages of Life," in C. G. Jung, *The Collected Works of C. G. Jung*, Vol. 8, *The Structure and Dynamics of the Psyche*, Pp. 387–403, Pantheon Books, New York, 1960.

41. KENISTON, K., "Youth as a Stage of Life," *Am. Scholar*, *39*:631–654, 1970.

42. KERR, M. D., "Psychohormonal Approach to the Menopause," *Modern Treatment*, *5*: 587–595, 1968.

43. KINSEY, A. C., *et al.*, *Sexual Behavior in the Human Female*, W. B. Saunders, Philadelphia, 1953.

44. KLERMAN, G. L., and LEVINSON, D. J., "Becoming the Director: Promotion as a Phase in Personal-Professional Development," *Psychiatry*, *32*:411–427, 1969.

45. LIEBERMAN, M. A., and COPLAN, A. S., "Distance from Death as a Variable in the Study of Aging," *Develop. Psychol.*, *2*: 71–84, 1970.

46. LOPATA, H., *Widowhood in an American City*, Schenkman Publishing Co., Cambridge, Mass., 1972.

47. LOWENTHAL, M. F., and CHIRIBOGA, D., "Transition to the Empty Nest," *Arch. Gen. Psychiat.*, *26*:8–14, 1972.

48. ———, *et al.*, *Aging and Mental Disorder in San Francisco: A Social Psychiatric Study*, Jossey-Bass, San Francisco, 1967.

49. MAOZ, B., DOWTY, N., ANTONOVSKY, A., and WIJSENBEEK, H., "Female Attitudes to Menopause," *Soc. Psychiat.*, *5*:35–40, 1970.

50. MASTERS, W. H., and JOHNSON, V. E., *Human Sexual Response*, Little, Brown, Boston, 1966.

51. MUNNICHS, J. M. A., *Old Age and Finitude: A Contribution to Psychogerontology*, Karger, Basel, 1966.

52. MYRDAL, A., and KLEIN, V., *Women's Two Roles: Home and Work*, 2nd ed., Routledge and Kegan Paul, London, 1968.

53. NEUGARTEN, B. L., "The Awareness of Middle Age," in Owen, R. (Ed.), *Middle Age*, British Broadcasting Co., London, 1967.

54. ———, "Dynamics of Transition of Middle Age to Old Age: Adaptation and the Life Cycle," *J. Geriatric Psychiat.*, *4*:71–87, 1970.

55. ——— (Ed.), *Middle Age and Aging: A Reader in Social Psychology*, University of Chicago Press, Chicago, 1968.

56. ———, *et al.*, *Personality in Middle and Late Life*, Atherton Press, New York, 1964.

57. ———, and GARRON, D. C., "Attitudes of Middle-aged Persons toward Growing Older," *Geriatrics*, *14*:21–24, 1959.

58. ———, and GUTMANN, D. L., "Age-Sex Roles and Personality in Middle Age: A Thematic Apperception Study," *Psychol. Monog.*, *72*, 1958.

59. ———, HAVIGHURST, R. J., and TOBIN, S. S., "The Measurement of Life Satisfaction," *J. Gerontol.*, *16*:134–143, 1961.

60. ———, and KRAINES, R. J., "Menopausal Symptoms in Women of Various Ages," *Psychosom. Med.*, *27*:266–273, 1965.

61. ———, KRAINES, R. J., and WOOD, V., "Women in the Middle Years," unpublished manuscript on file, Committee on Human Development, University of Chicago, 1965.

62. ———, and PETERSON, W. A., "A Study of the American Age-grade System," *Proceedings of the Fourth Congress of the International Association of Gerontology*, *3*:497–502, 1957.

63. ———, and WEINSTEIN, K., "The Changing American Grandparent," *J. Marriage & Fam.*, *26*:199–204, 1964.

64. ———, WOOD, V., KRAINES, R. J., and LOOMIS, B., "Women's Attitudes toward the Menopause," *Vita Humana* (now *Human Dev.*), 6:140–151, 1963.

65. OWEN, R. (Ed.), *Middle Age*, British Broadcasting Co., London, 1967.

66. PARKES, C. M., "Effects of Bereavement on Physical and Mental Health: A Study of the Medical Records of Widows," *Brit. Med. J.*, *2*:274–279, 1964.

67. PECK, R. F., and BERKOWITZ, H., "Personality and Adjustment in Middle Age," in Neugarten, B. L., *et al.*, *Personality in Middle and Late Life*, Atherton Press, New York, 1964.

68. REDLICH, F. C., and FREEDMAN, D. X., *The Theory and Practice of Psychiatry*, Basic Books, New York, 1966.

69. RIEGEL, K. F., "On the History of Psychological Gerontology," Working Paper No. 8, Center for Research on Conflict Resolution, History of Science Program, University of Michigan, Ann Arbor, 1971.

70. RILEY, M. W., FONER, A., *et al.*, *Aging and Society*, Vol. 1, *An Inventory of Research Findings*, Russell Sage Foundation, New York, 1968.

71. ———, RILEY, J. W., JOHNSON, M. E., *et al.* (Eds.), *Aging and Society*, Vol. 2, *Aging and the Professions*, Russell Sage Foundation, New York, 1969.

72. ROTHSTEIN, S. H., "Aging Awareness and Personalization of Death in the Young and Middle Years," Ph.D. diss., University of Chicago, 1967.

73. SODDY, K., *Men in Middle Life*, Tavistock, London, 1967.

74. SONKIN, L. S., and COHEN, E. J., "Treatment of the Menopause," *Modern Treatment*, *5*:545–563, 1968.

75. SUSSMAN, M. B., "The Help Pattern in the Middle Class Family," *Am. Sociol. Rev.*, *18*:22–28, 1957.

76. SZALITA, A. B., "Psychodynamics of Disorders of the Involutional Age," in Arieti, S. (Ed.), *American Handbook of Psychiatry*, Vol. 3, Basic Books, New York, 1966.

77. TROLL, L. E., "The Family of Later Life: A Decade Review," *J. Marriage & Fam.*, 263–290, 1971.

78. VEDDER, C. B., *Problems of the Middle Aged*, Pp. 45–64, Charles C Thomas, Springfield, Ill., 1965.

79. WARNER, W. L., and ABEGGLEN, J. C., *Occupational Mobility in American Business and Industry, 1928–1952*, University of Minnesota Press, Minneapolis, 1955.

80. WILSON, R. A., *Forever Feminine*, M. Evans and Co., New York, 1966.

81. ———, and WILSON, T. A., "The Fate of the Non-treated Post-menopausal Woman," *J. Am. Geriatric Soc.*, *11*:347–362, 1963.

CHAPTER 30

ADJUSTMENT TO RETIREMENT

Eugene A. Friedmann and Harold L. Orbach

Retirement and Industrial Society

RETIREMENT IS A CREATION of modern industrial society, an emerging pattern of social life that is without historical precedent. While prior socioeconomic systems have had varying numbers of older people, none has ever had the number or the proportion that are found in industrialized societies of the present. Nor have they experienced the phenomenon of withdrawal from work or other major roles based on considerations of age alone. Unlike other social role changes throughout the life span, which are marked by a series of role transitions and gradations from one more or less clearly defined social role to another, the retirement role is beset with a lack of socially defined positions and behaviors.[123] At best it is an emerging role in the process of being institutionalized. As a consequence considerations of adjustment to retirement cannot be viewed solely as the analysis of developmental variables operating in the context of relatively fixed social structural parameters. Rather, the process of adjustment must also be viewed in terms of the patternings and availabilities of social roles that are emerging for this life stage and the extent and the manner in which the individual perceives and accommodates himself to societal change.

Two perspectives are therefore required for the analysis of the transition to retirement, the *developmental* and the *historical*. The developmental perspective assumes generic processes of growth and change common to members of a group as they traverse the life span.[146] It posits established norms and role behaviors for each stage of the life cycle, and socialization processes that prepare the individual for and induct him into successive stages. The historical perspective focuses upon changes that occur in society's system of age role definitions *during* the life span of the individual and his generational cohort. These are factors that may transcend the usual boundaries of life cycle and socialization theory by adding a generational qualification to generic models of life cycle development. Thus, for example, if we were to observe political attitudes common to the aged of the 1960's that differed from those of younger age groups, we might infer, employing our developmental perspective, that these are changes that occur with aging. However, if historical inquiry reveals that the attitudes held by older persons in the 1960's were indeed similar to those acquired during their period of political attitude formation some 40 or 50 years earlier, then the change we have observed has not occurred to the individual as a consequence of the aging process; instead, it is a change in the larger societal system resulting in generational differences in political outlooks. In similar fashion, attitudes and adjustments to retirement can also be viewed from the perspective of developmental changes, which affect the individual's ability to accommodate himself to the role expectations and stresses of this period of life, and from an historical perspective, which views retirement as an emerging role for which earlier generations could not have acquired relevant definitions or expectations.[64,68,125]

Population Aging in Industrial Societies

The change from agriculture to manufacture as the basic mode of production that has been occurring worldwide over the past century and a half has set into motion a series of trends that are transforming the position of the aged in human society. The more advanced stages of industrialization have been accompanied by major shifts in the age composition of populations. The rapid increase of the population 65 and over has been a recent phenomenon. Preindustrial societies typically have less than 4 per cent of their population in the 65 and over age group. Declining fertility and increasing survival efficiency at younger ages, associated with the later stages of industrialization, has led to a rapid increase in the number and proportion of aged in the advanced industrial nations.[166,195,201]* In 1850 only one nation, France, had as much as 6 per cent of its population in the 65 and over age group; in 1900 only three countries—France, Sweden, and Norway—had as much as 7 per cent of their population in this age group. In 1969 only the advanced industrial societies of the world (comprising about one-fourth of the world's population) had 7 per cent or more in the 65 and over age bracket, with a proportion of 13 per cent or higher recorded for Belgium, France, the United Kingdom, Sweden, Norway, and East Germany.[196] In the United States population aging has been a twentieth-century phenomenon. The proportion of 65 and over has increased from 4 per cent in 1900 to 10 per cent in 1970; in numbers the increase has been from 3 million to 20 million in this period.[197] The aged in our population and in other industrial societies in this century have been increasing at over double the rate of the population as a whole.

Age as the Criterion for Retirement

Industrialization, and the increasing complexity of social organization that has accompanied it, has also led to a change in the basis of participation in the labor force. The taking of production out of the home into the factory established a separation of work life from the home and family life. The growth of large-scale corporate enterprise has rapidly reduced the proportion of self-employed, creating a labor force of "employees" for whom decisions such as the time and mode of withdrawal from work were no longer a matter of self-determination. The growth of specialization of tasks in mass production industries required the breaking down of complex operations into simpler components tied together

* Increased life span has not been a significant factor in this increase. Although life expectancy at birth has increased dramatically in the last century, life expectancy at age 50, for example, has increased only a couple of years. Thus, more persons are now living to the later stages of adulthood but, once having reached there, they do not have a substantially greater life expectancy than their counterparts 100 or 200 years ago.

into an assembly line production, and the introduction of new technologies and forms of industrial organization led to obsolescence of traditional skills and patterns of work relationships.

From the point of view of the older worker, this meant that remaining in the labor force was no longer determined by his ability to turn out a product as such. Rather, it depended upon factors external to him: that is, keeping pace with a production line whose rate of operation is usually set in terms of the capacities of younger workers, and learning new techniques where the training investments are reserved for younger workers.[21,109,110,206] The response of industrial societies to the increase in the proportion of the aged and to their difficulties in accommodating themselves to the new modes of production has increasingly led to a withdrawal of the older workers from the labor force. In the United States, for example, the proportion of older males in the labor force has declined from 65 per cent in 1900 to 45 per cent in 1950 and 26 per cent in 1970.[197]

Institutionalization of Pension Systems

The withdrawal of older workers from the labor force has also led to the emergence of institutionalized provision for their maintenance in retirement. In the past hundred years public and private pension and insurance programs specifically for the support of the retired workers have developed in industrial societies. These were independent of the "poor relief" patterns of support of the indigent aged that had been part of the Western tradition;[35] instead, these programs provided compensation based upon years of service rather than upon need per se. They were to emerge as an "earned right" and were to become instrumental in defining a retirement status as appropriate for the old worker.

Public pension programs, developed first for the military in the eighteenth century, were applied to police, firemen, and other public occupations characterized by hazardous service and then were gradually extended to include almost all categories of public and civil service.[36] In 1970 over 95 per cent of public employees in the United States were covered under public retirement programs. Private pension programs emerged in the latter part of the nineteenth century in a few of the new large-scale corporate enterprises (for example, railroads), but it was not until the last two decades when they became subject to collective bargaining that they were to effect a significant and still expanding portion of the labor force. In 1969 48 per cent of the U.S. labor force was included in private pension programs,[81] and federal income tax provisions have recently given impetus to the establishment of pension programs for the self-employed. The first modern social insurance laws making provision for the aged were enacted in Germany in 1883 and 1889. This national system provided for compulsory health insurance and disability, old age, and death benefits. By 1935 when the U.S. Social Security Act was passed, 20 countries had old age insurance systems and 23 had compulsory health insurance laws.[58] Social Security today is considerably broadened in scope as well as in coverage from the acts passed in the late 1930's. In 1940 only 7 per cent of the aged population was receiving retirement benefits under this legislation; in 1950 the figure was 17 per cent; and in 1968 it had expanded to 63 per cent of the population.[197] Further, its provisions had been expanded to include retirement, survivors, disability, and medical care insurance.

Emergence of Retirement Roles

These measures that institutionalized maintenance provisions for older workers removed from the labor force also usually established an age at which they became eligible to retire. In so doing, they were defining a new age role for human society, one that was to have more extensive and profound consequences than was intended by the mechanisms that brought it about.[44,52,123] The retirement role that is now being established as a normative expectation for workers in advanced industrial societies as yet does not have clearly defined functions or behaviors associated with it; nor is there a group of persons established in this role for a long enough period of time to serve as models for persons approaching this life stage.

Further, societal value systems themselves have failed to adjust to the changing significance of this new life stage and contain both contradictions and paradoxes. Thus, pensions and insurance provisions, which define retirement as a right and a deferred payment for service, still hold some of the stigma once attached to disability and indigency payments. A society that has for so long stressed only work-related values seems to be reluctant to accept without doubt or recrimination the possibility of acquiring a large leisure class in the form of the retired aged. Finally the aged themselves reflect this dilemma to the extent that they evaluate the merit of a status or activity in terms of its recognition in a market economy, rather than attaching intrinsic meanings and values to the leisure roles that the institutionalization of retirement will now permit.

Changing Orientations toward Work and Retirement

In historical perspective attitudes toward retirement have varied in response to the development of economic provisions for retirement and to the changing definition of the appropriate roles and age for retirement. In the United States the limited number of public and private pension programs that were in effect prior to 1930 usually were tied to declining or impaired work capacity as the condition for retirement. The few that specified a compulsory retirement age usually placed this at age 70,[36] again with the implication that the worker would not be a productive employee beyond this age. The definition of the retirement role implicit in these programs was essentially a negative one: the relinquishment of a major adult role because of the incapacitation of old age. The Depression of the 1930's, which forced the termination of most of these earlier plans, was also to lay the foundation for a federal old age and survivor insurance program through the introduction of the Social Security system in 1935. The legislation originally stipulated 65 as the age of retirement benefit eligibility. Although only a relatively small proportion of the labor force has been required to retire at 65, its inclusion as the age of eligibility has contributed significantly to its establishment as the normative age for retirement and, indirectly, to the acceptance of 65 as the point at which "old age" begins.

The decades of the 1950's and 1960's saw a significant expansion in Social Security coverage, a rapid increase in the number of people receiving retirement benefits, and the growth in supplemental pension programs in public and private employment. Persons drawing Social Security retirement benefits increased from 1.8 million in 1950 to 12.4 million in 1968; average monthly Social Security benefits received by retired workers increased from $44 to $99 in the same time period; and the number of persons drawing pension benefits under private supplementary plans increased from 0.4 million to 3.8 million.[197]

Studies at the beginning of the period indicated that more than half of all retired workers regarded retirement as an involuntary decision, giving physical disability or compulsory retirement requirements as the reason for their withdrawal from the labor force.[36,168,169] A check of auto and steel workers in the Detroit and Pittsburgh area in 1950 indicated that 60 per cent of those eligible for retirement were unwilling to accept retirement at age 65.[23] Other studies of the period in the United States and Great Britain suggested a widespread dissatisfaction with compulsory retirement requirements as well as a reluctance to accept 65 as the expected retirement age.[61,200] However, further analysis of workers' attitudes toward retirement during the period showed sharp variations according to income received or anticipated in retirement. Indications that opposition to retirement was essentially a concern over the consequences of reduced and inadequate income during this period rather than discontent about separation from work per se were suggested by studies such as those of Streib and Thompson.[179] According to their study, 63 per cent of the workers expecting incomes of $200 a month or more held favorable attitudes toward retirement, while only 35 per cent of the workers anticipating incomes under $100 a month viewed retirement favorably. In a study of employed industrial workers, Tuckman and Lorge[193] found that 59 per cent of the men and 73 per cent of the women studied were either planning to retire or would like to if they were financially able. Studies of retired workers similarly indicated a strong relationship between adequacy of retirement income and a favorable attitude toward retirement.[99] The spread of supplementary pension plans and the increase in coverage and adequacy of Social Security benefits for the retired over the past two decades, previously noted, have been accompanied by an accelerated withdrawal of the aged from the labor force. In 1950 45 per cent of the males 65 and over were in the labor force; the proportion has decreased steadily to 32 per cent in 1960 and 27 per cent in 1970.[59,197] More recent studies such as those of Palmore,[131] Reno,[141] and Shanas *et al.*[156] have shown a steady increase in the proportion of men in good health retiring voluntarily during the past two decades in the United States.

Retirement and the Meaning of Work

Although the most immediate consequence of retirement for the worker is economic, the termination of the work life is an event of considerable social and personal significance as well. In preindustrial societies work was inseparable from the total pattern of activity that survival or society dictated. It was basically a compelling activity necessary for survival in a harsh environment where withdrawal from work activities would represent the forfeiture of the means of survival. Or in more productive societies work activity was enmeshed in a network of major institutional functions with a single and usually hereditary status defining the individual's participation in a range of societal activities. Withdrawal from work responsibility then would be accompanied by a reduction or abandonment of the positions the person occupied in family and community as well.[122] In advanced industrial societies, however, work has emerged as a separate axis around which part of life's activities are organized, distinct from, although obviously not unrelated to, family, community, friendship, church, leisure, and the other axes of participation in modern life.

Despite the work-centered value systems that have characterized Western industrial societies and the interdependence between work activity and other areas of social participation, it is significant to note that as a separate sphere of life activity work does not seem to take precedence over other life areas for most workers. Dubin[45] found that most industrial workers in his survey sample failed to regard work as their "central life interest," nor did they perceive their job as the locus of their most important social experiences and relationships. A range of studies seem to suggest that job aspirations tend to peak in the middle years of adulthood for most workers, with variations according to skill and occupational level.[205] Studies of assembly workers in the auto industry put the peaking in their early thirties, noting that they look to other areas of their lives for achievement or personal gratification.[32,203] Studies of white-collar and professional occupations, however, put the aspiration peak in the fifties and tend to suggest that work remains a central life interest until late middle age at these occupational levels.[205] Palmer's[129] study of industrial workers in Norristown, Pennsylvania, indicated that the proportion of workers emphasizing job security aspects of work rose sharply around age 35, while the emphasis on achievement began to decline in this decade.

As a major but not necessarily dominant theme in the middle and later years of adult life, the job has come to have a variety of meanings for the worker, deriving from both the intrinsic significance that the performance of the task itself holds and a set of extrinsic significances that contribute to and structure his participation in other spheres of life activities. Viewed in terms of its intrinsic meanings, work represents a time-filling and purposeful activity, and it may also serve the worker in other ways as an important source of obtaining approved types of societal rewards and achievement, a significant component of his identity and concept of self, a creative outlet, a locus for friendship relations or for the gratification of other individual needs. Its extrinsic meanings would, of course, include the income-producing activity instrumental to the attainment of many life goals; but it may also serve as the basis of the worker's status and prestige or facilitate in other ways his ability to participate in groups or interests beyond that of work itself. Finally, for some it may also serve as a refuge from the stresses of participation in other areas of life activity. Studies of the meanings attached to the job suggest that no simple cultural theme fits all occupations and professions in industrial society, but that meanings attached to work tend to vary with the nature of the job itself and with the individual's place in the social structure.[122]

The concept of the meanings of work has been employed in the analysis of the transition to retirement to establish categories of "loss" experienced by the worker upon separation from job activity, to explore the differential significance of work among occupational levels, and to determine the relationship between the significance that work has for the individual and his attitude toward retirement. Friedmann and Havighurst,[54] in a study of employed men over 55 years of age in five occupations, found that intrinsic meanings of work varied markedly by occupational level. In the unskilled and semiskilled occupational groups, income as the primary or only factor that would be missed about the job was stressed to a far greater degree than in the skilled, white collar, or professional groups. Further, intrinsic meanings assigned by the lower skilled groups strongly stressed the routine activity and time-filling aspects of the job. In the skilled, white collar, and professional groups, the intrinsic meanings of work were increasingly stressed and displayed a more diverse content including a stressing of creativity, interpersonal relations, purposefulness, self-expression, and service to others among the several groups. Preference for retirement at age 65 was found to vary directly with occupational level, ranging from 68 per cent of the lower skilled workers to only 33 per cent of the professional group studied (physicians). A further analysis of individual respondents indicated a strong relationship between the desire to retire at 65 and the stressing of extrinsic meanings of work; intrinsic meanings characterized the group that wanted to continue past 65, although the relationship was not as strong. They note, however, that the relationship established cannot be interpreted to mean that retirement is not acceptable to persons who stress intrinsic meanings of work activity, since they found workers at all levels for whom intrinsic meanings were important who also accepted retirement. Rather the analysis of the full significance which work has come to have for the older worker at the time of his retirement can provide a more specific understanding of the losses experienced upon separation from it in retirement and the life meanings and satisfactions which have relevance for a successful adjustment to retirement.

Morse and Weiss[113] in a national sample of employed males aged 21 to 64 also indicated that manual workers stressed the sheer physical activity of the job as the factor they would miss about their jobs if they stopped working, regarding it as a necessary alternative to the boredom of idleness, while professional and white-collar workers tended to stress the sense of accomplishment, challenge, and intrinsic interest of the job. Although the study did not deal with the subject of retirement at age 65, it did pose a hypothetical question as to what they would do if they unexpectedly inherited enough money to live on. Men in the manual occupations showed a greater tendency to want to retire than white-collar or professional workers, with an increasing proportion of both groups opting for retirement in the upper age groups of the study. Some American studies have not consistently borne out this inverse relationship between occupational level and favorable attitude toward retirement.[59] Although they have not dealt directly with the significance of work per se, they show a direct relationship between amount or adequacy of income and satisfaction with retirement, with income adequacy increasing at the upper occupational levels.

Simpson, Back, and McKinney,[161] in a study of preretirement orientations of older males, found financial considerations to be the major determinant of a favorable or unfavorable outlook toward retirement for the semi-skilled workers. The middle stratum workers (clerical and sales, foremen and skilled workers) looked forward to retirement more than any other group in their study. Although they stressed intrinsic meanings of work to a somewhat greater extent and had a higher level of work commitment than the semiskilled workers, neither work orientations nor prospects of income loss proved to be related to a favorable anticipation of retirement. The upper occupational stratum workers (professionals and executives) had the highest percentage of unfavorable anticipations of retirement of any occupational level in the study and also stressed intrinsic meanings of work and indicated a high level of work commitment to a greater degree than any other group studied. Within this occupational level, differences in work orientation were the significant variables in distinguishing between favorable and unfavorable anticipations of retirement, with a stress upon intrinsic meanings of work and a high level of work commitment being related to unfavorable outlooks on retirement. But it should be noted that even among this subgroup of the upper occupational stratum, selected as a polar type for prediction of unfavorable attitude, over 40 per cent looked forward to retirement with a favorable anticipation.

A cross-national study of employed and retired workers in the United States, Denmark, and Britain[156] found the same relationship between occupational level and intrinsic meanings found in work as reported above for the American workers. But it indicated that American workers reported more positive satisfactions about both work *and* retirement than did their European counterparts. The Danish and English workers were much more likely to indicate that they would miss "nothing" about the job or "money only." The study also indicated a much larger proportion of Danish and English workers reporting no positive satisfactions to be found in retirement, when contrasted with American workers. What the study seems to suggest is a developing relationship between role expectations of retirement and those of work for the American worker: "In both Denmark and Britain men see the period of retirement as a time when one does nothing or when one rests after a lifetime of work. In the United States men see the retirement period as a time for activity, and Americans in retirement enjoy their 'free-time' activities or various leisure time pursuits. In fact, the data suggest that activity is so highly valued by older Americans that the pastimes of retirement take on the aspects of work." (p. 344)[156]

The Institutionalization of Retirement

The separation of work as a commitment and source of intrinsic satisfaction from other spheres of life activity is a product of industrialization. However, the extent to which work represents a dominant life theme for the individual is a culturally and historically specific phenomenon, rather than a universal reflecting a basic instinct of man.[102] The decade-by-decade sequence of the literature on orientations toward retirement strongly suggests the emergence of retirement as a normative expectation for the older worker within the last 30 years in the United States and an increasing acceptance of this new age role. The widespread opposition to retirement reported in studies during the 1940's and the early 1950's reflected an overwhelming concern over the consequences of serious financial deprivation associated with retirement in this period and the negative image of retirement that this had established.

That a major shift in outlook toward retirement among American workers occurred in the latter part of the 1950's and throughout the 1960's is evidenced by a sharp increase in the proportion of workers reporting favorable attitudes toward retirement and the substantial decline in the proportion of the labor force past age 65 remaining on the job. Further evidence is also provided by a growing trend toward acceptance of the early retirement (age 62) option that became available in the Social Security and selected private pension programs during the past decade.[7,62,75,128,137,141] By the middle of the 1960's American studies were consistently reporting a favorable outlook toward retirement among manual workers and many categories of white-collar workers provided they had an adequate income.[48,156,161,178]* The American experience suggests that work has come to have a variety of significances beyond that of furnishing income alone, and the specific significance it has varies with the occupational and skill levels. However, with the exception of selected executive and professional occupational groups,[48,54,159,161,178] work orientations are not related to the individuals' attitude toward retirement, but there are indications that the work orientations that have characterized American society in the past are now shaping the conceptions of retirement.[156] But the exception noted in the professional occupational levels is significant and deserves further discussion for two reasons. First, behavioral and social science researchers and practitioners concerned with retirement are themselves members of occupational groupings whose attitudes toward retirement are most influenced by the intrinsic meanings that they have attached to their jobs. Second, it suggests an added dimension of the work experience that has not been adequately analyzed in the investigations of adjustment of retirement—the degree of control that the job permits the individual in structuring both his work and nonwork activity patterns and time use.

* The growing relationship between institutionalization of retirement income provisions and its acceptance as a social role should not be allowed to obscure the extent of financial hardship among the currently retired. The institutionalization of measures for providing more adequate income support for the aged over the past two decades has reduced the proportion of the aged living below a government-defined poverty level from 35 percent in 1959 to 25 percent in 1968;[197] but it still leaves poverty as one of the major problems of the retired in the United States.

Work, Leisure, and the Uses of Time

The professions and the upper levels of the managerial ranks are among the few occupations in industrial society that are task-oriented. Economic return in these occupations is tied to the fulfillment of a task; the structuring of the time and activity patterns in the accomplishment of the task is largely the responsibility of the worker. In contrast, workers in most contemporary occupations are in effect selling their time in return for a salary or a wage; the structuring of the time commitment and activity patterns is no longer under the control of the individual but is imposed by the employer and the work organization. In preindustrial societies work was task-oriented and income was tied to the fulfillment of the task.[110] Men were able to structure their own working lives and the rhythm of their life patterns; the alternation between work and non-work activities was under their control as well. As the historian E. P. Thompson[183] notes, the "work pattern was one of alternate bouts of intense labour and of idleness wherever men were in control of their own lives" (p. 73).

Preindustrial man with his "natural time" did not have to be concerned with exactness in time allocation; the rhythm of life was marked by the seasons, the months, the days, the demands of family, friends, community, work, and leisure. These were interrelated demands and the time assigned among these by the individual could be varied according to his perception of their requirements. Activity units were the components of daily life patterns. The emerging factory system of the Industrial Revolution, however, depended upon the linking of the tasks of large numbers of workers into a single production process. As clocks and watches were developed in the early stages of Western industrialization, work time became the measure of work effectiveness, and work time was the unit for which the employee was paid.[2] Mature industrial societies are marked by "time thrift" and a separation between work time and other life spheres.[183]

Time has become the unit that the worker sells; it has also become a measure of value for nonwork activities as well. As the worker began to demand a shorter work week early in the century and as paid vacations were established as a work right, time was now being purchased by him for activities other than work. Leisure, which was once the possession of a privileged class not required to work, was now defined as "free time" in the life of the working man. It is time free from the obligations of work but, as Shanas's[156] study suggests, for the American worker the significance of nonwork activities is reckoned in the calculus of their time-filling values in addition to whatever intrinsic gratifications the tasks of leisure may have for the worker. Life is now paced by the clock, rather than by the task to be performed.

The problem of "filling time" in nonwork activities is not peculiar to the retired but is symptomatic of the fracture that has developed between work and nonwork spheres of activity in the life of industrial man. Maddox[102] notes that: "Off-work time, condemned in the nineteenth century as idleness, has achieved an increasingly respectable status. And work, which is still an important stabilizing force in contemporary life, has an uncertain future" (p. 126). A number of social science observers of industrial work forces have despaired of their being able to find satisfaction in the increasing amount of discretionary time being made available to them.[41,55,143] Their reasons deal variously with the absence of a leisure tradition, the prevailing norm that "leisure time" is well used if it advances work goals, the casting of free time in a worklike time use pattern, and the absence of a satisfactory mesh between work and nonwork life spheres. Wilbert Moore[111] suggests that a measure of satisfactory integration between the individual's work and nonwork lives might be measured in terms of a "perceived scarcity of time scale." Wilensky,[206] viewing the relationship in a more explicit developmental model, defines the work-nonwork relationship in terms of interlocking cycles of work, family, consumption, and participation. He notes two major points of disjunction and strain in the cycle of work-nonwork activities. The first is the disparity between the demands of the economic and family spheres in the life of the younger worker. The second occurs "toward the end when energy declines and abilities are blunted, the incongruity of activism as an ideological norm and apathy as the fact" (p. 231).

Considerations of the time-filling activity norms for nonwork time, the failure of an adequate leisure ethos to emerge to compensate for the declining significance of work as a central life interest, and the disjunction between work and nonwork activity are problems of the entire adult life cycle. Although they have not been applied analytically in the study of retirement, we might regard retirement as a special case of leisure use in adulthood, an ultimate extrapolation of the work-nonwork time use dilemma in the life of industrial man.

"Adjustment" to Retirement

Ideological Biases in Conceptions of Adjustment

The preceding discussion of the changing conceptions of the nature of work and retirement as instrumental and intrinsic activities or situations has raised the issue of changing ideological values attached to these activities or situations. This problem is magnified when we turn our attention to the issue of "adjustment" because of the varied and sometimes ambiguous and shifting meanings that have been given to this concept. As a result it is difficult to evaluate much of the research concerning "adjustment" to retirement without first classifying the kinds of considerations that have animated and dominated its development. Our first concern will be the meaning of adjustment and the interpretation of how this is best measured and evaluated. Following this, the changing ideological attitudes toward retirement will be explored. Finally an attempt will be made to provide an overview of the results of research in the light of these considerations.

Conceptions of Adjustment

The notion of adjustment, or more precisely of "good" or "functional" adjustment, has been widely employed in scientific disciplines, both natural and social. Implicitly or explicitly, these all rest upon assumptions of equilibrium of systems, whether physical, chemical, biological, social, or psychological. This assumption of equilibrium implies a movement toward the creation or restoration of a balance or harmony within or between such systems. However, in the absence of clearly specified basic framework parameters and the delineation of quantifiable variable relationships for such systems, speaking of "adjustment" as functional relationship depends upon an assumed analogy, for the conditions determining an equilibrium are not theoretically demonstrable. In this case criteria for evaluation of processes of adjustment must necessarily remain qualitative and subjective in character, resting on assumed valuations that can be highly ideological in nature, or else be arbitrarily derived from "common-sense" premises.

Adjustment in the social and psychological sciences has rested upon and been closely linked to the concept of change as involving growth by development, and generic processes of internal development have been assumed. Lacking framework parameters and objectively quantified relationships of system variables governing such processes, there has been an empirical quest for the measurable effects of postulated variables. Cross-sectional analyses are justified under such presuppositions. However, if the stability of the assumed system is itself questionable, that is, if it can be seen as an historically changing or evolving system, then the assumptions of cross-sectional analysis are inapplicable and longitudinal analysis is required to avoid confounding measurements of changing variable relationships for stable ones.

Concern with human aging in its social and psychological dimensions has grown directly out of concerns with human development as a generic process and with adjustment within such a framework.[177] The earliest work was largely organized around "problems of adjustment," and it gave rise to research seeking to determine empirically and measure the relevant adjustment variables and their correlates with a view to facilitating "good" adjustment.[30,136,198] Adjustment was conceptualized in three dimensions: (1) personal adjustment, (2) social adjustment, and (3) the interrelationship between personal and social adjustment. These focused upon psychological processes, sociological processes, and social psychological processes, respectively. Concern over personal adjustment tends to focus upon inner psychic events and the means of measuring these through essentially subjective data dependent upon individual responses, either to test instruments or life experiences. As Rosow[146] points out, this has involved conceptions of personality, psychological states and processes, self-images or evaluations, and a range of attitudinal concepts.[18,24,25,65,79,84,117,132,133,139] The social approach has concentrated on objectively identifiable patterns of behavior—activities, social roles and relationships, and organized life patterns of these—relating them to various psychological and societal correlates.[1,3,29,30,69,112,135,136] The social psychological approach has consistently sought to treat the question of adjustment as involving the interrelationship of both social and psychological processes, whether viewed as inherently inseparable (a transactional system) or as the outcome of two mutually interacting systems of variables.[9,16,40,118,120,121,178,179,208] It has typically attempted to integrate subjectively and objectively derived data rather than simply correlate them.[5,68,69,70,116,170,187,202]

However, it should be noted that these approaches have lacked base line criteria for adjustment standards as consequences of framework parameters and quantifiable system variable relationships. Rather, the latter have been empirically sought. As a result arbitrary premises about criteria standards for adjustment have been based upon presupposed values, theoretical analogies, and "common sense," often lacking any demonstrated normative base in empirical situations or clinical experience. The result has sometimes been a degree of circularity as when some measures of adjustment, using amount and variety of social activity as one criterion, found adjustment to be positively related to role activity and social interaction.[1,146] Another has been the imposition of developmental criteria as standards when their very existence and character have been the questions at issue. This has led to conflicting interpretations of the relative degree of consistency, stability, or change in personality development with age,[12] and to disagreements over what, in turn, "good adjustment" should consist of: consistency or change.[68,70,118,119,146,171] This obviously affects one's predisposed orientation to retirement adjustment. If one treats consistency as the basis of good adjustment, and retirement as effecting a major role change, one will be predisposed to view retirement negatively as a source of disruption or potential discontinuity. The source of part of the conflict between the two major approaches to retirement adjustment—the "activity" and the "disengagement" views—arises out of this type of opposed arbitrary assumption of criteria standards in a generic-developmental framework. The first postulates consistency and thus continuity of behavior as conducive to "good" adjustment; the latter postulates change—in this instance a decrease of activity—and thus reduction of social involvement as conducive to "good" adjustment.

Criteria of Adjustment

The major criteria used in judging retirement adjustment have been: (1) conceptions of personal happiness, indicated either by direct responses to questions about relative degree of happiness, or by indirect items or scales that act to measure correlative notions like "life satisfaction," "morale," "self-image," and other specific attitudinal feelings of a positive nature whose meaningful content is essentially a face-validity based on the common sense of ordinary usage; and (2) extent and degree of activity and social participation, largely indicated by reported activity and participation rather than by measures of observed activity and participation.

Rosow[146] points out that it is all too easy to confuse "happiness" as a social value with its usefulness as a scientific concept. One problem is that happiness as a value is exceedingly culture-bound, varying greatly between cultures, within subgroups of a given culture, and even between specific roles in a subculture. In addition, there is also evidence that happiness may be viewed as relative to given dimensions of age and sex: "As happy as a man (woman) of my age could be." In these terms it is of interest that research has found that most older persons rate themselves as happy, while at the same time indicating a steady, although irregular, decline in mean happiness ratings from young adulthood on, based on retrospective ratings by older subjects.[84] Similarly, cross-sectional data show a consistent and small decline in *degree* of happiness from ages 21 to 34, 35 to 44, 45 to 54, and 54 and older in the general population, while the rating of *happiness* is overwhelming: 95, 90, 86, and 82 per cent, respectively, consider themselves either pretty happy or very happy.[64] This might force us to treat all of adult life as a period of decreasing adjustment, if happiness were taken as an indicator of it.

While the same points can be applied to some correlative notions like "life satisfaction," they do not necessarily entirely apply to others such as "morale." However, the use of morale has created great difficulty because of the variety of scales developed to measure it, confusion about their conceptual content, the lack of consistency of scaling of items on such scales in different populations, conflict between scale scores and independent clinically based judges' morale ratings, and questions concerning the uni- or multidimensionality of the attitudes measured and whether they, in turn, tap more basic personality dimensions. Finally there are also questions about the social and cultural bias inherent in some of the items on morale scales that refer to happiness, unhappiness, satisfaction, and planning of activities—items that can be seen as individually discriminating between different socioeconomic and cultural groups in terms of their differential patterns of behavior, response, and historical experiences. As a consequence it is not surprising that the results of studies using measures of morale have produced interesting empirical data on the association between variable social situations and measured morale, and that there is often consistency in such data. Morale typically is not used to *explain* anything, but rather serves as an indicator, often very sensitive, of responses to changing social circumstances, and is thus interpreted as a sign of adjustment on a common-sense basis.[184–187] This is certainly unexceptionable, but in view of the finding that morale varies inversely and cumulatively with felt economic deprivation and poor health among retired men,[175] one can treat this at a direct factual level without involving the additional burdens of a concept of "adjustment."

The use of measures of activity and social interaction of various types to measure adjustment create two types of difficulties: (1) These discriminate again on socioeconomic and status differentials between different strata of the population and between different cultural subgroups. They tend to favor as adjusted the typical white middle-class, church-going, club-joining, socializing urban dweller and penalize the established, less organizationally active, lifelong patterns of the manual working class.[126,146] And they tend to ignore that only *meaningful* activity should contribute to an individual's adjustment. (2) The greater difficulty, however, arises from the fact that clinical judgments of "adjustment" or "good morale" as well as other measures independent of activity have often found types of "well-adjusted" older persons who are inactive and even isolates who may also be "happy" or "adjusted" *in their condition.*[139,208]

This has led to the concern with varying *types* or *kinds* of "adjustment" based on greatly different patterns of integration of social and psychological processes. The recognition of the relevance of life styles over the entire life span, developed by Reichard, Livson, and Peterson[139] and Williams and Wirths,[208] in part derived from Buhler,[18] cuts across attempts to develop adjustment measures that assume universally constant generic processes and tend to set absolute, unchanging criteria standards. If widely different patterns or styles of life can eventuate in "well-adjusted" individuals, judged via intersubjective ratings of qualified clinical observers or the felt expression of the individuals themselves, free of value biases for an "active" life, an "expressive" life, a "meaningful" life (in whose terms?), then "adjustment" can be treated as essentially a summary *descriptive* term, rather than a scientific concept, which can cover a variety of types of patterns of responses to situational conditions. Furthermore, if we can view these styles of life as *adaptive processes,* rather than as functional states, we no longer are put into the position of imposing or requiring the kinds of assumptions of constancy of state that the adjustment-equilibrium model demands. Adjustment can be treated as an empirically open question of adaptive processes suited to given situational contexts, which admit of a variety of equally valid alternative solutions. The emphasis turns to "adjustment" as *process* in a structural context where the dimensions of the context can change over time as a consequence of historical differences in socialization patterns and institutional structure. And total life history patterns of response and their organizational framework become the benchmark or standard for considering retirement adjustment.

Life History, Personality, and Types of Adjustment

Numerous suggestions for investigating adjustment from a life history perspective have been made in recent years, but their implications have not been fully taken up or translated into research meaningful for interpreting adaptation to the process of retirement. One reason for this is that they have not all clearly distinguished between developmental, generic processes of change and adaptation and historical, generational ones. Furthermore, most have been concerned with adjustment to old age, or the aging process, and have implicitly assumed that they would in consequence cover retirement adaptation. Or they have identified retirement adaptation as one aspect or facet of aging, or treated it as almost equivalent to a basic index of adjustment to aging.

This is in great measure because they continue to view the individual's adaptation as the unfolding of a generic, developmental process, albeit now with a variety of possible outcomes.

Buhler's[18] conception of "meaningful living," while emphasizing "the course of life," sees it as essentially set within a maturational (developmental) sequence of basic biophysical tendencies of life, which is co-determined by motivationally grounded goal-directed actions where acts of conscious self-determination toward a given objective serve to develop a continuous and coherent pattern of meaningfulness. While sociocultural influences are also included as co-determinants of development, they play a relatively lesser role somewhat on the order of filling content in a given form. Although Buhler allows for a variety of styles of outcome in retirement, both active and passive types of pursuits being equally valid as witnessed by her categorization of groups as "relaxed," "resigned," and "active," the emphasis is on a set of values that stress a self-conscious need-achievement orientation: "meaningful living is defined as living toward a fulfillment of life presupposing that a person has given some thought to what life is about and what he wants to live for" (p. 385). Although not specifically cast this way, Buhler's view could encompass generational differentiation rather than a purely generic-developmental one. This would involve some important changes, but since Buhler stresses total life continuity rather than simply the process of transition from middle to old age and retirement, she introduces the basis for evaluating the individual's adaptation in terms of his own behavior as a norm.

Peck,[132] following Erikson's developmental scheme, also emphasizes life history in terms of a set of universal developmental stages, creating given middle age and old age stages that he associates with specific ego problems in relation to processes of physical and physiological change. His scheme, however, makes assumptions concerning physical and sexual decline, and adjustment problems attendant upon them, that are today factually questionable. Peck's scheme is less open to amending its generic-developmental set of sequences to encompass generational change. It also views social events, such as retirement, as crises to be overcome—a specific value imposition growing out of his developmental equilibrium approach. The approach has not been empirically developed.

Reichard, Livson, and Peterson[139] emphasize *continuity* with past life style and long-term personality needs as the key to retirement adjustment. Thus, among their empirically derived personality types of older males, the well-adjusted types—the "mature," the "rocking-chair men," and "the armored"—as well as the poorly adjusted types—the "angry men" and the "self-haters"—all have specific types of adaptive processes that are lifelong and are by no means "to be explained solely, or even largely, on the basis of personality variations." While the small size of their sample precluded analysis of the interaction between personality types and social factors such as occupation and income, these empirically constructed types are free of many of the usual value biases encountered in many adjustment measures. Leaving open the influence of specific social factors, approaches such as this are amenable to generational-historical analyses as they leave open the actual empirical distributions in any given population at any given time as a consequence of varying social and institutional factors.

Gutmann,[65] utilizing data from studies of urban American, Mexican Indian, and American Indian men, develops three universal, generic-developmental types of reaction to internal personality crisis and explores their possible reaction to retirement: ". . . most normal men, regardless of their socio-cultural condition, experience in varying degrees, an internal crisis of relatively long duration, beginning in the late 40's, peaking around the middle 50's, and coming to some 'final' resolution in the early or mid-60's."

He sees this "masculine crisis of middle age" as arising out of internal origins, so no specific set of cultural events can be held responsible for initiating the opposing "active-productive" and "passive-receptive" modes or for bringing them into conflict. Gutmann posits a "return of the repressed," arguing that "the childish struggle between autonomy and dependence, between productivity and receptivity that had been settled earlier in favor of the autonomous, productive modes has to be re-fought."

The outcomes, or forms of crisis resolution, are pictured as taking a variety of types depending on individual history, temperament, and cultural factors. Three major ideal-type outcome possibilities—"counterdependence," "autoplastic autonomy," and "emphasized receptivity"—are probably to be found in all societies, with relative distribution discriminating different cultures. Each one of these postulated types relates differently to work and reacts differentially to the event and process of retirement. Gutmann[65] does not view American society as "particularly supportive of any of the likely patterns of crises resolution in later life. Viewed broadly, U.S. society supplies neither the age-status system nor the traditional patriarchal orientations most supportive to the counter-dependent outcome; rapid demographic and physical changes tend to undermine the ecological stability required by the autoplastic autonomy outcome; and family dispersion makes it much harder for emphasized-receptivity types to get the emotional supports they need. By and large, however, the difficulties are probably greatest for the counter-productive and the receptive types."

Gutmann's suggestive scheme, based upon TAT analyses of a selected number of men in three widely scattered cultures, exemplifies a generic-developmental approach that sees life history and social-cultural factors as largely providing the environment in terms of which inner universal processes work themselves out. In these terms life history becomes important only as providing the specific background. This hypothesis does not accord with the bulk of data derived from sociological and psychological research, using representative as well as selected samples, concerning the types of adjustments made by men in retirement (and it so far ignores women), but it does, as do other approaches stressing some aspect of life history, view the "retirement crisis" as the occasion, not the cause, of an adjustment problem that has its roots elsewhere and is manifested well before retirement. Since a data base is lacking for estimating the likely distribution of outcome types in the United States or other societies, the potential value of Gutmann's[65] approach to dealing with retirement adaptation remains unexplored. He does suggest, however, that:

> U.S. cultural values do tend to support the autoplastic autonomy style, and given sufficient income, such people can still wall off and furnish relatively private enclaves, and they can tend satisfying gardens within them. Senior citizen's clubs, communities, and special activity programs are also helping to create the required enclaves where elderly people of the autonomous persuasion can associate with their own kind in their own preserves and deplore the immorality of the young. For the affluent, autonomous aged, these may indeed be the "golden years." Post-retirement programs will probably have to deal mainly with casualties and potential casualties within the counter-dependent and emphasized-receptivity styles.

It would be useful to test Gutmann's predictions on suitable U.S. populations of retired people to see to what extent the original data, based on Kansas City residents in the 1950's, would retain validity in other contexts. Without some indication of the empirical distribution of various types, however, such an approach, however suggestive, remains entirely hypothetical, and the major question becomes precisely the determination of the occurrence of given types in a normal population. That is, the social cultural framework becomes the operative variable determining the actual nature of the problems of retirement adjustment.

Rosow[146] has analyzed the adjustment process in terms of a fourfold schema of continuity-discontinuity in role behavior, modified by positive-negative subjective impact of the change or lack of change on the individual. Continuity of life patterns between middle age and older age is viewed as an operational index of "good" adjustment; changes that eliminate previous negatively assessed aspects of life are regarded as contributing to "good" adjustment; and stable patterns that continue negatively assessed aspects are taken as contributing to "poor" adjustment. The four outcomes of occupational role change: (1) continuity of life pattern with positive subjective impact—voluntary employment, (2) continuity with negative subjective impact—involuntary employment, (3) discontinuity with positive subjective impact—voluntary retirement, and (4) discontinuity with negative subjective impact—involuntary retirement, provide an unbiased framework for a social psychological assessment of retirement adjustment that incorporates possible generational effects via the individual's assessment of change. As a consequence *prior attitude toward change* becomes a key variable. Research largely supports Rosow's model in that prior attitude to retirement consistently becomes perhaps the most important predictor variable for "good" adjustment with a variety of types of measures.[162,174,178,184–187]

Perhaps the most interesting example of an empirically generated social psychological approach is to be found in the work of Neugarten and her associates.[116–121] Using a generic-developmental approach that sought to relate changes in biology, psychology, and overt social behavior, these studies have shown that neither the "increased interiority of the personality" (as the increased inward turning of ego functions that projective tests indicated occurs with older age is characterized) nor the occurrence of a biologically timed major event (the female climacterium) had consistent age-associated effects on overt behavior or external aspects of personality. On the other hand, a normative network of age-graded behavioral expectations was found to operate as a system of social control, structuring individual perception of and reaction to the timing and occurrence of social events and providing an orderly and sequential time clock of major punctuation marks in the adult lifeline. Basically, normative expectations acquired through socialization were found to provide an age-graded framework to which behavior was oriented and in terms of which it was evaluated. Furthermore, the extent or degree of such age-graded norms increased with the respondents' age; that is, older persons had more extensive age-norm standards for appropriate behavior, while younger persons treated age-norm standards as less important. That situational and generational effects may be involved here is indicated by McKain's[94] finding that with respect to the question of "retirement marriages" by aged, widowed spouses, children had more rigid age-graded standards of appropriate behavior than their aged parents. Adjustment, then, can be treated in terms of the congruence of the individual's socially constrained, age-graded expectations of behavior with their actual occurrences. While Neugarten has explored *age-specific differences* in expectations along dimensions of social class, sex, and occupational role, this has largely remained within a generic-developmental perspective. That is, there is as yet no theoretical recognition of the influence of *generational differences* where these influences are systematically explored in terms of differential age-graded expectations of different age groups, conceived as distinctive cohorts.

Yet this remains perhaps the key consideration in exploring adjustment to retirement as the congruence of the adaptational responses of different generations of older persons with their role expectations. In this light inner personality variables cannot be treated as active apart from the cultural context since actual behavioral problems appear essentially as the consequence of the operation of social factors. And life styles or patterns of behavior cannot be separated from the situational and historical framework in which they are anchored.

Interestingly enough, the situational and generational influences of social values and expectations upon the perspectives of the scientific researchers who study retirement—possibly becoming sources of their own biases—remain to be taken into serious consideration, a point that Donahue, Orbach, and Pollak[44] and Rosow[146] have emphasized and that others are beginning to recognize as an important factor.[70] Therefore, we shall next turn to some of the more pervasive of the socially grounded myths concerning retirement adjustment to explore exactly how they have exerted an influence on the character of research and subtly continue to influence interpretations of research findings.

Myths of Retirement Adjustment

Pervasive cultural ideologies and biases have structured the perception of retirement adjustment as they have all matters associated with aging. They have done so both as general social myths and as incorporations into scientific viewpoints. The strong antipathy to aging or growing old, associated with the loss of vitality and virility, and to older persons in our society—the "youth cult"—has colored our perception of later life and of retirement in particular.

The association of aging with senescent decrements in abilities and functions has obviously animated the concern over continued vital activity as a means of demonstrating the absence of senescence. This has led to the uncritical acceptance of general empirical findings of such decremental losses in older persons. The long-established view that mental functioning declined inexorably with advancing age, based on general findings over more than a quarter of a century, has recently been shown to be the consequence of the failure to control for health as an independent variable. Despite a slowing down of the processes of intellectual functioning, when health status is controlled there is no clear evidence of any decrement in the character of intellectual functions due to aging *per se*.[11]

In a similar fashion retirement has been viewed as a crisis situation precipitating personal and social disorganization, familial disruption, organic illness, physical deterioration, mental illness, and even death.* Isolated clinical experience and general empirical findings, treated uncritically and ahistorically, have been employed to substantiate these views. The Cornell Study of Occupational Retirement, the most complete and only comprehensive longitudinal study in the United States, began with a series of theoretical hypotheses of the possible outcomes of retirement that posited entirely negative consequences of what was taken to be a critical discontinuity in life.[9] Little consideration was given to the possible positive features of a newly emerging social pattern. The effect of presumed pervasive social values regarding work and activity was treated as almost absolute and unchangeable. Historical fact and evidence were and still are being distorted, not necessarily intentionally, to fit uncritically held biases. One example: "In 1882, without pilot experimentation and without significant knowledge of the consequences, a major experiment *qua* social policy was introduced by Bismarck in Germany, in the wake of the industrial revolution."[25] This presumed description of the beginnings of retirement, given in 1966, could entirely ignore the facts that (1) the German retirement policy was adopted only after patchwork solutions to the problems of old age indigency via the traditional common law methods of poor relief, attempted for almost 50 years, had entirely broken down, threatening the social stability of large areas of urban, industrial Germany; (2) it was introduced to alleviate the real and desperate condition of the aged; (3) it was consciously modeled on the almost century-old system of military pensions and the 75-year-old system of civil service pensions in England, although not as generous in its benefits.[44]

Retirement in the United States is often treated as having originated as a conspiracy against the aged, as if older persons before 1937 enjoyed both free access to continued employment and relative economic and physical well-being, instead of desperate unemployment, economic misery, and physical ailment; as if it were foisted upon them instead of being in good measure the political and social response to the Townsend Movement, the most massive self-organization of the aged as

* Streib and Schneider[178] cite examples of this from the 1963 *Encyclopedia of Mental Health*. Dr. Douglas Orr writes, "Retirement is a mental health hazard under any circumstances." Discussing situations leading to reactive depressions, Dr. Felix Von Mendelssohn lists "forced retirement" along with such situations as "crippling accidents and diseases, totally hopeless situations . . . and loss of skills." According to Dr. Kenneth Soddy, "Many men find their most important identity in their career, which, being relinquished, establishes a strong tendency toward early breakdown and death" (p. 107). They also point out that in each of eighteen instances where the notion of retirement from work is meaningfully discussed, the depressive or negative effects are emphasized.

a political force that has occurred in our history.

Without ignoring the real problems retirement poses, it is of little benefit to treat these uncritically as inherent in the presumed nature of retirement, apart from the actual circumstances that surround the event of retirement, the process of retiring, and the character of retirement living. Reviewing the results of research on retirement demonstrates the consistent manner in which the blinding effect of biases have served to attribute to "retirement" consequences that are the result of poor health, inadequate income provisions, aspects of the individual's total life experiences, and general problems that face our society as a whole.

Adjustment to Retirement: Research Results

Physical Health and Morbidity

The assumption that retirement leads to a decline in physical well-being is based upon the presumed loss of sustaining patterns of the organism, which leads to or precipitates physiological collapse and even death in extreme cases. The abrupt disruption of these patterns by the event of retirement, the difficulties of adjusting to the new patterns in the process of retirement, and the new patterns in themselves all contribute. And it is a fact that many people die soon after retirement. *Post hoc, ergo propter hoc.*

However, research findings lend no support to this presumed retirement effect and, if anything, suggest its opposite: retirement benefits physical well-being.[44,96,105,115,150,178,186,193,194] Longitudinal analysis shows the same pattern of changes in health in both retired and nonretired members of the same sample cohorts.[178] Improvement in health status after retirement has been reported consistently in controlled studies.[44,105,150,193,194] Myer's[115] analysis of mortality rates in 1954 did show higher death rates than the general male population immediately after retirement for males who retired voluntarily, but his study also indicated that death rates for workers retired compulsorily were in agreement with actuarial expectancy. His conclusion was that people in poor health retire and that is why retirees die. Other studies reach similar conclusions.[105,194] Research into reasons for retirement have consistently shown that health is the major reason given for retirement,[44,156] and the 1968 Social Security Survey of Newly Entitled Beneficiaries has reported not only that 44 per cent of all newly retired beneficiaries give health as the major reason for retirement, but also that among those who retire early, 54 per cent report health as the major reason compared with 21 per cent of those who retire at age 65.[141] The inescapable conclusion is that it is often poor health, real or perceived, that is responsible for retirement, not the other way around. And that free of the burdens of work, itself often the cause of poor health status, retirement has no adverse effect on continued health, but more usually a beneficial one.[105,150]

Mental Health

The literature is full of clinical descriptions of the mental ailments of the retired. It was undoubtedly this factor that led the most comprehensive analysis of age and mental illness, the Langley-Porter Studies, to hypothesize retirement as a likely precipitant of mental illness in old age. But little evidence was found to support this view.[93,165]

> The factor which has recurred again and again in the consequential network is physical health . . . physical problems were seen as involved in the decision-making that led to the psychiatric ward in approximately two-thirds of the cases and precipitated the psychiatric illness in 10 per cent. . . . Even for those persons diagnosed as suffering from psychogenic disorders, studies of health history showed that physical problems often preceded the development of psychiatric symptoms (p. 68).[165]

Again, lifelong patterns of illness, both mental and physical, have often been overlooked as the basic causes of mental illness after retirement; and the interrelationship between lifelong patterns of health, economic circumstances, and the generic problems of advanced age, when carefully examined, lend little credence to retirement itself as a cause of mental illness. The bias toward the therapeutic effects of work, whose value in given circumstances is unquestionable, tends to ignore the sources of physical and mental illness in work in other circumstances. Here clinical experience must be viewed in the light of the overall empirical situation. Similarly the highly visible numbers and proportion of the aged in mental institutions can be seen to be the consequence of the more basic factors of physical health and economic circumstances, as well as the natural "aging" of young and middle-aged residents into aged residents.

Family Relationships

Retirement has also been described as a crisis time for family life and relationships, based on "theoretical" considerations, not actual knowledge.[173] This has involved two interrelated suppositions: (1) we are dealing with an emotionally charged, disruptive event that destroys prior continuity of life patterns; (2) the occupational role system largely controls the nature of the family role system; it is central to personal and family role identity; and its loss is "functionally" irreplaceable. This is implied by Burgess's[21] conception of retirement as leaving the man "imprisoned in a roleless role" (p.21). It is joined together in a somewhat melodramatic, but by no means atypical, position by Miller[107] (p. 78).

> The occupational identity of the individual establishes his position in the social system at large, allowing others to evaluate his status and role and providing a context within which his social activity can be interpreted. For example, the occupational identity of a male places him in appropriate relationship to other members of his family and supports his roles in that social system. Before retirement, the role of "husband" as mediated by his occupational identity results in high prestige and supports the various roles that the person is expected to assume in the family system. It would be extremely difficult to maintain the role of "head of the family" if an occupational identity were lacking. The occupational identity is that which provides the social substance by which other identities are maintained, various roles coordinated, and the appropriateness of social activity is substantiated. In other words, the retired person finds himself without a functional role which would justify his social future, and without an identity which would provide a concept of self which is tolerable to him and acceptable to others.

It would follow from this viewpoint that retirement should be totally disruptive of family life and relationships: loss of authority, respect, meaning, and function are predicated. It is interesting that this position that decries the impact on family life of retirement seems to overlook the fact that it has reduced family life to the status of a servile appendage of the world of work, robbing it of all independent meaning or value.

A similar "functional crisis" view has often been applied to the postparental period even before retirement, though its "shock" has often been viewed in terms of its effect on the wife. This has been based largely on clinical experiences with persons having so much difficulty that they resort to outside help. However, Deutscher's[42] study of postparental couples revealed no great sense of crisis in the overwhelming majority of cases over the "empty nest" stage. Indeed, the majority of couples saw it as a time of new freedoms: freedom from housework and from financial responsibilities, freedom to travel and explore interests, and finally freedom to be one's self for the first time since the children came along. So, too, studies of family relationships in retirement suggest no shock or crisis with retirement, but rather a high degree of continuity of activities and relationships and an enlarged and heightened relationship between spouses in many instances.*

It is an indicator of the pervasiveness of the "work is life" bias that the positive or enhancing potentialities of retirement for family life in terms of the time available for husband-wife or parent-child relationships free of the

* See references 20, 37, 76, 77, 82, 86, 87, 90, 145, 147, 172, 173 185, 209 in the Bibliography to this chapter.

demands of work, or the possible cushioning effect of the family on the "distressed" retired man, have been in the past largely ignored, not to mention the availability of family relationships as a primary outlet for the time, activity, and involvement presumed to be lost with cessation of work.

However, more recent research has begun to make clear the empirical facts concerning the adaptation of family relationships to retirement. Data arising out of the studies of the longitudinal Cornell Study of Occupational Retirement[172,173,178,185] show a high degree of continuity in family relationships, continued patterns of interaction, greater closeness between generations than prevailing views had foreseen, children viewing the impact of the father's retirement as largely not changing family relationships with those viewing change judging it positively, and little or no crisis in family life due to retirement. Streib[173] reports "a marked tendency for the retirees to place greater emphasis on the ties of affection between them and their adult children than the children do," a point also made in other studies.[148,155] The responses of the children of the retirees to particular items are of special interest. Replying to the question on how much family crisis was caused by the father's retirement, 71 per cent said it was no problem at all, 25 per cent said it was not serious, and only 4 per cent said it created a serious crisis. In response to a question asking if they felt retirement had brought their father closer to his immediate family, 68 per cent said there had been no change, 30 per cent said retirement had brought him closer to his family and 2 per cent said it had made him feel less close. Again, replying to a question asking whether the fact that the father had stopped working had been good or bad for the way members of the family got along with each other, 83 per cent reported no change, 15 per cent said it had been mostly for the good, and only 2 per cent said it had been mostly bad for the family. Streib's research, covering the first six years after retirement, does cover a somewhat better-than-average group with respect to income and occupation, and it does not claim to be an entirely representative national sample. But it is the best sample we have had in any American study of retirement, other than one restricted to a narrowly specific occupational group or region, and perhaps the best sample in any country.

Other studies have reported similar results concerning the affect of retirement on family relationships, especially concerning children.[20,155,156,210] And comparative studies of the Netherlands,[82] Italy, Germany, Poland, and the United States,[119,182] and of England, Denmark, and the United States,[156] have shown the important and often central role played by family relationships in retirement. Some of these have also demonstrated that differences in actual patterns of family relationships are often the consequence of situational conditions such as residential proximity, health, economic situation (possessing an automobile to enable easy travel), and even technical development in the society—universal as opposed to nonuniversal possession of a telephone for daily contact.[182] Continuity of family life is perhaps best illustrated by two patterns: greater parent affect in relationships with children, something that is clearly lifelong; and financial assistance or help continuing to be dominantly from parent to child, even in retirement.[155,156,173]

Social Interaction

In an analogous fashion to its supposed effect on family relationships, retirement has been linked to a serious decline in social interaction. The impact of retirement is seen as adversely inhibiting all types of social activity through the general process of removal of the individual from a functional location in society and its effect upon him, a dramatically reduced social life space and loss of social esteem. The general decline in activity noted with the older ages, taken as a sign of the process of aging, has been viewed as being spurred on by the event of retirement, and often the *process* of retirement has been almost assimilated to the social process of aging.

In one important sense this latter view lies at the core of the disengagement hypothesis: a mutually agreeable withdrawal from social interaction between the older person and society based on norms sanctioning such "retirement."[39,40] In the light of this, the disengagement approach, however, does not see the *event* of retirement as the initial cause of any crisis or traumatic situation, but rather as part of the resolution of problems intrinsic to the natural, that is, psychophysiological, processes of aging: "Retirement is society's permission to men to disengage"[40] (p. 146). Nevertheless, because the disengagement approach shares with the opposed "activity" point of view an equilibrium model of adjustment from one stable *state* to another, it views the *process* of retirement as potentially traumatic: "Because the abandonment of life's central roles—work for men, marriage and family for women—results in a dramatically reduced social life space, it will result in crisis and loss of morale unless different roles, appropriate to the disengaged state, are available"[40] (p. 215).

Thus, almost paradoxically, both the disengagement and the activity viewpoints see the *process* of retirement as initially maladjustive —one as consequence, the other as cause of a disturbed equilibrium. Their difference lies in their evaluation of the outcome. The disengagement perspective views the acceptance of a reduced level of social interaction in new roles as constituting positive adjustment; the activity perspective requires a continued high level of social interaction as necessary for optimal adjustment. Based on their respective theoretical assumptions—both cast in terms of a generic-developmental framework—each posits a different required new state of activity equilibrium as necessary in retirement, while sharing the almost identical characterization of the nature of the process as it is reflected in the individual's social interaction. As we have noted earlier, the disagreement rests on the assumed factor of consistency or change in equilibrium as constituting the universal nature of the developmental process.

Furthermore, both viewpoints see the preretirement state of equilibrium as similarly revolving around a single, basic type of work-role centrality for men, family-role centrality for women. They consequently overlook the possibility of different life history styles or patterns of social activity and social integration, and, in turn, different patterns of adaptation or change in which retirement has entirely different meanings and relationships to other forms of social interaction, in part dependent on the influence of social and situational factors. This has led to the search for a relatively universal set of relationships in patterns of activity, involvement, and satisfaction. It has also resulted in a tendency to emphasize sheer quantitative differences in social interaction patterns in place of evaluating the qualitative contribution these make to the individual's adaptation.[148]

It is not surprising, therefore, that research on the relationship between retirement and social interaction has led to seemingly inconsistent results when viewed as supportive of the disengagement or activity perspective.[3,68–70,80,82,87,99–101,112,119,134,135,145,148,170,178,182,202] Again, broad general findings based on cross-sectional data have shown, in different populations, under different social and situational conditions, and in different areas of social interaction, both patterns of increment and decrement of activity and satisfaction.[87,100,119,148,159,160,182] Maddox[100] has shown how the type of research design employed—gross cross-sectional analysis utilizing central tendencies as a basis for comparison—maximizes the probability of artifactual results and spurious conclusions. But it is now clear that whatever the complexities of the various relationships that can and do exist between retirement and social interaction of various types, there is no basis for the assumption that retirement causes or necessarily results in a constriction of life space and activity when the effects of health, socioeconomic status, situational and cultural factors, and lifetime patterns of adaptation are controlled.

Rosow[148] and Maddox[100] have distinguished the dimensions of personal friendships, formal participation in organizational settings, and noninterpersonal activity. Rosow has shown how situational factors such as the concentration of age peers varies systematically between working-class and middle-class retired persons in the extent of their friendship activities. Middle-class persons show less dependence on the immediate presence of peers for interpersonal activity, seeking out friends and activity over a wide geographical area, while working-class retirees are much more dependent on the immediate environment. For both, as the social role world contracts after retirement the local environment gains in importance as an integrating factor, but more so for the working-class person. Health, economic status, and marital status also act as influences here such that poor health, low economic status, and lack of a spouse serve to increase dependence on the immediate environment and thus on different kinds of social interaction.[164,202] Rosenberg's[145] study of retired men has demonstrated the same basic relationships between economic status, marital status, immediate environment, and social interaction: ". . . the transition to retirement is misconceived if it is thought of as a change to a condition lacking a central role. The role of neighbor becomes activated and replaces that of worker (for low economic status, working-class men)." Similar findings have been reported in cross-national studies of the United States, Germany, Poland, Italy, and the Netherlands.[87,119,182]

Finally the individual's evaluation of the satisfactions derived from his varying patterns of activity is also a function of his social status and lifelong patterns of adaptation. As Rosow[148] notes:

> Those who have little to do with neighbors are not invariably distressed, nor do those with high interaction with them necessarily feel satisfied. People's desires are one significant condition of their integration into friendship groups, and it is erroneous to regard such ties as a universal value, pervasively shared and uniformly enjoyed. This would be an uncritical view of a more complex matter (p. 104).

The same considerations have been shown to hold with respect to all types of social interaction.[87,101,119,182,202] What has become clear above all is that the effect of retirement on the individual's activity pattern is complex and varied, depending critically on his health, socioeconomic and marital status, ecological and social environment, and previous lifetime pattern of adaptation.

It seems that in the past we have been in the unfortunate situation of having the deviant, dramatic patterns presented as the typical ones. If one wishes to generalize one may do so on the basis of the modal patterns discovered, with the reservation that the heterogeneity of life styles cannot be overlooked. Streib and Schneider[178] summarize this issue cogently:

> . . . the assertion that, for men, retirement results in a sharp decline in social life space is not tenable for persons in many occupations. Work is . . . (not) an interesting and stimulating *social* experience for many persons who may tolerate the social side of work . . . but consider it unimportant in the long run. . . . It seems questionable whether the loss of work role may lead to a crisis for most older workers. Our longitudinal studies of the impact year point to the fact that retirement is usually not a crisis. . . . Cessation of work does not necessarily result in automatic disengagement in familial, friendship, neighborhood and other role spheres. It has been assumed by those who view retirement as a crisis that retirement was the precipitant for a series of retrenchments in role activities. This may occur in some instances—indeed it may be the only kind which comes to the attention of physicians, social workers, psychiatrists, clergymen, and other therapists—but these dramatic instances should not be used as the modal pattern (pp. 177–179).

❡ Attitudes toward Retirement and Satisfaction with the Retirement Experience

The General Picture

If one were to frame some broad generalizations concerning current attitudes toward retirement and satisfaction with it—recognizing all the pitfalls inherent in such a procedure and with the clear understanding that there are important and significant exceptions —one would have to conclude that most persons today have a generally positive attitude toward retirement as a future status; are more likely to strongly exhibit this attitude the higher their expected retirement income, the better their health, the greater their educational and occupational level and attainment, and the less they find work to be the major or only source of intrinsic satisfaction in life. This is the case regardless of their satisfaction with their work or with life in general. Furthermore, most retired persons adapt relatively well to their situation and find satisfaction in retirement to the degree that their attitudes and situations correspond with the preceding description.[7,37,62,67,74,75,92,137,138,156,174,177,182,184–187] There is good evidence that most Americans are coming to view retirement as a normally expected stage of life that follows a delimited period of work and one that has potentialities for its own intrinsic satisfactions.[37,67,124,156,177]

Satisfaction with Retirement

A Harris Poll[67] on retirement in 1965 concluded that: "Contrary to widely-held impressions, Americans do not contemplate retirement with deep doubts and fears. Instead of thinking that retirement means being put on the shelf, the majority of Americans see it as a chance to lead a different and not unpleasant life." In response to the question, "Has retirement fulfilled your expectations for a good life or have you found it less than satisfactory?" 61 per cent of a national survey said that retirement had fulfilled their expectations, 33 per cent found it less than satisfactory, and 6 per cent were not certain. Among the dissatisfied group 40 per cent gave financial problems as the reason for their attitude, 28 per cent gave poor health or disability, 10 per cent gave the loss of a spouse, and 22 per cent said they missed working. In effect, about one in five of the one-third dissatisfied retirees, or approximately 7 per cent of all retirees, gave a reason intrinsically related to retirement itself.

Similarly the more scientific area-probability samples studied by Shanas, *et al.*,[156] report 35 per cent of American retirees enjoying "nothing" in retirement, while 42 per cent of retirees in Britain and 50 per cent in Denmark express this attitude. When incapacity due to health is controlled, the proportion finding "nothing" enjoyed in retirement varies in direct relation to the existence or degree of incapacitation: from 27 to 55 per cent in the United States, 33 to 65 per cent in Britain, and 41 to 62 per cent in Denmark. "The more . . . physically limited, the more likely he is to find nothing in retirement that pleases him. . . . It is not so much their occupational backgrounds that determine what men enjoy in retirement. Rather it is their degree of capacity, their ability to get about, that influence their retirement attitudes" (p. 333).

The generally satisfactory adaptation to retirement made by most workers is highlighted by Streib and Schneider's[178] comparison of a number of preretirement and postretirement attitudes in their longitudinal study. In every case postretirement attitudes of the same groups of retirees, both men and women, were more positive toward the retirement experience than preretirement attitudes. Every one of their cohorts of retirees during the mid-1950's tended to overestimate the adverse effects of retirement before they retired. For example, 25 per cent of the men retiring in 1957 and 42 per cent of the men retiring in 1958 thought they would miss being with other people at work, but the proportions reporting they had missed being with other people at work after they retired were 16 per cent and 25 per cent, respectively. The comparable proportions for women in the same years were 63 and 56 per cent estimating they would often miss others at work compared to 29 and 28 per cent reporting they had so missed others after they were retired. Similar sizable differences were found between preretirement estimates and postretirement actualities with respect to often missing the feeling of doing a good job, wanting to go back to work, and worrying about not having a job. More direct evidence of the positive impact of the retirement experience is seen in the change in viewing "retirement as mostly good for a person." Each of four groups of men and of women retiring in different years showed increases in the proportion saying "retirement is mostly good for a person," and all but one of the eight groups experienced 20 to 45 per cent increases in the year after retirement. In addition, for those cohorts studied over a longer period of time, the proportions continue to increase, so that all of them finally show these magnitudes of change over four to six years, with an overall proportion of 75 to 87 per cent expressing this attitude. More generally, about one-third of both men and women report that their retirement has turned out better than they expected; a little less than two-thirds of both report that it has turned out about the way they expected it; and only 4 per cent of the men and 5 per cent of the women say it has turned out worse.

Changing Attitudes toward Retirement

Harris[67] also reported that two-thirds of American adults not yet retired think they are likely to retire at age 65 or earlier, while three-quarters of them would like to retire at age 65 or earlier. Although 22 per cent expected to retire at age 60 or under, 46 per cent would like to. While these expectations and desires are much more in evidence among younger adults—that is, among those 35 to 49 years of age, 25 per cent expect to retire at age 60 or under, while 53 per cent would like to, compared to 6 per cent and 23 per cent, respectively, of those 50 years of age and over—this is probably more a reflection of changed social attitudes than an expression of age-related attitudes since already retired men were not included and more than half of the men retiring and claiming Social Security benefits each year since 1962 have retired before age 65.[140]

Support for this interpretation comes from Katona's[74] studies in 1963 and 1966, which reported even stronger indications of early retirement plans and an increase in such plans among family heads in the labor force with annual incomes of $3,000 or more. In 1966, 43 per cent of family heads aged 35 to 44, 33 per cent aged 45 to 54, and 22 per cent aged 55 to 64 planned to retire before age 65, compared to 25 per cent, 23 per cent, and 21 per cent, respectively, so planning in 1963. Katona also found that 51, 54, and 48 per cent, respectively, of these same age groups looked forward to retirement with enthusiasm, while 17 per cent, 19 per cent, and 23 per cent dreaded retirement. Expected retirement income was positively related to planned early retirement. The proportion planning early retirement was about 30 per cent for those with expected retirement incomes ranging from under $2,000 to $3,999, rose to 40 per cent for those expecting $4,000 to $4,999, to over 50 per cent for those expecting $5,000 to $5,999, and to 57 per cent for those expecting $6,000.

Two major types of reasons—financial and health—were given for (1) either retiring early or planning to do so as opposed to (2) not retiring early or not planning to do so. Being able to afford to retire or poor health was linked with the first and not being able to afford to retire with the latter. In a like manner Reno,[140,141] analyzing data from the Social Security Administration's 1968 Survey of New Beneficiaries, reports that among men who claimed benefits at the earliest possible age and stopped working,* the proportion who did so willingly rose from 15 per cent of those with less than $1,000 in yearly retirement income to 75 per cent of those with $5,000 or more income. Twenty-five per cent of men retiring at age 62 did so willingly with a second pension and a median yearly income of $4,100. On the other hand, some 57 per cent of men retiring at age 62 gave health as the major reason, while at the same time 45 per cent had no second pension, did not want to retire, and had a median income of only $930 from Social Security benefits. While one cannot determine to what extent these last two categories are overlapping, there is no doubt that a sizable number of men in poor financial circumstances are forced to retire because of health. Although some of them may be somewhat satisfied in retirement because of the release from the burdens of work, many others undoubtedly are profoundly dissatisfied. Nonetheless, Reno reports that these men tend to file for retirement benefits as soon as they are qualified at the earliest age. However, it should be clear that these men's dissatisfaction is not due to retirement, but rather is a reflection of their poor health and income status. It is also apparent, as these and other studies

* Under Social Security men may file entitlement claims but not receive benefits because they continue to work. Another reason for filing in recent years is to qualify for Medicare benefits.

have shown,[131] that among those men retiring at age 62 with poor health and the lowest levels of retirement income, a large proportion have had lifelong patterns of low income due to irregular and disorderly work careers, often involving extensive periods of unemployment in part attributable to poor health and disability. For these men retirement hardly represents a significant change in an important or central occupational role, and one could question to what extent it is meaningful to properly speak of "retirement" for them. The continuing detailed studies of the Social Security Administration, including the recently commenced longitudinal survey of new beneficiaries, should provide us with the basis for a better understanding of these questions in the next few years.*

In the meantime problems like this emphasize the need to be cautious about broad generalizations and to consider carefully the complexities involved in understanding the special circumstances of distinct occupational, cultural, regional, racial, and sexual subgroups. Although focusing on the modal patterns restores a sense of balance to the melodramatic caricatures that have often been the consequence of focusing on extreme situations, it is precisely the complexity of the variety of relationships that exist between retirement attitudes, social circumstances, and retirement adaptation that demands attention. For regardless of the numerical size of subgroups, they may pose some of the severest problems that will demand public attention.

Occupational and Status Differences

A number of persistent differences in attitudes and satisfaction have been noted in specific occupational and status groupings. In a number of studies distinctions between blue-collar and white-collar workers have been reported. Stokes and Maddox[170] report higher levels of satisfaction in retirement for a small sample of Southern blue-collar workers (50 per cent) as opposed to white-collar workers (28 per cent). They find this to be related to the higher intrinsic satisfaction with work indicated by the white-collar workers in contrast to the blue-collar workers (78 per cent opposed to 38 per cent). However, their study also suggests that white-collar workers may experience a higher level of satisfaction in the long run since the blue-collar workers' level of satisfaction appears to decline with the passage of time while the white-collar workers' satisfaction shows some slight increment over time. Loether,[92] reporting on a large sample of Los Angeles civil service retirees, also finds a relationship between white-collar workers' higher job satisfaction and a less favorable attitude toward retirement. But despite the blue-collar workers' more favorable retirement attitudes and lower degree of job satisfaction, he reports a more satisfactory adjustment in retirement for white-collar workers and suggests their greater role flexibility as the explanation. He posits this as enabling them to make easier adaptations to the somewhat unstructured nature of retirement roles. On the other hand, Streib and Schneider,[178] with their longitudinal national sample, found a higher favorable orientation to retirement among professional workers than among clerical, skilled, semiskilled, and unskilled men, who showed little difference. Since favorable preretirement orientation was the most reliable predictor of satisfaction in retirement, they report higher levels of satisfaction among professionals in their study and the absence of the differences that Stokes and Maddox[170]report. Obviously their findings also are not congruent with Loether's.[92]

Automobile Workers and Early Retirement

Studies of automobile workers, as representative of highly organized mass production industrial workers, have indicated a strong and increasingly positive attitude toward re-

* One problem in particular is due to the fact that most studies of early retirement indicate a strong positive relationship with high retirement income, while studies of Social Security beneficiaries indicate, in addition, a stronger positive relationship to extremely low income and marginal relationship to the labor force. Perhaps these latter cases are by-passed in the other studies because their marginality excludes them from inclusion in the labor force.

tirement and a high degree of satisfaction with the retirement experience. Orbach[123,124] reported the growing tendency from 1953 to 1963 among automobile workers to retire before the compulsory age of retirement and the rising proportion electing to retire early, which had reached 35 per cent as early as 1962, almost three times the 12 per cent who waited until the compulsory age. The increases reflected a steady improvement in the retirement pension benefits and demonstrated a clearly instrumental attitude toward work. Barfield,[6] Barfield and Morgan,[7] and Pollman,[137,138] studying the massive movement to early retirement following extremely liberalized and financially improved early retirement provisions in 1964, reported satisfaction in retirement to be both pervasive and intense, with increasing enjoyment in retirement as the early retired auto worker moves further into the retirement period. Barfield[6] reports that 89 per cent of retirees believe their decision to retire when they did was right and that over three-quarters would advise others to retire at the same age as they did. Most report finding retirement about or exactly as expected. Of the 12 per cent Barfield found dissatisfied two to three years after retirement, half pointed to health problems as the primary cause of their dissatisfaction. Congruent with this, workers retired on disability pensions, who presumably had serious problems with their health, and those reporting declining health in recent years were much less likely to be satisfied with retirement than other retirees.

In all studies the basic factor in the decision to retire was the level of retirement income expected, and satisfaction with retirement life is substantially correlated with an adequate financial situation. Barfield reports, however, that this is less true for the most recent early retirees, who, he indicates, may be highly relieved to be free from the world of work and enjoying a relatively novel leisure-based life style. The special benefits of the automobile workers' pension arrangements make them somewhat of an "elite" group among industrial workers, and perhaps this leisure pattern can come into focus in the absence of financial problems. Barfield reports, for example, that over three-fourths of these early retirees have not been forced to dip into their savings; 20 per cent have in fact *increased* their savings since retirement. About two-thirds claim to be spending as much as or more than before retiring, and over three-fourths view their living standard as being at least as high as that enjoyed before retiring (and of those who have seen some deterioration in standard of living, about half still believe they have enough to live comfortably). Interestingly Barfield finds that those who have decreased their levels of social interaction, interest in world affairs, and amount of leisure activities—a relatively small proportion—report much lower levels of satisfaction with retirement life than those who have maintained the same levels of activity and those who have increased their activity levels. Part of the change in activity levels is related to health changes, but not all. However, some men were not in good health when they retired, and although they may not have perceived any change in health since then, they may have had to reduce their activities to maintain their health. At the same time one other factor, which has been reported in other studies of retirement satisfaction, was found to be important: retiring as planned or having to retire unexpectedly for whatever reason was highly related to stated satisfaction. Men who had to retire unexpectedly were much more likely to be dissatisfied. Barfield[6] concludes that, "In the final analysis, though, and inferring particularly from the apparent continuation of high levels of satisfaction among auto workers who have been out of the work environment for several years, it does not seem possible that the expressions of satisfaction in retirement on the part of most auto workers derive primarily from simple relief at having escaped a bad situation. We remain convinced that, for many people, the satisfaction of a life free from the demands of work are both pervasive and abiding" (pp. 45–46).

Another study of industrial workers in the oil industry in Texas finds early retirement to be substantially accepted by the overwhelming majority of men who chose it.[128] Those who were dissatisfied in some way and returned to work tended to be the unskilled, less educated, married, and nonwhite members of the study group who also had lower retirement incomes and financial reserves.

Civil Service Early Retirees

Messer[106] reported a high degree of satisfaction with retirement in a study of some 3,229 federal civil service early retirees, which is unusual because these workers were all between 55 and 60 years of age when they retired, with at least 30 years and up to 42 years of service. Sixty-eight per cent indicated they would certainly retire again under the same circumstances, 24 per cent probably would, and 6 per cent indicated they certainly would not. Among this select group of men and women, one-third had retired at age 55. Wanting to quit while still able to enjoy retirement was the most important reason given for retiring (23 per cent), followed by economic reasons such as seeking a better paid job, being better off with annuity plus outside earnings, wanting to qualify for Social Security (22 per cent), dissatisfaction with the job (18 per cent), health or family reasons (14 per cent). These retirees covered all categories of skill and professional levels in the federal service. Seventeen per cent reported living better than before retiring, 34 per cent about as well, 32 per cent not quite as well but all right, and 13 per cent reported having had to reduce their standard of living drastically to get by. Health did not seem to have been very much of a consideration in this group as one might expect from their young ages and type of work. Interestingly enough, former postal service employees were among the groups reporting the highest degree of satisfaction.

The rising rates of early retirement among a variety of public employees—teachers, policemen, firemen, skilled and semiskilled blue-collar workers, clerical employees, and members of the armed forces[97,98,160]—in addition to select groups of industrial workers like the automobile workers, has raised many questions about the nature of their future lives. Concern about boredom and adjusting to inactivity has given way to concern about their future financial viability. It is not yet clear to what extent various categories of workers will be able to feel financially secure in early retirement—ignoring for the moment the serious problem that large numbers of regular retirees who exist basically on Social Security benefits face—especially in view of inflationary trends in the economy. Some of these categories of early retirees will have qualified for generous retirement incomes, and many may have accumulated sufficient reserves to see them through their retirement years. If one were to judge by the early experience of the auto workers, where only about 17 per cent have engaged in remunerative employment since retirement, or by the Civil Service retirees, where about one-quarter have engaged in full-time employment with another 21 per cent working part-time or full-time temporarily, while about 44 per cent have not worked at all and 9 per cent only occasionally, then the overwhelming majority will probably not be dependent upon paid employment as a major source for financing their retirement. Many will for some time, or from time to time, seek some additional sources of income to maintain their desired level of living. Undoubtedly those in poor health will have the greatest difficulty since they will have additional expenses and be least likely to find employment to supplement their retirement income. For the present, however, the highly positive attitudes expressed indicate that for most of them, given an adequate retirement income, even quite early retirement poses no crisis or trauma but rather is seen as a valued privilege.

Professional Retirement

A number of studies of professional workers have also produced interesting findings, limited though they are in size and representativeness. Among the most interesting recent ones is Rowe's[149] study of academic scientists. Using a sample of 142 retired physical scientists and mathematicians from eleven of the largest American universities, and comparing his findings with Benz's[10] earlier study of retired academics and faculty and administrative officers of New York University, Rowe sought to determine how scientists would adapt to the circumstances of retirement given the implicit assumption of their high work orientation. Rowe[149] reports that, "Even though they generally do not plan for retirement beyond checking their annuity, retirement does not appear to be particularly disruptive for many of them. They tend to accept retirement as part of their 'life-cycle' and not especially disruptive to their happiness. Although they may perceive their research situation as less fruitful than their earlier years, their retirement situation is a relatively contented and independent one with opportunity to continue to 'engage' in science" (p. 118). Most continue a fairly active level of reading in their fields, some continue research, others attend meetings and professional gatherings, and some are able to continue to receive funds for studies and consulting work. Rowe found that the extent of continued involvement was related to the scientist's eminence in his field, which reflects the degree of socialization into the "ethos" of science as an activity independent of a specific working position. The scientist is oriented to his discipline and not to his "job," and this allows him to treat his job retirement as apart from his scientific commitment. Benz's[10] earlier study found a similar pattern among retired academics in general, although their degree of continued engagement was less than that of the scientists, while being much greater than that of the administrators who seem to lose touch with the "ethos" of the discipline. Rowe suggests that the identification with the discipline enshrines the continued search for knowledge as a transcending value. Just as with other retirees, however, problems of health and the death of a spouse are conditions that serve to lessen the scientists' activity and engagement with this quest for knowledge.

It is interesting that despite the common attitude that professionals, like farmers, want to work forever, few studies of specialized professional occupations have examined this assumption and sought to determine the extent of its validity as well as the differences between different categories of professionals. Rowe's study suggests that commitment to the scientific "ethos" may vary widely. Certainly some professions are more instrumental for their members and may lack the strong sense of scientific "ethos" of the physical scientist. Atchley's[4] study of work orientation among teachers found little evidence of a strong orientation to the profession among male teachers but a somewhat stronger orientation among women teachers. It would seem that the category of "professional" may be more a reflection of the professional's socioeconomic status than of his commitment insofar as it is related to work orientation and retirement attitude. Certainly Streib and Schneider's[178] findings that professionals have a more positive retirement orientation than other occupational groups in their study would indicate that very little is really known about the "professional" and retirement. Perhaps more research has been concerned with farmers, in the course of which the myth of the farmer's inability to stop working was laid to rest.[20,44,91,112,134,153]

Nonwhites in Retirement

Very few studies have been done on the adjustment of nonwhite groups to retirement. In most samples the number of blacks or other nonwhites have been too small to allow any specialized consideration. There has been some concern for ethnic aged studies in the last few years, but these have concentrated on the generic problems of aging among such groups. Grann's[60] study of retired black men suggests a positive attitude toward retirement, but it is difficult on the basis of this small study of one community to determine to what extent the major attitudes are those applicable to working-class poorer people in general. In addition, the relative newness of large numbers of blacks to the urban industrial environment in which retirement is a major feature of life may be a barrier. While there are many blacks among industrial and civil service retirees, there is no evidence that their attitudes would be much different from other persons of similar occupational, health, and income status. It is known, however, from Social Security data that blacks, especially from the South, are proportionately overrepresented among the very low income, reduced-benefit early retirees. But this would appear to be consistent with their having a low-income, irregular work history in the rural economy of the "old South."

Women in Retirement

Women are relatively neglected also, perhaps the most neglected in terms of the numbers involved. Although Streib and Schneider[178] compare men and women consistently in their study and find women workers in the 1950's to be generally less positively oriented toward retirement, less accepting of it, and consequently less satisfied with it, very few other studies have devoted much attention to women workers as a distinctive group.[4,38,87,119] In most instances they have been used as a basis for comparison with men, and almost no study has been devoted to women's attitudes and responses to retirement outside the context of their adjustment to aging per se.[44] Even given the extensive data available, almost no studies using Social Security survey information or program operation information have dealt with women workers as a special group. The fact that women overwhelmingly choose to retire under Social Security at the earliest possible ages for qualification has simply been taken as a reflection of their marginal role in the labor force in recent times. One would assume that this will soon be corrected since greater attention to women in the labor force has been evident for some time, even before the rise of women's liberation in the last few years.

The major exception (aside from Cottrell and Atchley's[38] tentative explorations) is Schneider's[151] study of the women workers in the Cornell Study of Occupational Retirement. Apart from indicating little change in health and general satisfaction with life as a consequence of retirement, Schneider found that single women workers, contrary to expectations, were more favorably oriented to retirement than married women workers and consistently expressed more satisfaction with their retirement role (91 per cent of the single women finding no difficulty in not working as compared to 65 per cent of the married women). He suggests that "homemaking" might be viewed as an activity rather than a role that single women may be favorably disposed to, as a pleasant change from office work or as an expression of their individuality, while the married woman, "If (she) is still working at sixty-five years of age, she is working for a good reason. Either she feels that she and her husband need the money, or she may enjoy work more than the tasks of homemaking" (p. 143).

Since the majority of persons living in retirement are women, the absence of studies devoted to comparisons of single, widowed, married, and divorced women workers and their relation to the event and process of retirement—both their own and that of their spouses and friends and neighbors—is a peculiarly unfortunate blind spot.

Preparation for Retirement and Retirement Satisfaction

One major consequence of the concern over the assumed negative consequences of retirement and over the relative openness or lack of structure of the retirement role has been a great deal of concern about planning for retirement.[44] Numerous programs have been developed over the past two decades by industry, labor groups, educational institutions, and private organizations. A large and extensive literature has been produced that includes many programs for educating workers preparing for retirement, and reports of such programs have been published and extensively discussed.[44] However, few research studies were designed or carried out to assess their value and impact until quite recently.[31,63,72] Thompson's[184] research concluded that for the achievement of satisfaction in retirement, "*the two most important factors are accurate pre-conception of retirement and a favorable pre-retirement attitude toward retirement.* Planning for retirement, which is often cited as a main objective of preretirement counseling programs, is shown to be of relatively less direct importance" (p. 43). Since he implied that planning may impede retirement adjustment among those who lack an accurate preconception of retirement, there has been much dispute over the value of preretirement counseling and other types of preparation for retirement other than the simple function of making available useful and otherwise not easily obtainable information. What has been at issue is the question of whether extensive programs really served the process of retirement adaptation for the individual, or whether they might be attempting to induce patterns of behavior that many workers were unable to assimilate because of the differing character of their own life styles.

Greene, *et al.*,[63] report on a complex and thorough experimental program of preretirement counseling, using a variety of types of companies and employees, with control groups not exposed to the programs. While they admit that the effects were difficult to measure, they do report a positive if weak association between exposure to the preretirement counseling program and retirement adjustment. Interestingly the factor of exposure to the program itself, rather than any specific characteristics or content aspects, seems to be important. There is some question about the lasting effects of the program, but evidence of any contribution, even if only at the beginning of the retirement process, suggests the value of the efforts that go into such programs.

Similarly, Hunter[72] reports important, though limited, positive results as the outcome of a longitudinal study of a program developed for automobile workers. This was a more extensive series of programmatic and informational meetings with specially designed reading materials and audiovisual aids. A positive association between participation in the program and a variety of measures of adjustment was reported in the first year of retirement. However, after the first year the differences between experimental and control subjects, although present, was diminished and lacked significance as measured by the test instruments. Hunter is careful to point out that the longer range implications of such programs are not known, nor is it entirely clear to what extent many of the positive results are the consequence of special factors outside of the program, such as self-selection of the volunteers who took the program and the preexistence of highly favorable attitudes toward retirement among selected groups of workers like the automobile workers in his program. Certainly the favorable benefit levels of the auto workers' pension system and the positive encouragement of the union are factors that have to be taken into consideration. If the program shows no measurable difference in the attitudes of experimental and control subjects after the first year, then the function of such programs may be so limited in comparison with Thompson's[184] claim of the preeminence of accurate conceptions of retirement and favorable preretirement attitudes that they may serve nothing more than an elaborate form of information communication that might be more readily achieved in other less costly and time-consuming fashion. On the other hand, although not measurable the highly favorable subjective response of the participants to much of the program might be viewed as indicating a useful form of adult socialization to the potentialities of continuing education for retired workers. To some extent, then, there may be great utility in having educational institutions develop and maintain such programs as part of a larger program of adult education that is lifelong and not just a seemingly last minute attempt to compensate for prior failings.[31]

Bibliography

1. ADAMS, D. L., "Correlates of Satisfaction among the Small Town Elderly," Ph.D. diss., University of Missouri, 1971.
2. ANDERSON, N., *Work and Leisure*, Free Press, New York, 1961.
3. ATCHLEY, R. C., "Retirement and Leisure Participation: Continuity or Crisis," *The Gerontologist*, *11*:13–17, 1971.
4. ———, "Retirement and Work Orientation," *The Gerontologist*, *11*:29–32, 1971.
5. BACK, K. W., and GUPTILL, C. S., "Retire-

ment and Self-Ratings," in Simpson, I. H., and McKinney, J. C. (Eds.), *Social Aspects of Aging*, Pp. 120–129, Duke University Press, Durham, N.C., 1966.

6. BARFIELD, R. E., *The Automobile Worker and Retirement: A Second Look*, Institute for Social Research, University of Michigan, Ann Arbor, Mich., 1970.
7. ———, and MORGAN, J. N., *Early Retirement: The Decision and the Experience*, Institute for Social Research, University of Michigan, Ann Arbor, Mich., 1969.
8. BARRON, M. L., "The Dynamics of Occupational Roles and Health in Old Age," in Anderson, J. E. (Ed.), *Psychological Aspects of Aging*, Pp. 236–239, American Psychological Association, Washington, D.C., 1956.
9. ———, STREIB, G. F., and SUCHMAN, E. A., "Research on the Social Disorganization of Retirement," *Am. Sociol. Rev.*, *17:* 479–482, 1952.
10. BENZ, M., "A Study of the Faculty and Administrative Staff Who Have Retired from New York University, 1945–1956," *J. Educ. Sociol.*, *32:*282–293, 1958.
11. BIRREN, J. E., "Increments and Decrements in the Intellectual Status of the Aged," *Psychiat. Res. Rep.*, *23:*207–214, 1968.
12. ——— (Ed.), *Relations of Development and Aging*, Charles C Thomas, Springfield, Ill., 1964.
13. BLAKELOCK, E., "A New Look at the New Leisure," *Admin. Sci. Quart.*, *4:*446–467, 1960.
14. BLAUNER, R., *Alienation and Freedom: The Factory Worker and His Industry*, University of Chicago Press, Chicago, 1964.
15. BRACEY, H. E., *In Retirement: Pensioners in Great Britain and the United States*, Louisiana State University Press, Baton Rouge, La., 1966.
16. BREEN, L. Z., "The Aging Individual," in Tibbitts, C. (Ed.), *Handbook of Social Gerontology*, Pp. 145–162, University of Chicago Press, Chicago, 1960.
17. ———, "Retirement-Norms, Behavior, and Functional Aspects of Normative Behavior," in Williams, R. H., Tibbitts, C., and Donahue, W. (Eds.), *Processes of Aging*, Vol. 2, Pp. 381–388, Atherton Press, New York, 1963.
18. BUHLER, C., "Meaningful Living in the Mature Years," in Kleemeier, R. W. (Ed.), *Aging and Leisure*, Pp. 345–388, Oxford University Press, New York, 1961.
19. BULTENA, G., "Life Continuity and Morale in Old Age," *The Gerontologist*, *9:*251–253, 1969.
20. ———, POWERS, E., FALKMAN, P., and FREDERICK, D., *Life after 70 in Iowa: A Restudy of Participants in the 1960 Survey of the Aged*, Sociology Report No. 95, Iowa State University, Department of Sociology and Anthropology, Ames, Iowa, 1971.
21. BURGESS, E. W., "Aging in Western Culture," in Burgess, E. W. (Ed.), *Aging in Western Societies: A Comparative Survey*, Pp. 3–28, University of Chicago Press, Chicago, 1960.
22. ———, COREY, L. G., PINEO, P. C., and THORNBURY, R. T., "Occupational Differences in Attitudes toward Aging and Retirement," *J. Gerontol.*, *13:*203–206, 1958.
23. *Business Week*. "Old Hands Snub Pensions," November 18, 1950.
24. BUTLER, R. N., "The Life Review: An Interpretation of Reminiscence in the Aged," *Psychiatry*, *26:*65–76, 1963.
25. ———, "Patterns of Psychological Health and Psychiatric Illness in Retirement," in Carp, F. (Ed.), *The Retirement Process*, Pp. 27–41, Public Health Service Publication No. 1778, U. S. Department of Health, Education and Welfare, National Institute of Child Health and Human Development, Washington, D.C., 1969.
26. CAIN, L. D., JR., "Aging and the Character of Our Times," *The Gerontologist*, *8:*250–258, 1968.
27. ———, "Life Course and Social Structure," in Faris, R. E. L. (Ed.), *Handbook of Modern Sociology*, Pp. 272–309, Rand McNally, Chicago, 1964.
28. CARP, F. M. (Ed.), *The Retirement Process*, Public Health Service Publication No. 1778, U. S. Department of Health, Education and Welfare, National Institute of Child Health and Human Development, Washington, D.C., 1969.
29. CAVAN, R. S., "Self and Role in Adjustment during Old Age," in Rose, A. M. (Ed.), *Human Behavior and Social Processes*, Pp. 526–536, Houghton Mifflin, Boston, 1962.
30. ———, BURGESS, E. W., HAVIGHURST, R. J.,

and GOLDHAMER, H., *Personal Adjustment in Old Age*, Science Research Associates, Chicago, 1949.

31. CHARLES, D. C., "Effect of Participation in a Pre-retirement Program," *The Gerontologist, 11* (No. 1, Part II): 24–28, 1971.
32. CHINOY, E., *Automobile Workers and the American Dream*, Doubleday, New York, 1955.
33. CHOWN, S., "Rigidity—A Flexible Concept," *Psychol. Bull., 56*:195–223, 1959.
34. CLAGUE, E., "Work and Leisure for Older Workers," *The Gerontologist, 11*:9–20, 1971.
35. CLARKE, H. J., *Social Legislation*, 2nd ed., Appleton-Century-Crofts, New York, 1957.
36. CORSON, J., and MCCONNELL, J., *Economic Needs of Older People*, Twentieth Century Fund, New York, 1956.
37. COTTRELL, F., *Technological Change and Labor in the Railroad Industry*, D. C. Heath, Lexington, Mass., 1970.
38. ———, and ATCHLEY, R. C., *Women in Retirement: A Preliminary Report*, Scripps Foundation, Oxford, Ohio, 1969.
39. CUMMING, E., "Further Thoughts on the Theory of Disengagement," *Internat. Soc. Sci. J., 15*:377–393, 1963.
40. ———, and HENRY, W. E., *Growing Old: The Process of Disengagement*, Basic Books, New York, 1961.
41. DE GRAZIA, S., *Of Time, Work and Leisure*, Twentieth Century Fund, New York, 1962.
42. DEUTSCHER, I., "Socialization for Postparental Life," in Rose, A. M. (Ed.), *Human Behavior and Social Processes*, Pp. 506–525, Houghton Mifflin, Boston, 1962.
43. DONAHUE, W., *et al.* (Eds.), *Free Time: Challenge to Later Maturity*, University of Michigan Press, Ann Arbor, Mich., 1958.
44. ———, ORBACH, H. L., and POLLAK, O., "Retirement: The Emerging Social Pattern," in Tibbitts, C. (Ed.), *Handbook of Social Gerontology*, Pp. 330–406, University of Chicago Press, Chicago, 1960.
45. DUBIN, R., "Industrial Workers' Worlds," *Soc. Prob., 3*:131–142, 1956.
46. ETENG, W. I. A., and MARSHALL, D. G., *Retirement and Migration in the North Central States: Comparative Analysis of Life and Retirement Satisfaction—Wisconsin, Florida and Arizona*, Population Series No. 20, University of Wisconsin, Department of Rural Sociology, Madison, Wisc., 1970.
47. FILLENBAUM, G. G., "A Consideration of Some Factors Related to Work after Retirement," *The Gerontologist, 11*:18–23, 1971.
48. ———, "On the Relationship between Attitude to Work and Attitude to Retirement," *J. Gerontol., 26*:244–248, 1971.
49. ———, "Retirement Planning Programs—At What Age, and for Whom?" *The Gerontologist, 11*:33–36, 1971.
50. ———, "The Working Retired," *J. Gerontol., 26*:82–89, 1971.
51. FLEMING, R. W., and MCGAUGHEY, R., *Civil Servants in Retirement*, University of Wisconsin, Industrial Relations Center, Madison, Wisc., 1952.
52. FRIEDMANN, E. A., "The Impact of Aging on the Social Structure," in Tibbitts, C. (Ed.), *Handbook of Social Gerontology*, Pp. 120–144, University of Chicago Press, Chicago, 1960.
53. ———, "The Work of Leisure," in Donahue, W., *et al.* (Eds.), *Free Time: A Challenge to Later Maturity*, University of Michigan Press, Ann Arbor, Mich., 1958.
54. ———, and HAVIGHURST, R. J., *The Meaning of Work and Retirement*, University of Chicago Press, Chicago, 1954.
55. FRIEDMANN, G., *The Anatomy of Work*, Free Press, New York, 1961.
56. GLENN, N. D., "Aging, Disengagement, and Opinionation," *Public Opinion Quart., 33*:17–33, 1969.
57. GLICK, P. C., and PARK, R., JR., "New Approaches in Studying the Life Cycle of the Family," *Demography, 2*:187–202, 1965.
58. GORDON, M., "Aging and Income Security," in Tibbitts, C. (Ed.), *Handbook of Social Gerontology*, University of Chicago Press, Chicago, 1960.
59. ———, "Work and Patterns of Retirement," in Kleemeier, R. (Ed.), *Aging and Leisure*, Oxford University Press, New York, 1961.
60. GRANN, L., "Social and Personal Adjustment of Retired Persons," *Sociol. & Soc. Res., 39*:311–316, 1955.
61. Great Britain Ministry of Pensions and Na-

tional Insurance, *Reasons for Retiring or Continuing Work*, H. M. Stationery Office, London, 1954.

62. GREENE, M. R., PYRON, H. C., MANION, V. V., and WINKELVOSS, H., *Early Retirement: A Survey of Company Policies and Retirees' Experiences*, University of Oregon, College of Business Administration, Eugene, Ore., 1969.
63. ———, *Preretirement Counseling, Retirement Adjustment, and the Older Employee*, University of Oregon, College of Business Administration, Eugene, Ore., 1969.
64. GURIN, G., VEROFF, J., and FELD, S., *Americans View Their Mental Health: A Nationwide Interview Study*, Basic Books, New York, 1960.
65. GUTMANN, D., *Ego Psychological and Developmental Approaches to the "Retirement Crisis" in Men*, unpublished paper, University of Michigan, Department of Psychology, Ann Arbor, Mich., n.d.
66. HAMOVITCH, M. B., "Social and Psychological Factors in Adjustment in a Retirement Village," in Carp, F. (Ed.), *The Retirement Process*, Pp. 115–126, Public Health Service Publication No. 1778, U. S. Department of Health, Education and Welfare, National Institute of Child Health and Human Development, Washington, D.C., 1969.
67. HARRIS, L., " 'Pleasant' Retirement Expected," *The Washington Post*, November 28, 1965.
68. HAVIGHURST, R. J., "Successful Aging," in Williams, R. H., Tibbitts, C., and Donahue, W. (Eds.), *Processes of Aging*, Vol. 1, Pp. 299–320, Atherton Press, New York, 1963.
69. ———, and ALBRECHT, R., *Older People*, Longmans, Green and Company, New York, 1953.
70. ———, NEUGARTEN, B. L., and TOBIN, S., "Disengagement and Patterns of Aging," in Neugarten, B. L. (Ed.), *Middle Age and Aging: A Reader in Social Psychology*, Pp. 161–172, University of Chicago Press, Chicago, 1968.
71. HERON, A., "Retirement Attitudes among Industrial Workers in the Sixth Decade of Life," *Vita Humana*, *6*:152–159, 1963.
72. HUNTER, W. W., *A Longitudinal Study of Retirement Education*, University of Michigan, Division of Gerontology, Ann Arbor, Mich., 1968.
73. KAPLAN, M., "Toward a Theory of Leisure for Social Gerontology," in Kleemeier, R. W. (Ed.), *Aging and Leisure*, Pp. 389–412, Oxford University Press, New York, 1961.
74. KATONA, G., *Private Pensions and Individual Saving*, Monograph No. 40, University of Michigan, Institute for Social Research, Ann Arbor, Mich., 1965.
75. ———, MORGAN, J. N., and BARFIELD, R. E., "Retirement in Prospect and Retrospect," in *Trends in Early Retirement*, Pp. 27–49, Occasional Papers in Gerontology, No. 4, Institute of Gerontology, University of Michigan-Wayne State University, Ann Arbor, Mich., 1969.
76. KERCKHOFF, A. C., "Family Patterns and Morale in Retirement," in Simpson, I. H., and McKinney, J. C. (Eds.), *Social Aspects of Aging*, Pp. 173–192, Duke University Press, Durham, N.C., 1966.
77. ———, "Husband-Wife Expectations and Reactions to Retirement," in Simpson, I. H., and McKinney, J. C. (Eds.), *Social Aspects of Aging*, Pp. 160–172, Duke University Press, Durham, N.C., 1966.
78. KLEEMEIER, R. W. (Ed.), *Aging and Leisure*, Oxford University Press, New York, 1961.
79. ———, "The Effect of a Work Program on Adjustment Attitudes in an Aged Population," *J. Gerontol.*, *6*:373–379, 1951.
80. ———, "Leisure and Disengagement in Retirement," *The Gerontologist*, *4*:180–184, 1964.
81. KOLODRUBETZ, W. W., "Trends in Employee-Benefit Plans in the Sixties," *Soc. Secur. Bull.*, *34*:21–34, 1971.
82. KOOY, G. A., and VAN'T KLOOSTER-VAN WINGERDEN, C. M., "The Aged in an Urban Community in the Netherlands," *Human Dev.*, *11*:64–77, 1968.
83. KREPS, J., "Employment Policy and Income Maintenance for the Aged," in McKinney, J. C., and de Vyver, F. T. (Eds.), *Aging and Social Policy*, Pp. 136–157, Appleton-Century-Crofts, New York, 1966.
84. KUHLEN, R. G., "Aging and Life-Adjustment," in Birren, J. E. (Ed.), *Handbook of Aging and the Individual*, Pp. 852–897, University of Chicago Press, Chicago, 1959.

85. ———, "Motivational Changes during Adult Years," in Kuhlen, R. G. (Ed.), *Psychological Backgrounds of Adult Education*, Pp. 77–113, Center for the Study of Liberal Education for Adults, Chicago, 1963.
86. Kutner, B., Fanshel, D., Togo, A. M., and Langner, T. S., *Five Hundred over Sixty: A Community Survey on Aging*, Russell Sage Foundation, New York, 1956.
87. Lehr, U., and Dreher, G., "Determinants of Attitudes toward Retirement," in Havighurst, R. J., Munnichs, J. M. A., Neugarten, B. L., and Thomae, H. (Eds.), *Adjustment to Retirement: A Cross-National Study*, 2nd ed., Pp. 116–138, Van Gorcum and Company, Assen, The Netherlands, 1970.
88. Lemon, B., Bengston, V. L., and Peterson, J., "Activity Types and Life Satisfaction in a Retirement Community: An Exploration of the Activity Theory of Aging," paper presented to the 8th International Congress of Gerontology, Washington, D.C., 1969.
89. Life Extension Foundation, *Retirement Study: Overall Summary of 1546 Questionnaires*, The Foundation, New York, 1956.
90. Lipman, A., "Role Conceptions and Morale of Couples in Retirement," *J. Gerontol.*, *16*:267–271, 1961.
91. Loeb, M., Pincus, A., and Mueller, J., *Growing Old in Rural Wisconsin*, University of Wisconsin, School of Social Work, Madison, Wisc., 1963.
92. Loether, H. J., "The Meaning of Work and Adjustment to Retirement," in Shostak, A. B., and Gomberg, W. (Eds.), *Blue-Collar World*, Pp. 517–525, Prentice-Hall, Inc., Englewood Cliffs, N.J., 1964.
93. Lowenthal, M. F., "Social Isolation and Mental Illness in Old Age," *Am. Sociol. Rev.*, *29*:54–70, 1964.
94. McKain, W. C., *Retirement Marriage*, Monograph No. 3, Storrs Agricultural Experiment Station, University of Connecticut, 1969.
95. McKinney, J. C., and De Vyver, F. T. (Eds.), *Aging and Social Policy*, Appleton-Century-Crofts, New York, 1966.
96. McMahan, C. A., and Ford, T. R., "Surviving the First Five Years of Retirement," *J. Gerontol.*, *10*:212–215, 1955.
97. McNeil, J. S., and Giffen, M. B., "Military Retirement: Some Basic Observations and Concepts," *Aerospace Med.*, *36*:25-29, 1965.
98. ———, "The Social Impact of Military Retirement," *Soc. Casework*, *46*:203–207, 1965.
99. Maddox, G. L., Jr., "Activity and Morale: A Longitudinal Study of Selected Elderly Subjects," *Soc. Forces*, *42*:195–204, 1963.
100. ———, "Fact and Artifact: Evidence Bearing on Disengagement Theory from the Duke Geriatrics Project," *Human Dev.*, *8*:117–130, 1965.
101. ———, "Persistence of Life Style among the Elderly: A Longitudinal Study of Patterns of Social Activity in Relation to Life Satisfaction," in *Proceedings of the 7th International Congress of Gerontology*, Vienna, *8*:309–311, 1966.
102. ———, "Retirement as a Social Event," in McKinney, J. S., and de Vyver, F. T. (Eds.), *Aging and Social Policy*, Pp. 117–135, Appleton-Century-Crofts, New York, 1966.
103. ———, and Eisdorfer, C., "Some Correlates of Activity and Morale among the Elderly," *Soc. Forces*, *40*:254–260, 1962.
104. Martel, M. U., and Morris, W. W., *Life after Sixty in Iowa: A Report on the 1960 Survey*, Iowa Commission for Senior Citizens, Iowa City, Iowa, 1961.
105. Martin, J., and Doran, A., "Evidence Concerning the Relationship between Health and Retirement," *Sociol. Rev.*, *14*:329–343, 1966.
106. Messer, E. F., "Thirty-Eight Years is a Plenty," in *Trends in Early Retirement*, Pp. 50–66, Occasional Papers in Gerontology, No. 4, Institute of Gerontology, University of Michigan-Wayne State University, Ann Arbor, Mich., 1969.
107. Miller, S. J., "The Social Dilemma of the Aging Leisure Participant," in Rose, A. M., and Peterson, W. A. (Eds.), *Older People and Their Social World*, Pp. 77–92, F. A. Davis Co., Philadelphia, 1965.
108. Moore, E. H., *The Nature of Retirement*, Macmillan, New York, 1959.
109. Moore, W. E., *Industrial Relations and the Social Order*, esp. Chapter 21, "The

Aged in Industrial Societies," Pp. 519–537, Macmillan, New York, 1951.

110. ———, *Industrialization and Labor: Social Aspects of Economic Development*, Cornell University Press, Ithaca, N.Y., 1951.

111. ———, *Man, Time and Society*, John Wiley, New York, 1963.

112. MORRISON, D. E., and KRISTJANSON, G. A., *Personal Adjustment among Older Persons*, Technical Bulletin No. 21, South Dakota State College, Agricultural Experiment Station, Brookings, S.D., 1958.

113. MORSE, N. C., and WEISS, R. R., "The Function and Meaning of Work and the Job," *American Sociol. Rev.*, *20*:191–198, 1955.

114. MURRAY, J. R., POWERS, E., and HAVIGHURST, R. J., "Personal and Situational Factors Producing Flexible Careers," *The Gerontologist*, *11* (No. 4, Part II): 4–12, 1971.

115. MYERS, R. J., "Factors in Interpreting Mortality after Retirement," *J. Am. Statistical A.*, *49*:499–509, 1954.

116. NEUGARTEN, B. L., "Adult Personality: Toward a Psychology of the Life Cycle," in Neugarten, B. L. (Ed.), *Middle Age and Aging: A Reader in Social Psychology*, Pp. 137–147, University of Chicago Press, Chicago, 1968.

117. ———, "A Developmental View of Adult Personality," in Birren, J. E. (Ed.), *Relations of Development and Aging*, Pp. 176–208, Charles C Thomas, Springfield, Ill., 1964.

118. ———, "Personality and the Aging Process," *The Gerontologist*, *12*:9–15, 1972.

119. ———, and HAVIGHURST, R. J., "Disengagement Reconsidered in a Cross-National Context," in Havighurst, R. J., Munnichs, J. M. A., Neugarten, B. L., and Thomae, T. (Eds.), *Adjustment to Retirement*, 2nd ed., Pp. 138–146, Van Gorcum and Company, Assen, The Netherlands, 1970.

120. ———, MOORE, J. W., and LOWE, J. C., "Age Norms, Age Constraints, and Adult Socialization," *Am. J. Sociol.*, *70*:710–717, 1965.

121. ———, and PETERSON, W. A., "The American Age-Grade System," in *Proceedings of the 4th Congress of the International Association of Gerontology*, Merano, Italy, *3*:497–502, 1957.

122. NOSOW, S., and FORM, W. H. (Eds.), *Man, Work and Society*, Basic Books, New York, 1962.

123. ORBACH, H. L., "Normative Aspects of Retirement," in Tibbitts, C., and Donahue, W. (Eds.), *Social and Psychological Aspects of Aging*, Pp. 53–63, Columbia University Press, New York, 1962.

124. ———, "Social and Institutional Aspects of Industrial Worker's Retirement Patterns," in *Trends in Early Retirement*, Pp. 1–26, Occasional Papers in Gerontology, No. 4, Institute of Gerontology, University of Michigan-Wayne State University, Ann Arbor, Mich., 1969.

125. ———, "Social Values and the Institutionalization of Retirement," in Williams, R. H., Tibbitts, C., and Donahue, W. (Eds.), *Processes of Aging*, Vol. 2, Pp. 389–402, Atherton Press, New York, 1963.

126. ———, and SHAW, D. M., "Social Participation and the Role of the Aged," *Geriatrics*, *12*:241–246, 1957.

127. ORZACK, L., and FRIEDMANN, E. A., *Work and Leisure Interrelationships*, paper presented at Fourth World Congress on Sociology, Milan, Italy, September 1959.

128. OWEN, J. P., and BELZUNG, L. D., "Consequences of Voluntary Early Retirement: A Case Study of a New Labor Force Phenomenon," *Brit. J. Indust. Rel.*, *5*:162–189, 1967.

129. PALMER, G., "Attitudes toward Work in an Industrial Community," *Am. J. Sociol.*, *63*:17–26, 1957.

130. PALMORE, E. B., "Differences in the Retirement Patterns of Men and Women," *The Gerontologist*, *5*:4–8, 1965.

131. ———, "Retirement Patterns among Aged Men: Findings of the 1963 Survey of the Aged," *Soc. Secur. Bull.*, *27* (No. 8): 3–10, 1964.

132. PECK, R., "Psychological Developments in the Second Half of Life," in Anderson, J. E. (Ed.), *Psychological Aspects of Aging*, Pp. 42–53, American Psychological Association, Washington, D.C., 1965.

133. PHILLIPS, B. S., "A Role Theory Approach to Adjustment in Old Age," *Am. Sociol. Rev.*, *22*:212–217, 1957.

134. PIHLBLAD, C. T., and ROSENCRANZ, H. A., *Retirement Status of Older People in a Small Town*, Interim Report, Vol. 3, Uni-

versity of Missouri, Department of Sociology, Columbia, Mo., 1968.

135. ———, *Social Adjustment of Older People in the Small Town*, Interim Report, Vol. 4, University of Missouri, Department of Sociology, Columbia, Mo., 1969.

136. POLLAK, O., *Social Adjustment in Old Age: A Research Planning Report*, Bulletin 59, Social Science Research Council, New York, 1948.

137. POLLMAN, A. W., "Early Retirement: A Comparison of Poor Health to Other Factors," *J. Gerontol.*, *26*:41–45, 1971.

138. ———, "Early Retirement: Relationship to Variation in Life Satisfaction," *The Gerontologist*, *11*:43–47, 1971.

139. REICHARD, S., LIVSON, F., and PETERSON, P. G., *Aging and Personality*, John Wiley, New York, 1962.

140. RENO, V., *Retirement Patterns of Men at OASDHI Entitlement*, Preliminary Findings from the Survey of New Beneficiaries, Report No. 2, U. S. Department of Health, Education and Welfare, Social Security Administration, Office of Research and Statistics, Washington, D.C., 1971.

141. ———, *Why Men Stop Working at or before Age 65*, Preliminary Findings from the Survey of New Beneficiaries, Report No. 3, U. S. Department of Health, Education and Welfare, Social Security Administration, Office of Research and Statistics, Washington, D.C., 1971.

142. RICHARDSON, A. H., "Alienation and Disengagement among the Very Aged," in *Proceedings of the 7th International Congress of Gerontology*, Vienna, *8*:539–545, 1966.

143. RIESMAN, D., *The Lonely Crowd*, Yale University Press, New Haven, 1950.

144. ———, and BLOOMBERG, W., JR., "Work and Leisure: Fusion or Polarity?" in Arensberg, C., *et al.* (Eds.), *Research in Industrial and Human Relations*, Pp. 69–85, Harper & Brothers, New York, 1957.

145. ROSENBERG, G. S., *The Worker Grows Old: Poverty and Isolation in the City*, Jossey-Bass, San Francisco, 1970.

146. ROSOW, I., "Adjustment of the Normal Aged," in Williams, R. H., Tibbitts, C., and Donahue, W. (Eds.), *Processes of Aging*, Vol. 2, pp. 195–223, Atherton Press, New York, 1963.

147. ———, "Intergenerational Relationships: Problems and Proposals," in Shanas, E., and Streib, G. (Eds.), *Social Structure and the Family*, Pp. 341–378, Prentice-Hall, Englewood Cliffs, N.J., 1965.

148. ———, *Social Integration of the Aged*, Free Press, New York, 1967.

149. ROWE, A. R., "The Retirement of Academic Scientists," *J. Gerontol.*, *27*:113–118, 1972.

150. RYSER, C., and SHELDON, A., "Retirement and Health," *J. Am. Geriatrics Soc.*, *17*: 180–190, 1969.

151. SCHNEIDER, C. J., "Adjustment of Employed Women to Retirement," Ph.D. diss., Cornell University, 1964.

152. SCOTCH, N. A., and RICHARDSON, A. H., "Characteristics of the Self-Sufficient among the Very Aged," in *Proceedings of the 7th International Congress of Gerontology*, Vienna, *8*:489–493, 1966.

153. SEWELL, W. H., RAMSEY, C. E., and DUCOFF, L. J., *Farmer's Conceptions and Plans for Economic Security in Old Age*, Research Bulletin 182, Wisconsin Agricultural Experiment Station, Madison, Wisc., 1953.

154. SHANAS, E., "Facts versus Stereotypes: The Cornell Study of Occupational Retirement," *J. Soc. Issues*, *14* (No. 2): 61–62, 1958.

155. ———, and STREIB, G. F., eds., *Social Structure and the Family: Generational Relations*, Prentice-Hall, Englewood Cliffs, N.J., 1965.

156. ———, TOWNSEND, P., WEDDERBURN, D., FRIIS, H., MILHØJ, P., and STEHOUWER, J., *Old People in Three Industrial Societies*, Atherton Press, New York, 1968.

157. SHEPPARD, H. L., "Discontented Blue-Collar Workers—A Case Study," *Monthly Labor Rev.*, *94*:25–32, 1971.

158. SIMMONS, L. W., *The Role of the Aged in Primitive Society*, Yale University Press, New Haven, 1954.

159. SIMPSON, I. H., BACK, K. W., and MCKINNEY, J. C., "Attributes of Work, Involvement in Society, and Self-Evaluation in Retirement," in Simpson, I. H., and McKinney, J. C. (Eds.), *Social Aspects of Aging*, Pp. 55–74, Duke University Press, Durham, N.C., 1966.

160. ———, "Continuity of Work and Retirement Activities, and Self-Evaluation," in

Simpson, I. H., and McKinney, J. C. (Eds.), *Social Aspects of Aging*, Pp. 106–119, Duke University Press, Durham, N.C., 1966.

161. ———, "Orientations toward Work and Retirement, and Self-Evaluation in Retirement," in Simpson, I. H., and McKinney, J. C. (Eds.), *Social Aspects of Aging*, Pp. 75–89, Duke University Press, Durham, N.C., 1966.

162. SIMPSON, I. H., and MCKINNEY, J. C. (Eds.), *Social Aspects of Aging*, Duke University Press, Durham, N.C., 1966.

163. SMITH, J., and MARSHALL, D. G., *Retirement and Migration in the North Central States: Two Planned Retirement Communities*, Population Series No. 23, University of Wisconsin, Department of Rural Sociology, Madison, Wisc., 1970.

164. SMITH, K. J., and LIPMAN, A., "Constraint and Life Satisfaction," *J. Gerontol.*, *27*: 77–82, 1972.

165. SPENCE, D. L., "Patterns of Retirement in San Francisco," in Carp, F. (Ed.), *The Retirement Process*, Pp. 63–76, Public Health Service Publication No. 1778, U. S. Department of Health, Education and Welfare, National Institute of Child Health and Human Development, Washington, D.C., 1969.

166. SPENGLER, J. J., "The Aging of Individuals and Populations," in McKinney, J. C., and de Vyver, F. T. (Eds.), *Aging and Social Policy*, Pp. 3–41, Appleton-Century-Crofts, New York, 1966.

167. STANFORD, E. P., "Retirement Participation in the Military," *The Gerontologist*, *11*: 37–42, 1971.

168. STECKER, M., "Beneficiaries Prefer to Work," *Soc. Secur. Bull.*, January 1951.

169. STEINER, P., and DORFMAN, R., *The Economic Status of the Aged*, University of California Press, Berkeley, Calif., 1957.

170. STOKES, R. G., and MADDOX, G. L., JR., "Some Social Factors on Retirement Adaptation," *J. Gerontol.*, *22*:329–333, 1967.

171. STREIB, G. F., "Disengagement Theory in Socio-Cultural Perspective," *Internat. J. Psychiat.*, *6*:69–76, 1968.

172. ———, "Family Patterns in Retirement," *J. Soc. Issues*, *14* (No. 2): 46–60, 1958.

173. ———, "Intergenerational Relationships: Perspectives of the Two Generations on the Older Parent," *J. Marriage & Fam.*, *27*:469–475, 1965.

174. ———, *Longitudinal Study of Retirement: Final Report to the Social Security Administration, Washington, D.C.*, Cornell University, Department of Sociology, Ithaca, N.Y., 1965.

175. ———, "Morale of the Retired," *Soc. Prob.*, *3*:270–276, 1956.

176. ———, "Participants and Drop-outs in a Longitudinal Study," *J. Gerontol.*, *21*: 200–209, 1966.

177. ———, and ORBACH, H. L., "Aging," in Lazarsfeld, P. F., Sewell, W. H., and Wilensky, H. L. (Eds.), *The Uses of Sociology*, Pp. 612–640, Basic Books, New York, 1967.

178. ———, and SCHNEIDER, C. J. (S.J.), *Retirement in American Society: Impact and Process*, Cornell University Press, Ithaca, N.Y., 1971.

179. ———, and THOMPSON, W. E., "Personal and Social Adjustment in Retirement," in Donahue, W., and Tibbitts, C. (Eds.), *The New Frontiers of Aging*, Pp. 180–197, University of Michigan Press, Ann Arbor, Mich., 1957.

180. TAIETZ, P., and LARSON, O. F., "Social Participation and Old Age," *Rural Sociol.*, *30*:229–238, 1956.

181. ———, STREIB, G. F., and BARRON, M. L., *Adjustment to Retirement in Rural New York State*, Bulletin 919, Cornell University Agricultural Experiment Station, Ithaca, N.Y., 1956.

182. THOMAE, H., "Cross-National Differences in Social Participation: Problems of Interpretation," in Havighurst, R. J., Munnichs, J. M. A., Neugarten, B. L., and Thomae, H. (Eds.), *Adjustment to Retirement: A Cross-National Study*, 2nd ed., Pp. 147–158, Van Gorcum and Company, Assen, The Netherlands, 1970.

183. THOMPSON, E. P., "Time, Work Discipline and Industrial Capitalism," *Past and Present: A Journal of Historical Studies*, *38*:56–97, 1967.

184. THOMPSON, W. E., "Pre-Retirement Anticipation and Adjustment in Retirement," *J. Soc. Issues*, *14* (No. 2): 35–45, 1958.

185. ———, and STREIB, G. F., "Meaningful Activity in a Family Context," in Kleemeier, R. W. (Ed.), *Aging and Leisure*, Pp.

177–212, Oxford University Press, New York, 1961.

186. ———, "Situational Determinants: Health and Economic Deprivation in Retirement," *J. Soc. Issues, 14* (No. 2): 18–34, 1958.

187. ———, ———, and KOSA, J., "The Effect of Retirement on Personal Adjustment: A Panel Analysis," *J. Gerontol., 15*:165–169, 1960.

188. TIBBITTS, C. (Ed.), *Handbook of Social Gerontology*, University of Chicago Press, Chicago, 1960.

189. ———, "Retirement Problems in American Society," *Am. J. of Sociol.*, 59:301–308, 1954.

190. ———, and DONAHUE, W. (Eds.), *Social and Psychological Aspects of Aging*, Columbia University Press, New York, 1962.

191. TISSUE, T., "Downward Mobility in Old Age," *Soc. Prob., 18*:67–77, 1970.

192. TRÉANTON, J.-R., "The Concept of Adjustment in Old Age," in Williams, R. H., Tibbitts, C., and Donahue, W. (Eds.), *Processes of Aging*, Vol. 1, pp. 292–298, Atherton Press, New York, 1963.

193. TUCKMAN, J., and LORGE, I., *Retirement and the Industrial Worker: Prospect and Reality*, Columbia University, Teachers College, Bureau of Publications, New York, 1953.

194. TYHURST, J. S., SALK, L., and KENNEDY, M., "Mortality, Morbidity and Retirement," *Am. J. Public Health, 47*:1434–1444, 1957.

195. United Nations, Department of Economic and Social Affairs, *The Aging of Populations and Its Economic and Social Implications*, Population Studies No. 26, United Nations, New York, 1956.

196. ———, Department of Economic and Social Affairs, *Demographic Yearbook, 1970*, United Nations, New York, 1971.

197. U.S. Department of Commerce, Bureau of Census, *Statistical Abstract of the United States, 1970*, U.S. Government Printing Office, Washington, D.C., 1971.

198. ———, "United States, General Population Characteristics, Advance Report," PC(V2)-1, February 1971.

199. U.S. Social Security Administration, *Soc. Secur. Bull.*, 33 (No. 4), 1970.

200. ———, Bureau of Old Age and Survivors Insurance, *More Selected Findings of the National Survey of Old Age and Retirement Insurance Beneficiaries, 1951*, Washington, D.C., 1954.

201. VALAORAS, V. G., "Young and Aged Populations," *Ann. Am. Acad. Pol. Soc. Sci., 316*:69–83, 1968.

202. VIDEBECK, R., and KNOX, A. B., "Alternative Participatory Responses to Aging," in Rose, A. M., and Peterson, W. A. (Eds.), *Older People and Their Social World*, Pp. 37–48, F. A. Davis Co., Philadelphia, 1965.

203. WALKER, C. R., and GUEST, R., *Man on the Assembly Line*, Harvard University Press, Cambridge, Mass., 1952.

204. WEBBER, I. L., "The Organized Social Life of the Retired," *Amer. J. Sociol.*, 59: 340–346, 1954.

205. WILENSKY, H., "Life Cycle, Work Situation and Participation in Formal Associations," in Kleemeier, R. W. (Ed.), *Aging and Leisure*, Pp. 213–242, Oxford University Press, New York, 1961.

206. ———, and LEBEAUX, C., *Industrial Society and Social Welfare*, Free Press, New York, 1965.

207. WILLIAMS, R. H., TIBBITTS, C., and DONAHUE, W. (Eds.), *Processes of Aging*, Vol. 2, Atherton Press, New York, 1963.

208. ———, and WIRTHS, C. A., *Lives through the Years*, Atherton Press, New York, 1965.

209. YOUMANS, E. G., "Family Disengagement among Older Urban and Rural Women," *J. Gerontol., 22*:209–211, 1967.

210. ———, "Objective and Subjective Disengagement among Older Rural and Urban Men," *J. Gerontol., 21*:439–441, 1966.

CHAPTER 31

OLD AGE

Robert N. Butler

FOR SOME, old age becomes a consummation of one's life; for others it is an occasion for grief, guilt, and despair; and for many it is complicated by severe socioeconomic circumstances—poverty, inadequate housing, insufficient medical services.[88] In this chapter we shall deal with the nature of old age, its normal conditions, and its potentiality for psychopathology. Definitive pathological conditions are dealt with in another volume of the *Handbook*.

"Many a man goes fishing all his life without realizing it is not the fish he is after," said Thoreau. In examining the nature of old age it is as perilous to avoid the philosophic as it is the social, psychological, economic, biological, and other perspectives.[12,32,71,85,96] Philosophic and religious considerations that are helpful in delineating the culmination of the life cycle range from the poetic insights of Ecclesiastes to the emphases upon self-awareness, absurdity, and despair in modern existentialism. These many perspectives bear upon our endeavors to provide therapeutic help as well as upon our understanding of old age.

Old age is one part of a continuing process of the life cycle and cannot be understood in a vacuum. The perspectives of both the life cycle and history are useful. The individual life cycle comprises successive stages and processes as well as altering modes of being, responsibilities, and tasks. The complex interplay of individual life cycles with socioeconomic, cultural, and historical conditions circumscribes a growing body of knowledge. Knowledge of life cycle features, such as characteristics of various stages, provides the psychotherapist and society with broad guidelines for appropriate support and participation of the elderly.[26]

The importance of history to understanding old age has two aspects. First, there is variation over time in the status and roles of old people. Second, there is the impact of the history of one's own times. The student of aging and the therapist should have an empathic understanding of the historic circumstances of an individual older person. To gain some insight into the impact of historic experience upon older people in the third quarter of the twentieth century, one should read such pertinent works as Barbara Tuchman's *The Proud Tower* and Frederick Lewis Allen's *The Big Change*.

Old people and children have become increasingly socially visible since the seventeenth century. The increased chances of survival, joined with certain socioeconomic conditions, has "unfolded" the life cycle, making its stages or phases prominent. Aristotle, Cicero, Shakespeare, Rousseau, and other philosophers and writers have considered the life cycle as a concept and have proposed various divisions and stages. The Swedish sociologist Ellen Key wrote of the twentieth century as the "century of the child." The French demographic and cultural historian Philippe Ariès, in his *Centuries of Childhood*, emphasized the social and historical evolution of the child in the last two centuries.[4] Victor Hugo, among others, observed the conspicuousness of old people in number and social significance from the turn of the nineteenth century. In the United States, in particular, social psychologists, sociologists, and psychologists have shown some interest in the nature of the life cycle. William James, Mead, Buhler,[19] Havighurst,[49] Benedek,[8] Erikson,[37] Pressey and Kuhlen,[78] and Neugarten[72] are among recent writers who have given attention to the nature of the life cycle with some particular emphasis upon old age and the transitional middle-age period. In 1922 G. Stanley Hall[48] published the first major American study of the psychology of old age. Rothschild, Grotjahn, Goldfarb, Weinberg, Linden, Busse, and Simon are a few American psychiatrists whose work has stimulated research into the psychodynamics of aging as well as into developing treatment approaches.

Nonetheless, the social and psychological sciences and the professions have not succeeded in keeping pace with demographic changes. As Gordon[43] has pointed out, sociology remains primarily concerned with class phenomena. Psychology has demonstrated great interest in child development in this century; the majority of psychological studies do not go beyond early adulthood. In most universities, courses in human development really mean child development, not the development of man throughout the course of life. The mass media have been somewhat more conscious of the significance of age-grading in human affairs, of stages of life, and of intergenerational conflict. Much has been made of the "generation gap," for example. However, studies of the so-called generation gap and comparative simultaneous studies of different age groups (such as of youth, the middle-aged, and old people) necessitate the recognition of three generic classes of variables, historical, generational, and life stage. First are historic changes accounting for any differences that are found, for example, different amounts of education. The second class concerns the parent-child relationship, the resolution of the oedipal conflict, the development of independence, and related matters. The third class of variables pertains to differences between one age group and another and to the processes or stages of life. For example, the roles, responsibilities, and preoccupations of a person in middle age must be compared to those of the adolescent or the old person.

How people live their lives, how they change in the course of time, and how they die describes a vast subject. Nonetheless, a comprehensive psychology ultimately requires a theory (and description) of the nature of the life cycle (or life cycles). There may be a number of different channels of human development rather than one. Special problems include the study of the subjective experience of the life cycle, including aging, changes of the body and the self-concept, and approaching death. Because of the intensive, long-term nature of psychoanalytic relationships, psychoanalysts could potentially provide special insights about the nature of the middle years, the experience of aging and disability, the denouement of character, and the crises of the creative personality moving through time from youth into the middle and later years.[22] However, psychoanalysis and other forms of intensive therapy have rarely been employed with older patients.

Old age cannot be seen as a static, fixed, unitary period, but must be seen as the resultant of a range of forces. In that sense the study of old age has particular value. It is only the elderly who can teach us about the nature of the life cycle as a whole, who can provide us with insights into the ultimate evolution of health—that is, characteristics of survival[6,7]—as well as of disorders. The elderly enable us to understand the nature of adaptation and survival in the wake of the many assaults made upon man as he moves through time. They can give us special, if poignant, insights into the nature of grief, despair, and depression.[76] If we understood these various subjects better—from our studies of the elderly—our understanding would have general applicability to other age groups. Our educational theories would be influenced by our better understanding of the old. Our treatment of grief and the possibility of suicide in the young would be helped by our understanding of these in the old. In the United States it is the elderly who account for one-fourth of all suicides.

Some General Social and Economic Characteristics of the Elderly[17,18]

In 1900, 4 per cent of the population was 65 years and older, and the average life expectancy at birth, combining both sexes, was 47 years of age. Life expectancy, of course, is influenced by high infant mortality. With improvements in general public health measures, including sanitation as well as reduced infant mortality, life expectancy has increased to an average of 71 years. The elderly now constitute nearly 10 per cent of the population. The aged group is growing at a faster rate than the general population.

In social terms old age is defined as 65 years and above. The choice of 65 derives from social legislation inaugurated by Chancellor Otto von Bismarck in 1883 in Germany. In terms of biology, aging begins with conception.

Of the 20 million Americans over 65, half are over 73. Gerontologists now think in terms of early and of advanced old age, 65 to 74 and 75 and above. Indeed, the age 65 is an obsolete cutoff point in terms of health, ability, and social status. Life expectancy at 65 is 15 years; for men it is 13 years, for women 16. More than 11 million of the elderly are women; more than eight million are men. More than 13,000 are over 100. Many old people do not think of themselves as old, some because of denial, others because of their excellent status. In 1970, for every 100 older persons over 65, 57 were women and 43 men. There were 135 older women to every 100 older men. The ratio increases from 120 at age 65 to 69 to more than 160 at age 85 and over. Throughout the world, wherever there are decent living conditions, low maternal mortality, and reasonably accurate statistics, women outlive men.

Most older men are married, most older women are widowed. There are almost four times as many widows as widowers. Marital status is extremely important to understanding the psychology and social situation of the elderly. The elderly woman, usually a widow, is often in a most precarious situation. Three times as many older women live alone or with nonrelatives as do older men. Once again we see the extent to which the older woman is especially isolated. Some 16,000 older women marry each year; some 35,000 older men do.

Seven of every ten older persons live in families. Nearly one-quarter live alone or with nonrelatives. Contrary to the general impression, only one in 20 older people live in an institution. Approximately 5 per cent live in old age homes, nursing homes, and a variety of other types of facilities. Put more positively, approximately 95 per cent of older people live in the community.

White persons make up less than 90 per cent of the total population, but they make up 92 per cent of the older population. This is because of the difference in life expectancy between the white and nonwhite populations. Although the black population dies at a greater rate throughout the course of life, once a black has survived to old age, he has a greater survival rate.

Educational attainment is another important parameter in describing the general characteristics of old age. Since "being educated" in the best sense is not equivalent to the amount of formal education, the meaning of the statistics is suspect. Half of the older population has had only elementary schooling or less, while half of those under 65 have had at least a high school education. Of every 100 older people, eight men and eight women 65 years and older had no schooling or less than five years and are therefore functionally illiterate. These educational disadvantages are pertinent to unemployment in old age and possibly to the utilization of one's internal resources in adapting to the solitude of old age. Approximately 6 per cent of old people are college graduates.

If one utilizes the highly stringent and conservative poverty index, one out of every four older people live below the poverty line. Thus, in 1969 4.8 million persons 65 and over lived in poverty. Although 10 per cent of the population, they constitute 20 per cent of America's poor. Eighty-five per cent of the aged poor are white. Between 1968 and 1969 the poor, 65 and over actually increased by 200,000. Many of the old who are poor became poor after becoming old.[88] If poverty is defined more realistically, as many as seven million of our 20 million elderly Americans live in poverty and severe deprivation. For instance, as of January 1971 ten million, that is, one-half of the elderly population of the United States, lived on less than $75 per week or some $10 per day. Twice as many aged blacks live in poverty as aged whites.

Social Security and Medicare have not met the income maintenance and medical needs of older people. The average Social Security earnings of older Americans in early 1971 was $118 per month. Medicare only covers approximately 45 per cent of all health needs and the older person has to make up the rest. Since the life expectancy for black men is 60.1, they often do not reap the benefits of either Social Security or Medicare.

Twenty-nine per cent of the income of the elderly still derives from their own earnings. Three of ten men and one of ten women over 65 are in the work force. This fact must be weighed against age discrimination in employment and the Social Security ceiling on earnings. These conspire to make participation in the labor market painfully difficult. Furthermore, educational level and technological and educational obsolescence contribute to reducing the potentiality of earnings. In 1900 about two-thirds of men over 65 years of age were in the labor market, but by 1970 only one-fourth were. On the other hand, the rate of employment of elderly females increased from about 8 per cent in 1900 to about 10 per cent in 1970. The aggregate income of the elderly is nearly $60 billion a year. Twenty-nine per cent is earned, 52 per cent comes from retirement, Social Security, and welfare programs, and the remainder comes from investments and contributions.

Older people, black and white, tend to live in central cities and in rural areas. There are particular concentrations of older people in the Midwest (where the young have left the farm), in New England, and in Florida (to which older people migrate). The majority of aged blacks reside in the South (60.8 per cent). States that have populations of older people greater than the national average include Florida (13.2 per cent), Iowa, Kansas, Maine, Massachusetts, Minnesota, Nebraska, New Hampshire, South Dakota, and Vermont. New York State has nearly two million older people. California, Illinois, and Pennsylvania also have over one million older people. New York City alone has approximately one million old people. Arizona, Nevada, Florida, Hawaii, and New Mexico showed the greatest percentage growth between 1960 and 1965.

The elderly are becoming an important political force.[35,74] Nearly 90 per cent of all older people are registered to vote, and nearly two-thirds of older people routinely do vote. These are much higher percentages of participation than many other age groups have. The bulk of American voters are in their forties. This is important from the psychological point of view because of the sense of powerlessness that older people feel. They have been displaced from the usual forms of participation in society, and there has been a growing restlessness and "militancy" for "senior power." With a sense of influence and power comes self-respect, important to any age group or category of individuals. However, at present the elderly have not organized themselves into an effective political force commensurate with their numbers and vote.

Psychology of Later Life

"Man as a work of art" and the search for "life-enhancing" situations, goals, people, and activities were lifelong concerns of the scholar and art critic Bernard Berenson.[10] But he also wrote, "from our conception, everything in existence is out to destroy us." His diaries written in his eighties and nineties delineate these feelings as one moves through the trajectory of the life cycle. On the one hand, there is growth and development and, on the other hand, involution and death, both occurring simultaneously. These seemingly contradictory processes can only be emotionally and intellectually reconciled (if they can be reconciled at all) by an understanding of one's relationships to one's immediate milieu, one's family, one's historic era, one's posterity, and to the generations that follow and the world left behind.

In old age, but beginning in middle age, there is an increasing interest in one's legacy. This is found in all socioeconomic strata of society. The poor are concerned with what happens to their children in their immediate world, the famous with their immortality. We see the sense of sponsorship of the young, we observe the desire to leave a mark, expressing a deepening personal and social interest. In various forms we may see philanthropy. This inner sensibility, of course, is not observed with equal intensity or frequency in all people as they move through the postmeridian period of life. Some grow increasingly eccentric, bitter, angry, and interior in their preoccupations. It is difficult to interpret these disparate psychological manifestations because of the variety of forces operating. For example, if one has two-flight dyspnea or lives on $54 a month or is wracked with pain from a lingering malignancy or chronic arthritis, one may be hypochondriacal, egocentric, cantankerous, indeed. Therefore, one may see exhibited the common stereotypes of old age that are really due to the inimical forces at work—or one may see interest in others and the outside world when circumstances are propitious.

Old people may fulfill the cultural "definition" of the elderly as dependent, resourceless, garrulous, forgetful, fretful, irritable. Some may be coercive, endeavoring to impose values, demand time. Many "characteristics" of old age, then, may be understood as reactions to participation in our society, the facts of disease and poor care, and the impact of poverty. Old age, of course, is also the congeries of outcomes of personalities that may have been shaped in a variety of ways, from early indulgence to acculturated self-reliance in childhood. Variations in personality depend upon both past and present conditions and can be misread. The scientist and the psychiatric practitioner must very carefully evaluate both the historical and contemporary circumstances of the older person in order to make an appropriate appraisal. Both the generalities of old age and the particularities of the individual older person must be evaluated.

We can see manifested in old age constructive, regenerative, and creative forces.[22,24,26] The sense of legacy or continuity may be seen in many forms, including arrangements for succession, willing of property, donation of the body, its organs or parts, counseling, teaching and sponsoring, and the autobiographical process. The interest in reviewing one's life at times leads to the creation of memoirs of considerable significance. We also observe the resolution of the problem of time (with the end of time panics and boredom) with the development of an appropriate valuation of time. With serenity or tranquillity there may develop a sense of historic perspective, the capacity to summarize and comment upon one's times and work as well as upon one's life.

It is common for students of psychology and personality to set up polarities. For instance, in his identification of the issues of middle life, Erikson[37] has contributed the idea of generativity, on the one hand, and stagnation, on the other. With respect to later life he has proposed the alternatives of ego integrity and despair. These polarities, while useful, are oversimplifications of extremely complex human experiences. Increasingly students of the middle and later years have a growing respect for the multidisciplinary approach. The Eriksonian description—with only two reference points beyond early adulthood—has not been satisfactorily demonstrated among various socioeconomic and cultural groups.

Cumming's and Henry's[33] theory of disengagement has become very popular in recent years. The disengagement theory, however, has been severely criticized and reevaluated by many. The process of giving up objects and taking on new ones is characteristic of the flow of life as a whole, if one takes the life cycle perspective. Young people, for example, must give up their parental attachments and seek new ones. However much difficulty the young may have in finding new object relations, this is certainly a problem of poignant dimensions for the elderly at times.

The disengagement theory of aging has justly been criticized because of the extent to which disengagement may be explained as a consequence of older people fulfilling our culture's role expectations of the aged, as a result of medical and psychiatric disease, and as a consequence of social adversities or socially induced withdrawal, for example, compulsory retirement. If disengagement is an inherent process peculiar to old age, as hypothesized by its authors, it remains to be definitively determined. Findings of the National Institute of Mental Health studies of human aging[13] contrasted with many of the common stereotypes of old age, such as rigidity and disengagement. Tobin and Neugarten[91] and Maddox[66] have presented data that challenge the theory, and Kleemeier[54] has summarized the issues lucidly. That disengagement is conducive to mental health is contrary to the hard data and the impressions of many sophisticated observers. Social valuation, roles, and interaction appear correlated with mental health.[1]

In populations such as the elderly, where changes are frequent, profound, and affect so many systems, bodily as well as social, the advantage of the multidisciplinary approach is undeniable in both the building and testing of theories. There is evidence of many patterns of behavior in old age and great diversity rather than uniformity.[79]

The problems of loss and of associated grief[82] comprise one of the essential issues of old age. The resolution of grief and efforts at restitution are crucial. A complex of factors influence the subjective experience, behavior, and adaptive level of older people. For purposes of convenience, these factors fall into two broad categories: extrinsic and intrinsic. Among the extrinsic factors are personal losses, or sometimes gains, of the marital partner, of children, or other loved and significant persons. Losses can cause isolation and extreme loneliness, and it may be extremely difficult to find substitutes. Anger and extreme rage in addition to grief and despair may be present.

Social losses of status and prestige occur in the absence of roles—the so-called rolelessness of which many writers have written. Socioeconomic adversities resulting from either a lowered income associated with retirement or the inflationary spiral related to the economic conditions of the country profoundly affect the older person. Many older people do not wish retirement, but it is increasingly universal and arbitrary and it occurs earlier than it did formerly. Altogether there may be a crushing sense of uselessness and a feeling of nonparticipation within the mainstream of society.

The intrinsic factors include those related to an individual's life history, personality, abilities, techniques for adjustment, and qualities for survival. Moreover, one's physical health is a critical determinant of one's well-being. Certain specific physical disabilities, including various diseases, such as cardiovascular and locomotor afflictions, markedly deplete the older person's energies. Particularly disabling are diseases of the integrative systems of the body—endocrine, vascular, and central nervous systems. Perceptual impairments, most especially hearing loss, may lead to depression and marked suspiciousness. Losses in sexual desire and capacity may be especially keenly felt. Organic brain disease that may be minimal or mild may, nonetheless, reduce one's adaptiveness. One may be affected by baffling "little strokes."[3]

The largely obscure, mysterious, and inexorable processes of aging per se may also be classified among the intrinsic factors. Very few phenomena have as yet been definitively established as functions of aging. Among these are losses in the speed of processes and of responses. For example, reaction time is slowed although it is also affected by disease states, depression, and blood pressure.[13]

An especially important intrinsic contribution to the psychology of aging is the subjective experiences of aging. There is the subjective awareness of the increasingly rapid passage of time. There is the approach of death to which some respond with fear, some with denial, and most with equanimity.[13] A few wish to die, and many increasingly welcome it, especially in terminal illness. There is a sense of release. I have had patients calmly put themselves to bed falsely thinking they are about to die. Some people consciously or unconsciously panic.

A 75-year-old man of a long-lived family who had been a major figure in American academic life and the author of a number of important books became quite depressed and fearful that he would run out of income before he died. He feared he would be unable to maintain the home of which he was so fond and properly take care of himself and his wife. After Professor J. had come for a number of visits he became more comfortable and revealed that he had almost a million dollars of available funds. In this case, he unconsciously desired to run out of money because of his fear of death. He wished to outsurvive his income.

Reactions to death depend upon one's sense of contribution to others and to those that follow as well as to the resolution of one's life experience. They also depend upon the openness and quality of one's immediate relationships. There is now an extensive literature on dying, including Eissler,[36] Glaser and Strauss,[40] and Kubler-Ross,[56] yet perhaps no one has delineated the problem of personal death so profoundly as Tolstoy.[92] The importance of honesty and the avoidance of isolation have been amply demonstrated. Modern-day secularization makes the process of mourning more difficult;[44] this is one aspect of the more general absence of rites de passage in our society.[94] The role of death as a disruptive force in social organization has been reviewed by Blauner.[15] The right to die remains a controversial subject.

The autobiographical process manifests itself in many ways throughout the life cycle, at times reflecting predominantly self-analytic or introspective qualities, at other times suggesting a need for self-documentation. A daily journal or diary is distinctly different from the retrospective memoir. In old age comes the salient process that I have called the life review,[24] prompted by the realization of approaching death and characterized by the progressive return to consciousness of past experiences and particularly the resurgence of unresolved conflicts that can be surveyed and integrated. If they are successfully reintegrated, the life review can give new and significant meaning to one's life and help prepare one for death, mitigating one's fears. This naturally occurring process has been found to correlate with adaptation.

A tentative hypothesis linking fragility with aging has been suggested as a result of the National Institute of Mental Health (NIMH) studies.[14] Yarrow observed that with each succeeding set of measurements over time, there was an increasing interdependence of the different variables. Thus, at a particular point, as one and perhaps only one variable began to fail, the whole "house of cards" might shatter. This is a variation of the threshold theory previously held by the multidisciplinary NIMH group: a number of factors operate within the older person; if any one of them or if a number of them collectively attain a certain intensity, there will be effects upon adaptation and survival.

People throughout life may be so preoccupied with questions that they don't realize how many answers they already have. Through their accumulated experience the elderly have a kind of knowledge and wisdom that may be overlooked both by themselves and by others. The creation of new, important, valued social roles for the elderly along with a sense of pleasure and celebration in life would provide important ingredients to adjustment to old age. Work and retirement patterns are important to mental health.[38]

Only a limited number of roles are presently available. There are modest programs to help the low-income elderly such as Foster Grandparents, Green Thumb, and Senior Aides as well as the SCORE (Service Corps of Retired Executives). Professional and scientific personnel are "shelved" despite tested ability and the desire to work.[83] Our society has chosen the lazy way out through arbitrary retirement rather than through the individualized approach.

It is particularly important to recognize that the nature of aging and the nature of the elderly are under constant change. Increasingly the elderly are younger, healthier, and better educated. The average age of admission into institutions has increased now to roughly 80 years old. Retirement, whose length now averages about 14 years, may average 25 years by the year 2000. It is apparent that educational and cognitive obsolescence is induced by our social and technological advances in the absence of continuing education. The scholar of human nature and the psychiatrist must have a vision of the changing characteristics of different age groups over time as a result of transforming social and technological conditions. They must be aware of varying patterns of health care and life expectancy. Newborn infants are not likely to be too different from historic era to historic era, depending, of course, upon basic health and sanitation. But those who have lived through a life of change may be quite surprisingly different from one historic period to another. Moreover, old people are a variegated group and not a homogeneous one. In fact, the standard deviations of various physiological measurements are greater in the elderly than in other age groups. Similarly there is greater variation in the character of old people. There are certain great "levelers," particularly massive brain disease, illiteracy, and poverty. But when those variables are held constant, the variation and uniqueness of human personality is obvious.

The significance of religion in the lives of old people may be seen in its social function and in the deeper sense of a need to acquire meanings about the nature of human existence and one's life and death.[90] Older people are not doctrinaire in their religious convictions. One finds beliefs in reunion, reincarnation, and ghosts, as well as in other personal, idiosyncratic religious ideas. These notions are found in our "advanced culture" as well as among so-called primitive societies.[86]

The old often need to make reparations for the past. They may undertake a variety of expiatory behaviors in order to resolve their profound sense of guilt regarding acts of omission as well as commission. The struggle for atonement demands respect and must not be treated by facile reassurance. Old people generally act as if they had free will and self-responsibility and not as if their behavior were determined by the historic conditions of their childhood. American old people in the third quarter of the century rely heavily upon their beliefs in self-reliance, independence, and pride. This helps account for their difficulty in seeking public assistance even when it is justified by poverty that occurred beyond their control. Old people resist seeking help although they have paid personal and property taxes over the years that would more than cover their cumulative welfare payments.

Suicide reaches its peak in men in their eighties. This partly reflects marginal social status or anomie. Some kill themselves to spare their families economic devastation when they have a lingering fatal illness. Others assert their right to die and request euthanasia. Old people, unlike young people, seem, on the whole, less fearful of death.

A liberation of the traditional patterns of the life cycle, in which education, work, and retirement are presently compartmentalized in that order, as well as increasing the flexibility of one's sense of identity and of roles, would be likely to shape more creative patterns in old age.[23] Older people harbor urgent desires to change identity—sometimes to the point of wishing to disappear—to move, or to find very different activities. These anti-identity forces are to be respected and should not be seen as pathological. Indeed, we see the dangers of excess identity vividly portrayed in Arthur Miller's *Death of a Salesman*.[68]

One of the most striking, important, and adaptive qualities of old age is the continuing presence of curiosity and surprise. This relatively rare type of enthusiasm reflects the successes of the individual in protecting himself from the more usual deformation of his essential human character.

The sense of consummation or of fulfillment in life is more common than recognized but not as common as possible. Both personal and social factors make sustained growth extremely difficult. Obstacles stand in the way of faith in one's self and in one's relationships to others.

Pertinent to our considerations is the problem of our contemporary psychiatric nosology. By strict adherence to the presumed predictive significance of psychopathology, it would be assumed that obsessive-compulsive and schizoid personalities, for example, are impairing at any age of life. However, in old age both of these may prove to be adaptive.[13] This changing adaptive value of personality and psychopathology may be seen in a number of forms. Because of the insulating aspect of his character, the schizoid individual may be protected from some of the painfulness of old age. The compulsive person may find his ritualistic and fastidious behavior useful in filling the void of retirement. Perlin[77] observed that dependent people may adjust better than independent persons in institutional settings. These three examples show how important it is to utilize psychiatric and personality concepts, not in a vacuum, but in the contemporary context that itself must be evaluated.

Some people hold grudges for a lifetime: hates may outlast love. Nonetheless, reconciliation of long-estranged relatives may occur when the old person confronts the prospect of death.

Much humor concerns changing sexuality in old age. As the result of various disease states, from senile vaginitis to prostatic disease, desire and activity may be affected. Changing aspects of marital relationships from alienation to boredom are influential. Under favorable circumstances of good health, and in good relationships, sexuality proceeds late into life. Once again chronological age per se, while important, may not have such overriding significance as it is often thought.

Older people tend to exploit or deny aging changes and to undertake counterphobic efforts to reassure themselves against aging and disability. Rigidity in old age may be a function of anxiety, as Atkin pointed out in 1941. Weinberg[95] has pointed out that old people may protectively exclude stimuli. This is seen in the older person who hears what he wishes to hear. Old people may not only exploit their disabilities in aging, however. Out of pride they may also refuse to acknowledge vulnerability and dependency. (Young invalids, too, may exploit their disabilities, especially if they fear their needs will not be met.) The dependent needs of the elderly, the reality of their reduced resources, has been stressed by Goldfarb,[41] who, in fact, thinks of age in terms of dependency rather than in terms of chronological passage of time.

In addition to health and social circumstances, personality—the enduring psychological features of an individual—ultimately and profoundly influence an older person's subjective experience, overt behavior, and level of adaptation. Given equal personal, social and bodily losses, one person may thrive while another fails. The most effective form of adaptation is that achieved through insight; the capacity to modify one's behavior in accordance with changing realistic circumstances and the willingness and ability to substitute available satisfactions for losses incurred. Another important feature is the development of an inner sense of the life cycle, definable as the sensation of the rhythm, variability, timing, and inevitability of changes. This is a profound awareness of process, maturation, obsolescence, and death. It is not morbid, but rather is a nonmorbid realization of the precious and limited quantity of life and how it changes. As counterpart of Hartman's concept of the average expectable environment, I have suggested the concepts of the average expectable life cycle as well as that of the *sense of the life cycle*.[26]

Not all older people display the various characteristics and qualities that have been observed in this chapter, but a sufficient number of older people do so, thus revealing the healthy normal trends related to the closing chapter of life.

Middle Age as Transition

Any understanding of old age requires consideration of the transitional period. In her studies of women Benedek[8,9] described parenthood and the climacterium as developmental phases. There are no biopsychological definitions of the later stages of the life cycle as there are for early childhood and adolescence. We rely primarily upon conventions, usually employing chronological age as the defining variable.

While in old age the autobiographical process manifests itself as the life review, in middle age it has the apparent purpose of stocktaking. It provides an opportunity to consider new possibilities and alternatives in order to reorganize commitments. Alternative possibilities range from fixed rigid closure and fatalism, on the one hand, to varieties of overexpansiveness, on the other. Another critical element in the middle years concerns the testing of one's personal, professional, and other commitments. This problem of fidelity underlies and includes the narrower question of marital fidelity. One's fidelity to one's neighbor and one's society is also a fundamental issue.

Still another important factor in the middle years—pointed out by Bernice L. Neugarten[72]—is that one begins to count backward from death rather than forward from birth. There may be rehearsal for widowhood. Growing awareness of the realities of aging and death also lead to body monitoring—another phrase of Neugarten's. A man may begin to envy his son's increasing sexuality. A middle-aged woman or man may make obscene or frenetic efforts to be youthful. Hostility toward or envy of the young may come to a head. Fantasies of rejuvenation may be frequent.

Middle age is often viewed as "the prime of life," but it is the period in which the sense of success or failure may deeply plague and frighten people of both sexes. The man is particularly preoccupied by his occupational and social status and the woman by her status as a parent and loved one. The so-called empty nest phenomenon may affect both the middle-aged man and woman as they see their children leave home and significant responsibilities change. If women continue to gain greater freedom over their lives (from the availability of day-care centers to changes in social, economic, and marital status), it will be interesting to see the effects in the middle years.

The menopause is commonly blamed for the development of psychiatric states in women. Old wives' tales particularly terrify women in this regard. Psychiatric disorders are "explained" by the menopause without deeply questioning why most women do not develop hospitalizable "involutional melancholia." There have been studies among women of various ages of the signs and symptoms regarded as menopausal. So-called menopausal symptoms have been found in adolescence.[70]

It is often said that in Western culture, particularly in the United States, childhood (and youth) is the central emphasis. There is much evidence to suggest, however, that it is the middle-aged who are the "command generation," both controlling and controlled by their pivotal position in the life cycle. Nearly 50 per cent of the U.S. population is under 25 and ten per cent over 65, leaving 40 per cent to bear the social, economic, and personal responsibilities for the two groups who are most vulnerable economically and in other respects to life's vicissitudes. Kluckhohn appraises a culture's values in terms of future orientation, youth orientation, and instrumentality. America, being especially pragmatic, leaves little room for the conceptual and contemplative, and yet these are values that the older person might potentially offer (if their whole life had not been otherwise prescribed). If the older person is valued by his culture and is supported in his opportunities to recollect, reminisce, counsel, and comment upon his experience, the final chapter of life may in the future offer the sense of dignity and self-respect that optimistic commentators hope for.

Changes with Age

There have been a variety of studies of changes in cognitive and psychomotor functions with age, generally indicating decline. Less frequently have there been studies of personality development and change in the adult years. Little has been specified about the continuity of personality over long time intervals. The elusive concepts of wisdom, experience, and judgment have been difficult for psychologists and psychiatrists to tackle. Long-term longitudinal studies of continuing samples are extremely difficult to obtain for various reasons. What few studies there are of cognitive abilities show the preservation and even the increase in intellectual functioning over the years. This is true of Owens's[75] studies of college alumni. Study of Terman's[89] group of gifted children followed into their late forties indicated that, with few exceptions, the superior child remained superior in adulthood.

When comparisons are made between one age group and another, decline is generally found but there are many methodological issues: survivor bias, historical change, educational level, health, and other differences. In the NIMH studies comparing healthy young controls and healthy old people, the old people did not show the declines of cerebral oxygen consumption, blood flow, and intellectual functioning that had been expected.[27] The long-term studies of community elderly by Busse[20] and his group are available. Aging and health in populations was the focus of one international symposium.[65]

Individual differences are remarkable. There are also marked discrepancies between the psychological, physiological, social, and chronological aspects of aging.

Psychoanalytic theory has had a tremendous influence upon our understanding of man, but it has given little attention to personality change after adolescence. Personality is regarded as having been fixed early and the immutability of character is assumed. Practicing psychoanalysts and psychotherapists are often pessimistic about change after middle life. Freud, himself very pessimistic about change after middle life, made most of his great contributions after he was 40 years of age.[22]

On the other hand, psychoanalytic constructs have been extrapolated forward in time and there is question about their application. Heinz Hartmann wrote of the "genetic fallacy." For instance, castration anxiety has been used to explain fear of death in older people, leaving out basic human concern with death per se in old age, which is also reinforced by specific, realistic, personal concerns about the effects of one's death upon other people. Another overused and misused construct has been regression. Among lay people the comparable overused and misused term is "second childhood."

Psychoanalytic theory, particularly ego psychology, is ripe for reappraisal in the light of investigations of the course of the life cycle.[11] Ego functioning may be studied by using elements that deplete the ego such as drugs and organic brain damage. Schuster[84] has reported on one 106-year-old man.

Kelly[53] studied 300 engaged couples first examined in the 1930's and retested in 1955 when they were in their forties. After correcting the correlations for attenuation, Kelly found that individual consistency was highest in the areas of vocational interest and values and was lowest in self-ratings. He observed considerable individual variation and concluded "our findings indicate that significant changes in the human personality may continue to occur during the years of adulthood."

Erikson's work, like Jung's, suggests a kind of passive acceptance of the inevitable. For example, in Erikson's[37] view the last stage manifests ego integrity at best; that is, inevitably "the acceptance of one's one and only life cycle, and of the people who have become significant in it, significant to it, is something that had to be; and that, by necessity, permitted of no substitutions. . . ." Buhler,[19] on the other hand, sees development in the life course more in terms of attainment of goals through growth. Krapf[55] has described atrophy of the ability to project oneself into the future as a feature of old age.

One major source for understanding the nature of old age is the writings of articulate, perceptive old people. Tolstoy's and Berenson's last diaries are remarkable.[10,93] The description by the author Eric Hodgins[50] of his "stroke" and the care he received (and didn't receive) should be read by all physicians.

Studies of individuals, especially creative people and people in political life, are important. Politicians and judges are relatively immune from retirement. Psychopathology, aging, and political behavior make for exciting study.[80]

The black aged experience a multiple jeopardy.[51,52] They are disadvantaged by poverty, suffering twice as much poverty as the white aged. Seventy-five per cent of aged blacks live in deteriorated housing. They are subject to greater incidence of cerebrovascular disease. Ironically, having been subject to prejudice, they may be better prepared for the prejudices operative against the elderly. Much more must be learned about the black elderly. The American Indians, Chicanos, Chinese-Americans, and Japanese-Americans also have yet to be studied in sufficient detail.

The family structure and relations of old people have been the subject of valuable research.[51,63,73,85] The subject of styles of grandparentage has been relatively rarely studied. Neugarten[73] has differentiated types: the formal grandparent, the fun-seeker, the surrogate parent, the reservoir of family wisdom, and the distant figure. Grandparentage is sometimes disdained by older people. They may feel exploited as baby-sitters by their children. Negative and conflictual relationships with their children may be revealed in their attitudes toward their grandchildren. On the other hand, grandparents and grandchildren may have very close relationships and enter into covert and secret struggles with the middle generation. Even outside of formal consanguineous relationships, the elderly and youth surprisingly often share values not appreciated by the middle generation.

Personal possessions, including one's home, one's pets, and familiar objects, are particularly significant to the elderly; heirlooms, keepsakes, photograph albums, old letters help support the older person in his environment, preserve his sense of continuity, aid his memory, and provide comfort. Fear of loss of possessions is not uncommon. In some there is excessive possessiveness, at times to the point of hoarding. The situation may create an understandable fear of change. What is to happen to one's belongings upon death is a frequent preoccupation.

The sense of elementality, of an appreciation for underlying perceptions of geometry, of color, and of place, is a particular feature of old age.[10] It may relate to the yearning in the middle-aged, usually not fulfilled, for simplification in one's life.

Uprooting especially affects older people.[2,16] Reactions to hospitalization, relocation, cataract operations, or transfer from one institution to another may create morbid changes, problems in adaptation, and mortality. Separation and depression are particularly critical in late life. Means of adapting to change include effective use of insight, denial and counterphobic maneuvers (to demonstrate one's capacity to overcome aging), compensatory activities, and acceptance.

Gutmann[47] has written, "psychological differences between age groups may reflect generational discontinuities—such as contrasting modes of early socialization between the generations—and may have nothing to do with intrinsic developmental 'programs.'" Thus, the developmental hypothesis of personality change is most strikingly tested by comparisons of different cultures. Gutmann's study of Mayan aging based on the Thematic Apperception Test (TAT) is a good illustration of the kind of data that can be gained by transcultural comparisons. Such cross-sectional and cross-cultural studies can be useful in determining the universal features of aging in all cultures as well as in suggesting useful social, economic, and cultural arrangements that might be of value in providing more adequate support for the aged in different cultures.[31]

Creativity may continue late into life; it does not invariably decline with age.[22,45,60] Sophocles, Michelangelo, Titian, Cervantes, Hals, Voltaire, Goethe, Humboldt, Franck, Hugo, Verdi, Tolstoy, Shaw, and Freud are but a few famous examples. Had Freud died before he was 40, we would hardly know him. He was 44 when his magnum opus, *The Interpretation of Dreams*, was published. He was 67 when *The Ego and the Id* appeared, containing his structural hypothesis with the concepts of id, ego, and superego. Factors that facilitate or impede the creative life must be studied. The autodidact or self-teacher exemplifies one creative process.

Grief and widowhood,[67] the nature of time and leisure,[34] late life alcoholism,[69] social isolation,[64] housing arrangements,[29,59,81] problems in retirement,[25,30,38] the need for guardianship and protective services while maximizing the preservation of decision making,[61] nursing homes,[42] psychological correlates with impending death,[62] prejudice against the elderly,[21] and a myriad of other specific subjects must be further explored in order to achieve a comprehensive picture of old age and of the potential vulnerabilities for emotional and mental disorders.[5,28,39,57,58] Old people are subjected to infantilization, patronization, blunt hostility, and disparagement. They are called crocks, biddies, old fogies, witches, and crones among other epithets. With increasing life expectancy likely through the control of major diseases such as cardiovascular disease (that is, arteriosclerosis) and malignancy, the so-called problems, but also the opportunities, of old age will be increasing. The human sciences and the therapeutic professions will be increasingly called upon to understand the basic nature, possibilities, and problems associated with early and advanced old age.[87] Public policies will have to change.[46]

Bibliography

1. ALBRECHT, R., "Social Roles in the Prevention of Senility," *J. Gerontol.*, *6*:380, 1951.
2. ALDRICH, K., and MENDKOFF, E., "Relocation of the Aged and Disabled: A Mortality Study," *J. Am. Geriatric Soc.*, *11*: 185–194, 1963.
3. ALVAREZ, W. C., "Cerebral Arteriosclerosis with Small Commonly Recognized Apoplexies," *Geriatrics*, *1*:189–216, 1952.
4. ARIES, P., *Centuries of Childhood: A Social History of Family Life*, Knopf, New York, 1962.
5. BAHN, A. K., *Outpatient Population of Psychiatric Clinics, Maryland 1958–59*, Public Health Monograph No. 65, 1965.
6. BARTKO, J. J., PATTERSON, R. D., and BUTLER, R. N., "Biomedical and Behavioral Predictors of Survival: A Multivariant Analysis," in Palmore, E. (Ed.), *Prediction of Life Span*, Heath, Lexington, Mass., 1971.
7. BEARD, B. B., *Social Competence of Centenarians*, University of Georgia Printing Department, Athens, 1967.
8. BENEDEK, T., "Climacterium: A Developmental Phase," *Psychoanal. Quart.*, *19*:1–27, 1950.
9. ———, "Parenthood as a Developmental Phase: A Contribution to Libido-Theory," *J. Am. Psychol. A.*, *7*:389–417, 1959.
10. BERENSON, B., *Sunset and Twilight: Diaries of 1947–58*, Harcourt, Brace & World, New York, 1963.
11. BEREZIN, M. A., Some Intrapsychic Aspects of Aging," in Zinberg, N. E., and Kaufman, I. (Eds.), *Normal Psychology of the Aging Process*, International Universities Press, New York, 1963.
12. BIRREN, J. E., *Handbook of Aging and the Individual*, University of Chicago Press, Chicago, 1959.
13. ———, BUTLER, R. N., GREENHOUSE, S. W., SOKOLOFF, L., and YARROW, M. R., *Human Aging: A Biological and Behavioral Study*, U.S. Public Health Service Monograph Publication No. 986, Washington, D.C., 1963, Paperback reprint, 1971.
14. ———, ———, ———, ———, and ———, "Reflections," in Granick, S., and

Patterson, R. D. (Eds.), *Human Aging II: An Eleven Year Biomedical and Behavioral Study*, U.S. Public Health Service Monograph, Washington, D.C., 1971.

15. BLAUNER, R., "Death of the Social Structure," *Psychiatry*, 29:378–394, 1966.
16. BLENKNER, M., "Environmental Changes and the Aging Individual," *The Gerontologist*, 7:2, 1967.
17. BROTMAN, H. B., *Facts and Figures on Older Americans*, U.S. Department of Health, Education and Welfare, Washington, D.C., March 1971.
18. ———, *Who Are the Aged: A Demographic View*, Institute of Gerontology, University of Michigan-Wayne State University, Ann Arbor, 1968.
19. BUHLER, C., "The Curve of Life as Studied in Biographies," *J. Appl. Psychol.*, 19:405–409, 1935.
20. BUSSE, E. W., *Therapeutic Implications of Basic Research with the Aged*, Institute of Pennsylvania Hospital, Strecker Monograph, Series No. 4, 1967.
21. BUTLER, R. N., "Ageism: Another Form of Bigotry," *The Gerontologist*, 9:243–246, 1969.
22. ———, "The Destiny of Creativity in Later Life," in Levin, S., and Kaharna, R. (Eds.), *Geriatric Psychiatry: Creativity, Reminiscing and Dying*, International Universities Press, New York, 1967.
23. ———, Hearings, Part 1, Subcommittee on Retirement and the Individual, Special Committee on Aging, U.S. Senate, U. S. Government Printing Office, Washington, D.C., 1967.
24. ———, "The Life Review: An Interpretation of Reminiscence in the Aged," *Psychiatry*, 26:65–76, 1963.
25. ———, and LEWIS, M. I., *Aging and Mental Health: Positive Psychosocial Approaches*, The C. V. Mosby Co., St. Louis, 1973.
26. BUTLER, R. N., "Toward a Psychiatry of the Life Cycle: Implications of Socio-psychologic Studies of the Aging Process for the Psychotherapeutic Situation," in Simon, A., and Epstein, L. J. (Eds.), *Aging in Modern Society*, Chap. 20, American Psychiatric Association, 1968.
27. ———, DASTUR, D., and PERLIN, S., "Relationships of Senile Manifestations and Chronic Brain Syndrome to Cerebral Circulation and Metabolism," *J. Psychiat. Res.*, 3:229–238, 1965.
28. ———, and SULLIMAN, L. G., "Psychiatric Contact with the Community-Resident, Emotionally-Disturbed Elderly," *J. Nerv. & Ment. Dis.*, 137:180–186, 1963.
29. CARP, F. M., *A Future for the Aged: Victoria Plaza and Its Residents*, University of Texas Press, Austin, 1966.
30. ———, *The Retirement Process*, Public Health Service Publication No. 1778, U.S. Government Printing Office, Washington, D.C., 1968.
31. CLARK, M., and ANDERSON, B. G., *Culture and Aging: An Anthropological Study of Older Americans*, Charles C Thomas, Springfield, Ill., 1967.
32. COMFORT, A., *The Process of Aging*, New American Library, New York, 1961.
33. CUMMING, E., and HENRY, W. E., *Growing Old: The Process of Disengagement*, Basic Books, New York, 1961.
34. DEGRAZIA, S., *Of Time, Work and Leisure*, Twentieth Century Fund, New York, 1962.
35. DONAHUE, W., and TIBBITTS, C. (Eds.), *Politics of Age*, University of Michigan Press, Ann Arbor, 1962.
36. EISSLER, K., *The Psychiatrist and the Dying Patient*, International Universities Press, New York, 1955.
37. ERIKSON, E. H., *Childhood and Society*, 2nd ed., Norton, New York, 1954.
38. FRIEDMAN, E. A., and HAVIGHURST, R. J., *The Meaning of Work and Retirement*, University of Chicago Press, Chicago, 1954.
39. GITELSON, M., "The Emotional Problems of Elderly People," *Geriatrics*, 3:135, 1948.
40. GLASER, B. G., and STRAUSS, A. L., *Awareness of Dying*, Aldine, Chicago, 1965.
41. GOLDFARB, A. I., "Patient-Doctor Relationship in Treatment of Aged Persons," *Geriatrics*, 19:18–23, 1964.
42. ———, "Prevalence of Psychiatric Disorders in Metropolitan Old Age and Nursing Homes," *J. Am. Geriatrics Soc.*, 10:77–84, 1952.
43. GORDON, M. M., "Social Class in American Sociology," *Am. J. Sociol.*, 55:262–268, 1949.
44. GORER, G., *Death, Grief and Mourning in*

Contemporary Britain, Grosset Press, London, 1965.

45. GREENLEIGH, L., "Timelessness and Restitution in Relation to Creativity and the Aging Process," *J. Amer. Geriatrics Soc.*, *8*:353–358, 1960.
46. Group for the Advancement of Psychiatry, *Toward a Public Policy on Mental Health Care of the Elderly*, Report No. 79, 1970.
47. GUTMANN, D., "Mayan Aging: A Comparative TAT Study," *Psychiatry*, *29*:246–259, 1956.
48. HALL, G. S., *Senescence, the Last Half of Life*, Appleton, New York, 1922.
49. HAVIGHURST, R. J., KUHLEN, R. G., and MCGUIRE, C., "Personality Development," *Rev. Educ. Res.*, *17*:333–344, 1947.
50. HODGINS, E., *Episode: Report on the Accident inside My Skull*, Atheneum, New York, 1964.
51. JACKSON, H., *Double Jeopardy—The Older Negro in America Today*, National Urban League, 1964.
52. JACKSON, J. J., "Social Gerontology and the Negro: A Review," *The Gerontologist*, *7*:169–178, 1967.
53. KELLY, E. L., "Consistency of the Adult Personality," *Am. Psychol.*, *10*:659–681, 1955.
54. KLEEMEIER, R. W., "Leisure and Disengagement in Retirement," *The Gerontologist*, *4*:180–184, 1964.
55. KRAPF, E. E., "On Aging," *Proc. Roy. Soc. Med.*, *46*:957–964, 1953.
56. KUBLER-ROSS, E., *On Death and Dying*, Macmillan, New York, 1969.
57. KUTNER, B., FANSHEL, D., TOGE, A. M., and LANGNER, T. E., *Five Hundred over Sixty: A Community Survey of Aging*, Russell Sage Foundation, New York, 1956.
58. LAWTON, M. P., "Gerontology in Clinical Psychology and Vice-Versa," *Aging & Human Dev.*, *1*:147–159, 1970.
59. ———, LIEBOWITZ, B., and CHAREN, H., "Physical Structure and the Behavior of Senile Patients Following Ward Remodeling," *Aging & Human Devel.*, *1*:231–239, 1970.
60. LEHMAN, H. C., *Age and Achievement*, Princeton University Press, Princeton, 1953.
61. LEHMANN, V., and MATHIASON, G., *Guardianship and Protective Services for Older People*, National Council on Aging Press, New York, 1963.
62. LIEBERMAN, M. A., "Psychological Correlates of Impending Death: Some Preliminary Observations," *J. Gerontol.*, *20*:181–190, 1965.
63. LINDEN, M. E., "The Older Person in the Family: Studies in Gerontologic Human Relations, VII," *Soc. Casework*, *37*:75, 1956.
64. LOWENTHAL, M. F., "The Relationship between Social Factors and Mental Health in the Aged," in Simon, A., and Epstein, L. J. (Eds.), *Aging in Modern Society*, American Psychiatric Association, 1968.
65. ———, and ZILLI, A., *Interdisciplinary Topics in Gerontology: Colloquium on Health and Aging of the Population*, Vol. 5, S. Karger, New York, 1969.
66. MADDOX, G. L., "Disengagement Theory: A Critical Evaluation," *The Gerontologist*, *4*:80–82, 103, 1964.
67. MARRIS, P., *Widows and Their Families*, Routledge and Kegan Paul, London, 1958.
68. MILLER, A., *Death of a Salesman*, Viking, New York, 1949.
69. MOOREHEAD, H. H., "Study of Alcoholism with Onset Forty-Five Years or Older," *Bull. N. Y. Acad. Med.*, *34*:99–106, 1958.
70. NEUGARTEN, B. L., " 'Menopausal Symptoms' in Women of Various Ages," *Psychosom. Med.*, *27*:266–273, 1965.
71. ——— (Ed.), *Middle Age and Aging: A Reader in Social Psychology*, University of Chicago Press, Chicago, 1968.
72. ———, *Personality in Middle and Later Life*, Atherton Press, New York, 1964.
73. ———, and WEINSTEIN, K. K., "The Changing American Grandparent," *J. Marriage & Fam.*, *26*:199–204, 1964.
74. ORIOL, W. E., "Social Policy Priorities: Age vs. Youth; The Federal Government," *The Gerontologist*, *10*:207–219, 1970.
75. OWENS, W. A., JR., Age and Mental Abilities: A Longitudinal Study," *Genet. Psychol. Monog.*, *48*:3–54, 1953.
76. PATTERSON, R. D., FREEMAN, L. C., and BUTLER, R. N., in Granick, S., and Patterson, R. D. (Eds.), *Human Aging II: An Eleven Year Biomedical and Behavioral Study*, U.S. Public Health Service Monograph, U.S. Government Printing Office, Washington, D.C., 1971.

77. PERLIN, S., "Psychiatric Screening in a Home for the Aged. I. A Followup Study," *Geriatrics, 13:*747–751, 1958.
78. PRESSEY, S. L., and KUHLEN, R. G., *Psychological Development through the Life Span*, Harper & Row, New York, 1957.
79. REICHARD, S., LIVSEN, F., and PATERSON, P. G., *Aging and Personality*, Wiley, New York, 1962.
80. ROGOW, A. A., *James Forrestal: A Study of Personality, Politics and Policy*, Macmillan, New York, 1964.
81. ROSOW, I., "Retirement Housing and Social Integration," in Tibbitts, C., and Donahue, W. (Eds.), *Social and Psychological Aspects of Aging*, pp. 327–340, Columbia University Press, New York, 1962.
82. SANDS, S. E., and ROTHSCHILD, D., "Sociopsychiatric Foundations for Theory of Reactions to Aging," *J. Nerv. & Ment. Dis., 116:*233, 1952.
83. SAVETH, E. N., *Utilization of Older Scientific and Professional Personnel*, National Council on Aging, New York, 1961.
84. SCHUSTER, D., "A Psychological Study of a 106-Year-Old Man," *Am. J. Psychiat., 109:*112, 1952.
85. SHANAS, E., and STREIB, G. F., *Social Structure and the Family: Generational Relations*, Prentice-Hall, Englewood Cliffs, N.J., 1965.
86. SIMMONS, L. W., *The Role of the Aged in Primitive Societies*, Yale University Press, New Haven, 1945.
87. STREHLER, B., "Long-Range Programs and Research Needs in Aging and Related Fields," Testimony, Hearings, Special Committee on Aging, U.S. Senate, 90th Congress, December 5 and 6, 1967, Part I, Pp. 1397–1414, U.S. Government Printing Office, Washington, D.C., 1965.
88. Task Force, Special Committee on Aging, U. S. Senate, *Economics of Aging: Toward a Full Share in Abundance*, U.S. Government Printing Office, Washington, D.C., 1969.
89. TERMAN, L. M., and ODEN, M. H., *The Gifted Group at Middle Life*, Stanford University Press, Stanford, 1959.
90. THOMPSON, P. W., "The Church and Its Role in the Promotion of Health in Older Persons," in *The Aging and the United Presbyterian Church*, United Presbyterian Church, Chicago, 1964.
91. TOBIN, S. S., and NEUGARTEN, B. L., "Life Satisfaction and Social Interaction with Aging," *J. Gerontol., 16:*244–246, 1961.
92. TOLSTOY, L. (1886), *The Death of Ivan Ilych*, Signet Classics, New York, 1960.
93. ———, *Last Diaries* (Ed. by STILMAN, L.), G. P. Putnam, New York, 1960.
94. VAN GENNEP, A., *The Rites of Passage*, University of Chicago Press, Chicago, 1960.
95. WEINBERG, J., "Personal and Social Adjustment," in Anderson, J. E. (Ed.), *Psychological Aspects in Aging*, American Psychological Association, Washington, D.C., 1956.
96. YARROW, M., BLANK, P., QUINN, O. W., YOUMANS, E., and STEIN, J., "Social Psychological Characteristics of Old Age," in *Human Aging: A Biological and Behavioral Study*, Chap. 14, Public Health Service Publication No. 986, 1953, Paperback reprint, 1971.

CHAPTER 32

LIFE'S MAJOR AND MINOR CRISES

Rose Spiegel

Overview of Crisis

Scope of Crisis

Crises are part of the vicissitudes of living—not only as painful stress, as pychopathology, but also as growth, challenge, excitement, and drama. Their range is wide, both in detail and in large patterns. They may erupt from extremes of emotional disorders outside the usual flow of life. They may arise from specific stress events that are predominantly situational but are frequent enough in the course of anyone's life to be within universal experience. Other universal crises are challenges occurring at transitional phases in the progression of epochs in the life of the individual. Still others are crises of confrontation between the individual and his culture, with a wide range of response. Crises may be heavily weighted to the obviously interpersonal or to the obviously intrapsychic or to the cultural. However, generally all these aspects are involved in some proportion. Actually a less obvious aspect may be the well-spring for the particular crisis and, in turn, may offer the more promising focus for therapy.

Crises have only recently been considered as an integral content appropriate for psychotherapy. Too often crisis has been considered to be outside the conventional boundaries of tightly defined psychiatric treatment. With a diversity of patients the therapist can hardly prevent crises from arising in their lives—nor in his own since he, too, is subject to vicissitudes and challenges in living. There is no mode or approach of psychotherapy, whether individual, family, or group, that does not at some time involve crisis. Crisis may be the moving force toward therapy. Indeed, some persons come to therapy because of their unrecognized involvement in crises in the lives of others. For instance, they may be bound to relationships with persons who are in crisis or who are crisis-prone and are demandingly dependent.

When crises do arise, whether in the course of therapy or as the precipitator of therapy in the first place, they generally occur in the context of specific life situations—of career, of marital or other interpersonal relationship. But these types of stress experience barely touch on the large tides of life that basically affect everyone and at times make the person seek help. However, under the ostensible specific events and situations of overt crises, the vulnerability often exists because of the deeper surge of these underlying critical phases. Not only these basic crises in the progression of life but also the crises of stress that we all share from cultural pressures may declare themselves situationally.

Emotional crises and crises in living may arise from such stark situations as the extreme of trauma, threat, and deprivation because of war and its aftermath, with which we were intimately familiar in the Nazi era and World War II, and which we are again forced to face in the crisis of values and the destruction resulting from the war in Vietnam. The current crisis of values, far from being a distant extreme, presses on us in everyday life. Cultural and societal action is no longer an abstraction, but an everyday issue of personal decision and position. Another variation in the role of the culture consists of those crises in living that the person experiences as generated by himself or another individual but that actually are rooted in the culture. That is, the culture often is crisis-making because of the values attached to the person, to his age, sex, color, achievement, "usefulness," affluence.

Other crises are more related to specific personal experience and to the intricacies of reaction formation that shape the personality of an individual. Somatic disturbances often precipitate psychological crises. Physical illness and damage in oneself, or in those significant to one, present a challenge in the areas of decisions directly related to the ailment, of threats to one's sense of identity, of the meaning the physical has to the person, and of psychological adaptation.

Still other upsurges of crisis are an accompaniment to life itself—life as the basic progression from birth through childhood, adolescence, the prime of maturity, and decline, and the challenges involved in coming to terms with the ultimate. Falling in love presents a crisis not only interpersonally but also intrapsychically. In career or social pursuits success and expansion often present not only a practical but also an intrapsychic crisis.

Besides this progression, this hidden pageantry in life itself, there are the prosaic, seemingly trivial everyday confrontations that, in miniature, test us in several directions: problem-solving of external events, resolving of interpersonal tensions and inner conflicts. These include the competitive and ethical issues of everyday life. Subtly these involve living with oneself in a self-esteem based on reality and with continuous maturing and growth of personality. Grace of living is involved; perhaps wisdom is achieved.

What Are the Core Characteristics of Crisis?

A crisis is characterized by a dilemma of decision or of direction to be taken in living, whose outcome for a beneficent or a destructive resolution is not always predictable, whether by the individual primarily facing the crisis, or by close participants, or even by those in a position simply to have a perspective on the crisis. This is true both in private lives and on the social scene. There are some crises, however, in which decisions are not within the power of the individual as far as the visible crisis is concerned. The inner crisis for him may involve his reaction to the visible crisis, to his lack of power to enter into its solution. Crisis is a threat to homeostasis. Both when the issue is a lack of power or simply a lack of equilibrium, the crisis needs to be met in terms of one's psychological adaptation in order to protect psychic integrity, to delimit stress.

It is the uncertainty of outcome, its potential for disappointment, even for disaster, that the protagonist in the crisis senses profoundly and with anxiety or even terror. The reactions and defenses that may fan out are legion. Anxiety is only one of that range and often not at all on a verbal cognitive level. The response may be a deep denial of crisis, may be depression, hyperactivity—even elation. Some even court crisis conditions, feeling they are most alive in the challenge of coping. At times the distinction between elation and an exhilarated courting of crisis cannot be drawn without a deep understanding of the individual. For some, crisis may be so threatening, the stress so inordinate, that an activation of flaws in personality may occur, to the extent that the "nervous breakdown" into psychosis ensues. In life crises, in particular, the ideal resolution is twofold: coping in appropriate practical terms and coming out on the other side of the crisis into the next phase in living with an enhancement of ego powers and of the whole personality.

To me the irreducible characteristic of crisis is the *unpredictability* of the favorable outcome. For whatever reasons the outcome is in doubt, generally for all concerned, including the person in the therapist role. Crisis has some of the urgency of emergency. However, in emergency disaster has already occurred; the action called for to meet the catastrophe is generally more clear than in crisis, although whether the coping measures will be effective remains in doubt. In crisis it is the wisdom of a particular decision that is obscure; in emergency it is the feasibility of the measure or the adequacy of the response to it. Emergency often calls for the mopping up of what the crisis has precipitated.

The *suspense* in crisis, often with the great intensity of the accompanying anxiety, may be an *unbearable stress* on the person or on others who are involved in it with him. That is, there is varying ability to endure the stress of unpredictability. This stress adds to the suffering in crisis. People vary in their stamina to face crisis, and the same person also does at different times, subject to what affects his stamina in general. The factual doubt inherent at a particular time in the crisis situation may be internalized as obsessional doubting. Not only do people vary in their personality reactions to crisis, but also in an overall sense they vary in the maturity they bring to crisis, in the preparation by life experience and individual endowment for fortitude and flexibility.

In some kinds of crises, particularly the crises of situation and events, the issue of problem-solving is involved, which underlies decision making. Thus there often is a convergence of the cognitive, the emotional, and the outer reality. Crisis often involves not only the individual but also other persons linked to him —the *central* individual—in his network of relationships, for instance, maritally and in the family, in work situations, in larger and larger groups, until with some individuals larger segments of society are involved, for which they generate crises in turn. The illness or the incapacity of a ruler or leader, particularly when its outcome is in doubt, is the classical example of the central crisis setting off spreading waves of secondary crises. Understandably the reassuring chant, now metaphor, "The king is dead. Long live the king!" For these reasons I have termed the central person in the crisis, whether he experiences it himself in his keenest awareness or whether he touches off intense crisis responses in others, the *storm center* in the crisis. This concept is useful in evaluating who does what in a crisis situation. Particularly it helps those involved in the crisis to realize their relationship and role both in generating and in coping. This clarification at times helps one not assume inappropriate responsibility and guilt.[15,16,18]

Crises have a quality of *transitoriness*. Indeed, Caplan,[2,3] in his work on crises, maintains that they have as a pattern a short-lived duration, that an *unresolved crisis*, such as in family situations, lasts about six weeks, generally changing its character of acuteness for chronicity. That is, either acute stress falls into a persistingly bad condition, with transformation of the sense of acuteness, or the situation takes still another form, with a realignment of its elements and of the relationship of the people involved in it.

However, although this time span offers a handy and useful formula, the divergences from it are important, and it would be a disservice to make the time factor the pivotal diagnostic criterion. A crisis situation sometimes persists far longer than the allotted span, eliciting the characteristic painful tension and desperation, and still remains unrecognized as crisis. For some individuals ignoring the crisis is an unaware attempt at homeostasis, in the passive hope that it will simply go away. And sometimes it does! Because of repression, denial, or lack of recognition, as the case may be, moving toward resolution may be deferred. Recognition of the critical situation makes the inchoate situation take shape as crisis.

Nor can a sharp time criterion be applied to transition points in development. In due course the next stage emerges, in better or worse form, whether in fullness or as a token marking the progression. We do know that immature individuals, while retaining a quality of immaturity, nevertheless experience some thrust to development on some level, by the sheer force of basic psychophysiological laws. Immaturity never really stays the force of aging.

Chronicity of a critical situation may depend not only on nonrecognition but on lack of direction for resolution or coping, on despair and a quiet giving up of hope. Although it may appear to be only a semantic distinction, it is useful to distinguish between a critical situation and crisis as follows: the latter involves the pressure and urgency that force confrontation. while *critical situation* conveys to me only one element of crisis.

The element of *choice*—its presence or absence, its feasibility or impossibility—demarcates a polarity in crisis. Vast crises in the physical world about us, for instance, earthquakes, leave no room for direct "choice." But they arouse intrapsychic and interpersonal crises in perturbations of anxiety and insecurity, particularly in children. We can see the parallel with raw war experiences, consisting of crisis situations whose direct solution is outside the range of choice and decision, and just because of this limitation force intrapsychic and interpersonal crises between the individual and society. The external crisis, whose outcome is not predictable and is potentially catastrophic, and which of itself is not open to choice and decision, moves the crisis into the intrapsychic world.

The characteristic *personality operations and styles* bear on the whole sweep of crisis, from inception and even provocation to the coping. The personality patterns of denial and nonrecognition have already been discussed. At the other extreme is a life style of relating to most problems in living as crises. Every semblance of the new and different is experienced as crisis, sometimes authentically, sometimes as an expression of manipulative dependence. This personality style at times is interesting in its sheer ingenuity, sometimes stressful and exhausting in its "hysterical" quality.[19]

One element in crisis is *stress*. The mobilization of forces and resources in coping with crisis involves stress. It involves both physical and emotional mobilization of energies and reading a level of readiness to act. This is illustrated by the contrast in the readiness of the unconditioned or unwarmed up athlete and the conditioned or warmed up one. In meeting stress and crisis a noncognitive learning is involved, and an alertness, which is opposite to passivity. The exception is when passivity itself is a way of coping and is adaptational in its function.

At a particular point in time stresses, with their demands on the total organism—here the personality—may be just too much, while at another point in time the same person can cope effectively. Where the person is at in the crisis response is important to sort out. For instance, there may be a confluence of stresses from different directions that cumulatively are too much, but in lesser combinations are manageable. This element needs to be explored and recognized. We are already familiar with the response to stress with hyperactivity, even elation and hypomanic states.

Because of these complexities it is suggested that we approach crisis in terms of its parameters:

1. The critical situation: interpersonal, intrapsychic, somatic, cultural-societal, or pertaining to forces in the external physical world? How did it come about? What is the dilemma? What is the potential for *choice*?
2. The persons involved and their role and reactions, ranging from the central person, perhaps as storm center, to the person who, formally or not, is in the helper role.
3. The degree of awareness and recognition of the existence of crisis, whether by the cen-

tral person, by others significant to him, or by those in a helping role, professional or informal. Who considers there is a crisis?

4. The personality and practical resources of each one involved in the crisis.
5. The external resources that may be called on to help cope with the crisis.

Developmental Transitions and Crises

All the world's a stage
And all the men and women merely players.
They have their exits and their entrances;
And one man in his time plays many parts,
His acts being seven ages.[12]

Development, the unfolding of maturation, should be seen, not as culminating and terminating with the prime of life, but as an arc of total life, going on to the final stages of change and adaptation, namely, aging and dying. Maturation does not fall into a simple, smooth linear graph, but involves transition stations that are critical points. These critical points are not necessarily points of crisis but are landmarks of vulnerability, which may become transformed into crisis, just as "critical situation" was distinguished from "crisis." More intensely these critical points are involved in normative crises, such as the normative "identity crises," which, according to Erikson,[6] is ascribed to the age of adolescence and young adulthood. Developmental transitions and critical points involve fulfillment of one phase and separation from its mode of experience and relatedness, followed by a new mode, with qualitative overlap with the past as well as with change.[5] In the early years the new epochs are involved with change into a different organization of the psyche. Characteristic also of developmental transitions and critical points are somatic forces, with correlated reverberations in the psyche that are outside of directed choice. One cannot directly will them to be otherwise, although the efforts of the medical and psychotherapy professions are directed to making choice possible through the mediation of still other forces. The somatic involvement in developmental transitions is dramatically prominent in adolescence, in pregnancy, and in the era of the menopause and male climacterium, accompanied by their psychological crises,[14] which varyingly may be ameliorated by medical as well as psychotherapeutic resources.

When these critical points and normative crises are subject to unfortunate visissitudes, pathological crises may arise. Birth itself is a crisis for the newborn as well as for the mother. The birth process is a physiological revolution, whose alternatives involve life and death, existence with a sound organism or a damaged one. Rank, in his *Trauma of Birth*, elevated aspects of the process of birth to an overall theory, which it is not necessary to discuss in his terms.[21] Indeed, there is separation from the mother, but for our purposes it is not necessary to follow Rank in taking this separation as the basic pattern for all separation, and "the temporary asphyxiation at birth" with its physiological symptoms as the fundamental model for anxiety. Suffice it to say that birth is a time of physiological crisis. Ronald Laing* reflected on the harshness of the birth experience, contributed to, according to his view, by the medical practice of hurried tying and severance of the umbilical cord and slapping the infant to hasten his breathing, thus curtailing the due and slower and less shocking changeover from the intrauterine world to the world of air and breathing.

The element of choice is an important and interesting theme to pursue throughout the developmental progression. It is clear that in infancy and early childhood choice is absent, and that with increasing maturity choice is possible on some psychological levels, which, in turn, indirectly influences the physiological, for better or worse. The emergence of the power of choice is part of development itself, and cultivating the art of choice is part of the art of living. In the very young the prototype of choice is close to the level of physiological protest against what violates him, and it

* In an unedited movie *Breathing and Running*, filmed by Peter Robinson, Inc., New York City, and exhibited at the William Alanson White Institute, 1970 (now edited).

blends with the power to rebel that soon begins to declare itself in more psychological form.

Outside of semantic development and the vicissitudes of either constitutional damage or illness from external sources, the crises that the very young are subject to usually are generated and inflicted by others. The upbringing of children, the various steps in acculturation of "good habits" involved in weaning, eating, and toilet training, are points of interpersonal interaction at physiological zones at different stages of maturation, as Sullivan[20] early pointed out. These are the everyday stress experiences, which, depending both on the child and the parent, may be severe enough to make for crises in the child's anxiety concerning mastery of skills and the "reflected appraisals" from the parents. The earliest experiences of anxiety, in terms of any of the major psychoanalytic theorists, may be considered one of the basic indicators of what the individual, particularly the young one, is experiencing as crisis. But it is only one.

In Sullivanian terms infantile developmental crises depend on somatic development and on the gratification of the budding psyche with its needs for tenderness and for communication. Thereby the infant is subject to crises of interpersonal relatedness that are outside of choice. Of that same early period of life before the age of nine months, Melanie Klein[8] proposed that there are critical "stations" of progression of intrapsychic development that have their respective characteristic modes of experiencing relationships. It is stressed that this progression of stations is fragile and vulnerable to regression. This vulnerability to regression is critical. Bowlby[1] has stressed separation from the mother in early infancy as a basic model of separation.

The various developmental stages of early childhood that are more clearly observable have their critical points of transition, which, whether for intrinsic reasons of miscarriage of development or for extrinsic reasons of illness or interpersonal mishap, may be subject to crisis. The outcome may be progression to the next stage of maturation, fixation, or regression.

The vulnerability to crisis is applicable as much to the Sullivanian concept of the epochs of maturation as to the Freudian concept of stages of development in terms of libido. In the Freudian frame of reference the working through of the Oedipus complex involves crisis. In Sullivan's concepts the phase of development of verbal communication in shared language is vulnerable and thereby a critical point. The crises arising from parental mishandling of communication here predispose the individual to obsessional misuse of language. There occurs also the critical period of enlarging relatedness to the outside world, typically beginning with school, playmates, and chums. Each enlargement bears with it a separation and a vulnerability. Unfortunate resolution of separation crises of what should be normal separation often takes its toll in school phobia.

Throughout childhood a wide range of vicissitudes occur from family tension, hostility, and indifference. The young are subject to anxiety and subsequent depression from separations that could have been handled better—particularly those of divorce. While the adult is undergoing the grave crises of serious sickness—physical or emotional and even death—the young one is immersed in his crises of reverberation, which he is virtually powerless to cope with, without outside help. Not only does the child suffer in the immediate experience, but also he may develop maladaptive modes of coping that persist outside the immediately damaging situation. For instance, in a London survey[9] it was revealed that children about age seven who were accident-prone came from homes with serious family problems of physical and psychiatric illness and marital discord. Other children in that study had responded to their stress with other kinds of symptoms such as asthma and enuresis. Indeed, one aspect of psychotherapy may be considered the undoing of the damage from crises of the past.

Adolescence is the grand crisis period, with the widening of personal, interpersonal, and societal horizons, the heightened meanings of genital sex, the establishment of more complete identity, the struggle with authority, the polarity of dependency-independence. The adolescent boy and the girl at puberty are flooded by the outer indications of change and by inner feeling states of which they have as yet no cognitive understanding. Their feelings include confusion, contentless anxiety, depression, and a sense of spinning in a world that others seem to understand more than they do. Adolescents frequently comment with surprise that others seem to "know" what they are experiencing more than they themselves do.

The full establishment of heterosexuality may involve a critical period of homosexual conflict, orientation, or experience.[11] "Critical period" is here used to avoid the particular denotations of "homosexual crisis."

For the girl a long-overlooked relationship of prime importance to her burgeoning femininity and sexuality concerns the role of her father. His acceptance of her as a woman and his anxiety, intimacy, or distancing from her at adolescence affect her from the point of view of sanction. But beyond the actual developmental period the separation from the father, whether initiated by the girl as a defense or by him, has been observed by the writer to be an important factor in the crises in the depressions of some women, particularly in the early forties and at the beginning of the menopause.[14] Indeed, crises of early life for which effective help was not given, and which then released in the later depressions, are often too simplistically attributed to the later developmental stage.

The *identity crisis* involves every transition in development, including middle age and old age, but it is most dramatic and is experienced as most affirmative in adolescence. One common element centers on the development of proper ego strengths, with new abilities, control, and sense of self. And in all the sense of self involves a new integration of the sexuality of that era. What is involved is not only sexuality but also the way the sexuality is related to internally. In discussing the identity crisis of adolescence, Erikson has stressed that crisis does not need to connote impending catastrophe, but rather a necessary turning point in development, with a marshaling of resources of growth and further differentiation. Both the core of the individual and the core of the community are involved.

It is an astonishing fact that in the last few years, I believe for the first time, adolescent turmoil has not been simply an internal affair, but has made its impact on society and the culture, forcing change in them. During their identity crises adolescents may manifest a range of disturbance—depression, confusion, anxiety, withdrawnness, various kinds of acting out. Involvement with society as a rebel, whether in authentic concern or as flight from coping within, is often associated with heightened questioning of personal, parental, and societal values—a questioning that deserves respect. The obscurity for many adolescents of life goals often enters into the rebellion and also into depression and despair. These and more variants generally fall within the range of the normative identity crisis. It is the my clinical impression that the adolescent for whom these large questions are answered despairingly or cynically often has recourse to a tough, cynical surface that hides depression, or to a dependence on drugs for quickie alleviation of depression and despair. Sometimes there occurs in adolescent turmoil immersion into clinically recognizable disorders, such as clinical depression and schizophrenic reactions. These carry a generally good prognosis, better than in the slower, less vivid process schizophrenia. Various suicidal syndromes occur in adolescence, from preoccupation with the possibility to completion of the act. Although the incitement to suicide is qualitatively similar through the various ages, still the sensitivity to a sense of worthlessness, to a failure to succeed, to a despair about being loved and accepted, particularly in a love affair, is enormous then. In my opinion the intensity of the despair predisposing to suicide is the outcome of a not yet established ego; love and its confirmation are experienced qualitatively as a young child and quantitatively with the power of youth, since the transition to mature strength in loving has not yet been made. Either the experience of falling in love is overwhelming, or the lack of that experience is a cause for low self-esteem—both potential for crisis.

The critical phase of adolescence involves accomplishment of the task of taking hold of life, of beginning to set terms to life, of developing choice. Adolescents, in contrast to young children, are far less at the mercy of adults, granting that no one is an island unto himself.

The establishment of maturity brings with it the added challenges both of loving and of affirming a mode of functioning in one's milieu. Loving in its mature sense involves not only intimacy and caring but also the intrapsychic process of making room within for the intimate other. In fact, falling in love and intimate loving present emotional crises that for some personalities, such as the schizoid, are very frightening. The threat of chaos and anxiety in the inner expansion of the psyche, the dread of vulnerability, of separation, rejection, and loss add a crisis quality to falling in love. Some, with an unhappy experience, take the position that safety lies in setting up barriers to loving.

Maturity—as well as adolescence—brings with it crises from both internal and external sources. The internal origin has to do directly with the progression of maturation. The external sources depend partly on what outer chance serves up to us, as well as on choices we make, often outside of awareness, which then bring their consequences. But there are some external events to which we are exposed simply because there is a high probability that they will happen. For example, as one gets older, in due course one loses one's parents, and one's children grow up and leave. Over these one has little choice but the deep choice of the mode of adaptation.

There are other time-linked critical events that are not biological but that have to do with the attitudes of others by which we are governed and over which we have no power, such as diminished employment opportunities in many fields after age forty-five and so on. It is at such critical points that create crises for individuals and their families that the counterforces of the community and protective laws can step in to compensate for the power the individual lacks.

Maturity brings with it the challenge of productivity and creativity, which may assume crisis intensity.

Maturity, marriage, and the birth and rearing of children bring critical situations, which have been discussed from the point of view of the child and are here to be viewed from the other end of the telescope. Pregnancy, particularly the first one for both parents, is a critical period, both physically and emotionally, in terms of the relationship between husband and wife. Like falling in love, it requires making room in the psyche for the developing baby, an internalizing of acceptance beyond the quality of an external, "objective" acceptance. How often this crisis period is passed through with flying colors is suggested by the extreme anguish and frustration when the crisis moves toward the catastrophic outcome. In my experience with one mother—of a schizoid personality with agitation and depression—in her first pregnancy, there was an extreme of impotent rage at the husband—soon divorced —for his "responsibility" in the pregnancy and at the baby, who was experienced as a hostile stranger invading her body and against whom she was helpless to withdraw as she longed to.

Still other emotional crises, particularly for the girl or woman, involve pregnancy. Important are the age and emotional era of life when it occurs, the relationship to the partner, and the socioeconomic situation. The decision for an abortion and the abortion itself (whether voluntary or spontaneous) are critical points, often followed by depression. Whether the abortion is performed under competent and socially permissible medical auspices or covertly and illegally affects the crisis situation, not only physically in terms of proper surgical service, but emotionally. The father of the abortive pregnancy is not immune to emotional crisis by any means.

For fathers, in their accompanying role throughout pregnancy, arise crises of loneliness, a sense of abandonment in favor of the unborn baby, reactivation of past critical experiences of rejection, as well as, affirmatively, the reinforcement of the love relationship and the fulfillment of masculinity.[4]

The crises of maturity, as has been said, include the situation of the growing up and departure of one's children. Parental overcontrol at this time is really a struggle against separation and abandonment by the children, a clinging dependence by the parent within the iron glove of his or her power. The reproach to the departing children, almost impossible to bear, is really allied to the classical depressive reaction of reproaching the loved one for failing to love.

Choice, the art of living and loving, and critical points in development converge in later maturity. Again the art of living involves cultivation of one's self, the capacity to love, the ability to be productive and even creative.

For women the time around the menopause is a crisis period. There is the well-known vulnerability to the depressions. The complexity of this period is compounded of physiological changes with their psychological reverberations, the dread of being unlovable and lonely, which, in turn, is contributed to by unresolved unfortunate vicissitudes in early love relationships and by cultural attitudes of undervaluing anyone of advancing years and of worshiping the cult of youth and immaturity. The hormonal crisis varies for different women, according to biochemical studies. For those for whom the physiological upheaval is experienced as critical, medical help is available—an outer resource for alleviating a (psycho) somatic crisis. The woman's struggles with this crisis period may involve the actual limitation of choice on a cultural basis, which calls for problem-solving on other levels. A particular woman may misread the attitudes, use the cultural attitudes as a cover-up for a withdrawal on her side. But, nonetheless, society plays a powerful role in limiting or challenging choice—which is part of what the women's liberation movement is about.

For men, at approximately this age, but generally ten years or so later, there is anxiety concerning waning sexual powers and symptoms in the genitourinary system. The apprehensions and depressive tendency of the middle age and aging crisis bring with them various attempts at solution. More particularly for men, partly because of society's greater permissiveness, solace is often sought frenetically in sexual relationships based on the need to test one's acceptability and to prove one's masculinity sexually. This period for men, as well as the rather earlier period for women, is depression-prone.

These periods of the crises of the middle years have the inner dimension of the implications of the flight of time, with unfulfilled aspirations in achievement and in personal relationships, but there is heavy responsibility on society for adding pressures of demands and devaluation to this vulnerability. A prominent role in our society, more so for men than for women, is acceptability for employment, and here society is harsh. For men diminished employment opportunity with the middle years creates, in addition to economic crisis, the crises of self-esteem, the feeling of worthlessness at being tacitly declared obsolescent. After the era of unwanted unemployment for some is ushered in the related one of retirement, which may be unwanted. If it is acceptable the challenge still is to self-esteem, productivity, and the art of living.

The challenge of old age brings deeper problems of one's acceptability to others as the extended family is vanishing, one's self-acceptance in the face of waning powers, and the dread of being helpless as a last indignity. Although one's life style plays a role in this time of crisis, so do X factors of health and constitution, and such situational factors as one's world of friends, relatives, and intimates. Obviously for many this is a period of powers shrunken below the level of self-sustaining independence, as society and the culture are recognizing and hopefully will deal with effectively and kindly.

Part of the adaptation to life as an ongoing development demands that each one as an individual and society in the large recognize dying and death, both of one's own person and of that of others (discussed in Chapter 33).

The Role of the Culture in Life's Crises

The "cultural" schools of psychoanalysis have stressed the role of culture and society in setting patterns of childrearing, shaping character, and establishing the values its members live by. And although the word "crisis" is rampant on the social scene, society's crisis-making force for the individual has hardly been faced in those terms. Society inculcates not only explicit values but also the standards for self-esteem and self-image. Standards for success and failure, one's self-expectations, and disappointment in career and status are intimately involved with cultural values and with the familiar periodic crises that few of us are immune too.

There are in our various subcultures actively punitive attitudes directed to individuals who don't conform—such as in love relationships with disapproved pairings of age, ethnic and religious differences, and sex. Those violating these rules are faced with group hostility, even if not by organized formal law, and they undergo crises of despair, sometimes unendurable and ending catastrophically. The hostility and power of a particular group—for example, the availability of housing, employment, and schooling on the basis of religion, ethnic origin, or sex—make for crises in individual lives and add heavy stress to the development of children and adolescents. Their relationship to parents, to the larger society, is pained. As a result the crisis may turn to depression, despair, rage, or violence.

Cultural attitudes have been changing at an accelerated pace within everyone's lifetime. Margaret Mead[10] has urged preparation for change in the culture's values and imperatives, to replace preparation in terms of the assumption that the culture's goals and values are enduring and unchanging. Some years ago, but still within the lifetime of many of us, the code, particularly for girls, was not to marry early, but after completing one's education, and to have two children to replace the parents. Then it was marry early and have a large family. Then it became more permissive about marriage, children in or out of wedlock, and small families, particularly because of apprehension about the population explosion.

The crisis for the middle-aged generation has been to find an orientation to these challenges and changes in sexual patterns, life styles, and the social emancipation of women. Middle-aged individuals, faced with setting up new relationships after a divorce or the death of a spouse, may experience an anchorless confusion, anxiety, depression, or acting out in a crisis of coming to new terms with a world they doubt that they made.

The flaws in the solution of employment opportunities after 45, particularly for higher-level business executives, makes for both personal and economic crises, and crises in family relationships. The attachment of a sense of individual worth to employment and employability, particularly for men, makes for a crisis in self-esteem, as by some mischance in society one becomes obsolescent. Certainly for members of an ethnic group with lower status in employment, who, regardless of youth, are not employed, despair, resentment, and even a propensity for violence make for crisis.

Finally large societal turbulence, such as war, economic depression, and political upheavals, as well as concentration camps in totalitarian states, makes for crises that, though widespread, still are experienced in the individual as a unit. Some of the crises in which society and the culture are involved allow for choice and an effective approach through problem-solving; still others involve pitifully little direct personal choice.

Coping with crisis involves recognition of what it's all about, problem-solving, ego strength, energy, resilience, self-assertiveness, and, at times, aggressiveness. It requires actual resourcefulness, appropriately pragmatic where outer reality is concerned. Ideally it requires a philosophy of life that includes values, a recognition of life's progression, and preparedness for change. It involves also a capacity to sublimate and to recognize that some crises concern relative trivia as part of the game of life.

Bibliography

1. BOWLBY, J., "Process of Mourning," in Gaylin, W. (Ed.), *The Meaning of Despair*, Science House, New York, 1968.
2. CAPLIN, G., *Principles of Preventive Psychiatry*, Basic Books, New York, 1964.
3. ———, HANSELL, N., JACOBSON, G. F., *Crisis Therapy*, San Francisco, American Psychiatric Association, May 11, 1970.
4. DEUTSCHER, M., "Some Clinical Observations on Expectant Fathers," paper presented at the Harry Stack Sullivan Society of the William Alanson White Institute, January 14, 1962.
5. ERIKSON, E. H., *Identity and the Life Cycle*, International Universities Press, New York, 1959.
6. ———, *Identity, Youth and Crisis*, Norton, New York, 1968.
7. FARBER, L. H., *The Ways of the Will*, Basic Books, New York, 1966.
8. KLEIN, M., "A Contribution to the Psychogenesis of Manic-Depressive States," in Gaylin, W. (Ed.), *The Meaning of Despair*, Science House, New York, 1968.
9. ———, "History of Family Problems Found in Children Who Suffer Accidents," *Medical Tribune*, December 8, 1971.
10. MEAD, M., "The Future as the Basis for Establishing a Shared Culture," *Daedalus*, Winter 1965, 135–155, 1965.
11. ROESLER, T., and DEISHER, R. W., "Youthful Male Homosexuality," *J.A.M.A.*, *219*: 1018–1023, 1972.
12. SHAKESPEARE, W., *As You Like It*, act II, scene 7, line 139 ff.
13. SMITH, W. G., "Critical Life Events and Prevention Strategies in Mental Health," *Arch. Gen. Psychiat.*, 25:103–110, 1971.
14. SPIEGEL, R., "Depressions and the Feminine Situation," in Goldman, G., and Milman, D. S. (Eds.), *Modern Woman*, Charles C Thomas, Springfield, Ill., 1972.
15. ———, "Management of Crises in Psychotherapy," in Goldman, G., and Stricker, G. (Eds.), *Practical Problems of a Private Psychotherapy Practice*. Charles C Thomas, Springfield, Ill., 1972.
16. ———, Panelist, Workshop on Crises in Psychotherapy, Chrzankowski, G., Chairman, Meeting, American Academy of Psychoanalysis, May 1969.
17. ———, "The Role of Father-Daughter Relationships in Depressive Women," in Masserman, J. H. (Ed.), *Science and Psychoanalysis*, pp. 105 ff., Grune & Stratton, New York, 1966.
18. ———, Seminars on Crises and Psychoanalysis, William Alanson White Psychoanalytic Institute.
19. ———, "Varieties of Crisis-Oriented Personalities," in preparation.
20. SULLIVAN, H. S., *Conceptions of Modern Psychiatry*, William Alanson White Psychiatric Foundation, Washington, D.C., 1947.
21. THOMPSON, C., *Psychoanalysis: Evolution and Development*, Hermitage House, New York, 1950.

CHAPTER 33

GRIEF AND BEREAVEMENT

Alberta B. Szalita

I follow'd rest; rest fled and soon forsook me;
I ran from grief; grief ran and overtook me.
—QUARLES, *Emblems* (1635)

THE SHADOW OF DEATH haunts man even when he does not think about it. With the expansion of consciousness characteristic of our age of awareness,[46,56] the transience of our existence is brought home with greater force, particularly since the survival of the race has become an ever present concern. Yet man manages to deny this ultimate reality—his mortality—with illusions of perpetual youth, reliance on advances in technology, and naive belief in magical solutions.

Only the giants of our globe, according to Gorki, are able to live as if death did not exist while living in fear of it. The average human being has the task of finding a midpoint between denial of death and obsessive fear of it. It is noteworthy that the greatest intellects as well as the mentally ill show a marked preoccupation with death. The former transform their obsession[58] into a creative quest for immortality; the latter suffer the anguish to the point of incapacitation. On the other hand, helping professionals, as well as average men, keep themselves busy avoiding confrontations with such questions as the impermanence of our existence.

Modern man does not place easy trust in religious systems to aid him in this dilemma; rarely does he believe in life in the hereafter. Grief, sorrow, and bereavement are inseparable companions of the human condition.

The health professions have recently addressed themselves more vigorously to the emotional impact of death and severe illness. Assistance to the dying patient and his family, as well as the rehabilitation of the physically injured or handicapped and those maimed by war or accidents, have finally emerged as central issues in modern psychiatric practice. And with good reason. Every patient soliciting help has many grounds for deep grief.

Bereavement has been defined as the "loss of a loved one and separation from others on whom we depend for sustenance, comfort, and security" (Erich Lindemann).[29] The data on bereavement are staggeringly abundant and diverse; its sources are manifold. It is necessary to restrict the scope of incidents of bereavement dealt with in this chapter. However, the reactions to the specific losses discussed—primarily, but not only, through death—may be usefully applied by the practitioner to other incidents.

Whenever an individual suffers a loss—be it his homeland, his language, an ideal, a treasured material possession, a part of his own body, or particularly a beloved person—he normally undergoes a period of grief and mourning of varying intensity until he recovers the energy he invested in the lost object. The process of mourning is very painful. It is a travail that reconciles him to the loss and permits him to continue his life with unimpaired vigor, or even with increased vitality. A similar process takes place when one is confronted with a disappointment, failure, the loss of a love object through rejection, divorce, abandonment, and the like.

Grief is a common human experience, for life does not spare any of us those events, major or minor, that cause sorrow. Normally grief, even over a very serious loss, is more easily resolved when shared; a person usually has some intimate friends or family members with whom he is able to share his pain. But beneath the surface of consciousness remains a core of deep sorrow that the bereaved person has to work through alone. Profound sorrow is silent.

In Agee's[2] novel, *A Death in the Family*, after the heroine's husband is suddenly killed in an automobile accident, her father warns her that it will take time for the loss to sink in, and then it will get worse. He continues:

> It'll be so much worse you'll think it's more than you can bear. Or any other human being. And worse than that, you'll have to go through it alone, because there isn't a thing on earth any of us can do to help, beyond blind animal sympathy.

What a grief-stricken person needs is "common sense," love of life, and what we are accustomed to call ego strength, that is, endurance, courage, and the ability to face up to things.

An individual who possesses the stamina for confronting a severe loss only rarely turns to a professional to aid him in going through this process. Those who seek help are sometimes very lonely people who have the sense to look for someone to share their grief with them, or they may display various aspects of the pathology of mourning, but the majority come for other reasons.

Pathology of Mourning

To paraphrase Parkes, grief may play as important a role in psychopathology as do inflammatory reactions in medicine.[31] The inability to engage in mourning after a loss gives rise to a number of pathological manifestations. These may be psychological, physiological, or mixed in nature. The psychological reactions range from chronic reactive depressions, agitated depressions similar to those observed in menopause or middle age, and emotional anesthesia (Minkowski, Paris*), to intractable melancholia. Addictions and acting out may also be precipitated by bereavement. In delayed mourning hypochondriasis plays a special role; it may be looked upon as a kind of strategy to master paranoid feelings that are linked with denial of the death.

On the physiological side there are to be found conversion symptoms, psychosomatic disturbances, and identification with the physical symptoms of the deceased. Some individuals seem to develop the same physical condition that caused the death. Somatic delusions have also been recorded.[39] Thus, a young man who lost his mother during adolescence and did not mourn her death developed the menopauselike symptoms that she presented prior to her death, including crying spells regularly accompanied by hot flashes. A woman whose late husband had suffered for years from Parkinson disease developed a Parkinson-like tremor, gait, and masklike face.

Reactions to bereavement are difficult to classify because they cover such a wide range of pathological manifestations. Peretz[39] dis-

* Personal communication.

tinguishes the following reactions to loss: (1) "normal" grief; (2) anticipatory grief; (3) inhibited, delayed, and absent grief; (4) chronic grief (perpetual mourning); (5) depression; (6) hypochondriasis and exacerbation of preexistent somatic conditions; (7) development of medical symptoms and illness; (8) psychophysiological reactions; (9) acting out (psychopathic behavior, drugs, promiscuity); and (10) specific neurotic and psychotic states.

Before focusing on some specific types of loss, I shall discuss the general psychodynamics of mourning. In order to complete successfully the process of mourning, one must have the capacity to form object relationships of a differentiated type and to deal with ambivalence.

The Mourning Process

Just as "being in love" is a prototype of human psychosis (Freud), so, too, are various manifestations of mourning. Both are common experiences in which the distinction between normality and abnormality is effaced. After all, there is nothing normal about nature. In reaction to the dead particularly, man still demonstrates an array of irrational beliefs and attitudes that fall within the range of normality. Some of these are culturally determined, some may be considered atavistic, handed down to us through folklore and, perhaps, through channels unknown to us. Thus, we often feel that the moment the soul leaves the body it is everywhere, sees everything, and is aware of our every thought. The first impulse is not to think anything bad nor to say anything derogatory about the dead. *De mortuis nil nisi bonum.*

The essential aspects of mourning are not clearly understood, particularly the kind of psychophysical pain[11,15] that accompanies the pangs of conscience; but to undergo the process effectively, one has to have the capacity to face the feelings of hopelessness and helplessness evoked by the finality of death. "Why this compromise by which the command of reality is carried out piecemeal should be so extraordinarily painful is not at all easy to explain," said Freud.[15] Some of it can be explained by the fear of the vengefulness of the spirit or ghost of the dead. A vengeful image of the dead only too often terrorizes those who had an intensely ambivalent relationship with the departed and who repressed or denied their hostile feelings, particularly obsessive, compulsive characters. But it would be a mistake to think that hysterically defended bereaved persons are free from such tortures.[4,11,16,19,30]

When the relationship was more or less mature and well integrated, the loss, however hard it may be to bear, has a different quality of pain. One may miss a person intensely, and feel the full impact of the loss—psychologically, economically, or in status, as in the case of a widow—and one may have feelings of slight remorse. Nevertheless, there is little need to expiate one's guilt, to justify one's past behavior, apologize to and appease the dead. One is able to deal more realistically with a real loss because of the relative absence of unconscious guilt.

I shall describe the process of mourning in a 34-year-old woman who lost her extended family—her parents, a brother three years older than herself, a sister two years younger, a grandfather, and a host of cousins, uncles, and aunts—during the German occupation of the Ukraine in World War II. Their neighbors rounded them up and shot them, under the supervision of the German invaders. The woman, a nurse, who lived thousands of miles away with her husband and three-year-old son, received word of the catastrophe in a letter from a friend of the family.

Although she had read and heard about Nazi atrocities, the thought that her family might be wiped out had never entered her mind. Such things might happen, but not to *her* family. Immediately after reading the letter, she felt stunned. "Something snapped in my head," she said. "It was as if a curtain had fallen down on a part of my brain and chopped off part of it."

Since she had not visited her childhood home for many years and had been out of communication with her family from the beginning of the war, news of the tragedy changed nothing in her life. And yet something changed drastically inside her. She became extremely fearful and irritable at home. At work she could take care of people "better than before," she said, "because I was able to shut out everything when I was with a patient; but at home I felt empty." She had previously read a great deal, but from the moment she learned about the extermination of her family, she was unable to concentrate on reading. She would sit in front of an open book as if submerged in deep thought, but would not be able to think about anything. To think meant to think about the fate of her family, which she was afraid to do. She knew, she said, that she had to follow in the footsteps of all the members of her family as they moved toward their collective grave, but she was unable to do so. Whenever she started to think about what had happened to them, she felt she would either lose her mind or kill herself. She avoided any discussion of the holocaust.

She became self-absorbed at home, unresponsive to the needs of her husband and child. For a time she felt incapable of being a good mother and entrusted the care of her son to a spinster sister-in-law, only to reclaim him later. Traveling to and from work she would often enter an empty subway car and scream at the top of her lungs. She was gripped by a kind of inner-directed attention, "as if I had to hold myself together and not think." The only time she allowed herself some pleasure, such as dining out or going to the theater, was when a stranger was there to enjoy it with her; then she could feel that she was doing something for another person. She entertained thoughts of divorcing her husband because she felt she had nothing to offer him. During this period she began to suffer from headaches, low back pain, and anxiety attacks during which she felt as though she would faint or die. She developed a number of phobias, including fears of crowds, elevators, high places, and also of going alone anywhere except to work. Meanwhile, she continued to work at peak efficiency. Thus, three years elapsed before she sought help.

In treatment, after a short preliminary period, she started to review her relationships, one after another, with her dead relatives. She would identify herself totally with whatever member of the family she was concentrating on, repeating his or her gestures, speaking in the same way, and behaving so as to please that member. Referring to this process as "going through purgatory or the last judgment," she said, "I have to give each of them an accounting of what we lived through together." For a long time she felt she had to expiate her guilt.

The patient had been brought up in a closely knit family whose members were not supposed to experience negative feelings, let alone utter them. In treatment she harbored suicidal thoughts when she became aware of her anger toward the dead for "abandoning" her and her envious attitudes toward her siblings; the uncovering of her death wishes for them so shocked her that she became even more desperate. When these wishes became clear, she felt that she was a danger to everyone close to her. She said that she was responsible for the deaths in her family; Hitler had simply carried out her wishes. She feared that her husband, son, and therapist would suffer a similar fate through her evil influence. She experienced a strong need to be punished. She burned herself while cooking or lighting a cigarette. In her dreams the deceased were alive and often angry with her. Whether awake or dreaming, she feared their retaliation.

Apparently each person she mourned was first sequestered in her mind, isolated as a unit of behavior, and then accounted to, as if still living, in order to obtain forgiveness. She conducted endless imaginary conversations with each of the departed before reaching the point where their deaths became a reality for her. Until then they maintained a life of their own within her.

Schematically one may distinguish the following stages in mourning:

1. *Complete identification with the deceased*—becoming like him and envisaging him as an authority, either totally benevolent or tyrannical, who directs the life of the sur-

vivor. Some persons stop at this level, maintaining a sort of symbiosis with the departed and fulfilling his wishes until the end of their days. This solution, achieved through unconscious mechanisms,[1] permits one to avoid guilt and responsibility for the death. The deceased is installed as a foreign body that governs the mourner's life.

2. *Splitting of the identification.* This takes place while reflecting on the relationship with the deceased, a process that leads to fragmentation of the "foreign body." The first consequence is a conflict between submission and rebellion.
3. A *detailed review of the relationship,* with a somewhat detached appraisal of one's own conduct toward the lost object. The self-evaluation encompasses a painful working through of myriads of minute elements and a complete scanning of one's life. There can be no glossing over in this process; shallowness is incompatible with mourning. The result of "digging in" is that one emerges as an integrated, enriched, and revitalized person.

In essence the process of mourning is not unlike the analysis of transference. The latter is also, in a sense, a kind of mourning as one struggles to give up dependence on parental figures and integrate the ambivalences of life. The issue is not so much the elimination of ambivalences as learning to live with them, for life is fraught with conflicts. There is good reason to equate effective living with the mastery of polyvalence.

What is particularly difficult about undergoing the process of mourning is the need to face the ultimate reality of life—*death*—with the full realization of one's helplessness to evade its finality. One dreams of eternal youth and denies one's own demise in countless ways, but real mourning affords no hiding place from this existential truth.

The rituals prescribed for funerals, which many include under the term "mourning," vary from culture to culture, but they express a common theme. They are all oriented toward appeasing and showing respect for the dead. These ceremonials tend to retard the mourning process and intensify the mood of bereavement, often artificially.* The intensification is accomplished through collective mourning, which initially serves to facilitate catharsis. Thereafter ensues a period in which the mourner is expected to conform to the restrictions imposed by societal or tribal customs, particularly for widows.

Collective mourning gives the bereaved an opportunity to shed tears, which they might not be able to do when they express their sorrow privately. Collective mourning reached its epitome following the assassinations of President John F. Kennedy, Martin Luther King, and Robert Kennedy. Never before the advent of television had an entire nation been joined together in expressions of bewilderment, sorrow, indignation. These successive cathartic experiences afforded some participants the opportunity to reflect on the transience of glory, or its acquisition through death, as well as for mourning the loss of their own loved ones.

Melancholia

Melancholia may be viewed as an extreme form of mourning, sharing its symptomatology —lugubriousness of mood, ritualistic rigidity, and many superstitions—but lacking the actual work of mourning as outlined above. Melancholia seems to be a sort of pseudomourning because of the inability to work through a situation, which, as already pointed out, is the essence of mourning as a dynamic process.

Instead, there is a clinging, demanding dependency,[48] an insistence on passive gratification that is impossible to meet. Particularly trying for the people around the melancholic

* An anecdote related by Rabbi Solomon Tarshansky of New York illustrates this artificiality. In many parts of the world it has been customary to hire mourners to keen at funerals. One *shtetl* boasted of a woman who excelled in that profession. She was renowned for her artistry in "carrying on" and inducing others to cry with her. But one day she refused to "perform" at the funeral of a wealthy townsman, and the disappointed customer asked her, "How come? Don't you need money any more?" She answered simply, "Today I cannot cry. My sister just died." With grief in her heart, she was unable to pretend sorrow.

person is the incessant complaining. Three W's—of Woe, Whine, and Wail—prevent the patient from taking stock of what has happened and what might be done about it. Thus, a closed circle is maintained, and this may be termed the *constipation of mourning*.

There are, however, transitional forms of mourning with similar symptomatology that are less intractable than melancholia. These forms include the agitated depressions of menopause,[52] in which it is impossible to trace the pathology to early introjects that were excluded from awareness. These exist throughout life, unchanged and unaffected by experience. The existence of such introjects, experienced as *not me*, is maintained mainly by repression and dissociation, reducing the integrative capacity of the ego and producing a latent tendency to depressive reactions. This accounts for a more or less chronic state of discontent with oneself as well as a tendency toward reaction-formation coupled with strong oral fixation. Under the influence of a new loss (object loss or any other), a decompensation takes place; anal defenses crumble and oral traits, such as envy, greed, avarice, and impatience become predominant. As Abraham[1] pointed out, envy may be oral and anal in character; it is usually both. The patient is preoccupied with what others have and devalues everything that he has or is. He blames others or fate for his deprivations.[48] The self-reproach is actually an accusation[15] directed at whoever deprived the patient of what was promised him at the cradle. Some individuals weep over real losses, others over lost opportunities or things they never had.[48]

Melancholic patients often present a caricature of the introjected, feared, malevolent person and are incapable of relating their behavior to any previous experience. Memory traces are lacking. If there is any recollection, it is only of undifferentiated feelings; there is an absence of a formulated thought about a loss. That is why it is important to inquire about early deaths and, in general, about those who were important in the patient's formative years.

The melancholic syndrome is perplexing in that not only is a painful experience retained, but also it seems to be actively intensified, with the persistence and effort worthy of a Sisyphus. Attempts to dissuade the patient are useless. The absurdity is accentuated by shameless self-vilifying ideation,[15] and the patient reacts impatiently to any effort to console him. He feels misunderstood, which indeed he is.

In instances where the ego is sufficiently intact and a connection can be established with an isolated introject, it is possible to obtain good psychotherapeutic results. It is important to find out as much as possible about the significant persons in the patient's early childhood in order to determine the source of the pathogenic influence. Of particular significance are his first encounters with death, whether of a human being or a pet. Although the patient's lamentations may strain the therapist's patience and foreclose his sympathy, the capacity for insight and the readiness to examine the relationship and confront the ambivalence with the therapist's help make it possible to integrate the not-me into the personality. This accounts for the new energy that becomes available after the completion of mourning.

Some of the above mentioned reactions are present during the process of mourning, but without the tenacity seen in melancholia. That is why it is essential that the therapist be familiar with all the manifestations of melancholia; otherwise it is impossible to sustain interest in helping the patient overcome the passive submission to pain, which teaches him nothing and maintains his restlessness, confusion, and perplexity. Patients invariably regress in proportion to the degree of their helplessness.

The pain of the melancholic well illustrates that suffering by itself does not generate wisdom. Protracted suffering exhausts the energy of the patient and the tolerance of the helper.

Emotional Anesthesia

Minkowski coined the term "emotional anesthesia"* to describe the feeling state of young survivors of the concentration camps.

* Personal communication.

When interviewed after World War II, they had no affective response to those who had perished around them, including those closest to them. One must differentiate the state of these survivors from total apathy; they were quite enterprising and were able to care for themselves and take advantage of any opportunity offered to further their immediate existence. Of course, there was an underlying depression.

In 1944, when I interviewed five children who survived the concentration camp at Majdanek, I was struck by the total absence of affect in them. For example, a nine-year-old boy described the death of his mother, who suffered from tuberculosis, in a matter-of-fact way; his face and voice betrayed no emotion whatsoever as he told his story:

> We all knew what she was going to do. She was going to throw herself into the river. She wanted to do it before but we wouldn't let her. This time we let her go. She went and drowned herself. Later in the day the Germans came to our home and killed everybody. I hid under the bed so they didn't see me. A Polish neighbor came to the house afterwards and took me home with her. She kept me as long as she could and then turned me over to the Germans. She had no choice.

Massive denial came to his rescue, permitting him to mobilize energy for survival.

Denial in Somatic Loss and Disability

Although actual loss, severe disablement, or disfigurement of a part of the body is not, as far as we know, accompanied by a specific mental constellation, clinical experience indicates that it stimulates so intense and massive a use of denial that this defense assumes a pathological quality.[39] A person with a severe physical handicap often denies it on a conscious as well as an unconscious level. Shame and fear of humiliation or pity are among other attitudes that are consistently observed in the severely handicapped.

For example, a young man with such severe paraplegia of the lower extremities that he could not walk without the aid of crutches, said during his first interview that he did not regard himself as a disabled person but as a "disembodied mind." Another young man uttered no complaint about the leg he had lost in an automobile accident—he thought he had resolved this problem by learning to use a prosthesis—but he solicited help in recovering his memory, which had begun to fail him after the accident, so that he could concentrate on his studies. A third patient, a paraplegic who was confined to a wheel chair and lacked control of his bladder and bowels, complained only about his stutter. If this speech defect could be overcome, he said that he would be a happy person.

Similar pronouncements have been made by many other patients with severe physical disabilities. As a rule they do not want to talk about their handicap or to have others confront them with it. The more intelligent the individual, the more intense his efforts to guard himself against any intrusion into this aspect of his experience.

The temporary or limited use of denial may be useful and even desirable for slowing down confrontation with a loss that is devastating to one's self-esteem or that threatens to shatter one's mental equilibrium. Used in this way, denial is most frequently associated with shame, humiliation, and guilt and indicates a proneness to dissociative processes. Before confrontation of the patient with the nature or full extent of the disability, it is *first* necessary to help him develop or consolidate the ego strength necessary to tolerate it.

On the other hand, the predominant and permanent resort to denial is always psychologically depleting and crippling. It suggests a commitment to a more or less double form of existence—that is, the incorporation of two distinctly separate images: one of a wholesome and healthy individual and the other of a disabled or disfigured self, which is, as a rule, actively hated, personified, and rejected. The stronger this conflict, the stronger is the subject's feeling of injustice, deprivation, and discrimination.

Before the patient can give up his denial of the disability, he must face his handicap. This, so far as we know, can be accomplished only through the laborious and painful process of mourning.[47,48,49] Since the ambivalence of the feelings seems to be stronger in the case of severe physical disability than in any other loss, the mourning is almost inevitably more prolonged and more painful, but it leads to reconciliation with the disability. Through this process a state of calm is achieved as one becomes capable of *making a distinction between oneself and the disability*. The disability then becomes depersonalized. In place of a split self-image, integration takes place; there emerges a unified personality with more or less limited equipment.

One should also bear in mind the deep narcissistic blow that physical injury inflicts on the ego. This is familiar to psychoanalysts who work with patients who have undergone mutilating operations, even when these affected only internal organs. The extent to which a physical defect disfigures a person is an important consideration. Particularly devastating to self-esteem are facial injuries.

In order to come to terms with the disability, the patient has to face the intense despair and grief that underlie his feelings toward his handicap; he also has to face an exceptionally intense castration anxiety. Despair nourishes a need for revenge. Most frequently there is no one to blame. Superstitions have given way to a scientific understanding of this predicament, as is the case in congenital disabilities. As Thomas Mann observes in *Royal Highness*, which deals with the life of a prince born with an atophied arm, the worst kind of misfortune is one for which no one is to blame.

The working through of a severe loss through the mourning process serves to diminish or alleviate shame, which invariably follows a loss of self-esteem. As with the physically handicapped or disfigured, shame may be markedly present in widows and widowers, in many of the dying aged, and in the parent who has lost a child. They feel stigmatized and ashamed, as if "marked by destiny," compared with more fortunate people. They also experience a sense of guilt, of being punished for their transgressions.

Grief has to be shared if it is to lead to hope. Only once have I heard a handicapped person express his grief spontaneously. It was soon after an accident in which he lost his right hand. "I miss my hand as one misses a sweetheart. I feel a yearning for it, as if it were a person." This was said by an immigrant peasant, whose right hand was his breadwinner. This I consider to be a normal reaction to a loss, one without recourse to denial.

The existent literature on the rehabilitation of the physically disabled[47,49] stresses the necessity of accepting the disability. However, there is nothing acceptable about a disability or the equally tragic loss of a beloved person; at best one may become dispassionate about it.

Loss of a Parent

There is a consensus that the loss of an important person cannot be mourned in early childhood since the child's ego is not equipped to deal with the task of separating itself from the object. As a rule the child continues to maintain a living representation of the deceased, usually an idealized image. The finality of death is inaccessible to a small child.[39] For example, after one mother had finished explaining to her four-year-old daughter that her father had "gone to heaven," the little girl asked, "When will daddy come back?"[2]

In treating adults who lost a parent in early childhood, one notices over and over again that the parent continues to live in their minds. In most cases two contrary and independent attitudes are maintained. One is an acknowledgment of the death and adaptation to it; the other is a denial of the death. In that context psychoanalytic literature refers to the splitting of the ego.[30]

A lawyer who had lost his father at the age of six did not begin to mourn him until he entered reanalysis 42 years later. He completed his first analysis ten years earlier without deep mourning for the parent. There was a mere acknowledgment of the death at the time. The father was talked about as a dead person. The patient reported how he recited the Kaddish by rote at the age of six without understanding a word of it and how, during the year he attended synagogue, older Jews admired him and whispered, "Poor orphan, so young." As they shook their bodies, he imitated their movements. Feeling accepted and at one with them gave him a sense of importance. He idealized his father and cherished his memory.

During the second analysis the patient began to confront himself with the meaning of his father's death and the realization that his father had never died for him. Very painful memories emerged and were expressed with tears, despair, and anger. Left with a "half-crazy" mother, he had to take care of her so that she wouldn't go completely mad. "I couldn't even go crazy," he cried out. "I had to control myself." When he went to work at the age of 11, his mother, he said, "kept waiting for me to come home, and if I came home late she would go out of her mind and abuse me both verbally and physically." In recalling the Kaddish, he said, "I wasn't praying to God; I was talking to my father. I told him, 'I do this for you. What are you going to do for me?'" The patient cried bitterly, "Father left me with *his job*—caring for mother." For years he had talked in this way to his father, even though his grandmother told him that his father was watching over him and taking care of him.

The patient had always gesticulated like an old Jew; many had noticed and commented on that behavior. Usually he responded to these remarks as to a pleasant joke, saying that imitating the old Jews he had met in the synagogue strengthened memories that preserved the bond with his dead father. In the reanalysis, during the period of working through oedipal guilt and the concomitant castration-anxiety, and also his relationship to his mother (then dead eight years), a curious identification with her behavior and style of speaking was noticeable.

In short, there is abundant evidence that a parent's death in childhood is accompanied by denial of the death, rigid avoidance of feelings that accompanied the event, unconscious identification and often idealization of the deceased parent, and fantasies of being reunited with him.

The Dying Patient (Anticipatory Mourning)

The dying patient has to face the ultimate reality—the loss of his own life.[39] Not infrequently, as he struggles to reconcile himself with the irreconcilable, he is surrounded by people waiting for his end to come. And often he feels guilty of imposing on them. Usually those around him, afraid of their own death, incapable of controlling their disturbed feelings over the imminent bereavement, or guilty over their own death wishes, avoid the dying patient in defense of their own feelings.* But if one wants to help the dying patient, one needs to be able to transcend this self-protective shield and become other-oriented, to alleviate his fear of abandonment and his aloneness, which are more frightening than death itself.

The absence of that other-oriented *quality of presence*[56] deprives the patient of the possibility of sharing his grief. Very often family and friends—and doctors—are unable to respond to this psychological need.[58] They "sign off" emotionally long before the death occurs. Their guilt and helplessness make the patient feel guilty, ashamed, envious, and bitter.

Even though they know they are going to die, human beings can hardly imagine the end of their own lives. Freud said: "In the unconscious, every one of us is convinced of his own immortality." A patient condemned to death by an incurable disease does not want to know he is dying. He wants the environment to deny death, and he hopes against hope that he

* Anticipatory grief of the dying patient is perhaps best described by Tolstoy in *The Death of Ivan Ilyich.*[50,58] It is an excellent study of a 45-year-old official, facing a lonely and agonizing death from cancer, whose wife and children are unwilling or unable to relieve his agony in the slightest measure.

will pull through, perhaps by some miracle. He wants to be given some hope.

A 52-year-old woman dying of cancer of the stomach wanted to talk about literature, to keep her mind on something else. Her husband and daughter avoided her because of their own fears and arranged for friends to visit her during her last weeks. When I made myself available to discuss anything she chose, including her illness, she evaded the topic. She did not want to hear that her life was endangered.

Other families begin the process of mourning long before the death occurs. In cases where the patient doesn't know that he is suffering from an incurable disease and those closest to him have to bear the burden alone, they are likely to come to the psychiatrist for help. He has to help them confront the situation, give them support and sometimes hope.

Conclusion and Treatment

Rarely if ever do patients mention difficulties in mourning as a presenting problem. Usually the bereaved complain about other difficulties, such as inability to concentrate on work, depression, disinterest in life, anxiety states, suicidal thoughts, and the like. As the classification cited above indicates, there are few psychiatric conditions that may not mask a delayed, unfinished, or absent mourning. Such a view may be challenged as an extreme oversimplification, for it virtually reduces the whole of psychiatry except schizophrenia to the pathology of mourning. Nevertheless, it is no oversimplification to assert that every human being has to face death in those close to him and his own death, and that the human condition is afflicted with sorrows, to which the schizophrenic is no less immune.

Hence, in addition to mastering the entire therapeutic armamentarium of modern psychiatry, every therapist has to develop skills in dealing with such matters. This makes an enormous demand on the psychiatrist. Since protracted complaining is frowned upon in our culture, those who indulge in it meet with pity, if not open contempt, even among the helping professions.

In order to develop appropriate responsiveness and to enable patients to undergo mourning when necessary, the therapist has to come to grips with his own fears of death and the defenses he has erected against them. Appropriate responsiveness springs from *empathy* and a *quality of presence*, both of which are *other*-oriented, and a readiness to face the countless forms of human unhappiness. As Bertand Russell[36] expressed it, "One needs, as the key to interpret alien experience, a personal knowledge of great unhappiness; but that is a thing which one need hardly set forth to seek, for it comes unasked" (p. 253).

Stefan Zweig,[62] with his deep understanding of human nature, is even more instructive in *Beware of Pity*. His hero found that once gained, an insight into human nature grows, in some mysterious way, so that "he to whom it has been given to experience vicariously even one single form of earthly suffering acquires by reason of this tragic lesson, an understanding of all its forms, even those most foreign to him, and apparently abnormal" (p. 78).

These thoughts point to suffering as the royal road to insight, wisdom, and empathy. It may be superfluous to observe that suffering per se does not generate wisdom, as is clearly demonstrated in cases of melancholia. But suffering that is accompanied by insight and gradual emancipation from narcissistic self-involvement leads to empathy that, in turn, contributes to the resolution of grief. With that comes a compassionate attitude toward others and a new commitment to life.

All of the bereaved need a great deal of compassion. But there are two kinds of compassion. One is accompanied by anger and pain—a form of self-protection and vulnerability that evokes angry feelings toward the object of compassion and wants him out of the way. The other is compassion free of self-centeredness—an empathic consideration for another person's feelings with a readiness to respond to his needs without having to shield or spare one's own sensibilities and without making someone else's suffering one's own burden.

Bibliography

1. Abraham, K., *Selected Papers*, Basic Books, New York, 1953.
2. Agee, J., *A Death in the Family*, Avon, New York, 1956.
3. Altschul, S., "Denial and Ego Arrest," *J. Am. Psychoanal. A., 16*:301–318, 1968.
4. ———, "Object Loss by Death of a Parent in Childhood," in "Panel on Depression and Object Loss," *J. Am. Psychoanal. A., 14*:142–153, 1940.
5. Arieti, S., "The Psychotherapeutic Approach to Depression," *Am. J. Psychother., 16*: 397–406, 1962.
6. Auclair, M., *Vers une vieillesse heureuse*, Editions du Seuil, Paris, 1970.
7. Bellak, L., *Manic-Depressive Psychosis and Allied Conditions*, pp. 220–229, Grune & Stratton, New York, 1952.
8. Bowlby, J., "Grief and Mourning in Infancy and Early Childhood," *Psychoanal. Study Child, 15*:9–52, 1960.
9. De Beauvoir, S., "Joie de vivre," *Harper's Magazine, 244*:33–40, 1972.
10. ———, *A Very Easy Death*, G. P. Putnam's, New York, 1966.
11. Fleming, J., and Altschul, S., "Activation of Mourning and Growth by Psychoanalysis," *Internat. J. Psychoanal., 44*:419–432, 1963.
12. Freud, Anna, *Normality and Pathology in Childhood: Assessments of Development*, International Universities Press, New York, 1957.
13. Freud, S. (1927), "Fetishism," in Strachey, J. (Ed.), *Standard Edition, Complete Psychological Works*, Vol. 21, pp. 149–157, Hogarth, London, 1961.
14. ——— (1924), "The Loss of Reality in Neurosis and Psychosis," in Strachey, J. (Ed.), *Standard Edition*, Vol. 19, pp. 183–187, Hogarth, London, 1961.
15. ——— (1905), "Mourning and Melancholia," in Strachey, J. (Ed.), *Standard Edition*, Vol. 14, pp. 243–286, Hogarth, London, 1964.
16. ——— (1925), "Negation," in Strachey, J. (Ed.), *Standard Edition*, Vol. 19, pp. 235–239, Hogarth, London, 1961.
17. ——— (1924), "Neurosis and Psychosis," in Strachey, J. (Ed.), *Standard Edition*, Vol. 19, pp. 149–157, Hogarth, London, 1957.
18. ——— (1940), "An Outline of Psychoanalysis," in Strachey, J. (Ed.), *Standard Edition*, Vol. 23, pp. 141–207, Hogarth, London, 1964.
19. ——— (1940), "Splitting the Ego in the Process of Defence," in Strachey, J. (Ed.), *Standard Edition*, Vol. 23, pp. 271–278, Hogarth, London, 1964.
20. Fromm-Reichmann, F., in *Selected Papers, Psychoanalysis and Psychotherapy* (Ed. by Bullard, D. M.), Chapters 4, 5, University of Chicago Press, Chicago, 1959.
21. Gibson, R. W., "The Family Background and Early Life Experience of the Manic-Depressive Patient," *Psychiatry, 21*:71–90, 1958.
22. Hartmann, H., "Comments on the Psychoanalytic Theory of the Ego," in *The Psychoanalytic Study of the Child*, Vol. 5, pp. 74–96, International Universities Press, New York, 1950.
23. Hilgard, J. R., Newman, M., and Fisk, F., "Strength of Adult Ego Following Childhood Bereavement," *Am. J. Orthopsychiat., 30*:788–798, 1960.
24. Jacobson, E., "Denial and Repression," *J. Am. Psychoanal. A., 5*:61–92, 1957.
25. Jankelevitch, V., *La mort*, Flammarion, Paris, 1966.
26. Klein, G. S., "Consciousness in Psychoanalytic Theory," *J. Am. Psychoanal. A., 7*:5–35, 1959.
27. Levin, S., "Panel Report: Depression and Object Loss," *J. Am. Psychoanal. A., 14*: 142–153, 1966.
28. ———, "The Psychoanalysis of Shame," *Internat. J. Psychoanal., 52*:355–362, 1971.
29. Lindemann, E., "Preface" to Langer, M., *Learning to Live as a Widow*, Gilbert Press, New York, 1957.
30. Miller, J. B. M., "Children's Reactions to the Death of a Parent: A Review of Psychoanalytic Literature," *J. Am. Psychoanal. A., 19*:697–719, 1971.
31. Parkes, C. M., "Bereavement and Mental Illness," *Brit. J. Med. Psychol.*, 38:1–26, 1965.
32. Piers, G., and Singer, M. B., *Shame and Guilt: A Psychoanalytic and a Cultural Study*, Bannerstone House, Springfield, 1953.

33. RANGELL, L., "The Nature of Conversion," *J. Am. Psychoanal. A.*, *7*:632–662, 1959.
34. ROCHLIN, G., "Dread of Abandonment," in *The Psychoanalytic Study of the Child*, Vol. 16, pp. 451–470, International Universities Press, New York, 1961.
35. RUITENBEEK, H. M., *Death: Interpretations*, Dell, New York, 1969.
36. RUSSELL, B., *The Autobiography of Bertrand Russell*, Little Brown, Boston, 1967.
37. ———, *Portraits from Memory*, Simon & Schuster, New York, 1951.
38. SCHELER, M., *Le sens de la souffrance*, Aubier, Paris, 1936.
39. SCHOENBERG, B., CARR, A. C., PERETZ, D., and KUTSCHER, A. H., *Loss and Grief: Psychological Management and Medical Practice*, Columbia University Press, New York, 1970.
40. SLATER, P., *The Pursuit of Loneliness: American Culture at the Breaking Point*, Saunders, Toronto, 1970.
41. SONTAG, S., *Against Interpretation*, Farrar, Straus & Giroux, New York, 1961.
42. SPERLING, S. J., "On Denial and the Essential Nature of Defense," *Internat. J. Psychoanal.*, *39*:25–39, 1958.
43. SPITZ, R. A., "Discussion of Dr. John Bowlby's Paper," in *The Psychoanalytic Study of the Child*, Vol. 15, pp. 85–95, International Universities Press, New York, 1960.
44. ———, *A Genetic Field Therapy of Ego Formation: Its Implications for Pathology*, International Universities Press, New York, 1959.
45. ———, and COBLINER, W. G., *The First Year of Life*, International Universities Press, New York, 1965.
46. SZALITA, A. B., "The Combined Use of Family Interviews and Individual Therapy in Schizophrenia," *Am. J. Psychother.*, *22*:419–430, 1968.
47. ———, "De fysisk handicappede," *Nordisk Psykiatrisk Tidsskrift*, *18*:479–484, 1964.
48. ———, "Deprived Families," *Bull. Fam. Ment. Health Clinic Jewish Fam. Serv.*, *1*:5–7, 1969.
49. ———, "Discussion," in Cayley, C., "Psychiatric Aspects of Rehabilitation," *Am. J. Psychother.*, *8*:538–539, 1954.
50. ———, "The Family in Literature and Drama," *Fam. Proc.*, *9*:99–105, 1970.
51. ———, "Further Remarks on the Pathogenesis and Treatment of Schizophrenia," *Psychiatry*, *15*:143–150, 1952.
52. ———, "Psychodynamics of Disorders of the Involutional Age," in Arieti, S. (Ed.), *American Handbook of Psychiatry*, Vol. 3, pp. 66–87, Basic Books, New York, 1966.
53. ———, "Psychotherapy and Family Interviews," in Arieti, S. (Ed.), *The World Biennial of Psychiatry and Psychotherapy*, Vol. 1, pp. 312–335, Basic Books, New York, 1970.
54. ———, "Reanalysis," *Contemp. Psychoanal.*, *4*:83–102, 1968.
55. ———, "Regression and Perception in Psychotic States," *Psychiatry*, *21*:53–63, 1958.
56. ———, "The Relevance of Family Interviewing for Psychoanalysis," *Contemp. Psychoanal.*, *8*:31–43, 1971.
57. ———, "Remarks on the Pathogenesis and Treatment of Schizophrenia," *Psychiatry*, *14*:295–300, 1951.
58. TOLSTOY, L., "The Life and Death of Ivan Ilyich," in Kamen, I. (Ed.), *Great Russian Stories*, Random House, New York, 1959.
59. TROYAT, H., *Fyodor Dostoyevsky*, Fayard, Paris, 1960.
60. WIENER, J. M., "Reaction of the Family to the Fatal Loss of a Child," and "Responses of Medical Personnel to the Fatal Illness of a Child," in Schoenberg, B., Carr, A. C., Peretz, D., and Kutscher, A. H. (Eds.), *Loss and Grief: Psychological Management and Medical Practice*, Chapters 6, 7, Columbia University Press, New York, 1970.
61. ZETZEL, E. R., "Anxiety and the Capacity to Bear It," *Internat. J. Psychoanal.*, *30*:1–12, 1949.
62. ZWEIG, S., *Beware of Pity*, Viking, New York, 1939.

CHAPTER 34

HELP IN THE DYING PROCESS

E. Mansell Pattison

DEATH has long been a taboo topic in American culture. But in a manner similar to the broaching of sexuality as a subject for investigation at the turn of the twentieth century, the decade of the 1960's has witnessed the broaching of death and dying. Research into this taboo area has presented to American culture and its medical profession the evidence of their neglect in a major human experience: the care of the dying person.

"There's nothing more I can do for you, you're going to die." So said the doctor, leaving the patient, the nursing staff, and the family to fend as best they could with one of the fundamental issues of life. In contrast, a general practitioner, Dr. Merrill Shaw,[78] himself soon to die of rectal cancer, wrote of his own dying process in a very different vein:

> The period of inactivity after a patient learns there is no hope for his condition can be a period of great productivity. I regard myself as fortunate to have had this opportunity for a "planned exit." Patients who have been told there is no hope need help with their apprehension. Any doctor forfeits his priesthood of medicine, if, when he knows his patient is beyond help, he discharges his patient to his own services. Then the patient needs his physician more than anyone else does. The doctor who says merely, "I'll drop in to see you once in awhile; there's nothing more I can do," is of no use to the patient. For the patient goes through a period of unnecessary apprehension and anxiety.

Research of the past decade on death and dying has now produced several thousand titles.[93] Thus, in this chapter we shall allude to major areas of research, but focus on the major clinical aspects of care of the dying.

Cultural Antecedents

As Feifel[24] has documented, the topic of death was culturally taboo in American culture until the past decade. That is not to say that death and dying were ignored, but rather what was repressed surfaced in the pornography of death.[32] The lurid, perverse, and seeming compulsive preoccupation with death and dying themes has been obvious, reflected in a movie like *The Loved One*—a parody on the Forest Lawn cemetery in Los Angeles.

Parsons and Lidz[64] observe that the Judeo-Christian heritage contributed greatly to the denial of death in Western civilization. Yet prior to the advent of modern medicine, the ravages of disease, war, and punishment continuously confronted people with death. The success of medical technology and the distant displacement of war has left many Americans with little direct contact with death until well into adult life. However, a number of factors converged to challenge our cultural denial. Medical advances prolonged life so that for the first time in our culture a major segment of aged remind us of the end of life. Medical advances prolong the lives of many who live among us while they die. New medical techniques of organ transplantation, dialysis, and radical medical and surgical therapies confront medicine and the culture with hard decisions about life and death.[91] Thus, we have a greater degree of decision regarding when we die, although we have no more capacity to influence the inevitability of death than before.

Perhaps even more fundamental is the crisis in meaning. The challenge to traditional values and their religious supports has left many facing an existential dilemma of providing purpose to keep on living. Thus, modern medicine becomes devoted to the preservation of life, while death is viewed as an intrusion into a scientific quest for eternal existence. The maintenance of life per se as the ultimate value is reflected in various science fiction fantasies, such as the movie *2001*, and the interest in cryogenic methods of preserving life. As an example one medical scientist[31] stated: "I, myself, tend to adhere to the concept of death as an accident, and therefore find it difficult to reconcile myself to it for myself or for others . . . people do not forgive themselves easily for having failed to save their own or others' lives."

Cultural and medical awareness of the problems of dying has not come directly to us. Our recent history begins perhaps with the studies of loss and bereavement heralded by Lindemann's[53] paper on grief in 1944 and re-emphasized by Engel's[20] distinction between healthy grief and pathological mourning. A more direct approach marked the popular exposés and analyses of American funeral practices.[7,36,57] Then, beginning in 1959 with the work of Feifel,[23] numerous books took up the topic of death, although usually in a philosophical or theoretical vein.[27,71,73,92] the problem of suicide and its prevention finally brought attention to problems of dying, rather than death itself, culminating in sociological studies on the context of dying.[29,30,97] But only recently have we come to address fully the problems of caring for the dying.[8,40,48,66,77,94,104] Thus, this evolution of research seems to reflect the difficulty in confronting death that professionals share with their culture.

Attitudes about Death

Our behavior throughout life is determined by our culture, and the same is no less true of our behavior in dying. Primitive cultures did not perceive death as a final biological state. The worlds of the gods and various concepts of continuing life provided a sense of continuum between this current life and extensions into other forms of existence. In contrast, classic Western cultures held a fatalistic view of death as the inevitable termination and destruction of existence.

At present we observe four different cultural attitudes toward death: death-denying, death-defying, death-desiring, and death-accepting. American culture has been generally death-denying. It has been suggested that physicians fail to provide assistance to the dying because of their unresolved conflicts about death. Feifel[22] reports that physicians tend to deny death more than other segments of our society. Although the choice of medicine as a career may partly reflect unconscious conflicts over, and denial of, death, the attitudes of physicians may be more determined by the general denial of death in American culture.

Psychiatrists, too, share in the same cultural denial, although they use more abstruse mechanisms. Thus, Wahl[95] observed:

It is interesting also to note that anxiety about death, when it is noted in the psychiatric literature, is usually described solely as a derivative and secondary phenomenon, often as a more easily endurable form of the "castration fear" . . . it is important to consider if these formulations also subserve a defensive need on the part of psychiatrists, themselves.

We can observe the threat of death in many psychotherapeutic situations: how a patient's threat of suicide cows the therapist; how the psychiatrist puts himself in physical danger with a dangerous patient without noting the real danger; the therapist's reluctance to allow a patient to expose his deepest threatening fantasies and psychotic thoughts that intimate the annihilation of personality. How often as a young psychiatrist I despaired of working with prisoners with a life sentence or facing death. How dead these men seemed to us with our bright anticipatory lives.

The denial of the anxiety of death is perhaps best exemplified in the avoidance of death among professionals. Most typically, in a hospital setting physicians and staff are very apprehensive about talking with dying patients—even more apprehensive about a psychiatrist's talking with a dying patient. They will suggest that the dying patient will become nervous, anxious, hurt, upset, injured. Yet Kubler-Ross[46] reports that only 2 per cent of 200 dying patients rejected an opportunity to discuss their dying, and she observed no adverse reactions in her interviews with dying patients. But she also reports that many professional personnel became upset watching interviews of dying patients. In my experience I have found most dying patients not only willing but desirous of discussing their illness and dying. To be sure they discuss these issues in guarded and partial ways at times, but they want to deal with their life at the moment. On the other hand, like Kubler-Ross, I have observed nurses become so anxious over a dying patient that they could not function, physicians cry out in despair during an interview, and psychiatrists angrily denounce the inhumaneness of talking with a patient about his dying. These observations suggest that the fears about the dying patient voiced by the personnel are projections of their own fears about handling the process of dying.[47]

The death-defying attitude is rooted in our traditional Judeo-Christian heritage. St. Paul sounds the keynote: "death is swallowed up in victory. O death, where is thy sting? O grave, where is thy victory?" (I Corinthians 15:54–55). A more poignant note is given by Dylan Thomas—"Death, Be Not Proud" and "Rage, rage against the dying of the light." Or consider those who have fought for causes, ideologies, families, or country, in defiance of the fact that they die in the doing.

The death-desiring attitude is perhaps less common in our culture, reflecting individual rather than cultural dynamics.[26] A desire to die may be a means to resolve life conflicts, or to kill the self for revenge or retaliation. The desire not to live may occur among the severely disabled, debilitated, or unhappy elderly, who seek release and escape from the misery of life. In instances of neurotic or psychotic fantasy, one may seek reunion with loved ones in the magical union of death. For example, in *Othello* and in *Aida*, the lovers will be reunited eternally in death.

In contrast to the above attitudes that sequester death apart from life, there are the death-accepting attitudes, in which death is seen as a part of life and an integral part of existence. Death as the concluding episode of one's life plan is eloquently described by Bertrand Russell:[74]

> An individual's human existence should be like a river—small at first, narrowly contained within its banks, and rushing passionately past boulders and over waterfalls. Gradually, the river grows wider, the banks recede, the waters flow more quietly, and, in the end, without any visible break, they become merged in the sea, and painlessly lose their individual being.

This view of death as the conclusion of a process does contain a certain romantic quality. As Schneidman[80] points out, however, this view of death tends to make accidental or early death a tragedy—the romantic life has been cut short!

A very different death-accepting attitude is seen in various existential philosophies, which posit death as a central issue in the manner with which we go about the process of living. For some existentialists dying does not refer to that one event of physiological end, but rather to the potential threat to one's own nonbeing.[15,16] Thus, Heidegger asserts that we are faced with a basic anxiety about our authentic potentiality for-Being-in-the-world.[28] If we do not face and resolve our existential death anxiety, we are not free to live. In a similar vein Norman O. Brown[10] argues in *Life Against Death* that neurosis arises from the incapacity to die. Once freed from avoiding death we can joyously embrace life.

These broad cultural attitudes toward death and dying have received less attention than the psychological and sociological factors. Some research has focused on the role of philosophical and religious attitudes in different patients' styles of coping with dying. This work has proven inconclusive thus far.[23,27,48,54,77] Patients with strong philosophical or religious convictions do not necessarily cope with their dying according to their intellectual affirmations. Clinically it appears that dying patients cope with the stress of dying in the same way they have previously coped with life stresses. Thus, emotional coping styles assume primacy over intellectual coping styles.

Meaning of Death throughout the Life Cycle

The meaning of death and the experience of dying varies throughout the life cycle. Thus, the clinician must approach the dying patient in terms of the patient's life stage.

For the infant and very young child dying is primarily the physical discomfort of the disease process and the treatment procedures. For the toddler and preschool child there is little concept of a personal death. For this child the primary fear is that of separation from the nurturant parents. Often, too, a child of this age may perceive illness as a punishment and hospitalization as parental rejection.[37,56] For the grade school child dying poses a disruption in his developing interpersonal relations and his growing sense of identity; to the adolescent dying poses a frustration and defeat of the newly competent youth, who is just coming to mastery of his life.[18,55]

For the active young adult dying poses a problem of narcissistic loss—loss of one's healthy body, of one's active, striving self-image, of one's investment in a growing family and life plan. Whereas for the middle-aged adult dying more often poses a concern for one's ongoing life involvement with loved ones and spouse. For example, a 50-year-old woman was quite agitated in the hospital because she feared she would die before she got home and saw that her house was in order, care for her children arranged, and her husband made comfortable about her departure.

For the aged dying faces one with the history, consequences, and meaning of one's life. As Willie Loman says in *Death of a Salesman*, "A man can't go out just the way he came in. He's got to amount to something." I am impressed by the perseverative reminiscence engaged in by residents in retirement homes. They are not yet ready to die because they have not yet come to peace with their life.[11] Erik Erikson[21] summed up this life crisis as that of integrity versus despair:

> It is acceptance of one's own and only life cycle and of the people who have become significant to it as something that had to be and that, by necessity, permitted of no substitutions . . . and that for him all human integrity stands and falls with the one style of integrity of which he partakes.

In addition, the elderly are faced with a large degree of social isolation, interpersonal deprivation, rejection for being deviant in our youthful society (i.e., old), and usually a high degree of "bereavement overload"—they are experiencing a rapid succession of losses of love relationships.

Death versus Dying

Death is not a problem of life. Death is not lived through. Although there are case reports of patients who have returned to life after experiencing a sense of death,[41,54,90] such events do not address the major issue of the dying process.

Death is not amenable to treatment. But the process of dying is very much a part of the person's life. When death is seen as an inevitable event coming soon, we may feel a sense of defeat and therefore ignore the period of living until the death event. This is the period of living-dying. The human dilemma of this living-dying process was brought home to me personally by a letter from a lady unknown to me:

Dear Dr. Pattison: Quite by accident I read your treatise on dying. Because I am so grateful for your guidance I am writing not only to thank you but to suggest that the article be made available to relatives who care for patients. . . . My husband has been treated for chronic glomeruli nephritis for nine years. For the past five years, he has had biweekly dialysis which equates to a living-dying stage of long duration. In these times, when there is no doctor-patient relationship in this type of indirect care, the entire burden of sharing the responsibility of death falls to the member of the family. . . . Your listing of the fears was so apparent when I read your paper, yet when my husband experienced them I was unprepared to see them or even acknowledge them. When a patient is accepted on a kidney program, he knows he is dying. Would it not be a kindness to the person caring for him to know his fears and how to help?

As said so clearly by his wife, the period of living-dying is most important to the patient, his family, and to professionals, and this dying process may last hours, days, weeks, months, even years.

The knowledge of his forthcoming death often produces a crisis in the life of the patient. During the period of acute crisis there is an increasing anxiety that will reach a peak of tolerance. No one can continue to function long at peak anxiety, and, therefore, the patient will call into play mechanisms to cope with the crisis of death knowledge. If the person successfully deals with this knowledge crisis, he can proceed in an integrated pattern of dying. But if the person does not deal with this crisis, his dying can become a disintegrative process with the use of many dysfunctional mechanisms. Thus, we can plot out what Straus and Glaser[86] call the "trajectory of dying" (Figure 34–1).

The knowledge of death as a crisis event can be analyzed in terms of five aspects of crisis:[62]

1. This stressful event poses a problem that by definition is insolvable in the immediate future. In this sense dying is the most stressful crisis because it is a crisis to which we bow, not solve.
2. The problem taxes one's psychological resources since it is beyond one's traditional problem-solving methods. One is faced with a new experience with no prior experience to fall back on, for although one lives amidst death, that is far different from one's own death.
3. The situation is perceived as a threat or danger to the life goals of the person. Dying interrupts a person in the midst of life; and even in old age it abruptly confronts one with the goals that one set in life.
4. The crisis period is characterized by a tension that mounts to a peak, then falls. As one faces the crisis of death knowledge, there is mobilization of either integrative or disintegrative coping mechanisms. There is a degree of diminishing anxiety as one approaches death. But the peak of acute anxiety usually occurs considerably before death.
5. The crisis situations awakens unresolved key problems from both the near and distant past. Problems of dependency, passivity, narcissim, identity, and more, may be reactivated during the dying process. Hence one is faced not only with the immediate dying process but also with the unresolved feelings from one's own lifetime and its inevitable conflicts.

This crisis of knowledge of one's own death may be experienced as an overwhelming, insuperable feeling of inadequacy—a potential dissolution of self. There is bewilderment, confusion, indefinable anxiety, and unspecified fear. Death confronts the person with a crisis to which there is seemingly no answer, and the ensuing anxiety makes it difficult to distinguish and cope with the various aspects of the dying process. But here lies the opportunity to intervene, for although we cannot deal with the ultimate problem of death, we can help the person to deal with the various parts of the process of dying.

Variations in Temporal Proximity of Aspects of Death

1. Ideal Proximity. (note termination of hope)

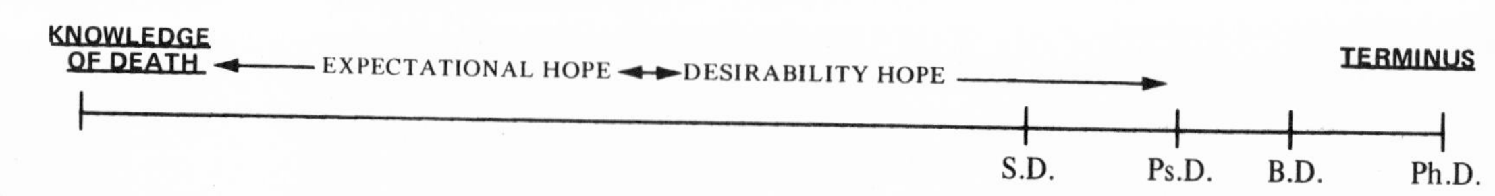

2. Social Rejection of Patient.

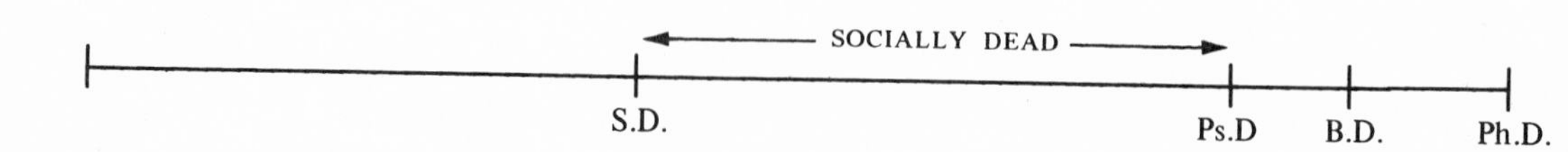

3. Social and Patient Rejection of Death.

4. Patient Rejection of Life.

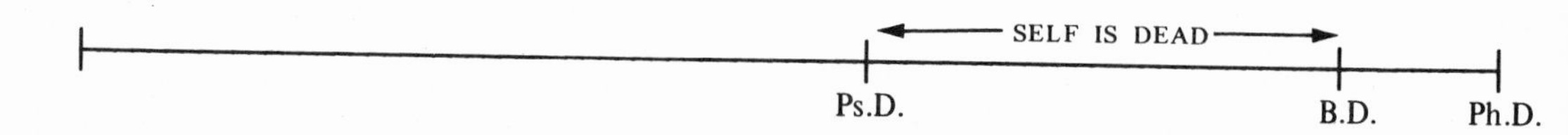

5. Social Rejection of Death with Artificial Maintenance.

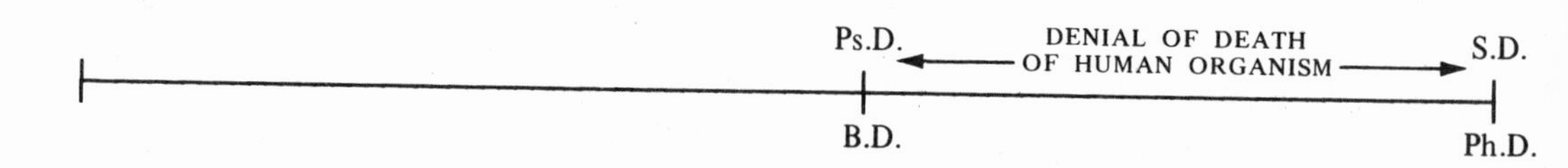

B.D. = Biological Death

Ph. D. = Physiological Death

Ps. D. = Psychic Death

S. D. = Sociological Death

Figure 34–1

By focusing upon the various aspects of the dying process, we can assist the dying person to resolve the crisis in a fashion that enhances his self-esteem, dignity, and integrity. The dying person can then take pride in having faced his crisis of dying with hope and courage.

Parts of the Dying Process

Now let us consider each of the fears that face the person in the experience of dying.

Fear of the Unknown

Freud[25] suggested that the unconscious does not recognize its own death but regards itself as immortal. "It is indeed impossible to imagine our own death; and whenever we attempt to do so, we can perceive that we are, in fact, still present as spectators."

In this view we fear the unknownness of death. On the other hand, the work of Lifton[52] and others suggest that death anxiety harkens back to the most primordial feelings of helplessness and abandonment.[6] The fear of the unknown is not the unknown death but the unknown of annihilation of self, of being, of identity.[35,38] Leveton[50] describes this sense of "ego chill" as "a shudder which comes from the sudden awareness that our non-existence is entirely possible."

This unknown cannot be processed within the self. Lifton[52] suggests that perhaps all of us must to some extent defend against this fear by a process of psychic closing off. Although this may not be totally effective, it may enable the person to focus on other aspects of his dying with which he can cope. Here a degree of denial and isolation may be quite appropriate to avoid unnecessary preoccupation. Janice Norton[60] describes this process clearly in a dying woman:

> She told me that her only remaining fear was that dying was strange and unknown to her, that she had never done it before. . . . She no longer worried about what was to happen to her after death. . . . She felt that she might be unnecessarily concerned with the actual process of death itself.

Fear of Loneliness

When one is sick there is a sense of isolation from oneself and from others, reinforced by the fact that others tend to avoid a sick person. This mutual withdrawal is even more evident when a person is dying. The isolation attendant to dying is not only a psychological phenomenon but is also a reflection of our urban culture and medical technology that has dehumanized and mechanized the dying process. No longer does our culture afford us the luxury of dying amidst our family and belongings; over 60 per cent of deaths now occur in the impersonal, isolated hospital room. One of the functions we have defined for hospitals in our society is that of dying places. Yet major sociological studies of hospital nursing and medical practices demonstrate that the hospital staff use a variety of denial mechanisms to preclude awareness of dying among themselves, the family, and the patient.[29,30,87] Therefore, most people experience their dying within a context of "closed awareness." As a consequence everyone engages in elaborate maneuvers to maintain the closed atmosphere, thereby preventing the staff, family, and patient from openly facing the dying process and responding in a humane and helpful manner.[29]

This closed context effectively isolates the patient from communication. In addition, family and staff begin to avoid the patient, visit less frequently, answer call lights more slowly, and minimize the duration of contact with the dying patient. The impact of all this isolation is a sense of human deprivation. As shown in many experiments the human deprived of contact with other humans quickly disintegrates and loses his ego integrity. For the dying person his isolation and human deprivation sets the stage for a type of anaclitic depression. This is not the depression of the loss of a loved one, but the loss of necessary human nurturance. Without this support one rapidly falls into the confusional syndrome of human deprivation that we term loneliness. It seems that this fear of loneliness is paramount when the person first faces the prospect of death and anticipates isolated rejection during dying.

Fear of Loss of Family and Friends

The process of dying confronts the person with the reality of losing one's family and friends through one's own death just as much as if they were dying. Hence there is a real object loss to be mourned and worked through. Rather than denying this real separation and preventing the grief work, it is possible for both the patient and his family to engage in "anticipatory grief work."[72] The completion of such grief work may allow the patient and his family to work out emotionally their mutual separation and part in peace. This is akin to the Eskimo custom of having a ritual feast of separation before the old person steps onto the ice floe and waves goodbye as he floats off to die in the sea; similarly, in the Auca tribe of South America, after a farewell ceremony, the old person leaves his clan to climb into a hammock to lie alone until he dies. Failure to recognize this real object loss may block the normal grief process and make it difficult for the dying person to distinguish between his own problem of death and the healthy process of grief that can be accomplished before death.

Fear of Loss of the Body

Since our bodies are so much a part of our self-image, when illness distorts our body, there is not only loss of function but a psychological sense of loss of self. This narcissistic blow to the integrity of the self may result in shame, feelings of disgrace and inadequacy, and loss of self-esteem. As before, the patient can be helped to mourn this loss actively and retain a sense of integrity in the face of separation from parts of oneself. Since we humans do not tolerate ambiguity well, it is more difficult to tolerate ambiguous distortion of body function. Patients will tolerate externally disfiguring disease better than internal disease because one can see clearly the loss of structure and function. Although external disfigurement may seem ugly, it poses less threat than the unknown and unspecified processes that the person cannot see and keep track of. Hence the failing heart or hidden cancer will provoke more anxiety than the external symptoms of disease.

In addition to the narcissistic loss, the patient may perceive his self as disfigured and unlovely, hence see himself as unlovable. Then the dying patient may try to hide his unlovely self from his loved ones, for fear that his family will likewise despise his ugly self, reject him, and leave him alone.

Fear of Loss of Self-Control

As debilitating disease progresses one is less capable of self-control. This is especially true when one's mental capacities are also affected. As shown in studies of brain-damaged persons, the actual functional deficit may be less problematic than the reaction to the perceived loss of control.

This problem is particularly acute in our society, which has placed strong emphasis on self-control, self-determination, and rationality. As a result most people in our culture become anxious and feel threatened by experiential states that pose loss of control or consciousness. This is reflected in our social ambivalence over the use of psychedelic drugs and alcohol, which produce states of diminished control and consciousness. In contrast to Eastern mystical experiential states, it is rare that Americans experience any sort of self-acceptable loss of control. Thus, when they come to the experience of dying, the loss of control of the body and the diminished sense of consciousness may create anxiety and fear and poses a threat to the ego. One is placed in a position of dependency and inadequacy so that in a sense the ego is no longer master of its own fate nor captain of the self.

Therefore, it is important to encourage and allow the dying person to retain whatever authority he can, sustain him in retaining control of daily tasks and decisions, avoid shaming for failure of control, and help the person to find reward and integrity in the exercise of self-determination available to him.

Fear of Pain

Although there are both cultural and individual differences in the response to pain, there is a more important distinction between pain and suffering. A certain level of awareness of self and one's body is a necessary precondition to suffering. This self-awareness may either enhance or diminish the sense of suffering. One may deal with suffering by providing temporary or partial oblivion to pain and hence diminish suffering. This is the typical medical response to pain in the dying patient. But this oblivion approaches and may be humanly indistinguishable from death. The other alternative is to diminish suffering through awareness and understanding. Bakan[3] suggests this is a more humanistic approach to suffering in that it retains human dignity and integrity, allowing the patient to understand his own pain and resolve his conflicts over it. This proposition is borne out clinically: pain relief is not merely a function of analgesic medication but is most influenced by the patient's own attitude toward his pain.[70,82] The fear of pain is not just a physical fear, but a fear of suffering, a fear of the unpleasant, of the unmanageable, of the unasked for. Senseless pain is perhaps intolerable. On the other hand, pain may be accepted and dealt with if that pain does not mean punishment or human suffering. People will not suffer long, but they will endure pain.

Fear of Loss of Identity

The loss of human contact, the loss of family and friends, the loss of body structure and function, the loss of self-control and total consciousness, all threaten the sense of one's identity. Human contacts affirm who we are, family contacts affirm who we have been, and contact with our own body and mind affirm our own being self.

We can see that the dying process faces the person with many threats to self-identity. How does one maintain identity and integrity in the face of these forces of dissolution? Bowers et al.[5] conclude that: "When life cannot be restored, then one can accept the fact with a meaning that gives dignity to his life, and purpose even to the process that is encroaching on his own vitality." Willie Loman, the salesman, speaks of his own death: "A man must not be allowed to fall into his grave as an old dog." It is not that we die, but rather how we die. The tasks are to retain self-esteem and respect for the self until death, to retain the dignity and integrity of the self through the process of living we call dying. If the person cannot sustain his ongoing sense of self, then he may fall prey to despair, the loss of self-esteem, the failure to respect oneself for what one has been.

One mechanism for the maintenance of integrity and identity comes from continuing respect and affirmation from the family and professional staff. This reaffirmation can continue to reflect to the dying person who he is.

Another aspect of identity comes from the sense of continuity, reinforcing one's identity through the maintenance and continuity of one's life via one's family and friends. One can see identity in one's children, life's work, and in the bequeathing of one's possessions to others. This is acted out in the leaving of a will and, in a more general sense, by leaving parts of one's body, such as in eye banks, bone banks, and the like. This personal sense of continuity was illustrated by a middle-aged man who was dying of lung cancer. I had spent much time with him during his dying and talked about both his life and my budding career. At one point the surgeons wanted to perform a biopsy. He refused unless I gave permission. I explained to him that the biopsy would not change his disease but might aid in my understanding of his disease. Then he was happy to comply, feeling he could give me part of himself that I would carry with me in my professional life. He had given me part of himself to remain with me after his death.

Still a third mechanism of identity maintenance that occurs is the desire for reunion with loved ones who have died before or who will die and join one. These reunion fantasies include the sense of return to the primordial mother figure as well as reunion with specific loved ones.[9] There will be reunion with one's parentage and one's progeny.[33] Hence one can place oneself at one point in the continuity of ongoing human relationships, of which man's death is merely a point in a more universal sense of existence.

Fear of Regression

Finally there is a fear of those internal instincts within oneself that pull one into retreat from the outer world of reality, into a primordial sense of being where there is no sense of time or space, no boundaries of self and others. Throughout the ego fights against this internal regression into selflessness. Freud called it Thanatos—the death instinct. Despite the metapsychological ambiguities of this concept, we can appreciate the phenomenological sense of this experience in the everyday universal experience of awakening in the morning. As the alarm rings we drowsily douse the noise, turn over, feel the immense weight of our sleep pulling us back into slumber. We luxuriate in the indefinite sense of our body boundaries, the relaxation of our awareness, the freedom from the demands and constrictions of the real world awaiting our awakening. And with exquisite pleasure we allow ourselves to float back off into a timeless, spaceless, selfless state of nonbeing.[59,81] Certain religious mystical experiences, psychedelic experiences, and body awareness exercises produce similar altered states of consciousness.[89] For the most part, however, most people in our culture encounter great difficulty in allowing themselves to enter these regressive states, much less experience such states as enjoyable rather than anxiety- and fear-producing.

For the dying person, especially as he approaches his terminal state with obtundation of body and mental awareness, such a sense of regression may be very frightening. He may fight against the regression to hold onto the concrete, hard reality-boundedness of himself. This may produce the so-called death agonies —the struggle against regression of the self.

In the attempt to attenuate this fear of regression, some ten years ago investigators began the experimental use of LSD and associated psychedelic agents.[17,61] Their rationale was to provide the dying patient with an ego-syntonic drug experience that was in itself pleasurable to the dying patient and might provide an anticipatory experience of acceptable ego regression. In the main these expectations were not consistently obtained, and there is little current interest in the procedures.

Nevertheless, the psychedelic experiments did serve to call attention to the problem of terminal regression and its management. Clinical concern here shifts from helping the patient face reality to helping the patient turn away from reality. With support and encouragement the patient can approach an acceptance and surrender to the process of renunciation of life and return to a sense of union with the world out of which he has sprung. Then psychic death is acceptable, desirable, and at hand.

Sequence of Dying

Above is sketched out a clinical outline of major events in the dying process of the patient. First, the patient is faced with the seemingly impossible crisis of knowledge of death, which threatens to overwhelm the self. Mankind seems to have always recognized that no one has the capability to face this crisis alone, for we develop cultural customs whereby we actually and literally help people to die. Given our interest, support, and guidance, the dying person can face death as an unknown with the realization that he cannot know, and instead he can turn to consider the processes of dying that he can know and deal with. If not deprived of human contact, he can learn to endure the inevitable degrees of separation without loneliness. He can face the loss of relatives, friends, and activities if he can actively mourn their loss where his grief is defined and accepted. He can tolerate the loss of body structure if others accept that loss with him. He can tolerate the loss of self-control if it is not perceived by himself and others as a shameful experience and if he can exercise control where feasible. He can tolerate pain if he can see the source of pain and define the nature of his suffering. He can retain dignity and self-respect in the face of the termination of his life cycle if he can place his life in perspective within his own personal history, family and human tradition. If this is accomplished, he can move toward an acceptable regression where the self gradually returns to a state of nonself.

In this sequence of dying I have focused on the major psychological conflicts that face the dying patient. Others have suggested that there are specific psychological phases through which the patient passes.[94] For example, Kubler-Ross[46] suggests five stages: denial and isolation, anger, bargaining, depression, and acceptance. However, in my own experience and review of the literature I am less impressed by the uniformity of dying experiences than by the divergency. Further I have observed a tendency among clinicians to ignore the actual conflicts and needs of the patient by viewing dying as a stereotyped process. Also I have observed that clinicians may push patients to handle dying the way the clinician might handle his own dying, or the clinician may push the patient to move through some preconceived set of stages of dying. Instead, clinical judgment requires an assessment of the conflicts and needs of the patient, which will vary with his age, values, coping mechanisms, family relationships, behavior of hospital staff, and other factors.

There is, however, a pattern of change over time in the trajectory between knowledge of death and death itself. Glaser and Straus[30] outline four types of "death expectations": (1) certain death at a known time; (2) certain death at an unknown time; (3) uncertain death, but a known time when the question will be resolved; (4) uncertain death, and an unknown time when the question will be resolved. Each of these expectational sets has a different effect upon the patient and his interaction with others. As the course of the illness becomes clear, that knowledge does force the patient and the staff to reckon with the impending death.

This expectational set plays an important role in the sense of hope. Stotland[85] makes a helpful distinction between two types of hope: hope as expectation and hope as desire. At the outset each dying patient holds hope for himself in terms of the expectation that he may not die. As that expectational set is dissolved, the patient may deny the inexorable process by clinging to expectational hope.

On the other hand, the patient may be assisted to change from expectational hope to desirable hope; that is, it would be good to live but it is not to be expected. This latter hope can play a role in the maintenance of integrity. However, as indicated in our discussion of regression, even this desirable hope might optimally give way to acceptable regression. For this reason Cappon[13] suggests that hope (in the desirability sense) should not cease until shortly before psychic death.

This leads us to a brief outline of four aspects of death itself. First is sociological death, that is, the withdrawal and separation between the patient and significant others. This sociological death often occurs days and weeks before terminus if the patient is left alone to die, or it may precede terminus as a gradual and seemingly imperceptible diminution. Second is psychic death, which occurs when the person accepts his death and regresses into himself. Such psychic death may proceed coincident with the diminution of actual psychic function. On the other hand, psychic death can occur earlier than terminus as in cases of voodoo death or in patients who predict their own death and refuse to continue living.[4,12,97] Third is biological death in which the organism as a human entity no longer exists, such as in irreversible coma, where the heart and lungs may continue to function with artificial support. And fourth is physiological death, which occurs when viable organ system function ceases. As illustrated in Figure 34–2 these four aspects of death are not necessarily closely related in

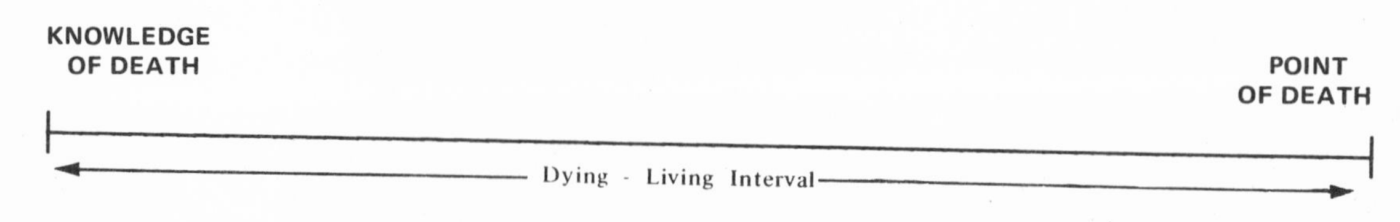

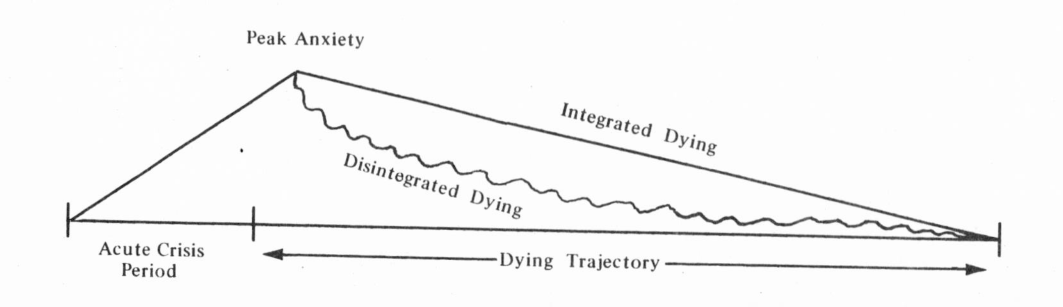

Figure 34–2

time, thus often producing many problems in patient care and dying decision making. Optimally our task in caring for the dying is to coordinate psychological and medical care so that all four aspects of death converge in close proximity.

Help in the Experience of Dying

If the dying person is provided the opportunity and assistance, it is possible for the experience of dying to be appropriately integrated into the process of living. Weisman and Hackett[99] suggest that we consider how to help a patient toward an appropriate death. They offer four criteria for an appropriate death that closely approximate the various conflicts discussed before:

1. Conflict is reduced.
2. Compatibility with the ego ideal is obtained.
3. Continuity of important relationships is preserved or restored.
4. Consummation of basic instincts and wishes of fantasy reemerge and are fulfilled.

The concept of an appropriate death is based on the clinical goal of adaptation rather than on an ideal goal. That is, we seek to help the patient integrate his dying into his life according to the style, meaning, and sequence of his life. Thus, the criteria of an appropriate death will be fulfilled in different ways for each patient. Each man's appropriate death will be different, but his dying will be appropriate to him.

Helping the Dying Patient

To assist the patient in his dying, certain patterns of assistance can be summarized in accordance with the conflicts and fears discussed above. We can share the responsibility for the crisis of knowledge of death, so that the patient can deal with the first impact of anxiety and bewilderment. We can clarify and identify with the patient the difference between death and dying and specify the processes of dying with which the patient can deal. These processes of dying are the realities of day-to-day life. We can make continued human contact available and rewarding. We can assist in working through the grief over the realistic losses of family and body image while retaining communication and meaningful relationships with those who will be lost. We can assume necessary body and ego function in the face of diminished self-control and pain, without incurring shame or depreciation for the person, thus helping him maintain integrity and self-respect. We can encourage the person to work out an acceptance of his life situation with dignity so that gradual regression can occur without conflict or guilt, while expectational hope can be transformed into desirable hope and relinquishment of self can be allowed.

Helping the Staff Who Care for the Dying

In most instances the psychiatrist does not assume major responsibility for the dying patient. Although there are occasions when there is need for brief psychotherapy or crisis intervention by mental health professionals, the more general task is to provide adequate consultation to medical and nursing staff, as well as educational and organizational assistance to enable the hospital as a social unit to deal more adequately with its function as a dying place.[39,102]

Because of the major organizational problems in dealing with dying, Glaser and Straus[30] offer four major recommendations for institutional change, each of which should be of concern to the psychiatrist:

1. Training for giving terminal care should be greatly amplified and deepened in schools of medicine and nursing.
2. Explicit planning and review should be given to the psychological, social, and organizational aspects of terminal care.
3. There should be explicit social psychological planning for phases of the dying trajectory that occur before and after residence in the hospital.
4. Medical and nursing personnel should encourage public discussion of issues that transcend professional responsibilities.

To provide such assistance to fellow professionals, it is incumbent that the psychiatrist first deal with his own conflicts and countertransferences. Numerous case reports indicate that the dying patient often abruptly confronts the psychiatrist, who becomes immobilized until he can resolve his own internal conflicts.[43,44,103] In his parable, *The League of Death*, Allen Wheelis[101] shows us we cannot help until we no longer fear and flee from death.

In addition, addressing the problems of the dying requires a shift in the usual therapeutic orientation. Typically we focus on the movement toward a fuller engagement of life in psychotherapy, but here we move into the unaccustomed orientation of disengagement.[75] The more traditional techniques of ongoing psychotherapy are less appropriate, and we must call upon our recently developed skills of brief psychotherapy.[51] In addition, work with the dying often calls for skills in family therapy, for the clinician must address the interpersonal aspects of the dying process. Finally the clinician should be able to utilize group therapy skills. In certain groups of patients with lingering fatal disease, the use of short-term group sessions may provide an effective means of modifying patient attitudes and behavior[1,65,79]

To enhance staff management of the dying requires the active involvement of the psychiatrist, therefore, not only with the dying but with those who care for the dying.[67] Weisman and Kastenbaum[100] have provided schematic procedures for what they call the "psychological autopsy." They propose that hospital staff hold an autopsy evaluation of the dying process, just as we currently conduct pathological autopsies. Through such a dispassionate scientific-medical method we may be able to assess distortions that occur in the dying process and consequently work out staff consensus for more effective management. Such procedures, or similar joint staff work on the problems of dying, may offer the vehicle for sound organizational improvement.

Helping the Family of the Dying

Since the majority of deaths occur in the hospital, the consulting psychiatrist and the staff may neglect the needs and relevance of the family because they are not part of the organizational structure. It is noteworthy that in the extensive literature on dying there is very little said about care for the family.[86] Yet they merit concern on two counts. First, they are integrally related to the manner in which the dying person will experience his dying. Second, they are themselves vulnerable and affected by the dying process, and they, too, may need assistance to live through the experience of dying in a healthy manner.[45] Just as with the patient, the family must face the anxiety and apprehension of the knowledge of the death crisis. They, too, must face the separation and loss, so that they can reciprocally work through the anticipatory grief of loss. They must make specific decisions in regard to children, home, finances, and belongings. And they must work out the meaning of their lives reconstituted without the dying person. In this arena we have generally neglected the contributions that the funeral director and the minister can make.[76] These people often have more intimate contact with the family than the medical staff does. Consultation, collaboration, and educational work with these community care agents should become a part of the overall pattern of care for the dying.

Preventive Aspects of Dying

At the outset I suggested that the problem of death is not solely an issue of management of the dying process, but involves the whole texture of human existence. A half century ago we learned that failure to reckon openly with sexuality gave rise to neurotic distortions of existence. In similar vein I have suggested that failure to reckon openly with death in our time may give rise to neurotic defenses against awareness of death in our lives.[14,69,84] If we can speak of sexual neuroses, it may be appropriate to speak of death neuroses.

Although there is little general agreement yet regarding the more general proposition of death neuroses, there is abundant clinical and experimental data on the impact of death and bereavement on personality adjustment.[88] For example, Moriarty[58] has recently summarized the effects of death in the family on subsequent personality development. He takes the position that experience with death in childhood poses serious trauma to the child, and, therefore, the child should be protected from exposure to death. This position, however, confuses psychic stress and psychic trauma. There is no doubt that death does impose a stress on the child, but that does not mean that such stress is necessarily traumatic. The psychic trauma may be an artifact of adult attitudes and adult reactions to death. Since we have not openly dealt with death as adults, it may be difficult to teach children how to deal appropriately with death.[2] In fact, Eissler[19] some fifteen years ago suggested that we should practice "orthonasia," that is, teach children about death as a part of life, so that children will learn how to incorporate healthy attitudes toward death into their coping repertoire and be adequately prepared to deal with death events in their own life cycle. Support for such preventive death education has been provided in recent publications.[34,42]

A related aspect of prevention bears on the grief and mourning process. The relationship between unresolved grief and neurotic depression syndromes has been amply discussed. This data have come primarily from patients with psychiatric symptomatology. Extensions of this process have been observed in recent epidemiological studies of bereavement. Such work demonstrates significant evidence of "bereavement morbidity." That is, there is an increased physical and emotional morbidity rate among the family and friends of the deceased.[63,68] Thus, there is a penumbra of psychopathological reactions that surround a person's death. Prevention of such "death contagion" presents us with a major challenge in community mental health. One example of preventive intervention along these lines is the widow-to-widow program, in which the widow and her family are contacted after the funeral by the local mental health team, which provides assistance in their grief work and guidance in reintegration into normal life patterns.[83] Other related preventive programs might include Parents without Partners programs, which provide a natural community group for single parents, and Big Brothers and Big Sisters programs, which provide parental surrogates for children who have lost a parent.[40] Such programs can provide support during the normal grief process, provide new human relationships that can partly replace the lost person, and provide guidance toward health integration.[47]

Summary

In contemporary culture death is viewed with a perverse and morbid curiosity. In the face of shifting pluralistic values, diminution of previously integrating religious values, and medical advances posing new types of death decision making, there are few reliable guidelines available to the layman and the professional in the management of the dying process. We are in the midst of a reevaluation of death in our culture. Death per se cannot be changed, but we can change the patterns of dying. Help in the dying process must focus on three areas: help to the dying patient, help to the hospital staff, and help to the person's family and friends. Since dying is a universal human experience, it merits the serious attention of the psychiatric profession. As with all human experience the clinician must first confront death within himself, and then he can begin to practice the high therapeutic art of helping people to die.

Bibliography

1. ADSETT, C. A., and BRUHN, J. G., "Short-term Group Psychotherapy of Post-Myocardial Patients and Their Wives," *Canad. Med. A. J.*, 99:577–584, 1968.
2. ANTHONY, S., *The Child's Discovery of Death: A Study in Child Psychology*, Harcourt Brace, New York, 1940.
3. BAKAN, D., *Disease, Pain, and Sacrifice: Toward a Psychology of Suffering*, University of Chicago Press, Chicago, 1968.
4. BARBER, T. X., "Death by Suggestion: A Critical Note," *Psychosom. Med.*, 23: 153–155, 1961.
5. BOWERS, M., *et al.*, *Counseling the Dying*, Thomas Nelson, New York, 1964.
6. BOWLBY, J., "Separation Anxiety: A Critical Review of the Literature," *J. Child Psychol., Psychiat. & Allied Disciplines*, 1: 251–275, 1960.
7. BOWMAN, L., *The American Funeral: A Study in Guilt, Extravagance and Sublimity*, Public Affairs Press, Washington, D.C., 1959.
8. BRIM, O. G., JR., *et al.*, *The Dying Patient*, Russell Sage Foundation, New York, 1970.
9. BRODSKY, B., "Liebestod Fantasies in a Patient Faced with a Fatal Illness," *Internat. J. Psychoanal.*, 40:13–16, 1959.
10. BROWN, N. O., *Life against Death*, Wesleyan University Press, Middletown, Conn., 1959.
11. BUTLER, R. N., "The Life Review: An Interpretation of Reminiscence in the Aged," *Psychiatry*, 26:65–76, 1963.
12. CANNON, W. B., "Voodoo Death," *Am. Anthropol.*, 44:169–181, 1942.
13. CAPPON, D., "The Dying," *Psychiat. Quart.*, 33:466–489, 1959.
14. CHADWICK, M., "Notes upon the Fear of Death," *Internat. J. Psychoanal.*, 10:321–334, 1929.
15. CHORON, J., *Death and Western Thought*, Collier, New York, 1963.
16. ———, *Modern Man and Mortality*, Macmillan, New York, 1964.
17. COHEN, S., "LSD and the Anguish of Dying," *Harper's Magazine*, 231:69–78, 1965.
18. EASSON, W. M., *The Dying Child: The Management of the Child or Adolescent Who Is Dying*, Charles C Thomas, Springfield, Ill., 1970.
19. EISSLER, K. R., *The Psychiatrist and the Dying Patient*, International Universities Press, New York, 1955.
20. ENGEL, G., "Is Grief a Disease?" *Psychosom. Med.*, 23:18–22, 1961.
21. ERIKSON, E. H., *Identity and the Life Cycle*, Psychological Monograhs, No. 1, International Universities Press, New York, 1959.

22. FEIFEL, H., "The Function of Attitudes towards Death," in Group for the Advancement of Psychiatry, *Death and Dying: Attitudes of Patient and Doctor*, New York, 1965.
23. ———, *The Meaning of Death*, McGraw-Hill, New York, 1959.
24. ———, "The Taboo on Death," *Am. Behav. Sci.*, *6*:66–67, 1963.
25. FREUD, S., "Thoughts for the Times on War and Death," in *Collected Papers*, Volume 4, Hogarth, London, 1915.
26. FRIEDLANDER, K., "On the Longing to Die," *Internat. J. Psychoanal.*, *21*:416–426, 1940.
27. FULTON, R. (Ed.), *Death and Identity*, John Wiley, New York, 1965.
28. GATCH, M. MCG., *Death: Meaning and Mortality in Christian Thought and Contemporary Culture*, Seabury Press, New York, 1969.
29. GLASER, B. G., and STRAUS, A. L., *Awareness of Dying*, Aldine, Chicago, 1966.
30. ———, and ———, *Time for Dying*, Aldine, Chicago, 1968.
31. GOLDFARB, A. I., "Discussion," in Group for the Advancement of Psychiatry, *Death and Dying: Attitudes of Patient and Doctor*, New York, 1965.
32. GORER, G., *Death, Grief and Mourning in Contemporary Britain*, Cresset, London, 1965.
33. GREENBERG, I. M., "An Exploratory Study of Reunion Fantasies," *J. Hillside Hospital*, *13*:49–59, 1954.
34. GROLLMAN, E. A. (Ed.), *Explaining Death to Children*, Beacon Press, Boston, 1967.
35. GROTJAHN, M., "Ego Identity and the Fear of Death and Dying," *J. Hillside Hospital*, *9*:147–155, 1960.
36. HABERSTEIN, R., and LAMERS, W. E., *The History of American Funeral Directing*, Bulfin, Milwaukee, 1962.
37. HAMOVITCH, M. B., *The Parent and the Fatally Ill Child*, Delmar, Los Angeles, 1964.
38. HEILBRUNN, G., "The Basic Fear," *J. Am. Psychoanal. A.*, *3*:447–466, 1955.
39. HICKS, W., and DANIELS, R. S., "The Dying Patient, His Physician, and the Psychiatric Consultant," *Psychosom.*, *9*:47–52, 1968.
40. HINTON, J. M., *Dying*, Penguin Books, Baltimore, 1967.
41. HUNTER, R. C. A., "On the Experience of Nearly Dying," *Am. J. Psychiat.*, *124*:122–126, 1967.
42. JACKSON, E. N., *Telling a Child about Death*, Meredith, Des Moines, 1966.
43. JOSEPH, F., "Transference and Countertransference in the Care of a Dying Patient," *Psychoanal.*, *49*:21–24, 1962.
44. KIRTLEY, D. D., and SACKS, J. M., "Reactions of a Psychotherapy Group to Ambiguous Circumstances Surrounding the Death of a Group Member," *J. Consult. & Clin. Psychol.*, *33*:195–199, 1969.
45. KLIMAN, G., *Psychological Emergencies of Childhood*, Grune & Stratton, New York, 1968.
46. KUBLER-ROSS, E., *On Death and Dying*, Macmillan, New York, 1970.
47. KUTSCHER, A. (Ed.), *But Not to Lose: A Book of Comfort for Those Bereaved*, Frederick Fell, New York, 1969.
48. ——— (Ed.), *Death and Bereavement*, Charles C Thomas, Springfield, Ill., 1969.
49. LANGER, M., *Learning to Live as a Widow*, Messner, New York, 1957.
50. LEVETON, A., "Time, Death, and the Ego-Chill," *J. Existentialism*, *6*:69–80, 1965.
51. LEWIN, K. K., *Brief Encounters: Brief Psychotherapy*, Warren H. Green, St. Louis, 1970.
52. LIFTON, R. J., "On Death and Death Symbolism: The Hiroshima Disaster," *Psychiatry*, *27*:191–210, 1964.
53. LINDEMANN, E., "Symptomatology and the Management of Acute Grief," *Am. J. Psychiat.*, *101*:141–148, 1944.
54. LUCAS, R. A., "Social Implications of the Immediacy of Death," *Can. Rev. Soc. and Anthropol.*, *5*:1–16, 1968.
55. MCCONVILLE, B. J., *et al.*, "Mourning Process in Children of Varying Ages," *Can. Psychiat. A. J.*, *15*:253–255, 1970.
56. MITCHELL, M. E., *The Child's Attitude to Death*, Schocken Books, New York, 1967.
57. MITFORD, J., *The American Way of Death*, Simon & Schuster, New York, 1963.
58. MORIARTY, D. M. (Ed.), *The Loss of Loved Ones: The Effects of a Death in the Family on Personal Development*, Charles C Thomas, Springfield, Ill., 1967.
59. NEEDLEMAN, J., "Imagining Absence, Non-Existence, and Death: A Sketch," *Review of Existential Psychology and Psychiatry*, *6*:230–236, 1966.

60. NORTON, J., "Treatment of a Dying Patient," in *The Psychoanalytic Study of the Child*, Vol. 18, pp. 541–560, International Universities Press, New York, 1963.
61. PAHNKE, W. N., and RICHARDS, W. H., "Implications of LSD and Experimental Mysticism," *J. Religion & Health*, 5:175–208, 1966.
62. PARAD, H. J. (Ed.), *Crisis Intervention: Selected Readings*, Family Service Association of America, New York, 1965.
63. PARKES, C. M., "Effects of Bereavement on Physical and Mental Health—A Study of the Medical Records of Widows," *Brit. Med. J.*, 2:274–279, 1964.
64. PARSONS, T., and LIDZ, V., "Death in American Society," in Sheidman, E. S. (Ed.), *Essays in Self-Destruction*, Science House, New York, 1967.
65. PATTISON, E. M., *et al.*, "Response to Group Treatment in Patients with Severe Chronic Lung Disease," *Internat. J. Group Psychother.*, 21:214–225, 1971.
66. PEARSON, L. (Ed.), *Death and Dying: Current Issues in the Treatment of the Dying Person*, Case Western Reserve University Press, Cleveland, 1969.
67. QUINT, J. C., *Nurse and the Dying Patient*, Macmillan, New York, 1967.
68. REES, W. D., and LUTKINS, S. G., "Mortality of Bereavement," *Brit. Med. J.*, 4: 13–20, 1967.
69. RHEINGOLD, J., *The Mother, Anxiety, and Death: The Catastrophic Death Complex*, Little Brown, Boston, 1967.
70. RIOCH, D., *et al.*, The Psychopathology of Death," in Simon, A. (Ed.), *The Physiology of Emotions*, Charles C Thomas, Springfield, Ill., 1961.
71. ROCHLIN, G., *Griefs and Discontents*, Little Brown, Boston, 1965.
72. ROSNER, A. A., "Mourning before the Fact," *J. Am. Psychoanal. A.*, 10:564–570, 1962.
73. RUITENBEEK, H. A. (Ed.), *Death: Interpretations*, Dell, New York, 1969.
74. RUSSELL, B., *Portraits from Memory*, Simon & Schuster, New York, 1956.
75. SAUL, L. J., "Reactions of a Man to Natural Death," *Psychoanal. Quart.*, 28:383–386, 1959.
76. SCHERZER, C. J., *Ministering to the Dying*, Prentice-Hall, Englewood Cliffs, N.J., 1963.
77. SCHNEIDMAN, E. S., "On the Deromanticization of Death," *Am. J. Psychother.*, 25: 4–17, 1971.
78. ———, "Suicide, Sleep and Death: Some Possible Interrelations among Cessation, Interruption and Continuous Phenomena," *J. Consult. Psychol.*, 28:95–106, 1964.
79. SCHOENBERG, B., *et al.* (Eds.), *Loss and Grief: Psychological Management in Medical Practice*, Columbia University Press, New York, 1970.
80. SHAW, M., "Dying of Cancer, 'Horror' Attitude Most Harmful," *Seattle Times*, March 24, 1954.
81. SHERE, E. S., "Group Therapy with the Very Old," in Kastenbaum, R. (Ed.), *New Thoughts on Old Age*, Springer, New York, 1964.
82. SHONTZ, F. C., and FINK, S. L., "A Psychobiological Analysis of Discomfort, Pain and Death," *J. Gen. Psychol.*, 60:275–287, 1959.
83. SILVERMAN, P. R., "The Widow-to-Widow Program," *Ment. Hyg.*, 53:333–337, 1969.
84. STEEN, M. M., "Fear of Death and Neurosis," *J. Am. Psychoanal. A.*, 16:7–31, 1968.
85. STOTLAND, E., *The Psychology of Hope*, Jossey-Bass, San Francisco, 1969.
86. STRAUS, A. L., and GLASER, B. G., *Anguish: A Case History of a Dying Trajectory*, The Sociology Press, Mill Valley, Calif., 1970.
87. SUDNOW, D., *Passing On: The Social Organization of Dying*, Prentice-Hall, Englewood Cliffs, N.J., 1969.
88. SWITZER, D. K., *The Dynamics of Grief: Its Source, Pain and Healing*, Abingdon Press, Nashville, 1970.
89. TART, C. T. (Ed.), *Altered States of Consciousness*, John Wiley, New York, 1969.
90. THURMOND, C. T., "Last Thoughts before Drowning," *J. Abnor. & Soc. Psychol.*, 38:165–184, 1943.
91. TORREY, E. F. (Ed.), *Ethical Issues in Medicine*, Little Brown, Boston, 1968.
92. TOYNBEE, A., *et al.*, *Man's Concern with Death*, McGraw-Hill, New York, 1969.
93. VERNICK, J. J., *Selected Bibliography on Death and Dying*, National Institute of Child Health and Human Development, Washington, D.C., 1971.

94. VERWOERDT, A., *Communication with the Fatally Ill*, Charles C Thomas, Springfield, Ill., 1966.
95. WAHL, C. W., "The Fear of Death," *Bull. Menninger Clin.*, *22*:214–223, 1958.
96. ——— *et al.*, *Helping the Dying Patient and His Family*, National Association of Social Workers, New York, 1960.
97. WALTERS, M. J., "Psychic Death: Report of a Possible Case," *Arch. Neurol. Psychiat.*, 52:84–85, 1944.
98. WEISMAN, A. D., "Misgivings and Misconceptions in the Psychiatric Care of the Terminal Patient," *Psychiatry*, *33*:76–81, 1970.
99. ———, and HACKETT, T., "Predilection for Death: Death and Dying as a Psychiatric Problem," *Psychosom. Med.*, 23:232–256, 1961.
100. ———, and KASTENBAUM, R., *The Psychological Autopsy*, Community Mental Health Journal Monograph Series, No. 4, 1968.
101. WHEELIS, A., *The Illusionless Man*, Norton, New York, 1966.
102. WODINSKY, A., "Psychiatric Consultation with Nurses on a Leukemia Service," *Ment. Hyg.*, *48*:282–287, 1964.
103. WORCESTER, A., *The Care of the Aged, The Dying and the Dead*, Charles C Thomas, Springfield, Ill., 1950.
104. WYLIE, H. W., JR., *et al.*, "A Dying Patient in a Psychotherapy Group," *Internat. J. Group Psychother.*, *14*:482–490, 1964.

PART FOUR

The Various Schools of Psychiatry

CHAPTER 35

THE PSYCHOBIOLOGICAL APPROACH

Wendell S. Muncie

ADOLF MEYER* (1866–1950) coined the term "psychobiology" to refer to a science of man which conceived that biography, with its mental functioning, was as truly biological as was physiology. Such a view naturally led to two further assumptions: (1) that the living man can only be studied as a whole person in action,† and (2) that this whole person represents an integrate of hierarchically arranged functions. Psychobiology offers a theory of personality organization and activity, and an attitude toward the approach to treatment of abnormalities of personality.

* For a good introduction to and a survey of Meyer's thinking and contributions in the fields of neurology, psychiatry, medical teaching, and mental hygiene, the interested reader is advised to turn to the introductions in the several volumes by, respectively, Louis Hausman, Sir David K. Henderson, Franklin G. Ebaugh, and Alexander H. Leighton.

For a more extensive and detailed treatise on treatment bearing largely the stamp of Meyer's influence, see references 6, 7, 21, 22, and 23.

† For a historical review of the development of the concept, see references 15 and 24.

The fundamental concept of psychobiology is that of integration. According to this concept, man is the indivisible unit of study, but this study can be approached from any of a number of hierarchically arranged levels: the physicochemical, the reflex, and other physiological systemic levels, and finally the psychobiological, that is, the activity of the whole person, as an item of biography, using economizing symbolizations binding together past, present, and future anticipations. This symbolizing activity, referred to as "mentation," is a specifically and characteristically human activity. The facts of any level of integration are included in the activity of the higher levels and are necessary for the complete understanding of the latter, but the facts of any level of integration are not to be understood as a summation of terms of the lower levels only. Briefly, the whole is greater than the sum of the parts. The activity at any level may be altered by change at either a higher or a lower level. This total activity of the individual may be sampled at any given time as a cross-sectional picture of the personality, or the personality can be studied longitudinally as a time-bound and changing phenomenon.

The old problem of "What is consciousness?" is replaced by the concept of mentation as a variety of action at the highest integrative level with varying degrees of perfection or completeness.

To complete the theory, the person-as-a-whole concept of necessity includes man in his society as a part of the whole. The study of the individual then inevitably merges with a study of his society, including the workable and less workable aspects of each item. Furthermore, the facts of human biology and societal structure force the attention for dynamic purposes onto the relationships (and attitudes) inherent in the child-family combine. There is abundant evidence, from normal as well as abnormal histories, of the lasting power of early acquired attitudes and the need for subsequent struggle to modify them.

These concepts are so simple and appear so self-explanatory that the scientific reader might be excused for expressing impatience with them for their simplicity and for their lack of any definition or rules as to "where to go from here." Meyer himself never supplied, or attempted to supply, detailed rules for the further amplification of study methods (except in regard to special topics of interest to him, such as schizophrenia, paranoid states, and the neurotic constitution). He stressed the selection of topics as the most vital job confronting the worker in this field. History shows how the selection of topics has changed in emphasis from one generation to another. And in no case can it be said that the topic has ever reached an exhaustive and authoritative working through.

To return to the simplicity of the concepts of psychobiology, let no one be misled by this appearance. There is no harder discipline to apply than that of working with holistic concepts; and perhaps none is less gratifying, because the gaps in knowledge are so omnipresent and glaring. It is no accident, therefore, that many scientific workers declare themselves "holists" but think and act otherwise, as psychophysical parallelists, or atomists, or other adherents of part-function philosophy.[5,13]

The significance of all this for treatment lies in the following facts:

1. Treatment rests on a sound view of the longitudinal behavior of the individual in his social setting, and in its precise sampling at any time in cross section. This has meaning for the matter of history taking.

2. Treatment may be instituted at any level of integration which can be shown to be involved in the origin of disorders of total functioning. This leads to multiple attacks on the most diverse problems in treatment, for example (*a*) simple psychotherapy as well as chemical attack on bromide delirium or (*b*) electroshock treatment as well as psychotherapy in depression.

Form and Content and Their Interrelations

Meyer often remarked that there seemed to be but a few ways in which people could react. This underscores the clinical fact that the diversity of life experience must finally be channeled into only a few varieties of behavioral expression, determined by the facts of biological personality organization. The emphasis in psychobiology, from its outset to the present, has been on the effort to elucidate the interrelations of life experience, objectively and subjectively viewed, and their biological means of expression. Actually, this means an effort at synthesis of the statistically valid descriptive generalities of mental disease (form) and the dynamic aspects (content) imparting meaning (that is, plausibility). This is that search which "psychosomatic medicine" has appropriated peculiarly to itself but which applies in all psychiatry if, indeed, not in all medicine.*

Common-sense observation leads to the conclusion that dynamics cannot be equated with causation. Meaning in illness may play a

* This is the monistic explanation Stanley Cobb recently underscored as the total of all psychosomatic effort.[5]

leading role or may figure in an incidental sense only. It cannot be said that a wholly satisfactory synthesis of form and content has as yet been achieved in regard to any mental-illness type, because of the inherent difficulties in the problem.

The facts of form were well, if not exhaustively, described by the early writers, culminating in the Kraepelinian systematizations. They deal with the phasic qualities, the tendency to recover versus the tendency to chronicity, to deterioration, with hallucinations, thinking disorders, disorders of affect, disorders of the body-image, and so forth. Yet there is considerable evidence for the belief that these items of form are themselves to a degree subject to cultural influence (content). This seems to be true after allowance is made for changes in "fashions" in diagnoses.

The forms of illness determine certain well-known content items and treatment necessities. For example, deep depression may be assumed to be associated with suicidal preoccupation and to demand adequate protection for the patient. Likewise, catatonic stupor assumes a pathological passivity and calls for caution in tube feeding.

Psychotherapy attempts to change the form of illness by attacking the content. Physical therapy—including the various forms of shock treatments and the use of stimulant, sedative, and tranquilizing drugs—attempts to change the content of illness through attack on the form. The shock treatments, to whatever degree they are justified through their pragmatic usefulness in shortening serious depressions and manic excitements or in ameliorating schizophrenic states of withdrawal or paranoid distortion, have left with many observers a note of regret at the massive shotgunlike attack, with a virtual dearth of intelligent understanding of their modes of action. We come somewhat closer to pinpointing our attack through modern chemical methods. The present rapidly developing selective psychopharmacology will eventually have to rest on sound knowledge of brain anatomy and physiology. At the moment, this is our best hope for attack on the "final common pathways" of form in the major "functional" psychoses.

It cannot be too often reiterated that the physical treatments are administered by people to people, and the full and final effect of the treatments will bear the imprint of the working relationship between them. In general, the physical treatments should be administered only when the ground has been prepared for them through demonstration by the patient of his belief in the physician's grasp of his problem. Under such conditions the treatments become a direct extension of the physician's influence in the situation.

Psychobiological theory interposes no objection to combined physicotherapy and psychotherapy. Accurate analysis of the factors at work should determine what treatment method should be the leading one and which one should serve in a more accessory capacity at any given time. All treatment has for its goals: (1) positively, the realization of that best potential of which the patient is capable, or at least willing to accept or to attempt; and (2) negatively, the avoidance of the introduction of factors that would leave the patient less able to deal with his life situation.

Treatment as Negotiation

In rereading Meyer's contributions to psychiatry, one is struck by the few discussions of detailed treatment methods.* The most vivid memory I have of his attitude to treatment does not appear in his collected works but derives from a statement made in a staff meeting late in his tenure at the Phipps Psychiatric Clinic. As I have reported this elsewhere,[22] it was substantially as follows: "The patient comes with his own view of his trouble; the physician has another view. Treatment consists of the joint effort to bring about that approximation of those views which will be the most effective and the most satisfying in the situation." This struck me forcibly at the time, for it laid down what, I recognized, had been our established working method at the

* The most detailed account, still a classic, is to be found in references 19 and 22.

clinic, but which had never been so aptly stated.

This succinctly asserts a cardinal principle: Treatment is a matter of negotiation of viewpoints and attitudes. This discards immediately old authoritarian views of treatment and uses, instead, mutual education through the elaboration of the material of the history as well as the working relationships existing between patient and physician to enlarge the area of negotiation. This view of treatment to me appears so basic, so elemental, and so self-evident—like much of Meyer's wisdom—that one can hear oneself saying impatiently, "Yes, of course. Now how to negotiate?" (That is, "How about the techniques?") Meyer was always interested in techniques, but he seems to have held the conviction that the great failures in psychiatry resulted more from a failure in basic attitudes than from a failure in techniques. Otherwise stated, if the basic attitudes were firmly established, every practitioner could be expected to develop, in time, those techniques commonly in good repute and to add his own variants depending on his own assets (and needs). Consequently, trainees under Meyer ended with the most diverse technical equipment, but all were touched to some extent by the simple basic elementals of treatment as noted above.

My observations over the years lead me to conclude that the concept of treatment as negotiation is basic for the best effort. I see this confirmed daily both in its observance and in its breach. Negotiation implies mutual respect and a willingness to give a sympathetic hearing to the other. It is especially the obligation of the physician to rid himself of any sense of justification for coercion which might arise from superior knowledge and faith in techniques. It is a humbling thought that, in some ways, the patient always knows more of himself than we ever will. If we can help in a more useful assembling of this self-knowledge, we will have served our purpose. The phrase "more useful assembling" has a certain teleological ring. We should not shrink from this fact, nor does its acknowledgment entitle us to any of the perquisites of omnipotence or omniscience.

To be condemned are enthusiastic parochially tinged injunctions to therapy addressed to a patient in no wise prepared for such well-meant advice. Treatment starts and ends with what is possible, and tries constantly to enlarge the area of the possible through patient understanding of the problem and communication to the sufferer of this expanding view, with the need for encouraging a greater participation on his part in an expanded goal. As purveyors of a service to sufferers, we must recognize that the patient has the inalienable right to determine the degree of his participation, and summary interference with his freedom of action in this regard can only be sanctioned when clear danger to himself or others is evident.

History Taking

How to get a history and what to do with it to alter the course of events constitute the fundamentals of therapy. We must start with the assumption that for every patient his own story is of special interest to him. This is inherent in the fact that he is suffering in some sense. He comes with a certain view of his problem, deriving from his own background and what the problem means to him. He deserves and must be given the fullest encouragement to develop the story as he sees fit and, in the process, to reveal progressively the motives at work as well as the form of the malady. The physician, for his part, must school himself to be a patient listener, and while the patient is talking and otherwise demonstrating his perturbation, must build up within his own mind a tentative view of the form of the illness and of its dynamic meaning, that is, of the items meaningful for provocation and continuation. The material then is sifted for the tentative allotments of etiology, and a "work sheet" of unfinished topics, or topics needing further elaboration, is set up. Subsequent sessions are concerned with filling in this material. This may be accomplished (1) by direct inquiry of the patient; (2) by use of devices for gaining access to material which, for adequate reasons in the preservation of the official version of the self, the patient has "forgotten" or relegated to a position of relative unimportance (using free association, analysis of slips of the tongue, dreams, daydreams, projective psychological test material)*; or (3) by inquiry from relatives or others with a legitimate interest in the patient's welfare. The material of (3) above is often absolutely vital to a full understanding of the case, especially in dealing with patients who act out in antisocial ways, but, more often, such material serves to arm the physician with a knowledge of specially sensitive items and to warn him away from brutal inquiry. It is good practice, in dealing with cooperative and voluntary patients, to secure permission for discussing the case with others. I do not mean that this must be a hard and fast rule, but in any case the physician needs to be prepared to justify such a move to the patient.

As I work, this turns out, in most instances, to be a matter of conversation, face to face, while encouraging the patient to express himself freely. In practice I do not put the patient under the obligation to tell all that passes through his mind. I assume that the willingness or the desire to "tell all" is itself a matter of growth of confidence. Furthermore, I respect the matter of privacy, a right dearly won and even more sorely pressed for its preservation. Neither do I feel barred from initiating inquiry, holding myself in readiness to justify my moves at all times.

I use interpretation of motives early in order to call attention to the main purpose of our collaboration—to bring a degree of plausibility into the story, because therein lies the opportunity to get purchase on the matter of significant meaning. Primarily, I use material of the expanded history, from the present and the past, to illustrate the personality trait in question which is in action at the moment. I bolster this with analysis of dream material for the purpose of pointing out to the patient that the material of our conversations finds its spontaneous corroboration in his own productions, in a setting not so clearly open to his defenses and to my suggestive influence.† I myself have never been able to see in dreams the superhighway to the understanding of personality. I reach my most valid interpretations from the analysis of historical material. But dreams offer the appearance of the spontaneous, unrehearsed production that carries its own note of validity for etiology and current expectations for therapy.

These remarks apply especially to the treatment of the psychoneuroses. I have not found the dreams of depressed patients helpful in treatment; they offer only simple corroboration in the same terms as the waking moods and preoccupations or a simple wish-fulfilling fantasy of well-being. In the schizoid or borderline schizophrenic, dreams may be useful to gauge the degree of tendency to panic and disintegration, and they can be best used to warn the physician against probing analysis, unless it is accompanied by massive personality support.

In order to keep the treatment related always to the complaint, that is, to the terms of the present, analysis of historical material for motives and attitudes is constantly interwoven with factual accounts of current activities. This serves to keep the patient alert to the obligation to use now what he has learned of his habitual tendencies and the opportunities for change in them, and keeps treatment from degenerating into that endless situation that allows the patient to say, in effect, "I'll change some day when I have come to know all there is to know about myself." Treatment should be carried on in an atmosphere that expects change consistent with the current understanding, with sympathetic understanding of failure and with support and encouragement to retrial.

* Projective tests may round out the clinical picture with added content details including the defensive mechanisms, or they may offer a personality profile wholly at variance with the clinical judgment. This dilemma is of utmost interest and needs further elucidation.

† This statement must be taken with the reservation that dream interpretation and the patient's acceptance of it are open to the same influences that interpret overt behavior and accept such interpretation.

The frequency of contacts with the patient is determined by:

1. The need to establish a sense of continuity in the search for effective plausible understanding, both in the patient and in the physician.
2. The degree of anxiety or other turmoil (for example, suicidal risk). The greater the turmoil, the more frequent must the visits be.
3. Certain practical considerations deriving from:
   a. Ease of accessibility.
   b. Cost of treatment.
   c. Case load of the physician.
4. A balance between the drive to develop a topic of inquiry and the need for time to elapse in which the significance of the development of the topic and data may be digested. On occasion, after a thorough working through of some aspect of the case, a prolonged therapeutic rest may be declared for the express purpose of discovering the patient's ability to put to use the new formulations.

History taking leads to certain conclusions about the severity of the disorder, its form, and the predominant mood and content. These items determine immediately certain practical points:

1. Does the patient need to be in a psychiatric hospital?
   a. Because of the suicidal risk, as in severe depressions.
   b. Because of the risk of asocial activity, as in manic states, paranoid states, etc.
   c. Because of the physical needs, as in stupors, deliriums, confusional states, organic syndromes.
   d. Because of the need for discipline and habit training, as in certain chronic schizophrenic states.
   e. Because of the need for certain physical treatments.
2. Can psychotherapy be the leading issue in treatment? This is the case in all neurotic states except the most severe, in the static personality disorders where antisocial acts are absent, and in the milder psychotic states of all sorts.
3. Can physical therapy—antidepressant, tranquilizing or sedative drugs, or electroshock—be profitably combined with psychotherapy? This will meet with diverse answers from many practitioners. Psychobiologists offer no theoretical objection to the combination, the decision resting on sound clinical judgment as to:
   a. The adequacy of psychotherapy alone.
   b. The degree of anxiety or other turmoil, that is, whether it is aiding or is impeding psychotherapy.
   c. Urgency for other reasons.

Treatment as Distributive Analysis and Synthesis

The central tool of therapy in psychobiological psychiatry was called by Adolf Meyer "distributive analysis and synthesis." The term attempted the description of the process by which the complaint was broken down into lines of inquiry (involving any integrative level and, of necessity, using the methods best adapted to study of the facts in question) combined with the attempt to reassess the factors at work in the form of alternative combination, that is, discovering alternative viewpoints and ways of working with more gratifying expectations.

The choice of methods to use in psychotherapy offers the widest range of variation. Meyer was not one to prescribe methods. His emphasis was on the use of any method in an experimental sense. Nevertheless, in actual practice this does not degenerate into the view that "anything works," that is, is admissible. Certain things are prerequisites for such work in this field:

1. The physician must have a profound respect for the patient and his efforts to get through this life with a maximum of gratification and a minimum of discomfort. It is a good thing to assume that the patient is doing as well as he can, considering all the circumstances of his heredity, his environment, and his own personality make-up. The physician's task is to help him arrive at a more useful effort.

2. The physician must have a high degree of sensitivity to the items of special emotional significance for the patient. This is enhanced by a broad understanding of (*a*) the infinite varieties of living, working, and playing; (*b*) the varieties of idealistic yearnings and their religious systematizations; (*c*) the wide differences between overt assertion and covert intention; and (*d*) the language of dreams and other "uncensored" productions, of the treasures of the ancients as expressed in mythology and their legends, and of the wisdom of philosophers, poets, essayists, dramatists, and artists of modern times.

3. The physician must have a sound knowledge of child development, of family structure and its variants, and of the accepted principles of personal and community mental hygiene as they apply to different ages and ethnic groups.

4. The physician must have understanding of his own personality assets and liabilities and of the degree of control of the latter. The physician must be aware of his own reactions to the patient and be able to deal honestly with them in the same manner that he expects his patient to do. On occasion, this may even require him freely to expose his own feelings to the patient, together with his methods of dealing with those feelings. Such self-exposure may alleviate the patient's anxiety in discovering that the physician also has the same sorts of problems with which he must contend. Customarily, however, it is not necessary for the physician to use such methods. He may safely allow himself to remain in the role allotted him by the patient—that of a person who knows how to live better than does the patient and who can help the patient find clues to his own better performance. Acknowledgement of the difficulties involved in mature living will be enough to indicate that the physician is not godlike in his perfection and has experienced his share of difficulties.

Distributive analysis begins with a consideration of the presenting complaint, the patient's or family's relatively naïve descriptive view of the trouble, with whatever concern for provocative factors can be elicited. This account commonly suffers in the latter aspect owing to (1) the patient's (or family's) innocence of sound knowledge of personality organization and operations and (2) the need to protect the self-esteem by throwing up defenses (to self and others) against the full exposure of provocative items in the nature of unacceptable personality traits or actions. There results, then, a crude effort at psychic homeostasis, where suffering is exhibited and at once is partially counterbalanced by defensive symptom formation. Both items operate through the common psychological mechanisms of repression, projection, denial, substitution, amnesia, conversion, psychosomatic display, and so forth.

As physicians, we learn to detect or to suspect such mechanisms selectively from the type of account given. Our further effort is directed toward bringing this material to the patient's awareness, to the degree that he can emotionally accept; that is, the analysis of motives and actions, conscious and less conscious, must proceed within the bounds set by the need to secure the safety of the patient. Depending on the type and severity of the illness, and conversely on the demonstrated degree of ego strength, this will vary from the most thoroughgoing exposure to positive efforts at aiding its repression and bolstering the ego in its efforts at reality adjustment.

Psychotherapy, as conceived of in psychobiology, begins with symptom analysis (the complaint). This leads quickly to an analysis of the motives involved, that is, motives of the moment, so to speak, with a consideration of compromises inherent in such symptom development—a compromise between the hurt sustained and the defense offered to it. This leads further to a study of the habitual attitudes and motives, that is, of long-term, "constitutional" personality assets and liabilities, their origins, their workings, and efforts at change.

Change in habitual attitudes is the final goal of psychotherapy and involves (1) understanding of the origins of the attitudes, their purposes, their relative usefulness as early developmental structures, and the point of departure to the present relatively obstructive role; (2) encouragement, indirectly by interpretation or more directly by suggestion and persuasion to attempt new activities based on new attitudes (In either case, the patient's willingness to try to change involves tentative acceptance of the interpretations of the attitudes, and trial-and-error efforts in the framework of the physician's understanding and approval.); and (3) the patient's freeing himself from the need for the physician's support and approval as a process of maturing, by his own efforts and those of the physician, and finally by the final acceptance of the validity of those formulations which he has derived wholly or in part from the therapy.

Direct counseling is kept at a minimum, in favor of the patient's discovery of his own prescription for living. This may not always be avoided, however, nor should one dodge the responsibility when it is clearly indicated. This need for categorical advice most commonly arises in dealing with immature characters—weak, indecisive, and lacking in drive. The danger with direct counseling, of course, lies in its furthering dependency needs in such weak characters. Yet this is a risk one must assume on occasion in order to get some initiative and direction into a rehabilitative program.

Psychotherapy Limited by Form of Illness

Experience teaches that certain aspects of the form of illness must be regarded as putting definite limits on the usefulness of psychotherapy, for example:

1. Depression of all but the most incidental neurotic sort is made worse by probing efforts at insight therapy. This must wait until the mood has lifted or until its effects can be counteracted by massive supportive therapy.

2. States of schizophrenic withdrawal and paranoid misinterpretation have resulted from serious insults to the self-esteem. This fact should impose a most cautious approach in psychotherapy. The patient's willingness to accept the physician as a friendly and sympathetic observer is at a minimum, and abrupt approaches are commonly brushed aside.

3. Where unchangeable and predominant reality factors are present, the opportunity for personality change may be minimal, since the latter does not operate in a vacuum but benefits from the presence of alternatives in the external field. There appear to be two strong dynamic determinants at work in the life history of any individual—the drive to autonomy and the drive to homonymy.[1] To put it briefly, everyone wants to be an individual, but not so much so as to be unpleasantly conspicuous; at the same time he wants to be like others (of his selection), but not so much so as to be lost in the crowd. How to be oneself and simultaneoulsy a member of the group is a challenge which each of us faces, and which each social aggregate faces with larger integrates. Experience shows that this problem is made all the more difficult when the individual suffers from an early imprint of patterns of behavior that deviate from socially acceptable norms, or when his society, chosen or forced on him by circumstances, deviates, in certain tendencies, too sharply from his personally acceptable norms. What the special forms are which conduce to major and to more minor disorders are not too clear at this point in our history, despite the appealing formulations of Freud, Adler, Jung, and others. In a very general sense, I see in the vicissitudes and fates of the contrasting developmental forces of free love (or approval) and contingent love (or approval), acting from the earliest days and months of the child's life, a basis for much diverse psychopathology. This sets the stage for the struggle toward autonomy and homonymy and for the attitudes to self and others. It contributes to the contrasts of ease and confidence or uneasiness and distrust. Sometimes we can get a fair documentation of these forces in actual historical material or in fantasies, and their exposure in a treatment setting of mutual respect, toleration, and inquiry can lead to alternate and pragmatically better attitudes (irrespective of the aspect of "diagnosis" in the case at hand).

Role Assignment*

In the course of taking the history and in its elaboration, it will become evident that the patient views the physician in a certain role. This is not a fixed matter; it changes from patient to patient and in any patient from time to time. It is almost certain that the kind of role has to do with (1) the frequency of visits, (2) the relatve dominance of direct conversation or of free association or other indirect method, and (3) the importance for the provocation of the illness of the preservation into the present of strong unresolved early relationships with parents, siblings, or other significant figures—unresolved in the sense that unhealthy childish attitudes are preserved (resting either on reality or fancy) and are all projected onto the physician. He is seen, then, not only as a physician but, in a sense, as that other person.

Treatment uses role taking to elicit the full-fledged emotional investment of the unhealthy attitude, but this is done patiently in order to stress the reality of the physician as such. Concentrated therapy tends to favor the emergence of such displaced and distorted role assignments. Fewer and wider-spaced visits tend to limit the role assignment to the realities, with the physician seen as the one who knows better how to live, as supporting and challenging the patient to learn and to use his insight in the development of newer and better attitudes and actual living habits.

Psychobiology uses all techniques as necessary, that is, as demanded by the nature of the complaint and the critical vision of the opportunity for change, but, unlike psychoanalysis, it does not view the development and resolution of the transference neurosis as a *sine qua non* for effective treatment. As I have outlined my working methods, so-called transference neuroses are kept at a minimum, but transference phenomena of lesser degree are abundantly-observable. These may be treated as incidental and passing items or may require analysis, depending on their strength. I do not shrink from their use, but I do not commonly encounter treatment situations which rest principally on such phenomena.

Suffering promotes the patient's cooperation in treatment, but too much suffering can become a road block. We therefore have to remain eternally sensitive to these contrasting needs: (1) the need to uncover the damaging personality aspects in their actual workings, thereby generating additional suffering as guilt, anxiety, and hostility directed against us or against self; and (2) the need to support the patient in his own efforts at betterment, encouraging the use of the known assets, stressing the compensating assets, and giving alternative interpretations to the excessively damaging ones he makes.

To prevent the turmoil from getting out of hand or to reduce it to the point where efforts at dynamic understanding may be undertaken, drug therapy with tranquilizers, antidepressants, sedatives or even electroshock therapy may be necessary.

Common sense tells us that the ways of getting through this life are judged as better or worse, acceptable or not acceptable, gratifying or humiliating, by the individual and by his entourage. In the most general terms this problem can be reduced to a variety of compromises between the drive to autonomy and the drive to homonymy.[1] In the last analysis our patients do invest us with the aura of having successfully accomplished this difficult feat ourselves, and of having a fund of information and skill from which they can also profit. It goes without saying that, as physicians, we should, in fact, justify this confidence to the degree that we have integrity in our performance. It is also basic that we do not attempt to impose our own solutions on others, but work toward the maximum realization of the patient's potentialities within the framework of the necessity of living with others who also have their needs.

In treatment as negotiation, we achieve our best performance to the degree that we can successfully deal with the following items:

* For an excellent description of role taking in therapy, see references 3 and 4.

(1) accurate recognition of the form of illness, that is, the kind of reaction exhibited (diagnosis); (2) the experiential material which affords the most plausible explanation of why this form of reaction has arisen and why it continues on (dynamic analysis); and (3) sorting out of the items wherever leverage may be applied in one way or another to change the forces at work (either in kind or degree), and the selection of method to be used for that purpose.

Failure in treatment may result from inadequate management of any one of these items, but I would single out as most basic the failure to achieve a working fusion of the concepts of form and of content in illness. Too many of us are too content with the understanding of only one or the other item. This means that those limitations and opportunities in treatment which are determined by the fusion or interaction of the two elements are not fully appreciated, leading to a premature choice of methods of working, and blocking the development in the negotiations of the fullest participation by the patient.

It is a cynical truism in medicine, and more so in psychiatry, that "anything works sometimes." To the extent that this is so, this means only that the wide variety of content items in experiential material and the wide variety of personality organizations lend themselves to a variety of negotiated compromises. It is not difficult to prove that what is meat for one is poison for another, or to show that any variety of personality organization has its strengths and its weaknesses.

Acceptance of the fact of variety does not imply acceptance of the truth of the cynical dictum above. If this were so, there would be no need to concern ourselves with the aspects of worst-bad-good-better-best. But since our patients concern themselves with this matter, then we must also (if for no better reasons).

The most common error arising from the inadequate recognition of the interrelationship of the form and content of illness is today to be noted in the management of those reactions loosely called "reactive depressions." The term is commonly applied to depressive reactions with gross provocation in immediate life experience acting on vulnerable personalities. The wealth of content material and the justified suspicion of deep underlying personality difficulties, together with the admixture of overt anxiety along with the depression, lead the inexperienced physician into the error of treating the illness as a neurotic problem, with a maximum of aggressive analysis and a minimum of ego support. This invariably has only one end—the patient is made more anxious and becomes compulsively wound up in futile efforts to extricate himself from the tangle. Experience should dictate, in such instances of well-established depression (form), that one should apply support (reassurance) liberally, based on the sound observation that such mood reactions do pass away. Analysis of the personality and its reaction to the immediate provocative life experience is offered as an elective process for the purpose of developing a sense of plausibility in the total reaction and for laying the groundwork for more extensive personality study at a later date, when the patient is able to face self-revelation without developing crippling loss of confidence, and if he chooses to pursue the matter further. By making the matter elective, in the reasonably sure expectation that the mood will pass anyway, the subject almost invariably responds to the support and can patiently collaborate in the search for ultimate causes, free from that frantic compulsion which ends in futility, deepened self-abasement, and even suicidal risk. There is no more necessary equipment for the therapist than to be sensitive to the nuances and degrees of depression and to judge whether it is safe to accelerate the processes of self-revelation.

Another error arises in the mistaken evaluation of derogatory delusions and hallucinations as being schizophrenic in the presence of deeply depressive affect. The safe rule of thumb is that when delusions and hallucinations are corroborative of the patient's acknowledged evaluation of himself, then they are to be viewed as affectively determined; when they are viewed by the patient in a sense of denial of his self-evaluation, they are schizophrenic. Treatment then is determined by the decision in the case.

Long-term treatment inevitably runs into sterile unproductive periods. These indicate (1) natural letdowns after more productive periods, (2) resistance to and evasion of topics needing discussion, or (3) the fact that a favorable plateau has been reached, and the patient is indirectly asking the physician to make the decision to let him go on his own for a period. In (1) the physician must school himself (and the patient) to ride out the inactive phase, knowing full well that activity will resume. In (2) evidences of the resistance and evasion become the topics for discussion, and commonly uncover topics of special sensitivity or hostile attitudes to the physician that must be worked through before effective work may be resumed. In (3) the physician's obligation is to bring into the open the question of the desirability of stopping treatment, for good or for the time being. The latter is, in effect, a therapeutic rest, designed to allow the patient an extended opportunity to gauge the effectiveness of his insights and new habit patterns without the support of the physician. A date for checking in can be made or left open, depending on the patient's own wishes, and his choice is a measure of his feeling of confidence and security.

My remarks so far have been concerned with the patient-physician relationship, but treatment is often a triangular affair, with the family or other interested environmental members being immediately affected. In the case of self-sustaining members of society, as is the case with most psychoneurotics, the "treatment" of the third member of the triangle may generally be left to the patient himself. Yet even here, a simple explanatory statement to the family, with the patient's permission, may be most helpful in furthering the goals of treatment. Specifically, the family's recurrent question, "What can I do to help?" needs a frank answer, even when the question is essentially self-serving and needs nothing more than the advice to "be yourselves." Patients do not live in a vacuum, and, by withholding any simple contact, we may do incalculable harm to the acceptance of treatment by the patient's family.

In psychoses, contact with family is commonly necessary for practical reasons, but, beyond this, the same need is clear in order to lay the foundations for the best acceptance of therapeutic goals and for a frank statement of the shortcomings of our methods and the limitations of our expectations. Such working with the family may be done with or without the patient being present, depending on the circumstances. The patient must know of our intention to see the family, and this may become a test of his ability to trust our efforts to protect his interests. He finally must be willing to trust us, and this may well be the first test of that fact.

Meyer stressed habit training as a primary aspect of treatment.[12,14,16,18] It is difficult to look back 50 years to a clear recognition of what the term meant then. What it has come to mean in the course of time is the organized effort on the part of the hospital community to enlist the patient's interest in the development of participative activities which will bring him into line with accepted social patterns. It is obvious that this becomes more meaningful the more he can understand his own difficulties, and as more opportunities open to him for alternative actions. So habit training today includes insight, understanding, and acceptance of a trial-and-error effort at behavior change in a social setting sympathetic to the patient's efforts.

The same general principles apply to the nonhospitalized patient, where his society is composed of his family and his work and play setting, and where he is expected to initiate and carry the greater burden of the trial-and-error effort at change. This is the psychobiological equivalent of "working through," with the emphasis on the actual performance rather than on the talking (or preliminary) phases.

To increase the degree and area of understanding is a first need in therapy. How best to accomplish this has led to great divergences in views. To bring into meaningful juxtaposition discrete items from varied aspects of behavior, and so to illustrate and to emphasize common behavior patterns (for good or for ill); to relate present behavior with childhood's precursors, including their residual emotional investments; to uncover material of fantasy or past memory long consigned to some degree of oblivion by personal need, and to relate this material to present behavior—these are some of the devices in current use. Meyer favored formulations based on generally available material, telling their own stories without need for translation into secondary symbolizations. In retrospect, he appears to have made too conservative an estimate of the number of those who could profitably pursue the exploration of the more recondite material,[11] for although there are still few who practice psychoanalysis, most psychiatrists are able to use much of the material so derived in that "detective work" Meyer thought was not generally to be achieved. In fact, the danger today is in the physician's preoccupation with the "unconscious" material, with a disastrous neglect of the readily available aspects of behavior. Much can be learned that is of dynamic import by the closest scrutiny of both the more and the less easily accessible material, and the best handling contrives a continuing interweaving of both sorts.

I have mentioned "plausibility" as the essence of the meaning of illness (content). The term carries an aura of naïve tentativeness which may be all that is necessary for the patient in the way of structural (theoretical) support to enable him to carry out trial-and-error efforts at personality change, with the active essential support of his physician. But the term need not mean only that. At the other end of the spectrum, it may carry a sense of conviction—to the patient if not to the physician. Conviction naturally makes for more full-fledged, less tentative performance, but not necessarily better performance. Premature jelling of conviction may be a real block to fullest understanding, and the physician must guard against contributing to this through premature and authoritative interpretations of historical material. I usually offer interpretations as speculation, for the purpose of arousing in the patient psychological curiosity and of opening the door to his own spontaneous meanings. The physician will use that frame of reference best suited to his own beliefs, derived from his own training experiences. I am not capable of interpretations that go far beyond the relatively easily visible symbolizations, and as I use dream material essentially to dramatize, through spontaneous illustration, the facts arrived at through more conventional conversational methods, this limited use suffices for the purpose. I am prepared for more extensive and searching use of such material by means not at my disposal, but it has always seemed to me that the more distant the symbolic reference, the more it had to be backed by authoritative and dogmatic theory, and the further it led away from the principle of treatment as the negotiated compromise to one of parochial coercion. I prefer to stay much closer to the more naïve meanings in the facts—meanings, as Meyer[14] pointed out, having close affinity with the problems and theories of sociology. I would not force my views on anyone, but for me the facts of intrafamily relationships seldom need to be translated into libidinous activities to tell their story. This already is parochial. But common sense can work with interpersonal and sociological relations, and for me this is the most fruitful means for increasing understanding in terms that are the closest to universal acceptance.

There is a tendency in our field for treatment procedures to become formalized. This applies to frequency of visits or treatments and to their total number. I am aware of the factor of experience in determining such items and of their usefulness to the patient in planning for the financial outlay for treatment, but the therapist should beware of becoming bound by tradition, for there is nothing more destructive of the principle of mutual participation in treatment. The goals will determine the outlook. For example, when one or two electroshock treatments have brought about a degree of volubility regarding meaningful material in a previously mute catatonic patient, the physician will have to make a decision at that point whether to use the new rapport for purposes of psychotherapeutic exploration or to give more electric shock in order to accomplish a forcible suppression of the material. Actually, in this instance the former may be attempted, keeping the latter in reserve for use if it is deemed advisable. I consider tentative use of electric shock in such an instance preferable to the rote application of, say, a series of ten to fifteen treatments.

As for psychotherapy of the psychoneuroses (at least those seen in office practice) far-reaching modifications—both in easing of symptoms and in change of attitudes and behavior expression—can be achieved by interviews spaced at weekly intervals and continuing for from six months to two years. It is not my experience that improvement under such circumstances can be properly called "transference cures," for, within the time alloted, the patient has had ample opportunity to demonstrate the worth of his new insights and to achieve that sense of self-sufficiency which does not need the physician's support. It is my belief that this derives from the fact that, from the first day of treatment, there is a tacit understanding that its purpose is to make the physician unnecessary. Every advance in understanding and management serves to underscore this aspect of the relationship. It is automatic with me to stress—not verbally but by my attitude—that I expect to do all I possibly can for the patient and that he will increasingly want to divest himself of my help. If this latter does not happen, his dependence becomes the object of examination. This is a matter for special concern in those patients who, from the earliest contacts, expect rejection at the therapist's hands. Without committing myself to interminable contacts, I do not hesitate to reassure the patient on this score, but at the same time I underline the objective of self-sufficiency.

The goal in treatment is to make the physician unnecessary to the patient. We must face the fact that this may not be possible. But even in such endless associations there should be no letup in the effort toward this goal. Encouragement to take vacation trips or even to move elsewhere, if the occasion arises, with assistance in finding for the patient in the new locality whatever help is necessary, will underscore the physician's determination not to let treatment lapse into parasitic symbiosis.

Meyer worked by appealing to the normally functioning part of his patient, and extended this area by removing hindrances arising from special complex content and by enlarging the opportunities for direct exploitation of the normal residual assets. The appeal was to the patient's spontaneity.[8,17] The physician must also be spontaneous. I can imagine no greater hindrance to effective rapport than the continual interposition between physician and patient of a curtain woven of petty rules and of calculatedly studied speech whose only effect is to arouse in the patient hatred of authority or obsequious abasement before omniscience, to say nothing of gaining us the popular reputation of being not quite normal! "Treatment" then becomes a cat-and-mouse game to be thoroughly condemned.*

As a physician, one may learn much about the patient's neurotic complexes by his reaction to the physician's spontaneous gestures of civility. For example, a woman, who had spent an hour expressing her craving for approval and her certainty that it would never be accorded her, exposed her problem dramatically when she could not accept the physician's gesture of helping her into her coat, and literally ran from the office. Much precious time in circuitous talking was avoided by direct use of this incident. Should the therapist have restrained his civilized spontaneity? I think not. The erection of utterly artificial barriers removes the patient that much farther from life itself and makes his return the harder.

Psychobiology as Objective Common Sense

Meyer referred to his psychiatric teaching and practice variously as genetic-dynamic, psychobiological, objective, and common sense. I have discussed the application of the terms "genetic-dynamic," and "psychobiological" in relation to his teaching and practice. By "objective," he stressed that people are judged by

* For example (from real life):
Newly admitted patient to the chief physician making rounds: "Good morning, Dr. ——."
Physician: "Why do you say that?"
Sequel: She recovered from her self-limited illness, but with contempt for this physician. And why not?

their behavior, and behavior is a matter for objective observation. This includes the corollary that full-fledged performance implies also subjective items which antedate and lead to overt behavior. It follows, then, that treatment should result in a change in objective behavior. In his hospital work he developed the use of ingenious charted checks on the behavior of the patient, showing at a glance the status of the patient at any given time and over the longer term. These, supplemented by nurses' notes, proved most valuable adjuncts to the physician's observation. For outpatients, such records are not practicable or necessary, but the principle of using the observations of those people who are near the patient is a device too little used today in deference to the concept of treatment as a closed circuit between patient and physician. Granted that such a reduction of the number of participants in treatment is generally adequate, there are instances where collateral objective observation is not only useful but necessary.

The term "objective" also refers to the quality of the quasi-experimental settings or procedures which can be devised for observing overt behavior and which will reduce to a minimum the variations introduced by the factor of the human observer. This is a field that has been studied experimentally in animals by Curt P. Richter and by W. Horsley Gantt, and the human subject by D. Ewen Cameron.[2] The reader is referred to their voluminous contributions for the details.

"Common sense" was a term which Meyer used with some relish and which has been badly misunderstood. Critics have wrongly assumed that he used the term in the naïve sense, leading one of them to say in rebuttal, "Psychiatry starts where common sense leaves off."

Meyer used the term in more than one sense:

1. He usually prefaced the term with the word "critical," to indicate that the common sense was that of persons entitled to be critical, that is, of those having an acquaintance with the subject.

2. Common sense was essentially a translation for the term "consensus," and as such represented Meyer's constant quest in theory, teaching, and practice for those items of agreed value.

3. In certain connotations, however, there was a flavor of the naïve in his awareness of the fact that the material of theory and practice of psychiatry, so dear to psychiatrists, was in no wise sacrosanct and, in fact, was subject to modification, acceptance, or rejection at the hands of the general public. He felt, therefore, that the main task of psychiatry was an educative one, that it must rise or fall on its demonstrated worth to the public generally. It is in this sense that the treatment situation reduces to a matter of negotiation.

It is in the search for consensus that Meyer's psychiatry exhibits its strength, for obviously in this search no method, theory, or experimental procedure can be ignored. It is clear that at any one institution or at any one person's hands, only a limited number of items of method, theory, or experiment can be put to test. But in the pooling of results, and in their critical evaluation, the consensus (common sense) can eventually be achieved. The objectivity of the results must be the final test of the value of the theory and method. This leaves Meyer's psychiatry as eclectic, free to use the method best suited to the dimensions and qualities of the problem and to the assets of the physician.

In indicating briefly how I work, I recognize this only as an individual sample of the free development of method under the terms of the Meyerian conceptions. I have reason to believe that many (maybe most) workers today employ the same general principles. A catalogue of the variations of method and their correlation with theory and with the personal assets and needs of the therapists would be a useful contribution to the goal of consensus—a goal not likely to be achieved, however, since we work in relative isolation. I was about to say "jealous isolation," but maybe "anxious isolation" would be nearer to the facts. My observations are offered for whatever they are worth, and in the spirit so well depicted by Adolf Meyer:

We do not so much aspire to eternity, but to leave, when we pass, the best opportunity for new times and new life. So it is with medicine. The goal of medicine is peculiarly the goal of making itself unnecessary; of influencing life so that what is medicine today will become mere common sense tomorrow, or at least with the next generation. The efforts of the worker of today become so assimilated in the common sense of tomorrow that it must be our pride to see that it has passed into the real objective nature of the world about us, no longer burdening our attention, but allowing us or those after us to do the same for ever new problems, with ever new achievements and satisfactions.[10]

Mental Hygiene

Treatment may end with the re-establishment of the personality state existing before the advent of illness. But psychiatrists, more than any other medical practitioners, hold dear the aim of deriving from the experience of illness the tools for guarding against recurrence or, indeed, to thwart other illness. In this learning-from-illness process, we depend wholly on psychotherapy to point the way to a more satisfying way of living—a personal mental hygiene. This is a complex matter at best, and the results are debatable, since it is difficult to see how any suitable test methods could be erected. In none of the functional diseases do we stand on that solid ground of mental hygiene which for the former alcoholic prescribes total abstinence. Nevertheless, each of us has assisted in the formulation of new attitudes which have been credited with bringing a lasting harvest of satisfaction in living. The problem is (1) to formulate the achievement in a way to permit statistical study and (2) to list the concrete therapeutic steps taken to consolidate the gains. This is an unfinished task confronting our generation, especially in regard to the major functional disorders—manic-depressive and schizophrenic reactions.

But a fully effective mental hygiene would envisage that of the group, the society in which the patient is but a member. This task is, perforce, a joint endeavor of all whose concern is organized society—general medicine, psychiatry, psychology, sociology, anthropology, law, religion, education, political science. I cannot see that any generally valid principles have been established applicable to all societies, or even to any one, but much experimentation and theorizing are in evidence, and serious concern is demonstrated for society's stake in the welfare of the individual as well as the individual's stake in the welfare of his society. Changing times are making for changing tools with which the individual expresses his parallel, and in some ways paradoxical, needs to autonomy and homonymy. New varieties of personality organizations develop, and new outcroppings of difficulties demand new efforts at correction. The one thing certain is that this process will never settle down in the foreseeable future into any static set of patterns which would simplify personal and social psychiatric theory and practice.*

Résumé

As with any significant figure, Meyer's mature teachings and achievements were somewhat prefigured in his involvement in, and reaction to, early family and cultural influences. His father was a Swiss Zwinglian minister and his uncle a practicing physician. His "psychobiology" was the effort to integrate the concern for the "spiritual" (mental) and "physical" represented by these family figures, and molded by the cohesive Swiss culture; but enriched by his involvement with the views of T. H. Huxley, Sherrington, Charcot, Forel, Kraepelin, Freud and his early associates, and particularly with the American school of pragmatism of William James and his followers.

Working in institutional settings, the state hospitals of Illinois, Massachusetts, and New

* For an account of current efforts in community mental hygiene, see reference 9.

York, he later was instrumental in the establishment of university teaching and treatment centers in New York and Baltimore. From these centers he saw the goals of psychiatry spread beyond the walls and the beginnings of a genuine community psychiatry developing.

For him the psychiatrist was a negotiator, working with the raw material of observable malfunction, with the purpose of assisting in the creation of a meaningful (that is, acceptably plausible) history up to the present; and from that creation the opening of new and better (that is, less threatening and more fulfilling and rewarding) options for the future. This goal is not unlike the creative artist's and avails itself of the most diverse theoretical views and methods. Always comes the ultimate test: Does it work?

His open-ended approach opposed rigid codification of theory and practice, for he, rightly, I think, saw that as the historical bane of our science. The search then centered on the consensus—the generally agreed on—with a welcome eye for the innovative.

Much of current theory and practice must be looked upon as dispensable at a date not too far distant. If I were to be asked what of the Meyerian tradition will likely live, I would single out the integration concept and treatment as negotiation.

Bibliography

1. ANGYAL, A., *Foundations for a Science of Personality*, Commonwealth Fund, Harvard, Cambridge, Mass., 1941.
2. CAMERON, D. E., *Objective Experimental Psychiatry*, 2nd ed., Macmillan, New York, 1941.
3. CAMERON, N., *The Psychology of Behavior Disorders*, Houghton Mifflin, Boston, 1947.
4. ———, and MAGARET, A., *Behavior Pathology*, Houghton Mifflin, Boston, 1951.
5. COBB, S., "Monism and Psychosomatic Medicine," *Psychosom. Med.*, *19*:177, 1957.
6. DIETHELM, O., *Treatment in Psychiatry*, 3rd ed., Charles C Thomas, Springfield, Ill., 1955.
7. HENDERSON, D. K., and GILLESPIE, R. D., *A Textbook of Psychiatry for Students and Practitioners*, 7th ed., Oxford, New York, 1950.
8. LEIGHTON, A., "Introduction to Collected Papers, Vol. IV," *Ment. Hyg.*, *4*:16, 1952.
9. LEMKAU, P., *Mental Hygiene in Public Health*, 2nd ed., McGraw-Hill, New York, 1955.
10. MEYER, A., "The 'Complaint' as the Center of Genetic-Dynamic and Nosological Teaching in Psychiatry," in Winters, E. E. (Ed.), *Collected Works of Adolf Meyer*, Vol. III, p. 1, Johns Hopkins Press, Baltimore, 1951.
11. ———, "A Discussion of Some Fundamental Issues in Freud's Psychoanalysis," in Winters, E. E. (Ed.), *Collected Works*, Vol. II, p. 604, Johns Hopkins Press, Baltimore, 1951.
12. ———, "Fundamental Conceptions of Dementia Praecox," in Winters, E. E. (Ed.), *Collected Works*, Vol. II, p. 432, Johns Hopkins Press, Baltimore, 1951.
13. ———, "Monismus als einheitlich kritrich geordneter Pluralismus," in Winters, E. E. (Ed.), *Collected Works*, Vol. III, p. 38, Johns Hopkins Press, Baltimore, 1951.
14. ———, "The Psychobiological Point of View," in Winters, E. E. (Ed.), *Collected Works*, Vol. III, pp. 429 ff., Johns Hopkins Press, Baltimore, 1951.
15. ———, *Psychobiology: A Science of Man*, (Tr. and Comp. by Winters, E. E., and Bowers, A. M.), Charles C Thomas, Springfield, Ill., 1958.
16. ———, "Remarks on Habit Disorganizations in the Essential Deterioration, and the Relation of Deterioration and the Psychasthenic, Neurasthenic, Hysterical and Other Constitutions," in Winters, E. E. (Ed.), *Collected Works*, Vol. II, p. 421, Johns Hopkins Press, Baltimore, 1951.
17. ———, "Spontaneity," in Winters, E. E. (Ed.), *Collected Works*, Vol. IV, p. 460, Johns Hopkins Press, Baltimore, 1952.
18. ———, "Thirty-five Years of Psychiatry in the United States from Present Outlook," in Winters, E. E. (Ed.), *Collected Works*, Vol. II, p. 1, Johns Hopkins Press, Baltimore, 1951.
19. ———, "The Treatment of Paranoic and

Paranoid States," in Winters, E. E. (Ed.), *Collected Works*, Vol. II, p. 517, Johns Hopkins Press, Baltimore, 1951.

20. Muncie, W., "Psychobiologic Therapy," *Am. J. Psychotherapy*, 7:225, 1943.
21. ———, *Psychobiology and Psychiatry*, 2nd ed., Mosby, St. Louis, 1948.
22. ———, "Treatment in Psychobiologic Psychiatry: Its Present Status," in Fromm-Reichmann, F., and Moreno, J. L. (Eds.), *Progress in Psychotherapy*, Grune, New York, 1956.
23. Strecker, E. A., Ebaugh, F. G., and Ewalt, J. R., *Practical Clinical Psychiatry*, 6th ed., Blakiston, New York, 1947.
24. Syz, H., "The Concept of the Organism-as-a-Whole and Its Application to Clinical Situations," *Human Biol.*, 8:489, 1936.

CHAPTER 36

THE ORGANISMIC APPROACH

Kurt Goldstein*

ORGANISMIC PSYCHOTHERAPY may seem a paradoxical term. For the organismic approach in general—from which we here consider the problem of psychotherapy—the concept is basic that there are no separate apparatus or mechanisms determining the activity of a living being. The organism is considered a unit, and all behavior—normal and pathological—is an expression of the different ways in which the organism functions in its totality. The organization of this unit depends on the task with which the organism is confronted and with which it must come to terms. How this is achieved is certainly based on the organism's structure, but ultimately it is determined by the basic trend of organismic life. Any behavior, normal or pathological, can be understood only if we consider it as an expression of the trend of the organism to realize all its capacities in harmony, in other words, its nature. The degree to which this realization is fulfilled is dependent upon the relationship between the organism's capacities and the demands of the outer and inner world, that is, on how much the organism can come to terms with them (p. 197).[6]

What appears to be the effect of the function of a part of the organism corresponds—considered from the organismic approach—to the activity of this unit in a definite organization, by which the organism comes to terms, as best he can, with the demands. In this organization, the process in one part is in the foreground and represents the figure of the figure-ground organization that underlies every performance of the organism (p. 109),[6] whereas the activity in the rest of the organism represents the ground belonging to the definite figure which appears, on face value, as *the* reaction to the demand.

We are interested here in the consequences that the organismic point of view has on psychotherapy. How can one justify a therapy that considers *one* part-process of the organism, the psychological, as all-important as psychotherapy pretends; in other words, how

* Deceased.

is psychotherapy possible from this point of view? Before trying to answer this question, we must consider at least briefly what the so-called psychophysical relationship, especially the psychological influence on physical phenomena, can mean, restricting our discussion to the problem: Does our approach provide a concept on which psychotherapy can be based?

The Psychophysical Problem

The discussion of the psychophysical problem has been undertaken in the past, particularly by philosophers and psychologists—without, however, many fruitful results. At the beginning of this century, it began to attract the special interest of physicians when it became evident that not only so-called psychological but also some somatic conditions—such as asthma or hypertonia—could be improved by psychological therapy, and thus it became necessary to decide whether we should apply a psychological or a physical method, or both, in a given case. It was this decision that demanded a search for a better clarification of the psychophysical relationship.

It is natural that the psychotherapists, by their success, were induced to ascribe a primacy to the psychological phenomena. Here, experiences with hypnosis and with the application of Freudian ideas were of special significance. There was a time when the "psychological" was considered so all-important that psychotherapeutic treatment even of bodily diseases was inaugurated, as, for instance, by Groddeck,[12] whose ideas attracted considerable attention after publication of his book, *Das Ich und das Es* (*The Ego and the It*). Even if this extreme point of view did not find much acceptance in the therapy of physical diseases, the particular evaluation of the "psychological" is reflected in a number of prescriptions in psychotherapy, especially in psychoanalysis—for example, the strict demand to divert, as much as possible, the attention of the neurotic patient from a somatic interpretation of symptoms, and others which will be discussed later from the organismic point of view. It also had a considerable influence on the practical physician's concept of the role the "psychological" plays in the development of disease and in therapy. Certainly, the somatically minded physician had never denied the significance of the psychological in the development of disease. He was well aware of the relevancy of the mind for what is going on in the body in disease, and he appraised the implication of psychic phenomena. He never considered them irrelevant epiphenomena, as philosophers and psychologists had often done. Indeed, he attributed to the mental aspect a special domain, separated from the somatic and only secondarily connected with it, corresponding to the general natural-science concept that the organism is constructed out of parts that only secondarily are connected with each other.

The observation of symptoms and the effect of psychological or physical stimulation on each other suggested a mutual relationship of separate processes. But consideration of the phenomena from the organismic point of view reveals that this relationship becomes understandable only if one relates it to the activity of the whole organism—influenced on the one hand by psychological, on the other by physical stimulation, which in the organism are never isolated processes. They are made to appear thus only by the use of isolating abstract consideration. Therefore, when we speak of psychological or physiological phenomena, we should be clear from the outset that these words represent only imperfect descriptions of the facts, that they refer only to the "figures" in the present process of the whole organism. They represent data which can be evaluated in their significance for the behavior of the individual only when we consider them in their functional belongingness to the present organization and activity of the whole individual organism.

From this point of view, it follows that psychological and physiological processes are determined by the same laws. This is not because the laws are equal in two different fields; rather, they are the laws of the function of the organism as a whole, which appear in the same way in the two groups of phenomena. In other words, we are not justified in speaking of parallel processes—neither of isomorphism in the sense of Koehler.

In this conceptual framework, understanding a mental or physical condition that we call sickness means (1) determining the significance of the psychic or physical processes in the development of the condition; (2) determining the role that psychic or physical phenomena play within the totality of the clinical picture and the experience of the patient; and (3) determining by which means—whether the psychological or the physiological or both—the abnormal condition can be brought back to the "norm," how the patient can best regain his health—that is, become able again to realize his nature to the highest possible degree.

Sickness from the Organismic Point of View

Before one can apply these concepts to the procedure in psychotherapy, one must define the meaning of sickness within the organismic framework. Sickness is not simply any modification of the structure or functioning of the organism, nor is it a loss of definite psychic or somatic performances. Differently considered from a differently determinable "norm," such a modification may be an anomaly, but the individual with this anomaly may neither appear, nor feel, sick. (p. 249)[6] The individual becomes sick if the condition brings the organism into a state of disorder—into catastrophe—so that he is no longer able to realize the capacities inherent in his nature—at least, to a degree that life still appears to be worth living.

The objectively verifiable changes of special functions, bodily or psychological, are the expressions of this state of disorder of the organism—of the fact that the normal, adequate relationship between the organism and the demands made upon it, which is the presupposition for the realization of the organism's capacities according to its nature, no longer exists.

The disordered function—the catastrophic condition—is revealed in the disordered behavior—that is, in different symptoms—and is accompanied by the experience of anxiety. The anxiety is generated not by the experiences of failure brought about by the actual damage, psychological or material, of the organism, but by the experience of danger to the realization of that individual's nature which is produced by the failure. (pp. 291 ff.)[6] The danger need not even be real; anxiety occurs also if the individual only imagines that he is no longer able to realize his nature.

Therapy from the Organismic Point of View

From this characterization of "being sick," it is understandable that breakdown can be the effect of very different events, bodily or psychological—in other words, any condition producing such disorder that the realization of the individual nature becomes essentially impossible.

Furthermore, the outstanding symptom need not be directly related to the cause of the disease; the patient, however, assuming that it is, and suffering from it, may demand its elimination, particularly if the symptom consists in unbearable pain or anxiety. Obviously, one might be inclined to respond to this appeal, particularly if alleviation of the symptom might make the patient more responsive to the real therapeutic procedure. Reducing pain or anxiety by medication, shock treatment, and so forth, is justified however *only if one explains to the patient that the elimination of pain does not represent the real treatment* and might conceivably delay the final improvement.

Any attempt to reduce disturbing symptoms—that is, symptomatic therapy—demands careful consideration of the effect this may have on the self-realization of the patient. This consideration requires evaluation of the premorbid personality, the character of the patient, his goals for further life, what can be expected from more intensive treatment in respect to greater or lesser restitution of the personality, and what that will mean for his future life. Whether or not this procedure proves to be useful or harmful is determined largely by the patient's capacity for understanding the physician's intention. This capacity will differ, depending upon the degree of the mental defect caused by the underlying disease.

I want to discuss, for purposes of illustration, a condition in which the defect is very severe—as, for example, the case of brain-damaged individuals.

Because frequent catastrophic conditions produce severe disturbances in brain-injured individuals, it is obviously necessary to change their environment so that it no longer makes demands upon them that they cannot meet. In doing this, we are only imitating in treatment what would normally occur passively; for we observe that this change in milieu seems to develop spontaneously after a certain time, even without treatment. This modification occurs because the patient, by withdrawal from the world around, eliminates a number of stimuli, including those producing catastrophe. The defect remains the same (as special examinations reveal), but the patient is in a more ordered condition and able to perform many undisturbed activities which were not possible for him previously. (For detailed observations of this change of behavior, see Ref. 6, pp. 35 ff.) This change of behavior is the result of an adaptation on the part of an individual, due to the trend of the organism to realize its nature—in a brain-injured individual, to stick to the preserved capacities, that is, the only thing he is able to do. He cannot avoid demands voluntarily. These demands are eliminated by this sticking to what he can do.

In terms of the organismic concept of sickness, we can say that the patient is in a "healthier" state once he has achieved this new balance between damaged function and limited environment—and there is no doubt that he also feels healthier. This state does not represent normalcy, however; it goes along with more or less outspoken restrictions of the individual's capacities, of his nature; a consequent shrinkage of the patient's present world as compared with his world as it existed before. Life may be more secure, but one can assume that, if the individual were aware of the restrictions placed upon it, he would not consider it still worth living.

Because of his mental defect, however, a severely brain-injured individual does not recognize this shrinkage of the world and his personality, especially when he lives in a custodial environment that allows the patient to get as much personal satisfaction as he needs. He may not recognize that this "custody" excludes him, in a high degree, from normal communion with his fellow men. As a matter of fact, should he become aware of his factual position—something that can happen easily if he is approached by someone who does not realize his vulnerability in this respect—this awareness alone hurls him back into catastrophe. The occurrence of such shocks during treatment—when retraining demands that the patient be confronted with tasks he cannot fulfill—is avoided more or less successfully by the transference situation (see later).

Thus, some kind of self-realization is achieved here, in spite of the restrictions imposed by the protective mechanism.

The situation is different, of course, when we deal with patients who are aware of the restriction of their world as a consequence of the protection against distress and pain. For mentally normal persons suffering from severe bodily diseases, or for neurotics and psychotics, living under such restrictions may create either temporary or insurmountable problems. A patient with a severe heart failure, for example, may be able to bear restrictions—such as those imposed by the need for bed rest—not only because he feels that they will mean an improvement in his heart disturbances but also because he is able to realize his needs to a considerable degree in spite of them, and because he hopes the restrictions will be only transient.

On the other hand, the person suffering from a chronic bodily illness may not always be able to bear the restrictions imposed by the procedure necessary to avoid pain. If such a patient were to become convinced that this condition of living was to be permanent, he might reach the conclusion that suicide is the only way out—the only means of protecting himself against the horrifying affliction of not being able to carry on with tasks that are, to him, essential; the only way of escaping the perpetual catastrophes and anxiety and, particularly, the exclusion from his world. Then we meet the apparent paradox (from our point of view, a logical conclusion) that an individual prefers death to a life so shrunken that it appears to him no longer suitable for realizing his true nature.

The situation becomes particularly complicated in neurotics. Although the patient may, for a certain time, live comfortably with the protective and defense mechanisms developed during the early years of life against conflicts and anxieties, when confronted with new conflicts—particularly the new external and internal demands which arise during puberty—he may begin to feel unbearably restricted because it becomes impossible for him to realize his nature in these circumstances. There is only one way out of the dilemma if he wants to avoid suicide, if he wants to "exist." He has to learn to bear some conflicts, some suffering and anxiety, voluntarily. The choice lies between this and the unbearable restrictions. If the patient is able to make this choice, he may still suffer, but he will no longer feel sick; that is, although he may be somewhat disordered and stricken by anxiety, he is, at least, able to realize his essential capacities to a considerable degree, thus regaining health.

Health, in this framework, is not an objective condition which can be understood by the methods of natural science alone. It is, rather, a state related to a lofty mental attitude by which the individual has to value what is essential for his life. "Health" appears thus as a value; its value consists in the individual's capacity to actualize his nature to a degree that, for him at least, is essential. "Being sick" appears as a loss or diminution of the value of self-realization, of existence. The central aim of "therapy"—in cases in which full restitution does not occur spontaneously or is not possible at all—appears to be a transformation of the patient's personality in such a way as to enable him to make the right choice; this choice must bring about a new orientation which is adequate enough to his nature to restore the sense that life is worth while.

Generally, it is demanded that psychotherapy avoid value judgments. As far as the therapist's attitude toward the failures of the patient is concerned, this demand is correct. The therapist must not impose his own values upon the patient; but that does not mean that the problem of value has to, or even can, be avoided totally. Freud, for example, believed that therapy should be based on scientific methods and concepts alone. "All that is outside of science is delusion, particularly religion." Whether or not Freud's attitude was free of value judgments is debatable, since a positive belief in natural science alone is in itself based upon a value judgment. Freud's stress on the significance of pleasure as a driving force in man was based on his special estimation of it for normal life, on the value he saw in the relief of tension.

How efficient an individual is in making the aforementioned choice and in enduring conflict and anxiety depends upon various factors.

It depends first, upon the structure of the premorbid personality—particularly the nature of his "inborn character." Here the intrinsic *courage* of the individual is of paramount importance (cf. Ref. 6, p. 306). Therapy has to make him aware of his character, able to accept the limitations which belong to his nature, to recognize life's value in spite of them, and to see a possibility for self-realization. Therapy may help him to learn that it is possible to meet the conflicts with "fear" rather than overwhelming "anxiety."[8]

Second, the capacity for choice depends upon whether the totality of the personality is involved in pathology or an essential part of it has remained normal. This difference shows up when one compares neurotics with patients suffering from organic brain defects or from schizophrenia. In the latter cases, the use of the abstract capacity, which is prerequisite for exercising choice, is reduced. Under these conditions, the patient can only try out the possible ways of behavior which may best bring about "order" and satisfactory use of his capacities; the brain-injured does it by sticking to what he is able to do in the protective environment; the schizophrenic by withdrawing more or less completely from the world and building up his own world—by using his preserved capacity for abstraction.

Finally, the choice is dependent upon past experiences and their influence on the patient's current condition, particularly with regard to how much they interfere with solving the current conflict.

How can we help the neurotic patient to find the new orientation of his personality which will bring about the condition of health?

Our first task is to help the patient in his search for the causes which have previously produced disorder and anxiety. The conflict with which we are concerned is always a current conflict. There is no doubt, however, that the current conflict also depends upon the aftereffects of previous experiences, physical and psychic alike. Here the protective mechanisms that were developed in childhood to protect the individual against anxiety can have a disastrous aftereffect. Their persistence shows, as we have already mentioned, in some traits of the neurotic's behavior which themselves produce conflicts. Thus, unearthing of previous events and experiences is of paramount significance for psychotherapy. But the material which comes to the fore in the utterances of the patient has to be scrutinized and used with the greatest care. What can be uncovered at present does not at all correspond to the real previous experiences—not even to the fantasies which, although they may have played a great role in childhood, may not be effective in the present conflicts.

Discussion of this important point in detail would involve the consideration of a number of complex problems. I should like, however, at least to point them out. First, there is the problem of the essential difference in the structure between the experiences of the infant and those of the adult. This makes it difficult for the latter to recall the previous experiences. Recollection presupposes a similarity between the situation of the organism at the time of the experience and its condition when remembrance has to take place. As I have explained on another occasion (cf. Ref. 5, p. 317), the feelings and attitudes which are predominant in childhood usually cannot be re-experienced by the adult because they cannot as a rule be made conscious; they belong to a level of awareness which cannot be authentically exposed in the psychotherapeutic situation willingly (see later).

Second, during the years of development, the aftereffects of childhood experiences undergo systematic modifications produced by the maturation of the personality and by the cultural influences under which the child grows up. Schachtel[15] has provided us with important insights into this complex phenomenon. Furthermore, aftereffects of previous experiences normally become effective (or not effective) only according to their significance (or non-significance) for self-realization in the present.

I doubt whether repression as it was described by Freud plays an essential role in forgetting in childhood. Much of what is called repression is, I believe, the effect of the modification of the child's behavior brought about by the personality changes of maturation and by influences from the outer world. These factors create new patterns which determine the behavior of the organism. Elimination of some previous experiences (called repression) occurs when the maturing organism readapts itself to a new environment and gains new patterns, of which those that appear to be "repressed" actually are no longer a part. The former reactions have not been repressed; rather, they cannot be remembered because they are no longer part of the attitudes of later life and, therefore, cannot become effectve. They can be revived or recalled under definite conditions, conditions similar to those under which they originated, such as, for example, the psychotherapeutic situation, in free associations, and in dreams; but what now comes to the fore as recollection is not an authentic reconstruction of the child's original experience. Overlooking this difference has led to many mistakes of interpretation.

The ambiguity of language creates particular difficulties in the interpretation of the adult's description of "childhood experiences." The same word may have different meanings in different situations; this statement is true, generally, but it is especially true of the way a child uses words. Just as adults and infants experience objects in ways that differ in principle, so also may adults and children use the same words to describe totally different experiences. The patient must use the language of the adult when he refers to previous experience, but this language is particularly unfit to describe the childhood experience because it is—as a rule—built to conform to the demands of the objective adult world. This language, unfortunately, is inadequate to describe the feelings, attitudes, etc., which are predominant in infancy and childhood, even when the feelings are recalled more or less clearly. Thus, when the patient speaks of father, mother, child, sexuality, and so forth, the therapist must remember that the words may not necessarily convey an accurate impression of what was actually going on in the child.

Finally, the therapist must remember that recollection is often impeded by the anxiety and catastrophes which arise from the patient's growing awareness of the dangerous conflict implicit in some experiences. The patient fights against relinquishing his previously acquired protective mechanisms. Overcoming this resistance is one of the most important functions of psychotherapy, for two reasons: first, only when the resistance has been dissipated can the patient become aware of his conflicts; second, it is particularly through the treatment of resistance that the patient gains insight into the psychic processes underlying his conflicts.

Recalling dangerous experiences is made easier for the patient if he is protected against the anxiety attached to the recollection. This protection is achieved through the development of the transference relationship. Since Freud's earliest formulations, transference has been considered a tool essential in the treatment of neurotics. In my experience, it is equally essential in the treatment and retraining of patients with organic brain lesions. In such cases, the retraining situation itself produces catastrophic conditions so frequently, even in the everyday life of the patient, that all retraining must begin with the development of transference.

The transference problem is significant for all forms of therapy and we therefore have reason to discuss it here from the organismic point of view.[10] I have come to the conclusion that transference is effective in the treatment of all diseases, organic or functional; secondly, it always has the same basic character, which is modified somewhat in the different aspects of sickness. The similarity is understandable inasmuch as we are dealing with the same dynamic problem in all conditions of sickness—the individual's reaction to unbearable conflicts and restrictions. Whether the causes of the conflict lie in an organic or a psychological defect is not relevant to the central issue.

In mentally normal patients with chronic somatic diseases, the helpful aspect of the patient-physician relationship may lie in the patient's confidence in the capacities of the physician, in his reputation as a skilled and honest man. Indeed, only if a deeper mutual relationship has been established—if the patient believes that the physician is as deeply involved as he himself is—will the patient continue treatment in the not uncommon cases where the therapy does not seem to lead to improvement, or where the symptoms increase. The development of that deeper relationship becomes imperative in the treatment of patients with whom, because of a defect of abstract attitude, a normal understanding is not possible, or is possible only to a restricted degree. With such cases as brain-injured patients, for example, the physician may not even be able to acquire enough information for retraining the patient through the usual examinations, because these for the most part require some degree of abstraction. The physician will have still more difficulty in evaluating the significance of the defect for the patient's future because, in order to evaluate the patient's potential for improvement and for the eventual achievement of the highest possible degree of self-realization, the physician must determine how much of the pattern which developed after injury as a protective mechanism against catastrophe can be eliminated, how much must not be touched because of the unbearable catastrophe that might ensue. Such an evaluation requires deep insight into the patient's previous personality, his aims, hopes, fears, and conflicts—and ways of handling them—as well as insight into the changes produced by the brain damage. Only through this insight will the physician be able to help an individual who, lacking capacity for abstraction, cannot grasp the meaning of the procedure and is unable to check directly whether the instructions he has to follow will be useful for him, and who further has to learn not to be afraid of making mistakes. This presupposition of all the effects of retraining is still more important when the patient is later expected to use what he has learned without being able to understand whether it is correct or not. The patient will only be able to meet such demands when he is convinced that not only is the physician capable of helping him but he can be trusted absolutely.

Such a conviction cannot be acquired in the usual way of communication, but it can be built up. It originates from communication on that level of consciousness which I have distinguished as level of awareness (cf. Ref. 4, p. 311) which is preserved in the patient in spite of the defect in abstraction. In this way, direct and immediate relations through common activities, feelings, and attitudes become effective in building up a state of solidarity, a state which I call communion. This state exists also between normal individuals. Without it normal mutual understanding is not possible (cf. Ref. 9). It usually exists beside, and embedded in, the level of so-called consciousness in which abstract attitude plays a predominant role; it normally originates in our voluntary act of giving ourselves over to it. The brain-injured individual is not able to develop such a relationship voluntarily because to do so presupposes abstract attitude. Communion can originate for him only in the milieu, and this must be created by the physician.

But it is not enough that the physician create the environment out of which communion may develop; he must also participate in the communion. This demands a deeply sympathetic attitude toward the patient. The physician must see him as a human being like himself with whom he can live in spite of the fact that the patient is deprived of essential human capacities. Only if he achieves this kind of countertransference will the physician be able to communicate with the patient and behave in such a way that the patient not only feels protected against the occurrence of catastrophes but understands the significance of the physician's procedure for using the psychic capacities still at his disposal for his existence in the future.

Before discussing the problem of transference in neurotics and schizophrenics, I would like to make a general remark concerning the phenomenon. There may be doubt as to whether what we describe as transference in organic patients is identical with the usual concept of transference, and whether we are justified in considering it as only a modification corresponding to the difference of the condition in neurosis and psychosis. In order to answer the question, we must consider the situation in these diseases somewhat in detail.

Our experiences with organic patients have taught us the particularly important fact that a state of transference can develop in an individual with a defect of abstraction. This brought us insight into the difficulties and the possibilities of developing transference in schizophrenics.

Freud thought that the treatment of schizophrenics through psychoanalysis would scarcely be possible because the development of transference in these patients is made difficult, even unlikely, by their narcissism—an observation which holds true if one tries to use the methods he found useful in neuroses.

To explain why the establishment of transference is so difficult in schizophrenia, it would be necessary to give a detailed account of the mental condition in this illness. Opinions on this subject have changed with the times. Different mental defects—lack of attention, disturbance of apperception, weakness or narrowing of consciousness—have been considered as explanations of the variety of symptoms in schizophrenia. I have tried to understand schizophrenic behavior as a change of personality, approaching it as I did behavior defects in organic brain damage (see Ref. 8, p. 17). To my knowledge, Storch was one of the first to emphasize the change of personality in its totality by stressing the abnormal concreteness of the schizophrenic. But his assumption, to which mine corresponds, met with little approval in later interpretations of schizophrenia. Only after careful investigation of schizophrenic behavior by means of performance tests such as those used for study of impairment of abstraction in organic patients did the problem of total personality change in schizophrenia begin to attract attention (see Ref. 11).

The Russian psychiatrist Vigotski, following the organismic concept of the defect in organic patients and using the procedures of investigation initiated in our studies, demonstrated impairment of abstraction and abnormal concreteness as characteristics of schizophrenia. His findings were confirmed by the studies of Hanfmann and Kasanin,[13] Bolles and myself,[2] and others. Vigotski[16] spoke of disturbance in abstract thinking. We considered the anomaly of thinking as *one* expression of the change of the total personality, an assumption which was agreed to by Hanfmann in so far as she believes that the intellectual and emotional disturbances are probably only two manifestations of one basic change.

This impairment of abstraction might suggest that schizophrenia is an organic disease. But such an assumption did not seem appropriate. Looking for another explanation of the symptoms, I did a careful study of the phenomena in the concrete behavior of schizophrenics and compared these with the phenomena in the behavior of organic patients. The results were illuminating: the comparison showed essential differences between the two groups.

It is not necessary to enter into a detailed discussion of these differences in order to understand what these findings indicated about the characteristic change of the schizophrenic personality and its significance for the development of transference. It may be sufficient to mention the main differences. The pattern which the organic patient shows in his concrete behavior can be understood as being due more or less to disintegration of sensory, motor, or mental processes. They show the characteristics of the dedifferentiation of function typical of all organic damages. The schizophrenic, on the other hand, develops characteristic individual patterns in his concrete procedure which reveal influences from the patient's ideas, feelings, etc. All this pointed to qualitative differences in the origin of the impairment of abstraction. This origin would be found in consideration of this defect in its relation to the totality of the schizophrenic picture, its development, its mental features, etc., which cannot be given here. It may be sufficient to point to the disturbance of abstraction in relation to one symptom of schizophrenia which is generally considered outstanding—the withdrawal of the patient from the world—and to consider why the schizophrenic withdraws. Could the impairment of abstraction be considered a means to guarantee this withdrawal? In this respect, again, our experience with brain-injured patients became important, for we have learned that, through their concreteness, a great number of demands made by our world which he cannot fulfill and which send him into a state of anxiety are eliminated. The organization of our normal world shows a greater dependence on the individual's capacity for abstraction than one is usually aware of. Is the concreteness of the schizophrenic a means of avoiding dangers which arise for him out of our world, dangers based on conflicts between him and our world which may lead him to catastrophe and anxiety? There is general agreement that experiences in early infancy play an essential role in the development of schizophrenia. Sullivan stated that the damage to the interpersonal relationship between infant and mother is of great significance for the development of schizophrenia. I consider the situation as one in which the original organic unity between infant and mother is disrupted by birth. Catastrophes, anxiety, and hindered normal development may ensue if the disruption is not repaired.[7] In this stage, the new unity between infant and world must be built up at a psychophysical level. To what extent this can be achieved depends upon the conditions of the environment, particularly on the behavior of the mother or other significant persons. Not only must the various needs of the infant—corresponding to his developing capacities—be adequately met, but, even more important, the disrupted communion between infant and world (particularly the mother) must be restored. Otherwise catastrophes may occur which the infant is unable to bear.

The infantile organism reacts to catastrophe by escape, for the organism at this stage of development is not able to build up other protective mechanisms. (In this stage, the organism can be compared in this respect to brain-injured patients with severe impairment of abstraction.) The result is that the infant tends to withdraw from the world, particularly the private, personal environment. Persons who later become schizophrenic often show symptoms of this tendency—shyness, suspicion, anxiety, withdrawal—at an early age, before the disease, under the pressure of special conflicts, breaks out.

From this point of view, the abnormal concreteness appears as a secondary phenomenon, a protective mechanism against unbearable danger and anxiety. (In principle, my interpretation of schizophrenic withdrawal agrees with Arieti's concept.[1]) The assumption that the danger arises from the world of personal relations, and that the patient tries to withdraw because of the anxiety stimulated by it, makes a number of peculiarities in the patient's behavior understandable. In this respect I must mention that, if lack of abstraction is a protective mechanism, the withdrawal will be utilized only, or particularly, in situations which are dangerous for the patient; it will be less evident when there is no danger. We see that the patient does not always manifest the failures arising out of concreteness. In the same way, it is understandable that the child who is potentially schizophrenic may develop normal intellectual capacities. One has the impression that, although the individual develops his intellectual capacities, they too represent a kind of protective mechanism, because intellectualization involves no personal relationship. In other words, since, as Abraham and later Federn have stressed, the schizophrenic does not always show behavior symptoms of withdrawal, we can conclude that the patient is not *impaired* in his abstract attitude as is a brain-injured patient, but that he does not use it in dangerous situations.

Since the patient sees the personal world as dangerous, it becomes understandable that it is difficult, and may even be impossible, to develop transference. The patient wants to avoid any communication; he resists, sometimes violently, if his conflicts are touched, because—as Federn has shown—he knows his conflicts. The use of language, important for the development of any relationship and for establishing transference in neuroses, can only be a hindrance in treating schizophrenics. We know cases which show clearly that the schizophrenic does not want to understand our language and changes his own language in such a way that we cannot understand him.

If we want to establish contact with the patient, we must avoid all topics which require abstract attitude; we must proceed in a concrete, direct way, in careful consideration of the patient's ideas, desires, tendencies, etc., and avoid all conflicts as much as possible. The physician's behavior must make the patient feel that there is not so much difference between his world and our world, that he is not so much in opposition to the latter. Thus, he does not have to be afraid of us. Only then may contact with the physician become less dangerous. The patient may thus give up his withdrawal, at least in certain situations, and even be ready to talk about his problems and to accept the help of the therapist.

This is a very crude description of the difficulties in the development of transference in schizophrenics, but it may highlight the essential points necessary for understanding and helping overcome them. It is an extremely difficult job. It requires not only knowledge, endurance, and courage but also deep devotion to one's work. I believe the successes of Klaesi, Frieda Fromm-Reichmann, Rosen, and others to be the result of such procedure. When

Rosen stresses the necessity of almost continuous proximity and attention to the patient, this corresponds to our concept of the basis of transference in schizophrenics. Frieda Fromm-Reichmann's[5] description of procedure indicates that it is similar even in details to that we apply in treatment of organic patients.

From our point of view it is understandable that one of the main points of the clinical setting of psychoanalysis—the physician sitting behind the patient who lies on the couch—is contraindicated. Only by looking at each other can patient and physician come as close to one another as is necessary. Free association should be avoided, as it is apt to increase rather than reduce the disturbance in thinking. The physician should make the patient understand why his behavior, which might have been necessary before, is no longer necessary when he has a closer contact with our world. That means a *certain neglect of the contents*, which have often been put too much in the foreground. Frieda Fromm-Reichmann stresses that: "The actual role of the therapeutic use of the contents of the schizophrenic manifestations has undergone considerable change." Much more important is the genesis of the dynamics which determine the contents of the schizophrenic productions.

In respect to the problem of contents, I am in agreement with Ferenczi, Reich, Rank, F. Alexander, Frieda Fromm-Reichmann—all of whom emphasized that it is not necessary to reach all "repressed" experiences. Alexander says that eliciting memories by free association may be less the cause of the therapeutic progress than its result. I do not wish to imply that recollections of previous experiences and conflicts are not important; I believe, however, the attitudes are often more important than the contents, a fact stressed first by Max Friedemann.[3]

Freud considered the transference in neurosis a spontaneous occurrence, an expression of the neurosis of the patient, of his pathological desire for an intensive active relationship with the analyst. It is regarded especially as an expression of the drive of repetition compulsion. The patient feels forced to experience again the difficulties of the relationship between infant and parents, by which procedure he shifts his affect against his parents to his analyst. This displacement enables the patient to become aware of the conflicts and to learn to cope with them under the protection of the therapist. Thus, transference neurosis appears as necessary for treatment. In the development of the theory, the physician's active role in the development of transference was increasingly stressed.

It is the first purpose of the therapy of neuroses to bring to light material the patient cannot remember under normal conditions but the knowledge of which is important for the understanding of the origin of the symptoms; the second is to help the patient regain his health. Often the first purpose was considered the most significant. No matter how correct that may be, the second purpose seems to have at least the same, and perhaps greater importance: to help the patient transform his personality in such a way that he will be able not only to get rid of old conflicts but to handle new ones. What should the patient-physician relationship be like in order to fulfill these tasks?

In my attempt to answer this question I would like to start from MacAlpine's[14] description of the structure of the analytic situation, despite the fact that not all analysts will agree with her and some may consider her description exaggerated. When we wish to understand a phenomenon, an extreme appearance often shows its structure particularly clearly. MacAlpine states that it is impossible for the patient to live in the setting to which he is exposed by the analytic technique. This setting forces him to regress to an infantile level—he responds to the deprivation of object relations by the situation through curtailing the "conscious ego function," thus giving himself over to infantile reactions and attitudes determined by the pleasure principle. The author even goes so far as to assume that in this situation the patient loses not only object relations but the objective world, all his actions in and out of the analytic sessions being imbued with infantile reactions. I would like to stress that this state may also exist more or less apart from the analytic sessions. In other words, the whole personality, the patient's whole life is involved. I cannot discuss here the questions of whether this description fits the condition of transference in general, whether it corresponds to Freud's concept, whether this condition develops spontaneously or is the effect of neurotic trends, whether it is more or less unwillingly produced by the analyst or develops independently out of the whole situation. I would like to confine myself to some essential remarks.

The first concerns the problem of *regression*. On another occasion[7] I came to the conclusion that there is no justification for assuming such a regression from the phenomena which gave rise to its assumption. They can be described without it and even in a better, less biased way. I have tried to show that the similarity between the behavior of an infant and a grownup in the situation described becomes understandable as an expression of the process of isolation with which we are dealing in both conditions: in infancy, due to the lack of development of the conscious behavior, and, in the situation in which we are interested here, due to the artificial state of mind the patient is in. Isolation abnormally strengthens the phenomena belonging to the level of awareness (p. 312).[4] To this state belong such experiences as attitudes, feelings, deviations from the normal functions, particularly a prevalence of directly stimuli-determined reactions which produce elimination of the disturbing effect of stimulation and which are experienced as release of tension—in other words, reactions corresponding to the pleasure principle, ambivalent reactions, etc.; they are all due to modifications of functions characteristic of *behavior in isolation in general*, not only of that in infancy. Closer observation shows further that the similarities to infant reactions are merely superficial. The contents of the behavior may differ essentially from that of the infant, and when they appear similar—for example, in verbal expressions—the words, as closer analysis will show, may mean something very different from what they meant in childhood. The tendency to compare the behavior of adults in pathological conditions to that of children or animals has produced great confusion in the interpretation of symptoms in organic pathology and is only too apt to do the same in the interpretation of neuroses. Patients show deviations from normal conscious behavior above all because the different techniques which Freud has proposed—free association, report of dreams and daydreams, etc.—are particularly apt to produce a state of "isolation."

Even if it were possible to bring the adult into an infantile state, which, as MacAlpine says, concerns the whole personality, we wonder whether this condition would be useful for therapy. We could not be sure whether what the patient utters is important for his present conflict and his neurosis. Furthermore, how could anything be gained in such a state but an idea of what happened to him before? How can he realize what it may mean to him *today*, in a mental state in which just that capacity is reduced by which this decision could be made? One may think that some improvement could be achieved by a form of acting out of partial conflicts which come to the fore. But is that real improvement? I think acting out has no positive value for the cure, however important it may be for release of tension in certain stages of treatment and for facilitation of further therapeutic procedures. It is decidedly unsuited to eliminating the present conflict because it cannot help the patient reach an adjustment to those remaining conflicts which cannot be eliminated; the patient cannot acquire the new orientation which is an essential part of psychotherapy in cases where *restitutio ad integrum* is not possible, or in neuroses where such a restitution is hardly ever achieved.

Important as the aftereffects of previous—particularly infantile—experiences may be for the development of the present neurosis, the conflicts with which we deal in neuroses can be understood only if we consider the total situation in which the individual is *now*. The patient must not only become aware of previous conflicts and the anxiety connected with them but must also understand their origin, in particular that they were unavoidable at the time of infancy but that they no longer apply to the present situation, however much he may feel the anxiety related to them. He must realize the present conflict and its significance for his self-realization now and in the future, he must see which conflicts can be eliminated and which cannot. How could the patient, in a state of transference as described by MacAlpine, find a new attitude toward himself and toward the world, and make his choice according to a value system which corresponds to his total present situation? The patient will be able to do that only if he can face the problems he has to deal with—in other words, if his attitude toward them changes from anxiety into fear. It is one of the paramount tasks of psychotherapy to help the patient to accomplish this change, and for that he must make use of his abstract capacity which is preserved in the normal part of his personality. That could not be achieved in an infantile state.

One might think that this transformation of the personality could take place in a second stage of therapy, after the infantile material has first been brought to the fore. But such a distinction between two separate stages does not correspond to the facts. Some infantile conflicts, emotions, and attitudes related to definite experiences may be remembered and lived through with emotions and intellectual insight. But during one and the same period of treatment the patient passes alternately through different stages—always the one which corresponds to the tasks he has to fulfill—just as does a normal person. Sometimes ambivalent emotional reactions are quite incorrectly called childish, i.e., regressed. They likewise belong to normal, adult life situations. If such a state of "regression" occurs in treatment, many of the difficulties of abnormal attitudes, aggressiveness, or love toward the therapist may arise. The dependence on the analyst may become so strong that one could fairly speak of a special kind of neurosis with the characteristic fixation and ambivalence. This situation should be avoided, and this can best be achieved through the organization of the patient-physician relationship which corresponds essentially to the presupposition for treatment of organic patients and schizophrenics: the patient-physician communion.

I consider such a relationship the presupposition also of the development of a transference neurosis. Without it the patient will not be willing to regard his contact with the therapist as a father-son relationship. On the other hand, I do not consider it necessary that an outspoken transference neurosis take place, although I do not mean that recollection of the parental relationship and a correspondingly ambiguous attitude toward the physician will not or should not occur at all.

I find it gratifying that my experiences correspond to those of some well-known psychoanalysts, for example, Franz Alexander when he says: "The emphasis is no longer on transference neurosis, the transference relationship becomes the axis of therapy," and when he further stresses that "the therapist should always be in control of the transference neurosis, avoiding a more extensive neurosis and restricting the growth of it to those facets which reflect the conflict."

The structure of the state of communion, in my opinion the basis of all treatment, leads us to understand why the difficulties with which one is often confronted in transference neurosis either do not occur in this condition or occur mildly. One of the significant causes of the usual difficulties is the fact that the patient is isolated from the physician, who, so to say, remains out of the game. Consequently, the patient is always afraid to lose the therapist's protection and reacts to the situation in a "primitive" way, as we all react in anxiety—that is, he adopts the pleasure principle and clings to the physician by all available means.

It is not only the person of the therapist on whom he depends abnormally but the idea that improvement will come from the outside, from a kind of powerful God or doctrine which the physician represents and which alone can cure him. This is expressed by the patient's attempt to use analytic interpretation and terminology, a practice which can prevent him from seeing the facts. He cannot consider discontinuing treatment; he cannot accept another therapist; he will consider the slightest deviation from what he has learned with disbelief. In other words, he does not become free; he does not learn to master his conflicts himself—the requisite for regaining health.

These difficulties can only be avoided if, from the very beginning, the relationship is arranged so that the patient experiences it as a common enterprise of himself and the physician, in which the latter is leading only because he has learned how to handle difficult problems, but which will be successful only when the patient shows good will and participates in the procedure.

This feeling of a common enterprise presupposes the development of communion. Only in this condition can the communication take place which is necessary to make the patient aware that his problems are not alien to the therapist, that they are common more or less to all human beings, that—however different the symptoms may be by which they are recognizable—they arise basically from the disruption of the mutual relationship between him and others, the basis of all human existence. He learns further that human existence —self-realization—always necessitates some sacrifice, which need not be taken as an expression of a positive value but as the price man has to pay for being an individual.

This valuable experience of the significance of mutual human relationship, which he has realized in the transference situation, he will take away with him when, later on, he has to live without direct contact with the therapist. He will no longer need the physician as a person; the mutual relationship between patient and physician—whether or not they later meet again—will never cease. This experience is important in that it can shorten the time of treatment, but perhaps as important is the fact that it is still effective after the treatment has ended.

If one considers the development of the state of communion a first requisite in therapy, many psychoanalytic procedures which have been treated rather like sacraments are affected. I will mention primarily the attitude toward the use of the couch as one of the most disputed procedures. Organismic technique states that the patient should not lie down on the couch before development of communion. Further, our method of approach influences the number of sessions per week, the duration and cost of the treatment, the relationship between the patient and the physician outside the consulting room, and after the end of the treatment, the relationship of the physician to the patient's relatives, and his communication with them, treatment of both husband and wife—not together but at the same time— when family problems, particularly sexual ones, are involved.

The principle of communion as a prerequisite of treatment does not imply that the patient and the physician play the same part.

With organic patients, the physician is the guide and careful observer during the whole treatment—and afterward in so far as he helps the patient to organize life in the future. But the patient must not play a passive role; he must understand that he has to be active, must learn to bear difficulties for the sake of the best form of self-realization.

In the case of schizophrenia, the physician should at first obtain complete control, but so imperceptibly that the patient will be encouraged by experiences to cooperate and to participate in the attempt to overcome the difficulties. It is important that the physician alternate in his attitude toward the patient— remaining passive when he feels that his activity touches conflicts the patient is not yet ready to bear, and becoming more active when that is possible. He should at the same time be in close contact with, and keep his distance from the patient. Important as it is that the patient feel his closeness to the physician, the latter must not show more affection than the patient is able to bear at that moment.

In the case of neurosis, the patient should be induced to participate in the procedure from the beginning. When free associations are used, particularly in relation to dreams, it should be done only periodically; long periods of free associations should be avoided, the patient should again and again be brought back to reality, and a synthesis of the results achieved should be attempted. The relationship of the patient to the physician should never be merely passive; a friendship should develop which might last after treatment during long periods of life.

Let me conclude with some general remarks

about the phenomenon of communion (cf. Ref. 7). It is not easy to describe what is understood by this term. One must experience the state in order to realize what it represents. It is an example of the frequently discussed "We" experience—or Buber's "I-Thou" experience—which we undergo, as has been said, without an act of reflection. One could call it the experience of a unity of individuals—a unity which does not eliminate them as such but on the contrary promotes their full development. It disentangles the individual from many irrelevant experiences and from many conflicts of the past and present, it makes him free to realize the essentials of life in general and of his individual existence in particular, the basis of self-realization.

We consider communion as the presupposition for every successful treatment, precisely because, in such a situation, we are dealing with an expression of one of the fundamentals of human existence, the possibility of understanding and accepting each other. The union is based on the normal drive in man to help and to be helped out of which originates the mutual concern and thus the guaranty of self-realization in the highest possible degree for the particular individual and the "other."

Bibliography

1. ARIETI, S., *Interpretation of Schizophrenia*, Brunner, New York, 1955.
2. BOLLES, M., and GOLDSTEIN, K., "A Study of Impairment of Abstract Behavior in Schizophrenic Patients," *Psychiatric Quart.*, *12*:42, 1938.
3. FRIEDMANN, M., "Die Ueberlastung des Unbewussteir in der Psychoanalyse," *Allg. Ztschr. f. Psychotherapie*, *1*:65, 1929.
4. FROMM-REICHMANN, F., *Principles of Intensive Psychotherapy*, University of Chicago Press, Chicago, 1950.
5. ———, "Transference Problems in Schizophrenia by Psychoanalytic Psychotherapy," *Psychiatry*, *11*:3, 1948.
6. GOLDSTEIN, K., *Der Aufbau des Organismus*, Haag Nijhoff, 1934 (English tr.: *The Organism*, American Book Co., New York, 1939).
7. ———, "The Concept of Health, Disease and Therapy," *Am. J. Psychotherapy*, *8*:745, 1954.
8. ———, "Methodological Approach to the Study of Schizophrenic Thought Disorder," in Kasanin, J. S. (Ed.), *Language and Thought in Schizophrenia*, University of California Press, Berkeley, 1944.
9. ———, "The Smiling of the Infant and the Problem of Understanding the 'Other,'" *J. Psychol.*, *44*:175, 1957.
10. ———, "Transference in the Treatment of Organic and Functional Nervous Disorders," *Internat. Congr. Psychotherapy*, Zurich, 1954.
11. ———, and SCHEERER, M., "Abstract and Concrete Behavior," *Psychol. Monogr.*, *53* (2):239, 1941.
12. GRODDECK, G., *Das Ich und das Es*, Nervous and Mental Disease Pub. Co., New York, 1928.
13. KASANIN, J. S. (Ed.), *Language and Thought in Schizophrenia, Collected Papers*, University of California Press, Berkeley, 1944.
14. MACALPINE, I., "The Development of Transference," *Psychoanal. Quart.*, *19*:50, 1950.
15. SCHACHTEL, E. G., "On Memory, etc.," *Psychiatry*, *10*(1):1, 1947.
16. VIGOTSKI, L., "Thought in Schizophrenia," *Arch. Neurol. & Psychiat.*, *31*:1063, 1934.

CHAPTER 37

CLASSICAL PSYCHOANALYTIC SCHOOL

A. The Theory of the Neuroses

Marvin G. Drellich

Introduction

In 1945 Fenichel published a 703-page volume, *The Psychoanalytic Theory of Neurosis*.[20] This encyclopedic treatise quickly became a standard reference since it contained an exhaustive survey of the classical psychoanalytic literature to that date. The subsequent 25 years have produced an unending stream of publications in this area, many of which are of superior merit and a few are truly seminal. Obviously a summary of the classical psychoanalytic theory of neuroses in one concise chapter must be regarded as merely an extended outline.

This outline will attempt an overview of the major theories that are the core concepts in contemporary classical psychoanalysis. The highlights of the historical evolution of these theories will be traced, and those concepts that are most extensively and effectively used by contemporary psychoanalysts will be indicated.

Preliminary Definitions

The terms "neuroses" and "psychoneuroses" are used here as essentially synonymous. They refer to the group of psychiatric illnesses characterized by prominent symptoms that have no significant somatic origin. The symptoms include disturbances of feelings (anxiety, depression, guilt), disturbances of thought (obsessions), and disturbances of behavior (compulsions and phobic inhibitions), all of which are experienced as alien to the comfort and well-being of the individual. There are no prominent disturbances of the sense of reality as are present in the psychoses.

Classical psychoanalysis is somewhat more difficult to define. The definition given here will be entirely satisfactory only to the author and perhaps to a small number of like thinkers. This definition will be substantially but not completely satisfactory to a majority of those who consider themselves classical psychoanalysts. A small number of classical psychoanalysts may hold major disagreements with the definition.

Classical psychoanalytic theory is defined as the theory or group of theories of human psychological development, functioning, and behavior that were formulated by Sigmund Freud and progressively evolved over the 50 years (1888–1938) of his creative scientific life. Classical theory today embraces nearly all of Freud's theoretical concepts as they were orginally defined by him and as they have been modified, clarified, and extended both by him and by numerous others who are closely associated with this body of knowledge. Classical theory includes all five of the metapsychological hypotheses or points of view (dynamic, economic, topographic, genetic, and structural) that constitute the major building blocks of Freud's comprehensive view of human behavior and psychological activity. The five hypotheses are not judged to be equal in clinical usefulness, in internal consistency, or in survival value. Rapaport[74] has subjected each metapsychological theory to careful scrutiny and has made predictions about which are most likely to be retained, which will undergo major modifications, and which may be entirely discarded as psychoanalytic theory continues to be responsive to influences from clinical observations, from the psychological insights derived from nonclassical psychoanalytic sources and from the nonanalytic behavioral sciences. Significant criticism of the economic and topographic points of view has come from some analysts who otherwise subscribe to the classical theories.[6,15,45,61,76,81]

Classical psychoanalytic theory retains most of the original vocabulary of Freud's formulations, although a few modifications, refinements, and elaborations of terms have occurred as the science has evolved. Where major theoretical modifications have been introduced or original concepts have been formulated, they have been included in the classical theory to the extent that the new ideas are: (1) consistent with clinical psychoanalytic observation, (2) consistent with the main body of classical concepts, and (3) formulated in terms of and integrated into one or more of the five metapsychological points of view. Greenson[47] has described the elements of classical psychoanalytic practice and technique. His concepts appear to be consonant with the statement of classical theory described herein.*

It must be emphasized that classical psychoanalytic theory is in no sense a static entity. It has undergone progressive development during and after Freud's lifetime. His nuclear theories evolved piecemeal and were formulated and published over a span of 33 years (1893–1926) as his clinical experience increased, as his technique became more penetrating, as his understanding became more refined, and as existing theories proved incapable of explaining fully the complexities of the neurotic processes he was exploring. As each new theory was developed he made radical changes in some of the concepts. His final theories of sexuality, anxiety, and repression are strikingly different from his initial concepts. Where possible he attempted to retain his existing theories, integrate the newer with the older concepts, and increase the comprehensiveness of the whole.

* A definition of a classical psychoanalyst along organizational lines was considered but did not prove to be satisfactory. The majority of the members of the American Psychoanalytic Association may be classified as classical psychoanalysts, but a substantial minority appear not to be of this persuasion. A significant minority of the members of the American Academy of Psychoanalysis subscribe to classical theory as do a large number of classically trained, nonmedical psychoanalysts who are associated with local and national nonmedical psychoanalytic organizations and training institutions.

Formative Influences on Freud's Theories of the Neuroses

In 1886 at the age of 32, Freud began the practice of neurology in Vienna. A substantial number of his patients suffered from what was then called hysteria, a broadly inclusive term applied to several disorders that would currently be called psychoneuroses. He discovered that his preparatory education in medicine, psychiatry, and neurology did not equip him to understand or to treat effectively these disorders.

Freud's Medical and Neurological Education

He had attended medical school for eight years (1873–1881). During five of those years (1876–1881) he undertook study and research on the histology of the nervous system under Ernst Brücke, the great neuroanatomist and physiologist. Brücke was a leading exponent of the "Helmholtz School of Medicine," which held that all functions of the human organism could be ultimately explained in exclusively physical terms. Brücke taught that the reflexes are the basis of all activities of the nervous system, including all mental processes.[4,16,59]

Freud was profoundly impressed with Brücke as a person as well as a teacher, and he worked for an additional year (1882) in Brücke's laboratory after obtaining his medical degree. From 1882 to 1885 he was house physician at the General Hospital in Vienna where he rotated through several services, including internal medicine, dermatology, surgery, psychiatry, and neurology. For more than a year he served in the department of nervous diseases under Franz Scholz, developer of subcutaneous injections. For five months he worked in the psychiatric clinic under Theodor Meynert, who taught that all normal and abnormal mental processes could be related to the structure of the brain and the reflex activity of the nervous system. During the entire three years of hospital service, Freud was permitted to use Meynert's laboratory of cerebral anatomy where the young house physician carried out research on human brain anatomy. As a result of his years of research in neurophysiology under Brücke and his study of neuroanatomy and pathology in Meynert's laboratory, Freud was appointed Lecturer in Neuropathology at the University of Vienna Medical School in 1885, shortly before completing his hospital service.

A penultimate formal preparation for the practice of neurology occurred when Freud requested and was given a grant to travel to Paris for study with Jean Martin Charcot, the internationally known neurologist. Charcot was not primarily concerned with neuroanatomy or neurophysiology. He was a clinician whose emphasis was on describing and demonstrating the multitude of symptoms of hysteria. During the six months that Freud studied in Paris, Charcot demonstrated that hysterical symptoms could be removed or induced by direct suggestion to patients who were under hypnosis. He taught that hysterical symptoms were neither imagined nor feigned. He asserted that ideas alone could cause hysterical symptoms to occur, especially when these ideas constituted a psychic trauma. Clearly these psychological theories as explanations for mental disorders were in direct conflict with the exclusively physical, neuroanatomical theories of Brücke and Meynert.

Private Practice

When Freud finally entered into the private practice of neurology, he found that there were few effective treatment methods for his patients who suffered from hysteria and other psychiatric disorders. Initially he used the then current techniques such as hydrotherapy, electrotherapy, massage, and the Weir-Mitchell rest cure.[78] Late in 1887 he began to use hypnosis to remove symptoms by direct suggestion. About a year later he began to use hypnosis to achieve catharsis, that is, uncovering of painful forgotten thoughts and ideas and release or discharge of the associated intense feelings. This mode of using hypnosis he had learned from Joseph Breuer.

The Influence of Joseph Breuer

Joseph Breuer was a prominent Viennese physician whom Freud had met when both were doing research at Brücke's institute. Breuer, 14 years older, was initially a revered senior colleague but gradually became a fatherly friend who referred many cases when Freud entered into private practice in 1886. Several years earlier (1881–1882) Breuer had treated a young woman (Anna O.) for various hysterical symptoms. When he recounted the remarkable story of her illness, treatment, and apparent cure to Freud, it had a profound impact on Freud's ideas concerning the causes and treatment of neuroses.

Breuer described how the patient's bizarre hysterical symptoms, paralysis, contractures, and ocular disturbances could be traced back, under hypnosis, to their origins, that is, to "traumatic" events in her past life. When these events were remembered and the associated emotions were re-experienced, the patient had a dramatic disappearance of her symptoms. She herself called it a "talking cure." Breuer and Freud called it "catharsis," and it is currently known as abreaction. Breuer's assumption was that hysterical symptoms are caused by traumatic experiences that are too painful to be remembered. These traumata reappear in a symbolic form as hysterical symptoms.

In 1888 an article entitled "Hysteria" was published in Villaret's *Medical Handbook*. This unsigned essay has been fairly conclusively attributed to Freud.[29] In it he attempted to bring together the anatomical views of Brucke and Meynert with his growing recognition of the role of psychical factors, his inheritance from Charcot and Breuer. He asserted that "hysteria is based wholly and entirely on physiological modifications of the nervous system," but he added that conditions of life, including functional sexual problems, which have a "high psychical significance," play a part in the etiology of neurosis and that psychical traumas are frequent incidental causes of hysteria.

In 1889 he went to Nancy, France, for three weeks to perfect his hypnotic techniques in the clinic of Ambroise Liébeault and Hippolyte Bernheim. He returned to Vienna where for several years he continued to use hypnosis to relieve symptoms by direct suggestion and also to uncover the traumatic causes and effect a cathartic cure of hysterical symptoms.

The years 1888 to 1898 were Freud's most creative,[77] and after several false starts he produced the fundamental concepts on which have been built the classical psychoanalytic theories of the neuroses.

The Earliest Psychoanalytic Theories of the Neuroses

Freud continued to treat patients with hypnosis, waking concentration, and finally free association, a technique that he evolved in the years 1892–1893. He had gathered an enormous amount of data on the current and past events in the lives of his patients who suffered from psychoneurotic disorders. He could not be content with the sketchy and relatively superficial psychological concepts of Charcot, Breuer, or Bernheim, so he began to formulate his own more complex and penetrating theories based largely on inferences drawn from his clinical observations. Several preliminary theories were formulated but never published because he judged them to be unsatisfactory. These false starts have been preserved through his correspondence and his preliminary drafts that were sent to Wilhelm Fliess, his sole confidant for many years.[38]

In the first publications that can properly be called psychoanalytic, we find diverse concepts that were later to be integrated into an overall theory of the neuroses. Most important is the now universally recognized concept that there are elements in the human mind, ideas, thoughts, memories, and associated feelings that are *unconscious*, that is, entirely outside of conscious awareness. These unconscious elements are not easily retrieved or brought to consciousness. They are not, however, entirely inconsequential. Unconscious psychical elements (only later were they more precisely described as drives, defenses, and so forth) are able to find expression through the symptoms of neuroses.

A second crucial observation was that *sexual factors* are the most important of the unconscious psychical elements. The detailed formulation of the role of sexuality in the etiology of the neuroses was to take another decade for Freud to work out.

Role of Sex in the Etiology of Neuroses

In 1888 Freud had said that "conditions related functionally to sexual life play a great part in the etiology of hysteria (as of all neuroses)."[29] Here he was referring to adult sexual life or to the sexuality of adolescence or young adulthood. In 1893 he published with Breuer a preliminary communication based on Breurer's Anna O. and several of Freud's own cases.[13] Their expanded publication[14] in 1895 made explicit the importance of traumatic sexual experiences occurring in adolescence or young adult life for the etiology of the neuroses.

Freud's publications in this decade include several concepts that were to find prominence in his later systematic formulations. In "The Neuro-Psychoses of Defense" (1894) the concept of *defense* is extensively discussed for the first time.[35] He described the *repression* of unacceptable sexual ideas and the *conversion* of excitation into "something somatic". The concept of defense, also called "the theory of repression," is the cornerstone of the classical dynamic hypothesis. The discovery of unconscious psychical elements and the recognition of unconscious defenses are nuclear in the theory of *intrapsychic conflict*. The conflict theory is the fundamental dynamic hypothesis and is indispensable to all psychodynamic theories.

In this paper he went beyond the theory of sexual trauma and introduced the idea that masturbation plays a crucial role in the etiology of neurasthenia and coitus interruptus is the pathogenic factor in anxiety neurosis. These theories were eventually to be abandoned in favor of more sophisticated theories of sexuality, but there is another theme introduced at this time (1894–1895)[37] that was to have lasting significance: the first formulation of Freud's theory of *anxiety*.

He described a "quota of affect," which indicated a quantity of feeling or emotion, and used it synonymously with "a sum of excitation," which indicated a quantity of energy. He asserted that anxiety neurosis occurs when a sum of undischarged affect or excitation, specifically sexual excitation, is transformed into anxiety. He called this condition an "actual neurosis" and distinguished it from the psychoneuroses. The actual neurosis is an essentially toxicological concept in that dammed up libido is transformed into anxiety. The psychoneuroses (hysteria, obsessions, phobias) were already being explained in psychodynamic terms.

The toxic theory of anxiety was to be dramatically modified by Freud in later years.[30,82] Anxiety was to be recognized as the outcome of an intrapsychic conflict. The revised anxiety theory was to become a central element in all psychodynamic theories as, indeed, it is today. The idea of a "quota of affect" or a "sum of excitation" was destined to find a place in his economic hypothesis where it has some acceptance today.

In 1896 Freud began to refer to his method and theory as *psychoanalysis*.[28] He advanced his sexual theory in dramatic terms.[27] He asserted that psychical defenses are in every instance directed against a sexual experience of a traumatic nature. Further, the traumatic sexual experience always occurred before the age of puberty and was, in fact, the seduction of the child by the child's father. In cases of hysteria the patient as a child was a passive participant in the seduction, while in the obsessional neuroses the child played an active role in the seduction.

The seduction theory was not destined to be retained for long. As early as September 1897 Freud had privately expressed doubts about the seduction theory.[59] He had formulated this theory from the histories of many patients who claimed to have remembered such seductions. He could not support the implication that there were so many perversely seductive fathers to account for the legions of hysterical and obsessional patients. Freud was discouraged by this reversal and despaired of ever finding the connection between sexuality and neuroses. From these ruins, according to Strachey,[78] "Freud became aware of the part played by fantasy in mental events and this opened the door to the discovery of infantile sexuality and the Oedipal Complex." He came to realize that his patients' "seductions" occurred only in their fantasies, which compelled him to recognize (1) that sexuality is a significant element in earliest childhood, and (2) that infantile sexuality is the consequence of unconscious *instinctual drives** that remain psychically active throughout life.

In 1897 Freud began his systematic self-analysis by recording and associating to his own dreams.[32] He sent much of this material to Fliess in Berlin, who supplied friendly encouragement for this undertaking, but appears to have contributed no substantive insights or creative interpretations. By March 1899 the daily self-analysis was discontinued, although he continued to analyze his dreams intermittently thereafter.

Most of the substance of his self-analysis forms the core of his monumental work, *The Interpretation of Dreams* (1900). Freud always regarded the dream book as his major creative work. This volume, along with his *Three Essays on the Theory of Sexuality* (1905), ushered in a substantially new and integrated theory of psychoanalysis, a theory that has been expanded and modified but remains in many respects intact today as the heart of the classical psychoanalytic theory.

The Five Classical Psychoanalytic Hypotheses

Classical psychoanalytic theory currently consists of five distinct but overlapping hypotheses or points of view. The foregoing historical discussion has indicated some of the intellectual antecedents that found their way into the five theories. Each hypothesis represents an attempt to comprehend and explain the organization and functioning of the human mind or psychic apparatus. Each theory, viewed separately, attempts to be inclusive and thereby explain the motives and vicissitudes of normal as well as neurotic thinking and behavior.

None of the five theories is, in fact, inclusive, and no claim has been made for the completeness of any or all. Indeed, these concepts are most meaningful and explanatory when taken together, that is, when they are considered to supplement and clarify each other. These five theories are to a greater or lesser extent metaphorical concepts. Classical psychoanalytic clinicians and theoreticians will generally use all five as conceptual tools, but, as will be shown, the dynamic, genetic, and structural theories appear to be the most extensively used in recent years.

No attempt will be made here to outline a complete summary, much less a statement of classical theory in all its aspects. Several volumes exist that present the entire scope of classical theory: psychopathology of neuroses and psychoses, concepts of normal and abnormal development, the theoretical basis of psychoanalytic technique, the psychology of dreams and wit, and applied psychoanalysis.[10,20,33,55,80] Here the hypotheses will be described to the extent that they bear upon our understanding of the etiology and psychopathology of the neuroses.

The Dynamic Hypothesis

The dynamic point of view is concerned primarily with the interaction of the largely unconscious factors within the mind. The most important interaction is the conflict of psychical elements within the individual; in fact, this theory has also come to be called the theory of *intrapsychic conflict.*

The recognition of the existence of unconscious mental activity is a central element in this theory. Freud's earliest discoveries of the

* In the United States classical psychoanalysts generally use the term "instinctual drive" to translate Freud's word "triebe." British psychoanalysts, including James Strachey, translator and editor of the *Standard Edition* of Freud's works, translate "triebe" as instinct. In this essay the American usage will be followed.

psychological factors in mental illness, discoveries that emerged from his encounters with Charcot, Breuer, and Bernheim and from his own clinical work, involved the bringing into consciousness of psychical elements that were in the patient's mind but were previously entirely excluded from conscious awareness. He was first concerned with uncovering the traumatic events of the patient's earlier life. These traumata had left the patients with painful and unacceptable memories that had to be put out of conscious awareness by means of a psychical *defense*. The concept of defense, which he called the "theory of repression," implies a dynamic interaction of opposing forces in the mind.

What then are the elements involved in these dynamic interactions or conflicts of intrapsychic forces? At first Freud believed that memories of sexual traumata were the elements that were repressed, but his sexual theories were then sidetracked. His concern with adult sexual practices, masturbation, coitus interruptus, and abstinence were toxicological, not psychological, theories and had little significance for the theory of intrapsychic conflict. His short-lived theory of childhood seduction was within the psychological realm, but more important, it led to his discovery of *infantile sexuality* and his formulation of the role of *sexual instinctual drives* in the dynamic psychological processes of the mind.

The dynamic hypothesis then is concerned with the conflict between the instinctual drives that "propel the organism toward gratification"[47] and the defenses, the counterforces that oppose the expression and gratification of these impulses.

For many years the instinctual drives were considered to be exclusively sexual in nature. The instinctual drives were described as "the demands that the body makes upon the psychological apparatus."[31] These demands produce within the psyche a *psychical representation* of the instinctual drive. Any fantasy or overt action that directly or indirectly may serve to express or discharge the instinctual drive is called a *drive derivative*.

In 1920 Freud[24] published a controversial volume in which he postulated the existence of a death instinct, one part of which was an aggressive instinctual drive. The death instinct has had very little acceptance in the psychoanalytic community, but the recognition of the aggressive instinctual drives and of the defenses against them has become an integral part of the dynamic point of view.[12,46,54] It should be noted that the aggressive drives are never spoken of as having a somatic source or origin but rather are judged to be entirely psychological. Nevertheless, it is possible to speak of an idea that is the psychical representation of the aggressive drive and also to speak of an aggressive instinctual drive derivative.

The defenses or counterforces oppose the discharge or expression of the sexual and aggressive drives because the individual has learned from a very early age to associate many instances of drive discharge with pain. The pain includes distressing feelings (anxiety, guilt, shame), physical pain, and danger to one's safety, security, or even survival.

From the earliest years of childhood the individual has encountered external opposition to his immediate gratification of instinctual drives. Prohibitions and threats from parents, society, and all other external sources, whether explicit or implicit, are perceived and remembered. This leads to the unconscious conviction that the direct expression of drive impulses will lead to serious consequences—pain, retaliation, or punishment. The instinctual demands upon the psyche are continual but not necessarily constant in intensity. In circumstances where external factors stimulate either sexual or aggressive responses, the internal demands are even more insistent. (Quantitative questions will be discussed in the section on the economic hypothesis.)

The opposition to the expression of the instinctual drives is initially the perceived external threats of pain or danger. Early in life the threats become internal, generate a sense of greater or lesser danger, and lead to the internal defenses which are set in opposition to the awareness and the undisguised expression of the instinctual drives. Fear of annihilation, fear of castration, fear of isolation, and even fear of being overwhelmed by unbearably intense excitation are among the unconscious fears that motivate the defenses.

The variety and vicissitudes of the defenses were not systematically worked out until after the formulation of the structural theory. Within the dynamic theory it is simply stated that the expectation of pain or injury that is associated with the gratification of sexual or aggressive drive derivatives serves to mobilize intrapsychic defenses against these impulses.[11,79]

It must be emphasized that the foregoing description was at one time considered to be characteristic of *all* mental activity, normal and abnormal. Now there has been a growing tendency to accept Hartmann's[51] views on the existence of conflict-free mental activities, but it remains a widely accepted viewpoint that *most* mental activity and subsequent behavior, be they normal or pathological, involve some degree of intrapsychic conflict.

One of the important implications of the dynamic viewpoint is the principle of *psychic determinism.* Brenner[10] states, "in the mind, as in physical nature about us, nothing happens by chance or in a random way. Each psychic event is determined by the ones which preceded it." There is a cause or causes for every mental event, conscious and unconscious. The final overt manifestation, the dream, the joke, the slip of the tongue, the choice of job, the choice of spouse, the largest and smallest decision, and, for our immediate concern, the manifest symptom or symptoms of the psychoneuroses, all are determined. All are the result of the many unconscious psychic processes that interact and conflict and thereby determine the final overt manifestations, be they normal or pathological.

The principle of the dynamic functioning of the mind—the unconscious intrapsychic conflict—for all its simplicity has been included in all four subsequent psychoanalytic hypotheses. Each new approach included a place for the continuing, active, unconscious intrapsychic conflict between instinctual drives and defenses. The structural concepts will be shown to be the most direct, though very elaborate and sophisticated, development of the dynamic concepts.

The conflict theory was the first in which Freud tried to develop a consistent and enduring preliminary statement about the psychopathology of the neuroses. The symptoms of neurosis are seen to arise out of intrapsychic conflict. The sexual and aggressive instinctual drives have intensity (see economic theory) and phase-specific qualities (see genetic theory). The defenses, which for many years were thought of as repression, direct exclusion from consciousness, have been recognized to include many simple and complex psychic activities (see structural hypothesis). The wide variety of defenses that come to be used singly or in concert to cope with instinctual drive demands, and with internal and external prohibitions, have a great deal to do with the attainment of mental health, the qualities of one's personality or character, and the occurrence of psychiatric illness. The foregoing discoveries are actually an elaborate fleshing out on the sturdy skeleton of the dynamic hypothesis.

The Economic Hypothesis

The antecedents of the economic hypothesis can be found in the influence of Brücke's, Meynert's, and Freud's early work in neuroanatomy and neurophysiology. Brücke was the model of scientific integrity, discipline, and dedication to truth. These principles found a prominent place in Freud's standards of personal and scientific conduct and persisted long after he had departed from Brücke's neuroanatomical theories of mental life.

Brücke believed that organisms are products of the physical world, systems of atoms moved by forces according to the principle of the conservation of energy. He held that all mental processes were reducible to physical reflexes of the nervous system. Moreover, all activities of the nervous system were based on a quantity of excitation, originating in the system, which travels along the nerve fibers and collects or accumulates at a larger place in the channel due to summation of stimuli.[4]

As late as 1895 Freud[38] had prepared a draft, "Project for a Scientific Psychology," that he hoped would serve as a comprehensive psychology for neurologists. He tried to explain his burgeoning psychological insights in terms that were consonant with the current neurophysiological theories. He tried to "represent psychical processes as quantitatively determined states of specifiable material particles." He described the occupation of a neuron by a quantity of energy as the *cathexis* (*besetzung*). Above all, he was concerned with the "distribution and circulation of quantities of energy within the hypothetical brain structure."[69]

He never published the "Project" because he was on the threshold of moving much further away from the neuroanatomical model of the mind. It should be noted, however, that concern with quantities of energy and cathexis was to appear again in his substantially psychological formulations, especially the economic hypothesis.

The economic theory is concerned primarily with the quantity and movement of psychic energy within the psychic apparatus. Indeed, the economic theory has been called an energy transfer theory.[69] It is one logical development of the dynamic theory in that it tries to explain the outcome of intrapsychic conflict in quantitative terms. It is concerned with the intensity of instinctual drives, the strength of psychic defenses, and the internal and external factors that affect the quantity of their energies.

The first systematic statement of the economic point of view is in Chapter 7 of *The Interpretation of Dreams.* He postulated the existence of a hypothetical psychic energy that has a quantity and displaceability. Psychic energy was introduced to explain the internal operations of his first (topographic) model of the psychic apparatus. Psychic energy is alluded to in many of his subsequent clinical and theoretical publications, culminating in the thoroughgoing restatement and refinement in his 1915 series on metapsychology.[31,41,43]

It must be repeated at the outset that psychic energy is a hypothetical concept, an abstraction that refers to an entity that is neither real nor measurable in physical terms. Psychic energy has been postulated and retained in classical psychoanalysis as the most effective way to understand, explain, and communicate observations about the vicissitudes of the instinctual drives and the opposing forces.

The instinctual drive, defined as the demand that the body makes upon the psyche, is conceptualized as producing a hypothetical central stimulation. This excitation or tension is considered to be constantly present but undergoes fluctuations in intensity. The increase and decrease of levels of intensity are described in terms of levels or quantities of psychic energy. The tension is entirely unconscious, and it is said to make demands upon the psychic apparatus in order that actions may be initiated that would permit the tension to be reduced. When instinctual tension increases it is labeled *unpleasure*; when appropriate action intervenes to reduce the tension it is labeled *pleasure*. This *pleasure-unpleasure principle* is an essential element in the economic theory; however, it must be clarified that the term "pleasure" is not used here in the narrow sense of enjoyment or delight, but rather it indicates a satisfaction or gratification that is the consequence of the lessening of unpleasure.

An instinctual drive is considered to have (1) a source, the organ or tissue from which it arises; (2) an intensity, the quantity of psychic energy that may accumulate (increasing tension) or discharge (decreasing tension); (3) an aim, in all instances satisfaction or gratification; and (4) an object, the person or thing in the external world that may be acted upon or related to in such a manner that the interaction will effectively discharge or reduce the inner tension. Actually the psychic energy is said to be directed toward, attached to, or invested upon the psychical representation of a person or thing. This investment is called the *cathexis* of the object.

An instinctual drive invests the psychical representation of the external object with increasing cathexis. This impels the individual to initiate action toward the real object in order to achieve gratification. Prohibitions and fears may oppose the motor activity that would permit gratification. Indeed, prohibitions may even oppose being conscious of the wish to act. This opposition to action or to the wish to act is called countercathexis. Countercathexis is seen to require a quantity of psychic energy to maintain inhibition of action and unawareness of the drive wish. Here is the dynamic concept of *repression*, restated in economic terms as the countercathexis directed against awareness of the drive wish.

As a result of repression and inhibition, drive tension will persist or increase and will seek alternate channels for discharge. This involves a *displacement* of instinctual energy, an important concept to explain many behavioral phenomena in economic terms. A hysterical symptom may be described as the alternate channel or pathway that symbolically discharges the drive tension but does so in a disguised manner that conceals the true nature of the drive wish. Indirect, incomplete discharge is thus possible while the drive wish remains repressed. The same symptom may also symbolically exact the punishment (pain) that is unconsciously associated with gratification of the forbidden instinctual drive.

Drives may undergo other vicissitudes besides repression and displacement. A drive may be reversed into its opposite, as when an *active* instinctual wish may be opposed and find an acceptable pathway of discharge through a *passive* involvement with the object. *Sadistic* impulses may be opposed and lead to a *masochistic* cathexis of the object. The object itself may be changed. A young man may have persistent excessive sexual wishes directed toward his mother, who had been both sexually provocative and prohibiting. These oedipal wishes will be unconscious and unexpressed, but they may be directed toward another female who in any way can be unconsciously associated with his mother. Such change of object will often be a determinant of the choice of the love object in adult life. When the original drive wish was very strongly opposed, the opposition may also be transferred to the new object, and sexual inhibitions (impotence) may be the outcome of any sexual acts directed toward the woman identified with his mother. When still stronger opposition to oedipal wishes occurs, it may lead to avoidance of all women as sexual objects, and the man may be aware only of sexual wishes toward men. These clinical illustrations are necessarily one-dimensional and cannot be taken as complete explanations of impotence, homosexuality, or any other symptom or behavior pattern. All symptoms and behavior are *overdetermined*; that is, they are the outcome of many psychic elements, drives, and defenses in prolonged interaction. In the foregoing clinical examples the elements mentioned were only a few of the many forces that combined to produce the symptom.

In 1905 Freud[42] introduced the term "libido," which he described as the expression of sexual instinctual energy. At that time he spoke only about the sexual instinct, postulated no other instinctual drives, and described the opposition to discharge as coming from internal censors or counterinstinctual forces. In 1910 he introduced the concept of *ego instincts*, which he identified as the self-preservative instincts and hence the repressive forces.[39] After that he referred to the conflict between sexual instincts (libido) and ego instincts. In 1920 he again revised his instinctual drive theory. Here he introduced the concept of the *aggressive instincts*, that is, instinctual drives directed toward the destruction of the object.[24] (See section on dynamic theory for additional comments on the acceptance of the aggressive instinctual drives.)

The wide acceptance of the concept of aggressive drives has given classical psychoanalysis a *dual theory of instinctual drives.* This theory postulates two fundamental drives: (1) the sexual drive, whose energy is called libido and which is judged to have a somatic source; and (2) the aggressive drive, whose energy has not been given an acceptable name and which has not been judged to have a somatic source. The two drives may be in conflict—that is, they may vie for priority of expression—but in most instances the two drives are fused. This means that the object representation is cathected and drive wishes initiate actions that serve to gratify both drives simultaneously. All behavior, normal and pathological, is determined by the conflicts between the fused drives and the counterforces. Obviously in any specific instance one or the other drive may be the more prominent instinctual element.

If the economic hypothesis seems somewhat obscure, that is precisely Freud's own evaluation of these concepts. He asserted that "the instincts are the most obscure element of psychological research"[24] and that this theory must undergo constant alteration with advances in knowledge.[31] Although many objections[15,61] have been raised over the years to the energic concept, it has also received some thoughtful and persuasive support.[65] Brenner's[10] comments seem to be representative of the current classical position; he wrote, "it cannot be emphasized too strongly that the division of drives that we use is based on clinical grounds and will stand or fall on these grounds alone."

At this time there remains a distinct need for a theory that deals with quantification, increase and decrease of tensions, intensity of drives, strength of defenses, thresholds, displacement, and discharge. These quantitative formulations, while hypothetical and abstract, are, nevertheless, the most effective frame of reference in which to organize the quantitative variables of psychology and behavior.

The Topographic Hypothesis

This is the name that Freud gave to his first published model of the mental apparatus. In Chapter 7 of *The Interpretation of Dreams* he described the psyche as consisting of three systems: the conscious (Cs), the preconscious (Pcs), and the unconscious systems (Ucs). He theorized a linear, spatial relationship between these systems. The Cs included the perceptual apparatus that received external stimuli and responded with a hypothetical excitation that could flow from one system to another. In the absence of internal opposition the flow of excitation will reach the motor apparatus, and action will ensue directed toward discharging the excitation.

He made it clear that "psychical topography has nothing to do with anatomy,"[43] and the psychical systems do not correspond to any location in the brain. The Ucs consisted of the instinctual drive representations that seek to discharge their cathexes, that is, to move their excitation into the Pcs and Cs and thence to impel the motor apparatus to discharge the excitation. Unconscious processes are timeless, have little regard for external reality, and press relentlessly for satisfaction. The mode of functioning of the Ucs is called the *primary process*, the tendency to discharge without delay and without awareness or regard for reality factors or internal interference. Primary process includes a readiness to displace its energies on to any object representation that permits immediate gratification; hence it is said to have a *mobility of cathexes*. In contrast, the secondary process, the mode of functioning of mental content in Pcs and Cs, takes into consideration the external realities, prohibitions, appropriateness of objects, and the need for delay or postponement of gratification.

The topographic theory provides a model of the mind within which the energy of instinctual drives moves as it presses for discharge. The psychic energy becomes attached to ideas and representations of external objects, and it encounters the censoring counterforces that are "located" in the Pcs. These forces function to control or bind the mobile cathexes that seek to move into the Pcs and Cs from their origin in the Ucs.

The topographic theory has been summarized here very briefly because in its original spatial or hydraulic form it has not retained its usefulness for a substantial number of classical psychoanalysts. The decline of this model has been accelerated since 1923 when Freud postulated the more sophisticated structural model of the psychic apparatus.[26] Freud never intended the structural model to replace the topographic one. In his later publications he continued to use both models in his formulations. Nevertheless, the topographic vocabulary has undergone changes as it continued to be used by classical psychoanalysts.

Current theory retains the concept that mental content may be fully conscious and in immediate awareness; it may be preconscious and relatively easily brought into consciousness, or it may be unconscious and therefore recoverable with considerable difficulty because of the opposing repressive forces. In this sense the terms are no longer used to designate systems but rather to describe qualities of mental content. As adjectives the words find almost unanimous acceptance.

Bellak[7] has advanced a still more precise approach in asserting that degrees of consciousness-unconsciousness must lie on a continuum. Mental content may occupy (figuratively, not spatially) any position or degree along this continuum. Obviously the three systems of the topographic model do not provide the means of distinguishing the many degrees of consciousness-unconsciousness of mental content. Gill,[45] after a thoroughgoing examination of the topographic model, concluded "there should not be a topographic point of view in addition to a structural one." Arlow and Brenner[6] concurred: "there is essentially little in the topographic model . . . which cannot be more satisfactorily explained by the structural model." Lewin,[63] on the other hand, supports the topographic terminology when discussing dreams, but he finds the structural theory more valuable when discussing the neuroses.

Primary process has been retained as a clinical term to describe a certain kind of thinking. It is primitive, irrational, wishful thinking that is dominated by emotions and is relentlessly pleasure-seeking. Primary process thinking is characteristic of the normal infant; it is more prominent in dreams and can be seen to be dominant in the psychoses. Secondary process thinking is characteristic of the more mature, rational, integrated adult personality. It takes into account the external realities, judges the appropriateness of objects and opportunities for satisfaction, and delays gratification in accordance with these considerations.

The Genetic Hypothesis

The concepts that comprise the genetic theory are widely known and are popularly associated with Freud because they include many of his most creative insights into human sexuality. Actually the genetic hypothesis is first and foremost a developmental theory, specifically a *theory of psychosexual development.* According to Hartmann and Kris,[52] "Genetic propositions describe how any condition under observation has grown out of an individual's past and extended throughout his total life span."

Genetic theory is concerned with the regular, predictable phases of development, with the tasks and problems that are inevitably encountered in each phase, and with the attempt to solve these problems in the course of maturation. Particular attention is directed toward the child's psychosexual development, the characterisitcs of the sexual instinctual drives at specific phases of the maturational process. The formal publication of the genetic theory occurred in 1905,[42] but it represents the culmination of 15 years of investigation and several unsuccessful theories of the significance of sexuality, especially infantile sexuality, for the genesis of the neuroses. His earlier theories of sexual traumata, unhealthy adult sexual practices, and childhood sexual seduction all had to be abandoned (see discussion of earliest psychoanalytic theories). From the ashes of the seduction theory came the discovery of the sexual fantasies of the child, the concept of infantile sexuality, and the timetable of psychosexual development as described in his *Three Essays on the Theory of Sexuality.*

In many of its aspects the genetic theory is based on and dependent on the economic hypothesis. Attention to the origins, quantity, displacements, and vicissitudes of sexual instinctual energy (libido) are an integral part of the developmental scheme; hence the term "libido theory" is used as synonymous with psychosexual theory.

Freud began by postulating that the sexual instinctual drive is present in all persons from the earliest days of life. The drive originates from somatic sources and makes demands on the psychic apparatus for discharge of the accumulating sexual energies or libido. The major somatic source of libido appears to shift in a regular, predictable pattern in the course of development. These major sources are called the *libidinal* or *erogenous zones.*

In earliest infancy and roughly until the age

of 18 months, the infant's libidinal tensions originate primarily from the mouth, lips, and tongue. In addition to being the chief source of libidinal tension, this oral zone is also the primary organ for gratification of this tension. The only way the infant can relieve tension (unpleasure) in the oral area is by regular opportunities to engage in vigorous sucking and later biting and chewing. The prominence of the oral zone as the source of libidinal tension and the organ of achieving pleasure has logically led to calling the first year to 18 months of life the *oral phase* of psychosexual development.

At approximately 18 months the major source of libidinal tension gradually shifts to the *anal zone*. The anus and surrounding area make up the chief erogenous zone. Unpleasure is experienced as the tension in the anal area associated with the accumulation of feces in the rectum and the conscious urge to defecate. Pleasure occurs in the anal area when defecation occurs and anal tension is relieved. Later in the *anal phase* the retention of feces and controlled defecation come to be experienced as pleasurable, and the feces themselves become an object of the child's interest.

Sometimes in the third year of life the chief erogenous zone again shifts, this time to the genitals, the penis in boys and the clitoris in girls. This is the *phallic phase* of development, when the phallic organs are the sources of libidinal tension and the organs for achieving pleasure through infantile masturbation. It appears that pleasure is also achieved by seeing the genitals of others (voyeuristic component instinct), showing one's genitals to others (exhibitionistic component instinct), and urinating (urethral erotism). The phallic phase lasts for about three years, but it must be understood that this is in no sense an infantile equivalent of adult genital sexuality.

It is essential that we take into consideration the role of *object relations* in psychosexual development. In the oral phase the first object of the sexual drive is the infant's own body, especially his own mouth. Soon he begins to have a hazy awareness of things outside himself. At first it is an awareness of the mother, the feeding person, or more accurately the visually or tactilely perceived breast, which acquires a psychic representation (memory trace) and becomes cathected as an essential object for gratification. In the anal phase the mother seeks to begin toilet training. She takes a position in opposition to the child's spontaneous pleasure of defecation. She is now an object who interferes with gratification. She seeks to assume control of the anal activity and introduces the issue of discipline. Control and discipline are issues that are destined to have far-reaching consequences on the child's ultimate character structure.

The child becomes concerned with such questions as: Who has control of my body and its functions? Must I conform, obey, and surrender control to mother, or may I somehow defy her and retain control of where and when I defecate? What is the punishment for defiance? What are the rewards for conformity? None of these occur as verbal questions but rather as nonverbal concerns of the preverbal child.

In the phallic phase an event of momentous importance occurs, coincidental with the shift to the dominance of the genital erogenous zone. This event is the *Oedipus complex*, which occurs inexorably during the phallic phase, and therefore this period is also referred to as the *oedipal phase* of object relations.

The child's sexual interest in the parent of the opposite sex and his rivalrous hostile wish to displace the parent of the same sex are too well known to bear detailed discussion. This is a normal developmental event; it does not necessarily lead to pathology, sexual or otherwise, in later life. Depending on the quality of the mother-child relationship, the father-child relationship, and the parents' relationship to each other, there is precipitated in the child's mind the elements (memories, attitudes, beliefs, convictions, fears, prohibitions) that will shape the character of his or her adult heterosexual life.

It must be made clear that the stages of psychosexual development do not come to an abrupt end. Under all circumstances there remain elements of the oral, anal, and phallic libidinal drives that remain active and become incorporated in the adult psychic life and behavior. Under favorable circumstances these *partial instinctual drives* will serve to enhance and enrich the adult personality and sexuality. Under unfavorable circumstances the persistent partial drives or the defenses against them will produce character peculiarities, rigidities, or, as we shall see, psychoneurotic symptoms.

Another necessary clarification is that the word "sexual" as used in this theory and all classical psychoanalytic theories does not have the narrow meaning by which it is ordinarily defined. Freud and classical psychoanalysts use the word "sex" as a very broad and inclusive term to indicate all psychical and behavioral processes associated with pleasure-seeking. Pleasure is associated with discharge of instinctual drive tension. There remains, regretfully, an adamant refusal on the part of many critics of psychoanalysis to understand or to remember the psychoanalytic use of the word "sex," and they base their criticism on the assumption that "sex" can only refer to adult genital sexual activity.

In order to understand the value of the genetic hypothesis for the theory of the neuroses, it is necessary to consider the phenomena of *fixation* and *regression*.[20] Fixation indicates an arresting of development, wherein characteristic elements of an early phase persist to an excessive degree in later stages of development. Fixation is likely to occur when there have been unusual frustrations or, less often, excessive gratifications during a given stage. When there have been unusual frustrations the unfulfilled drives of that stage remain unchanged and active in seeking the phase-specific satisfactions into later phases of life. An exaggerated preoccupation and pursuit of oral or anal gratifications, either directly or indirectly, are examples of the consequence of fixations at these stages.

An excess of gratification during a given phase may cause the individual to renounce this level with great reluctance and defer moving on to the challenges and satisfactions of the next phase, which may seem so much more difficult to achieve. It follows that the most favorable outcome of any phase of development occurs when there has been an optimal balance between gratifications and frustrations at that time.

Regression is seen to occur when there already exists a fixation or predisposition to fixation that had not yet produced phase-specific symptoms. At times of great stress, trauma, or frustration in a later stage of development, the individual may regress to symptoms, behavior, or defenses that are specific for the phase of fixation. In a sense a regression is either the process whereby a latent or potential fixation becomes manifest or the process whereby a minor or moderate overt fixation becomes more flagrant and dominates current behavior.

The Structural Hypothesis

In 1923 Freud worked out a new model of the mental apparatus.[26] This theory was expected to have far more explanatory potential than did the topographic model. This expectation has clearly been realized. The structural hypothesis represents the most widely accepted statement of the functional divisions of the human mind. It has been proven to have enormous usefulness and an almost inexhaustible potential for refined formulations based on clinical observations.

The structural theory divides the mind into three functional divisions, the *id*, the *ego*, and the *superego*. These are neither physical structures nor physical divisions. They correspond to no physical locations in the brain. The mental processes that are "grouped" together within each "structure" have essentially similar or closely related functions, relatively constant objectives, and consistent modes of operation. Beres[8] has warned against the danger of reifying or concretizing the psychic structures. Id, ego, and superego are abstract conceptual metaphors; they must not be treated as concrete entities. These structures are nothing more than functional systems; they can only be defined by the functions of the elements in each structure.

The id includes the psychic representations of the sexual and aggressive instinctual drives. Its contents are totally unconscious. It includes "the sum total of 'wishes' which are the resultant of certain perceptions and memories of the earlier gratification of basic physiological needs."[68]

The ego is the structure that includes the widest variety of functions. It is well to remember that Freud used the word "ego" from the beginning of his scientific career.[73] He used it loosely and at various times he meant it to refer to the self, the defensive agency of the mind, and the seat of the self-preservative instinct. With the formulation of the structural theory the word "ego" now has assumed and retained a firm meaning, and the casual, unfocused use of the term has all but disappeared.

The ego includes all the mental elements that regulate the interaction between the instinctual drives and the demands of the external world. This includes the perception of the needs of the individual (physical and psychological needs) and the characteristics and demands of the environment. The ego coordinates and regulates these disparate elements. It strives to achieve maximum gratification of instinctual wishes while maintaining the individual in a realistic relationship to the external environment.

It is impossible to say too much about the ego because the major thrust of classical psychoanalytic theory over the past 40 years has been in the direction of *ego psychology*. The latter is the conceptualization of nearly all mental activities in terms of the functions of the ego as it mediates between instinctual drives, external demands, and superego judgments. Ego psychology is concerned with understanding the many *defense mechanisms* that are available to cope with and protect against threats, real or imagined, from internal and external sources. Many symptoms of psychoneurotic illness are recognized as reflecting specific defensive operations that have been evolved by the individual to protect against the demands of the instinctual drives. Anna Freud[22] is credited with the major role in describing the important defense mechanisms in the repertory of the ego.

Repression, the first defense mechanism to be recognized as such, is the exclusion of an idea and its associated feelings from consciousness. It may exclude a thought that was once conscious but was too painful and had to be banished from conscious awareness. It may keep ideas or inclinations from ever reaching conscious awareness.

Isolation is the defense whereby ideas are split off from the feelings that are associated with and appropriate to them. An obsessional idea such as "my brother is going to have an accident" may be permitted to reach consciousness if it is isolated from its associated feelings of wishfulness and guilt. The thought without feelings is experienced as an alien, intrusive idea that has no real connection with the thinker.

Reaction formation maintains repression by replacing an unacceptable drive derivative with its opposite. A person may remain entirely unaware of anal erotic impulses to make a great mess by maintaining a conscious, active concern with order, neatness, and cleanliness.

Displacement refers to the tendency to direct an unacceptable instinctual wish away from the original object and on to an essentially neutral or less threatening object. A young woman is unaware of her hatred toward her mother but feels an often inexplicable resentment toward older women in positions of authority.

Projection is the defense whereby a painful or unacceptable impulse or idea is attributed to someone else. A young woman who cannot allow herself to recognize sexual feelings toward her married employer will be convinced that he is sexually interested in her and that he is communicating his interest in many indirect ways.

Undoing is a reparative defense whereby the individual makes amends for having thought or acted in an unacceptable, guilt-producing way. An unscrupulous businessman will make generous charitable contributions to his church on Sunday and thereby permit himself to resume his dishonest business practices on Monday. He "paid the price" and "wiped his slate clean."

Turning against the self is a variety of displacement whereby the instinctual impulse, usually aggressive in nature, is displaced from the original external hated object and directed toward the self as self-hatred, self-accusation, and self-deprecation. This is an important mechanism in depression and masochistic character neuroses.

Denial is a very primitive defense whereby the individual remains unaware of certain tangible, visible aspects of external reality that would be painful to acknowledge. The persistent feeling that the death of a loved one has not occurred and that the deceased is still alive is an obvious instance of denial.

Rationalization occurs when a person convinces himself that he is carrying out or avoiding an action for some neutral or acceptable reason in order to remain unaware of the unacceptable instinctual drive derivative that is the actual but unconscious motive.

Identification is far more than a defense mechanism. It is a ubiquitous aspect of maturation and learning through which a child becomes like another person, usually a loved or feared parent or parent substitute. By identification the child acquires some of the values, morality, mannerisms, behavioral style, and even the pathological symptoms of the object. Identification is used as a defense as a way of coping with separation or loss of the love object.

It must be stressed that defense is not to be equated with pathology. All persons need and use a variety of defense mechanisms without signs of mental disturbance. What is significant is the specific repertory of defenses, their flexibility and appropriateness, which determine a person's character structure and manifest symptoms of psychoneurosis when it occurs.

The ego includes many mental elements that are unconscious as well as almost all that are conscious. Remembering, concentrating, decision making, judgment, intellectual activities of all kinds, planning, and learning are among the functions of the ego. Initiation and control of motor activity and the development and maintenance of relations with others (object relations) are also within the province of the ego. Reality testing is an obvious ego function, as is "regression in the service of the ego," a phenomenon whereby the artistic or creative person can draw upon repressed mental content in the service of the creative process.

The third psychic structure, the superego, is actually a highly specialized aspect of the ego. It is the structure that represents the ethical and moral attitudes, the readiness to feel guilt, the unconscious roots of the sense of conscience. The superego content is the result of internalization of the moral and ethical standards of society, especially as these values have been communicated and to some extent interpreted and modified by one's parents.

One of the virtues of the structural hypothesis is that it meshes very effectively with the dynamic and genetic points of view. The three psychic structures are not considered to be fixed or immutable; rather they are viewed as developing under the influence of inborn constitutional factors and the individual's unique developmental experiences. *Ego development*, especially, has received a great deal of study and is recognized as proceeding parallel with and complementary to libido development.

Ego development has not been as thoroughly worked out as has libido development, in part because the structural hypothesis was a later arrival on the theoretical scene, but primarily because ego development is a far more complex matter. It can only be understood in terms of the development of the many ego functions: defense, thinking, dealing with reality, object relations, regulation and control of instinctual drives, the various autonomous functions (to be discussed in the next section), and the organizing or integrative function of the ego. Only a brief sketch of ego development is possible here; for additional details see Hartmann and Kris,[52] Hartmann, Kris, and Loewenstein,[53] and Jacobson.[57]

The earliest pre-ego process is believed to be the *stimulus barrier*.[44] This barrier protects the infant from the multitude of external stimuli that constantly impinge upon him and would produce an intolerable overstimulation were it not for the dampening by the stimulus barrier. The barrier serves to protect against the "traumatic situation,"[30,33] the situation in which the organism is flooded with an excess of stimulation that it is helpless to escape or discharge. It is further postulated that an innate pre-ego perceptual capacity is present at birth. At the outset this perception may simply be able to experience and differentiate painful states (increasing drive tension) and pleasurable states (decreasing tension).

The ego and the id develop from an *undifferentiated matrix* of instinctual energy in the earliest weeks of life. As the infant begins to endure the first painful consequences of contact with the external world (for example, when he is not fed at the first signs of hunger), there is an instantaneous attempt to relieve this tension by a hallucination. This hallucination consists of memory traces of previous satisfactions (being fed) and the fragmentary perceptions of the external object (mother) that had been associated with previous gratifications. This is the beginning of what will later be fantasy formation. Hallucinatory fulfillment is only momentary because the actual tension of hunger persists and increases. Next an overt motor-affective response ensues; the infant begins to cry and thrash his arms and legs about. When the mother responds promptly by feeding and fondling the infant, there is a reduction of drive tension, a sense of gratification, and in the earliest months the infant falls asleep.

This event includes the earliest perceptions, memory traces, and object representations. Here, too, is the beginning of the sense of differentiation of the *self* and the *object representation.* In the earliest months it is postulated that during or after gratification the differentiation is not maintained, and a sense of *fusion* of self and object ensues. This blissful fusion persists until hunger or other painful tension occurs to disturb the ecstatic, intrapsychic union.[64] The fusion of self and object is a precursor of the mechanism of identification.

It is the frustration and the failure of the hallucinatory wish fulfillment that provide the impetus for the activation of the ego functions of consciousness, differentiation of self, and object and motor affective responses. The latter have an impact on the environment (are "successful" in bringing about feeding) and also contribute to the perception of one's body and body parts. Primitive perception of reality and focusing of attention occur with this event. The hallucinatory wish fulfillment is a primary process response for instant gratification. The failure of hallucination and the sequence of subsequent activities are secondary process responses. A series of similar events produces memory traces that are the basis for *learning from experience.*

Every day brings frustrations and unpleasurable perceptions that evoke ego responses that are to a greater or lesser extent successful in achieving gratification. The ego functions become more refined and additional functions are activated as different tensions arise. Differentiation of self and mother becomes more distinct, and secondary process thinking, with ability to endure longer delays in gratification (frustration tolerance), becomes more prominent.

Before the end of the first year the infant has some sense of the success of his motor-affective responses in evoking consistently gratifying responses from the mother. With this comes the capacity to be awake and alone for short periods of time with the expectation that mother will be available and responsive when needed. This growing sense of *object constancy* has lasting impact on the development of *basic trust.*[18] It also produces a sense of one's effective influence upon the external world, an important ingredient for feelings of *self-esteem.* The repeated frustrations, gratifications, and separations in the second and third years are part of what Mahler calls the *separation-individuation* phase of ego development.

The crucial question of anxiety is related to these first three years.[30] The traumatic situation of flooding the organism with excess, undischargeable excitation has already been described. The excitation may come from external sources or from instinctual drives. As development progresses, the likelihood of flooding from external sources is lessened, but the danger of flooding by undischargeable drive tensions remains prominent. The earliest instances of increasing drive tension (hunger) and the absence of mother are probably recorded as unified memory traces; that is, the extended absence of mother and the mounting drive tension are "remembered" as inseparably associated events. Later the mere absence of mother may evoke the association with increasing drive tension. Before the tension reaches the level of flooding, the ego diverts a small quantity of instinctual energy to produce the subjective feeling of anxiety. This *anxiety serves as a signal* to warn that the traumatic situation is imminent and to mobilize the executive functions of the ego. Anxiety as a signal of danger is one of the first defensive operations of the ego and is fated to have the most lasting consequences.

The anxiety signal is evoked by all subsequent perceptions of danger from external or internal sources, whether these perceptions are real or imagined. The role of anxiety in the genesis of psychoneurotic symptoms cannot be overestimated. In all instances where an unacceptable instinctual impulse threatens to overwhelm the ordinary defenses, the anxiety signal is evoked and new or intensified defenses are set into operation in order to maintain repression. Throughout life the persistence of anxiety in the absence of external danger, the commonest symptom of all psychoneuroses, is a clear indication that the ego unconsciously perceives a continuing danger of instinctual drive derivatives reaching conscious thought or action. The theoretical and clinical publications on the psychoanalytic theory of anxiety are numerous and of high caliber.[9,21,49,70,72,82]

Separation anxiety appears to be the earliest occurrence of the signal function. In the oedipal phase a new surge of phallic instinctual drive tension introduces new dangers. First, the possibility exists that the sexual interest in the parent will erupt into conscious thought or action, and then the parent of the same sex will retaliate against such expressions of sexuality. The punishment is unconsciously conceptualized as castration, that is, loss of the "offending" organ that is simultaneously, the source of the excitation and also the organ of expression of the forbidden impulse. In the face of *castration anxiety* the child takes in the real or imagined parental prohibitions and threats, the psychic representation of which forms the nucleus of a separate structure, the superego.

Development and elaboration of ego functions proceeds into later childhood, where motor, intellectual, and social skills are evolved as the child must meet new demands of the external world. In adolescence the capacity for efficient regulation and control of sexual drives is developed in response to the increased libidinal demands of this period. The ability to restrain effectively and to express appropriately the sexual drive and the capacity to relinquish excessive self-love (narcissism) in favor of object love are ego activities that must be developed in adolescence. The maturation of ego functions continues into adult life, especially in the areas of work and love.

Classical Metapsychology after Freud, The Adaptive Hypothesis

One outgrowth of the increasing concern with ego psychology has been the special attention that has been paid to normal development and nondefensive ego functions, the so-called *adaptive hypothesis*. Since this area is not immediately concerned with the neuroses it will be mentioned in the briefest way.

In 1939 Hartmann took up the problem of adaptation in ego psychology.[50] In this and subsequent works he postulated a *conflict-free* area of ego activities. He referred to development of conflict-free ego functions as *autonomous ego development*.[51] Many ego functions that arise independent of the drive-defense conflict are described as *primary autonomous ego functions*. Further, he went on to describe certain ego functions that originate in the drive-defense conflict but become detached from conflict and come to function in the service of the ego. These he called *secondary autonomous ego functions*.

Autonomous ego functions operate with neutralized libido or neutralized aggressive energy and include aspects of perception, memory, motility, and the stimulus barrier thresholds. Taking an evolutionary approach, Hartmann postulated that the primary autonomous ego apparatuses are adaptive elements whose functions have evolved as preparedness for coping with an "average expectable environment."

Many of Hartmann's theoretical concepts are accepted in the mainstream of classical psychoanalytic theory. Somewhat less widely accepted but clearly on the threshold of acceptance are the theoretical innovations of Erikson.[18,19] In the adaptational spirit he has formulated a timetable of psychosocial development. He has emphasized the interdependence of libido development, ego development, and social experience, not only in childhood, but throughout the entire life cycle. Erikson describes an *epigenetic* sequence of developmental phases, parallel to libido and ego development. The epigenetic concept emphasizes the inevitable, universal ground plan. There are specific social developmental tasks and challenges occurring at predictable developmental phases of social adaptation. Each person develops unique solutions to the phase-specific social tasks.[60]

It is noteworthy that neither Hartmann nor Erikson made any attempt to integrate their theories into a unified adaptive point of view.[81] Perhaps each regarded his formulations as preliminary. Certainly these most recent additions to classical theory will require further clinical observation and evaluation, but Rapaport has judged them both to have a high likelihood of survival.[74]

The General Theory of the Neuroses

The foregoing summary of psychoanalytic metapsychology is in no sense complete. It is intended to provide an organized statement of the major theories that permit the formulation of a general theory of the neuroses in classical psychoanalytic terms. It is apparent that this formulation draws upon all five metapsychological points of view to the extent that each system permits meaningful and consistent generalizations that can be based, first and last, on clinical observations. Much of what follows has been discussed in the sections on the five hypotheses. Here the implications for the genesis of neuroses will be brought together.

Intrapsychic Conflict

The most important element in the psychoanalytic theory of the neuroses is the concept of intrapsychic conflict. The conflict is considered to be between an unconscious instinctual impulse that presses to reach consciousness and an opposing force or forces. To be more precise, the unconscious impulse is a mental derivative (an unconscious fantasy or wish) of an instinctual drive (either sexual or aggressive). The derivative is seeking to enter consciousness as a conscious thought or through overt physical action. The opposition has been attributed to countercathexis in the economic theory, to a censor in the topographic theory, and to the defensive functions of the ego in the structural theory. This opposition occurs because of the real or apparent dangers that are associated with the emergence of the forbidden instinctual derivative into consciousness. Many prohibitions derive from reality-adaptive factors, such as the needs to survive, to remain physically intact, to avoid pain, to assure stable and gratifying object relations, all of which operate through the ego functions of perception, learning from experience, memory, and capacity for anticipation. Other prohibitions derive from moral forces, which are said to be superego demands upon the ego to maintain repression.

Unconscious intrapsychic conflict is a universal occurrence in all persons, healthy or psychologically disturbed. Conflict is postulated to begin sometime in the first year of life when the earliest ego functions appear to develop and are called upon to oppose the immediate and indiscriminate discharge of the inborn instinctual drives.

Psychoanalytic theory is concerned with quantitative factors in the conflict between drive derivatives and repressive forces. The quantitative factors are inferred from clinical observations of the result of the conflict, and there is no intent to suggest that such quantities are measurable in any concrete sense. When the strength of the drive is greater than the repressive forces, the drive derivative will reach conscious awareness as a thought, idea, or fantasy, or as a physical action that is directed toward direct gratification. The strength of the instinctual drives is judged to be increased at critical times in psychosexual development. The oral, anal, and phallic instinctual drives are maximal at the phases of development that correspond to the dominance of each of these erogenous zones. At puberty the instinctual drives are considered to be strengthened by the psychophysiological changes of this period. The instinctual drives are judged to be weakened at times of physical illness and in later life.

When the repressive forces are sufficient to prevent the emergence of the instinctual drive into consciousness, *repression* is said to occur. Repression cannot be expected to be permanent; the instinctual drives continue to exert a pressure toward discharge, and the repressive functions must exert a continuous counterforce in order to maintain repression. Such continuing unconscious conflict is ubiquitous in all persons at all times. The outcome of the conflict will have consequences for the character and mental health of the individual.

Stable Outcome of Conflict

The most favorable outcome occurs when there is a stable relationship between drive and defense, and the conflict has little or no intrusive impact on conscious waking life, although it may be detected in dreams. During sleep the defensive functions are somewhat relaxed or reduced; the unacceptable instinctual drives are relatively stronger and threaten to reach consciousness. In this situation the dream functions to provide a hallucinatory gratification (wish fulfillment) of the instinctual drive derivative, while the residual defensive forces operate to disguise the direct nature of the wish fulfillment lest the dreamer be awakened by confrontation with the drive derivative in its direct, undisguised form.

Another consequence of the stable relationship between drive and defense is the development of the individual's *character*. The variety of defenses and tactics that the ego has employed to admit, repel, or modify the instinctual drive demands are unique for each individual. A relatively fixed pattern of defenses is established for each person, and these defenses shape the regularly recurring tactics for coping with internal conflicts. Depending upon the specific experiences of psychosexual development (traumas, frustrations, gratifications) and upon the effectiveness of ego development (variety, strength, and flexibility of defenses), the individual may come to adulthood with a "normal" character structure or with a *character neurosis*. A normal character structure involves the capacity for a degree of direct gratification at realistically appropriate times and places, a successful use of modified drive gratification (sublimation), and the relative flexibility of defensive patterns. In the character neuroses the individual is burdened with a very narrow spectrum of defensive options. The limited number of defenses are endowed with excessive strength and interfere with direct drive gratification even when such satisfaction would be appropriate. Successful sublimation of drives is limited, and the constricted but intense defenses produce rigid, stereotyped patterns of behavior.

Neurotic Outcome of Conflict

When there is a significant disturbance in the stable equilibrium between the instinctual drives and the repressive forces, particularly when the strength of the instinctual drive threatens to overcome the defenses, the development of psychoneurotic symptoms is a likely outcome. The symptom occurs as a result of a *compromise formation* as follows: (1) the drive derivative and its associated unconscious fantasy wishes find a channel that permits an incomplete gratification; (2) the repressing forces impose a highly disguised form upon the gratifying process in order to make the specific drive derivative consciously unrecognizable; and (3) the superego imposes an element of suffering onto the drive expression in order to disguise the fact that gratification of unacceptable drive derivatives has indeed occurred. The superego-imposed suffering also functions as a self-punishment that is unconsciously introduced to ward off the expected external punishment for drive gratification. These dynamics explain why the symptom is often called *the return of the repressed.*

The psychoneurotic symptom, whatever may be its external form, can be explained as the outcome of an intrapsychic conflict that has resulted in a compromise between partial expression of the drive, profound disguise of final expression of the drive, and the imposition of conscious suffering as punishment for drive gratification. In the section on the special theory of the neuroses (see below), specific psychoneurotic symptoms will be discussed in the framework of the theory of conflict.

Psychosexual Development, Ego Development, and Neurotic Conflict

The foregoing description of the psychoneurotic outcome of the intrapsychic conflict has included brief comments on the strength and character of instinctual drives and defensive forces. Here there must be a more detailed consideration of the elements that influence the strength and character of the drives and the ego functions. These influences are regularly traced back to childhood developmental experiences, especially the course of infantile psychosexual development and the course and outcome of ego development. Psychosexual development and ego development occur simultaneously, and each is profoundly influenced by the other; both are also influenced by the child's interaction with the significant persons in his immediate external environment. (See sections above on psychosexual and ego development.)

Freud's discovery of the role of childhood experiences and especially the role of infantile sexuality in the genesis of the neuroses is firmly held today and is conceptualized in the framework of the genetic hypotheses. Fixation at a particular phase of psychosexual development is an arresting of development where the partial instinctual drive of that phase is said to persist with an excessive intensity into the later stages. There is a concomitant arresting of ego development wherein the ego defenses that are a permanent feature of the phase of fixation retain an excessive prominence in the repertoire of ego functions in later life. Excessive frustration is the chief basis for fixation, although in some instances excessive gratification will also contribute. Fixation requires that extraordinary or unusually strong ego defenses be mobilized in order to contain the intense, persistent partial instinctual drive. Such diverting of defensive energy to oppose persistent drives of an earlier phase (which should have achieved relative stability) will impair the ego's energy and versatility in coping with the partial impulses of the current phase.

Regression occurs in later life in response to stress, trauma, and frustration; it involves reverting back to the libidinal drive derivatives and also to the prominent ego defenses that are characteristic of the specific phase of fixation.

The Special Theory of the Neuroses

The extended discussion of classical psychoanalytic theory is a necessary preliminary for a presentation of the psychogenesis of the psychoneuroses. It will be helpful, before proceeding, to reiterate a highly condensed statement of the genesis of psychoneurotic symptoms: (1) Unconscious intrapsychic conflict between instinctual drive and defense is ubiquitous. (2) Some of the functions of the ego are: (a) to maintain repression of unacceptable drives, (b) to permit the expression of acceptable drive derivatives, and (c) to arrange compromises through which unacceptable drives may find expression in a disguised form. (3) One form of compromise is the psychoneurotic symptom. Its manifest elements are determined by the nature of the drive, the form of the disguise imposed on it, the defense involved in maintaining the disguise, and the addition of painful elements as punishment for drive discharge. (4) Anxiety is the *sine qua non* that signals the possible or imminent emergence of repressed impulses. The anxiety signal usually causes an increased intensity of defenses as well as evoking new defenses to assure that repression is maintained and thereby anxiety is reduced or eliminated. (5) In this last sense it is convenient to speak of defenses against anxiety as a shorthand for defenses against the instinctual drive whose threatened emergence produces an anxiety signal.

Anxiety Reaction

Anxiety reaction is the psychoneurosis characterized by chronic, free-floating anxiety and episodes of acute intense anxiety. Earlier this was called *anxiety neurosis* and *actual neurosis* and was considered to be caused by a damming up of libido and direct transformation of libido into anxiety. This was a purely physiological or toxic theory and has now been replaced by a psychological theory. The distressing anxiety is caused by the failure of all the defenses to maintain effective stable control over the instinctual drives. Patients with anxiety reaction may have phobic, obsessional, compulsive, or depressive symptoms. They are likely to have passive-dependent personality traits, sexual and work inhibitions, and occasional paranoid feelings, but neither the neurotic symptoms nor the neurotic character traits are sufficient to stabilize the conflict, so the patient is rarely free of anxiety.

Persistent feelings of helplessness are indicators of traumatic experiences that date from the earliest developmental phase and recur in later phases. Separation anxiety and fear of being overwhelmed by one's one feelings derive from traumata in the oral phase. Fear of losing control of oneself and fear of having no control over the diffusely dangerous external environment result from anal phase difficulties. Sexual inhibitions and gender identity problems are traced to persistent oedipal phase problems.

Throughout early development the child was subjected to a heightened influx of stimuli from severe traumatic experiences (external sources) or from unusual degrees of infantile frustration (internal source). Available channels for discharge of excitation are likely to have been reduced because of extraordinary parental neglect or prohibitions.

There is some question whether the acute anxiety attack is not only a sudden signal of danger but may also be a diffuse, nonspecific discharge or safety valve for the internal excitation.[21,72] This is unsettled, but if it is a discharge phenomenon as well as a signal, it is unsuccessful because little or no relief of chronic anxiety is experienced after the acute anxiety attack.

Phobic Reaction

Phobic reaction is the psychoneurosis that has also been called *anxiety hysteria*. The patient has one or more prominent phobias. A phobia is an intense feeling of anxiety that is evoked by a specific, ordinarily neutral place or thing called the phobic object or situation. The patient will experience mounting anxiety as he approaches or comes into contact with the phobic situation or object.

A patient with a bridge phobia will have intense anxiety as he comes near a bridge and as he crosses it. As he drives off the bridge there is a sharp decline in anxiety, and when he can stay away from bridges he may be relatively free of anxiety.

The phobic object or situation symbolically represents a forbidden or painful instinctual impulse; it simultaneously represents an external object or situation that is likely to increase the intensity of the forbidden or dreaded instinctual drive. By the mechanism of *displacement* the anxiety becomes diverted from the internal drive and from the realistically threatening external stimulus and attaches to the otherwise neutral phobic object. To the extent that the patient can *avoid* the phobic object he remains generally free of anxiety.

Displacement and avoidance are the definitive defense mechanisms in phobias, but they are only partially successful. When the phobic object cannot be avoided the anxiety returns. Furthermore, the phobic object is frequently a thing of uncanny fascination to the phobic patient. The fascination occurs because the phobic object, while ostensibly neutral, contains symbolic qualities that cryptically reveal the feared object and the feared instinctual drive. The cryptic representation functions in much the same manner as the manifest dream disguises, but it also reveals or leads to the revealing of the latent dream wish.[63]

The phobic object represents a *temptation*, an unacceptable sexual or aggressive impulse, and simultaneously represents the *punishment* believed to occur if the impulse were to be acted upon. *Claustrophobia*, the fear of closed places and crowds, illustrates some of these dynamics. A man who slept in his parents' bedroom for more than eight years had frequently overheard and observed parental intercourse. This recurring *primal scene* is a trauma in that it floods the youngster with sexual excitation for which there is no discharge channel. His fear of closed places in later life was traced in part to these oedipal origins. The closed place represents first the bedroom from which he couldn't escape. It also represents the wish to intrude into the parental sexual act, not only to separate his parents but also to become a participant in the sexual act. In some unspecific way he sought to relieve the painful undischargeable sexual excitation by participation in the parents' sexual activity. Finally his adult fear of being smothered and actual difficulty in breathing in the claustrum represent the punishment for the incestuous wishes.

Phobias may originate from earlier phase traumas. A young woman had had a talented, dramatic, overstimulating mother whose behavior toward the daughter was highly inconsistent. The little girl had experienced unpredictable, impulsive attentions and equally unpredictable indifference from her mother. The attentions included singing, dancing, endless talking, reading aloud, all accompanied by dramatic hugging and other physical contact. The girl was excited, delighted, and also terrified by these attentions. When her mother was preoccupied and unresponsive the child was initially relieved, but later she sought out, with unpredictable success, more interaction with her mother. As an adult she had claustrophobic symptoms especially pronounced during sexual intercourse. With her husband in the superior position and manifesting great sexual excitement and activeness, the patient would develop feelings of being overwhelmed and smothered by him. Analysis established that the phobia represented the wish to achieve primitive oral gratification with her mother, to achieve a psychical sense of fusion with the "loving," nourishing mother. The fear represented the memory trace of the overwhelming, overstimulating mother who, she believed, was capable of actually enveloping and incorporating her. The claustrum was the wished for mother who simultaneously loved and threatened to devour her and also included punishment for the patient's aggressive feelings, which were responses to the mother's inconsistency and unpredictable rejections.

Conversion Reaction

Conversion reaction was more common in Freud's time when it was known merely as *hysteria*, but it is by no means rare today. The most prominent feature is the *conversion* of instinctual drive derivatives into one or more apparently somatic symptoms. In the absence of physical disease the patient may have spasms, paralyses, weakness, anesthesia, paresthesia, coughing, shortness of breath, pains, or any other symptom ordinarily associated with somatic illness.

The patient is likely to be unsophisticated about psychological matters and is naïvely unable to see through the often tissue-thin disguises that are intended to conceal the instinctual impulses that are both expressed and punished by the symptom. For many years the conversion reaction was considered to be the result of a phallic phase fixation. The symptom represented the disguised gratification of forbidden oedipal wishes as well as the symbolic retribution. The woman who is unable to swallow is discovered to be expressing her fellatio fantasy directed toward her father. Weakness, paralysis, or pains in the hands are expressive of (1) masturbatory impulses associated with unconscious oedipal fantasies, and (2) destructive impulses toward the parent of the same sex who "interferes" with oedipal gratification.

The hysterical patient invariably shows a pronounced preoccupation with sexual matters, transparent seductiveness as well as an apparent sexual "innocence," strong sense of sexual morality and sexual inhibitions, all of which are immature sexual responses and are consistent with the concept of oedipal phase fixation.

In the past two decades clinical evidence points toward a primary role for oral fixations and an important but secondary role for oedipal fixations.[67] The patient shows gross immaturity, constant seeking of love and attention, and minimal capacity to tolerate frustration. The prominent defense mechanisms of the hysterical personality, identification, denial, and fantasy formation, are that which are regularly evoked during the normal development in the oral phase. By *identification* the hysteric, who is highly suggestible, takes on as his own the symptoms and behavior of those who are closest to him. Identification by oral incorporation is normally the first type of object relationship, but for the hysteric it persists into later phases and even into adult life as a mode of relating to others. One consequence is the absence of a distinct sense of self and an ease in changing to whatever behavior and response appear to promise instant gratification of needs.

Fantasy formation is another prominent ego activity in hysterics. The fantasies are not mature rehearsals for real experiences but rather are wishful gratifications. The most unrealistic wishful fantasies are retained against all the evidence of experience; this is because *denial* of external realities is a regularly encountered ego defense.

The oedipal phase is not coped with in a satisfactory fashion because the prehysteric child brings to this phase exaggerated oral dependency strivings and ego functions that were more suited to cope with oral phase issues. Small wonder that the adult heterosexual life has many infantile features. It betrays strivings for anaclitic gratifications and oral phase ego defenses in what is only superficially adult genital activity.

Obsessive-Compulsive Neurosis

Obsessive-compulsive neurosis is a widely recognized disorder. The patient is troubled by persistent obsessional thoughts, usually of a painful, guilt-producing, or otherwise distressing nature. The obsessions will often dominate his thinking, will interfere with full participation in intellectual, social, sexual, and work activities, and, above all, will be experienced as alien and intrusive. The compulsions are the stereotyped recurring acts and rituals that the patient feels an irresistible urge to carry out. Anxiety is felt if the compulsion is not performed, and temporary relief of anxiety occurs when the act is carried out in the "proper" fashion and the "correct" number of times. Checking over and over again to see that the door is locked, that the gas is turned off, or that the baby is still breathing are obvious examples. Dressing or undressing in a fixed sequence, counting rituals, compulsive touching, or avoidance of touching are also typical.

The obsessive-compulsive is likely to have a characteristic life style that does not seem alien to him but may be a trial to others. He is a serious, conscientious, hardworking person who seems to be more comfortable as a follower rather than as a leader. He is careful to obey the law and follow the wishes and orders of others. He is honest, orderly, thrifty, inflexible, unimaginative, overly intellectual, and literal. These traits describe the individual who is said to have an anal personality.[2,58]

Psychoanalytic investigation has traced the genetic roots of obsessive-compulsive neurosis to disturbances of development in the anal stage. One of the major issues in this phase has to do with the child's struggle to retain *control* over the evacuation of the bowels and over the pleasure associated with expulsion and retention. The struggle includes the mother's attempt to institute toilet training; she attempts to impose her will as to where and when the evacuations will occur. When toilet training has been harsh, rigid, and carried out in an atmosphere of fear and threats, the child will submit out of fear and will manifest an apparent compliance while concealing profound rage, resentment, and the need to find opportunities for disguised defiance. The prototype of all later disciplinary, conforming experiences is established here.

The child will be burdened with unusually strong aggressive and destructive drives, which can rarely be allowed into consciousness or expressed directly lest he experience intense anxiety and guilt. Disguised expressions of the aggressiveness and defiant and oppositional impulses are reflected in many character traits and actions. The development of genital sexuality is impaired by the intrusion of anal-sadistic and aggressive drive derivatives into the adult sexual situation. Sometimes it is the defenses against these drives that are most obvious in the form of inhibitions of sexual performance.

The defenses that are prominent in opposition to anal-sadistic drive derivatives include *reaction formation*, changing impulses into their opposite. Rage and resentment are expressed as exaggerated kindness and considerateness; tendencies toward messiness and disorder are outwardly seen in neatness, orderliness, and cleanliness. *Undoing* is a technique that allows defiance, rage, and negativism to be expressed with less disguise provided a severe penalty, punishment, or compensatory corrective action is arranged for to undo the original destructive action. *Isolation* of feelings from ideas permits the obsessive-compulsive to have grossly destructive obsessional fantasies and keep them unreal and less painful because they are isolated from their associated feelings. *Magic thinking* finds a function with these patients and is the basis of many of the private rituals and fantasied pacts with fate. *Regression* to anal erotic forms of pleasure and to primitive, anal phase ego activities is a response to new stress or traumata in later life.

The *compulsions* themselves serve a defensive function to channel off the aggressive energies into "approved," non-guilt-producing, outwardly harmless activities. All defenses are directed toward maintaining unawareness of aggressive tendencies and preventing any overt, undisguised aggressive or destructive acts. The unconscious "knowledge" of his intense rage requires that the obsessive-compulsive never allows himself to lose control of himself. Freedom, spontaneity, and impulsive behavior are associated with loss of control of oneself and the unleashing of one's wide-ranging murderousness and destructiveness.

The foregoing discussion of the genesis and dynamics of several psychoneuroses is far from exhaustive and is intended only to illustrate in a preliminary way how the concepts of classical psychoanalytic theory may begin to be applied to specific diagnostic entities. The reader is referred to the separate chapters in this *Handbook* that are devoted to the individual psychoneurotic syndromes for further details.

Bibliography

1. ABRAHAM, K. (1925), "Character Formation on the Genital Level of Libido Development," in *Selected Papers*, Hogarth, London, 1949.
2. ——— (1921), "Contribution to the Theory of the Anal Character," in *Selected Papers*, Hogarth, London, 1949.
3. ——— (1924), "Influence of Oral Erotism on Character Formation," in *Selected Papers*, Hogarth, London, 1949.
4. AMACHER, P., *Freud's Neurological Education and Its Influence on Psychoanalytic Theory*, International Universities Press, New York, 1965.
5. ARLOW, J. A., "Conflict, Regression and Symptom Formation," *Int. J. Psychoanal.*, *44*:12–22, 1963.
6. ———, and BRENNER, C., *Psychoanalytic Concepts and the Structural Theory*, International Universities Press, New York, 1964.
7. BELLAK, L., "The Unconscious," in *Conceptual and Methodological Problems in Psychoanalysis*, *Ann. N.Y. Acad. Sc.*, *76*: 1066–1097, 1959.
8. BERES, D., "Structure and Function in Psychoanalysis," *Int. J. Psychoanal.*, *46*:53–63, 1965.
9. BLAU, A., "A Unitary Hypothesis of Emotion 1. Anxiety, Emotions of Displeasure and Affective Disorders," *Psychoanal. Quart.*, *24*:75–103, 1955.

10. BRENNER, C., *An Elementary Textbook of Psychoanalysis*, International Universities Press, New York, 1955.
11. ———, "The Nature and Development of the Concept of Repression in Freud's Writings," in *The Psychoanalytic Study of the Child*, Vol. 12, pp. 18–46, International Universities Press, New York, 1957.
12. ———, "The Psychoanalytic Concept of Aggression," *Int. J. Psychoanal.*, 52:137–144, 1971.
13. BREUER, J., and FREUD, S. (1893), "On the Psychical Mechanism of Hysterical Phenomena: Preliminary Communication," in Strachey, J. (Ed.), *Standard Edition*, Vol. 2, Hogarth, London.
14. ———, and ——— (1895), *Studies on Hysteria*, Basic Books, New York, 1957.
15. COLBY, K. M., *Energy and Structure in Psychoanalysis*, Ronald Press, New York, 1955.
16. ELLENBERGER, H. F., *The Discovery of the Unconscious*, Basic Books, New York, 1970.
17. ENGEL, G., "Anxiety and Depression-Withdrawal. The Primary Affects of Unpleasure," *Int. J. Psychoanal.*, 43:89–97, 1962.
18. ERIKSON, E. H., *Childhood and Society*, Norton, New York, 1950.
19. ———, *Identity and the Life Cycle*, International Universities Press, New York, 1959.
20. FENICHEL, O., *The Psychoanalytic Theory of Neurosis*, Norton, New York, 1945.
21. FLESCHER, J., "A Dualistic Viewpoint on Anxiety," *J. Am. Psychoanal. A.*, 3:415–446, 1955.
22. FREUD, A. (1936), *The Ego and the Mechanisms of Defense*, International Universities Press, New York, 1946.
23. FREUD, S. (1925), *An Autobiographical Study*, in Strachey, J. (Ed.), *Standard Edition*, Vol. 20, Hogarth, London.
24. ——— (1920), *Beyond the Pleasure Principle*, in Strachey, J. (Ed.), *Standard Edition*, Vol. 18, Hogarth, London.
25. ——— (1924), "The Economic Problem of Masochism," in Strachey, J. (Ed.), *Standard Edition*, Vol. 19, Hogarth, London.
26. ——— (1923), *The Ego and the Id*, in Strachey, J. (Ed.), *Standard Edition*, Vol. 19, Hogarth, London.
27. ——— (1896), "Further Remarks on the Neuro-Psychoses of Defense," in Strachey, J. (Ed.), *Standard Edition*, Vol. 3, Hogarth, London.
28. ——— (1896), "Heredity and the Aetiology of the Neurosis," in Strachey, J. (Ed.), *Standard Edition*, Vol. 3, Hogarth, London.
29. ——— (1888), "Hysteria," in Strachey, J. (Ed.), *Standard Edition*, Vol. 1, Hogarth, London.
30. ——— (1926), *Inhibitions, Symptoms and Anxiety*, in Strachey, J. (Ed.), *Standard Edition*, Vol. 20, Hogarth, London.
31. ——— (1915), "Instincts and Their Vicissitudes," in Strachey, J. (Ed.), *Standard Edition*, Vol. 14, Hogarth, London.
32. ——— (1900), *The Interpretation of Dreams*, in Strachey, J. (Ed.), *Standard Edition*, Vols. 4 & 5, Hogarth, London.
33. ——— (1916–1917), *Introductory Lectures on Psychoanalysis*, in Strachey, J. (Ed.), *Standard Edition*, Vols. 15 & 16, Hogarth, London.
34. ——— (1905), *Jokes and Their Relations in the Unconscious*, in Strachey, J. (Ed.), *Standard Edition*, Vol. 8, Hogarth, London.
35. ——— (1894), "The Neuro-Psychoses of Defense," in Strachey, J. (Ed.), *Standard Edition*, Vol. 3, Hogarth, London.
36. ——— (1933), *New Introductory Lectures on Psychoanalysis*, in Strachey, J. (Ed.), *Standard Edition*, Vol. 22, Hogarth, London.
37. ——— (1885), "On the Grounds for Detaching a Particular Syndrome from Neurasthenia under the Description 'Anxiety Neurosis,'" in Strachey, J. (Ed.), *Standard Edition*, Vol. 3, Hogarth, London.
38. ——— (1885), "Project for a Scientific Psychology," in *The Origins of Psychoanalysis*, Basic Books, New York, 1954.
39. ——— (1910), "The Psychoanalytic View of Psychogenic Disturbance of Vision," in Strachey, J. (Ed.), *Standard Edition*, Vol. 11, Hogarth, London.
40. ——— (1901), *The Psychopathology of Everyday Life*, in Strachey, J. (Ed.), *Standard Edition*, Vol. 6, Hogarth, London.
41. ——— (1915), "Repression," in Strachey, J. (Ed.), *Standard Edition*, Vol. 14, Hogarth, London.
42. ——— (1905), *Three Essays on the Theory of Sexuality*, in Strachey, J. (Ed.),

Standard Edition, Vol. 7, Hogarth, London.

43. ——— (1915), "The Unconscious," in Strachey, J. (Ed.), *Standard Edition*, Vol. 14, Hogarth, London.

44. GEDIMAN, H. K., "The Concept of Stimulus Barrier: Its Review and Reformulation as an Adaptive Ego Function," *Int. J. Psychoanal.*, *52*:243–258, 1971.

45. GILL, M. M., *Topography and Systems in Psychoanalytic Theory*, International Universities Press, New York, 1963.

46. GILLESPIE, W. H., "Aggression and Instinct Theory," *Int. J. Psychoanal.*, *52*:155–160, 1971.

47. GREENSON, R. R., "The Classical Psychoanalytic Approach," in Arieti, S. (Ed.), *American Handbook of Psychiatry*, Basic Books, New York, 1959.

48. ———, "The Origin and Fate of New Ideas in Psychoanalysis," *Int. J. Psychoanal.*, *50*:503–516, 1969.

49. ———, "Phobia, Anxiety and Depression," *J. Am. Psychoanal. A.*, *7*:663–674, 1959.

50. HARTMANN, H. (1939), *Ego Psychology and the Problem of Adaptation*, International Universities Press, New York, 1958.

51. ———, *Essays on Ego Psychology*, International Universities Press, New York, 1964.

52. ———, and KRIS, E., "The Genetic Approach in Psychoanalysis," in *The Psychoanalytic Study of the Child*, Vol. 1, pp. 11–30, International Universities Press, New York, 1945.

53. ———, ———, and LOEWENSTEIN, R. M., "Comments on the Formation of Psychic Structure," in *The Psychoanalytic Study of the Child*, Vol. 2, pp. 11–38, International Universities Press, New York, 1946.

54. ———, ———, and ———, "Notes on the Theory of Aggression," in *The Psychoanalytic Study of the Child*, Vol. 3, pp. 9–36, International Universities Press, New York, 1949.

55. HENDRICKS, I., *Facts and Theories of Psychoanalysis*, Alfred A. Knopf, New York, 1939.

56. JACOBSON, E., "The Affects and Their Pleasure-Unpleasure Qualities in Relation to the Psychic Discharge Processes," in Loewenstein, R. M. (Ed.), *Drives, Affects, Behavior*, International Universities Press, New York, 1953.

57. ———, "The Self and the Object World," in *The Psychoanalytic Study of the Child*, Vol. 9, pp. 75–127, International Universities Press, New York, 1954.

58. JONES, E., "Anal Erotic Character Traits," in *Papers on Psychoanalysis*, Balliere, Tyndall and Cox, London, 1948.

59. ———, *The Life and Work of Sigmund Freud*, Vol. 1, Basic Books, New York, 1953.

60. KAYWIN, L., "An Epigenetic Approach to the Psychoanalytic Theory of Instincts and Affects," *J. Am. Psychoanal. A.*, *8*:613–658, 1960.

61. KUBIE, L. S., "The Fallacious Use of Quantitative Concepts in Dynamic Psychology," *Psychoanal. Quart.*, *16*:507–518, 1947.

62. LAMPL-DE GROOT, J., "Symptom Formation and Character Formation," *Int. J. Psychoanal.*, *44*:1–11, 1963.

63. LEWIN, B. D., "Phobic Symptoms and Dream Interpretation," *Psychoanal. Quart.*, *21*:295–322, 1952.

64. ———, *The Psychoanalysis of Elation*, Norton, New York, 1950.

65. LUSTMAN, S. L., "Introduction to Panel on the Use of the Economic Viewpoint in Clinical Psychoanalysis," *Int. J. Psychoanal.*, *50*:95–102, 1969.

66. MAHLER, M., "On the Significance of the Normal Separation-Individuation Phase," in Schur, M. (Ed.), *Drives, Affects, Behavior*, Vol. 2, International Universities Press, New York, 1965.

67. MARMOR, J., "Orality in the Hysterical Personality," *J. Am. Psychoanal. A.*, *1*:656–671, 1952.

68. MOORE, B. E., and FINE, B. D. (Eds.), *A Glossary of Psychoanalytic Terms and Concepts*, American Psychoanalytic Association, New York, 1968.

69. PUMPIAN-MINDLIN, E., "Propositions Concerning Energetic-Economic Aspects of Libido Theory: Conceptual Models of Psychic Energy and Structure in Psychoanalysis," in Bellak, L. (Ed.), *Conceptual and Methodological Problems in Psychoanalysis*, *Ann. N.Y. Acad. Sc.*, *76*:1038–1065, 1959.

70. RANGELL, L., "A Further Attempt to Resolve the 'Problem of Anxiety,'" *J. Am. Psychoanal. A.*, *16*:371–404, 1968.

71. ———, "The Scope of Heinz Hartmann," *Int. J. Psychoanal.*, 46:5–30, 1965.
72. ———, "A Unitary Theory of Anxiety," *J. Am. Psychoanal. A.*, 3:389–414, 1955.
73. RAPAPORT, D., "A Historical Survey of Psychoanalytic Ego Psychology," in Erikson, E. H., *Identity and the Life Cycle*, International Universities Press, New York, 1959.
74. ———, *The Structure of Psychoanalytic Theory*, International Universities Press, New York, 1960.
75. SCHAFER, R., "An Overview of Heinz Hartmann's Contributions to Psychoanalysis," *Int. J. Psychoanal.*, 51:425–446, 1970.
76. SILVERBERG, W. V., *Childhood Experience and Personal Destiny*, Springer, New York, 1952.
77. STEWART, W. A., *Psychoanalysis: The First Ten Years, 1888–1898*, Macmillan, New York, 1967.
78. STRACHEY, J., "The Emergence of Freud's Fundamental Hypotheses," in Strachey, J. (Ed.), *Standard Edition*, Vol. 3, pp. 62–70, Hogarth, London, 1962.
79. VAN DER LEEUW, P. J., "On the Development of the Concept of Defense," *Int. J. Psychoanal.*, 52:51–58, 1971.
80. WAELDER, R., *Basic Theory of Psychoanalysis*, International Universities Press, New York, 1960.
81. YANKELOVICH, D., and BARRETT, W., *Ego and Instinct*, Random House, New York, 1970.
82. ZETZEL, E. R., "The Concept of Anxiety in Relation to the Development of Psychoanalysis," *J. Am. Psychoanal. A.*, 3:369–388, 1955.

B. The Theory of Psychoanalytic Technique

Ralph R. Greenson

Introduction

This presentation is an attempt to outline the theoretical foundation upon which psychoanalytic practice and technique are based. There is a very special reciprocal relationship between theory and technique in psychoanalysis. Every change in psychoanalytic theory is followed by corresponding changes in technique; every technical rule can be considered valid only if it can be rooted in a specific piece of psychoanalytic theory (Anna Freud).[19]

Psychoanalysis is the only type of psychotherapy that attempts to remove the causes of neuroses. The vast majority of psychoanalysts believe that psychoanalytic technique and theory stands or falls on the psychoanalytic theory of neurosis (Walder).[71] In recent years a number of psychoanalysts have attempted to widen the field of application of psychoanalytic technique and have described the analytic treatment of delinquents, borderline cases, and psychotics. Aichhorn,[1] Eissler,[11] Stone,[70] Arieti,[3] Searles,[63] and Wexler,[72] among others, have contributed to this endeavor, but all of them acknowledge the deviations in technique from classical psychoanalysis. Others, like Rosenfeld,[62] Boyer and Giovacchini,[8] and Arlow and Brenner,[5] maintain it is possible to treat schizophrenics with the traditional psychoanalytic method.

I shall limit this essay to the theoretical foundations of classical psychoanalytic technique. I shall also touch upon some unanalytic procedures that are used regularly in psychoanalytic treatment, but that do not violate classical psychoanalytic theory. I hope it will then become clear why Freudian psychoanalysis, without basic modifications, is not generally suitable for the treatment of borderline and psychotic patients.

The Historical Development of Psychoanalytic Therapy

I believe the quickest way of ascertaining what is essential in psychoanalytic technique is to take a birds-eye view of its historical development, noting the major changes in the technical procedures and the therapeutic processes. I am using the term "technical procedure" to refer to a course of action, a tool, an instrumentality, undertaken by the patient or the therapist. Free association, suggestion, and interpretation are examples of technical procedures. A "therapeutic process" refers to an interrelated series of psychic events within the patient, a continuity of psychic reactions that have a remedial aim or effect. Therapeutic processes are usually instigated by the technical procedures; examples include the recapturing of memories, emotional catharsis, transference, and the obtaining of insight.

Psychoanalytic technique was not suddenly discovered or invented. It evolved gradually as Freud struggled to find a way to help his neurotic patients. Although Freud disclaimed any enthusiasm for therapeutic results, it was his therapeutic intent that led to the discovery of psychoanalysis. The changes in his technique were never abrupt nor complete. There would be a shift of emphasis, a change in the order of importance assigned a given procedure or a therapeutic process, but Freud was loath to discard completely his old ideas. Nevertheless, it is possible to delineate different phases in the development and significance of different technical procedures and therapeutic processes.

The early psychoanalytic concepts were discovered by Freud in the ten years between 1888 and 1898. Stewart[67] has written a carefully documented account of the sequence of discoveries during this period. *The Studies on Hysteria*, written by Breuer and Freud,[9] can be regarded as the beginning of psychoanalysis. In the section of that book called *Preliminary Communication*, published originally in 1893, Breuer described the treatment of Anna O., a patient who overcame her own amnesia and in sudden spells spontaneously hypnotized herself and spoke out of her unconscious mind. Not all patients were amenable to hypnosis, and Freud described how he gradually began to circumvent this difficulty by putting his patients in a state of concentration by the use of pressure on the forehead. The aim of the treatment of hysterical patients at that time was to get the patient to remember a traumatic experience that had been cut off from consciousness and that had resulted in a strangulated state of the accompanying affects. Freud and Breuer[9] maintained that "*Each individual hysterical symptom immediately and permanently disappeared when we had succeeded in bringing clearly to light the memory of the event by which it was provoked and its accompanying affect, and when the patient had described that event in the greatest possible detail and had put the affect into words*" (p. 6).

At this point the therapeutic process was the abreacting and remembering of a traumatic event, and the therapeutic procedure was hypnosis or suggestion. The fact that certain patients, like Frau Emmy von N., fought against being hypnotized revealed to Freud the presence of *resistance*. This force, he realized, was the same force that kept the pathogenic ideas from becoming conscious. In the section on *The Psychotherapy of Hysteria*, written by Freud[26] alone and first published in 1895, he realized that the patient's resistance was a defense, their "not knowing" was, in fact, a "not wanting to know" (pp. 269–270).

According to Jones,[50] Freud gradually gave up hypnosis, suggestion, and pressing between 1892 and 1896 and relied instead on *free association*. Hints of this are already mentioned in 1889 in the case of Emmy von N. and later in the treatment of Elizabeth von R. The procedure of free association became known as the fundamental rule of psychoanalysis. It has remained the basic and unique method of communication for patients in psychoanalytic treatment. Other means of communication occur during the course of psychoanalysis but they are secondary, preparatory, and not typical for psychoanalysis. Freud[25] described this method as follows:

> Without exerting any other kind of influence, he invites them to lie down in a comfortable attitude on a sofa, while he himself sits on a chair behind them outside their field of vision. He does not even ask them to close their eyes, and avoids touching them in any way, as well as any other procedure which might be reminiscent of hypnosis. The session thus proceeds like a conversation between two people equally awake, but one of whom is spared every muscular exertion and every distracting sensory impression which might divert his attention from his own mental activity. . . . In order to secure these ideas and associations he asks the patient to "let himself go" in what he says, "as you would do in a conversation in which you were rambling on quite disconnectedly and at random." [pp. 250–251]

Freud also realized that the personal influence of the physician could be of great value and suggested that the therapist act as an elucidator, a teacher, and a father confessor. However, he also became aware that under certain conditions the patient's relation to the physician can become "disturbed," a factor that turns the patient-physician relationship into the "worst obstacle" we can come across.

Thus Freud had discovered the phenomena of resistance and transference, but they were then considered essentially obstacles to the work. His main objective was to achieve affective abreaction and to recover traumatic memories. Transference reactions and resistances were hindrances to be circumvented or overcome.

The second phase of psychoanalytic technique began with Freud's[28] *Interpretation of Dreams* published in 1900. In the famous seventh chapter of that book Freud had described the different forms, qualities, and properties that distinguish conscious from unconscious psychic phenomena. He designated two sets of governing principles: the primary and secondary processes. By recognizing the occurrence of condensation, symbol formation, reversal, and displacement, Freud discovered the importance of a new technical procedure: interpretation. Freud now realized that the structure of a neurosis was too complex to be dealt with merely by symptom removal.

The Fragment of an Analysis of a Case of Hysteria,[24] known in psychoanalytic circles as the Dora case, was essentially a clinical example of how dream interpretation could be used as a technical tool. It was written immediately after the *Interpretation of Dreams* in 1901 but for various reasons was delayed in publication. In it Freud stated that psychoanalytic technique had been completely revolutionized (p. 12). He no longer tried to clear up each symptom, one after the other. By allowing the patient to do free association and to choose the subject matter of the hour, the analyst would then be able to interpret the meaning of the patient's material and resistances and thus undo the repressions, leading to filling in the gaps of memory.

It was in the Dora case that Freud[24] first stated that, "Transference, which seems ordained to be the greatest obstacle to psychoanalysis, becomes its most powerful ally, if its presence can be detected each time and explained to the patient" (p. 117). In a postscript to that case Freud described how the patient had broken off treatment because he had failed to analyze the multiple transference elements that interfered with the treatment situation.

The major shift in the theory of the therapeutic process was the emphasis on making the unconscious conscious by means of interpreting the patient's associations, including the resistances and the transference reactions. However, the overriding importance of the transference and the transference resistances was still not fully recognized. The recovery of memories had now become the major goal, and emotional catharsis was allocated to a secondary position.

The next major changes in technique took place in the years 1911–1915 when Freud published a series of six technical papers. In *The Dynamics of the Transference*[21] he attempted to answer why transference can be the greatest obstacle and also the greatest ally of the analyst. Psychoanalysis does not produce transference; it only provides an arena for it to appear. Our patients all suffer from frustrations in their sexual and erotic life and therefore come to psychoanalysis with a transference readiness. During treatment the patient will regress to protect and hold onto his childhood fantasies, now displaced onto the person of the analyst. The analyst's task is to permit both the loving and hateful transference reactions to occur and to resolve them by interpretation, by making them conscious to the patient.

Part of the last paragraph of *The Dynamics of Transference* is of such importance that it deserves to be quoted:

> Just as happens in dreams, the patient regards the products of the awakening of his unconscious impulses as contemporaneous and real; he seeks to put his passions into action without taking any account of the real situation. The doctor tries to compel him to fit these emotional impulses into the nexus of the treatment and of his life-history, to submit them to intellectual consideration and to understand them in the light of their psychical value. This struggle between the doctor and the patient, between intellect and instinctual life, between understanding and seeking to act, is played out almost exclusively in the phenomena of transference. It is on that field that the victory must be won—the victory whose expression is the permanent cure of the neurosis. It cannot be disputed that controlling the phenomena of transference presents the psycho-analyst with the greatest difficulties. But it should not be forgotten that it is precisely they that do us the inestimable service of making the patient's hidden and forgotten erotic impulses immediate and manifest. For when all is said and done, it is impossible to destroy anyone in absentia or in effigie.

The next paper published later that same year described the ideal attitude of the analyst. He should listen with evenly hovering attention and should try to avoid gratifying the patient's infantile wishes by reacting like an opaque mirror to the patient, reflecting back only what the patient displaces onto him. The analyst's personal views and reactions would only interfere with the analysis of the transference reactions.[38]

The paper "On Beginning the Treatment"[33] is noteworthy for our purposes because in it Freud stated that the psychoanalyst should not make important interpretations to the patient until a proper rapport had been established between patient and analyst. How this is accomplished Freud stated as follows:

> To ensure this, nothing need be done but to give him time. If one exhibits a serious interest in him, carefully clears away the resistances that crop up at the beginning and avoids making certain mistakes, he will of himself form such an attachment and link the doctor up with one of the imagos of the people by whom he was accustomed to be treated with affection. It is certainly possible to forfeit this first success if from the start one takes up any standpoint other than one of sympathetic understanding, such as a moralizing one, or if one behaves like a representative or advocate of some contending party (Pp. 139, 140).

I believe this is the first description of the origin and importance of what later came to be known as the therapeutic or working alliance (Zetzel,[74] Stone,[68] Greenson).[48]

Some important additions to psychoanalytic technique were added in "Remembering, Repeating and Working Through.[39] Interpretations are never thoroughly effective if given only once. They have to be repeated because all neurotic patients have a "compulsion to repeat." Patients will act out in the analytic situation what they cannot remember; therefore, the transference is singularly valuable. Finally in a successful analysis the patient will replace his previous neurosis with a "transference neurosis." This means that eventually the analyst will become the most important person in the patient's life, and all the patient's neurotic symptoms and attitudes will revolve around the analyst. It is the resolution of the transference neurosis that insures a successful analysis.

If we review this phase of development in the history of psychoanalytic technique, we see that the central and most important technical procedures are aimed at facilitating the development of a regressive transference neurosis. In this way the patient relives within the analytic situation what he is unable to remember. By a nonintrusive, humane atttitude, and by repeatedly and consistently interpreting the transference and resistance, the psychoanalyst is able to resolve the transference neurosis and help the patient break away from his neurotic past.

The last major addition to the technique of Freudian psychoanalysis was the emphasis on producing alterations in the ego. This was first mentioned in the twenty-seventh lecture of *The Introductory Lectures* (p. 455),[29] but the nature of the alteration and how it could be effected was not described. In *Inhibitions, Symptoms and Anxiety*, published in 1926, Freud[27] returned to the concept of "defensive processes" and considered repression only one of several defensive maneuvers. The ego uses the defenses, which exert an *anticathexis*, a counterforce, in order to keep some dangerous impulse, feeling, or thought out of consciousness. As a consequence there are alterations in the ego. For example, an attitude of excessive cleanliness may be used as a reaction formation against an unconscious tendency to enjoy dirtiness. The defenses operate through the ego, and our major therapeutic efforts are aimed directly at the ego. Anna Freud's[17] *The Ego and the Mechanisms of Defense*, published in 1936, was the first attempt to place ego psychology in the forefront of psychoanalytic theory as well as to systematize the various mechanisms of defense.

In *The Ego and the Id* Freud[22] stated that by analyzing the transference reactions and resistances, psychoanalysis had become "an instrument to enable the ego to achieve a progressive conquest of the id" (p. 56). Later he wrote that the therapeutic efforts of psychoanalysis are intended "to strengthen the ego, to make it more independent of the super-ego, to widen its field of perception and enlarge its organization, so that it can appropriate fresh portions of the id. Where id was, there ego shall be" (p. 80). Again in *Analysis Terminable and Interminable* Freud[20] stated: "The business of the analysis is to secure the best possible psychological conditions for the functions of the ego; with that it has discharged its task" (p. 250).

The Psychoanalytic Theory of Neurosis

I shall only touch on some of the highlights of the theory of neuroses and psychoses as they relate to psychoanalytic therapy. This subject has been covered in detail in Part A of this chapter.

The Conflict Theory of Neurosis

Beginning in 1894 in "The Neuro-Psychoses of Defence" Freud[31] postulated that the neuroses are the result of a conflict between an instinctual drive and a defense. This theory is mentioned at many different points in Freud's writings, but it is often obscured by the fact that in his early writings the only instinctual drive he described clinically was the sexual drive, and he used the term "repression" for the concept of defense mechanism in general. In his *Three Essays on Sexuality*[35] he very clearly describes his theory of the etiology of hysteria: "The character of hysterics shows a degree of sexual repression in excess of the normal quantity, an intensification of resistance against the sexual instinct (which we have already met with in the form of shame, disgust and morality), and what seems like an instinctive aversion on their part to any intellectual consideration of sexual problems. As a result of this, in especially marked cases, the patients remain in complete ignorance of sexual matters right into the period of sexual maturity" (p. 164).

Perhaps Freud's[34] most unequivocal statement on the conflict theory can be found in *On the History of the Psycho-Analytic Movement*. His point of view will be clarified if one replaces the term "repression" with the concept of mechanisms of defense.

> The theory of repression is the corner-stone on which the whole structure of psycho-analysis rests. It is the most essential part of it; and yet it is nothing but a theoretical formulation of a phenomenon which may be observed as often as one pleases if one undertakes an analysis of a neurotic without resorting to hypnosis. In such cases one comes across a resistance which opposes the work of analysis and in order to frustrate it pleads a failure of memory. The use of hypnosis was bound to hide this resistance; the history of psycho-analysis proper, therefore, only begins with the new technique that dispenses with hypnosis. The theoretical consideration of the fact that this resistance coincides with an amnesia leads inevitably to the view of unconscious mental activity which is peculiar to psycho-analysis and which, too, distinguishes it quite clearly from philosophical speculations about the unconscious. It may thus be said that the theory of psycho-analysis is an attempt to account for two striking and unexpected facts of observation which emerge whenever an attempt is made to trace the symptoms of a neurotic back to their sources in his past life: the facts of transference and of resistance. Any line of investigation which recognizes these two facts and takes them as the starting point of its work has a right to call itself psycho-analysis, even though it arrives at results other than my own. But anyone who takes up other sides of the problem while avoiding these two hypotheses will hardly escape a charge of misappropriation of property by attempted impersonation, if he persists in calling himself a psycho-analyst (p. 16).

It was only with the development of the structural point of view that it was possible to describe the psychoanalytic theory of neurosis succinctly. I believe Fenichel's[15] approach that all psychoneuroses are relative traumatic neuroses is a valuable one for understanding the formation of psychoneuroses in general. Psychoanalysis maintains that the psychoneuroses are based on unresolved *unconscious* conflicts between id impulses (instinctual drives) seeking discharge and ego defenses that are attempting to ward off the impulses' direct discharge or access to consciousness. These conflicts eventually lead to a damming up of the instinctual drives. The ego, which gives rise to the mechanisms of defense, becomes drained of its energies and is eventually overwhelmed. As a result involuntary discharges occur that manifest themselves clinically as the symptoms of the neurosis.

The superego plays a more complicated role in the neuroses. It may enter the conflict on the side of the ego and may make the ego feel guilty even for thoughts and fantasies of instinctual satisfaction. On the other hand, the superego's self-reproaches may become regressively reinstinctualized so that the self-reproaches take on a drivelike quality. All parts of the psychic apparatus take part in the formation of neurotic symptoms.

The external world also plays an important role in the causation of a psychoneurosis, but it does so only when it becomes allied with the ego, id, or superego. The external world may represent and mobilize some instinctual temptation. Then all situations that are reminiscent of the instinctual temptation may trigger an eruption of symptoms. The external world may be reacted to as a superego, and people or situations will be avoided because they stir up guilt and shame. What seems to begin as a conflict with the external world turns out to have become an internal conflict between the ego, the id, and the superego.

It should be remembered that Freud[30] believed every adult neurosis is built on an infantile neurosis and will be re-experienced in the transference neuroses. It was once thought that all children go through some form of infantile neurosis; some overcome it, "grow out of it," while others develop a neurosis in later life (Fenichel).[13] Anna Freud[18] has stated that perhaps our more modern, flexible methods of upbringing lead to other less well-defined developmental disorders in the children of today.

The key factor in understanding the pathogenic outcome of the neurotic conflicts is the ego's need constantly to expend its energies in attempting to keep the dangerous or forbidden drives from gaining access to consciousness or motility. Ultimately this leads to a relative insufficiency of the ego, and disguised derivatives of the original neurotic conflicts will overwhelm the depleted ego and break through into consciousness and behavior. These distorted and disguised involuntary discharges manifest themselves as the symptoms of the psychoneurosis. I would like to cite a relatively simple clinical example from my book on technique (Greenson).[47]

Some years ago a young woman, Mrs. A., came for treatment accompanied by her husband. She complained that she was unable to leave her house alone and felt safe only with her husband. In addition, she complained of a fear of fainting, a fear of dizziness, and a fear of becoming incontinent. Mrs. A.'s symptoms had begun quite suddenly some six months earlier while she was in a beauty parlor.

The analysis, which lasted several years, revealed that the actual trigger for the outbreak of the patient's phobias was the event of having her hair combed by a male beautician. We were able eventually to uncover the fact that at that moment she was reminded of her father combing her hair when she was a little girl. The reason she had gone to the beauty parlor that day was her pleasurable expectation of seeing her father, who was to visit the young married couple for the first time since their marriage. He was to stay in their home and she was filled with great delight, consciously. However, unconsciously, she was full of guilt feelings for loving her father and for her predominantly unconscious hostility toward her husband.

The apparently innocuous event of having her hair combed stirred up old incestuous longings, hostilities, guilt, and anxiety. To put it briefly, Mrs. A. had to be accompanied by her husband in order to be sure he had not been killed by her death wishes. Also his presence protected her from acting out sexually. The fears of fainting, of dizziness, and of incontinence were symbolic representations of losing her moral balance, losing her self-control, soiling her good character, humiliating herself, and falling from her high position. The young woman's symptoms had links to the pleasurable body sensations of childhood as well as to infantile punishment fantasies.

I believe one can formulate the events as follows: the combing of her hair stirred up repressed id impulses which brought her into conflict with her ego and superego. Despite the absence of obvious neurotic symptoms prior to the outbreak of the phobias, there were indications that her ego already was relatively depleted and her id lacked adequate discharge possibilities. Mrs. A. had had difficulty in sleeping for years, nightmares, and inhibitions in her sexual life. As a consequence the fantasies mobilized by the hair combing increased the id tensions to a point where they flooded the infantile defenses of the ego and involuntary discharges took place, eventuating in acute symptom formation (Pp. 19, 20).

The Psychoanalytic Theory of Psychosis

Freud's writings on the psychoses were not consistent and were often obscure and even contradictory. Yet a careful reading of his work does reveal that he felt there were *qualitative* differences between the neuroses and the psychoses. In *The Introductory Lectures*[29] he distinguishes between patients who can form a *transference neurosis* during psychoanalytic treatment and those who cannot because they suffer from a *narcissistic neurosis* (pp. 420–423). Earlier in the Schreber case Freud[37] described how Schreber's "Subjective world had come to an end since his withdrawal of love from it." In his paper on *The Unconscious*[32] Freud stated very clearly, "In the case of schizophrenia, on the other hand, we have been driven to the assumption that after the process of repression the libido that has been withdrawn does not seek a new object, but retreats into the ego; that is to say, that here the object-cathexes are given up and a primitive objectless condition of narcissism is re-established" (pp. 196–197).

I believe that Wexler[72] is particularly clear on the subject of the psychoses, and I would like to quote from his most recent paper. Referring to Freud's remarks above, he says:

By this statement, Freud did not mean only the withdrawal from external reality objects. He specified repeatedly that "in schizophrenia, this flight consisted in withdrawal of instinctual cathexis from the points which represent the unconscious presentation of the object." Here was the really crucial differentiation between neurosis and psychosis. Here was the real basis for the clinical experience of inner and outer world destruction. Here was the central theoretical construct by which one could understand that if the ego was a "precipitate of abandoned object-cathexes," then the dissolution of those representations must necessarily lead to the psychic disasters of schizophrenia (p. 93).

I believe these theoretical formulations based on the clinical findings in psychoses indicate how dangerous the classical psychoanalytic procedure can be for such patients. Lying on a couch with the analyst out of sight and doing free association can mean a loss of contact with reality for the schizophrenic patient. For these patients the nontransference, real relationship is far more important. Only when there is sufficient structure building would consistent interpretation be indicated (Greenson and Wexler).[49]

The Metapsychology of Psychoanalysis

The basic aim of psychoanalytic therapy is to undo the unresolved conflicts that are the cause of the neurosis. To understand the patient's pathology and the interactions between the patient and the analyst during therapy, we attempt to study these phenomena from five basic theoretical points of view. These are grouped together as the metapsychology of psychoanalysis and consist of the topographical, dynamic, economic, genetic, and structural points of view. In actual practice we only analyze our patient's productions partially and fragmentarily in a given interval of time. Nevertheless, if we do succeed in working through all our insights, we realize that we have, in fact, utilized all five metapsychological approaches. Freud's metapsychological writings are scattered throughout his writings and are neither systematic nor complete. Here I shall only attempt to give a working definition of these concepts. For a more comprehensive survey the reader is referred to Fenichel,[15] Rapaport and Gill,[61] and Arlow and Brenner.[4]

The earliest point of view Freud postulated was the *topographical* one. In Chapter 7 of *The Interpretation of Dreams* he described the different modes of functioning that are characteristic for conscious and unconscious phenomena. The primary process holds sway over unconscious material, and the secondary process directs conscious phenomena. Unconscious material has only one aim—discharge. There is no sense of time, order, or logic, and contradictions may coexist without nullifying one another. Condensation and displacement are other characteristics of the primary process. Designating a psychic event as conscious or unconscious implies more than merely a difference in quality. Archaic and primitive modes of functioning are characteristic of unconscious phenomena.

The *dynamic point of view* considers all mental phenomena to be the result of the interaction of forces, some working in unison and others working concurrently in opposition to one another. An example of this is the following: A young man greets an attractive guest sitting at the family dinner table by saying: "How nice to see you, Dolores. Every time you come it is like a *breast* of fresh air. Oh, I mean it's like a breath of *flesh* air." The first sentence expresses the young man's pleasure in a socially acceptable way. The slip of saying breast instead of breath indicates a breakthrough of sexual feelings toward the woman. He tries to repair this breach but succeeds only partially. He represses the breast, but fresh gets turned into flesh. It is only a partial victory for the antisexual forces within him. The wish to expose and to hide his sexual desires is vividly demonstrated in his slips. The dynamic point of view is the basis for all hypotheses concerning instinctual drives and defenses, ego interests, neurotic conflicts, symptom formation, ambivalence, and overdetermination.

The *economic point of view* concerns the distribution, transformation, and expenditures of psychic energies. Such concepts as the ego's ability or inability to cope, sublimation, sexualization, aggressivization, and binding are based on this hypothesis. An example of economics can be seen in the case of Mrs. A., whom I described above. Before the outbreak of the patient's phobias she was in a state of damned up instinctual tensions, but her ego was still able to carry out its defensive functions adequately enough so that Mrs. A. could function without obvious symptoms. At the point of her father's visit the hair combing brought back sexual and romantic memories from the past. In addition, it increased her hostility to her husband. Mrs. A.'s ego could not cope with this new influx of id strivings seeking discharge. The instinctual impulses broke through in feelings of fainting, dizziness, and incontinence. This led to a phobia about leaving her house unaccompanied by her husband.

The *genetic point of view* deals with the origin and development of psychic phenomena. It deals not only with how the past lives on in the present, but also why a specific solution was used in certain conflicts. This includes the biological-constitutional factors as well as the experiential ones. An example would be a male patient who uses excessive submissiveness to avoid conflict with a strong man. This was his mother's way of dealing with his father, which he identified with.

The *structural point of view* assumes the psychic apparatus can be divided into several persisting function units. This was Freud's[22] last major theoretical contribution. The hypothesis that the psychic apparatus is made of an ego, id, and superego is based on the concept of psychic structure. It is implied whenever we talk of interstructural conflicts like symptom formation or intrastructural processes like the ego's synthetic function.

The Psychoanalytic Situation

At this point I believe we are ready to explore the three components of the psychoanalytic situation—the patient, the psychoanalyst, and the setting—and ask ourselves what does each contribute to the process of being analyzed and how do these three essential elements interact with one another. We can now state the aim of psychoanalytic therapy more succinctly. *The analyst attempts to resolve the patient's neurotic conflicts by reuniting with the conscious ego those portions of the id, superego, and unconscious ego that had been excluded from the maturational processes of the healthy remainder of the total personality.*

The Components of Classical Psychoanalytic Technique

The Patient's Production of Material

Free Association

The psychoanalyst approaches the unconscious by using the derivatives of the unresolved conflicts. Derivatives are "half-breeds" that are not conscious and yet are highly organized in accordance with the secondary process; they are accessible to the conscious ego and can be put into coherent language (Freud).[32] Psychoanalysis requires that the patient try to approximate free association, the so-called basic rule, in order to facilitate the communication of derivatives. Derivatives appear also in dreams, slips, symptoms, and acting out.

The patient is asked to try, to the best of his ability, to let things come up in his mind and to say them without regard for logic or order; he is to report things even if they seem trivial, shameful, or impolite. By letting things come to mind a regression in the service of the ego takes place, and derivatives of the unconscious ego, id, and superego tend to come to the surface. The patient moves from strict secondary-process thinking in the direction of the primary process. It is the analyst's task to analyze these derivatives for the patient.

Usually this is attempted only after the preliminary interviews have been concluded, and we have decided the patient seems to have the capacities needed to work in the analytic situation. We want to be reasonably certain that the patient has the resilience in his ego functions to oscillate between the more regressive ego functions as they are needed in free association and the more advanced ego functions required for understanding the analyst's interventions and resuming his everyday life at the end of the hour.

The patient may report events from his daily life or past history in addition to reporting his free associations. It is characteristic for psychoanalysis to ask the patient to include his associations as he recounts any other happening in his life. Free association has priority over all other means of producing material in the analytic situation.

However, free association, like any other tool of psychoanalysis, may be misused by the patient or the psychoanalyst (A. Freud).[19] This does not mean such tools are outdated and should be replaced or downgraded as suggested by Lacan,[55] Alexander and French,[2] or Marmor.[58] I agree with Anna Freud who said that the tools of any trade should be periodically inspected, revised, and sharpened. Every psychoanalyst should be familiar with the most common misuses of our technical rules. Some schizophrenic patients cannot stop free associating. The error in such an instance would be in asking such a patient to enter into an analytic situation.

THE TRANSFERENCE REACTIONS

Psychoanalysis is distinguished from all other therapies by the way it promotes the development of the transference reactions and by how it attempts systematically to analyze transference phenomena. By transference we refer to a special kind of relationship toward a person; it is a distinctive type of object relationship. The main characteristic is the experience of feelings for a person that do not befit that person and that actually apply to another. Essentially a person in the present is reacted to as though he were a person in the past. Transference is a repetition, a new edition of an old object relationship (Freud).[36] A displacement has taken place; impulses, feelings, and defenses pertaining to a person in the past have been shifted onto a person in the present. It is primarily an unconscious phenomenon, and the person reacting with transference feelings is in the main unaware of the distortion.

Transference may consist of any of the components of an object relationship; that is, it may be experienced as feelings, drives, wishes, fears, fantasies, attitudes, and ideas or defenses against them. The people who are the original sources of transference reactions are the meaningful and significant people of early childhood (S. Freud,[21] A. Freud).[17] Transference occurs in analysis and outside of analysis, in neurotics, in psychotics, and in healthy people. All human relations contain a mixture of realistic and transference reactions (Fenichel).[14]

Transference reactions are always inappropriate. They may be so in the quality, quantity, or duration of the reaction. One may overreact or underreact, or one may have a bizarre reaction to the transference object. The transference reaction is unsuitable in its current context; but it was once an appropriate reaction to a past situation. Just as ill-fitting as transference reactions are to a person in the present, they fit snugly to someone in the past.

> For example, a young woman patient reacts to my keeping her waiting for two or three minutes by becoming tearful and angry, fantasying that I must be giving extra time to my favorite woman patient. This is an inappropriate reaction in a thirty-five-year-old intelligent and cultured woman, but her associations lead to a past situation where this set of feelings and fantasies fit. She recalls her reactions as a child of five waiting for her father to come to her room to kiss her good night. She always had to wait a few minutes because he made it a rule to kiss her younger sister good night first. Then she reacted by tears, anger, and jealousy fantasies—precisely what she is now experiencing with me. Her reactions are appropriate for a five-year-old girl, but obviously not fitting for a thirty-five-year-old woman. The key to understanding this behavior is recognizing that it is a repetition of the past, i.e., a transference reaction. [Greenson,[47] pp. 151–153]

It is important to recognize that in transference reactions the patient tends to repeat instead of to remember; and in this sense transference is always a resistance in regard to the function of memory. However, by repeating, by re-enacting the past, the patient does make it possible for the past to enter into the treatment situation. Transference repetitions bring into the analysis material that might otherwise be inaccessible. If properly handled the analysis of transference will lead to memories, reconstructions, and insight, and an eventual cessation of the repetition.

There are many ways of classifying the various clinical forms of transference reactions. The most commonly used designations are the positive and the negative transference. The positive transference refers to the different forms of loving and the negative transference implies some variety of hatred toward the analyst. It should be borne in mind that all transference reactions are essentially ambivalent. What appears clinically is only the surface.

For transference reactions to take place in the analytic situation, the patient must be willing and able to risk some temporary regression in terms of ego functions and object relations. The patient must have an ego capable of temporarily regressing to transference reactions, but this regression must be partial and reversible so that the patient can be treated analytically and still live in the real world. People who do not dare regress from reality and those who cannot return readily to reality, like the psychotics, are poor risks for psychoanalysis. Freud[29] divided the neuroses into two groups on the basis of whether a patient could develop and maintain a relatively cohesive set of transference reactions and still function in the analysis and in the external world. Patients with a "transference neurosis" could do this, while patients suffering from a "narcissistic neurosis" could not.

Freud[39] also used the term "transference neurosis" to describe that intensity and extent of transference reactions in which the analyst and the analysis have become the central interest in the patient's emotional life, and the patient's major neurotic conflicts are relived in the analytic situation. All the important features of the patient's neurosis will be relived or re-enacted in the analytic situation (Freud).[24]

Psychoanalytic technique is geared to insure the maximal development of the transference neurosis. The relative anonymity of the analyst, his nonintrusiveness, the so-called rule of abstinence, and the "mirrorlike" behavior of the analyst all have the purpose of preserving a relatively uncontaminated field for the budding transference neurosis (Fenichel,[14] Greenacre,[45] Gill).[41] The transference neurosis is an artifact of the analytic situation; it can be undone only by the analytic work. It serves as a transition from illness to health.

On the one hand, the transference neurosis is the most important vehicle for success in psychoanalysis; on the other, it is the most frequent cause of therapeutic failure (Freud,[21,39] Glover).[43] The transference neurosis can be resolved only by analysis; other procedures may change its form, but will only perpetuate it (Gill).[41]

Psychoanalysis is the only form of psychotherapy that attempts to resolve the transference reactions by systematically and thoroughly analyzing them. In some briefer or diluted versions of psychoanalysis one does so only partially and selectively. Thus one might analyze only the hostile transference when it threatens to disrupt the treatment, or one analyzes only as deeply as required for the patient to be able to work in the therapeutic situation. In such cases there is always a residual of unresolved transference reactions after the treatment is completed. This implies that there is some unanalyzed neurosis left unchanged.

In the *antianalytic* forms of psychotherapy the transference reactions are not analyzed but gratified and manipulated. The therapist assumes the role of some past figure, real or fantasied, and gratifies some infantile wish of the patient's. The therapist might act like a loving or encouraging parent, or like a punishing moralist, and the patient might feel some temporary improvement or even be "cured." But these "transference cures" are fleeting and last only as long as the idealized transference to the therapist is untouched (Fenichel,[15] Greenson).[47]

The Working Alliance and the Real Relationship

As important as the unfolding and interpreting of the transference reactions are, they are not sufficient for producing lasting changes in the patients. In order for a neurotic patient to work effectively in the analytic situation he must establish and maintain a working or therapeutic alliance with the analyst.[46,47,48] Zetzel[74] and Stone[68] were among the first to stress this aspect of the psychoanalytic situation. The core from which the working alliance is derived is the real, nontransference relationship between the analyst and patient (Greenson and Wexler).[49]

Ever since Freud's Dora case of 1905, psychoanalysts have made the analysis of transference the major focus of psychoanalytic technique. This development has reached such proportions that Kleinian psychoanalysts consider all interactions between the patient and his analyst as transference or countertransference and would make interpretation the only correct intervention. "Orthodox" Freudians often recognize that personal interactions other than transference may occur but tend to treat them as irrelevant or trivial, at least in their writings. They even acknowledge that interventions other than strictly defined interpretations may at times be necessary, but these are mainly considered "parameters" and are to be used sparingly and then eliminated (Eissler).[10,12] On the whole both groups ignore the subject.

Over the years, however, a number of psychoanalysts, too heterogeneous to be classified, have taken a growing interest in what may be broadly termed the working alliance and the nontransference interactions that take place in the course of psychoanalytic treatment. As stated previously, transference is the experiencing of impulses, feelings, fantasies, attitudes, and defenses with respect to a person in the present that do not appropriately fit that person but are a repetition of responses originating in regard to significant persons of early childhood, unconsciously displaced onto persons in the present. Transference is an indiscriminate, nonselective repetition of the past that ignores or distorts reality and is inappropriate (Greenson).[47]

The very fact that the concept of transference has, over the years, come to have this rather precise meaning implies that it was technically and theoretically necessary to differentiate it from other reactions that are relatively transference-free. Anna Freud, in a recent personal communication on the subject of differentiating between transference and nontransference relationships, had the following to say: 'I have always learned to consider transference in the light of a distortion of the real relationship of the patient to the analyst, and, of course, that the type and manner of distortion showed up the contributions from the past. If there were no real relationship this idea of the distorting influences would make no sense."

All object relations consist of some elements of repetition from the past, but the so-called real, the nontransference, relationship differs from transference in the degree of relevance, appropriateness, accuracy, and immediacy of what is expressed. Furthermore, nontransference responses are basically readily modifiable by both internal and external reality; they are adaptive and realistic.

The terms "transference," "nontransference," "transference-free," and "real relationships" must be considered as relative and overlapping. All transference contains some germs of reality, and all real relationships have some transference elements. All object relationships consist of different admixtures and blendings of transference and nontransference components. Nevertheless, I feel it is important to draw some clear-cut distinction between them. For this purpose a clinical example may serve better than abstract definitions.

A young man, Kevin, in the fifth year of analysis, told me hesitantly after I had made an interpretation that he had something to say that was very difficult for him. He had been about to skip over it when he realized he had been doing just that for years. Taking a deep breath, he said: "You always talk a bit too much. You tend to exaggerate. It would be much easier for me to get mad at you and say you're cockeyed or wrong or off the point or just not answer. It's terribly hard to say what I mean because I know it will hurt your feelings."

I believe the patient had correctly perceived some traits of mine, and it was indeed somewhat painful for me to have them pointed out.

I told him he was right on both counts, but I wanted to know why it was harder for him to tell it to me simply and directly as he had just done than to act in an angry fashion. He answered that he knew from experience that I would not get upset by an exhibition of temper since that was obviously his neurosis and I wouldn't be moved by it. Telling me so clearly about my talking too much and exaggerating was a personal criticism and that would be hurtful. In the past he would have been worried that I might retaliate in some way, but he now knew it was not likely. Besides, he no longer felt my anger would kill him.

Here the difference between transference and nontransference reactions becomes clear. The patient had correctly perceived some characteristics of his analyst's way of working and had also quite realistically predicted that it would be painful for the analyst to have them pointed out. These are nontransference phenomena; they are contemporaneous, appropriate, and realistic. His earlier fantasies about a potentially retaliatory anger that might kill him were historically rooted carry-overs from his childhood anxieties, inappropriate exaggerations, and therefore transference distortions. The patient had developed a good working alliance in relation to his temper outbursts at the analyst, but this alliance could not maintain itself when it came to more realistic criticism. This only developed in his fifth year of analysis (Greenson and Wexler).[49]

It might be well at this point to clarify the relationship between the working alliance, transference, and the real relationship. The working alliance is the nonneurotic, rational, reasonable rapport that the patient has with his analyst and that enables him to work purposefully in the analytic situation despite his tranference impulses (Zetzel,[74] Stone,[68,69] Greenson).[48] The patient and the psychoanalyst contribute to the formation of the working alliance. The patient's awareness of his neurotic suffering and the possibility of help from the analyst impel him to seek out, and work in, the analytic situation. The positive transference, the overestimation and overevaluation of the psychoanalyst, may also be a powerful ally, but it is treacherous. Above all, the reliable core of the working alliance is to be found in the real or nontransference relationship between the patient and the analyst. Transference reactions, whether loving or hateful, from the most infantile to the most mature, eventally lead to idealization, sexualization, or aggressivization and become important sources of resistance in the end. The analyst's participation will be discussed in the section on the analyst's contributions to the analytic situation.

The Resistances

Resistance means opposition. All those forces within the patient that oppose the procedures and processes of analysis—that is, that hinder the patient's free association, that interfere with the patient's attempts to remember and to gain and assimilate insight, that operate against the patient's reasonable ego and his wish to change—all of these forces are to be considered resistance (Freud).[28] Resistance may be conscious, preconscious, or unconscious, and may be expressed by means of emotions, attitudes, ideas, impulses, thoughts, fantasies, or actions. Resistance is in essence a counterforce in the patient, operating against the progress of the analysis, the analyst, and the analytic procedures and processes. Freud[21] had already recognized the importance of resistance in 1912 when he stated: "The resistance accompanies the treatment step by step. Every single association, every act of the person under treatment must reckon with the resistance and represents a compromise between the forces that are striving towards recovery and the opposing ones" (p. 103).

Resistance opposes the patient's reasonable ego, defending the neurosis, the old, the familiar, and the infantile from exposure and change. It may be adaptive. The term "resistance" can be equated with all the defensive operations of the mental apparatus as they are evoked in the analytic situation. The defenses are processes that safeguard against danger and pain and are to be contrasted to the instinctual activities, which seek pleasure and discharge. In the psychoanalytic situation the defenses manifest themselves as resistances. Freud used the terms synonymously throughout most of his writings. The function of defense is originally and basically an ego function, although every kind of psychic phenomenon may be used for defensive purposes.[44]

Freud distinguished among several types of resistances, but I believe it is safe to state that no matter what its origin may be, for a psychic phenomenon to be used for defensive purposes, it must operate through the ego. This is the rationale for the technical rule that the analysis of resistance should begin with the ego. Resistance is an operational concept; it is nothing new that is created by the analysis; the analytic situation only becomes the arena for these forces of resistance to show themselves.

The Analyst's Contribution

In classical psychoanalysis a great number of therapeutic procedures are employed in varying degrees. It is characteristic of all techniques that are considered analytic that their major aim is to increase the patient's insight about himself. Some procedures do not add insight per se, but strengthen those ego functions that are required for gaining understanding. For example, abreaction may permit a sufficient discharge of instinctual tension so that a beleaguered ego will no longer feel imminently endangered. The more secure ego is now able to observe, think, remember, and judge, functions it had lost in the acute anxiety state. Insight now becomes possible. Abreaction is one of the *nonanalytic* procedures that is frequently used in psychoanalytic treatment; it is often an indispensable prerequisite for insight.

The *antianalytic* procedures are those that block or lessen the capacity for insight and understanding. The use of any measure or course of action that diminishes the ego functions of observing, thinking, remembering, and judging belongs in this category. Some obvious examples are the administering of certain drugs and intoxicants, quick and easy reassurances, infantile transference gratifications, diversions, and so forth.

The most important analytic procedure is interpretation; all others are subordinated to it both theoretically and practically. All analytic procedures are steps that lead to an interpretation or make an interpretation effective (E. Bibring,[7] Gill,[41] Menninger).[59]

The term "analyzing" is a shorthand expression that refers to typical insight-furthering techniques. It usually includes four distinct procedures: confrontation, clarification, interpretation, and working through. Before discussing these procedures I believe it would be helpful to consider the dynamics of the treatment situation in general and also explore the question of how an analyst listens.

The Dynamics of the Treatment Situation

The treatment situation mobilizes conflicting tendencies within the patient. It would be valuable to survey the alignment of the forces within the patient in the analytic situation (see Freud,[33] pp. 142–144). I shall begin by enumerating those forces that are on the side of the psychoanalyst and the psychoanalytic processes and procedures:

1. The patient's neurotic misery, which impels him to work in the analysis, no matter how painful.
2. The patient's conscious rational ego, which keeps the long-range goals in view and comprehends the rationale of the therapy.
3. The id, the repressed, and their derivatives; all those forces within the patient seeking discharge and tending to appear in the patient's productions.
4. The working alliance, which enables the patient to cooperate with the psychoanalyst despite the coexistence of opposing transference feelings.
5. The deinstinctualized positive transference, which permits the patient to overvalue the competence of the analyst. On the basis of little evidence the patient will accept the analyst as an expert. The erotic positive transference may also induce the patient to work temporarily, but that is far more un-

reliable and prone to turn into its opposite.

6. The rationale superego, which impels the patient to fulfill his therapeutic duties and obligations. Menninger's[59] "contract" and Gitelson's "compact" express similar ideas.
7. Curiosity and the desire for self-knowledge, which motivate the patient to explore and reveal himself.
8. The wish for professional advancement and other varieties of ambition.
9. Irrational factors, such as competitive feelings toward other patients, getting one's money's worth, the need for atonement and confession, all of which are temporary and unreliable allies of the psychoanalyst.

All the forces listed above influence the patient to work in the analytic situation. They differ in value and effectiveness and change during the course of treatment.

The forces within the patient opposing the analytic processes and procedures may be broken down as follows:

1. The unconscious ego's defensive maneuvers, which provide the models for the resistance operations.
2. The fear of change and the search for security, which impel the infantile ego to cling to the familiar neurotic patterns.
3. The irrational superego, which demands suffering in order to atone for unconscious guilt.
4. The hostile transference, which motivates the patient to defeat the psychoanalyst.
5. The sexual and romantic transference, which leads to jealousy and frustration and ultimately to a hostile transference.
6. Masochistic and sadistic impulses, which drive the patient to provoke a variety of painful pleasures.
7. Impulsivity and acting-out tendencies, which impel the patient in the direction of quick gratifications and against insight.
8. The secondary gains from the neurotic illness, which tempt the patient to cling to his neurosis.

These are the forces that the analytic situation mobilizes in the patient. As one listens to the patient, it is helpful to have this rather simplified division of forces in the back of one's mind (Greenson).[47]

How the Analyst Listens

It might seem unnecessarily pedantic to set down in writing how a psychoanalyst should listen. Yet clinical experience has taught us that the way a psychoanalyst listens is just as unique and complex a procedure as doing free association is for the patient. Here only an outline will be sketched as a preliminary briefing.

The analyst listens with three aims in mind:

1. To translate the productions of the patient into their unconscious antecedents. The patient's thoughts, fantasies, feelings, behavior, and impulses have to be traced to their unconscious predecessors.
2. The unconscious elements must be synthesized into meaningful insights. Fragments of past and present history, conscious and unconscious, must be connected so as to give a sense of continuity and coherence in terms of the patient's life.
3. The insights so obtained must be communicable to the patient. As one listens one must ascertain what uncovered material will be constructively utilizable by the patient.

Clinical experience has suggested a few basic guidelines in order to accomplish these divergent aims (Freud,[38] pp. 111–117). One listens with evenly suspended, evenly hovering, free-floating attention. One does not make a conscious attempt to remember. The analyst will remember the significant data if he pays attention and if the patient is not stirring up the analyst's own transference reactions. Nonselective, nondirected attention will tend to rule out one's own special biases and will allow the analyst to follow the patient's lead. From the evenly suspended, free-floating position the analyst can oscillate and make blendings from among his free associations, empathy, intuition, introspection, problem-solving thinking, theoretical knowledge, and so forth (Ferenczi,[16] Sharpe).[64]

All activities that interfere with the capacity to make the oscillations described above are to be avoided. An analyst should not take any notes if this interferes with his free-floating listening. Word-for-word notes are obviously contraindicated since that would distort his main purpose. The analyst is primarily an understander and a conveyer of insight. He is not essentially a recorder or a collector of research data (Berezin).[6] In order to listen effectively one must also pay attention to one's own emotional responses since these responses often lead to important clues. Above all, the analyst must be alert to his own tranference and resistance reactions since they can impede or help his understanding of the patient's productions.

The analytic situation is essentially a therapeutic one. The analyst is to administer insight and understanding for therapeutic purposes. He listens to gain insight, and he listens from a position of free-floating attention, with restrained emotional responses, with compassion, and with patience. All other scientific pursuits have to be put aside if he is to perform his complicated tasks effectively.[47]

Analyzing the Patient's Material

CONFRONTATION

The first step in analyzing a psychic phenomenon is confrontation. The phenomenon in question has to be made evident to the patient's *conscious* ego. For example, before I can interpret the reason a patient may have for avoiding something, I first have to get him to acknowledge that he is avoiding something. Eventually the patient himself will recognize this, and it will be unnecessary for me to do so. However, before any further analytic steps are taken it must be certain that the patient recognizes the psychic phenomenon within himself that we are attempting to analyze.

CLARIFICATION

Confrontation leads to the next step, clarification. Usually these two procedures blend together, but I find it valuable to separate them because there are instances where each of them cause distinct problems. Clarification refers to those activities that aim at placing the psychic phenomenon being analyzed in sharp focus. The significant details have to be dug out and carefully separated from extraneous matter. The particular variety or pattern of the phenomenon in question has to be singled out and isolated.

Let us take a simple example. I confront a patient, Mr. N., with the fact that he is resisting and he recognizes that it is indeed so, he does seem to be running away from something. The patient's further associations may then lead in the direction of revealing why he is resisting or what he is resisting. Let us take the former instance. The resistant patient's associations lead him to talk of various events of the past weekend. Mr. N. went to a P.T.A. meeting at his daughter's school and felt abashed by the presence of so many wealthy-appearing parents. This reminds him of his childhood and how he hated to see his father attempt to ingratiate himself with his wealthy clients. His father was a tyrant in his dealings with his employees and an "ass-kisser" with the rich. He was afraid of his father until he left home to go to college. Then he developed a contempt for him. He still has a feeling of contempt for him, but he doesn't show it. After all, it would serve no purpose, his father is too old to change. His father must be getting close to sixty, his hair is almost all white, "whatever is left of it." The patient becomes silent.

I had the impression that Mr. N.'s associations were pointing to certain feelings he had about me and it was those feelings which had caused him to be resistant in the early part of the hour. I also felt that this probably had to do with contempt and, more precisely, the patient's fear of expressing his contempt for me directly. When the patient became silent, I said that I wondered if he didn't feel some contempt for another white-haired man. The patient's face flushed and his first response was to say: "I suppose you think I was talking about you. Well, it's just not true. I don't feel any contempt for you—why should I? You treat me very well—most of the time. I have no idea how you treat your family or your friends. But, that's none of my business. Who knows, maybe you are one of those men who steps on the little guy and makes up to the 'big shots.' I don't know and I don't care."

At that moment I pursued the point. I replied that I felt he was relieved not to know how I really behaved outside the hour. If he knew he might feel contempt and he would be afraid to express it to me directly. Mr. N. was silent for a few seconds and answered that if he imagined me doing something contemptible, he wouldn't know what to do with the information. This reminded him of an occasion a few weeks back. He had been in a restaurant and heard a man's angry voice belaboring a waiter. For a fleeting instant the voice sounded like mine and the back of the man's head resembled mine. He was relieved a few moments later to see that it wasn't true.

It was now possible to point out to the patient that he was trying to avoid feeling contempt for me because if he were to do so, he would be afraid of expressing it, just as he had with his father. It was this specific complex pattern of emotional responses that had to be singled out for clarification before one could go on with the further analysis of his resistances. [Greenson,[47] pp. 38–39]

Interpretation

The third step in analyzing is interpretation. This is the procedure that distinguishes psychoanalysis from all other psychotherapies because in psychoanalysis interpretation is the ultimate and decisive instrument. Every other procedure prepares for interpretation or amplifies an interpretation and may itself have to be interpreted. To interpret means to make an unconscious phenomenon conscious. More precisely it means to make conscious the unconscious meaning, source, history, mode, or cause of a given psychic event. This usually requires more than a single intervention. The analyst uses his own unconscious, his empathy and intuition, as well as his theoretical knowledge, for arriving at an interpretation. By interpreting we go beyond what is readily observable and we assign meaning and causality to a psychological phenomenon. We need the patient's responses to determine the validity of our interpretation (E. Bibring,[7] Fenichel).[15]

The procedures of clarification and interpretation are intimately interwoven. Very often clarification leads to an interpretation that leads to a further interpretation (Kris).[52]

Let me return to the clinical excerpt from the treatment of Mr. N. to illustrate these points. The patient is aware that he is resisting facing something, but he is unaware of the feelings and toward whom they are directed. His associations to his contempt for his father impel me to clarify this further, and I point out that he is *afraid* of feeling contempt and having to express it to another white-haired man. The fear of expressing contempt is a clarification; the introduction of "another white-haired man" is an indirect transference interpretation. Mr. N.'s immediate response is a vehement denial: "I suppose you think I was talking about you. Well, it's just not true. . . . I have no idea how you treat your family or your friends. But, that's none of my business. . . . I don't know and I don't care." This response is so intense that I feel I have touched something inside him that is on the verge of becoming conscious.

I then interpret that he is relieved not to know how I treat people in the outside world because if he were to feel contempt he would be afraid to express it to me directly. Mr. N. then recalls an experience in a restaurant when he heard an angry voice belaboring a waiter and for a flash he thought of me. His relief that it was not me indicates that he would have felt contempt for such a man. Mr. N. thus validated the correctness of my clarification and interpretation. I was then able to connect his feelings in the hour to me to the same specific pattern he had felt to his father —a deepening of the interpretation.

Working Through

The fourth step in analyzing is working through. Working through refers to a complex set of procedures and processes that occur after an insight has been given. Working through makes it possible for an insight to lead to change (Greenson).[46] It refers in the main to the repetitive, progressive, and elaborate explorations of the resistances that prevent an insight from leading to change. In addition to the broadening and deepening of the analysis of resistances, reconstructions are also of particular importance. A variety of circular processes are set in motion by working through in which insight, memory, and behavior change influence each other (Kris).[53,54]

In an hour, some two weeks after the session reported above, Mr. N. reports a fragment of a dream. All he can remember is that he is waiting for a red traffic light to change when he feels that someone has bumped into him from behind. He rushes out in fury and finds out, with relief, it was only a boy on a bicycle. There was no damage to his car. The associations led to Mr. N.'s love of cars, especially sport cars. He loved the sensation, in particular, of whizzing by those fat old expensive cars. The expensive cars seem so sturdy, but they fall apart in a few years. The little sports car of his can outrun, outclimb, outlast the Cadillacs, the Lincolns, and the Rolls Royces. He knows this is an exaggeration, but he likes to think so. It tickles him. This must be a carry-over from his athletic days when he loved to be the underdog who defeated the favorite. His father was a sports fan who always belittled my patient's achievements. His father always hinted that he had been a great athlete, but he never substantiated it. He was an exhibitionist, but Mr. N. doubted whether his father really could perform. His father would flirt with a waitress in a cafe or make sexual remarks about women passing by, but he seemed to be showing off. If he were really sexual, he wouldn't resort to that. [Greenson,[47] p. 40]

It seemed clear to me that the patient's material concerns comparing himself with his father in terms of sexual competence. It also deals with people who pretend to be what they are not. Up until this point in the analysis the patient had no memories concerning the sexual activities of his parents. In the hour just described Mr. N. stated that he felt sure his mother disliked sex and his father made use of prostitutes. When I pointed out that this could be a wish-fulfilling fantasy, Mr. N. fell into a long silence.

In the following hour Mr. N. admitted he was furious with my interpretation, but toward the end of the session he acknowledged that his sex life with his wife was "all screwed up." He was unable to look at his wife the day after they had sexual relations because he felt she must abhor his sexual behavior.

In the course of the next several weeks Mr. N. became aware of the fact that alongside of his wish that his wife become sexually free, he had contempt for her when she was sexually excited. This was connected to a childhood screen memory of seeing his mother wink slyly at his father when they saw two dogs copulating in the street. This material was followed by memories of childhood concerning his mother being repulsive to him when he detected her menstruating. This insight was followed by several weeks of complete sexual avoidance of his wife.

During the next several months Mr. N. raged at the hypocrisy of most married adults, "You and your wife included." It was more honest to pay cash for sex than to buy sex with marriage and expensive homes. I interpreted this as contempt covering over an unconscious envy of adults who had a good sex life. Mr. N. reacted with sullen anger for several days. Only gradually was he able to realize that he was reacting like a little boy who begrudged his parents a good sex life. Finally he was able to accept the notion that he could let his parents have their own sex life, and he was perfectly free to enjoy the sex in his own bedroom.

All of this work took a period of six months beginning with the dream of being bumped in his sports car at a red light by a boy on a bicycle. This was not the end of Mr. N.'s sexual problems, there were many back-and-forth movements, but progress continued. Eventually the theme of homosexuality and aggression entered the picture, and there were varieties of regression and progression. I also want to stress that Mr. N. himself did a good deal of the analytic work of working through outside the analytic hour (Stewart).[66]

Working through is the most time-consuming element in psychoanalytic therapy. Only rarely does insight lead very quickly to a change in behavior; and then it is usually transitory or remains isolated and unintegrated. Ordinarily it requires a great deal of time to overcome the powerful forces that resist change and to establish lasting structural changes. The interesting relationship between the work of mourning and working through, the importance of the repetition compulsion and the death instinct, may be pursued in greater detail in the writings of Freud,[20,27,39] Fenichel,[14] Greenacre,[44] Kris,[53,54] Novey,[60] and Greenson.[46]

The four steps outlined above represent a schematized version of what is implied by the concept of analyzing a psychic event. All these steps are necessary, but some may be done spontaneously by the patient, particularly the confrontation or part of the clarification. These steps do not follow in the exact order described, since each procedure can produce new resistances that will have to be taken up first. Or an interpretation can precede a clarification and can facilitate a clarification of a given phenomenon. An additional variable is the fact that the imponderables of everyday life can intrude into the patient's life and take precedence for psychoeconomic reasons over everything else that is going on in the analysis. Nevertheless, confrontation, clarification, interpretation, and working through are the four basic procedures that the analyst performs when he analyzes.

There are two further important processes that play a role in the analyst's contribution to the psychoanalytic situation. I am referring here to the countertransference and the real relationship.

Countertransference

We call it *countertransference* when a psychoanalyst reacts to a patient as if the *patient* were a significant person in the *analyst's* past. Countertransference is a transference reaction of an analyst to a patient, a parallel, a counterpart, to transference as it occurs in patients. The counter in countertransference means analogue, duplicate of, like the counter in counterpart. I stress this point because some authors use the term as though the only countertransference reactions are the psychoanalyst's reactions to the patient's transference reactions. This is not the case. The psychoanalyst may develop countertransference reactions based on the patient's physical or emotional qualities, the patient's psychological history and experiences, and so forth. What is of importance is that countertransference reactions in the analyst, if unrecognized by him, can lead to persistent misunderstanding or mistreatment of the patient.

I can give a simple and brief example. After several years of treatment I suddenly realized to my dismay that I had been giving a young woman patient 55 to 65 minutes each session. I had been completely oblivious of this until I wondered why the patient had so few hostile reactions to me. I then realized this was due to a countertransference reaction. I did some free association about it and was able to trace it back to my feelings toward a certain member of my family. I then started giving the patient the regular 50 to 55 minutes, which she eventually recognized. She questioned me if it were not true that I was giving her less time than I had previously. I admitted that I had realized I had been giving her extra time and I was now correcting my error. She was very curious about the cause of my error. I replied that giving her extra time had not been deliberate, but that I believed my personal unconscious reasons did not belong in her analysis. Then we analyzed her many fantasies to my previous behavior as well as to my asserting a right to privacy and the inequality of the analytic situation. It was not long before the pent-up hostility came out and the analysis proceeded to greater depths.

Countertransference reactions, particularly if they are mild, controllable, and recognizable by the analyst, may be valuable clues to the goings on in the patient that are escaping the notice of the analyst. For example, the first sign you may have of a patient's wish to annoy you is a sense of feeling annoyed for no apparent reason. Finally these few remarks should not be concluded without stating that all psychotic patients and all primitive reactions in our patients will stir up countertransference reactions in ourselves. Winnicott's[73] brief paper on *Hate in the Countertransference* and Searles's[63] textbook are excellent reference sources.

The Real Relationship

Just as the patient reacts to his analyst with other than transference distortions, so does the analyst react to the patient as a real person. Our basic technical tool is interpretation, and the analyst's relative incognito and muted personal responses facilitate the maximal development of the transference reactions that we analyze. Nevertheless, I, along with a growing number of analysts, contend that the technique of "only interpreting" will stifle and distort the development of the patient's transference neurosis and block his capacity to develop realistic object relationships.

The analyst can contribute to the patient's working alliance and nontransference relationship by his consistent and unwavering pursuit of insight, plus his concern, respect, and care for the totality of the patient's personality, sick and healthy. The analyst must help the patient's beleaguered ego distinguish between what is appropriate and distorted, correct and false, realistic and fantastic in regard to his reactions to people, above all, toward his psychoanalyst. He must not fall prey to the vocational hazard of one-upsmanship.

The transference reactions are the vehicle that enables the patient to bring the warded off, inaccessible material into the analytic situation. The working alliance makes it possible for the patient to understand the analyst's insights, review and organize interpretations and reconstructions, and finally integrate and assimilate the material of the analysis. The basis for the working alliance is the capacity for relatively conflict-free ego functioning and the ability, to some degree, to form a real, nontransference relationship to the analyst.

In the clinical instance of Kevin, given above, confirmation by the analyst of the patient's judgments put the analyst for the moment on the side of the patient's observing, realistic ego. In this the analyst supports the working alliance in its efforts to overcome the experiencing ego, previously flooded by neurotic transference. Acknowledging that the patient was right in his criticism was certainly unanalytical. What is more important, however, is that the procedure is not antianalytic. It does advance the analysis, it helps reality testing.

I would like to quote from a paper on this subject:

> All patients, whether neurotic, borderline, or psychotic, have transference reactions in and out of the therapeutic situation. It is our belief that only those patients are analysable who have the capacity for transference-free relationships as well. This is necessary to "get into" analysis. Patients who lack this capacity for transference-free relating require preparatory psychotherapy. This means they need to be helped to build an object relationship based on reliable and predictable perceptions, judgments, and responses. They require more than interpretation and insights. Even most of our neurotic patients, at different periods of the analysis (for example, at the height of the transference neurosis), may require such additional measures. While exact prescriptions for building or strengthening a "real" object relationship in the analytic situation cannot be given, some general guidelines may prove helpful.
>
> The most important and most difficult ingredient to describe is the creation of a productive analytic atmosphere. This should consist of a sense of serious purpose, of dedicated work, of discipline and restraint on the part of the psychoanalyst. Yet this atmosphere must also contain indications of the analyst's humanitarian concern and respect for the patient's predicament. . . . The analyst has to explore and probe into sensitive and intimate areas, and insights should be given with precision, directness and frankness, yet with full awareness of the patient's vulnerability and exposure. The analyst is a physicianly person who must be able to administer painful insights without unnecessary sugar-coating or damaging delays (see Greenson,[47], Chapter 4.2).
>
> We have found it beneficial to explain every new or strange procedure or measure to the patient so that he understands why we work in a certain way. An important rule of thumb we have found useful in promoting the non-transference reactions is the frank admission of any and all errors of technique, whether they be due to countertransference reactions, faulty interpretations, or shortcomings in the analyst's personality or character. The timing of the admission of error and the issue whether one expresses regret verbally or by tone are too complex to be discussed in this limited presentation.
>
> All our patients, to varying degrees, doubt their judgment perceptions, and worthiness. If we "only interpret" or "only analyze" we unintentionally leave them with the impression that their reactions were "'merely" repetitions of their infantile past, and that their behaviour was immature, wrong or crazy. If part of our therapeutic aim is to increase the patient's healthy ego functions and capacity for object relations, it is important that we confirm those aspects of his behaviour which indicate healthy functioning. By ignoring those undistorted aspects of the patient's productions we unwittingly imply that his realistic reactions are unimportant, hardly worthy of comment, and that all that matters is understanding the

unconscious meaning of his behaviour. . . . Beyond that, many of our patients need the experience of feeling in ways that "they are right." They need the experience of having their appropriate ego functions and object relationships acknowledged and respected by the analyst's proper "handling" of both the transference and non-transference phenomena. Structure building occurs not only as a result of dissolution through interpretation but by positive recognition and dealing with the patient's most effective levels of performance. [Greenson and Wexler, pp. 36–37][49]

To conclude this summary on the importance of the working alliance and the real relationship:

> With this in mind, we want to state our basic propositions: To facilitate the full flowering and ultimate resolution of the patient's transference reactions, it is essential in all cases to recognize, acknowledge, clarify, differentiate, and even nurture the non-transference or relatively transference-free reactions between patient and analyst. The technique of "only analysing" or "only interpreting" transference phenomena may stifle the development and clarification of the transference neurosis and act as an obstacle to the maturation of the transference-free or "real" reactions of the patient. Central as the interpretation of transference is to psychoanalytic therapy, and about this there can be no question, it is also important to deal with the non-transference interactions between patient and analyst. This may require non-interpretive or non-analytic interventions but these approaches are vastly different from anti-analytic procedures. [Greenson and Wexler, pp. 27–28][49]

The Analytic Setting

The term "analytic setting" refers to the physical framework and the routines of psychoanalytic practice that form an integral part of the process of being psychoanalyzed. Although it is true that one or another of these elements may be altered without making psychoanalysis impossible, it is also true that the analytic "atmosphere" does influence the various processes that do occur in psychoanalytic treatment. We also know that transference reactions take place spontaneously in neurotic human beings who are not in psychoanalytic therapy. Yet experience has shown that the classical analytic setting does facilitate and maximize the appearance of all the different transference reactions.

Until relatively recently most psychoanalysts stressed the overwhelming importance of the patient's past history and of the analyst's attitude of relative neutrality, incognito, and passivity as the factors determining the course of the transference reactions. Although this is still essentially valid, we do recognize today that certain elements in the analytic setting and procedures may promote or hinder these developments. The papers of Macalpine,[57] Greenacre,[45] Lewin,[56] Spitz,[65] and Stone[68] have been of particular value in illuminating the significance of the analytic setting for the evolution of the various transference reactions.

Greenacre[45] stressed that the circumstance of two people meeting together repeatedly and alone for a long period of time makes for an intensity of emotional involvement. The fact that one is troubled and relatively helpless and the other expert and offering help facilitates an uneven, "tilted" relationship, with the troubled one tending to regress to some form of infantile dependency. The routine of having the patient lie on the couch also contributes to the regression in a variety of ways. The reclining position is a carry-over from the days of hypnosis and a modification of the attempt to put the patient to sleep (Lewin,[56] Khan).[51] The diminution of the external stimuli, the fact that the patient does not see the analyst, that the analyst is relatively silent, and that there is no physical or visual contact between them, also furthers a sleeplike state (Macalpine,[57] Spitz).[65]

Spitz[65] emphasized other elements that push the patient in the direction of objectlessness. The patient is lying down and therefore is lower than the analyst sitting upright behind him, the patient's locomotion and bodily movements are restricted, and he speaks but he cannot see to whom. It is Greenacre's[45] contention that this combination of elements recapitulates the matrix of the mother-child relationship of the first months of life. Free association itself is an invitation to regress toward the primary process and the dream (Macalpine,[57] Lewin).[59] It resembles as well the prattling of a child in that we ask the patient to say everything without discrimination and without responsibility (Spitz).[65]

The analyst's routine behavior also contributes to the regressive pull of the analytic setting. His muted emotional responses, relative incognito, and generally nongratifying attitude regarding the patient's neurotic wishes all expedite the transference neurosis (Macalpine,[57] Spitz).[65] The circumstance that the analyst is a treater of the sick, a therapist, also activates the many infantile antecedents of the doctor in the patient's fantasy life.

Many components of the setting described above that further the regression toward the infantile neurosis also contribute to the formation and maintenance of the working alliance. All procedures that become predictable make for a relative sense of security; and if they are perceived as having a therapeutic intent, they will create a feeling of trust, which is the core of the working alliance. Security and trust make it possible for the patient to allow himself to regress just as they give him the courage to risk discarding a neurotic defense and trying a new form of adaptation. The analyst's unflagging pursuit of insight and understanding, his respect and protection of the rights, potentials, and dignity of the patient, his concern and compassion, and his frank and thoughtful commitment to relieving the patient's neurotic misery should be part of the analytic atmosphere.

It is characteristic for many of the processes of psychoanalysis eventually to mobilize ambivalence. The insatiable instinctual hunger of the neurotic patient can turn even the analyst's gratificatory attitudes into a frustration; the patient's mistrust may twist the analyst's therapeutic concern into a form of rejection and the analyst's patience into indifference. The crux of the matter is the relative strength of the patient's reasonable ego in regard to his id, to his superego, and to the external world at a given moment. The patient's working relationship to the analyst is dependent upon these factors.

The imponderables of everyday life can play a decisive role. Despite the fact that the analytic setting is of importance in the therapeutic equation, it cannot replace psychoanalytic technique: the art of interpretation and the skill in relating to a human being. It must also be remembered with all humility that even with the best technique it requires a goodly amount of time to overcome the formidable tyranny of the neurotic patient's past and his compulsion to repeat (Greenson).[47]

Bibliography

1. AICHHORN, A., *Wayward Youth*, Viking Press, New York, 1945.
2. ALEXANDER, F., and FRENCH, T. M., *Psychoanalytic Therapy*, Ronald Press, New York.
3. ARIETI, S. (Ed.), *American Handbook of Psychiatry*, Vol. 1, pp. 419–566, Basic Books, New York, 1959.
4. ARLOW, J. A., and BRENNER, C., *Psychoanalytic Concepts and the Structural Theory*, International Universities Press, New York, 1964.
5. ———, and ———, "The Psychopathology of the Psychoses: A Proposed Revision," *Int. J. Psychoanal.*, 50:5–14, 1969.
6. BEREZIN, M., "Note Taking during the Psychoanalytic Session," *Bull. Phila. A. Psychoanal.*, 7:96–101, 1957.
7. BIBRING, E., "Psychoanalysis and the Dynamic Psychotherapies," *J. Am. Psychoanal. A.*, 2:745–770, 1954.
8. BOYER and GIOVACCHINI, cited in Arlow, J. A., and Brenner, C., *Int. J. Psychoanal.*, 50, 1969.
9. BREUER, J., and FREUD, S. (1893–1895), *Studies on Hysteria*, in Strachey, J. (Ed.), *Standard Edition*, Vol. 2, Hogarth, London, 1955.
10. EISSLER, K. R., "The Effect of the Structure of the Ego on Psychoanalytic Technique," *J. Am. Psychoanal. A.*, 1:104–143, 1953.
11. ———, "Ego-Psychological Implications of the Psychoanalytic Treatment of Delinquents," in *The Psychoanalytic Study of the Child*, Vol. 5, pp. 97–121, International Universities Press, New York, 1950.
12. ———, "Remarks on Some Variations in Psycho-Analytical Technique," *Int. J. Psychoanal.*, 39:222–229, 1958.

13. FENICHEL, O., *Neurotic Acting Out. Collected Papers of Otto Fenichel*, Vol. 2, pp. 296–304, Norton, New York, 1954.
14. ———, *Problems of Psychoanalytic Technique*, Psychoanalytic Quarterly, Albany, 1941.
15. ———, *The Psychoanalytic Theory of Neurosis*, Norton, New York, 1945.
16. FERENCZI, S., *The Elasticity of Psycho-Analytic Technique: Final Contributions to the Problems and Methods of Psycho-Analysis*, pp. 87–101, Basic Books, New York, 1955.
17. FREUD, A., *The Ego and the Mechanisms of Defense*, International Universities Press, New York, 1936.
18. ———, "Mutual Influences in the Development of Ego and Id: Introduction to the Discussion," in *The Writings of Anna Freud*, Vol. 4, pp. 230–244, International Universities Press, New York, 1945–1956.
19. ———, "Problems of Technique in Adult Analysis," in *The Writings of Anna Freud*, Vol. 4, pp. 377–406, International Universities Press, New York, 1945–1956.
20. FREUD, S. (1937), *Analysis Terminable and Interminable*, in Strachey, J. (Ed.), *Standard Edition*, Vol. 23, pp. 209–253, Hogarth, London.
21. ——— (1912), *The Dynamics of Transference*, in Strachey, J. (Ed.), *Standard Edition*, Vol. 12, pp. 97–108, Hogarth, London.
22. ———, *The Ego and the Id*, in Strachey, J. (Ed.), *Standard Edition*, Vol. 19, pp. 3–66, Hogarth, London.
23. ——— (1914), *Fausse Reconnaissance (Deja Raconte) in Psycho-Analytic Treatment*, in Strachey, J. (Ed.), *Standard Edition*, Vol. 13, pp. 201–207, Hogarth, London.
24. ——— (1901), *Fragment of an Analysis of a Case of Hysteria*, in Strachey, J. (Ed.), *Standard Edition*, Vol. 7, pp. 3–122, Hogarth, London.
25. ——— (1903), "Freud's Psychoanalytic Procedure," in Strachey, J. (Ed.), *Standard Edition*, Vol. 7, pp. 249–254, Hogarth, London.
26. ——— (1896), "Further Remarks on the Neuro-Psychoses of Defence," in Strachey, J. (Ed.), *Standard Edition*, Vol. 3, pp. 159–185, Hogarth, London.
27. ——— (1926), *Inhibitions, Symptoms, and Anxiety*, in Strachey, J. (Ed.), *Standard Edition*, Vol. 20, pp. 77–175, Hogarth, London.
28. ——— (1900), *The Interpretation of Dreams*, in Strachey, J. (Ed.), *Standard Edition*, Vols. 4 & 5, Hogarth, London.
29. ——— (1916–1917), *Introductory Lectures on Psycho-Analysis*, in Strachey, J. (Ed.), *Standard Edition*, Vols. 15 & 16, Hogarth, London.
30. ——— (1918), "Lines of Advance in Psycho-Analytic Therapy," in Strachey, J. (Ed.), *Standard Edition*, Vol. 17, pp. 157–168, Hogarth, London.
31. ——— (1894), "The Neuro-Psychoses of Defence," in Strachey, J. (Ed.), *Standard Edition*, Vol. 3, pp. 43–68, Hogarth, London.
32. ——— (1915), "Observations on Transference-Love," in Strachey, J. (Ed.), *Standard Edition*, Vol. 12, pp. 157–171, Hogarth, London.
33. ——— (1913), "On Beginning the Treatment," in Strachey, J. (Ed.), *Standard Edition*, Vol. 12, pp. 121–144, Hogarth, London.
34. ——— (1914), *On the History of the Psycho-Analytic Movement*, in Strachey, J. (Ed.), *Standard Edition*, Vol. 14, pp. 3–66, Hogarth, London.
35. ——— (1904), "On Psychotherapy," in Strachey, J. (Ed.), *Standard Edition*, Vol. 7, pp. 257–268, Hogarth, London.
36. ——— (1905), "Psychical (or Mental) Treatment," in Strachey, J. (Ed.), *Standard Edition*, Vol. 7, pp. 283–302, Hogarth, London.
37. ——— (1911), *Psycho-Analytic Notes on an Autobiographical Account of a Case of Paranoia (Dementia Paranoides)*, in Strachey, J. (Ed.), *Standard Edition*, Vol. 12, pp. 3–82, Hogarth, London.
38. ——— (1912), "Recommendations to Physicians Practising Psycho-Analysis," in Strachey, J. (Ed.), *Standard Edition*, Vol. 12, pp. 109–120, Hogarth, London.
39. ——— (1914), "Remembering, Repeating, and Working-Through," in Strachey, J. (Ed.), *Standard Edition*, Vol. 12, pp. 145–156, Hogarth, London.
40. ——— (1913), *Totem and Taboo*, in Strachey, J. (Ed.), *Standard Edition*, Vol. 13, pp. 1–161, Hogarth, London.
41. GILL, M. M., "Psychoanalysis and Explora-

tory Psychotherapy," *J. Am. Psychoanal. A.*, *2*:771–797, 1954.

42. ———, *Topography and Systems in Psychoanalytic Theory*, Psychological Issues, Monogr. 10, International Universities Press, New York, 1963.
43. GLOVER, E., *The Technique of Psycho-Analysis*, International Universities Press, New York, 1955.
44. GREENACRE, P., "Re-evaluation of the Process of Working Through," *Int. J. Psychoanal.*, *37*:439–444, 1956.
45. ———, "The Role of Transference: Practical Considerations in Relation to Psychoanalytic Therapy," *J. Am. Psychoanal. A.*, *2*: 671–684, 1954.
46. GREENSON, R. R., "The Problem of Working Through," in Schur, M. (Ed.), *Drives, Affects, Behavior*, Vol. 2, pp. 277–314, International Universities Press, New York, 1965.
47. ———, *The Technique and Practice of Psychoanalysis*, International Universities Press, New York, 1967.
48. ———, "The Working Alliance and the Transference Neurosis," *Psychoanal. Quart.*, *34*:155–181, 1965.
49. ———, and WEXLER, M., "The Non-Transference Relationship in the Psychoanalytic Situation," *Int. J. Psychoanal.*, *50*:27–39, 1969.
50. JONES, E., *The Life and Work of Sigmund Freud*, Vol. 1, Basic Books, New York, 1953.
51. KHAN, M. M. R., "Dream Psychology and the Evolution of the Psycho-Analytic Situation," *Int. J. Psychoanal.*, *43*:21–31, 1962.
52. KRIS, E., "Ego Psychology and Interpretation in Psychoanalytic Therapy," *Psychoanal. Quart.*, *20*:15–30, 1951.
53. ———, "On Some Vicissitudes of Insight in Psycho-Analysis," *Int. J. Psychoanal.*, *37*: 445–455, 1956.
54. ———, "The Recovery of Childhood Memories in Psychoanalysis," in *The Psychoanalytic Study of the Child*, Vol. 11, pp. 54–88, International Universities Press, New York, 1956.
55. LACAN, J. M., "Commentaires sur des textes de Freud," *Psychanalyses*, Vol. 1, pp. 17–28, 1953–1955.
56. LEWIN, B. D., "Dream Psychology and the Analytic Situation," *Psychoanal. Quart.*, *24*:169–199, 1955.
57. MACALPINE, I., "The Development of the Transference," *Psychoanal. Quart.*, *19*: 501–539, 1950.
58. MARMOR, J., "Limitations of Free Association," *Arch. Gen. Psychiat.*, *22*:160–165, 1970.
59. MENNINGER, K. A., *Theory of Psychoanalytic Technique*, Basic Books, New York, 1958.
60. NOVEY, S., "The Principle of 'Working Through' in Psychoanalysis," *J. Am. Psychoanal. A.*, *10*:658–676, 1962.
61. RAPAPORT, D., and GILL, M. M., "The Points of View and Assumptions of Metapsychology," *Int. J. Psychoanal.*, *40*:153–162, 1959.
62. ROSENFELD, H., *Psychotic States*, International Universities Press, New York, 1965.
63. SEARLES, H. F., *Collected Papers on Schizophrenia and Related Subjects*, International Universities Press, New York, 1965.
64. SHARPE, E. F., *The Technique of Psycho-Analysis: Collected Papers on Psycho-Analysis*, pp. 9–106, Hogarth, London, 1950.
65. SPITZ, R. A., "Transference: The Analytical Setting and Its Prototype," *Int. J. Psychoanal.*, *37*:380–385, 1956.
66. STEWART, W. A., "An Inquiry into the Concept of Working Through," *J. Am. Psychoanal. A.*, *11*:474–499, 1963.
67. ———, *Psychoanalysis: The First Ten Years*, Macmillan, New York, 1967.
68. STONE, L., *The Psychoanalytic Situation*, International Universities Press, New York, 1961.
69. ———, "The Psychoanalytic Situation and Transference," *J. Am. Psychoanal. A.*, *15*: 3–58, 1967.
70. ———, "The Widening Scope of Indications for Psychoanalysis," *J. Am. Psychoanal. A.*, *2*:567–594, 1954.
71. WAELDER, R., *Basic Theory of Psychoanalysis*, International Universities Press, New York, 1960.
72. WEXLER, M., "Schizophrenia: Conflict and Deficiency," *Psychoanal. Quart.*, *9*:83–99, 1971.
73. WINNICOTT, D. W., "Hate in the Counter-Transference," *Int. J. Psychoanal.*, *30*:69–74, 1949.
74. ZETZEL, E. R., "Current Concepts of Transference," *Int. J. Psychoanal.*, *37*:369–376, 1956.

CHAPTER 38

THE ADLERIAN AND JUNGIAN SCHOOLS

A. Individual Psychology

Heinz L. Ansbacher

The most important question of the healthy and the diseased mental life is not whence? but, whither? . . . In this whither? the cause is contained.
ALFRED ADLER

THE ADLERIAN SCHOOL of psychiatry, represents a unique theory of personality, psychopathology, and psychotherapy and corresponding practices, which Adler named Individual Psychology.

1. It is consistently humanistic, rejecting analogies from physics, chemistry, or animals (except anthropomorphized animals from fables), while stressing man's striving to overcome difficulties as an aspect of the general evolutionary principle.
2. It is consistently holistic, regarding man as an individual in the sense of being indivisible and unique as well as inextricably embedded in larger systems of his fellow men.
3. It focuses on what is specific to man, his cognitive ability for abstract behavior,[68,69] for creating fictions,[109] and for anticipating future events.
4. It regards human creativity with its freedom of choice as decisive, and heredity and environmental factors as subordinated to it. Animals in their natural habitat, instinctually determined, are not subject to mental disorders.
5. Man requires values as criteria for choice. While the choice is invariably in the direction of a goal of success, what constitutes success is individually determined.
6. Functional mental disorders are based on mistaken schemata of apperception and mistaken ways of living guided by unsuitable goals of success—mistaken life styles. These are not in the patient's awareness but can be inferred from his actions and their

consequences. In psychotherapy the patient's cognitive misconceptions and mistaken goals are pointed out to him together with alternatives, thus confronting him with new choice situations.

7. Individual Psychology is pragmatic[114] rather than positivistic, accepting such alternative concepts and assumptions as are therapeutically valuable and rejecting those associated with pathogenicity. This does not mean that Adler was blind to all the existing pathologies, only that he preferred to regard them as avoidable mistakes rather than as something innate. He was deliberately an optimist from the realization that pessimism is virtually a negation of the work of psychotherapy.

Adler's contribution is to have accepted many time-honored and newer philosophical and scientific humanistic conceptions and to have forged these into an original theory of psychopathology and system of psychotherapy.

Alfred Adler: Development and Systematic Position

Alfred Adler was born in 1870, in a suburb of Vienna, received his medical degree at the University of Vienna in 1895, and subsequently practiced medicine there. From 1902 to 1911 he was associated with Freud in the initial group that became the Vienna Psychoanalytic Society. He then developed his own school of Individual Psychology. In the 1920's he founded a chain of child guidance centers in Vienna. He began traveling to the United States in 1926 and settled there in 1935. In 1937 he died on a lecture tour in Aberdeen, Scotland. Four biographical accounts of Adler by associates have been published,[43,66,90,105] and one by a detached psychiatrist historian, Henri Ellenberger,[61] which is also the best documented.

Constancy

Adler was consistently guided by the idea of social progress and melioration. As a small child he had decided to become a physician "to overcome death" (p. 199).[10] His first publications were on social medicine, with references to Rudolf Virchow, the great nineteenth-century research and public health physician, liberal, and champion of the poor. As a student he had become interested in socialism and later read before the Freudian circle a paper on "The Psychology of Marxism."[12] He introduced the name Individual Psychology with Virchow's definition of individual as "a unified community in which all parts cooperate for a common purpose" (p. iv).[11] After World War I Adler condemned the Bolshevik terror,[5] wrote a passionate defense against the notion of collective guilt,[6] and, in a handbook on active pacifism, denounced personal power over others as a false ideal to be replaced by one of social interest.[17] One of his last papers was on "The Progress of Mankind" (pp. 23–28).[21] Adler's crowning theoretical achievement was the concept of communal feeling (*Gemeinschaftsgefühl*), and his outstanding contribution to practice was counseling before a group. In view of such positive orientation toward the community of man, he had a positive regard for religion as having always pointed to "the necessity for brotherly love and the common weal" (p. 462),[10] and he appreciated the concept of God as "the dedication of the individual as well as of society to a goal which rests in the future and which enhances in the present the driving force toward greatness by strengthening the appropriate feelings and emotions" (p. 460).[10]

Change

Adler's changes were in his theoretical formulations—in the direction away from a mechanistic-causalistic model of man and toward a humanistic-finalistic model.

When Adler wrote in 1907 about organ inferiority and compensation[10] it was a step toward an organismic orientation,[1] although he was still causalistic in his expression, and his concept of motivation was essentially one of homeostasis through emphasis on the central as against the peripheral and autonomic nervous system. When he introduced in 1908 the concept of the aggression drive[10] he took again a step away from elementarism and toward holism in that this was the result of a confluence of several drives, although he still spoke in terms of a drive psychology. When in 1910 he introduced "inferiority feeling"[10] he brought in the concept of the self with its subjectivity and creativity since the feeling was not in a one-to-one relationship to actual conditions. When in 1910–1912 Adler introduced the "masculine protest" and the "will to power"[10] these were decisive steps in replacing a causalistic drive psychology by a finalistic value psychology.

When Adler first wrote about communal feeling or social interest, *Gemeinschaftsgefühl*, it was almost in opposition to self-interest, a dualism quite foreign to a holistic theory. Only in the late 1920's did he clarify that this was a cognitive function, "an innate potentiality which must be consciously developed" (p. 134).[10] During this period Adler also introduced the term "life style," superseding some previous terms, a truly holistic, humanistic conception previously used by the philosopher Wilhelm Dilthey and the sociologist Max Weber.[33]

Writings

Adler's psychological writings extend from 1907 until his death in 1937. It is during the second half and increasingly during the last quarter of this span that he achieved the sophistication and comprehensiveness on which his present-day relevancy is largely based, although the important foundations were laid during the first period.

Most of the books of the second period are available as paperbacks,[8,14,15,18,19,23,24] and articles after his last book have been collected into a volume[21] that includes also his essay on religion and a complete bibliography of some 300 titles. Of his earlier works[11,20,25] important excerpts are to be found in a book of selections from his writings.[10] Another volume[13] consists of 28 papers from professional journals up to 1920, mainly on psychopathology and psychotherapy.

Relationship to Freud

Freud formed his original circle in 1902 by inviting four younger men to meet with him one evening a week to discuss problems of neurosis. One of these was Adler, 14 years Freud's junior.[74] This group developed into the Vienna Psychoanalytic Society, and Adler eventually became its president and co-editor of one of its journals—just a year before his resignation in 1911.

Freud considered Adler to have been his pupil, which Adler consistently denied.[37] He would admit essentially only that "I profited by his mistakes" (p. 358).[10] This position is supported by Ellenberger[61] who states, "Adler seems to have used Freud largely as an antagonist who helped him . . . by inspiring him in opposite ways of thought" (p. 627). Ellenberger advises that in order to understand Adler the reader "must temporarily put aside all that he learned about psychoanalysis and adjust to a quite different way of thinking" (p. 571). The difference is that between a physicalistic-causalistic[73,79] and a humanistic-finalistic way of thinking.

Systematic Position

Because of the physical contiguity, similarity of subject matter, and Freud's seniority, Adler's Individual Psychology was usually classified as a variant of psychoanalytic theory and therapy. In recent years, however, better classifications have been introduced. Adler has been designated as "the ancestral figure of the 'new social psychological look'" (p. 115);[72] as among those advancing a "pilot" rather than a "robot," view of man, where man is largely master of his fate (p. 597);[63] as probably the first among the "cognitive change theorists of psychotherapy" (p. 357),[108] which include Albert Ellis, Adolph Meyer, Fred C. Thorne, George A. Kelly, Rollo May, Viktor Frankl, and O. Hobart Mowrer; as the first among the "third force humanistic psychologists" (p. ix);[83] as advancing a "fulfillment model" rather than a "conflict model" of personality (pp. 17–19);[81] and as representing the philosophy of the Enlightenment.[61] In these various designations Adler is always found in a group opposed to Freud. In sum they support the statement that Adler originated a system of psychotherapy in which a mechanistic medical model of the functional disorders was replaced, not by resorting further to the natural sciences, but by aligning itself with the humanities or human studies (*Geisteswissenschaften*) as described by Wilhelm Dilthey and Eduard Spranger,[10] while keeping well aware of the somatic aspects.

Methodology

Methodologically Adler's approach is what we have called phenomenological operationalism.[35] On the phenomenological, subjective side Adler[10] held: "More important than disposition, objective experience and environment, is the [individual's] subjective evaluation of these. Furthermore, this evaluation stands in a certain, often strange relation to reality" (p. 93). Adler was convinced that "a person's behavior springs from his opinion" (p. 182). "Individual Psychology examines the attitudes of an individual" (p. 185).

On the operational, objective side Adler's[21] principle was, "By their fruits ye shall know them" (pp. 64, 283), that is, by overt behavior and its consequences. In this respect Individual Psychology comes close to behaviorism, although the two differ widely in their respective concepts of human nature. In contrast to other subjectivistic approaches and psychoanalysis, one will not find in Adlerian literature such terms as real self, primary processes, inner forces, latent states, inner conflict, emotions that the individual has to "handle," and many others, because they are like reifications of abstractions and cannot be operationalized.

From this it follows that Individual Psychology is not a *depth* psychology, in the sense that something substantive can be found lurking within the individual if you only dig deeply enough. Rather it is a *context* psychology, in the sense that the meaning of a specific form of behavior can be determined by regarding it in its larger concrete context of which the individual himself is likely not to be aware. By the same token Individual Psychology is a concrete and idiographic science, more concerned with arriving at the lawfulness of the individual case than arriving at general principles, which is the emphasis of the nomothetic approaches.

Personality Theory

Aiming for a humanistic, nonreductionistic, and pragmatically valuble model of man and personality theory, Adler borrowed from philosophy and the other humanities. Thus his conception is in accord with the broad stream of human development found in the other social sciences, in daily life, and in history—aiming toward a better life for all, greater freedom, and greater humaneness.[84]

Man's Creativity—Style of Life

Adler presupposes that the human organism is a unified whole that is not completely determined by heredity and environment, but, once brought into existence, develops the capability of influencing and creating events, as evidenced by the cultural products all around us, beginning with language. Adler[21] quoted from Pestalozzi (1746–1827): "The environment molds man, but man molds the environment" (p. 28), a sentence later also used by Karl Marx.

Heredity and environment merely supply the raw material that the individual uses for his purposes. To quote Adler[21] again: "The important thing is not what one is born with, but what use one makes of that equipment" (p. 86). To understand this one must assume "still another force: the creative power of the individual" (p. 87).

Adler thus advocated, in fact, a "Third Force" psychology, stressing human self-determination, while psychoanalysis essentially stressed heredity, and behaviorism stressed environment.[31] This means in practice that the last two look for objective causes in the past to *explain* behavior, while Adlerian psychology looks for the individual's intentions, purposes, or goals, which are of his own creation, to *understand* behavior.

Creativity is the essential part of Adler's model of man. Its criterion is the capacity to formulate, consciously or most often unknowingly, a goal of success for one's endeavors and to develop planful procedures for attaining the goal, that is, a life plan under which all life processes become a self-consistent organization, the individual's style of life.[33] Only in the feeble-minded is such purposeful creative power absent.[21]

Striving to Overcome: Goal of Success

A unitary concept of man such as Adler's requires one overall dynamic principle. This Adler derived from the processes of growth and expansion that are common to all forms of life. Having the capacity to anticipate the future and having a range of freedom of choice, man develops values and personal ideals, that is, mental constructs[76] or fictions,[109] that serve him as criteria of choice in his movement through life toward a goal of success. The movement takes the form of a striving from relative minus to relative plus situations —from inferiority to a goal of superiority.

Regarding the question whether the inferiority feeling or the goal-striving is primary, Adler does not give a clear answer. But we hold strongly that the latter must be primary because where there is no prior conception of what we should and want to be, there can be no inferiority feeling. Most of us have no inferiority feelings about not being able to speak Chinese because in our environment this is not likely to find inclusion in our goal or image of success. But everybody's goal of success includes wanting to be a worthy human being. "The sense of worth of the self shall not be allowed to be diminished." Adler[10] called this "the supreme law" of life (p. 358).

Movement from inferiority to superiority is a dialectical conception in which Adler was undoubtedly influenced by Nietzsche. Adler[11] found: "Nietzsche's 'will to power' . . . includes much of our understanding" (p. ix); he considered Nietzsche "one of the soaring pillars of our art" (p. 140) and credited him with "a most penetrating vision" (p. 24).[15] Just as will to power meant for Nietzsche not domination over others but the dynamics toward self-mastery, self-conquest, self-perfection,[75] so for Adler[10,21] power meant overcoming difficulties, with personal power over others representing only one of a thousand types, the one likely to be found among patients.

Social Interest

Adler's most important concept, and the one most specific to him, is communal feeling (*Gemeinschaftsgefühl*). It is often translated as social interest, meaning not an interest in the other as an object for one's own purposes, but "an interest in the interests" of the other. It includes also the conception of being attuned to the universe in which we live.[32]

Adler's holistic emphasis saw man not only as an indivisible whole but also as a part of larger wholes—his family, community, humanity, our planet, the cosmos. Man lives within a social context from which he meets the various life problems of occupation, of sex, and of society in general—all social problems.

Social interest as a conception is based on the simple assumption that we have a natural aptitude for acquiring the skills and understanding to live under the conditions into which we are born as human beings. Human nature "includes the possibility of socially affirmative action" (p. 35),[15] a conception quite different from conformity and superego, leaving room for social innovation, even through rebellion. If such an aptitude were not present in man, how would social living and the creation of the various cultures ever have started? This aptitude, however, needs to be consciously trained and developed. A developed social interest is the criterion for mental health. When the actions belie words of social interest, as is so often the case with neurotics—the "yes-but" of the neurosis[10]—the actions are taken to speak louder than the words.

Social interest is then direction-giving. The

direction it gives to action is toward synergy of the personal striving with the striving of others. It is on the socially useful side, in line with the interests of others. All failures in life, on the other hand, have in common a striving for a goal of success that has only personal meaning—and thus becomes in the long run unsatisfactory even to the individual himself.

Social Interest-Activity Typology

Adler eventually added to his basic dynamics of overcoming a second dimension, beyond that of social interest, namely, the individual's degree of activity. This led him to a fourfold typology that, however, does not quite correspond to the four categories created by the two dimensions. The types are: high social interest, high activity—the ideally normal; low social interest, high activity—the ruling type, tyrants, delinquents; low social interest, low activity—the getting type, expecting from and leaning on others, and the avoiding type, found in neuroses and psychoses; with the category high social interest, low activity left unrepresented by a type (pp. 167–169).[10]

Approach to Biography and Literary Criticism

"Every individual," Adler[10] stated, "represents both a unity of personality and the individual fashioning of that unity. The individual is thus both the picture and the artist . . . of his own personality, but as an artist . . . he is rather a weak . . . and imperfect human being" (p. 177).

This then became Adler's approach to characters in literary works. The great authors succeed in creating their characters in similar fashion as characters in real life "create" themselves. Adler venerated and admired the great writers "for their perfect understanding of human nature," and considered any attempt to explain artistic creations by tracing them to assumed underlying "causes" as being profane and desecrating.[96] Adlerian literary criticism is concerned with understanding the basic "mistake" in the life style of the tragic hero, or in the case of an actual biography, perhaps of a great man, understanding what acts of constructive overcoming of difficulties became decisive. Recent studies along these lines have dealt with *Hamlet*,[82] the casket scenes from *The Merchant of Venice*,[78] *Oedipus Rex*,[39] Somerset Maugham,[44] and Ben Franklin.[80] There are also numerous earlier studies.

Theory of Psychopathology

Everyone lives in a world of his own construction, in accordance with his own "schema of apperception." There are great individual variations in the opinion of one's own situation and of life and the world in general, involving innumerable errors. While the "absolute truth" eludes us, we can discern between greater and lesser errors. The former are more in accordance with "private sense" (also private intelligence or private logic) and characteristic of mental disturbance, the latter, with "common sense" (H. S. Sullivan's "consensus"), an aspect of social interest, and characteristic of mental health (pp. 253–254).[10]

It was Kant (1724–1804), with whose writings Adler was well acquainted, who had originally observed, "The only feature common to all mental disorders is the loss of common sense (*sensus communis*), and the compensatory development of a unique, private sense (*sensus privatus*)."[34] Interestingly, the Latin *communis*, in addition to meaning common and general, also means equal and public.

The Patient's Creativity

The patient's creative process, involving more serious errors than is ordinarily the case, has brought him to the predicament in which he finds himself. His "symptoms" are further creations, his own "arrangements" (pp. 284–286)[10] to serve as excuses for not meeting his life problems. They assure him freedom from responsibility. In trying to convey an excuse they are forms of communication. Since the organism is a unified whole, the autonomic functions can become part of these arrangements, the basis for psychosomatic medicine. In this sense Adler[10] spoke of organic symptoms as "organ dialect" (p. 223), actually symbolic acts rather than symptoms.

In most cases the patient's circumstances were conducive to mistaken constructs. As Adler[10] stated, "Every neurotic is partly right" (p. 334). But the Adlerian school does not accept these adverse circumstances as absolutely binding. Difficulties can be overcome in one way or another. Thus the patient is "right" in that there were "traumatas" and all sorts of "frustrations" in his life, which can easily be construed as adverse "causes." But he is only "partly right" in that he was not obligated to construct his life in the inexpedient way in which he did. Others with similar experiences constructed their lives differently.

Conflict and Emotions Seen Holistically and Teleologically

On the basis of its holistic orientation Adlerian theory does not recognize any internal dualities and antitheses resulting in antithetical unconscious impulses, or in onslaughts from an unconscious on the conscious. It does not recognize any "intrapersonal" conflicts, only "interpersonal" conflicts arising from the opposition of the patient's private sense to the common sense. "Individual Psychology is not the attempt to describe man in conflict with himself. What it describes is always the same self in its course of movement which experiences the incongruity of its life style with the social demands" (p. 294).[10]

What is often described as "ambivalence" is considered the use of seemingly antithetical means to arrive at the same end, as a trader sometimes buys and sometimes sells, but always for the same end of making money. The related concepts of doubt and indecision are also arrangements of the patient, serving one unrecognized goal—to maintain the status quo.

Emotions are not understood as in conflict with rationality, but in the service of the hidden purposes of the individual. Thus anxiety supports the creation of a distance between the person and his tasks of life in order to safeguard the self-esteem when there is fear of defeat. "Once a person has acquired the attitude of running away from the difficulties of life, this may be greatly strengthened . . . by the addition of anxiety" (p. 276).[10] Lack of social interest always being involved in pathology, "the anxious person . . . also feels himself forced by necessity to think more of himself and has little left over for his fellow man" (p. 277).[10] Anxiety neurosis and all kinds of phobias serve the purpose of blocking the way and thus cover up the simple fear of personal defeat. Actually all neurotic symptoms develop out of an effort to conceal "the hated feeling of inferiority" (p. 304).[10] "The emotions are accentuations of the character traits . . . they are not mysterious phenomena. . . . They appear always where they serve a purpose corresponding to the life method or guiding line of the individual" (p. 227).[10]

The Pampered Life Style

The predisposing condition for pathology is the pampered style of life. This is more likely to be developed by individuals who as children (1) have actually been pampered, although it is often found also in those who (2) have been unwanted or neglected, or (3) suffered from physical handicaps (organ inferiorities)—the three overburdening childhood situations. The pampered life style is ultimately the individual's own creative response to which he is by no means obligated by the situation. The pampered life style is characterized by leaning on others, always expecting from them, attempting to press them into one's service, evading responsibility, and blaming circumstances or other people for one's shortcomings, while actually feeling incompetent and insecure. "The pampered life style" eventually replaced Adler's original term of "the neurotic disposition."

Also, "We must always suspect an opponent," according to Adler,[15] "and note who suffers most because of the patient's condition. . . . There is always this element of concealed accusation" (p. 81). This accusation "secures some triumph or at least allays the fear of defeat," not in the light of common sense, of course, but in accordance with the patient's private logic (p. 80).[15]

There is then always a degree of self-deception to the extent that the patient makes himself believe he is not to be blamed because he is not responsible. This is, of course, a general tendency. But "when the individual helps it along with his devices, then the entire content of life is permeated by the reassuring, anesthesizing stream of the life-lie which safeguards the self-esteem" (p. 271).[10] Later Adler most often used the term "self-deception" instead of "life-lie." This idea was taken up many years later by Sartre in his concept of "bad faith," *mauvaise foi*.[107]

Unity of Mental Disorders

In keeping with a unitary dynamic theory, Adler presented essentially a unitary theory of mental disorders. These are not considered as different illnesses, but the outcomes of mistaken ways of living by discouraged people, people with strong inferiority feelings and unrealistically high and rigid compensatory goals of personal superiority, and people with insufficiently developed social interest. "What appear as discrete disease entities are only different symptoms which indicate how one or the other individual considers that he would dream himself into life without losing the feeling of his personal value" (p. 300).[10]

"Naturally, anyone, who stands for the unity and uniform structure of the psychoneuroses," Adler[10] observed as early as 1909, "will have to explain each particular case individually" (p. 301). At the same time Adler recognized the commonly observed symptom categories, and in the following we shall give his views on some of these.

Compulsion Neurosis

As Freud used the hysteric as the paradigm of the neurotic, so Adler, in fact, considered compulsion neurosis as the prototype of all mental disorders. One of the early Adlerians, Leonhard Seif, had noted that "one could call virtually any neurosis a compulsion neurosis" (p. 138).[21] The following is a summary description adapted from Adler. (1) A striving for personal superiority is diverted into easy channels. (2) This striving for an exclusive superiority is encouraged in childhood by excessive pampering. (3) Compulsion neurosis occurs in the face of actual situations where the dread of a blow to vanity through failure leads to a hesitating attitude. (4) These means of relief, once fixed upon, provide the patient with an excuse for failing. (5) The construction of the compulsion neurosis is identical with that of the entire life style. (6) The compulsion does not reside in the compulsive actions themselves, but originates in the demands of social living that represent a menace to the patient's prestige. (7) The life style of the compulsion neurotic adopts all the forms of expression that suit its purpose and rejects the rest. (8) Feelings of guilt of humility, almost always present, are efforts to kill time in order to gain time (pp. 135–137).[21]

Even the psychoses share, according to Adler,[21] characteristics of the compulsion neuroses. "Compulsive symptoms may border on manic-depressive insanity or schizophrenia and resolve themselves into one or the other" (p. 137). "All three groups are variants of a single condition: an extreme superiority complex and [confrontation with social] tasks which call for more social interest than the patient has" (p. 138).

The book by Leon Salzman on *The Obsessive Personality*[98] fits exceedingly well within the Adlerian framework. The author considers himself close to Rado, Horney, Alexander, Sullivan, Strauss, Goldstein, and Bonime, among others.

Depression, Suicide, and Mania

Depression was for Adler[13] "a remarkable artistic creation (*Kunstwerk*), only that the awareness of creating is absent and that the patient has grown into this attitude since childhood." It is actually "the endeavor through anticipation of one's ruin to force one's will upon others and to preserve one's prestige." The depressed patient makes "a formidable weapon" out of his weakness "to gain recognition and to escape responsibility"

(p. 239).* "The most prominent weapon . . . consists in complaints, tears and a sad, dejected mood" (p. 250).*

"It is always a question of effect upon the environment" (p. 251).[13] Such paradoxical use of weakness to gain control over others, that is, the attempt to control others without accepting the responsibility for doing so, has more recently been recognized and designated by the term "paradoxical communication" (p. 17).[71]

A particular person in the patient's environment may be considered the "opponent." As we have seen above, Adler regarded this, in general, as a most useful principle for understanding a patient's dynamics. But it was in the context of depression that he originally recommended "to raise the question of the 'opponent' " (p. 236).[13]

Regarding suicide, Adler proposed in 1910 that it had a social intention like depression. Talking about adolescent suicide, he considered it an "act of revenge," in which "one's own death is desired, partly to cause sorrow to one's relatives, partly to force them to appreciate what they have lost in the one whom they have always slighted. . . . In later years, . . . a teacher, a beloved person, society, or the world at large is chosen as the object of this act of revenge."[36] Adler[11] soon added that neurosis, in general, is "a self-torturing device for the purpose of raising the self-esteem and troubling the immediate environment" (p. 412) and that suicide is similar. The potential suicide "hurts others by dreaming himself into injuries or hurting himself," when confronted with an exogenous problem for which his social interest is insufficient (p. 252).[21] "The 'other' is probably never lacking. Usually it is the one who suffers most by the suicide" (p. 251).[21]

Farberow and Shneidman in their *The Cry for Help*[62] have offered an opportunity to compare the Adlerian understanding of suicide with that of other schools by having one case of attempted suicide discussed from various viewpoints. The viewpoints of Freud, modified psychoanalysis, Jung, Adler (by the present writer), Sullivan, Horney, George Kelly, and Carl Rogers are each presented in a separate chapter.

The manic state that often accompanies depression is interpreted by Kurt Adler[27] as follows: "Mania is a frantic effort by the patient to force success in the service of his goal of superiority. . . . In his overcompensation of his inferiority feelings . . . he appears to take literally the 'all' in the 'all or nothing' proposition so typical of the neurotic. . . . Both the manic and the depressed never really believe in themselves, do not appreciate others, and are always eager to exploit others for their own purposes. Both negate reality by the use of delusion about their prophetic gift: one, by foreseeing that everything will be wonderful and that he can do anything, the other, that everything will be dismal and that he can do nothing" (p. 60).

Schizophrenia

The Adlerian theory of schizophrenia assumes an abysmally low self-esteem that is, so to speak, balanced by an extravagant reified goal of superiority, such as to be Jesus Christ or Napoleon. This can be maintained only "when the individual has, by losing all interest in others, also lost interest in his own reason and understanding" (p. 128).[15]

The characteristic hallucinations, connected with the role of superiority that the patient has created for himself, "arise always when the patient wants something unconditionally, yet at the same time wants to be considered free from responsibility" (p. 317).[10] The hallucination is a trick to make subjective impulses appear as something objective. "The coercion toward irresponsibility prevents the will from being guided by objective determiners and replaces these by apparently strange voices and visions" (p. 259).[11] The life style of the schizophrenic along these lines has been concisely described by Kurt Adler.[29]

This conception is quite similar to that later arrived at by Ludwig Binswanger. The area of agreement could be described as follows: "From a strong feeling of inferiority the schizophrenic throws away colorful human

* Translation modified from the original.

weakness for a soaring fiction which can be maintained only at the cost of reality, human contact, and the whole shared world which gives existence its deep meaning."[110] This is also similar to Harry Stack Sullivan's position.

Recently an Adlerian book on schizophrenia has been published. Its author, Bernard Shulman,[102] questions the role of genetic or environmental factors "as direct linear 'cause' of schizophrenia," and holds instead that "a teleological factor must be present, namely, a set of *personal values* which are largely self-determined and which 'call forth' the psychosis" (p. 8). In this sense schizophrenia is not only a reaction but also "an action, a decision, a choice" (p. xi). On this basis Shulman gives many practical examples of treatment that could be followed by any therapist, regardless of his theoretical orientation.

Perversion and Crime

Adler[10] summarized the common factors in all sexual perversions (homosexuality, sadism, masochism, masturbation, fetishism, and so forth) as early as 1917 in the following:

1. Every perversion is an expression of increased psychological distance between man and woman.
2. The perversion indicates a revolt against the normal sexual role, and is an unconscious trick to depreciate the normal sexual partner and to enhance one's own self-esteem.
3. Some animosity against the normal sexual partner is always evident.
4. Perversions in men are compensatory attempts in the face of the overrated power of women; in women, in the face of the assumedly stronger male.
5. Perversions develop in persons who generally are oversensitive, excessively ambitious, and defiant. They are likely to be egocentric, distrustful, and domineering, have little inclination to "join in the game," whether with men or women. Their social interest is greatly limited (p. 424).

Adler[10] stressed that the consistent exclusion of the other sex is a matter of self-training. "No sexual perversion without preparation. . . . Each person has formed it for himself; he has been directed to it by the psychological constitution he has himself created, although he may have been misled into it by his inherited physical constitution which makes the deviation easier for him" (p. 424). A contemporary Adlerian exposition of homosexuality has been provided by Kurt Adler.[28]

Regarding crime, while Adler wrote quite extensively on the life style and treatment of the criminal, and crime prevention,[21,24] the basic theory can be stated briefly. The criminal, like other failures, fails in social interest (p. 411).[10] But unlike the others he displays a certain degree of activity, albeit on the useless side of life; he can cooperate, though, only with his kind (p. 413);[10] and he is likely to develop "a cheap superiority complex" (p. 414).[10] He is extreme in attempting to free himself from responsibility, always looking "for reasons that 'force' him to be a criminal" (p. 413).[10] How well the Adlerian formulations fit especially the cases of the various assassins of American presidents has been shown in a discussion by James P. Chaplin.[45]

Process of Psychotherapy

Psychotherapy is the endeavor to help the patient reconstruct his assumptions and goals in line with greater social usefulness. "The fault of construction is discovered and a reconstruction is accomplished" (p. 22).[8] This is done largely through extending social interest toward the patient, getting him to see his goal of personal superiority stemming from hidden inferiority feelings, and encouraging his actions on the socially useful side. Thereby his behavior will be modified despite all objective adversities, including those from the past that in any event cannot be altered. The process is one of cognitive reorganization, "by a correction of the faulty picture of the world, and the unequivocal acceptance of a mature picture of the world" (p. 333).[10]

Psychotherapy is an "artistic" (p. 192),[21] in the sense of creative, task in which the therapist brings his own creativity to bear to influence that of the patient. It consists in the art of imparting the understanding the therapist has gained of the patient to him, to make him see the mistake in his life style—giving him insight—and the alternatives available to him.

Three or four phases of psychotherapy have variously been identified and described.[2,3,55,92,113] However, these must be understood not as distinct units following each other neatly in time, but as components to be found in any of the many incidents during the course of treatment as well as the treatment as a whole. Not all the phases are always represented. As we see it, the phases are: (1) establishing and maintaining a good relationship with the patient; (2) gathering data from the patient to understand him, to have source material for interpretation, for conceptualizing his life style; (3) interpreting the data; (4) provoking therapeutic movement, change of behavior.

But before turning to a description of these phases, we should like to consider briefly improvement without insight and somatotherapy. According to Adler,[10] the criterion of success of treatment is objective. "As soon as the patient can connect himself with his fellow men on an equal and cooperative footing, he is cured" (p. 347). Therefore, since mental disorder is a phenomenon of problem-solving in a situation, a patient may improve through merely a change of his situation or renewed interest in others.[15] Adler[10] gives the example of a burglar who becomes a good citizen. "Perhaps he is growing older and fatter . . . his joints are stiff and he cannot climb so well: burglary has become too hard for him" (p. 418). Or, since the patient somehow unknowingly fell into making the more erroneous choice, he may also improve under certain circumstances without knowing how this came about. It is the change in behavior that counts; insight is not absolutely necessary.[55,93] This is the point where Individual Psychology and behavior therapy meet.

Regarding somatic treatment techniques, let us remember that Adlerian theory had its origin in observing the variety of compensatory responses instigated by organ inferiorities; also that it is a pragmatic theory, rather than one attempting to establish absolutes. Thus Adlerian psychiatrists welcome the help they find in modern somatic techniques. "Drugs and electroshock therapy," according to Kurt Adler,[27] "probably cause a break in the constant, intensive preoccupation of the patient with his prestige strivings and morbid delusions" (p. 64). From similar reasoning Alexandra Adler[4] states that drug therapy "may result in a more positive response to work, and an increased interest in human contact." However, the success of such treatment "depends upon experience with and interest in the management of the whole personality of the patient."

Establishing and Maintaining a Good Relationship

"Psychotherapy," according to Adler,[10] "is an exercise in cooperation and a test of cooperation" (p. 340). To this Dreikurs[55] adds, "Therapeutic cooperation requires an alignment of goals" (p. 65). Goals and interests of patient and therapist must not clash. "The first rule is to win the patient; the second is never to worry about your own success" (p. 341).[10] In addition to friendliness, an important way to win the patient is to make him feel understood, whereby one also wins his respect.[57] This feeling is generated by interpretation of the patient's behavior in a way that is new and plausible to him.

To assure the continued cooperation of the patient, it is necessary to be tactful and avoid dogmatic statements. Adler[8] referred to Benjamin Franklin in recommending the use of such phrases as "perhaps," "probably," or "possibly" when making proposals to patients. In the same vein Dreikurs[55] offers interpretations with such phrases as, "Would you like me to tell you?" "Could it be?" "Are you willing to listen?" (p. 274). Sometimes the surprise element is helpful in maintaining a fruitful relationship,[52] and Adler[21] recommended "to have a series of dramatic illustrations at one's disposal" (p. 201).

The therapist must also know that the patient may want to depreciate him as he has done with others, in order to raise his own self-esteem.[10] Resistance is an expression of this depreciation tendency. The patient may praise the therapist or express great expectations as a build-up to be followed by an all the greater letdown as a form of depreciation. The therapist must "take the wind right out of the patient's sails!" (p. 338).[10]

Gathering Data

The purpose of the psychological exploration is to arrive at a self-consistent conceptualization of the patient's style of life, with emphasis on his mistaken goal and methods of striving for it. The exploration is not extended beyond this point. Not believing in the "causal" significance of past events per se, the Adlerian considers their recollection as active "arrangements" by the patient, and takes them as significant samples of his life style. In this respect the Adlerian approach differs from the Sullivanian one with which it has much in common.[38] It considers a complete exploration of the past unnecessary, and should be quite timesaving by comparison.

The Adlerian emphasis is on concrete events and actions as well as on the patient's "private world." The idea is to get a representative picture of the patient, with his private views in the total context of his concrete social system, past and present. Thus the Adlerian therapist is interested in the primary family constellation in which the life style emerged, the patient's early recollections, his dreams, and so forth, but also in the actual time and circumstances under which his problem developed, as well as his present concrete social and occupational situation and problems.

To complete the picture the Adlerian is likely to follow any statement by the patient of what happened to him with the question, "And what did you do?"[57] Often the request is made to describe a typical day.[58] Most importantly, the question is asked, "What would you do if you were well?"[10,55]

The areas to be explored are roughly outlined in an interview guide.[10,19] There is also a guide for exploring the family constellation.[103]

Interpretation

The therapist listens to the patient dialectically;[97] that is, he asks himself what opposite could be paired with a certain statement. This is based on the assumption of the self-deception of the patient mentioned before. The patient sees and recognizes only that part of the situation that is consistent with his life style, and thus in this sense he is actually unconscious "even when he is conscious" (p. 217).[22] "While he regards one point, we must look at the other. He looks at his obstacles; we must look at his attempt to protect his fictive superiority and rescue his ambition" (p. 199).[21] When the patient speaks of his generosity, the therapist may understand an accusation of stinginess against others.

The therapist synthesizes the two aspects into an inference regarding the patient's possible intention and goal. Thus the answer to the question, "What would you do if you were well?" leads to the interpretation that this activity may be exactly the one from which the patient is excusing himself by his symptom.

The dialectics may show to the patient more directly alternatives of action open to him; he is urged further into these by the therapist pointing to the paradoxes that the patient created by overlooking the part of the situation that does not suit his life style. For example, when the wife complains, "My husband comes home late at night," the question, "And what do you do?" may elicit the answer, "I scold him." From this the therapist gives her the insight that she "is not merely a victim . . . but a most active participant" (p. 269).[55] This reply is a "therapeutic paradox" (pp. 184–185)[71] in that it contains a reproach in telling her that she is not as innocent as she tried to appear, and at the same time an encouragement in telling her that she actually is capable of taking the initiative. Such interpretation comes close to being a confrontation since it is likely to move the patient toward change, if this is further suggested. In the above case the wife may from now on receive her husband quite differently when he comes home late, which, in turn, confronts him with a new situation.

Adler was quite aware of the phenomenon described today as paradoxical communication,[112] when he quoted Socrates, "Young man of Athens, your vanity peeps from the holes in your robe" (p. 232).[10]

Interpretations are also given from concretizing or operationalizing a statement, or in reference to the consequences of an action. Adler would operationalize a complaint about indecision or doubt by stepping back and forth, actually remaining in the same place, and from this inferring that the meaning of indecision is the hidden intention to preserve the status quo. It is similar with Adler's[21] acceptance of Nietzsche's interpretation of guilt feelings as mere wickedness: "The patient is demonstrating virtue and magnanimity," while, in fact, doing nothing to remedy the situation (p.137).

A book on psychological interpretation by Leon H. Levy[77] is very much in accord with the Adlerian position. The author holds that interpretation does not "uncover" any new "facts" hitherto hidden in the unconscious, but rather brings "an alternate frame of reference" to bear to facilitate change. The author acknowledges his intellectual debt to George A. Kelly and Julian B. Rotter, both close to Adler.

Confrontations and Directives

Confrontation in Adlerian psychotherapy is a technique particularly calculated to provoke therapeutic movement. When an interpretation is followed by a question challenging the patient to take a stand, it becomes a confrontation.[101] These questions are most often calculated to make the patient face the concrete reality, the common sense. For example, a middle-aged man who had been in psychoanalytic treatment before he came to Adler told him he suffered from an unresolved Oedipus complex, whereupon Adler[35] confronted him with: "Look here, what do you want of the old lady?"

The confrontation is designed to get a commitment from the patient to make a choice on his own. Thus really any interpretation can become a confrontation. In the example of the wife in the previous section, presenting the option in the form of a question—"Will you continue to scold him although it does not help?"—would have made it into a confrontation.

At times directives are given. When these refer to the symptom, they are most often the paradoxical encouragement of the symptom as Haley[71] described it, namely, "in such a way that the patient cannot continue to utilize it (p. 55)." When a patient complained, "There is nothing I like doing," Adler[10] would direct him to "refrain from doing anything you dislike" (p. 346–347). To a patient characterized by indecision and finally asking, "What shall I do?" Adler[113] would say, "Do for a few months more what you have been doing! Above all, don't do anything rash!" (p. 101). This is what Viktor Frankl[64] calls "paradoxical intention," and Dreikurs,[55] after Erwin Wexberg, calls "antisuggestion," that is, to practice the very thing that one had been fighting against. Antisuggestion specifically is not limited to the symptom but applicable to any behavior. To return once more to the wife of the drinking husband, one might say, "Go right ahead scolding him each time. But it won't make you feel any better."

Child Psychotherapy

Treatment of children in the Adlerian literature is concerned with disturbing behavior rather than psychotic abnormality. It does not differ from adult treatment in that it involves giving the child an understanding of the goals of his behavior, showing him how he is, in fact, behaving by pointing out the consequences of his behavior on others, and encouraging him to conceive and choose alternative ways, that would lead to socially desirable successes. Dreikurs[56,60] has distinguished four goals of misbehavior in children: to gain attention, power, or revenge, or avoid defeat by withdrawal. Teachers in classroom situations have found this distinction especially helpful.

It is quite possible to give children the necessary understanding through simple terms, sometimes with gestures. Adler[10] thought that if he did not succeed in explaining to a child the roots of his mistakes, "I can be sure that I have blundered in interpreting his situation or in describing it to him" (p. 397). The preferred form of therapy is counseling before a group or in the classroom situation or smaller group of peers (see below).

¶ Illustrations of the Adlerian Approach

To demonstrate the Adlerian approach to understanding and therapy, we shall in the following give a concrete example of dream interpretation and the interpretation of an early recollection.

Dream Interpretation

William D. Dement,[49] the dream physiologist, briefly reports the following four dreams: "One subject in our laboratory in a single night ran the gamut from being with 'two hippopotamuses in a millpond' through a 'taffy pull in the Soviet Embassy' to 'hearing Handel's *Messiah* sung by a thousand-voice chorus in this beautiful cathedral,' back to 'writing at my desk' " (p. 308).

He gives this as an example of "the wildly unpredictable nature of dream content," even in dreams from a single night, and finds, "The fundamental determinants of dream content remain cloaked in obscurity" (pp. 308, 309). For the Adlerian the search for determinants would be a "pseudoproblem" since he accepts "the nearly limitless possibilities of the creative power" (p. 777)[16] of the individual. However, while specific dream content is unpredictable, it is not unintelligible. Since the Adlerian assumes dreams to be attempted solutions of current problems facing the dreamer, in line with his life style in general, one should be able to "guess" from the dream content what kind of person the dreamer would be. Thus challenged, the present writer wrote to Dement: "The common denominator in these dreams is bigness, strength, activity and a pleasant feeling tone. Hence we are willing to 'predict' with a considerable degree of confidence that the dreamer in waking life shows great activity, buoyancy, and optimism, with perhaps some grandiosity and manic traits. He is also a cultured person and interested in music. The dreams maintained his frame of mind for the next day's work. One might 'predict' further, that dreams on subsequent nights would still carry this person's mark of confidence and optimism" (H. L. Ansbacher, personal communication to W. C. Dement, March 29, 1967). To this Dement replied: "The individual in question is essentially as you describe him" (William C. Dement, personal communication, April 3, 1967).

If the author of these dreams were a patient, the feeling tone would probably be somewhat different. But if the content would otherwise be the same, we might guess that the patient would be manic or depressed. We could use the dreams to help us make him see his basic mistake, namely, the expectation of associating only with the big and the ultimate, thereby perhaps missing out on solving the daily problems of life—provided other data would point in the same direction. The dreams could help in giving the patient insight into what he is in fact doing and in clarifying to him the available alternatives of which he had been unaware.

An excellent presentation of the Adlerian view on dream interpretation has been provided in a chapter by Bernard Shulman,[100] and the book by Walter Bonime[42] has been widely acclaimed by Adlerians.

Early Recollections

For the Adlerian a recollection is essentially a response, an action. For him it is important that the individual selected this particular incident as memorable, and how he acts in it. The earliest recollections are especially important. In Adler's[24] words, "They represent the individual's judgment, 'even in childhood, I was such and such a person,' or, 'even in childhood, I found the world like this' " (p. 75). A brief description of the technique is to be found in all of Adler's books. Among contemporary descriptions an earlier one is by Harold H. Mosak[85] and a recent one by Verger and Camp.[111]

As an illustration we are citing the earliest recollection of the great German author, playwright, and poet, Goethe, giving first Freud's causalistic interpretation and then an Adlerian finalistic interpretation, following Paul Rom's[95] account. Goethe's recollection was: "One fine afternoon, when everything was quiet in the house I was amusing myself with my pots and dishes . . . and not knowing what to do next, I hurled one of my toys into the street. . . . [Neighbors] who saw my delight at the fine crash it made, and how I clapped my hands for joy, cried out, 'Another!' Without delay I flung out a pot, and as they went on calling for more, by degrees the whole collection . . . were dashed upon the pavement. My neighbors continued to express their approbation, and I was highly delighted to give them pleasure."

Freud[65] reports that when he first read this story he was merely puzzled. But years later he found among patients who were jealous of their younger siblings recollections of throwing things out of the window as a symbolic gesture of getting rid of the rival. "The new baby must be *thrown out*, through the window, perhaps because he came through the window," brought by the stork. Thus Freud furnishes a "causal" explanation that is at the same time elementaristic and generalizing by giving a single "element" a general symbolic meaning. Actually in Goethe's case sibling rivalry could barely be supported, and if it were, it could not tell us anything particularly characteristic of Goethe.

The Adlerian "finalistic" understanding notes every detail of the story and places it in its larger context. We would then say, when Goethe as a small boy was bored one time, he "experimented" by throwing a dish out of the window and enjoyed the fine crash it made. The adult Goethe actually devoted considerable time to scientific investigation. But the boy Goethe enjoyed even more the applause from neighbors, being "delighted to give them pleasure." The impression we receive is that of an active, independent child who is at first investigative and then sees himself as contributing to the pleasure of people outside the family in the "big" world. He is giving a show and is so much carried away by the applause that he stops at nothing.

If this were the recollection of a patient his complaint would undoubtedly be related to the life style that is here expressed in prototypical form. It might then be helpful to show him that he apparently thinks he must do something extraordinary, sensational, and receive applause from an audience, and that for him no price is too high to achieve this goal. We might then point out that this is a quite unrealistic goal, which led him to his present predicament, as we assume it did. If our proposition rings true to the patient, we have to this extent added to his insight by giving him new concepts that could become the starting point for a reconstruction, or further strengthen a reconstruction already in progress.

Group Process and Group Approaches

Individual Psychology is not individualistic. On the contrary, by regarding the individual as inextricably socially embedded and by considering the major life problems such as to require a well-developed social interest for their solution, Individual Psychology is very much a social psychology. For Adler[10] the function of psychotherapy was "a belated assumption of the maternal function" (p. 341), and this, in turn, was (1) to "give the child his first experience of a trustworthy fellow being"; and (2) to "spread this trust and friendship until it includes the whole of our human society" (p. 373). This definition is quite consistent with that of mental health as the presence of developed social interest and common sense (in contrast to private intelligence).

With this orientation Adlerian theory is so keyed to the group process that it brings the group factor even into individual therapy, namely, in the sense that "the therapist appears as the representative for the human community" (p. 90).[55] When more than one person is brought into the therapy situation this may be seen as a concretization of the "common sense," described as "the pooled intelligence of the social group" (p. 19).[16]

Here it is understood that the group "must be based on healthy social values. . . . Otherwise . . . social validation within the group would result in 'socially shared autisms.'"[93] Adlerians have been suspicious of recent encounter groups insofar as they do not seem to offer enough assurance for the prevalence of "common sense" and social interest. According to Kurt Adler,[26] "These groups foster mainly catharsis . . . and very often . . . overt depreciation of others. . . . There is no sensitivity about the feeling of others in most of these sensitivity groups" (p. 116). A poorly led sensitivity group "reinforces the neurotic behavior of the self-centered person by existing mainly to have him experience his own sensations, talk about them, and attack others, . . . in a fit of 'honesty'" (p. 67).[89]

Adler originated a form of group approach that is in a sense the most daring and has, to our knowledge, been practiced only by Adlerians. It is the treatment of children before a group of observers, the children coming in with, or shortly after, their parents or teachers. The group was at first thought of as having only the function of a training seminar for teachers in how to deal with difficult children. But a second function soon became apparent in that the group actually facilitated therapy. The observers by their mere presence embodied the common sense, as "witnesses" so to speak. The children realized "that 'no man liveth unto himself alone,' and that the mistakes of every individual affect many lives and are of public concern" (p. 491).[7] The work of these counseling centers is described in a volume by Adler and his associates[9] and has been carefully reported and evaluated by Madelaine Ganz.[67] There were over 30 such centers in operation when they were closed by the Austrian fascists in 1934.

Today this type of open community mental health work is carried on primarily under the leadership of Rudolf Dreikurs in numerous Family Education Centers. The technique is described in various books[58,60] and has become particularly teachable through a series of video tapes of actual counseling sessions before classes of graduate students.[53,57]

Within the framework of social interest Adlerians have used any form of therapy beyond the one-to-one ratio, since it always means for the individual patient an increased representation of the common sense. We are referring here to multiple psychotherapy,[59] family therapy,[50] outpatient treatment and therapeutic social clubs,[40,41] psychodrama and action therapy,[47,87,88,99,104] milieu therapy,[91] conventional group therapy,[46,48] as well as educational group counseling.[52] There is a mimeographed collection of papers on group therapy by Dreikurs.[54] An earlier symposium on Adlerian techniques, group and otherwise, was edited by Kurt Adler and Danica Deutsch,[30] while a recent survey of techniques has been offered in a volume edited by Arthur Nikelly.[86]

Adler recognized the necessity for the widest application of basic psychotherapeutic principles, and on the basis of this realization pioneered in the use of nonprofessionals—parents, teachers, and peers—as psychotherapeutic and prophylactic agents.[70] In this way Adlerians, starting with educational counseling centers, have carried psychology into the classroom, originally in Vienna[106] and presently in the United States and elsewhere.[51,56]

Finally we should like to mention the relationship of Individual Psychology to pastoral counseling. Although detached from any organized religion, Individual Psychology is intrinsically attractive to the religious counselor through its concept of man and its therapeutic aim, which is to get the individual to relinquish his self-boundedness and turn to the larger world, "so that he will play his role harmoniously in the orchestral pattern of society" (p. 399).[10] A recent symposium was concerned with this relationship.[94]

Adlerian Organization

Since Adler's days there has been an International Association of Individual Psychology. It consists today of member groups in Austria,

Brazil, Denmark, France, Germany, Great Britain, Greece, Israel, Holland, Italy, Switzerland, and the United States, publishes the *Individual Psychology News Letter*, and holds scientific meetings every three years.

The American Society of Adlerian Psychology sponsors the *Journal of Individual Psychology* and the *Individual Psychologist* and holds annual scientific meetings. Teaching and training are offered by the Alfred Adler Institutes of New York, Chicago, and Minneapolis. New York also has an Alfred Adler Mental Hygiene Clinic. Graduate training in Adlerian counseling is offered at the Universities of Arizona, Oregon, Vermont, and West Virginia. There are furthermore some 25 regional Adlerian associations in the United States and Canada that sponsor one or more family counseling centers and arrange for study groups for parents and teachers.

Bibliography

1. ADLER, ALEXANDRA, "The Concept of Compensation and Overcompensation in Alfred Adler's and Kurt Goldstein's Theories," *J. Ind. Psychol.*, 15:79–82, 1959.
2. ———, "Individualpsychologie (Alfred Adler)," in Frankl, V. E., von Gebsattel, V. E., and Schultz, J. H. (Eds.), *Handbuch der Neurosenlehre und Psychotherapie*, pp. 221–268, Urban & Schwarzenberg, Munich, 1957–1959.
3. ———, "Individual Psychology: Adlerian School," in Harriman, P. L. (Ed.), *Encyclopedia of Psychology*, pp. 262–269, Philosophical Library, New York, 1946.
4. ———, "Modern Drug Treatment and Psychotherapy," *J. Ind. Psychol.*, 15:79–82, 1959.
5. ADLER, ALFRED, "Bolschewismus und Seelenkunde," *Internationale Rundschau, Zurich*, 4:597–600, 1918.
6. ———, *Die andere Seite: eine massenpsychologische Studie über die Schuld des Volkes*, Leopold Heidrich, Vienna, 1919.
7. ———, "A Doctor Remakes Education," *Survey Graphic*, 58:490–495, 1927.
8. ——— (1930), *The Education of Children*, Henry Regnery, Chicago, 1970.
9. ———, *et al.*, *Guiding the Child on the Principles of Individual Psychology*, Greenberg, New York, 1930.
10. ———, *The Individual Psychology of Alfred Adler: A Systematic Presentation in Selections from His Writings* (Ed. by Ansbacher, H. L., and Ansbacher, R. R.), Basic Books, New York, 1956.
11. ——— (1912), *The Neurotic Constitution*, Moffat, Yard, New York, 1917.
12. ——— (1909), "On the Psychology of Marxism," in Nunberg, H., and Federn, E. (Eds.), *Minutes of the Vienna Psychoanalytic Society*, Vol. 2, 1908–1910, pp. 172–174, International Universities Press, New York, 1967.
13. ——— (1920), *The Practice and Theory of Individual Psychology*, Littlefield, Adams, Totowa, N.J., 1969.
14. ——— (1930), *The Problem Child: The Life Study of the Difficult Child as Analyzed in Specific Cases*, Dutton, Capricorn Books, 1963.
15. ——— (1929), *Problems of Neurosis: A Book of Case Histories*, Harper & Row Torchbooks, New York, 1964.
16. ———, "Psychiatric Aspects Regarding Individual and Social Disorganization," *Am. J. Sociol.*, 42:773–780, 1937.
17. ——— (1928), "The Psychology of Power," *J. Ind. Psychol.*, 22:166–172, 1966.
18. ——— (1929), *The Science of Living*, Doubleday-Anchor, New York, 1969.
19. ——— (1933), *Social Interest: A Challenge to Mankind*, Dutton, Capricorn Books, 1963.
20. ——— (1907), *Study of Organ Inferiority and Its Psychical Compensation*, Nervous and Mental Diseases Publishing Co., New York, 1917.
21. ———, *Superiority and Social Interest: A Collection of Later Writings* (Ed. by Ansbacher, H. L., and Ansbacher, R. R.), Northwestern University Press, Evanston, 1964.
22. ——— (1912), *Ueber den nervösen Charakter*, 4th ed., Bergmann, Munich, 1928.
23. ——— (1927), *Understanding Human Nature*, Fawcett, New York, 1969.
24. ——— (1931), *What Life Should Mean to You*, Dutton, Capricorn Books, 1958.
25. ———, and FURTMÜLLER, C. (Eds.),

Heilen und Bilden: Aerztlichpädagogische Arbeiten des Vereins für Individualpsychologie, Reinhardt, Munich, 1914.

26. ADLER, K. A., "Adlerian View of the Present-Day Scene," *J. Ind. Psychol.*, *26*:113–121, 1970.
27. ———, "Depression in the Light of Individual Psychology," *J. Ind. Psychol.*, *17*:56–67, 1961.
28. ———, "Life Style, Gender Role, and the Symptom of Homosexuality," *J. Ind. Psychol.*, *23*:67–78, 1967.
29. ———, "Life Style in Schizophrenia," *J. Ind. Psychol.*, *14*:68–72, 1958.
30. ———, and DEUTSCH, D. (Eds.), *Essays in Individual Psychology: Contemporary Applications of Alfred Adler's Theories*, Grove Press, New York, 1959.
31. ANSBACHER, H. L., "Alfred Adler and Humanistic Psychology," *J. Humanistic Psychol.*, *11*:53–63, 1971.
32. ———, "The Concept of Social Interest," *J. Ind. Psychol.*, *24*:131–149, 1968.
33. ———, "Life Style: A Historical and Systematic Review," *J. Ind. Psychol.*, *23*:191–212, 1967.
34. ———, "Sensus Privatus versus Sensus Communis," *J. Ind. Psychol.*, *21*:48–50, 1965.
35. ———, "The Structure of Individual Psychology," in Wolman, B. B. (Ed.), *Scientific Psychology*, pp. 340–364, Basic Books, New York, 1965.
36. ———, "Suicide as Communication: Adler's Concept and Current Applications," *J. Ind. Psychol.*, *25*:174–180, 1969.
37. ———, "Was Adler a Disciple of Freud?" *J. Ind. Psychol.*, *18*:126–135, 1962.
38. ANSBACHER, R. R., "Sullivan's Interpersonal Psychiatry and Adler's Individual Psychology," *J. Ind. Psychol.*, *27*:85–98, 1971.
39. ATKINS, F., "The Social Meaning of the Oedipus Myth," *J. Ind. Psychol.*, *22*:173–184, 1966.
40. BIERER, J., "The Day Hospital: Therapy in a Guided Democracy," *Ment. Hosp.*, *13*:246–252, 1962.
41. ———, and EVANS, R. I., *Innovations in Social Psychiatry: A Social Psychological Perspective through Dialogue*, Avenue Publishing, London, 1969.
42. BONIME, W., *The Clinical Use of Dreams*, Basic Books, New York, 1962.
43. BOTTOME, P., *Alfred Adler: Portrait from Life*, Vanguard, New York, 1957.
44. BURT, F. D., "William Somerset Maugham: An Adlerian Interpretation," *J. Ind. Psychol.*, *26*:64–82, 1970.
45. CHAPLIN, J. P., "The Presidential Assassins: A Confirmation of Adlerian Theory," *J. Ind. Psychol.*, *26*:205–212, 1970.
46. CORSINI, R. J., *Methods of Group Psychotherapy*, McGraw-Hill, New York, 1957.
47. ———, *Roleplaying in Psychotherapy: A Manual*, Aldine, Chicago, 1966.
48. ———, and ROSENBERG, B., "Mechanisms of Group Psychotherapy: Processes and Dynamics," *J. Abnorm. & Soc. Psychol.*, *15*:406–411, 1955.
49. DEMENT, W. C., "Psychophysiology of Sleep and Dreams," in Arieti, S. (Ed.), *American Handbook of Psychiatry*, Vol. 3, pp. 290–332, Basic Books, New York, 1966.
50. DEUTSCH, D., "Family Therapy and Family Life Style," *J. Ind. Psychol.*, *23*:217–223, 1967.
51. DINKMEYER, D., and DREIKURS, R., *Encouraging Children to Learn: The Encouragement Process*, Prentice-Hall, Englewood Cliffs, N.J., 1963.
52. ———, and MURO, J. J., *Group Counseling: Theory and Practice*, Peacock, Itasca, Ill., 1971.
53. DREIKURS, R., *Counseling the Adolescent: Guidebook* (Ed. by Peterson, J. A., and Peterson, N. M.), Vermont Educational Television, University of Vermont, Burlington, 1971.
54. ———, *Group Psychotherapy and Group Approaches: Collected Papers*, Alfred Adler Institute, Chicago, 1960.
55. ———, *Psychodynamics, Psychotherapy, and Counseling: Collected Papers*, Alfred Adler Institute, Chicago, 1967.
56. ———, *Psychology in the Classroom*, 2nd ed., Harper & Row, New York, 1968.
57. ———, *Understanding Your Children: Study Guidebook* (Ed. by Peterson, J. A., and Peterson, N. M.), Vermont Educational Television, University of Vermont, Burlington, 1969.
58. ———, CORSINI, R. J., LOWE, R., and SONSTEGARD, M., *Adlerian Family Counseling: A Manual for Counseling Centers*, University of Oregon Press, Eugene, 1959.

59. ———, SHULMAN, B. H., and MOSAK, H. H., "Patient-Therapist Relationship in Multiple Psychotherapy," *Psychiat. Quart.*, 26:219–227, 590–596, 1952.

60. ———, and SOLTZ, V., *Children: The Challenge*, Duell, Sloan & Pearce, New York, 1964.

61. ELLENBERGER, H. F., "Alfred Adler and Individual Psychology," in *The Discovery of the Unconscious*, pp. 571–656, Basic Books, New York, 1970.

62. FARBEROW, N. L., and SHNEIDMAN, E. S. (Eds.), *The Cry for Help*, McGraw-Hill, New York, 1961.

63. FORD, D. H., and URBAN, H. B., *Systems of Psychotherapy*, John Wiley, New York, 1963.

64. FRANKL, V. E., "Paradoxical Intention: A Logotherapeutic Technique" (1960), in *Psychotherapy and Existentialism: Selected Papers on Logotherapy*, pp. 143–163, Simon & Schuster, Clarion Book, New York, 1968.

65. FREUD, S. (1917), "A Childhood Recollection from *Dichtung und Wahrheit*," in *Collected Papers*, Vol. 4, pp. 357–367, Hogarth, London, 1925.

66. FURTMÜLLER, C., "Alfred Adler: A Biographical Essay," in ALFRED ADLER, *Superiority and Social Interest*, pp. 309–394, Northwestern University Press, Evanston, 1964.

67. GANZ, M., *The Psychology of Alfred Adler and the Development of the Child* (1935), Routledge & Kegan Paul, London, 1953.

68. GOLDSTEIN, K., "Functional Disturbances in Brain Damage," in Arieti, S. (Ed.), *American Handbook of Psychiatry*, Vol. 1, pp. 770–794, Basic Books, New York, 1959.

69. ———, "Notes on the Development of My Concepts," *J. Ind. Psychol.*, 15:5–14, 1959.

70. GUERNEY, B. G., JR., "Alfred Adler and the Current Mental Health Revolution," *J. Ind. Psychol.*, 26:124–134, 1970.

71. HALEY, J., *Strategies of Psychotherapy*, Grune & Stratton, New York, 1963.

72. HALL, C. S., and LINDZEY, G., *Theories of Personality*, John Wiley, New York, 1957.

73. HOLT, R. R., "A Review of Some of Freud's Biological Assumptions and Their Influences on His Theories," in Greenfield, N. S., and Lewis, W. C. (Eds.), *Psychoanalysis and Current Biological Thought*, pp. 93–124, University of Wisconsin Press, Madison, 1965.

74. JONES, E., *The Life and Work of Sigmund Freud*, Vol. 2, Basic Books, New York, 1955.

75. KAUFMANN, W., *Nietzsche: Philosopher, Psychologist, Antichrist*, 3rd ed., Princeton University Press, Princeton, N.J., 1968.

76. KELLY, G. A., *The Psychology of Personal Constructs*, Norton, New York, 1955.

77. LEVY, L. H., *Psychological Interpretation*, Holt, Rinehart & Winston, New York, 1963.

78. LICKORISH, J. R., "The Casket Scenes from the Merchant of Venice: Symbolism or Life Style," *J. Ind. Psychol.*, 25:202–212, 1969.

79. LOWRY, R., "Psychoanalysis and the Philosophy of Physicalism," *J. Hist. Behav. Sc.*, 3:156–167, 1967.

80. MCLAUGHLIN, J. J., and ANSBACHER, R. R., "Sane Ben Franklin: An Adlerian View of His Autobiography," *J. Ind. Psychol.*, 27: 189–207, 1971.

81. MADDI, S. R., *Personality Theories: A Comparative Analysis*, Dorsey Press, Homewood, Ill., 1968.

82. MAIRET, P., "Hamlet as a Study in Individual Psychology," *J. Ind. Psychol.*, 25:71–88, 1969.

83. MASLOW, A. H., *Toward a Psychology of Being*, 2nd ed., Van Nostrand, Princeton, N.J., 1968.

84. MATSON, F. W., *The Broken Image: Man, Science and Society*, Doubleday-Anchor, New York, 1966.

85. MOSAK, H. H., "Early Recollections as a Projective Technique," *J. Proj. Tech.*, 22:302–311, 1958.

86. NIKELLY, A. G. (Ed.), *Techniques for Behavior Change: Applications of Adlerian Theory*, Charles C Thomas, Springfield, Ill., 1971.

87. O'CONNELL, W. E., "Adlerian Psychodrama with Schizophrenics," *J. Ind. Psychol.*, 19:69–76, 1963.

88. ———, "Psychotherapy for Everyman: A Look at Action Therapy," *J. Existent.*, 7:85–91, 1966.

89. ———, "Sensitivity Training and Adlerian

Theory," *J. Ind. Psychol.*, *27*:65–72, 1971.

90. ORGLER, H., *Alfred Adler: The Man and His Work*, Dutton, Capricorn Books, New York, 1965.
91. PAPANEK, E., "Re-education and Treatment of Juvenile Delinquents," *Am. J. Psychother.*, *12*:269–296, 1958.
92. PAPANEK, H., "Alfred Adler," in Freedman, A. M., and Kaplan, H. I. (Eds.), *Comprehensive Textbook of Psychiatry*, pp. 320–327, Williams & Wilkins, Baltimore, 1967.
93. ———, "Psychotherapy without Insight: Group Therapy as Milieu Therapy," *J. Ind. Psychol.*, *17*:184–192, 1961.
94. "Religion and Individual Psychology," *J. Ind. Psychol.*, *27*:3–49, 1971.
95. ROM, P., "Goethe's Earliest Recollection," *J. Ind. Psychol.*, *21*:189–193, 1965.
96. ———, and ANSBACHER, H. L., "An Adlerian Case or a Character by Sartre?" *J. Ind. Psychol.*, *21*:32–40, 1965.
97. RYCHLAK, J. F., *A Philosophy of Science for Personality Theory*, Houghton Mifflin, Boston, 1968.
98. SALZMAN, L., *The Obsessive Personality: Origins, Dynamics and Therapy*, Science House, New York, 1968.
99. SHOOBS, N. E., "Individual Psychology and Psychodrama," *J. Ind. Psychol.*, *12*:46–52, 1956.
100. SHULMAN, B. H., "An Adlerian View," in Kramer, M., *et al.* (Eds.), *Dream Psychology and the New Biology of Dreaming*, pp. 117–137, Charles C Thomas, Springfield, Ill., 1969.
101. ———, "Confrontation Techniques in Adlerian Psychotherapy," *J. Ind. Psychol.*, *27*:167–175, 1971.
102. ———, *Essays in Schizophrenia*, Williams & Wilkins, Baltimore, 1968.
103. ———, "The Family Constellation in Personality Diagnosis," *J. Ind. Psychol.*, *18*:35–47, 1962.
104. ———, "A Psychodramatically Oriented Action Technique in Group Psychotherapy," *Group Psychother.*, *13*:34–39, 1960.
105. SPERBER, M., *Alfred Adler, oder das Elend der Psychologie*, Fritz Molden, Vienna, 1970.
106. SPIEL, O., *Discipline without Punishment: An Account of a School in Action* (1947), Faber & Faber, London, 1962.
107. STERN, A., "Existential Psychoanalysis and Individual Psychology," *J. Ind. Psychol.*, *14*:38–50, 1958.
108. SUNDBERG, N. D., and TYLER, L. E., *Clinical Psychology*, Appleton-Century-Crofts, New York, 1962.
109. VAIHINGER, H., *The Philosophy of "As If": A System of the Theoretical, Practical and Religious Fictions of Mankind* (1911), Harcourt, Brace, New York, 1925.
110. VAN DUSEN, W., and ANSBACHER, H. L., "Adler and Binswanger on Schizophrenia," *J. Ind. Psychol.*, *16*:77–80, 1960.
111. VERGER, D. M., and CAMP, W. L., "Early Recollections: Reflections of the Present," *J. Consult. Psychol.*, *17*:510–515, 1970.
112. WATZLAWICK, P., BEAVIN, J. H., and JACKSON, D. D., *Pragmatics of Human Communication: A Study of Interactional Patterns, Pathologies, and Paradoxes*, Norton, New York, 1967.
113. WEXBERG, E., *Individual Psychological Treatment* (1927), 2nd ed., (Revised & annotated by Shulman, B. H.), Alfred Adler Institute, Chicago, 1970.
114. WINETROUT, K., "Adler's Psychology and Pragmatism," *J. Ind. Psychol.*, *24*:5–24, 1968.

B. Analytical Psychology

Joseph L. Henderson and J. B. Wheelwright

The Unconscious

JUNG DIVIDES the concept of the unconscious in two parts, *the personal unconscious* and the *collective unconscious*.

We can distinguish a *personal unconscious*, comprising all the acquisitions of personal life, everything forgotten, repressed, subliminally perceived, thought, felt. But, in addition to these personal unconscious contents, there are other contents which do not originate in personal acquisitions but in the inherited possibility of psychic functioning in general, i.e., in the inherited structure of the brain. These are the mythological associations, the motifs and images that can spring up anew anytime anywhere, independently of historical tradition or migration. [p. 485][8]

In his explorations of the unconscious Jung was always very much concerned with the role of consciousness, which he frequently refers to as *the conscious*. Thus he maintains there is a relationship or a dissociation of the conscious and the unconscious. The unconscious may, in the case of a relationship, appear to be complementary to the conscious, filling out or completing what is found lacking there. In the case of dissociation the unconscious is exaggeratedly opposed to the contents of the conscious (for example, the story of Dr. Jekyll and Mr. Hyde). But,

The functional relation of the unconscious processes to consciousness may be described as *compensatory*, since experience shows that they bring to the surface the subliminal material that is constellated by the conscious situation, i.e., all those contents which could not be missing from the picture if everything were conscious. The compensatory function of the unconscious becomes more obvious the more one-sided the conscious attitude is; pathology furnishes numerous examples of this. [p. 485][8]

This compensatory function of the unconscious to the conscious is the cornerstone of Jung's psychology as a basis for psychotherapy. Unlike Freud's psychoanalysis, which seeks to retrieve repressions from the unconscious, Jung's empirical approach to the unconscious makes possible the inclusion by the conscious of many images and emotions that have never previously been experienced consciously.

We know from experience, too, that sense perceptions which, either because of their slight intensity or because of the deflection of attention, do not reach conscious *apperception*, none the less become psychic contents through unconscious apperception, which again may be demonstrated by hypnosis, for example. Finally, experience also teaches that there are unconscious psychic associations—mythological *images*—which have never been the object of consciousness and must therefore be wholly the product of unconscious activity. [p. 840][8]

The Objective Psyche

These basic observations led Jung in later years to speak less of the unconscious versus the conscious and to refer to their potential for unity, for which he adopted the term "the objective psyche." Whitmont[12] summarizes this development as follows:

> Jung has suggested the term *objective psyche* for that totality of the psyche which generates concepts and autonomous image symbols. Hence the ego-centered, subjective consciousness is a partial rather than a complete manifestation of the psyche. In the views of the psyche which were prevalent until Jung's studies became known, psychological functioning was a meaningful organization only in and through the activity of the ego. The drives themselves which constitute Freud's id were regarded as merely irrational, chaotic and senseless, not even related to a balance which keeps the organism alive but only striving to satisfy their own innate needs. Any meaning to be attached to the psychic organism could therefore be viewed solely in terms of ego rationality. The unconscious was quasi-attached to the ego as a general receptacle for that which the ego must repress because it is culturally or personally unacceptable. The psyche was thus "my" psyche, a part of my subjectiveness.
>
> The term *objective psyche* replaces and enlarges the earlier concept of the *collective unconscious* originally used by Jung to denote a dimension of the unconscious psyche that is of an *a priori*, general human character, rather than merely the precipitate of personal repressed material. Because this term gave rise to many confusions and misinterpretations—such as the seeeming advocacy of collectivity or of a mass psyche—he substituted the term *objective psyche* in later writings.
>
> The objective psyche exists independently of our subjective volition and intent. It operates independently of the ego, but can be experienced and comprehended to a limited extent by the ego. That which, lacking understanding, we would view as merely chaotic imaginations, urges and impulses, can disclose meaning when we are capable of interpreting its image manifestations symbolically.

Complexes and Archetypes

The term "complex" was initially used to describe certain emotionally toned reactions to typical happenings or persons, which were causally conditioned by early childhood experiences (traumatic or otherwise). However, it came to be seen, even during the period of Jung's association with Freud, that complexes not only are personal, ego-centered reactions but also conform to certain collective representations (for example, the Oedipus complex). This fact led Jung to postulate a nonpersonal factor in the formation of complexes. Quite early Jung[8] states:

> The work of the Zürich school gives careful and detailed records of the individual material. There we find countless typical formations which show obvious analogies with mythological formations. These parallels have proved to be a new and exceedingly valuable source for the comparative study of delusional systems. It is not easy to accept the possibility of such a comparison, but the only question is whether the materials to be compared are really alike or not. It may also be objected that pathological and mythological formations are not directly comparable. This objection cannot be raised *a priori*, since only careful comparison can show whether a real parallelism exists. At present all we know is that both are fantasy-structures which, like all such products, are based essentially on the activity of the unconscious. Experience must show whether the comparison is valid. The results so far obtained are so encouraging that further research along these lines seems to me very well worth while. [Pp. 187–188]

Further research did prove that the comparison was valid, and the use of the words "type" and "typical" in this early passage shows that Jung was reaching for a term that could embrace specific observations. The term he finally adopted conveys a sense of the inevitability of all new discoveries. At first he favored a term already in circulation, *primordial image* (borrowed from Burkhart), which had the advantage of suggesting something ancient, eternal, and creative. This did justice to the mythological formations, but it failed to account for some element that seemed to come from all plants or animals to repeat identical patterns of behavior as part of a development process. The word "*archetype*" was accordingly adopted, which was observed to combine patterns of both image and behavior in its configuration.[5]

Therefore, Whitmont[12] rightly observes:

> The term *complex* denotes the basic structural element of the objective psyche, and the central element of the complex is the *archetype*. We shall see more clearly how complexes manifest themselves if we again turn to an actual case.
>
> Complexes therefore operate not only as sets of inner tendencies and drives, but also as expectations, hopes and fears concerning the outward behavior of people and objects. Philosophically speaking, since all our perceiving occurs in terms of our psychological predispositions we may regard all perceptions as projections upon the object, the "thing in itself," but in our clinical usage we limit the term to those situations in which the reality perception is distorted by the compelling power of a constellated complex or archetype. [p. 57]

Whitmont[12] further tells us:

> Jung saw in every complex two aspects. The first he called a *shell* of the complex, the other *the core*. The shell is that surface which immediately presents itself as the peculiar reaction pattern dependent upon a network of associations grouped around a central emotion and individually acquired, hence of a personal nature.

The *core* of the complex is represented by its archetypal content, which frequently suggests a mythological theme, and "Apparently the energic charge of the complex which accounts for its disturbing field effect originates, not in the personal layer . . ."[12] but in the mythological core.

The Archetypal Image

Jung[8] describes the archetypal image as: "having the psychological character of a fantasy idea and never the quasi-real character of an hallucination, i.e., it never takes the place of reality, and can always be distinguished from sensuous reality by the fact that it is an 'inner image' " (p. 442).

The Psychoid Pole

In his later work Jung seems to have recognized that he had overemphasized the importance of the archetypal image at the expense of the equally important archetypal behavior pattern. In a paper "On the Nature of the Psyche," he recapitulated, and for the first time, gave full value to the correspondence or complementarity existing between the archetypal image and the pattern of behavior. In this paper Jung uses the visual image of the spectrum to illustrate how human consciousness, in any individual sense, mediates between the instinct, or the "psychoid" pole of experience, and the archetypal image at the opposite pole. Instinct is likened to the infrared area of the spectrum, the archetypal image to the ultraviolet area. The intermediate yellow area is then the meeting place, or place of blending, where the archetype is subjectively experienced as a whole. Instinct gives reality to the image; the image gives meaning to the instinct.

Thus, in harmony with the most advanced biological research, we begin to see more clearly as time goes on how environment is to be regarded as a function of the organism, just as much as the organism is to be regarded as a function of its environment. Man's capacity for bringing about certain basic changes in the archetypal patterns presented to his imagination, over and above the instinctually predetermined psychoid area of his being, becomes a challenge of unlimited ethical and spiritual consequences.

The Symbol

The archetypes have sometimes been called "organs of the unconscious," and as such they maintain some kind of psychic metabolism that is seen in symbolic activity, usually of a visual nature:

> . . . the symbol provides the mode of manifestation by which the archetype becomes discernible. . . . Consequently one can *never* encounter the "archetype as such" *directly* but only indirectly in a symbol, or in a *complex* or a *symptom*. . . . Hence any statement about the archetype is an "inference."[12]

A thorough, well-documented study of symbol formation may be found in G. Adler's case presentation of a woman patient in *The Living Symbol.*[1]

Ego, Persona, Shadow

Jung[8] wrote: "By ego I understand a complex of ideas which constitutes the center of my field of consciousness and appears to possess a high degree of continuity and identity. Hence I also speak of an *ego-complex*" (p. 425). The ego in a mature person stands between and must mediate between the external (objective) and the internal (subjective) worlds. In order to adapt to the external world, the person has to play one or more roles without his personality becoming dissociated: "He puts on a *mask*, which he knows is in keeping with his conscious intentions, while it also meets the requirements and fits the opinions of society. . . . This mask . . . I have called the *persona* which is the name for the masks worn by actors in antiquity. . ." (p. 425).[8]

In relation to the subjective inner world the ego encounters its shadow aspect, its own weakness and self-doubt, and so it usually has an unpleasing, at times unacceptable, appearance, often carrying repressed emotions or thoughts. But it is not always so negative; in fact, it may appear to contain just those characteristics that the persona lacks and that the ego needs to bring into consciousness to balance the one-sidedness of the persona. This accounts for the personal aspect of the shadow, but, since the shadow stands at the doorway leading to the collective unconscious, it has an archetypal aspect as well, as represented in Satan or other demonic figures.

Anima and Animus

In contrast to the shadow, the anima-animus has no personal connections with the ego. It is always purely archetypal in character, providing a "soul image."

> Just as the *persona* (v. Soul), or outer attitude, is represented in dreams by images of definite persons who possess the outstanding qualities of the persona in especially marked form, so in a man the soul, i.e., anima or inner attitude, is represented in the unconscious by definite persons with the corresponding qualities. Such an image is called a "soul-image." Sometimes these images are of quite unknown or mythological figures. With men the anima is usually personified by the unconscious as a woman; with women the animus is personified as a man.
>
> For a man, a woman is best fitted to be the real bearer of his soul-image because of the feminine quality of his soul; for a woman it will be a man. Wherever an impassioned, almost magical, relationship exists between the sexes, it is invariably a question of a projected soul-image. Since these relationships are very common, the soul must be unconscious just as frequently—that is, vast numbers of people must be quite unaware of the way they are related to their inner psychic process. Because this unconsciousness is always coupled with complete identification with the persona, it follows that this identification must be very frequent too. [Pp. 470–471][8]

The Self

Jung describes the self as an archetype that stands in the greatest contrast with the ego. The ego is small, partial, personal; the self is infinitely and indefinably larger, all-encompassing or central, and impersonal. He says:

> I therefore distinguish between the ego and the self since the ego is only the subject of my con-

sciousness, while the self is the subject of my total psyche, which also includes the unconscious. In this sense the self would be an ideal entity which embraces the ego. In unconscious *fantasies* the self often appears as supraordinate or ideal personality, having somewhat the relationship of Faust to Goethe or Zarathustra to Nietzsche.

As an empirical concept, the self designates the whole range of psychic phenomena in man. It expresses the unity of the personality as a whole. [Pp. 425, 460][8]

This concept caused some controversy in Jungian circles. In one sense Jung seems to imply that the self is the totality that encompasses all psychic activity, at other times that it is simply the central archetype.[10]

Neumann, Edinger, and Fordham[4] have elaborated Jung's concept of the self. In differing ways their concepts stress the primary nature of the self as an archetypal container for the child-mother pair in the first year of life. Rising out of this unconscious matrix by a process of deintegration,[4] the ego develops by stages. In later life, beginning in adolescence, the ego learns again to relate to the self progressively by establishing an ego-self axis that allows the ego to relate to this numinous content at the same time maintaining a polite distance from it. Without this axis the ego would be in danger of succumbing to an identification with the self (megalomania), or becoming alienated and nihilistic, or falsely protected by an unyielding atheism.

Jung maintains that in the second half of life the ego necessarily must relinquish its exclusive dominion over the psyche and give precedence to the fateful direction of life in accordance with the self. This frequently leads to some form of religious conviction wherein the evil power inherent in the archetypal shadow may be redeemed through growth of individual consciousness and faith (pistis).

Individuation

This type of growth invariably leads to symbol formation as previously mentioned, and this means learning to balance opposite tendencies hitherto unreconciled, as, for instance, allegiance to the individual versus allegiance to society.

Individuation is a natural necessity inasmuch as its prevention by a levelling down to collective standards is injurious to the vital activity of the individual. Since *individuality* is a prior psychological and physiological datum, it also expresses itself in psychological ways. Any serious check to individuality, therefore, is an artificial stunting. It is obvious that a social group consisting of stunted individuals cannot be a healthy and viable institution; only a society that can preserve its internal cohesion and collective values, while at the same time granting the individual the greatest possible freedom, has any prospect of enduring vitality. As the individual is not just a single, separate being, but by his very existence presupposes a collective relationship, it follows that the process of individuation must lead to more intense and broader collective relationships and not to isolation. [Pp. 448–449][8]

Psychological Types

It is enormously helpful to the therapist to be able to estimate the abilities and limitations of his patients, in terms of their possible behavior and adaptation. And it is essential that he speaks to them in a language that they understand. To talk intuitively to a factual man, or intellectually to a woman who lives through feeling, is a waste of breath.

Many vexed marriage situations revolve around this business of types. Our research shows that in a series of over a thousand subjects, the overwhelming majority have married their polar opposites, although for friends they tend to pick similar types. It is a little startling to think of most of us marrying people that we would never pick as friends.

Jung's idea of individuation is closely related to types. As long as we are content to let somebody else carry our introverted side or our feeling, we remain relatively unconscious and undeveloped. And he believes that growth involves a constant increase of consciousness—that is, the incorporation into our conscious personalities of aspects of our psyche that have hitherto lain in the unconscious. This shift is usually carried out by projections that are identified and consciously reclaimed, as in the analysis of the transference.

Jung's function types represent the four principal ways of adapting to people, things, and situations. These functions work under the aegis of the habitual attitude type—that is, introversion or extroversion. They are especially useful because they are derived from normal individuals and apply to any class of person—high or low, educated or uneducated, complicated or simple—and to either sex.

Extroversion and introversion are attitudes and represent specific direction in which psychic energy, or libido, can flow. Extroversion is a flowing of energy toward the outer world; introversion is a flow of energy toward the inner world. The extrovert tends to explain things from the point of view of environment, seeing a fact produced in a person as coming from without. An unconscious extrovert values the outer object and fears his own inner self.

The extrovert is at ease in the outer world—with objects, people, and situations. His attitude toward the object is romantic and adventurous. He is likely to look on any subjective activity as morbidly introspective. When he deals with his unconscious material he is ill at ease, as an introvert is in the outside world—feeling his way with caution, reserve, and fear, as though dealing with an uncanny power.

He has "a constant tendency to appeal for interest and to produce impressions upon his milieu."[8] Or he has an exaggerated intimacy with those around him and a tendency to adjust to his surroundings through imitation. The more neglected the subjective, introverted side, the more primitive and infantile it becomes. When the extrovert talks about himself, he seems naïve or superficial to the introvert.

In the introvert the energy flows away from the object to the subject. Unlike the extrovert the introvert's subjective reaction to the outer stimulus is the most important thing. He abstracts from his environment whatever he needs to satisfy his inner processes. He may be shy, taciturn, impenetrable.

Subordinate to introversion or extroversion are the four methods of adaptation. These are sides of the personality that Jung called functions. Most of us have one or two developed functions, and the other two or three are relatively unavailable and lie in the unconscious. In assessing the developed functions it is easiest to find how an individual works out of an impasse—does he think his way out, does he wait for a hunch or intuition, does he use his observation or sensation to find a loophole, or does he try to solve the problem with feeling, that is, through relationship.

The superior and inferior functions are equally potent in any individual. But he can run the superior one, whereas the inferior one runs him. For example, a thinking type can direct his thinking, but he is very vulnerable on the feeling side. His feelings are easily hurt and may come up volcanically and take him over.

According to Jung, people gather data with their perceptive functions—sensation or its opposite, intuition. He calls these functions irrational. The data is then processed by the rational, assessing (judging) functions. These are thinking and feeling.

Sensation

Sensation is a perceptive function. The extroverted types are realists. They can retain a great many objective, unrelated facts. They are constantly experiencing the concrete world, and the more extroverted they are, the less they assimilate these experiences. Their thinking is factual, and entering the unconscious is often hard for them, as they tend to be object-bound.

When it is differentiated introverted sensation is highly tuned and spiritual, not being limited to the actual physical sensation. It is perhaps the most inarticulate function of all and is best expressed in color or form. Jung says, "Introverted sensation has a preferential, objective determination, and those objects which release the strongest sensation are decisive for the individual's psychology."

An introverted sensation woman said to me, "I go into the outside world as long as I get subjective reactions to it. When they cease to come, I go away by myself and boil these past reactions down to abstractions. Once set down on paper, or in paint, and put away, I feel satisfied and am ready to meet the world again."

Intuition

Intuition is perception via the unconscious. For the extrovert it is directed toward outer objects and for the introvert toward inner ones. It appears to be an attitude of expectation as it is concerned with seeing possibilities and having hunches. It works best when there is nothing to go on and is an excellent function for a pioneer. It works when there are no facts, no moral support, no proved theories—only possibilities. The intuitive is transiently interested in objects and facts, and then only as stepping stones.

Extroverted intuitives are particularly able to ferret out the potential in people and to foster its growth. For that reason they do well as stage and movie directors and as educators. Their thinking is speculative.

The introverted intuitive draws from the deepest layer of his unconscious, which is common to humanity. It is a particularly useful function in psychology as this is pioneering work and deals with intangibles. Having vision, he avoids the pitfall of the sensation type, which is to get bogged down in a welter of facts and details. This is the predominant type among Jungian analysts.[3]

Thinking

There are two kinds of thinking—the kind that derives from objective data and the kind that may be traced to a subjective source. This stamps the latter type with a subjective inner direction. It is the development of a vague inner image or idea, which has a mythological quality. It has little to do with objective fact, and is only accepted by the extroverted world when it is adjusted to outer facts.

Extroverted thinking is an intellectual reconstruction of concrete actuality or generally accepted ideas, and is concerned with promoting them. The gauge by which thinking can be considered extroverted or introverted is this: Where does it lead back to? Does it go back to generally established ideas, external facts (as with an engineer), or does it remain an abstraction and return to the subject—as with many philosophers? Actually both points of view are essential for balanced thinking. Introverted abstractions save extroverted thought from a materialistic or repetitious fate. Equally extroverted thought saves the introvert thinker from an abstraction that has no relation to the accepted traditional world.

The introverted thinker is often inarticulate because he is forever trying to present the image that comes to him from the unconscious, which usually does not tally with objective facts.

When the introverted thinker is exposed to an objective situation he becomes timid and anxious or aggressive. When he presents his ideas he makes no transitions, but throws them out as they are with little realization of his audience's reactions. He is not aware that because his ideas are clear to him they are not necessarily clear to others. However, this type may have the power to create ideas that do not yet exist, although certain apparently unrelated facts may be known. Mendeleev achieved this in his construction of the table of atomic weights. He left blank spaces for substances that have since been discovered and been found to fit, as he predicted.

Feeling

As I see it this type chooses its friends on the basis of character rather than interest. According to Jung, the type is commonest among women, where it is generally considered normal. It occurs less often in men; when it does it is often a problem. This is because such men do not conform to the stereotyped notion that all men are thinkers.

This is the type that makes value judgments at their best. Extroverted feeling values are traditional and generally valid, being determined by social standards. To quote Jung:[8] "The function is designed not to upset the general feeling situation." But when feeling becomes overexaggerated, the subjective, unrealized egotistical attitude creeps in and makes it untrustworthy, cold, and material. At its worst it becomes vicious, putting other people in a bad light, while appearing blameless and even worthy itself.

The feeling type has very definite likes and dislikes—an appraising quality that cuts like a knife and is orderly and consistent. This is what makes feeling a rational function.

The extroverted feeling man is often found in the ministry, in psychology, and in society, but in most other walks of life he is under a severe handicap. A man in our society is supposed to be a thinking creature, and it is difficult for him to maintain his masculinity against this social prejudice. He may be able to think, but only when it supports his feeling.

The introverted feeling type is very unlike the extrovert. His feeling works from an inner premise—he is almost hostile to the object; he may be inaccessible and silent. He must protect himself from the outside world, and in order to do so, sometimes belittles it or depreciates it. His feeling is more intensive than extensive, because it is not drained off by an easy adjustment to the outside world.

His premise is a vague inner image. It resembles the concept of the introverted thinking type. In order to get his image across, an artistic talent is helpful. This type tends to break with traditional values, as its premise is very original, literally original, deriving from the basic historical patterns of the mind.

J. B. W.

Jungian Therapy

Jungian concepts cover so wide a field of psychic experience that their use in therapy becomes considerably complicated. Their value lies in the flexibility of any therapy inspired by them. Jung himself used to say he had no method of psychotherapy and that each analyst had to create his own method. However, there are certain guidelines that may tell us how a Jungian analysis may proceed. From the beginning Jungian therapy presupposed the inclusion of the relevant techniques used in Freud's psychoanalysis or Adler's Individual Psychology. Gerhard Adler,[2] a follower of Jung, quotes Jung's own point of view. According to this:

> . . . he distinguishes four different stages of analysis, each requiring a special technical approach; the first stage of "confession" (or the cathartic method); the second stage of "elucidation" or "interpretation" (in particular the interpretation of the transference, thus being very near to the "Freudian" approach); the third stage of "education" (the adaptation to social demands and needs, thus most nearly expressing the standpoint of Alfred Adler); and finally what he calls the stage of "transformation" (or "individuation"), in which the patient discovers and develops his unique individual pattern, the stage of "Jungian" analysis proper.
>
> These stages are not meant to represent either consecutive or mutually exclusive stages of treatment, but different aspects of it, which interpenetrate and vary according to the needs of the particular patient and the therapeutic situation. Thus, treatment has to be undogmatic, flexible, and adjusted to the needs of the individual patient, and this specification is one of the main tenets of analytical psychology. [p. 338]

In addition to this schematic conception, Jung's "constructive" method makes possible a subjective approach to the material presented that allows both therapist and patient (analysand) to work together, including their subjective response to the dream or other fantasy material at hand. This establishes the basis for a method of exploring the archetypal unconscious as well as the personal unconscious. This is the method of amplification used together with personal free associations.

> Amplification gives the widest context for interpretation because it opens the way for a confrontation between the historical remnants (the archaic heritage) and the immediate needs of the personal psyche. The archetypal symbols can then be accepted or rejected by the individual's own choice; this is of the greatest importance in depth psychology, for no one can experience the archetypal images without being temporarily fascinated, terrified, or possessed by them. The free associations and the amplifications, when properly handled by analyst and patient, gradually reduce the undesirable power of these images and render them accessible to consciousness as organs of healing. What can be integrated remains; what

s dangerous or unacceptable falls back into the unconscious, whence it may reappear later, when ego-consciousness is ready to receive and integrate it. [p. 6][5]

This method goes hand in hand with a conscious recognition and continual reassessment of the transference and countertransference between analyst and analysand.

Among such prospective unconscious material the primordial archetypal images are of particular significance, and it is on this that the specific approach of analytical psychology is focussed. Applied to the interpretation of transference phenomena this means that underneath what appears as a merely personal transference relationship, archetypal, transpersonal images are active. Every intense experience of a personal nature will also actualize the corresponding archetypal image. In other words, every actual experience of, say, father or mother consists of a complex blend of two components; the parents as such and the archetypal image projected onto them. The personal experience acts as the evoking factor for the archetypal image (Neumann), and the two together in their interpenetration for the image. The archetypal aspect must never be overlooked in interpreting the unconscious processes in general (Adler, G.), and in the transference relationship in particular.

Thus the transference, as well as dealing with repressed infantile conflicts, also aims at raising into consciousness the archetypal, transpersonal substratum of personal experience.

Regarding the attitude to countertransference we come to a fundamental point in the theory and practice of analytical psychology. Here I must make it quite clear from the start that I shall use the term "countertransference" in a positive sense, as indicating the analyst's constructive subjective reaction arising from his own unconscious activated in the analytical relationship (which would perhaps be better described as the "analytic field"). As such it is an inevitable, necessary and indeed desirable instrument of treatment. This constructive countertransference has, of course, to be most decisively distinguished from such undesirable countertransference manifestations as unconscious identifications and projections due to the analyst's unanalyzed neurotic complexes and leading to harmful unconscious involvements—in which case further analysis of the analyst is clearly indicated. [p. 340][2]

A Jungian analysis is conducted by establishing a vis-à-vis, conversational situation between therapist and analysand. This insures that the therapist will remain alert and participate humanly in the analysand's experience. The therapist may then be in a position to observe significant nonverbal forms of communication, and may even implement this by making available colored pencils or clay with which to form images of nonverbal experiences. Some therapists have a separate room in which they keep a sand table, in which figures and designs may be placed. This may be used as a projective technique, or as a device to help overcome a resistance, or simply as an experience for its own sake. It was originally used by Margaret Lowenfeld in London and adapted by Dora Kalff (Zürich). It is especially useful in child therapy, but has been extended in recent years to adolescent and adult communications.

The Jungian School

J. B. Wheelwright,[11] the retiring president of the International Association of Analytical Psychologists, states:

I think it is important to say, at the outset, that Jung did not intend or want to compete with Freud. It was his conviction that an analyst could not transcend himself. He said, "Philosophical criticism has helped me to see that every psychology—my own included—has the character of a subjective confession . . . it is only by accepting this as inevitable that I can serve the cause of man's knowledge to man."[8] In short, he attempted to abstract and generalize his truth—not the truth.

In the broadest terms, Freud focussed on sexuality, Adler focussed on power, and Jung focussed on growth, which he called individuation. With these central concerns went attitudes and values that many people have found ego-syntonic. It is a matter of different, not better or worse approaches. As time goes on, the barriers between the schools seem to be dissolving. Frieda Fromm-Reichmann said, "The goal of analysis is self-realization." Few Jungians would quarrel with that statement.

It is worth remembering that Jung saw the dangers in organization. He was well aware that rigidity in institutions is a constant danger and that a man who is identified with an orthodoxy is inaccessible. One cannot communicate with an institution. He once told me that he supposed an organization was necessary, but that he thought it should be as disorganized as possible.

Jung addressed himself to two previously neglected aspects of life. With his concepts of the animus and the anima, he placed women on an equal footing with men—no better, no worse. The other aspect was aging and the second half of life. He was primarily always interested in whatever age he happened to be. As he lived to be a very old man, he learned and formulated a lot about old age. And, unlike many geriatric specialists, he did not regard it as a combination of illness and an economic liability to society.

There has been no basic change in Jungian theory coming from any member or group of the Jungian School. Modifications have been suggested by Neumann, Fordham, Edinger, but there remains a basic conformity to the basic concepts. There has been considerable borrowing and learning from other schools—from the ego psychologists, such as Erikson, and the neo-Freudian, Melanie Klein, chiefly represented by Fairbairn and Winnicott in England. A friendly relationship exists between Jungians and certain members of the school of existential analysis (*daseinsAnalyse*), for example, Medard Boss and Rollo May. Adler[2] sees the Jungian School as comprising three different Jungian approaches to therapy: "The 'orthodox' approach tends to keep Jung's concepts 'pure' and virtually unchanged, with the accent of its practical work on archetypal interpretation (the 'synthetic-constructive' method)." An unorthodox approach is seen in those analysts, chiefly in London, who have tended to revive the old Freudian reductive method of interpretation in their practical work. Finally there is ". . . a solid center group firmly linked to Jung's teachings . . . but accepting modifications in the light of further experience and using in their work a combination of reductive and constructive interpretation."[2]

Jungians differ among themselves concerning the value of group therapy and the use of multiple analysis (that is, analysis with more than one analyst), especially in training analyses. The Zürich Institute requires multiple analysis; the London group discourages it; the New York and San Francisco groups have no formal requirements, but are friendly to multiple analysis when it seems indicated.

Bibliography

1. ADLER, G., *The Living Symbol*, Bollingen Series 63, Princeton, 1961.
2. ———, "Methods of Treatment in Analytical Psychology," in Wolman, B. B. (Ed.), *Psychoanalytic Techniques*, Basic Books, New York, 1967.
3. BRADWAY, K., "Jung's Psychological Types," *J. Analyt. Psychol.*, 9, no. 2: pp. 129–137, 1964.
4. FORDHAM, M., *Children as Individuals*, C. G. Jung Foundation, New York, 1970.
5. HENDERSON, J. L., *Thresholds of Initiation*, Wesleyan University Press, Middletown, Conn., 1947.
6. JACOBI, J., *Complex, Archetype and Symbol*, Bollingen Series 56, Princeton, 1959.
7. JUNG, C. G., "Psychogenesis of Mental Disease," in *Collected Works*, pp. 187–188, Bollingen Series 3, Princeton, 1960.
8. ———, *Psychological Types*, Bollingen Series 20, Princeton, 1971.
9. ———, "The Structure and Dynamics of the Psyche," in *Collected Works*, pp. 159–234, Bollingen Series 6, Princeton, 1947.
10. PERRY, J. W., "Reconstitutive Process in Psychopathology of the Self," *Ann. N.Y. Acad. Sc.*, 96:853–876, 1962.
11. WHEELWRIGHT, J. B., 1963.
12. WHITMONT, E. C., *The Symbolic Quest: Concepts of Analytical Psychology*, Basic Books, New York, 1970.

CHAPTER 39

BRITISH PSYCHOANALYTIC SCHOOLS

A. The Kleinian School

R. E. Money-Kyrle

Melanie Klein and Kleinian Psychoanalytic Theory

MELANIE KLEIN was born in Vienna in 1882 and died in London in 1960. She had originally intended to study medicine at the Vienna University and would have done so, had not an early marriage intervened. However, years later during World War I, she had a second opportunity to recapture her old interest in a new form. She came in contact with Freud's work, recognized what she felt she had been looking for and, from then on, dedicated herself to it. She started her training with Sándor Ferenczi during the war and, after the armistice, continued it with Karl Abraham. Both encouraged her to specialize in the analysis of children, at that time almost a new field. (Later she also analyzed adults and, at the end of her life, was largely engaged in training analyses.)

One of her early patients was a very silent child. She tried giving him toys, discovered she could interpret his play as if it had been verbal associations, and so found herself in possession of a new implement of psychoanalytic research. The results of her research with this implement, which she began to publish in a long series of papers and a few books, were regarded by some as departures from Freud and are still often criticized as such. Others, including her own teacher Abraham, till his death in 1926, welcomed them as important contributions to analytic insight and therapeutic power. She herself always saw her work as rooted in Freud's and a development of it, which inevitably also involved some modifications.

Since most of the ideas she introduced had their source in her early papers and were gradually developed and clarified by her in later writings, it is not easy to pinpoint them by single references; but a short bibliography of her main publications is given at the conclusion of this chapter. What follows here is an attempted summary of her contributions to theory, although no summary of the work of such an original thinker can do justice to her thought.

To begin with, a word about two distinctive qualities of Melanie Klein's views on technique. First, it is probably true that she developed Freud's conception of transference analysis into "pure transference analysis," a movement which, in particular, involved the discarding of all forms of reassurance, on the one hand, and educational pressure, on the other, with both children and adults. She felt these could only blur what should be analyzed, namely, the transference picture of the analyst as it emerges and changes in the patient's mind. Second, she always tried to direct her interpretations at whatever seemed to be the patient's main anxieties at any given time. Once, at the beginning of her practice, she was herself alarmed by the amount of anxiety she seemed to be arousing in a child patient by this means. But Abraham advised her to persist, and she found that by so doing she was best able to relieve the anxiety the analysis was evoking. After this experience, she never had further doubts about the correctness of her approach.

Coming now to Melanie Klein's contributions to theory, these can be listed under the following heads: early stages of the Oedipus complex and superego formation, early operation of introjective and projective mechanisms in building up the child's inner world of fantasy, the concepts of paranoid-schizoid and depressive positions, a clarification of the difference between two sorts of identification, introjective and projective, and, lastly, the importance of a very early form of envy.

Perhaps the most far-reaching of these, in its effects on theory and practice, is her concept of a paranoid-schizoid position in early infancy, followed a little later by a depressive one. Both presuppose her acceptance of Freud's basic concept of ambivalence, of conflict between love and hate—ultimately of the life and death instincts. In the first position, this ambivalence expresses itself mainly in mental acts of splitting and projection. Thus, in her view, the infant's ambivalence toward the breast, loving when satisfied, hating when frustrated, causes him to divide this object into two: a "good" breast containing projected love that is felt to love the child, and a "bad" one containing projected hate that is felt to hate him. Both become internalized, making it possible for him to feel alternately supported and attacked from within himself. Moreover, the reprojection of these inner objects onto the external breast, and their further reintrojection, set up benign or vicious spirals leading to increasing well-being on the one hand, or an increasing sense of persecution on the other.

In particular, the persecutory feelings aroused by splitting and projection of hate are often dealt with by further splitting as a defense, and this can develop into a terrifying sense of mental disintegration.

Although such states of mind are, in her view, characteristic of earliest infancy, she spoke of them as belonging to a "position" rather than a phase—a word which, on the one hand, avoids the implication that the infant is always split and persecuted, and on the other, by its spatial analogy, suggests an attitude that can be abandoned and again adopted at any time.

As distinct from this position, Melanie Klein held that a very different one begins to be at least temporarily adopted in the second quarter of the first year, when the infant is integrated enough to relate himself to his mother as a whole person. Whereas in the earlier position the anxiety is centered on his own survival, in the later one it centers more on the survival of his good objects, both inside and outside himself. And what, in the last analysis, he fears is that his own destructive and greedy impulses will destroy, or have destroyed, the good breast—an anxiety that may be consciously expressed in later childhood as the fear that his mother, or father, or both may die. Thus, in Kleinian terminology, depression connotes a state of sadness allied to mourning and should be distinguished from such other feelings as the sense of worthlessness or of hopeless confusion that is often mixed with it. In Melanie Klein's view, because some persecutory feelings are always involved, depression is never observed in isolation.

In Kleinian theory, the depressive position is the main hurdle in development. Surmounting it involves the acceptance of responsibility for the damage in the inner world (sometimes also in the outer), followed by mental acts of reparation. But if this acceptance is too painful to be borne, various defenses come into operation. Of these the most usual are a regression to the paranoid-schizoid position, or a swing into a manic state, in which either the extent of the inner damage or its importance is denied.

If correct, this theory of early positions of development must be expected to throw light on the psychoses of adults. And it has been applied with an encouraging degree of success to the treatment of some of these disorders by several members of her school, as well as by others influenced by it. Among the pioneers in this field, three of her pupils should be mentioned—Bion, Rosenfeld and Segal—whose example has since been followed by several others.

Although Melanie Klein's ideas about the origin of the superego and early stages of the Oedipus complex began to develop before she had formulated her concept of paranoid-schizoid and depressive positions, the former may be retrospectively regarded as elaborations of the latter. Thus, in her view, the good and bad breasts internalized in the paranoid-schizoid position are forerunners of the superego. In the depressive position, they become more integrated; and, in the developed superego, they contribute to its dual character as friendly mentor and implacable judge.

Meanwhile, of course, the fact that the child has two parents related to each other exerts its influence on his internal objects. Almost from the beginning of her work, Klein believed she had found evidence of the presence of an Oedipus complex at a far earlier age than had previously been thought possible. The child's rivalry in a triangular situation seemed to begin as early as the oral stage, so that his father could be internalized as an object denying him the breast. At the same time, there would also be a "good" father, split off from the bad one, to be internalized as the donor of, or fused with, the good breast.

Moreover, the two parents could be felt as a combination that either supports the child or frustrates him. Indeed, both the paranoid-schizoid and depressive positions with regard to the breast reappear with regard to this concept of "combined parents." In the first position, both a friendly and a hostile combination are felt to exist, and become internalized. In the second position, when these two opposite aspects of the combined parents are more integrated, the child is depressed because his fantasy attacks on the bad ones are felt to have damaged the good ones, too.

It will be seen that, in Melanie Klein's view, the developed Oedipus complex and superego formation discovered by Freud have a long and complex prehistory. This configuration Freud conceived of, in the first instance, as a kind of jealous internal father-god, who maintained in his sons the taboos on incest and parricide and tended generally to inhibit sexuality. But Freud seems to have been well aware that much more remained to be discovered about it—for instance, about what form it took in women, whether and how a mother imago entered into its composition, and about its kindly aspects, which he considered a source of consolation through humor. Melanie Klein did not discard Freud's concept; she accepted it, worked backward from it, and believed she had contributed to tracing it to its source. No Kleinian would claim that this task is even yet fully accomplished.

Another of Freud's concepts, which she also worked on, was that of identification. Its use in his *Totem and Taboo* does seem to imply that he had two kinds of identification in mind. But the distinction was not very clear; and because identification resulting from introjection was already well recognized, the possibility of identification by projection tended to be lost sight of, till Melanie Klein gave it a name, "projective identification."

This concept, as used by her school, appears in two main contexts. In the first place, it helps to explain a number of pathological conditions. There are, for example, certain megalomanic states (observable in smaller degree in otherwise normal people) in which a projection of part of the self into someone (often the analyst) standing for an admired parent is followed by an elated sense of identification with him. Or, a similar state of elation seems to result from an intrapsychic projective identification of the ego into the superego. But such forceful penetration is usually felt either to injure its object or turn it into an enemy, and then the outcome is a claustrophobic sense of being imprisoned in a depressed or persecutory interior. But this is not all; for, as Rosenfeld has pointed out, the reinternalization of an object felt to have been injured or made hostile by projective identification can result in depressive or persecutory hypochondria. He has also traced confusional states to the same basic cause: The patient does not know who or where he is because in fantasy he is inside someone else.

In the second place, the concept of projective identification is used to explain the emotional affect some patients may produce in an analyst. When this affect appears to exceed what can be explained in terms of countertransference, Kleininians believe it to be a manifestation of the most primitive means by which a baby can communicate emotions to its mother and, if they are disagreeable emotions, can experience relief by so doing. If the initial motive is to "evacuate" distress—and distressed patients in need of their next session often dream of needing a lavatory—the angry infant can soon use the same mechanism as an attack designed to distress the mother.

As to the means by which the projection is brought about, I would suppose the baby—or the patient in analysis—to be equipped with a phylogenetically prehuman, and ontogenetically preverbal, capacity to express feeling through behavior. If so, it must also be supposed that mothers are phylogenetically equipped to understand it. Indeed, in Bion's view, one of the important characteristics of a good mother is an uninhibited capacity to do just this. And, of course, the same applies to analysts. But, as a rule, personal difficulties must be overcome before an analyst can expose himself, without too much anxiety, to the peculiar stresses that sensitivity to a very ill patient's projections seem to involve.

It was through her interest in aggressive forms of projective identification occurring in analysis that Melanie Klein reached her concept of a very early form of envy. For some patients behave toward an analyst as if they wished to destroy any sense of superior equanimity they may suppose him to possess. Moreover, since their dreams often seem to indicate that they feel they do so by projecting their own fecal product into an otherwise admired object to render it worthless, and they do this on occasions when other patients might have felt love and gratitude, she inferred that it expressed a very primitive form of envy directed toward the good breast felt to contain every desired quality that the baby feels he lacks. In this, envy, which aims at the destruction of goodness, is to be sharply distinguished from greed, which aims at appropriating it. Everyone knows, of course, that envy is universal in the human species, and appears to be constitutionally stronger in some people than in others. Freud has also familiarized us with the concept of penis envy in women—a term that includes both the greedy desire to steal the penis and the envious desire to belittle it. That the purely envious component in this could have a forerunner in envy of the breast has seemed unacceptable to some; but many analysts have since found the concept indispensable in overcoming certain hitherto intractable difficulties with patients—in particular, with patients who display a marked negative therapeutic response. Here the aim is to make the patient aware that he is envious, and also to expose the many delusions about the supposedly carefree happiness of other people that his envy causes him to form and that, in turn, increase it. For, although the amount of constitutional envy possessed by any individual seems to be unalterable, much can be done to expose and correct the way it distorts his beliefs.

Enough has been said, perhaps, to give

some idea of Melanie Klein's theoretical contributions to psychoanalysis. Of these, the central role must be allotted to her concept of a depressive position arising when the infant is sufficiently integrated both to mourn and to feel responsible for the destruction of his good objects in his own inner world of fantasy. The therapeutic aim of those who agree with her is first to analyze the defenses against the reexperience of this position in analysis, and, by so doing, also to reintegrate the split-off parts of the self, including the destructive elements responsible for the depression, in order that they can be brought under the control of the rest of the personality and used ego-syntonically. So far as this is achieved, it also brings about a better integration of those internal objects that have remained, as it were, unaltered forerunners of the superego, removing much of the superego's bizarre severity and giving it more the character of a friendly mentor. (These views on the central role of the depressive position, in fact, largely determined her technique of pure transference analysis. For she believed that any departure involving reassurance prevented, or at least delayed, the working through of this position, and could, therefore, actually be dangerous.)

Of course, the extent to which the depressive position can be worked through in the way described is always limited. But it is the aim of Kleinian analysis and as such has sometimes been criticized as moralistic. That well-integrated people tend to be more "moral," if this means having a greater sense of mature responsibility, seems to be a fact. But I do not think this result was anticipated, nor is any moral pressure put on patients to develop in any particular direction. That, in a successful analysis, a patient does develop in the described direction seems to be purely the result of the analytic process.

If a reason is sought, I would suppose it to lie in the conditions of our racial past which, if it favored the development of aggressive impulses, also favored the development of a cooperative type of man who could harness them for social ends. It would seem that, if freed from psychotic and neurotic disabilities, he tends automatically to develop in this way.

Further Development of the Kleinian School

Melanie Klein not only developed a number of psychoanalytic theories, which extended and in some cases modified Freud's work, from which she always took her departure; she also founded a school and, toward the end of her life, was much concerned about its definition and its future. It began as a group of colleagues, mainly in England, who were most influenced by her and who supported her in controversy, soon included analysts who had been trained by her, and then second-generation analysts trained by these, and so on. But naturally there were some disagreements and defections, and this raised the question of who was to be called a Kleinian.

To this question probably no wholly precise answer can be given. Certainly there are now a large number of analysts who understand, accept, and apply all Melanie Klein's theories. But these shade off into others who, although they understand and accept most of her work, do not understand or accept all of it. Then there is the complication of those who are good at theory but perhaps lack the insight to be really good analysts, and conversely of those who have the insight but are confused about the theory. Moreover, analysis is a growing science, and many Kleinian analysts have developed theories of their own. Usually these are extensions of Mrs. Klein's views, of which she almost certainly would have approved. But one cannot always be sure even at this early date, and, of course, the uncertainty will increase with time as yet newer theories are developed. This is not important in itself except for the purpose of defining Kleinians, since only the truth or falsehood of theories is what matters.

The Kleinian School, therefore, has no clear-cut boundary, but its core consists of those who feel most inspired by gratitude for Melanie Klein's work. Moreover, the School now has groups in several different countries.

It is impossible for any one writer to be wholly fair to all his colleagues, and in the account that follows allowance must be made for my being little acquainted with the work of some of them that may in no sense be inferior to the work of others with whom I am in closer touch. Moreover, there is not space for more than a selection, so what follows is not a summary of the original work of Kleinians, but a summary of some samples of this work. Nor, indeed, is it possible to give more than a sketchy account of the work of any one analyst.

Among Melanie Klein's initial collaborators were Susan Isaacs,[13-15] Joan Riviere,[23-28] Nina Searl,[30-35] and others who analyzed children by her methods. This early group throve under the sympathetic protection, until the time of his death, of Ernest Jones, who wrote the Preface to the special number of the *International Journal of Psychoanalysis* (1952) brought out on the occasion of Klein's seventieth birthday. The group also included her early pupils, and of those who survived to carry on her work, the best known names (in alphabetical order) are probably Bion,[7-12] Rosenfeld,[29] and Segal,[36-42] each of whom have made important original contributions.

Bion was first known for his work on groups during World War II. Here one of his major contributions was the hypothesis (invented to explain his actual experiences) that all groups meet under the influence of unconscious basic assumptions. Thus there could be the dependent group under the unconscious assumption that it exists to be dependent on some kind of god, the fight-flight group assumed to be there for the purpose of fighting or fleeing, and the pairing group assumed to exist for lovemaking. Moreover, the work group, under the inspiration of a conscious purpose, is always liable to come under the influence of one or other of these more archaic or, indeed, psychotic fantasies from the unconscious. Bion also records interesting observations about the way in which an unorganized group will choose its leader—often its illest member. Since Bion believed that each discipline should begin by working out its own concepts for itself, these concepts are not directly Kleinian, but they are by no means incompatible with Kleinian theory.

After the war, however, he returned to psychoanalytic practice and included a significant proportion of psychotics among his patients. These he analyzed on strictly Kleinian lines and soon began to publish his results in a number of papers and books. His theoretical works include a theory of thinking that proceeds from a distinction between normal and pathological projective identification. At the beginning of his life the infant projects his troubles into the breast and gradually reabsorbs (introjects) an object that can understand and deal with them inside himself. But if there is an excess of envy in the baby, or of resistance in the mother to understanding him, or a combination of both factors, the projective identification becomes an attack both upon the breast and upon the infant's own dawning capacity to understand, which should have been derived from it. This, in very rough outline, is Bion's view of the origin of schizophrenia. Much of his work, however, has been devoted less to theory than to technique. Thus he elaborated a grid to help the analyst in his task of recognizing the analytic significance of his patients' material and the exact moment at which they were ripe for an interpretation. In still more recent years he has stressed the importance of the analyst's freeing himself from preconceptions by "forgetting" his theories, his desires, and his patients before he sees them, so that each session acquires something of the freshness of an initial interview. If he does this, relevant theories and memories about his patients will be more likely to float back into his mind as required and will not forcibly intrude to distort his unbiased perception of the material.

Meanwhile, Herbert Rosenfeld and Hannah Segal were the first to apply a strict Kleinian technique to the analysis of the schizophrenic. That is to say, they renounced all attempts to play any definite role, either positive or negative, in his life, did not attempt to educate him, and confined themselves as far as possible to analyzing the transference exactly as they would in analyzing a child or a neurotic adult. The difference lay only in the nature of the transference itself, which does differ markedly from those met with in classical analyses in that it is a psychotic transference. For example, because of the psychotic's excessive use of projective identification, the analyst's transference role is largely that of the person or object with which the patient feels confused. Moreover, since the psychotic's object relations are mainly to part-objects (breast, penis, and so forth), it is mainly as a part-object that his analyst will appear to him.

Mrs. Klein had already suggested that the schizophrenic is someone who cannot tolerate the depressive position, when whole-object relations begin, and for this reason regresses to the paranoid-schizoid position. For the same reason his object relationships remain at, or regress to, the part-object level; and as Hannah Segal was able to show, this is also a reason for his massive use of projective identification by which process he feels he can put his depression into the analyst. Another motive is usually destructive envy.

Many general contributions to the understanding of schizophrenia are shared between Bion, Rosenfeld, and Segal, who all approached the problem from the same Kleinian angle. But apart from these Rosenfeld is probably best known for his work on differentiating between various types of confusional states, and Segal for isolating a presymbolic form of thinking, characteristic of the schizophrenic and of the schizoid part of more normal people, in which symbol and object symbolized are concretely identified. For example, to the normal or neurotic musician playing the violin may symbolize masturbation, but to the psychotic it *is* masturbation and, therefore, he cannot do it in public. Segal is also the author of an extremely lucid and condensed exposition of Kleinian analysis.[37]

Among those who have done outstanding work on the application of Kleinian theory in the social field, special mention should be made of Elliott Jaques.[16–18] Thus, for example, in his "Social Systems as a Defence against Persecutory and Depressive Anxiety"[18] he shows how the functions of a working group he studied in a factory were structured at the fantasy level in such a way as to drain off the "bad" impulses, and consequent persecutory and depressive anxieties, that might otherwise have impaired the work of the factory as a whole. But in so doing it impaired its efficiency as a work group of managers and operatives designed to work out a new method of payment.

In a number of other books, less obviously though still in fact under the influence of Kleinian ideas, Jaques has evolved a method by which the value of any job can be assessed in terms of the amount of "discretionary responsibility" the job demands. If this method could be generally agreed on as right and fair (which it has been in a number of factories), it could obviously be used as an acceptable basis for differential payments, and much argument and strife could be avoided.

Among the younger Kleinians Donald Meltzer[19–21] has an outstanding place. He has made important discoveries, for example, about the role of early anal masturbation in causing a baby, who feels himself to be deserted, to become thoroughly confused between breast and bottom and between his own personality and that of his mother. Meltzer also has an unusually clear grasp of the stages through which an analysis should pass and in which stage a given patient is at a given time. Much of this he has recorded in his book *The Psycho-Analytic Process.*[20]

Among those whose work has been more exclusively in the applied field may be mentioned Adrian Stokes,[43–46] not a practicing analyst, who has written much on art, and myself,[22] who has contributed to ethics and politics.

It will be clear by now, as I said at the outset, that only a sketchy account of the work of only a few members of the Kleinian School has been given. The real work of any analyst is done in the consulting room, and no one else, except possibly those patients who have collaborated most successfully in it, is in a position to assess it adequately. This is where insights gained in the analyst's own analysis are tested again, and where new insights are conceived and tested in their turn. The work is arduous and fraught with difficulties and dangers; and in order to endure these without undue strain, the analyst needs to have acquired a strong sense of the value of his own analysis and so of analysis in general. This, and the sense of being able to convey it in varying degrees to his patients, is necessary and sufficient for his peace of mind; and since these feelings are of the same form as the sense of having had good parents and of oneself being a good parent in turn, it is a basic satisfaction—perhaps never perfectly achieved. The desire to write papers and books, whether case histories in which new insights are recorded, or theories in which they are generalized or applied in other fields, would seem to be more complex. Some may do it simply because they have a facility for this kind of work, others from ambition or because they generously want their colleagues to share their new discoveries, or from a mixture of all these motives and more. But although writing about analysis can never convey an adequate impression of how the author does analysis, it is only those who write who are likely to be assessed at all.

Melanie Klein was a very generous writer who believed her discoveries to be important (but not herself) and wished to share them as soon as she could. I think many of her pupils have inherited this motive for publication. But it is impossible to say how much of the theory that they (and their pupils) have accumulated since her death is strictly Kleinian in the sense defined at the beginning of this article. Moreover, the unconscious has a fluidity about it that is quite foreign to conscious verbal thought, so that it is less easy to be sure whether anything said about it, even by oneself, is true or more probably a half-truth, and, of course, the difficulty is greater when assessing what is said by someone else. Nevertheless, a body of theory is accumulating in the Kleinian School, and it is mainly self-consistent. It is expected to go on accumulating, and new models (or theories) that express it better are likely to be invented. The foundation of the Kleinian School is Freudianism extended and modified by Melanie Klein; and this will always be so however much it may in time be hidden under the expanding structure of new theory built upon it.

It may be worth concluding with a note on a certain kind of stress to which all permanent groups are occasionally subject. I have in mind what Bion, who first drew attention to it analytically, has called the confrontation of the mystic with the Establishment. Bion himself discusses various outcomes, in one of which the group is disrupted or the mystic, or innovator, "loaded with honours and sunk without a trace." But perhaps this tragic outcome results from the group paying too much attention to the supposed originality of the mystic or the supposed conservativism of the Establishment, and not enough to an investigation of the degree of objective truth to be found on the one side or the other. It is to be hoped that the Kleinian School will be able to mobilize sufficient objectivity to deal more successfully with such crises as and when they should arise.

Bibliography

1. KLEIN, MELANIE, *The Psycho-Analysis of Children*, Hogarth, London, 1932.
2. ———, *Contributions to Psycho-Analysis, 1921–1945*, Hogarth, London, 1948.
3. ———, *et al.*, *Developments in Psycho-Analysis*, Hogarth, London, 1952.
4. ——— (Ed.), *New Directions in Psycho-Analysis*, Tavistock, London, 1955.
5. ———, *Envy and Gratitude*, Tavistock, London, 1957.
6. ———, *Our Adult World and Other Essays*, Heinemann, London, 1963.
7. BION, W. R., *Attention and Interpretation*, Tavistock, London, 1970.
8. ———, *Elements of Psycho-Analysis*, Heinemann, London, 1963.
9. ———, *Experience in Groups*, Tavistock, London, 1961.
10. ———, *Learning from Experience*, Heinemann, London, 1962.
11. ———, *Second Thoughts*, Heinemann, London, 1967.
12. ———, *Transformation*, Heinemann, London, 1965.
13. ISAACS, S., "The Nature and Function of Phantasy," *Developments in Psycho-Analysis*, Hogarth, London, 1952.
14. ———, "Privation and Guilt," *Int. J. Psychoanal.*, *10*, 1929.

15. ———, *Social Development in Young Children*, Routledge, London, 1933.
16. JAQUES, E., "Death and the Mid-Life Crisis," *Int. J. Psychoanal.*, *46*, 1965.
17. ———, "Psycho-Analysis and the Current Economic Crisis," in *Psycho-Analysis and Contemporary Thought*, Hogarth, London, 1958.
18. ———, "Social Systems as a Defence against Persecutory and Depressive Anxiety," in Klein, M., *et al.* (Eds.), *New Directions in Psycho-Analysis*, Tavistock, London, 1955.
19. MELTZER, D., "The Differentiation of Somatic Delusions from Hyperchondria," *Int. J. Psychoanal.*, *45*, 1964.
20. ———, *The Psycho-Analytical Process*, Heinemann, London, 1967.
21. ———, "The Relation of Anal Masturbation to Projective Identification," *Int. J. Psychoanal.*, *47*, 1964.
22. MONEY-KYRLE, R. E., *Man's Picture of His World*, Duckworth, London, 1961.
23. RIVIERE, J., "Jealousy as a Mechanism of Defence," *Int. J. Psychoanal.*, *17*, 1932.
24. ———, "The Negative Therapeutic Reaction," *Int. J. Psychoanal.*, *17*, 1936.
25. ———, "On the Genesis of Psychological Conflict in Earliest Infancy," in *Developments in Psycho-Analysis*, Hogarth, London, 1952.
26. ———, "Symposium on Child Analysis," *Int. J. Psychoanal.*, *8*, 1927.
27. ———, "The Unconscious Phantasy of an Inner World Reflected in Examples from Literature," and "The Inner World in Ibsen's *Master Builder*," in Klein, M., *et al.* (Eds.), *New Directions in Psycho-Analysis*, Tavistock, London, 1955.
28. ———, "Womanliness as a Masquerade," *Int. J. Psychoanal.*, *10*, 1929.
29. ROSENFELD, H., *Psychotic States*, Hogarth, London, 1965.
30. SEARL, N., "Danger Situations of the Immature Ego," *Int. J. Psychoanal.*, *10*, 1929.
31. ———, "The Flight to Reality," *Int. J. Psychoanal.*, *10*, 1929.
32. ———, "A Note on Depersonalization," *Int. J. Psychoanal.*, *13*, 1933.
33. ———, "A Paranoic Mechanism as Seen in the Analysis of a Child," Abstract in *Int. J. Psychoanal.*, *9*, 1928.
34. ———, "Play, Reality and Aggression," *Int. J. Psychoanal.*, *14*, 1933.
35. ———, "The Role of the Ego and Libido in Development," *Int. J. Psychoanal.*, *11*, 1930.
36. SEGAL, H., "Fear of Death: Notes on the Analysis of an Old Man," *Int. J. Psychoanal.*, *39*, 1958.
37. ———, *Introduction to the Work of Melanie Klein*, Heinemann, London, 1964.
38. ———, "A Necrophilic Phantasy," *Int. J. Psychoanal.*, *34*, 1953.
39. ———, "A Note on Schizoid Mechanisms Underlying Phobic Formation," *Int. J. Psychoanal.*, *35*, 1957.
40. ———, "Notes on Symbol Formation," *Int. J. Psychoanal.*, *38*, 1957.
41. ———, "A Psycho-Analytical Approach to Aesthetics, in Klein, M., *et al.* (Eds.), *New Directions in Psycho-Analysis*, Tavistock, London, 1955.
42. ———, "Some Aspects of the Analysis of a Schizophrenic," *Int. J. Psychoanal.*, *31*, 1950.
43. STOKES, A., *Greek Culture and the Ego*, Tavistock, London, 1958.
44. ———, *The Invitation in Art*, Tavistock, London, 1968.
45. ———, *Painting and the Inner World*, Tavistock, London, 1963.
46. ———, *Three Essays on the Painting of Our Time*, Tavistock, London, 1961.

B. Psychoanalytic Object Relations Theory: The Fairbairn-Guntrip Approach

Harry Guntrip

Forerunners of Object Relations Theory

Fairbairn was the first psychoanalyst to work out a full-scale, systematic object relations theory of our psychic life as persons growing in relationships. But it would be a mistake to associate object relations theory exclusively with his name. He would have disapproved of any attempt to create yet another school of theory. Fairbairn was no sectarian, but a philosophically, scientifically, and artistically educated man, "seeing life steadily and seeing it whole," but, above all, following the clues provided by the actual pressure of the patient's experiences on the necessarily limited *first* attempt of Freud to create a *psychobiology* of human living. Thus he wrote:

The clinical material from which the whole of my special views are derived may be formulated from the general proposition that libido is not primarily pleasure-seeking, but object-seeking. The clinical material on which this proposition is based may be summarized in the protesting cry of a patient to this effect—"You're always talking about my wanting this and that desire satisfied: but what I really want is a father"[6] (p. 137).

Fairbairn observed that after the introduction of the superego concept, Freud failed to make the necessary modification of his biological libido theory that this new object relational concept demanded. In fact, object relations theory is the development of the personal aspect of Freud's theory, as distinct from the physical or biological aspect, and it was really there from the start as soon as Freud abandoned as impossible to achieve his first attempt to formulate his findings in neurophysiological terms. Thereafter, the physical and the truly psychological, the biological and the personal, alternated between being confused and being distinguished at every stage of Freud's theoretical development. Object relations theory represents the ultimate emergence to the forefront of the personal and properly psychological part of his first formulations. This became marked when, after World War I, he turned his attention more and more to ego problems. In his last book, *An Outline of Psychoanalysis*,[10] the title of the last and unfinished chapter is "The Internal World," a thoroughly object relational concept.

Freud was the pioneer who said the first, not the last word, but his work ended at a point that makes it remain a tremendous stimulus to further thinking. Theoretical developments beyond Freud began with Freud himself, for he was always outthinking his own earlier provisional hypotheses. There were earlier critical object relational studies of Freud's biological psychology from outside the psychoanalytic movement, a healthy sign of the impact psychoanalysis was making. Ian Suttie of the Tavistock Clinic in Britain, in *The Origins of Love and Hate*,[25] rejected instinct theory and held that deep-seated fears, creating a "taboo on tenderness" in personal relations, were the basis of neurosis. In America Harry Stack Sullivan, founder of the Washington School of Psychiatry and the William Alanson White Institute of Psychiatry and Psychoanalysis in New York, made an even more sustained and systematic development of the same kind, allowing for a biological substrate of the personality as the raw material of a self or person, and then devoting himself to the study of all types of psychosis and psychoneurosis in their social settings and in terms of interpersonal relationships. This kind of theoretical development was bound to take place. Inside the psychoanalytic movement itself we may even see it adumbrated by Sandor Ferenczi, whose work at that earlier date really implied an object relations point of view, with more personal methods of treatment. It was unfortunate that Ferenczi was before his time. Freud was not ready for this development and disavowed Ferenczi, but his work has been recently justified by his pupil, the late Michael Balint, in that important book, *The Basic Fault*.[1] He traced the need of the very ill patient to regress as far back as the basic fault, the original failure of secure ego development at the beginning. This can only be remedied, not by a search for gratifications of instinctive needs (q.v. Fairbairn's patient) that could only become an endless series of makeshift temporary solutions, but by a search for recognition as a self, a person, by the analyst, an experience that can be the starting point for a new beginning in personality development. In this view the personal object relational need transcends the biological needs.

Thus object relational thinking is very far from being synonymous with Fairbairn's systematic elaboration of it, and even Fairbairn owed his primary stimulus to the work of Melanie Klein. At this point we must recognize that one of the accidents of history made an enormous difference to the way psychoanalysis developed in America and Britain. Political unrest in Europe scattered the original Freudian circle, and two outstanding psychoanalytic thinkers, Melanie Klein and Heinz Hartmann, settled respectively in Britain and America. As a result the main body of psychoanalytic thought in the two countries showed, over the next few decades, marked divergencies, which are now, one hopes, becoming more closely studied and better understood. In America the interpersonal relations theory of Sullivan continued to live a vigorous life in such thinkers as Karen Horney, Erich Fromm, Clara Thompson, and others, but Hartmann's[15] influence became predominant in the psychoanalytic societies. He pointed out how Freud wavered between system ego and person ego concepts, and both came down firmly on the system ego, and the structural analysis of the biopsyche into the id (the it, impersonal biological energies, instincts), the ego (the I), and the superego (conscience). Hartmann, however, allowed the superego concept (which, as an internalization of parental authority, is clearly object relational) to fall into the background, and he dealt mainly with the system ego and its dual functions of controlling id drives and adapting to the external world. Adaptation is a biological term, and for Hartmann psychoanalysis was a biological science, which he strove to develop toward a rapprochement with general psychology (which, in behaviorism, has become steadily more mechanistic and impersonal). One can admire Hartmann's intellectual power and clarity without agreeing with him, and I feel that the concept of adaptation is wholly inadequate to ego psychology, or any account of human beings as persons. Frequently our highest human functioning leads to a refusal to adapt and to preparedness to sacrifice life itself in the service of our values.

It is by comparison with this type of psychoanalytic theory that the object relations theory can be most clearly expounded. In Britain the great turning point in theoretical development was the work of Melanie Klein, which is the subject of Chapter 38A. I need therefore only pick out that aspect of it that inspired the fresh thinking of Fairbairn, Winnicott, and others. Fairbairn acknowledged that it was her paper on "Manic-Depressive States" (1935)[18] that made the all-important impact on him; and Winnicott told me that it was his training analyst, James Strachey, who said to him: "As a pediatrician you must hear what Mrs. Klein is saying." Even though Strachey as a classical analyst could not accept all her views, he felt that what she was saying was of great importance. Melanie Klein did not herself, I believe, recognize how radical was her development beyond Freud, mainly because Freud himself had adumbrated it. She accepted his theory of instincts of sex and aggression, even making more of the death instinct than Freud himself did. Her use of projection and introjection is absolutely essential for psychoanalysis, but it is its implications for her theory that concern us here. For Klein the infant's greatest fear is of his own death instinct, and he introjects (takes into his inner mental world) the good breast as a defense. But then he feels the good object is endangered inside, so he projects the death instinct into the external breast, only to find himself faced with an external bad object, which he introjects to control it. Thus *he has now built up in his unconscious a whole internal world of good and bad objects* with which he can be seen, in his fantasy and dream life, to be carrying on active relationships. Whatever we think of this instinct theory, the important result is that this concept of instinct has now become really *superfluous for psychology. The unconscious has been reinterpreted,* described no longer biologically in terms of a seething cauldron of id drives, but psychologically as an internal world of ego-object relationships. Thus by projection and introjection human beings live in two worlds at once, the inner mental world and the external material world, and constantly confuse the two together. It is the business of psychoanalysis to expose this confusion by interpretation of the transference of the patient's inner relationships on to the analyst. Klein's work is a development, not of the biological element of Freud's instinct theory, but of the personal, object relational element of his superego concept, the result of the relationships that develop between the child and his parents. As Hartmann allowed the superego concept to fall into the background while giving a very impersonal account of the ego, so Klein allowed the ego concept to fall too much into the background while giving a highly object relational account, based on the concept of introjection and the superego, of the hitherto biologically described unconscious.

At this point W. R. D. Fairbairn[6] came to the rescue of the Ego and worked out in a detailed way, "A Revised Theory of the Psychoses and Psychoneuroses" (Ch. 2). We may pause to ask how it was possible for two such different lines of thought as Hartmann's and Klein's to develop. It is not simply that one is right and the other wrong. It is that the subject matter is extremely complex, and it was not for a long time that the major problems involved began to be clarified. For that we have had to wait finally for further developments in the philosophy of science, which were not available in Freud's time, with his pre-Einsteinian, Helmholtzian inheritance. Freud was by education a physical scientist and by natural genius an intuitive psychologist with deep insights into the subjective experience of human beings. His intellectual problem was to determine in what sense his explorations of subjective human experience could be a "science." Was it a science at all?

His determination to make it scientific led him to seek to base it first on neurology and then on biology, but, in fact, although it could not be seen at that time, he was forcing a consideration of the question: "What is science when it can include at one end atomic physics, and at the other the exploration of the inner, unconscious, motivated life of personal selves in relationships?" All possibilities had to be explored: Hartmann explored in one direction; Melanie Klein and Fairbairn in the other, more personal direction.

Freud worked with the concept of psychic reality, but the object relations theory has given new depth and certainty to this concept, revealing psychoanalysis as a true *psycho*dynamic science. Bowlby[3] rejects Freud's concept of psychic energy on the ground that all energy is physical. My criticism of it would be that with Freud it was not yet a truly psychological concept, but physical energy with a psychic label on it. True psychic energy is motivational energy, in which a man's values can energize a lifetime of devoted labor. Bronowski's[4] view that man is both a machine and a self, and that there are two different kinds of knowledge, knowledge of the machine, which is physical science, and knowledge of the self, would have been a godsend to Freud. Bronowski finds knowledge of the self in literature, but that is only one area in which knowledge of the self is expressed. Its systematic conceptualization is a scientific task and produces psychodynamic science. This I believe psychoanalysis, especially in its form as object relations theory, to be: the science of human beings as persons developing in the medium of personal relationships, past and present. This view of psychological science is also supported by the view of Medawar,[20] leaning heavily on Karl Popper, on the structure of knowledge. Expounding the "hierarchical model of the structure of knowledge" as rising tier by tier from its ground floor in physics, he rejects reductionism, writing: "Many ideas belonging to a sociological level of discourse make no sense in biology, and many biological ideas make no sense in physics." We may add that neither can psychological ideas be reduced to the lower levels of any other science, and since the Person is the highest product of evolution known to us, the irreducible science of the nature of man as personal must crown the edifice of scientific knowledge. It will be the conceptual basis of the healing art of psychoanalytic therapy.

These developments in the philosphy of science are of extreme importance for understanding the implications not only of Freud's original concept of psychic reality but also of Fairbairn's development of it. Freud's Oedipus complex, however much it is represented as an instinct phenomenon, represents the object relations of the child and parents as persons, and this is psychic reality. When Sullivan and the American Culture Pattern school shifted the emphasis from the biological to the sociological, studying the fate of the individual as a person in his social milieu, they were exploring psychic reality, subjective personal experience. When in the 1930's and 1940's Melanie Klein elaborated Freud's concept of the superego into a full-scale analysis of the internal psychic world and its developmental processes, she was exploring an endopsychic phenomenon, which used the biological raw material of living, but grew wholly out of the quality of the child's relationships with the parents, that is, psychic reality. Sullivan closely approached Klein's theory here in his view of parataxic distortion. In the words of Clara Thompson, [26] "Interpersonal relations as understood by Sullivan refer to more than what actually goes on between two or more factual people. There may be 'fantastic personifications' such as for instance the idealization of a love-object. . . . One may also endow people falsely with characteristics taken from significant people in one's past. An interpersonal relationship can be said to exist between a person and any one of these more or less fantastic people, as well as between a person or group evaluated without distortion" (Pp. 215–216). Here is an unmistakable account of purely psychic reality, but neither Sullivan nor Klein saw as yet that this would demand a complete redevelopment of ego psychology. This was Fairbairn's major step forward and rested on his concept of the dynamic structure of psychic reality.

The Evolution of Fairbairn's Ego Analysis

Ernest Jones, in his foreword to Fairbairn's book, *Object Relations Theory of the Personality,*[6] wrote: "Instead of starting as Freud did, from stimulation of the nervous system proceeding from excitation of various erotogenous zones and internal tension arising from gonadic activity, Dr. Fairbairn starts at the centre of the personality, the ego, and depicts its strivings and difficulties in its endeavour to reach an object where it may find support. . . . This constitutes a fresh approach in psychoanalysis" (p. v).

To emphasize the ego and its search for security, which it can find only by dealing satisfactorily with its bad objects and maintaining reliable relations with its good objects, is to bring the whole problem of object relationships into the very center of the psychoanalytic inquiry. It also lifts psychoanalysis above psychobiology to the level of a true *psycho*dynamic theory—that is, a theory of the person, not simply of the organism—and gives full meaning to Freud's term "psychic reality," with psychoanalysis having the status of psychodynamic science. With Melanie Klein instinct theory still held the place that should have been held by the ego, for, as Fairbairn later pointed out, instincts are the instincts of, or properties of a person-Ego since modern science does not now separate energy and structure (or id and ego). However, Klein regarded the phenomena of internal object relations as illustrating the vicissitudes of instincts rather than the struggles of an ego in search of security. Nevertheless, it is evident that the whole drift of psychoanalytic theory was toward placing object relationships at the very heart of the psychodynamic problem. We shall see that gradually, at the hands of Fairbairn and Winnicott, the problem of the person-Ego underwent a subtle but highly important development. It became more than E. Jones's "endeavour to reach an object where it may find support," more than a search for security; nothing less than the ultimate need for self-discovery, for self-development, for the realization and growth of the potential ego's full possibilities in relationship with other persons. If the term "security" is to be used, it must imply secure possession of one's own full selfhood, and this may involve the possession of the inner strength to face up to external insecurity, persecution, with an overriding determination to be true to one's real self. Winnicott entitled one of his most important books *The Maturational Processes and the Facilitating Environment,*[29] and if the infant has a genuinely facilitating environment for long enough at the beginning of life, he can withstand the pressures of very unfacilitating environments in later life. On the nature of the ego, I accept Fairbairn's view of its absolute fundamental centrality and importance for psychoanalytic theory.

Fairbairn's original contribution began with his 1940 paper on "Schizoid Factors in the Personality."[6] Prior to that his writings fall into two groups, 1927–33 and 1933–40. In the first period he began as a fully orthodox exponent of the classic psychoanalytic instinct theory. But as early as 1931, in a paper[6] presented at the British Psychoanalytical Society, he showed unmistakably his basic concern with ego analysis. In his patient's dreams various aspects of her ego or total self appeared clearly personified and differentiated as the little girl, the mischievous boy, the martyr, and the critic. He compared this with Freud's structural theory of id, ego, and superego and made one of his most important statements:

> The data provided by the case . . . leave no doubt about the existence of functioning structural units corresponding to the ego, the id, and the super-ego, but the same data seem equally to indicate the impossibility of regarding these functioning structural units as *mental entities.* . . . Perhaps the arrangement of mental phenomena into functioning structural units is the most that can be attempted by psychological science. At any rate it would appear contrary to the spirit of modern science to confer the status of entity upon "instincts"; and in the light of modern

knowledge an instinct seems best regarded as a characteristic dynamic pattern of behaviour. [p. 218]

In Fairbairn's patient's personifications the critic would clearly be a superego phenomenon, the little girl and the mischievous boy would be personalized id phenomena, and the martyr might well be the ego caught, as Freud said, between id and superego pressures. This subsequently led Fairbairn to redefine the id in personal or ego terms. At that time he still used the id concept, but he came to see that it is an impersonal and nonpsychological term, and that everything in human psychology must be presented as an aspect of ego functioning.

It was not, however, until Fairbairn had absorbed the work of Melanie Klein from 1933 to 1940 that he could work out the full-scale revision of psychoanalytic theory demanded by the bringing of personal object relations into the center of the picture. Klein's analysis of the internalized or psychic object enabled Fairbairn to proceed to a radical ego analysis. Hitherto the ego had been treated as a superficial phenomenon "on the surface of the id" (Freud), developed for the control and adjustment of impulse to the demands of outer reality. It was more a mechanism than a real self. The term "self" should connote the dynamic center or heart of the personality, its basic unity in health. *An internalized object is itself an experience of the ego or self.* Fairbairn realized that Melanie Klein's theory of the external object as split in the course of psychic internalization into a variety of internal objects, good and bad, involved parallel splits in the experiencing ego, since the ego is libidinally attached in different ways to the different aspects of the original external object. How real is the internal object and how much it is a part of the total ego or self are clear in such a dream as that of a sensitive married woman too easily made to feel guilty. She dreamed that as she was walking along the street, a tall, dark, stern-faced woman followed her wherever she went, keeping an eye on her. It was her mother from whom, however, she had been parted by marriage for over ten years. Sometimes these internalized dream objects can acquire momentary hallucinatory reality in a half-awake state, as when a man woke to see a small boy dart across the room and disappear up the chimney, and realized he was dreaming. Still more striking is the unwitting acting out of a dream, as with a patient whose mother repeatedly beat her as a child, so that in her forties when she began analysis, she was still having nightmares of being beaten by her mother. But when very emotionally disturbed she would beat herself, and as she was doing this in one session I said, "You must be terrified being punched like that." She stopped, stared at me, and said "I'm not being hit. I'm the one doing the hitting." I commented "You are both." She was acting out her split ego, part self and part bad mother.

Freud himself, in his last unfinished *Outline of Psychoanalysis*,[10] had taken his stand on the view that ego splitting is not confined to the psychoses but is universally present in the psychoneuroses as well. Fairbairn's long-established concern about ego analysis enabled him to draw out the implications of Klein's work and of Freud's last statement. He summarized them thus: "Psychology may be said to resolve itself into a study of the relationships of the individual to his objects, whilst, in similar terms, psychopathology may be said to resolve itself more specifically into a study of the relationships of the ego to its internalized objects" (p. 60).[6]

The distinctions Fairbairn made within the overall whole of the complex ego structure are thus not entities (like parts of a machine) but processes, differing but simultaneous reactions of a whole personal ego dealing with his environment of complex personal objects. Some of these reactive processes are so fundamental that they become habitual and relatively enduring characteristics of the whole person, especially those based on the infant's and small child's reactions to parents, and can be used to describe the relatively enduring structure of the psychic self in its dealings with its object world, its human and cultural environment.

The Relevance of the Schizoid Problem

The revision of psychoanalytic theory proposed by Fairbairn had not only *theoretical* roots in his primary concern about ego analysis, and in Melanie Klein's theory of internal objects, but also an all-important *clinical* root in his study of schizoid states in psychoanalytic therapy. This must be examined before his structural theory can be further developed. *The schizoid problem goes deeper than depression*, which is, at least in the classic conception of it, a phenomenon of the moral aspect of growth. "Moral" here means the capacity to feel for the object. Melanie Klein adopted an object relational view of the stages of development in infancy when she distinguished between the paranoid position (internal persecutory bad objects) and the depressive position (anxiety over internal endangered good objects), and later added Fairbairn's concept of schizoid to describe the basic emotional position as paranoid-schizoid. Fairbairn accepted this view of the two basic psychopathological states. The classic psychoanalytical concept was that the infant psyche developed beyond the autoerotic and narcissistic level of ego libido, which was its original *objectless* condition. From an original primary identification with the object, the infant grew, through physical birth and psychic growth, to a capacity to differentiate the object from himself. Thus object libido arose as the ability to feel for others (primarily the mother). Then when the infant experiences the good object and the bad object as aspects of one and the same mother, he is caught in an ambivalent love-hate relationship and guilt. He can neither hate for loving, nor love for hating, and guilt paralyzes him. This is a psychology of impulse, and Freud's scheme of endopsychic structure—id, ego, and superego—conceptualizes the analysis of depression, as is shown by the fact that it arose out of his analysis of melancholia and obsessional neurosis. The id, held to be the origin of antisocial, destructive impulses of sex and aggression, cannot be adequately managed by the ego, and the superego develops as an internalized version of the authoritarian parent, to help the ego in its struggles to master id impulses. I have retraversed this familiar ground because it is essential to the understanding of Fairbairn's work that it should be closely compared with Freud's on these two main points of impulse psychology and endopsychic structure.

Fairbairn realized that the schizoid patient is not primarily concerned with the control of impulses in object relationships, a secondary matter, but in the end with whether he has a sufficiently real ego to be capable of forming object relationships at all. He finds object relationships so difficult, not merely because he has dangerous impulses, but because he has a weak, undeveloped ego; he is infantile and dependent because at bottom his primary maternal object failed to treat him as real, to "love him for his own sake, as a person in his own right." In his 1940 paper Fairbairn[6] described the schizoid tendency to treat people as less than persons, as things, part-objects, "breasts" to be used, as a result of the breast-mother's inability to give the spontaneous and genuine expressions of affection that would make her a real person to her baby, robbing him of the chance to feel himself becoming a real person for her. Such an infant grows up only able to use, not really able to relate to, people. His inner unsureness of his own reality as an ego is likely to be shown by role playing, exhibitionism, with little real communication, taking rather than giving, fearing to give since it may feel like a loss or self-emptying. He may seek to bridge the gulf between himself and others by "thinking" rather than "feeling," by impersonal intellectualization with greater investment in theories and creeds than in real people. It is in the struggle to overcome this basic weakness that his impulses become antisocial.

Fairbairn at first regarded the schizoid's withdrawal from objects as due to his fear that his unsatisfied needs, which the object has failed to meet, had become so greedy and devouring that his love had become even more dangerous than his hate. This is clearly met with in analysis, but is only halfway to the more complete explanation toward which Fairbairn's work developed as he discarded impulse psychology; namely, that the final problem is an infantile ego too weak to be able to cope with the outer world, because he is already split in his growing emotional life by the inconsistency of his primary parental objects, and becomes a prey to loss of internal unity, radical weakness, and helplessness. While still partly struggling to deal with the outer world, he also partly withdraws from it and becomes detached, out of touch, "introverted" (Jung), finding refuge in an internal fantasy world. This is not a problem of impulse control, but of ego splitting, and it calls for a different type of theory of endopsychic structure. In fact, Freud provided a model for this in his theory of the superego, described at first as "a differentiating grade of the ego." Here is the beginning of the conceptualization of ego splitting in a structural theory. Fairbairn[6] wrote: "What manifests itself on the surface as a divorce between thought and feeling must be construed as the reflection of a split between (1) a more superficial part of the ego representing its higher levels and including the conscious, and (2) a deeper part of the ego representing its lower levels, highly endowed with libido, the source of affect" (p. 21). He later developed a more systematic theory of endopsychic structure, based on the analysis, not of depression, but of the schizoid problem. Winnicott later suggested describing the basic split as between "a true self put away in cold storage," when it cannot find a nourishing environment, and "a false self on a conformity-basis" to cope with the external world; a stimulating if not complete description of the total problem.

Psychobiology and Instinct Theory

Before Fairbairn's structural theory is outlined, it is well to state explicitly his attitude to biological psychology and instinct theory. Naturally he does not ignore biology; he accepts the existence of biological factors as providing the raw material of personality, much as Sullivan[23] refers to the biological substrate. But Fairbairn discarded the concept of instincts as biological entities existing outside the psychological ego, which they are then supposed to invade. He regards a human being as whole from the start, and personality as developing, not by integration of separate elements, but by internal psychic differentiation within the whole, under the impact of experience of the external world. Although a biological and psychological aspect are distinguishable in theory as different levels of scientific abstraction and conceptualization, in reality they are aspects of a unity. They should not be confused or mixed, as in psychobiological theories, nor can psychology be reduced to biology. In a letter to me on March 15, 1962 Fairbairn wrote: "I do not consider that a psycho*biological* view is valid at any level of abstraction. Psychological and biological are both valid at their appropriate levels of abstraction, but to my mind a *psycho*biological view is not valid because it confuses two quite separate disciplines." Freud's theory of structure is a mixture of a biological id, or matrix of drive energies, and a psychological ego and superego as control systems. The id is impersonal. The ego and superego are personal and properly psychological concepts. The person includes the organism, but in dealing with the person on the psychological level of abstraction, we are thinking on a higher, more comprehensive level than when dealing with the biological organism. Organism is a wider concept than person. There exist organisms without personality, but not personality without organism. Thus the statement "Person includes organism" is not reversible, and the personal cannot be dragged down to and accounted for on the organic level.

Roughly speaking, the organism accounts for potentiality, primary energies, raw material, the genes that determine the hereditary constitution, the phenomena of maturational stages from infancy to old age, the physiological appetites (for body maintenance), the neurophysiological mechanisms for sensory perception and action, all the complex machinery the person needs for dealing with the external world. In nonpersonal organisms the idea of instinct as drive entity may be more meaningful, but its meaning is vague and not very useful. In the human being there are no fixed instincts determining the functioning of the personality, in Freud's sense of sex and aggression, but a growing personality possessing biologically based energies for action, which operate in ways determined by the state of the personal whole ego in the medium of object relations. The activity of these biological energies expresses the condition of the ego; the ego is not in the power of fixed instincts. Freud's sex is one of the appetites, like eating, drinking, excreting, breathing, and so on, although sex is the appetite most easily involved in human relations. It arises out of biochemical conditions within the organism, but may then be inhibited or overstimulated, or left to function normally by the whole person-ego. Aggression cannot be treated as comparable with sex. It is a phenomenon parallel to anxiety. Freud very late on changed his definition of anxiety from "damned up sexual tension" to "an ego reaction to threat." Anxiety and aggression, fear and anger, flight and fight, are the twin "ego reactions to threat," not just to bodily existence but ultimately and more important to the personality as such.

Thus Fairbairn held that the infant is oral because he is immature, not immature because oral. If an adult is adequately genitally sexual, it is because he is mature, not vice versa. *The development of the individual personality takes place in the medium of personal object relationships*, beginning with the mother-child relation. *This is not a biological but a psychodynamic phenomenon*, and this is the proper study of psychoanalysis. Fairbairn for some time preferred the adjective "instinctive" to the noun "instinct" as a safeguard against the tendency to reify biological potentialities on the psychological level, but the term becomes increasingly meaningless for his theory. We deal with a psychodynamic ego using its biological endowments in the conduct of personal object relations, in quest of security. "Security" here refers not merely to material security, bodily self-preservation. Subpersonal organisms presumably have only that aim. The pursuit of security in the sense of mere economic provision is only an extension of that primitive aim. "Psychodynamic security" means "security of the personality as such." It could be better termed, as by Fairbairn, the quest of reality "as a person in one's own right," of personal significance and stability, of the capacity to maintain oneself as a meaningful member of persons in relationship. *Psychodynamic security is only achieved by adequate ego growth*, initiated by what Winnicott calls "good enough" personal (parent-child) relationships from infancy onward.

This is exactly what schizoid personalities are found to lack. They suffer as persons from what R. D. Laing[19] calls "ontological insecurity." The deeper one goes into their mental makeup analytically, the more they are found to be experiencing themselves as empty, worthless nonentities—meaningless, futile, isolated, lonely, and aimless. They experience a craving for close contact for security's sake, that is, dependent, which at the same time they fear because they feel they can do nothing effective to realize it or accept it. Thus Fairbairn's work involves a decisive shift of the center of gravity in psychoanalysis from impulse theory, guilt, and depression to the failure of ego development and the schizoid problem, as well as the sheer primitive fears that blocked the infant's growth emotionally and forced his withdrawal into himself. He wrote: "I have become increasingly interested in the problems presented by patients displaying schizoid tendencies. . . . The result has been the emergence of a point of view which, if well-founded, must have far-reaching implications for both psychiatry in general and for psychoanalysis in particular . . . a considerable revision of prevailing ideas regarding the nature and aetiology of schizoid conditions. . . . Also a recasting and reorientation of the libido theory and modification of various classical psychoanalytical concepts" (p. 28).[6] This calls for a new conceptualization of endopsychic structure. K. Colby[5] showed the inadequacy of Freud's pioneer theory of id, ego, and superego to account for present-day enlarged clinical knowledge, but the model he proposed is too frankly mechanistic to meet our needs. Fairbairn provided a model that is consistently psychodynamic and fully personal.

The Theory of Endopsychic Structure

Fairbairn's theory of endopsychic structure is based on his theory of dynamic structure. He pointed out that Freud's view is based on a dualistic separation of energy (id) and structure (ego and superego), which is Newtonian and Helmholtzian, but not in accord with modern scientific theories. Thus Bertrand Russell[21] tells us that when a proton and an electron collide they do not split into more "things," but disappear into energy, which, he says, "at any rate is not a 'thing.'" Fairbairn[9] postulates a "pristine, unitary whole ego at birth," possessing its own energy and developing its own internal structure as a result of its earliest experiences in object relations. I would modify his view of its nature in one respect. It would be more accurate to say that at birth there is a "pristine, unitary whole human psyche with ego potential" that immediately begins to grow into a developing self, a person-ego. Its developmental fate depends on its finding a loving, supportive, facilitating, and especially maternal environment in which to grow. There is no impersonal id; all is ego, and development could proceed as a stable, unified, steadily enriched growth of the pristine ego if the infant experienced only absolutely good object relations. Good object relations simply promote good ego development. Bion[2] says that in the infant "good experience is simply digested." This, however, is impossible in practice, there being no perfect parents and no absolutely reliable external circumstances. The infant's experience is a mixture of good and bad, satisfaction and deprivation, free self-expression and frustration. The parent who is at one time good is at another bad—inevitably, from the infant's point of view, and often in actual fact. In the struggle to cope with his mixed experience and difficult outer world, *the infant goes through a series of spontaneous psychic maneuvers that result in his external objects coming to be represented by internal psychic counterparts* (which Melanie Klein called "internal good and bad objects") for their easier management. Broadly the infant seeks to see his outer world as good when it first becomes intolerable by the expedient of taking its bad aspects into his inner mental world to deal with them there. Bion[2] says that bad experience cannot be digested and absorbed, but only projected if not retained as a foreign body. This, however, does not solve his problems. Rather it tends to result in an unrealistic idealization of his *real objects* (for example, a patient who said at the first session "I have the best mother on earth," who turned out to be the major source of all her problems), and the creation *inside* himself of what Fairbairn called a "fifth column of internal persecutors." This procedure is the beginning of the process of schizoid withdrawal, and ego splitting into an external reality self and an internal reality self. Unless the infant's real-life object relations are good enough to keep him in genuine touch with his outer world, he becomes more and more dominated by fear and retreat into himself, and he loses contact with external reality in a flight from life into his inner world of fantasy objects.

This is not, however, adequately described simply as fantasy, for it becomes an enduring feature of his psychic life and develops as the unconscious structural pattern of his personality. (Structure per se is unconscious and only becomes knowable through active functioning.) Fairbairn has shown how Klein's object splitting is paralleled by ego splitting, and he systematizes the multiplicity of internal object relations revealed in dreams, symptoms, and disturbed external human relationships, reducing them to three main groups that represent the fundamental pattern of our endopsychic structure. Freud's scheme was really an early adumbration of this, and it is remarkable how through the centuries attempts to analyze the constitution of the human mind have all conformed to a threefold pattern; for example, Plato's chariot with a charioteer (of reason) and two steeds, the many-headed beast of fleshly lusts, and the lion, courage, the fighting principle, (that is, Freud's sex and aggression); also the familiar "body, mind, and spirit," which is in principle the same as Freud's id, ego, and superego. For Fairbairn, however, we must start with a primitive whole ego as yet undeveloped at birth, which becomes differentiated or split into three aspects that then function as lesser egos in opposition to each other, because of the self-contradictory nature of the reactions evoked in the infant by inconsistencies in his primary objects. Thus the primary unity of human nature is lost, disintegrated in internal civil war.

Fairbairn's structural pattern is threefold: (1) An infantile *Libidinal Ego* (cf. Freud's id), in a state of dissatisfaction and frustration, is related to an internal bad object, which excites but never satisfies the child's basic needs, and which Fairbairn calls the *Exciting Object.* Thus L.E.-E.O. embodies the experiences of the baby, insofar as he is deprived of adequate parental love. One male patient dreamed of a man following a woman who constantly retreated. (2) An infantile *Anti-Libidinal Ego* (a constituent of Freud's superego), an aspect of the infantile psyche in which the baby, not being able to secure a good object relation with the unsatisfying parents, is driven back on identifying with them as *Rejecting Objects* (an internal bad object forming a further constituent in Freud's superego). He is thus turned against his own libidinal needs. A female patient dreamed that she was a little girl feeling frightened, and she saw me in a room and thought, "If I can get to him I'll be safe," and she began to run to me. But another girl strode up and smacked her face and drove her away, her own Anti-Libidinal Ego at one with the punishing mother who was always saying "Don't bother me." Thus Anti-L.E.-R.O. embodies the experience of the deprived infant insofar as he sides with the critical, angry, denying parent against himself. His anger against his bad objects is turned back against himself in an attempt to suppress those of his needs that they will not meet. *The combination of Anti-L.E. and internal R.O. is a more precise formulation of the sadistic aspect of the superego,* and accounts for internal self-persecution of the L.E., for which reason Fairbairn at first called the Anti-L.E. the "internal saboteur" (cf. the patient who beat herself as her mother beat her). (3) The good or understanding aspects of the parents are left to form the *Ideal Object,* which is projected back into the real-life parent, causing idealized overevaluation of the parents in the external world. The I. O. is a still further constituent of the Freudian superego, accounting for its moral rather than sadistic aspect. The Ideal Object is related to by the *Central Ego* (the Freudian ego), the conscious self of everyday living. Thus C.E.-I.O. embodies the experiences of the child insofar as he seeks to preserve emotionally undisturbed good relationships with his parents in the outer world.

Fairbairn discarded Freud's definition of libido as pleasure-seeking and regarded *libido as object-seeking, the primary life drive to object relations and ego growth.* He regarded libido as having priority over aggression, which arises as a reaction of intensified self-assertion in the face of frustration. *Libido is the energy of all three subegos into which the primary psyche or nascent whole ego is split* as it develops in an environment of disturbing human relations. Even the Anti-L.E., has a libidinal basis, for this self-persecuting function of "aggression turned against the self" arises out of the infant's need to maintain object relations even with bad objects, as a result of which he is involved in identification with their negative attitudes toward himself.

Fairbairn's theory in its bare essentials may be summarized thus: (1) The ego, a pristine psychosomatic whole at birth, becomes split or loses its natural unity as a result of early bad experience in object relationships. (2) Libido is the primary life drive of the psychosomatic whole, the energy of the ego's search for good relationships, which make good ego growth possible. Energy and structure are not separated as in Freud's id and ego, which are replaced in object relations theory by Fairbairn's dynamic structure and ego splitting. (3) Aggression is the natural defensive reaction to a thwarting of the libidinal drive, which makes it parallel to Freud's second definition of anxiety as a reaction to a threat of the ego. (4) The structural ego pattern that emerges when the pristine ego or psychic unity is lost conforms to a threefold pattern of ego splitting and of internal ego-object relations: (a) L.E.-E.O., the primary needy natural self left unsatisfied; (b) Anti-L.E.-R.O., the angry infant employing his aggression to stifle his own needs as weaknesses; and (c) C.E.-I.O., the practical, conformist, conscious self seeking to get by as tolerably as possible in real life, repressing emotional experience into the unconscious (as in a and b), and unrealistically idealizing the parents he cannot do without, since bad objects are better than none. (5) Fairbairn regarded Freud's oral, anal, and genital stages of development as unsatisfactory, for mouth, anus, and genitals are biological organs used by the person to make relationships in both natural and disturbed ways. The anal phase he regarded as nonexistent normally, unless an obsessional mother forces it on the child in cleanliness training. The child is always using excretory organs. That the earliest infantile ego is markedly a "mouth ego" is simply due to his immaturity, and later on genitals may be used maturely or immaturely according to the state of the ego. Fairbairn therefore proposed, as the three stages of development, (a) *immature dependence* in infancy, (b) *a transitional phase* on the way to (c) *mature dependence*, or the relationship of equals on an adult level. In the early infantile phase of immature dependency, he regarded the schizoid and the depressive "positions" (Klein) as the two ultimate psychopathological states of internal bad object relations. Before we deal with that we must look at psychotherapy.

Psychotherapy

Psychotherapy is naturally based by Fairbairn on object relations theory. Since bad objects make the child ill, only a good object relation can make him well, that is, give him a belated opportunity to undergo an ego-maturing growth in a therapeutic relationship with an analyst he discovers at last (by working through transference problems) to be reliable, understanding, and concerned to enable him to find his own true self. Repression is carried out, not on instincts, but on internal bad objects, and the parts of the ego related to them, the ultimate internal bad object states and deepest psychic disasters being depression and the schizoid condition. The psychoneuroses arise out of a variety of attempted defenses against internal bad objects, which the patient can only discard when the analyst has become a sufficiently good external object to him. This, however, is not simply a matter of the analyst being a genuine good object in reality—reliable, understanding, and truly caring. The patient's difficulty is that he cannot trust or believe that can be true. The analyst will not come to be experienced by the patient as a therapeutic good object with whom he can regrow his personality in security, unless the analyst can help him to relive and outgrow his internal bad object relations in the transference. The analyst must prove capable of reliably surviving all the patient's projections of internal bad objects on to him, thereby bringing the patient through to an undistorted realistic relationship in which he can find his *own* true self.

Winnicott distinguishes two levels of treatment, "oedipal analysis" for the problems of later childhood, and "Management analysis" where problems go down as deep as the mother-infant relationship. By this he implies that with such deep problems the analyst must accept and support the infantile dependence of the patient. At such depth, as Balint says, the patient may not be able to accept interpretations as interpretations but only as attacks. Winnicott's unrivaled experience as a psychoanalytic pediatrician gives particular value to his views on the mother-baby relation. In "The Location of Cultural Experience"[28] he says that the experience of relationship is deeper and stronger than the experience of the satisfaction of instinctive needs. "The rider must ride the horse, not be run away with." The mother's "primary maternal preoccupation" with the baby (which develops during pregnancy and only fades, in the healthy mother, as the baby grows securely independent of her) gives her a knowledge of the baby's needs that no other person can have. Gradually the securely mothered baby develops a mental image of mother that, if undamaged, comes to allow the baby to tolerate her absence for a certain time, and the gap can be bridged by the transitional object, the cuddly toy that represents mother, the first symbol of relationship. But if the mother is absent from the baby too long, his mental image of her is lost, and with it "whatever of ego structure has begun to develop." This is the basis of "madness." This analysis raises the problem of regression to which we must finally turn.

The Problem of Regression

In the light of clinical problems the whole foregoing analysis points toward and requires one further step for its completion. Broadly speaking four groups of psychoneurotic conditions are recognized; hysteric, phobic, obsessional, and nonpsychotic paranoid states. Fairbairn regards these as corresponding to the four possible arrangements of good and bad, external (real) and internalized (psychic) objects, as related to by the split ego. *Obsessional states* represent the effort to maintain total internal control over all good and bad objects regarded as internalized, and over the suffering Libidinal Ego. In *phobic states* the suffering Libidinal Ego takes flight from bad objects to safe good ones, all of them projected into and seen as part of outer reality. In neurotic *paranoid states* the ego treats the good object as internal and identifies with it, while its bad objects are projected into the outer world, and the suffering Libidinal Ego hates them there. In *hysteric states* the opposite policy is pursued. The good objects are seen as projected into the outer world where the suffering Libidinal Ego can appeal to them for help against its bad objects, which are regarded as internal persecutors. One patient for a period changed regularly month by month from a hysteric to a paranoid attitude toward me and toward everyone. When paranoid his bodily health was perfect but everyone was against him, and I was only treating him to get fees out of him. When hysteric his body was full of aches and pains while he sought frantically for friends in his external world, and I was his one great hope.

At deeper levels depression and schizoid states take us into the region of borderline and psychotic cases. *Depression* is the paralysis of the suffering Libidinal Ego by guilt under the accusatory persecution of internal bad objects and the Anti-Libidinal Ego. The *schizoid state* takes us deepest of all, arising in its extreme form out of the flight from all bad objects, both internal and external, and, indeed, from all object relations, into the depths of the unconscious. There are varying degrees of seriousness in schizoid reactions mixed with all the other psychoneuroses and psychoses, but at its worst the search for a solution can involve one other form of illness that so far has not been fitted into the psychodynamic conceptual scheme, *regression* to infantile dependence in search of Balint's "new beginning," a chance to be psychically "born again." Whatever its degree this running backward to earlier levels of experience in search of security is a schizoid withdrawal from the world of bad object experience. Present-day realities are experienced as intolerable, mostly because the internal bad object world is projected on to them or they play on and reinforce real external bad object relations. It is well to bear in mind Freud's caution that we cannot raise anything out of the unconscious purely by analysis, but must wait until real life stirs it up. But the deepest schizoid withdrawal and the profoundest regression into apathy, exhaustion, and extreme infantile dependence are an escape from an intolerably bad internal world. One such apathetic patient acted out in the night while fast asleep scenes of being burned on the back with a hot iron by her psychotic mother. I was present on two such occasions when her husband rang me in the night, and I discovered scars on her back that he had not known were there. Gradually the patient became able to remember these scenes on waking and to work through them in sessions, and she lost her suicidal impulses.

How far can regressive schizoid withdrawal go? Clearly it can go as far as a fantasied and unconsciously experienced flight back into the womb, and many myths, dreams, and illness reactions represent just that. Such cases require Winnicott's "management." One patient, during the analysis of a prolonged hysteric phase, dreamed that she could not cope with adult life because she had a hungry baby under her apron clamoring for food. She produced a hysteric conversion symptom of an acute pain in her right forearm, which she nursed like a baby. She worked through this phase and became markedly schizoid, aloof, silent, and out of touch. She then had a prolonged fantasy of a dead or sleeping baby buried alive in her womb, which led on to a vivid dream of opening a steel drawer and finding inside it a live baby, staring with wide-open, expressionless eyes because there was nothing to see. This suggested to me that there is one last ultimate split in the ego, *in the infantile L.E. itself, into a clamoring, orally active L.E. (hysteric), and a deeply withdrawn, passive L.E. (schizoid).* This I have called the *Regressed Ego*, and it would account for a wide range of phenomena, including compulsive sleep, exhaustion, feelings of nonentity, the sense of having lost some part of the self, the strange isolation of feeling out of touch—in fact, all the marked schizoid states.

Fairbairn wrote to me that this concept accounted for phenomena that he had not hitherto been able to fit into the conceptual scheme, and he regarded it as the logical development and completion of his theory. Furthermore, I had gone beyond him at this point. Whereas he had treated depression and the schizoid state as equally ultimate psychic disasters, I had treated the schizoid state as deeper than depression, and he agreed that this was so. The patient cited who oscillated between hysteric and paranoid phases progressed into a suicidal depression, dreaming of a man pointing at him and saying "You are the guilty man." When finally this did not yield to orthodox analysis, and I said to him: "I don't think you are depressed in the accepted sense of the term, but seriously afraid of life, retreating from it, and trying to force yourself back by a sense of guilt," he produced at once the classic schizoid feeling of a sheet of plate glass between him and the world and said that as soon as he got home, he had, since his breakdown, gone straight to bed and curled up under the clothes. Fairbairn's work makes it clear that, psychotherapeutically, oedipal analysis is sufficient for many patients, but for others radical therapeutic success will only be achieved when the problems of schizoid regression are solved. It also shows that aggression is not the ultimate factor that it was classically assumed to be. In the last analysis it arises out of the desperate struggle of a radically weakened schizoid ego to maintain itself in being at all. As one patient said: "When I'm very frightened, I can only keep going at all by hating."

Bibliography

1. BALINT, M., *The Basic Fault*, Barnes & Noble, New York, 1968.
2. BION, W. R., *Learning from Experience*, Heinemann, London, 1962.
3. BOWLBY, J., *Attachment and Loss*, Vol. 1, *Attachment*, Basic Books, New York, 1969.
4. BRONOWSKI, J., *The Identity of Man*, Doubleday, New York, 1965.
5. COLBY, K., *Energy and Structure in Psychoanalysis*, Ronald Press, New York, 1955.
6. FAIRBAIRN, R., *Object Relations Theory of the Personality*, Basic Books, New York, 1952.
7. ———, "Observations on the Nature of Hysterical States," *Brit. J. M. Psychol.*, *28*:144–156, 1955.
8. ———, "On the Nature and Aims of Psychoanalytical Treatment," *Int. J. Psychoanal.*, *29*:374–385, 1958.
9. ———, "Synopsis of an Object-Relations Theory of the Personality," *Int. J. Psychoanal.*, *44*:224–225, 1963.
10. FREUD, S., *An Outline of Psychoanalysis*, Hogarth, London, 1949.
11. GUNTRIP, H., *Healing the Sick Mind*, Allen & Unwin, London, 1964.
12. ———, *Personality Structure and Human Interaction*, International Universities Press, 1961.

13. ———, *Psychoanalytic Theory, Therapy, and the Self*, Basic Books, New York, 1971.
14. ———, *Schizoid Phenomena, Object-Relations and the Self*, International Universities Press, New York, 1968.
15. HARTMANN, H., *Ego Psychology and the Problem of Adaptation*, International Universities Press, New York, 1958.
16. ———, *Essays on Ego Psychology*, International Universities Press, New York, 1964.
17. KHAN, M., "An Essay on Dr. M. Balint's Researches," *Int. J. Psychoanal.*, 50:237–248, 1969.
18. KLEIN, M., "A Contribution to the Psychogenesis of Manic-Depressive States," in *Contributions to Psychoanalysis, 1921–1945*, International Universities Press, New York, 1948.
19. LAING, R. D., *The Divided Self*, Tavistock, London, 1962.
20. MEDAWAR, P. B., *Induction and Intuition in Scientific Thought*, American Philosophical Society, 1969.
21. RUSSELL, B., *History of Western Philosophy*, Allen & Unwin, London, 1946.
22. SULLIVAN, C. T., *Freud and Fairbairn: Two Theories of Ego-Psychology* (Preface by Fairbairn, W.), Doylestown Foundation, Doylestown, Pa. (available from the Director of Research).
23. SULLIVAN, H. S., *The Interpersonal Theory of Psychiatry*, Norton, New York, 1955.
24. SUTHERLAND, J. D., "Object-Relations Theory and the Conceptual Model of Psychoanalysis," *Brit. J. M. Psychol.*, 36:109–120, 1963.
25. SUTTIE, I., *The Origins of Love and Hate*, Kegan Paul, London, 1935.
26. THOMPSON, C., *Psychoanalysis: Evolution and Development*, Nelson, New York, 1952.
27. WINNICOTT, D. W., *The Family and Individual Development*, Tavistock, London, 1965.
28. ———, "The Location of Cultural Experience," in Winnicott, D. W. (Ed.), *Playing and Reality*, Basic Books, New York, 1971.
29. ———, *The Maturational Processes and the Facilitating Environment*, International Universities Press, New York, 1965.
30. WISDOM, J. O., "Fairbairn's Contribution on Object Relationship, Splitting and Ego-Structure," *Brit. J. M. Psychol.*, 36:145–160, 1963.

CHAPTER 40

AMERICAN NEO-FREUDIAN SCHOOLS

A. The Interpersonal and Cultural Approaches

Earl G. Witenberg

A THEORY is an attempt to organize a chaotic universe of observations into a comprehensible whole, and the form it takes is largely determined by the modes of thought habitual to men of its time. Thus, the belief in individualism of the eighteenth and nineteenth centuries is reflected in Freud's view of the individual in a fateful struggle with his instincts, and the advance of the biological and physical sciences left their imprint on his structuralization of the libidinal impulses and on his conceptualization of the power of human drives in terms of energy quanta. Similarly all psychiatric theories that depict man as an organism in interaction show evidences of a new mode of thinking that began to become persuasive at the turn of the century. In physics events were seen to be relational; man in crowds was seen to be more than man alone; and society was appreciated as more than a compact entered into by discrete individuals for their mutual benefit. Psychiatry as a study of the reciprocal processes between people could not be long in coming, and the line of its development may be seen in the work of Durkheim, Cooley, Mead,[15] Peirce, Dewey, Lewin, and Sapir.[20] It reached conceptualization in the work of Harry Stack Sullivan,[22,23] with its influence clearly revealed in the writings of Erich Fromm,[6–13] Frieda Fromm-Reichmann,[14] Karen Horney, Clara Thompson,[24,25,26] and, indeed, of the whole field of psychiatry, child development, social psychology, and even of historical research.

Harry Stack Sullivan

Sullivan has been characterized by Redlich and Freedman as "America's most original modern psychiatrist." A colleague of his, William Silverberg, has stated that the only person he knew who could have done what Freud had done if Freud hadn't already done it was Sullivan.

Sullivan made one of the most comprehensive statements describing man as a biologically rooted but socially interacting organism. What gives Sullivan's work a flavor peculiarly different from that of other workers in the field are two elements that are, perhaps, specific to American thinking. Whether the influence is direct or not, it is clear that the pragmatic climate of America, noted by de Tocqueville more than a century ago and conceptualized by Peirce, James, and Dewey, has been influential in his approach. Sullivan's thinking is empirical rather than rationalistic. He is more concerned with developing formulations that will facilitate his purpose—the development of an effective therapy—than with theoretical "systems" of a high order of abstraction. He is interested in language and linguistics in a way no previous psychiatrist has been. His close association with E. Sapir, an anthropologist interested in linguistics and communication, enriched Sullivan's formulations. The scrutiny he gives to the development of language and symbols in the reciprocal processes between the parent or culture-surrogate and the child is the most telling in any of the psychoanalytic theories. This empiricism utilizes the approach known as operationalism. His formulations are generally stripped bare of abstractions in order that they may approach testability, and if some of the color and richness of human life seems to be lost in the process, at least what is there is hopefully verifiable by means of observation.

What Sullivan did, to a degree more than any other psychiatrist of his time, was to take note of the rising belief that, as Fichte put it, "the 'I' is not a fact but an act," and to test that belief in a clinical situation. In Sullivan's view the person we commonly refer to as an individual is the result of an interplay among physiological, psychobiological, and situational factors. Physiological factors include, among others, native endowment, nutrition, disease, and physical injury. Under psychobiological factors are those that inhibit or facilitate the evolution of the person, his education and acculturation, the development of his perceptions, his emotional sensitivity to events and to people. From the first mysterious contagion between child and mother to the last personal interchange of the old man at the moment of death, the human person is a being in process—not a fact but an act. To illustrate the dynamic nature of this view, we can contrast the common mode of speech when speaking of the person who exhibits much anxiety (or love, or hate) with the mode of Sullivan's frame of reference. The common way of thinking, which sees man as a possessor of traits, speaks of "an anxious person," whereas Sullivan might say that he is a person who under such and such circumstances experiences anxiety with others. And by others Sullivan would include fantasied persons and various distortions of the other. Situational factors include those actual interpersonal opportunities that reflect the interplay of the culture and its participants, the changing institutional setting of life, and the opportunity for new experience.

Leston Havens has remarked that Sullivan has "secretly influenced all of American psychiatry." Not so secret is his influence on such writers as Arieti,[1,2] Burnham,[3] Chrzanowski, Green, Tower, Rioch,[19] Weigert, Searles, Spiegel,[21] Stanton, Schwartz, and Will.[27]

The Basic Needs: Satisfaction and Security

TENSION OF NEEDS

Crucial in the development of a personality is the nature of the interaction between the parent and the infant in regard to the tension of needs or drives. From birth onward the infant experiences tensions of needs. There are tensions caused by physiological needs, both general and zonal; there is the tension of the need for tenderness; there are also tensions caused by the needs for curiosity, focal attention, activity, and mastery. Also there is the innate capacity for the tension of anxiety. Sullivan conceptualized tension as an intrinsic potential for action of organisms consisting of energy transformation so that the organism may maintain itself, that it may maintain a balance among various subsystems or organ systems, and it may maintain an equilibrium between its internal and external environments. Human experience may be said to consist of the enduring influence of this tensional history on the present and the future.

General needs are needs for food, breathing, and sleep. The zonal areas are oral, anal, vestibulo-kinesthetic, retinal, and genital. All these needs arise in the internal physicochemical milieu and are experienced centrally as tension. As a result of such tension the infant makes certain movements, such as crying, at first toward the universe at large and later toward the more specific agents who, experience has shown, may relieve his tension. If his efforts are successful in producing actions by others that result in the satisfaction of his needs, his tension abates, and he experiences the state that Sullivan terms "euphoria." Early experiences of such tension may be caused by anoxia, with increase in respiratory ventilation as a means of reducing tension, or by hypoglycemia, with the resulting tension causing sucking movements as a signal to the mothering person that a need requires satisfaction. The entire collection of tensions that requires intervention of a mothering person may be summated as the infant's need for "tenderness."

It can be seen that the appeasement of these primary needs requires interaction with others. From the moment of birth the parents are the agents through whose acts the child's needs are either satisfied or denied, and in the process of such satisfaction or denial, one or another mode of relationship is established. It is in the evolution of the relationship with the mothering one and others during the course of development that the self-dynamism is organized. If the "satisfactions" can be achieved without the development of significant anxiety in the infant, the basis for "security" is established, and the self-system that develops is one that has the interpersonal competence to achieve the satisfaction of its needs without loss of security. If, on the other hand, the early experiences through which the self-system is developed lead to a self that lacks the interpersonal competence to secure the satisfaction of its primary needs without the repeated experience of anxiety or its substitute states, the person is said to lack security and to be mentally ill. Thus, the need to satisfy these needs or drives leads to interaction with others, and this interaction leads to the patterning of a self-system that is interpersonally competent or incompetent, secure or insecure.

The Response to Needs

There must be a conjunction between the infant's needs and the emotional state of the mothering person if the signs that signal the needs are to be heeded. To this end Sullivan postulates that the activities of the infant that signal the presence of tension within him cause a state of tension within the mothering one that, in turn, is experienced as tenderness. Thus, the tension produced by needs in the infant induces a state of tension in the mother that, felt as tenderness, causes her to take steps to satisfy the infant's needs. This is the sequence of events that occurs in the case of the fortunate infant whose signals of tension to the mother result in the feeling of tenderness uncomplicated by anxiety. The infant finds his needs satisfied without the production of anxiety within himself, and the result is the achievement of security.

The Significance of Anxiety

The results of the infant's signaling of his needs are not always quite so fortunate. If those needs upon which life itself depends are not satisfied, he, of course, does not survive. But there is yet another possibility fraught with disastrous consequences for him. His signals may result in the satisfaction of his bodily needs, but under emotional circumstances of such a kind that grave damage is done to his sense of security, and as a result to the development of his personality. Here Sullivan takes up the question of anxiety, which is basic in the development of the distortions of personality known as neurosis and psychosis.

In the tension of needs the disequilibrium is physiological—there may be, for example, oxygen deficiency, dehydration, or hypoglycemia. The state of anxiety, on the other hand, results from an emotional interchange between one person and another, in this case the mothering one and the infant. If the mothering one is anxious for whatever reason (that is, for reasons totally unrelated or totally related to the signals of the infant), there may be some interference with the satisfaction of his needs, but by and large they will be satisfied. They will be satisfied, however, in an emotional atmosphere that will cause anxiety in the infant. The mechanism by means of which the infant's signals may cause anxiety in the mother may be readily imagined and will, of course, depend upon the existing personality of the mother and her state of mind at the moment. A mother beset with anxiety about a current marital problem will be anxious in the interchange with the infant. A mother with low self-esteem in regard to her womanliness will find her baby's needs threatening, since they cause her to face the very area of her anxiety. But the means by which the anxiety of the mother can cause anxiety in the infant are less easy to explain. If the infant is newborn with an as yet poorly developed perceptual apparatus, it may be difficult to understand by what means he will perceive the anxiety of his mother. In order to explain such a perception, Sullivan postulates the existence between the infant and the mother of a quality called "empathy." By this he means the emotional contagion by which the infant may become aware of the emotional state of his mother without the mediation of any of the sense organs.

Given anxiety in the mother, the infant becomes anxious and experiences the state Sullivan terms insecurity. Thus, lack of security is the result of interpersonal deficit as the presence of security is the result of interpersonal adequacy. The presence of the tension of needs is a sign of physiological disequilibrium; the presence of the tension of anxiety is a sign of interpersonal difficulty. As satisfaction is to the tension of needs, security is to the tension of anxiety. The perception of this tension of anxiety causes the infant to become uncomfortable and to signal this discomfort, but for the infant there is no relief because the mothering one has no way of relieving it. Severe psychopathology results from intense anxiety early in life. Severe anxiety does not convey any information about experience. It erases any experience or occurrence that may have preceded or accompanied it. The most severe form of anxiety, that associated with the contents of the not-me personification, is experienced as uncanny emotion, as awe. Dread, loathing, and horror are terms associated with the state. Less intense anxiety permits a gradual realization of the interpersonal circumstances under which it occurs.

The problem for both infant and adult is not simply devising means for satisfying the tension of needs; the means for doing this alone are usually at hand. The usual problem is to alter activity in the direction of lesser rather than greater anxiety, of greater rather than lesser security. To summarize: the drives and the ensuing tensions of the individual require the cooperation of others for their satisfaction. This makes the attempt to satisfy needs an interpersonal process, and depending upon the interpersonal competence with which it is carried on, it results in security or insecurity.

The Development of the Self-System

The self-system develops out of the interpersonal experiences the individual has with others in trying to relieve the tension of his needs or drives. He expresses his tension in interpersonal situations; and as a result he experiences feeling states through empathy, he notices facial expressions, voice tones, body tensions, and gestures of various sorts in the other person, and he is the recipient of more or less direct statements of the reaction of the other person to his needs. Raw events are not directly experienced. How and what is experienced is essentially a symbolic process that is different at different stages of development.

The earliest and most primitive form of experience of the young infant is experience in the prototaxic mode. This consists of discrete, total, unrelated impressions. There is no separation of self from environment and no differential localization of the source of the impressions. It is what William James calls the world of the infant, "blooming, buzzing confusion." This is the hypothecated state of experience for the young infant. Later on in life it may be seen during the acute phase of some severe mental disorders—functional or chemical (as in the untoward effects of psychedelic drugs).

The young infant gradually acquires a rudimentary form of perception called prehension. It is not the mothering one who is prehended first but the nipple. This is a complex image with a broad reference. There is a beginning capacity for foresight. The infant prehends crying as action that will lead to the relief of distress (this is the beginning of appreciation of the "magical potency" of vocal behavior). He also begins to prehend distinctions, say, between "good" and "bad" nipple; the former brings relief of distress, the latter brings additional distress. The development of this ability to identify differences in the environnment may be construed to be the first step toward experience in the parataxic mode.

Another step on the way is the integration of data provided by more than one of the zones of interaction, for example, visual and auditory. There is no logical or causal connection to the parts into which the original global experience is broken. The parts are connected by association; there is a more or less clear appreciation of relationships among experiences—relationships of coincidence and concomitance, of similarity and difference. The crying infant who needed to be picked up bodily before he would stop crying now will cease the crying when he hears the mother's footsteps or feels her hand on his cheek. This form of experience in the parataxic mode, this experience by concomitance and association, is the primordial form of knowing (as is the prototaxic when it recurs in major disorders). It may influence and color cognitive functioning into adulthood. The parataxic mode is the primary mode until the child learns the shared meanings of language. At this point there is the development of the syntaxic mode of experience, and consensual validation is available; the relationships are logical and causal here, not private, idiosyncratic, or governed by anxiety.

Related Appraisals and Personifications

Late in infancy the training for socialization is begun, and for the first time the infant becomes aware of the *appraisal* of himself by his parents. If a particular act is met with anxiety on the part of the parents, the infant takes his parents' reaction as an appraisal of himself and his worth. He begins to organize himself in terms of such reflected appraisals originally experienced in the parataxic mode. Impulses in himself that call forth anxiety in his parents, and hence in himself, are organized into a concept of himself that he personalizes as "bad-me"; while those impulses that do not result in anxiety but result in euphoria are organized into a concept of "good-me." Since those impulses, acts, thoughts, emotions, and fantasies that are part of the "bad-me" concept are regularly associated with the unpleasant emotion of anxiety, the individual becomes alert to all performances that result in a disturbing emotion in his parents and hence in himself. He tunes himself, as it were, to their emotional wavelength and by focusing on that in himself that may or may not provoke anxiety, he less clearly notices other experiences of himself. As Sullivan puts it, he "selectively inattends" to those aspects of experience or of himself that are issues for his parents, and in so doing may fail to experience either the creative or destructive aspects of his experience or personality. But the failure to notice such aspects of experience or of the self is not a profound disturbance; if pointed out by an accepting person, the inattended part of the experience may be recovered. Selective inattention serves, then, to control consciousness in those situations that might provoke anxiety. However, some activities may be met with such intense anxiety or anger that selective inattention is insufficient for psychic safety. For such processes the method is dissociation. Dissociated aspects of the self cannot come into awareness through ordinary conscious human experience, though they may reveal themselves during crucial periods such as adolescence and menopause in night terrors, panics, and so forth. The patterns that the person develops are his self-system. This serves to limit his anxiety by controlling his awareness. It thereby allows him to function with an apparent security, but it limits his opportunities to experience the new, the novel and limits the amount of syntaxic experience. With the advent of language "good-me" and "bad-me" become fused as "me."

"Not-me" is not integrated into the self. When the acts or even the very existence of the child result in total anxious disapproval on the part of the parents, the child is exposed to a major psychic disaster. He experiences such overwhelming anxiety that the boundaries of the self are eliminated, a state to which Sullivan applied the term "not-me." In crucial periods of later life such as adolescence, menopause, the beginning of a schizophrenic episode, in periods of overwhelming anxiety, or in certain nightmares, the "not-me" is experienced as the uncanny emotions of awe, dread, and loathing.

The Process of Socialization

During the latter part of infancy the child begins to make repeatable noises. He has been surrounded from birth with the linguistic signs and symbols of his parents and others; he experiences them in the prototaxic and parataxic modes. The child responds to these sounds actively. He also "rehearses" them subvocally and thereby begins the reverie processes that continue through life. Most thinking occurs preverbally. His first attempts at vocalization bear little relationship to the sounds heard. He thereby develops a personal language primarily in the parataxic mode; this autistic speech for some may persist into adulthood and under circumstances of extreme anxiety may become apparent. It may also become obvious in the production of poets who have access to this personal language.

If a particular sound is met with tenderness, it will be reiterated. If it is met with anxiety or displeasure, it will disappear or become covert to add to the reverie processes. The sounds that will be reiterated are, of course, chosen by the parents as agents of their own desire and of the culture. This explains why the magic word "mama" is so often the first word chosen or alleged to be chosen by the infant for the initiation of speech. Speech is the first type of communication in the syntaxic mode. The process of verbal communication makes possible the learning of social patterns since this variety of communication is less susceptible to the emotionally idiosyncratic distortions in relationship between parent and child. Consensual validation now becomes possible. The infant is now a child and has the opportunity to learn what a word means from more than one source.

The child is now at the threshold of socialization. His signals no longer bring unqualified cooperation from his parents. He begins to learn his social role via the anxiety gradient. If the degree of anxiety provoked in him is not too great, it serves as an educational force guiding him in learning required behavior. Thus, the thwarting of a need results in anxiety that teaches how and when society permits his needs to be satisfied. The child is now on the way to becoming a civilized person. If the child is faced with great anxiety, tension will increase, and the child has the choice of facing disintegration of his patterns of organization, resorting to attempts at sublimation, or of attempting to discharge his needs symbolically in dreams or substitutive activity. With advancing age, learning by reward and punishment begins to play a more active role in the social process. The child begins to discriminate authority figures and authority situations. If the means of enforcing the parents' authority are appropriate and the issues on which the authority is exerted are consonant with the culture, this will lead to a healthy and necessary discrimination of what sort of behavior is acceptable and what is not. To be appropriate the means must be adequate enough to be effective, not degrading to the child, and not punishment for punishment's sake alone. If the demands of the authority figures are consonant with the culture, if the child observes that the demands are approximately the same as those made by his parents on others and on themselves, he will not feel the precepts of the authorities as arbitrary or unjust. If, however, the authority is irrational and driven by anxiety, or if the modes of punishment become ends instead of means, the child may begin to protect himself by deceiving himself and his parents, by concealing what is going on within him. He may use verbalisms to avoid punishment and learn the use of "as if" performances to deceive the authority, deceiving himself in the process.

If the child's signals for the need for tenderness, approval, or affection are met with marked anxiety, being made fun of, being taken advantage of, or being hurt, instead of the tenderness he requires, what Sullivan calls the "malevolent transformation" takes place. This is a basic confusion in the relation of stimulus and response, much like turning on the hot water faucet and receiving a cascade of ice cold water. In the shower it may result only in a mistrust of plumbing, but in an area as vital as the need for tenderness, approval, and affection, it results in much more distressing consequences. It results in the conviction that one lives among enemies and can expect no satisfaction of needs outside oneself. An additional consequence of the development of the malevolent transformation is that it vitiates the trust in others so necessary for the progressive experience of unthwarted personality development. It also results in others being repelled by the child's attitude so that the child is never able to benefit from experience with potentially kind and friendly people.

The nature of the exercise of authority by the parent over the child may be significantly influenced by the gender of the parent in question. The parent of the same gender as the child is likely to have more patterned responses in his relationship with the child. The father "knows" how a little boy should be reared with a certainty he would never assume for that mysterious creature, a little girl. His own unfulfillment may seek fulfillment in his son. Since he feels his ideas on how his son should be raised as certitude, he is less open to observing the child as he really is and hence less likely to be responsive to the child's actual needs. The result will be the exercise of authority that is fixed and irrational, and in marked contrast to that of the parent of the opposite sex. Given the tendency to the exercise of irrational and excessive authority by the parent of the same sex and excessive tenderness, in contrast, by the parent of the opposite sex, the result may be behaviorally identical to Freud's description of the Oedipus complex.

Reflection and Revision

As the child advances to grammar school age, the mode of relationship idiosyncratic to his own family may, for the first time, be compared to the varieties of relationship of those outside the family. This becomes not only possible but necessary if humiliation, disapproval, or punishment by the new figures in his environment—his classmates and teachers—is to be avoided. For the first time the peculiarities and limitations of his own family are open to comparison and remedy. He comes under the influence of new authorities, teachers, recreational directors, and others whose manner of exerting authority and responding to, say, defiance may be different from his parents, particularly to the extent that his parents do not reflect the culture. At the same time the child meets others with a variety of personalities and learns to accommodate to them. In discussion with his peers, he has the opportunity to check upon the habits, values, and reactions of the significant persons of his past and present life, and by comparing them with other adults, he may come to make some value judgments about them. In questioning for the first time the infallibility of their judgments, he may revise some opinions of himself that have come from reflected appraisals. During this period competition is natural, and with proper intervention by social and school authorities, the child learns compromise and cooperation if he has not already experienced both with his brothers and sisters.

The importance of this period in the development of the personality cannot be overemphasized. It is a time for reflection and for revision. For the first time the child reflects upon the nature of his parents and, as a result, may bring them down to life size. The altering of his view of his parents makes for revision in his view of himself. It also is a time for learning social subordination and social accommodation. Juveniles learn how to relate to authorities; they learn how many slight differences in living there are; they learn acceptable and unaccceptable ways of being. They learn that what is valued outside the home may be different from what is valued inside the home. They learn from authorities other than the parents and the siblings; they also learn from each other.

So important is this time for reflection and revision, and so necessary the association with peers that makes this possible, that Sullivan states that children isolated from other children by the circumstances of geography invent realistic imaginary playmates. These playmates can have their source, however, only in personifications of the self or in storybook characters.

The opportunity for reflection and revision during this period of the child's life may be denied to him with fateful consequences for the rest of his life. If his learning of socialization patterns in the home has been very deviant from the patterns sanctioned by the community, the child is thrust into a world that will not accept him. In addition to the effect of the resulting social ostracism (disastrous enough), he loses the opportunities that may result from experiences with his peers. He will compete, cooperate, and learn social differences defectively. If his parents disparage the children to whom he is attracted, his peers will then lose their function of providing a new view of himself and his parents. A further loss will be the opportunity for a corrective revision of his self-system to accept differences, to cooperate, and to feel secure under many circumstances.

THE NEED FOR INTIMACY

During a brief period before the onset of adolescence termed preadolescence, the individual develops a great need for intimacy. This marks the emergence of the capacity for love. For Sullivan, love is a state of affairs in which the maintenance and enhancement of the satisfaction and security of another one is as important as one's own satisfaction and security. At this particular state there is a pairing, and a chum of the same sex is selected for purposes of intimacy. There is a sharing of details of living, including those private ones never previously shared. The individual then has the opportunity to see himself through the eyes of one who is like himself, who comes from the same age segment of culture, and who does not represent the authorities of society. Since the chum is perceived to be much like oneself with a community of interest and intent, the learning of collaboration proceeds. Because the individual now sees himself through the eyes of another, this period is one in which fantastic ideas about oneself may be corrected. Preadolescence is a wonderful opportunity to learn consensual validation in a way not before possible. It also is a time to learn the nature of intimacy with its ups and downs. While homosexuality may occur, it is not the homosexual phase described by Freud, but homosexuality as an instrumentality of intimacy, a sharing of sexual "secrets." The less intimate but more leveling relationships within groups or gangs that occur at this period provide further opportunity for remedying personal distortions that may have developed. If the child is not prepared to enter this phase because of previous major disturbances, he experiences a deficit that cannot be made up later in life.

In preadolescence loneliness first assumes the power of a major integrating force. It motivates the person to move toward another, even though the movement is attended by anxiety. Driven by loneliness, one may move through the distress of anxiety to the rewards of relationship.

THE INTEGRATION OF LUST, INTIMACY, AND SECURITY

Adolescence is ushered in by the first distinct appearance of a clearly sexual need. The tension of this need is manifested by covert processes of frankly sexual content and by awkward and misdirected approaches to persons of the opposite sex. There is a shift from the chum, and the gang diminishes in importance except as a way of gaining information about this new lust drive. It is a period dominated by three needs that require major orientations in adaptation; serious damage to the personality occurs if the reorientations in adaptation are not successfully accomplished. The basic need for security continues; the need for intimacy, begun in preadolescence, requires a shift from the partner of the same sex to one of the opposite sex; and the need for lustful satisfaction, requiring the collaboration of another person, makes its first appearance. The need for intimacy with someone of the opposite sex, heretofore a stranger, may result in the shaking of the sense of security if the movement toward the other is done with an ineptness that provokes the reflected appraisal of disdain. If the drive for lustful satisfaction collides with a fear of intimacy of any sort, a turning toward the self, in the form of increased self-absorbing masturbation, which is not a reverie preparation for heterosexual activity, may occur. If the need for lustful satisfaction is not accompanied by a shift from the preadolescent intimacy with a chum of the same sex to an adolescent need for intimacy with a friend of the opposite sex, one of a number of unfortunate consequences may occur: fantasies of a homosexual character, coupled with security operations to prevent their being discovered; conscious homosexual reveries associated with severe anxiety; a lifelong search for the "ideal" woman or man; or a homosexual way of life. Each of these is an attempt to solve the problem of possessing a need for lustful satisfaction with an accompanying inability to dare to seek intimacy with a person of the opposite sex.

It thus may be seen that the need for lustful satisfaction may be either an integrating or disintegrating dynamism depending upon the ability or the inability of the adolescent to achieve intimacy with a person of the opposite sex without the loss of security or self-esteem. The needs that drove the infant and child into relation with others are problems that have been more or less competently solved by this time. But the adolescent now finds himself driven by a tension of need for lust that requires a degree of intimacy that he has not experienced with a new, and as yet, strange person, who is in some ways comfortingly like himself and in other ways terrifyingly, and yet happily, different. If he is able to accomplish the integration between his need for lustful satisfaction and his need for security and his need for intimacy with a girl, he creatively organizes three aspects of himself. If he cannot accomplish the integration, he has no choice but to exist in a world of partial satisfaction—now of lust, now of security, now of intimacy, never of all three. The successful experiencing of this epoch after negotiating the other developmental eras makes for true adulthood. The person is now free to grow in interpersonal situations. His experience is syntaxic. He knows himself as he is. He can then effectively utilize whatever opportunities, personal or social, there are available to him. He is able to choose rather than having only one course. The orientation in living, begun hopefully during the juvenile stage, reaches its full development. This orientation is described by Sullivan in the following way:

> One is oriented in living to the extent to which one can formulate, or can be easily led to formulate (or has insight into), data of the following types: the integrating tendencies (needs) which customarily characterize one's interpersonal relations; the circumstances appropriate to their satisfaction and relatively anxiety-free discharge; and the more or less remote goals for the approximation of which one will forego intercurrent opportunities for satisfaction or the enhancement of one's prestige.[22]

Clinical Application

The interpersonal approach to the difficulties that arise from defects in the developmental history may usefully be applied clinically by dividing the contemporary personality into three components: the active, waking self, the part of the self not immediately accessible to awareness, and the period spent in sleep. To be studied in these states of the personality are the operations of the dynamisms, those methods of the self-system for decreasing anxiety. Among them are sublimation, obsession, selective inattention, hypochondria, algolagnia, the paranoid condition, the dynamism of "emotion" (fear, anger, rage, hatred, grief, guilt, pride, conceit, envy, and jealousy), dissociation, and the schizophrenic dynamism. These dynamisms are used to a greater or lesser degree by all people; the so-called normal person uses the entire repertory at one time or another. In mental disorder the person places extreme dependence on only one dynamism and, as a consequence, effaces a large part of the range of his personality.

The use of selective inattention and dissociation as dynamisms for avoiding anxiety has already been described. Selective inattention is troublesome only when it excludes from awareness what is relevant in an interpersonal situation. Sublimation is similar to Freud's concept; it is the unwitting substitution of a socially acceptable behavior pattern that partially satisfies the tension of a need for an unacceptable behavior that causes anxiety in collision with the self-system. It begins in infancy. A similarly common example is the use of the obsessional dynamism. It begins early in life (with the advent of language) and shows itself as a preoccupation with certain words, phrases or sentences that superficially seem to be communicative but that are used autistically. The words, phrases, or sentences are plucked from a childhood situation in which they were used to propitiate, placate, or mollify a parent or other authority figure. These verbalisms are a ritualistic performance whose purpose is to avoid other experiences that would produce anxiety. This magical performance allays anxiety with remarkable effectiveness. Sleep is usually undisturbed since there is conflict between needs and anxiety. The verbalisms or rituals substitute for anxiety. The use of the dynamism thus serves to maintain a shaky security in a frightening interpersonal field.

The theory of interpersonal relations was developed by Sullivan out of direct study and treatment of both hospitalized and ambulatory patients over a 30-year period. His contribution to the practice and theory of treatment is one of Sullivan's major achievements. It would not be possible, within the scope of this present chapter, to cover adequately these studies. The best sources are in his "Clinical Studies in Psychiatry"[22]; Arieti,[1] Will[25], Spiegel,[21] and Burnham[3] have all extended and modified this approach.

THE THERAPEUTIC SITUATION

In psychiatric and psychoanalytic thought it has long been customary to think of people as possessed of impulses, traits, and goals operating under the forces of constructs such as id, ego, and superego, but the fact is that human beings think, feel, and act as organisms responsive to an environment. We learn nothing about the human being except as we observe and experience him in an interpersonal field. If we conclude that a man is an "angry person," we are simply using an obscuring shorthand for what we have experienced; namely, that in a series of interpersonal operations, he has reacted with anger more frequently than we deem appropriate. If we have had the opportunity to experience another human being over a long period of time and in a sufficient variety of situations, we may attempt to characterize his personality on the basis of his predominant modes of interaction. But it is important to realize that we don't describe him from the point of view of the observer alone. The so-called observer, the psychoanalyst in the present example, is himself part of the interpersonal field, both as subject and object. His very presence in the field of the patient either in fact or in fantasy affects the so-called object of his observation. Since the observer, too, is human, he is affected by the communications of the patient. This conception of the potentially complex interaction between patient and therapist may be disquieting, for it certainly postulates a situation far more complex than the anonymous, unaffected therapist observing the patient, as supposed by classical psychoanalysis. But this view, at the same time that it renders the situation more complex, puts at the disposal of the therapist more of the relevant data, and it frees him from the inhibition imposed by classical theory of the use of his own emotional reactions in treatment.

If we visualize a consulting room containing but one patient and one psychoanalyst, from the point of view of field theory we see at once that it is rather crowded. For the patient the room is peopled with at least the following: the psychoanalyst as he is, fantastic personifications of people (the psychoanalyst distorted by the patient), potent representatives of once significant people, storybook people, and the institutions of the culture such as church and school to the extent that they are manifested in people significant to the patient. If the consulting room does not yet seem sufficiently crowded, we must remember that there are also representatives of the same listing for the analyst, though we would hope that the circumstances of his life and the remedial effect of his own psychoanalysis have made him aware of their impact on the treatment situation.

Sullivan's view that the psychoanalyst is a participant-observer, rather than the detached observer of classical psychoanalysis, permits him to become aware of data that might otherwise be unavailable to him. Let us suppose that the patient should state that he does not like to go out of doors because he does not like to have people staring at him. If the psychoanalyst is aware of himself as a participant in the present field of the patient, it will occur to him that the next, though perhaps unexpressed, thought of the patient is that the analyst is staring at him for much the same reason he believes others do. As a result the analyst will understand the patient's anxiety or anger and make a relevant comment. He will strive to understand the presuppositions held by the patient in seeking treatment, for these will distort the psychoanalyst into a special role, be it magic helper, kind father, or the like. He will be aware of his own emotional reactions to the patient's manner and communications, first because they may give him a clue about the patient's operations, and second because his feelings, whether expressed overtly or covertly, will necessarily cause shifts in the interpersonal field in which the patient is operating.

Even the process of the initial history taking, which Sullivan believed should be extensive, is part of the interpersonal experience for the patient. It is actually the onset of therapy. Sullivan recommended that the analyst be formal, frank, and direct and that he inform the patient what he knows about him from referring and collateral sources. This procedure may allay certain anxieties in the patient about what the psychiatrist already thinks of him, as he observes the psychiatrist relating what he already knows. It also serves as an opportunity to clarify any misinformation the psychiatrist has gathered. But most important in this stage of history taking is the clarification of the reasons the patient seeks a therapeutic collaboration at this moment in his life. The goal of the second stage of history taking, the reconnaissance, is to obtain a social sketch of the patient and his family. During this, data on his relationship with the psychiatrist will be noted (for example, the first parataxic distortion of the psychiatrist will probably be with an adult other than the parent), and tentative experiences with free association may be provided whenever the patient meets blind spots in his recollections. During the detailed inquiry the psychiatrist is particularly alert to those areas and epochs of life in which anxiety is or was a prominent feature, and he seeks to gain an understanding of the security operations that the patient has habitually used. In taking such a history the significant phenomena that arise in each epoch of life are used as the skeleton of the inquiry, and an attempt is made to determine the types of security operations used at each epoch. In infancy what was the security operation used when the need for tenderness was frustrated? During the epoch of childhood was there the ability to enter into cooperative play with others, and, if not, in what way did he cope with this inability? Are there memories or remnants of parataxic or autistic language? If socialization—as social subordination or social accommodation—was omitted during the juvenile era, what substituted for the deficiency? Did the patient have a chum of the same sex during preadolescence? How did he handle the need for intimacy, the need for lustful satisfaction, and the need for security with a person of the opposite sex during adolescence? And now in adulthood does he have self-respect and respect for others? Does he have freedom for personal initiative? Does he have sexual performance of an adequate kind? If not, what seems to get in the way? In addition to these, his toilet training, speech habits, eating habits, competitive relationships, ambition, and so forth, are investigated. From the data secured during such a survey the analyst attempts to make a statement clarifying the issues that seem to him to have brought the patient to treatment, and by this time these may have become quite other from the reasons initially noted by the patient himself. Some clarification may be suggested concerning the underlying anxiety, and it may be possible to note some of the security operations used by the patient. Finally the goals of the collaboration are outlined in general terms.

It follows from this survey of Sullivan's theory of personality (however incomplete) that the exploration of the patient's difficulties involves an inquiry into what went wrong at each era of his life. What went wrong at any era will influence each subsequent era. If the appropriate experience was lacking, the analyst attempts to determine the reason for its omission, and how the attendant anxiety was handled. The red thread that leads to an understanding of the patient's difficulties in living is the analyst's sensitivity to the presence of anxiety, either in its overt form or in one of its many guises. If the patient is pursuing a line of thought and suddenly shifts to another, the analyst presumes that the patient would have experienced anxiety had he pursued the original thought. He therefore notes two things: first, the moment of shift and the content of thought that was interrupted in order that the red thread of anxiety may again be taken up; second, the kind of security operation that was used to escape from the anxiety-laden path that was being pursued. He notes and follows the course of anxiety not because he is primarily interested in anxiety itself but because the presence of anxiety and the means used to avoid it lead to the possibility of discovering those areas of the self-system that have been lost to awareness through the operations of selective inattention or dissociation. For this is the essence of psychoanalysis, the rediscovery of the self-system in its entirety, to the end that with the bringing the "bad-me" and "not-me" into awareness, they may be seen as archaic and irrelevant organizations and may be exorcised as personal demons that no longer need exist or be the ultimate source of anxiety against which one adopts various elaborate security devices.

One of the great virtues of Sullivan's approach to therapy and theory is that it is operational and pragmatic. It avoids the dichotomy between theory and practice that is seen in systems that are constructs of a high order of abstraction and that must, at the same time, deal with human beings and their everyday lives. Its postulates are few, and its elasticity is great. It provides no final answers. But what it does with a disarming simplicity is to attempt to organize the myriad facts about the personality of the human being into a useful and malleable whole. There are weaknesses in the formulations in the areas of perception, cognition, and learning. These areas have been and are under investigation by the investigators listed above. The history of science shows that no theory could do more than organize these data and show the way for additional areas to be developed, and perhaps guide improvements that will eventually make the theory inadequate.

Erich Fromm

In the work of Erich Fromm we find the insights resulting from the fusion of the individual approach of psychoanalysis and the group perspective of the social sciences. His conceptualizations are rooted in history, political theory, and philosophy as well as in clinical observations. Perhaps the area in which Fromm has had his greatest impact upon psychoanalytic thought is in his critique of the impact our contemporary social fabric has upon the personality. Writing in the tradition of dissent, with psychoanalytic experience in three segments of Western culture, he ranges over the currents of contemporary culture with a persuasive precision. He combines the findings of psychoanalysis with a scientific analysis of society and culture.

Man's Hope and Man's Fate

The problems of man that go beyond his relation to his fellow men and to his culture constitute one of Fromm's primary interests. These problems arise because man is unique in the animal kingdom. He is an organism yet he is aware of himself. His fate is to be a creature of organic nature; his hope is that his self-awareness can lead him to transcend his passive "creatureliness." In approaching the problem Fromm states certain underlying hypotheses. Man is a creature who is relatively free of instinctual regulation in the process of adaptation to the surrounding world. He has new and unique qualities differentiating him from other animals: he is self-aware; he remembers the past; he visualizes the future; he is capable of symbolic activity. What we see as progress results from the very conflict of his existence, not from innate drive; for conflict creates problems that demand solution, and each solution contains contradictions that must be faced and solved at new levels. Man, an animal, is set apart from nature, yet he is a part of nature and subject to its physical laws. This split in man's nature leads to dichotomies—contradictions that he can react to but can never annul. His reaction to these existential dichotomies are dependent upon his character and culture. For Fromm the most fundamental dichotomy is that between life and death (in this instance physical, not psychological, death). All knowledge about death leads man to realize that death cannot be escaped and hence must be accepted. It is a meaningful part of life, but at the same time it is a defeat of life. As a result of man's mortality he can never achieve all his possibilities. Man is alone and aware of himself as a separate unique entity, and yet he needs to be related to his fellow man.

In addition to his existential dichotomies, man is faced with dichotomies of his own making. These are the historical dichtomies. For example, contemporary man has the technical means for vast material production, yet he is unable to use them exclusively for his own welfare. These dichtomies are potentially solvable through man's courage and wisdom (including self-knowledge); their contradictions call man to attempt their solution. It is this uncertainty that impels man to "unfold his powers." Only by facing the truth, by avoiding the siren voices of easy external solutions can he find meaning in his life through the unfolding of his powers. Reason, love, and productivity are faculties peculiar to man, and the development of these is necessary for a full life.

Man, the animal, is aware of himself. This very self-awareness in the body of an animal creates a tremendous sense of separateness and fright. He therefore looks for some unity and for some meaning. Responding with his mind, his feelings, and his actions, man passionately strives to give meaning to his existence. This need for orientation of himself in the world and devotion to his own existence has led him to devise answers varying widely in content and form: among these are totemism and animism, nontheistic systems of faith, philosophic systems, and monotheistic ones. Within this frame of reference are secular systems, involving, for example, strivings for success and prestige, conquest and submission, that can be explained on the basis of this need for orientation and devotion. This need is the most powerful source of energy in man. Neuroses and irrational strivings can be understood as the individual's attempt to fulfill his need for orientation and devotion. Fromm sees neurosis as a personal religion differing mainly from organized religion in its individual nonpatterned characteristics. The need for orientation and devotion is common to all men, but the particular contents of the individual solution vary with differences in value and character.

Individuation in the Social Context

While men are alike in that they share the human situation and the existential dichotomies, they are unique in their way of solving historical dichotomies. The attempt by the members of a society to solve their historical dichotomies determines the patterning of personality in that society. Within the framework of social patterning there is ample range for individual variation. Fromm stresses the importance of warmth and encouragement in early childhood in the developing of personality. He points out that the character structure of the parents will largely determine the manner in which the child experiences weaning, toilet training, and sex. The atmosphere, attitude, and overall climate of feeling, training, and indoctrination in these matters counts most in childhood for personality development, rather than the nature of routine used.

The child at birth makes no distinction between himself and the environment. It is months before the infant can separate himself from his surroundings. For Fromm the ties of the infant to the environment are primary, necessary, and enjoyable. As the infant's ability to manipulate his environment grows, his confidence and independence increase. These are fostered by the competence induced in education and training. If the process of individuation takes place at the same rate as the normal growth in competence and strength, there is no difficulty. Since there is often a lag in one or the other, feelings of isolation and of helplessness grow. With the anxiety attendant on this lag, the child, and later the adult, attempts to recapture the primary emotional ties. This process is one of the mechanisms by which the individual "escapes from freedom of the self." Man's fate and his hope is that he can never recapture the primary ties.

Along with others Fromm holds that the family is the psychic agency of society. Through its training the family makes the child want to do what he has to do. Through the agency of the family a core character structure is formed for each society or culture. The adult personality is a highly interwoven complex of inborn endowment, early experience in the family, and later experience in a social group. A constant interplay between society, the family, and the individual shapes character structure. While class mobility, for example, will not change the character structure of the parents, attitudinal changes will be communicated to the children. Fromm emphasizes the constant interplay. No factor is considered separately; the whole constellation—the parent-child situation, the child-child relationship, the child-school interplay, and the child-external authority situation later on—has to be understood.

Varieties of Relatedness

Man has to be related to things and to people in order to live. The man who has a recognition of his true self will have a productive orientation to living. The person who has enjoyed good early relationships will have respect and love for himself, will be able to cherish and love others, and will be able to use his capacities in fruitful work. This is the productive way of living.

Too often man's relationship to the outside world is not one of loving cooperation but rather one of symbiosis. He escapes from his feelings of isolation and loneliness by entering into relationships of reciprocal dependence. The slave needs the master; the master needs the slave. Each needs the other to avoid feelings of loneliness and isolation. Each places self-aggrandizement above people or ideals.

Major social changes may occur for a multiplicity of reasons quite remote from the character of the individual. For example, changing technological and economic forces will alter the situation with which the individual has to cope. As a result of this, his character will change. Capitalism, originating independently of psychological forces, fostered the rise of the Calvinist conscience; the growth of capitalism was made possible by the institutionalization of the psychological attitudes whose development it found necessary.

Fromm sees living as following two kinds of relatedness to the outside world—that of acquiring and assimilating things and that of relatedness to people. The orientation by which the person relates himself to the world is the core of his character. By permitting the individual to act consistently and reasonably, his character provides the basis for his adjustment to society. In discussing character types Fromm differentiates between productive and nonproductive orientation. Man, beset by the sense of separateness and fright engendered by his self-awareness, looks for a unity or meaning and does so by either progressing (productive) or regressing (nonproductive). The extreme of regression is for man to become a nonreflective animal and thereby be free of the problems of awareness and reason. This man cannot live creatively, cannot create. He, nonetheless, does not want to be, cannot be, completely passive and nonproductive. He wants to transcend life; he wants to make an imprint on the world. He can transcend his creature status by destroying and thereby triumph over life. This is another mechanism to escape from freedom of the self. There are three main ways of escaping from freedom—automaton conformity, destructiveness, and authoritarianism. There is no wholly productive or wholly nonproductive orientation in real life; there is some of each in everyone.

Varieties of Character Types

Receptive Orientation

The receptive orientation is typified by the person who feels that all that is good or necessary is outside of himself. He needs to be loved and yet cannot love. He makes no effort to gather information on his own; he looks for answers from others. He is looking for a "magic helper." God will help him, but he does not believe that this is due to his own works. He is optimistic and helpful but is anxious when the "source of supply" is threatened. He is dependent on authorities for knowledge and support. Alone and helpless in making decisions or taking responsibility, he relies on the people about whom he has to make the decisions. He dreams of being fed as synonymous with being loved. This type of character structure is most prevalent in societies so structured so that one group may exploit another. Feudalism and the institution of slavery are examples. In America the receptive orientation manifests itself in the need to conform, to please, to succeed without effort. Twentieth-century man is the eternal suckling for Fromm, taking in cigarettes, drinks, knowledge. This character type is clinically identical with the oral receptive type of Freud and Abraham. The difference in the concept of origin is that Fromm sees the attitude as being inculcated by the nature of the culture and the family (that is, one gets things passively), which then secondarily is applied to erogenous zones. In Freud the zonal need is primary.

The Hoarding Orientation

The hoarding orientation is based on the premise that security for the individual depends upon what he can save or own. He feels there is nothing new under the sun. He tries to possess rather than to love others. He is obsessively orderly and rigid. Fear of death and destruction mean more to him than life and growth. Intimacy is a threat. One gets things by hoarding. Fromm has shown how this character type was typical of the bourgeois economy of the eighteenth and nineteenth centuries.

The Exploitative Orientation

The basic premise for the exploitative type is the same as for the receptive type—the source of all good is external to the individual. He cannot produce. He takes what he can by cunning and thievery. He is pessimistic, suspicious, and angry. He "steals" his ideas from others. The robber baron of feudal times, the "adventure" capitalist of the eighteenth century, and the Nazi clique are examples of this orientation.

The Marketing Orientation

The marketing orientation is one of Fromm's original contributions to psychoanalytic characterology. This character type is typical of our time and could only come into being in a highly organized (bureaucratized) capitalistic society. The existence of a marketplace that places value and measure upon both people and things is the main condition for its development and perpetuation. In this type there is a loss of person-to-person valuation of the individual. Purely personal qualities have no value in themselves. They are valuable only as commodities to the extent that they are valuable to others. The goal is to sell one's self.

It is no longer sufficient to have integrity, ingenuity, skill, and knowledge. As there are fashions in material things, there are fashions in the desirability of personality types communicated by the mass media. In this time of social mobility and constantly broken ties between the individual and his neighbor, employer, or colleague, there is an increasing alienation of man from his own deeper feelings—an increasing superficiality accentuating man's isolation and loneliness.

The receptive and exploitative orientations enable man to relate in terms of himself through symbiosis. The hoarding orientation enables man to relate in terms of his own feelings by withdrawal or destructive means. The marketing orientation leaves man alone, alienated from his own feelings and from his fellow man. Fromm thinks this type of personality is more common in America because of its highly developed corporate structure; as capitalism becomes more developed through bureaucratization and the heightened influence of anonymous authority, subliminal and widespread, this type will become prevalent in both Eastern and Western Europe. This type of society requires millions to live with a serious defect without becoming ill. Man is an automaton under these circumstances.

THE NECROPHILIC ORIENTATION

This is typified by the person who is attracted to death, decay, illness, to all that is not alive, to the inorganic and not the organic. It is not to be confused with the sexual perversion with the same name. It derives from Unamuno's description of a general in the army of General Franco; this man had a favorite motto—"Long live death!" It may be regarded as a malignant form of the "hoarding orientation" or the "anal character" of Freud. Hitler is the best known example of this. During World War I he was found in a trancelike state gazing at the decayed corpse of a soldier, and he had to be dragged away from it. Hitler's end was really what he unconsciously wished. Consciously he wanted to save Germany; unconsciously he was working for its destruction. Behavior in a social context determines any type; needless cruelty and destruction of millions of persons are manifestations of the necrophilic type. Unfortunately there are a number of people like this in society. Man, isolated and helpless, wants to transcend himself. Searching for unity and meaning he overcomes automaton passivity by becoming destructive if he is blocked from becoming productive.

THE PRODUCTIVE ORIENTATION

The productive character is exemplified by the man who is able to use his own powers and to realize his own potentialities. He is free and not dependent on someone who controls his powers. He produces what he wants relatively independently of others. He is at one with his powers. All the expressions of his being are authentic; they are genuinely his and not put into him by an outside influence such as a newspaper. He is active in his work, in feeling, in thinking, and in relationship with people. Through love he can unite himself with the world and at the same time retain the separateness and integrity of his own self.

In actuality we always deal with blends of these orientations. A person of a receptive orientation will relate to one with an exploitative orientation; this is because they both need closeness to the other, in contrast to the distance from the other typical of the hoarding orientation. If the method of relating is totally nonproductive, the ability to accept, take, save, or exchange turns into the craving to receive, exploit, hoard—the dominant ways of acquisition. Loyalty, authority, and assertiveness in the productive character becomes submissiveness, domination, and withdrawal in the nonproductively oriented relationship. In other words, any orientation has a positive or negative aspect, dependent upon the degree of productiveness in the individual.

The Social Character

There is an intimate relationship between the nature of a society and the nature of character types that predominate in that society. A society requires a predominance of human beings whose character structures are consistent with its institutions. It finds this in what

Fromm calls the social character. The social character is the nucleus of character structure held in common by the members of a culture. It serves the function of molding and channeling human energy in a manner that facilitates the functioning of its society. But it is not as though man were a blank sheet of paper upon which society writes its text, for it is inherent in the nature of man that he strives for happiness, harmony, love, and freedom, meaning, and unity. Since such strivings are basic, universal, and extracultural, to speak of the socioeconomic circumstances as molding the character of man is to give only one side of the interaction. The other side of the interaction is the impact of these primary strivings of individual men upon the structure of the society and the direction in which it moves. Thus, the social character tends to be conservative and stabilizing, while that which is contributed by the impact of the individual tends to be catalytic and changing.

The inculcation of the social character into the character of each individual is begun in childhood, and, of course, the parents are the main agents utilized by the society for this task. Since the parents are hardly aware of their role and are simply doing what is natural for them, the social character may perhaps more accurately be said to be insinuated rather than inculcated into the individual. The person receives training in the social character in all of his interactions, including those associated with oral, anal, and phallic activity, since in early life these are the zones through which he makes frequent and immediate contact with his parents. Thus, in a social organization that requires precision, accuracy, and dependability in its human subjects for efficient utilization of its factories and institutions, the bowel habits taught will be those associated with regularity and dependablity, not because the parents are in a state of reaction formation against their own anal eroticism, but because the culture has taught them the rewards that stem from preciseness, accuracy, and dependability and the sanctions that stem from their absence.

Fromm vividly describes some of the disruptive effects upon the individual and his society of the great changes that have occurred in capitalistic organization and technology during the twentieth century. Science and our knowledge of nature have become more abstract and more distant from the experience of everyday human beings. The act and the object of the act are alien to the doer; now the act often has a life of its own. The man who kills a hundred or a thousand people through the pushing of a button cannot react emotionally to his act in so abstract a world, though the same man might experience feelings of disturbed consciousness were he to injure one helpless person. Man has become a thing as a result of being dependent upon powers outside of himself. He has become alienated from his state, from his work, from the things he consumes, from his fellow man, and from himself. Conformity, not variety, is the order of the day. It is not that he is in such great danger of becoming a creature of arrogant and raucous authority. From such hazards he is protected by the fact that their demands are overt. His greatest danger is his unknowing response to the autonomous, invisible authorities everywhere about him, so that he becomes a puppet without being aware of the strings that determine his every movement. Because large-scale production requires mass consumption, the individual has been taught to expect that every desire can be satisfied and no wish frustrated. Reason has been replaced by intelligence; ethics have been replaced by fairness; work is equated with toil rather than joy.

Fromm feels that by the use of knowledge gained in the fields of psychoanalysis, economics, sociology, politics, and ethics, men may become aware of the crippling effects of their society; this is one of the first steps on the road to social sanity. As a result man may be able to reassert himself so that he again occupies the central place in his own life. He will no longer be an instrument of economic aims; he will no longer be estranged from himself, his fellow man, or from nature. But such a change must occur in all areas, not simply in the political area, or the economic area, or in a spiritual rebirth alone, or in an alteration in sexual attitudes. Societal reform must get at the roots of the difficulty.

Basic to Fromm's approach to the ills of man and his society are two premises: the first is the concept that human beings have certain needs that are primary and undetermined by culture; the second is that it is possible to know these needs here and now. He believes that the great philosophical and religious leaders of the past—Moses, Jesus, Buddha, Lao-Tse, Ikhnaton, Socrates, and others—described what are more or less the same norms, though they had little knowledge of or influence over each other. He states he has observed the same strivings for peace, harmony, love, and solidarity in his patients as was described by the great thinkers of the past.

From his experience with patients, from his knowledge of the functioning of past and contemporary societies, from his study of religious and philosophical thinkers, Fromm has endeavored to define precisely what is the character of the mentally healthy person. This is the productive, nonalienated person. He relates to the world lovingly and uses his reason to grasp reality objectively. He experiences himself as a unique individual entity and at the same time feels at one with his fellow man. He is not subject to irrational authority but accepts willingly the rational authority of conscience and reason. He is in the process of being born as long as he is alive, and he considers the gift of life the most precious he has. The capacity for this is inherent in every man and will assert itself as long as socioeconomic and cultural conditions permit.

Treatment: Process and Goals

As yet Fromm has not written directly about treatment techniques and practices. However, he views the analytic process as a reciprocal relationship between the patient and the analyst. Within this context the analyst responds to the patient's communication with what he feels and hears, even if it is different from what the patient said or intended to say. His aim is to arrive at the patient's unconscious processes as they are going on in the patient at that time. As a humanist the analyst knows what the patient is experiencing from his own experience; he can thereby communicate this knowledge in an accepting, nonjudgmental way. Of course, the analysis of dreams, of transference, and of resistance phenomena plays an essential part in the therapeutic endeavor. Treatment is designed to increase the self-responsibility and self-activity of the patient; dependency is not fostered. There is a never-ending struggle for self-understanding that goes on after the formal analysis.

Fromm has drawn a sharp distinction between psychoanalysis where adjustment to one's culture is the aim of the cure and his humanistic psychoanalysis where realization of one's human potentialities and individuality, which must transcend one's particular culture, is the goal. For Fromm adjustment is a person's ability to be comfortably what the culture requires although he lose his unique individuality in the process. In the traditional approach no universal human norms are postulated, but a kind of social relativism that assumes that adjustment to the criteria of the extant culture is the appropriate goal of treatment. In Fromm's approach the truly creative, productive, and life-loving unconscious forces will be brought to full awareness. The goal, then, is the transformation of the personality from its culture-bound state to its full human range.

Writing with the insights of a psychoanalyst and with the perspectives of a social scientist, Erich Fromm has extended the dimensions of psychoanalysis. In terms of his clinical experience and observations of Western culture, Fromm has described the essential relationship between the character orientations held in common by the members of a society and the society as a social organization. He has shown how the social, economic, and political organization of prewar Germany was related to the growth of the authoritarian character, and has described the relationship between the development of the marketing character and contemporary capitalistic society.

He has postulated the existence of basic human needs that are valid for all men in all cultures, and in doing this he has substituted normative needs for the relativistic needs that had long been the hallmark of much social thought. Since our culture is deficient in its ability to satisfy these basic human needs, Fromm is an eloquent and active critic of our social values and organization. Given potentially loving and creative man, who is in constant danger of being deformed by a society deficient in sanity, Fromm contends that it is imperative that the psychoanalyst be concerned not with his patient's ability to adjust to his culture but rather with his necessity to transcend it. This requires an understanding of all unconscious processes so that the individual may become free. His socio-individual point of view has permitted Fromm to cast a fresh and vivid light on the questions of love, incest, motherhood, narcissism, character, and dream theory and practice. Erich Fromm has made such serious criticisms of our style of life that the discussion of the issues he has raised will prove to be as important to our survival and our future as any of our current concerns.

Bibliography

1. ARIETI, S., *Interpretation of Schizophrenia*, Brunner, New York, 1955.
2. ———, *The Intrapsychic Self*, Basic Books, New York, 1967.
3. BURNHAM, D. L., *et al.*, *Schizophrenia and the Need/Fear Dilemma*, University Free Press, Inc., New York, 1969.
4. COHEN, M. B., "Countertransference and Anxiety," *Psychiatry*, 15:231, 1952.
5. ———, *et al.*, "An Intensive Study of Twelve Cases of Manic-Depressive Psychosis," *Psychiatry*, 17:103, 1954.
6. FROMM, E., *The Art of Loving*, Harper, New York, 1956.
7. ———, *Beyond the Chains of Illusion*, Trident Press, Pocket Books, New York, 1962.
8. ———, *The Crisis of Psychoanalysis*, Holt, Rinehart & Winston, New York, 1970.
9. ———, *Escape from Freedom*, Rinehart, New York, 1941.
10. ———, *The Forgotten Language*, Rinehart, New York, 1951.
11. ———, *Man for Himself*, Rinehart, New York, 1947.
12. ———, *Psychoanalysis and Religion*, Yale University Press, New Haven, 1950.
13. ———, *The Sane Society*, Rinehart, New York, 1955.
14. FROMM-REICHMANN, F., *Psychoanalysis and Psychotherapy*, University of Chicago Press, Chicago, 1959.
15. MEAD, G. H., *Mind, Self and Society*, University of Chicago Press, Chicago, 1934.
16. MULLAHY, P. (Ed.), *The Contributions of Harry Stack Sullivan*, Hermitage Press, New York, 1952.
17. ———, *Oedipus Myth and Complex*, Hermitage Press, New York, 1948.
18. ——— (Ed.), *A Study of Interpersonal Relations*, Grove Press, New York, 1950.
19. RIOCH, J. M., "The Transference Phenomenon in Psychoanalytic Pyschiatry," *Psychiatry*, 6:147, 1943.
20. SAPIR, E., *Language: An Introduction to the Study of Speech*, Harcourt Brace Jovanovich, New York, 1971.
21. SPIEGEL, R., "Communication with Depressive Patients," *Contemp. Psychoanal.*, 2: 30, 1965.
22. SULLIVAN, H. S., *Collected Works of Harry Stack Sullivan*, 2 vols., Norton, New York, 1953.
23. ———, *Personal Psychopathology*, William Alanson White Psychiatric Foundation, Washington, D.C., 1965.
24. THOMPSON, C., "From and Sullivan," in Fromm-Reichmann, F., and Moreno, J. L. (Eds.), *Progress in Psychotherapy*, Grune & Stratton, New York, 1956.
25. ———, *Psychoanalysis*, Hermitage House, New York, 1950.
26. ———, "Transference and Character Analysis," in Mazer, M., and Witenberg, E. (Eds.), *An Outline of Psychoanalysis*, Modern Library, New York, 1955.
27. WILL, O., "Schizophrenia and the Psychotherapeutic Field," *Contemp. Psychoanal.*, 1:1–29, 110–135, 1964.
28. WITENBERG, E. (Ed.), *Interpersonal Explorations in Psychoanalysis: New Directions in Theory and Practice*, Basic Books, New York, 1973.
29. ———, and CALIGOR, L., "The Interpersonal Approach to Treatment with Particular Emphasis on the Obsessional" in Wolman, B. (Ed.), *Psychoanalytic Techniques*, Basic Books, New York, 1967.

B. The School of Karen Horney

Isidore Portnoy

Freud and Horney

For more than 15 years Karen Horney worked with, taught, and contributed to the traditional theories and techniques of Freudian psychoanalysis in which she had received her psychoanalytic education. Her first doubts about Freud's theories arose in response to his views on feminine psychology.[6,11] Her own experience and the influence of Erich Fromm and others led to a comprehensive evaluation of Freud's concepts.[11] In her view his basic and imperishable contributions, constituting the "groundwork of psychoanalytic theory and method,"[17] were his concepts of unconscious forces, psychic determinism, the role of inner conflict in a dynamic theory of personality, the meaningfulness of dreams, the role of anxiety in neurosis, and the crucial importance of childhood in human development. In the realm of therapy she valued most his emphasis on the necessity of bringing unconscious forces into consciousness in order to bring about basic change; the importance of recognizing and dealing with the patient's resistance; the recognition that, in spite of resistance, unconscious forces continue to operate and find expression in inadvertent symbolic behavior, dreams, free associations, and patterns of relating to the analyst. Although Horney differed with Freud on many aspects of these concepts, she remained convinced that essentially they constituted the "common base" uniting all schools of psychoanalysis. Horney rejected Freud's instinctivistic orientation, including the concepts of the libido and the death instinct; the repetition compulsion; the concepts of innate destructiveness[5] and dualities; his belief that the scientific attitude excludes moral valuation; and, most of all, his pessimistic philosophy in which ". . . man is doomed to dissatisfaction whichever way he turns. He cannot live out satisfactorily his primitive instinctual drives without wrecking himself and civilization. He cannot be happy alone or with others. He has but the alternative of suffering himself or making others suffer" (p. 377).[9]

Culture and Neurosis

Horney viewed man as a social being who can only develop his humanity in a cultural milieu. She saw neurosis as a particular form of human development "generated not only by incidental individual experiences, but also by the specific cultural conditions under which we live" (p. 8).[10] She noted the fallacy of each culture in assuming its own attitudes and values to be the standard of normality for all times and places. She recognized the infinite variety of ways in which cultures differ—in their definitions of health and illness, in the fears they generate and the defenses they provide against those fears, in their attitudes toward all aspects of human relating and striving.

Although there are gross inconsistencies in this regard, the increasing openness of our society, the emphasis on the uniqueness and worth of the individual, and the stress placed on self-expression and self-fulfillment are important positive values furthering self-realization. Among the neurotogenic aspects of our society she focused on three sets of factors. The first includes those conditions that foster neurotic development by creating feelings of helplessness, insecurity, potential hostile tension, and emotional isolation. She emphasized particularly the role of competitiveness, which begins in the economic field and brings "the germs of destructive rivalry, disparagement, suspicion, and begrudging envy into every human relationship" (p. 173).[11] Economic exploitation, inequality of rights and opportunities, and the overemphasis on success breed feelings of insecurity, impair initiative and self-confidence. A second set of factors relates to contradictions in our culture that are the anlage for the development of neurotic conflict. Thus our culture emphasizes the importance of success and winning in competition, while it places equal emphasis on brotherly love and humility. Needs and expectations are unceasingly stimulated, while many individuals are frustrated in the fulfillment of these needs. Freedom is greatly stressed in the face of the great number of existing restrictions. The third set of factors relates to those achievements and qualities of character that our culture rewards and that come to constitute elements in the neurotic individual's idealized image.

We are all the children of our culture. We benefit and suffer to various degrees from the growth-furthering and growth-obstructing aspects of our culture. The difference between the healthy and the neurotic individual is always one of degree. The difference between healthy and neurotic forces is, however, qualitative and basic. The neurotic individual is one in whom neurotic forces predominate and interfere seriously with his self-fulfillment, particularly in his relating to himself and others. In Horney's view he is one who has experienced injurious cultural influences "in an accentuated form, mostly through the medium of childhood experiences. We might call him a stepchild of our culture" (p. 290).[10]

¶ The Real Self and Self-Realization

As Horney evolved her own psychoanalytic concepts, her optimistic and humanistic philosophy came to occupy a central position. Her optimism was not predictive but an expression of her belief "in the inner dignity and freedom of man and in the constructiveness of the evolutionary forces inherent in man."[8] She viewed the real self as the dynamic core of the human personality, "the central inner force, common to all human beings and yet unique in each, which is the deep source of (healthy) growth" (p. 17),[9] the self "we refer to when we say that we want to find ourselves" (p. 158).[9] She saw the real self as the source of our capacities for experiencing and expressing our alive and spontaneous feelings; for evolving our own values and making choices based on them; for taking responsibility for our own actions and the consequences of them. It is the source of genuine integration, a natural process in which all aspects of the individual function harmoniously and without serious inner conflict, giving the individual a solid sense of his realness, his wholeness, his identity. The constructive forces of the real self unfold and develop in the process of healthy growing that Horney[9] called self-realization. This process does not essentially aim at the development of special abilities, but it involves, "in a central place," the development of good human relationships with oneself and with others. It is not a static goal, but a direction and a process.

In relation to others self-realization encompasses spontaneous moving toward others in giving and receiving affection; healthy altruism, discriminate trust, cooperation, empathy, and sympathy; the joy of meaningful involvement with others in play, in work, in sex, in the sharing of mutual interests and ideas; the democratic ideals of mutuality and respect for the individuality and rights of others. Healthy moving against others includes the ability and the freedom to oppose, to accept and enjoy healthy friction, to exercise rational authority, to meet challenges and threats to genuine convictions and interests. Healthy moving away from others includes natural striving toward autonomy, freedom, independence; the ability to stand alone, to accept one's separateness; to enjoy "the meaningful solitude"[13] that is essential to the process of living creatively.

In relation to oneself self-realization includes the striving to find, to own, to develop one's own identity; genuine self-interest; living in accord with values derived from what Horney[9] called "a morality of evolution" (p. 13), in terms of which we consider moral that which enhances our healthy being and growing, and we consider immoral that which obstructs such growing. It would include an individual's developing "for his well functioning, . . . both the vision of possibilities, the perspective of infinitude, *and* the realization of limitations, of necessities, of the concrete" (p. 35).[9]

In the area of work self-realization means the ability to commit one's attention and energies wholeheartedly to tasks one has undertaken either out of desire or genuine necessity. It includes the capacity for enjoying congenial and creative work and the varied pleasures of working alone and with others; the development of special talents; the wish to tap the deep wells of creativeness that exist in all, as evidenced in dreaming.

Finally the process of self-realization moves the individual into the realm of involvement in and concern with broader issues relating to the community and the world at large. Here man accepts his place in the world and takes responsibility for his role in creating his culture as well as being created by it.

The Neurotic Character Development

Basic Anxiety and the Search for Safety

In Horney's concept every neurosis is a character neurosis, a developmental process that begins in childhood, as an expression of the child's efforts to cope with a human environment that he experiences as inimical to him, and that evolves into increasingly disturbed relationships to others and to himself. The process of healthy growing in a young child can only take place if he feels loved and accepted as the person he actually is; if his individuality is respected and he is given the encouragement and guidance he needs to express and develop his own truly unique being. "If he can thus grow *with* others, in love and in friction, he will also grow in accordance with his real self" (p. 18),[9] on the basis of an inner core of security and self-confidence. Instead, the significant persons in his environment often relate to him predominantly in terms of their own compulsive, egocentric neurotic needs, are indifferent or hostile to his legitimate needs and wishes, offer "love" that is smothering and guilt inducing; "guidance" that is coercive and exploitative. Particularly damaging to the child are the parents' neurotic conflicts expressed in inconsistent treatment, alternating between indulgence, admiration, and idealization of the child, on the one hand, and unrealistic expectations, hostility, subtle or overt undermining and disparagement, on the other hand. It is always the impact of the whole human environment, and especially the balance of constructive and destructive influences in that environment, that is decisive. Where these influences are mainly injurious and threatening to the child, he develops feelings of uncertainty, hostility, and precariousness, a crucial triad of feelings of helplessness, isolation, and of being surrounded by a potentially hostile world. As these feelings pervade the child's whole experience and perception of the world around him, they come to constitute what Horney[10] called the basic anxiety—the nutritive soil in which the neurotic structure develops. Horney[1] distinguished the basic anxiety from the *Angst der Kreatur*, which is an expression in philosophic terms of man's awareness of his actual helplessness toward such inevitable processes as aging and death. While both involve an awareness of greater powers, only in basic anxiety are those powers experienced as hostile. The practical importance of this distinction lies in the fact that, since basic anxiety is not an inevitable aspect of the human condition, its significant reduction is one of the major goals of psychoanalytic therapy.

Basic anxiety compels the child to make efforts to allay his feelings of precariousness by inhibiting his true feelings and adopting strategic patterns of relating in a search for safety. At first these *ad hoc* strategies have some flexibility, the child being able to move back and forth between clinging, opposing, and withdrawing. Depending on the intensity of the basic anxiety, however, these strategies become increasingly rigid, indiscriminate, compulsive character trends, each fostering the growth of its own needs, qualities, sensitivities, inhibitions, and values. The natural movement toward others becomes a compulsive pattern of compliance. The natural movement against others becomes a compulsive pattern of aggression. The natural movement away from others becomes a compulsive pattern of detachment.[13] An invariable aspect of developing compulsiveness is that these drives, like all neurotic drives, are in all their essentials unconscious.

The particular neurotic pattern that the child adopts depends on his temperament and the contingencies of his human environment. *The compliant child* is one whose early years were spent "*under the shadow* of somebody" (p. 222),[9] in a family in which affection could be obtained from an adored or self-sacrificing mother, a benignly despotic father, a preferred or domineering sibling, but only at the price of submission and subordination of himself. This child "accepts" the helplessness in his basic anxiety and seeks safety by gaining the love, approval, and protection of the powerful persons in his environment. He develops increasing needs for physical closeness and intimacy, feels anxious and excluded when he is alone or when hostility arises in him or toward him. Hence he inhibits in himself all that could arouse hostility in others, including any critical, ambitious, or assertive tendencies. Sex may constitute an indispensable proof that he is needed and loved. He cultivates in himself qualities of lovableness and sensitiveness to the wishes of others, becomes appeasing, ingratiating, compulsively appreciative and considerate. He tends to take the blame readily and to judge himself by the opinions of others.

The aggressive child is one whose human environment was characterized by gross neglect, contemptuous, hypocritical, or brutal treatment, subtle or overt. Initially he may have attempted to gain security through affection and closeness, but when these efforts fail he feels increasingly rejected and humiliated, "accepts" the hostility of the world around him, turns his back on the whole area of affection, and strives for safety through opposing, excelling, dominating, and getting recognition. He cultivates in himself qualities of shrewdness, toughness, and resourcefulness, and he worships strength and will power. He abhors softness, fearfulness, and needing others, which he experiences as weakness, although he also needs others to provide the recognition and submission without which he feels anxious and lost. Sex may be important to him as a means of controlling and conquering.

The detached child is one whose human environment was characterized by "cramping influences" (p. 275)[9] that were either so subtle or so powerful that rebellion was not possible. The family atmosphere was one of tightness, with implicit or explicit demands on the child that he fit in at all costs. He was allowed little room for individuality and privacy. Such a child will often first go through periods of compliance and aggression, but eventually he "accepts" the emotional isolation in his basic anxiety and attempts to achieve safety and keep his inner life intact by emotionally withdrawing from others. His primary aims are to be independent, never to need anyone, to feel as little as possible for or against others. His fear of bondage is so powerful that he has to squelch feelings and wishes that might move him toward others. He can find outlets for his need to love in the realm of nature, art, and animals, as well as in intellectual pursuits where he can maintain his imperturbability and serenity. Sex may be a bridge to others, but he will cross it only if he can leave his feelings behind. If he feels that his secret inner world is being invaded, he reacts with anxiety and retreats to the ivory tower in which he can feel safe, unique, superior.

Basic Conflict and the Search for Inner Unity

What is crucial about the compulsive drives for safety is that, whatever the presenting defensive orientation, the other two continue to operate, less visibly but just as compulsively. The result is that the neurotic individual is driven by three diametrically opposed, irreconcilable patterns of relating to others of which he is essentially unaware and needs to remain unaware. This is the basic conflict, the dynamic core of the neurotic character structure.[13] Neither alternative in the basic conflict represents what the individual genuinely wishes. Nor can he renounce either of the alternatives because of their defensive function. The poignancy of all neurotic conflicts derives from the natural need for inner unity, the threat to such unity due to the incompatibility of the opposing forces, the underlying anxiety, and the increasing weakness in the individual's integrating powers resulting from the degree of alienation from the real self that has already taken place.

An individual threatened by basic conflict must devote his energies to defenses that will provide him with a sense of being unified.[13] These neurotic solutions, actually pseudosolutions, aim not at resolution of conflict but at getting rid of awareness of conflict. One solution is to make one of the basic moves—compliance, aggression, or detachment—predominant and to repress the others. Another solution is to attempt to keep conflict out of operation by moving away from the two more active forces, compliance and aggression, and reinforcing detachment. The individual may also attempt to experience one or other aspect of his inner conflict as occurring outside himself. By this radical process of externalization he no longer experiences the conflict as being within himself. It now appears in his consciousness as being a conflict between himself and his environment. In addition to these solutions and the solution of the idealized image, he resorts also to "auxiliary approaches to artificial harmony" (p. 131)[13] to bolster his defensive structure. These include blind spots, compartmentalization, rationalization, excessive self-control, arbitrary rightness, elusiveness, and cynicism.

It is axiomatic in Horney's thinking that all neurotic positions are maintained out of stringent inner necessity, come to have great subjective value for the individual, and have consequences that are seriously obstructive to healthy being and growing. The major consequences of unresolved conflict and of the whole neurotic development up to this point are a beginning alienation from self, a shift of his center of gravity to the outside, increasing vulnerability and shaky equilibrium, a great propensity for developing anxiety, and a feeling of hopelessness "with its deepest root in the despair of ever being wholehearted and undivided" (p. 183).[13] "He no longer knows where he stands or 'who' he is" (p. 21).[9] Lacking a feeling of self-confidence and a sense of his own identity is a particularly serious disability as the individual finds himself in the stormy period of adolescence. What he needs is a solution that will not only help him avoid the experience of conflict but will also provide a "substitute" for self-confidence and a sense of identity. At this point he turns for help to the uniquely human capacity of imagination and in so doing initiates a shift of the neurotic development to the intrapsychic, to the area of his relationship to himself.

Self-Idealization and Self-Actualization

The individual creates in his imagination an idealized image,[13] a fantasy of all that he unconsciously believes he is, could be, or ought to be, a flawless image of godlike perfection. The focus may be on saintliness, the absolute of goodness; on omnipotence, the absolute of power and invulnerability; or on omniscience, the absolute of knowledge. It consists of elements in part fictitious, in part actual qualities greatly exaggerated—weaknesses that have been distorted to appear glorious, inconsistencies that have been made to appear harmonious. The idealized image is a shining illusion, the first step in a process of self-idealization that has the most far-reaching significance for the person's life. In the next step he attempts to identify with the image as an idealized self. This idealized self comes to have for the neurotic individual more reality than his actual self. His center of gravity shifts back to himself, to his idealized self, not his real self.

Imagination does not suffice, and the individual now is driven to express, to prove his idealized self in his actual living. Together self-idealization and self-actualization constitute a *search for glory*. Self-actualization includes activity that derives from neurotic ambition, spurring the individual to achieve power and prestige; vindictiveness, driving the individual toward revenge for humiliations suffered and toward a feeling of triumph over others; and neurotic perfectionism.[9]

Neurotic Claims

Self-actualization inevitably involves the development in the individual of the feeling that he is *entitled* to have his neurotic needs fulfilled. In these irrational claims he unconsciously expects and demands that the world treat him as if he factually were his idealized self. Some of these claims, in neurotic as well as in psychotic individuals, are frankly grandiose. Others are in themselves reasonable as wishes and understandable as needs. It is the feeling of entitlement that makes them irrational. The individual justifies them consciously on various grounds—such cultural grounds as being a parent or a child; on grounds of actual or imagined special merit; on grounds of love, helplessness, suffering, "justice"—each supporting his magical expectations. Claims play an important role in the self-actualizing process, and any frustration of claims produces acute reactions of rage, righteous indignation, self-pity, and abused feelings. Less acute but more serious reactions to unfulfilled claims are a deep sense of envy, a feeling of being singled out by others and by life for deprivation, while others are invariably seen as more fortunate. In making claims on others and shifting responsibility to them, however, the individual in fact weakens his incentive and ability to assert and make efforts for his own legitimate rights.[9]

The Tyranny of the Should

The final means by which the individual attempts to actualize his idealized self, the one most decisive for his future development, consists of strenuous attempts to mold himself into a state of perfection. These attempts lead to the development of an inner dictatorship in which the individual is driven to fulfill, absolutely and immediately, all that, in terms of his idealized values, he should be, should feel, should think, should do. Disregarding feasibility, his general human and specific individual limitations, the efforts required, the arithmetic of reality, the individual "accepts" the premise that nothing should be impossible for him.

The particular "shoulds" depend on the specific direction of the neurotic development, but in all neurotic individuals the range of the inner dictates is seen to be far more vast than cultural dictates or the dictates of rigid political or religious systems. The focus is on flaws that must be erased and problems that must be removed through will power. Actually the individual's concern is increasingly with the *appearance* of perfection, not its actuality. This inevitably leads to the development of numerous pretenses constituting a façade of the qualities he idealizes. Where the shoulds are concerned with moral perfection, the result is actually to produce an immoral counterfeit of genuine moral concern. The individual is often caught in a particularly painful dilemma when his shoulds conflict—for example, when he feels he should fulfill all that others expect of him at the same time as he demands that he should be totally independent. Because of the enormously coercive character of the inner dictates, it is always necessary to externalize them, either actively (making his perfectionistic demands on others) or passively (experiencing others making the demands on him). In this way he avoids becoming conscious of the real nature and full impact of these dictates. In doing so, however, he has introduced another factor that further impairs his already disturbed human relationships by making him hypercritical of others or hypersensitive to their criticism. Since the demand is often made that superiority should be achieved effortlessly, he has great difficulty making the efforts that would in fact help him achieve more. The most serious consequence of the shoulds is the impairment of the individual's spontaneity. He is not free to feel his true feelings, but must unconsciously persuade himself that he feels what his shoulds dictate. In this way the inner dictates constitute the single greatest force producing alienation from the self.[9]

Neurotic Pride

To the degree that the neurotic individual succeeds in his self-idealizing and self-actualizing, satisfying what Maimonides called the "thirst for glory," he experiences neurotic pride. Where healthy self-esteem derives from what the individual factually is and does, neurotic pride is essentially false pride, since it is invested not only in imagined or exaggerated assets and achievements, but also in aspects of the neurotic character structure about which he has no choice and that are, indeed, destructive to him and to others. "The development of pride is the logical outcome, the climax, and consolidation initiated with the search for glory" (p. 109).[9] There is ample evidence in his fantasies and dreams and in all aspects of his living of the great subjective value that neurotic pride has for the individual. Its great failing, however, is its extreme vulnerability. It can easily be hurt by others, producing intense feelings of humiliation when they do not fulfill his claims; or it can be hurt by himself, producing intense feelings of guilt or shame when he does not fulfill his shoulds. The catastrophic effect of hurt pride may be experienced in intense anxiety, depression, rage, psychosomatic and acute psychotic episodes. Instead of these, what may show on the surface are evidences of narcotizing measures, for example, alcohol, drugs, compulsive eating, sexual promiscuity, and so forth. To avoid hurt pride becomes a major concern, and the individual is driven to avoid people, situations, thoughts, feelings, and activities that might result in hurt pride. He must also institute measures for restoring his pride when it has been injured. Revenge is the most effective and ubiquitous of these measures. By triumphing vindictively over others, turning the tables on them, hitting back harder, he not only gets revenge but also vindicates his pride.[18] He may restore his pride by withdrawing his interest from activities or persons, including himself, who have hurt his pride. His "decision" not to try rather than to risk failure may produce great restrictions in his life. He may restore pride by means of humor, denial, blaming others, reinforcing his claims on them, and refusing to take responsibility for defects in himself, attributing them to his neurosis or his "unconscious," as if these were not aspects of himself.[9]

Self-Hate

When the individual is confronted by his failure to fulfill inner dictates, he experiences self-hate. From the perspective of the idealized self to which his center of gravity has shifted, he is bound to hate and despise his actual self. To this hatred is added the vindictive rage of the proud self that impotently needs the actual person for attaining glory and, therefore, feels betrayed by him. No matter how strenuous his attempts to keep shortcomings from awareness, in his depths the neurotic individual does perceive them, and the intensity of his self-loathing is often dramatically expressed in dreams. Because of his increasing alienation from himself, he has little feeling of kinship or sympathy for himself with which to cope with the onslaught. Furthermore, the severity of his self-condemnation is itself invested with pride and thus serves to maintain self-glorification. Opposed as are neurotic pride and self-hate, the fact that both in different ways serve the interest of the search for glory led Horney to group them together as the major aspects of *the pride system.*

Self-hate may be expressed as relentless unconscious demands on oneself whose hostile threatening quality greatly increases the severity of the shoulds; merciless self-accusations in which the verdict of guilty arises from the failure to achieve godlike perfection, so that the neurotic individual spends so much of his life in a courtroom in futile self-defense; self-contempt, directed mainly against any striving for improvement or achievement. Self-contempt includes active drives to undermine self-confidence by belittling, disparaging, doubting, discrediting, and ridiculing oneself. It often appears as an unconscious sabotaging drive in the course of analysis. Self-hate may take the form of self-frustration in which the active intent is to frustrate any hopes, any strivings for pleasure. The feeling is that he does not deserve and has no right to anything better. Nowhere is the neurotic dilemma more poignant than in this picture of an individual whose claims express his feeling entitled to everything, while his self-hate denies him the right to anything. Self-hate may take the form of self-tormenting in a spirit of vindictive glee. This may be actively externalized in sadistic behavior to others or passively externalized in masochistic behavior. He is both the torturer and the tortured, his sadism and masochism being but two sides of one coin. Finally self-hate may be expressed in self-destructiveness, in which the unconscious intent is to bring about the person's physical, psychic, or spiritual destruction.

In this crescendo of terror the shift is increasingly toward active drives against the despised self. The real nature of self-hate is now clear. In his compulsive self-disparaging comparisons with others, for example, the neurotic individual does not loathe himself because he is inferior to others. The truth is that he feels inferior to others *because he hates himself.* Horney viewed this state of inner terror as man being at war with himself. For his own survival he needs to defend himself by externalization, by increasing his claims on others, by soaring in his imagination to greater heights of self-glorification in neurotic and psychotic distortion of his own reality, and by investing still more energy in the fulfillment of his shoulds. Horney viewed this great human tragedy as analogous to the devil's pact, expressed in religion, mythology, and the literature of many ages and many peoples (Goethe's *Faust*, Wilde's *The Picture of Dorian Gray*, Balzac's *The Magic Skin*). In each a human being in distress reaches for the absolute and godlike and in return gives up his soul and ends in hell. The hell in the devils' pact is the agony of self-hate. The loss of soul is the quieter and deeper despair of alienation from self. The search for glory has led the individual to become a hated stranger to himself (p. 375).[9]

Alienation from the Self

Every step in the neurotic development moves the individual away from the total actuality of himself and the deeper reality of his real self. The compulsiveness of his drives for safety and his solutions to basic conflict produce the first step in the alienating process. Externalization cuts the individual off from major aspects of his being. Every element in the search for glory leads him further from himself. The need to avoid hurt pride and to restore pride make him withdraw interest from himself. Above all, the shoulds foster his alienation in rigidly imposing on him a whole set of counterfeit feelings while making spontaneity taboo. Self-hate introduces active moves against the self into the alienating process. It should be noted that each step in the neurotic development not only increases alienation but also could not have been taken without alienation already being present.

Two forms of alienation were described by Horney. The first is alienation from the actual self. Symptomatically this may be expressed in feelings of unreality and depersonalization, hysterical phenomena, and the like. More chronic and serious is the "paucity of inner experiences"[14] present to different degrees in all neuroses. In severe degrees the person lives as if in a fog, cut off from the turbulent inner life that is not accessible to his consciousness but may appear in his dreams. Feelings of emptiness, nothingness, and boredom may be experienced directly or may be evidenced in the person's efforts to fill the emotional void with activity, food, and sex. The person may respond with anxiety and dread to this feeling of emptiness and nothingness (a quite different phenomenon from what some existentialists have seen as being part of the human condition), but he also dreads feelings of aliveness, since numbing has acquired the important function of keeping him from awareness of conflict, of unfulfilled shoulds, and of self-hate.

At the core of the alienating process is the second type of alienation, alienation from the real self. The effects of this process are seen in the impairment of spontaneous feelings, the lesser availability of energies for self-realization, the weakening of natural directive and integrating powers, the impairment of the capacity for choice and for taking responsibility for oneself. Of all the consequences of the neurotic development, alienation from the real self was viewed by Horney as the most serious.[9]

Intrapsychic Conflicts and Solutions

The pride system gives rise to two major intrapsychic conflicts. The first of these, central inner conflict, is the most comprehensive of all since it involves all of the forces of the pride system opposed to all of the constructive forces of the real self. However, since self-realization has been obstructed to a great degree, the opposing forces are not sufficiently equal in strength to bring this conflict into the open. This occurs during therapy and will be discussed under that heading.

The second major conflict is within the pride system itself, between all the forces driving the individual toward self-glorification and all the forces driving him toward self-extinction. It arises from the individual's efforts to identify *in toto* with both the idealized and the despised self. The individual relieves his inner tension by active alienating moves away from himself; by externalization of his inner experiences, which can become severe enough to produce a consistent pattern of externalized living; by psychic fragmentation; by belief in the supremacy of the mind, a type of fragmentation in which the mind, like a magic ruler for whom everything is possible, becomes an unrelated onlooker of oneself; and by automatic control that comes into play when the individual feels the neurotic structure to be on the verge of disintegration. Since the attempt to solve intrapsychic conflict through these measures does not provide sufficient pseudointegration, the individual resorts to the more radical and comprehensive solution of "streamlining" (p. 190)[9] or withdrawing from the inner battle. Horney viewed these solutions as a possible basis for a psychoanalytic typology, but preferred to consider them directions of development.

The first of these solutions is the *expansive solution* (p. 191),[9] in which the individual identifies with his proud self and attempts to keep from awareness the existence of his despised self. Three subdivisions of this solution were described by Horney. The first is narcissism, in which the individual unconsciously feels identified with his idealized image. To the extent that others fulfill his claims and expectations, he can for a time maintain an unquestioned belief in his own superiority and uniqueness. The second subdivision is perfectionism, in which the individual identifies with his superior standards, his shoulds. He must rely heavily on remaining unaware of unfulfilled shoulds. The third subdivision is arrogant vindictiveness, in which the individual identifies with his pride, arrogates to himself all powers and "rights," and denies them to others. Vindictiveness is the crucial motivating force in the sadistic trends that characterize this picture. It protects him against the hostility of others and, through externalization, defends him from his particularly harsh self-hate. His self-hate is directed toward his spontaneous as well as neurotic moves toward others and the self-effacing trends that he must keep from awareness. Vindictiveness is his main way of restoring hurt pride and is itself glorified by him. He uses it to intimidate others into fulfilling his claims. The excitement and thrill that he experiences in his pursuit of a vindictive triumph may make this the dominant passion of his life. In the interpersonal phase of the neurotic development, aggression was his main solution to basic conflict. His idealized image focuses on omnipotence, invulnerability, and inviolability. He is predominantly active in his externalizing. His reactions to frustrated claims are chiefly rage and militant rightness. The dominant principle in the expansive solution is the individual's drive for mastery of himself and of the world about him through will power and intellect.

The second major solution is *self-effacement* (p. 214).[9] Here the individual identifies with his despised self and attempts to blot out all aspects of his real self and of his idealized self. His solution to basic conflict was compliance. His idealized image focuses on goodness, kindness, unselfishness, and lovableness. He externalizes passively both his shoulds and his self-hate and is, therefore, extremely dependent on others for reassurance. Helplessness and suffering come to have indispensable functions for him. They serve as the basis for his claims, and they are the means by which he secretly expresses vindictiveness and attempts to control others. Suffering is unconsciously glorified, and he feels proud of his unique martyrdom. His helplessness and passivity make him as weak in defending himself against inner attack as against outer attack. He not only "accepts" his guilt but also wallows in it and resists efforts to lessen it, since it may be his only remaining defense against self-destruction. Belittling himself and idealizing others, he can live for others, and his striving for their love is the dominant motif in his life. He often finds his supreme fulfillment in erotic love. Loving affirms for him the qualities of his idealized self; being loved means to him redemption and purification, a reprieve from the verdict of guilt. In this morbid dependency (p. 239)[9] he is able to lose himself in ecstatic feelings, to surrender, to merge with the arrogant and vindictive partner whom he usually chooses and whose ruthlessness he secretly envies. Only such a partner can break his pride through degradation and enable him to surrender totally. In this union he secretly reunites with the pride that he has had to disown in himself and that he can now live out vicariously through his partner. In his sexual life this solution may produce a severe degree of sexual masochism.

The third solution to intrapsychic conflict is *neurotic resignation* (p. 259).[9] Here the individual attempts to solve his conflict by giving up his search for glory as well as his striving for self-realization. He attempts to become a nonparticipant, a detached onlooker in his own life, to achieve peace by severely restricting his needs and wishes. Whatever self-effacing and expansive trends remain in the picture are divested of their active elements, the drives for love and mastery. This individual solved his basic conflict through detachment. In his idealized image he glorifies self-sufficiency, independence, serenity. He externalizes passively and chafes under the restrictions and coercions that he experiences as coming from others. The dominant appeal in this solution is *freedom from* active striving, involvement, and commitment. He resists planning toward goals, and his resistance against effort and change seriously obstructs him in analysis. Neurotic resignation may be evident in a pattern of persistent resignation, in which increasing inertia involves thinking and feeling as well as acting; rebellious resignation, in which there is some liberation of energies; and shallow living, in which the severely alienated and resigned person moves to the periphery of life, loses his sense of essentials, and fills his life with ceaseless distractions, automaton behavior, and drifting. Even at this advanced point in the deteriorating process, the forces and feelings of the real self remain alive, but they are heard in dreams as distant voices and seen as pale shadows.

Conclusion

While Horney was concerned with the way in which the neurotic character development originated, she was more interested in clarifying the processes that perpetuated this development. In her view vicious circles provide the inner dynamism that furthers and perpetuates neurosis. Thus compulsive defenses against anxiety and solutions to conflict produce alienation and weakening of natural integration, requiring increasing reliance on the pseudointegration provided by neurotic solutions. The shoulds produce alienation and weakening of the directive power of the real self, requiring greater reliance on the shoulds for direction. The more severe the alienation, the more the need to cling to the sense of identity provided by the idealized self, the more severely alienated the individual then becomes. The more pride, the greater the self-hate, the more stringent is the need to reinforce and cling to the glorified self. The more he shifts responsibility to others in defending against self-hate, the weaker and more remote his real self becomes, the more vulnerable he is to the onslaught of self-hate. A final and particularly important vicious circle relates to the effect of the search for glory on the neurotic individual's human relationships. Pride makes him egocentric, makes him view others in terms of his externalizations and relate to others on the basis of his claims. Because of these distortions, the increased vulnerability, and the great proliferation of unconscious fears, the pride system essentially reinforces the basic anxiety that initiated the neurotic development.

Horney felt that "neuroses represent a peculiar kind of struggle for life under difficult conditions" (p. 11).[11] These conditions were seen to be at first external and interpersonal, and later intrapsychic, each eventually reinforcing the other. In her dynamic formulations both healthy and neurotic character structures are seen as processes in which forces operating in the here and now support and oppose each other in complex and changing patterns of interaction. What Horney said about morbid dependency applies equally to the whole neurotic structure. "We cannot hope to understand it as long as we are unreconciled to the complexities of human psychology and insist upon a simple formula to explain it all" (p. 258).[9] In Horney's holistic approach no aspect of the neurotic structure can be truly understood except in relation to all other aspects, to the whole person, and to the world around him. This viewpoint characterizes not only her work[28] but also the work of other members of her group.[20,21,23]

❰ The Therapeutic Process

Improving the effectiveness and possibilities of psychoanalytic therapy was always a central concern for Horney. Her first paper was on therapy,[16] and one of her major contributions was her study of the possibilities and limitations of self-analysis.[15] In her lectures on technique,[1–4,22,24–27,29,30] she emphasized basic therapeutic principles and avoided specific rules, believing that each analyst would naturally evolve his own patterns of being with and working with his patients. She was characteristically flexible about the use of the couch, the frequency of sessions, and so forth.

In the Horneyan approach the goal of analysis is to bring about a basic change in the direction and quality of a person's life, a shift of energies from self-actualization to self-realization.[25] "*The road of analytic therapy is the road to reorientation through self-knowledge*" (p. 341).[9] Therapy is viewed as a complex cooperative process of working toward increasing awareness, taking place in the matrix of a unique, evolving human relationship. Both insight and relationship are viewed as essential to helping the patient outgrow his neurosis.[7] The patient comes to treatment motivated usually by the wish to relieve suffering (mostly the suffering caused by unresolved conflict, hurt pride, and the frustration of neurotic needs) and by the unconscious expectation that the analyst will, through the magic of love, reason, or will, rid him of the painful consequences of his neurotic drives while reinforcing the drives themselves. Behind this may lie a deeper concern about his being blocked in many areas of his life, which may serve as a constructive incentive for analysis.

The patient's most important task in the analytic work is to express all of his experiences—feelings, thoughts, sensations, memories, fantasies—as freely and honestly as possible as they arise in the session. The capacity for such free association[24] is always limited at first, but evolves as the analysis progresses. All that the patient experiences is an expression of his present character and is potentially important for developing insight. Of special interest are his *dreams*, which Horney viewed as creative expressions of the individual's attempts to face his inner conflicts and solutions in a search for inner unity.[29] In dreams externals submerge, and he is closer to the truth of himself, which he expresses in poetic, symbolic form. This truth includes rejected and hidden aspects that he needs to deny in consciousness. It may involve his feelings about the analytic relationship that he cannot otherwise acknowledge. Perhaps most important of all, it is in dreams that the alive forces of the real self may still be visible in spite of severe psychopathology. *Childhood memories* play an important part in the analytic process. The patient may use them obstructively, attempting to establish simple cause-and-effect connections between childhood and the present. This effort constitutes a massive detour to avoid facing and taking responsibility for owning and changing what he is now. On the positive side childhood memories help him to understand the course of his development and to develop a feeling of connectedness with himself. This may be an area in which he can touch the alive core of himself, experiencing a poignancy and depth of feeling that he may for a long time be unable to experience in the symbols of the present. Getting a clearer view of the way in which his problems began can lessen his judgmental self-accusations and move him to self-acceptance and compassion. The patient's *patterns of relating to the analyst* eventually constitute the most important area for developing insight. The analytic relationship is not viewed as a repetition of the past. It reflects all that the patient is in the present, both healthy and neurotic. It is the most "here and now" experience during the analytic session, and whatever pertains to it, therefore, has the greatest impact, both for furthering or hindering the progress of the analyis.

The analyst understandably has the greater responsibility in the analytic work.[19] Measuring the relative strength of obstructive and constructive forces in the patient, he makes the decision as to whether analysis is advisable and feasible for the particular patient.[2] He must recognize from the start that, while his professional knowledge and psychological acumen often enable him to know a good deal *about* the patient rather quickly, he cannot really know this particular person without himself struggling toward deeper, more meaningful understanding.[22] Having himself experienced the analytic process and having moved substantially toward self-realization enable him to use his whole being as his primary tool in his work. His first function is to create for the patient an atmosphere of nonjudgmental acceptance. His aim is to be open to the patient, to all that the patient brings in verbally and nonverbally, and to what the patient omits. He observes, listens to, senses, thinks about, and feels not only what is going on in the patient but also what is going on in himself, including his own free associations, healthy responses, and neurotic reactions.[4] The analyst's activity is measured not as much by what he does or says as by the degree and quality of his being with the patient and himself in the sessions. It is this, together with his professional skills, that enables him to recognize themes emerging in the patient's associations. His nonverbal and verbal responses to the patient may take the form of exploratory questions that encourage and stimulate more associations, questions and observations concerning the patient's emotional responses, tentative uncovering and revealing interpretations, comments that express recognition of constructive efforts and offer encouragement and support, and, in the later stages of analysis, philosophic help.[27] It is particularly in this area that the analyst conducts the analysis, acts as a guide who is more familiar with the terrain, can see what is more available for opening up and penetrating, what the patient can tolerate, and what can best move the analysis forward.

Another major function of the analyst is to work with the patient toward removing blockages.[30] Horney preferred this term to resistance because it is more neutral and allows inclusion of the retarding effects of the analyst's participation. The same forces that obstruct the patient's development also serve as retarding forces in the analysis. While therapy aims at undermining neurotic positions, each position is essential to the stability and survival of the whole structure, and anxiety, manifest or covert, is the characteristic reaction when any aspect of the structure is threatened. The analyst respects the patient's defenses and understands their positive value, since the patient must defend what has become the basis for his existence until he has some new ground to stand on. His defenses operate against his experiencing many of his feelings, against his developing awareness (particularly of the compulsiveness and irrationality of his neurotic solutions), and against his changing. His defense of the status quo involves most urgently his need to ward off awareness of inner conflict and of self-hate, the latter also involving the need to deny the real nature of his pretenses and illusions. Blockages often appear in sessions in the form of the patient's ignoring, minimizing, discarding, pseudo-accepting, or attacking interpretations, intellectualizing, shifting to episodes in which he was the victim in the past, defensive self-recriminations, attacking the analyst and analysis, and so forth. Equally important in obstructing the progress of the analysis are the inevitable unconscious efforts the patient makes to fulfill his neurotic needs in the analytic relationship. To varying degrees he is bound to see the analyst in the light of his own pride and self-hate and of the distortions arising from his externalizing. The patient inevitably feels his defenses threatened, his pride hurt, his claims frustrated. He may also be perceiving and responding to actual mistreatment, to what the analyst's neurotic needs bring into the relationship and the work. The balance of healthy and neurotic motivation in the analyst is a decisive factor in the progress of the analysis.[1] His periodic efforts to evaluate change in the patient[26] and in himself help him to become aware of major blockages.

Finally the analyst has the important function of giving the patient human help. This does not mean that he interferes with the patient's developing autonomy by playing the role of adviser or surrogate parent. What it means is that the analyst gives the patient needed emotional support by his feeling with and for the patient, his striving to understand him, his recognition of the patient's possibilities for healthy growth, and his wish for the patient to become freer for living a creative and joyous life.

The Disillusioning Process

Horney called the first phase of the analytic work a disillusioning process since it deals mainly with the undermining of the neurotic individual's two powerful illusions—one maintaining that he is his superhuman idealized self, the other maintaining that he is his subhuman despised self. For Horney "working through" did not mean tracing a present problem to its infantile roots. In her orientation it means gaining ever deeper and broader consciousness of neurotic trends in all their aspects and dimensions. The patient begins with awareness of a symptom. As the analysis moves forward he is able to see some aspect of his neurotic character, at first briefly and vaguely. Returning again and again to the problem, his first "insight" is intellectual, partial, general. As defenses against feeling and awareness are worked on he begins to see and feel his needs operating in specific situations and in more and more areas of his life—in the past and present, in relation to others, to himself, and to the analyst, in the organic and sexual aspects of his being, in dreams, and in fantasies. He comes to feel the compulsive, irrational, insatiable quality of his drives, the grip they have on him, the functions they serve, the connections between them and his whole being, and the obstructive effects, particularly of the pride system, on his whole life. The last constitutes an appeal to his self-interest and can strengthen in him the incentive to work toward relinquishing pride. He begins to question the reality of his illusions and pretenses. Do claims really bring him what he wants or needs? Does deadening his feelings really make him less vulnerable? This process of reality testing goes hand in hand with a process of value testing in which he weighs more honestly the worth of what he gains from his neurotic drives. What, for example, has he gained by vindictively going to pieces and was it worth the price? Thus the logical outcome of the disillusioning process is a reorientation of values.

In addition to what he gains from his individual insights, deep and genuine self-knowledge leads to increasing self-acceptance. Feelings of helplessness and hopelessness diminish as the individual gains strength and hope from struggling to become more aware of himself, more connected with himself, more able to face himself squarely instead of denying and evading. As he is able to exercise some capacity for free and rational choice he *experiences* the difference between being an active force in his own life and being driven; between living affirmatively and being always on the defensive. It is liberating and exciting to feel the difference, for example, between the natural wish for sexual fulfillment and the addictive use of sex in the service of neurotic needs and solutions. As his alienation lessens and he becomes less proud and self-hating, he feels less vulnerable, less isolated, less hostile. He recognizes that his being trapped is not inevitable. Taking back externalizations enables him to see others in less distorted ways. This includes the analyst whom he can begin to see as an ally and friend in his further struggle. Through all this there is the incalculable value of sharing his inner being with another person who believes that "nothing human is alien" to him or his patient. In these ways the analytic work has helped to mobilize incentives for further growth.

Mobilizing the Constructive Forces of the Real Self

"*The therapeutic value of the disillusioning process lies in the possibility that with the weakening of the obstructive forces, the constructive forces of the real self have a chance to grow*" (p. 348).[9] The analyst has recognized the patient's constructive forces from the start. Wherever possible he has attempted to identify expressions of spontaneity and aliveness, of the patient's deep longing to return to himself, of his wondering about the reality and truth of himself, all of these appearing often in childhood memories and in dreams. Early in analysis direct interpretations concerning these forces may arouse self-hate. With the pride system weakened, however, the analyst can more directly encourage the patient to see how little he has consulted his own wishes, determined his own directions in life, taken responsibility for himself.

Resolution of Conflict

With the lessening of alienation and the mobilizing of constructive forces, the integrating powers of the real self become stronger and the patient is capable of coming to grips with his inner conflicts. He moves from occasional glimpsing of inconsistencies, contradictions, and patterns of ambivalence, to becoming aware of the deep rifts within him, to daring to stay with his conflicts as they appear in more areas of his being. Facing and staying with conflict is a painful but strengthening and liberating experience. It leads finally to an increasing relinquishing of the major neurotic solutions. What comes into focus now is the most comprehensive conflict of all—the central inner conflict—between the constructive forces moving the individual toward self-realization and the weakened but still powerful drives toward self-actualization. It is now the analyst's major task to keep this conflict in the foreground, to identify it with the patient, to provide the support needed in this turbulent period. The patient's constructive moves are followed by negative therapeutic reactions, repercussions in which there is an upsurge of self-hate, of anxiety, of neurotic defenses, and of narcotizing measures. It is essential that the analyst communicate to the patient his understanding that these are painful but encouraging evidences of growth. As the turmoil subsides the patient can enjoy the experience of becoming a real person. He is ready to become his own analyst and has acquired the incentives and tools for working toward his further evolution, a task that Horney called our "prime moral obligation" and "in a very real sense, (our) prime moral privilege" (p. 15).[9]

Bibliography

*1. AZORIN, L. A., "The Analyst's Personal Equation," *Am. J. Psychoanal., 17*:34, 1957.

*2. CANTOR, M. B., "The Initial Interview," *Am. J. Psychoanal., 17*:39, 1957; *17*:121, 1957.

*3. ———, "Mobilizing Constructive Forces," *Am. J. Psychoanal., 27*:188, 1967.

*4. ———, "The Quality of the Analyst's Attention," *Am. J. Psychoanal., 19*:28, 1959.

5. HORNEY, K., "Culture and Aggression," *Am. J. Psychoanal., 20*:130, 1960.

6. ———, *Feminine Psychology*, Norton, New York, 1967.

7. ———, "Human Nature Can Change: A Symposium," *Am. J. Psychoanal., 12*:67, 1952.

8. ———, "The Individual and Therapy," in "Psychoanalysis and the Constructive Forces in Man: A Symposium," *Am. J. Psychoanal., 11*:54, 1951.

9. ———, *Neurosis and Human Growth*, Norton, New York, 1950.

10. ———, *The Neurotic Personality of Our Time*, Norton, New York, 1937.

11. ———, *New Ways in Psychoanalysis*, Norton, New York, 1939.

12. ———, "On Feeling Abused," *Am. J. Psychoanal., 11*:5, 1951.

13. ———, *Our Inner Conflicts*, Norton, New York, 1945.

14. ———, "The Paucity of Inner Experiences," *Am. J. Psychoanal., 12*:3, 1952.

15. ———, *Self-Analysis*, Norton, New York, 1942.

16. ———, "The Technique of Psychoanalytic Therapy," *Am. J. Psychoanal., 28*:3, 1968.

17. ———, "Tenth Anniversary," *Am. J. Psychoanal., 11*:3, 1951.

18. ———, "The Value of Vindictiveness," *Am. J. Psychoanal., 8*:3, 1948.

19. ———, "What Does the Analyst Do?" in Horney, K. (Ed.), *Are You Considering Psychoanalysis?* Norton, New York, 1946.

20. ———, *et al., Advances in Psychoanalysis*, Norton, New York, 1964.

21. ———, *et al., New Perspectives in Psychoanalysis*, Norton, New York, 1965.

*22. METZGER, E. A., "Understanding the Patient as the Basis of All Technique," *Am. J. Psychoanal., 16*:26, 1956.

23. RUBINS, J. L., "Holistic (Horney) Psychoanalysis Today," *Am. J. Psychother., 21*:198, 1967.

*24. SHEINER, S., "Free Association," *Am. J. Psychoanal., 27*:200, 1967.

*25. SLATER, R., "Aims of Psychoanalytic Therapy," *Am. J. Psychoanal., 16*:24, 1956.

*26. ———, "Evaluation of Change," *Am. J. Psychoanal., 20*:3, 1960.

*27. ———, "Interpretations," *Am. J. Psychoanal., 16*:118, 1956.

28. WEISS, F. A., "Karen Horney. A Bibliography," *Am. J. Psychoanal., 14*:15, 1954.

*29. WILLIG, W., "Dreams," *Am. J. Psychoanal., 18*:127, 1958.

*30. ZIMMERMAN, J., "Blockages in Therapy," *Am. J. Psychoanal., 16*:112, 1956.

*Horney's lectures on psychoanalytic technique, listed under their editors and compilers.

C. The Cognitive-Volitional School

Silvano Arieti

The neo-Freudian schools headed by Sullivan, Fromm, and Horney have added dimensions to the original Freudian theory and have attempted to curtail the conceptions that have not passed the test of time. They have continued to ignore or minimize, however, some aspects of the psyche that may even be the most important at a human and social level of development. These are the aspects that enable man to be a symbolic animal and a center of will.

A stress on cognition and volition does not imply that affects are not major agents in human conflict and in conscious or unconscious motivation. It implies, however, that at a human level all feelings, except the most primitive, are consequent to *meaning* and *choice*. In their turn they generate new meanings and choices. Simple levels of physiopsychological organization, such as states of hunger, thirst, fatigue, need for sleep, and a certain degree of temperature, sexual urges, or relatively simple emotions, such as fear about one's physical survival, are powerful dynamic forces. They do not include, however, the motivational factors that are possible only at higher levels of cognitive development.

Although motivation can be understood as a striving toward pleasure or avoidance of unpleasure, gratification of the self or the self-image (and not necessarily of one's instincts or biological needs) becomes the main motivational factor at a conceptual level of development. Concepts like inner worth, personal significance, mental outlook, appraisals reflected from others, attitudes toward ideals, aspirations, capacity to receive and give acceptance, affection, and love are integral parts of the self and of the self-image, together with the emotions that accompany these concepts. To think that these emotional factors, which are sustained by complicated cognitive processes, are only displacements or rationalizations that cover more primitive instinctual drives is a reductionistic point of view.

Moreover, feelings which theoretically stand on their own, become involved with symbolic processes, which give them special meaning and involve them in intricate networks of motivation. A typical example is sexual life, which obviously cannot be considered only from a sensuous or instinctive point of view. Sexual gratification or deprivation become involved with such concepts as being accepted or rejected, desirable or undesirable, loved or unloved, lovable or unlovable, capable or incapable, potent or impotent, normal or abnormal. Thus sexual gratification or deprivation become phenomena that affect the whole self-image.

Conation (especially volition) has been almost completely ignored in psychoanalysis, and, in general, psychology has been studied predominantly from the point of view of external behavior. But with the exception of a relatively small number of automatic or involuntary movements, human external behavior is preceded by conscious or unconscious cognitive processes and mechanisms of choice.

Undoubtedly the previous schools of psychoanalysis have dealt with cognitive and conative phenomena since it is impossible to explore the human mind without doing so. However, they have dealt with them reluctantly and inadequately. The Freudian school interprets any symbolic phenomenon as a derivative or substitute of instinctual drives. The cultural and interpersonal schools, which are more closely related to the cognitive-volitional approach, have rightly stressed that one becomes a person mainly by virtue of relations with other human beings and not predominantly by virtue of inborn instinctual drives. They do not indicate, however, the symbolic and volitional mechanisms by which relations with other human beings take place. They also do not indicate that the sequence of external influences is integrated by intrapsychic mechanisms, so that it becomes personal history and part of the inner self. In other words, they do not show how the external influences and the intrapsychic mechanisms by which these external influences are integrated become that part of the human psyche that in the various terminologies has been called "inner life," "psychic reality,""intrapsychic self," and the like.

The combination of all the conditions affecting the individual (biological, interpersonal, environmental, and sociocultural) confers a unique status on each person. However, according too the cognitive-volitional approach, individuality is not exclusively the algebraic sum of all the factors or exclusively an emerging new form resulting from a chance constellation of all the previous or present conditions. There is also a margin of autonomy, which, although small in an overwhelmingly deterministic world, confers a special status on the human condition. This element has to be taken into account; as a matter of fact, the main aim of mental health is to facilitate the increase of this margin.

Whereas Sullivan adheres to an operational point of view, a cognitive-volitional approach does not. Operationalism, according to its strongest exponent, Bridgman,[31] advocates that "all concepts of which we cannot give an adequate account in terms of operations" no longer be used. Most of what originates in the inner life is thus debunked as nonoperational because it cannot be subjected to a set of operations or experiments that permit verification. Only what comes from the external environment can be operational, even if we embellish it with the word "interpersonal." A purely operational frame of reference remains behavioristic and is forced to deny or minimize the intrapsychic. It is doubtful that Sullivan, in his profound and insightful studies of schizophrenia, could confine himself to an operational level. Theoretically, however, the Sullivanian system is founded on conceptions deriving from English empiricism, American pragmatism, the teachings of the anthropologist Sapir, and of the social psychologist George Herbert Mead. On the other hand, the cognitive-volitional approach is related to conceptions deriving from Giambattista Vico, Immanuel Kant, Ernst Cassirer, and Susanne Langer. The cognitive-volitional approach retains the very important interpersonal dimension of the Sullivanian school, included, however, in the frame of reference of symbolism and choice derived from the just mentioned cultural influences.

Whereas some schools (behavioristic, behavior therapy, conditioned response, aversion therapy, and so forth) are interested in studying and altering man's behavior and capacity for adaptation, most psychoanalytic and psychotherapeutic schools are interested in studying and changing the inner self, even if it is more difficult to do so. The premise of psychoanalysis and psychodynamic psychotherapy is that if you change the inner self, sooner or later a change, and a more reliable one, will occur also in the external behavior and capacity for adaptation.

As Guntrip[35] has pointed out, historically psychoanalysis is the first science to illustrate the existence of a psychic reality that is distinct from the reality of the external world. It is one of Freud's great achievements to have demonstrated the existence of this psychic reality as an entity in its own right, an entity that is alterable, but at the same time highly resistant to change.

According to the cognitive-volitional school, inner life, or inner reality, may represent, substitute for, distort, enrich, or impoverish the reality of the external world. Although this inner reality, too, has constant exchanges with the environment, it has an enduring life of its own. It becomes the essence of the individual. Its organization is what we call the "inner self."

Inner reality is the result of a continuous re-elaboration of past and present experiences. Its development is never completed throughout the life of man, although its greatest rate of growth occurs in childhood and adolescence. It is based on the fact that perceptions, thoughts, feelings, actions, and other psychological functions do not cease completely to exist once the neuronal mechanisms that mediated their occurrence have taken place. Although they cannot be retained as they were experienced, their effects are retained as various components of the psyche. Freud wrote that in mental life nothing that has once been formed can perish.

From approximately the ninth month of life the child internalizes: he retains as inner objects mental representations of external objects, events, relations, and the feelings associated with these psychological events. Inner objects acquire a relative independence from the correspondent external stimuli that elicited them. They progressively associate and organize in higher constructs. The integration of all these intrapersonal and interpersonal factors gives origin to a potentially infinite psychological universe, the universe of symbols. Merely from the point of view of survival, symbolic function is not so important as other physiological functions, for instance, hypothalamic regulation; but from the specific point of view of human psychology and psychopathology, it is much more important.

Some psychoanalytic schools have tried, although not very successfully, to include in their theories the psychodynamic and psychostructural aspects of high symbolism. Melanie Klein,[43] for instance, recognizes that internalized object relations become permanent features of inner life. For her, however, these mental incorporations correspond to oral incorporation. She sees the formation of the psyche in a theoretical framework that retains Freud's oral, anal, and genital stages. She believes that these stages unfold much earlier in life than Freud had postulated, and therefore even more than Freud she is compelled to neglect cognitive forms that develop after the first year of life. Klein repeatedly refers to unconscious fantasies but does not indicate the cognitive features or the media that sustain these fantasies. It is difficult to visualize how in the three-month-old baby they can consist of ideas, thoughts, images, feelings of hopelessness, feelings of abandonment, and so forth. Although Klein has correctly stressed the importance of man's inner world, she has been quite nebulous in her description of the structure and functioning of this inner world. Fairbairn,[32] too, stressed the importance of the endopsychic structure and its relevance to object relations, but he did not examine the cognitive elements of this structure. The classic psychoanalytic school has studied internalization, but with a few recent exceptions it has studied them predominantly from the energetic or economic point of view.

Classic psychoanalysis has dealt with the highest levels of the psyche in its formulation of the ego and superego. The superego is seen as an institution that imposes renunciation by requesting a repression of instinctual life. This formulation does not require an inquiry into cognitive processes, which in Freudian theory are generally attributed to the ego. However, most ego psychologists follow the leadership of Hartmann, Kris, and Loewenstein,[37] and Rapaport[50] in considering these processes only as carriers of conflicts that originate elsewhere. As a matter of fact, these authors have called these cognitive functions autonomous and conflict-free areas and do not see them as sources of, or direct participants in, the conflict. For the cognitive school they are not conflict-free but active participants or generators of the conflict. Some may believe that the cognitive approach deals only with what in classic psychoanalysis pertains to the ego and superego and ignores the primitive or id psychology. This impression is incorrect. In the present frame of reference the Freudian id psychology is also seen as originated and mediated to a large extent by cognitive mechanisms. Contrary to the other neo-Freudian schools, the cognitive approach does not reject the concept of the primary process, which is seen predominantly as a cognitive system operating according to *primary cognition*. I personally believe that without such a concept or equivalent ones it is impossible to explain the structure and meaning of dreams, psychoses, or any kind of severe pathology. Some theorists (especially those belonging to the self-actualization schools, such as Fromm, Rogers, Horney, and Maslow) do not include in their system anything pertaining to the primary process. Consequently their contributions, although offering numerous important insights about some aspects of man and about mild psychopathology, are of much less value in understanding severe psychopathology and dream work.

It is important at this point to mention two important psychologists whose works in the field of cognition are illustrated in other chapters of this volume: Jean Piaget and Heinz Werner. Piaget's contributions are very significant, but they have not received adequate consideration, mainly because they are difficult to integrate with a psychodynamic study of man. They neglect affect as much as classic psychoanalytic studies neglect cognition, and do not deal with motivation, unconscious processes, and conflicts of forces. They are more illustrations of a process of cognitive maturation and adaptation to environmental reality than a representation of intrapsychic life. The cognitive functions, as described by Piaget, seem really autonomous and conflict-free, as the ego psychologists have classified them. In spite of this convergence of views, all attempts to absorb Piaget's contributions into classic psychoanalysis—including the attempt made by Odier[47]—have failed. The only contributions of Piaget that could be reconciled with classic psychoanalysis are those he made very early in his career when he was still under the influence of the psychoanalytic school.[48] For instance, his concept of the child's egocentrism is related to the psychoanalytic concept of the child's feeling of omnipotence.

Werner's contributions have been neglected more than Piaget's. And yet, perhaps even more than Piaget's, they are pertinent to psychiatric studies as they take into consideration pathological conditions.[56,57,58] Like the works of Piaget they do not make significant use of the concepts of the unconscious, primary process, and motivation.

Contributions of Other Authors

An increasing number of psychoanalysts have made important contributions within the cognitive frame of reference. For lack of space I shall mention only three of them.

Expanding on the work of psychoanalytic writers such as Bibring, Rapaport, and Arieti, Aaron Beck has formulated a theory of cognition, affect, and psychopathology. The core of his theory is that the conceptualization of situations determines the individual's emotional responses. In psychopathological syndromes these conceptualizations are frequently distorted and lead to the other symptoms that are characteristic of the disorders. Idiosyncratic cognitive schemas underlie the neurotic patient's systematic bias in interpreting and integrating his experiences.[26]

The individual's appraisal of an event, and specifically of its effect on his personal domain, determines his emotional response. Particular cognitions are chained to particular affects. For example, an appraisal of loss leads to sadness, gain to euphoria, threat to anxiety. Anger is produced when an individual attributes causality for an event he perceives as both unpleasant and unjustified to an external object. Thus, if he perceives a criticism as unwarranted, he will respond with anger toward the offender. If he accepts the criticism and focuses on his own loss of self-worth, he is likely to experience sadness.

In psychopathological syndromes the patient interprets his environment in an idiosyncratic way. His interpretation is dominated by cognitive schemata or patterns, acquired developmentally, that are activated under conditions of stress. The schema give rise to cognitions that appear "automatic" and plausible to the patient, yet violate rules of logic and reality testing and resist change. The affects typical of psychopathological syndromes are based on these distorted cognitions. A "snowballing" effect may result when the affect is, in turn, interpreted according to the faulty cognitive schema.

Psychopathological syndromes are each associated with a particular type of cognitive distortion. In depression the individual takes a negative view of his world, his future, and himself (the "cognitive triad"); his thinking centers around the theme of loss.[28] The manic tends to perceive life experiences in terms of an appreciation of his domain. Anxiety is related to themes of personal danger, the anger of the paranoid to ideas of persecution or abuse, hysteria to ideas of somatic dysfunction. Phobias are characterized by an expectation of harm in specific situations. In Beck's theory obsessions are seen as recurrent thoughts, compulsions as related self-commands.

Beck's cognitive therapy comprises a wide range of techniques aimed at altering the idiosyncratic cognitive schema and cognitive distortions that underlie the overt symptom or behavior problem. The cognitive therapist trains the patient to recognize the idiosyncratic cognitions that intervene between an event and his affective reaction. He tries to achieve "distancing," or a way of viewing events objectively, to prepare the patient to use rules of logic and reality testing, and to correct systematic errors such as arbitrary inference, overgeneralization, magnification, and cognitive deficiencies.[27] Feedback from experimental work (Loeb, Beck, and Diggory)[46] led Beck to introduce structured success experiences into the therapy of depressed patients.

Jospeh Barnett maintains that the cognitive element is central to the understanding of the dynamics of the obsessional neuroses and of character in general. He proposed that the obsessional neuroses were rooted in a basic cognitive fault, contrary to the assumptions of most theories that fears of affect or instinctual drives are central to the pathology. For Barnett[23] the basic problem of the obsessional is avoidance of knowing, or an abnormal need to maintain innocence. The ambiguity of the obsessional's early family experience creates severe damage to his self-esteem. His self-contempt dictates that he avoid knowing his impact on others or their impact on him in order to avoid the exposure he fears in the implications of his interactions. He "maintains innocence" with special cognitive processes described by Barnett. The obsessional's apparent emotional poverty is the result of "affective implosion," a mechanism in which affect is internally erupted in order to disintegrate inferential processes of thought that might organize those aspects of experience that would be threatening to self-esteem.

Later Barnett[21] redefined cognition as a broad form of experiential knowing that included both thought and affect as parts of a continuum rather than as opposing functions. Character was defined in relation to characteristic modes of organizing experience on this continuum and according to the way affect and thought were integrated. Character pathology was then related to those extremes of explosion and implosion of affects that interfered with knowing and meaning in characteristic ways. Obsessional, hysteric, paranoid, depressive, and impulsive character disorders were defined in relation to the cognitive dynamics.

According to Barnett,[22] the obsessional's need to maintain innocence leads to pervasive feelings of shame and fears of exposure. His aggression, therefore, is most often covert. It is aggression by omission rather than by commission. The peculiar split dependency of the obsessional, in which he has been expected to perform in impersonal areas of living, leads to his performance orientation, which increases competitiveness and interpersonal hostility.

The characteristics of obsessional sexuality were similarly seen to derive from these cognitive dynamics.[23] The primary symptom of obsessional sexuality is seen to be mechanization of sex. Competent and stereotyped, it is designed to avoid the exposure and interpersonal intimacy risked by spontaneity and freedom of expression. Shame and fears of exposure lead to a *secret life*, in which experiences involved in the patient's systems of innocence are kept encapsulated and isolated from his restricted cognitive functioning. The secret life exists in varying degrees of awareness and may be kept isolated in fantasy or acted out. Withholding and fears of exposure may lead to such symptoms as impotence, retarded orgasm, premature ejaculation, or compulsive genital activities. Most recently Barnett[25] has formulated therapeutic techniques in the treatment of the obsessional that center around the concept of cognitive repair.

Jules Bemporad[29] has also studied depression from a predominantly cognitive point of view; that is, as a result of a false view of one's abilities as well as a false view of other people's expectancies and reactions. It is a reaction on the part of specific individuals to their conceptions of certain situations: loss, abandonment, continued pain, and so forth. Depression is based more on a cognitive view of the situation than on metapsychological events. In therapy the analyst must help the patient to alter his views.

In what follows I will present my own cognitive-volitional approach, borrowing from my previous writings.

The Primordial or Presymbolic Self

The primordial self is a human psyche that has not yet reached the symbolic-volitional level of development. It includes primitive feelings, learning directly associated with perceptual stimuli, and behavior that is automatic or consequent to simple conditioning without intermediary complex mechanisms. Most of these functions start at birth or in the first few months of life and persist throughout the life of the individual.

Feeling is a characteristic unique to the animal kingdom and is the basis of psychological life. Feeling is unanalyzable in its essential subjective nature and defies any attempt toward a noncircular definition. Synonyms of "feeling" that are often used are "awareness," "subjectivity," "consciousness," "experience," "felt experience." Although each of these terms stresses a particular aspect, all refer to subjective experience.

Transmission of information from one part of the organism to another exists even without subjective experience. For instance, the important information transmitted through the spinocerebellar tracts never reaches the level of awareness. As long as information is transmitted without awareness, the organism is not too different from an electronic computer or a transmitter or transformer of data. When any change in the organism is accompanied by awareness, a new phenomenon emerges in the cosmos—experience. Awareness and experience introduced the factor psyche.

As Freud clarified, and as we shall consider later in this chapter, some psychological functions lose the quality of awareness and become unconscious. However, *if in phylogeny some functions had not become endowed with awareness, the psyche would not have emerged*, and the physiology of the nervous system would consist only of unconscious neurological functions.

The most primitive forms of felt experiences are simple sensations and perceptions such as pain, temperature, perception, hearing, vision, thirst, hunger, olfaction, and taste. These sensations and perceptions can be considered in two main ways: (1) as subjective experiences that occur in the presence of particular somatic states, for instance, a specific state of discomfort that we call pain; (2) as functions mirroring (or producing analogs of) aspects of reality.[5]

We encounter here a basic dichtomy. On

one hand is the awareness of a particular state of the body or part of the body; that is, the awareness of an inner status of the organism, the experience as experience. On the other hand is the function of mirroring reality, a function that generally expands into numerous ramifications that deal with cognition. The importance of these two components varies a great deal in the various types of perceptions. The experience of inner status is very important in the perception of pain, hunger, thirst, temperature, and less important in other perceptions such as touch, taste, smell. In auditory and visual perceptions the experience of a change of inner status plays a minimal role. These perceptions make the animal aware of what happens in the external world and become the foundation of cognition.

Elsewhere[7] I have described the various experiences of inner status (sensations, perceptions, physiosensations such as hunger, thirst, fatigue, sleepiness, sexual urges, other instinctual experiences) and how from all of them we can abstract feelings of pleasure and unpleasure. Motivation becomes connected with the awareness of what is pleasant (and to be searched for) and what is unpleasant (and to be avoided).*

I have also tried to show how not only the simple experiences of inner status, such as sensations, but also all emotions or affects can be included in the category of feeling. They are experienced within the organism. They are felt experiences. From all of them the motivational characteristics of pleasure and unpleasure can be abstracted. It is not without reason that in English the word "feeling" has a connotation so vast as to include simple sensations as well as high-level affects.

Emotions can be divided into several ranks or categories. The simplest (protoemotions or first order emotions), which can occur already in the primordial self, are of at least five types: (1) tension, a feeling of discomfort caused by different situations such as excessive stimulation, hindered physiological or instinctual response; (2) appetite, a feeling of expectancy that accompanies a tendency to move toward, contact, grab, or incorporate an almost immediately attainable goal; (3) fear, an unpleasant, subjective state that follows the perception of danger and elicits a readiness to flee; (4) rage, an emotion that follows the perception of a danger to be overcome by fight, not flight; (5) satisfaction, an emotional state resulting from gratification of physical needs and relief from other emotions.

In a general sense we can say that protoemotions (1) are experiences of inner status that cannot be sharply localized and that involve the whole or a large part of the organism; (2) either include a set of bodily changes, mostly muscular and humoral, or retain some bodily characteristics; (3) are elicited by the presence or absence of specific stimuli that are perceived by the subject as related in a positive or negative way to its safety and comfort; (4) become important motivational factors and to a large extent determine the type of external behavior of the subject; (5) in order to be experienced require a minimum of cognitive work. For instance, in fear or rage a stimulus must be promptly recognized as a sign of present or imminant danger.

What we have described under the headings of simple sensations, physiological and instinctual functions, perception, nonsymbolic learning, and protoemotions enable the animal organism (human or infrahuman) to survive and adjust to the environment. The effect of all these functions tends to be immediate or almost immediate. If they unchain a delayed reaction, the delay ranges only from a fraction of a second to a few minutes. Protoemotions are not experienced immediately, as are such simple sensations as pain or thirst. They require some cognitive work. However, this cognitive work is presymbolic or, in some cases, symbolic to a rudimentary degree. Presymbolic cognition includes perception and simple learning, which have been intensely investigated by experimental psychologists. It also includes sensorimotor intelligence, which has been accurately studied by Piaget in the first year and a half of life.

The motivational organization, based on the

* For the intricate relation between feeling, causality, and motivation see Chapter 2 of reference 7.

physiological, instinctual, or elementary emotional states we have mentioned, unchains in man, too, powerful dynamic forces, but it does not include all the psychological factors that affect him. Because of the impact of symbolic cognition, man's needs, desires, purposes, and conflicts go far beyond physiological-protoemotional motivation.

Most of the functions mentioned in this section do not lead to the formation of inner constructs unless associated with other psychological mechanisms. However, it would be inaccurate to state that they have no role in the making of inner reality. In the human being the memory and recall of these experiences become connected with higher level functions, especially after the acquisition of language. Feelings and protoemotions are also part of inner reality as long as they are experienced.

Protoemotions have even a greater role as potentialities in that they remain as affective or primary predispositions toward a given type of personality when they are not well balanced by other emotions. There are some human beings in whom fear (later changed into anxiety) is the predominant emotion. Depending on its interaction with other emotions and on the prevailing type of their interpersonal relations, these people eventually may become detached (that is, prone to withdraw from frightening stimuli), compliant (prone to placate the source of fear), or simply remain anxious. People in whom rage prevails tend to become aggressive and hostile. People in whom appetite is the principal emotion tend to become hedonistic. When tension predominates, the individual is likely to be hypochondriacal and more interested in his body than in the external environment. When satisfaction is the principal protoemotion, the person's predominant outlook is conservative, centered on the status quo.

In the sphere of volition the primordial self remains at a rudimentary level. The subject (animal or infant) can direct his movements in goal-directed forms of behavior; possibly he has already many internal representations of actions (exocepts or engrams). However, he does not have the ability to choose. His behavior is determined by instincts, conditioned reflexes, operant conditioning, or equivalent mechanisms.

After the first year of life the mechanisms of the primordial self remain the prevailing ones only in the seriously mental defective. The normal child very soon in life shows precursors of the volitional faculty in his playful activity and his capacity to imitate.[18] The inability to will makes the young child totally dependent on adults. It is not so in subhuman animals, although they remain at a presymbolic level even in their adult life. Comparative psychologists have reported that when animals are kept together in a certain environment, some assume a dominant and some a submissive role. Presumably this role is determined by a preponderance of a protoemotion and by the learning of a type of behavior that fits the environmental circumstances. Among these environmental circumstances the type of behavior of the other animals and consequent interplay are very important. The preponderance of a protoemotion and of some kinds of presymbolic learning and ways of relating constitute the temperament.

For expository reasons we have so far made no reference to evolvement in time and to interpersonal relations. Actually the presymbolic stage of childhood can be divided into many substages. In the first six to eight weeks of life the infant is predominantly a bundle of proprioceptive and interoceptive sensations. Then he becomes more and more involved with sensory perceptions and the simplest forms of learning. First, the presence of the mother as the overpowering environmental object and, second, the development of locomotion help the little child to shift the focus of his awareness from his body to his immediate environment. The predominant figure in his environment is that of the mother (or mother substitute), with whom he establishes a *primary bond*. By this term I mean an elementary, all-embracing social relation that is sustained not by symbolism but by immediately given sensory and affective components.

Symbolic Cognition and Volition

The development of new psychological functions—imagery and language—change enormously the child's relation with his environment. Although at this level of maturation the cortical centers of language are ready to begin to function, it is necessary for the child to be in contact with a human environment to transform his babbling and environmental sounds and noises into meanings. He learns to connect them with things, events, or special states of the organism. That is why children who are deaf or without human contacts cannot learn symbolic processes in normal ways. Consensual validation, that is, the recognition that a given sound has the same or a related effect on the mother and on himself, is an absolute necessity to trigger off verbal symbolization in the child. When the function of language is chiefly denotative, consensual validation is easy; this is "daddi" and this is "mamme." From the very beginning, however, language goes beyond its purely denotative functions. In the course of evolving toward maturity, language and other symbolic functions can be classified as pertaining to three categories of cognition: primary, secondary, and tertiary.

The designations primary and secondary derive from Freud's[34] original formulation of the primary and secondary processes in Chapter 7 of *The Interpretation of Dreams*. To quote Jones:[38]

> Freud's revolutionary contribution to psychology was not so much his demonstrating the existence of an unconscious, and perhaps not even his exploration of its content, as his proposition that there are two fundamentally different kinds of mental processes, which he termed primary and secondary. . . .

Freud gave the first description of these two processes, but tried to differentiate the particular laws or principles that rule the primary process only. He called the primary process primary because, according to him, it occurs earlier in the ontogenetic development and not because it is more important than the secondary process. Freud elucidated very well two mechanisms by which the primary process operates: namely, the mechanisms of displacement and of condensation. However, after this original breakthrough he did not make other significant discoveries in the field of cognition. This arrest of progress is to be attributed to several factors. First, Freud became particularly interested in the primary process as a carrier of unconscious motivation. Second, inasmuch as he interpreted motivation more and more in the function of the libido theory, the primary process came to be studied predominantly as a consumer of energy.

The Freudian school as a rule has continued to study the primary process almost exclusively from an economic point of view. Its main characteristic would be the fact that it does not bind the libido firmly but allows it to shift from one investment to another.[19] Some Freudians, however, for instance, Schur,[51] reassert the preponderantly cognitive role of the primary process.

I am also particularly concerned with the cognitive functions of the primary process; namely, what I call "primary cognition." Primary cognition prevails for a very short period of time early in life as a normal aspect of development. In most cases it is almost immediately overlapped by secondary cognition so that it is difficult to retrieve it in pure forms even in the young child. Primary cognition also prevails and is easier to detect in those mental mechanisms (1) that are classified in classic psychoanalysis as belonging to the id, such as dream work; (2) in the early stages of what Werner[57] called the "microgenetic process"; and (3) in psychopathological conditions. Its most typical forms occur in advanced stages of schizophrenia.[3,6,9,13,14,15]

Secondary cognition consists predominantly of conceptual thinking; most of the time it follows the laws of logic and inductive and deductive processes. Tertiary cognition occurs in the process of creativity and generally consists of specific combinations of primary and secondary forms of cognition. It is doubtful whether we can distinguish three categories of volition corresponding to the three types of cognition. Certainly there are mature and immature forms of making choices, as we shall see in the following sections.

The Primary Self

I call *primary self* a self that functions predominantly with the mechanism of primary cognition and certain forms of immature volition. The primary self is a theoretical construct because it can never be observed clinically in its entirety. We may reconstruct it from observations made in early life, imagery, memories, dreams, and especially from what we find in severe psychopathology, particularly in schizophrenia. A stage of development dominated by primary cognition never exists in pure form in normal conditions because the child who has outgrown the stage of the primordial self soon functions according to *secondary cognition* and not primary cognition. If the social bond with the mother and the other significant adults unfolds properly, the child accepts promptly from them secondary cognition, and his potentiality for individual, private, primary ways is to a large extent suppressed. Samples of these primary cognitions, however, exist in every normal child and even in adults. These samples multiply in psychopathological conditions, especially in schizophrenia, and in dreams.

The primary self develops the capacity to use symbols not only in order to communicate with other human beings but also to internalize the external world and thus to build an inner reality. This internalization may start toward the eighth or ninth month of life and may precede the onset of language. At first psychological internalization occurs through images.* An *image* is a memory trace that assumes the form of a representation. It is an internal quasi-reproduction of a perception that does not require the corresponding external stimulus in order to be evoked. Indeed, the image is one of the earliest and most important foundations of human symbolism, if by symbol we mean something that stands for something else that is not present. Whereas previous forms of cognition and learning permitted an understanding based on the immediately given or experienced, from now on cognition will rely also on what is absent and inferred. For instance, the child closes his eyes and visualizes his mother. She may not be present, but her image is with him; it stands for her. The image is obviously based on the memory traces of previous perceptions of the mother. The mother then acquires a psychic reality that is not tied to her physical presence.

Image formation is the basis for all higher mental processes; it starts in the second half of the first year of life. It introduces the child into that inner world that I have called "phantasmic."[7] It enables the child not only to re-evoke what is not present but also to retain an affective disposition for the absent object. For instance, the image of the mother may evoke the feelings that the child experiences toward her. The image thus becomes a substitute for the external object. It is actually an inner object, although it is not well organized. It is the most primitive of the inner objects if, because of their sensorimotor character, we exclude motor engrams from the category of inner objects. When the image's affective associations are pleasant, the evoking of the image reinforces the child's longing or appetite for the corresponding external object. The image thus has a motivational influence in leading the child to search out the actual object, which in its external reality is still more gratifying than the image. The opposite is true when the image's affective associations are unpleasant: the child is motivated not to exchange the unpleasant inner object for the corresponding external one, which is even more unpleasant.

Imagery soon constitutes the foundation of inner psychic reality. It helps the individual not only to understand the world better but also to create a surrogate for it. Moreover, whatever is known or experienced tends to become a part of the individual who knows and experiences. Thus *cognition can no longer be considered only a hierarchy of mechanisms, but also an enduring psychological content*

* See Chapter 5 of reference 7.

*that retains the power to affect its possessor, now and in the future.**

The child who has reached the level of imagery is now capable of experiencing not only such simple emotions as tension, fear, rage, and satisfaction, as he did in the first year of life, but also anxiety, anger, wish, perhaps in a rudimentary form even love and depression, and, finally, security. Anxiety is the emotional reaction to the expectation of danger, which is mediated through cognitive media. The danger is not immediate, nor is it always well defined. Its expectation is not the result of a simple perception or signal. At subsequent ages the danger is represented by complicated sets of cognitive constructs. At the age level that we are discussing now, it is sustained by images. It generally refers to a danger connected with the important people in the child's life, mother and father, who may punish or withdraw tenderness and affection. At this age anger is also rage sustained by images. Wish is also an emotional disposition that is evoked by the image of a pleasant object. The image motivates the individual to replace the image with the real object of satisfaction. Depression can be felt only at a rudimentary level at this stage, if by depression we mean an experience similar to the one that the depressed adult undergoes. At this level depression is an unpleasant feeling evoked by the image of the loss of the wished object and by the experience of displeasure caused by the absence of the wished object. Love at this stage remains rudimentary. For the important emotion, or emotional tonality, called after Sullivan's security, I must again refer the reader to another publication.[7]

The child does not remain for a long time at a level of integration characterized exclusively by sensorimotor behavior, images, simple interpersonal relations, and the simple emotions that we have mentioned. Higher levels impinge almost immediately so that it is impossible to observe the phantasmic level in pure culture. Nevertheless, we can recognize and abstract some of its general characteristics.

* For a study of the phenomenology of images and the formations of their derivatives—paleosymbols—see Chapter 5 of reference 7.

Images, of course, remain as psychological phenomena for the rest of the life of the individual. However, at a stage during which language does not exist or is very rudimentary, they play a very important role. Unless initiated, checked, or corrected by subsequent levels of integration (secondary process), they follow the rules of the primary process. They are fleeting, hazy, vague, shadowy, cannot be seen in their totality, and tend to equate the part with the whole. For instance, if the subject tries to visualize his kitchen, now he reproduces the breakfast table, now a wall of the room, now the stove. An individual arrested at the phantasmic level of development would have great difficulty in distinguishing images and dreams from external reality. He would have no language and could not tell himself or others, "This is an image, a dream, a fantasy; it does not correspond to external reality." He would tend to confuse psychic with external reality, almost as a normal person does when he dreams. Whatever was experienced would become true for him by virtue of its being experienced. Not only is consensual validation from other people impossible at this level, but even intrapsychic or reflexive validation cannot be achieved. The phantasmic level of young children is characterized by what Baldwin[20] called "adualism," or at least by difficult dualism: lack of the ability to distinguish between the two realities, that of the mind and that of the external world. This condition may correspond to what orthodox analysts, following Federn,[33] called "lack of ego boundary."

Another important aspect that the phantasmic level shares with the sensorimotor level of organization is the lack of appreciation of causality. The individual cannot ask himself why certain things occur. He either naïvely accepts them as just happenings or he expects things to take place in a certain succession, as a sort of habit rather than as a result of causality or of an order of nature. The only phenomenon remotely connected with causation is a subjective or experiential feeling of expectancy derived from the observation of repeated temporal associations.

The *endocept* is a mental construct representative of a level intermediary between the phantasmic and the verbal. At this level there is a primitive organization of memory traces, images, and motor engrams (or exocepts). This organization results in a construct that does not tend to reproduce reality as it appears in perceptions or images; it remains nonrepresentational. In a certain way the endocept transcends the image, but inasmuch as it is not representational, it is not easily recognizable. On the other hand, it is not an engram (or exocept) that leads to prompt action. Nor can it be transformed into a verbal expression; it remains at a preverbal level. Although it has an emotional component, most of the time it does not expand into a clearly felt emotion.

The endocept is not, of course, a concept. It cannot be shared. We may consider it a disposition to feel, to act, to think that occurs after simpler mental activity has been inhibited. The awareness of this construct is vague, uncertain, and partial. Relative to the image, the endocept involves considerable cognitive expansion; but this expansion occurs at the expense of the subjective awareness, which is decreased in intensity. The endocept is at times experienced as an "atmosphere," an intention, a holistic experience that cannot be divided into parts or words—something similar to what Freud called "oceanic feeling." At other times there is no sharp demarcation between endoceptual, subliminal experiences and some vague protoexperiences. On still other occasions strong but not verbalizable emotions accompany endocepts.

For the evidence of the existence of endocepts and for their importance in adult life, dreams, and creativity, the reader is referred elsewhere.* In children endocepts remain in the forms of vague memories that will affect subsequent periods of life. In adult life they often evoke memories expressed with mature language that was not available to the child when the experiences originally took place.

Endoceptual experiences exist even when the child has already learned some linguistic expressions—expressions, however, that are too simple to represent the complexities of these experiences. To avoid misinterpretations I wish to repeat at this point that the acquisition of language (that is, the verbal level) overlaps the endoceptual, phantasmic, and, to a small degree, toward the end of the first year of life, even the sensorimotor (or exoceptual) level.

Preconceptual Levels of Thinking

It is beyond the purpose of this chapter to study the child's acquisition of language (see Chapter 16 of this volume). It is also outside the scope of this chapter to study the experience of high-level emotions, which presuppose verbal symbols. I am referring to the mature experience of depression, hate, joy, and derivative emotions.† From the acquisition of language (naming things) to a logical organization of concepts, various substages follow one another so rapidly and overlap in so many multiple ways that it is very difficult to retrace and individualize them. These intermediary stages are more pronounced and more easily recognizable in pathological conditions.

Some of the stages, which some authors call "prelogical" and which I call "paleological" (or following ancient logic),[6,15] use a type of cognition that is irrational according to our usual logical standards. However, paleological thinking is not haphazard, but susceptible of being interpreted as following an organization or logic of its own. A considerable aspect of paleological thinking can be understood in accordance with Von Domarus's[54] principle, which (in a formulation I have slightly modified) states: Whereas in mature cognition or secondary cognition identity is accepted only on the basis of identical subjects, in paleological thinking identity is accepted on the basis of identical predicates. In other publications[2,7] I have illustrated the relations among part-perception, paleological thinking and some psychological mechanisms reported by ethologists, for instance, Tinbergen.[53]

* Reference 7, especially Chapter 6.

† See Chapter 7 of reference 7.

Paleological cognition occurs for a short period of time early in childhood, from the age of one to three. It is difficult to recognize because it is, in most instances, overlapped by secondary cognition. Here are a few examples. An 18-month-old child is shown pictures of different men. In each instance he says, "Daddy, daddy." It is not enough to interpret this verbal behavior of the child by stating that he is making a mistake or that his mistake is owing to lack of knowledge, inadequate experience of the world, or inadequate vocabulary. Obviously he makes what we consider a mistake; however, even in the making of the mistake he follows a mental process. From perceptual stimuli he proceeds to an act of individualization and recognition. Because the pictures show similarities with the perception of his daddy, he puts all these male representations into one category: they are all daddy or daddies. In other words, the child tends to make generalizations and classifications that are wrong according to a more mature type of thinking. Obviously there is in this instance what to the adult mind appears a confusion between similarity and identity. Children tend to give the role of an identifying or essential predicate to a secondary detail, attribute, part, or predicate. This part is the essential one to them either because of its conspicuous perceptual qualities or because of its association with previous very significant experiences. Levin[44] reported that a 25-month-old child was calling "wheel" anything that was made of white rubber, as, for example, the white rubber guard that was supplied with little boys' toilet seats to deflect urine. The child knew the meaning of the word "wheel" as applied, for example, to the wheel of a toy car. This child has many toy cars whose wheels, when made of rubber, were always of white rubber. It is obvious that an identification had occurred because of the same characteristic, namely, white rubber.*

Young children soon become aware of causality and repeatedly ask why. At first causality is teleological: events are believed to occur because they are willed or wanted by people or by anthropomorphized forces.

We should not conclude that young children must think paleologically; they only have a propensity to do so. Unless abnormal conditions (either environmental or biological) make difficult either the process of maturation or the process of becoming part of the adult world, this propensity is almost entirely and very rapidly overpowered by the adoption of secondary process cognition. Moreover, they may still deal more or less realistically with the environment when they follow the more primitive type of nonsymbolic learning, which permits a simple and immediate understanding. In secondary process cognition the individual learns to distinguish essential from nonessential predicates and develops more and more the tendency to identify subjects that are indissolubly tied to essential predicates.

At these preconceptual levels of cognition we can recognize two opposite tendencies: one, characterized by the propensity to proceed toward higher levels; the other, characterized by the propensity to return to lower forms, for instance, the image. In dreams and in schizophrenia the second tendency prevails: thinking is transformed into perceptions, as in hallucinations and oneiric images. Thus we have thoughts represented in personal symbols or what seem to be metaphors.[7]

At the stage of the primary self volition, too, occurs in primitive forms. Instinctive, automatic, and conditioned behavior persists; but we have also voluntary behavior that follows one of two patterns: wish fulfillment or obedience. In wish fulfillment the individual enacts the sequence of movements that he believes will lead him to gratify his wish. Flight is also included in this category because to escape danger is a wish. The obedient individual puts into effect whatever he senses the adult wants him to do. I have illustrated these mechanisms elsewhere,[18] and only a brief account can be repeated here.

Will starts with a "no," a "no" that the little

* This confusion between identity and similarity reacquires prominence in some psychopathological conditions. It has been studied intensely in schizophrenia by Von Domarus[54] and later by Arieti.[6,15]

child is not able to say but is able to enact upon his own body. The child enacts a "no" when he controls his sphincters in spite of the urge to defecate and to urinate. Now he has a choice: he may allow his organism to respond automatically in a primitive way or not. It is important to point out that although the child has developed neurologically the ability to inhibit these functions, he still has the urge to allow them to occur: it is easier and more comfortable to defecate when the rectum is distended by feces or to urinate when the bladder is full. But during toilet training he learns not to do so because he understands that another person (generally the mother) does not want him to. Thus in early childhood behavior acquires a new dimension when it becomes connected with the anticipation of how other people will respond to that behavior. Any activity ceases to be just a movement, a physiological function, or a pleasure-seeking mechanism on the part of the child; it acquires a social dimension and thereby becomes an *action*. The first enacted "no" is also the first enacted "yes"; "yes" to mother, "no" to oneself. Thus to will, in its earliest forms, means to will what other people want. As a matter of fact, the child learns to will an increasingly large number of actions that are wanted by his mother. In other words, he learns "to choose" as his mother and the other adults around him want him to choose. The child senses an attitude of imperativeness in the surrounding adults. The adult is experienced as the *imperative other*, a person who gives a command. The child senses that he has to obey in order to please or at least not to displease.

At first, the command is not internalized. The child obeys only if the adult is present. In the absence of the adult the wish or the primitive wish retains supremacy. When later the child introjects the command (that is, he obeys even when his mother is absent), another phenomenon occurs, not always but in several instances, that induces him to disobey even when his mother is present. He becomes *negativistic*. He says "no" to many suggestions of his mother. He does not want to eat, to be dressed, undressed, washed, and so forth. As I wrote elsewhere,[18] "We must not look upon this almost universal phenomenon as a period of naughtiness children go through or a difficult stage parents have to tolerate. All the 'no's' constitute a big 'no' to extreme compliance, to the urge to be extremely submissive to the others or to accept the environment immediately. They constitute a big 'yes' to the self. The submissive tendencies are always very strong because it is through them that the child becomes receptive to the environment and allows himself to experience influences that nourish him and make him grow. Negativism is a healthy correction. The no period is a decisive turn of the will. Control now no longer involves only body functions, like the inhibitions of reflexes or elementary physiological behavior. It is interference with behavior willed by others and the beginning of a liberation from the influence or suggestion of others. It is the first spark of that attitude that later on will lead the mature man to fight for his independence or to protect himself from the authoritarian forces that try to engulf him."

Early Images of the Mother and of the Self

The randomness of experience is more and more superseded by the gradual organization of inner constructs. These constructs continually exchange some of their components and increase in differentiation, rank, and order. A large number of them, however, retain the enduring mark of their individuality. Although in early childhood they consist largely of the cognitive forms that we have described (images, endocepts, paleological thoughts) and of their accompanying feelings (from sensations to emotions), they become more and more complicated and difficult to analyze. Some of them have powerful effects and have an intense life of their own; and yet at the stage of our knowledge we cannot give them an anatomical location or a neurophysiological interpretation. They may be considered the very inhabitants of inner reality. The two most important ones in the preschool age, and the only two that we shall describe here, are the image of the mother and the self-image.*

In normal circumstances the mother as an inner object will consist of a group of agreeable images: as the giver, the helper, the assuager of hunger, thirst, cold, loneliness, immobility, and any other discomfort. She becomes the prototype of the good inner object. The negative characteristics of the mother play a secondary role that loses significance in the context of the good inner object. In some pathological conditions the mother becomes a malevolent object, and an attempt is made to repress this object from consciousness.[10,12]

Much more difficult to describe in early childhood is the self-image. This construct will be easier to understand in later developmental stages. At the sensorimotor level the primordial self probably consists of a bundle of relatively simple relations between feelings, kinesthetic sensations, perceptions, motor activity, and a partial integration of these elements. At the phantasmic level the child raised in normal circumstances learns to experience himself not exclusively as a cluster of feelings and of self-initiated movements, but also as a body image and as an entity having many kinds of relations with other images, especially those of the parents. Inasmuch as the child cannot see his own face, his own visual image will be faceless—as, indeed, he will tend to see himself in dreams throughout his life. He wishes, however, to be in appearance, gestures, and actions like people toward whom he has a pleasant emotional attitude or by whom he feels protected and gratified. The wish tends to be experienced as reality, and he believes that he is or is about to become like the others or as powerful as the others. Because of the reality value of wishes and images, a feeling results that in psychoanalytic literature has been called a "feeling of omnipotence."

In the subsequent endoceptual and paleological stages the self-image will acquire many more elements. However, these elements will continue to be integrated so that the self-image will continue to be experienced as a unity, as an entity separate from the rest of the world. The psychological life of the child will no longer be limited to acting and experiencing, but will include also observing oneself and having an image of oneself.

In a large part of psychological and psychiatric literature a confusion exists between the concepts of self and of self-image. In this section we shall focus on the study of the self-image.† Also in a large part of the psychiatric literature the self and the consequent self-image are conceived predominantly in a passive role. For instance, Sullivan has indicated that the preconceptual and first conceptual appraisals of the self are determined by the relationships of the child with the significant adults. Sullivan[52] considers the self (and self-image) as consisting of reflected appraisals from the significant adults: the child would see himself and feel about himself as his parents, especially the mother, see him and feel about him. The self cannot be seen, however, merely as a passive reflection. The mechanism of the formation of the self cannot be compared to the function of a mirror. If we want to use the metaphor of the mirror, we must specify that we mean an activated mirror that adds to the reflected images its own distortions, especially those distortions that at an early age are caused by primary cognition. The child does not merely respond to the environment. He integrates experiences and transforms them into inner reality, into increasingly complicated structures. He is indeed in a position to make a contribution to the formation of his own self.

The self-image may be conceived as consisting of three parts: body image, self-identity, and self-esteem. The body image consists of

* We must warn the reader about a confusion that may result from the two different meanings given to the word "image" in psychological and psychiatric literature. The word "image" is often used, as we did in a previous section of this chapter, in reference to the simple sensory images that tend to reproduce perceptions. This word also refers to those much higher psychological constructs or inner objects that represent whatever is connected with a person; for instance, in this more elaborate sense, the image of the mother would mean a conglomeration of what the child feels and knows about her.

† The vaster concept of the self will be more accurately dealt with in a subsequent section of this chapter.

the internalized visual, kinesthetic, tactile, and other sensations and perceptions connected with one's body. The body is discovered by degrees, and also the actions of the body on the not-self are discovered by degrees. The body image eventually will be connected with belonging to one of the two genders. Self-identity, or personal identity or ego identity, depends on the discovery of oneself not only as continuous and as same but also as having certain definite characteristics and a role in the group to which the person belongs.

Self-esteem depends on the child's ability to do what he has the urge to do, but it is also connected with his capacity to avoid doing what the parents do not want him to do. Later it is connected also with his capacity to do what his parents want him to do. His behavior is explicitly or implicitly classified by the adults as bad or good. Self-identity and self-esteem seem thus to be related, as Sullivan has emphasized, to the evaluation that the child receives from the significant adults. However, again this self-evaluation is not an exact reproduction of the one made by the adults. The child is impressed more by the appraisals that hurt him the most or please him the most. These partial, salient appraisals and the ways they are integrated with other elements will make up the self-image.

Primitive Organization of Images and Constructs

In the primary self, images, such as the ones discussed in the previous section, and cognitive constructs do not follow the organization found in more mature stages of development. Often the image of a person is only built out of some salient features, those that impress or affect the child more, and does not include all features or even the most important ones. The child does not respond equally to all appraisals and roles attributed to him. Those elements that hurt him more and, in rare cases, that please him more stand out disproportionally. Thus the self, although related to the external appraisals, is not a reproduction of them. At times it may even be what appears to us a caricaturelike representation of these appraisals. Moreover, the self is constituted of all the defenses that are built to cope with the disturbing appraisals and their distortions. In schizophrenia we find that often the early self-image was a grotesque representation of the individual. Unfortunately, in turn, the self-image does change the individual and makes him become similar to this inner construct.

In normal conditions the image of the mother is made up of only the predominant, positive characteristics of the mother, and the negative ones are overlooked. However, when the mother has definite and very marked negative characteristics (anxiety, hostility, detachment, callousness), the child becomes particularly sensitized to these characteristics. He becomes aware only of them because they are the parts of the mother that hurt and to which he responds deeply. He ignores the others. He responds to parts, not to the whole. His use of primary process cognition results in his achieving only partial or distorted awareness. In schizophrenia the patient early in life often responded only to the pronounced negative parts of the mother, and the result was a transformation of the mother into a monstrous image.

Why do most people respond more to or become more aware of negative parts of the surrounding world, especially the interpersonal world? The answers to this question are only hypothetical. Inasmuch as the negative parts or characteristics hinder adaptation that may be dangerous from the point of view of survival, evolution has probably favored a stronger response to them. We react more vigorously to pain than to pleasure and to sorrow than to joy. Although we find this phenomenon in most people, part-perception becomes particularly pronounced in pathological cases, and especially in schizophrenia. Part-awareness is generally followed by other types of organization that I call protopathic or spontaneous.[8] In this type of organization any experience may add to a previous construct, not because of logical consistency, but either by contiguity, conditioned reflex mechanisms, repetition, part-perception, and so forth, or because an overpowerful emotion or motivation affects the mental process. For instance, if a person has been experienced by the individual as very frightening, any subsequent action of this person, even a benevolent one, may tend to be interpreted or experienced as frightening. Every relation with that person may thus accrue to the original negative image. Similarly an overpowering sexual or aggressive motivation may be reinforced by any stimulus whatsoever emanating from the source of that motivation.

These mechanisms of the primary self continue to exist even later in life when they could be replaced by the more mature ones of the secondary self. Thus a 19-year-old girl, reported elsewhere, misinterpreted or saw in a peculiar slant whatever involved the stepmother.[8] Quite often she interpreted paleologically whatever the stepmother said or did. In other words, the spontaneous organization of the input coming from the stepmother was predominantly organized according to the rules of primary process cognition. Her inner need to hate the stepmother was stronger than her respect for the demands of reality, and the patient succumbed to the seduction of the spontaneous organization.

Dispositions, attitudes, and communications of adults involving particular persons or situations quite often follow primitive constructs built early in life. At times these constructs are applied to entire categories, such as all elderly or maternal women, authorities, siblings, and so forth. We may state that the individual identified a person in his present life with one in the past in a transferential modality. Whatever terminology we follow, we must recognize that we deal with cognitive structures.

Another characteristic of primitive constructs is that they have an enduring or persistent quality even when they should be superseded by more appropriate constructs. What is organized early in life not only undergoes dissolution with great difficulty but also tends to inhibit the organization of more mature or subsequent constructs by which it could be replaced. This applies to normal constructs as well as to pathological ones. This is not to say that previous learning prevents later learning. For instance, we all know that the knowledge of arithmetic does not prevent the learning of algebra; as a matter of fact, it is a prerequisite. However, the learning of a second language later in life is in some respects inhibited by the knowledge of the first language. The dissolution of primary constructs, which have been built with preconceptual mechanisms and methods of spontaneous organization, constitutes a difficult task, also because it has to be actualized through the methods of the secondary self, which are often alien to primitive constructs.

It is easy to understand that primitive constructs about the self, parents, siblings, and other people have an important bearing on the affective attitude of the individual and on the realm and scope of his motivation. To the extent that the appraisal of life is effectuated by the individual through the mechanisms of the primary self, his consequent emotions and motivations appear distorted to a mature observer. To the extent that these primitive motivations (originated at the level of the primordial and of the primary self) contrast with those of the secondary self, they bring about conflict in the individual.

The Secondary Self

The secondary self is organized predominantly in accordance with secondary cognition. It is beyond the scope of this chapter to describe the stages intermediary between early childhood and mature adulthood. We shall consider only the role of concepts. As Vygotsky[55] has illustrated, conceptual thinking starts early in life, but it is in adolescence that it acquires prominence. Conceptual life is a necessary and very important part of mature life. Many authors[48,56,58,59] have made important studies of the mechanisms involved in the formation of concepts and of concepts as psychological forms. In this chapter I shall instead stress their content. This position is a departure from what I have done in reference to less mature forms of cognition.[3,6,9,15] In fact, in psychiatric studies, especially in such conditions as schizophrenia in which severe pathology is found, it is important to study not only content but also form; it is crucial to understand not only what the individual experiences but how he experiences it. Is he perceiving in terms of parts or wholes? Is he using images, endocepts, or paleological cognition? How are these cognitive modalities varying during the course of the illness or even during the course of a single therapeutic session? What is the meaning of such variety of forms? On the other hand, the psychiatrist's and analyst's main interest in concepts resides in determining how their content psychodynamically affects human life.

In a large part of psychiatric, psychoanalytic, and psychological literature concepts are considered static, purely intellectual entities, separate from human emotions and unimportant in psychodynamic studies. I cannot adhere to this point of view. Concepts and organized clusters of concepts become depositories of emotions and also originators of new emotions. They have a great deal to do with the conflicts of man, his achievements and his frustrations, his states of happiness or of despair, his anxiety or his security.[2] They become the repositories of intangible feelings and values. Not only does every concept have an emotional counterpart, but also concepts are necessary for high emotions. In the course of reaching adulthood emotional and conceptual processes become more and more intimately interconnected. It is impossible to separate the two. They form a circular process. The emotional accompaniment of a cognitive process becomes the propelling drive not only toward action but also toward further cognitive processes. Only emotions can stimulate man to overcome the hardship of some cognitive processes and lead to complicated symbolic, interpersonal, and abstract processes. On the other hand, only cognitive processes can extend indefinitely the realm of emotions. As I have illustrated elsewhere,[7] some very important human emotions could not exist without a conceptual foundation. For instance, depression should not be confused with lower feelings, which require no cognitive counterpart at all or only nonsymbolic learning. I am referring to the state of deprivation, discomfort, or anaclitic frustration of lower animal forms or human babies. Depression requires an understanding of the meaning of loss (actual or symbolic) and a state of despair (which follows a belief that what is lost cannot be retrieved). The importance of this understanding is not recognized because it is based on cognitive processes that often become almost immediately unconscious (see below). The conceptual presuppositions to mature love, to symbolic anxiety, to hate (as distinguished from rage or anger) have been described elsewhere.[7]

Reification of concepts (that is, the assumption that concepts faithfully correspond to external reality) is considered by science to be an invalid procedure. It is obvious that concepts do not correspond to external reality, nor do they represent reality adequately most of the time. Nevertheless, they do have an enduring psychological life or a reality of their own as psychological constructs.

Even more criticized and reputed unscientific is the reification of emotions or feelings in general. Certainly thoughts and feelings do not easily submit to the rigor and objectivity of a stimulus-response psychology. However, to dismiss all studies of thoughts and feelings as mystical is to dismiss most of man as mystical. Thoughts and feelings make up what is most valuable in man. If this essential part of man requires methods of study that do not correspond to those of standardized science, we must be ready to accept unusual methods of inquiry.

Even what I said about the relative lack of importance of concepts as forms needs clarification. Concepts, too, undergo organization of increasing order, rank, and level and become components or organized conceptual constructs, whose grammar and syntax we do not yet know. Undoubtedly future studies will reveal the structure of these so far obscure organizations and configurations.

From a psychiatric and psychoanalytic point of view, the greatest importance of concepts resides in the fact that to a large extent they come to constitute the self-image.[2] When this development occurs the previous self-images are not completely obliterated. They remain throughout the life of the individual in the form of minor components of the adult self-image or as repressed or suppressed forms. In adolescence, however, concepts accrue to constitute the major part of the self-image. As we already mentioned at the beginning of this chapter, worth, personal significance, mental outlook, more mature evaluations of appraisals reflected from others, attitudes toward ideals, aspirations, capacity to receive and give acceptance, affection, and love are integral parts of the self and of the self-image, together with the emotions that accompany these concepts. These concepts and emotions that constitute the self are generally not consistent with one another, in spite of a prolonged attempt made by the individual to organize them logically.

To complicate the picture is the fact that many conceptual forms are fused or mixed with more primitive ones (paleological thoughts, endocepts, spontaneous organizations, and so forth) so that often the apparently logical conceptual overlay hides the irrational inner structure. One of the aims of therapy is to distinguish and disentangle the irrational from the rational.

The motivation of the human being varies according to the various levels of development. When higher levels emerge, motivations originated at lower levels do not cease to exist. At a very elementary sensorimotor level the motivation consists of obtaining immediate pleasure and avoiding immediate displeasure by gratification of bodily needs. When imagery emerges, either phylogenetically or ontogenetically, the individual becomes capable of wishing something that is not present and is motivated toward the fulfillment of his wishes. He will continue to be wish-motivated in more advanced stages of primary cognition, such as the paleological stage. As we have already mentioned, although the motivation can always be understood as a search for, or as an attempt to retain, pleasure and avoid unpleasure, gratification of the self becomes the main motivational factor at a conceptual level of development. Certainly the individual is concerned with danger throughout his life: immediate danger, which elicits fear, and a more distant or symbolic danger, which elicits anxiety. However, whereas at earlier levels of development this danger is experienced as a threat to the physical self, at higher levels it is many times experienced as a threat to an acceptable image of the self. In some instances sexual deprivation is acceptable at an instinctual level but not at a conceptual level, where it becomes connected with many negative meanings and unpleasant emotions. Many psychological defenses are devices to protect the self or the self-image. Here are a few examples. A woman leads a promiscuous life; she feels she is unacceptable as a person, but as a sexual partner she feels appreciated. The hypochondriac protects his self by blaming only his body for his difficulties. The suspicious person and the paranoid attribute to others shortcomings or intentions that they themselves have. These examples could easily be multiplied. They represent cognitive configurations that lead the patient to feelings, ideas, and strategic forms of behavior that make the self-image acceptable or at least less unacceptable. Neurotic behavior is to a large extent based on these particular defensive cognitive configurations, which often become unconscious and are applied automatically. Often the patient has learned to apply these configurations in situations in which they were appropriate. Later, because of his lack of security or ability to discriminate, he has extended their sphere of applicability. In most cases important cognitive configurations are completely repressed because ungratifying or inconsistent with one's cherished self-image. *Contrary to what is often believed, repression and suppression from awareness apply not only to primitive strivings and to the contents of primary cognition, but also to the content of high conceptual ideation.*

In some psychoneuroses, such as phobic conditions and obsessive-compulsive syndromes, the self is also, in specific situations, protected by some use of primary cognition. In the schizophrenic psychosis the self is defended by an extensive use of primary cognition, which is not corrected or counterbalanced by secondary cognition. Various forms of displacement and transformation of a cognitive construct into another that is less disturbing (such as in regression, fixation, paleological thinking) occur in dreams and in many pathological conditions.

As described elsewhere,[7] no anticipation of the future is possible without symbolic processes. In order to feed his present self-esteem and maintain an adequate self-image, the young individual has, so to speak, to borrow from his expectations and hopes for the future. It is when a present vacillating self-esteem cannot be supported by hope and faith in the future that severe psychopathological conditions may develop.[10,12]

Learning has a very important role in the organization of inner reality. However, we must specify that the content of a learning experience becomes part of inner reality when it becomes integrated with the rest of the psyche and has an effect that transcends the original experience. Undoubtedly there are intermediary stages and special relations between learning experiences and components of inner reality. I have illustrated some of them elsewhere.[16]

My views on the role of primary and secondary cognition in the symptomatology, psychodynamics, and psychotherapy of schizophrenia and psychotic depression are reported in my chapters appearing in Volume 3 of this *Handbook*.

General Aspects of the Self

From a general point of view the self (and in particular, the inner self) can be examined from five different aspects: (1) representational function, (2) subjectivity, (3) potentiality, (4) integration, and (5) desubjectivization. These aspects are so interrelated that we cannot understand any of them without taking into consideration the others. For didactical reasons only we shall discuss them separately.

The representational function is the formation of inner constructs that represent objects or events of the external world. These representations are not analogical reproductions of what they stand for.

The second aspect of the self refers to the fact that what is objective or objectivizable becomes subjective, is appropriated by the individual as a subjective experience, acquires a subjective reality, and becomes part of the individual himself. This subjectivization is not adequately accounted for by psychological or psychiatric authors who give exclusive or almost exclusive importance to the environment. This subjectivization is a phenomenon as difficult to understand as the whole mind-body problem. The intensity of a subjective construct does not correspond to the external objective event or stimulus to which it is related. It depends, as we have already mentioned, on the mechanisms triggered off in its formation, on the selections of other constructs with which it is integrated, and on the processes used in such integration. The subjective aspect is particularly evident when we study sensations and emotions, but emotions accompany and transform cognitive constructs and, in turn, are transformed by them.

The potentiality of the self can be seen (1) in the way it affects the behavior of the individual in relation to other people, himself, and the world in general; (2) in the way it affects itself. In other words, the self feeds on the external world as well as on itself. Potentiality is also connected with the very important function of will or volition, to which we shall devote a subsequent section of this chapter.

The term "integration" refers to a large number of psychological mechanisms by virtue of which all functions of the psyche are related to one another and synthesized into higher ranks of organization. We have outlined a few of them, but most of them are still unknown.

Desubjectivization is the fifth property and in a certain way the opposite of subjectivity. It refers also to a group of phenomena that tend to decrease the private or subjective aspects of the experience. We have seen that in subhuman animals and during the first year and a half of human life the inner self exists only in rudimentary form. In the growing and grown human being, however, it becomes so intense that during the evolution of Homo sapiens mechanisms developed that had the purpose of decreasing its intensity and prominence. These various mechanisms have been described, under various terminologies, in psychological, psychiatric, and psychoanalytic literatures. We have already mentioned some of these mechanisms as the adoption of special cognitive configurations or special types of cognition. Others should be considered under the heading of desubjectivization:

1. Decrease of affective or sensuous content, for instance, by the mechanisms of denial, reaction formation, undoing, blunting of affect, depersonalization, alienation, hysterical anesthesia, and so forth.
2. Suppression, or more or less voluntary removal, of some psychological content from the focus of attention or of consciousness. This content goes into a state of quiescence, like a language or skill that is not used.
3. Repression, or removal of psychological content from consciousness. The study of this mechanism is the main topic of classic Freudian psychoanalysis.

These mechanisms alter and complicate, but do not eliminate, inner life. Desubjectivization is not necessarily a useful procedure. Although from the point of view of the whole human race it may have statistically useful survival effects, it often has undesirable consequences on the individual. As a matter of fact, it is the aim of psychoanalysis to make conscious again what became unconscious. Other writers and I believe that this psychoanalytic procedure is not therapeutically sufficient in most instances. Therapy must also aim at reintegrating harmoniously with the rest of the self what was restored to consciousness.

Volition

In the previous section we spoke of the potentiality of the individual. We have to stress now an aspect of it related to volition, that faculty of man that has been neglected by psychoanalysts even more than cognition. A full analysis of the unfolding of mature volition is given elsewhere.[18]

Generally at the age of four orders start to be internalized. The child feels he *must* act in a certain way. Why must the mother and father bring about an internalization of the command? At a very early age the child must follow orders at a level of complexity that he cannot understand. How can a year-old child understand that he should not break a glass, that he should not eat two big ice creams? The parental order is a substitution for the understanding that the immature child lacks. A gap between the required action and the understanding of it is unavoidable: *the comprehension-action gap.*

Behaviorists have stressed that the child learns to obey by operant conditioning, through punishment and reward. Actually in the human the process, although retaining some characteristics of conditioning, is much more complicated because it is connected with expanding symbolism. First of all, punishment and reward are a *neutralization* of the child's act. The punishment received, let us say, after eating the second ice cream, neutralizes the pleasure experienced in eating it. The reward, on the contrary, is a compensation for the sustained deprivation of the second ice cream. Mother offers a different pleasantness: her approval, her keeping the child on her lap, hugging him, and so forth. In the second place punishment and reward are symbolic or anticipatory of the consequences of the child's act. Thus the punishment that the mother inflicts on the child who has eaten the second ice cream is symbolic of the bad effect that eating excessive ice cream will cause. The immediate punishment takes the place of the painful condition that the bad act, if allowed to be repeated several times, would eventually cause. Conversely, the reward (as approval, tenderness, or love on the part of the mother) is symbolic of the state of welfare and happiness that characterizes a life resulting from doing the right thing. One of the functions of the command, therefore, is to anticipate the value of the act in the distant or near future.

The child who knows he has done something considered wrong by the adults generally feels he deserves to be punished. Unless he is punished the psychological equilibrium will not be restored. The bad deed has produced a rupture, a gap, a state of tension or disequilibrium (remotely reminiscent of the state of a conditioned dog that is prevented from responding). Punishment will re-establish the equilibrium and may even be welcome. If punishment is considered the natural thing to restore the individual, a feeling of guilt will ensue until punishment is received.

Conditioning plus the actual and symbolic value of reward and punishment, however, are not enough to explain the transformation of an external command into an internal one. The person, parent, adult, or leader, who gives the command has a quality that I call the *imperative attitude*. This imperative attitude is experienced in its totality as a benevolent and cooperating force, although it has attributes that may not seem positive. These attributes are the properties of magnetizing, electrifying, hypnotizing, having charismatic qualities, charming, seducing, fascinating, intruding, possessing, capturing, enchanting, and so forth. Imperative attitude cannot be replaced by any of these terms because it contains an element of the meanings of each of these terms.

The person, generally the child, who responds to this imperative attitude, experiences a tendency to obey, what philosophers call a sense of "oughtness." You ought to do what the commanding person wants you to do and you must do it, no longer because he wants you to do it, but because you must want to do it. An external power no longer obligates you; now it is an inner power, what I call *endocratic power*. If you do not do what you must do, you will no longer incur only external punishment but also *guilt feeling*.

The sense of "oughtness" is partially experienced unconsciously, and in a certain way constitutes the Freudian superego; but to a large extent it is also experienced consciously and manifests itself as a sense of duty. Often the parents and other authorities engender in the individual not just endocracy but an excessive amount of it, what I call the *endocratic surplus*.

It is difficult for the individual to exert his will since his wishes of instinctual and/or conceptual derivation or endocratic surplus predispose him to act in primitive or prescribed ways. The individual is thus not *completely* free to choose or act. He has only a margin of free will. Will is not a function fully given to man, but a mechanism of partial determination that permits a striving toward autonomy, freedom, and creativity. The less disturbed the individual is, the larger is his margin of will and the less is he at the mercy of deterministic forces. The mature man is the one who has the largest possible margin of free will; on the other extreme is the catatonic patient who has lost his will almost completely.[6,17] I have described elsewhere[18] the various vicissitudes of the will.

A large part of psychological life consists of the relations between one's own will and the will of others. In optimal conditions, for instance, in the relation between the child and the loving, nonneurotic mother, a state of communion is approximated. The primary social bond is transformed into a feeling of basic trust and subsequently of security. Unfortunately, however, often the will of others is experienced as a power that restricts, distorts, or deflects one's personal will. The will of others may remain an external power, and the individual develops special mechanisms for dealing with it: compliance, submission, rebellion, hostility, detachment, and paranoid attitudes. It may be internalized and become an internal power (endocracy) that may strenghten the individual's will but also cripple, deform it, or paralyze it. This power (external or internal) affects every personal relation and disturbs it to such a point that a state of trust or communion between two or more people is no longer possible. When two people are together, an unequal distribution of power—that is, unequal ability to exert one's will—tends to develop unless strong measures are taken to maintain the equilibrium. This need to dominate may disturb the relation between parent and child, husband and wife, siblings, teacher and pupil, employer and employee, and so on. A relation that is meant to be based on love, affection, learning, or cooperation becomes corrupted by power seeking—most of the time implemented not just by conscious mechanisms but also, and in many cases predominantly, by unconscious maneuvers. Society generally sanctions the more common unequal distributions of power, which may thus remain unchallenged for thousands of years, or until liberation movements occur.

In the field of psychoanalysis and psychology Alfred Adler is the author who first realized the importance of power in human life and who advanced meaningful hypotheses about its origin. In my recent book[18] I discuss the merits and limitations of Adler's theory. Biological, economic, historical, and political conditions greatly determine unequal distributions of power. However, I feel that man would be able to overcome the effects of these conditions and establish an equal distribution of power if other psychological factors would not prevent him.

Because of the expansion of his conceptual processes and of a philosophy of life that he comes to build in his contacts with other human beings, man becomes aware that a discrepancy exists between the way he sees himself and the way his thinking makes him visualize what he could be. He is always short of what he can conjecture; he can always conceive a situation better than the one he is in. When he sees himself as less than what he would like to be, he believes others, too, are dissatisfied with him. He faces a theoretical infinity of space, time, things, and ideas that he can vaguely visualize but cannot master. On the other hand, he becomes aware of his finitude. He knows he is going to die, and that the range of the experiences he is going to have is limited. He cannot be better than he is capable of being, and he cannot enjoy more than a certain amount of food and sex. Being able to conceive the infinite, the immortal, the greater and greater, he cannot accept his limitation. He may feel betrayed by his own nature and may desperately search for ways to overcome his condition. It often seems to him that a way to expand the prerogatives of his life is by invading the life of others. He feels his life will be less limited if he takes away the freedom of others, if he makes others work for him, submit to him, give up their will for his own. Thus instead of accepting his limitations and helping himself and his fellow men within the realm of these limitations, he develops conscious and unconscious conceptualizations that make him believe that he can by-pass his finitude and live more by making others less alive. The network of rationalizations that ramifies from these conceptualizations and assumptions is described in more detail in my recent book.[18]

The Tertiary Self

The tertiary self is the human being in the process of creativity. The individual uses a special form of cognition, the tertiary process, which consists of specific combinations of primary and secondary forms of cognition. The important topic of creativity cannot be dealt with in this essay. The reader is referred to my chapter on creativity in Volume 6 of this *Handbook* and to my book *The Intrapsychic Self.*[7]

Some Principles of Psychopathology

Unless hindered by adverse biological, intrapsychic, or environmental conditions, the individual tends to live in accordance with a mode of life congruous with a high level of symbolism and profound emotional involvement. The preponderance of adverse emotions like severe anxiety, depression, hostility, detachment, aggression, guilt and related feelings, conflict, frustration, hindrance in the exercise of one's will, and unequal distribution of power may require the adoption of mechanisms that reduce or modify the psychological faculties of the individual. These mechanisms become particularly pronounced when they occur in a state of unconsciousness. The mechanisms occurring in schizophrenia and depression will be described in detail in may chapters appearing in Volume 3 of this *Handbook*.

More general mechanisms were described in *The Intrapsychic Self* (Chapter 13).[7] Here I shall mention two important processes that in various degrees and forms occur in several psychiatric syndromes: *somatization* and *cognitive regression.* Somatization is the transformation of a psychological difficulty or disorganization into a physiological or organic one. This mechanism, which Freud referred to as "the mysterious leap from the psychological to the physical," occurs in disorders generally called psychosomatic. Typical examples are gastric ulcer, colitis, hypertension, and bronchial asthma.

Some authors consider psychosomatic disorders only those in which "the leap" occurs through the intervention of the autonomic nervous system. Others (for instance, Arieti[11] and by implication, Jung)[39] also consider psychosomatic some disorders in which the psychological factors engender changes directly in the central nervous system. Sexual dysfunctions are well-known psychosomatic disorders. Less known is a particular type of somatization called *eroticization.* Many usually nonerotic actions of the individual and of others acquire an erotic meaning and may become capable of eliciting an erotic feeling in the patient. The mechanism is generally the opposite of the one described in Freudian literature. It is the anxiety, or conflict, or other abnormal states that confer the erotic quality. The erotic quality is a reductive mechanism, or regression. It may be heterosexually or homosexually oriented. Thus a heterosexual man, unable to relate at a mature interpersonal level, may interpret any action or gesture of a woman as erotogenic. As a matter of fact, he actually responds to them with an erotic feeling. The insecure man, who is not a full-fledged homosexual but who has some homosexual tendencies, often goes through the following sequence: he perceives himself as inadequate, inferior to another man. He thus feels that he must submit to him. The submission is eroticized and becomes a homosexual feeling.

Cognitive regression is the adoption of an immature form of symbolism or thinking in order to support rationalizations, prejudices, irrational wishes, and delusions. The most typical examples of regression occur in schizophrenia, where cognition often assumes the paleological modality of the primary process to a limited degree. Cognitive regression occurs also in the everyday living of normal people. In summary, we could say that when the human psyche cannot advance or at least function adequately, it tends to escape from mature meanings and choices.

Culture and Human Conflict

Several neo-Freudian schools have emphasized the role of society and culture in determining human conflict. Portnoy[49] has described the role that Horney has attributed to culture in human neurosis, and Witenberg[60] has done so in relation to Fromm.

The cognitive-volitional approach stresses that the emotional and social bonds existing among people are determined not only by physical proximity and biological needs but also by the cognitive ways they use and the choices they make. A cultural milieu is essential for the development of man (see also Whorf).[59] At the same time that society permits man to survive and grow, it may undermine his basic functions and aspirations and enhance psychological difficulties in many ways. We shall discuss this topic here only in relation to some aspects of present Western culture.

Western culture presents many conflictful areas that are internalized by people and therefore transformed into personal conflicts. Contradictory conceptualizations are very common. At the same time that the individual is asked to feel brotherly love for his neighbor, he is urged to compete with him. At the same time that he is taught to observe sexual restrictions, he is exposed to strong sexual stimulation. Although he is taught to be parsimonious, he is exposed to the inducement of a consumeristic society. Whereas he is told to believe that men are created equal, he sees artificial inequalities being established and maintained. Whereas he is supposed to cherish freedom, his freedom is secretly or openly infringed upon in a thousand ways. Whereas he is supposed to follow his own initiative, he is conditioned more and more, with more refined and therefore almost unconscious methods of reinforcement.

Freud was very skillful in uncovering the ways by which the individual represses in order to escape the punishment of society, but he did not become concerned with how society represses and transmits that repression to the individual.

As I wrote elsewhere,[18]

> The individual has a double burden to repress: his own and that of society. How does society repress? By teaching the individual not to pay attention to many facts (selective inattention); by masquerading the real value of certain things; by giving an appearance of legality and legitimacy to unfair practices; by transmitting ideas and ideals as absolute truths without any challenge or search for the evidence on which they are supposed to be based; by teaching certain habits of living, etc. The defenses against objectionable wishes which Freud described in the individual (for instance, repression, reaction-formation, isolation, and rationalization) can be found in society, too [Pp. 45–46].

Cognitive psychoanalysis discloses that cultural introjections become part of deep, unconscious levels of the psyche and operate with little participation of the conscious self. Because of his scientific achievements, modern man has gained more freedom from his biological needs but is in danger of becoming a slave of culture and society. I have described three major negative ways by which modern society affects the whole psyche and especially the will of the individual: primitivization, endocratic surplus, and deformation of the self.

Primitivization consists of all the mechanisms and habits that foster the primitive functions of the psyche at the expense of high-level functions. Prominent among these mechanisms are exaggerated decontrol of the sexual and aggressive drives, craving for immediate satisfaction, and return to magic and shamanism.

Endocratic surplus is excessive introjection of authorities and of false and self-perpetuating principles. Endocratic surplus may reduce the individual to blind obedience and enhance the occurrence of tyranny.

Deformation of the self has been brought about by the application of the scientific model to all aspects of life, the actual reduction of people's feelings and ideas to numbers, mass production, and consumerism. By being manipulated or persuaded in hidden ways, the individual acquires the habit of reacting, not acting. His will becomes atrophic while he retains the illusion of freedom. Reaction is confused with spontaneity, promiscuity with romance, intrusion into one's privacy with sincerity and comradeship. The self is deformed and tends to be alienated, and addictive drugs are used to combat alienation.

If we try to interpret these three adverse sociopsychological factors in reference to Freudian terminology, we can say that primitivization expands the id at the expense of the other parts of the psyche; deformation of the self warps the ego; and endocratic surplus overburdens and distorts the superego. Freud, of course, was aware of the important impact of society on the individual's psyche, but was not aware of these three mechanisms. He felt that society acted predominantly by summoning the service of the superego for the repression of the id. The superego was an ally of society and of the ego insofar as it controlled the undesirable aspects of the id. Endocratic surplus of social origin was not seen by Freud as constricting the will of the individual and promoting tyranny. For him sexual repression was the ultimate inhibiting force, and the search for power was not important enough to be considered.

Freud conceived of the ego as that part of the psyche that deals with reality; but he did not see how "reality"—that is, the social environment—may warp the ego. Also because of the Victorianism in which he was brought up, he could not conceive that society itself at times promotes eruption of the primitive id.

Cultural conceptualizations become *cognitive domains* or *assumptions*, which often act unconsciously. Together with people working in other fields (such as sociology, politics, social psychology, economics, and history) the psychoanalyst has an important role in revealing how these cognitive domains affect the individual and society. Thus psychoanalysis, which was criticized for limiting its sphere of operation to individual therapy of a restricted class of people, acquires a new function: advocacy.* Already a host of writers, working in psychoanalysis and related fields, have made outstanding contributions in this area (Coles,[30] Halleck,[36] Keniston,[40,41,42] Lifton).[45]

* I heard the term "advocacy psychiatry" for the first time in a speech delivered by Robert Lifton to the American Academy of Psychoanalysis in New York City on December 4, 1971.

Bibliography

1. ADLER, A., *Understanding Human Nature*, Greenberg, New York, 1927.
2. ARIETI, S., "Conceptual and Cognitive Psychiatry," *Am. J. Psychiat.*, *122*:361–366, 1965.
3. ———, "Contributions to Cognition from Psychoanalytic Theory," in Masserman, J. (Ed.), *Science and Psychoanalysis*, Vol. 8, pp. 16–37, Grune & Stratton, New York, 1965.
4. ———, "Creativity and Its Cultivation: Relation to Psychopathology and Mental Health," in Arieti, S. (Ed.), *American Handbook of Psychiatry*, Vol. 3, pp. 720–741, Basic Books, New York, 1966.
5. ———, "The Experiences of Inner Status," in Kaplan, B., and Wapner, S. (Eds.), *Perspectives in Psychological Theory*, International Universities Press, 1960.
6. ———, *Interpretation of Schizophrenia*, 1st ed., Brunner, New York, 1955. 2nd ed., completely revised and expanded, Basic Books, New York, 1974.
7. ———, *The Intrapsychic Self: Feeling, Cognition, and Creativity in Health and Mental Illness*, Basic Books, New York, 1967.
8. ———, "The Meeting of the Inner and the External World in Schizophrenia, Everyday Life and Creativity," *Am. J. Psychoanal.*, *29*:115–130, 1968.
9. ———, "The Microgeny of Thought and Perception," *Arch. Gen. Psychiat.*, *6*:454–468, 1962.
10. ———, "New Views on the Psychodynamics of Schizophrenia," *Am. J. Psychiat.*, *124*:453–458, 1968.
11. ———, "The Possibility of Psychosomatic Involvement of the Central Nervous System in Schizophrenia," *J. Nerv. & Ment. Dis.*, *123*:324–333, 1956.
12. ———, "The Psychodynamics of Schizophrenia: A Reconsideration," *Am. J. Psychother.*, *22*:366–381, 1968.
13. ———, "Schizophrenia: The Manifest Symptomatology, the Psychodynamic and Formal Mechanisms," in Arieti, S. (Ed.), *American Handbook of Psychiatry*, 1st ed., Vol. 1, pp. 445–484, Basic Books, New York, 1959.
14. ———, "Schizophrenic Thought," *Am. J. Psychother.*, *13*:537, 1959.
15. ———, "Special Logic of Schizophrenic and Other Types of Autistic Thought," *Psychiatry*, *11*:325–338, 1948.
16. ———, "The Structural and Psychodynamic Role of Cognition in the Human Psyche," in Arieti, S. (Ed.), *The World Biennial of Psychiatry and Psychotherapy*, Vol. 1, Basic Books, New York, 1970.
17. ———, "Volition and Value: A Study Based on Catatonic Schizophrenia," *Comprehensive Psychiat.*, *2*:74–82, 1961.
18. ———, *The Will to Be Human*, Quadrangle, New York, 1972.
19. ARLOW, J. A., "Report on Panel: The Psychoanalytic Theory of Thinking," *J. Am. Psychoanal. A.*, *6*:143–153, 1958.
20. BALDWIN, J. M. (1929), quoted by Piaget, J., *The Child's Conception of Physical Causality*, Harcourt, Brace, New York, 1930.
21. BARNETT, J., "Cognition, Thought and Affect in the Organization of Experience," in Masserman, J. (Ed.), *Science and Psychoanalysis*, Vol. 12, pp. 237–247, Grune & Stratton, New York, 1968.
22. ———, "On Aggression in the Obsessional," *Contemp. Psychoanal.*, *6*:48–57, 1969.
23. ———, "On Cognitive Disorders in the Obsessional," *Contemp. Psychoanal.*, *2*:122–134, 1966.
24. ———, "Sexuality in the Obsessional Neuroses," in Witenberg, E. G. (Ed.), *Interpersonal Explorations in Psychoanalysis*, Basic Books, New York, 1972.
25. ———, "Therapeutic Intervention in the Dysfunctional Thought Processes of the

Obsessional," *Am. J. Psychother.*, June 1972.

26. BECK, A. T., "Cognition, Affect and Psychopathology," *Arch. Gen. Psychiat.*, *24*:495–500, 1971.
27. ———, "Cognitive Therapy: Nature and Relation to Behavior Therapy," *Behav. Ther.*, *1*:184–200, 1970.
28. ———, *Depression: Clinical, Experimental, and Theoretical Aspects*, Hoeber, New York, 1967.
29. BEMPORAD, J. R., "New Views on the Psychodynamics of the Depressive Character," in Arieti, S. (Ed.), *The World Biennial of Psychiatry and Psychotherapy*, Vol. 1, Basic Books, New York, 1970.
30. COLES, R., *Children of Crisis*, Dell, New York, 1968.
31. BRIDGMAN, P. W., *The Logic of Modern Physics*, Macmillan, New York, 1928.
32. FAIRBAIRN, W. R. D., *An Object-Relations Theory of the Personality*, Basic Books, New York, 1954.
33. FEDERN, P., *Ego Psychology and the Psychoses*, Basic Books, New York, 1952.
34. FREUD, S., *The Interpretation of Dreams*, Basic Books, New York, 1960.
35. GUNTRIP, H., *Personality Structure and Human Interaction*, International Universities Press, New York, 1961.
36. HALLECK, S. L., *The Politics of Therapy*, Science House, New York, 1971.
37. HARTMANN, H., KRIS, E., and LOEWENSTEIN, R. M., "Comments on the Formation of Psychic Structure," in *The Psychoanalytic Study of the Child*, Vol. 2, pp. 11–38, International Universities Press, New York, 1946.
38. JONES, E., *The Life and Work of Sigmund Freud*, Vol. 1, Basic Books, New York, 1953.
39. JUNG, C. G. (1903), *The Psychology of Dementia Praecox*, Nervous and Mental Disease Monograph Series, New York, 1936.
40. KENISTON, K., *The Uncommitted: Alienated Youth in American Society*, Harcourt, Brace & World, New York, 1965.
41. ———, *Young Radicals: Notes on Committed Youth*, Harcourt, Brace & World, New York, 1968.
42. ———, *Youth and Dissent: The Rise of a New Opposition*, Harcourt Brace Jovanovich, New York, 1971.
43. KLEIN, M., *Contributions to Psycho-Analysis*, Hogarth, London, 1948.
44. LEVIN, M., "Misunderstanding of the Pathogenesis of Schizophrenia, Arising from the Concept of 'Splitting,'" *Am. J. Psychiat.*, *94*:877, 1938.
45. LIFTON, R., *Death in Life: Survivors of Hiroshima*, Random House, New York, 1967.
46. LOEB, A., BECK, A. T., and DIGGORY, J. C., "Differential Effects of Success and Failure on Depressed and Nondepressed Patients," *J. Nerv. & Ment. Dis.*, *152*:106–114, 1971.
47. ODIER, C., *Anxiety and Magic Thinking*, International Universities Press, New York, 1956.
48. PIAGET, J., *The Child's Conception of Physical Causality*, Harcourt, Brace, New York, 1930.
49. PORTNOY, I., "The School of Karen Horney," Ch. 40 B of this volume.
50. RAPAPORT, D., "The Structure of Psychoanalytic Theory," in *Psychological Issues*, Vol. 2, International Universities Press, New York, 1960.
51. SCHUR, M., *The Id and the Regulatory Principles of Mental Functioning*, International Universities Press, New York, 1966.
52. SULLIVAN, H. S., *Conceptions of Modern Psychiatry*, Norton, New York, 1953.
53. TINBERGEN, N., *The Study of Instinct*, Clarendon, Oxford, 1951.
54. VON DOMARUS, E., "The Specific Laws of Logic in Schizophrenia," in Kasarin, J. S. (Ed.), *Language and Thought in Schizophrenia: Collected Papers*, University of California Press, Berkeley, 1944.
55. VYGOTSKY, L. S., *Thought and Language*, M.I.T. Press, Cambridge, 1962.
56. WERNER, H., *Comparative Psychology of Mental Development*, International Universities Press, New York, 1956.
57. ———, "Microgenesis and Aphasia," *J. Abnorm. & Soc. Psychol.*, *52*:347, 1956.
58. ———, and KAPLAN, B., *Symbol Formation: An Organismic-Developmental Approach to Language and the Expression of Thought*, John Wiley, New York, 1963.
59. WHORF, B., *Language, Thought and Reality*, John Wiley, New York, 1956.
60. WITENBERG, E. G., "The Interpersonal and Cultural Approaches," Ch. 40 A of this volume.

CHAPTER 41

MISCELLANEOUS PSYCHOANALYTIC APPROACHES

Simon H. Nagler

THIS CHAPTER is devoted to the consideration of a diverse group of psychoanalysts who made significant contributions to psychoanalysis and to the eclectic structure of modern psychiatry, although they did not create major systems of theory and practice. The work of Wilhelm Reich, Sandor Ferenczi, Otto Rank, and Sandor Rado had a common origin in orthodox Freudian psychoanalysis. For the most part their contributions were through modifications of or even deviations from this original source. Sandor Ferenczi, however, was perhaps second only to Freud in his basic contributions to the structure of analytic theory and practice. But of special importance in this group is Wilhelm Reich, who became in his time a very controversial figure, even a tragic and ill-fated one. His ultimate activity was far afield from the established boundaries of psychoanalysis and psychiatry. Like Otto Rank he did not consider himself a psychoanalyst in his final phase.

Wilhelm Reich (1897–1957)

Biographical Sketch

Wilhelm Reich was born in Austria in 1897 of Jewish parentage. A brilliant person of extraordinary vitality, he was restless, moody, and hypersensitive. He received his M.D. degree in 1922 from the University of Vienna. During his undergraduate years he became intensely interested in psychoanalysis (1919) and was given the unusual privilege of joining the Vienna Psychoanalytic Society while but a student. After graduation he continued his psychiatric studies for two years with Julius von Wagner-Jauregg and Paul Schilder. Because of his enthusiasm and enormous capacity for work, Reich quickly rose to importance within the young analytic movement in Vienna. From 1924 to 1930 he was director of the Vienna Seminar for Psychoanalytic Therapy, which was a training institute. He became involved in the highly charged political atmosphere of the period and studied Marxian theory for its implications for the social causation of mental illness. He joined the Austrian Socialist party, but left it for the Communist party in 1928, in a continuing attempt to reconcile Marxian and Freudian concepts.

Reich's relationship with Freud is of great interest. Reich was said to have been a "favorite son" and a most brilliant assistant. Their estrangement began early in 1927, but its basis is shrouded in controversy. Some maintain that it was Reich's efforts to synthesize Marxian and Freudian concepts that caused the conflict, and that Freud considered Reich a political fanatic. Others attribute the alienation to Reich's insistence on the sexual origin of every neurosis. Reich himself attributed their difficulties to theoretical differences about the social implications of psychoanalysis and to the professional jealousy of colleagues like Paul Federn.[12] On the other hand, Freud, commenting on Reich's paper on the masochist character, said that Reich had developed his theory "in the service of the Bolshevist party." Reich maintained later that he was never a political participant as such but that he was interested in the mental health needs of the workers. He helped to open clinics where information was available on birth control, childrearing, sex education, and so forth. In 1929 he made a visit to the Soviet Union, where he was disappointed with its bourgeois, moralistic attitudes toward sex and especially toward childhood and adolescent sexuality.

Suffice it to say, Reich's political attitudes and activities made him increasingly *persona non grata* to the members of the Vienna Psychoanalytic Society. In 1930 he went to Berlin for a personal analysis with Sandor Rado. He joined the German Communist party there. In discussions with Erich Fromm, Siegfried Bernfeld, and Otto Fenichel, he continued his effort to synthesize Freud and Marx. He established a publishing house, the Sexpol-Verlag, in 1931 and issued pamphlets on sex education for children and adolescents. He extended his critique of bourgeois sexual morality, contending that economically determined sexual taboos lead to sexual repression and submission to the authoritarian family and state, which breed mental illness and political totalitarianism. The rise of fascism and the reaction of the masses led in 1933 to his theoretical work, *The Mass Psychology of Fascism.*[36]

Although the International Psychoanalytic Press had published Reich's first two books, including *The Function of the Orgasm,*[33] it rejected his epochal *Character Analysis* (*Die Charakteranalyse*), published by the Sexpol Verlag in 1933. At the International Congress held at Lucerne in August 1934, he either was expelled or resigned. The year before the German Communist party had expelled Reich because his politicalization of sex was compromising, especially in light of the rising power of fascism. He realized at this point that he was no longer thinking within the psychoanalytic framework. In 1933 he went to Copenhagen, expecting to function as a teaching and training analyst, but the International Psychoanalytic Association rejected his request for permission to do so. He was asked to desist from training analysts, giving public lectures, or even publishing. After six months in Denmark the authorities revoked Reich's right of residence in the country.

He settled in Malmö, Sweden, in September 1933, but in less than a year his permission to live in Sweden was revoked, because "the combination of sex, psychoanalysis and politics was too much for the authorities to absorb."[31] He then went to Oslo, Norway, where at the Psychological Institute of the University of Oslo, he began long-delayed experiments on the bioelectrical nature of anxiety and sexuality. During 1934–1935 he sought to measure the biological excitation in the erogenous zones by means of an oscillometer. He claimed to have demonstrated the rise of skin potential with pleasure and its fall with anxiety. The major target of his later work was the elucidation of the nature of the energy involved. He was evolving thus the further development of his highly significant character analysis, still within the Freudian framework, into vegetotherapy, based on the function of the autonomic or vegetative nervous system. This was to eventuate in the so-called *psychiatric orgone therapy*, his final therapeutic formulation.

In the course of his experiments in Oslo Reich claimed to have discovered the "bions," energy vesicles created from organic matter and thus demonstrating the process of biogenesis. These bions he later related to carcinogenesis. But of greatest importance in this period was his "discovery" of *orgone* or life energy, which he came to believe was of cosmic distribution.

The first publication in 1937 of his work on bions unleased a violent newspaper campaign by biologists and psychiatrists and by both political camps. The vituperation ranged from "the quackery of psychoanalysis" through "the Jewish pornographer" to "God Reich creates life." Reich refused to defend himself against what he termed "the emotional plague." He finally left Norway in August 1939, when he received a contract through friends in the United States at the New School for Social Research as an Associate Professor of Medical Psychology. He taught for two years, giving lectures on the "Biological Aspects of Character Formation," and a seminar on the "Psychological Approach to Psychosomatic Research." In 1940 he bought property in Rangeley, Maine, which was to be the site of his research activities for many years and his eventual burial place at Orgonon. In the fall of 1940 he took the step that was finally to bring him into disastrous conflict with the law. He built the first orgone energy accumulator for use in experiments with cancerous mice. It was based on the principle that organic materials attract and absorb orgone while metals attract and repel the energy. The accumulator was first used on human subjects in December 1940.

This development brought Reich deeper into the realm of physics. He even arranged a meeting with Einstein, who at first was supposedly impressed, but later rejected Reich's discoveries. He reacted bitterly, terming Einstein's action the result of a communist-inspired plot. Soon after Pearl Harbor Reich was arrested as an enemy alien, but was soon released. In the summer of 1942 the construction of the Orgone Institute was begun in Maine. Many psychologists and psychiatrists sought instruction in Reich's character analytic vegetotherapy, later to be termed psychiatric orgone therapy. Teachers of gymnastics and dancers were highly intrigued by Reich's concepts of muscular armoring.

In 1946–1947 there began an attack on Reich's ideas in the major magazines and newspapers, culminating in an investigation by the Food and Drug Administration of the orgone accumulator. In 1950 Reich made plans for and initiated the so-called Oranur Experiment, which terminated in near disaster. Reich hoped in a Messianic fashion to eliminate the effects of the atom bomb by a threefold approach using orgone energy to neutralize the effect of the bomb, heal radiation sickness, and immunize mankind against nuclear radiation. Since the Atomic Energy Commission would put no restrictions on him, he was able to obtain the necessary isotopes. He expected that high concentrations of orgone would neutralize nuclear energy, but instead presumably the orgone got more excited, resulting in enormous Geiger counter readings. Many of the assistants and students developed radiation sickness, and the laboratories had to be evacuated.

In spite of his obvious disorganization Reich continued to work and write in a highly polemical manner. In 1951 he had a very severe heart attack. Nonetheless, the next year Reich started a new phase in his work that he termed Core (cosmic orgone engineering). It dealt with cosmic phenomena, outer space, weather conditions, drought, and rainmaking. Toward the end of 1952 many of his co-workers and students left because they were unwilling to pursue this disturbing course of development in Reich's ideas. He was convinced that the President and powerful officials of the air force were aware of his work and were protecting him by planes overhead, that spacemen knew how to use orgone energy to run their ships, and so forth. In the spring of 1953 his *The Murder of Christ*[37] appeared. Its thesis is that Christ represents life, which the "armored" man cannot tolerate and so must repeatedly kill it in all spheres of human existence—family, politics, science, and religion.

In 1954 the Federal Food and Drug Administration sought an injunction against Reich and his Foundation at Orgonon. He refused to defend himself. The decree of injunction was granted, and in May 1956 he was convicted of disobeying it and sentenced to two years' imprisonment. At Danbury (Connecticut) Penitentiary he was considered paranoid, and the psychiatrist suggested his transfer to the federal penitentiary at Lewisburg (Pennsylvania) for treatment. Here, however, he was declared legally sane and competent, although his wife and even sympathetic psychiatrists concurred in the conclusion that he had paranoid ideas. He died in prison on November 3, 1957 as the result of myocardial insufficiency. It is hardly to the credit of the psychoanalytic movement that so brilliant an early leader should have been allowed to die in a prison rather than in a hospital, which he obviously deserved. As Mary Higgins[12] eloquently wrote: "That Wilhelm Reich . . . should die in a federal penitentiary is shocking, that those who cared were helpless, and that there were many who knew and who did not care, is tragic."

The Psychoanalytic Period

THE FUNCTION OF THE ORGASM

Reich's life and work is sharply divided into two distinct parts by the publication of *Character Analysis* in 1933. It was a milestone in psychoanalytic characterology and perhaps even more in analytic psychotherapy. Because of its social emphasis along with its biological basis, it was an important forerunner of the culturalist orientation in analysis. It related libidinal energies, character structure, and social order in one coherent formulation: "the characterological anchoring of the social order."[32] Only *Character Analysis* and his prior work are acceptable to the official psychiatric world.

Before the publication of this important work, Reich had made another controversial contribution in the analytic vein with his *The Function of the Orgasm*.[33] Reich considered his discovery of the true nature of "orgastic potency" to have been the first link in the chain leading to his discovery of orgone in 1936 and thence to physical and psychiatric orgone therapy. In this study Reich[19] sought the basis for unsatisfactory analytic results. He concluded that the reason was the failure of the patient to achieve a satisfactory genital life. The disturbance in genital function was not merely another neurotic symptom. In fact, it is *the* neurotic symptom, and the neurosis is the result of a genital disturbance in the form of orgastic impotence. The disturbance in orgasm leads to a damming up of sexual energy, and this undischarged sexual energy is the energy source of neurosis. (This is apparently a return to Freud's earliest toxicological theory of anxiety as due to blocked libido.)

This stasis of sexual energy does not occur only in the "actual" neurosis (Freud), but in all psychic disorders.[38] The goal of analytic therapy then becomes the establishment of orgastic potency with the elimination of sexual stasis. Psychic conflict and sexual stasis are directly related. The central conflict of every neurosis is the sexual child-parent relationship; the content of the neurosis is the historical experiential material of this relationship. Sexual stasis through social inhabition leads to fixation on the parents, and "the pathogenicity of the Oedipus complex . . . depends on whether or not there is a physiologically adequate discharge of sexual energy."[33] This depends on a healthy sex life with a beloved partner of the opposite sex, with whom there is a complete, even convulsive, discharge of sexual energy. This capacity Reich termed *orgastic* potency.

Orgasm is an essential function for the total psychosomatic organism since through this function of sexuality that part of bioenergy not used in other activities must be dissipated for the health and happiness of the individual.[19] Unused bioenergy interferes with the freedom of thought, feeling, and action, leading even to life-inimical behavior in the defensive nature of the character structure or personality. Different character traits are related in a unitary defense against instinctual forces, dangerous emotions, and the external world. This defense Reich named the *character armor*, which has its origin in childhood, during which instinctual drives are frustrated by fear of punishment. At the center of the character is the oedipal conflict, with its inevitable frustration.

The social order determines the character structure it requires for its activity and survival. In the preface to *Character Analysis* Reich[32] defined the task of scientific characterology to be the discovery of "the means and mechanisms by way of which social existence is transformed into psychic structure and with that into ideology. . . . The character structure is a crystallization of the sociological process of a given epoch."

As a result of threats in childhood the individual develops characteristic defensive ways of behaving manifested early both in analysis and in daily life, as the *how* rather than the *what*, the *manner* rather than the *content* of activity. It is these traits such as argumentativeness, distrust, and the like, that function as resistance in analysis. This insight led to Reich's technique of character resistance analysis. The entire neurotic character becomes manifest in treatment as a condensed rigid and inflexible defense mechanism. Muscular rigidities and attitudes are important parts of this armoring.

Consequently behind every symptom neurosis there is a neurotic character, and since, therefore, every analysis deals with character resistance, every analysis is a character analysis. This was a revolutionary concept at the time of its formulation. The origin of the difficulty is the thwarting of children and adolescents through antisexual, authoritarian attitudes that have their roots in the prevailing social ideologies, however much these claim religious, philosophical, and moral foundations. The suppression of natural sexual urges leads to secondary drives, chiefly sadomasochistic, which thus have their roots in repressive social forces, political parties, and the like.

In the course of his clinical work seeking to release energies bound up in the character armor and muscular armoring, Reich discovered the *orgasm* reflex. But even before this he had described the events of the orgasm in his orgasm formula: mechanical tension, bioelectric charge, bioelectric discharge, relaxation. Reich believed that sexuality and anxiety formed the basic antithesis of negative existence.

Character Structure and Character Analysis

Although Freud and Abraham wrote briefly on character, Reich[32] was first to give major significance to character structure, formulating a consistent theory. His basic concept is that character is a defensive structure, an armoring of the ego. It functions as resistance in analysis, and the measuring of this character resistance can be recognized without infantile material. At the beginning of an analysis, character resistance predominates. For each case there is "only one technique, which has to be derived from its individual structure."

There are two stages in each analysis. The first is "education to analysis by analysis," being for the most part an approach to the character structure. The second stage begins after much of the character resistance has been eliminated. This phase concentrates on the liberated infantile material, seeking to bring about the genital libidinal fixation. Although Horney and others considered the first character analytic phase of treatment as the chief therapeutic goal, Reich considered it as preparatory to the working through of the repressed infantile material in a true Freudian manner.

Reich delineated several types of pathological character formation, that is, the hysterical, the compulsive, the phallic-narcissistic, and the masochistic. The core of the personality structure is the incestuous wishes and the manner of their resolution. The rigidity of the ego is due to many factors: identification with the frustrating person, anxiety aroused by the frustration leading to motoric inhibition manifest in the character structure, and so forth. The armoring that results from the manner of the conflict solution becomes the source of later neurotic conflicts and symptom neuroses. The nature of the conflict solution depends on the intensity of fear aroused by threats of punishment, the amount of instinctual satisfaction permitted, the personality of the parents, and other factors. Reich stressed this conflict solution as the source of the latent negative transference so deadly to progress in the analysis.

He sought to distinguish a symptom from a character neurosis. The symptom is rarely as well rationalized as the character trait, and is always ego-alien and of a much simpler construction. The more a symptom is analyzed, the closer is the approach to the characterological reaction basis. Since the character armor uses instinctual energies in the production of reaction formations and other neurotic defenses, it binds anxiety. It is the threatened release of anxiety that calls forth the renewed activity of the narcissistic character defense. The character armor was produced in infancy for the same defensive purposes that character resistance functions in analysis, that is, to avoid pain and anxiety.

Reich established definite rules for his technique of character analysis. He warned against too early interpretation of deeply repressed material and against inconsistent and unsystematic interpretation. He stressed the recognition of character resistance and its early dissolution. The patient must learn that neurosis can only be avoided by evolving a personality with the capacity for sexual and social freedom.

Finally the character structure depends on the level of psychosexual development at which the personality was most decisively shaped and also on the degree of sexual freedom attained. The genital or healthy character utilizes orgasm and sublimation to deal with anxiety; the neurotic character employs pregenital means of gratification and reaction formation.

The hysterical character is fixated at the genital level with little inclination toward sublimation or reaction formation and great tendency toward bodily behavior that discharges sexual tension. The compulsive personality is fixated at the anal-sadistic level due to rigid toilet training. There is a regression to the phase of anal interest and hostility with the concomitant diminution of genitality. Intense reaction formations in the latency period shape the final rigid personality.

The phallic-narcissistic character is overtly aggressive and hostile, being devoid of reaction formation defense. This type is attractive sexually although orgastically impotent. The phallus is in the service of hate, not love, as a result of the rejection of phallic display when the individual was a child seeking attention and love. This resulted in identification with the rejecting object. This character type seeks to degrade or destroy the woman for whom he has contempt. There is none of the passivity of the anal stage.[32]

Reich repudiated the Freudian death instinct, especially as an explanatory concept for masochism in all its manifestations. Instead, he regarded masochism as a result of inhibited exhibitionism. The masochistic personality is the outcome of the repression of exhibitionistic tendencies during the genital stage. The masochist tends to be self-damnatory and querulous, with a compulsion to torment self and others. Some of these individuals develop masochistic perversions. Reich believed that the source of the problem lay in a fear of pleasurable excitation, resulting in an inhibition of the sensation of pleasure and its conversion into pain. In truth the masochist secretly has inordinate demands for love and a very low frustration tolerance. Thus the masochist appears giving, but this is really the projection of his own insatiability.

Postanalytic Investigations

BIOENERGETICS

Reich undoubtedly considered his postanalytic investigations, writings, and discoveries to be his most important and most enduring work. During his work on the orgasm he was concerned with the nature of the energy involved, considering it probably bioelectrical. Later in his character analytic work he recognized that the muscular rigidities were "the somatic side of the process of repression and the basis for its continued existence."[33] He noted that in treatment these rigidities and spasms gave way to spontaneous, soft, and harmonious movements in the rhythm of breathing. If these movements were not impeded, they would quicken and even eventuate in a reflex bodily convulsion, which he termed the *orgasm reflex*. Reich devoted the major part of his efforts to the clarification of the energy question involved.

In 1939 in Oslo he discovered energy vesicles, the *bions*, while working with heated inorganic substances like coal dust and sand. He had reasoned that if the orgasm formula was characteristic of life, it might also throw light on biogenesis. Under the microscope he noted that these mysterious bions organized into cells or protozoa.[19] Since most of the bions originated from distintegrated organic material, Reich speculated on the origin of cancer cells in this manner. He made numerous experiments to substantiate this hypothesis.

While working with the bions in 1939 in Oslo, Reich discovered that they emitted a previously undescribed energy, which he called *life energy* or *orgone*. He later came to consider this energy as cosmic and related to cosmic radiation. He claimed to have demonstrated the existence of this mysterious energy visually, microscopically, and thermically. This was the bioenergy he had sought since student days. Its liberation had been his constant therapeutic aim, and so he now called his technique *orgone therapy*. Under the new science of *orgonomy*, Reich now distinguished psychological or psychiatric orgone therapy (formerly character analytic vegetotherapy) and physical orgone therapy, which employed the orgone accumulator and other orgone devices. It was the orgone accumulator that later led to his imprisonment.

In 1945 he reported the experimental production of *primary biogenesis*, that is, the development of life from earth and water. This was the so-called Experiment XX.[31] Live cells and protozoa were supposedly produced from "sterilized bionous water through freezing and thawing." In 1950 he began experiments on the previously observed antagonism between orgone and radioactivity. As noted above, the outcome of the Oranur Experiment was disastrous since many of his assistants developed radiation sickness and the building and area became radioactive for years. Along with these endeavors in biophysics beyond the realm of psychiatry and psychology, Reich considered the formation of hurricanes and tornadoes, planned rainmaking expeditions, and postulated that the origin of galaxies might result from the confluence of two orgone streams (cosmic superimposition).[19] He thought that a hazy black atmosphere around Orgonon was a "deadly orgone," which he termed DOR, and he sought to remove it by a contraption of metallic tubes called a cloudbuster.

His fate and the scientific consideration of his work in bioenergetics must inevitably cast a pall over the preoccupations of his postanalytic period. But there is little question that in his character analysis he decisively influenced psychoanalytic theory and practice and pointed the way for many to follow.

Sandor Ferenczi (1873–1933)

Biographical Sketch

Sandor Ferenczi, brilliant contributor to the psychoanalytic movement, was born in Hungary in 1873 of Polish parentage.[14] His father's bookshop was the early source of his continued interest in all art forms. He studied medicine in Vienna, but acquired an intense interest in psychological matters. After receiving his medical degree in 1894, he spent a year in military service and then became a staff physician in a municipal hospital in Budapest. As a result of being required to treat numerous prostitutes there, he became interested in sexual pathology and thus ultimately in nervous and mental disease. He was a neurologist and also served as the psychiatric consultant to the Royal Court of Justice in Budapest. He had a genuine human kindness and a respect for people.

He became involved with psychoanalysis through reading Freud's *The Interpretation of Dreams*, and in 1908 he began a personal analysis with the master himself. He rapidly became one of Freud's favorites and a central figure in the development of psychoanalysis. He accompanied Freud and Jung on their historic visit in 1909 to Clark University in the United States. In 1910 he proposed the formation of the International Psychoanalytic Association in which he was active until his death. With the aid of Sandor Rado he organized the Hungarian Psychoanalytic Society in 1913. He was appointed professor of psychoanalysis at the University of Budapest in 1919, the first time such a title was bestowed.

He was a tireless worker, and his contributions together with those of Freud and Abraham form the cornerstone of psychoanalysis. Freud called Ferenczi's writings on theory and technique "pure gold." *The International Journal of Psychoanalysis* was largely his creation in 1920. He wrote a fundamental paper on male homosexuality, and his delineation in 1913 of the "Stages in the Development of the Sense of Reality" was an epochal early contribution to the development of ego psychology. In 1924 he published *Thalassa: A Theory of Genitality*, an attempt to correlate biological and psychological phenomena through a new scientific methodology he called "bioanalysis." Freud considered this work to be the "boldest application of psychoanalysis that was ever attempted."

Difficulties for Ferenczi within the movement began with his papers on active therapy, which Jones called a "striking new departure in technique." This new approach was further elaborated in his book on *The Development of Psychoanalysis*, written in conjunction with Otto Rank in 1925. It dealt with problems in psychoanalytic technique and especially with the role of activity in therapy. F. Alexander and E. Glover, influential figures in the psychoanalytic establishment, were quite critical of Ferenczi's innovations. He himself considered active therapy only an adjunct to "real" analysis, and he wrote a paper on the "Contra-Indications to Active Psychoanalytic Technique" in 1925. However, additional papers on active therapy between 1930 and 1933 further widened the gulf between him and the orthodox posture. Jones indicated there was a real break with Freud in 1929, but other authors dispute this. It seems, nonetheless, that at the time of his death in 1933 he was somewhat isolated from the psychoanalytic movement. Freud wrote to Oskar Pfister that Ferenczi's death was a "distressing loss," and that his memory would be preserved for a long time by "some of his work, his genital theory, for instance." Pfister[15] has called him Freud's "distinguished champion." Clara Thompson wrote that Ferenczi "never ceased until his death in trying to win Freud's approval," but "their friendship was severely shaken, never to be restored on the same basis, by (his) last paper, 'Confusion of Tongues between Adults and the Child.' "[8]

Contributions to the Theory and Techniques of Psychoanalysis

Even in his brilliance and originality, Sandor Ferenczi remained for the most part within the orthodox analytic framework to whose structure he had significantly contributed. He deviated only in his modifications of therapeutic technique, but despite his repudiation of his collaboration with Rank in 1924,[9] he eventually alienated Freud.[2] Therapy had always been the center of his psychoanalytic interest.

His contributions may best be discussed by consideration of the contents of the three volumes of his selected papers.[2,3,4] The first volume, *Sex in Psychoanalysis*,[4] covers the years from 1908 to 1914 and reveals his originality as well as his matchless exposition of Freud's ideas. Aside from papers on dreams and symbolism, this volume contains three significant contributions: (1) to Freud's notion of the masochistic attitude of the subject as basic in hypnosis; (2) to the knowledge about the development of the sense of reality; and (3) to the understanding of the nature of homosexuality.

In the paper "Introjection and Transference" Ferenczi[4] favored the view that in hypnosis and suggestion the psychic work is done chiefly by the subject and not by the hypnotist. The mechanism at work is the Freudian one of transference, in which unconscious, repressed impulses and affects are tranferred on to persons or objects in the outer world. The "parental complexes" are the most important in transference, and "repressed infantile impulses of the hypnotized persons" are transferred on to the authoritative person of the hypnotist. Moreover, unconscious sexual ideas are the basis of the sympathetic capacity to be hypnotized or subject to suggestion. Fear and love are basic to the hypnotic techniques, and these emotions are certainly associated with both parental figures. The hypnotist must be able to inspire in the subject "the same feelings of love or fear, the conviction of infallibility as those with which his parents inspired him as a child." Finally Ferenczi described several striking cases that confirmed Freud's views that hypnotic credulity and pliancy are rooted "in the masochistic component of the sexual instinct."

An important forerunner in the development of ego psychology is Ferenczi's[4] paper on the "Stages in the Development of the Sense of Reality." In this study he set out "to learn something new about the development of the ego from the pleasure to the reality principle, since it seemed to be probable that the replacement . . . of the childhood megalomania by the recognition of the power of natural forces composes the essential content of the development of the ego." Ferenczi demarcated four stages in this development: (1) the period of unconditional omnipotence, (2) the period of magical, hallucinatory omnipotence, (3) the period of omnipotence by the help of magic gestures, and (4) the period of magic thoughts and magic words.

Feelings of inferiority that are so prominent in the neurotic do not contradict the "almost incurable megalomania of mankind," for these feelings are only a reaction to the underlying sense of omnipotence. And the striving for power is a return of the repressed in a yearning for the former effortless omnipotence. In narcissism or self-love we may always retain the illusion of omnipotence. In this essay discussing the choice of the neurosis, Ferenczi suggests that the mechanism of the neurosis is determined by the stage of development of the sense of reality that the individual was in at the time of the determining inhibition of the sexual drive—for example, hysteria is determined by regression to the stage of magic gestures.

In 1911 Ferenczi[4] delivered a fundamental paper on "The Nosology of Male Homosexuality (Homoerotism)." He distinguished the character structures and psychodynamics of the active and passive types of homosexuals or, as he preferred, homoerotics (a term that emphasizes the psychological rather than the biological aspect of the impulse). Only the passive male homosexual is truly an "invert" since he feels like a woman in all ways of life; he is thus a "subject homoerotic." The active homosexual, on the other hand, feels like "a man in every respect," except that of object choice, so that he is "an object-homoerotic." The passive type rarely seeks psychological help since his form of satisfaction suits him without any inner conflict. On the other hand, the active or object homoerotic is beset with conflict and seeks help.

Analytically the subject or passive homosexual reveals evidences of inversion in his earliest history, even the presence of an inverted oedipal wish, that is, to replace the mother in his father's favor. Ferenczi considered this type of homosexual incurable by analysis. The active homosexual, on the other hand, has normal oedipal fantasies, is intellectually precocious, and has strong anal erotism. He often has a history of severe punishment for early heterosexual impulses, resulting in dread of the woman. Ferenczi considered this type to be an obsessional neurotic, in whom the homosexual impulse is the compulsion symptom. He notes that considering the homosexual a neurotic was in opposition to Freud's dictum that the neurosis is the negative of perversion, but this was only an apparent contradiction. Actually it is still a perversion in the service of neurosis. Ferenczi considered the active homosexual to be the most common and to be on the increase; possibly, he suggests, this is due to the return of the abnormally repressed homosexual instinctual component in civilized man.

The second volume of the selected papers was published in 1926 as *Further Contributions to the Theory and Technique of Psycho-Analysis.*[3] The papers fall into two main groups, one with a medical and sexual outlook, the other devoted to problems of technique. Among the latter papers is the very significant one on "The Further Development of an Active Therapy in Psychoanalysis." This was the beginning of the development that led to the final rift with Freud. It was Ferenczi's intention "to speed up the analytic technique by so-called 'active' measures." Classic procedure was predominantly passive. Taking a hint from Freud's insistence that the phobic expose himself to the painful situation, Ferenczi introduced active measures like forcing the patient to renounce pleasurable activities, heightening anxiety, and producing new memories that accelerated the analysis. This was analyzing in a condition of abstinence, produced by "sytematic issuing and carrying out of commands and prohibitions."

These auxiliary measures were not intended to replace the main activity of the analysis, the search for unconscious and infantile material. "In requiring what is inhibited, and inhibiting what is uninhibited, we hope for a fresh distribution of the patient's psychic . . . energy that will further the laying bare of repressed material." Ferenczi considered that he had overworked Rank's suggestion to set a limit to the duration of the analysis. He corrects this in his paper on "Contra-Indications of the Active Technique."[3] Experience has shown him that the desired detachment from the analyst does not always occur upon the predeterminated dismissal of the patient. However, he considers it a distinct analytic advance to follow Rank's suggestion to "regard *every* dream, *every* gesture, *every* parapraxis, *every* aggravation or improvement in the condition of the patient as above all an expression of transference and resistance."

In 1925 together with Rank he wrote *The Development of Psychoanalysis,*[9] which was a summary of technical innovations, including "active" therapy. But when the theoretical implications became clear in the smoke of the controversy aroused, Ferenczi repudiated his basic alignment with his friend Rank. Nonetheless, Ferenczi's greatest contributions to psychoanalysis remain in the area of technique.

Volume 3 of the selected papers appeared in English in 1955 as *Final Contributions to the Problems and Methods of Psychoanalysis.*[2] From 1928 to 1933 Ferenczi wrote several very important papers on "relaxation" therapy as further modifications of his concept of active therapy. In the paper "The Principles of Relaxation and Neocatharsis" he introduces "the principle of indulgence," encouraging the patient to greater freedom and mobility. There are thus two opposite methods, one producing tension by frustration, the other encouraging relaxation through indulgence. He told analysts to be more humble toward patients, even to encourage the patient in his free expression of anger toward the analyst. He encouraged the analyst to soften the analytic atmosphere. Some neurotics have remained at the level of the child and require more than orthodox treatment. They need "to be adopted and to partake for the first time in their lives of the advantages of a normal nursery."

In other controversial papers in this third volume, like "Confusion of Tongues between Adults and the Child," Ferenczi increased the gulf between him and his orthodox colleagues. He challenged the emphasis on heredity and constitution, he attacked the Godlike, holier-than-thou stance of the analyst, he emphasized the real problems between parents and children. He shifted the Freudian emphasis from the father to the mother, being perhaps the first to emphasize the crucial nature of the mother-child relationship. He boldly said children have problems because of bad parents, and patients have difficulties because of analysts with inadequately analyzed problems. He stressed the need for thorough analysis of the analyst and for greater humility and honesty of the therapist. In the largest sense he sought to humanize the analytic procedure. Clara Thompson wrote of him that he was "all unwittingly a prophet."

In concluding this discussion of Ferenczi's far-ranging interests and contributions, one must refer to his ideas on sexuality, which were stimulated by his early work with prostitutes as a staff physician. He translated Freud's *The Three Essays on the Theory of Sexuality* into Hungarian. His own ideas were finally published in 1924 in *Thalassa: A Theory of Genitality.*[5] It was lauded by Freud and many consider it Ferenczi's masterpiece. It links ontogeny with phylogeny in the sex life of mankind. Intrauterine life is a re-enactment of the oceanic existence of earlier life forms. Emergence on to dry land was a trauma in the sense of Rank's concept of birth trauma. Man is plagued by a nostalgic longing to return, not merely to his mother's womb, but also to the ancient life at the bottom of the sea. Symbols reveal these primeval connections. *Thalassa* is an impressive and fascinating example of the work of a creative mind with enormous erudition. It illuminates wide biological vistas. Ferenczi was the first to formulate the twofold purpose of the orgasm, that is, to discharge all emotional tensions as well as sexual tensions. This was an important psychoanalytic contribution. In the allotted space it would not be possible merely to list his many insights.

Otto Rank (1884–1939)

Biographical Note

Otto Rank was born in Vienna on April 22, 1884, third child in a prosperous middle-class Jewish family (Rosenfeld).[13] His life falls readily into four distinct periods: birth to meeting Freud in 1906; secretary of Freud's "Committee," 1906 to 1926; years in Paris, 1926 to 1934; 1934 to his death on October 31, 1939, about five weeks after Freud's death in London.

He loved the theater and considered becoming an actor, but he had a technical school education when he met Freud in 1906 through an introduction by Alfred Adler. Freud was so impressed by Rank's manuscript, *Der Kunstler*, that he encouraged the young man of 22 to go to the university and "to devote himself to the non-medical side of psychoanalytic investigation." He became secretary to Freud's "Committee," rapidly appearing as, in Havelock Ellis's words, "perhaps the most brilliant and clairvoyant of Freud's disciples." He was an avid reader in philosophy, psychology, literature, ethics, aesthetics, art, and history, and he obtained his Ph.D. degree by 1912. He was always interested in the artist, the hero, and the problem of creativity. Nietzsche was a great influence on him, and from Schopenhauer he derived the concept of "*will*," which was the core of his psychology.

He was actively involved with Freud's inner circle for a decade after Adler and Jung had severed their personal and theoretical bonds with the master. During the prestigious years within the charmed circle Rank was secretary to the Vienna Psychoanalytic Society, helped to establish and edit the journal *Imago* (1912–1924), and edited the *International Journal of Psychoanalysis.* He also continued to write widely on psychoanalysis as related to literature, art, and mythology.

Rank was becoming increasingly disenchanted with the lengthy and inflexible analytic therapeutic procedure. His best friend in the inner circle was Sandor Ferenczi, who also was critical of the standard Freudian therapeutic process. They both questioned the inflexible approach and especially the ever lengthening duration of the treatment due chiefly to Freud's insistence on intensive exploration without due regard for the welfare of the patient. Ferenczi was experimenting with a more actively directive approach, while Rank was manipulating the therapeutic situation as to physical arrangement of therapist and patient, setting a time limit to the analysis, and so forth. These problems were extensively considered in 1924 in a book written with Ferenczi, *The Development of Psychoanalysis.*[9] This publication marked the beginning of Rank's movement beyond the pale of Freudian psychoanalysis. Ferenczi, however, repudiated this book when he correctly perceived its theoretical implications.

Nonetheless, Rank was unprepared for and rudely shocked by the storm precipitated by his publication of *The Trauma of Birth*[28] in 1924 without prior discussion with the "Committee." He was not defended by Freud against the attack, since the latter considered the theory of birth trauma to be a challenge to the Oedipus complex as a primary explanatory principle in psychoanalysis. At that point Rank went to Paris to allow the storm to dissipate, but this effectively marked the end of his membership in the Vienna group.

Rank remained in Paris from 1926 to 1934. But beginning in 1927 he made frequent trips to the United States where he had acquired loyal followers, especially in the field of social work, whose development and direction he strongly influenced. He was associated chiefly with the Pennsylvania School of Social Work in Philadelphia, lecturing and holding seminars. He also taught and held seminars in New York City. In 1930 his recognition of the importance of will psychology for therapy was introduced at such a seminar. His friend and close adherent, Jessie Taft,[41] wrote of this period that "the 'will' focus liberated Rank finally from his Freudian past, from the biological, developmental details of family history as the core of analytic procedure, and from the old psychoanalytic terminology." During this period he wrote *The Technique of Psychoanalysis*,[27] the second and third volume of which were published in English translation as *Will Therapy*.[29] He also published during his stay in Paris a three-volume work on *Genetic Psychology*,[23] the third volume of which appeared in English translation as *Truth and Reality*.[29]

In 1934, mindful of the gathering political clouds in Europe, he came to New York, which was to be his final abode, although he traveled frequently to Philadelphia and other cities to lecture and conduct seminars. In New York he taught several courses at the Graduate School for Jewish Social Work. He also worked as a psychotherapist, often to the point of exhaustion. *Beyond Psychology*,[21] published posthumously by friends in 1941, was written during this period. After several illnesses he died in New York on October 31, 1939 from an overwhelming infection.

Introduction

Not only was Otto Rank one of the founders of psychoanalysis, but also he was a forerunner of ego psychology and of the neo-Freudian, culturalist modifications of Karen Horney and Clara Thompson. He introduced an approach to feminine psychology diametrically opposed to the derogatory attitude of the Freudian concept of the woman as a castrated male. His formulations would meet no criticism from the women's liberation movement today, for he took psychoanalysis from a male, father-oriented system to a female, mother-oriented construct.

Rank's major contributions may be summarized under three headings: (1) the concept of birth trauma; (2) the concept of "will" and the psychology of personality; (3) "will" or relationship therapy.

The Concept of Birth Trauma[13, 28]

Paradoxically Rank's break with classic psychoanalysis was the result of his brilliant elaboration of Freud's conception of birth trauma as the first condition of anxiety during the initial separation from the mother.[13] Following Freud's lead, Rank came to "recognize in the birth trauma the ultimate biological basis of the psychical . . . and a fundamental insight into the nucleus of the unconscious. . . ." Furthermore, Rank sought to demonstrate that man's entire psychical development reveals the importance of the birth trauma and the continually recurring attempts to master it.[28]

Because of this original trauma, separation becomes the most dreaded of human experiences, and the central human conflict is the desire to return to the womb, which wish arouses anxiety. Weaning is thus another separation from the mother rather than a frustration of orality. And genital impulses, at least for the male, are a desire to return to the mother. "The female, on the other hand, can achieve this goal only through identification with her father or brothers or her own child."[29]

Analytic experience convinced Rank that resolution of the transference meets its strongest resistance "in the form of the earliest infantile fixation on the mother." The analysis allows the patient to separate from the mother more successfully than in the past. Indeed, "the analysis finally turns out to be a belated accomplishment of the incompleted mastery of the birth trauma."[28] Patients of both sexes make a mother transference. It is the mother libido, as it existed in the prenatal physiological connection between mother and child, that must be resolved analytically.

Why did Freud oppose the development of his own concept? Rank concluded that his work threatened the patriarchal bias of classic psychoanalysis and the primacy of the Oedipus complex as an explanatory concept. In fact, as his ideas developed, Rank gave up his original effort to establish his theory biologically, as Freud had, and sought confirmation in a psychocultural interpretation of the mother-child relationship as the basic scheme of all human relationships. Obviously Rank's formulation made amends for Freud's weakest point, that is, his psychology of women, or better, the absence of such in his masculine-centered psychological system. Some students of culture consider this bias of Freud's a contribution to the development of our present destructive, selfish, and competitive society with its dehumanizing forces. If this be so, Rank's neglected emphasis could be an antidote to the social poisons that abound.

The Psychology of Personality and the Concept of Will

Rank's personality theory is built on a Freudian base from which it never completely deviated, certainly not to the extent that the psychologies of Jung and Adler did. Jung's constructive attitude and Adler's social emphasis influenced Rank's thinking, but his final formulations are distinctly his own and not a mere composite of these diverse influences. Unlike Freud he was not inclined to system building. In fact, Rank considered the Freudian attempt at universalization of theory as essentially deceptive. His later writings stressed the doctrine of psychological relativism. He insisted on the existence of psychologies rather than a psychology, on theories of personality rather than a theory.[21,29] He considered Freudian formalization of treatment as an obstacle to the evolution of a dynamic and creative therapy.

Central to Rank's psychology and theory of personality is his concept of "will," which stands in opposition to the patently passive and mechanistic psychoanalytic formulations. As a result of this emphasis on will, Rank was able to project a more positive, creative, and flexible view of personality structure and human behavior. The use of the somewhat discredited construct of will was perhaps unfortunate, but it was necessary to Rank's more active view of personality function.

Rank's will psychology is distinctly an ego psychology in which "will (is) a positive guiding organization and integration of self which utilizes creatively, as well as inhibits and controls the instinctual drives."[29] Furthermore, Rank[20] wrote: "For me the problem of willing, in a philosophical sense of the word, had come to be the central problem of the whole question of personality, even of all psychology." The voluntaristic concept of will emphasized the factors of choice, responsibility, and autonomy, which were largely ignored in Freudian thought.

The will is not merely a mediator between instinct (id) and society (superego), but a dynamic psychocultural entity with roots in biology and the sociocultural milieu. "Not only is the individual ego naturally the carrier of higher goals, even when they are built on external identifications, it is also the temporal representative of the cosmic primal force no matter whether one calls it sexuality, libido or id. . . . the ego . . . (is) the autonomous representative of the will and ethical obligation in terms of a self-constituted ideal."[29] Rank equates the will to the Freudian wish.

The personality is conceptualized largely in terms of impulse, emotion, inhibition, and will.

Inhibition arises from the fear and anxiety originally associated with the trauma of birth. Childhood is largely devoted to the mastery of this original anxiety. Inhibition is thus a basic and autonomous intrapsychic process, rather than a secondary imposition from the environment as set forth in the Freudian notion of repression. On a more conscious level inhibition appears in the form of denial, both being an expression of negative will. As forms of resistance in analysis, they are to be transformed into positive expressions of will. Stubbornness, willfulness, and disobedience are forms of negative counterwill.

"Impulse is conceived of as a dynamic aspect of will. Its neutrality is in opposition to the biological notion of instinct. To Rank impulse is a concept between instinctive drive and social conditioning, and its formulation permits a more unified view of innate and acquired factors in the structure of personality as determinants of behavior."[13] Rank recognized basic drives other than sexuality. He especially emphasized the creative nature of will in the form of the creative impulse, not the result of sublimated sexuality. The creative will principle gradually came to have central importance in his thought analogously with the spiritual libido tendency in Jung's system.[13]

Rank[29] rejected the Freudian "over-evaluation of the power of the unconscious impulsive life in men, and . . . the under-evaluation of his conscious willing ego." He considered the entire gamut of human emotions to be essentially a phenomenon of consciousness. It reaches its highest form as an instrument of observation and knowledge of itself—self-consciousness. The latter strongly influences the superego "in terms of the self-constructed ideal formation and . . . a creative sense of the outer world." So that in the last analysis, to Rank,[29] psychology can only be "a psychology of consciousness."

Individuation and Personality Types

Rank was greatly concerned with the individualization of each unique personality, which development he conceived of in terms of birth symbolism, as an "evolution from blind impulse through conscious will to self-conscious knowledge . . . (the) continued result of births, rebirths and new births."[29] The individual goes through three phases. In the first stage he wills for himself what previously has been determined by parental and social demands and his biological urges. The second period is characterized by a conflict between the will and counterwill, in which process of self-creativity the individual evolves his own ideals, standards, and aims. In the final stage there exists a truly autonomous ego.

The "average" or normal individual has remained at the first stage. There is a relative unity of personality functions, but there is little creativity. Conformity is the ideal. The second stage, that of conflict, gives rise to the neurotic type, with the tendency to self-criticism, feelings of inferiority and guilt. Estranged from the ideals of society, he is, nonetheless, unable to create his own values since that would depend on his own self-acceptance. In general, this personality functions in an oppositional manner.

The final stage of individuation culminates in the artistic or creative type, the highest integration of will and spirit. Creativity is its expression. Such an individual can accept his own ideals and values without the compulsion to impose them on others. To be one's self is the aim and ideal, neither driven by impulses nor neurotically restrained by a superego. Such an individual epitomizes the autonomy of the ego with a self-constituted ideal. Repeatedly Rank decried the fact that Freudianism deprived the personality of its inherent awareness and potential autonomy with responsibility, creativity, and ethics.

In the creative individual the expression of will predominates, whereas inhibition does so in the neurotic. The artistic type reconciles separation or individuation and the need for union in a constructive manner, but the neurotic is frustrated, incapable of integration of the conflictual trends in life, yet is unable to choose the average way out. He becomes involved in trivia to avoid independent behavior or is either compliant or rebellious. The neurotic fears life and preserves himself by a compulsive relationship with others. But the neurotic may become the antisocial or criminal type, which Rank delineated as a special personality outcome. In this detached psychopathic type impulse dominates behavior. He has a fear of death, and union with another threatens his individuality.

It is apparent that the three special personality types (creative, neurotic, and antisocial) arise as the result of an imbalance of the personality-organizing principles of will, inhibition, and impulse. The normal individual is more harmoniously integrated than these special types, but also less creative. However, Rank relates the neurotic to the artistic, the former being a miscarriage of the artistic temperament. The neurotic, therefore, is a failure in creativity rather than in normal development. This conception is a reflection of Rank's objection to the psychoanalytic derivation of creativity on the basis of sublimation.

Will or Relationship Therapy

The heart of Rank's psychotherapy is the concept of relationship, in which he emphasized the emotional dynamics experienced within the analytic situation as the essential therapeutic agent. Contrariwise, he minimized the value of intellectual learning, making the unconscious conscious, and theoretical insight. The therapy must be a genuinely creative experience for both the patient and therapist, albeit in a patient-centered approach, in which the individual is to be understood from himself and not from theoretical presuppositions. The analytic situation is focused on the present and on what is new in the therapeutic activity rather than on what is old and repetitious of past patterns of behavior. There is greater concern for the form of reaction than for the specific content, especially that dealing with the infantile period.

Of special significance is the device of end setting, enabling the analyst to shorten the therapy and, by setting the "time limit," to effect a more gradual solution of the pivotal problem of separation, which Rank considered so significant in therapy and general life adjustment because of his mother-centered interpretation of anxiety and dependence. This procedure combats excessive dependence on the therapist.

Reactions in treatment depend not only on the patterns of the patient but also on what is new in the therapeutic relationship and on the personality of the therapist. Since the neurotic is more akin to the artistic type of personality, the therapist might best be of the creative type himself to aid the patient toward self-realization.

Guilt is a nuclear problem of the neurosis, indeed, of personality development in general, and of every human relationship. It is of particular importance in therapy because the patient cannot accept the help he seeks without developing guilt feelings. Therapy can only succeed if the relationship in it is effective in making "it possible for the sufferer to take from the other what he needs emotionally without getting guilt feelings."[29]

The goal of therapy, therefore, is to help the patient to accept his individuality and will without guilt. In this light resistance is not an obstacle in treatment but rather a negative will manifestation, which is a positive striving toward independence and is to be encouraged and directed rather than to be eliminated. What is therapeutically effective "is the same thing that is potent in every relationship between two human beings, namely, will."[29] No therapeutic progress is possible without recognition of the positive nature of resistance.

Even hostility in the analysis may be an expression of counterwill directed against dependency. Of course, such hostility is to be utilized to aid the patient to overcome his life fear and to face separation and individuation. The recognition of primary fear of life and the fear of death is of great importance in will therapy. Growing independence in treatment may arouse these fears and success may intensify moralistic guilt. The Rankian therapist supports movement toward independence and constantly subtly assures the patient that he can be loved without fear of dominance. But too much love or acceptance may arouse death fears, since the neurotic both seeks and fears close union. However, the increased fear of death (reunion with the mother) may intensify the drive toward independence, a healthy cycle resulting. The therapist's early disclosure to the patient of his death fears helps to minimize the vacillation between trust and fear.

Rank's therapy, like his psychology, has a prominent social orientation, stressing the interdependent relation of the individual and the social order in which self-realization is sought. Society is not merely an obstacle to libidinal strivings but also a necessary condition for expression. Constructive interrelation of the individual and society, rather than their opposition in the Freudian sense, is the keynote of individuation according to Rank. The therapist encourages a realistic partialization in behavior toward a positive and eventually creative expression of will, accepting and even applauding the manifestations of will within mutually understood limits, and consistently rejecting the inappropriate, repetitive roles assigned to him from the past. In the terminal phase of therapy the analyst "is transformed from assistant ego to assistant reality." The neurotic views reality as hostile and painful. The average individual can use reality therapeutically, something the neurotic can only attain in therapy. "In the last analysis therapy can only strive for a new attitude toward the self, a new valuation of it in relation to the past, and a new balancing in relation to and by means of, present reality."[29]

Sandor Rado (1890–1972)

Biographical Notes

Sandor Rado was born on January 8, 1890 in Hungary. He became acquainted with Freud's writings in 1910 through a paper by Sandor Ferenczi. Excited by these new ideas he changed his career direction to study medicine at the University of Budapest and to specialize in psychiatry and psychoanalysis. While pursuing his analytic training (1911–1915), he became a member of Ferenczi's informal study group and in time his warm friend. He was strongly influenced by this pioneer, who was an intimate and favorite of the master himself.[16]

In 1913 Rado went to Vienna to meet Freud and to hear him lecture on the interpretation of dreams. During this year he aided Ferenczi in organizing the Hungarian Psychoanalytic Society. Significantly his first paper before this Society was on the contributions of psychoanalysis to biological problems, thus indicating early in his career the very attitude that was later to alienate Freud, namely, Rado's insistence that "psychoanalysis must seek to win its logical place in the system of medical sciences or else . . . float in mid-air."[18]

He undertook a personal analysis in 1922 with Karl Abraham in Berlin. This procedure was then becoming accepted as a necessary part of analytic training. He soon became a faculty member in the Berlin Psychoanalytic Institute, of which Abraham was the director. In 1924 Freud appointed Rado executive editor of the important psychoanalytic journals, the *Zeitschrift* and *Imago*.

Being aware of the political direction developing in Germany, he emigrated to the United States in the mid-1930's to become the educational director of the New York Psychoanalytic Institute, to whose establishment along the lines of its Berlin counterpart he had earlier contributed. This move further alienated Freud, who apparently was less alert to the political atmosphere and who was already unsympathetic to Rado's insistence that psychoanalysis remain exclusively a part of medicine. In 1939 Rado repudiated Freud's psychobiological libido theory, but called for a psychoanalysis based "on our established biological knowledge of man."

As many of his papers indicate, Rado was intensely interested in psychiatric and psychoanalytic education. Because of theoretical differences he ceased teaching at the New York Psychoanalytic Institute in 1941, and from 1944 to 1955 he was the first director of the newly established Psychoanalytic Clinic for Training and Research, which was part of of Psychiatry. He lived in retirement in New York City until his death on May 14, 1972. the department of psychiatry of the College of Physicians and Surgeons of Columbia University. From 1958 to 1967 he was professor of psychiatry and dean of the New York School

Introduction

During his many years of clinical and educational experience within the classic Freudian framework, Rado became increasingly aware of the deficiencies of this theoretical structure in light of scientific methodology and biological knowledge. He deplored the use of numerous undefinable analytic concepts and an unscientific vocabulary without generally established definitions. In fact, his newly introduced terminology is an outstanding feature of his "system." In seeking a remedy for this situation, he became, as Robert Heath[11] recently wrote, "an insufficiently recognized pioneer among the few psychoanalysts who attempted to base psychodynamic formulations on sound physiologic data." This direction was really a return to Freud's original psychobiological orientation before his progress toward a more personal object-relational ego psychology. Rado's work programmatically was based on three goals: the rigorous application of scientific method to psychoanalysis, a thorough re-examination and reevaluation of Freud's theories, and finally, "a complete resystemization of psychodynamics." It culminated in *adaptational psychodynamics,* a unified conceptual system with a holistic view of man achieved by combining the complementary introspective and inspective techniques of observation. Continually aware of the mind-body problem, Rado sought to evolve a psychoanalytic theory related to brain physiology and biological evolutionary principles.

Adaptational Psychodynamics

This systematic revision of classical Freudian principles sought to encompass "a solid mass of empirical findings" in a less abstract manner based on a hierarchical theoretical structure, whose central concept is *adaptation* in the basic evolutionary survival sense. Behavior is evaluated in terms of its usefulness for the preservation of the individual and his species; it is always motivated and is necessarily regulated.

Although Rado recognized the pivotol position of pleasure in the regulation of behavior, he did not accept Freud's generalization, in the libido theory, that all desire for pleasure is sexual in origin. In a further critique Rado[18] considered that in Freudian instinct theory the role of the emotions in behavior, healthy or deranged, is indistinct and even obscure, for they appear "as one of the manifestations of hypothetic instincts rather than as elementary facts of clinical observation." Instead of finding motivation in the activity of hypothetical energies cathecting equally imaginary psychic structures in a tripartite mind, adaptational psychodynamics seeks for the source of human motivation in the emotions mobilizing the organism to fulfill its survival needs.

Influenced by Walter B. Cannon, Rado[17] divided the emotions into two categories: "the emergency emotions based on present pain or the expectation of pain, such as fear, rage, retroflexed rage, guilty fear, and guilty rage; and the welfare emotions based on present pleasure or the expectation of pleasure, such as pleasure, desire, affection, love, joy, self-respect and pride." Adopting Ferenczi's formulation, Rado defined adaptive measures as autoplastic (changes in the organism) and alloplastic (changes produced in the environment). These adaptations are accomplished through learning, creative imagination, and goal-directed activity.

The individual's needs are of two classes, as Sullivan indicated. There are "aboriginal needs" for safety, growth, repair, and reproduction; and there are social or cultural needs for security, self-esteem, and self-realization. Need is a theoretical construct, manifesting itself in feelings, thoughts, and impulses as motivated forces. The individual becomes aware of these forces through pleasurable or painful tensions, causing appropriate reactions toward successful control and resultant satisfaction.

The integration of behavior is achieved by the mind. This concept had been further removed from the brain by Freud's unconsious mind. Rado sought to rectify this unscientific situation by considering the brain's activity as partly "self-reporting" and partly "nonreporting." Consciousness or awareness is the result of the self-reporting process. It is known to us by introspection and it is psychodynamic. The process of reporting is open to the methods of inspection and is physiological. Nonreporting activity can carry over into the self-reporting range and enter awareness. Conscious activity regularly passes into the nonreporting area, as in memory storage. Most of brain activity is nonreporting but, nonetheless, motivational, although its meaning is not accessible to ordinary measures of inspection. Meanings may be inferred as from the material of free association.

In Rado's comprehensive terminology *psychodynamic cerebral system* refers to (1) the entire self-reporting range of brain activity, and (2) to the range of nonreporting activity accessible to psychodynamic investigation and also to physiological methods. And "the foremost objective of adaptational psychodynamics is to discover the mechanisms by which the psychodynamic cerebral system accomplishes its integrative task."[17]

The integrative activity of this system is based on four hierarchical ordered units reflecting phylogenesis. In ascending order these units are the hedonic level, and the levels of brute emotions, emotional thought, and unemotional thought. The hedonic or lowest level is governed by the pleasure principle based on the expectation that what is good for survival will be signaled by pleasure, while pain announces a threat to one's organic integrity. This is the level of "hedonic self-regulation," and the recognition of the primacy of this regulatory factor is "a cornerstone of the theoretic structure of adaptational psychodynamics." It corresponds to the protozoa level in the evolutionary scale. There is little thought for the future. The organism merely moves toward pleasure and away from noxious influences, which is obviously of great survival value. Although pain-pleasure physiology is dominant at this level, it also operates in higher levels of integration. Mechanisms for the avoidance of painful stimulation exist on all levels; Rado referred to these as "riddance mechanisms." Repression is a psychological riddance mechanism; coughing, sneezing, and vomiting are physiological examples of riddance behavior.

The emergency and welfare emotions control the levels of brute emotions and emotional thought. The level of preverbal brute emotion parallels the metazoan stage in evolution. Fear and rage appear at this level and function, like pain, adaptively as emergency emotions. The next higher level, that of emotional thought, is a tempered development of brute emotion. Brute fear and rage are moderated, becoming apprehension and angry thought. While higher cortical function is apparent, it is still inferior to the ability of unemotional thought to discriminate and interpret its environment. At this level there are further differentiation and refinements of the basic emotions. Learning is increased along with a lengthening of the period of dependence. Reasoning, symbolism, trial-and-error logic, and control by reward and punishment develop. But the individual is selective, not objective, seeking to rationalize the controlling emotion.

The highest level is unemotional thought, crowned by reason, common sense, objectivity, and the potential for intelligent, self-disciplined behavior. Culture and the survival of civilization depend on the increasing function of this level of human evolution.

The behavior of the entire organism is integrated by any one of these levels of organization or usually by combinations of them. This integration may be nonreporting as well as self-reporting. The passage from the former to the latter is governed by a precautionary "pain barrier." Messages from the individual to his social environment are controlled by a similarly monitory "social pain guard."

On the pinnacle of the psychodynamic integrative apparatus, Rado[17] postulates the "action self." "Of proprioceptive origin . . . it then integrates the contrasting pictures of total organism and total environment that provide the basis for the selfhood of the conscious organism. . . . These integrations . . . represent highly complex organizations composed of sensory, intellectual, emotional and motor components." At first the individual considers his willed behavior as all-powerful. In time this sense of omnipotence yields to reality testing, and the individual distinguishes between a realized self and an idealized self. Moreover, a self-moderating and self-judging mechanism, the conscience, arises as the precipitate of parental and other authority. Conscience is governed by the principle of obedience and serves most to control rage and defiance.

This model of man and his behavior appears to neurologize and mechanize him. Rado[17] hastens to reassure us that "man is not a computing machine; his emotional needs must be met if he is to function in a stage of health and to prosper. . . . Hedonic control is of the essence of the biologic organism. . . . If (it) could be removed, the residual entity would be neither human nor an organism."

Classification of Behavior Disorders

Based on this theoretical system, Rado[17] offered a new classification of behavior disorders, which he defined as "disturbances of psychodynamic integration that significantly affect the organism's adaptive life performance, its attainment of utility and pleasure." Owing to our great ignorance of brain physiology, normal and disordered, we must content ourselves with the elucidation of the psychodynamic aspect of etiology.

Safety is the organism's prime survival concern. The emergency control function of the psychodynamic cerebral system accounts for disordered reactions. Overproduction of emergency emotions (rage, fear, guilty fear, and guilty rage) prevents the organism's effective handling of the demands of daily life. These excessive reactions themselves become a disorganizing source of threat from within. Such failures of emergency control Rado termed "emergency dyscontrol," the simplest form of behavior disorder, which is a factor in the more complex disorders. The classification is hierarchical, in ascending complexity of pattern and mechanism.

Class I. *Overreactive disorders*

1. Emergency dyscontrol: the emotional outflow, the riddance through dreams, the phobic, the inhibitory, the repressive, and the hypochrondriac patterns, the gainful exploitation of illness.
2. Descending dyscontrol: the overflow of emergency emotions into the various organ systems, producing the psychosomatic disorders.
3. Sexual disorders: Failures and impairment of standard sexual activity. Impotence, frigidity, fetishism, sadomasochistic behavior, homosexuality, fire setting and shoplifting as sexual equivalents.
4. Social overdependence: search for a substitute parent, compulsive competition, avoidance of competition, self-destruction, defiant behavior.
5. Common maladaptation: combinations of groups 3 and 4.
6. The expressive pattern: expressive elaboration of common maladaptations, such as ostentatious self-presentation, dreamlike interludes, expressive complication of incidental disease, and the like.
7. The obsessive pattern: obsessive elaboration of common maladaptations, such as broodings, rituals, tics, stammering, bedwetting, nail-biting, grinding of teeth in sleep, and the like.
8. The paranoid pattern: nondisintegrative elaboration of common maladaptations; includes hypochondriacal, self-referential, persecutory, and grandiose behavior.

Class II. *Mood cyclic disorders*

Cycles of depression, elation, and alternate cycles: cycles of minor elation, of depression masked by elation, of preventive elation.

Class III. *Schizotypal disorders*

1. Compensated schizo-adaptation.
2. Decompensated schizo-adaptation (pseudoneurotic schizophrenia).
3. Schizotypal disorganization with adaptive incompetence (disorganization of action self).

Class IV. *Extractive disorders*

The ingratiating and extortive patterns of transgressive behavior.

Class. V. *Lesional disorders (organic)*

Class VI. *Narcotic disorders*
Pattern of drug dependence.

Class VII. *Disorders of war adaptation (war neurosis)*

Rado also contributed special formulations on schizotypal organization, depression, homosexuality, and psychotherapy.

Schizotypal Organization

Rado always stressed the etiological point of view. He proposed the heuristic concept of "pathogenic phenotypes," and he accepted the genetic origin of schizophrenia. An individual burdened by such an heritage is a "schizophrenic genotype" or, in short, a "schizotype."[16] The traits of this type are "schizotypal organization" with manifestations termed "schizotypal behavior."

The schizotypal organization is due to two basic kinds of damage to the psychodynamic cerebral system. There is a diminished capacity for pleasure and distorted awareness of the individual's own body. The basis for these disturbances is unknown, but Rado suggested that a "molecular disease" (Linus Pauling) of genetic origin will be found to account for them. Since pleasure is of crucial psychodynamic importance ("pleasure is the source and fulfillment of life and death is its problem"), its deficiency disturbs the total operation of the integrative system, resulting in an expansion of the emergency emotions and a corresponding diminution of the welfare emotions. Furthermore, since the proper function of the action self depends on proprioceptive information, a deficiency in this function is crucially damaging. The two schizotypal deficiencies render the action self subject to the fragmentation described clinically. This, in turn, may be the chief source of the sense of inferiority and the excessive fear of death.

The individual's response to this genetic damage is the creation of a compensatory form of adaptation, characterized by "extreme overdependence, operational replacement in the integrative apparatus, and a scarcity economy of pleasure." There is also a great increase in magical craving, which manifests an extreme lack of self-reliance. Operationally the schizotypal uses cold intelligence to compensate for lack of warm emotions. The success of the entire compensatory adjustment rides on the balance of the schizotype's assets and liabilities, such as degree of genetic damage, intelligence, talent, and socioeconomic class. An unfavorable balance may generate tensions beyond the individual's adaptive capacity, and he may then decompensate to a pseudoneurotic schizophrenic level, and ultimately, if not immediately, to a stage of disintegrated schizotypal behavior with disorder of thought and behavior.

Depression

In line with his concept of pathogenic genotypes, Rado postulated a "class of mood cyclic phenotypes," characterized, among other things, by a frequent occurrence of depressive spells. Depression is a special form of emergency dyscontrol in which the patient has sustained a great loss or behaves in an "as if" manner. He reacts with the opposing emotions of rage and guilty fear. The latter prevails while most of the coercive rage is unrepressed and turned against the self in self-punishment and remorse, which are but a façade for his bitterness and wounded pride. This is the result of the puncturing of the postulated pain barrier at the beginning of the depression. Freud considered loss of self-esteem the result of the depression. Rado and other authors consider the loss of the love object as representing the essential loss of self-esteem. Others see this not as a loss of self-esteem but as a loss of self-confidence in one's ability to master the environment. It is noteworthy that clinically the depressed patient is converted from a mature, self-reliant adult to a frightened, helpless child. Recovery occurs only with the patient's regaining of the capacity for true pleasure, not the spurious type that is the result of "the mere illusion of control."

Homosexuality

Rado severely criticized the Freudian theory of bisexuality. The genital pleasure function is just part of the entire "pleasure organization."[17] The assumption of a universal constitutional homosexual component as an inevitable result of bisexuality is erroneous. Numerous and diverse aspects of interpersonal relations have been assigned to "unconscious homosexuality." The facile assumption of a homosexual constitutional component has stalled the investigation of what such an "element" could really be. Therapeutically the patient has been reduced to despair by the notion of a struggle with an innate predisposition.

The basic problem in the field of genital psychopathology, according to Rado, is to find the reason for the individual's application of aberrant stimuli to standard genital apparatus. He considered the chief causal factor to be anxiety that forces the ego-action system to seek a "reparative adjustment." He emphasized the essential heterosexuality of man, based on biological foundations and the truly universal social institutionalization of marriage with its early imprinting on the child. The resultant movement toward heterosexuality may be damaged in the child by parental threats. Rado concludes that "the psychodynamics underlying the behavior of the homosexual has many patterns," but the basic dynamic is "fear of the genitals of the opposite sex."

Psychotherapy

Rado considered psychotherapy to be applied psychodynamics, a "problem of controlled intercommunications." The classical psychoanalytic technique fosters the patient's childlike emotional dependence. In a word, it operates at the "parentifying level." There is no counterbalancing therapeutic force impelling the patient to deal more successfully with his actual life situation. "To overcome repressions and thus be able to recall the past is one thing; to learn from it and be able to act on the new knowledge, another." The therapist must counter the patient's attempt to "parentify" him. To succeed the therapist "must first bolster up the patient's self-confidence on realistic grounds." Interpretation has a dual function, to modify "the patient's present life performance, his present adaptive task," and also to regulate behavior within the therapy. Rado stressed the need to minimize undue regression in the therapy and to prevent evasion of the current adaptive tasks. Nonetheless, Rado insisted that his adaptational technique seeks "the Freudian goal of total reconstruction."

Rado, it is conceded, has undertaken one of the most systemic revisions of standard psychoanalytic theory and practice. But he remained true to Freud's early tie to biology and physiology, and was basically concerned with the machinery of personal existence and insufficiently with its quality, with the mechanisms of behavior rather than with "the meaningful personal experience that is the essence of the personal self."

Bibliography

1. BAKER, E. F., *Man in the Trap*, Macmillan, New York, 1967.
2. FERENCZI, S., *Final Contributions to the Problems and Methods of Psychoanalysis*, Basic Books, New York, 1955.
3. ———, (1926), *Further Contributions to the Theory and Technique of Psychoanalysis*, Hogarth, London, 1950.
4. ———, *Sex in Psychoanalysis: Contributions to Psychoanalysis*, Brunner, New York, 1950.
5. ———, (1924), *Thalassa: A Theory of Genitality*, Psychoanalytic Quarterly, New York, 1938.
6. ———, *Versuch einer Genitaltheorie*, Internationaler Psychoanalytischer Verlag, Vienna, 1924.
7. ———, ABRAHAM, K., SIMMEL, E., and JONES, E., *Psychoanalysis and the War Neuroses*, International Psychoanalytic Press, London, 1921.
8. ———, and HOLLOS, S., *Psychoanalysis and the Psychic Disorder of General Paresis*, Nervous and Mental Diseases Publishing Co., New York, 1925.
9. ———, and RANK, O., *The Development of Psychoanalysis*, Nervous and Mental Diseases Publishing Co., New York, 1925.
10. GUNTRIP, H., *Psychoanalytic Theory, Ther-*

apy, and the Self, Basic Books, New York, 1971.

11. Heath, R. G., *Perspectives for Biological Psychiatry*, in Cancro, R. (Ed.), *The Schizophrenic Syndrome*, Vol. 1, Brunner/Mazel, New York, 1971.
12. Higgins, M., and Raphael, C. M. (Eds.), *Reich Speaks of Freud*, Noonday Press, New York, 1967.
13. Karpf, F. B., *The Psychology and Psychotherapy of Otto Rank*, Philosophical Library, New York, 1953.
14. Lorand, S., "Ferenczi, Sandor," in Eidelberg, L. (Ed.), *Encyclopedia of Psychoanalysis*, Free Press, New York, 1968.
15. Pfister, O., *Psychoanalysis and Faith*, Basic Books, New York, 1963.
16. Rado, S., *Adaptional Psychodynamics: Motivation and Control*, Science House, New York, 1969.
17. ———, *Psychoanalysis of Behavior. Collected Papers, 1922–1956*, Grune & Stratton, New York, 1956.
18. ———, *Psychoanalysis of Behavior, Collected Papers*, Volume 2, *1956–1961*, Grune & Stratton, New York, 1962.
19. Raknes, O., *Wilhelm Reich and Orgonomy*, Penguin Books, Baltimore, 1971.
20. Rank, O., *Art and Artist, Creative Urge and Personality Development*, Alfred A. Knopf, New York, 1932.
21. ———, *Beyond Psychology*, Haddon Craftsmen, Camden, N.J., 1941.
22. ———, *Das Inzest-Motiv in Dichtung und Sage*, Franz Deuticke, Leipzig and Vienna, 1912.
23. ———, *Grundzuge einer Genetischen Psychologie auf Grund der Psychoanalyse der Ich-Struktur*, Vol. 1, *Genetische Psychologie*, Vol. 2, *Gestaltung und Ausdruck der Personlechkeit*, Vol. 3, *Wahrheit und Wisklichkeit*, Franz Deuticke, Leipzig and Vienna, 1927–1929.
24. ———, *Modern Education, A Critique of Its Fundamental Ideas*, Alfred A. Knopf, New York, 1932.
25. ———, *The Myth of the Birth of the Hero*, Nervous and Mental Diseases Publishing Co., 1914.
26. ———, *Psychology and the Soul*, University of Pennsylvania Press, Philadelphia, 1950.
27. ———, *Technik der Psychoanalyse*, Vol. 1, *Die Analytische Situation (The Analytic Situation)*, Vol. 2, *Die Analytische Reaktion (The Analytic Reaction)*, Vol. 3, *Die Analyse des Analytikers (The Analyst and His Role)*, Franz Deuticke, Leipzig and Vienna, 1926, 1929, 1931.
28. ———, *The Trauma of Birth*, Brunner, New York, 1952.
29. ———, *Will Therapy and Truth and Reality*, Alfred A. Knopf, New York, 1950.
30. ———, and Sachs, H., *The Significance of Psychoanalysis for the Mental Sciences*, Nervous and Mental Diseases Publishing Co., New York, 1916.
31. Reich, I. O., *Wilhelm Reich: A Personal Biography*, Avon Books, New York, 1970.
32. Reich, W., *Character Analysis*, Orgone Institute Press, New York, 1949.
33. ———, *The Discovery of the Orgone: The Function of the Orgasm*, Noonday Press, New York, 1971.
34. ———, *The Invasion of Compulsory Sex-Morality*, Farrar, Straus & Giroux, New York, 1971.
35. ———, *Listen, Little Man*, Noonday Press, New York, 1971.
36. ———, *The Mass Psychology of Fascism*, Farrar, Straus & Giroux, New York, 1970.
37. ———, *The Murder of Christ*, Noonday Press, New York, 1971.
38. ———, *Selected Writings: An Introduction to Orgonomy*, Farrar, Straus & Giroux, New York, 1960.
39. ———, *The Sexual Revolution*, Farrar, Straus & Giroux, New York, 1969.
40. Sachs, H., *Freud, Master and Friend*, Harvard University Press, Cambridge, 1944.
41. Taft, J., *Otto Rank*, Julian Press, New York, 1958.

CHAPTER 42

THE EXISTENTIAL SCHOOL

James L. Foy

Introduction

THE EXISTENTIAL SCHOOL in contemporary psychiatry is hardly a formal school at all, having no institutional arrangements, no curriculum, and no certification of candidates. The school is more like a movement or loose confederation of practitioners and theoreticians who share a common approach to the human person and to the varied distortions of human potential. The actual common denominator of this broadly based group of psychiatrists and psychotherapists would be their grounding in existential thinking, a major trend in modern philosophy that has left its imprint on many aspects of twentieth-century culture.

Ordinarily we do not think of academic philosophy as having an intimate connection with the developments in psychiatry over the past 100 years. Idealism, pragmatism, vitalism, logical positivism, and linguistic analysis have all contributed to the climate in which psychiatry has matured as a science and an art; however, these philosophical endeavors have not influenced theory or practice in specific ways. As a matter of fact, psychiatrists have tended to be wary of the abstractions and controversies in philosophical thought, perhaps because they have been busy promulgating abstractions and controversies of their own. For example, a major philosophical issue, the mind-body problem, is generally given brief treatment in modern medicine and psychiatry, and it is dismissed too quickly as solved by a superficial nod in the direction of such concepts as interactionism, psychophysical parallelism, or epiphenomenalism. Careful and prolonged questioning of the mind-body problem will result in a tangle of unsettling and contradictory answers, which are so often swept under some convenient conceptual rug.

Freud as a builder of psychological theory was also a philosopher of sorts, a particularly inconsistent and even eccentric philosopher. It is remarkable that Freudians have been able to accommodate psychoanalytic theory to such divergent philosophies as neo-Kantian idealism, epiphenomenalism, and mechanistic materialism. Existential philosophy offers the example of a consistent and systematic philosophical penetration of psychiatric theory and practice. This is the only instance where a rigorous philosophical method and viewpoint has had a decisive influence on the formation of a school of psychiatry, with its own constructs and clinical applications.

If existential philosophy has left its mark on psychiatry this is only one of many impacts it has had upon the entire edifice of modern culture. Its influence has been conspicuous in all the arts, humanities, social sciences, religion and theology. It has proposed a critique of reductionistic and dehumanizing tendencies in contemporary science and technology, and at the same time it has revitalized the older humanistic tradition with ideas open to change and the expansion of human possibilities. In politics existential thinking has been welcomed in both conservative and radical wings. The career of Jean-Paul Sartre gives testimony to an existentialist's participation in all the controversies of our time: political, social, artistic, and intellectual.

Some Basic Definitions

Existentialism as a term almost defies definition. As an "ism" or ideology, the noun form is universally rejected by existential thinkers, who remain suspicious of global systems and catchall labels. The adjective form, existential, is used more widely and seems a necessary generic term to be applied to thinkers, psychiatrists, and psychotherapists who share an existential orientation. Existential then pertains to a viewpoint on man that recognizes the unique place and presence of individual human existence and the centrality of human action and freedom within human existence. "Existence" is a word strictly reserved for man and never applied to the being of other nonhuman things. Warnock[84] describes the existential thinker as promoting human freedom, the possibilities of action based on choice, and the primacy of an ethical commitment. Existential philosophy is a practical philosophy. Barrett[4] offers a similar definition. Copleston[22] emphasizes the dramatic quality in the existential viewpoint: the isolation in freedom of a self-transcending subject, exposed and threatened but choosing who he is to become under the staggering limits of his human condition. Olson,[62] in his useful introduction to the existential mode of thought, indicates the principal themes related by the philosophers, and, I might add, by the psychiatrists as well. These key themes are: values, anxiety, reason and unreason, freedom, authenticity, the other person, and death.

Phenomenology is a term closely related to the existential approach to man, his situation, and his world. Phenomenology is the method of the existential thinker, the disciplined and rigorous manner in which he describes and elucidates human existence. Phenomenology is defined as a science of the subjective, the descriptive analysis of subjective process, the analysis of things as they appear to human consciousness. To the phenomenologist consciousness is not some vague container or receptacle of representations; it is strictly conceived as an act of intentionality, a process of connecting and relating subject to object, a subject pointing to, intending an object. All existential thinkers acknowledge their debt to the founder of phenomenology as a philosophical discipline. This seminal philosopher was Edmund Husserl, a genius of the century who developed from logician to pure philosopher to existentialist in his late years. By far the best exposition of phenomenology and the phenomenological movement is the two-volume work by Spiegelberg.[74]

Ontology is defined as the theory of being as being, and it is the fundamental enterprise of philosophical speculation. The term is more or less synonymous with metaphysics, although the existentialists prefer ontology since it leads back to the ground of being rather than beyond material substances. Ontology always refers to the most basic of sciences. Ontological, therefore, is the adjective form that pertains to the basic structures of being. Ontic is an adjective form that pertains to particular beings or entities and the facts about them.

Anthropology is a term used by the existentialists in a way that is quite different from its common usage in this country, where it is more or less synonymous with ethnology or the science of human culture. In European philosophy anthropology is defined, in very literal fashion, as the theory of man and his essential human structures. The compound term "philosophical anthropology" clarifies this concept for the British or American reader.

Dasein is a German word that one finds often enough untranslated in the existential texts. This term means literally there-being and it is sometimes rendered in English as human existence or human reality. It could also be rendered as being-here-and-now, keeping some of the human immediacy of the original German. Daseinanalysis, therefore, is defined as the descriptive analysis and elucidation of a particular human existence, and as such is a term often employed by existential psychiatrists.

Philosophical Perspectives

Existential thinking and more recently existential phenomenology are products of European intellectual history. The original sources come from writers in Scandinavia, Russia, Germany, France, Italy, and Spain. British and American philosophy, until quite recently, was unreceptive to this style of thinking. An important exception is William James,[34] whose lectures on religious experience have an existential theme and temper. The roots of existential thought lie deep in history, and their outline may be discerned in works by the pre-Socratic Greeks, Augustine, Dante, Pascal, and certain German Romantics. The mid-nineteenth century affords the time and place for the full, bold statements of existential thought in revolt against the hypocrisies, systems, and utopias of that age. Dostoevsky's short novel *Notes from the Underground* is a disturbing overture to existential revolt.[39] The novelist, who was always a perceptive psychologist, dissects the dehumanizing effects of the advancing industrial age, so enthralled with scientism, progress, and the crystal palaces of the great exhibitions. Dostoevsky asserts an alarming and crude version of human freedom and responsibility in the face of mass society, conformity, and scientific naturalism.

Friedrich Nietzsche is another more forceful example of the European intellectual in existential revolt against the received ideas and complacencies of his time. Philosopher, poet, and psychologist, he became a specialist in unmasking the self-deceptions that enabled his contemporaries to live in a divided world, a world split between ideal and reality, religion and ungodliness. In a series of cruel and unmerciful books he spoke for a ruthless sincerity and a facing up to the epidemic phoniness of modern life.

Kierkegaard

By far the most important single innovator within nineteenth-century existential thought was Sören Kierkegaard, who died in 1855 after a short brilliant career as literary gadfly, serious philosopher, and religious thinker in his native Denmark. Kierkegaard's voluminous works were rediscovered in the early years of this century and translated into all European languages. His influence as the father of existential thinking has been persuasive and profound, and his influence continues to grow in the human sciences, such as psychiatry, because of the provocative and heuristic value of his work. This is not the place to survey his prolific and varied output; however, his role as an insightful psychologist of human reality deserves special attention. In an early work like *Either/Or*,[41] a miscellany of fiction and commentary, Kierkegaard analyzes the pleasure principle in the individual's everyday life. His exposition of the erotic and its subtle effect upon many areas of behavior, his discussion of techniques of self-deception, defensive posturing, and hidden motivations, all show him to be a depth psychologist very much in advance of his contemporaries. His repeated excursions into self-analysis indicate that he had a clinician's grasp of "identity crisis" and the unconscious workings of conscience. These self-analyses also provide a theoretical framework for understanding neurotic defenses and syndromes. In many books he returns to an analysis of anxiety, boredom, indecision, the absurd, commitment, and the lack of it.

Acting out of religious conviction and suppositions, Kierkegaard encounters human reality in its strictly individual and personal form, and not in its abstract form as man in general. He always writes about *my* existence in all its concrete and irreplaceable uniqueness. His self-psychology receives its most thorough elucidation in his book on despair, *Sickness unto Death.*[15] This masterful existential analysis of depression is widely applied in existentially oriented psychotherapy. After discussing the possibility of a despair that is unconscious or the despairing unconsciousness of having a self, Kierkegaard goes on to elaborate in detail two conscious forms: (1) despair at not willing to be oneself, the despair of weakness; and (2) despair at willing despairingly to be oneself in defiance and rigidity. Although his own "therapy" is firmly based on religious experience and recognition of an eternal self, these clinical insights into the torments of the despairer's emotional life lend themselves to a natural psychotherapeutic strategy and procedure.

Kierkegaard's insistence upon the primacy of existence, subjectivity, personal development, and human limits places him as a mentor to all twentieth-century existential thinkers. His legacy is in the unsparing manner in which he faced his own complex fate and inwardness and out of the struggle bequeathed intellectual and spiritual riches to later generations.

French and German Existential Thinkers

Kierkegaard's existential point of view has taken firm roots in philosophical reflections by French and German writers during the first half or more of this century. Gabriel Marcel's independent, sensitive, humanistic inquiry into a range of existential themes has attracted attention in psychiatry but has gained little influence. He has lectured several times in the United States, delivering the William James Lectures at Harvard in 1961, which summarized his life's work.[51] Marcel has had an affiliation with American philosophy, particularly with Royce and Hocking. His work should be better known by psychiatrists and psychotherapists, since his emphasis on the intersubjective and participation coincides with familiar existential preoccupations of therapists. Marcel has reflected deeply on the human situation, on the body as relation to the world, on paternity, on creative fidelity, and on suicide.

Albert Camus and Jean-Paul Sartre are the most widely known figures in the French school. It is only in his earliest writings that Camus can be considered an existential thinker, and his essays *The Myth of Sisyphus*[19] are important as psychological explorations. The same holds true, of course, in regard to his early novels and plays. Sartre also is famous for his versatility as the author of novels, stories, dramas, autobiography, criticism, and philosophical texts. His evolution from ontological existentialist to existential Marxist provides evidence of both consistency and inconsistency in his thought. His lengthy and difficult text, *Being and Nothingness,* appeared in France in 1943. This book has had extraordinary influence on existential psychiatry, perhaps for the reason that it is read for its rich psychological analyses rather than for its strict ontological speculations.[66] Sartre contributes splendid phenomenological descriptions of the body, sexuality, desire, sadomasochism, hostile situations, character defenses, which he terms "bad faith," and most pointedly an extended analysis of human freedom and responsibility. Sartre's later Marxist-oriented revisionism had had little influence on psychiatry; however, the Scottish psychiatrist R. D. Laing[45] has shown a continued responsiveness to Sartre's work.

The last important existential philosopher from France is Maurice Merleau-Ponty, who died in 1961, in midcareer. He was associated with Sartre for a time and like him was deeply indebted to Husserl and Heidegger. Merleau-Ponty was trained in psychology and was a serious student of psychopathology and psychoanalysis, all of which highly recommends him to the existential movement in psychiatry, where his reputation has steadily grown since his death. His early work, *Phenomenology of Perception,*[58] an analysis of perception and the role of the body-subject, will likely have expanding influence in psychiatric and psychosomatic theory. His appreciation of brain-damaged states and their psychopathology and the problems of psychoanalysis[25] are thoroughly informed and underlie an astute, meaningful existential interpretation. More than the other existential thinkers, Merleau-Ponty speaks the clinical language of the psychiatrist, while retaining his position as a gifted and disciplined philosopher.

In Germany the first scholar to respond to Kierkegaard's message was Karl Jaspers, the psychiatrist and philosopher, who died in 1969 after a long, honorable academic career. He spent only seven years in clinical psychiatry at Heidelberg before joining the Faculty of Philosophy, although that period enabled him to publish his *General Psychopathology*, which has had enormous prestige and influence in continental European psychiatry. Curiously enough, although this often revised text is phenomenological in method, it does not contain the writer's absorption with his philosophy of existence, the most significant contribution by him to the ideas of our time.[36] As a philosopher Jaspers has consistently concentrated upon existential man: his world views, his boundary situations, his encounter with truth, his search for the limits of technique and theory, his encompassing transcendental possibilities. All human sciences will continue to react to Jaspers' critique and questioning. His work has a lasting power, and as Stierlin[76] has pointed out, behind some of his biases, an indifference to Freud, for example, Jaspers had a tough mind that warned against reducing man to theory, or worse, objectifying him through technological manipulation. In this regard Jaspers'[35] critical appraisal of psychotherapy is worth reading for its wisdom about goals, levels of communication, limits, and practice.

Martin Buber and Paul Tillich are two existential thinkers who, though born in Germany, spent their later careers in other countries, where their influence on the mental health professions has been considerable. Buber, a Jew, was an early immigrant to Palestine where he taught until the end of his life. Mystical thought, theology, education, and philosophical anthropology were the cornerstones of his productive career. His earliest essay, *I and Thou*,[16] is a classic existential text and paves the way for all his subsequent writings. Buber's relation to psychiatry is very relevant and has been summarized by Farber,[23] an American psychoanalyst who first responded to the philosopher's work. In 1957 Buber visited the United States and lectured at the Washington School of Psychiatry on themes related to psychotherapy.[17] Tillich, a theologian by vocation, came from Germany to the United States where he taught at Harvard and Chicago. As an existential thinker he had a broad knowledge of and intense interest in psychology and psychotherapy. His popular book, *The Courage to Be*,[79] is a Kierkegaardian reflection upon anxiety, despair, individualization, and acceptance. Tillich has had a major influence upon existential psychiatry in this country through his own writing, lecturing, and through his association with the psychoanalyst Rollo May.[55]

Heidegger

All of the authors cited above have been affected by the elaborate and profound existential philosophy of the German thinker Martin Heidegger. His *magnum opus, Being and Time*,[31] has dominated the entire movement since its publication in 1927. This work is the single most informing text for existential developments in clinical psychiatry, psychotherapy, and psychoanalysis. It has stimulated interpretation, misinterpretation, admiration, and reaction. It has provided an extensive, if difficult, vocabulary of new terms. It has introduced a wealth of ideas, some of which have already passed into the domain of cliché, as in the popular usage of the word "openness". It has been both praised and damned for its acknowledged heaviness and difficulty. Its author has gone on to other philosophical interests, but this book, actually an incomplete work, has been recognized as the bible of the existential school.

Hannah Arendt[3] and others have attested to the importance of Heidegger as a thinker of profundity and originality. His reputation has risen to rival that of Husserl's, for whom he was an assistant and later the successor in the Chair of Philosophy at the University of Freiburg. Heidegger's ambitions to overthrow traditional metaphysics and recover the structures of being will not necessarily illuminate his significance for the human sciences. Rather it is his creative thinking about the being of human being, the analytics of Dasein, that challenges us to rethink our own theory and practice.

Human existence is a standing out, an emergence of being or a becoming. Dasein always implies the personal pronoun: "I am," "you are." Dasein is essentially its own *possibility*, but its being is not given for once and for all, to unfold automatically like things in the organic world. Human existence discloses itself, questions itself, and chooses itself in its becoming, and it can be won or lost, it can be gained or forfeited. This is what Heidegger means by the concepts authentic and inauthentic existence: existence owned or existence disowned. We can understand another human being only as we see what he is moving toward, what he is becoming by fulfilling himself or depleting himself. As we will see over and over again, existential thinking returns to the possibilities of being human in human beings.

For Heidegger the fundamental existential constitution of Dasein is being-in-the-world, which must be grasped as a unitary phenomenon. That is to say, human existence does not rest inside a world, but rather it is given within a world, it is together with a world; and spatiality is something derived from the primary condition of being-in-the-world. The consequences of this original constitution are considerable, since this establishes an attack upon the traditional dichotomy between subject and object, an attack that rejects the Cartesian split that has afflicted psychology and psychiatry for centuries. Being-in-the-world supplants all closed models; be it a black box with input and output, an inside brain with representations of what is out there, or a pure psychic subjectivity. Human existence is always and already mixed up with a world. There is a worldishness about our very existence. We are our horizons, our landscapes, our abodes. Nature and the things out there in the environment come later and appear as afterthoughts to our immediate presence within a world.

The analysis of the existential world or worlds becomes a central task for basic psychiatry, and it has been pursued by the clinicians we will discuss below. Heidegger himself distinguishes the *Umwelt* and the *Mitwelt*. The *Umwelt* is the first and a nearest world of the body and its proximate situation. The *Mitwelt* is the with-world of coexistence, Dasein's being with others, which determines its own hazards and opportunities. There is the everyday world of the everyday self alongside of everybody else. This is the ordinary world of the "they" as in "they say it's going to rain." The impersonal self-world relation is the commonplace condition of all of us and leads directly to an inauthentic realm of conformity, anonymity, and facelessness. Contrasted with this is the condition of coexistence in openness and the co-presence of Dasein to Dasein. This is the realm of existential encounter and authentic meeting face to face, where our own greatest possibilities are discovered in our solicitous being with others. The concept of encounter, so important in psychiatry and psychotherapy, is derived from an ontological formulation. First and foremost we have our being in a world of others.

Human existence has a peculiarly outgoing or foregoing structure, because our becoming is never finished until the moment we die. To live is to take ownership of the project of our personal existence. According to common experience an aim is something I have before me, something ahead of me. When I set a specific aim for myself, for example, the goal of climbing Mt. Everest, I conceive of it as a possibility that I may or may not achieve sometime in the future. Until then I let this possibility, in advance, determine all the steps I take here and now: I undergo a rigorous training, I expose myself to hardship and danger, I bend my energies toward organizing my expedition, collecting equipment, recruiting guides—and all this for the sake of, or to take care of, a possibility that may never be realized but on whose outcome I have staked my life's efforts.

Remarkable as the pursuit and perhaps the achievement of such a goal is, for Heidegger what is even more remarkable is how man must be to be capable of heading out toward a goal at all. For this he must be able to project himself forward into a future, to take aim at as yet completely "nonexistent" things and events, and to take direction from them for what should be done here and now. Above all, he must be able to understand himself, not only in what he was and what he is, but also in what he can become—and thus move toward himself, so to speak, clothed in his own possibilities. In other words, man is able to transcend, to go out beyond himself as he already is to the full range of the possibilities of his being. This is the unique concept of being that Heidegger reveals in human existence and to which he gives the name care. The temporal meaning of this fore-throw structure of care is clearly the future. For Heidegger the future is the primary marker and mode of time.[42]

The past is the domain of the being I already am and have been. This is the fact of my being, marked by a thrownness into the world, a "has been" left to myself and stuck with my own facticity. The present is the moment as I fall captive to the world, giving myself away to everyday busyness in time that claims my inauthentic, disowned self.

For Heidegger existence as care is actually the unity of time—future, past, and present. Man does not merely exist in time like things; on the contrary, man originates or brings himself to ripeness as time. So much of Heidegger's thought is a meditation on existence, care, and time, locating identical structures in each dimension.

If Dasein is a self-disclosing, forward-moving, care-taking, future-directed being, there is one possibility that always beckons each human existence, and that master possibility is death. Dasein is being-toward-death, my death that is only mine to die. In his analysis of death Heidegger tends to see the ontological structure as the most authentic and proper possibility for human reality. This is not a pessimistic philosophy, but rather a living reminder of the seriousness of living itself. Resoluteness is the corollary of being-toward-death, and it calls us and rededicates us to authentic existence, to take care of our being. Heidegger would say that no one can take someone else's death away from him. "They" say that death will come someday, but not yet, and the power and certainty of it are denied. As soon as man is born, he is old enough to die; nevertheless, no one is so old that he does not possess a future.

Space does not permit an account of Heidegger's analyses of dread and guilt, which he approaches as ontologically obligatory for human existence. These analyses should not be confused with neurotic anxiety and neurotic guilt. Dread as a basic human mood is elucidated in Heidegger's essay, "What is Metaphysics?"[32] The imposing contribution of this body of thought is a thoroughgoing hermeneutics of human existence, and this penetrating interpretation seems to be a by-product of Heidegger's project of investigating the question of being. The existential school has assimilated, applied, and, in some instances, elaborated on this complex and rich interpretation.

Existential Aspects of Basic Psychiatry

The development of the phenomenological school in psychology has a bearing on the emergence of the existential approach, since the phenomenological method is fastened upon by nearly all existentialists. Phenomenology is also identified with Gestalt psychology and the now classical investigations of perception. As a method of dwelling on an exhaustive description of psychological phenomena and withholding explanatory hypotheses, phenomenology was later applied to social psychology[50] and to the psychology of personality. In his Terry Lectures, entitled *Becoming*, Allport[2] made an important contribution in bringing an existential viewpoint to the study of persons against a background of pheomenological orientation. Other American psychologists, Maslow[52] and Rogers,[64] made like-minded approaches, and recently their work has stimulated a lively movement of humanistic psychology with a wide variety of applications in the clinical field, from encounter groups to body awareness exercises.

While these developments in American psychology have taken place in the past several decades, existential phenomenology has had a much longer and more powerful effect upon European psychology and psychiatry. This fact was brought vividly to the attention of Americans with the very significant publication in 1958 of the book *Existence*,[56] which has received wide attention. The editors of the book, May, Ellenberger, and Angel, gathered together essays on basic approaches and clinical case studies, translated into English for the first time. This book focused American attention on the contributions of the psychiatrists Binswanger, Minkowski, and Straus.

Basic psychiatry has been genuinely informed and instructed through the critical work of Erwin Straus, whose career has bridged European and American circles. He is indebted to Husserl and phenomenology because of his reworking of the concept of the lived world, the world of firsthand experience. Straus claims Heidegger as his ally in his similar, constitutive revelation of being-in-the-world. In an early book Straus[78] undertakes a critical examination of stimulus-response psychology and the Pavlovian school. His critique is supported by detailed phenomenological investigations of sense experience and body movement. Because of his lifelong interest in the senses and movement, he has completed studies of hallucinations, anosagnosia, catatonic states, and a variety of neurological disorders. His mastery of the nuances of the clinical problems involved is exceptional. Straus's[77] selected papers reveal his existential orientation, especially his paper "The Upright Posture," a brilliant elucidation of the "standing" position and the presence of human existence. At this point in time his teaching has a moderating influence upon a technologically slanted and mechanistic trend in American psychiatry. Straus's reputation among a group of young American philosophers is growing through his regularly held conferences in Lexington, Kentucky, addressed to pure and applied phenomenology.

A basic science of psychiatry is not likely to mature from the existential example and approach; instead, the movement seems to offer a style and content for criticizing scientific doctrines of man, mind, and behavior. Benda[6] and Serban[71] offer a critical look at Freudian theory that relates to Binswanger's[7] earlier views. Existential psychology has been called a third force, an alternative between Freud's natural-scientific model and Skinner's behavioristic model. Indeed, the concept of model applied to man is anathema to existential thought. Van Kaam[82] has attempted a more ambitious survey of the foundations for an anthropological phenomenology in psychology based on existential theory, but unfortunately his work has received only minimal attention.

Existential Aspects of Clinical Psychiatry

Mention should be made of two early interpreters of existential thinking to American clinicians. Wolff[88] published a number of case studies from his own practice that placed extraordinary emphasis on crisis, values, and goals. His method included dream analysis. Sonnemann[73] was one of the first to present Binswanger's work to non-German readers, and his neglected scholarly book, *Existence and Therapy*, is rich with references to the European existential tradition and has a fine grasp of the historical situation. His avoidance of case material might explain the book's neglect among psychiatrists.

Existential Case Studies

A reading of clinical case studies is recommended as the first approach to existential psychiatry. The material now available is extensive and useful. These cases tell us the "what" of psychopathology, rather than explaining the "why" of symptoms and syndromes. Van Den Berg's[81] case of an obsessional and phobic young man illustrates the clinician's methodical description of his patient's world, bodily experiences, relations with others, spatiality, and temporality. Descriptions of lived space and lived time in psychopathological states are hallmarks of all these studies. Laing's[45] cases of schizoid states and schizophrenia are now fairly well known in psychiatry. Boss[13] has published studies of a variety of sexual perversions. He has also written on the existential analysis of dreams.[12] Binswanger's case studies of mania and schizophrenia will be commented on separately. Cases by Minkowski, Von Gebsatell, and Kuhn are available in the collection *Existence*.[56]

Against the background of his clinical practice of psychoanalysis, Farber[24] has collected material on anxiety neurosis, depression, hysteria, and schizophrenia. His articles contain an existential re-examination of the psychopathology of will. The dislocations, perversions, and promotions of will and willfulness are delineated in neurotic problems and some areas of scientific interest, such as sex research. Farber's analyses of the uses, misuses, and abuses of the modern will have influenced May,[54] who expands on their relevance for contemporary values and attitudes regarding sexuality and love. An excellent overview of sex and existence has been presented by Van Kaam.[83]

Lynd,[49] a sociologist, has written an astute and clinically valuable study of shame, personality, and an existential interpretation of identity. Her work is filled with clinical examples from literature, a practice that is common to many of the existentialists. Her treatment of the sense of identity provides a commentary on and a revision of Erikson's familiar concept. Her work is useful to psychotherapists willing to explore existential dimensions of their endeavors.

Binswanger

Ludwig Binswanger, the Swiss psychiatrist, scholar, and philosopher who died in 1965, is the exemplary clinician of the existential movement. His teaching, his books and articles, his administration of a mental hospital, his example, all serve to illustrate his formative role and enduring influence as an existential thinker. Born into a family tradition that included many prominent psychiatrists over several generations, he worked at the sanatorium at Kreuzlingen, on the shores of Lake Constance. Binswanger is the most thorough psychiatric interpreter of Heidegger from the period of *Being and Time*.[31] He has called his particular reading of this text a "fruitful misunderstanding," and, indeed, his response to Heidegger is more creative than imitative. Binswanger's psychiatry cannot be understood on the basis of Heidegger alone. There are equal elements of phenomenology (descriptions of lived time and lived space), Kierkegaard's self psychology, Buber's dialogue between I and Thou, not to mention essential ingredients from Bleuler and Freud.[28]

Binswanger was a lifelong friend of Freud, with whom he had important theoretical differences. He was an early champion of psychoanalysis in Switzerland and delivered an important address at Freud's eightieth birthday celebration. It was after Freud's death that he systematically developed his ideas on existential analysis, although he was working out his position from the early 1930's.[70] In spite of open difference Freud never broke off warm personal relations with Binswanger, as was so often the case with Freud and his dissident disciples.

Binswanger's contributions include his critical-theoretical writings and a number of long case histories that contain the application of his ideas to the study of lives, rendered in an elegant and admirable style. From Heidegger he develops the notion that man is an essentially temporal being, that man exists to the degree that he stands facing his own future, and that out of this relation to a future he orders his present and connects it with his past. Binswanger observed in his patients a core of trouble connected with the loss of direction and meaning in their lives, which fundamentally meant that they no longer stood in meaningful relation to time itself—they could not bind it "care-fully" and harmoniously together.

Elaborating on the individual and personal application of Dasein as being-in-the-world and the meaning structure of care as it actually manifests itself in a particular human being's daily life, Binswanger searches out the world design of his patient. The patient's expression and transformation of care is analyzed for his existential worlds. The first and nearest world of the *Umwelt* and the world with others in direct relations, the *Mitwelt*, have been discussed above. Binswanger adds the *Eigenwelt*, the world of self in its continuity, consistency, and identity. Psychopathology is possible because the world design of the patient is constricted, simplified, distorted, or depleted. Pathology is always a loss of world content, a loss of the complexity of contexts of reference. Illness is viewed as the overpowering of Dasein by one world design.[7]

Binswanger's clinical case studies are, for the most part, histories and Daseinanalyses of manic and schizophrenic persons. In a number of articles on mania he first worked out the manic mode of being-in-the-world.[10] In the flighty and surface world of the manic patient the future holds everything, and he propels himself headlong into a future brimming over with too many possibilities. His care is strewn before him recklessly, and his existence is squandered among a fleeting chaos of things and faces. Manic phenomena lend themselves readily to an existential interpretation of temporality and being-in-the-world, and Binswanger returns frequently to these clinical examples. His case studies of schizophrenia are more complex, "The Case of Ellen West"[8] being the longest and most detailed of them. It has achieved the status of a classic case, but more than a clinical history, it is a human narrative of great depth and poignancy. "Ellen West" was not treated by Binswanger himself. She had been examined, treated, and psychoanalyzed by a number of eminent specialists. Binswanger reconstructed the study from exhaustive reports, diagnostic evaluations, treatment summaries, her diaries, poems, and letters, all of which he collected after her death. The analysis defies any brief summary; however, this document demonstrates how Binswanger supplements psychoanalytic interpretation with a basic existential interpretation of the patient's tortuously divided world design, a world split irrevocably between an upper, ethereal one and a lower, earthy one. It is the existential dilemma of the obsessive-compulsive person pushed to extreme limits and beyond, into psychosis. Binswanger's other cases,[7] "Lola Voss," for example, deserve wider readership. These human narratives are his finest achievement.

> Existential analysis is not a psychopathology, nor is it clinical research nor any kind of objectifying research. Its results have first to be recast by psychopathology into forms that are peculiar to it. . . . psychopathology would be digging its own grave were it not always striving to test its concepts of functions against the phenomenal contents to which these concepts are applied and to enrich and deepen them through the latter. Additionally, existential analysis satisfies the demands for a deeper insight into the nature and origin of psychopathological symptoms. If in these symptoms we recognize "facts of communication" —namely, disturbances and difficulties in communication—we should do our utmost to retrace their causes—retrace them, that is, to the fact that the mentally ill live in "worlds" different from ours. . . . Thus we also comply here with a *therapeutic* demand.[9]

Binswanger's work was first brought to the notice of American psychiatrists by Weigert,[86] who presented his analysis of the ontological structures of care and love. Weigert has also recommended the work of Max Scheler[67] to psychiatrists. Scheler was an existential sociologist and philosopher of the first quarter of this century, whose book on sympathy and fellow feeling has had a persistent influence on Binswanger and psychoanalysts from a number of schools. Binswanger has received commentaries by Blauner[11] and Kahn[38] and a more critical examination by Arieti and Schmidl,[69] who detect a therapeutic nihilism in his Daseinanalysis and a rivalry with psychoanalysis. Stierlin[75] also suspects the therapeutic usefulness of a trend toward dedifferentiation in Daseinanalysis, with its undermining of unconscious processes, transference phenomena, and any convergence with the social sciences. In answer to these objections Binswanger repeatedly claimed that his work was an elucidation of basic forms of existence, an infrastructure, upon which psychoanalysis or other systems of explanatory hypotheses could build their theory and procedures for treatment.

Existential Aspects of Psychotherapy

If existential psychiatry does not propose an encompassing theory or prescribe techniques for the conduct of treatment, nevertheless, it does hold significant implications for psychotherapy. Existentially minded psychotherapists have come from many traditions: Freudian, neo-Freudian, and Adlerian. Methods and techniques such as free association, dream analysis, attention to self-systems and interpersonal relations are not abandoned, but rather are taken up in an approach to the patient that has special therapeutic consequences. Flexibility and individualization within the therapeutic situation is given primary emphasis, and diagnostic labeling is supplanted by more extensive descriptions of where the patient is at and what he commences with at the outset of psychotherapy.

The therapeutic consequences of an existential approach are outlined by Werner Mendel.[57] He delineates six guidelines, each of which would seem to indicate a specific strategy to be adopted by the psychotherapist. His first point is that when it comes to the past of a given human being, history is constantly rewritten. The past, as it is reconstructed from memories, can never be recovered in some global, permanent form of how it really was with this person. All of us revise our past as we go along, since memory itself is dependent upon the present situation and, often enough, upon one's orientation toward the future. It is to be noted that much current thinking in historical studies recognizes revisionism as a necessary task for each generation. Personal histories likewise require their own revision, a looking backward again and again to sum up the relevant past. A second time-related point insists upon the influence that the future has and will continue to have upon the outcome of therapy. The future is before the patient in his plans, expectations, anticipations, and chosen goals.

Third, there is a readiness to dwell on conscious material, to let it speak for itself without applying formulas to reduce it to something else, to something underneath. This means that a phenomenological attitude is maintained toward accounts of symptoms, events, fantasies, and dreams. Manifest content is patiently described before turning to a hidden latent content. A fourth point is that the reality of the therapeutic encounter between therapist and patient is recognized, without reducing every vicissitude in the relationship to manifestations of transference or countertransference.

Fifth, in existential therapy there is an important distinction made between human saying and human doing. A person puts more of his existence into what he does than into what he says. Convergence or divergence between saying and doing is attended to in the therapy. Sixth, the point is made that decision and action are integral to the process of therapy, and are not to be conceived as interminably delayed until the work of therapy is accomplished. Action, more than reflection upon past action, is the central theme of the ongoing process.

In *Psychology and the Human Dilemma* Rollo May[55] has collected his articles on existentialism and related problems, and in this book he discusses a number of issues involving psychoanalysis, analytical psychotherapy, and existential thought. He continues a mode of inquiry initiated by Tillich.[80] May's work has continued to enlarge on the context and meaning of psychotherapy, with particular attention to its goals. He has some cautionary things to say about the irrationalist aberration in some existential promotions of immediacy and here-and-now experience, a tendency found in the recent American interest in Zen Buddhism.

May has been sensitive to the charge that

the contemporary psychotherapist is often an agent of the culture, a technician whose job is to pave the way to adjustment and conformity to group norms and societal pressures. He locates a powerful remedy for the techniques of adjustment in an existentially oriented therapy that undercuts objectification and manipulation of the patient and, at the same time, seeks a goal in the fulfillment of personal existence. May would insist that the goal of any psychotherapeutic endeavor should be the patient's coming to himself, experiencing his existence as both his limits and his possibilities, accepting his anxiety and guilt and constructing upon them a responsible commitment to his unique potentialities. May and Colm[21] have been responsive to the encountering context of psychotherapy. Coexistence, presence, and openness underlie the actual relationship between patient and therapist. There are intuitive dimensions to the encounter that are reciprocal and deserve as much attention as transference and countertransference. In a deeper sense perhaps transference and countertransference are made possible because of the encountering ground upon which both participants meet. There is nothing sentimental about the encounter, nor is it susceptible to a neat intellectual formulation.[37] The reality of encounter is the most neglected subject in the conventional psychotherapies; while, on the other hand, encounter is the most misrepresented subject in the unconventional psychotherapies, with their frantic appeals to nudity, touching, and stroking as means of enhancing it.

Existential approaches to psychotherapy have been influenced by other practitioners and authors. Victor Frankl,[27] an Austrian who survived the Nazi concentration camps, has popularized his logotherapy in several books. Logotherapy is based upon what Frankl calls the human will to meaning, described as an existential *a priori* to all psychotherapeutic aims. His work recovers the importance of values to all forms of psychotherapy. The loss of the meaning of life is not a unitary concept, but rather a reflection of a set of specific values the patient has abandoned, misplaced, or neglected to articulate. The therapist's role is not to hand over values to his patient, but to be a detective in the patient's search for his own values. Jordan Scher,[68] an American editor and therapist, has written extensively on the existential approach. His ontonalysis is indebted to Binswanger and Frankl, but is stamped with his own broad clinical and social interests and points in the direction of an existential social psychiatry.

Among American psychoanalysts Weigert[86] and Weisman[87] have shown a willingness to assimilate existential views into neo-Freudian and Freudian theory and practice. The Swiss psychoanalyst Medard Boss,[14] on the other hand, has published a lengthy and radical existential demolition of analytic theory, while attempting to preserve the trappings of analytic procedure and method. Boss takes his cue from the later work of Heidegger. His book, *Psychoanalysis and Dasein-analysis*, has four parts: an exposition of Heidegger's analytics of Dasein, a re-evaluation of basic psychoanalytic theory in the light of this, a discussion of neurotic syndromes with some excellent case material, and an informative section on the impact of Daseinanalysis on orthodox analytic techniques. Boss seems to carry out the therapeutic program of Binswanger's existential analysis, under the banner of an idiosyncratic psychoanalysis.

Other psychotherapies have their existentialists; however, they seem to be minority voices. Group psychotherapy, which is quite naturally suited to an existential orientation, has been reformulated by many authors who favor some kind of experiential learning. Mullan[61] expounds this view in a number of papers. Communications theory and family psychotherapy have the Palo Alto group among their early pioneers, some of whose members present an existential viewpoint.[85] Child psychotherapists of an existential persuasion are represented in a collection of clinical articles edited by Moustakas.[60] Burton[18] has built on an existential base in elaborating his humanistic psychotherapy.

These developments in the psychotherapies are all of recent vintage, and the future holds possibilities for more existential inroads into theory and practice. The most important change existential approaches have brought about in the psychotherapies has been the loosening of rigid, doctrinaire concepts and procedures.

Bibliography

1. ALLERS, R., *Existentialism and Psychiatry*, Charles C Thomas, Springfield, Ill., 1961.
2. ALLPORT, G. W., *Becoming*, Yale University Press, New Haven, 1955.
3. ARENDT, H., "Martin Heidegger at Eighty," *New York Review 17*, No. 6:50–54.
4. BARRETT, W., *What is Existentialism?* Grove Press, New York, 1964.
5. BENDA, C. E., *The Image of Love*, Free Press, New York, 1961.
6. ———, "What is Existential Psychiatry?" *Am. J. Psychiat.*, *123*:288–296, 1966.
7. BINSWANGER, L., *Being-in-the-World*, Basic Books, New York, 1963.
8. ———, "The Case of Ellen West," in May, R., Angel, E., and Ellenberger, H. F. (Eds.), *Existence*, Basic Books, New York, 1958.
9. ———, "The Existential Analysis School of Thought," in May, R., Angel, E., and Ellenberger, H. F. (Eds.), *Existence*, Basic Books, New York, 1958.
10. ———, "On the Manic Mode of Being-in-the-World," in Straus, E. W. (Ed.), *Phenomenology: Pure and Applied*, Duquesne University Press, Pittsburgh, 1964.
11. BLAUNER, J., "Existential Analysis," *Psychoanal. Rev.*, *44*:51–64, 1957.
12. BOSS, M., *The Analysis of Dreams*, Philosophical Library, New York, 1958.
13. ———, *Meaning and Content of Sexual Perversions*, Grune & Stratton, New York, 1949.
14. ———, *Psychoanalysis and Daseinanalysis*, Basic Books, New York, 1963.
15. BRETALL, R. (Ed.), *A Kierkegaard Anthology*, Modern Library, New York, 1946.
16. BUBER, M., *I and Thou*, Scribner's, New York, 1958.
17. ———, *The Knowledge of Man*, Harper & Row, New York, 1965.
18. BURTON, A., *Modern Humanistic Psychotherapy*, Jossey-Bass, San Francisco, 1967.
19. CAMUS, A., *The Myth of Sisyphus*, Alfred A. Knopf, New York, 1955.
20. COLM, H., *The Existentialist Approach to Psychotherapy with Adults and Children*, Grune & Stratton, New York, 1966.
21. ———, "The Therapeutic Encounter," *Rev. Existent. Psychol. & Psychiat.*, *2*:137–159, 1965.
22. COPLESTON, F., *Contemporary Philosophy*, Burns & Oates, London, 1956.
23. FARBER, L. H., "Martin Buber and Psychiatry," *Psychiatry*, *19*:109–120, 1956.
24. ———, *The Ways of the Will*, Harper Colophon Books, New York, 1968.
25. FISHER, A. L. (Ed.), *The Essential Writings of Merleau-Ponty*, Harcourt, Brace & World, New York, 1969.
26. FOUCAULT, M., *Madness and Civilization*, Random House, New York, 1965.
27. FRANKL, V. E., *Man's Search for Meaning*, Washington Square Press, New York, 1963.
28. FRIEDMAN, M., "Phenomenology and Existential Analysis," *Rev. Existent. Psychol. & Psychiat.*, *9*:151–168, 1969.
29. ———, *The Worlds of Existentialism*, Random House, New York, 1964.
30. HARPER, R., *Existentialism*, Harvard University Press, Cambridge, 1948.
31. HEIDEGGER, M., *Being and Time*, Harper & Row, New York, 1962.
32. ———, *Existence and Being*, Henry Regnery, Chicago, 1949.
33. HUSSERL, E., *Phenomenology and the Crisis of Philosophy*, Harper & Row, New York, 1965.
34. JAMES, W., *The Varieties of Religious Experience*, New American Library, New York, 1958.
35. JASPERS, K., *The Nature of Psychotherapy*, University of Chicago Press, Chicago, 1965.
36. ———, *Philosophy of Existence*, University of Pennsylvania Press, Philadelphia, 1971.
37. JOHNSON, R. E., *Existential Man: The Challenge of Psychotherapy*, Pergamon Press, New York, 1971.
38. KAHN, E., "An Appraisal of Existential Analysis," *Psychiat. Quart.*, *31*:203–227, 417–444, 1957.
39. KAUFMANN, W., *Existentialism from Dostoevsky to Sartre*, World Publishing Co., Cleveland, 1956.

40. KEEN, E., *Three Faces of Being: Toward an Existential Clinical Psychology*, Appleton-Century-Crofts, New York, 1970.
41. KIERKEGAARD, S., *Either/Or*, 2 vols., Doubleday-Anchor. New York, 1959.
42. KING, M., *Heidegger's Philosophy*, Macmillan, New York, 1964.
43. KOCKELMANS, J. J. (Ed.), *Phenomenology*, Doubleday, New York, 1967.
44. KORS, P. C., "The Existential Moment in Psychotherapy," *Psychiatry*, *24*:153–162, 1961.
45. LAING, R. D., *The Divided Self*, Quadrangle Books, Chicago, 1960.
46. ———, "Minkowski and Schizophrenia," *Rev. Existent. Psychol. & Psychiat.*, *3*: 195–207, 1963.
47. LAWRENCE, N., and O'CONNOR, D. (Eds.), *Readings in Existential Phenomenology*, Prentice-Hall, Englewood Cliffs, N.J., 1967.
48. LUIJPEN, W. A., *Existential Phenomenology*, Duquesne University Press, Pittsburgh, 1960.
49. LYND, H. M., *On Shame and the Search for Identity*, Science Editions, New York, 1961.
50. MACLEOD, R. B., "The Phenomenological Approach to Social Psychology," in Kuenzli, A. E. (Ed.), *The Phenomenological Problem*, Harper, New York, 1959.
51. MARCEL, G., *The Existential Background of Human Dignity*, Harvard University Press, Cambridge, 1963.
52. MASLOW, A. H., *Toward a Psychology of Being*, Van Nostrand, Princeton, 1962.
53. MAY, R. (Ed.), *Existential Psychology*, Random House, New York, 1961.
54. ———, *Love and Will*, Norton, New York, 1969.
55. ———, *Psychology and the Human Dilemma*, Van Nostrand, Princeton, 1967.
56. ———, ANGEL, E., and ELLENBERGER, H. F. (Eds.), *Existence*, Basic Books, New York, 1958.
57. MENDEL, W. M., "Introduction to Existential Psychiatry," *Psychiat. Dig.* (Nov. 1964) 23–34.
58. MERLEAU-PONTY, M., *Phenomenology of Perception*, Routledge & Kegan Paul, London, 1962.
59. ———, *The Structure of Behavior*, Beacon Press, Boston, 1963.
60. MOUSTAKAS, C. (Ed.), *Existential Child Therapy*, Basic Books, New York, 1966.
61. MULLAN, H., and SANGILULIANO, I., "Interpretation as Existence in Analysis," *Psychoanal. & Psychoanalyt. Rev.*, *45*:52–73, 1958.
62. OLSON, R. G., *An Introduction to Existentialism*, Dover, New York, 1962.
63. PLESSNER, H., *Laughing and Crying*, Northwestern University Press, Evanston, 1970.
64. ROGERS, C., *On Becoming a Person*, Houghton Mifflin, Boston, 1961.
65. RUITENBEEK, H. M. (Ed.), *Psychoanalysis and Existential Philosophy*, Dutton, New York, 1962.
66. SARTRE, J. P., *Being and Nothingness*, Philosophical Library, New York, 1956.
67. SCHELER, M., *The Nature of Sympathy*, Routledge & Kegan Paul, London, 1954.
68. SCHER, J. M., "Intending, Committing and Crisis in Ontoanalysis," *Existent. Psychiat.*, *6*:347–354, 1967.
69. SCHMIDL, F., "Psychoanalysis and Existential Analysis," *Psychoanal. Quart.*, *29*:344–354, 1960.
70. ———, "Sigmund Freud and Ludwig Binswanger," *Psychoanal. Quart.*, *28*:40–58, 1959.
71. SERBAN, G., "Freudian Man vs. Existential Man," *Arch. Gen. Psychiat.*, *17*:598–607, 1967.
72. ———, "The Theory and Practice of Existential Analysis," *Behav. Neuropsychiat.*, *1*:8–14, 1969.
73. SONNEMANN, U., *Existence and Therapy*, Grune & Stratton, New York, 1954.
74. SPIEGELBERG, H., *The Phenomenological Movement*, 2 vols., Martinus Nijhoff, The Hague, 1965.
75. STIERLIN, H., "Existentialism Meets Psychotherapy," *Phil. & Phenomenol. Res.*, *24*: 215–239, 1963.
76. ———, "Karl Jaspers (1883–1969)," *Rev. Existent. Psychol. & Psychiat.*, *9*:78–86, 1968.
77. STRAUS, E. W., *Phenomenological Psychology*, Basic Books, New York, 1966.
78. ———, *The Primary World of Senses*, Free Press, New York, 1963.
79. TILLICH, P., *The Courage To Be*, Yale University Press, New Haven, 1952.
80. ———, "Existentialism and Psychotherapy," *Rev. Existent. Psychol. & Psychiat.*, *1*:8–16, 1961.

81. VAN DEN BERG, J. H., *The Phenomenological Approach to Psychiatry*, Charles C Thomas, Springfield, Ill., 1955.
82. VAN KAAM, A., *Existential Foundations of Psychology*, Image Books, New York: 1969.
83. ———, "Sex and Existence," *Rev. Existent. Psychol. & Psychiat.*, 3:163–181, 1963.
84. WARNOCK, M., *Existentialism*, Oxford University Press, London, 1970.
85. WATZLAWICK, P., BEAVIN, J., and JACKSON, D. D., *Pragmatics of Human Communication*, Norton, New York, 1967.
86. WEIGERT, E., *The Courage to Love*, Yale University Press, New Haven, 1970.
87. WEISMAN, A. D., *The Existential Core of Psychoanalysis*, Little, Brown, Boston, 1965.
88. WOLFF, W., *Values and Personality*, Grune & Stratton, New York, 1950.
89. WYSS, D., *Depth Psychology*, Norton, New York, 1966.

CHAPTER 43

THE BEHAVIOR THERAPY APPROACH

Joseph Wolpe

MOST PSYCHIATRIC THERAPY is directed at overcoming habitual or repetitious unadaptive behavior. All habits depend on the specifics of neural organization, which is ordinarily shaped by physiological development, learning, or both. Unadaptive habits, however, sometimes result from physiological "distortions" due to lesions of the nervous system or to biochemical abnormalities, such as circulating toxins, or to disorders of endocrine secretion. It is a reasonable presumption that when an unadaptive habit has come into existence through learning, it should be erasable by learning. Experimentally established knowledge of the learning process and of the factors influencing it has generated a range of new procedures for modifying or eliminating unadaptive learned habits; and this is what is known as conditioning therapy or *behavior therapy*.[85,86] All unadaptive habits based on learning are proper targets for behavior therapy. Its main applications have been to the neuroses, to other learned habits that do not fit the definition of neuroses, and to those unadaptive habits of schizophrenics that are based on learning.

Immediately it must be noted that the unadaptive learned habits amenable to behavior therapy have for centuries been treated by other methods, often successfully. This means that when soothsayers or priestly counselors and, later, psychoanalysts, nondirective therapists, or Christian Scientists have achieved success, a learning process relevant to the unadaptive habits concerned must have gone on. Since the efforts of these healers were never aimed at the stimulus-response bonds central to the offending habits, it must be concluded that whatever habit changes they achieved were due to inadvertent and unprogrammed events that occurred during their interactions with their patients. In behavior therapy a neurotic anxiety response habit can usually be systematically eliminated by counteracting the anxiety response by another emotion incompatible with it. In *any* kind of therapy, if a patient happens to have fairly strong nonanxious emotional responses to his therapist, it may be supposed that neurotic anxieties tending to be aroused by relevant verbal stimuli will sometimes be inhibited by the ongoing therapist-evoked emotion—thus weakening the anxiety habit in much the same way as when "therapeutic" emotion is deliberately counterposed to anxiety in behavior therapy.

If the behavioristic view of the nature of neurosis is correct, it follows that treatment will be both less time-consuming and more effective if it adds to the inadvertent interpersonal therapeutic effects, the effects obtainable by the deliberate employment of methods based on learning principles. This deduction has received a good deal of empirical support both clinically (Wolpe,[86] Moore,[50] Sloane, *et al.*)[72] and in laboratory contexts (Lang, Lazovik, and Reynolds,[37] Paul,[58] Paul and Shannon,[59] Paul,[57] Mealiea and Nawas).[48]

The Clinical Field of Behavior Therapy

Among the habits based upon learning that are the natural targets of behavior therapy, the commonest are the following:

The Neuroses

A neurosis is defined as a persistent unadaptive habit that has been acquired in one or a succession of anxiety-generating situations.[86] Anxiety is almost always a central component of neurotic habits. While the central role of anxiety is quite obvious in a great many neurotic patients—those with anxiety neuroses, phobias, tension states, interpersonal anxiety, and "free-floating" anxiety—there are others in whom its presence is obscured by symptoms secondary to it. The patient's main complaint may be of sexual inadequacy, homosexuality, transvestism, asthma, migraine, peptic ulceration, obsessional behavior, stuttering, or blushing, or it may take the form of a "character neurosis"—among other possibilities. Yet when such cases are subjected to behavior analysis—an analysis of the relations between neurotic responses and their antecedents—it almost invariably turns out that anxiety is their crux. The stutterer, with only rare exceptions, is found to stutter only in social situations that make him anxious; the asthmatic or the peptic ulcer patient has somatic attacks in situations that raise his emotional tension.* When women are frigid or men are impotent, it is usually because of anxieties that are evoked by features of sexual situations. Most homosexuals have either interpersonal anxiety toward people in general, with an emphasis on females, or specific anxiety evoked by women in contexts of physical intimacy. Similarly anxiety is the wellspring of kleptomania, adult tantrums, and other manifestations of "character neurosis," as well as of many depressions (Wolpe)[84] and of most obsessions and compulsions, the patient characteristically becoming anxious when he does not perform his rituals. The only kind of neurotic case in which anxiety is consistently not found is the classical hysteria with *la belle indifference.*

Unadaptive Learned Habits Other than Neuroses

Many unadaptive learned habits seen in otherwise normal people do not conform to the definition of neurosis given above in one way or another, usually because they have not originated in anxiety-generating situations. Such habits are most often encountered in young children, common examples being enuresis nocturna, encopresis, nail-biting, thumbsucking, tantrums, aggressiveness, and disruptive classroom behavior. Adult examples are habitual tardiness, frittering away of time instead of settling down to assignments, and "localized" reactions, such as a retching response to dentures in the mouth (Stoffelmayr)[74] and some cases of anorexia nervosa (Bachrach, Erwin, and Mohr,[5] Scrignar).[66]

* It should be noted that a great many cases of asthma, migraine, and other "psychosomatic" illnesses have purely somatic etiologies.

The Learned Habits of Schizophrenics

I believe that schizophrenia is basically an organic illness (for example, Gottlieb et al.,[26] Rubin).[62] Some of its manifestations, for example, associational gaps in thinking (see Chapman),[18] appear to be a direct function of the biological abnormality. But other schizophrenic behavior seems to be the result of the organic state's predisposing to the acquisition through learning of bizarre habits. These latter habits are modifiable by the use of learning procedures.

Paradigms Commonly Employed in Behavior Therapy

Positive Reinforcement

Owing to the influence of Skinner,[70,71] positive reinforcement has come to be the most widely studied and dissected paradigm in the whole field of conditioning. Any response is rendered more probable if it is followed by *reinforcement* in the form of food, water, sex, money, affection, or anything else that is "rewarding" to the individual organism. Eventually, of course, an asymptote, or maximum habit strength, is reached. This relationship—the strengthening of habits by rewards—has been demonstrated with great constancy in simple organisms as well as in man. Typically the experiments have involved musculoskeletal behavior; and this is also true of most of the therapeutic applications of positive reinforcement (Ayllon and Azrin,[3] Schaefer and Martin).[65]

Negative reinforcement is a term applied to the special case where the "reward" derives from the *removal* of some disturbing stimulus, such as continuous electrical stimulation, or any other source of stress.

Extinction

Extinction is the progressive diminution of the probability of a response that is repeatedly elicited without reinforcement, that is, without being followed by any kind of "reward" as defined above. The habit-weakening thus procurable is, in general, just as reliable as the habit-strengthening with reinforcement. But it can be impeded by certain factors—notably if there was intermittency in the reinforcements by which the habit was established. As Thomas[76] has noted, resistance to extinction is frequently high in clinical cases because the behavior to be extinguished often has a long history of intermittent reinforcement. Therefore, once treatment has begun, the cessation of reinforcement must be complete.

Conditioned Inhibition Based on Reciprocal Inhibition

A considerable number of methods implement this paradigm, which has had its main applications in eliminating unadaptive autonomic habits, especially anxiety. *Reciprocal inhibition* (Sherrington)[69] is the activation of a reaction incompatible with that expressing the habit to be weakened. The model for reciprocal inhibition therapy was derived (Wolpe)[86] from observations on the treatment of experimental neuroses—habits of severe anxiety induced by shocking an animal in a small cage—habits as strangely persistent as the neuroses of man. So severe is the anxiety conditioned to the experimental environment that the neurotic animal will not eat there even if he is starving. But the animal will eat in a place remotely similar to that in which he was shocked and where there is only weak anxiety; and repeated helpings of food there diminish this anxiety to zero. Environments increasingly similar to the experimental environment are similarly treated, until finally even the experimental cage loses its ability to elicit any anxiety.

While feeding has also been used for the gradual deconditioning of human neuroses (Jones)[33] other anxiety-inhibiting responses have been found to be more widely applicable.

A notable example of reciprocal inhibition applied to emotions other than anxiety is *aversion therapy*. Here an unpleasant response, such as produced by a strong electrical stimulus to a limb or by suggested nauseation (Cautela),[16] is made to inhibit unadaptive pleasant excitation—aroused by a fetishistic object, for example.

Retroactive Inhibition

Much unadaptive behavior results from people having acquired objectively incorrect cognitive associations. For example, a patient who has lapses of memory because of pervasive anxiety may believe that these are an indication of advancing intellectual deterioration, to which belief additional anxiety is quite a natural response. It is necessary to attach the awareness of these lapses to the idea that they are the result of anxiety and without organic significance. In other words, the wrong cognitive response must be replaced by the correct one. The therapist must repeatedly and forcefully provide the new association. Each time the new response is evoked, there is an inhibition of the tendency to perform the old response and a weakening of its habit. This process, known as retroactive inhibition, is the basis of ordinary forgetting. As Osgood[52] has pointed out, it is an instance of reciprocal inhibition.

Conditioned Inhibition Based on Transmarginal Inhibition

In recent years there have been many reports (for example, Malleson,[44] Frankl,[23] Stampfl and Levis,[73] Wolpin and Raines,[88] Wolpe)[85] of the overcoming of neurotic fears by exposing patients either in reality or in imagination, usually for prolonged periods (10 to 60 minutes), to strongly anxiety-arousing stimuli. Many of these patients have improved markedly, and some rapidly. The most widely accepted label for procedures of this kind is *flooding*.

If a conditioned stimulus is administered to an animal at various intensities, it is usually found that as stimulus intensity rises the strength of the response increases until it reaches an asymptote—that is, remains at its top level no matter how much stronger the stimulus becomes. But sometimes, after the response has reached maximum strength, its evocation paradoxically becomes weaker as the intensity of stimulation is further increased. Pavlov[60] used to call this kind of inhibition of response *transmarginal inhibition*. By analogy with reciprocal inhibition I have proposed that habit strength may be decreased if the response has undergone transmarginal inhibition; and this may be the mechanism of change by flooding (p. 192).[85]

¶ The Clinical Examination

There is a rather widespread belief among people outside the field that the behavior therapist plunges into conditioning techniques after a superficial scanning of his cases. It is imagined, for example, that a patient who complains of fears of crowds will be subjected to desensitization (see below) to crowd stimuli without ado. In point of fact, a careful investigation is especially necessary in preparation for behavior therapy because the therapist can only proceed rationally if he has a clear and exact knowledge of the relations between the unadaptive responses and the stimuli that trigger them. First, the therapist makes a comprehensive inventory of the patient's complaints, that is, the reactions he finds uncomfortable, distressful, or disadvantageous. Then he takes a complete history of each complaint, starting from the circumstances of its onset, and following it through its vicissitudes up to the present time. It is frequently found that a neurotic reaction has become worse with the passage of time, generally because the power to evoke it has spread to a widening range of cues on the basis of second order or multiple order conditioning (Wolpe).[86] For example, a woman with a fear of crowds began to manifest a secondary conditioned fear of movie houses and other public buildings following an occasion when a movie house unexpectedly filled with students while she was sitting there.

Because the patient's current stimulus-response relationships will be the target of therapy, it is to them that the most searching examination is given. Since anxiety is, as stated above, usually the central constituent of neurotic reactions, the greatest amount of attention is generally given to identifying its eliciting stimuli. But in the many cases that do not present themselves as overt anxiety problems, such as frigidity or stuttering, the first task is to try to establish whether anxiety is a precondition of the presenting syndrome—as it usually is. Then the antecedents of the anxiety will, in turn, be determined.

After the patient's presenting reactions have been explored, the therapist takes a general history of his life and background. He inquires about early relationships with parents and siblings, special relationships that may have existed with grandparents, religious background and training, traumatic experiences, and neighborhood life. Next he surveys the patient's education—whether he enjoyed school, how well he did academically, whether he took part in sports, whether he made friends, and whether there were any stressful relationships. The patient is asked when he left school and what he then did and how he got on at each institution or place of employment, both occupationally and with respect to personal relationships.

The patient's sex life is then traced from the time of his first awareness of erotic arousal. He is asked about the practice of masturbation and his beliefs and emotional attitudes toward it, and about his love relationships, dwelling upon the important emotional involvements. Each relationship, in turn, right up to the present time, is examined in detail, both socially and sexually.

After the anamnesis the patient is given several inventories to fill out. The Willoughby Schedule (Willoughby)[81] consists of 25 questions that yield information on a five-point scale about common areas of neurotic activity and about general emotional sensitivity. The Fear Survey Schedule (Wolpe and Lang)[87] lists 108 stimulus situations to which a fear is unadaptive and to which response is made on a five-point scale. The Bernreuter Self-Sufficiency Inventory provides information with regard to dependency habits.

Unless the patient has been referred for treatment by a physician (and sometimes even then), a medical examination should be arranged if there is any suggestion at all that organic disease may have a role in his illness. A particularly common indication for this is a history of episodic attacks of anxiety to which no constant stimulus antecedents can be related. Frequently such episodes turn out to be based upon functional hypoglycemia (Conn and Seltzer,[20] Salzer).[64]

Behavior Therapy Techniques in Particular Syndromes

Neurotic Syndromes

In the majority of neurotic patients we find combinations or assemblages of syndromes. As previously stated, anxiety is at the crux of the great bulk of them, so that very often, sooner or later, the treatment is that described below for the common phobias and allied reactions. Therefore, this must receive especially full coverage. But quite frequently a particular syndrome may deflect the course of treatment in special ways. We shall therefore consider treatment in relation to various syndromes. It should be noted that in certain complex interactions of responses and in cases resistant to routine techniques, considerable variation from the procedures described may take place.

The Common Phobias and Allied Reactions

The distinguishing feature of a phobia is that the patient responds with anxiety to a particular class of stimuli to which there is no rational motor or verbal response. The stimulus classes comprising the classical phobias involve easily defined stimulus configurations, such as crowds, animals, enclosed spaces, heights, and darkness. However, numerous stimuli that are outside the ambit of those classically associated with phobias—perhaps because they are less easy to define—can also be the antecedents of anxiety reactions according to the above-mentioned criterion, such stimuli as feelings of losing control, hearing expressions of disapproval or of praise, or seeing people quarreling. Because in a stimulus-response sense there is no difference between the latter group of reactions and the classical phobias, they are treated by the same methods.

Systematic Desensitization. Systematic desensitization is the most widely used technique for the treatment of phobias and related reactions. In its general outlines it is similar to the method that was used for treating experimental neuroses, alluded to above. In the standard procedure the emotional effects of deep muscle relaxation take the place of those of feeding as the means of inhibiting anxiety; and the anxiety-evoking stimuli are presented to the patient's imagination instead of to his perception.

Relaxation is taught beforehand by a short version of the method of Edmund Jacobson,[30,31] two to four muscle groups being dealt with during a 10 to 15 minute period at each of several sessions. During these sessions the stimulus controls of the patient's anxiety responses are also studied. Suppose, for example, that the patient responds with anxiety to being looked at by others. It may be found that the anxiety increases according to the number of people looking at him or according to their demeanor, their age, or their sex, or according to the vulnerability to criticism that he feels he incurs by some feature of the behavior that he is performing at that time—such as its presumptive awkwardness.

Whatever the relevant dimension, a list of situations involving it is made up and ranked according to the amount of disturbance they elicit. Such a ranked list is called a *hierarchy*. The items on the list may differ from each other by so simple a factor as number or by complex factors that need in each case to be specified. The following is an example of a hierarchy providing such specification on the theme of "rejection." The items are ranked in *descending* order of their anxiety-producing effects.

1. My apology for a blunder is not accepted by a friend.
2. My invitation to my apartment for dinner or drinks is refused by a friend.
3. I speak to a peer and he does not seem to hear.
4. A project important to me is criticized by peers.
5. I am left out of plans or not invited.
6. I am spoken to by a peer in a tone of voice sharper than the speaker uses for somebody else present.
7. Nobody remembers my birthday.
8. My greeting to an acquaintance in the street is not returned.

Another preliminary to systematic desensitization is to acquaint the patient with the *subjective anxiety scale*. The therapist says, "Think of the worst anxiety you have ever had and call it 100; then think of being absolutely calm and call that 0. On the scale from 0 to 100 you can at any time estimate your ongoing anxiety. And if your anxiety level changes, you can say by how much."

The technique of systematic desensitization is as follows. The patient reclines in his own chosen position in an adjustable armchair. He is asked to close his eyes and to relax as he has been instructed. His anxiety level must be down to 0 or very close to it for an adequate counteranxiety autonomic state to have been achieved. If it is not close to 0, it is necessary to find out why and to act accordingly. For example, there may be pervasive ("free-floating") anxiety. Then one may give the patient one or more single, full-capacity inhalations of a mixture of 35 per cent oxygen and 65 per cent carbon dioxide to reduce the anxiety (Wolpe).[85] Lacking facilities for this, one may try the administration of a tranquilizing drug an hour before the session. One reason why muscle relaxation may fail to calm is that it happens to activate a fear of "letting go." In that case another anxiety-inhibiting agent must be used.

Having assured himself that the patient's anxiety level is close to 0, the therapist says: "Your eyes being closed, I am going to ask you to imagine a number of scenes. You will imagine them very clearly. The moment the image that I suggest is clearly formed, indicate it by raising your left index finger one inch." After a matter of seconds the finger will go up. The therapist waits about five seconds; then says, "Stop imagining that scene"; and then, "By how much did that scene increase your anxiety level?" The patient will reply with a number on the subjective anxiety scale. Thereafter, he is enjoined to relax for about 20 seconds before being asked to imagine the same scene again.

In treating the hierarchy given above, starting with the weakest item (8), I spoke as follows: "Walking in the street on a pleasant Monday morning, you are approaching a man to whom you spoke for 10 minutes at a party the previous Saturday night. You smile in greeting, but he walks past with a blank stare." When the patient's finger rose to indicate that he was imagining the scene, I let a period of about six seconds pass before requesting the cessation of imagining. I then asked the patient by how much the anxiety level rose during the visualization. He said "15 units." When the scene was presented a second time after a period of relaxation, the anxiety rose only 8 units, the third time 3 units, and the fourth time 0. Then I began presentations of the next scene on the hierarchy.[7] "It is 5:00 P.M. on your birthday; you have spoken to half a dozen of your friends and nobody has remembered." This was presented repeatedly according to the same rules that applied to the first scene. It is worth noting that the mention of a specific time (5:00 P.M.) introduced a gradient of anxiety responses *within* that particular item: there was increasing anxiety with increasing lateness and with increasing numbers of unremembering friends. Eventually the patient came to be able to imagine without any anxiety the strongest item in the hierarchy (1).

What makes this procedure worthwhile is that there is a transfer of the change of reaction from each imaginary situation to the corresponding real situation. Nor is the change confined to the specific social interactions that figure in the treatment, but to a whole array of similar situations. In other words, if the hierarchy has covered the ground adequately, the patient should no longer be disturbable by any rejections to which disturbance is unrealistic and unnecessary. Plainly, if the changes noted in the sessions did not transfer themselves to the real situations, this whole therapeutic enterprise would be a waste of time.

Some Variants of Systematic Desensitization. Systematic desensitization is, generally speaking, a convenient and economical procedure, which, as Paul[56] has noted, is the first in the history of psychotherapy to have been vindicated by controlled experiments. Further economy may sometimes be sought by treating patients in groups (Paul and Shannon,[59] Goldstein and Wolpe).[25] Nonetheless, we do not take it for granted that the standard procedure is optimal; and recent observations (Lang, Melamed and Hart)[38] make it likely that beneficial modifications will evolve before long.

Aside from the issue of general efficacy, standard systematic desensitization is not a viable method for certain kinds of patients. There are those who do not respond fearfully when they imagine things that they find fearful in the real world. For them it is necessary to present the relevant stimuli perceptually—either concretely or in photographic form. This kind of technical variation is called *desensitization in vivo*. An extension of the use of *in vivo* exposures that is often strikingly effective is *modeling* (Bandura).[7] This involves the patient witnessing a fearless model engaging in progressively more intimate interactions with the phobic object and later, under guidance, making gradual approaches himself.

There are a good many patients in whom relaxation cannot be effectively used to inhibit anxiety. Some of these are unable to relax because of a conditioned fear of "letting go." Fortunately there are several anxiety-inhibiting agents that can replace relaxation in a desensitization program (Wolpe).[85] One of them seems to invoke external inhibition (Pavlov).[60] The patient is asked to imagine a scene from a hierarchy, and when he indicates by raising his finger that the image is clear, a *weak* electrical stimulus is administered very rapidly, two or three times, through electrodes attached to his forearm. This can apparently compete with the anxiety, for in many cases the anxiety elicited by the scene progressively decreases, with lasting effect. Scenes are given six or eight times per minute. Another method employs imagined evocations of other emotions—pleasurable emotions, such as may be aroused by sexual scenes, skiing, or floating on the waves, or whatever else evokes hedonic arousal in the individual. A questionnaire by Cautela and Kastenbaum[17] is often helpful in identifying possible stimuli to such arousal.

Flooding. This mode of treatment is in a way the antithesis of desensitization. Instead of the subject being exposed briefly to carefully graded "doses" of fear-evoking stimuli, he is exposed insistently to very high, though not necessarily maximal, doses for prolonged periods—up to 30 minutes and sometimes more. There is no doubt that this strategy is often effective.

AGORAPHOBIA

It has been customary to class agoraphobia among the classical phobias because the patient responds with anxiety in relation to a well-defined class of stimuli—usually physical distance from a place of safety or the relative inaccessibility of a "safe" person. However, it is only in a minority of agoraphobics that the level of anxiety is controlled purely by distance from "safety." Much more often the distance or accessibility of a trusted person is the occasion for the anxiety rather than its immediate stimulus antecedent. Behavior analysis may show that what the patient really fears is a bodily catastrophe; for example, he may perceive a heart attack as imminent when he feels a certain kind of pain in the chest. If the pain is in actuality brought on by a recurrent phenomenon such as gas in the gastrointestinal tract, the patient has frequent attacks of anxiety; and the farther away from possible help he is when it arises, the more anxious he feels. Therapy will focus on the pain in the chest. It will first be necessary to insure that there is no heart disease by sending the patient to an internist. Occasionally negative findings will entirely reassure the patient. But more often the anxiety response to the pain persists even after the patient is intellectually convinced that his heart is normal. Systematic desensitization to the pain, usually on a dimension based on distance from the center of the chest, will be necessary to effect recovery. (Neurotic fears of illness are often encountered without agoraphobia.)

The commonest cases of agoraphobia occur in unhappily married women, low in self-sufficiency, after several years of marriage. If a woman with normal self-sufficiency is dissatisfied with her husband, she usually takes active steps to indicate how she would like him to change; and if no amelioration occurs, she is able to consider divorce. The woman with low self-sufficiency cannot contemplate divorce because the threat of being left on her own is extremely frightening to her. There is a powerful impulse to separate from the husband; but its projected consequences make it too fearful to translate into action. The physical situation of being alone seems to be a generalization of the aloneness implied in the wished-for divorce, and thus also evokes fear. In such cases the first necessity is to work on the marriage, to try to make it satisfactory if possible. A measure that is always necessary to this end is to train the patient in assertive behavior. If the marriage cannot be improved, it should be dissolved. Whatever happens to the marriage, the treatment of the case cannot be regarded as complete until the patient has also overcome her fear of being alone (shown by her low self-sufficiency); this can usually be accomplished by a combination of assertive training and systematic desensitization.

Recent reports by Marks and Boulougeris[45,46] and a well-controlled experiment by Gelder and Gath[24] strongly suggest that flooding may turn out to be a particularly effective method in the treatment of agoraphobia.

TIMIDITY

Timid people are inhibited from expressing themselves adequately to others because of unadaptive fears of people that have been conditioned in them. The solution sometimes lies in desensitization; but this does not often fully fit the bill since the subject is also deficient in ways of speech and other motor behavior toward others. A mode of treatment that brings about change in both emotional and motor habits is assertive training. The patient is instructed to express his previously inhibited emotions in all suitable situations.

The situations that are relevant are varied and often numerous. The patient, by reason of his interpersonal fears, may be unable to complain about poor service in a restaurant, to contradict friends with whom he disagrees, to get up and leave a social situation that has become boring, to chastise a subordinate, or to express affection, appreciation, or praise. As a first step he has to be made to realize (if he does not already) that giving voice to his feelings in such situations is appropriate. He must also understand that it is anxiety that has been standing in the way, and that in the act of expressing the other emotions, the anxiety will each time to some extent be inhibited and its habit strength diminished.

Change in behavior must be prescribed in detail. To illustrate this let us take the example of an everyday situation in which a patient has indicated he has difficulty: he cannot bring himself to request a waiter to take back unsatisfactory food. The therapist discusses with him the situation where he has ordered a steak done rare, which is the way he likes it, and it has arrived done to a frazzle. The conversation proceeds as follows:

DOCTOR: What would you do?

PATIENT: I wouldn't say anything to the waiter and would feel irritated, but would eat the steak.

DR.: Why would you not speak up to the waiter?

P.: Because I wouldn't like to antagonize him.

DR.: But you would be perfectly within your rights to do so. You have not got what you have ordered and what you will be paying for. You should ask him to take that steak back and bring you another, which he will probably do; but if he is disagreeable, you should complain to the manager.

Very frequently, on the basis of such instructions, the patient becomes gradually more capable of expressing himself adequately in an increasing number of situations. Often, however, it proves helpful to give him some practice in the consulting room. The therapist takes the role of the "other person" and the patient "plays himself," so that his errors can be corrected and improving behavior shaped. This is known as *behavior rehearsal.*

Although many patients can be instructed to carry out changed behavior of this kind "across the board," in those who have a great deal of anxiety, it is necessary to grade the tasks in a way that parallels desensitization. As Salter,[63] the pioneer of assertive techniques, has expressed it, "Therapy should begin where the patient's level of inhibition is lowest." In some cases preliminary desensitization may be needed—for example, to the idea of being or seeming "aggressive" (Wolpe).[85]

It should be noted that assertive training incorporates two of the change processes outlined early in this chapter. There is reciprocal inhibition of anxiety by the expression of anger, annoyance, affection, or whatever other emotion is germane to the situation; and there is positive reinforcement of new modes of motor and verbal response, the reinforcement coming from whatever is rewarding in consequence—whether it be the achievement of control of a social interplay, or getting the rare steak that one has ordered.

MALE SEXUAL INADEQUACY

For the patient with erectile incompetence or premature ejaculation, a particularly convenient and effective instrument of reciprocal inhibition of anxiety is sexual emotional arousal. In all its phases short of ejaculation the male sexual response is mainly a parasympathetic function. Anxiety interferes with the sexual response because sympathetic arousal inhibits parasympathetic function. As a result penile erection may be impaired, or the normal overflow into the sympathetic from parasympathetic sexual activation, which is the trigger to ejaculation, may be activated too soon. In order to employ the sexual response successfully for the purpose of overcoming sexual anxiety, it is necessary at all times to insure that it is "stronger" than the anxiety to which it is opposed.

Thus, in using the sexual response as an anxiety inhibitor, the first requirement is to ascertain at what point in the sexual approach anxiety begins and what factors increase it. Perhaps the man begins to feel anxiety the moment he enters the bedroom, or perhaps it is when he is lying in bed in the nude next to his wife. The basic idea of the treatment is explained to him: he is to extend his sexual approach in slowly graded steps, never allowing himself to advance beyond a stage that arouses slight anxiety in him. If, for example, he feels anxiety to a minimal extent just lying next to his wife in bed, he is permitted to advance no further until the anxiety has entirely dissipated. This will usually occur after two or three sessions. Then he can go on to the next stage—perhaps fondling her breasts, caressing her thighs, or lying on top of her without attempting intromission. When this, in turn, ceases to evoke anxiety he may be permitted to approximate the penis to the clitoris. After this he may be allowed a small degree of entry, and then increasing degrees—after which increasing movement. The precondition for advancing beyond a stage is always the disappearance of all anxiety from it.

The cooperation of the wife is obviously crucial to such a program. She has to know that her husband's problem must be handled gently and that she must not make him anxious by mocking or goading him to achieve any particular level of performance. Though this may mean her enduring a good deal of frustration, she may hope to reap the rewards of her patience eventually.

The details of treatment naturally vary from case to case. Male sex hormones are occasionally indispensable (Miller, Hubert, and Hamilton,[49] Wolpe)[85] and tranquilizing drugs also have a place. A procedure that is frequently of great value has been described by Semans.[67] The wife is asked to manipulate the penis to a point just short of ejaculation and then to stop. After an interval she does this again, and may repeat it several times. The effect is to increase the latency to ejaculation, which may rise to a half-hour or more.

In an average of about eight weeks on this schedule, about two-thirds of the patients have been found to achieve completely normal sexual function, and another one-sixth to become reasonably satisfying to their partners.

FEMALE SEXUAL INADEQUACY

Where female sexual inadequacy (frigidity) is the problem, two kinds of cases must be distinguished. In "essential" frigidity, there is lack of response to all males, while in "situational" frigidity it is to a particular male—the patient's husband—or perhaps to a class of males, or to a particular environment. In situational frigidity the solution requires changing the situation or else extricating the patient from it.

Some cases of essential frigidity have an organic basis. One encounters women who do not recall ever having known sexual arousal and give no history of distressing sexual experiences that might have led to the conditioned inhibition of sexual responses. In some a hormonal cause is found that can sometimes be corrected. In others frigidity is due to some pathological condition that makes coitus painful—perhaps a very thick hymen or an inflammatory lesion. I once saw a woman who had been psychoanalyzed for four years for vaginal spasms that were due to local ulceration. A gynecological examination is mandatory in every case of frigidity in which there is any possibility of a physical disease.

But in most cases essential frigidity is a matter of conditioned inhibition. Some women are either absolutely or relatively frigid owing to previous experiences that have attached negative, usually anxious, feelings to sexual stimuli. In others the frigidity is the result of traumatic experiences related to sexual behavior. Sometimes the relevant experience is nothing more than having been frightened in the act of masturbation.

The treatment of essential frigidity depends upon the individual case. Where there has been faulty indoctrination, it is necessary to remove misconceptions about sex and to reeducate the woman concerning sexual activity. Having done this, one is almost always still left with a negative emotional response and an accompanying behavioral inadequacy. Typically this is treated by systematic desensitization (Wolpe).[85] Brady[10] has reported particularly good results with desensitization in which relaxation is assisted by sodium methohexital.

Homosexuality

Experience in the Behavior Therapy Unit suggests that homosexuality depends on one or more of the following three factors: (1) conditioned anxiety reactions toward women in contexts of physical closeness; (2) interpersonal anxiety of the kind that calls for assertive training, with peak reactivity toward females; and (3) positive erotic approach conditioning to men. Aversive conditioning is an appropriate treatment only for the last factor. If either or both kinds of anxiety conditioning are present, treatment should be directed to them first; and it will sometimes be found that aversive conditioning has become needless because sexual interest is transferred from males to females. The most favored technique incorporating aversive conditioning is that of Feldman and MacCulloch[22] in which approach attitudes to females are conditioned coordinately with aversion to males.

Another method of producing aversive effects that can be applied to homosexuality—as well as to other conditions—was introduced by Cautela[15] under the name "covert sensitization." The patient is asked to visualize the stimulus to which a negative attitude is to be conditioned, and at the same time aversive imagery is suggested that evokes nauseation, disgust, or some other unpleasant response.

Other Neurotic Syndromes

As stated above, the treatment of neuroses is mainly a matter of identifying and deconditioning unadaptive anxiety responses. A number of technical variations have emerged for the treatment of particular syndromes besides those discussed above. A few of them will be briefly mentioned.

Brady[11,12] has descrized the use of a micrometronome resembling a hearing aid for the treatment of stuttering. A variety of methods have been used in cases of obsessive-compulsive behavior, including positive reinforcement (Wetzel,[80] Bailey and Atchinson),[6] thought-stopping (Yamagami),[89] and covert sensitization (Cautela,[16] Wisocki).[82] Fourteen cases of various sexual perversions have been reported by Marquis[47] as having been successfully treated by "orgasmic reconditioning" in which chosen fantasies are carefully programmed to precede masturbatory orgasms. Serber[63] has overcome transvestism and other sexual deviations by the use of "shame"—having the patient undress, in the case of transvestism, in the presence of an audience. While reactive depression is generally treatable as an offshoot of anxiety (Wolpe),[84] in some cases reinforced activity has a key therapeutic role (Burgess).[13]

Pure Learned Unadaptive Habits That Are Not Neurotic

Unadaptive learned habits are excluded from the definition of neuroses when they have not originated in anxiety-generating situations. The most common of these habits are seen in children, in such forms as thumb-sucking, nail-biting, tantrums, enuresis nocturna, and encopresis. Occasionally such habits persist into adult life. But there are also other members of this class of habits that arise during adolescence or later. Some cases of anorexia nervosa seem to develop on a basis of instrumental conditioning and without relation to anxiety (see Bachrach, Erwin, and Mohr)[5] in contrast to the kind of case described by Lang.[36] A particularly common unadaptive habit of adolescent origin is distractibility from study (for example, Sulzer,[75] Wolpe).[85]

All of these habits call for treatment on principles of instrumental conditioning—positive or negative reinforcement, or aversion therapy. Kimmel and Kimmel[34] treated enuresis nocturna by encouraging the child to drink freely at all times and rewarding prolongation of *diurnal* retention of urine. On the first training day the parent waits for the child's first report of a desire to urinate. The child is asked to hold it in for, say, five minutes, being promised a reward of cookies, candy, or soda pop, etc., depending upon his likes and dislikes. When the necessary time has expired, the promised reward is given and the child permitted to urinate. The same essential procedure is followed whenever the child has a urinary urge, and as it becomes clear that the initial time requirement has become easy for him, it is increased by a few minutes. The increase must always be small in order to avoid failure or refusal to cooperate. In a few days the time can be increased to as much as 30 minutes. This may require an increase in the amount of reward also. All three of the cases reported by Kimmel and Kimmel stopped bed-wetting within 14 days after the beginning of the treatment, and none of them showed more than a single lapse during a one-year followup.

In recent years increasing use has been made of social reinforcers for children's unadaptive habits. Hart *et al.*,[28] have described deconditioning of the excessive crying behavior of two nursery school children. The teachers were told to distinguish between crying due to pain or other appropriate causes and "operant crying." They were instructed to ignore each child's operant cries, not going to him, speaking to him, or even looking at him while he was crying (except for an initial glance to assess the situation). If a particular child was close to a teacher when he began to cry, she was to turn her back and walk away or busy herself with another child. Within five days after introduction of the program, operant crying decreased from between five and ten times per morning to less than two. At this point, in order to substantiate the hypothesis that the operant crying of these children was truly a function of social reinforcement, the therapists decided to try to reinstate the crying behavior by instructing the teachers to give attention to every approximation to a cry, such as whimpering and sulking. The base line rate of crying was soon reestablished. After four days' reintroduction of the policy of removal of attention from crying, the behavior was practically eliminated.

As an adult example we may summarize a case reported by Sulzer.[75] A graduate student complained of an inability to persist in reading long enough to maintain high grades in his courses. In his living room he could concentrate for only 20 to 30 minutes before becoming so restless that he could not continue reading. He would join his wife watching television in the bedroom or read some nonschool material. He would rarely be able to resume study that evening. A program was devised for the patient to study according to new rules in one of the university reading rooms. He was to go to the library daily for a week and each day read six pages of assigned material, no more and no less. He was not to study at home. The average time per day spent studying in the first week was less than 20 minutes, but the student had some difficulty in restraining himself from reading further. The therapist expressed approval to him for following the program exactly. Each following week the daily reading material was increased by two pages until, after several months, it reached 60 pages daily. Later, measures were adopted to enable the student to study at home.

A few further examples can be only briefly mentioned. Risley and Hart[61] have reported the alteration of nonverbal behavior in preschool children by reinforcing related verbal behavior. Patterson[54] recounts the virtual elimination of hyperactive classroom behavior in a nine-year-old child by reinforcing attentive behavior by candy, having found that most of the "hyperactivity" consisted of such inappropriate behavior as talking, pushing, hitting, pinching, looking about the room and out of the window, moving about the room, and squirming and tapping at the desk. Barrish *et al.*[8] and Bushell *et al.*,[14] have described the effective use of individual behavioral contingencies with group consequences rebounding on the individual. The treatment of encopresis by various kinds of positive reinforcement has been described by Neale,[51] Madsen,[43] Tomlinson,[77] and Edelman.[21] The use of behavior modification for youthful offenders in a group setting is described by Cohen and Filipczak,[19] and the treatment of autistic children by Lovaas, Schaffer, and Simmons.[42] Homme[29] has shown how favored activities may be used as reinforcers on the basis of the Premack principle. An overview of procedures in the home and the classroom has been provided by Patterson.[55]

In order to make behavior modification widely available, it is necessary to train teachers, social workers, and nursing aides as dispensers of reinforcement (Guerney,[27] Jehu,[32] Patterson,[55] Birnbrauer *et al.*).[9] Lindsley,[39] who has for a long time argued that a massive effort in behavior modification with children cannot be successfully implemented by psychotherapists, even with the help of teachers, and that parents must be brought into the therapeutic team, has recently provided a brief account of his program for parents.

Modifying the Learned Behavior of Schizophrenics

As previously stated, schizophrenia is essentially an organic disease, but may predispose patients to acquire a variety of unadaptive habits on the basis of learning. These learned habits, as would be expected, are susceptible to modifications by learning. The possibility of this was first put forth by Lindsley.[40] Much of the clinical pioneering was performed by Ayllon and his co-workers.[1,2,4] A case described by Ayllon[1] illustrates his typical ingenious retraining procedures. The patient, a 47-year-old female schizophrenic who had been hospitalized for nine years, had a number of bizarre habits, including wearing 25 pounds of clothing simultaneously, overeating, stealing the food of other patients, and hoarding towels. All of these were successfully overcome by reinforcement procedures. The towel hoarding habit was treated by getting the nurses progressively to add more and more towels to the patient's room. At first she was delighted and arranged them in neat piles. After two or three weeks, when their number was in the region of 600, they became unmanageable; and the patient would beg the nurses not to bring any more. After about six weeks she herself began taking the towels out of her room, and they did not come back. She ultimately removed practically all of them, remaining, like other patients, with one or two. Evidently at a certain stage the massiveness of the hoard had become aversive, so that removal of the towels provided negative reinforcement. In the course of a year's continuous observation, she never again reverted to towel hoarding.

To carry out behavior modification programs as exemplified by the above case requires specially trained ward personnel. Once they have been employed in reasonable numbers, it becomes possible to utilize their services widely in special operant wards, a rapidly growing number of which have been established in mental hospitals throughout the United States in the past decade. The training of such personnel has thus become increasingly important. It has been greatly aided by two lucid texts, by Schaefer and Martin[65] and by Ayllon and Azrin.[3]

The efforts at behavior modification that go on in operant wards are not confined to special programs for individual patients. Selected behavior of all the members of a group may be programmed for change in a particular direction. The implementation of group programs has been facilitated by the use of tokens (Ayllon and Azrin)[3] that the patients receive for the performance of desired behavior and that they can subsequently spend to obtain objects or privileges of their own choice. Wholesale changes of a socializing kind are frequently procured, including regular attendance at suitable occupations, improved cleanliness, and cooperation with the hospital staff.

More specific group targets of operant conditioning procedures in wards include reducing urinary incontinence (Wagner and Paul),[79] controlling overeating (Upper and Newton),[78] and decreasing pill-taking behavior (Parrino *et al.*).[53]

The Place of Behavior Therapy in Psychiatry

Once it is realized that unadaptive behavior must either be the result of abnormal anatomical or physiological states or else have been acquired by learning, it is but a small step to the conclusion that the most logical approach to the latter category is by making use of experimentally established knowledge of the factors that influence learning. This is what behavior therapy does. However, as pointed out at the beginning of this chapter, emotional learning based upon the spontaneous emotional responses of the patient to the therapist benefits many neurotic patients, no matter what the orientation of the therapist. It seems also that experienced and sympathetic therapists of many kinds often influence their patients to change their behavior toward reasonable assertiveness on the basis of common sense. But it can scarcely be doubted that the deliberate application of solidly established principles is a surer way to success than any rule of thumb.

The results so far obtained by behavior therapy of the neuroses quite strongly support the foregoing. Uncontrolled comparisons (such as, Wolpe)[83] have shown behavior therapy to produce markedly favorable results much more rapidly than psychoanalysis does (on the criteria suggested by Knight).[35] A large range of controlled studies (such as, Paul,[58] Moore,[50] and Lomont *et al.*)[41] have uniformly indicated superior efficacy for specific behavior therapy techniques; and Sloane *et al.*,[72] have found, in a carefully controlled comparison of behavior therapy with psychoanalytically oriented therapy, that behavior therapy achieves significantly more personality change in areas outside the target symptom than psychoanalytically oriented therapy. Furthermore it is quite evident that the opinion of psychoanalytically oriented psychiatrists that change procured by conditioning cannot be reliable because it leaves the "dynamic roots" of neurosis untouched, and that relapse or symptom substitution is to be expected, has not been borne out (Wolpe,[86] Paul).[58] Neither relapse nor symptom substitution is seen in patients whose neuroses have been overcome by skillful behavior therapy based on competent behavior analyses. In every one of the rare cases that have displayed a renewal of symptoms, we have found a clear history of exposure to new anxiety-conditioning experiences.

Bibliography

1. AYLLON, T., "Intensive Treatment of Psychotic Behavior by Stimulus Satiation and Food Reinforcement," *Behav. Res. Ther.*, *1*:53–61, 1963.
2. ———, and AZRIN, N. H., "Reinforcement and Instructions with Mental Patients," *J. Exp. Anal. Behav.*, *7*:327–331, 1964.
3. ———, and ———, *The Token Economy: A Motivational System for Therapy and Rehabilitation*, Appleton-Century-Crofts, New York, 1968.
4. ———, and HAUGHTON, E., "Control of the Behavior of Schizophrenic Patients by Food," *J. Exp. Anal. Behav.*, *5*:343–352, 1962.
5. BACHRACH, A. J., ERWIN, W. J., and MOHR, J. P., "The Control of Eating Behavior in an Anorexic by Operant Conditioning Techniques," in Ullman, L., and Krasner, L. (Eds.), *Case Studies in Behavior Modification*, Holt, Rinehart & Winston, New York, 1965.
6. BAILEY, J., and ATCHINSON, T., "The Treatment of Compulsive Handwashing Using Reinforcement Principles," *Behav. Res. Ther.*, *7*:327–329, 1969.
7. BANDURA, A., *Principles of Behavior Modification*, Holt, Rinehart & Winston, New York, 1969.
8. BARRISH, H. H., SAUNDERS, M., and WOLF, M. M., "Good Behavior Game: Effects of Individual Contingencies for Group Consequences on Disruptive Behavior in a Classroom," *J. App. Behav. Anal.*, *2*:119–224, 1969.
9. BIRNBRAUER, J. S., BURCHARD, J. D., and BURCHARD, S. N., "Wanted: Behavior Analysts," in Bradfield, R. H. (Ed.), *Behavior Modification: The Human Effort*, Dimensions, San Rafael, 1970.
10. BRADY, J. P., "Brevital-Relaxation Treatment of Frigidity," *Behav. Res. Ther.*, *4*:71–77.
11. ———, "Metronome-Conditioned Speech Retraining for Stuttering," *Behav. Res. Ther.*, *2*:129–150, 1971.
12. ———, "Studies on the Metronome Effect on Stuttering," *Behav. Res. Ther.*, *7*:197–204, 1969.
13. BURGESS, E. P., "The Modification of Depressive Behaviors," in Rubin, R. D., and

Franks, C. M. (Eds.), *Advances in Behavior Therapy*, Academic Press, New York, 1968.

14. BUSHELL, D., WROBEL, P. A., and MICHAELIS, M. L., "Applying 'Group' Contingencies to the Classroom Study Behavior of Preschool Children," *J. App. Behav. Anal.*, *1*:55–61, 1968.
15. CAUTELA, J., "Covert Sensitization," *Psychol. Rep.*, *20*:459, 1967.
16. ———, "Treatment of Compulsive Behavior by Covert Sensitization," *Psychol. Rec.*, *16*:33–41, 1966.
17. ———, and KASTENBAUM, R., "A Reinforcement Survey Schedule for Use in Therapy, Training, and Research," *Psychol. Rep.*, *20*:1115–1130, 1967.
18. CHAPMAN, J., "The Early Symptoms of Schizophrenia," *Brit. J. Psychiat.*, *112*: 225–251, 1966.
19. COHEN, H. L., and FILIPCZAK, J. A., *A New Learning Environment*, Jossey-Bass, San Francisco, 1971.
20. CONN, J. W., and SELTZER, H. S., "Spontaneous Hypoglycemia," *Am. J. Med.*, *19*: 460–478, 1955.
21. EDELMAN, R. I., "Operant Conditioning Treatment of Encopresis," *J. Behav. Ther. Exp. Psychiat.*, *2*:71–73, 1971.
22. FELDMAN, M. P., and MACCULLOCH, M. J., "Aversion Therapy in the Management of Homosexuals," *Brit. Med. J.*, *1*:594, 1967.
23. FRANKL, V., "Paradoxical Intention: A Logotherapeutic Technique," *Am. J. Psychother.*, *14*:520, 1960.
24. GATH, D., and GELDER, M. G. A., *Treatment of Phobias—Desensitization versus Flooding*, paper delivered to the Department of Psychiatry, Temple University Medical School, November 8, 1971.
25. GOLDSTEIN, A., and WOLPE, J., "Behavior Therapy in Groups," in Kaplan, H. I., and Sadock, B. J. (Eds.), *Comprehensive Group Psychotherapy*, Williams & Wilkins, Baltimore, 1971.
26. GOTTLIEB, J. S., FROHMAN, C. E., TOURNEY, G., and BECKETT, P. G. S., "Energy Transfer Systems in Schizophrenia: Adenosine-triphosphate (ATP)," *Arch. Neurol. Psychiat.*, *81*:504–510, 1959.
27. GUERNEY, B. G., *Psychotherapeutic Agents: New Roles for Nonprofessionals, Parents and Teachers*, Holt, Rinehart & Winston, New York, 1969.
28. HART, B. M., ALLEN, K. E., BUELL, J. S., HARRIS, F. R., and WOLF, M. M., "Effects of Social Reinforcement on Operant Crying," *J. Exp. Child Psychol.*, *1*:145–153, 1964.
29. HOMME, L. E., "Control of Coverants, The Operants of the Mind," *Psychol. Rec.*, *15*: 501–511, 1965.
30. JACOBSON, E., *Progressive Relaxation*, University of Chicago Press, Chicago, 1938.
31. ———, *Self-Operations Control*, J. B. Lippincott, Philadelphia, 1964.
32. JEHU, D., "The Role of Social Workers in Behavior Therapy," *J. Behav. Ther. Exp. Psychiat.*, *1*:5–15, 1970.
33. JONES, M. C., "A Laboratory Study of Fear: The Case of Peter," *J. Genet. Psychol.*, *31*: 308–311, 1924.
34. KIMMEL, H. D., and KIMMEL, E., "An Instrumental Conditioning Method for the Treatment of Enuresis," *J. Behav. Ther. Exp. Psychiat.*, *1*:121–123, 1970.
35. KNIGHT, R. P., "Evaluation of the Results of Psychoanalytic Therapy," *Amer. J. Psychiat.*, *98*:434, 1941.
36. LANG, P. J., "Behavior Therapy with a Case of Nervous Anorexia," in Ullman, L. P., and Krasner, L. (Eds.), *Case Studies in Behavior Modification*, Holt, Rinehart & Winston, New York, 1965.
37. LANG, P. J., LAZOVIK, A. D., and REYNOLDS, D., "Desensitization, Suggestibility and Psychotherapy," *J. Abn. Psychol.*, *70*:395–402, 1965.
38. LANG, P. J., MELAMED, B. G., and HART, J., "A Psychophysiological Analysis of Fear Modification Using an Automated Desensitization Procedure," *J. Abn. Psychol.*, *76*: 221–234, 1970.
39. LINDSLEY, O. R., "An Experiment with Parents Handling Behavior at Home," in Fargo, G. A., Behrens, C., and Nolan, P. (Eds.), *Behavior Modification in the Classroom*, Wadsworth Publ. Co., Belmont, Calif., 1970.
40. ———, "Operant Conditioning Methods Applied to Research in Chronic Schizophrenia," *Psychiat. Res. Rep.*, *5*:118–139, 147–153, 1956.
41. LOMONT, J. F., GILNER, F. H., SPECTOR, N. J., and SKINNER, K. K., "Group Assertion Training and Group Insight Therapies," *Psychol. Rep.*, *25*:463–470, 1969.
42. LOVAAS, O. I., SCHAFFER, B., and SIMMONS,

J. Q., "Building Social Behavior in Autistic Children Using Electric Shock," *J. Exp. Stud. Pers.*, *1*:99–109, 1965.

43. MADSEN, C. H., "Positive Reinforcement in the Toilet Training of a Normal Child: A Case Report," in Ullman, L. P., and Krasner, L. (Eds.), *Case Studies in Behavior Modification*, Holt, Rinehart & Winston, New York, 1965.
44. MALLESON, N., "Panic and Phobia," *Lancet*, *1*:225–227, 1959.
45. MARKS, I. M., and BOULOUGERIS, J. C., "Implosion (Flooding)—A New Treatment for Phobias," *Brit. Med. J.* (June 1969) *1*:721–723.
46. ———, ———, and MARSET, P., "Superiority of Flooding (Implosion) to Desensitization for Reducing Pathological Fear," *Behav. Res. Ther.*, *9*:7–16, 1971.
47. MARQUIS, J. N., "Orgasmic Reconditioning: Changing Sexual Object Choice through Controlling Masturbation Fantasies," *J. Behav. Ther. Exp. Psychiat.*, *1*:263–271, 1970.
48. MEALIEA, W. L., and NAWAS, M. M., "The Comparative Effectiveness of Systematic Desensitization and Implosive Therapy in the Treatment of Snake Phobia," *J. Behav. Ther. Exp. Psychiat.*, *2*:85–94, 1971.
49. MILLER, N. E., HUBERT, E., and HAMILTON, J., "Mental and Behavioral Changes Following Male Hormone Treatment of Adult Castration Hypogonadism and Psychic Impotence," *Proc. Soc. Exp. Biol. Med.*, *38*:538, 1938.
50. MOORE, N., "Behavior Therapy in Bronchial Asthma: A Controlled Study," *J. Psychosom. Res.*, *9*:257–276, 1965.
51. NEALE, D. H., "Behavior Therapy and Encopresis in Children," *Behav. Res. Ther.*, *1*:139–149, 1963.
52. OSGOOD, C. E., "Meaningful Similarity and Interference in Learning," *J. Exp. Psychol.*, *38*:132, 1946.
53. PARRINO, J. J., GEORGE, L., and DANIELS, A. C., "Token Control of Pill-Taking Behavior in a Psychiatric Ward," *J. Behav. Ther. Exp. Psychiat.*, *2*:181–185, 1971.
54. PATTERSON, G. R., "An Application of Conditioning Techniques to the Control of a Hyperactive Child," in Ullman, L. P., and Krasner, L. (Eds.), *Case Studies in Behavior Modification*, Holt, Rinehart & Winston, New York, 1965.
55. PATTERSON, G. R., "Behavioral Intervention Procedures in the Classroom and the Home," in Bergin, A. E., and Garfield, S. L. (Eds.), *Handbook of Psychotherapy and Behavior Change*, John Wiley, New York, 1970.
56. PAUL, G. L., "Behavior Modification Research: Design and Tactics," in Franks, C. M. (Ed.), *Behavior Therapy: Appraisal and Status*, McGraw-Hill, New York, 1969.
57. ———, "Inhibition of Physiological Response to Stressful Imagery by Relaxation Training and Hypnotically Suggested Relaxation," *Behav. Res. Ther.*, *7*:249–256, 1969.
58. ———, *Insight versus Desensitization in Psychotherapy*, Stanford University Press, Stanford, 1966.
59. ———, and SHANNON, D. T., "Treatment of Anxiety through Systematic Desensitization in Therapy Groups," *J. Abn. Psychol.*, *71*:124–135, 1966.
60. PAVLOV, I. P., *Conditioned Reflexes* (Tr. by Anrep, G. V.), Liveright, New York, 1927.
61. RISLEY, T. R., and HART, B., "Developing Correspondence between the Non-Verbal and Verbal Behavior of Preschool Children," *J. App. Behav. Anal.*, *1*:267–281, 1968.
62. RUBIN, L. S., "Pupillary Reflexes as Objective Indices of Autonomic Dysfunction in the Differential Diagnosis of Schizophrenic and Neurotic Behavior," *J. Behav. Ther. Exp. Psychiat.*, *1*:185–194, 1970.
63. SALTER, A., *Conditioned Reflex Therapy*, Creative Age, New York, 1949.
64. SALZER, H. M., "Relative Hypoglycemia as a Cause of Neuropsychiatric Illness," *J. Nat. Med. A.*, *58*:12, 1966.
65. SCHAEFER, H. H., and MARTIN, P. L., *Behavioral Therapy*, McGraw-Hill, New York, 1969.
66. SCRIGNAR, C. B., "Food as the Reinforcer in the Outpatient Treatment of Anorexia Nervosa," *J. Behav. Ther. Exp. Psychiat.*, *2*:31–36, 1971.
67. SEMANS, J. H., "Premature Ejaculation: A New Approach," *South Med. J.*, *49*:353, 1956.
68. SERBER, M., "Shame Aversion Therapy," *J. Behav. Ther. Exp. Psychiat.*, *1*:213–215, 1970.
69. SHERRINGTON, C. S., *Integrative Action of*

the Nervous System, Yale University Press, New Haven, 1906.

70. SKINNER, B. F., *The Behavior of Organisms*, Appleton-Century-Crofts, New York, 1938.
71. ———, *Science and Human Behavior*, Macmillan, New York, 1953.
72. SLOANE, R. B., CRISTOL, A. H., YORKSTON, N. J., WOLPE, J., and FREED, H., *Psychotherapy and Behavior Therapy for Neuroses: A Comparative Study*, in print.
73. STAMPFL, T. G., and LEVIS, D. J., "Essentials of Implosive Therapy: A Learning-Theory-Based Psychodynamic Behavioral Therapy," *J. Abn. Psychol.*, *72*:496–503, 1967.
74. STOFFELMAYR, B. E., "The Treatment of a Retching Response to Dentures by Counteractive Reading Aloud," *J. Behav. Ther. Exp. Psychiat.*, *1*:163–164, 1970.
75. SULZER, E. S., "Behavior Modification in Adult Psychiatric Patients," in Ullman, L. P., and Krasner, L. (Eds.), *Case Studies in Behavior Modification*, Holt, Rinehart & Winston, New York, 1965.
76. THOMAS, E. J., "Selected Sociobehavioral Techniques and Principles: An Approach to Interpersonal Helping," *Soc. Work*, *13*: 12–16, 1968.
77. TOMLINSON, J. R., "The Treatment of Bowel Retention by Operant Procedures: A Case Report," *J. Behav. Ther. Exp. Psychiat.*, *1*:83–85, 1970.
78. UPPER, D., and NEWTON, J. G., "A Weight-Reduction Program for Schizophrenic Patients on a Token Economy Unit: Two Case Studies," *J. Behav. Ther. Exp. Psychiat.*, *2*:113–115, 1971.
79. WAGNER, B. R., and PAUL, G. L., "Reduction of Incontinence in Chronic Mental Patients: A Pilot Project," *J. Behav. Ther. Exp. Psychiat.*, *1*:29–37, 1970.
80. WETZEL, R., "Use of Behavioral Techniques in a Case of Compulsive Stealing," *J. Consult. Psychol.*, *30*:367–374, 1966.
81. WILLOUGHBY, R. R., "Some Properties of the Thurston Personality Schedule and a Suggested Revision," *J. Soc. Psychol.*, *3*:401, 1932.
82. WISOCKI, P. A., "Treatment of Obsessive-Compulsive Behavior by Covert Sensitization and Covert Reinforcement: A Case Report," *J. Behav. Ther. Exp. Psychiat.*, *1*:233–239, 1970.
83. WOLPE, J., "The Discontinuity of Neurosis and Schizophrenia," *Behav. Res. Ther.*, *8*: 179–187, 1970.
84. ———, "Neurotic Depression: Experimental Analog, Clinical Syndromes, and Treatment," *Am. J. Psychother.*, *25*:362–368, 1971.
85. ———, *The Practice of Behavior Therapy*, Pergamon Press, New York, 1969.
86. ———, *Psychotherapy by Reciprocal Inhibition*, Stanford University Press, Stanford, 1958.
87. ———, and LANG, P. J., *A Fear Survey Schedule*, Educational and Industrial Testing Service, San Diego, 1969.
88. WOLPIN, M., and RAINES, J., "Visual Imagery, Expected Roles and Extinction as Possible Factors in Reducing Fear and Avoidance Behavior," *Behav. Res. Ther.*, *4*:25–37, 1966.
89. YAMAGAMI, T., "The Treatment of an Obsession by Thought-Stopping," *J. Behav. Ther. Exp. Psychiat.*, *2*:133–135, 1971.

PART FIVE

Contributions to Psychiatry from Related Fields

CHAPTER 44

PSYCHIATRY AND PHILOSOPHY

Abraham Edel

THE RELATIONS OF philosophy and psychiatry have sometimes involved dialogue, sometimes confrontation, often illumination. But on the whole there has been little of what would itself be most desirable, cooperation on shared problems.

That there are many problems in common is clear enough from current literature. An obvious one lies in the field of value theory. Moral philosophy directly seeks firm bases for guiding practice, while psychiatry not only seeks scientific grounds for judging mental health but also is pressed to show that its conception of mental health does not smuggle in current social standards. Other common problems concern conceptions of the human being and methods of studying human behavior. Philosophy is still enmeshed in disputes about whether the methods that have proved so successful in the physical sciences can be extended to human action, and psychiatry here is beset with a variety of schools and methodological approaches. Again, other common problems are rooted in metaphysical presuppositions, especially about the relation of body and mind. Frequently these presuppositions are cast as reality claims—for example, in psychiatry that the unconscious is a reality and not a fiction, or that human reality lies in the phenomenological field. Philosophers have the heritage of reality claims in their traditional schools of materialism, idealism, dualism, phenomenalism, and the rest. How contemporary philosophy deals with these conflicts may suggest how to analyze psychiatric concepts in this domain. In this chapter I should like to approach these shared problems following the order indicated and then conclude with some reflections on the nature of the philosophical activity involved. After all, there has been controversy not only about what philosophy's role is in relation to psychiatry but also among philosophers themselves about what philosophy is up to.

Problems in the Theory of Value

Influential views in twentieth-century philosophy have made a sharp separation of fact and value (such as, G. E. Moore).[47] This separation has often been helpful in detecting disguised norms or values masquerading as factual judgments. But elevating it to a philosophic dogma instead of recognizing it as a distinction relative to important contexts has made a mystery of the relations of the scientific and the ethical. Major energies in philosophy and science are consequently required to remove ethical judgments from the status of sheer fiat and to make them responsible to knowledge. Let us illustrate these problems from the viewpoint of both disciplines.

Medicine and psychiatry both employ concepts of illness and health, of needs, of what is normal and abnormal. The clearest root values in medicine are the undesirability of pain and of the inability to function. These get expanded and articulated into an ideal of health as we acquire knowledge of the physiological conditions of the kind of functioning that minimizes pain and breakdown. A major form of this articulation is in terms of an account of men's needs (Edel,[15] ch. 6). The logic of "need" involves a value reference; for example, even tissue needs imply that if they are not satisfied (e.g., water, salt) then certain forms of breakdown inimical to survival will result. It is important to note that such knowledge may extend into *social* conditions for given states of society, so that a medical critique of social institutions becomes possible. Thus, the ideal of public health makes possible a moral critique of some urban conditions, such as the overcrowding that is productive of epidemics or economic arrangements that engender dangerous pollution. Such judgments are still *medical*, but as critiques they are only *partial*. For they cannot determine that other values may not overweigh the risks involved: they cannot decide whether to risk a 5 per cent increase in the incidence of cancer to assure a tremendous increase in electrical power; they cannot determine simply on the ground that death is a medical evil that there should be no cultural arrangements allowing an individual to commit suicide or to sacrifice his health to research. But the fact that medical critiques of value are partial does not belie their ability to go as far as they can on the basis of the knowledge that has been acquired. In short, medical knowledge imposes constraints on value judgments, it helps render them more determinate, but it cannot settle them completely for all kinds of cases.

At present psychiatry is in a less favorable position than medicine to help such moral judgment because the scope of its established knowledge is less extensive. But where its knowledge exists we can see the same process of joining factual information to basic value criteria to yield a complex ideal of mental health. There is no sharp discontinuity between the general ideal of health and that of mental health. Where psychology utilizes knowledge of brain damage to criticize the use of certain drugs and additives, it is operating with the minimal value criteria given above. Where it brings its knowledge of the conditions that beget such phenomena as hysterical blindness or psychic impotence to bear on a critique of interpersonal relations, it is extending the medical criterion of the undesirability of pain at least to the undesirability of being anxiety-ridden.

Such knowledge and value judgment are often generalized through the concepts of normal and abnormal behavior and development. Controversy without end has centered around these concepts. Some insist that their scientific meaning must be cast as a statistical norm: to judge the normal as better than the abnormal simply smuggles in conformity to currently dominant value patterns. Others argue contrariwise that normality means more than statistical distribution, for a statistically dominant behavior (say, mass hysteria or scapegoating) can be abnormal. In this view the judgment of abnormality refers to the *causal* analysis of the behavior (Wegrocki,[62] cf. Devereux).[11] A definite model is implicit. A human being is assumed to have certain energies aiming in a given direction, achievement of which would be reckoned a good. A distortion process takes place, whether through blocking or other interference, such that the original aim, *which is still being sought*, is not achieved, and a substitute compromise direction is taken, which is productive of intense anxiety discoverable in the quality of present experience and its affects. The behavior in this substitute direction is thus declared abnormal. A comparably more complex analysis involving stages is offered for abnormal development, for example, in Erikson's[17] stages of man.

Seen in this way the concept of normality joins in a definite pattern components of statistical distribution, causality, and value. But it does not determine where that kind of pattern is to be found and where a distinction of normal and abnormal is relevant. Thus, it may hold adequately for the understanding of extreme alcoholism; but whether it holds for all homosexuality or only some forms (as well as some forms of heterosexuality) is still debatable. General attacks on the concept of mental health (such as by Szasz)[60] as bootlegging social values would then be equivalent to denying that there is adequate causal knowledge of the processes involved, as well as an adequate analysis of the precise values involved. Extended philosophical analysis of these problems is required, but the shift should definitely be such as to spotlight the kind of knowledge that would justify the psychiatric conception and to ask whether we have it and if it is attainable and where it applies, rather than wholesale argument for one or another general model throughout the whole area.

If philosophy can help psychiatry in this way, psychiatry can repay the debt in many different ways. Thus, moral philosophy long remained stuck with the assumption that a man can get a correct account of his own basic values only by introspection. When two men thus find themselves in ultimate value disagreement—one respects all human beings, the other only his own group or race and regards the rest of humanity as simply a means —there is no further basis of judging between them (Edel,[15] ch. 3). Here psychiatry, with the concept of mechanisms of defense, opened up the possibility that what is held in consciousness as an ultimate value may be serving quite different intrapsychic functions; most dramatically in reaction formations in which the very rigor with which a value is held in consciousness (cleanliness, helpfulness, nonviolence) may in the specific cases reflect the strength of the opposing impulse. These concepts do not settle issues of moral conflict but open up the possibility of going beyond introspective ultimacy. Again psychoanalytic theory has had marked impact on moral psychology in its theories of the source and operations of conscience and of the moral emotions generally (shame, guilt, pride, and so forth; cf. Fromm,[28] Lewis,[41] Lynd,[43] Piers and Singer[51]). For here, too, it has strengthened the interest in genetic accounts and removed the apparent intuitive authority that older moral philosophies assigned to these all too human processes. Such contributions, which are numerous and multiple in their impact, do not mean basing ethics on science, but bringing science to bear in refining and deepening ethical awareness of its presuppositions.

❡ Man and His Powers

How to understand man and his behavior is the second set of shared problems. Philosophy is involved in metaphysical considerations of the mind-body problem (Hook),[30] in epistemological exploration of the sources of knowledge, and in methodological analyses of psychology, the social sciences, and history. Psychiatry is plunged directly into these problems, for as part of medicine it is an art or technology of therapy of the human being. Indeed, if we take "psychiatry" literally as the cure of the psyche, and understand the psyche (as Aristotle[2] did in the classical foundations of psychology) as the manifold functions of man from moving and eating at one extreme to sensing and abstract thinking at the other, we might even say that medicine is a part of psychiatry!

Given such breadth, both philosophy and psychiatry are compelled to work out some model of the human being. Suggestions are sought from the many fields whose data and methods are brought to bear upon the inquiry. For example, philosophy developed the idea of determinism in the context of the growing physical sciences, tried it out on the realm of human action, and then grappled with the consequences for traditional notions of freedom and initiative in morality and in history. As philosophers became self-conscious about the relation of their models to the different fields of knowledge, they fashioned systematically the philosophy of social science as a branch of their work. Similarly psychiatry, in developing a theory of therapy, draws upon a host of sciences from biochemistry and pharmacology to psychology, and to disciplines not organized as systematic science, such as history (biography), cultural anthropology, even literature (phenomena of symbolism), and so forth. Moreover, in the selection of its sources —in what it follows as well as what it ignores —psychiatry often enshrines specific models of man that are made the basis of its directives in therapy. Thus, psychiatry at a given time may have a particular primary focus corresponding to the model of man that it employs and to the area of human knowledge that has impinged strongly upon it.

Of course these interrelations are never one-way. The development of special fields is itself influenced by current general models of man. In this process philosophy generalizes the model from one field, suggesting it thereby to others. But the central philosophical contribution here is to criticize rigorously the structure and presuppositions of the models.

The common concerns of philosophy and psychiatry may be illustrated in three presently controversial issues. The first is the problem of primary focus—whether a model of man should be cast in terms that are intra-individual, interpersonal, group and institutional, or historical-evolutionary. The second is the demand for seeing man not merely as fashioned and determined, but as an active being, decisional and responsible. The third is the methodological crisis precipitated by the current revolt against the scientific models, which in effect challenges traditional modes of inquiry in the psychological and social sciences.

The notion of primary focus is suggested by a comparison of schools in both philosophy and psychiatry. Philosophical issues about egoism and individualism, the relation of the individual to others and to society as a whole, and about the bases and objects of loyalty and obligation rest on presuppositions about what selves are, how they are related, and what community and society involve (Dewey).[13] When philosophy looks to psychology to clarify these foundations, it finds the same positions embedded in different schools. Hence a mutual effort at clarification and the search for evidence is required.

In psychiatry each of the schools seems committed to its own primary focus. The Freudian view focuses on the internal economy of the individual and the career of instinctual demands and energies. Other people are objects assessed by the extent to which they satisfy these demands. Freud[25] even sought to extend this view to group life, tracing the changes in libidinal ties as needs for the larger group organization emerged. In general, this is a broad reductive program based on an intrapsychic focus. It was somewhat modified by ego psychology, which, exploring the conflict-free bases of ego development (Rapaport),[53] showed how other people can be perceived on a nondemand basis. In such approaches as Sullivan's,[59] although there is careful attention to the self-system or the experiential feedback from the functioning of ego machinery, the primary focus seems to shift to interpersonal relations in the explanation of personal development. An existentialist view like Martin Buber's[8] proceeds more directly because it is not concerned with the machinery of development, and focuses on the basic dialogue or direct relation of men. In such approaches both what goes on within the individual and what goes on in society are now dependent phenomena reflecting the character of interpersonal relations. Buber thus treats both the individual pursuit of pleasure and the emphasis on collective institutions as a retreat from the primary I-Thou relation. With an interpersonal focus even the character of a man's relation to the cosmos (as in religion, for example) might be seen as a reflection of the type and quality of interpersonal relations. Focus on groups and institutions, as in sociological approaches, sees the individual in his whole development as the intersection of roles in the ongoing life of the society. At times it threatens to reduce the individual to a point of intersection of social demands, and intrapsychic factors are seen as simply necessary conditions to provide raw materials for the social shaping. Finally a historical focus sets all problems of understanding man in terms of a changing process in which the character of institutions, roles, interpersonal relations, and even the quality of intrapsychic events can be understood only in a framework of directional social movement. Thus, for example, the character of human aggressive behavior is taken neither as an inherent property of the id nor as a special feature of interpersonal orientation, but as an internalized reflection of a predatory capitalism at its particular stage of historical development.

The conflict of focuses here quite transparently reflects the different psychological schools and social and historical disciplines. In the long run these solutions must be empirical —which model of man will prove more fruitful in inquiry and in therapeutic guidance. But there are also possibilities of integrated models, not simply piling model on model in an intellectual compromise. But most of all, as we shall see shortly, to have related the models to the different disciplines in the study of man can make us sensitive to their limited and possibly changing character. For it scarcely makes scientific sense to invest all theoretical hope in one particular stage of development of one among many modes of regarding the human being.

Our second illustration—the demand for providing a more active conception of man—touches a powerful philosophical tendency in recent thought. It is basically a revolt against a deterministic conception of man, which itself had grown as the scientific approach moved on from physics and biology into the psychological and social sciences. Different philosophical strategies were employed. Some simply took a moral stand and reiterated the demand for free will (Berofsky)[6] as the necessary condition for human responsibility (echoing Kant's dictum that "I could have acted otherwise" is a presupposition of any blame or guilt or obligation). Others fashioned a new quasi dualism between "action" and "behavior" (Kenny)[35] in which the former already entails initiation by the self; in a similar way the existentialists insisted that choice is pervasive in human life and any disguising of it is inauthentic or bad faith (Sartre).[56] Still others adopted the sharp distinction between the perspective of the spectator and that of the agent, hoping to show that the legitimate determinism of the spectator's view is misapplied in the agent's own understanding of his decision.

Parallel strategies are found in psychiatric theory with the same end in view—that of restoring a kind of active and responsible dignity to the human being. The Freudian model is criticized as reducing the individual to a passive product of causal drives and forces. It is argued that the patient should not be shaped, reconditioned, or readjusted, but his activity should be elicited with perhaps blocks removed or the social situation restructured to allow the emergence of constructive capacities. A new respect is even proposed for the active forces at work in neurosis, since this represents an effort within a given life to work out a unique convergence of problems.

The controversies over what should be the primary focus generally remain within a scientific tradition. In contrast, the activist conception of man is often cast in antiscientific terms: a possibly legitimate ethical ideal is made the basis for ruling scientific approaches out of bounds in the study of man. Accordingly there is no way of avoiding a basic philosophical reckoning on these questions. Such a reckoning obviously has already led us into the basic methodological controversies about the character and limits of scientific inquiry into human beings. These controversies cannot, of course, be settled here, but it is important at least to keep them from simply recapitulating outworn issues and using outworn pictures of science.

In the first place the picture of science and its method has undergone great transformation in the twentieth century (Nagel).[48] Instead of the sharp contrast of quantitative science and qualitative man with its dualism of matter and spirit, the logic of measurement has shown that types of order constitute a whole range of varying strength and degree; the quest for order cannot be barred from any domain on dogmatic grounds, but what kind of order will be found is an empirical issue. Similarly, instead of the old Laplacean world determinism, there is simply the search for what kinds of determinisms may be discoverable. Probabilities have made inroads on fixed laws; the physical determinist image has been supplemented if not wholly supplanted by the picture of evolutionary changes with complex patterns of interaction over time; attempts to understand process over time as well as to control application where different systems have to be invoked has led to operational systems approaches in which placeholders remain open to receive readings that are not necessarily anticipated or predicted but that will thereafter aid in control within a given delimited field. And such changes hold for the physical sciences and their related technologies, for example, meteorology, space science, geology; they are not novel issues when they become typical in the study of man.

The growth of the social sciences and their methods has done much to diminish the dogmatic picture of science as a whole. Although behaviorist reduction programs are still widespread, there is nothing to prevent the responsible incorporation of phenomenological description within scientific psychologies. The relation of different levels of phenomena in the data of scientific work remains an open question for inquiry. We are no longer committed to the old set of levels spun off from the traditional distinction of sciences. No reduction program can be ruled out as a logical possibility; but none is guaranteed simply by being propounded.

The distinction of spectator and agent perspectives is indeed significant for some intellectual purposes. But it cannot be absolutized, since there can be an observer's study of the agent at work. How far prediction is possible here is an empirical issue; although philosophical attempts have been made to render decision logically inaccessible to scientific study, they have not been wholly successful. At best what has been shown is that decision can always be looked at from the agent's perspective in the context of action. Thus, the occurrence of action and decision need not constitute absolute breaks in the web of scientific inquiry, nor need they require a wholly different methodology.

This is, of course, a summary statement of the aim of a scientific view that has expanded its self-picture in the light of the growth of contemporary philosophy and contemporary knowledge. It would require an extended elaboration. Nor does it regard itself as a final picture, for there are points at which questions of general direction are not yet resolved and can only await the progress of inquiry. For example, we cannot tell how far there will emerge an integrated model of man to supplant the multiplicity of models that now exist. Sometimes there is an attempt to prejudge this from the nature of model-making as selective and abstractive; it is said that the use of a model is a way of looking sharply at one phase of existence that shuts out other phases. But if the result is simply different selective pictures, we cannot bar the possible discovery of a unifying theory to cover them all; and if it is a conflicting picture in which the viewings are incompatible and would distort each other's perspective, there may be a discoverable theory that would permit corrections. (Laughing and crying, or observing and acting, may be incompatibles, but not only do we shuttle rapidly from one to the other but a unified theory does not seem out of reach.) Even more, types of integration may come in different patterns: the victory of one model by successful reduction of the rest, the discovery of fresh terms in which the others are systematized, a minimally comprehensive scheme in which the elements of each of the present models would be expanded to allow place for variables from the others, and so forth. Certainly the rapid growth of the physical sciences and of their modes of interrelation shows that there can be no antecedent dogmatic decision about the forms that interrelation will take.

Within psychiatry there has been ample discussion about the extent to which inquiry is scientific. No question seems to arise about experiments in physiology and pharmacology, but the situation of psychoanalysis is vigorously argued both ways (cf. Hook,[31] Kubie,[38] Pumpian-Mindlin,[52] Rapaport[54]). Actually there need be no single general answer. A full investigation could separate the different components of method. It is one question whether a psychoanalytic session can be regarded as a scientific experiment (just as one can ask about a classroom hour of teaching). It is quite another to ask whether there are low-level laws—for example, "Subjects who draw the head as the last feature of their picture usually show disturbance in interpersonal relations" (Machover)[44]—or whether appeal to childhood experience to explain adult behavior is to be construed as a kind of temporal "action at a distance" or posits an intervening continuum in principle capable of being traced (neural traces, muscular residues). Thus, there need be no contradiction between using childhood traumas as explanatory and a principle like Allport's[1] "functional autonomy," in which the cause is always of the present, or Lewin's[40] principle of "contemporaneity."

It is to be noted that the growth of social psychiatry and the use of cultural materials add fresh dimensions to the scientific understanding of mental illness (cf. Opler).[49] But this brings with it some of the same methodological controversies. For example, the phenomenological approach may interpret culture largely as a system of symbols and use it to look for basic symbols of the human condition and the structure of the life-world. The scientific approach will want to see the genetic basis of symbols, how energies took such form and what causal relations underlie symbolic relations. On a hoped-for integrated model of man, the meaning of the symbol or the human act as symbolic can be revealed only in the full cultural, institutional, and historical development of the people, if it has a generic hold, and in the biography of the individual, if it is idiosyncratic. Universal symbols are possible on both approaches, but in the scientific they require a causal explanation of invariance.

In general, while methodological principles that are accepted guide inquiry, the methodologies themselves have presuppositions about what the world and man are like. A comprehensive view of the relations of psychiatry and philosophy cannot be without some reckoning of such basic outlook.

Body, Mind, and "Reality"

Contemporary psychiatry, psychology generally, and philosophy are all characterized by a revolt against dualism, particularly in its Cartesian form (such as, Ryle).[55] In psychiatry at the moment, the attack is spearheaded by the phenomenological approach; in psychology the most militant is a reductive behaviorism; in philosophy the revolt often takes the form of a functionalist theory of mind, or else some theory of the identity of mind and body. Very often in all of these the debate takes the form of a reality claim. Traditional philosophy has been beset by metaphysical controversies about whether matter or mind is the ultimate reality. Behaviorist psychology not merely reacted against introspection by making the study of behavior the sole permissible experimental method, but occasionally proclaimed behavior the sole reality. Psychiatry not merely finds itself arguing that the unconscious is as real as the conscious, but even has to defend the conscious against the restrictive claims of the organic. For example, in a symposium on "Integrating the Approach to Mental Disease" (Kruse),[37] we find an exchange in which Alexander says that nostalgic feelings stimulate the parasympathetic nervous system and Gerard objects, "That is where you introduce a gremlin, feelings do not stimulate neurons," and Alexander replies, "It is observed that the secretion decreases when the nostalgic mood wanes." Psychiatry in particular has had the burden of spanning the dualistic gap, to provide some notion of the meaning of an act or to deal with purposive categories in diagnosing from acts. Otherwise, it would simply be left with the "mysterious leap from the mind to the body," for example, in studying conversion phenomena (cf. Deutsch).[10]

Philosophy is the clearest place in which to see what reality claims amount to. For traditional metaphysical systems were cast in terms of reality claims, each offering its own candidate for the post of the ultimately real. Of course, "reality" is an elusive term. Sometimes it is used in the hospitable all-comprehensive sense in which it is bestowed on everything; sometimes it is relegated, as in Kantian philosophy,[34] to an unknowable beyond anything specific. In the vast mass of metaphysical theories, however, the candidates are specifically characterized: macro-objects of ordinary life, micro-objects of physical science (such as, atoms), consciousness and its constituents, minds, God, and universals or ideal objects. Positivistic philosophy denies any meaning to "reality" except as emotive (cf. Ayer).[3] But if we pay attention to the way the candidates behave and the policies they adopt when they are thought (by their supporters) to have won out, there is a definite pattern in the controversies. Concepts of reality can be seen to embody programs of explanation and reduction: those entities are assigned reality that are most effective in an explanatory mode for phenomena, and those entities are dispensed with that can be understood as functions of the accepted basic entities in their operation. So a naturalism dispenses with God and with substantive minds, as well as interpreting universals as modes of organization. (Dewey;[12] cf. James).[33] A physical materialism (such as, Hobbes)[29] installs the microentities that make a system of the most advanced physics of the day. A Berkeleian idealism[5] prunes reality into constructs of sensory building blocks. Even a Kantian relegation of reality to noumena, neither empirically nor rationally reachable, is carrying out a task—it is openly devised as a way to reconcile morality with seemingly destructive scientific presuppositons. All these and other concepts of reality (such as, Hume's[32] phenomenalism) are products of interaction between specific inquiries in the context of their typical problems at a given stage of the growth of knowledge and the results achieved at that stage, together with the methods of inquiry that have proved successful. That is why the growth of science, the success of its methods, and its expansion to all areas of inquiry have given such great strength to a scientific naturalism and its concept of reality. But the consequence has been that the competition among concepts of reality has been simply transferred to the differences resulting from the uneven development of the many sciences. This is clear even in the history of materialism itself, which has taken first dominant physical-mechanical form, then biological form, then historical form; this has involved reconciling the methods of physics, evolutionary biology, and sociohistorical disciplines, a task by no means completed (cf. Sellars).[57] Such intrascientific conflicts are the major sources of differing concepts of reality today—except for attempted philosophical reinterpretations of the scientific enterprise itself. Outright irrational mysticisms are outworn, and attempts to hem in science with no-trespass signs are too busy moving the signs as the forbidden territory is overrun by fresh discoveries.

If we look at Cartesian dualism in this light, then the center of gravity is considerably shifted both in the interpretation of Descartes' work and in the battles over the Freudian unconscious and the contemporary phenomenological critiques. Let us explore this briefly.

The many-sided Cartesian metaphysics of the seventeenth century is often blamed for having enshrined the sharp separation of body and mind (Descartes),[9] carrying with it the sharp separation of method of inquiry in the physical and psychological sciences—the mathematical treatment of the quantitative and the introspective treatment of the qualitative. In this way it generated the honorific battle between an objective, external, publicly observable natural world and an inner, subjective, inaccessible world. But in debating such issues one is likely to neglect the scientific context of Descartes' scheme. The strength of his dualism lay in isolating physics from the interference of the mental, the religious, and the teleological-metaphysical, so that it could go ahead with mathematical methods; Descartes cannot be berated for the troubles this brought to a barely existing psychology. No more can Galileo be blamed for his choice; as Philip Frank has somewhere pointed out, it lay not between a simple and a complex physics, but between a simple physics with a complex theology and a complex physics with a simple (traditional) theology! Having made his choice, Descartes then projected the mechanistic program of explaining matter and motion, animal life and all biological functions, even emotional reactions—all the way to the threshold of consciousness—in material or physical terms. But he put a barrier of triple-plated steel between all this and consciousness itself. Although philosophical criticism could compel refinement in his categories, discover puzzles and paradoxes in the theory of interaction of body and mind, and work out alternative schemes, so long as the growth of physics was protected, its explanatory domain enlarged, and so long as psychology could manage to get along either by separate introspective procedures or by correlating mental phenomena to physical-organic events as bases, the schema was not likely to be abandoned.

In Freud's conceptual scheme we do not have a reconstruction of the Cartesian, but an expansion of the mental side in such a way as to threaten the established borders. In short, he added to consciousness, which had monopolized the domain of the mental, the unconscious, which in a well-behaved Cartesianism should be interpreted in physical terms. The reverberations still continue. There are programs of reduction that maintain that everything that can be said in terms of the unconscious can be said equally well without it. Others attempt to apportion the expanded domain of phenomena that Freud revealed between the strictly organic, operating on a mechanical-physical model, and the strictly conscious, operating on a finalistic or purposive model; they berate the unconscious as a strange and confused hybrid of the mechanical and the purposive (Peters).[50] We need not enter into the multiplicity of arguments—logical, empirical, metaphysical, valuational—that have centered about the unconscious (cf. Edel).[14] But it is important to note that Freud's installation of the unconscious in the domain of mental phenomena was not uncritical. His panegyric of the unconscious in *The Interpretation of Dreams* (Freud)[26] as the true psychic reality is probably just a touch of sibling rivalry with consciousness. More typical is the care to be found, for example, in the 1915 paper on "The Unconscious." Here Freud[27] differentiates an unconscious idea, which continues after repression as an actual formation, from an unconscious affect, which cannot exist since an affect is a process of discharge. The paper shows that he is grappling within a whole system of existent ideas about neurological processes to work out a concept that will help explain the range of phenomena that he was so insightfully reinterpreting. The advance of the problem since then has followed a typical scientific path, for it consists less in answering the question in the terms in which it was asked than in weakening, and even transcending, the sharp contrasts and vigorous dichotomies with which it began. The phenomena of the unconscious expand to yield a whole continuum from unawareness of vegetative and neurological processes, through unconscious automatization in the learning process, forgetting, unconscious expressive movement, through gradations of fantasy, daydream and other quasi-dream phenomena, dreams, and so on (Bellak).[4] They are differentiated, of course, by the types of conflict that they express and in which they play a part. On the other side of the old divide, consciousness stretches out to bridge the gap by incorporating phenomena of peripheral awareness, the unnoticed that is capturable in memory, subliminal perception outside of awareness but affecting awareness (Klein).[36] Indeed, the whole range from deepest unconscious to most explicit conscious begins to be captured in a designedly all-embracing concept of *registration* of experience, which minimizes automatic commitment to the older entities. The problem of the mental and the physical may yet remain, but the battle of the conscious and the unconscious has deservedly abated, except for the schools of the psychological establishments.

Moreover, a similar process can be suggested for the conflict of the physical-causal and purposive models of explanation in these domains. The objection is far from fatal that the Freudian unconscious operates purposively (as in the idea of repressed desires for a goal) and yet interacts causally (as psychic energies pushing for an outlet). It shows rather that there is an intermediate domain in human affairs between the clearly purposive and the clearly mechanical, in which there are partial purposes, purpose in the making, precursor phenomena to full-fledged purposes. It is a challenge to the psychologist to refine his two traditional models rather than, having enshrined them, to deny the existence of discovered embarrassing borderlands.

In such processes of scientific advance coupled with such modes of philosophical analysis, we get beyond the sheer conflict of reality claims and come to understand how the mantle of "reality" is the reward waiting for variables that will turn out to occupy strategic roles in explanation. It is important to note the same lesson with respect to principles as well as concepts. Take, for example, what is increasingly asserted in textbooks of Freudian psychology as a basic law or presupposition—the principle of *psychic determinism.** A philosopher reading such formulations is often puzzled. Does psychic determinism mean the determinism of psychic phenomena (an old materialistic view) or rather the *psychic* determinism of psychic phenomena? If the latter, does it refer to *all* psychic phenomena? (But some psychic consequences come from brain damage. And if psychic phenomena can have organic consequences as in conversion symptoms and psychosomatic illness, why not the reverse?) If the principle simply means that many psychic phenomena spring from the unconscious segment of the mental, then it would no longer be a basic principle but a secondary consequence. On the other hand, if we look to the proposed principle in the context of the problems from which it emerged, and the way it functions, we see it in a quite different light. Thus, the principle may have operated first as a suggestion of general direction, turning attention from the kinds of explanations in the medical textbooks of Freud's day to proposed psychological explanations—of dreams, parapraxes, and so forth. Or again, in the therapeutic situation it may serve as a rule of procedure—never to accept the patient's attempt to close an avenue by saying "I just *happened* to think of it (feel that way, etc.)." Or the principle may sum up successful lessons in a variety of types of cases where a psychological explanation edged out a physiological one. For example, Freud explains a déjà vu phenomenon in terms of repressed wishes utilizing a past experience as signal. This is probably more successful than, say, Bergson's physiological conjecture that occasionally there is a break in a present experience so that the completing second part already sees the first part with a sense of the past. Or again, the principle of psychic determinism may function in a theoretical attempt to broaden the concept of a "psychological phenomenon" by insisting that where a mental phenomenon is physiologically initiated (for example, conscious pain by a brain tumor), it be not left in isolation but be explored in terms of the total behavioral response of the individual, or his phenomenological patterning of the experience, or the impact within it of his personality structure, and so on. All such contexts to which the principle of psychic determinism may be attached should, of course, be distinguished and analyzed; but in none of them does it seem to emerge as a general formulable principle capable of mustering its own evidence or antecedently required as a general truth.

In general, it is not surprising, given the strain that the Freudian approach and its career have imposed on the dualistic conceptual scheme, that there should be many

* There is not complete uniformity in this. For example, while Brenner[7] and Monroe[46] give the principle a basic role, Fenichel[19] does not even include "psychic determinism" in his extremely rich index!

contemporary attempts in psychiatry at large-scale conceptual revision. Current phenomenological psychiatry is perhaps the most drastic. It has all the earmarks of a wholesale revolt against the established conceptions, and like many powerful revolutions it passes readily into a new imperialistic dogmatism. Aware of the way in which dualism fashioned all theoretical constructs on the assumption of the distinctness of body and mind, it wants to avoid asking questions about experience in terms of the constructs. It thus wishes to separate the directly experiential from interpretations in terms of psychological and physical theory.

Phenomenology generally is characterized by the demand for initial description of the direct field of experience. In this respect its clearest scientific example is Gestalt psychology and its study of the visual field. The field is strictly accepted in its own terms in the beholding; the relations of its parts (for example, figure and ground) and the configurations of its elements are discerned and described. The question of the relation of the phenomenological qualities to the physical bases of vision and the physical constitution of the objects beheld, as well as to psychological reactions of the beholder, is a secondary matter of correlation *after* the phenomenological job is done. In phenomenological approaches to psychiatry (cf. Van Den Berg),[61] emphasis likewise falls first on the doctor's shifting his attitude from merely being a spectator to somehow entering into the patient's world as the patient sees and lives it. Hence phenomenological case description tends to be rich and sympathetic, to discover fine nuances in the patient's world.* One would have expected, however, that in the light of the initial program, the phenomenologically oriented scientist would have gone on from his enriched phenomenological base to look for phenomenological-physical and phenomenological-psychological relations. Instead, because he takes the "Lebenswelt" of the individual to be the real, because he mistrusts the scientific world view and its attempt to trace the continuity of man and the natural world, because he suspects traditional approaches to man and psychology to be residues of dualism, the phenomenological psychiatrist remains within the phenomenological domain. This approach seeks instead to develop the second main task of philosophical phenomenology, that is, to lay bare the essential structure of the real and to find it in phenomenological terms. It is assumed that everyday living is definitely structured and that the basic form is invariant for human beings. This would contrast with a sociocultural view that regards form as expressive of function and function as depending on the institutional and cultural tasks to be found in the lives of social groups. However, if the level of search for structure is general enough, it may be that sociocultural invariants and phenomenological structures converge. But differences in method would still remain. Thus, where an evolutionary naturalism would seek the consequences of the shift from animal posture to man's upright position, a phenomenological approach would pass over genetic and developmental considerations to give (such as, Straus)[58] instead a view of the centrality of standing up in relation to the I and my world and its transcending character in which is embedded motility and a contrapositive sensory relation to the rest of the world. Yet since Straus refers to it as rising up against gravity, perhaps a more outright study of phenomenological-physical relations would furnish a richer comparative phenomenological description of new experiences of weightlessness beyond the gravitational field!

By refusing to seek "external" relations and remaining within the phenomenological reality, the phenomenological approach is also compelled to find the marks of normality within the field. It cannot appeal to different lines of causation to distinguish the normal from the abnormal, as is done in the scientific conception. The phenomenological answer is that each man's life-world has a discoverable norm, the disintegration of which constitutes sickness. The claim for an external standard—comparable to physical measurements demonstrating the illusory character of a visual

* This is true also for existentialist psychiatry (cf. May, Angel, and Ellenberger).[45]

presentation—is met with the general argument that the scientist's measuring is itself within the phenomenological world. In many respects, then, the approach is like Berkeleian idealism, which, having reduced all material objects to sensations, goes on to distinguish the abiding sensory order (God's ideas) from flickering copies in the human mind. It may be that in the long run Lukács's rebuke to Scheler had a point: when Scheler said we could do a phenomenological study even of the devil, provided that we bracketed the question of his reality, Lukács[42] retorted that when the study was done, Scheler would open the brackets, and there would be the devil standing before him!

On the whole the constructive elements in the phenomenological approach, in contrast to its attempt to become a self-enclosed system, are capable of being integrated in a scientific approach, especially if the broader conception of science considered above is kept in mind. The growing knowledge of the physical and social conditions for the existence of consciousness and its forms seems too well rooted to justify a return to a philosophical idealism.

What Philosophy Does

My concluding aim is to render explicit the conception of the work of philosophy involved in the previous discussion of its relation to psychiatry. Before going on to this, however, I should like to criticize a quite different mode of relation utilized by some philosophers who were especially interested in psychoanalytic theory. They saw the origins of basic philosophizing as lying in the disguised emotional quest for security against death, for subtle modes of expressing aggression, and for projecting fears and desires. Lazerowitz[39] analyzed detailed metaphysical puzzles in such terms; for example, the Heraclitean "everything changes" means emotionally that everything dies, and yet dying is lived through since changing is living, and so the doctrine serves as an emotional phoenix ever rising from the ashes. J. O. Wisdom[63] saw the unconscious origin of Berkeley's idealism in an interpretation of matter as a poison within the system to be gotten rid of at all costs. Feuer[20,21,22,23,24] formulated a general account of the projective base of philosophical work and sought evidence in specific analysis of dreams and ideas in case studies of Descartes, Spinoza, and Kant.

I have elsewhere attempted an evaluation of genetic accounts of philosophical doctrines to determine where they add to the meaning of the doctrine and where they show only specific functions (psychological or social) that the doctrine may have also served (Edel).[16] Thus, Berkeley may have had the emotional attitude to matter that Wisdom speaks of, but his doctrine also has sound arguments against a particular conception of matter (the Lockean), embodies an analysis of verification for scientific statements that gives a critique of Newton, and serves to fight atheism on behalf of the then established religion! Only if other tasks—philosophical, scientific, cultural, and social—are deemed empty, can one set the projective as the core. If, however, the thesis is that *all* philosophy has these psychological elements, then fresh distinctions would have to be drawn within the psychoanalytic materials themselves between the kind of philosophizing that expresses projective and neurotic elements and the kind that is realistic, faces problems clearly, and so forth. It follows, therefore, that a sound sense of what understanding psychoanalytic theory can bring to the view of philosophical activity should not brand philosophy generally as illusory, but rather should sharpen the criteria for a more realistic contribution of philosophy to the understanding of the world and man.

This whole view of philosophy as emotional expression was really cashing in on a contemporary uncertainty about what philosophy is doing. It reflected a state in which philosophy had isolated itself from the contexts of its problems in the growth of the many fields of human inquiry and the needs of human direction in practice. Not being simply empirical, and no longer permitted to legislate a priori about the ultimate nature of reality, philosophy could then only be seen as engaged in either linguistic analysis of emotive expression or personal decisional commitment.

In the conception of philosophizing implicit in this study, I have followed a quite different path. Philosophical ideas and principles do not constitute an isolated domain developed and certified apart from the realms of their application. They are intellectual products closely related to the problems and developments of human inquiries, and to the tasks of cultural patterning and social guidance. They thus have embedded presuppositions that reflect the stage of scientific and social development at given times. To understand the philosophical ideas involves seeing them operating in the different fields, unpacking the presuppositions as well as the traditional linguistic patterns and established and emerging human purposes embedded in them. Philosophizing is the continual task of clarifying and understanding the conceptual network that is being employed in ordinary life and in specialized inquiries, comparing and correlating the shape that it takes in different areas, evaluating the extent to which it advances or impedes the goals and helps solve the problems of these inquiries. Such clarifying activity is not, however, merely lighting up the conceptual schemes; evaluating involves being continually critical along many different dimensions. In this way philosophy is often altering, even subverting, traditional foundations. Old categories with embedded presuppositions, in terms of which whole areas have been organized and inquiry guided, may thus be shaken or even shattered. New categories—for philosophy may be inventive, sometimes sensing what is needed, sometimes picking up a theme from some special field and generalizing it—point to fresh modes of inquiry. Indeed, philosophy in its grander traditional formulations has served as an intellectual workshop in which categories and models taking rise in one area of inquiry, or one phase of human life, have been generalized, given stricter logical form, and tried speculatively on other or even all areas of inquiry and human action, whether descriptive or explanatory or normative and regulative. In this sense philosophy has a constant and indispensably creative role in which it serves at minimum as an intellectual broker, at maximum as intellectual inventor of possible and possibly productive structures. In the contemporary world, with the rapidity of change in all fields, it is not surprising that there will be an awareness of intellectual turning points at which conceptual schemes need reconstruction. For example, a philosopher of science warns a congress of psychologists that the philosophical formulations that were relied upon in the development of a narrow operationalism in psychology have now been abandoned by the philosophers of science themselves (Feigl).[18] Ideally, one might hope, such a reckoning by the philosophers would include what kind of psychology a strict operationalism had encouraged and what kinds of results it got in its psychological work.

Such a philosophical interpretation of conceptual schemes and their application can do much to alleviate the common evil suffered in both philosophy and psychiatry—the hardening of school conflicts. Psychiatry has been perhaps more adversely affected, but only because the immediacy of therapeutic work exerts a greater pressure on it, while philosophy can wait at leisure for school diseases to run their course! Both have to learn or relearn the experimental character of intellectual constructions and become more fully aware of how the great dichotomies of a particular period in the development of knowledge, such as the mind-body cleavage, have permeating scope, and how their reconsideration in the light of advancing knowledge and reflection requires thoroughgoing intellectual reconstruction, and how this cannot be done without wide cooperative reflection on the philosophical and scientific sides together.

Bibliography

1. ALLPORT, G. W., *Personality*, Ch. 7, Holt, New York, 1937.
2. ARISTOTLE, *On the Soul* (Translated by Hett, W. S.), Harvard University Press, Cambridge, 1957.

3. AYER, A. J., *Language, Truth and Logic*, 2nd ed., Dover, New York, 1950.
4. BELLAK, L., *et al.*, "Conceptual and Methodological Problems in Psychoanalysis," *Ann. N.Y. Acad. Sci.*, 76:1087, 1096, 1959.
5. BERKELEY, G. (1710), *Principles of Human Knowledge*, Liberal Arts Press, Indianapolis.
6. BEROFSKY, B. (Ed.), *Free Will and Determinism*, Harper & Row, New York, 1966.
7. BRENNER, C., *An Elementary Textbook of Psychoanalysis*, Doubleday-Anchor, New York.
8. BUBER, M., *I and Thou* (Tr. by Smith, R. G.), T. and T. Clark, Edinburgh, 1937.
9. DESCARTES, R. (1641), *Meditations*, Liberal Arts Press, Indianapolis.
10. DEUTSCH, F. (Ed.), *On the Mysterious Leap from the Mind to the Body*, International Universities Press, New York, 1959.
11. DEVEREUX, G., "Normal and Abnormal: The Key Problem of Psychiatric Anthropology," in *Some Uses of Anthropology: Theoretical and Applied*, pp. 23–48, The Anthropological Society of Washington, Washington, D.C., 1956.
12. DEWEY, J., *Experience and Nature*, Norton, New York, 1929.
13. ———, *Human Nature and Conduct*, Modern Library, New York, 1930.
14. EDEL, A., "The Concept of the Unconscious: Some Analytic Preliminaries," *Phil. of Sci.*, 31:18–33, 1964.
15. ———, *Ethical Judgment: The Use of Science in Ethics*, Free Press Paperback, New York, 1964.
16. ———, *Method in Ethical Theory*, Ch. 11, Bobbs-Merrill, Indianapolis, 1963.
17. ERIKSON, E. H., *Childhood and Society*, Ch. 7, Norton, New York, 1950.
18. FEIGL, H., "Philosophical Embarrassments of Psychology," *Am. Psychol.*, 14:115–128, 1959.
19. FENICHEL, O., *The Psychoanalytic Theory of Neurosis*, Norton, New York, 1945.
20. FEUER, L. S., "Anxiety and Philosophy: The Case of Descartes," *Am. Imago*, 20:411–449, 1963.
21. ———, "The Bearing of Psychoanalysis upon Philosophy," *Phil. & Phenomenol. Res.*, 19:323–340, 1959.
22. ———, "The Dream of Benedict de Spinoza," *Am. Imago*, 14:225–240, 1957.
23. ———, "The Dreams of Descartes," *Am. Imago*, 20:3–26, 1963.
24. ———, "Lawless Sensations and Categorial Defenses: The Unconscious Sources of Kant's Philosophy," in Hanly, C., and Lazerowitz, M. (Eds.), *Psychoanalysis and Philosophy*, International Universities Press, New York, 1971.
25. FREUD, S. (1921), *Group Psychology and the Analysis of the Ego*, Bantam Books, New York.
26. ——— (1900), *The Interpretation of Dreams*, Basic Books, New York, 1965.
27. ——— (1915), "The Unconscious," in *Collected Papers*, Vol. 4, Hogarth, London, 1934.
28. FROMM, E., *Man for Himself*, Rinehart, New York, 1947.
29. HOBBES, T. (1651), *Leviathan*, Book I, Liberal Arts Press, Indianapolis.
30. HOOK, S. (Ed.), *Dimensions of Mind*, New York University Press, New York, 1960.
31. ——— (Ed.), *Psychoanalysis, Scientific Method and Philosophy*, New York University Press, New York, 1959.
32. HUME, D. (1758), *An Inquiry Concerning Human Understanding*, Liberal Arts Press, Indianapolis.
33. JAMES, W., *Pragmatism*, Longmans, Green, New York, 1907.
34. KANT, I. (1783), *Prolegomena to Any Future Metaphysics*, Liberal Arts Press, Indianapolis.
35. KENNY, A., *Action, Emotion and Will*, Routledge and Kegan Paul, London, 1963.
36. KLEIN, G. S., "Consciousness in Psychoanalytic Theory: Some Implications for Current Research in Perception," *J. Am. Psychoanal. A.*, 7, 1959.
37. KRUSE, H. D. (Ed.), *Integrating the Approaches to Mental Disease*, p. 68, Hoeber-Harper, New York, 1957.
38. KUBIE, L., "The Scientific Problems of Psychoanalysis," in Wolman, B. B., and Nagel, E. (Eds.), *Scientific Psychology*, Ch. 17, Basic Books, New York, 1965.
39. LAZEROWITZ, M., *The Structure of Metaphysics*, Routledge and Kegan Paul, London, 1955.
40. LEWIN, K., *Principles of Topological Psychology*, McGraw-Hill, New York, 1936.
41. LEWIS, H., *Shame and Guilt in Neurosis*, International Universities Press, New York, 1971.

42. LUKÁCS, G., "Existentialism," in Sellars, R. W., McGill, V. J., and Farber, M. (Eds.), *Philosophy for the Future: The Quest of Modern Materialism*, p. 574, Macmillan, New York, 1949.
43. LYND, H. M., *On Shame and the Search for Identity*, Harcourt Brace World, New York, 1958.
44. MACHOVER, K., *Personality Projection in the Drawing of the Human Figure*, p. 40, Charles C Thomas, Springfield, Ill., 1949.
45. MAY, R., ANGEL, E., and ELLENBERGER, H. F. (Eds.), *Existence: A New Dimension in Psychiatry and Psychology*, Basic Books, New York, 1958.
46. MONROE, R. L., *Schools of Psychoanalytic Thought*, Holt, New York, 1955.
47. MOORE, G. E., *Principia Ethica*, University of Cambridge Press, Cambridge, 1903.
48. NAGEL, E., *The Structure of Science*, Harcourt Brace World, New York, 1961.
49. OPLER, M. (Ed.), *Culture and Mental Health*, Macmillan, New York, 1959.
50. PETERS, R. S., *The Concept of Motivation*, Routledge and Kegan Paul, London, 1958.
51. PIERS, G., and SINGER, M. B., *Shame and Guilt*, Charles C Thomas, Springfield, Ill., 1953.
52. PUMPIAN-MINDLIN, E. (Ed.), *Psychoanalysis as Science*, Stanford University Press, Stanford, 1952.
53. RAPAPORT, D. (Ed.), *Organization and Pathology of Thought*, Columbia University Press, New York, 1951.
54. ———, *The Structure of Psychoanalytic Theory*, International Universities Press, New York, 1960.
55. RYLE, G., *The Concept of Mind*, Hutchinson's University Library, London, 1949.
56. SARTRE, J. P., *Being and Nothingness* (Tr. by Barnes, H.), Part 1, ch. 2 and Part 4, ch. 1, Philosophical Library, New York, 1956.
57. SELLARS, R. W., MCGILL, V. J., and FARBER, M. (Eds.), *Philosophy for the Future: The Quest of Modern Materialism*, Macmillan, New York, 1949.
58. STRAUS, E. W., "Psychiatry and Philosophy," in Natanson, M. (Ed.), *Psychiatry and Philosophy*, Springer-Verlag, New York, 1969.
59. SULLIVAN, H. S., *The Interpersonal Theory of Psychiatry*, Norton, New York, 1953.
60. SZASZ, T. S., *The Myth of Mental Illness*, Hoeber-Harper, New York, 1961.
61. VAN DEN BERG, J. H., *The Phenomenological Approach to Psychiatry*, Charles C Thomas, Springfield, Ill., 1955.
62. WEGROCKI, H. G., "A Critique of Cultural and Statistical Concepts of Abnormality," *J. Abnorm. & Soc. Psychol.*, *34*:166–178, 1939.
63. WISDOM, J. O., *The Unconscious Origin of Berkeley's Philosophy*, Hogarth, London, 1953.

CHAPTER 45

PSYCHIATRY AND MORAL VALUES

Paul Ricœur

ANY INVESTIGATION that would undertake to cover the entire field of problems posed by psychiatry with regard to ethics would unavoidably lose itself in generalities—not only because the problem and the schools of thought that claim to be part of psychiatry are innumerable, but also because the ethical implications themselves are of such a diverse nature that they are practically incomparable. This is why we have deliberately chosen to limit this chapter to one branch of psychiatry, psychoanalysis, and to one author, Freud. There are two reasons for this: first, it is Freud's work that exercises the greatest influence on contemporary culture at the popular as well as the scientific level of discussion; second, his work permits us to pose the problem of the relations between psychiatry and ethics in the most radical terms. At first glance the Freudian analysis of morality appears to be a traumatic negation of traditional moral beliefs. But the real problems, those that surpass ordinary banality, only take shape beyond this first shock. When we no longer resist, when we no longer seek to justify ourselves, we discover what is essential—namely, that we must not ask psychiatry and psychoanalysis for an alternative answer to unchanged questions, but for a new manner of asking moral questions.

A preliminary question is worth consideration: is psychiatry, and, above all, Freudian psychoanalysis, competent to deal with ethics? Someone might object that Freud's writings on art, morals, and religion constitute the extension of individual psychology to collective psychology and, beyond psychological phenomena, to a domain where psychiatry is no longer competent, the highest realm of human existence. Certainly it was during the last part of his life that Freud's great texts about culture accumulated: *The Future of an Illusion* (1927); *Civilization and Its Discontents* (1930); *Moses and Monotheism* (1937–1939). But it is not a question of a belated extension from analytic experience to a general theory of culture. Already in 1908 Freud had written "Creative Writers and Daydreaming." *Delu-*

sions and Dreams in Jensen's "Gradiva" dates from 1907; *Leonardo da Vinci and a Memory of His Childhood* from 1910; *Totem and Taboo* from 1913; "Thoughts for the Times on War and Death" from 1915; "The 'Uncanny'" from 1919; "A Childhood Recollection from *Dichtung und Wahrheit*" from 1917; "The Moses of Michelangelo" from 1914; *Group Psychology and the Analysis of the Ego* from 1921; "A Neurosis of Demonical Possession in the Seventeenth Century" from 1923; and "Dostoevsky and Parricide" from 1928. The great "intrusions" into the domains of aesthetics, sociology, morality, and religion are strictly contemporary with texts as important as *Beyond the Pleasure Principle, The Ego and the Id*, and, above all, the "Papers on Metapsychology."

The truth of the matter is that these works are not just "applied" psychoanalysis, but psychoanalysis pure and simple.

How is this possible? What justifies psychoanalysis in speaking from its very beginning about art, ethics, and religion, not as a secondary extension of its task, but in conformity with its original intention?

The question is all the more legitimate in that the first intersection between psychoanalysis and a general theory of culture precedes all the works we have just cited and dates from the first interpretation of the Greek myth of Oedipus in a letter to Fliess of May 31, 1897: "Another presentiment tells me, as if I knew it already—although I do not know anything at all—that I am about to discover the source of morality." He clarifies this discovery in a second letter (October 15, 1897): "Only one idea of general value has occurred to me. I have found love of the mother and jealousy of the father in my own case, too, and now believe it to be a general phenomenon of early childhood. . . . If that is the case, the gripping power of *Oedipus Rex*, in spite of all the rational objections to the inexorable fate that the story presupposes, becomes intelligible, and one can understand why later fate dramas were such failures. . . . but the Greek myth seizes on a compulsion which everyone recognizes because he has felt traces of it in himself. Every member of the audience was once a budding Oedipus in fantasy, and this dream-fulfillment played out in reality causes everyone to recoil in horror, with the full measure of repression which separates his infantile from his present state." In one fell swoop Freud claims to have found the interpretation for a private dream and a public myth. From its very beginning psychoanalysis is both a theory of neurosis and a theory of culture.

Once again, how is this possible?

The principal answer is as follows. The object of psychoanalysis is not human desire as such—by which we mean wishes, libido, instinct, and eros (all these words having a specific signification in their specific contexts)—but human desire as understood in a more or less conflictual relation with a cultural world, whether this world is represented by parents —especially by the father—or by authorities, by anonymous external or internal prohibitions, whether articulated in discourse or incorporated in works of art or social, political, and religious institutions. In one way or another the object of psychoanalysis is always *desire plus culture*. This is why Freud does not extend concepts that could have first been elaborated within a sort of neutralized cultural framework to cultural realities. Whether we consider *The Interpretation of Dreams* or the *Three Essays on the Theory of Sexuality*, the instinctual level is confronted from the very beginning by something like censorship, "dams," prohibitions, ideals. The nuclear figure of the father is merely the system's center of gravity. And even when we claim to isolate a human instinct, or a genetic phase of that instinct, we reach it only in the expressions of this instinct at the level of linguistic or prelinguistic signs and nowhere else. Analytic experience itself, insofar as it is an exchange of words and silences, of speaking and listening, belongs to what we can call the order of signs, and as such becomes part of that human communication on which culture reposes. There is a psychoanalytic institution in the proper sense of the word from the codification of the therapy session right up to the organization of psychoanalytic societies.

For these historical and systematic reasons psychoanalysis is the theory of *the dialectic between desire and culture.* Consequently no human phenomenon is foreign to it to the degree that all human experience implies this dialectic.

The result of the unified structure of psychoanalytic theory is that it does not approach ethics as an isolated problem, but as a particular aspect of *culture,* itself considered as a whole. Psychoanalysis is a global theory that touches culture itself as a totality. The originality of Freudianism consists entirely in this. And it is by way of a global theory of culture that psychoanalysis takes up the phenomenon of morality.

An "Economic" Model of the Phenomenon of Culture

What is "culture"? Let us first say negatively that there is no question here of opposing civilization and culture to each other. This refusal to use a distinction that seems likely to become classic is itself very enlightening. There is not, on the one hand, a utilitarian enterprise to dominate the forces of nature that would be civilization and, on the other hand, a disinterested, idealistic undertaking to realize values that would be culture. This distinction, which can make sense from a point of view other than that of psychoanalysis, no longer holds as soon as we decide to approach culture from the point of view of a balance sheet of libidinal investments and counterinvestments.

This economic interpretation dominates all Freudian considerations about culture.

From this point of view the first phenomenon to be considered is coercion, because of the *repression of instincts* that it implies. It is on this note that *The Future of an Illusion* opens. Culture, Freud notes, began with the prohibition of the oldest desires: incest, cannibalism, murder. And yet coercion does not constitute the whole of culture. *Illusion,* whose future Freud is examining, finds its place in a larger cultural task of which prohibition is merely the outer manifestation. Freud delineates the problem with three questions: To what point can we *diminish* the burden of the instinctual sacrifices imposed on man? How to *reconcile* them with those renouncements that are ineluctable? How, beyond this, to offer individuals satisfying *compensations* for these sacrifices? These questions are not, as we might first believe, about culture; rather they constitute culture itself. What is in question in the conflict between prohibition and instinct is this triple problematic: the diminution of the instinctual burden; reconciliation with the ineluctable; and compensation for sacrifice.

But only an economic interpretation can make sense of this task. Here we reach the unitary point of view that not only holds together all Freud's essays on art, morality, and religion, but also connects "individual psychology" and "collective psychology," and roots them in "metapsychology."

This economic interpretation of culture is displayed in two moments, especially well-illustrated in *Civilization and Its Discontents.* First, there is everything that we can say without recourse to the death instinct. Then there is what we cannot say without making this instinct intervene. Short of this point of inflection that opens it to the tragic within culture, the essay advances with a calculated simplicity. Culture's economy appears to coincide with what we could call a general "erotics." The goals sought by the individual and those that animate culture appear as sometimes converging, sometimes diverging figures of the same Eros. It is the same "erotics" that binds group together and that brings an individual to look for pleasure and flee suffering—the triple suffering that the world, his body, and other men inflict upon him. Culture's development is, as is the growth of an individual from infancy to adulthood, the fruit of Eros and Ananke, of love and work. It is the fruit of love more than of work, however, because the necessity to be united in work to exploit nature is insignificant compared to the libidinal tie that unites individuals into a single social body.

It seems then that it is the same Eros that animates the search for individual happiness and that wants to unite men into ever vaster groups. But the paradox quickly appears: as a struggle against nature, culture gives men power heretofore conferred on the gods, but this resemblance to the gods leaves man unsatisfied: the discontent of civilization. . . . Why is this? We could undoubtedly account for certain tensions between the individual and society solely on the basis of this general "erotics," but we cannot account for the grave conflict that makes culture tragic. It is easy, for example, to explain that family ties resist being expanded to larger groups. For every adolescent the passage from one circle to the other necessarily appears as rupturing the oldest and the narrowest tie. We understand, too, that something about feminine sexuality resists this transfer of the privately sexual to the libidinal energies of the social tie. We can go even further in the direction of conflicting situations without encountering radical contradictions. Culture, we know, imposes sacrifices of enjoyment on all sexuality—prohibition of incest, censorship of infantile sexuality, supercilious channeling of sexuality into the narrow ways of legitimacy and monogamy, imposition of the imperative to procreate, and so forth. But however painful these sacrifices may be and as inextricable as these conflicts may be, they still do not constitute true antagonism. We can even say that, on the one hand, the libido resists with all its inertial force the task that culture imposes upon it to abandon all its previous positions, and, on the other hand, that the libidinal tie of society so feeds on the energy deducted from sexuality as to menace it with atrophy. But all this is so little "tragic" that we might even dream of a sort of truce or settlement between the individual libido and the social tie.

So the question arises again: why does man fail to be happy? Why is man unsatisfied insofar as he is a cultural being?

It is here that the analysis changes direction: consider what is laid down for man, an absurd commandment, love his neighbor as himself; an impossible demand, love his enemies; a dangerous order that squanders love, rewards the wicked, and leads to loss for anyone imprudent enough to apply it. But the truth that is hidden behind this *unreasonable imperative* is the unreasonableness of an instinct that escapes a simple erotics:

> The element of truth behind all this, which people are so ready to disavow, is that men are not gentle creatures who want to be loved, and who at most can defend themselves if they are attacked; they are, on the contrary, creatures among whose instinctual endowments is to be reckoned a powerful share of aggressiveness. As a result, their neighbor is for them not only a potential helper or sexual object, but also someone who tempts them to satisfy their aggressiveness on him, to exploit his capacity for work without compensation, to use him sexually without his consent, to seize his possessions, to humiliate him, to cause him pain, to torture and to kill him. *Homo homini lupus.* [p. 111][1]

The instinct that so perturbs the relation of man to man and that requires society to rise up as an implacable dispenser of justice is, as we have recognized, the death instinct, the primordial hostility of man for man.

With the introduction of the death instinct the whole economy of the essay is recast. While the "social erotic" could consistently appear to be the *extension* of the sexual erotic, either as a displacement of object or sublimation of goal, the division of Eros and death in two on the plane of culture can no longer appear as the extension of a conflict that could be better understood on the plane of the individual. On the contrary, it is the tragic in culture that serves as the privileged revelator of an antagonism that remains silent and ambiguous at the level of an individual life and psyche. Certainly Freud had forged his doctrine of the death instinct as early as 1920 (*Beyond the Pleasure Principle*), without accentuating the *social* aspect of aggressivity, and within an apparently *biological* framework, but it remained something of an adventurous speculation, despite experimental support for the theory (repetition neurosis, infantile play, the tendency to relive painful episodes, and so forth). In 1930 Freud saw more clearly that the death instinct remained a *silent* instinct "in" the living being and that it only became *manifest* in its social expression of aggressivity and destruction. It is in this sense that we said above that the interpretation of culture becomes the revelator of the antagonism of instincts.

Thus in the second half of the essay we see a sort of rereading of the theory of instincts beginning from their cultural expression. We understand better why the death instinct is, in the psychological scheme of things, both an unavoidable inference and an unassignable experience. We never grasp it but in conjunction with Eros. Eros utilizes it by diverting it from one person onto another. It is mingled with Eros when it takes the form of sadism, and we surprise it working against the individual himself through masochistic satisfaction. In short, it only betrays itself when mixed with Eros, sometimes doubling the object libido, sometimes overloading the narcissistic ego libido. It is unmasked and revealed as anticulture. Thus there is a progressive revelation of the death instinct across the three levels, biological, psychological, and cultural. Its antagonism becomes less and less silent to the extent that Eros serves first to unite the individual to himself, then the ego to its object, and finally individuals into ever larger groups. As it is repeated from level to level, the struggle between Eros and death becomes more and more manifest and attains its complete meaning only at the level of culture:

> This aggressive instinct is the derivative and the main representative of the death instinct which we have found alongside Eros and which shares world-dominion with it. And now, I think, the meaning of the evolution of civilization is no longer obscure to us. It must present the struggle between Eros and Death, between the instinct of life and the instinct of destruction, as it works itself out in the human species. This struggle is what all life essentially consists of, and the evolution of civilization may therefore be simply described as the struggle for life of the human species. And it is this battle of the giants that our nurse-maids try to appease with their lullaby about Heaven." [p. 122][1]

But this is not all, for in the last chapters of *Civilization and Its Discontents*, the relation between psychology and the theory of culture is completely inverted. At the beginning of this essay it was the libido's economy, borrowed from the metapsychology, that served as guide in the elucidation of the phenomenon of culture. Then with the introduction of the death instinct, the interpretation of culture and the dialectic of instincts were seen to refer to one another in a circular movement. The sense of guilt is introduced, in effect, as the "means" by which civilization holds aggressivity in check. The cultural interpretation is pushed so far that Freud[1] can affirm that the express intention of his essay was "to represent the sense of guilt as the most important problem in the development of civilization" (p. 134) and to show, moreover, why the progress of civilization must be paid for by a loss of happiness due to the reinforcement of this feeling. He cites the famous words of Hamlet in support of this conception: "Thus Conscience does make cowards of us all . . ."

If, therefore, the sense of guilt is the specific means by which civilization holds aggressivity in check, it is not surprising that *Civilization and Its Discontents* should contain the most developed interpretation of this feeling, whose dynamics, however, are fundamentally psychological. But the psychology of this feeling is only possible if we begin with an "economic" interpretation of culture. From the point of view of individual psychology, the sense of guilt appears to be merely the effect of an internalized, introjected aggressivity that the superego has taken over in the form of conscience and that it turns back against the ego. But its whole "economy" only appears when the need for punishment is placed within a cultural perspective: "Civilization, therefore, obtains mastery over the individual's dangerous desire for aggression by weakening and disarming it and by setting up an agency within him to watch over it, like a garrison in a conquered city" (pp. 123–124).[1]

Thus the economic, and, if we may say so, the structural interpretation of the sense of guilt depends upon a cultural perspective, and it is only within the framework of the structural interpretation that the diverse partial genetic interpretations elaborated during different periods by Freud concerning the murder of the primeval father and the instituting of remorse can be situated and understood. Considered by itself, this explanation remains somewhat problematic because of the contingency that it introduces into the history of a feeling that elsewhere is presented as having the characteristics of "fatal inevitability" (p. 132).[1] However, the contingent character of this development, as it is reconstituted by the genetic explanation, is attenuated as soon as the genetic explanation itself is subordinated to the structural, economic interpretation:

> Whether one has killed one's father or has abstained from doing so is not the really decisive thing. One is bound to feel guilty in either case, for the sense of guilt is an expression of the conflict due to ambivalence, of the eternal struggle between Eros and the instinct of destruction or death. This conflict is set going as soon as men are faced with the task of living together. So long as the community assumes no other form than that of the family, the conflict is bound to express itself in the Oedipus complex, to establish the conscience and to create the first sense of guilt. When an attempt is made to widen the community, the same conflict is continued in forms which are dependent on the past; and it is strengthened and results in a further intensification of the sense of guilt. Since civilization obeys an internal erotic impulsion which causes human beings to unite in a closely-knit group, it can only achieve this aim through an ever-increasing reinforcement of the sense of guilt. What began in relation to the father is completed in relation to the group. If civilization is a necessary course of development from the family to humanity as a whole, then—as a result of the inborn conflict arising from ambivalence, of the eternal struggle between the trends of love and death—there is inextricably bound up with it an increase of the sense of guilt, which will perhaps reach heights that the individual finds hard to tolerate. [Pp. 132–133][1]

Examining these two texts has not yet told us anything specific about ethics, but a framework has been assembled wherein the ethical problem can be placed in new terms drawn from the economic function of culture considered as a whole. We can say two contrary things about this theory of culture. On the one hand, to the degree that all processes of culture are viewed from the economic point of view, we can say that psychoanalysis is a reductive theory. We will consider this interpretation at the end of this essay. But we must also say in an inverse sense that the supremacy of the economic point of view could only be established by the intermediary of an interpretation of cultural phenomena that gives a voice, an expression, a language to those forces that by themselves are mute. The conflicts between instincts that are at the root of these cultural phenomena can only be approached, in effect, within the cultural sphere where they find an indirect expression. The economics passes through a hermeneutic.

The Economy of Ethical Phenomena

It is now possible to deal directly with the interpretation of moral phenomena in Freudian theory. By understanding them in a new way, psychoanalysis can, in effect, change our very "lived" moral experience. But as we said at the beginning of this chapter, when psychoanalysis turns its gaze toward morality, it is received as trauma and aggression by the uninitiated. Let us therefore cross this wasteland in Freud's company.

We will consider successively the clinical-descriptive, the genetic-explanatory, and finally the economic-theoretical, where we will rejoin the level attained directly in the preceding analysis of the global phenomenon of culture.

1. If we limit ourselves to the properly descriptive, Freud's discovery about morality consists essentially in applying to ethical phenomena the instruments that had proved themselves in the description of pathological phenomena such as obsessional neurosis, melancholy, and masochism. This allows us to extend concepts forged in the clinic such as cathexis, repression, and defense mechanisms, to this new order of phenomena. Morality then appears as annexed to the pathological sphere. But to assure this extension of descriptive concepts forged in contact with dreams and neuroses, it was necessary to extend the unconscious character of the sphere of the repressed to that sphere of repressing that

Freud calls the *superego*. This is why Freud adds a new topography (id, ego, superego) to his first topography (unconscious, preconscious, conscious) that allows him to account for the fundamentally unconscious character of the processes by which the agency of repression itself is constituted. The new agencies required to take up the ethical phenomenon are not so much places as roles in a personology. Ego, id, and superego are expressions that denote the relation of the personal to the anonymous and suprapersonal in the founding of the ego. The very question of the *ego* is a new question with respect to the question of consciousness treated in *The Interpretation of Dreams*. To become an ego is different from becoming conscious, that is, lucid, present to oneself, and attentive to reality. Rather, becoming an ego concerns the alternative of being dependent or autonomous. This is no longer a phenomenon of perception (either internal or external perception), but of strength and weakness, that is, of mastery. According to the title of one of the chapters of *The Ego and the Id*, the second topography has its end in "The Ego's Relations of Dependence" (Chapter 5). These relations of dependence are master-slave relations: dependence of the ego on the id; dependence of the ego on the world; dependence of the ego on the superego. Through these alienated relations a personology is outlined. The role of the ego, carried by the personal pronoun, is constituted in relation to the anonymous, the sublime, and the real.

These new considerations, which are not contained in the trilogy unconscious-preconscious-conscious, may be introduced in a properly descriptive fashion. What in effect, from a properly clinical point of view, is the superego? Freud gives a very revealing synonym for it in the third chapter of *The Ego and the Id*. He says, "ego ideal *or* superego." *The New Introductory Lectures on Psychoanalysis*[6] are more specific: "But let us return to the superego. We have alloted to it the functions of self-observation, of conscience and of [maintaining] the ideal" (p. 66).

From observation Freud designates this division of self experienced as a feeling of being observed, watched, criticized, condemned: the superego manifests itself as an eye and a regard.

Conscience, in turn, designates the strictness and cruelty of this experience. It resists our actions like Socrates' demon, which says "No," and condemns us after the action. Thus not only is the ego watched, but also it is mistreated by its inner and superior other. We need not emphasize that these two traits of observations and condemnation are in no way borrowed from a Kantian style of reflection on the condition of the good will and the *a priori* structure of obligation, but from clinical experience. This split between the observer and the rest of the ego is revealed in a greatly exaggerated way in the delusion of being observed, and melancholy declares its cruelty.

As for the ideal it is described as follows: the superego "is also the vehicle of the ego ideal by which the ego measures itself, which it emulates and whose demand for ever greater perfection it strives to fulfill" (pp. 64–65).[6] At first glance it may seem that no pathological model presides over this analysis. Is it not a question here of moral aspiration, of the desire to conform to, of forming oneself in the image of, of having the same content as a model? The preceding text does permit such an analysis. But Freud is always more attentive to the character of constraint than to the spontaneity of the responses that the ego gives to the demands of the superego. Moreover, placed with the two preceding traits, this third characteristic takes on a coloration that we can readily call pathological in the clinical and the Kantian sense of the word. Kant spoke of the "pathology of desire"; Freud speaks of the "pathology of duty" and its modes of observation, condemnation, and idealization.

The pathological approach reveals the initial situation of morality as alienated and alienating. A "pathology of duty" is just as instructive as a pathology of desire. In the final analysis it is no more than a prolongation of the latter. In effect, the ego oppressed by the superego is in a situation vis-à-vis this internal stranger analogous to the ego confronted by the pressure of its desires. In terms of the superego we are "foreign" to ourselves. Thus Freud[6] speaks of the superego as an "internal foreign territory" (p. 57).

We must not ask of psychoanalysis what it cannot give: the origin of the ethical problem, its founding principle; but what it can give: the source and the genesis of this problem. The difficult problem of identification has its roots here. The question is how can I become myself, beginning from another, such as the father? The advantage of a thought that begins by rejecting the primordial character of the ethical ego is that it displaces our attention to the process of interiorization by which the external becomes internal. That way not only the proximity with Nietzsche is discovered, but also the possibility of a confrontation with Hegel and his concept of the doubling of consciousness by which it becomes self-consciousness. Certainly by rejecting the primordial character of the ethical phenomenon, Freud can only encounter morality as the humiliation of desire, as prohibition and not as aspiration. But the limitation of his point of view is the counterpart of its coherence. If the ethical phenomenon first appears in a wounding of desire, it is justifiable by a general erotics, and the ego, prey to its diverse masters, again falls under an interpretation bound up with an economics.

Such is the clinical description of the moral phenomenon. This description, in turn, calls for an explanation that can only be genetic. If, in effect, moral reality presents characteristics so markedly inauthentic, it must be treated as derived and not as original. "Since [the superego] goes back to the influence of parents, educators and so on, we learn still more of its significance if we turn to those who are its sources" (p. 67). This declaration from the *New Introductory Lectures*[6] is a good expression of the function of genetic explanation in a system that does not recognize either the original character of the *cogito* or its ethical dimension. Genetic explanation takes the place of a transcendental foundation.

It would be fruitless to argue that Freudianism in its basic intention is anything other than a variety of evolutionism or moral geneticism. In every case, study of the texts allows us to affirm that beginning dogmatically, Freudianism does not cease to render its own explanation more problematic to itself to the extent that it carries it out.

For one thing the proposed genesis does not constitute an exhaustive explanation. The genetic explanation reveals a source of authority —the parents—that only transmits a prior force of constraint and aspiration. The text cited above continues, "A child's superego is in fact constructed on the model not of its parents but of its parents' sugerego; the contents which fill it are the same and it becomes the vehicle of tradition and all the time-resisting judgments of value which have propagated themselves in this manner from generation to generation."[6] Therefore, it would be fruitless to seek a full justification for moral judgments within genetic explanation. Their source is somehow given in the world of culture. The genetic explanation only circumscribes the earliest phenomenon of authority without really exhausting it.

2. Genetic explanation depends on the convergence of ontogenesis and phylogenesis, in other words, on the convergence of the psychoanalysis of the infant and that of primitive societies.

One thing that strikes every reader of Freud's first writings is the lightning character of his discovery of the Oedipus complex, which was simultaneously recognized as being both an individual drama and the collective destiny of humanity, both a psychological fact and the source of morality, both the origin of neurosis and the origin of culture. The Oedipus complex receives its intimately personal character from the discovery that Freud made through his own self-analysis. But at the same time its general character is suddenly glimpsed in the background of this individual experience. If his self-analysis unveils the striking effect, the compulsive aspect of the Greek legend, the myth, in return, attests to the fatality that adheres to the individual experience. Perhaps it is within this global intuition of a coincidence between an individual experience and universal destiny that we must look for the real motivation (which no anthropological investigation could exhaust) of all the Freudian attempts to articulate the ontogenesis, the individual's secret, in terms of the phylogenesis, our universal destiny. The scope of this universal drama is apparent from the beginning. It is attested to by the extension of the interpretation of *Oedipus Rex* to the personage of Hamlet: if "the hysterical Hamlet" hesitates to kill his mother's lover, it is because within him lies "the obscure memory that he himself had meditated the same deed against his father because of his passion for his mother" (p. 224).[7] This is a brilliant and decisive comparison, for if Oedipus reveals the aspect of destiny, Hamlet reveals the aspect of guilt attached to this complex. It was not by accident that as early as 1897, Freud[7] was citing Hamlet's words, "Thus conscience does make cowards of us all . . ." on which he remarks, "His conscience is his unconscious feeling of guilt."

Now what gives the individual's secret a universal destiny and an ethical character, if not the passage through institutions? The Oedipus complex is the dream of incest when "incest is antisocial and civilization consists in a progressive renunciation of it" (p. 210). Thus the repression that belongs to everyone's history of desire coincides with one of the most formidable cultural institutions, the prohibition of incest. The Oedipus complex poses the great conflict between civilization and instincts that Freud never stopped commenting on from "'Civilized' Sexual Morality and Modern Nervous Illness" (1908) and *Totem and Taboo* (1913), to *Civilization and Its Discontents* (1930) and *Why War?* (1933). Repression and culture, intrapsychical institution and social institution, coincide in this exemplary point.

Can phylogenesis be carried beyond ontogenesis? We might think so from reading *Totem and Taboo* (we are thinking of the section dealing with taboos: Chapter 1, "The Horror of Incest," and Chapter 2, "Taboo and Emotional Ambivalence"). As is well known, the kernel of his explanation is constituted by putting together the prohibition of incest as established by anthropology and the Oedipus complex as it comes from clinical study of obsessional neurosis. But, in truth, *Totem and Taboo* only provides the occasion for a psychoanalytic interpretation of anthropology in which psychoanalysis rediscovers what it already knew, although now on the scale of human history.

The guiding thread of the analogy between the history of an individual and the history of the species is furnished by the structural kinship between taboo and neurotic obsessions. The former functions as a collective neurosis and the latter functions as an individual taboo. Four characteristics assure this parallel: "(1) the fact that the prohibitions lack any assignable motive; (2) the fact that they are maintained by an internal necessity; (3) the fact that they are easily displaceable and that there is a risk of infection from the prohibited object; and (4) the fact that they give rise to injunctions for the performance of ceremonial acts" (pp. 28–29).[10] But the most important reason for putting these two together is constituted by the analysis of emotional ambivalence. The taboo is both attractive and repulsive. This double affectivity of desire and fear strikingly illumines the psychology of temptation and recalls Saint Paul, Saint Augustine, Kierkegaard, and Nietzsche. Taboo puts us in a place where the forbidden is attractive because it is forbidden, where the law excites concupiscence: "the basis of taboo is a prohibited action, for performing which a strong inclination exists in the unconscious" (p. 32).[10] The primitive clearly presents the psychic life's ambivalence. What finally appears in fear is the force of desires and the "indestructibility and insusceptibility to correction which are attributes of unconscious process" (p. 70).[10] Because he is like a child, the savage reveals in a fantastic exaggeration what only appears to us in the very dissimulated and attenuated figure of the moral imperative, or in the distortions of obsessional neurosis. Emotional ambivalence appears, then, as the common ground of taboo conscience (and its remorse), on the one hand, and the moral imperative as it has been formalized by Kant, on the other.

But if Freud was to derive conscience from emotional ambivalence, he had to assume the prior existence of some authoritative social figures or agencies. The father figure in the Oedipus complex and the passage from biological relations to "group kinship" in totemic organization require an already existing authority. *Totem and Taboo* clarifies the emotional expression of this authority more than its ultimate origin. The psychology of temptation to which the theme of emotional ambivalence belongs only makes more evident the lack of an original dialectic of desire and law. What is left unspoken in these two chapters is the existence of institutions as such.

In order to fill this gap Freud had to posit a real Oedipus complex at the beginning of mankind, an original parricide whose scar all subsequent history bears.

We will not consider here the details of the Freudian myth of the first murder of the father figure, which brings into play not only an old-fashioned anthropological apparatus but also a reconstruction of totemism itself that surpasses the phenomenon of taboo properly speaking. But at the completion of this reconstitution of origins, the problem of institutions reappears in all its force. In mythical terms how could the prohibition against "fratricide" arise from a "parricide"? In unmasking the father figure in the alleged totem, Freud has only made more acute the problem that he wanted to resolve, namely, the ego's adoption of external prohibitions. Certainly without the horde's jealousy of the father there is no prohibition, and without the "parricide" there would be no stopping of the jealousy. But these two ciphers of jealousy and parricide are still ciphers for violence. Parricide puts a stop to jealousy, but what prevents the repetition of the crime of parricide? This was already Aeschylus's problem in the *Oresteia*, as Freud is quick to acknowledge. Remorse and obedience in retrospect of the crime allow us to speak of a contract with the father, but this only explains at most the prohibition against killing, not the prohibition against incest. That requires another contract, a convenant between the brothers. By this pact they decide not to repeat their jealousy of the father; they renounce that violent possession that was the motive for the murder. "Thus the brothers had no alternative, if they were to live together, but—not, perhaps, until they had passed through many crises—to institute the law against incest, by which they all alike renounced the women whom they desired and who had been their chief motive for despatching their father" (p. 144).[10] And a little further on: "In thus guaranteeing one another's lives the brothers were declaring that no one of them must be treated by another as their father was treated by them all jointly. They were precluding the possibility of a repetition of their father's fate. To the religiously-based prohibition against killing the totem was now added the socially-based prohibition against fratricide" (p. 146).[10] With this renunciation of violence under the spur of discord, we are given all that is necessary for the birth of institutions. The true enigma of law is fratricide, not parricide. With the symbol of the pact among the brothers, Freud has met the true requirement for analytical explanation of the problem of Hobbes, Spinoza, Rousseau, and Hegel: the change from war to law. The question is whether that change still belongs to an economics of desire.

3. We are now ready to take the last step, that is, integrating the clinical description and the genetic explanation in an economic point of view such as we have presented at the beginning of this essay at the level of the global phenomenon of culture.

What is an economic explanation of morality? Its task is to account for what has until now remained external to desire as a "differentiation" of the instinctual substratum; in other words, to make the historical process of the introjection of authority correspond to an economic process of distribution of cathexes. It is this differentiation, this modification of instincts, that Freud calls the "superego." In these terms this new economic theory is much more than a translation of a collection of clinical, psychological, and anthropological material into a conventional language. It is charged with resolving a hitherto insoluble problem on both the descriptive and the historical planes. The fact of authority has constantly appeared as the presupposition of the Oedipus complex as applied to either an individual or a collectivity. Authority and prohibition must be introduced in order to pass from individual or collective prehistory to the history of the adult and the civilized person. The entire effort of the new theory of agencies is to inscribe authority within the history of desire, to make it appear as a "difference" within desire. The institution of the *superego* is the response to this perplexity. The relationship between the genetic and economic points of view is therefore reciprocal. It is a question of putting the Oedipus event and the advent of the superego into relation and of stating this relation in economic terms.

One important concept plays a decisive role in accomplishing this: the concept of identification. We can follow its development from the *Three Essays on the Theory of Sexuality* (more precisely the section added in 1915),[9] where identification is compared to idealization; then in the article "Mourning and Melancholia,"[4] where identification is conceived as a reaction to the loss of the beloved object through internalization of the lost object; to *Group Psychology and the Analysis of the Ego*,[3] where the intersubjective character of identification comes to the fore: "Identification is known to psychoanalysis as the earliest expression of an emotional tie with another person." This is how Chapter 7 entitled "Identification," begins. Not only the relation to another as a model is emphasized, but this relation itself divides into a wish *to be like* and a wish *to have* and *to possess*. "It is easy to state in a formula the distinction between an identification with the father and the choice of the father as an object. In the first case one's father is what one would like to *be*, and in the second he is what one would like to *have*. The distinction, that is, depends upon whether the tie attaches to the subject or to the object of the ego. The former kind of tie is therefore already possible before any sexual object-choice has been made. It is much more difficult to give a clear metapsychological representation of the distinction. We can only see that identification endeavours to mould a person's ego after the fashion of the one that has been taken as a model" (p. 106).[3] Freud never more vigorously expressed the problematic and the nonproblematic character of identification.

Freud puts these properly economic discoveries together in the synthesis of *The Ego and the Id*. The question that dominates its third chapter is: how can the superego, which from a historical point of view stems from parental authority, derive its energies from the id according to an economic point of view? How can the internalization of authority be a differentiation of intrapsychical energies? The intersecting of these two processes, belonging to two different planes from a methodological point of view, explains how what is sublimation from the point of view of effects, and introjection from the point of view of method, can be likened to "regression" from an economic point of view. This is why the problem of "replacement of an object-cathexis" by an identification is taken in its most general sense as a kind of algebra of placements, displacements, and replacements. So presented, identification appears as a postulate in the strong sense of the term, a demand that we must accept from the beginning. Consider the following text:

> When it happens that a person has to give up a sexual object, there quite often ensues an alteration of his ego which can only be described as a setting up of the object inside the ego, as it occurs in melancholia; the exact nature of this substitution is as yet unknown to us. It may be that by this introjection, which is a kind of regression to the mechanism of the oral phase, the ego makes it easier for the object to be given up or renders that process possible. It may be that this identification is the sole condition under which the id can give up its objects. At any rate the process, especially in the early phases of development, is a very frequent one, and it makes it possible to suppose that the character of the ego is a precipitate (*Niederschlag*) of abandoned object-cathexes and that it contains the history of those object-choices. [p. 29][2]

The abandonment of the object of desire, which initiates sublimation, coincides with something like a regression. This is a regression, if not in the sense of a temporal regression to a previous stage of organization of the libido, at least in the economic sense of a regression from object libido to the narcissistic libido, considered as a reservoir of energy. In effect, if the transformation of an erotic object choice into an alteration of the ego is really a method of dominating the id, the price that must be paid is as follows. "When the ego assumes the features of the object, it is forcing itself, so to speak, upon the id as a love-object and is trying to make good the id's loss by saying, 'Look, you can love me too—I am so like the object!' " (p. 30).[2]

We are now prepared for the generalization that will henceforth dominate the problem: "The transformation of object-libido into narcissistic libido which thus takes place obviously implies an abandonment of sexual aims, a desexualization—a kind of sublimation, therefore. Indeed, the question arises, and deserves careful consideration, whether this is not the universal road to sublimation, whether all sublimation does not take place through the mediation of the ego, which begins by changing sexual object-libido into narcissistic libido and then, perhaps, goes on to give it another aim" (p. 30).[2]

Freud's whole effort from here on is to make the identification with the father of individual and collective prehistory a part of the theoretical schema of identification by abandoned object cathexes. We will not consider his theoretical elaboration of this since it no longer concerns the ethical incidences of psychoanalysis. It should suffice to have shown, on the plane of doctrine, the convergence among (1) a clinical description of morality, (2) a genetic explanation of this information, and (3) an economic explanation of the processes implied by this genesis.

Ethics and Psychoanalysis

The preceding analyses lead us to the threshold of the crucial question: can we speak of a psychoanalytic ethics? The answer must be frankly and clearly negative if by ethics we mean a prescribing of duties, either old or new. But this negative response to a question that has not itself been affected by psychoanalysis does not exclude our asking whether its critique of morality does not imply a new way of thinking about ethics.

But first we must consider the negative response. That psychoanalysis prescribes nothing follows first from its theoretical status, then from its discoveries concerning morality, and finally from its character insofar as it is a therapeutic technique.

First, the theoretical status of psychoanalysis prevents it from becoming prescriptive. The Freudian interpretation of culture, taken overall, and of ethics considered in particular, implies a limitation of a certain kind. Psychoanalytic explanation, we have seen, is essentially an economic explanation of the moral phenomenon. Its limit results from its very project of understanding culture from the point of view of its emotional cost in pleasures and pains. Therefore, we cannot expect anything else from this enterprise than a *critique of authenticity*. Above all, we cannot ask it for what we might call a *critique of foundations*. This is a task for another method, another philosophy. Psychoanalysis as such is limited to unmasking the falsifications of desire that inhabit the moral life. We have not founded a political ethic, or resolved the enigma of power, because we have discovered—as, for example, in *Group Psychology and the Analysis of the Ego*—that the tie to the chief mobilizes an entire libidinal cathexis with a homosexual characteristic. Nor have we resolved the enigma of the authority of values when we have discerned the father figure and identification with him as fantastic as it is real in the background of the moral and social phenomena. The legitimation of a phenomenon such as power or value is something else. So is what we make of the emotional cost of experience, the sum of pleasures and pains in our lives. Because psychoanalysis cannot pose the question of moral legitimation, it must limit itself to a sort of empty marking of the place of a phenomenon as important as that of sublimation, in which an axiological point of view is mixed with an economic one. In sublimation, in effect, an instinct is working on a higher level, although we must say that the energy invested in new objects is the same energy that was formerly invested in a sexual object. The economic point of view only accounts for this connection, not for the new value promoted by this process. One postpones the difficulty by speaking of socially acceptable goals and objects, for social utility is a cape of ignorance thrown over the problem of value raised by sublimation.

Thus psychoanalysis cannot deal with the problem of value, legitimation, or radical origin, because its economic point of view is only economic. Its force is that of suspicion, not justification or legitimation, and still less that of prescription.

Second, the discoveries of psychoanalysis about morality prohibit it from moralizing. In a sense close to that of Nietzsche in the *Genealogy of Morals*, the exploration of conscience's archaisms reveals that man is wrongly accused in the first place. This is why it is fruitless to ask psychoanalysis for an immediate ethic without conscience first having changed its position as regards itself. Hegel saw this before Nietzsche and Freud. In criticizing the "moral vision of the world" in Chapter 6 of *The Phenomenology of Mind*, he denounced the "judging conscience" as denigrating and hypocritical. It should recognize its own finitude, its equality with the judged conscience, so that the "forgiveness of sins" might be possible as knowledge of a reconciling self. But in distinction to Nietzsche and Hegel, Freud does not accuse accusation. He understands it and in understanding it he makes its structure and strategy public. An ethic where the cruelty of the superego would yield to the severity of love is possible in this direction, but first it would be necessary to learn in depth that the catharsis of desire is nothing without the catharsis of the judging conscience.

The fundamentally nonethical character of psychoanalysis results not only from its theoretical status, or even from its discoveries concerning morality, but also from its *technique* in that it is therapeutic. This therapy implies in principle the neutralization of the moral point of view. In the essay entitled "Remembering, Repeating and Working-Through," Freud insists that psychoanalysis is not just, or not even principally, a purely intellectual interpretation, but *work* against resistances and a "handling of" the forces released by transference. Not only has psychoanalytic explanation an economic character, but also treatment itself is an economic operation. This economic work Freud calls *Durcharbeiten*: "This working-through of the resistances may in practice turn out to be an arduous task for the subject of the analysis and a trial of patience for the analyst. Nevertheless it is a part of the work which effects the greatest changes in the patient and which distinguishes analytic treatment from any kind of treatment by suggestion" (Pp. 155–156).[8] In another essay, "On Beginning the Treatment," Freud rigorously attaches this handling of resistances to the handling of transference: the name "psychoanalysis" applies only "if the intensities of the transference have been used for the overcoming of resistances" (p. 143).[5]

This struggle against resistances and by means of transferences leads us to the decisive insight that the sole ethical value that is thereby brought into play is *veracity*. If psychoanalysis is a technique, it is not included in the cycle of techniques of domination; it is a technique of veracity. What is at stake is self-recognition and our itinerary runs from misunderstanding to recognition. In this regard it has its model in the Greek tragedy *Oedipus Rex*. The fate of Oedipus is to have already killed his father and married his mother. But the drama of recognition begins beyond this point, and this drama consists entirely in his recognition of the man whom earlier he had cursed: "I am that man. In a sense I always knew it, but in another sense I didn't; now I know who I am."

Beyond this what can the expression "technique of veracity" signify? First, that it takes place entirely on the plane of speech. Therefore, we are faced with a strange technique. It is a technique according to its work character and its commerce with emotional energies and mechanisms belonging to the economy of desire. But it is a unique technique in the sense that it only attains or handles these energies through and across the effects of meaning, across a work of speech. From then on what is in question in analysis is the access to true discourse, which is certainly something different from social adaptation, talk of which hastens to overthrow the scandal of psychoanalysis and make it socially acceptable. For who knows where a true discourse may lead as regards the established order, that is, for the established disorder?

If, therefore, veracity is the sole ethical value implied by its analytic technique, psychoanalysis is bound to practice, as regards every other ethical value, what we could call a "suspension" of ethics. But an ethic reduced to veracity is still something. It contains the seed for new attitudes issuing from the end of dissimulation.

Certainly the vulgarization of psychoanalysis tends to draw a sort of babble about everybody's libido from this disoccultation, which has nothing to do with *working through*, with the *work* of truth. The vulgarization of the results of psychoanalysis, apart from its technique and its work, even tends to induce reductive schemes and to authorize saying the first thing to come to mind about all the eminent expressions of culture: "Now we know that all the works of culture are nothing but, or nothing other than. . . ." Psychoanalysis in this sense reinforces what Max Weber called "disenchantment" (*Entzäuberung*). But this is the price modern culture must pay to have a better understanding of itself. Whether we like it or not, psychoanalysis has become one of the media through which our culture seeks to understand itself. And it is unavoidable that we should only become aware of its signification through the truncated representations that are allowed by the narcissism of our resistances. Misunderstanding is the necessary path to understanding.

This same misunderstanding inclines popular consciousness to look for a system of justifications for moral positions in vulgarized psychoanalysis, positions that have not undergone its questioning in their depths, even though psychoanalysis wanted to be precisely a tactic for unmasking every justification. Thus some want it to ratify education without restraints—because neurosis comes from repression—and see in Freud a discreet apologetic for and camouflaging of a new epicureanism; others, taking their stand on the theory of stages of maturation and integration, and on the theory of perversions and regressions, utilize it to the profit of traditional morality—did not Freud define culture as the sacrificing of instincts? Once set off on this way, nothing stops us from psychoanalyzing psychoanalysis itself: did not Freud publicly provide a "bourgeois" justification for the discipline of monogamy, while secretly providing the "revolutionary" justification for orgasm? But the conscience that poses this question, and that attempts to enclose Freud within this ethical either-or, is a conscience that has not undergone the critical test of psychoanalysis.

The Freudian revolution is its diagnostic technique, its cold lucidity, its laborious search for truth. It is a mistake to attempt to change its science into moralizing; a mistake to ask ourselves whether it is the scientist or the Viennese "bourgeois" trying to justify himself who speaks of perversion and regression; a mistake to suspect it—in order to blame it or commend it—of sliding under the diagnosis of the libido to approval of an unacknowledged epicureanism, when it really only turns the unpitying gaze of science on the sly conduct of moral man. Here is our misunderstanding: we listen to Freud as if he were a prophet, although he speaks like an unprophetic thinker. He does not bring us a new ethic, but he changes the conscience of those for whom the ethical question remains open. He changes our conscience by changing our knowledge of it and by giving us the key to some of its ruses. Freud can indirectly change our ethics because he is not directly a moralist.

For my part I would say that Freud is too *tragic* a thinker to be a moralist. Tragic in the sense of the Greeks. Instead of turning us toward heartrending options, he makes us look at what he himself calls the "hardness of life," following the German poet Heine. He teaches us that it is difficult to be human. If from time to time he seems to be pleading for the diminution of instinctual sacrifice through an easing of social prohibitions or for an acceptance of this sacrifice in the name of the reality principle, it is not because he believes that some sort of immediate diplomatic action is possible between the clashing agencies. Rather he waits for a total change of consciousness that will proceed from a wider and better articulated understanding of the human tragedy, without worrying about drawing its ethical consequences too soon. Freud is a tragic thinker because human situations for him are unavoidably conflictual situations. Lucid understanding of the necessary character of these conflicts constitutes, if not the last word, at least the first word of a wisdom that would incorporate the instruction of psychoanalysis. It is not by accident that Freud—naturalist, determinist, scientist, child of the Enlightenment—kept returning to the language of tragic myths: Oedipus and Narcissus, Eros, Ananke, Thanatos. We must assimilate this tragic knowledge to reach the threshold of a new ethic, which we should stop trying to derive directly from Freud's works, an ethic that would be prepared slowly and at length by the fundamentally nonethical instruction of psychoanalysis. The self-awareness that psychoanalysis offers modern man is difficult and painful because of the narcissistic humiliation it inflicts on us—but at this price, it rejoins that reconciliation whose law was pronounced by Aeschylus: "wisdom comes only through suffering."

Bibliography

1. FREUD, S., *Civilization and Its Discontents*, in Strachey, J. (Ed.), *Standard Edition*, Vol. 21, Hogarth, London.
2. ———, *The Ego and the Id*, in Strachey, J. (Ed.), *Standard Edition*, Vol. 19, Hogarth, London.
3. ———, *Group Psychology and the Analysis of the Ego*, in Strachey, J. (Ed.), *Standard Edition*, Vol. 18, Hogarth, London.
4. ———, "Mourning and Melancholia," in Strachey, J. (Ed.), *Standard Edition*, Vol. 14, p. 249, Hogarth, London.
5. ———, "On Beginning the Treatment," in Strachey, J. (Ed.), *Standard Edition*, Vol. 12, pp. 121–144, Hogarth, London.
6. ———, *New Introductory Lectures on Psychoanalysis*, in Strachey, J. (Ed.), *Standard Edition*, Vol. 22, pp. 64–65, Hogarth, London.
7. ———, *Origins of Psychoanalysis*, Basic Books, New York, 1954.
8. ———, "Remembering, Repeating, and Working-Through," in Strachey, J. (Ed.), *Standard Edition*, Vol. 12, pp. 145–156, Hogarth, London.
9. ———, *Three Essays on the Theory of Sexuality*, in Strachey, J. (Ed.), *Standard Edition*, Vol. 7, p. 198, Hogarth, London.
10. ———, *Totem and Taboo*, in Strachey, J. (Ed.), *Standard Edition*, Vol. 13, pp. 1–161, Hogarth, London.

CHAPTER 46

PSYCHIATRY AND RELIGION

Kenneth E. Appel

PSYCHIATRY is the study of emotional and mental disturbance and illness. Religion is one's system of devotions, reverences, allegiances, and practices—whether avowed or implicit, conscious or unconscious. Psychiatry deals with illness, its treatment and prevention. Religion is concerned with the development of the spiritual aspects of personality and the enrichment of personal and social life. Ideally it should help the tolerance and endurance of pain and suffering, the maintenance of health, and the prevention of illness. Standards and values, the need to belong, the need for togetherness, the need to feel worthwhile and of value, the desire to be cared for—all of these are important in human life. They are the concerns of psychiatry and are related to religion as well.

Religion has many aspects[11] and a varied history.[19] The various parts of the personality are given importance by different persons and by different faiths. Reason, belief, faith, ritual, and church memberships are variously emphasized by different leaders as essential aspects of religion. With some the intellect is stressed, and such people require explanations, demonstrations, reasons, and logic, in part following St. Thomas. Others emphasize emotion, belief, conviction, and dogma. In these they find security, on these they are dependent, and so Schleiermacher believed a feeling of dependency was the essence of religion. For Pascal hope was a cornerstone. James wrote of the will to believe. St. Augustine stressed the will rather than the intellect. G. F. Moore thought the essence of religion was the conservation of social values. Santayana believed in the beauty of religion. Others emphasized rituals, ceremonies, and sacraments. Rules, taboos, and magic have been considered very important by some. Group belonging and togetherness have been emphasized by others. Protestantism finds mediation unnecessary and has stressed the individual's direct access to God.[12,32,41] Catholicism emphasizes help from the group, from authority, from history, from tradition.[8,34,37,42] Judaism exalts and insists on social justice.[16,21,24,27] "Righteousness exalteth a nation."

The scientists and philosophers have by no means dispensed with religion, as is so often thought to be the case.[22,31,38] Einstein writes of awe before the great unknown and reverence for the harmony and beauty that exists in the world of nature. Bertrand Russell, in his Free Man's Worship, although disassociating himself from traditional religion, writes eloquently of stoic struggle and devotion to duty in the face of the destructive forces of nature and man, which most people would find was essentially religious. Rabbi Kagan quotes from the Talmud that even if a man denies God and yet lives according to him, he is acceptable to God. Many religious psychiatrists feel that Freud, notwithstanding his writings about the neurotic obsessiveness of religious beliefs and practices, was religious in the sincerity, integrity, and persistence of his devotion and research.[20,29,42]

The existentialists bring psychiatry, religion, and philosophy together.[14,25,33] Efforts at the denial of religion, assertions that it has been eliminated, affirmations of the reign of reason and the hegemony of science have not banished anxiety, as existentialists point out. People cannot rely on reason and science and the inevitability of progress to bring about security and relief from fear and anxiety. The self-assurance of Victorian thinking and Newtonian mechanics has been shaken by quantum mechanics and by nuclear, astro, and mathematical physics. The present was called "The Age of Anxiety"[4,24] long before Sputnik appeared. The existentialists, from Kierkegaard to Tillich, speak of anxiety as being involved in actualizing any possibility, in creativity, and in the realization of selfhood. They speak of the inevitability of anxiety, its confrontation in the development of individuation, freedom, and responsibility. For Tillich anxiety is the reaction to the threat of nonbeing, the threat of meaninglessness in one's existence. The capacity to bear anxiety is a measure of selfhood. Mowrer[26] speaks of integrated behavior as the capacity to bring the future into the psychological present. Allport writes of *becoming* "as an integration of the past into the present." Niebuhr says man, in contrast to animal, sees the future contingencies and anticipates their perils. Anxiety is the concomitant of freedom and finiteness.

There are devotions, reverences, thoughts, and principles of life, action, and behavior that contribute to effectiveness and health. Religion, beneath its many varieties, consists of constructive devotions and practices. There are forces at work in patients that are not determined by what we see, touch, weigh, smell, or hear. People seem to get well because of relationships with other human beings or under the influence of ideals or in devotion to causes. Relationships with other human beings or causes are transpersonal values. Such relationships involve a dedication to the constructive forces in life. Faith, hope, love, and justice are such forces. They have been manifested in the medical profession at its best and are related to religion. The doctor who engages in an active and tireless search to heal his patient is exemplifying these constructive forces, and the patient somehow comes to identify with him and becomes inspired to join the search. Inspiration and aspiration promote health. They are factors that cannot be neglected in medicine and psychiatry. A devoted and dedicated physician or priest is able to tap and mobilize resources for growth and life in the individual (eros, agape).

Psychiatry has been related to religion both historically and ideologically. Throughout the Middle Ages there were islands of constructive humanity—love and kindness at Monte Cassino and other places. For a long time in the Middle Ages the primitive demonological conception of illness held sway, however, particularly in connection with the mentally ill. "Thou shalt not suffer a witch to live" seemed to be a religious injunction that Judeo-Christian religion followed with regard to the treatment of mental conditions. Protestant theology, perhaps following Calvin, emphasized the inborn sinfulness of man. In Judaism there is an optimistic view about man's ability to overcome sin. Judaism knows about Yetzer Hara and Yetzer Hatov, the struggle between good and evil, but there is the belief that Yetzer Hatov, the good, will be victorious. Catholicism offers the weight of tradition, authority, and exercises (penances) to help man overcome his destructive impulses. Protestantism leaves man more to his own inner devices—suffering and punishment—to effect contrition and salvation. Traditions and some religious faiths are often great allies in the struggle to overcome the destructive forces that appear in neuroses, psychoses, and character disorders. Ideologically St. Paul stated succinctly the problem of conflict with which much psychiatric thinking is concerned: "For the good that I would, I do not; but the evil which I would not, that I do." In the Judeo-Christian scriptures there are many statements such as the following that deal with the problems of suffering, illness, psychology, and healing:

Because he has set his love on me,
I will deliver him . . .
He will call on me and I will answer him,
I will be with him in trouble,
I will deliver him . . .
God is our refuge and strength,
A very present help in trouble
Therefore will we not fear . . .

PSALMS 91 AND 46

Love suffers long, is kind . . .
Takes no offense . . .
Is not joyful over wrong,
But is joyful with the truth;
Overlooks all . . .
Hopes all, endures all.

1 CORINTHIANS 13

For centuries religion and medicine have been closely related. Psychiatry is a branch of medicine, and its psychotherapeutic methods at times have approximated closely those of religion. Science and religion assumed distinctive roles in society as time went on, but they continued to share common goals. In order to clarify this belief, in 1947 the Group for the Advancement of Psychiatry made a statement that they believed in the dignity and integrity of the individual and that the major goal of treatment was the attainment of social responsibility. They recognized the crucial significance of the home and its influence on the individual and the problem of ethical training. They emphasized the important role religion can play in bringing about an improved emotional, moral, and physical state. Methods of psychiatry help patients achieve healthy emotional lives so that they may live in harmony with society and its standards. Psychiatry does not actually conflict with the morals and ethics of religion, but frees the person to assert the morality and ethics that stem from a pattern of conscious affirmation, reasoned devotions, and accepted behavior rather than from a compulsive, fear-based pattern of behavior. These psychiatrists wrote that there was no real conflict between psychiatry and religion. It was felt that in practice competent psychiatrists would be guided by these principles. Psychiatry confirms the fact that beliefs and devotions affect not only physiological functions of the body, as Dr. Wolff and others have shown, but also the social functions of the individual, as indicated by the behavior of people.

There has been a growing realization that the fields of religion and psychiatry have a common concern. Evidence of this appears in the many organizations that are bringing religion and psychiatry into closer relationship. Prominent among these movements have been the American Foundation of Religion and Psychiatry and the Academy of Religion and Mental Health. A program has existed for years that trains divinity students in clinical experience. It has afforded theological students firsthand experience with illness in both general and mental hospitals; students can learn more effectively to perform their traditional function in the extremes of human suffering. Efforts are being made to have clergymen become more cognizant of the roots of emotional problems. Psychiatrists need to become more familiar with the basic religious thinking of people today. One hundred million people in the United States are church members. What do they believe? How does it affect their points of view, their outlooks on life, their mental health? In psychiatric practice one sees results of the crippling accentuation of illness when punitive and rigid religious and moral concepts are held. But religion that is supportive, wholesome, courage-inspiring and that does not make impossible demands on human nature is an important factor in health. The fundamental beliefs and values of the patient cannot be casually dismissed or explained away by the psychiatric reduction of all values to infantile fixations and fears of father.

Psychotherapy should not be an exercise of

intellectual understanding alone but an experience of growth—hence a living relationship. The process of reorganization that is called psychotherapy does not always take place consciously; much of it is automatic and unconscious. New experience is needed to change the balance of the destructive exaggerations of fear, anger, resentment, and guilt, to oppose previous noxious experience. New experience is indicated more often than surgery or drugs. Such corrective experience is psychotherapy. It is the influence one human being has on another. It is not just discussion, explanation, or the development of insight. It is not merely abreaction or ventilation. It is often identification with a constructive, heuristic, experimental approach on the part of the physician. The patient can identify with this sympathetic prospecting of the physician, and this process of identification is very important. The patient gradually sees the physician tackling difficult problems, no matter how hopeless they seem, without becoming disturbed. The physician asks himself what is the procedure to follow, what is the thing to do, what are the possibilities, what has contributed to this? Through these questions he establishes a certain amount of identification with the patient. He moves beyond the patient's immediate problems and explores possibilities, searching for what can be done in the situation. No situation is usually so terrible that something cannot be done. The patient gradually takes over. Doctors who are patient, understanding, considerate, interested, and eager to aid will be able to help patients *feel* this and respond constructively. This is really tapping the growth or love impulses, or eros, of the patient—of the patient's id. When standards are deficient and defective, therapy consists in helping the individual develop new guiding principles or ego ideals.[7,15] This can come through identification with "causes," with friends, or with the therapist. The relationship with the therapist is vital for the patient, and perhaps the most health-producing factor is the spiritual quality of maturity, care, and supportiveness on the part of the psychiatrist. Religion, too, is concerned with what is deeper than the intellectual, with forces within and beyond the individual. In experiences with prayer groups and healing services, there is contact with forces within and beyond that has helped people to realign their personalities in the direction of health.

Psychiatrists should have a sympathetic understanding of the religious thinking and feeling of the times and be ready to use these resources that are available. This will offer a creative, hopeful, forward orientation in therapy that will serve as a positive supplement to the intellectual or reductive type of therapy so prevalent today. A brilliant psychiatrist was unable to help a patient because he ignored the patient's religious beliefs, orientation, and background. Another psychiatrist made light of the guilt feelings of a man in his late fifties. He thought that by minimization he would be able to help his patient. This psychiatrist did not recognize the need for suffering and atonement. Superficial advice and adjurations cannot remove the paradigms of sin, suffering, and atonement that have become so much a part of our Judeo-Christian civilization. Religion can be used constructively and reassuringly even when there is severe illness. I think of cases of schizophrenia, true depressions, and reactive depressions in which making use of the religious resources and background of the patients helped them not only to endure their suffering but also to come through their illness with stronger personalities than before they became ill. Collaboration of clergy and psychiatrists is often most helpful.

Various types of psychotherapy have apparently produced the same percentage of cures. Figures from faith-healing centers report percentages of cures in certain conditions similar to figures reported by psychiatrists, as does Alcoholics Anonymous. There is comparable effectiveness. Therefore, there must be something basic beneath intellectual formulations and theoretical frames of reference. There are limits to analytic thinking that James and Bergson pointed out long ago. I believe that the basic factor is probably the emotional relationship between the therapist and the patient, the care, love, devotion, and interest on the part of the therapist, and the faith the patient has in him. When St. Augustine went to Milan as a teacher of rhetoric, he was dissatisfied with rationalism and reason in the conduct of life. In Milan he met Ambrose, Bishop of Milan. Augustine recounts in his *Confessions*, "That man of God received me as a father, and showed me Episcopal kindness on my coming. Henceforth I began to love him, at first, indeed not as a teacher of the truth." That is to say, the primary, constructive, real integrating force of this great teacher (a shaper of Western thinking, of Western religion) was his human relationship with the kindly Ambrose. This historic evidence is important in our reflections on the relationship of religion to psychiatry.

Knowledge by itself is not the salvation of the individual nor a guarantee of the survival of society. This is one of the myths that has developed, as a narcissistically consoling compensation, through several centuries. Two world wars, with unheard of barbarity and the continuance of nightmarish violence all over the world, have not yet jolted humanity hard enough for it to recognize that service, activation, cooperation, participation, involvement, and sharing with others positive personal and social values are essential not only for health and happiness but also for survival. And this is the eternal lesson of religion, whatever its form. Psychiatry and religion can cooperate and supplement one another in helping people with emotional and mental difficulties and illness. Psychiatry is no substitute for religion. Religion is no substitute for psychiatry. They both contribute to the health and fulfillment of the individual.

Bibliography

1. Appel, K. E., "Anxiety Problems within Cultural Settings," in Masserman, J. H., and Moreno, J. L. (Eds.), *Progress in Psychotherapy*, Grune, New York, 1957.
2. ———, "Do's and Dont's in Psychotherapy," presented at Omaha-Midwest Clinical Society, November 6, 1957, to be published.
3. ———, "The Present Challenge of Psychiatry," *Am. J. Psychiat.*, 3, 1954.
4. Auden, W. H., *The Age of Anxiety*, Random House, New York, 1947.
5. Barnett, L., *The Universe and Dr. Einstein*, Harper, New York, 1948.
6. Boisen, A. T., *Religion in Crisis and Custom*, Harper, New York, 1955.
7. Boulding, K., *The Image*, University of Michigan Press, Ann Arbor, 1956.
8. Braceland, F. J., "Clinical Psychiatry Today and Tomorrow," in Braceland, F. J. (Ed.), *Faith, Reason and Modern Psychiatry*, Kenedy, New York, 1955.
9. Doniger, S. (Ed.), *Religion and Human Behavior*, Association Press, New York, 1954.
10. Einstein, A., *Ideas and Opinions*, Crown, New York, 1954.
11. Ginsburg, S. W., "Religion: Man's Place in God's World," in Herma, H., and Kurth, G. M. (Eds.), *Elements of Psychoanalysis*, World Publishing Co., Cleveland, 1950.
12. Heuss, Rev. J., "Our Imperfections," in McCauley, L., and McCauley, E. (Eds.), *A Treasury of Faith*, Dell, New York, 1957.
13. Hiltner, S., *Self-Understanding*, Scribner's, New York, 1951.
14. Hubben, W., *Four Prophets of Our Destiny*, Macmillan, New York, 1952.
15. Jung, C. G., *Modern Man in Search of a Soul*, Harcourt, New York, 1934.
16. Kagan, H. E., "Psychiatry and Religion," Minnie K. Landsberg Memorial Foundation, Cleveland, 1952.
17. Kegley, C. W., and Bretail, R. W. (Eds.), *The Theology of Paul Tillich*, Macmillan, New York, 1952.
18. Kew, C. E., and Kew, C. J., *You Can Be Healed*, Prentice-Hall, Englewood Cliffs, N.J., 1953.
19. Lamprecht, S. P., *Our Religious Traditions*, Harvard University Press, Cambridge, 1950.
20. Lee, R. S., *Freud and Christianity*, Wyn, New York, 1949.
21. Liebman, J. S. (Ed.), *Psychiatry and Religion*, Beacon Press, Boston, 1948.
22. Long, E. L., Jr., *Religious Beliefs of American Scientists*, Westminster, Philadelphia, 1951.
23. McCauley, L., and McCauley, E. (Eds.), *A Treasury of Faith*, Dell, New York, 1957.
24. May, R., *The Meaning of Anxiety*, Ronald, New York, 1950.

25. MICHALSON, C. (Ed.), *Christianity and the Existentialists*, Scribner's, New York, 1956.
26. MOWRER, O. H., *Psychotherapy*, Ronald Press, New York, 1953.
27. NOVECK, S. (Ed.), *Judaism and Psychiatry*, Basic Books, New York, 1956.
28. OATES, W. E., *Religious Factors in Mental Illness*, Association Press, New York, 1955.
29. OPPENHEIMER, J. R., *Science and the Common Understanding*, Simon & Schuster, New York, 1954.
30. OUTLER, A. C., *Psychotherapy and the Christian Message*, Harper, New York, 1954.
31. STACE, W. T., *Religion and the Modern Mind*, Lippincott, Philadelphia, 1952.
32. STEERE, D. V., *On Listening to Another*, Harper, New York, 1955.
33. TILLICH, P., *The Courage to Be*, Yale University Press, New Haven, 1952.
34. VANDERVELDT, J. H., and ODENWALD, R. P., *Psychiatry and Catholicism*, McGraw-Hill, New York, 1952.
35. WHITE, E., *Christian Life and the Unconscious*, Harper, New York, 1955.
36. ———, *The Way of Release: For Souls in Conflict*, Marshall, Morgan & Scott, London, 1947.
37. WHITE, V., *God and the Unconscious*, Harvill Press, London, 1952.
38. WHITEHEAD, A. N., *Science and the Modern World*, Macmillan, New York, 1925.
39. WISE, C. A., *Pastoral Counseling, Its Theory and Practice*, Harper, New York, 1951.
40. ———, *Psychiatry and the Bible*, Harper, New York, 1956.
41. ———, *Religion in Illness and Health*, Harper, New York, 1942.
42. ZILBOORG, G., "Some Denials and Affirmations of Religious Faith," in Braceland, F. J. (Ed.), *Faith, Reason and Modern Psychiatry*, Kenedy, New York, 1955.

Kenneth E. Appel
James R. MacColl, III

LIBERAL THINKING in religion and theology, from Harnack, Bultmann, Bonhoeffer, Niebuhr, Tillich, and Buber, has progressed up through Harvey Cox and Malcolm Boyd. Even participation in religious rituals and services by the young and lay people has been developed. Teilhard and Pope John have contributed in part to this development.

Unconventionality in behavior and dress, even in the most intimate behavior of men and women, coeducational dormitories in educational institutions, communal living without the benefit of the marriage ceremony, all these changes in acting and custom create great strains and disruptions in traditional family living.

War, crime, violence, drugs, unemployment, and race relations contribute to social unrest, disruption and tragedy. Church membership falls off. There are decreased financial contributions. Applications for traditional training in the ministry decline. Dissatisfaction with new experiments in church services and rituals are seen. The older generation and the young develop nonunderstanding. The increased number of elderly create new problems in living and loneliness in society. There is more need for new, progressive, gradual, and tolerable accommodations in society. Religion has always pointed out toward a new life—new potentials in life, the realization of change, and experiment toward creative accommodation. All these changes and conflicts cause new and unexpected tensions and produce new substantive contents and activities for medicine and religion. Used understandingly, the inspirational and cohesive forces of religion can help new adjustments both physiologically, psychologically, and sociologically. Thus new emphases in religion (both in theory and practice) and in medicine call for new imaginative and resourceful collaboration.

Hippies, vagabonds, unconventionals, rebels, and drug addicts permeate our society. All sorts of methods are being experimented with in meeting these unconventionalities, irregularities, and disturbances. Efforts are being made to tap the positive and viable forces in these people. For example, many hippies show a great deal of frankness in discussing their feelings, attitudes, and actions. One does not meet with the deception, the hypocrisy, and the protective façade that amounted to deception in trying to reach and talk with many young people as in former days. A great deal of *helpfulness* exists among these groups—helping one another. Dr. Peabody long ago wrote that the care of the patient started with caring emotionally for the patient.

The youth festival at Woodstock, New York, produced remarkably little injury and violence.

Experiments with folk music in religious services are tried. There are folk masses. English may be substituted for Latin. There is great interest in Indian and Oriental thinking and religion with its many practices of meditation and contemplation.

The little Haiku lines of Japanese poetry picture aspects of nature that emphasize beauty. Tagore speaks of this poetry's function as fulfilling the needs of people when there are heartaches and personal wounds. There is a sharing of beauty and the friendliness of nature seen in the seasons, the birds, and the flowers that do not hurt, humiliate, or maim, but that are healing.

So, although religion in many ways has become secular and interest in traditional practice and beliefs has lessened, there still remain positive aspects that have always been emphasized in the heart of religion. Religion has become ecumenical. There is much collaborative caring, loyalty, and helping in the community among young people that are religious.

Malcolm Boyd has been able to develop new relationships and fellowship with the unconventional and the outcasts. He has been able to obtain positive responses through his approaches of caring, kindness, and compassion—not formal, not blame, not critical, not rejecting, not derogatory, not dominating, but *accepting* of contacts, communication. He does not exclude and reject people and has been able to work with those whose conduct and impulses one does not customarily approve or condone. There is much writing about God is dead, following Nietzsche and Bishop Robinson, yet the essential creativeness and fellowship that exist remain as positive forces in human nature and the community. Leo Rosten's *Trumpet for Reason*, James Michener's *America vs. America* and *The Drifters* all tackle these problems from different points of view.

Although there has been much writing derogating religion in "sophisticated" psychiatry, a recent book by Thomas A. Harris, entitled *I'm OK—You're OK*,[7] has an important chapter on moral values. He writes:

> Persons are important in that they are all bound together in a universal relatedness which *transcends* their own personal existence. . . . There is the rationale of the position I'm OK—You're OK. Through this position only are we persons instead of things. Returning man to his rightful place of personhood is the theme of redemption, or reconciliation, or enlightenment, central to all of the great world religions. The requirement of this position is that we are responsible to and for one another, and this responsibility is the ultimate claim imposed on all men alike. The first inference we can draw is Do Not Kill One Another. [P. 220, 223]

The new thinking about religion has resulted in many new emphases and practices among the clergy and laity of Judeo-Christian persuasion. There is a strong re-entry into social, economic, industrial, and political problems. The motivating factors are to bring about attitudinal and institutional change.

1. a. Politics: the Vietnam war, the draft, conscientious objectors.
   b. Industry: scrutinizing of corporations whose large interests in other countries (Africa) are racial.
   c. Ecclesiastical: clergymen running for elective political office and opposing discriminating racial practices within the institutional religious and social establishment.
   d. The turmoil within the American Psychiatric Association and the American Medical Association being carried on by professionals and nonprofessionals with concern for social change.
2. The development of the so-called new breed of clergyman, who does not act or think parochially or pastorally in the traditional sense. His prime concern is effecting change, hitting at the causes of our problems; he is far less concerned with treating victims. One result of this is the emergence of more specialized forms of ministry.
   a. Institutional and industrial ministries, nonstipendiary ministries—men earning their living by secular employment.
   b. Development of paraprofessionals and medical clergy teams.
3. The emerging relevancy of religion for youth, for example, Jesus freaks.
   a. Large enrollment of students taking religious courses in colleges.
   b. Clustering of theological seminaries of different denominations in several large cities.
4. Emerging new roles for women in the professional ranks of the church.
   a. Ordination.
   b. Greater responsibility in church governing bodies.

All these changes are closely linked with social psychiatry, for psychiatry is concerned not only with the adjustment of the individual to society but in lending its knowledge and skills to bring about an environment that will provide a better medium for happy living and a healthier development of the individual.

Writings can be helpful in the form of *bibliotherapy* and even excerpts from literature.

From *The Understanding Heart*:[1]

> The happiness of life is made up of minute fractions—the little soon forgotten charities of a kiss or smile, a kind look, a heartfelt compliment, and the countless infinitesimals of pleasureable and genial feeling.
>
> SAMUEL TAYLOR COLERIDGE

> Happiness, I have discovered, is nearly always a rebound from hard work. It is one of the follies of men to imagine that they can enjoy mere thought, or emotion, or sentiment! As well try to eat beauty! For happiness must be tricked! She loves to see men at work. She loves sweat, weariness, self-sacrifice. She will be found not in palaces but lurking in cornfield and factories and hovering over littered desks: she crowns the unconscious head of the busy child. If you look up suddenly from hard work, you will see her—but if you look too long she fades sorrowfully away.
>
> DAVID GRAYSON

> Good manners is the art of making those people easy with whom we converse. Whoever makes the fewest persons uneasy is the best bred in the company.
>
> JONATHAN SWIFT

> The web of our life is of a mingled yarn, good and ill together; our virtues would be proud if our faults whipped them not; and our crimes would despair if they were not cherished by our virtues.
>
> WILLIAM SHAKESPEARE

From *Quiet Thoughts*:[10]

Living by Grace

Grace is doing for another being kindnesses he doesn't deserve, hasn't earned, could not ask for, and can't repay. Its main facets are beauty, kindness, gratitude, charm, favor, and thankfulness. Grace offers man what he cannot do for himself. The unwritten creed of many is that God is under obligation to them, but grace suggests that we are under obligation to God. To live in that consciousness is to live by grace. Living by grace is costly; it means sharing. It has no meaning apart from a spirit of self-sacrifice that prompts the soul to think more of giving than of receiving, of caring for others rather than for one's self.

Power to Become

To criticize or to find fault with someone (often) is to fail to see that person in his full possibilities. It is to see his many weaknesses rather than his many strengths. It is an attempt, albeit unconsciously and usually unsuccessfully, to get the other person to conform to our way of thinking. This alienates. The irony is that the critic himself is usually (or often) the one who is blind. To accept people as they are and for what they are, to place confidence in them and to encourage them, is to help them become better than they are. To treat people as if they were what they ought to be is to help them to become what they are capable of becoming. Within every person is the capacity to become something greater than he now is. It is possible for each of us to become better and to help others to become what they ought to be.

The Fellowship of Those Who Care
How grateful we are for churches, hospitals, museums, symphonies and civic centers—to mention only a few of the philanthropic endeavors which are made possible because somebody cares enough to support them. It is not irrelevant to ask whether these organizations are made possible because of us or in spite of us. Are we, as someone has suggested, part of the problem or part of the solution? The world moves forward and progress is made because there are those who care. People who are trying to lift the level of living of those around them soon discover that they have ties which bind them together. This worldwide fellowship of those who care transcends language, color and nationality. The world moves forward because of those who build noble projects and support them.

Joseph Fort Newton's[11] writings are sound psychologically and most helpful religiously. For example, these headings: Live a Day at a Time (p. 24), Forgive Yourself (p. 88), What Can I Do? (p. 104).

There is a collection of quotations by Lewis C. Henry[11] that presents excellent thoughts for contemplation. For example:

Self-Love
He was like a cock who thought the sun had risen to hear him crow.

GEORGE ELIOT

He that falls in love with himself will have no rivals.

BENJAMIN FRANKLIN

Self-love is the greatest of all flatterers.

LA ROCHEFOUCAULD

To love one's self is the beginning of a life-long romance.

WILDE

Kindness
Have you had a kindess shown?
Pass it on;
'Twas not given for thee alone,
Pass it on;
Let it travel down the years,
Let it wipe another's tears,
'Till in Heaven the deed appears—
Pass it on.

REV. HENRY BURTON

Their cause I plead—plead it in heart and mind;
A fellow-feeling makes one wondrous kind.

DAVID GARRICK

Art thou lonely, O my brother?
Share thy little with another!
Stretch a hand to one unfriended,
And thy loneliness is ended.

JOHN OXENHAM

Selfishness is the only real atheism; aspiration, unselfishness, the only real religion.

ZANGWILL

Bibliography

1. BACHELDER, L., *The Understanding Heart*, Peter Pauper Press, Mount Vernon, N.Y., 1966.
2. BERMAN, R., *American in the Sixties*, Free Press, New York, 1968.
3. BOYD, M., *Are You Running with Me, Jesus?* Avon Books, New York, 1965.
4. ———, *Book of Days*, Fawcett, New York, 1968.
5. ———, *Free to Live, Free to Die*, Signet Books, New York, 1967.
6. FORD, B. (Ed.), *The Modern Age*, Penguin Books, Baltimore, 1961.
7. HARRIS, T. A., *I'm OK—You're OK*, Harper & Row, New York, 1969.
8. HENRY, L. C., *Best Quotations for All Occasions*, Fawcett, New York, 1966.
9. HOPKINS, J. (Ed.), *The Hippie Papers*, Signet Books, New York, 1968.

10. McElroy, P. S., *Quiet Thoughts*, Peter Pauper Press, Mount Vernon, N.Y., 1964.
11. Newton, J. F., *Everyday Religion*, Abingdon Press, New York, 1950.
12. Roszak, T., *The Making of a Counter-Culture*, Doubleday, New York, 1969.
13. Scaduto, A., *The Beatles*, Signet Books, New York, 1968.
14. Toffler, A., *Future Shock*, Bantam Books, New York, 1970.
15. Tynan, K., *Oh! Calcutta!* Grove Press, New York, 1969.

Jack Bemporad

The twentieth century could probably be characterized in many ways, but certainly, above all, as the age of secularization. The one assumption that seems to find acceptance in all parts of modern Western civilization today is that man can function adequately without faith.

Yet people throughout the Western world seem to be gripped by a curious malaise. It can be seen in the sudden popularity of the Jesus movement, of Far Eastern religions and mysticism, and in the rage for astrology; it is evident in our literature, which has come to be dominated by a parade of antiheroes; it manifests itself in the increasing mechanization of our art and in the rising tide of drug use that seems to cut across all ages and classes. Certainly it is evident in the burgeoning crime rates everywhere and the ever-increasing ranks of the emotionally disturbed. Life, for modern technological man at any rate, seems to have become one long quest for meaning; a journey in search of values that seems to grow more intense as the descriptive is substituted for the normative in all areas of human existence.

It is within this context that I will try to explore the relationship between Judaism and psychiatry, for these two disciplines could serve as an example of the tension that exists today between faith and science. While they have many common characteristics, they differ in some views on the nature of man, a fact that has had wide-ranging implications not only for the treatment of the mentally disturbed but also for the emotional and spiritual lives of all men in our time.

Some of these differences are inherent in the very nature of the two spheres of thought. Psychiatry is a branch of medicine whose goal is the understanding of the human psyche and the health of the mentally and emotionally disturbed. Its main task is to care for and treat individuals seeking or in need of the help of a psychiatrist. Psychiatry is generally recognized as an important method of treating emotional illness. Also, in spite of differences and disagreements, psychiatrists strive to be guided by the norms and standards of the scientific method in their theory and practice. It is important also to note parenthetically that many psychiatrists maintain that there is no strict relationship between psychiatric theory and practice. Thus most therapists are eclectic and pragmatic in their approach to patients.

Judaism is primarily a religion, with the distinction that it is the religion of an historic people who have survived intact as a cultural group from ancient times until the present. As a religion it affirms certain fundamental beliefs that are neither empirically derived nor subject to any empirical test. Judaism affirms the existence and unity of God, has a prescribed ritual that makes both moral and ceremonial demands, and subscribes to certain basic ideas about the being and nature of God, creation, revelation, and redemption. It has a specific view on the nature of man and the good life.

Judaism appeals to experience to validate its beliefs and maintains that these beliefs enable one to interpret and give significance to life. If there are experiences that seem to contradict or be antagonistic to these fundamental beliefs, for example, the reality of evil in the world, Judaism attempts to re-establish and justify the faith that these experiences seem to threaten. Psychiatry, on the other hand, is bound primarily by the standards of scientific method, and its duty is to follow the facts wherever they lead.

Prior to the sixteenth century scientific and religious philosophical affirmations were in harmony. Both religion and classical philosophy placed at the center of importance the spiritual and intellectual life and emphasized the objective character of man's values. When physics emerged in the sixteenth century, only those things that could be mathematically deduced became real and certain, and mathematics became the only true method for investigation of the laws of nature.

The Galilean description of nature was purely mechanistic, without reference to either values or the purpose of existence. Whereas previously values and purpose rested in a natural order of the universe, they now came to rest solely in man. Soon this mechanism extended itself to the domain of the self, and man found himself an anomaly with respect to nature. The rise of neo-Darwinism in the modern era brought with it the dimming of man's spiritual importance. Whereas in the classical and medieval world man had viewed himself as the image of God and as the culmination of the order and hierarchy of the objective value structure of the cosmos, the man of the modern scientific world could view himself only as a forlorn, homeless, unintelligible entity in a mechanical universe.

Science generalizes, idealizes, and abstracts from our overall experience only those things that can be mathematicized or that fit some theoretical formulation, and thus ignores the greater part of experience. It errs in that it judges only those experiences that natural science can deal with as real and all other experience as subjective and illusory, a view that allows no place for religion. Nowhere is this exclusion more apparent than in some psychiatric views of religion. When psychiatry has chosen to follow the reductionist tendencies of the natural sciences, it has been led to an inevitable clash with religious ideas.

Although there are certain statements in Jewish tradition that would seem to show psychiatric insights (such as the admonition in Leviticus 19 that one should rebuke his neighbor rather than repress his anger and thereby be led to sin) and numerous psychiatric perspectives that seem to be connected with religious affirmation (for example, the works of Jung and Frankl), psychiatry has, for the most part, chosen the path of the natural sciences.

Certain psychiatrists, notably Freud[6] and his followers, have tried to explain religion as the "universal, obsessional neurosis of humanity" or described it as "comparable to a childhood neurosis" (p. 68) that ". . . mankind will overcome as a neurotic phase, just as so many children grow out of their similar neuroses" (p. 71).

The reductionism implicit in this approach, which has been followed by many psychiatrists, not only fails to take religion seriously and accord it its proper sphere but also suffers from the etiological fallacy that the origin of something disproves its value. Freud believed that religion remained in a primitive state and, unlike other areas of human intellectual pursuit, did not keep pace with the general development of Western culture or attain the heights of the arts and science. It is simply prejudice on Freud's part to affirm that religion can only be infantile neurosis or nothing. Why should religion be the only exception to the development of cultural creative awareness? That it is not has been amply demonstrated by Silvano Arieti[1] in one of his recent works, *The Intrapsychic Self.* In one of the best arguments against the reductionist position, Arieti traces the psychological development of religion through the ages and convincingly shows that Freudian reductionism is mistaken in its assertion that religion has failed to evolve and develop parallel to other advances in civilization.

In his discussion of religious and mystical experience, Arieti asserts that the appearance of the supernatural was the result of a special application of teleological causality, by which the world is interpreted as a place in which every act is willed. This presupposes the presence of a person or personified entity to do the willing. This "immanence of the divine in nature" extends to the whole world what a child experiences in his early personal relationships with his mother or father, when every act or object is willed by significant adults in his life.

Arieti summarizes those anthropological and social studies of religion that show that religion, while not having its origins in childhood neuroses, does have its beginnings in childhood emotions. The evolution of religious ideas began, he states, "with the creation of momentary dieties" who represented to primitive man the animated and personified manifestation of every feeling, object, or condition of his life. Just as his parents willed his comforts and discomforts, so primitive man invested his surroundings with willing entities, although they were usually invisible and were felt to reside in the object or activity they represented.

"The idea of physical causality did not develop in primitive man," according to Arieti. "However, as he became aware of the precarious nature of his existence, the feelings of hope and trust which he earlier associated with his mother were experienced in a larger context."

Thus religion can be seen from its origin as a set of cognitive constructs that prolong hope in the survival of the individual and of the small social group to which he belongs. Later, of course, hope is expanded further and embraces the survival of the tribe, state, nation, or human race; or it is focused on eternal survival (immortality) of at least part of the individual (soul) or on general human progress. Faith comes to mean two things: not only belief in the existence of the divinity, but also trust or confidence. Religion is thus not just a way of interpreting the world, but of hoping.

The immediacy and simplicity of primitive life, Arieti continues, was particularly conducive to the conception of numerous gods. Whereas in early primitive life each object or function had a separate god, the same god later became responsible for many objects or actions. Whereas the dieties were originally conceived as residing in the activity or objects of the natural world, they now became abstracted.

This development represented the first great transformation in religion—the move from idolatry to paganism—and while its impetus is unknown and lost in history, its significance was very great. Arieti states, "The principle of teleologic causality still applied (i.e., the world is this way because the gods willed it this way), [but] the gods are seen as more and more separate from the reality of man." They became "a third reality," different from either the reality of human life or the reality of art.

The process of abstraction, however, did not proceed beyond the physical level; that is, it separated the gods from things. It remained for Judaism to effect the even greater revolution, the change from paganism to monotheism. This religious revolution, Arieti says, "later adopted by the whole western world, consisted of further removals from concretization. Whereas primitive religions tend toward the primitive mechanisms of concretizations, paleologic transformations and phantasmic ideation, Judaism, as a rule, emerges as a revolt against these trends. According to biblical account, Abraham is the first man who had the revelation that God is one and invisible. What some consider revelation, others consider insight. Of course, we have no historical proof of the existence of Abraham, the first

Jew. His myth, however, represents the beginning of a very long trend in Judaism against religious concretization and other primitive mechanisms . . . the Hebrew God loses human characteristics and becomes far superior to any other reality. The third reality is conceived as having preceded any other reality; as a matter of fact, as having created anything else which is existing."

The continued development of man was accompanied by still further developments in monotheistic religion, in which religion became inseparably fused with moral life. This was known as the prophetic period, and it was distinctive in that the third reality now became not only precedent to existence and superior to other realities but also the moral standard and guidance for man. All religious history to the present day, including the rise of Christianity, has simply been an extension of the prophetic mandate.

Arieti's outline of the conceptual development of psychological life from prehistoric to modern times shows that religion, far from remaining in a primitive and childish state, paralleled and finally became the foundation for much scientific, intellectual, and artistic achievement. He further refutes the Freudian attack on religion as a useless, neurotic device by showing that faith is a means not only of understanding the universe and one's place in it but also of hoping, drawing strength for the future, and fighting despair. The function of religion as emotional sustenance and inspiration has been ignored by Freud, but its importance cannot be discounted.

Freud's[6] reductionist fallacy is further coupled with a mistaken understanding of the relationship between faith and knowledge throughout history. He maintained that "the more the fruits of knowledge become accessible to men, the more widespread is the decline of religious belief, at first only the obsolete and objectionable expressions of the same and then of its fundamental assumptions also."

Freud maintains that it is knowledge or science that does away with paganism and idolatrous religion. However, as a matter of historic fact, this was not the case. The most significant transformation in religion was from polytheism and paganism to ethical monotheism, and history shows that it was not science that did away with this idolatry and polytheism, but prophetic religion.

The rejection of the mythical is the essence of prophetic religion. It does away with the gods in nature and by doing so, changes man's relationship to God. Prophetic religion is characterized by the idea that man's relationship to God is moral in character; it believes that God makes a moral demand on man. God is understood as pure spirit, separate from the natural world, and no images or representations of Him are possible.

As long as the gods were conceived as forces of nature (gods of fertility, rain, harvesting), worship was aimed at appeasing these natural forces. But when God is seen as beyond nature, then man's relationship to God is not one of appeasement but one of ethical obligation. Perhaps the most explicit statement of this opposition occurs in the Book of Micah. It is the year 701 B.C. and Jerusalem is surrounded, her doom imminent. The people are confused and do not know what to do. How should one propitiate God to avoid the calamity? How shall one appease or influence Him? "Shall one come before God with burnt offerings, with calves of a year old? Will the Lord be pleased with thousands of rams, with ten thousand rivers of oil? Shall one give his first-born for his transgressions, the fruit of his body—will the sacrifice of children appease God's anger?" Micah's answer echoes down through the centuries. "It hath been told thee, O man, what is good and what the Lord doth require of thee. But to do justly, love mercy and walk humbly with Thy God." All the pre-exilic prophets fought against sacrificial appeasement and asserted that God wanted justice and righteousness.

When God is seen as transcendent, as beyond nature, then man's relation to God can be a spiritual one. God is set over against nature as the only true Being, who stands in relation to man as an imperative to action. The contrast between God as a force in nature and God as a transcendent spiritual being is clearly illustrated by an incident in the life of Elijah, the prophet. Elijah went up upon the mountain to seek God and was confronted by a shattering wind and then an earthquake and after that a fire. But God was not to be found in any of these, for God was not a force of nature. After all these natural forces there came a still, small voice. The still, small voice, the inner voice that is the conscience of man, this was Elijah's communication with God.

Once God was seen as a transcendent, unique spiritual being, then the concepts of man and nature became fundamentally different. First of all, the concept of a unified and transcendent God gave rise to the concept of man as transcending nature. Man became not only a natural but also a spiritual being. This can be seen in the creation story. God creates man as a "thou" having a special place in the universe. Eichrodt[4] describes the difference between man and nature quite rightly when he states,

> Man is not simply a piece of nature . . . the earlier account of creation ascribes the clear boundary between man and the animals which prevents man from finding his complement and completion in the subhuman creation, to the effects of man's independent spiritual nature by which he is set on God's side. In man's destiny as a being made in the image of God, the priestly thinker, however, brings together the sayings about man's special place in the creation and gives pregnant utterance to the thought that man cannot be submerged in nature or merged in the laws of the cosmos, so long as he remains true to his destiny. The creator's greatest gift to man, that of the personal I, necessarily places him in analogy with God's being at a distance from nature.

The idea of one transcendent God transformed not only the concept of man but also the concept of nature, and far from being vulnerable to scientific advance, actually made science possible. Since nature was no longer full of gods, it finally became possible to act meaningfully in nature, for there were no more forces to be appeased. Instead of propitiating nature, men could now act to understand and transform it. This is a crucial point, for it made moral action possible.

It likewise made science possible, for as long as nature was full of gods, all separate powers, it could never be consistently apprehended. But with the concept of God as a transcendent creator and the universe as his creation, the idea of cosmos became possible. The world became the matrix of creation, the arena in which one acted to actualize the ideal. Man realized that he had the power to act and transform the universe rather than be helpless before it.

I have gone into such detail on this point because I want to indicate the original creative perspective of the Biblical view, which is totally missed by Freud and many other psychiatrists who deny any upward growth or development in the sphere of religion. What Freud presumes to be a scientific view on the nature of man is, in fact, a type of psychological Hobbesianism. He states, "Insecurity of life, and equal danger for all, now united men into one society, which forbids the individual to kill and reserves to itself the right to kill in the name of society the man who violates this prohibition. This, then, is justice and punishment." Freud takes as a premise the Hobbesian transformation of the summum bonum into the summum malum, a violent death that must be avoided. Man must be protected against his drives by the state. There is a growing despair in Freud, culminating in the view that sees civilization as the veneer that is incapable of ever really doing much against man's destructive instincts.

Fromm,[7] in his *Psychoanalysis and Religion*, tries to defend Freud but also to distinguish between two types of religion—humanistic and authoritarian. Yet he continues the reductionist tendencies of Freud by maintaining, à la Feuerbach, that God is a projection of man. He states, "God is not a symbol of power over man, but of man's own powers," or again, "God is the image of man's higher self, a symbol of what man potentially is or ought to become." There is a bifurcation between the good and bad aspects of man's nature, each projected onto a fictitious being, namely, God: if we are humanists, it is a loving God, and if we are authoritarian, then it is an authoritarian God. In any case God is made in man's image and is a projection of man. It is interesting to note that Fromm reverses himself on this fundamental issue when he begins to discuss idolatry and monotheism. He states,

We forget that the essence of idolatry is not the worship of this or that particular idol, but is a specifically human attitude. This attitude may be described as a deification of things, of partial aspects of the world and man's submission to such things in contrast to an attitude in which his life is devoted to the realization of the highest principles of life, those of love and reason, *to the aim of becoming what he potentially is, a being made in the likeness of God.* It is not only pictures in stone and wood that are idols. Words can become idols, and machines can become idols; leaders, the state, power and political groups may also serve. Science and the opinion of one's neighbors can become idols, and God has become an idol for many.*

Fromm here implicitly accepts Maimonides' definition of idolatry as the worship of the created, and in trying to define idolatry he inverts his own theory by saying that man must become what he "potentially is, a being made in the likeness of God." But isn't God, according to Fromm, merely a projection of man? How then can man be an image of God, when God is an image or projection of man? It is because what Fromm defines as God is, indeed, an idol, (that is, something man creates or projects) that he then has no ground for defining idolatry and thus contradicts himself.

The basic question here then is not whether psychiatry and psychology may not have valuable contributions to make to the psychology of religion, but whether they are limiting their judgment to psychological phenomena only or are also making assertions about the ontological status of the reference of these phenomena.

This issue has been clearly raised by Martin Buber[2] in his reply to Jung. Buber criticizes Jung for maintaining that "God does not exist independent of the human subject," and states that the controversial question is therefore this: "Is God merely a psychic phenomenon or does he also exist independently of the psyche of men?" (pp. 133–134).

Buber[2] continues, "the distinction which is here in question is thus not that between psychic and non-psychic statements, but that between psychic statements to which a super psychic reality corresponds and psychic statements to which none corresponds. The science of psychology, however, is not authorized to make such a distinction; it presumes too much, it injures itself if it does so. The only activity that properly belongs to the science of psychology in this connection is a reasoned restraint. Jung does not exercise such a restraint when he explains that God cannot exist independent of man" (p. 135).

There is no universal agreement about whether Jung rejected the ontological status of a transcendent diety. My own inclination is to side with Buber. (A careful treatment of this issue is to be found in Avis M. Dry, *The Psychology of Jung.*)[3]

Although the psychiatrist may claim to follow scientific procedure, there is one area where he is very close to a religious position, and that goes to the very essence of the nature of religion in general and Judaism in particular.

Judaism claims that man is by nature a religious animal; he feels awe, a sense of mystery, dependence, and wonder. Indeed, he is ready to worship, to exalt, to deify. The problem, of course, is that what man worships most often is not God, but as Hosea says, the works of his hands or the projection of his fears. The crucial task for Judaism is how man can transcend his natural status. How can he become a moral and spiritual being? The religious life for Judaism is the process whereby man can observe certain practices and precepts that will transform the natural into the moral. Thus by having a doctrine of the good life, Judaism makes specific assertions about man's nature, his good, and so forth.

Now in practice psychiatry acts similarly. It is not value-free. It tries to help its patients to live "better lives." Therefore, implicitly it ac-

* My italics.

cepts man's freedom, the capacity to change and the capacity to redirect one's life. Also psychiatry, by and large, believes in the uniqueness and individuality as well as the dignity of man. Here it borrows from the religious tradition and most specifically from Biblical teaching. The belief in freedom is the belief in moral responsibility, and such responsibility requires the formation of character.

But what constitutes character? Is it a necessary agent or a creative agent? The determinist sees character as a link or series of links in a chain of natural causes, existing on an equal level with past and present externalities. But character is not so much the result of circumstances as one's integration or conception of circumstances. It is man's conception of circumstances that enables him to select certain of them as important and others as meaningless.

Further, this selection cannot be considered determined if it also functions as a unifying factor—that is, in terms of an ideal that transcends actual circumstances. Character should be seen as a constant remolding and redefining of the facts; not a static entity, but a dynamic, creative, self-transcending process. Deterministic explanations cannot explain that element of character or self-consciousness that synthesizes the diverse effects that impinge upon man. In fact, it is this very process that makes it possible for psychiatry to function; the individual capacity to integrate one's experience with an ideal and achieve new and higher insights brings about internal change.

That psychological concept of man that embraces a deterministic position, one that makes him no more than a high form of animal bound to his desires and determined by prior causes, has led us to that most dreadful of views on man: the automaton. This is the view that man can be manipulated by external pressures and temptations. It has been in large part responsible for the rise of modern totalitarianism and the use of the propaganda and brainwashing techniques, which have gained such currency today. It is this view that is alien both to psychiatry and religion.

It is not that man should abhor determinism, for without determinism there can be no willing or action; it is simply that self-determination is what he seeks to achieve. Man is a self-conscious being, and this quality essentially distinguishes him from the rest of nature and mediates the circumstances and causes that impinge upon him. Thus his goals and ideals are included in his action on circumstances, and there is a constant redefinition of the constituents involved in the act of choice. The seeking of self-determination is what Arieti calls the will to be human. When man reflects on circumstances, he is really mentally rehearsing his action and considering what its effects would be. The more self-aware he is, the more his reflection will be to the point and thus bring about a change in his environment in terms of his aim and goals.

We can then say that man's character is the presentation to self-consciousness of what man believes, knows, and wills. It is the creative, continuing process that is constantly being reformulated in accord with an idea of the good that man is trying to actualize. Rather than being a static given, to be combined as a mere sum of contingencies compelling action, it is itself actualized in the act. The element of dissatisfaction is here to the point: man is aware of a distinction between what he is at present and what he would like to be. Change comes from the fact of an idea or ideal goal that is put before man by man.

Judaism was the first religious system to affirm that freedom contravened the tyranny of fate. Unlike the philosophy of the ancients, which could never separate good and evil from their interrelationships with fate, Judaism has always asserted the value of repentance—not only for the self-healing of the soul, but also in order to transform the future and transcend the past. Judaism affirms that historical reality is incomplete and, therefore, redeemable. The past may have taken place in fact, but its meaning is not decided until the end of history. If a past act causes us to reconsider and change our life, then it can redeem us and regenerate us.

The first people to talk seriously about repentance were the prophets. The prophets said that it was the end goal that makes the past and present meaningful; in other words, they gave us the idea of the Messianic. Repentance is genuine because the world is in the making; the prophets discovered this and thus discovered freedom. A man who can repent is a man who is free. This is the key meaning of the story of Jonah. Unlike the stories of Noah and of Sodom and Gomorrah, where the righteous are saved and the wicked are destroyed by the decree of God, the meaning of Jonah is that God maintains: The people of the city have repented, they have put on sackcloth and ashes. They have changed their ways, and if they have changed their ways they force me to change my ways. The doctrine of repentance came with the prophets, and it meant that man's action can affect God so that he could annul His decrees and thus eliminate the power of fate. It became man's task to bring God into the world, to bring the kingdom of God about on earth, and thus it made man free.

Moreover, it is this doctrine of repentance that confirms the great psychological value of religion, for it asserts that repentance must be an emotional and spiritual experience in order to be truly redeeming. Rather than being simply an intellectual or secondary process, the formation of character is really possible only through primary process mechanisms, and these mechanisms have always found one of their chief vehicles and supports to be religion.

This point has been well defended by Arieti[1] in *The Intrapsychic Self*. Refuting Freud's point of view as it was expressed in *The Future of an Illusion*, Arieti asks,

> Must we share this view that religion is based on unconscious, primitive and irrational psychic processes? Must we accept the idea that in some of its beliefs religion is a form of collective schizophrenia and in some of its practices a form of obsessive-compulsive psychoneurosis, different from the psychiatric syndromes because it is socially acceptable?
>
> These are possibilities that we cannot easily dismiss just because we do not like them. However, another point of view, not mentioned by Freud, is suggested by our study of the aesthetic process. We have seen that one of the bases of the work of art is the use of the primary process, yet few would deny that art has value and merit. But aesthetic methods, results and values are certainly different from those obtained by pure secondary-process mechanisms.
>
> In a similar way, we may state that religious methods, results and values are different from those derived from pure secondary processes. The religious value consists in giving people faith in the survival of man and man's ideals. Religion becomes an incentive to greatness of the spirit. It offers new insights . . . which open up new dimensions of understandings and feelings. These new dimensions, although they are abstractions constructed at higher psychological levels, need the support of lower mechanisms. We should not make the common mistake of considering the new insights and aims of religion irrational or primitive, just because they are partially founded upon primary process mechanisms.

The adopted primary process, Arieti continues, not only reinforces secondary processes, but often makes insights possible that are in opposition to prevailing attitudes or conventions of the historical period in which they take place, and that therefore might be repressed. A good example of this was the prophets of the Old Testament, who needed a mystical experience to support their mission of bringing the message of justice and love to the people.

"The mystical experience," he states, "transforms uncertainty into certitude, confusion into clarity, hesitation and cautiousness into courage and determination. The mystical experience becomes a revelation and is accepted as a reality by the subject. Therefore of all creative processes, it is the one that seems closest to the psychotic experience. It is a loss of reality, but as we have seen, it is a loss that helps open up new dimensions of reality. Furthermore, it is experienced as a 'third reality,' a reality distant from the first in its attributes and yet always in contact with the first."[1]

Although Arieti concedes that the value of the mystical experience for humanity could be denied on the basis of all the wars, persecutions, hate, and prejudices that have resulted from "illuminations and revelations," he replies, "Results such as these are obviously not to be valued. It must be said that when religion leads to these results it fails, just as bad poetry fails—except that in the failure of religion the consequences are much more harmful. On the other hand, good religious insights become norms for generations to come. By promoting the survival of man or of his ideals, they are recognized as valid by the high levels of the psyche."[1]

He concludes his argument by showing that modern ethical culture has eliminated the supernatural and replaced rite with ethical behavior alone. "Faith in survival is thus based only on the ethical behavior of man and on secondary-process mechanisms. Whether the religious needs of man can be fulfilled without also resorting to the primary-process is a debatable question."[1]

Therefore, we can reaffirm with Arieti that religion, as a quest for meaning, is not an abstract or intellectual pursuit alone, nor is it a reversion to childish emotions and needs. Rather it lies at the very depths of the human self once he begins to ask himself the question of the why of existence. The quest for religion begins when man searches for the meaning of his existence, when he seeks the purpose and significance of his life, and when he judges himself by terms that transcend his finite self.

It is as W. P. Montague[8] so beautifully said:

Religion as we shall conceive it is the acceptance of neither a primitive absurdity nor of a sophisticated truism, but of a momentous possibility—the possibility namely that what is highest in spirit is also deepest in nature, that the ideal and the real are at least to some extent identified, not merely evanescently in our own lives, but enduringly in the universe itself. If this possibility were an actuality, if there truly were at the heart of nature something akin to us, a conserver and increaser of values, and if we could not only know this and act upon it, but really feel it, life would suddenly become radiant. For no longer should we be alien accidents in an indifferent world, uncharacterized by-products of the blindly whirling atoms; and no longer would the things that matter most be at the mercy of the things that matter least.

Bibliography

1. ARIETI, S., *The Intrapsychic Self*, Basic Books, New York, 1967.
2. BUBER, M., *Eclipse of God*, Harper & Row, New York, 1952.
3. DRY, AVIS M., *The Psychology of Jung*, Methuen, London, 1961.
4. EICHRODT, W., *Man in the Old Testament*, Allenson, Naperville, Ill., 1951.
5. FREUD, S., *Civilization and Its Discontents*.
6. ———, *The Future of An Illusion*, Liveright, New York, 1949.
7. FROMM, E., *Psychoanalysis and Religion*, Yale University Press, New Haven, 1950.
8. MONTAGUE, W., *Belief Unbound*, Yale University Press, New Haven, 1930.

John W. Higgins*

THE SUBJECT MATTER under the heading of psychiatry and religion† is most diverse. This variety cannot be decried when it reflects the richness and complexity of the interaction between the two. The differences, however, are also products of factors which limit any approach to the topic. These include the grasp of information pertinent to each sphere, the individual viewpoint from which man and the universe are understood and their relations conceived, and sundry personal biases. The interplay of these factors can produce a maze of quarrels, confusion, and tangents which obstructs scholarly efforts to delineate the problem and to capitalize on the questions posed.

As a guide through this maze, interrelations may be divided into three gross levels: (1) the level of conflict, (2) the level of pacification, and (3) the level of argument. These distinctions are simplifications, since the transition from one to the other may actually be gradual and continuous. Also, the following discussion of them is highly condensed and intended as a representative survey rather than as an exhaustive summary.

The Level of Conflict

It is meant here to describe those essays which are principally characterized by a spirit of hostile attack. Attitudes toward religion, arising from an individual's own experiences, are often rife with prejudice, overt or covert. As a result, quite before any issues are identified, attack and defense may begin. At best, the essential issues have not always remained clear. They have sometimes been obscure because an undeveloped state of knowledge in either sphere has not allowed clear identification of viewpoints. At other times, false issues have grown out of erroneous conceptions of the established doctrines of either or both fields.

A complete survey of all conflicts would be unwieldy, but the following are nuclear issues capable of various translations.

World View

Many controversies ostensibly about psychiatry and religion are more properly seen as fundamental differences of outlook toward the nature of man and the universe. Such differences are properly the subject of philosophical discourse. Unfortunately, empirical data and

* The brilliant young author of this chapter died suddenly February 23, 1968. Only 47 years of age, he was a rare combination of a skillful psychoanalyst and a deeply spiritual man and was able to integrate his knowledge and his social concerns without jeopardizing his identity as a physician. The editors have done well to let his chapter stand as is, for no one with his particular combination of skills has yet appeared on the horizon to replace him.—Francis J. Braceland.

† "Religion" will generally mean, here, the body of beliefs, practices, and doctrines of the Christian tradition: by "psychiatry" will be meant especially, although not exclusively, the body of facts and theories about human behavior which have evolved primarily from the psychological treatment of mental illness. The more particular radiation of interest is from the Roman Catholic Church on the one hand and Freudian psychoanalysis on the other. This statement is not made in a restrictive sense but is simply to make explicit the influences which affect the selection of certain examples, and to some degree inevitably affect the view of the whole topic.

behavioral theories have become not only bludgeons but also victims in what are essentially philosophical arguments. As Philip[19] comments, "It is possible to maintain either that science can only give a partial interpretation of the truth, or to claim that no truth is knowable apart from a scientific approach" (p. 129). Excursions into this kind of conflict are facilitated by a denial that psychiatric theory rests on any preconceived system of ethics and the assertion that it arrives at one by empirical evidence. Zilboorg,[22] addressing this position, has said, "Psychoanalysis . . . found itself able to go along officially without moral values not because it rejected these values but because it carried them implicitly and inherently as everything human carries them" (p. 49). It can be asserted that the truth or fallacy of a belief in God does not depend on scientific demonstration. Nevertheless, it can be anticipated that conflicts centering around world view will continue.[3]

Even when a kind of agreement about the validity of religion is reached, divergency is fully possible. Illustrating this are the definitions of religion by two authorities writing sympathetically about the interrelations of the two spheres. Of religion, Dempsey,[4] a psychologist and Capuchin priest, says it is that

> which treating of God or the absolute, of man's relations to God, comprehends a series of speculative doctrines about the nature and activities of God and the nature and destiny of man, a series of precepts which direct behavior, and an external mode of worship by which the human being as a psychosomatic and social entity gives expression to his inner sentiments and convictions. [p. 34]

Erich Fromm[9] defines religion as "any system of thought and action shared by a group which gives the individual a frame of orientation and an object of devotion" (p. 21). Those who could agree with one of these might well quarrel with the other. The more or less orthodox undoubtedly would find neither acceptable. If respect, although not necessarily belief, is given to the opponent's first principles, at least the lines of reasoning can be followed and the conclusions weighed.

The careful student will beware of misidentification of psychiatry as a religious or ethical system. He should also beware of rejecting all the psychiatric concepts of an individual with whose ethical system he disagrees.

Differences among Schools of Psychiatric Thought

Certainly there is disagreement about the best way to comprehend and treat mental illness; neither is there usually an optimal degree of clarity about the essentials of the theories in the other schools. This expectable state of affairs produces attacks against or briefs for one or another school. Sometimes such arguments are miscarried into discussions about religion and psychiatry.

To select one example, in their evaluation of the philosophical and religious implications of psychoanalysis, Vanderveldt and Odenwald[21] include attacks on the libido theory and the theory of the universality of the Oedipus complex in neuroses, and deal in a critical manner with other concepts such as resistance and transference (ch. 9). In the succeeding chapter they propose their own preferred method of treatment (psychological or psychosynthesis) and imply that it is more acceptable within Christian belief. The arguments about the psychological matters are clearly capable of being waged outside the arena of psychiatry and religion. That, for instance, dealing with psychosynthesis was discussed long ago by Freud (p. 394).[8] The malleability of psychoanalytic theory according to one's preference in the service of "reconciling" it to religious belief can be observed by a perusal of such monographs as those by Dempsey,[4] Gemelli,[10] Lee,[13] and Nuttin,[17] to name a few.

The remedies for such conflicts are clear thinking and an adequate grasp of the psychiatric theory in question.

Reductionist Theories and Pansexualism

The most manifold aspects of behavior and ideation have been demonstrated to be determined in varying degrees and ways by the forces of instinctual drives. This explanation of behavior by tracing it back to instinctual sources has been applied to religion. In the therapeutic situation this has unquestionably been a fruitful avenue of individual inquiry; religious symbols, rituals, and beliefs are fully capable of being invested with neurotic and psychotic processes. The promulgation of such discoveries has caused chagrin among some who seem to accept them as casting doubt on the validity of religion. Yet, the fact remains that these psychiatric findings do not introduce something new to religion. A review of the writings of the mystics, such as St. John of the Cross, will show that religious contemplation for centuries has been concerned with illuminating (and diminishing) the covert self-satisfying aims of religious practice.

The attack proper on religion occurs when, by reductionism, it is proposed wholly to explain some or all of its aspects. Caution on this very practice was expressed by Freud:[7] "If psychoanalysis deserves any attention, then—without prejudice to any other sources or meanings of the concept of God, upon which psychoanalysis can throw no light—the paternal element in that concept must be a most important one" (p. 147). Despite this statement, the general sense of Freud's writings about religion is expressive of his atheism, which he presumably felt to be supported by his discoveries. The religious person will maintain that questions of religion cannot be fully dealt with only in the psychiatric frame of reference, but require also the historical, theological, and other frames. (For a more thorough, critical review of Freud's major explicit statements about religion, reference is made to the monograph by Philip.)[19]

Reactions to the concept of reductionism are often voiced under the heading of "pansexualism." The implication is that "psychoanalysis explains everything by sex." Strictly speaking, this is not so. Furthermore, development of psychoanalytic ego psychology emphasizes other considerations in addition to the drives in comprehending behavior (see Hartmann[11]). If the notion of pansexualism is generously accepted in its loose usage, it should carry with it an appreciation of pregenital sexuality. The moral acceptibility of this connotation is soundly argued by Parcheminey.[18] It should be remembered, "moreover, what psychoanalysis called sexuality was by no means identical with the impulsion toward a union of the two sexes or toward producing a pleasurable sensation in the genitals; it had far more resemblance to the all-inclusive and all-preserving Eros of Plato's Symposium" (p. 169).[6]

Determinism and Free Will

The almost endless debates around psychic determinism and free will arise from oversimplified, if not erroneous, conceptions of what the Church has to say regarding free will and about that which is called determinism in psychiatry. Part of the difficulty appears to have arisen out of equating free will with indeterminism. The complexity of this issue is explored by Dempsey.[4] A study of the viewpoints of religion and psychiatry regarding these suggests that they are not wholly contradictory concepts. For example, Pope Pius XII,[20] in discussing the effect of unconscious instinctual drives, said: "That these energies may exercise pressure upon an activity does not necessarily signify that they compel it" and their force is not to be regarded "as a kind of fatality, as a tyranny of the affectual impulse streaming forth from the subconscious and escaping completely from the control of the conscience and of the soul." From the side of psychiatry, R. P. Knight[12] writes:

> Determinism does not say that causal factors of the past, nor even of recent past, compel a neurotic course in an individual for the rest of his life. It says merely that the individual's total makeup and probable reactions at any given moment are strictly determined by all the forces, early and late, external and internal, past and present which have played and are playing on him.[12]

The similarities in these authoritative viewpoints allow for further investigation without rancor.

The Role of the Therapist

The absence of a concise, universally applicable description of psychotherapy is of legitimate concern to psychiatric research and training in general. This deficiency has allowed the incursion of various notions into psychiatry-religion considerations, a few of which are notable. A general fear is that some moral damage will be dealt the patient. It would be invalid to suggest that *no* "harm" (measured in psychological, social, or religious terms) could come to a patient from contact with a physician who may assume a position of great direct or indirect influence on him. All psychiatrists must always ask themselves to recognize this and to be particularly aware of differences in world view. The psychiatrist must maintain the right of inquiry, but the closer his treatment approaches indoctrination to his points of view, the farther it gets from good psychiatric treatment. (Novey[16] has given an excellent description of an analyst's actual therapeutic approach to problems presented as connected with religion.)

Two allegations about psychotherapy are more common than others. One is that psychiatrists promote a "do as thou wilt," quasi-Rabelaisian attitude. A rejoinder of Freud's[5] to this would probably be subscribed to by most:

> It is out of the question that part of the analytic treatment should consist of advice to "live freely"—if for no other reason because we ourselves tell you that a stubborn conflict is going on in the patient between libidinal desires and sexual repression, between sensual and ascetic tendencies. This conflict is not resolved by helping one side to win a victory over the other. [p. 375]

A second idea, related to the first, is that psychiatrists believe all guilt to be "neurotic" in basis and therefore to be expunged, thereby leaving no inner force of conscience. There are intricate issues here which would require a separate treatise (for example, Zilboorg[23]). This question has been dealt with in a general way by, for example, distinguishing between guilt and guilt feelings—reasonable versus excessive guilt—real guilt versus neurotic guilt. Although these efforts may not have produced the final answer, most psychiatrists would disclaim that they function as "forgivers." Yet they would maintain that their patients often present hypertrophied senses of guilt which must be dealt with in treatment. How to go about this is a technical matter depending upon the theoretical orientation and the goal of the therapy.

The Level of Pacification

Here is meant those interrelations which are especially characterized by the making or maintaining of "peace." There are undoubted fruits to be borne in this atmosphere, such as mutual educational activities, the promotion of treatment facilities, the organization of study groups, etc. There are also inherent hazards, since the attainment of peace may appear so desirable an end that premature and even false solutions of issues may be seized upon and the pursuit of further knowledge thereby hindered. Consequently, it is well briefly to examine some of the ways by which "calm" is obtained.

Dilution of Concepts

There is less likelihood of conflict with nonadherents, the less dogmatic a religion or school of psychiatric thought is. How "liberal" any group can become and still maintain its identity is a matter which its members must settle for themselves.

Denial of Conflict

Insensitivity to issues of the sort discussed earlier ensures a kind of placidity. Somewhat similarly, general statements of principles can suggest an air of agreement which disappears in practice. For example, it can be held that psychiatrists deal with the mind; the clergy, with the soul. Another position is "there can be no conflict between truths." If such statements are taken as keynotes for future endeavors, they are acceptable. It is inaccurate to view them as defining an actual state of affairs. Especially to expedite mutually agreed-upon activities, a deliberate decision to avoid conflictful issues may be made. In the early stages of study groups and colloquia, this is not uncommon.

Selective Acceptance

Concepts of religion or a school of psychiatric thought may be accepted in whole or part, based on how little they seem to threaten the concepts of the other. This can be a valid step, assuming it proceeds from optimal understanding, and especially if it does not mask a conflict of essential issues. Partnerships based on common opposition to another religious or psychiatric system should be approached cautiously.

A psychiatric theory which does not particularly concern itself with motives can appear to leave such matters wholly to religion or its equivalent, and therefore seem "safe" to religion. A psychiatry based on a rationalistic psychology can seem more consonant with certain religions than one which includes an interest in and a certain respect for the irrational. Related to this is an antipathy for the unfamiliar in psychiatric terminology and methods of approach, which is conducive to an acceptance of schools which employ more familiar ones. Religious adherents may become greatly attracted to psychiatric writings if they specifically include religiouslike terms. Probably the best known examples of this are to be found in Jung. Recent reappraisals of Jung from the theological viewpoint support the notion that initial enthusiastic acceptance can be at least uncritical.[4,10,17] When the interrelations between religion and psychiatry are primarily characterized by either conflict or pacification, the possibility should be borne in mind of a blurring of the boundaries between the two, or misidentification of issues.

The Level of Argument

The level of argument means that interaction whose aim is fostering the growth of knowledge. Although in certain areas of each there may be little of immediate legitimate concern to the other, a number of interests of psychiatry and religion overlap or appear to overlap. In a general but simplified sense, this overlapping is the concern for the behavior of man, his motivations, ideals, and limits, and his relations with other men. Each deals with such matters within different frames of reference—how different will depend on the psychiatric theory and religion in question. This difference in frame of reference has often led to the assumption that the concepts of the one would be either inimical or useless to the other. Consequently, there has been the tendency for either to lay exclusive claim to a particular domain and more or less to neglect what the other had to say except to engage in conflict, as previously described.

Arising from and also perpetuating such separateness has been a difference in terminology which has effectively obscured the fact that, upon study, a number of the concepts, some of the ways of formulating problems, and some of the approaches to solution are strikingly similar. The clear identification of such areas requires that the specialist be acquainted with and have a reasonably sound grasp of the pertinent considerations in the other field.

Some in religion are prepared to look to psychiatry for new approaches to a variety of problems; for example, Bishop Marling[15] has blocked out some problems in moral theology, in hagiography, in the clarification of mystical phenomena, and in questions relating to the vocation to the religious life (see also Aumann[1]). Modern psychiatry has not yet significantly tapped the fund of information available under the heading of "religion," about the human condition, ranging from inspirational writings about the aims of living to highly systematized doctrines about the organization of behavior. To name only one set of the latter, there is a rich store of writing concerning the virtues. Those who are prepared to become acquainted with this literature will find potentially useful data for speculative research on the adaptive functions of the ego. The utilization of such information need not be linked to acquiescence to theological belief. At the end of a paper in which he explores in more detail some of the interplay suggested here, Mailloux[14] sounds the keynote of the level of argument: ". . . our knowledge of man is still fragmentary, and . . . only the joint efforts of closely related disciplines, representing widely diversified methodological approaches, can justify our hope for the attainment of a synthesis that will be satisfying to our minds and illuminating for our actions."

This level is clearly the most diffcult to reach. In some ways not enough is known even to formulate questions clearly. If prejudice and ignorance are overcome, psychiatry and religion can represent one of those areas where true division provides an opening to new depths of knowledge.

Appendix

Since the author's untimely demise many events have transpired, but none seem to have materially changed the problems about which he discoursed in this chapter. Presently the nations of the world are in conflict or in fear of it; our own nation is in an ugly mood and hatred and destruction are rampant. Both religion and psychoanalysis are having identity crises, and this is not the time, nor is it the overall mood, for this chapter to be rewritten with any promise of permanency. The same problems that concerned Dr. Higgins under his various subheadings are still being debated and unfortunately are still unsettled.

Just as unfortunately neither religion nor psychiatry have escaped the general unrest. Catholicism has been depleted and saddened by the desertion of some brilliant minds from its clerical ranks. Psychoanalysis is fractionated and in some places denigrated as attention is more and more being focused heavily upon sociological problems as the prime cause of illness.

Most distressing of all is a movement—its shadow presently no bigger than a man's hand—that already can be designated as antipsychiatric. Surfacing during the student uprising in France in 1968, it has received impetus from the writings of Foucault and the works of Laing and Cooper in England. Some traces of it are appearing on our own West Coast and undoubtedly it will spread. Just as surely it will eventually disappear as man's common sense and reasonableness once again assert themselves, but meanwhile they will cause some havoc. It is wise, therefore, to let Dr. Higgins's chapter stand as it is, for he covers the essentials to which we must continually turn our attention.

FRANCIS G. BRACELAND

Bibliography

1. AUMANN, J., "Sanctity and Neurosis," p. 267 in Ref. 2.
2. BRACELAND, F. J., *Faith, Reason and Modern Psychiatry*, Kenedy, New York, 1955.
3. DAY, F., "The Future of Psychoanalysis and Religion," *Psychoanal. Quart.*, 13:84, 1944.
4. DEMPSEY, P. J. R., *Freud, Psychoanalysis, Catholicism*, Henry Regnery, Chicago, 1956.
5. FREUD, S. (1916), *A General Introduction to Psychoanalysis* (Tr. by Riviere, J.), Garden City, New York, 1943.
6. ——— (1925), "The Resistances to Psychoanalysis," in *Collected Papers*, Vol. 5, p. 163, Basic Books, New York, 1959.
7. ——— (1913), "Totem and Taboo," in Strachey, J. (Ed.), *Standard Edition*, Vol. 13, Hogarth, London, 1955.
8. ——— (1919), "Turnings in the Ways of Psychoanalytic Therapy," in *Collected Papers*, Vol. 2, pp. 392–402, Basic Books, New York, 1959.
9. FROMM, E., *Psychoanalysis and Religion*, Yale University Press, New Haven, 1950.
10. GEMELLI, A., *Psychoanalysis Today*, Kenedy, New York, 1955.
11. HARTMANN, H., *Ego-Psychology and the Problem of Adaptation*, International Universities Press, New York, 1958.
12. KNIGHT, R. P., "Determinism, 'Freedom,' and Psychotherapy," *Psychiatry*, 9:251, 1946; reprinted in Knight, R. P., and Friedman, C. R. (Eds.), *Psychoanalytic Psychiatry and Psychology*, International Universities Press, New York, 1954.
13. LEE, R. S., *Freud and Christianity*, Wyn, New York, 1949.

14. MAILLOUX, N., "Psychology and Spiritual Direction," in Braceland, F. J. (Ed.), *Faith, Reason and Modern Psychiatry*, p. 247, Kenedy, New York, 1955.
15. MARLING, J. M., "Opportunities for the Catholic Psychiatrist," *Am. Ecclesiastical Rev.*, *135*:73, 1956.
16. NOVEY, S., "Utilization of Social Institutions as a Defense Technique in the Neuroses," *Int. J. Psychoanal.*, *38*:2, 1957.
17. NUTTIN, J. (Tr. by Lamb, G.), *Psychoanalysis and Personality*, Sheed, New York, 1953.
18. PARCHEMINEY, G., "The Problem of Ambivalence," in Bruno De Jesus-Marie, P. (Ed.), *Love and Violence*, Sheed, New York, 1954.
19. PHILP, H. L., *Freud and Religious Belief*, Rockliff, London, 1956.
20. POPE PIUS XII, "Discourse to the Delegates of Fifth International Congress of Psychotherapy and Clinical Psychology," *Acta Apostolica Sedis*, *45*:278, 1953; English tr., *Linacre Q.*, *20*:97, 1953; excerpts in Furlong, Francis P., "Peaceful Coexistence of Religion and Psychiatry," *Bull. Menninger Clin.*, *19*:210, 1955.
21. VANDERVELDT, J. H., and ODENWALD, R. P., *Psychiatry and Catholicism*, McGraw-Hill, New York, 1952.
22. ZILBOORG, GREGORY, "Psychoanalysis and Religion," *Atlantic Monthly*, *183*:47, 1949.
23. ———, "The Sense of Guilt," *Proc. Inst. for Clergy on Problems in Pastoral Psychology*, Fordham, New York, 1955.

Eilhard Von Domarus*

THE TOPIC OF PSYCHOTHERAPY and Oriental religions is formidable and ideally presupposes a wide knowledge of Western science and Oriental philosophy. But the task facing the West of coming to a greater understanding of the East has become imperative in this era, and, therefore, any attempt that may contribute to such understanding should be welcomed, despite the newness and challenge of some of the concepts it presents.

What do we want? What does the Orient want? What do we, with our scientific-psychotherapeutic methods, hope to achieve? At what does the Orient, with its spiritual-metaphysical speculations, aim? Further, what is the scientific-historical method itself? What is the spiritual-metaphysical method?

The scientific method proceeds from particulars to particulars. We observe a great many wolves being predatory and we make the inductive, "universal" statement that "all wolves are predatory"; from this proposition we deduce that any particular wolf we may see thereafter is also predatory. Such a universal proposition, however, refers to particulars only—not to the experience of a totality involving wolves as components.

The spiritual method teaches that behind the world of seen particulars there is an unseen world which regulates all visible particulars. If we learn to experience this spiritual world of impersonal, moral laws, we can know all the particulars in their significance for the total at once. Wolves are thus seen as necessary part-events of the economy of nature, of the world, and of the cosmos.

* Deceased.

The scientific method is also historical. Darwin's discovery of the natural origin of the species (1859) and Freud's discovery of the psychosexual or psychosocial development of man (1893 and following) may be considered the climax of the scientific-historical method. By this empirical method of science and technology, man attempts to make himself master of his environment and thus feels challenged to make his environment, for his purposes, progressively better and better.

The spiritual method is little interested in particulars, or in race and individual histories. From time out of mind, the Orient believed that the world is a moral universe—a totality, a unity—following a predestined, impersonal order, that man and every creature in it has to play its role in this tragicomedy, and that the environment existed only as the stage for the never-ceasing appearance of the actors.

Thus the scientific Occident heeded almost exclusively its environment, leaving a man's emotions in a sordid, psychoneurotic condition; and the Oriental philosopher heeded his own personal moral, spiritual development, leaving his environment in a sordid, unhealthy condition.

Both Occident and Orient started with the conviction that pain and suffering and disease, even death, were perverse phenomena of nature—they all ought not be. The Occident assumed that all suffering could be eliminated by "progress," by changing the environment of man (including his body and later also his mind) and making it favorable, so that the adjustment of man to it would be easy. The result was that man became a servant of his servants (his machines and his money) and was plagued with a feeling of meaninglessness. What started with optimism is about to end in pessimism.

The Orient assumed that suffering in nature was unavoidable. If there are differences between things, the gain of the one entailed the loss for another: when a spider joyfuly eats up a fly caught in its web, the fly is subjected to the agonies of dying. But, according to the Orient, this world is not primarily a "natural world"; it is, above all, a moral order. When a man plays his role in this world of moral tragicomedy well, he may be spared further agonies; and a man who no longer lives in delusion about himself and the world, and who uses his knowledge with compassion to become a "liberated" soul, will not be born again into this relative world (by way of reincarnation) but will have passed beyond all pain and pleasure, birth and death. Thus the Orient starts with pessimism and ends in optimism.

For the Occident, religion does not seem to be a necessity, but the scientific-historical method does. For the Orient the scientific method seemed to be of little value, but religion always was considered a practical necessity. All this is about to change. Before Freud physicians believed that organic disease may cause emotional distress; after Freud it was seen that emotional distress may cause organic disease. The subtle emotions came to be recognized as causally more important than the gross, material organs. While Freud never followed up this line of thought into metaphysics, his discovery opened up the way toward acknowledgment of the "subtle" as of paramount significance for practical living. Conversely, in the Orient a new understanding of the importance of healthy living, which alone can make a sane religion possible, gained ground in the Ramakrishna-Vivekananda movement, in which compassion is considered more important for a man's salvation than abstract knowledge of the "absolute."

If we ask ourselves what is the soil in which Occidental psychoanalysis and the Oriental's religious desire for liberation from worldly misery evolved, we find that in both cases the man of utter truthfulness—the theoretical scientist and the contemplative saint—found the solution. And the cause of disease and misery was found by both to be the unhealthy (though socially conditioned) living of the individual. The Occident came to discover the conflict arising from undue demands of a traditional superego upon a weakened ego by descriptive, empirical analysis of individual cases, using the method of free association and scientific dream interpretation (interpreting dreams in terms of the life pattern of the individual as he had acquired it from childhood).[4] The Orient came to discover the conflict between the desire for wealth, lust, compulsive duty and the longing for liberation by the Yoga method of concentration and meditation and metaphysically oriented dream evaluation (by which the dream state is considered as important as the waking).

In other words, the problems were the same. But the Occident used the method of free association and scientific dream interpretation (leading to the discovery of dynamic mechanisms of the unconscious), and the Orient used the mystical-religious method of meditation, including metaphysical dream evaluation (leading to the discovery of a transcendent spiritual realm).

To make these statements more explicit, we shall proceed by first discussing Hinduism and Buddhism, and then Taoism and Confucianism.

Hinduism

According to orthodox Hinduism, the aims of a man are either worldly or philosophical-religious. When a man wants to succeed in a worldly way, he may concentrate his energies on the acquisition of wealth (artha), on sensuous indulgence (kama), or on such moral living as will bring him rewards on earth or in a hereafter (dharma). These three aims of worldly living are called together "Trivarga." If, on the other hand, after many reincarnations, after many trials and tribulations, successes and failures, and the sufferings connected with satiety, a man finally aims at liberation (moksha) from the bonds of his existence altogether, he pursues the path of "Apavarga." Let us consider first the ways of Trivarga, following fairly closely the descriptions of Heinrich Zimmer.[16]

Trivarga

ARTHA (WEALTH)

If a man wishes to become wealthy and powerful, he has to follow the path of Matsya-Nyaya, the path described in the Artha Sastras as the Law of the Fishes. This is the law of savage, competitive living among the creatures of the ocean, and, when pursued by men, leads to an imitation of the pitiless laws of nature, unalleviated by moral and religious scruples. Seven ways are described, known as Saman, Danda, Dana, Bheda, Maya, Upeksha, and Indra-Jala. Since for sane religious living a knowledge of the laws of irreligious living—in order to cope with them—is as important as the decision to live a religious life, these seven ways will be considered briefly, as they form the contrasting background against which the desire for religious living arose.

Saman. Saman is the show of courteous behavior intended to conceal hostility, of surface friendliness for the purpose of manipulating someone.

Danda. Danda is "bulldozing," with or without violence. Once your neighbor has been charmed, you may (yourself fully armed) attack him. It is always characterized by aggression and is a logical development from Saman. Manu, the legendary Hindu lawgiver, says: "For the increase of a kingdom, Saman and Danda are the chief aims." Nevertheless, there are auxiliary methods that are important for the "Dark Secrets of the Crooked Way" (p. 123)[16]—or for psychoneurotic, psychopathic living.

Dana. Dana is the method of flattery and bribery, of giving gifts to one whom you like, because he is weak, and not to another whom you dislike, because he is strong.

Bheda. Bheda is the method of divide and rule, of giving gifts to the weaker, ingratiating him to you and separating him from his stronger neighbor, your enemy.

Maya. Maya is the method of external probity, of displaying moral superiority, while under its cover one follows the Law of the Fishes. Maya, which makes much use of name-calling, may also be called the method of anathematizing.

Upeksha. Here a man does not acknowledge the truth when it would inconvenience him. For example, he ignores an accident to which he is a witness because his help would entail a time-consuming court appearance. Upeksha means that we don't follow moral demands by denying an obligation. We are like the man who wore bells on his ears and, whenever he was about to hear something disagreeable, shook his head and heard nothing but his own tune. Upeksha is the method par excellence of the fundamentalist who brooks no disturbance of his preconceived interests but rings his bells every time he hears something disagreeable.

Indra-Jala. This last way of crooked living is employed especially during wartime. It is human creativeness in the service of hatred, the unabashed deception of one's enemy by the use of dummies, spies, misleading statements, etc.

The "Seven Dark Secrets of the Crooked Way" have been described here because it is the conviction of psychotherapists, as well as of religious thinkers, that nothing can be overcome unless it is first consciously and explicitly known. They obviously correspond psychoanalytically to the description of the possessive, sadistic personality of our time that primarily makes use of the superego-ridden masochists to pursue its autocratic ends.

What is the cure for these conditions? First, individual and/or political diagnosis needs to be made; second, its etiology found; then its prognosis; and finally its therapy needs to be outlined. In ancient times the cure was hoped for in the appearance of the Cakra-Vartin, the Great World Savior, the Universal King. He was conceived of as all-knowing, all-good, and all-powerful—but always in a worldly sense. He corresponded possibly to the Superman of Nietzsche. He embodied the collective imagery of his people in such a way that they believed that with him they might achieve the millenium, without him only disaster would overtake them. This "ideal" contrasts sharply with the ideal of the saint, who only wishes to show the right way of living in order to save himself from his own propensities toward Matsya-Nyaya living, as well as to help others to do likewise. Not different is the way of psychotherapy: the therapist tries to show the "castrating" patient that such living will destroy him, as well as the people of his environment, and that in his own interest healthy living is preferable to living according to the Law of the Fishes. Religion does not enter yet.

KAMA

Ramakrishna taught that the greatest obstacles to a man's internal progress and final absolution lay with his lust and greed. In terms of modern psychotherapy that would mean that man's sufferings come primarily from sexual and economic maladjustments. The latter stem from his inherited animal propensities and his desire to continue as an animal in civilized society, following the Law of the Fishes. His sexual maladjustments are also due to this inheritance—his desire to experience sensual excitements without end and, by his early conditioning, also without end of "variation." Hindu philosophy describes the path of man's sensuous desires and their consequences in the Kama Sutras, in which four stages are distinguished: jambha, moha, stambha, and vasa.*

Jambha is the opening of oneself to sensuous desires and then letting oneself be infatuated by the opposite sex. Although the immediate experience of jambha is one of being overcome, actually it is only when our nature permits us to be overcome that such infatuation will occur. If one is not sexually disposed, no such infatuation will take place. That is to say, the cause of the infatuation (contrary to common-sense experience) lies in the experiencer of the infatuation rather than in the "object" of infatuation.

Moha means confusion. Once infatuated, confusion sets in.

Stambha signifies the paralysis of the will—man's will is now stupefied by his infatuation. It imperceptibly leads on to the final stage.

Vasa. The infatuated man is now humbled, wholly subject to his desires. According to this description, the sensuality of the man himself deprives him of his will rather than the object of his infatuation.

In Hindu philosophy jambha has been compared with the flowers that are arrows, moha with a bow, stambha with the bait, and vasa

* The above and following paragraphs have been drawn in part from Heinrich Zimmer. For a fuller description see Ref. 16.

with the hook that catches the fish that has taken the bait. This is an accurate description of the masochistic man who permits himself to be mistreated by a sadistic woman—or, generally, for any "falling" in love. And for the man of philosophy any use of the will except for understanding and liberation is, in Oriental opinion, a falling away from the upward, anagogic path of religion, to pursue which is man's highest aim.

Dharma

Man's Trivarga includes the three worldly aims of wealth and possession (or artha), pleasure and sensuousness (or kama), and finally, on a higher stage of development, the fulfillment of his duty (or dharma). Hinduism asserts that every man determines his own destiny. His deeds form his character and hence have consequences for him (rather than only for others), and these "fruits" are reaped by him either in this or in a future life. Society brings to every man's attention what his duties are, what role he is expected to play in the society in which he is born, according to the merits (or demerits) of his previous incarnations. He may be born as a sudra (common man), as a vaisya (merchant, businessman), as a kshatriya (warrior, politician), or as a brahmin (philosopher, etc.). If he plays his role well in his stratum of society and follows his own dharma, he will acquire merits.

The Hindu mind had little empirical interest in psychology or sociology. The individual developing into a person dominated by artha, kama, or dharma did so because the metaphysically determined wheel of existence made a man what he found himself to be, although as a result of his former actions (which, in turn, appear to be metaphysically determined). The primary interest was in the transcendent, which for the nonphilosophical person remains forever unconscious. For many purposes transcendent means unconscious (or subconscious), but the transcendent is conceived as the metaphysical supraindividuality, whereas the unconscious (or subconscious) is the realm of obscure psychological motivation.

The importance of metaphysics, the science of the unfathomable, for psychology is reemerging, as is indicated by the increasing interest of psychotherapists in religion. Likewise, the recognition of the importance of the subconscious psychology for mysticism is lucidly expressed by Vivekananda (1863–1902):

> Psychology is the science of science. . . . To control the mind you must go deep down into the subconscious mind, classify and arrange in order all the different impressions, thoughts, etc., stored up there, and control them. This is the first step. By the control of the subconscious mind you get control of the conscious.[14]

It is indeed true that psychoanalysis is the best way ever devised to get to know the intimate workings of one's mind—by "archeologically" exploring its development through different infantile stages to maturity as reactions to the parental environment. Thus infantile sucking, biting, anal, phallic, latency, and pubertal phases are shown to exist in each individual. This psychosexual (or, better, psychosocial) development, if left to develop relatively undisturbed by understanding parents, leads to (genital) maturity with the development of a societally determined superego, whereas, when disturbed, it leads to sadomasochism, etc. The latter stages obviously correspond to the man living in terms of artha and kama rather than a healthy dharma.

Is living by dharma, then, the end-all of human development? The answer of Oriental religion is No. It says that man's mind develops still further—not owing to its own laws, but as enlarged and purified by love, compassion, and the voice of conscience. Here is the parting of the ways of orthodox psychoanalysis and metaphysically oriented psychotherapy. We are confronted with the age-old question: on what do ultimate, eschatological truths depend? Do they depend on our "psychology" or does our psychology depend on them?

Apavarga (moksha)

The role of an individual in Hindu society depends on the position or caste in which he is born, which, in turn, is dependent upon his deeds in previous incarnations. What caste he is born into will largely determine the "superego" that he acquires. If he is no longer born as a common man but in one of the three castes, he will ordinarily fulfill his dharma first as a student (brahmacharya), a householder (garhasthya), a retired forest dweller (vanaprastha), and God-realized hermit (sannyas). Here in the liberated sannya we see definitely emerge man's ultimate goal on earth, his Apavarga.

When discussing the laws of Trivarga, we were considering the ways by which a man might adjust himself to his environment. But the religious mind does not rest here. As a matter of fact, even in the earliest times, ultimate, so-called eschatological questions preoccupied the writers of the Vedic hymns, not to speak of the philosophical parts of the sacred writings. The *Svetasvatara Upanishad*, for example, asks the following questions:

> Whence are we born? Whereby do we live? Whither do we go? Oh ye who know Brahman [the Impersonal Reality], tell us at whose command do we abide here, whether in pain or in pleasure. Should time or nature or necessity or chance or the elements be considered as the cause, or He who is called Purusha, the Man, the Supreme Spirit?[8]

According to the existentialists, modern man suffers from a sense of meaninglessness.[13] In the midst of wealth (artha) and comfort (kama) and a collective materialistic security of conformity (dharma), he finds no peace within himself, and anxiety is the most characteristic symptom of our age. Indeed, it is precisely because Western man, at the close of the Middle Ages, shelved (but had not solved) eschatological questions that he became materialistically so successful. But what does this avail if the man himself did not change? And it is precisely because he did not pay attention to his inner life that he developed into the frantic or bored "meaningless" creature of psychoneurotic anxiety.

Psychoanalysis is the empirical, historical approach to this problem. To an extent never even hoped for in previous generations, it has solved many an urgent problem and is, indeed, preparing man for a new form of living. But, among analysts, at least Freud and his followers did not tackle eschatological questions. It is here that metaphysics in general and Hindu metaphysics in particular become of increasing relevance.

The Hindu claims that the experience of a "personal ego" as an ultimate reality is the basis of all false living. It gives rise to innumerable ambitions, all eventually failing ever to be satisfied because there is no ultimate reality to this personal ego, this "I" of daily language. It is the inscrutable power of "false seeing," of the Divine Maya, that conjures up this ego, and ever afterward man is caught in the meshes of "relativity," until, by the growth of the metaphysical sense, he is allowed to discover, or rather to experience directly, the absolute "I," the Brahman, the ultimate source of all existence. And he finds that the *Ding an sich* behind his subjectivity, the Atman, is the same as the *Ding an sich* behind objectivity, the Brahman[7]—that his real I is not the personal I of the relative world but the impersonal I of Brahman. Indeed, if we ask ourselves, "What are we intended for here?" the Hindu answer is, and has always been, "To discover our identity with the Absolute." Says Ramakrishna: ". . . One has attained Perfect Knowledge if one believes in God as sporting as man."[9]

It is impossible to do justice here to the metaphysical achievements of Hinduism and their bearing upon psychotherapy. Psychologically the West, with its empirical-historical approach to all problems, including those of the mind, has added to our knowledge of the mind, and psychoanalysis is the most formidable method yet discovered for the penetration of the mind's workings. But this by no means implies that psychoanalysis has substituted for the Yoga systems (the disciplines of Hinduism) and that the Yoga method of meditation and contemplation has become obsolete. Modern psychotherapy and Hinduism are not mutually exclusive alternatives but, properly understood, supplement each other.

Association and scientific dream interpretation are designed to cure man of psychoneurotic and psychotic fantasies; mediation and metaphysical world interpretation are designed to show man his role and significance in the cosmic play of things, leading to his liberation. Obviously only a healthy mind can possibly live up to the demands of such a metaphysics. If a man believes he can do without psychology, he will continue a life of self-deception; and, if he believes he can live a life without metaphysics, he will inevitably lead a life of dharma at best, or else become an autocrat living unconsciously by the Law of the Fishes. This contradicts Freudianism and orthodox analysis—just as the empirical, psychological approach to the mind will be difficult to acquire for an orthodox Hindu. Such facts, however, should not deter an earnest seeker after truth from integrating both methods into his life and personality.

Buddhism

In contrast to Hindu metaphysics stands "religions" that teach no metaphysics, no God, no soul. These schools (atheism and Buddhism), in contrast to the orthodox Asian schools, have comparatively less instructive significance for us here, as they are more closely related to the Western agnostic way of thinking.

Generally speaking, religions either emphasize the transcendent mystery of the self or the brotherhood of man. The great appeal that Buddha's teaching had for the masses was its emphasis on the brotherhood of man, and (contrary to Western opinion) its fundamental optimism. Buddhism teaches that all is impermanent, and that there is no independent reality behind this phenomenal world of suffering. All living is suffering, an anomaly, a disease; but the cause of the disease can be known—it lies in man's desires for worldly living. These desires can be and are removed by the Eightfold Noble Path of Buddha—right views, right aspiration, right speech, right conduct, right vocation, right thoughts, right remembering of Buddha's teaching (or right mindfulness), and right contemplation.

Buddha thus lays no stress on metaphysics; he proceeds like an empirical physician, making a diagnosis of the illness, finding its etiology and prognosis, and proceeding to therapy. Buddha's therapy leads to the extinction (or nirvana, "blowing out") of a man's worldly desires and the reabsorption of his individuality in the "wheel of woes."

Buddha lived from about 560 to 480 B.C. Buddhism attained its peak in India under King Asoka (about 274–237 B.C.) and was eliminated from Indian soil about A.D. 1250. In the meantime Buddhist monks had spread his gospel far and wide—into Ceylon, Tibet, and China, and to Japan where it gave rise to a special school known as Zen Buddhism.

What is it that makes Buddhism so appealing to the modern materialistic mind (rather than Hindu vedanta)? The interest possibly lies even more in its doctrine of dependent origination than in its emphasis upon the brotherhood of man. In this doctrine Buddha teaches: (1) that all things are conditioned; (2) that all things are for that reason impermanent, subject to change—as all things must change when the conditions upon which they depend change; (3) that for that same reason there can be nothing permanent behind the world of phenomena—no soul in man, no God of the universe, no teachable "metaphysics" that can be positively known; and (4) that the only continuity there is, is by way of the inexorable law of the wheel of woe.

We see here, in this doctrine of dependent origination, or mutual conditioning, besides which there is nothing—no great cause, no God, no soul—very much the attitude of the scientist of today who also says, "That being so, this is so; when that ceases to be, this ceases to be; when that comes to be, this comes to be; and there is nothing besides that which I empirically know—there is no great cause, there is no permanent substance behind all this relativity." There is no room here for speculations; it leaves man meaningless (except that he wishes to escape the wheel of misery), just as modern science leaves man meaningless. Concerning metaphysics, a modern analyst says similarly of a musician-patient: "He never tried in his search after that transcendental and supernatural secret of the Absolute and did not recognize that the great secret of the transcendental . . . is that it does not exist."[2]

My opinion contrasts sharply with this. Being an analyst, fully cognizant of the irreplaceable value of psychotherapy for gaining a true and verifiable understanding of one's mind, I nevertheless believe, from my own experiences, that life cannot be fully understood without metaphysics and cannot be satisfactorily lived without religion. I believe that Hinduism basically offers more to the modern, objective mind than does Buddhism.

Hinduism declares that life cannot be comprehended except as a play, an orderly divine play; that man depends not only on cosmic but also on supracosmic forces, whatever they may be called; that the world is not only relative but absolute; that man is really God manifesting himself as man. This conception of man as God-in-man (or God-as-man) sets man free to use his ego as a representative of the universal "I," and releases his ego from the continuous sway of its own ever doubting mind (his personal consciousness); or the sway of his id (his organic consciousness), which, through greed or lust, drives him into a life of artha or kama; or the sway of the superego (his collective consciousness), which drives him into a life of compulsive duty (or dharma). There is no better way to prepare oneself to live according to such grasp of reality than psychoanalysis.

Taoism

In China two movements similar to Hinduism and Buddhism originated, and at about the same time—Taoism, founded by Lao-Tse (604–531 B.C.), and Confucianism, founded by Kung Fu-Tse (557–479 B.C.). In the following exposition we shall freely follow Edwin A. Burtt.[2]

Tao Teh Ching is the main book of Taoism—Teh meaning virtue; and Tao, being, in fact, an untranslatable word, may be rendered as "the cosmic way" rather than the "ordinary common way" of man. The ordinary way of man is to continue his animal propensities into social conduct. Power, prestige, and possessions (Karen Horney) form his ambition, but not living with the Tao. What is the Tao? It is the metaphysical conviction that man "belongs" to the law of the universe, will come to grief if he selfishly tries to assert an anticosmic, separate existence, but will be reunited with the whole, the Tao, if he will yield his egoism so that the whole (Tao) may be preserved. As water seeks the lowest places but maintains the highest trees, so man, following the laws of natural simplicity, will be humble and unassuming and "feed" the tree of all life by unselfish action. Obviously Lao-Tse taught in terms of his time the eventual misery of the psychoneurotic and, in contrast to this, the happiness of the Teh, the man of healthy living. He taught it in terms of a metaphysics of Tao and Teh rather than in terms of psychoanalysis and psychotherapy.

Confucianism

The second great autochthonous religion of China was founded by Kung Fu-Tse. We usually connect the name of Kung Fu-Tse, or Confucius, with ritualistic, obsessive-compulsive behavior. But Kung Fu-Tse himself was primarily religious, not moralistic. He did, however, claim that only by the discovery of moral righteousness and order would we be able to discover the cosmic order of Yang (heaven), Yin (earth), and their inseparable cooperation. In Confucianism moral life "comes first," is first to be achieved, and it is after that that we discover the morality of the universe and its order; our moral "divinity" reveals to us "God's" divinity, not vice versa. In a sense this procedure is scientific-empirical and therefore close to psychoanalysis. In psychoanalysis it is the discovery of the psychosocial development that prepares a man for finding the divine. In Confucianism it is the discovery of the moral law that prepares a man to discover the divine "unity" of the "universe."

Here again we find that, in contrast to the restless world of the materialist, there is an emphasis upon order—not just a factual but a moral order. The two following quotations from Kung Fu-Tse show the closeness of the basic scientific attitude of Confucianism with the modern psychotherapeutic procedure:

> . . . To discover a systematic order in the world depends upon the achievement of a harmonious order in oneself. . . . If there is integrity within, a unified cosmos is discoverable without; if the moral harmony of a mature personality is in the process of realization within, one finds himself part of a universe in which all things are moving toward the goal of full growth and development.[2]

Here we find the gist of a metaphysically oriented psychotherapy and a psychotherapeutically oriented religion: first discover within yourself the laws of health, and then you will be able to live a sane life in an insane world—and you will discover that what seemed an insane world is really a world progressing toward sanity and morality, which always existed but had to be discovered by your first living a sane life yourself.

In summary, it may be said that it is impossible to do justice to the role that Oriental religion might play in complementing modern scientific endeavors. The difficulty in realizing this importance lies in acquiring enough understanding of both approaches to combine them sensibly in one's own mind. It may be mentioned that that was also the case when the West tried to combine its religion with scientific discoveries. But the attempt to see the unity of purpose in all living may be in progress and foreshadow a new era.

From the standpoint of psychotherapeutic thinking, the question eventually will arise: is there nothing more to the human than his id, his ego, his superego, and his environment? Or do we experience an inner force, transcendental to all this, that, though we remain unconscious of it most of the time, makes itself known at certain times as the voice of conscience and (not to be confused with the demands of the superego) reveals what role we should assume in a situation otherwise seemingly insoluble? Western religious thinkers are now again trying to develop a religion centered on a loving ego rather than a punitive superego, but in nontranscendental terms.[6] However valuable such attempts may be, we must ask ourselves again and again, as did the Hindu sage: "Whereby do we live? Whither do we go? At whose command do we abide here?" And the very attempt to answer these questions will lead us sooner or later to seek out what answer Oriental religion can provide and whether it will help to free us from the widespread thralldom of meaninglessness.

Bibliography

1. AKHILANANDA, SWAMI, *Mental Health and Hindu Psychology*, Harper, New York, 1951. [Reviewed in *Am. J. Psychother.*, 6(4):772, 1952.]
2. BURTT, E. A., *Man Seeks the Divine*, Harper, New York, 1957.
3. ——— (Ed.), *The Teachings of the Compassionate Buddha*, New American Library, New York, 1955.
4. GUTHEIL, E. A., *Handbook of Dream Analysis*, Liveright, New York, 1951.
5. HIRIYANNA, M., *The Essentials of Indian Philosophy*, G. Allen, London, 1949.
6. LEE, R. S., *Freud and Christianity*, J. Clarke, London, 1948.
7. MULLER, F. M., *Indian Philosophy, Vedanta and Purva-Mimamsa*, Vol. II, p. 59, Susil Gupta, India, 1952.
8. ———, *The Six Systems of Indian Philosophy*, Vol. I, p. 8, Susil Gupta, India, 1951.
9. NIKHILANANDA, SWAMI, *Ramakrishna, Prophet of New India*, Harper, New York, 1948.
10. PRABHAVANANDA, SWAMI (Tr. by Isherwood, C.), *The Bhagavad-Gita*, Harper, New York, 1944; reprinted in paperbound ed., New American Library, New York, 1954.
11. RADHAKRISHNAN, S., *The Hindu View of Life*, G. Allen, London, 1927.
12. REIK, T., *The Haunting Melody*, p. 344, Farrar, Straus, New York, 1953.
13. TILLICH, P., *The Courage to Be*, Yale, New Haven, 1952.
14. VIVEKANANDA, SWAMI, *Complete Works, Advaita Ashrama; Mayavati*, Vol. VI, pp. 27, 30, Almora, Himalayas, 1947.
15. ———, *The Yogas and Other Works*, Ramakrishna-Vivekananda Center, New York, 1953.
16. ZIMMER, H., *Philosophies of India*, Campbell, Joseph (Ed.), Bollingen Series, Pantheon, New York, 1951.

CHAPTER 47

LITERATURE AND PSYCHIATRY

Leon Edel

THE COMMON GROUND of literature and psychiatry is the world of irrational being: that is, the study of humans prone to a wildness of the imagination beyond the experiences of everyday life and "the usual"—and an acting out of this irrationality. As psychiatry has addressed itself from the first—from ancient times—to the study and treatment of mental states, so literature has depicted them with great imaginative power. We need only remind ourselves of the passionate and prophetic Cassandra and the horror of her visions; the violence of Lear's rage at the moment of the disintegration of his world; the sleepwalking and madness of Lady Macbeth; or the eternal question of Hamlet's sanity. Nor should we forget Dostoevsky's "idiot" or, in modern times, William Faulkner's empathic portrayal of a mental defective. These are literary cases that not only reveal an intense and powerful observation of mental states; they also represent remarkable intuitive understanding by artists of the workings of the distraught mind. As such, literature has provided verbal pictures of the very stuff—the human stuff—in which psychiatry deals.

Traditionally poets have been considered mad. In their transcendent visions, and in their use of symbolic language, they seem to talk in fables and mysteries. They have been regarded, like Cassandra, as irrational but also as gifted with extraordinary insight. Out of this was born, long ago, the observation of the "daemonic" in man. The individual as one "possessed," facing priests and doctors who must drive out the devils, is a familiar figure in the old dramas, both religious and secular.

Literature has helped establish mythic archetypes, that is, supreme examples, for varieties of mental being and has made psychiatrists aware of mental states beyond those they encounter in daily practice. Medical literature, moreover, contains examples of healers who have themselves been imaginative writers and have recorded for us the mysteries of madness, whether in the sparse annals of primitive societies, the lore of witchcraft and demonology, the occult of the Middle Ages, or the theories of mental being during the Age of Reason—when the "unreasoning" began to be shut away from society. We must remember that *The Rake's Progress* ends up in Bedlam. It was perhaps no accident also that the celebrated Jean Martin Charcot, teacher of both

Freud and William James, should have been an authority on demonology, or that Jung should have conducted studies into the history of alchemy, or that William James, all his life, pursued psychical research in a scientific way.

Journey into the Unconscious

Writers have always felt some sense of mystery in their creations. They have recognized that a subliminal or unconscious self presides over the material they put down on paper; they have known that at a given moment they seem to be able to bring out of dim recesses whole trains of lost associations and memories. They have been aware of their power to take the clutter of life and bring some kind of beautiful order into it. In the East, centuries ago, Buddha spoke of the mind and of the imagination as consonant with life itself. He said, "All we are is the result of our thoughts. It is founded on our thoughts, made up of our thoughts."[13] If we accept these words—and modern psychology tends to bear them out—we find ourselves well on the road to the discoveries of Sigmund Freud and what he spoke of as the unconscious. To be sure, Buddha was speaking of conscious thought. But I suspect he was also speaking of everything that comes into the mind, even the irrational; and I do not think he was saying *Cogito ergo sum*—"I think, therefore I am"—the celebrated dictum of Descartes that gave so pronounced a stamp to the Age of Reason. Descartes referred wholly to rational men: he repudiated the guidance of the senses. The Age of Reason distinctly felt that man's senses tend to mislead, that they interfere with and prevent the observing of reality.

The Romantic Movement, as we know, rejected the enthronement of Reason. Within that movement we can discern a great "inward turning" that ultimately created the intellectual and artistic climate for Freud's discoveries. Goethe argued that the novel is valid only when it portrays man's inmost thoughts; no other form of literature, he said, can more effectively give us access to the inner modalities of a human being. Blake, a precursor of romanticism both in his poetry and in his drawings, found symbols for inner states and cultivated a personal mythology. Coleridge, in glimpsing peripheral states of consciousness (aided by his addiction to opium), could speak of man's "flights of lawless speculation" and man's "modes of inmost being." Rousseau in seeking childhood experience to understand himself, Schlegel and Jean Paul in their quest for the laws of man's nature, Balzac in his recognition that there was an "undiscovered world of psychology," or Hawthorne in his awareness of the "topsy-turvy" world of dreams—all these writers were fascinated by the life of their imagination, its contradictions and ambivalences, its mythic landscapes quite as real and often more real than the human landscape outside themselves in which they moved. Coleridge writes of the "stuff of sleep and dreams" adding, "but Reason at the Rudder,"[2] and Lamb characterizes the difference between night dream and the fantasies of the artist by saying that artists dream "being awake." So, too, writers as different as Dostoevsky, Strindberg, Ibsen, Dickens, Henry James, and Conrad knew before the systematic observations of Freud and his charting of the dream world that they lived with old submerged and disguised realities of their lives, tissued into a fabric of personal mythology. Dickens stumbled on the fact that hypnosis could touch hidden depths and, long before Charcot and Freud, tried in his amateur way to give aid to a mentally ill friend by hypnotizing her. He wrote a letter on dreams in which he argued that these never deal directly with daytime experiences. He explained how they are elaborate transformations—all this half a century before Freud's book on dreams. "If I have been perplexed during the day in bringing out the incidents of a story," he wrote, "I find that I dream at night, never by any chance of the story itself, but perhaps of trying to shut a door that *will* fly open, or to screw something tight that *will* be loose, or to drive a horse on some very important journey, who unaccountably becomes a dog and can't be urged along." He added: "I sometimes think that the origin of all fable and allegory, the very first conception of such fictions, may be referable to this class of dreams."[3]

Between the times of Dickens's insights and the publication of Freud's book on dreams, there occurred a series of movements that were the logical consequences of romanticism—a bursting of the bonds of the rational and the observed—as if the Western world were preparing for the journey into the unconscious. "Realism" and "naturalism" had had their day, but to this period belongs (in literature) the development of the Symbolist Movement guided by the poets Baudelaire, Mallarmé, and Rimbaud, who understood symbolic metaphor and, like their fellows of the paint and brush, sought the *impression* of things, the sensory world, and all that this afforded by way of intuition and subjective experiences. At a later stage, in the realm of psychology and philosophy, Bergson and William James explored in a more rational way psychological or human time, and man's feelings for the occult. No one so far as I know has traced the "climate" in which Freud worked on his dream studies; but we know that in all the arts the "inward-turning" was taking place, and in the so-called decadent movement of the 1890's symbolism had led to curious adventures in synesthesia. Freud's *Interpretation of Dreams* of 1900 brought these currents into focus; it was a key book, like Darwin's *Origin of the Species*. From the period just before the Freudian work, and in the immediate decades that followed, we can date "the modern movement" in art and the direct insemination of surrealism by Freud's explorations of the unconscious after World War I. Surrealism established a conscious link between literature and psychoanalysis when André Breton journeyed to Vienna and dedicated his first book to Sigmund Freud. Author of the surrealist manifestoes, Breton would for years seek to find verbal equivalents for the inchoate stuff of the unconscious, through automatic writing, through recording of consciousness at hypnagogic moments, and through a persistent attempt to tap primary processes. The modern movement in painting would dramatize the visual counterpart of literary surrealism in its quest for the symbolic abstractions of dream material, as in the work of Dali or Chagall.

Interdisciplinary Problems

What value does literature have for psychiatry, or psychiatry for literature, beyond the contribution of each to the culture of the practitioner? There is no question in literature of attempting diagnosis; quite obviously therapy is not involved. Characters described in literature are figments of their author's imagination. We can never know Hamlet's blood pressure, nor his blood chemistry; nor can we usefully say he was paranoid or schizoid or manic-depressive. Literature can, however, suggest to psychiatry systems of value, dimensions of human personality, evidence of the way in which structures of words are part of symbol and myth. Psychiatry on its side offers literary criticism and literary biography valuable guidelines and insights into the psychology of thought, language, and imagination.

The area most common to literature and psychiatry is the exploration of the unconscious, initiated in particular in the therapy of Freud. Psychoanalysis brings psychiatry much closer to literature—and correspondingly, literature to psychiatry—by the very fact that it is concerned with the imaginative faculty of man. Man's way of dreaming, thinking, behaving—here we can say the two disciplines establish a remarkable kinship. In literature this study involves the exploration of verbal forms and structures—the novel, the play, the poem—in which the artist embodies personal and social symbols and myths. In psychoanalysis personal symbolism and personal mythology is studied privately between psychiatrist and patient, for therapeutic ends. Literary theory and literary ideas as well as the substance of fiction, poetry, and drama offer much material by which the psychiatric horizons may be widened. Once we begin to think in terms of man's unconscious, literature and psychiatry become sister disciplines.[8]

Freud's studies in literature were revolutionary: his examination of a minor German novel, *Gradiva*, was an excursion into fictional dream work, and even his speculations on Leonardo,[7] which he so carefully qualified because he was extrapolating from meager data, had within them useful guidance for students of literature. If Freud extended our grasp of Aristotelian catharsis, he went beyond to see how far art represents a territory between wish-denying reality and wish-fulfilling fantasy. This was of enormous help to the biography of art: it made possible for artists a deeper sense of the unconscious promptings in their own work. However, Freud recognized that psychoanalysis has its limitations in such application: it has yet to explain the genesis of art; and he carefully placed this question among the unsolved mysteries of human experience.

Psychiatry and Biography

Literature and psychiatry achieve their most relevant mutual irradiation in the field of biography, that is, the writing of lives of individuals who have aroused in the world a particular curiosity. A corollary field is the study of the process by which such individuals have been enabled to create. We might roughly speak of two categories of biography: writing of "quiet" lives, that is, of individuals who assert themselves by a kind of physical passivity and by tremendous imaginative action; and writing of those who are physically active and seek the world, rather than withdrawing from it in order to verbalize it. Literary biography is concerned mainly with individuals at their writing desk. It has particular value for interdisciplinary study because we find in it the verbalization of so much of man's inner landscape. Human curiosity usually leads us to seek the personality of the poet once we have been excited by his poem—or by his novel or play. The literary work suggests mysteries; it offers, as T. S. Eliot puts it, the "objective correlative"[6] of the poem's creating imagination. We understandably speculate about the poem's immediate source, the human vitality and the mind, the dream-making symbols and myths that have gone into its production. The biography that celebrates a statesman or a general may be concerned with certain of his imaginative ideas, but fundamentally it describes an individual in action—in battle, or in parliament, in a hurly-burly of public life—as against the private life of the writer.

In speaking of the "objective correlative" we recognize that a work of art is in reality indissoluble. It is impossible to recover the intricate threads of memory, association, craft, tradition (and much more) out of which the artistic structure has come into being. Biographers also are forced to consider questions of privacy, even when they are not bound by the oath of Hippocrates; they know that they cannot map all the stages of creativity. The artist scarcely wishes to offer his life history as appendage to his created work. At best all we can hope to do is obtain glimpses of the imagination in action. But precisely by seeking the particular symbolic landscape or imagination, and the particular myth that these symbols express, the psychiatrically oriented biographer is writing the only kind of biography that can be said to have validity. Biographers of James Joyce, who have accepted that writer's own legend and myth (which was that of a world hostile to Joyce), have simply chronicled the illusions by which the Irish writer lived and defended himself. With the knowledge of what life patterns are, and how defenses function, it is possible to assert that a psychiatric biography of Joyce would reveal precisely the opposite of his own legend—that Joyce was extremely hostile to the world. His formidable aggressivity, his explosive thrust, can be read in every line he set down. Privacy of a writer is not invaded in this kind of biography, for our concern is not with his little daily doings (although these, too, may illustrate character and personality), but his whole imaginative being *as he himself expressed it.* A whole generation, misled by *A Portrait of the Artist as a Young Man*, has sympathized with the protagonist, Stephen Dedalus, in his conflict between the flesh and the devil, and the manner in which this is told in the novel; and its readers have been so dazzled by the virtuosity of the writing that they fail to see that the young man of this book is callow and possessed of an unpleasant and distinctly hostile nature. The two disciplines enable us to understand with greater certainty both the aesthetic and the psychological qualities of this autobiographical work and to recognize that however much Joyce shuffled the facts of his own boyhood and youth, he expressed their truest meaning, the deepest modes of his being. This kind of biography, far from being gratuitous, helps us to arrive at truth, the truth of which Goethe wrote when he attempted the story of his own life—even though he characteristically deceived himself in telling his personal story, as all writers must. The biographer, working within the truths of criticism and psychoanalysis, can unmask the self-deception.

Dr. Phyllis Greenacre,[9] in her study of Charles Darwin entitled *The Quest for the Father*, draws a valuable distinction between the roles of the psychoanalyst who wishes to write biography and the psychoanalyst engaged in therapy. It is apparent, she writes, "that the psychoanalytic biographer approaches the study of his subject from vantage points precisely the opposite of those of the psychoanalytic therapist. The latter works largely through the medium of his gradually developing and concentrating relationship with the patient who is seeking help and accepts the relationship for this purpose. The personal involvement and neutrality of the therapist permit the patient to be drawn almost irresistibly into reproducing, toward the analyst, in only slightly modified forms, the attitudes (and even their specific content) which have given rise to his difficulties. In this setting, the analyst can help the patient to become feelingly aware of the nature of his difficulties and to achieve a realignment of the conflict-driven forces within him. Psychoanalysis as a *technique* is distinctly for therapeutic purposes, and is not generally useful for investigating the personality structure of the individual who is in a good state of balance." From this, Dr. Greenacre is led to define, both for the psychoanalyst and for the literary biographer, the precise difference between the writer who works with the living subject and the writer who works from documents. In contrast to the "biography" of the patient formed out of direct confrontation and the transference situation, the "psycholiterary" biographer, if one dares to combine the two, "approaches his subject almost wholly by avenues which are unavailable in therapeutic technique. He has no direct contact with his subject, and there is no therapeutic aim. He amasses as much material from as many different sources as possible. Lacking the opportunity to study the subject's reactions through the transference neurosis, he must scrupulously scrutinize the situations from which the source material is drawn, and assess the personal interactions involved in it. Further, the study is made for the purpose of extending analytic [and we might insert also literary] knowledge and is not sought by the subject."[9]

This helps us clear the ground. It suggests to literature that the Boswell type of biography may need to be reexamined, for the Boswells know their subjects personally and the direct consequence of such relationships between subject and author requires close scrutiny. One would have to examine, for instance, the large biography that Lawrance Thompson wrote of Robert Frost; or that planned by Carvell Collins of William Faulkner; or the one proposed by Richard Goldstone of Thornton Wilder—in the light of their years of intimacy and observation of their subjects. Boswell, worshiping Johnson, is extremely vulnerable as an objective biographer, and there can be no doubt that his formidable work is a biography of countertransference. On this subject Freud long ago offered us valuable advice. Undertaking a biographical speculation in his Leonardo,[7] he remarked that biographers in many cases "have chosen their hero as the subject of their studies because—for reasons of their personal emotional life—they have felt a special affection for him from the very first. They then devote their energies to a task of idealization, aimed at enrolling the great man among the class of their infantile models—at reviving in him, perhaps, the child's idea of the father. To gratify this wish they obliterate the individual features of their subject's physiognomy; they smooth over the traces of his life's struggle with internal and external resistances; and they tolerate in him no vestige of human weakness or imperfection. They thus present us with what is in fact a cold, strange, ideal figure, instead of a human being to whom we might feel ourselves distantly related. That they should do this is regrettable, for they thereby sacrifice truth to an illusion, and for the sake of their infantile phantasies abandon the opportunity of penetrating the most fascinating secrets of human nature."[7]

In an earlier paper I pointed out that Freud omitted the kind of biographer who uses his subject as an expression of anger and hate rather than of love for the "infantile model."[4] This explains the school of "debunking" biography, which arose with Lytton Strachey and resulted in the wholesale smashing of Victorian idols. It simply replaced adulation with hostility, excessive praise with extreme mockery.

The best, the most objective, biography is the one that recognizes the unpleasant as well as the admirable characteristics of genius; that is, the biography that uses its evidence in a "scientific" way. Boswell's stated desire to show us his subject, warts and all, conforms to the scientific ideal of most biographers. Yet their countertransference soon blurs the lines and alters the image, as Freud predicted. What are biographers to do? One hardly can propose they undergo a prolonged psychoanalysis; but they might seek the counsel of the psychiatrist, and recognize that it is precisely here that interdisciplinary relations should be cultivated. It can make the difference between a work created in total worship or in total malice. On his side the psychiatrist who feels a need to write a biography rather than a case history should call on his colleagues in literature to teach him something about literary tradition, proper saturation in the materials, and the cultivation, insofar as his talents permit, of literary art. Such a psychiatrist, we would hope, has himself been sufficiently analyzed to be able to avoid the dangers of countertransference. But there are many instances of superficiality in both disciplines. We can find literary biographies that have modeled themselves on Erik Erikson without fully understanding the relation of art to the unconscious; and psychoanalytical biographies that have ignored the modalities of the creative act. And then psychiatrists often fail to translate their special language into the discourse of everyday life. They write, after all, in the language of their profession, rather than in that of the unindoctrinated reader. A reader without psychoanalytical orientation is asked to understand assumptions, concepts, conclusions, known to anyone who has been exposed to psychiatry and psychoanalysis, but which can only baffle the uninitiated. I have always given as an example of this a paragraph in the first chapter of Ernest Jones's life of Freud[11] in which he writes of Freud as a child during the impending birth of a sibling:

> Darker problems arise when it dawned on him that some man was even more intimate with his mother than he was. Before he was two years old, for the second time, another baby was on the way, and soon visibly so. Jealousy of the intruder, and anger for whoever had seduced his mother into such an unfaithful proceeding, were inevitable. Discarding his knowledge of the sleeping conditions in the house, he rejected the unbearable thought that the nefarious person could be his beloved and perfect father.

This doubtless has meaning in psychoanalysis, but is it valid biography? Surely it never "dawns" on a two year old that some man "is even more intimate with his mother than he was." As I have had occasion to remark, such a precocity would be ready for the couch and would not need play therapy. In a word this passage is written in nonbiographical language. Jones is reading into the consciousness and awareness of the child material that according to Freudian theory exists (we may speculate) in his unconscious. The use of words such as "jealousy" and "seduction" in relation to a two-year-old, even if that child be as gifted as Sigmund Freud, disturbs the reader's sense of verisimilitude. A more skilled biographer would say that Freud, years later, may have felt that in his infantile experience he had undergone a period of disturbance, or a sense of having been set aside, on the advent of another child into the family. And even though he had felt, in the intuitive ways of childhood, that somehow his father was involved in the event, he had concealed this thought that his parent, who seemed to him an ideal and powerful figure, would deprive him of the place he occupied in the very center of his mother's life. In some such way the psychoanalytical ideas can be translated into the language of everyday life.

In my work as a biographer I have found that Freud's observations on slips of the pen can be of enormous help in reading the manuscripts of writers. In the old days these slips would have been ignored; today they can be of extraordinary use in uncovering hidden process. In writing the life of Henry James I first began to pay attention to the unusual sibling rivalry between the novelist and William, his psychologist brother, when I noticed that Henry had made an error in giving his brother's birth date, in a book entitled *Notes of a Son and Brother*. Such errors, I suppose, are not always significant; nevertheless, Freud alerted us to them, and I was led from this slip of the pen to its context, and ultimately to observing that what had always been described as an unusually affectionate relation between two men of genius had in it profound subterranean rivalries. The novelist nearly always chose second sons as his heroes (William was fifteen months older) and usually relegated older brothers to wars, where they were killed off, or sent to mental institutions; and upon the younger rival then fell the mantle and guardianship of family name and honor. The image the world originally had of Henry and William James was that of peers in their creative kingdoms. What was important was to see the many levels of this relationship: the overt affection and concern of the two, their roles in the cultural life of America and Europe, their consistent development, yet each working out his own destiny within a frame of covert infantile emotion, transformed by the novelist into fiction, and often great art, and transformed by the psychiatrist—for William James had taken a medical degree—into observation of human behavior. In order to give an objective picture, I told the story not only from Henry's side (he being my subject) but from William's as well.[5] In this way I offered an interpersonal study. And since what occurred between these brothers was an age-old phenomenon of which psychiatry is well aware, I translated the entire situation into the Jacob and Esau story of the Old Testament, extending the relation I was examining by bringing out its mythic and archetypal qualities.

The most common accusation leveled at the use of psychiatry in literary criticism and biography is that it is "reductive." It is argued that we take the life of a genius and deal with its pathology, or try to show that certain "conditioning" produced the poetic inspiration. The genesis of poetry can be called irrelevant, for the poetry itself represents the essence of the genius. The genesis of any work can be of no concern to a critic; it belongs to the study of creative process. But there is more to be said about the "reductive" nature of any psychological or biographical probing of a writer and his work. If the psychoanalytical student of literary creativity spends his time (as I have had occasion to observe)[4] snorkeling around the base of the iceberg seeking to see what is submerged without looking at the glittering exposed mass visible to the world, then, indeed, the process is reductive. We can always find id explanations for this or that part of a talented life and say that an individual's "orality" or his "cannibalism" has made him keep his large grasp on life in order that he may consume it. The early essays in "applied" psychoanalysis are filled with this kind of inquiry. Nevertheless, it is the visible shape of genius that has encountered the world, and it is the relationship between the submerged and the visible that should be the focus of our study. The inquiry of Thomas Mann in *The Magic Mountain* into man's way of passing through illness into health is an example in fiction of the nonreductive method: that is, the recognition that it is not necessarily the bellyache that inspires the poem, but the poet's determination to overcome the bellyache. Most art of the transcendent kind represents a drive to health, not to illness; and while one could find illness in the lives of many of our greatest artists—Proust's allergies, Virginia

Woolf's depression, Joyce's pathological defenses—what is truly striking is the heroic resistance of their art to death and annihilation, the age-old will to perpetuate the human spirit and the human consciousness. Approached in this way, the psychoanalytical inquiry can be life-enhancing rather than life-reductive.

On the other hand, we know that pathology exists in certain kinds of artists, and that their work reflects illness rather than a drive to health. Psychoanalysis, and the larger field of psychiatry, could do a great deal to help literary criticism understand the nature of pathology. There must be found ways of educating literary criticism to understand that some of the elements of art that may seem highly idiosyncratic are in reality products of alienation and madness, derived from man's destructive side. If we could say with greater assurance that the truly "sick" work of art is profoundly pathological, it might not be praised for the wrong reasons. This would require a large measure of delicacy and insight; and I am certain that most critics today would balk at such instruction, feeling the dangers of "censorship" and recalling the Hitlerian formulization of "bad" art and "good" art. It is not difficult to see the pathology of a book like *Mein Kampf* and to recognize how contagious such pathology can be. But it is more difficult to deal with the writing of the avant-garde, as it calls itself today, which confuses psychedelic experience with natural experience and dwells on the excremental and the obscene in ways that degrade life and only record profound morbidity. In this one must distinguish with great caution between the need for education and for evaluation while making clear censorship is not intended.

Creative Process

The study of creative process need not be gone into in any detail here; a large literature is available on the subject.[1] What we can say is that psychiatry and psychoanalysis has opened up new avenues of insight into the ways in which individuals of genius are capable of transforming life experience into works of art. What is not sufficiently realized is that any direct "imitation" of life is essentially an act of journalism; and that the imaginative transformation is an artistic act involving to a much greater degree man's sensory—that is, his aesthetic—faculties. In the study of the creative process we attempt to enter—however tentatively and always with great caution—the landscape of a given imagination by examining the works brought forth by that imagination. This is the counterpart in literary study to the psychoanalytic examination of dreams, which constantly use man's symbol-making capacities. Works of art are much more than dream: they are dream that has been given conscious shape by the verbal, the rhythms and patterns, the color and tactile senses of the artist,[14] depending on his medium.[12] Some psychiatrists have tended to regard a novel or a poem as if it were a dream. To do this is to ignore tradition, influence, structure, form, the saturation of the writer or painter with the art of the past, and the ways in which the gifted individual mobilizes knowledge and affect within his creative power.

"Creative process" is thus a large term; any study of it requires delicate probing. We must keep in mind always that at best we can arrive only at some crude *schema* or map of our explorations. Nevertheless, literary criticism has performed many subtleties of explication, and when to such criticism is joined the awareness of certain kinds of mental progress, it becomes possible for us to engage in the delicate and humane adventure of exploring a poetic landscape. How delicate and complex the study of creative process invariably is may be judged by Silvano Arieti's[1] bold attempt to define the relations between primary and secondary process and his suggestion that a tertiary or innovative process must also be discerned. The poetic work combines similarities through symbol and metaphor to arrive at the new. The body of knowledge available to us in Freud, Jung, Kris,[12] Arieti,[1] Schachtel,[14] Greenacre,[9] to choose but a few names out of an immense bibliography, makes possible a cross-fertilization of disciplines by which we can arrive at revelations of form and watch the transformations embodied in the literary work. Literary criticism itself is unaware of how much it has learned since Freud in its study of the iconography of literature, although much of this learning suffers from lack of exact psychological and psychiatric knowledge. Literary critics, borrowing from psychiatry, tend to confuse their speculation about manifest content with imperfect understanding of possible unconscious elements. And then they tend to regard symbols as having fixed meanings when they draw on certain symbolic explanations in books devoted to psychiatry. The critic, moreover, must be careful in his disassembling of certain creative elements not to fragment the work of art.

Thoreau long ago recognized that a poem is a piece of "very private history, which unostentatiously lets us into the secret of man's life," and Henry James observed that "the artist is present in every page of every book from which he sought so assiduously to eliminate himself."

In the wake of Freud there have been important exponents of "applied psychoanalysis" who met with varying success. Among them we might mention Jung, Rank, Jones, Sachs, Pfister, Kris, Alexander, Fromm, Greenacre, and in the camp of literature such diverse figures as Cazamian, Badouin, Bachelard, Graves, Edmund Wilson, Maud Bodkin, Trilling, and others. The writings of Jung have had particular appeal to literary criticism because of his study of the nature of myth and archetype. His search for parallels between primordial images and fantasies and contemporary dream material touched the wellspring of poetic experience and his theory of the "collective unconscious," while wholly speculative, had in it a viable attempt to examine the nature of archetypal symbols and fantasies. A striking adaptation of some of his ideas is to be found in the critical theorizing of Northrop Frye, particularly in his *Antomy of Criticism* (1957) where he observes that "literature, conceived as . . . a total imaginative body is in fact a civilized, expanded and developed mythology."

To sum up: psychiatry and literature share common ground—in the interest of both disciplines—in the expression of man's behavioral variousness; in capturing aberrations and idiosyncrasies of thought; in the study of the ways in which the artist projects himself through literary forms. Where in older times these forms were regarded as impersonal, we know them today as embodying and encapsulating the intimate fantasies of the imagination through intricate uses of memory and association and "learned" reactions. The verbal forms of expression may be of help to psychiatry in its pursuit of the physical data; but both disciplines share the pursuit of the biographical record, whether biography as art or biography as case history. The creative imagination and the artist's dream work can take psychiatry beyond diagnosis and therapy in offering projections of extraordinary cases of a highly individual kind. The disciplines are enabled thus to work toward a better knowledge of human creativity when it takes the form we describe as "genius," and to extend thereby the potentials of human creativity.

Bibliography

1. ARIETI, S., *The Intrapsychic Self*, Basic Books, New York, 1967.
2. COLERIDGE, S. T., *Notebooks* (Ed. by Coburn, K.), Pantheon, New York, 1957.
3. DICKENS, C., *Selected Letters* (Ed. by Dupee, F. W.), Farrar, Straus, and Cudahy, New York, 1960.
4. EDEL, L., "The Biographer and Psychoanalysis," *Internat. J. Psychother.*, *42*:4–5, 1961.
5. ———, *The Life of Henry James*, 5 vols., Lippincott, New York and Philadelphia, 1953–1972.
6. ELIOT, T. S., "Hamlet," in *Selected Essays*, new ed., Harcourt, Brace, New York, 1950.
7. FREUD, S., "Leonardo da Vinci," in Strachey, L. (Ed.), *Standard Edition*, Vol. 11, pp. 107, 130, Hogarth, London, 1957.
8. FROMM, E., *The Forgotten Language*, Reinhart, New York, 1951.

9. GREENACRE, P., *The Quest for the Father*, International Universities Press, New York, 1963.
10. JONES, E., *Hamlet and Oedipus*, Doubleday, New York, 1955.
11. ———, *Sigmund Freud: Life and Work*, Basic Books, New York, 1960.
12. KRIS, E., *Psychoanalytic Explorations in Art*, International Universities Press, New York, 1952.
13. ROSS, N. W., *Three Ways of Asian Wisdom*, Simon & Schuster, New York, 1966.
14. SCHACHTEL, E., *Metamorphosis*, Basic Books, New York, 1959.

CHAPTER 48

PSYCHIATRY AND HISTORY

Bruce Mazlish

Psychiatry per se has been defined as the medical study, diagnosis, treatment, and prevention of mental illness. Since most of the individuals and groups studied by historians cannot be classified as "mentally ill," and are certainly not amenable to treatment, psychiatry would appear to have little applicability to history, even as a diagnostic aid. Only when psychiatry itself is broadened beyond mental illness into a kind of general psychology can it, even in principle, become available for significant use by historians and social scientists.

Presumably such broadening has taken place in the development of psychoanalysis by Freud and his followers. As one standard history of psychiatry has put it, "In our century a scientific revolution has taken place: psychiatry has come of age. . . . This advancement . . . became possible only after Freudian discoveries transformed psychiatry and penetrated general medical thought."[3]

Freudian psychoanalysis has two outstanding features that make it highly attractive for the historian who attempts to use it in his own work. First, although originating in psychiatry, it claims to be a general psychology whose observations apply as much to normal as to abnormal people, to mentally healthy as well as to ill personalities. Discovering unconscious mental processes, and the "laws" that hold good in that realm, in the course of offering treatment and therapy to patients, Freud extended his findings to nonpatients: in short, to all humanity. Historians can feel comfortable with such a conclusion.

Second, Freudian psychoanalysis is itself an "historical" science. That is, many of its procedures and methodological assumptions are similar to historical ones. In fact, it has often been noted that psychoanalysis deals with *personal history*. In any case, as Hans Meyerhoff[40] has so well illustrated, both psychoanalysis and history deal with materials from the past, seek to "reconstruct" a pattern of events from fragmentary data, offer an "explanation" based on the totality of this reconstruction rather than on general laws, and are essentially retrodictive rather than predictive disciplines.

Nevertheless, with all their similarities psychoanalysis differs from history in one essential aspect: it does claim to be a generalizing science. Although its explanations are not offered in lawlike formulas, but rather in terms of a holistic reconstruction, psychoanalysis approaches its materials with a general theory of its own (initially derived, of course, from clinical observations), while history does not. The postulate of an unconscious, the dynamics involved in repression, resistance, and transference, the mechanisms of defense and adaptation utilized by the ego, all these and many other processes made familiar to us by psychoanalysis constitute an effort at a systematized science.

As a new kind of science or, to pitch the claim lower, as a discipline offering a more cogent and systematic way of understanding personality, it can give the historian a means other than his mere intuition or common-sense psychology by which to explain the motives of historical individuals or groups. And since motive is a key factor in much historical explanation, the historian finds himself more and more drawn to the use of psychoanalysis in his work. When to this is added psychoanalysis' concern for unconscious as well as conscious thought processes, which is unavailable to any other psychological or psychiatric school, we can see why historians have concentrated almost exclusively on some variant of psychoanalysis when they have come to "apply" psychology to history.

In its "applied" form the conjoining of psychoanalysis and history is nowadays frequently referred to as "psychohistory." A number of scholars are unhappy with the term—it seems, for example, to exclude other social scientists—and the names psycho-social science and psycho-social history have been suggested; but psychohistory appears to be gaining general if reluctant acceptance. In any case the new discipline, or interdiscipline, is misleadingly thought of as the mere "application" of psychoanalysis to history, for, to a great extent, psychohistory turns to the sociological, demographic, and economic fields as well for its materials and theories. Ideally, too, psychohistory allows for the "application" of history to psychoanalysis, in an effort to reexamine the validity and variability of the latter's concepts and theories in an historical context. In short, though emerging out of the application of psychoanalysis to history, psychohistory claims to be a true fusion and intermingling of the two.

Freud's Work in History

Freud himself pioneered the application of psychoanalysis to history. Apparently around 1910 his interest in the subject came first to flood tide, culminating in the publication of *Leonardo Da Vinci* in that year and *Psychoanalytic Notes on an Autobiographical Account of a Case of Paranoia* (*The Case of Schreber*) in 1911.

A word about the latter first. It is usually not mentioned in the context of Freud's historical reconstructions, but it definitely should be. In the summer of 1910 (thus, just after completing his Leonardo manuscript), Freud's attention was caught by Schreber's *Memoirs* (published in 1903). By the end of the year he had finished his new manuscript. In this work Freud made his first analysis of an actual "historical" figure on the basis of an extensive autobiographical document, thus anticipating a fundamental procedure of psychohistorians working on life histories (for example, Erikson[14] using Gandhi's *Autobiography* in *Gandhi's Truth*). He also gave an instance of how one could use such materials, penetrating through their censored and distorted nature. While Schreber was essentially a "patient," that is, mentally ill, Freud showed how to work with materials removed from the clinical situation itself.

Leonardo, of course, to which almost all attention has been directed in this context, also appealed to a "memoir," but this was only a short childhood memory, inserted almost accidentally into Leonardo's scientific notebooks. The memory concerning what Freud called vultures, added to various books on Leonardo (such as Merezhkovsky's study and Scognamiglio's monograph on Leonardo's youth) and to Leonardo's own paintings, served as the documentary basis for Freud's historical analysis. It embodies, as James Strachey[49] observes, not only the first but "the last of Freud's large-scale excursions into the field of biography" (though we might wish to qualify this statement by adding the controversial work, in collaboration with William Bullitt, on Woodrow Wilson).

Freud's psychobiographical study of Leonardo is too well known to need summarizing here. Unlike the Schreber that followed, it dealt with a dead person as well as a historically famous one. It has also aroused a good deal of controversy. For example, Freud made the mistake of accepting as the translation for *nibbio*, Italian for "kite," the German word for "vulture," and then proceeded to offer recondite myths about vultures to confirm his general analysis. So, too, as the eminent art historian Meyer Schapiro[47] has pointed out, Freud took as particular to Leonardo's paintings (for example, the raised finger of John the Baptist) what was, in fact, common to all the iconography of the period. In his *Leonardo Da Vinci*, however, K. R. Eissler[12] has sought to respond to these and other criticisms, and the interested reader must be referred to that magisterial book. For our purposes we need not pass final judgment on Freud's work here, but merely note that it was a pioneering effort (along with similar attempts at psychological biographies by Hitschmann, Sadger, Stekel, and so forth) to establish the possibility of psychoanalytic study of historical figures, illustrating at the same time the numerous difficulties and problems attending such an effort.

It must be mentioned here, nevertheless, that the limitations of Freud's approach are highlighted in the last work in this genre to which he lent his name. *Thomas Woodrow Wilson: A Psychological Study*[18] by Freud and William Bullitt first appeared in 1967, long after Freud's death. It purported to be a collaboration begun around 1932 and finished by the end of Freud's life in 1939 (though held back from publication at that time). Most commentators are willing to agree that the Introduction is by Freud and that the rest is based on conversations Freud had with Bullitt; they also feel that Bullitt, in the writing, distorted some of Freud's views and removed all the subtlety of his interpretations. Nevertheless, Freud must be held partly responsible for a work that treats its subject—a highly creative political leader—as if he were *nothing but* a clinical patient. By its emphasis on the pathological, the book shows how such efforts to apply psychoanalysis to history can lead to sheer reductionism.

The weaknesses of the *Woodrow Wilson* book must not, however, obscure the fact that Freud not only discovered the science of psychoanalysis but also pioneered its application to historical materials. One other work of his, the essay on *A Seventeenth-Century Demonological Neurosis* (1923), should be mentioned since it inspired a whole host of further researches concerning witchcraft, millenarian movements, and so forth.[8,9,10] In sum, whatever the particular and understandable lapses in Freud's historical work, he opened the way for others to learn by his work and to carry on further in the direction he had set.

Disciples and Developments

Most of the initial attempts to carry on Freud's work were undertaken by his disciples, who, while professionally trained in psychoanalysis, had, not surprisingly, only amateur interests in history. Thus, quite naturally, they always teetered on the edge of reductionism. The next thing to note is that the shift in emphasis in psychoanalytic theory, from concentration on id processes to ego and superego processes, and from infantile sexuality to adolescence or adulthood, seemed a necessary prelude to further advances in the direction of psychohistory. We can best illustrate these two points by brief comments on some of the disciples and developments in psychoanalysis that followed Freud.

Jung and His Influence

In his deviations from Freud, Jung offered an alternate set of terms, although often for the same data (for example *anima* and *animus* instead of bisexuality), and stressed the collective unconscious, especially as manifested in archetypes and symbols. In his own work Jung carried out recondite investigations into the history of alchemy, mandala symbolism, and so forth. Although these studies are "historical" in nature, they cannot be viewed as steps toward psychohistory itself. In principle the stress on the collective unconscious might be useful to historians as they struggle to help create a group psychology applicable to the analysis of group phenomena; but in practice nothing of significance along these lines has emerged. Similarly Jungian analysis might be applied to individual life histories; and it has been done so significantly in at least one book, Arnold Künzli's *Karl Marx*.[27] Generally, however, historians have not resorted to the Jungian approach when they have used psychoanalytic concepts and theories.

Adler and His Influence

Adler's theories of inferiority and superiority, and especially of overcompensation, and his stress on considering the individual in his social setting has had more resonance than Jung's work among historians and social scientists. Partly this may be because social scientists are much concerned with issues of power and therefore welcomed an alternative to Freud's emphasis on the sex drive. In any case, as early as 1930 the political scientist Harold D. Lasswell[30,31] pioneered in the study of "psychopathology and politics," to take the name of one of his books. In his work Lasswell used life-history material from patients in mental hospitals and from volunteers with no obvious mental pathology, both, however, involved in politics, to establish a classification of "political types" (for example, agitators, administrators, and so forth) and to try to "uncover the typical subjective histories of typical public characters." Emerging from this data with the postulate that politicians were in search of power and that their "most important private motive is a repressed and powerful hatred of authority," Lasswell developed his famous formula "p } d } r = P, where p equals private motives; d equals displacement onto a public object; r equals rationalization in terms of public interest; P equals the political man; and } equals transformed into."

Exactly how the displacement of private affects upon public objects takes place has turned out to be a more involved problem than was originally thought, with the intervening links difficult to trace in the case of actual, functioning politicians. Nevertheless, Adler's influence, mediated through Lasswell's formulations, has continued to inspire political scientists and, to a lesser degree, historians as well. Thus, Alexander and Juliette George,[21] in their exemplary study, *Woodrow Wilson and Colonel House*, weave much of their interpretation around the notion that Wilson's repressed hostility toward his father found displaced expression in many of his political struggles, where his "taste for achievement and power" could find satisfaction. Needless to say, the taste for power was rationalized in terms of a lofty idealism: Lasswell's "rationalization in terms of public interest." The Georges conclude that in Wilson's political behavior, "power was for him a compensatory value, a means of restoring the self-esteem damaged in childhood" (p. 320).

In another direction, staying close to the Adlerian notions, but applying them to analysis of group phenomena rather than individual life histories, one can mention the highly suggestive book by O. Mannoni, *Prospero and Caliban: The Psychology of Colonization*.[35,36] Mannoni analyzes the interaction of two different personality types—the Western colonizer and the native colonized—as highlighted by the uprising in Madagascar in 1947. He sees the colonizer as asserting superiority to overcome his fears of inferiority and projecting his own unconscious fears onto the natives. The latter, in turn, have a dependency complex, derived from their cult of the dead, that conditions their relations to the colonizers. The contact of these two personality types, and their mutual incomprehension, makes for the colonial experience, which Mannoni seeks to analyze in detail. Although he claims that "if one had to reduce the psychological theory to one system, I believe one could do it by applying the ideas of Karl Abraham, and especially of Melanie Klein" (p. 33),[36] it is clear that in his borrowing "from various schools of psychology," Mannoni leans heavily on Adler and his notions of inferiority and superiority, especially for the colonizers, though putting these notions very much to his own particular usages.

Sullivan and Fromm

Elsewhere in this Volume (Chapter 40A), mention is made of a line of development seen in Durkheim, Cooley, Mead, Lewin, and Sapir that "reached conceptualization in the work of Henry Stack Sullivan, with its influence clearly revealed in the writings of Erich Fromm, Frieda Fromm-Reichmann, Karen Horney, Clara Thompson [and others] . . ." (p. 843). What was this development, and what were its consequences for psychohistory? We shall take Sullivan and Fromm as the prototypes for what was involved.

For our purposes we need only highlight a few of Sullivan's emphases to illustrate certain developments in psychoanalysis. Sullivan stressed the following: (1) man must be viewed primarily as a socially interacting organism, although he is biologically rooted; (2) man is "not a fact but an act," that is, he develops and changes in a continuous process; and (3) his psychic states, for example, anxiety, are the result primarily of interpersonal relations (which are determined largely by his particular society and its socialization processes) rather than intrapsychic conflict.

Clearly Sullivan's shifts in emphasis from the classic Freudian position favored the study of man in society and developing over time, in contrast to the analysis of an individual in a relative vacuum. As such it would seem to be congenial to the work of historians. Certainly it influenced other analysts, such as Fromm and Horney, to explore the way particular societies created particular character types, for example, a "marketing character," or a "neurotic personality of *our time* (my italics)." Strangely enough, however, Sullivan's developments seem to have had little *direct* influence on historians per se, although his work undoubtedly affected the climate of opinion in which they worked. Perhaps this was because, in spite of its differing conceptual stresses, it really offered historians no tools or operational theories separate from the orthodox Freudian ones with which to work.

The outstanding example of history psychoanalytically informed along the lines of Sullivan's thinking was the work of an analyst, not an historian: Erich Fromm's *Escape from Freedom.*[20] The influence of this wide-ranging book has been rather extraordinary. Published in 1941, and obviously influenced by the Nazi phenomenon of the time, the book has enjoyed numerous reprintings.

Fromm conveniently states both his intention and his thesis at the very beginning of his work. He intends the book to be part of a broad study "concerning the character structure of modern man and the problems of the interaction between psychological and sociological factors." The Sullivanian overtones are clear and later openly acknowledged in various places (although it must be noted that by 1955 Fromm, in his *Sane Society,* turned against Sullivan). Fromm's thesis is that "modern man, freed from the bonds of pre-individualistic society, which simultaneously gave him security and limited him, has not gained freedom in the positive sense of the realization of his individual self; that is, the expression of his intellectual, emotional and sensuous potentialities. Freedom, though it has brought him independence and rationality, has made him isolated and, thereby, anxious and powerless. This isolation is unbearable and the alternative he is confronted with are [*sic*] either to escape from the burden of his freedom into new dependencies and submission, or to advance to the full realization of positive freedom which is based upon the uniqueness and individuality of man" (p. viii).

Fromm has been influenced by a number of different sources, and a few of them need to be remarked upon. First, Fromm had studied sociology (receiving a Ph.D. from Heidelberg in 1922), rather than history, before entering psychoanalytic training. Thus, he had professional competence in at least those two fields. In *Escape from Freedom* he borrowed heavily from sociological theories concerning man's alienation from modern industrial society, and one detects heavy echoes of Tönnies's division of Gemeinschaft and Gesellschaft, Durkheim's anomie, and Max Weber's general analysis of capitalist society and values. (Incidentally a splendid psychohistorical analysis of Weber, bearing on exactly the issues propounded in Fromm's thesis, is Arthur Mitzman's *The Iron Cage.*)[41] However, Karl Marx seems to be the outstanding influence. Passages in *Escape from Freedom* seem to read almost as quotations from *The Communist Manifesto*, as when Fromm talks of how capitalism "helped to sever all ties between one individual and the other and thereby isolated and separated the individual from his fellow men."

To his sociology, strongly Marxist-colored, Fromm adds psychoanalysis, heavily tinted by Sullivanian hues. He begins, however, by postulating a "drive for freedom" that is rooted in the individual's necessary "emergence from a state of oneness with the natural world." Fromm describes this earlier state as involving "primary ties," which, although affording security and a feeling of belonging, must be broken. The result is that the individual now feels his freedom as isolation, as "a curse." Two resolutions are open to him: he may turn to authority and slavishly submerge himself in a group, that is, "escape from freedom"; or he may embrace the "one possible, productive solution for the relationship of individualized man with the world: his active solidarity with all men and his spontaneous activity, love and work, which unite him again with the world, not by primary ties but as a free and independent individual."

Fromm does not state his insights merely in sociological and psychoanalytic terms; he places them in the context of an historical analysis. In a long chapter on the Reformation, he tries to show how the "capitalist" individual broke his "primary ties" during a specific historical period. Similarly, in a chapter on the psychology of Nazism, he seeks to show how in the twentieth century the escape from freedom into authoritarianism took specific shape in Germany. In short, he offers a sort of psychological history of modern times.

As psychological *history* his work paints with a broad brush in a way that might leave many historians filled with misgivings. For example, Fromm asserts without much real use of hard historical data that medieval man, in spite of many dangers, "felt himself secure and safe." This hardly accords with other views of the medieval period, where anxiety seems endemic. If 1348–1349 is still "medieval" (and Fromm makes no effort to be precise), then one must reckon with the psychic consequences of the Black Death, as William L. Langer so eloquently reminds the historian in his ringing invitation to the application of psychology to history, "The Next Assignment."[29] Fromm seems also to assume a "middle class" in the medieval period; most historians would judge this as present-minded. On a broader issue, "escape from freedom" in the twentieth century seems less related to highly developed liberal capitalist societies, such as Great Britain and the United States, than to latecomers to capitalism, e.g., Germany; to incipiently industrialized countries, e.g., Italy; or to backward and underdeveloped countries, e.g., Czarist Russia. Such questions suggest that Fromm's work deals more with sociological categories than with concrete historical data, and historians have accordingly resisted following it.

As psychology and sociology, which is what Fromm himself primarily intended his book to be, it has been more successful in instigating further work. Fromm's psychoanalytic interpretation of the escape from freedom into authoritarianism as being rooted in sadomasochistic strivings has found its echo in such large-scale investigations as *The Authoritarian Personality*, by T. Adorno *et al*,[2] and in specific studies such as William Blanchard's[5] book on Jean-Jacques Rousseau. Fromm's chapter on "The Psychology of Nazism" has anticipated a flood of studies on Nazism, Nazi anti-Semitism, and Adolf Hitler.[1,19,28,48,51] His attention to the "person who gives up his individual self and becomes an automaton, identical with millions of other automatons around him," points directly to David Riesman's *The Lonely Crowd.*[45]

Thus, whatever its own limitations as history and therefore psychohistory, *Escape from Freedom* has been a seminal book in inspiring related studies. Keeping steadily in mind the injunctions of Sullivan's version of psychoanalysis, Fromm has sought to deal with individuals as interacting with other individuals in a social and historical setting. Above all, he has shown others how to avoid sheer reductionism, where everything becomes translated into psychology. As Fromm comments, "Nazism is a psychological problem, but the psychological factors themselves have to be understood as being molded by socio-economic factors; Nazism is an economic and political problem, but the hold it has over a whole people has to be understood on psychological grounds." An insight such as this, worked out in terms of actual data, as Fromm has attempted it in *Escape from Freedom*, tries to give historical life to the changes in emphasis brought to psychoanalytic theory by Sullivan and his co-workers. It also opens the way for a truer fusion of psychoanalysis, sociology, and history.

W. Reich and Marcuse

A brief word must be added about some contributions spiritually related to Fromm's efforts. Indeed, Wilhelm Reich, whose work has unexpectedly come into prominence recently (see, for example Robinson's *The Freudian Left*),[46] predates Fromm. Reich's contributions to psychoanalysis carried Freud's theories to their two extremes. On one side Reich stressed the biological, that is, the libido, which he tried to measure quantitatively in biopsychic energy, practically reducing the sexual to the merely genital. On the other side Reich emphasized the social, insisting on the unique importance of social and historical factors in psychic development. Thus, in his theory of "character neurosis" Reich focused attention, not on particular symptoms, but on the patient's total character structure, seen as the result of his entire personal and societal history.

In his major contribution to "historical" studies, *The Mass Psychology of Fascism* (1933),[44] Reich turned to Marxism as the key to the social factors and tried to fuse Marx and Freud. In this book Reich attempted to delineate an authoritarian character structure, brought into being as a result of bourgeois economic and social developments. Suggestive, the book is generally not judged successful; and it is still almost unknown to most historians.

Herbert Marcuse is a nonanalyst who has followed in the footsteps of Reich and Fromm, trying to synthesize the work of Marx and Freud. His *Eros and Civilization* (1955)[37] represents the work of a philosopher and a political theorist and makes no appeal to clinical evidence. However, Marcuse holds fast to the Freudian emphases on childhood and on sexual repression and accuses the neo-Freudians (such as Fromm) of watering down or ignoring the fundamentals of psychosexual development. In his very difficult book Marcuse attempts to place repression in an historical dimension and to show that sexual repression under capitalism is *surplus* repression, that is, the equivalent of Marx's surplus value. He also analyzes the "performance principle" as operating in the service of capitalism by desexualizing the pregenital erogenous zones. (Thus, Marcuse is here also criticizing Reich's emphasis on genitality.) Although accepting the necessity of a bare minimum of repression, Marcuse seems to look with favor upon a return to "polymorphous perversity." In a non-capitalist society sexual repression would no longer be essential to insure social repression and economic exploitation.

Norman O. Brown,[6] in his brilliant excursion into metapsychology, *Life Against Death*, eschews Marcuse's Marxism, but carries even further his eulogy of "polymorphous perversity." The end of repression would mark, it seems, man's release from the nightmare of history. In the last part of his book, it should be added, Brown presents specific studies in anality, especially as it has manifested itself in the Protestant Era.

Ego Psychology and Erikson

Almost all of the post-Freudian developments mentioned above have been more contributions to the philosophy of history, or to metapsychology, rather than the actual application of psychoanalysis to the traditional materials with which historians have worked, that is, precise documents relating to specific individuals and events. With the work of Erik H. Erikson,[13–17] a "revolution" in history is occurring, marked by the use of the term "psychohistory." Not since Freud himself has the impact on history been so great.

Sullivan and his school helped prepare the way for Erikson, but it is primarily the developments in ego psychology, associated with Freud himself, his daughter Anna Freud, Heinz Hartmann, David Rapaport, and others, that opened the way in theory for Erikson's work. As is well known, attention was now centered on the interrelationship of id, ego, and superego processes, and stress placed on the defensive and adaptive functions of the ego. Normality and creativity became as interesting and valid as psychopathology and breakdown, and the personality was seen more as a functioning whole than as a bundle of neuroses; hence reductionism was more easily avoided.

With these inspirations, to which he contributed, Erikson turned to the elaboration of what has come to be called psychohistory. In *Childhood and Society*,[13] which has become practically a handbook in the field, he outlined in simple, clear terms his "Theory of Infantile Sexuality." Here he tried to show how id, ego, and superego processes interrelate during all the stages of psychosexual development; they are, in short, corresponding processes. Next he deals with the orthodox stages of oral, anal, phallic, and genital in terms of what he calls "zones, modes, and modalities," thus freeing them from a predominantly biological orientation. Implicit, too, in this essay, though more fully developed in the later chapter, "Eight Ages of Man," are Erikson's stages of development, ranging through infancy, early childhood, play age, school age, adolescence, young adult, adulthood, and mature age, where the individual is presented with such antinomies as "trust versus mistrust," "autonomy versus shame," "initiative versus guilt," and so on.[16] Although such stages carry with them the danger of being applied mechanically, they offer, if correctly viewed, merely a useful schema of psychosexual development. In any case, throughout his work, Erikson strives to show how the biologically given stages are elaborated upon by culture, with varied and different results.

Much of *Childhood and Society* is devoted to exemplifications of Erikson's theories in relation to specific case studies: anthropological, as in the study of the Yurok and Sioux Indians; historical, as in the studies of American, German, and Russian national character. In *Young Man Luther*[17] and in *Gandhi's Truth*[14] Erikson really practiced what he preached and gave full-scale examples of what he intended by psychohistory (though at first he did not use the term). Thus, much of Erikson's effect on historians has resulted from the fact that he united theory and practice to an unusual degree, and in a way that they could see themselves following.

Erikson's successful inspiration of a number of historians may be attributed to some of the following factors. First, his psychoanalytic theories, giving due weight to ego and superego processes, allowed him to take seriously historical materials as telling us what, in fact, was the cultural content of these processes. Second, to understand the historical he used actual historical materials—letters, autobiographies, and similar documentary materials—rather than resorting to large-scale sociological, and generally Marxist, theories; in fact, it is anthropology rather than sociology that has had the greatest influence on Erikson. Third, he studied his historical materials closely (though some historians disagree with the way he does this), adding to the usual historian's insight his own psychohistorical methods derived from a secure base in clinical data; that is, the same analysis of psychological processes, such as projection, displacement, and so forth, are applied rigorously and with great insight to the historical documents. Fourth, he concerned himself with problems of historical method and has shown an unusual awareness of problems of evidence and inference, and objectivity and subjectivity, for example, of transference or countertransference phenomena as manifested in the historian himself.[15] For these and similar reasons many historians have felt themselves at home with Erikson, or at least willing to learn from him. He does not violate their method and materials, but rather adds a new dimension to them.

A few of Erikson's psychohistorical theories ought to be mentioned here. In *Young Man Luther* and *Gandhi's Truth* he calls attention to the way in which the great leader, solving his own problems—primarily an identity crisis—offers a solution also to the problems of many others in his time and society. The identity crisis, then, has become a heuristic way of looking at leaders and groups, and a number of historians and political scientists have been inspired in their work by this notion.[43] By choosing religious leaders who, in addition, became political leaders, Luther and Gandhi, Erikson has also tried to bridge the gap between psychoanalytic analyses of religious phenomena and of political phenomena; incidentally he has modified Freud's view of religion as merely an illusion and tried to treat it rather as a valid way of symbolically and emotionally ordering the world. Finally, by stressing the concept of mutuality—which he stretches from the mother's initial relation to the child (where a failure here can mean schizophenia) all the way to Gandhi's theory of *satyagraha*, or nonviolence—Erikson has sought to indicate the sort of therapy that psychohistorical studies can offer to mankind.

Such theories as those above, however, give little indication of the actual impact of Erikson's work on historians. The major impact comes not from such large-scale ideas but from Erikson's precise and detailed application of psychoanalysis to history, *and* of history to psychoanalysis. It is his fusion of the two in psychohistory that has opened up a whole new field of endeavor, actively being pursued today by a growing number of historians and other social scientists.[23,25,26,32,33]

Problems and the Future

Psychohistory is in process of becoming a flourishing field. Starting with Freud's work, it has drawn inspiration from the contributions of some of his disciples, whose efforts rely heavily on shifts in emphasis in psychoanalytic theory; we have touched briefly and selectively on some of these developments in order to indicate, without any pretense at complete coverage, the general lines of evolution. Now we need to consider what lies ahead in the way of both problems and promise.

One problem clearly is in the area of training. A few analysts are turning to graduate work in history; there seems no inherent problem here except time, money, and inclination. Most workers in psychohistory from the psychoanalytic side, however, will presumably "pick up" their history from private reading and study, with all the attendant dangers of amateurism and superficiality. From the social science and history side, the dangers of amateurism and superficiality in "picking up" psychoanalysis by private reading would seem even greater. However, some of the psychoanalytic institutes have now started to give courses in psychoanalysis specifically for social scientists; it remains to be seen how appropriate these courses will be. (In addition, one ought to note the possibility of collaboration between individuals trained in history and in psychiatry.)

Ought the psychohistorian to have had a full analysis himself? Or at least psychotherapy? A good deal of contact with actual clinical cases? Should he himself have the experience of treating a few patients in therapy, under supervision? Is adequate funding available to the social scientist for these kinds of experience? Such problems as these, and related ones, are not to be taken lightly. Concerted efforts to define what is optimum training in psychohistory is needed. Fortunately such efforts are now being at least talked about.

Much thought must also be given to certain methodological problems. For example, what sorts of materials lend themselves adequately to psychohistorical interpretation? With the psychohistorian's subjects generally dead, or at least out of reach, the clinical analyst's resort to free association and dream analysis, for example, is simply not available. Are letters, memoirs, autobiographies, accounts by contemporaries, and other such documentary remains sufficient evidence for psychohistorical interpretations? Next, how can these interpretations be verified?

A related problem is the role of the psychohistorian's own personality in his interpretations. Historians have long recognized that political or social biases, national or ethnic commitments, may unduly distort a particular piece of work. Now psychohistory seeks to call attention to the role of the historian's personality as well in affecting his interpretation; and this will necessarily be intensified in *psychohistorical* interpretation. Again Erik Erikson, as remarked earlier, has pioneered in trying to work out a careful consideration and analysis of the role played by what can be called transference and countertransference in the historian's interpretation of his documents.[15]

Even if we assume that a psychohistorical interpretation is possible on the basis of the documents, of what consequence is such an interpretation for *historical explanation*? For example, what is the connection between Adolf Hitler's anti-Semitism, explained as it may be on personal grounds, and the actual political decision to exterminate six million Jews? To link the two we need a general historical explanation, and this may not be easy. In general, a "great man"—a Luther or a Hitler, a Gandhi or a Stalin—can fairly readily be linked to a major event that he has helped create: the Reformation, Nazism, nonviolence, or the 1936 Purge; and this is surely one reason why life histories have attracted the first real efforts in psychohistory. Nevertheless, the general problem of fitting psychohistory into general historical explanation remains.

These and related problems are gradually being subjected to careful scrutiny by historians and other social scientists.[22,33,38,39,50,52] Of a different order, however, from these methodological problems is the problem of how one moves from individual psychology and individual life histories to group psychology and group history. Freud's own contributions to group psychology were extremely tentative. Yet most of what historians are concerned with falls into the realm of group phenomena. How can psychohistory resolve this basic dilemma?

One solution is to proceed with the effort to link the life histories of great leaders with the mass phenomena they seemed to have evoked. In so doing we seek to understand more about the society and history of a period in terms of how it helped create the given individual—Erikson's emphasis on "a convergence in all three processes [somatic, ego, and social]"—and then vice versa: how the great individual helps change his society.

Another procedure is to investigate more closely, and in psychohistorical terms, the history of the family as the intermediate and nuclear group shaping the individual. The family, of course, is itself a changing entity, though it may be analyzed in terms of certain presumably universal constants, for example, the oedipal conflict. To state this incidentally is to raise the methodological problem of the universal applicability of psychoanalytic concepts rooted in a nineteenth-century European context; and one part of psychohistory's task is to shed light on this problem by, for example, reexamining the meaning and significance of such concepts as the oedipal complex. At the moment the best work on this particular subject has been done by an anthropologist, Anne Parsons,[42] but psychohistorians should soon have more to say about it and related concepts. Meanwhile, however, basic work on the history of the family is seriously under way, with Philippe Ariès's *Centuries of Childhood*[4] outstanding in this regard.

As we have suggested, psychohistorical studies of groups and group phenomena are probably the most difficult and certainly the least developed aspect of the new discipline. Lacking a sound group psychology to "apply," historians have been hard put to respond effectively to William Langer's "Next Assignment."[2] Norman Cohn's *Warrant for Genocide*[9] is a pioneering effort to understand an episode in collective psychopathology. Various political scientists have been trying to talk about group identity.[24] Historians have sought to analyze, for example, the eighteenth-century American religious experience known as "The Great Awakening,"[7] and an historian and a sociologist have collaborated in an effort to understand the development of autonomy in the modern world, especially in relation to revolutionary activity.[53] Examples such as these, chosen largely for illustrative purposes, suggest a wide diversity of views and approach to the problem of group history.

The task for psychohistory seems, therefore, to develop each of its parts—life history, family history, and group history—to integrate them thoroughly, and to do so with a keen awareness of the numerous methodological problems involved. In developing the group-history aspect, additional support from psychoanalysis, psychology, and psychiatry would seem essential; as in the past, changes in emphasis in psychoanalytic theory prepared the way for advances in psychohistory. In the future, however, such advances would seem to be most realistically expected from the collaboration of psychiatrists and historians and social scientists—a collaboration that would take the form of a true fusion of thought and effort, reflected in professional training, and manifesting itself in important steps forward in a partially autonomous field, psychohistory.

Bibliography

1. ACKERMAN, N., and JAHODA, M., *Anti-Semitism and Emotional Disorder: A Psychoanalytic Interpretation*, Harper, New York, 1950.
2. ADORNO, T., *et al.*, *The Authoritarian Personality*, Harper & Row, New York, 1950.
3. ALEXANDER, F. G., and SELESNICK, S. T., *The History of Psychiatry*, Harper & Row, New York, 1966.
4. ARIÈS, P., *Centuries of Childhood* (Tr. by Baldick, R.), Knopf, New York, 1962.
5. BLANCHARD, W. H., *Rousseau and the Spirit of Revolt: A Psychological Study*, University of Michigan Press, Ann Arbor, 1967.
6. BROWN, N. O., *Life against Death*, Wesleyan University Press, Middletown, Conn., 1959.
7. BUSHMAN, R., *From Puritan to Yankee: Character and the Social Order in Connecticut, 1690–1765*, Harvard University Press, Cambridge, 1967.
8. COHN, N., *The Pursuit of the Millennium*, Secker & Warburg, London, 1957.
9. ———, *Warrant for Genocide*, Harper & Row, New York, 1966.
10. DEMOS, J., "Underlying Themes in the Witchcraft of Seventeenth-Century New England," *Am. Hist. Rev.*, 75:1311–1326, 1970.
11. EDINGER, L., *Kurt Schumacher: A Study in Personality and Political Behavior*, Stanford University Press, Stanford, 1965.
12. EISSLER, K. R., *Leonardo Da Vinci: Psychoanalytic Notes on the Enigma*, International Universities Press, New York, 1961.
13. ERIKSON, E. H., *Childhood and Society*, Norton, New York, 1950.
14. ———, *Gandhi's Truth*, Norton, New York, 1969.
15. ———, "On the Nature of Psycho-Historical Evidence: In Search of Gandhi," *Daedalus*, 97:695–730, 1968.
16. ———, "The Problem of Ego Identity," *J. Am. Psychoanal. A.*, 4:56–121, 1956.
17. ———, *Young Man Luther*, Norton, New York, 1958.
18. FREUD, S., and BULLITT, W. C., *Thomas Woodrow Wilson, Twenty-Eighth President of the United States: A Psychological Study*, Houghton Mifflin, Boston, 1967.
19. FRIEDLANDER, S., *L'Antisémitisme Nazi. Histoire d'une psychose collective*, Editions du Seuil, Paris, 1971.
20. FROMM, E., *Escape from Freedom*, Rinehart, New York, 1941.
21. GEORGE, A. L., and GEORGE, J. L., *Woodrow Wilson and Colonel House*, John Day Co., New York, 1956.
22. GREENSTEIN, F. I., *Personality and Politics*, Markham Publishing Co., Chicago, 1969.
23. HUNT, D., *Parents and Children in History: The Psychology of Family Life in Early Modern France*, Basic Books, New York, 1970.
24. ISAACS, H., "Group Identity and Political Change: The Houses of Muumbi," paper presented to Psychohistorical Study Group, American Academy of Arts and Sciences, October 16, 1971.
25. KENISTON, K., *The Uncommitted: Alienated Youth in American Society*, Harcourt, Brace & World, New York, 1960.
26. ———, *Young Radicals: Notes on Committed Youth*, Harcourt, Brace & World, New York, 1968.
27. KÜNZLI, A., *Karl Marx: Eine Psycho-graphie*, Europe Verlag, Wien, 1966.

28. KURTH, G., "The Jew and Adolf Hitler," *Psychoanal. Quart.*, *16*:11–32, 1947.
29. LANGER, W. L., "The Next Assignment," *Am. Hist. Rev.*, *63*:283–304, 1958.
30. LASSWELL, H., *Power and Personality*, Norton, New York, 1948.
31. ———, *Psychopathology and Politics*, University of Chicago Press, Chicago, 1930.
32. LIFTON, R. J., *Death in Life: Survivors of Hiroshima*, Random House, New York, 1967.
33. ———, *History and Human Survival*, Random House, New York, 1970.
34. ———, *Revolutionary Immortality: Mao Tse-tung and the Chinese Cultural Revolution*, Random House, New York, 1968.
35. MANNONI, O., *Prospero and Caliban: The Psychology of Colonization* (Trans. by Powesland, P.), Frederick A. Praeger, New York, 1956.
36. MARCUSE, H., *Eros and Civilization*, Beacon Press, Boston, 1955.
37. MASON, P., *Prospero's Magic*, Oxford University Press, London, 1962.
38. MAZLISH, B., "Clio on the Couch: Prolegomena to Psycho-History," *Encounter*, *31*: 46–54, 1968.
39. ——— (Ed.), *Psychoanalysis and History*, rev. ed., Grosset & Dunlap, New York, 1971.
40. MEYERHOFF, H., "On Psychoanalysis and History," *Psychoanalysis and the Psychoanalytic Review*, 49, 1962.
41. MITZMAN, A., *The Iron Cage: An Historical Interpretation of Max Weber*, Knopf, New York, 1970.
42. PARSONS, A., "Is the Oedipus Complex Universal? The Jones-Malinowski Debate Revisited and a South Italian 'Nuclear Complex,'" *Psychoanal. Study Soc.*, *3*:278–328, 1964.
43. PYE, L., *Politics, Personality, and Nation Building: Burma's Search for Identity*, Yale University Press, New Haven, 1962.
44. REICH, W. (1933), *The Mass Psychology of Fascism*, 3rd ed., Orgone Press, New York, 1946.
45. RIESMAN, D., *The Lonely Crowd*, Yale University Press, New Haven, 1950.
46. ROBINSON, P. A., *The Freudian Left*, Harper & Row, New York, 1969.
47. SCHAPIRO, M., "Leonardo and Freud: An Art-Historical Study," *J. Hist. Ideas*, *17*: 147–178, 1956.
48. SIMMEL, E. (Ed.), *Anti-Semitism: A Social Disease*, International Universities Press, New York, 1946.
49. STRACHEY, J., "Editor's Note," in *Standard Edition, Complete Psychological Works*, Vol. 11, p. 60, Hogarth, London, 1953.
50. STROUT, C., "Ego Psychology and the Historian," *Hist. & Theory*, *7*:281–297, 1968.
51. WAITE, R., "Adolf Hitler's Anti-Semitism: A Study in History and Psychoanalysis," in Wolman, B. B. (Ed.), *The Psychoanalytic Interpretation of History*, Basic Books, New York, 1971.
52. WEHLER, H. U., "Zum Verhältnis von Geschichtswissenschaft und Psychoanalyse," *Historische Zeitschrift*, *208*:529–554, 1969.
53. WEINSTEIN, F., and PLATT, G. M., *The Wish to Be Free*, University of California Press, Berkeley, 1969.

CHAPTER 49

THE STUDY OF LANGUAGE IN PSYCHIATRY

The Comparative Developmental Approach and Its Application to Symbolization and Language in Psychopathology

Bernard Kaplan

SINCE THERE HAS BEEN but a limited number of distinct generative ideas in the history of thought,[104] it is not surprising that some kind of comparative developmental orientation has been with us from the beginnings of Western reflection. Wherever investigators have been struck by different modes of adjustment to the environment—whether between species, within a species, or even within a single individual under diverse conditions—and have sought some criterion for stratifying these variegated modes of being-in-the-world, a comparative developmental approach has been operative, at least in rudimentary form. Readers interested in earlier formulations relevant to comparative developmental systems are referred to the following:[1–2,13,15,17,34,37,39,41,43,47,49–52,58–59,67,72–73,75–76,78–80,86–87,93,103–106,108,110–111,113,117,124–125,128,131,135,142,145,147].

The approaches to comparative developmental phenomena by such men as Freud,[126] Piaget,[120,121] Wallon,[160–162] and Werner[163–168] have much in common when contrasted with those of self-styled "behavior theorists" and others whose orientations are essentially agenetic and noncomparative; nevertheless, there are basic divergences in presuppositions among the above-mentioned developmental positions, and these differences preclude either a syncretism or a treatment of any one of the positions as if it were interchangeable with any of the others. For various reasons the comparative-developmental approach of Heinz Werner has been chosen for exposition here. This choice rests on a number of considerations: at least in the United States, Werner is the one most generally identified with comparative developmental psychology; Werner has been the one most directed toward a comparative and developmental analysis of the formal aspects of pathological symbolization and linguistic expression vis-à-vis representation and expression in normal individuals; finally, the author, having worked closely with Werner for more than fifteen years, is most familiar with and most competent to present his position and its application to problems of symbolization and language in pathology.

Werner's Comparative Developmental Approach

Every key "catchword," A. O. Lovejoy reminds and warns us, has been invested with multiple and sometimes antithetical meanings —not merely in the writings of different thinkers, but also in the works of the same thinker at different times. The terms "comparative" and "developmental" are clearly such catchwords. Encrusted with multiple connotations deriving from their use in a variety of disciplines,[17,75,106] and further compromised by their diverse employments within a single discipline,[86] they are each likely, exposed to the kind of analysis for which Lovejoy was justly famous, to unfurl at least a dozen different meanings, including a number at odds with each other. It is not germane to attempt to lay bare these multiple meanings here. What is relevant is that different conceptions may underlie the usage of a key term, even in the writings of the same thinker. In the latter case, this is sometimes due to a lack of differentiation, a fusion of a number of distinguishable conceptions; sometimes, to a development from an initially diffuse to a later more articulated notion; sometimes, to "regression" or dedifferentiation. Often a change ensues as one becomes clearer about one's subject matter. Sometimes, under the pressure of criticism or of questioning from without, the thinker seeks to maintain the legitimacy of his enterprise by modifying his concepts to remedy defects, remove inconsistencies, or clarify obscurities.

Werner used some key terms, particularly "development" and its derivatives, in several different ways; some of the usages diverged considerably from the more customary ones; and some of these divergences were explicitly intended to maintain the legitimacy of the Wernerian undertaking: the establishment of a comprehensive developmental approach to all life phenomena, unsullied by questionable and obsolescent presuppositions that had undermined earlier attempts at such an enterprise, that continued to vitiate kindred contemporary points of view, and that had partly infected Werner's early writings, despite his explicit rejection of them.

Ingredient in these observations is the warning that one must not regard Werner's comparative developmental approach as if it had sprung forth, fully formed and immutable, at a particular time, or as if it had not since undergone modification. Furthermore, the position as advanced in Werner's classic work *Comparative Psychology of Mental Development* is not identical with that taken, jointly with the present author, in *Symbol Formation*; nor is the position sketched in that work identical with the one toward which we were working at the time of Werner's death. The approach changed over time; hopefully, it also developed. Although space precludes the tracing of all of the vicissitudes, the main outlines of the shift in conceptualization may be briefly presented.

Early Formulation of the Concept of Development

Initially, in Werner's writings the status of the concept of development was unclear. Although the meaning of the term was relatively unambiguous—*"an increasing differentiation of parts* and *an increasing subordination or hierarchization*"—it was uncertain whether it was to be taken as designating an "empirical law," a generalization derived from an unbiased analysis of the character of changes actually manifested in a wide variety of processes, or as an heuristic principle, a way of looking at phenomena in the "interest of reason."

In a context in which there was a general belief in a cosmic law of "progressive development," immanent and efficacious in the actual course of history,[34,106] an acceptance of a law of "orthogenesis," ingredient in the emergence of new species over time, and an affirmation of a law of "recapitulation,"[74] governing sequential changes in human ontogenesis, such a conflation of usages would probably have gone unchallenged, and perhaps even unnoticed. However, in the light of historical,[106] biological,[29] anthropological[65,74] and epistemological criticism,[43] it became obvious that if "development" were to retain its connotation of sequential changes in a system, yielding novel increments both in structure and mode of operation,[115] it would have to be elevated to an ideal status and be distinguished from actual history or evolution. One could not presume that historical, evolutionary, or, by extension, any kind of change over time, for example, change with age, was, by the fact itself, developmental.

One had to posit what one meant by development, take a thus-defined developmental progression as a standard or "ideal of natural order,"[158] and then determine to what extent and through what factors or means historical, evolutionary, ontogenetic, and other changes conformed to or deviated from such an ideal progression. With this way of regarding development, one could introduce an *orthogenetic principle*,[167] without being bedeviled by biologists, anthropologists, and others proclaiming censoriously that neither evolution nor culture change necessarily reveals such an orthogenesis.

To reflect this changed status of the concept of development, it was re-formulated not only so that it entailed orthogenesis, but also so that its empirical applicability was left bracketed: *Insofar as development occurs in a process under consideration, there is a progression from a state of relative undifferentiatedness to one of increasing differentiation and hierarchic integration.* Such a formulation did not commit one to the view that any process was exclusively or predominantly a developmental one; at the same time, it allowed one to examine every process to determine the extent to which it revealed features of increasing differentiation and of hierarchic integration over time. Furthermore, by omitting references to a particular time-scale, and hence, allowing for different time-scales for different processes,[51] it permitted one to apply developmental conceptualization to culture history, to the individual's life career, and to the "microgenesis" of a particular percept or thought.[10,63,65]

Recent Modifications of the Concept of Development

Recently[85,86] even the new formulation of the concept of development was seen as too time-bound for the comprehensive comparative psychology of mental development such as Werner had envisaged and had taken initial steps to realize. It still did not seem to permit a comparison and developmental ordering of the behavior of groups who were contemporaneous (contemporary scientific man and nonliterate man) without invoking the palpably unwarranted assumption that one of these types of mentality was arrested at an earlier period in actual history or was a throwback to such an earlier period in a curiously conceived anthropogenesis. It precluded a comparison and developmental ordering of child and psychotic behavior without invoking the palpably absurd thesis that the psychotic had regressed to an earlier phase in the ontogenetic process,[156] that he had become a child once again. It still seemed to bar a comparison and developmental ordering of the modes of functioning of higher primates, of adult members of nonliterate societies, and of children in a technologically advanced society without invoking the palpably untenable Meckel-Haeckel "biogenetic law"[131] and its even more unwarranted codicil "recapitulation theory."[65,74] Finally, it did not permit a comparison and developmental ordering of the modes of functioning of adult human beings in the different worlds that they inhabit (for example, the dream world, the fantasy world, the everyday practical world, the aesthetic world, the scientific-theoretical world) without assuming that these worlds had emerged successively in time, in a curiously conceived evolution of consciousness.

To justify the comparative developmental approach, it appeared that one had, paradoxically, to formulate the concept of development so that it would not be limited to processes unfolding in a particular entity over time, but would also apply to the atemporal relationship of one pattern of organization or mode of functioning to others.[86] Development had to be conceived in terms of an ideal sequence of organizations, of systems of transaction, and of modes of adaptation, irrespective of their actual locus in our unilinear time scheme. Only in this way, it appeared, was it possible to encompass the range of phenomena Werner sought to encompass, without lapsing into questionable or untenable assumptions.

Such an idealization or "essentializing" of development does not mean that the concept is rendered inapplicable to a single system taken as changing over time. The main consequence of the progressive attempt to render the concept of development context-free is that *development* becomes a manner of looking at phenomena, a new way of representing phenomena,[158] rather than merely a particular phenomenon in itself. Or perhaps more clearly stated, the consequence is that it leads one to see everything in terms of "development": development (increasing differentiation and hierarchic integration) is looked for not only in the formation of personality, but also in the formation of a percept; not only with regard to the conception of self, but also with regard to conceptions of space, time, number, and causality; not only with regard to individual behavior, but also with regard to cultures taken as organic unities; not only with respect to time-bound series, but also with respect to patterns of organization or to modes of functioning, regardless of time of occurrence.

A comparative developmental approach pertains to any and all aspects of behavior susceptible to analysis and ordering in terms of the very general concepts of differentiation and integration, concepts that require specification in the diverse contexts to which they were applied. It pertains to whatever can be construed as a functional whole, a system, an organized totality, whatever can be viewed in terms of part-whole and means-end relationships. Its range extends from functional subsystems within an organism to transpersonal patterns of objectified mind, for example, linguistic systems, technologies, and so on. Assuming a developmental progression as in Toulmin's terms, "an ideal or natural order,"[158] developmental psychology focuses throughout on the immanent rules, the modes of operation, revealed in the functioning of actual systems. Its aims are to articulate and distinguish systems, to which Werner sometimes referred as "genetic levels,"[163] in terms of their specific principles of organization, to order such systems according to the degree to which they reveal differentiated and hierarchically integrated functioning, and to determine the conditions or constraints that militate for or against the realization of ideal development.

Explication of Some Major Concepts

Having characterized the status of the concept of development and of its defining "orthogenetic principle," one may introduce some ancillary notions further to specify the concept of development and to clarify its application.

Primitivity

The much-abused concept of "primitivity"[21,71,97–99,167] is often employed in developmental analysis. It should be recognized that to the degree that the concept of development is logically disentangled from chronology, so too is the concept of "primitivity" freed from its bondage to time.[85] Just as the developmental status of a mode of functioning is determined not by its time of occurrence, but by organizational characteristics ingredient in it, so also the primitivity of a mode of functioning is determined formally rather than temporally. Moreover, it should be emphasized that the developmental status of a mode of functioning, primitive or advanced, is a relative matter, depending upon the other systems with which it is compared and contrasted. Thus to speak of the mentality of a young child as primitive vis-à-vis that of an adult, to characterize the mentality of a psychotic as primitive contrasted with that of a normally adapted individual, to describe the functioning of a member of a nonliterate group as primitive compared with that of a scientifically imbued member of a Western society, or to assess the mode of functioning typical of the dream state as primitive relative to that of the alert, waking state, in no way entails an identification of infantile, psychotic, nonliterate, and oneiric mentation. Nor does it imply that the same causes or motives underlie the diverse forms of primitive mentation.[47]

FORMAL PARALLELISM

The "law of recapitulation," advanced by Haeckel and Stanley Hall, and adopted by Freud,[65,74] is repudiated by the Wernerian comparative developmental approach.[163] In its place the concept of formal parallelism is introduced. This concept suggests a comparability (with regard to general organizational features) of different domains or theoretically constructed series. Indeed, this notion follows from the application of developmental conceptualization to diverse domains. In the domain of animal life, in the career of the human being, in the realm of socio-cultural organization, and so on, one would expect that forms of life, modes of being-in-the-world, types of consciousness, or patterns of organization would lend themselves to an ordering in terms of degree of differentiation and hierarchic integration.[163] Therefore, the value of the concept of formal parallelism is heuristic: alerting one to material, situational, and efficient-causal differences in the various domains, it nevertheless suggests similarities in organization and prompts one to look in one domain for analogues to phenomena in other domains.[22,25,27,30,82,120,161,163–168]

POLARITIES IN ORTHOGENESIS

In comparing, contrasting, and ordering modes of organization, developmental theorists have found it useful to particularize the orthogenetic principle, and abstractly, to distinguish within a circumscribed organism-environment system the ends of the organism or the functions toward which it is directed, the means by which it executes its functions, the consequent structure of its transactions with its surrounds, and its capacity to maintain its integrity and to adapt itself to internal and external vicissitudes. With regard to these various aspects, prescinded from the total organism-environment transaction, pairs of polar concepts are employed to specify particular developmental progressions. Thus, a developmental progression with respect to *functions* or ends may be characterized as a movement from *interfused* to *subordinated*; in the former, ends or goals, susceptible to being distinguished, are not sharply differentiated, and in the latter, functions are differentiated and hierarchized, with drives and momentary motives subordinated to more central, long-range goals. The progression in *means* is characterized as a movement from *relative syncresis* to *discreteness*; for example, in the former such means of coping with the environment as perceiving and remembering, wishing and acting, and so forth are more or less undifferentiated from each other (as in the dream state), and in the latter, they are distinguished and "freely combinable." With regard to the *structure* of a behavioral act or to the outcome of an organism-environment transaction (for example, a drawing, a tool, an utterance) one may speak of a progression from *diffuse* to *articulate*; genetically primitive acts or act-products are relatively global, with little internal articulation, and developmentally advanced acts or products are segmented, with clearcut parts subordinated to the goal or unity of action. Concerning the capacity of the organism to maintain its integrity and to adapt itself to inner and outer vicissitudes, one may use the complementary polarities: *rigid-flexible* and *labile-stable*; a rigid mode of organization, and hence a genetically more primitive one, is one in which the organism is incapable of altering or modifying its response despite marked changes in the environment demanding such alteration or modification; a flexible system is one in which the capacity for such modification is present. Correspondingly, a labile system is one in which the organism is pulled from its course or goal by minor variations in its surroundings or by slight disruptions within, and a stable system is one in which the organism has the capacity to retain its integrity and adaptation despite such variations or disruptions.

In sum, a more primitive mode of organismic functioning, irrespective of the factors that may be invoked to account for its manifestation (for example, neurological immaturity, brain damage, extreme anxiety), is one that may be characterized as showing a greater interfusion of functions, a greater syncresis of means or operations, a more diffuse structurization of acts or act-products, and a greater rigidity and lability in relation to changing inner and outer conditions.

Multiple Modes of Functioning and Levels of Organization

The concepts of multiple modes of functioning and levels of organization are related.[17,32,61] Although the comparative-developmental theorist takes as his initial tasks, methodologically, the description of theoretically isolated modes of functioning ("ideal types") and the ordering of these modes in terms of the orthogenetic principle, he has the additional aim of utilizing these abstracted modes of functioning to describe changes in organization in the course of a life career or in varied circumstances at any one period in the life career. Individuals observed in the course of human ontogenesis do not manifest a single mode of functioning, but rather reveal multiple modes of functioning, ranging from quite primitive to more advanced. Empirically, changes in organization as a function of age are at least up to early adulthood, consonant with the orthogenetic principle; that is, there is a developmental progression in ontogenesis. This, however, should not lead one to believe that higher modes of functioning simply replace lower or more primitive ones. The biologically mature organism is characteristically constituted by different levels of organization and will under varied circumstances (for example, drowsiness, intoxication, anxiety, impoverishment of the environment) reveal more primitive modes of adjustment, although capable of higher levels of functioning. These concepts may enable one to clear up a widespread misunderstanding, mainly by anthropologists, of the developmental position. This misunderstanding has been due, partly, to infelicitous formulations by developmentally oriented psychologists and psychopathologists. Developmental theorists do not distinguish types of men, but rather distinguish multiple modes of functioning. There is no normal man capable only of primitive mentation. Likewise, there is no man incapable of primitive mentation. However, the actual conditions of existence, the society into which an individual is born, and so on, may promote primitive levels of organization or may allow advanced ones as customary ways of being. Developmental theorists fully accept the doctrine of "the psychic unity of mankind." However, they do not confuse unity with homogeneity or capability with actuality.

Psychopathology

In dealing with psychopathology, as with actual ontogenesis, a comparative-developmental approach does not focus indiscriminately on the multiplicity of changes that occur; its concern with formal or organizational features of functioning leads it, rather, to concentrate on part-whole, structure-function, and/or means-end relationships. On the assumption that pathology entails some degree of "primitivization of mentality," it expects to find, in pathological individuals, a *dedifferentiation* and *disintegration* of functioning. It is preferable to speak of disintegration rather than delamination ("peeling off of layers"), because in actual ontogenesis higher levels of activity, as they emerge, are not merely grafted on lower levels, but also modify lower levels of functioning; hence lower level integrative mechanisms may be dissolved or transvalued and may not recoup their earlier status or potency, if released from higher level regulation and control.

In studying psychopathology, one is oriented toward the interfusion of ends or goals. Ends, distinguished in the normal individual or in the patient, premorbidly, would be expected to merge, with each activity being overdetermined; for example, the goal of securing esteem or love may not be distinguished from the goal of securing nutrition. One is oriented toward the syncresis of means distinguished at higher levels of organization; the differentiation of desiring, imagining, remembering, perceiving, judging, overt acting, so important for establishing the cardinal distinction between self and nonself and for separating the different "spheres of reality" in which the normal is capable of living, would be expected to collapse. One would also look for a diffuseness in acts or act-products, relative to the normal or premorbid condition, that is, a failure to distinguish parts and wholes, things and attributes, container and contained, the literal and the metaphorical, and so on. Finally one would be oriented in psychopathology toward manifestations of the loss of hierarchic integration, the rigidity and lability of the organism vis-à-vis its environment, revealed in such phenomena as stereotype, perseveration, stimulus-boundedness, sudden shifts from one sphere of reality to another (for example, from the communal Lebenswelt to the autistic fantasy); the complete segregation of activities and subsystems, each operating with unchecked local autonomy; the unregulated and uncontrolled incursion of activities into domains from which they had been excluded through higher level controls, for example, personforming activities intruding into the domain of object-formation, with the consequent personification of things, or, conversely, object-forming activities intruding into the domain of person-formation, with the consequent apprehension of persons as manipulable things; and, most important, the *loss of ideality*, the inability to sustain ideal relations, or the tendency to concretize the abstract, to collapse similarity into identity, to confuse the symbol and the referent, and so on.

All of these phenomena of dedifferentiation and disintegration, variously explained, have been observed as characteristic of psychopathology by psychiatrists of all persuasions. The loss of ideality is at the core of what Goldstein[69,70] has described as "the loss of the abstract attitude" and what Arieti has described in terms of "Von Domarus' principle."[5] Since the mode of analysis characterizing the comparative developmental approach to psychopathology is discussed and exemplified by Arieti elsewhere in this series of volumes, the reader is referred there.[7] Other discussions and applications of this approach, or closely related ones, may be found in the following:[3–11,22,25,27,33,36,64,66,68,112,119,122,136–140,149,150,156,166,168]. (For a masterful general exposition of the absence of ideality in primitive thought the reader is referred to Cassirer.[45])

One may cite the following remarks by the psychiatrist, Harold Searles[137]: "From a phenomenological viewpoint, schizophrenia can be seen to consist essentially in an impairment of both integration and differentiation which . . . are but opposite faces of a unitary process. From a psychodynamic viewpoint as well, this malfunctioning of integration-differentiation seems pivotal to all the bewilderingly complex and varied manifestations of schizophrenia, and basic to the writings on schizophrenia by Bleuler, Federn, Sullivan, Fromm-Reichman, Hill, and other authorities in the field" (p. 261). There is only one modification that may be made in Searles's statement: for a developmental psychologist, his remarks would apply to other forms of psychopathology as well as to schizophrenia.

Symbolization and Language

A persistent doctrine in Western thought, reinforced greatly in biology and psychology by Darwinism, has been the "law of continuity."[92,104] This law, dogmatically rather than methodologically maintained, has led some psychologists explicitly to deny or implicitly to ignore fundamental differences among species and, a fortiori, any basic differences as to modes of functioning within a species or an individual. In many instances, it has been tacitly assumed that one mode of functioning could be reduced to another if the former could, in some way, be "derived" from the latter.

As essential feature of the comparative developmental approach, as has been noted above, is its emphasis on multiple modes of functioning and on different levels of organization,[17,32,61] not only with regard to different species, but also with regard to the domain of human behavior. Different forms of life have different modes of transaction with the selfsame physical environment, irrespective of how these differences have come about. This is also the case for the human being, not only in the course of ontogenesis, but also in the adult form, under various conditions.

One may grossly distinguish three general modes of transaction, each entailing different "worlds," "*Umwelten*" or "behavioral environments" for the organisms engaged in them. These are: (1) reflex-reaction to physical energies; (2) practical goal-directed action upon or toward presentations qua signals; (3) reflective or detached knowing about objects and events.[168] The human being is capable of existing in all of these worlds. However, his distinctive world is the last one. Living in this world is rendered possible by the distinctive capacity for symbolization, which is among the human being's biological endowments, and the pervasive presence of one or more forms of that universal instrumentality, language, which is an essential feature of the human being's normal (social) environment.[159]

Explication of Concepts

Since "symbolization" and "language" are among those catchwords about whose protean character Lovejoy has warned us, it is important to clarify how these terms are to be used in the present context. Sometimes identified, and even when distinguished often treated as co-ordinate, symbolization and language are here taken to belong to different categories: hence the distinction between symbolization as a capacity or activity and language as an instrumentality. This distinction permits one to highlight the following: As an activity, symbolization may exploit a number of instrumentalities, of which language is only one;[16,23,26–27,92,161,168] as an instrumentality, language may enter into a number of distinguishable activities of which symbolization is only one[119] (pp. 264ff.).

SYMBOLIZATION

One of the difficulties in discussing symbolization in a work directed mainly toward psychiatrists and clinical psychologists is that the activity of symbolization has often been taken in their literature as a manifestation of primitivity among the young and of pathology among the old. Due mainly to an acceptance of the "dogma of immaculate perception," and its corollary, a copy theory of knowledge, symbolization has implied "distortion," or a failure to designate things as they are. With the increasing recognition that all knowing transcends the so-called sensory given has come the realization that symbolization is requisite for all of the higher manifestations of man's nature[18,41,92,168,170] and, in fact, enters into the very constitution of his world of objects.[40]

The essence of the activity of symbolization —and this core is clearest when advanced manifestations of that activity are in play[168] —is representation, in a relatively circumscribed medium, of some organismic experience that would otherwise be ineffable and incogitable. Thus, in the sense that "metaphor" ultimately signifies the use of some aspect of experience to represent something other than itself, symbolization at any level involves "radical metaphorizing."[42,84]

As in the case with other human activities, symbolization must be posited as syncretically fused with other acts at lower levels of functioning; at such levels, there is no sharp separation among such activities as desiring, doing, perceiving, imagining, remembering, representing, and so on. It is an anachronistic misnomer to designate any of these activities as they operate in the global gruel of primordial functioning by the same discrete designations which they only half-legitimately warrant even at more advanced levels of functioning.

Due to this syncresis characterizing primitive levels of functioning, all the phenomena that are recognized as intangible, ideal, or "subjective" at higher levels are immersed in and experienced on the same plane as the concretely "objective" products of perceptual-motor action. In Cassirer's terms the "law of concrescence" operates[45] with ideal significations like the part-whole relationship, relationship of resemblance, and so on, not yet emergent from or collapsed into material and efficacious identity; that is, part equals whole, what is "like" is identical to, and so forth (pp. 64ff.). What would at higher levels be in a symbolizing relationship to something else does not represent but *is* that something else.

At higher levels of functioning, with the individuation and articulation of distinct activities and with the correlative stabilization of the domains of the subjective and objective, symbolizing becomes relatively autonomous, and the self becomes aware of its symbolic activity. The individual is capable of a distinction between vehicle and referent and can recognize the differences between the activity of symbolizing, the work of symbol-formation, and the outcome of the activity and the work —the symbol.

LANGUAGE

As is the case with any other socially shared instrumentality, a language[27,57,62] possesses functional potentialities and structural complexities that are fully apprehended and articulated by the individual only in the later phases of ontogenesis and only as they are practically exploited by the individual in the course of his vital and intellectual activities. As is the case with other instrumentalities, the functional structures of a language are grasped in use before they are articulated for and by reflective thought. Like other instrumentalities, a language is not only influenced by the activities which it subserves, but it also shapes those activities. Finally, as other instrumentalities, a language, capable of the most refined uses, may be only grossly exploited or aberrantly employed. It will be clear that in this conception of language, speech and language are not equivalent terms. Speech is an activity that may occur without the use of language;[66,70,77,94,119] language, on the other hand, is an instrumentality that may be understood and used without the activity of speech.[66,70,77,95,119]

It should be reasonably clear from this sketch why symbolizing is here regarded as independent of the instrumentality of language, although the individual subsequently appropriates and exploits this socially shared instrumentality as the principal means of representing, and hence objectifying and communicating, his thinking and feelings. This is not to overlook the above-mentioned fact that the instrumentality exploited by an individual in carrying out an activity is not only shaped and guided, but also shapes and guides, that activity, and the agent who executes it.

From an individual point of view, the immediate functions or usages of language are the individual's activities and tendencies which language may subserve. To be sure, there is no clear consensus as to what these are; their number and kind seem to vary with the investigator and his principal area of inquiry, as well as with his penchant for specification;[20,31,62,109,118,119] but one may follow the aphasiologist, A. Ombredane, and distinguish the affective, the ludic (play), the practical, the representational, and the dialectical[119] (pp. 264ff.).

These usages are increasingly social in character, increasingly detached from the exigencies of practical life, increasingly entail the distinctive features of the linguistic instrumentality to the exclusion of other means of realization.

Affective Usage. One using language *affectively*, an activity which must be sharply distinguished from referring to affective states representatively, uses principally intonational and rhythmic features of the linguistic instrumentality. He may use also exclamations, interjections, and curses, denuded of lexical significance. Insofar as he uses words having a circumscribed significance in the social code, he uses them without regard for their customary meaning and mainly to exhibit his feelings. Such words may then suffer what Arieti[5] has designated as a "reduction of connotation power" and a corresponding increase in "verbalization" (pp. 211, 215). The same verbal sign may be applied to quite different situations by virtue of an affective equivalence among the situations. The lack of concern for the linguistic code reveals itself in forms of utterance that are agrammatical and approximate jargon, in the extreme. As Ombredane puts it: "Distinctions of declension and conjugation are effaced, the sentence is simplified in the extreme, rejoining in its structure the eminently elliptical infantile sentence, where the copulas, the morphemes are omitted, where juxtaposition is substituted for subordination, where words follow each other in the psychological order of the ideas rather than in the grammatical order of the language" (p. 268).[119] In this usage, one scarcely requires an interlocutor and is not concerned with an object of reference.

Ludic Usage. One using language ludically is concerned primarily with the rhythmic and echoic features of the linguistic instrumentality. There is a play with sounds, words, and phrases, with relative or total disregard for the semantic values of these forms. Relationships of assonance and alliteration predominate over semantic relations. Here again, there is an absence of connotation power and an emphasis on verbalization. In engaging in the ludic usage, "the individual abandons himself to an unreflective and facile verbalization which can admit neither the constraints of meaning nor those of the grammatical code" (p. 269).[119] "The successive moments of this verbalization are determined . . . by the force of mechanical and musical connections: phonetic and verbal assimilations, alliterations, assonances, annominations, reduplications, etc., whence it results that ludic verbalization tends regularly to stereotypy" (p. 270).[119] Here again, one scarcely requires an auditor and is not concerned with an object of reference.

Practical Usage. Language is used practically when it is employed to facilitate ongoing action and when it pre-eminently involves a primary face-to-face group. What characterizes the practical usage and distinguishes it most markedly from the representative usage is the centrality of a perceptually shared situation or of at least a situation presumed known to all. The primary goal is the prompt adaptation of action to circumstances when the circumstances are present to the individuals involved. "It follows that practical language is characterized by the extreme reduction of representative elements and by the maximum development of suggestive, excitatory, or inhibitory elements" (p. 271).[119] Such a use of language is quite elliptical, with a predominance of imperatives without any specification of the object. Characteristically, in this usage, an auditor, a particular one, is involved, and the semantic values of words and phrases do play some role, although, typically, much of the meaning is supplied by non-linguistic context.

Representative Usage. With the representative use of language, one is no longer bound to a perceptually shared context or to one that is presupposed by the interlocutors but is free to refer to absent and counterfactual states of affairs. It is with this usage that symbolization truly meets language. That which in the practical attitude is taken for granted is given a linguistic articulation through the grammatical and lexical resources of the linguistic code. Detached from contextual supports, one must create everything linguistically. "Hence the necessity of defining the setting, the persons, the relations of the actions in time, hence the necessity of marking presence or absence, indicating aim and instrument, explicating the chain of facts and the organization of reasons" (p. 273).[119] In this usage, one cannot neglect the subtle grammatical features of the linguistic instrument. To use Sapir's formulation, one seeks to give each of the elements in the flow of language "its very fullest conceptual value" (p. 14).[133]

It must not be thought that representative

usage entails literality or that it simply subserves the conveying of information. It also serves concretely to depict intangible or complex notions, to represent affective states, and so on. It is with the representative usage that true metaphor comes into existence: that is, where the duality between what is literally stated and what is meant is clearly maintained.[42,88] It will be understood, in this connection, that it is with representative usage that the individual is capable of maintaining a "categorial attitude" toward objects and events. It is obvious that the representative usage of language is generally brought into play when there is an object of reference; it also entails an auditor, but now a more "generalized other."

Dialectical Usage. The dialectial usage involves an analytical attitude toward the linguistic instrumentality itself. It is oriented toward the discovery of the rules immanent in the language, and is further concerned with either shaping the ordinary language or constructing artificial languages for specific functions. Although this usage is customarily tied up with logical and scientific concerns, it may be regarded more generally as an orientation toward the rules involved in any employment of language. Hence, the student of poetry, seeking to make explicit the syntax of poetry, is also engaged in a dialectical activity.[14,54,107,116,148]

As noted, the emergence and differentiation of these usages, and hence the carving out of specialized means for actualizing them, go hand in hand with an increasing expropriation of previously untapped resources of the language. Linguistic features, unnecessary for the mere expression of pleasure or displeasure, are grasped and internalized (linked with symbolization and speech) in order to indicate why, and about what, one is emotionally exercised; in order to guide, with some precision, the behavior of others; in order to represent absent or ideal states of affairs to impersonal and remote addressees.

In this connection, it cannot be overemphasized that all of the usages coexist at higher levels of development: an "advanced usage" (for example, representation) does not simply replace an "inferior usage" (for example, practical handling of language). Rather, as each usage, or each "attitude," is manifested, it becomes progressively differentiated from the others, and then becomes integrated with them in the varied contexts of human functioning. As representation emerges, it becomes progressively distinguished from the other "inferior" usages in the means it employs, each attitude exploiting certain aspects of language for its distinctive actualization. In all usages, nonlinguistic activities (for example, body movements) and the contexts of utterance are also distinctively employed.

It is also important to emphasize that these usages are manifested on different levels of functioning: affectivity expressed in exclamations and in curse words is quite different from affectivity expressed in sarcasm; the ludic handling of the linguistic medium in lallation is markedly different from that involved in a witty play on words; the representative activity of the young child is not on the same level as the representative activity of the poet.

Symbolization and Language Usage in Psychopathology

Prejudice and Approach

There are two prejudices that have often interfered with an adequate comparative developmental approach to symbolization and language usage in psychopathology; one of these prejudices is parochial and of relatively recent origin, the other more ancient and more pervasive.

The first prejudice is that advanced by a number of orthodox psychoanalysts[83] who would limit what they call "true symbolism" to a process of unwitting realization of unconscious meanings in sensuous or tangible form. For this group, language used to represent tion would not be symbolic; only language revealing the repressed wishes of the individstates of affairs and to communicate informaual in disguised form would have that status.

This prejudice is clearly ungrounded and has been recognized as baseless even by some Freudian theorists.[90,132] One need not go to the opposite extreme, as some have suggested,[18] and deny any symbolic status to such "distorted" expressions of the unconscious. Instead of opting for one or another form of the activity of symbolization as the "true" one, a developmental orientation should lead one to recognize that symbolization is an activity occurring on different levels, ranging from the most primitive, where it is syncretically fused with perception and action, to the most advanced, where it comes into its own.

The second prejudice is a more subtle and insidious one and more difficult to uproot since it constitutes part and parcel of the Western rationalist-intellectual tradition. Indeed, it has played and continues to play an important role in many of the developmental theories advanced in Western societies, including, to some extent, early comparative-developmental theory. This prejudice, often implicit, would have it that the aim of development, whether in the social group or in the individual, is the pre-eminence of a scientific-technological orientation in every domain of human life. It is therefore led to assess all performances and to evaluate their "developmental status" in terms of this putative aim. Characteristically those who espouse this conception of development regard symbolic activity and language usages that do not subserve the communication of information in precise and exact linguistic symbols as, by the very nature of the case itself, primitive or pathological or, at the very least, "regressed in the service of the ego." For many of them, no matter how they mask it, art, play, poetry, and religious symbolism and language are either neurotic symptoms, emotional expressions devoid of intellectual content, or embodiments of inferior forms of cognition, to be superseded in ontogenesis or societal evolution by fully articulated, unequivocal discourse, representing scientific conceptualization and impersonal communication of thought.

Once again, it is neither a fact of history nor an induction from the study of ontogenesis that the aim of social and individual change is a progression toward an exclusively scientific-literal orientation in all spheres of activity. It is an unacknowledged bias of considerable magnitude to assume that any individual, normal or psychotic, is seeking always to express himself with due regard for the rules of formal logic. Without denying the usefulness of such an assumption as a "fiction" of the investigator, introduced in the "interest of reason," one must strongly question any thesis that would attribute to a person a prelogical, paralogical, or paleological process of thought on the basis of stretching all behavioral products to the Procrustean bed of logical analysis.

The hazards of using a logical standard in reconstructing or attempting to explain the underlying processes of individuals from their behavioral products (linguistic and nonlinguistic) have been pointed out recently by a number of investigators, concerned mainly with establishing the non-inferiority and autonomy of poetry and play.[14,96,100,101,127,153,169] These hazards have also been highlighted by the much-maligned Levy-Bruhl, the source and origin of the notion of prelogical mentality[55,97–99,114] Levy-Bruhl early maintained that the behavior of nonliterate peoples could not be attributed to faults or defects in their logical processes,[55] but was due to a different (namely, "affective")orientation toward experience from the one which governs Western man qua scientist. More recently,[99] he stressed that the character of the thought processes of nonliterate man had been misrepresented by himself and others in that they had not distinguished clearly between physical impossibility and logical incompatibility; that someone believes that a thing which is A may also be B is not an indication of a defect in reasoning but is due to a different conception of reality; similarly, it is neither a logical fallacy nor a manifestation of primitive logic to believe in bilocation, although such a belief is untenable in the Western scientific conception of the physical world.

These points could be glossed over in the present context were it not for the fact that the same issue has arisen concerning the interpretation of pathological language usage. The problem has been well posed by one of the leading investigators of schizophrenic language, Maria Lorenz. She writes: "The evaluation of the thought processes of a patient through language when evidence is obtained from experimental situations . . . often appears at variance with the impression obtained from a spontaneous talk with the patient. . . . Certain kinds of demands, implicit in test situations, seem to precipitate reactions of irritability, frustration, defensiveness, which quickly lead to resistance, negativisms at times, and nearly always a stubborn clinging to the individual's inflexible mode of viewing the world." She continues: ". . . quite a different form of language is utilized when expression of inwardly experienced states takes precedence over the communication of facts of judgment. . . . The whole area of expressive use of language, poetry, would appear in a sadly illogical, paralogical light if criticized on the basis of logic. To criticize poetry on the basis of logic misses the point of poetry. To criticize the thought of a schizophrenic patient on the basis of logic when he does not assume a reasoning attitude is often to miss completely the alternative meaning of his response. The pathology may lie less in an inability to think logically than in an overemphasis and inflexibility of other modes of thought (p. 608).[102] Similar arguments have been advanced by one of the leading European investigators of pathological symbolization and language, Jean Bobon.[20-27]

It is important to disentangle some of the points at issue here. There is no controversy concerning the radical gap that separates the psychotic's mode of being-in-the-world from that of the normal wide-awake individual. Nor need there be an argument over the possibility, or even limited usefulness, of using logico-linguistic norms for an analysis of a patient's performance, so long as one does not impute to the patient logical errors or mistakes when he is not oriented toward the logico-linguistic representation of thought. What is questioned here is the legitimacy of assuming that a patient is necessarily thinking paralogically or paleologically when he uses tropes in his utterances. (Since, in the main, Lorenz' criticism seems to be directed toward Arieti's use of the Von Domarus principle, it is only fair to say that Arieti has typically shown himself to be aware of the need for caution in inferring underlying thought processes directly from isolated linguistic utterances.)

Although Werner in practice was not entirely free of the rationalist prejudice, he generally recognized the error of interpreting primitive experience in terms of logicolinguistic schemata (p. 23).[163] But even beyond that, it should be clear that the orthogenetic principle of current comparative developmental theory does not entail an exclusively scientific-technical orientation or any other orientation as an inherent and exclusive aim of a developmental process. From a developmental point of view, it is not the submergence of all but a single orientation that constitutes genetic advance, but rather the differentiation and perfection of all of the orientations, and their harmonious integration in the functioning of the individual.

One consequence of this conception is that one must be wary of speaking of primitivity or pathology of symbolic activity or of language usage solely on the basis of isolated productions of individuals.[101] To make a determination of primitivity or pathology, one must include a consideration of the demands of the situation, the intentionality of the individual issuing the product, the relevance of the production to the individual's ends, and so on. Specifically, no act or act-product removed from its functional context[144] is primitive or pathological. Such terms ultimately refer to means-ends or form-function relationships, not to external forms taken in themselves. Neither the blurring of contours or distortion of perspective (in a painting), nor the play on words (in intended wit), nor the personification of time (in a sonnet) are primitive or pathological in themselves, no more so than the use of words for abstract thought is primitive or a "regression in the service of the ego," because such words were originally representative of undifferentiated concrete thinking. To beat swords into ploughshares is not primitive or aberrant unless one is still intent on waging war. This point is so important to stress because developmental theorists, Werner and I included, have occasionally written and sometimes also thought that a means or activity was primitive in itself, irrespective of the function for which it was deployed.

These considerations suggest that one must be extremely cautious in drawing conclusions as to the specific processes or mechanisms culminating in acts or act-products (bodily movements, actions, utterances, paintings, and so on) of individuals, normal or pathological. The same end-product may have a quite different meaning and mode of formation in different individuals or in the same individual at different times. For example, to consider an action by one of Bleuler's patients, an individual may make the movements of a shoemaker to represent to another what a shoemaker does, or to magically incarnate one's ancient lover, a shoemaker. To adapt an utterance by one of Rosenfeld's[129] patients, one may remark, "The Russians *were* our allies" (p. 459) to convey factual information, because one wishes to allude consciously to the dangers inherent in relying on the permanent friendship of anyone, or because one does not distinguish in thought and experience between the "betrayal" of the allies by the Russians and the "betrayal" of oneself by a therapist. Inferences as to processes and mechanisms require a knowledge of the mental status of the individual, a knowledge of the orientation or attitude that the individual has adopted, an awareness of the context of the act, and so on.

It should be noted that the author, in citing illustrations from literature to exemplify the application of the comparative developmental approach to symbolization and language, has assumed the legitimacy of the descriptions and inferences drawn by the various investigators quoted. In any case, since the citations will be merely illustrative, they may serve their function even if the specific inferences are open to question.

Clarification, Elaborations, and Illustrations

When one seeks to characterize an activity amenable to realization in varied forms, it is generally preferable to describe it in terms of its more mature form than in terms of a rudimentary manifestation where its specific features are likely to be obscured. Symbolization has been characterized as the capacity to represent, that is, the ability to take items of experience or to intend materials of the environment to exemplify or mean something other than themselves. As such, the ability to symbolize is purely formal; prescinded from its natural ties with one or another "posture of the mind"[123] and one or another mode of giving form to the flux of impressions, the activity does not itself determine what is symbolized, what material is used as a symbol, or the manner in which the symbol and its significate are related for the symbolizer. These aspects of symbolic activity are determined in great measure by the "posture of the mind," attitudes, orientations, and modes of organizing experience.

In order to discuss symbolic activity in psychopathology one has first to consider those attitudes and ways of forming a world that one finds in schizophrenia and in related disorders. Before turning in this direction, however, one may note that there are pathological cases in which it can perhaps be said that it is primarily the activity of symbolization or is at least the work of symbol formation that is disturbed and impaired; by symbol formation is here meant a specialization of symbolic activity in which one takes or shapes properties of a particular medium (for example, sounds, lines, body movements; later, objects and word-forms) to represent something other than themselves (for example, objects, concepts, propositions). Without going into details or without introducing the requisite refined distinctions and qualifications, one may regard some of the aphasias, apraxias, and agnosias as reflecting such a relatively direct impairment of symbolization. Thus, one may find the inability to transform heard sounds into words, although the patient can still entertain concepts; or one may find that a patient is capable of getting the wordform but is unable to go from it to the concept it normally represents; again, one may find patients who are capable of propositional thought but who are unable to re-present their thoughts in language, producing paraphasic and asyntactic utterances, and so on.

Such cases clearly reflect the dedifferentiation and disintegration of functioning that characterizes all pathological primitivization. In an essay of broader scope, they would have to be included—for their own sake, as well as for purposes of comparison and contrast—with phenomena of psychopathology proper. Important as they are, however, they cannot be considered here. The reader is referred to the writings of some of the aphasiologists.[53,66,69–70,77,119] One might also call attention here to works by psychologists,[56] philosophers,[46] and linguists[81,82] who have approached the problems of aphasia and kindred disorders from other perspectives.

In returning to disturbances of symbolization in psychopathology proper, one need not dwell on a point which every serious student of such phenomena has emphasized; that is, disturbances of symbolization in psychiatric cases are not simply consequences of an impairment in the capacity to represent per se, but are, rather, part and parcel of a more profound and pervasive disruption of the sentiments, attitudes, and ways of giving form to experience that the activity of symbolizing subserves. The patient's entire interpersonal and intrapsychic life undergoes at least some degree of dedifferentiation and disintegration.

One may characterize the world that an extremely disturbed patient establishes, the "reality" in which he is immersed, as a filmy flatland, devoid of those crucial distinctions between what there is and what is merely appearance, what is substantial and what is ideal, what is felt or imagined and what is taking place. Here all experiences are on a par; one no longer has control over his various intentionalities and is no longer able to allocate his monentary impressions into domains of subjective and objective, that is, the seen, the fantasied, the thought, and the performed. A patient may momentarily recover these distinctions, but then without warning he finds himself back in the flatland again. There is thus an unregulated incursion of the primordial affective-mythopoetic mode of functioning into domains from which it had been excluded in the course of normal ontogenesis, a mode in which intensity of experience and in which affective relevance are the sole or main determinants of the "real" and "objective," and hence one to which reasoning, critical analysis, and control are essentially alien.[45,60,91,98] Phenomena familiar to us from our own circumscribed oneiric and hypnagogic modes of functioning become the standard phenomena in the waking lives of the more disturbed patients.

In a world formed through the affective-mythopoetic mode of functioning, symbolization must be radically altered. Organically involved in the development of mentality and integral to the individual's socialization, symbolic activity now becomes syncretically fused with feeling-acting-perceiving and loses its distinctive status as the means of representing the ideal, the intangible, and the remote.

Let us first distinguish the major constituents of those situations (symbol-situations) in which symbols, both linguistic and nonlinguistic, are characteristically employed. In well-articulated symbol-situations, one may distinguish at least the following components: the *addressor*, or one who uses symbols (in part, at least) to communicate; the *addressee*, or one to whom the communication is addressed; the *intention*, or that which the addressor wishes to communicate to the addressee; the *referent*, that object or state of affairs to which the addressor wishes to call the addressee's attention; the *context*, or situation in which the communication takes place; the *scene*, or locus of the referent, insofar as the referent is not part of the context; and, finally, the *medium*, the means (one or more) by which the addressor conveys his intentions and/or represents his referents to the addressee.

In both the relatively undifferentiated mentality of early childhood and in the dedifferentiated and disintegrating mentality of psychotics, these distinguishable aspects of symbol-situations are far less articulated from each other and, paradoxically, far less integrated with each other than is the case for normal adults.

For the normal adult, through a complex

process of socialization and quasi-autonomous intellectual development, in both of which symbolic activity plays an enormous role, the world has become a diversified and stratified realm. There are people in it who are distinct from himself and from inert objects. They live their own lives. Some are close to him; the overwhelming majority are unknown to him and unknowing as well as uncaring of him. He may represent them individually or collectively, to himself, and in this way make them part of his thought world; he thus can conceive of himself as capable of acting with respect to them, addressing and communicating to them. However, throughout he is aware that even those closest to him do not share his memories and are not privy to his feelings, fantasies, and fleeting thoughts. If he wishes to make these known, he must express them by means that allow him to communicate them to others, being aware as he does so of the varying distances of his addressees from his personal life and experiences.

Through his multiple transactions with his social and, in large measure, socially defined physical environment, transactions in which, to an incalculable degree, his capacity for symbolization is essential, the normal adult comes to represent to himself a domain of social objects, all not only distinct from persons, but also distinguished among themselves. Almost automatically, he is able to articulate his global experiences and to segregate his feelings, hopes, and fears about these objects from the objects themselves and to perceive and classify them in ways that other members of his community are likely to categorize them. Thus he can distinguish his intentions toward the objects or distinguish the personal meanings and attitudes that he has in regard to them from the objects themselves as socially shared referents. In the main, he locates these objects in a causal-pragmatic-functional network, one which he shares with other normal adults. Nevertheless, he is aware that objects and events of concrete experience can be viewed in other ways, that is, aesthetically, religiously, and so on. These ways of viewing are, however, sharply distinguished from the pragmatic-causal, and he does not confuse the relations thus established among objects and events with their causal relationships.

The normal adult is able, through the indissociable interplay of socialization, semiautonomous intellectual development, and symbolic activity, to entertain also a world of ideal objects, occupying a different "place" from that pre-empted by the everyday things and events with which he has direct and immediate commerce. Through an activity of symbolically mediated hypostatization, he can and does build up in the matrix of society purely intellectual objects and relations, which concrete objects and events are taken to exemplify. These, as well as the particular things of experience, can become the objects of his thought and reference as well as of the thought and reference of others.

In his communication to others, the adult takes account not only of his addressee, but also of the context, socially-symbolically defined, in which his communication takes place.[144] Normally, he adapts his communication to this context, referring to it when pertinent, using it as one component of his medium of communication when germane, disregarding it to the extent that it is irrelevant to that about which he wishes to communicate. Especially in the latter situations, he makes a sharp distinction between context and scene and recognizes the necessity to provide an ideal, symbolically delineated locus for the subject matter of his communication, enabling his addressees to share with him the nonpresent situation to which his symbolic utterance pertains.

To stabilize and define the transitory impressions which are experienced by him, to locate them in one or another region of his world, and to think and communicate about them, especially when they have vanished, the normal adult uses chiefly the linguistic instrumentality, which he shares as a common medium with others. Early in ontogenesis, this instrument, or better, the rules governing its use, are internalized and integrated with his innate capacity for vocalization; this capacity, subordinated to thought and symbolization, is thus transformed into speech. In early ontogenesis, however, speech is syncretically fused with context, action, and private images and is immersed in the child's affective, ludic, and practical life; it is used mainly as a means of expressing feelings and demands, as a thing with which to play, or as a device which substitutes for action in the attempt to control and regulate the behavior of others. Furthermore, language is there assimilated to an interfused, syncretic, diffuse, labile, and relatively rigid mentality and cannot have the value of representing stable thoughts and articulated concepts of which the child is as yet incapable.

Speech can be used in these ways by the adult in special circumstances and with particular addressees, but he is also capable of using the medium in an ideal manner to represent objects and events, to symbolize how he feels about phenomena, and so on. To do this adequately, he must at the minimum be cognizant of and respect the communal and autonomous values of the various parts of the medium, not only the referential values of lexicon and syntax, but the expressive values as well. Even in poetry,[107,116,148] where he may be interested in the aesthetic properties of linguistic sounds, he must recognize that he can "alter the sounds of words no farther than the [common] sense would follow,"[107] on pain of excommunication. Archibald MacLeish has succinctly put it: "If you want the sound of *lurk* instead of *lark* in your sonnet you can write it down but your bird will disappear. If you want to play sonorous games with *l'amour, la mort,* and *la mer* you may: but you will still have love, death, and the sea on your hands with no possibility of escape. . . ."[107] In sum, the principal medium of representation and communication is, for the normal adult, recognized as autonomous and interpersonal; a structure with values distinct from his own actions, thoughts, feelings and associations, and one that cannot be manhandled in an idiosyncratic and arbitrary way.

It will be recognized that the young child participates in symbol-situations of quite a different kind from those in which the adult is capable of engaging. The child dwells in a world of a limited number of vaguely differentiated addressees from whom he, himself, is not yet articulated; is governed primarily by affective, ludic, and practical intentions; is concerned with relatively few referents, and these highly charged and infused with personal meanings, which he is unaware are not shared by others; is restricted to a few concrete contexts; does not clearly take into account the differences between his contexts of communication and the scenes in which his referents are located and is, furthermore, often incapable of representing these scenes if requested to do so; and, finally, does not sharply distinguish the medium of language from his other media, from his context of utterance, and from his affect and action. For a detailed treatment of the ontogenetic progression, the reader is referred to *Symbol Formation.*[168]

Again, in psychopathology, there is a tendency toward dedifferentiation and disintegration of symbol-situations as a whole and of all the constituents of such situations. This does not mean that the psychotic regresses to childhood. In the dissolution of his functioning he almost invariably carries with him residues of social and intellectual attainments, mastered at higher levels of functioning than the child is capable of reaching. Moreover, in less extreme cases at least, the psychotic often manifests such higher levels of functioning, even if only transitorily, sporadically, and outside executive control. Again, as interfused, syncretic, diffuse, labile, inflexible, and even unintegrated, as the child's functioning may be, he is not disintegrated in his activity. The primitivity of the psychotic, it must be reiterated, is of a different kind than the primitivity of the child, just as it is of a different kind than the socially adaptive primitivity of men in technologically backward societies.

In discussing the character of psychopathological symbol-situations, one need not elaborate either on the fusion between the patient and his momentary addressee or on the lack of differentiation among addressees. These closely related phenomena have almost invariably been observed in schizophrenia and related disorders. In the first instance, the patient tends to feel fused with, incorporated within, or threatened by invasion from, the one to whom he communicates. In the second instance, the addressee is not grasped as a distinct, stable, and socially determinate contemporary, but is a diffusely interwoven composite of remembered, feared, desired, "need-relevant" persons; in extreme cases there is that radical autism in which even the unstable linkage of the patient with such composites dissolves. Underlying these phenomena from a formal or structural point of view is the syncresis of activities (feeling, wishing, perceiving, and so forth) and the consequent psychophysical undifferentiatedness (lack of distinction between the ideal and the substantial-concrete) that characterize the most primitive levels of functioning.

One need not dwell overlong on the differentiation and disintegration of the patient's intentions, that is, his attitudes, purposes, and meanings with regard to objects and to others in symbol situations. These are invariably fused and ambivalent, frequently unknown to him either before or after he has expressed them, often manifested in an involuntary and uncontrolled manner, and sometimes experienced as unrelated to himself and infused into him by malevolent others. Rarely is he oriented toward an impersonal, factual communication about neutral states of affairs or oriented toward representing his feelings and wishes to another; in the main, his posture is egocentric-affective, and he is prompted unwittingly to express or enact his diffusely felt rage, fear, love, or desire to control the objects of his world, and the like. Since he has little control over his attitudes, they are liable to be both labile and inflexible; for example, Searles[136] reports one patient who suddenly paused in the midst of vicious paranoid tirades against him to ask him in a calm and friendly manner for a light for her cigarette (p. 543).

That the patient's relations to his world of objects and events, his actual and potential referents, undergo dissolution, likewise requires little commentary or illustration. The factors in play here are, in large measure, the same as those that enter into the patient's relation to his addressees. Due to a syncresis of wishing, remembering, imagining, perceiving, and so forth, and an impairment of critical analytic and synthetic operations, the patient is often unable to articulate his momentary impressions in a manner so as to shape and categorize objects and events in social-consensual terms. Rather he senses and defines impressions in terms of idiosyncratic-affective categories. Such tendencies toward construing impersonal events in personal-emotional terms are especially illuminating in those cases where the critical faculties are still operative but have become to some extent dissociated from percept formation. For example, Alberta Szalita refers to one of her patients, who reported, "I went to visit a recent acquaintance of mine. . . . We had dinner together. . . . After dinner, the hostess served coffee. When I raised my eyes as I was reaching for the cup of coffee, her face looked different than before. I felt that my sister was handing me the cup. I had to move closer to check whether it was my sister or not." Another of Szalita's[156] patients, looking at the ceiling of her office, claimed to have seen a witch there moving her arms. He later remarked, "You need not tell me that there is no witch on the ceiling—I know that as well as you do. But I really felt it" (p. 59).

There are many related phenomena in the schizophrenic's affective-mythopoetic construction of reality: the unwitting transformation of feeling states into things and concrete happenings, the substantialization of thoughts, the equation of parts with wholes, attributes with things, and so on. The reader is referred to the writings of the close students of schizophrenia.[3–11,36,124,134,136–141,149–150,155–157] One may also once again refer the reader to the works of Levy-Bruhl[55,97–99] and Cassirer;[42,44,45] despite the fact that these authors do not directly concern themselves with pathological cases, their discussions of the principles governing the formation of an affective-mythopoetic world are clearly relevant. Both have the further merit of avoiding the "naïve realism," which would take as given to primitive mentality those distinctions that are established only at higher levels of functioning.

The dedifferentiation and disintegration of the relation between patient and context is of the same order as that between the patient and addressee and between the patient and referents. It is noteworthy that, on one hand, the patient often cannot exclude the context, even when it is irrelevant to the communication situation; on the other hand, that he frequently takes no cognizance of the socially defined context in expressing himself. Thus, with regard to the lack of differentiation, one of McGhie and Chapman's[112] patients remarked, "My concentration is very poor. I jump from one thing to another. If I am talking to someone they only need to cross their legs or scratch their heads and I forget what I am saying. I think I could concentrate better with my eyes shut"[112] (p. 104). Another patient remarked, "I can't concentrate. It's diversion of attention that troubles me. I am picking up different conversationsl It's like being a transmitter. The sounds are coming through to me but I feel my mind cannot cope with anything. It's difficult to concentrate on any one sound. It's like trying to do two or three different things at one time" (p. 104).

With regard to the tendency of the patient to be oblivious to or to be dissociated from the socially defined context, the following case is illustrative: one of Cameron, Freeman, and McGhie's patients[36] "would occasionally spring to her feet, with her face convulsed, and scream obscenely. The content of these comments was usually to the effect that a fat old woman was in bed having sexual intercourse with a man who did not belong to her. She, the patient, was not going to continue to bring home her pay-packet to keep them in this situation—and she was not going to scrub the floors either" (p. 273). Such dissociation may be due in part to the tendency of the patient unwittingly to equate the affectively relevant scene in which the "referent" is located to the present, socially-defined context. It is also in many cases due to the loss of hierarchic integration, and hence it is the tendency of the autistic patient to blurt out involuntarily whatever he feels irrespective of the present context.

Striking in many cases of psychopathology is the disintegration and dedifferentiation of the relationship between patient and the scene in which the referent of the communication would normally be located. This dissolution is one of the major determinants of the bizarre appearance of patients' expressions, even if these are comprised of well-formed sentences. In normal persons such settings may either be justifiably assumed to be known by others or may be symbolically (ideally) established. Due to the pathological person's syncretic mentality, however, all ideal relationships tend to disintegrate. The consequence is that the patient becomes immersed in the scene and conflates it with his current context. The normal person's temporary immersion in affective memories and fantasies is an approximation to this kind of situation.

The strikingly altered relationships between addressor and medium (or media) of representation are part and parcel of the more pervasive dedifferentiation and disintegration of functioning which marks the disturbed individual's relation to all of the other constituents of symbol situations and, indeed, to all of the other events in his life.

Due to the syncresis of his mental operations and to the profound impairment of his capacity to maintain purely ideal relationships (including those of symbol to intention and symbol to referent), the severely disorganized patient often apprehends the communal symbol systems (for example, language, conventional gestures, pictures) in an affective-mythopoetic way. He thus does not treat them as autonomous of himself, with relatively fixed values, and subject to stable rules of usage enjoined on all; instead he tends to assimilate them to his magical-austistic universe and to endow them with idiosyncratic-emotional significance. Construing his world with a relatively unstratified mentality, he often has difficulty in distinguishing between the conventional values of symbols and his personal wishes, fears, images, and uncontrolled associations. Furthermore, the loss of ideality[35] bars him from distinguishing items of his experience as things and actions. Thus words and gestures which the normal person would take as having merely representational values may, for the patient, be experienced as incarnate objects and efficacious actions; and ordinary objects and actions may be infused with a "mystical" significance and be perceptually transformed in terms of that significance. Such interpenetration and loss of ideality sometimes leads to the patient either to refusing to use symbols or to dismembering words as he would things that threaten him; analogously, actual or magical destructive activities may be carried out against ordinary things that are invested with malevolent significance, including as Szasz notes,[157] one's own body. Such interpenetration allows the patient to construct his own forms and to imbue these with a significance that they have for him alone, although he may feel that this significance is obvious to anyone; hence, in part, the emergence of neologisms, neomorphisms, glossolalias, and the like.

These processes of dedifferentiation and of the correlative disintegrative processes may lead also, in the extreme, to a radical dissociation between the patient and the communal symbols systems. He may find it difficult to channel his diffuse-affective experiences into the conventional linguistic forms. He may experience an enormous gap between his thought-feelings and his utterances. His own utterances and productions, themselves, may appear to him alien, external, or thrust into him from without. Sometimes they will be totally incomprehensible to him. As the rules governing the different usages of a medium (for example, language) interpenetrate—rules internalized in the course of ontogenesis and operative in the production and comprehension of symbols in their varied functions—and, as hierarchic control diminishes, the patient's utterances may become dystaxic or agrammatic or may verge on verbigeration. Within the microgenesis of a single utterance, he may sometimes be pulled by the external phonaesthetic features of words, sometimes by their syntagmatic relations (for example, horse runs), sometimes by their paradigmatic relations (for example, cow-horse, cow-calf), and so on. The final outcome may be a word-salad.

It is not possible here to illustrate all of these phenomena, but examples of several of them may be presented. It will be observed that these examples often reveal more than one of the phenomena. Consider the idiosyncratic use of communal symbol systems. It is unlikely that Rosenfeld's patient, a severely disturbed schizophrenic, used the statement "The Russians *were* our allies" to impart factual information or to speak metaphorically or allegorically; rather it appears that he gave this utterance the personal value. "A person who may appear to be your friend for a while can turn against you, and you might do that to me." One may say in this connection that, although one cannot conclusively rule out an awareness of an allegorical intention,[101] it seems likely here that the patient infused his vaguely sensed horror of being betrayed into the apparently neutral utterance. Such phenomena characteristically occur in early phases of the genetic actualization of thought ("microgenesis"),[10,63,165] but are normally barred from overt expression.

In another instance of the infusion of personal meanings into conventional symbols, one of Bobon's patients drew the eye of a fish, which he believed not only gave him access to his past states, but also which he felt would also be efficacious in allowing others access to unknown realms.[24] Again, for Mme. Sechehaye's patient, Renée, the drawing of a circle with a point in it was both plurisignificant and profound in personal meaning[141]; the point in the circle signified both a process of disintegrating into nothingness and a feeling that in this process one would rediscover mother (p. 983).

To illustrate the transformations of "linguistic forms" into efficacious actions, one may mention Schilder's patient,[134] who believed she could destroy objects by words, an act she neologically designated as "bumping off" ("*bumbse ab*"). One may consider also another of Rosenfeld's patients who felt that whenever his analyst made an interpretation he literally put himself into the patient's mind.[130] For another example, one may take Bobon and Roumeguere's patient, Antoine, who, persecuted by a mass of invisible living corpuscles, used "words" (and gestures) as efficacious actions to dispel or control these malevolent entities.[28] For the patient, the spoken word was a power in itself; in pronouncing it, one perturbed whatever it designated for the utterance of the word automatically unsettled the *elementaux* that corresponded to it. The patient, himself, remarked: ". . . each word represents the material thing: it is a power, it is the stuff in question . . . you say the name of a city and you sense that the atmosphere of the city has changed . . . when you say the name of a person, you influence him in the same way; my name influences me in a certain way when it is pronounced, how it is pronounced, and by whom . . ." (p. 818). Underlying all of these instances is an interpenetration and fusion of meaning (thought), referent, and "symbol."

To illustrate the transformations of gestures into magically efficacious actions, one may again refer to Bobon and Roumeguere's patient.[28] Antoine would stop all influx of aggressive elements against his person by turning his back to this influx, arms dangling and palms turned toward the rear. He would purify himself by allowing his arms to hang, palms turned toward his body, fingers spread. A rotary movement of his body and elevation of his head, with or without concomitant torsion of the trunk, constituted an infallible attack (pp. 816ff.).

Again, symbol-realism often underlies the avoidance and/or dismemberment of "linguistic forms." A striking example of this phenomenon is provided by Bobon's[25] patient Joseph, who admitted that he amputated and deformed certain words, even to the point of unrecognizability, because their use was mysteriously charged with unlucky influence (pp. 361ff.). "Certain words should not be pronounced," he said, "because they are revolting . . . because there are always words which attract bad things." Thus, this patient would use "tection" instead of "protection" because "tection is protection in the good sense. I take half of the word because protection is the bad meaning."

A fascinating example of the steps in the construction of a progressively complicated neologism (and a corresponding neomorphism) is presented by Stuchlick and Bobon.[152] Their patient attempted to "synthesize his ideas" both in drawings and in words. In one instance he started out with the discrete notions and drawings of a fish (*poisson*), a maiden (*pucelle*), a pacifier (*suçon*), a caterpillar (*chenille*), a cow (*vache*), and a locomotive (*machine*). He next joined together, in drawings and "words," pairs of these referents: "*poicelle*," "*sucelle*," "*sucenille*," "*vachenille*," "*mache*." Then triplets: "*poisucelle*," "*sucelille*," "*suvachenille*," "*machenille*." Finally, he constructed a conglomerate of fragments of all the drawings and names: "*poisucevamachenille*."[151]

There are many other phenomena in the disturbed person's use of communal media and in construction of his own vehicles of symbolization that would further reveal the tendencies toward dedifferentiation and disintegration characterizing psychopathology.

The brief outline here of some aspects of the comparative-developmental approach to psychopathology of symbolic activity may be filled out and supplemented by the reader through a perusal of *Symbol Formation*. The reader is also referred to the following:[6–7, 137, 139–140, 146, 149–150] Explicit mention should be made of the significant works of Piro[122] and Bobon and others,[22–28, 151] works that are relatively unknown in America.

Normal Analogues to Symbolization and Language Uses in Psychopathology

This chapter should not be concluded without, at least, a brief discussion of phenomena occurring naturally in the everyday lives of normal adults that bear a remarkable resemblance to the handling of symbolization and language in psychopathology.

Dreams

There is the dream. Freud has provided a classic treatment of oneiric phenomena. In general, he has approached these phenomena from a psychodynamic point of view, although he presents an extensive discussion of formative factors in the structuralization of dreams. Unfortunately, however, he and his followers have tended to convey the impression that the latent contents or the dream thoughts are initially more or less discrete, lexico-syntactically organized patterns which are subsequently operated upon by acts of condensation, displacement, and the like, to produce the manifest content of the dream. Whether or not Freud actually intended to maintain that the outcome of analysis is temporally prior in the formation of the dream, such a notion must be rejected by those holding to a comparative developmental viewpoint. Interfusion, syncresis, and diffuseness precede articulation and discreteness in the microgenesis as well as in the ontogenesis of explicit thought.

Due to the brilliance of Freud's work and to the spread of a psychodynamic orientation, the outstanding monograph by Kraepelin on speech disturbances in dreams[89] has been generally overlooked in psychiatric circles. In this monograph, Kraepelin analyzes speech in dreams from primarily a formal point of view and highlights the similarities between such speech and the speech of severely disturbed schizophrenics. For a brief treatment of Kraepelin's work the reader is referred to *Symbol Formation*. Bobon[22] presents a detailed analytical summary of Kraepelin's main points.

Hypnagogic Phenomena

Of equal importance to the study of dreams is the investigation of hypnagogic phenomena. As one knows, Silberer[143] believed that these phenomena were susceptible to quasi-experimental control and that they could thus provide an excellent way of examining in slow motion the manner in which thoughts are given form on relatively primitive levels of organization. Silberer's views on the formation of symbols, including his tacit belief that one is not limited to the expression of a circumscribed sphere of contents either in dreams or in hypnagogic states, accord closely with those maintained by comparative developmental theorists. It may be noted that, in the main, Silberer was more concerned with imaginal representation than with linguistic forms in hypnagogic states.

The author, in the course of writing this paper, adopted Silberer's procedure, but was oriented toward such "linguistic forms." In one instance, he dozed off as he was thinking of those very narrow views of cognition which observe the thinking process from a remote vantage point. This "thought" was realized in an image of a long road that at the same time looked like a pencil telescope; there was someone looking through it. At the same time the thought "It's a tunnel potential" was uttered. In the hypnagogic state, there was a vague feeling that the author wanted to say "tunnel vision" and was aware that "potential" was somewhat tangential to what he was trying to say.

In another instance the author had just read a passage in a work where a cautious alienist had discussed a theme to the effect that one could not be very sure concerning the nature of thought organization in schizophrenia in the absence of experimental work. Earlier in the day the author's mother-in-law had arrived with many pieces of soap for his youngest son. Earlier, too, a colleague, noted for cautious experimentation and stringent criticisms of any conclusions not based on experiment, had been given a birthday party attended by the author. As the author dozed off, thinking about the objectivity of "clean" experiments and about the difficulty of getting impeccable information about pathological thought, he found himself hearing his colleague say "It's unjective to throw soap" in a tone which suggested that the colleague was once again railing against his bête noir.

These illustrations may suggest that the processes of symbol formation and that the genetic actualization of the transformation of "thoughts" into words may well benefit from a more thorough examination of those hypnagogic states where one can partly witness the formation of a symbolic expression "not answering the aim/and that unbodied figure of the thought/that gave 't surmised shape."

Attempts to approach primitive levels in the formation of symbols in normal adults in a somewhat more orthodox, but still far from clean, experimental fashion are discussed in detail in the author's joint work with Heinz Werner, *Symbol Formation*.

Bibliography

1. ABRAMS, MEYER H. *The Mirror and the Lamp*. New York: Oxford University Press, 1955.
2. ALLPORT, GORDON. *Becoming*. New Haven, Conn.: Yale University Press, 1955.
3. ARIETI, SILVANO. "Special Logic of Schizophrenic and Other Types of Autistic Thought," *Psychiatry*, 4 (1948), 325–338.
4. ———. "Autistic Thought," *Journal of Nervous and Mental Diseases*, 3 (1950), 288–303.
5. ———. *Interpretation of Schizophrenia*. New York: Robert Brunner, 1955.
6. ———. "Some Aspects of Language in Schizophrenia," in Heinz Werner, ed., *On Expressive Language*. Worcester, Mass.: Clark University Press, 1955.
7. ———. "Schizophrenia: The Manifest Symptomatology, the Psychodynamic and Formal Mechanisms," in Silvano Arieti, ed., *American Handbook of Psychiatry*, Vol. 1. New York: Basic Books, 1959.
8. ———. "The Loss of Reality," *Psychoanalysis and the Psychoanalytic Review*, 48 (1961), 3–24.
9. ———. "Volition and Value: A Study Based on Catatonic Schizophrenia," *Comprehensive Psychiatry*, 2 (1961), 74–82.
10. ———. "The Microgeny of Thought and Perception," *Archives of General Psychiatry*, 6 (1962), 454–468.
11. ———. "Studies of Thought Processes in Contemporary Psychiatry," *The American Journal of Psychiatry*, 120 (1963), 58–64.
12. BALKEN, EVA RUTH. "Psychological Researches in Schizophrenic Language and Thought," *Journal of Psychology*, 16 (1943), 153–176.
13. BARBU, ZEVEDIE. *Problems of Historical Psychology*. New York: Grove Press, 1960.
14. BARFIELD, OWEN. (1927) *Poetic Diction*. London: Farber and Farber, 1951.
15. BERGIN, THOMAS G., and MAX H. FISCH, eds., *The New Science of Giambattista Vico*. New York: Doubleday Anchor, 1961.
16. BERNHEIMER, RICHARD. *The Nature of Representation: A Phenomenological Inquiry*. New York: New York University Press, 1961.
17. BERTALANFFY, LUDWIG VON. *Modern Theories of Development*. London: Oxford University Press, 1933.
18. ———. "On the Definition of the Symbol," in Joseph Royce, ed., *Psychology and the Symbol*. New York: Random House, 1965.
19. ———. "General Systems Theory and Psychiatry," in Silvano Arieti, ed., *American Handbook of Psychiatry*, Vol. 3. New York: Basic Books, 1966.
20. BION, W. R. "Language and the Schizophrenic," in Melaine Klein, et al., eds., *New Directions in Psychoanalysis*. New York: Basic Books, 1957.
21. BOAS, FRANZ. *The Mind of Primitive Man* (1911). New York: Collier Books, 1963.
22. BOBON, JEAN. *Introduction historique à l'étude des néologismes et des glossolalies en psychopathologie*. Paris: Masson, 1952.
23. ———. "Psychopathologie de l'expression plastique (mimique et picturale)," *Acta Neurologica et Psychiatrica Belgica*, 11 (1955), 923–929.
24. ———. "Symbolisme et metamorphoses du poisson dans l'oeuvre picturale d'un schizophrene," *La Vie Medicale* (1956), 40–45.
25. ———. "Les pseudo-glossolalies ludiques et magiques," *Journal Belge de Neurologie et de Psychiatrie*, 47 (1947), 219–238, 327–395.
26. ———. "Contribution à la psychopthologie de l'expression plastique, mimique et picturale: les 'neomimismes' et les 'neomorphismes,'" *Acta Neurologica et Psychiatrica Belgica*, 57 (1957), 1031–1067.
27. ———. *Psychopathologie de l'expression*. Paris: Masson, 1962.
28. ———, and PIERRE ROUMEGUERE. "Du geste et du dessin magiques au néographisme conjuratoire," *Acta Neurologica et Psychiatrica Belgica*, 57 (1957), 815–829.
29. BOYD, WILLIAM C. "The Contributions of Genetics to Anthropology," in Sol Tax, ed., *Anthropology Today: Selections*. Chicago: Phoenix Books, 1962.
30. BRUNSWIK, EGON. "Ontogenetic and Other Developmental Parallels to the History of

Science," in H. M. Evans, ed., *Men and Moments in the History of Science.* Seattle, Washington: University of Washington Press, 1959.

31. BUHLER, KARL. *Sprachtheorie.* Jena, Germany: Gustav Fisher Verlag, 1934.
32. BUNGE, MARIO. "Levels: A Semantical Preliminary," *Review of Metaphysics*, 13 (1960), 396–406.
33. BURNHAM, DONALD L. "Misperception of Other Persons in Schizophrenia," *Psychiatry*, 19 (1956), 283–303.
34. BURY, JOHN B. *The Idea of Progress.* London: Macmillan, 1932.
35. CAIRNS, DORIAN. "The Ideality of Verbal Expressions," *Philosophy and Phenomenological Research*, 39 (1941), 453–462.
36. CAMERON, JOHN L., et al. "Clinical Observations on Chronic Schizophrenia," *Psychiatry*, 19 (1956), 271–281.
37. CASSIRER, ERNST. *Das Erkenntnisproblem in der Philosophie und Wissenschaft der neueren Zeit.* 3 vols. Berlin: B. Cassirer, 1906, 1907, 1920.
38. ———. (1910) *Substance and Function.* Chicago: Open Court, 1923.
39. ———. (1932) *The Philosophy of the Enlightenment.* Boston: Beacon Press, 1955.
40. ———. "Le langage et la construction du monde des objets," *Journal de Psychologie*, 30 (1933), 18–44.
41. ———. *An Essay on Man.* New Haven, Conn.: Yale University Press, 1944.
42. ———. *Language and Myth.* New York: Dover, 1946.
43. ———. *The Problem of Knowledge.* New Haven, Conn.: Yale University Press, 1950.
44. ———. *Philosophy of Symbolic Forms*, Vol. 1. New Haven, Conn.: Yale University Press, 1953.
45. ———. *Philosophy of Symbolic Forms*, Vol. 2. New Haven, Conn.; Yale University Press, 1955.
46. ———. *Philosophy of Symbolic Forms*, Vol. 3. New Haven, Conn.: Yale University Press, 1957.
47. ———. *The Logic of the Humanities.* New Haven, Conn.: Yale University Press, 1961.
48. CHAPMAN, LOREN J., et al. "Regression and Disorders of Thought," *Journal of Abnormal and Social Psychology*, 63 (1961), 540–545.
49. CLIVE, GEOFFREY. "Notes toward a Topography of the Irrational since the Enlightenment," *Journal of Existential Psychiatry*, 15 (1963), 177–204.
50. COLLINGWOOD, R. G. *The Idea of History* (1946). New York: Oxford University Press, 1956.
51. ———. *The Idea of Nature.* London: Oxford University Press, 1945.
52. CORNFORD, FRANCIS M. *Plato's Theory of Knowledge.* New York: Liberal Arts Press, 1957.
53. CRITCHLEY, MACDONALD. "The Evolution of Man's Capacity for Language," in Sol Tax, ed., *Evolution after Darwin*, Vol. 2. Chicago: University of Chicago Press, 1959.
54. DAVIE, DONALD. *Articulate Energy: An Enquiry into the Syntax of English Poetry.* New York: Harcourt Brace & Co., 1958.
55. DAVY, GEORGES. "La psychologie des primitifs d'après Levy-Bruhl," *Journal de Psychologie*, 27 (1930), 112–176.
56. DELACROIX, HENRI. *Le Langage et la pensée.* Paris: Alcan, 1930.
57. DE LAGUNA, GRACE. *Speech, Its Function and Development.* New Haven, Conn.: Yale University Press, 1927.
58. DEUTSCH, KARL. "Mechanism, Organism, and Society: Some Models in Natural and Social Science," *Philosophy of Science*, 18 (1951), 230–252.
59. DILTHEY, WILHELM. *Pattern and Meaning in History*, H. P. Richman, ed., New York: Harper & Brothers, 1962.
60. ELIADE, MIRCEA. *The Sacred and the Profane.* New York: Harcourt Brace & Co., 1959.
61. FEIBLEMAN, JAMES K. "Theory of Integrative Levels," *British Journal of the Philosophy of Science*, 5 (1954), 59–66.
62. FIRTH, JAMES R. "Modes of Meaning," in James R. Firth, ed., *Papers in Linguistics 1934–1951.* London: Oxford University Press, 1957.
63. FLAVELL, JOHN, and JURIS DRAGUNS. "A Microgenetic Approach to Perception and Thought," *Psychological Bulletin*, 54 (1957), 197–217.
64. FREEMAN, THOMAS. "On the Psychopathol-

ogy of Schizophrenia," *Journal of Mental Science*, 106 (1960), 925–937.

65. GEERTZ, CLIFFORD. "The Growth of Culture and the Evolution of Mind," in Jordan Scher, ed., *Theories of the Mind.* Glencoe, Ill.: The Free Press, 1962.
66. GELB, ADHEMAR. "Remarques générales sur l'utilisation des données pathologiques pour la psychologie et la philosophie du langage," *Journal of Psychologie*, 30 (1933), 403–429.
67. GLASS, BENTLEY, OWSEI TEMKIN, and WILLIAM L. STRAUSS, JR. *Forerunners of Darwin.* Baltimore, Md.: The Johns Hopkins Press, 1959.
68. GOLDBERGER, EMANUEL. "The Id and the Ego: A Developmental Interpretation," *Psychoanalytic Review*, 44 (1957), 235–288.
69. GOLDSTEIN, KURT. *The Organism.* New York: American Book Company, 1939.
70. ———. *Language and Language Disturbances.* New York: Grune & Stratton, 1948.
71. ———. "Concerning the Concept of 'Primitivity,'" in Stanley Diamond, ed., *Culture in History: Essays in Honor of Paul Radin.* New York: Columbia University Press, 1960.
72. GULLEY, NORMAN. *Plato's Theory of Knowledge.* New York: Barnes and Noble, 1964.
73. GURWITSCH, ARON. "Approach to Consciousness," *Philosophy and Phenomenological Research*, 15 (1955), 303–319.
74. HALLOWELL, A. IRVING. "The Recapitulation Theory of Culture," in A. Irving Hallowell, ed., *Culture and Experience.* Philadelphia: University of Pennsylvania Press, 1955.
75. HARRIS, DALE B., ed., *The Concept of Development.* Minneapolis, Minn.: University of Minnesota Press, 1957.
76. HAZARD, PAUL. *The European Mind: 1680–1715* (1935). Cleveland, Ohio: World Publishing Co., 1963.
77. HEAD, HENRY. *Aphasia and Kindred Disorders.* 2 vols. New York: Macmillan, 1926.
78. HENEL, HEINRICH. "Type and Protophenomenon in Goethe's Science," *PMLA*, 71 (1956), 651–669.
79. HOBHOUSE, LEONARD T. *Development and Purpose.* London: Macmillan, 1913.
80. HUGHES, H. STUART. *Consciousness and Society* (1958). New York: Vintage Books, 1961.
81. JAKOBSON, ROMAN. "Aphasia as a Linguistic Problem," in Heinz Werner, ed., *On Expressive Language.* Worcester, Mass.: Clark University Press, 1955.
82. ———. *Kindersprache, Aphasie und allgemeine Lautgesetze.* Uppsala, Sweden: Almqvist, 1941.
83. JONES, ERNEST. "The Theory of Symbolism," in Ernest Jones, ed., *Papers on Psychoanalysis.* Boston: Beacon Press, 1961.
84. KAPLAN, BERNARD. "Radical Metaphor, Aesthetic and the Origin of Language," *Review of Existential Psychology and Psychiatry*, 2 (1962), 75–84.
85. ———. "Developmental Aspects of the Representation of Time." Unpublished paper read at Fourth Annual Meeting of the New England Phychological Association, November 13, 1964.
86. ———. "Meditations on Genesis." Unpublished paper presented to the Boston Psychoanalytic Society, February 24, 1965.
87. KLUBACK, W. *Wilhelm Dilthey's Philosophy of History.* New York: Columbia University Press, 1956.
88. KONRAD, HEDWIG. *Étude sur la métaphore.* Paris: Vrin, 1958.
89. KRAEPELIN, EMIL. *Über Sprachstörungen im Traum.* Leipzig, Germany: Engelmann, 1906.
90. KUBIE, LAWRENCE. "The Distortion of the Symbolic Process in Neurosis and Psychosis," *Journal of American Psychoanalytic Association*, 1 (1953), 59–86.
91. KUNTZ, PAUL G. "Mythical, Cosmic and Personal Order," *Review of Metaphysics*, 16 (1963), 718–748.
92. LANGER, SUSANNE K. *Philosophy in a New Key.* Cambridge, Mass.: Harvard University Press, 1942.
93. LAVINE, THELMA. "Knowledge as Interpretation: An Historical Survey," *Philosophy and Phenomenological Research*, 10 (1950), 526–540, 11 (1950), 80–103.
94. LENNEBERG, ERIC. "Language Disorders in Childhood," *Harvard Educational Review*, 34 (1964), 152–177.
95. ———. "Understanding Language without Ability to Speak: A Case Report," *Jour-*

nal of Abnormal and Social Psychology, 65 (1962), 419–425.

96. LEVI, ALBERT W. *Literature, Philosophy and the Imagination*. Bloomington, Ind.: Indiana University Press, 1962.
97. LEVY-BRUHL, LUCIEN. *How Natives Think*. London: George Allen & Unwin, 1926.
98. ———. *Primitive Mentality*. London: George Allen & Unwin, 1923.
99. ———. *Les carnets de Levy-Bruhl*. Paris: Presses Universitaires de France, 1949.
100. LÉVI-STRAUSS, CLAUDE. *La Pensée sauvage*. Paris: Plon, 1962.
101. LORENZ, MARIA. "Expressive Behavior and Language Patterns," *Psychiatry*, 18 (1955), 353–366.
102. ———. "Problems Posed by Schizophrenic Language," *Archives of General Psychiatry* 4 (1961), 603–610.
103. LOVEJOY, ARTHUR O. *Revolt Against Dualism*. London: George Allen & Unwin, 1930.
104. ———. *The Great Chain of Being*. Cambridge, Mass.: Harvard University Press, 1936.
105. ———. *Essays in the History of Ideas*. Baltimore, Md.: The Johns Hopkins Press, 1948.
106. LOWITH, KARL. *Meaning in History*. Chicago: University of Chicago Press, 1949.
107. MACLEISH, ARCHIBALD. *Poetry and Experience*. Boston: Houghton Mifflin, 1960.
108. MAGNUS, RUDOLF. *Goethe as a Scientist* (1906). Translated by Heinz Norden. New York: Collier Books, 1961.
109. MALINOWSKI, BRONISLAW. "The Problem of Meaning in Primitive Languages," in C. K. Ogden and I. A. Richards, *The Meaning of Meaning*. New York: Harcourt Brace, 1923.
110. MANUEL, FRANK. *Shapes of Philosophical History*. Stanford, Calif.: Stanford University Press, 1965.
111. MARCUSE, HERBERT. *Reason and Revolution* (1941). Boston: Beacon Press, 1960.
112. MCGHIE, ANDREW, and JAMES CHAPMAN. "Disorders of Attention and Perception in Early Schizophrenia," *British Journal of Medical Psychology*, 34 (1961), 103–115.
113. MEAD, GEORGE HERBERT. *Movements of Thought in the Nineteenth Century*. Chicago: University of Chicago Press, 1936.
114. MEYERSON, IGNACE. "Review of Levy-Bruhl, Lucien, La Mentalité primitive," *L'Année Psychologique*, 23 (1922), 214–222.
115. NAGEL, ERNST. "Determinism and Development," in Dale Harris, ed., *The Concept of Development*. Minneapolis, Minn.: University of Minnesota Press, 1957.
116. NOWOTTNY, WINIFRED. *The Language Poets Use*. London: Athlone Press, 1962.
117. NUYENS, FRANCISCUS JOHANNES. *L'Evolution de la psychologie d'Aristotle*. Louvain, Belgium: Institut Supérieure de Philosophie, 1948.
118. OGDEN, C. K., and I. A. RICHARDS. *The Meaning of Meaning*. New York: Harcourt Brace, 1923.
119. OMBREDANÉ, ANDRE. *L'aphasie et l'élaboration de la pensée explicite*. Paris: Presses Universitaire de France, 1951.
120. PIAGET, JEAN. *Introduction a l'epistemologie génétique*. 3 vols. Paris: Presses Universitaire de France, 1950.
121. ———. *Play, Dreams and Imitation*. New York: W. W. Norton, 1951.
122. PIRO, S. *Semantica del linguaggio schizofrenico*. Naples: Acta Neurologica Policlinico, 1958.
123. PRICE, KINGSLEY. "The Work of Art and the Postures of the Mind," *Review of Metaphysics*, 12 (1959), 540–569.
124. RAMZY, ISHAK. "From Aristotle to Freud," *Bulletin of Menninger Clinic*, 20 (1956), 112–123.
125. RAPAPORT, DAVID. "Dynamic Psychology and Kantian Epistemology," Unpublished paper, 1947.
126. ———. "Psychoanalysis as a Developmental Psychology," in Bernard Kaplan and Seymour Wapner, eds., *Perspectives in Psychological Theory: Essays in Honor of Heinz Werner*. New York: International Universities Press, 1960.
127. REID, LOUIS ARNAUD. *Ways of Knowledge and Experience*. London: George Allen & Unwin, 1961.
128. RIESE, WALTHER. "The Pre-Freudian Origins of Psychoanalysis," in *Science and Psychoanalysis*. New York: Grune & Stratton, 1958.
129. ROSENFELD, HERBERT. "Transference Phenomena and Transference Analysis in an Acute Catatonic Schizophrenic Patient," *International Journal of Psychoanalysis*, 33 (1952), 457–464.

130. ———. "Considerations Regarding the Psychoanalytic Approach to Acute and Chronic Schizophrenia," *International Journal of Psycho-Analysis*, 35 (1954), 135–140.
131. RUSSELL, EDWARD A. *Form and Function: A Contribution to the History of Animal Morphology*. London, John Murray, 1916.
132. RYCROFT, CHARLES. "Symbolism and Its Relationship to the Primary and Secondary Processes," *International Journal of Psycho-Analysis*, 37 (1956), 137–146.
133. SAPIR, EDWARD. *Language*. New York: Harcourt Brace, 1921.
134. SCHILDER, PAUL. *Wahn und Erkenntnis*. Berlin: Springer, 1914.
135. SCHILPP, PAUL A., ed., *The Philosophy of Ernst Cassirer*. Evanston, Ill.: Library of Living Philosophers, 1949.
136. SEARLES, HAROLD F. "Integration and Differentiation in Schizophrenia," *Journal of Nervous and Mental Disease*, 129 (1959), 542–550.
137. ———. "Integration and Differentiation in Schizophrenia: An Overall View," *British Journal of Medical Psychology*, 32 (1959), 261–281.
138. ———. *The Nonhuman Environment: Its Normal Development and in Schizophrenia*. New York: International Universities Press, 1960.
139. ———. "Schizophrenic Communication," *Psychoanalysis and the Psychoanalytic Review*, 48 (1961), 3–50.
140. ———. "The Differentiation between Concrete and Metaphorical Thinking in the Recovering Schizophrenic Patient," *Journal of the American Psychoanalytic Association*, 10 (1962), 22–49.
141. SECHEHAYE, MARGUERITE. "'Affects' et besoins frustrés vus à travers les dessins d'une schizophrène," *Acta Neurologica et Psychiatrica Belgica*, 57 (1957), 972–992.
142. SHUTE, CLARENCE. *The Psychology of Aristotle*. New York: Columbia University Press, 1941.
143. SILBERER, HERBERT. "Report on a Method of Elicting and Observing Symbolic Hallucination Phenomena," in David Rapaport, ed., *Organization and Pathology of Thought*. New York: Columbia University Press, 1951.
144. SLAMA-CAZACU, TATIANA. *Langage et contexte*. 'S-Gravenhage: Mouton, 1961.
145. SPENCER, HERBERT. "The Comparative Psychology of Man," *Mind*, 1 (1876), 8–20.
146. SPIEGEL, ROSE. "Specific Problems of Communication in Psychiatric Conditions," in Silvano Arieti, ed., *American Handbook of Psychiatry*, Vol. 1. New York: Basic Books, 1959.
147. STACE, W. T. *The Philosophy of Hegel* (1923). New York: Dover, 1955.
148. STANKIEWICZ, EDWARD. "Problems of Emotive Language," in Thomas Sebeok et al., eds., *Approaches to Semiotics*. The Hague: Mouton, 1964.
149. STORCH, ALFRED. *Das archaisch-primitive Erleben und Denken der Schizophrenen*. Berlin: Springer, 1922.
150. ———. "Die Welt der beginnenden Schizophrenie und die archaische Welt," *Zeitschrift für die gesamte Neurologie und Psychiatrie*, 127 (1930), 799–810.
151. STUCHLIK, JAROSLAV. "Contribution à la psychopathologie de l'expression verbale: les néophasies et les néographies," *Acta Neurologica et Psychiatrica Belgica*, 57 (1957), 1004–1030.
152. ———, and JEAN BOBON. "Les 'druses écrites et dessiniées (Kontaminationen, blendings); pathogene de certain néomorphismes." *Acta Neurologica et Psychiatrica Belgica*, 50 (1960), 529–550.
153. SUTTON-SMITH, BRIAN. "Piaget on Play: A Critique," *Psychological Review*, 17 (1966), 104–110.
154. SZALITA-PEMOW, ALBERTA. "Remarks on Pathogenesis and Treatment of Schizophrenia," *Psychiatry*, 14 (1951), 295–300.
155. ———. "Further Remarks on the Pathogenesis and Treatment of Schizophrenia," *Psychiatry*, 15 (1952), 143–150.
156. ———. "Regression and Perception in Psychotic States," *Psychiatry*, 21 (1958), 53–63.
157. SZASZ, THOMAS. "The Psychology of Bodily Feelings in Schizophrenia," *Psychosomatic Medicine*, 19 (1957), 11–16.
158. TOULMIN, STEPHEN. *The Philosophy of Science*. London: Hutchinson University Library, 1953.
159. VENDRYES, JOSEPH. "Le Caractère social du langage et la doctrine de F. de Saussure,"

Journal de Psychologie, 18 (1921), 617–624.

160. WALLON, HENRI. "De l'Expérience concrète à la notion de causalité et à la represéntation-symbole," *Journal de Psychologie*, 29 (1932), 112–144.
161. ———. *De l'Acte à la pensée*. Paris: Flammarion, 1942.
162. ———. *Les Origines de la pensée chez l'enfant*. Paris: Presses Universitaires de France, 1947.
163. WERNER, HEINZ. *Comparative Psychology of Mental Development* (1940). 3d ed. New York: International Universities Press, 1957.
164. ———. "Change of Meaning: A Study of Semantic Processes Through the Experimental Method," *Journal of General Psychology*, 50 (1954), 181–208.
165. ———. "Microgenesis and Aphasia," *Journal of Abnormal and Social Psychology*, 52 (1956), 347–353.
166. ———. "The Concept of Development from a Comparative and Organismic Point of View," in Dale B. Harris, ed., *The Concept of Development: An Issue in the Study of Human Behavior*. Minneapolis, Minn.: University of Minnesota Press, 1957.
167. ———, and BERNARD KAPLAN. "The Developmental Approach to Cognition: Its Relevance to the Psychological Interpretation of Anthropological and Ethnolinguistic Data," *American Anthropologist*, 58 (1956), 866–880.
168. ———. *Symbol Formation*. John Wiley & Sons: New York, 1963.
169. WHEELWRIGHT, PHILIP. *The Burning Fountain*. Bloomington, Ind.: Indiana University Press, 1954.
170. WHITEHEAD, ALFRED N. *Symbolism: Its Meaning and Effect*. New York: Macmillan, 1927.

CHAPTER 50

MATHEMATICS AND CYBERNETICS

Anatol Rapoport

THE FUNDAMENTAL CONTRIBUTION of mathematics to science has been to provide a precise and contentless language in which to describe events, to formulate generalizations, and to deduce consequences of assumptions. Precision and independence from content are interdependent. The vocabulary of everyday language depends on the way perceptions and concepts are organized; for instance, on the particular way objects are classified or relations among them are interpreted. In attaching names to objects, properties, or actions, we fix the categories in which we think. These categories are of necessity too crude to capture the infinite variety of events that constitute "objective reality." Thus a content-bound language may impose a structure on our perceptions of the world and on the abstract concepts we form, and this structure may or may not correspond to the structure of reality.

Because mathematical language is contentless, that is, totally abstracted from perceptions, its structure is entirely transparent. In the exact (mathematicized) sciences the structure of a mathematical theory is constantly compared with the structure of a portion of the world under study. Mathematics itself, however, is concerned with the structure of relations independent of empirical content.

As an example consider the equation (a mathematical statement) relating the area of a circle to its radius, $A = \pi R^2$. It says that whatever be the radius of a circle, the ratio of the area to the square of the radius is always constant, equal approximately to 3.1415926. The statement is actually a composite of a potentially infinite number of statements, since it specifies the magnitude of the area of *all* possible circles. Assuming that the radius can be specified with infinite precision, the area can also be specified with infinite precision, because the number π can be calculated with infinite precision. However, the statement cannot refer to anything in the empirically observable world, because there are no perfect circles and because physical measurements cannot be made with infinite precision.

The statement refers only to idealized objects in an idealized mathematical world.

The scientific revolution of the seventeenth century was a consequence of a discovery that certain real events could be *almost* precisely described by idealized mathematical models, in the first instance, the motions of heavenly bodies and the behavior on physical bodies subjected to specified forces under controlled conditions. Thus the first mathematicized science was born—mechanics.

The scientific revolution of the seventeenth century is generally recognized as the impetus that stimulated the Industrial Revolution of the eighteenth century and consequently the immense social changes that came in its wake. To appreciate the significance of this impetus fully, it is necessary to recognize the *conceptual* impact of mathematicized science. First, the world of matter appeared to be governed by physical laws. These laws, however, could no longer be stated as metaphysical principles like "Nature abhors a vacuum," "There is no effect without a cause," or "All things consist of substance and form." Physical laws are invariably stated as mathematical equations—relations among quantities—and the quantities themselves represent results of specified measurements, that is, concrete operations with meter sticks, balances, clocks, thermometers, barometers, potentiometers, and the like. Implied in each physical law are predictions of what will be observed under *specified* conditions. Both the conditions and the observations having been specified as quantities, that is, readings on instruments, the truth or the falsehood of the assumptions can in principle be determined by independent observers. Thus *philosophical* arguments about the validity of generalizations become irrelevant. In the final analysis the truth of an assertion becomes a matter of objective verification of observations. Therefore, the first result of the scientific revolution was that of fixating the specific meaning of "truth" in the context of scientific discourse, making it independent of pronouncements of authority, of speculations couched in verbal arguments, or metaphysical concepts.

Second, mathematical language has greatly expanded the scope of *deduction*. Deduction is a process by means of which, assuming the truth of some assertions, we can assert the truth of other assertions. Syllogistic reasoning is an example of deduction applied to assertions involving class inclusion. For instance, assuming that no A is B and some C are A, we can conclude that some C are not B. Rules of mathematical operation vastly expand the range of deductive reasoning. Thus, from mathematical equations expressing physical laws, a vast number of other quantitative relations can be deduced by chains of mathematical reasoning. Empirical verification of the deduced relations corroborates the validity of the laws. Empirical falsification of the deductions necessitates a search for the roots of the discrepancy. At times it is discovered that certain conditions had not been taken into account. At times the formulation of the laws is modified to bring them into closer correspondence to reality.

In this way science has changed fundamentally the old conception of knowledge as a collection of insights of wise men to be absorbed by studying texts. Scientific knowledge revealed itself as constantly growing and constantly being revised in the light of new observations and new interpretations of what is observed. The most important single factor effecting this change has been the adoption of mathematics as the language of the exact sciences. Assertions in that language leave no doubt about what is asserted (and consequently what must be done to test the assertions) and, moreover, bind the assertions into logical interdependence, the organic structure of scientific theories.

The role of the exact sciences in technology is obvious. Their role in the development of scientific medicine is no less apparent. Diagnostic procedures have come to depend more and more on refined observations made possible by instruments and laboratory procedures. In fact, scientific diagnosis is largely formulated in quantitative terms: temperatures, blood pressures, concentrations of substances in body fluids, shapes of electrocardiograms and electroencephalograms. Chemotherapy and physiotherapy are extensions of chemical and physical technology to medicine. Genetic etiology of diseases is discovered by statistical techniques. Effectiveness of drugs and other forms of therapy is evaluated by statistical inference. Indeed, the bulk of contemporary scientific medicine stems from the conception of the living body as a material system and of its living process as a complex network of physical and chemical processes that preserve a certain dynamic balance. The balance can in principle be described by certain limits within which the parameters of the process may vary. Disease can be defined in terms of deviations from these limits. If the deviations are reversible the "normal state" can be restored. Otherwise, death eventually occurs, which means that the dynamic processes that characterize the living organism can no longer be re-established.

To the extent that psychiatry is rooted in knowledge of organic structure and function, the same methods and conceptualizations apply to its findings. Neural anatomy and histology, neurophysiology, biochemistry, and genetics have all contributed to scientific psychiatry and so have demonstrated the relevance of the contributions of mathematics. No less important are the contributions of statistics (a branch of applied mathematics), an indispensable tool in studying gross trends and in evaluating results of therapeutic procedures on populations of patients.

In short, wherever psychiatry is concerned with physical events or with assessment of causes and effects on a gross scale, mathematics (including statistics) contributes to it as it does to any other science.

The Mind-Body Problem

The cleavage that still persists between psychiatry and other branches of science, including scientific medicine, is rooted in the Cartesian mind-body dualism as reflected, for instance, in the distinction between "organic" and "functional" mental disorders. From the standpoint of the essentially materialistic world outlook embodied in at least classical natural science, the mind-body dichotomy is not essential. It is seen not as a reflection of a dualism of reality but simply as an idea induced by our direct introspective knowledge of our state of consciousness, which seems different from the sort of knowledge we obtain through our senses about the external world. For the materialist, "mind," "consciousness," and so forth are only aspects of material events; for instance, nervous activity perceived "from the inside" as it were, rather than from the outside. From this point of view "thoughts," "concepts," "memories," "emotions," and the like are assumed to be the subjective aspects of objectively observable events, in principle describable in physiological terms.

The qualification "in principle" frees the adherent of this view from the necessity of demonstrating its validity in each specific instance. He is content to search for physiological correlates of mental activity, and whenever he finds apparent correlates, he is satisfied that the discovery corroborates the basic reductionist assumption.

The question remains of what constitutes a correlate of mental activity. Some phenomena clearly deserve the name; for instance, reports by individuals of thoughts, feelings, and the like reproducibly evoked by stimulating specific areas of the brain (as in experiments performed on patients undergoing brain surgery). Other evidence is obtained from ablation experiments on animals, where reproducible behavioral changes are effected. Here, since we have no access to the animals' mental state via reports, the corroboration of the reductionist hypothesis must depend on a tacitly assumed linkage between mental activity (not directly observable) and behavior (directly observable). Since, however, the materialist takes this linkage for granted, he is satisfied that reproducible correlations between anatomical structures and physiological events, on the one hand, and behavior patterns, on the other, corroborates the identification of "mind" with material events. The task of reduction, accordingly, becomes that of disclosing the "mapping" of neural events upon behavioral events.

The simplest "mappings" of this sort go back to the discovery of the reflex arc. A large advance is associated with Pavlov's discovery of the conditioned reflex. Thereby the extreme flexibility of behavior patterns of higher animals appeared explainable in principle. Objections to what appeared to be a mechanistic conception of behavior (and, by implication, of mental activity) revolved around the so-called purposefulness or "goal-directedness" of animal behavior, which, it was said, eluded all explanations based on mechanical models. The argument is similar to that of vitalists, who would subsume all living processes (not only behavior) under "goal-directed" ones, to be clearly differentiated from mechanical (not goal-directed) processes characteristic of the nonliving world.

It is true that classical physical science expelled goal-directedness from its conceptual repertoire. However, the conception of instantaneous local "causality," devoid of teleological components, is not confined to classical mechanics. It pervades all mathematicized physical science. Processes governing chemical reactions and the propagation of electromagnetic waves are typically formulated in differential equations, which relate magnitudes of variable quantities to their rates of change. The solutions of these equations are time courses of the variables. Thus, if the totality of these magnitudes and the relations among them are taken to be the description of a system, then each instantaneous state of the system is, in a way, the "cause" of the immediately succeeding state. "Causality," then, when analyzed completely, turns out to be acting "here and now" without reference to future states or "goals."

A detailed examination of *some* aspects of the living process showed that they could be explained in terms of obeying known physical and chemical laws. In particular, the early contentions of the vitalists that the energetics of living processes cannot be derived from the law of conservation of energy proved to be groundless. Also many of the regulating physiological processes, which keep temperatures, concentrations of substances, and so forth within certain limits, turned out to be the effects of homeostasis, the preservation of nonequilibrium steady states. It has been shown that nonliving systems can also be regulated by homeostasis as long as they are permeable to exchanges of matter and energy with the environment (open systems).

A much more serious difficulty in the way of extending the mechanistic paradigm to apply to living systems is the conspicuously goal-directed nature of gross behavior. Only if such apparently purposeful behavior could be exhibited in a system where nothing but established physical laws were known to operate, could the mechanistic conception be extended to the behavior of living organisms in relation to the outside world.

The problem of vindicating the wider applicability of the mechanistic view of nature (to include at least some aspects of living behavior) became linked with the problem of constructing machines that would exhibit purposeful behavior. The actual construction of such machines was spurred on by other than philosophical motives. The need was for machines that could transcend the limitations of the human brain so as to guide the performance of other machines that transcended the limitations of the human muscle.[11] This need is being met by modern automation technology. The brilliance of this technological achievement, however, should not obscure the importance of its philosophical implications, namely, a corroboration (not a proof, of course!) of the conjecture that the behavior of organisms can be explained in terms of known physical laws.

Servomechanisms

Machines capable of what appears to be goal-directed behavior are called *servomechanisms*. The branch of technology dealing with their construction developed especially rapidly during and since World War II and has been christened *cybernetics*. Figuratively speaking, cybernetics deals with the "intelligence" of machines. The engines of the precybernetic era had no "intelligence" to speak of. Vehicles had to be steered, guns had to be aimed; power had to be turned on or off by human operators as conditions demanded. Even in the early days of the industrial era, however, certain simple cybernetic devices were known. Steam engines had governors that controlled the speed of the flywheel by automatic action triggered by a critical speed. In the thermostat, another familiar device, the source of heat is turned off when the column of mercury in the thermometer reaches a critical height and turned on when it sinks below it. When the rudder of a ship is set in a certain position, the ship will eventually assume a prescribed course, since there will be a torque on the hull as long as the ship is *not* on the prescribed course.

These examples illustrate the fundamental principle of cybernetics, namely, the utilization of *error* in correcting the error. Every machine is designed to respond in prescribed ways (emit certain outputs) to given conditions in the environment (inputs). In servomechanisms the performance of the machine itself, or rather the comparison between its performance and some prescribed end state, serves as an input. In a way a servomechanism can be viewed as a machine that keeps asking "How am I doing?" Through a system of closed loops, called feedback loops, a servomechanism responds not only to the environment but also to its response to the environment, to the response to the previous response, and so on. This circularity of responses creates the impression that a servomechanism is guided by a preset "goal" and so simulates the purposeful behavior of a living organism.

Similarity is a symmetric relation. If servomechanisms can be said to behave in some ways like living organisms, then living organisms can be said to behave in some ways like servomechanisms. Once this analogy is noticed, new methods of investigation suggest themselves in psychology and in the behavioral sciences in general. For the theory of cybernetics, linked with rich engineering experience, gives rise to concepts, hypotheses, and conjectures that often can be translated in behavorial terms.

To take an example, consider the concept of the transfer function, central in system engineering. An engineer's system is designed to give a prescribed output to each of the inputs to which it is sensitive. For instance, the input may be the image made by the path of an airplane and the output an appropriate aiming of the antiaircraft gun. The motion of the plane is described in terms of its instantaneous position, the instantaneous rate of change of position (velocity), the rate of change of the rate of change (acceleration), and so on ad infinitum. Clearly there is a limit on how rapidly the output can change appropriately. It would be difficult for a gun weighing several tons to follow the motions of a swallow. The inertia of the gun is one limitation; another is the speed with which information inputs can be processed. The capacity of a servomechanism to respond to inputs is determined by its transfer function, which depends, in turn, on a system of interconnections of its parts, an analogue of a "nervous system" processing the inputs and translating them into outputs. The structure of this "nervous system" is, of course, completely known to the designer of the machine. Indeed, his task is to design servomechanisms with prescribed transfer functions or else to calculate the characteristics of a transfer function capable of achieving the purpose for which the servomechanism is designed.

Consider now the inverse problem: given the performance of a servomechanism, to infer the structure of its "nervous system." This problem, called the "black box" problem, is central to the task of a physiological psychologist, who seeks to infer at least the general features of a nervous system that could account for some observed behavior pattern.

As a rule inverse problems are harder than direct ones, and their solutions often are not unique. That is, a great many arrangements can give the same transfer function, so that even its precise determination does not give much information about the underlying structure. Thus it would be hopeless to try to infer the vast collection of servomechanistic arrangements in a human brain by noting correspondences between stimuli and responses arbitrarily chosen, or chosen for their supposed importance in human behavior. In some situations, however, the transfer function itself is an object of interest, the determination of which depends on our ability to describe the inputs and the outputs in precise mathematical terms. When this *can* be done such situations are singled out for study, not because they are necessarily behaviorally important, but because they are analyzable by the methods at our disposal and so can serve as stepping stones in the development of the theory. In the mathematicized sciences the choice of a problem is, of necessity, often guided by tractability.

So-called tracking problems are of this sort. Their investigation was motivated partly by the need to understand the performance of the human component in man-machine systems, but their theoretical tractability was an additional impetus. The usual tracking experiment involves the task of following a target by moving a lever. The input (the motion of the target) is fully describable in terms of superimposed simple motions. Thus the complexity of the input is a controllable quantity. The output (the tracking motions of the subject) are likewise analyzable. From the mathematical relations between the input and the output, the subject's transfer function can be determined. This knowledge is useful to the engineer designing a man-machine system. It also provides theoretical leverage for the black box problem. On the basis of the inferred transfer function, the neurophysiologist can at least make guesses about structural features in the nervous system that can account for the transfer function.

There have been suggestions for using cybernetic methods in diagnostic procedures. Already in the earliest formal treatment of the subject, N. Wiener[40] called attention to the similarity between certain kinds of nervous pathology and servomechanism malfunctioning, particularly the oscillations accompanying the loss of motor control. The corresponding causes of the malfunctioning in servomechanisms being known or inferable, it appeared to Wiener that such knowledge might be transferable to the neurological situation. The work of L. Stark and T. N. Cornsweet[33] on the servomechanistic analysis of the pupil reflex is an example. If a sinusoidally varying light intensity impinges on an eye, the pupil will respond by periodic contractions and dilations. This is essentially a "tracking" task. From these oscillations the corresponding transfer function has been computed. As the gain of the system (the decrease of intensity due to the contraction divided by the increase of applied intensity) is increased past a certain threshold, the system becomes unstable, and the pupil oscillates at its "natural" frequency. This frequency is calculated from the transfer function and turns out to be about 72 cycles per minute. The actually observed "natural" frequencies in human subjects ranged from 62 to 80 cycles per minute (in some 80 subjects). But in 70 pupils of patients with multiple sclerosis, these oscillations averaged only 41 cycles per minute.

Since oscillations are clinical manifestations of a wide variety of neurological diseases (tremor, ataxia, clonus, nystagmus) and since servomechanistic analysis leads to specific neurological hypotheses, Wiener's early conjecture concerning the diagnostic value of the cybernetic approach may be a valuable guide to research on the working of the nervous system.

Information Theory

The central concept underlying the technology of the First Industrial Revolution had been that of energy. The primary function of an engine is to utilize a source of energy, such as fuel, to do work, to move masses of matter, for instance. The central concept underlying the (cybernetic) technology of the Second Industrial Revolution is that of information. The primary function of a servomechanism is to process inputs to convert them into appropriate outputs. This is also the central problem in the technology of telecommunication.

Information theory (or, more properly, the mathematical theory of communication) deals with that which is carried by signals abstracted from what the signals are made of or signify. A signal can be sent by producing an air disturbance, or an electrical disturbance, or a light. The method of sending the signal is not important in information theory. What is important is what "knowledge" the signal conveys, or rather the quantitative aspect of that knowledge. The most fundamental idea in information theory is that the "amount of information" depends not on *what* is said in a message but on what *could* have been said. In much the same way the probability of an event is associated not with the event itself but rather with the whole context in which the event could occur. Thus the probability of drawing a particular ball from an urn depends on the number of balls in the urn. Hence it is not the ball in question that determines the probability but the number of balls that *could* be drawn. Similarly the amount of information in a message is not defined unless one can specify all possible messages from which the message in question is selected. This is the meaning of the "amount of information" in the mathematical theory of communication. The meaning is made precise by abstracting totally from the content of the messages. For example, the telegraph operator is not concerned (indeed, his professional ethics do not allow him to be concerned) with the content or meaning of the messages he sends. To him the messages are only sequences of signals. The amount of information in a message is calculated in terms of the *a priori* probability of that message being selected from all the possible messages that could have been sent

This concept of the "amount of information" allows the telecommunication engineer to design efficient and economical equipment for transmitting expected information loads over channels. In this context information becomes something that flows over channels, much like power flows over power lines or oil flows through pipes. It makes sense to speak of capacities and volumes of flow and efficient "packing methods" (which in telecommunication become "coding") quite in the same way that one speaks of the flow of traffic, people, or material goods where efficiency depends on scheduling or packaging. These considerations apply to all forms of telecommunication, telephone, radio, and television.

Here, then, is still another view of the nervous system—that of a telecommunicative device. It is that, of course, in the literal sense and has long been recognized as such. What information theory has achieved is to have created a powerful and precise language for describing the performance of telecommunication devices.[31] This language is now used as a tool in constructing theories of nervous function,[6,18] which, it is hoped, can be extended to theories of behavior.

The formulation of the essentials of information theory has inspired numerous psychological experiments based on the central concepts of the theory. In these experiments the individual is treated as a "channel" whose overall characteristics—for example, channel capacity—are to be estimated. One way of doing this is by pumping information through the individual, that is, by making him a link in a telecommunication channel, a transducer. For example, if the individual is required to respond differentially to each of a collection of signals presented in random sequence, his channel capacity is expected to put an upper limit on the rate and on the accuracy of his performance. (Information theory also deduces the mathematical relation between rate and accuracy.) Now the amount of information per signal can be varied at will by varying the number of signals from which selections are made, by varying the relative frequencies with which the different signals are sent, and by varying the sequential probabilities of the signals. Thus it is possible to have the same average amount of information per signal in several different situations, involving different numbers of signals, different relative frequencies, or different sequential probabilities. The conception of the individual as a link in a communication channel suggested that his performance should be determined by the rate of information flow rather than by the particular way this rate is achieved. Early experiments on choice reaction times provided some corroboration for this hypothesis.[16] Of equal or even greater importance, however, were the discrepancies that could not be accounted for by the channel model. These led to the design of more refined experiments and to a more detailed analysis of reaction times, which revealed some distinct inadequacies of the information theory approach and advanced alternative interpretations of the experimental results.[18] Thus in its very failure information theory served in a constructive role *as a point of departure* for a theory of information processing in the nervous system.

Automata

In addition to its central role in telecommunication, information theory is also important in the theory of automata, a class of machines to which the high-speed computers belong. For what is called the "memory" of a digital computer is simply a storage facility for information, a reservoir, to which information is shunted to be recovered when needed. In cybernetics, too, information theory ideas are important. The "conditionality of response" of a piece of automatic equipment, that is, the complexity of instructions it can "understand," is also measurable in information units. The "intelligence" of machines thus becomes a measurable quantity, just as in the early days of technological evolution, mechanical advantage and, later, horsepower were standard evaluative units. The ability of computing machines to perform not only arithmetical calculations but also complex logical operations has induced their classification (partially in jest, one supposes) as "thinking machines." There is no question, of course, that in some respects automata simulate the thought process. Again, turning the simile around, we might ask whether in some respects our thinking organs may not function on the principle of computing machines.

W. S. McCulloch and W. Pitts[19] showed that a model of the functional logical processes, as denoted by the operations of symbolic logic, is entirely consistent with certain simplified assumptions concerning the interaction of neurons.* Suppose we picture a neuron as a unit that can exist in only one of two possible states—"firing" and "nonfiring." (This is not factually correct, of course, but is an idealized version of the "all-or-none" law.) Suppose further that the firing of a neuron is occasioned by the impingement on its dendrites or cell body of the summed activities of other neurons, transmitted via axones to the terminal buttons. Let the threshold of firing of a neuron be defined as the minimal number of active terminal buttons sufficient to fire it. Finally suppose that some of the terminal buttons are inhibitory; that is, their activity subtracts from rather than adds to the firing potential impinging on the neuron. These characteristics are sufficient to represent any conditionality of response of any neuron or set of neurons by an appropriate arrangement of excitatory and inhibitory connections. It follows not only that one-to-one stimulus-response relations can be represented in an idealized nervous system (this could be done already with the old "telephone switchboard" models), but also the dependencies of responses on "inner states," be they interpreted as memories, accidental associations, or random fluctuations, can be included.

Following the completely abstract "logical" model of the nervous system, several "engineering models" of neurons were proposed and built in connection with experiments simulating the activity of neurons or neural nets.[15,20,34] The engineering models, in turn, stimulated theoretical analysis of information processing in elements assumed to have the characteristics of "real" neurons, for example, membrane potential, absolute and relative refractoriness, and so forth.[1]

The behavior of automata has been shown to be capable of far greater variability and flexibility than had been imagined. Modern computing machines do not just perform specified listed operations in order; they are capable of "making decisions," as is evident from the programs that guide their operations;

* This work was anticipated in the context of electrical switching circuits by C. E. Shannon.[30] Subsequently J. Von Neumann gave an extensive and lucid exposition of the theory in a general context.[37]

for example, "Add column 6 to column 13, compare the result with the last entry in column 2; if the sum is greater, extract square root of column 10, otherwise proceed to step 7, etc." Computing machines can solve logical problems such as this one: If bandits don't drink beer only if the sun shines and the moon is in the first quarter, and if, whenever the sun shines, shrimps cannot whistle unless the moon is either in the second or third quarters, and if bandits drink beer at the same time when shrimps do not whistle only when gosphers go skating in the moon's last quarter, it being understood that when the last-mentioned does not occur it does not mean that bandits cannot drink beer if shrimps whistle or that shrimps must whistle if bandits do not drink beer; what may or must be the phase of the moon when gophers go skating on a cloudy day while the shrimps remain silent?

Certainly some of the thinking we do is of this type (though not as complicated). It had been almost taken for granted, until the theory of automata showed otherwise, that "thinking" is necessarily a different sort of activity from what machines are able to do. Indeed, machines had been habitually looked upon as strong but stupid. In some circles an argument rages about whether the technology of automata has refuted this view, whether computers "really" think. The theory of automata has shown that once we have described the thinking process with sufficient precision, we can build an automaton to simulate it. Nor does the simulated thought process need to be rigid. For the *rules* of inference can be made to change in consequence of the automaton's "experience." The "lifelike" character of automata equipped with simple servomechanistic regulatory units and just one or two "motivation" mechanisms has been dramatically demonstrated. A "turtle" that persists in following white lines randomly drawn on the floor seems to have an "aim in life." It seems even more lifelike when it is observed to run to electrical outlets to get recharged as its batteries threaten to run down, and even more so when it changes its behavior patterns after being "spanked."[38]

None of these demonstrations is sufficient to change the minds of those who insist that "machines can't think." It is always possible to keep revising the definition of "thinking" so as to keep it in the residual area of what has *not* yet been successfully simulated. But this sort of procedure may be a rationalization of an aversion to equating men with machines rather than the discovery of the basic difference between men and machines.

Turning to the possibility of applying the theory of automata (as it pertains to logical operations) to a theory of specific nervous activity involved in thinking, we find that the practical difficulties are enormous. Whereas, in the case of cybernetic and information theory approaches, it was possible to ferret out gross concepts reasonably applicable to nervous regulatory and signal-transmitting activity (for example, transfer function, channel capacity, and so forth), we find that, in viewing the nervous system as an automaton, we must postulate the existence of specific units and specific relations among them. The basis of the theory is the correspondence between the fundamental logical operations and certain arrangements of relays. These are arrangements corresponding to the logical operations of "and," "or," "implies," "not," and the like. In this way every logical function consisting of binary variables (propositions and their negations) and logical operations can be mapped (not uniquely, though!) upon certain networks of relays, which one may interpret as "neurons." Even if the real neurons obligingly acted in every way like those automaton units, it would be all but inconceivable with our present techniques to identify the particular arrangement responsible for even a modest range of behavior patterns of a living organism. At best this is possible in the simplest instances. For example, B. Hassenstein and W. Reichhardt[13] have studied the responses of a beetle to stimuli impinging upon the separate contiguous facets of its complex eye with the aid of a neural model essentially of the McCulloch-Pitts type. The responses were sufficiently simple so that they can be fully analyzed; yet they contain sufficient conditionality to necessitate an apparatus more complex than a simple aggregate of reflex pathways.

Specifically the beetle responds with rotations of its body to various patterns of stimulus incidence, depending on (1) the order of stimulation of contiguous facets, (2) the relative intensity of the successive stimuli, and (3) the time interval between the presentations. To account for all these aspects of behavior (remarkably consistent), Hassenstein and Reichhardt have postulated the simplest conceivable arrangement of automaton units, of which only a few are required to serve each pair of facets. As a consequence of this arrangement the prediction is made and verified that stimuli impinging consecutively upon facets separated spatially by more than one facet do not interact with each other. By and large the model is mainly an explanatory one; that is, its theoretical significance is confined to a schematization of neural elements to account in the simplest way for observed behavior. The model thus serves as a *possible* solution to a black box problem.

Naturally one expects rather more from a model. If, for example, the postulated arrangements were identified anatomically or at least indirectly by further consequences not observed in the preliminary investigations, the theoretical force of the model would have been greatly enhanced. On the other hand, the conceptual value of automaton models is not to be underestimated. It is instructive to note how "much" can be done with only a few "neurons." "Much" is put in quotes advisedly. Richness of conditionality of response is to be distinguished from ordinary complexity of response. A response may be marvelously complex in the sense of having many components and yet not necessitate any complicated neurological mechanism. This would be so if each step in the sequence were rigidly determined by the preceding step. To be sure, neural connections would be required to link the steps sequentially. But there would be no need of information-processing and decision-making units. It is the *conditionality* of behavior that necessitates complex automation, behavior described in terms of "if so, then so, unless so, in which case so, provided this or that but not both . . ." and so on. The few hundred neurons of the ant must be sufficient to provide it with all the conditionality at its disposal. This relatively small number reflects the circumstances that, although the behavior of the ant may seem quite complex, the conditionality of its behavior patterns must be rather small compared to that of animals with enormously larger numbers of neurons.

Mathematical Theories of Neural Nets

It appears, therefore, that the weakest link in the application of automaton theory to the anatomy and physiology of the brain is the specificity of automaton models. True, for any pattern of behavior of any prescribed conditionality, an automaton to simulate it can be in principle constructed. But if the model is simply a translation from logical propositions to networks of relays (as it is in the McCulloch-Pitts model), the loss of a single unit may radically change the entire behavior pattern of the automaton. It is inconceivable that such sensitivity to single units (neurons in this case if the analogy applies) should characterize the living brain. We are constantly impressed by the plasticity and adaptability of living behavior. Specific failures traceable to specific excisions are still exceptions rather than the rule. On the other hand, building in sufficient alternative connections to forestall every conceivable specific failure would probably necessitate more neurons than are available in the largest brains.

Another feature that distinguishes living behavior from that of precisely constructed automata is the "approximate" character of the former. Actions of living organisms are not mathematically precise; nor are they necessarily the most direct and efficient. They are "adequate," with wide error margins. Moreover they are often recognized as responses to "fuzzy" stimuli. An object is recognized by a higher animal as "itself" from different visual angles, in different orientations, and at different distances, a circumstance emphasized in

Gestalt psychology. These synthesizing and abstracting functions of the nervous system cannot be accounted for by assuming simple one-to-one correspondences between elementary stimuli impinging on specific elements and determinate responses of the latter.

Attempts to simulate Gestalt phenomena are reflected in the construction of networks of elements designed to recognize *patterns*, that is, gross features of events regardless of perturbations. Examples of this approach are found in the work of Rosenblatt,[27] D. Rutovitz,[28] R. Narasimhan,[23] and many others. Work along these lines is clearly inspired by the ambitions of automation technology. One can well imagine a typist-automaton that takes oral dictation; that is, is able to recognize words regardless of accent, vocal characteristics, or speech peculiarities of the person dictating. Also the theoretical spinoffs of these investigations may be considerable. Constructing pattern-recognizing automata may suggest ideas about how living organisms synthesize information carried by impinging stimuli.

The immense plasticity of living behavior has led several workers concerned with the theory of the nervous system to attempt to construct "probabilistic" (statistical, stochastic) models. Here connections or stimulus-response relations are not specified, but only their probabilities. Experiences of the organism (learning, metabolic changes, and so forth) are supposed to operate on these probabilities. From these probabilities one infers only gross aspects of behavior, not its details, and the variability of behavior—its continuous rather than discrete character—can be attributed to statistical fluctuations in the "functional structure" of the system. To cite an analogy, the general outline of a fountain persists, but it is not rigid, nor does it depend on the path of each individual water drop.

A convenient starting point of a probabilistic model of a nervous system is a "random net," formally defined as a collection of nodes (neurons) among which the synaptic connections are indicated only as probabilities. Statistical computations then give the gross connectivity characteristics of such a net; for example, the expected number of paths between an arbitrary pair of neurons, the expected number of neurons so many synaptic connections removed from each neuron, and so forth. Given such gross statistical features and certain assumed laws of synaptic transmission, the activity of such a net, resulting from some initial input, can also be calculated. For example, given the probability distribution of the number of axons emanating from each neuron and the probability distribution of their targets, the "critical input" can be established, one that if exceeded results in the spread of excitation through the net and if not exceeded results in a dying away of the initial excitation.[2,26]

A physical demonstration of a systematic, even systematically modifiable, behavior of a servomechanism with a randomly connected "nervous system" is provided by the "homeostat," which illustrates the so-called principle of *ultrastability*.[3,4] The stability of a servomechanism depends, of course, on its connections. If the connections of a thermostat, for example, were reversed, it would become unstable: rising temperature would result in even more heat from the furnace until something would "give." But if a servomechanism switched its connections whenever some variable exceeded a certain limit, it would have ultrastability. In the case of a thermostat we could initially connect the leads randomly. If we happened to make it unstable, the rise of temperature would switch the connections and make it stable, after which the connections would no longer be switched, because a critical temperature would not be exceeded. In the homeostat, whenever certain voltages are exceeded, the connections are randomly shuffled until stability is achieved. This property, besides insuring stability, even enables the homeostat to exhibit simple learning behavior. If a certain response pattern is "punished" (by increasing voltages beyond the tolerated limits), connections will be switched until the right ones for the situation are found. The principle of learning thus exhibited is that of random search and fixation on the correct response.

The theory of probabilistic automata also underlies much of the work of pattern recognition.[24,27]

In short, concepts derived from cybernetic technology have been a rich source of ideas in theories of neural structure and function,[17,32] which, it is hoped, will strengthen the still tenuous links between physiology and theories of mental phenomena.

Mathematical Linguistics and Psycholinguistics

Let us now see how a computer would solve the above-mentioned logical problem, involving bandits, shrimps, and gophers. The "givens" of the problem must first be stripped of all semantic meaning (which only interferes with the reasoning). Then it can be presented in a language the computer can "understand," the language of two-valued symbolic logic—essentially a branch of mathematics where the variables can assume only either of two values, 0 (representing "false") and 1 representing "true"). The variables, symbolized by letters, stand for propositions. For instance, b stand for "bandits drink beer"; s for "the sun is shining"; m_i for the "moon is in the i-th quarter"; g for "gophers go skating." The denials of the propositions are symbolized by corresponding letters with bars over them. For instance, $\bar{b}$ stands for "bandits don't drink beer." Besides the symbols representing propositions, the language of symbolic logic contains symbols representing relations among propositions. These are "$\wedge$," meaning "and"; "$\vee$," meaning "and/or"; "$\rightarrow$," meaning "implies"; and parentheses for punctuating sentences. We can now represent the entire information given in the problem by the following "sentences" written in the language of symbolic logic:

$$\underline{\bar{b}} \rightarrow \underline{s} \wedge \underline{m_1}$$
$$\underline{s} \rightarrow (\underline{w} \rightarrow \underline{m_2} \vee \underline{m_3})$$
$$\underline{\bar{b}} \wedge \underline{w} \rightarrow \underline{g} \quad m_4$$

The computer is programmed to perform certain operations on the symbols in accordance with specified rules. The result of these operations leads to the following sentence:

$$\underline{g} \wedge \underline{\bar{s}} \wedge \underline{\bar{w}} \rightarrow \underline{m_4}$$

which, retranslated into English, says "When gophers go skating while the sun is not shining, and the shrimps do not whistle, the moon must be in the fourth quarter," the required solution.

We can say, therefore, that to solve the problem the computer must be presented with it in a language it "understands." The "grammar" of that language (in this case the rules of operation of symbolic logic) is built into the computer, and this is what we mean by saying that the computer "understands" it.

One of the problems attacked by computer technology was that of automatic translation. Automatic translation would be simple if sentences could be translated from one language to another word by word. For then the only "rules" that would have to be programmed would be those that link each word in one language with its equivalent in another. As is well known, however, the problem is vastly complicated, not only by the fact that most words have more than one meaning but also by the fact that grammars of even closely related languages are different. On the other hand, substituting whole sentences will not do, since the number of possible sentences is potentially infinite. (It is safe to assume that the sentence you are now reading has never been spoken or written before.) Thus, the problem of automatic translation is that of giving a complete description of a grammar of a natural language in a language accessible to a computer. The immense difficulty of this task has now been realized.

Although automatic translation still seems to be a thing of the distant future, the "theoretical spin-offs" of the associated problems have been considerable. The attention of linguists (who are only incidentally or not at all interested in automatic translation) has been turned to one of the most challenging and, possibly, one of the most important problems of human psychology: to describe rigorously (not intuitively) the internalized grammatical rules that enable a human being, only a few years after birth, to produce and comprehend a practically unlimited number of sentences in his native language.[9] "Comprehension" in this context means much more than associating words with their referents; for language is much more than assigning labels to objects or situations.

Mathematical linguistics is, in part, concerned with the construction of rigorous theories of grammar. "Mathematical" in this context does not mean "quantitative," as it does in classical physical science. Here, the relevant branches of mathematics (for example, set theory, symbolic logic, the theory of semigroups, etc.) deal not with measurable quantities but only with rigorous rules of symbolic transformations. The term "mathematical" is justified in view of the definition of mathematics as a contentless language of rigorous description and deduction. To put it another way, mathematical linguistics is concerned with the abstract relational "framework" of language, which determines meaning by "shaping" the content that is poured into it.

Another mathematical approach to language behavior is via statistical linguistics. The verbal output of an individual, or of a population of individuals such as a speech community, can be viewed as a vast number of minute sequentially produced units, for example, words selected from the lexicon of a language. Because of their large size, these collections exhibit certain statistical regularities. The smaller the units, the less are the statistical characteristics of these large samples dependent on content. For example, in large samples of printed English the relative frequencies of the letters of the alphabet are very nearly the same, regardless of source. The frequencies of larger units (for example, words and phrases) will, of course, be more dependent on the source or content. Nevertheless, certain statistical features common to all large corpuses can be abstracted also on these levels. G. K. Zipf[43] particularly stressed the repeated observation of the following relation. Let the different words in a large corpus (a sample of verbal output) be ordered in the order of the frequency of their occurrence, so that rank 1 ($r = 1$) is assigned to the most frequently occurring word, rank 2 ($r = 2$) to the next most frequently occurring word, etc. To each rank corresponds the actual frequency of occurrence in that corpus, denoted by f. Then in all large corpuses the product $f \times r^{\gamma}$ is approximately constant, where γ is a number close to 1 and, in almost all cases, somewhat larger than 1. This rank-frequency relation (in other contexts the rank-size relation) was observed in a great many widely disparate situations and was attributed by Zipf to an underlying universal law, which he called the principle of least effort. Zipf's justification of the law and its consequences was often extremely vague and cannot be considered as a significant theoretical contribution. Nevertheless, the basic idea—that of examining the "statistical profile" of verbal outputs—has remained fruitful. The point is that these statistical profiles are determined by certain parameters (indices) and can serve as a basis of objective comparisons. Thus Zipf noted that the verbal output of schizophrenics is characterized by unusually large values of γ, the characteristic parameter of the rank-frequency relation.

Comparison of statistical profiles involving more than just rank-frequency relations has been used in determining the authorship of texts. In fact, there are cases on record where disputes concerning the authorship of texts have been decided by such comparisons. The validity of these methods depends on the circumstance that, although an individual may well exercise voluntary control over detailed actions, *in the large* his patterns of behavior, including his verbal outputs, are much more determined by habits, predispositions, and the like. Thus the statistical profile of an individual's verbal output "reveals" his identity in the same way as his handwriting or the spectral characteristics of his voice. In a way the verbal output is a "secretion"; therefore, it seems reasonable to develop methods of analyzing this secretion parallel to those developed in scientific medicine for analyzing physical secretions. The implications for psychiatry are obvious.

Content analysis is essentially an extension

of statistical methods to include the "semantic" features of a verbal output. Part of its task is the development of coding techniques, which map "meanings" on objectively identifiable units. These techniques require considerable competence in the subject matter of the verbal output undergoing analysis. However, after the coding procedure has been designed and the "content" translated into a statistical profile, analysis becomes entirely objective. For instance, the statistical profiles of two or more outputs or their trends over time can be compared in the same way as spectra of different light sources. In this way "hard" content analysis can be used to supplement the conclusions of "soft" content analysis, which depends on intuitive conjectures of the analyst (for example, interpretation of dreams, literary or musical criticism), and to put the theories of the latter to scientifically objective tests.

Examples of the application of content analysis, both hard and soft, ranging from analysis of international crises to shifts of emphasis in grade school readers, can be found in Gerbner, *et al.*[12]

In the "semantic differential," an instrument based on factor analysis techniques, the object is to construct the "semantic space" of a subject or of a population of subjects. The theory is based on the observation that a great many adjectives can be characterized by the connotations they evoke on three principal axes: a value axis, along the good-bad scale; a potency axis, along the big-little or strong-weak scale; an activity axis, along the active-passive scale. Moreover, a great many other words, especially those with strong emotional overtones, can also be so characterized. Thus, from a subject's associative responses to a set of "concepts," a (connotative) "semantic space" can be constructed, in which each concept appears in a definite position, determined by its three coordinates on the three axes. The semantic differential has been used in comparative studies of such semantic spaces characterizing different individuals, populations of individuals of different cultural backgrounds, and the same individuals at different times. For instance, of particular interest to psychiatrists may be a study, undertaken by Osgood, Suci, and Tannenbaum,[25] involving a comparison of the semantic spaces associated with the three components of a "split personality" ("The Three Faces of Eve").

Exploration of Ideas: Opportunities and Dangers

Progress in science depends essentially on successful generalizations that unite apparently disparate phenomena into unified theoretical schemes. The evolution of physical science illustrates this process most clearly. The law of conservation of energy, first established in classical mechanics, was later extended to unite mechanics with thermodynamics. Electrical and magnetic phenomena were united in electrodynamics and extended to include all forms of radiant energy. Statistical mechanics revealed the deep connection between information and entropy via the mathematical expressions of the "amount of order" (or disorder). There are also dangers lurking behind attempted generalizations guided by metaphorical instead of rigorous mathematical analogizing. Every model is, of course, an analogy. What makes a model heuristically useful is its conception as a point of departure rather than of arrival. Unfortunately the richly suggestive ideas of mathematicized theories are often used as explanatory props rather than as raw material for constructing testable hypotheses. This is probably inevitable as long as in many lines of inquiry "theory" continues to be understood as a collection of mental images or figures of speech, which, it is somehow felt, harmonize with intuitive feelings of what constitutes an "explanation." To take an example at random, Freud's "hydraulic" model of psychodynamics is a "theory" in that sense. In essence many sociological theories are collections of definitions, that is, invitations to organize experience in a particular way. Although this is not the way the term "theory" is used in natural science, it would be rash to consider the construction of such theories altogether useless. After all, the organization of thought along certain lines is often a prerequisite of any progress toward insight. Metaphorical theorizing is not of itself necessarily misleading. It can become misleading when hazy notions are coupled with precise-sounding terminology. The use of the latter may give an impression (to the theorizer himself, as well as to his audience) that precision has been achieved when, in fact, concepts that are precise in proper contexts have been muddled by metaphorical transformations of meanings.

The Homeostasis Metaphor

The concept of homeostasis was formulated in the context of physiological regulation by W. B. Cannon.[8] In such regulations homeostatic mechanisms operate so as to keep certain variables (concentrations, pressures, temperatures) of the organism's "internal environment" within certain limits of tolerance. Homeostasis is also a central principle of cybernetics, since the regulation activity of servomechanisms can be described in the same terms as the regulation of the physiological processes. In general, homeostasis operates on the principle of feedback. In negative feedback the "restoring force" is always opposite to the error, so that the variable in question tends to some equilibrium value. In positive feedback the error is self-enhancing, so that either a variable increases without bound or oscillations of ever increasing amplitude result. Homeostasis in system engineering is attained by a proper arrangement of feedback loops. (Positive feedback loops also have their place, where it is required that the system pass quickly from one steady state to another.)

As long as the variables represent real measurable quantities and the network of influences among them is actually observed or specifically assumed, one may speak of homeostasis in a great variety of situations; for instance, in engineering, where the variables are voltages or tensions, or water levels; in physiology, as described above; in ecology, where the variables may be populations, gene frequencies, etc., and where the "forces" are statistical trends, which, of course, do not have the physical characteristics of forces but have similar mathematical properties. One can, in the same spirit, speak of homeostasis in economics, where the variables are prices, interest rates, trade volumes, etc. In these instances the concept of homeostasis is, indeed, a unifying principle of several widely disparate areas. It is a "general systems" principle par excellence.

When the operational meaning of the variables is lost sight of and replaced by intuitive notions, the terminology associated with homeostasis becomes at best metaphorical. The models become paraphrases of impressions and cease to have theoretical significance, as this significance is understood in "hard" science. The various models of behavior, personality, and society, couched in terms borrowed from theories of homeostasis, give the illusory impression that powerful and rigorous methods are being applied to the study of man.*

To speak of the defense mechanisms of the individual, or of the mutual impact of political systems or cultures, in the language of homeostasis may be subjectively enlightening, but there is no way of knowing whether such enlightenment is any different from the sort experienced by philosophers who, in the days before the advent of physical science (and often afterward), "explained" the physical, the biological, and the social universes by picturing them as manifestations of metaphysical laws that reflect no more than grandiose verbiage.

The Information-Entropy Metaphor

There is a link between information theory and thermodynamics that carries a tantalizing suggestion of being of prime importance for theoretical biology, along with all the dangers of speculative promiscuity. The formal resemblance of the mathematical expression for the average amount of information per signal to

* For an extensive critique of the use of the homeostasis concept in psychology, see, for example, H. Toch and A. H. Hastorf.[36]

the expression for the entropy of a physical system, as calculated in statistical mechanics, was noted by N. Wiener[40] and C. E. Shannon,[31] who laid the foundations of cybernetics and information theory, respectively. The definition of entropy is highly technical: very roughly speaking, entropy can be taken as a measure of disorder present in a system. It is this disorder in the motions of molecules that makes it impossible to convert heat energy *fully* into mechanical energy without other changes accompanying the process. Stated in another way (as the famous Second Law of Thermodynamics), in a system completely isolated from its environment, the total entropy can never decrease; it keeps increasing until the system is in thermodynamic equilibrium. When this state is attained the heat energy of the system can no longer be converted into "useful work."

When thermodynamic considerations first began to be applied to biological systems, some biologists forgot the important qualification "isolated from its environment" and argued that living systems violated the Second Law (an argument for vitalism!) since such systems tended, at least in their development, toward "greater organization," rather than toward chaos as the Second Law demands. Since no living system is isolated from its environment, the argument rested on a non sequitur. At any rate E. Schroedinger[29] pointed out that life must "feed on negative entropy," by which is meant simply that organisms must ingest substances rich in "free energy," in other terms, low in entropy.* In metabolism this free energy becomes "degraded"; that is, entropy increases, and this surplus of entropy, dumped in excretion upon the outside world, "pays" for the decreases in entropy (increased organization) that the organism effects within itself. In this way entropy-lowering life processes can go on without the Second Law being violated in the end result.

So far the argument has been presented in thermodynamic terms, and its relevance for information theory is far from evident. A clear connection can be found, however, in an early paper by L. Szilard.[35] Szilard analyzed the operation performed by Maxwell's demon, a hypothetical creature posited by James Clerk Maxwell in 1869. The demon is supposed to be able to "see" the molecules of an enclosed volume of gas, mechanically and thermally isolated from the environment (an isolated system). By sorting them he can "increase the order" in the system and so lower its entropy in apparent violation of the Second Law. By considering the simplest possible system of this sort, consisting of a single molecule, Szilard was able to analyze completely the nature of the demon's intervention. He showed that if the Second Law does hold, the demon himself (being part of the isolated system) must suffer an increase in entropy that at least compensates for the decrease he effects in the rest of the system. This conclusion is simply a logical consequence of the assumption that the Second Law does hold. The remarkable feature of Szilard's analysis is the exact quantitative relation between the "amount of information" that the demon must utilize in his operation and the resulting decrease of entropy. He showed that in utilizing one bit† of information the demon lowers the entropy of the system (excluding himself) by $k \log_e 2$ ergs per degree, where k, the so-called Boltzmann's constant, is 1.37×10^{-16} ergs per degree. Thus a transformation factor connecting entropy and information was established, analogous to the transformation factor connecting a unit of work and a unit of heat, discovered almost a century earlier. In 1951 L. Brillouin[7] was able to show that the demon must indeed suffer at least the prescribed increase of entropy, regardless of the method he uses in determining the position or the velocity of a molecule.

The implications of these theoretical results for events in biological systems *on the molecular level* are now being actively investigated.[10,21] There the connection between information and entropy is quite clear: what

* Schroedinger's fortunate phase has become a byword and has made facile speculation in biological thermodynamics fashionable. The basic idea, however, was already formulated 20 years earlier.[22]

† The "bit" is a unit of information, the amount conveyed in a decision between two equally probable alternatives.

appears in the language of gross thermodynamics as entropy (units: energy over temperature) appears in the statistical formulation (the mechanical basis of thermodynamics) as information (units: pure numbers, logarithms of probabilities). Trouble arises when results are extrapolated in attempts to apply the concepts to information in its vernacular meaning. It is taken for granted by many writers that the quantitative information measure can be applied to the *content* of communications, so that the number of bits in this chapter, for example can be stated with as much precision as the weight of the paper it is printed on. In a way this is true but irrelevant to the informative content of the chapter. Quantity of information is defined with reference to the *statistical* properties of the source from which the signals are chosen. To be sure, the "information" of any verbal output can be measured in this way by reference to the statistical distributions of its units—say, letters, or phonemes, or words—but only by disregarding the meaningful content of the corpus. Thus one bit of information is gained by someone who is told the outcome resulting from a toss of a fair coin; one bit was gained by Paul Revere when he saw two lights appear in the tower of the Old North Church. What the last-mentioned "bit of information" meant for the American Revolution is irrelevant from the point of view of information theory. Indeed, the amount of information conveyed by a meaningless scramble of randomly selected letters is actually greater than that conveyed by a meaningful text of the same length, because the random selection is subjected to fewer statistical constraints. From the point of view of telecommunication this makes sense, because it would take more channel capacity to transmit random combinations of signals at a given rate than statistically constrained combinations. From the point of view of the recipient, however, who considers that he gets information when he is "informed," the statistical definition makes no sense.* Therefore, no operational meaning can be assigned to a statement such as "A has received so many bits of information and has thereby lowered his entropy (increased his internal order) by so many units." The statement can acquire meaning only if it is shown just how A has utilized this information and how the decrease of entropy was compensated by an increase elsewhere. However, the seductive power of metaphors is great, as evidenced by an abundance of loose talk about the relation of "entropy" and "information" in human affairs. So far extrapolations of the information-entropy identity to regions where receiving information means being informed have dissolved into vague and, one suspects, sterile speculations.

* In strictly limited contexts some progress in constructing a theory of semantic information has been made.[5]

Do Machines "Think"?

Simulation of "thought" by machines has raised some questions of philosophical and ethical import. Crudely put, the fundamental question is whether it is proper to ascribe "thought" to machines or, conversely, to picture man as a complex machine. Inevitably the posing of these questions is charged with affect. Answers in the affirmative seem to some to imply a denial of man's humanity, while others see in the erasure of distinction between living and nonliving systems another step toward the unification of science and toward the abandonment of anthropomorphism—a continuation of a maturing process instigated first by the heliocentric theory and later by the theory of evolution.

It is possible to by-pass the emotional overtones of these questions by a careful distinction between different meanings of "thinking." There are some things that information-processing machines can demonstrably do; for instance, solve logical and mathematical problems and exercise control over physical processes. At one time it was thought inconceivable that inanimate systems might be capable of performing these apparently "intelligent" tasks. There are also some things that presently existing information-processing machines cannot do. However, the limitations are more difficult to *spell out* than the achievements. The difficulty is that, once the limitations are specifically spelled out, ideas are suggested on how to overcome them. It has been said that a computer cannot compose a poem or a quartet. Promptly computers were programmed to compose "poems" and "quartets." The objection that these products are not really works of art, *because* they have been programmed, can be met by a powerful challenge, namely, to distinguish the machine-made products from some contemporary examples of man-made ones.

Arguments to the effect that what goes on in computers is not "thought" because the processes are "preprogrammed" are not conclusive. The analogy between information processing by man and by machine rests on the assumption that the processes in man's nervous system are *also* preprogrammed, namely, by the structure of the nervous system and its physiological state at a given moment. The admittedly vast difference in complexity between the two kinds of processes is not sufficient reason for dismissing the analogy. Nor are arguments about "free will," supposedly possessed by man but not by the machine, relevant to the issue, if what is wanted is evidence to resolve it one way or another. Our conviction of having "free will" stems from a metaphysical (or religious) position or is induced by introspection. Metaphysical positions are impermeable to evidence. Introspection is accessible only to the introspecting subject; hence there is no way of knowing whether the machine does or does not "introspect."

There remains only the *ethical* basis for distinguishing between man and machine. The real meaning of the questions "Do machines think?" or "Are men machines?" is embodied in another question: "Shall our attitudes toward men and machines be similar or different?" This question is obviously value-oriented and should be frankly posed and recognized as such. It has substantial ethical import in a civilization where the lives of human beings are to a large extent organized by work in the services of machines. Comparing men to machines does deny man's humanity in the sense of turning attention to man as an instrument: "Machines can in principle do everything men can do." This sort of comparison turns attention away from man's *intrinsic* worth, which, unlike his instrumental worth, resides not in what he can do but in what he is, namely, man.[41] The fact that we communicate with other human beings *without* knowing analytically how this is done; the fact that we ascribe consciousness to other human beings, not on the basis of "evidence" (we have no access to another being's consciousness), but intuitively, by identifying with them, puts relations between human beings into a unique category. Insistence on the uniqueness of these relations is a manifestation of certain values, and the adherence to these values is the only significant meaning underlying the refusal to identify human thought with automated information processing.

❨ Conclusions

Physical science, with its formidable methodological machinery in which controlled experiment, induction, and mathematical deduction are meshed, has nourished the life sciences almost from their inception. As the methods of physical science are becoming extended to areas where not matter and energy but organization and information processing are of central interest, the basis for integration becomes even firmer and a hope emerges of extending the integration to include those aspects of the life process that have been considered absolutely *sui generis*, aspects involving "psychical" rather than physical events. Such an integration would lead to the final dissolution of the mind-body duality in the context of scientific investigations.

Those who attempt to realize such integration borrow from these latest developments of physical science their methods, their ideas, and their language. The conditions for a fruitful extension of method are explicit. Mathematicized science deals with exactly specifiable structural relations. Whenever such structural relations can be unambiguously defined (in terms of observations, operations, or mathematical manipulations), the method of mathematical deduction can serve as a powerful tool of theory construction. Discrepancies between theory and observation serve to initiate the cyclical process of hypothesis-deduction-verification-new hypothesis. Therefore, initial accuracy of assumed relations is not essential; only the unambiguous specification of variables and relations is a prerequisite for extending the mathematical method to new areas.

When such specification cannot be made, the heuristic value of the ideas immanent in mathematicized science may still remain. Therefore, rather than attempt explicit mathematical modeling, some behavioral scientists seek to adapt the general ideas emerging from mathematical analysis to theories of behavior. The value of such adaptations is an open question. They may be enlightening or they may be misleading. To illustrate take the so-called uncertainty principle of atomic physics. The principle sets limits to the precision with which the position and the momentum of a particle can be simultaneously measured. As such it is a principle of theoretical physics and nothing else. However, the principle has philosophical implications. One implication has to do with the failure of strict causality on certain submicroscopic levels of events. This had led to contentions that the uncertainty principle "proves" the existence of "free will," largely a play on words, "free will" being the verbal antithesis of "strict causality." The irrelevance of such conclusions to science need hardly be pointed out. However, there is another implication suggested by the uncertainty principle, namely, that events may be affected by being observed. These effects have been long felt to operate in psychology. As stated in quantum mechanics, the uncertainty principle is exact and explicit. It singles out pairs of so-called complementary quantities, position and momentum being one such pair, energy and time another. A specified amount of precision in determining one member of the pair introduces a specified minimum amount of uncertainty in the other. The principle could be of genuine heuristic value in psychology if analogous complementary pairs were sought and discovered. For an example of a rigorous treatment of the uncertainty principle in the context of signal detection, see C. W. Helstrom.[14]

In short, an idea is scientifically fruitful if it serves to stimulate thinking that leads to discoveries. Such thinking may well start with consideration of analogies, provided they are not merely suggested by metaphorical use of language but are rooted in some aspect of reality.

The line between fruitful and sterile ideas is hard to draw. Some wild speculations of today may contain the germs of fundamental theoretical formulations of tomorrow. A typical sample of rather free-wheeling theorizing about the wider implications of cybernetics, ranging from a comparison of human and automated chess playing to an analysis of conscience and liberty, is contained in a volume published to commemorate Norbert Wiener's seventieth birthday in November 1964.[42] Wiener died in March of that year, and the book came out as a memorial volume.

On occasion some theorizers have simply borrowed the *language* of modern developments in the exact sciences. The most serious dangers of speculative promiscuity are rooted in this practice. Neologisms being, for some reason, more distasteful to physical scientists than to others, physical scientists tend to adapt common words to highly technical usage. The everyday connotations of these terms remain. For instance, "information," "feedback," "stability," "redundancy," "noise," etc., terms common in system engineering and cybernetics, are not entirely unrelated to the meanings of the corresponding common usage words. But common usage words are also heavy usage words; that is to say, they are rich in marginal, metaphorical meanings. It is here that the tendency to "theorize" by juggling words in their various contexts is greatest among those who are impressed but not disciplined by the spirit of the exact sciences. Whereas in information theory "noise" is defined as precisely as "heat" is in physics, in psychological speculations spiked with cybernetic terminology, "noise" often assumes a range of meanings stretching from the noise of traffic to the disturbances in the mental processes of a psychiatric patient.

The high prestige of science in a society dominated by technology can and has been utilized by quacks and charlatans to exploit the gullible. A notorious result of this practice was the "dianetics" fad that swept the United States in the 1950's. Dianetics was an amalgam of vulgarized notions lifted from psychoanalysis and a mumbo jumbo potpourri of terms common in cybernetics. It was offered as a sure-fire, cheap method of psychotherapy guaranteed not only to cure mental and emotional disorders but also to raise the intelligence of customers to genius levels. No better example is needed to demonstrate the seductiveness of scientific terminology in the role of word magic.

"Scientism" (simulation of scientific rigor by the misuse of technical terms) is especially harmful in psychiatry, which is concerned with events and conditions that elude precise objective analysis. By creating the impression that a new, powerful arsenal of concepts is being applied, scientism detracts from the important aspect of psychiatry as an art, where intuitive insights and empathetic understanding continue to be indispensable. The most important contributions of mathematics and cybernetics are not so much to the practice of the healing art as to the scientific infrastructure that underlies our understanding of the life processes, including mental activity and behavior.

Behavioral scientists and psychiatrists who feel that mathematics, cybernetics, and allied subjects have something of value to contribute to their area of concern will do well to draw from those fields of knowledge something of their discipline as well as inspiration. To the extent that this is done, the transplant of ideas may bear fruit. After all, the ideas of cybernetics are themselves transplants. Their origin was in biological science. They are the "organismic" ideas that had been banished from classical mechanics and whose absence in physical science has been so eloquently deplored by A. N. Whitehead.[39] The early exclusion of "organismic" concepts from physical science was justified. The soil of early physics could not have nourished these concepts. Only when this soil was sufficiently enriched could the seeds sprout. Cybernetics, rooted in the physical sciences, is the result. Now the young shoots may be ready for transplantation back to the biological and behavioral sciences where they belong. The only question is whether the present soil in those areas can nourish them properly.

Bibliography

1. AFANASYEVA, L. G., and PETUNIN, YU. I., "Teoreticheskiy Analog Obrabotki Impulsnoy Informatsii Neyronom" ("Theoretical Analysis of the Processing of Impulse Information by a Neuron"), *Kibernetika*, 3: 74–81, 1971.
2. ALLANSON, J. T., "Some Properties of a Randomly Connected Neural Network," in Cherry, C. (Ed.), *Information Theory*, Academic Press, New York, 1956.
3. ASHBY, W. R., *Design for a Brain*, John Wiley, New York, 1952.
4. ———, *Introduction to Cybernetics*, John Wiley, New York, 1956.
5. BAR-HILLIEL, Y., "Semantic Information and Its Measures," *Tr. 10th Conference on Cybernetics*, Macy Foundation, New York, 1955.
6. BRAZIER, M. A. B., "How Can Models from Information Theory Be Used in Neurophysiology," in Fields, W. S., and Abbott, W. (Eds.), *Information Storage and Neural Control*, Charles C Thomas, Springfield, Ill., 1963.
7. BRILLOUIN, L., "Maxwell's Demon Cannot Operate: Information and Entropy, I," *J. Appl. Phys.*, 22:334, 1951.
8. CANNON, W. B., *The Wisdom of the Body*, Norton, New York, 1939.
9. CHOMSKY, N., *Syntactic Structures*, Mouton, The Hague, 1957.
10. ECHOLS, H., "Genetic Control of Protein Synthesis," in Fields, W. S., and Abbott, W. (Eds.), *Information Storage and Neural Control*, Charles C Thomas, Springfield, Ill., 1963.
11. GERARD, R. W., "Instruments and Men," *Instruments*, 18:10, 1945.

12. GERBNER, G., HOLSTI, O. R., KRIPPENDORFF, K., PAISLEY, W. J., and STONE, P. J. (Eds.), *The Analysis of Communication Content*, John Wiley, New York, 1969.
13. HASSENSTEIN, B., and REICHHARDT, W., "Systemtheoretische Analyse der Zeit-, Reihenfolgen- und Vorzeichenauswertung bei der Bewegungsperzeption des Rüsselkäfers Chlorophanus," *Ztschr. Naturforsch.*, *11*b: 513, 1956.
14. HELSTROM, C. W., "Detection Theory and Quantum Mechanics (I)," *Information and Control*, *10*:254; (II), *Information and Control*, *13*:156, 1968.
15. HILTZ, F. F., "Analog Computer Simulation of a Neural Element," *Tr. Inst. Radio Engrs.*, BME–9, *1*:12–20, 1962.
16. HYMAN, R., "Stimulus Information as a Determinant of Reaction Time," *J. Exper. Psychol.*, *45*:188, 1953.
17. JOHN, E. R., "Neural Mechanisms of Decision Making," in Fields, W. S., and Abbott, W. (Eds.), *Information Storage and Neural Control*, Charles C Thomas, Springfield, Ill., 1963.
18. KORNBLUM, S., "Sequential Determinants of Information Processing in Serial and Discrete Choice Reaction Time," *Psychol. Rev.*, *76*:113, 1969.
19. MCCULLOCH, W. S., and PITTS, W., "A Logical Calculus of the Ideas Immanent in Nervous Activity," *Bull. Math. Biophys.*, *5*:114, 1943.
20. MAZZETTI, P., MONTALENTI, G., and SOARDO, P., "Experimental Construction of an Element of a Thinking Machine," *Kybernetik*, *1*:170, 1962.
21. MEDVEDEV, Z. A., "Oshibki Reproduktsii Nukleinovykh Kislot, Belkov i Ikh Biologicheskoye Znachenie" ("Errors of Reproduction of Nucleic Acids and Nucleoproteins and Their Biological Significance"), *Problemy Kibernetiki*, *9*:241, 1963.
22. MEYERHOF, O., "Thermodynamik des Lebensprozesses," *Handbuck der Physik*, *9*: 238, 1926.
23. NARASIMHAN, R., "Labeling Schemata and Syntactic Description of Pictures," *Information and Control*, *7*:151, 1964.
24. NASU, M., and HONDA, H., "Fuzzy Events Realized by Finite Probabilistic Automata," *Information and Control*, *12*:284, 1968.
25. OSGOOD, C. E., SUCI, G. J., and TANNENBAUM, P. H., *The Measurement of Meaning*, University of Illinois Press, Urbana, 1957.
26. RAPOPORT, A., "Ignition Phenomena in Random Nets," *Bull. Math. Biophys.*, *14*:35, 1952.
27. ROSENBLATT, F., "The Perceptron: A Probabilistic Model for Information Storage and Organization in the Brain," *Psychol. Rev.*, *65*:386, 1958.
28. RUTOVITZ, D., "Pattern Recognition," *J. Roy. Statistical Soc.*, Ser. A., *129*:504, 1966.
29. SCHROEDINGER, E., *What is Life?* Macmillan, New York, 1945.
30. SHANNON, C. E., "A Symbolic Analysis of Relay and Switching Circuits," *Tr. Amer. Inst. Elec. Engrs.*, *57*:713, 1938.
31. ———, and WEAVER, W., *The Mathematical Theory of Communication*, University of Illinois Press, Urbana, 1949.
32. SPERRY, R. W., "Orderly Function with Disordered Structure," in Von Foerster, H., and Zopf, G. W. (Eds.), *Principles of Self-Organization*, Pergamon, New York, 1962.
33. STARK, L., and CORNSWEET, T. N., "Testing a Servoanalytic Hypothesis for Pupil Oscillations," *Science*, *127*:588, 1958.
34. STEINBUCH, K., *Automat und Mensch*, Springer Verlag, Berlin, 1961.
35. SZILARD, L., "Ueber die Entropieverminderung in einem thermodynamischen System bei Eingriffen intelligenter Wesen," *Ztschr. Physik.*, *53*:840, 1929.
36. TOCH, H., and HASTORF, A. H., "Homeostasis in Psychology," *Psychiatry*, *18*:81, 1955.
37. VON NEUMANN, J., "The General and Logical Theory of Automata," in Jeffress, L. A. (Ed.), *Cerebral Mechanisms in Behavior: The Hixon Symposium*, John Wiley, New York, 1955.
38. WALTER, W. G., *The Living Brain*, Norton, New York, 1953.
39. WHITEHEAD, A. N., *Science and the Modern World*, Pelican Books, New York, 1948.
40. WIENER, N., *Cybernetics*, John Wiley, New York, 1948.
41. ———, *God and Golem Inc.*, M.I.T. Press, Cambridge, 1964.
42. ———, and SCHADÉ, J. P. (Eds.), *Progress in Biocybernetics*, Vol. 2, Elsevier, Amsterdam, 1965.
43. ZIPF, G. K., *Human Behavior and the Principles of Least Effort*, Addison-Wesley, Cambridge, 1949.

CHAPTER 51

GENERAL SYSTEM THEORY AND PSYCHIATRY

Ludwig von Bertalanffy

The Quandary of Modern Psychology

In recent years the concept of system has gained increasing influence in psychology and psychopathology. Numerous investigations have referred to general system theory or to some part of it;[2,5,7,13,15,57,62,110,116,127,129,133,142,160] Gordon W. Allport[6] ended the re-edition of his classic with "Personality as System"; Karl Menninger[130] based his system of psychiatry on general system theory and organismic biology; Rapoport[143] even spoke of the "epidemic-like popularity in psychology of open systems." The comprehensive works by Grinker[90] and Gray, Duhl, and Rizzo,[86] presenting general system theory and psychiatry (or unified theory in Grinker's term) in a broad frame of general considerations and specific psychiatric questions and applications, are indispensable for this study. With special gratification the present writer may cite the agreement of the two deans of American psychiatry. If there be a third revolution in psychiatry (after the behavioristic and psychoanalytic), says Grinker,[90] it is in the development of general (systems) theory; and Karl Menninger[128] honored von Bertalanffy as "one of his most influential teachers." The question arises why this trend has appeared.*

Systems thinking in psychiatry is part of a global reorientation that extends over the spectrum of intellectual life. It essentially is the search for new "paradigms" in scientific thinking, to use Thomas Kuhn's[112] poignant expression, after the paradigm of classical mechanistic thinking, which started with the scientific revolution of the sixteenth and seventeenth centuries, had reached its boundaries, and its limits as a scientific method, a theory, and a world view became apparent.

* The reader is referred to Grinker's article, "The Relevance of General Systems in Psychiatry" in Volume 6 of this *Handbook*. As is well known, Grinker's efforts in the field go back to conferences he started in 1951. The fact that the present contribution and that of Grinker were written in the same spirit, but independently, may lead to some overlapping, but hopefully also to further elucidation.

In the sciences from physics to the biological and social sciences and the humanities, the paradigm of an analytical-elementalistic-summative approach reached its limits wherever problems circumscribed by notions like "system," "wholeness," "teleology," and the like appeared and demanded new ways of thinking.[30] This was the case in physics as the limitations of classical theory were discovered; in the life sciences with the innumerable problems of order and organization of parts and processes in the living organism; in psychology with the problems of personality; in the social sciences with the problems of organizations both natural (family, tribe, and the like) and formal (an army or bureaucracy). Similarly technology transcended the traditional fields of (mechanical, electric, chemical, etc.) engineering and had to meet both in its "hardware" and "software" with essentially new requirements of communication and control,[175] man-machine systems, system analysis of industrial, commercial, economic, ecological, military, and political problems up to the social problems and international relations. And the surfeit of social criticism, new philosophies, counterculture, and social utopias, in its motivation and in often grotesque ways, equally expresses the discontent with the world view of yesterday and the search for a new one.

Such need was especially felt in psychology and psychiatry. American psychology in the first half of the twentieth century was dominated by the concept of the reactive organism, or, more dramatically, by the model of man as a robot. This conception was common to all major schools of American psychology, classical and neobehaviorism, learning and motivation theories, psychoanalysis, cybernetics, the concept of the brain as a computer, and so forth. According to a leading personality theorist:

> Man is a computer, an animal, or an infant. His destiny is completely determined by genes, instincts, accidents, early conditionings and reinforcements, cultural and social forces. Love is a secondary drive based on hunger and oral sensations or a reaction formation to an innate underlying hate. In the majority of our personological formulations there are no provisions for creativity, no admitted margins of freedom for voluntary decisions, no fitting recognitions of the power of ideals, no bases for selfless actions, no ground at all for any hope that the human race can save itself from the fatality that now confronts it. If we psychologists were all the time, consciously or unconsciously, intending out of malice to reduce the concept of human nature to its lowest common denominator, and were gloating over our successes in so doing, then we might have to admit that to this extent the Satanic spirit was alive within us.[137]

The tenets of robot psychology have been extensively criticized in the works by Allport,[3,6] Matson,[124] Koestler,[106,107,109] Bertalanffy,[41] and others. The theory, nevertheless, remained dominant for obvious reasons. The concept of man as a robot was both an expression of and a powerful motive force in industrialized mass society. It was the basis for behavioral engineering in commercial, economic, political, and other advertising and propaganda; the expanding economy of the "affluent society" could not subsist without such manipulation. Only by manipulating humans ever more into Skinnerian rats, robots, buying automata, homeostatically adjusted conformers and opportunists (or bluntly speaking, into morons and zombies) can this great society follow its progress toward an ever increasing gross national product. As a matter of fact,[97] the principles of academic psychology were identical with those of the "pecuniary conception of man."

* *

Since the present article was first written (1964), a number of fashions in psychology and psychiatry have come and gone without, however, essentially changing the predominant "robot" or "zoomorphic" model of man. It may be helpful for the present exposition briefly to enumerate the major currents that, partly with sensational success, have appeared in the intervening period.

1. *Ethology*, the comparative study of animal behavior, was broadly used for the zoomorphic theory, that is, the reduction of human to animal modes of behavior. It is obvious that all too much of human behavior has biological roots; and few periods of human history have more vividly experienced the bestiality of man under the thin veneer of so-called civilized society. It is not a

new discovery that, as man is an animal, more specifically an anthropoid ape in his anatomy, histology, biochemistry, physiology, and so forth, he also shares many behavioral mechanisms with his animal forebears and relatives. Study of his "biological drives," sex and aggression in particular, obviously is urgent in order to recognize and, if possible, to educate them. However, a "reductionist" theory, the contention that man is "nothing but" a naked ape, was certainly not the intention of the pioneers of ethology, who, like Lorenz,[119,120] emphasized man's uniqueness expressed by obvious facts like culture, tradition, history, and the like. Such reticence was alien to works of a sensationalist nature, which,[10,135,164] on the basis of often most specious arguments,[45,70,78,115,125] derived great popular success from the zoomorphic doctrine. Probably this success originated in the masochism of contemporary society thus finding alleviation of guilt feelings. Modern atrocities, criminality, and the like are more easily excused and tolerated when they come from man's biological, irresistible "aggressive drives." Similarly sex research and manuals for sexual practice belong to the same trend to discard what specifically "human" may be left in this technical and commercial but otherwise inhuman age.*

2. Under the banner of a "third force" in psychology (versus behaviorism and psychoanalysis), a new *"humanistic" psychology* was introduced.[59,60] Its leaders, such as Maslow, Charlotte Bühler, Matson, and others, exerted a thoroughly admirable influence by emphasizing the specifics of human psychology, the necessity of considering the healthy not the sick as the basic model, the investigation of the human life course, the emphasis (as against the supposedly solely normal, utilitarian behavior of the average American in commercialized society) on self-realization, "peak experience," "being cognition," and so forth. The reaction against the emotional emptiness of our society is equally understandable. Soon, however, the movement submitted to commercialism. Encounter groups and the "Human Potential Movement" became an industry run by practitioners (called "trainers" in a significant and revealing appellation) with highly questionable credentials.[101] While in part using respectable techniques of group therapy, "humanistic psychology" became big business that, with T-groups, sensitivity training, nude marathons, and the like, offered a way out of the boredom of affluent society and a shortcut to an emotional "high," with sometimes devastating results.[8] At the same time the alleged "humanism" became "zoomorphic" in a somewhat modified way. Salvation was sought in the "group," and consequently the individual was reduced to the lowest common denominator, becoming an "undifferentiated and diffused region in a social space," his self obliterated by manipulative psychology and the techniques of social engineering.[105] In this somewhat roundabout way commercialism and dehumanization, deplored as the worst outgrowth of industrial mass society, were reaffirmed by voluntary, well-paying customers.[101]

3. The well-known illnesses of present society were frontally attacked by the advocates of the counterculture[148] and Consciousness III.[145] The criticism of corporate society and the psychological wasteland of our times was appropriate enough. The remedies proposed: drugs, rock music, beards, bell-bottom trousers, commune living, and exotic religions—and this is the rather complete list of what the counterculture has to offer for saving humanity—were juvenile and silly. Not only rock festivals and students' protests but also the counterculture as a whole seems on the wane, after a surprisingly short life span for a worldwide "revolution" with highest aspirations to remodel society.

4. And here the circle closes. For apparently the latest major development (as of February 1972) is the success (with supposedly 200,000 copies sold) of Skinner's recent book, *Beyond Freedom and Dignity*,[156] which is the revival or rather the reiteration of old-fashioned behaviorism. It disregards that in the meantime animal experimentation[102] has demonstrated that even rodent behavior in wildlife situations does not follow the conditioning scheme. Conventional learning theory presently seems to apply to laboratory artifact (positive reinforcement in the Skinner cage, classical conditioning in nonsense-syllable learning, and similar techniques in advertising) but neither to natural animal behavior nor to the normal psychological development of the child.[45] Behavior therapy seems successful in certain pathology, especially bed-wetting, but it is more dubitable whether the same principles apply to the education of Einsteins, Mozarts, and even of ordinary citizens. Nevertheless, behaviorism came back with a vengeance and a high measure of intolerance with Skinner's most recent work.

* It would seem that no previous time needed to learn lusty sex "from the book." The frescoes in Pompeii's *lupanar* were professional advertising rather than visual aids in sex education.

There is no need to enter into a discussion of *Freedom and Dignity* that is of an essentially philosophical or possibly verbal nature. But it would appear that Skinner has never seen a Gothic cathedral, or even the skyscrapers of New York, never heard music from Beethoven's Ninth to the cheapest rock hit, never thought about his own laboratory, books, and university—and never made the somewhat trivial observation that rats, pigeons, and apes just don't do any such things. It is well in its place to look at the animal world for the first beginnings of language, use of tools, tradition, and the like. But human psychology cannot possibly ignore that the world of culture (of symbolic activities, to use this writer's phraseology) is something new, an emergent that cannot be reduced to the levels of conditioning and learning theory.*

Such a survey of the past eight years is useful because it shows that the kaleidoscopically changing fashions and fads in psychology (and psychiatry) actually did not alter the basic presuppositions or paradigm of American psychology. The patient on the couch, the rat in the Skinner box, the stickleback aggressively defending his territory, the T-group undergoing sensitivity training (preferably in the state of nudity), and the drug experience certainly are rather different "models of man." But they agree in the basic paradigm, namely, the neglect or "bracketing out" of what is specifically human; the consequent reduction of human to animal behavior; further, the environmentalism seeing human behavior as a product of outer factors (such as childhood experience, reinforcement, group training, the implements of the counterculture, as the prevailing theory may be), but never seeing specifically human or individual factors; and the resulting manipulation by psychoanalysis or conditioning or sensitivity training or folk music and drugs. A new paradigm is demanded to effectuate a "revolution" in this and other sciences and in practical life and society as well.

* According to Skinner,[156] culture "is a set of contingencies of reinforcement" (p. 182). This may well be true for American popular culture where an entertainer draws some $30,000 for an evening, or a boxer a couple of millions. But how this statement may apply to a Mozart whose "reinforcement" mainly was getting tuberculosis and being seated at the lackey's table or, for that matter, to any creative person—including even university professors, who certainly would do better applying their IQs to the used car or other business—is no less wondrous to this writer than Mozart's work itself.

The enormous threat contained in Skinner's latest work is that his is not a program or project to undo *Freedom and Dignity*, but a description of what is widely realized in the thought control exerted by the mass media, television, commercial society, and politics. The question whether or not the "controllers" consciously followed the academic theory of passive and operant conditioning is inconsequential, although one would suspect that they often do.[51]

* *

Modern society, provided a large-scale experiment in manipulative psychology. If its principles are correct, conditions of tension and stress should lead to an increase of mental disorder. On the other hand, mental health should be improved when basic needs for food, shelter, personal security, and so forth are satisfied; when repression of infantile instincts is avoided by permissive training; when scholastic demands are reduced so as not to overload a tender mind; when sexual gratification is provided at an early age, and so on.

The behavioristic experiment led to results contrary to expectation. World War II—a period of extreme physiological and psychological stress—did not produce an increase in neurotic[140] or psychotic[118] disorders, apart from direct shock effects such as combat neuroses. In contrast, the affluent society produced an unprecedented number of mentally ill. Precisely under conditions of reduction of tension and gratification of biological needs, novel forms of mental disorder appeared as existential neurosis, malignant boredom, and retirement neurosis,[1] that is, forms of mental dysfunction originating not from repressed drives, from unfulfilled needs, or from stress but from the meaninglessness of life. There is the suspicion[17,43] (although not substantiated statistically) that the recent increase in schizophrenia may be caused by the "other-directedness" of man in modern society. And there is no doubt that in the field of character disorders, a new type of juvenile delinquency has appeared; crime not for want or passion, but for the fun of it, for "getting a kick," and born from the emptiness of life.[9,91] As Erich Fromm[79] recently asserted, boredom is "the illness of the age" and the root of its violence in war and crime.

Thus theoretical as well as applied psychology was led into a malaise regarding basic principles. This discomfort and the trend toward a new orientation were expressed in many different ways, such as in the various neo-Freudian schools, ego psychology, personality theories (Murray, Allport), the belated reception of European developmental and child psychology (Piaget, Werner, Charlotte Bühler), the "new look" in perception, self-realization (Goldstein, Maslow), client-centered therapy (Rogers), phenomenological and existential approaches, sociological concepts of man (Sorokin),[158] and others. In the variety of these currents there is one common principle: to take man not as a reactive automaton or robot but as *an active personality system.*

Therefore, the reason for the current interest in general system theory appears to be that it is hoped that it may contribute toward a more adequate conceptual framework for normal and pathological psychology.

Why Systems Research?

In the past few decades scientific developments have taken place that can be subsumed under the general title of "systems research."[28,29] They concern a broad front in the scientific endeavor encompassing biology, psychology, behavioral and social science, technology, and other sciences. Although differing in theoretical structure, models, mathematical methods, and so forth, these developments are similar in their motives and aims. Representative of these new disciplines are general system theory, cybernetics, information theory, game and decision theory, and others. These theoretical approaches are paralleled by developments in applied science arising from the increasing complexities in technology, automation, and society in general, such as systems engineering, operations research, and the like. In recent years academic programs and job denominations have appeared that go under the name of "systems research" (or some variant) and are novel in comparison to traditional specialties. (For an introduction into the field, the following works are suggested: von Bertalanffy,[30,41] Klir,[103] Buckley,[58] Gray *et al.*,[86] Miller,[132] Rapoport.)[144]

The emergence of a "system science" is based on three major considerations:

1. Up to recent times physics was the only "exact" science, that is, the only science permitting explanation, prediction, and control within a highly developed conceptual (mathematical) framework. With the rise of biological, behavioral, and social sciences, the need for similar theoretical constructs became apparent. Simple application of physics does not suffice for this purpose. Hence a *generalization* of scientific concepts became necessary.

2. The encounter with biological, behavioral, and social problems has shown that traditional science cannot account for many aspects that are predominant in these fields. Interaction in multivariable systems, organization, differentiation, self-maintenance, goal-directedness, and the like are of fundamental importance in biological, behavioral, and social phenomena. These aspects cannot be bypassed by declaring them to be "unscientific" or "metaphysical" by decree of a physicalistic and obsolete metaphysics. Hence generalization of scientific concepts implies the *introduction of new categories.*

3. Such expanded and generalized theoretical constructs are *interdisciplinary*; that is, they transcend the traditional compartments and are applicable to phenomena in different divisions of science.

These developments are comparatively novel and largely provide "explanation on principle"[94] rather than detailed explanations and predictions. However, the same was, and still is, true of the great theories of Darwin and Freud. And in the present status of psychology and psychopathology the need is for new ways of conceptualization to permit recognition of problems and aspects that previously were overlooked or were intentionally excluded.

General System Theory and Cybernetics

Within the present context cybernetics, in its formulation as homeostasis, and general system theory, in its application to dynamic systems, are of special interest. The relation of both theories is not always well understood, and cybernetics is sometimes identified with general system theory. Hence a clarification is in order.[30,41]

The basic model of *cybernetics*[104,167,175] is the feedback scheme:

Stimulus → Message / Message → Response

Complex *feedback arrangements* found in modern servomechanisms and automation, as well as in the organism, can be resolved into aggregates of feedback circuits of this type. Applied to the living organism, the feedback scheme is called "homeostasis";[63] at least this is the common usage of a term that can be given different meanings.[165] Homeostasis is the ensemble of regulations that maintain variables constant and direct the organism toward a goal, and are performed by feedback mechanisms; that is, the result of the reaction is monitored back to the "receptor" side so that the system is held stable or led toward a target or goal. The simplest illustration is thermoregulation both by the familiar thermostat and in the warm-blooded organism; a large number of physiological and behavioral regulations are controlled by feedback mechanisms of sometimes extraordinary complexity.

General system theory[30,52] pertains to principles that apply to systems in general. A system is defined as a complex of components in mutual interaction. General system theory contends that there are principles of systems in general or in defined subclasses of systems, irrespective of the nature of the systems, of their components, or of the relations or "forces" between them. System principles may be expressed in mathematical models, may often be simulated by electronic or other analogues; they have been applied in numerous fields of pure and applied science. Concepts and principles of system theory are not limited to material systems, but can be applied to any "whole" consisting of interacting "components," as especially practical applications in systems engineering[92,93] show.

A case particularly important for the living organism is that of *open systems*,[23,30,39] that is, systems maintained in the exchange of matter with the environment, by import and export and the building up and breaking down of components. Open systems, compared to closed systems of traditional physics, show singular characteristics. An open system may attain a *steady state* in which it remains constant, but in contrast to conventional equilibriums, this constancy is one of continuous exchange and flow of component material. The steady state of open systems is characterized by *equifinality*; that is, in contrast to equilibriums in closed systems, which are determined by initial conditions, the open system may attain a time-independent state that is independent of initial conditions and determined only by the system parameters. Open systems show *thermodynamic characteristics* that are apparently paradoxical. According to the Second Law of Thermodynamics, the general course of physical events (that is, in closed systems) is directed toward increasing entropy, leveling down differences and states of maximum disorder. In open systems the import of "negative entropy" is possible with the transfer of matter. Hence such systems can maintain themselves in states of high improbability and at a high level of order and complexity; they may even advance toward increasing order and differentiation, as is the case in development and evolution.

General systems and cybernetics are applicable to certain ranges of phenomena. In some cases either model may be applied, and the equivalence of description in the "languages" of cybernetics (feedback circuits) and dynamical system theory (interactions in a multivariable system) can be shown. We note that no scientific model is monopolistic; each may reproduce, more or less successfully, certain aspects of reality. The present chapter is limited to general system theory in the narrower sense, excluding cybernetics, information theory, etc., in their possible applications to psychiatry. Within the space available only a small selection of key concepts, annotated by examples relevant to psychiatry or psychological theory, can be presented.

System Concepts in Psychopathology

General system theory has its roots in the organismic conception in biology. On the European continent this was developed by the present author[35] in the 1920's, with parallel developments in the Anglo-Saxon countries (Whitehead, Woodger, Coghill, and others) and in psychological Gestalt theory (W. Köhler). It is interesting to note that Eugen Bleuler[54] followed with sympathetic interest this development in its early phase. A similar development in psychiatry was represented by Goldstein.[83] Somewhat later the homeostasis principle became recognized in physiology through Cannon's[63] work. Organismic biology introduced the concept of the organism as an open system, which led to important expansions of physical theory, especially in thermodynamics. A further generalization was the proposal of an interdisciplinary "general system theory."[49] In a somewhat parallel way developments in communication engineering,[155] computers, and servomechanisms led to cybernetics[115] as an interdisciplinary field. The proposal of general system theory entailed the unexpected discovery that similar trends were active in many fields of the behavioral and social sciences. These tendencies joined in the formation of the Society for General Systems Research (1954), which since has tried to serve as a unifying agency for such studies.[52]

Like other fundamental conceptions (for example, the atomic theory, the machine theory of organism, the cyclic theory of history, the positivistic theory in philosophy of science), the modern system concept has a long history.[31] In biology and medicine one may trace it to Claude Bernard and the Paracelsian-Hippocratic tradition; in psychology and philosophy, to the Leibnitzian tradition,[3,4] Nicholas of Cusa,[36] and even further. It appears, however, that the idea of a science of systems could emerge only at the present state of scientific development.

Organism and Personality

In contrast to physical forces such as gravity or electricity, the phenomena of life are found only in individual entities called organisms. Any organism is a system, that is, a dynamic order of parts and processes standing in mutual interaction.[40] Similarly psychological phenomena are found only in individualized entities that in man are called personalities. "Whatever else personality may be, it has the properties of a system" (p.109).[6]

The "molar" concept of the psychophysical organism as a system contrasts with its conception as a mere aggregate of "molecular" units such as reflexes, sensations, brain centers, drives, reinforced responses, traits, factors, and the like. Psychopathology clearly shows mental dysfunction as a system disturbance rather than as a loss of single functions. Even in localized traumata (for example, cortical lesions) the ensuing effect is impairment of the total action system, particularly with respect to higher and, hence, more demanding functions. Conversely, the system has considerable regulative capacities.[53,82,114]

The Active Organism

"Even without external stimuli, the organism is not a passive but an intrinsically active system. Reflex theory has presupposed that the primary element of behavior is response to external stimuli. In contrast, recent research shows with increasing clarity that autonomous activity of the nervous system, resting in the system itself, is to be considered primary. In evolution and development, reactive mechanisms appear to be superimposed upon primitive rhythmic-locomotor activities. The stimulus (i.e., a change in external conditions) does not *cause* a process in an otherwise inert system; it only *modifies* processes in an autonomously active system" (p. 133 ff).[26]

The living organism maintains a disequilibrium called the steady state of an open system, and thus it is able to dispense existing potentials or "tentions" in spontaneous activity or in response to releasing stimuli; it even advances toward higher order and organization. The robot model considers response to stimuli, reduction to tensions, re-establishment of an equilibrium disturbed by outside factors, adjustment to environment, and the like as the basic universal scheme of behavior. The robot model, however, only partly covers animal behavior and does not cover an essential portion of human behavior at all. The insight into the primary immanent activity of the psychophysical organism necessitates a basic reorientation that can be supported by any amount of biological, neurophysiological, behavioral, psychological, and psychiatric evidence.[41]

Autonomous activity is the most primitive form of behavior;[40,65,98,99,153,169] it is found in brain function[96] and in psychological processes. The discovery of activating systems in the brain stem[20,95,123] has emphasized this fact. Natural behavior encompasses innumerable activities beyond the S-R scheme, from exploring, play, and rituals in animals[153] to economic, intellectual, aesthetic, and religious pursuits to self-realization and creativity in man. Even rats seem to "look" for problems,[95] and the healthy child and adult are going far beyond the reduction of tensions or gratification of needs in innumerable activities that cannot be reduced to primary or secondary drives.[6] All such behavior is performed for its own sake, deriving gratification ("function pleasure," after K. Bühler) from the performance itself.

For similar reasons complete relaxation of tensions, as in sensory deprivation experiments, is not an ideal state but is apt to produce insufferable anxiety, hallucinations, and other psychosislike symptoms. Prisoner's psychosis, acerbation of symptoms in the closed ward, and retirement and weekend neuroses are related clinical conditions attesting that the psychophysical organism needs an amount of tension and activity for healthy existence. It appears that a proper distance between both understimulation and overstimulation should be maintained; the effects of the latter are called "culture shock" owing to the nervous overload in a rapidly changing society.[166]

It is a symptom of mental disease that spontaneity is impaired. The patient increasingly becomes an automaton or S-R machine, pushed by biological drives, obsessed by needs for food, elimination, sexual gratification, and so on. The model of the passive organism is a quite adequate description of the stereotype behavior of compulsives, of patients with brain lesions, and of the waning of autonomous activity in catatonia and related psychopathology. But by the same token this emphasizes that normal behavior is different.

Homeostasis

Many psychophysiological regulations follow the principle of homeostasis. Its limitations[72,165] have been aptly summarized by Charlotte Bühler:[62]

> In the fundamental psychoanalytic model, there is only one basic tendency, that is toward need gratification or *tension reduction*. . . . Present-day biological theories emphasize the "spontaneity" of the organism's activity which is due to its built-in energy. The organism's autonomous functioning, its "drive to perform certain movements" is emphasized by Bertalanffy. . . . These concepts represent *a complete revision of the original homeostasis principle* which emphasized exclusively the tendency toward equilibrium.[62]

In general, the homeostasis scheme is not applicable (1) to dynamic regulations, that is, regulations not based upon fixed mechanisms but taking place within a system functioning as a whole (for example, regulative processes after brain lesions), (2) to spontaneous activities, (3) to processes whose goal is not reduction but building up of tensions, and (4) to processes of growth, development, creation, and the like. We may also say that homeostasis is inappropriate as an explanatory principle for those human activities that are non-utilitarian, that is, not serving the primary needs of self-preservation and survival and their secondary derivatives, as is the case with many cultural manifestations. The evolution of Greek sculpture, Renaissance painting, or German music had nothing to do with adjustment or survival because they are of symbolic rather than biological value[32,48] (compare below). But even living nature is by no means merely utilitarian.[40]

The principle of homeostasis has sometimes been inflated to a point where it becomes silly. The martyr's death at the stake is explained[56] "by abnormal displacement" of his internal processes so that death is more "homeostating" than continuing existence; the mountain climber is supposed to risk his life because "losing valued social status may be more upsetting."[159] Such examples show to what extremes some writers are willing to go in order to save a scheme that is rooted in economic-commercial philosophy and sets a premium on conformity and opportunism as ultimate values. It should not be forgotten that Cannon,[63] eminent physiologist and thinker that he was, is innocuous of such distortions; he explicitly emphasized the "priceless unessentials" beyond homeostasis.

The homeostasis model is applicable in psychopathology because nonhomeostatic functions, as a rule, decline in mental patients. Thus Karl Menninger[130] was able to describe the progress of mental disease as a series of defense mechanisms, settling down at ever lower homeostatic levels until mere preservation of physiological life is left. Arieti's[17] concept of progressive teleological regression in schizophrenia is similar.

Differentiation

"Differentiation is transformation from a more general and homogeneous to a more special and heterogeneous condition" (p. 19).[69] "Wherever development occurs it proceeds from a state of relative globality and lack of differentiation to a state of increasing differentiation, articulation and hierarchic order."[170]

The principle of differentiation is ubiquitous in biology, the evolution and development of the nervous system, behavior, psychology, society, and culture. We owe to Werner[169] the insight that mental functions generally progress from a syncretic state, where percepts, motivation, feeling, imagery, symbols, concepts, and so forth are an amorphous unity, toward an ever clearer distinction of these functions. In perception the primitive state seems to be one of synesthesia (traces of which are left in the human adult and which may reappear in schizophrenia, mescaline, and LSD experience) out of which visual, auditory, tactual, chemical, and other experience are separated. In animal and a good deal of human behavior, there is a perceptual-emotive-motivational unity; perceived objects without emotional-motivational undertones are a late achievement of mature, civilized man. The origins of language are obscure; but insofar as we can form an idea, it seems that "holophrastic" (W. Humboldt)[169] language and thought, that is, utterances and thoughts with a broad aura of associations, preceded separation of meanings and articulate speech. Similarly the categories of developed mental life, such as the distinction of "I" and objects, space, time, number, causality, and so forth, evolved from a perceptual-conceptual-motivational continuum represented by the "paleological" perception of infants, primitives, and schizophrenics.[11,17,141,169] Myth was the prolific chaos from which language, magic, art, science, medicine, mores, morals, and religion were differentiated.[66]

Thus "I" and "the world," "mind" and "matter," or Descartes's "*res cogitans*" and "*res extensa*" are not a simple datum and primordial antithesis. They are the final outcome of a long process in biological evolution, mental development of the child, and cultural and linguistic history, wherein the perceiver is not simply a receptor of stimuli but in a very real sense *creates* his world.[56,64,80,124] The story can be told in different ways,[6,34,37,66,77,131,141,169] but there is general agreement that differentiation arose from an "undifferentiated absolute of self and environment".[21] The animistic experience of the child and the primitive (persisting still in Aristotelian philosophy), the "physiognomic" outlook,[169] the experience of "we" and "Thou" (still much stronger in Oriental than in Western thinking),[108] empathy, and so forth were steps on the way until Renaissance physics eventually "discovered inanimate nature."[151] "Things" and "self" emerge by a slow build-up of innumerable factors of gestalt dynamics, of learning processes, and of social, cultural, and linguistic determinants; the full distinction between "public objects" and "private self" is certainly not achieved without naming and language, that is, processes at the symbolic level; and perhaps this distinction presupposes a language of the Indo-Germanic type.[172]

In psychopathology and especially schizophrenia all these primitive states may reappear by way of regression and in bizarre manifestations; bizarre because there are arbitrary combinations of archaic elements among themselves and with more sophisticated thought processes. On the other hand, the experience of the child, savage, and non-Westerner, though primitive, nevertheless forms an organized universe. This leads to the next group of concepts to be considered.

Centralization and Related Concepts

"Organisms *are* not machines; but they can to a certain extent *become* machines, congeal into machines. Never completely, however; for a thoroughly mechanized organism would be incapable of reacting to the incessantly changing conditions of the outside world" (pp. 17 ff.).[40] The *principle of progressive mechanization* expresses the transition from undifferentiated wholeness to higher function, made possible by specialization and "division of labor"; this principle implies also loss of potentialities in the components and of regulability in the whole.

Mechanization frequently leads to establishment of *leading parts,* that is, components dominating the behavior of the system. Such centers may exert "trigger causality"; that is, in contradistinction to the principle, *causa aequat effectum,* a small change in a leading part may by way of *amplification mechanisms* cause large changes in the total system. In this way a *hierarchical order* of parts or processes may be established.[30,174] These concepts hardly need comment except for one debated point.

In the brain as well as in mental function, centralization and hierarchical order are achieved by stratification,[81,117,122,149] that is, by superimposition of higher "layers" that take the role of leading parts. Particulars and disputed points are beyond the present survey. However, one will agree that, in gross oversimplification, three major layers or evolutionary steps can be distinguished. These are the evolution of (1) the paleencephalon, "old brain" or brain stem, in lower vertebrates, (2) the neencephalon (cortex), evolving from reptiles to mammals, and (3) certain "highest" centers, especially the motoric speech (Broca's) region and the large association areas that are found only in man. Concurrently there is an anterior shift of controlling centers, for example, in the apparatus of vision from the colliculi optici of the mesencephalon (lower vertebrates) to the corpora geniculata lateralia of the diencephalon (mammals) to the regio calcarina of the telencephalon (man).

In some way parallel is stratification in the mental system of personality. Again in extreme oversimplification, this may be circumscribed as the domain of instincts, drives, emotions, the primeval "depth personality"; that of conscious perception and voluntary action; and that of the specific activities characteristic of man, called "symbolic" in Western science, and the "secondary signal system" in Russian. Somewhat different is Arieti's intrapsychic organization of primary, secondary, and tertiary processes.[14]

Thus it is clear that stratification exists both in the brain and in mental processes, and that these correspond in some way, but the particulars present great difficulties. The neurophysiological meaning of a small portion of neural processes (of the cortex jointly with the arousal system), being "conscious" while the majority is not, is completely unknown. The Freudian distinction of id, ego, and superego is certainly insufficient; especially so because the Freudian id (or unconscious) comprises only limited aspects, and disregarded its creative side, which was already emphasized by pre-Freudian authors.[173] The "unconscious" is not only a cellar to put in what has been "repressed" but also the fountainhead from which "creative" processes—in science, art, religion, presumably even evolution—arise. Unfortunately this is not widely known to American psychoanalysts; one may guess that the development of neo-Freudian thought and practice would have been different if the fact was recognized that the Freudian is but one version of the theory of the "unconscious." Furthermore, the "unconscious" comprises both the lowest intrapsychic level ("primary process," "animal drives," "instinct," and the like) and, paradoxically, the highest (variously named "oceanic feeling," mystical, "peak" experience, Consciousness III, and so forth).

Thus stratification in its neurophysiological and psychological aspects is a fact, but it leaves many problems whose exploration would widely exceed the frame of the present article. In any case it is certainly incorrect when Anglo-Saxon authors refuse stratification for being "Philosophical"[71] or insist that there is no fundamental difference between the behavior of a rat and that of man.[157] Such an attitude simply ignores elementary zoological facts. Moreover, stratification is indispensable for understanding psychiatric disturbances.

Among the consequences of the stratified hierarchy of both the brain and mental function is a dismal one. It was expressed by the present author[22,41] as follows:

> Man is characterized by the massive development of the cerebral cortex and the specific regions mentioned; while no comparable development is recognizable in the lower strata of his brain. (N.B., The hypothalamic regions are less highly differentiated in man than in lower mammals and monkeys.) This presumably is the reason why man's evolution is almost exclusively on the intellectual side. The ten billion neurons of the cortex made possible the progress from stone axes to airplanes and atomic bombs, and from primitive mythology to quantum theory. However, there is no corresponding development on the instinctual side. For this reason man's moral instincts have hardly improved over those of the chimpanzee.

Unfortunately this applies to all utopian hopes for man's betterment, from the preaching of the great religions to the Enlightenment's faith in reason to the nineteenth-century belief in progress and to Consciousness III. There is, quite simply, no anatomical substratum for the expected improvement.

This conception has been elaborated by Koestler[107] and MacLean[109] in the doctrine of the "three brains of man" to which we refer for detailed information on this important aspect.

Regression

The psychotic state is sometimes said to be a "regression to older and more infantile forms of behavior." This is incorrect; already E. Bleuler noted that the child is not a little schizophrenic but a normally functioning, though primitive, being. "The schizophrenic will regress to, but not integrate at a lower level; he will remain disorganized" (p. 475).[17] Regression is essentially disintegration of personality, that is, dedifferentiation and decentralization. Dedifferentiation means that there is not a loss of meristic functions, but rather a reappearance of primitive states (syncretism, synesthesia, paleological thinking, and so forth). Decentralization is, in the extreme, functional dysencephalization in the schizophrenic.[17] Splitting of personality, according to E. Bleuler, in milder form neurotic complexes (that is, psychological entities that assume dominance), disturbed ego function, weak ego, and so forth, similarly indicate loosening of the hierarchical mental organization.

Boundaries

Any system as an entity that can be investigated in its own right must have boundaries, either spatial or dynamic. Strictly speaking, spatial boundaries exist only in naïve observation, and all boundaries are ultimately dynamic. One cannot exactly draw the boundaries of an atom (with valences sticking out, as it were, to attract other atoms), of a stone (an aggregate of molecules and atoms that mostly consist of empty space, with particles in planetary distances), or of an organism (continually changing matter with environment).

In psychology the boundary of the ego is both fundamental and precarious. As already noted, it is slowly established in evolution and development and is never completely fixed. It originates in proprioceptive experience and in the body image, but self-identity is not completely established before the "I", "Thou", and "it" are named. Psychopathology shows the paradox that the ego boundary is at once too fluid and too rigid. Syncretic perception, animistic feeling, delusions and hallucinations, and so on, make for insecurity of the ego boundary; but within his self-created universe the schizophrenic lives "in a shell," much in the way animals live in the "soap bubbles" of their organization-bound worlds (von Uexküll).[153] In contrast to the animal's limited "ambient," man is "open to the world" or has a "universe"; that is, his world widely transcends biological bondage and even the limitations of his senses. To him "encapsulation" (Royce)[150]—from the specialist to the neurotic, and, in the extreme, to the schizophrenic—sometimes is a pathogenic limitation of potentialities. These are based in man's symbolic functions.

Symbolic Activities

"Except for the immediate satisfaction of biological needs, man lives in a world not of things but of symbols."[22,41] We may also say that the various symbolic universes, material and nonmaterial, that distinguish human cultures from animal societies are part, and easily the most important part, of man's behavior system. It can be justly questioned whether man is a rational animal; but he certainly is a symbol-creating and symbol-dominated being throughout.

Symbolism is recognized as the unique criterion of man by biologists,[41,98] physiologists of the Pavlovian school ("secondary signal system"),[121] psychiatrists,[9,17,82] and philosophers.[66,113] It is not found even in leading textbooks of psychology and most recent behavioristic work[156] in consequence of the predominant robot philosophy. But it is precisely because of symbolic functions that "motives in animals will not be an adequate model for motives in man" (p. 221)[6] and that human personality is not finished at the age of three or so, as Freud's instinct theory assumed.

The definition of symbolic activities will not be discussed here; the author has attempted to do so elsewhere.[22,37,41] It suffices to say that probably all notions used to characterize human behavior are consequences or different aspects of symbolic activity. Culture or civilization; creative perception in contrast to passive perception (Murray, G. W. Allport), objectivation of both things outside and the self,[163] ego-world unity,[138] self-reflexiveness;[45] abstract against concrete stratum;[82] having a past and future, "time-binding," anticipation of future; true (Aristotelian) purposiveness,[38] intention as conscious planning;[6] dread of death, suicide; will to meaning,[73] interest as engaging in self-gratifying cultural activity,[6] idealistic devotion to a (perhaps hopeless) cause, martyrdom; "forward trust of mature motivation";[6] self-transcendence; ego autonomy, conflict-free ego functions; "essential" aggression;[25] conscience, superego, ego ideal, values, morals, dissimulation, truth, and lying—these are very different formulations or aspects, but all stem from the root of creative symbolic universes and therefore cannot be reduced to biological drives, psychoanalytic instincts, reinforcement of gratifications, or other biological factors. The distinction between *biological values* and *specific human values* is that the first concern the maintenance of the individual and the survival of the species, the latter always concern a symbolic universe.[40,41,48]

Consequently mental disturbances in man, as a rule, involve disturbances of symbolic functions. Kubie[111] appears to be correct when, as a "new hypothesis" on neuroses, he distinguished "psychopathological processes which arise through the distorting impact of highly charged experiences at an early age" from those "consisting in the distortion of symbolic functions." Frankl's[74] distinction of somatogenic, psychogenic, and noogenic neuroses should be generally accepted. Disturbances in schizophrenia also are essentially at the symbolic level and are able to take many different forms: loosening of associational structure, breakdown of the ego boundary, speech and thought disturbances, concretization of ideas, desymbolization, paleological thinking, and others. We refer to Arieti's[17] and Goldstein's[82] discussions.

The conclusion (which is by no means generally accepted) is that mental illness is a *specifically human phenomenon.* Animals may behaviorally show (and for all we know by empathy experience) any number of perceptual, motoric, and mood disturbances, hallucinations, dreams, faulty reactions, and the like. Animals cannot have the disturbances of symbolic functions that are essential ingredients of mental disease. In animals there cannot be disturbance of ideas, delusions of grandeur or of persecution, etc., for the simple reason that there are no ideas to start with. Similarly "animal neurosis" is only a partial model of the clinical entity.[42]

This is the ultimate reason why human behavior and psychology cannot be reduced to biologistic notions like restoration of homeostatic equilibrium, conflict of biological drives, unsatisfactory mother-infant relationships, and the like. Another consequence is the culture dependence of mental illness both in symptomatology and epidemiology. To say that psychiatry has a physio-psycho-sociological framework is but another expression of the same fact.

For the same reason human striving is more than self-realization; it is directed toward objective goals and realization of values,[72–75] which mean nothing else than symbolic entities that in a way become detached from their creators.[22,37,41] Perhaps we may venture a definition. There may be conflict between biological drives and a symbolic value system; this is the situation of psychoneurosis. Or there may be conflict between symbolic universes, or loss of value orientation and experience of meaninglessness in the individual; this is the situation when existential or "noogenic" neurosis arises. Similar considerations apply to "character disorders" like juvenile delinquency that, quite apart from their psychodynamics, stem from the breakdown or erosion of the value system. Among other things culture is an important psychohygienic factor.[32,48]

The System Concept in Psychopathology

Having gone through a primer of system-theoretical notions, we may summarize that these appear to provide a consistent framework for psychopathology.

Mental disease is essentially a disturbance of system functions of the psychophysical organism. For this reason isolated symptoms or syndromes do not define the disease entity.[43] Look at some classical symptoms of schizophrenia. "Loosening of associational structure" (E. Bleuler) and unbridled chains of associations; quite similar examples are found in "purple" poetry and rhetoric. Auditory hallucinations; "voices" told Joan of Arc to liberate France. Piercing sensations; a great mystic like St. Teresa reported an identical experience. Fantastic world constructions; those of science surpass any schizophrenic's. This is not to play on the theme "genius and madness"; but it is apt to show that not single criteria but integration makes for the difference.

Psychiatric disturbances can be neatly defined in terms of system functions. In reference to *cognition*, the worlds of psychotics, as impressively described by writers of the phenomenological and existentialist schools,[126] are "products of their brains." But our normal world also is shaped by emotional, motivational, social, cultural, and linguistic factors, amalgamated with perception proper. Illusions and delusions, and hallucinations at least in dreams, are present in the healthy individual; the mechanisms of illusion even play an important role in constancy phenomena, without which a consistent world image would be impossible. The contrast between normality[139] and schizophrenia is not that normal perception is a plane mirror of reality "as is," but that schizophrenia has subjective elements that run wild and that are disintegrated.

The same applies at the symbolic level. Scientific notions, such as the earth running with unimaginable speed through the universe or a solid body consisting mostly of empty space interlaced with tiny energy specks at astronomical distances, contradict all everyday experience and "common sense" and are more fantastic than the "world designs" of schizophrenics. Nevertheless, the scientific notions happen to be "true"; that is, they fit into an integrated scheme.

Similar considerations apply to *motivation.* The concept of spontaneity draws the borderline. Normal motivation implies autonomous activity, integration of behavior, plasticity in and adaptability to changing situations, free use of symbolic anticipation, decision, and so forth. This emphasizes the hierarchy of functions, especially the symbolic level superimposed upon the organismic. Hence besides the organismic principle of "spontaneous activity" the "humanistic" principle of "symbolic functions" must be basic in system-theoretical considerations.

Hence the answer whether an individual is mentally sound or not is ultimately determined by whether he has *an integrated universe consistent within the given cultural framework.*[43] So far as we can see this criterion comprises all phenomena of psychopathology as compared with normality and leaves room for the culture dependence of mental norms. What may be consistent in one culture may be pathological in another, as cultural anthropologists[19] have shown.

This concept has definite implications for psychotherapy. If the psychophysical organism is an active system, occupational and adjunctive therapies are an obvious consequence; evocation of creative potentialities will be more important than passive adjustment. If these concepts are correct, more important than "digging the past" will be insight into present conflicts, attempts at reintegration, and orientation toward goals and the future, that is, symbolic anticipation. This, of course, is a paraphrase of recent trends in psychotherapy, which thus may be grounded in "personality as system." If, finally, much of present neuroses are "existential," resulting from the meaninglessness of life, then "logotherapy" (Frankl),[73,75] that is, therapy at the symbolic level, will be in place.

It therefore appears that—without falling into the trap of "nothing but" philosophy and disparaging other conceptions—a system theory of personality provides a sound basis for psychology and psychopathology.

Conclusion

System theory in psychology and psychiatry is not a dramatic denouement or new discovery, and if the reader has a *déjà vu* feeling, we shall not contradict him. On the other hand, it should be recognized that the "model of man" in systems terms is totally different from the still widely dominant "robot model" of neobehaviorism and other modern currents as enumerated in the beginning of the chapter. It was our intention to show that the system concept in this field is not speculation, is not an attempt to press facts into the strait jacket of a theory that happens to be in vogue, and has nothing to do with the "mentalistic anthropomorphism" so feared by behaviorists. Nevertheless, the system concept is a radical reversal with respect to robot theories, leading to a more realistic (and incidentally more dignified) image of man. Moreover, it entails far-reaching consequences for the scientific world view that can only be alluded to in the present outline:

1. The system concept provides a theoreti-

cal framework that is *psychophysically neutral.* Physical and physiological terms such as action potentials, chemical transmission at synapses, neural network, and the like are not applicable to mental phenomena, and even less can psychological notions be applied to physical phenomena. System terms and principles like those discussed can be applied to facts in either field.

2. The mind-body problem cannot be discussed here, and the author has to refer to other investigation.[33,34] We can only summarize that the Cartesian dualism between matter and mind, objects outside and ego inside, brain and consciousness, and so forth is incorrect in the light both of direct phenomenological experience and of modern research in various fields; it is a conceptualization stemming from seventeenth-century physics that, even though still prevailing in modern debates,[100,152] is obsolete. In the modern view science does not make metaphysical statements, whether of the materialistic, idealistic, or positivistic sense-data variety. It is a conceptual construct to reproduce limited aspects of experience in their formal structure. Theories of behavior and of psychology should be similar in their formal structure or isomorphic. Possibly systems concepts are the first beginning of such "common language" (compare Piaget and Bertalanffy).[162] In the remote future this may lead to a "unified theory"[173] from which eventually material and mental, conscious and unconscious aspects could be derived (L. Whyte).

3. Within the framework developed the problem of *free will or determinism* also receives a new and definite meaning. It is a pseudoproblem, resulting from confusion of different levels of experience and of epistemology and metaphysics. We *experience* ourselves as free, for the simple reason that the category of causality is not applied in direct or immediate experience. Causality is a category applied to bring order into objectivated experience that is reproduced in symbols. Within the latter we try to *explain* mental and behavioral phenomena as causally determined and can do so with increasing approximation by taking into account ever more factors of motivation, by refining conceptual models, and so forth. Will is not *determined*, but is *determinable*, particularly in the machinelike and average aspects of behavior, as motivation researchers and statisticians know. However, causality is not metaphysical necessity, but is one instrument to bring order into experience, and there are other "perspectives"[27] of equal or superior standing.

4. Separate from the epistemological question is the moral and legal question of *responsibility.* Responsibility is always judged within a symbolic framework of values as accepted in a society under given circumstances. For example, the M'Naughten rules that excuse the offender if "he cannot tell right from wrong" actually mean that the criminal goes unpunished if his symbolic comprehension is obliterated and hence his behavior is determined only by "animal" drives. Killing is prohibited and is punished as murder within the symbolic framework of the ordinary state of society, but is commanded (and refusal of command is punished) in the different value frame of war.

Some Current System-Theoretical Issues in Psychiatry

It would be difficult to review briefly the numerous publications in "system-theoretical psychiatry. It may be useful, however, to enumerate a few major problem areas. Once more we refer to Grinker's presentation in another volume of this *Handbook.*

1. The study of the intrapsychic self,[14] that is, of the human psyche in its totality, is essentially a systems approach. So are Menninger's vital balance[130] and unitary concept of mental illness.[129] In contrast to behavioristic and Freudian theories (the latter essentially are limited to the primary process), Arieti's "types" or rather levels of cognition (primary as in the child, primitive, and schizophrenic; secondary in conceptual thinking; tertiary in creative processes) is a pioneering attempt toward a conceptual construction of the whole human psyche. Arieti's[16] persistent effort to introduce "cognition" into psychiatry (as he

justly complains, "almost completely ignored by classical psychoanalysts"), corresponds with what has been discussed here under the label of symbolic activities. Combined with available insight into the evolution of the brain, this is perhaps as far as we may presently go. The task of the future (apart from further refinement) may be in establishing the isomorphism of the neurophysiological and mental aspects by means of a "psychophysically neutral" general system theory.[33,34]

2. The problem of schizophrenia remains at present unsolved, and the enormous literature reflects a state of confusion. Recent investigators such as Grinker[88] and Arieti[12] (cf. also Bertalanffy)[45] agree that it must be approached not in a monocausal way (for example, as a biochemical disturbance, as either genetic or environmental, as a result of psychodynamics, of double bind, and so forth), but in a system approach taking into account many interacting levels and factors.

3. According to what has been said above, learning theory in its conventional form, that is, based on classic or operant conditioning (positive/negative reinforcement), requires a thorough overhauling.[45] It is not a model for animal behavior in the normal "wild" state as adaptive behavior is learned here without positive reinforcement and long-time repetition.[41,102] It covers human learning in certain respects (so that teaching machines are well in place to a certain extent), but does not cover learning by "insight into meaning." At present no adequate theory covering the aspects mentioned appears to exist; but this is an urgent desideratum not only for theoretical psychology but also for psychiatry, in view of the involvement of learning processes in neurosis and psychosis, the application of behavior therapy and its limits, and so forth.

4. The systems approach appears to be particularly fruitful in family and group therapy and community psychiatry, where it has been widely used.[86]

5. The same applies to "borderline" fields that in part are in the domain of psychiatry and mental health service. Such are court psychiatry and juvenile delinquency;[146,147] health service in general, which, in its present chaotic state and in view of the many "variables" involved, obviously requires a systems approach not only in the way of formal and learned programs but also in practical implementation;[155,168] medical education[87] and educational administration;[134] functionalism in sociology,[69] political science,[176] and other fields.

6. The system approach is tacitly implied in a good deal of psychiatric practice. It has not produced a new wonder cure (fortunately one may say). But the two essential insights—(a) that mental illness and therapy are not monocausal, but are processes in an enormously complex, interacting, multilevel whole; and (b) this whole or system essentially is not passive, robotlike, or an environment-dependent S-R machine, but is an active system, whose potentialities and activities should be employed both in normal life and in the therapeutic process—formulate a reorientation in psychiatry that is followed by many practitioners and finds a theoretical framework in system theory.

7. The above also partly answers a complex of questions that endanger the very existence of psychiatry as a medical specialty and that are epitomized in Szasz's[161] "myth of mental illness." It is obvious that mental illness and psychiatry far transcend conventional science and the "medical model" because they are largely on the "symbolic" level about which (physicalistic) science tells us nothing. Furthermore, psychotherapy has far expanded into the sphere of problems and conflicts that arise in a complex society and previously were handled by a nonmedical adviser, the wise friend, the teacher, the priest, and the like. This is apparent already in Roger's rebaptizing of patients as "clients," more recently in T-groups and allied forms of "psychotherapy" for the supposedly healthy. In anxiety, marital problems, and the psychological and behavioral abnormalities resulting therefrom, it is or may be extremely hard to say whether they are moral and value problems or constitute a medical case; consequently, whether relief should be sought in social (practically speaking, financial and social) help beyond the doctor's control, or else in pills, professional or nonprofessional psychotherapy, group ther-

apy, and the like. There is hardly a doubt that the notion of "mental illness" has been blown up far beyond legitimate medical science and practice, for example, when considering alcoholism, drug addiction, and juvenile delinquency as "illness" rather than misbehavior beyond current social and legal rules. The question whether or not they are "psychiatric" remains most precarious as is shown by conflicting testimonies of respectable psychiatrists in any court case. Furthermore, there is something basically wrong with a specialty suffering from the "neuropsychiatric split"[16] when the supposedly identical "illness" is treated either by brain surgery or by "soft talk"; comparable to warfare that combines napalm and intercontinental missiles with prayers and malediction of the enemy (as, paradoxically, is the case in our "enlightened" age). On the other hand, there obviously is "mental illness" as there is organic illness of various sorts, and it justifies the specialty of psychiatry in the same way as other specialties of medicine are justified.

8. Possibly systems theory may play a unifying role in psychology. It is a major objection raised against psychology (and by implication, psychiatry) that it lacks consistent development and the cumulative nature that is characteristic of science.[105] In any legitimate science results—empirical and theoretical—when once established, remain so and are accumulated in a continual evolution. Galileo or Mendel are still uncontested authorities, however much quantum mechanics and molecular genetics have transcended their discoveries. In psychology, in contrast, we seem to see kaleidoscopically changing fashions. It is an arena of contesting "schools," theories convincing only to their author and his pupils, or limited to a particular professor, university department, or movement, and too many ephemeral and sensationalist fads.

This situation appears somewhat less hopeless from a system-theoretical view. Many seemingly different "systems" in psychology are descriptions of essentially the same facts in different "languages" as it were, or different aspects of such facts. To give an example, in developmental psychology we find the Piagetian, Wernerian, Brunnerian, and other "schools." Analysis would show, however, that they are complementary rather than contradictory; that is, they present essentially similar models or paradigms in different languages (similarly as the same mathematical structure can be expressed in an equation or a graph, or the same physical facts can be expressed in the languages of classical thermodynamics and of statistical mechanics). General system theory, because of its abstract nature, may be the best approach to a "common language" that unifies psychological theories and makes psychology into a science fulfilling the requirement of cumulation of established statements.

Humanistic General System Thinking

In summary, we may, with Gray,[84] emphasize that there exists a mechanistic and an organismic trend within systems theory. The first is understandably connected with technological developments such as control theory, cybernetics, system analysis in commerce and industry, and so forth. A systems (that is, multivariable) approach is obviously imperative to deal with the complex problems in modern society; it carries the danger of making the human individual ever more into a small wheel of the social "megamachine."[55,136] On the other hand, organismic-humanistic system theory is, according to Gray,[84,85] characterized by what he calls the "five Bertalanffian principles," namely, (1) the organismic systems or nonreductionist approach, emphasizing the wholeness of the organism and its accessibility to scientific method, contrasted with the elementaristic and summative approach of conventional science; (2) the principle of the active organism in contradistinction to the reactive organism, the robot or S-R scheme; (3) the emphasis on the specificities of human compared with animal psychology and behavior subsumed under the notion of symbolic activities; (4) the principle of anamorphosis, that is, the trend toward higher order or organization in contrast to the entropic trend in ordinary physical processes, which is made possible by the open-system nature of the living organism and manifest in creativity and its manifold manifestations, ranging from evolution in its nonutilitarian aspects to behavior in play and exploratory activities and to the highest human creativity and culture; and (5) as a consequence of the latter, the introduction of specifically human and suprabiological values into the scientific world view.

There is obviously an inverse relationship between rigor and broadness of problems and answers. Problems of control theory or biophysics are amenable to answers in technical (mathematical) language, while human concerns can be discussed only in everyday language (although formalization is possible, for example, modern linguistic shows). But we would be amiss, especially in psychology and psychiatry, in disregarding fundamental "systems" properties and principles discussed in informal ways. In this sense the broad conception of "humanistic" system theory is or will be, we believe, indispensable in arriving at a broader understanding of man and the "human condition" than was provided by previous approaches.

Bibliography

1. ALEXANDER, F., *The Western Mind in Transition: An Eye-witness Story*, Random House, New York, 1960.
2. ALLPORT, F., *Theories of Perception and the Concept of Structure*, John Wiley, New York, 1955.
3. ALLPORT, G. W., *Becoming: Basic Considerations for a Psychology of Personality*, Yale University Press, New Haven, 1955.
4. ———, "European and American Theories of Personality," in David, H., and von Bracken, H. (Eds.), *Perspectives in Personality Theory*, Tavistock, London, 1957.
5. ———, "The Open System in Personality Theory," *J. Abnorm. & Soc. Psychol.*, *61*: 301–310, 1960.
6. ———, *Pattern and Growth in Personality*, Holt, Rinehart and Winston, New York, 1961.
7. ANDERSON, H., "Personality Growth: Conceptual Considerations," in David, H., and von Bracken, H. (Eds.), *Perspectives in Personality Theory*, Tavistock, London, 1957.
8. "APA Weighs Pros, Cons of Encounter Groups," *Psychiat. News*, May 1970.
9. APPLEBY, L., SCHER, J., and CUMMINGS, J. (Eds.), *Chronic Schizophrenia*, Free Press, Glencoe, Ill., 1960.
10. ARDREY, R., *The Territorial Imperative*, Dell, New York, 1968.
11. ARIETI, S., "Contributions to Cognition from Psychoanalytic Theory," in Masserman, G. (Ed.), *Science and Psychoanalysis*, Vol. 8, Grune & Stratton, New York, 1965.
12. ———, "Current Ideas on the Problem of Psychosis," in *Problems of Psychosis*, International Colloquium on Psychosis, Montreal, November 5–8, 1969.
13. ———, *Interpretation of Schizophrenia*, Brunner, New York, 1955.
14. ———, *The Intrapsychic Self*, Basic Books, New York, 1967.
15. ———, "The Microgeny of Thought and Perception," *Archiv. Gen. Psychiat.*, *6*: 454–468, 1962.
16. ———, "The Present Status of Psychiatric Theory," *Am. J. Psychiat.*, *124*:1630–1639, 1968.
17. ———, "Schizophrenia," in Arieti, S. (Ed.), *American Handbook of Psychiatry*, Vol. 1, Basic Books, New York, 1959.
18. ———, "The Structural and Psychodynamic Role of Cognition in the Human Psyche," in Arieti, S. (Ed.), *World Biennial of Psychiatry and Psychotherapy*, Vol. 1, Basic Books, New York, 1970.
19. BENEDICT, R. (1934), *Patterns of Culture*, Mentor Books, New York, 1946.
20. BERLYNE, D. E., *Conflict, Arousal, and Curiosity*, McGraw-Hill, New York, 1960.
21. ———, "Recent Developments in Piaget's Work," *Brit. J. Educ. Psychol.*, *27*:1–12, 1957.
22. BERTALANFFY, L. VON, "A Biologist Looks at Human Nature," *Scientific Monthly*, *82*:33–41, 1956. (Reprinted in Daniel, R. (Ed.), *Contemporary Readings in Psychology*, Houghton Mifflin, Boston, 1959; *Reflexes to Intelligence: A Reader in Clinical Psychology*, Glencoe, Free Press, 1959.)

23. ———, *Biophysik des Fliessgleichgewichts* (Tr. by Westphal, W.), Vieweg, Braunschweig, Germany, 1953.
24. ———, "Body, Mind and Values," in Laszlo, E., and Wilbur, J. B., *Human Values and the Mind of Man*, Gordon and Breach, London, 1971.
25. ———, "Comments on Aggression," *Bull. Menninger Clin.*, 22:50–57, 1958. (Reprinted in Sarason, I. G. (Ed.), *Psychoanalysis and the Study of Behavior*, Van Nostrand, New York, 1965.)
26. ———, *Das Gefüge des Lebens*, Teubner, Leipzig, 1937.
27. ———, "An Essay on the Relativity of Categories," *Phil. Sci.*, 22:243–263, 1955.
28. ———, "General System Theory," *Main Currents in Modern Thought*, 11:75–83, 1955.
29. ———, "General System Theory—A Critical Review," *General Systems*, 7:1–20, 1962.
30. ———, *General System Theory: Foundations, Development, Applications*, rev. ed., George Braziller, New York, 1973.
31. ———, "History and Status of General Systems Theory," in Klir, G. (Ed.), *Trends in General System Theory*, John Wiley, New York, 1971.
32. ———, "Human Values in a Changing World," in Maslow, A. (Ed.), *New Knowledge in Human Values*, Harper, New York, 1959.
33. ———, "Mind and Body Re-examined," *J. of Humanistic Psychol.*, 6:133–138, 1966.
34. ———, "The Mind-Body Problem: A New View," *Psychosom. Med.*, 24:29–45, 1964. (Reprinted in Matson, F. W., and Montague, A. (Eds.), *The Human Dialogue*, Free Press, New York, 1967.)
35. ——— (1928), *Modern Theories of Development* (Tr. by Woodger, J. H., 1933), Harper Torchbooks, New York, 1962.
36. ———, *Nikolaus von Kues*, G. Müller, Munich, 1928.
37. ———, "On the Definition of the Symbol," in Royce, J. (Ed.), *The Symbol: An Interdisciplinary Symposium*, Random House, New York, 1964.
38. ———, "An Outline of General System Theory," *Brit. J. Phil. Sci.*, 1:134–165, 1950.
39. ———, "Principles and Theory of Growth," in Nowinski, W. W. (Ed.), *Fundamental Aspects of Normal and Malignant Growth*, Elsevier, Amsterdam, 1960.
40. ——— (1949), *Problems of Life*, Harper Torchbooks, New York, 1960.
41. ———, *Robots, Men and Minds: Psychology in the Modern World*, George Braziller, New York, 1967.
42. ———, "The Significance of Psychotropic Drugs for a Theory of Psychosis," World Health Organization, *AHP*, 2, 1957.
43. ———, "Some Biological Considerations on the Problem of Mental Illness," in Appleby, L., Scher, J., and Cumming, J. (Eds.), *Chronic Schizophrenia*, Glencoe, Free Press, 1960. (Reprinted in *Bull. Menninger Clin.*, 23:41–51, 1959.)
44. ———, "Some Considerations on Growth in Its Physical and Mental Aspects," *Merrill-Palmer Quart.*, 3:13–23, 1956.
45. ———, "System, Symbol and the Image of Man," in Galdston, I. (Ed.), *The Interface between Psychiatry and Anthropology*, Brunner/Mazel, New York, 1971.
46. ———, "Theoretical Models in Biology and Psychology," in Krech, D., and Klein, G. (Eds.), *Theoretical Models and Personality Theory*, Duke University Press, Durham, 1952.
47. ———, "The Theory of Open Systems in Physics and Biology," *Science*, 111:23–29, 1950.
48. ———, "The World of Science and the World of Value," *Teachers College Record*, 65:496–505, 1964. (Reprinted in Bugenthal, J. F. T. (Ed.), *Challenges of Humanistic Psychology*, McGraw-Hill, New York, 1967; Shoben, E. J., Jr., and Goldberg, S. (Eds.), *Problems in Contemporary Education*, Scott, Foresman and Co., Glenview, Ill., 1968.)
49. ———, "Zu einer allgemeinen Systemlehre," *Blätter für Deutsche Philosophie*, 18:3–4, 1945.
50. ———, Hempel, C., Bass, R., and Jonas, H., "General System Theory: A New Approach to Unity of Science," *Human Biol.*, 23:302–361, 1951.
51. ———, with Hilgartner, G. A., and Koch, S., "Author-Reviewers Symposia: Robots, Men and Minds, L. von Bertalanffy," *Phil. Forum*, 9:301–329, 1971.
52. ———, with Rapoport, A., *General Systems*, Yearbook of the Society for General Systems Research, Society for General

Systems Research, Washington, D.C., 16 vols. since 1956.

53. BETHE, A., "Plastizität und Zentrenlehre," in Bethe, A. (Ed.), *Handbuch der normalen und pathologischen Physiologie*, Vol. 15, Springer, Berlin, 1931.

54. BLEULER, E., *Mechanismus-Vitalismus-Mnemismus*, Springer, Berlin, 1931.

55. BOGUSLAW, W., *The New Utopians*, Prentice-Hall, Englewood Cliffs, N.J., 1968.

56. BRUNER, J., "Neural Mechanisms in Perception," in Solomon, H. (Ed.), *The Brain and Human Behavior*, Williams & Wilkins, Baltimore, 1958.

57. BRUNSWICK, E., "Historical and Thematic Relations of Psychology to Other Sciences," *Scientific Monthly*, 83:151–161, 1956.

58. BUCKLEY, W., *Modern Systems Research for the Behavioral Scientist*, Aldine, Chicago, 1967.

59. BUGENTHAL, J. F. T. (Ed.), *Challenges of Humanistic Psychology*, McGraw-Hill, New York, 1967.

60. BÜHLER, C., "Basic Theoretical Concepts of Humanistic Psychology," *Am. Psycholog.*, 26:378–386, 1971.

61. ———, *Psychologie im Leben unserer Zeit.*, Knaur, Munich and Zurich, 1962.

62. ———, "Theoretical Observations about Life's Basic Tendencies," *Am. J. Psychother.*, 13:561–581, 1959.

63. CANNON, W., *The Wisdom of the Body*, Norton, New York, 1932.

64. CANTRIL, H., "A Transaction Inquiry Concerning Mind," in Scher, J. (Ed.), *Theories of the Mind*, Free Press, New York, 1962.

65. CARMICHAEL, L. (Ed.), *Manual of Child Psychology*, 2nd ed., John Wiley, New York, 1954.

66. CASSIRER, E., *The Philosophy of Symbolic Forms*, 3 vols., Yale University Press, New Haven, 1953–1957.

67. COWDRY, E., *Cancer cells*, 2nd ed., W. B. Saunders, Philadelphia, 1955.

68. "Crime and Criminologists," *The Sciences*, 2:1–4, 1963.

69. DEMERATH, N. J. III, and PETERSON, R. A. (Eds.), *System, Change and Conflict. A Reader on Contemporary Sociological Theory and the Debate over Functionalism*, Free Press, New York, 1967.

70. ELLIS, H. F., "The Naked Ape Crisis," *New Yorker*, 1968.

71. EYSENCK, H., "Characterology, Stratification Theory, and Psychoanalysis: An Evaluation," in David, H., and von Bracken, H. (Eds.), *Perspectives in Personality Theory*, Tavistock, London, 1957.

72. FRANKL, V. E., "Das homöostatische Prinkip und die dynamische Psychologie," *Ztschr. f. Psychother. Med. Psychol.*, 9:41–47, 1959.

73. ———, *From Death-Camp to Existentialism*, Beacon Press, Boston, 1959.

74. ———, "Irrwege seelenärztlichen Denkens (Monadologismus, Potentialismus und Kaleidoskopismus)," *Nervenarzt*, 31:385–392, 1960.

75. ———, *The Will to Meaning: Foundations and Applications of Logotherapy*, World, New York, 1969.

76. FREEMAN, G., *The Energetics of Human Behavior*, Cornell University Press, Ithaca, 1948.

77. FREUD, S. (1920), *A General Introduction to Psychonanalysis*, Permabooks, New York, 1953.

78. FRIED, M. H., Review of *Men in Groups*, *Science*, 165:883–884, 1969.

79. FROMM, E., "The Erich Fromm Theory of Aggression," *New York Times Magazine*, February 27, 1972.

80. GEERTZ, C., "The Growth of Culture and the Evolution of Mind," in Scher, J. (Ed.), *Theories of the Mind*, Free Press, New York, 1962.

81. GILBERT, A. R., "On the Stratification of Personality," in David, H., and von Bracken, H. (Eds.), *Perspectives in Personality Theory*, Tavistock, London, 1957.

82. GOLDSTEIN, K., "Functional Disturbances in Brain Damage," in Arieti, S. (Ed.), *American Handbook of Psychiatry*, Vol. 1, Basic Books, New York, 1959.

83. ———, *The Organism*, American Book Co., New York, 1939.

84. GRAY, W., "The Contributions of Ludwig von Bertalanffy to Modern Psychiatry," in Laszlo, E. (Ed.), *The Relevance of General Systems Theory. Papers presented to Ludwig von Bertalanffy on His 70th birthday*, Braziller, New York, 1972.

85. ———, "Ludwig von Bertalanffy's General System Theory as a Model for Human-

istic System Science," *XIII International Congress of the History of Science*, Subsection "History of System Analysis," Moscow (USSR), August 18–24, 1971.

86. ———, DUHL, F., and RIZZO, N., *General Systems Theory and Psychiatry*, Little, Brown, Boston, 1969.

87. GRINKER, ROY R., SR., "Biomedical Education as a System," *Arch. Gen. Psychiat.*, *24*:291–297, 1971.

88. ———, "An Essay on Schizophrenia and Science," *Arch. Gen. Psychiat.*, *20*:1–24, 1969.

89. ———, "Goals for the Future of American Psychiatry," *Mount Sinai J. Med.*, *38*: 226–242, 1971.

90. ———, *Toward a Unified Theory of Human Behavior*, 2nd ed., Basic Books, New York, 1967.

91. HACKER, F. J., "Juvenile Delinquency," Hearings before the U.S. Senate Subcommittee Pursuant to S. Res. 62, June 15–18, 1955, U.S. Govt. Printing Office, Washington, D.C., 1955.

92. HALL, A., *A Methodology for Systems Engineering*, Van Nostrand, Princeton, 1962.

93. ———, and FAGEN, R., "Definition of System," *General Systems*, *1*:18–28, 1956.

94. HAYEK, F., "Degrees of Explanation," *Brit. J. Phil. Sc.*, *6*:209–255, 1955.

95. HEBB, D. O., "Drives and the C.N.S. (Conceptual Nervous System)," *Psycholog. Rev.*, *62*:243–254, 1955.

96. ———, *The Organization of Behavior*, John Wiley, New York, 1949.

97. HENRY, J., *Culture against Man*, Random House, New York, 1963.

98. HERRICK, C., *The Evolution of Human Nature*, Harper Torchbooks, New York, 1956.

99. HOLST, E. VON, "Vom Wesen der Ordung im Zentralnervensystem," *Naturwissenschaften*, *25*:625–631, 641–647, 1937.

100. HOOK, S. (Ed.), *Dimensions of Mind*, Collier Books, New York, 1961.

101. HOWARD, J., *Please Touch: A Guided Tour of the Human Potential Movement*, McGraw-Hill, New York, 1970.

102. KAVANAU, J. L., "Behavior of Captive White-footed Mice," in Willems, E. P., and Raush, H. (Eds.), *Naturalistic Viewpoints in Psychological Research*, Holt, Reinhart and Winston, New York, 1969.

103. KLIR, G. (Ed.), *Trends in General Systems Theory*, John Wiley, New York, 1971.

104. KMENT, H., "The Problem of Biological Regulation and Its Evolution," *General Systems*, *4*:75–82, 1959.

105. KOCH, S., "Stimulus/Response," *Psychology Today*, *3*, 1969.

106. KOESTLER, A., *The Act of Creation*, Hutchinson, London, 1964.

107. ———, *The Ghost in the Machine*, Macmillan, New York, 1967.

108. ———, *The Lotus and the Robot*, Hutchinson, London, 1960.

109. ———, and SMYTHIES, J. R. (Eds.), *Beyond Reductionism. The Alpbach Symposium*, Macmillan, New York, 1970.

110. KRECH, D., "Dynamic Systems as Open Neurological Systems," *Psychol. Rev.*, *57*: 283–290, 1950.

111. KUBIE, L., "The Distortion of the Symbolic Process in Neurosis and Psychosis," *J. Am. Psychoanal. A.*, *1*:59–86, 1953.

112. KUHN, T. S., *The Structure of Scientific Revolutions*, University of Chicago Press, Chicago, 1962, 2nd ed., 1970.

113. LANGER, S. (1942), *Philosophy in a New Key*, Mentor Books, New York, 1948.

114. LASHLEY, K. S. (1929), *Brain Mechanisms and Intelligence*, Hafner Publishing Co., New York, 1964.

115. LEIBOWITZ, L., "Desmond Morris is Wrong about Breasts, Buttocks, and Body Hair," *Psychology Today*, February 1970.

116. LENNARD, H. L., and BERNSTEIN, A., *The Anatomy of Psychotherapy*, Columbia University Press, New York, 1960.

117. LERSCH, P., and THOMAE, H .(Eds.), *Handbuch der Psychologie*, Vol. 4, *Persönlichkeitsforschung und Persönlichkeitstheorie*, Hogrefe, Göttingen, 1960.

118. LLAVERO, F., "Bemerkungen zu einigen Grundfragen der Psychiatrie," *Nervenarzt*, *28*:419–420, 1957.

119. LORENZ, K., *On Aggression*, Harcourt, Brace & World, New York, 1963.

120. ———, "Rats, Apes, Naked Apes, Kipling, Instincts, Guilt, The Generations and Instant Copulation. A Talk with Konrad Lorenz," *New York Times Magazine*, July 5, 1970.

121. LURIA, A., *The Role of Speech in the Regulation of Normal and Abnormal Behavior*, Pergamon Press, New York, 1961.

122. LUTHE, W., "Neuro-humoral Factors and Personality," in David, H., and von Bracken, H. (Eds.), *Perspectives in Personality Theory*, Tavistock, London, 1957.
123. MAGOUN, H., *The Waking Brain*, Charles C Thomas, Springfield, Ill., 1958.
124. MATSON, F. W., *The Broken Image*, George Braziller, New York, 1964.
125. ———, Review of *Men in Groups*, *Psychology Today*, December 1969.
126. MAY, R., ANGEL, E., and ELLENBERGER, H. (Eds.), *Existence: A New Dimension in Psychiatry and Psychology*, Basic Books, New York, 1958.
127. MEIR, A., "General System Theory, Developments and Perspectives for Medicine and Psychiatry," *Arch. Gen. Psychiat.*, *21*:302–310, 1969.
128. MENNINGER, K., "The Psychological Aspects of the Organism under Stress. Regulatory Devices of the Ego Under Major Stress," *J. Am. Psychoanal. A.*, *2*:67–106, 280–310, 1954. Reprinted in *General Systems*, *2*:142–172, 1957.
129. ———, ELLENBERGER, H., PRUYSER, P., and MAYMAN, M., "The Unitary Concept of Mental Illness," *Bull. Menninger Clin.*, *22*:4–12, 1958.
130. ———, MAYMAN, M., and PRUYSER, P., *The Vital Balance*, Viking, New York, 1963.
131. MERLOO, J., *The Rape of the Mind*, World, New York, 1956.
132. MILLER, J. G., "Living Systems: Basic Concepts," in Gray, W., Duhl, F. J., and Rizzo, N. D. (Eds.), *General Systems Theory and Psychiatry*, Little, Brown, Boston, 1969.
133. ———, "Towards a General Theory for the Behavioral Sciences," *Am. Psychol.*, *10*:513–531, 1955.
134. MILSTEIN, M. M., and BELASCO, J., *Educational Administration and the Behavioral Sciences: A Systems Perspective*, Allyn and Bacon, Boston, 1972.
135. MORRIS, D., *The Naked Ape*, McGraw-Hill, New York, 1968.
136. MUMFORD, L., *The Myth of the Machine*, Harcourt, Brace & World, New York, 1967.
137. MURRAY, H. A., "The Personality and Career of Satan," *J. Soc. Issues*, *18*:36–54, 1962.
138. NUTTIN, J., "Personality Dynamics," in David, H., and von Bracken, H. (Eds.), *Perspectives in Personality Theory*, Tavistock, London, 1957.
139. OFFER, D., and SABSHIN, M., "The Concept of Normality," Ch. 8 of this volume.
140. OPLER, M. K., *Culture, Psychiatry and Human Values*, Charles C Thomas, Springfield, Ill., 1956.
141. PIAGET, J., *The Construction of Reality in the Child*, Basic Books, New York, 1959.
142. PUMPIAN-MINDLIN, E., "Propositions Concerning Energetic-economic Aspects of Libido Theory," *Ann. N.Y. Acad. Sc.*, *76*:1038–1052, 1959.
143. RAPAPORT, D., "The Structure of Psychoanalytic Theory," *Psychol. Issues*, 2, 1960, Mono. 6.
144. RAPOPORT, A., "Systems Analysis," in Sills, D. L. (Ed.), *International Encyclopedia of the Social Sciences*, Vol. 15, Macmillan and Free Press, New York, 1968.
145. REICH, C. A., *The Greening of America*, Bantam Books, New York, 1971.
146. RIZZO, N. D., "The Court Clinic and Community Mental Health—Systems Theory in Action," in *The Seventh Annual John W. Umstead Series of Distinguished Lectures*, North Carolina Department of Mental Health, February 5–6, 1970.
147. ———, "Recent Applications of General System Theory in Schools and Courts," *XIII International Congress of the History of Science*, Subsection "History of Systems Analysis," Moscow (U.S.S.R.), August 18–24, 1971.
148. ROSZAK, T., *The Making of a Counter-Culture*, Doubleday, New York, 1969.
149. ROTHACKER, E., *Die Schichten der Persönlichkeit*, 3rd ed., Barth, Leipzig, 1947.
150. ROYCE, J. R., *The Encapsulated Man*, Van Nostrand, New York, 1964.
151. SCHAXEL, J., *Die Grundzüge der Theorienbildung in der Biologie*, 2nd ed., Fischer, Jena, 1922.
152. SCHER, J. M., (Ed.), *Theories of the Mind*, Free Press, New York, 1962.
153. SCHILLER, C. H. (Ed. and Trans.), *Instinctive Behavior*, Methuen, London, 1957.
154. SHANNON, C. E., and WEAVER, W., *The Mathematical Theory of Communication*, University of Illinois Press, Urbana, Ill., 1949.
155. SHELDON, A., BAKER, F., and MCLAUGHLIN,

C. P. (Eds.), *Systems and Medical Care*, M.I.T. Press, Cambridge, 1970.

156. SKINNER, B., *Beyond Freedom and Dignity*, Alfred A. Knopf, New York, 1971.

157. ———, "The Flight from the Laboratory," in Marx, M. (Ed.), *Theories in Contemporary Psychology*, Macmillan, New York, 1963.

158. SOROKIN, P., "Reply to My Critics," in Allen, P. (Ed.), *Pitirim A. Sorokin in Review*, Duke University Press, Durham, 1963.

159. STAGNER, R., "Homeostasis as a Unifying Concept in Personality Theory," *Psychol. Rev.*, *58*:5–17, 1951.

160. SYZ, H., "Reflections on Group—or Phylo-Analysis," *Acta Psychotherapeutica*, *11*: 37–38, 1963.

161. SZASZ, T., *The Myth of Mental Illness*, Hoeber-Harper, New York, 1961.

162. TANNER, J., and INHELDER, B. (Eds.), *Discussions on Child Development*, Vol. 4, Tavistock, London, 1960.

163. THUMB, N., "Die Stellung der Psychologie zur Biologie. Gedanken zu L. von Bertalanffy's Theoretischer Biologie," *Zentralblatt Psychotherapie*, *15*:139–149, 1943.

164. TIGER, L., *Men in Groups*, Random House, New York, 1969.

165. TOCH, H. H., and HASTORF, A. H., "Homeostasis in Psychology: A Review and Critique," *Psychiatry*, *18*:81–91, 1955.

166. TOFFLER, A., *Future Shock*, Random House, New York, 1970.

167. WAGNER, R., *Das Regelproblem in der Biologie*, Thieme, Stuttgart, 1954.

168. WERLEY, H. H., "Health Research and the Systems Approach," Symposium given March 1, 1971, Center for Nursing Research, Wayne State University, Detroit, Michigan.

169. WERNER, H. (1940), *Comparative Psychology of Mental Development*, International Universities Press, New York, 1957.

170. ———, "The Concept of Development from a Comparative and Organismic Point of View," in Harris, D. (Ed.), *The Concept of Development*, University of Minnesota Press, Minneapolis, 1957.

171. ———, and KAPLAN, B., *Symbol Formation*, John Wiley, New York, 1963.

172. WHORF, B. L., *Language, Thought, and Reality. Selected Writings of B. L. Whorf*, ed. by John Carroll, John Wiley, New York, 1956.

173. WHYTE, L., *The Unconscious before Freud*, Basic Books, New York, 1960.

174. ———, WILSON, A. G., and WILSON, D., *Hierarchical Structures*, Elsevier, New York, 1969.

175. WIENER, N., *Cybernetics*, John Wiley, New York, 1948.

176. YOUNG, O. R., *Systems of Political Science*, Prentice-Hall, Englewood Cliffs, N.J., 1968.

PART SIX

Classification and Assessment of Psychiatric Conditions

CHAPTER 52

CLASSIFICATION AND NOMENCLATURE OF PSYCHIATRIC CONDITIONS

Henry Brill

Introduction

It is traditional to complain about the short-comings and the illogical nature of the various systems of nomenclature and classification; yet this writer knows of no psychiatry that can get along without them. Great psychiatrists at least from the time of Pinel onward have expressed such dissatisfactions, and today the complaints and the suggested solutions are at least as numerous and more sophisticated than ever before.[22] Some authorities advocate an abandonment of all labeling on the grounds that there is no such thing as mental disorder,[8] while others insist that there is only one type of mental disorder whose variations are individual, infinite, and not to be further classified. Still others would replace all existing systems with new synthetic ones based on completely novel approaches, many of which rely on computer-based analyses of quantitative data.[21]

Nevertheless, it is fair to say that for all practical purposes clinical psychiatry continues to place its reliance on regular periodic revisions of existing classifications. Such revisions seek to incorporate advances in psychiatry and to accommodate to changing views, as well as to develop technical improvement in the methods of classification. Another aim of recent years has been to work toward a convergence among all existing psychiatric systems, and much progress has been made in this direction through the World Health Organization's eighth edition of the *International Classification of Diseases* (ICD-8),[15] and further progress is expected in the next edition, which is now in preparation (ICD-9). The problem is far less difficult than might be expected because, in spite of the great divergencies on specifics that Stengel[29] outlined, the major classifications that are now in active use all derive from a common psychiatric history and share a common scientific literature, and thus they are very similar in basic structure.

Psychiatric classification and nomenclature is often portrayed as a long series of arbitrary inventions that have been created and destroyed in endless succession. This is distinctly not true. It is a matter of common knowledge that such terms as mania, melancholia, and paranoia were already in common use in classic Greek and Roman times.[2] Other terms such as neurosis and neurasthenia were added in the course of time, and although much was tried and abandoned, what has survived the centuries has a vigor that speaks of some real usefulness in what is perhaps the most pragmatic and empirical of all medical undertakings, the treatment of the mentally ill.

The arrangement of individual disorders into major classes, namely, the psychoses, neuroses, character disorders, and the mental deficiencies, is likewise of long development as is the clear distinction between the organic and the functional disorders. Most of this has been accomplished during the last 150 years, and all of this is common to the major classifications worldwide.

Another factor that tends to bring the various systems into alignment is that they must all meet the test of practicability in actual operation. Six essential requirements may be listed for a clinical classification to become generally acceptable:

1. It should be as simple as possible and practical for application under field conditions; highly complex terminology or overelaborate recording procedures are a serious bar to widespread use.
2. It should lend itself to the various operations necessary for public health statistics. It must, therefore, be constructed to comply with technical statistical requirements—a reminder that diagnosis is more than a personal matter between a patient and his physician.
3. A glossary defining each term in the classification is essential. In arriving at such definitions, it is well to note Stengel's[29] position that it is futile to indulge in "last ditch battles" about an exact definition of a type of neurosis or a subgroup of schizophrenia. In the present state of our knowledge such definitions must be taken more as conventions than as absolute truths, although they do have operational significance and, if generally accepted, provide a valuable medium of communication. ICD-8 has not hitherto had such a glossary, but the World Health Organization is now preparing one.
4. The system should allow maximum comparability between its terms and those of the major psychiatric classifications.
5. The psychiatric classification should be gathered into one list and not scattered through a general manual of medical diagnoses.
6. A good index of all psychiatric terms such as is found in ICD-8 is a valuable part of any classification.

Let us now examine some of the definitions of important terms that are used in describing classifications generally because these can be a source of difficulty. We can then turn to an examination of some of the general and theoretical objections and criticisms that have been raised with respect to current psychiatric classifications, including DSM-II and ICD-8.

Definition of Terms

It might be expected that the field of classification would have a well-standardized terminology, but as Crowson[9] points out this is not the case. Even at the most expert level one finds considerable variation in usage of such basic terms as classification, systematics, nosology, and taxonomy. The word "classification" may, for example, refer to the process of classifying or to its product. "Systematics" is used by some writers as a synonym for classification, while others reserve the term to describe the general science and theory of classification.

For purposes of this chapter the following definitions have been adopted:

1. *Nomenclature:* This is a system of the names in a scientific classification. Ideally the terms of a nomenclature are rigorously defined and specific and do not overlap in meaning. Thus a nomenclature can be distinguished from a terminology, which is a general collection of all terms used in a technical field.

The aim of a nomenclature is to promote stability and uniformity in scientific naming. It should be noted that these are names, not of things, but of concepts; thus schizophrenia is, strictly speaking, the name of a concept of a disorder, not of the disorder itself. The APA nomenclature,[6] in general, recognizes only one primary term for each condition, although some synonyms are mentioned. In ICD-8 there is also only one primary term for each classified condition; but a large number of synonyms are listed as inclusion terms, and to further assist in defining the various entities ICD-8 also lists for them a series of exclusion terms, names of conditions that are similar but different enough to be excluded from the rubric. The nomenclature also includes names of groups of conditions such as the neuroses and psychoses, as well as subdivisions of various conditions such as "schizophrenia catatonic excited" and "schizophrenia catatonic withdrawn."

2. *Classification:* This is an orderly arrangement of names into a hierarchical system with successively higher orders of generalization. In psychiatry the basic elements are such names as hysteria and catatonic schizophrenia. The next higher level consists of names of groups of disorders such as schizophrenia and manic-depressive psychoses. The next level is made up of more generalized groupings such as psychoses, neuroses, character disorders, and the like.

3. *Natural and Artificial Classifications:* A classification may be either natural or artificial. If it reflects some deeper underlying pattern or reality, it is called "natural" or Aristotelian. An example is the current classification of animals and plants that reflects the principles of Darwinian evolution. In psychiatry the classification of organic brain syndromes of known etiology is generally recognized as a "natural" system.

An artificial classification, as the term implies, is purely arbitrary and synthetic and is developed for utilitarian purposes. An example in psychiatry is the classification of patients by pattern of behavior. A natural classification must be discovered; an artificial one is invented. As a result the members of the first group will share points of resemblance and characteristics other than those required to assign them to the group, but this is not true of members grouped in an artificial system.[9] One of the most difficult questions in modern psychiatry is whether the classification of the so-called functional psychoses should be treated as natural or artificial, and opinion is strongly divided on this point. The current classification of personality and behavior disorders is generally considered to be an artificial one. It is important to note that artificial systems can serve and have served important scientific purposes and that some systems that began as artificial conventions were subsequently found to reflect underlying natural laws. The classification of plants and animals by Linnaeus himself had no theoretical basis until it was later provided by Darwin, and Metchnikoff's periodic table also began as a purely "artificial" arrangement of elements by atomic weight. In psychiatry the identification of psychosis with pellagra and general paresis preceded any knowledge of underlying causes.

4. *Diagnosis:* This term has many connotations, but essentially it refers to nosology and classification of medical disorders. Some require that a diagnosis must include a knowledge of etiology, or at the very least a well-defined somatic demonstrable pathology. But these statements are really objections to the principle of artificial classification, and they are based on the assumption that if a classification cannot be proved to be a natural one it must be considered artificial and therefore it is not a diagnosis in any medical or scientific sense. Such a limitation has not been customary in medicine or psychiatry, and it is a matter of medical history that the present level of medical diagnosis was achieved through a series of successively better approx-

imations that began as purely artificial designations and only much later emerged as such entities as vitamin deficiencies, endocrine disorders, and the various specific infections. It is noteworthy that the term "diagnosis" was used for these disease names long before their nature was understood.

5. *Nosology, Taxonomy, Systemics, Classification:* Specialists make various distinctions among these terms, but they are often interchanged in a confusing manner. For our purposes they can be considered essentially interchangeable, although nosology is usually applied in medical fields, taxonomy to biology, and systemics is a term of broad connotations that include the more abstract aspects of the science and philosophy of classification.[27]

Dialectic and Debate

All of the concepts in classification and nomenclature are continually tested in the debate that has become traditional in this field.

One of the simplest and most recent criticisms is that psychiatric classification and nomenclature is mere pejorative labeling.[28] This criticism indicates that such labeling is in itself an improper procedure, but yet it seems to use the very technique that it condemns by attaching a label of "mere labeling" to the practice of psychiatric diagnosis and classification. If one examines matters more closely, however, he finds that this statement is often linked to an attack on the validity of current categories of mental disorder. This, in turn, is based on a variety of arguments; one of the most common is well stated by Lorr, Klett, and McNair[21] as follows: The psychiatric syndrome is "not a class concept . . . but should represent instead a continuous quantitative variable measurable in terms of degree." The assumption seems to be that where a continuity can be traced from one condition to another the two cannot be really different, or briefly stated, that continuity among things signifies identity. This position seems to ignore the possibility that the continuity can be due to an overlapping of normal and abnormal states as in the case of body weight, blood pressure, hemoglobin measurements, physical stature, and basal metabolic rates.

An even more challenging application of the argument based on continuity is the statement that mental illness does not exist altogether because one can find all possible transitional states between mental health and mental illness. The argument is that this situation is peculiar to psychiatry, and, therefore, the medical model that applies in other specialties is not appropriate for psychiatry.[8] The fallacy in this logic is the mistaken assumption that there is no continuity between physical health and physical illness. As a British health publication[25] states, ". . . in strictly scientific or technological terms there is no sharp distinction between a healthy and a diseased state in an individual. For a vast range of biochemical and physical observations . . . there is a continuous distribution curve for the population as a whole . . . there is no sharp discontinuity." Thus what was was introduced as the distinctive and therefore disqualifying characteristic of psychiatric classification is, in fact, not distinctive, but is shared by fields in medicine where the medical model is not questioned.

Nor are these problems in classification limited to the medical field, because they are, in fact, found in the biological sciences generally, and it is worth noting that the theory of classification and nomenclature has perhaps had its most intensive development in relation to botany and zoology. In this connection it is most instructive to read such works as that of Crowson[9] who says, "The species is no exception to the rule that the concepts and categories employed in natural history are never susceptible to precise rigorous or final definition; any scientist who is not content to operate with more or less vague and inexact basic principles and ideas is temperamentally unsuited to the study of natural history." Reading further in this and similar works on biological classification we find that other problems that have been discussed as if they were specific to psychiatry are generally encountered in classification of other biological data, and finally that psychiatry shares a number of other controversies about classification with botany and zoology. Crowson,[9] speaking about biology, objects to "a classification which is limited in its basis to characters which can be counted or measured," and the reasons that he presents will be familiar to any psychiatrist who has been concerned with classification in that field. He also calls attention to an "academic trend over the last fifty years . . . of a progressive deprecation of the importance of systematics [the general name of the science of classification], and it has produced the effect that young (recently graduated) botanists and zoologists (have) less real systematic knowledge than at any time in the last hundred years. A projection . . . would suggest the virtual disappearance of systematic content from academic botany and zoology courses." All of which is quite familiar in the field of psychiatry as are the reasons he gives for thinking that the trend will be halted, and this includes the indispensability of classification for practical purposes.

His chapter on "numerical taxonomy" will also be familiar to those interested in psychiatric classification. This chapter is a vigorous attack on the principle (which he traces to American sources) that only a mechanized, computer-produced classification is valid, and his chapter on the noncongruence principle restates the fact that different classificatory characters are rarely coincident in their distributions. His comment on the "splitters" who would create endless subcategories and the "lumpers" who would go to the opposite extreme is also directly applicable to psychiatry. In brief then it would seem that we must be careful in discussing the problems of psychiatric nomenclature and classification to distinguish between those that are specific to psychiatry, and may reasonably be expected to find a remedy within that discipline, and those that are or seem to be inherent in the classification of all biological data, and are far less likely to do so.

Other issues that are regularly raised with respect to psychiatric nomenclature and classification are displacements from social and political controversies and do not really relate to psychiatric systematics as such. Here one may class the arguments that psychiatric classification can be misused to hospitalize the rejected, to label them in such a way as to express the bias of society, and to stigmatize them and that it can be used as a punishment for behavior unacceptable to the dominant society.[8,22,28] Still other arguments and objections appear to represent sheer dialectic, and here one may classify the statement that classification is traditional and should on that account be discarded. This statement ignores the historical fact that attacks on classification are also highly traditional in psychiatry and date back at least to the time of Pinel. Whether one wishes to condemn a practice merely on the grounds that it is traditional is, of course, a political and not a scientific question.

Finally we come to those criticisms that arise out of actual experience in the application of psychiatric classification. These are quite specific, and we shall see the important role that they play in the development of such systems when we examine the classification of the American Psychiatric Association (DSM II) and of the World Health Organization (ICD-8).

The American Psychiatric Association's Diagnostic and Statistical Manual and ICD-8, The International Classification of Diseases

The 1968 classification of the American Psychiatric Association will now be reviewed in some detail and will be compared with Section V, the psychiatric section of ICD-8, the World Health Organization's classification.[15] The APA classifications began as a national system, but if we include the earlier versions it has probably had more extensive actual use than any other system in the history of psychiatry, having been generally applied in the United States and also used in a number of other countries in North and South America. Stengel[29] pointed this out in his masterly review of national systems.

Among the advantages of DSM II[7] are its glossary* and its convertibility to the World Health Organization's system since it uses the same code numbers and, for the most part, uses the same or similar terms. This has become steadily more important as the WHO system has been more widely adopted in Europe and elsewhere.

The changes in DSM II as compared with DSM I are quite extensive and in many cases represent a return to previous terminology. These changes resulted in part from a serious bilateral effort to develope convergence between the APA classification and that of the WHO. The changes were also responsive to a series of general criticisms in this country that DSM I had moved too far from previously established classification systems. The next decennial revision of the International Classification is now getting under way, but current indications are that this revision will not be extensive and that the alignment with DSM II will not be disturbed. Nevertheless, some changes are to be expected because some problems have become apparent during use; new conditions and new requirements have developed and new data of classificatory significance have emerged, as will be shown further on in this chapter.

As already noted the terms in DSM II are generally the same as those in ICD-8, but differences remain, and these have been identified by marking with an asterisk DSM II items not found in ICD-8 and closing between squared brackets ICD-8 items listed in DSM II but "to be avoided" in actual use. A second significant difference between the two systems is the order of listing of some of the items or groups of items. For example, DSM II lists the forms of mental retardation first, while ICD-8 lists the organic brain syndromes first. The Arabic numerals that comprise the numbering system of ICD-8 have been retained in DSM II, but their sequence has been broken by the rearrangements of the items of classification. Thus the APA list begins with the 310–315 series (the mental retardations), followed by the 290–294 series (the psychoses associated with organic brain syndromes). Then comes the series number 309 (nonpsychotic organic brain syndromes), followed, in turn, by series 295–298 (psychoses not attributed to physical conditions listed previously).

In ICD-8 the terms are listed in numerical order starting with item 290 (senile and presenile dementia) and ending with item 315 (unspecified mental retardation). Several unused numbers left at the end of the ICD-8 psychiatric series have been used for additional terms in the APA Manual, namely, 316–318 (conditions without manifest psychiatric disorder and nonspecific conditions) and 319 (nondiagnostic terms).

To create an overall numerical sequence that corresponds to the order of its own presentation, DSM II has identified eleven main groups of items with Roman numerals, thus it opens with item I (mental retardation) and closes with XI (nondiagnostic terms for administrative use).

All of these changes were adopted in order to make the APA system (DSM II) compatible with that of the WHO (ICD-8) and interconvertible with it, and at the same time have it be acceptable in American practice. The history of the collaborative efforts between the APA committee and the WHO representatives that preceded the publication of DSM II is fully described in the opening pages of that publication.

Following is a brief and modified version of the current APA classification as listed in DSM II under the general title of "The Diagnostic Nomenclature."

I. Mental retardation (310–315)

The primary categories listed are borderline (310), mild (311), moderate (312), severe (313), profound (314), and unspecified (315), and each of these is to be followed by an additional phrase identified by a decimal digit specifying one of ten broad categories of associated conditions of etiological or pathogenic nature, as follows:

.0 infection or intoxication
.1 trauma or physical agent

* The World Health Organization is now preparing a glossary for ICD-8, which will go far toward making this document more effective. The glossary should be available by 1974.

.2 disorders of metabolism, growth, or nutrition
.3 gross brain disease (postnatal)
.4 disease and conditions due to (unknown) prenatal influence
.5 chromosomal abnormality
.6 prematurity
.7 following major psychiatric disorder
.8 psychosocial (environmental) deprivation
.9 other conditions

When known the specific associated physical condition is specified as an additional diagnosis. The fourth digit is also used to identify subdivisions of various major rubrics as in the case of the alcoholic psychoses 291.1, 291.2, etc. DSM II adds a second decimal place to this basic four-digit system of ICD-8, and this creates additional categories by subdividing a rubric as in the case of 309.13* and 309.14.*

II. Organic brain syndromes
(disorders caused by or associated with impairment of brain tissue function)

II–A. Psychoses associated with organic brain syndromes (290–294)

290 Senile and presenile dementia
291 Alcoholic psychosis
.0 delirium tremens
.1 Korsakov's psychosis (alcoholic)
.2 other alcoholic hallucinosis
.3 alcohol paranoid state (alcoholic paranoia)
.4* acute alcohol intoxication*
.5 alcoholic deterioration*
.6* pathological intoxication*
.9 other alcoholic psychosis
292 Psychosis associated with intracranial infection
.0 general paralysis
.1 other syphilis of central nervous system
.2 epidemic encephalitis
.3 other and unspecified encephalitis
.9 psychosis with other intracranial infection
293 Psychosis associated with other cerebral condition
.0 cerebral arteriosclerosis
.1 other cerebrovascular disturbance
.2 epilepsy
.3 intracranial neoplasm
.4 degenerative disease of the central nervous system
.5 brain trauma
.9 other cerebral condition
294 Psychosis associated with other physical condition
.0 endocrine disorder
.1 metabolic or nutritional disorder
.2 systemic infection
.3 drug or poison intoxication (other than alcohol)
.4 childbirth
.8 other undiagnosed physical condition and unspecified
.9 [psychosis with unspecified physical condition]

II–B. Nonpsychotic organic brain syndromes (OBS) (309)

309 Non-psychotic organic brain syndromes associated with physical conditions
.0 nonpsychotic OBS with intracranial infection
[.1 nonpsychotic OBS with drug, poison, or systemic intoxication]
.13* alcohol* (simple drunkenness)
.14* other drug, poison, or systemic intoxication*
.2 brain trauma
.3 circulatory disturbance
.4 epilepsy
.5 disturbance of metabolism, growth, or nutrition
.6 senile or presenile brain disease
.7 intracranial neoplasm
.8 degenerative disease of central nervous system
.9 with other physical condition

III. Psychosis not attributed to physical conditions listed previously (295–298)

295 Schizophrenia
.0 simple
.1 hebephrenic
.2 catatonic
.3 paranoid
.4 acute episode
.5 latent
.6 residual
.8 schizo-affective
.8* childhood,*
.90* chronic undifferentiated*
.99* other*

296 Major affective disorders
- .0 involutional melancholia
- .1 manic-depressive illness, manic
- .2 depressed
- .3 circular
- .8 other

297 Paranoid states
- .0 paranoia
- .1 involutional paranoid state
- .9 other paranoid state

298 Other psychoses
- 0. psychotic depressive reaction
- [.1 reactive excitation]
- [.2 reactive confusion]
- [.3 acute paranoid reaction]
- [.9 reactive psychosis unspecified]

IV. Neuroses (300)

300 Neuroses
- .0 anxiety
- .1 hysterical neurosis
- [.13* conversion type, .14* dissociative type]
- .2 phobic
- .3 obsessive-compulsive
- .4 depressive
- .5 neurasthenic
- .6 depersonalization
- .7 hypochondriacal
- .8 other

V. Personality disorders and certain other nonpsychotic mental disorders (301–304)

301 Personality disorders
- .0 paranoid
- .1 cyclothymic
- .2 schizoid
- .3 explosive
- .4 obsessive-compulsive
- .5 hysterical
- .6 asthenic
- .7 antisocial
- .81* Passive-aggressive*
- .82* inadequate*
- .89* other of specified types*
- [.9 unspecified personality disorder]

302 Sexual deviations
- .0 homosexuality
- .1 fetishism
- .2 pedophilia
- .3 transvestitism
- .4 exhibitionism
- .5* voyeurism*
- .6* sadism*
- .7* masochism*
- .8 other sexual deviation

303 Alcoholism
- .0 episodic excessive drinking
- .1 habitual excessive drinking
- .2 alcohol addiction
- .9 other alcoholism

304 Drug dependence

Drug dependence of several distinct types are listed, namely, those due to natural and synthetic drugs of morphinelike action; the barbiturate group; other hypnotics, sedatives, or tranquilizers, cocaine, cannabis; other psychostimulants; and hallucinogens.

VI. Psychophysiological disorders (305)

Ten main topographical subtypes are listed. They include skin, musculo-skeletal, respiratory, etc.

VII. Special symptoms (306)

This is a list of symptoms most often found in child psychiatry, although most of the terms are not limited to any age group (speech disturbance, specific learning disturbance, tic, enuresis, encopresis, etc.).

VIII. Transient situational disturbances (307)

DSM II lists five types under this heading—those of infancy, childhood, adolescence, adult life, and late life. ICD-8 differs in that it lists only the last three types at this point and places the first two under behavior disorders of childhood.

IX. Behavior disorders of childhood and adolescence (308)

The hyperkinetic, withdrawing, overanxious, runaway, unsocialized, and group delinquent reactions of childhood or adolescence are individually listed under this category, which in DSM II differs from the parallel category of ICD-8 (behavior disorders of childhood) since the latter also includes the adjustment reactions of infancy and childhood.

X. Conditions without manifest psychiatric disorder and nonspecific conditions (316*–318*)*

This is a heterogeneous series of terms that include marital maladjustment, social and occupational maladjustment, and dyssocial behavior. The ICD-8 calls this category "social maladjustment without manifest psychiatric disorder."

XI. Nondiagnostic terms for administrative use (319*)*

This is a series of terms such as "diagnosis deferred, boarder, experiment only." In ICD-8 such terms are listed under a section "special conditions and examinations without sickness."

Comment on the Classification

The main strong points and the chief criticisms of the specific categories of DSM II may be listed as follows.

Organic brain syndromes. All the so-called organic mental disorders are listed here according to etiology in an order that is largely standard for somatic disorders generally (genetic and prenatal influence, infections, intoxication, trauma, circulatory disturbance, and so forth).

This is the least controversial division of the classification. Here one finds the classical association of etiology, pathology, and pathological physiology or illness. On the psychic level an organic syndrome has been identified, and this includes defects of sensorium, lability of emotion, and disorders of judgment, volition, and conduct. The method of classification is logical, flexible, and comprehensible.

The APA terminology has now returned to a more complete alignment with the older literature. DSM I's distinction between "acute" and "chronic" has been dropped or very much subordinated, and the organic disorders have been condensed into a single etiological list. In addition, the names of the disorders have been shortened and simplified; "chronic brain syndrome associated with central nervous system syphilis, meningo encephalitic type" has been replaced by the older term "psychosis with general paralysis," and "acute brain syndrome associated with alcohol intoxication" has now been replaced by such terms as "delirium tremens," "acute alcohol intoxication," and so forth. The general term "organic brain syndrome" still is retained in the overall designation of psychoses associated with cerebral pathology, but it has been almost eliminated from the designations of the individual forms appearing only as the abbreviation OBS in the nonpsychotic group (number 309). It will be noted that this category has been moved up to follow immediately after 294, psychoses with other physical condition, instead of coming immediately before mental retardation as it does in ICD-8.

As stated above, this group is the least controversial one in the classification, but this is based on the relative ease with which the syndromes can be defined and not on a real understanding of pathogenesis. For example, the mechanism of hallucinations and delusions in the senile or the paretic are still no better understood than are those of a schizophrenic, and while the delusions and hallucinations of organic cases are usually of a different pattern than those of functional cases, yet they may be quite indistinguishable as often occurs in the amphetamine psychoses. This observation has led some to feel that the psychic phenomena of all types of psychosis are identical and of a different order than the somatic ones, and this opinion brings to mind the Cartesian philosophy that mind and body operate on different planes. This dualism may reflect also a deep religious feeling that equates the mind with the soul and views the soul as incorruptible. Such dualism is not usually enunciated in explicit form, but it is often implicit in discussions about the organic syndromes. Organic defect states, for example, are intuitively accepted as due to brain damage or loss of brain function, as in the case of senile dementia, but such states are often seen as different in nature from the organic psychoses with secondary systems even if the behavior problems are identical in both instances. This has lead to endless futile rhetoric about the real meaning of the word "psychosis," a discussion that is founded on the fallacy that this term has some intrinsic and inherent meaning other than the arbitrary significance assigned by usage, custom, and the definitions of professional bodies. Insofar as such discussions distract attention from the real needs of the patient and center everything on the meaning of a word, they are worse than futile because the word "psychosis" does not have a clear and rigorous definition, a characteristic that it shares with other essential words such as "illness" and "health."

A question has recently been raised with respect to the separation of the psychotic and the nonpsychotic forms of OBS on the ground that one should not make such a basic distinction between cases of lesser and greater severity where the basic illness is the same. Other questions have also been raised, but it now seems that changes in the OBS groups are likely to be limited to those based on new discoveries such as that which may move Jakob-Creuzfelt Disease into the category of infectious disorders from the presenile dementias.

Mental deficiency. The very brief listing previously allocated to mental deficiency has been replaced by a condensed version of a special classification of these disorders that gathers them all in one series that includes those of known and of unknown etiology. In DSM II this is the first major division, while it is the last in ICD-8. This part of the classification now appears to be one of the most satisfactory sections, whereas it was previously one of the least acceptable.

Psychoses (functional). This group now consists of schizophrenia, the affective disorders, and the paranoid states. Also listed and marked as "not for American use" are the reactive psychoses, which play an important role in ICD-8 and are widely recognized in European psychiatry. Whether these last represent a nosological entity remains a fundamental issue in classification and one that future research will have to solve; at this time opinion remains firm on both sides of the question.[6,11]

For a long time critics have claimed that the classifications listed too many varieties of schizophrenia, but in spite of much discussion it was not possible to abbreviate the list and the number of divisions remains at ten or more.

The five main subgroups, namely, the catatonic, hebephrenic, paranoid, simple, and childhood types are generally accepted, and the main differential diagnosis is between the affective disorders and schizophrenia; the differential response to somatic therapies continues to provide indications that the distinction is significant.

During the preparation of DSM II criticism against this part of the APA classification centered particularly on the acute undifferentiated, chronic undifferentiated, schizo-affective, and residual types, and these are again the focus of considerable discussion, but for the moment they remain either in DSM II or ICD-8 or in both. Another criticism had to do with the use of the term "reaction" as applied to schizophrenia in DSM I. This was interpreted by many as implying a knowledge that we do not have of etiology, and it was even looked upon as propaganda in favor of a hypothesis about the mode of origin of the symptoms. As a result the term "reaction" that was also widely used elsewhere in DSM I has been replaced almost everywhere in DSM II by terms such as "disorder" or "illness," and this change does not appear likely to be reversed. Europeans in particular tend to believe that many schizophrenic cases are of endogenous origin and felt that "reaction" implied that all cases are exogenous; hence they considered the term insufficiently neutral for a classification.

It is interesting to note that although much of the discussion with respect to this issue was carried on in terms of exogenous causes of functional disorders contrasted with endogenous origins, Lewis[20] has pointed out in a very incisive paper that this distinction is itself vague and does not really clarify the theoretical question.

Major affective disorders. These continue to be a problem in nosology. Recent experience with treatment seems to indicate there is a significant difference between cases that suffer only depressive or manic attacks and those who have both types of syndromes in the course of time. A distinction is thus being made between monopolar and bipolar affective disorders, and this may find its way into the next revision.

In addition, the long-standing debate about the validity of the entity of involutional melancholia still continues, and it seems no nearer to solution than it was in Kraepelin's time. The paranoid forms are now classed with the paranoid states and no longer grouped with the affective disorders, but so many depressed cases show paranoid elements that issues about their proper classification continue to be raised from time to time. Finally one cannot but continue to be concerned about the great disparity in the proportion of cases classified as schizophrenic in the United States as compared with Britain and Scandinavia. Not all of this is due to differences in classification standards, but important differences do exist and efforts continue toward eliminating them through improved nosology.

Neuroses. Called the psychoneurotic disorders in DSM I, this category has taken on increasing importance with the growth of outpatient psychiatry. Long the foundation of psychoanalytic practice and theory, these illnesses as a group have been as well recognized as other major psychiatric divisions, but the subdivisions have remained somewhat vague and subject to shifts from time to time that do not show any clear line of evolution in any specific direction. The "psychasthenia" of Janet, for example, once included the phobias and the obsessive-compulsive reactions, and in the *American Handbook of Psychiatry*, Volume 3, neurasthenia is combined with hypochondriasis.

Virtually all psychiatrists who treat neurotics readily acknowledge that in this class mixed syndromes predominate, that there is a considerable tendency for symptoms to shift from one category to another, and that lack of fundamental knowledge obviously restricts our efforts at classification. This uncertainty goes so far as to leave considerable doubt about the demarcation between the neuroses and other functional disorders, including the psychophysiological disorders, the psychoses, and certain personality disorders, since many cases of personality disorder have neurotic elements and some cases seem to show a transition from neurosis to psychosis. Finally the hysterical psychosis, long rejected in American nosology, is now again being seriously reconsidered.

Among the subdivisions of the neuroses, the obsessive-compulsive syndrome and the phobias (anxiety hysteria in the Manual) are all relatively stable in the various classifications. As a group the neuroses are distinguished from psychoses by absence of change of the basic personality and by lack of delusions or hallucinations and, more recently, by differential response to therapy, since as a class they do not respond strikingly to somatic treatments as do many psychoses, while some forms are far more suitable for psychotherapy than are the psychoses. Curiously enough, limited psychosurgery remains the treatment of last resort for intractable cases of obsessive-compulsive neurosis that do not respond well to psychotherapy.

In DSM II the hysterical neurasthenic and hypochondriacal neuroses reappear after having been dropped in DSM I, and this reflects the general state of uncertainty about the subdivisions of this highly prevalent form of disorder. The other major change is the introduction of the new term "depersonalization neurosis" (or syndrome).

Personality disorders and certain other non-psychotic mental disorders. This major group has long been a source of active controversy and confusion. DSM II has abandoned the attempt of DSM I to separate the so-called personality pattern disturbances from personality trait disturbances; it now groups them together as personality disorders and adds to the title the very significant words "and certain other nonpsychotic mental disorders." This broader term must, however, be seen as a temporary expedient because new restrictions on the term "mental disorder" are emerging in social psychiatry, where such problems play an important role. There is, for instance, much doubt whether homosexuality necessarily constitutes a form of mental disorder, and the nature of the problem in some of the other forms of sexual deviation is also under debate.

Questions have been raised also about the appropriateness of listing alcoholism or drug dependence per se as a mental disorder since some feel that these may occur in the absence of mental disorder unless one includes alcohol or drug-seeking behavior as mental disorder by definition. No one, of course, can deny that mental disorder is often a cause or effect of dependence.

Omission of the rubric "sociopathic personality disturbance" in DSM II has been the expression of a similar trend to purge the classification of items that label deviant behavior in and of itself as a form of mental disorder. ICD-8 contains essentially the same categories under "other nonpsychotic mental disorders" and faces the same problems.

There is much less question about personality disturbances that are lifelong, fixed, relatively mild, and have some of the qualities of a major psychiatric syndrome without actually being a manifestation of such illness. Among these types DSM II lists the schizoid, cyclothymic, and paranoid personality types, as well as the compulsive and the passive-aggressive varieties. However, the way in which these conditions have been grouped, and the terms used for such grouping, are still open to controversy. Although the manifest content of these controversies relates to principles of nosology, there can be no doubt that the issue of stigma injects heat into the discussion, and it is with respect to those who are seen as socially deviant in behavior that the controversy is sharpest. In these cases medico-legal issues are involved, and some of the forms of personality disorder are seen quite as frequently in correctional facilities as in psychiatric installations. In many such cases the controversy has to do with which of two unwilling types of facilities, namely, jails or mental hospitals, will have to deal with a given case.

Among the suggestions for improvement of this category are (1) a simplification of the list, (2) removing sexual deviations, drug dependence, and perhaps antisocial personality to more neutral positions in the classification. As noted above, progress was made in this direction when "sociopathic personality disturbance" was eliminated as the name of a group of entities whose only common denominator was conflict with established codes of behavior and with various legal sanctions, and further developments of this type may be expected.

The Psychophysiological autonomic and visceral disorders. Also known as the psychosomatic disorders, this category is now firmly entrenched in European as well as American medicine, and in both DSM II and ICD-8 the subdivisions are listed in the standard anatomical sequence of the general medical nosology. These disorders have some resemblance to the neuroses but are distinguishable. They are characterized by disorders of function of various organs that are considered to have important emotional elements. In DSM I they were designated "psychophysiological autonomic visceral disorders," but this was considered to be ponderous and contaminated by etiological assumptions. Therefore, the more neutral term "psychophysiological disorder" was substituted in DSM II. ICD-8 has adopted the concept under the name "physical disorders of presumably psychogenic origin."

Special symptoms. In DSM I this was a subgroup listed under "personality disorders." It has now been giving a more independent status and remains a necessary listing, but it will undoubtedly undergo further change and rearrangement together with the two groups that we shall discuss next, since all three are used extensively in child psychiatry where efforts at restructuring of the nosology are already well advanced.[5,17]

Transient situational personality disturbances. Under this head are grouped a rather heterogeneous collection of terms, many of which are also used in child and adolescent psychiatry. The reactive element is stressed, and the effect has been heightened by removing the term "reaction" from most of the other parts of the Manual. As the title indicates, one essential characteristic of the transient group is its good prognosis; another is the absence of specific symptoms belonging to other types of psychiatric disorder. It has been argued that many of these syndromes may last for years, leaving the term "transient" open to debate. Furthermore, this category fails to satisfy child psychiatrists, although no generally accepted replacement has yet been developed.

Behavior disorders of childhood and adolescence. This is a new group in DSM II characterized partly by the fact that the duration of the disturbance lies between that of the transient situational disturbances and the psychoses, neuroses, and personality disorders. The patterns of behavior are also fairly well defined, and the rubric is represented in ICD-8 but in considerably abbreviated form. This category, together with the two preceding ones, appears likely to be much affected by the previously mentioned nosological work now being done in child psychiatry.

Conditions without manifest psychiatric disorder and nondiagnostic terms for administrative use. These two major groups include various conditions likely to be encountered in psychiatric practice and provide a means for indicating that although the cases were seen in a psychiatric setting they were not considered to manifest psychiatric pathology. Included in the first group are various forms of maladjustment and dyssocial behavior, while various housekeeping entries such as "examination only" or observation are found in the second. This category is not found in Section V of ICD-8, which uses no code numbers above 315, while DSM II uses numbers 316, 317, 318, and 319 for this purpose. These categories are considered useful even though their validity has been questioned on the ground that their inclusion in a psychiatric classification implies psychiatric pathology, but its purpose is altogether the reverse: namely, to indicate affirmatively that no psychiatric pathology was diagnosed.

Code Numbers

Like other modern classification systems, DSM II identifies each entity by a name and also by a corresponding code number. With certain relatively minor exceptions the names and their code numbers in this list are the same as the ones in ICD-8. If one examines the list it will, however, be noted that the DSM II series is not presented in numerical order. Thus the list is headed by the 310–315 series (the mental retardations), and these are followed by the 290–294 series (the psychoses associated with organic brain syndromes), and these are followed by Section 309, the nonpsychotic organic brain syndromes, after which comes a section containing items numbered 295–298. The Arabic numerals represent the order of items as they are listed in Chapter V, the Psychiatric Section of ICD-8. The irregularities found in the sequence of code numbers in DSM II resulted from rearrangements that were made in an attempt to conform the system to American practices without losing the interconvertibility of the two systems. The sequence of presentation of major categories in DSM II is identified by the Roman numerals I–XI inclusive superimposed on the Arabic numerals that identify each name in the list.

The addition of such serial numbers to names for purposes of better identification is almost universal in our society and is exemplified by Social Security numbers, credit card numbers, and hospital identification numbers. Such numbers render identification more accurate and facilitate statistical and control operations, especially with computer technology. When applied to a classification system such serial numbers have additional advantages in that they can be used to reflect the hierarchical structure of the classification; decimal places of decreasing magnitude are attached to subdivisions of decreasing importance, while the higher values are attached to the major subdivisions. The use of decimal values to introduce subcategories is illustrated in the way DSM II has added the subcategories 302.5 (voyeurism), 302.6 (sadism), and 302.7 (masochism) under the major ICD-8 heading of 302 (sexual deviations), which lists specifically only four forms, ending with 302.4 (exhibitionism). Just as it is possible to add items by expanding the numerical listing, it is possible to drop items that are not locally acceptable without breaking the overall classification pattern, and this has been done with respect to 298.1 (reactive excitation), 298.2 (reactive confusion), 298.3 (acute paranoid reaction), and 298.9 (reactive psychoses unspecified), all of which are marked "to be avoided in the U.S." This was done because of a fundamental difference of opinion about the so-called reactive psychoses, which are not widely recognized in the United States, but are fully accepted as valid in Europe, especially in France and the Scandinavian countries.

Another use of the code number system in DSM II is to identify modifying phrases. These are designated by a fourth digit that can be used to specify additional characteristics of a syndrome such as acute or chronic (.x1 or .x2), or mild, moderate, or severe (.x6, .x7, .x8). Thus acute psychosis with brain trauma in DSM II would be identified by the code number 293.51, and the chronic form would have the code 293.52.

Finally it may be noted that the ICD classification system allots three digits for designation of major disease categories and a fourth digit, a decimal digit, for specification of additional details within each category. DSM II adds a fifth digit (in the next decimal place) to provide for qualifying phrases and for other purposes.

In spite of this flexibility both systems provide for only one disorder or disability in each coding, while in clinical practice psychiatric disorders often occur not separately but in interacting combinations with each other. Alcoholism may, for example, occur in combination with schizophrenia, manic-depressive illness, or mental deficiency; similarly schizophrenia may be combined with epilepsy. A very few combined disorders such as propf-schizophrenia have been given specific names, but these have not achieved general recognition, and for practical purposes the practice of the past has been to select one diagnosis on the ground that it is the underlying, the presenting, or the most serious one. This was felt to be necessary because there are no names for most of the possible permutations of such disorders, and if they were created it would not be feasible or useful to deal statistically with such a large number of entities. The use of multiple diagnoses would solve this problem, but this was long impractical for similar reasons. Computer technology has now removed many of these limitations, and DSM II states that multiple diagnoses should be used where indicated, and this Manual for the first time "encourages the recording of such diagnosis as alcoholism or mental retardation separately," with the caution that no more conditions should be diagnosed than are needed to account for the clinical picture.

In summary, one may say of the code numbers that they may be completely ignored by the clinician, but a fuller understanding of the nature of psychiatric classification as a system may be gained by mastering the relatively simple principles on which it is based.

¶ Nosology of This Handbook

This *Handbook* is based essentially on the current classification, but it shows many variations and such variations occur regularly in textbooks and at virtually all levels of psychiatric communication. In part they are required in order to maintain a continuity with the psychiatric literature of the past; in part they simply illustrate what has been clearly enunciated by Jaspers,[16] namely, that no classification is equally suitable for all purposes, and thus for different situations different classifications may be quite appropriate. For example, the literature on psychiatric states precipitated by the stress of war, or toxic, exhaustive, infectious conditions tends to emphasize the stressing factors. In so doing it often departs from the formal classification systems under which one might assign many of these cases to such categories as schizophrenia, depression, or neurosis. The difference from ordinary classification is further intensified by the use of specific terms such as "war neurosis" that are not found in the currently accepted nomenclature. Such variations, however, will only rarely cause any problems, and for practical purposes the context will enable one to translate the terms into those of the standard classification with no difficulty. The feasibility of such a procedure can be seen from the fact that it is still the practice in some countries for each psychiatrist to enter the diagnoses of his cases in the classification of his choice. These terms are then gathered in a central statistical bureau and translated into the standard system by technical personnel. This is considered by no means ideal, but it is said to be quite practical and even gives reasonably satisfactory results.

For a textbook the alternative to such flexibility of presentation would be to resurvey all that had been done in the past and attempt to recast the entire literature with every significant change of nomenclature and classification, even though revisions of the formal classification now occur quite regularly at ten-year intervals. Even more difficult is the problem of new categories that have not yet been formally placed in the classification, as is the case with the monopolar affective disorders. Such new developments obviously must be given recognition long before the official classification can incorporate them. All of this means that deviations from the formal classification and nomenclature must be acknowledged as frequent and, for many purposes, necessary. Such deviations do not, however, constitute a repudiation of the accepted system, nor do they diminish the need for such a system to be used for normal clinical records from which public health data must be developed.

Conclusion

While they are vigorously challenged on academic and even political grounds, current systems of classification remain virtually unchallenged in clinical usage, and in one or another form are utilized throughout the world. They all derive from the same evolutionary process and stem from a common psychiatric literature; this is clearly reflected in their structure. The main categories of psychosis, neurosis, character or conduct disorder, and mental deficiency are everywhere to be found; the leading entities such as schizophrenia, manic-depressive disorder, and so forth are also easily recognized. The division of etiology into functional and organic appears unchallenged, and with the passage of time the list of organic causes grows steadily. We have no certain knowledge of how many of these categories were merely invented, and are thus artificial, and how many are natural like paresis, which was discovered. Many difficulties of a scientific nature remain to be overcome before this question can be answered for the functional disorders; even in the case of the so-called organic cases the age-old body-mind dilemma appears to block any real understanding of how a given physical pathology is transmuted into the corresponding psychic disorder. This issue has been raised most recently with respect to the role of drug abuse in psychiatric disorder; however, it remains without answer even in the case of the amphetamine psychoses, where the cause-effect relation appears quite clear-cut. The irregularity of response pattern, the lack of anatomical or biochemical specificity, and the incomplete correlation between the extent of somatic insult and the psychic response is not different here than it is in relation to cerebral arteriosclerosis or senility, but it shows that our clearest concepts of etiology are severely limited when they are closely examined and that the identification of a somatic agent is but one step toward the understanding of psychiatric disorder.

In spite of these and other limitations, progress has been made within the framework of the existing systems. Criticism of psychiatric classification has always been active; it is one of the most traditional parts of psychiatry and has been a safeguard against dogmatism, but one doubts that even the most vigorous current critics would wish to sweep away all classification and all "naming" or "labeling." At least I do not know of any active psychiatric service that operates without such "labels." Doubts about the validity of the medical model are frequently expressed and have had considerable recent attention, but they are by no means new, as will be seen from such little known older literature as that of the philosopher Kant[18] on psychiatric classification and that of Jerome Gaub,[26] who preceded him by many years. On the other hand, it is probably correct to say that a large number of mental health professionals expect that sooner or later the computer or the laboratory will open up great new vistas in this field; this view is shared by some who reject the medical model and expect that it will be replaced by a social model of mental disorder. For some this expectation is so strong that there is a real danger that it could lead them into seeing the event before it occurs; such premature acceptance of a pseudoadvance could, if it involved influential individuals, create considerable confusion in psychiatry. At this time, however, it seems that the great preponderance of the professional world is still inclined to wait for firm evidence that any proposed replacement of the current psychiatric classification will be superior in actual performance to what we now have, and that conviction will come from performance rather than from debate. Indications are that progress will be relatively slow and will be built on the foundations of previous work rather than on revolutionary new principles. Everyone, of course, hopes for a spectacular breakthrough, but at the moment none seems to be in prospect. It also seems clear that the various systems of classification and nomenclature that now are used in various parts of the world will continue to converge toward a common form, providing a better international language for psychiatry, a better basis for public health studies, and, last but not least, a better basis for treatment.

Bibliography

1. ADAMS, H. B., "Mental Illness or Interpersonal Behavior?" *Am. Psychol.*, 19, 1964.
2. AURELIANUS, C. (Ed. by Drabkin, I. E.), *On Acute Diseases and Chronic Diseases*, University of Chicago Press, Chicago, 1950.
3. BAN, T. A., and LEHMANN, H. E., *Experimental Approaches to Psychiatric Diagnosis*, Charles C Thomas, Springfield, Ill., 1971.
4. BURGESS, L. G. (Ed.), *Current Medical Information and Terminology*, 4th ed., American Medical Association, Chicago, 1971.
5. Committee on Child Psychiatry of the Group for the Advancement of Psychiatry, *Psychopathological Disorders in Childhood: Theoretical Considerations and a Proposed Classification*, Volume 6, Report No. 62, June 1966.
6. Committee on Nomenclature and Statistics, *Diagnostic and Statistical Manual of Mental Disorders* (DSM I), American Psychiatric Association, Washington, D.C., 1952.
7. ———, *Diagnostic and Statistical Manual of Mental Disorders* (DSM II), American Psychiatric Association, Washington, D.C., 1968.
8. COOPER, D., *Psychiatry and Anti-Psychiatry*, Tavistock, London, 1967.
9. CROWSON, R. A., *Classification and Biology*, Atherton, New York, 1970.
10. EVERITT, B. S., "Cluster Analysis," *Brit. J. Psychiat.*, *120*:143–145, 1972.
11. FAERGEMAN, P. M., *Psychogenic Psychoses*, Butterworths, London, 1963.
12. FEINSTEIN, A. R., "Boolean Algebra and Clinical Taxonomy," *New Eng. J. Med.*, *269*: 929–938, 1963.
13. FLECK, S., "Labelling Theory," *N.Y. State District Branch Bull.*, May 1969, p. 1.
14. HOWELLS, J. G., *Nosology of Psychiatry*, Special Report to the Society of Clinical Psychiatrists, Claver Press, Ipswich, 1970.
15. *International Classification of Diseases* (ICD-8), 1965 revision, World Health Organization, Geneva, 1967.
16. JASPERS, K., *General Psychopathology*, University of Chicago Press, Chicago, 1963.
17. JENKINS, R. L., and COLE, J. O., *Diagnostic Classification in Child Psychiatry*, Psychiatric Research Reports of the American Psychiatric Association, Report No. 18, Washington, D.C., October 1964.
18. KANT, I., *The Classification of Mental Disorders* (Tr. and ed. by Sullivan, C. T.), Doylestown Foundation Paper, Doylestown, Pa., 1964.
19. KATZ, M. M., COLE, J., and BARTON, W. E., *The Role and Methodology of Classification in Psychiatry*, Public Health Service Publication No. 1584, National Institute of Mental Health, Washington, D.C., 1968.
20. LEWIS, A., "'Endogenous' and 'Exogenous': a Useful Dichotomy?" *Psychol. Med.*, *1*: 191–196, 1971.

21. LORR, M., KLETT, J. C., and MCNAIR, D. M., *Syndromes of Psychosis*, Pergamon Press, New York, 1963.
22. MENNINGER, K., MAYMAN, M., and PRUYSER, P., *The Vital Balance*, Viking, New York, 1963.
23. MORIYAMA, I. M., "The Classification of Disease, a Fundamental Problem," *J. Chronic Dis.*, 2:462–470, 1960.
24. ÖDEGARD, Ö., "Reactive Psychoses," *Acta Psychiatrica Scandinavica*, Suppl. 203:23–28 (undated).
25. Office of Health Economics, *Prospects in Health*, Paper No. 37, White Crescent Press, London, 1971.
26. RATHER, L. J., *Mind and Body in Eighteenth Century Medicine, a Study Based on Jerome Gaub's De regime mentis*, University of California Press, Berkeley, 1965.
27. SAVORY, T., *Naming the Living World*, John Wiley, New York, 1962.
28. SCHEFF, T. J., "Schizophrenia as Ideology," *Schizophrenia Bull.*, National Clearing House for Mental Health Information, no. 2, Fall 1970.
29. STENGEL, E., "Classification of Mental Disorders," *Bull. W.H.O.*, 21:601–663, 1959.
30. VEITH, I., *Hysteria: The History of a Disease*, University of Chicago Press, Chicago, 1965.

CHAPTER 53

THE PSYCHIATRIC INTERVIEW

Ian Stevenson

THE PSYCHIATRIC INTERVIEW as practiced in most American psychiatric facilities has undergone a marked change during the past 60 years. Formerly a question-and-answer type of interview satisfied the requirements of psychiatric interviewing, as it did and still does satisfy those of medical history-taking with regard to exclusively physical illnesses. But the modern psychiatric interview, although it includes questions, puts much more emphasis on a free-flowing exchange between the psychiatrist and the patient. This alteration in our practice has resulted from changes in the kinds of information we want about patients and in our ideas of how we can best obtain this information. We also have learned the limitations of verbal communications. We now notice not only what the patient says but also his manner of saying it, for this may show what his words conceal. And we have learned that, when two people talk together, what they say depends not only upon what they want to tell each other but also upon what they think about each other. In what follows I shall discuss first the information a psychiatrist usually wishes to obtain in an initial interview, next how the psychiatrist's relationship with the patient influences what the patient tells him, then the psychiatrist's optimal attitude, and finally some techniques that can increase the yield of an interview.

Both the theory and technique of psychiatric interviewing receive attention from American research psychiatrists, although not as much as they should. We may hope that from their efforts will emerge changes fully as great as those that the last 60 years have brought. This will require, among other things, that each of us challenge constantly his own habits and remain unwilling to practice, for the rest of his lifetime, only whatever his teachers have taught him.

I shall discuss the psychiatric interview chiefly with regard to the initial evaluation of a patient. Sometimes we can achieve this in one interview, but quite often we need several. Moreover, the initial interview or interviews should blend with the psychiatric examination. Chapter 54 discusses the psychiatric examination and the methods of including part of the examination in the psychiatric interview and of making a transition from the interview to the more definitive examination.

What the Psychiatrist Wants to Learn

The psychiatrist should obtain first what the patient usually most wants to give, namely, a description of his symptoms and the story of their onset and progress. After this the importance of life stresses in precipitating mental illness requires a detailed review of the patient's current environment. The discussion of this can lead easily into talk about the patient's early environment and thence toward his family history. From this may naturally follow an account of the patient's own earlier life—his personal history.

Most psychiatrists understand the importance of eliciting this material in initial interviews, and only two items deserve further emphasis. First, much importance must be attached to a detailed account of the patient's symptoms. We should try to imagine what the patient has experienced and now experiences. We should try to see the world as he sees it, but we can do this only if we let him talk to us in great detail. Moreover, many psychological symptoms require study not only as direct experiences of the patient but also with regard to the purpose they serve the patient in adapting to other people or to other forces within himself. In short, we enter into detail so that we may know both what functions are disturbed and how these functions relate to others. Second, the study of the patient's current environment must be emphasized. Although we all recognize the importance of major life stresses in precipitating mental illnesses, we often neglect, to our and the patient's disadvantage, the careful study of how the patient lives. Only by entering into his daily life, as it were, can we come to appreciate the subtle but cumulatively powerful relationships between the patient and others close to him. And usually only such an appreciation will permit us to dissect the respective contributions of the patient and those around him to the strain he experiences.

While listening to the history, the psychiatrist should not only attend to the bare facts of peoples, places, and events as chronicled by the patient; he must also study the meaning of these events for the patient and the attitudes that the patient showed to them then and, if those have changed, the attitudes he now shows toward them. In studying attitudes the psychiatrist must include, in addition to the patient's words, observations of the patient's emotions.

We have also another important reason for observing the patient's emotions as he talks. The psychiatric interview begins and includes much of the psychiatric examination. The patient's recital of his complaints and his history contributes valuable data about the illness. But that illness is a product (in most instances) of the action of stresses on sensitivities. The psychiatric interview should therefore study the special sensitivities and vulnerabilities of the patient. As the patient talks, the psychiatrist should scan him and his remarks for signs that certain events or topics are of special importance to him. The signals that reveal such events or topics deserve a brief review.

One may ask the patient directly about the events, people, and thoughts that bother him most. More often than is usually done, we should ask for this information directly. At the same time we should remember the frequent, almost invariable, inability of patients to give a frank and complete answer to direct questions. In studying physical illnesses we can ask patients about the occurrence of nausea bloody stools, or swollen feet and usually expect reasonable and valuable answers. But we cannot ask a patient to tell us about his marriage, his parents, or his employer and expect that the words he returns us can alone contain all we need to know. Several factors are responsible for this difference. In the first place our society strongly emphasizes the importance of other persons having a good opinion of us. For psychiatric patients this becomes especially important, since they usually think poorly of themselves and have become doubly dependent upon approval by other people. When a patient finds himself in a psychiatrist's office, he has additional reasons for winning and preserving the favorable opinion of the psychiatrist. Consequently with his words he attempts to portray himself (unless he is very depressed or self-effacing) as a person who is in all respects lovable and "normal." Second, even the patient with the greatest candor has within himself large and important aspects of mind and behavior that lie quite outside his awareness. With the best will in the world he cannot tell us what he does not know about himself. And finally, even if he knew much more than he does, words would still furnish only a feeble channel for the communication of life's richest experiences, both of suffering and of happiness.

The psychiatrist needs to remember also that what the patient tells him about some past event or experience, even a rather recent one, may not correspond closely with the facts, if they could be ascertained, or the memories of other persons, if they are interviewed about the same events. Investigations of memories have shown them to be much less stable than was at one time thought true.[1,18] A person's account of his past given at one time may differ markedly from his account of the same events given at another time.[31] And some events are remembered more accurately than others.[16,25]

Despite these limitations of verbal communications and memories, the psychiatrist can use certain valuable clues provided by the patient to guide him toward at least some of what he wants to learn. These clues lie in the various signs of emotion shown by the patient as he talks, for our most important experiences become bound to emotions, or, more accurately, they become important because they affect us deeply.

Emotions show themselves in many and sometimes unexpected ways. The patient's arrangement and manner of presenting his verbal statements reveal much. The psychiatrist should note what the patient says first (both at the beginning of an interview and subsequently in response to questions), what he talks about most, what he returns to many times, and what he omits or glides over quickly. Thus the psychiatrist needs to learn what the patient especially wants to talk about and what he especially wants to avoid talking about. Unusual speed of speech, hesitations, blockings, amnesias, and confusions all deserve attention as signs of emotion and, hence, clues to the significance of events or topics. The order of the patient's remarks deserves attention, and especially the connections of thoughts associated in one sentence or adjoining ones. Verbal associations betray affective links.

The psychiatrist should notice changes in the pitch and timbre of the patient's voice as he talks. Such changes express alterations in the tensions of skeletal muscles, of which many other indications can appear in the patient's face and limbs. Accordingly, the psychiatrist should watch for the play of emotion in the patient's face, in the posture of his body, and in the movements and gestures of his limbs.

Changes in the patient's viscera deserve equal attention, for emotions affect the autonomic nervous system as markedly as the central nervous system. Physiological investigations have shown the occurrence of many important visceral changes during emotional disturbances. Not many of these lie exposed to the unaided eye of the interviewing physician. Nevertheless, he may notice changes in the patient's breathing and in his heart rate, observed perhaps in the beat of the carotid artery in the neck. He can notice flushing and pallor of the face and sometimes perspiration. The patient's mouth may dry up, or tears may glisten in his eyes. During an interview emotional changes may bring on (and sometimes remove) the patient's symptoms. Thus palpitations may occur, or a headache may disappear. A patient with a psychophysiological skin reaction may scratch a tender spot on the skin when the conversation touches something tender in his mind. Each patient has his own special mode of expressing his emotions, almost as characteristic as his gait or his fingerprints. Some patients, for example, rub their eyes, others glance swiftly away from the interviewer, and still others swallow whenever they experience anxiety. The psychiatrist should watch the patient for characteristic traits especially in the early phases of the interview, partly because the patient is then usually most anxious and partly because the psychiatrist can use what he then learns to identify moments of anxiety later in the interview.

The identification of an important emotion only begins its study by the psychiatrist. With techniques described later he should try to open a further discussion of the topic that has evoked the emotion, although he may often defer this to a more appropriate time. In that discussion he wants to learn in what way this topic is important to the patient and how it became so. Exactly what thoughts does the patient have about the event, person, or topic that causes these strong emotions? The psychiatrist cannot consider that his study of an emotion is complete unless he has elicited from the patient the details of the accompanying thoughts. For this he returns again to the patient's words through which alone (outside art) the patient can communicate his thoughts.

The detection of emotion during the psychiatric interview contributes to the examination of the patient, which, as already mentioned, starts at the beginning of the interview and, indeed, cannot and should not be separated from it. During the interview the psychiatrist has ample opportunities to examine other aspects of the patient's mental functioning, as described in Chapter 54.

A final purpose of the psychiatric interview is the evaluation of the patient's readiness for psychiatric treatment and efforts to improve this when necessary. Since this properly belongs to psychiatric treatment, it is mentioned here without further discussion. But the psychiatrist should remember it during his interview. Although his assigned tasks may resemble those of a juggler keeping five balls in the air, unless the psychiatrist can include in his technique a study and strengthening of the patient's motivation for treatment, he may conduct a superb interview that leads to nothing.

The Physician-Patient Relationship

As mentioned earlier, the patient's wish to tell his story is frequently obstructed by his wish to win and preserve the psychiatrist's approval and assistance. This interference is experienced by all patients to some degree. But each patient varies from every other one in the experiences that have led to this shielding of himself and to other behavioral patterns. And each psychiatrist differs from all others in his capacity to stimulate or reduce such patterns in his patients.

When the patient was a child, like everyone else he learned from experiences what to expect that his parents (and other people) would do. He then generalized many of these expectations, first learned with particular persons, to guide his behavior with other persons. Sometimes his generalizations guided him correctly, at other times inappropriately. A dog conditioned to respond to a sound with a frequency of 512 cycles per second may respond (unless carefully trained) to a range of sound, say, between 475 and 550 cycles per second. The more careful and prolonged the conditioning, the more discriminating will be the dog's response to different stimuli. But his discrimination may weaken under stress or without proper reinforcement. In much the same way humans may discriminate poorly as well as correctly. They may respond to physicians as if they were duplicates of their parents. Such misperceptions on the part of a patient never occur first with regard to the psychiatrist; on the contrary, they have happened often before and have contributed importantly to the patient's difficulties with other people. But the psychiatrist should especially notice how the patient perceives him, first, because he can study this directly instead of depending upon observations of other people and, second, because the patient's perceptions of the psychiatrist furnish important clues to his difficulties with other people.

The more closely the psychiatrist resembles the significant persons of the patient's earlier life, the more he will be likely to evoke the behavior in which they trained the patient. (The frequency of 512 cycles per second stimulates the conditioned dog, mentioned above, to the greatest extent, even though he may respond to a lesser extent to other frequencies.) Suppose that the psychiatrist, after studying in advance a portrait of the patient's father or, better still, a moving picture sequence, should carefully disguise himself in appearance and manner to resemble the patient's father. We could hardly blame the patient for responding to the psychiatrist-actor as if somehow his father had wandered into the psychiatrist's office and sat behind his desk. After a moment of initial surprise the patient would engage in conversation, so he would believe, with his father. Now suppose that the disguise has been arranged very poorly, that, in fact, the psychiatrist has put on a mustache like the father's but does not shave his head to a similar baldness or imitate the father's gruff voice or smoke cheap cigars. If then the patient still acts as if the psychiatrist is his father, the psychiatrist would have important evidence of poor discrimination.

By partially resembling earlier persons in the patient's life, the psychiatrist may stimulate the conditioned responses of his patients in many ways. Each deserves brief mention here and much attention in the interviews. First, as already mentioned, the psychiatrist's physical appearance influences the patient's responses. The psychiatrist's sex and age, especially, but other features of appearance hardly less, strongly guide the patient's thinking about what it will be useful or safe to reveal. Second, the patient responds to the social role of the psychiatrist as he conceives it. In this he mingles his concept of the role of the physician. Two features usually blend. Physicians have authoritative roles in our culture, with power to recommend and execute drastic treatments or to commit to certain hospitals. This aspect of our work leads the patient to confuse us with policemen, sergeants, judges, teachers, and, most important of all, with fathers. But physicians also have a role of succoring the sick and weak; in this connection a patient frequently achieves a mental montage of a physician and his own mother. Third, our behavior may also stimulate in the patient patterns of behavior laid down in earlier experiences. Some of this behavior derives from our professional work. We ask questions and so we may remind the patient of his mother, who always asked her little boy pressing questions, sometimes requiring painful answers, when he came home from school. However, some of what we do our work does not require and may indeed be better off without it. Thus suppose we, like the patient's mother, have an inordinate preoccupation with sex, and we question the patient excessively about this. He may then react strongly, although not necessarily irrationally.

In all these various ways the psychiatrist may evoke behavioral patterns in the patient that can partly, or sometimes entirely, interfere with that part of him that perceives the psychiatrist as a helpful expert to whom he should tell his story. Patients vary greatly in their capacities to see the psychiatrist as he is and to avoid confusing him with other people. If the psychiatrist is to study the patient's discrimination, he not only must attend to the patient's behavior but also must learn as much as he can about himself. If the patient falsely attributes a mustache to the psychiatrist, the psychiatrist can only evaluate the possible misperception in this if he can recall whether he himself has shaved during the last few days. He must know what he himself brings as stimuli into the interview. He must remember that the patient responds both to what the psychiatrist does and to what he is.

In this connection it is worth mentioning that even when different interviewers adopt a somewhat uniform approach in the conduct of an interview, they may have markedly different effects on different patients.[7,14] In drawing attention to this fact, I am not recommending the adoption of a uniform style in interviewing. This would be as undesirable as it would be infeasible. But I do exhort the interviewer to become as much aware as he can of his own behavior with patients and its differing effect on different patients.

A physician-patient relationship is clearly not fixed or capable of permanent description. It is a shifting complex of behavior that includes changes in both patient and physician. The patient does not necessarily continue in his misperceptions of the psychiatrist, and his speed of correcting them furnishes an important point of prognostic value. During their further contacts psychiatrist and patient have the opportunity to correct their initial and frequently false categorizations of each other. If first impressions repel, they may discover—with the ever fresh pleasure this brings—that each is, after all, rather a pleasant person once one gets to know him a little. More often first impressions attract, because each shows socially conventional behavior. In a new situation our behavior at first tends to conform to the social roles we believe the situation assigns to us. Afterward, closer acquaintance may bring to the fore traits at first concealed. For with growing intimacy there emerge various patterns of behavior learned in the less uniformly structured experiences of the family. Thus it happens that after a time the psychiatrist does something or fails to do something that frustrates one of the patient's expectations of him, or he may offend the patient in many such ways. These events he must also study carefully.

The usual initial positive attraction of psychiatrist and patient for each other is largely sustained by their fantasies of what each can expect from the other. When the fantasies yield to closer inspection, and when at the same time intimate behavior begins to replace more formal behavior, the relationship may weaken. At this point one factor alone saves most physician-patient relationships from dissolving. In the time taken for the patient's irrational expectations of him to collapse, the psychiatrist has a chance to show, one should not say to display, his real professional competence. Then, as the patient learns that the psychiatrist is not what he first thought him to be—perhaps a doting mother or an eternally patient father—he may discover that as a helpful physician the psychiatrist can now contribute even more than the mother or father. This transition from a tenuous relationship based on fantasy to a firm one based on an experience of competence demands that the psychiatrist offer the patient something considerably more than he can find in ordinary social intercourse. The following sections of this chapter offer suggestions concerning the content of this "something."

The importance of the physician-patient relationship in influencing what the patient will tell the psychiatrist and what the psychiatrist should tell the patient requires that the psychiatrist constantly evaluate this relationship. He should note how readily the patient talks and all other behavior of the patient toward him. Psychiatrists notice minutiae of social conduct—for example, punctuality, hesitancy in smoking, deference in going through doors—that would and should be overlooked or not noticed at all in other situations. But in an interview psychiatrists should observe all these items of behavior as clues to the attitudes that such behavior expresses. The psychiatrist should also help the patient to use any opportunity that arises to state what he thinks of the psychiatrist. In initial interviews most patients cannot achieve much candor in such comments. The psychiatrist can usually expect conventional formulas. But often, and even in guarded remarks, the patient may say something revealing and relevant. In drawing out the patient's thoughts about ourselves, if we press the patient artificially we will usually only increase his conformity to socially acceptable platitudes. Natural opportunities will arise, however, that we can exploit. If the patient has referred himself or chosen the psychiatrist from among several of whom he has heard, we can ask him, "Why did you select me to consult?" If he generalizes about physicians or psychiatrists, we can say, "Are you including me in that?" I shall discuss later the special value and importance of discussing the patient's thoughts about the psychiatrist whenever the patient seems to become unusually anxious.

The Optimal Attitude and Behavior of the Psychiatrist

We should often ask ourselves in what ways we can be of more use to our patients than even their best friends can be. The difference may lie principally in the *degree* to which we show a friend's helpful qualities and, most of all, in the tenacity and patience that permit us (and the best of friends) to sustain a relatively stable relationship with another person over a long period of time. In addition, four other qualities for the psychiatrist—interest, acceptance, detachment, and flexibility—are recommended.

For his task the psychiatrist certainly requires interest in the patient and in his difficulties. This interest can include to a degree the biologist's curiosity about the wonders of living organisms, yet we cannot allow ourselves to become so preoccupied with the details of morbid anatomy and physiology that we lose interest in the whole patient. Our specialty particularly concerns itself with the responses of the whole man. Our interest should be in the patient and for the patient; it should not pursue, disguised as diagnostic fervor, our own special predilections and curiosities. We rarely can entirely prevent these from interfering with our guidance of interviews, but we can at least strive to become aware of the ways in which our interest in ourselves may mingle harmfully with our interest in our patients. The interest we show in patients should include, and chiefly derive from, an attempt to understand them. Our limited success in this task may matter less than our efforts to try and to improve. We know that a fumbling medical student may learn much from a patient in a psychiatric interview. At present there is so little difference in skill between the worst and the best of us that we must rank the wish to understand as hardly less important than any understanding we achieve. At any rate patients respond well to both. Finally our interest should always include attention to the assets of the patient as well as to his deficiencies and difficulties. To this aspect of our interest patients also respond favorably, and with it we may help them to tell us more freely about their sufferings.

The psychiatrist should next try to reach a capacity for complete acceptance of his patients. Our profession does not ask that we approve all that our patients do or abandon our own ethical principles in favor of moral relativism. But we do improve our skill when we can accept patients unreservedly, regardless of what they may say or do that would be quite offensive in another context. Just how offensive people can be, the psychiatrist has a better chance than anyone to learn. But he also can learn more easily just how important it is to all of us to gain and hold the affection of others despite our sortcomings. Here we can often be of more help than the patient's family and friends. Because they frequently have become alienated by his behavior or their own, so that the patient believes himself to be without the friends we all need, we should have the deepest reservoirs of kindness.

If the psychiatrist does surpass the performance of family and friends in this regard, he often owes his success to the cultivation of a third quality required in his work. We may call it detachment, separating this sharply from the aloofness with which it has sometimes been confused. Because we live outside the circle of the patient's family and friends, we are not so closely—and hence so emotionally—involved in the patient's difficulties as they are. What the patient does cannot affect us so much. It should affect us somewhat, or we would not want to help him or be capable of doing so, but it must not affect us to the extent that the strength of our emotions disturbs our judgment of the patient in the manner that the strength of his emotions has disturbed his judgment. His anxiety prevents him from thinking clearly. He needs a less troubled mind to help him correct his misperceptions and faulty reasoning. Here again we can establish maxims more easily than we can follow them, and for this reason among many, psychiatrists should know themselves as well as they can.

Every internist taking a history and performing a physical examination finds that he omits less if he follows a routine order of procedure. The psychiatrist's study of his patient should be equally thorough and usually must be longer. But the psychiatrist cannot afford to impose a rigid form on his interviews and examinations. Although careful to think and ask about everything that might relate to the patient's symptoms or difficulties, he should not expect always to learn things in the same order. Nor should he expect ever to learn the same things in every interview, for different symptoms require different emphases in the discussions. Lack of space prohibits a review here of some of the common variations in interviewing that occur with, for example, patients who have depressions, hypochondriasis, schizophrenia, anxiety states, and psychophysiological reactions. For these variations alone, flexibility becomes another desirable attribute of a successful interviewer, but he also needs this quality especially to reduce the resistances within patients that often prevent their talking freely about many important topics. Some patients can talk easily about their wives but dare not discuss their parents. Others may pour out a cataract of information about their parents and close up like a bank vault when the psychiatrist inquires about their wives. Many varieties occur in such resistances, but the physician can nearly always count on finding some. Fortunately time helps the psychiatrist. Talking itself predisposes the patient to further talking. If the physician yields at first to the patient's reluctance to talk about certain subjects and lets him discuss others, he may thus prepare him eventually to return to the previously avoided material. This is not to say that the patient should be permitted to seize and retain control of the interviews. On the contrary, the physician should preserve guidance throughout and, if necessary, make his guidance explicit to the patient, but he should not use his skill and power to confront the patient prematurely with subjects that are seriously disturbing. This can trouble and even shatter the developing positive attraction of the patient for the physician. The flow of the patient's remarks is sometimes delicately balanced between the wish for help and the fear of injury at the hands of those to whom he gives his confidence. If he experiences painful emotions too much and too early, his expectation that he could be hurt may be confirmed (not unreasonably), even though the interviewer said nothing intended to hurt him. We all turn away from pain and often also from those with whom we associate the pain, even when they have tried to help us. And so the psychiatrist should let things come gently and naturally, perhaps learning this lesson from skillful obstetricians.

The Technique of Interviewing

Arrangements for the Interview

In any interview stimuli reach the patient not only from the physician but also from the entire setting in which it takes place. The psychiatrist will find worthwhile a study of the setting of his interviews even if, and perhaps especially if, he cannot change the setting. Privacy and reasonable comfort for the patient and the physician are absolutely essential. A separate room best assures complete privacy, but not if telephones ring and secretaries run in and out. A public ward, with its chatter and other hubbub, gives more privacy than a semiprivate room. Bright precinct-station lights should not blind the patient as he talks. The physician and patient should preferably sit so that each can look at the other without having to do so continuously if they prefer not. The psychiatrist should reserve enough time for a satisfactory interview. In the present state of our knowledge anyone who does not keep at least 45 minutes or an hour for an interview identifies himself as practicing some psychiatric formula that does not include listening to patients. Brief interviews may have their place in medicine, surgery, and even in certain authoritative and directive psychotherapies, but they have no relevance to diagnostic and therapeutic psychiatric interviewing at its best. (A later section will discuss reasons for this.) Moreover, one interview, even of the length suggested, rarely suffices for a thorough exploration of the patient's difficulties, and the psychiatrist will usually have to arrange for several further meetings.

The psychiatrist should always make notes during or after an interview. Apart from the value of having some record of the talk, the process of making notes passes the material through the mind of the psychiatrist again and thereby adds to his study of it. If he makes his notes during the interview, he should be reasonably certain that the note taking does not interfere with his own spontaneity. Some psychiatrists can pass this test, others cannot. And he should also be certain that the note taking does not trouble the patient. About this he should not necessarily expect to hear from his patients, many of whom will communicate their objections indirectly rather than with words.

In connection with notes and records I shall refer briefly to the use of questionnaires in eliciting a medical history.[3,4,20] Questionnaires can be filled out by the patient before an interview, perhaps in the waiting room. They often save time and they provide a valuable check for completeness of the survey of the patient's history and condition. They cannot, however, substitute for the interview, and this for at least two reasons. First, the psychiatrist cannot usually observe the patient's emotional responses as he fills out the questionnaire, and these provide important clues to the feelings and events of importance. (He may watch for signs of emotion as he discusses the questionnaire later, but this reduces the time saved, and in any case the strong emotions usually only come to expression during a fairly free conversation, not in response to questions.) Second, the psychiatric interview has other purposes than that of gathering information. It should provide the beginning of a trustful relationship in which psychiatrist and patient collaborate for the improvement of the patient's condition. Since questionnaires cannot replace interviews their main value at present lies in research and sometimes in supplementing the interview by assuring comprehensiveness of the topics covered.

Starting the Interview

As patient and psychiatrist meet, the initiative lies with the psychiatrist. He should introduce himself, lead the patient to his office, offer him a chair, and start the conversation. One can begin well enough with a brief introductory statement such as, "I know about you only the little that Dr. X told me. So it would be best for you to tell me in your own words what troubles you." After this the physician should usually remain silent until the patient's first responsive flow has dried up. He can soon tell whether the patient can talk freely or needs additional help. If the patient does need help the physician should give it promptly, not letting him bathe in the sweat of tense silences. Sometimes the patient does not know what he should give in the way of a history. Since psychiatrists do ask for kinds of information different from that required by internists and surgeons, the patient may simply need a little guidance. Sometimes the patient's anxiety mounts so high that it blocks his free expression. In that case the physician can channel the conversation into something less painful to the patient. Often he can reduce the patient's anxiety by asking questions that free the patient of the fear that he will say too much and of the responsibility for giving emphasis to important topics. Later the patient may relax enough to talk freely. If such measures fail, often the psychiatrist should ask the patient about his anxiety and should suggest possible origins of it in order to encourage further expression. He can say, for example, "You seem frightened. Can you tell me what makes you so?" If the patient still blocks, the physician can suggest, "Perhaps you are afraid of how I will react to the things you may want to tell me about. Is that so?" The patient may then respond by verbalizing the origins of his immediate anxiety and can then continue with other parts of the interview.

Once the patient has begun to talk, the physician's task consists in helping him to talk freely and in guiding him to speak about the most relevant topics. These will be discussed separately, although in an interview they naturally intermingle.

How to Help the Patient Talk Freely

If the physician has a strong interest in his patients, he can influence most of them to talk freely, because everyone talks better to an interested listener than to a bored and reluctant one. The awareness of the psychiatrist's interest reinforces the patient's wish to talk and his conviction that the psychiatrist merits his confidences. It may often be difficult to listen without interrupting. The psychiatrist's other medical training frequently impels him to intrude a question about a date or place so that he is sure to know all the data. Or something the patient says may infect him a little with the patient's anxiety or depression. Then he can quite unconsciously deflect the patient from such sensitive topics (for him more than for the patient perhaps)by asking the patient about something else. Each little interruption in itself may seem trivial, and usually is, but each adds to a cumulative effect on the patient that tells him, "The doctor wants something from me. What is it? How can I tell him what he wants to know?" When patients become occupied in gving us the information they think we want, they can easily forget to tell us what they want and need to say, of which we as yet know nothing. Every time we let the patient talk as he wishes, we encourage him to say something else that, perhaps up until that moment, he thought he ought never to confide in anyone.

In addition to deflecting the patient's line of thought, the interruptions by the psychiatrist also tell the patient more about the psychiatrist. There are advantages to the patient's knowing rather little about the psychiatrist; the less he knows, the less he can censor what he says in accordance with the assumed attitudes of the psychiatrist. This may make for a freer revelation of the patient himself.

Should the psychiatrist then always say and do nothing as the patient talks? Certainly not. He should say and do whatever becomes necessary to sustain the patient's flow and to guide it. Silence may suffice, or it may not. Sooner or later some further responses become necessary or additionally helpful. In offering these the psychiatrist may move from silence toward levels of increasing activity, each designed to emphasize to the patient a little more strongly his wish to hear more. Thus grunts of "uh-uh" and leaning forward expectantly stimulate the patient a little more, or sometimes much more, than silence. If such gestures prove inadequate, the psychiatrist can questioningly repeat the last word or phrase of something the patient has said. After this come gentle urgings such as, "What happened then?" "Go ahead," and "I'd like to hear some more about that." Should these fail, and assuming that the patient knows in general what he should talk about, his anxiety toward the psychiatrist has probably interfered too greatly. The psychiatrist should then bring this into the discussion directly, help the patient verbalize it, and, if necessary, apply appropriate reassurance. Thus he can begin by saying, "something makes it hard for you to talk to me about this matter. Can you tell me what it is?" Often the patient will respond satisfactorily to such leads. If not, the psychiatrist should suggest possible misperceptions of him by the patient, such as those mentioned above in connection with reducing initial anxiety.

He can say, for example, "Perhaps you are afraid of what I will think of you?" If all such efforts to loosen the patient's tongue fail, the psychiatrist's task usually includes pointing out to the patient his share of responsibility for their difficulties in talking. The psychiatrist might say, for example, "We have to work together on this, I'm sure you know. It's a collaboration between us, and I can do little for you unless you can tell me more about yourself." At this point the psychiatrist may learn of the patient's distrust about the privacy of his communications. On this matter and other similar doubts, the psychiatrist should provide firm reassurance based on actual performance. He should not, for example, assure the patient that what he learns from the patient goes no further and then schedule an interview with the patient's parents without the patient's knowledge.

With this repertoire of techniques increasing serially in stimulating the patient to talk, when should the psychiatrist use his influence? I believe he usually needs to increase his activity in the following circumstances: to show his interest, to reduce the patient's anxiety, to encourage the patient's emotional expression, to control garrulity and irrelevance, and to channel the interview toward topics of the greatest importance. I will defer discussion of the last two of these to a section on guiding the interview, but the first three pertain to helping the patient talk freely.

Some psychiatrists have more interest in their patients than they show. I think young psychiatrists are especially liable to make this error when they mistakenly apply in initial interviews the silence that is conventional and sometimes helpful in certain psychotherapeutic techniques. In attempting to stay out of the patient's way, a psychiatrist may say so little as to give the patient the impression he is mute. Patients have been known to leave some psychiatrists after one or two interviews because they do not understand these psychiatrists' unresponsiveness and become alienated by it. Most patients have already received training by internists and surgeons in the question-and-answer method of history-taking. They may misinterpret excessive silence on the part of the psychiatrist as simply incompetence. Moveover, previously important persons have often communicated aloofness, indifference, disapproval of, or even anger toward the patient by means of silence. The patient may confuse the psychiatrist with these persons, and if so the interview can perish, or it can become unnecessarily uncomfortable for the patient as well as less productive, since anxiety interferes with thinking and with expression. It makes sense, therefore, for the psychiatrist to remain silent if he can and needs to do no more, but also to offer freely whatever signs of interest the patient seems to require. He can easily insert such additional communications of interest often enough with nods of the head, with "Uh-uhs," or with simple words such as, "Surely," "Naturally," "Of course," and "I see." Words matter less than attitudes. With a friendly attitude we will find the right words, expressing them in a gentle speech and with a kind face. The psychiatrist should also offer, from time to time, more explicit signs of his understanding of what the patient did or felt with remarks such as, "I can see how hard that must have been for you," or "That must have made you feel better." Remarks of this kind should articulate what the patient has rather clearly expressed and should not influence him to agree, against his own knowledge, with the psychiatrist's interpretation of events. When the psychiatrist does not understand what a particular experience meant for the patient, he should usually inquire further, but when he does understand, if he will occasionally echo what the patient says he can lubricate the interview.

Experience will teach the psychiatrist the level of anxiety proper with each patient for a flowing interview. When a patient's anxiety becomes too great, the physician should try to reduce it by some of the techniques mentioned earlier. As already mentioned, excessive anxiety during an interview usually derives from misperceptions of the psychiatrist as being more menacing than he is. Anxiety felt by the patient with regard to other persons drives him to talk, while anxiety felt toward the psychiatrist blocks his talking. The psychiatrist should generally try to reduce or keep minimal the patient's anxiety toward him in initial interviews. Certainly he should note it and may subsequently wish to allow its full exposure, but if the patient becomes very anxious with regard to the psychiatrist before a strong attachment has developed, he may block harmfully or fail to return. Since the patient nearly always hungers for the psychiatrist's approval, his anxiety toward the psychiatrist can often be easily reduced by encouraging and praiseful remarks with regard to the patient's exposition of his difficulties. For example, the psychiatrist can say, at a moment when the patient hesitates and looks inquiringly at him, "Go ahead, you're doing very well. Keep going the way you were."

Yet we need to remember also that anxiety can run too low in an interview. Physician and patient can unwittingly exclude the patient's anxiety from expression and agree that he is much better than he (or a referring physician) thought he was. This comes about when the human wish to reduce human suffering urges the psychiatrist to offer reassurance prematurely. In doing this the psychiatrist deprives himself of the opportunity of tracing the patient's anxiety to its specific origins. For example, suppose a patient says, "Doctor, I think I am going crazy." To this the psychiatrist can immediately reply, "Oh, no you're not. You don't have the symptoms." More useful remarks would be either, "What do you mean by 'crazy'?" or "What makes you think you are going crazy?" To such questions the patient may then answer with details of his anxious thoughts. It then turns out, perhaps, that he thought he was going crazy because his memory has faltered recently and an aunt who died in a mental hospital also complained of this at one time. Further inquiries remind the patient that she was, as a matter of fact, an aunt by marriage. To such specific details the psychiatrist can then provide specific reassurance. The best reassurance comes from understanding and explanation. Patients can usually distinguish reassurance based on careful inquiry and explanation from shallow statements to the effect that "everything is going to be all right." Their ability to penetrate our weaknesses in this respect provides another reason for avoiding premature reassurance. Such reassurance can seal off further exposures of the patient's anxiety. He may think to himself, "Why should I tell my troubles to someone who minimizes them all as my family does?" Moreover, premature reassurance, when the patient does accept it, tends to promote the patient's excessive dependence on the psychiatrist. If we say, "Everything is going to be all right" (and there may be times and places when we should), we should realize that we have thereby accepted responsibility for their being so. When we insist that the patient join us in a careful exploration of his symptoms and difficulties, we communicate firmly to him our expectation that he will also share responsibility for his getting well.

We can control the amount of anxiety in the patient rather well by changes in the amount of talking we do. As the patient talks more and the psychiatrist less, the patient's anxiety tends to increase, at least initially, although after catharsis it may decrease again. As much as possible the psychiatrist should talk to modify the patient's anxiety, not his own. To do this he needs to remember that patients often tolerate silences rather well and frequently use them to think before speaking. A patient occupied in telling his story may not even notice silences, and sometimes does not seem even to notice the interviewer. But if a patient uses a silence to delete some repellent thoughts, he usually becomes aware of the silence, and his anxiety mounts. Then, if the psychiatrist has not prematurely spoken in order to ease his own tension, the patient will speak to reduce it in himself.

A common dissimilarity between the interviews of interested amateurs, such as sensitive internists, and experienced psychiatrists exists in the differing extents to which they permit, encourage, and facilitate the expression of their patient's emotions. This being so, we may ask why we psychiatrists encourage the free expression of emotions. We do it first because, as I mentioned earlier, emotions give importance to an experience and at the same time communicate that importance to other people. They should also communicate its importance to the person himself. And this they do when the emotions become strong enough. But often patients have not expressed themselves freely to other people. Consequently the related emotions may recede somewhat, and the patient may think himself untroubled by them. Talking brings the emotions to the surface, and if they become strong enough the patient may be astonished by the extent to which he has been affected. Patients frequently comment on this with remarks such as, "I never cry when I think about these things at home, but when I come here and talk I seem to cry all the time." This illustrates Sir Charles Sherrington's comment that in motor activity talking lies midway between thinking and acting. And it brings us to an additional reason for encouraging the patient's expression of emotions—the therapeutic benefit to him. Although this subject properly belongs to therapy rather than to diagnostic interviewing, the psychiatrist can remind himself that initial interviews begin therapy by observing the simultaneous benefit for both diagnosis and therapy of the patient's freely expressing strong emotions. Moreover, the relief usually experienced by the patient cements his attachment to the psychiatrist and makes the patient eager to talk more at the next interview.

This does not always happen. Sometimes patients recoil in anger or guilt when they find they have talked too freely and shown some emotion they previously condemned and imagined they could not experience. A patient may resent the psychiatrist's hearing him criticize his parents perhaps for the first time, or seeing him cry, or eliciting the confession of some wickedness. One cannot easily predict which patients will react in this way. Fortunately the best safeguards lie within the patients, for those who are most likely to be hurt by too rapid a release of emotions are those who are most inhibited in the first interviews. They will require several or many interviews before they talk freely. But the psychiatrist should still observe the patient's reaction to the interview itself and notice whether the patient shows concern about the things he says and the emotions he displays. Within the patient's tolerance the psychiatrist should encourage the patient to express his emotions fully. Weak emotions, like mild pain, are often of doubtful significance, but strong emotions tell both psychiatrist and patient alike that they are working in relevant subjects.

Some of the chief techniques for encouraging the patient's expression of emotions have already been mentioned. The physician should sustain and show his interest over at least forty-five minutes or an hour. In brief interviews the patient rarely has time to overcome his almost invariable initial reserve. In a ten-minute interview discussion of the weather may take five; in a fifty-minute interview one can give five to the weather and still do much besides. In addition, emotions cut grooves for related thoughts of the same theme, which, in turn, bring stronger emotions to the surface. The longer one talks about a particular subject, the more emotion accompanies the evoked thoughts. Fully developed emotions usually occur only in longer interviews, because shorter ones do not permit this self-fueling of emotions to occur.

Beyond the requirements of showing interest and allowing plenty of time, the physician can further increase the patient's emotional expression by careful attention to some additional technical points. These are emphasizing detail in the patient's narration, reinforcing the patient's emotion by communicating understanding of his feelings, and naming the experienced emotion.

When we tell others about a past experience, we partially relive the events we tell and partially experience again the emotions we then had. The extent to which we feel again the old emotions depends upon the vividness of reliving. Simple, uneducated people easily slip into a present-tense style of narration in which they seem almost completely to relive what they describe. More educated and more controlled patients, on the other hand, tend to talk in the past tense or to confine themselves to general statements. The psychiatrist should press the patient to provide specific examples of what he says. For example, if the patient says, "My father was always mean to me," the psychiatrist should ask, "Do you remember that? What do you remember?" He should frequently ask, "Can you give me an example of that?" or "Such as what?" Questions of this type oblige the patient to focus on specific events and, at least partially, to relive them. Moreover, the discipline of documenting general statements contributes to the patient's understanding of his own misperceptions. Once the patient has begun to tell about an incident, the psychiatrist can easily heighten the portrayal of detail by interjecting questions that ask for further details such as, "What happened then?" "What did your father say to that?" and "What did you do after you left the house?" After a little guidance of this kind the patient will continue to give detail on his own, partly because he knows what the psychiatrist wants and partly because he begins to experience the relief of catharsis, which usually only comes with vivid retelling.

Remembering the influence of the audience on any speaker, the psychiatrist can increase the patient's emotional expression by showing understanding of his emotions and attitudes in the events narrated. This does not need to include or imply an endorsement of the patient's behavior; rather it implies an awareness that what he then did was natural for him at the time. Remarks (offered in a questioning way) such as, "So you felt no one was on your side," and "At that point you thought your father was trying to control you," can tell the patient that he at last has someone to talk to who can understand him, and so he will want to talk more.

Patients frequently come close to the expression of strong emotions without quite permitting themselves to reach it spontaneously. Frequently fears of the psychiatrist's reaction to strong emotions inhibit them. When a tear moistens the patient's eye, the psychiatrist can profitably tell the patient he has noticed the emotion with a remark such as, "I can see it makes you sad to talk about this." Such a statement says to the patient, as it were, "It's all right to cry here. Go ahead." And frequently such little remarks will help the patient to cry or experience other strong emotions. The psychiatrist gains nothing if he runs too far ahead of the patient in using this technique. Many patients have great difficulty in acknowledging and showing anger. If the psychiatrist too rapidly confronts such a patient with a name such as "anger" or "rage" for these emotions, the patient may shrink back in horrified denial that he could house such feelings within himself. In that case, however, the psychiatrist does not need to retreat all the way. If he finds himself ahead of the patient and encounters denial, he could still say, "Well, of course, I could be mistaken, but I think nearly everyone in your situation would have been annoyed at what happened to you." This provides the patient with a hint of the acceptability of some anger that he may later wish to use.

Although I have emphasized the importance of the patient's talking freely, the psychiatrist should retain general control of the interview. Free talking does not mean unlimited free association. The right of the patient to say what he wants does not convey also the right to babble on tediously about irrelevant matters. The psychiatrist has the privilege and even the duty of curtailing circumstantiality and garrulity. But before he does so he should first ask himself (and perhaps the patient) why the patient behaves in this way. There are many reasons, and it is worth finding out which applies. Sometimes the irrelevant chatter results from a long-standing inability to think clearly, a form of mental deficiency. Sometimes it indicates failure of memory, with the patient substituting an appearance of remembering details for accuracy of recall. Sometimes the patient talks about something else in order to postpone talking about some more affecting topic, or to conceal it altogether. This commonly happens in the description of hypochondriacal complaints in which the patient, by focusing the attention of himself and everyone else on his heart or stomach, withdraws it from his marriage or disastrous financial predicament. Sometimes with such excessive talk the patient tries to communicate covertly something that he thinks about himself but cannot or dare not articulare explicitly, or of which he may even be unaware. The patient who offers unnecessary detail may never have thought that his affairs seem less important to other people than to himself. Or when a patient recounts details of his previous illnesses and operations in uninvited detail, he may really want us to know in this way how much he has suffered and needs our sympathy.

Before cutting off the patient, or while cutting him off, the psychiatrist should usually inquire about the excessive talk. He can say, for example, "I notice you spend a lot of time telling me about your past illnesses. I can see that they are important to you, but I don't think I understand why. Can you tell me how they are important to you at this time?" If such inquiries prove futile to stem the flow of the patient's irrelevancies, the psychiatrist can then move gently, but if need be also firmly, to deflect the patient. He can say, for example, "Perhaps later we can come back to what you are talking about. But since our time is limited, I wish you would tell me about so and so." This brings us to the various techniques for channeling the interview toward significant topics.

Guiding the Interviewer Toward Significant Topics

As in his encouragement of the patient's talking freely, the psychiatrist should guide the interview covertly when possible and only secondarily with more open directions. Often he can use the devices mentioned previously for showing greater interest in a topic of special importance that the patient only mentions. Thus he can channel the patient into another topic without the patient's becoming aware of his influence. But the psychiatrist should be aware of it. He should know that he is guiding the patient, and for a definite reason.

All psychiatrists should study carefully reports of experiments that have shown the profound influence on other persons of systematic utterances (by an experimenter) of such simple sounds as "Uh-huh." Such interjections have been found to influence the number of plural words spoken by a subject told to say all the words that come to his mind. As the experimenter gives an "Uh-huh" after each plural word, the subject, even without any awareness of being influenced, tends to increase the number of plural words he says.[15] An even greater effect occurs when the subject judges that the experimenter means to communicate approval by his "Uh-huh."[22] Similar experiments have shown that such interjected "Uh-huhs" can influence subjects to give more emotional responses during an interview[27] and to vary the types of memories recalled.[26] Now a patient always knows, or thinks, that the psychiatrist wants something, and he usually wants to satisfy the psychiatrist much more than experimental subjects want to satisfy psychologists. Consequently, if the psychiatrist interjects his "Uh-huhs" unconsciously and pursues his special interests in sex, religion, money, or something else, the patient will almost certainly go along with him. Both may find the hour enjoyable, but it may be unrewarding because of a one-sided emphasis on their favorite topic.

If the psychiatrist's subtlest signs of increased interest do not guide the patient to talk more about some significant object, then he may direct the patient more openly. He should exploit as much as possible the associations and references already provided by the patient. For example, suppose the patient says, "My headaches are getting worse every day, and my wife says she can't stand it much longer." The psychiatrist can catch the patient's reference to his wife and inquire, "What does your wife say about your headaches?" This broaches the subject of the patient's marriage, and other inquiries and information naturally follow. Sometimes the psychiatrist should not interrupt the patient in order to pursue an association at that time. This can interfere with the patient's flow toward something equally important. But the psychiatrist can make a mental note of the patient's remark and return to it later. He can say, for example, "You mentioned five minutes ago that your wife couldn't stand your headaches. Will you tell me some more about that?" By using the patient's own references and associations, an experienced psychiatrist can sometimes conduct an entire and thorough history-taking interview without ever himself introducing a new topic. Since the patient seems always to be elaborating further on what he himself first brought out, he cannot reasonably believe that the psychiatrist has forced him to talk of things he did not mention.

Even with the most skillful use of indirect techniques, the psychiatrist will at times have to ask questions, bring up new topics, or inquire directly for further details. Although, as already mentioned, the psychiatrist should usually defer questions about dates, places, and details of events omitted in the patient's initial story, eventually he should ask for whatever facts he believes necessary to satisfy the requirements of a thorough history.

Before he does so, however, he should remember that questions frequently introduce errors into histories. It has been shown that spontaneously given accounts of events include fewer errors than accounts elicited with interrogation.[18,28] This occurs for the simple reason that most people cannot bear to say "I don't remember" or "I don't know," and they are therefore inclined to answer questions with some information even when they are unsure of its accuracy. Patients, who are eager to obtain help and who often imagine that they must qualify for this help by pleasing the interviewer, have a special vulnerability to this tendency.

When the interviewer does ask questions, his attention to careful phrasing of them proves rewarding. Slight differences in wording can greatly influence the patient and his responses. If our "Uh-huhs" can tell the patient what we want to hear, our explicit questions provide a much more forceful and sometimes harmful guidance to the patient. The questions asked should provide the fewest possible clues to the answers expected and the least possible channeling of the answers. Most desirable are "open" questions that ask about a topic in general and to which the patient must reply with one of several sentences. The least desirable questions are leading questions to which the patient can answer "Yes" or "No" and then remain silent. Compare, for example, the differing values of asking the patient, "Do you and your wife quarrel often?" and "Tell me about your marriage." The first question, apart from its abruptness, which can offend, may evoke a simple "No" from the patient and nothing else, unless irritation. The second question invites and almost obliges the patient to reply with a sentence or more. Moreover, it does not confine the patient in his reply to the present time. The psychiatrist can learn much from noting what the patient selects to talk about first in answer to such a question. To illustrate this important principle further, an exercise for a psychiatrist who wishes to improve his technique is to arrange opposite each other in a list closed questions and more valuable open ones. For example, one can ask a patient "Was the pain severe?" but a better question would be, "What was your illness like?" "Did you miss your daughter when she married?" will yield less than, "How did you feel when your daughter married?" We can ask, "Do you have a bad temper?" but we can improve on this by saying instead, "How is your temper?" I do not mean to proscribe all leading questions focused sharply on a specific point, but these should come after more general open questions have given the patient an opportunity to answer freely without the suggestions and guidance of leading questions.

In asking questions that broach new topics, tact and timing reward the interviewer for the extra care they require. Careful phrasing of questions can greatly improve their yield. For example, in talking to an unemployed patient one should avoid asking, "Have you been on welfare often?" Instead, one can say more usefully "Have you had much trouble finding work?" Or, to illustrate further, one can unnecessarily offend a patient by asking, "Have you quit many jobs?" The patient would give the same and more information if asked, "What has led to your various changes of jobs?"

The state of the physician-patient relationship should influence our timing of questions and opening of topics to which the patient may be sensitive. As the patient and psychiatrist become more attached to each other, the patient feels freer to disclose more of himself, and the psychiatrist feels freer to ask him to do so. We can ask questions in the last five minutes of an interview that we could not ask in the first five, and we can ask questions in the fifth interview that would have been inappropriate in the first.

We can make many questions less painful by embedding them, as it were, in a matrix of other questions to which the patient is less sensitive. Thus one can lead a woman patient fairly easily to talk about sexual intercourse if one inquires about this at the end of a series of questions on pregnancies. In asking about further pregnancies the physician may naturally inquire whether the patient's sexual relations have been satisfactory and, if not, why not. Similarly one can ask questions about impairment of memory right after asking about the effects of the patient's illness on his vital functions such as sleep and appetite. With the question placed in this context, the patient is much less likely to believe that the psychiatrist thinks he is "crazy" than if a question about memory confronts him abruptly as a new topic.

Do not, however, confuse tact with timidity. The psychiatrist should never hesitate, out of feeling for the sensitivity of the patient, to ask a question that is necessary for thorough evaluation. In talking to a depressed patient, for example, the psychiatrist should discover whether the patient has had suicidal thoughts and the likelihood of his acting on these thoughts. Often he can learn about such thoughts indirectly, but when he cannot, then he should pose questions directly. A question firmly asked will usually elicit a more direct answer than one offered hesitantly.

The psychiatrist should try to avoid offering gratuitous comments and interpretations that can trip the patient as he tries to tell his story. Instead, he should try to offer simple questions that, while asking for more information, encourage the patient to talk further. For example, suppose the patient says, "I feel I need affection and can't get it." One might respond to this with, "Well, we all need affection, and you're not alone in this." A much more useful response would be, "What interferes with your getting affection?" This second comment reassures the patient that he needs affection, but it also inquires further about what he himself may do to deprive himself of it. Or again a patient may say, "I'm afraid I may lose control of myself." To this one could reply with, "Would that be bad?" but an even better response would be, "What do you think would happen if you did?" Or as a final illustration, suppose a patient says, "I was afraid of my parents as a child." The psychiatrist could answer reassuringly, "Yes, many children are afraid of their parents." A more productive answer, however, would be, "What about them made you afraid?"

Ending Interviews

When patients do express emotions freely, we should give them some warning of the end of an interview before it closes. This permits the patient to regain some calmness before leaving the office. About five minutes in advance one can say something like, "I can see that all this is extremely important to you, and we need to talk about it some more. But our time for today will soon be up, and we will have to postpone the rest."

I find it helpful always to ask the patient at the end of diagnostic interviews if he has anything further he would like to bring out or has any questions he would like to ask. In these final moments patients frequently reveal some matter of great importance to them. Previously anxiety prevented their reaching these subjects, but as they see the interview closing, they often decide to risk the exposure. Usually time does not then permit a full discussion, but the psychiatrist can defer this until the next interview.

Much of the best work of interviews occurs after psychiatrist and patient have separated. The patient (and a good psychiatrist also) goes on thinking about the subjects of the interview. New associations and often new emotions come to the surface and provide additional material at the next interview. The psychiatrist can usefully ask patients on parting to think further about the things discussed and to note these additional thoughts. Such instructions often stimulate patients who have shown marked resistance to psychological explorations. In the interview itself their great anxiety frequently prevents their talking or even thinking freely, and they often present defensive and obviously incorrect denials of symptoms and attitudes for which abundant evidence exists in other signs. After the interview and away from the psychiatrist, many of these patients relax and then begin to think constructively about the topics discussed. At the same time the image of the psychiatrist becomes less awesome. After a few hours or days of rumination the patient may eagerly welcome a second interview and may talk much more freely.

At the end of any initial interview the psychiatrist should discuss with the patient plans for further interviews or for treatment. Often the patient will press him for an immediate diagnostic opinion. The psychiatrist may then have to explain that he will need further interviews and perhaps other examinations and tests before offering an evaluation of the patient's illness. He can usually include some initial reassurance covering what he knows up to that point. He should avoid blanket reassurance that he may afterward have to revise, and he should avoid offering prematurely a diagnostic opinion or recommendations for treatment. But always he should tell the patient what he plans to do next. Attention to such details of courtesy and cooperation greatly aids the transition from initial and diagnostic interviews to treatment.

Few single interviews sufficiently reveal the patient's difficulties for the purposes of the thorough evaluation that sound practice requires. Not many healthy people can pass from strangership to intimacy with another person in any hour, or even in several. So we should not expect this of anxious or otherwise troubled patients. Therefore, we must turn to additional interviews and to additional informants, both of which are nearly always desirable. With the patient's consent (rare exceptions to this occurring in the cases of irrational, psychotic patients or young children) we should interview important relatives of the patient, so that we may benefit from their often quite different perceptions of the patient and his illness. The discrepancies between the patient's account of himself and that of a relative frequently astonish us and also show us how differently people appear to different observers. Our psychiatric interviews can improve if we frequently remind ourselves of their significant limitations in giving us the information we need.

Bibliography

1. BARTLETT, F., *Remembering: A Study in Experimental and Social Psychology*, Cambridge University Press, Cambridge, 1954.
2. BIRD, B., *Talking with Patients*, Lippincott, Philadelphia, 1955.
3. BRODMAN, K., ERDMANN, A. J., LORGE, I., and WOLFF, H. G., "The Cornell Medical Index: An Adjunct to Medical Interview," *J.A.M.A.*, *140*:530–534, 1949.
4. ———, "The Cornell Medical Index-Health Questionnaire: II. As a Diagnostic Instrument," *J.A.M.A.*, *145*:152–157, 1951.
5. DEUTSCH, F., "The Associative Anamnesis," *Psychoanal. Quart.*, 8:354–381, 1939.
6. ———, and MURPHY, W. F., *The Clinical Interview*, Vols. 1 & 2, International Universities Press, New York, 1955.
7. DIMASCIO, A., BOYD, R. W., GREENBLATT, M., and SOLOMON, H. V., "The Psychiatric Interview: A Sociophysiologic Study," *Dis. Nerv. System*, *16*:4–9, 1955.
8. DUNBAR, F., "Psychosomatic Histories and Techniques of Examination," *Am. J. Psychiat.*, 95:1277–1305, 1939.
9. FINESINGER, J. E., "Psychiatric Interviewing: 1. Some Principles and Procedures in Insight Therapy," *Am. J. Psychiat.*, *105*:187–195, 1948.
10. FROMM-REICHMANN, F., *Principles of Intensive Psychotherapy*, Ch. 5, "The Initial Interview," University of Chicago Press, Chicago, 1950.
11. GARRETT, A. M., *Interviewing: Its Principles and Methods*, Family Welfare Association of America, New York, 1942.
12. GILL, M., NEWMAN, R., and REDLICH, F. C., *The Initial Interview in Psychiatric Practice*, International Universities Press, New York, 1954.
13. GLIEDMAN, L. H., GANTT, W. H., and TEITELBAUM, H. A., "Some Implications of Conditional Reflex Studies for Placebo Research," *Am. J. Psychiat.*, *113*:1103–1107, 1957.
14. GOLDMAN-EISLER, F., "Individual Differences between Interviewers and Their Effect on Interviewees' Conversational Behavior," *J. Ment. Sc.*, 98:660–671, 1952.
15. GREENSPOON, J., "The Reinforcing Effect of Two Spoken Sounds on the Frequency of Two Responses," *Am. J. Psychol.*, 68:409–416, 1955.
16. HAGGARD, E. A., BREKSTAD, A., and SKARD, A. G., "On the Reliability of the Anamnestic Interview," *J. Abnorm. Soc. Psychol.*, *61*:311–318, 1960.
17. HENDRICKSON, W. J., COFFER, R. H., JR., and CROSS, T. N., "The Initial Interview," *A.M.A. Arch. Neurol. & Psychiat.*, *71*:24–30, 1954.
18. HUNTER, I. M. L., *Memory: Facts and Fallacies*, Penguin Books, Harmondsworth, England, 1957.
19. IMBODEN, J. B., "Brunswick's Theory of Perception: A Note on Its Applicability to Normal and Neurotic Personality Functioning," *A.M.A. Arch. Neurol. & Psychiat.*, 77:187–192, 1957.
20. KANNER, I. F., "Programmed Medical History-Taking with or without Computer," *J.A.M.A.*, *207*:317–321, 1969.
21. LEVY, D. M., "Modifications of the Psychiatric Interview," *Am. J. Psychother.*, *18*: 435–451, 1964.
22. MANDLER, G., and KAPLAN, W. K., "Subjec-

tive Evaluation and Reinforcing Effect of a Verbal Stimulus," *Science*, *124*:582–583, 1956.

23. MEARES, A., *The Medical Interview*, Charles C Thomas, Springfield, Ill., 1958.
24. MENNINGER, K. A. (with MAYMAN, M., and PRUYSER, P. W.), *A Manual for Psychiatric Case Study*, 2nd ed., Grune & Stratton, New York, 1962.
25. PYLES, M. L., STOLZ, H. R., and MACFARLANE, J. W., "The Accuracy of Mothers' Reports on Birth and Developmental Data," *Child Development*, *6*:165–176, 1935.
26. QUAY, H., "The Effect of Verbal Reinforcement on the Recall of Early Memories," *J. Abnorm. & Soc. Psychol.*, *59*:254–257, 1959.
27. SALZINGER, K., and PISONI, S., "Reinforcement of Verbal Affect Responses of Normal Subjects during the Interview," *J. Abnorm. & Soc. Psychol.*, *60*:127–130, 1960.
28. STERN, W., *Allgemaine Psychologie auf personalistischer Grundlage*, Martinus Nijhoff, Haag, 1935 (Eng. Trans. by Spoerl, H. D.), *General Psychology from the Personalistic Standpoint*, Macmillan, New York, 1938.
29. STEVENSON, I., *The Diagnostic Interview*, 2nd ed., Harper & Row, New York, 1971.
30. SULLIVAN, H. S., *The Psychiatric Interview*, Norton, New York, 1954.
31. TOBIN, S. S., and ETIGSON, E., "Effect of Stress on Earliest Memory," *Arch. Gen. Psychiat.*, *19*:435–444, 1968.
32. WHITEHORN, J. C., "Guide to Interviewing and Clinical Personality Study," *Arch. Neurol. & Psychiat.*, *52*:197–216, 1944.

CHAPTER 54

THE PSYCHIATRIC EXAMINATION

Ian Stevenson and William M. Sheppe, Jr.

LIKE THE PSYCHIATRIC INTERVIEW, the psychiatric examination has changed considerably in the past 60 years. Once largely restricted to the examination of severe mental disorganizations, it has now become an extensive study of the whole personality, with special emphasis on thought contents and their accompanying emotions. Moreover, psychiatric examinations are no longer restricted to patients with severe mental illnesses. Although naturally varying in emphasis and detail with different patients, some psychiatric examination should now form part of every careful medical examination.

Introduction

More specifically the psychiatric examination may uncover significant data under any or all of the following headings:

1. Signs of psychological disturbances that are expressions of the patient's major illness, for example, organic brain syndromes or schizophrenic reactions.
2. Psychological factors that are (partially) causative of a major physical illness or that exacerbate it, for example, anxiety in a patient with peptic ulcer, essential hypertension, or angina pectoris.
3. Psychological reactions to the presence of another illness, for example, anxiety about his condition experienced by a patient with congestive heart failure.
4. Psychological factors that interfere with the patient's cooperation in treatment, for example, a denial of illness or a resentment of authority as represented by the physician.

In addition to studying psychopathology, the psychiatric examination should also encompass the individual characteristics and assets of the patient. Unfortunately the preoccupation of physicians with ill health sometimes leads them to neglect the healthy aspects of their patients. Yet the outcome of an illness is influenced fully as much by the patient's assets and strengths as by the nature or apparent severity of his symptoms or illness. Without some awareness of these assets the physician cannot expect to promote healing by his efforts.

Although modern psychiatric theory emphasizes the adaptations that patients make to life stresses, it is possible to take a harmfully one-sided view of the patient's symptoms by considering them exclusively as defenses. Such narrowness can lead to such absurdities as the explanation of a schizophrenic psychosis as an "escape" from difficulties in living. The schizophrenic patient becomes psychotic because the stress to which he is subjected disorganizes his mental machinery. With this there may come some relief, but it is certainly secondary. Similarly a young man about to go to college may develop an incapacitating dermatitis a few days before his departure. The illness prevents him from attending college, which he did not want to do. But he did not become ill in order to avoid going to college; rather his dermatitis was an accompaniment and an effect of the anxious thoughts connected with going to college. Many symptoms have adaptive or defensive value, others do not.

Therefore, psychiatric symptoms may be: (1) disorganizations or disorders directly produced by a lesion or acquired during a stress; (2) efforts to compensate for these disorders; or (3) efforts to counteract the lesion or stress.

The psychiatrist should remember that his examination may not elicit adequately representative data. This point deserves brief discussion. In a manner quite impossible for the physical state to do, the psychological state can vary widely with changes in the external environment. Two physicians may obtain totally different evaluations of the same patient's mental state. The patient may have different feelings and attitudes toward each; he responds differently to the two different examiners, and consequently they observe different data in their examinations. For example, if the patient's attitude toward one physician carries unpleasant tension, this tension may impair his mental efficiency. A patient of greater than average intelligence may thus appear intellectually subnormal when his mental efficiency is momentarily so reduced. Another physician, evoking a different response from the patient, might conclude that the patient was highly intelligent. These discrepancies can occur even when the patient consciously wishes to cooperate with both physicians. Although the patient's attitude toward the physician depends partly on the approach of the physician himself, it is also influenced by the distorted perceptions of the physician by the patient, who may misperceive him as resembling some similar past significant person more than he actually does. In addition, the physician usually is a stranger and, in this instance, a person with some authority to make important decisions relative to the patient. Any observer of human behavior, by his mere presence, may modify the very behavior that he would like to observe.

A difference in the evaluation of the mental state may arise not only from differing attitudes of the patient toward different observers; equally the patient's emotional state may be changed by some other person or event, the effect of which may linger for many hours or days and may still be affecting him at the time of the examination.

Furthermore, the patient's anxiety may be aroused by the examination itself, altogether apart from the thoughts he has about the examiner. Some of this anxiety may be rational; for example, he may know that the physician may make important recommendations about him (such as admission to a hospital) as a result of the examination. Or the anxiety may derive from quite irrational ideas about the examination; for example, the patient may believe that the physician has been hired by the FBI to trick him. Whatever its origin, anxiety about the examination can markedly alter the patient's mental status as observed during it.

The occurrence of important variations leading to lack of reliability in psychiatric examinations is not just a surmise. A number of tests of the reliability of psychiatric examinations have shown clearly that psychiatrists do not yet have anything like as fine a tool as they need in their current procedures of psychiatric examination. An awareness of these limitations may help to reduce them, and we shall therefore briefly review some investigations of the reliability of the psychiatric examination.

In considering reliability we must first acknowledge very different incidences of the diagnostic categories assigned to supposedly comparable groups of patients in different countries and even in different American hospitals. Although the reported incidence of schizophrenia in North America is two to three times the incidence reported for Europe, there is little evidence for believing that the true incidence rates would differ so widely if judged by the same physicians utilizing the same criteria. Studying the records of New York State psychiatric hospitals, Hoch and Rachlin[11] found marked variations in the assignment of the diagnosis "manic-depressive psychosis" over a period of some years. They also reported marked differences in the incidence of this diagnosis between New York and California State hospitals over the same period. Differences of the magnitude reported were almost certainly not due to differences in the incidence of "real disease" over the period studied or in these two states. It is much more probable that the difference derived from variations in methods of examination and in diagnostic criteria.

The observations of Temerlin[35] indicate rather clearly that clinicians' diagnoses may be markedly influenced by suggestion and the theoretical bias of the group or professional milieu in which they practice. Other studies suggest that psychiatrists and other mental health professionals tend to perceive members of different social classes quite differently and hence to observe and emphasize different features of behavior in patients coming from social classes different from their own.[12,27]

When different examiners conduct independent interviews, one source of unreliability lies in the different ways in which the examiners conduct the interview. Two equally skilled examiners may elicit markedly different data from separate interviews with the same patient. Even when two examiners conduct a joint examination of the same patient, their individual perceptions and formulations of the identical data may lead to considerable disagreement. Ash's[1] studies of reliability of categorizations based on data derived from a conference method of examining patients are illustrative. Agreements for major diagnostic categories among paired psychiatrists ranged from 57.6 per cent to 67.4 per cent and, for specific diagnostic categories, from only 31.4 per cent to 43.5 per cent. These results suggest a rather low order of agreement.

If psychiatrists have cause for embarrassment in the rather low reliability of their examinations, they may derive some consolation from noting that equally large errors in history-taking for physical symptoms may occur. Cochrane, *et al.*,[3] observed that different physicians recorded markedly different incidences of pulmonary symptoms in coal miners interviewed under similar circumstances and presumed to have the same actual incidence of the symptoms that were inquired into. Serious discrepancies have even been found in the actual physical examination of patients and interpretations of laboratory examinations by different examiners.[6,7,8,10] Summaries of such errors by Johnson[13,14] and Kilpatrick[15] provide some solace in suggesting that we are not alone among the medical specialties in the impreciseness of our current diagnostic skills. Perhaps the honest recognition of the limited nature of our current knowledge and clinical skills is essential to the eventual development of a more precise science of human behavior. Progress may eventually come with the development of more uniform systems of examination[30,31,32,33] the use of similar check lists by different examiners, which will insure collection of more comparable data,[21,34] and the further development of partially quantitative rating scales for the assessment of subtle changes in human behavior.[18]

In the meantime, however, the psychiatrist should remember what we now know about the various causes of discrepancies between his examinations of the same patient at different times and between his examination at any time and that of a colleague at the same time or another time. He needs to remember that the patient's mental state at the time of an examination may be importantly influenced by: (1) alterations in the patient's psychological state induced by recent events; (2) the attitude of the patient toward the examining physician; (3) the attitude of the patient toward the examination itself; (4) the attitude of the examining physician toward the patient.

Thus discepancies in the observations of different physicians do not always reflect different percipiences in the physicians. They can arise from changes in the patient's psychological state between one examination and another. The physician should therefore make his first examination as complete as possible, but he should also supplement his observations with the evaluations of other individuals, for example, nurses, occupational therapists, aides, members of the patient's family, who have had an opportunity to observe the patient in varying settings. And he should himself make repeated observations of the patient's behavior, if possible, in different settings. Like the history of physical examination, the psychiatric examination should be a continuing process in which the physician is ever observant of new data.

The psychiatric examination, as well as the physical examination, requires evaluation by two standards. In some respects all healthy human beings resemble each other, and gross deviations from statistical normality may be safely labeled abnormal. In evaluating such deviations the physician relies upon a broad knowledge of human nature and human behavior in the culture in which he and the patient live. But he must also remember that in some respects everyone is unique. For these aspects the patient himself furnishes the standard of reference. The physician should therefore try to compare the patient's present condition with his premorbid level of function. Only in this way can he accurately evaluate the patient's mental state.

Sometimes the patient himself can reliably contribute information about his "base line." However, the physician rarely knows the patient's dependability in this regard, at least initially, and so should check the subject's statements with those of others who have known him closely. The patient's relatives and friends can often provide valuable information about the extent of deviation from his usual personality.

Before presenting an outline of the psychiatric examination we think it important to remind our readers that psychological states have a wholeness the existence of which we can obscure by our habit of examining part functions. We cannot avoid examining parts, but we can reduce the attendant errors by constantly remembering the interdependence of the parts on each other and the quality that is given to each part by its membership in the whole. For example, we sometimes differentiate intelligence from personality, as if these two words denoted exclusive functions. Actually the qualities included in the concept of intelligence, far from being separate, form a major part of personality. Personality may be defined as the sum of the habitual reactions of a person to external events. Perceptions, intellectual functions, and affective states all influence these reactions. We may usefully abstract for our discussions the processes of perception, integration, and response; in life, however, they merge inextricably. Almost any item of behavior may illustrate the instantaneous blending and the impossibility of really separating these processes. For example, suppose the patient angrily accuses the physician of not being sufficiently interested in him. He reveals his misperception of the physician as being similar to his neglectful father and dissimilar to his adoring mother. At the same time he shows his inadequate understanding of the physician's role and usefulness and of the physician's genuine interest on his behalf. Such understanding might have corrected, to some extent, his initial false perception. He also shows how his own comfort depends upon the interest and affection of other persons, and how threatened he feels when this is lessened. Finally his anger shows the pattern of his response to this anxiety and his lack of inhibition of the expression of his anger.

A mental examination reveals order or disorders of function. The physician does not examine a mental organ directly in the same manner that he can examine a physical organ, for example, when he auscultates the heart and lungs. His examination of the mental state cannot even be as direct as the examination of nervous tissue. In the latter case the nerves are not seen and rarely are even felt. Disturbances of their function are usually deduced from noting changes in the activity of other organs, for example, the muscles. Similarly in the mental examination the physician infers disturbances by noting abnormalities in the way the patient perceives, integrates, and responds to external events. He cannot observe the mental processes directly. He must deduce their condition from the patient's verbal statements and from observations of his other behavior.

Before describing the psychiatric examination itself we wish to emphasize the importance of placing it within a comprehensive examination of the whole patient. This should include a careful history of the patient's physical health together with a physical examination and all indicated laboratory tests. The interrelationships of psychiatric disorders and physical ones are often subtle and easily overlooked. Each type of disorder may mimic or conceal one of the other type. For example, an important percentage of patients diagnosed as psychoneurotic have been found later to have significant physical illnesses that accounted for their symptoms.[5,9,28] A large number of brain tumors[19,24,25,29] and other diseases of the brain[2,37] may present as "obvious" psychiatric syndromes, and their proper treatment may be overlooked in the absence of careful assessment of the patient's physical condition. The psychiatrist cannot count on the patient's leading him to the diagnosis of physical illness. Indeed, patients with psychiatric disorders often deny the presence of major physical illnesses that other persons would have complained about and sought treatment for much earlier.[38] In addition to the aid in actual diagnosis that physical examinations afford, they may also help in the assessment of the gravity and potential dangerousness of the patient's condition. In this connection electroencephalograms have proven increasingly helpful in distinguishing patients inclinced to express aggressive impulses violently.[26,36]

Outline of the Psychiatric Examination

We shall next present an outline of the psychiatric examination, arranged in headings for easier memorization. Then we shall discuss individual parts of the examination. We offer this outline without any conviction that it surpasses all others. The value of an outline lies chiefly in the prevention of omissions in the examination, and one outline may accomplish this as well as another. Certainly we should use no outline rigidly in our own thinking and even less in our examination of the patient. Each examination should differ from every other, according to the needs of the patient and physician and other circumstances. However, our advocacy of flexibility does not condone casual or incomplete examinations. As mentioned earlier, details and emphasis will vary from one case to another, but in every instance the psychiatrist should consider, at least in his own mind, each of the major headings of his outline. Most errors arise from omissions, relatively few from faulty observations. Sometimes circumstances prevent as adequate or as complete an examination as we wish or the patient's condition requires. When this occurs we should record the deficiency in our notes and never fail to remedy it as soon as conditions change.

The outline presented here has three main sections that correspond, so to speak, to the afferent, central, and efferent portions of a reflex arc. However, because the patient presents his emotions and behavior first to the observation of the examiner, we have placed these first. The physician can often complete this part of the examination during his interview with the patient. Through transitions that we shall mention later, he can then extend the interview into the more detailed examination of the other parts of the outline.

Outline of Psychiatric Examination

General Observations
- Circumstances and setting of examination
- General description of the patient

Emotions and Behavior
- Emotions
- Behavior, with special emphasis on potential for destructive behavior

Central Organizing Processes
- Intelligence, including memory
- Thought processes
- Thought contents, including self-concept and insight

Perceptions
- Misperceptions
- Illusions
- Hallucinations
- Attention
- Orientation

General Observations

In addition to the usual note about the date and place of the examination, the physician should also make some mention of the actual setting in which the examination was conducted. As previously noted, human behavior may differ markedly in different surroundings and under varying circumstances. Strange surroundings, or those that might be interpreted as threatening, may induce such severe anxiety as to obscure other important aspects of the patient's psychological state. The setting and circumstances of the examination therefore acquire their importance from the effect they might have on the patient. Any unusual reasons for making an examination—for example, at the request of a court—other than the patient's request for medical assistance should also be noted.

The physician should next attempt something akin to a novelist's sketch of the patient, using a photograph of words to convey a meaningful picture of the patient as a person. As Francis Peabody[23] once stated aptly, ". . . a 'clinical picture' is not just a photograph of a man sick in bed: it is an impressionistic painting of the patient surrounded by his home, his work, his relations, his friends, his joys, sorrows, hopes and fears" (p. 15). By way of illustration such a description may include the general appearance, demeanor, and clothing of the patient. A female patient who enters the office with hair unkempt, little or no make-up, no stockings, and badly scuffed shoes immediately communicates something about her attitude toward the social norms of dress. What this format means for her (for example, defiance of social customs, hostility to men, or lowered estimate of herself) will only emerge later, but its occurrence should be noted immediately.

To this general description the physician should also add a note concerning the patient's initial attitude toward the examiner, the interview, and the examination, his expectations about these, and his degree of cooperativeness.

Emotions and Behavior

EMOTIONS

If the psychiatric interview is sufficiently complete and ranges widely enough to permit full expression of the patient's emotions in connection with the topics of major significance to him, the emotional content can often be quite adequately evaluated during the course of the ordinary history-taking interview. Chapter 53 having discussed this extensively, we shall not here fully consider this aspect of the psychiatric examination.

Dominant Emotions. The physician should observe the dominant emotions of the patient, their intensity and duration, and their appropriateness to the patient's immediate situation and thought content. The physician should know whether a dominant emotion has recently arisen or is habitual, and hence more properly called a mood. Sometimes he may watch emotions come and go during an interview and thus gain information about the capacity of the patient to return to his habitual feeling state; at other times he may need to make special inquiries on this point. Thus he may obtain considerable information about the lability or fixity of the patient's emotional life. He should also learn about the sensations experienced during a given emotion. Some persons have only mild changes of sensations during emotions such as fear and anger; others may undergo what amounts to a severe physiological storm.

Events acquire meanings through the quality of our experiences. As different experiences give different meanings to outwardly similar events, we must always learn the meaning of events for the patient. When we observe a strong emotion in the patient, we know that we have reached something important for him, but why it is important we learn only from getting him to tell us about his thoughts and feelings. As the patient is frequently a poor witness of his own thoughts and feelings, we can often help him and ourselves to a better understanding of his emotions by learning what situations evoke the various emotions in him.

Since much human behavior and a great deal of pschopathology derive from efforts to control or suppress the expression of strong emotions, the study of the patient's various mechanisms for dealing with emotions forms a most important part of every psychological evaluation. Does the patient handle emotions such as anxiety in an integrative and constructive fashion? To what extent have the patient's efforts at controlling or concealing emotions succeeded, and to what extent have they failed? If they have failed, to what extent has the patient's mental organization become impaired?

To summarize and illustrate, if the patient is angry, the physician needs to know what has provoked his anger, why this angers him, how he feels, what he shows, and what he does when he is angry.

When the psychiatrist notices an important durable emotion, such as a depressed or elated mood, he should carefully evaluate the degree of mood change and its present or potential connection with irrational and perhaps dangerous behavior, for example, suicide in depressed patients or wild financial schemes in elated ones.

Appropriateness of Emotions. In evaluating the appropriateness of emotion to a given situation, the physician should remember the following modifying factors:

Cultural Differences in Emotional Expression. Wide variations in patterns of emotional expression occur. To the Scandinavian the Italian may appear wildly emotional, while the Italian may think the Scandinavian cold and unfeeling. The physician must therefore know the patient's customary or cultural pattern of emotional expression before evaluating its appropriateness.

Suppression and Concealment of Emotion. When the patient feels ashamed of his emotions or has other motives for their concealment, his emotional reactions to stressful situations may seem inappropriate.

Masking of Other Emotions by a Dominant Emotion. Any strong and lasting emotion can dominate the experience of the patient so as to exclude other emotions from expression in situations that might ordinarily evoke them. For example, a patient extremely anxious or depressed will not smile or cry when stimulated by events that would previously have made him do so. A general decline in all emotional expression, known as apathy, can similarly exclude ordinary emotional expression.

Displacement of Emotional Expression Toward an Inappropriate Person or Object. A person may inhibit the discharge of a strong emotion toward the person stimulating it and later discharge a similar emotion toward someone else who perhaps precipitates a slighter degree of this emotion but receives the full force of the accumulated earlier emotion. Misperceptions of other people may also lead to inappropriate expectations of them and hence to inappropriate emotional reactions. Such misperceptions probably account for most inappropriate emotional responses. Their frequency emphasizes the importance of observing the roles that the patient assigns to himself and to others. Significant clues to these perceptions may often be found in the patient's emotional reactions to the examining physician.

Inappropriateness (apparent or real) of emotion to thought content may arise in a number of ways.

Unawareness of Actual Thoughts. The psychiatrist may be unaware of the patient's actual thoughts. First, the patient may conceal his thoughts. Second, he may reveal them but the physician may not understand the special significance that they have for the patient. A patient with a schizophrenic reaction may smile or laugh as he mentions the death of his father; however, we may find that the father was a tyrant whose death brought joyful relief to the patient. Third, the patient's thoughts may move so quickly that the physician may observe the occurrence of an emotion but fail to detect its accompanying thought. Apparently inappropriate emotional expression also occurs in schizophrenia as a result of disrupted associations. In talking about something that is quite serious, the patient may be observed to smile or laugh in a manner quite inappropriate to the topic under discussion. Upon being questioned about this, however, the patient may state that an amusing thought entered his mind at the time he smiled. Thus again the emotion expressed in the smiling may have been quite appropriate for the thought. The abnormality lay in the intrusion of this thought at this time, a result of the disorder in association that is a characteristic feature of this condition.

Dissociation Between Thoughts and Emotions. The central abnormality here is not that meanings have changed but that thoughts no longer bring the usual accompanying emotions. This kind of dissociation can often be observed in dreams in which, for example, situations that would be quite alarming or quite ludicrous to the waking consciousness fail to evoke anxiety or laughter. When this condition is found in the waking state, it strongly suggests a schizophrenic disorder.

Neurophysiological Disorder in the Expression of Emotions. In certain brain diseases, such as pseudobulbar palsy, so many cortical neurons are injured that an interference occurs with the modulation of emotional expression. Accordingly, a slightly amusing thought brings a shower of laughter, and a sad thought may precipitate tears. The patient usually knows that his emotions are excessive and inappropriately expressive of his thoughts, but he cannot control these reactions.

BEHAVIOR

Emotion and behavior overlap and merge and are to a considerable degree interrelated. Behavior generally refers to action and expression with the whole body or major parts of it. Behavior may express the same motives as those that give rise to the conscious experience and expression of emotions. On the other hand, a given attitude may not enter conscious awareness as a felt emotion but, nevertheless, may reach expression indirectly in behavior. For example, if a wife slights her husband, he may not experience conscious anger, but if he forgets his wife's birthday the next day, we might be justified in attributing this forgetting to a hostile motive provoked by his wife's remark. Or the husband, still without showing any anger, may bring into play techniques for making his wife feel guilty. He thus retaliates not openly but by a much more subtle hurting of his wife.

In evaluating the expressive and communicative significance of motor activity, it should be noted that some motor activity is a concomitant of a central neural state. The communicative aspect of this kind of activity is incidental and only medical. It may tell the physician about a central disorder, but it is not created for this purpose. Other motor activity has communicative intention but is not symbolically expressive. And still other motor activity communicates symbolically to those who understand the symbols. Thus we may observe motor activity of the hand in a kind of gradient of communicativeness, as follows:

1. Tremors of the hands associated with organic diseases of the brain, for example, Parkinson's Disease. The symptom is a concomitant of a disturbance in the central nervous system, having only a medical communicative value.
2. Tremors of the hands associated with anxiety states. No communication is "intended," but some may result if the tremor is noticed by those who can interpret its significance.
3. Tapping of fingers on a table during partially inhibited anger. Here the motor activity is partly an accompaniment of heightened psychological tension but also may have, and be intended to have, communicative value to other persons.
4. Banging the fist against a table. In this act the patient communicates symbolically his anger and his desire to hit another person.

The physician should note the patient's general level of physical tension. Is his activity random and unorganized, or is it channeled into some constructive project? Is he aware of periods of hyperactivity or hypoactivity, and do these disturb him in any way? Is he prone to act out or express his frustrations and anxieties directly through various behavioral outlets? If so are these techniques successfully adaptive or self-defeating? Can he communicate readily with words, or does he communicate chiefly with nonverbal behavior?

The behavior of the patient with other people and the effect he has upon them are of major importance in making a comprehensive psychological evaluation. One should note, for example, those persons with whom he shows anxiety, those toward whom he feels superior, those of whom he stands in awe, and those to whom he comes for help. In a similar vein does the patient evoke pity, anxiety, anger, attention, friendliness, or aloofness in most of the people he encounters? If so is this a habitual, chronic pattern or one of recent development?

Any study of the patient's interpersonal relations should certainly not be limited to simple observations of his interaction with different persons. Although the patient's behavior with different people varies, he is a unique person with certain needs that dominate his behavior and give it a more or less consistent pattern. One basis for the concept of personality is that there are habits and patterns in behavior that give some predictability to each person's responses to his environment. Therefore, the physician should try to detect the dominant traits of the patient, or his recurring patterns of activity, as expressed in his behavior with different people. Such traits as ingratiation, aggressiveness, cautiousness, defensiveness, and irritability will occur as leitmotivs over and over again in the patient's behavior with many different people. The major generalizations that he has made from his past experiences thus become expressed in the recurrent traits of his behavior in relations with other persons.

Potentiality for Destructive Behavior. The patient's potential for destructive behavior, whether directed toward himself or others, should be considered and recorded. Although determinations of suicidal and homicidal risks are paramount, other more subtle forms of life-threatening behavior, such as chronic alcoholism in a patient with known liver damage, drug addiction, pathological risk-taking, and persistent accident proneness, should also be considered.

Since depressed patients have the highest incidence of suicide, it is appropriate to consider the detection of suicidal risk in connection with depression. However, since suicidal behavior is frequently encountered in many other diagnostic groups, such as chronic alcoholism and schizophrenia, a consideration of suicidal risk should not be limited to the depressive group. Helpful data for this evaluation may be derived from observation of the patient's nonverbal behavior, dominant emotion, and thought content. The examiner should also interrogate the patient directly, albeit gently, about any thoughts of suicide he may entertain.

The possibility that an occasional patient may constitute a risk to others should also be kept in mind. Individuals with clinical signs of organic brain damage and a history of episodic dyscontrol and violent rage constitute a special risk. Other patients showing highly organized delusions of persecution, relatively intact intellectual processes, high levels of energy and aggression, weak self-controls, and a history of responding to anxiety with overt aggression deserve special note. Unfortunately most patients who fall within this category are rarely referred for psychiatric help until they are involved in serious legal difficulty.

Potentials for destructive behavior, whether it be suicide or compulsive kleptomania, can only be evaluated and recorded in terms of probability. For purposes of concise recording, some clinicians[34] have recommended a three-point continuum or scale on which the behavioral potential is noted to be either absent or minimal, definitely present or a major risk. Cohen, *et. al.*,[4] have developed a 14-point check list for suicide risk that takes little time to fill out and assures that the main features of the patient's condition that increase the risk of suicide are all considered. Whether check lists and scales are actually employed in evaluating the potential for destructive behavior, it is essential that a systematic assessment of such potentiality be made in every patient whose condition suggests its presence.

Central Organizing Processes

Under central processes we include all those mental processes that occur between stimulus and response; that is, everything mental that contributes to the organization and utilization of experience. For purposes of study and discussion the processes may be divided into a number of topics, such as intelligence, thought processes, and thought content.

INTELLIGENCE

Intelligence is a term applied to a number of mental components and processes, including speed and accuracy of thinking, richness of thought content, capacity for complex thinking, and the ability to manipulate thoughts and objects. Standardized tests of intelligence administered by clinical psychologists can often provide useful information to the psychiatrist. Sometimes, however, poor cooperation on the part of the patient or other factors result in test results indicating an intelligence lower than that actually possessed by the patient. Even when he has the aid of a clinical psychologist and always when he has not, the psychiatrist should make his own appraisal of the patient's intelligence. In evaluating the patient's intelligence he should take into account the history and also the following factors:

Vocabulary and Range of Information. Both of these functions may be demonstrated during the course of history-taking interviews or other parts of the examination. Care should be taken to avoid confusion between intelligence and formal education.

Memory. The accuracy and extent of the patient's memory for the more remote past can usually be determined from his account of childhood and earlier life given in the history-taking interviews. Recent memory will usually be revealed in the patient's account of the events of the immediate past, such as the circumstances just prior to the examination. The importance of memory in behavior requires the careful examination of the patient with regard to any defect discovered. Do gaps of memory begin and end abruptly? Are they confined to one period or scattered over the whole span of life? Do they include unpleasant events only or pleasant and unpleasant ones indiscriminately? Is the defect of memory due to lack of attention to the environment or to poor retention or recall? Does the patient know he has a poor memory? Does he fill in gaps in memory with confabulations or compensate with garrulity?

Judgment. By judgment is meant a person's ability to use all the resources of his intelligence in solving constructively the problems that he meets. Although behavior supplies the only really reliable guide to judgment, some information about the patient's problem-solving techniques and the kind of decisions he makes may be derived by posing for him certain hypothetical dilemmas.

THOUGHT PROCESSES

Although the processes of thought are chiefly evaluated through the patient's speech, other indicators, such as drawings and special tests, may be employed. In general, one may assume a rather close correlation between the patient's speech and his thought content. Certainly speech expresses thought, but it often does so imperfectly. Some patients, such as those who are shy or uncooperative, may hide behind their few words an initially unsuspected clarity and richness of thought. And a smaller number of patients may acquire a verbal facility that successfully conceals blurred concepts and other limitations of thinking.

Speed of Thought. The speed of the thought processes communicates valuable information about the patient. Some people think faster than others all their lives. This greater speed of thought (often accompanied by greater speed of speech) may reflect superior intelligence rather than the mood of the person. The physician needs to know whether there has been a change in speed of thought and speech from what is habitual for the patient. If so this usually indicates a change in mood, since in elation thoughts and flow of speech are accelerated, and in depression they are retarded.

Besides changes in mood other circumstances may alter the speed of thoughts and speech. For example, in delirioid states and in schizophrenic disorders thoughts ordinarily held unconscious may invade consciousness. The patient may react to this increased imagery by attempts to describe it or to respond to it. The resulting increase in speed of speech may therefore indicate an increase in imagery rather than elation.

Accuracy and Clarity of Thought and the Capacity for Higher Forms of Thinking. As the patient talks, the physician should study the organization of his productions. Various degrees of disorganization may be noted. The patient may talk in grammatical, well-constructed sentences, but these may bear little relation to each other, giving a lack of coherence to the whole; or individual sentences may be poorly constructed, disorganized, or asyndetic. Words may be fragmented, or new words may be invented. Such neologizing may result in words such as "bastitute," "incarsterate," and "mental teleprosy," whose etymology and meaning are reasonably clear. The origin and meaning of other neologisms are often obscure, although they nearly always become intelligible through careful study of the patient and his past experiences.

The relation of individual sentences to each other may indicate some disturbance in associations. When associations are disturbed so that irrelevant thoughts intrude into consciousness, the disturbance is usually reflected in the speech. The interference may be manifest only as a blocking, so that the flow of speech becomes temporarily interrupted. This occurs commonly in states or moments of anxiety. The psychiatrist should note the topic about which the patient is talking at the time of blocking or hesitancy. The affect associated with the topic acts like sand in the mental machinery. Sometimes the intruding thoughts are themselves expressed, giving rise to more or less disorganization of speech. When every thought seems to be verbalized and all filtration of thoughts apparently suspended, the physician should suspect a psychotic disorder. A complete disinhibition of the expression of thought, that is, true free association, does not occur in a first interview or mental examination unless a severe mental disorder, for example, a dissociation, is present.

The examiner should note the ability of the patient to pursue a goal idea. When speech is accelerated, as in elation, thoughts often skip from one idea to another in what is appropriately called a flight of ideas. This rapid succession of ideas may result in the expression of no central theme whatever, so that although the patient may appear to have some goal idea in mind, he never reaches it. The physician should note the patient's ability to pursue a line of thought during distracting stimuli. Can he continue a theme after he has been interrupted?

The examiner should try to understand the reasons for a patient's inability to reach a goal idea. Is his stream of thought interrupted by intruding thoughts or by extraneous stimuli? If intruding thoughts interrupt him, are they repetitive, obsessional thoughts or the products of loosened association? Or does the patient have one dominant thought content that crowds out all others?

The meaning of circumstantiality, in which the patient floods the psychiatrist with irrelevant detail, also deserves attention. Has the patient never learned to focus his thoughts, or does this symptom arise as a response to some anxiety? The patient may tell his story in great detail to win the psychiatrist's approval for "his side," or he may do this to avoid talking about something he wishes to deny or conceal. He may give much detail to evoke sympathy for his suffering, or he may be trying to substitute a wealth of other detail for important gaps in memory. Only careful study can clarify the distinctions between these motivational factors and disorganization of thinking due to organic brain disease.

The examination of thought processes should include study of the patient's capacity for conceptual thinking. Under this heading the physician should consider and, if need be, test the patient's ability to perform abstractions.

Rigidity of Thought Processes. The human tendency to form habits affects not only behavior but thought itself, which is a foundation of behavior. Many serious difficulties in living occur because the patient cannot entertain alternative explanations of events or cannot consider alternative responses to them. Earlier associations and generalizations become hardened so that they cannot be broken down to make way for new associations or more appropriate generalizations. Under this heading one should, therefore, consider the extent to which the patient's mind is open to new ideas and thoughts. Is it closed to explanations and interpretations different from those he has worked out for himself? Or, on the other hand, is he abnormally suggestible and excessively liable to pick up the thoughts of others?

THOUGHT CONTENT

Central Themes. From observations during the history-taking interview the physician will usually have obtained much information about the main themes or central preoccupations of the patient's thoughts. However, judicious direction of the interview may be necessary in order to uncover central themes that are being avoided or unconsciously concealed. The examiner may be able to quantify somewhat the different values attached by the patient to different subjects by noting how frequently each comes up in the course of conversation. On the other hand, that which is obviously avoided or left unsaid frequently reveals as much about the patient as his overt verbalizations. Because affect is usually directly related to thoughts, central themes of thought should be correlated with the dominant affect.

Abnormalities of Thought Content. Under this heading one should consider such symptoms as fixed ideas, phobias, obsessional ideas, delusions, and excessively unrealistic fantasies. The degree of distortion of thought content and the tenacity with which the patient maintains his distortions should be noted. To what extent does the abnormality of thought content influence the patient's behavior? Are his abnormal thoughts of recent origin or habitual with him? Do they create further anxiety, or is he blandly content in his current situation?

Self-Concept. The set of thoughts that the patient entertains about himself and that en bloc constitutes his self-concept is of the highest importance. What does he think about himself? Does he exaggerate or minimize either his assets or his deficiencies? Is his self-concept congruent with the way other people see him? Does he feel that he has undergone change or is in some way not his "usual self"? Does his self-concept match his ideal of himself? How would he like to change?

Insight. Insight in this connection means the patient's understanding of his illness and the attendant circumstances. Some patients are unaware that they are ill; others feel sick without understanding what is going on; still others have some grasp of their illness but feel no responsibility for it, even blaming it altogether on others. The highest form of insight allows knowledge of how one's behavior affects others and implements constructive steps toward appropriate modifications. The physician should note the patient's feeling of responsibility for his situation, his awareness of how his behavior is affecting others, and his curiosity about his illness and its treatment.

Perceptions

Disorders of perception occur in a wide variety of forms and in various degrees of severity. The patient may incorrectly perceive either himself or the external environment. The physician should often inquire first about simple variations in the intensity of perceptions, for example, hyperacousia, myopia, anesthesias, hyperalgesia, and so forth. Then he should inquire about the more severe disorders of perception. These may range, in approximate order of severity, from misperceptions of other people, through illusions and hallucinations, to states of disorientation and confusion. Intermediate disorders of perception, such as the hallucinations found in schizophrenic psychoses, seem to arise from a combination of motivational influences and the disruptive effects of strong anxiety on mental processes. The more severe disorders of perception, such as confusion and disorientation, are usually noted in the presence of serious interference with cerebral function such as that encountered in toxic states or structural disease of the brain.

Misperceptions

Simple misperceptions of the mildest degree consist of mistaking the role of another person, that is, expecting different behavior from him than he can reasonably be expected to show. These errors are usually based on past experiences, but they are also influenced by strong affects, especially fear. An example would be the patient's expectation that the psychiatrist will always treat him with tenderness and indulgence, just as his mother did when he was a child. To note misperceptions the psychiatrist should study carefully the patient's comments on different people and his behavior with them. He should notice especially, since he can observe it most easily, the patient's behavior with the psychiatrist.

Illusions or Misidentifications

In the simple misperceptions mentioned above, the patient does not actually misidentify objects and people but only misunderstands their roles. In illusions perceptual misidentification of objects and people occurs. The patient then may mistake the psychiatrist for, say, his brother. The examiner should avoid confusing illusions and delusions. In a delusion, which is a false belief, the patient might think that the psychiatrist was his brother disguised skillfully to look like the psychiatrist.

Hallucinations

Hallucinations are sensory perceptions occurring in the absence of external stimuli. The following aspects of hallucinatory phenomena should be noted:

1. Sensory modality in which they occur.
2. Content of hallucinations, including resemblances of voices or visual appearances to those of persons known to the patient.
3. Circumstances associated with their occurrence or with fluctuations in their intensity.
4. The patient's insight into their occurrence and his reaction to them. Does he communicate about them freely or does he conceal their occurrence?

Attention

By attention is meant the patient's ability to focus his consciousness on a part of the field of awareness. It may be compared to focusing vision on a part of the visual field. The examiner should consider attention with regard to the following aspects:

1. General direction, that is, toward outside stimuli or toward inner thoughts.
2. Concentration. Can the patient's attention be diverted from its main focus to other stimuli? Is it too easily diverted so that he cannot concentrate? Sometimes we can distinguish active attention, that which the patient directs himself, and passive attention, that which occurs without his interest or willed participation. A combination of inwardly directed and fixed attention is a feature of dissociated states, in which the patient attends largely or only to his own stream of thoughts and responds less or not at all to outside stimuli. He is then said to be "out of contact" with the examiner. This can exist in all degrees and many variations.
3. Clarity of attention. Clarity of attention may become impaired as during toxic interferences with cerebral functioning. This is often referred to as a clouding of consciousness. Clouding of consciousness is not necessarily a feature of dissociated states. Clouding of consciousness can exist in all degrees from slight impairment to states of severe confusion merging into coma.

Orientation

In the more severe disorders of perception, orientation becomes impaired. However, orientation may be preserved in the presence of other severe disorders, such as hallucinations and dissociated states. Orientation is usually studied under the following headings:

1. Time—knowledge of the hour, day, date, month, season, and year;
2. Place—knowledge of one's location and spatial orientation.
3. Person—knowledge of one's identity.

Of the above three aspects of orientation the sense of time seems to be the most fragile and is usually the first to become disorganized as organic brain syndromes develop. Orientation for person is the most stable aspect of the three. It is worth noting also that the patient may give a correct statement about the place he is when he can no longer locate the place properly in relation to other areas. Thus he may answer in response to a question about orientation for place that he is in, say, the "University Hospital," but be unable to describe how he can go from the hospital to his home. At a later stage of disorientation the patient loses orientation for place as well as that for location. Sometimes, we also consider under orientation:

4. Situation—knowledge of why one is in a certain place.

This includes correctly identifying the people around one and their functions. Defects of orientation for situation are usually delusions. Sometimes the phrase "paranoid disorientation" is used for them. The patient, for example, knows that he is in a hospital ward but insists that it has secretly been changed into a prison and that the doctors are guards.

The Technique of the Psychiatric Examination

The Approach to the Examination

We cannot and should not sharply separate the psychiatric interview and psychiatric examination. The psychiatric examination begins the moment patient and psychiatrist meet. As the patient talks during the history-taking interview, the psychiatrist should attempt to evaluate his emotional state, his intelligence, his thought contents, or any other function of the patient's personality that may reveal itself. As the psychiatrist's skill in mental examination increases, he will find that he can conduct more and more of his examination during the history-taking interview, leaving ever smaller portions for some specific inquiries at the end of the interview or for another occasion. Thus the experienced examiner can often guide a history-taking interview so as to carry out a rather complete survey of the patient's mental state. On the other hand, errors can arise from careless reliance upon the history-taking interview as a substitute for a thorough examination, particularly in the hands of the inexperienced. Such deficiencies may occur if the psychiatrist fails to have the interview range over an area broad enough to provide all the data necessary for an adequate and comprehensive evaluation. The psychiatrist should not think that if he just keeps the patient talking, all the important data will simply fall into his ears. A great deal of essential material may be overlooked unless it is specifically sought, either by skillful direction of the interview or by specific and thoughtful questioning. In the course of such inquiry the examiner may use the data of an earlier part of the same interview or of another interview to guide his questions so that they will be most appropriate and most tactful.

Any change of emphasis in the interview toward more detailed or specific examination should be made as subtly as possible and preferably without the patient's even being aware of such a transition. The examiner who inadvertently makes the patient feel that he is being "tested" or interrogated as a witness may succeed only in confirming the patient's worst expectations about himself and the nature of psychiatric therapy. A clumsy approach may intensify the patient's anxiety so that he becomes unable to cooperate in the examination to any significant degree. Tactfulness in examination, apart from sparing the patient discomfort and additional anxiety, may therefore add considerably to the information actually obtained.

The transition to more detailed examination of the patient can sometimes be made smoothly by an appearance of investigating the patient's complaints more thoroughly. For example, the psychiatrist can say, "You mentioned a few minutes ago that your memory was not working as well as it used to. I would like to ask you a few questions so we can see just how much trouble you have been having." Abnormal thought contents such as delusions can often be studied ostensibly as part of a more detailed review of the patient's history.

However, some topics of examination do not lend themselves to such exploration, and these the examiner will nearly always need to present with a brief explanatory comment. His questions should also tactfully attribute maximal function to the patient. The following questions illustrate such introductions.

"I suppose you keep pretty good track of time, don't you?" The patient answers, and the psychiatrist continues, "Would you know what the date is today?"

"Do you read the newspapers and keep up with current events?" The patient answers, and the physician continues, "In that case I am sure you'd know who the President is. Can you tell me?"

"You said you went to the eighth grade, I think. Would you mind if I asked you a few questions in arithmetic? They are quite simple."

"You said your memory was pretty good. Do you mind if I see just how good it is?"

"You've told me a good deal about yourself. Now would you mind if I asked you some more specific questions about yourself and how your mind works?"

If the patient makes no objections to the inquiries, the psychiatrist need volunteer no further explanation. If, on the other hand, the patient seriously hesitates, becomes uncooperative, or asks what is going on, he is entitled to a reasonable explanation. The psychiatrist may explain that, as the patient has evidently had some psychological symptoms (reference being made in passing to these), additional information is needed to find out what can be done to help. The examination should never be presented to the patient as a "test" in which what he does may be compared to what someone else does. Such an analogy may evoke memories of failures in school and anxiety-laden thoughts of inferiority, and such feelings may reduce mental functioning further or destroy cooperation. Instead, the examination should be explained to the patient as an effort to compare what he does now with what he once did, and with what he is believed capable of doing when he is at his best. But it is altogether preferable if the examination can be accomplished with sufficient skill so that elaborate explanations are unnecessary.

The Importance of Not Humiliating or Tiring the Patient

Anyone can be chagrined by noting an inadequacy of function, especially a mental function, in relation to an assumed standard or to one's previous performance. The examiner should remember this especially with patients who are already somewhat aware of a loss of mental function as, for example, those with arteriosclerotic or other brain diseases. In such instances a few questions may quickly bring out some loss of intellectual capacity. The patient may well have managed to compensate partially for these defects by denial; if he is confronted with them through formal testing or insistent interrogation, however, the defensive denial may be torn from him with catastrophic results. Anxiety so aroused may impair function still more and can lead to further anxiety in a vicious cycle. Inquiries should therefore be pushed only far enough to reveal information needed for the evaluation and eventual well-being of the patient. As always the emphasis should be upon the welfare of the patient and not upon the completion of a routine medical form. Although the examination must be complete and eventually recorded in an organized fashion, the data derived from a skillfully conducted interview are seldom obtained in the final written form. Indeed, any attempt to adhere rigidly to a specific outline will often impair the initial relationship between examiner and patient to such a degree as to obviate the very purpose of the psychological examination.

For this reason we follow with interest rather than with approval recent developments in constructing a uniform psychiatric examination in which each patient is asked exactly the same questions and presumably in the same manner even by different examiners.[30,31,32,33] Such uniform interviews may have value for certain types of research, but they seem to us likely to interfere too much with the spontaneous demonstrations of interest and warmth (to the right amount!) that the psychiatrist should provide both to help the patient and to facilitate communication. In a different class, however, are check lists that the psychiatrist can fill out immediately *after* the examination.[21,34] These do not interfere with the flow of the interview or examination, and they are most helpful in reducing important omissions in the examination.

THE IMPORTANCE OF NONVERBAL BEHAVIOR

We wish to emphasize also that the words of the patient often form a relatively small portion of all the data used in evaluating the mental state. Observations of other forms of communication, such as posture, gestures, facial expressions, or other signs of emotions and thoughts, can, in many instances, provide even more significant information than the patient's overt verbalizations. The long-suffering hypochondriac's secret twinkle of enjoyment as she describes the effects of her many illnesses upon her family may supply an invaluable clue about the pathogenesis and meaning of her neurosis. The heavy-lidded, lusterless eye of the depressive frequently bespeaks more sorrow than his words could hope to encompass. The skillful examiner, like the good pediatrician, is exceptionally sensitive to that vast sphere of human interaction known as nonverbal communication. As in all forms of communication, however, specific components may be misleading, especially where their function is that of defense, denial, or masking. The smiling depressive who is secretly planning the "perfect" suicide may illustrate the latter all too tragically.

THE IMPORTANCE OF ASKING THE PATIENT ABOUT HIS CONDITION

In our zeal to observe the patient we can sometimes forget that he has the best opportunity of anyone to observe himself. We should try to enlist him as a collaborative observer. Certainly gaps and distortions will occur in his observations, as in ours, but if we omit to ask him directly what he has noticed, how he feels, and what he thinks, we lose some of the most valuable material we can obtain. The following questions may illustrate ways of inviting the patient's opinion of his condition: "What do you think about your condition?" "What would you say about your mood?" "How do you think your mind works now compared to when you were well (or compared to a year ago)?"

Special Points of Technique

EXAMINING EMOTIONS

Chapter 53 included a description of the signals whereby one person communicates emotions to another, which we will not repeat here. As mentioned above, the examiner should always invite the patient to describe his emotions directly. Defensive concealment of emotions may cause discrepancies between what the psychiatrist observes in other signs and what the patient himself reports. Sometimes the patient may not understand the psychiatrist's question. To the question, "How are your spirits?" the patient may reply, "Fine," not because he feels well, but because he does not understand what the word "spirits" means in this connection. Questions about emotions and moods may need to be asked in several different ways before the patient understands.

In making inquiries about depth of depression and suicidal intentions, it is well to begin with the thought content of the depressive mood and work gradually upward toward the possible suicidal thoughts. This gives the patient a chance to volunteer whatever he will about suicidal thoughts and may also produce less of a shock when the psychiatrist finally asks direct questions. The frequency and importance of suicide require us to ask questions bearing on it whenever its possibility arises, as it does in most depressed patients and in many others.

We suggest the following group of questions as one series that can move the inquiry gradually toward the subject of suicide:

"How low do you get when you are depressed?"
"What do you think about when you are depressed?"
"Do you ever think that life is not worth living?"
"Do you ever wish you could die?"
"Have you ever thought of killing yourself?"

"What did you think you would do?"
"What would happen after you died?"
"Why does that seem the only solution?"

In evaluating mood and suicidal intentions the patient's verbal comments in response to the above questions furnish only a part and sometimes a small part of all the relevant data. The patient's speech, his motor attitudes and movements, his appetite and sleep, and his other verbalizations, for example, of hypochondriacal ideas or ideas of hopelessness, all constitute important information for the evaluation of his mood.

Examining Central Processes

Examining Memory. Skillful guidance of the interview will often permit an accurate evaluation of both recent and remote memory. When it does not, more direct inquiries and examinations are indicated. These can begin with a question such as, "Have you had any trouble with your memory?" This gives the patient a chance to furnish further information. The psychiatrist may then continue with, "Do you mind if I ask you some questions to see just how good your memory is?" The patient assenting, specific questions that test the patient's remote and recent memory can be employed. For example, in testing remote memory the patient may be asked the years of his birth, of his graduation from school, of his marriage, and of the births of his children. The psychiatrist may then wish to refine his examination by inquiring about items less likely to be well remembered, that is, less important events in the patient's life, such as the year of a change of job, or events in the nation's history through which the patient has lived.

Recent memory can be tested by questions pointed toward current situations or events just past. For example, the patient may be asked what he did the day before, where he lives, what he ate for breakfast, or who has visited him. If the examiner thinks the patient's memory markedly impaired, he may ask the patient to tell him his (the physician's) name. Assuming they were introduced at the beginning of the interview, this may test the patient's memory over a period of half an hour or more. Additional tests of recent memory can be accomplished by (1) showing the patient some objects, for example, a key, comb, coin, and watch, and asking him half an hour later to say what these were; (2) telling the patient a simple story and asking him to repeat it with all remembered detail; or (3) speaking a series of random digits to the patient and asking him to repeat them forward and backward. A healthy person of average memory should be able to repeat accurately a series of eight to nine digits forward and six to seven digits backward.

Recognition nearly always exceeds recall in disorders of memory. Consequently in examining memory the physician may first test the patient's recall for a given item and then, if he fails, offer it to him in a series of other items. For example, the physician can first ask a patient who has visited him. If the patient says he thinks someone visited but he cannot remember whom, the physician can next inquire: "Was it your wife, sister, brother, aunt, cousin?" With such assistance from the physician the patient may remember the correct name.

Examining Thought Processes. Much of the data concerning the patient's thought processes will emerge during the history-taking interview as the psychiatrist observes the patient's behavior and studies his responses to specific questions. Notwithstanding these ample sources of information, the psychiatrist should often ask the patient himself what he thinks about his thought processes. Suitable questions might include the following: "Does your mind work well, or as well now as it ever has?" "Do you have any difficulty in controlling your thoughts?" "Can you concentrate your thoughts now?"

Where doubt remains after the above observations or where diagnosis requires a more refined understanding of thought processes, the psychiatrist should proceed to more detailed testing. Sometimes he can accomplish this with a few simple questions, but sometimes he will need more elaborate tests administered by himself or a clinical psychologist.

The ability to manipulate mathematical symbols may be tested by asking the patient simple arithmetical problems such as: "Do you remember what 12 times 12 is?" or "Would you subtract 8 from 56 for me?" The subtraction of 7 from 100 serially provides an excellent test of calculation. Upon completing this exercise the patient may be asked to divide 7 into 100 with a view to noting whether the result agrees with that of the test of subtraction and, if not, whether the patient notices this discrepancy. Thus one test measures attention, the speed of thought processes, and the ability to detect errors and correct them. Moreover, when the test is timed one patient's performance can be compared with that of another and also with his own performance on later occasions.

Sometimes calculations can be tested by asking the patient questions about his personal life that require him to calculate. For example, the psychiatrist may ask the patient naturally enough during the history-taking for the dates of various important events, such as the patient's marriage or the birth of his first child. Later in the interview he can ask the patient how old he was at the time of these events. This method has special value because it can be applied without the patient's being aware of any testing, which can hardly be the case when he is asked to subtract 7 from 100 serially. The subtler method can be used first, the more open method later if the first suggests a need for further examination of this mental function.

The psychiatrist will usually have had occasion during the earlier parts of the interview to evaluate the patient's ability for conceptual thinking. He should note whether the patient can form and express concepts clearly. When important impairments in concept formation are suspected, this capacity can be examined more definitively by testing the patient's ability to abstract.

Abstraction may be tested by asking the patient to compare and state the differences or similarities between different paired objects. Pairs commonly used are an orange and an apple, a tree and a fly, a dwarf and a child, a man and a dog, a lie and a mistake, a church and a theater. In offering such objects to test abstraction, the psychiatrist must know that the patient is familiar with them. Otherwise he may test only the patient's range of information. The patient may also be given a series of words such as *high, cold, dark, near*, and so forth, and asked to supply their opposites or their synonyms.

The ability to generalize may be evaluated by asking the patient to give the meaning of well-known proverbs, such as: "The proof of the pudding is in the eating" or "A rolling stone gathers no moss." To these proverbs the patient may reply with several different types of abnormal responses. In a correct response the patient should interpret the proverb as a generalization of human experience. A patient with a disorder of thinking may fail to make the transition from the concrete example of the proverb to its general application to human behavior. Sometimes the patient may "explain" the first proverb by another proverb without interpreting the first proverb, or he may become engrossed in thinking about and discussing the images of the proverb. Thus to the proverb, "A new broom sweeps clean," a patient may reply, "Well, that's right, when you get a new broom it will sweep cleaner." The patient may also see personal references in the proverb. For example, to the proverb, "It never rains but it pours," he may reply, "That's the way it is with me. Every time I go out in the rain it pours." The intelligence of the patient must always be considered in evaluating the response to such proverbs. A person of low intelligence or poor education, however, usually gives no answer at all to the proverb, whereas a schizophrenic patient is more apt to supply a concrete answer, one with personal reference, or one with bizarre associations mixed in.

Examining Thought Content. The psychiatrist should probe the patients' thought content indirectly, encouraging him to talk freely and to disclose his important thoughts without being questioned specifically. Should this maneuver fail, the examiner should not hesitate to ask direct questions about the patient's central preoccupying thoughts. He can say, for example, "Can you give me an idea of the main things you think about?" "What do you think about most?" "What sort of thoughts come into your mind?" "Do you have any thoughts that keep coming back to you?" "Do you have any thoughts you can't get rid of?" "Do you have any special daydreams or fantasies?"

Many patients do not understand the word "fantasy" and may even be unfamiliar with the expression "daydream"; therefore, the psychiatrist may find it helpful to ask the patient such questions as: "What things do you imagine yourself doing in the future?" "What would you ask for if you could have three wishes granted?" "Do you often imagine how you would like things to be in the future?"

If the patient has not exposed his thoughts freely in the course of the history-taking interviews, he may not do so any more readily in response to the direct questions suggested above; the latter certainly do not guarantee a frank answer. On the contrary, direct questions may merely intensify defensiveness. When the psychiatrist does ask direct questions, he will have to judge each time to what extent the patient has been able to expose his thoughts. The detailed examination of irrational ideas and delusions requires simply an extension of the principles already outlined. The psychiatrist must gradually evoke the patient's trust so that he can confide without fear. In doing so he must often travel the narrow path between endorsement of the patient's delusions and rejection of them as totally preposterous or "imaginary."

Note should be taken of the ease with which the patient talks about his irrational ideas or delusions. When a patient holds his beliefs critically and still tries to test them, he may be unwilling to confide them in a physician. One does not ordinarily communicate an unusual idea to another person unless he either believes strongly in its truth or feels assured of the other person's acceptance of him. Therefore, delusions that carry strong conviction are usually readily revealed. On the other hand, skillful inquiry may be needed to discover maturing but unripe delusions. Material suspected of delusional elaboration must be approached tactfully and with an assumption of the reasonableness of the patient. If a critical, challenging attitude is adopted, the patient may avoid the subject altogether or at least conceal the existence of any idea, such as an incipient delusion, that he thinks the psychiatrist may doubt.

Some psychiatrists avoid all interrogation of psychotic patients about delusions or hallucinations on the ground that such questions can harmfully crystallize ideas that the patient still holds in solution, as it were. We have already advocated an examination in which the patient is guided to deliver the necessary data in a conversational way, with as much avoidance as possible of direct questions. However, sometimes questions are helpful and essential to the clarification of the patient's symptoms. We believe the manner of questioning is much more important than the questions themselves in promoting the patient's delusions or otherwise injuring him. And certainly the psychiatrist should carefully avoid any suggestion of forcing the patient to talk of what he wants to keep to himself.

When the patient does confide delusional material to the psychiatrist, the latter should also ascertain to what extent he has revealed his thoughts to other people. The psychiatrist should know whether the patient considers the psychiatrist a special person and whether he has delusional ideas about the psychiatrist. Failure to note the latter may seriously impair the validity of the entire examination.

Patients often spontaneously verbalize expressions of apparent insight that have little relevance to a sound understanding or little application in their subsequent behavior. Consequently it is often helpful and necesary to ask direct questions focusing more sharply on insight. However, as with responses to other direct questions, the patient's answers will take their place with other data that may prove more important in evaluating insight. The physician may ask such questions as: "How do you think you became ill?" "What do you think is the matter?" "What brought all this on?" "What do you think are the causes of your illness?"

EXAMINING PERCEPTIONS

Examining Illusions and Hallucinations. When inquiring about illusions and hallucinations, the examiner should start with questions about experiences resembling those anyone can have. Because illusions and hallucinations are unusual experiences, the patient may have difficulty in recognizing what he has experienced or in describing it in everyday language. He may need to be offered a variety of questions before he understands what is being asked.

The following examples of introductory questions may prove useful:

Have you ever had something like a dream when you thought you were awake?

Have you ever had anything you might call a vision or seen an apparition?

Have you ever thought you heard people talking when no one was around?

Do you ever seem to confuse your thoughts and someone else's thoughts, as if someone might be putting things into your mind or even saying things to you?

When the patient answers such questions affirmatively, or when his behavior clearly indicates that he is hallucinating at the time of the examination, the psychiatrist should push the inquiry with further questions designed to learn as much as possible about such experiences. He can ask, for example: "What do you see (hear)?" "Who seems to be doing this (saying these things)?" "What is happening?" "Why are they doing this?"

The patient's complaints will usually include depersonalization and estrangement when these have occurred importantly in the illness. However, when he does not volunteer information about these symptoms that the psychiatrist nevertheless suspects from other data, they should be inquired about directly. One may ask, for example: "Have you ever felt as if you were not yourself, or as if you had changed so you couldn't recognize yourself?" "Did you ever think that everything around you seemed different or strange or unfamiliar?"

Examining Attention, Concentration, and Clarity of Consciousness. Important defects of concentration will usually appear in the interviews without special testing. However, it may be desirable to test the patient's ability to hold his attention on some task, and this can be done by asking him to subtract 7 serially from 100. If the patient cannot subtract he may be asked to repeat groups of digits of increasing length. Ability to subtract serial 7's from 100 provides a useful and semiquantitative index for comparing changes in the patient's condition from day to day as a delirium clears or becomes worse. Performance can be timed and notes made of errors, of whether the patient notices his mistakes and tries to correct them, and of any wandering of his attention from the task. If this test is used often, the patient's speed may improve through learning. Varying the test by asking the patient to start at 101 or 102 can reduce whatever distortion this might bring.

Examining Orientation. Much can be learned about the patient's orientation simply by careful attention to his remarks during the interview. If these do not provide adequate data, the physician should always turn toward more direct questions. Orientation may therefore be evaluated in several different ways:

1. Inferences made from listening to the patient's spontaneous references to people, places, and events during the earlier parts of the interview.
2. Questions designed to test orientation, but not asking directly about this. The patient may be asked where he was before coming to the office (or hospital), and whether he had difficulty finding the office. His reply may reveal whether he is oriented for place. Simple inquiries about recent events demonstrate whether he is oriented for time. Other questions may similarly test orientation for situation and person.
3. Direct questions about orientation. The psychiatrist should use these only after he has tried the other approaches, unless the patient seems obviously disoriented; in such cases he may proceed immediately to direct questioning. However, he should not hesitate to use direct questions when any doubt remains after use of the indirect approaches mentioned above.

The psychiatrist should usually begin with inquiries that suggest less extreme disorders and proceed from these to test for complete disorientation. For example, a first level of inquiry might include the following questions:

Do you keep track of time fairly well? Do you know what day (date, month, year) it is?

Where do you live? How do you get there from here?

Why did you come here?

A second and potentially more shocking level of questions might include the following:

Do you know where you are now? What place is this?

Do you know who I am? What is your name?

What time of day is it now?

A partially delirious patient may be disoriented without knowing this. Questions such as the above may alarm him by exposing the severity of his condition. This additional anxiety may actually increase his confusion. After such questions the psychiatrist should therefore always offer some appropriate reassurance. And if the patient is not oriented, the psychiatrist should attempt to furnish orientation to the patient. He should say who he (the psychiatrist) is, where the patient is, briefly why he is where he is, and what he may expect next, together with any other information that he thinks the patient can usefully assimilate in his condition.

Concluding the Examination

Even an initial or diagnostic interview must have some discernible value to the patient. The nuclear components of all therapy, the decrement of anxiety through a relationship of basic trust and the increment of self-esteem, are just as vital to the initial interview as they are to long-term treatment.

No matter how limited his remarks, the examiner should express, by word or gesture, some understanding of the patient's problems and his appreciation for his cooperation during the interview. If at all possible he should conclude the interview with some clarifying statement about the patient's immediate future. If the examination has been skillfully conducted, the opportunity for ventilation in an uncritical, nonjudgmental setting may be quite anxiety relieving and integrative in itself. Every effort should be made to increase the patient's sense of wholeness and dignity in an often insecure and troubled world. Trite generalities and superficial platitudes are often more alienating than therapeutic. Clarification and reassurance, medicine's most common aids, must always be realistic and based on an accurate, sincere evaluation of the facts, including those that the patient may well have overlooked. Good medical practice requires the active participation and intelligent cooperation of the patient; communicating and enlisting the latter are primarily responsibilities of the skilled examiner.

Appendix
Examination of Inaccessible Patients

Many psychiatric patients are inaccessible for effective verbal communication. This occurs when patients are mute (for example, in catatonia and severe depression), excessively garrulous (for example, in manic psychoses and some senile psychoses), or excited (for example, in schizophrenic excitements). The inaccessibility of different patients of this group varies greatly, and it also varies in the same patient from time to time. Indeed, few patients are totally verbally uncommunicative. The psychiatrist should try to discover under what circumstances the patient becomes communicative with him or with others. Certain persons or certain remarks of the psychiatrist himself may open up the patient's verbal communications. These successful stimuli should be carefully noted in order to utilize them in further developing contact with the patient.

While waiting for the availability of such stimuli, the psychiatrist may still learn much from observing a patient who does not talk usefully or talk at all. Such a patient may signal much about himself with the movements of his face and body or with his gestures or other behavior. A sudden grimace, a single glance, or a series of body movements may reveal more than a thousand words.

Facial Expression

Both the habitual expression of the face and changes in expression should be noted. The hangdog face of the depressive and the fatuous grin of the hebephrenic communicate immediately to the observant. The eyes especially deserve attention. The poets have referred to the eyes as "the doorway to the soul." In a more mundane but practical sense an awareness of ocular expression can be of immense assistance to the astute clinician. Patients whose life experiences have conditioned them to distrust verbal communication, for example, certain schizophrenic patients, may be quite sensitive and adept in this sphere.

Bodily Activity

Posture, gait, and other motor activity may be informative. The despairing slouch of the depressive and the proud, defensive posture of the paranoid vividly communicate important attitudes. The patient may employ such communicative signals as grunts, grimaces, or gestures with his hands. Motor expressions of hostility and anger often tell much about the thoughts of a mute patient. A seemingly disinterested catatonic may revealingly clench his fist or shift his position when his defensive isolation is challenged too aggressively. Seductive posturing and unconscious gestures may portray attitudes that could not otherwise be communicated.

Response to Stimulation

The patient's responses to environmental changes, verbal greetings, or direct physical stimulation should be noted and recorded. The examiner may also observe the patient's reactions when personal topics relative to his home or family are discussed within his hearing. Will he cooperate with such routine requests as suggestions that he open his eyes or change his position? Does he disregard noxious stimuli such as glaring sunlight or an obviously uncomfortable posture? Does he let his cigarette burn his hand? Many mute and seemingly unresponsive schizophrenic patients can be brought into activity by throwing them a ball and engaging in direct motor activity with them. Such stimulation apparently causes them to leave their inner world temporarily and enter into contact with others.

Other Behavior

The spontaneous behavior of the patient should be noted, especially with respect to his responses to persons who approach him. What does he do when he is left alone? Does he make any effort to communicate his needs? Does he accept cigarettes or food when offered? Is he incontinent? Does he feed himself? Does he preserve some concern for his appearance or neglect this altogether? The psychiatrist might note, for example, whether a mute female patient continues to comb her hair or use lipstick; such behavior usually indicates a degree of self-esteem and some continuing interest in others.

❨ Bibliography

1. ASH, P. "The Reliability of Psychiatric Diagnosis," *J. Abnorm. & Soc. Psychol., 44:* 272–276, 1949.
2. CHAMBERS, W. R., "Neurosurgical Conditions Masquerading as Psychiatric Diseases," *Am. J. Psychiat., 112:*387–389, 1955.
3. COCHRANE, A. L., CHAPMAN, P. J., and OLDHAM, P. D., "Observers' Errors in Taking Medical Histories," *Lancet, 1:*1007–1009, 1951.
4. COHEN, E., MOTTO, J. A., and SEIDEN, R. H., "An Instrument for Evaluating Suicide Potential: A Preliminary Study," *Am. J. Psychiat., 122:*886–891, 1966.

5. COMROE, B. I., "Follow-Up Study of 100 Patients Diagnosed as 'Neurosis,'" *J. Nerv. & Ment. Dis.*, *83*:679–684, 1936.
6. DAVIES, L. G., "Observer Variation in Reports on Electrocardiograms," *Brit. Heart J.*, *20*:153–161, 1958.
7. DERRYBERRY, M., "The Reliability of Medical Judgments on Malnutrition," *Public Health Report*, *53*:263–268, 1938.
8. FLETCHER, C. M., "The Clinical Diagnosis of Pulmonary Emphysema—An Experimental Study," *Proc. Roy. Soc. Med.*, *45*:577–584, 1952.
9. FRIESS, C., and NELSON, M. J., "Psychoneurotics Five Years Later," *Am. J. Med. Sc.*, *203*:539–558, 1942.
10. GARLAND, L. H., "The Problem of Observer Error," *Bull. N.Y. Acad. Med.*, *36*:570–584, 1960.
11. HOCH, P., and RACHLIN, H. L., "An Evaluation of Manic-Depressive Psychosis in the Light of Follow-Up Studies," *Am. J. Psychiat.*, *97*:831–843, 1941.
12. HOLLINGSHEAD, A. B., and REDLICH, F. C., *Social Class and Mental Illness*, John Wiley, New York, 1958.
13. JOHNSON, A. M. L., *The Anatomy of Judgment*, Basic Books, New York, 1960.
14. ———, "Observer Error. Its Bearing on Teaching," *Lancet*, *2*:422–424, 1955.
15. KILPATRICK, G. S., "Observer Error in Medicine," *J. Med. Educ.*, *38*:38–43, 1963.
16. KRETSCHMER, E., *A Textbook of Medical Psychology*, 2nd ed., Hogarth, London, 1952.
17. LEWIS, N. D. C., *Outlines of Psychiatric Examinations*, 3rd ed., New York State Department of Mental Hygiene, Albany, 1943.
18. LYERLY, S. B., and ABBOTT, P. S., *Handbook of Psychiatric Rating Scales (1950–1964)*, Public Health Service Publication No. 1495, National Institute of Mental Health, Bethesda, Md., 1966.
19. MCINTYRE, H. D., and MCINTYRE, A. P., "The Problem of Brain Tumor in Psychiatric Diagnosis," *Am. J. Psychiat.*, *98*:720–726, 1942.
20. MAYER-GROSS, W., SLATER, E., and ROTH, M., "Examination of the Psychiatric Patient," *Clinical Psychiatry*, 2nd ed., Ch. 2, Cassell, London, 1960.
21. MEIKLE, S., and GERRITSE, R., "A Comparison of Psychiatric Symptom Frequency under Narrative and Check List Conditions," *Am. J. Psychiat.*, *127*:379–382, 1970.
22. MENNINGER, K. A. (with MAYMAN, M., and PRUYSER, P. W.), "The Psychological Examination," *A Manual for Psychiatric Case Study*, 2nd ed., Ch. 3, Grune & Stratton, New York, 1962.
23. PEABODY, F., *The Care of the Patient*, Harvard University Press, Cambridge, 1927.
24. POOL, J. L., and CORRELL, J. W., "Psychiatric Symptoms Masking Brain Tumor," *J. Med. Soc. N.J.*, *55*:4–9, 1958.
25. RUBERT, S. L., and REMINGTON, F. B., "Why Patients with Brain Tumors Come to a Psychiatric Hospital: A 30-Year Survey," *Psychiat. Quart.*, *37*:253–263, 1963.
26. SAYED, Z. A., LEWIS, S. A., and BRITTAIN, R. P., "An Electroencephalographic and Psychiatric Study of Thirty-Two Insane Murderers," *Brit. J. Psychiat.*, *115*:1115–1124, 1969.
27. SHADER, R. I., BINSTOCK, W. A., OHLY, J. I., and SCOTT, D., "Biasing Factors in Diagnosis and Disposition," *Comprehensive Psychiat.*, *10*:81–89, 1969.
28. SLATER, E. T. O., and GLITHERO, E., "A Follow-Up of Patients Diagnosed as Suffering from 'Hysteria,'" *J. Psychosom. Res.*, *9*:9–13, 1965.
29. SONIAT, T. L. L., "Psychiatric Symptoms Associated with Intracranial Neoplasms," *Am. J. Psychiat.*, *108*:19–22, 1951.
30. SPITZER, R. L., "Immediately Available Record of Mental Status Exam: The Mental Status Schedule Inventory," *Arch. Gen. Psychiat.*, *13*:76–78, 1965.
31. SPITZER, R. L., FLEISS, J. L., BURDOCK, E. I., and HARDESTY, A. S., "The Mental Status Schedule: Rationale, Reliability, and Validity," *Comprehensive Psychiat.*, *5*:384–395, 1964.
32. SPITZER, R. L., FLEISS, J. L., ENDICOTT, J., and COHEN, J., "Mental Status Schedule: Properties of Factor-Analytically Derived Scales," *Arch. Gen. Psychiat.*, *16*:479–493, 1967.
33. SPITZER, R. L., FLEISS, J. L., KERNOHAN, W., LEE, J. C., and BALDWIN, I. T., "Mental Status Schedule: Comparing Kentucky and New York Schizophrenics," *Arch. Gen. Psychiat.*, *12*:448–455, 1965.
34. STEVENSON, I., *The Psychiatric Examination*, Little, Brown, Boston, 1969.

35. TEMERLIN, M. K., "Suggestion Effects in Psychiatric Diagnosis," *J. Nerv. & Ment. Dis.*, *147*:349–353, 1969.
36. TREFFERT, D. A., "The Psychiatric Patient with an EEG Temporal Lobe Focus," *Am. J. Psychiat.*, *120*:765–771, 1964.
37. WAGGONER, R. W., and BAGCHI, B. K., "Initial Masking of Organic Brain Changes by Psychic Symptoms," *Am. J. Psychiat.*, *110*: 904–910, 1954.
38. WHITEHORN, J. C., "Guide to Interviewing and Clinical Personality Study," *Arch. Neurol. & Psychiat.*, *52*:197–216, 1944.
39. WINGFIELD, R. T., "Psychiatric Symptoms That Signal Organic Disease," *Virginia Med. Monthly*, *94*:153–157, 1967.

CHAPTER 55

THE PSYCHOLOGICAL EXAMINATION

Alan K. Rosenwald

Introduction

Several paths could be taken in describing the nature of the psychological examination. One approach would emphasize its validity, reliability, standardization procedures, and other methodological issues drawn from the studies that involve the use of the major psychological tests in clinical settings. Indeed, it is essential for the psychological examiner to consider whether what he says about a given individual can be substantiated from sound scientific data or merely reflects his subjective fancy.* As one can see from the extensive bibliography cited, the literature on diagnostic and personality testing of the past ten years has been reviewed with the intent of giving the reader who wishes to go further an extensive sampling of such studies.† This chapter will make limited use of the bibliography.

Another approach that might be taken is the creation of a test manual. This would give brief elementary instructions on how to give intelligence, personality, and neuropsychological tests, how to record the results, and how to interpret their meanings.‡ This has a resemblance of rationality but, I feel, runs the risk of trying to provide the reader with skills in "Ten Easy Lessons" when much more is demanded. I do propose to describe some of the more widely used tests, illustrate the nature of the psychological examination by presenting psychological test data, and discuss

* *Research in Clinical Assessment*, edited by E. I. Megargee,[57] is illustrative of a number of volumes that deal persuasively with problems in validating clinical methods and describe approaches that may be made to such problems. It is beyond the scope of this chapter to consider in detail the multiplicity of procedures and concerns that characterize these studies.

† I am indebted to Mr. John Hamilton, graduate student from the University of Chicago, who spent many hours summarizing reports of studies done since 1960. The bibliography is intended to provide a representative sample of studies directly and indirectly related to an understanding of the psychological examination. In addition to the sampling of the last decade, basic references are also included.

‡ *Physician's Guide to Psychiatric Tests*, a monograph prepared by Schering in 1965, is such a manual.

how one derives meaning from such data and communicates the findings. In doing this I have assumed that my readers will largely be psychiatrists plus a smaller number of other mental health specialists who might wish to see what goes under the aegis of the psychological examination.

Referral Procedures

Let's start at the beginning when a person is referred for a psychological study. Referrals should be made when the referrer seeks an answer about the individual whom he is treating or evaluating that does not appear readily or easily obtainable from his clinical interaction with the patient. He feels that a separate opinion will benefit his pursuit of understanding. For example, the referrer seeks to confirm his tentative opinion that a patient is suffering from schizophrenia. He wants an independent opinion derived from a different set of data than that developed from his interview material and behavioral observations. Another major reason for referrals is the referrer's judgment that the psychological examination will give a special source of information that is not readily available. The usual example of this would be a more objective measure of the level of intelligence. In fact, the psychologist is seen as having the tools to evaluate affairs of the intellect. Personality assessment is less likely to be regarded as the elusive bailiwick of the psychologist. One often hears such statements as "give me an hour or two of interview and I don't need psychologicals." Evidence exists that the inferences developed from psychological test data are frequently quite different from those obtained from a series of interviews and prove to be of inestimable value in understanding and planning for the patient. Referrals may also be made for didactic purposes when a conference is to be held to demonstrate the functioning of each of the mental health disciplines. They may be requested with the intent of studying an individual prior to beginning and following the end of a particular therapeutic procedure.

Once the referral has been made, the psychologist seeks to utilize those psychological tests that will most adequately answer the question or questions that have been asked. Tests are selected on the basis of what functions they are designed to reveal and how well they work. Psychologists are constantly concerned with the validity and reliability of their testing instruments. Test results are often affected significantly by the interaction between patient and examiner. Many factors contributing to the nature of this patient-examiner interaction may affect performance. The patient wants to know the significance of the psychological examination. To what end is the study being done? What good will it do him? Will it help him get out of the hospital? Will the results keep him out of the armed services or effect a discharge from the armed services? Medication may adversely affect the performance so the psychologist must know if the patient is being medicated. The examiner may be concerned about his own time and try to speed things up. He may dislike examining a particular patient and be too brusque in his relationship with him. Other illustrations could be given. I feel that the patient should be told the purpose of the examination although at times its purpose is clearly apparent to the examinee. Questions are asked frequently about how the patient should be prepared for the fact that he is to be given a series of psychological tests. He should be told that as part of the evaluation procedure, a psychological study is being done in order to learn about him by a different method than that to which he has already been exposed. The study is to be done to develop the most effective procedures to plan for his treatment both while he is a patient of the institution and when he leaves. General statements of this kind should be made rather than emphasizing any unusual significance to the examination. All too frequently the subject is led to believe that if he does "well" he may expect good things to happen such as leaving the hospital or transfer to a better setting. In turn, he may be terrified because he has been led to believe that if he does "poorly," dire consequences will follow. No matter how objective the instructions or the explanation of the use of the results, the examiner should always try to maintain or re-establish the proper climate so as to get optimum results.

Illustrative Examples of Referral Procedures

By way of illustrating the referral process, let me turn to recent examples. L. K., a young married woman of 24 came into the outpatient clinic. She had a history of smoking pot and once used LSD. Since her use of LSD, she experienced flashbacks of a "bad trip." Her husband, she said, had no flashbacks even though he had taken drugs identical in kind and quantity to hers. She had also tried peyote and morning-glory seeds. She said she no longer used drugs, but had all the negative experiences that she had when she was actually using them. She had seen several psychiatrists and psychologists for help in overcoming her hallucinations, "spaced-out experiences," and "my mental confusion." She had quit all the therapists because either they viewed her problem as "psychological only," thereby dismissing her feelings that there must be an organic component resulting from actual brain damage, or they saw the problem as organic, but either wouldn't or didn't do anything about it. At any rate she had come to still another setting for help with her problem. She was referred to the psychology service for aid in understanding her difficulties and evaluating the degree and nature of the psychopathology.

A brief interview prior to the start of the psychological examination, coupled with other data derived from the outpatient service, yielded little or no evidence that L. K. had sustained brain damage as a result of her use of drugs, nor was there anything in her previous history that suggested organic involvement. This does not exclude the possibility that she had previously experienced an acute neurophysiological disturbance. The only thing that could be established was that her functioning at present did not reflect impairment due to brain damage. Occupationally she had been both an actress and a model for TV commercials. She had also functioned as a schoolteacher in urban and suburban schools.

Let us turn to another referral of an entirely different sort. S. M., a 58-year-old black man, was referred for a psychological study to evaluate his potential to return to work. He had a stable work history, having worked eight years in the meat-packing industry, and having left there when the plant sharply reduced its activites. In his present employment he functioned as a driver who was responsible for organizing truck loads and dispensing partial loads to various individuals; this required reloading what was left in the truck. He became depressed and confused on the job, apparently being unable to keep up with the time demands of rapid loading, and developed a sense of failure. He was hospitalized, given shock therapy, recovered, and returned to work. Several months later he had similar symptoms and was referred to two psychiatrists, the one who had previously treated him and a consultant psychiatrist. The former thought the man could return to work at his old job; the latter did not. My task was to make an independent recommendation on the basis of a psychological study. His company was interested in securing advice in vocational planning. They were aware of the stability of his work history and wanted to continue to employ him, provided there was some assurance that he could work, if not at his former level, at least at some kind of work. They were willing to reassign him to another area of the plant if it were indicated.

Motivational Factors of Patient toward Psychological Study

One of the concerns of the psychological examination is what the individual being examined seeks to accomplish. The literature is replete with "faking bad" and "faking good," in which the individual tries to appear as sick or as well as the circumstances warrant. The more obvious the intent of the examination, the easier it is to manipulate. Although some objective personality tests such as the Minnesota Multiphasic Personality Inventory have validity scales designed to detect the "faking" individual, still we are interested in the personalities we can extract from individuals who try to dissimulate. Therefore, a desirable quality of a test is that it should be difficult to fake. Consider, for example, the Rorschach test. Assume an individual decides to obfuscate responses on the Rorschach. He gives bizarre content on the grounds that this will really insure his pathology. However, we get data from the Rorschach other then the content of his responses, so he can say some pretty strange things and still appear relatively intact. I was asked to serve as a judge in studies of posthypnotic suggestion that the hypnotized subject had sustained severe damage to his brain and he should demonstrate his intellectual impairment on a battery of psychological tests. The records were presented to the judges to be evaluated for the degree of organicity, and nothing was known about the fact that the subjects were hypnotized. Even though the content of responses given was bloody and anatomical, the formal structure of the Rorschach was quite intact on individuals who were not brain damaged. Formal structure refers to such things as areas of blot used, accuracy and complexity of percepts, and so forth.

Let us look at the individual who seeks to put himself in a bad light (faking bad). Perhaps it is not even a question of faking. The individual decides that it is appropriate to place no checks on his perceptions. If he sees something, he will give it. This is in contrast to a more wary individual who does not subscribe freely to the examiner's injunction to tell everything that he sees as soon as he sees it. We see this illustrated frequently by a high percentage of sexual responses. Judging from his remarks, the examinee seems prone to think that this is the only significant parameter of the test. He is more likely to think the goodness and badness of the content is the pathology. Not so.*

Description of a Battery of Psychological Tests

Rorschach Ink Blot Test

The Rorschach consists of a standard series of ten ink blots. Each blot is reproduced by itself on cardboard. Five of the blots are gray, two are gray and red, and three are multicolored. The task of the examinee is to tell what he sees on each of the ten cards. The examiner records verbatim, or as close to verbatim as possible, what the examinee says. Upon completion of the series inquiry is made about each response so that the examiner knows where the response is located (whole ink blot, commonly or infrequently used parts of the blot, or space within or surrounding the ink blot). Further inquiry is used to evoke the determinants that entered into the response (shape, color, shading, tridimensionality, perception of human figures as though moving).† Each response should be scored according to location, determinant(s) used, and category of content. To illustrate, a person gives as his first response to Card 1, "bat." The inquiry indicates that he has used the whole

* This has been a source of difficulty in courtrooms. The psychologist may be asked what the defendant has seen, for example, a bat. The defense attorney may point out that he, too, sees a bat. "Does this make him crazy, too?" he will ask. This is why expert testimony in the courtroom should be an opinion based on the psychological data, not a reporting of the psychological data itself. After all, it is evident that similarity of content, or even exact equivalence, between prisoner and defense attorney establishes only that the men may share communal experiences.

† There are a number of scoring systems that have been elaborated by workers with the Rorschach. The scores mentioned above are patterned primarily after Rorschach[70] and Beck.[4] Other systems used in this country are Klopfer, Hertz, Piotrowski[66] and Rapaport,[68] to mention the most prominent ones. It is not my concern here to involve the reader in a discussion of the variations of systems. No system is capable of reflecting all the nuances of all the responses one gets.

ink blot, and the bat was determined only by shape. The scoring would be W (whole), F+ (form, accuracy acceptable), A (content category), and P. A particular response given to designated areas is scored P (popular). P responses are the most frequently given responses in the test series. P is an index of intellectual conformity, an individual's ability to recognize the conventional standards by which members of society govern their ordinary day-to-day behavior. An inadequate number of P responses suggests intellectual alienation from society.

The utility of the Rorschach resides not only in the content of the responses given but also in the formal quality of the responses themselves. One of the basic concepts that Rorschach evolved and designated is the Erlebnistypus (Experience Type or Experience Balance). The EB, as it is commonly designated, is a ratio between the number of responses that involve color (hue) and the number of responses in which humans and animals are perceived in movements that involve actions considered as characteristically human. M examples are "People dancing in rhythm," and "Two witches going at it at the top, holding Tommy guns." The number, quality, and cards on which they occur are all variables used in the evaluation of the meaning of M. M responses are scored M+ or M—. The + or — refers to the form quality of the percept. Color responses are scored FC+ or FC—, CF+ or CF—, with + or — again standing for the quality of the form or shape. FC is scored when the response is determined primarily by shape and secondarily by color; CF when the response is dominated by color and form is secondary to the percept; C when the response is based on the color values. The color sum (C sum), one half of the Experience Balance (EB), is derived in the following manner: Each FC response is one-half unit of color, each CF response is one unit, and each C response one and one-half units. The C sum is the arithmetical total of all responses involving the use of color. M is simply the absolute number of M responses and is reported on the left-hand side of the ratio; C sum is reported on the right. The Experience Balance represents a ratio between the individual's subjective life and his affective responsivity. M responses appear to reflect more the internal processes of an individual, which help us to understand how he perceives his world, whereas C responses tell us something about how he adopts to stimulation that he perceives as lodged in the environment. The Experience Balance is complex and cannot be considered in detail.

There are responses that involve the primary use of form and the secondary use of shading (FY), shading dominant over form (YF), and shading only (Y). An example of FY is "horrid bug," of YF is "rain clouds," and of Y is "death." Other determinants in the Rorschach within the Beck scoring system are responses involving the use of Vista (V). "A path leading up a hillside on top of which is a church" is an example of FV. "A canyon," in which the three-dimensional quality dominates the form, would be scored VF, and an "unexplainable void," in which depth alone was experienced, would be scored V. There are responses that involve texture. "A soft bearskin rug," in which form dominates the feeling of texture, would be scored FT. "A piece of satin," in which the texture dominated the form, would be scored TF. "A feeling of roughness or scaliness" without regard to form would be scored T. A single response can involve the use of several determinants such as form, color, and movement, but we will not consider here the complexities of such scoring. The test is suitable for all ages and populations. Frequently it has been used in ethnological studies to remarkable advantage.

Thematic Apperception Test (TAT)

The TAT is one of the most widely used projective techniques. Originally developed at the Harvard Psychological Clinic by Henry A. Murray[63,64] and his co-workers, it is used in a wide variety of settings to get an individual's thought content and to explore his basic needs or wishes, to discover the obstacles that seem to thwart such needs, and to discern attitudes he has toward his family, peers, and self. The test is considered suitable for girls, boys, adult females, and adult males. Adult is defined as age 14 and over. On the backs of the cards are numbers, some of which are followed by letters. For example, 1, 2, 3BM, 3GF, 12M, 12F, 12BG, 13MF, 13B, 13G, 13MF, etc. Those cards with numbers only are to be used for adults and children of both sexes. Those with letters are to be used as follows: B, boys only; G, girls only; BG, boys and girls; M, adult males only; F, adult females only; MF, adult males and females. Twenty cards were available for each of the four groups (boys, girls, adult males, adult females). The entire set contains 31 cards. The person is asked to tell a short story with a beginning, middle, and end. He is asked to say what the characters in the picture might be thinking or feeling. Current practice is usually to use less than 20 cards and to select a basic core of cards that is given to all examinees. Some examiners prefer using two basic sets, one for males, the other for females. Other examiners select cards on the basis of what they believe are key problems. What is significant is that there is much variation in the set of cards used as well as in the total number, and that this varies from clinician to clinician. No such leeway is permitted with the Rorschach; all ten cards must be used, and the order in which they are given is always the same.

There are many ways one can approach the analysis of the TAT. Many workers in the field work with TAT stories by seeking the dominant conflicts, their resolution, the mood of the story, the interaction between the story characters, the significant elements of the picture left out or distorted, the characters or situations introduced into the story that are not present in the picture, the hero of the story with whom the storyteller seems to identify, and a number of other variables, the number depending on which particular scoring system or TAT theoretician you wish to follow.* To illustrate, Card 1 shows a young boy seated with a violin lying on a desk or table. A typical story concerns itself with the boy who has to practice his violin, usually coerced by a maternal figure. As a result of his compliance he is then able to go outside and play baseball with his friends. Another characteristic result of being forced into patterns of practice is a rise to Olympian heights and becoming an internationally renowned violinist like Jascha Heifetz or Yehudi Menuhin.

One of the most tempting things to do with the TAT is to translate the material as though it were autobiographical. Sometimes we are really dealing with essential autobiography, but if the material is that well controlled by the patient then we would do better by requesting his actual autobiography. In some sense we have to try and distinguish between the overt content of the story and some of the latent or covert qualities of the patient. This proves very difficult for the psychological examiner. He must recognize that the instructions ask the patient to tell a short story and to say what the characters are thinking or feeling. Then in a quixotic maneuver the examiner seems to be saying that the stories that are told are to serve almost exclusively as a vehicle by which to describe the personality of the storyteller.†

This is undoubtedly a true reflection of what does happen, yet one must recognize that a certain portion of the material has relatively limited value in the sketching of personality dynamics. We have to learn what the usual expectations are for a particular picture. There have been attempts to derive normative data for the TAT cards.[16] The examiner must be sensitive to the normative responses so that he avoids the pitfall of making a subtle interpretation about a story when he is working with a plot that is a common response to the picture. In working with the TAT, most clinicians do not have a rigorous set of norms that they follow in deriving their interpretations, but they do have an appreciation of the usual stories, endings, thoughts, and feelings so that

* Murray, Bellak, Henry, Tomkins, and Stein are but a few of the individuals who have worked to develop rationales for scoring and interpretation.

† One needs only to survey the field of world literature to see that this technique of analysis is a frequent occurrence. Dostoevsky, Henry James, Dickens, and Ibsen have all been studied psychologically by the content of their writings and the literary style that they employ.

deviations from the usual expectations are noted. Interpretations of stories are derived both from the individual's awareness of the conventional plot and from the individual's deviances from this common core. This is a mandate that all too often is ignored. Essentially we must attend both to an individual's capacity to recognize the most obvious elements of a situation and to his ability to make the appropriate response. This is an index of his awareness of social norms, a reflection of his sensitivity to the expected patterns of thought. At the same time we are also interested in his ability to be unique, to depart from the traditional modes of thought or conscious awareness and move in the direction of individualism.

Wechsler Adult Intelligence Scale (WAIS)

The WAIS is the most widely used individual intelligence test for evaluating the intellectual functions of adults. The WAIS is made up of 11 subtests, six of which constitute the Verbal Scale, five the Performance Scale. The Verbal Scale consists of Information, Comprehension, Arithmetical Reasoning, Digit Span, Similarities, and Vocabulary. Information seeks to establish the range of general knowledge possessed by an individual. Comprehension measures verbal judgment as applied to everyday problems and an individual's understanding of the nature of his society. Arithmetical Reasoning consists of a series of arithmetic problems, all of which are to be solved without pencil and paper. Digit Span consists of repeating progressively longer series of numbers that are read to the subject. There is also a series that he repeats backward from the order in which it has been given. Similarities is a subtest designed to measure an individual's ability to think abstractly. One is asked to state the similarity between two things ranging in complexity from "orange" and "banana" to "Praise" and "'Punishment." Vocabulary asks for a definition of words. The Performance Scale is composed of Picture Completion, Picture Arrangement, Object Assembly, Block Design, and Digit Symbol. Picture Completion demands that the individual select the most important item missing from the pictures shown. Picture Arrangement consists of a series of pictures that must be rearranged from the order given so that they make the most sensible sequence. Object Assembly is like a simple jigsaw puzzle in which the individual assembles several common objects. In block Design one must copy designs reproduced in a booklet by using blocks that are red, red and white, and white. Digit symbol consists of filling in symbols that stand for the numbers one to nine; the task is to fill in as many spaces below the numbers as quickly as one can.

The Performance Scale has time limits, and bonus points may be given for rapid performance that is accurate as well. The Verbal Scale, except for Arithmetical Reasoning, has no time limits. The test administration usually takes somewhat over an hour. Wechsler has also created scales for preschool and primary age children aged four to six and a half (Wechsler Preschool and Primary Scale of Intelligence) and a test for older children aged five through fifteen (Wechsler Intelligence Scale for Children).

Bender-Gestalt

We turn next to the Bender-Gestalt Test. This test was developed by Lauretta Bender in 1938, patterned after designs developed by Max Wertheimer. The individual's task is to copy nine geometric designs (Gestalt figures). Because of the nature of these designs and the test instructions, we derive some awareness of how the individual perceives the designs, how he copies them, and how he organizes the designs in relationship to each other. This latter feature is evident because the individual is given as much paper as he likes to reproduce the designs. The instructions are simply that the basic nature of the task is to reproduce the designs presented him. The way he proceeds, the location of the designs on the paper, the rate of speed, and so forth are left to the discretion of the examinee. He knows that there is more than one design, but he is not told the number of designs he is expected to copy. This test is used both to assess the malfunctioning due to neural injury and to tell us something about personality functions.

Proverbs

Proverbs have been used in psychiatric settings for many years to evaluate an individual's ability to abstract, to generalize from the given instance of the proverb. A fair amount of evidence suggests that schizophrenics have difficulty in assuming an abstract attitude. For example, if the proverb used is "To the boiling pot, the flies come not," difficulty in abstraction would be reflected in the following response: "That's true. If a pot has boiling water, the fly is a smart insect and doesn't come to the pot because if he got too close to the water, his wings might get saturated or he would burn to death. So he stays away." Contrast that with an acceptable abstraction, "The carrion of society are not attracted to centers of activity," or "Trouble does not strike where progress is being made."

Take another proverb, "The fairer the paper, the fouler the blot." A concrete response would be, "If you have a good grade of paper, the worse the blot will be." A typical abstract reply might be, "The finer the character, the more the slightest indiscretion shows." Here is one final proverb to illustrate the method: "The ripest fruit falls first." Illustrative replies are, "The best things are the most quickly gone," "The best things are the most readily available," and "As is the pattern of growth, development, and resultant productivity of nature, so also people expand their horizons as they grow and attain maturity." Although it is tempting to discuss the significance of these replies, they are presented only to illustrate the variability of responses.

Psychological Studies

When reference to test protocols is made, it is important to keep in mind that these psychological examinations were done to meet the needs of the referrer and were not selected to illustrate a particular psychodynamic process. They were not selected for their uniqueness but as recent records obtained in a clinic. They are working examples of what may be done in a psychological study.

Psychological Study of L. K.

The behavior of L. K. during the examination was characterized by an apparent interest in the examination itself. She was seen on three occasions, mainly because she had scheduled a job interview the first time that resulted in a shorter session than intended. Before the second session she called to say she would be late, so that only the final session remained unaffected by time considerations.

This slender and rather pretty young woman showed a grace of movement. A certain wide-eyed disingenuous smile was one of her significant characteristics. She apparently hid behind the façade of this fixed smile as she demonstrated when she told of having feelings of violence toward me, even though at the same time she was telling me that she liked me. However, she showed no overt signs of distress. She understood that the purpose of the examination was to evaluate her for treatment, either for individual psychotherapy or for marital or couple therapy with her husband.* She seemed to accept the psychological examination as a necessary prelude to treatment, but it was never patently clear whether she sought to appear to advantage or disadvantage. In the main I assume the latter because the kind of messages she chose to communicate were ones such as the feelings of violence toward me engendered within her or feelings of disintegration. At the same time she wanted to share these experiences with me, so that we could have material that would enable us to help her from the morass into which she felt she had descended.

Let us turn to her performance on the tests. After three seconds exposure to Card 1 in the Rorschach she says, "Death was my first response." She goes on, "It looks like some horrid bug. It's got terrible pincers, native envi-

* An equivalent psychological study was done on her husband to evaluate him for therapy.

ronment is South America. There aren't many of them here but it's deadly."

This was what she reported on Card 1, and she returned the card after only 32 seconds. The interpretation of the Rorschach rests on the totality of responses given to all ten cards. So it was with L. K. However, to illustrate the process of interpretation, let's start with the one response and see where this might take us. The response was given rapidly; not only was three seconds the shortest reaction time it took to get the first response from any of the ten cards, but also 32 seconds was the shortest period of time a card was held by the patient until she had completed responding. We know that there are two popular responses (P) on Card 1. The whole blot (W) is seen as a butterfly, bat, or moth. A central detail (D) is seen as the hips and legs, most often of a woman. She gave neither of these responses. One inference that we start to make is that she does not take time to "rally her defenses," to compose herself so as to take an objective stance. She misses P responses. More than that, we note that she gives a highly impressionistic response, "death," which immediately carries with it an ominous aura. The inquiry clarifies what has suggested death. "Covered with black, I associated blackness with death." So now we become aware that the individual is giving us a highly symbolic response. Granted that it is arbitrary and not bound by objective considerations, she becomes aware of the black and bursts out, "Death was my first response."* This leads to a series of inferences, which become strengthened or modified dependent upon the totality of her responses.

Her response to blackness, darkness, suggests an overwhelming anxiety, a sense of panic, feelings of terror about the unknown, a generalized phobic reaction. Note that it is important that her responses to black shadings are not strongly bound by a consideration of objective form. Devoid of finite boundaries, the response to black carries with it the impression of not being able to see, terrors of the night, fear of the dark. We note a continuation of this response to black in her second and last response. Here, however, we see her moving into control, taking charge. We have a "horrid bug." This bug, however, has shape. She is aware of a form, its deadly and horrid qualities again being derived from blackness. Both clinical and research studies combine to tell us that strong reactions to black suggest strong undefined fears, phobic reactions, overwhelming sensitivity to superstitions. (This young woman turns out to have a strong belief in witches, ESP, and astrological phenomena). For the time being let us leave the Rorschach and turn to her responses on proverbs.

To the proverb, "To the boiling pot, the flies come not," L. K. replies, "Where there is constructive activity, evil does not appear." To the proverb, "The fairer the paper, the fouler the blot," L. K. responds, "The purer the person, the more obvious and sad the sin against his character."

There are two things to be noted about her replies. "Boiling pot" is frequently defined as activity; L. K. adds "constructive." "The flies come not" is interpreted often as "hangers-on," "laggards," and so on; she interprets "evil." At this point one sets up the possibility that it is important for this individual to place a premium on achievement and that when it is not consummated, inimical forces outside one's control may interfere with the pursuit of an individual's goals. Virtue cannot exist in the midst of adversity. Here it becomes important to note that we are talking about the possible interpretations. The probability of correctness of our inference rests on similar instances that bolster our initial hypothesis. We derive some additional support from "The fairer the paper, the fouler the blot." "The fairer the paper" is equated with "the purer the person." This emphasis on purity suggests virtue at its highest, a premium on goodness. "The fouler the blot" means "the more obvious and sad the sin against his character." The indiscretions committed are clear to everyone. Sad is gratuitous. Perhaps L. K. laments what has befallen her. Sin may be what the individual has done to himself or it might be what has been done to him. At least we are becoming aware that L.

* Actually she did not burst out. She gave the response in no way significantly different from the way she communicated her other responses.

K. places a premium on her virtues and sees that her actions or the actions of others will prove to be her undoing. L. K.'s response to "The ripest fruit falls first" is "The best and the most desirable are the first to be pulled down." Again we have the theme of that which is good, or pure, or meritorious being eroded from without. There are two statements that one can make about L. K. The first would be that there are unusually strong narcissistic elements in her life style. She appears absorbed with her own virtues, morality, and perfection. At the same time she sees herself as very vulnerable to the assaults of others. She is victimized by the rapacious quality of the unspecified others. Alternately L. K. may also be saying, "I have brought ruination to my own life because of my sins and evil deeds." Furthur study is necessary.

To Card 1 in the TAT she tells the following story: "This little fellow is a very shy child. He doesn't get along well with other children. He doesn't relate. He doesn't know how to. His parents are pretty well off and they push, particularly the mother. This child should belong to every organization in the school. He should . . . also why doesn't he have friends over to the house, bring some friends home? He has just reached the age where he would be allowed to be in the grammar school band. He didn't want to be . . . not because he didn't like music and didn't want to play an instrument but because he would be thrown up against the other children. His mother insists. They bought him the finest violin they could find—not a Strad . . . but a very good violin. She thinks violins are so elegant. We see him here after the first day of band practice. Music open that his instructor gave him and his violin. He is thinking he made so many mistakes he can't possibly go back the next day. Everybody made so many mistakes but his were the worst of all. Because of alphabetical seating he is placed next to the most pretty girl in school in the band. This is so upsetting he doesn't know how to deal with it. He knows she must hate him. It is so embarrassing. Well, he says to himself, 'I could go on to school and I could do my work and explain things to mother at home the best I could, but I can't sit in that band. That's too much. I'd rather be dead.' Then he thinks of his Uncle Joe and Uncle Joe is kind of a free spirit . . . roams around the countryside doing a bit of this and that and Uncle Joe never seemed to force him into anything he didn't want to do. He always affirmed him. Any little success he had, Uncle Joe made a great big thing out of. 'That's it' . . . he decides. 'Tonight, when they have all gone to bed, I'm going to run away and live with Uncle Joe.' Let's say . . . finish it there and he comes out O.K. at the end."

The other card is 3GF. "Some of you might judge me severely for what I am about to do, but let me tell you and you will be able to understand if you can't accept. You see, I once had a fine, intelligent mind. Personality that people said sparkled, physical beauty, and a husband that loved me. Now I have none of that. It all started with the drugs. It happened so fast I didn't have time to fight back. One day, there I was at a party smoking pot, dropping acid, laughing my head off, then, the next day thrown into limbo . . . hopelessly lost, unable even to understand a telephone number to call a doctor for help. Oh, I got help all right. Everyone tried. I think I tried hardest of all but none of it worked. I've destroyed my brain, the most valuable of all my assets. I disgust everyone but most of all myself. It's time to end this ridiculous fairy tale that some miracle is going to make it better. There is a razor in there. It won't take long and people tell me it's not very painful. If you who are listening can't understand, then at least allow me the dignity of my decision. This is not a call for help. This is a decision. So be it."

We find that there are no perceptual distortions in the story that L. K. tells about Card 1. She clearly sees this as a young boy with a violin who has his sheet music in front of him. True to the empirical tradition she brings in the coercive mother. From here on, however, to the end of the story, the plot becomes highly individualized. The mother figure is berated mercilessly. She, the maternal figure, is concerned only about achievement and socialization, and these values of hers are described as clearly dominant over all others. The patient perceives this figure as controlling, the determiner of what the child should or should not do, denying him his independence. One may question if she talks about herself, especially since the hero has the fate of being cast into crisis by the prettiest girl in school. The hero can't stand this and seeks surcease from all this turmoil by going with Uncle Joe, who provides the escape route from such travail.

What the patient wants to tell us about herself may contradict what she actually revealed about herself. The first such inference that I would make is that the patient has a need to "psychologize," to become an analyst of personality dynamics. She has a need to talk about how "a personality is born." The ready assignment of many psychological qualities in the picture suggests that L. K. may attribute a number of qualities to others that are hers rather than intrinsic to the development of the story.

Let us turn to Card 3GF for further illustration of the process. "Some of you might judge me severely for what I am about to do, but let me tell you and maybe you will be able to understand if you can't accept." This is the first thought that L. K. expresses. Most often individuals tell their stories in the third person; L. K. moves in the direction of first person. This use of the autobiographical genre enables her to move to the center of the stage and engage in a soliloquy. She almost immediately alerts us (me, the examiner) that she has a dramatic message that she wishes to communicate. She warns me to listen. She says, "understand me before you judge harshly." Once again we are confronted by the distinction we must make between L.K.'s levels of awareness as to what she wants to communicate and what she has communicated without being aware of it. The overt message is that because of the damage she has done to herself as a result of taking drugs, and in spite of numerous efforts to help her, including her own, she can't overcome the difficulties. So suicide is the only path open. This certainly is what she wishes to tell us. All hope is gone, she is irreversibly damaged, and killing oneself is not too painful. However, we must look beyond the plot and inquire about both the structure and the elements used to weave the plot.

It becomes clear that there is a coherence to the story. It flows clearly from beginning to end. One almost gets the impression that it has been rehearsed and that given the appropriate stimulus she responds automatically with her story. However, it becomes important to note that this suicide has many social parameters to it. She tells us that, "La commedia è finita," as does Canio in *Pagliacci*. The dramatic overtones are certainly there. She has discussed the weapon for suicide, and the razor has not been found wanting as an instrument of destruction. More than that, we note what she once had: "Personality that sparkled, people said, physical beauty and a husband that loved me." Thus L. K. describes herself as being fully in the arena of society. She was a luscious object to behold, both intellectually and physically. The egocentricity and the narcissism of L. K. are evident. The world revolves around her—nay, she is almost the world, to which or to whom others should pay homage. Again one must not be too eager to say she knows that she is talking exclusively about herself. It is at this juncture of interpretation that we begin to leave the manifest content of the obvious storytelling of L. K. She is not nearly as aware of what she is communicating, that she is egocentric, dramatic, and controlling. These are qualities that she does not use to describe herself. They arise from the manifest content of the data. She is described as controlling since she puts people on guard so that they don't ignore her. She denies that this is a "call for help. This is a decision."

She seeks to give messages that have a psychiatric flavor to them and then retracts the intent of the message. Further, it is important to note the one specific character introduced is a husband who is brought in to demonstrate that "one loved her"—even as "everyone tried to help." So that people focus on her—to do things for her. Again one must ask about the need to tell a story devoted to self-destruction with no hope. Why does she seem dedicated to the principle that all is futile? Why does she assume her brain is destroyed? Why can't she move in the direction of telling a story in the third person that describes an individual experiencing despair over her indiscretions who subsequently overcomes her difficulties?*

An analysis of all her stories on the TAT indicates that much of what she tells is overly dramatic and, in fact, even has a humorous touch. Card 4 involving male-female interaction, with the female seen as restraining or attempting to control the male, follows immediately the suicidal story told by L. K. She begins, "'You can't go out there, Bill,' the young woman said, scratching the sweaty arm of her husband." Later, "'I don't care what he said he was supposed to do,' the angry Latino gritted his teeth, 'Don't give him no right to rape my woman and treat me like dirt.'" It becomes clear that the versatility of L. K. permits her to tell a suicidal theme one moment and the next to obviously caricature man-woman relationships.

Perhaps, then, one of the things that we can infer is that this individual has great difficulty in dealing with intense feelings. There is a mocking, a denial, an exaggeration, but the overwhelming quality that emerges is the simulated intense affect with the impression that drama will win out. In general, the responses on the Rorschach that mirrored affective relationships were equally controlled. She is controlled but lacks warmth. L. K. gives more than an average number of P (popular) responses. What might this above average number of popular responses signify? First, the individual demonstrates a clear recognition of social standards, the commonplace patterns of behavior that account for so much of our daily lives. She can be described as quite aware of and potentially responsive to her environs. Second, it can also be argued that although she is overtly compliant and knowledgeable about the expectation of others, she needs to know "all the angles" so that she can be on the outlook for anything that deviates from the expected, the usual. More than that, she needs to recognize the conventional so she can clearly deviate from it. This is derived from other data that establish her staunch negativism. The result is that we are drawn to the possibility not only that L. K. is very much in touch with the "real world," but also that her awareness may lead her to be suspicious of situations that don't meet conventional standards. Alternatively she needs to know all the "right answers" so she can deviate significantly from conventional patterns of behavior when the need arises.

We are in a better position to further expand the meaning of a relatively high number of P, as we return to the meaning of the Experience Balance (EB). Her response to color suggests a *pro forma* recognition of emotionality. She goes through the motions of making the proper gestures of feeling, showing pleasure or pain as the situation demands. However, the quality of the color responses suggests that she has difficulty in generating a genuine emotional experience, and that there is a forced shamlike quality to the dramatic portrayal of her feelings. Here we look to the FC responses themselves. Take one example as illustrative of the way L. K. produces color responses. She sees orange hats as part of a percept. The hat is a quite acceptable form. It is an orange hat because the color is reddish orange. If the color had been blue it would have been a blue hat.

A more "genuine" FC response would be one in which the shape and color blend harmoniously and when the color is specific for the percept. Because of its shape and characteristic green, a pine tree is an illustration of a harmonious FC+ response. L. K.'s responses are more forced where the color does not seem intrinsically related to the form. This, then, leads to the supposition that L. K. is controlled, that she gives evidence for trying to be adaptive, but that she would have some difficulty in having any genuine feelings. She is much more controlled than her dramatic

* Again it should be noted that L. K. perceives the significant elements of the story and grasps the basic mood of the picture. The TAT Manual describes 3GF, "A young woman is standing with downcast head, her face covered with her right hand. Her left arm is stretched forward against a wooden door."

qualities might lead one to guess. The environment doesn't stimulate but oppresses her. She seems to anticipate calamities, the nature of which remain vague, that might befall her and are abundant and omnipresent. There is no serenity—only bleakness and feelings of panic. It becomes clear that what she experiences is very much at variance with the artificial nature of the positive feelings she is able to communicate. Yet her controls are significantly better than her clinical history would suggest. In spite of a pervasive anxiety she is able to create a façade of controlled behavior. While she experiences feelings of depression, she maintains a sense of humor. While she experiences nothing that is positive, she has a vigorous, lively imagination. Her Rorschach responses that involve humans are active, accurate percepts. They have an impulsive quality to them and mirror a strong wish for exhibitionism. They enable her to take a strong stand in pursuit of goals or to override obstacles that thwart her. She seems to be a creature of aggressive impulses, both outer and self-directed. To use an often overworked phrase, she acts out her impulses but has misgivings about some of her actions. The depressive reaction she experiences is a feeling of futility and abandonment of support; it does not seem to have its origin in guilt derivatives.

Psychological Study of S. M.

If time had permitted, a more complete study would have been done on S. M. He was seen on two occasions, each time for a period of three hours. His test behavior contributed to the length of the examination. He was very deliberate, pausing for protracted periods of time before he would respond. He could not permit himself to say, "I don't know," and made bad guesses when he didn't know the correct response. He checked the accuracy of his responses and having checked them, he frequently rechecked all over again. He verbalized at length when he was given questions where brevity of response was indicated.

S. M.'s performance on visual-motor tests is at the low end of the average range for his age group.* He shows a mild motor retardation. He is also at the low end of the average range on the two tests that measure visual discrimination and awareness of the sequence of simple social events. His performance on these two tests is qualitatively poor. For example, when asked to identify the missing object from the flag of the United States, he replies, "color blue, and no holder for the flag." (This, though the flag is in black and white.) The correct reply is "not enough stars." This introduction of color, when it is not present in any of the other pictures, is an irrelevancy that is all too characteristic of his thinking. He is better on verbal tests of intelligence than on visual-motor tasks. One wonders whether S. M. could have done better in the performance scale if not restrained by time limitations, but an analysis of his performance indicates that such was not the case. When given the opportunity to complete tasks in which he had exceeded the time limits, he was rarely able to do so. So even though he was overly deliberate, his score seemed related more to an inability to do tasks rather than a failure due to time limitations. His lowest score on the verbal scale was on similarities. This test consists of 13 items that receive scores ranging from zero to two. On the first item, "In what way are an orange and a banana alike?" a two-point response would be "fruit." S. M. says, "Each has a peeling." To the second item in the series, "In what way are a coat and dress alike," S. M. replies, "They both fit over the body," being unable to come up with the more abstract response, "clothing or wearing apparel." Out of a possible score of 26 on this test, he gets six. More important is that he never gave a two-point response. His replies on the similarities fail to come up to an abstract level of response, but he does not get involved in the personal symbolism or bizarre responses so characteristic of schizophrenic

* There is always a legitimate concern about cultural bias of such scales when used with groups who may differ significantly from the standardization group. In this instance S. M. has been in Chicago for 40 years, so that he is not as alienated from standardization groups as he might have been if he were from the rural South of the 1930's. Nevertheless, one must exercise caution.

thought. The comprehension subtest, rather than being viewed as an exclusive test of judgment, can be seen as composed of three parts. The first part concerns what one should do in simple social situations. For example, "What is the thing to do if you find an envelope in the street that is sealed, addressed, and has a new stamp?" The second part seeks to explore the individual's awareness of how society is organized. An example is "Why should people pay taxes?" On these two parts he performs quite adequately. He knows the appropriate response in simple social situations and has an adequate awareness of the structure of society. The third part, three proverbs, is a measure of verbal concept formation. Earlier, responses to proverbs were noted as measures of the ability to think abstractly. To the proverb, "Strike while the iron is hot," he replies, "That would mean bend while it's hot . . . if it's iron." To the proverb, "One swallow doesn't make a summer," he responds, "That would mean it wouldn't complete the summer. It's incomplete." So we infer that S. M.'s thinking is literal and inflexible. He has great difficulty in drawing things to an end and is quite circuitous in both behavior and thought. Whatever the disturbance, it doesn't seem to have invaded his immediate memory function as measured by the digit span subtest. This was his top score on the verbal scale.

On the Bender-Gestalt he does not meet the criteria that are specified as characteristic of patients who are suffering from intracranial damage. Although he has one or two of the signs, his performance does not suggest any serious difficulties. However, it should be noted that while he maintains his compulsive orientation, his organization and planning ahead are not nearly as precise as he desires. What then are we to conclude? The probability is that S. M. has shown a significant decrease in efficiency related to processes of aging, possibly aggravated by the use of electric shock therapy. However, the Bender-Gestalt helps us to reaffirm that S. M. does not suffer from an acute disturbance. More than that, data drawn from the Rorschach permit the inference that this is not an action-oriented man, but one who seeks to lead a relatively isolated, subjective life. He is self-contained and distrustful of his environment. Overall one must conclude that there are changes in this individual that appear chronic and suggestive of a moderate impairment in intellectual function characteristic of an individual with diffuse brain damage. Also his character structure is such that overtly he might appear to be much more stable than he really is.

The psychological data of L. K. and S. M. have served to demonstrate how one clinician pursues the process of interpretation. The psychological data were, for the most part, verbatim records of how L. K. and S. M. responded to the tasks given them. This enables other clinicians to work with the same set of data so inconsistencies in interpretation can be discussed and reconciled. Some psychological tests permit relatively little leeway in the interpretation of the data they provide, but many tests lend themselves to a variety of interpretations. This variation may reflect skills, personal style, theoretical orientation, the amount of experience of the examiner, and so forth. However, given the same set of data, it is possible to demonstrate a remarkable consistency between major conclusions drawn by independent, sophisticated examiners.

Discussion

Before summarizing the nature of the psychological examination both as presented in the foregoing and coupled with subjective impressions on the intrinsic nature of the psychological examination, I should like to point out what has been left out from consideration. I note this not to apologize but to emphasize the complexity, variety, and sheer number of psychological tests. Minimal reference was made to the examination of children. A prodigious amount of work has been done on infant and preschool populations to explore intellectual potential as well as to expand our knowledge about personality development. The whole area of achievement testing, which is an integral part of so many school systems, was not discussed. Remarks on the psychological examination were confined to individual examinations of adults, so that group tests were not discussed.

Some tests were mentioned only briefly. Draw-a-person tests and their relationship to intelligence and personality characteristics were not mentioned.* Materials for this test consist simply of paper and pencil. The individual is asked to draw a person. After completion of the drawing he is asked to draw a person of the opposite sex. This is the basic test, although there are other variations of the drawing task such as when the individual is asked to draw a house, tree, and person or to draw a person in the rain. In general, the test is considered useful for examining the intellectual level of the patient, for detecting organic signs or assessing personality characteristics with special reference to the body image concept. Neither were objective personality tests discussed for that matter. The best known test of this kind is the Minnesota Multiphasic Inventory. The MMPI consists of 550 statements that the individual checks as true or false depending on whether they are characteristic of him. There is a "cannot say" if he can't decide whether the item is true or false. The items are personality statements such as "I work under a great deal of tension" or "the sight of blood neither frightens me nor makes me sick." Items were developed empirically from major diagnostic groups that reflected the range of feelings and ideational experiences characteristic of patients in these groups. The test has been well standardized, and interpretation rests on the profile of scores obtained from responses to the test items. Items are scored according to the diagnostic group or groups that most appropriately reflect the nature of the item. Some reference was made to the neuropsychological tests, and the performance of S. M. was evaluated in terms of possible diffuse cortical brain damage that could affect his intellectual functioning. Yet it is not possible to consider all the concept formation tests developed by Goldstein, Hanfman,[33] Kasanin, and others. Halsted, Reitan,[69] and others have devoted their energies to deriving measures of intellectual impairment due to cortical dysfunction.

Perhaps my concern about what I haven't said should be tempered by a remark by Ralph Nickleby in Charles Dickens', *Nicholas Nickleby*, "Of all fruitless errands, sending a tear to look after a day that is gone, is the most fruitless."

What, then, is a psychological examination? I believe passionately that when one is asked to do a psychological study, one must be able to say much about the personality of a given individual. The referrer need not be concerned with the techniques that the psychologist uses, but he should be able to expect valid clinical information that augments his own observations. The clinical process is more than the particular tests we give. When one gives intelligence tests he not only gets a representative sample of intellectual functions, but he also gets attributes of personality. It is easy to see the personality characteristics of a 61-year-old female patient who, when asked the question, "How far is it from Paris to New York," replied "3,240 miles and Lindbergh was the first man to make a solo flight and he landed at LeBourget Field in Paris. Oh, my God, I can't remember the name of the field in New York that he departed from on May 20, 1927." When projective personality tests like the Rorschach and TAT are being used, we should not limit ourself to the description of the emotional style, degree of anxiety, and defense against anxiety but talk also about intellectual characteristics. No one who has sat through a Rorshach that involves nothing but responses consisting of animal percepts escapes the conclusion that the individual's intellectual horizon is extremely limited.

What I should like to emphasize in closing is that the psychological examiner uses the test battery as a means of weaving together the intellectual and emotional life of an individual. I do not mean to minimize the separate tests; rather I propose to maximize their use in deriving the composite picture of an individual. The psychological report should describe the examinee so that he emerges from the written pages as an individual who has certain

* Hammer has an entire volume devoted to *The Clinical Application of Projective Drawings.*

needs and wishes, barriers that block them from being consummated, and certain reactions to such obstacles. One should be aware of the level of intelligence and the intellectual style he uses. The penultimate value of the psychological study must rest in its ability to describe the major personality characteristics of a person in graphic, meaningful terms. The ultimate aim is that the reader of the report understands the patient with sufficient clarity to be able to anticipate the kinds of behavior that will appear.

Bibliography

1. ANASTASI, A., "Psychology, Psychologist and Psychological Testing," *Am. Psychol.*, *22:* 297–307, 1967.
2. APPELBAUM, S. A., "Half-Hidden Influences on Psychological Testing and Practice," *J. Proj. Tech. & Pers. Ass.*, *29:*128–133, 1965.
3. BAKER, G., "A Therapeutic Application of Diagnostic Test Results," *J. Proj. Tech. & Pers. Ass.*, *28:*3–8, 1964.
4. BECK, S. J., *Rorschach's Test*, Vol. 1, *Basic Processes*, 3rd ed., Grune & Stratton, New York, 1961.
5. BENJAMIN, J. D., "A Method for Thinking and Evaluating Formal Thinking Disorders in Schizophrenia," in Kasanin, J. S. (Ed.), *Language and Thought in Schizophrenia*, pp. 65–88, University of California Press, Berkeley, 1944.
6. BLATT, S. J., and ALLISON, J., "Methodological Considerations in Rorschach Research: The W Response as an Expression of Abstractive and Integrative Strivings," *J. Proj. Tech.*, *27:*267–278, 1963.
7. BREGER, L., "Psychological Testing: Treatment and Research Implications," *J. Consult. & Clin. Psychol.*, *32:*176–181, 1968.
8. BRIM, O. G., JR., "American Attitudes toward Intelligence Tests," *Am. Psychol.*, *20:*125–130, 1965.
9. BRUHN, J. C., CHANDLER, B., and WOLF, S., "A Psychological Study of Survivors and Nonsurvivors of Myocardial Infarction," *Psychosom. Med.*, *31:*8–20, 1969.
10. BUTCHER, J. N., and TELLEGEN, A., "Objections to MMPI Items," *J. Consult. Psychol.*, *30:*527–534, 1966.
11. CAMP, B. W., "WISC Performance in Acting-Out and Delinquent Children with and without EEG Abnormality," *J. Consult. Psychol.*, *30:*350–353, 1966.
12. CARLSON, R., "Where is the Person in Personality Research," *Psychol. Bull.*, *75:*203–219, 1971.
13. CRONBACH, L. J., "Statistical Methods Applied to Rorschach Scores: A Review," *Psychol. Bull.*, *46:*393–429, 1949.
14. DE VOS, G. A., "A Quantitative Approach to Affective Symbolism in Rorschach Responses," *J. Proj. Tech.*, *16:*133–150, 1952.
15. EBEL, R. L., "Must All Tests Be Valid?" *Am. Psychol.*, *16:*640–647, 1961.
16. ERON, L. K., "Frequencies of Themes and Identifications in the Stories of Schizophrenic Patients and Non-Hospitalized College Students," in Megargee, E. I. (Ed.), *Research in Clinical Assessment*, pp. 440–450, Harper & Row, New York, 1966.
17. FINE, R., "The Case of El: The MAPS Test," *J. Proj. Tech.*, *25:*483–489, 1961.
18. FISHER, S., and CLEVELAND, S. E., *Body Image and Personality*, Van Nostrand, Princeton, 1958.
19. ———, and FISHER, R. L., "A Projective Test Analysis of Ethnic Subculture Themes in Families," *J. Proj. Tech.*, *24:*366–369, 1960.
20. FISKE, D. W., "Homogeneity and Variation in Measuring Personality," *Am. Psychol.*, *18:*643–652, 1963.
21. FONDA, C. P., "The White-Space Response," in Rickers-Ovsiankina, M. A. (Ed.), *Rorschach Psychology*, pp. 80–105, John Wiley, New York, 1960.
22. FORER, B. R., "The Case of El: Vocational Choice," *J. Proj. Tech.*, *25:*371–375, 1961.
23. FREEMAN, E. H., FEINGOLD, B. G., GORMAN, F. J., and SCHLESINGER, K., "Psychological Variables in Allergic Disorders: A Review," *Psychosom. Med.*, *26:*543–576, 1964.
24. FULKERSON, S. C., and BARRY, J. R., "Methodology and Research on the Prognostic Use of Psychological Tests," *Psychol. Bull.*, *58:*197–204, 1961.
25. GARFIELD, S. L., "The Clinical Method in Personality Assessment," in Wepman, J. W., and Heine, R. W. (Eds.), *Concepts*

of Personality, pp. 474–502, Aldine, Chicago, 1963.

26. GILL, M. M. (Ed.), *The Collected Papers of David Rapaport*, Basic Books, New York, 1967.
27. GOFF, A. F., and PARKER, A. W., "Reliability of the Koppitz Scoring System for the Bender Gestalt Test," *J. Clin. Psychol.*, 25:407–409, 1969.
28. GOUGH, H. G., ROZYNKO, V. V., and WENK, E. A., "Parole Outcome as Predicted from the CPI, the MMPI and a Base Expectancy Table," *J. Abnorm. Psychol.*, 70:432–441, 1965.
29. GRIFFITH, R. M., "Rorschach Water Percepts: A Study in Conflicting Results," *Am. Psychol.*, 16:307–311, 1961.
30. GUERTIN, W. H., FRANK, G. H., LADD, C. E., and RABIN, A. I., "Research with the Wechsler Intelligence Scales for Adults," *1955–1960 Psychol. Bull.*, 59: 1–26, 1962.
31. HALL, L. P., and LA DRIERE, L., "Patterns of Performance on WISC Similarities in Emotionally Disturbed and Brain-Damaged Children," *J. Consult. & Clin. Psychol.*, 33:357–364, 1969.
32. HAMMER, E. H., *Clinical Applications of Projective Drawings*, Charles C Thomas, Springfield, Ill., 1958.
33. HANFMANN, E., and KASANIN, J., "Conceptual Thinking in Schizophrenia," *Nerv. & Ment. Dis. Monogr.*, 67, 1942.
34. HOFFER, A., and OSMOND, H., "A Card Sorting Test Helpful in Making Psychiatric Diagnosis," *J. Neuropsychiat.*, 2:306–331, 1961.
35. HOLT, R. R., "Clinical and Statistical Prediction: A Reformulation and Some New Data," in Megargee, E. I. (Ed.), *Research in Clinical Assessment*, pp. 657–671, Harper & Row, New York, 1966.
36. ———, and HAVEL, J. A., "A Method for Assessing Primary and Secondary Processes in the Rorschach," in Rickers-Ovsiankina, M. A. (Ed.), *Rorschach Psychology*, pp. 263–315, John Wiley, New York, 1960.
37. HOLTZMAN, W., "Recurring Dilemmas in Personality Assessment," *J. Proj. Tech. & Pers. Ass.*, 38:144–150, 1964.
38. HOOKER, E., "The Case of El: A Biography," *J. Proj. Tech.*, 25:252–267, 1961.
39. HUTT, M. L., *The Hutt Adaptation of the Bender-Gestalt Test*, Grune & Stratton, New York, 1969.
40. JACOBS, M. A., KNAPP, P. H., ROSENTHAL, S., and HASKELL, D., "Psychologic Aspects of Cigarette Smoking in Men: A Clinical Evaluation," *Psychosom. Med.*, 32:469–485, 1970.
41. KAHN, M. W., "Psychological Test Study of a Mass Murderer," *J. Proj. Tech.*, 24:148–160, 1960.
42. KAHN, R. L., and FINK, M., "Prognostic Value of Rorschach Criteria in Clinical Response to Convulsive Therapy," *J. Consult. Psychol.*, 5:242, 1960.
43. KAPLAN, M. L., HIRT, M. L., and DURITZ, R. M., "Psychological Testing: Comprehensive Psychiatry," 8:299–309, 1967.
44. KARON, B. P., "Reliability: Paradigm or Paradox, with Especial Reference to Personality Tests," *J. Proj. Tech. & Pers. Ass.*, 30: 223–227, 1966.
45. KAYE, J. D., "Percept Organization as a Basis for Rorschach Interpretation," *Brit. J. Proj. Psychol. & Pers. Study*, 14:7–15, 1969.
46. KNUDSEN, A. K., GORHAM, D. R., and MOSELEY, E. C., "Universal Popular Responses to Inkblots in Five Cultures: Denmark, Germany, Hong Kong, Mexico, and U.S.A., *J. Proj. Tech. & Pers. Ass.*, 30: 135–142, 1966.
47. KRAUS, J., "A Combined Test for the Diagnosis of Organic Brain Condition: Predictive Validity Based on Radiographic and EEG Criteria," *J. Abnorm. Psychol.*, 75:187–188, 1970.
48. LERNER, B., "Rorschach Movement and Dreams: A Validation Study Using Drug-Induced Dream Deprivation," *J. Abnorm. Psychol.*, 71:75–86, 1966.
49. LEVENTHAL, T., GLUCK, M. R., ROSENBLATT, B. P., and SLEPIAN, H. J., "The Utilization of the Psychologist-Patient Relationship in Diagnostic Testing," *J. Proj. Tech.*, 26:66–80, 1962.
50. LEVIN, R. B., "An Empirical Test of the Female Castration Complex," *J. Abnorm. Soc. Psychol.*, 71:181–188, 1966.
51. LITTLE, K. B., and SHNEIDMAN, E. S., "Congruencies among Interpretations of Psychological Test and Anamnestic Data," in Megargee, E. I. (Ed.), *Research in Clinical Assessment*, pp. 574–611, Harper & Row, New York, 1966.

52. LOVELL, V. R., "The Human Use of Personality Tests: A Dissenting View," *Am. Psychol.*, *22:*383–393, 1967.
53. MACHOVER, K., *Personality Projection in the Drawings of the Human Figure*, Charles C Thomas, Springfield, Ill., 1949.
54. MARKEL, N. N., "Relationships between Voice-Quality Profiles and MMPI Profiles in Psychiatric Patients," *J. Abnorm. Psychol.*, *74:*61–66, 1969.
55. MASLING, J. M., "The Influence of Situational and Interpersonal Variables in Projective Testing," *Psychol. Bull.*, *57:*65–85, 1960.
56. ———, and HARRIS, S., "Sexual Aspects of TAT Administration," *J. Consult. & Clin. Psychol.*, *33:*166–169, 1969.
57. MEGARGEE, E. I. (Ed.), *Research in Clinical Assessment*, Harper & Row, New York, 1966.
58. MEEHL, P. E., *Clinical Versus Statistical Prediction*, University of Minnesota Press, Minneapolis, 1954.
59. ———, "When Shall We Use Our Heads Instead of the Formula?" in Megargee, E. I. (Ed.), *Research in Clinical Assessment*, pp. 651–657, Harper & Row, New York, 1966.
60. MEYER, M. M., "The Case of El: Blind Analysis of the Tests of an Unknown Patient," *J. Proj. Tech.*, *25:*375–382, 1961.
61. MOOS, R. H., "Sources of Variance in Response to Questionnaires and in Behavior," *J. Abnorm. Psychol.*, *74:*405–412, 1969.
62. MOYLAN, J. S., and APPLEMAN, W., "Passive and Aggressive Responses to the Rorschach by Passive-Aggressive Personalities and Paranoid Schizophrenics," *J. Proj. Tech.*, *24:*17–21, 1960.
63. MURRAY, H. A., "Commentary on the Case of El," *J. Proj. Tech.*, 25:404–411, 1961.
64. ———, *Explorations in Personality*, Oxford University Press, New York, 1938.
65. MURSTEIN, B. I., *Theory & Research in Projective Techniques*, John Wiley, New York, 1963.
66. PIOTROWSKI, Z. A., *Perceptanalysis*, Macmillan, New York, 1957.
67. RABIN, A. I., *Projective Techniques in Personality Assessment*, Springer, New York, 1968.
68. RAPAPORT, D., GILL, M. M., SCHAFER, R., and HOLT, R. R. (Eds.), *Diagnostic Psychological Testing*, International Universities Press, New York, 1968.
69. REITAN, R. M., "Psychological deficit," *Ann. Rev. Psychol. Rev.*, *13:*415–444, 1962.
70. RORSCHACH, H., *Psychodiagnostics*, 5th ed., Hans Huber, Bern, 1921.
71. ROSENTHAL, D. (Ed.), *The Genain Quadruplets: A Study of Heredity and Environment in Schizophrenia*, Basic Books, New York, 1963.
72. SATTLER, J. M., and THEYE, F., "Procedural, Situational and Interpersonal Variables in Individual Intelligence Testing," *Psychol. Bull.*, *68:*347–360, 1967.
73. SCHNEIDMAN, E. S., "The Case of El: Psycological Test Data," *J. Proj. Tech.*, 25: 131–154, 1961.
74. ———, "The Logic of El: A Psychological Approach to the Analysis of Test Data," *J. Proj. Tech.*, *25:*390–403, 1961.
75. SCHUBERT, J., "Rorschach Protocols of Asthmatic Boys," *Brit. J. Proj. Psychol. & Pers. Study*, *14:*16–22, 1969.
76. SHAKOW, D., "The Nature of Deterioration in Schizophrenia," *Nerv. & Ment. Dis. Monogr.*, *70:*1–88, 1946.
77. SILVER, A. W., "TAT & MMPI Psychopath Deviant Scale Differences between Delinquent and Nondelinquent Adolescents," *J. Consult. Psychol.*, *27:*370, 1963.
78. SIMPSON, R., "Study of the Comparability of the WISC and WAIS," *J. Consult. & Clin. Psychol.*, *34:*156–158, 1970.
79. SINGER, M. I., "Comparison of Indicators of Homosexuality on the MMPI," *J. Consult. & Clin. Psychol.*, *34:*15–18, 1970.
80. SULLIVAN, P. F., and ROBERTS, L. K., "Relationship of Manifest Anxiety to Repression-Sensitization on the MMPI," *J. Consult. & Clin. Psychol.*, *33:*763–764, 1969.
81. TALLENT, N., "Clinical Psychological Testing: A Review of Premises, Practices and Promises," *J. Proj. Tech. & Pers. Ass.*, *29:*418–435, 1965.
82. TYLER, L., "Psychological Assessment and Public Policy," *Am. Psychol.*, *25:*264–266, 1970.
83. WAITE, R. R., The Intelligence Test as a Psychodiagnostic Instrument," *J. Proj. Tech.*, *25:*90–102, 1961.
84. WALKER, A. M., RABLEN, R. A., and ROGERS, C., "Development of a Scale to Measure Process Changes in Psychotherapy," *J. Clin. Psychol.*, *16:*79–85, 1960.
85. WANDERER, Z. W., "Validity of Clinical Judg-

ments Based on Human Figure Drawings," *J. Consult. & Clin. Psychol.*, 33:143–150, 1969.

86. WECHSLER, D., *The Measurement and Appraisal of Adult Intelligence*, 4th ed., Williams & Wilkins, Baltimore, 1958.
87. WEINER, I. B., *Psychodiagnosis in Schizophrenia*, John Wiley, New York, 1966.
88. WERKMAN, S. L., and GREENBERG, E. S., "Personality and Interest Patterns in Obese Adolescent Girls," *Psychosom. Med.*, 29:72–79, 1967.
89. WIGGINS, J. S., "Strategic Method, and Stylistic Variance in the MMPI," *Psychol. Bull.*, 59:224–242, 1962.
90. WOLL, J., "Traditional and Contemporary Views of Psychological Testing," *J. Proj. Tech. & Pers. Ass.*, 27:359–369, 1963.
91. ZIEGLER, F. J., KRIEGSMAN, S. A., and RODGERS, D. A., "Effect of Vasectomy on Psychological Functioning," *Psychosom. Med.*, 28:50–63, 1966.
92. ZULLIGER, H., *The Behn-Rorschach Test*, Hans Huber, Bern, 1956.

CHAPTER 56

SOCIAL WORK

Milton Wittman

It will be the purpose of this chapter* to reflect four main themes: an overview of the emergence of social work and its relation to psychiatry and mental health over the last century; a review and analysis of significant events influencing social work in the past decade; a discussion of social work as a professional resource in mental health; and an assessment of the main issues in social work practice as it relates to mental health.

When stripped to the barest essentials, the goals of the professions of social work and psychiatry can be said to be derived from a mutual concern for the continued well-being of the individual and for the preservation and enhancement of the health-generating capabilities of the family, community, and society in general. The antecedents of the two callings have deep roots in antiquity, in the Middle Ages, and in the colonial period of American history. The Biblical injunction to tithe for the welfare of the widow, the orphan, the stranger implies a concern for the economically deprived that translates in modern times to the guaranteed annual income, still distant in any terms satisfactory to social workers. It was the fate of the mentally ill or disordered to be regarded either as holy or demon-possessed, with only painfully slow emergence of more humanitarian efforts at care and treatment as the science of medicine emerged from the Dark Ages and men like Benjamin Rush in America, Philippe Pinel in France, and William Tuke in England introduced immense changes in the care and treatment of the mentally ill.[28] In colonial times the almshouse was the traditional community resource for the indigent and disabled citizen, if he was not auctioned to the lowest bidder who offered to provide care at the least cost. Almshouses were administered by overseers of the poor under pauper laws derived from the Elizabethan Poor Law of 1601. The need for more specialized care was recognized with the founding of the first state mental hospital at Williamsburg, Virginia, in 1773, but these early hospitals were frequently nothing more than almshouses under another name.

A number of authors have described the

* This chapter reflects the opinions of the writer and does not represent the policy of the National Institute of Mental Health, Alcohol, Drug Abuse, and Mental Health Administration, U.S. Department of Health, Education and Welfare.

The writer wishes to acknowledge the assistance of Anne L. Martin in bibliographic research. Margaret Daniel provided useful comments on the manuscript.

growth and development of institutions for the provision of care for the mentally ill during the nineteenth century.[20,26,70,75] Perhaps one of the most significant contributions to improvement of care for the mentally ill was made by Dorothea Lynde Dix, a pioneer in reform who was active in mid-century. Albert Deutsch[28] describes at some length her vast contributions to the cause of the mentally ill. She was responsible for changes in the programs for the care of mental patients in 20 states and in foreign countries as well. She was directly responsible for the founding or enlarging of 32 mental hospitals in the United States and overseas, including the Government Hospital for the Insane in Washington, D.C. (now known as St. Elizabeth's Hospital). It is worth noting that Miss Dix was successful in securing passage through the Congress in 1854 of a bill that would have allocated federal land to the support of "the indigent insane." This bill was vetoed by President Franklin Pierce in a message that contained a flat denial of federal responsibility for provision of resources for the care of the mentally ill.[70] It took almost a full century to bring about a complete reversal in the philosophy of government, which culminated in President Harry S Truman's signing of the National Mental Health Act in 1946.

Social workers of the nineteenth century were concerned with problems of poverty and social pathology. They took the lead in moving states toward taking responsibility for the more orderly structure of social services in the nineteenth century. This tended to be in the form of institutions derived from the almshouse of the colonial period. The range of public and other eleemosynary institutions included not only hospitals for the insane but also institutions for orphans, the mentally retarded, and aged persons.[75] These organizations for congregate care were built with the best of intentions to provide a place where food, clothing, and housing, the essentials of life, could be provided in a central location and at the lowest cost to the community. Frequently these institutions were built out in the country adjacent to farm property so that the residents, or "inmates," could be employed.

American social work in the nineteenth century was concerned with the organization of charity and the extension of welfare benefits to the poor and disabled members of the population. The early charity organization societies were intended to bring together the various existing charitable institutions so that they could provide service in a more systematic manner. Borrowing on a model already established in England, the first Charity Organization Society was established in Buffalo, New York, in 1877. The problems of poverty in the big cities were addressed mainly by the social settlements established along the model of Hull House in Chicago. Here Jane Addams established a service unit in an old home in the midst of the immigrant area. The concept of service established by Jane Addams at the turn of the century was one of forthright advocacy for the dependent and immigrant populations. Hers was one of the early social service centers that offered multiple services to the community. These services included educational activities for adults and children, counseling resources for people with problems, and promotional activities to correct injustices in the community. Jane Addams was an early supporter of the labor movement in this country, aggressive efforts toward achieving women's rights, correction of child labor abuses, and improvement of the human lot through social legislation.

The end of the century saw the beginning of professional social work education in the establishment of the New York School of Social Work in 1898.[22] Social services of the pattern offered by the typical charity organization society were the model for voluntary effort in behalf of the urban poor.[90] The first decade of the twentieth century saw the establishment of medical social work in Boston and psychiatric social work in New York City.[58,65] In 1906 the addition of a trained worker, Miss E. H. Horton, to Manhattan State Hospital under the auspices of the State Charities Aid Association marked the first effort to employ a trained person in social services related to the mentally ill.[28] The development of social work as a profession brought early recognition of the remedial nature of this occupational group. The emergence of social case work was noted with the publication in 1917 of *Social Diagnosis* by the Russell Sage Foundation. Miss Mary Richmond, an early pioneer in social case work, very soon emphasized the need to consider the broad social planning aspects as well as the individual treatment issues involved in dealing with human problems.[71]

The period of World War I gave birth to a school for social work devoted entirely to the training of psychiatric social workers (Smith College). It also gave impetus to the development of the mental health movement, a field in which social workers became much interested ever since Clifford Beers wrote his stirring account of his experiences as a patient in *A Mind That Found Itself*. An additional impetus came from the establishment of the first child guidance clinic in Chicago, involving collaborative work between social workers, psychologists, and psychiatrists. The need to provide service for mentally ill soldiers and later for veterans provided an incentive for social work to move substantially into the area of social services for the mentally ill. The first social services for veterans were established under the auspices of the U.S. Public Health Service and then transferred to the Veterans' Administration. This was the period that saw the emergence of the American Association of Schools of Social Work in 1919 and the American Association of Social Workers in 1921. The former was concerned with standards for education of members of the profession. In 1926 the American Association of Psychiatric Social Workers (AAPSW) began with a requirement for rigid training patterns that would assure that the graduate social worker would have an adequate orientation in mental health and psychiatric content. This type of instruction was introduced in a number of schools in the 1920's. Following World War I psychoanalysis came into vogue and was introduced in schools of social work. This event preoccupied the field for 20 years. Some have felt that the dominant role of psychoanalytic theory retarded the emergence of advocacy as a role for social work.[17]

During this period the Commonwealth Fund financed a series of child guidance clinic demonstrations in several American cities. Social workers participated in the team activities in these demonstration clinics, many of which still exist today. Another important innovation developed by the Commonwealth Fund was that of school social work. Psychiatric social workers found themselves very much involved in social work in the schools because of the possibility for preventive work with children at an early stage in their lives. The Commonwealth Fund fostered the development of psychiatric social work not only in the United States but also in England.[85] A number of English social workers came to the United States for training and returned home to establish training for psychiatric social work, first at the London School of Economics and Political Science in the late 1920's and then later in other universities throughout the United Kingdom.

Of all the vulnerable population groups requiring social services, the mentally ill seem foremost in this respect. The nature of mental disorder is such that it causes serious problems not only for the individual but also for the family. The unpredictable nature of schizophrenia and the devastating effect it can and does have on the immediate family is cause for social concern. The same is true of mental retardation or childhood mental disorders, which are so frequently beyond the scope of parents' ability to deal with without expert help. Psychiatric social workers moved very quickly to develop working relationships with psychoanalysts and psychiatrists in the provision of individual and group services for the mentally ill.[5] Social workers also pioneered in the development of foster care as a means of treatment for the mentally ill. Borrowing on ideas for home care developed in Belgium and elsewhere, New York State and later Maryland pioneered at an early stage in the movement of patients into foster homes in the community as the alternative to hospital care.[26] These foster homes were the predecessors to the halfway houses of today, which provide for group care for patients on an extramural basis.

A study by Lois M. French[35] published in

1940 surveyed the field of psychiatric social work and recorded the distribution of social workers in mental hospitals and child guidance clinics. It also revealed the numbers in training at the time and reviewed the educational programs available to them. This was a period of organization for psychiatric social workers, and they developed a powerful professional organization that played a most useful role in the ultimate development of community mental health services and of wartime social services for mentally ill soldiers. An ample literature describes both of these developments.[13,58]

Major changes occurred in the 1930's and 1940's when the major national and international events of the Great Depression and World War II visibly affected the field of social work.[20,22] The Great Depression of 1931–1939 saw a large number of social workers drawn into public assistance and public child welfare. The public social services became a major arena for the practice of social work. Similarly the impact of World War II provided new challenges to the field.[22] Both on the home front and in the military services social workers were called upon to do more to become involved with problems of individual and social pathology. Social workers in the American Red Cross carried on the work of their predecessors who were field agents for the U.S. Sanitary Commission during the American Civil War. The social workers were very closely related to problems of military families left behind and to crisis situations. In the military services case work and group work were the typical methods used in direct services for military personnel. A most significant development took place through the encouragement and reinforcement of psychiatrists outside and in the service.[54] The result of the introduction of social services into the military establishment was a creation of a group of social work officers, with career lines for enlisted men as well in the role of social work technician. The most far-reaching development took place in the U.S. Army, beginning with psychiatric social work at first, but extending to include medical social work and eventually developing as a program of community services. A somewhat similar development has taken place in the U.S. Air Force. The Navy never developed a uniformed military social service program, and as a result it is without the benefit of an infrastructure of social services for its personnel.

The pattern of development of social services for veterans provided for both institutional and community services during the postwar period. The Veterans' Administration has responsibility for care, treatment, and rehabilitation of over 100,000 hospital patients (more than one-half of these are mental patients). There is an acute need for personal and community social services for veterans and their families.[7] The VA now employs over 2,000 qualified social workers as part of the manpower devoted to the provision of services to veterans.

Some of the research conducted by the Veterans' Administration has had significant import for the practice of social work and psychiatry. For example, an outplacement study, which provided for a large-scale evaluation of all patients from 16 VA hospitals, found that 16 per cent of the medical, surgical, and neurological patients and 50 per cent of the psychiatric patients could have been cared for in a nonhospital setting if "appropriate services and/or living care situations were or could be made available."[61] It is quite possible that similar studies in other mental and general medical and surgical hospitals would reveal somewhat similar findings. This places the important responsibility on the professional staff to insure that patients are properly evaluated and are not retained in institutional situations longer than is necessary. In the mental health field the development of halfway houses and intermediate institutions has done much to curtail the lengthened hospitalization of patients beyond the period actually needed.

The period following World War II brought about profound change in the direction of social work and the mental health field. During the war the ferment began that led by 1955 to the amalgamation of five separate social work organizations into a single united national organization called the National Association of

Social Workers (NASW). The immediate effect of the experience gained from the problems of mental illness among soldiers, coupled with the unrest emerging from the dismal record of the mental hospital system as a means of providing care, led to the passage on July 3, 1946 of the National Mental Health Act (Public Law 79–487). The National Mental Health Act made matching funds available for the expansion of state mental health services, with particular emphasis on outpatient care.[32] Funds were made available on a much expanded basis for support of mental health research and training. The availability of teaching grants and stipends for students permitted the training of greatly increased numbers of psychiatrists, clinical psychologists, psychiatric social workers, and psychiatric nurses for work in mental health. A number of related projects advanced work on educational standards in the core professions.

At Dartmouth College in 1949 a group of practitioners, educators, and psychiatric social workers on the faculties of schools of social work met to review the status of training for psychiatric social work. A system of communication was established among these educators that led to the introduction of more modern teaching methods and to the adaptation of content from psychiatry and the social sciences into the social work curriculum.[6]

The publication in 1952 of the AAPSW Study, *Social Work in Psychiatric Hospitals and Clinics*,[11] permitted an overview of where social workers were to be found and what they were doing in the mental health field. During this period the state mental health authorities were established in all 53 states and territories, with state-level planning and administrative staff dedicated to the improvement and expansion of services throughout the state. There was a period of reassessment of mental hospital care and its relation to community mental health. It was during the 1950's that the existence of case work as a dominant social work method was emphatically challenged. A significant growth took place among practitioners and educators who were prepared to offer the group method or community organization method in social work practice. It was also a period in which the field of social work education moved toward assessment and evaluation. The publication of the Hollis-Taylor report led to the establishment of the Council on Social Work Education, which became the standard-setting body governing preparation for the field of social work and social welfare.[44] The curriculum policy activity of this organization led eventually to the declaration of the end of specialization in social work training at the master's level. By 1959 it was agreed that all specialization training would end, and that the master's degree would be regarded as a generic concentration, which would only provide for a specialization by method rather than field of practice.

The Council on Social Work Education in the late 1950's undertook a massive national curriculum study. The result was publication of 13 volumes dealing with several areas of the social work curriculum and with general issues affecting social work education. The study gave a strong impetus to the development of undergraduate social work education. One of the major recommendations was that there be exploration of the continuum from undergraduate to graduate education in social work.[15]

During the same period a National Joint Commission on Mental Illness and Mental Health was established by the Mental Health Study Act of 1955. After five years of survey and research the final report of the Joint Commission was published as a volume entitled *Action for Mental Health*.[1] This report was released in the early years of the Kennedy administration, and by 1963 new legislation was passed that established the community mental health center as the focal point for the delivery of community mental health services. It should be noted that during the 1950's the advent of drug therapy and its widespread use led to a significant decline in hospital populations. The number of people in hospitals began to decline at the rate of 5 to 7 per cent a year. The result was an increasing demand for community care facilities. The Mental Health Centers Act provided for a planning period with the allocation of funds for this purpose to each of the states and territories. The states' plans permitted a restructuring of resources to include a complex of hospital and community treatment and rehabilitation facilities as a single network. Many states moved quickly to pass their own community services acts as a means of providing a legislative framework for the restructuring of the delivery of mental health services.[50] As the new community mental health centers came into being through the availability of construction funds and staffing grants, new modes of practice began to develop. A considerable literature now exists on the early stages of the operation of community mental health centers.[14,21,52,62] The role of social work in these centers is significant. Social workers participate not only in the treatment services, which in many cases still operate along traditional lines, but also in the outreach and community organization functions of the community mental health center. An early study of professional staff functions in mental health centers indicated that over half of staff time was directly related to patient care.[36] Outreach functions were in their early stages.

Significant Developments in the 1960's

During the 1960's a number of highly significant events took place in social work education. The development of a new curriculum policy statement in 1962 provided a reorientation of master's level education in social work. Increased prominence was given to development of the social work methods of group work and community organization. The curriculum policy moved toward a much more integrated curriculum involving human growth and behavior and content on the social environment. Expanded content in social welfare policy and services was added. The guidelines provided for flexibility of curriculum organization and structure. A number of schools of social work conducted curriculum studies that led to the reorganization of program content in social work education.

Under the aegis of the Council on Social Work Education, a five-year period of consultation service intended to foster the development of new schools of social work in unserved areas resulted in the addition of a number of graduate schools of social work in states that had hitherto been without social work education. These were mainly Southern states such as South Carolina, Arkansas, and Alabama. In addition, unserved regions in other states such as western Michigan, central Kentucky, and northeastern Pennsylvania also developed educational resources.

Perhaps the most significant shift in social work education came from the rising recognition of the place in the curriculum of community organization and social policy. Several influences arising from social and economic developments in the 1960's contributed to this situation. One was the rise of the movement toward comprehensive health and mental health services and the introduction of national legislation calling for regional planning and for the development of community health planning as a collaborative effort to organize health services for more effective and efficient delivery at the local level. A second major influence was the introduction of the poverty program in 1964. This massive "war on poverty" resulted in a host of federally funded endeavors aimed at creating "equal opportunity" for the poor and underprivileged in American society. Community action programs were introduced in most major urban areas and in many rural areas that had never before had a consciously directed effort to look at the causes of poverty and to take some action about them.[72] As inadequate education was seen as a major deterrent to upward mobility, the Head Start program was introduced as a means of upgrading the educational prospects for children in families where deprivation was the characteristic mode of life. Lastly a new modality of service for local delivery was developed in the neighborhood service center. These centers were established to serve as a basis for community action programs and other types of social services in the community. An outstanding example of such a center was the Mobilization for Youth program in New York City, which served as a model for a research training and service agency based in a poverty area and making use of new careers and paraprofessional personnel drawn from the indigenous population in the community.[10]

By the end of the 1960's a new network of services was spread through the community, calling for entirely new modes of professional behavior. Social work and social workers were profoundly influenced by these developments. This was particularly true in the mental health field.[33,93]

During this period case work as a method for delivery of service came under fire as being too narrowly related to the individual and his personal problems and not sufficiently related to problems of society and the welfare of the community as a whole.[49] Social case work has been described as a "problem-solving process"[64] and as "psychosocial therapy"[45] by leading theorists in the field. The activist atmosphere of the 1960's led to greater concern for factors influencing the entire social system.[29,82] Students and minority groups pressed for changes. Consequently much more of the curriculum was devoted to studying the social forces influencing society and the means of changing social policy. There was considerable increase in the proportion of students enrolled for field instruction in community organization and in group work.[24] In addition, there were increasing numbers enrolled in social work educational programs that provided for training in integrated methods, bringing together case work, group work, and community organization as a single integrated social work practice method.[9] The pioneering efforts in this regard began on the West Coast and have influenced social work education throughout the country.

Schools of social work have responded to this challenge by developing learning and teaching centers built principally around a single field instruction agency, or a cluster of agencies, where a variety of experiences can be made available to the student in individual treatment, group work or group therapy activities, and community organization.[56]

The development of Medicare and Medicaid as an extension of the poverty program was intended to aid the aged and those who are in economic need for any reason to receive the full range of medical care. The application of Medicare and Medicaid to problems of mental health and mental illness has not been easy. One problem is the lack of adequate facilities throughout the country to provide the required care; the second is the problem of articulation of these programs with the existing health and welfare establishments in order to provide a good coordinated effort.[88]

A particularly important influence on the field of social work in the past decade has been the advent of the community mental health centers program. By 1970 there were 425 such centers throughout the country, serving catchment areas of from 75,000 to 150,000 population. As Bertram S. Brown[19] has pointed out, these facilities will need to plan for interaction with the proposed health maintenance organizations. Social workers have taken the initiative in helping develop new roles for consultation and education,[67] for emergency services,[59] and for community planning.[48] Social workers have also taken the initiative in the introduction of group work and group therapy into mental health settings.[60,76] One of the early studies of social work practice in hospitals and communities demonstrated that about 12 per cent of social workers were conducting some form of group activity in their hospitals and clinics.[11] The role of social work in community mental health in stimulating citizen participation is increasing, and much work has been done to help organize consumers. More social workers are found in institutions involved in delivery of mental health services.[58] It has been found that aggressive work in the community can frequently avoid hospitalization of mental patients.[83] This takes careful work with the patient, with his family, and with the community resources that provide the support system.[30]

In summary, it could be said that the decade of the 1960's was a period of development and consolidation of the practice of psychiatric social work beyond the clinical concept of personal treatment on an individual or group basis. The role of social work as the communication link between the patient or client and the community continued to be regarded as essential to total service for patients. Many social workers took an important part in the revolution in mental health, which found some mental hospitals serving more frequently as diagnostic and short-term treatment facilities as an adjunct to their former status as long-term care institutions. Unfortunately there remains a hard core of mental patients who have become so adapted to the culture of the institution that it is difficult to interrupt what has become a satisfying social situation. In addition, for many of these patients their families have disappeared over the years, and there would be no way to arrange for community care except through foster care or similar arrangements. The use of more home care facilities and extended community care has permitted the prevention of hospitalization or the reduction of its length and the provision of services closer to the client's home or in his own home. The development of the community mental health center had by the end of the 1960's pointed a new way to service provision at the local level.[62]

Social Work as a Professional Resource in Mental Health

French's[35] 1940 study outlined the characteristics and deployment of psychiatric social workers at that time. Social workers were found in mental hospitals, in mental health clinics, in child guidance clinics, in school social work programs, and in a wide variety of services where their training and background permitted the application of knowledge about human growth and development to psychosocial problems of individuals and families. Psychiatric social workers were envisioned as team members making a specific contribution to the diagnosis, treatment, and rehabilitation of mental illness. These support and treatment dimensions have continued to the present.[58] Social workers are found in a wide variety of mental health resources, including the new community mental health centers.

The psychiatric social worker provides service throughout the cycle of illness and recovery by being available from the point of intake into the clinic or hospital through ultimate recovery. An important initial screening function takes place at intake. Situations involving serious interpersonal conflict or parent-child dysfunction in social relationships require prompt action in terms of interpretation to the sick person or to the concerned relative. In emergency services the social worker frequently initiates a plan for the temporary care of children or for counseling and advice to a spouse who is left without his or her mate. When mental breakdown requiring hospitalization occurs, these pressures are met through the development of alternate care within the childrens' own home with the assistance of relatives or homemakers or through alternate means of care that can be provided through child welfare facilities in the community. The advent of drug therapy has frequently meant that it is possible for the affected parent to be returned home fairly soon. This is another point at which the psychosocial evaluation is materially aided through the participation of the psychiatric social worker. During the early stages of assessment and evaluation of psychiatric illness, the social data are compiled and made part of the total assessment of the factors that lead to treatment plans and to continuing evaluation of the patient's progress and his capability for being returned to his home environment. The better equipped communities now have a number of alternatives available that materially affect the possibilities for convalescence and complete recovery. These are favorably influenced by the availability of well-staffed day-care centers, halfway houses, and opportunities for partial hospitalization, which make it possible for the patient to resume employment using the hospital or intermediate care facility as a base. The social worker participates in planning for the continuum of care so that, as the patient moves toward control of himself and appropriate use of his capabilities, he is able to establish a stable relationship with his home and community. The social worker frequently is used as a resource for information regarding the service system of any given community.

In the child guidance clinic the psychiatric

social worker, in addition to performing in the intake and evaluation function, carries main responsibility for intensive treatment of parents and children in a collaborative team relationship with other members of the mental health service constellation. In addition to individual case work treatment, the social worker frequently provides group therapy, a modality that is now found in nearly every psychiatric resource. The responsibility for group therapy may be carried by an individual staff member or by a team working toward specific objectives with a selected group of patients or relatives. Group therapy may be conducted in the community with relatives of mental patients as a means of providing support and interpretation and as a means of assisting in the resumption of normal living when the patient returns home.

The provision of direct treatment by psychiatric social workers is a responsibility undertaken in the light of the long-standing tradition that social workers are equipped to undertake intensive treatment. Case illustrations describing the interaction of social worker and client in social and mental health resources are found in the case work literature. The Grayson Case in Perlman,[64] for example, describes work with a male veteran receiving service in a privately supported psychiatric clinic (pp. 207–222).

Case work involves direct application of knowledge about psychological and social behavior in the interaction with individuals and families with problems of adaptation to life stress and interpersonal conflict. The case work method is one of the primary tools available to the psychiatric social worker in the practice of direct work with clients and patients. It is the most typical of the methods used by social workers providing direct services.[17] In addition to the application of case work and group work methods in mental health, many social workers are now being recruited because of their community organization skills. The community organization social worker is involved in developing citizen participation through advisory committees and similar structures. He also carries on linkage and liaison relationships with relevant social and health agencies. His function on the mental health team is critical to the development of community relationships so importantly linked to the education and consultation functions of community mental health.

The psychiatrist, psychologist, nurse, and social worker are seen as co-workers in service delivery administration and educational activities in the care, treatment, and rehabilitation of the mentally ill. Characteristically they are engaged in collaborative work on preventive activities as well. Parent education on child development is one frequent area for collaborative work. Another involves cooperative activities in the public school system with teachers and guidance personnel. Social workers initiate outreach activities that bring services directly to the community. A program of this type is described by Briar and Miller[17] (pp. 237–238). In this case social workers work from a specially equipped van that is operated directly on the streets of an urban poverty area. There is growing interest in such experimentation with extended social services.

Trends and Implications for Future Practice of Social Work

There has been a considerable evolution in the role and function of social work in mental health. This change has reflected newer attitudes and applications of knowledge and skill throughout the structure of human services in the United States. In addition to manning the established agencies, social workers have moved into community mental health services from the very beginning. With the exception of psychiatric nurses they tend to rank highest in terms of numbers employed in mental health centers.[36] There are emerging a number of descriptive patterns of the work conducted by community mental health centers. The variety of programs offered can be seen from a report on the provision of mental health training for public welfare personnel.[67] The NIMH has periodically reported on the status of work in community mental health centers, including reference to community involvement and the use of paraprofessional personnel.[62] One of the more unique aspects of mental health center operations is that involving the emergency services. These are intended to prevent entrance of patients into the typical mental hospital system.[34] Active use is made of home care, foster care, and day care as means of providing alternatives to hospitalization.[4] here is also an extension of social service and mental health consultation in connection with outreach services.[21] These have been found particularly helpful in consulting around problems of illegitimacy in the community.[79] Some of the newer trends have involved the use of behavior modification as part of the treatment armamentarium available to social workers.[84]

During recent years social work has penetrated deeply into such insitutions as the public schools,[47] the military services,[27] and the Veterans' Administration, where they have staffed community as well as mental health treatment services. Veterans' Administration social workers have experimented with outreach programs dealing with the physically as well as the mentally disabled.[7] Social workers with mental health preparation have been particularly useful in suicide prevention centers,[31] in alcoholism service centers,[51,74] and in narcotic treatment programs. Some have conducted long-range research in drug addiction dealing with the social factors involved in narcotics addiction.[18] Other social workers have developed programs in the fields of geriatrics[42] and of mental retardation.[68] In a number of institutions a wide range of group work and group therapy services have developed.[60] The participation by social workers in such programs has increased considerably during the last decade.

It could be said that a wide range of skills is brought to community mental health services by social workers prepared at the master's and doctoral levels. It is possible that the numbers of baccalaureate level workers will increase considerably as more use is made of manpower with less than full training. The number of community mental health workers trained at the associate of arts level will also be increasing as means are found for deploying personnel with a wide range of skills and capabilities. Several studies have been conducted and a number of reports exist on work being done by paarprofessionals in the mental health field.[41,80] It is beginning to be seen that this group of mental health workers is essential to the offering of a complete mental health service in any community and in any institution. A number of special training programs have developed to help prepare paraprofessional workers in the mental health field.[80]

It can be anticipated that the number of patients in the older type mental hospitals will be changing in keeping with the trend away from custodial care toward treatment in the community.[40] Social work will be playing an increasingly important role in the development of extension and outreach services. Indeed, there will be a considerable development of social work in connection with prevention as well.[63,69,93] The capabilities of social work are particularly related to outreach and community organization aspects of mental health. If it is true that mental illness can affect no single individual without some lateral impact on his family, then it is true that attention needs to be given to the social factors in mental illness.[43] A continuum of care is needed from the time of onset to recovery and afterward. As is now well known, the prospects for success and treatment are much enhanced if the patient and his family already have some general conception of what mental illness is about and of what resources are available to provide for care, treatment, and rehabilitation of an individual who may become mentally ill. Not only is it necessary for citizens of all economic levels to know about mental illness and mental health resources, but also it is important for individuals employed in the human services to have this knowledge and background. Public education has been undertaken nationally and locally by the National Association for Mental Health, the present-day national organization deriving from the old National Committee for Mental Hygiene founded by Clifford Beers. Education of the health services and human services professions tends to be undertaken by members of the mental health establishment of all disciplines. Social workers are found in medical schools and in service establishments. Eleanor Ireland has indicated[46] the range and nature of social work participation in the training of child psychiatrists.[46] From another aspect Imena A. Handy[42] has demonstrated the effect of social work in the care of mentally ill geriatric patients.

Social workers have undertaken a number of educational efforts directed to the several professions, while also they are concerned with the deepening of mental health content in sensitive areas of human existence. There has been increasing interest, for example, in the area of human sexuality in relation to social living. Harvey L. Gochros and his associates[38] have undertaken to develop content on human sexuality in social work education and also to communicate more knowledge in this area to the mental health and social service professions and to the general public. Additional experimentation has attempted to relate existentialism to human treatment.[89] This implies the adaptation of philosophic reasoning to psychosocial therapy.

One of the problems in expanding resources for the care of the mentally ill is the massive ignorance about the nature of mental illness and its response to treatment. As a result there have been painfully slow developments in the coverage of mental illness by health insurance. A British visitor in the mid-1960's, Richard M. Titmuss,[86] reviewed the relationship of social policy to economic progress in 1966. His main theme was that the allocation of world resources has yet to be devoted to the fullest extent possible to the interests of the disadvantaged.[86] A breakthrough of major consequence occurred when mental health care was brought under Medicare and Medicaid.[88] The limitations of both of these programs still leave much to be desired in terms of universal coverage for the economically deprived and for older people who need mental health services. One of the secondary gains from the introduction of Medicare payments has been the highlighting of staffing shortages both in hospitals and in extended care facilities. The shortage extends not only to social work personnel but also to other trained personnel. These shortages have been periodically highlighted in terms of care of the mentally ill. It is obvious that the physically ill and disabled need psychosocial services just as much as those who are mentally ill. A number of authors have highlighted the shortages of personnel at several levels.[2,3,8] Melvin A. Glasser has touched on the major issues involved in the funding of mental health services through national health insurance. He proposes offering "health security" that resembles what is now available under law in terms of Social Security, with adequate funding; he calls for a national drive toward improvement of the standards of health care and the distribution of health care and services more generally throughout the population.[37]

There is a move within the field of social work to strengthen and further develop the clinical practice area of the field. One symptom of this development is the growth during the past decade of the number of practitioners who have entered the field of private practice, full- or part-time. These social workers may be employed in group practice, they may work collaboratively with a number of other disciplines, or they may operate as independent private practitioners on a fee-for-service basis. Social workers have, therefore, promoted a system of registration or licensing throughout the country, seeking legislation that will establish a legal basis for professional standards for the practice of social work. It can be expected that, just as psychology has obtained recognition as a treatment profession in several states, social work will in time receive similar status in the community. Margaret A. Golton,[39] in her review of private practice in social work, indicates the first licensing of social workers began in 1952. There are four states that now require registration certification or licensing, and in a number of other states chapters of the National Association of Social Workers are pressing for the legal recognition of social work private practice. As Wilbert E. Moore[57] points out, the move toward autonomous practice is one symbol of the achievement of full professional status. He indicates, moreover, that this is frequently a cause for role strain among the various professions.

Social work has its own critics who periodically call for re-examination of the objectives and purposes of the profession. John B. Turner[87] has pointed to the need to "avoid professional obsolescence" and proposes that the profession shift its manpower from preoccupation with traditional welfare institutions to increasing utilization in other institutions having an impact on the social structure. He sees the need to base planning on consumer-identified needs and to look at basic causes of social and psychological problem behavior. Social work, along with the other health professions, needs to look at its basic educational doctrines, the array of services, and the means by which these can be extended to the total population.

In the early 1970's the nation is in an inexorable crush from the pressures deriving from rapid industrialization, automation, and economic disequilibrium. At a peak of scientific development the fact remains that important segments of the American community are threatened with economic insecurity and insufficient provision of adequate means for maintaining life and health. There is poverty in the inner city and in the rural areas of the country. There are vastly insufficient manpower resources to provide the health and mental health services needed to improve the mental health of individuals, families, and communities.[94] The complexities of the mental health manpower picture have been fully described.[12,25] The physical problem involved in the shift of personnel from mental hospitals to community centers is tremendously complex and defeating. The problem is how to assure adequate provision of social and mental health services for patients remaining in the hospital and for those being moved to, or being supported in, the community during the transitional period. It is apparent from the report submitted by Rehin and Martin[55] that even the British face perplexing decisions. In considering the shift to community mental health services, Pascal Scoles[77] has described some of the interesting problems that arise in the provision of care for the chronic mental patient in an urban community. He refers to a return to the "older ways in social work," a phrase that alludes to the return to concern for the patient through home visits and through direct activities referring to employment, job satisfaction, improved use of social and health resources, and other means for providing concrete services that tend to be overlooked very frequently in the intensive psychotherapy phase of rehabilitation. The application of common sense to the organization and delivery of mental health services frequently does require direct service provision that cannot occur without some flexibility in outlook on the part of the provider of services, and without an orientation to the operation of the network of community systems, which function imperfectly with reference to individual needs.[91] Rockmore and Conklin[73] have demonstrated what can be done at the state level to provide for use of a state office as a referral agency. This is seen as a somewhat different interpretation of the role of an administrator in a psychiatric social service.[73]

As one projects ahead, the gradually increasing U.S. population (now well over 200 million) and the lack of significant change in the incidence of mental illness and disorder as a result of preventive or treatment measures lead one to speculate that the problems of providing adequate services will remain a pressing issue in the political, economic, and social environment for many years to come. In fact, it is quite likely that the dominating issue for the remainder of the twentieth century will be how to assure that threats to human existence can be resolved. Whether these exist in the form of war, poverty, hunger, ecological danger, or social deprivation, the aims of social work and social or community psychiatry are parallel. They are dedicated not only to the remedial efforts involved in social and psychiatric treatment but also to the elimination of causes for social and mental breakdown. It is the ultimate objective of social work and social psychiatry to project a social and economic system that will make appropriate services available throughout the life cycle of the individual. In addition, social work has as a primary objective the ultimate abolition of poverty and the reduction of social pathology to its absolute minimum. This can only be done by restructuring community resources so that they provide basic services in the immediate environment of the individual.[92] Curriculum development in social work education needs to take social changes into account to remain relevant.[66] There is a great deal currently known about prevention and early intervention that suggests that health resources ought to be available in every community regardless of economic level.[69] There is also the notion of early intervention that can occur around crisis situations. These arise in entrance into the school system. They arise in day-care facilities and in nursery schools. They arise in industrial and commercial establishments that provide for the daily livelihood of people. It could be projected that counseling services, if available in sufficient quality and quantity, could do much to alleviate problems of drug addiction, alcoholism, and family breakdown. The need for better interpersonal understanding is cardinal and lies behind most parental conflict and marital disequilibrium.

The resources of the nation need to be mustered to provide the types of health and social services that will assure that every child born in the country can expect to receive adequate food, clothing, and housing,[23] and should be able, in moving into maturity, to obtain first-rate educational and health services as well.[86] The vast need for such assistance in the black community has been underscored by Whitney M. Young, Jr.,[95] but his admonition on needs for the country could be extended to every category of ethnic group and to the general population as well. Social work is an inherent part of the mental health movement and is closely related to the service development in the American community. The pattern of participation in delivery of mental health services has reflected vast changes over the past decades. More social workers have moved to administrative, teaching, consultative, and research positions.[78] Social workers also have moved into social policy positions in governmental and voluntary structures. A strong continuity of interest in direct treatment services has manifested itself throughout the century. Social work clinicians hold firmly to the responsibility for direct involvement in behavior change so vital to social and emotional competence. One-fifth of the members of the National Association of Social Workers in 1969 were employed in or expressed an interest in psychiatric and mental health services.[81] The place of social work in mental health is one of essential involvement in the several levels of service, training, and research directed toward the well-being of people.

Bibliography

1. *Action for Mental Health: Final Report of the Joint Commission on Mental Illness and Health*, Basic Books, New York, 1961.
2. ADLER, J., and TROBE, J. L., "The Obligations of Social Work Education in Relation to Meeting Manpower Needs at Differential Levels in Social Work," *Child Welfare, 47*:346–350, 1968.
3. ALBEE, G. W., *Mental Health Manpower Trends*, Basic Books, New York, 1959.
4. ALBINI, J. L., "The Role of the Social Worker in an Experimental Community Mental Health Clinic: Experiences and Future Implications," *Com. Ment. Health J., 4*: 111, 1968.
5. ALLEN, F. H., "The Influence of Psychiatry on Social Case Work," in *Proceedings of the National Conference of Social Work*, University of Chicago Press, Chicago, 1935.
6. American Association of Psychiatric Social Workers, *Education for Psychiatric Social Work: Proceedings of The Dartmouth Conference*, New York, 1950.
7. ANDERSON, D. M., "Veterans Services," in *Encyclopedia of Social Work*, pp. 1513–1518, National Association of Social Workers, New York, 1971.
8. ARNHOFF, F. N., RUBINSTEIN, F. A., and SPEISMAN, J. C., *Manpower for Mental Health*, Aldine, Chicago, 1969.
9. BARTLETT, H. M., *The Common Base of Social Work Practice*, National Association of Social Workers, New York, 1970.
10. BECK, B. M., "A Social Work Approach to a Mental Health Objective," in Magner, G. (Ed.), *Leadership Training in Mental Health*, pp. 61–71, National Association of Social Workers, New York, 1970.
11. BERKMAN, T. D., *Practice of Social Workers*

in Psychiatric Hospitals and Clinics, American Association of Psychiatric Social Workers, New York, 1953.

12. BETTIS, M. C., and ROBERTS, R. E., "The Mental Health Manpower Dilemma," *Ment. Hyg.*, 53:163–175, 1969.
13. BEVILAQUA, J. J., and MORGAN, R. W., "Military Social Work," in *Encyclopedia of Social Work*, pp. 851–855, National Association of Social Workers, New York, 1971.
14. BINDMAN, A. J., and SPIEGEL, A. D., *Perspectives in Community Mental Health*, Aldine, Chicago, 1969.
15. BOEHM, W. W., *Objectives for the Social Work Curriculum of the Future*, Council on Social Work Education, New York, 1959.
16. ———, "Social Psychiatry and Social Work," *Canad. Psychiat. A. J.*, 12:29–42, 1967.
17. BRIAR, S., and MILLER, H., *Problems and Issues in Social Casework*, Columbia University Press, New York, 1971. See also Borenzweig, H., "Social Work and Psychoanalytic Theory: An Historical Analysis," *Soc. Work*, 16:7–16, 1971.
18. BRILL, L., and LIEBERMAN, L., *Authority and Addiction*, Little, Brown, Boston, 1969.
19. BROWN, B. S., "Community Mental Health: The View from Fund City," *The Social Welfare Forum, 1971*, Columbia University Press, New York, 1971.
20. BRUNO, F. J., *Trends in Social Work as Reflected in the Proceedings of the National Conference of Social Work: 1874–1946*, Columbia University Press, New York, 1948.
21. COHEN, L. M., *et al.* (Eds.), *Patients in Programs: At Area C Community Mental Health Center*, Department of Human Resources, Washington, D.C., 1971.
22. COHEN, N. E., *Social Work in the American Tradition*, The Dryden Press, New York, 1958.
23. *Crisis in Child Mental Health: Report of the Joint Commission on Mental Health of Children*, Harper & Row, New York, 1969.
24. Council on Social Work Education, *Statistics on Social Work Education: 1970*, Council on Social Work Education, New York, 1971.
25. COWNE, L. J., "Approaches to the Mental Health Manpower Problem: A Review of the Literature," *Ment. Hyg.*, 53:176–187, 1969.
26. CRUTCHER, H. B., *Foster Home Care for Mental Patients*, The Commonwealth Fund, New York, 1944.
27. DAVIS, J. A., "Outpatient Group Therapy with Schizophrenic Patients," *Soc. Casework*, 52:172–178, 1971.
28. DEUTSCH, A., *The Mentally Ill in America*, Columbia University Press, New York, 1946.
29. DUMONT, M., *The Absurd Healer: Perspectives of a Community Psychiatrist*, Science House, New York, 1968.
30. EVANS, P. P., and BRACHT, N. F., "Meeting Social Work's Challenge in Community Mental Health," *Ment. Hyg.*, 55:295–297, 1971.
31. FEIDEN, E. S., "One Year's Experience with a Suicide Prevention Service," *Soc. Work*, 15:26–31, 1970.
32. FELIX, R. H., "A Comprehensive Mental Health Program," in *Mental Health and Social Welfare*, Columbia University Press, New York, 1961.
33. FISCHER, J., "Portents from the Past: What ever happened to Social Diagnosis?" *Int. Soc. Work*, 13:18–28, 1970.
34. FLOMENHAFT, K., KAPLAN, D. M., and LANGSLEY, D. G., "Avoiding Psychiatric Hospitalization," *Soc. Work*, 14:38–45, 1969.
35. FRENCH, L. M., *Psychiatric Social Work*, The Commonwealth Fund, New York, 1940.
36. GLASSCOTE, R. M., and GUDEMAN, J. E., *The Staff of the Mental Health Center: A Field Study*, Joint Information Service of the American Psychiatric Association and National Association of Mental Health, Washington, D.C., 1969.
37. GLASSER, M. A., "Mental Health, National Health Insurance, and the Economy," *Hosp. & Com. Psychiat.*, 23:17–22, 1972.
38. GOCHROS, H. L., and SCHULTZ, L. G., *Human Sexuality and Social Work*, Association Press, New York, 1972.
39. GOLTON, M. A., "Private Practice in Social Work," in *Encyclopedia of Social Work*, pp. 949–955, National Association of Social Workers, New York, 1971.
40. GREENBLAT, M., YORK, R. H., and BROWN, E. L., *From Custodial to Therapeutic Patient Care in Mental Hospitals*, Russell Sage Foundation, New York, 1955.
41. GROSSER, C., HENRY, W. E., and KELLEY,

J. G., *Nonprofessionals in the Human Services*, Jossey-Bass, San Francisco, 1969.

42. HANDY, I. A., "Social Work Services to the Mentally Ill Geriatric Patient," *J. Am. Geriat. Soc.*, *17*:1145–1146, 1969.
43. HOLLINGSHEAD, A. B., and REDLICH, F. C., *Social Class and Mental Illness*, John Wiley, New York, 1958.
44. HOLLIS, E. V., and TAYLOR, A. L., *Social Work Education in the United States*, Columbia University Press, New York, 1951.
45. HOLLIS, F., *Casework: A Psychosocial Therapy*, Random House, New York, 1964.
46. IRELAND, E., "The Social Worker's Contribution to Training in Child Psychiatry," *J. Child Psychol. & Psychiat. & Allied Disciplines*, *8*:99–104, 1967.
47. JOHNSON, A., *School Social Work: Its Contribution to Professional Education*, National Association of Social Workers, New York, 1962.
48. KAHN, A. J., *Studies in Social Policy and Planning*, Russell Sage Foundation, New York, 1969.
49. KLEIN, P., *From Philanthropy to Social Welfare*, Jossey-Bass, San Francisco, 1968.
50. KNEE, R. I., and LAMSON, W. C., "Mental Health Services," in *Encyclopedia of Social Work*, pp. 802–812, National Association of Social Workers, New York, 1971.
51. KRIMMEL, H., *Alcoholism: Challenge for Social Work Education*, Council on Social Work Education, New York, 1971.
52. LAMB, H. R., HEATH, D., and DOWNING, J. J., *Handbook of Community Mental Health Practice*, Jossey-Bass, San Francisco, 1969.
53. LEVINE, R. A., "Consumer Participation in Planning and Evaluation of Mental Health Services," *Soc. Work*, *15*:41–46, 1970.
54. MAAS, H. S. (Ed.), *Adventure in Mental Health*, Columbia University Press, New York, 1951.
55. MARTIN, F. M., and REHIN, G. F., *Towards Community Care*, Berridge and Company, London, 1969.
56. *Modes of Professional Education*, Tulane University School of Social Work, New Orleans, 1969.
57. MOORE, W. E., *The Professions: Roles and Rules*, Russell Sage Foundation, New York, 1970.
58. NACMAN, M., "Social Workers in Mental Health Services" in *Encyclopedia of Social Work*, pp. 822–828, National Association of Social Workers, New York, 1971.
59. NASH, K. B., "Social Work in a University Hospital: Commitment to Social Work Teaching in a Psychiatric Emergency Division," *Arch. Gen. Psychiat.*, *22*:332–337, 1970.
60. National Association of Social Workers, *Use of Groups in the Psychiatric Setting*, New York, 1960.
61. *Outplacement Study*, Department of Medicine and Surgery, U.S. Veterans Administration, Washington, D.C., September 1965.
62. OZARIN, L. D., FELDMAN, S., and SPANER, F. E., "Experience with Community Mental Health Centers," *Am. J. Psychiat.*, *127*:88–92, 1971.
63. PARAD, H. J. (Ed.), *Crisis Intervention: Selected Readings*, Family Service Association of America, New York, 1965.
64. PERLMAN, H. H., *Social Casework: A Problem-Solving Process*, University of Chicago Press, Chicago, 1957.
65. PHILLIPS, B., "Social Workers in Health Services," *Encyclopedia of Social Work*, pp. 565–574, National Association of Social Workers, New York, 1972.
66. PINS, A. M., "Changes in Social Work Education and Their Implications for Practice," *Soc. Work*, *16*:5–15, 1971.
67. POLLEY, G. W., MCALLISTER, L. W., OLSON, T. W., and WILSON, K. P., "Mental Health Training for County Welfare Social Work Personnel: An Exercise in Education and Community Organization," *Com. Ment. Health J.*, *7*:29–38, 1971.
68. POND, E. M., and BRODY, S. A., *Evolution of Treatment Methods at a Hospital for the Mentally Retarded*, Research Monograph No. 3, State of California, Department of Mental Hygiene, 1965.
69. *Public Health Concepts in Social Work Education*, Council on Social Work Education, New York, 1962.
70. PUMPHREY, R. E., and PUMPHREY, M. W., *The Heritage of American Social Work*, Columbia University Press, New York, 1961.
71. RICHMOND, M. E., *Social Diagnosis*, Russell Sage Foundation, New York, 1917.
72. RIESSMAN, F., and POPPER, H. I., *Up From Poverty: New Career Ladders for Nonprofessionals*, Harper & Row, New York, 1968.

73. ROCKMORE, M. J., and CONKLIN, J. J., "A Blending of Two Roles: The Administrator-Psychiatric Social Worker," *State Gov.*, Summer 1971.

74. ROOT, L. E., "Social Casework with the Alcoholic and His Family," in Catanzaro, R. (Ed.), *Alcoholism: The Total Treatment Approach*, pp. 208–222, Charles C Thomas, Springfield, Ill., 1968.

75. ROTHMAN, D. J., *The Discovery of the Asylum*, Little, Brown, Boston, 1971.

76. SCHEIDLINGER, S., "Therapeutic Group Approaches in Community Mental Health," *Soc. Work*, *13*:87–95, 1968.

77. SCOLES, P., "The Chronic Mental Patient: Aftercare and Rehabilitation," in *Social Work Practice*, *1969*, pp. 61–75, Columbia University Press, New York, 1969.

78. SHORE, M. F., and MANNINO, F. V., *Mental Health and the Community: Problems, Programs, and Strategies*, Behavioral Publications, New York, 1969.

79. SIGNELL, K. A., "Mental Health Consultation in the Field of Illegitimacy," *Soc. Work*, *14*:67–74, 1969.

80. SOBEY, F., *The Nonprofessional Revolution in Mental Health*, Columbia University Press, New York, 1970.

81. STAMM, A. M., "NASW Membership: Characteristics, Deployment, and Salaries," *Personnel Information*, *12*, 1969.

82. STEINER, C., "Radical Psychiatry: Principles," *Radical Ther.*, *2*:3, 1971.

83. STUBBLEBINE, J. M., and DECKER, J. B., "Are Urban Mental Health Centers Worth It?" *Am. J. Psychiat.*, *127*:84–88, 1971.

84. THOMAS, E. J. (Ed.), *Behavioral Science for Social Workers*, Free Press, New York, 1967.

85. TIMMS, N., *Psychiatric Social Work in Great Britain, 1939–62*, Routledge and Kegan Paul, London, 1964.

86. TITMUSS, R. M., "Social Policy and Economic Progress," *The Social Welfare Forum, 1966*, Columbia University Press, New York, 1966.

87. TURNER, J. B., "In Response to Change: Social Work at the Crossroads," *Soc. Work*, *13*:7–15, 1968.

88. U. S. Social Security Administration, Office of Research and Statistics, *Financing Mental Health Care Under Medicare and Medicaid*, U. S. Government Printing Office, Washington, D.C., 1971.

89. WEISS, D., "Social Work as Encountering," *J. Jewish Communal Service*, *46*:238–245, 1970.

90. WILENSKY, H. L., and LEBEAUX, C. N., *Industrial Society and Social Welfare*, Free Press, New York, 1965.

91. WITTMAN, F. D., "Alcoholism and Architecture; the Myth of Specialized Treatment Facilities," in Mitchell, W. J. (Ed.), *Environmental Design: Research and Practice*, Proceedings of the EDRA 3/AR 8 Conference, University of California, Los Angeles, 1972.

92. WITTMAN, M., "The Social Welfare System: Its Relation to Community Mental Health," in Golann, S. E., and Eisdorfer, C. (Eds.), *Handbook of Community Mental Health*, Appleton-Century-Crofts, New York, 1972.

93. ———, "The Social Worker in Preventive Services," in *The Social Welfare Forum*, pp. 136–147, Columbia University Press, New York, 1962.

94. ———, "Social Work Manpower in the Health Services," *Am. J. Pub. Health*, *55*:393–399, 1965.

95. YOUNG, W. M., *To Be Equal*, McGraw-Hill, New York, 1964.

NAME INDEX

Note: Bold face figures indicate chapter pages.

SUBJECT INDEX